Building Construction Handbook

Building Construction Handbook

FREDERICK S. MERRITT

Consulting Engineer, Syosset, N.Y.

Third Edition

McGRAW-HILL BOOK COMPANY

New York St. Louis San Francisco Auckland Düsseldorf
Johannesburg Kuala Lumpur London Mexico Montreal
New Delhi Panama Paris São Paulo Singapore
Sydney Tokyo Toronto

Library of Congress Cataloging in Publication Data

Merritt, Frederick S ed.
Building construction handbook.

(McGraw-Hill handbooks)
Includes bibliographical references.
1. Building—Handbooks, manuals, etc. I. Title.
TH151.M4 1975 690'.02'02 75-6553
ISBN 0-07-041520-X

234567890 KPKP 784321098765

The editors for this book were Harold B. Crawford and Robert Braine, the designer was Naomi Auerbach, and the production supervisor was George Oechsner. It was set in Caledonia by Bi-Comp, Inc.

It was printed and bound by The Kingsport Press.

Contents

SECTION 2. BUILDING MATERIALS by A. G. H. Dietz 2–1

Part 1. Masonry Materials

BRACING

FASTENERS

CONNECTIONS

STEEL ERECTION

PAINTING

FIRE PROTECTION FOR STRUCTURAL STEEL

SECTION 8. WOOD CONSTRUCTION by Maurice J. Rhude 8–1

SECTION 9. WINDOWS by Phillip M. Grennan 9–1

SECTION 23. VERTICAL TRANSPORTATION by Frederick S. Merritt 23–1

SECTION 24. SURVEYING FOR BUILDING CONSTRUCTION by R. S. Brackett 24–1

SECTION 25. ESTIMATING BUILDING CONSTRUCTION COSTS by E. D. Lowell 25–1

SECTION 26. CONSTRUCTION MANAGEMENT by Robert F. Borg 26–1

SECTION 27. SPECIFICATIONS by Joseph F. Ebenhoeh, Jr. 27–1

SECTION 28. INSURANCE AND BONDS by C. S. Cooper 28–1

Index follows Section 28.

Contributors

THOMAS H. BOONE *(Retired) Formerly Building Research Division, National Bureau of Standards, Washington, D.C.* (SEC. 14)
ROBERT F. BORG *President, Kreisler Borg Florman Construction Company, Scarsdale, N.Y.* (SEC. 26)
REGINALD S. BRACKETT *Bay Shore, N.Y.* (SEC. 24)
C. S. COOPER *Bradentown, Fla.* (SEC. 28)
WILLIAM E. DIAMOND II *Associate, Smith, Hinchman and Grylls Associates, Inc., Detroit, Mich.* (SEC. 1)
A. G. H. DIETZ *Professor, Department of Building Engineering, School of Architecture and Planning, Massachusetts Institute of Technology, Cambridge, Mass.* (SEC. 2)
JOSEPH F. EBENHOEH, JR. *Senior Associate, Albert Kahn Associates, Inc., Detroit, Mich.* (SEC. 27)
F. E. FAHY *Bethlehem, Pa.* (SEC. 7)
DR. JACOB FELD *Consulting Engineer, New York, N.Y.* (SEC. 4)
C. C. FISHBURN *Tucson, Ariz.* (SEC. 12)
JOHN E. FITZGIBBONS *Chemical Engineer, Standardization Division, U.S. Naval Construction Battalion, Davisville, R.I.* (SEC. 14)
PHILIP M. GRENNAN *Consulting Engineer, Rockville Center, N.Y.* (SEC. 9)
JOHANNA C. GUDAS *Editor-Publisher, American Roofer and Building Improvement Contractor, Oak Park, Ill.* (SEC. 15)
TYLER G. HICKS *International Engineering Associates, New York, N.Y.* (SEC. 21)
E. S. HOFFMAN *Chief Structural Engineer, Engineers Collaborative, Chicago, Ill.* (SEC. 5).
RICHARD A. HUDNUT *Product Standards Coordinator, Builders Hardware Manufacturers Association, New York, N.Y.* (SEC. 16)
HUGH B. KIRKMAN *Consulting Engineer, Charleston, S.C.* (SEC. 20)
E. D. LOWELL *Chief Estimator, Kaiser Engineers, Oakland Calif.* (SEC. 25)
FREDERICK S. MERRITT *Consulting Engineer, Syosset, N.Y.* (SECS. 3, 10, 13, and 23)
RAYMOND V. MILLER *Director of Construction, Rider College, Lawrenceville, N.J.* (SEC. 16).
T. H. QUINLAN *Senior Associate, Seelye, Stevenson, Value & Knecht, New York, N.Y.* (SEC. 18)
MAURICE J. RHUDE *President, Sentinel Structures, Peshtigo, Wisc.* (SEC. 8)
PAUL F. RICE *Technical Director, Concrete Reinforcing Steel Institute, Chicago, Ill.* (SEC. 5)

E. B. J. ROOS *Hamlin, Pa.* (SEC. 18)
HENRY J. STETINA *Consulting Engineer, Jenkintown, Pa.* (SEC. 6)
RALPH TOROP *Chief Engineer, Forman Air Conditioning Company, New York, N.Y.* (SEC. 19)
CHARLES J. WURMFELD *Consulting Engineer, New York, N.Y.* (SEC. 22)
LYLE F. YERGES *Consulting Engineer, Downers Grove, Ill.* (SEC. 17)
ALAN H. YORKDALE *Director, Engineering and Research, Brick Institute of America, McLean, Va.* (SEC. 11)

Preface

The Third Edition of the *Building Construction Handbook* is virtually a new book. It contains several new sections and almost complete revisions of previous sections to include important new developments. Many of the revisions were necessary because of drastic changes in the specifications for design and construction with commonly used construction materials.

In preparing this edition, we maintained the same objectives as for the previous editions: We set out to provide in a single volume information that would be of greatest usefulness to everyone concerned with building design and construction, and especially to those who have to make decisions affecting building materials and construction methods. We wanted it to meet the needs of owners, architects, consulting engineers of various specialties, plant engineers, builders, general contractors, subcontractors of various trades, material and equipment suppliers, manufacturers, financiers, building inspectors, construction labor, and many others, all of whom may have different problems and different interests in building design and construction. As before, we were faced with an extremely difficult problem in selecting subject matter since there was such a wealth of material available that each section could readily be expanded into a thick handbook.

We decided to adopt the same solution that gained ready acceptance for the previous editions:

- The handbook is comprehensive, but each topic is treated as briefly as clarity permits.
- Information incorporated is of a nature that should be valuable in making decisions—characteristics of building materials and installed equipment, essentials of stress analysis and structural design, recommended construction practices and why they are used, cost estimating, and construction management.
- Frequent reference is made to other sources where additional authoritative, detailed information can be obtained.
- Each section is written for the nonspecialist in the field, on the theory that the specialist prefers to seek answers to his problems in a more detailed text dealing exclusively with his own field. Emphasis is placed on fundamentals rather than on tables of design data.

• Tables of design data, building codes, standard specifications, and similar material that may be obtained easily from trade associations, technical societies, and government agencies are referred to but not reprinted in this book.

• The practical approach is stressed throughout. Methods are presented that are as simple and short as possible.

Some of the new sections in the third edition were added to present new developments in design and construction. Other new sections resulted from contributions of new authors, who brought a fresh viewpoint, new ideas, and additional information to the treatment of their subjects. In addition, the other authors took advantage of the opportunity opened up by publication of a new edition to update and improve almost all parts of the book and to add material on other new developments. To make space for the new material without allowing the handbook to grow into a cumbersome size, obsolete and less important information in the earlier editions has been deleted. As a result, the Third Edition is almost entirely new.

Some of the new sections resemble the older ones, at least in title. Section 1, Business, Art, and Profession of Architecture, like the former Section 1, deals with the responsibilities of architects. But the new section puts more emphasis on how to obtain new design contracts and execute them profitably. Completely rewritten Sections 5 through 8 reflect the changes that have occurred in specifications and construction practices in recent years. Section 17, Acoustics, emphasizes basic principles of sound control and presents information on sound measurement. Section 25, Estimating Building Construction Costs, presents the fresh viewpoint and efficient methods of a new author. Similarly, Section 26, Construction Management, introduces the new ideas and practices of a new contributor.

Other sections are completely new to the handbook. Section 11, Engineered Brick Masonry, presents rational design of brick walls, including their use as shear walls. This presentation supplements the treatment of masonry walls in the revised Section 10, Walls, Partitions, and Doors, which deals with design and construction of nonengineered masonry walls. Section 20, Fire Protection, offers practical methods for prevention of fire damage and for insuring life safety if fires occur.

Many significant changes have been made in the sections that have been retained. Most noteworthy are the addition of structural dynamics and an introduction to the finite-element method to Section 3, Stresses in Structures; the slurry trench method for constructing foundations and prestressed anchorages for sheeting to Section 4, Soil Mechanics and Foundations; and new glass materials and new methods of installation to Section 9, Windows.

The authors and the editor hope that you will find this edition even more useful than the previous one.

Frederick S. Merritt

Building Construction Handbook

Section 1

Business, Art, and Profession of Architecture

WILLIAM E. DIAMOND II, AIA

Associate, Smith, Hinchman & Grylls Associates, Inc., Detroit, Mich.

The practice of architecture is a business, an art, and a profession. An architect, however, is primarily a professional. He is a qualified expert, licensed to practice. His functions are discharged primarily for the welfare of other than his own interests. His profession, when it reaches its highest potential, is not only artful in a figurative sense, but is a literally and historically accepted art form. Functioning within an economic frame of reference, he is inevitably required to exercise good business sense both in the delivery of his service and in his own behalf.

1-1. Business of the Architect. An architect is a businessman as well as a professional. His business, large or small, has many of the requirements of any business. To perform, it must be profitable.

When a business happens to be the profession of architecture, the profit requirement takes on special significance. Architects must obtain and manage the resources required to perform services well. Without good business sense and proper business practices, an architect may find himself restricted in his ability to provide the best services to his clients. An architect who settles for too little fee, who does not keep his office costs in order, or who does not manage his own processes well is constrained in his ability to be a good professional. It is a wise client who understands and incorporates similar concerns in selection of an architect and fee negotiation.

Architects have a responsibility to employees to provide the most favorable working conditions and compensation the marketplace will permit. If they fail to obtain or manage fees properly, they will not be able to discharge that responsibility.

Obtaining Commissions and Making Agreements. Architects obtain commissions through reputation, personal contacts, and respect generated through their practices and community and personal activities. The agreements reached in securing work are always centered around the architectural services to be provided and the fees required to cover those services.

The major concern of an architect once he has been selected is to arrive at an accurate assessment of the scope of the work to be performed. The nature of the project, the degree of professional involvement, and the skills required must be considered in arriving at an equitable fee arrangement. Types of fees that may be used are:

Percentage of construction cost of project.
Fee plus reimbursement of expenses.
Multiple of direct personnel expense.
Stipulated or lump sum.
Salary.

For a typical project requiring standard services, the *percentage-of-construction-cost fee* is a safe standard. Years of experience with the relationship between the scope of architectural services required for various sizes of standard construction contracts provide a basis for such rule-of-thumb fee agreements.

For projects where atypical work or services are required, other arrangements are more suitable. For example, for projects where the scope of work is indefinite a *cost-plus fee* is often best. It permits work to proceed on an as-authorized basis, without undue gambling for either party to the contract. Under such an agreement, the architect is reimbursed for his costs, and also receives an agreed-on fee for each unit of effort he expends on the project. Special studies, consultations, investigations, and unusual design efforts are often performed under such an arrangement.

For atypical projects where the scope can be clearly defined, a *lump-sum fee* is often appropriate. In all cases, however, if the architect is a good businessman, he will know his own costs and will be able to project accurately the scope of work required to accomplish fixed tasks. He should take care, for the protection of his client, his staff, and his own interests, that he does not settle for fees that do not cover his own costs adequately. Otherwise, the client's interests will suffer, and his own financial stability will be undermined.

Fee agreements should be accompanied by a well-defined understanding in the form of a written agreement between architect and owner.

Architect As a Manager. The architect chosen for a project manages several aspects of the project simultaneously. He manages his own internal resources, he manages the processes that deliver service to the owner, and through that service he manages the programs of owner needs through the facility process to the creation of a built environment. The requirement that the architect be a capable businessman is, therefore, far reaching. Furthermore, changes in the construction industry place increasing demands on this aspect of architecture.

The need for good business sense, and a thorough knowledge of the architect's costs is reinforced by the need to manage his fees throughout the project. Allocation, commitment, and monitoring of the expenditure of resources are of critical importance to the success or failure of every project. Only when these are properly done can the correct services, the right advice, the appropriate design, and the best documents be delivered to clients.

Once the resources required to deliver services are insured, the architect must provide management skills to see that these services are timely, well coordinated, accurate, and well related to the client's needs. This is especially important on large projects, in large offices, or when dealing with the architect's employees and consultants. The best talent must be secured, appropriately organized, directed, and coordinated to insure that the project receives well-integrated and well-directed professional service.

Finally, the object of the entire endeavor is to produce the facility the owner needs, within his budget and on schedule. The architect's resources and the services provided are in many ways only a tool in managing the construction process for the benefit of the owner. The architect's management of materials and technology and his relationships with the owner and contractors will account in good measure for the success of the project.

Running an Architectural Practice. As a businessman, an architect is faced with acquiring a staff, advancing outstanding staff members, and removing unacceptable workers. He must keep records of his business expenses, file tax returns, provide employee benefits, distribute and account for profits, and keep accurate cost records for project planning and to satisfy government requirements. He must meet legal requirements for practice as an individual or as a corporation. In many of these areas, he is assisted by experts. It is impossible for an architect to practice effectively or successfully without a thorough understanding and complete concern for the business of architecture.

1-2. Art of the Architect. It is the art of the architect that is the most visible aspect of his practice; yet, it is the most difficult to discuss objectively. Historically, architecture has been regarded as one of the fine arts, and has been defined as the art of building well. No course in art history would be complete without a thorough discussion of the edifices of the past. But not all architects are artists, and few buildings can be termed architecture in the strict sense of the word.

While some architects are regarded as artistic, the general public is not likely to apply the term *artist* to the architect who lives down the street. This is in part due to a narrow attitude about art and a broad view of architects and their services.

In the Renaissance period of history, many great artists were equally skilled in various arts. Painting, sculpture, architecture, physics, and anatomy were such arts. Leonardo da Vinci is a well-known example of this generality. Since da Vinci's time, knowledge has broadened and specialization has taken over. The arts have been separated from the sciences: one considered objective, the other subjective.

Art and the artist have become more specialized as well. The fine artist is not part of the daily experience of the general public. His work is the subject of special shows and exhibits. Few homes or buildings are graced with original works of a true artist. Buildings on the other hand have become more technical, more scientific in their design and construction.

This discussion is not intended to imply that art has diminished in the concern of architects. They are no less concerned today about the esthetic, subjective aspects of the environment they create than before. But their concerns have become broader. They no longer practice only for a wealthy few. They no longer design only churches, palaces, and monuments. They are no longer restricted by a very limited technology or relatively unrestricted by budget. Finally, the construction process is no longer solely the work of highly skilled independent artisans and craftsmen.

With all these changes, the art of the architect has also changed. Architects no longer are able to lavish human labor on endless surface areas. Social and economic changes have dictated that no one man's productive life may be totally consumed by the carving or setting of tile for a few square yards of wall. Buildings may not be constructed from basic materials worked with great skill. Instead, mass-produced materials and equipment are fitted together artfully to form the built environment. Facilities are constructed for all elements in the society, with an ever-widening and changing technology, and within stringent budgetary requirements.

Architecture is simpler, cleaner, yet more diverse, and in most ways just as artful as in the past. This does not mean that every building can be called architecture in the strictest sense, any more than in the past. Too many builders are building too much. It does mean, however, that a broader cross section of the built environment receives the services of architects, and thus some esthetic concern. Often, the results are less than perfect to the purist; yet it cannot be said that the art is gone from architecture. To the contrary, the best work of the best architects is technically superb and still a work of art in the traditional sense.

So architecture is still an art; an art, however, inseparable from technology. The divisions of art and science have not been and cannot be imposed on architecture.

1-3. Profession of the Architect. The term "professional" is broadly applied in regular usage. It is often used to imply a high degree of skill and capability in a given field of endeavor. It also may indicate that the person to whom the term is applied discharges his functions fairly and with great concern for the interest of the people and communities with whom he deals. Finally, professional in its narrowest usage differentiates a group of vocational pursuits that are other than business or labor.

Architects are professionals in the latter sense, through choice, education, experience, and license. When they discharge responsibilities well, they lay claim to the term in the fullest sense of the word. Their responsibilities to be professional are not only traditional, but also a legal requirement.

Licensed Architects. The most basic professional credential of an architect is his license in each state in which he practices. Licensing is intended to insure that adequate skills and knowledge are applied to protect the health, safety, and welfare of those affected by the practice of the profession, and to make certain that anyone who represents himself as an architect may be relied on to observe basic requirements for design and construction of buildings. Licensing insures that an owner seeking services, that the public occupying a building, and that architects in practice are all fairly protected from the unqualified, the incompetent, and the unethical.

Obtaining a license to practice places special obligations on architects, to the public, to clients, and to the architectural profession. Consequently, a candidate, to receive his license, must meet educational and experience requirements. He must be recommended by other architects under whose supervision he has worked. He must pass a series of examinations that test his abilities in functional design and basic engineering, his knowledge of construction and professional practice, and his understanding of the requirements and conduct of the profession.

A typical architect receives 5 to 6 years of college-level education, and obtains either a professional degree at the Bachelor's level or a Master's degree. This is typically combined with 3 to 4 years of office experience before he is eligible to take license examinations. After this period, it is usual for the architect to serve an additional period of internship in a firm other than his own before he feels worthy of offering his services directly to the public.

Professional Ethics. The American Institute of Architects has formulated the following basic principles for guidance of architects:

Advice and counsel constitute the service of the profession. Given in verbal, written, or graphic form, they are normally rendered in order that buildings with their equipment and the areas about them, in addition to being well suited to their purposes, well planned for health, safety, and efficient operation and economical maintenance, and soundly constructed of materials and by methods most appropriate and economical for their particular uses, shall have a beauty and distinction that lift them above the commonplace. It is the purpose of the profession of architecture to render such services from the beginning to the completion of a project.

The fulfillment of that purpose is advanced every time the architect renders the highest quality of service he is capable of giving. Particularly, his drawings, specifications, and other documents should be complete, definite, and clear concerning his intentions, the scope of the contractor's work, the materials and methods of construction to be employed, and the conditions under which the construction is to be completed and the work paid for. The relation of the architect to his client depends on good faith. The architect should explain the exact nature and extent of his services and the conditional character of estimates made before final drawings and specifications are complete.

The contractor depends on the architect to guard his interests as well as those of the client. The architect should condemn workmanship and materials that are not in conformity with the contract documents, but it is also his duty to give every reasonable aid toward a complete understanding of those documents so that mistakes may be avoided. He should not call on a contractor to compensate for oversights and errors in the contract documents without additional consideration. An exchange of information between architects and those who supply and handle building materials is commended and encouraged.

The architect, in his investments and in his business relations outside his profession, must be free from financial or personal interests that tend to weaken or discredit his standing as an unprejudiced and honest adviser, free to act in his client's best interests. The use of free engineering services offered by manufacturers, jobbers of building materials, appliances, and equipment, or contractors may be accompanied by an obligation that can become detrimental to the best interest of the owner.

The architect may offer his services to anyone on the generally accepted basis of commission, salary, or fee, as architect, consultant, adviser, or assistant, provided he rigidly maintains his professional integrity, disinterestedness, and freedom to act.

He should refrain from associating himself with, or allowing the use of his name by, any enterprise of questionable character.

Architects should work together through their professional organizations to promote the welfare of the physical environment, and should share in the interchange of technical information and experience.

The architect should seek opportunities to be of service in civic affairs, and, to the best of his ability, advance the safety, health, and well-being of the community in which he resides by promoting appreciation of good design, the value of good construction, and the proper placement of facilities and the adequate development of the areas about them.

The architect should take action to advance the interests of his employees, providing suitable working conditions for them, requiring them to render competent and efficient services, and paying them adequate and just compensation therefor. The architect should encourage and sponsor those who are entering the profession, assisting them to a full understanding of the functions, duties, and responsibilities of the architectural profession.

Every architect should contribute toward justice, courtesy, and sincerity in his profession. In the conduct of his practice, he must maintain a wholly professional attitude toward those he serves, toward those who assist him in his practice, toward his fellow architects, and toward the members of other professions. His daily performance should command respect to the extent that the profession will benefit from the example he sets both to other professionals and to the public in general.

Standards of Behavior (Established by the American Institute of Architects and mandatory for its membership, AIA Document No. 330). An architect is remunerated for his services solely by his professional commission, salary, or fee, and is debarred from any other source of compensation in connection with the works and duties entrusted to him.

An architect may propose to a possible client the service he is able to perform, but shall not submit free sketches except to an established client.

An architect shall not falsely or maliciously injure the professional reputation, prospects, or business of a fellow architect. He shall not attempt to supplant another architect after definite steps have been taken by a client toward his employment, nor shall he undertake a commission for which another has been previously employed until he has determined that the original employment has been definitely terminated.

An architect who has been engaged or retained as professional adviser in a competition, cannot, if the competition is abandoned, be employed as architect for this project.

An architect shall avoid exaggerated, misleading, or paid publicity. He shall not take part, nor give assistance, in obtaining advertisements or other support toward meeting the expense of any publication illustrating his works, nor shall he permit others to solicit such advertising or other support in his name.

An architect shall represent truthfully and clearly to his prospective client or employer his qualifications and capabilities to perform services.

An architect shall above all serve and promote the public interest in the effort to improve human environment, and he shall act in a manner to bring honor and dignity to the profession of architecture. He shall conform to the registration laws governing the practice of architecture in any jurisdiction in which he practices.

BASIC AND TRADITIONAL SERVICES

A typical project begins at the moment a need for facilities is identified by a potential owner. This is true whether the facility will be a modest home or a massive multimillion-dollar complex.

Among his first acts, the owner selects an architect. Establishment of a basic program or statement of requirements, along with budgets and time schedules, follow close behind. With these in hand, the intense work of the architect begins. He designs a building, estimates its cost, and reviews his work with the owner. On approval of the design, the architect incorporates his design into contract documents to be bid on by contractors. He presides over bidding, assists in selection of a bidder, and aids in establishing contracts between owner and contractor. Finally, the architect oversees construction, interprets contract documents when necessary, and assists the contractor in his dealings with the owner. At the conclusion of the work, the architect tries to insure that all contracts have been completed, and in accordance with plans and specifications, and that all contractual obligations have been satisfied by both parties to the contract.

1-4. Selection of Architect. Proper selection of an architect for a project is critical to the project, and should not be mishandled or performed on the wrong basis. The services of an architect are expertise, advice, and counsel; therefore, the owner's selection process should be directed at determining the following factors:

Professionalism. The ability of the architect to perform his duties fairly and in a professional manner is of major importance. His reputation and actions during the selection process are particularly important indicators. Those who simply appear concerned about obtaining the commission without arriving at a full understanding of the owner's needs and demonstrating the ability to perform the necessary services should be avoided. The standing in the community and reputation among previous clients of the architect to be selected are important.

Overall Abilities. While licensing laws establish reliable minimum standards of competence, ability varies. Professional ability is the only competition for work that is to the benefit of the owner. In this area, previous work is the guide (unless a design competition is held).

Background and Experience. Of importance in selecting the appropriate professional is any peculiarity of the commission. Some architects may have special talents or greater experience in certain areas. In some cases, knowledge or understanding of local or special conditions may be critical to the project.

One of the most common mistakes made by owners is solicitation of proposed fees as a basis of selection. Most reputable architects, believing the procedure to be unprofessional and counter to the best interest of the client, will not compete on the basis of fees. Because the direct costs of most architects are the same, differences in proposed fees reflect varying degrees or qualities of service. Since the value of a finished project is determined more by the services the architect provides than by any other single factor, it is critical to the owner that the degree or quality of service be based on a clear and detailed understanding of the commission. A wise owner selects an architect on the basis of his abilities. Negotiations for fees should be undertaken after the architect has been selected and after detailed discussions with the architect. In consideration of total project value, this is the only sensible procedure and may be relied on to produce the most efficient services.

In the final analysis, the value of architectural services to an owner depends on the effectiveness of the architect. If the right services are obtained, the benefit can be expected to be maximized. These services are the direct result of the owner's selection processes.

1-5. Building Program. The program, representing the results of early basic, critical decisions, is a statement of requirements for the facility to be built and is associated with a budget. The architect for the project often is called on to assist the owner in defining his needs in physical terms. (This assistance is not normally included in typical fee curves used as a guide.) Functions are translated into length, width, and height, and become rooms. Related functions become areas of buildings. These are not designed solutions, but requirements for design to satisfy.

For complex buildings, experts or consultants in facility planning may be engaged. Hospitals, schools, and industrial installations are only a few of such complex cases. For these, the requirements of a complex set of physical and functional needs must be reconciled into clear requirements for a building design.

In many cases, a basic view of the kind of construction is developed during programming, simple or complex, unusual or typical. Areas of special concern are often identified for early work. For example, unusual environmental requirements may dictate an intense effort in the area of mechanical equipment.

During this early stage, the architect is called on to determine which requirements beyond the direct needs of the client must be satisfied. Codes and zoning requirements are the most common such issues. The architect must early identify restrictions and requirements that may affect the design and construction of the building.

1-6. Building Codes. The impact of building codes runs throughout a project. But early attention to these regulations inevitably simplifies the work.

Municipal building codes are ordinances essentially for provision of minimum construction requirements to protect public health, safety, and welfare.

While there is growing acceptance of model and national standards in building codes, various local agencies and officials modify and interpret requirements in various ways. Early review not only of the local code, but also discussions with local officials, are advisable. These officials typically attempt to apply the code in a locally significant but systematic way. Their concerns and attitudes, if understood early, can be a progressive addition to the program requirements. If, on the other hand, design is accomplished without a code review, the codes can become an impediment to the project.

In order that the art and science of building be advanced, most building codes and code officials permit individual initiative to improve design, materials, assembly, and equipment by means that comply with basic principles and standards. In an age of rapidly advancing technology, it is most important that there be legal provision for performance-type codes, properly safeguarded approvals or revisions by boards of standards, separate boards of appeals, and adoption of current versions of nationally recognized standards by reference—all without the necessity of tedious and costly legislative procedures at frequent intervals.

Four model codes have been widely adopted nationally or regionally. They are under constant study and periodic revision, by or with the advice of experts. The codes and the organizations that sponsor them are:

National Code—American Insurance Association.

Uniform Building Code—International Conference of Building Officials (ICBO).

Southern Standard Building Code—Southern Building Code Conference.

Basic Building Code—Building Officials and Code Administrators International, Inc. (BOCA).

The American Insurance Association, ICBO, and BOCA also publish abbreviated codes suitable for use in small communities.

The basic advantages of model codes are that their use materially reduces the cost to any municipality of writing its own code, and their use promotes nationwide uniformity in building regulations, based on sound standards developed by experts. Nevertheless, communities rarely adopt model codes without some modifications. While architects must understand local modifications and applications, a modified model code is much easier to deal with than a completely different document in each locality.

Electrical and plumbing installations and other building service equipment are subject to regulation for which model codes are also available:

National Electrical Code—National Fire Protection Association.

National Plumbing Code—Coordinating Committee for National Plumbing Code, jointly supported by U.S. Department of Commerce and Housing and Home Finance Agency.

See also Art. 1-8.

1-7. Zoning. Although zoning regulates building construction in a positive but general way, it is generally regarded as a control separate from building codes. Zoning, though, like building codes, is based on the need for general protection of public health, safety, and welfare. But zoning is primarily a regulation of land use, in terms of types of occupancy, building height, and density of population and activity.

Zoning is essentially a legal tool and administrative method of putting into current effect certain features of a comprehensive or master plan. Many of these features are conveniently represented on a zoning map, approved by the local legislature. (A future land-use plan, which is the basic and most important single part of a master plan, must not be confused with a zoning map, which is the current legal and immediately practicable version.) The zoning map is subject to change in the course of years, in the direction of the future land-use plan.

See also Art. 1-8.

1-8. Occupational Safety and Health Laws. Under federal and state laws, minimum standards of health and safety have been established for practically all occupations. The requirements are imposed mainly on employers and, in that they affect physical conditions, these regulations are an important consideration in the design of buildings where workers are employed. Such standards include lighting requirements, toilet-fixture counts, ventilation, and noise levels, to mention only a few.

Zoning ordinances and building codes are principally enforced by two methods, granting of building permits and inspection of construction work in progress, leading to issuance of certificates of occupancy when completed buildings satisfy all regulations. These allow examination of building plans and work under way by local officials charged with administration of these regulations. Conformance is assured through review, discussion, agreement, inspection, and permit granting. With federal regulations, however, the procedure is different. While advice is available, official review of plans does not occur. Enforcement is often undertaken only after worker complaint. An inspection is then made. If the complaint is upheld, the employer may be liable to a fine without sufficient time to correct the deficiency. While inspectors may be reasonable, the procedure places the employer, who is often the building owner, and the architect in a difficult position. Because official review and consent are not possible in assessing the specific application of the various regulations, and because of the legal liability inherent in a violation of federal occupational and safety laws, the architect and the owner must exercise great care to comply with those regulations.

1-9. Design Stage. Once the basic requirements of a project have been established, including the owner's requirements, budget, and schedules, as well as zoning, building codes, and other environmental patterns, design of a building can begin. The central objective of the design process is to arrive at the proposed structure that will most appropriately meet the owner's needs for physical space. The requirements may be unusual in some respects, or relatively straightforward. In all cases, schedules and costs are major considerations.

The design process is a complex interaction of the requirements of the owner; the experience, knowledge, and ability of the architect; and the site and surrounding environment of the proposed structure. The process is one of making tentative decisions in response to parts of the problem, and testing for validity in comparison with solutions to other parts and other requirements (Table 1-1). The final solution must be a well-integrated whole, and not just an assembly of individual answers to various portions of the problem.

Table 1-1. Summary of Design Phase Activities

Conference with client to outline project, its general purposes, general plan, feasibility, location, general type of construction and equipment; to discuss the probable time required to build, and to estimate approximate costs
Development of the background of the project as a basis and format for contract documents
Visits to the site and siting studies
Diagrammatic and blocked-out studies of elements of the project (schematics)
Examination of laws, ordinances, codes, standards, rules, and regulations of governmental authorities
On approval of schematics by the client, preparation of preliminary drawings—plans, sections, elevations (design development)
Preparation of outline specifications, describing construction, materials, equipment, and basic mechanical and electrical systems
Preparation of preliminary estimates of the costs of the work
Detailing, preparation of contract documents, and final cost estimate

Owner review of the proposed solution from the earliest decisions to a complete schematic design is essential. The proposals of the architect must be examined to determine that the owner's needs have been properly understood and integrated, and that budgets and other requirements are being maintained.

Costs. The initial requirements of the owner must include some budget figure. The key word is *budget.* A budget is a value judgment or willingness to pay. An owner places a value on his new facility that he is willing to pay; this is his budget figure.

The architect prepares cost estimates. His early estimates are based on a rough approximation of the cost of a facility not yet designed but which can approximate the owner's requirements. The key word is *estimate.*

The comparison with budget is important. When problems regarding cost occur, it is generally the result of a situation in which the owner finds his willingness to pay, or budget, considerably lower than the architect's estimate of probable cost for meeting project requirements. On the assumption that the owner's requirements are clearly understood by the architect, the owner may face a critical set of decisions:

Budget the same funds for reduced or modified requirements?

Increase the budget?

Proceed with the objective that detail design efficiency will make the comparison more favorable?

Cancel the project?

Some combination of the first three actions is the usual response. It is important, however, for both the owner and the architect to understand clearly that the setting of requirements and establishing of budgets is the ultimate responsibility of the owner. Designing the facility and accurate cost estimating are the responsibility of the architect.

Most disputes arising between an architect and owner usually flow out of a failure to respect these responsibilities by one or both parties. The architect must design the building to meet owner requirements and not to satisfy his own desires. He has a responsibility to keep the owner informed as to the probable cost implications of each change in requirements, and at each stage in the design process. The owner, on the other hand, should understand that his requirements must be clearly communicated and that changes in the project will affect costs.

The architect, though responsible for making accurate estimates, cannot guarantee his estimates. While he can anticipate, he cannot control the cost of materials, equipment, labor, or other conditions in the construction market that affect costs. He can be expected, to be responsible for designing efficiently and within the budget. He can also be expected to bring considerable design and management leverage to bear on costs.

Schematic Designs. The schematic design stage should be completed with site plans and floor plans worked out, basic building massing determined, facades laid out in a general way, and basic materials and the physical systems (structural, mechanical, electrical) selected. A formal submission to the owner is normally made at completion of the schematic phase, complete with a review of estimated construction costs and items that will be incorporated in the contract documents peculiar to the owner or the project.

Design Development. On approval by the owner of the schematic design, the project moves into the design development phase. The emphasis moves from concern for over-all relationships, functions, and massing to more technical concerns for the constructability of the building. While technology is never beyond the consideration of the architect, it is during this stage that he concentrates on the impact of technology on the over-all solution. The architect's esthetic concerns are not reduced during this period; rather, they are shifted from function, mass, and space to surface and detail.

It is possible, during the design development stage, that changes will occur in the basic design. For example, the design development of technology, materials, or building systems may lead to a problem or to an opportunity previously unforeseen in earlier steps. This is, however, an exception rather than a general rule. The main purpose of design development is to bring the facility into clearer focus and to a higher level of resolution.

The basic structural, mechanical, and electrical systems will be generally engineered and laid out, typical exterior walls will be designed, typical construction details worked out, and the junctions of typical materials considered. While the facility will not yet be completely detailed, much of the uncertainty will be resolved.

Outline Specifications. Drawings describing the facility in some detail are prepared, and outline specifications established. These specifications are not written for contract purposes, but to record for review the basic decisions that will be incorporated in the contract specifications. The outline specifications tell the owner what materials and special conditions of construction and special requirements of the contract are proposed for the work.

At this point, the architect prepares another formal cost estimate. He is in a position to refine his former estimate to a great degree, and can expect that it will have a much higher degree of accuracy than that for schematics. (See also Art. 1-10.)

1-10. Contract Documents. With a fully approved design (Art. 1-9) and the authorization of the owner, the architect next prepares contract documents, including working drawings. The main elements of contract documents are:

Drawings
Specifications
General conditions
Supplementary conditions
Owner-contractor agreement

(See also Arts. 26-4 and 26-15 and Sec. 27.)

While drawings are prepared to permit a contractor to construct the building, the phase during which these drawings are made has a greater significance. The architect, at this phase, is preparing documents as a basis for a legal contract between the owner and one or more contractors. These documents center around the drawings, and the task of preparing the drawings consumes most of the architect's resources for the phase.

But there are other critical parts of the total effort. Detailed specifications will also be prepared. In addition, the basic legal conditions (general and supplementary) will be determined. Aside from defining the parties to the contract, and their relationships, these documents describe such concerns as the rights, duties, and responsibilities of all parties during the construction period, insurance requirements, methods of payments, procedures for settling disputes, and ways of dissolving the contract. Additionally, requirements peculiar to the owner or the facility are spelled out. Provisions for such items as site utilities, temporary structures, and temporary heat for winter construction may be detailed in various locations in the documents. (See also Art. 26-5.)

Simultaneously with creation of the legal documents, fine detailing of the building proceeds and the last level of design decision making occurs. Every condition of material, surface, joint, and construction must be considered and defined so as to be clear to the contractor. The drawings and specifications must be complete in defining the building.

The contract documents are the basis for bidding, construction contract, and construction. The architect's responsibilities to the owner for clarity and accuracy are thus extended to cover the contractor, who will work from the documents. These documents must be clear, correct in every detail, and without conflicting information or requirements.

Contractors who bid, contract, and work from the architect's documents rely on his abilities. Any mistakes the architect may make can prove costly to a contractor as well as to the owner.

Final Cost Estimate. On completion of the contract documents, the architect prepares his final and most accurate estimate of construction cost. It is intended to confirm previous estimates, incorporate more detailed knowledge of the building, reflect owner-requested changes, and assess the construction market at the time of bidding. (See also Sec. 25 and Art. 26-5.)

Bidding and Contract Award. With authorization of the owner, the project is put on the construction market. In some cases, a single contractor may be asked to negotiate a contract directly. This is particularly appropriate for small work or where unique abilities or knowledge peculiar to a specific contractor are important.

For private work, a selected list of prequalified bidders may be invited to bid competitively for the work. The list is usually prepared with contractor reputation, financial stability, and construction capabilities in mind.

For public work, bidding is open to all licensed contractors. In this case, bids are requested by advertisement, and plans and specifications are made generally available to all bidders.

Under all conditions, bidding must be handled fairly and equitably. All contractors must be treated even-handedly, and given the same information at about the same time. Bids must be received simultaneously, and must be in precisely the same form for fair comparison. The low bidder may not always be the successful bidder, but the criteria of selection must be uniformly and fairly applied without partiality. (See also Arts. 26-6, 27-4, and 27-11.)

The preceding activities are summarized in Table 1-2.

Table 1-2. Summary of Contract Documents, Bid, and Award

(Prepared as the basis for taking bids and establishing a contract between owner and contractor)

Working drawings:
- Architectural drawings
- Structural drawings
- Mechanical drawings
- Electrical drawings
- Site development and other special engineering drawings

Specifications:
- Type and quality of materials
- Desired finishes
- Manner of construction

General conditions:
- Responsibilities of parties to the contract
- Insurance requirements, bonds, methods of payment, and other conditions supplementary to drawings and specifications

Approvals:
- Assistance of the architect in obtaining approval of governmental agencies, when required

Bidding:
- Architect furnishes stipulated number of copies of plans and specifications to the client (and to contractors for bidding, if directed to do so)
- Architect may issue "addenda" to change, modify, or clarify the contract documents to be bid
- Architect may provide assistance in preparation and letting of contracts for construction
- Prequalification of bidders and advertisement of bids
- Proposal form for bid submission, to insure fair competition
- *For public works,* "notice to contractors," usually advertised in an official publication, containing:
  - Brief description of building
  - Where proposals will be received
  - Date and time for receipt of proposals
  - Where contract documents may be examined and obtained
  - Deposit required for securing copies of drawings and specifications
- "Instructions to bidders" with requirements regarding submission of bids
- *For all public and some private work,* a certified check or bidder's bond is required, to insure that the successful bidder will sign a contract for the bid price
- When the architect presides over the bid opening, he must insure simultaneous receipt of bids and proper tabulation for impartial comparison

Award of contract:
- On review of all bids, the architect is generally responsible for the recommendation of award and preparation of the contract between client and successful bidder

1-11. Architect's Role during Construction. Construction of a building results from fulfillment of a contract (or contracts) between an owner and one or more contractors. The architect's responsibility is to insure fair performance of this contract by both parties. His prime functions in this regard are interpretation of drawings and specifications and establishment and enforcement of standards of acceptability for the work.

Administration. Traditional duties of the architect during construction include administration of the contract as the agent of the owner. The architect seeks to insure that the contractor fulfills the terms of the contract documents. The architect, however, also has a serious responsibility to the contractor, which requires interpretation of contract documents fairly and without prejudice to the contractor and assisting the contractor in obtaining payment.

The architect does not traditionally become involved in the conduct of the work or determine the methods of construction. He does, however, have a general responsibility to see that the construction site is maintained in a safe condition and that all terms of the contract are observed. Among his diverse concerns, he has a responsibility to the owner to see that the contractor discharges his financial obligations to subcontractors and suppliers during execution of the contract, so that no financial irregularities complicate the project.

At the outset of construction, the architect has many duties. Among them, he advises the owner of his responsibilities in regard to property insurance, to protect the property of the contractor (including the work) from fire or "acts of God." The architect's dealings with the contractor usually center around approval of subcontractor lists, contractor's schedule of values, and proposed progress schedule to determine the value of in-place work as a basis of payment.

Construction Supervision. Once work is under way, the architect inspects construction, as necessary, to check progress of work, assure compliance with contract requirements, and guide the contractor in interpretation of the contract documents for full performance of his contract.

The architect's supervision should cover:

Checking of bench marks for building location.

Examination of soil for footings after excavation. (Tests may be necessary.)

Inspection of connections with public utilities.

Determination of need for additional subsoil drainage.

Verification of accuracy of placing base plates for steel columns.

Checking concrete forms for alignment and bracing.

Inspection of concrete reinforcing and electrical conduits, and of location of sleeves.

Check on quality of concrete.

Determination of adequacy of weather protection.

Check of wiring and piping for conformance with contract documents.

Check of partition layout and plastering grounds.

Check of finish materials for proper installation and compliance with contract documents.

Similar checks on other items required by plans and specifications.

Shop-drawing Checking. Throughout the construction period, there are important specific activities the architect may perform that keep the project running smoothly. To assure that the intent of design is carried out, and to assist the contractor in interpreting the documents, the architect checks shop drawings submitted by contractors. His checking must be performed with care and dispatch, to prevent upset of construction schedules. Where the contractor is unclear either in his shop drawings or in any other area, the architect may prepare full-size detail drawings or other supplementary drawings necessary to assure compliance with contract requirements.

Payments to Contractors. From a financial point of view also, the architect has serious responsibilities. He issues certificates to actuate owner payment to the contractor as prescribed in the agreement (usually on a monthly basis). In ascertaining the amount to be paid under any certificate, the architect should assure that:

Value of work done has been accurately ascertained.

Amount of the retained percentage has been deducted (usually 10%).

Total amount of change orders approved by the owner to date, and the correct total of previous payments are shown.

No reasons exist for withholding payment; that is, there is no defective work, no evidence indicating the probable filing of claims, no failure of the contractor to make payments to subcontractors or suppliers, no reasonable doubt that the contract can be completed for the balance unpaid, no notice of damage to another contractor.

On determination that the work has been completed, the architect issues a "certificate of payment" to the contractor, sending him the contractor's and owner's copies. The contractor sends the owner's copy to the owner for payment. If required, the contractor should certify that bills have been paid and that his payment has been received.

Valuation of work forming the basis of a certificate of payment should be regarded as one of the architect's more serious duties. Failure to correctly value the work will do some injury to the owner or the contractor. In past court cases, where a loss has fallen on an owner or contractor because of an inaccurate valuation, the architect has been held liable for the full amount of the loss.

Change Orders. During the contract period, reason may arise to make changes in the work. These are accomplished by issuing documents to the contractor describing the desired changes. Such changes may include owner deletion or addition of equipment, adjustments for unforeseen site conditions, or errors or omissions in the documents.

Once the proposed change has been reviewed by the contractor, he makes a proposal for the work. The architect checks the extra cost or credit, and issues bulletins and change orders, as necessary, for any modification of contract.

Issuance of orders for changes in the contract, however, must be authorized in writing by the owner. (See also Arts. 26-15 and 26-16.)

Project Completion Activities. As the project nears completion, the architect's attention is turned to wrapping up work and agreements in an orderly and well-coordinated manner. This is accomplished through final acceptance of the work and issuance of the final certificates. The normal procedure is as follows:

1. The contractor sends the architect a statement claiming substantial completion.
2. If the architect concurs, he sends a statement of substantial completion to the owner with copies to the contractor.
3. A semifinal inspection is made by the architect in the presence of the owner's project representative.
4. The contractor submits to the architect for approval all required guarantees, certificates of inspection, and bonds.
5. Lists of required corrections are made, one for each trade, with provision for indicating satisfactory correction by checking or punching.
6. Final inspection is made by the architect and the owner.
7. If everything is acceptable, a certificate of completion is sent to the contractor with copies to the owner.
8. All required operating instructions are submitted to the architect for approval.
9. Complete keying schedule with master keys, submaster keys, room keys, and any special keying are delivered to the architect for approval.
10. The contractor submits an affidavit that all bills have been paid, with any exceptions noted and covered by releases of liens.
11. The architect sends owner's and contractor's copies of final certificate of payment to the contractor.
12. The owner makes final payment and notifies the architect.
13. The architect sends the owner permit to occupy, with copies to the contractor.

Before issuing a final or semifinal certificate of payment, the architect should be sure that:

Degree of completion of work called for by the contract has been reached.

Any period of time required to elapse between the issuance of the semifinal and final certificate for payment has elapsed.

Owner's interests have been protected in respect to the filing of liens, usually by a release of liens and affidavit.

Accounts between owner and contractor have been adjusted, including original contract sum, additions, and deductions as included on change orders, cash allowances, deductions for uncorrected work, and deduction for liquidated damages.

Any written guarantees, such as those for roofing, have been deposited with the architect or owner, as well as certificate of occupancy and all certificates of inspection, such as for boilers or electrical work, or by public authorities.

1-12. Liability. In general, an architect, in discharging basic and traditional services for a project, is liable for losses that would not have occurred had he used *reasonable care and skill.* Legally, this means the degree of care and skill currently furnished by members of the profession generally, rather than by its most outstanding members.

This, however, does not mean that an architect can expect to get by with just average performance. His reputation rides on his performance, and reputation is the most important means of obtaining commissions. As a professional, every architect must act in the best interests of his client. It is not an accident that only through discharging this responsibility an architect insures his own success.

The construction process entails commitment of sizable resources by owners and contractors. The architect is instrumental in the proper commitment of these funds. His main function is to protect the owner. The architect does not, however, guarantee the work of the contractor.

In general, the contract drawings serve as a record of the instructions given by the architect. If losses result from the contractor's failure to follow these instructions, the liability is clearly the contractor's rather than the architect's. If, on the other hand, losses are due to faulty design, as indicated by the contract documents, the liability is clearly the architect's.

In the case of foundations, it is generally impracticable after completion of a building either to determine whether or not the contract documents have been followed or to check the soundness of the design. It is difficult to uncover the work, and because of lack of data regarding the character of the soil, checking the design would have little meaning. For these reasons, settlement of foundations is usually taken as prima facie evidence of neglect or lack of skill on the part of the architect. The architect, therefore, is generally held to be liable for damage resulting from a foundation failure.

Another definite liability of the architect is for any losses that result from errors in his certificates for payment to the contractor.

Any architect may delegate authority to his employees. But by doing so, he does not escape responsibility for their acts. As a professional and a businessman, he must be expected to choose competent employees and associates and to supervise their work closely.

1-13. Architect and Consultants. In the performance of any services, an architect may retain consultants to supplement his own skills, to deliver a more complete and well-coordinated service to his clients. Such consultants include structural, mechanical, and electrical engineers, and specialists in specific building types.

In all cases, the architect's legal responsibilities to the owner remain firm. He is fully responsible for the services he delivers. The consultants, in turn, are responsible to the architect. Following this principle, the architect is responsible to his clients for an error in structural engineering made by his structural consultant. The architect may claim damages in a separate action if he is sued, but the responsibility to his client is his alone. Consequently, it is wise for the architect and his clients to evaluate the expertise the architect may have in supervising others before retaining consultants in other areas of responsibility.

1-14. Professional Engineers. Integral with the design of any facility is a series of systems concerning which the architect has basic knowledge and some design skill. These systems are the structural, mechanical, electrical, and site-paving and drainage systems. On small, simple projects, the architect may choose to design the entire project himself. On larger projects, however, he usually relies on the design capabilities of specialists in these engineering areas.

In some architect offices, these skills are retained in much the same fashion that an owner retains an architect, through a professional services contract. In other, generally larger offices, the staff may include professionals in one or more of these disciplines.

By whatever means they are employed for a project, these specialists play an indispensable role in design and construction. They are expected to perform in just as professional a fashion as the architects with whom they collaborate.

The major distinctions between architects and engineers run along generalist and specialist lines. The generalists are ultimately responsible for the work. It is for this reason that an architect is generally employed by an owner. On some special projects, however, where one of the engineering specialties becomes the predominant feature (dams, power plants, sewage treatment, certain research or industrial installations), an owner may select an engineering professional to assume responsibility for design and construction. On other projects, it is the unique and imaginative contribution of the engineer that may make the most significant total impact. The over-all strength of a dynamic, exposed structure, the sophistication of complex lighting systems, or the quiet efficiency of a well-designed mechanical system may prove to be the major source of owner pride in his facility. In any circumstance, the responsibilities of the professional engineer for competence and contribution are just as great and important to the project as those of the architect.

1-15. Design by Team. For many reasons, building design and construction tend to become increasingly complex and sophisticated, and sizes of typical projects to become larger. As a result, design and construction processes tend to become too difficult to be controlled by a single individual and require instead the efforts of a closely integrated team of diverse professionals, each of whom is preeminent in his area of expertise and identifies with the total project.

Design by team places extraordinary demands on the team members. These members may be architects, engineers, construction specialists, financial experts, planners, statisticians, building-type consultants, owner representatives, or other experts. Three rules govern their performance:

1. The field of expertise of each team member must be distinct and clear. The member must be sufficiently demanding and expert in his field to command respect of his teammates, and he must be given complete and final professional authority in his field.

2. Team members must be concerned more about the total project than about their own field of expertise. They must be willing to contribute to all areas of the work. They must establish evaluation criteria for their own fields based on total-project requirements rather than just on the standards of their disciplines. They must be sensitive to the fields of the other professionals.

3. Team members must be able to identify conditions where project demands and standards of their disciplines coincide (an opportunity), and to reconcile the conflicts (a process of design optimization).

Team members generally find that they must tackle some design problems individually and others in concert with one or more teammates. The most difficult problems are often those that must be solved by a group. If team effort is successful and creative in developing solutions, a higher-quality project than the contribution of a single designer is produced. But if the team members develop solutions by compromise, the project is "designed by committee," with poor results likely.

Team design, therefore, has the potential of rising above the work of a single designer. It also has the potential to go the other way. The outcome depends on the quality of the professionals employed and the manner in which they are organized and operate. Leadership, in particular, can play an important role in developing the potential of the team.

ADDITIONAL SERVICES

Beyond traditional services, architects often are called on to extend themselves into areas where, because of basic knowledge, skill, and experience, they may provide unique professional services to clients. Such services may be delivered as part of basic services, or on a completely separate basis.

1-16. Nontraditional Services. Most architects may provide certain of these services. Site selection, program definition, and various kinds of feasibility studies are good examples. In other instances, particular experience, a specialized practice, or a unique association with other professionals may prepare a practitioner to provide services that could not normally be obtained from an architect. Hospital consultation (with a specialist in health facilities) and computer scheduling and cost control (from a computer-oriented architectural firm) are examples in this area.

The architect should be a generalist, problem solver, businessman, and manager. It is these characteristics, in combination with in-depth experience in some specific areas, that often lead to selection of an architect for performance of hybrid tasks where intra- or interdisciplinary tradeoffs are required to determine broad issues, make complex decisions, or produce over-all plans.

Architects have been called on to plan hospitals, manage large construction projects, inventory national resources, select sites, and plan universities. It is important to understand, however, that the training and licensing of an architect do not generally qualify all architects for such work. It is instead the unique combination of generalist background with specialized experience that produces such a consultant.

In whatever area an architect undertakes to provide service, whether through his own skills or through the skills of subcontractors, his responsibilities to perform competently as a professional and in a legal sense are undiminished.

MAJOR CONDITIONS AFFECTING ARCHITECTURAL PRACTICE

Changes set in motion with the industrial revolution continue to bring major shifts in construction practices. With these changes, the architect and his services also change.

1-17. Economic Factors. A building owner views his facilities in economic terms first and foremost. As real estate, facilities have investment value that must be related to cost, return, and other economic issues. The trend toward more owner sophistication in economic decision making regarding facilities will continue. This means that the clients architects serve will become increasingly cost conscious and will expect these concerns to be dealt with more expertly. For the architect, this is a challenge, both as a professional and a businessman, to develop greater skill in cost control through design and involvement with construction.

Costs of material and labor in the construction industry rise continuously. This means that new facilities are usually more expensive to construct than in the past, often not only from an absolute point of view but relative to everything else. For a large project of several years' duration, costs may rise substantially while construction proceeds. A contractor bidding a lump sum for a multimillion-dollar project with the expectation of providing materials and labor several years ahead and at costs that cannot be predicted must cover such risks. He cannot bid "tight." He must arrange to receive funds from the contract to cover such risks. The contract sum always includes these costs.

1-18. Changing Construction Market. As costs rise, skills employed in the construction industry have tended to decline. Skilled craftsmen capable of artistic interpretation have become scarce and, in effect, have been replaced by laborers who perform rote construction chores.

At the same time, the trend has been for large general contractors to become removed from direct involvement in current technology or in actual performance of construction work and to become more involved as managers of the process. Those involved in technology and its changes are suppliers, manufacturers, and subcontractors, who may have no direct relationship to the owner and the architect from a contractual point of view.

These changes in construction in a time of increasing economic pressure on owners, especially owners of large projects, place greater stress on architects to manage the costs and technology associated with construction but reduce leverage via traditional processes.

1-19. Systems. Development of systems methods based on computer technology has impacted the design and construction of facilities. Construction scheduling, cost-control techniques, accounting, information-retrieval systems, graphics, program modeling, and simulation—all computer-assisted—are often used in the facility production process. These techniques and the capabilities they make possible have an enormous impact on facility planning, design, and construction.

Building Systems. Architects currently seldom detail stone and wood trim for buildings. Instead, they select products that incorporate the details most suitable for their buildings. Manufacturers or producers design such items along with their products. Windows, doors, wall surfaces, toilet enclosures, plumbing accessories, masonry units are usually designed not by an architect for a project but by a manufacturer. Mass production, high labor costs, and development of manufacturing processes and materials heretofore not available have contributed to these changes in architectural practice.

Similarly, engineers can select more comprehensive products. Air-conditioning systems of increasingly larger sizes are becoming available with a greater variety of equipment and controls. Larger structural systems are available on a preengineered basis. Integrated ceiling systems, including lighting, air conditioning, ceiling surface, and ceiling structure, with the ability to adapt to wall systems, are produced by several manufacturers on a total-concept basis.

Also, contractors and suppliers in mechanical, electrical, and structural areas are becoming more specialized.

All these factors have contributed to development of a phenomenon known as *building systems.* A building is usually thought of as an assembly of systems, such as structural; exterior skin; heating, ventilating, and air conditioning; plumbing; power, signal and lighting; walls; and ceilings. In many instances, these physical systems conform to the area of one, or of a group of, contractors, building tradesmen, or suppliers who function in a discrete set of construction activities.

Building systems have always existed within facilities. For instance, structural systems and wall systems have occurred in all buildings. But these distinctions have become less academic and more real. The continuing increase in complexity of buildings; in specialization of design roles; and in specialization and sophistication of manufacturers, suppliers, and contractors has made the boundaries between the various systems within buildings more clearly defined. In most cases, the subcontracting of work to a general contractor and various specialist contractors has recognized and followed these lines. Consequently, many building systems can be obtained as a package or each one can be viewed as a clearly separate portion of the total building.

Systems Building. The existence of well-defined and generally accepted building systems has led to another level in the systems process. Contracts are sometimes awarded by an owner to a contractor for one or more of the building systems. This level of recognition of the existence of discrete parts of the total facility is indicative of a phenomenon called *systems building.*

Systems building, while relying on the definitions of building systems, is not "hardware" oriented. It is instead a management process that manipulates the design and construction of the facility through the design and installation of the various building systems.

It is possible, for example, to buy a structural system of known capacity and dimensions, a preengineered air-conditioning system, a completely designed exterior-wall system, and a ceiling and interior-partitioning system, and to combine them uniquely with other parts of the building designed in a more traditional way. By this method, it is possible to place on a unique foundation, as a total facility, a building that has not been designed or built before.

Systems building is applicable to a number of circumstances. Commercial, industrial, and governmental owners often have a continuing building program that requires repetitive erection of a single-facility prototype in various locations. Such owners are not interested in a unique design and construction effort for substantially the same facility each time a new site is acquired. The process consumes needless time. Furthermore, economic efficiency in design and construction cannot be achieved with such a traditional one-at-a-time process. Under the general heading of systems building, a broad range of options are open to such owners (Table 1-3).

Table 1-3. System Buildings Options for Facility Acquisition

Specific design for an individual building (traditional approach)
Specific systems designed for a group building program
Application or assembly of predesigned systems, or preconstructed or commercially available systems, buildings, or components

Design of a unique building or building part is not always necessary. Preengineered or premanufactured buildings or building parts are available. Schools, for example, often acquire additional classroom space by the purchase and integration of such structures or components into a larger complex of complete buildings.

Facilities may also be acquired via specific systems designed for a group building program with direct participation from systems suppliers. (Such a program would call for erection of a large number of basically similar buildings over a period of time.)

An extreme example of systems building might have facilities acquired through the lease and assembly of preconstructed buildings and building systems components. This approach could meet temporary and urgent needs. Furthermore, in a second phase, leased structures could be replaced in phased erection of either uniquely designed or system-designed permanent structures. The options and combinations associated with systems building are endless.

Thus, systems building generates choices other than one-at-a-time design for one-at-a-time, on-site construction. In the examples cited, a fairly broad definition of systems building and building systems has been utilized. The most significant efforts in this regard have been in more comprehensive product design, systems design for special building programs, and, in the factory, assembly of building components and totally integrated building parts. Housing, schools, and hotel design and construction have been in the forefront of systems building. (See also Art. 26-24.)

1-20. Multiple-contract Construction. Under traditional procedures, a general contract is awarded to a single contractor for the total project. He, in turn, assumes complete responsibility for the total work and is obligated to deliver the completed project for the contract price. To accomplish this, he establishes subcontracts between himself and other contractors and suppliers who provide most of the actual facility.

Under multiple-contract construction, however, prime contracts are established directly between an owner and the suppliers and subcontractors who might normally deal with a general contractor.

Prime contracts directly between the owner and the contractor or supplier provide greater flexibility in design of a facility. Design of some parts of a building can be combined with purchase and installation of predesigned and premanufactured items, and with products that are especially designed for a specific application by the manufacturer or supplier. Thus, the owner and architect can obtain the most competitive and appropriate preengineered systems, or develop and acquire unique products through direct contact with suppliers. Traditionally designed and constructed portions of the building can also be contracted for separately in one or more packages.

The interface between multiple contracts must be carefully defined and managed. One method of achieving this goal is assignment of the contracts to a general contractor once the unique design or most appropriate system has been obtained. Other methods are discussed in Art. 1-22.

The award of multiple contracts for a single facility may occur simultaneously. Reasons for simultaneous multiple contracts include advantages of dealing directly with the system supplier and financial leverage. Award of several contracts of moderate size instead of one large one has a number of beneficial effects:

Bidding is more diverse. A large project may be beyond the financial capability of many otherwise qualified contractors. With the total project reduced to several relatively small contracts, more contractors can bid competitively. Where the contracts are written to follow trade or system specialties, bids can be taken on each of the parts of the facility, and total project cost assembled from a combination of low bids. A general contract for the same project, in contrast, would provide a single low bid that would not necessarily be comprised of the lowest cumulative total of all subcontract prices.

With budgets, design, estimates, contract documents, and bids established for several parts of the projects, instead of one large lump sum, it is easier to make design tradeoffs throughout the entire process to control overall costs. If the structural contract is over the budget, it is feasible to reduce the cost of carpeting and drapes to compensate and thus maintain the over-all project budget. This can also be accomplished under a single release contract. But the complete contract structure imposed on an entire project, and more detailed bidding information, make tight cost control a direct and strong result of multiple contracts.

Multiple contracts are not always awarded simultaneously in the same manner as if the total project were covered by a single contract. When some staggering of contracts occurs, the object is usually to reduce the total time required to design and construct the project. (See also Arts. 1-21 and 26-6.)

1-21. Accelerated Design and Construction. The culmination of most of the procedures discussed in this article, accelerated design and construction rely on the traditional services of architects and require both systems concepts and multiple contracts. The traditional process of design and construction and the roles and responsibilities of the various parties need not be profoundly changed.

In the traditional process, the entire facility moves stage by stage through the entire process (programming, design, design development, contract documents, bid and award of contracts, construction, and acceptance of completed project).

With any form of accelerated design and delivery, these phases remain substantially the same. But the various building systems or subsystems move through them at different times and result in multiple contracts released at different times.

For any project, basic building siting is determined early in the design process. Therefore, at an early date in design, a contract can be let for rough site work.

Similarly, basic structural decisions can be made before all the details of the building are established, and permit early award of foundation and structural contracts. Under such circumstances, construction can be initiated very early in the design process, instead of at the end of a lengthy design and contract preparation period. Months and even years can be taken out of the traditional project schedule. When the systems building techniques discussed in Art. 1-19 are employed, purchase of preengineered, commercially available building systems can be integrated into the accelerated design and construction process, reducing time even more drastically.

The major requirements for a project in which design and construction occur simultaneously are threefold:

1. Accurate cost estimating to maintain project budgets.
2. Full understanding of the construction process so that design decisions and contract documents for each building system or subsystem are completed in a fashion that meets the requirements of an ongoing construction process.
3. Tight management of the construction process with feedback into the design process to maintain clear definition of the required contract packages. Over-all project cost control and over-all project construction responsibilities, including management of the interface of independent prime contracts, must also be undertaken by a construction manager.

Often, the major purpose of accelerated design and construction is to reduce the effect of rapidly increasing construction costs during the extended project design and construction period. For large projects extending several years, contractors and subcontractors quote costs for materials and labor that will be installed several years later. In most cases, the costs associated with such work are uncertain. Bidding for such work, especially when it is of large magnitude, must be very conservative. Accelerated design and construction, however, bring all the financial benefits of shortened project duration and reduce the effect of cost escalation. Also, bidding can be closer to the actual costs, thus reducing bidding risk to the contractor. The combination of early bidding, shortened contract duration, reduced escalation, smaller packages, and greater number of bidders can produce large savings in over-all construction costs.

A major objection to accelerated design and construction is that project construction is initiated prior to obtaining bids for the total project and prior to complete assurance that the total project budget can be maintained. In this regard, the reliability of estimating becomes even more critical. It is the experience of most owners and architects involved with multiple contracts, however, that such contracts, bid one at a time, can be readily compared with a total budget breakdown and thus provide safeguards against budget overruns. The ability to design, bid, and negotiate each contract as a separate entity provides optimum cost control.

For accelerated design and construction programs to work effectively, a new role is required. This is the role of the professional construction manager (Art. 1-22).

1-22. Construction Managers. While a construction manager is necessary on any multiple-contract project, he is most fully utilized in accelerated design and construction. He is a professional who does not contract for performing the actual work. Rather, he provides services during the design and construction process. He must provide the direction and strategy necessary for composing contract packages for award. He must schedule the construction processes and communicate the requirements that continuing construction places on the design process. He must manage the processes of construction in the field. He must provide the leadership required to solve field problems and coordinate the work of the several contractors.

Since all contractors contract directly with the owner, the construction manager has no direct financial interest in the work. He acts only as the agent of the owner in the same professional capacity as the architect.

The construction manager may be a qualified architect, engineer, or contractor. It is not safe to assume, however, that all architects, engineers, or contractors possess the thorough knowledge of design and construction and the management skills necessary to discharge this unique role successfully. No matter who plays the role, however, a close team relationship between design and construction management leaders is critical. The process is continuous and interacting. Its success depends on the correct flow of information, decisions, and documents.

Use of construction management is not restricted to accelerated design and construction programs. On any multiple-contract project, for example, the owner may elect to employ a construction manager or management agency to maintain coordination between the work of the several contractors and to assume responsibility for certain project-wide construction tasks such as site maintenance. Regardless of the type of project or method of awarding contracts, the construction manager must be expert in construction and in management techniques. He must be retained on a professional basis with his fees earned a result of his service to the owner and not a result of construction performed. (See also Arts. 26-1 and 26-6.)

CONSTRUCTION WITHOUT PROFESSIONAL SERVICES

There are numerous occasions on which an owner does not choose to employ an architect as his agent. Such instances include the small-builder house and employment of a design-build firm for more substantial facilities. When the owner employs a professional as his agent, he secures an expert in design and construction committed to acquiring the most suitable facility for the owner's needs and budget. When an owner acquires facilities without use of such services, he deals directly with those whose business it is to provide a built facility for profit. With the exception of some residential construction, an architect is involved in the process. His client, however, is the package builder and not the owner.

1-23. Package Builders. There are several kinds of package builders. Their main attraction is that they will, for a fixed and often guaranteed cost, deliver to the owner a designed and constructed facility in a single package. Often, such a builder will acquire property, arrange financing, and include the entire project under one contract. Such services vary in degree of control and participation that the owner is afforded. In some cases, the owner may be offered one or more stock prototypes. In other cases, the design of the facility may be specific for the building to be erected. (See also Arts. 1-24 to 1-26, 26-1, and 26-6.)

1-24. Home Builders. Residential builders typically offer several models with which their forces are familiar. The potential owner is offered a few choices at reasonable prices. He must shop for the model that best suits his budget and his needs.

There are local, capable, and sensitive builders who take pride in providing the best quality at a price the market will bear. Such builders may or may not employ an architect. In either case, they are to be sought out and valued by the owner seeking a modest builder home.

1-25. Design-and-Build Companies. Designer-builders offer to design and construct a facility for a fixed lump-sum price, as a total contract. They will bid competitively to provide this service or will provide free design services prior to owner commitment to the project and as a basis for negotiation. Their design work is not primarily aimed at cost-performance tradeoffs, but at reduced cost for acceptable quality.

The design-build approach to facilities is best employed when an owner requires relatively straightforward facilities and does not require the ability to participate in detail decision making regarding the various building systems and materials. This does not mean that he has no control over these items. On the contrary, he is often permitted a wide range of selection. But the range of choices is affected by the fixed-cost restraints imposed by the designer-builder and accepted by the owner. When the facilities required are within the range of relatively standard industry-wide prototypes, this restriction may have little significance.

A common misconception regarding design-build is that poor-quality work inevitably results. While there is a general benefit to the builder for reductions in material and labor costs, the more reputable designer-builder may be relied on to deliver a building well within industry standards.

Facilities where higher quality systems, more sensitive design needs, or atypical technical requirements occur deserve the services of an independent design professional. The designer-builder or package builders, however, may offer a desirable service for more straightforward projects. (See also Arts. 26-1 and 26-6.)

1-26. Turn-key Developers. The term *turn key* is often used interchangeably with design-build (Art. 1-25). The major distinction in meaning is that turn key implies more complete independence on the part of the turn-key builder.

In turn-key projects, the major input of the owner is a document outlining his requirements. Bids are usually taken wherein the turn-key developer submits design and cost proposals to meet these requirements. Once a developer is selected, the owner sells his property to the developer or authorizes its purchase from a third party under option. From this time on, the owner has little or no participation in the project. Instead, he guarantees that he will buy the project back from the developer, as a package, at completion.

As with design-build companies, the professional designer is on the developer team and is not an agent of the owner. Once over-all requirements are established, detail decisions are made by the turn-key developer. The owner deals only with the final product. This means he has less control than with design-build or other similar procedures.

Section 2

Building Materials

PROFESSOR ALBERT G. H. DIETZ
Massachusetts Institute of Technology

Part 1. MASONRY MATERIALS*

Masonry materials, in the widest sense of the term, include all the inorganic, nonmetallic materials of construction. These encompass a wide range of materials, from insulating blocks of lightweight concrete to slate tile roofs. Only the most important of these will be presented in this section, and attention will be focused as far as possible on those characteristics of the materials that affect their utilization, strength, durability, volume change, etc.

2-1. Cementitious Materials. Cementitious materials include the many and varied products that may be mixed with water or another liquid to form a paste that may or may not have aggregate added to it. The paste (mortar or concrete if aggregate is used) is temporarily plastic and may be molded or deformed. But later, it hardens or sets to a rigid mass.

There are many varieties of cements and numerous ways of classification. One of the simplest classifications is by the chemical constituent that is responsible for the setting or hardening of the cement. On this basis, the silicate and aluminate cements, wherein the setting agents are calcium silicates and aluminates, constitute the most important group of modern cements. Included in this group are the portland, aluminous, and natural cements.

Limes, wherein the hardening is due to the conversion of hydroxides to carbonates, were formerly widely used as the sole cementitious material, but their slow setting and hardening are not too compatible with modern requirements. Hence, their principal function today is to plasticize the otherwise harsh cements and add resilience

* Part 1 was originally written by James A. Murray (deceased), former Associate Professor of Materials, Massachusetts Institute of Technology, for the first edition of the handbook.

to mortars and stuccoes. Use of limes is beneficial in that their slow setting promotes healing, the recementing of hairline cracks.

Another class of cements is comprised of calcined gypsum and its related products. The gypsum cements are widely used in interior plaster and for fabrication of boards and blocks; but the solubility of gypsum prevents its use in construction exposed to any but extremely dry climates.

Oxychloride cements constitute a class of specialty cements of unusual properties. Their cost prohibits their general use in competition with the cheaper cements; but for special uses, such as the production of sparkproof floors, they cannot be equaled.

Masonry cements or mortar cements are widely used because of their convenience. While they are, in general, mixtures of one or more of the above-mentioned cements with some admixes, they deserve special consideration because of their economies.

THE SILICATE AND ALUMINATE CEMENTS

2-2. Portland Cements. Portland cements are the most common of the modern hydraulic cements. They are made by blending a carefully proportioned mixture of calcareous and argillaceous materials. The mixture is burned in a rotary kiln at a temperature of about 2700°F to form hard nodulized pellets, called clinker. The clinker is ground with a retarder (usually rock gypsum) to a fine powder, which constitutes the portland cement.

Since cements are rarely used in construction without addition of aggregates to form mortars or concretes, properties of the end products are of much greater importance to the engineer than the properties of the cement from which they are made. Properties of mortars and concretes are determined to a large extent by the amount and grading of aggregate, amount of mixing water, and use of admixtures.

One property of mortars and concretes—durability—is greatly affected by the chemical composition of the cement used. It is therefore pertinent to consider the composition of the several types of cement and the effect of composition on the use of the cement.

The composition of portland cement is now usually expressed in terms of the potential phase compounds present, the tricalcium silicate (C_3S), dicalcium silicate (C_2S), tricalcium aluminate (C_3A), and tetracalcium aluminum ferrite (C_4AF). Portland cements are made in five types, distinction between the types being based upon both physical and chemical requirements. Some requirements of the American Society for Testing and Materials Specification for Portland Cement (C150-73) are given in Table 2-1. Other requirements of the specification, such as autoclave expansion, time of set, and magnesia content, which are constant for all types of cement, are omitted from the table.

Type I, general-purpose cement, is the one commonly used for many structural purposes. Chemical requirements for this type of cement are limited to magnesia and sulfur trioxide contents and loss on ignition, since the cement is adequately defined by its physical characteristics.

Type II is a modified cement for use in general concrete where a moderate exposure to sulfate attack may be anticipated or where a moderate heat of hydration is required. These characteristics are attained by placing limitations on the C_3S and C_3A content of the cement. Type II cement gains strength a little more slowly than Type I but ultimately will achieve equal strength. It is generally available in most sections of the country and is preferred by some engineers over Type I for general construction.

Type III cement attains high early strength. In 3 days, strength of concrete made with it is practically equal to that made with Type I or Type II cement at 28 days. This high early strength is attained by finer grinding (although no minimum is placed upon the fineness by specification) and by increasing the C_3S and C_3A content of the cement. Type III cement, however, has high heat evolution and therefore should not be used in large masses. It also has poor sulfate resistance. Type III cement is not always available out of building materials dealers' stocks but may be obtained by them from the cement manufacturer on short notice.

Table 2-1. Chemical and Physical Requirements for Portland Cement

Type	I	II	III	IV	V
Name	General-purpose	Modified	High early	Low-heat	Sulfate-resisting
C_3S, max. %	....	....	....	35	
C_2S, min, %	....	....	....	40	
C_3A, max. %	....	8	15	7	5
Fineness, specific surface, sq cm per g, avg, min	1,600	1,600	....	1,600	1,600
Compressive strength, psi, mortar cubes of 1 part cement and 2.75 parts graded standard sand after:					
3 days					
Standard	1,800	1,500	3,500	....	1,200
Air-entraining	1,450	1,200	2,800		
7 days					
Standard	2,800	2,500	....	1,000	2,200
Air-entraining	2,250	2,000			
28 days					
Standard	....	....	....	2,500	3,000

Type IV is a low-heat cement, which has been developed for mass concrete construction. Normal Type I cement, if used in large masses that cannot lose heat by radiation, will liberate enough heat during the hydration of the cement to raise the temperature of the concrete as much as 50 or 60°F. This results in a relatively large increase in dimensions while the concrete is still soft and plastic. Later, as the concrete cools after hardening, shrinkage causes cracks to develop, weakening the concrete and affording points of attack for aggressive solutions. The potential-phase compounds which make the largest contribution to the heat of hydration are C_3S and C_3A; so the amounts of these that are permitted to be present are limited. Since these compounds also produce the early strength of cement, the limitation results in a cement that gains strength relatively slowly. This is of little importance, however, in the mass concrete for which this type of cement is designed.

Type V is a special portland cement intended for use when high sulfate resistance is required. Its resistance to sulfate attack is attained through the limitation on the C_3A content. It is particularly suitable for structures subject to attack by liquors containing sulfates, such as sea water and some other natural waters.

Both Type IV and Type V cements are specialty cements; they are not normally available from dealer's stock but are usually obtainable for use on a large job if arrangements are made with the cement manufacturer in advance.

For use in the manufacture of air-entraining concrete, agents may be added to the cement by the manufacturer, thereby producing "Air-Entraining Portland Cements." These cements are available as Types IA, IIA, and IIIA.

Most cements will exceed the requirements of the specification as shown in Table 2-1 by a comfortable margin.

2-3. Aluminous Cements. Aluminous cements are prepared by fusing a mixture of aluminous and calcareous materials (usually bauxite and limestone) and grinding the resultant product to a fine powder. They are composed primarily of calcium aluminates with a small proportion of dicalcium silicate or of dicalcium alumina silicate (C_2AS).

Aluminous cements are characterized by their rapid-hardening properties and the high strength developed at early ages. Table 2-2 (adapted from F. M. Lea "Chemistry of Cement and Concrete," St. Martin's Press, Inc., New York) shows the relative strengths of 4-in. cubes of 1:2:4 concrete made with normal portland, high-early-strength portland, and aluminous cements.

Since a large amount of heat is liberated with rapidity by aluminous cement during hydration, care must be taken not to use the cement in places where this

Table 2-2. Relative Strengths of Concretes Made from Portland and Aluminous Cements

Days	Compressive strength, psi		
	Normal portland	High-early portland	Aluminous
1	460	790	5,710
3	1,640	2,260	7,330
7	2,680	3,300	7,670
28	4,150	4,920	8,520
56	4,570	5,410	8,950

heat cannot be dissipated. It is usually not desirable to place aluminous-cement concretes in lifts of over 12 in., otherwise the temperature rise may cause serious weakening of the concrete.

Aluminous cements are much more resistant to the action of sulfate waters than are portland cements. They also appear to be much more resistant to attack by water containing aggressive carbon dioxide or weak mineral acids than the silicate cements. Their principal use is in concretes where advantage may be taken of their very high early strength or of their sulfate resistance, and where the extra cost of the cement is not an important factor.

2-4. Natural Cements. Natural cements are formed by calcining a naturally occurring mixture of calcareous and argillaceous substances at a temperature below that at which sintering takes place. The American Society for Testing and Materials Specification for Natural Cement (C10) requires that the temperature shall be no higher than is necessary to drive off the carbonic acid gas. Since natural cements are derived from naturally occurring materials and no particular effort is made to adjust the composition, both the composition and properties vary rather widely. Some natural cements may be almost the equivalent of portland cement in properties; others are much weaker. Natural cements are principally used now in masonry mortars and as an admixture in portland-cement concretes.

2-5. Hydraulic Limes. Hydraulic lime is made by calcining a limestone containing silica and alumina to a temperature short of incipient fusion so as to form sufficient free lime (CaO) to permit hydration and at the same time leaving unhydrated sufficient calcium silicates to give the dry powder its hydraulic properties (see American Society for Testing and Materials Specification C141).

Because of the low silicate and high lime contents, hydraulic limes are relatively weak. They find their principal use in masonry mortars.

LIMES

2-6. Quicklimes. When limestone is heated to a temperature in excess of 1700°F, the carbon dioxide content is driven off and the remaining solid product is quicklime. It consists essentially of calcium and magnesium oxides plus impurities such as silica, iron, and aluminum oxides. The impurities are usually limited to less than 5%. If they exceed 10%, the product may be a hydraulic lime.

Two classes of quicklime are recognized, high-calcium and dolomitic. A high-calcium quicklime usually is considered as one containing less than 5% magnesium oxide. A dolomitic quicklime usually contains from 35 to 40% magnesium oxide. A few quicklimes are found that contain from 5 to 35% magnesium oxide and are called magnesian limes.

The outstanding characteristic of quicklime is its ability to slake with water. When quicklime is mixed with from two to three times its weight of water, a chemical reaction takes place: The calcium oxide combines with water to form calcium hydroxide, and sufficient heat is evolved to bring the entire mass to a boil. The resulting product is a suspension of finely divided calcium hydroxide (and magnesium hydroxide or oxide if dolomitic lime is used) in water. On cooling, the semifluid mass stiffens to a putty of such consistency that it may be shoveled or carried in a hod. This slaked quicklime putty is the form in which the material is used in construction. Quicklime should never be used without thorough slaking.

The yield of putty will vary, depending upon the type of quicklime, its degree of burning, slaking conditions, etc., and will usually be from 70 to 100 cu ft of putty per ton of quicklime. The principal use of the putty is in masonry mortars, where it is particularly valuable because of the high degree of plasticity or workability it imparts to the mortar. It is used at times as an admixture in concrete for the purpose of improving workability. It also is used in some localities as finish-coat plaster where full advantage may be taken of its high plasticity.

2-7. Mason's Hydrated Lime. Hydrated limes are prepared from quicklimes by addition of a limited amount of water. After hydration ceases to evolve heat, the resulting product is a fine, dry powder. It is then classified by air-classification methods to remove undesirable oversize particles and packaged in 50-lb sacks. It is always a factory-made product, whereas quicklime putty is almost always a job-slaked product.

Mason's hydrated limes are those hydrates suitable for use in mortars, base-coat plasters, and concrete. They necessarily follow the composition of the quicklime. High-calcium hydrates are composed primarily of calcium hydroxide. Normal dolomitic hydrates are composed of calcium hydroxide plus magnesium oxide.

Plasticity of mortars made from normal mason's hydrated limes (Type N) is fair. It is better than that attained with most cements but not nearly so high as that of mortars made with an equivalent amount of slaked quicklime putty.

The normal process of hydration of a dolomitic quicklime at atmospheric pressure results in the hydration of the calcium fraction only, leaving the magnesium oxide portion substantially unchanged chemically. When dolomitic quicklime is hydrated under pressure, the magnesium oxide is converted to magnesium hydroxide. This results in the so-called "special" hydrates (Type S), which not only have their magnesia contents substantially completely hydrated but also have a high degree of plasticity immediately on wetting with water. Mortars made from Type S hydrates are more workable than those made from Type N hydrates. In fact, Type S hydrates are nearly as workable as those made from slaked quicklime putties. The user of this type of hydrate may therefore have the convenience of a bagged product and a high degree of workability without having the trouble and possible hazard of slaking quicklime.

2-8. Finishing Hydrated Limes. Finishing hydrated limes are particularly suitable for use in the finishing coat of plaster. They are characterized by a high degree of whiteness and of plasticity. Practically all finishing hydrated limes are produced in the Toledo district of Ohio from dolomitic limestone. The normal hydrate is composed of calcium hydroxide and magnesium oxide. When first wetted, it is no more plastic than Type N mason's hydrates. It differs from the latter, however, in that, on soaking overnight, the finishing hydrated lime develops a very high degree of plasticity, whereas the mason's hydrate shows relatively little improvement in plasticity on soaking.

Finishing hydrated lime is also made by the pressure method, which forms a Type S hydrate. This material, similar in characteristics (except color) to the Type S mason's hydrate, is also used for finish-coat plaster but does not have quite as satisfactory characteristics for plastering as does Type N finishing hydrate.

GYPSUM CEMENTS

2-9. Low-temperature Gypsum Derivatives. When gypsum rock ($CaSO_4 \cdot 2H_2O$) is heated to a relatively low temperature, about 130°C, three-fourths of the water of crystallization is driven off. The resulting product is known by various names such as hemihydrate, calcined gypsum, and first-settle stucco. Its common name, however, is **plaster of paris.** It is a fine powder, usually white. When mixed with water, it sets rapidly and attains its strength on drying. While it will set under water, it does not gain strength and ultimately, on continued water exposure, will disintegrate.

Plaster of paris, either retarded or unretarded, is used as a molding plaster for preparing ornamental plaster objects or as gaging plaster, which is used with finishing hydrated lime to form the smooth white-coat finish on plaster walls. The unretarded plaster of paris is used by manufacturers to make gypsum block, tile, wallboard, and plaster board.

When plaster of paris is retarded and mixed with fiber such as sisal, it is marketed under the name of hardwall plaster or cement plaster. (The latter name is misleading, since it does not contain any portland cement.) Hardwall plaster, mixed with water and with from 2 to 3 parts of sand by weight, is widely used for base-coat plastering. In some cases wood fiber is used in place of sand, making a "wood-fibered" plaster.

Special effects are obtained by combining hardwall plaster with the correct type of aggregate. With perlite aggregate, a lightweight plaster is obtained; with pumice or some other aggregates, acoustical plasters are produced.

Gypsum plasters, in general, have a strong set, gain their full strength when dry, do not have abnormal volume changes, and have excellent fire-resistance characteristics. They are not well adapted for use under continued damp conditions or intermittent wet conditions.

2-10. High-temperature Gypsum Derivatives. When gypsum rock is heated to a high temperature, all the water is driven off and anhydrous calcium sulfate results. This material will not set. But addition of a small amount of alum to the rock before calcination acts as an accelerator, and the resulting product, known as **Keene's cement,** sets hard. It forms a more durable plastering material than hardwall plaster and is much more resistant to water. Hence it is extensively used as a plastering material in bathrooms and kitchens. It also has good fire-retarding properties. Cost of Keene's cement, however, is appreciably higher than that of hardwall plaster; consequently, its use has been somewhat limited except for special finishes.

MISCELLANEOUS CEMENTS

2-11. Oxychloride Cements. Lightly calcined magnesium oxide mixed with a solution of magnesium chloride forms a cement known as magnesium oxychloride cement, or **Sorel cement.** It is particularly useful in making flooring compositions in which it is mixed with colored aggregates. Floors made of oxychloride cement are sparkproof and are more resilient than floors of concrete.

Oxychloride cement has very strong bonding power and may be used with greater quantities of aggregate than are possible with portland cement because of its higher bonding power. It also bonds well with wood and is used in making partition block or tile with wood shavings or sawdust as aggregate. It is moderately resistant to water but should not be used under continually wet conditions.

2-12. Masonry Cements. Masonry cements, or—as they are sometimes called—mortar cements, are intended to be mixed with sand and used for setting brick, tile, stone, etc. They may be any one of the hydraulic cements already discussed or mixtures of them in any proportion.

Many commercial masonry cements are mixtures of portland cement and pulverized limestone, often containing as much as 50 or 60% limestone. They are sold in bags containing from 70 to 80 lb, each bag nominally containing a cubic foot. Price per bag is commonly less than that of portland cement, but because of the use of the lighter bag, cost per ton is higher than that of portland cement.

Since there is no specification as to chemical content and physical requirements of masonry cement, specifications are quite liberal; some manufacturers vary the composition widely, depending upon competition, weather conditions, or availability of materials. Resulting mortars may vary widely in properties.

AGGREGATES

2-13. Desirable Characteristics of Aggregates. Aggregate is defined as "inert material which when bound together into a conglomerated mass by a matrix, forms concrete, mastic, mortar, plaster, etc." [American Society for Testing and Materials (ASTM) designation C58.]

As such, the term aggregate includes the normal sands, gravels, and crushed stone used for making mortar, plaster, and concrete. It also includes the lightweight materials, such as slag, vermiculite, and pumice, used to prepare lightweight plasters and concretes. It does not include finely divided or pulverized stone or sand sometimes used with cement that is more properly termed an admixture or filler.

As in the case of cements, properties of aggregates are of less importance than properties of mortars or concretes made from aggregates. The most important property of an aggregate is its soundness, or its dimensional stability. Concrete may be unsound if made with a sound aggregate if not properly proportioned, handled, and placed; but no amount of precaution will enable sound concrete to be made with unsound aggregate.

Cleanliness, strength, and gradation are important in any aggregate. For most purposes, an aggregate should be free of clay, silt, organic matter, and salts. Such cleanliness is usually readily obtainable if facilities are available for washing the aggregate; but bank sands and gravels may contain clay and loam unless special precautions are taken. Strength is important but is infrequently measured, reliance being placed to a large extent upon the performance of existing structures as being evidence of adequate strength. Grading, while important, is under the control of the producer, who can usually adjust his operation to meet the grading requirements of any normal specification.

2-14. Coarse Aggregates. Coarse aggregates comprise that portion of the aggregate retained on a No. 4 sieve (4.75 mm). Normal coarse aggregate is composed of gravel or crushed stone and for structural concrete is usually required to be fairly uniformly graded from a maximum size of about 2 in. down to $\frac{3}{16}$ in. It should, of course, be clean and strong as well as sound.

ASTM Standard C125 defines gravel as:

1. Granular material predominantly retained on a No. 4 sieve and resulting from natural disintegration and abrasion of rock or processing of weakly bound conglomerate; or

2. That portion of an aggregate retained on a No. 4 sieve and resulting from natural disintegration and abrasion of rock or processing of weakly bound conglomerate.

The first definition applies to an entire aggregate, in a natural condition or after processing; the second to a portion of an aggregate.

Crushed stone is "the product resulting from the artificial crushing of rocks, boulders or large cobblestones, substantially all faces of which have resulted from the crushing operation." (See ASTM Standard C125.) As a result of its processing, it is angular in contour. It should be roughly cubical in shape for use in concrete but frequently contains flat or needle-shaped particles that may be a source of weakness in the concrete. Crushed stone is usually of a single rock species and is not so heterogeneous as gravel. It may contain unsound impurities of the same characteristics as found in gravel.

Blast-furnace slag is used as a coarse aggregate in some localities where it is economically available. It is lighter in weight than gravel and crushed stone but normally is not considered a lightweight aggregate.

2-15. Fine Aggregates. Fine aggregate is that portion of aggregate finer than a No. 4 sieve. It is usually graded fairly uniformly from No. 4 to No. 100 in size. Unless otherwise mentioned, fine aggregate is usually considered to be sand, the product of the natural disintegration and abrasion of rock or processing of completely friable sandstone. The fine product obtained from crushing of stone is known as "stone sand," while that from crushing of slag is known as "slag sand." Sand for use in concrete must contain negligible amounts of clay, coal, and lignite, material finer than a No. 200 sieve, shale, alkali, mica, coated grains, soft and flaky particles, and organic impurities.

2-16. Lightweight Aggregates. Lightweight aggregates include perlite, exfoliated vermiculite, pumice, scoria, tuff, blast furnace slag, fly ash, diatomite, and expanded clay, shale, and slate. When mixed with cement, the resulting concrete also is light in weight and has low thermal conductivity. Strength of the concrete will be almost inversely proportional to the weight of the aggregate, although comparisons are not always easily made because of the difficulty of making concrete mixes that have the type of aggregate as the sole variable.

ASTM Specifications C330, C331, and C332 require that lightweight aggregates for structural and insulating concrete and concrete masonry units be composed predominantly of cellular and granular organic material. For structural concrete and masonry, dry loose weight should not exceed 70 lb per cu ft for fine aggregates, 55 lb per cu ft for coarse aggregates, and 65 lb per cu ft for combined fine and coarse aggregates. C332 divides aggregates for insulating concretes into two groups. Group I contains aggregates, such as perlite and vermiculite, prepared by expanding products. Dry loose weight of the perlite should be between 7.5 and 12 lb per cu ft, and of the vermiculite, between 6 and 10 lb per cu ft. These aggregates generally produce concretes weighing from 15 to 50 lb per cu ft, the thermal conductivity of which may range from 0.45 to 1.50. Group II contains aggregates, such as pumice, scoria, and tuff, prepared by processing natural materials, and aggregates prepared by expanding, calcining, or sintering products. For Group II, dry loose weight should not exceed 70 lb per cu ft for fine aggregates, 55 lb per cu ft for coarse aggregates, and 65 lb per cu ft for combined fine and coarse aggregates. The aggregates generally produce concretes weighing from 45 to 90 lb per cu ft, the thermal conductivity of which may range from 1.05 to 3.00.

MORTARS AND CONCRETES

2-17. Mortars. Mortars are composed of a cementitious material, fine aggregate, sand, and water. They are used for bedding unit masonry, for plasters and stuccoes, and with the addition of coarse aggregate, for concrete. In this article, consideration will be given primarily to those mortars used for unit masonry and plasters.

Properties of mortars vary greatly, being dependent on the properties of the cementitious material used, ratio of cementitious material to sand, characteristics and grading of the sand, and ratio of water to solids.

Packaging and Proportioning. Mortars are usually proportioned by volume. A common specification is that not more than 3 cu ft of sand shall be used with 1 cu ft of cementitious material. Difficulty is sometimes encountered, however, in determining just how much material constitutes a cubic foot: A bag of cement (94 lb) by agreement is called a cubic foot in proportioning mortars or concretes, but an actual cubic foot of lime putty may be used in proportioning mortars. Since hydrated limes are sold in 50-lb bags, each of which makes somewhat more than a cubic foot of putty, weights of 40, 42, and 45 lb of hydrated lime have been used as a cubic foot in laboratory studies; but on the job, a bag is frequently used as a cubic foot. Masonry cements are sold in bags containing 70 to 80 lb, and again a bag is considered a cubic foot.

Workability is an important property of mortars, particularly of those used in conjunction with unit masonry of high absorption. Workability is controlled by the character of the cement and amount of sand. For example, a mortar made from 3 parts sand and 1 part slaked lime putty will be more workable than one made from 2 parts sand and 1 part portland cement. But it will also have much less strength when tested by the conventional cube test. By proper selection or mixing of cementitious materials, a satisfactory compromise may usually be obtained, producing a mortar of adequate strength and workability.

Water retention—the ratio of flow after one minute standard suction to the flow before suction—is used as an index of the workability of mortars. A high value of water retention is considered desirable for most purposes. There is, however, a wide variation in water retention of mortars made with varying proportions of cement and lime and with varying limes. ASTM C270 requires mortar mixed to an initial flow of 100 to 115, as determined by the method of ASTM C109, to have a flow after suction of at least 70%.

Strength of mortar is frequently used as a specification requirement, even though it has little relation to the strength of masonry. (See, for example, ASTM Specifications C270 and C476.) The strength of mortar is affected primarily by the amount of cement in the matrix. Other factors of importance are the ratio of sand to cementing material, curing conditions, and age when tested.

Volume change of mortars constitutes another important property. Normal volume change (as distinguished from unsoundness) may be considered as the shrinkage during early hardening, shrinkage on drying, expansion on wetting, and changes due to temperature.

After drying, mortars expand again when wetted. Alternate wetting and drying produces alternate expansion and contraction, which apparently continues indefinitely with portland-cement mortars.

Coefficient of Expansion. Palmer measured the coefficient of thermal expansion of several mortars ("Volume Changes in Brick Masonry Materials," *Journal of Research of the National Bureau of Standards,* Vol. 6, p. 1003). The coefficients ranged from 0.38×10^{-5} to 0.60×10^{-5} for masonry-cement mortars; from 0.41×10^{-5} to 0.53×10^{-5} for lime mortars, and from 0.42×10^{-5} to 0.61×10^{-5} for cement mortars. Composition of the cementitious material apparently has little effect upon the coefficient of thermal expansion of a mortar.

High-Bond Mortars. When polymeric materials based on polyvinylidene chloride are added to mortar, greatly increased bonding, compressive, and shear strength result. To obtain this high strength, the other materials, including sand, water, Type I or III portland cement, and a workability additive, such as pulverized ground limestone or marble dust, must be of quality equal to that of the ingredients of standard mortar. The high strength of the mortar makes it possible to apply appreciable bending and tensile stresses to masonry, resulting in thinner walls and making possible the prelaying of single-wythe panels that can be hoisted into place.

2-18. Normal Concrete. Concrete is a mixture of portland cement, fine aggregate, coarse aggregate, and water, which is temporarily fluid and capable of being poured, cast, or molded, but which later sets and hardens to form a solid mass. Normal concrete is here considered to be concrete made with sand as fine aggregate and with conventional aggregates, such as gravel or crushed stone, and to which no air-entrainment admixture has been added. It is to be distinguished from air-entrained concrete, which has some air deliberately added and from lightweight concretes made with special lightweight aggregates or with a large air content.

The ultimate consumer of concrete is interested in three major properties of concrete—adequate strength, adequate durability, and minimum cost. In the attainment of these three objectives, the concrete designer may wish to vary the proportions of ingredients widely. The principal variations which may be imposed are:

1. The water-cement ratio.
2. The cement-aggregate ratio.
3. The size of coarse aggregate.
4. The ratio of fine aggregate to coarse aggregate.
5. The type of cement.

With varying ingredients, concretes may be made with strengths as low as 1,500 psi or more than 8,000 psi at 28 days. Durability may vary from almost complete durability under the most severe exposure to none at all unless the concrete is protected from weather. Material costs may vary from very expensive to very cheap.

Workability. This property of fluid concrete is of great importance. Workability is much talked about but is poorly defined and poorly measured. It involves the concepts of fluidity, cohesiveness, adhesiveness, plasticity, and probably others. A definition sometimes used is "ability to flow without segregation." Workability is enhanced by high cement factors and high sand-to-coarse-aggregate ratios; but these factors also involve high cost and may involve undesirable, high-volume change. The concrete designer must strike a balance between all factors involved and must frequently allow less workability than he wants in order to obtain concrete that is satisfactory from some other aspect.

Density of fluid concrete made with conventional aggregates will approximate 150 lb per cu ft. This value may be used in calculating strength of formwork and density of set concrete.

Setting and hardening of concrete are caused by a chemical reaction between the constituents of the cement and the water. This reaction is accompanied by the evolution of considerable heat. Normally, in structural concrete, this heat is removed by conduction, radiation, or convection about as rapidly as it is generated; but in large masses, the heat cannot be dissipated rapidly and may be sufficient to raise the temperature of the mass by 50 to 60°F. In such cases, Type IV cement may be used to keep the temperature rise as low as possible and thereby prevent delayed cracking.

Strength of concrete is conventionally measured by testing in compression cylinders 6 in. in diameter and 12 in. high.

The prime factor affecting strength is the water-cement (*W/C*) ratio. This is commonly expressed as a ratio by weight but may be expressed as gallons

Table 2-3. Relation between *W/C Ratio and Compressive Strength**

Net W/C by Weight	*28-day Strength, Psi*
0.40	5,500
0.42	5,200
0.44	5,000
0.46	4,800
0.49	4,400
0.53	4,000
0.58	3,600
0.62	3,300
0.67	2,900
0.71	2,600
0.75	2,400

* Water-cement.

Table 2-4. Relation between Age and Compressive Strength, Psi

Age	Type I cement	Type II cement	Type III cement
3 days	2,100	2,100	3,400
7 days	3,200	3,000	4,400
28 days	4,500	4,500	5,500
90 days	5,000	5,400	6,100
1 year	5,600	5,800	6,700

Table 2-5. Relation between Cement Content and Strength of Concrete

Cement Content, Lb per Cu Yd	*Avg 28-day Strength, Psi*
350	2,000
400	2,700
450	3,300
500	3,800
550	4,300
600	4,800
650	5,200

Table 2-6. Effect of Curing Temperature on Relative Strength of Concrete

Curing Temp, °F	*Relative 28-day Strength 70°F = 100*
40	78
55	98
70	100
85	101
100	103
115	105

of water per sack of cement. Table 2-3 shows the relation between the *W/C* ratio and 28-day compressive strength of normal concrete as used by the Bureau of Reclamation ("Concrete Manual," Government Printing Office, Washington, D.C.).

The strength is affected by the age of the concrete when tested, as shown in Table 2-4.

Strength of concrete is affected by the cement content, as shown in Table 2-5.

Strength of concrete is diminished by curing at low temperatures, as is shown in Table 2-6.

Other factors have a bearing on the strength of a particular concrete but may usually be compensated for in mix design. Strength of a concrete made from one aggregate may be somewhat less than that made from a second aggregate, but a slight decrease in water-cement ratio of the former concrete may make the two identical. Similarly, variations in strength due to grading of aggregate and to the type or brand of cement may be compensated in the mix design.

Volume Change. This property is of importance equal to that of strength. Volume changes that occur while concrete is in a plastic state are normally not of great importance in mass concrete, but in reinforced concrete they may be. The result of settlement, sedimentation, or bleeding may be to cause the concrete to settle away from the underside of the reinforcement, thereby impairing bond of the concrete to the reinforcing steel. This may be avoided by using a workable nonbleeding concrete, obtainable by using less water, finer cement, pozzolanic admixtures, or air entrainment.

Volume changes that occur after hardening are apt to result in cracking, which weakens the concrete and makes it susceptible to early disintegration by attack by frost or by aggressive solutions.

Concrete is subject to expansion and contraction with wetting and drying. The drying shrinkage of concrete is slightly more than ½ in. per 100 ft. This, in magnitude, is equivalent to a thermal change of about 83°F.

Coefficient of Expansion. The thermal coefficient varies with the cement content, water content, type of aggregate, and other factors, but averages 0.0000055 in. per in. per degree Fahrenheit ("Concrete Manual," Bureau of Reclamation, Government Printing Office, Washington, D.C.).

Modulus of Elasticity. The stress-strain curve for concrete is curved over its length. The portion from zero load to about 40% of the ultimate load, however, is sufficiently straight that a secant modulus of elasticity may be determined for this portion. The modulus of elasticity, calculated in this manner, is about 1,000 times the compressive strength in pounds per square inch.

Creep. When concrete is subjected to a load, it initially undergoes an elastic deformation, the magnitude of which is proportional to the load and the elasticity of the concrete. If load application continues, however, deformation increases gradually with time. This deformation under load is termed creep or plastic flow. The creep continues for an indefinite time, measurements having been made for periods in excess of 10 years. It proceeds at a continuously diminishing rate and approaches some limiting value. The amount of creep at this limiting value appears to be from one to three times the amount of the elastic deformation. Usually, more than 50% of the ultimate creep takes place within the first 3 months after loading. Upon unloading, an immediate elastic recovery takes place, followed by a plastic recovery of lesser amount than the creep on first loading.

2-19. Air-entraining Concretes. All normal concrete contains some air, usually between 0.5 and 1.0%. By the addition of a small amount of an air-entraining agent, the amount of air contained in the concrete may be materially increased to 10% or more (by volume) if desired. This extra air is present in the form of small, discrete air bubbles, which have a marked effect upon the workability, durability, and strength of the concrete.

Since addition of air decreases strength of concrete, it is necessary to compensate for this decrease by decreasing the water-cement ratio or decreasing unit water content, or both. Even with compensating adjustments of this nature, it is usually impossible to maintain the design strength of the concrete if more than 6% air is present. Consequently, air contents are usually specified to be about 4%, with 6% air as a maximum.

Air-entrained concrete has a lower unit weight than normal concrete. A concrete that would weigh 150 lb per cu ft without air will weigh only 144 with 4% entrained air.

Addition of air to concrete makes a great improvement in workability. If air is added to a normal concrete of 2-in. slump without compensating adjustments, an air-entrained concrete with a slump of 6 to 8 in. will result. By removal of some of the water, thereby increasing the water-cement ratio, the slump may be brought back to 2 in., but the resulting air-entrained concrete is still much more workable and more cohesive and has less bleeding and segregation than the normal concrete of equivalent slump. This improvement in workability persists, even if some of the sand is withdrawn, as is common practice.

Durability of air-entrained concrete exposed to severe weathering is much superior to that of normal concrete. This has been adequately demonstrated by many laboratory and field studies which have been reported in recent years.

The other properties of air-entrained concrete such as rate of development of strength, volume change, and elasticity appear to be identical with those of normal concrete, if allowance is made for the weakening effect of the air and the additional workability attained through its incorporation.

2-20. Lightweight Concretes. Appreciable savings in weight and accompanying savings in steel costs and other costs are often obtainable through the use of lightweight concretes. Often, these are concretes wherein the normal heavy aggregates have been replaced with lightweight aggregates (see Art. 2-16).

Lighter concretes, in general, have higher absorptions, lower strengths, lower elastic moduli, lower thermal conductivity, and higher shrinkage than heavier concretes. The designer of lightweight concrete must base his design on specific tests of the particular aggregate under consideration rather than upon average data obtainable from tables and charts.

Other types of lightweight concretes are produced by various methods.

Expanded Concretes. One of these methods is to mix a concrete or mortar with a treated aluminum powder. The reaction of the aluminum with calcium hydroxide liberated by the cement evolves hydrogen, which causes the plastic concrete to expand. The expansion may be as much as 100% of the unexpanded volume. The cement sets at about the time the expansion is completed. Hence the set concrete remains in the expanded form. Concretes weighing from 75 to 100 lb per cu ft are produced in this manner. The upper surface of the concrete may be irregular if there has been irregularity in distribution of the aluminum powder.

Foamed Concrete. Lightweight concrete can be made by mixing cement and fine filler, such as pulverized stone, silica flour, or fly ash, with a stabilized foam. By proper proportioning of the volumes of foam and solids, concretes weighing as little as 10 lb per cu ft have been produced. These materials have very low strength but are well adapted for production of insulated and fire-resistant roof slabs and other members where strength is of little importance.

Nailable Concrete. Concrete into which nails can be driven and which will hold the nails firmly is produced by mixing equal volumes of cement, sand, and pine sawdust with enough water to give a slump of 1 to 2 in. This concrete, after moist curing for 2 days and drying for 1 day, has excellent holding power for nails.

2-21. Concrete Masonry Units. A wide variety of manufactured products are produced from concrete and used in building construction. These include such items as concrete brick, concrete block or tile, concrete floor and roof slabs, precast lintels, and cast stone. These items are made both from normal dense concrete mixes and from mixes with lightweight aggregates.

Properties of the units vary tremendously—from strong, dense load-bearing units used under exposed conditions to light, relatively weak, insulating units used for roof and fire-resistant construction.

Many types of concrete units have not been covered by adequate standard specifications. For these units, reliance must be placed upon the manufacturer's specifications. Requirements for strength and absorption of concrete brick and block have been established by the American Society for Testing and Materials. A summary of these requirements is presented in Table 2-7.

Manufactured concrete units are subject to the advantage (or sometimes disadvantage) that curing is under the control of the manufacturer. Many methods of curing are used, from simply stacking the units in a more or less exposed location to curing under high-pressure steam. The latter method appears to have considerable merit in reducing ultimate shrinkage of the block. Easterly found shrinkages ranging from 1/4 to 3/8 in. per 100 ft for concretes cured with high-pressure steam ("Shrinkage and Curing in Masonry Units," *Journal of the American Concrete Institute,* Vol. 23, p. 393). These values are about one-half as great as those obtained with normal atmospheric curing. Easterly also obtained values for the moisture movement

Table 2-7. Summary of ASTM Specification Requirements for Concrete Masonry Units

	Compressive strength, min, psi		Moisture content for Type I units, max, % of total absorption (average of 5 units)			Moisture absorption, max, pcf (average of 5 units)		
			Avg annual relative humidity, %			Oven-dry weight of concrete, pcf		
	Avg of 5 units	Individual min	Over 75	75 to 50	Under 50	125.1 or more	105.1 to 125.1	Under 125.1
Concrete building brick, ASTM C55:								
U-1, U-II (high strength severe exposures)	3,500	3,000	...	...	...	10	10	10
P-1, P-II (general use, moderate exposures)	2,500	2,000	...	...	...	13	14–17	18
G-1, G-II (backup or interior use)	1,500	1,250						
Linear shrinkage, %:								
0.03 or less	....	....	45	40	35			
0.03 to 0.045	....	....	40	35	30			
Over 0.045	....	....	35	30	25			
Solid, load-bearing units, ASTM C145:								
U-I, U-II (unprotected exterior walls below grade or above grade exposed to frost)	1,800	1,500	...	...	...	10	11–14	15
P-I, P-II (protected exterior walls below grade or above grade exposed to frost)	1,800	1,500	...	...	...	13	14–17	18
G-I, G-II (general use, no frost)	1,200	1,000						
Linear shrinkage, %: (Same as for brick)								
Hollow, load-bearing units, ASTM C90:								
N-I moisture-controlled units (general use)	1,000	800	...	...	...	13	15	18
S-I moisture-controlled units (above grade, weather protected)	700	600	...	...	...	...	....	20
Linear shrinkage, %: (Same as for brick)								
Hollow, non-load-bearing units, ASTM C129	350	300						
Linear shrinkage, %: (Same as for brick)								

in test blocks that had been cured with high-pressure and high-temperature steam. He found expansions of from ¼ to ½ in. per 100 ft on saturating previously dried specimens.

BURNED-CLAY UNITS

Use of burned-clay structural units dates from prehistoric times. Hence durability of well-burned units has been adequately established through centuries of exposure in all types of climate.

Modern burned-clay units are made in a wide variety of size, shape, color, and texture to suit the requirements of modern architecture. They include such widely diverse units as common and face brick; hollow clay tile in numerous shapes, sizes, and designs for special purposes; ceramic tile for decorative and sanitary finishes, and architectural terra cotta for ornamentation.

Properties of burned-clay units vary with the type of clay or shale used as raw material, method of fabrication of the units, and temperature of burning. As a consequence, some units, such as salmon brick, are underburned, highly porous, and of poor strength. But others are almost glass hard, have been pressed and burned to almost eliminate porosity, and are very strong. Between these extremes lie most of the units used for construction.

2-22. Brick. Brick have been made in a wide range of sizes and shapes, from the old Greek brick, which was practically a 23-in. cube of 12,650 cu in. volume, to the small Belgian brick, approximately $1\frac{3}{4} \times 3\frac{3}{8} \times 4\frac{1}{2}$ in. with a total volume of only 27 cu in. The present common size in the United States is $2\frac{1}{4} \times 3\frac{3}{4} \times 8$ in., with a volume of $67\frac{1}{2}$ cu in., and a weight between $4\frac{1}{2}$ and 5 lb. Current specification requirements for strength and absorption of building brick are given in Table 2-8 (see ASTM Specification C62). Strength and absorption of brick from different producers vary widely.

Table 2-8. Physical Requirements for Clay or Shale Brick

Grade	Compressive strength, flat, min, psi		Water absorption, 5-hr boil, max—%		Saturation* coefficient, max—%	
	Avg of 5	Indi-vidual	Avg of 5	Indi-vidual	Avg of 5	Indi-vidual
SW—Severe weathering	3,000	2,500	17.0	20.0	0.78	0.80
MW—Moderate weathering	2,500	2,200	22.0	25.0	0.88	0.90
NW—No exposure	1,500	1,250	No limit	No limit	No limit	No limit

* Ratio of 24-hr cold absorption to 5-hr boil absorption.

Thermal expansion of brick has been measured by Ross ("Thermal Expansion of Clay Building Brick," *Journal of Research of the National Bureau of Standards,* Vol. 27, p. 197). He obtained values ranging from 0.0000017 per °F for a sample of fire-clay brick to 0.0000069 per °F for a surface-clay brick. The average coefficient for 80 specimens of surface-clay brick was 0.0000033; that for 41 specimens of shale brick was 0.0000034. These results are of the same order of magnitude as those obtained by Palmer, who found a variation of from 0.0000036 to 0.0000085 ("Volume Changes in Brick Masonry Materials," *Journal of Research of the National Bureau of Standards,* Vol. 6, p. 1003).

Palmer studied the expansion of brick due to wetting and obtained wetting expansions varying from 0.0005 to 0.025%.

The thermal conductivity of dry brick as measured by several investigators ranges from 1.29 to 3.79 Btu/hr/sq ft/°F/in. The values are increased by wetting.

2-23. Structural Clay Tile. Structural clay tile are hollow burned-clay masonry units with parallel cells. Such units have a multitude of uses: as a facing tile for interior and exterior unplastered walls, partitions, or columns; as load-bearing tile in masonry constructions designed to carry superimposed loads; as partition tile for interior partitions carrying no superimposed load; as fireproofing tile for protection of structural members against fire; as furring tile for lining the inside of exterior walls; as floor tile in floor and roof construction; and as header tile, which are designed to provide recesses for header units in brick or stone-faced walls.

Two general types of tile are available—side-construction tile, designed to receive its principal stress at right angles to the axis of the cells, and end-construction tile, designed to receive its principal stress parallel to the axis of the cells.

Tile are also available in a number of surface finishes, such as opaque glazed tile, clear ceramic-glazed tile, nonlustrous glazed tile, and scored, combed, or roughened finishes designed to receive mortar, plaster, or stucco.

Requirements of the appropriate ASTM specifications for absorption and strength of several types of tile are given in Table 2-9 (see ASTM Specifications C34, C56, C57, C212, and C126 for details pertaining to size, color, texture, defects, etc.). Strength and absorption of tile made from similar clays but from different sources and manufacturers vary widely.

Table 2-9. Physical Requirement Specification for Structural Clay Tile

Type and grade	Absorption, %		Compressive strength, psi (based on gross area)			
			End-construction tile		Side-construction tile	
	Avg of 5 tests	Individual max	Min avg of 5 tests	Individual min	Min avg of 5 tests	Individual min
Load-bearing (ASTM C34):						
LBX	16	19	1,400	1,000	700	500
LB	25	28	1,000	700	700	500
Non-load-bearing (ASTM C56):						
NB	......	28				
Floor tile (ASTM C57):						
FT 1	......	25	3,200	2,250	1,600	1,100
FT 2	......	25	2,000	1,400	1,200	850
Facing tile (ASTM C212):						
FTX	7 (max)	9				
FTS	13 (max)	16				
Standard	......	...	1,400	1,000	700	500
Special duty	......	...	2,500	2,000	1,200	1,000
Glazed units (ASTM C126)	......	...	3,000	2,500	2,000	1,500

LBX. Tile suitable for general use in masonry construction and adapted for use in masonry exposed to weathering. They may also be considered suitable for direct application of stucco.

LB. Tile suitable for general use in masonry where not exposed to frost action, or in exposed masonry where protected with a facing of 3 in. or more of stone, brick, terra cotta, or other masonry.

NB. Non-load-bearing tile made from surface clay, shale, or fire clay.

FT 1 and FT 2. Tile suitable for use in flat or segmental arches or in combination tile and concrete ribbed-slab construction.

FTX. Smooth-face tile suitable for general use in exposed exterior and interior masonry walls and partitions, and adapted for use where tile low in absorption, easily cleaned, and resistant to staining are required and where a high degree of mechanical perfection, narrow color range, and minimum variation in face dimensions are required.

FTS. Smooth or rough-texture face tile suitable for general use in exposed exterior and interior masonry walls and partitions and adapted for use where tile of moderate absorption, moderate variation in face dimensions, and medium color range may be used, and where minor defects in surface finish, including small handling chips, are not objectionable.

Standard. Tile suitable for general use in exterior or interior masonry walls and partitions.

Special duty. Tile suitable for general use in exterior or interior masonry walls and partitions and designed to have superior resistance to impact and moisture transmission, and to support greater lateral and compressive loads than standard tile construction.

Glazed units. Ceramic-glazed structural clay tile with a glossy or satin-mat finish of either an opaque or clear glaze, produced by the application of a coating prior to firing and subsequently made vitreous by firing.

Foster presents data on the modulus of elasticity of tile showing a range in modulus from 1,620,000 to 6,059,000 psi for tile from 18 different sources ("Strength,

Absorption and Freezing Resistance of Hollow Building Tile," *Journal of the American Ceramic Society*, Vol. 7, No. 3, p. 189).

Data for thermal expansion and for wetting expansion of tile apparently are not available, but since tile are made from the same clays as brick, the values previously cited for brick should be applicable to tile (see Art. 2-22). Some data are available on thermal conductivity of individual tile; but since the conductivity of the whole wall is more important than that of the units, these data will be considered under the properties of assemblages (Sec. 18).

2-24. Ceramic Tiles. Ceramic tile is a burned-clay product used primarily for decorative and sanitary effects. It is composed of a clay body on which is superimposed a decorative glaze.

The tile are usually flat but vary in size from about ½ in. square up to about 6 in. or even larger. Their shape is also widely variable—squares, rectangles, and hexagons are the predominating forms, to which must be added coved moldings and other decorative forms. They are not dependent upon the color of the clay for their final color since they are usually glazed. Hence they are available in a complete color gradation from pure whites through pastels of varying hue to deep solid colors and jet blacks.

Properties of the base vary somewhat, particularly absorption, which ranges from almost zero up to about 15%. The glaze is required to be impervious to liquids and should not stain, crack, or craze.

Ceramic tile are applied on a solid backing, either masonry or plaster, by means of a mortar or adhesive. They are usually applied with the thinnest possible mortar joint; consequently accuracy of dimensions is of greatest importance. Since color, size, and shape of tile are the important features in these tile, reliance must be placed upon the current literature of the manufacturer for information on these characteristics; general summaries are of little value.

2-25. Architectural Terra Cotta. The term "terra cotta" has been applied for centuries to decorative molded-clay objects whose properties are similar to brick. The shapes are molded either by hand or by machine and fired in a manner similar to brick. Terra cotta is frequently glazed to produce a desired color or finish. This introduces the problem of cracking or crazing of the glaze, particularly over large areas. Structural properties of terra cotta are similar to those of brick.

BUILDING STONES

Principal building stones generally used in the United States are limestones, marbles, granites, and sandstones. Other stones such as serpentine and quartzite are used locally but to a much lesser extent. Stone, in general, makes an excellent building material, if properly selected on the basis of experience; but the cost may be relatively high.

Properties of stone are not under the control of the manufacturer. He must take what nature has provided, and therefore, he cannot provide the choice of properties and color available in some of the manufactured building units. The most the stone producer can do is to avoid quarrying certain stone beds that have been proved by experience to have poor strength or poor durability.

2-26. Strength of Stone. Data on the strength of building stones are presented in Table 2-10, summarized from the extensive work of D. W. Kessler and his associates (*U.S. National Bureau of Standards Technical Papers*, No. 123, B. S. Vol. 12; No. 305, vol. 20, p. 191; No. 349, Vol. 21, p. 497; *Journal of Research of the National Bureau of Standards*, Vol. 11, p. 635; Vol. 25, p. 161). The data in Table 2-10 pertain to dried specimens. Strength of saturated specimens may be either greater or less than that of completely dry specimens. In a comparison of 90 specimens of marble tested in compression both dry and saturated, it was found that the strength of 69 saturated marbles was less than that of the same marbles when dry by amounts ranging from 0.3 to 46.0% and averaging 9.7%. On the other hand, 21 marbles showed an increase in strength on soaking; the increase ranged from 0.04 to 35.4% and averaged 7.6%.

The modulus of rupture of dry slate is given in Table 2-10 as ranging from 6,000 to 15,000 psi. Similar slates, tested wet, gave moduli ranging from 4,700

Table 2-10. Strength Characteristics of Commercial Building Stones

Stone	Compressive strength, psi, range	Modulus of rupture, psi, range	Shear strength, psi, range	Tensile strength, psi, range	Elastic modulus, psi, range	Toughness		Wear resistance	
						Range	Avg	Range	Avg
Granite	7,700–60,000	1,430–5,190	2,000–4,800	600–1,000	5,700,000–8,200,000	8–27	13	43.9–87.9	60.8
Marble	8,000–50,000	600–4,900	1,300–6,500	150–2,300	7,200,000–14,500,000	2–23	6	6.7–41.7	18.9
Limestone	2,600–28,000	500–2,000	800–4,580	280–890	1,500,000–12,400,000	5–20	7	1.3–24.1	8.4
Sandstone	5,000–20,000	700–2,300	300–3,000	280–500	1,900,000–7,700,000	2–35	10	1.6–29.0	13.3
Quartzite	16,000–45,000	...	...	...	...	5–30	15		
Serpentine	11,000–28,000	1,300–11,000	...	800–1,600	4,800,000–9,600,000	...	...	13.3–111.4	46.9
Basalt	28,000–67,000	...	...	...	...	5–40	20		
Diorite	16,000–35,000	...	...	...	...	6–38	23		
Syenite	14,000–28,000								
Slate	...	6,000–15,000	2,000–3,600	3,000–4,300	9,800,000–18,000,000	10–56	...	5.6–11.7	7.7
Diabase	...	...	...	...	...	6–50	19		
Building limestone	...	...	...	...	...	3–8	4.4		

to 12,300 psi. The ratio of wet modulus to dry modulus varied from 0.42 to 1.12 and averaged 0.73.

2-27. Specific Gravity, Porosity, and Absorption of Stone. Data on the true specific gravity, bulk specific gravity, unit weights, porosity, and absorption of various stones have been obtained by Kessler and his associates. Some of these data are given in Table 2-11.

2-28. Permeability of Stone. Permeability of stone varies with type of stone, thickness, and driving pressure that forces water through the stone. Table 2-12 presents data by Kessler for the more common stones at three different pressures ("Permeability of Stone," *U.S. National Bureau of Standards Technical Papers*, No. 305, Vol. 20, p. 191). The units of measurement of permeability are cubic inches of water that will flow through a square foot of a specimen ½ in. thick in 1 hr.

Table 2-11. Specific Gravity and Porosity of Commercial Building Stones

Stone	Specific gravity		Unit weight, lb per cu ft	Porosity, %	Absorption, %	
	True	Bulk			By weight	By volume
Granite	2.599–3.080	2.60–3.04	157–187	0.4–3.8	0.02–0.58	0.4–1.8
Marble	2.718–2.879	2.64–2.86	165–179	0.4–2.1	0.01–0.45	0.04–1.2
Limestone	2.700–2.860	1.87–2.69	117–175	1.1–31.0	....	6–15
Slate	2.771–2.898	2.74–2.89	168–180	0.1–1.7	0.00–1.63	0.3–2.0
Basalt	....	2.9–3.2				
Soapstone	....	2.8–3.0				
Gneiss	....	2.7–3.0				
Serpentine	....	2.5–2.8	158–183			
Sandstone	....	2.2–2.7	119–168	1.9–27.3	....	6–18
Quartzite	....	....	165–170	1.5–2.9		

Table 2-12. Permeability of Commercial Building Stones
(cu in. per sq ft per hr for ½-in. thickness)

Pressure, psi	1.2	50	100
Granite	0.6–0.8	0.11	0.28
Slate	0.006–0.008	0.08–0.11	0.11
Marble	0.06–0.35	1.3–16.8	0.9–28.0
Limestone	0.36–2.24	4.2–44.8	0.9–109
Sandstone	4.2–174.0	51.2	221

Table 2-13. Coefficient of Thermal Expansion of Commercial Building Stones

Stone	*Range of Coefficient*
Limestone	$(4.2–22) \times 10^{-6}$
Marble	$(3.6–16) \times 10^{-6}$
Sandstone	$(5.0–12) \times 10^{-6}$
Slate	$(9.4–12) \times 10^{-6}$
Granite	$(6.3–9) \times 10^{-6}$

2-29. Thermal Expansion of Stone. Thermal expansion of building stones as measured by Kessler and his associates is given in Table 2-13. It shows that limestones have a wide range of expansion as compared with granites and slates.

The effect of repeated heating and cooling on marble has been studied. A marble that had an original strength of 9,174 psi, after 50 heatings to 150°C, had a strength of 8,998 psi—a loss of 1.9%. After 100 heatings to 150°C, the strength was only 8,507 psi, or a loss of 7.3%. The latter loss in strength was identical with that obtained on freezing and thawing the same marble for 30 cycles.

Wheeler measured the permanent expansion that occurs on repeatedly heating marble ("Thermal Expansion of Rock at High Temperatures," *Transactions of the Royal Society of Canada,* Vol. 4, sec. 3, p. 19). Using specimens 20 cm in length, he obtained total expansions of over 2 mm and permanent expansions of over 1 mm.

2-30. Freezing and Thawing Tests of Stone. Kessler made freezing and thawing tests of 89 different marbles ("Physical and Chemical Tests of Commercial Marbles of U.S.," *U.S. National Bureau of Standards Technical Papers,* No. 123, vol. 12). After 30 cycles, 66 marbles showed loss of strength ranging from 1.2 to 62.1% and averaging 12.3% loss. The other 23 marbles showed increases in strength ranging from 0.5 to 43.9% and averaging 11.2% increase.

Weight change was also determined in this same investigation to afford another index of durability. Of 86 possible comparisons after 30 cycles of freezing and thawing, 16 showed no change in weight, 64 showed decreases in weight ranging from 0.01 to 0.28% and averaging 0.04% loss, while 6 showed increases in weight ranging from 0.01 to 0.08% and averaging 0.04%.

GYPSUM PRODUCTS

2-31. Gypsum Wallboard. Gypsum wallboard consists of a core of set gypsum surfaced with paper or other fibrous material firmly bonded to the core. It is designed to be used without addition of plaster for walls, ceilings, or partitions and provides a surface suitable to receive either paint or paper (see also Sec. 13.)

Wallboards are extensively used in "dry-wall" construction, where plaster is eliminated. They are available with one surface covered with aluminum or other heat-reflecting type of foil. Other wallboards have imitation wood-grain or other patterns on the exposed surface so that no additional decoration is required.

2-32. Gypsum Lath. Gypsum lath, or as it was formerly called, gypsum plasterboard, is similar to gypsum wallboard in that it consists of a core of set gypsum surfaced with paper. The paper for wallboard, however, is treated so that it is ready to receive paint or paper, while that for gypsum lath is specially designed or treated so that plaster will bond tightly to the paper. In addition, some lath provides perforations or other mechanical key devices to assist in holding the plaster firmly on the lath. It is also available with reflective foil backing (see also Sec. 13.)

2-33. Gypsum Sheathing Board. Gypsum sheathing boards are similar in construction to wallboards, except that they are provided with a water-repellent paper surface. They are commonly made in only ½-in. thickness, and with a nominal width of 24 or 48 in. They are made with either square edges or with V tongue-and-groove edges. Sheathing boards also are available with a water-repellent core.

2-34. Gypsum Partition Tile or Block. Gypsum tile or block are used for non-load-bearing partition walls and for protection of columns, elevator shafts, etc., against fire. They are made from set gypsum in the form of blocks that are 12 in. in height, 30 in. in length, and in thicknesses varying from 1½ to 6 in. The 1½- and 2-in. thicknesses are commonly used as furring block and are solid. The 3-, 4-, 5-, and 6-in. thicknesses are provided with round, elliptical, or rectangular cores parallel to the long dimension of the block to reduce weight.

2-35. Gypsum Plank. A precast gypsum product used particularly for roof construction is composed of a core of gypsum cast in the form of a plank and usually with tongue-and-groove metal edges and ends. The planks are available in two thicknesses—a 2-in. plank, which is 15 in. wide and 10 ft long, and a 3-in. plank which is 12 in. wide and 30 in. long. Planks are reinforced with wire fabric. (See ASTM C377.)

GLASS AND GLASS BLOCK

2-36. Glass for General Use. Glass is so widely used for decorative and utilitarian purposes in modern construction that it would require an encyclopedia to list all the varieties available. Clear glass for windows and doors is made in varying thicknesses or strengths, also in double layers to obtain additional thermal insulation. Safety glass, laminated from sheets of glass and plastic, or made with embedded wire reinforcement, is available for locations where breakage might be hazardous. For ornamental work, glass is available in a wide range of textures, colors, finishes, and shapes. See also Art. 9-13.

For many of its uses, glass presents no problem to the materials engineer. It is used in relatively small sheets or sections, which are frequently free-floating in their supports; therefore, expansion is not a problem. Most kinds of glass are highly resistant to weathering so that durability of the glass itself is not a problem, although the weathertightness of the frame may be a serious problem. Much glass is used in interiors for esthetic reasons where it is not subject to the stresses imposed upon many of the other masonry materials. An exception to this general condition is found in the case of glass block.

2-37. Glass Block. Glass block are made by first pressing or shaping half blocks to the desired form, then fusing the half blocks together to form a complete block. It is usually 3⅞ in. thick and 5¾, 7¾, or 11¾ in. square. The center of the block is hollow and is under a partial vacuum, which adds to the insulating properties of the block. Corner and radial block are also available to produce desired architectural effects.

Glass block are commonly laid up in a cement or a cement-lime mortar. Since there is no absorption by the block to facilitate bond of mortar, various devices are employed to obtain a mechanical bond. One such device is to coat the sides of the block with a plastic and embed therein particles of sand. The difficulty in obtaining permanent and complete bond sometimes leads to the opening up of mortar joints; a wall of glass block, exposed to the weather, may leak badly in a rainstorm unless unusual precautions have been taken during the setting of the block to obtain full and complete bond.

Glass block have a coefficient of thermal expansion which is from 1½ to 2 times that of other masonry. For this reason, large areas of block may expand against solid masonry and develop sufficient stress so that the block will crack. Manufacturers usually recommend an expansion joint every 10 ft or so, to prevent building up of pressure sufficient to crack the block. With adequate protection against expansion and with good workmanship, or with walls built in protected locations, glass-block walls are ornamental, sanitary, excellent light transmitters, and have rather low thermal conductivity.

ASBESTOS-CEMENT PRODUCTS

Asbestos-cement products are formed from a mixture of portland cement and asbestos fiber that has been wetted and then pressed into a board or sheet form. Organic fiber is added in some cases to promote resiliency and ease of machining. Curing agents, water-repellent admixtures, pigments, mineral granules, and mineral fillers may or may not be added, depending upon the particular end use of the product and its desired appearance.

2-38. Properties of Asbestos-Cement. Since these materials are used in sheet form and frequently under exposed weather conditions, the properties of most importance are flexural strength, deflection under load while wet, and absorption. The requirements of ASTM Specifications C220, C221, C222, and C223 for asbestos-cement products are given in Table 2-14.

Type U, or general-utility flat sheets, are suitable for exterior and interior use, having sufficient strength for general utility and construction purposes and suitable where high density, glossy finish, and lowest absorption are not essential. Sheets of this type are easy to cut and work, are bendable, and need not be drilled for nailing when thickness is ¼ in. or less (ASTM Specification C220).

Type F, or flexible, high-strength flat sheets, are suitable for exterior and interior use where a combination of strength and density, flexibility, smooth surface, and low absorption are desired. Sheets of this type are readily cuttable and workable and need not be drilled for fastening when thickness is ¼ in. or less (ASTM C220).

Corrugated sheets are designed to provide weather-exposed surfaces for roofs, walls, and other elements of buildings and other structures, and for decorative and

Table 2-14. Specification Requirements for Asbestos-Cement Products

	Thickness, in.	Size of test specimen, in.	Length of test span, in.	Minimum flexural strength, lb	Average deflection, in. at maximum load		Maximum water absorption, % by weight
					Parallel with length	Parallel with width	
Flat sheets							
Type F......	1/8	6 × 12	10	20	0.45	0.35	25
	3/16	6 × 12	10	50	0.30	0.20	25
	1/4	6 × 12	10	90	0.20	0.15	25
	3/8	6 × 12	10	190	0.12	0.10	25
	1/2	6 × 12	10	360	0.08	0.06	25
	5/8	6 × 12	10	560	0.05	0.03	25
Type U......	3/16	6 × 12	10	35	0.20	0.15	30
	1/4	6 × 12	10	65	0.15	0.10	30
	3/8	6 × 12	10	145	0.10	0.08	30
	1/2	6 × 12	10	260	0.07	0.05	30
	5/8	6 × 12	10	400	0.04	0.02	30
Corrugated sheets							
Type A......	9/32 min	12.6 × 36	30	500 per ft	....		25
Type B......	9/64 min	12.6 × 36	30	200 per ft	....		25
Roofing shingles. .	0.150 min						
Stronger direction....	....	6 × 12	10	27	0.15		25
Weaker direction....	....	6 × 12	10	22	0.15		25
Siding	0.150 min						
Stronger direction....	....	6 × 12	10	25	0.15		30
Weaker direction....	....	6 × 12	10	20	0.15		30

other purposes (ASTM C221). Corrugated sheets supplied under C221 may be Type A (standard) or Type B (lightweight). Type A sheets are about 3/8 in. thick and weigh about 4 lb per sq ft. They are designed as a structural roofing and siding sheet, and are suitable for use over sheathing or for direct application over girts and purlins. For structural roofing, spans should not exceed 4½ ft, and for siding, 5½ ft c to c supports. Type B sheets are about 3/16 in. thick and weigh about 2 lb per sq ft. They are designed as a roofing and siding sheet for use over sheathing, or for direct application on girts and purlins. For roofing, spans should not exceed 2½ ft, and for siding, 3½ ft c to c supports.

Roofing shingles are designed to provide weather-exposed surfaces on roofs of buildings (ASTM C222). They are available in a number of shapes and patterns.

Siding shingles and clapboards are designed to provide weather-exposed sidewall surfaces of buildings (ASTM C223).

Part 2. WOOD, METALS, AND OTHER MATERIALS

WOOD

2-39. Mechanical Properties of Wood. Because of its structure, wood has different strength properties parallel and perpendicular to the grain. Tensile, bending, and compressive strengths are greatest parallel to the grain and least across the grain, whereas shear strength is least parallel to the grain and greatest across

Table 2-15. Strength of Some Commercially Important Woods Grown in the U.S.*
(Results of tests on small, clear specimens in the green and air-dry condition)

Commercial and botanical name of species	Moisture content, %	Specific gravity	Static bending			Compression parallel to grain		Compression perpendicular to grain—fiber stress at proportional limit, psi	Shear parallel to grain—max shearing strength, psi	Hardness: Load required to embed a 0.444 in. ball to ½ its diameter, lb	
			Fiber stress at proportional limit, psi	Modulus of rupture, psi	Modulus of elasticity, psi	Fiber stress at proportional limit, psi	Max crushing strength, psi			End	Side
Ash, commercial white (*Fraxinus* sp.)	43	0.54	5,300	9,500	1,400	3,360	4,060	860	1,350	1,010	940
	12	0.58	8,900	14,600	1,680	5,580	7,280	1,510	1,920	1,680	1,260
Beech (*Fagus grandifolia*)	54	0.56	4,300	8,600	1,380	2,550	3,550	670	1,290	970	850
	12	0.64	8,700	14,900	1,720	4,880	7,300	1,250	2,010	1,590	1,300
Birch (*Betula* sp.)	62	0.57	4,400	8,700	1,560	2,640	3,510	550	1,160	910	850
	12	0.63	10,100	16,700	2,070	6,200	8,310	1,250	2,020	1,660	1,340
Chestnut (*Castanea dentata*)	122	0.40	3,100	5,600	930	2,080	2,470	380	800	530	420
	12	0.43	6,100	8,600	1,230	3,780	5,320	760	1,080	720	540
Cypress, southern (*Taxodium distichum*)	91	0.42	4,200	6,600	1,180	3,100	3,580	500	810	440	390
	12	0.46	7,200	10,600	1,440	4,740	6,360	900	1,000	660	510
Douglas fir (coast region) (*Pseudotsuga taxifolia*)	36	0.45	4,800	7,600	1,550	3,410	3,890	510	930	510	480
	12	0.48	8,100	11,700	1,920	6,450	7,420	910	1,140	760	670
Douglas fir ("Inland Empire" region) (*Pseudotsuga taxifolia*)	42	0.41	3,600	6,800	1,340	2,460	3,240	500	870	530	470
	12	0.44	7,400	11,300	1,610	5,520	6,700	950	1,190	720	630
Elm, American (*Ulmus americana*)	89	0.46	3,900	7,200	1,110	1,920	2,910	440	1,000	680	
	12	0.50	7,600	11,800	1,340	4,030	5,520	850	1,510	1,110	
Elm, rock (*Ulmus racemosa*)	48	0.57	4,600	9,500	1,190	2,970	3,780	750	1,270	980	
	12	0.63	8,000	14,800	1,540	4,700	7,050	1,520	1,920	1,510	
Gum, black (*Nyssa sylvatica*)	55	0.46	4,000	7,000	1,030	2,490	3,040	600	1,100	790	640
	12	0.50	7,300	9,600	1,200	3,470	5,520	1,150	1,340	1,240	810

Gum, red (*Liquidambar styraciflua*)	81	0.44	3,700	6,800	1,150	2,230	2,840	460	1,070	630	520
	12	0.49	8,100	11,900	1,490	4,700	5,800	860	1,610	950	690
Gum, tupelo (*Nyssa aquatica*)	97	0.46	4,200	7,300	1,050	2,690	3,370	590	1,190	800	710
	12	0.50	7,200	9,600	1,260	4,280	5,920	1,070	1,590	1,200	880
Hemlock, eastern (*Tsuga canadensis*)	111	0.38	3,800	6,400	1,070	2,600	3,080	440	850	500	400
	12	0.40	6,100	8,900	1,200	4,020	5,410	800	1,060	810	500
Hemlock, western (*Tsuga heterophylla*)	74	0.38	3,400	6,100	1,220	2,480	2,990	390	810	520	430
	12	0.42	6,800	10,100	1,490	5,340	6,210	680	1,170	940	580
Hickory, pecan (*Hicoria* sp.)	68	0.59	5,300	9,900	1,380	3,810	4,320	980	1,260	1,274	1,308
	12	0.65	9,100	16,300	1,780	6,360	8,280	2,040	1,770	1,930	1,820
Hickory, true (*Hicoria* sp.)	57	0.65	6,100	11,300	1,570	3,650	4,570	1,080	1,360		
	12	0.73	10,900	19,700	2,180	.....	8,970	2,310	2,140		
Honey locust (*Gleditsia triacanthos*)	63	0.60	5,600	10,200	1,290	3,320	4,420	1,420	1,660	1,440	1,390
	12	....	8,800	14,700	1,630	5,250	7,500	2,280	2,250	1,860	1,580
Larch, western (*Larix occidentalis*)	58	0.48	4,600	7,500	1,350	3,250	3,800	560	920	470	450
	12	0.52	7,900	11,900	1,710	5,950	7,490	1,080	1,360	1,110	760
Maple, sugar (*Acer saccharum*)	58	0.56	5,100	9,400	1,550	2,850	4,020	800	1,460	1,070	970
	12	0.63	9,500	15,800	1,830	5,390	7,830	1,810	2,330	1,840	1,450
Oak, red (*Quercus* sp.)	80	0.57	4,400	8,500	1,360	2,590	3,520	800	1,220	1,050	1,030
	12	0.63	8,400	14,400	1,810	4,610	6,920	1,260	1,830	1,490	1,300
Oak, white (*Quercus* sp.)	70	0.59	4,700	8,100	1,200	2,940	3,520	850	1,270	1,110	1,070
	12	0.67	7,900	13,900	1,620	4,350	7,040	1,410	1,890	1,420	1,330
Pines, southern yellow:											
Longleaf (*Pinus palustris*)	63	0.54	5,200	8,700	1,600	3,430	4,300	590	1,040	550	590
	12	0.58	9,300	14,700	1,990	6,150	8,440	1,190	1,500	920	870
Shortleaf (*Pinus echinata*)	81	0.46	3,900	7,300	1,390	2,500	3,430	440	850	410	440
	12	0.51	7,700	12,800	1,760	5,090	7,070	1,000	1,310	750	690
Poplar, yellow (*Liriodendron tulipifera*)	64	0.38	3,400	5,400	1,090	1,930	2,420	330	740	390	340
	12	0.40	6,100	9,200	1,500	3,550	5,290	580	1,100	560	450
Redwood (virgin) (*Sequoia sempervirens*)	112	0.38	4,800	7,500	1,180	3,700	4,200	520	800	570	410
	12	0.40	6,900	10,000	1,340	4,560	6,150	860	940	790	480
Spruce, eastern (*Picea sp.*)	46	0.38	3,300	5,600	1,110	2,120	2,600	290	710	390	340
	12	0.40	6,500	10,100	1,440	4,160	5,590	590	1,070	630	490

* From U.S. Forest Products Laboratory.

the grain. Except in plywood, the shearing strength of wood is usually governed by the parallel-to-grain direction.

The compressive strength of wood at an angle other than parallel or perpendicular to the grain is given by the following formula:

$$C_\theta = \frac{C_1 C_2}{C_1 \sin^2 \theta + C_2 \cos^2 \theta} \qquad (2\text{-}1)$$

in which C_θ is the strength at the desired angle θ with the grain, C_1 is the compressive strength parallel to grain, and C_2 is the compressive strength perpendicular to the grain.

Increasing moisture content reduces all strength properties except impact bending, in which green wood is stronger than dry wood. The differences are brought out in Table 2-15. In practice, no differentiation is made between the strength of green wood in engineering timbers because of seasoning defects that may occur in timbers as they dry and because large timbers normally are put into service without having been dried. This is not true of laminated timber, in which dry wood must be employed to obtain good glued joints. For laminated timber, higher stresses can be employed than for ordinary lumber. In general, compression and bending parallel to the grain are affected most severely by moisture, whereas modulus of elasticity, shear, and tensile strength are affected less. In practice, tensile strength is taken equal to the bending strength of wood.

In Table 2-15 are summarized also the principal mechanical properties of the most important American commercial species.

Values given in the table are average ultimate strengths. To obtain working stresses from these, the following must be considered: (1) Individual pieces may vary 25% above and below the average. (2) Values given are for standard tests that are completed in a few minutes. Over a period of years, however, wood may fail under a continuous load about $\frac{9}{16}$ that sustained in a standard test. (3) The modulus of rupture of a standard 2-in.-deep flexural-test specimen is greater than that of a deep beam. In deriving working stresses, therefore, factors of 75%, $\frac{9}{16}$, 0.9 (for an average-depth factor), and $\frac{3}{5}$ (for a safety factor) are applied to the average ultimate strengths to provide basic stresses, or working stresses, for blemishless lumber. These are still further reduced on the basis of knots, wane, slope of grain, shakes, and checks to provide working stresses for classes of commercial engineering timbers. (See Sec. 8 for engineering design in timber.)

2-40. Hygroscopic Properties of Wood. Because of its nature, wood tends to absorb moisture from the air when the relative humidity is high, and to lose it when the relative humidity is low. Moisture imbibed into the cell walls causes the wood to shrink and swell as the moisture content changes with the relative humidity of the surrounding air. The maximum amount of imbibed moisture the cell walls can hold is known as the fiber-saturation point, and for most species is in the vicinity of 25 to 30% of the oven-dry weight of the wood. Free water held in the cell cavities above the fiber-saturation point has no effect upon shrinkage or other properties of the wood. Changes in moisture content below the fiber-saturation point cause negligible shrinkage or swelling along the grain, and such shrinkage and swelling are normally ignored; but across the grain, considerable shrinkage and swelling occur in both the radial and tangential direction. Tangential shrinkage (as in flat-cut material) is normally approximately 50% greater than radial shrinkage (as in edge-grain material).

Table 2-16 gives average-shrinkage values in the radial and tangential directions for a number of commercial species as they dry from the fiber-saturation point (or green condition) to oven-dry. For intermediate changes in moisture content, shrinkage is proportional; e.g., if the fiber-saturation point is assumed to be 28% (average), a change in moisture content from 21 to 14% would cause shrinkage one-quarter as great as the shrinkage from 28% to zero.

Table 2-17 gives measured average moisture contents of wood in buildings in various parts of the United States.

Table 2-16. Shrinkage Values for Commercially Important Woods Grown in the United States.

Species	Shrinkage (% of dimension when green) from green to oven-dried to 0% moisture (test values)		
	Radial	Tangential	Volumetric
Ash, commercial white	4.6	7.5	12.8
Basswood	6.6	9.3	15.8
Beech	5.1	11.0	16.3
Birch	6.9	8.9	16.3
Cedar:			
Eastern red	3.1	4.7	7.8
Northern white	2.1	4.7	7.0
Port Orford	4.6	6.9	10.1
Southern white	2.8	5.2	8.4
Western red	2.4	5.0	7.7
Cherry, black	3.7	7.1	11.5
Chestnut	3.4	6.7	11.6
Cypress, southern	3.8	6.2	10.5
Douglas fir:			
Coast region	5.0	7.8	11.8
"Inland Empire" region	4.1	7.6	10.9
Rocky Mountain region	3.6	6.2	10.6
Elm:			
American	4.2	9.5	14.6
Rock	4.8	8.1	14.1
Gum:			
Black	4.4	7.7	13.9
Red	5.2	9.9	15.0
Hemlock:			
Eastern	3.0	6.8	9.7
Western	4.3	7.9	11.9
Hickory:			
Pecan	4.9	8.9	13.6
True	7.3	11.4	17.9
Larch, western	4.2	8.1	13.2
Maple:			
Red	4.0	8.2	13.1
Sugar	4.9	9.5	14.9
Oak:			
Red	4.3	9.0	14.8
White	5.4	9.3	16.0
Pine:			
Loblolly	4.8	7.4	12.3
Longleaf	5.1	7.5	12.2
Northern white	2.3	6.0	8.2
Ponderosa	3.9	6.3	9.6
Shortleaf	4.4	7.7	12.3
Sugar	2.9	5.6	7.9
Western white	4.1	7.4	11.8
Poplar, yellow	4.0	7.1	12.3
Redwood	2.6	4.4	6.8
Spruce:			
Eastern	4.3	7.7	12.6
Engelmann	3.4	6.6	10.4
Walnut, black	5.2	7.1	11.3

* From U.S. Forest Products Laboratory.

Table 2-17. Measured Moisture-content Values for Various Wood Items*

Use of lumber	Moisture content (% of weight of oven-dry wood) for					
	Dry Southwestern states		Damp Southern coastal states		Remainder of the United States	
	Avg	Individual pieces	Avg	Individual pieces	Avg	Individual pieces
Interior finish woodwork and softwood flooring	6	4–9	11	8–13	8	5–10
Hardwood flooring	6	5–8	10	9–12	7	6–9
Siding, exterior trim, sheathing, and framing	9	7–12	12	9–14	12	9–14

* From U.S. Forest Products Laboratory.

2-41. Commercial Grades of Wood. Lumber is graded by the various associations of lumber manufacturers having jurisdiction over various species. Two principal sets of grading rules are employed: (1) for softwoods, and (2) for hardwoods.

Softwoods. Softwood lumber is classified as dry, moisture content 19% or less; and green, moisture content above 19%.

According to the American Softwood Lumber Standard, softwoods are classified according to use as:

Yard lumber: lumber of grades, sizes, and patterns generally intended for ordinary construction and general building purposes.

Structural lumber: lumber 2 in. or more in nominal thickness and width for use where working stresses are required.

Factory and shop lumber: lumber produced or selected primarily for remanufacturing purposes.

Softwoods are classified according to extent of manufacture as:

Rough lumber: lumber that has not been dressed (surfaced) but has been sawed, edged, and trimmed.

Dressed (surfaced) lumber: lumber that has been dressed by a planing machine (for the purpose of attaining smoothness of surface and uniformity of size) on one side (S1S), two sides (S2S), one edge (S1E), two edges (S2E), or a combination of sides and edges (S1S1E, S1S2, S2S1E, S4S).

Worked lumber: lumber that, in addition to being dressed, has been matched, shiplapped or patterned:

> *Matched lumber:* lumber that has been worked with a tongue on one edge of each piece and a groove on the opposite edge.
>
> *Shiplapped lumber:* lumber that has been worked or rabbeted on both edges, to permit formation of a close-lapped joint.
>
> *Patterned lumber:* lumber that is shaped to a pattern or to a molded form.

Softwoods are also classified according to nominal size:

Boards: lumber less than 2 in. in nominal thickness and 2 in. or more in nominal width. Boards less than 6 in. in nominal width may be classified as strips.

Dimension: lumber from 2 in. to, but not including, 5 in. in nominal thickness, and 2 in. or more in nominal width. Dimension may be classified as framing, joists, planks, rafters, studs, small timbers, etc.

Timbers: lumber 5 in. or more nominally in least dimension. Timber may be classified as beams, stringers, posts, caps, sills, girders, purlins, etc.

Actual sizes of lumber are less than the nominal sizes, because of shrinkage and dressing. In general, dimensions of dry boards, dimension lumber, and timber less than 2 in. wide or thick are 1/4 in. less than nominal; from 2 to 6 in. wide or thick, 1/2 in. less, and above 6 in. wide or thick, 3/4 in. less. Green lumber less than 2 in. wide or thick is 1/32 in. more than dry; from 2 to 4 in. wide or thick, 1/16 in. more, 5 and 6 in. wide or thick, 1/8 in. more, and 8 in. or above in width and thickness, 1/4 in. more than dry lumber. (See Table 8-2.) There are exceptions, however.

Yard lumber is classified on the basis of quality as:

Select: lumber of good appearance and finishing qualities.
- Suitable for natural finishes.
  - Practically clear.
  - Generally clear and of high quality.
- Suitable for paint finishes.
  - Adapted to high-quality paint finishes.
  - Intermediate between high-finishing grades and common grades, and partaking somewhat of the nature of both.

Common: lumber suitable for general construction and utility purposes.
- For standard construction use.
  - Suitable for better-type construction purposes.
  - Well adapted for good standard construction.
  - Designed for low-cost temporary construction.
- For less exacting construction purposes.
  - Low quality.
  - Lowest recognized grade must be usable.

Structural lumber is assigned modulus of elasticity values and working stresses in bending, compression parallel to grain, compression perpendicular to grain, and horizontal shear in accordance with ASTM procedures. These values take into account such factors as sizes and locations of knots, slope of grain, wane, and shakes or checks, as well as such other pertinent features as rate of growth and proportions of summerwood.

Factory and shop lumber is graded with reference to its use for doors and sash, or on the basis of characteristics affecting its use for general cut-up purposes, or on the basis of size of cutting. The grade of factory and shop lumber is determined by the percentage of the area of each board or plank available in cuttings of specified or of given minimum sizes and qualities. The grade of factory and shop lumber is determined from the poor face, although the quality of both sides of each cutting must be considered.

Hardwoods. Because of the great diversity of applications for hardwood both in and outside the construction industry, hardwood grading rules are based upon the proportion of a given piece that can be cut into smaller pieces of material clear on one or both sides and not less than a specified size. Grade classifications are therefore based on the amount of clear usable lumber in a piece.

Special grading rules of interest in the construction industry cover hardwood interior trim and moldings, in which one face must be practically free of imperfections and in which Grade A may further limit the amount of sapwood as well as stain. Hardwood dimension rules, in addition, cover clears, which must be clear both faces; clear one face; paint quality, which can be covered with paint; core, which must be sound on both faces and suitable for cores of glued-up panels; and sound, which is a general-utility grade.

Hardwood flooring is graded under two separate sets of rules: (1) for maple, birch, beech, and pecan; and (2) for red and white oak. In both sets of rules, color and quality classifications range from top-quality to the lower utility grades. Oak may be further subclassified as quarter-sawed and plain-sawed. In all grades, top-quality material must be uniform in color, whereas other grades place no limitation on color.

Shingles are graded under special rules, usually into three classes: numbers 1, 2, and 3. Number 1 must be all edge grain and strictly clear, containing no sapwood. Numbers 2 and 3 must be clear to a distance far enough away from the butt to be well covered by the next course of shingles.

2-42. Destroyers and Preservatives. The principal destroyers of wood are decay, caused by fungus, and attack by a number of animal organisms of which termites, carpenter ants, grubs of a wide variety of beetles, teredo, and limnoria are the principal offenders. In addition, fire annually causes widespread destruction of wood structures.

Decay will not occur if wood is kept well ventilated and air-dry or, conversely, if it is kept continuously submerged so that air is excluded.

Most termites in the United States are subterranean and require contact with the soil. The drywood and dampwood termites found along the southern fringes of the country and along the west coast, however, do not require direct soil contact and are more difficult to control.

Table 2-18. Resistance of Various Heartwoods to Decay

Heartwood durable even when used under conditions that favor decay:	Cedar, Alaska Cedar, eastern red Cedar, northern white Cedar, Port Orford Cedar, southern white Cedar, western red Chestnut Cypress, southern Locust, black Osage-orange Redwood Walnut, black Yew, Pacific
Heartwood of intermediate durability but nearly as durable as some of the species named in the high-durability group:	Douglas fir (dense) Honey locust Oak, white Pine, southern yellow (dense)
Heartwood of intermediate durability:	Douglas fir (unselected) Gum, red Larch, western Pine, southern yellow (unselected) Tamarack
Heartwood between the intermediate and the nondurable group:	Ash, commercial white Beech Birch, sweet Birch, yellow Hemlock, eastern Hemlock, western Hickory Maple, sugar Oak, red Spruce, black Spruce, Engelmann Spruce, red Spruce, Sitka Spruce, white
Heartwood low in durability when used under conditions that favor decay:	Aspen Basswood Cottonwood Fir, commercial white Willow, black

Teredo, limnoria, and other water-borne wood destroyers are found only in salt or brackish waters.

Various wood species vary in natural durability and resistance to decay and insect attack. The sapwood of all species is relatively vulnerable; only the heartwood can be considered to be resistant. Table 2-18 lists the common species in descending order of resistance. Such a list is only approximate, and individual pieces deviate considerably.

Preservatives employed to combat the various destructive agencies may be subdivided into oily, water-soluble salts, and solvent-soluble organic materials. The principal oily preservatives are coal-tar creosote and creosote mixed with petroleum. The most commonly employed water-soluble salts are zinc chloride, chromated zinc chloride, sodium fluoride, copper salts, mercuric salts, and a variety of other materials that are often sold under various proprietary names. The principal solvent-soluble organic materials are chlorinated phenols, such as pentachlorphenol.

Preservatives may be applied in a variety of ways, including brushing and dipping, but for maximum treatment, pressure is required to provide deep side-grain penetration. Butts of poles and other parts are sometimes placed in a hot boiling creosote or salt solution, and after the water in the wood has been converted to steam, they are quickly transferred to a cold vat of the same preservative. As the steam condenses, it produces a partial vacuum, which draws the preservative fairly deeply into the surface.

Pressure treatments may be classified as full-cell and empty-cell. In the full-cell treatment, a partial vacuum is first drawn in the pressure-treating tank to withdraw most of the air in the cells of the wood. The preservative is then let in without breaking the vacuum, after which pressure is applied to the hot solution. After treatment is completed, the individual cells are presumably filled with preservative. In the empty-cell method, no initial vacuum is drawn, but the preservative is pumped in under pressure against the back pressure of the compressed air in the wood. When the pressure is released, the air in the wood expands and forces out excess preservative, leaving only a coating of preservative on the cell walls.

Retentions of preservative depend on the application. For teredo-infested harbors, full-cell treatment to refusal may be specified, with a range from 16 to 20 lb per cu ft of wood. For ordinary decay conditions and resistance to termites and other destroyers of a similar nature, the empty-cell method may be employed with retentions in the vicinity of 6 to 8 lb of creosote per cu ft of wood. Salt retentions generally range in the vicinity of 1½ to 3 lb of dry salt retained per cu ft of wood.

Solvent-soluble organic materials, such as pentachlorphenol, are commonly employed for the treatment of sash and door parts to impart greater resistance to decay. This is commonly done by simply dipping the parts in the solution and then allowing them to dry. As the organic solvent evaporates, it leaves the water-insoluble preservative behind in the wood.

These organic materials are also employed for general preservative treatment, including fence posts and structural lumber. The water-soluble salts and solvent-soluble organic materials leave the wood clean and paintable. Creosote in general cannot be painted over, although partial success can be achieved with top-quality aluminum-flake pigment paints.

Treatment against fire consists generally of salts containing ammonium and phosphates, of which mono-ammonium phosphate and di-ammonium phosphate are widely employed. At retentions of 3 to 5 lb of dry salt per cu ft, the wood does not support its own combustion, and the afterglow when fire is removed is short. A variety of surface treatments is also available, most of which depends on the formation of a blanket of inert-gas bubbles over the surface of the wood in the presence of flame or other sources of heat. The blanket of bubbles insulates the wood beneath and retards combustion.

2-43. Glues and Adhesives. A variety of adhesives is now available for use with wood, depending on the final application. The older adhesives include animal glue, casein glue, and a variety of vegetable glues, of which soybean is today the most important. Animal glues provide strong, tough, easily made joints, which, however, are not moisture-resistant. Casein mixed with cold water, when properly formulated, provides highly moisture-resistant glue joints, although they cannot be called waterproof. The vegetable glues have good dry strength but are not moisture-resistant.

The principal high-strength glues today are synthetic resins, of which phenol formaldehyde, urea formaldehyde, resorcinol formaldehyde, and melamine formaldehyde are the most important. Phenol, resorcinol, and melamine provide glue joints that are completely waterproof and will not separate when properly made even on boiling. Urea formaldehyde provides a glue joint of high moisture resistance, although not quite so good as the other three. Phenol and melamine require application of heat, as well as pressure, to cure the adhesive, whereas urea and resorcinol can both be formulated to be mixed with water at ordinary temperatures and hardened without application of heat above room temperature. Waterproof plywood is commonly made in hot-plate presses with phenolic or melamine adhesive. Resorcinol is employed where heat cannot be applied, as in a variety of assembly operations and the manufacture of laminated parts like ships' keels, which must have the maximum in waterproof qualities. Epoxide resins provide strong joints.

An emulsion of polyvinyl acetate serves as a general-purpose adhesive, for general assembly operations where maximum strength and heat or moisture resistance are not required. This emulsion is merely applied to the surfaces to be bonded, after which they are pressed together and the adhesive is allowed to harden.

2-44. Plywood. As ordinarily made, plywood consists of thin sheets, or veneers, of wood glued together. The grain is oriented at right angles in adjacent plies. To obtain plywood with balance—that is, which will not warp, shrink, or twist unduly—the plies must be carefully selected and arranged to be mirror images of each other with respect to the central plane. All plies are at the same moisture content at the time of manufacture. The outside plies or faces are parallel to each other and are of species that have the same shrinkage characteristics. The same holds true of the cross bands or inner plies, and finally of the core or central ply itself. As a consequence, plywood, as ordinarily made, has an odd number of plies, the minimum being three. Matching the plies involves thickness, the species of wood with particular reference to shrinkage, equal moisture content, and making certain that the grain in corresponding plies is parallel.

Principal advantages of plywood over lumber are its more nearly equal strength properties in length and width, greater resistance to checking, greatly reduced shrinkage and swelling, and resistance to splitting.

The approach to equalization of strength of plywood in the various directions is obtained at the expense of strength in the parallel-to-grain direction; i.e., plywood is not so strong in the direction parallel to its face plies as lumber is parallel to the grain. But plywood is considerably stronger in the direction perpendicular to its face plies than wood is perpendicular to the grain. Furthermore, the shearing strength of plywood in a plane perpendicular to the plane of the plywood is very much greater than that of ordinary wood parallel to the grain. In a direction parallel to the plane of the plywood, however, the shearing strength of plywood is less than that of ordinary wood parallel to the grain, because in this direction rolling shear occurs in the plywood; i.e., the fibers in one ply tend to roll rather than to slide.

Depending on whether plywood is to be used for general utility or for decorative purposes, the veneers employed may be cut by peeling from the log, by slicing, or today very rarely, by sawing. Sawing and slicing give the greatest freedom and versatility in the selection of grain. Peeling provides the greatest volume and the most rapid production, because the logs are merely rotated against a flat knife and the veneer is peeled off in a long continuous sheet.

Plywood is classified as interior or exterior, depending on the type of adhesive employed. Interior-grade plywood must have a reasonable degree of moisture resistance but is not considered to be waterproof. Exterior-grade plywood must be completely waterproof and capable of withstanding immersion in water or prolonged exposure to outdoor conditions.

In addition to these classifications, plywood is further subclassified in a variety of ways depending on the quality of the surface ply. Top-quality is clear on one or both faces, except for occasional patches. Lower qualities permit sound defects, such as knots and similar blemishes, which do not detract from the general utility of the plywood but detract from its finished appearance.

2-45. Wood Bibliography

Forest Products Laboratory, Forest Service, U.S. Department of Agriculture: "Wood Handbook," Government Printing Office, Washington, D.C.

Hunt, G. M., and G. A. Garratt: "Wood Preservation," McGraw-Hill Book Company, New York.

National Hardwood Lumber Association, Chicago, Ill.: "Rules for the Measurement and Inspection of Hardwood Lumber, Cypress, Veneer, and Thin Lumber."

National Lumber Manufacturers Association, Washington, D.C.: "National Design Specifications for Stress-grade Lumber and Its Fastenings."

U.S. Department of Commerce, National Bureau of Standards, Washington, D.C.: American Softwood Lumber Standard, Voluntary Practice Standard PS20; Douglas Fir Plywood, Commercial Standard CS 45; Hardwood Plywood, Commercial Standard CS 35.
Western Wood Products Association, Portland, Ore.: "Western Woods Use Book."

FERROUS METALS

The iron-carbon equilibrium diagram in Fig. 2-1 shows that, under equilibrium conditions (slow cooling) if not more than 2.0% carbon is present, a solid solution of carbon in gamma iron exists at elevated temperatures. This is called austenite. If the carbon content is less than 0.8%, cooling below the A_3 temperature line causes transformation of some of the austenite to ferrite, which is substantially pure alpha iron (containing less than 0.01% carbon in solution). Still further cooling to below the A_1 line causes the remaining austenite to transform to pearlite—the eutectoid

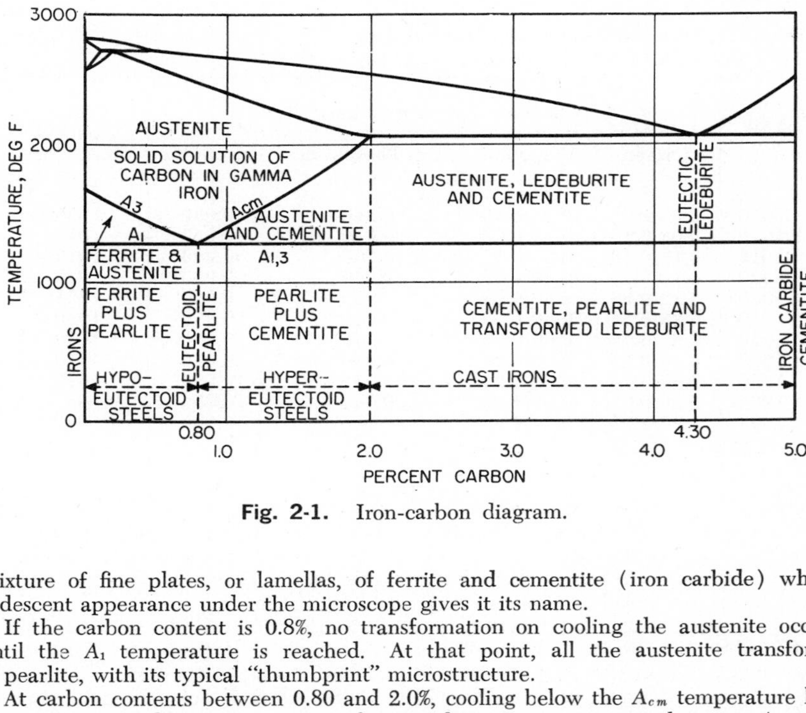

Fig. 2-1. Iron-carbon diagram.

mixture of fine plates, or lamellas, of ferrite and cementite (iron carbide) whose iridescent appearance under the microscope gives it its name.

If the carbon content is 0.8%, no transformation on cooling the austenite occurs until the A_1 temperature is reached. At that point, all the austenite transforms to pearlite, with its typical "thumbprint" microstructure.

At carbon contents between 0.80 and 2.0%, cooling below the A_{cm} temperature line causes iron carbide, or cementite, to form in the temperature range between A_{cm} and $A_{1,3}$. Below $A_{1,3}$, the remaining austenite transforms to pearlite.

2-46. Steels. The plain iron-carbon metals with less than 0.8% carbon content consist of ferrite and pearlite and provide the low (0.06 to 0.30%), medium (0.30 to 0.50%), and high carbon (0.50 to 0.80%) steels called hypoeutectoid steels. The higher-carbon or hypereutectoid tool steels contain 0.8 to 2.0% carbon and consist of pearlite and cementite. The eutectoid steels occurring in the vicinity of 0.8% carbon are essentially all pearlite.

2-47. Irons. Metals containing substantially no carbon (several hundredths of 1%) are called irons, of which wrought iron, electrolytic iron, and 'ingot" iron are examples.

Wrought iron, whether made by the traditional puddling method or by mixing very low carbon iron and slag, contains a substantial amount of slag. Because it contains very little carbon, it is soft, ductile, and tough, and like low-carbon ferrous metals generally, is relatively resistant to corrosion. It is easily worked. When broken, it shows a fibrous fracture because of the slag inclusions. "Ingot" iron is a very low carbon iron containing no slag, which is also soft, ductile, and tough.

Above 2.0% carbon content is the region of the cast irons. Above the $A_{1,3}$ temperature, austenite, the eutectic ledeburite and cementite occur; below the $A_{1,3}$ temperature, the austenite transforms to pearlite, and a similar transformation of the ledeburite occurs.

When the silicon content is kept low, and the metal is cooled rapidly, *white cast iron* results. It is hard and brittle because of the high cementite content. White cast iron as such has little use; but when it is reheated and held a long time in the vicinity of the transformation temperature, then cooled slowly, the cementite decomposes to ferrite and nodular or temper carbon. The result is black-heart *malleable iron*. If the carbon is removed during malleabilization, white-heart malleable iron results.

Table 2-19. Standard Steels for Hot-rolled Bars
(Basic open-hearth and acid bessemer carbon steels)

AISI No.	Chemical composition limits, %				Corresponding SAE No.
	Carbon	Manganese	max Phosphorus	max Sulfur	
C1006	0.08 max	0.25/0.40	0.040	0.050	1006
C1010	0.08/0.13	0.30/0.60	0.040	0.050	1010
C1015	0.13/0.18	0.30/0.60	0.040	0.050	1015
C1020	0.18/0.23	0.30/0.60	0.040	0.050	1020
C1025	0.22/0.28	0.30/0.60	0.040	0.050	1025
C1030	0.28/0.34	0.60/0.90	0.040	0.050	1030
C1040	0.37/0.44	0.60/0.90	0.040	0.050	1040
C1050	0.48/0.55	0.60/0.90	0.040	0.050	1050
C1070	0.65/0.75	0.60/0.90	0.040	0.050	1070
C1085	0.80/0.93	0.70/1.00	0.040	0.050	1085
C1095	0.90/1.03	0.30/0.50	0.040	0.050	1095

If the silicon content is raised, and the metal is cooled relatively slowly, *gray cast iron* results. It contains cementite, pearlite, ferrite, and some free carbon, which gives it its gray color. Gray iron is considerably softer and tougher than white cast iron and is generally used for castings of all kinds. Often, it is alloyed with elements like nickel, chromium, copper, and molybdenum.

At 5.0% carbon, the end product is hard, brittle iron carbide or cementite.

2-48. Standard Steels. The American Iron and Steel Institute and the Society of Automotive Engineers have designated standard compositions for various steels including plain carbon steels and alloy steels. AISI and SAE numbers and compositions for several representative hot-rolled carbon-steel bars are given in Table 2-19.

Most of the steel used for construction is low- to medium-carbon, relatively mild, tough, and strong, fairly easy to work by cutting, punching, riveting, and welding. Table 2-20 summarizes the most important carbon steels, cast irons, wrought iron, and low-alloy steels used in construction as specified by the ASTM. This includes structural steels, bolt steels, various cast metals, and reinforcing steels.

Prestressed concrete imposes special requirements for reinforcing steel. It must be of high strength with a high yield point and minimum creep in the working range. ASTM specifications require the mechanical properties in Table 2-21.

2-49. Heat-treatment and Hardening of Steels. Heat-treated and hardened steels are sometimes required in building operations. The most familiar heat-treatment is annealing, a reheating operation in which the metal is usually heated to the austenitic range and cooled slowly to obtain the softest, most ductile state.

Cold working is often preceded by annealing. Annealing may be only partial, just sufficient to relieve internal stresses that might cause deformation or cracking, but not enough to reduce markedly the increased strength and yield point brought about by the cold working, for example.

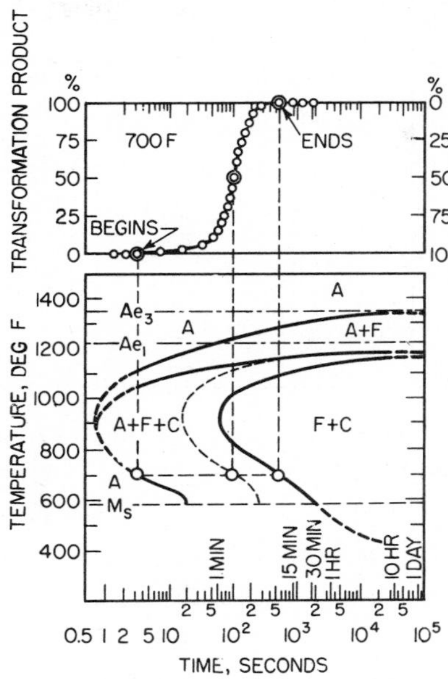

Fig. 2-2. Typical isothermal transformation curve at a single temperature (*top*) and complete isothermal transformation diagram (*bottom*), showing effects of heat-treatment. (*"Metals Handbook," American Society for Metals.*)

Heat-treatments are best exemplified by isothermal transformation curves, also called S curves and *TTT* (transformation-temperature-time) curves. Figure 2-2 is typical. It shows that above A_{e3} austenite is stable, between A_{e1} and A_{e3} some austenite transforms to ferrite, but below A_{e1} various transformations occur at various rates depending on temperature. A specimen cooled to above the "knee" at 900°F shows proeutectoid transformation to ferrite (between the first two curves) followed by transformation to pearlite (ferrite plus carbide) in the second zone, until transformation is complete. At the knee, transformation is extremely rapid, and the proeutectoid formation of ferrite disappears. A specimen cooled rapidly to below the knee and held at a lower temperature transforms completely to ferrite plus carbide in a strong, hard form called bainite. If cooled quickly to below the M_S temperature and held, the transformation product is the very hard, strong, but relatively brittle form known as martensite.

In plain carbon steels, the knee of the curve is extremely critical because of the short transformation time. It may be impossible to cool the centers of heavy parts fast enough to get below the knee, and hardenability is therefore limited. The addition of alloying elements almost always shifts the knee to the right, that is, makes transformation more sluggish and therefore increases the hardenability of the metal. Many alloys, such as several stainless steels, remain austenitic at ordinary temperatures; others can easily be made martensitic throughout, but still others merely become ferritic and cannot be hardened.

Often, a hard surface is required on a soft, tough core. Two principal casehardening methods are employed. For *case carburizing,* the low- to medium-carbon steel is packed in carbonaceous materials and heated to the austenite range. Carbon

Table 2-20. ASTM Requirements for Structural, Reinforcing, and Fastening Steels

	ASTM specification	Tensile strength, min, ksi	Yield point, min, ksi	Elongation in 8 in., min, %	Elongation in 2 in., min, %	Bend test, ratio of bend diameter, in., to specimen thickness, in.				
						0–¾	¾–1	1–1½	1½–2	Over 2
Structural steel	A36	58–80	36	20	23	½	1½	2	2½	3
High-strength structural steel	A440									
High-strength, low-alloy, structural Mn-V steel.	A441	63–70	42–50	18	21	1	1½	2	2½	3
High-strength, low-alloy, structural steel	A242	63–70	42–50	18	21	1	1½	2	2½	3
High-strength, low-alloy columbium-vanadium steels . . .	A572	60–80	42–65	20–15	24–18	Depends on grade				
High-strength, low-alloy structural steel	A588	63–70	42–50	18	21	1	1½	2	2½	3
						0–1	1–2½	Over 2½		
High-yield-strength, quenched and tempered alloy steel	A514	105–135	90–100	. . . .	17–18	2	3	4		
Structural steel	A529	60–85	42	19	. . . .	1				
High-strength steel bolts	A325	105	77	. . . .	14					

High-strength, alloy steel bolts	A490	140-150	115-130	. . . .	14	
Carbon-steel forgings, Cl	A235	66	33	. . . .	22-23	
Heavy-gage, hot-rolled sheets	A245	45-55	25-40	20-14	27-15	
Carbon-steel sheet and strip	A570	45-58	25-42	20-12	27-13	
Bolts and nuts, machine	A307	60	. . .	. . . .	18	
Sheetpiling.	A328	70	38.5	17	. . . .	
Cast steel, 65-35, annealed	A27	65	35	. . . .	24	
High-strength cast steel, 80-50	A148	80	50	. . . .	22	
						90° bend test; ratio of pin diameter to specimen diameter
Reinforcing steel for concrete:						
Billet-steel bars . . .	A615					
Grade 40	. . . .	70	40	7-11	9 (No. 14-18)	Under No. 6: 3 Nos. 6, 7, 8: 4 Nos. 9, 10, 11: 5
Grade 60	. . . .	90	60	7-9	9 (No. 9-18)	Under No. 6: 4 Nos. 6, 7, 8: 5 Nos. 9, 10, 11: 6
Grade 75	. . . .	100	75	5 (No. 11-14)	6 (No. 11-18)	No. 11: 8
Rail-steel bars	A616					
Grade 50	. . . .	90	50	$\frac{1,000,000}{\text{tens. str.}}$		
Grade 60	. . . .	90	60	$\frac{1,000,000}{\text{tens. str.}}$		
Cold-drawn wire . .	A82	70-80	56-70			W7 or under, 180°: 1 over W7, 180°: 2

Table 2-21. ASTM Requirements for Prestressing Wire

Material	ASTM specification No.	Tensile strength, ksi	Minimum yield point, ksi
Stress-relieved strand for prestressed concrete:	A416		
Grade 250	....	250	85% of breaking strength, 1% extension
Grade 270	....	270	
Stress-relieved wire for prestressed concrete:	A421		
BA	....	240	192, 1% extension
WA	....	235–250	188–200, 1% extension

diffuses into the surface, providing a hard, high-carbon case when the part is cooled. For *nitriding*, the part is exposed to ammonia gas or a cyanide at moderately elevated temperatures. Extremely hard nitrides are formed in the case and provide a hard surface. Cast iron is frequently cast against metal inserts in the molds. The cast metal is chilled by the inserts and forms white cast iron, whereas the rest of the more slowly cooled metal becomes gray cast iron.

2-50. Steel Alloys. Plain carbon steels can be given a great range of properties by heat-treatment and by working; but addition of alloying elements greatly extends those properties or makes the heat-treating operations easier and simpler. For example, combined high tensile strength and toughness, corrosion resistance, high-speed cutting, and many other specialized purposes require alloy steels. However, the most important effect of alloying is the influence on hardenability. The most important alloying elements from the standpoint of building, and their principal effects, are summarized below:

Aluminum restricts grain growth during heat-treatment and promotes surface hardening by nitriding.

Chromium is a hardener, promotes corrosion resistance (see stainless steels, below), and promotes wear resistance.

Copper promotes resistance to atmospheric corrosion and is sometimes combined with molybdenum for this purpose in low-carbon steels and irons. It strengthens steel and increases the yield point without unduly changing elongation or reduction of area.

Manganese in low concentrations promotes hardenability and nondeforming, nonshrinking characteristics for tool steels. In high concentrations, the steel is austenitic under ordinary conditions, is extremely tough, and work-hardens readily. It is therefore used for teeth of power-shovel dippers, railroad frogs, rock crushers, and similar applications.

Molybdenum is usually associated with other elements, especially chromium and nickel. It increases corrosion resistance, raises tensile strength and elastic limit without reducing ductility, promotes casehardening and improves impact resistance.

Nickel boosts tensile strength and yield point without reducing ductility; increases low-temperature toughness, whereas ordinary carbon steels become brittle; promotes casehardening; and in high concentrations improves corrosion resistance under severe conditions. It is commonly associated with chromium (see stainless steels, below). *Invar* contains 36% nickel.

Silicon strengthens low-alloy steels; improves oxidation resistance; with low carbon yields transformer steel, because of low hysteresis loss and high permeability; in high concentrations provides hard, brittle castings, resistant to corrosive chemicals, useful in plumbing lines for chemical laboratories.

Sulfur promotes free machining, especially in mild steels.

Titanium prevents intergranular corrosion of stainless steels by preventing grain-boundary depletion of chromium during such operations as welding and heat-treatment.

Table 2-22. Effects of Alloying Elements in Steel*

Element	Solid solubility		Influence on ferrite	Influence on austenite (hardenability)	Influence exerted through carbide		Principal functions
	In gamma iron	In alpha iron			Carbide-forming tendency	Action during tempering	
Aluminum (Al)	1.1 % (increased by C)	36 %	Hardens considerably by solid solution	Increases hardenability mildly, if dissolved in austenite	Negative (graphitizes)	. . . .	1. Deoxides efficiently 2. Restricts grain growth (by forming dispersed oxides or nitrides) 3. Alloying element in nitriding steel
Chromium (Cr)	12.8 % (20 % with 0.5 % C)	Unlimited	Hardens slightly; increases corrosion resistance	Increases hardenability moderately	Greater than Mn; less than W	Mildly resists softening	1. Increases resistance to corrosion and oxidation 2. Increases hardenability 3. Adds some strength at high temperatures 4. Resists abrasion and wear (with high carbon)
Cobalt (Co)	Unlimited	75 %	Hardens considerably by solid solution	Decreases hardenability as dissolved	Similar to Fe	Sustains hardness by solid solution	1. Contributes to red hardness by hardening ferrite
Manganese (Mn)	Unlimited	3 %	Hardens markedly; reduces plasticity somewhat	Increases hardenability moderately	Greater than Fe; less than Cr	Very little, in usual percentages	1. Counteracts brittleness from the sulfur 2. Increases hardenability inexpensively
Molybdenum (Mo)	3 % ± (8 % with 0.3 % C)	37.5 % (less with lowered temp)	Provides age-hardening system in high Mo-Fe alloys	Increases hardenability strongly (Mo > Cr)	Strong; greater than Cr	Opposes softening by secondary hardening	1. Raises grain-coarsening temperature of austenite 2. Deepens hardening 3. Counteracts tendency toward temper brittleness 4. Raises hot and creep strength, red hardness 5. Enhances corrosion resistance in stainless steel 6. Forms abrasion-resisting particles
Nickel (Ni)	Unlimited	10 % (irrespective of carbon content)	Strengthens and toughens by solid solution	Increases hardenability mildly, but tends to retain austenite with higher carbon	Negative (graphitizes)	Very little in small percentages	1. Strengthens unquenched or annealed steels 2. Toughens pearlitic-ferritic steels (especially at low temperature) 3. Renders high-chromium iron alloys austenitic

Table 2-22. Effects of Alloying Elements in Steel (*Continued*)

Element	Solid solubility		Influence on ferrite	Influence on austenite (hardenability)	Influence exerted through carbide		Principal functions
	In gamma iron	In alpha iron			Carbide-forming tendency	Action during tempering	
Phosphorus, (P)	0.5 %	2.8 % (irrespective of carbon content)	Hardens strongly by solid solution	Increases hardenability	Nil	....	1. Strengthens low-carbon steel 2. Increases resistance to corrosion 3. Improves machinability in free-cutting steels
Silicon (Si)	2 % ± (9 % with 0.35 % C)	18.5 % (not much changed by carbon)	Hardens with loss in plasticity (Mn < Si < P)	Increases hardenability moderately	Negative (graphitizes)	Sustains hardness by solid solution	1. Used as general-purpose deoxidizer 2. Alloying element for electrical and magnetic sheet 3. Improves oxidation resistance 4. Increases hardenability of steel carrying nongraphitizing elements 5. Strengthens low-alloy steels
Titanium (Ti)	0.75 % (1 % ± with 0.20 % C)	6 % ± (less with lowered temp)	Provides age-hardening system in high Ti-Fe alloys	Probably increases hardenability very strongly as dissolved. The carbide effects reduce hardenability	Greatest known (2 % Ti renders 0.50 % carbon steel unhardenable)	Persistent carbides probably unaffected. Some secondary hardening	1. Fixes carbon in inert particles *a.* Reduces martensitic hardness and hardenability in medium-chromium steels *b.* Prevents formation of austenite in high-chromium steels *c.* Prevents localized depletion of chromium in stainless steel during long heating
Tungsten (W)	6 % (11 % with 0.25 % C)	33 % (less with lowered temp)	Provides age-hardening system in high W-Fe alloys	Increases hardenability strongly in small amounts	Strong	Opposes softening by secondary hardening	1. Forms hard, abrasion-resistant particles in tool steels 2. Promotes hardness and strength at elevated temperature
Vanadium (V)	1 % (4 % with 0.20 % C)	Unlimited	Hardens moderately by solid solution	Increases hardenability very strongly, as dissolved	Very strong (V < Ti or Cb)	Max for secondary hardening	1. Elevates coarsening temperature of austenite (promotes fine grain) 2. Increases hardenability (when dissolved) 3. Resists tempering and causes marked secondary hardening

* "Metals Handbook," American Society for Metals.

Table 2-23. Stainless Steels—ASTM A276-70T

ASTM No.	Carbon, max %	Chromium, %	Nickel, %	Other elements, max %	Tensile strength, ksi		Yield, point, ksi		Elongation % in 2 in.		Reduction area, %	Drawing or stamping	Remarks*	
					A†	B‡	A†	B‡	A†	B‡	A†		General properties	Welding properties
302	0.15	17-19	8-10	Mn 2, Si 1	75	95-125	30	45-100	40	12-28	50	Very good	A readily fabricated stainless steel for decorative or corrosion-resistant use	Very good; tough welds
302B	0.15	17-19	8-10	Si 2-3, Mn 2	75	.....	30	.....	40	.....	50	....	Silicon added to increase resistance to scaling at high temperatures	Fairly good. May crack; not recommended
304	0.08	18-20	10.5	Mn 2, Si 1	75	95-125	30	45-100	40	12-28	50	....	A low-carbon 18-8 steel, weldable with less danger of intercrystalline corrosion	
308	0.08	19-21	10-12	Mn 2, Si 1	75	.....	30	.....	40	.....	50	....	For use when corrosion resistance greater than that of 18-8 is needed	
309	0.20	22-24	12-15	Mn 2. Si 1	75	.....	30	.....	40	.....	50	Good	For use at elevated temperature, combining high scaling resistance and good strength	Good
309S	0.08	22-24	12-15	Mn 2, Si 1	75	.....	30	.....	...	.....	...	....	Low carbon permits welded fabrication with a minimum of carbide precipitation	
310	0.25	24-26	19-22	Mn 2, Si 1	75	.....	30	.....	40	.....	50	Good	Similar to 25-12 stainless, with higher nickel content for greater stability at welding temperatures	Good
316	0.08	16-18	10-14	Mo 2-3, Mn 2, Si 1	75	95-125	30	45-100	40	12-28	50	Good	Superior resistance to chemical corrosion	Very good
321	0.08	17-19	9-12	Ti 5 × C min, Mn 2, Si 1	75	.....	30	.....	40	.....	50	Good	An 18-8 type, stabilized against intercrystalline corrosion at elevated temperatures	Very good
347	0.08	17-19	9-12	Mn 2, Si 1, Cb-Ta 10 × C, min	75	.....	30	.....	40	.....	50	Good	A stabilized 18-8 steel for service at elevated temperatures	Very good

* Metals Handbook, American Society for Metals. † A, annealed. ‡ B, cold-finished.

Tungsten, vanadium, and *cobalt* are all used in high-speed tool steels, because they promote hardness and abrasion resistance. Tungsten and cobalt also increase high-temperature hardness.

The principal effects of alloying elements are summarized in Table 2-22.

2-51. Stainless Steels. Stainless steels of primary interest in building are the wrought stainless steels of the austenitic type. The austenitic stainless steels contain both chromium and nickel. Total content of alloy metals is not less than 23%, with chromium not less than 16% and nickel not less than 7%. Table 2-23 summarizes the composition and principal properties of these metals.

Austenitic stainless steels are tough, strong, and shock-resistant, but work-harden readily; so some difficulty on this score may be experienced with cold working and machining. These steels can be welded readily but may have to be stabilized (e.g., AISI types 321 and 347) against carbide precipitation and intergranular corrosion due to welding unless special precautions are taken. These steels have the best high-temperature strength and resistance to scaling of all the stainless steels.

Types 303 and 304 are the familiar 18-8 stainless steels widely used for building applications. These and Types 302 and 316 are the most commonly employed stainless steels. Where maximum resistance to corrosion is required, such as resistance to pitting by sea water and chemicals, the molybdenum-containing types 316 and 317 are best.

For resistance to ordinary atmospheric corrosion, some of the martensitic and ferritic stainless steels, containing 15 to 20% chromium and no nickel, are employed. The martensitic steels, in general, range from about 12 to 18% chromium and from 0.08 to 1.10% carbon. Their response to heat-treatment is similar to the plain carbon steels. When chromium content ranges from 15 to 30% and carbon content is below 0.35%, the steels are ferritic and nonhardenable. The high-chromium steels are resistant to oxidizing corrosion and are useful in chemical plants.

2-52. Welding Ferrous Metals. General welding characteristics of the various types of ferrous metals are as follows:

Wrought iron is ideally forged but may be welded by other methods if the base metal is thoroughly fused. Slag melts first and may confuse unwary operators.

Low-carbon iron and steels (0.30%C or less) are readily welded and require no preheating or subsequent annealing unless residual stresses are to be removed.

Medium-carbon steels (0.30 to 0.50%C) can be welded by the various fusion processes. In some cases, especially in steel with more than 0.40% carbon, preheating and subsequent heat-treatment may be necessary.

High-carbon steels (0.50 to 0.90%C) are more difficult to weld and, especially in arc welding, may have to be preheated to at least 500°F and subsequently heated between 1200 and 1450°F. For gas welding, a carburizing flame is often used. Care must be taken not to destroy the heat-treatment to which high-carbon steels may have been subjected.

Tool steels (0.80 to 1.50%C) are difficult to weld. Preheating, postannealing, heat-treatment, special welding rods, and great care are necessary for successful welding.

Welding of structural steels for buildings is usually governed by the American Welding Society Structural Welding Code D1.1, the American Institute of Steel Construction Specification for the Design, Fabrication and Erection of Structural Steel for Buildings, or a local building code. AWS D1.1 specifies tests to be used in qualifying welders and types of welds. The AISC Specification and many building codes require, in general, that only qualified welds be used and that they be made only by qualified welders.

Structural steels to be welded should be of a weldable grade, such as A36, A441, A572, A588, or A514. These steels may be welded by shielded metal arc, submerged arc, gas metal arc, flux-cored arc, electroslag, electrogas, or stud-welding processes.

Shielded-metal-arc welding fuses parts to be joined by the heat of an electric arc struck between a coated metal electrode and the material being joined, or base metal. The electrode supplies filler material for making the weld, gas for shielding the molten metal from the air, and flux for refining this metal.

Submerged-arc welding fuses the parts to be joined by the heat of an electric arc struck between a bare metal electrode and base metal. The weld is shielded from the air by flux. The electrode or a supplementary welding rod supplies metal filler for making the weld.

Gas-metal-arc welding produces fusion by the heat of an electric arc struck between a filler-metal electrode and base metal, while the molten metal is shielded by a gas or mixture of gas and flux. For structural steels, the gas may be argon, argon with oxygen, or carbon dioxide.

Electroslag welding uses a molten slag to melt filler metal and surfaces of the base metal and thus make a weld. The slag, electrically conductive, is maintained molten by its resistance to an electric current that flows between an electrode and the base metal. The process is suitable only for welding in the vertical position. Moving, water-cooled shoes are used to contain and shape the weld surface. The slag shields the molten metal.

Electrogas welding is similar to the electroslag process. The electrogas process, however, maintains an electric arc continuously, uses an inert gas for shielding, and the electrode provides flux.

Stud welding is used to fuse metal studs or similar parts to other steel parts by the heat of an electric arc. A welding gun is usually used to establish and control the arc, and to apply pressure to the parts to be joined. At the end to be welded, the stud is equipped with a ceramic ferrule, which contains flux and which also partly shields the weld when molten.

Preheating before welding reduces the risk of brittle failure. Initially, its main effect is to lower the temperature gradient between the weld and adjoining base metal. This makes cracking during cooling less likely and gives entrapped hydrogen, a possible source of embrittlement, a chance to escape. A later effect of preheating is improved ductility and notch toughness of base and weld metals and lower transition temperature of weld. When, however, welding processes that deposit weld metal low in hydrogen are used and suitable moisture control is maintained, the need for preheat can be eliminated. Such processes include use of low-hydrogen electrodes and inert-arc and submerged-arc welding.

Weldability of structural steels is influenced by their chemical content. Carbon, manganese, silicon, nickel, chromium, and copper, for example, tend to have an adverse effect, whereas molybdenum and vanadium may be beneficial. To relate the influence of chemical content on structural steel properties to weldability, the use of a carbon equivalent has been proposed. One formula suggested is

$$C_{eq} = C + \frac{Mn}{4} + \frac{Si}{4} \tag{2-2}$$

where C = carbon content, %
Mn = manganese content, %
Si = silicon content, %

Another proposed formula includes more elements:

$$C_{eq} = C + \frac{Mn}{6} + \frac{Ni}{20} + \frac{Cr}{10} - \frac{Mo}{50} - \frac{V}{10} + \frac{Cu}{40} \tag{2-3}$$

where Ni = nickel content, %
Cr = chromium content, %
Mo = molybdenum content, %
V = vanadium content, %
Cu = copper content, %

Carbon equivalent appears to be related to the maximum rate at which a weld and adjacent base metal may be cooled after welding without underbead cracking occurring. The higher the carbon equivalent, the lower will be the allowable cooling rate. Also, the higher the carbon equivalent, the more important use of low-hydrogen electrodes and preheating becomes. (O. W. Blodgett, "Design of Welded Structures," The James F. Lincoln Arc Welding Foundation, Cleveland, Ohio.)

2-53. Corrosion of Iron and Steel. Corrosion of ferrous metals is caused by the tendency of iron (anode) to go into solution in water as ferrous hydroxide and displace hydrogen, which in turn combines with dissolved oxygen to form more water. At the same time, the dissolved ferrous hydroxide is converted by more oxygen to the insoluble ferric hydroxide, thereby allowing more iron to go into solution. Corrosion, therefore, requires liquid water (as in damp air) and oxygen (which is normally present dissolved in the water).

Protection against corrosion takes a variety of forms:

Deaeration. If oxygen is removed from water, corrosion stops. In hot-water heating systems, therefore, no fresh water should be added. Boiler feedwater is sometimes deaerated to retard corrosion.

Coatings

1. *Paints.* Most paints are based on linseed oil and a variety of pigments, of which red lead, oxides of iron, lead sulfate, zinc sulfate, graphite, aluminum, and various hydrocarbons are a few. No one paint is best for all applications. Other paints are coatings of asphalt and tar.

2. *Metallic.* Zinc applied by hot dipping (galvanizing) or hot powder (sherardizing), hot tin dip, hot aluminum dip, and electrolytic plates of tin, copper, nickel, chromium, cadmium, and zinc. A mixture of lead and tin is called terneplate. Zinc is anodic to iron and protects, even after the coating is broken, by sacrificial protection. Tin and copper are cathodic and protect as long as the coating is unbroken but may hasten corrosion by pitting and other localized action once the coating is pierced.

3. *Chemical.* Insoluble phosphates, such as iron or zinc phosphate, are formed on the surface of the metal by treatment with phosphate solutions. These have some protective action and also form good bases for paints. Black oxide coatings are formed by treating the surface with various strong salt solutions. This coating is good for indoor use but has limited life outdoors. It provides a good base for rust-inhibiting oils.

Cathodic Protection. As corrosion proceeds, electric currents are produced as the metal at the anode goes into solution. If a sufficient countercurrent is produced, the metal at the anode will not dissolve. This is accomplished in various ways, such as connecting the iron to a more active metal like magnesium (rods suspended in domestic water heaters) or connecting the part to be protected to buried scrap iron and providing an external current source such as a battery or rectified current from a power line (protection of buried pipe lines).

2-54. Ferrous Metals Bibliography

American Iron and Steel Institute, New York, N.Y.:
"Carbon Steels, Chemical Composition Limits," "Constructional Alloys, Chemical Composition Limits"; "Steel Products Manuals," secs. 4, 6, 7, 16, 18, 21, 28.
American Society for Testing and Materials, Philadelphia, Pa.: "Standards."
American Society for Metals, Cleveland, Ohio: "Metals Handbook."

ALUMINUM AND ALUMINUM-BASED ALLOYS

Pure aluminum and aluminum alloys are used in buildings in various forms. High-purity aluminum (at least 99% pure) is soft and ductile but weak. It has excellent corrosion resistance and is used in buildings for such applications as bright foil for heat insulation, roofing, flashing, gutters and downspouts, exterior and interior architectural trim, and as pigment in aluminum-based paints. Its high heat conductivity recommends it for cooking utensils. The electrical conductivity of the electrical grade is 61% of that of pure copper on an equal-volume basis and 201% on an equal-weight basis.

Aluminum alloys are generally harder and stronger than the pure metal. Furthermore, pure aluminum is difficult to cast satisfactorily, whereas many of the alloys are readily cast.

2-55. Aluminum-alloy Designations. The alloys may be classified: (1) as cast and wrought, and (2) as heat-treatable and non-heat-treatable. Alloys are heat-treatable if the dissolved constituents are less soluble in the solid state at ordinary temperatures than at elevated temperatures, thereby making age-hardening possible.

Table 2-24. Basic Temper Designations for Wrought Aluminum Alloys*

-F As fabricated. This designation applies to the products of shaping processes in which no special control over thermal conditions or strain-hardening is employed. For wrought products, there are no mechanical property limits.

-O Annealed (wrought products only). This designation applies to wrought products that are fully annealed to obtain the lowest strength condition.

-H Strain-hardened (wrought products only). This designation applies to products that have their strength increased by strain-hardening, with or without supplementary thermal treatments to produce some reduction in strengths. The H is always followed by two or more digits.

-W Solution heat-treated. An unstable temper applicable only to alloys that spontaneously age at room temperature after solution heat-treatment. This designation is specific only when the period of natural aging is indicated: for example, W ½ hr.

-T Thermally treated to produce stable tempers other than F, O, or H. This designation applies to products that are thermally treated, with or without supplementary strain-hardening, to produce stable tempers. The T is always followed by one or more digits:

Basic Tempers. The first digit following the H or T indicates the specific combination of basic operations, as follows:

-H1 Strain-hardened only. This designation applies to products that are strain-hardened to obtain the desired strength without supplementary thermal treatment. The number following this designation indicates the degree of strain-hardening.

-H2 Strain-hardened and partially annealed. This designation applies to products that are strain-hardened more than the desired final amount and then reduced in strength to the desired level by partial annealing. For alloys that age soften at room temperature, the H2 tempers have the same minimum ultimate tensile strength as the corresponding H3 tempers. For other alloys, the H2 tempers have the same minimum ultimate tensile strength as the corresponding H1 tempers and slightly higher elongation. The number following this designation indicates the degree of strain-hardening remaining after the product has been partially annealed.

-H3 Strain-hardened and stabilized. This designation applies to products that are strain-hardened and whose mechanical properties are stabilized by a low-temperature thermal treatment that causes slightly lowered tensile strength and improved ductility. H3 is applicable only to those alloys that, unless stabilized, gradually age-soften at room temperature. The number following this designation indicates the degree of strain-hardening before the stabilization treatment.

Degree of strain-hardening. The digit after H1, H2, and H3 indicates degree of strain-hardening. Zero represents annealing, while 8 indicates tempering to produce an ultimate tensile strength equivalent to that achieved by a cold reduction (temperature during reduction not to exceed 120°F) of about 75% following a full anneal. Tempers producing ultimate tensile strengths at equal intervals between that of the 0 temper and that of the 8 temper are designated by 1 to 7. Numeral 9 designates tempers that produce a minimum ultimate tensile strength exceeding that of the 8 temper by 2.0 ksi or more.

-T1 Cooled from an elevated-temperature shaping process and naturally aged to a substantially stable condition. This designation applies to products for which the rate of cooling from an elevated-temperature shaping process, such as casting or extrusion, is such that their strength is increased by room-temperature aging.

-T2 Annealed (cast products only). This designation applies to cast products that are annealed to improve ductility and dimensional stability.

-T3 Solution heat-treated and then cold-worked. This designation applies to products that are cold-worked to improve strength, or in which the effect of cold work in flattening or straightening is recognized in mechanical property limits.

-T4 Solution heat-treated and naturally aged to a substantially stable condition. This designation applies to products that are not cold-worked after solution heat-treatment, or for which the effect of cold work in flattening or straightening may not be recognized in mechanical property limits.

Table 2-24. Basic Temper Designations for Wrought Aluminum Alloys* *(Continued)*

-T5	**Cooled from an elevated-temperature shaping process and then artificially aged.** This designation applies to products that are cooled from an elevated-temperature shaping process, such as casting or extrusion, and then artificially aged to improve mechanical properties or dimensional stability or both.
-T6	**Solution heat-treated and then artificially aged.** This designation applies to products that are not cold-worked after solution heat-treatment, or for which the effect of cold work in flattening or straightening may not be recognized in mechanical property limits.
-T7	**Solution heat-treated and then stabilized.** This designation applies to products that are stabilized to carry them beyond the point of maximum strength to provide control of some special characteristics.
-T8	**Solution heat-treated, cold-worked, and then artificially aged.** This designation applies to products that are cold-worked to improve strength, or for which the effect of cold work in flattening or straightening is recognized in mechanical property limits.
-T9	**Solution heat-treated, artificially aged, and then cold-worked.** This designation applies to products that are cold-worked to improve strength.
-T10	**Cooled from an elevated-temperature shaping process, artificially aged, and then cold-worked.** This designation applies to products that are artificially aged after cooling from an elevated-temperature shaping process, such as casting or extrusion, and then cold-worked to further improve strength.
	Treatment Variations. Additional digits, the first of which may not be zero, may be added to designations T1 through T10 to indicate a variation in treatment that significantly alters the characteristics of the product.

* Aluminum Association

When heat-treated to obtain complete solution, the product may be unstable and tend to age spontaneously. It may also be treated to produce stable tempers of varying degree. Cold working or strain hardening is also possible, and combinations of tempering and strain hardening can also be obtained.

Because of these various possible combinations, a system of letter and number designations has been worked out by the producers of aluminum and aluminum alloys to indicate the compositions and the tempers of the various metals. Wrought alloys are designated by a four-digit index system. 1xxx is for 99.00% aluminum minimum. The last two digits indicate the minimum aluminum percentage. The second digit represents impurity limits. (EC is a special designation for electrical conductors.) 2xxx to 8xxx represent alloy groups in which the first number indicates the principal alloying constituent, and the last two digits are identifying numbers in the group. The second digit indicates modification of the basic alloy. The alloy groups are

Copper	2xxx
Manganese	3xxx
Silicon	4xxx
Magnesium	5xxx
Magnesium and silicon	6xxx
Zinc	7xxx
Other elements	8xxx
Unused series	9xxx

Cast alloys have not been brought into the same digit system, and are designated by numbers, which may have letter prefixes to indicate modifications, e.g., 295 and B295. Casting alloys may be sand or permanent-mold alloys.

Among the wrought alloys, the letters F, O, H, W, and T following the letter S indicate various basic temper designations. These letters in turn may be followed by numerals to indicate various degrees of treatment. Temper designations are summarized in Table 2-24.

The structural alloys generally employed in building are 2014-T6, 6061-T6, and 6063-T6. Architectural alloys often used include 3003-H18, 5005-H16, 5050-0, and 6463-T6.

2-56. Clad Aluminum. Pure aluminum is generally more corrosion-resistant than its alloys. Furthermore, its various forms—pure and alloy—have different solution potentials; that is, they are anodic or cathodic to each other, depending on their relative solution potentials. A number of alloys are therefore made with centers or "cores" of aluminum alloy, overlaid with layers of metal, either pure aluminum or alloys, which are anodic to the core. If galvanic corrosion conditions are encountered, the cladding metal protects the core sacrificially.

2-57. Corrosion Resistance and Corrosion Protection of Aluminum. Although aluminum ranks high in the electromotive series of the metals, it is highly corrosion-resistant because of the tough, transparent, tenacious film of aluminum oxide which rapidly forms on any exposed surface. It is this corrosion resistance that recommends aluminum for building applications. For most exposures, including industrial and seacoast atmospheres, the alloys normally recommended are adequate, particularly if used in usual thicknesses and if mild pitting is not objectionable.

Pure aluminum is the most corrosive-resistant of all and is used alone or as cladding on strong-alloy cores where maximum resistance is wanted. Of the alloys, those containing magnesium, manganese, chromium, or magnesium and silicon in the form of $MgSi_2$ are highly resistant to corrosion. The alloys containing substantial proportions of copper are more susceptible to corrosion, depending markedly on the heat-treatment.

Certain precautions should be taken in building. Aluminum is subject to attack by alkalies, and it should therefore be protected from contact with wet concrete, mortar, and plaster. Clear methacrylate lacquers or strippable plastic coatings are recommended for interiors, and methacrylate lacquer for exterior protection during construction. Strong alkaline and acid cleaners should be avoided and muriatic acid should not be used on masonry surfaces adjacent to aluminum. If aluminum must be contiguous to concrete and mortar outdoors, or where it will be wet, it should be insulated from direct contact by asphalts, bitumens, felts, or other means. As is true of other metals, atmospheric-deposited dirt must be removed to maintain good appearance.

Electrolytic action between aluminum and less active metals should be avoided, because the aluminum then becomes anodic. If aluminum must be in touch with other metals, the faying surfaces should be insulated by painting with asphaltic or similar paints, or by gasketing. Steel rivets and bolts, for example, should be insulated. Drainage from copper-alloy surfaces onto aluminum must be avoided. Frequently, steel surfaces can be galvanized or cadmium-coated where contact is expected with aluminum. The zinc or cadmium coating is anodic to the aluminum and helps to protect it.

2-58. Finishes for Aluminum. Almost all finishes used on aluminum may be divided into three major categories in the system recommended by The Aluminum Association: mechanical finishes, chemical finishes, and coatings. The last may be subdivided into anodic coatings, resinous and other organic coatings, vitreous coatings, electroplated and other metallic coatings, and laminated coatings.

In The Aluminum Association system, mechanical and chemical finishes are designated by M and C, respectively, and each of the five classes of coating is also designated by a letter. The various finishes in each category are designated by two-digit numbers after a letter. The principal finishes are summarized in Table 2-25.

2-59. Welding and Brazing of Aluminum. Weldability and brazing properties of aluminum alloys depend heavily on their composition and heat-treatment. Most of the wrought alloys can be brazed and welded, but sometimes only by special processes. The strength of some alloys depends on heat-treatment after welding. Alloys heat-treated and artificially aged are susceptible to loss of strength at the weld, because weld metal is essentially cast. For this reason, high-strength structural alloys are commonly fabricated by riveting or bolting, rather than by welding.

Brazing is done by furnace, torch, or dip methods. Successful brazing is done with special fluxes.

Table 2-25. Finishes for Aluminum and Aluminum Alloys.

Type of finish	Designation*
Mechanical finishes:	
As fabricated	M1*Y*
Buffed	M2*Y*
Directional textured	M3*Y*
Nondirectional textured	M4*Y*
Chemical finishes:	
Nonetched cleaned	C1*Y*
Etched	C2*Y*
Brightened	C3*Y*
Chemical conversion coatings	C4*Y*
Coatings:	
Anodic	
General	A1*Y*
Protective and decorative (less than 0.4 mil thick)	A2*YZ*
Architectural Class II (0.4–0.7 mil thick)	A3*Y*
Architectural Class I (0.7 mil or more thick)	A4*Y*
Resinous and other organic coatings	R*XY*
Vitreous coatings	V*XY*
Electroplated and other metallic coatings	E*XY*
Laminated coatings	L*XY*

* *X, Y,* and *Z* represents digits from 0 to 9.

2-60. Aluminum Bibliography

The Aluminum Association, New York, N.Y.: "Aluminum Standards and Data," and "Designation Systems for Aluminum Finishes."

Van Lancker, M.: "Metallurgy of Aluminum Alloys," John Wiley & Sons, Inc., New York.

COPPER AND COPPER-BASED ALLOYS

Copper and its alloys are widely used in the building industry for a large variety of purposes, particularly applications requiring corrosion resistance, high electrical conductivity, strength, ductility, impact resistance, fatigue resistance, or other special characteristics possessed by copper or its alloys. Some of the special characteristics of importance to building are ability to be formed into complex shapes, appearance, and high thermal conductivity, although many of the alloys have low thermal conductivity and low electrical conductivity as compared with the pure metal.

In the following paragraphs, brief comments will be made respecting the properties and uses of the principal copper and copper-based alloys of interest in building. The principal properties are summarized in Table 2-26.

2-61. Copper. The excellent corrosion resistance of copper makes it suitable for such applications as roofing, flashing, cornices, gutters, downspouts, leaders, fly screens, and similar applications. For roofing and flashing, soft-annealed copper is employed, because it is ductile and can easily be bent into various shapes. For gutters, leaders, downspouts, and similar applications, cold-rolled hard copper is employed, because its greater hardness and stiffness permit it to stand without large numbers of intermediate supports.

Copper and copper-based alloys, particularly the brasses, are employed for water pipe in buildings, because of their corrosion resistance. Electrolytic tough-pitch copper is usually employed for electrical conductors, but for maximum electrical conductivity and weldability, oxygen-free high-conductivity copper is used.

When arsenic is added to copper, it appears to form a tenacious adherent film, which is particularly resistant to pitting corrosion. Phosphorus is a powerful deoxidizer and is particularly useful for copper to be used for refrigerator tubing and other applications where flaring, flanging, and spinning are required. Arsenic and phosphorus both reduce the electrical conductivity of the copper.

For flashing, copper is frequently coated with lead to avoid the green patina formed on copper that is sometimes objectionable when it is washed down over adjacent surfaces, such as ornamental stone. The patina is formed particularly in industrial atmospheres. In rural atmospheres, where industrial gases are absent, the copper normally turns to a deep brown color.

Principal types of copper and typical uses are (see also Table 2-26):

Electrolytic tough pitch (99.90% copper) is used for electrical conductors—bus bars, commutators, etc.; building products—roofing, gutters, etc.; process equipment—kettles, vats, distillery equipment; forgings. General properties are high electrical conductivity, high thermal conductivity, and excellent working ability.

Deoxidized (99.90% copper and 0.025% phosphorus) is used, in tube form, for water and refrigeration service, oil burners, etc.; in sheet and plate form, for welded construction. General properties include higher forming and bending qualities than electrolytic copper. They are preferred for coppersmithing and welding (because of resistance to embrittlement at high temperatures).

2-62. Plain Brass. A considerable range of brasses is obtainable for a large variety of end uses. The high ductility and malleability of the copper-zinc alloys, or brasses, make them suitable for operations like deep drawing, bending, and swaging. They have a wide range of colors. They are generally less expensive than the high-copper alloys.

Grain size of the metal has a marked effect upon its mechanical properties. For deep drawing and other heavy working operations, a large grain size is required, but for highly finished polished surfaces, the grain size must be small.

Like copper, brass is hardened by cold working. Hardnesses are sometimes expressed as quarter hard, half hard, hard, extra hard, spring, and extra spring, corresponding to reductions in cross section during cold working ranging from approximately 11 to 69%. Hardness is strongly influenced by alloy composition, original grain size, and form (strip, rod, tube, wire).

Brass compositions range from higher copper content to zinc contents as high as 40% or more. Brasses with less than 36% zinc are plain alpha solid solutions; but Muntz metal, with 40% zinc, contains both alpha and beta phases.

The principal plain brasses of interest in building, and their properties are (see also Table 2-26):

Commercial bronze, 90% (90.0% copper, 10.0% zinc). Typical uses are forgings, screws, weatherstripping, and stamped hardware. General properties include excellent cold working and high ductility.

Red brass, 85% (85.0% copper, 15.0% zinc). Typical uses are dials, hardware, etched parts, automobile radiators, and tube and pipe for plumbing. General properties are higher strength and ductility than copper, and excellent corrosion resistance.

Cartridge brass, 70% (70.0% copper, 30.0% zinc). Typical uses are deep drawing, stamping, spinning, etching, rolling—for practically all fabricating processes—cartridge cases, pins, rivets, eyelets, heating units, lamp bodies and reflectors, electrical sockets, drawn shapes, etc. General properties are best combination of ductility and strength of any brass, and excellent cold-working properties.

Muntz metal (60.0% copper, 40.0% zinc). Typical uses are sheet form, perforated metal, architectural work, condenser tubes, valve stems, and brazing rods. General properties are high strength combined with low ductility.

2-63. Leaded Brass. Lead is added to brass to improve its machinability, particularly in such applications as automatic screw machines where a freely chipping metal is required. Leaded brasses cannot easily be cold-worked by such operations as flaring, upsetting, or cold heading. Several leaded brasses of importance in the building field are the following (see also Table 2-26):

High-leaded brass (64.0% copper, 34.0% zinc, 2.0% lead). Typical uses are engraving plates, machined parts, instruments (professional and scientific), nameplates, keys, lock parts, and tumblers. General properties are free machining and good blanking.

Forging brass (60.0% copper, 38.0% zinc, 2.0% lead). Typical uses are hot forgings, hardware, and plumbing goods. General properties are extreme plasticity when hot and a combination of good corrosion resistance with excellent mechanical properties.

Table 2-26. Copper and Copper Alloys*

Alloy name	Electrolytic tough pitch	Deoxidized	Commercial bronze, 90 %	Red brass, 85 %	Cartridge brass, 70 %	Muntz metal	High-leaded brass	Forging brass	Architectural bronze
Working properties:									
1. Cold-working	1. Excellent	1. Excellent	1. Excellent	1. Excellent	1. Excellent	1. Fair	1. Poor	1. Fair	1. Very poor
2. Hot-working	2. Excellent	2. Excellent	2. Excellent	2. Good	2. Good	2. Excellent	2. Poor	2. Excellent	2. Excellent
3. Machining	3. Fair	3. Poor	3. Poor	3. Poor	3. Fair	3. Good	3. Excellent	3. Good	3. Good
4. Welding†	4. deoxidized copper preferred	4. ‡	4. ‡	4. ‡	4. ‡	4. ‡	4. Not recommended	4. Not recommended	4. Not recommended
5. Soldering	5. Excellent	5. Excellent	5. Excellent	5. Excellent	5. Excellent	5. Excellent	5. Excellent	5. Good	5. Excellent
6. Polishing	6. Excellent	6. Excellent	6. Excellent	6. Excellent	6. Excellent	6. Excellent	6. Excellent	6. Excellent	6. Excellent
Young's modulus of elasticity, psi (000,000 omitted)	17	17	17	17	16	15	15	15	14
Avg coefficient of thermal expansion, per °F (68 to 570°F)	0.0000098	0.0000098	0.0000102	0.0000104	0.0000111	0.0000116	0.0000113	0.0000115	0.0000116
Electrical conductivity, % IACS at 68°F	100 min	85 annealed	44	37	28	28	26	27	28
Thermal conductivity, Btu, per sq ft, per ft, per hr, per °F, at 68°F	226	196	109	92	70	71	67	69	71

	Electrolytic tough pitch		Deoxidized		Commercial bronze, 90 %		Red brass, 85 %		Cartridge brass, 70 %		Muntz metal		High-leaded brass		Forging brass		Architectural bronze	
	Hard	Soft	Hard	Soft	Hard	Soft	Hard	Soft	Hard	Soft	Hard	Soft	Hard	Soft	Hard	Soft*	Hard	Soft¶
Tensile strength, 1,000 psi:																		
Sheet	50	32	50	32	61	37	70	40	76	47	80	54	74	49	...	...	...	
Rod	48	32	...	...	55	40	57	40	70	48	75	54	...	...	...	52	...	60
Tube	55	32	55	32	...	38	70	40	78	47	74	56						
Elongation, % in 2 in.:																		
Sheet	6	45	5	45	5	45	5	47	8	62	10	45	7	52				
Rod	16	55	...	...	20	50	23	55	30	65	20	50	...	...	...	45	...	30
Tube	8	45	8	45	...	50	8	55	8	65	10	50						
Yield strength, 1000 psi:																		
0.5 % extension under load:																		
Sheet	47	8	45	10	54	10	57	12	63	15	60	21	60	17				
Rod	45	8	...	...	48	10	52	10	52	16	55	21	...	...	...	20	...	
Tube	52	8	50	10	...	12	58	12	64	15	55	23	...	...	...	...	...	20
0.2 % offset																		
Sheet	47	8	45	10	56	10	61	12	74	15	66	21	66	17				
Rod	45	8	...	...	48	10	54	10	53	16	59	21	...	...	...	20	...	20
Tube	52	8	50	10	...	12	61	12	70	15	60	23						
Rockwell hardness:																		
Sheet	50B	40F	50B	40F	70B	53F	77B	59F	82B	64F	85B	80F	80B	68F				
Rod	47B	40F	...	...	60B	55F	75B	55F	80B	65F	80B	80F	...	...	...	78F	...	65B
Tube	60B	40F	60B	40F	...	57F	77B	60F	82B	64F	80B	82F						

Table 2-26. Copper and Copper Alloys* (*Continued*)

Alloy name	Admiralty		Manganese bronze (A)		Nickel silver, 18% (A), deep drawing		Nickel silver, 7½% key stock		Cupronickel, 30%		Silicon bronze (high-silicon bronze, A)		Aluminum silicon bronze		Phosphor bronze, 5% (A)		Phosphor bronze, 8% (C)	
Working properties:																		
1. Cold-working	1. Good		1. Poor		1. Excellent		1. Poor		1. Excellent		1. Good		1. Poor		1. Excellent		1. Excellent	
2. Hot-working	2. Poor		2. Excellent		2. Fair		2. Poor		2. Fair		2. Excellent		2. Excellent		2. Poor		2. Poor	
3. Machining	3. Fair		3. Good		3. Fair		3. Excellent		3. Fair		3. Fair		3. Excellent		3. Fair		3. Fair	
4. Welding†	4. ‡		4. ‡		4. ‡		4. Non-leaded preferred		4. ‡		4. ‡		4. ‡		4. ‡		4. ‡	
5. Soldering	5. Good		5. Excellent		5. Excellent		5. Excellent		5. Excellent		5. Excellent		5. Fair		5. Excellent		5. Excellent	
6. Polishing	6. Good		6. Excellent		6. Excellent		6. Excellent		6. Good		6. Good		6. Excellent		6. Excellent		6. Excellent	
Young's modulus of elasticity, psi (000,000 omitted)	16		15		18		17		22		17		14		16		16	
Avg. coefficient of thermal expansion, per °F (68 to 570°F)	0.0000112		0.0000118		0.0000090		0.0000083		0.0000090		0.0000100		0.0000092		0.0000099		0.0000101	
Electrical conductivity, % IACS at 68°F	25		24		6		10		4.6		7		7		15		13	
Thermal conductivity, Btu, per sq ft, per ft, per hr, per °F, at 68°F	64		61		19		29		17		21		22		40		36	
	Hard	Soft	Hard	Soft	Hard	Soft	Hard	Soft	Hard	Soft	Hard§	Soft	Hard§	Soft	Hard	Soft	Hard	Soft
Tensile strength, 1,000 psi:																		
Sheet	90	50	...	...	85	58	85	54	82	54	94	60	...	...	100	47	112	55
Rod	....	...	84	65	...	...	...	...	...	...	92	58	...	80	75			
Tube	100	53	...	...	...	...	...	...	80	60								
Elongation, % in 2 in.:																		
Sheet	5	55	...	...	3	40	14	38	4	35	8	60	...	...	4	64	3	70
Rod	....	...	19	33	...	...	...	...	...	...	22	60	...	20	25			
Tube	3	65	...	...	...	...	...	...	5	45								
Yield strength, 1000 psi:																		
0.5% extension under load:																		
Sheet	70	18	...	...	74	25	73	18	76	10	58	25	...	...	80	19	75	
Rod	....	...	60	30	...	...	...	...	...	...	55	22	...	43	65			
Tube	80	22	...	...	...	...	...	...	75	25								
0.2% offset																		
Sheet	76	18	...	...	80	25	82	18	78	10	83	25	...	...	91	19	100	
Rod	....	...	...	30	...	...	...	...	...	...	...	22	...	43	70			
Tube	90	22	...	...	...	...	...	...	77	25								
Rockwell hardness:																		
Sheet	90B	25B	...	...	87B	85F	83B	33B	85B	40B	93B	82F	...	...	95B	73F	93B	75F
Rod	....	...	90B	65B	...	...	...	...	...	...	90B	60B	...	80B	80B			
Tube	95B	75F	...	...	...	...	...	...	82B	80F								

* Revere Copper and Brass, Inc.
† By fusion methods: conductivity of the coppers makes resistance welding (spot and seam) impractical. Coppers, however, may be resistance-brazed by a patented method.
Basis of rating: sheet, hard—0.040 in. stock, previously rolled 4 B & S numbers hard (37.5% area reduction).
Basis of rating: rod, hard—1 in. and under diameter, previously drawn through 25 to 35% area reduction.
‡ For welding—gas-shielded-arc processes preferred.
¶ As extruded.
§ Hard, for rod, bolt temper, extra hard.

Architectural bronze (56.5% copper, 41.25% zinc, 2.25% lead). Typical uses are handrails, decorative moldings, grilles, revolving door parts, miscellaneous architectural trim, industrial extruded shapes (hinges, lock bodies, automotive parts). General properties are excellent forging and free-machining properties.

2-64. Tin Brass. Tin is added to a variety of basic brasses to obtain hardness, strength, and other properties which would otherwise not be available. Two important alloys are (see also Table 2-26):

Admiralty (71.0% copper, 28.0% zinc, 1.0% tin, 0.05% arsenic). Typical uses are condenser and heat-exchanger plates and tubes, steam-power-plant equipment, chemical and process equipment, and marine uses. General properties are excellent corrosion resistance, combined with strength and ductility.

Manganese bronze (58.5% copper, 39.2% zinc, 1.0% iron, 1.0% tin, 0.3% manganese). Typical uses are forgings, condenser plates, valve stems, and coal screens. General properties are high strength combined with excellent wear resistance.

2-65. Nickel Silvers. These are alloys of copper, nickel, and zinc. Depending on the composition, they range in color from a definite to slight pink cast through yellow, green, whitish green, whitish blue, to blue. A wide range of nickel silvers is made, of which only two typical compositions will be described (see also Table 2-26). Those that fall in the combined alpha-beta phase of metals are readily hot-worked and therefore are fabricated without difficulty into such intricate shapes as plumbing fixtures, stair rails, architectural shapes, and escalator parts. Lead may be added to improve machining.

Nickel silver, 18% (A), (65.0% copper, 17.0% zinc, 18.0% nickel). Typical uses are hardware, architectural panels, lighting, electrical and plumbing fixtures. General properties are high resistance to corrosion and tarnish, malleable, and ductile. Color: silver-blue-white.

Nickel silver, 7½% leaded key stock. Typical uses are keys, products requiring machinability, lock washers, cotter pins, and fuse clips. It has good machinability but is preferred non-leaded for bending and drawing.

2-66. Cupronickel. Copper and nickel are alloyed in a variety of compositions of which the high-copper alloys are called the cupronickels. A typical commercial type of cupronickel contains 30% nickel (Table 2-26):

Cupronickel, 30% (70.0% copper, 30.0% nickel). Typical uses are condenser tubes and plates, tanks, vats, vessels, process equipment, automotive parts, meters, refrigerator pump valves. General properties are high strength and ductility, resistance to corrosion and erosion. Color: white-silver.

2-67. Silicon Bronze. These are high-copper alloys containing percentages of silicon ranging from about 1% to slightly more than 3%. In addition, they generally contain one or more of the four elements, tin, manganese, zinc, and iron. A typical one is high-silicon bronze, type A (see also Table 2-26):

High-silicon bronze, A (96.0% copper, 3.0% silicon, 1.0% manganese). Typical uses are tanks—pressure vessels, vats; weatherstrips, forgings. General properties are corrosion resistance of copper and mechanical properties of mild steel.

2-68. Aluminum Bronze. Like aluminum, these bronzes form an aluminum oxide skin on the surface, which materially improves resistance to corrosion, particularly under acid conditions. Since the color of the 5% aluminum bronze is similar to that of 18-carat gold, it is used for costume jewelry and other decorative purposes. Aluminum-silicon bronzes (Table 2-26) are used in applications requiring high tensile properties in combination with good corrosion resistance in such parts as valves, stems, air pumps, condenser bolts, and similar applications. Their wear-resisting properties are good; consequently, they are used in slide liners and bushings.

2-69. Tin Bronze. Originally and historically, the bronzes were all alloys of copper and tin. Today, the term bronze is generally applied to engineering metals having high mechanical properties and the term brass to other metals. The commercial wrought bronzes do not usually contain more than 10% tin because the metal becomes extremely hard and brittle. When phosphorus is added as a deoxidizer, to obtain sound dense castings, the alloys are known as phosphor bronzes. The two most commonly used tin bronzes contain 5 and 8% tin. Both have excellent cold-working properties (Table 2-26).

2-70. Copper Bibliography
American Society for Testing and Materials, Philadelphia Pa.: "Standards."
Revere Copper and Brass, Inc., New York: "Revere Copper and Copper Alloys."

LEAD AND LEAD-BASED ALLOYS

Lead is used primarily for its corrosion resistance. Lead roofs 2,000 years old are still intact. Exposure tests indicate corrosion penetrations ranging from less than 0.0001 in. to less than 0.0003 in. in 10 years in atmospheres ranging from mild rural to severe industrial and seacoast locations. Sheet lead is therefore used for roofing, flashing, spandrels, gutters, and downspouts.

2-71. Applications of Lead. Because the green patina found on copper may wash away sufficiently to stain the surrounding structure, lead-coated copper is frequently employed. Three classes recognized by ASTM are listed in Table 2-27.

Table 2-27. Classes of Lead-coated Copper

Class	Weight of lead coating, lb per 100 sq ft, total, coated both sides	
	Min	Max
Class A standard (general utility)	12	15
Class B heavy	20	30
Class C extra heavy	40	50

Lead pipe for the transport of drinking water should be used with care. Distilled and very soft waters slowly dissolve lead and may cause cumulative lead poisoning. Hard waters apparently deposit a protective coating on the wall of the pipe and little or no lead is subsequently dissolved in the water.

Principal alloying elements used with building leads are antimony (for hardness and strength) and tin. But copper, arsenic, bismuth, nickel, zinc, silver, iron, and manganese are also added in varying proportions.

Soft solders consist of varying percentages of lead and tin. For greater hardness, antimony is added, and for higher-temperature solders, silver is added in small amounts. ASTM Standard B32 specifies properties of soft solders.

Low-melting alloys and many bearing metals are alloys of lead, bismuth, tin, cadmium, and other metals including silver, zinc, indium, and antimony. The fusible links used in sprinkler heads and fire-door closures, made of such alloys, have a low melting point, usually lower than the boiling point of water. Yield (softening) temperatures range from 73 to 160°F and melting points from about 80 to 480°F, depending on the composition.

2-72. Lead Bibliography
American Society for Metals, Cleveland, Ohio: "Metals Handbook."
American Society for Testing and Materials, Philadelphia, Pa.: "Standards."

NICKEL AND NICKEL-BASED ALLOYS

Nickel is used mostly as an alloying element with other metals, but it finds use in its own right, largely as electroplate or as cladding metal. Among the principal high-nickel alloys are Monel and Inconel. The nominal compositions of these metals are given in Table 2-28.

2-73. Properties of Nickel and Its Alloys. Nickel is resistant to alkaline corrosion under nonoxidizing conditions but is corroded by oxidizing acids and oxidizing salts. It is resistant to fatty acids, other mildly acid conditions, such as food processing and beverages, and resists oxidation at temperatures as high as 1600°F.

Monel is widely used in kitchen equipment. It is better than nickel in reducing conditions like warm unaerated acids, and better than copper under oxidizing condi-

Table 2-28. Composition of Nickel Alloys

Content	Nickel alloy, low-carbon 201	Nickel alloy 200	Monel 400	Inconel 600	70-30 cupro-nickel	90-10 cupro-nickel
	ASTM A265	ASTM B162	ASTM B127	ASTM B168	ASTM B171	ASTM B171
Carbon	0.02	0.15	0.3	0.15		
Manganese	0.35	0.35	1.25	1.0	1.0 max	1.0 max
Sulfur	0.01	0.01	0.024	0.015		
Silicon	0.35	0.35	0.5	0.5		
Chromium	....	....	....	14–17		
Nickel	99 min	99 min	63–70	72 min	29–33	65 min
Copper	0.25	0.25	Remainder	0.5	65 min	86.5 min
Iron	0.40 max	0.40 max	2.5 max	6–10	0.70 max	0.5–2.0
Lead	....	....	....	.....	0.05 max	0.05 max
Zinc	....	....	....	.....	1.0	1.0

tions, such as aerated acids, alkalies, and salt solutions. It is widely used for handling chlorides of many kinds.

Inconel is almost completely resistant to corrosion by food products, pharmaceuticals, biologicals, and dilute organic acids. It is superior to nickel and Monel in resisting oxidizing acid salts like chromates and nitrates but is not resistant to ferric, cupric, or mercuric chlorides. It resists scaling and oxidation in air and furnace atmospheres at temperatures up to 2000°F.

2-74. Nickel Bibliography

International Nickel Co., New York: "Nickel and Nickel Alloys."

Hoerson, Albert, Jr.: "Nonferrous-clad Plate Steels," Chap. 13 in A. G. H. Dietz, "Composite Engineering Laminates," M.I.T. Press, Cambridge, Mass.

PLASTICS

The synonymous terms plastics and synthetic resins denote synthetic organic high polymers, all of which are plastic at some stage in their manufacture. Plastics fall into two large categories—thermoplastic and thermosetting materials.

Thermoplastics may be softened by heating and hardened by cooling any number of times. Thermosetting materials are either originally soft or liquid, or they soften once upon heating; but upon further heating, they harden permanently. Some thermosetting materials harden by an interlinking mechanism in which water or other by-product is given off, by a process called condensation; but others, like the unsaturated polyesters, harden by a direct interlinking of the basic molecules without any by-product's being given off.

Most plastics are modified with plasticizers, fillers, or other ingredients. Consequently, each base material forms the nucleus for a large number of products having a wide variety of properties. This section can only indicate generally the range of properties to be expected.

Because plastics are quite different in their composition and structure from other materials, such as metals, their behavior under stress and under other conditions is likely to be different from other materials. Just as steel and lead are markedly different and are used for different applications, so the various plastics materials—some hard and brittle, others soft and extensible—must be designed on different bases and used in different ways. Some plastics show no yield point, because they fail before a yield point can be reached. Others have a moderately high elastic range, followed by a highly plastic range. Still others are highly extensible and are employed at stresses far beyond the yield point.

More than many other materials, plastics are sensitive to temperature and to the rate and time of application of load. How these parameters influence the

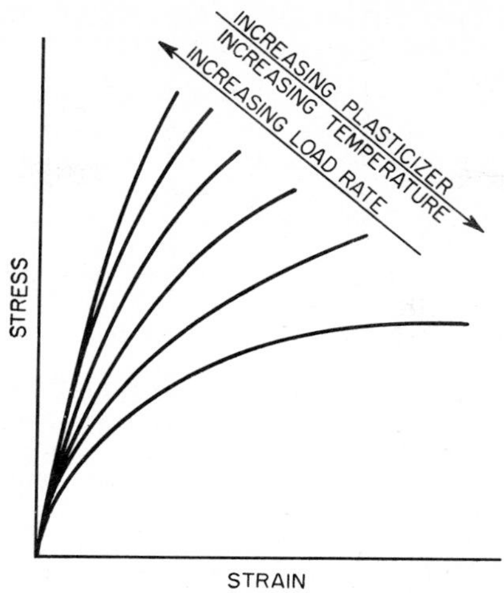

Fig. 2-3. Stress-strain diagram shows the influence of temperature, plasticizer, and rate of loading on behavior of plastics.

properties is indicated in a general way in Fig. 2-3, which shows that for many plastics an increase in temperature, increase in plasticizer content, and decrease in rate of load application mean an increase in strain to fracture, accompanied by a decrease in maximum stress. This viscoelastic behavior, combining elastic and viscous or plastic reaction to stress, is unlike the behavior of materials which are traditionally considered to behave only elastically.

2-75. Fillers and Plasticizers. Fillers are commonly added, particularly to the thermosetting plastics, to alter their basic characteristics. For example, wood flour converts a hard, brittle resin, difficult to handle, into a cheaper, more easily molded material for general purposes. Asbestos fibers provide better heat resistance; mica gives better electrical properties; and a variety of fibrous materials, such as chopped fibers, chopped fabric, and chopped tire cords, increase the strength and impact properties.

Plasticizers are added to many thermoplastics, primarily to transform hard and rigid materials into a variety of forms having varying degrees of softness, flexibility, and strength. In addition, dyes or pigments, stabilizers, and other products may be added.

2-76. Molding and Fabricating Methods for Plastics. Both thermosetting and thermoplastic molding materials are formed into final shape by a variety of molding and fabricating methods:

Thermosetting materials are commonly formed by placing molding powder or molded preform in heated dies and compressing under heat and pressure into the final infusible shape. Or they are formed by forcing heat-softened material into a heated die for final forming into the hard infusible shape.

Thermoplastics are commonly formed by injection molding; that is, by forcing soft, hot plastic into a cold die, where it hardens by cooling. Continuous profiles of thermoplastic materials are made by extrusion. Thermoplastic sheets, especially transparent acrylics, are frequently formed into final shape by heating and then blowing to final form under compressed air or by drawing a partial vacuum against the softened sheet.

Foamed plastics are employed for thermal insulation in refrigerators, buildings, and many other applications. In buildings, plastics are either prefoamed into slabs, blocks, or other appropriate shapes, or they are foamed in place.

Prefoamed materials, such as polystyrene, are made by adding a blowing agent and extruding the mixture under pressure and at elevated temperatures. As the material emerges from the extruder, it expands into a large "log" that can be cut into desired shapes. The cells are "closed"; that is, they are not interconnecting and are quite impermeable.

Foamed-in-place plastics are made with pellets or liquids. The pellets, made, for example, of polystyrene, are poured into the space to be occupied, such as a mold, and heated, whereupon they expand and occupy the space. The resulting mass may be permeable between pellets. Liquid-based foams, exemplified by polyurethane, are made by mixing liquid ingredients and immediately casting the mixture into the space to be occupied. A quick reaction results in a foam that rises and hardens by a thermosetting reaction. When blown with fluorocarbon gases, such foams have exceptionally low thermal conductivities.

All the plastics can be machined, if proper allowance is made for the properties of the materials.

Plastics are often combined with sheet or mat stocks, such as paper, cotton muslin, glass fabric, glass filament mats, nylon fabric, and other fabrics, to provide laminated materials in which the properties of the combined plastic and sheet stock are quite different from the properties of either constituent by itself. Two principal varieties of laminates are commonly made: (1) High-pressure laminates employing condensation-type thermosetting materials, which are formed at elevated temperatures and pressures. (2) Reinforced plastics employing unsaturated polyesters and epoxides, from which no by-products are given off, and consequently, either low pressures or none at all may be required to form combinations of these materials with a variety of reinforcing agents, like glass fabric or mat.

In the following articles, the principal varieties of plastics are briefly described and their principal fields of application indicated. In Table 2-29, ranges of values to be expected of specimens made and tested under specific conditions, usually those of the American Society for Testing and Materials, are given. The method of fabrication and specific application of a particular plastic may severely modify these properties.

2-77. Thermosetting Plastics. *Phenol Formaldehyde.* These materials provide the greatest variety of thermosetting molded plastic articles. They are used for chemical, decorative, electrical, mechanical, and thermal applications of all kinds. Hard and rigid, they change slightly, if at all, on aging indoors but, on outdoor exposure, lose their bright surface gloss. However, the outdoor-exposure characteristics of the more durable formulations are otherwise generally good. Phenol formaldehydes have good electrical properties, do not burn readily, and do not support combustion. They are strong, light in weight, and generally pleasant to the eye and touch, although light colors by and large are not obtainable because of the fairly dark-brown basic color of the resin. They have low water absorption and good resistance to attack by most commonly found chemicals.

Epoxy and Polyester Casting Resins. These are used for a large variety of purposes. For example, electronic parts with delicate components are sometimes cast completely in these materials to give them complete and continuous support, and resistance to thermal and mechanical shock. Some varieties must be cured at elevated temperatures; others can be formulated to be cured at room temperatures. One of the outstanding attributes of the epoxies is their excellent adhesion to a variety of materials, including such metals as copper, brass, steel, and aluminum.

Polyester Molding Materials. When compounded with fibers, particularly glass fibers, or with various mineral fillers, including clay, the polyesters can be formulated into putties or premixes that are easily compression- or transfer-molded into parts having high impact resistance.

Melamine Formaldehyde. These materials are unaffected by common organic solvents, greases, and oils, as well as most weak acids and alkalies. Their water absorption is low. They are insensitive to heat and are highly flame-resistant, depending on the filler. Electrical properties are particularly good, especially resistance to arcing. Unfilled materials are highly translucent and have unlimited color possibilities. Principal fillers are alpha cellulose for general-purpose compounding; minerals to improve electrical properties, particularly at elevated temperatures; chopped fabric to afford high shock resistance and flexural strength; and cellulose, mainly for electrical purposes.

Table 2-29. Selected Properties of Plastics*

Property	ASTM test method	ABS Acrylonitrile-butadiene-styrene	PMMA Acrylic	CA, CAB, CAP, CN, CP, EC Cellulosics	EP Epoxies	FEP, PCTFE, PTFE, PVF Fluoro-plastics	MF Melamine-formal-dehyde	PA Nylon Polyamide	PF Phenol-formal-dehyde
Tensile strength, psi	D638–D651	4000–8000	7000–11,000	2000–9000	4000–30,000	2000–7000	5000–13,000	7000–35,000	3000–18,000
Elongation, %	D638	2–300	2–10	5–100	0.5–70	80–300	0.30–0.90	10–320	0.13–2.25
Tensile Modulus, 10^6 psi	D638	0.23–1.03	0.35–0.50	0.065–0.60	0.001–3.04	0.05–0.30	1.2–2.4	0.11–1.80	0.25–5.00
Compressive strength, psi	D695	7000–22,000	11,000–19,000	2000–36,000	1000–40,000	1700–10,000	20,000–45,000	6700–24,000	10,000–70,000
Compressive modulus, 10^6 psi	D695	0.17–0.39	0.37–0.46	. . . .	. . . .	to 0.12	. . . .	0.185–0.248	. . . .
Flexural yield strength, psi	D790	5000–27,000	12,000–17,000	2000–16,000	1000–60,000	7400–9300	9000–23,000	No break to 17,500	4000–60,000
Flexural modulus, 10^6 psi	D790	0.20–1.30	0.39–0.47	. . . .	. . . .	to 0.20	. . . .	0.14–1.14	to 2.4
Hardness, Rockwell	D785	R75–M100	M80–M105	R34–R125	M80–M120	R25–95 (Shore) D50–D80	M110–M125	R108–E75	M37–E101
Impact strength, ft-lb per in. notch	D256	1.0–10	0.3–0.5	0.4–8.5	0.2–10	3.0 to no	0.24–6	1.0–5.5	0.2–18
Thermal conductivity, Btu per sq ft per in. per °F	C177	1.3–2.3	1.2–1.7	1.1–2.3	1.2–8.7	0.9–1.7	1.9–4.9	1.5–2.5	0.9–6.4
Thermal expansion, 10^{-6} per °F	D696	39–73	28–50	44–111	3–55	25–66	11–25	7–83	14–33
Resistance to heat, continuous, °F		140–230	140–200	115–220	200–550	300–550	210–400	175–400	200–550
Burning rate, in. per min	D635	Slow to self-extinguishing	Slow	Self-extinguishing to very fast	Slow to non-burning	None to self-extinguishing	Nonburning to very slow	Self-extinguishing to slow burning	None to slow
Effect of sunlight	. . . .	None to slight yellowing	None	Slight to discoloration, embrittlement	None to slight	None to slight bleaching	Slight to darkening	Slight discoloration	Darkens
Clarity	. . . .	Translucent to opaque	Excellent to opaque	Transparent to opaque	Transparent to opaque	Transparent to opaque	Translucent to opaque	Translucent to opaque	Transparent to opaque
Machining qualities	. . . .	Good to excellent	Fair to excellent	Good to excellent	Poor to excellent	Excellent	Fair to good	Fair to excellent	Poor to good
24-hr water absorption, 1/8-in. thickness, %	D570	0.2–0.45	0.3–0.4	0.8–7.0	0.08–4.0	0.00–0.04	0.08–0.80	0.4–1.5	0.1–2

Table 2-29. Selected Properties of Plastics (*Continued*)

Property	PC Poly-carbonate	Polyesters	PE Poly-ethylene	PP Poly-propylene	PS, SAN, SBP, SRP Polystyrene	SI Silicones	UF Urea-formal-dehyde	UP Urethanes	PVAc, PVAl, PVB, PVC, PVCAc, PVFM
Tensile strength, psi	8000–20,000	800–50,000	1000–5500	2900–9000	1500–20,000	800–35,000	5500–13,000	175–10,000	500–9000
Elongation, %	0.9–1.30	0.5–310	15–1000	2–700	0.75–80	to 100	0.5–1.0	10–1000	2–450
Tensile modulus, 10^6 psi	0.35–1.85	0.3–2.0	0.014–0.18	0.1–0.9	0.15–1.4	0.0009–3.0	1.0–1.5	0.01–1.0	0.05–0.6
Compressive strength, psi	12,500–19,000	12,000–50,000	to 5500	3700–8000	4000–22,000	100–18,000	25,000–45,000	20,000	1000–22,000
Compressive modulus, 10^6 psi	0.3–0.45	. . . .	to 0.15	to 0.3	to 0.53	. . . .	. . . .	0.004–0.1	to 0.6
Flexural yield strength, psi	13,500–30,000	8000–80,000	to 7000	5000–11,000	5000–26,000	to 35,000	10,000–18,000	– to 9000	to 17,000
Flexural modulus, 10^6 psi	0.34–1.20	to 2.0	to 0.35	0.125–0.825	to 1.8	. . . .	1.3–1.6	0.01–0.35	to 0.4
Hardness, Rockwell	M70–R118	60(Barcol)–E98	D30(Shore)–R15	R30–R110	R50–E60	40(Shore)–M95	M110–M120	20A(Shore)–M28	10A(Shore)–M85
Impact strength, ft-lb per in. notch	1.2–17.5	0.2–16.0	0.5–2.0 to no break	0.5–20.0	0.25–11.0	to 15	0.25–0.40	5 to flexible	0.4–20, impact strength varies with type and amount of plasticizer
Thermal conductivity, Btu per sq ft per in. per °F	0.7–1.5	1.2–7.2	2.3–3.6	0.6–1.2	0.3–1.0	1.0–3.8	2.0–2.9	0.5–2.1	0.9–20
Thermal expansion, 10^{-6} per °F	10–37	7–56	56–195	16–57	19–117	4–167	12–20	56–112	28–195
Resistance to heat, continuous, °F	250–275	250–450	180–275	190–320	140–220	400–>600	170	190–250	120–210
Burning rate, in. per min	Self-extinguishing	Slow to non-burning	Slow to self-extinguishing	Slow to non-burning	Slow to non-burning	None to slow	Self-extinguishing	Slow to self-extinguishing	Slow to self-extinguishing
Effect of sunlight	Slight color change	None to slight yellowing, embrittlement	Unprotected crazes fast, weather resistance available	Unprotected crazes fast, weather resistance available	Slight yellowing	None to slight	Pastels, gray	None to yellowing	Slight
Clarity	Transparent to opaque	Transparent to opaque	Transparent to opaque	Transparent to opaque	Excellent to opaque	Clear to opaque	Transparent to opaque	Clear to opaque	Transparent to opaque
Machining qualities	Fair to excellent	Poor to excellent	Fair to excellent	Fair to good	Fair to good	Fair to good	Fair	Fair to excellent	Poor to excellent
24-hr water absorption, ⅛-in. thickness, %	0.07–0.20	0.01–1.0	<0.01–0.06	<0.01–0.05	0.03–0.6	<0.2	0.4–0.8	0.02–1.5	0.02–3.0

* "Modern Plastics Encyclopedia."

Alkyd. These are customarily combined with mineral or glass fillers, the latter for high impact strength. Extreme rapidity and completeness of cure permit rapid production of large numbers of parts from relatively few molds. Because electrical properties, especially resistance to arcing, are good, many of the applications for alkyd molding materials are in electrical applications.

Urea Formaldehyde. Like the melamines, these offer unlimited translucent to opaque color possibilities, light-fastness, good mechanical and electrical properties, and resistance to organic solvents as well as mild acids and alkalies. Although there is no swelling or change in appearance, the water absorption of urea formaldehyde is relatively high, and it is therefore not recommended for applications involving long exposure to water. Occasional exposure to water is without deleterious effect. Strength properties are good, although special shock-resistant grades are not made.

Silicones. Unlike other plastics, silicones are based on silicon rather than carbon. As a consequence, their inertness and durability under a wide variety of conditions are outstanding. As compared with the phenolics, their mechanical properties are poor, and consequently glass fibers are added. Molding is more difficult than with other thermosetting materials. Unlike most other resins, they may be used in continuous operations at 400°F; they have very low water absorption; their dielectric properties are excellent over an extremely wide variety of chemical attack; and under outdoor conditions their durability is particularly outstanding. In liquid solutions silicones are used to impart moisture resistance to masonry walls and to fabrics. They also form the basis for a variety of paints and other coatings capable of maintaining flexibility and inertness to attack at high temperatures in the presence of ultraviolet sunlight and ozone. Silicone rubbers maintain their flexibility at much lower temperatures than other rubbers.

2-78. Thermoplastic Resins. Materials under this heading in general can be softened by heating and hardened by cooling.

Acrylics. In the form of large transparent sheets, these are used in aircraft enclosures and building construction. Although not so hard as glass, they have perfect clarity and transparency. They are the most resistant of the transparent plastics to sunlight and outdoor weathering, and they have an optimum combination of flexibility and sufficient rigidity with resistance to shattering. A wide variety of transparent, translucent, and opaque colors can be produced. The sheets are readily formed to complex shapes. They are used for such applications as transparent windows, outdoor and indoor signs, parts of lighting equipment, decorative and functional automotive parts, reflectors, household-appliance parts, and similar applications. They can be used as large sheets, molded from molding powders, or cast from the liquid monomer.

Polyethylene. In its unmodified form, this is a flexible, waxy, translucent plastic maintaining flexibility at very low temperatures, in contrast to many other thermoplastic materials. The heat-distortion point of the low-density polyethylenes is low; these plastics are not recommended for uses above 150°F. Newer, high-density materials have higher heat-distortion temperatures. Some may be heated to temperatures above 212°F. The heat-distortion point may rise well above 250°F for plastics irradiated with high-energy beams. Unlike most plastics, this material is partly crystalline. It is highly inert to solvents and corrosive chemicals of all kinds at ordinary temperatures. Usually low moisture permeability and absorption are combined with excellent electrical properties. Its density is lower than that of any other commercially available nonporous plastic. When compounded with black pigment, its weathering properties are good. It is widely used as a primary insulating material on wire and cable and has been used as a replacement for the lead jacket in communication cables and other cables. It is widely used also as thin flexible film for packaging, particularly of food, and as corrosionproof lining for tanks and other chemical equipment.

Two principal forms of polyethylene exist: low-density, or standard, and high-density, or linear. The latter is somewhat more dense than the former, has greater strength and stiffness, withstands somewhat higher temperatures, and has a more sharply defined softening-temperature range.

Polypropylene. This polyolefin is similar in many ways to its counterpart, polyethylene, but is generally harder, stronger, and more temperature-resistant. It finds a great many uses, among them complete water cisterns for water closets in plumbing systems abroad.

Teflon. A polytetrafluoroethylene that finds uses in building calling for resistance to extreme conditions, or for applications requiring low friction. In steam lines, for example, supporting pads of Teflon permit the line to slide easily over the pad as expansion and contraction with changes in temperature cause the line to lengthen and shorten. The temperatures involved have little or no effect. Other low-friction applications include, for example, seats for trusses. Mechanical properties are only moderately high, and reinforcement may be necessary to prevent creep and squeeze-out under heavy loads.

Polytetrafluoroethylene. When fluorine replaces hydrogen, the resulting plastic is a highly crystalline linear-type polymer, unique among organic compounds in its chemical inertness and resistance to change at high and low temperatures. It has an extremely low dielectric-loss factor. In addition, its other electrical properties are excellent. Its outstanding property is extreme resistance to attack by corrosive agents and solvents of all kinds. At temperatures well above 500°F, it can be held for long periods with practically no change in properties, except loss in tensile strength. Service temperatures are generally maintained below 480°F. It is not embrittled at low temperatures, and its films remain flexible at temperatures below −100°F. It is difficult to mold because it has no true softening temperature. Because of this, a modified form in which one chlorine is substituted for fluorine is employed. This is *polymonochlorotrifluoroethylene.* Like the silicones, these fluorocarbons are difficult to wet; consequently, they have high moisture repellence and are often used as parting agents, or where sticky materials, such as candy, must be handled.

Polyurethane. This plastic is finding a number of applications in building. As thermal insulation, it is used in the form of foam, either prefoamed or foamed in place. The latter is particularly useful in irregular spaces. When blown with fluorocarbons, the foam has an exceptionally low *K*-factor and is, therefore, widely used in thin-walled refrigerators. Other uses include field-applied or baked-on clear or colored coatings and finishes for floors, walls, furniture, and casework generally. The rubbery form is employed for sprayed or troweled-on roofing, and for gaskets and calking compounds.

Polyvinylfluoride. This has much of the superior inertness to chemical and weathering attack typical of the fluorocarbons. Among other uses, it is used as thin-film overlays for building boards to be exposed outdoors.

Polyvinyl Formal and Polyvinyl Butyral. Polyvinyl formal resins are principally used as a base for tough, water-resistant insulating enamel for electric wire. Polyvinyl butyral is the tough interlayer in safety glass. In its cross-linked and plasticized form, polyvinyl butyral is extensively used in coating fabrics for raincoats, upholstery and other heavy-duty moisture-resistant applications.

Vinyl Chloride Polymers and Copolymers. These materials vary from hard and rigid to highly flexible. Polyvinyl chloride is naturally hard and rigid but can be plasticized to any required degree of flexibility as in raincoats and shower curtains. Copolymers, including vinyl chloride plus vinyl acetate, are naturally flexible without plasticizers. Nonrigid vinyl plastics are widely used as insulation and jacketing for electric wire and cable because of their electrical properties and their resistance to oil and water. Thin films are used for rainwear and similar applications, whereas heavy-gage films and sheets are widely used for upholstery. Vinyl chlorides are used for floor coverings in the form of tile and sheet because of their abrasion resistance and relatively low water absorption. The rigid materials are used for tubing, pipe, and many other applications where their resistance to corrosion and action of many chemicals, especially acids and alkalies, recommends them. They are attacked by a variety of organic solvents, however. Like all thermoplastics, they soften at elevated temperatures.

Vinylidene Chloride. This material is highly resistant to most inorganic chemicals and to organic solvents generally. It is impervious to water on prolonged immersion, and its films are highly resistant to moisture-vapor transmission. It can be sterilized, if not under load, in boiling water. Mechanical properties are good. It is not recommended for uses involving high-speed impact, shock resistance, or flexibility at subfreezing temperatures.

Polystyrene. Polystyrene formulations constitute a large and important segment of the entire field of thermoplastic materials. Numerous modified polystyrenes provide a relatively wide range of properties. Polystyrene is one of the lightest of the presently available commercial plastics. It is relatively inexpensive, easily molded, has good dimensional stability, and good stability at low temperatures; it is brilliantly clear when transparent and has an infinite range of colors. Water absorption is negligible even after long immersion. Electrical characteristics are excellent. It is resistant to most corrosive chemicals, such as acids, and to a variety of organic solvents, although it is attacked by others. Polystyrenes as a class are considerably more brittle and less extensible than many other thermoplastic materials, but these properties are markedly improved in copolymers. Under some conditions, they have a tendency to develop fine cracks, known as craze marks, on exposure, particularly outdoors. This is true of many other thermoplastics, especially when highly stressed.

Nylon. Molded nylon is used in increasing quantities for impact and high resistance to abrasion. It is employed in small gears, cams, and other machine parts, because even when unlubricated they are highly resistant to wear. Its chemical resistance, except to phenols and mineral acids, is excellent. Extruded nylon is coated onto electric wire, cable, and rope for abrasion resistance. Applications like hammerheads indicate its impact resistance.

2-79. Cellulose Derivatives. Cellulose is a naturally occurring high polymer found in all woody plant tissue and in such materials as cotton. It can be modified by chemical processes into a variety of thermoplastic materials, which, in turn, may be still further modified with plasticizers, fillers, and other additives to provide a wide variety of properties. The oldest of all plastics is cellulose nitrate.

Cellulose Acetate. This is the basis of safety film developed to overcome the highly flammable nature of cellulose nitrate. Starting as film, sheet, or molding powder, it is made into a variety of items such as transparent packages and a large variety of general-purpose items. Depending on the plasticizer content, it may be hard and rigid or soft and flexible. Moisture absorption of this and all other cellulosics is relatively high, and they are therefore not recommended for long-continued outdoor exposure. But cellulose acetate film, reinforced with metal mesh, is widely used for temporary enclosures of buildings during construction.

Cellulose Actate Butyrate. The butyrate copolymer is inherently softer and more flexible than cellulose acetate and consequently requires less plasticizer to achieve a given degree of softness and flexibility. It is made in the form of clear transparent sheet and film, or in the form of molding powders, which can be molded by standard injection-molding procedures into a wide variety of applications. Like the other cellulosics, this material is inherently tough and has good impact resistance. It has infinite colorability, like the other cellulosics. Cellulose acetate butyrate tubing is used for such applications as irrigation and gas lines.

Cellulose Nitrate. One of the toughest of the plastics, cellulose nitrate is widely used for tool handles and similar applications requiring high impact strength. The high flammability requires great caution, particularly in the form of film. Most commercial photographic film is cellulose nitrate as opposed to safety film. Cellulose nitrate is the basis of most of the widely used commercial lacquers for furniture and similar items.

LAMINATES CONTAINING PLASTICS

2-80. High-pressure Laminates. Laminated thermosetting products consist of fibrous sheet materials combined with a thermosetting resin, usually phenol formaldehyde or melamine formaldehyde. The commonly used sheet materials are paper, cotton fabric, asbestos paper or fabric, nylon fabric, and glass fabric. The usual form is flat sheet, but a variety of rolled tubes and rods is made.

Decorative Laminates. These high-pressure laminates consist of a base of phenolic resin-impregnated kraft paper over which an overlay of printed paper or wood veneer is applied. Over all this is laid a thin sheet of melamine resin. When the entire assemblage is pressed in a hot-plate press at elevated temperatures and pressures, the various layers are fused together and the melamine provides a completely transparent finish, resistant to alcohol, water, and common solvents. This material is widely used for tabletops, counter fronts, wainscots, and similar building applications. It is customarily bonded to a core of plywood to develop the necessary thickness and strength. In this case, a backup sheet consisting of phenolic resin and paper alone, without the decorative surface, is employed to provide balance to the entire sandwich. When a thin sheet of aluminum is placed directly under the decorative sheet, it carries heat rapidly away from heat sources like burning cigarettes and prevents scorching.

2-81. Reinforced Plastics. These are commonly made with phenolic, polyester, and epoxide resins combined with various types of reinforcing agents, of which glass fibers in the form of mats or fabrics are the most common. Because little or no pressure is required to form large complex parts, rather simple molds can be employed for the manufacture of such things as boat hulls and similar large parts. In buildings, reinforced plastics have been rather widely used in the form of corrugated sheet for skylights and side lighting of buildings, and as molded shells, concrete forms, sandwiches, and similar applications.

These materials may be formulated to cure at ordinary temperatures, or they may require moderate temperatures to cure the resins. Customarily, parts are made by laying up successive layers of the glass fabric or the glass mat and applying the liquid resin to them. The entire combination is allowed to harden at ordinary temperatures, or it is placed in a heated chamber for final hardening. It may be placed inside a rubber bag and a vacuum drawn to apply moderate pressure, or it may be placed between a pair of matching molds and cured under moderate pressure in the molds.

The high impact resistance of these materials combined with good strength properties and good durability recommends them for building applications. When the quantity of reinforcing agent is kept relatively low, a high degree of translucence may be achieved, although it is less than that of the acrylics and the other transparent thermoplastic materials.

Table 2-30 presents some engineering properties of several types of reinforced plastics.

2-82. Plastics Bibliography

American Society for Testing and Materials, "Standards."

"Modern Plastics Encyclopedia," Plastics Catalog Corp., New York.

Dietz, A. G. H.: "Plastics for Architects and Engineers," M.I.T. Press, Cambridge, Mass.

Skeist, I.: "Plastics in Building," Van Nostrand Reinhold Company, New York.

PORCELAIN-ENAMELED PRODUCTS

Porcelain-enameled metal is used for indoor and outdoor applications because of its hardness, durability, washability, and color possibilities. For building purposes, porcelain enamel is applied to sheet metal and cast iron, the former for a variety of purposes including trim, plumbing, and kitchen fixtures, and the latter almost entirely for plumbing fixtures. Most sheet metal used for porcelain enameling is steel—low in carbon, manganese, and other elements. Aluminum is also used for vitreous enamel.

2-83. Porcelain Enamel on Metal. Low-temperature softening glasses must be employed, especially with sheet metal, to avoid the warping and distortion that would occur at high temperatures. To obtain lower softening temperatures than would be attainable with high-silica glasses, boron is commonly added. Fluorine may replace some of the oxygen, and lead may also be added to produce easy-flowing brilliant enamels; but lead presents an occupational health hazard.

Composition of the enamel is carefully controlled to provide a coefficient of thermal expansion as near that of the base metal as possible. If the coefficient of the enamel is greater than that of the metal, cracking and crazing are likely

Table 2-30. General Properties of Some Reinforced Plastics*

Property	Polyester			Epoxy		Phenolic		
	Glass mat	Glass cloth	Asbestos felt mat	Glass mat	Glass cloth	Glass mat	Glass cloth	Asbestos felt mat
Specific gravity	1.35–2.3	1.50–2.1	1.6–1.9	1.8–2.0	1.9–2.0	1.7–1.9	1.8–1.95	1.7–1.9
Reinforcement content, % by weight	35–45	60–67	50–80	40–50	65–70	45–50	60–65	60–70
Tensile strength, psi	20,000–25,000	30,000–70,000	30,000–60,000	14,000–30,000	20,000–60,000	5,000–20,000	40,000–60,000	40,000–60,000
Compressive strength, psi	15,000–50,000	25,000–50,000	30,000–50,000	30,000–38,000	50,000–70,000	17,000–26,000	35,000–40,000	45,000–55,000
Flexural strength, psi	10,000–40,000	40,000–90,000	50,000–70,000	20,000–26,000	70,000–100,000	10,000–60,000	65,000–95,000	50,000–90,000
Impact strength, ft-lb per in. of notch (½ × ½ in. notched bar, Izod test)	2–10	5–30	2–8	8–15	11–26	8–16	10–35	1–6
Water absorption, % (24 hr, ⅛ in. thickness)	0.01–1.0	0.05–0.5	0.01–0.1	0.05–0.95	0.08–0.7	0.1–1.2	0.07–0.9	0.01–0.1
Heat resistance, °F (continuous)	300–350	300–350	300–450	330–500	330–500	350–500	350–500	350–600
Burning rate	Slow to self-extinguishing	Slow to self-extinguishing	None	Slow to self-extinguishing	Slow to self-extinguishing	None	None	None
Volume resistivity, ohms per cu cm (at 50% RH and 73°F)	10^{14}	10^{14}	6.6×10^{8}	3.8×10^{15}	3.8×10^{15}	7×10^{12}	7×10^{12}	10^{10}–10^{12}
Arc resistance, sec	120–180	60–120	138	125–140	100–110	40–150	20–130	120–200

* "Modern Plastics Encyclopedia."

to occur, but if the coefficient of the enamel is slightly less, it is lightly compressed upon cooling, a desirable condition because glass is strong in compression.

To obtain good adhesion between enamel and metal, one of the so-called transition elements used in glass formulation must be employed. Cobalt is favored. Apparently, the transition elements promote growth of iron crystals from base metal into the enamel, encourage formation of an adherent oxide coating on the iron, which fuses to the enamel, or develop polar chemical bonds between metal and glass.

Usually white or colored opaque enamels are desired. Opacity is promoted by mixing in, but not dissolving, finely divided materials possessing refractive indexes widely different from the glass. Tin oxide, formerly widely used, is now largely displaced by less expensive and more effective titanium and zirconium compounds. Clay adds to opacity. Various oxides are included to impart color.

Most enameling consists of a ground coat and one or two cover coats fired on at slightly lower temperatures; but one-coat enameling of somewhat inferior quality can be accomplished by first treating the iron surface with soluble nickel salts.

The usual high-soda glasses used to obtain low-temperature softening enamels are not highly acid-resistant and therefore stain readily and deeply when iron-containing water drips on them. Enamels highly resistant to severe staining conditions must be considerably harder; i.e., have higher softening temperatures and therefore require special techniques to avoid warping and distorting of the metal base.

Interiors of refrigerators are often made of porcelain-enameled steel sheets for resistance to staining by spilled foods, whereas the exteriors are commonly baked-on synthetic-resin finishes, such as those incorporating melamine and urea formaldehyde.

2-84. Porcelain Bibliography

Andrews, A. I.: "Enamels," The Twin City Printing Co., Champaign, Ill.

Norton, F. H.: "Elements of Ceramics," pp. 183–189, Addison-Wesley Publishing Company, Cambridge, Mass., 1952.

Kingery, W. D.: "Introduction to Ceramics," John Wiley & Sons, Inc., New York.

RUBBER

Rubber for construction purposes is both natural and synthetic. Natural rubber, often called crude rubber in its unvulcanized form, is composed of large complex molecules of isoprene.

2-85. Synthetic Rubbers. The principal synthetic rubbers are the following:

GR-S is the one most nearly like crude rubber and is the product of styrene and butadiene copolymerization. It is the most widely used of the synthetic rubbers. It is not oil-resistant but is widely used for tires and similar applications.

Nitril is a copolymer of acrylonitrile and butadiene. Its excellent resistance to oils and solvents makes it useful for fuel and solvent hoses, hydraulic-equipment parts, and similar applications.

Butyl is made by the copolymerization of isobutylene with a small proportion of isoprene or butadiene. It has the lowest gas permeability of all the rubbers and consequently is widely used for making inner tubes for tires and other applications in which gases must be held with a minimum of diffusion. It is used for gaskets in buildings.

Neoprene is made by the polymerization of chloroprene. It has very good mechanical properties and is particularly resistant to sunlight, heat, aging, and oil; it is therefore used for making machine belts, gaskets, oil hose, insulation on wire cable, and other electrical applications to be used for outdoor exposure, such as building and glazing gaskets and roofing.

Sulfide rubbers—the polysulfides of high molecular weight—have rubbery properties, and articles made from them, such as hose and tank linings and glazing compounds, exhibit good resistance to solvents, oils, ozone, low temperature, and outdoor exposure.

Silicone rubber, which is discussed also in the section on plastics, when made in rubbery consistency forms a material exhibiting exceptional inertness and temperature resistance. It is therefore used in making gaskets, electrical insulation, and similar products that maintain their properties at both high and low temperatures.

Additional elastomers, which are sometimes grouped with the rubbers, include polyethylene, cyclized rubber, plasticized polyvinyl chloride, and polybutene.

2-86. Properties of Rubber. In the raw state, rubbers are generally quite plastic, especially when warm, have relatively low strength, are attacked by various solvents, and often can be dissolved to form cements. These characteristics are necessary for processing, assembling, and forming but are not consistent with the strength, heat stability, and elasticity required in a finished product. The latter desirable properties are obtained by vulcanization, a chemical change involving the interlinking of the rubber molecules and requiring incorporation of sulfur, zinc oxide, organic accelerators, and other ingredients in the raw rubber prior to heat-treatment.

Other compounding ingredients do not enter into the chemical reaction but add considerably to the mechanical properties of finished rubber. This is particularly true of carbon black. Antioxidants retard the rate of deterioration of rubber on exposure to light and oxygen. The great variety of materials entering into various rubber compounds therefore provides a wide range of properties. In addition, many rubber products are laminated structures of rubber compounds combined with materials like fabric and metals.

The two principal categories of rubber are hard and soft. Hard rubber has a high degree of hardness and rigidity produced by vulcanization with high proportions of sulfur (ranging as high as 30 to 50% of the rubber). In soft rubber, however, sulfur may range as low as 1 to 5% and usually not more than 10%.

Principal properties of rubber compounds of interest in building are the following:

Age Resistance. Rubber normally oxidizes slowly on exposure to air at ordinary temperatures. Oxidation is accelerated by heat and ozone. Hard rubber ages less rapidly than soft, and most of the synthetics age at somewhat the same or greatly retarded rates compared with natural rubber.

Compressibility. With the exception of sponge rubber, completely confined rubber, as in a tight container, is virtually impossible to compress.

Elongation. Soft vulcanized compounds may stretch as much as 1,000% of the original length, whereas the elongation of hard rubber ranges usually between 1 and 50%. Even after these great elongations, rubber will return practically to its original length.

Energy Absorption. The energy-storing ability of rubber is about 150 times that of spring-tempered steel. The resilient energy-storing capacity of rubber is 14,600 ft-lb per lb as compared with 95.3 for spring steel.

Expansion and Contraction. The volumetric changes caused by temperature are generally greater for rubber than for metals. Soft vulcanized-rubber compounds, for example, have thermal coefficients of expansion ranging from 0.00011 to 0.00005, whereas hard vulcanized-rubber compounds range from 0.00004 to 0.000015, as compared with coefficients of aluminum of 0.000012, glass of 0.000005, and steel of 0.000007.

Friction. The coefficient of friction between soft rubber and dry steel may exceed unity. But when the surfaces are wet, the friction coefficient drops radically and may become as low as 0.02 in water-lubricated rubber bearings.

Heat Resistance. Vulcanized-natural-rubber compounds are normally limited in use to temperatures of 150 to 200°F, whereas some of the synthetic-rubber compounds like the silicones may be used at temperatures as high as 450°F.

Light Resistance. The natural, GR-S, and nitryl rubber compounds under tension are likely to crack when exposed to sunlight. Other rubbers are highly resistant to similar attack. Sunlight discolors hard rubber somewhat and reduces its surface electrical resistivity.

Gas Permeability. The polysulfide and butyl compounds have very low rates of permeation by air, hydrogen, helium, and carbon dioxide, in contrast with the natural and GR-S compounds, whose gas permeability is relatively high. A butyl inner tube loses air only about 10% as rapidly as a natural rubber tube.

Strength. The tensile strength based on original cross section ranges from 300 to 4,500 psi for soft-rubber stocks, and from about 1,000 to 10,000 psi for hard rubber. Under compression, soft rubber merely distorts, whereas true hard rubber can be subjected to 10,000 to 15,000 psi before distorting markedly.

2-87. Laminated Rubber. Rubber is often combined with various textiles, fabrics, filaments, and metal wire to obtain strength, stability, abrasion resistance, and flexibility. Among the laminated materials are the following:

V Belts. These consist of a combination of fabric and rubber, frequently combined with reinforcing grommets of cotton, rayon, steel, or other high-strength material extending around the central portion.

Flat Rubber Belting. This laminate is a combination of several plies of cotton fabric or cord, all bonded together by a soft-rubber compound.

Conveyor Belts. These, in effect, are moving highways used for transporting such material as crushed rock, dirt, sand, gravel, slag, and similar materials. When the belt operates at a steep angle, it is equipped with buckets or similar devices and becomes an elevator belt. A typical conveyor belt consist of cotton duck plies alternated with thin rubber plies; the assembly is wrapped in a rubber cover, and all elements are united into a single structure by vulcanization. A conveyor belt to withstand extreme conditions is made with some textile or metal cords instead of the woven fabric. Some conveyor belts are especially arranged to assume a trough form and made to stretch less than similar all-fabric belts.

Rubber-lined Pipes, Tanks, and Similar Equipment. The lining materials include all the natural and synthetic rubbers in various degrees of hardness, depending on the application. Frequently, latex rubber is deposited directly from the latex solution onto the metal surface to be covered. The deposited layer is subsequently vulcanized. Rubber linings can be bonded to ordinary steel, stainless steel, brass, aluminum, concrete, and wood. Adhesion to aluminum is inferior to adhesion to steel. Covering for brass must be compounded according to the composition of the metal.

Rubber Hose. Nearly all rubber hose is laminated and composed of layers of rubber combined with reinforcing materials like cotton duck, textile cords, and metal wire. The modern laminated rubber hose can be made with a large variety of structures. Typical hose consists of an inner rubber lining, a number of intermediate layers consisting of braided cord or cotton duck impregnated with rubber, and outside that, several more layers of fabric, spirally wound cord, spirally wound metal, or in some cases, spirally wound flat steel ribbon. Outside of all this is another layer of rubber to provide resistance to abrasion. Hose for transporting oil, water, wet concrete under pressure, and for dredging purposes is made of heavy-duty laminated rubber.

Vibration Insulators. These usually consist of a layer of soft rubber bonded between two layers of metal. Another type of insulator consists of a rubber tube or cylinder vulcanized to two concentric metal tubes, the rubber being deflected in shear. A variant of this consists of a cylinder of soft rubber vulcanized to a tubular or solid steel core and a steel outer shell, the entire combination being placed in torsion to act as a spring. Heavy-duty mounts of this type are employed on trucks, buses, and other applications calling for rugged construction.

2-88. Rubber Bibliography

American Society for Testing and Materials, Philadelphia, Pa.: "Standards."

ASPHALT AND BITUMINOUS PRODUCTS

Asphalt, because of its water-resistant qualities and good durability, is used for many building applications to exclude water, to provide a cushion against vibration and expansion, and in similar applications.

2-89. Dampproofing and Waterproofing Asphalts. For dampproofing (mopped-on coating only), and waterproofing (built-up coating of one or more plies) three types of asphalt are recognized (ASTM Specification D449):

Type A, an easily flowing, soft, adhesive, "self-healing" material for use underground or wherever similar moderate temperatures are found.

Type B, a less susceptible asphalt for use aboveground where temperatures do not exceed 125°F.

Type C, for use aboveground where exposed on vertical surfaces to direct sunlight or other areas where temperatures exceed 125°.

Softening ranges are, respectively, 115 to 145°F, 145 to 170°F, and 180 to 200°F.

2-90. Roofing Asphalts and Pitches. For constructing built-up roofing, four grades of asphalt are recognized (ASTM D312): Type 1, for inclines up to 1 in. per ft; Type II, for inclines up to 3 in. per ft; Type III, for inclines up to 6 in. per ft; and Type IV, suited for relatively steep slopes, generally in areas with relatively high year-round temperatures. Types I through IV may be either smooth or surfaced with slag or gravel. Softening ranges: 135 to 150°F, 160 to 175°F, 180 to 200°F, and 205 to 225°F, respectively.

Coal-tar pitches for roofing, dampproofing, and waterproofing are of two types (ASTM D450): Type A, for gravel or slag-surfaced roofing on inclines up to 3 in. per ft (nailed) or 1 in. per ft (not nailed), and for dampproofing and waterproofing above ground; Type B, for dampproofing and waterproofing under moderate temperature conditions below ground. Softening ranges are, respectively, 140 to 155°F, and 120 to 140°F.

2-91. Roofing Felts. For built-up waterproofing and roofing, three types of membranes are employed: felt (ASTM D226, D227), asbestos felt (ASTM D250), and cotton fabrics (ASTM D173). Felts are felted sheets of vegetable or animal fibers or mixtures of both, saturated with asphalt or with coal tar conforming to ASTM D312 and D450.

Standard asphalt felts weigh 15 or 30 lb per square (100 sq ft), and standard coal-tar felts weigh 15 lb per square.

Asbestos felts, weighing 15 or 30 lb per square, are asphalt-saturated felts containing at least 85% asbestos fiber.

Cotton fabrics are open-weave materials weighing at least 3½ oz per sq yd before saturating, with thread counts of 24 to 32 per in. The saturants are either asphalts or coal tars. The saturated fabric must weigh at least 10 oz per sq yd.

2-92. Roll Roofing. Asphalt roll roofing, shingles, and siding consist basically of roofing felt, first uniformly impregnated with hot asphaltic saturant and then coated on each side with at least one layer of a hot asphaltic coating and compounded with a water-insoluble mineral filler. The bottom or reverse side, in each instance, is covered with some suitable material, like powdered mica, to prevent sticking in the package or roll.

Granule-surfaced roll roofing (ASTM D249) is covered uniformly on the weather side with crushed mineral granules, such as slate. Minimum weight of the finished roofing must be 80 to 83 lb per square (100 sq ft), and the granular coating must weigh at least 18.5 lb per square.

Roll roofing (ASTM 224), surfaced with powdered talc or mica, is made in two grades, 65 and 55 lb per square, of which at least 18 lb must be the surfacing material.

2-93. Asphalt Shingles. There are two standard types: Type I, uniform thickness; Type II, thick butt. Average weights must be 95 lb per square (100 sq ft). For Type I, the weather-side coating must weigh 23.0 lb per square; for Type II, 30.0 lb per square. The material in these shingles is similar to granule-surfaced roll roofing.

2-94. Asphalt Mastics and Grouts. Asphalt mastics used for waterproofing floors and similar structures, but not intended for pavement, consist of mixtures of asphalt cement, mineral filler, and mineral aggregate, which can be heated at about 400°F to a sufficiently soft condition to be poured and troweled into place. The raw ingredients may be mixed on the job or may be premixed, formed into cakes, and merely heated on the job (ASTM D491).

Bituminous grouts are suitable for waterproofing above or below ground level as protective coatings. They also can be used for membrane waterproofing or for bedding and filling the joints of brickwork. Either asphaltic or coal-tar pitch materials of dampproofing and waterproofing grade are used, together with mineral aggregates as coarse as sand (ASTM D170, D171).

2-95. Bituminous Pavements. Asphalts for pavement (ASTM D946) contain petroleum asphalt cement, derived by the distillation of asphaltic petroleum. Various grades are designated as 40-50, 50-60, 60-70, 85-100, 120-150, and 200-300 depending upon the depth of penetration of a standard needle in a standard test (ASTM D5).

Emulsions range from low to high viscosity and quick- to slow-setting (ASTM D977).

2-96. Asphalt Bibliography

American Society for Testing and Materials, Philadelphia, Pa.: "Standards."

MASTIC SEALERS

Mastic sealers or calking compounds are employed to seal the points of contact between similar and dissimilar building materials that cannot otherwise be made completely tight. Such points include glazing, the joints between windows and walls, the many joints occurring in the increasing use of panelized construction, the copings of parapets, and similar spots.

The requirements of a good mastic sealer or calking compound are: (1) good adhesion to the surrounding materials, (2) good cohesive strength, (3) elasticity to allow for compression and extension as surrounding materials retract or approach each other because of changes in moisture content or temperature, (4) good durability or the ability to maintain their properties over a long period of time without marked deterioration, and (5) no staining of surrounding materials such as stone.

2-97. Calking Compounds. Most calking compounds have been composed of a variety of drying oils combined with fillers such as ground limestone, asbestos fiber, chalk, silica, and similar materials. For extreme outdoor exposure conditions, rubber-based elastomeric compounds, including those based on polysulfide and silicone rubbers, are commonly used. They have proved durable when properly applied in a variety of outdoor applications, including use as building sealants.

In the oil-based compounds, the oils are given a variety of treatments depending upon their ultimate uses. Standard procedure is to treat the oils by either or both of the following procedures:

1. *Polymerization.* The oil is raised to a temperature of 500 to 600°F to promote an increase in molecular size, which at the same time changes the viscosity of the oil. Oils so treated have good package stability and increased resistance to absorption into the capillaries of masonry materials.

2. *Oxidation.* The oil is raised to a temperature of 200 to 300°F, and air is blown through the hot oil. Oxygen is absorbed and increases the molecular size and the viscosity of the oil. Oxidized oils have a tendency to continue to oxidize or "body" even while in storage in sealed drums. Oxidized oils resist the tendency to be drawn into the capillaries of masonry and other porous surfaces even more than the polymerized oils and are therefore chosen for calking compounds to be applied directly to porous unprimed masonry. Their package stability is not so good as that of the polymerized oils. Oxidized oils are rapid film formers and tend to skin over quickly either in the package or after being applied on the building.

In addition to the foregoing treatments, it is possible to separate the slow-drying and fast-drying portions of linseed, fish, and other oils, and thereby to obtain faster- or slower-hardening mastics. Materials based on synthetic resins such as the polybutenes of various molecular weights have opened new possibilities in the formulation of mastics. Some of these remain flexible at temperatures as low as −65°F.

Four basic types of consistencies are required for the various types of building applications:

1. Knife-consistency glazing compounds are used to glaze lights of glass in window frames. Whereas standard putties consist of about 10% raw linseed oil and 90% ground calcium carbonate, the newer glazing compounds employ about 14 to 15% mixed raw and bodied oils with oil-absorbing pigments such as asbestos fiber and other fibrous silicates in addition to ground calcium carbonate or marble dust. Reactive pigments such as lead carbonate are used to promote flexible skin formation.

2. Hand- or tool-consistency compounds are formulated to be highly cohesive and more adhesive than the knife glazing compounds. Their consistency is something

like that of modeling clay and they are formulated with bodied oils only. Asbestos fiber, amorphous calcium carbonate whiting, and some crystalline whiting in the form of ground calcium carbonate are the usual pigments.

3. Gun-consistency calking compounds are formulated entirely from bodied oil much of which is oxidized. Pigmentation consists of asbestos fibers, silicate fibers of shorter sizes, ground clays, and sometimes ground calcium carbonate in crystalline form.

4. Paddle or spray mastics are thin consistencies like extremely heavy paints and are specifically designed for use in narrow joints.

Because mastic sealers are based on drying oils that eventually harden in contact with the air, the best joints are generally thick and deep, with a relatively small portion exposed to the air. The exposed surface is expected to form a tough protective skin for the soft mass underneath, which in turn provides the cohesiveness, adhesiveness and elasticity required. Thin shallow beads cannot be expected to have the durability of thick joints with small exposed surface areas.

Rubber-based compounds generally are composed of two components, base material and catalyst. These two are mixed when they are to be used and, once mixed, proceed to harden by intermolecular linking and must therefore be put in place before they stiffen too much. At ordinary temperatures the "pot life," or period during which they are workable, is approximately 6 to 7 hr. Heating shortens the pot life; cooling lengthens it. One-component rubber-based compounds, which do not require mixing on the job, also are available. They cure on exposure to the atmosphere, for example, by reaction with moisture in the air.

Various consistencies and hardnesses are possible. The soft formulations (Shore A hardness 10 to 20) show elongations of about 300 to 350%, whereas the heavier gun consistencies may show elongations of 90 to 150%.

The best of the rubber-based formulations show little change in hardness, elongation, and adhesiveness with age. Manufacturers recommend them particularly for uses involving large differential movements where maximum adhesion must at the same time be maintained, as in large double-glass units set in metal frames, curtain walls in which metal sheets undergo marked changes in dimension with changes in temperature, and similar applications.

For setting large sheets of glass and similar units, setting or supporting spacer blocks of rubber are often combined with gaskets of materials such as vulcanized synthetic rubber and are finally sealed with the elastomeric rubber-based mastics or glazing compounds.

2-98. Mastic Sealers Bibliography

Cook, J. P.: "Construction Sealants and Adhesives," John Wiley & Sons, Inc., New York.

Damusis, A.: "Sealants," Van Nostrand Reinhold Company, New York.

COATINGS

Protective and decorative coatings generally employed in building are the following:

Oil Paint. Drying-oil vehicles or binders plus opaque and extender pigments.

Water Paint. Pigments plus vehicles based on water, casein, protein, oil emulsions, and rubber or resin latexes, separately or in combination.

Calcimine. Water and glue, with or without casein, plus powdered calcium carbonate and any desired colored pigments.

Varnish. Transparent combination of drying oil and natural or synthetic resins.

Enamel. Varnish vehicle plus pigments.

Lacquer. Synthetic-resin film former, usually nitrocellulose, plus plasticizers, volatile solvents, and other resins.

Shellac. Exudations of the lac insect, dissolved in alcohol.

Japan. Solutions of metallic salts in drying oils, or varnishes containing asphalt and opaque pigments.

Aluminum Paint. Fine metallic aluminum flakes suspended in drying oil plus resin, or in nitrocellulose.

2-99. Drying-oil Vehicles or Binders. Most vehicles contain drying oils. These are vegetable and animal oils that harden or "dry" by absorbing oxygen. The most important drying oils are:

Linseed oil derived from flaxseed by crushing, cooking, and pressing. Like other drying oils, this raw oil hardens slowly and is therefore usually treated by "boiling" or blowing. Boiling consists of heating at elevated temperatures to promote polymerization, which is accompanied by increased body and viscosity and increased readiness to react with oxygen. Blowing consists of heating to moderate temperature and blowing air through the oil to cause bodying and increased viscosity by oxidation.

Dehydrated castor oil, made by heating under vacuum with catalysts to remove water. Natural castor oil is not a drying oil; but dehydrated castor oil is excellent, with good flexibility, adhesion, and rapid-drying characteristics.

Fish oil, used mostly in cheap paints where softness and aftertack are not undesirable. The oil must be refined, as by refrigeration, to remove nonhardening constituents.

Soybean oil, obtained from the seeds of the soya plant, has excellent flexibility and is nonyellowing but dries slowly and is usually combined with tung, perilla, or linseed oil.

Tung or chinawood oil, derived from the tung tree nut, is especially waterpoof, fast-drying, and durable. It is usually used with less reactive oils, such as linseed or soya. In the raw state, it tends to wrinkle on hardening and is used for wrinkle finishes; otherwise it is treated or combined with other oils.

Other oils include oiticica, perilla, chia, hempseed, peanut, corn, poppyseed, safflower, sunflower, and walnut.

An important class of vehicles consists of **alkyd resins** combined with drying oils. A considerable number of alkyds can be reacted with a wide variety of drying oils in combinations ranging from mostly alkyd to mostly oil, to provide a large choice of alkyd-oil combinations useful for numerous air-drying and heat-hardening finishes.

Dryers are catalysts that hasten the hardening of drying oils. Most dryers are salts of heavy metals, especially cobalt, manganese, and lead, to which salts of zinc and calcium may be added. Iron salts, usable only in dark coatings, accelerate hardening at high temperatures. Dryers are normally added to paints to hasten hardening, but they must not be used too liberally or they cause rapid deterioration of the oil by overoxidation.

Thinners are volatile constituents added to coatings to promote their spreading qualities by reducing viscosity. They should not react with the other constituents and should evaporate completely. Commonly used thinners are turpentine and mineral spirits; i.e., derivatives of petroleum and coal tar.

2-100. Pigments for Paints. Pigments may be classified as white and colored, or as opaque and extender pigments. The hiding power of pigments depends on the difference in index of refraction of the pigment and the surrounding medium—usually the vehicle of a protective coating. In opaque pigments, these indexes are markedly different from those of the drying-oil vehicles; in extender pigments, they are nearly the same. The comparative hiding efficiencies of various pigments must be evaluated on the basis of hiding power per pound and cost per pound. The relative hiding power per pound of various important white pigments, in descending order, is approximately as follows: rutile titanium dioxide, anatase titanium dioxide, zinc sulfide, titanium calcium, titanium barium, zinc sulfide barium, titanated lithopone, lithopone, antimony oxide, zinc oxide.

Principal white pigments are zinc oxide, lithopone, titanated lithopone, zinc sulfide, titanated zinc sulfide, zinc sulfide–barium, zinc sulfide–calcium, zinc sulfide–magnesium, antimony oxide, titanium dioxide (anatase and rutile), titanium-barium, titanium-calcium, and titanium-magnesium.

Zinc oxide is widely used by itself or in combination with other pigments. Its color is unaffected by many industrial and chemical atmospheres. It imparts gloss and reduces chalking but tends to crack and alligator instead.

Zinc sulfide is a highly opaque pigment widely used in combination with other pigments.

Titanium dioxide and extended titanium pigments are relative newcomers but have forged into the forefront because of their high opacity and generally excellent properties. Various forms of the pigments have different properties. For example, anatase titanium dioxide promotes chalking, whereas rutile inhibits it.

Colored pigments for building use are largely inorganic materials, especially for outdoor use, where the brilliant but fugitive organic pigments soon fade. The principal inorganic colored pigments are:

Metallic. Aluminum flake or ground particle, copper bronze, gold leaf, zinc dust.

Black. Carbon black, lampblack, graphite, blue basic lead sulfate (gray), vegetable black, and animal blacks.

Earth colors. Yellow ocher, raw and burnt umber, raw and burnt sienna, reds, and maroons.

Blue. Ultramarine, iron (Prussian, Chinese, Milori).

Brown. Mixed ferrous and ferric oxide.

Green. Chromium oxide, hydrated chromium oxide, chrome greens.

Maroon. Iron oxide, cadmium.

Orange. Molybdated chrome orange.

Red. Iron oxide, cadmium red, vermilion.

Yellow. Zinc chromate, cadmium yellows, hydrated iron oxide.

Extender pigments are added to extend the opaque pigments, increase durability, provide better spreading characteristics, and reduce cost. The principal extender pigments are silica, china clay, talc, mica, barium sulfate, calcium sulfate, calcium carbonate, and such materials as magnesium oxide, magnesium carbonate, barium carbonate, and others used for specific purposes.

2-101. Resins for Paints. Natural and synthetic resins are used in a large variety of air-drying and baked finishes. The natural resins include both fossil resins, which are harder and usually superior in quality, and recent resins tapped from a variety of resin-exuding trees. The most important fossil resins are amber (semiprecious jewelry), Kauri, Congo, Boea Manila, and Pontianak. Recent resins include Damar, East India, Batu, Manila, and rosin. Shellac, the product of the lac insect, may be considered to be in this class of resins.

The synthetic resins, in addition to the alkyds already considered under drying oils (Art. 2-99), are increasing in importance, especially for applications requiring maximum durability. Among them are phenol formaldehyde, melamine formaldehyde, urea formaldehyde, epoxides, urethanes, silicones, polyvinyl chlorides, and acrylics.

Phenolics in varnishes are used for outdoor and other severe applications on wood and metals. They are especially durable when baked.

Melamine and urea find their way into a large variety of industrial finishes like automobile and refrigerator finishes.

Epoxides bond well to a variety of surfaces, including metals, and are used for tough, durable, and transparent or pigmented coatings.

Silicones are used when higher temperatures are encountered than can be borne by the other finishes.

Urethanes, polyvinyl chlorides, and acrylics provide tough, durable films for indoor and outdoor use.

2-102. Coatings Bibliography

Burns, R. M., and Bradley, W.: "Protective Coatings for Metals," Van Norstrand Reinhold Company, New York.

Martens, C. R., "The Technology of Paints, Varnishes and Lacquers," Van Nostrand Reinhold Company, New York.

Section 3

Stresses in Structures

FREDERICK S. MERRITT

Consulting Engineer, Syosset, N.Y.

STRESS AND STRAIN

Loads are the external forces acting on a structure. Stresses are the internal forces that resist them. Depending on the manner in which the loads are applied, they tend to deform the structure and its components—tensile forces tend to stretch, compressive forces to squeeze together, torsional forces to twist, and shearing forces to slide parts of the structure past each other.

If the structure and its components are so supported that, after a very small deformation occurs, no further motion is possible, they are said to be in equilibrium. Under such circumstances, internal forces, or stresses, exactly counteract the loads.

3-1. Static Equilibrium. Several useful conclusions may be drawn from the state of static equilibrium: Since there is no translatory motion, the sum of the external forces must be zero; and since there is no rotation, the sum of the moments of the external forces about any point must be zero.

For the same reason, if we consider any portion of the structure and the loads on it, the sum of the external and internal forces on the boundaries of that section must be zero. Also, the sum of the moments of these forces must be zero.

In Fig. 3-1, for example, the sum of the forces R_L and R_R needed to support the roof truss is equal to the 20-kip load on the truss (1 kip = 1 kilopound = 1,000 lb = 0.5 ton). Also, the sum of the moments of the external forces is zero about any point; about the right end, for instance, it is $40 \times 15 - 30 \times 20 = 600 - 600$.

In Fig. 3-2 is shown the portion of the truss to the left of section *AA*. The internal forces at the cut members balance the external load and hold this piece of the truss in equilibrium.

Generally, it is convenient to decompose the forces acting on a structure into components parallel to a set of perpendicular axes that will simplify computations. For example, for forces in a single plane—a condition commonly encountered in building design—the most useful technique is to resolve all forces into horizontal

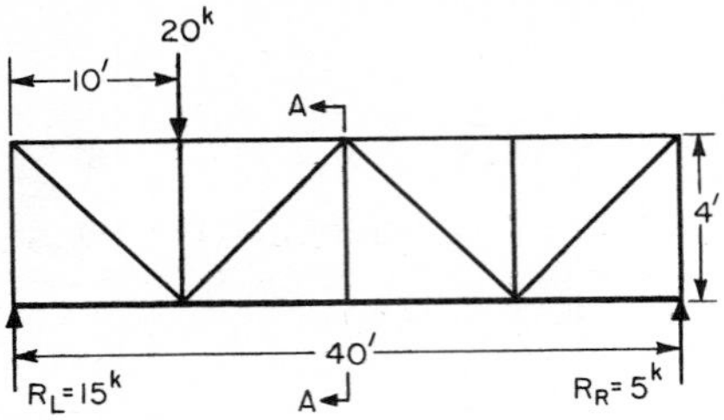

Fig. 3-1. Truss in equilibrium under load. Upward-acting forces equal downward.

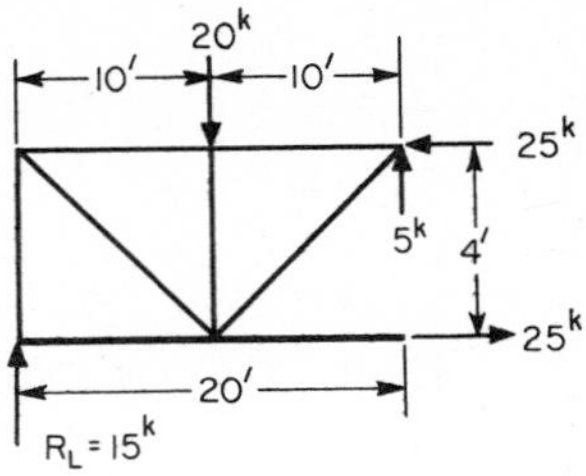

Fig. 3-2. Section of truss kept in equilibrium by stresses in its components.

and vertical components. Then, for a structure in equilibrium, if H represents the horizontal components, V the vertical components, and M the moments of the components about any point in the plane,

$$\Sigma H = 0 \qquad \Sigma V = 0 \qquad \text{and} \qquad \Sigma M = 0 \tag{3-1}$$

These three equations may be used to evaluate three unknowns in any nonconcurrent coplanar force system, such as the roof truss in Figs. 3-1 and 3-2. They may determine the magnitude of three forces for which the direction and point of application already are known, or the magnitude, direction, and point of application of a single force.

Suppose, for the truss in Fig. 3-1, the reactions at the supports are to be computed. Taking moments about the right end and equating to zero yields $40 R_L - 30 \times 20 = 0$, from which left reaction $R_L = 600/40 = 15$ kips. Equating the sum of the vertical forces to zero gives $20 - 15 - R_R = 0$, from which the right reaction $R_R = 5$ kips.

3-2. Types of Load. External loads on a structure may be classified in several different ways. In one classification, they may be considered as static or dynamic.

Static loads are forces that are applied slowly and then remain nearly constant. One example is the weight, or dead load, of a floor or roof system.

Dynamic loads vary with time. They include repeated and impact loads.

Repeated loads are forces that are applied a number of times, causing a variation in the magnitude, and sometimes also in the sense, of the internal forces. A good example is a bridge crane.

Impact loads are forces that require a structure or its components to absorb energy in a short interval of time. An example is the dropping of a heavy weight on a floor slab, or the shock wave from an explosion striking the walls and roof of a building.

External forces may also be classified as distributed and concentrated.

Uniformly distributed loads are forces that are, or for practical purposes may be considered, constant over a surface area of the supporting member. Dead weight of a rolled-steel I beam is a good example.

Concentrated loads are forces that have such a small contact area as to be negligible compared with the entire surface area of the supporting member. A beam supported on a girder, for example, may be considered, for all practical purposes, a concentrated load on the girder.

Another common classification for external forces labels them axial, eccentric, and torsional.

An axial load is a force whose resultant passes through the centroid of a section under consideration and is perpendicular to the plane of the section.

An eccentric load is a force perpendicular to the plane of the section under consideration but not passing through the centroid of the section, thus bending the supporting member (see Arts. 3-33, 3-35, and 3-78).

Torsional loads are forces that are offset from the shear center of the section under consideration and are inclined to or in the plane of the section, thus twisting the supporting member (see Art. 3-78).

Also, building codes classify loads in accordance with the nature of the source. For example:

Dead loads include materials, equipment, constructions, or other elements of weight supported in, on, or by a building, including its own weight, that are intended to remain permanently in place.

Live loads include all occupants, materials, equipment, constructions, or other elements of weight supported in, on, or by a building and that will or are likely to be moved or relocated during the expected life of the building.

Impact loads are a fraction of the live loads used to account for additional stresses and deflections resulting from movement of the live loads.

Wind loads are maximum forces that may be applied to a building by wind in a mean recurrence interval, or a set of forces that will produce equivalent stresses. Mean recurrence intervals generally used are 25 years for structures with no occupants or offering negligible risk to life, 50 years for ordinary permanent structures, and 100 years for permanent structures with a high degree of sensitivity to wind and an unusually high degree of hazard to life and property in case of failure.

Snow loads are maximum forces that may be applied by snow accumulation in a mean recurrence interval.

Seismic loads are forces that produce maximum stresses or deformations in a building during an earthquake, or equivalent forces.

For design loads, use probable maximum loads or the loading specified in the local building code, whichever is larger. In the absence of a local building code, use the loads specified in American National Standard "Building Code Requirements for Minimum Design Loads in Buildings and Other Structures," A58.1, and "Standard for Tents, Grandstands and Air-Supported Structures Used for Places of Assembly," Z20.3, latest editions, American National Standards Institute, 1430 Broadway, New York, N.Y. 10018. Alternatively, Tables 3-1 and 3-2 may be used. Also, see Art. 3-84 for seismic loads.

3-3. Unit Stress. For convenience in evaluating the load-carrying ability of a member of a structure, it is customary to express its strength in terms of the maximum unit stress, or internal force per unit of area, that it can sustain without suffering structural damage. Then, application of a safety factor to the maximum unit stress determines a unit stress that should not be exceeded when design loads are applied to the member. That unit stress is known as the allowable stress or working stress.

To ascertain whether a structural member has adequate load-carrying capacity, the designer generally has to compute the maximum unit stress produced by design loads in the member for each type of internal force—tensile, compressive, or shearing—and compare it with the corresponding allowable unit stress.

When the loading is such that the unit stress is constant over a section under consideration, the stress may be obtained by dividing the force by the area of the section. But in general, the unit stress varies from point to point. In that case, the unit stress at any point in the section is the limiting value of the ratio of the internal force on any small area to that area, as the area is taken smaller and smaller.

3-4. Unit Strain. Sometimes in the design of a structure, unit stress may not be the prime consideration. The designer may be more interested in limiting the deformation or strain.

Deformation in any direction is the total change in the dimension of a member in that direction.

Unit strain in any direction is the deformation per unit of length in that direction.

When the loading is such that the unit strain is constant over a portion of a member, it may be obtained by dividing the deformation by the original length of that portion. In general, however, the unit strain varies from point to point in a member. Like a varying unit stress, it represents the limiting value of a ratio (Art. 3-3).

Table 3-1. Minimum Design Live, Wind, and Snow Loads

A. Uniformly Distributed Live Loads, Psf, Impact Included[a]

Occupancy or use	Load	Occupancy or use	Load
Assembly spaces:		Marquees	75
Auditoriums[b] with fixed seats	60	Morgue	125
Auditoriums[b] with movable seats	100	Office buildings:	
Ballrooms and dance halls	100	Corridors above first floor	80
Bowling alleys, poolrooms, similar recreational areas	75	Files	125
		Offices	50
Conference and card rooms	50	Penal institutions:	
Dining rooms, restaurants	100	Cell blocks	40
Drill rooms	150	Corridors	100
Grandstand and reviewing-stand seating areas	100	Residential:	
		Dormitories:	
Gymnasiums	100	Nonpartitioned	60
Lobbies, first-floor	100	Partitioned	40
Roof gardens, terraces	100	Dwellings, multifamily:	
Skating rinks	100	Apartments	40
Bakeries	150	Corridors	80
Balconies (exterior)	100	Hotels:	
Up to 100 sq ft on one- and two-family houses	60	Guest rooms, private corridors	40
		Public corridors	80
Bowling alleys, alleys only	40	Housing, one- and two-family:	
Broadcasting studios	100	First floor	40
Catwalks	30	Storage attics	80
Corridors:		Uninhabitable attics	20
Areas of public assembly, first-floor lobbies	100	Upper floors, habitable attics	30
		Schools:	
Other floors same as occupancy served, exept as indicated elsewhere in this table		Classrooms	40
		Corridors	80
		Shops with light equipment	60
Fire escapes:		Stairs and exitways	100
Multifamily housing	40	Handrails, vertical and horizontal thrust, lb per lin ft	50
Others	100		
Garages:		Storage warehouse:	
Passenger cars	50	Heavy	250
Trucks and buses	*c*	Light	125
Hospitals:		Stores:	
Operating rooms, laboratories, service areas	60	Retail:	
		Basement and first floor	100
Patients' rooms, wards, personnel areas	40	Upper floors	75
		Wholesale	100
Kitchens other than domestic	150	Telephone equipment rooms	80
Laboratories, scientific	100	Theaters:	
Libraries:		Aisles, corridors, lobbies	100
Corridors above first floor	80	Dressing rooms	40
Reading rooms	60	Projection rooms	100
Stack rooms, books and shelving at 65 pcf, but at least	150	Stage floors	150
		Toilet areas	40
Manufacturing and repair areas:			
Heavy	250		
Light	125		

[a] For live loads of 100 psf or less on a member carrying a loaded area of 150 sq ft or more, other than roofs, garages, or places of assembly, design live loads may be reduced at the rate of 0.08% per sq ft of supported area, but not more than $23(1 + D/L)\% < 60\%$, where D = dead load, psf, and L = live load, psf, on member. No reductions may be made for live loads exceeding 100 psf, except that design live loads on columns, piers, and walls supporting such loaded areas exceeding 150 sq ft, including storage areas, parking spaces, and places of assembly, may be reduced a maximum of 20%.

[b] Including churches, schools, theaters, courthouses, and lecture halls.

[c] Use American Association of State Highway and Transportation Officials highway lane loadings.

Table 3-1. Minimum Design Live, Wind, and Snow Loads (*Continued*)

B. Concentrated Live Loads[d]

Location	Load, lb
Elevator machine room grating (on 4-sq in. area)	300
Finish, light floor-plate construction (on 1-sq in. area)	200
Garages:	
Passenger cars:	
Manual parking (on 20-sq in. area)	2,000
Mechanical parking (no slab), per wheel	1,500
Trucks, buses (on 20-sq in. area), per wheel	16,000
Office floors (on area 2.5 ft square)	2,000
Roof-truss panel point over garage, manufacturing, or storage floors	2,000
Scuttles, skylight ribs, and accessible ceilings (on area 2.5 ft square)	200
Sidewalks (on area 2.5 ft square)	8,000
Stair treads (on 4-sq in. area at center of tread)	300

[d] Use instead of live load, except for roof trusses, if concentrated loads produce greater stresses or deflections. Add impact factor for machinery and moving loads: 100% for elevators, 20% for light machines, 50% for reciprocating machines, 33% for floor or balcony hangers. For craneways, add a vertical force equal to 25% of maximum wheel load; a lateral force equal to 10% of the weight of trolley and lifted load, at the top of each rail; and a longitudinal force equal to 10% of maximum wheel loads, acting at top of rail.

C. Wind Pressures for Design of Framing, Vertical Walls, and Windows of Ordinary Rectangular Buildings, Psf[e]

Height zone, ft above curb	Exposures								
	A[f]	B[g]	C[h]	A[f]	B[g]	C[h]	A[f]	B[g]	C[h]
	Coastal areas, N.W. and S.E. United States[i]			Northern and central United States[j]			Other parts of United States[k]		
0–50	20	40	65	15	25	40	15	20	35
51–100	30	50	75	20	35	50	15	25	40
101–300	40	65	85	25	45	60	20	35	45
301–600	65	85	105	40	55	70	35	45	55
Over 600	85	100	120	60	70	80	45	55	65

[e] For winds with 50-year recurrence interval. For computation of more exact wind pressures, see ANSI Standard A58.1-1972.
[f] Centers of large cities and very rough, hilly terrain.
[g] Suburban areas, towns, city outskirts, wooded areas, and rolling terrain.
[h] Flat, open country; open, flat coastal belts, and grassland.
[i] 110-mph basic wind speed.
[j] 90-mph basic wind speed.
[k] 80-mph basic wind speed.

3-5. Hooke's Law. For many materials, unit strain is proportional to unit stress, until a certain stress, the **proportional limit,** is exceeded. Known as Hooke's law, this relationship may be written as

$$f = E\epsilon \qquad \text{or} \qquad \epsilon = \frac{f}{E} \tag{3-2}$$

where f = unit stress
ϵ = unit strain
E = modulus of elasticity

Table 3-1. Minimum Design Live, Wind, and Snow Loads *(Continued)*

D. Snow Loads on Sloping Roofs, Psf[l]

Basic load, psf[m]	Shed roof				Gable roof[n]				
	Roof angle with horizontal, deg								
	0–30	40	50	60	0–20	30	40	50	60
5	5	5	0	0	5	5	5	5	0
10	10	5	5	0	10	10	10	5	5
15	10	10	5	5	10	15	10	10	5
20	15	10	10	5	15	20	15	15	5
30	25	20	10	5	25	30	25	20	10
40	30	25	15	10	30	40	30	25	10
50	40	30	20	10	40	50	40	30	15
60	50	35	25	10	50	60	45	40	15

[l] For snow loads of uniform depth with 50-year mean recurrence interval and snow load coefficient of 0.8 adjusted for slopes. For mountainous regions, snow load should be based on analysis of local climate and topography. For computation of more accurate snow loads, see ANSI Standard A58.1-1972.

[m] 0–5 psf—Southern and southwestern states, except for mountainous regions.
10 psf—Central and northwestern states.
20–30 psf—Middle Atlantic states and southern Great Lakes area.
40–50 psf—Northern Great Lakes area and New England states.
60–80 psf—Northern New England.

[n] Investigate roof fully loaded and alternatively with only one slope loaded.

E. Minimum Design Loads for Materials

Material	Load, lb per cu ft	Material	Load, lb per cu ft
Aluminum, cast	165	Gravel, dry	104
Bituminous products:		Gypsum, loose	70
Asphalt	81	Ice	57.2
Petroleum, gasoline	42	Iron, cast	450
Pitch	69	Lead	710
Tar	75	Lime, hydrated, loose	32
Brass, cast	534	Lime, hydrated, compacted	45
Bronze, 8 to 14% tin	509	Magnesium alloys	112
Cement, portland, loose	90	Mortar, hardened:	
Cement, portland, set	183	Cement	130
Cinders, dry, in bulk	45	Lime	110
Coal, anthracite, piled	52	Riprap (not submerged):	
Coal, bituminous or lignite, piled	47	Limestone	83
Coal, peat, dry, piled	23	Sandstone	90
Charcoal	12	Sand, clean and dry	90
Copper	556	Sand, river, dry	106
Cork, compressed	14.4	Silver	656
Earth (not submerged):		Steel	490
Clay, dry	63	Stone, quarried, piled:	
Clay, damp	110	Basalt, granite, gneiss	96
Clay and gravel, dry	100	Limestone, marble, quartz	95
Silt, moist, loose	78	Sandstone	82
Silt, moist, packed	96	Shale, slate	92
Sand and gravel, dry, loose	100	Tin, cast	459
Sand and gravel, dry, packed	110	Water, fresh	62.4
Sand and gravel, wet	120	Water, sea	64
Gold, solid	1,205	Zinc	450

Table 3-2. Minimum Design Dead Loads

Walls	*Psf*
Clay brick	
High-absorption, per 4-in. wythe	34
Medium absorption, per 4-in. wythe	39
Low-absorption, per 4-in. wythe	46
Sand-lime brick, per 4-in. wythe	38
Concrete brick	
4-in., with heavy aggregate	46
4-in., with light aggregate	33
Concrete block, hollow	
8-in., with heavy aggregate	55
8-in., with light aggregate	35
12-in., with heavy aggregate	85
12-in., with light aggregate	55
Clay tile, loadbearing	
4-in.	24
8-in.	42
12-in.	58
Clay tile, nonloadbearing	
2-in.	11
4-in.	18
8-in.	34
Furring tile	
1½-in.	8
2-in.	10
Glass block, 4-in.	18
Gypsum block, hollow	
2-in.	9.5
4-in.	12.5
6-in.	18.5
Partitions	*Psf*
Plaster on masonry	
Gypsum, with sand, per in. of thickness	8.5
Gypsum, with lightweight aggregate, per in.	4
Cement, with sand, per in. of thickness	10
Cement, with lightweight aggregate, per in.	5
Plaster, 2-in. solid	20
Metal studs, plastered two sides	18
Wood studs, 2 × 4-in.	
Unplastered	3
Plastered one side	11
Plastered two sides	19
Concrete Slabs	*Psf*
Stone aggregate, reinforced, per in. of thickness	12.5
Cinder, reinforced, per in. of thickness	9.25
Lightweight aggregate, reinforced, per in. of thickness	9
Insulation	*Psf*
Cork, per in. of thickness	1.0
Foamed glass, per in. of thickness	0.8
Glass-fiber bats, per in. of thickness	0.5
Polystyrene, per in. of thickness	0.08
Urethane, 2-in.	1.2
Vermiculite, loose fill, per in. of thickness	0.05
Floor Finishes	*Psf*
Asphalt block, 2-in.	24
Cement, 1-in.	12
Ceramic or quarry tile, 1-in.	12
Hardwood flooring, ⅞-in.	4
Plywood subflooring, ½-in.	1.5
Resilient flooring, such as asphalt tile and linoleum	2
Slate, 1-in.	15
Softwood subflooring, per in. of thickness	3
Terrazzo, 1-in.	12
Wood block, 3-in.	4

Table 3-2. Minimum Design Dead Loads (*Continued*)

Floor Fill	*Psf*
Cinders, no cement, per in. of thickness	5
Cinders, with cement, per in. of thickness	9
Sand, per in. of thickness	8
Waterproofing	*Psf*
Five-ply membrane	5
Glass	*Psf*
Single-strength	1.2
Double-strength	1.6
Plate, 1/8-in.	1.6

Wood joists, double wood floor, joist size	Psf	
	12-in. spacing	16-in. spacing
2 × 6	6	5
2 × 8	6	6
2 × 10	7	6
2 × 12	8	7
3 × 6	7	6
3 × 8	8	7
3 × 10	9	8
3 × 12	11	9
3 × 14	12	10

Ceilings	*Psf*
Plaster (on tile or concrete)	5
Suspended metal lath and gypsum plaster	10
Suspended metal lath and cement plaster	15
Roof and Wall Coverings	*Psf*
Asbestos-cement corrugated or shingles	4
Asphalt shingles	2
Composition:	
3-ply ready roofing	1
4-ply felt and gravel	5.5
5-ply felt and gravel	6
Copper or tin	1
Corrugated iron	2
Sheathing (gypsum), ½-in.	2
Sheathing (wood), per in. thickness	3
Slate 3/16-in.	7
Wood shingles	3
Masonry	*Pcf*
Cast-stone masonry	144
Concrete, stone aggregate, reinforced	150
Ashlar:	
Granite	165
Limestone, crystalline	165
Limestone, oölitic	135
Marble	173
Sandstone	144

Hence, when the unit stress and modulus of elasticity of a material are known, the unit strain can be computed. Conversely, when the unit strain has been found, the unit stress can be calculated.

3-6. Constant Unit Stress. The simplest cases of stress and strain are those in which the unit stress and strain are constant. Stresses due to an axial tension or compression load or a centrally applied shearing force are examples; also an evenly applied bearing load. These loading conditions are illustrated in Figs. 3-3 to 3-6.

For the axial tension and compression loadings, we take a section normal to

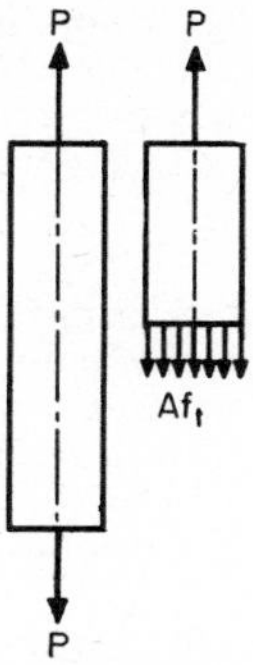

Fig. 3-3. Tension member.

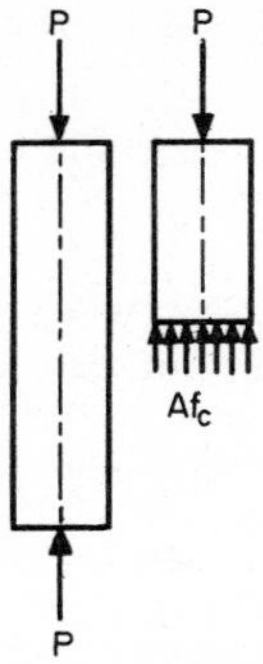

Fig. 3-4. Compression member.

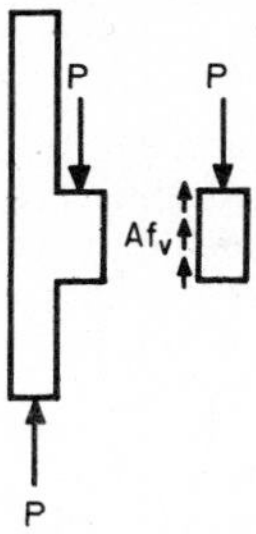

Fig. 3-5. Bracket in shear.

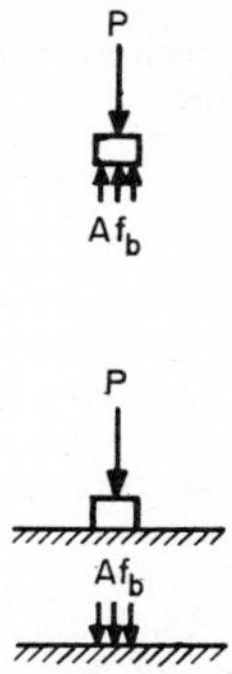

Fig. 3-6. Bearing load.

the centroidal axis (and to the applied forces). For the shearing load, the section is taken along a plane of sliding. And for the bearing load, it is chosen through the plane of contact between the two members.

Since for these loading conditions, the unit stress is constant across the section, the equation of equilibrium may be written

$$P = Af \tag{3-3}$$

where P = load
f = a tensile, compressive, shearing, or bearing unit stress
A = cross-sectional area for tensile or compressive forces, or area on which sliding may occur for shearing forces, or contact area for bearing loads

For torsional stresses, see Art. 3-78.

The unit strain for the axial tensile and compressive loads is given by the equation

$$\epsilon = \frac{e}{L} \tag{3-4}$$

where ϵ = unit strain
e = total lengthening or shortening of the member
L = original length of the member

Applying Hooke's law and Eq. (3-4) to Eq. (3-3) yields a convenient formula for the deformation:

$$e = \frac{PL}{AE} \tag{3-5}$$

where P = load on the member
A = its cross-sectional area
E = modulus of elasticity of the material

[Since long compression members tend to buckle, Eqs. (3-3) to (3-5) are applicable only to short members.]

While tension and compression strains represent a simple stretching or shortening of a member, shearing strain represents a distortion due to a small rotation. The load on the small rectangular portion of the member in Fig. 3-5 tends to distort it into a parallelogram. The unit shearing strain is the change in the right angle, measured in radians.

Modulus of rigidity, or shearing modulus of elasticity, is defined by

$$G = \frac{v}{\gamma} \tag{3-6}$$

where G = modulus of rigidity
v = unit shearing stress
γ = unit shearing strain

It is related to the modulus of elasticity in tension and compression E by the equation

$$G = \frac{E}{2(1 + \mu)} \tag{3-7}$$

where μ is a constant known as Poisson's ratio.

3-7. Poisson's Ratio. Within the elastic limit, when a material is subjected to axial loads, it deforms not only longitudinally but also laterally. Under tension, the cross section of a member decreases, and under compression, it increases. The ratio of the unit lateral strain to the unit longitudinal strain is called Poisson's ratio.

For many materials, this ratio can be taken equal to 0.25. For structural steel, it is usually assumed to be 0.3.

Assume, for example, that a steel hanger with an area of 2 sq in. carries a 40-kip (40,000-lb) load. The unit stress is 40,000/2, or 20,000 psi. The unit tensile strain, taking the modulus of elasticity of the steel as 30,000,000 psi, is 20,000/30,000,000, or 0.00067 in. per in. With Poisson's ratio as 0.3, the unit lateral strain is -0.3×0.00067, or a shortening of 0.00020 in. per in.

3-8. Thermal Stresses. When the temperature of a body changes, its dimensions also change. Forces are required to prevent such dimensional changes, and stresses are set up in the body by these forces.

If α is the coefficient of expansion of the material and T the change in temperature, the unit strain in a bar restrained by external forces from expanding or contracting is

$$\epsilon = \alpha T \tag{3-8}$$

According to Hooke's law, the stress f in the bar is

$$f = E\alpha T \tag{3-9}$$

where E = modulus of elasticity

3-9. Strain Energy. When a bar is stressed, energy is stored in it. If a bar supporting a load P undergoes a deformation e the energy stored in it is

$$U = \tfrac{1}{2}Pe \tag{3-10}$$

This equation assumes the load was applied gradually and the bar is not stressed beyond the proportional limit. It represents the area under the load-deformation curve up to the load P. Applying Eqs. (3-2) and (3-3) to Eq. (3-10) gives another useful equation for energy:

$$U = \frac{f^2}{2E} AL \tag{3-11}$$

where f = unit stress
E = modulus of elasticity of the material
A = cross-sectional area
L = length of the bar

Since AL is the volume of the bar, the term $f^2/2E$ indicates the energy stored per unit of volume. It represents the area under the stress-strain curve up to the stress f. Its value when the bar is stressed to the proportional limit is called the **modulus of resilience.** This modulus is a measure of the capacity of the material to absorb energy without danger of being permanently deformed and is of importance in designing members to resist energy loads.

Equation (3-10) is a general equation that holds true when the principle of superposition applies (the total deformation produced by a system of forces is equal to the sum of the elongations produced by each force). In the general sense, P in Eq. (3-10) represents any group of statically interdependent forces that can be completely defined by one symbol, and e is the corresponding deformation.

The strain-energy equation can be written as a function of either the load or the deformation.

For axial tension or compression:

$$U = \frac{P^2L}{2AE} \qquad U = \frac{AEe^2}{2L} \tag{3-10a}$$

where P = axial load
e = total elongation or shortening
L = length of the member
A = cross-sectional area
E = modulus of elasticity

For pure shear:

$$U = \frac{V^2L}{2AG} \qquad U = \frac{AGe^2}{2L} \tag{3-10b}$$

where V = shearing load
e = shearing deformation
L = length over which deformation takes place
A = shearing area
G = shearing modulus

For torsion:

$$U = \frac{T^2L}{2JG} \qquad U = \frac{JG\phi^2}{2L} \tag{3-10c}$$

where T = torque
ϕ = angle of twist
L = length of shaft
J = polar moment of inertia of the cross section
G = shearing modulus

For pure bending (constant moment):

$$U = \frac{M^2L}{2EI} \qquad U = \frac{EI\theta^2}{2L} \tag{3-10d}$$

where M = bending moment
θ = angle of rotation of one end of the beam with respect to the other
L = length of beam
I = moment of inertia of the cross section
E = modulus of elasticity (see also Art. 3-55)

For beams carrying transverse loads, the strain energy is the sum of the energy for bending and that for shear.

STRESSES AT A POINT

Tensile and compressive stresses are sometimes referred to also as normal stresses, because they act normal to the cross section. Under this concept, tensile stresses are considered as positive normal stresses and compressive stresses as negative.

3-10. Stress Notation. Suppose a member of a structure is acted upon by forces in all directions. For convenience, let us establish a reference set of perpendicular coordinate x, y, and z axes. Now let us take at some point in the member a small cube with sides parallel to the coordinate axes. The notations commonly used for the components of stress acting on the sides of this element and the directions assumed as positive are shown in Fig. 3-7.

For example, for the sides of the element perpendicular to the z axis, the normal component of stress is denoted by f_z. The shearing stress v is resolved into two components and requires two subscript letters for a complete description. The first letter indicates the direction of the normal to the plane under consideration. The second letter indicates the direction of the component of the stress. Considering the sides perpendicular to the z axis, the shear component in the x direction is labeled v_{zx} and that in the y direction v_{zy}.

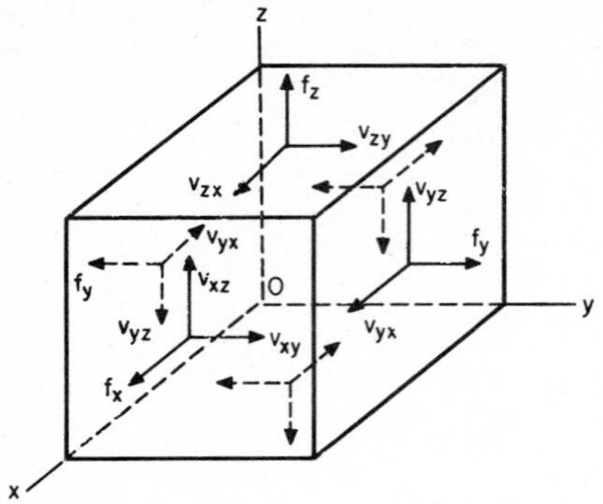

Fig. 3-7. Stresses in a rectangular coordinate system.

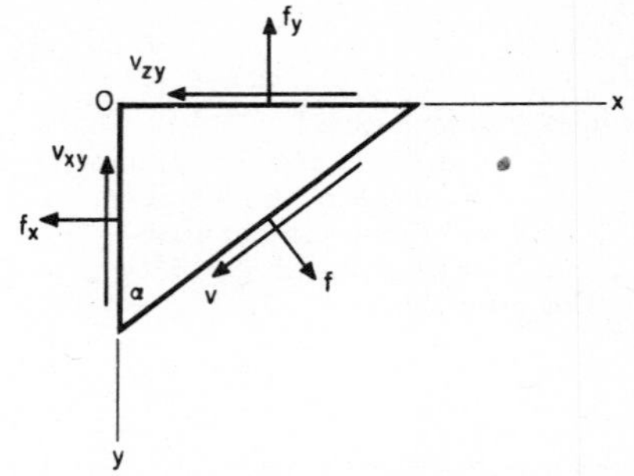

Fig. 3-8. Stresses at a point on a plane inclined to the axes.

3-11. Stress Components. If, for the small cube in Fig. 3-7, moments of the forces acting on it are taken about the x axis, considering the cube's dimensions as dx, dy, and dz, the equation of equilibrium requires that

$$v_{zy}\,dx\,dy\,dz = v_{yz}\,dx\,dy\,dz$$

(Forces are taken equal to the product of the area of the face and the stress at the center.) Two similar equations can be written for moments taken about the y axis and the z axis. These equations show that

$$v_{xy} = v_{yx} \qquad v_{zx} = v_{xz} \qquad \text{and} \qquad v_{zy} = v_{yz} \tag{3-12}$$

In words, the components of shearing stress on two perpendicular faces and acting normal to the intersection of the faces are equal.

Consequently, to describe the stresses acting on the coordinate planes through a point, only six quantities need be known. These stress components are f_x, f_y, f_z, $v_{xy} = v_{yx}$, $v_{yz} = v_{zy}$, and $v_{zx} = v_{xz}$.

If the cube in Fig. 3-7 is acted on only by normal stresses f_x, f_y, and f_z, from Hooke's law and the application of Poisson's ratio, the unit strains in the x, y, and z directions, in accordance with Arts. 3-5 and 3-6, are, respectively,

$$\begin{aligned} \epsilon_x &= \frac{1}{E}[f_x - \mu(f_y + f_z)] \\ \epsilon_y &= \frac{1}{E}[f_y - \mu(f_x + f_z)] \\ \epsilon_z &= \frac{1}{E}[f_z - \mu(f_x + f_y)] \end{aligned} \tag{3-13}$$

where μ = Poisson's ratio

If only shearing stresses act on the cube in Fig. 3-7, the distortion of the angle between edges parallel to any two coordinate axes depends only on shearing-stress components parallel to those axes. Thus, the unit shearing strains are (see Art. 3-6)

$$\gamma_{xy} = \frac{1}{G}v_{xy} \qquad \gamma_{yz} = \frac{1}{G}v_{yz} \qquad \text{and} \qquad \gamma_{zx} = \frac{1}{G}v_{zx} \tag{3-14}$$

3-12. Two-dimensional Stress. When the six components of stress necessary to describe the stresses at a point are known (Art. 3-11), the stress on any inclined plane through the same point can be determined. For the case of two-dimensional stress, only three stress components need be known.

Assume, for example, that at a point O in a stressed plate, the components f_x, f_y, and v_{xy} are known (Fig. 3-8). To find the stresses for any plane through the z axis, take a plane parallel to it close to O. This plane and the coordinate planes form a triangular prism. Then, if α is the angle the normal to the plane makes with the x axis, the normal and shearing stresses on the inclined plane, obtained by application of the equations of equilibrium, are

$$f = f_x \cos^2 \alpha + f_y \sin^2 \alpha + 2v_{xy} \sin \alpha \cos \alpha \tag{3-15}$$
$$v = v_{xy}(\cos^2 \alpha - \sin^2 \alpha) + (f_y - f_x) \sin \alpha \cos \alpha \tag{3-16}$$

3-13. Principal Stresses. A plane through a point on which stresses act may be assigned a direction for which the normal stress is a maximum or a minimum. There are two such positions, perpendicular to each other. And on those planes, there are no shearing stresses.

The directions in which the normal stresses become maximum or minimum are called principal directions and the corresponding normal stresses principal stresses.

To find the principal directions, set the value of v given by Eq. (3-16) equal to zero. The resulting equation is

$$\tan 2\alpha = \frac{2v_{xy}}{f_x - f_y} \tag{3-17}$$

If the x and y axes are taken in the principal directions, v_{xy} is zero. Consequently, Eqs. (3-15) and (3-16) may be simplified to

$$f = f_x \cos^2 \alpha + f_y \sin^2 \alpha \tag{3-18}$$
$$v = \tfrac{1}{2} \sin 2\alpha (f_y - f_x) \tag{3-19}$$

where f and v are, respectively, the normal and shearing stress on a plane at an angle α with the principal planes and f_x and f_y are the principal stresses.

Pure Shear. If on any two perpendicular planes only shearing stresses act, the state of stress at the point is called pure shear or simple shear. Under such conditions, the principal directions bisect the angles between the planes on which these shearing stresses occur. The principal stresses are equal in magnitude to the unit shearing stresses.

3-14. Maximum Shearing Stress. The maximum unit shearing stress occurs on each of two planes that bisect the angles between the planes on which the principal stresses act. The maximum shear is equal to one-half the algebraic difference of the principal stresses:

$$\max v = \frac{f_1 - f_2}{2} \tag{3-20}$$

where f_1 is the maximum principal stress and f_2 the minimum.

3-15. Mohr's Circle. The relationship between stresses at a point may be represented conveniently on Mohr's circle (Fig. 3-9). In this diagram, normal stress f and shear stress v are taken as coordinates. Then, for each plane through the point, there will correspond a point on the circle, whose coordinates are the values of f and v for the plane.

To construct the circle given the principal stresses, mark off the principal stresses f_1 and f_2 on the f axis (points A and B in Fig. 3-9). Tensile stresses are measured to the right of the v axis and compressive stresses to the left. Construct a circle with its center on the f axis and passing through the two points representing the principal stresses. This is the Mohr's circle for the given stresses at the point under consideration.

Suppose now, we wish to find the stresses on a plane at an angle α to the plane of f_1. If a radius is drawn making an angle 2α with the f axis, the coordinates of its intersection with the circle represent the normal and shearing stresses acting on the plane.

Mohr's circle can also be plotted when the principal stresses are not known but the stresses f_x, f_y, and v_{xy}, on any two perpendicular planes, are. The procedure is to plot the two points representing these known stresses with respect to the

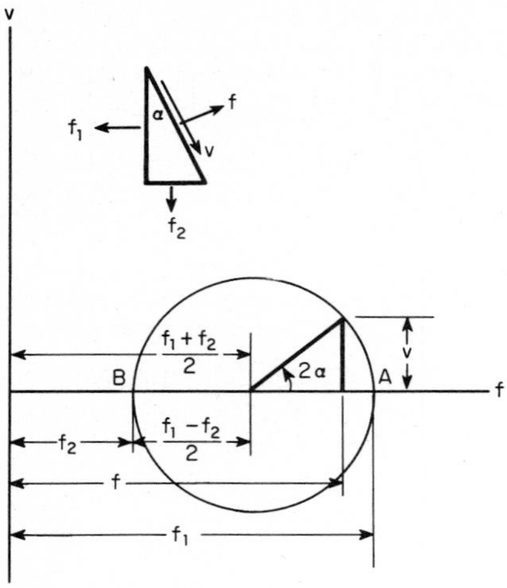

Fig. 3-9. Mohr's circle for stresses at a point—constructed from known principal stresses.

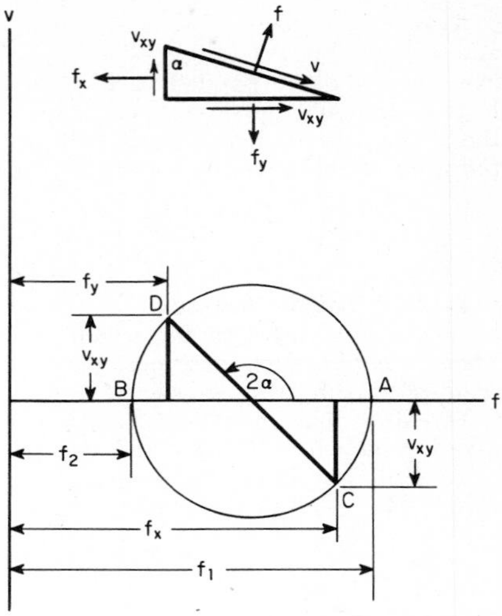

Fig. 3-10. Stress circle constructed from two known normal positive stresses, f_x and f_y, and a shear v_{xy}.

f and *v* axes (points *C* and *D* in Fig. 3-10). The line joining these points is a diameter of Mohr's circle. Constructing the circle on this diameter, we find the principal stresses at the intersection with the *f* axis (points *A* and *B* in Fig. 3-10).

For more details on the relationship of stresses and strains at a point, see Timoshenko and Goodier, "Theory of Elasticity," McGraw-Hill Book Company, New York.

STRAIGHT BEAMS

Beams are the horizontal members used to support vertically applied loads across an opening. In a more general sense, they are structural members that external loads tend to bend, or curve.

3-16. Types of Beams. There are many ways in which beams may be supported. Some of the more common methods are shown in Figs. 3-11 to 3-16.

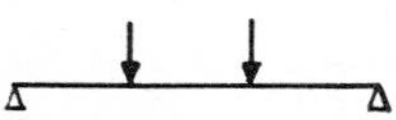

Fig. 3-11. Simple beam.

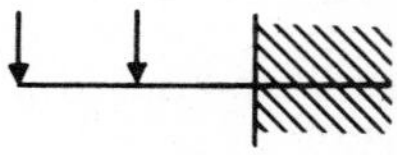

Fig. 3-12. Cantilever beam.

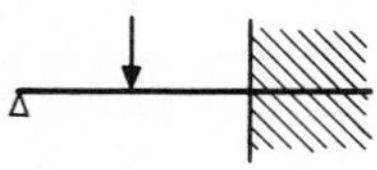

Fig. 3-13. Beam with one end fixed.

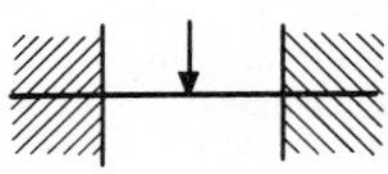

Fig. 3-14. Fixed-end beam.

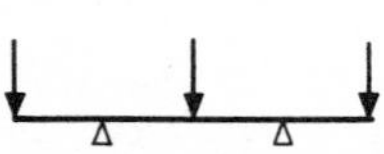

Fig. 3-15. Beam with overhangs.

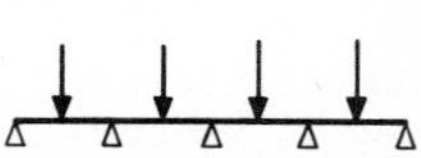

Fig. 3-16. Continuous beam.

The beam in Fig. 3-11 is called a simply supported, or **simple beam.** It has supports near its ends, which restrain it only against vertical movement. The ends of the beam are free to rotate. When the loads have a horizontal component, or when change in length of the beam due to temperature may be important, the supports may also have to prevent horizontal motion. In that case, horizontal restraint at one support is generally sufficient.

The distance between the supports is called the **span.** The load carried by each support is called a **reaction.**

The beam in Fig. 3-12 is a **cantilever.** It has only one support, which restrains it from rotating or moving horizontally or vertically at that end. Such a support is called a **fixed end.**

If a simple support is placed under the free end of the cantilever, the beam in Fig. 3-13 results. It has one end fixed, one end simply supported.

The beam in Fig. 3-14 has both ends fixed. No rotation or vertical movement can occur at either end. In actual practice, a fully fixed end can seldom be obtained. Some rotation of the beam ends generally is permitted. Most support conditions are intermediate between those for a simple beam and those for a fixed-end beam.

In Fig. 3-15 is shown a beam that overhangs both its simple supports. The overhangs have a free end, like a cantilever, but the supports permit rotation.

When a beam extends over several supports, it is called a **continuous beam** (Fig. 3-16).

Reactions for the beams in Figs. 3-11, 3-12, and 3-15 may be found from the equations of equilibrium. They are classified as **statically determinate beams** for that reason.

The equations of equilibrium, however, are not sufficient to determine the reactions of the beams in Figs. 3-13, 3-14, and 3-16. For those beams, there are more unknowns than equations. Additional equations must be obtained based on the deformations permitted; on the knowledge, for example, that a fixed end permits no rotation. Such beams are classified as **statically indeterminate.** Methods for finding the stresses in that type of beam are given in Arts. 3-55, 3-57, 3-58 to 3-69, and 3-105 to 3-108.

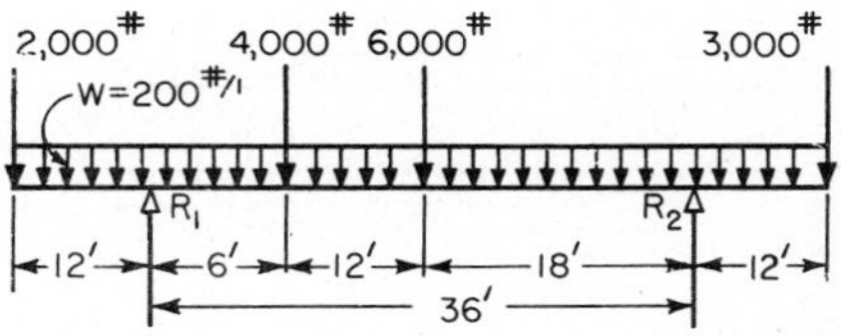

Fig. 3-17. Beam with overhangs loaded both uniformly and with concentrated loads.

3-17. Reactions. As an example of the application of the equations of equilibrium (Art. 3-1) to the determination of the reactions of a statically determinate beam, we shall compute the reactions of the 60-ft-long beam with overhangs in Fig 3-17. This beam carries a uniform load of 200 lb per lin ft over its entire length and several concentrated loads. The supports are 36 ft apart.

To find reaction R_1, we take moments about R_2 and equate the sum of the moments to zero (clockwise rotation is considered positive, counterclockwise, negative):

$$-2{,}000 \times 48 + 36R_1 - 4{,}000 \times 30 - 6{,}000 \times 18 + 3{,}000 \times 12 - 200 \times 60 \times 18 = 0$$

$$R_1 = 14{,}000 \text{ lb}$$

In this calculation, the moment of the uniform load was found by taking the moment of its resultant, which acts at the center of the beam.

To find R_2, we can either take moments about R_1 or use the equation $\Sigma V = 0$. It is generally preferable to apply the moment equation and use the other equation as a check.

$$3{,}000 \times 48 - 36R_2 + 6{,}000 \times 18 + 4{,}000 \times 6 - 2{,}000 \times 12 + 200 \times 60 \times 18 = 0$$

$$R_2 = 13{,}000 \text{ lb}$$

As a check, we note that the sum of the reactions must equal the total applied load:

$$14{,}000 + 13{,}000 = 2{,}000 + 4{,}000 + 6{,}000 + 3{,}000 + 12{,}000$$

$$27{,}000 = 27{,}000$$

3-18. Internal Forces. Since a beam is in equilibrium under the forces applied to it, it is evident that at every section internal forces are acting to prevent motion. For example, suppose we cut the beam in Fig. 3-17 vertically just to the right of its center. If we total the external forces, including the reaction, to the left of this cut (see Fig. 3-18*a*), we find there is an unbalanced downward load of 4,000 lb. Evidently, at the cut section, an upward-acting internal force of 4,000 lb must be present to maintain equilibrium. Again, if we take moments of the external forces about the section, we find an unbalanced moment of 54,000 ft-lb. So there must be an internal moment of 54,000 ft-lb acting to maintain equilibrium.

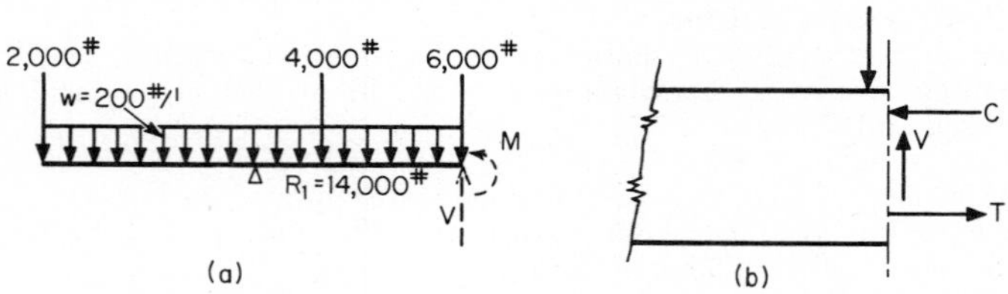

Fig. 3-18. Sections of beam kept in equilibrium by internal stresses.

This internal, or resisting, moment is produced by a couple consisting of a force C acting on the top part of the beam and an equal but opposite force T acting on the bottom part (Fig. 3-18*b*). The top force is the resultant of compressive stresses acting over the upper portion of the beam, and the bottom force is the resultant of tensile stresses acting over the bottom part. The surface at which the stresses change from compression to tension—where the stress is zero—is called the **neutral surface.**

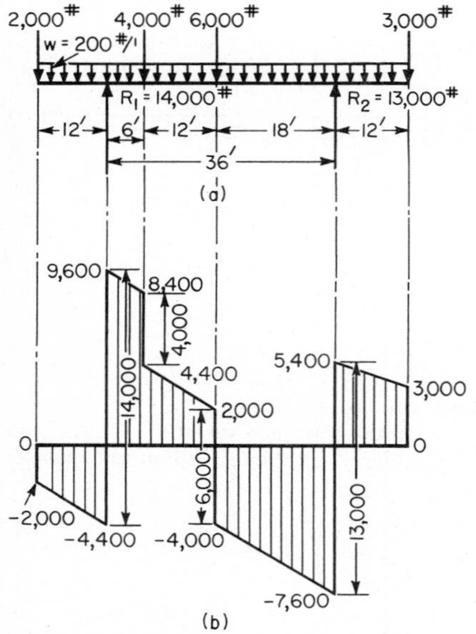

Fig. 3-19. Shear diagram for beam of Fig. 3-17.

3-19. Shear Diagrams. The unbalanced external vertical force at a section is called the shear. It is equal to the algebraic sum of the forces that lie on either side of the section. Upward-acting forces on the left of the section are considered positive, downward forces negative; signs are reversed for forces on the right.

A diagram in which the shear at every point along the length of a beam is plotted as an ordinate is called a shear diagram. The shear diagram for the beam in Fig. 3-17 is shown in Fig. 3-19*b*.

The diagram was plotted starting from the left end. The 2,000-lb load was plotted downward to a convenient scale. Then, the shear at the next concentrated load—the left support—was determined. This equals $-2{,}000 - 200 \times 12$, or $-4{,}400$ lb. In passing from just to the left of the support to a point just to the right, however, the shear changes by the magnitude of the reaction. Hence, on the right-hand side of the left support the shear is $-4{,}400 + 14{,}000$, or 9,600 lb. At the next concentrated load, the shear is $9{,}600 - 200 \times 6$, or 8,400 lb. In passing the 4,000-lb load, however, the shear changes to $8{,}400 - 4{,}000$, or 4,400 lb. Proceeding in this manner to the right end of the beam, we terminate with a shear of 3,000 lb, equal to the load on the free end there.

It should be noted that the shear diagram for a uniform load is a straight line sloping downward to the right (see Fig. 3-21). Therefore, the shear diagram was completed by connecting the plotted points with straight lines.

Shear diagrams for commonly encountered loading conditions are given in Figs. 3-30 to 3-41.

3-20. Bending-Moment Diagrams. The unbalanced moment of the external forces about a vertical section through a beam is called the bending moment. It is equal to the algebraic sum of the moments about the section of the external forces that lie on one side of the section. Clockwise moments are considered positive, counterclockwise moments negative, when the forces considered lie on the left of the section. Thus, when the bending moment is positive, the bottom of the beam is in tension.

A diagram in which the bending moment at every point along the length of a beam is plotted as an ordinate is called a bending-moment diagram.

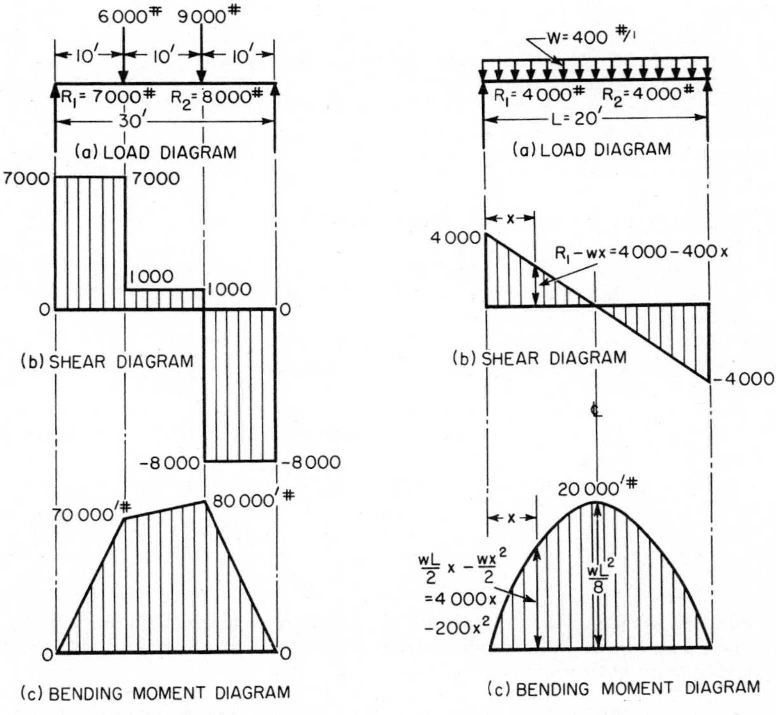

Fig. 3-20. Shear and moment diagram for beam with concentrated loads.
Fig. 3-21. Shear and moment diagram for uniformly loaded beam.

Figure 3-20*c* is the bending-moment diagram for the beam loaded with concentrated loads only in Fig. 3-20*a*. The bending moment at the supports for this simply supported beam obviously is zero. Between the supports and the first load, the bending moment is proportional to the distance from the support, since it is equal to the reaction times the distance from the support. Hence the bending-moment diagram for this portion of the beam is a sloping straight line.

The bending moment under the 6,000-lb load in Fig. 3-20*a* considering only the force to the left is 7,000 × 10, or 70,000 ft-lb. The bending-moment diagram, then, between the left support and the first concentrated load is a straight line rising from zero at the left end of the beam to 70,000, plotted to a convenient scale, under the 6,000-lb load.

The bending moment under the 9,000-lb load, considering the forces on the left of it, is $7{,}000 \times 20 - 6{,}000 \times 10$, or 80,000 ft-lb. (It could have been more easily obtained by considering only the force on the right, reversing the sign convention: $8{,}000 \times 10 = 80{,}000$ ft-lb.) Since there are no loads between the two concentrated loads, the bending-moment diagram between the two sections is a straight line.

If the bending moment and shear are known at any section of a beam, the bending moment at any other section may be computed, providing there are no unknown forces between the two sections. The rule is:

The bending moment at any section of a beam is equal to the bending moment at any section to the left, plus the shear at that section times the distance between sections, minus the moments of intervening loads. If the section with known moment and shear is on the right, the sign convention must be reversed.

For example, the bending moment under the 9,000-lb load in Fig. 3-20*a* could also have been obtained from the moment under the 6,000-lb load and the shear to the right of the 6,000-lb load given in the shear diagram (Fig. 3-20*b*). Thus, $80{,}000 = 70{,}000 + 1{,}000 \times 10$. If there had been any other loads between the two concentrated loads, the moment of these loads about the section under the 9,000-lb load would have been subtracted.

Bending-moment diagrams for commonly encountered loading conditions are given in Figs. 3-30 to 3-41. These may be combined to obtain bending moments for other loads.

3-21. Moments in Uniformly Loaded Beams. When a beam carries a uniform load, the bending-moment diagram does not consist of straight lines. Consider, for example, the beam in Fig. 3-21*a*, which carries a uniform load over its entire length. As shown in Fig. 3-21*c*, the bending-moment diagram for this beam is a parabola.

The reactions at both ends of a simply supported uniformly loaded beam are both equal to $wL/2 = W/2$, where w is the uniform load in pounds per linear foot, $W = wL$ is the total load on the beam, and L is the span.

The shear at any distance x from the left support is $R_1 - wx = wL/2 - wx$ (see Fig. 3-21*b*). Equating this expression to zero, we find that there is no shear at the center of the beam.

The bending moment at any distance x from the left support is

$$R_1x - wx\left(\frac{x}{2}\right) = \frac{wLx}{2} - \frac{wx^2}{2} = \frac{w}{2}x(L - x)$$

Hence:

The bending moment at any section of a simply supported, uniformly loaded beam is equal to one-half the product of the load per linear foot and the distances to the section from both supports.

The maximum value of the bending moment occurs at the center of the beam. It is equal to $wL^2/8 = WL/8$.

3-22. Shear-Moment Relationship. The slope of the bending-moment curve for any point on a beam is equal to the shear at that point; i.e.,

$$V = \frac{dM}{dx} \tag{3-21}$$

Since maximum bending moment occurs when the slope changes sign, or passes through zero, maximum moment (positive or negative) occurs at the point of zero shear.

After integration, Eq. (3-21) may also be written

$$M_1 - M_2 = \int_{x_2}^{x_1} V\,dx \tag{3-22}$$

This equation indicates that the change in bending moment between any two sections of a beam is equal to the area of the shear diagram between ordinates at the two sections.

3-23. Moving Loads and Influence Lines. One of the most helpful devices for solving problems involving moving loads is an influence line. Whereas shear and moment diagrams evaluate the effect of loads at all sections of a structure, an influence line indicates the effect at a given section of a unit load placed at any point on the structure.

For example, to plot the influence line for bending moment at some point *A* on a beam, a unit load is applied at some point *B*. The bending moment at *A* due to the unit load at *B* is plotted as an ordinate to a convenient scale at *B*. The same procedure is followed at every point along the beam and a curve is drawn through the points thus obtained.

Actually, the unit load need not be placed at every point. The equation of the influence line can be determined by placing the load at an arbitrary point and computing the bending moment in general terms (see also Art. 3-57).

Suppose we wish to draw the influence line for reaction at *A* for a simple beam *AB* (Fig. 3-22*a*). We place a unit load at an arbitrary distance of xL from *B*. The reaction at *A* due to this load is $1\ xL/L = x$. Then, $R_A = x$ is the equation of the influence line. It represents a straight line sloping upward from zero at *B* to unity at *A* (Fig. 3-22*a*). In other words, as the unit load moves across the beam, the reaction at *A* increases from zero to unity in proportion to the distance of the load from *B*.

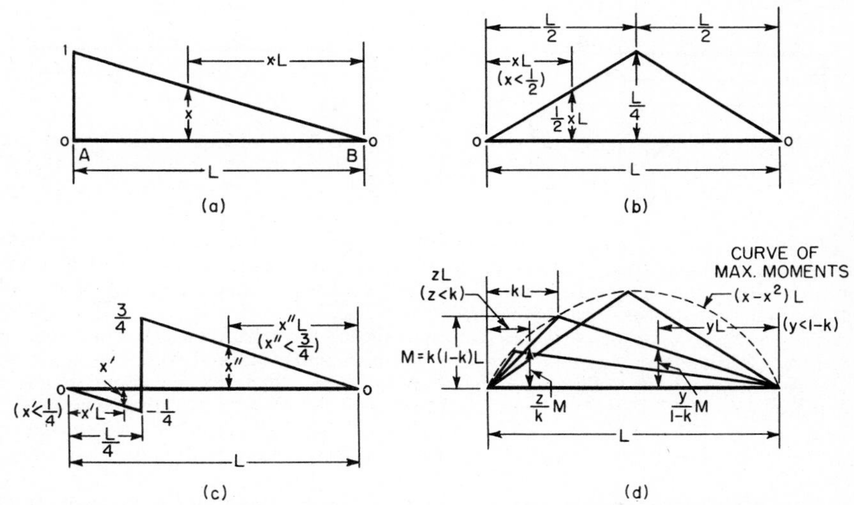

Fig. 3-22. Influence lines for (*a*) reaction at *A*; (*b*) midspan bending moment; (*c*) quarter-point shear; and (*d*) bending moments at several points in a beam.

Figure 3-22*b* shows the influence line for bending moment at the center of a beam. It resembles in appearance the bending-moment diagram for a load at the center of the beam, but its significance is entirely different. Each ordinate gives the moment at midspan for a load at the corresponding location. It indicates that, if a unit load is placed at a distance xL from one end, it produces a bending moment of $\frac{1}{2}\ xL$ at the center of the span.

Figure 3-22*c* shows the influence line for shear at the quarter point of a beam. When the load is to the right of the quarter point, the shear is positive and equal to the left reaction. When the load is to the left, the shear is negative and equal to the right reaction.

The diagram indicates that, to produce maximum shear at the quarter point, loads should be placed only to the right of the quarter point, with the largest load at the quarter point, if possible. For a uniform load, maximum shear results when the load extends from the right end of the beam to the quarter point.

Suppose, for example, that the beam is a crane girder with a span of 60 ft. The wheel loads are 20 and 10 kips, respectively, and are spaced 5 ft apart. For maximum shear at the quarter point, the wheels should be placed with the 20-kip wheel at that point and the 10-kip wheel to the right of it. The corresponding ordinates of the influence line (Fig. 3-22*c*) are $\frac{3}{4}$ and $\frac{40}{45} \times \frac{3}{4}$. Hence, the maximum shear is $20 \times \frac{3}{4} + 10 \times \frac{40}{45} \times \frac{3}{4} = 21.7$ kips.

Figure 3-22*d* shows influence lines for bending moment at several points on a beam. It is noteworthy that the apexes of the diagrams fall on a parabola, as shown by the dashed line. This indicates that the maximum moment produced at any given section by a single concentrated load moving across a beam occurs when the load is at that section. The magnitude of the maximum moment increases when the section is moved toward midspan, in accordance with the equation shown in Fig. 3-22*d* for the parabola.

3-24. Maximum Bending Moment. When there is more than one load on the span, the influence line is useful in developing a criterion for determining the position of the loads for which the bending moment is a maximum at a given section.

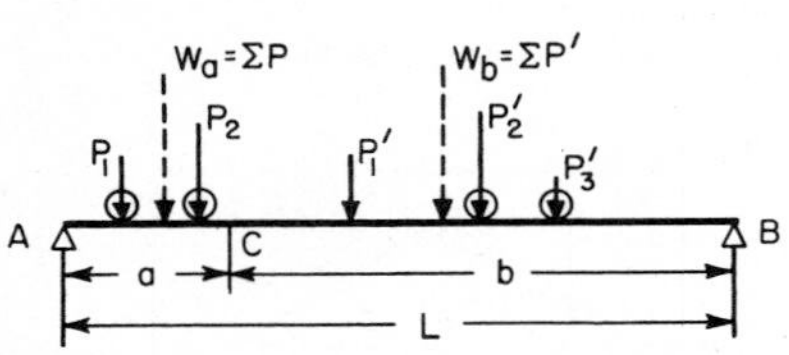

Fig. 3-23. Moving loads on simple beam *AB* placed for maximum moment at *C*.

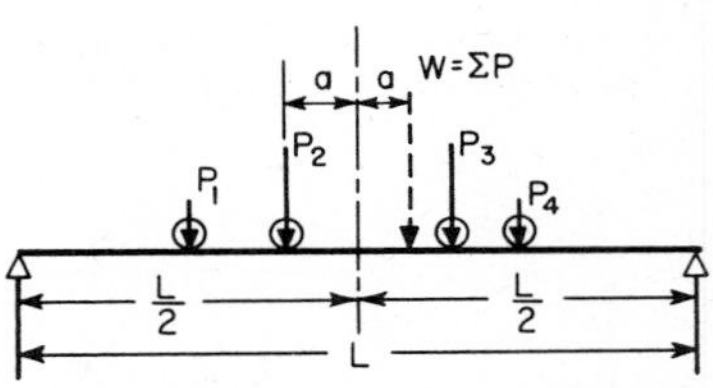

Fig. 3-24. Moving loads placed for largest maximum moment in a simple beam.

Maximum bending moment will occur at a section *C* of a simple beam as loads move across it when one of the loads is at *C*. The proper load to place at *C* is the one for which the expression $W_a/a - W_b/b$ (Fig. 3-23) changes sign as that load passes from one side of *C* to the other.

When several loads move across a simple beam, the maximum bending moment produced in the beam may be near but not necessarily at midspan. To find the maximum moment, first determine the position of the loads for maximum moment at midspan. Then shift the loads until the load P_2 (Fig. 3-24) that was at the center of the beam is as far from midspan as the resultant of all the loads on the span is on the other side of midspan. Maximum moment will occur under P_2.

When other loads move on or off the span during the shift of P_2 away from midspan, it may be necessary to investigate the moment under one of the other loads when it and the resultant are equidistant from midspan.

3-25. Bending Stresses in a Beam. To derive the commonly used flexure formula for computing the bending stresses in a beam, we have to make the following assumptions.

1. The unit stress on any fiber of a beam is proportional to the unit strain of the fiber.
2. The modulus of elasticity in tension is the same as that in compression.
3. The total and unit axial strain of any fiber are both proportional to the distance of that fiber from the neutral surface. (Cross sections that are plane before bending remain plane after bending. This requires that all fibers have the same length before bending; thus, that the beam be straight.)
4. The loads act in a plane containing the centroidal axis of the beam and are perpendicular to that axis. Furthermore, the neutral surface is perpendicular to the plane of the loads. Thus, the plane of the loads must contain an axis

of symmetry of each cross section of the beam. (The flexure formula does not apply to a beam loaded unsymmetrically.)

5. The beam is proportioned to preclude prior failure or serious deformation by torsion, local buckling, shear, or any cause other than bending.

Equating the bending moment to the resisting moment due to the internal stresses at any section of a beam yields

$$M = \frac{fI}{c} \tag{3-23}$$

M is the bending moment at the section, f is the normal unit stress on a fiber at a distance c from the neutral axis (Fig. 3-25), and I is the moment of inertia of the cross section with respect to the neutral axis. If f is given in pounds per square inch (psi), I in in.^{4}, and c in inches, then M will be in inch-pounds. Generally, c is taken as the distance to the outermost fiber. See also Arts. 3-26 and 3-27.

3-26. Moment of Inertia. The neutral axis in a symmetrical beam is coincidental with the centroidal axis; i.e., at any section the neutral axis is so located that

$$\int y\,dA = 0 \tag{3-24}$$

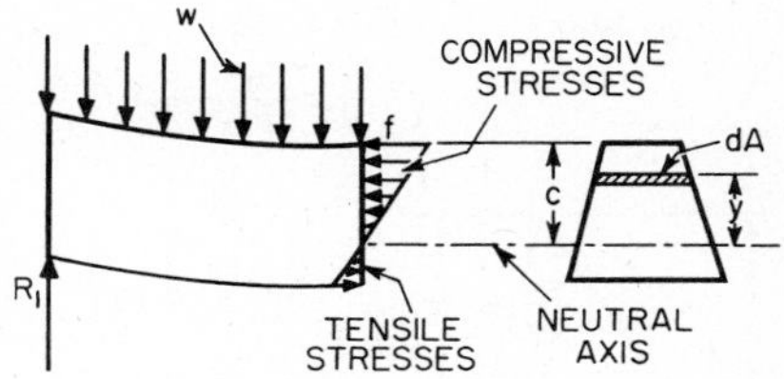

Fig. 3-25. Unit stresses due to bending on a beam section.

where dA is a differential area parallel to the axis (Fig. 3-25), y is its distance from the axis, and the summation is taken over the entire cross section.

Moment of inertia with respect to the neutral axis is given by

$$I = \int y^2\,dA \tag{3-25}$$

Values of I for several common types of cross section are given in Fig. 3-26. Values for structural-steel sections are presented in "Steel Construction Manual," American Institute of Steel Construction, New York. When the moments of inertia of other types of sections are needed, they can be computed directly by application of Eq. (3-25) or by breaking the section up into components for which the moment of inertia is known.

If I is the moment of inertia about the neutral axis, A the cross-sectional area, and d the distance between that axis and a parallel axis in the plane of the cross section, then the moment of inertia about the parallel axis is

$$I' = I + Ad^2 \tag{3-26}$$

With this equation, the known moment of inertia of a component of a section about the neutral axis of the component can be transferred to the neutral axis of the complete section. Then, summing up the transferred moments of inertia for all the components yields the moment of inertia of the complete section.

When the moments of inertia of an area with respect to any two rectangular axes are known, the moment of inertia with respect to any other axis passing through the point of intersection of the two axes may be obtained through the use of Mohr's circle as for stresses (Fig. 3-10). In this analog, I_x corresponds with f_x, I_y with f_y, and the **product of inertia** I_{xy} with v_{xy} (Art. 3-15).

$$I_{xy} = \int xy\,dA \tag{3-27}$$

The two perpendicular axes through a point about which the moments of inertia are a maximum and a minimum are called the principal axes. The products of inertia are zero for the principal axes.

3-27. Section Modulus. The ratio $S = I/c$ in Eq. (3-23) is called the section modulus. I is the moment of inertia of the cross section about the neutral axis and c the distance from the neutral axis to the outermost fiber. Values of S for common types of sections are given in Fig. 3-26.

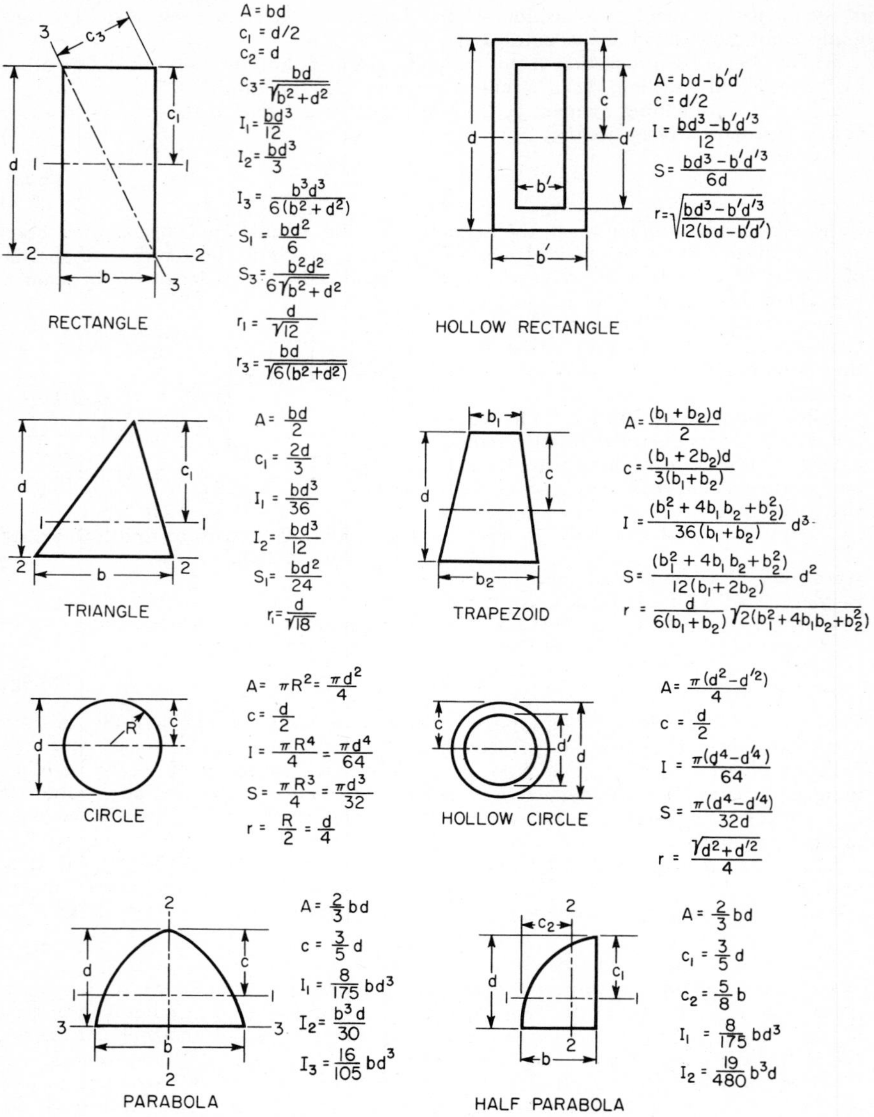

Fig. 3-26. Geometric properties of sections.

3-28. Shearing Stresses in a Beam. The vertical shear at any section of a beam is resisted by nonuniformly distributed, vertical unit stresses (Fig. 3-27). At every point in the section, there is also a horizontal unit stress, which is equal in magnitude to the vertical unit shearing stress there [see Eq. (3-12)].

At any distance y' from the neutral axis, both the horizontal and vertical shearing unit stresses are equal to

$$v = \frac{V}{It} A'\bar{y} \tag{3-28}$$

where V = vertical shear at the cross section
t = thickness of beam at distance y' from neutral axis
I = moment of inertia about neutral axis
A' = area between the outermost fiber and the fiber for which the shearing stress is being computed
$\bar{y}$ = distance of center of gravity of this area from the neutral axis (Fig. 3-27)

For a rectangular beam, the maximum shearing stress occurs at mid-depth. Its magnitude is

$$v = \frac{12V}{bd^3b}\frac{bd^2}{8} = \frac{3}{2}\frac{V}{bd}$$

That is, the maximum shear stress is 50% greater than the average shear stress on the section. Similarly, for a circular beam, the maximum is one-third greater than the average. For an I beam, however, the maximum shearing stress in the web is not appreciably greater than the average for the web section alone, assuming the flanges to take no shear.

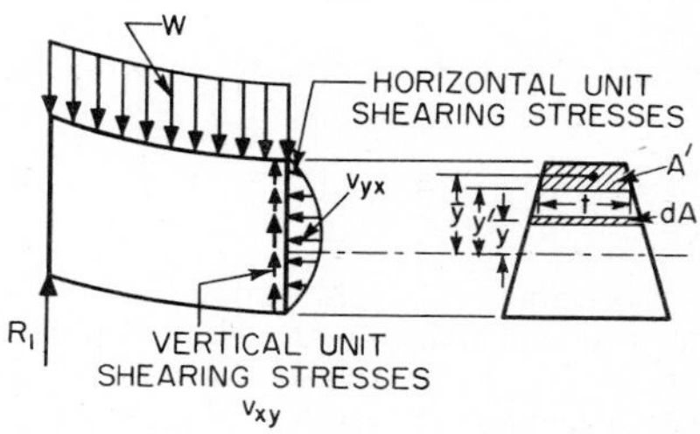

Fig. 3-27. Unit shearing stresses on a beam section.

3-29. Combined Shear and Bending Stress. For deep beams on short spans and beams made of low-tensile-strength materials, it is sometimes necessary to determine the maximum stress f' on an inclined plane due to a combination of shear and bending stress—v and f, respectively. This stress f', which may be either tension or compression, is greater than the normal stress. Its value may be obtained by application of Mohr's circle (Art. 3-15), as indicated in Fig. 3-10, but with $f_y = 0$, and is

$$f' = \frac{f}{2} + \sqrt{v^2 + \left(\frac{f}{2}\right)^2} \tag{3-29}$$

3-30. Stresses in the Plastic Range. When the bending stresses exceed the proportional limit and stress no longer is proportional to strain, the distribution of bending stresses over the cross section ceases to be linear. The outer fibers deform with little change in stress, while the fibers not stressed beyond the proportional limit continue to take more stress as the load on the beam increases. (J. O. Smith and O. M. Sidebottom, "Inelastic Behavior of Load-carrying Members," John Wiley & Sons, Inc., New York.)

Modulus of rupture is defined as the stress computed from the flexure formula [Eq. (3-23)] for the maximum bending moment a beam sustains at failure. It is not the actual maximum unit stress in the beam but is sometimes used to compare the strength of beams.

3-31. Beam Deflections. When a beam is loaded, it deflects. The new position of its longitudinal centroidal axis is called the **elastic curve.**

At any point of the elastic curve, the radius of curvature is given by

$$R = \frac{EI}{M} \tag{3-30}$$

where M = bending moment at the point
E = modulus of elasticity
I = moment of inertia of the cross section about the neutral axis

Since the slope dy/dx of the curve is small, its square may be neglected, so that, for all practical purposes, $1/R$ may be taken equal to d^2y/dx^2, where y is the deflection of a point on the curve at a distance x from the origin of coordinates. Hence, Eq. (3-30) may be rewritten

$$M = EI\frac{d^2y}{dx^2} \tag{3-31}$$

To obtain the slope and deflection of a beam, this equation may be integrated, with M expressed as a function of x. Constants introduced during the integration must be evaluated in terms of known points and slopes of the elastic curve.

Equation (3-31), in turn, may be rewritten after one integration as

$$\theta_B - \theta_A = \int_A^B \frac{M}{EI}\,dx \tag{3-32}$$

in which θ_A and θ_B are the slopes of the elastic curve at any two points A and B. If the slope is zero at one of the points, the integral in Eq. (3-32) gives the slope of the elastic curve at the other. It should be noted that the integral represents the area of the bending-moment diagram between A and B with each ordinate divided by EI.

The **tangential deviation** t of a point on the elastic curve is the distance of this point, measured in a direction perpendicular to the original position of the beam, from a tangent drawn at some other point on the elastic curve.

$$t_B - t_A = \int_A^B \frac{Mx}{EI}\,dx \tag{3-33}$$

Equation (3-33) indicates that the tangential deviation of any point with respect to a second point on the elastic curve equals the moment about the first point of the M/EI diagram between the two points. The **moment-area method** for determining the deflection of beams is a technique in which Eqs. (3-32) and (3-33) are utilized.

Suppose, for example, the deflection at mid-span is to be computed for a beam of uniform cross section with a concentrated load at the center (Fig. 3-28).

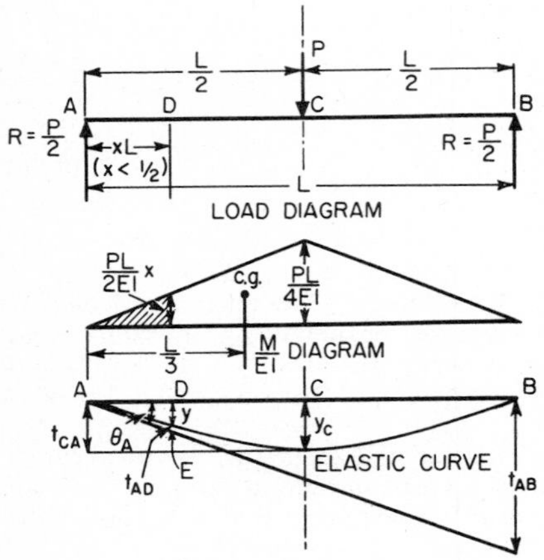

Fig. 3-28. Elastic curve for a simple beam.

Since the deflection at midspan for this loading is the maximum for the span, the slope of the elastic curve at the center of the beam is zero; i.e., the tangent is parallel to the undeflected position of the beam. Hence, the deviation of either support from the midspan tangent is equal to the deflection at the center of the beam. Then, by the moment-area theorem [Eq. (3-33)], the deflection y_c is given by the moment about either support of the area of the M/EI diagram included between an ordinate at the center of the beam and that support.

$$y_c = \frac{1}{2}\frac{PL}{4EI}\frac{L}{2}\frac{2}{3}\frac{L}{2} = \frac{PL^3}{48EI}$$

Suppose, now, the deflection y at any point D at a distance xL from the left support (Fig. 3-28) is to be determined. Referring to the sketch, we note that the distance DE from the undeflected position of D to the tangent to the elastic curve at support A is given by

$$y + t_{AD} = xt_{AB}$$

where t_{AD} is the tangential deviation of D from the tangent at A and t_{AB} is the tangential deviation of B from that tangent. This equation, which is perfectly general for the deflection of any point of a simple beam, no matter how loaded, may be rewritten to give the deflection directly:

$$y = xt_{AB} - t_{AD} \tag{3-34}$$

But t_{AB} is the moment of the area of the M/EI diagram for the whole beam about support B. And t_{AD} is the moment about D of the area of the M/EI diagram included between ordinates at A and D. Hence.

$$y = x\frac{1}{2}\frac{PL}{4EI}\frac{L}{2}\left(\frac{2}{3}+\frac{1}{3}\right)L - \frac{1}{2}\frac{PLx}{2EI}xL\frac{xL}{3} = \frac{PL^3}{48EI}x(3 - 4x^2)$$

It is also noteworthy that, since the tangential deviations are very small distances, the slope of the elastic curve at A is given by

$$\theta_A = \frac{t_{AB}}{L} \tag{3-35}$$

This holds, in general, for all simple beams regardless of the type of loading.

The procedure followed in applying Eq. (3-34) to the deflection of the loaded beam in Fig. 3-28 is equivalent to finding the bending moment at D with the M/EI diagram serving as the load diagram. The technique of applying the M/EI diagram as a load and determining the deflection as a bending moment is known as the **conjugate-beam method.**

The conjugate beam must have the same length as the given beam; it must be in equilibrium with the M/EI load and the reactions produced by the load; and the bending moment at any section must be equal to the deflection of the given beam at the corresponding section. The last requirement is equivalent to requiring that the shear at any section of the conjugate beam with the M/EI load be equal to the slope of the elastic curve at the corresponding section of the given beam. Figure 3-29 shows the conjugates for various types of beams.

Deflections for several types of loading on simple beams are given in Figs. 3-30 to 3-35 and for overhanging beams and cantilevers in Figs. 3-36 to 3-41.

When a beam carries a number of loads of different types, the most convenient method of computing its deflection generally is to find the deflections separately for the uniform and concentrated loads and add them up.

For several concentrated loads, the easiest solution is to apply the reciprocal theorem (Art. 3-57). According to this theorem, if a concentrated load is applied to a beam at a point A, the deflection it produces at point B is equal to the deflection at A for the same load applied at B ($d_{AB} = d_{BA}$).

Suppose, for example, the midspan deflection is to be computed. Then, asume each load in turn applied at the center of the beam and compute the deflection at the point where it originally was applied from the equation of the elastic curve given in Fig. 3-33. The sum of these deflections is the total midspan deflection.

Another method for computing deflections of beams is presented in Art. 3-55. This method may also be applied to determining the deflection of a beam due to shear.

3-32. Combined Axial and Bending Loads. For stiff beams, subjected to both transverse and axial loading, the stresses are given by the principle of superposition if the deflection due to bending may be neglected without serious error. That is, the total stress is given with sufficient accuracy at any section by the sum of the axial stress and the bending stresses. The maximum stress equals

$$f = \frac{P}{A} + \frac{Mc}{I} \tag{3-36}$$

where P = axial load
A = cross-sectional area
M = maximum bending moment
c = distance from neutral axis to outermost fiber at the section where maximum moment occurs
I = moment of inertia of cross section about neutral axis at that section

When the deflection due to bending is large and the axial load produces bending stresses that cannot be neglected, the maximum stress is given by

$$f = \frac{P}{A} + (M + Pd)\frac{c}{I} \tag{3-37}$$

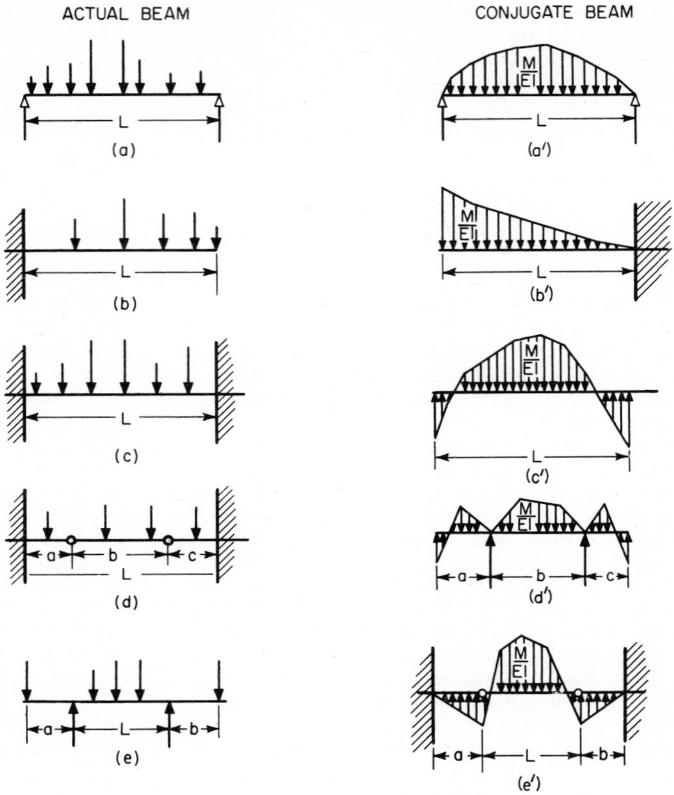

Fig. 3-29. Conjugate beams.

where d is the deflection of the beam. For axial compression, the moment Pd should be given the same sign as M, and for tension, the opposite sign, but the minimum value of $M + Pd$ is zero. The deflection d for axial compression and bending can be obtained by applying Eq. (3-31). (S. Timoshenko and J. M. Gere, "Theory of Elastic Stability," McGraw-Hill Book Company, New York; Friedrich Bleich, "Buckling Strength of Metal Structures," McGraw-Hill Book Company, New York.) However, it may be closely approximated by

$$d = \frac{d_o}{1 - (P/P_c)} \tag{3-38}$$

where d_o = deflection for the transverse loading alone
P_c = the critical buckling load π^2EI/L^2 (see Art. 3-71)

3-33. Eccentric Loading. An eccentric longitudinal load in the plane of symmetry produces a bending moment Pe where e is the distance of the load from the centroidal axis. The total unit stress is the sum of the stress due to this moment and the stress due to P applied as an axial load:

$$f = \frac{P}{A} \pm \frac{Pec}{I} = \frac{P}{A}\left(1 \pm \frac{ec}{r^2}\right) \tag{3-39}$$

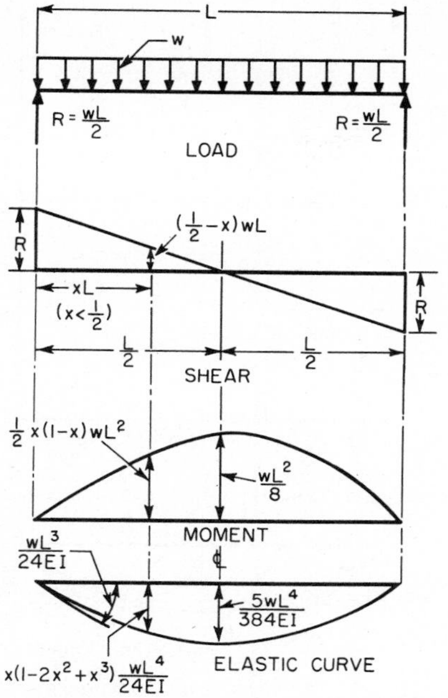

Fig. 3-30. Uniform load over the full span of a simple beam.

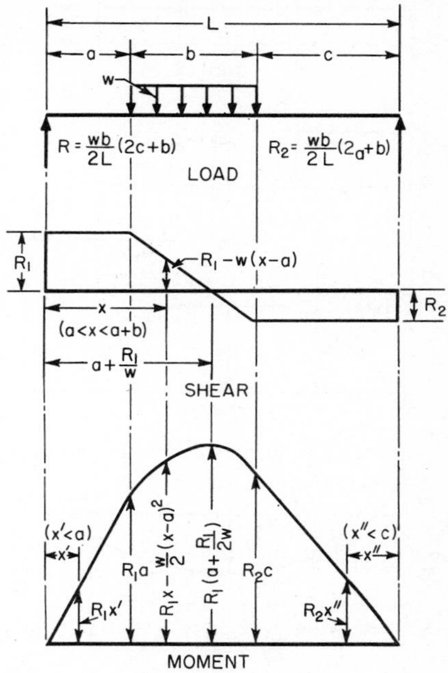

Fig. 3-31. Uniform load over part of a simple beam.

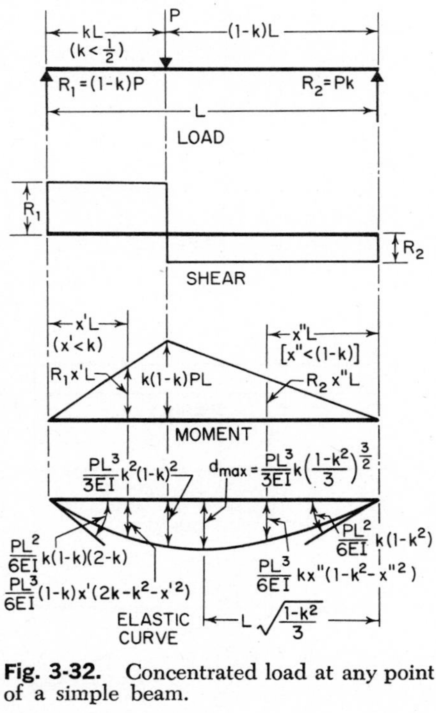

Fig. 3-32. Concentrated load at any point of a simple beam.

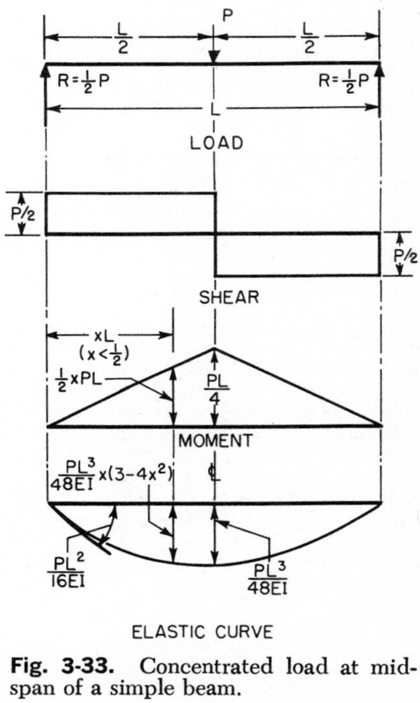

Fig. 3-33. Concentrated load at midspan of a simple beam.

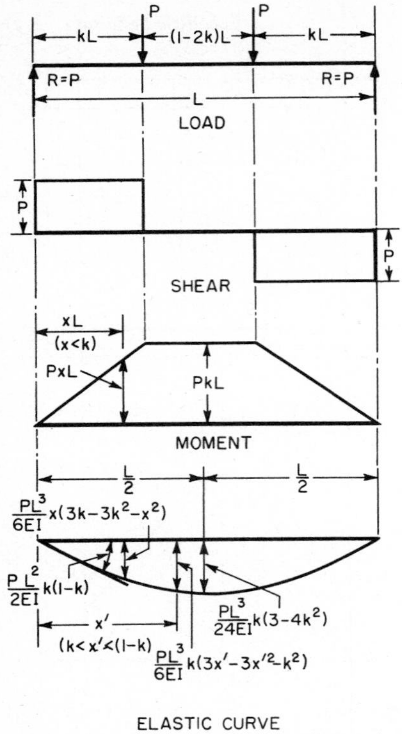

Fig. 3-34. Two equal concentrated loads on a simple beam.

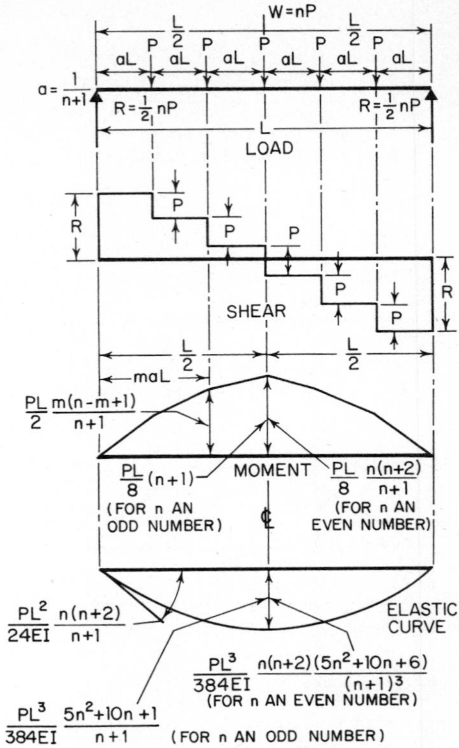

Fig. 3-35. Several equal concentrated loads on a simple beam.

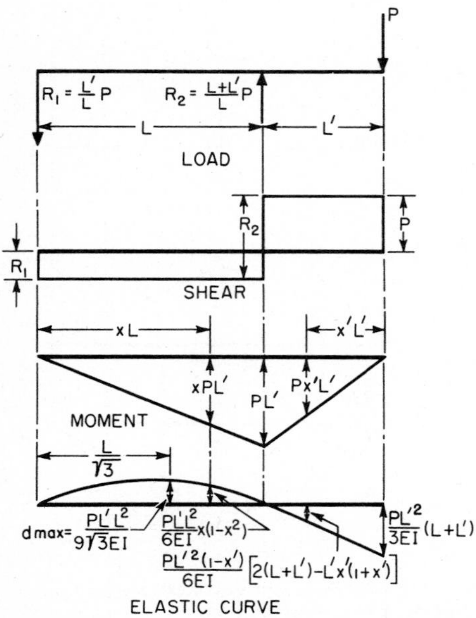

Fig. 3-36. Concentrated load on a beam overhang.

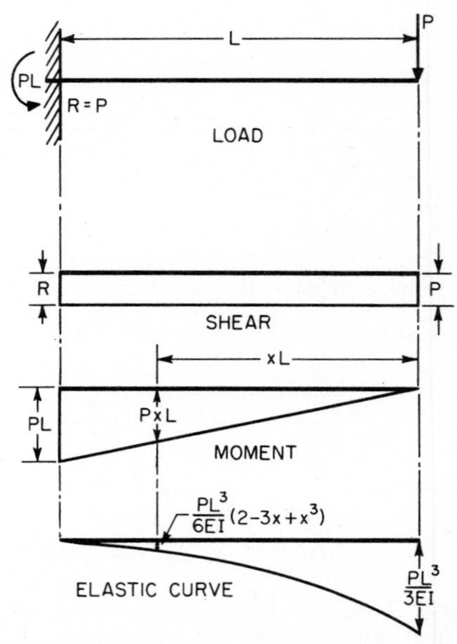

Fig. 3-37. Concentrated load on the end of a cantilever.

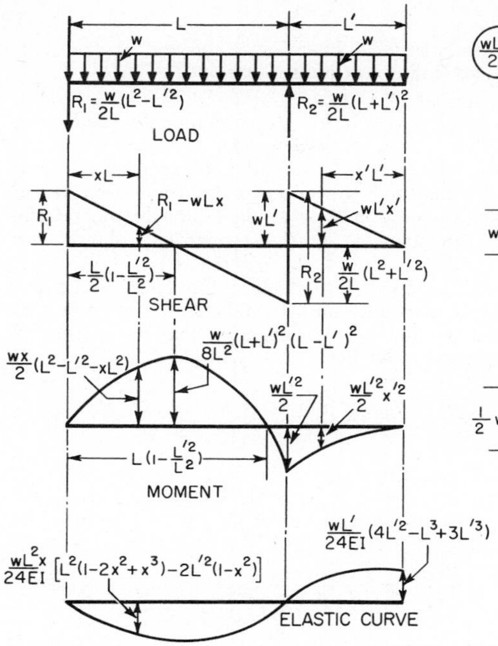

Fig. 3-38. Uniform load over the full length of a beam with overhang.

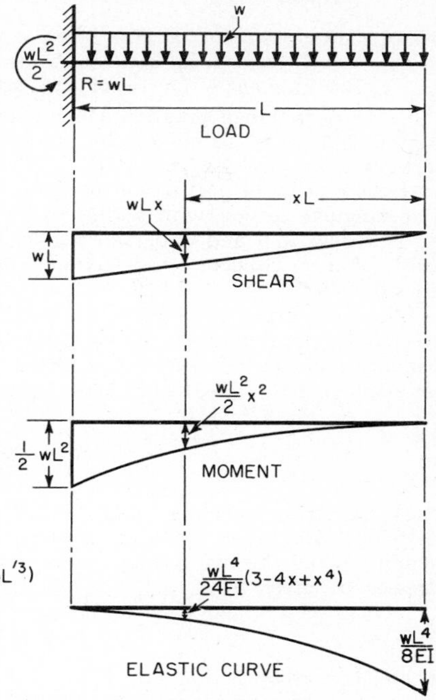

Fig. 3-39. Uniform load over full length of a cantilever.

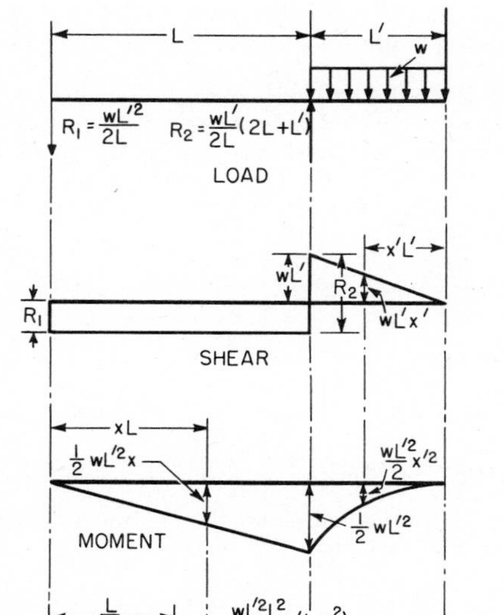

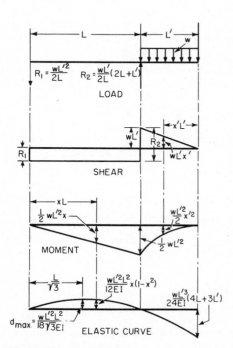

Fig. 3-40. Uniform load on a beam overhang.

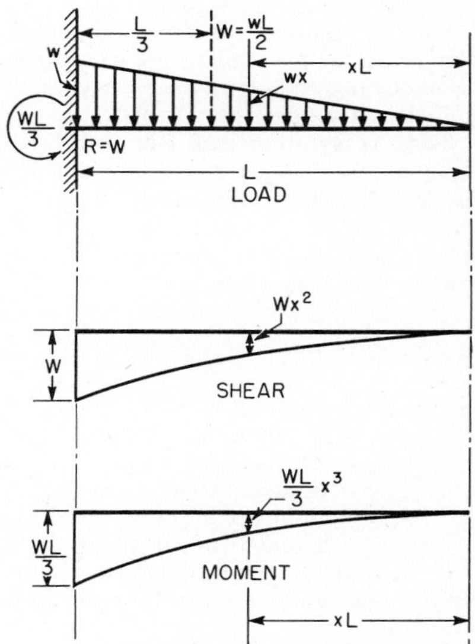

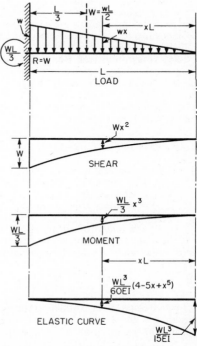

Fig. 3-41. Triangular loading on a cantilever.

where A = cross-sectional area
c = distance from neutral axis to outermost fiber
I = moment of inertia of cross section about neutral axis
r = **radius of gyration** which is equal to $\sqrt{I/A}$

Figure 3-26 gives values of the radius of gyration for some commonly used cross sections.

If there is to be no tension on the cross section, e should not exceed r^2/c. For a rectangular section with width b and depth d the eccentricity, therefore, should be less than $b/6$ and $d/6$; i.e., the load should not be applied outside the middle third. For a circular cross section with diameter D the eccentricity should not exceed $D/8$.

When the eccentric longitudinal load produces a deflection too large to be neglected in computing the bending stress, account must be taken of the additional bending moment Pd, where d is the deflection. This deflection may be computed by employing Eq. (3-31) or closely approximated by

$$d = \frac{4eP/P_c}{\pi(1 - P/P_c)} \tag{3-40}$$

P_c is the critical buckling load, π^2EI/L^2 (see Art. 3-71).

If the load P does not lie in a plane containing an axis of symmetry, it produces bending about the two principal axes through the centroid of the cross section. The stresses are given by

$$f = \frac{P}{A} \pm \frac{Pe_xc_x}{I_y} \pm \frac{Pe_yc_y}{I_x} \tag{3-41}$$

where A = cross-sectional area
e_x = eccentricity with respect to principal axis YY
e_y = eccentricity with respect to principal axis XX
c_x = distance from YY to outermost fiber
c_y = distance from XX to outermost fiber
I_x = moment of inertia about YY
I_y = moment of inertia about XX

The principal axes are the two perpendicular axes through the centroid for which the moments of inertia are a maximum or a minimum and for which the products of inertia are zero.

3-34. Unsymmetrical Bending. Bending caused by loads that do not lie in a plane containing a principal axis of each cross section of a beam is called unsymmetrical bending. Assuming that the bending axis of the beam lies in the plane of the loads, to preclude torsion (see Art. 3-25), and that the loads are perpendicular to the bending axis, to preclude axial components, the stress at any point in a cross section is given by

$$f = \frac{M_xy}{I_x} \pm \frac{M_yx}{I_y} \tag{3-42}$$

where M_x = bending moment about principal axis XX
M_y = bending moment about principal axis YY
x = distance from point for which stress is to be computed to YY axis
y = distance from point to XX axis
I_x = moment of inertia of the cross section about XX
I_y = moment of inertia about YY.

If the plane of the loads makes an angle θ with a principal plane, the neutral surface will form an angle α with the other principal plane such that

$$\tan \alpha = \frac{I_x}{I_y} \tan \theta \tag{3-43}$$

3-35. Beams with Unsymmetrical Sections. In the derivation of the flexure formula $f = Mc/I$ (Art. 3-25), the assumption is made that the beam bends, without twisting, in the plane of the loads and that the neutral surface is perpendicular to the plane of the loads. These assumptions are correct for beams with cross sections symmetrical about two axes when the plane of the loads contains one of these axes. They are not necessarily true for beams that are not doubly symmetrical. The reason is that in beams that are doubly symmetrical the bending axis coincides with the centroidal axis, whereas in unsymmetrical sections the two axes may be separate. In the latter case, if the plane of the loads contains the centroidal axis but not the bending axis, the beam will be subjected to both bending and torsion.

The **bending axis** may be defined as the longitudinal line in a beam through which transverse loads must pass to preclude the beam's twisting as it bends. The point in each section through which the bending axis passes is called the **shear center,** or center of twist. The shear center is also the center of rotation of the section in pure torsion (Art. 3-77). Its location depends on the dimensions of the section.

If a beam has an axis of symmetry, the shear center lies on it. In doubly symmetrical beams, the shear center lies at the intersection of the two axes of symmetry and hence coincides with the centroid.

For any section composed of two narrow rectangles, such as a T beam or an angle, the shear center may be taken as the intersection of the longitudinal center lines of the rectangles.

For a channel section with one axis of symmetry, the shear center is outside the section at a distance from the centroid equal to $e(1 + h^2A/4I)$, where e is the distance from the centroid to the center of the web, h is the depth of the channel, A the cross-sectional area, and I the moment of inertia about the axis of symmetry. (The web lies between the shear center and the centroid.)

Locations of shear centers for several other sections are given in Friedrich Bleich, "Buckling Strength of Metal Structures," Chap. III, McGraw-Hill Book Company, New York.

CURVED BEAMS

Structural members, such as arches, crane hooks, chain links, and frames of some machines, that have considerable initial curvature in the plane of loading are called curved beams. The flexure formula of Art. 3-25, $f = Mc/I$, cannot be applied to them with any reasonable degree of accuracy unless the depth of the beam is small compared with the radius of curvature.

Unlike the condition in straight beams, unit strains in curved beams are not proportional to the distance from the neutral surface, and the centroidal axis does not coincide with the neutral axis. Hence the stress distribution on a section is not linear but more like the distribution shown in Fig. 3-42*c*.

3-36. Stresses in Curved Beams. Just as for straight beams, the assumption that plane sections before bending remain plane after bending generally holds for curved beams. So the total strains are proportional to the distance from the neutral axis. But since the fibers are initially of unequal length, the unit strains are a more complex function of this distance. In Fig. 3-42*a*, for example, the bending couples have rotated section AB of the curved beam into section $A'B'$ through an angle $\Delta\, d\theta$. If ϵ_o is the unit strain at the centroidal axis and ω is the angular unit strain $\Delta\, d\theta/d\theta$, then the unit strain at a distance y from the centroidal axis (measured positive in the direction of the center of curvature) is

$$\epsilon = \frac{DD'}{DD_o} = \frac{\epsilon_o R\, d\theta - y\Delta\, d\theta}{(R - y)\, d\theta} = \epsilon_o - (\omega - \epsilon_o)\frac{y}{R - y} \tag{3-44}$$

where R = radius of curvature of centroidal axis

Equation (3-44) can be expressed in terms of the bending moment if we take advantage of the fact that the sum of the tensile and compressive forces on the section must be zero and the moment of these forces must be equal to the bending moment M. These two equations yield

$$\epsilon_o = \frac{M}{ARE} \qquad \text{and} \qquad \omega = \frac{M}{ARE}\left(1 + \frac{AR^2}{I'}\right) \tag{3-45}$$

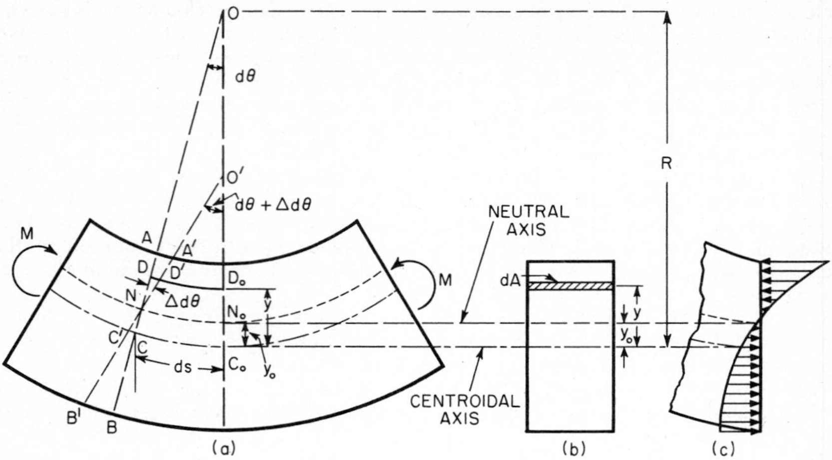

Fig. 3-42. Stresses in a curved beam.

where A is the cross-sectional area, E the modulus of elasticity, and

$$I' = \int \frac{y^2\,dA}{1 - y/R} = \int y^2 \left(1 + \frac{y}{R} + \frac{y^2}{R^2} + \cdots\right) dA \tag{3-46}$$

It should be noted that I' is very nearly equal to the moment of inertia I about the centroidal axis when the depth of the section is small compared with R, so that the maximum ratio of y to R is small compared with unity. M is positive when it decreases the radius of curvature.

Since the stress $f = E\epsilon$, we obtain the stresses in the curved beam from Eq. (3-44) by multiplying it by E and substituting ϵ_o and ω from Eq. (3-45):

$$f = \frac{M}{AR} - \frac{My}{I'}\frac{1}{1 - y/R} \tag{3-47}$$

The distance y_o of the neutral axis from the centroidal axis (Fig. 3-42) may be obtained from Eq. (3-47) by setting $f = 0$:

$$y_o = \frac{I'R}{I' + AR^2} \tag{3-48}$$

Since y_o is positive; the neutral axis shifts toward the center of curvature.

3-37. Curved Beams with Rectangular Sections. Taking the first three terms of the series in Eq. (3-46) and integrating, we obtain for a rectangular cross section

$$I' = I\left(1 + \frac{3}{5}\frac{c^2}{R^2}\right) \tag{3-49}$$

where I is the moment of inertia about the centroidal axis and c is one-half the depth of the beam. The stresses in the outer fibers are given by

$$\begin{aligned} f_{\max} &= -\frac{Mc}{I}\left(\frac{I}{I'}\frac{R}{R - c} - \frac{c}{3R}\right) \\ f_{\min} &= +\frac{Mc}{I}\left(\frac{I}{I'}\frac{R}{R + c} + \frac{c}{3R}\right) \end{aligned} \tag{3-50}$$

It is noteworthy that the factor by which Mc/I is multiplied in Eq. (3-50) is a function of c/R and independent of the other dimensions of the beam.

3-38. Curved Beams with Circular Section. Integrating the series in Eq. (3-46) for a circle and taking the first three terms of the result, we obtain

$$I' = I\left(1 + \frac{1}{2}\frac{c^2}{R^2}\right) \tag{3-51}$$

where I = moment of inertia about centroidal axis
c = one-half depth of beam.

The stresses in the outer fibers are given by

$$\begin{aligned} f_{\max} &= -\frac{Mc}{I}\left(\frac{I}{I'}\frac{R}{R-c} - \frac{c}{4R}\right) \\ f_{\min} &= +\frac{Mc}{I}\left(\frac{I}{I'}\frac{R}{R+c} + \frac{c}{4R}\right) \end{aligned} \tag{3-52}$$

For this section, too, the factor by which Mc/I is multiplied is a function of c/R and independent of the other dimensions of the beam.

3-39. Curved I or T Beams. If Eq. (3-47) is applied to I or T beams or tubular members, it may indicate circumferential flange stresses that are much lower than will actually occur. The error is due to the fact that the outer edges of the flanges deflect radially. The effect is equivalent to having only part of the flanges active in resisting bending stresses. Also, accompanying the flange deflections, there are transverse bending stresses in the flanges. At the junction with the web, these reach a maximum, which may be greater than the maximum circumferential stress. Furthermore, there are radial stresses (normal stresses acting in the direction of the radius of curvature) in the web that also may have maximum values greater than the maximum circumferential stress.

A good approximation to the stresses in I or T beams is presented in Seely and Smith, "Advanced Mechanics of Materials," John Wiley & Sons, Inc., New York. In brief, the authors contend that, for circumferential stresses, Eq. (3-47) may be used with a modified cross section, which is obtained by using a reduced flange width. The reduction is calculated from $b' = \alpha b$, where b is the length of the portion of the flange projecting on either side from the web, b' is the corrected length, and α is a correction factor determined from equations developed by H. Bleich. α is a function of b^2/rt, where t is the flange thickness and r the radius of the center of the flange:

$b^2/rt =$	0.5	0.7	1.0	1.5	2	3	4	5
$\alpha =$	0.9	0.6	0.7	0.6	0.5	0.4	0.37	0.33

When the parameter b^2/rt is greater than 1.0, the maximum transverse bending stress is approximately equal to 1.7 times the stress obtained at the center of the flange from Eq (3-47) applied to the modified section. When the parameter equals 0.7, that stress should be multiplied by 1.5, and when it equals 0.4, the factor is 1.0. In Eq. (3-47), I' for I beams may be taken for this calculation approximately equal to

$$I' = I\left(1 + \frac{c^2}{R^2}\right) \tag{3-53}$$

where I = moment of inertia of modified section about its centroidal axis
R = radius of curvature of centroidal axis
c = distance from centroidal axis to center of the more sharply curved flange

Because of the high stress factor, it is advisable to stiffen or brace curved I-beam flanges.

The maximum radial stress will occur at the junction of web and flange of I beams. If the moment is negative; that is, if the loads tend to flatten out the beam, the radial stress is tensile, and there is a tendency for the more sharply curved flange to pull away from the web. An approximate value of this maximum stress is

$$f_r = -\frac{A_f}{A}\frac{M}{t_w c_g r'} \tag{3-54}$$

where f_r = radial stress at junction of flange and web of a symmetrical I beam
A_f = area of one flange
A = total cross-sectional area
M = bending moment
t_w = thickness of web
c_g = distance from centroidal axis to center of flange
r' = radius of curvature of inner face of more sharply curved flange

3-40. Axial and Bending Loads on Curved Beams. If a curved beam carries an axial load P as well as bending loads, the maximum unit stress is

$$f = \frac{P}{A} \pm \frac{Mc}{I} K \tag{3-55}$$

where K is a correction factor for the curvature [see Eqs. (3-50) and (3-52)]. The sign of M is taken positive in this equation when it increases the curvature, and P is positive when it is a tensile force, negative when compressive.

3-41. Slope and Deflection of Curved Beams. If we consider two sections of a curved beam separated by a differential distance ds (Fig. 3-42), the change in angle $\Delta\, d\theta$ between the sections due to a bending moment M and an axial load P may be obtained from Eq. (3-45), noting that $d\theta = ds/R$.

$$\Delta\, d\theta = \frac{M\, ds}{EI'}\left(1 + \frac{I'}{AR^2}\right) + \frac{P\, ds}{ARE} \tag{3-56}$$

where E is the modulus of elasticity, A the cross-sectional area, R the radius of curvature of the centroidal axis, and I' is defined by Eq. (3-46) [see also Eqs. (3-49), (3-51), and (3-53)].

If P is a tensile force, the length of the centroidal axis increases by

$$\Delta\, ds = \frac{P\, ds}{AE} + \frac{M\, ds}{ARE} \tag{3-57}$$

The effect of curvature on shearing deformations for most practical applications is negligible.

For shallow sections (depth of section less than about one-tenth the span), the effect of axial forces on deformations may be neglected. Also, unless the radius of curvature is very small compared with the depth, the effect of curvature may be ignored. Hence, for most practical applications, Eq. (3-56) may be used in the simplified form:

$$\Delta\, d\theta = \frac{M\, ds}{EI} \tag{3-58}$$

For deeper beams, the action of axial forces, as well as bending moments, should be taken into account; but unless the curvature is sharp, its effect on deformations may be neglected. So only Eq. (3-58) and the first term in Eq. (3-57) need be used. (S. Timoshenko and D. H. Young, "Theory of Structures," McGraw-Hill Book Company, New York.) See also Arts. 3-86 to 3-89.

GRAPHIC-STATICS FUNDAMENTALS

A force may be represented by a straight line of fixed length. The length of line to a given scale represents the magnitude of the force. The position of the line parallels the line of action of the force. And an arrowhead on the line indicates the direction in which the force acts.

Forces are concurrent when their lines of action meet. If they lie in the same plane, they are coplanar.

3-42. Parallelogram of Forces. The resultant of several forces is a single force that would produce the same effect on a rigid body. The resultant of two concurrent forces is determined by the parallelogram law:

If a parallelogram is constructed with two forces as sides, the diagonal represents the resultant of the forces (Fig. 3-43*a*).

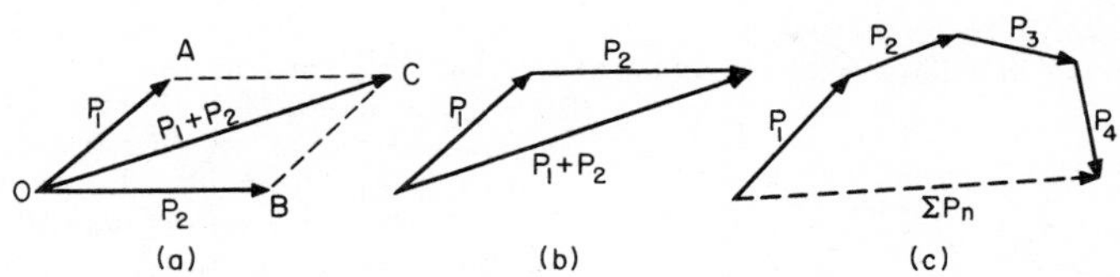

Fig. 3-43. Addition of forces by (*a*) parallelogram law; (*b*) triangle construction; and (*c*) polygon construction.

The **resultant** is said to be equal to the sum of the forces, sum here meaning, of course, addition by the parallelogram law. Subtraction is carried out in the same manner as addition, but the direction of the force to be subtracted is reversed.

If the direction of the resultant is reversed, it becomes the **equilibrant,** a single force that will hold the two given forces in equilibrium.

3-43. Resolution of Forces. To resolve a force into two components, a parallelogram is drawn with the force as a diagonal. The sides of the parallelogram represent the components. The procedure is: (1) Draw the given force. (2) From both ends of the force draw lines parallel to the directions in which the components act. (3) Draw the components along the parallels through the origin of the given force to the intersections with the parallels through the other end. Thus, in Fig. 3-43*a*, P_1 and P_2 are the components in directions *OA* and *OB* of the force represented by *OC*.

3-44. Force Polygons. Examination of Fig. 3-43*a* indicates that a step can be saved in adding the two forces. The same resultant could be obtained by drawing only the upper half of the parallelogram. Hence, to add two forces, draw the first force; then draw the second force beginning at the end of the first one. The resultant is the force drawn from the origin of the first force to the end of the second force, as shown in Fig. 3-43*b*. Again, the equilibrant is the resultant with direction reversed.

From this diagram, an important conclusion can be drawn: **If three forces meeting at a point are in equilibrium, they will form a closed force triangle.**

The conclusions reached for addition of two forces can be generalized for several concurrent forces: To add several forces, P_1, P_2, P_3, . . . , P_n, draw P_2 from the end of P_1, P_3 from the end of P_2, etc. The force required to close the force polygon is the resultant (Fig. 3-43*c*).

If a group of concurrent forces are in equilibrium, they will form a closed force polygon.

3-45. Equilibrium Polygons. When the forces are coplanar but not concurrent, the force polygon will yield the magnitude and direction of the resultant but not its point of application. To complete the solution, the easiest method generally is to employ an auxiliary force polygon, called an equilibrium, or funicular (string), polygon. Sides of this polygon represent the lines of action of certain components of the given forces; more specifically, they take the configuration of a weightless string holding the forces in equilibrium.

In Fig. 3-44*a* are shown three forces P_1, P_2, and P_3. The magnitude and direction of their resultant *R* are determined from the force polygon in Fig. 3-44*b*. To construct the equilibrium polygon for determining the position of the resultant, select a point *O* in or near the force polygon. From *O*, which is called a **pole,** draw rays OA, OB, OC, and OD to the extremities of the forces (Fig. 3-44*b*). Note that consecutive pairs of rays form closed triangles with the given forces, and therefore, they are components of the forces in the directions of the rays. Furthermore, each of the rays OB and OC, drawn to a vertex of the polygon formed by two given forces, represents, in turn, two forces with the same line of action but acting in opposite directions to help hold the given forces in equilibrium. Ray OA is a component not only of P_1 but also of R; similarly, OD is a component of both P_3 and R, and with OA holds R in equilibrium.

If in Fig. 3-44*a*, where the given forces are shown in their relative positions, lines are drawn from each force in the direction of the rays forming their components, they will form a funicular polygon. For example, starting with a convenient point on P_1 (Fig. 3-44*a*), draw a line parallel to OB until it intersects P_2. This represents the line of action of both force BO, which helps hold P_1 in equilibrium, and OB, which helps hold P_2 in equilibrium. Through the intersection with P_2, draw a line parallel to OC until it intersects P_3. Through this last intersection, draw a line parallel to DO, one component of R, and through the starting point on P_1, draw a line parallel to OA, the second component of R, completing the funicular polygon. These two lines intersect on the line of action of R.

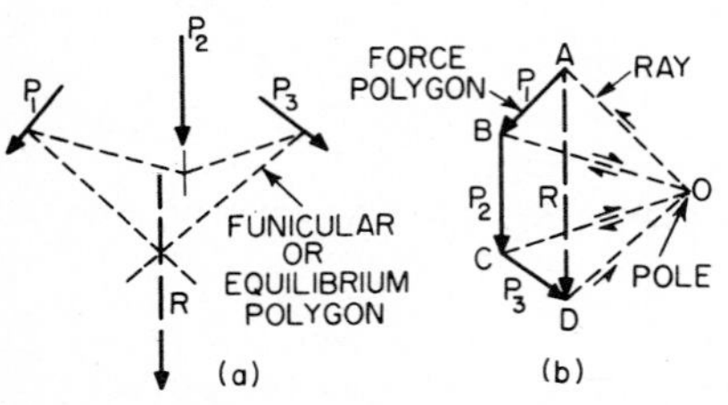

Fig. 3-44. Force and equilibrium polygons for a system of forces.

3-46. Beam and Truss Reactions by Graphics. The equilibrium polygon can also be used to find the reactions of simple beams and trusses supporting vertical loads. First, a force polygon is constructed to obtain the magnitude and direction of the resultant of the loads, which is equal in magnitude but opposite in direction to the sum of the reactions. Second, rays are drawn to the vertexes of the polygon from a conveniently located pole. These rays are used to construct all but one side of an equilibrium polygon. The closing side is the common line of action of two equal but opposite forces that act with a pair of rays already drawn to hold the reactions in equilibrium. Therefore, draw a line through the pole parallel to the closing side. The intersection with the resultant separates it into two forces, which are equal to the reactions sought.

For example, suppose the reactions were to be obtained graphically for the beam (or truss) in Fig. 3-45*a*. As a first step, the force polygon $ABCD$ is constructed, a pole selected, and rays drawn to the extremities of the forces (Fig. 3-45*b*). Since the loads are parallel, the force polygon is a straight line. The sum of the reactions, in this case, is equal to the length of the line AD.

Next, the equilibrium polygon is constructed. Starting with a convenient point on the line of action of R_1, draw a line oa parallel to ray OA in Fig. 3-45*b* and locate its intersection with the line of action of P_1 (Fig. 3-45*d*). Through this intersection, draw ob, a line parallel to OB to the intersection with P_2, then a line oc parallel to OC, and finally a line od parallel to OD, which terminates on the line of action of R_2. Draw the closing line oe of the polygon between this last intersection and the starting point of the polygon. The last step is to draw through the pole (Fig. 3-45*b*) OE, a line parallel to oe, the closing line of the equilibrium polygon, cutting the force-polygon resultant at E. Then, $DE = R_2$, and $EA = R_1$. (See Art. 3-17 for an analytical method of computing reactions.)

3-47. Shear and Moment Diagrams by Graphics. The shear at any section of a beam is equal to the algebraic sum of the loads and reactions on the left of the section, upward-acting forces being considered positive, downward forces negative. If the forces are arranged in the proper order, the shear diagram may be obtained directly from the force polygon after the reactions have been determined.

For example, the shear diagram for the beam in Fig. 3-45*a* can be easily obtained from the force polygon $ABCDE$ in Fig. 3-45*b*. The zero axis is a line parallel to the beam through E. As indicated in Fig. 3-45*c*, the ordinates of the shear

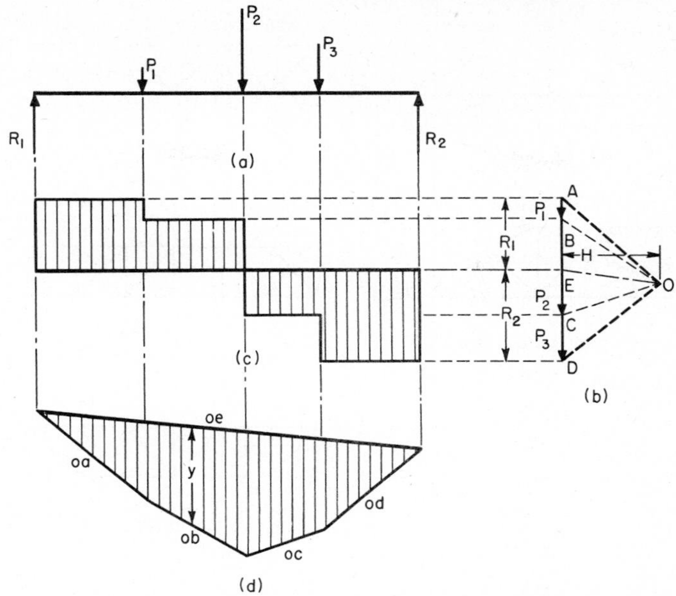

Fig. 3-45. Shear and moment diagrams obtained by graphic statics for (*a*) loaded beam; (*b*) force polygon; (*c*) shear diagram; and (*d*) equilibrium polygon and bending-moment diagram.

diagram are laid off, starting with R_1 along the line of action of the left reaction, by drawing lines parallel to the zero axis through the extremities of the forces in the force polygon. (See Art. 3-19 for an analytical method of determining shears.)

The moment of a force about a point can be obtained from the equilibrium and force polygons. In the equilibrium diagram, draw a line parallel to the force through the given point. Measure the intercept of this line between the two adjacent funicular-polygon sides (extended if necessary) that originate at the given force. The moment is the product of this intercept by the distance of the force-polygon pole from the force. The intercept should be measured to the same linear scale as the beam and load positions, and the pole distance to the same scale as the forces in the force polygon.

As a consequence of this relationship between the sides of the funicular polygon, each ordinate (parallel to the forces) multiplied by the pole distance is equal to the bending moment at the corresponding section of the beam or truss. In Fig. 3-45*d*, for example, the equilibrium polygon, to scale, is the bending-moment diagram for the beam in Fig. 3-45*a*. At any section, the moment equals the ordinate y multiplied by the pole distance H. (See Art. 3-20 for an analytical method.)

ROOF TRUSSES

A truss is a coplanar system of structural members joined together at their ends to form a stable framework. Neglecting small changes in the lengths of the members due to loads, the relative positions of the joints cannot change.

Three bars pinned together to form a triangle represent the simplest type of truss. Some of the more common types of roof trusses are shown in Fig. 3-46.

The top members are called the upper chord; the bottom members, the lower chord; and the verticals and diagonals, web members.

The purpose of roof trusses is to act like big beams, to support the roof covering over long spans. They not only have to carry their own weight and the weight

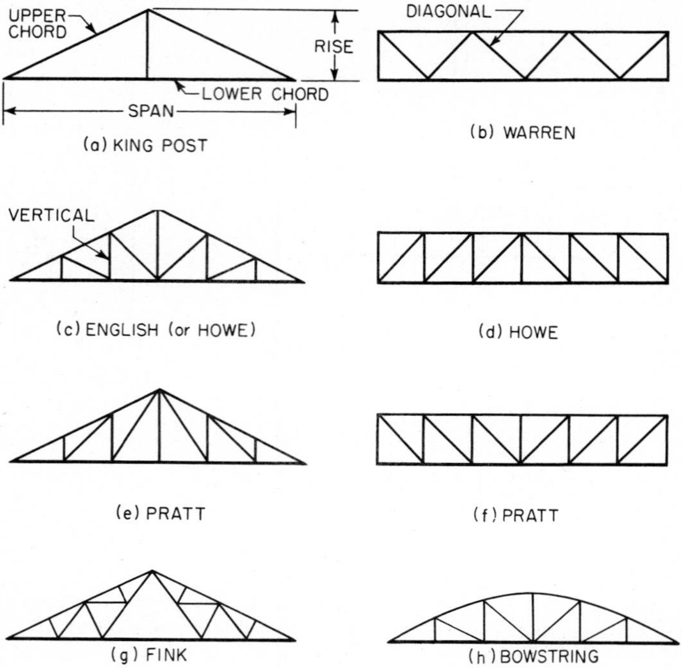

Fig. 3-46. Common types of roof trusses.

of the roofing and roof beams, or purlins, but cranes, wind loads, snow loads, suspended ceilings, and equipment, and a live load to take care of construction, maintenance, and repair loading. These loads are applied at the intersection of the members, or panel points, so that the members will be subjected principally to direct stresses.

3-48. Method of Sections. A convenient method of determining the stresses in truss members is to isolate a portion of the truss by a section so chosen as to cut only as many members with unknown stresses as can be evaluated by the laws of equilibrium applied to that portion of the truss. The stresses in the cut members are treated as external forces. Compressive forces act toward the panel point and tensile forces away from the joint.

Suppose, for example, we wish to find the stress in chord AB of the truss in Fig. 3-47*a*. We can take a vertical section XX close to panel point A. This cuts not only AB but AD and ED as well. The external 10-kip (10,000-lb) loading and 25-kip reaction at the left are held in equilibrium by the compressive force C in AB, tensile force T in ED, and tensile force S in AD (Fig. 3-47*b*). The simplest way to find C is to take moments about D, the point of intersection of S and T, eliminating these unknowns from the calculation.

$$-9C + 36 \times 25 - 24 \times 10 - 12 \times 10 = 0$$

from which C is found to be 60 kips.

Similarly, to find the stress in ED, the simplest way is to take moments about A, the point of intersection of S and C:

$$-9T + 24 \times 25 - 12 \times 10 = 0$$

from which T is found to be 53.3 kips.

On the other hand, the stress in AD can be easily determined by two methods. One takes advantage of the fact that AB and ED are horizontal members, requiring AD to carry the full vertical shear at section XX. Hence we know that the vertical component V of $S = 25 - 10 - 10 = 5$ kips. Multiplying V by sec θ (Fig. 3-47b), which is equal to the ratio of the length of AD to the rise of the truss ($^{15}/_{9}$), S is found to be 8.3 kips. The second method—presented because it is useful when the chords are not horizontal—is to resolve S into horizontal and vertical components at D and take moments about E. Since both T and the horizontal component of S pass through E, they do not appear in the computations, and C already has been computed. Equating the sum of the moments to zero gives $V = 5$, as before.

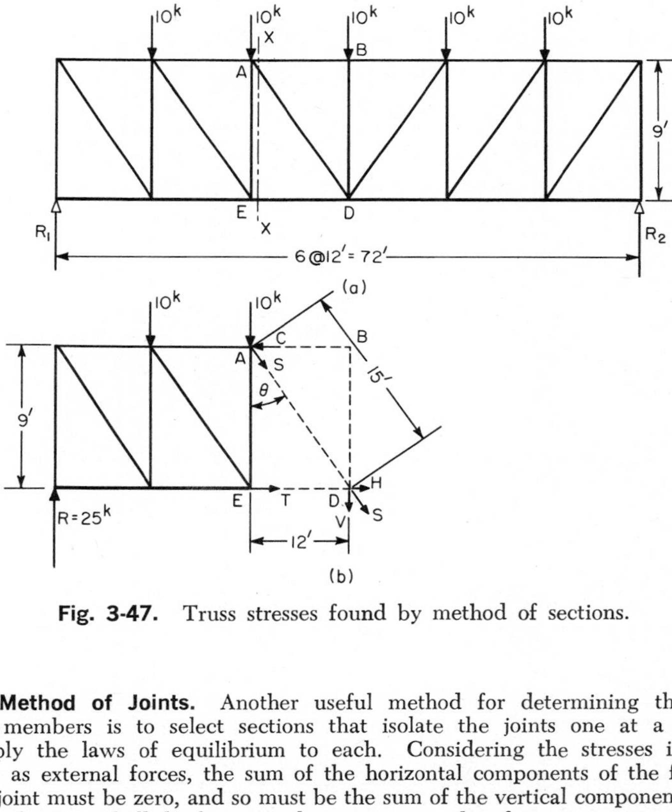

Fig. 3-47. Truss stresses found by method of sections.

3-49. Method of Joints. Another useful method for determining the stresses in truss members is to select sections that isolate the joints one at a time and then apply the laws of equilibrium to each. Considering the stresses in the cut members as external forces, the sum of the horizontal components of the forces acting at a joint must be zero, and so must be the sum of the vertical components. Since the lines of action of all the forces are known, we can therefore compute two unknown magnitudes at each joint by this method. The procedure is to start at a joint that has only two unknowns (generally at the support) and then, as stresses in members are determined, analyze successive joints.

Let us, for illustration, apply the method to joint 1 of the truss in Fig. 3-48a. Equating the sum of the vertical components to zero, we find that the vertical component of the top chord must be equal and opposite to the reaction, 12 kips (12,000 lb). The stress in the top chord at this joint, then, must be a compression equal to $12 \times {}^{30}/_{18} = 20$ kips. From the fact that the sum of the horizontal components must be zero, we find that the stress in the bottom chord at the joint must be equal and opposite to the horizontal component of the top chord. Hence the stress in the bottom chord must be a tension equal to $20 \times {}^{24}/_{30} = 16$ kips.

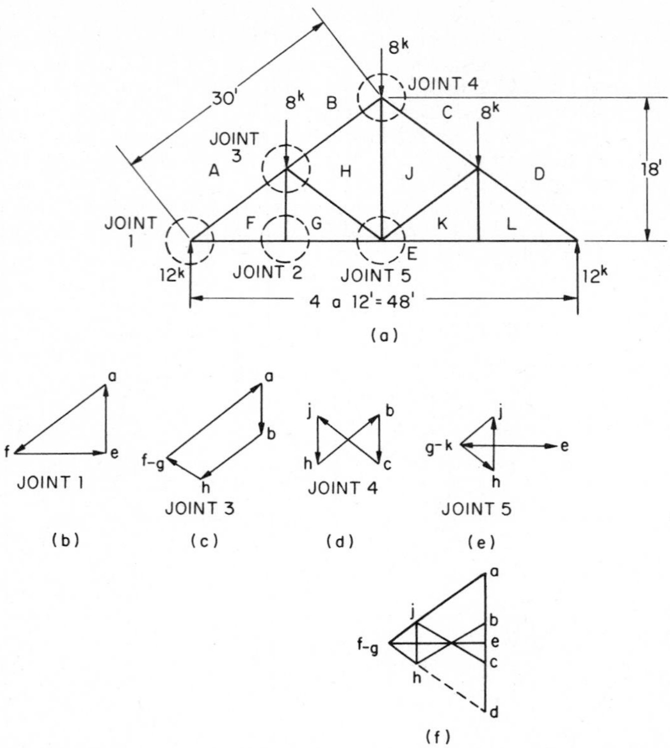

Fig. 3-48. Method of joints applied to a roof truss, leading to a graphical solution with a Maxwell diagram (*f*).

Moving to joint 2, we note that, with no vertical loads at the joint, the stress in the vertical is zero. Also, the stress is the same in both bottom chord members at the joint, since the sum of the horizontal components must be zero.

Joint 3 now contains only two unknown stresses. Denoting the truss members and the loads by the letters placed on opposite sides of them, as indicated in Fig. 3-48*a*, the unknown stresses are S_{BH} and S_{HG}. The laws of equilibrium enable us to write the following two equations, one for the vertical components and the second for the horizontal components:

$$\Sigma V = 0.6S_{FA} - 8 - 0.6S_{BH} + 0.6S_{HG} = 0$$
$$\Sigma H = 0.8S_{FA} - 0.8S_{BH} - 0.8S_{HG} = 0$$

Both unknown stresses are assumed to be compressive; i.e., acting toward the joint. The stress in the vertical does not appear in these equations, because it was already determined to be zero. The stress in FA, S_{FA}, was found from analysis of joint 1 to be 20 kips. Simultaneous solution of the two equations yields $S_{HG} = 6.7$ kips and $S_{BH} = 13.3$ kips. (If these stresses had come out with a negative sign, it would have indicated that the original assumption of their directions was incorrect; they would, in that case, be tensile forces instead of compressive forces.)

3-50. Bow's Notation. The method of designating the loads and stresses used in the previous paragraph is the basis of Bow's notation. Capital letters are placed in the spaces between truss members and between forces. Each member and load is then designated by the letters on opposite sides of it. For example, in Fig. 3-48*a*, the upper chord members are *AF*, *BH*, *CJ*, and *DL*. The loads are

AB, *BC*, and *CD*, and the reactions are *EA* and *DE*. Stresses in the members generally are designated by the same letters but in lower case.

3-51. Graphical Analysis of Trusses. The method of joints may also be used with graphical techniques. Figure 3-48*b*, *c*, *d*, and *e* shows how the polygon of force may be applied at each joint to determine the two unknown stresses. Since each joint is in equilibrium, the loads and the stresses in each member must form a closed polygon. The known forces are constructed first. Then, from the origin and the end point, respectively, a line is drawn parallel to the line of action of each of the two unknown stresses. Their intersection determines the magnitude of the unknown stresses.

This solution, as well as the preceding ones, presumes that the reactions are known. They may be computed analytically or graphically, with load and funicular polygons, as explained in Arts. 3-17 and 3-46.

Examination of Fig. 3-48*b*, *c*, *d*, and *e* indicates that each stress occurs in two force polygons. Hence the graphical solution can be shortened by combining the polygons. The combination of the various polygons for all the joints into one stress diagram is known as a **Maxwell diagram.**

The procedure consists of first constructing the force polygon for the loads and reactions and then applying the method of joints, employing each stress as it is found for finding the stresses in the next joint. To make it easy to determine whether the stresses are compression or tension, loads and reactions should be plotted in the force polygon in the order in which they are passed in going clockwise around the truss. Similarly, in drawing the force polygon for each joint, the forces should be plotted in a clockwise direction around the joint. If these rules are followed, the order of the space letters indicates the direction of the forces—compressive stresses act toward the joint, tensile stresses away from the joint. Going around joint 1 clockwise, for example, we find the reaction *EA* as an upward-acting vertical force *ea* (*e* to *a*) in Fig. 3-48b; the top-chord stress *af* acting toward the joint (*a* to *f*), and the bottom-chord stress *fe* acting away from the joint (*f* to *e*). Hence *af* is compressive, *fe* tensile.

To construct a Maxwell diagram for the truss in Fig. 3-48*a*, we lay off the loads and reactions in clockwise order (Fig. 3-48*f*). The force polygon *abcde* is a straight line because all the forces are vertical. To solve joint 1, a line is drawn through point *a* in Fig. 3-48*f* parallel to *AF* and a second line through point *e* parallel to *FE*. Their intersection is point *f*. Going around this triangle in the same order as the loads and members are encountered in traveling clockwise around joint 1, we find *ea*, the reaction, as an upward force; going from *a* to *f*, we move toward the joint, indicating that *af* is a compressive stress; and going from *f* to *e*, we move away from the joint, indicating that *fe* is a tensile stress.

At joint 2, the stress in the vertical is zero. Hence point *g* coincides with point *f*.

To solve joint 3, we start with the known stress *af* and proceed clockwise around the joint. To complete the force polygon, we draw a line through point *b* parallel to *BH* and a line through *g* parallel to *GH*. Their intersection locates point *h*. Similarly, the force polygon for joint 4 is completed by drawing a line through *c* parallel to *CJ* and a line through *h* parallel to *JH*.

Wind loads on a truss with a sloping top chord are assumed to act normal to the roof. In that case, the load polygon will be an inclined line or a true polygon. The reactions are computed generally on the assumption either that both are parallel to the resultant of the wind loads or that one end of the truss is free to move horizontally and therefore will not resist the horizontal components of the loads. The stress diagram is plotted in the same manner as for vertical loads after the reactions are found.

Some trusses are complex and require special methods of analysis. For methods of solving these, see Norris and Wilbur, "Elementary Structural Analysis," McGraw-Hill Book Company, New York, or T. Au, "Elementary Structural Mechanics," Prentice-Hall, Inc., Englewood Cliffs, N.J. These references also present a graphical method for obtaining the deflections of a truss. An analytical method is given in Art. 3-56.

GENERAL TOOLS FOR STRUCTURAL ANALYSIS

For some types of structures, the equilibrium equations are not sufficient to determine the reactions or the internal stresses. These structures are called **statically indeterminate.**

For the analysis of such structures, additional equations must be written based on a knowledge of the elastic deformations. Hence methods of analysis that enable deformations to be evaluated in terms of unknown forces or stresses are important for the solution of problems involving statically indeterminate structures. Some of these methods, like the method of virtual work, are also useful in solving complicated problems involving statically determinate systems.

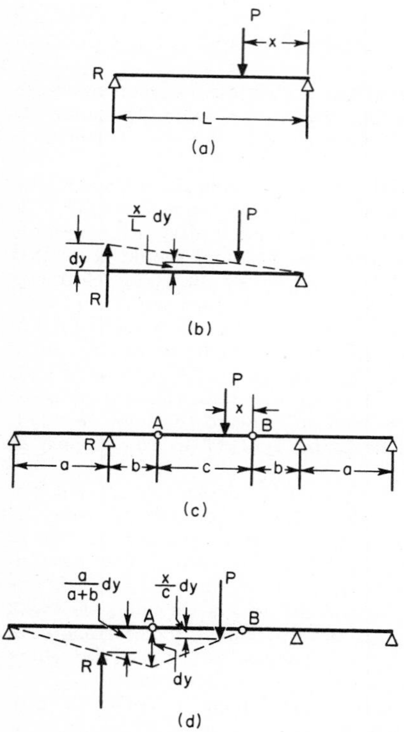

Fig. 3-49. Virtual work applied to determining a simple-beam reaction (*a*) and (*b*) and the reaction of a beam with suspended span (*c*) and (*d*).

3-52. Virtual Work. A virtual displacement is an imaginary small displacement of a particle consistent with the constraints upon it. Thus, at one support of a simply supported beam, the virtual displacement could be an infinitesimal rotation $d\theta$ of that end but not a vertical movement. However, if the support is replaced by a force, then a vertical virtual displacement may be applied to the beam at that end.

Virtual work is the product of the distance a particle moves during a virtual displacement by the component in the direction of the displacement of a force acting on the particle. If the displacement and the force are in opposite directions, the virtual work is negative. When the displacement is normal to the force, no work is done.

Suppose a rigid body is acted upon by a system of forces with a resultant R. Given a virtual displacement ds at an angle α with R, the body will have virtual work done on it equal to $R \cos \alpha \, ds$. (No work is done by internal forces. They act in pairs of equal magnitude but opposite direction, and the virtual work done by one force of a pair is equal but opposite in sign to the work done by the other force.) If the body is in equilibrium under the action of the forces, then $R = 0$ and the virtual work also is zero.

Thus, the principle of virtual work may be stated: **If a rigid body in equilibrium is given a virtual displacement, the sum of the virtual work of the forces acting on it must be zero.**

As an example of how the principle may be used to find a reaction of a statically determinate beam, consider the simple beam in Fig. 3-49*a*, for which the reaction R is to be determined. First, replace the support by an unknown force R. Next, move that end of the beam upward a small amount dy as in Fig. 3-49*b*. The displacement under the load P will be $x\,dy/L$, upward. Then, by the principle of virtual work, $R\,dy - Px\,dy/L = 0$, from which $R = Px/L$.

The principle may also be used to find the reaction R of the more complex beam in Fig. 3-49*c*. The first step again is to replace the support by an unknown force R. Next, apply a virtual downward displacement dy at hinge A (Fig. 3-49*d*). The displacement under the load P will be $x\,dy/c$, and at the reaction R will be $a\,dy/(a+b)$. According to the principle of virtual work, $-Ra\,dy/(a+b)$

$+ Px\ dy/c = 0$, from which $R = Px(a + b)/ac$. In this type of problem, the method has the advantage that only one reaction need be considered at a time and internal forces are not involved.

When an elastic body is deformed, the virtual work done by the internal forces is equal to the corresponding increment of the strain energy dU, in accordance with the principle of virtual work.

Assume a constrained elastic body acted upon by forces $P_1, P_2, \ldots$, for which the corresponding deformations are $e_1, e_2 \ldots$. Then, $\Sigma P_n\ de_n = dU$. The increment of the strain energy due to the increments of the deformations is given by

$$dU = \frac{\partial U}{\partial e_1} de_1 + \frac{\partial U}{\partial e_2} de_2 + \cdots$$

In solving a specific problem, a virtual displacement that is most convenient in simplifying the solution should be chosen. Suppose, for example, a virtual displacement is selected that affects only the deformation e_n corresponding to the load P_n, other deformations being unchanged. Then, the principle of virtual work requires that

$$P_n\ de_n = \frac{\partial U}{\partial e_n} de_n$$

This is equivalent to

$$\frac{\partial U}{\partial e_n} = P_n \tag{3-59}$$

which states that the partial derivative of the strain energy with respect to any specific deformation gives the corresponding force.

Suppose, for example, the stress in the vertical bar in Fig. 3-50 is to be determined. All bars are made of the same material and have the same cross section. If the vertical bar stretches an amount e under the load P, the inclined bars will each stretch an amount $e \cos \alpha$. The strain energy in the system is [from Eq. (3-10a)]

$$U = \frac{AE}{2L}(e^2 + 2e^2 \cos^3 \alpha)$$

and the partial derivative of this with respect to e must be equal to P; that is

$$P = \frac{AE}{2L}(2e + 4e \cos^3 \alpha)$$

$$= \frac{AEe}{L}(1 + 2 \cos^3 \alpha)$$

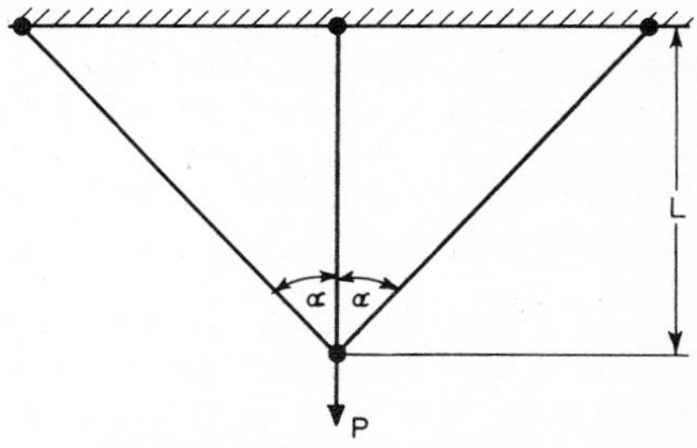

Fig. 3-50. Indeterminate truss.

Noting that the force in the vertical bar equals AEe/L, we find from the above equation that the required stress equals $P/(1 + 2 \cos^3 \alpha)$.

3-53. Castigliano's Theorem. It can also be shown that, if the strain energy is expressed as a function of statically independent forces, the partial derivative of the strain energy with respect to one of the forces gives the deformation corresponding to that force. (See Timoshenko and Young, "Theory of Structures," McGraw-Hill Book Company, New York.)

$$\frac{\partial U}{\partial P_n} = e_n \tag{3-60}$$

This is known as Castigliano's first theorem. (His second theorem is the principle of least work.)

3-54. Method of Least Work. If displacement of a structure is prevented, as at a support, the partial derivative of the strain energy with respect to that supporting force must be zero, according to Castigliano's first theorem. This establishes his second theorem:

The strain energy in a statically indeterminate structure is the minimum consistent with equilibrium.

As an example of the use of the method of least work, we shall solve again for the stress in the vertical bar in Fig. 3-50. Calling this stress X, we note that the stress in each of the inclined bars must be $(P - X)/2 \cos \alpha$. With the aid of Eq. (3-10*a*), we can express the strain energy in the system in terms of X as

$$U = \frac{X^2 L}{2AE} + \frac{(P - X)^2 L}{4AE \cos^3 \alpha}$$

Hence, the internal work in the system will be a minimum when

$$\frac{\partial U}{\partial X} = \frac{XL}{AE} - \frac{(P - X)L}{2AE \cos^3 \alpha} = 0$$

Solving for X gives the stress in the vertical bar as $P/(1 + 2 \cos^3 \alpha)$, as before (Art. 3-52).

3-55. Dummy Unit-load Method. In Art. 3-9, the strain energy for pure bending was given as $U = M^2L/2EI$ in Eq. (3-10*d*). To find the strain energy due to bending stress in a beam, we can apply this equation to a differential length dx of the beam and integrate over the entire span. Thus,

$$U = \int_0^L \frac{M^2\,dx}{2EI} \tag{3-61}$$

If M represents the bending moment due to a generalized force P, the partial derivative of the strain energy with respect to P is the deformation d corresponding to P. Differentiating Eq. (3–61) under the integral sign gives

$$d = \int_0^L \frac{M}{EI} \frac{\partial M}{\partial P}\,dx \tag{3-62}$$

The partial derivative in this equation is the rate of change of bending moment with the load P. It is equal to the bending moment m produced by a unit generalized load applied at the point where the deformation is to be measured and in the direction of the deformation. Hence, Eq. (3-62) can also be written

$$d = \int_0^L \frac{Mm}{EI}\,dx \tag{3-63}$$

To find the vertical deflection of a beam, we apply a dummy unit vertical load at the point where the deflection is to be measured and substitute the bending moments due to this load and the actual loading in Eq. (3-63). Similarly, to compute a rotation, we apply a dummy unit moment. (See also Art. 3-31.)

As a simple example, let us apply the dummy unit-load method to the determination of the deflection at the center of a simply supported, uniformly loaded beam of constant moment of inertia (Fig. 3-51*a*). As indicated in Fig. 3-51*b*, the bending moment at a distance x from one end is $(wL/2)x - (w/2)x^2$. If we apply a dummy unit load vertically at the center of the beam (Fig. 3-51*c*), where the vertical deflection is to be determined, the moment at x is $x/2$, as indicated in Fig. 3-51*d*. Substituting in Eq. (3-63) gives

$$d = 2 \int_0^{L/2} \left(\frac{wL}{2} x - \frac{w}{2} x^2\right) \frac{x}{2} \frac{dx}{EI} = \frac{5wL^4}{384EI}$$

As another example, let us apply the method to finding the end rotation at one end of a simply supported, prismatic beam produced by a moment applied at the other end. In other words, the problem is to find the end rotation at B, θ_B, in Fig. 3-52*a*, due to M_A. As indicated in Fig. 3-52*b*, the bending moment at a distance

x from B due to M_A is $M_A x/L$. If we applied a dummy unit moment at B (Fig. 3-52c), it would produce a moment at x of $(L - x)/L$ (Fig. 3-52d). Substituting in Eq. (3-63) gives

$$\theta_B = \int_0^L M_A \frac{x}{L} \frac{L - x}{L} \frac{dx}{EI} = \frac{M_A L}{6EI}$$

To determine the deflection of a beam due to shear Castigliano's theorems can be applied to the strain energy in shear

$$V = \iint \frac{v^2}{2G} \, dA \, dx$$

where v = shearing unit stress
G = modulus of rigidity
A = cross-sectional area

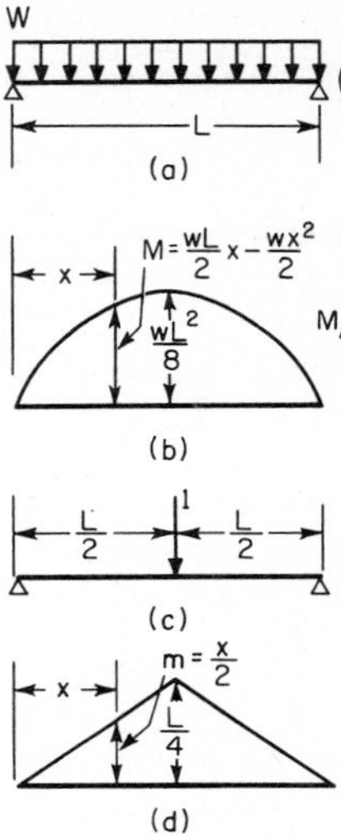

Fig. 3-51

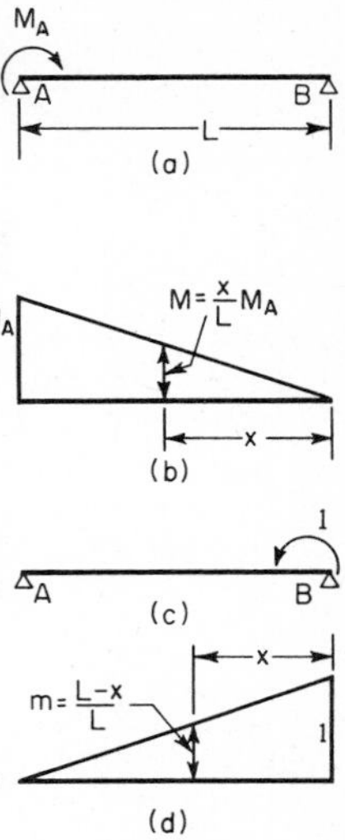

Fig. 3-52

Fig. 3-51. Dummy unit-load method applied to a uniformly loaded beam (a) to find midspan deflection; (b) moment diagram for the uniform load; (c) unit load at midspan; (d) moment diagram for unit load.

Fig. 3-52. End rotation due to end moment (a) by dummy unit-load method; (b) moment diagram for end moment; (c) unit moment applied at beam end; (d) moment diagram for unit moment.

3-56. Truss Deflections. The dummy unit-load method may also be adapted for the determination of the deformation of trusses. As indicated by Eq. (3-10a), the strain energy in a truss is given by

$$U = \sum \frac{S^2 L}{2AE} \tag{3-64}$$

which represents the sum of the strain energy for all the members of the truss. S is the stress in each member due to the loads. Applying Castigliano's first theorem and differentiating inside the summation sign yield the deformation:

$$d = \sum \frac{SL}{AE} \frac{\partial S}{\partial P} \tag{3-65}$$

The partial derivative in this equation is the rate of change of axial stress with the load P. It is equal to the axial stress in each bar of the truss u produced by a unit load applied at the point where the deformation is to be measured and in the direction of the deformation. Consequently, Eq. (3-65) can also be written

$$d = \sum \frac{SuL}{AE} \tag{3-66}$$

To find the deflection of a truss, apply a dummy unit vertical load at the panel point where the deflection is to be measured and substitute in Eq. (3-66) the stresses in each member of the truss due to this load and the actual loading. Similarly, to find the rotation of any joint, apply a dummy unit moment at the joint, compute the stresses in each member of the truss, and substitute in Eq. (3-66). When it is necessary to determine the relative movement of two panel points, apply dummy unit loads in opposite directions at those points.

It is worth noting that members that are not stressed by the actual loads or the dummy loads do not enter into the calculation of a deformation.

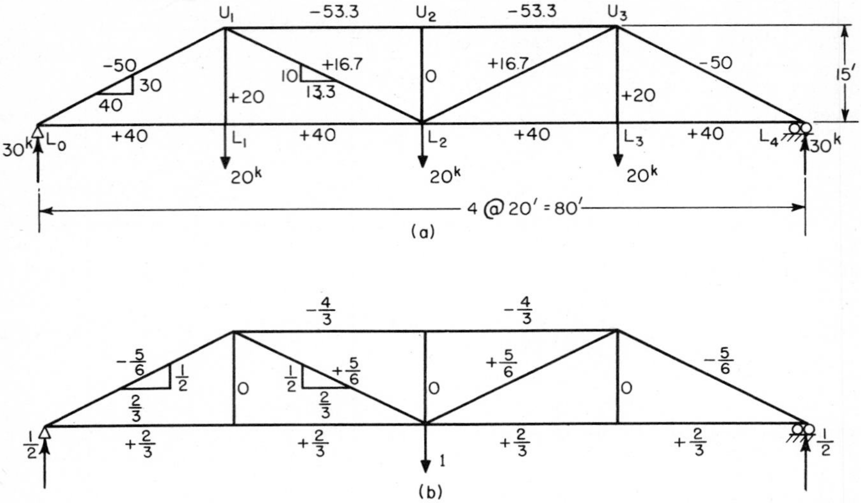

Fig. 3-53. Dummy unit-load method applied to a loaded truss (*a*) to find midspan deflection; (*b*) unit load applied at midspan.

As an example of the application of Eq. (3-66), let us compute the deflection of the truss in Fig. 3-53. The stresses due to the 20-kip load at each panel point are shown in Fig. 3-53*a*, and the ratio of length of members in inches to their cross-sectional area in square inches is given in Table 3-3. We apply a dummy unit vertical load at L_2, where the deflection is required. Stresses u due to this load are shown in Fig. 3-53*b* and Table 3-3.

Table 3-3. Deflection of a Truss

Member	L/A	S	u	SuL/A
L_0L_2	160	+40	+2/3	4,267
L_0U_1	75	−50	−5/6	3,125
U_1U_2	60	−53.3	−4/3	4,267
U_1L_2	150	+16.7	+5/6	2,083
				13,742

$$d = \frac{SuL}{AE} = \frac{2 \times 13{,}742{,}000}{30{,}000{,}000} = 0.916 \text{ in.}$$

The computations for the deflection are given in Table 3-3. Members not stressed by the 20-kip loads or the dummy unit load are not included. Taking advantage of the symmetry of the truss, we tabulate the values for only half the truss and double the sum. Also, to reduce the amount of caculation, we do not include the modulus of elasticity E, which is equal to 30,000,000, until the very last step, since it is the same for all members.

3-57. Reciprocal Theorem. Consider a structure loaded by a group of independent forces A, and suppose that a second group of forces B are added. The work done by the forces A acting over the displacements due to B will be W_{AB}.

Now, suppose the forces B had been on the structure first, and then load A had been applied. The work done by the forces B acting over the displacements due to A will be W_{BA}.

The reciprocal theorem states that $W_{AB} = W_{BA}$.

Some very useful conclusions can be drawn from this equation. For example, there is the reciprocal deflection relationship: **The deflection at a point *A* due to a load at *B* is equal to the deflection at *B* due to the same load applied at *A*. Also, the rotation at *A* due to a load (or moment) at *B* is equal to the rotation at *B* due to the same load (or moment) applied at *A*.**

Another consequence is that deflection curves may also be influence lines to some scale for reactions, shears, moments, or deflections (**Muller-Breslau principle**). For example, suppose the influence line for a reaction is to be found; that is, we wish to plot the reaction R as a unit load moves over the structure, which may be statically indeterminate. For the loading condition A, we analyze the structure with a unit load on it at a distance x from some reference point. For loading condition B, we apply a dummy unit vertical load upward at the place where the reaction is to be determined, deflecting the structure off the support. At a distance x from the reference point, the displacement is d_{xR} and over the support the displacement is d_{RR}. Hence $W_{AB} = -1\ (d_{xR}) + Rd_{RR}$. On the other hand, W_{BA} is zero, since loading condition A provides no displacement for the dummy unit load at the support in condition B. Consequently, from the reciprocal theorem,

$$R = \frac{d_{xR}}{d_{RR}}$$

Since d_{RR} is a constant, R is proportional to d_{xR}. Hence the influence line for a reaction can be obtained from the deflection curve resulting from a displacement of the support (Fig. 3-54). The magnitude of the reaction is obtained by dividing each ordinate of the deflection curve by the displacement of the support (see also Art. 3-23).

Similarly, the influence line for shear can be obtained from the deflection curve produced by cutting the structure and shifting the cut ends vertically at the point for which the influence line is desired (Fig. 3-55).

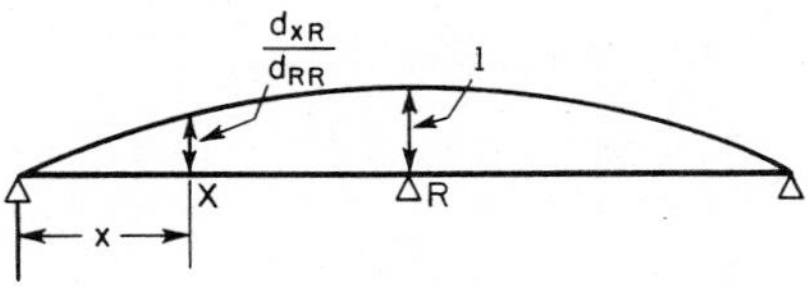

Fig. 3-54. Reaction-influence line for a continuous beam.

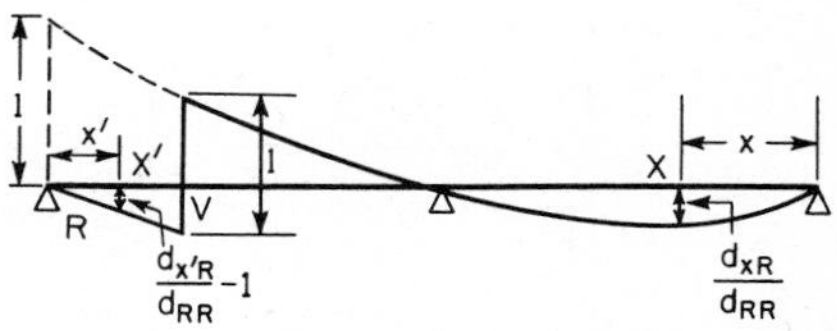

Fig. 3-55. Shear-influence line for a continuous beam.

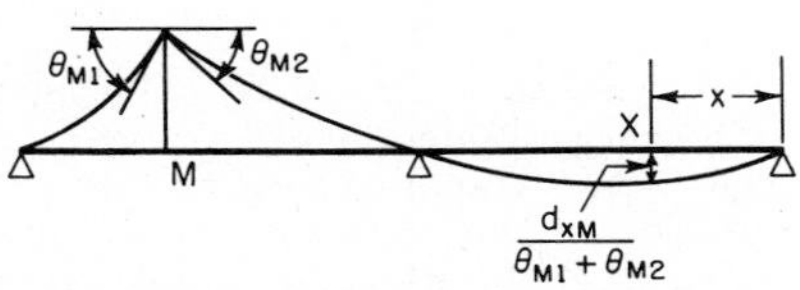

Fig. 3-56. Moment-influence line for a continuous beam.

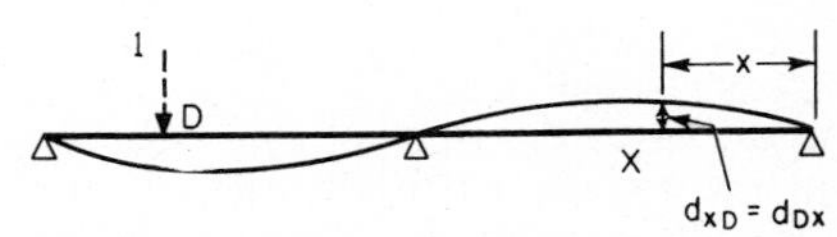

Fig. 3-57. Deflection-influence line for a continuous beam.

The influence line for bending moment can be obtained from the deflection curve produced by cutting the structure and rotating the cut ends at the point for which the influence line is desired (Fig. 3-56).

And finally, it may be noted that the deflection curve for a load of unity at some point of a structure is also the influence line for deflection at that point (Fig. 3-57).

CONTINUOUS BEAMS AND FRAMES

Fixed-end beams, continuous beams, continuous trusses, and rigid frames are statically indeterminate. The equations of equilibrium are not sufficient for the determination of all the unknown forces and moments. Additional equations based on a knowledge of the deformations of the member are required.

Hence, while the bending moments in a simply supported beam are determined only by the loads and the span, bending moments in a statically indeterminate member are also a function of the geometry, cross-sectional dimensions, and the modulus of elasticity.

3-58. General Method of Analysis for Continuous Members. Continuous beams and frames consist of members that can be treated as simple beams, the ends of which are prevented by moments from rotating freely. Member *LR* in the continuous beam *ALRBC* in Fig. 3-58*a*, for example, can be isolated, as shown in Fig. 3-58*b*, and the elastic restraints at the ends replaced by couples M_L and M_R. In this way, *LR* is converted into a simply supported beam acted upon by transverse loads and end moments.

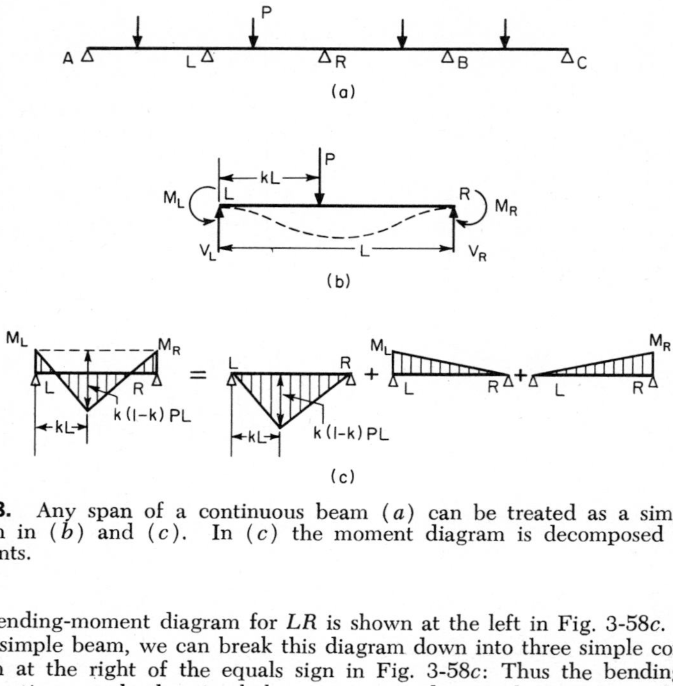

Fig. 3-58. Any span of a continuous beam (*a*) can be treated as a simple beam, as shown in (*b*) and (*c*). In (*c*) the moment diagram is decomposed into basic components.

The bending-moment diagram for *LR* is shown at the left in Fig. 3-58*c*. Treating *LR* as a simple beam, we can break this diagram down into three simple components, as shown at the right of the equals sign in Fig. 3-58*c*: Thus the bending moment at any section equals the simple-beam moment due to the transverse loads, plus the simple-beam moment due to the end moment at *L*, plus the simple-beam moment due to the end moment at *R*.

Obviously, once M_L and M_R have been determined, the shears may be computed by taking moments about any section. Similarly, if the reactions or shear are known, the bending moments can be calculated.

A general method for determining the elastic forces and moments exerted by redundant supports and members is as follows: Remove as many redundant supports or members as necessary to make the structure statically determinate. Compute for the actual loads the deflections or rotations of the statically determinate structure in the direction of the forces and couples exerted by the removed supports and members. Then, in terms of these forces and couples, compute the corresponding deflections or rotations the forces and couples produce in the statically determinate structure (see Arts. 3-55 and 3-31). Finally, for each redundant support or member write equations that give the known rotations or deflections of the original structure in terms of the deformations of the statically determinate structure.

For example, one method of finding the reactions of the continuous beam *ALRBC* in Fig. 3-58*a* is to remove supports *L*, *R*, and *B* temporarily. The beam is now simply supported between *A* and *C*, and the reactions and moments can be computed from the laws of equilibrium. Beam *AC* deflects at points *L*, *R*, and *B*, whereas we know that continuous beam *ALRBC* is prevented from deflecting at these points by the supports there. This information enables us to write three equations in terms of the three unknown reactions that were eliminated to make the beam statically determinate.

So we first compute the deflections d_1, d_2, d_3 of simple beam *AC* at *L*, *R*, and *B* due to the loads. Then, in terms of the unknown reactions R_1 at *L*, R_2 at *R*, R_3 at *B*, we compute the deflection at *L* when *AC* is loaded with only the three reactions. This deflection is the sum $-y_{11}R_1 - y_{12}R_2 - y_{13}R_3$, where y_{ij} represents the deflection at support i due to a unit load at support j. The first equation is obtained by noting that the sum of this deflection and the deflection at *L* due to the loads must be zero. Similarly, equations can be written making the total deflection at *R* and *B*, successively, equal to zero. In this way, three equations with three unknowns are obtained, which can be solved simultaneously to yield the required reactions.

$$\begin{aligned} d_1 &= y_{11}R_1 + y_{12}R_2 + y_{13}R_3 \\ d_2 &= y_{21}R_1 + y_{22}R_2 + y_{23}R_3 \\ d_3 &= y_{31}R_1 + y_{23}R_2 + y_{33}R_3 \end{aligned} \tag{3-67}$$

See also Arts. 3-105 to 3-108.

3-59. Sign Convention. For moment distribution, the following sign convention is most convenient: A moment acting at an end of a member or at a joint is positive if it tends to rotate the joint clockwise, negative if it tends to rotate the joint counterclockwise. Hence, in Fig. 3-58, M_R is positive and M_L is negative.

Similarly, the angular rotation at the end of a member is positive if in a clockwise direction, negative if counterclockwise. Thus, a positive end moment produces a positive end rotation in a simple beam.

For ease in visualizing the shape of the elastic curve under the action of loads and end moments, bending-moment diagrams will be plotted on the tension side of each member. Hence, if an end moment is represented by a curved arrow, the arrow will point in the direction in which the moment is to be plotted.

3-60. Fixed-end Moments. A beam so restrained at its ends that no rotation is produced there by the loads is called a fixed-end beam, and the end moments are called fixed-end moments. Fixed-end moments may be expressed as the product of a coefficient and *WL*, where *W* is the total load on the span *L*. The coefficient is independent of the properties of other members of the structure. Thus, any member can be isolated from the rest of the structure and its fixed-end moments computed.

Fixed-end moments may be determined conveniently by the moment-area method or the conjugate-beam method (Art. 3-31).

Fixed-end moments for several common types of loading on beams of constant moment of inertia (prismatic beams) are given in Figs. 3-59 to 3-62. Also, the curves in Fig. 3-64 enable fixed-end moments to be computed easily for any type of loading on a prismatic beam. Before the curves can be entered, however, certain characteristics of the loading must be calculated. These include $\bar{x}L$, the location of the center of gravity of the loading with respect to one of the loads; $G^2 = \Sigma b_n^2 P_n/W$, where b_nL is the distance from each load P_n to the center of gravity of the loading (taken positive to the right); and $S^3 = \Sigma b_n^3 P_n/W$. (See case 9, Fig. 3-63.) These values are given in Fig. 3-63 for some common types of loading.

The curves in Fig. 3-64 are entered with the location *a* of the center of gravity with respect to the left end of the span. At the intersection with the proper

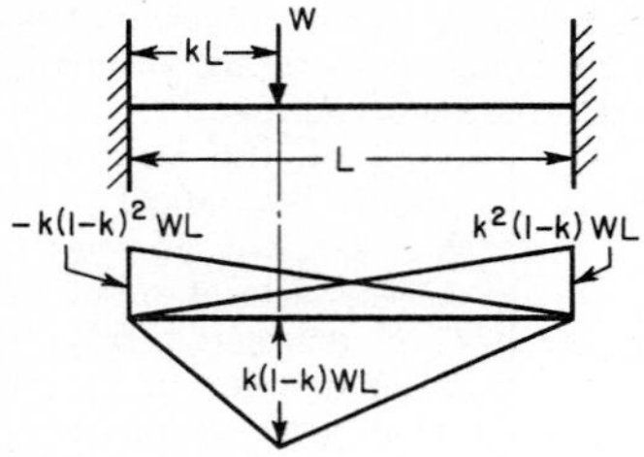

Fig. 3-59. Concentrated load at any point of a fixed-end beam.

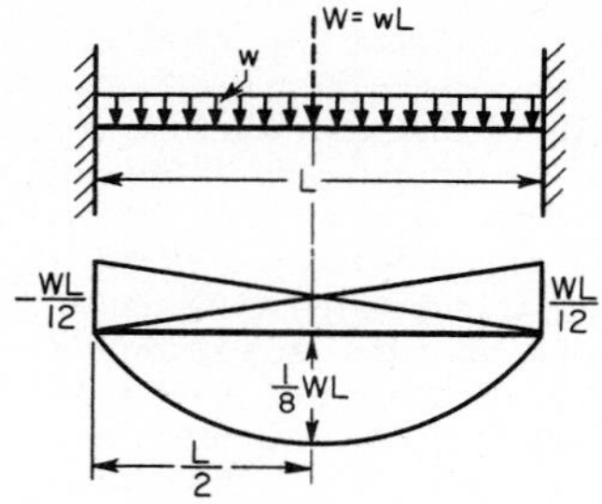

Fig. 3-60. Uniform load on a fixed-end beam.

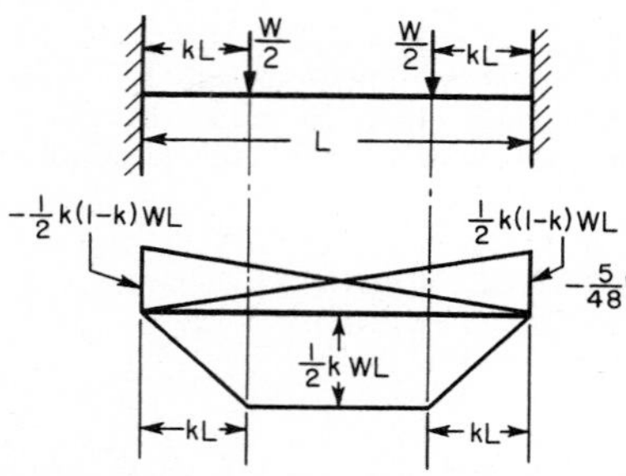

Fig. 3-61. Two equal concentrated loads on a fixed-end beam.

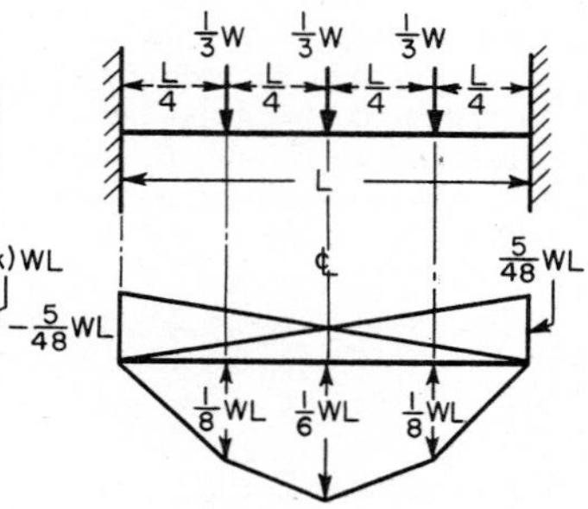

Fig. 3-62. Several equal concentrated loads on a fixed-end beam.

G curve, proceed horizontally to the left to the intersection with the proper S line, then vertically to the horizontal scale indicating the coefficient m by which to multiply WL to obtain the fixed-end moment. The curves solve the equations:

$$m_L = \frac{M_L{}^F}{WL} = G^2[1 - 3(1 - a)] + a(1 - a)^2 + S^3 \qquad (3\text{-}68a)$$

$$m_R = \frac{M_R{}^F}{WL} = G^2(1 - 3a) + a^2(1 - a) - S^3 \qquad (3\text{-}68b)$$

where $M_L{}^F$ is the fixed-end moment at the left support and $M_R{}^F$ at the right support.

As an example of the use of the curves, find the fixed-end moments in a prismatic beam of 20-ft span carrying a triangular loading of 100 kips, similar to the loading shown in case 4, Fig. 3-63, distributed over the entire span, with the maximum intensity at the right support.

Case 4 gives the characteristics of the loading: $y = 1$; the center of gravity is $0.33L$ from the right support, so $a = 0.667$; $G^2 = 1/18 = 0.056$; and $S^3 = -1/135 = -0.007$. To find $M_R{}^F$, we enter Fig. 3-64 with $a = 0.67$ on the upper scale at the bottom of the diagram, and proceed vertically to the estimated location of the intersection of the coordinate with the $G^2 = 0.06$ curve. Then, we move horizontally to the intersection with the line for $S^3 = -0.007$, as indicated by the dash line in Fig. 3-64. Referring to the scale at the top of the diagram, we find the coefficient m_R to be 0.10. Similarly, with $a = 0.67$ on the lowest scale, we find the coefficient m_L to be 0.07. Hence, the fixed-end moment at the right support is $0.10 \times 100 \times 20 = 200$ ft-kips, and at the left support $-0.07 \times 100 \times 20 = -140$ ft-kips.

3-61. Fixed-end Stiffness. To correct a fixed-end moment to obtain the end moment for the actual conditions of end restraint in a continuous structure, the

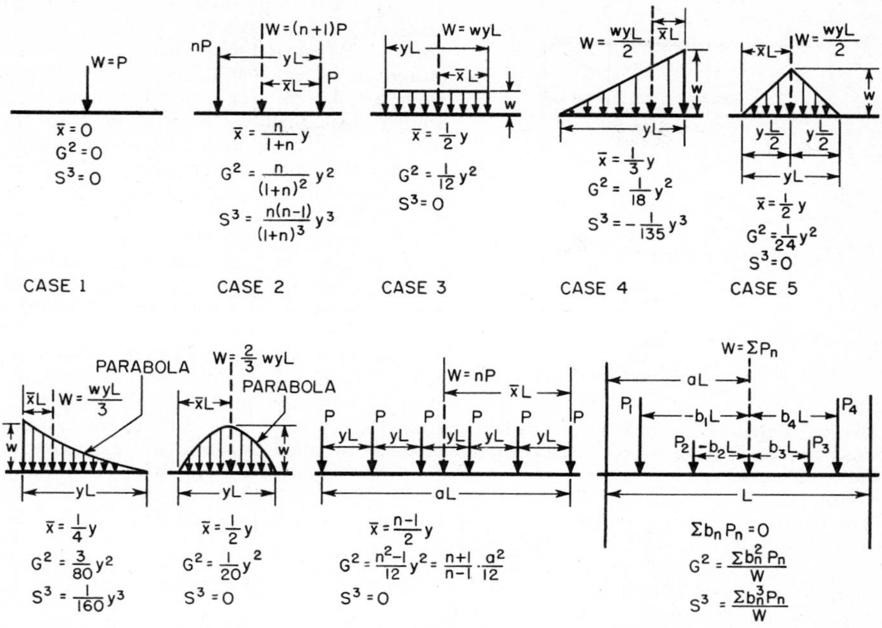

Fig. 3-63. Characteristics of loadings.

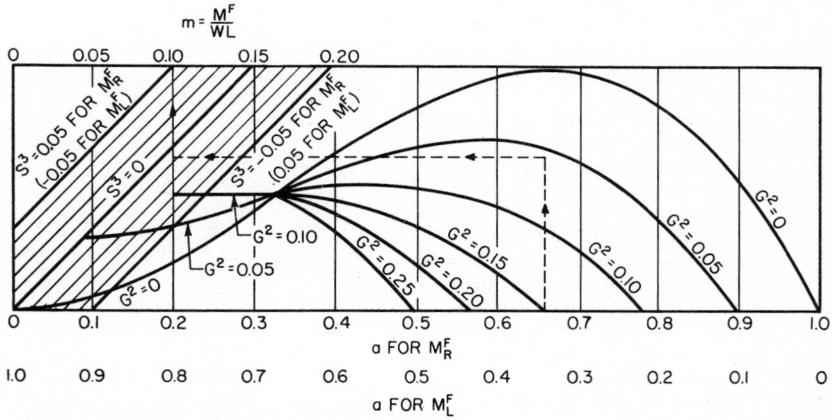

Fig. 3-64. Chart for fixed-end moments due to any type of loading.

end of the member must be permitted to rotate. The amount it will rotate depends on its stiffness, or resistance to rotation.

The fixed-end stiffness of a beam is defined as the moment required to produce a rotation of unity at the end where it is applied, while the other end is fixed against rotation. It is represented by K_R^F in Fig. 3-65.

For prismatic beams, the fixed-end stiffnesses for both ends are equal to $4EI/L$, where E is the modulus of elasticity, I the moment of inerta of the cross section

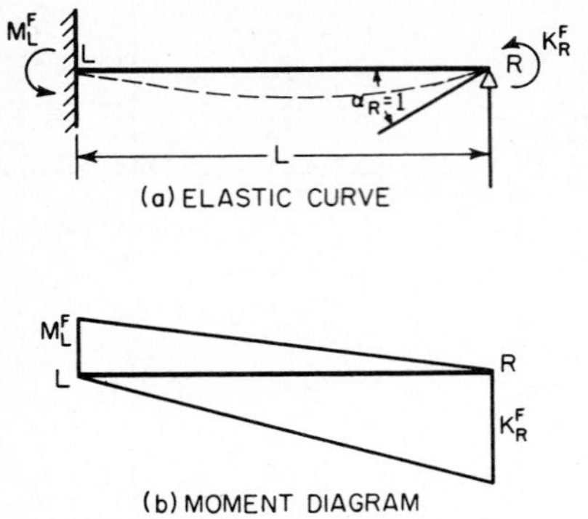

Fig. 3-65. Fixed-end stiffness.

about the centroidal axis, and L the span (generally taken center to center of supports). When deformations are not required to be calculated, only the relative value of K^F for each member need be known; hence, only the ratio of I to L has to be computed. (For prismatic beams with a hinge at one end, the actual stiffness is $3EI/L$, or three-fourths the fixed-end stiffness.)

For beams of variable moment of inertia, the fixed-end stiffness may be calculated by methods presented later in this section or obtained from prepared tables. See Art. 3-62.

3-62. Fixed-end Carry-over Factor. When a moment is applied at one end of a beam, a resisting moment is induced at the far end if that end is restrained against rotation (Fig. 3-65). The ratio of the resisting moment at a fixed end to the applied moment is called the fixed-end carry-over factor C^F.

For prismatic beams, the fixed-end carry-over factor toward either end is 0.5. It should be noted that the applied moment and the resisting moment have the same sign (Fig. 3-65); i.e., if the applied moment acts in a clockwise direction, the carry-over moment also acts clockwise.

For beams of variable moment of inertia, the fixed-end carry-over factor may be calculated by methods presented later in this section or obtained from tables such as those in the "Handbook of Frame Constants," Portland Cement Association, Skokie, Ill., and J. M. Gere, "Moment Distribution Factors for Beams of Tapered I-section," American Institute of Steel Construction, New York.

3-63. Moment Distribution by Converging Approximations. The frame in Fig. 3-66 consists of four prismatic members rigidly connected together at O and fixed at ends A, B, C, and D. If an external moment U is applied at O, the sum of the end moments in each member at O must be equal to U. Furthermore, all members must rotate at O through the same angle θ, since they are assumed to be rigidly connected there. Hence, by the definition of fixed-end stiffness, the proportion of U induced in the end of each member at O is equal to the ratio of the stiffness of that member to the sum of the stiffnesses of all the members at the joint.

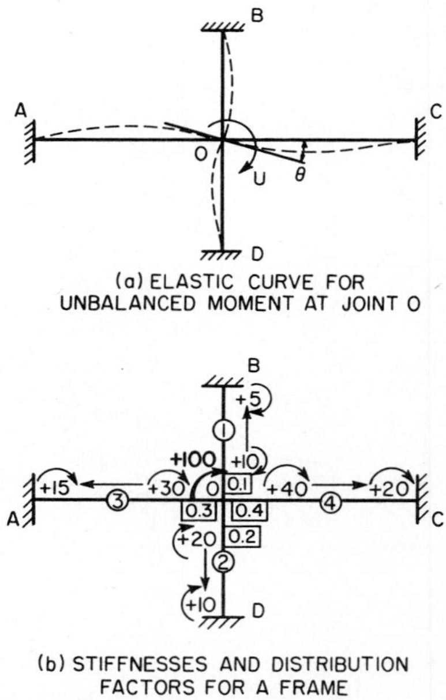

Fig. 3-66. Moments in a simple frame.

Suppose a moment of 100 ft-kips is applied at 0, as indicated in Fig. 3-66*b*. The relative stiffness (or I/L) is assumed as shown in the circle on each member. The distribution factors for the moment at O are computed from the stiffnesses and shown in the boxes. For example, the distribution factor for OA equals its stiffness divided by the sum of the stiffnesses of all the members at the joint: $3/(3+2+4+1) = 0.3$. Hence, the moment induced in OA at O is $0.3 \times 100 = 30$ ft-kips. Similarly, OB gets 10 ft-kips, OC 40 ft-kips, and OD 20 ft-kips.

Because the far ends of these members are fixed, one-half of these moments are carried over to them. Thus $M_{AO} = 0.5 \times 30 = 15$; $M_{BO} = 0.5 \times 10 = 5$; $M_{CO} = 0.5 \times 40 = 20$; and $M_{DO} = 0.5 \times 20 = 10$.

Most structures consist of frames similar to the one in Fig. 3-66, or even simpler, joined together. Though the ends of the members are not fixed, the technique employed for the frame in Fig. 3-66*b* can be applied to find end moments in such continuous structures.

Before the general method is presented, one short cut is worth noting. Advantage can be taken when a member has a hinged end to reduce the work of distributing moments. This is done by using the true stiffness of a member instead of the fixed-end stiffness. (For a prismatic beam, the stiffness of a member with one end hinged is three-fourths the fixed-end stiffness; for a beam with variable I, it is equal to the fixed-end stiffness times $1 - C_L^F C_R^F$.) Naturally, the carry-over factor toward the hinge is zero.

When a joint is neither fixed nor pinned but is restrained by elastic members connected there, moments can be distributed by a series of converging approximations. All joints are locked against rotation. As a result, the loads will create fixed-end moments at the ends of every loaded member. At each joint, a moment equal to the algebraic sum of the fixed-end moments there is required to hold

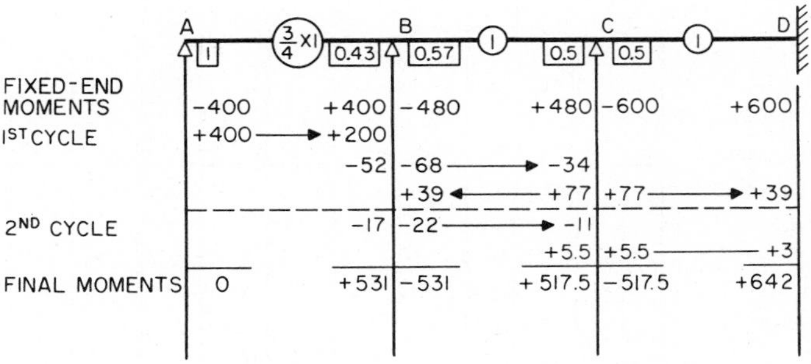

Fig. 3-67. Moments distribution by converging approximations.

it fixed. Then, one joint is unlocked at a time by applying a moment equal but opposite in sign to the moment that was needed to prevent rotation. The unlocking moment must be distributed to the members at the joint in proportion to their fixed-end stiffnesses and the distributed moments carried over to the far ends.

After all joints have been released at least once, it generally will be necessary to repeat the process—sometimes several times—before the corrections to the fixed-end moments become negligible. To reduce the number of cycles, the unlocking of joints should start with those having the greatest unbalanced moments.

Suppose the end moments are to be found for the continuous beam *ABCD* in Fig. 3-67. The I/L values for all spans are equal; therefore, the relative fixed-end stiffness for all members is unity. However, since A is a hinged end, the computation can be shortened by using the actual relative stiffness, which is ¾. Relative stiffnesses for all members are shown in the circle on each member. The distribution factors are shown in boxes at each joint.

The computation starts with determination of fixed-end moments for each member (Art. 3-60). These are assumed to have been found and are given on the first line in Fig. 3-67. The greatest unbalanced moment is found from inspection to be at hinged end *A*; so this joint is unlocked first. Since there are no other members at the joint, the full unlocking moment of +400 is distributed to *AB* at *A* and one-half of this is carried over to *B*. The unbalance at *B* now is +400 − 480 plus the carry-over of +200 from *A*, or a total of +120. Hence, a moment of −120 must be applied and distributed to the members at *B* by multiplying by the distribution factors in the corresponding boxes.

The net moment at B could be found now by adding the entries for each member at the joint. However, it generally is more convenient to delay the summation until the last cycle of distribution has been completed.

The moment distributed to BA need not be carried over to A, because the carry-over factor toward the hinged end is zero. However, half the moment distributed to BC is carried over to C.

Similarly, joint C is unlocked and half the distributed moments carried over to B and D, respectively. Joint D should not be unlocked, since it actually is a fixed-end. Thus, the first cycle of moment distribution has been completed.

The second cycle is carried out in the same manner. Joint B is released, and the distributed moment in BC is carried over to C. Finally, C is unlocked, to complete the cycle. Adding the entries for the end of each member yields the final moments.

3-64. Continuous Frames. In practice, the problem is to find the maximum end moments and interior moments produced by the worst combination of loading. For maximum moment at the end of a beam, live load should be placed on that beam and on the beam adjoining the end for which the moment is to be computed. Spans adjoining these two should be assumed to be carrying only dead load.

For maximum midspan moments, the beam under consideration should be fully loaded, but adjoining spans should be assumed to be carrying only dead load.

	A			B			C			D		E
1. RELATIVE STIFFNESS		$K^F=1$			$K^F=1$			$K^F=1$			$K^F=1$	
2. DISTRIBUTION FACTOR	1/3		1/4	1/4		1/4	1/4		1/4	1/4		1/3
3. F.E.M. DEAD LOAD	—		+91	−37		+37	−70		+70	−59		—
4. F.E.M. TOTAL LOAD	−172	+99	+172	−78	+73	+78	−147	+85	+147	−126	+63	+126
5. CARRY-OVER	−17 →	+11	↙ +29	−1 →	+1	↙ −2	−11 →	+7	↙ +14	−21 →	+13	↙ +7
6. ADDITION	−189	+18	+201	−79	−1	+76	−158	+9	+161	−147	+5	+133
7. DISTRIBUTION	+63		−30	−30		+21	+21		−4	−4		−44
8. MAX. MOMENTS	−126	+128	+171	−109	+73	+97	−137	+101	+157	−151	+81	+89

Fig. 3-68. Moments in a continuous frame.

The work involved in distributing moments due to dead and live loads in continuous frames in buildings can be greatly simplified by isolating each floor. The tops of the upper columns and the bottoms of the lower columns can be assumed fixed. Furthermore, the computations can be condensed considerably by following the procedure recommended in "Continuity in Concrete Building Frames," Portland Cement Association, Skokie, Ill., and indicated in Fig. 3-68.

Figure 3-68 presents the complete calculation for maximum end and midspan moments in four floor beams AB, BC, CD, and DE. Building columns are assumed to be fixed at the story above and below. None of the beam or column sections is known to begin with; so as a start, all members will be assumed to have a fixed-end stiffness of unity, as indicated on the first line of the calculation.

On the second line, the distribution factors for each end of the beams are shown, calculated from the stiffnesses (Arts. 3-61 and 3-63). Column stiffnesses are not shown, because column moments will not be computed until moment distribution to the beams has been completed. Then the sum of the column moments at each joint may be easily computed, since they are the moments needed to make the sum of the end moments at the joint equal to zero. The sum of the column moments at each joint can then be distributed to each column there in proportion to its stiffness. In this problem, each column will get one-half the sum of the column moments.

Fixed-end moments at each beam end for dead load are shown on the third line, just above the heavy line, and fixed-end moments for live plus dead load on the fourth line. Corresponding midspan moments for the fixed-end condition also are shown on the fourth line, and like the end moments will be corrected to yield actual midspan moments.

For maximum end moment at *A*, beam *AB* must be fully loaded, but *BC* should carry dead load only. Holding *A* fixed, we first unlock joint *B*, which has a total-load fixed-end moment of +172 in *BA* and a dead-load fixed-end moment of —37 in *BC*. The releasing moment required, therefore, is $-(172 - 37)$, or —135. When *B* is released, a moment of $-135 \times 1/4$ is distributed to *BA*. One-half of this is carried over to *A*, or $-135 \times 1/4 \times 1/2 = -17$. This value is entered as the carry-over at *A* on the fifth line in Fig. 3-68. Joint *B* is then relocked.

At *A*, for which we are computing the maximum moment, we have a total-load fixed-end moment of —172 and a carry-over of —17, making the total —189, shown on the sixth line. To release *A*, a moment of +189 must be applied to the joint. Of this, $189 \times 1/3$, or 63, is distributed to *AB*, as indicated on the seventh line of the calculation. Finally, the maximum moment at *A* is found by adding lines 6 and 7: $-189 + 63 = -126$.

For maximum moment at *B*, both *AB* and *BC* must be fully loaded, but *CD* should carry only dead load. We begin the determination of the moment at *B* by first releasing joints *A* and *C*, for which the corresponding carry-over moments at *BA* and *BC* are +29 and $-(+78 - 70) \times 1/4 \times 1/2 = -1$, shown on the fifth line in Fig. 3-68. These bring the total fixed-end moments in *BA* and *BC* to +201 and —79, respectively. The releasing moment required is $-(201 - 79) = -122$. Multiplying this by the distribution factors for *BA* and *BC* when joint *B* is released, we find the distributed moments, —30, entered on line 7. The maximum end moments finally are obtained by adding lines 6 and 7: +171 at *BA* and —109 at *BC*. Maximum moments at *C*, *D*, and *E* are computed and entered in Fig. 3-68 in a similar manner. This procedure is equivalent to two cycles of moment distribution.

The computation of maximum midspan moments in Fig. 3-68 is based on the assumption that in each beam the midspan moment is the sum of the simple-beam midspan moment and one-half the algebraic difference of the final end moments (the span carries full load but adjacent spans only dead load). Instead of starting with the simple-beam moment, however, we begin with the midspan moment for the fixed-end condition and apply two corrections. In each span, these corrections are equal to the carry-over moments entered on line 5 for the two ends of the beam multiplied by a factor.

For beams with variable moment of inertia, the factor is $\pm 1/2[(1/C^F) + D - 1]$ where C^F is the fixed-end carry-over factor toward the end for which the correction factor is being computed and *D* the distribution factor for that end. The plus sign is used for correcting the carry-over at the right end of a beam, and the minus sign for the carry-over at the left end. For prismatic beams, the correction factor becomes $\pm 1/2(1 + D)$.

For example, to find the corrections to the midspan moment in *AB*, we first multiply the carry-over at *A* on line 5, —17, by $-1/2(1 + 1/3)$. The correction, +11, is also entered on the fifth line. Then, we multiply the carry-over at *B*, +29, by $+1/2(1 + 1/4)$ and enter the correction, +18, on line 6. The final midspan moment is the sum of lines 4, 5, and 6: $+99 + 11 + 18 = +128$. Other midspan moments in Fig. 3-68 are obtained in a similar manner.

3-65. Moment-influence Factors. In certain types of framing, particularly those in which different types of loading conditions must be investigated, it may be convenient to find maximum end moments from a table of moment-influence factors. This table is made up by listing for the end of each member in the structure the moment induced in that end when a moment (for convenience, +1,000) is applied to every joint successively. Once this table has been prepared no additional moment distribution is necessary for computing the end moments due to any loading condition.

For a specific loading pattern, the moment at any beam end M_{AB} may be obtained from the moment-influence table by multiplying the entries under *AB* for the various joints by the actual unbalanced moments at those joints divided by 1,000, and summing (see also Art. 3-67 and Table 3-4).

3-66. Deflection of Supports. For some framing, it is convenient to know the effect of a deflection of a support normal to the original position of a continuous beam. But the moment-distribution method is based on the assumption that such movement of a support does not occur. However, the method can be modified to evaluate the end moments resulting from a support movement.

The procedure is to distribute moments, as usual, assuming no deflection at the supports. This implies that additional external forces are exerted at the supports. These forces can be computed. Then, equal and opposite forces are applied to the structure to produce the final configuration, and the moments that they produce are distributed as usual. These moments added to those obtained with undeflected supports yield the final moments.

To apply this procedure, it is first necessary to know the fixed-end moments for a beam with supports at different levels. In Fig. 3-69*a*, the right end of a beam with span L is at a height d above the left end. To find the fixed-end moments, we first consider the left end hinged, as in Fig. 3-69*b*. Noting that a line connecting the two supports makes an angle approximately equal to d/L (its tangent) with the original position of the beam, we apply a moment at the hinged end to produce an end rotation there equal to d/L. By the definition of stiffness, this moment equals $K_L^F d/L$. The carry-over to the right end is C_R^F times this.

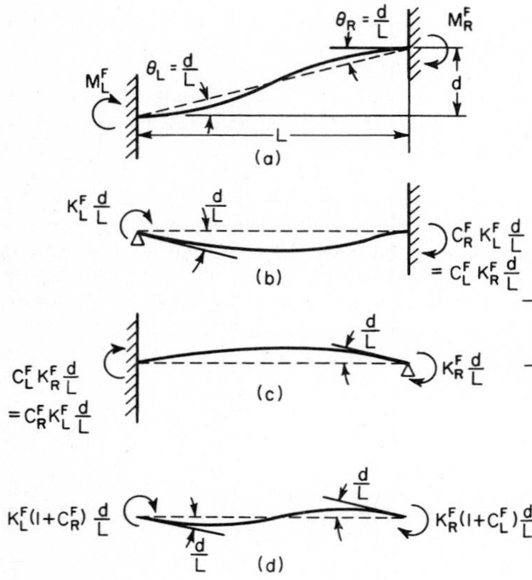

Fig. 3-69. Moments due to deflection of a fixed-end beam.

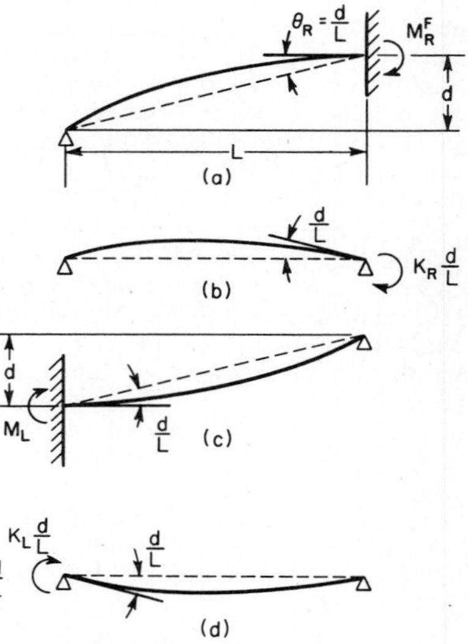

Fig. 3-70. Moments due to deflection of a beam with one end fixed, one end hinged.

By the law of reciprocal deflections (Art. 3-57), the fixed-end moment at the right end of a beam due to a rotation of the other end is equal to the fixed-end moment at the left end of the beam due to the same rotation at the right end. Therefore, the carry-over moment for the right end in Fig. 3-69*b* is also equal to $C_L^F K_R^F d/L$ (see Fig. 3-69c). By adding the end moments for the loading conditions in Fig. 3-69*b* and *c*, we obtain the end moments in Fig. 3-69*d*, which is equivalent to the deflected beam in Fig. 3-69*a*:

$$M_L^F = K_L^F(1 + C_R^F)\frac{d}{L} \tag{3-69}$$

$$M_R^F = K_R^F(1 + C_L^F)\frac{d}{L} \tag{3-70}$$

In a similar manner, the fixed-end moment can be found for a beam with one end hinged and the supports at different levels (Fig. 3-70):

$$M^F = K\frac{d}{L} \tag{3-71}$$

where K is the actual stiffness for the end of the beam that is fixed. For prismatic beams, this value is three-fourths that for fixed-end stiffness; for beams of variable moments of inertia, it is equal to the fixed-end stiffness times $(1 - C_L{}^F C_R{}^F)$.

3-67. Procedure for Sidesway. The problem of computing sidesway moments in rigid frames is conveniently solved by the following method:

1. Apply forces to the structure to prevent sidesway while the fixed-end moments due to loads are distributed.
2. Compute the moments due to these forces.
3. Combine the moments obtained in steps 1 and 2 to eliminate the effect of the forces that prevented sidesway.

Suppose the rigid frame in Fig. 3-71 is subjected to a 2,000-lb horizontal load acting to the right at the level of beam BC. The first step is to compute the moment-influence factors (Table 3-4) by applying moments of +1,000 at joints B and C, assuming sidesway prevented. Since there are no intermediate loads on the beams and columns, the only fixed-end moments that need be considered are those in the columns due to lateral deflection of the frame caused by the horizontal load. This deflection, however, is not known initially. So we assume an arbitrary deflection, which produces a fixed-end moment of $-1{,}000M$ at the top of column CD. M is an unknown constant to be determined from the fact that the sum of the shears in the deflected columns must be equal to the 2,000-lb load. The same deflection also produces a moment of $-1{,}000M$ at the bottom of CD [see Eqs. (3-69) and (3-70)].

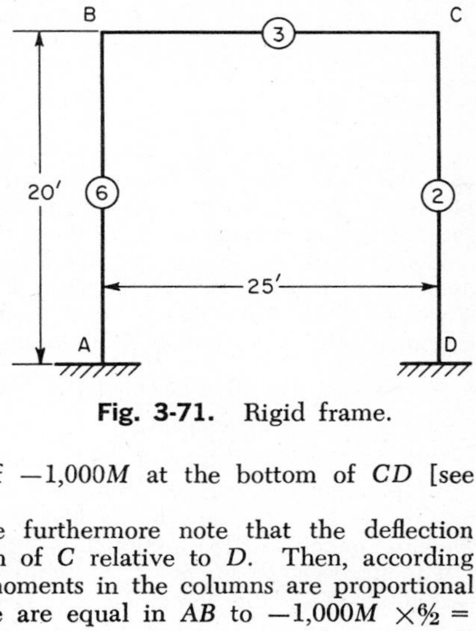

Fig. 3-71. Rigid frame.

From the geometry of the structure, we furthermore note that the deflection of B relative to A is equal to the deflection of C relative to D. Then, according to Eqs. (3-69) and (3-70), the fixed-end moments in the columns are proportional to the stiffnesses of the columns and hence are equal in AB to $-1{,}000M \times 6/2 = -3{,}000M$. The column fixed-end moments are entered in the first line of Table 3-5, which is called a moment-collection table.

Table 3-4. Moment-influence Factors for Fig. 3-71

Member	+1,000 at B	+1,000 at C
AB	351	−105
BA	702	−210
BC	298	210
CB	70	579
CD	−70	421
DC	−35	210

In the deflected position of the frame, joints B and C are unlocked. First, we apply a releasing moment of $+3{,}000M$ at B and distribute it by multiplying by 3 the entries in the column marked "+1,000 at B" in Table 3-4. Similarly,

Table 3-5. Moment-collection Table for Fig. 3-71

Remarks	*AB*		*BA*		*BC*		*CB*		*CD*		*DC*	
	+	−	+	−	+	−	+	−	+	−	+	−
Sidesway, *FEM*	......	3,000*M*	......	3,000*M*	......	......	......	....	....	1,000*M*	....	1,000*M*
B moments	1,053*M*	......	2,106*M*	......	894*M*	......	210*M*	....	....	210*M*	....	105*M*
C moments	......	105*M*	......	210*M*	210*M*	......	579*M*	....	421*M*	......	210*M*	
Partial sum	1,053*M*	3,105*M*	2,106*M*	3,210*M*	1,104*M*	......	789*M*	....	421*M*	1,210*M*	210*M*	1,105*M*
Totals	......	2,052*M*	......	1,104*M*	1,104*M*	......	789*M*	....	....	789*M*	....	895*M*
For 2,000-lb load	......	17,000	......	9,100	9,100	......	6,500	....	....	6,500	....	7,400
4,000-lb load, *FEM*	......	......	......	......	......	12,800	3,200					
B moments	4,490	......	8,980	......	3,820	......	897	....	....	897	....	448
C moments	336	......	672	......	......	672	......	1,853	....	1,347	....	672
Partial sum	4,826	......	9,652	......	3,820	13,472	4,097	1,853	....	2,244	....	1,120
No-sidesway sum	4,826	......	9,652	......	......	9,652	2,244	....	....	2,244	....	1,120
Sidesway *M*	......	4,710	......	2,540	2,540	......	1,810	....	....	1,810	....	2,060
Totals	120	......	7,110	......	......	7,110	4,050	....	....	4,050	....	3,180

a releasing moment of $+1{,}000M$ is applied at C and distributed with the aid of Table 3-4. The distributed moments are entered in the second and third lines of Table 3-5. The final moments are the sum of the fixed-end moments and the distributed moments and are given in the fifth line.

Isolating each column and taking moments about one end, we find that the overturning moment due to the shear is equal to the sum of the end moments. We have one such equation for each column. Adding these equations, noting that the sum of the shears equals 2,000 lb, we obtain

$$-M(2{,}052 + 1{,}104 + 789 + 895) = -2{,}000 \times 20$$

from which we find $M = 8.26$. This value is substituted in the sidesway totals in Table 3-5 to yield the end moments for the 2,000-lb horizontal load.

Suppose now a vertical load of 4,000 lb is applied to BC of the rigid frame in Fig. 3-71, 5 ft from B. Tables 3-4 and 3-5 can again be used to determine the end moments with a minimum of labor:

The fixed-end moment at B, with sidesway prevented, is $-12{,}800$, and at C $+3{,}200$. With the joints locked, the frame is permitted to move laterally an arbitrary amount, so that in addition to the fixed-end moments due to the 4,000-lb load, column fixed-end moments of $-3{,}000M$ at B and $-1{,}000\ M$ at C are induced. Table 3-5 already indicates the effect of relieving these column moments by unlocking joints B and C. We now have to superimpose the effect of releasing joints B and C to relieve the fixed-end moments for the vertical load. This we can do with the aid of Table 3-4. The distribution is shown in the lower part of Table 3-5. The sums of the fixed-end moments and distributed moments for the 4,000-lb load are shown on the line "No-sidesway sum."

The unknown M can be evaluated from the fact that the sum of the horizontal forces acting on the columns must be zero. This is equivalent to requiring that the sum of the column end moments equals zero:

$$-M(2{,}052 + 1{,}104 + 789 + 895) + 4{,}826 + 9{,}652 - 2{,}244 - 1{,}120 = 0$$

from which $M = 2.30$. This value is substituted in the sidesway total in Table 3-5 to yield the sidesway moments for the 4,000-lb load. The addition of these moments to the totals for no sidesway yields the final moments.

This procedure enables one-story bents with straight beams to be analyzed with the necessity of solving only one equation with one unknown regardless of the number of bays. If the frame is several stories high, the procedure can be applied to each story. Since an arbitrary horizontal deflection is introduced at each floor or roof level, there are as many unknowns and equations as there are stories.

The procedure is more difficult to apply to bents with curved or polygonal members between the columns. The effect of the change in the horizontal projection of the curved or polygonal portion of the bent must be included in the calculations ("Gabled Concrete Roof Frames Analyzed by Moment Distribution," Portland Cement Association, Skokie, Ill.). In many cases, it may be easier to analyze the bent as a curved beam (arch).

3-68. Single-cycle Moment Distribution. In the method of moment distribution by converging approximations, all joints but the one being unlocked are considered fixed. In distributing moments, the stiffness and carry-over factors used are based on this assumption. However, if actual stiffnesses and carry-over factors are employed, moments can be distributed throughout continuous frames in a single cycle.

Formulas for actual stiffnesses and carry-over factors can be written in several simple forms. The equations given in the following text were chosen to permit the use of existing tables for beams of variable moment of inertia that are based on fixed-end stiffnesses and fixed-end carry-over factors, such as the "Handbook of Frame Constants," Portland Cement Association, Skokie, Ill.

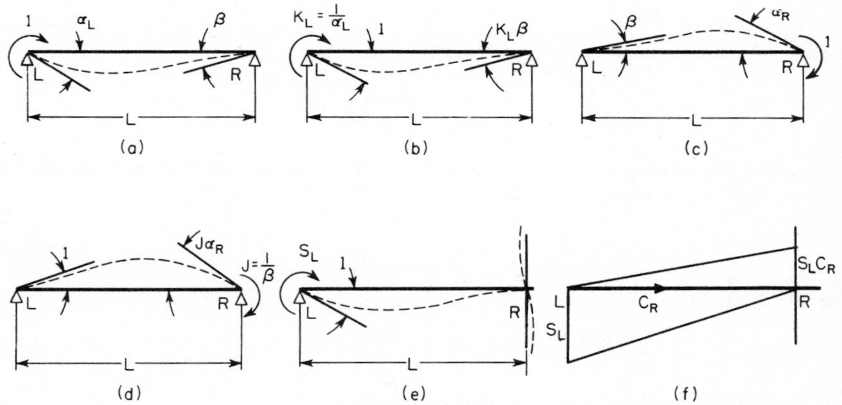

Fig. 3-72. End rotations of simple beams.

Considerable simplification of the formulas results if they are based on the simple-beam stiffness of members of continuous frames. This value can always be obtained from tables of fixed-end properties by multiplying the fixed-end stiffness given in the tables by $(1 - C_L^F C_R^F)$, in which C_L^F is the fixed-end carry-over factor to the left and C_R^F is the fixed-end carry-over factor to the right.

To derive the basic constants needed, we apply a unit moment to one end of a member, considering it simply supported (Fig. 3-72*a*). The end rotation at the support where the moment is applied is α, and at the far end, the rotation is β. By the dummy-load method (Art. 3-55), if x is measured from the β end,

$$\alpha = \int_0^L \frac{x^2}{EI_x}\,dx \qquad (3\text{-}72)$$

$$\beta = \int_0^L \frac{x(L-x)}{EI_x}\,dx \qquad (3\text{-}73)$$

in which I_x = moment of inertia at a section a distance of x from the β end
E = modulus of elasticity

Simple-beam stiffness K of the member is the moment required to produce a rotation of unity at the end where it is applied (Fig. 3-72*b*). Hence, at each end of a member, $K = 1/\alpha$.

For prismatic beams, K has the same value for both ends and is equal to $3EI/L$ For haunched beams, K for each end can be obtained from tables for fixed-end stiffnesses, as mentioned previously, or by numerical integration of Eq. (3-72).

While the value of α, and consequently of K, is different at opposite ends of an unsymmetrical beam, the value of β is the same for both ends, in accordance with the law of reciprocal deflections (compare Fig. 3-72*a* and *c*). This is also evident from Eq. (3-73), where $L - x$ can be substituted for x without changing the value of the integral.

Now, if we apply a moment J at one end of a simple beam to produce a rotation of unity at the other end (Fig. 3-72*d*), this moment will be equal to $1/\beta$ and will have the same value regardless of the end at which it is applied. As can be shown, K/J is equal to the **fixed-end carry-over factor.**

J is equal to $6EI/L$ for prismatic beams. For haunched beams, it can be computed by numerical integration of Eq. (3-73).

Actual stiffness S of the end of an unloaded span is the moment producing a rotation of unity at the end where it is applied when the other end of the beam is restrained against rotation by other members of the structure (Fig. 3-72*e*).

The bending-moment diagram for a moment S_L applied at the left end of a member of a continuous frame is shown in Fig. 3-72*f*. As indicated, the moment carried over to the far end is S_LC_R, where C_R is the carry-over factor to the right. At L, the rotation produced by S_L alone is S_L/K_L, and by S_LC_R alone is $-S_LC_R/J$. The sum of these angles must equal unity by definition of stiffness:

$$\frac{S_L}{K_L} - \frac{S_LC_R}{J} = 1$$

Solving for S_L and noting that $K_L/J = C_L^F$, the fixed-end carry-over factor to the left, we find the formula for the stiffness of the left end of a member:

$$S_L = \frac{K_L}{1 - C_L^F C_R} \qquad (3\text{-}74)$$

For the right end:

$$S_R = \frac{K_R}{1 - C_R^F C_L} \qquad (3\text{-}75)$$

For prismatic beams, the stiffness formulas reduce to

$$S_L = \frac{K}{1 - C_R/2} \qquad S_R = \frac{K}{1 - C_L/2} \qquad (3\text{-}76)$$

where $K = 3EI/L$

When the far end of a prismatic beam is fully fixed against rotation, the carry-over factor equals ½. Hence, the fixed-end stiffness equals $4K/3$. This indicates that the effect of partial restraint on prismatic beams is to vary the stiffness between K for no restraint and $1.33K$ for full restraint. Because of this small variation, an estimate of the actual stiffness of a beam may be sufficiently accurate in many cases.

Restraint R at the end of an unloaded beam in a continuous frame is the moment applied at that end to produce a unit rotation in all the members of the joint. Since the sum of the moments at the joint must be zero, R must be equal to the sum of the stiffnesses of the adjacent ends of the members connected to the given beam at that joint.

Furthermore, the moment induced in any of these other members bears the same ratio to the applied moment as the stiffness of the member does to the restraint. Consequently, *end moments are distributed at a joint in proportion to the stiffnesses of the members.*

Actual carry-over factors can be computed by modifying the fixed-end carry-over factors. In Fig. 3-72*e* and *f*, by definition of restraint, the rotation at joint R is $-S_LC_R/R_R$, which must be equal to the rotation of the beam at R due to the moments at L and R. The rotation due to S_LC_R alone is equal to S_LC_R/K_R, and the rotation due to S_L alone is $-S_L/J$. Hence $-S_LC_R/R_R = S_LC_R/K_R - S_L/J$. Solving for C_R

and noting that $K_R/J = C_R^F$, the fixed-end carry-over factor to the right, we find the actual carry-over factor to the right:

$$C_R = \frac{C_R^F}{1 + K_R/R_R} \tag{3-77}$$

Similarly, the actual carry-over factor to the left is

$$C_L = \frac{C_L^F}{1 + K_L/R_L} \tag{3-78}$$

Approximate Carry-over Factors. In analyzing a continuous beam, we generally know the carry-over factors toward the ends of the first and last spans. Starting with these values, we can calculate the rest of the carry-over factors and the stiffnesses of the members. However, in many frames, there are no end conditions known in advance. To analyze these structures, we must assume several carry-over factors.

This will not complicate the analysis, because in many cases it will be found unnecessary to correct the values of C based on assumed carry-over factors of preceding spans. The reason is that C is not very sensitive to the restraint at far ends of adjacent members. When carry-over factors are estimated, the greatest accuracy will be attained if the choice of assumed values is restricted to members subject to the greatest restraint.

A very good approximation to the carry-over factor for prismatic beams may be obtained from the following formula, which is based on the assumption that far ends of adjacent members are subject to equal restraint:

$$C = \frac{\Sigma K - K}{2(\Sigma K - \delta K)} \tag{3-79}$$

where ΣK = sum of K values of all members at joint toward which carry-over factor is acting

K = simple-beam stiffness of member for which carry-over factor is being computed

δ = a factor that varies from zero for no restraint to $\frac{1}{4}$ for full restraint at far ends of connecting members

Since δ varies within such narrow limits, it affects C very little.

Estimation Example. To illustrate the estimation and calculation of carry-over factors, the carry-over factors and stiffness in the clockwise direction will be computed for the frame in Fig. 3-73*a*. Relative I/L, or K values, are given in the circles.

A start will be made by estimating C_{AB}. Taking $\delta = \frac{1}{8}$, we apply Eq. (3-79) at B with $K = 3$ and $\Sigma K = 5$ and find $C_{AB} = 0.216$, as shown in Fig. 3-73*a*. We can now compute the stiffness S_{AB} employing Eq. (3-76): $S_{AB} = 3/(1 - 0.108) = 3.37$. Noting that $R_{AD} = S_{AB}$, we can use the exact formula [Eq. (3-78)] to obtain the carry-over factor for DA: $C_{DA} = 0.5/(1 + 6/3.37) = 0.180$. Continuing around the frame in this manner, we return to C_{AB} and recalculate it by Eq. (3-78), obtaining 0.221. This differs only slightly from the estimated value. The change in C_{DA} due to the new value of C_{AB} is negligible.

If a bending moment of 1,000 ft-lb were introduced at A in AB, it would induce a moment of 1,000 C_{AB} = 221 ft-lb at B; 221 × 0.321 = 71 ft-lb at C; 71 × 0.344 = 24 ft-lb at D; 24 × 0.180 = 4 ft-lb at A, etc.

Beam Analysis. To show how moments in a continuous beam would be computed, the end moments will be determined for the beam in Fig. 3-73*b*, which is identical with the one in Fig. 3-67 for which moments were obtained by converging approximations. Relative I/L, or K values, are shown in the circles on each span. Since A is a hinged end, $C_{BA} = 0$. $S_{BA} = 1/(1 - 0) = 1$. Since there is only one member joined to BC at B, $R_{BC} = S_{BA}$, and $C_{CB} = 0.5/(1 + 1/1) = 0.250$. With this value, we compute $S_{CB} = 1.14$. To obtain the carry-over factors for the opposite direction, we start with $C_{CD} = 0.5$, since we know D is a fixed end. This enables us to compute $S_{CD} = 1.33$ and the remainder of the beam constants.

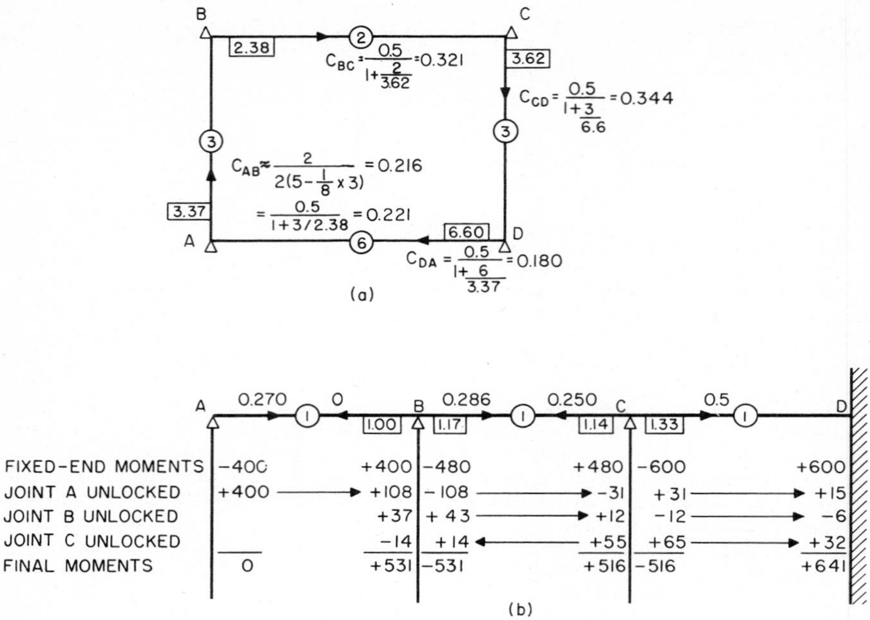

Fig. 3-73. Moments in (*a*) a quadrangular frame and (*b*) a continuous beam.

The fixed-end moments are given on the first line of calculations in Fig. 3-73*b*. We start the distribution by unlocking *A* by applying a releasing moment of +400. Since *A* is unrestrained, the full 400 is given to *A* and $400 \times 0.270 = 108$ is carried over to *B*. If several members had been connected to *AB* at *B*, this moment, with sign changed, would be distributed to them in proportion to their stiffnesses. But since only one member is connected at *B*, the moment at *BC* is −108. Next, $-108 \times 0.286 = -31$ is carried over to *C*. Finally, a moment of +15 is carried to *D*.

Then, joint *B* is unlocked. The unbalanced moment of $400 - 480 = -80$ is counteracted with a moment of +80, which is distributed to *BA* and *BC* in proportion to their stiffnesses (shown in the boxes at the joint). *BC*, for example gets

$$80 \times \frac{1.17}{1.17 + 1.00} = 43$$

The carry-over to *C* is $43 \times 0.286 = 12$, and to *D*, −6.

Similarly, the unbalanced moment at *C* is counteracted and distributed to *CB* and *CD* in proportion to the stiffnesses shown in the boxes at *C*, then carried over to *B* and *D*. The final moments are the sum of the fixed-end and distributed moments.

Special Cases. Advantage can be taken of certain properties of loads and structures to save work in distribution by using carry-over factors as the ratio of end moments in loaded members. For example, suppose it is obvious, from symmetry of loading and structure, that there will be no end rotation at an interior support. The part of the structure on one side of this support can be isolated and the moments distributed only in this part, with the carry-over factor toward the support taken as C^F.

Again, suppose it is evident that the final end moments at opposite ends of a span must be equal in magnitude and sign. Isolate the structure on each side of this beam and distribute moments only in each part, with the carry-over factor for this span taken as 1.

3-69. Method for Checking Moment Distribution. End moments computed for a continuous structure must be in accordance with both the laws of equilibrium and the requirements of continuity. At each joint, therefore, the sum of the moments must be equal to zero (or to an external moment applied there), and the end of every member connected there must rotate through the same angle. It is a simple matter to determine whether the sum of the moments is zero, but further calculation is needed to prove that the moments yield the same rotation for the end of each member at a joint. The following method not only will indicate that the requirements of continuity are satisfied but also will tend to correct automatically any mistakes that may have been made in computing the end moments.

Consider a joint O made up of several members OA, OB, OC, etc. The members are assumed to be loaded and the calculation of end moments to have started with fixed-end moments. For any one of the members, say OA, the end rotation at O for the fixed-end condition is

$$0 = \frac{M_{OA}{}^F}{K_{OA}} - \frac{M_{AO}{}^F}{J_{OA}} - \phi \tag{3-80}$$

where $M_{OA}{}^F$ = fixed-end moment at O
$M_{AO}{}^F$ = fixed-end moment at A
K_{OA} = simple-beam stiffness at O
J_{OA} = moment required at A to produce a unit rotation at O when the span is considered simply supported
ϕ = simple-beam end rotation at O due to loads

For the final end moments, the rotation at O is

$$\theta = \frac{M_{OA}}{K_{OA}} - \frac{M_{AO}}{J_{OA}} - \phi \tag{3-81}$$

Subtracting Eq. (3-80) from Eq. (3-81) and multiplying by K_{OA} yields

$$K_{OA}\theta = M_{OA} - M_{OA}{}^F - C_{AO}{}^F M'_{OA} \tag{3-82}$$

in which the fixed-end carry-over factor toward O, $C_{AO}{}^F$, has been substituted for K_{OA}/J_{OA}, and M'_{OA} for $M_{AO} - M_{AO}{}^F$. An expression analogous to Eq. (3-82) can be written for each of the other members at O. Adding these to Eq. (3-82), we obtain

$$\theta \Sigma K_O = \Sigma M_O - \Sigma M_O{}^F - \Sigma C_O{}^F M'_O \tag{3-83}$$

Equating the value of θ obtained from Eqs. (3-82) and (3-83) and solving for M_{OA}, we determine the equations that will check the joint for continuity:

$$M_{OA} = M_{OA}{}^F + C_{AO}{}^F M'_{OA} - m_{OA} \tag{3-84}$$

$$m_{OA} = \frac{K_{OA}}{\Sigma K_O}(-\Sigma M_O + \Sigma M_O{}^F + \Sigma C_O{}^F M'_O) \tag{3-85}$$

Similar equations can be written for the other members at O by substituting the proper letter for A in the subscripts.

If the calculations based on these equations are carried out in table form, the equations prove to be surprisingly simple (see Tables 3-6 and 3-7).

For prismatic beams, the terms $C^F M'$ become ½ the change in the fixed-end moment at the far end of each member at a joint.

Examples. Suppose we want to check the beam in Fig. 3-73*b*. Each joint and the ends of the members connected are listed in Table 3-6, and a column is provided for the summation of the various terms for each joint. K values are given on line 1, the end moments to be checked on line 2, and the fixed-end moments on line 3. On line 4 is entered one-half the difference obtained when the fixed-end moment is subtracted from the final moment at the far end of each member. $-m$ is placed on line 5 and the corrected end moment on line 6. The $-m$ values are obtained from the summation columns by adding line 2 to the negative of the sum of lines 3 and 4 and distributing the result to the members of the joint in proportion to the K values. The corrected moment M is the sum of lines 3, 4, and 5.

Table 3-6. Moment-distribution Check
(No deflection at supports)

Line No.	Term	Joint A		Joint B			Joint C			Joint D	
		AB	Sum	BA	BC	Sum	CB	CD	Sum	DC	Sum
1	K	1	1	1	1	2	1	1	2	1	∞
2	M	0	0	+600	−600	0	+450	−450	0	+700	
3	M^F	−400	−400	+400	−480	−80	+480	−600	−120	+600	
	First Cycle										
4	$\frac{1}{2}M'$	+100	+100	+200	−15	+185	−60	+50	−10	+75	
5	$-m$	+300	+300	−53	−52	−105	+65	+65	+130	0	
6	M	0	0	+547	−547	0	+485	−485	0	+675	
	Second Cycle										
7	$\frac{1}{2}M'$	.....	.....	+200	+2	+202	−33	+37	+4	+57	
8	$-m$	.....	.....	−61	−61	−122	+58	+58	+116	0	
9	M	.....	.....	+539	−539	0	+505	−505	0	+657	

Assume that a mistake was made in computing the end moments for Fig. 3-73*b*, giving the results shown on line 2 of Table 3-6 for the fixed-end moments on line 3. The correct moments can be obtained as follows:

At joint B, the sum of the incorrect moments is zero, as shown in the summation column on line 2. The sum of the fixed-end moments at B is −80, as indicated on line 3. For BA, $\frac{1}{2}M'$ is obtained from lines 2 and 3 of the column for AB: $\frac{1}{2} \times (0 + 400) = +200$, which is entered on line 4. The line 4 entry for BC is obtained from CB: $\frac{1}{2} \times (450 - 480) = -15$. The sum of the line 4 values at B is therefore $200 - 15 = +185$. Entered in the summation column, this is then added to the summation value on line 3, the sign is changed and the number on line 2 (in this case zero) added to the sum, giving −105, which is noted in the summation column on line 5. The values for BA and BC on line 5 are obtained by multiplying −105 by the ratio of the K value of each member to the sum of the K values at the joint; that is, for BA, $-m = -105 \times \frac{1}{2} = -53$. The corrected moment for BA, the sum of lines 3, 4, and 5, is $+400 + 200 - 53 = +547$. The other corrected moments are found in the same way and are shown on line 6.

The new end moments differ considerably from the moments on line 2, indicating that those moments were incorrect. A comparison with Fig. 3-73*b* shows that the new moments are nearer to the correct answer than those on line 2. Even closer results can be obtained by repeating the calculations using the moments on line 6 as demonstrated in the second cycle in Table 3-6.

Equations (3-80) to (3-85) can be generalized to include the effect of the movement d of a support in a direction normal to the initial position of a span of length L:

$$K_{OA}\theta = M_{OA} - M_{OA}{}^F - C_{AO}{}^F M'_{OA} + K_{OA}\frac{d}{L} \tag{3-86}$$

$$M_{OA} = M_{OA}{}^F + C_{AO}{}^F M'_{OA} - K_{OA}\frac{d}{L} - m_{OA} \tag{3-87a}$$

$$m_{OA} = \frac{K_{OA}}{\Sigma K_O}\left(-\Sigma M_O + \Sigma M_O{}^F + \Sigma C_O{}^F M'_O - \Sigma K_O \frac{d}{L}\right) \tag{3-87b}$$

For each span with a support movement, the term Kd/L can be obtained from the fixed-end moment due to this deflection alone; for OA, for example, by multiplying moment M_{OA}^F by $(1 - C_{AO}^F C_{OA}^F)/(1 + C_{OA}^F)$. For prismatic beams, this factor reduces to ½. Equation (3-86) is called a **slope-deflection equation.**

In Table 3-7, the solution for the frame in Fig. 3-71 is checked for the condition in which a 4,000-lb vertical load was placed 5 ft from B on span BC. The computations are similar to those in Table 3-6, except that the terms $-Kd/L$ are included for the columns to account for the sidesway. These values are obtained from the sidesway fixed-end moments in Table 3-5. For BA, for example, $Kd/L = ½ \times 3{,}000M$, with $M = 2.30$, as found in the solution. The check indicates that the original solution was sufficiently accurate for a slide-rule computation. If line 7 had contained a different set of moments, the shears would have had to be investigated again. A second cycle could be carried out by distributing the unbalance to the columns to obtain new Kd/L values.

Table 3-7. Moment-distribution Check

(Frame with sidesway)

Line No.	Term	Joint *A*		Joint *B*			Joint *C*			Joint *D*	
		AB	Sum	*BA*	*BC*	Sum	*CB*	*CD*	Sum	*DC*	Sum
1	K	6	∞	6	3	9	3	2	5	2	∞
2	M	+120	. . .	+ 7,110	− 7,110	0	+4,050	−4,050	0	−3,180	
3	M^F	0	. . .	0	−12,800	−12,800	+3,200	0	+3,200	0	
4	$-Kd/L$	−3,450	. . .	−3,450	0	− 3,450	0	−1,150	−1,150	−1,150	
5	$+½M'$	+3,555	. . .	+60	+425	+485	+2,845	−1,590	+1,255	−2,025	
6	$-m$	0	. . .	+10,510	+ 5,255	+15,765	−1,983	−1,322	−3,305	0	
7	M	+105	. . .	+ 7,120	− 7,120	0	+4,062	−4,062	0	−3,175	

BUCKLING OF COLUMNS

Columns are compression members whose cross-sectional dimensions are relatively small compared with their length in the direction of the compressive force. Failure of such members occurs because of instability when a certain load (called the critical, or **Euler load**) is equaled or exceeded. The member may bend, or buckle, suddenly and collapse.

Hence the strength of a column is not determined by the unit stress in Eq. (3-3) ($P = Af$) but by the maximum load it can carry without becoming unstable. The condition of instability is characterized by disproportionately large increases in lateral deformation with slight increase in load. It may occur in slender columns before the unit stress reaches the elastic limit.

3-70. Stable Equilibrium. Consider, for example, an axially loaded column with ends unrestrained against rotation, shown in Fig. 3-74. If the member is initially perfectly straight, it will remain straight as long as the load P is less than the critical load P_c. If a small transverse force is applied, it will deflect, but it will return to the straight position when this force is removed. Thus, when P is less than P_c, internal and external forces are in stable equilibrium.

3-71. Unstable Equilibrium. If $P = P_c$ and a small transverse force is applied, the column again will deflect, but this time, when the force is removed, the column will remain in the bent position (dash line in Fig. 3-74). The equation of this elastic curve can be obtained from Eq. (3-31):

$$EI\frac{d^2y}{dx^2} = -P_c y \tag{3-88}$$

in which E = modulus of elasticity

I = least moment of inertia

y = deflection of the bent member from the straight position at a distance x from one end

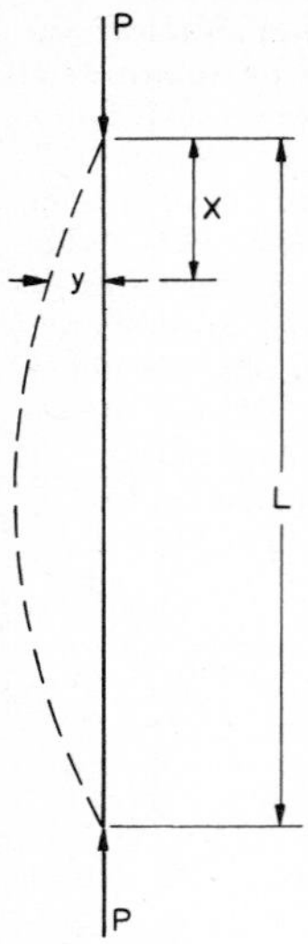

Fig. 3-74. Buckling of a long column.

This assumes, of course, that the stresses are within the elastic limit. Solution of Eq. (3-88) gives the smallest value of the Euler load as

$$P_c = \frac{\pi^2 EI}{L^2} \tag{3-89}$$

Equation (3-89) indicates that there is a definite finite magnitude of an axial load that will hold a column in equilibrium in the bent position when the stresses are below the elastic limit. Repeated application and removal of small transverse forces or small increases in axial load above this critical load will cause the member to fail by buckling. Internal and external forces are in a state of unstable equilibrium.

It is noteworthy that the Euler load, which determines the load-carrying capacity of a column, depends on the stiffness of the member, as expressed by the modulus of elasticity, rather than on the strength of the material of which it is made.

By dividing both sides of Eq. (3-89) by the cross-sectional area A and substituting r^2 for I/A (r is the radius of gyration of the section), we can write the solution of Eq. (3-88) in terms of the average unit stress on the cross section:

$$\frac{P_c}{A} = \frac{\pi^2 E}{(L/r)^2} \tag{3-90}$$

This holds only for the elastic range of buckling; i.e., for values of the slenderness ratio L/r above a certain limiting value that depends on the properties of the material. For inelastic buckling, see Art. 3-73.

3-72. Effect of End Conditions. Equation (3-90) was derived on the assumption that the ends of the column are free to rotate. It can be generalized, however, to take into account the effect of end conditions:

$$\frac{P_c}{A} = \frac{\pi^2 E}{(kL/r)^2} \tag{3-91}$$

where k is the factor that depends on the end conditions. For a pin-ended column, $k = 1$; for a column with both ends fixed, $k = \frac{1}{2}$; for a column with one end fixed and one end pinned, k is about 0.7; and for a column with one end fixed and one end free from all restraint, $k = 2$.

3-73. Inelastic Buckling. Equations (3-89) to (3-91), having been derived from Eq. (3-88), the differential equation for the elastic curve, are based on the assumption that the critical average stress is below the elastic limit when the state of unstable equilibrium is reached. In members with slenderness ratio L/r below a certain limiting value, however, the elastic limit is exceeded before the column buckles. As the axial load approaches the critical load, the modulus of elasticity varies with the stress. Hence Eqs. (3-89) to (3-91), based on the assumption that E is a constant, do not hold for these short columns.

After extensive testing and analysis, prevalent engineering opinion favors the Engesser equation for metals in the inelastic range:

$$\frac{P_t}{A} = \frac{\pi^2 E_t}{(kL/r)^2} \tag{3-92}$$

This differs from Eqs. (3-89) to (3-91) only in that the tangent modulus E_t (the actual slope of the stress-strain curve for the stress P_t/A) replaces the modulus of elasticity E in the elastic range. P_t is the smallest axial load for which two equilibrium positions are possible, the straight position and a deflected position.

3-74. Column Curves. Curves obtained by plotting the critical stress for various values of the slenderness ratio are called column curves. For axially loaded, initially straight columns, the column curve consists of two parts: (1) the Euler critical values, and (2) the Engesser, or tangent-modulus critical values.

The latter are greatly affected by the shape of the stress-strain curve for the material of which the column is made, as shown in Fig. 3-75. The stress-strain curve for a material, such as an aluminum alloy or high-strength steel, which does not have a sharply defined yield point, is shown in Fig. 3-75*a*. The corresponding column curve is drawn in Fig. 3-75*b*. In contrast, Fig. 3-75*c* presents the stress-strain curve for structural steel, with a sharply defined point, and Fig. 3-75*d* the related column curve. This curve becomes horizontal as the critical stress approaches the yield strength of the material and the tangent modulus becomes zero, whereas the column curve in Fig. 3-75*b* continues to rise with decreasing values of the slenderness ratio.

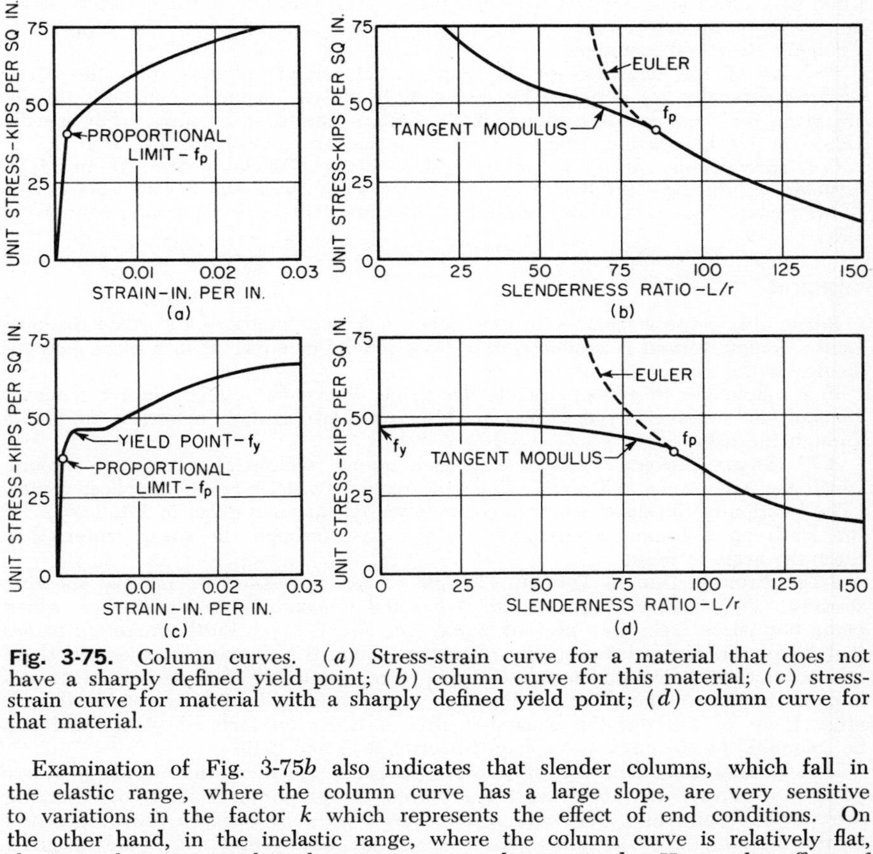

Fig. 3-75. Column curves. (*a*) Stress-strain curve for a material that does not have a sharply defined yield point; (*b*) column curve for this material; (*c*) stress-strain curve for material with a sharply defined yield point; (*d*) column curve for that material.

Examination of Fig. 3-75*b* also indicates that slender columns, which fall in the elastic range, where the column curve has a large slope, are very sensitive to variations in the factor k which represents the effect of end conditions. On the other hand, in the inelastic range, where the column curve is relatively flat, the critical stress is relatively insensitive to changes in k. Hence the effect of end conditions on the stability of a column is of much greater significance for long columns than for short columns.

3-75. Local Buckling. A column may not only fail by buckling of the member as a whole but as an alternative, by buckling of one of its components. Hence, when members like I beams, channels, and angles are used as columns or when sections are built up of plates, the possibility of the critical load on a component (leg, half flange, web, lattice bar) being less than the critical load on the column as a whole should be investigated.

Similarly, the possibility of buckling of the compression flange or the web of a beam should be looked into.

Local buckling, however, does not always result in a reduction in the load-carrying capacity of a column. Sometimes, it results in a redistribution of the stresses enabling the member to carry additional load.

3-76. Behavior of Actual Columns. For many reasons, columns in structures behave differently from the ideal column assumed in deriving Eqs. (3-89) and (3-92). A major consideration is the effect of accidental imperfections, such as nonhomogeneity of materials, initial crookedness, and unintentional eccentricities of the axial load, since neither field nor shopwork can be perfect. These and the effects of residual stresses usually are taken into account by a proper choice of safety factor.

There are other significant conditions, however, that must be considered in any design rule: continuity in frame structures and eccentricity of the axial load. Continuity affects column action in two ways. The restraint at column ends determines the value of k, and bending moments are transmitted to the column by adjoining structural members.

Because of the deviation of the behavior of actual columns from the ideal, columns generally are designed by empirical formulas. Separate equations usually are given for short columns, intermediate columns, and long columns. For specific materials—steel, concrete, timber—these formulas are given in Secs. 5 to 8.

For more details on column action, see F. Bleich, "Buckling Strength of Metal Structures," McGraw-Hill Book Company, New York, 1952; and S. Timoshenko and J. M. Gere, "Theory of Elastic Stability," McGraw-Hill Book Company, New York, 1961.

TORSION

Forces that cause a member to twist about a longitudinal axis are called torsional loads. Simple torsion is produced only by a couple, or moment, in a plane perpendicular to the axis.

If a couple lies in a nonperpendicular plane, it can be resolved into a torsional moment, in a plane perpendicular to the axis, and bending moments, in planes through the axis.

3-77. Shear Center. The point in each normal section of a member through which the axis passes and about which the section twists is called the shear center. (The location of the shear center in commonly used shapes is given in Art. 3-35.) If the loads on a beam, for example, do not pass through the shear center, they cause the beam to twist.

3-78. Stresses Due to Torsion. Simple torsion is resisted by internal shearing stresses. These can be resolved into radial and tangential shearing stresses, which being normal to each other also are equal (see Art. 3-11). Furthermore, on planes that bisect the angles between the planes on which the shearing stresses act, there also occur compressive and tensile stresses. The magnitude of these normal stresses is equal to that of the shear. Therefore, when torsional loading is combined with other types of loading, the maximum stresses occur on inclined planes and can be computed by the methods explained in Arts. 3-12 and 3-15.

If a circular shaft (hollow or solid) is twisted, a section that is plane before twisting remains plane after twisting. Within the proportional limit, the shearing stress at any point in a transverse section varies with the distance from the center of the section. The maximum shear occurs at the circumference and is given by

$$v = \frac{Tr}{J} \tag{3-93}$$

where T = torsional moment
r = radius of section
J = polar moment of inertia

Polar moment of inertia is defined by

$$J = \int \rho^2 \, dA \tag{3-94}$$

where ρ = radius from shear center to any point in section
dA = differential area at the point

In general, J equals the sum of the moments of inertia about any two perpendicular axes through the shear center. For a solid circular section, $J = \pi r^4/2$.

Within the proportional limit, the angular twist for a circular bar is given by

$$\theta = \frac{TL}{GJ} \tag{3-95}$$

where L = length of bar
G = shearing modulus of elasticity (see Art. 3-6)

If a shaft is not circular, a plane transverse section before twisting does not remain plane after twisting. The resulting warping increases the shearing stresses in some parts of the section and decreases them in others, compared with the shearing stresses that would occur if the section remained plane. Consequently, shearing stresses in a noncircular section are not proportional to distances from the shear center. In elliptical and rectangular sections, for example, maximum shear occurs on the circumference at a point nearest the shear center.

For a solid rectangular section, this maximum may be expressed in the following form:

$$v = \frac{T}{kb^2d} \tag{3-96}$$

where b = short side of the rectangle
d = long side
k = a constant depending on the ratio of these sides:

d/b =	1.0	1.5	2.0	2.5	3	4	5	10	∞
k =	0.208	0.231	0.246	0.258	0.267	0.282	0.291	0.312	0.333

(S. Timoshenko and J. N. Goodier, "Theory of Elasticity," McGraw-Hill Book Company, New York.)

If a thin-shell hollow tube is twisted, the shearing force per unit of length on a cross section (shear flow) is given approximately by

$$H = \frac{T}{2A} \tag{3-97}$$

where T = torsional moment
A = area enclosed by mean perimeter of wall of tube

And the unit shearing stress is given approximately by

$$v = \frac{H}{t} = \frac{T}{2At} \tag{3-98}$$

where t = thickness of tube

For a rectangular tube with sides of unequal thickness, the total shear flow can be computed from Eq. (3-97), and the shearing stress along each side from Eq. (3-98).

For a narrow rectangular section, the maximum shear is very nearly equal to

$$v = \frac{T}{\frac{1}{3}b^2d} \tag{3-99}$$

where b = short dimension of the rectangle
d = long side

This formula can also be used to find the maximum shearing due to torsion in members, such as I beams and channels, made up of thin rectangular components. Let $J = \frac{1}{3}\Sigma b^3d$, where b is the thickness of each rectangular component and d the corresponding length. Then, the maximum shear is given approximately by

$$v = \frac{Tb}{J} \tag{3-100}$$

In this equation, b is the thickness of the web or the flange. The maximum shearing stress is at the center of one of the long sides of the rectangular part that has the greatest thickness (Seely and Smith, "Advanced Mechanics of Materials," John Wiley & Sons, Inc., New York).

STRUCTURAL DYNAMICS

Article 3-2 noted that loads can be classified as static or dynamic and that the distinguishing characteristic was the rate of application of load. If a load is applied slowly, it may be considered static. Since dynamic loads may produce stresses and deformations considerably larger than those caused by static loads of the same magnitude, it is important to know reasonably accurately what is meant by slowly.

A useful definition can be given in terms of the natural period of vibration of the structure or member to which the load is applied. If the time in which a load rises from zero to its maximum value is more than double the natural period, the load may be treated as static. Loads applied more rapidly may be dynamic. Structural analysis and design for such loads are considerably different from and more complex than those for static loads.

In general, exact dynamic analysis is possible only for relatively simple structures, and only when both the variation of load and resistance with time are a convenient mathematical function. Therefore, in practice, adoption of approximate methods that permit rapid analysis and design is advisable. And usually, because of uncertainties in loads and structural resistance, computations need not be carried out with more than a few significant figures, to be consistent with known conditions.

3-79. Properties of Materials under Dynamic Loading. In general, mechanical properties of structural materials improve with increasing rate of load application. For low-carbon steel, for example, yield strength, ultimate strength, and ductility rise with increasing rate of strain. Modulus of elasticity in the elastic range, however, is unchanged. For concrete, the dynamic ultimate strength in compression may be much greater than the static strength.

Since the improvement depends on the material and the rate of strain, values to use in dynamic analysis and design should be determined by tests approximating the loading conditions anticipated.

Under many repetitions of loading, though, a member or connection between members may fail because of "fatigue" at a stress smaller than the yield point of the material. In general, there is little apparent deformation at the start of a fatigue failure. A crack forms at a point of high stress concentration. As the stress is repeated, the crack slowly spreads, until the member ruptures without measurable yielding. Though the material may be ductile, the fracture looks brittle.

Some materials (generally those with a well-defined yield point) have what is known as an **endurance limit.** This is the maximum unit stress that can be repeated, through a definite range, an indefinite number of times without causing structural damage. Generally, when no range is specified, the endurance limit is intended for a cycle in which the stress is varied between tension and compression stresses of equal value. For a different range, if f is the endurance limit, f_y the yield point, and r the ratio of the minimum stress to the maximum, then the relationship between endurance stresses is given approximately by

$$f_{\max} = \frac{2f}{(1 - r) + (f/f_y)(1 + r)} \tag{3-101}$$

A range of stress may be resolved into two components—a steady, or mean, stress and an alternating stress. The endurance limit sometimes is defined as the maximum value of the alternating stress that can be superimposed on the steady stress an indefinitely large number of times without causing fracture. If f is the endurance limit for completely reversed stresses, s the steady unit stress, and f_u the ultimate tensile stress, then the alternating stress may be obtained from a relationship of the type

$$f_a = f\left(1 - \frac{s^n}{f_u}\right) \tag{3-102}$$

where n lies between 1 and 2, depending on the material.

Design of members to resist repeated loading cannot be executed with the certainty with which members can be designed to resist static loading. Stress concentrations may be present for a wide variety of reasons, and it is not practicable to calculate their intensities. But sometimes it is possible to improve the fatigue strength of a material or to reduce the magnitude of a stress concentration below the minimum value that will cause fatigue failure.

In general, avoid design details that cause severe stress concentrations or poor stress distribution. Provide gradual changes in section. Eliminate sharp corners and notches. Do not use details that create high localized constraint. Locate unavoidable stress raisers at points where fatigue conditions are the least severe. Place connections at points where stress is low and fatigue conditions are not severe. Provide structures with multiple load paths or redundant members, so that a fatigue crack in any one of the several primary members is not likely to cause collapse of the entire structure.

Fatigue strength of a material may be improved by cold-working the material in the region of stress concentration, by thermal processes, or by prestressing it in such a way as to introduce favorable internal stresses. Where fatigue stresses are unusually severe, special materials may have to be selected with high energy absorption and notch toughness.

(J. H. Faupel, "Engineering Design," John Wiley & Sons, Inc., New York; C. H. Norris et al., "Structural Design for Dynamic Loads," McGraw-Hill Book Company, New York; W. H. Munse, "Fatigue of Welded Steel Structures," Welding Research Council, 345 East 47th Street, New York, N.Y. 10017; Almen and Black, "Residual Stresses and Fatigue in Metals," McGraw-Hill Book Company, New York.)

3-80. Natural Period of Vibration. A preliminary step in dynamic analysis and design is determination of this period. It can be computed in many ways, including by application of the laws of conservation of energy and momentum or Newton's second law of motion, $F = M(dv/dt)$, where F is force, M mass, v velocity, and t time. But in general, an exact solution is possible only for simple structures. Therefore, it is general practice to seek an approximate—but not necessarily inexact—solution by analyzing an idealized representation of the actual member or structure. Setting up this model and interpreting the solution require judgment of a high order.

Natural period of vibration is the time required for a structure to go through one cycle of free vibration, that is, vibration after the disturbance causing the motion has ceased.

To compute the natural period, the actual structure may be conveniently represented by a system of masses and massless springs, with additional resistances provided to account for energy losses due to friction, hysteresis, and other forms of damping. In simple cases, the masses may be set equal to the actual masses; otherwise, equivalent masses may have to be computed (Art. 3-84). The spring constants are the ratios of forces to deflections.

For example, a single mass on a spring (Fig. 3-76*b*) may represent a simply supported beam with mass that may be considered negligible compared with the load W at midspan (Fig. 3-76*a*). The spring constant k should be set equal to the load that produces a unit deflection at midspan; thus, $k = 48EI/L^3$, where E is the modulus of elasticity, psi; I the moment of inertia, in.^{4}; and L the span, in., of the beam. The idealized mass equals W/g, where g is the acceleration due to gravity, 386 in. per sec^2.

Also, a single mass on a spring (Fig. 3-76*d*) may represent the rigid frame in Fig. 3-76*c*. In that case, $k = 2 \times 12EI/h^3$, where I is the moment of inertia, in.^{4}, of each column and h the column height, in. The idealized mass equals the sum of the masses on the girder and the girder mass. (Weight of columns and walls is assumed negligible.)

The spring and mass in Fig. 3-76*b* and *d* form a one-degree system. The **degree of a system** is determined by the least number of coordinates needed to define the positions of its components. In Fig. 3-76, only the coordinate y is needed to locate the mass and determine the state of the spring. In a two-degree system, such as one comprising two masses connected to each other and to the ground

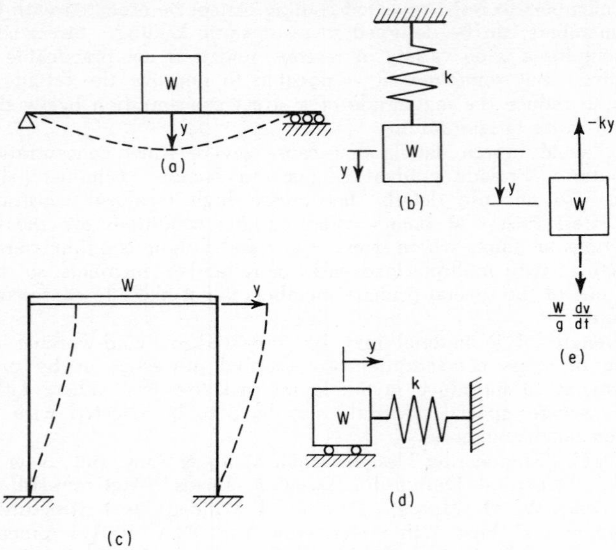

Fig. 3-76. Mass on weightless spring (b) or (d) may represent the motion of a beam (a) or a rigid frame (c) in free vibration.

by springs and capable of movement in only one direction, two coordinates are required to locate the masses.

If the mass with weight W, lb, in Fig. 3-76 is isolated, as shown in Fig. 3-76*e*, it will be in dynamic equilibrium under the action of the spring force $-ky$ and the inertia force $(d^2y/dt^2)(W/g)$. Hence, the equation of motion is

$$\frac{W}{g}\frac{d^2y}{dt^2} + ky = 0 \tag{3-103}$$

where y = displacement of mass, in., measured from rest position

Equation (3-103) may be written in the more convenient form

$$\frac{d^2y}{dt^2} + \frac{kg}{W}y = \frac{d^2y}{dt^2} + \omega^2 y = 0 \tag{3-104}$$

The solution is

$$y = A \sin \omega t + B \cos \omega t \tag{3-105}$$

where A and B are constants to be determined from initial conditions of the system, and

$$\omega = \sqrt{\frac{kg}{W}} \tag{3-106}$$

is the natural circular frequency, radians per sec.

The motion defined by Eq. (3-105) is harmonic. Its natural period, sec, is

$$T = \frac{2\pi}{\omega} = 2\pi\sqrt{\frac{W}{gk}} \tag{3-107}$$

Its natural frequency, cps, is

$$f = \frac{1}{T} = \frac{1}{2\pi}\sqrt{\frac{kg}{W}} \tag{3-108}$$

If, at time $t = 0$, the mass has an initial displacement y_0 and velocity v_0, substitution in Eq. (3-105) yields $A = v_0/\omega$ and $B = y_0$. Hence, at any time t, the mass is completely located by

$$y = \frac{v_0}{\omega} \sin \omega t + y_0 \cos \omega t \tag{3-109}$$

The stress in the spring can be computed from the displacement y.

In multiple-degree systems, an independent differential equation of motion can be written for each degree of freedom. Thus, in an N-degree system with N masses, weighing $W_1, W_2, \ldots, W_N$ lb, and N^2 springs with constants k_{rj} ($r = 1, 2, \ldots, N$; $j = 1, 2, \ldots, N$), there are N equations of the form

$$\frac{W_r}{g}\frac{d^2y_r}{dt^2} + \sum_{j=1}^{N} k_{rj}y_j = 0 \qquad r = 1, 2, \ldots, N \tag{3-110}$$

Simultaneous solution of these equations reveals that the motion of each mass can be resolved into N harmonic components. They are called the fundamental, second, third, etc., harmonics. Each set of harmonics for all the masses is called a **normal mode** of vibration.

There are as many normal modes in a system as degrees of freedom. Under certain circumstances, the system could vibrate freely in any one of these modes. During any such vibration, the ratio of displacement of any two of the masses remains constant. Hence, the solutions of Eqs. (3-110) take the form

$$y_r = \sum_{n=1}^{N} a_{rn} \sin \omega_n(t + \tau_n) \tag{3-111}$$

where a_{rn} and τ_n are constants to be determined from the initial conditions of the system and ω_n is the natural circular frequency for each normal mode.

To determine ω_n, set $y_1 = A_1 \sin \omega t$; $y_2 = A_2 \sin \omega t$ . . . . Then, substitute these and their second derivatives in Eqs. (3-110). After dividing each equation by $\sin \omega t$, the following N equations result:

$$\begin{aligned}
\left(k_{11} - \frac{W_1}{g}\omega^2\right) A_1 + k_{12}A_2 + \cdots + k_{1N}A_N &= 0 \\
k_{21}A_1 + \left(k_{22} - \frac{W_2}{g}\omega^2\right) A_2 + \cdots + k_{2N}A_N &= 0 \\
\cdots\cdots\cdots\cdots\cdots\cdots\cdots\cdots\cdots & \\
k_{N1}A_1 + k_{N2}A_2 + \cdots + \left(k_{NN} - \frac{W_N}{g}\omega^2\right) A_N &= 0
\end{aligned} \tag{3-112}$$

If there are to be nontrivial solutions for the amplitudes $A_1, A_2, \ldots, A_N$, the determinant of their coefficients must be zero. Thus,

$$\begin{vmatrix}
k_{11} - \frac{W_1}{g}\omega^2 & k_{12} & \cdots & k_{1N} \\
k_{21} & k_{22} - \frac{W_2}{g}\omega^2 & \cdots & k_{2N} \\
\cdots & \cdots & \cdots & \cdots \\
k_{N1} & k_{N2} & \cdots & k_{NN} - \frac{W_N}{g}\omega^2
\end{vmatrix} = 0 \tag{3-113}$$

Solution of this equation for ω yields one real root for each normal mode. And the natural period for each normal mode can be obtained from Eq. (3-107).

If ω for a normal mode now is substituted in Eqs. (3-112), the amplitudes A_1, A_2, . . . , A_N for that mode can be computed in terms of an arbitrary value, usually unity, assigned to one of them. The resulting set of modal amplitudes defines the **characteristic shape** for that mode.

The normal modes are mutually orthogonal; that is,

$$\sum_{r=1}^{N} W_r A_{rn} A_{rm} = 0 \tag{3-114}$$

where W_r is the rth mass out of a total of N, A represents the characteristic amplitude of a normal mode, and n and m identify any two normal modes. Also, for a total of S springs

$$\sum_{s=1}^{S} k_s y_{sn} y_{sm} = 0 \tag{3-115}$$

where k_s is the constant for the sth spring and y represents the spring distortion.

When there are many degrees of freedom, this procedure for analyzing free vibration becomes very lengthy. In such cases, it may be preferable to solve Eqs. (3-112) by numerical, trial-and-error procedures, such as the Stodola-Vianello method. In that method, the solution converges first on the highest or lowest mode. Then, the other modes are determined by the same procedure after elimination of one of the equations by use of Eq. (3-114). The procedure requires assumption of a characteristic shape, a set of amplitudes A_{r1}. These are substituted in one of Eqs. (3-112) to obtain a first approximation of ω^2. With this value and with $A_{N1} = 1$, the remaining $N - 1$ equations are solved to obtain a new set of A_{r1}. Then, the procedure is repeated until assumed and final characteristic amplitudes agree.

Because even this procedure is very lengthy for many degrees of freedom, the Rayleigh approximate method may be used to compute the fundamental mode. The frequency obtained by this method, however, may be a little on the high side.

The Rayleigh method also starts with an assumed set of characteristic amplitudes A_{r1} and depends for its success on the small error in natural frequency produced by a relatively larger error in the shape assumption. Next, relative inertia forces acting at each mass are computed: $F_r = W_r A_{r1}/A_{N1}$, where A_{N1} is the assumed displacement at one of the masses. These forces are applied to the system as a static load and displacements B_{r1} due to them calculated. Then, the natural frequency can be obtained from

$$\omega^2 = \frac{g \sum_{r=1}^{N} F_r B_{r1}}{\sum_{r=1}^{N} W_r B_{r1}^2} \tag{3-116}$$

where g is the acceleration due to gravity, 386 in. per sec^2. For greater accuracy, the computation can be repeated with B_{r1} as the assumed characteristic amplitudes.

When the Rayleigh method is applied to beams, the characteristic shape assumed initially may be chosen conveniently as the deflection curve for static loading.

The Rayleigh method may be extended to determination of higher modes by the Schmidt orthogonalization procedure, which adjusts assumed deflection curves to satisfy Eq. (3-114). The procedure is to assume a shape, remove components associated with lower modes, then use the Rayleigh method for the residual deflection curve. The computation will converge on the next higher mode. The method is shorter than the Stodola-Vianello procedure when only a few modes are needed.

For example, suppose the characteristic amplitudes A_{r1} for the fundamental mode have been obtained and the natural frequency for the second mode is to be computed. Assume a value for the relative deflection of the rth mass A_{r2}. Then, the shape with the fundamental mode removed will be defined by the displacements

$$a_{r2} = A_{r2} - c_1 A_{r1} \tag{3-117}$$

where c_1 is the participation factor for the first mode.

$$c_1 = \frac{\sum_{r=1}^{N} W_r A_{r2} A_{r1}}{\sum_{r=1}^{N} W_r A_{r1}^2} \tag{3-118}$$

Substitute a_{r2} for B_{r1} in Eq. (3-116) to find the second-mode frequency and, from deflections produced by $F_r = W_r a_{r2}$, an improved shape. (For more rapid convergence, A_{r2} should be selected to make c_1 small.) The procedure should be repeated, starting with the new shape.

For the third mode, assume deflections A_{r3} and remove the first two modes:

$$a_{r3} = A_{r3} - c_1 A_{r1} - c_2 A_{r2} \tag{3-119}$$

The participation factors are determined from

$$c_1 = \frac{\sum_{r=1}^{N} W_r A_{r3} A_{r1}}{\sum_{r=1}^{N} W_r A_{r1}^2} \qquad c_2 = \frac{\sum_{r=1}^{N} W_r A_{r3} A_{r2}}{\sum_{r=1}^{N} W_r A_{r2}^2} \tag{3-120}$$

Use a_{r3} to find an improved shape and the third-mode frequency.

For some structures with mass distributed throughout, it sometimes is easier to solve the dynamic equations based on distributed mass than the equations based on equivalent lumped masses. A distributed mass has an infinite number of degrees of freedom and normal modes. Every particle in it can be considered a lumped mass on springs connected to other particles. Usually, however, only the fundamental mode is significant, though sometimes the second and third modes must be taken into account.

For example, suppose a beam weighs w lb per lin ft and has a modulus of elasticity E, psi, and moment of inertia I, in.^{4}. Let y be the deflection at a distance x from one end. Then, the equation of motion is

$$EI \frac{\partial^4 y}{\partial x^4} + \frac{w}{g} \frac{\partial^2 y}{\partial t^2} = 0 \tag{3-121}$$

(This equation ignores the effects of shear and rotational inertia.) The deflection y_n for each mode, to satisfy the equation, must be the product of a harmonic function of time $f_n(t)$ and of the characteristic shape $Y_n(x)$, a function of x with undetermined amplitude. The solution is

$$f_n(t) = c_1 \sin \omega_n t + c_2 \cos \omega_n t \tag{3-122}$$

where ω_n is the natural circular frequency and n indicates the mode, and

$$Y_n(x) = A_n \sin \beta_n x + B_n \cos \beta_n x + C_n \sinh \beta_n x + D_n \cosh \beta_n x \tag{3-123}$$

where

$$\beta_n = \sqrt[4]{\frac{w \omega_n^2}{EIg}} \tag{3-124}$$

For a simple beam, the boundary (support) conditions for all values of time t are $y = 0$ and bending moment $M = EI\, \partial^2 y/\partial x^2 = 0$. Hence, at $x = 0$ and $x = L$, the span length, $Y_n(x) = 0$ and $d^2Y_n/dx^2 = 0$. These conditions require that

$$B_n = C_n = D_n = 0 \qquad \beta_n = \frac{n\pi}{L}$$

to satisfy Eq. (3-123). Hence, according to Eq. (3-124), the natural circular frequency for a simply supported beam is

$$\omega_n = \frac{n^2\pi^2}{L^2}\sqrt{\frac{EIg}{w}} \tag{3-125}$$

The characteristic shape is defined by

$$Y_n(x) = \sin\frac{n\pi x}{L} \tag{3-126}$$

The constants c_1 and c_2 in Eq. (3-122) are determined by the initial conditions of the disturbance. Thus, the total deflection, by superposition of modes, is

$$y = \sum_{n=1}^{\infty} A_n(t)\sin\frac{n\pi x}{L} \tag{3-127}$$

where $A_n(t)$ is determined by the load (see Art. 3-82).

Equations (3-122) to (3-124) apply to spans with any type of end restraints. Figure 3-77 shows the characteristic shape and gives constants for determination of

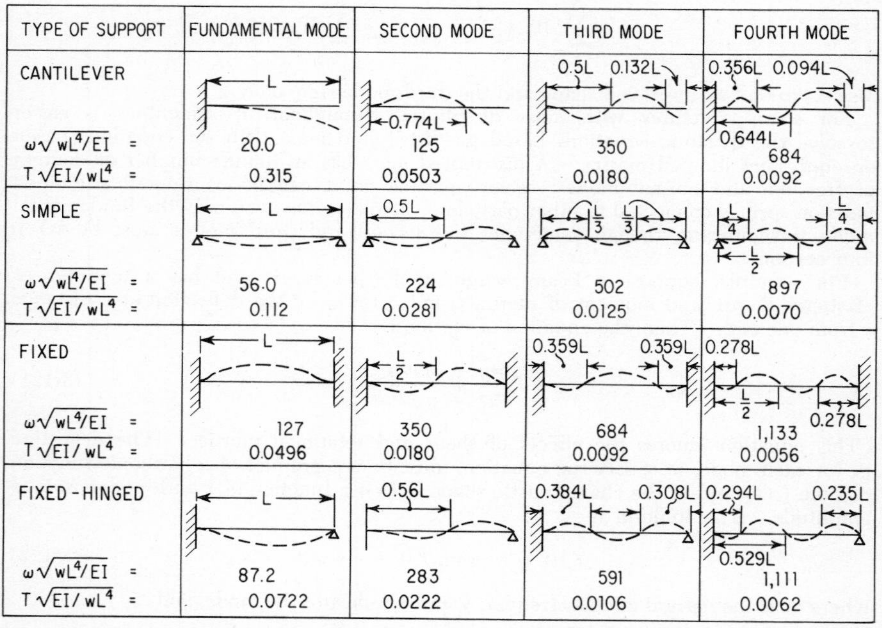

TYPE OF SUPPORT		FUNDAMENTAL MODE	SECOND MODE	THIRD MODE	FOURTH MODE
CANTILEVER		L	0.774L	0.5L, 0.132L	0.356L, 0.094L, 0.644L
	$\omega\sqrt{wL^4/EI}$ =	20.0	125	350	684
	$T\sqrt{EI/wL^4}$ =	0.315	0.0503	0.0180	0.0092
SIMPLE		L	0.5L	L/3, L/3	L/4, L/4, L/2
	$\omega\sqrt{wL^4/EI}$ =	56.0	224	502	897
	$T\sqrt{EI/wL^4}$ =	0.112	0.0281	0.0125	0.0070
FIXED		L	L/2	0.359L, 0.359L	0.278L, L/2, 0.278L
	$\omega\sqrt{wL^4/EI}$ =	127	350	684	1,133
	$T\sqrt{EI/wL^4}$ =	0.0496	0.0180	0.0092	0.0056
FIXED-HINGED		L	0.56L	0.384L, 0.308L	0.294L, 0.235L, 0.529L
	$\omega\sqrt{wL^4/EI}$ =	87.2	283	591	1,111
	$T\sqrt{EI/wL^4}$ =	0.0722	0.0222	0.0106	0.0062

Fig. 3-77. Coefficients for computing natural circular frequencies and natural periods of vibration of prismatic beams.

natural circular frequency ω and natural period T for the first four modes of cantilever, simply supported, fixed-end, and fixed-hinged beams. To obtain ω, select the appropriate constant from Fig. 3-77 and multiply it by $\sqrt{EI/wL^4}$. To get T, divide the appropriate constant by $\sqrt{EI/wL^4}$.

To determine the characteristic shapes and natural periods for beams with variable cross section and mass, use the Rayleigh method. Convert the beam into a lumped-mass system by dividing the span into elements and assuming the mass of each element to be concentrated at its center. Also, compute all quantities, such as deflection and bending moment, at the center of each element. Start with an assumed characteristic shape and apply Eq. (3-116).

Methods are available for dynamic analysis of continuous beams. (G. S. Rogers, "An Introduction to the Dynamics of Framed Structures," John Wiley & Sons, Inc., New York; D. G. Fertis and E. C. Zobel, "Transverse Vibration Theory," The Ronald Press Company, New York.) But even for beams with constant cross section, these procedures are very lengthy. Generally, approximate solutions are preferable.

(J. M. Biggs, "Introduction to Structural Dynamics," McGraw-Hill Book Company, New York; N. M. Newmark and E. Rosenblueth, "Fundamentals of Earthquake Engineering," Prentice-Hall, Englewood Cliffs, N.J.)

3-81. Impact and Sudden Loads. Under impact, there is an abrupt exchange or absorption of energy and drastic change in velocity. Stresses caused in the colliding members may be several times larger than stresses produced by the same weights applied statically.

An approximation of impact stresses in the elastic range can be made by neglecting the inertia of the body struck and the effect of wave propagation and assuming that the kinetic energy is converted completely into strain energy in that body. Consider a prismatic bar subjected to an axial impact load in tension. The energy absorbed per unit of volume when the bar is stressed to the proportional limit is called the **modulus of resilience.** It is given by $f_y^2/2E$, where f_y is the yield stress and E the modulus of elasticity, both in psi. Below the proportional limit, the stress, psi, due to an axial load U, in.-lb, is

$$f = \sqrt{\frac{2UE}{AL}} \tag{3-128}$$

where A is the cross-sectional area, sq in., and L the length of bar, in.

This equation indicates that energy absorption of a member may be improved by increasing its length or area. Sharp changes in cross section should be avoided, however, because of associated high stress concentrations. Also, uneven distribution of stress in a member due to changes in section should be avoided. For example, if part of a member is given twice the diameter of another part, the stress in the larger portion is one-fourth that in the smaller. Since the energy absorbed is proportional to the square of the stress, the energy taken per unit of volume by the larger portion is therefore only one-sixteenth that absorbed by the smaller. So despite the increase in volume due to doubling of the diameter, the larger portion absorbs much less energy than the smaller. Thus, energy absorption would be larger with a uniform stress distribution throughout the length of the member.

If a static axial load W would produce a tensile stress f' in the bar and an elongation e', in., then the axial stress produced when W falls a distance h, in., is

$$f = f' + f'\sqrt{1 + \frac{2h}{e'}} \tag{3-129}$$

if f is within the proportional limit. The elongation due to this impact load is

$$e = e' + e'\sqrt{1 + \frac{2h}{e'}} \tag{3-130}$$

These equations indicate that the stress and deformation due to an energy load may be considerably larger than those produced by the same weight applied gradually.

The same equations hold for a beam with constant cross section struck by a weight at midspan, except that f and f' represent stresses at midspan and e and e', midspan deflections.

According to Eqs. (3-129) and (3-130), a sudden load ($h = 0$) causes twice the stress and twice the deflection as the same load applied gradually.

For very long members, the effect of wave propagation should be taken into account. Impact is not transmitted instantly to all parts of the struck body. At first, remote parts remain undisturbed, while particles struck accelerate rapidly to the velocity of the colliding body. The deformations produced move through the struck body in the form of elastic waves. The waves travel with a constant velocity, fps,

$$c = 68.1\sqrt{\frac{E}{\rho}} \tag{3-131}$$

where E = modulus of elasticity, psi
ρ = density of the struck body, lb per cu ft

If an impact imparts a velocity v, fps, to the particles at one end of a prismatic bar, the stress, psi, at that end is

$$f = E\frac{v}{c} = 0.0147v\sqrt{E\rho} = 0.000216\rho cv \tag{3-132}$$

if f is in the elastic range. In a compression wave, the velocity of the particles is in the direction of the wave. In a tension wave, the velocity of the particles is in the direction opposite the wave.

In the plastic range, Eqs. (3-131) and (3-132) hold, but with E as the tangent modulus of elasticity. Hence, c is not a constant and the shape of the stress wave changes as it moves. The elastic portion of the stress wave moves faster than the wave in the plastic range. Where they overlap, the stress and irrecoverable strain are constant.

(The impact theory is based on an assumption difficult to realize in practice—that contact takes place simultaneously over the entire end of the bar.)

At the free end of a bar, a compressive stress wave is reflected as an equal-tension wave, and a tension wave as an equal-compression wave. The velocity of the particles there equals $2v$.

At a fixed end of a bar, a stress wave is reflected unchanged. The velocity of the particles there is zero, but the stress is doubled, because of the superposition of the two equal stresses on reflection.

For a bar with a fixed end struck at the other end by a moving mass weighing W_m lb, the initial compressive stress, psi, is

$$f_o = 0.0147v_o\sqrt{E\rho} \tag{3-133}$$

where v_o is the initial velocity of the particles, ft per sec, at the impacted end of the bar and E and ρ the modulus of elasticity, psi, and density, lb per cu ft, of the bar. As the velocity of W_m decreases, so does the pressure on the bar. Hence, decreasing compressive stresses follow the wave front. At any time $t < 2L/c$, where L is the length of the bar, in., the stress at the struck end is

$$f = f_o e^{-2\alpha t/\tau} \tag{3-134}$$

where e = 2.71828, α is the ratio of W_b, the weight of the bar, to W_m, and $\tau = 2L/c$.

When $t = \tau$, the wave front with stress f_0 arrives back at the struck end, assumed still to be in contact with the mass. Since the velocity of the mass cannot change suddenly, the wave will be reflected as from a fixed end. During the second interval, $\tau < t < 2\tau$, the compressive stress is the sum of two waves moving away from the struck end and one moving toward this end.

Maximum stress from impact occurs at the fixed end. For α greater than 0.2, this stress is

$$f = 2f_o(1 + e^{-2\alpha}) \tag{3-135}$$

For smaller values of α, it is given approximately by

$$f = f_o\left(1 + \sqrt{\frac{1}{\alpha}}\right) \tag{3-136}$$

Duration of impact, time it takes for the impact stress at the struck end to drop to zero, is approximately

$$T = \frac{\pi L}{c\sqrt{\alpha}} \tag{3-137}$$

for small values of a.

When W_m is the weight of a falling body, velocity at impact is $\sqrt{2gh}$, when it falls a distance h, in. Substitution in Eq. (3-133) yields $f_o = \sqrt{2EhW_b/AL}$, since $W_b = \rho AL$ is the weight of the bar. Putting $W_b = \alpha W_m$; $W_m/A = f'$, the stress produced by W_m when applied gradually, and $E = f'L/e'$, where e' is the elongation for the static load, gives $f_o = f' \sqrt{2h\alpha/e'}$. Then, for values of α smaller than 0.2, the maximum stress, from Eq. (3-136), is

$$f = f'\left(\sqrt{\frac{2h\alpha}{e'}} + \sqrt{\frac{2h}{e'}}\right) \tag{3-138}$$

For larger values of α, the stress wave due to gravity acting on W_m during impact should be added to Eq. (3-135). Thus, for α larger than 0.2,

$$f = 2f'(1 - e^{-2\alpha}) + 2f'\sqrt{\frac{2h\alpha}{e'}}\,(1 + e^{-2\alpha}) \tag{3-139}$$

Equations (3-138) and (3-139) correspond to Eq. (3-129), which was developed without taking wave effects into account. For a sudden load, $h = 0$, Eq. (3-139) gives for the maximum stress $2f'(1 - e^{-2\alpha})$, not quite double the static stress, the result indicated by Eq. (3-29). (See also Art. 3-82.)

(S. Timoshenko and J. N. Goodier, "Theory of Elasticity," McGraw-Hill Book Company, New York; S. Timoshenko and D. H. Young, "Engineering Mechanics," McGraw-Hill Book Company, New York; D. D. Barkan, "Dynamics of Bases and Foundations," McGraw-Hill Book Company, New York; J. H. Faupel, "Engineering Design," John Wiley & Sons, Inc., New York.)

3-82. Dynamic Analysis of Simple Structures. (See also Arts. 3-79 to 3-81.) As noted in Art. 3-80, an approximate solution based on an idealized representation of an actual member or structure is advisable for dynamic analysis and design. Generally, the actual structure may be conveniently represented by a system of masses and massless springs, with additional resistances to account for damping. In simple cases, the masses may be set equal to the actual masses; otherwise, equivalent masses may be substituted for the actual masses (Art. 3-84). The spring constants are the ratios of forces to deflections (see Art. 3-80).

Usually, for structural purposes, the data sought are the maximum stresses in the springs and their maximum displacements and the time of occurrence of the maximums. This time is generally computed in terms of the natural period of vibration of the member or structure, or in terms of the duration of the load. Maximum displacement may be calculated in terms of the deflection that would result if the load were applied gradually.

The term D by which the static deflection e', spring forces, and stresses are multiplied to obtain the dynamic effects is called the **dynamic load factor.** Thus, the dynamic displacement is

$$y = De' \tag{3-140}$$

And the maximum displacement y_m is determined by the maximum dynamic load factor D_m, which occurs at time t_m.

One-degree Systems. Consider the one-degree-of-freedom system in Fig. 3-78*a*. It may represent a weightless beam with a mass weighing W lb applied at midspan and subjected to a varying force $F_of(t)$, or a rigid frame with a mass weighing

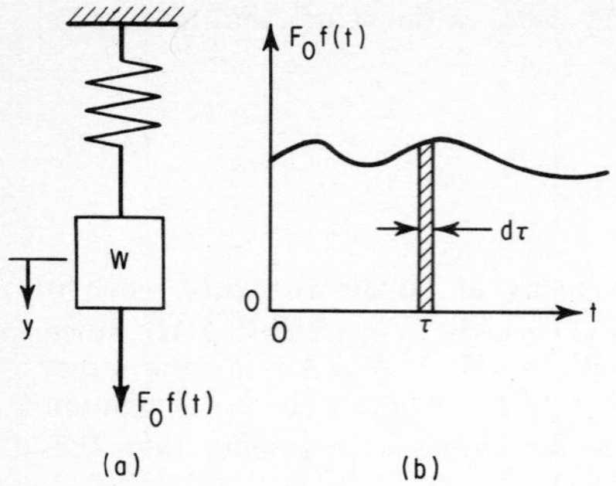

Fig. 3-78. One-degree system acted on by varying force.

W lb at girder level and subjected to this force. The force is represented by an arbitrarily chosen constant force F_o times $f(t)$, a function of time.

If the system is not damped, the equation of motion in the elastic range is

$$\frac{W}{g}\frac{d^2y}{dt^2} + ky = F_o f(t) \qquad \text{(3-141)}$$

where k is the spring constant and g the acceleration due to gravity, 386 in. per sec². The solution consists of two parts. The first, called the complementary solution, is obtained by setting $f(t) = 0$. This solution is given by Eq. (3-105). To it must be added the second part, the particular solution, which satisfies Eq. (3-141).

The general solution of Eq. (3-141), arrived at by treating an element of the force-time curve (Fig. 3-78*b*) as an impulse, is

$$y = y_o \cos \omega t + \frac{v_o}{\omega} \sin \omega t + e'\omega \int_0^t f(\tau) \sin \omega(t - \tau)\, d\tau \qquad \text{(3-142)}$$

where y = displacement of mass from equilibrium position, in.
y_o = initial displacement of mass ($t = 0$), in.
$\omega = \sqrt{kg/W}$ = natural circular frequency of free vibration
k = spring constant = force producing unit deflection, lb per in.
v_o = initial velocity of mass, in. per sec
$e' = F_o/k$ = displacement under static load, in.

A closed solution is possible if the integral can be evaluated.

Assume, for example, the mass is subjected to a suddenly applied force F_o that remains constant (Fig. 3-79*a*). If y_o and v_o are initially zero, the displacement y of the mass at any time t can be obtained from the integral in Eq. (3-142) by setting $f(t) = 1$:

$$y = e'\omega \int_0^t \sin \omega(t - \tau)\, d\tau = e'(1 - \cos \omega t) \qquad \text{(3-143)}$$

The dynamic load factor $D = 1 - \cos \omega t$. It has a maximum value $D_m = 2$ when $t = \pi/\omega$. Figure 3-79*b* shows the variation of displacement with time. For values of D_m for some other types of forces, see Table 3-8.

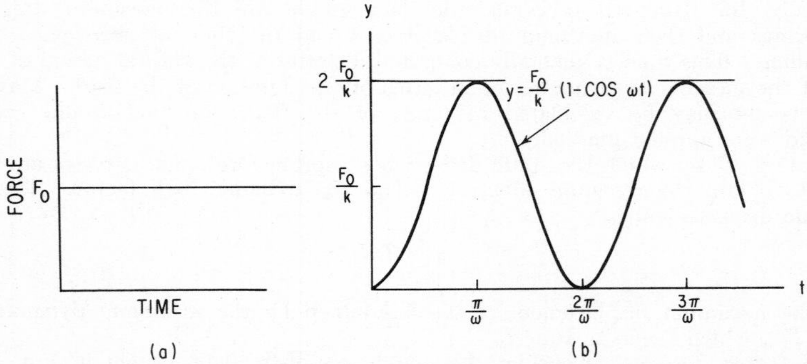

Fig. 3-79. Harmonic vibrations (*b*) result when constant force (*a*) is applied to an undamped one-degree system such as the one in Fig. 3-78*a*.

Multidegree Systems. A multidegree lumped-mass system may be analyzed by the modal method after the natural frequencies of the normal modes have been determined (Art. 3–80). This method is restricted to linearly elastic systems in which the forces applied to the masses have the same variation with time. For other cases, numerical analysis must be used.

In the modal method, each normal mode is treated as an independent one-degree system. For each degree of the system, there is one normal mode. A natural frequency and a characteristic shape are associated with each mode. In each mode, the ratio of the displacements of any two masses is constant with time. These ratios define the characteristic shape. The modal equation of motion for each mode is

$$\frac{d^2A_n}{dt^2} + \omega_n{}^2A_n = \frac{gf(t)\sum_{r=1}^{j} F_r\phi_{rn}}{\sum_{r=1}^{j} W_r\phi_{rn}{}^2} \tag{3-144}$$

where A_n = displacement in the nth mode of an arbitrarily selected mass
ω_n = natural frequency of the nth mode
$F_rf(t)$ = varying force applied to the rth mass
W_r = weight of the rth mass
j = number of masses in the system
ϕ_{rn} = ratio of the displacement in the nth mode of the rth mass to A_n
g = acceleration due to gravity

We define the modal static deflection as

$$A'_n = \frac{g\sum_{r=1}^{j} F_r\phi_{rn}}{\omega_n{}^2\sum_{r=1}^{j} W_r\phi_{rn}{}^2} \tag{3-145a}$$

Then, the response for each mode is given by

$$A_n = D_nA'_n \tag{3-145b}$$

where D_n = dynamic load factor.

Since D_n depends only on ω_n and the variation of force with time $f(t)$, solutions for D_n obtained for one-degree systems also apply to multidegree systems. Thus, the dynamic load factor for each mode may be obtained from Table 3-8. The total deflection at any point is the sum of the displacements for each mode, $\Sigma A_n\phi_{rn}$, at that point.

Beams. The response of beams to dynamic forces can be determined in a similar way. The modal static deflection is defined by

$$A'_n = \frac{\int_0^L p(x)\phi_n(x)\,dx}{\omega_n{}^2\dfrac{w}{g}\int_0^L \phi_n{}^2(x)\,dx} \tag{3-145c}$$

where $p(x)$ = load distribution on the span [$p(x)f(t)$ is the varying force]
$\phi_n(x)$ = characteristic shape of the nth mode (see Art. 3-80)
L = span length
w = uniformly distributed weight on the span

The response of the beam then is given by Eq. (3-145b), and the dynamic deflection is the sum of the modal components, $\Sigma A_n\phi_n(x)$.

Nonlinear Responses. When the structure does not react linearly to loads, the equations of motion can be solved by numerical analysis if resistance is a unique function of displacement. Sometimes, the behavior of the structure can be represented by an idealized resistance-displacement diagram that makes possible a solution in closed form. Figure 3–80a shows such a diagram.

Elastic-plastic Responses. Resistance is assumed linear ($R = ky$) in Fig. 3-80*a* until a maximum R_m is reached. After that, R remains equal to R_m for increases in y substantially larger than the displacement y_e at the elastic limit. Thus, some portions of the structure deform into the plastic range. Figure 3-80*a*, therefore, may be used for ductile structures only rarely subjected to severe dynamic loads. When this diagram can be used for designing such structures, more economical designs can be produced than for structures limited to the elastic range, because of the high energy-absorption capacity of structures in the plastic range.

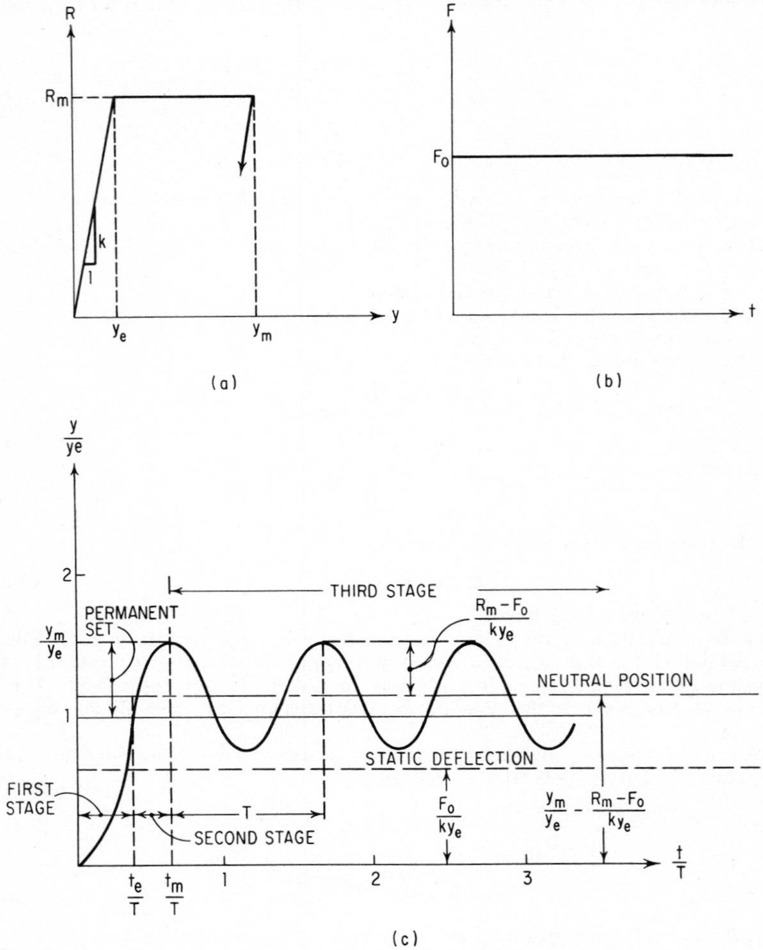

Fig. 3-80. Response in the plastic range of a one-degree system with resistance characteristics indicated in (*a*) to a constant force (*b*) is shown in (*c*).

For a one-degree system, Eq. (3-141) can be used as the equation of motion for the initial sloping part of the diagram (elastic range). For the second stage, $y_e < y < y_m$, where y_m is the maximum displacement, the equation is

$$\frac{W}{g}\frac{d^2y}{dt^2} + R_m = F_of(t) \qquad \text{(3-146}a\text{)}$$

For the unloading stage, $y < y_m$, the equation is

$$\frac{W}{g}\frac{d^2y}{dt^2} + R_m - k(y_m - y) = F_o f(t) \tag{3-146b}$$

Suppose, for example, the one-degree undamped system in Fig. 3-78*a* behaves in accordance with the bilinear resistance function of Fig. 3-80*a* and is subjected to a suddenly applied constant load (Fig. 3-80*b*). With zero initial displacement and velocity, the response in the first stage ($y < y_e$), according to Eq. (3-143), is

$$y = e'(1 - \cos \omega t_1) \tag{3-147a}$$

$$\frac{dy}{dt} = e'\omega \sin \omega t_1 \tag{3-147b}$$

Equation (3-143) also indicates that y_e will be reached at a time t_e such that $\cos \omega t_e = 1 - y_e/e'$.

For convenience, let $t_2 = t - t_e$ be the time in the second stage; thus, $t_2 = 0$ at the start of that stage. Since the condition of the system at that time is the same as at the end of the first stage, the initial displacement is y_e and the initial velocity $e'\omega \sin \omega t_e$.

The equation of motion is

$$\frac{W}{g}\frac{d^2y}{dt^2} + R_m = F_o \tag{3-146c}$$

The solution, taking into account initial conditions after integrating, for $y_e < y < y_m$ is

$$y = \frac{g}{2W}(F_o - R_m)t_2^2 + e'\omega t_2 \sin \omega t_e + y_e \tag{3-148}$$

Maximum displacement occurs at the time

$$t_m = \frac{W\omega e'}{g(R_m - F_o)} \sin \omega t_e \tag{3-149}$$

and can be obtained by substituting t_m in Eq. (3-148).

The third stage, unloading after y_m has been reached, can be determined from Eq. (3-146*b*) and conditions at the end of the second stage. The response, however, is more easily found by noting that the third stage consists of an elastic, harmonic residual vibration. In this stage the amplitude of vibration is $(R_m - F_o)/k$, since this is the distance between the neutral position and maximum displacement, and in the neutral position the spring force equals F_o. Hence, the response can be obtained directly from Eq. (3-143) by substituting $y_m - (R_m - F_o)/k$ for e', because the neutral position, $y = y_m - (R_m - F_o)/k$, occurs when $\omega t_3 = \pi/2$, where $t_3 = t - t_e - t_m$. The solution is

$$y = y_m - \frac{R_m - F_o}{k} + \frac{R_m - F_o}{k} \cos \omega t_3 \tag{3-150}$$

Response in the three stages is shown in Fig. 3-80*c*. In that diagram, however, to represent a typical case, the coordinates have been made nondimensional by expressing y in terms of y_e and the time in terms of T, the natural period of vibration.

Figure 3-81 shows some idealized force-time functions. When actual conditions approximate these, maximum displacement and time of its occurrence may be obtained from Table 3-8, which was calculated on the assumption of a bilinear resistance-displacement relationship (Fig. 3-80*a*). For each type of load, as load duration or time of rise varies, Table 3-8 lists the dynamic load factors in the elastic range and the ratio $R_m/F_o = y_e/e'$ in the plastic range. Values in the plastic range are given for several values of y_m/y_e.

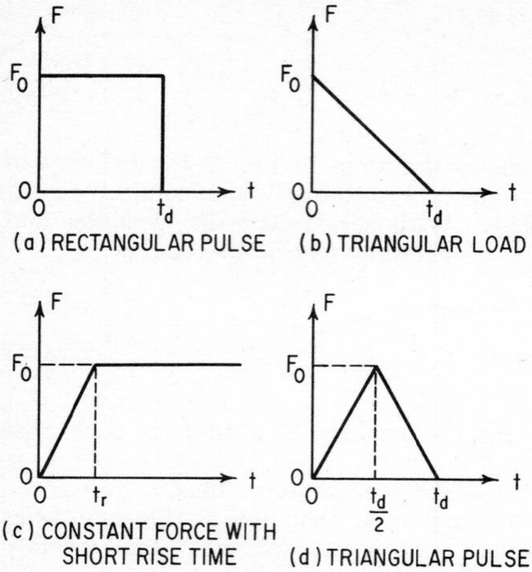

Fig. 3-81. Idealized force-time relationships.

(J. M. Biggs, "Introduction to Structural Dynamics," McGraw-Hill Book Company, New York; G. L. Rogers, "Dynamics of Framed Structures," John Wiley & Sons, Inc., New York; D. G. Fertis and E. C. Zobel, "Transverse Vibration Theory," The Ronald Press Company, New York; N. M. Newmark and E. Rosenblueth, "Fundamentals of Earthquake Engineering," Prentice-Hall, Englewood Cliffs, N.J.)

3-83. Resonance and Damping. Damping in structures, due to friction and other causes, resists motion imposed by dynamic loads. Generally, the effect is to decrease the amplitude and lengthen the period of vibrations. If damping is large enough, vibration may be eliminated.

When maximum stress and displacement are the prime concern, damping may not be of great significance for short-time loads. These maximums usually occur under such loads at the first peak of response, and damping, unless unusually large, has little effect in a short period of time. But under conditions close to resonance, damping has considerable effect.

Resonance is the condition of a vibrating system under a varying load such that the amplitude of successive vibrations increases. Unless limited by damping or changes in the condition of the system, amplitudes may become very large.

Two forms of damping generally are assumed in structural analysis, viscous or constant (Coulomb). For viscous damping, the damping force is taken proportional to the velocity but opposite in direction. For Coulomb damping, the damping force is assumed constant and opposed in direction to the velocity.

Viscous Damping. For a one-degree system (Arts. 3-80 to 3-82), the equation of motion for a mass weighing W lb and subjected to a force F varying with time but opposed by viscous damping is

$$\frac{W}{g}\frac{d^2y}{dt^2} + ky = F - c\frac{dy}{dt} \tag{3-151}$$

where y = displacement of the mass from equilibrium position, in.
k = spring constant, lb per in.
t = time, sec
c = coefficient of viscous damping
g = acceleration due to gravity = 386 in. per sec²

Let us set $\beta = cg/2W$ and consider those cases in which $\beta < \omega$, the natural circular frequency [Eq. (3-106)], to eliminate unusually high damping (overdamping). Then, for initial displacement y_o and velocity v_o, the solution of Eq. (3-151) with $F = 0$ is

$$y = e^{-\beta t}\left(\frac{v_o + \beta y_o}{\omega_d}\sin \omega_d t + y_o \cos \omega_d t\right) \tag{3-152}$$

where $\omega_d = \sqrt{\omega^2 - \beta^2}$ and $e = 2.71828$. Equation (3-152) represents a decaying harmonic motion with β controlling the rate of decay and ω_d the natural frequency of the damped system.

When $\beta = \omega$

$$y = e^{-\omega t}[v_o t + (1 + \omega t)y_o] \tag{3-153}$$

Table 3-8. Maximum Response and Occurrence Time for Short-time Loads*

Rectangular Pulses (Fig. 3-81*a*)

t_d/T	Elastic response		$y_m/y_e = 1.5$		$y_m/y_e = 2.0$		$y_m/y_e = 5.0$		$y_m/y_e = 10.0$	
	D_m	t_m/T	R_m/F_o	t_m/T	R_m/F_o	t_m/T	R_m/F_o	t_m/T	R_m/F_o	t_m/T
0.1	0.6	0.3	0.5	0.3	0.4	0.4	0.2	0.5	0.2	0.7
0.2	1.2	0.4	0.8	0.4	0.6	0.4	0.4	0.5	0.3	0.8
0.3	1.6	0.4	1.2	0.4	1.0	0.5	0.5	0.6	0.4	0.8
0.4	1.9	0.5	1.4	0.5	1.1	0.5	0.7	0.7	0.5	0.9
0.5	2.0	0.5	1.5	0.6	1.2	0.6	0.8	0.8	0.6	0.9
1.0	2.0	0.5	1.5	0.6	1.3	0.8	1.1	1.1	0.9	1.3
10.0	2.0	0.5	1.5	0.6	1.3	0.8	1.1	1.7	1.1	3.0

Triangular Loads (Fig. 3-81*b*)

t_d/T	Elastic response		$y_m/y_e = 1.5$		$y_m/y_e = 2.0$		$y_m/y_e = 5.0$		$y_m/y_e = 10.0$	
	D_m	t_m/T	R_m/F_o	t_m/T	R_m/F_o	t_m/T	R_m/F_o	t_m/T	R_m/F_o	t_m/T
0.1	0.3	0.3	0.2	0.3	0.2	0.4	0.1	0.5		
0.2	0.6	0.3	0.4	0.4	0.3	0.4	0.2	0.6	0.1	0.8
0.3	0.8	0.4	0.6	0.4	0.5	0.5	0.3	0.7	0.2	0.9
0.4	1.0	0.4	0.7	0.4	0.6	0.5	0.4	0.7	0.3	0.9
0.5	1.2	0.4	0.9	0.4	0.7	0.6	0.4	0.7	0.3	0.9
1.0	1.5	0.5	1.1	0.5	1.0	0.6	0.7	0.8	0.5	1.0
5.0	2.0	0.5	1.4	0.6	1.3	0.6	1.0	1.4	0.9	2.0
10.0	2.0	0.5	1.4	0.7	1.3	0.7	1.1	1.5	1.0	2.5

Constant Force with Short Rise Times (Fig. 3-81*c*)

t_r/T	Elastic response		$y_m/y_e = 1.5$		$y_m/y_e = 2.0$		$y_m/y_e = 5.0$		$y_m/y_e = 10.0$	
	D_m	t_m/T	R_m/F_o	t_m/T	R_m/F_o	t_m/T	R_m/F_o	t_m/T	R_m/F_o	t_m/T
0.1	2.0	0.5	1.5	0.8	1.3	1.0	1.1	1.7	1.1	3.5
0.5	1.7	0.8	1.3	1.0	1.2	1.1	1.1	2.6	1.0	6.0
1.0	1.0	1.0	1.0	1.0	1.0	1.0	1.0	1.0	1.0	1.0
1.5	1.2	1.4	1.0	2.2	1.0	3.0	1.0	3.0		
2.0	1.0	2.0	1.0	2.0	1.0	2.0	1.0	2.0		

Triangular Pulses (Fig. 3-81*d*)

t_d/T	Elastic response		$y_m/y_e = 1.5$		$y_m/y_e = 2.0$		$y_m/y_e = 5.0$		$y_m/y_e = 10.0$	
	D_m	t_m/T	R_m/F_o	t_m/T	R_m/F_o	t_m/F	R_m/F_o	t_m/F	R_m/F_o	t_m/F
0.1	0.3	0.3	0.2	0.3	0.2	0.4				
0.2	0.6	0.4	0.5	0.4	0.4	0.4	0.2	0.6	0.2	0.6
0.3	1.0	0.4	0.7	0.4	0.5	0.4	0.3	0.6	0.2	0.9
0.4	1.2	0.5	0.8	0.5	0.7	0.5	0.4	0.7	0.3	0.9
0.5	1.3	0.5	0.9	0.6	0.8	0.6	0.5	0.7	0.3	1.0
1.0	1.6	0.7	1.1	0.7	1.0	0.8	0.7	1.0	0.5	1.3
5.0	1.1	2.7	1.0	3.0	0.9	3.1	0.8	3.5	0.8	3.9
10.0	1.0	5.0	1.0	5.5	0.9	6.0	0.9	6.3	0.8	7.0

* Source: U.S. Army Corps of Engineers, "Design of Structures to Resist the Effects of Atomic Weapons," Manual EM 1110-345-415.

which indicates that the motion is not vibratory. Damping producing this condition is called critical, and the critical coefficient is

$$c_d = \frac{2W\beta}{g} = \frac{2W\omega}{g} = \sqrt{\frac{kW}{g}} \tag{3-154}$$

Damping sometimes is expressed as a percent of critical (β as a percent of ω).

For small amounts of viscous damping, the damped natural frequency is approximately equal to the undamped natural frequency minus $\frac{1}{2}\beta^2/\omega$. For example, for 10% critical damping ($\beta = 0.1\omega$), $\omega_d = \omega[1 - \frac{1}{2}(0.1)^2] = 0.995\omega$. Hence, the decrease in natural frequency due to damping generally can be ignored.

Damping sometimes is measured by **logarithmic decrement,** the logarithm of the ratio of two consecutive peak amplitudes during free vibration.

$$\text{Logarithmic decrement} = \frac{2\pi\beta}{\omega} \tag{3-155}$$

For example, for 10% critical damping, the logarithmic decrement equals 0.2π. Hence, the ratio of a peak to the following peak amplitude is $e^{0.2\pi} = 1.87$.

The complete solution of Eq. (3-151) with initial displacement y_o and velocity v_o is

$$y = e^{-\beta t}\left(\frac{v_o + \beta y_o}{\omega_d} \sin \omega_d t + y_o \cos \omega_d t\right) + e' \frac{\omega^2}{\omega_d} \int_0^t f(\tau) e^{-\beta(t-\tau)} \sin \omega_d (t - \tau)\, d\tau \tag{3-156}$$

where e' is the deflection that the applied force would produce under static loading. Equation (3-156) is identical to Eq. (3-142) when $\beta = 0$.

Unbalanced rotating parts of machines produce pulsating forces that may be represented by functions of the form $F_o \sin \alpha t$. If such a force is applied to an undamped one-degree system, Eq. (3-142) indicates that if the system starts at rest the response will be

$$y = \frac{F_o g}{W}\left(\frac{1/\omega^2}{1 - \alpha^2/\omega^2}\right)\left(\sin \alpha t - \frac{\alpha}{\omega} \sin \omega t\right) \tag{3-157}$$

And since the static deflection would be $F_o/k = F_o g/W\omega^2$, the dynamic load factor is

$$D = \frac{1}{1 - \alpha^2/\omega^2}\left(\sin \alpha t - \frac{\alpha}{\omega} \sin \omega t\right) \tag{3-158}$$

If α is small relative to ω, maximum D is nearly unity; thus, the system is practically statically loaded. If α is very large compared with ω, D is very small; thus, the mass cannot follow the rapid fluctuations in load and remains practically stationary. Therefore, when α differs appreciably from ω, the effects of unbalanced rotating parts are not too serious. But if $\alpha = \omega$, resonance occurs; D increases with time. Hence, to prevent structural damage, measures must be taken to correct the unbalanced parts to change α, or to change the natural frequency of the vibrating mass, or damping must be provided.

The response as given by Eq. (3-157) consists of two parts, the free vibration and the forced part. When damping is present, the free vibration is of the form of Eq. (3-152) and is rapidly damped out. Hence, the free part is called the **transient response,** and the forced part, the **steady-state response.** The maximum value of the dynamic load factor for the steady-state response D_m is called the **dynamic magnification factor.** It is given by

$$D_m = \frac{1}{\sqrt{(1 - \alpha^2/\omega^2)^2 + (2\beta\alpha/\omega^2)^2}} \tag{3-159}$$

With damping, then, the peak values of D_m occur when $\alpha = \omega\sqrt{1-\beta^2/\omega^2}$ and are approximately equal to $\omega/2\beta$. For example, for 10% critical damping,

$$D_m = \frac{\omega}{0.2\omega} = 5$$

So even small amounts of damping significantly limit the response at resonance.

Coulomb Damping. For a one-degree system with Coulomb damping, the equation of motion for free vibration is

$$\frac{W}{g}\frac{d^2y}{dt^2} + ky = \pm F_f \qquad (3\text{-}160)$$

where F_f is the constant friction force and the positive sign applies when the velocity is negative. If initial displacement is y_o and initial velocity is zero, the response in the first half cycle, with negative velocity, is

$$y = \left(y_o - \frac{F_f}{k}\right)\cos \omega t + \frac{F_f}{k} \qquad (3\text{-}161)$$

equivalent to a system with a suddenly applied constant force. For the second half cycle, with positive velocity, the response is

$$y = \left(-y_o + 3\frac{F_f}{k}\right)\cos \omega\left(t - \frac{\pi}{\omega}\right) - \frac{F_f}{k} \qquad (3\text{-}162)$$

If the solution is continued with the sign of F_f changing in each half cycle, the results will indicate that the amplitude of positive peaks is given by $y_o - 4nF_f/k$, where n is the number of complete cycles, and the response will be completely damped out when $t = ky_oT/4F_f$, where T is the natural period of vibration, or $2\pi/\omega$.

Analysis of the steady-state response with Coulomb damping is complicated by the possibility of frequent cessation of motion.

(J. P. Den Hartog, "Mechanical Vibrations," McGraw-Hill Book Company; L. S. Jacobsen and R. S. Ayre, "Engineering Vibrations," McGraw-Hill Book Company; D. D. Barkan, "Dynamics of Bases and Foundations," McGraw-Hill Book Company; W. C. Hurty and M. F. Rubinstein, "Dynamics of Structures," Prentice-Hall, Englewood, Cliffs, N.J.)

3-84. Approximate Design for Dynamic Loading. (See also Arts. 3-79 to 3-83.) Complex analysis and design methods seldom are justified for structures subject to dynamic loading because of lack of sufficient information on loading, damping, resistance to deformation, and other factors. In general, it is advisable to represent the actual structure and loading by idealized systems that permit a solution in closed form.

Whenever possible, represent the actual structure by a one-degree system consisting of an equivalent mass with massless spring. For structures with distributed mass, simplify the analysis in the elastic range by computing the response only for one or a few of the normal modes. In the plastic range, treat each stage—elastic, elastic-plastic, and plastic—as completely independent; for example, a fixed-end beam may be treated, when in the elastic-plastic stage, as a simple supported beam.

Choose the parameters of the equivalent system to make the deflection at a critical point, such as the location of the concentrated mass, the same as it would be in the actual structure. Stresses in the actual structure should be computed from the deflections in the equivalent system.

Compute an assumed shape factor ϕ for the system from the shape taken by the actual structure under static application of the loads. For example, for a simple beam in the elastic range with concentrated load at midspan, ϕ may be chosen, for $x < L/2$, as $(Cx/L^3)(3L^2 - 4x^2)$, the shape under static loading, and C may be set equal to 1 to make ϕ equal to 1 when $x = L/2$. For plastic conditions (hinge at midspan), ϕ may be taken as Cx/L, and C set equal to 2, to make $\phi = 1$ when x = $L/2$.

For a structure with concentrated forces, let W_r be the weight of the rth mass, ϕ_r the value of ϕ at the location of that mass, and F_r the dynamic force acting on W_r. Then, the equivalent weight of the idealized system is

$$W_e = \sum_{r=1}^{j} W_r \phi_r^2 \tag{3-163}$$

where j is the number of masses. The equivalent force is

$$F_e = \sum_{r=1}^{j} F_r \phi_r \tag{3-164}$$

For a structure with continuous mass, the equivalent weight is

$$W_e = \int w \phi^2 \, dx \tag{3-165}$$

where w is the weight in lb per lin ft. The equivalent force is

$$F_e = \int q \phi \, dx \tag{3-166}$$

for a distributed load q, lb per lin ft.

The resistance of a member or structure is the internal force tending to restore it to its unloaded static position. For most structures, a bilinear resistance function, with slope k up to the elastic limit and zero slope in the plastic range (Fig. 3-80a), may be assumed. For a given distribution of dynamic load, maximum resistance of the idealized system may be taken as the total load with that distribution that the structure can support statically. Similarly, stiffness is numerically equal to the total load with the given distribution that would cause a unit deflection at the point where the deflections in the actual structure and idealized system are equal. Hence, the equivalent resistance and stiffness are in the same ratio to the actual as the equivalent forces to the actual forces.

Let k be the actual spring constant, g acceleration due to gravity, 386 in. per sec², and

$$W' = \frac{W_e}{F_e} \Sigma F \tag{3-167}$$

where ΣF represents the actual total load. Then, the equation of motion of an equivalent one-degree system is

$$\frac{d^2y}{dt^2} + \omega^2 y = g \frac{\Sigma F}{W'} \tag{3-168}$$

and the natural circular frequency is

$$\omega = \sqrt{\frac{kg}{W'}} \tag{3-169}$$

The natural period of vibration equals $2\pi/\omega$. Equations (3-168) and (3-169) have the same form as Eqs. (3-104), (3-106), and (3-141). Consequently, the response can be computed as indicated in Arts. 3-80 to 3-82.

Whenever possible, select a load-time function for ΣF to permit use of a known solution, such as those in Table 3-8.

For preliminary design of a one-degree system loaded into the plastic range by a suddenly applied force that remains substantially constant up to the time of maximum response, the following approximation may be used for that response:

$$y_m = \frac{y_e}{2(1 - F_o/R_m)} \tag{3-170}$$

where y_e is the displacement at the elastic limit, F_o the average value of the force, and R_m the maximum resistance of the system. This equation indicates that for purely elastic response, R_m must be twice F_o; whereas, if y_m is permitted to be large, R_m may be made nearly equal to F_o, with greater economy of material.

For preliminary design of a one-degree system subjected to a sudden load with duration t_d less than 20% of the natural period of the system, the following approximation can be used for the maximum response:

$$y_m = \frac{1}{2} y_e \left[\left(\frac{F_o}{R_m} \omega t_d \right)^2 + 1 \right] \tag{3-171}$$

where F_o is the maximum value of the load and ω the natural frequency. This equation also indicates that the larger y_m is permitted to be, the smaller R_m need be.

For a beam, the spring force of the equivalent system is not the actual force, or reaction, at the supports. The real reactions should be determined from the dynamic equilibrium of the complete beam. This calculation should include the inertia force, with distribution identical with the assumed deflected shape of the beam. For example, for a simply supported beam with uniform load, the dynamic reaction in the elastic range is $0.39R + 0.11F$, where R is the resistance, which varies with time, and $F = qL$ is the load. For a concentrated load F at midspan, the dynamic reaction is $0.78R - 0.28F$. And for concentrated loads $F/2$ at each third point, it is $0.62R - 0.12F$. (Note that the sum of the coefficients equals 0.50, since the dynamic-reaction equations must hold for static loading, when $R = F$.) These expressions also can be used for fixed-end beams without significant error. If high accuracy is not required, they also can be used for the plastic range.

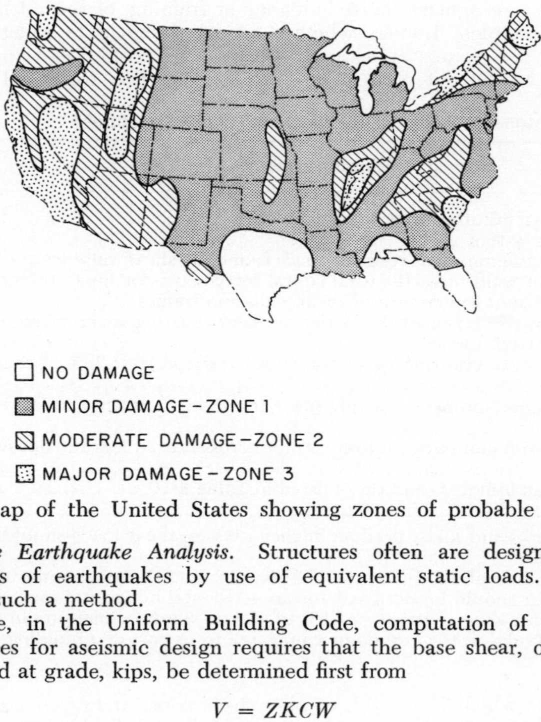

Fig. 3-82. Map of the United States showing zones of probable seismic intensity.

Approximate Earthquake Analysis. Structures often are designed to resist the dynamic forces of earthquakes by use of equivalent static loads. Many building codes specify such a method.

For example, in the Uniform Building Code, computation of equivalent static, horizontal forces for aseismic design requires that the base shear, or total horizontal force developed at grade, kips, be determined first from

$$V = ZKCW \tag{3-172}$$

Z provides for variation in design forces with changes in the probability of seismic intensity throughout the country. $Z = 0.25$ for zone 1; $Z = 0.5$ for zone 2; and $Z = 1$ for zone 3 (Fig. 3-82).

W, in general, is the total dead load, kips. For storage and warehouse occupancies, however, W should be taken as the dead load plus 25% of the live load. There are other cases, however, where designers would find it prudent and realistic to include a portion of the live load in W.

$C = 0.05/\sqrt[3]{T}$ is called the seismic coefficient. T is the fundamental period of vibration of the structure, sec. For a building as a whole, C need not exceed 0.1. For one- and two-story buildings, it may be taken as 0.1.

T may be computed from

$$T = \frac{0.05h_n}{\sqrt{D}} \tag{3-173}$$

where h_n = height, ft, of uppermost level above base

D = plan dimension, ft, in direction parallel to applied forces

More appropriate values of T may be used if they can be substantiated by technical data. Note, however, that C is not sensitive to small changes in T. If the lateral-force resisting system is a moment-resisting space frame designed to withstand all the lateral forces, and is not enclosed or adjoined by more rigid elements, T may be taken as $0.1N$, where N = number of stories above grade.

Space frames, as defined in the Uniform Building Code, are three-dimensional structural systems composed of interconnected members, other than bearing walls, and laterally supported to function as a complete, self-contained unit, with or without the aid of horizontal diaphragms or floor bracing systems. Such vertical-load-carrying frames may be considered moment-resisting if the members and joints are capable of resisting design lateral forces by bending moments.

K takes into account the potential for inelastic energy absorption in moment-resisting frames. It also recognizes the redundancy of framing, or second line of defense, present in most complete frames, whether or not designed to resist lateral loads. Buildings that do not possess at least a complete vertical-load-carrying space frame are penalized by assignment of a high K. Table 3-9 lists suggested values for K.

Table 3-9. K for Aseismic Design of Buildings and Other Structures*

Framing system	K
Building framing systems other than those listed below	1.00
Building with a box system	1.33
Buildings with a ductile, moment-resisting space frame and shear walls designed so that:	
Frames and shear walls resist the total lateral force in proportion to their rigidities, taking into account interaction of shear walls and frames	
Shear walls acting independently of the moment-resisting space frame resist the total required lateral force	
The ductile, moment-resisting space frame can resist at least 25% of the required lateral force	0.80
Buildings with a ductile, moment-resisting space frame with capacity to resist the total required lateral force	0.67
Elevated tanks plus full contents, on four or more cross-braced legs and not supported by a building†	3.00
Structures other than buildings and those listed in Table 3-10	2.00

* Where prescribed wind loads produce higher stresses, those loads should be used instead of the seismic loads.

† The minimum value of KC in Eq. (3-172) is 0.12, and the maximum value need not exceed 0.25. Framing should be designed for an accidental horizontal torsion due to the lateral shear acting with an eccentricity of 5% of the maximum dimension in plan at the level of the shear. For tanks on the ground or connected to or part of a building, see Table 3-10.

Box systems, for which $K = 1.33$, employ shear walls or braced frames subjected primarily to axial stresses to resist the required lateral forces. Such systems are considered to lack a complete vertical-load-carrying space frame.

Note, however, that a building with shear walls or braced frames may be assigned $K = 0.80$ if it also has a ductile moment-resisting space frame capable of resisting at least 25% of the lateral force. In that case, the shear walls, acting independently of the space frame, must be able to withstand the total required lateral force.

A *ductile, moment-resisting space frame* of structural steel is a space frame capable of developing plastic hinges at connections of beams and girders to columns.

In zones 2 and 3, each such connection should be capable of developing the full plastic capacity of the beam or girder at the connection, unless analysis or tests indicate that a smaller capacity would be adequate. To insure adequate ductility where beam flange area is reduced, for example, by bolt holes, plastic hinges should not be permitted to occur at such locations unless the ratio of ultimate strength to yield strength of the steel exceeds 1.5. In addition, special precautions must be taken to prevent buckling; for example, width-thickness and depth-thickness ratios must satisfy plastic-design criteria. And the effective lengths used in determining the slenderness ratios of columns should ignore any assistance provided by shear walls and braced frames in resisting lateral forces. Furthermore, the capacity of butt welds in tension should be verified by nondestructive testing.

For ductile, moment-resisting space frames of reinforced concrete, the Uniform Building Code places stringent requirements on member sizes, amount and location of reinforcement, concrete strength, anchorage of reinforcement, and beam-column joints.

A building whose sole resistance to lateral forces is provided by a ductile, moment-resisting space frame may be assigned $K = 0.67$. Before assigning a K value of 1.00 or less, however, designers should ascertain that action or failure of more rigid elements surrounding or adjoining the frame will not impair its ability to carry vertical or lateral loads.

Buildings often may qualify for more than one K value, for example, when they have different framing systems in perpendicular directions. Note, however, that if $K = 1.33$ is required in one direction, it also will be required in the other.

For elevated tanks, independent of buildings, the relatively high value $K = 3$ is assigned because of their relatively poor performance in earthquakes. This high value is especially desirable for water tanks because of their importance to communities in case of fire after an earthquake.

Structures other than buildings are assigned $K = 2$. This relatively high value reflects a general lack of redundancy, so that little yielding can be tolerated without failure. This value also assumes absence of nonstructural elements that can contribute materially to damping.

Building components should be designed for seismic forces given by

$$F_p = ZC_pW_p \tag{3-174}$$

(This equation has no bearing on determination of lateral forces to be applied to the building as a whole.)

Values of the seismic coefficient C_p are given in Table 3-10. W_p, in general, is the weight of the component. For floors and roofs acting as diaphragms, however, W_p is the tributary load from that story. And for storage and warehouse occupancies, W_p should be taken as the dead load plus 25% of the floor live load.

Distribution of Seismic Loads. Seismic forces are assumed to act at each floor level on vertical planar frames, or bents, extending in the direction of the loads. The seismic loads are distributed in proportion to the tributary weights, or masses carried by the bents. Thus, the seismic design force at any floor or roof level is proportional to that portion of W, used in Eq. (3-172), that is located at or assigned to the level.

Total horizontal shear at any level should be distributed to the bents of the lateral-force distribution system in proportion to their rigidities. The distribution should, however, take into account the rigidities of horizontal bracing and diaphragms (floors and roofs). In lightly loaded structures, for example, diaphragms may be sufficiently flexible to permit independent action of the lateral-force resisting bents. A strong temblor could cause severe distress in frames and diaphragms if relative rigidities were not properly evaluated.

The total horizontal shear, kips, on a structure is V, given by Eq. (3-172). The Uniform Building Code recommends that the portion of V to be assigned to the top of a multistory structure be computed from

$$F_t = 0.004V\left(\frac{h_n}{D_s}\right)^2 \tag{3-175}$$

where h_n = height, ft, of top of structure above ground level
D_s = plan dimension, ft, of the vertical lateral-force resisting system in the direction of the seismic forces

F_t need not be more than $0.15V$ and may be taken as zero when $h_n/D_s < 3$. Equation (3-175) recognizes the influence of higher modes of vibration as well as deviations from straight-line deflection patterns, particularly in tall buildings with relatively small dimensions in plan.

Table 3-10. C_p for Aseismic Design of Components of Buildings and Other Structures

Components	Direction of force	C_p
Exterior bearing and nonbearing walls, interior bearing walls and partitions, interior nonbearing walls and partitions over 10 ft high, masonry or concrete fences over 6 ft high*	Normal to flat surface	0.20
Cantilever parapet and other cantilever walls, except retaining walls	Normal to flat surface	1.00
Exterior and interior ornamentations and appendages	Any direction	1.00
When connected to or part of a building: towers, tanks, towers and tanks plus contents, chimneys, smokestacks, and penthouses	Any direction	0.20†
When resting on the ground: tank plus effective mass of its contents	Any direction	0.10
Floors and roofs acting as diaphragms‡	Any direction	0.10
Connections for exterior panels or other elements attached to or enclosing the exterior§	Any direction	2.00
Connections for prefabricated structural elements other than walls, with force applied at center of gravity of assemblage¶	Any horizontal direction	0.30

* Deflection of walls or partitions under a force of 5 psf applied perpendicular to them shall not exceed 1/240 the span for walls with brittle finishes or 1/120 the span for walls with flexible finishes.

† For any building, when $h_n/D \geq 5$, increase C_p by 50%.

‡ Apply a minimum value of C_p of 0.10 to load tributary from that story unless a greater value of C is required for Eq. (3-172).

§ Precast, nonbearing, nonshear wall panels or other elements attached to or enclosing the exterior shall accommodate movements of the structure resulting from lateral forces or temperature changes. Connections and panel joints shall allow for a relative movement between stories of at least ¼-in. or twice the story drift under wind or seismic forces, whichever is greater. Connections should have sufficient ductility and rotation capacity to preclude fracture of concrete or brittle failures at or near welds. Movement in the plane of a panel for story drift may be provided by sliding connections with slotted or oversize holes, or by bending of steel.

¶ Use dead load plus 25% of the floor live load for W_p in storage and warehouse occupancies.

The code also recommends that the seismic force F_x, to be assigned to any level at a height h_x, ft, above the ground be calculated from

$$F_x = (V - F_t)\frac{w_x h_x}{\sum_{i=1}^{n} w_i h_i} \tag{3-176}$$

where w_x = portion of W located at or assigned to level x
h_x = height, ft, of level x above ground level
w_i = portion of W located at or assigned to level i
h_i = height, ft, of level i above ground level
n = number of levels in structure

For one- and two-story buildings, however, a uniform distribution of seismic loading should be used, because of their relatively large stiffness.

Overturning. The equivalent static lateral forces applied to a building at various levels induce overturning moments. At any level, the overturning moment equals the sum of the products of each force and its height above that level. The overturning moments acting on the base of the structure and in each story are resisted by axial forces in vertical elements and footings.

At any level, the increment in the design overturning moment should be distributed to the resisting elements in the same proportion as the distribution of shears to those elements. Where a vertical resisting element is discontinued, the overturning moment at that level should be carried down as loads to the foundation.

(J. M. Biggs, "Structural Dynamics," McGraw-Hill Book Company, New York; G. L. Rogers, "Dynamics of Framed Structures," John Wiley & Sons, Inc., New York; U.S. Army Corps of Engineers, "Design of Structures to Resist the Effects of Atomic Weapons," Manual EM 1110-345-415; N. M. Newmark and E. Rosenblueth, "Fundamentals of Earthquake Engineering," Prentice-Hall, Englewood Cliffs, N.J.; S. Okamoto, "Introduction to Earthquake Engineering," Halsted Press, New York.)

3-85. Vibrations. In general, vibrations in a building are objectionable from the viewpoint of physical comfort, noise, protection of delicate apparatus, and danger of fatigue failures (Figs. 3-83 and 3-84). The simplest ways to eliminate them are to isolate vibrating machinery, install supports that transmit a minimum of energy, install shock absorbers, or insert mufflers and screens to absorb noise.

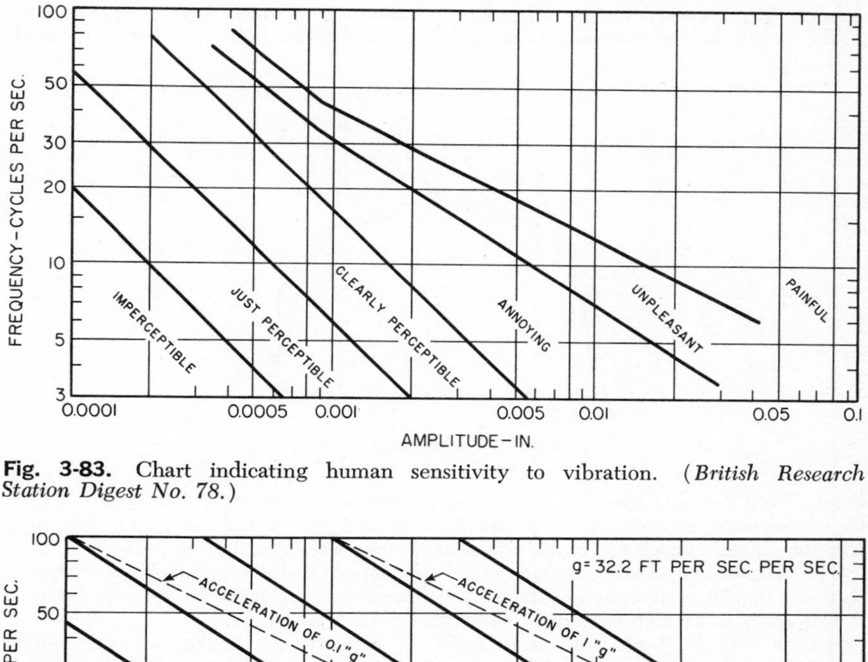

Fig. 3-83. Chart indicating human sensitivity to vibration. (*British Research Station Digest No. 78.*)

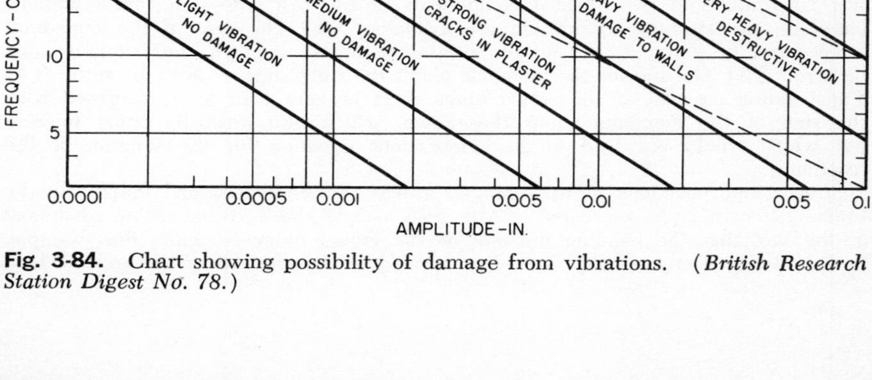

Fig. 3-84. Chart showing possibility of damage from vibrations. (*British Research Station Digest No. 78.*)

Sometimes, machines with rotating parts can be made to vibrate less by balancing the rotating parts to reduce the exciting force or the speed can be adjusted to prevent resonance. Also, the masses of the supports can be adjusted to change the natural frequency so that resonance is avoided.

The surfaces of a foundation subjected to vibration should be shaped to preclude the possibility of the vibrations being reflected from them back into the interior in a direction in which a stress buildup can occur. Such a foundation should be made of a homogeneous material, since objectionable stresses may occur in the region of an inhomogeneity. The foundation should be massive—weight comparable with that of the machine it supports—and should be isolated by an air gap or insulating material from the rest of the building.

("Vibration in Buildings," Building Research Station Digest No. 78, Her Majesty's Stationery Office, London, June, 1955; D. D. Barkan, "Dynamics of Bases and Foundations," McGraw-Hill Book Company, New York.)

STRESSES IN ARCHES

An arch is a curved beam, the radius of curvature of which is very large relative to the depth of the section. It differs from a straight beam in that: (1) loads

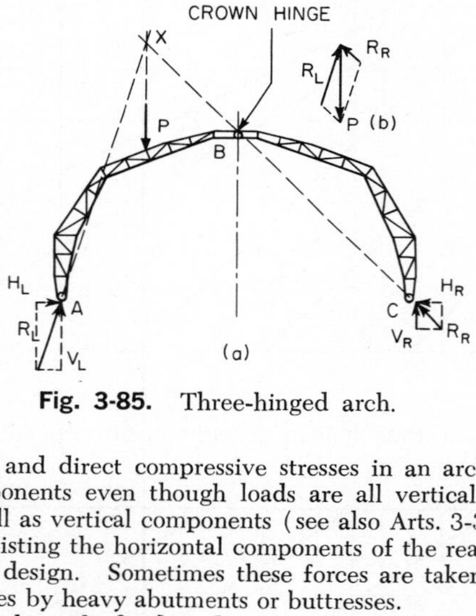

Fig. 3-85. Three-hinged arch.

induce both bending and direct compressive stresses in an arch; (2) arch reactions have horizontal components even though loads are all vertical; and (3) deflections have horizontal as well as vertical components (see also Arts. 3-36 to 3-41).

The necessity of resisting the horizontal components of the reactions is an important consideration in arch design. Sometimes these forces are taken by tie rods between the supports, sometimes by heavy abutments or buttresses.

Arches may be built with fixed ends, as can straight beams, or with hinges at the supports. They may also be built with a hinge at the crown.

3-86. Three-hinged Arches. An arch with a hinge at the crown as well as at both supports (Fig. 3-85) is statically determinate. There are four unknowns—two horizontal and two vertical components of the reactions—but four equations based on the laws of equilibrium are available: (1) The sum of the horizontal forces must be zero. (2) The sum of the moments about the left support must be zero. (3) The sum of the moments about the right support must be zero. (4) The bending moment at the crown hinge must be zero (not to be confused with the sum of the moments about the crown, which also must be equal to zero but which would not lead to an independent equation for the solution of the reactions).

Stresses and reactions in three-hinged arches can be determined graphically by utilizing the principles presented in Arts. 3-43 to 3-47 and 3-51 and taking advantage of the fact that the bending moment at the crown hinge is zero. For example, in Fig. 3-85*a*, a concentrated load P is applied to segment AB of the arch. Then, since the bending moment at B must be zero, the line of action of the reaction at C must pass through the crown hinge. It intersects the line of action of P at X. The line of action of the reaction at A must also pass through X. Since P is equal to the sum of the reactions, and since the directions of the reactions have thus been determined, the magnitude of the reactions can be measured from a parallelogram of forces (Fig. 3-85*b*). When the reactions have been found, the stresses can be computed from the laws of statics or, in the case of a trussed arch, determined graphically.

3-87. Two-hinged Arches. When an arch has hinges at the supports only (Fig. 3-86), it is statically indeterminate, and some knowledge of its deformations is

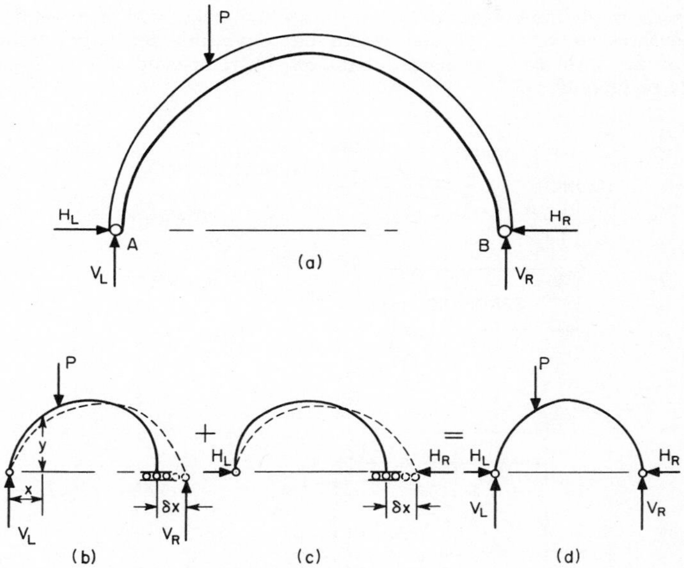

Fig. 3-86. Two-hinged arch.

required to determine the reactions. One procedure is to assume that one of the supports is on rollers. This makes the arch statically determinate. The reactions and the horizontal movement of the support are computed for this condition (Fig. 3-86*b*). Then, the magnitude of the horizontal force required to return the movable support to its original position is calculated (Fig. 3-86*c*). The reactions for the two-hinged arch are finally found by superimposing the first set of reactions on the second (Fig. 3-86*d*).

For example, if δx is the horizontal movement of the support due to the loads, and if $\delta x'$ is the horizontal movement of the support due to a unit horizontal force applied to the support, then

$$\delta x + H\,\delta x' = 0 \tag{3-177}$$

$$H = -\frac{\delta x}{\delta x'} \tag{3-178}$$

where H is the unknown horizontal reaction. (When a tie rod is used to take the thrust, the right-hand side of the first equation is not zero, but the elongation of the rod HL/AE.) The dummy unit-load method [Eq. (3-63)] can be used to compute δx and $\delta x'$:

$$\delta x = \int_A^B \frac{My}{EI}\,ds - \int_A^B \frac{N\,dx}{AE} \tag{3-179}$$

where M = moment at any section due to loads
N = normal thrust on cross section
A = cross-sectional area of arch
y = ordinate of section measured from A as origin, when B is on rollers
I = moment of inertia of section
E = modulus of elasticity
ds = differential length along axis of arch
dx = differential length along horizontal

$$\delta x' = -\int_A^B \frac{y^2}{EI}\,ds - \int_A^B \frac{\cos\alpha\,dx}{AE} \tag{3-180}$$

where α = the angle the tangent to axis at the section makes with horizontal

Unless the thrust is very large and would be responsible for large strains in the direction of the arch axis, the second term on the right-hand side of Eq. (3-179) can usually be ignored.

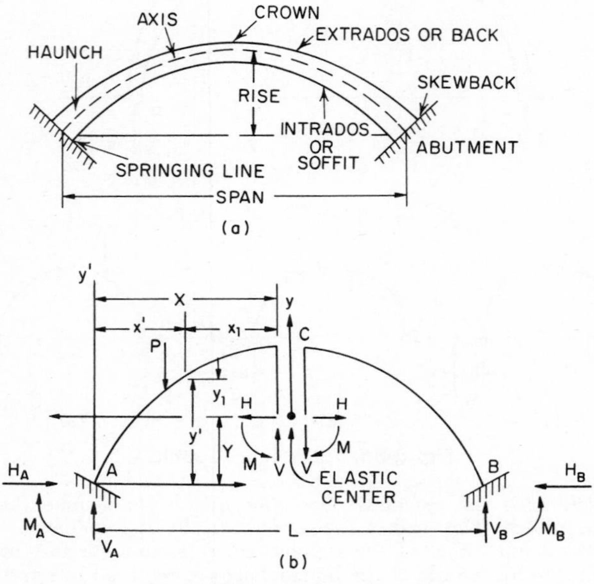

Fig. 3-87. Fixed-end arch.

In most cases, integration is impracticable. The integrals generally must be evaluated by approximate methods. The arch axis is divided into a convenient number of sections and the functions under the integral sign evaluated for each section. The sum is approximately equal to the integral. Thus, for the usual two-hinged arch,

$$H = \frac{\sum_A^B (My\,\Delta s/EI)}{\sum_A^B (y^2\,\Delta s/EI) + \sum_A^B (\cos\alpha\,\Delta x/AE)} \tag{3-181}$$

3-88. Fixed Arches. An arch is considered fixed when rotation is prevented at the supports (Fig. 3-87). Such an arch is statically indeterminate; there are six reaction components and only three equations are available from conditions of equilibrium. Three more equations must be obtained from a knowledge of the deformations.

One way to determine the reactions is to consider the arch cut at the crown, forming two cantilevers. First, the horizontal and vertical deflections and rotation produced at the end of each half arch by the loads are computed. Next, the deflection components and rotation at those ends are determined for unit vertical force, unit horizontal force, and unit moment applied separately at the ends. These deformations, multiplied, respectively, by V, the unknown shear; H, the unknown horizontal thrust at the crown; and M, the unknown moment there, yield the deformation caused by the unknown forces at the crown. Adding these deformations algebraically to the corresponding one produced by the loads gives the net movement of the free end of each half arch. Since these ends must deflect and rotate the same amount to maintain continuity, three equations can be obtained in this manner for the determination of V, H, and M. The various deflections can be computed by the dummy unit-load method [Eq. (3-63)], as demonstrated in Art. 3-87 for two-hinged arches.

The solution of the equations, however, can be simplified considerably if the center of coordinates is shifted to the elastic center of the arch and the coordinate axes are properly oriented. If the unknown forces and moments V, H, and M are determined at the elastic center, each equation will contain only one unknown. When the unknowns at the elastic center are determined, the shears, thrusts, and moments at any point on the arch can be found by application of the laws of equilibrium.

Determination of the location of the elastic center of an arch is equivalent to finding the center of gravity of an area. Instead of an increment of area dA, however, an increment of length ds multiplied by a width $1/EI$ must be used. (E is the modulus of elasticity, I the moment of inertia.) Since, in general, integrals are difficult or impracticable to evaluate, the arch axis usually is divided into a convenient number of elements of length Δs and numerical integration is used, as described in Art. 3-87. Then, if the origin of coordinates is temporarily chosen at A, the left support of the arch in Fig. 3-87b, and if x' is the horizontal distance to a point on the arch and y' the vertical distance, the coordinates of the elastic center are

$$X = \frac{\sum_A^B (x' \Delta s/EI)}{\sum_A^B (\Delta s/EI)} \qquad Y = \frac{\sum_A^B (y' \Delta s/EI)}{\sum_A^B (\Delta s/EI)} \tag{3-182}$$

If the arch is symmetrical about the crown, the elastic center lies on a normal to the tangent at the crown. In that case, there is a savings in calculations by taking the origin of the temporary coordinate system at the crown and measuring coordinates parallel to the tangent and the normal. To determine Y, the distance of the elastic center from the crown, Eq. (3-182) can be used with the summations limited to the half arch between crown and either support. For a symmetrical arch also, the final coordinate system should be chosen parallel to the tangent and normal to the crown.

After the elastic center has been located, the origin of a new coordinate system should be taken at the center. For convenience, the new coordinates x_1, y_1 may be taken parallel to those in the temporary system. Then, for an unsymmetrical arch, the final coordinate axes should be chosen so that the x axis makes an angle α, measured clockwise, with the x_1 axis such that

$$\tan 2\alpha = \frac{2\sum_A^B (x_1 y_1 \Delta s/EI)}{\sum_A^B (x_1^2 \Delta s/EI) - \sum_A^B (y_1^2 \Delta s/EI)} \tag{3-183}$$

The unknown forces H and V at the elastic center should be taken parallel, respectively, to the final x and y axes.

For a coordinate system with origin at the elastic center and axes oriented to satisfy Eq. (3-183),

$$H = \frac{\sum_A^B (M'y\,\Delta s/EI)}{\sum_A^B (y^2\,\Delta s/EI)}$$

$$V = \frac{\sum_A^B (M'x\,\Delta s/EI)}{\sum_A^B (x^2\,\Delta s/EI)} \qquad (3\text{-}184)$$

$$M = \frac{\sum_A^B (M'\,\Delta s/EI)}{\sum_A^B (\Delta s/EI)}$$

where M' is the average bending moment on each element due to loads. To account for the effect of an increase in temperature t, add $EctL$ to the numerator of H, where c is the coefficient of expansion and L the distance between abutments.

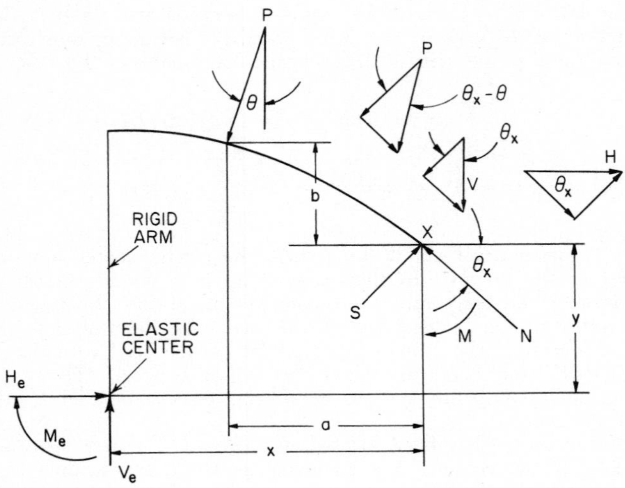

Fig. 3-88. Portion of an arch, including the elastic center.

(S. Timoshenko and D. H. Young, "Theory of Structures," McGraw-Hill Book Company, New York; S. F. Borg and J. J. Gennaro, "Modern Structural Analysis," Van Nostrand Reinhold Company, Inc., New York; G. Winter and A. H. Nilson, "Design of Concrete Structures," McGraw-Hill Book Company, New York; V. Leontovich, "Frames and Arches," McGraw-Hill Book Company, New York.)

3-89. Stresses in Arch Ribs. When the reactions have been found for an arch (Arts. 3-86 to 3-88), the principal forces acting on any cross section can be found by applying the equations of equilibrium. For example, consider the portion of a fixed arch in Fig. 3-88, where the forces at the elastic center H_e, V_e, and M_e are known and the forces acting at point X are to be found. The load P, H_e, and V_e may be resolved into components parallel to the axial thrust N and the shear S at X, as indicated in Fig. 3-88. Then, by equating the sum of the forces in each direction to zero, we get

$$\begin{aligned} N &= V_e \sin \theta_x + H_e \cos \theta_x + P \sin (\theta_x - \theta) \\ S &= V_e \cos \theta_x - H_e \sin \theta_x + P \cos (\theta_x - \theta) \end{aligned} \tag{3-185}$$

And by taking moments about X and equating to zero, we obtain

$$M = V_e x + H_e y - M_e + Pa \cos \theta + Pb \sin \theta \tag{3-186}$$

The shearing unit stress on the arch cross section at X can be determined from S with the aid of Eq. (3-28). The normal unit stresses can be calculated from N and M with the aid of Eq. (3-36).

In designing an arch, it may be necessary to compute certain secondary stresses, in addition to those caused by live, dead, wind, and snow loads. Among the secondary stresses to be considered are those due to temperature changes, rib shortening due to thrust or shrinkage, deformation of tie rods, and unequal settlement of footings. The procedure is the same as for loads on the arch, with the deformations producing the secondary stresses substituted for or treated the same as the deformations due to loads.

THIN-SHELL STRUCTURES

A structural shell is a curved surface structure. Usually, it is capable of transmitting loads in more than two directions to supports. It is highly efficient structurally when it is so shaped, proportioned, and supported that it transmits the loads without bending or twisting.

A shell is defined by its middle surface, halfway between its extrados, or outer surface, and intrados, or inner surface. Thus, depending on the geometry of the middle surface, it might be a type of dome, barrel arch, cone, or hyperbolic paraboloid. Its thickness is the distance, normal to the middle surface, between extrados and intrados.

3-90. Thin-shell Analysis. A thin shell is a shell with a thickness relatively small compared to its other dimensions. But it should not be so thin that deformations would be large compared with the thickness.

The shell should also satisfy the following conditions: Shearing stresses normal to the middle surface are negligible. Points on a normal to the middle surface before it is deformed lie on a straight line after deformation. And this line is normal to the deformed middle surface.

Calculation of the stresses in a thin shell generally is carried out in two major steps, both usually involving the solution of differential equations. In the first, bending and torsion are neglected (membrane theory, Art. 3-91). In the second step, corrections are made to the previous solution by superimposing the bending and shear stresses that are necessary to satisfy boundary conditions (bending theory, Art. 3-96).

3-91. Membrane Theory for Thin Shells. Thin shells usually are designed so that normal shears, bending moments and torsion are very small, except in relatively small portions. In the membrane theory, these stresses are ignored.

Despite the neglected stresses, the remaining stresses are in equilibrium, except possibly at boundaries, supports, and discontinuities. At any interior point, the number of equilibrium conditions equals the number of unknowns. Thus, in the membrane theory, a thin shell is statically determinate.

The membrane theory does not hold for concentrated loads normal to the middle surface, except possibly at a peak or valley. The theory does not apply where boundary conditions are incompatible with equilibrium. And it is inexact where there is geometric incompatibility at the boundaries. The last is a common condition, but the error is very small if the shell is not very flat. Usually, disturbances of membrane equilibrium due to incompatibility with deformations at boundaries, supports, or discontinuities are appreciable only in a narrow region about each source of disturbance. Much larger disturbances result from incompatibility with equilibrium conditions.

To secure the high structural efficiency of a thin shell, select a shape, proportions, and supports for the specific design conditions that come as close as possible to satisfying the membrane theory. Keep the thickness constant; if it must change, use a gradual taper. Avoid concentrated and abruptly changing loads. Change curvature gradually. Keep discontinuities to a minimum. Provide reactions that are tangent to the middle surface. At boundaries, insure, to the extent possible, compatibility of shell deformations with deformations of adjoining members, or at least keep restraints to a minimum. Make certain that reactions along boundaries are equal in magnitude and direction to the shell forces there.

Means usually adopted to satisfy these requirements at boundaries and supports are illustrated in Fig. 3-89. In Fig. 3-89*a*, the slope of the support and provision for movement normal to the middle surface insure a reaction tangent to the middle surface. In Fig. 3-89*b*, a stiff rib, or ring girder, resists unbalanced shears and transmits normal forces to columns below. The enlarged view of the ring girder

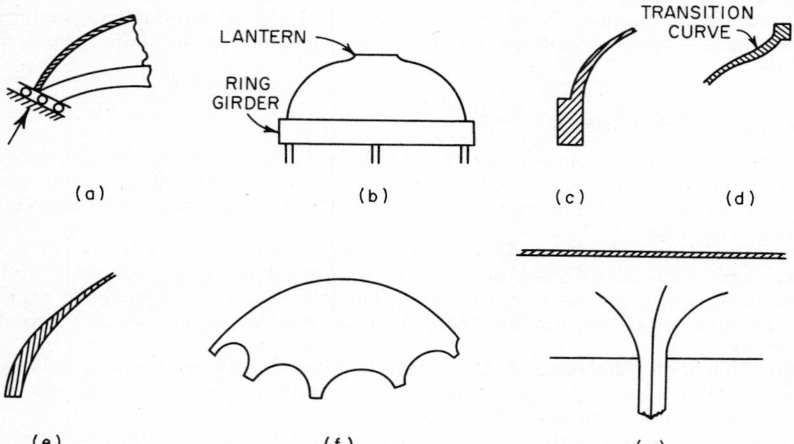

Fig. 3-89. Special provisions made at supports and boundaries of thin shells to meet requirements of the membrane theory include (*a*) a device to insure a reaction tangent to the middle surface; (*b*) stiffened edges, such as the ring girder at the base of a dome; (*c*) gradually increased shell thickness at a stiffening member; (*d*) a transition curve at changes in section; (*e*) a stiffening edge obtained by thickening the shell; (*f*) scalloped edges; and (*g*) a flared support.

in Fig. 3-89*c* shows gradual thickening of the shell to reduce the abruptness of the change in section. The stiffening ring at the lantern in Fig. 3-89*d*, extending around the opening at the crown, projects above the middle surface, for compatibility of strains, and connects through a transition curve with the shell; often, the rim need merely be thickened when the edge is upturned, and the ring can be omitted. In Fig. 3-89*e*, the boundary of the shell is a stiffened edge. In Fig. 3-89*f*, a scalloped shell provides gradual tapering for transmitting the loads to the supports, at the same time providing access to the shell enclosure. And in Fig. 3-89*g*, a column is flared widely at the top to support a thin shell at an interior point.

Even when the conditions for geometric compatibility are not satisfactory, the membrane theory is a useful approximation. Furthermore, it yields a particular solution to the differential equations of the bending theory.

3-92. Membrane Forces in Shells of General Shape. By applying the equilibrium conditions to a small element of the middle surface, one obtains the membrane forces. The sides of the element are acted upon by normal forces and shears, as indicated in Fig. 3-90. For simplification, the equilibrium equations are usually expressed in terms of the projection of these forces on the xy plane. Thus:

$$\bar{N}_x = N_x \frac{\cos \phi}{\cos \theta} \qquad \bar{N}_y = N_y \frac{\cos \theta}{\cos \phi} \qquad \bar{T} = T \tag{3-187}$$

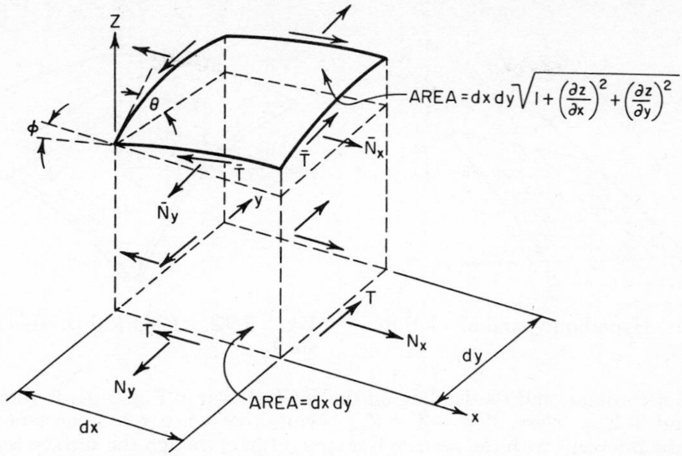

Fig. 3-90. Shears and normal forces acting on an element of the middle surface of a thin shell in the membrane theory.

The forces act per unit of length of the edges of the element and over the full thickness of the shell, and hence are called unit forces. The subscripts indicate the coordinate axis they parallel.

Assume that the loading on the element per unit area in the xy plane is given by its components X, Y, and Z. Then, for equilibrium in the directions of the three coordinate axes:

$$\frac{\partial N_x}{\partial x} + \frac{\partial T}{\partial y} + X = 0 \qquad \frac{\partial N_y}{\partial y} + \frac{\partial T}{\partial x} + Y = 0$$
$$\frac{\partial}{\partial x}\left(N_x \frac{\partial z}{\partial x} + T\frac{\partial z}{\partial y}\right) + \frac{\partial}{\partial y}\left(N_y \frac{\partial z}{\partial y} + T\frac{\partial z}{\partial x}\right) + Z = 0 \tag{3-188}$$

where $\tan \phi = \partial z/\partial x$ and $\tan \theta = \partial z/\partial y$. These equations can be reduced to a single equation by introduction of a stress function $F(x,y)$ such that

$$N_x = \frac{\partial^2 F}{\partial y^2} - \int X\,dx \qquad N_y = \frac{\partial^2 F}{\partial x^2} - \int Y\,dy \qquad T = -\frac{\partial^2 F}{\partial x\,\partial y} \tag{3-189}$$

where the lower limits of the integrals are constants and the upper limits are x and y, respectively.

The equilibrium equation for the general case of a thin shell then is

$$\frac{\partial^2 F}{\partial x^2}\frac{\partial^2 z}{\partial y^2} - 2\frac{\partial^2 F}{\partial x\,\partial y}\frac{\partial^2 z}{\partial x\,\partial y} + \frac{\partial^2 F}{\partial y^2}\frac{\partial^2 z}{\partial x^2} = P$$
$$P = -Z + X\frac{\partial z}{\partial x} + Y\frac{\partial z}{\partial y} + \frac{\partial^2 z}{\partial x^2}\int X\,dx + \frac{\partial^2 z}{\partial y^2}\int Y\,dy \tag{3-190}$$

Hence, the membrane forces are determined when this equation is solved for F. (S. Timoshenko and S. Woinowsky-Krieger, "Theory of Plates and Shells," McGraw-Hill Book Company, New York.)

3-93. Membrane Forces in a Hyperbolic-paraboloid Shell. Equation (3-190) of Art. 3-92 can be used to determine the membrane forces in a thin shell with a hyperbolic-paraboloid middle surface. Assume, for example, that the equation of the middle surface is

$$z = cxy$$

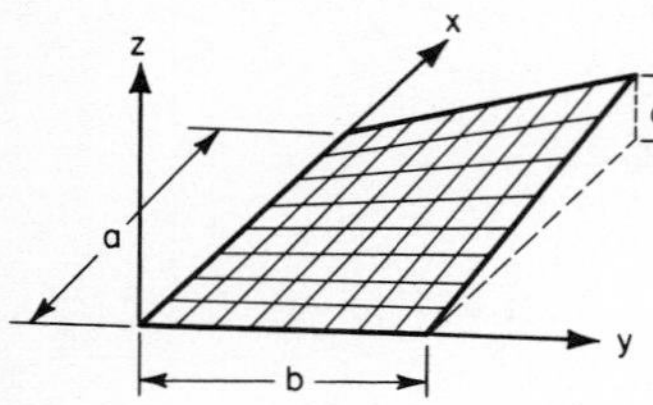

Fig. 3-91. Hyperbolic-paraboloid thin shell.

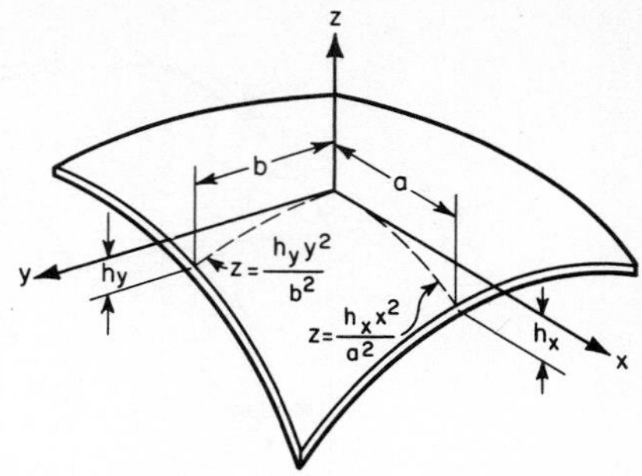

Fig. 3-92. Elliptical-paraboloid thin shell.

where c is a constant, and the loading on the shell, shown in Fig. 3-91, is vertical, say, a snow load, $-Z_s$. Then, $P = -Z = Z_s$. (Note that when x is a constant or y is a constant, the intercept with the surface is a straight line, though the surface is curved.)

Differentiation of the surface equation gives

$$\frac{\partial^2 z}{\partial x^2} = \frac{\partial^2 z}{\partial y^2} = 0 \qquad \frac{\partial^2 z}{\partial x\, \partial y} = c$$

Substitution of these derivatives in Eqs. (3-190) and (3-189) yields

$$-2 \frac{\partial^2 F}{\partial x\, \partial y} c = 2Tc = -Z = Z_s$$

and this gives

$$T = \frac{Z_s}{2c} \tag{3-191}$$

Furthermore, N_x can be any function of y, and N_y any function of x.

Suppose the shell to be stiffened along its edges by a diaphragm with negligible resistance normal to its plane. Then, $N_x = \bar{N}_x = 0$ and $N_y = N_y = 0$, since the edges are free of normal forces.

Assume now the load to be the weight of the shell, a constant $-Z_w$ per unit area of the surface. The area of the horizontal projection of an element of the surface equals $1/(1 + c^2x^2 + c^2y^2)^{1/2}$ times the area of the element. Then, the horizontal projections of the unit forces are

$$T = \frac{Z_w}{2c}(1 + c^2x^2 + c^2y^2)^{1/2}$$

$$N_x = -\frac{Z_w y}{2} \log_e [cx + (1 + c^2x^2 + c^2y^2)^{1/2}] + f_1(y) \tag{3-192}$$

$$N_y = -\frac{Z_w x}{2} \log_e [cy + (1 + c^2x^2 + c^2y^2)^{1/2}] + f_2(x)$$

where $f_1(y)$ is an arbitrary function of y and $f_2(x)$ an arbitrary function of x; logarithms are taken to the base e. The actual unit forces $\bar{N}_x$ and $\bar{N}_y$ can be calculated by applying Eq. (3-187) to Eq. (3-192), noting that $\tan \phi = -cy$ and $\tan \theta = -cx$. (S. Timoshenko and S. Woinowsky-Krieger, "Theory of Plates and Shells," McGraw-Hill Book Company, New York.)

3-94. Elliptical-paraboloid Membrane Forces. Equations (3-190) of Art. 3-92 also can be used to obtain membrane forces in an elliptical paraboloid (parabolas in vertical planes, ellipses in horizontal)

$$z = \frac{h_x x^2}{a^2} + \frac{h_y y^2}{b^2}$$

rectangular in plan (Fig. 3-92). The solution usually is given as a Fourier series, summed for $n = 1, 3, 5, \ldots, \infty$.

Suppose the shell to be subjected to a uniform load p over its horizontal projection. Assume that forces normal to the boundaries vanish. Then, the unit forces, acting per unit of length over the thickness of the shell, are

$$\bar{N}_x = -\frac{pb^2}{2h_x}\left(\frac{a^4 + 4h_x{}^2x^2}{b^4 + 4h_y{}^2y^2}\right)^{1/2}\left[1 + \frac{4}{\pi}\sum_n (-1)^{(n+1)/2}\frac{\cosh(n\pi x/c)}{n\cosh(n\pi a/c)}\cos\frac{n\pi y}{2b}\right]$$

$$\bar{N}_y = \frac{2pa^4}{\pi h_x b^2}\left(\frac{b^4 + 4h_y{}^2y^2}{a^4 + 4h_x{}^2x^2}\right)^{1/2}\sum_n (-1)^{(n+1)/2}\frac{\cosh(n\pi x/c)}{n\cosh(n\pi a/c)}\cos\frac{n\pi y}{2b} \tag{3-193}$$

$$T = \frac{2pa^2}{\pi h_x}\sum_n (-1)^{(n+1)/2}\frac{\sinh(n\pi x/c)}{n\cosh(n\pi a/c)}\sin\frac{n\pi y}{2b}$$

where $c = 2a(h_y/h_x)^{1/2}$

The series for T does not converge at the corners. Actually, bending moments and torsion will exist in those regions, counteracting the huge membrane shears.

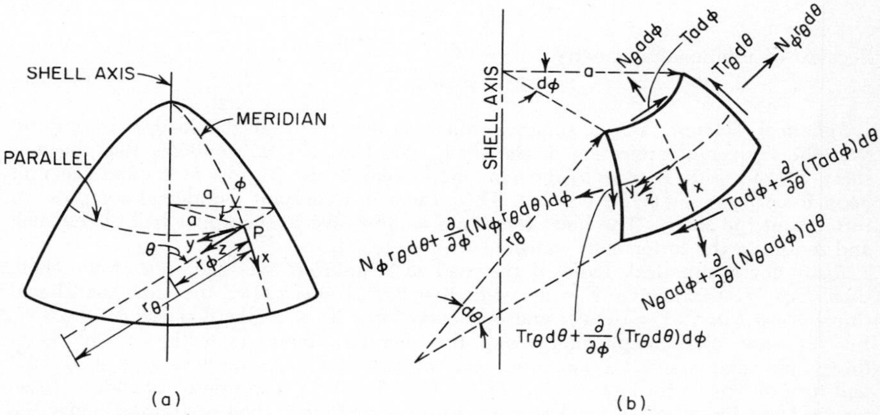

Fig. 3-93. Shell of revolution. (*a*) Coordinate system used in analysis; (*b*) unit forces acting on an element.

3-95. Membrane Forces in Shells of Revolution. In a shell of revolution, the middle surface results from the rotation of a plane curve about an axis, called the "shell axis." A plane through the axis cuts the surface in a meridian. A plane normal to the axis cuts the surface in a circle, called a parallel.

The membrane forces in a shell of revolution usually are taken parallel to the x, y, z, axes of a cartesian coordinate system at each point of the middle surface. The x axis is tangent to the meridian, the y axis is tangent to the parallel, and the z axis lies in the direction of the normal to the surface (Fig. 3-93).

A given point P on the surface is determined by the angle θ between the shell axis and the normal through P, and by the angle ϕ between the radius through P of the parallel on which it lies and a fixed, reference direction. Let r_θ be the radius of curvature of the meridian, and let r_ϕ, the length of the shell normal between P and the shell axis, be the radius of curvature of the normal section at P. If a is the radius of the parallel through P, then $a = r_\phi \sin\theta$.

Consider the element of the middle surface of the shell in Fig. 3-93*b*. Its edges are subjected to unit normal and shear forces, acting per unit of length of edge over the thickness of the shell. Assume that the loading on the element per unit of area is given by its X, Y, Z components. Then, the equations of equilibrium are

$$\frac{\partial}{\partial \theta}(N_\theta r_\phi \sin\theta) + \frac{\partial T}{\partial \phi} r_\theta - N_\phi r_\theta \cos\theta + X r_\theta r_\phi \sin\theta = 0$$
$$\frac{\partial N_\phi}{\partial \phi} r_\theta + \frac{\partial}{\partial \theta}(T r_\phi \sin\theta) + T r_\theta \cos\theta + Y r_\theta r_\phi \sin\theta = 0 \qquad (3\text{-}194)$$
$$N_\theta r_\phi + N_\phi r_\theta + Z r_\theta r_\phi = 0$$

Symmetrical Loading. When the loads also are symmetrical about the shell axis, Eqs. (3-194) take a simpler form. If a is the radius of a parallel and R the resultant of the total vertical load on the shell above that parallel, then the first equation can be written in the alternative form:

$$N_\theta = -\frac{R}{2\pi a}\sin\theta = -\frac{R}{2\pi r_\phi}\sin^2\theta \qquad (3\text{-}195a)$$

Substitution of this in the third equation yields

$$N_\phi = \frac{R}{2\pi r_\theta}\sin^2\theta - Z r_\phi \qquad (3\text{-}195b)$$

Because of rotational symmetry

$$T = 0 \qquad (3\text{-}195c)$$

Spherical Domes. For a spherical shell, $r_\theta = r_\phi = r$. If the load p is uniform over the horizontal projection of the shell, then Eqs. (3-195) indicate that the unit shear $T = 0$; with $R = \pi a^2 p$, the unit meridional thrust $N_\theta = -pr/2$; and the unit hoop force $N_\phi = -(pr/2)\cos 2\theta$. Thus, there is a constant meridional compression throughout the shell. The hoop forces are compressive in the upper half of the shell and tensile in the lower half, vanishing at $\theta = 45°$.

Again for a spherical dome, if the load w is uniform over the area of the shell, then Eqs. (3-195) give $T = 0$; with $R = 2\pi r^2(1 - \cos\theta)w$, the unit meridional thrust $N_\theta = -wr/(1 + \cos\theta)$; and unit hoop force $N_\phi = wr[1/(1 + \cos\theta) - \cos\theta]$. In this case, the compression along the meridian increases with θ. The hoop forces are compressive in the upper part of the shell, become zero at $\theta = 51°\ 51'$ and turn to tension for larger values of θ. Usually, at the lower dome boundary, these tensile hoop forces are resisted by a ring girder, and since shell and girder under the membrane theory will have different strains, bending stresses will be imposed on the shell.

When there is an opening around the crown of the dome, as in Fig. 3-89*b*, the upper edge may be thickened or reinforced with a ring girder to resist the hoop forces. If $2\theta_0$ is the angle of the opening and P the vertical load per unit of length of the compression ring, then

$$N_\theta = -wr\frac{\cos\theta_0 - \cos\theta}{\sin^2\theta} - P\frac{\sin\theta_0}{\sin^2\theta}$$
$$N_\phi = wr\left(\frac{\cos\theta_0 - \cos\theta}{\sin^2\theta} - \cos\theta\right) + P\frac{\sin\theta_0}{\sin^2\theta} \qquad (3\text{-}196)$$

Conical Shells. For a conical shell, parallel circles are located by the coordinate s, measured from the vertex along a generator, instead of the angle θ, and the unit meridional force is denoted by N_s (Fig. 3-94). The equilibrium equations are

$$\frac{\partial}{\partial s}(N_s s) + \frac{\partial T}{\partial \phi}\frac{1}{\cos\theta} - N_\phi + Xs = 0$$
$$\frac{\partial N_\phi}{\partial \phi}\frac{1}{\cos\theta} + \frac{\partial}{\partial s}(Ts) + T + Ys = 0 \qquad (3\text{-}197)$$
$$N_\phi + Zs\cot\theta = 0$$

For a vertical load p uniform over the horizontal projection of a conical shell, Eqs. (3-197) give

$$T = 0 \qquad N_s = -\frac{ps}{2}\cot\theta \qquad \text{and} \qquad N_\phi = -ps\frac{\cos^3\theta}{\sin\theta} \tag{3-198}$$

Suppose that the cone has an opening at the top at $s = s_0$, and the shell is subjected to a vertical uniform load w over its area. Then,

$$T = 0 \qquad N_s = -\frac{s^2 - s_0^2}{2s}\frac{w}{\sin\theta} \qquad \text{and} \qquad N_\phi = -ws\frac{\cos^2\theta}{\sin\theta} \tag{3-199}$$

For a horizontal load q uniform over the vertical projection of the cone with lantern,

$$T = -q\frac{s^3 - s_0^3}{3s^2}\sin\phi$$

$$N_s = -\frac{qs}{2}\left[\cos\theta - \frac{1}{3\cos\theta} - \frac{s_0^2}{s^2}\left(\cos\theta - \frac{1}{\cos\theta}\right) - \frac{s_0^3}{s^3}\frac{2}{3\cos\theta}\right]\cos\phi \tag{3-200}$$

$$N_\phi = -qs\cos\theta\cos\phi$$

Barrel Arches. For a cylindrical shell with horizontal axis, a point on the surface is located by its distance s along a generator from one end and by the angle ϕ which

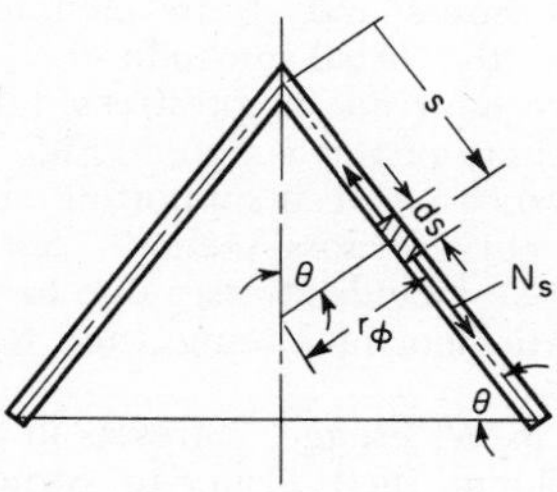

Fig. 3-94. Conical thin shell.

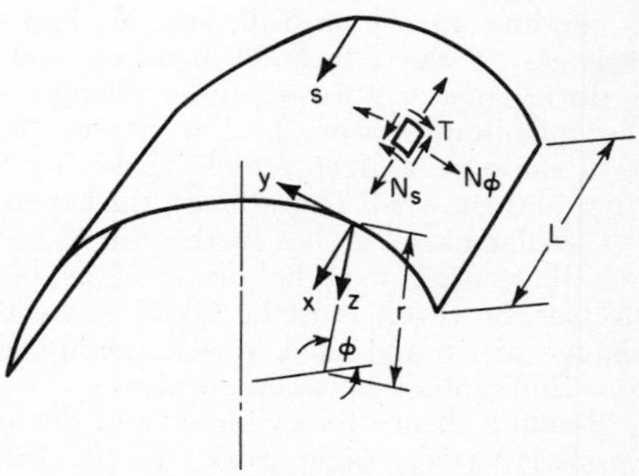

Fig. 3-95. Cylindrical thin shell.

the radius of curvature at the point makes with the horizontal (Fig. 3-95). The equilibrium equations are

$$\frac{\partial N_s}{\partial s}r + \frac{\partial T}{\partial \phi} + Xr = 0 \qquad \frac{\partial N_\phi}{\partial \phi} + \frac{\partial T}{\partial s}r + Yr = 0 \qquad N_\phi + Zr = 0 \tag{3-201}$$

The third equation indicates that $N_\phi = 0$ when $Z = 0$. Hence, for uniform vertical loads over the horizontal projection of the shell area, vertical reactions vanish along the longitudinal edges of a cylindrical shell if the tangent to the directrix at the edges is vertical. Shells with a semicircular, semielliptical, or cycloidal directrix have this property.

Equations (3-201) can be used to demonstrate an unusual property of parabolic vaults. The equation of the parabola can be rewritten from the common form $u = v^2/2r_0$, where r_0 is the radius of curvature at the vertex ($\phi = \pi/2$), into the more convenient form $r = r_0/\sin^3\phi$. Now, for a vertical load p uniform over the horizontal projection, $X = 0$, $Y = -p\sin\phi\cos\phi$, and $Z = p\sin^2\phi$. Then, solution of Eqs. (3-201) yields $T = 0$ and $N_s = 0$, so that loads are transmitted to the supports only in the ϕ direction. $N_\phi = -pr_0/\sin\phi$. Thus, under this type of a loading, a parabolic vault behaves like a parabolic arch. The same results are obtained with a catenary-shaped vault under vertical loading uniform over the shell area.

For a semicircular barrel arch, r is a constant. If the vertical load w is uniform over the shell area, $X = 0$, $Y = w\cos\phi$, and $Z = w\sin\phi$. Then, Eqs. (3-201) give $T = -w(L - 2s)\cos\phi$, $N_s = -w(s/r)(L - s)\sin\phi$, and $N_\phi = -wr\sin\phi$. Note that no supports are needed along the longitudinal edges, since N_ϕ is zero there, but provision must be made to resist the shear. Such provision, usually in the form of an edge member, generally creates bending stresses due to incompatibility of deformations at the shell boundary. The shears transmit the loads to the edges $s = 0$ and $s = L$, where supports must be provided.

Other Shapes. For membrane stresses in a variety of shells, see W. Flügge, "Stresses in Shells," Springer-Verlag, New York; D. P. Billington, "Thin Shell Concrete Structures," McGraw-Hill Book Company, New York; and S. Timoshenko and S. Woinowski-Krieger, "Theory of Plates and Shells" McGraw-Hill Book Company, New York.

3-96. Bending Theory for Thin Shells. When equilibrium conditions are not satisfied or incompatible deformations exist at boundaries, bending and torsion stresses arise in the shell. Sometimes, the design of the shell and its supports can be modified to reduce or eliminate these stresses (Art. 3-91). When the design cannot eliminate them, provision must be made for the shell to resist them.

But even for the simplest types of shells and loading, the stresses are difficult to compute. In bending theory, a thin shell is statically indeterminate; deformation conditions must supplement equilibrium conditions in setting up differential equations for determining the unknown forces and moments. Solution of the resulting equations may be tedious and time-consuming, if indeed solution is possible.

In practice, therefore, shell design relies heavily on the designer's experience and judgment. The designer should consider the type of shell, material of which it is made, and support and boundary conditions, and then decide whether to apply a bending theory in full, use an approximate bending theory, or make a rough estimate of the effects of bending and torsion. (Note that where the effects of a disturbance are large, these change the normal forces and shears computed by the membrane theory.) For domes, for example, the usual procedure is to use as a support a deep, thick girder or a heavily reinforced or prestressed tension ring, and the shell is gradually thickened in the vicinity of this support (Fig. 3-89*c*).

Circular barrel arches, with ratio of radius to distance between supporting arch ribs less than 0.25, may be designed as beams with curved cross section. Secondary stresses, however, must be taken into account. These include stresses due to volume change of rib and shell, rib shortening, unequal settlement of footings, and temperature differentials between surfaces.

Bending theory for cylinders and domes is given in W. Flügge, "Stresses in Shells," Springer-Verlag, New York; D. P. Billington, "Thin Shell Concrete Structures," McGraw-Hill Book Company, New York; S. Timoshenko and S. Woinowsky-Krieger, "Theory of Plates and Shells," McGraw-Hill Book Company, New York; "Design of Cylindrical Concrete Shell Roofs," Manual of Practice No. 31, American Society of Civil Engineers.

3-97. Stresses in Thin Shells. The results of the membrane and bending theories are expressed in terms of unit forces and unit moments, acting per unit of length over the thickness of the shell. To compute the unit stresses from these forces and moments, usual practice is to assume normal forces and shears to be uniformly distributed over the shell thickness and bending stresses to be linearly distributed.

Then, normal stresses can be computed from equations of the form

$$f_x = \frac{N_x}{t} + \frac{M_x}{t^3/12} z \tag{3-202}$$

where z = distance from middle surface
t = shell thickness
M_x = unit bending moment about an axis parallel to direction of unit normal force N_x

Similarly, shearing stresses produced by central shears and twisting moments may be calculated from equations of the form

$$v_{xy} = \frac{T}{t} \pm \frac{D}{t^3/12} z \tag{3-203}$$

where D = twisting moment

Normal shearing stresses may be computed on the assumption of a parabolic stress distribution over the shell thickness:

$$v_{xz} = \frac{V}{t^3/6}\left(\frac{t^2}{4} - z^2\right) \tag{3-204}$$

where V = unit shear force normal to middle surface

For axes rotated with respect to those used in the thin-shell analysis, use Eqs. (3-15) and (3-16) to transform stresses or unit forces and moments from the given to the new axes.

FOLDED PLATES

A folded-plate structure consists of a series of thin planar elements, or flat plates, connected to one another along their edges. Usually used on long spans, especially for roofs, folded plates derive their economy from the girder action of the plates and the mutual support they give one another.

Longitudinally, the plates may be continuous over their supports. Transversely, there may be several plates in each bay (Fig. 3-96). At the edges, or folds, they may be capable of transmitting both moment and shear or only shear.

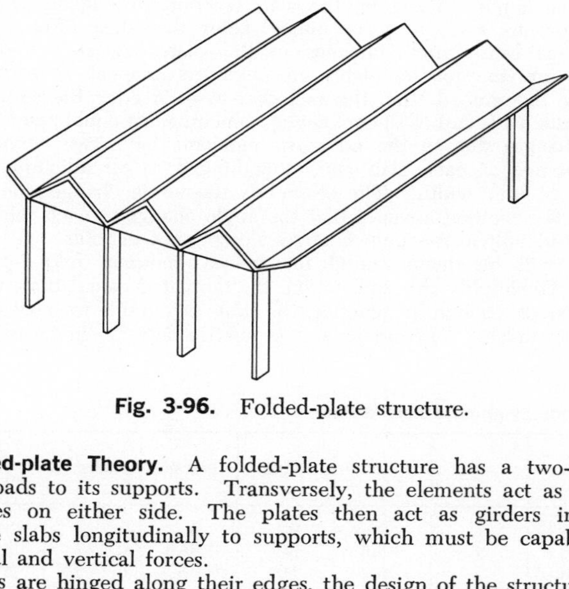

Fig. 3-96. Folded-plate structure.

3-98. Folded-plate Theory. A folded-plate structure has a two-way action in transmitting loads to its supports. Transversely, the elements act as slabs spanning between plates on either side. The plates then act as girders in carrying the load from the slabs longitudinally to supports, which must be capable of resisting both horizontal and vertical forces.

If the plates are hinged along their edges, the design of the structure is relatively simple. Some simplification also is possible if the plates, though having integral edges, are steeply sloped or if the span is sufficiently long with respect to other dimensions that beam theory applies. But there are no criteria for determining when such simplification is possible with acceptable accuracy. In general, a reasonably accurate analysis of folded-plate stresses is advisable.

Several good methods are available (D. Yitzhaki, "The Design of Prismatic and Cylindrical Shell Roofs," North Holland Publishing Company, Amsterdam, available in the United States from W. S. Heinman Books, 400 East 72d Street, New York, N.Y.; "Phase I Report on Folded-plate Construction," Proceedings Paper 3741, *Journal of the Structural Division, American Society of Civil Engineers*, December, 1963; and A. L. Parme and J. A. Sbarounis, "Direct Solution of Folded Plate Concrete Roofs," Portland Cement Association, Skokie, Ill.). They all take into account the effects of plate deflections on the slabs and usually make the following assumptions:

The material is elastic, isotropic, and homogeneous. The longitudinal distribution of all loads on all plates is the same. The plates carry loads transversely only by bending normal to their planes and longitudinally only by bending within their planes. Longitudinal stresses vary linearly over the depth of each plate. Supporting members, such as diaphragms, frames, and beams, are infinitely stiff in their own planes and completely flexible normal to their own planes. Plates have no torsional stiffness normal to their own planes. Displacements due to forces other than bending moments are negligible.

Regardless of the method selected, the computations are rather involved; so it is wise to carry out the work by computer or, when done manually, in a well-organized table. The Yitzhaki method (Art. 3-99) offers some advantages over others in that the calculations can be tabulated, it is relatively simple, it requires the solution of no more simultaneous equations than one for each edge for simply supported plates, it is flexible, and it can easily be generalized to cover a variety of conditions.

3-99. Yitzhaki Method for Folded Plates. Based on the assumptions and general procedure given in Art. 3-98, the Yitzhaki method deals with the slab and plate systems that comprise a folded-plate structure in two ways. In the first, a unit width of slab is considered continuous over supports immovable in the direction of the load (Fig. 3-97*b*). The strip usually is taken where the longitudinal plate stresses are a maximum. Secondly, the slab reactions are taken as loads on the plates, which now are assumed to be hinged along the edges (Fig. 3-97*c*). Thus, the slab reactions cause angle changes in the plates at each fold. Continuity is restored by applying to the plates an unknown moment at each edge. The moments can be determined from the fact that at each edge the sum of the angle changes due to the loads and to the unknown moments must equal zero.

The angle changes due to the unknown moments have two components. One is the angle change at each slab end, now hinged to an adjoining slab, in the transverse strip of unit width. The second is the angle change due to deflection of the plates. The method assumes that the angle change at each fold varies in the same way longitudinally as the angle changes along the other folds.

The method will be demonstrated for the symmetrical folded-plate structure in Fig. 3-97*a*. To simplify the explanation, the tabulated calculations will be broken up into several steps, though, in practice, the table is suitable for continuous calculation from start to finish. Dimensions and geometric data are given in the following table.

Geometric Data for Symmetric Folded Plate

Plate	h, in.	t, in.	a, in.	ϕ, deg	cos ϕ	tan ϕ	A, sq in.	W, lb per ft
1 and 6	48	7	0	90	0	. . . . .	326	350
2 and 5	108	3	93.5	30	0.866	0.577	324	337.5
3 and 4	108	3	106.4	10	0.985	0.1763	324	337.5

In the table, h is the depth in inches, t the thickness in inches, a the horizontal projection of the depth in inches, and A the cross-sectional area of a plate in square inches. ϕ is the angle each plate makes with the horizontal, in degrees. W is the vertical load in pounds per linear foot of plate and thus also the total load in pounds on a 12-in.-wide slab.

Step 1. Compute the loads on a 12-in.-wide transverse strip at mid-span. For the edge beam, $W = 350$ lb per lin ft. For a load of 37.5 lb per sq ft on the plates, each slab in the strip carries 337.5 lb.

Step 2. Consider the strip as a continuous slab supported at the folds (Fig. 3-97*b*), and compute the bending moments by moment distribution. To take advan-

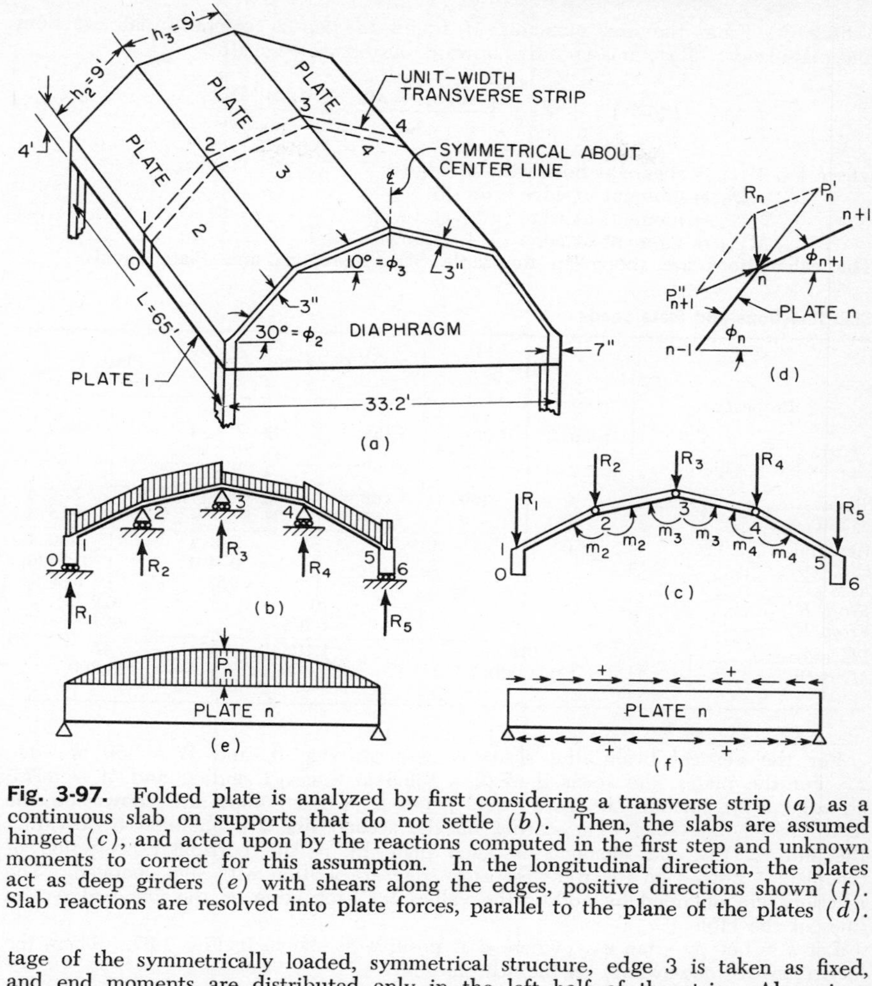

Fig. 3-97. Folded plate is analyzed by first considering a transverse strip (a) as a continuous slab on supports that do not settle (b). Then, the slabs are assumed hinged (c), and acted upon by the reactions computed in the first step and unknown moments to correct for this assumption. In the longitudinal direction, the plates act as deep girders (e) with shears along the edges, positive directions shown (f). Slab reactions are resolved into plate forces, parallel to the plane of the plates (d).

tage of the symmetrically loaded, symmetrical structure, edge 3 is taken as fixed, and end moments are distributed only in the left half of the strip. Also, since edge 1 is free to rotate, the relative stiffness of plate 2 is taken as three-fourths its actual relative stiffness to eliminate the need for distributing moments to edge 1. The computations are shown in the table Slab on Rigid Supports.

Slab on Rigid Supports

Property	Plate 1		Plate 2		Plate 3	
	Edge 0	Edge 1	Edge 1	Edge 2	Edge 2	Edge 3
Stiffness	...	...	...	¾ × 1	1	Fixed
Ratio	...	...	...	0.428	0.572	0
M^F, in.-lb	0	0	0	3,945	−2,996	2,996
				−406	−543	−272
M, in.-lb	0	0	0	3,539	−3,539	2,724

Step 3. From the end moments M found in step 2, compute slab reactions and plate loads. Reactions (positive upward) at the nth edge are

$$R_n = V_n + V_{n+1} + \frac{M_{n-1} + M_n}{a_n} - \frac{M_n + M_{n+1}}{a_{n+1}} \tag{3-205}$$

where V_n, V_{n+1} = shears at both sides of edge n
M_n = moment at edge n, in.-lb
M_{n-1} = moment at edge $(n - 1)$, in.-lb
M_{n+1} = moment at edge $(n + 1)$, in.-lb

The calculations are shown in the table Slab Reactions and Plate Loads.

Slab Reactions and Plate Loads

Property	Plate 1		Plate 2		Plate 3	
	Edge 0	Edge 1	Edge 1	Edge 2	Edge 2	Edge 3
Shears	0	350	168.8	168.8	168.8	337.5
$\pm\Sigma M/a_n$	0	0	−37.8	37.8	7.7	−15.4
Reaction	0	481		383.1		322.1
k	...	...	......	0.401		0.3526
R/k	...	...	......	957		913
From R_{n-1}	0		0		−972	
From R_n	481		1,105		927	
Plate load	481		1,105		−45	
M, ft-lb	254,000		584,000		−23,800	

For the vertical beam, the shear is zero at edge 0, and $W = 350$ at edge 1. For the plates, the shear is $W/2 = 168.8$ at edges 1 and 2, and $W = 337.5$ at edge 3 to include the load from plate 4. These values are shown on the first line of the table. The entries on the second line are obtained by dividing the sum of the end moments in each plate by a, the horizontal projection of the plate. Entries on the right for each edge are written with sign changed. The reactions are obtained by adding the values for each edge on the first and second lines of the table.

Let $k = \tan \phi_n - \tan \phi_{n+1}$, where ϕ is positive as shown in Fig. 3-97. Then, the load (positive downward) on the nth plate is

$$P_n = \frac{R_n}{k_n \cos \phi_n} - \frac{R_{n-1}}{k_{n-1} \cos \phi_n} \tag{3-206}$$

(Figure 3-97 shows the resolution of forces at edge n; $n - 1$ is similar.) Equation (3-206) does not apply for the case of a vertical reaction on a vertical plate, for R/k is the horizontal component of the reaction.

In the table, the two components of Eq. (3-206) are shown on separate lines after the values of R/k, the second component being entered first, with sign changed. The components are summed to obtain the plate loads in pounds per linear foot.

Step 4. Calculate the midspan (maximum) bending moment in each plate. In this example, each plate is a simple beam and $M = PL^2/8$ ft-lb, where L is the span in feet. The moments for each plate are entered in the table after the plate loads. These moments are assumed to vary parabolically along the edges.

Step 5. Determine the free-edge longitudinal stresses at midspan. In each plate, these can be computed from

$$f_{n-1} = \frac{72M}{Ah} \qquad f_n = -\frac{72M}{Ah} \tag{3-207}$$

where f is the stress in psi, M the moment in ft-lb from step 4, and tension is taken as positive, compression as negative.

Step 6. Apply a shear to adjoining edges to equalize the stresses there. Compute the adjusted stresses by converging approximations, similar to moment distribution. To do this, distribute the unbalanced stress at each edge in proportion to the reciprocals of the areas of the plates, and use a carry-over factor of $-\frac{1}{2}$ to distribute the stress to a far edge. Edge 0, being a free edge, requires no distribution of the stress there. Edge 3, because of symmetry, may be treated the same, and distribution need be carried out only in the left half of the structure. The calculations are shown in the table Plate Stresses for Slab Loads

Plate Stresses for Slab Loads

Property	Plate 1		Plate 2		Plate 3	
	Edge 0	Edge 1	Edge 1	Edge 2	Edge 2	Edge 3
$1/A$ ratio	.....	0.491	0.509	0.5	0.5	
Free-edge f	1,133	−1,133	1,200	−1,200	−49	49
Distribution	−573	1,145	−1,188	594		
			−139	278	−279	139
	34	−68	71	−36		
			−9	18	−18	9
	2	−5	4	−2		
				1	−1	
Adjusted f	596	−61	−61	−347	−347	197

Step 7. Compute the midspan edge deflections. In general, the vertical component δ, in inches, can be computed from

$$\frac{E}{L^2}\,\delta_n = \frac{15}{k_n}\left(\frac{f_{n-1} - f_n}{a_n} - \frac{f_n - f_{n+1}}{a_{n+1}}\right) \tag{3-208a}$$

where E = modulus of elasticity, psi

$k = \tan\phi_n - \tan\phi_{n+1}$, as in step 3

The factor E/L^2 is retained for convenience; it is eliminated by dividing the simultaneous angle equations by it. For a vertical plate, the vertical deflection is given by

$$\frac{E}{L^2}\,\delta_n = \frac{15(f_{n-1} - f_n)}{h_n} \tag{3-208b}$$

The calculations are shown in the table Edge Rotations for Plate Loads.

Edge Rotations for Plate Loads

Property	Plate 1		Plate 2		Plate 3	
	Edge 0	Edge 1	Edge 1	Edge 2	Edge 2	Edge 3
$\pm\Delta f/(a \text{ or } h)$	.....	13.70	3.06	5.11	−5.11	−5.11
$E\delta_n/L^2$	.....	205.5		305.6		−435
$\mp E\Delta\delta/L^2a$	.....	....	....	1.07	6.95	−13.90
$E\theta_P/L^2$	.....	....	....	8.02		−13.90

On the first line of the table are entered the results of dividing the difference of the stresses at the edges of each plate by the depth for the vertical edge beam and by the horizontal projection of the depth for the other plates. The results for the top edge of a plate are entered with sign changed. The second line, giving $E\delta_n/L^2$, is obtained from the first line by multiplying by 15 for the edge beam or by adding the entries for each plate, and then multiplying by $15/k_n$.

Step 8. Compute the midspan angle change θ_P at each edge. This can be determined from

$$\frac{E}{L^2}\theta_P = -\frac{\delta_{n-1} - \delta_n}{a_n} + \frac{\delta_n - \delta_{n+1}}{a_{n+1}} \tag{3-209}$$

by continuing the calculations of step 7 in the table. The third line of the table is obtained by subtracting the successive entries on the second line and then dividing by the horizontal projection of the plate. The left-hand entry at each edge is written with sign changed. The fourth line, giving $E\theta_P/L^2$, is the sum of the third-line entries for each edge.

Step 9. To correct the edge rotations with a symmetrical loading, apply an unknown moment of $+1{,}000m_n \sin(\pi x/L)$, in-lb (positive when clockwise) to plate n at edge n and $-1{,}000m_n \sin(\pi x/L)$ to its counterpart, plate n' at edge n'. Also, apply $-1{,}000m_n \sin(\pi x/L)$ to plate $(n+1)$ at edge n and $+1{,}000m_n \sin(\pi x/L)$ to its counterpart; x is the distance along an edge from the end of a plate. (The sine function is assumed to make the loading vary longitudinally in approximately the same manner as the deflections.) At midspan, the absolute value of these moments is $1{,}000m_n$.

The 12-in.-wide transverse strip at midspan, hinged at the supports, will then be subjected at the supports to moments of $1{,}000m_n$. Compute the rotations thus caused in the slabs from

$$\frac{E}{L^2}\theta''_{n-1} = \frac{166.7h_n m_n}{L^2 t_n^3}$$
$$\frac{E}{L^2}\theta''_n = \frac{333.3m_n}{L^2}\left(\frac{h_n}{t_n^3} + \frac{h_{n+1}}{t^3_{n+1}}\right) \tag{3-210}$$
$$\frac{E}{L^2}\theta''_{n+1} = \frac{166.7h_{n+1}m_n}{L^2 t^3_{n+1}}$$

The results of this calculation are shown in the table Edge Rotations, Reactions, and Plate Loads for Unknown Moments.

Edge Rotations, Reactions, and Plate Loads for Unknown Moments

	Plate 1		Plate 2		Plate 3	
	Edge 0	Edge 1	Edge 1	Edge 2	Edge 2	Edge 3
$E\theta''/L^2$ for m_2	...	...	...	$0.631m_2$		$0.316m_2$
$E\theta''/L^2$ for m_3	...	...	...	$0.158m_3$		$0.631m_3$
R for m_2	...	$-10.70m_2$		$20.08m_2$		$-18.76m_2$
R for m_3	...	...	...	$-9.38m_3$		$18.76m_3$
k	...	...	...	0.401		0.3526
R/k for m_2	...	...	...	$50.3m_2$		$-53.2m_2$
R/k for m_3	...	...	...	$-23.4m_3$		$53.2m_3$
P for m_2	$-10.70m_2$		$58.1m_2$		$-105.2m_2$	
P for m_3	0		$-27.1m_3$		$77.8m_3$	
M for m_2	$-4{,}580m_2$		$24{,}840m_2$		$-45{,}100m_2$	
M for m_3	0		$-11{,}620m_3$		$33{,}300m_3$	

Because edge 1, in effect, is hinged in the actual structure, the correction moments need be applied only at edges 2 and 3. And the angle changes in the slabs due to these moments need be computed only at those edges. Results are given in terms of m_2, the moment at edge 2, and m_3, the moment at edge 3.

Step 10. Compute the slab reactions and plate loads due to the unknown moments. The reactions are

$$R_{n-1} = -\frac{1{,}000m_n}{a_n} \qquad R_n = 1{,}000m_n\left(\frac{1}{a_n} + \frac{1}{a_{n+1}}\right) \qquad R_{n+1} = -\frac{1{,}000m_n}{a_{n+1}} \tag{3-211}$$

The plate loads are

$$P_n = \frac{1}{\cos\phi_n}\left(\frac{R_n}{k_n} - \frac{R_{n-1}}{k_{n-1}}\right) \tag{3-212}$$

The calculations are shown as a continuation of the table of step 9.

Step 11. Assume that the loading on each plate is $P_n \sin(\pi x/L)$ (Fig. 3-97*e*), and calculate the midspan (maximum) bending moment. For a simple beam,

$$M = PL^2/\pi^2 \text{ ft-lb}$$

The moments also are shown in the table in terms of m_2 and m_3.

Step 12. Using Eq. (3-207), compute the free-edge longitudinal stresses at midspan. Then, as in step 6, apply a shear at each edge to equalize the stresses. Determine the adjusted stresses by converging approximations. The results of this calculation are shown in the table Plate Stresses for Unknown Moments.

Plate Stresses for Unknown Moments

Stress	Plate 1		Plate 2		Plate 3	
	Edge 0	Edge 1	Edge 1	Edge 2	Edge 2	Edge 3
Free-edge f, for m_2	$-20.5m_2$	$20.5m_2$	$51.1m_2$	$-51.1m_2$	$-92.6m_2$	$92.6m_2$
Adjusted f, for m_2	$-31.2m_2$	$41.9m_2$	$41.9m_2$	$-66.3m_2$	$-66.3m_2$	$79.5m_2$
Free-edge f, for m_3	......	......	$-23.8m_3$	$23.8m_3$	$68.5m_3$	$-68.5m_3$
Adjusted f, for m_3	$9.2m_3$	$-18.4m_3$	$-18.4m_3$	$41.4m_3$	$41.4m_3$	$-54.9m_3$

Step 13. Compute the vertical component of the edge deflections at midspan from

$$\frac{E}{L^2}\delta_n = \frac{144}{\pi^2 k_n}\left(\frac{f_{n-1} - f_n}{a_n} - \frac{f_n - f_{n+1}}{a_{n+1}}\right) \tag{3-213a}$$

or for a vertical plate from

$$\frac{E}{L^2}\delta_n = \frac{144(f_{n-1} - f_n)}{\pi^2 h_n} \tag{3-213b}$$

The results of this calculation are shown in the table Edge Rotations for Unknown Moments. The procedure is the same as step 7.

Edge Rotations for Unknown Moments

Rotation	Plate 1		Plate 2		Plate 3	
	Edge 0	Edge 1	Edge 1	Edge 2	Edge 2	Edge 3
$\pm\Delta f/(a \text{ or } h)$ for m_2	.....	$-1.522m_2$	$1.155m_2$	$1.368m_2$	$-1.368m_2$	$-1.368m_2$
$E\delta/L^2$, for m_2	.....	$-22.2m_2$		$92.0m_2$		$-113.2m_2$
$\mp E\Delta\delta/L^2a$, for m_2	.....	......	......	$1.222m_2$	$1.927m_2$	$-3.854m_2$
$E\theta'/L^2$, for m_2	.....	......	......	$3.149m_2$		$-3.854m_2$
$\pm\Delta f/(a \text{ or } h)$ for m_3	.....	$0.574m_3$	$-0.640m_3$	$-0.904m_3$	$0.904m_3$	$0.904m_3$
$E\delta/L^2$, for m_3	.....	$8.4m_3$		$-56.2m_3$		$74.8m_3$
$\mp E\Delta\delta/L^2a$, for m_3	.....	......	......	$-0.691m_3$	$-1.230m_3$	$2.460m_3$
$E\theta'/L^2$, for m_3	.....	......	......	$-1.921m_3$		$2.460m_3$

Step 14. Using Eq. (3-209), determine the midspan angle change θ' at each edge. The computations are given in the continuation of the table of step 13 and follow the procedure of step 8.

Step 15. At each edge, set up an equation by putting the sum of the angle changes equal to zero. Thus, after dividing through by E/L^2: $\theta_P + \theta'' + \Sigma\theta' = 0$. Solve these simultaneous equations for the unknown moments.

The sum of the angles for edge 2 is

$$8.02 + 0.631m_2 + 0.158m_3 + 3.149m_2 - 1.921m_3 = 0$$

The sum of the angles for edge 3 is

$$-13.90 + 0.316m_2 + 0.631m_3 - 3.854m_2 + 2.460m_3 = 0$$

The solution is $m_2 = -0.040$ and $m_3 = 4.45$.

Step 16. Determine the actual reactions, loads, stresses, and deflections by substituting for m_2 and m_3 the values just found. For example, the final longitudinal stress at edge 0 is $596 - 31.2(-0.040) + 9.2(4.45) = 638$ psi. Similarly, the final longitudinal stress at edge 1 is -145 psi; at edge 2, -160 psi; and at edge 3, -50 psi.

Step 17. Compute the shear stresses. The shear stress at edge n in pounds (Fig. 3-97f) is

$$T_n = T_{n-1} - \frac{f_{n-1} + f_n}{2} A_n \tag{3-214}$$

In the example, $T_0 = 0$; so the shears at the edges can be obtained successively, since the stresses f are known. The calculations are shown in the table Shearing Stresses.

For a uniformly loaded folded plate, the shear stress S, psi, at any point on an edge n is approximately

$$S = \frac{2T_{max}}{3Lt}\left(\frac{1}{2} - \frac{x}{L}\right) \tag{3-215}$$

with a maximum at plate ends of

$$S_{max} = \frac{T_{max}}{3Lt} \tag{3-216}$$

The edge shears computed with Eq. (3-216) are given in the table.

The shear stress, psi, at middepth (not always a maximum) is

$$v_n = \left(\frac{3P_nL}{2A_n} + \frac{S_{n-1} + S_n}{2}\right)\left(\frac{1}{2} - \frac{x}{L}\right) \tag{3-217}$$

and has its largest value at $x = 0$:

$$v_{max} = \frac{0.75P_nL}{A_n} + \frac{S_{n-1} + S_n}{4} \tag{3-218}$$

The middepth shears also are given in the table.

Shearing Stresses

Stress	Plate 1		Plate 2		Plate 3	
	Edge 0	Edge 1	Edge 1	Edge 2	Edge 2	Edge 3
$-(f_{n-1} + f_n)A_n/2$	...	-83,500		49,500		34,000
Midspan edge shear, lb	0	-83,500		-34,000		0
Edge shear, psi, $x = 0$	0	-61.2	-143	-58	-58	0
$0.75PL/A$	69.9		148.3		46.0	
0.25 edge shear	0	-15.3	-35.8	-14.5	-14.5	0
Middepth shear, psi	54.6		98.0		31.5	

For more details, see D. Yitzhaki and Max Reiss, "Analysis of Folded Plates," Proceedings Paper 3303, *Journal of the Structural Division, American Society of Civil Engineers,* October, 1962.

CATENARIES

3-100. Stresses in Cables. When a cable is suspended from two points of support, the end reactions have a horizontal as well as a vertical component. If the cable carries only vertical loads, the horizontal component of the tension at any point in the cable is equal to the horizontal component of the end reaction. The vertical components of the reactions can be computed by taking moments about the supports in the same way as for beams, provided the cable supports are at the same level. When the supports are not at the same level, another equation derived from a knowledge of the shape of the cable is required.

Generally, it is safe to assume that the thickness of the cable is negligible compared with the span. Hence, bending stresses may be neglected, and the bending moment at any point is equal to zero. As a consequence, the shape assumed by the cable is similar to that of the bending-moment diagram for a simply supported beam carrying the same loads. Stresses in the cable are directed along the axis.

At the lowest point of a cable, the vertical shear either is zero or changes sign on either side of the point.

The horizontal component of the reaction can be computed from the fact that the bending moment is zero at a point on the cable at which the sag is known (usually the low point).

For example, suppose a cable spanning 30 ft between supports at the same level is permitted to sag 2 ft in supporting a 12-kip load at a third point. Taking moments about the supports, we find the vertical components of the reactions equal to 8 and 4 kips, respectively. Setting the bending moment at the low point equal to zero, we can write the equation $8 \times 10 - 2H = 0$; from which the horizontal component H of the reaction is found to be 40 kips. The maximum tension in the cable then is

$$T = \sqrt{8^2 + 40^2} = 40.8 \text{ kips}$$

(F. S. Merritt, "Structural Steel Designers' Handbook," McGraw-Hill Book Company, New York.)

WIND AND SEISMIC STRESSES IN TALL BUILDINGS

Buildings must be designed to resist horizontal forces as well as vertical loads. In tall buildings, the lateral forces must be given particular attention, because if they are not properly provided for, they can cause collapse of the structure.

Horizontal forces are generally taken care of with X bracing, shear walls, or wind connections. **X bracing,** usually adopted for low industrial-type buildings, transmits the forces to the ground by truss action. **Shear walls** are vertical slabs capable of transmitting the forces to the ground without shearing or buckling; they generally are so deep in the direction of the lateral loads that bending stresses are easily handled. **Wind connections** are a means of establishing continuity between girders and columns. The structure is restrained against lateral deformation by the resistance to rotation of the members at the connections.

The continuous rigid frames thus formed can be analyzed by the methods of Arts. 3-63 to 3-68, 3-79 to 3-85 or Arts. 3-105 to 3-108. Generally, however, approximate methods are used.

It is noteworthy that for most buildings even the "exact" methods are not exact. In the first place, the forces acting are not static loads, but generally dynamic; they are uncertain in intensity, direction, and duration. Earthquake forces, usually assumed as a percentage of the weight of the building above each level, act at the base of the structure, not at each floor level as is assumed in design, and accelerations at each level vary nearly linearly with distance above the base. Also, at the beginning of a design, the sizes of members are not known, so the exact resistance to lateral deformation cannot be calculated. Furthermore, floors, walls, and partitions help resist the lateral forces in a very uncertain way.

3-101. Portal Method. Since an exact analysis is impossible, most designers prefer a wind-analysis method based on reasonable assumptions and requiring a minimum of calculations. One such method is the so-called "portal method."

It is based on the assumptions that points of inflection (zero bending moment) occur at the midpoints of all members and that exterior columns take half as much shear as do interior columns. These assumptions enable all moments and shears throughout the building frame to be computed by the laws of statics.

Consider, for example, the roof level (Fig. 3-98*a*) of a tall building. A wind load of 600 lb is assumed to act along the top line of girders. To apply the

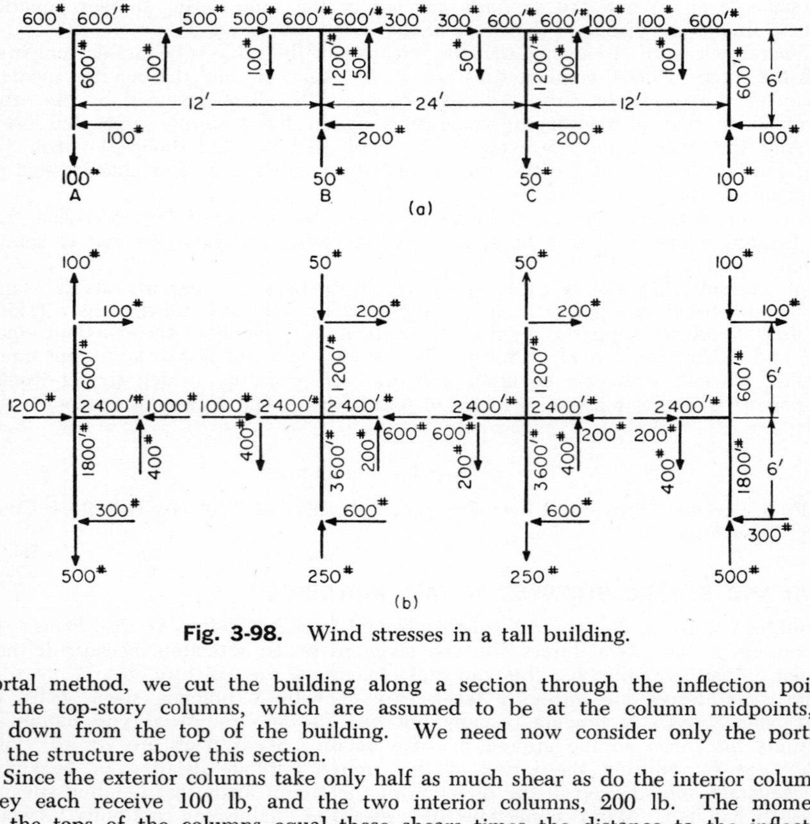

Fig. 3-98. Wind stresses in a tall building.

portal method, we cut the building along a section through the inflection points of the top-story columns, which are assumed to be at the column midpoints, 6 ft down from the top of the building. We need now consider only the portion of the structure above this section.

Since the exterior columns take only half as much shear as do the interior columns, they each receive 100 lb, and the two interior columns, 200 lb. The moments at the tops of the columns equal these shears times the distance to the inflection point. The wall end of the end girder carries a moment equal to the moment in the column. (At the floor level below, as indicated in Fig. 3-98*b*, that end of the end girder carries a moment equal to the sum of the column moments.) Since the inflection point is at the midpoint of the girder, the moment at the inner end of the girder must be the same as at the outer end. The moment in the adjoining girder can be found by subtracting this moment from the column moment, because the sum of the moments at the joint must be zero. (At the floor level below, as shown in Fig. 3-90*b*, the moment in the interior girder is found by subtracting the moment in the exterior girder from the sum of the column moments.)

Girder shears then can be computed by dividing girder moments by the half span. When these shears have been found, column loads can be easily computed from the fact that the sum of the vertical loads must be zero, by taking a section around each joint through column and girder inflection points. As a check, it should be noted that the column loads produce a moment that must be equal to the moments of the wind loads above the section for which the column loads were computed. For the roof level (Fig. 3-98*a*), for example, $-50 \times 24 + 100 \times 48 = 600 \times 6$.

3-102. Cantilever Method. Another wind-analysis procedure that is sometimes employed is the cantilever method. Basic assumptions here are that inflection points are at the midpoints of all members and that direct stresses in the columns vary as the distances of the columns from the center of gravity of the bent. The assumptions are sufficient to enable shears and moments in the frame to be determined from the laws of statics.

For multistory buildings with height-to-width ratio of 4 or more, the Spurr modification is recommended ("Welded Tier Buildings," U.S. Steel Corp.). In this method, the moments of inertia of the girders at each level are made proportional to the girder shears.

The results obtained from the cantilever method generally will be different from those obtained by the portal method. In general, neither solution is correct, but the answers provide a reasonable estimate of the resistance to be provided against lateral deformation. (See also *Transactions of the ASCE*, Vol. 105, pp. 1713–1739, 1940.)

ULTIMATE STRENGTH OF DUCTILE FLEXURAL MEMBERS

When an elastic material, such as structural steel, is loaded with a gradually increasing load, stresses are proportional to strains up to the yield point. If the material, like steel, also is ductile, then it continues to carry load beyond the yield point, though strains increase rapidly with little increase in load (Fig. 3-99*a*).

Similarly, a beam made of an elastic material continues to carry more load after the stresses in the outer fibers reach the yield point. However, the stresses will no longer vary with distance from the neutral axis; so the flexural formula [Eq. 3-23)] no longer holds. However, if simplifying assumptions are made, approximating the stress-strain relationship beyond the elastic limit, the load-carrying capacity of the beam can be computed with satisfactory accuracy.

3-103. Theory of Plastic Behavior. For a ductile material, the idealized stress-strain relationship in Fig. 3-99*b* may be assumed. Stress is proportional to strain until the yield-point stress f_y is reached, after which strain increases at a constant stress.

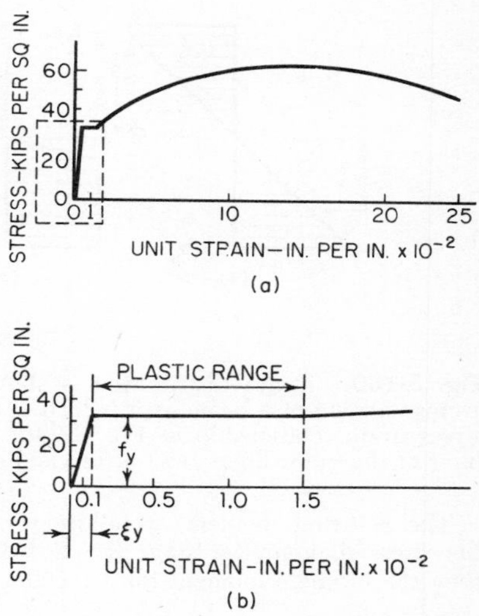

Fig. 3-99. Stress-strain relationship for a ductile material generally is similar to the curve shown in (*a*). To simplify plastic analysis, the portion of (*a*) enclosed by the dash lines is approximated by the curve in (*b*), which extends to the range where strain hardening begins.

For a beam of this material, the following assumptions will also be made: plane sections remain plane, strains thus being proportional to distance from the neutral axis; properties of the material in tension are the same as those in compression; its fibers behave the same in flexure as in tension; and deformations remain small.

Strain distribution across the cross section of a rectangular beam, based on these assumptions, is shown in Fig. 3-100*a*. At the yield point, the unit strain is ϵ_y and the curvature ϕ_y, as indicated in (1). In (2), the strain has increased several times, but the section still remains plane. Finally, at failure, (3), the strains are very large and nearly constant across upper and lower halves of the section.

Corresponding stress distributions are shown in Fig. 3-100*b*. At the yield point, (1), stresses vary linearly and the maximum is f_y. With increase in load, more and more fibers reach the yield point, and the stress distribution becomes nearly constant, as indicated in (2). Finally, at failure, (3), the stresses are constant across the top and bottom parts of the section and equal to the yield-point stress.

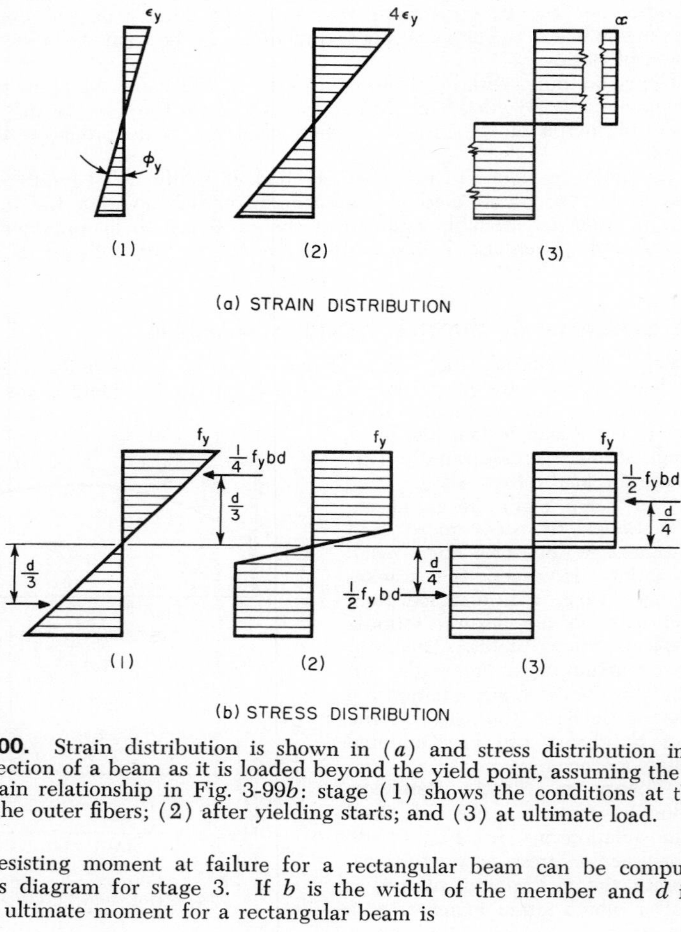

Fig. 3-100. Strain distribution is shown in (*a*) and stress distribution in (*b*) for a cross section of a beam as it is loaded beyond the yield point, assuming the idealized stress-strain relationship in Fig. 3-99*b*: stage (1) shows the conditions at the elastic limit of the outer fibers; (2) after yielding starts; and (3) at ultimate load.

The resisting moment at failure for a rectangular beam can be computed from the stress diagram for stage 3. If b is the width of the member and d its depth, then the ultimate moment for a rectangular beam is

$$M_P = \frac{bd^2}{4} f_y \qquad (3\text{-}219)$$

Since the resisting moment at stage 1 is $M_y = f_y bd^2/6$, the beam carries 50% more moment before failure than when the yield-point stress is first reached in the outer fibers.

A circular section has an M_P/M_y ratio of about 1.7, while a diamond section has a ratio of 2. The average wide-flange rolled-steel beam has a ratio of about 1.14.

The relationship between moment and curvature in a beam can be assumed to be similar to the stress-strain relationship in Fig. 3-99*b*. Curvature ϕ varies linearly with moment until $M_y = M_P$ is reached, after which ϕ increases indefinitely at constant moment. That is, a plastic hinge forms.

This ability of a ductile beam to form plastic hinges enables a fixed-end or continuous beam to carry more load after M_P occurs at a section, because a redistribution of moments takes place. Consider, for example, a uniformly loaded fixed-end beam. In the elastic range, the end moments are $M_L = M_R = WL/12$, while the midspan moment M_C is $WL/24$. The load when the yield point is reached in the outer fibers is $W_y = 12M_y/L$. Under this load, the moment capacity of the ends of the beam is nearly exhausted; plastic hinges form there when the moment equals M_P. As load is increased, the ends then rotate under constant moment and the beam deflects like a simply supported beam. The moment at midspan increases until the moment capacity at that section is exhausted and a plastic hinge forms. The load causing that condition is the ultimate load W_u since, with three hinges in the span, a link mechanism is formed and the member continues to deform at constant load. At the time the third hinge is formed, the moments at ends and center are all equal to M_P. Therefore, for equilibrium, $2M_P = W_uL/8$, from which $W_u = 16M_P/L$. Since for the idealized moment-curvature relationship, M_P was assumed equal to M_y, the carrying capacity due to redistribution of moments is 33% greater.

3-104. Upper and Lower Bounds for Ultimate Loads. Methods for computing the ultimate strength of continuous beams and frames may be based on two theorems that fix upper and lower limits for load-carrying capacity.

Upper-bound Theorem. A load computed on the basis of an assumed link mechanism will always be greater than or at best equal to the ultimate load.

Lower-bound Theorem. The load corresponding to an equilibrium condition with arbitrarily assumed values for the redundants is smaller than or at best equal to the ultimate loading—provided that everywhere moments do not exceed M_P.

Equilibrium Method. This method, based on the lower-bound theorem, usually is easier for simple cases. The steps involved are: (1) Select redundants that if removed would leave the structure determinate. (2) Draw the moment diagram for the determinate structure. (3) Sketch the moment diagram for an arbitrary value of each redundant. (4) Combine the moment diagrams, forming enough peaks so that the structure will act as a link mechanism if plastic hinges are formed at those points. (5) Compute the value of the redundants from the equations of equilibrium, assuming that at the peaks $M = M_P$. (6) Check to see that there are sufficient plastic hinges to form a mechanism and that M is everywhere less than or equal to M_P.

Consider, for example, the continuous beam *ABCD* in Fig. 3-101*a* with three equal spans, uniformly loaded, the center span carrying double the load of the end spans. Assume that the plastic moment for the end spans is k times that for the center span ($k < 1$). For what value of k will the ultimate strength be the same for all spans?

Figure 3-101*b* shows the moment diagram for the beam made determinate by ignoring the moments at B and C, and the moment diagram for end moments M_B and M_C applied to the determinate beam. Figure 3-101*c* gives the combined moment diagram. If plastic hinges form at all the peaks, a link mechanism will exist.

Since at B and C the joints can develop only the strength of the weakest beam, a plastic hinge will form when $M_B = M_C = kM_p$. A plastic hinge also will form at the center of span BC when the midspan moment is M_P. For equilibrium to be maintained, this will occur when $M_P = wL^2/4 - \frac{1}{2}M_B - \frac{1}{2}M_C = wL^2/4 - kM_P$, from which

$$M_P = \frac{wL^2}{4(1 + k)}$$

Maximum moment occurs in spans AB and CD where $x = L/2 - M/wL$—or if $M = kM_P$, where $x = L/2 - kM_P/wL$. A plastic hinge will form at this point when the moment equals kM_P. For equilibrium, therefore,

$$kM_P = \frac{w}{2}x(L - x) - \frac{x}{L}kM_P = \frac{w}{2}\left(\frac{L}{2} - \frac{kM_P}{wL}\right)\left(\frac{L}{2} + \frac{kM_P}{wL}\right) - \left(\frac{1}{2} - \frac{kM_P}{wL^2}\right)kM_P$$

This leads to the quadratic equation:

$$\frac{k^2M_P{}^2}{wL^2} - 3kM_P + \frac{wL^2}{4} = 0$$

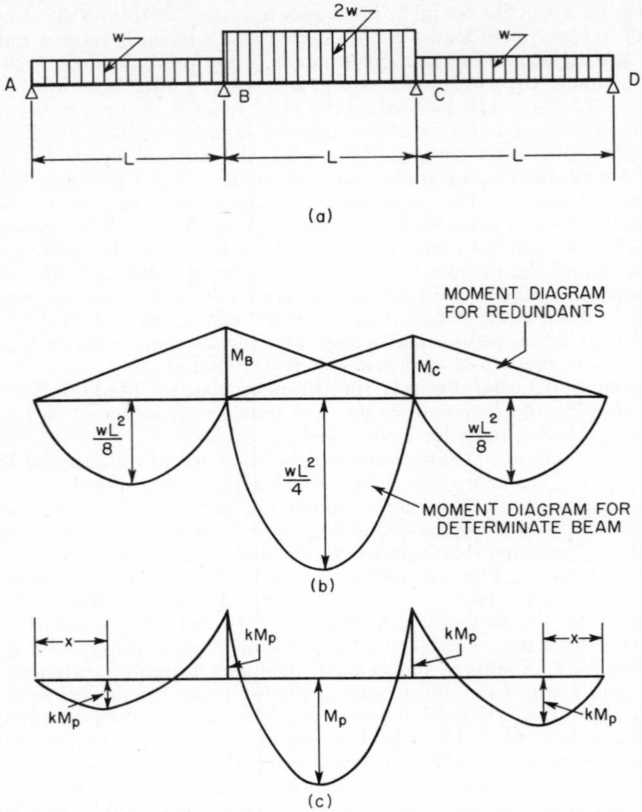

Fig. 3-101. Continuous beam shown in (a) carries twice as much uniform load in the center span as in the side span. In (b) are shown the moment diagrams for this loading condition with redundants removed and for the redundants. The two moment diagrams are combined in (c), producing peaks at which plastic hinges are assumed to form.

When the value of M_P previously computed is substituted in this equation, it becomes

$$7k^2 + 4k = 4 \qquad \text{or} \qquad k(k + 4/7) = 4/7$$

from which $k = 0.523$. And the ultimate load is

$$wL = \frac{4M_P(1 + k)}{L} = 6.1 \frac{M_P}{L}$$

Mechanism Method. This method, based on the upper-bound theorem, requires the following steps: (1) Determine the location of possible hinges (points of maximum moment). (2) Pick combinations of hinges to form possible link mechanisms. Certain types are classed as elementary, such as the mechanism formed when hinges are created at the ends and center of a fixed-end or continuous beam, or when hinges form at top and bottom of a column subjected to lateral load, or when hinges are formed at a joint in the members framing into it, permitting the joint to rotate freely. If there are n possible plastic hinges and x redundant forces or moments, then there are $n - x$ independent equilibrium equations and $n - x$ elementary mechanisms. Both elementary mechanisms and possible combinations of them must be investigated. (3) Apply a virtual displacement to each possible mechanism in turn and compute the internal and external work (Art. 3-52). (4) From the equality of internal and external work, compute the critical load for each mechanism. The mechanism with the lowest critical load is the most probable, and its load is the ultimate, or limit load. (5) Make an equilibrium check to ascertain that moments everywhere are less than or equal to M_P.

As an example of an application of the mechanism method, let us find the ultimate load for the rigid frame of constant section throughout in Fig. 3-102*a*. Assume that the vertical load at midspan is equal to 1.5 times the lateral load.

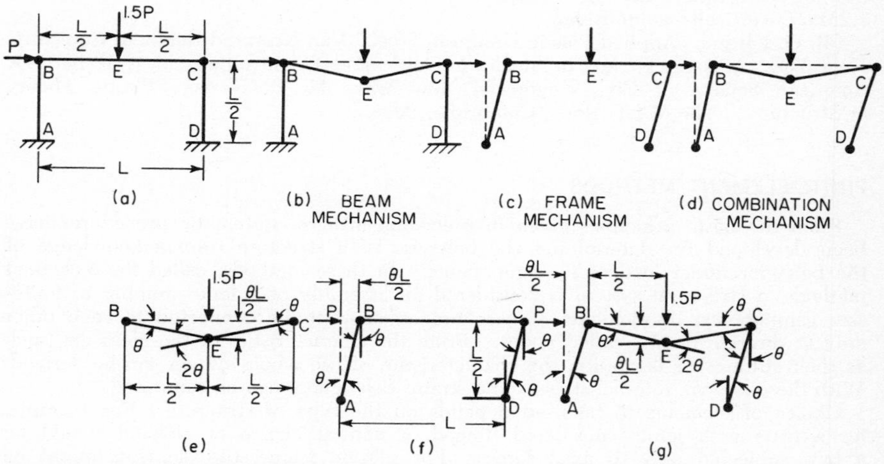

Fig. 3-102. Ultimate-load possibilities for a rigid frame of constant section with fixed bases.

Maximum moments can occur at five points—*A*, *B*, *C*, *D*, and *E*—and plastic hinges may form there. Since this structure has three redundants, the number of elementary mechanisms is $5 - 3 = 2$. These are shown in Fig. 3-102*b* and *c*. A combination of these mechanisms is shown in Fig. 3-102*d*.

Let us first investigate the beam alone, with plastic hinges at *B*, *E*, and *C*. As indicated in Fig. 3-102*e*, apply a virtual rotation θ to *BE* at *B*. The center deflection, then, is $\theta L/2$. Since *C* is at a distance of $L/2$ from *E*, the rotation at *C* must be θ and at *E*, 2θ. The external work, therefore, is $1.5P$ times the midspan deflection, $\theta L/2$, which equals $\frac{3}{4}\theta PL$. The internal work is the sum of the work at each hinge, or $M_P\theta + 2M_P\theta + M_P\theta = 4M_P\theta$. Equating internal and external work: $\frac{3}{4}\theta PL = 4M_P\theta$, from which $P = 5.3M_P/L$.

Next, let us investigate the frame mechanism, with plastic hinges at *A*, *B*, *C*, and *D*. As shown in Fig. 3-102*f*, apply a virtual rotation θ to *AB*. The point of application of the lateral load *P* will then move a distance $\theta L/2$, and the external work will be $\theta PL/2$. The internal work, being the sum of the work at the hinges, will be $4M_P\theta$. Equating internal and external work, $\theta PL/2 = 4M_P\theta$, from which $P = 8M_P/L$.

Finally, let us compute the critical load for the combination beam-frame mechanism in Fig. 3-102*g*. Again, apply a virtual rotation θ to *AB*, moving *B* horizontally and *E* vertically a distance of $\theta L/2$. The external work, therefore, equals

$$\frac{\theta PL}{2} + \frac{3}{4}\,\theta PL = \frac{5\theta PL}{4}$$

The internal work is the sum of the work at hinges *A*, *E*, *C*, and *D* and is equal to $M_P\theta + 2M_P\theta + 2M_P\theta + M_P\theta = 6M_P\theta$. Equating internal and external work, $5\theta PL/4 = 6M_P\theta$, from which $P = 4.8M_P/L$.

The combination mechanism has the lowest critical load. Now an equilibrium check must be made to insure that the moments everywhere in the frame are less than M_P. If they are, then the combination mechanism is the correct solution.

Consider first the length EC of the beam as a free body. Taking moments about E, noting that the moments at E and C equal M_P, the shear at C is computed to be $4M_P/L$. Therefore, the shear at B is $1.5 \times 4.8M_P/L - 4M_P/L = 3.2M_P/L$. By taking moments about B, with BE considered as a free body, the moment at B is found to be $0.6M_P$. Similar treatment of the columns as free bodies indicates that nowhere is the moment greater than M_P. Therefore, the combination mechanism is the correct solution, and the ultimate load for the frame is $4.8M_P/L$ laterally and $7.2M_P/L$ vertically at midspan.

(R. O. Disque, "Applied Plastic Design in Steel," Van Nostrand Reinhold Company, New York; "Plastic Design in Steel—A Guide and Commentary, M & R No. 41, American Society of Civil Engineers, New York; M. R. Horne, "Plastic Theory of Structures," The M.I.T. Press, Cambridge, Mass.)

FINITE-ELEMENT METHODS

From the basic principles given in preceding articles, systematic procedures have been developed for determining the behavior of a structure from a knowledge of the behavior under load of its components. In these methods, called finite-element methods, a structural system is considered an assembly of a finite number of finite-size components, or elements. These are assumed to be connected to each other only at discrete points, called nodes. From the characteristics of the elements, such as their stiffness or flexibility, the characteristics of the whole system can be derived. With these known, internal stresses and strains throughout can be computed.

Choice of elements to be used depends on the type of structure. For example, for a truss with joints considered hinged, a natural choice of element would be a bar, subjected only to axial forces. For a rigid frame, the elements might be beams subjected to bending and axial forces, or to bending, axial forces, and torsion. For a thin plate or shell, elements might be triangles or rectangles, connected at vertices. For three-dimensional structures, elements might be beams, bars, tetrahedrons, cubes, or rings.

For many structures, because of the number of finite elements and nodes, analysis by a finite-element method requires mathematical treatment of large amounts of data and solution of numerous simultaneous equations. For this purpose, the use of high-speed computers is advisable.

The mathematics of such analyses is usually simpler and more compact when the data are handled in matrix form. Matrix notation is especially convenient in indicating the solution of simultaneous linear equations. For example, suppose a set of equations is represented in matrix notation by $\mathbf{AX} = \mathbf{B}$. Multiplication of both sides of the equation by inverse $\mathbf{A}^{-1}$ yields $\mathbf{A}^{-1}\mathbf{AX} = \mathbf{A}^{-1}\mathbf{B}$. Since $\mathbf{A}^{-1}\mathbf{A} = \mathbf{I}$, the identity matrix, and $\mathbf{IX} = \mathbf{X}$, the solution of the equations is given by $\mathbf{X} = \mathbf{A}^{-1}\mathbf{B}$. (F. S. Merritt, "Modern Mathematical Methods for Engineers," McGraw-Hill Book Company, New York.)

The methods used for analyzing structures generally may be classified as force (flexibility) or displacement (stiffness) methods.

In analysis of statically indeterminate structures by force methods, forces are chosen as redundants, or unknowns. The choice is made in such a way that equilibrium is satisfied. These forces are then determined from the solution of equations that insure compatibility of all displacements of elements at each node. After the redundants have been computed, stresses and strains throughout the structure can be found from equilibrium equations and stress-strain relations.

In displacement methods, displacements are chosen as unknowns. The choice is made in such a way that geometric compatibility is satisfied. These displacements are then determined from the solution of equations that insure that forces acting at each node are in equilibrium. After the unknowns have been computed, stresses and strains throughout the structure can be found from equilibrium equations and stress-strain relations.

In choosing a method, the following should be kept in mind: In force methods, the number of unknowns equals the degree of indeterminancy. In displacement methods, the number of unknowns equals the degrees of freedom of displacement at nodes. The fewer the unknowns, the fewer the calculations required.

3-105. Force-displacement Relations. Assume a right-handed cartesian coordinate system, with axes x, y, z. Assume also at each node of a structure to be analyzed a system of base unit vectors, $\mathbf{e}_1$ in the direction of the x axis, $\mathbf{e}_2$ in the direction of the y axis, and $\mathbf{e}_3$ in the direction of the z axis. Forces and moments acting at a node are resolved into components in the directions of the base vectors. Then, the forces and moments at the node may be represented by the vector $P_i\mathbf{e}_i$, where P_i is the magnitude of the force or moment acting in the direction of $\mathbf{e}_i$. This vector, in turn, may be conveniently represented by a column matrix $\mathbf{P}$. Similarly, the displacements—translations and rotation—of the node may be represented by the vector $\Delta_i\mathbf{e}_i$, where Δ_i is the magnitude of the displacement acting in the direction of $\mathbf{e}_i$. This vector, in turn, may be represented by a column matrix $\boldsymbol{\Delta}$.

To conserve space in the following, column vectors, when given in terms of their components, will be represented by their transposes (rows interchanged with columns), represented by a superscript T. Thus, the vector $P_i\mathbf{e}_i$ will be represented by the row matrix $\mathbf{P}^T = [P_1\ P_2\ \ldots\ P_6]$. Also, vectors and matrices will be indicated by boldface symbols.

For compactness, and because, in structural analysis, similar operations are performed on all nodal forces, all the loads, including moments, acting on all the nodes may be combined into a single column matrix $\mathbf{P}$. Similarly, all the nodal displacements may be represented by a single column matrix $\boldsymbol{\Delta}$.

If the independent loads and displacements are listed in the appropriate order in each matrix, a matrix equation relating them can be written:

$$\boldsymbol{\Delta} = \mathbf{FP} \tag{3-220}$$

For an elastic structure, $\mathbf{F}$ is the flexibility matrix. It is square (has the same number of rows as columns) and symmetric ($\mathbf{F} = \mathbf{F}^T$). A typical element F_{ij} of $\mathbf{F}$ gives the deflection of a node in the direction of displacement Δ_i when a unit force acts at the same or another node in the direction of force P_j. The jth column of $\mathbf{F}$, therefore, contains all the nodal displacements when one force P_j is set equal to unity and all other independent forces are zero.

Multiplication of both sides of Eq. (3-220) by $\mathbf{F}^{-1}$ yields $\mathbf{F}^{-1}\boldsymbol{\Delta} = \mathbf{F}^{-1}\mathbf{FP} = \mathbf{IP} = \mathbf{P}$, or

$$\mathbf{P} = \mathbf{K}\boldsymbol{\Delta} \tag{3-221}$$

where $\mathbf{K}$ = inverse of flexibility matrix = $\mathbf{F}^{-1}$

For an elastic structure, $\mathbf{K}$ is the stiffness matrix. It too is square and symmetric. A typical element K_{ij} of $\mathbf{K}$ gives the force at a node, in the direction of load P_i, that results when the same or another node is given a unit displacement in the direction of displacement Δ_j. The jth column of $\mathbf{K}$, therefore, contains the nodal forces that produce a unit displacement of the node at which Δ_j occurs and in the direction of Δ_j, but no other nodal displacements throughout the structure.

For some structures, the flexibility or stiffness matrix can be constructed from the definition. In other cases, the matrix may have to be synthesized from the flexibility or stiffness matrices of finite elements comprising the structures. The procedures are explained in the following articles.

3-106. Matrix Force (Flexibility) Method. In this article, the structure is assumed to consist of n finite elements connected only at the nodes. The flexibility matrices of the elements are assumed known. Right-handed cartesian coordinate axes, x, y, z, are chosen for the structure, with base unit vectors $\mathbf{e}_i$ parallel to these axes at each mode, as in Art. 3-105. The flexibility matrix for the whole structure can be constructed from the element flexibility matrices, as will be demonstrated.

For the ith element, nodal forces and displacements are related by

$$\boldsymbol{\delta}_i = \mathbf{f}_i\mathbf{S}_i \qquad i = 1, 2, \ldots n \tag{3-222}$$

where $\boldsymbol{\delta}_i$ = matrix of displacements of the nodes of the ith element in directions of base vectors

$\mathbf{f}_i$ = flexibility matrix of the ith element

$\mathbf{S}_i$ = matrix of forces, moments, torques acting at nodes of the ith element in directions of base vectors

For compactness, this relationship between nodal displacements and forces for each element can be combined into a single matrix equation applicable to all the elements:

$$\boldsymbol{\delta} = \mathbf{fS} \tag{3-223}$$

where $\boldsymbol{\delta}$ = matrix of all nodal displacements for all elements

$\mathbf{S}$ = matrix of all forces acting at the nodes of all elements

$$\mathbf{f} = \begin{bmatrix} \mathbf{f}_1 & 0 & \cdots & 0 \\ 0 & \mathbf{f}_2 & \cdots & 0 \\ \cdot & \cdot & \cdots & \cdot \\ 0 & 0 & \cdots & \mathbf{f}_n \end{bmatrix} \tag{3-224}$$

Element forces $\mathbf{S}$ can be determined from the loads $\mathbf{P}$ at the nodes to meet equilibrium requirements at the nodes:

$$\mathbf{S} = \mathbf{a}_0\mathbf{P} \tag{3-225}$$

where $\mathbf{a}_0$ is a matrix of influence coefficients. The jth column of $\mathbf{a}_0$ contains the element forces when one load $P_j = 1$, and all other loads are set equal to zero.

From Eqs. (3-223) and (3-225),

$$\boldsymbol{\delta} = \mathbf{fa}_0\mathbf{P} \tag{3-226}$$

From energy relations, it can be shown that

$$\boldsymbol{\Delta} = \mathbf{a}_0^T\boldsymbol{\delta} \tag{3-227}$$

where $\boldsymbol{\Delta}$ = matrix of displacements at the nodes of the structure

$\mathbf{a}_0^T$ = transpose of $\mathbf{a}_0$ = matrix $\mathbf{a}_0$ with rows and columns interchanged

Finally, from Eqs. (3-226) and (3-227) comes the relationship that permits computation of the flexibility matrix of the whole structure from the flexibility matrices of its elements:

$$\boldsymbol{\Delta} = \mathbf{a}_0^T\mathbf{fa}_0\mathbf{P} \tag{3-228}$$

Since also, by Eq. (3-220), $\boldsymbol{\Delta} = \mathbf{FP}$,

$$\mathbf{F} = \mathbf{a}_0^T\mathbf{fa}_0 \tag{3-229}$$

Equations (3-220) and (3-222) to (3-229) can be used to compute all forces and displacements in a statically determinate structure.

If the structure is statically indeterminate, however, $\mathbf{a}_0$ cannot be developed from the equilibrium equations alone. Forces and displacements, though, can be determined from the following equations:

First, the structure is reduced, made statically determinate by replacing some constraints with redundant forces, represented by a matrix $\mathbf{X}$. All element forces $\mathbf{S}$ are related to loads $\mathbf{P}$ at the nodes of the reduced structure by

$$\mathbf{S} = \mathbf{a}_0\mathbf{P} + \mathbf{a}_1\mathbf{X} \tag{3-230}$$

where $\mathbf{a}_0$ = matrix of influence coefficients for loads acting on reduced structure

$\mathbf{a}_1$ = matrix of influence coefficients for redundants acting on reduced structure

The ith column of $\mathbf{a}_1$ contains the element forces resulting from a unit load applied in the direction of $\mathbf{X}_i$ at the node where $\mathbf{X}_i$ acts. Substitution of Eq. (3-230) in Eq. (3-223) yields

$$\boldsymbol{\delta} = \mathbf{fS} = \mathbf{f}(\mathbf{a}_0\mathbf{P} + \mathbf{a}_1\mathbf{X}) \tag{3-231}$$

Generally, the displacements associated with redundants are zero. Hence, by Eq. (3-227),

$$\mathbf{a}_1^T\boldsymbol{\delta} = \mathbf{0} \tag{3-232}$$

Solving for the redundants from Eqs. (3-231) and (3-232) yields

$$\mathbf{X} = -\mathbf{D}^{-1}\mathbf{D}_0\mathbf{P} \tag{3-233}$$

where $\mathbf{D}_0 = \mathbf{a}_1^T\mathbf{f}\mathbf{a}_0$
$\mathbf{D}^{-1}$ = inverse of matrix $\mathbf{D}$
$\mathbf{D} = \mathbf{a}_1^T\mathbf{f}\mathbf{a}_1$

Substitution of the known values of the redundants in Eq. (3-230) permits computation of the element forces from

$$\mathbf{S} = (\mathbf{a}_0 - \mathbf{a}_1\mathbf{D}^{-1}\mathbf{D}_0)\mathbf{P} \tag{3-234}$$

Methods of developing flexibility matrices of finite elements are given in Art. 3-108.

(H. I. Laursen, "Matrix Analysis of Structures," McGraw-Hill Book Company, New York; O. C. Zienkiewicz, "The Finite Element Method in Engineering Science," McGraw-Hill Book Company, New York; C. S. Desai and J. F. Abel, "Introduction to the Finite Element Method," Van Nostrand Reinhold Company, New York; M. F. Rubinstein, "Structural Systems—Statics, Dynamics and Stability," Prentice-Hall, Englewood Cliffs, N.J.)

3-107. Matrix Displacement (Stiffness) Method. In this article, the structure is assumed to consist of n finite elements connected only at the nodes. The stiffness matrices of the elements are assumed known. Right-handed cartesian coordinate axes, x, y, z, are chosen for the structure with base unit vectors $\mathbf{e}_i$ parallel to these axes at each node, as in Art. 3-105. Forces and displacements are resolved into components along $\mathbf{e}_i$. The stiffness matrix for the whole structure can be constructed from the element stiffness matrices, as will be demonstrated.

For the ith element, nodal forces and displacements are related by

$$\mathbf{S}_i = \mathbf{k}_i\boldsymbol{\delta}_i \qquad i = 1, 2, \ldots n \tag{3-235}$$

where $\mathbf{S}_i$ = matrix of forces, including moments and torques, acting at the nodes of the ith element
$\mathbf{k}_i$ = stiffness matrix of the ith element
$\boldsymbol{\delta}_i$ = matrix of displacements of the nodes of the ith element

For compactness, this relationship between nodal displacements and forces for each element can be combined into a single matrix equation applicable to all the elements:

$$\mathbf{S} = \mathbf{k}\boldsymbol{\delta} \tag{3-236}$$

where $\mathbf{S}$ = matrix of all forces acting at the nodes of all elements
$\boldsymbol{\delta}$ = matrix of all nodal displacements for all elements

$$\mathbf{k} = \begin{bmatrix} \mathbf{k}_1 & 0 & \cdots & 0 \\ 0 & \mathbf{k}_2 & \cdots & 0 \\ \cdots & \cdots & \cdots & \cdots \\ 0 & 0 & \cdots & \mathbf{k}_n \end{bmatrix} \tag{3-237}$$

Element nodal displacements $\boldsymbol{\delta}$ can be determined from the displacements $\boldsymbol{\Delta}$ of the nodes of the structure to insure geometric compatibility:

$$\boldsymbol{\delta} = \mathbf{b}_0\boldsymbol{\Delta} \tag{3-238}$$

where $\mathbf{b}_0$ is a matrix of influence coefficients. The jth column of $\mathbf{b}_0$ contains the element nodal displacements when the node where Δ_j occurs is given a unit displacement in the direction of Δ_j, and no other nodes are displaced.

From Eqs. (3-236) and (3-238),

$$\mathbf{S} = \mathbf{k}\mathbf{b}_0\boldsymbol{\Delta} \tag{3-239}$$

From energy relationships, it can be shown that

$$\mathbf{P} = \mathbf{b}_0^T\mathbf{S} \tag{3-240}$$

where $\mathbf{P}$ = matrix of all the loads acting at the nodes of the structure
$\mathbf{b}_0^T$ = transpose of $\mathbf{b}_0$ = matrix $\mathbf{b}_0$ with rows and columns interchanged

Finally, from Eqs. (3-239) and (3-240) comes the relationship that permits computation of the stiffness matrix of the whole structure from the stiffness matrices of its elements:

$$\mathbf{P} = \mathbf{b}_0{}^T\mathbf{k}\mathbf{b}_0\mathbf{\Delta} \tag{3-241}$$

Since also, by Eq. (3-221), $\mathbf{P} = \mathbf{K\Delta}$,

$$\mathbf{K} = \mathbf{b}_0{}^T\mathbf{k}\mathbf{b}_0 \tag{3-242}$$

With the stiffness matrix known, the nodal displacements can be computed from the nodal loads. Multiplication of both sides of Eq. (3-221) by $\mathbf{K}^{-1}$ gives

$$\mathbf{\Delta} = \mathbf{K}^{-1}\mathbf{P} \tag{3-243}$$

This permits member forces to be computed from Eq. (3-239).

For structures with a large number of nodal displacements, inversion of the stiffness matrix $\mathbf{K}$ directly may be impracticable. The size of $\mathbf{K}$, however, often can be reduced by taking advantage initially of conditions at nodes where loads do not act in the directions of permissible displacements.

Let $\mathbf{X}$ be the matrix of the unknown displacements at those nodes. Then, as in Eq. (3-238),

$$\boldsymbol{\delta} = \mathbf{b}_0\mathbf{\Delta} + \mathbf{b}_1\mathbf{X} \tag{3-244}$$

where $\mathbf{b}_0$ = matrix of influence coefficients for displacements of nodes where loads act in directions of displacements

$\mathbf{b}_1$ = matrix of influence coefficients for displacements of nodes where loads do not act in directions of displacements

The jth column of $\mathbf{b}_1$ contains the element nodal displacements when the node where X_j occurs is given a unit displacement in the direction of X_j, and no other nodes are displaced. Substitution of Eq. (3-244) in Eq. (3-236) yields

$$\mathbf{S} = \mathbf{k}(\mathbf{b}_0\mathbf{\Delta} + \mathbf{b}_1\mathbf{X}) \tag{3-245}$$

Since $P_j = 0$ for those displacements not associated with loads, Eq. (3-240) indicates that

$$\mathbf{b}_1{}^T\mathbf{S} = \mathbf{0} \tag{3-246}$$

Solving for the unknown displacements from Eq. (3-245) and (3-246) gives

$$\mathbf{X} = -\mathbf{B}^{-1}\mathbf{B}_0\mathbf{\Delta} \tag{3-247}$$

where $\mathbf{B}_0 = \mathbf{b}_1{}^T\mathbf{k}\mathbf{b}_0$

$\mathbf{B}^{-1}$ = inverse of matrix $\mathbf{B}$

$\mathbf{B} = \mathbf{b}_1{}^T\mathbf{k}\mathbf{b}_1$

Substitution for $\mathbf{X}$ in Eq. (3-244) gives the element nodal displacements:

$$\boldsymbol{\delta} = (\mathbf{b}_0 - \mathbf{b}_1\mathbf{B}^{-1}\mathbf{B}_0)\mathbf{\Delta} = \mathbf{b}_2\mathbf{\Delta} \tag{3-248}$$

where $\mathbf{b}_2 = \mathbf{b}_0 - \mathbf{b}_1\mathbf{B}^{-1}\mathbf{B}_0$

From Eqs. (3-245), (3-247), and (3-248), the element member forces are given by

$$\mathbf{S} = \mathbf{k}\mathbf{b}_2\mathbf{\Delta} \tag{3-249}$$

And from Eq. (3-240),

$$\mathbf{P} = \mathbf{b}_0{}^T\mathbf{k}\mathbf{b}_2\mathbf{\Delta} \tag{3-250}$$

Since also, by Eq. (3-221), $\mathbf{P} = \mathbf{K\Delta}$, the stiffness matrix for the whole structure is given by

$$\mathbf{K} = \mathbf{b}_0{}^T\mathbf{k}\mathbf{b}_2 \tag{3-251}$$

This matrix may be considerably smaller than that given by Eq. (3-242). As before, nodal displacements can be computed from Eq. (3-243) after $\mathbf{K}$ has been inverted.

Methods of developing stiffness matrices of finite elements are described in Art. 3-108.

(See bibliography at the end of Art. 3-106.)

3-108. Element Flexibility and Stiffness Matrices. As indicated in Arts. 3-106 and 3-107, the relationship between independent forces and displacements at nodes of finite elements comprising a structure is determined by flexibility matrices $\mathbf{f}_i$ [Eq. (3-222)] or stiffness matrices $\mathbf{k}_i$ [Eq. (3-235)] of the elements. In some cases, the components of these matrices can be developed from the defining equations.

The jth column of a flexibility matrix of a finite element, for which we will use the symbol $\mathbf{f}$ without subscript in this article, contains all the nodal displacements of the element when one force S_j is set equal to unity and all other independent forces are set equal to zero.

The jth column of a stiffness matrix of a finite element, for which we will use the symbol $\mathbf{k}$ without subscript in this article, consists of the forces acting at the nodes of the element to produce a unit displacement of the node at which δ_j occurs and in the direction of δ_j but no other nodal displacements of the element.

Bars with Axial Stress Only. As an example of the use of the definitions of flexibility and stiffness, consider the simple case of an elastic bar under tension applied by axial forces P_i and P_j at nodes i and j, respectively (Fig. 3-103). The bar might be the finite element of a truss, such as a diagonal or a hanger. Connections to other members are made at nodes i and j, which can transmit only forces in the directions i to j or j to i.

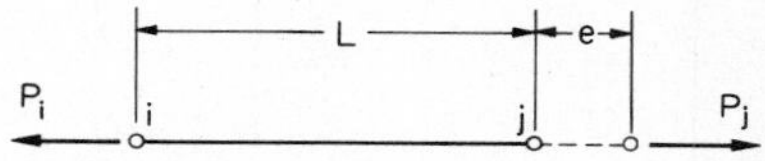

Fig. 3-103. Elastic bar in tension.

For equilibrium, $P_i = P_j = P$. Displacement of node j relative to node i is e. From Eq. (3-5), $e = PL/AE$, where L is the initial length of the bar, A the bar cross-sectional area, and E the modulus of elasticity. Setting $P = 1$ yields the flexibility of the bar,

$$f = \frac{L}{AE} \tag{3-252}$$

Setting $e = 1$ gives the stiffness of the bar,

$$k = \frac{AE}{L} \tag{3-253}$$

Beams with Bending Only. As another example of the use of the definition to determine element flexibility and stiffness matrices, consider the simple case of an elastic beam in bending applied by moments M_i and M_j at nodes i and j, respectively (Fig. 3-104a). The beam might be a finite element of a rigid frame. Connections to other members are made at nodes i and j, which can transmit moments and forces normal to the beam.

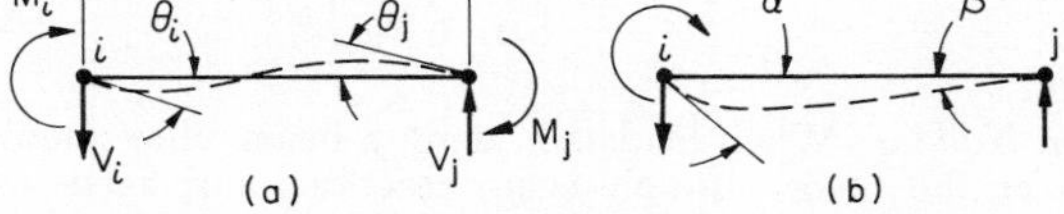

Fig. 3-104. Beam subjected to end moments and shears.

Nodal displacements of the element can be sufficiently described by rotations θ_i and θ_j relative to the straight line between nodes i and j. For equilibrium, forces $V_j = -V_i$ normal to the beam are required at nodes j and i, respectively, and $V_j = (M_i + M_j)/L$, where L is the span of the beam. Thus, M_i and M_j are the only independent forces acting. Hence, Eqs. (3-222) and (3-235) can be written for this element as

$$\boldsymbol{\theta} = \begin{bmatrix} \theta_i \\ \theta_j \end{bmatrix} = \mathbf{f} \begin{bmatrix} M_i \\ M_j \end{bmatrix} = \mathbf{fM} \tag{3-254}$$

$$\mathbf{M} = \begin{bmatrix} M_i \\ M_j \end{bmatrix} = \mathbf{k} \begin{bmatrix} \theta_i \\ \theta_j \end{bmatrix} = \mathbf{k}\boldsymbol{\theta} \tag{3-255}$$

The flexibility matrix **f** then will be a 2×2 matrix. The first column can be obtained by setting $M_i = 1$ and $M_j = 0$ (Fig. 3-104*b*). The resulting angular rotations are given by Eqs. (3-72) and (3-73). For a beam with constant moment of inertia I and modulus of elasticity E, the rotations are $\alpha = L/3EI$ and $\beta = -L/6EI$. Similarly, the second column can be developed by setting $M_j = 0$ and $M_i = 1$.

The flexibility matrix for a beam in bending then is

$$\mathbf{f} = \begin{bmatrix} \dfrac{L}{3EI} & -\dfrac{L}{6EI} \\ -\dfrac{L}{6EI} & \dfrac{L}{3EI} \end{bmatrix} = \frac{L}{6EI} \begin{bmatrix} 2 & -1 \\ -1 & 2 \end{bmatrix} \tag{3-256}$$

The stiffness matrix, obtained in a similar manner or by inversion of **f**, is

$$\mathbf{k} = \begin{bmatrix} \dfrac{4EI}{L} & \dfrac{2EI}{L} \\ \dfrac{2EI}{L} & \dfrac{4EI}{L} \end{bmatrix} = \frac{2EI}{L} \begin{bmatrix} 2 & 1 \\ 1 & 2 \end{bmatrix} \tag{3-257}$$

Beams Subjected to Bending and Axial Forces. For a beam subjected to nodal moments M_i and M_j and axial forces P, flexibility and stiffness are represented by 3×3 matrices. The load-displacement relations for a beam of span L, constant moment of inertia I, modulus of elasticity E, and cross-sectional area A are given by

$$\begin{bmatrix} \theta_i \\ \theta_j \\ e \end{bmatrix} = \mathbf{f} \begin{bmatrix} M_i \\ M_j \\ P \end{bmatrix} \qquad \begin{bmatrix} M_i \\ M_j \\ P \end{bmatrix} = \mathbf{k} \begin{bmatrix} \theta_i \\ \theta_j \\ e \end{bmatrix} \tag{3-258}$$

In this case, the flexibility matrix is

$$\mathbf{f} = \frac{L}{6EI} \begin{bmatrix} 2 & -1 & 0 \\ -1 & 2 & 0 \\ 0 & 0 & \eta \end{bmatrix} \tag{3-259}$$

where $\eta = 6I/A$, and the stiffness matrix is

$$\mathbf{k} = \frac{EI}{L} \begin{bmatrix} 4 & 2 & 0 \\ 2 & 4 & 0 \\ 0 & 0 & \psi \end{bmatrix} \tag{3-260}$$

where $\psi = A/I$

Loads between Nodes. When loads act along a beam, they should be replaced by equivalent loads at the nodes—simple-beam reactions and fixed-end moments, both with signs reversed. The equivalent forces should be used in constructing $\mathbf{a}_0$ for Eqs. (3-225) and (3-230) for determining redundants; but the fixed-end moments should not be included in $\mathbf{a}_0$ in Eq. (3-234) for computing element forces. When the element forces are obtained from Eq. (3-239) or (3-249), the final element forces then are determined by adding the fixed-end moments and simple-beam reactions to the solution of Eq. (3-239) or (3-249).

General Methods for Computing Flexibility. Each component f_{ij} of a flexibility matrix equals a displacement at a node due to a unit force at the same or another node. When the matrix for a finite element cannot readily be constructed from this definition, other means must be used.

For some elements, the components of the flexibility matrix may be obtained by integrating over the whole volume of the element

$$f_{ij} = \int_V \boldsymbol{\sigma}^T \boldsymbol{\varepsilon} \, dV \tag{3-261}$$

where $\boldsymbol{\sigma}^T$ = transpose of the matrix of the stresses produced in the element by a unit force at the node where displacement f_{ij} occurs and in the direction of that displacement

$\boldsymbol{\varepsilon}$ = matrix of the unit strains produced in the element by a unit force at the same or another node

dV = differential volume

For use in the integral, note that for a three-dimensional, homogeneous, elastic material with modulus of elasticity E, shearing modulus G, and Poisson's ratio μ, stresses $\boldsymbol{\sigma}$ and unit strains $\boldsymbol{\varepsilon}$ in the direction of x, y, z Cartesian coordinate axes are related by

$$\boldsymbol{\varepsilon} = \boldsymbol{\phi\sigma} \tag{3-262}$$

$$\boldsymbol{\varepsilon}^T = [\epsilon_x \quad \epsilon_y \quad \epsilon_z \quad \gamma_{xy} \quad \gamma_{xz} \quad \gamma_{yz}] \tag{3-263a}$$

$$\boldsymbol{\sigma}^T = [\sigma_x \quad \sigma_y \quad \sigma_z \quad \tau_{xy} \quad \tau_{xz} \quad \tau_{yz}] \tag{3-263b}$$

$$\boldsymbol{\phi} = \frac{1}{E}\begin{bmatrix} 1 & -\mu & -\mu & 0 & 0 & 0 \\ -\mu & 1 & -\mu & 0 & 0 & 0 \\ -\mu & -\mu & 1 & 0 & 0 & 0 \\ 0 & 0 & 0 & \frac{E}{G} & 0 & 0 \\ 0 & 0 & 0 & 0 & \frac{E}{G} & 0 \\ 0 & 0 & 0 & 0 & 0 & \frac{E}{G} \end{bmatrix} \tag{3-263c}$$

See Arts. 3-10 to 3-12.

For some elements, the components of the flexibility matrix may be obtained by differentiating twice the strain energy U in the element due to forces at the nodes:

$$f_{ij} = \frac{\partial^2 U}{\partial S_i \, \partial S_j} \tag{3-264}$$

where S_i = force acting at the node where displacement f_{ij} occurs and in the direction of that displacement

S_j = force acting at the same or another node

Flexibility of Thin Triangular Element. For an elastic, triangular element with thickness t very small compared with other dimensions, the terms in Eq. (3-262) become

$$\boldsymbol{\varepsilon}^T = [\epsilon_x \quad \epsilon_y \quad \gamma_{xy}] \tag{3-265a}$$

$$\boldsymbol{\sigma}^T = [\sigma_x \quad \sigma_y \quad \tau_{xy}] \tag{3-265b}$$

$$\boldsymbol{\phi} = \frac{1}{E}\begin{bmatrix} 1 & -\mu & 0 \\ -\mu & 1 & 0 \\ 0 & 0 & \frac{E}{G} \end{bmatrix} \tag{3-265c}$$

See Arts. 3-10 to 3-12.

Assume that forces S_i act only at the vertices (nodes) of the triangle and in the directions of the x and y axes. Of the six possible forces that can act on the element, only the three forces S_1, S_2, S_3, shown in Fig. 3-105, are independent. The other three can be computed from these from equilibrium conditions. The independent forces can be formed into a column matrix **S** for loads on the element. This matrix

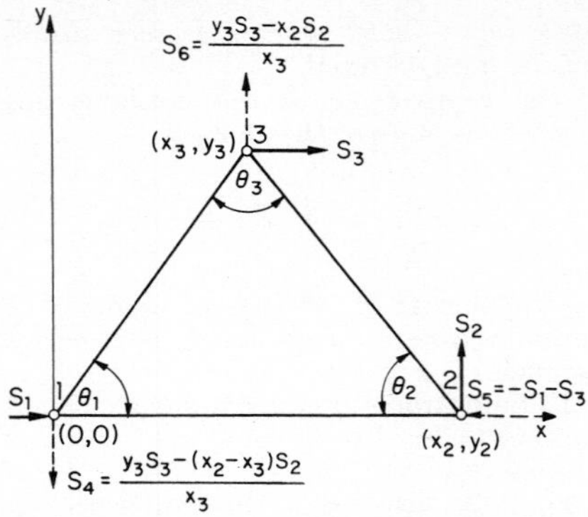

Fig. 3-105. Forces acting at nodes of thin triangular element.

is related to the nodal displacements $\boldsymbol{\delta}$ of the element by Eq. (3-223), $\boldsymbol{\delta} = \mathbf{fS}$, where $\mathbf{f}$ is the flexibility matrix of the element. In this case, $\boldsymbol{\delta}$ also is a column vector with three components. Thus, $\mathbf{f}$ is a 3×3 matrix.

A key assumption will now be made: *Stresses in the element are constant.*

This places constraints on the displacements and forces of the structure composed of such elements. As a result, equilibrium or geometric compatibility may be violated, and the idealized structure may then be stiffer or more flexible than the actual structure. The approximation may be improved by such refinements as selecting an element with more nodes (triangle with nodes at vertices and midpoints of sides) or a smaller element. (The former choice usually requires fewer calculations for about the same degree of accuracy.)

For constant σ_x, σ_y, and τ_{xy}, the stresses and nodal forces are related in terms of the nodal coordinates by

$$\boldsymbol{\sigma} = \boldsymbol{\alpha}\mathbf{S} \tag{3-266}$$

$$\boldsymbol{\alpha} = \frac{2}{t}\begin{bmatrix} -\dfrac{1}{y_3} & 0 & \dfrac{x_3 - x_2}{x_2 y_3} \\ 0 & -\dfrac{1}{x_3} & \dfrac{y_3}{x_2 x_3} \\ 0 & 0 & \dfrac{1}{x_2} \end{bmatrix} \tag{3-267}$$

From energy relations, it can be shown that for integration over the volume of the element

$$\int_V \boldsymbol{\varepsilon}^T\boldsymbol{\sigma}\, dV = \boldsymbol{\delta}^T\mathbf{S} \tag{3-268}$$

Substituting Eq. (3-266) in Eq. (3-268) and transposing matrices result in the relation between strains and nodal displacements:

$$\boldsymbol{\delta} = \int_V \boldsymbol{\alpha}^T\boldsymbol{\varepsilon}\, dV \tag{3-269}$$

By Eqs. (3-262) and (3-266), $\boldsymbol{\varepsilon} = \boldsymbol{\phi\sigma} = \boldsymbol{\phi\alpha}\mathbf{S}$. Hence, Eq. (3-269) can be written

$$\boldsymbol{\delta} = \left[\int_V \boldsymbol{\alpha}^T\boldsymbol{\phi\alpha}\, dV\right]\mathbf{S} \tag{3-270}$$

But also, $\boldsymbol{\delta} = \mathbf{fS}$. Hence, the flexibility matrix is given by

$$\mathbf{f} = \int_V \boldsymbol{\alpha}^T\boldsymbol{\phi\alpha}\, dV = \int_V \boldsymbol{\alpha}^T\boldsymbol{\phi\alpha} t\, dA = \frac{x_2 y_3}{2} t\, \boldsymbol{\alpha}^T\boldsymbol{\phi\alpha} \tag{3-271}$$

Substitution for α from Eq. (3-267) and for ϕ from Eq. (3-265c) gives, with $E/G = 2(1+\mu)$,

$$\mathbf{f} = \frac{2}{Et}\begin{bmatrix} \dfrac{x_2}{y_3} & -\dfrac{\mu x_2}{x_3} & \dfrac{x_2 - x_3}{y_3} + \dfrac{\mu y_3}{x_3} \\ -\dfrac{\mu x_2}{x_3} & \dfrac{x_2 y_3}{x_3^2} & \dfrac{\mu(x_3 - x_2)}{x_3} - \dfrac{y_3^2}{x_3^2} \\ \dfrac{x_2 - x_3}{y_3} + \dfrac{\mu y_3}{x_3} & \dfrac{\mu(x_3 - x_2)}{x_3} - \dfrac{y_3^2}{x_3^2} & \dfrac{(x_3 - x_2)^2}{x_2 y_3} + \dfrac{2y_3(\mu x_2 + x_3)}{x_2 x_3} + \dfrac{y_3^3}{x_2 x_3^2} \end{bmatrix} \tag{3-272}$$

Flexibility Matrix for Rotated Axes. If the orientation of the coordinate axes for Eq. (3-272) differs from that of the axes chosen for the whole structure, the flexibility matrix and nodal forces and displacements must be transformed accordingly. Assume that the triangular element is rotated counterclockwise through an angle θ relative to the coordinate axes of the structure. Then, new nodal forces S in the directions of those axes and the original nodal forces

$$\mathbf{S}_o = [S_1 S_4 S_2 S_5 S_3 S_6]^T$$

are related by

$$\mathbf{S}_o = \mathbf{RS} \tag{3-273}$$

where **R** is a diagonal matrix with elements

$$\mathbf{R}_{ij} = \begin{bmatrix} \cos\theta & \sin\theta \\ -\sin\theta & \cos\theta \end{bmatrix} \quad \text{for } i = j \qquad \mathbf{R}_{ij} = \mathbf{0} \text{ for } i \neq j \tag{3-274a}$$

From Eq. (3-273), the relation between original nodal forces and new nodal forces can be expressed as

$$[S_1 S_2 S_3]_o^T = \mathbf{r}[S_1 S_2 S_3]^T \tag{3-274b}$$

The transformed flexibility matrix **f** then is given in terms of the flexibility matrix $\mathbf{f}_o$ of Eq. (3-272) by

$$\mathbf{f} = \mathbf{r}^T \mathbf{f}_o \mathbf{r} \tag{3-275}$$

General Methods for Computing Stiffness. Each component k_{ij} of a stiffness matrix equals a force at a node when a unit displacement occurs only at the same or another node. When the matrix for a finite element cannot readily be developed from this definition, other means must be used.

For some elements, the components of the stiffness matrix can be obtained by integrating over the whole volume of the element

$$k_{ij} = \int_V {}^T\boldsymbol{\sigma}\, dV \tag{3-276}$$

where $\boldsymbol{\varepsilon}^T$ = transpose of the matrix of unit strains produced in the element when the node at which force k_{ij} acts is given a unit displacement in the direction of the force while no other displacements occur

$\boldsymbol{\sigma}$ = matrix of the stresses in the element when a unit displacement occurs at the same or another node while no other displacements occur

dV = differential volume

See Arts. 3-10 to 3-12.

For some elements, the components of the stiffness matrix can be obtained by differentiating the strain energy U in the element because of nodal displacements:

$$k_{ij} = \frac{\partial^2 U}{\partial \delta_i \partial \delta_j} \tag{3-277}$$

where δ_i = displacement of the node where k_{ij} acts, in the direction of k_{ij}
δ_j = displacement of the same or another node

Stiffness of Thin Triangular Element. For an elastic triangular element with modulus of elasticity E, shearing modulus G, Poisson's ratio μ, and thickness t very small relative to other dimensions, stresses $\boldsymbol{\sigma}$ are related to strains $\boldsymbol{\varepsilon}$ by

$$\boldsymbol{\sigma} = \boldsymbol{\kappa}\boldsymbol{\varepsilon} \tag{3-278}$$

$$\boldsymbol{\sigma}^T = [\sigma_x \sigma_y \tau_{xy}] \tag{3-279a}$$

$$\boldsymbol{\varepsilon}^T = [\epsilon_x \epsilon_y \gamma_{xy}] \tag{3-279b}$$

$$\boldsymbol{\kappa} = \frac{E}{1 - \mu^2}\begin{bmatrix} 1 & \mu & 0 \\ \mu & 1 & 0 \\ 0 & 0 & \dfrac{1-\mu}{2} \end{bmatrix} \tag{3-279c}$$

See Arts. 3-10 to 3-12.

Assume that forces $\mathbf{S}_i$ act only at the vertices (nodes) of the triangle and in the directions of the x, y axes. A displacement δ_i in the direction of each axis can occur at each node, as shown in Fig. 3-106. Hence, six nodal displacements can be chosen.

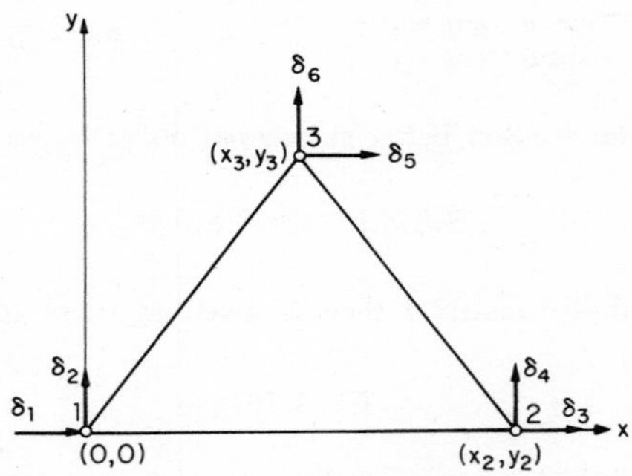

Fig. 3-106. Displacements at nodes of thin triangular element.

They can be formed into a displacement matrix $\boldsymbol{\delta}$. This matrix is related to the matrix $\mathbf{S}$ of nodal forces by Eq. (3-235), $\mathbf{S} = \mathbf{k}\boldsymbol{\delta}$, where $\mathbf{k}$ is the stiffness matrix of the element.

A key assumption will now be made: *Unit strains in the element are constant.* (This places constraints on the displacements and forces, as discussed in connection with the development of the flexibility matrix of a triangular element.) Because

$$\boldsymbol{\varepsilon}^T = \left[\frac{\partial \delta_x}{\partial x} \quad \frac{\partial \delta_y}{\partial y} \quad \frac{\partial \delta_x}{\partial y} + \frac{\partial \delta_y}{\partial x}\right] \tag{3-280}$$

where $\boldsymbol{\varepsilon}^T$ = transpose of the matrix of unit strains in the element
δ_x = component of displacement within the element in the direction of x axis
δ_y = component of displacement within the element in the direction of y axis

this assumption is equivalent to requiring that displacements vary linearly within the element. In matrix form, this relation can be written:

$$\begin{bmatrix} \delta_x \\ \delta_y \end{bmatrix} = \mathbf{Xc} = \begin{bmatrix} 1 & x & y & 0 & 0 & 0 \\ 0 & 0 & 0 & 1 & x & y \end{bmatrix} [c_1 c_2 c_3 c_4 c_5 c_6]^{\mathrm{T}} \qquad (3\text{-}281)$$

where c_i = constant. The strains then, by Eq. (3-280), are given by

$$\boldsymbol{\varepsilon} = \mathbf{Cc} = \begin{bmatrix} 0 & 1 & 0 & 0 & 0 & 0 \\ 0 & 0 & 0 & 0 & 0 & 1 \\ 0 & 0 & 1 & 0 & 1 & 0 \end{bmatrix} [c_1 c_2 c_3 c_4 c_5 c_6]^{\mathrm{T}} \qquad (3\text{-}282)$$

Nodal displacements can be obtained from Eq. (3-281) on substitution of the nodal coordinates (Fig. 3-106):

$$\boldsymbol{\delta} = \begin{bmatrix} 1 & 0 & 0 & 0 & 0 & 0 \\ 0 & 0 & 0 & 1 & 0 & 0 \\ 1 & x_2 & 0 & 0 & 0 & 0 \\ 0 & 0 & 0 & 1 & x_2 & 0 \\ 1 & x_3 & y_3 & 0 & 0 & 0 \\ 0 & 0 & 0 & 1 & x_3 & y_3 \end{bmatrix} \begin{bmatrix} c_1 \\ c_2 \\ c_3 \\ c_4 \\ c_5 \\ c_6 \end{bmatrix} \qquad (3\text{-}283)$$

Solving for the constants yields

$$\mathbf{c} = \frac{1}{x_2 y_3} \begin{bmatrix} x_2 x_3 & 0 & 0 & 0 & 0 & 0 \\ -y_3 & 0 & 0 & y_3 & 0 & 0 \\ x_3 - x_2 & 0 & -x_3 & 0 & x_2 & 0 \\ 0 & x_2 y_3 & 0 & 0 & 0 & 0 \\ 0 & -y_3 & 0 & y_3 & 0 & 0 \\ 0 & x_3 - x_2 & 0 & -x_3 & 0 & x_2 \end{bmatrix} \begin{bmatrix} \delta_1 \\ \delta_2 \\ \delta_3 \\ \delta_4 \\ \delta_5 \\ \delta_6 \end{bmatrix} \qquad (3\text{-}284)$$

Substitution of Eq. (3-284) in Eq. (3-282) develops the relation between unit strains and nodal displacements:

$$\boldsymbol{\varepsilon} = \boldsymbol{\beta\delta} \qquad (3\text{-}285)$$

$$\boldsymbol{\beta} = \frac{1}{x_2 y_3} \begin{bmatrix} -y_3 & 0 & 0 & y_3 & 0 & 0 \\ 0 & x_3 - x_2 & 0 & -x_3 & 0 & x_2 \\ x_3 - x_2 & -y_3 & -x_3 & y_3 & x_2 & 0 \end{bmatrix} \qquad (3\text{-}286)$$

From energy relations, it can be shown that for integration over the whole volume of the element

$$\int_V \boldsymbol{\sigma}^{\mathrm{T}} \boldsymbol{\varepsilon} \, dV = \mathbf{S}^{\mathrm{T}} \boldsymbol{\delta} \qquad (3\text{-}287)$$

Substituting Eq. (3-285) in (3-287) and transposing matrices results in the relation between stresses and nodal forces:

$$\mathbf{S} = \int_V \boldsymbol{\beta}^{\mathrm{T}} \boldsymbol{\sigma} \, dV \qquad (3\text{-}288)$$

By Eqs. (3-278) and (3-285), $\boldsymbol{\sigma} = \boldsymbol{\kappa\varepsilon} = \boldsymbol{\kappa\beta\delta}$. Hence, Eq. (3-288) can be written as

$$\mathbf{S} = \left[\int_V \boldsymbol{\beta}^{\mathrm{T}} \boldsymbol{\kappa\beta} \, dV \right] \boldsymbol{\delta} \qquad (3\text{-}289)$$

But also, $\mathbf{S} = \mathbf{k}\boldsymbol{\delta}$. Consequently, according to Eq. (3-289), the stiffness matrix for the triangular element is given by

$$\mathbf{k} = \int_V \boldsymbol{\beta}^{\mathrm{T}} \boldsymbol{\kappa\beta} \, dV = \int_V \boldsymbol{\beta}^{\mathrm{T}} \boldsymbol{\kappa\beta} t \, dA = \frac{x_2 y_3}{2} t \boldsymbol{\beta}^{\mathrm{T}} \boldsymbol{\kappa\beta} \qquad (3\text{-}290)$$

Substitution for β from Eq. (3-286) and for κ from Eq. (3-279*c*) gives

$$\mathbf{k} = \frac{Et}{2(1-\mu^2)\,x_2y_3}\begin{bmatrix} y_3^2 + x_{23}^2\lambda & x_{23}y_3(\mu+\lambda) & -y_3^2 + x_3x_{23}\lambda & y_3(x_3\mu - x_{23}\lambda) & -x_2x_{23}\lambda & -x_2y_3\mu \\ x_{23}y_3(\mu+\lambda) & x_{23}^2 + y_3^2\lambda & -y_3(x_{23}\mu - x_3\lambda) & x_3x_{23} - y_3^2\lambda & -x_2y_3\lambda & -x_2x_{23} \\ -y_3^2 + x_3x_{23}\lambda & -y_3(x_{23}\mu - x_3\lambda) & y_3^2 + x_3^2\lambda & -x_3y_3(\mu+\lambda) & -x_2x_3\lambda & x_2y_3\mu \\ y_3(x_3\mu - x_{23}\lambda) & x_3x_{23} - y_3^2\lambda & -x_3y_3(\mu+\lambda) & x_3^2 + y_3^2\lambda & x_2y_3\lambda & -x_2x_3 \\ -x_2x_{23}\lambda & -x_2y_3\lambda & -x_2x_3\lambda & x_2y_3\lambda & x_2^2\lambda & 0 \\ -x_2y_3\mu & -x_2x_{23} & x_2y_3\mu & -x_2x_3 & 0 & x_2^2 \end{bmatrix} \tag{3-291}$$

where $\lambda = (1-\mu)/2$

$x_{23} = x_2 - x_3$

Stiffness Matrix for Rotated Axes. If the orientation of the coordinate axes for Eq. (3-291) differs from that of the axes chosen for the whole structure, the stiffness matrix and nodal displacements must be transformed accordingly. Assume that the triangular element is rotated counterclockwise through an angle θ relative to the coordinate axes of the structure. Then, new nodal displacements $\boldsymbol{\delta}$ in the direction of those axes are related to the original nodal displacements $\boldsymbol{\delta}_o$ used in determining Eq. (3-291) by

$$\boldsymbol{\delta}_o = \mathbf{R}\boldsymbol{\delta} \tag{3-292}$$

where $\mathbf{R}$ is the diagonal matrix with elements given by Eq. (3-274*a*). The transformed stiffness matrix $\mathbf{k}$ then is given in terms of the flexibility matrix $\mathbf{k}_o$ of Eq. (3-291) by

$$\mathbf{k} = \mathbf{R}^T\mathbf{k}_o\mathbf{R} \tag{3-292}$$

(See bibliography at the end of Art. 3-106.)

Section 4

Soil Mechanics and Foundations

JACOB FELD, Ph.D.
Consulting Engineer, New York, N.Y.

Two hundred years ago, John Muller wrote in his "Treatise Containing the Practical Part of Fortifications" (A. Muller, London, 1755):

"As the foundations of all buildings, in general, are of utmost importance, in respect to the strength and duration of the work, we shall enter into all the most material particulars, which may happen in different soils, in order to execute works with all the security possible, because many great buildings have been rent to pieces, and some have fallen down, for want of having taken proper care in laying the foundation."

Muller made five recommendations, which are still pertinent:

1. Use proper augers to bore holes from 10 to 15 ft deep to discover the nature of the soil and its hardness.
2. If the soil is gravel or hard stiff clay, no sinking will occur.
3. If the soil is not uniform, or loose sand, lay a grate of timbers crossed two ways, sometimes boarded over with 3-in. wood planks.
4. If the soil is first soft and hard below, then drive piles and lay a grate above them; drive a few piles to determine the proper length before cutting all the piles.
5. Sometimes a bed of gravel or clay has soft mud below, and it is dangerous to drive piles; then build a timber deck and let it sink to bearing, a method used in Russia and Flanders with good results.

TYPES AND NORMAL QUALITIES OF SOILS

The geologist defines a soil as an altered rock. The engineer defines a soil as the material which supports a structure.

The geologist distinguishes among three soil classes:

1. Primitive—formed by the mechanical and chemical decomposition of a rock.
2. Derived—a mixture of primitives with changes resulting from water and air action.
3. Decayed—a mixture of minerals with vegetable- and animal-decay refuse.

4-1. Physical Properties of Soils. The engineer needs to know, for complete identification of a soil: the size of grains, gradation, particle shapes, grain orientation, chemical make-up, and colloidal and dust fractions. Even then, the physical properties can be made to vary over a large range with slight chemical additives or electrochemical controls. The importance of shape in controlling the ratio of surface areas to volume is readily seen from the following example of fractures from a cube:

Shape	Total volume	Surface area of all pieces
Cube	1	6.0
Cubes cut into two slices	1	8.0
Cube cut diagonally	1	8.8
Cube cut into three slices	1	10.0
Cube cut into three tetrahedrons	1	14.5

Where surface properties are important, grain shape becomes at least equal in influence to size gradation. Normally, an important characteristic is the relative positioning of the grains within the soil body, since this controls the resistance to internal displacement and is at least a qualitative measure of shear and compression strengths.

There have been many attempts to codify all soils into classes of similar and recognizable properties. As more information is collected on soil properties, the systems of classification become more elaborate and complicated. One difficulty is the attempt to use similar classifications for different uses; for instance, a system applicable to highway design has little value when the problem is basically that of building foundations.

Structurally, the soil problem can be solved if the physical action can be predicted and deviations therefrom restricted within desired limits. Most, if not all, subsurface failures occur when a soil, expected to act in accordance with the laws governing solid behavior, actually becomes viscous, plastic, or even liquid. Such change may be temporary; but the release of internal stresses, while the soil is in the nonsolid phase, causes sufficient displacements to crack the superstructure.

Soils are peculiar in their structural behavior in that, while static equilibrium can be shown to exist along a series of closely proximate points, yet failure may occur along surfaces enclosing such investigated points. Because of the lack of isotropic structure, surplus strength at one point will seldom transfer to adjacent points that require a little extra strength to prevent failure. Recognition of these inherent characteristics of soil will take the mystery out of many observed incidents.

4-2. States of Matter Affecting Soil Behavior. Soils can exhibit solid, viscous, plastic, or liquid action; if the true state can be predicted, the structural design can be prepared accordingly.

Solids are materials having constant density, elasticity, and internal resistance, little affected by normal temperature changes, moisture variations, and vibration below seismic values. Deformation by shearing takes place along two sets of parallel planes, with the angle between the sets constant for any material and independent of the nature or intensity of the external forces inducing shearing strain. The harder and more brittle the material, the more this angle differs from a right angle.

If tension is taken as negative compression, Hartmann's law states that the acute angle formed by the shear planes is bisected by the axis of maximum compression, and the obtuse angle by the axis of minimum compression (usually tension). Shearing planes do not originate simultaneously and are not uniformly distributed.

The angle of shear is related to the ultimate compression and tension stresses by Mohr's formula:

$$\cos \theta = \frac{f_1 - f_2}{f_1 + f_2} \tag{4-1}$$

i.e., the cosine of the angle of shear equals the ratio of the difference of the normal stresses to the sum. When lateral pressures are applied externally simultaneously with loads, the angle of shear increases; i.e., the material becomes less brittle.

These basic properties of solids can be used in designs affecting soils only if the soils remain solid. When changes in conditions modify soil structures so that they are not solid, these properties are nonexistent, and a new set of rules governs the actions. Most soils will act like solids, but only up to a certain loading. This limit loading depends on many external factors, such as flow of moisture, temperature, vibration, age, and sometimes the rate of loading.

No marked subdivision exists between liquid, plastic, and viscous action. These three states have the common property that volume changes are difficult to accomplish, but shape changes occur continuously. They differ in the amount of force necessary to start motion. There is a minimum value necessary for the plastic or viscous states, while a negligible amount will start motion in the liquid state. Upon cessation of the force, a plastic material will stop motion, but liquid and viscous materials will continue indefinitely, or until counteracting forces come into play. Usually, division between solid and plastic states is determined by the percentage of moisture in the soil. This percentage, however, is not a constant but decreases with increase in pressure on the material. Furthermore, the entire relationship between moisture percentage and change of state can be deranged by adding chemicals to the water.

Liquids retained in a vessel are almost incompressible. In water-bearing soils, therefore, if movement or loss of water can be prevented, volume change and settlement will be avoided. Water loss can be prevented by both physical and chemical changes in the nature of the water.

4-3. Soil Moisture. If the water content of soil is chiefly film or adsorbed moisture, the mass will not act as a liquid. All solids tend to adsorb or condense upon their surfaces any liquids (and gases) with which they come into contact.

The kind of ion, or metal element, in the chemical make-up of a solid has a great influence on how much water can be adsorbed. Ion-exchange procedures for soil stabilization and percolation control are therefore an important part of soil mechanics.

The effect of temperature on soil action may be explained by the reduction of film thickness with rise of temperature. This is comparable with the thinning out of glue holding a granular mass together. Soil failures are most common in the spring months after warm rains, when the strong water films are weakened by temperature rise and by accretion of more water.

Water films are tougher than pore water. (Terzaghi, in 1920, stated that water films less than two-millionths of an inch thick are semisolid in action; they will not boil or freeze at normal temperatures.) Consequently, saturated soil freezes much more easily than damp soil, and ice crystals will grow by sucking free moisture out of the pores. A sudden thaw will then release large concentrations of water, often with drastic results.

When liquids evaporate, they first change to films, requiring large thermal increment for the change from film to vapor. The strong binding power of water film is the explanation for the toughness of beaches, for the ebbing tides pull out free moisture.

Compaction of soils artificially or by natural causes is best carried out under fairly definite moisture contents, since redistribution of soil grains to a closely packed volume cannot be accomplished without sufficient moisture to coat each particle. The film acts as a lubricant to permit easier relative movements and by its capillary tension holds the grains in position. Obviously, finer grains require more water for best stabilization than coarser materials.

4-4. Resistance of Soils to Pressure. Before 1640, Galileo distinguished among solids, semifluids, and fluids. He stated that semifluids, unlike fluids, maintain their condition when heaped up, and if a hollow or cavity is made, agitation causes the filling up of the hollow; whereas in solids, the hollow does not fill up. This is an early description of the property known as the natural slope of granular materials—an easily observed condition for clean dry sands, but a variable slope for soils containing clays with varying moisture percentages. The angle of natural slope should not be confused with the angle of internal friction, although many writers have followed Woltmann (1799), who, in translating Coulomb's papers, made that mistake.

The fundamental laws of friction were applied to soils by Coulomb. He recognized that the resistance along a surface of failure within a soil is a function of both the load per unit area and the surface of contact.

One factor determining resistance of a soil is the coefficient of friction; friction is a resistance independent of the area loaded, but directly proportional to the intensity of loading. A second factor is cohesion, which is a characteristic of the material. Mayniel, in 1808, describes these factors as: "Perfect friction, as in sand, caused by the intercogging of particles; cohesion, a reunion of masses, like a glueing together; imperfect friction, the rubbing or rolling of particles over each other, due entirely to their own individual weights."

4-5. Soil Properties Important in Engineering. The normally important soil properties are:

1. *Weight or Density*. The amount of solid material in a unit volume is called the "dry weight." Usually, it is taken as a measure of solidity or compressive strength. The ratio of actual weight as found in nature or in a prepared soil to the corresponding weight of an artificially compacted sample is the measure of density relative to the standard optimum. Of the various standards used, the technique developed by Proctor was first accepted but has been replaced by a modification known as the AASHO test.

2. *Elastic Modulus*. This is of value only where the soil acts as a solid, with complete rebound on release of load. Most soils do act in this fashion over limited ranges of load application and favorable external conditions. Reliance on elasticity is limited by the possibility of unfavorable change in conditions during the life of a structure.

3. *Internal Resistance*. This is probably the most important soil property. There are many methods for its determination. Unfortunately, results obtained vary with the method. In a perfectly dry, graded sand of similarly shaped grains, shear resistance is a definite value; but in other soils, the value varies, being larger at the beginning of motion than after the first slip has started. Variations in shear value can result from temperature, chemicals in the soil, moisture, direction of moisture flow, vibration, and time of application or release of loading. In practice, it is necessary to estimate the worst possible set of conditions and use the corresponding value. Internal resistance is a combination of frictional and cohesive forces. Both can vary, but usually not in parallel relationships.

4. *Internal Friction*. Coulomb pure friction corresponds to simple shear in the theory of elasticity. It can never exceed the value of internal resistance. Many designers use it as the total shear resistance—an assumption also made in most earth-pressure formulas.

5. *Cohesion*. This is the maximum tensile value of the soil. It is a complicated coordination of many factors, such as gravitational grain attraction, colloidal adhesion of the grain covering, capillary tension of the moisture films, atmospheric pressure where gases in the void space have been dissolved or adsorbed, electrostatic attraction of charged surfaces, and many others.

6. *Dilatancy or volume change* as a result of applied external forces. This, too, is a complicated property, since it is difficult to separate true dilatancy from shrinkage (sometimes expansion) as a result of variation in fluid content. Dilatancy is a reversible phenomenon; unfortunately, so is the effect of fluid change. If either action is disregarded, prediction of movements of structures on volume-changing soils can lead to inaccurate values.

7. *Poisson's Ratio*. This measures the lateral effects produced by a linear strain,

in a continuous series of values of reversible sign. With a value of unity, as occurs in a fluid, any force is transferred undiminished in all directions, simultaneously in a pure liquid and with a lag in other fluids.

There are other minor properties sometimes of importance. The vegetable-refuse content of a soil may affect the fixity of any properties induced by treatment. For instance, soils very heavy in rotted vegetation, containing tannic acids, are not suitable for cement stabilization. Soils with large limestone-dust content can be weakened by water flow through the mass, to say nothing of the disintegrating effect from percolating sewage and other waste liquors.

IDENTIFICATION, SAMPLING, AND TESTING OF SOILS

To permit the application of previously collected experience to new problems, a standard system of soil identification is necessary; then classification of a soil follows the determination of physical properties in accordance with standardized testing procedures. Testing for soil properties or reactions to applied loads involves both laboratory and field procedures.

4-6. Soil Identification. Field investigations to identify soils can be made by surface surveying, aerial surveying, or geophysical or subsurface exploratory analysis. Complete knowledge of the geological structure of an area permits definite identification from a surface reconnaissance. Tied together with a mineralogical classification of the surface layers, the reconnaissance can at least recognize structures of some soils. The surveys, however, cannot alone determine soil behavior unless identical conditions have previously been encountered.

Up to recent times, most foundation designs were based on assumed properties deduced from surface observations. Geological maps and detailed reports, useful for this purpose, are available for most populated areas in the United States. They are issued by the U.S. Department of Agriculture, U.S. Geological Survey, and corresponding state offices.

Old surveys are of great value in locating original shore lines and stream courses, as well as existence of surface-grade changes. Correlation between geological names of exposed soils and the engineering terminology is not a difficult matter.

A complete site inspection is a necessary adjunct to the data collected from maps and surveys and will often clarify the question of uniformity. Furthermore, inspection of neighboring structures will point up some possible difficulties.

Aerial surveying has been developed to the extent where very rapid determination of soil types can be made at low cost over very large areas. Stereoptic photographic data, correlated with standard charts, identify soil types from color, texture, drainage characteristics, and ground cover.

Soil Classification. The Corps of Engineers and the Bureau of Reclamation have adopted a Unified Soil Classification (Table 4-1), which sets definite criteria for naming soils and lists them in fixed divisions based on grain size and laboratory tests of physical characteristics.

4-7. Subsoil Exploration. Geophysical or seismic exploration of subsoils is a carry-over from practice standardized in oil-field surveys to determine discontinuities in soil structure. The principles involved are the well-known characteristics of sound-wave transmission, reflection, and refraction in passing through materials of different densities. The method charts time and intensity of sound-wave emergence at various points, induced by the explosion of a submerged charge.

A similar technique uses the variation in electric conductivity of various densities and discontinuities in layer contacts.

These methods can give a good qualitative picture of certain soil properties, depths of layers, and depth to solid rock. They cannot be expected to give more than an average or statistical picture of conditions or to give definite identification of the kinds of soils. However, they are an excellent guide to programing a complete exploratory investigation, since the locations of changes and discontinuity in soil layers cannot otherwise be learned.

Borings. Usually, exploratory investigation of subsurface soil conditions is by borings. If properly made, these will portray the depth and thickness of each

Table 4-1. Unified Soil Classification Including Identification and Description[a]

Major division	Group symbol	Typical name	Field identification procedures[b]	Laboratory classification criteria[c]	
A. Coarse-grained soils (more than half of material larger than No. 200 sieve)[d]					
1. Gravels (more than half of coarse fraction larger than No. 4 sieve)[e]					
Clean gravels (little or no fines)	GW	Well-graded gravels, gravel-sand mixtures, little or no fines	Wide range in grain sizes and substantial amounts of all intermediate particle sizes	$D_{60}/D_{10} > 4$ $1 < D_{30}^2/D_{10}D_{60} < 3$ D_{10}, D_{30}, D_{60} = sizes corresponding to 10, 30, and 60% on grain-size curve	
	GP	Poorly graded gravels or gravel-sand mixtures, little or no fines	Predominantly one size, or a range of sizes with some intermediate sizes missing	Not meeting all gradation requirements for GW	
Gravels with fines (appreciable amount of fines)	GM	Silty gravels, gravel-sand-silt mixtures	Nonplastic fines or fines with low plasticity (see ML soils)	Atterberg limits below *A* line or PI < 4	Soils above A line with 4 < PI < 7 are borderline cases, require use of dual symbols
	GC	Clayey gravels, gravel-sand-clay mixtures	Plastic fines (see CL soils)	Atterberg limits above *A* line with PI > 7	
2. Sands (more than half of coarse fraction smaller than No. 4 sieve)[e]					
Clean sands (little or no fines)	SW	Well-graded sands, gravelly sands, little or no fines	Wide range in grain sizes and substantial amounts of all intermediate particle sizes	$D_{60}/D_{10} > 6$ $1 < D_{30}^2/D_{10}D_{60} < 3$	
	SP	Poorly graded sands or gravelly sands, little or no fines	Predominantly one size, or a range of sizes with some intermediate sizes missing	Not meeting all gradation requirements for SW	
Sands with fines (appreciable amount of fines)	SM	Silty sands, sand-silt mixtures	Nonplastic fines or fines with low plasticity (see ML soils)	Atterberg limits above *A* line or PI < 4	Soils with Atterberg limits above *A* line while 4 < PI < 7 are borderline cases, require use of dual symbols
	SC	Clayey sands, sand-clay mixtures	Plastic fines (see CL soils)	Atterberg limits above *A* line with PI > 7	

Table 4-1. Unified Soil Classification Including Identification and Description (*Continued*)

Information required for describing coarse-grained soils:

For undisturbed soils, add information on stratification, degree of compactness, cementation, moisture conditions, and drainage characteristics. Give typical name; indicate approximate percentage of sand and gravel; maximum size, angularity, surface condition, and hardness of the coarse grains; local or geological name and other pertinent descriptive information; and symbol in parentheses. Example: *Silty sand*, gravelly; about 20% hard, angular gravel particles, ½ in. maximum size; rounded and subangular sand grains, coarse to fine; about 15% nonplastic fines with low dry strength; well compacted and moist in place; alluvial sand; (SM).

Major division	Group symbol	Typical names	Identification procedure[f]			Laboratory classification criteria[c]
			Dry strength (crushing characteristics)	Dilatancy (reaction to shaking)	Toughness (consistency near PL)	
B. Fine-grained soils (more than half of material smaller than No. 200 sieve)[d]						
Silts and clays with liquid limit less than 50	ML	Inorganic silts and very fine sands, rock flour, silty or clayey fine sands, or clayey silts with slight plasticity	None to slight	Quick to slow	None	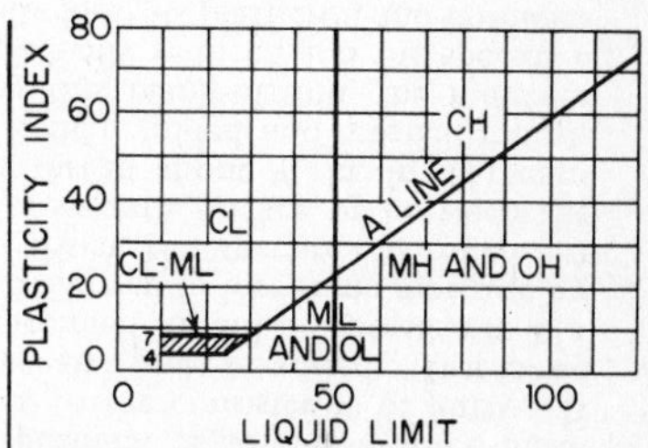 Plasticity chart for laboratory classification of fine-grained soils compares them at equal liquid limit. Toughness and dry strength increase with increasing plasticity index (PI)
	CL	Inorganic clays of low to medium plasticity, gravelly clays, sandy clays, silty clays, lean clays	Medium to high	None to very slow	Medium	
	OL	Organic silts and organic silty clays of low plasticity	Slight to medium	Slow	Slight	
Silts and clays with liquid limit more than 50	MH	Inorganic silts, micaceous or diatomaceous fine sandy or silty soils, elastic silts	Slight to medium	Slow to none	Slight to medium	
	CH	Inorganic clays of high plasticity, fat clays	None to very high	None	High	
	OH	Organic clays of medium to high plasticity	Medium to high	None to very slow	Slight to medium	
C. Highly organic soils						
	Pt	Peat and other highly organic soils	Readily identified by color, odor, spongy feel, and often by fibrous texture			

Table 4-1. Unified Soil Classification Including Identification and Description (*Continued*)

Field identification procedures for fine-grained soils or fractions[g]:

Diiatancy (reaction to shaking)

After removing particles larger than No. 40 sieve, prepare a pat of moist soil with a volume of about ½ cu in. Add enough water if necessary to make the soil soft but not sticky.

Place the pat in the open palm of one hand and shake horizontally, striking vigorously against the other hand several times. A positive reaction consists of the appearance of water on the surface of the pat, which changes to a livery consistency and becomes glossy. When the sample is squeezed between the fingers, the water and gloss disappear from the surface, the pat stiffens, and finally it cracks or crumbles. The rapidity of appearance of water during shaking and of its disappearance during squeezing assist in identifying the character of the fines in a soil.

Very fine clean sands give the quickest and most distinct reaction, whereas a plastic clay has no reaction. Inorganic silts, such as a typical rock flour, show a moderately quick reaction.

Dry strength (crushing characteristics)

After removing particles larger than No. 40 sieve, mold a pat of soil to the consistency of putty, adding water if necessary. Allow the pat to dry completely by oven, sun, or air drying, then test its strength by breaking and crumbling between the fingers. This strength is a measure of character and quantity of the colloidal fraction contained in the soil. The dry strength increases with increasing plasticity.

High dry strength is characteristic of clays of the CH group. A typical inorganic silt possesses only very slight dry strength. Silty fine sands and silts have about the same slight dry strength but can be distinguished by the feel when powdering the dried specimen. Fine sand feels gritty, whereas a typical silt has the smooth feel of flour.

Toughness (consistency near PL)

After particles larger than the No. 40 sieve are removed, a specimen of soil about ½ cu in. in size is molded to the consistency of putty. If it is too dry, water must be added. If it is too sticky, the specimen should be spread out in a thin layer and allowed to lose some moisture by evaporation. Then, the specimen is rolled out by hand on a smooth surface or between the palms into a thread about ⅛ in. in diameter. The thread is then folded and rerolled repeatedly. During this manipulation, the moisture content is gradually reduced and the specimen stiffens, finally loses its plasticity, and crumbles when the plastic limit (PL) is reached.

After the thread crumbles, the pieces should be lumped together and a slight kneading action continued until the lump crumbles.

The tougher the thread near the PL and the stiffer the lump when it finally crumbles, the more potent is the colloidal clay fraction in the soil. Weakness of the thread at the PL and quick loss of coherence of the lump below the PL indicate either organic clay of low plasticity or materials such as kaolin-type clays and organic clays that occur below the *A* line.

Highly organic clays have a very weak and spongy feel at PL.

Table 4-1. Unified Soil Classification Including Identification and Description (*Continued*)

Information required for describing fine-grained soils:

For undisturbed soils, add information on structure, stratification, consistency in undisturbed and remolded states, moisture, and drainage conditions. Give typical name; indicate degree and character of plasticity; amount and maximum size of coarse grains; color in wet condition; odor, if any; local or geological name and other pertinent descriptive information; and symbol in parentheses. Example: *Clayey silt,* brown; slightly plastic; small percentage of fine sand; numerous vertical root holes; firm and dry in place; loess; (ML).

---

[a] Adapted from recommendations of Corps of Engineers and U.S. Bureau of Reclamation. All sieve sizes United States standard.

[b] Excluding particles larger than 3 in. and basing fractions on estimated weights.

[c] Use grain-size curve in identifying the fractions as given under field identification.

For coarse-grained soils, determine percentage of gravel and sand from grain-size curve. Depending on percentage of fines (fractions smaller than No. 200 sieve), coarse-grained soils are classified as follows:

Less than 5% fines . . . . GW, GP, SW, SP
More than 12% fines . . . . GM, GC, SM, SC
5% to 12% fines. . . . . Borderline cases requiring use of dual symbols

Soils possessing characteristics of two groups are designated by combinations of group symbols; for example, GW-GC indicates a well-graded, gravel-sand mixture with clay binder.

[d] The No. 200 sieve size is about the smallest particle visible to the naked eye.

[e] For visual classification, the ¼-in. size may be used as equivalent to the No. 4 sieve size.

[f] Applicable to fractions smaller than No. 40 sieve.

[g] These procedures are to be performed on the minus 40-sieve-size particles (about 1/64 in.).

For field classification purposes, screening is not intended. Simply remove by hand the coarse particles that interfere with the tests.

soil layer; its color and texture; resistance to penetration of a fixed area of plunger under standardized impact or static forces; depth of the free-water surface and depth of rock.

Since it is impossible to guarantee that all rock encountered is bedrock, some additional core borings are necessary when subsurface conditions vary. The minimum depth of a core should be 5 ft, but if the recovery indicates rock nature or seams not consistent with the local geology, the cores should be at least 10 ft long.

Many foundation-construction difficulties, often of very expensive nature, result from either unreliable methods of taking borings or from too much reliance on extrapolation of the results. After all, except in uniform soil bodies, of which there are practically none, a boring can only be expected to depict truly the condition at the boring location. Hence, choosing the number and location of borings calls for good judgment. Many municipal and state codes fix a minimum number for each project.

Unfortunately, except in large projects, discontinuities of layers do not become known until the boring rig has been removed from the site, making the cost of additional borings quite high. Without exception, however, that cost is a small part of either the wastage in overdesign or the cost of corrective revisions necessary during foundation construction; additional borings should be made wherever a discontinuity is indicated. The usual plan is to spot a boring midway between a pair showing inconsistent results; then if the new boring is not a fair average of the two, placing the next one midway between the two showing the greatest variation, and so on. In this way, the location of changes in condition can be closely fixed and overconservative estimates eliminated.

The New York City Building Code gives minimum requirements for number and depth of borings as follows:

Borings or test pits shall show the nature of the soil in at least one location in every 2,500 sq ft of building area, and for buildings supported on piling of such type or capacity that load tests are required, one boring in every 1,600 sq ft of building area. Required number of borings for buildings having a plan area in excess of 10,000 sq ft may be reduced by half if underlaid by uniform layers of good soil, and may be eliminated where sound rock exists as a bearing stratum.

The borings or test pits shall be carried sufficiently into good bearing material to establish its character and thickness. For structures more than one story in height, dwellings more than two stories high, and structures having an average area load exceeding 1,000 psf, there shall be at least one boring in every 10,000 sq ft of building area, and at least one boring per building, carried down to a depth of 100 ft below curb level.

Where rock is encountered in borings, cores shall be taken for at least 5 ft, or further to obtain a recovery of over 35% from 5 ft of penetration. However, for piles or drilled-in caissons of more than 80-ton capacity bearing on rock, the cores shall be 10-ft minimum length.

Unless hard sound rock, medium hard rock, or intermediate rock is encountered, borings shall extend below the deepest part of the excavation as necessary to satisfy the more restrictive of the following requirements: 10 ft in hardpan, 15 ft for one- and two-family residences, and 25 ft in other cases; into material with bearing capacity of fine sand or better when upper layers are of poorer material; depth at which vertical stress caused by the proposed construction is reduced to 10% or less of the original vertical stress at this depth due to the weight of the overburden.

In every built-up community, there should be a central public depository of subsurface boring records. A compilation of boring results plotted on a city street map is invaluable. In areas where consistent soil conditions exist, a map record of previous investigations can well permit one to dispense with any further exploration.

The tendency in recent years is toward more careful, and therefore more expensive, boring methods. Except for the test pit, the older methods (auger and wash borings) are now no longer accepted as true indications of soil conditions. The auger mixes up all the soils; the wash boring deletes the finer fractions and often completely revises the make-up of the soils.

Test Pits. The test pit is a method of showing what exists but, of course, is limited as to depth. Test pits are especially useful where boring records are not consistent with predictions from surface and geological records.

Boring Methods. The usual boring procedure is to drive a pipe into the ground, remove the soil to the lower extremity either by auger or washing methods, then insert a sampling tube, which is pressed into the ground to permit recovery of a true piece of the soil at that depth. (There has been some standardization of equipment dimensions and penetration forces, although outside barrels used vary from 1½ to 4 in. and inside samplers from 1 to 2½ in. ID.) Record is kept of the number of blows with a falling weight required to force the casing down each foot. (It is noteworthy that resistance decreases whenever the inside of the casing is cleared out.) Record is also made of the resistance of the sampling spoon to penetration, usually

Table 4-2. Correlation of Boring-spoon Results

Descriptive term	Number of blows per ft			
	A	*B*	*C*	*D*
	For sand (medium)			
Loose	15	10	25	15 or less
Medium compact	16–30	11–25	25–45	
Compact	30–50	26–45	45–65	16 or 50
Very compact	Over 50	Over 45	Over 65	50 or more
	For clay			
Very soft	To 3	To 2	To 5	To 2
Soft	4–12	3–10	5–15	3–10
Stiff	12–35	10–25	15–40	11–30
Hard	Over 35	Over 25	Over 45	Over 30

A. 2-in. spoon, 148-lb hammer, 30-in. fall.
B. 2-in. spoon, 300-lb hammer, 18-in. fall.
C. 2-in. spoon, 150-lb hammer, 18-in. fall.
D. 2½-in. spoon, 300-lb hammer, 18-in. fall.

NOTE: Coarser soil requires more blows; finer material fewer blows. A variation of 10% in the weight of the hammer will not materially affect values in the table. (Compiled by T. V. Burke.)

for a depth of 1 ft. Here, the thickness and shape of the cutting edge are important factors.

Some correlation has been attempted between the resistances of various sized spoon samplers under various impact pressures. The New York City Department of Housing and Buildings has issued a table of such correlation as a guide, together with an indication of the density of the soil as measured by the resistance to penetration (see Table 4-2). The table also shows the code definition of soil compaction.

Where laboratory tests are to be made on the samples recovered, both to identify the soil better and to determine its physical properties, special methods are necessary, such as "undisturbed sampling." The soil sample is taken with an ultra-thin, noncorrosive metal shell, which is immediately sealed upon removal from the casing to avoid all changes in moisture or density of the sample.

Rock borings or drillings are made by rotating a hard steel tube with a bit, or lower end, specially treated to cut the rock along an annular ring. Bits are surfaced with some extra-hard, tough alloy, such as tungsten carbide, or with commercial diamond chips known as "borts." High-pressure water jets are always necessary to dissipate the frictional heat. The lower tube section, or recovery barrel, contains part of the rock freed by the annular cutting. In the best of ledge rock, not over 80% of the depth cored will be found in the recovered core. This is reduced by the existence of seams to some 50%. In soft rock, the total of all pieces left in the core may be only 10 to 20% of the drilled length.

Where orientation of seams is an important factor, such as in the problem of determining amount and direction of rock anchors, the driller should be instructed to mark the compass orientation of the recovery barrel and paint a corresponding mark on the recovered core. The log should also note the number of pieces making up each length of recovery, usually 5 ft, and not only the percent of length recovered. A report of 60% recovery in a core has little meaning by itself, since it may be a single piece with mud seams at each end, or a dozen or more pieces with narrow seams scattered throughout the length. If the orientation is known, a three-dimensional picture of seam structure can be prepared from a set of borings, with better prediction of strength of the rock. Excavations in rock with sloping seams

Table 4-3. Usual Laboratory Soil Tests*

Test	Soil to which adaptable	General use of results
Specific gravity	All	Void-ratio determination
Grain size	Granular	Estimate permeability, frost action, compaction, shear strength
Grain shape	Fine granular	Estimate shear strength
Liquid and plastic limits	Cohesive	Compressibility, compaction
Water content	Cohesive	Compressibility, compaction
Void ratio	All	Compressibility, strength
Unconfined compression	Cohesive	Estimate shear strength

NOTE: The last two tests must be of undisturbed samples. Mineralogical identification techniques in various clays have been developed by T. W. Lambe and R. T. Martin, Massachusetts Institute of Technology, and are reported excellent for soil description and usually reliable indicators of behavior for clay soils.

* Adapted from Sowers and Sowers, "Introductory Soil Mechanics and Foundations," The Macmillan Company, New York.

require different amounts of protection in different faces, and such preinformation permits proper preparation of protective measures, rather than the usual emergency operations after excavation is well progressed.

4-8. Soil Tests. Laboratory, and to a smaller extent field, testing of soils has developed into a complicated maze of interrelated tests, with many forms of apparatus. A summary of these tests was prepared by the American Society for Testing Materials Committee D-18, and published as "Procedure for Testing Soils," July, 1950. Many of the test techniques are applicable only to certain restricted soil groups and give misleading data if otherwise used. Table 4-3 lists the usual laboratory tests.

Where important structures are founded on rather uncertain soil types and some estimate of long-time settlements is desired, or where more than one soil type is found under a structure, more detailed testing is warranted. Laboratory determinations of unconfined compression, direct or simple shear, triaxial shear, and consolidation rate and extent are also required.

In-place soil tests are more popular in Europe than in the United States. One method is to drive a casing to the desired depth, and after the soil is cleaned out, the exposed bottom is tested by statically loading cone-shaped plungers or tight-fitting bearing plates. In another test, called the *Vane shear test,* a rod is inserted into the ground. The rod has two to four vertical steel plates. The torque necessary to start rotation and to maintain rotation of the rod is determined. These values are measures of the shear resistance and friction and are quite reliable in soft and fine-grained soils. Under the European techniques, resistances measured at various depths are used to determine not only soil-bearing values but also pile-friction resistances, from which are obtained required pile lengths for desired load capacity.

4-9. Allowable Bearing and Load Tests on Soils. Use of load tests for determining bearing values is as old as the art of foundation design. The techniques described centuries ago have only been changed by the use of equipment to prepare the test area and to handle the static loads.

Loads may be applied by jacks reacting against a loaded truck or weight of castings set up on a frame. Recently, the trend is toward jacking against a beam held in place by screw anchors. Now available for this purpose is an aluminum truss, which reduces the weights to be handled and expedites the making of load tests in isolated sites.

There is no completely accepted formula for determination of safe bearing value of a soil from the results of a load-bearing test. For equal settlement, average bearing values depend on the shape, size, and rigidity of the footing. Choice of bearing value without relation to settlement is meaningless. Furthermore, bearing

Table 4-4. Allowable Soil Bearing Pressures

Soil material	Tons per sq ft	Notes
Hard sound rock	60	No unusual seam structure
Medium hard rock	40	
Intermediate rock	20	
Soft rock	8	
Hardpan	12	Well cemented
Hardpan	8	Poorly cemented
Gravel soils	10	Compact, well graded
Gravel soils	8	Compact with more than 10% gravel
Gravel soils	6	Loose, poorly graded
Gravel soils	4	Loose, mostly sand
Sand soils	3 to 6	10% of blow count on sampling spoon per ft penetration
Fine sand	2 to 4	
Clay soils	5	Hard
Clay soils	2	Medium
Silt soils	3	Dense
Silt soils	1½	Medium
Fills and soft soils		By load test only

capacity as a contact pressure at the footing base is not necessarily proof of safety, since failure by local or general shear within the soil may be the true criterion to be considered. The distribution of contact pressure, usually assumed as uniform over the entire area, is also variously assumed, as shown in Fig. 4-2 (Art. 4-18).

Better correlation between load tests and safe bearing values can be expected in granular soils than in clays and silts.

Under constant unit load on clay soils, settlements will vary almost directly with the diameter of the test loading device. For a desired settlement in clay soils, bearing value will vary with the density. F. L. Plummer has estimated a value of 1 ton per sq ft if the density is 100 lb per cu ft and 5 tons per sq ft if the density is 140 lb per cu ft. The Krey method of plastic equilibrium, based on an assumed circular section of failure below the footing, shows the bearing value as directly proportional to the width of the footing, if cohesion is disregarded.

Allowable soil bearing pressures without tests for various soils are given in Table 4-4 for normal conditions. These basic bearing pressures may be increased when the base of the footing is embedded beyond normal depth. Rock values may be increased by 10% for each foot of embedment beyond 4 ft in fully confined conditions, but the values may not exceed twice the basic values.

If load tests are required, bearing pressures should be limited to values such that the proposed construction will be safe against failure of the soil under 100% overload.

4-10. Settlement. Settlement of structures must always be expected. Measurement is usually by surveying methods, referring the elevations of fixed points to a bench mark sufficiently remote so as not to be affected by the loaded structure. For differential settlements within a structure, a simpler leveling device consists of water-filled tubes with gage glasses.

For approximate settlements under various loadings at different ages, a soil-loading test made with an undisturbed sample in a triaxial-compression machine—the lateral loads representing restraint within the soil—gives results which should be used as a general guide only. In saturated soils, the load applied will be entirely carried by the water until it has enough time to flow out. The settlement at any given time after load application is

$$S_t = qS \tag{4-2}$$

where q is the percentage of the consolidation with time as shown by the laboratory test and

$$S = \frac{h(e_i - e_f)}{1 + e_i} \tag{4-3}$$

where h = depth of layer
e_f = final voids ratio
e_i = initial voids ratio.

A fairly accurate indication of future behavior can be obtained from this type of test and analysis, but quantitive results closely fitting field measurements over long periods of time should not be expected.

SOIL CONSOLIDATION AND IMPROVEMENT

Soil as an engineering material differs from stone, timber, and other natural products in that it can be improved and altered to conform to desired characteristics. Up to recent years, the usual studies were for improvement of bearing capacity.

4-11. Methods for Improving Soil Bearing Capacity. In normal soil conditions, increasing the depth of the footing is the simplest method of increasing bearing capacity, especially if the concrete is poured directly against the soil at the bottom of the pit. The counterbalancing of lateral displacements by the surcharge of the unexcavated soil surrounding a pit markedly improves the capacity of the soil to carry load. There are exceptions to this rule, especially if the pit is carried down to below the free-water level in clay or silt, where the opposite effect will be found.

Drainage is another well-known method to improve physical characteristics. Where soil is possibly moved by subterranean water flow, open-joint tile drains should be placed just above the footing base to lead away free water before it can move or soften the bearing strata.

A compressible soil is improved by covering with a porous granular layer—sand, gravel, or crushed rock—and by impressing or blending it into the natural material, forming a new and much stronger soil stratum. In the Spezia, Italy, docks, built in a tidal marsh of great thicknesses of soft silt, it was found that a cushion layer of sand could be hydraulically deposited to form a base for very heavy masonry docks and quay walls. The interlocking of fine and porous-grained materials, often found in nature as hardpans, can be artificially produced.

Dry density of a soil is a fair measure of its strength. Therefore, to improve a soil, reduce the void volume. This was first accomplished, centuries ago, by driving short wood piles into soft ground—not as piles to carry loads to lower strata but to compress or consolidate the upper layer and make it available for the support of loading. About 1800, it was found that, if the wood piles, usually only 6 to 10 ft in length and driven manually, were retracted and the open space filled with friable material, such as sand or limestone dust, the improvement of bearing value was at least equal to the effect of the wood piles. Further, by ramming the sand in position, additional compaction of the soil was accomplished. Later, when pile-driving machinery became available, the holes were prepared by ramming, and the space was filled with concrete. In this way, there originated the sand pile for soil consolidation and the cast-in-place concrete pile for load transfer by friction along the surfaces in contact.

Extremely weak soils can be used for the support of buildings. On a sanitary fill on marshland, for example, very light consolidation by bulldozers of a 12-in. layer of fine cinders formed a blanket strong enough to support one-story frame dwellings and light-gage metal arch huts, as well as access roads and subsurface utilities. Similarly, a three-story masonry-wall and timber-floor housing project was placed on a 30-ft depth of domestic ashes, filling in an abandoned clay pit. Based upon some extensive load tests, a spread-footing design for a maximum of 750 psf added load was found to be an economical solution. For many structures on similar soils, on the other hand, pile foundations have been found unsatisfactory because of later consolidation of the soft soils and movements of the piles.

S. J. Johnson, at the 1952 Soil Stabilization Conference held at Massachusetts Institute of Technology, differentiated between temporary procedures to permit construction of foundations and permanent improvements of the soil. Among temporary procedures are control of ground-water levels by sumps and pumps, well points, or electro-osmosis and the surrounding of excavations by sheeting.

Sumps to lower water-table levels by pumping, placed not too close to the slopes of an excavation to avoid local slides, are effective only if the soil has a high coefficient of permeability; i.e., greater than 0.1 cm per sec, although in very dense, well-graded soils good results are achieved with medium coefficients. Glossop and Skempton have reported good results with wells where the soil has 25% finer than 0.05-mm grain size (silt smaller than No. 200 mesh), while Terzaghi and Peck state that the limit is 10% of 0.05-mm grain size.

Electro-osmosis as a control of moisture is feasible only in fine-grained soils and is still in the experimental stage for use where economical methods are necessary.

Sheetpiling as an enclosure to resist unbalanced pressures is an important procedure for temporary control of soil strength.

4-12. Compaction of Soils. Permanent soil alteration is accomplished usually by physical means. Most soils can be compacted or densified by the application of pressure sufficient to reduce the water content. The effect of water in soils is unimportant if less than 3 to 7% of the soil grains will pass a No. 200 mesh screen.

Compaction by heavy rubber-tired wheels will show change in density to depths as much as 4 ft, especially if large tire pressures (100 psi) are used.

Surface vibration is of value only for local uses. Deep methods of soil compaction, such as the "Vibrofloat" machinery, can compact loose sands to substantial depths over limited areas but are not suitable for clays, silts, and marls. Even explosives are restricted to free-draining soils, if proper consolidation is desired.

Since 1941, the use of artificially stabilized fills for building foundations has become fairly common. Such fills must be made under careful moisture control with due regard to the local rainy season. Densities should be 100% of optimum for sandy soils and about 97% for cohesive soils. Since usually the settlements will be less than for the local soils, special care must be taken at the limits of the fills, and proper slip and articulated joints must be provided in the buildings.

4-13. Densification of Soils. Loose soils can be grouted for densification by adding cement (often mixed with fly ash or bentonite to provide better mobility) into the pores, or with silt intrusion, chemical or asphalt emulsion. These procedures replace part of the void content, which is partly or wholly filled with moisture, with some more solid or colloidal filler.

Prestressing of soil by temporary overloading, usually simultaneously with sand drains or wicks to release surplus water, has been used in road and airport designs and is also applicable to larger building projects.

The opposite problem is the control of heaving from frost entering the soil or from the expansion of dried-out clay when water seepage is available. These conditions are serious; the frost difficulty can be eliminated only by carrying the footings below the deepest frost penetration; the solution of the heaving-soil problem is not so simple and is still not completely solved.

Freezing of soils to unexpected depths below cold-storage rooms, as well as the complete drying out of soils under boiler settings and similar hot-process equipment, result in serious foundation displacements. These locations require special insulation blankets with free-drainage qualities, so that soil volumes will remain constant.

Many of the troubles with soil behavior can be eliminated if proper densification is obtained. The cost of obtaining density by mechanical means is roughly proportional to the compactive effort employed. However, different compaction methods must be used with different soils, and only the proper methods are of value. For instance, vibration can be very helpful with sands but may be useless or harmful with clays. It is therefore necessary to identify the soils being operated on, and after determination of the possibilities of property change by different construction procedures has been made, a choice of the proper method can be based on the costs of each procedure.

As a rough approximation, in large-scale works, 95% of optimum density can be obtained at a cost of about 4 cents per cubic yard. This is reported as the experience of the U.S. Bureau of Reclamation on Embankments and Foundations by Walker and Holtz.

Proper gradation of soil to be compacted is desirable. Under present techniques, the maximum possible density is more easily obtained with fine-grained soils; larger percentages of grain sizes above ¼ in. reduce the density obtained with a constant compactive effort.

Modern sheepsfoot rollers weigh 15 tons and will compact a 10-ft strip with 640 psi on each "foot" to a depth of 6 in. below the surface. The tendency in recent years has been to use heavy rubber-tired rolling equipment for compaction, with some extremely heavy "supercompactors" for the larger operations. These units with 80 to 100 psi tire pressures and 25,000-lb wheel loads have proved effective in compacting 18-in. layers of gravelly soil with fewer passes than were required by the use of sheepsfoot rollers. None of these machines can provide compaction to the top surface, which must be prepared either with smooth-surface rollers or by blading off the top layer.

Field density determinations are used for control and acceptance of compaction operations. In minor work, a simple test is to run a light automobile over the surface being compacted and swing the car around in a fairly sharp curve. Tire marks show if the soil is too moist, and dust will fly off the tires if the soil needs more sprinkling.

Overcompaction is dangerous where later changes in temperature and moisture may cause serious heaving. The density increase of soils improves their strength, but too high a compression will produce an incipient unstable internal structure when the soils become wet.

Some experimental work reported by W. Kjellman of Sweden deals with cases where soil layers are compacted by covering with a membrane and exhausting the air below the membrane to apply a surcharge. One form of membrane used in actual work consisted of a sand cover on a clay soil. Similar densification and stability have been noted where wellpoint suction creates a partial vacuum in the soil.

More difficulty must be expected in obtaining uniform density in glacial-drift deposits or blends of granular soils than in the finer soils. Using the modified Proctor test, of a 10-lb weight falling 18 in. and with 25 blows per layer, the following results were obtained with two natural glacial deposits and a blend of the same:

Deposit	Optimum moisture, %	Max dry density, lb per cu ft
Brown sand with trace of silt	6.9	120.6
Brown silt with trace of sand	8.1	122.1
50% of each	9.5	127.6

Note that the blended soil required a higher moisture content but gave a considerably higher density than either of the natural deposits.

SPREAD FOOTINGS

The most economical foundation support for a load is that which transmits the load most directly to the soil. That makes the spread footing for an isolated column load the most frequent type of foundation.

4-14. Bearing Capacity with Spread Footings. Sometimes, the complete solution of this subject is given in one simple equation:

$$A = \frac{P}{S} \tag{4-4}$$

where area of footing needed A equals the load to be carried P divided by the unit bearing value of the soil S. Unfortunately, this equation has no real meaning; nor is it even approximately correct without some limitations on each factor.

Actually, the equation should read: The area of a footing required to comply with the limitation of a desired amount of settlement of the column carried—for a known distribution of loading on the base of the footing and for known elastic characteristics of the footing—approximately equals the actual average load divided by the average bearing of the soil. But this relationship holds only for footings of the same shape and size, under the same unit loading, for a definite subsidence in the desired time period.

The purpose of a footing is to support a definite assigned portion of a building. Since all footings will settle, i.e., the load imposed will cause a displacement, all footings in a perfect design will be displaced equally, both horizontally and vertically.

4-15. Cause of Settlement of Footings. Since there are on record many cases of foundation failures but very few footing failures, it must be concluded that the ground on which the footings rest is the usual cause of the trouble.

The displacement of the load carried can be one or several of the following movements:

1. Compression of the footing material elastically.
2. Deflection of the footing unit elastically.
3. Compression of the soil directly below the footing elastically.
4. Vertical yield of the soil directly below the footing because of horizontal transfer of stress, with consequent horizontal strain—elastic and plastic—possibly not equally in all directions.
5. Vertical yield of the soil due to change in character of the soil, such as squeeze of water content under pressure, change in volume at contact planes of different materials, change in chemical constituents of soil or rock.

It should be kept in mind that items 3, 4, and 5 may be caused not only by the direct loading but also by loadings on adjacent footings. Also, the loading at the bottom of a footing is not uniform over its entire area.

The action of loads on soil is pictured by Hogentogler as a series of interrelated phases, chiefly:

1. The load compresses the soil directly below it.
2. The compressed soil bulges laterally.
3. The bulge compresses the side soil areas.
4. The side soil areas resist and distort, sometimes showing definite surface bulges.

Fehr and Thomas found ripples in sand surfaces beyond loaded test areas.

4-16. Amounts of Settlement. From data on settlement of structures that have been accumulated it can be safely concluded that:

1. In soft rock, hardpan, and dense glacial deposits, settlements of from ¼ to ⅜ in. must be expected and will occur almost immediately under loads up to 10 tons per sq ft.
2. In dry sands, settlements of ½ to ¾ in. must be expected under loads from 3 to 4 tons per sq ft and will occur almost immediately.
3. In wet sands, settlement of ¼ to ½ in. will occur immediately under loads from 2 to 3 tons per sq ft and will increase to ¾ in. in a period of months, especially if the sand is drained.

4. In dense clays, usually mixed with gravel or sand, settlements under loads of 3 to 4 tons per sq ft will be ½ in. immediately and will increase to about 2 in. in a period of 2 years.

5. In soft clays, settlements under 1 to 2 tons per sq ft will be 1 in. immediately and will increase to 3 in. in 3 years. The increase will continue with time. Values of 6 to 9 in. are not unusual.

These values are for small areas. The settlement of large areas in granular soils will not be uniform, chiefly because the base pressures are not uniform. Furthermore, edge shear on a footing provides some resistance to settlement. The settlement of large areas in plastic soils will not be uniform, because the base pressures are not uniform and the effect of plastic flow is much less in the interior regions.

4-17. Consolidation of Subsoil. The settlements in Art. 4-16 are for isolated footings. The total settlement of a structure will be increased by two factors:

1. Consolidation of the subsoil by the added load.
2. Effect of adjacent loads on the footing in question.

The relative importance of subsoil consolidation can be evaluated from Terzaghi's statement that 80% of settlement comes from the compression of soil within a depth one and a half times the width of the footing. (Cummings computed the unit stresses at various depths under different base-pressure distributions and concluded that the stress at a depth of twice the footing width is independent of the base-pressure distribution.)

The effect of adjacent loads is a more serious factor. For example, F. Kögler (*Harvard Soil Mechanics Conference*, Vol. 3, p. 66, 1936) describes, as the reason for the tipping of large tanks toward each other, the greater consolidation of the soil between tanks due to the addition of compression from several loadings, as against the lesser compression on the outside of the cluster of tanks. In tall buildings founded on clays, the interior footings will settle more than those along the street lines. The seriousness of building adjacent footings at different levels, with resultant unequal restraint of the subsoil under the footings, is proved by cracked walls and tipped structures. The effect becomes most serious when continuous footings under walls are built with stepped subgrades.

A related cause of trouble is the upheaval of the bottom of excavations due to lateral pressure of the banks, causing a decrease of soil density during construction. In clays, such as those in Texas, excavated pits are immediately covered with a mat of concrete, at least 3 in. thick. In water-bearing areas, the excavation is kept flooded to prevent unbalanced pressures on the subgrade.

The effect of uplift due to changing water levels, as occurs along many streams as well as along tidal waters, must be considered when the design has footings at different depths. Along shores, very light bungalows on spread footings will move with the tides.

4-18. Pressure Distribution under Spread Footings. Even though the settlement characteristics of a soil are determined, the actual expected settlement cannot be even approximated unless the distribution of pressure at the base of the footing is known. Without such information, the design of a footing is impossible.

In the past, concrete footings were made very rigid. Prior to 1910, for example, footings 8 ft square were usually made about 4½ ft deep (Fig. 4-1); combined footings for two columns were layers of steel-grillage beams, embedded in concrete. The rigidity so obtained forces a uniform indenture into the soil; a uniform deformation meant a uniform soil resistance. The pressure distribution, therefore, was different from that under a reinforced-concrete footing.

For flexible (i.e., appreciably elastic deformable) footings, the base pressure is a simple function of the elastic curve for any soil loaded below the permanent set or flow value. This is another way of saying that additional settlement is directly proportional to the additional load.

Kögler, in 1936, stated that the base-distribution pressure must be assumed as either uniform, parabolic with zero at the edges, or bell-shaped with a constant minimum along the edges and a maximum under the center of the footing area (see Fig. 4-2*a*).

A. E. Cummings (*Transactions of the ASCE*, 1935) stated that the pressure-distribution curve is a direct function of the elastic curve formed by the footing. There are two general cases: (1) a parabolic distribution for a rigid footing on a dense sand, and (2) a double-bulge shape, with maxima near the quarter points for a less rigid footing on a rigid or elastic material. There is a great difference in the footing design under the two cases (see Fig. 4-2*b*).

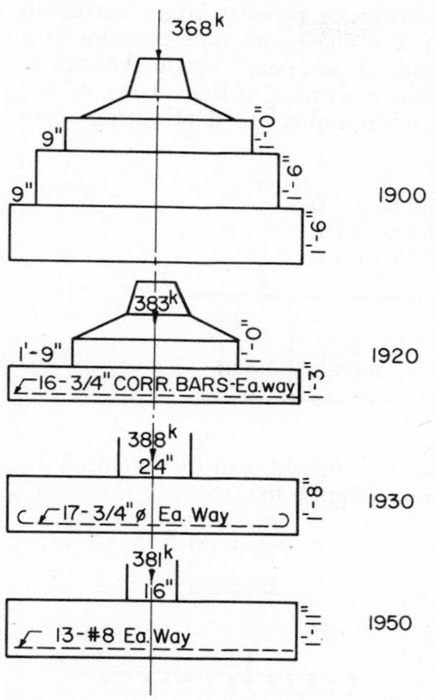

Fig. 4-1. 8-ft 2-in. square footing on 3-ton-capacity soil as designed at various times.

Unit bearing capacity for equal settlement varies with the size and shape of footing; it is greater for smaller areas and smaller for round areas. W. S. Housel (*University of Michigan Research Bulletin* 13, 1929) reported on tests of four round plates and four square plates, each shape with areas of 1, 2, 4, and 9 sq ft. He concluded that for equal settlement, the bearing value P equals the area A times the strength of the pressure bulb m plus the perimeter S times the edge shear resistance of the soil n. Thus, $P = Am + Sn$.

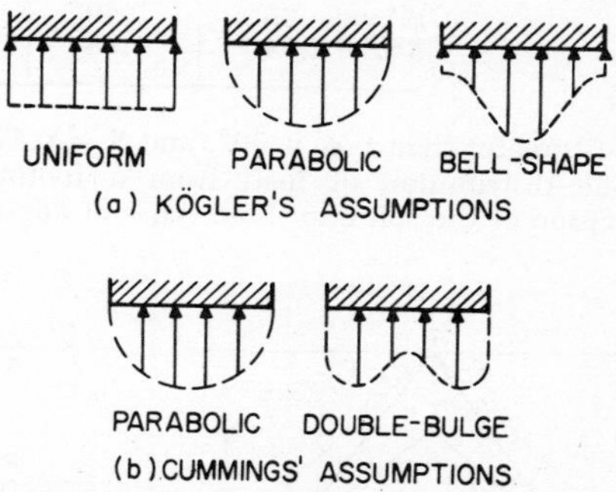

Fig. 4-2. Some assumed distributions of loadings at the base of a spread footing.

If m and n are assumed constant for any one soil, two load tests on different areas are sufficient for a complete solution.

According to A. W. Skempton (1951 London Building Congress), clays with zero internal friction, but with cohesive strength, have the ultimate bearing value given by $(cN + p)$, where c is the tested cohesion or shear strength of the soil, p is the total overburden pressure at foundation level, and N is a factor dependent upon the shape of the footing. If R is the ratio of depth of the bottom to width of footing, values of N are:

When R is	0	0.25	0.5	0.75	1.0	1.5	2.0	2.5	3.0	4.0+
For circular footings . .	6.2	6.7	7.1	7.4	7.7	8.1	8.4	8.6	8.8	9.0
For strip footings	5.1	5.6	5.9	6.2	6.4	6.8	7.0	7.2	7.4	7.5

The ultimate bearing value should be divided by the chosen factor of safety to obtain a safe unit value of the clay.

For the purpose of footing design, knowledge of the pressure distribution in the soil below the contact area is necessary when soft layers are encountered that may fail under the combination of stresses from various loads. Or it may be needed for estimating the total expected settlement of the structure. In weak soils, the initial settlement either compresses the material to a denser and less yielding one, or continuous settlements occur, with only one result—ultimate failure.

In 1885, Boussinesq published a solution for the distribution of stress in an elastic material caused by a point load on a horizontal surface. This solution is applicable approximately to pressure distribution in the deeper soil layers under footings. The vertical stress at any point in the mass equals $3Pz^3/2\pi R^5$, where z is vertical depth, P is the load, and R is the radius vector. In 1935, A. E. Cummings recommended that, in place of a constant exponent 3 for z in the numerator, an amount varying from 3 for hard dry clays to 6 for sands be used.

Kögler modified the Boussinesq distribution by substituting for the curve average straight lines, somewhat steeper than the curve, with a somewhat larger maximum stress vertically below the load (Fig. 4-3*a*). If S is maximum unit pressure at a depth z according to Kögler, and t maximum depth of permanent set in the soil, ϕ' the slope from the vertical to the point of zero stress according to Boussinesq, ϕ'' the slope according to Kögler, then (the area above a horizontal at any depth being equal to the load P):

For loose sand:	$\phi' = 35$ to $40°$	$t = 6.0$ to 3.5 ft
For medium sand:	$\phi' = 40$ to $55°$	$t = 3.5$ to 2.7 ft
For dense sand:	$\phi' = 35$ to $65°$	$t = 2.7$ to 1.5 ft

ϕ' =	40°	50°	60°	75°	90°
ϕ'' =	35°	40°	45°	50°	55°
S^2/P =	2.04	1.34	0.96	0.65	0.478

If z is greater than t, ϕ' is 90°, and $S = 0.478P/z^2$.

The distribution of load from a footing into the subsoil can be pictured by a division of the soil into three types of affected areas (Fig. 4-3*b*):

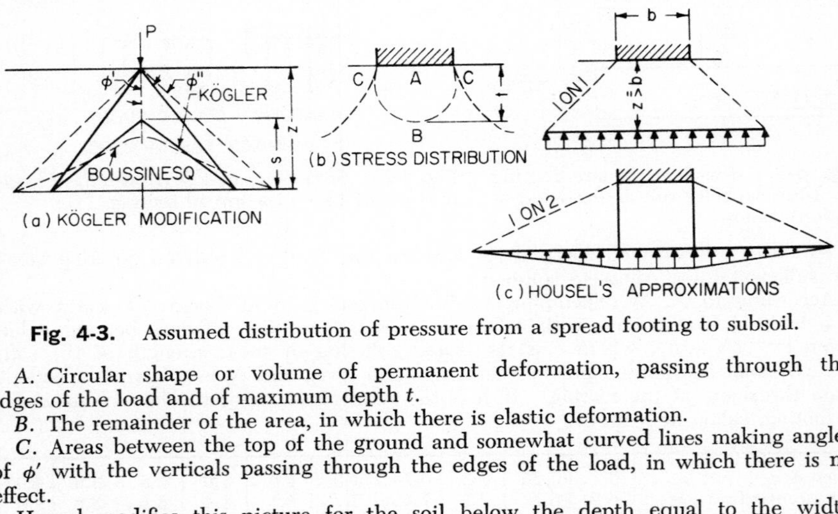

Fig. 4-3. Assumed distribution of pressure from a spread footing to subsoil.

A. Circular shape or volume of permanent deformation, passing through the edges of the load and of maximum depth t.

B. The remainder of the area, in which there is elastic deformation.

C. Areas between the top of the ground and somewhat curved lines making angles of ϕ' with the verticals passing through the edges of the load, in which there is no effect.

Housel modifies this picture for the soil below the depth equal to the width of the loaded area by the assumption that the pressure is uniformly distributed over the width within 1:1 slope lines from the edges of the footing—or as an alternate, is uniformly distributed over the width of the footing and then reduces to zero at the intersection with a 1:2 slope line from the edge of the footing (Fig. 4-3*c*).

4-19. Footing Loadings. The load to be used for footing design must be at least the dead load, including the weight of all fixed partitions, furniture, finishes, etc. However, it is doubtful whether the live load has any effect on settlements, except in such structures as grandstands, where the live-load increment may be over 150 psf during the minutes before and after a touchdown.

Where city building codes are enforced, the loads used in foundation design usually are fixed by a code. Two requirements are common:

1. Footings to be designed to sustain full dead load and a reduced summation of live loads on all floors, the percentage of reduction increasing with the number of stories, but seldom exceeding 50%. Pressures under the footings are limited by listed maximum unit pressures for various classes of soils.

2. Footings to be designed as in 1, but the areas of the footings are to be proportioned for equal soil pressure under dead load only.

In method 2, unit base pressures under dead plus live load are greater for interior footings than for wall footings. Since wall footings are usually rectangular—and if at street lines, on soil less affected by adjacent loads—the opposite condition would be more conducive to equal settlements. In method 1, since the computed live loads seldom, if ever, exist, the actual base pressures on the interior footings are less than those on the exterior footings.

4-20. Effect of Pressure Distribution on Footing Design. The rigid solution of the stresses in a footing for a pressure distribution on the base that is a function of the settlement is many times indeterminate. As compared with a uniform distribution, the effect of the variable distribution on the computed bending moments in a single loaded footing (whether square, round, or strip shape) may be large, but the effects on computed shear, as well as unit shear and bond stresses in the concrete, may be disregarded from both the point of safety as well as economy. Where bending moment controls the design, as well as in combination footings of either the mat, cantilever, or combined types, the departure from uniform distribution may affect both safety and economy.

Why, then, have there not been many structural failures of footings constructed from designs based on the assumption of uniform distribution? The usual explanations point to the surplus strength in the materials (factors of safety) as well as the nonexisting live loads included in the design. However, since unequal settlements probably occur, even though the total base pressures may be less than computed, another explanation is necessary. It is found in the theory of "limit design": In a closed system of forces and restraints, local failure cannot occur, because the continuity of the mass causes a transfer of load to the spots which can best take the load.

4-21. Footing Design. Despite the uncertainties of pressure distribution under footings, an approximate solution for some simple footings may be developed, the method being necessary only where large moments control the design. Where allowable pressures are above 3 tons per sq ft, uniform pressure distribution may be safely assumed in footing design [see Eq. (4-4), Art. 4-14].

The simplest case is a strip footing carrying a uniform continuous line load. The total load per foot on the footing w is equal to the summation of unit pressure below the footing. The shape of the loading curve is assumed similar to the deflection curve of the footing. The maximum unit pressure is $(F + A + B)$; the minimum is $(F + A)$, where F is the unit weight of the footing, A the minimum pressure, exclusive of footing weight, and B the maximum change in pressure from a uniform distribution due to deflection of the footing (Fig. 4-4).

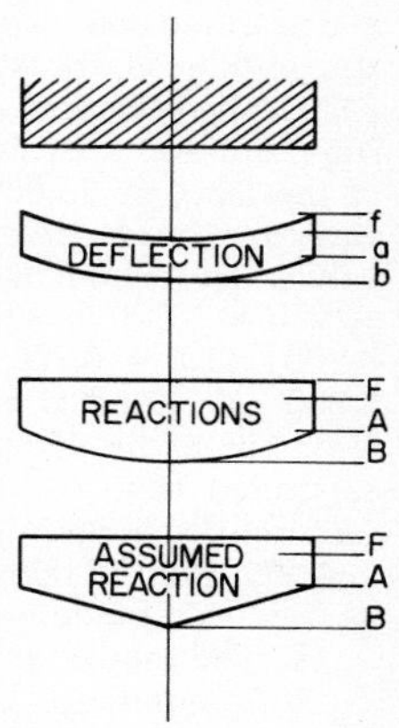

Fig. 4-4. Soil reaction on a strip footing.

First, compute the maximum deflection of the footing as a cantilever, assuming the load w uniformly distributed: $b = wL^4/8EI$, where L is the cantilever length, E the modulus of elasticity, and I the moment of inertia based on an assumed section, footing weight being disregarded temporarily. Next, assume that the ratio of this deflection to the maximum settlement is the same as the ratio of the pressure for b to maximum soil pressure. Draw the resulting base-pressure curve, taking into account the fact that the area between the pressure curve and the zero axis must equal the total load. Assume that the deflection curve is a straight line to simplify the computation (assumed reaction diagram, Fig. 4-4). Then, repeat the process using the new pressure curve in place of the assumed uniform distribution. Generally, the first approximation is close enough for economy.

Taking a specific example to illustrate the method, we shall compute the pressures for a strip footing 10 ft wide and 20 in. deep, which carries a wall load of 60,000 lb per lin ft, the soil being such that the average pressure, 6,000 psf, will cause a settlement of ¼ in.

Under uniform distribution, w is 6,000 psf; E for concrete will be taken as 2,000,000 psi; I is $12 \times 20^3/12$, or 8,000 in.^{4}.

$$b = \frac{wL^4}{8EI} = \frac{6{,}000 \times 5^4 \times 12^3}{8 \times 2{,}000{,}000 \times 8{,}000} = 0.05 \text{ in.}$$

$$B = \frac{0.05}{0.25} \times 6{,}000 = 1{,}200 \text{ psf}$$

Since the area of the pressure curve must equal 60,000 lb; i.e.,

$$\frac{L}{2}(2A + B) = P$$

$$A = 6{,}000 - \frac{1{,}200}{2} = 5{,}400 \text{ psf}$$

Footing weight of 250 psf must be added to these pressures. Base pressures then vary from 5,650 psf at the edge to 6,850 psf at the center.

For use in the next step, note that deflection of a cantilever under triangular load, with zero at the free end, is $WL^3/15EI$ where W is the total load. Assume the pressures just computed as the load. The new approximate deflection then is $12^3[5{,}400 \times 5^4/8EI + 1{,}200 \times 5^4/(15 \times 2EI)]$, or practically 0.05 in., differing less than 5% from the first approximation. Moments and shears can be computed from the pressures already obtained, with sufficient accuracy (see also Art. 5-69).

For a square footing with a point load, similar assumptions can be made. In addition, assume that the deflection of a corner is $wL^4/8EI$ and that of the middle of a side, two-thirds as much, so that the deflected footing is dish-shaped.

For a footing 10 × 10 ft by 30 in. thick carrying a load of 600,000 lb on a soil that settles ¼ in. under the average load, 6,000 psf, the deflections under a uniform 6,000 psf load are computed to be 0.050 and 0.033 in. at the corners and middle of the sides, respectively. The corresponding vertical ordinates of the dish-shaped loading curve are $0.05 \times 6{,}000/0.25 = 1{,}200$ and 800 lb (departure from uniform distribution). Consider half a quarter section, carrying an eighth of the load, or 75,000 lb on an area of 12.5 sq ft, and assume straight-line pressure variation from center to corners and to center of edges. The "loss" in bearing value, represented by a pyramidal volume with 5-ft altitude, is $(800 + 1{,}200)/2 \times 5/3 \times 5 = 8{,}300$ lb. Unit pressure at the center of footing, therefore, is $(75{,}000 + 8{,}300)/12.5 = 6{,}700$ psf. At the middle of the edge, the pressure is $6{,}700 - 800 = 5{,}900$ psf. And at the corner, the pressure is $6{,}700 - 1{,}200 = 5{,}500$ psf. To these units are to be added the weight of the footing, or 375 psf. Moments can then be computed from this distribution, and will show some saving over the assumption of uniform distribution (see also Arts. 5-75 to 5-77).

4-22. Helpful Hints for Footing Design. Several practical considerations may be used as a guide in design and construction:

1. The more rigid the footing, the more uniform the pressure on the base.
2. Nonuniform distribution may be disregarded in footings resting on soils having a bearing value over 3 tons per sq ft.
3. On similar soils, with the same loads per unit area, square footings settle more than rectangular footings.
4. On the same soils, for the same shaped footings under the same average load intensity, the larger the footing, the greater the settlement.
5. Footings on the exterior of a building, not adjacent to other loads, will settle less than those affected by adjacent loads.
6. Footings will probably settle again when adjacent soil areas are loaded.
7. Bearing-value tests are reliable only when site conditions are duplicated exactly.

8. Flexible footings may be designed for smaller bending moments than are given by the assumption of uniform distribution.

9. Footing failures are seldom structural failures.

10. Uniform settlements of buildings can be obtained by taking into consideration: actual loadings to be expected; greater soil bearing value of exterior locations; provision for slip joints for settlement due to adjacent future loadings; effect of difference in levels of footings in close proximity; changes in portions of the soil area due to water, weathering, and vibration; and differences in bearing value of differently shaped footings.

11. Maximum soil pressure under centers of footings is no criterion of expected settlements, because of the distribution of load due to rigidity.

12. Bearing pressures of soil and expected settlements deduced from load tests in the field or laboratory tests on undisturbed samples have no value if the conditions in the construction work are not considered.

13. Exposure of soils during excavation may modify the soil structure and change all the characteristics.

14. All structures will settle, and provision must be made for the expected settlement by keeping them free of adjacent structures that have moved into a stable position.

15. Lateral resistance in soils must not be disturbed or reduced below that necessary to balance the lateral stresses resulting from the added loadings.

COMBINED FOUNDATIONS, MATS, AND RAFTS

In many cases, it is not possible or economical to provide a footing centrally located under each column and wall load. There are two corrective procedures normally used for such cases: combination of loads for one footing or imposition of balancing loads to neutralize the eccentric relative positions of load and footing. Two or more loads may be supported on a single footing, and in some locations an entire building may be constructed on a mat or raft. Balancing of loads can be accomplished by means of structural beams acting as levers (often called "pan handles" or "pump handles") between column loads, or by imposition of counterbalancing moment resistance in the form of rigid frames.

Entire building areas may be covered by a single raft, or mat, where soils have extremely low bearing values at desired settlements and where the weight of the structure must sometimes be partly or wholly balanced by removal of soil weight to avoid soil failure.

4-23. Design of Combined Footings. For isolated footings the form of soil-resistance distribution may not be a determining factor in the economy of structural design, since the bending moment of the pressures about the centrally located column usually does not control the design of either the concrete or the reinforcing. The opposite condition holds, however, when the superimposed loads are not single and centrally located. And since very little is known concerning the true distribution of soil resistance at the contact face between soil and concrete, the usual design procedure is to assume uniform distribution, with sufficient soil resistance to counterbalance eccentricity of loading.

Once a design of this type is prepared, analysis of the deflections resulting from a uniform distribution (not to be carried to extreme precision because the accuracy of deflection computations for concrete structures is questionable) can be used to revise the shape of the soil-reaction curve. The resulting moments will usually be found less than when computed for the uniform-reaction assumption.

For this computation, the soil-reaction modulus, or relationship between gross deformation of the soil and bearing value, must be known. This can be taken from the results of a load-bearing test, but the effect of edge shears should be considered.

When more than one load is supported on a footing, the center of gravity of the loads should coincide with the geometric center of the footing. In addition, axes of symmetry through the column loadings should coincide with axes dividing the footing into equal areas. Since column loads are seldom of constant value, and since the variation in uncertain loadings (live load, impact, wind, etc.) is different even for adjacent columns, these requirements can only be approximately met.

When the soil is highly sensitive to load changes, combined footings should be avoided, if possible, since irreversible tipping may cause structural damage. A simple example is to stand on a plank on soft soil and shift the weight from one leg to another.

For combined footings, a study must be made of different possible load combinations, and the shape of the footing should be adjusted so that there will be pressure over the entire base area at all times. Any curling of the footing, especially in water-bearing soils, will permit softening of part of the base area. Unequal settlements must result (see also Art. 5-72).

4-24. Mat Footings. The application of the **Moseley principle** of least resistance—if a system of reactions statically stable can be shown to exist to counteract all external forces, then the structure will be safe if designed for such reactions—permits some economical modifications of the assumption of uniform base pressures. This is especially true where the footings are extended to provide support for interior walls in addition to adjacent columns. One assumption is to allot to an area symmetrically located about each column the load of that column, the area being determined by the full soil bearing value. In that way the soil resistances are in the form of rectangular loads with blank spaces between, and the computed bending moments are considerably reduced. In large mats, the same assumption will result in strips of loading under the lines joining the columns. Some mats are constructed as such strips, with thinner slabs closing the "no-load" gaps.

One danger in the construction of mats or rafts, often overlooked, should be noted. Since they are usually built on soft, saturated soils, where moisture removal means shrinkage and settlement, the absorption of moisture by the concrete mat and dissipation by evaporation within cellar spaces must be prevented. Even with "good" concrete, such moisture travel can be appreciable, since the temperature gradient will always be upward. A waterproof layer preventing moisture loss is a wise precaution.

4-25. Strapped Footings. Used to support two columns, strapped footings consist of a pad under each column with a strap between them. Where soil resistance is assumed only under the footings the strap connections (levers), often built on disturbed soil, or even a flexible form liner, are more economical than combined footings. The usual assumptions are that base pressure is uniform under each pad, or bearing area, and that a rigid beam connects the two column loads, so that the center of gravity of the bearing areas coincides with the center of gravity of the column loads (see also Art. 5-73).

Any attempt to modify the assumption of uniform distribution results in complex detail founded on little fact; however, any distribution that reduces the eccentricity between load and resistance shows considerable return in economy. Before starting an analysis based on nonuniform distribution, one should carefully consider the reliability of deflection and rigidity computations for reinforced concrete, because the only justification for nonuniform soil distribution is a relationship of soil resistance and deformation.

PILES AND PILE DRIVING

A pile is defined as a structural unit introduced into the ground to transmit loads to lower soil strata or to alter the physical properties of the ground. It is of such shape, size, and length that the supporting material immediately underlying the base of the unit cannot be manually inspected. This definition does not limit the type of materials making up the pile, the manner of insertion into the ground, or the loads to be transferred to lower strata. However, where manual inspection of the bottom is possible, as in shallow pits or sufficiently wide, deep caissons, these types of construction are not classified as piles.

Piles transfer load to lower strata by frictional resistance along the surface in contact with the soil and by direct bearing into a compressed volume at or near the bottom of the pile. The portion of the load carried by friction and by end bearing is known between limits.

The maximum frictional resistance between soil and pile surface is a definite limit to the amount of load that can be transferred by friction.

The bearing values of soils as normally determined, either by empirical classification or by load test, cannot be used to evaluate the end bearing resistance of a pile. The confined character of the soil considerably increases its bearing value and a compressed volume of soil at the pile tip acts as an integral part of the base, of unknown total area. Apparently, the total bearing area at the pile tip has a definite relationship to the area required to transfer that part of the pile load not taken up by skin friction.

4-26. Pile Deterioration. Any analysis of load transfer and determination of pile load-carrying value is based on the premise that the pile structure will not change. That danger comes from many directions; the pile may rust away or rot, and exterior conditions, such as soil acidity, fluctuation in water levels, and loss of lateral support in the soil encasing the piles must be carefully considered and precautionary measures taken.

The rusting of steel surfaces in the ground (the average corrosion loss is 0.003 in. per year) cannot harm the load value of a pile, unless the mill scale becomes loose and surface friction is seriously affected. The life of steel H and sheet piling is far beyond the normal life of a building.

4-27. Timber Piles. Continuously kept wet, timber piles are quite permanent. But a narrow band of alternately wet and dry condition, sometimes caused by seasonal water-level fluctuations, will soon be apparent in pile failures. Where many timber-pile foundations exist, a large-scale control of the water-table level by always maintaining full level, as is done at Amsterdam, Holland, is a necessary expense.

Almost every kind of timber can be used for a pile. Some species take the shock of driving better than others, and the species have different static compression values. Prime-grade sticks cannot be expected to be sold for piles; yet it is reasonable to insist on timber cut above the ground swell, free from decay, from unsound or grouped knots, from wind shakes and short reversed bends. The maximum diameter of any sound knot should be one-third the diameter of the pile section where the knot occurs, but not more than 4 in. in the lower half of pile length or more than 5 in. elsewhere. All knots should be trimmed flush with the body of the pile, and ends should be squared with the axis.

Piles should have reasonably uniform taper throughout the length and should be so straight that a line joining the centers of point and butt should not depart from the body of the pile. All dimensions should be measured inside the bark, which may be left on the pile if untreated but must be removed before treatment with any preservative. The diameter at any section is the average of the maximum and minimum dimensions; piles not exactly circular in section should not be rejected.

No timber pile should have a point less than 6 in. in diameter, except for temporary use. Where the point is reinforced with a steel shoe, the minimum dimension is at the upper end of such shoe. No untreated timber piles should be used unless the cutoff or top level of the pile is below permanent water-table level. This level must not be assumed higher than the invert level of any sewer, drain, or subsurface structure, existing or planned, in the immediate vicinity, or higher than the water level at the site resulting from the lowest drawdown of wells or sumps.

Where timber piles may be partly above water level, the pile should be pressure-treated to a final retention of not less than 12 lb of creosote per cu ft of wood. Preservative should be a grade 1 coal-tar creosote oil, with penetration by the empty-cell process, both in accordance with Federal specifications No. TT-W-571-b or similar standard.

The tops of all creosoted piles as cut off should be below ground level and the cutoff section should be treated with three coats of hot creosote oil. Creosoted piles above ground level require some replenishment of preservative with age, as well as protection against fire damage.

Of the various timber species, cedar, western hemlock, Norway pine, spruce, or similar kinds are restricted to an average unit compression at any section of 600 psi; cypress, Douglas fir, hickory, oak, southern pine, or similar kinds are allowed a value of 800 psi. A maximum load of 20 tons is commonly allowed on timber piles with 6-in. points, and 25 tons where the point is 8 in. or more, subject, of course, to compliance with indicated or tested driving resistance and maximum stresses.

Timber piles are seldom used as end bearing piles, but there is no real objection to such use, if properly installed.

4-28. Steel H Piles. Where heavy resistances exist in layers that overlie the bearing depths, steel H piles will often take the punishment of the heavy driving much better than any other type. After insertion in the ground, the soil gripped between the web and inside faces of the flanges becomes an integral part of the pile, so that frictional resistance is measured along the surface of the enclosing rectangle and not along the metal surface of the section.

To prevent local crippling during driving, the section should have flange projections not exceeding 14 times the minimum thickness of metal in either web or flange and with total flange width not less than 85% of the depth of the section. All metal thicknesses should be at least 0.40 in.

Other structural-steel sections, or combinations of sections, having flange widths and depths of at least 10 in. and metal thickness of at least ½ in. may also be used—usually in combination with steel sheeting. Most rolling mills issue special sections for use as piles.

The load-carrying value of a steel pile is usually restricted to a maximum unit stress of 12,600 psi at any cross section but not more than 35% of the yield stress.

4-29. Concrete Piles. Precast piles are most suitable for large projects. The necessity for accurate length determination and the added cost of cutoffs, plus the high cost of equipment to make and handle the piles, tend to make precast piles more expensive than the various cast-in-place types.

Precast piles must be reinforced for local stresses during lifting and driving and must be cured to full strength before use. Development of prestressed-concrete piles, some with the prestressed rods removable after the pile has been driven, improves economical possibilities for precast piles, since the cost of reinforcing is at least half the total cost of a precast pile before driving.

Cast-in-place concrete piles come in many variations, in which concrete fills premolded cavities in the soil. The cavity for a pile can be formed by removal of the soil or by forcing the volume displaced into the adjacent soil, as is done by timber or precast-concrete piles. The soil cavity can be made by auger or casing carried to a desired depth and filled with concrete. Volume displacement is accomplished by driving a thin, metal, closed-end shell stiffened with a retractable mandrel or a heavier cylindrical metal shell, which is retracted as the concrete filling is installed.

Retracted shells are usually heavy steel pipe, sometimes with reinforced cutting edges and with driving rings welded at the top to provide a grip for retraction.

In some types, the concrete is compressed by a vibrating or drop hammer and tends to squeeze out laterally below the casing to form annular expansions. These greatly increase the bearing value of the pile.

Whether the shell is left in place or not, one advantage of cast-in-place piles is the possibility of examining the cavity before placing the concrete, to check depth of penetration and variation from plumb or desired batter position.

Usually, the empirical resistance taken as a measure of pile value is indicated by the downward movement during casing driving, before any concrete is placed. It is therefore often claimed that the completed piles have much higher values than expected. The reasoning is that the concrete filling causes an expansion of the shell—or even with the shell removed, of the soil-contact cylinder—which must increase the frictional support value of the pile.

Shoes are sometimes integral with the shells; there are also several types of special shoes such as the flat boot, pointed castings of 60 and 45° cones, ball-pointed ribbed caps, inverted paraboloid, precast-concrete "button," and flat plates welded to the piles.

For all cast-in-place piles, care must be exercised to clean the cavity of all foreign matter and to fill the entire volume with concrete. If water enters the shell, tremie tubes may be necessary for pouring the concrete.

The structural concrete pile is designed as a short concrete column, with a maximum unit stress of 25% of the 28-day cylinder strength in compression for working-stress design or with proper load factors, an assumed 5% eccentricity of loading, and $\phi = 0.70$ ($\phi = 0.75$ where a permanent steel casing of ¼ in. or greater wall thickness is used) for ultimate-strength design. Reinforcing steel is evaluated as for reinforced concrete columns. Metal shells in which concrete is cast and which acts with the concrete as part of a pile, may, with an ⅛-in. allowance for loss in thickness due to corrosion, be stressed to 35% of the yield stress but not to more than 12,600 psi.

4-30. Pile Splices. Piles are often made up of special sections of similar or different structural elements. Such composite piles are usually evaluated as equal to the weaker section. The splice must be designed to hold the sections together during driving, as well as to resist later effects from the driving of adjacent piles or from lateral loads. (Timber-to-concrete splice details and tests were given by H. D. Dewell, in *Civil Engineering*, September, 1938, p. 574.) Timber pile sleeves consist of a tight-fitting steel pipe section into which the pile ends are driven; sometimes drift pins or through bolts are used to hold the assembly together. Steel rolled sections and pipe piles are often extended by a full-strength weld. Concrete piles can be extended, but the detail is extremely expensive and should be employed in emergency cases only.

4-31. Soil Limitations on Piles. The structural pile can be relied upon within the limitations normally set on the materials. Maximum pile loading is more dependent on the nature of the soils into which the pile is driven than on the materials comprising the pile.

Structures on piles installed in unstable soil strata, which may be subject to lateral movements, should be braced by batter piles or completely tied with structural members having both tensile and compressive resistance in the plane of the pile caps.

Piles installed in soils that exhibit considerable subsidence and consolidation during driving, or where great depths of unconsolidated silts and mucks are penetrated, must be investigated for possible future loads when the consolidation occurs and the weight of these upper layers hangs on the piles. Several suggestions offered to avoid such overloading have been tried but found wanting. Among them are artificial lubrication of the pile, or surrounding the upper section with a loose paper or metal sleeve, to eliminate friction. It is true that the maximum weight that can hang on the pile is equal to the surface frictional resistance. However, the lateral strength of long piles driven through unconsolidated layers is so low that insertion of adjacent piles, weight of construction equipment, and unequal consolidations often cause lateral displacement of the piles. In these cases, immediate lateral staying of the piles is advisable, even before construction of the pile caps and bracing system between caps.

4-32. Reused Piles. Reuse of piles where structures are demolished, or even for new loading conditions, should be restricted to places where complete data are available on length and driving conditions of each pile. Where a group of piles is to be reused for new loadings, their supporting value should be restricted to 75% of rated capacity. Additional piles driven for the same structure should be of similar length and type as those being reused and should also be restricted to 75% of rated capacity.

4-33. Minimum Sizes of Piles. The usual liberal restrictions on pile sections are a minimum 6-in.-diameter point, with a minimum 8-in.-diameter butt at cutoff for a tapered pile. Piles of uniform section are limited to a minimum 8-in. diameter, or 7½-in. side if not circular.

Actually, if the piles can satisfy stress restrictions as short columns, there is no justification for any dimensional limitations. Generally, such small sizes as given here are suitable only for end-bearing piles and have too little surface to develop sufficient skin friction.

4-34. Pile Spacing. Since, basically, the purpose of piles is to transfer loads into soil layers, pile spacing should be a function of the load carried and the capacity of the soil to take over that load.

Piles bearing on rock or penetrating into rock should have a minimum spacing center to center of twice the average diameter, or 1.75 times the diagonal of the pile, but not less than 24 in. This spacing will not overload the rock, even if the bearing piles are fully loaded and no load relief occurs from skin friction.

All other piles, transferring a major part of the load to the soil by skin friction, should have a minimum spacing center to center of twice the average diameter, or 1.75 times the diagonal of the pile, but not less than 30 in. For a pile group consisting of more than four piles, such minimum spacing must be increased by 10% for each interior pile, to a maximum of 40%; i.e., to 42-in. spacing, if the principal support of the piles is in soil with less than 6 tons per sq ft bearing value. Figure 4-5 gives the pile spacing for various groups where spacing is governed by soil value.

Studies of the load value of groups, as related to an individual-pile value, show definite reduction in average pile value as the cluster increases in size. From results of a large-scale test program of wood piles, the approximate formula was deduced that the value of a pile is reduced by one-sixteenth for each other pile within normal spacing (Fig. 4-5). Increased spacing tends to eliminate the difference and give the piles nearly their full values.

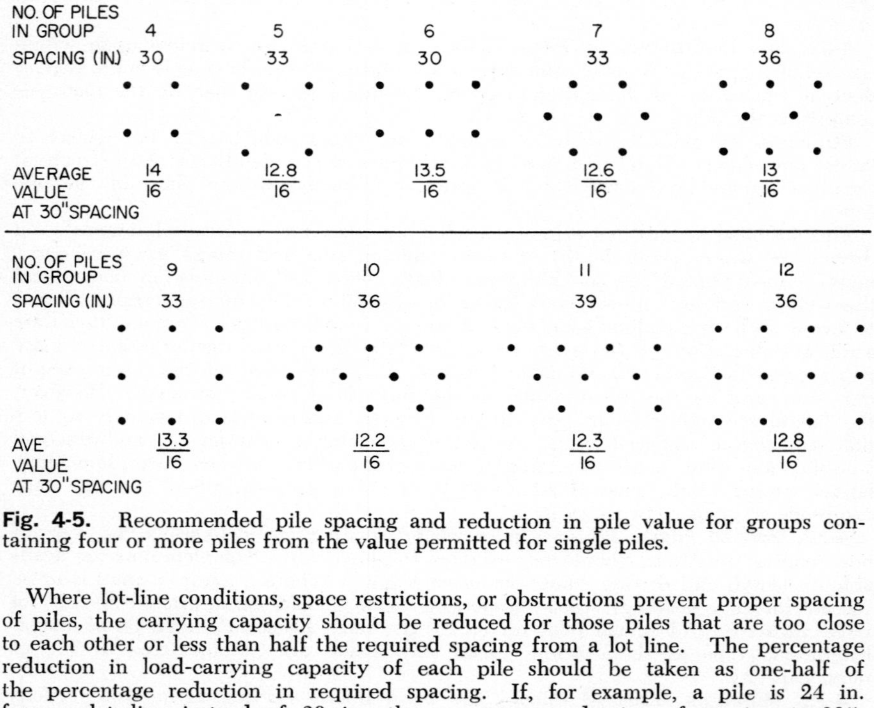

Fig. 4-5. Recommended pile spacing and reduction in pile value for groups containing four or more piles from the value permitted for single piles.

Where lot-line conditions, space restrictions, or obstructions prevent proper spacing of piles, the carrying capacity should be reduced for those piles that are too close to each other or less than half the required spacing from a lot line. The percentage reduction in load-carrying capacity of each pile should be taken as one-half of the percentage reduction in required spacing. If, for example, a pile is 24 in. from a lot line instead of 30 in., the percentage reduction of spacing is 20%; so pile capacity is reduced 20%/2, or 10%. Such reduction in values is sometimes economical where closer spacing permits the placing of a pile group below the columns and eliminates eccentricity straps or combined footings.

4-35. Pile Embedment. Piles require a definite grip in resistant soil. It is a wise precaution to insist on at least 10 ft of embedment in natural soil of sufficient stiffness to warrant a presumptive bearing value of at least 2 tons per sq ft. The embedment should be below grade or such elevation as will not be affected by future excavations.

Each pile must be braced against lateral displacement by rigid connection to at least two other piles in radial directions not less than 60° apart. Where fewer than three piles are in a group, or where extra lateral stability is made necessary by the nature of loads or softness of soil layers, concrete ties or braces must be provided between caps. Such bracing should be considered as struts, with minimum dimensions of one-twentieth the clear distance between pile caps, but not less than 8 in. All bars should be anchored in the caps to develop full tension value. A continuous reinforced-concrete mat 6 in. or more in thickness, supported by and anchored to the pile caps, or in which piles are embedded at least 3 in., may be used as the bracing tie, if the slab does not depend on the soil for the direct support of its own weight and loads.

Pile caps are designed as structural members distributing the load into the piles, with no allowance for any load support on the soil directly below the pile cap.

4-36. Tolerances for Piles. In designing a cap, each pile is considered a point load concentrated at the center of the cutoff cross section. However, only in ideal soil conditions and with perfect equipment will the piles all be inserted vertically and exactly in the planned position so that this assumption is true. Some practical rules of acceptance of piles not perfectly positioned must control every pile job:

A tolerance of 3 in. in any direction from the designated location should be permitted without reduction in load capacity—provided that no pile in the group will then carry more than 110% of its normal permitted value, even though the average loadings of the piles are much less. The distribution of column loads into the individual piles should be determined either analytically or graphically from a plotted graph of the surveyed locations of the cutoff centers. Where any later resurvey is expected, such centers should be marked by nails or chisel cuts.

Where any pile is overloaded, either additional balancing piles should be added to the group, or structural levers should be added for load redistribution or connection with adjacent groups.

Computation of actual pile loads can be simplified by plotting the pile and column locations to scale, then locating the center of gravity of the piles graphically and using the line between this point and the center of the column loads as an axis of symmetry. Multiplying pile values by the offsets of all piles from the normal axis drawn through the center of gravity of the piles gives a total moment equal to the eccentricity moment. Assume a linear variation for reduction or increase of pile value with distance from the axis. If the maximum pile value is just beyond the 110% limit, a second approximation should be tried, relocating the center of gravity of the piles on a weighted basis, using the values from the first solution. This will slightly reduce the eccentricity and may be sufficient to permit approval of the pile group.

Where piles are installed out of plumb, a tolerance of 2% of the pile length, measured by plumb bob inside a casing pile or by extending the exposed section of solid piles, is permissible. A greater variation requires special provision to balance the resultant horizontal and vertical forces, using ties to adjacent column caps where necessary. For hollow-pipe piles, which are driven out of plumb or where obstructions have caused deviations from the intended line, the exact shape of the pile can be determined by the use of a magnetic shape-measuring instrument, or an electronic plumb bob. After the deflection curve of a pile is plotted, consecutive differentiation gives the slope, moment, and shears acting on the pile, from which the reduction in allowable load, if any, can be determined.

4-37. Pile-driving Hammers. Equipment and methods for installing piles should be such as to produce satisfactory piles in proper position and alignment. This involves the proper choice of equipment for the size and weight of pile, expected load value, and type of soil to be penetrated. Generally, heavier piles require heavier hammers. Piles also may be driven with vibratory hammers.

Experience with wood piles indicates that a lighter hammer than required by some building codes should be used to reduce breakage of piles, certainly in areas where soil layers change suddenly in resistance to penetration. (T. C. Bruns, "Don't Hit Timber Piles Too Hard," *Civil Engineering*, December, 1941.) Although the usual argument that a big spike should not be driven with a tack hammer is true, so is the opposite statement that a long slender finishing nail should not be driven with a sledge. The usual pile-driving hammers used on building foundations are listed in Table 4-5.

Table 4-5. Pile-driving Hammers and Rated Values

1. Drop Hammers
Vulcan sizes from 500- to 3,000-lb weight of hammer, dropping 12 to 48 in.

Hammer type	Rated energy, ft-lb	Ram, lb	Stroke, in.
2. Single-acting Steam Hammers			
McKiernan-Terry No.:			
S3	9,000	3,000	36
S5	16,250	5,000	39
S10	32,500	10,000	39
S14	37,500	14,000	32
S20	60,000	20,000	36
Warrington-Vulcan No.:			
O20	60,000	20,000	36
O16	48,750	16,250	36
O14	42,000	14,000	36
O10	32,500	10,000	39
O8	26,000	8,000	39
O6	19,500	6,500	36
OR	30,200	9,300	39
0	24,300	7,500	39
1	15,000	5,000	36
2	7,260	3,000	29
3. Double-acting Steam or Air Hammers*			
Super-Vulcan No.:			
18C	3,600 at 150 s/m	1,800	10½
30C	7,260 at 133 s/m	3,000	12½
50C	15,100 at 120 s/m	5,000	15½
65C	19,500 at 117 s/m	6,500	15½
80C	24,450 at 111 s/m	8,000	16½
140C	36,000 at 103 s/m	14,000	15½
200C	50,200 at 98 s/m	20,000	15½
DGH900	4,000	900	10
McKiernan-Terry No.:			
5	1,000 at 300 s/m	200	7
6	2,500 at 275 s/m	400	8¾
6.5	3,200 at 280 s/m	600	8¾
7	4,150 at 225 s/m	800	9½
9-B-3	8,750 at 145 s/m	1,600	17
10-B-3	13,000 at 105 s/m	3,000	19
11-B-3	19,150 at 95 s/m	5,000	19
C826	24,000 at 90 s/m	8,000	18
C3	9,000 at 130 s/m	3,000	16
C5	16,000 at 110 s/m	5,000	18
C8	26,000 at 80 s/m	8,000	20
Industrial Brownhoist No.:			
Short	6,000 at 150 s/m	1,600	16
Long	9,650 at 110 s/m	1,900	24
Union Ironworks No.:			
00	54,900 at 85 s/m	6,000	36
0A	22,050 at 90 s/m	5,000	21
0	19,850 at 110 s/m	3,000	24
1	13,100 at 130 s/m	1,850	21

Table 4-5. Pile-driving Hammers and Rated Values (*Continued*)

Hammer type	Rated energy, ft-lb	Ram, lb	Stroke, in.
1A	10,020 at 120 s/m	1,600	18
1½ A	8,680 at 125 s/m	1,500	
2	5,755 at 145 s/m	1,025	16
3	3,660 at 160 s/m	700	14
3A	4,390 at 150 s/m	820	13½
4	2,100 at 200 s/m	370	12
Raymond Pile Co. No.:			
65C (steam)	19,500 at 117 s/m	6,500	16
80C (steam)	24,450 at 111 s/m	8,000	16½
65C (hydraulic)	19,500	6,500	16
4. Diesel Hammers			
Syntron No.:			
3,000	3,000 at 110 s/m	1,200	33
5,000	5,200 at 110 s/m	2,100	34
9,000	9,000 at 105 s/m	3,800	34
16,000	16,000 at 84 s/m	5,400	48
Delmag No.:			
D5 (single acting)	9,100	1,100	98
D12 (single acting)	22,500	2,750	98
D22 (single acting)	39,800	4,850	98
McKiernan-Terry No.:			
DE10 (single acting)	9,900	1,100	108
DE20 (single acting)	18,800	2,000	113
DE30 (single acting)	30,100	2,800	129
DE40 (single acting)	43,000	4,000	129
Link-Belt Speeder No.:			
312 (partial double acting)	18,000 maximum	3,855	31
520 (partial double acting)	30,000 maximum	5,070	43
180 (partial double acting)	8,100	1,725	38
105 (partial double acting)	7,500	1,445	35
440 (double acting)	18,200	4,000	37
5. Vibratory Hammer			
Bodine–Guild (U.S.A.)	2–50 hp		
Vibrodriver (France)			
PTC-B	2–17 hp		
	2–35 hp	9,000 weight	
Taylor-Woodrow (English)	8 rams at 200-ton thrust		
Foster	10 models: 12 hp to 240 hp		

*s/m = strokes per minute.

4-38. Jetting Piles into Place. Other methods of pile insertion include the Franki drop weight for pulling a pipe casing into the ground by acting on a compressed gravel or concrete plug in the bottom of the casing, or water-jet loosening of the soil allowing the pile to drop by gravity or under the weight of a hammer resting on the butt.

Jetting is not necessary or advisable in soils other than fairly coarse dense sands. Even there, it must be remembered that jetting is excavation of soil; it has been so decided by the New York State Supreme Court, and all legal restrictions concerning excavations along a property line are available protections for the adjacent owner.

Jetting should be stopped at least 3 ft above the final pile-tip elevation and the piles driven at least 3 ft to the required resistance. The condition of adjacent completed piles should be checked, and if loss of resistance is found, all piles should be redriven.

Use of jacks against static loads (in the form of existing structural mass or temporary setup of weights) is usually limited to adding piles for additional support during underpinning operations or within completed buildings.

4-39. Load Value of Piles. There are three general methods of evaluating the safe carrying capacity of a pile:

1. From the indicated resistance to the impact of a moving hammer.
2. From the net settlement under a static load.
3. From the computed shear resistance over the embedded pile surface area.

Recent tendency is to give greater reliability to the load-test method and restrict the use of pile-driving formulas to the lower pile-load values and in areas of uniform soil conditions.

In the use of impact formulas, minimum necessary data are (1) the energy per blow—usually given by the manufacturer for a definite number of blows per minute, and for steam or air hammers, at definite pressures on the piston—and (2) the net penetration of the pile into the ground under a blow.

The net penetration is the gross penetration under impact less the rebound. It is measured along a scale on the fixed portion of the pile-driver frame.

For many of the pile formulas, additional weights of pile and other parts of the hammer, conditions at the surfaces of impact and other empirical constants also must be evaluated. In any case, the formulas can apply only to ideal conditions; penetration measurements are not reliable if made when a pile head is broomed or shows plastic flow, if new cushion blocks are used, and if frictional resistance hinders the free fall of the hammer. Note, too, that for piles driven to large batters, the true impact is the vertical component of the rated value of the hammer.

Many pile-driving formulas have been proposed; all have some disadvantages, and none can be used blindly. To correlate a static bearing value with the resistance to an impact—as some formulas attempt to do—is basically impossible; that the results have little correlation with load-test values is therefore not surprising. Some of the more common formulas are shown in Table 4-6.

Where no soft layers underlie firm soil layers (as would be shown by erratic reports on casing resistance), and where the expected maximum pile value does not exceed 30 tons, any of the common formulas can be used, with the proper constants for the corresponding hammer type.

The minimum impact blow should be related to the load value: a minimum of 15,000 ft-lb for 25- to 30-ton pile values, 12,000 ft-lb for 21 to 25 tons, 8,000 ft-lb for 15 to 20 tons. Restrictions on hammer weights based on pile diameter do not have any logical basis, as they require too heavy a hammer for a wood pile, with consequent breakage during driving, and permit too small a hammer for an 8-in. steel pile, which may carry 30 or more tons.

The New York City Building Code recommends that minimum penetration resistances for timber piles be determined by the *Engineering News* formula (see Table 4-6), but for all other types of piles, the values should be at least those given in Table 4-7.

When the allowable pile load is to be determined by load tests, subsurface investigations must first prove that the conditions in the area are fairly uniform. For each 15,000 sq ft of building area or part thereof, three piles should be driven—distributed over the area—close to points where borings were made. A complete pile-driving log record should be kept. Driving should be carried to resistances indicated by dynamic formula for the desired load value. If the lengths of the piles or their driving records (blows per foot) vary more than 20%, additional piles should be driven until agreement is found.

Two piles in each area should be load-tested to 200% of the desired load. The load should be applied in eight equal increments, each to remain until the settlement is constant within 0.001 ft for 2 hr. The total load should remain until the settlement

Table 4-6. Common Pile-driving Formulas

General energy-resistance formula (Redtenbacher, 1859):

$$R = \frac{AE}{L}\left[-S \pm \sqrt{S^2 + WH\frac{(W + Pn^2)2L}{(W + P)AE}}\right]$$

where W = weight of hammer, lb
H = height of drop of W, in.
R = ultimate pile value, lb (no factor of safety)
S = pile penetration per blow, in.
E = modulus of elasticity of pile
A = cross-sectional area of pile, sq in.
P = weight of pile, lb
L = length of pile, in.
n = coefficient of restitution due to impact—zero for nonelastic soil; unity for elastic soil

The Redtenbacher formula, as used in Europe, takes $n = 0$.

If $n = 1$, this formula becomes $R = WH/(S + C)$, where C is a constant. This is the Wellington, or *Engineering News*, formula.

Boston Pile Code formulas:

For drop hammer: $$R' = \frac{3WH'}{S + K}\frac{W}{W + P}$$

For single-acting hammer: $$R' = \frac{3.6WH'}{S + K}\frac{W}{W + P}$$

For double-acting hammer: $$R' = \frac{4F}{S + K}\frac{W}{W + P}$$

where H' = height of fall, ft
S = average pile penetration for last 5 blows, in.
R' = allowable pile value, lb
F = actual hammer-blow energy, ft-lb
$K = \frac{3R'L}{SAE} + C$, in.
C = 0.05 for wood pile or cap
K = constant for various type and sizes of piles

Hiley formula often used in Britain (1930):

$$R = \frac{YWH}{S + \frac{1}{2}C}$$

where Y = efficiency of blow transmission, always less than unity
C = sum of energy losses due to compression of driving head, of pile, and of ground

is constant within 0.001 ft for 48 hr. After that, the load should be removed in four equal decrements, with intervals of not less than 1 hr. The rebound should be measured after each removal and also 24 hr after the load is entirely removed.

Based on the plotted record of the test, the maximum allowable pile load is the smaller of the following: one-half the load causing a net settlement of 0.01 in. per ton of load, or one-half the load causing a gross settlement of 1 in.

In the subsequent driving of piles in that area, using the same equipment on the same type of pile, under similar operating conditions, all piles showing the same or greater resistance to penetration as was shown by the piles tested should be allowed the average values obtained from the load tests. Any piles that stop at higher levels and are one-third or more shorter than the average length of the tested piles should be rejected or load-tested.

Where piles are installed by jacking against static loads, the safe carrying capacity may be taken as 50% of the maximum jacking load required at final penetration. If the jacking load is held constant for a period of 10 hr, the carrying capacity can be taken as two-thirds the jacking load. In temporary underpinning work, piles may be used to support the full jacking load, but if the piles are to remain as permanent supports, not more than two-thirds the jacking load should be allowed.

Only piles receiving their support entirely from frictional resistances along their surface can be evaluated by summing the resistances in each layer penetrated. Such frictional unit resistances must be determined either from undisturbed samples taken within each soil layer or by means of a plunger cone or vane-type investigation.

Table 4-7. Minimum Driving Resistance and Minimum Hammer Energy for Other Than Timber Piles

Pile capacity, tons	Hammer energy, ft-lb†	Minimum driving resistance, blows per ft*				
		Friction piles	Piles bearing on hardpan	Nondisplacement piles bearing on decomposed rock	Displacement piles bearing on decomposed rock	Piles bearing on rock
Up to 20	15,000	19	19	48	48	5 blows per ¼ in. (minimum hammer energy of 15,000 ft-lb)
	19,000	15	15	27	27	
	24,000	11	11	16	16	
30	15,000	30	30	72	72	
	19,000	23	23	40	40	
	24,000	18	18	26	26	
40	15,000	44	50	96	96	
	19,000	32	36	53	53	
	24,000	24	30	34	34	
50	15,000	72	96	120	120	
	19,000	49	54	80	80	
	24,000	35	37	60	60	
	32,000	24	25	40	40	
60	15,000	96		240	240	
	19,000	63		150	150	
	24,000	44		100	100	
	32,000	30		50	50	
70 and 80	19,000		5 blows per ¼ in. (minimum hammer energy of 15,000 ft-lb)	5 blows per ¼ in. (minimum hammer energy of 19,000 ft-lb)		
	24,000					
	32,000					
100						
Over 100						

**Final driving resistance* shall be the sum of tabulated values plus resistance exerted by nonbearing materials. The driving resistance of nonbearing materials shall be taken as the resistance experienced by the pile during driving but which will be dissipated with time and may be approximated as the resistance to penetration of the pile recorded when the pile has penetrated to the bottom of the lowest stratum of nominally unsatisfactory bearing material or to the bottom of the lowest stratum of clay or silt soils, but only where such strata are completely penetrated by the pile.

Sustained driving resistance: Where piles are to bear on soft rock or hardpan, the minimum driving resistance shall be maintained for the last 6 in., unless a higher sustained driving resistance requirement is established by load test. Where piles are to bear in weaker soil classes, the minimum driving resistance shall be maintained for the last 12 in. unless load testing demonstrates a requirement for higher sustained driving resistance. No pile need be driven to a resistance to penetration, blows per ft, more than twice the resistance indicated in this table, nor beyond the point at which there is no measurable net penetration under the hammer blow.

The tabulated values assume that the ratio of total weight of pile to weight of striking part of hammer does not exceed 3.5. If a larger ratio is to be used, or for other conditions for which no values are tabulated, the driving resistance shall be as approved by the commissioner.

For intermediate values of pile capacity, minimum requirements for driving resistance may be determined by straight-line interpolation.

†The hammer energy indicated is the rated energy.

Any attempt to combine theoretical end bearing with such surface friction should be discouraged. Not enough is known about the possibility of combined action; even with friction alone, the unproved assumption is made that the shearing strains in all soil layers are equal, or remain constant up to failure; otherwise a summation of the resistances is not permissible.

The maximum unit shear coefficient at any depth cannot exceed the internal coefficient of shear of the soil, because failure will occur at the pile surface or within the soil adjacent to it.

The surface of shear resistance is cylindrical for all piles, the diameter of the cylinder for a battered or stepped pile being substantially the diameter at the butt of the pile. Noncylindrical shapes are more efficient friction piles.

As an approximation, subject to check by field determination, the maximum friction resistance may be taken as 10% of the normally safe bearing value for each soil encountered. In no case should a pile value be permitted above 50% of the accumulated computed surface-skin frictional resistance.

4-40. Pile Loads. For all types of piles, the design must safely support the maximum combination of the following loads:

1. All dead loads and the weight of the pile cap and superimposed loads.
2. All live loads, reduced as permitted because of floor area or number of stories in the building.
3. Lateral force and moment reactions, including the effect of eccentricity, if any, between the column load and the center of gravity of the pile group.
4. That amount of the vertical, lateral, and moment reactions resulting from wind and seismic loads in excess of one-third of the respective reactions totaled from 1, 2, and 3.

The maximum load permitted on a vertical pile is the allowable pile load, as determined by formula, test, or computation, applied concentrically in the direction of its axis. No lateral loads in excess of 2,000 lb per pile shall be permitted on a vertical pile, unless actual test shows that the pile will resist double the lateral load with less than ½-in. lateral movement at the ground surface. If batter piles are used to take lateral forces, the resultant loads along the axis of the pile should not exceed the allowable pile value.

Piles are considered as columns fixed at a point from 5 to 10 ft below the surface; the lower value can be used whenever the soils at the top are compact. Allowable unit stresses in any pile section must not be exceeded. The total pile load is considered as existing in all end bearing piles without relief to a depth of 40 ft and with a maximum reduction of 25% if the pile is longer.

For a friction pile, the entire load is assumed to exist at the section located at two-thirds the embedded length of the pile, measured from the top.

4-41. Bearing Piles. Piles driven to a resistance providing end bearing in rock or extremely tough and dense layers directly overlying bedrock are called bearing piles. For one type, an open casing is used, which can be cleaned out for visual inspection of the material at the bottom of the pile. Where exposure of bedrock is possible and the steel pipe is driven into the rock to ultimate resistance—denoted by less than ¼-in. net penetration under five blows of a hammer having a rated energy in foot-pounds of at least 25% of the desired pile value in pounds—the maximum pile value can be taken as 80% of the structural value of the combined steel and concrete section. The limit usually is 200 tons.

Piles of solid section similarly driven are similarly evaluated but usually are further restricted to a maximum value of 120 tons if driven into rock or 80 tons if the point stops above the rock surface. Such values cannot be used, unless checked by load tests; a maximum value of 40 tons should be set on end bearing piles evaluated by formula only.

Special pile forms may be used to obtain a very high allowable load. One frequently used is known as a **drilled-in caisson,** a proprietary name, or a **caisson pile**. It is formed starting with a heavy steel pipe with a reinforced cutting edge, which is driven into bedrock and then cleaned out. A socket is drilled or chopped into the rock at the bottom of the pipe. Next, the pipe is filled with concrete. For higher-load values, the pile may be reinforced in the same way as a spiral-reinforced concrete column or with a structural-steel shape that extends into the socket.

The load-carrying capacity is the structural value of the pile as a column, but may not exceed the load transfer from the socket to the rock. This load is the sum of the allowable bearing value on the bottom of the socket plus the bond of the concrete to the rock along the surface area of contact in the depth of the socket. Not economical for small loads, such piles have been used to sustain loads of 3,000 tons, requiring reinforcing with the heaviest rolled-steel column sections with plates welded to them.

The high cost of the steel pipe shell warrants retraction during concreting. Often, when the shell is to be retracted, a thin shell liner is inserted to act as a form for the concrete.

4-42. Friction Piles. Piles not driven to a bearing on hard ground or rock are known as friction piles and are assumed to have no end bearing. When the piles stop in soils underlain by compressible layers, a careful analysis of the imposed loading on such layers is necessary, to determine what settlements are to be expected from consolidation. Usually, piles evaluated by impact formula only are restricted to a 30-ton value, with further restriction to 25 tons or less for wood piles.

Steel tubes 18 in. in diameter have been used as friction piles for 100 and 200 tons, but in some cases, it was evident that the soils could not take such load concentration in groups of four to six piles and that resistances dissipated with time, even when the piles were not loaded.

Bearing value of friction piles is affected by vibrations transferred to the piles. Tests under 500 to 850 vibrations per minute showed additional settlements in the range of 0.2 to 0.5 in. in silt and sand.

CAISSONS AND SHAFTS

Before the invention of heavy construction equipment, the only way that a foundation could be carried to considerable depths through soft soils was by slow, manual excavation within a protective construction. To a large degree, this development was a carry-over from experience gathered in well-sinking operations. (A parallel development in more recent times is the drilled-caisson procedure modeled on oil-well-exploration methods.) The sheeted and braced deep-water well logically became a construction shaft for a deep footing. The ancient masonry ring enclosure in a well, which was successively underpinned as the excavation progressed, as developed in many parts of North Africa, the Babylonian area of influence, and in India, is still copied, with only slight modifications, in very deep shafts.

4-43. Open Shafts for Foundations. Shafts are constructed to great depths, usually through saturated soft soil by the following methods:

1. Excavation, followed by sheeting the desired outline using braced horizontal or vertical members of wood, steel, or precast concrete.
2. Excavation, following the insertion of vertical sheeting, introducing bracing at the desired depths, when reached. This type of construction usually is called a **cofferdam.**
3. Method 2 sometimes is modified by setting up a completely braced shell, or **caisson,** which sinks from its weight (aided, if necessary, by temporary surcharge loadings) as the excavation undercuts the shell.
4. **Pneumatic caissons** are used for excavations carried to depths at which it would be impossible with ordinary caissons to control the flow of water or soil through the bottom. They are capped to provide a top seal and work is done under artificially applied air pressure, sufficient to counterbalance hydrostatic pressure. Entrance to the working chamber is through double-gated locks.
5. Resistance to sinking of a caisson limits the depth to about 100 ft; so the **nested caisson** was introduced. Lower sections of this type of caisson fit within the completed ring and can be jacked down against the resistance of the upper section. With this procedure, a three-section caisson has reached 400-ft depths.
6. Similar depths can be reached by freezing the ground around the volume to be excavated.

4-44. Soil Resistance When Sinking Caissons. In all the sinking methods, frictional soil resistance governs depth of penetration. The design of the sheeting and bracing depends on the pressure of the soil during sinking or during excavation within the enclosure.

In all types of sinking methods, where additional bearing area is required when the shaft reaches the desired depth, the base area can be extended by undercutting if the soil at that depth can be retained sufficiently to install the concrete base. In heavy clays, the cutting can be done without any special protection. In clays that squeeze or displace laterally on excavation of a free surface, the cost of sheeting and bracing to withstand such movement is prohibitive; the shaft must be carried down full size. In granular materials, sloping sheeting can be driven through the bottom of a caisson to cover and protect the conical section of enlargement.

Resistance to vertical penetration of a rigid body in any soil indicates that, during the period when there is actual relative motion, the soil is in a viscous state, and the nature of the material forming the rigid body has little effect on the resistance. The depth of cover also has little effect on the amount of resistance per unit area of buried surface. Apparently, the tackiness and adhesion of the soil to the caisson, plus the shear resistance of any projections on the surface, are always greater than the internal resistance of the soil, and therefore the lesser value governs the total resistance to be expected. Large additions of water or slurry along the interface will considerably reduce the resistance, especially in soils where cohesion to the caisson is a major factor in total internal resistance. Continuity of movement is desirable, to eliminate "freezing" of the interface that is caused by dissipation of enough free liquid to change the soil state from viscous to solid. (Jacob Feld, International Soil Mechanics Congress, Paper No. Vc5, vol. 7, Rotterdam, 1948.)

In addition to surface resistance, there is also the bearing resistance of any surfaces at the base of the moving caisson, which must be overcome by load to the point of failure; i.e., the amount of unit pressure must be enough to make the soil act as a nonsolid. From a summary of many caisson operations, the load required to overcome surface resistance has been found to range approximately as follows, in pounds per square foot of buried surface:

Dense clay, sometimes mixed with fine sand	700–800
Tight sand, clay, and silt mixtures	500–600
Saturated sands	400–500
Soft clay and wet sand mixtures	200–300

4-45. Pressure on Sheeting. Soil pressure on sheeting that is inserted into the ground without permitting soil expansion or movement depends on the ratio of width or diameter of excavation to depth.

In any type of soil, there is no lateral pressure at a certain depth, which depends on the width of the excavation. For derivation of theoretical earth-pressure formulas, see *Journal of the American Concrete Institute,* Vol. 16, pp. 441–451, 1945, from which Table 4-8 is reproduced.

It is noteworthy that zero unit pressure occurs in normal granular soils at a depth about six times the width of the pit or shaft. If surface resistance was a friction as normally encountered in solids, it could not restrict the depth of penetration, since the frictional value is a direct function of the pressure acting against the sheeting.

4-46. Types of Sheeting. Simplest pit sheeting consists of square-edge wood boards, successively placed one at a time below previously braced boards, following careful hand cutting of the soil to closely simulate the shape of the boards. When soil conditions do not permit such shaping and where the soil will not maintain a vertical face long enough to permit insertion of the next sheet, any one of several louver-type sheets may be used.

The basic requirement of **louver sheeting** is provision of packing in back of the sheets, to correct irregularities of excavation and to eliminate open voids in back of the sheeting. The simplest device is to interpose small blocks between adjacent sheets and thereby leave packing gaps. One patented unit provides a recessed gap in each sheet, which eliminates the handling of small blocks. An efficient method is to cut a diagonal from one edge of a board and tack it to the opposite edge, forming a parallelogram, so that an 8-in. board will cover 10 in. (if of 2-in stock). The sharp lower edge permits digging into the bottom of the pit and rotating the board into position against the earth. The upper level protects the gap opening and simplifies the packing operation. Bracing of the

Table 4-8. Soil Pressures in Pits

Angle of internal resistance	Ratio of width to depth: Zero unit pressure	Max unit pressure	Zero total pressure	For a pit 5 ft wide unit pressure is zero at a depth of (ft)
0	0	0	0	∞
5	0.054	0.108	0.036	93
10	0.098	0.196	0.065	51
15	0.126	0.252	0.084	40
20	0.147	0.294	0.098	34
25	0.159	0.318	0.106	31
30	0.167	0.334	0.111	30
35	0.168	0.336	0.112	30
40	0.165	0.330	0.110	30
45	0.159	0.318	0.106	31
50	0.149	0.298	0.099	34
55	0.135	0.270	0.090	37
60	0.120	0.240	0.080	52
65	0.104	0.208	0.069	48
70	0.084	0.168	0.056	60
75	0.065	0.130	0.043	77
80	0.042	0.084	0.028	119
85	0.024	0.048	0.016	209

pit sides can be accomplished in several ways, all copied from the methods used in making wooden boxes, and based on the support of each side by the two adjacent sides.

The **Chicago caisson** is an open circular pit lined with vertical plank staves, which are braced internally by steel rings wedged against the staves. The excavation is carried down for the depth equal to the length of staves—4 to 6 ft—and usually two bracing rings are inserted for each section. This type is suitable chiefly for easily cut clays that will stand unbraced for several hours and where free-flowing water or running silt is not encountered. This method has been used for many of the large structures in Chicago, to depths as great as 200 ft. Sometimes, enlargements are made at the bottom into the bearing stratum.

Where softer soils do not permit the use of box sheeting or lagging, a steel cylinder is sunk as a cutoff of soil and water until it reaches required soil bearing layers and can be excavated. If the bottom cannot be sealed off, excavation is under water and the concrete is poured by tremie method.

4-47. Gow Caisson. Developed in the Boston area, this caisson consists of a nest of steel cylinders from 8 to 16 ft high. Each cylinder is 2 in. smaller in diameter than the preceding. The caisson is driven down by impact or jacking as excavation proceeds. Since no internal bracing is required, the excavation is easily performed, and the cylinders can be withdrawn as concreting progresses.

4-48. Slurry Trench Walls. Use of a bentonite-slurry-filled trench dug to or into rock, and later displacement of the slurry by tremie concrete, is often an economical method for foundation-wall construction. Walls, usually a minimum of 24 in. thick, can be constructed by this method in any soil condition. Reinforcing cages may be inserted into the slurry before concreting. Good bond between the steel and the concrete has resulted.

Reuse of the slurry, a specially homogenized machine mix, necessitates dividing the length of the wall into sections. On some projects, leakage problems have caused trouble. In the slurry-wall constructions for the Budapest subway, the entire inside face was covered with ⅜-in. steel plates welded to a continuous sheet to prevent leakage. On other jobs, grouting of the joints was found necessary.

Application of the slurry method to deep cellars, such as those of the World Trade Center in New York, has shown that the method can compete economically with normal cofferdam construction. For such large areas, interior bracing frames are too expensive. Anchors drilled into the exterior rock are usually provided to replace the temporary bracing.

DRILLED-SHAFT FOOTINGS

As a development of power drilling for water and oil wells, there is a group of drilled-shaft procedures for obtaining economical exposures of suitable bearing layers.

4-49. Cored Shafts. In normal plastic soil, sometimes interspaced with limestone or sand seams, a spinning auger-bottom bucket is rotated by a square- or oblong-section Kelly bar to excavate a circular shaft 24 to 72 in. in diameter. The operation is rapid and simple. When the bottom is exposed, a sawtooth-bottom steel shell is rapidly spun and a cut made, until the desired hardness of material is reached, or else an underreamer is used to cut an enlargement to the desired area. The concrete pier is cast without forms. Skin friction is depended on to carry some of the load, although normally the base pressures are satisfactory. The concrete core may be reinforced to increase carrying capacity.

This development is chiefly used in Texas and Southern California, where sandy loams or clays can be cut and left standing for inspection of the bottom and concreting done without bracing. In the limestone areas near Dallas, cores or the bottom exposures are taken for testing and base-pressure values are correlated with the compressive strength of core samples.

Cored shafts have been carried to depths over 80 ft, but the economical range is about 40 ft.

The time and power used to lift each bucket of soil, and the special rigs required for the deep holes, make other methods quite competitive. For depths of about 20 ft where soft rock or equivalent can be reached, however, the cored caisson foundation will be found competitive in cost with a spread footing on the upper soil.

When running water is encountered, the shafts must be kept full of water during excavation to prevent caving of the side walls. If in addition, seams of fine sand or silt are cut, the shaft must be enclosed with a steel shell, usually with a serrated bottom edge for cutting into the layer that permits sealing off the running material. To prevent binding, the top of the shell has two or more outriggers, or recessed gripping cuts, for turning the shell part of a revolution every hour or so. Within the shell, the coring operation continues normally. As concrete is placed, the shell is retracted.

4-50. Montee Caisson. Used for deep foundations, this consists of a reinforced steel shell, with cutting edge of serrated hardened plate, which is connected to a large motor for continuous rotation. The action is similar to the circular saw used in some large granite quarries to remove cores of 8 and 10 ft in diameter, to serve as weakening planes for freeing the rock sheets. To maintain completely fluid conditions, a slurry is fed into the cylinder to float the interior soil outward under the cutting edge and up along the outer surface of the shell.

The difficulty encountered in the use of the Montee caisson comes chiefly from rock surfaces encountered on a tilt. Well-bedded boulders or riprap are easily cut through, but a loose boulder may damage the cutting edge. When uneven rock is engaged, it is difficult to get a complete seating and seal to the rock. The apparatus requires a large investment; so the method cannot be considered except for large projects. A French development, by Solatanche, oscillates the cutting shell and has proved more reliable in getting deep caissons to desired depths.

4-51. Driven Shafts. In a similar group of methods, soil in shafts is extruded by displacement, and piers are concreted in the voids so formed. One method that has been used for light structures on partly consolidated sanitary fill is to insert a steel flat-pointed 12×12 in. mandrel into the ground with a light pile hammer to a depth of 4 ft and place concrete in the voids. The consolidation from squeezing the 4 cu ft of soil into the immediate adjacent area may be sufficient strengthening to provide acceptable support.

Franki Piles. The Franki extrusion "pile" is similar. A steel shell is set into a pit; the bottom is filled with run-of-bank gravel or dry concrete to a depth of two or three diameters; and the fill then is pushed into the soil by dropping a heavy ram inside the shell. The gravel packs and grips the shell sufficiently to pull it along through almost any type of soil. On reaching a desired depth, rapid impact with the ram breaks up the plug and extrudes it into the soil in the shape of an inverted mushroom. Concrete is added before the plug material has entirely cleared the shell. As the additional concrete is extruded, the shell is withdrawn. The result is a very rough-surfaced cylinder with a number of annular fins projecting into the soil. At the bottom is a mushroom-shaped expansion. Under load test, these piles carry much more load than is indicated by the area of the mushroom times the normally presumptive bearing value of the soil.

Drilled Injection Piles. This is a small-diameter concrete shaft. It is constructed by drilling into the soil with an auger, followed by simultaneous extraction of the auger and filling of the cavity below with a cement grout injected through the hollow shaft of the auger. Friction piles of this type have been used. The method also has been used for building continuous bulkhead cutoffs, similar to tight sheeting.

EARTH PRESSURES ON WALLS

In all soil-mechanics and foundation problems, four types of earth pressures are involved:

1. Active earth pressure. This exerts horizontal and vertical components against any structure that impedes the tendency of the earth to fall, slide, or creep into its natural state of equilibrium.
2. Passive earth resistance. This pressure is mobilized by the tendency of a structure to compress the earth.
3. Vertical load reaction of soil above a structure that has been built into the earth or which is covered with earth.
4. Vertical pressure distribution through a soil from an imposed loading

Each of these pressures is dependent on many of the physical properties of the soil, as well as the relative rigidity of soil and structure. The most important properties of the soil appear to be density and coefficient of internal resistance, the latter being a direct factor in the ratio of active pressure to the weight of soil above the level considered. Some typical values for various soil types are given in Table 4-9.

4-52. Active Earth Pressure. This can safely and economically be determined by the following rules:

1. Horizontal component of active earth pressure for any material for a normal-type wall is closely given by the general wedge theory for the case of a vertical wall and substantially horizontal fill. The wall is assumed to be backfilled by usual construction methods and not so rigid that a small rotational movement (of the magnitude of 0.001 its height) cannot occur. The necessary rotation to mobilize the internal friction of the backfill is equivalent to an outward movement of ¼ in. at the top of a 20-ft wall.

The horizontal component of lateral pressure E_a is closely given for vertical walls and horizontal fills by the formula

$$E_a = \tfrac{1}{2}wH^2 \tan^2\left(45° - \frac{\phi}{2}\right) = \tfrac{1}{2}wH^2K_a \tag{4-5}$$

where w = average unit weight of fill, lb per cu ft
H = height of fill, ft
ϕ = angle whose tangent is coefficient of internal resistance of the fill

The value of $\tan^2(45° - \phi/2)$ is often designated by the symbol K_a. Table 4-9 gives average values for w and $\tan \phi$. Table 4-10 gives values of $\tan^2(45° - \phi/2)$ for various values of ϕ.

Table 4-9. Typical Soil Weights and Internal Resistance

Soil type	Average weight, w, lb per cu ft	Coefficient of internal resistance, $\tan \phi$
Soft flowing mud	105–120	0.18
Wet fine sand	110–120	0.27–0.58
Dry sand	90–110	0.47–0.70
Gravel	120–135	0.58–0.84
Compact loam	90–110	0.58–1.00
Loose loam	75–90	0.27–0.58
Clay	95–120	0.18–1.00
Cinders	35–45	0.47–1.00
Coke	40–50	0.58–1.00
Anthracite coal	52	0.58
Bituminous coal	50	0.70
Ashes	40	0.84
Wheat	50	0.53

Table 4-10. Active and Passive Pressure Coefficients

Values of $K_a = \tan^2\left(45° - \frac{\phi}{2}\right) = \frac{1 - \sin\phi}{1 + \sin\phi}$

and of $K_p = \tan^2\left(45° + \frac{\phi}{2}\right) = \frac{1 + \sin\phi}{1 - \sin\phi}$

ϕ	$\tan \phi$	K_a	K_p
0	0	1.00	1.00
10	0.176	0.70	1.42
20	0.364	0.49	2.04
25	0.466	0.41	2.47
30	0.577	0.33	3.00
35	0.700	0.27	3.69
40	0.839	0.22	4.40
45	1.000	0.17	5.83
50	1.192	0.13	7.55
60	1.732	0.07	13.90
70	2.748	0.03	32.40
80	5.671	0.01	132.20
90	∞	0	∞

2. The general wedge-theory formulas (taking into account friction along the surface of a wall) may be used for evaluation of the horizontal component for all other conditions of wall batters and sloping fills. However, comparison with experimental results indicates that the results are somewhat too small for negative surcharges and somewhat too large for positive surcharges, but the differences are no greater than 10%. It is quite accurate enough, taking into account the uncertainties of conditions as to actual slope, to use a table of ratios, referred to the simplified case of vertical wall and horizontal fill (see Table 4-11).

3. The vertical component of lateral pressure, in all cases, is such that the resultant pressure forms an angle with the normal to the back of the wall equal to the angle of wall friction. However, under no condition can this angle exceed the angle of internal friction of the fill.

4. The pressure of soils that, because of lack of drainage and because of their nature, may become fluid at any time—whether such fluid material is widespread

Table 4-11. Lateral Pressure Ratios for General Conditions

For Horizontal Fill

*Wall Slope**	%
+1:6	107
+1:12	103
Vertical	100
−1:12	95
−1:6	90

For Vertical Walls

ϕ =	40°	30°	20°
Fill Slope†		%	
+10°	123	110	109
0	100	100	100
−10°	74	83	88

NOTE: Percentages apply to values given by Eq. (4-5) for the case of vertical wall and horizontal fill and for a specific value of ϕ. In computing the above values, the angle of friction between the wall and fill is assumed equal to ϕ, but not greater than 30°.

* + represents a slope away from the fill; − represents a slope toward the fill.

† + represents a slope uphill away from the wall; − represents a downhill slope.

or only a narrow layer against the wall—is the same as hydrostatic pressure of a liquid having the same density as that soil.

5. The pressure of submerged soils is given by Eq. (4-5) with the weight of material reduced by buoyancy (for the solid fraction of the soil only) and the coefficient of internal friction evaluated for the submerged condition; in addition, full hydrostatic pressure of water must be included. For granular materials, submergence affects the coefficients of internal and wall friction very little; submergence changes silt materials to liquids.

6. The pressure of fills during saturation and prior to complete submergence and the pressure during drainage periods are affected by the rapidity of water movement. Drainage produces a slight temporary decrease in pressure from normal. Submergence produces an expansion of the fill with consequent increase in pressure. Such variations will not occur if adequate provision is made for drainage; if such provision is not made, the soil may become submerged and the pressure should be computed as for a submerged soil.

7. The pressure of granular fills may be increased about 10% by earthquake and other vibrations. The increase remains for a time and slowly disappears. Silt materials and some kinds of clay (thixotropic clays) under the action of earthquakes or vibrations by heavy trucking may become liquid.

8. The pressure of fills varies directly with temperature and normally decreases with age. Both factors can be safely disregarded in the design of abutments and walls.

9. Surface loading of the fill increases the lateral pressure on the wall. A trapezoidal pressure distribution—obtained by adding to the ordinary fill pressure the pressure computed from Eq. (4-5) for a depth of earth equivalent in weight to the surface loading or surcharge—may be assumed. Although not in complete accord with recent experimental work, this assumption is considered safe for walls because of the unlikely occurrence of full surcharge over the entire area of influence. For liquid fills, even though drained but remaining fluid, the moisture in the pores transmits the full weight of the surcharge to the wall. Hence, in such cases,

that part of the horizontal pressure on the wall due to the surcharge equals the full weight of the surcharge at all depths.

10. The resultant horizontal pressure acts a distance from the base equal to 0.33 to 0.45 of the height of the wall. Liquid and negative-sloped fill pressures act at 0.33 of the height. Theoretical points of application of surcharge fills are listed in Table 4-12. Except for liquid and negative-sloped fill pressures, the point of application should be assumed no lower than 0.375 of the height, and higher if the surcharge ratio so requires. This assumption is on the safe side sufficiently to compensate for any effect of vibration, age, and local fill compaction. The location of the resultant affects wall design much more than corresponding accuracy in amount of pressure.

11. The pressure in pits and bins is not given by the wedge theory unless a correction for side-wall friction is made, in which case actual field observations of pressures are closely checked. (See Table 4-8.)

Table 4-12. Pressures from Surcharges on Fills

Ratio of surcharge equivalent to fill height	Ratio of height of resultant to fill height	Total pressure ratio (theoretical)	Total pressure ratio (Spangler)*
0	0.33	100	100
0.1	0.36	120	105
0.2	0.38	140	120
0.3	0.40	160	150
0.4	0.41	180	210
0.5	0.42	200	290
0.6	0.42	220	
0.7	0.42	240	
0.8	0.43	260	
0.9	0.43	280	
1.0	0.43	300	

* The Spangler ratios are based on data shown in Fig. 10, p. 63, *Highway Research Board Proceedings,* 1938, assuming normal unsurcharged pressure equal to 16 h^2, the extrapolation being from a single surcharge load (100 psf).

12. Many attempts have been made to derive a consistent formula for the pressure of cohesive soils, including a reduction factor for the cohesive resistance. Empirical evaluation indicates a formula of the form:

$$K = 0.75 - \frac{q}{wH} \tag{4-6}$$

where K = hydrostatic pressure ratio to be used in Eq. (4-5) for K_a
w = unit weight of soil, lb per cu ft
H = depth of excavation of fill, ft
q = the average unconfined compressive strength of soil samples, psf

(See chart by R. S. Knapp and R. B. Peck, *Engineering News-Record,* Nov. 20, 1941.)

4-53. Passive Earth Pressure. The horizontal component of passive (or maximum) resistance to pressure before failure (termed also "passive pressure") is often computed from the formula

$$E_p = wH^2 \tan^2\left(45° + \frac{\phi}{2}\right) = \tfrac{1}{2}wH^2K_p \tag{4-7}$$

where w = average unit weight of fill, lb per cu ft
H = fill height, ft
ϕ = angle whose tangent is the coefficient of internal resistance of the fill
K_p = $\tan^2 (45° + \phi/2)$

Experimental work, especially with sheetpiles in sand, shows that actual values of passive resistance are larger than those given by Eq. (4-7). For the design of sheetpiling embedded in sandy materials, a maximum value of passive resistance equal to twice E_p may be recommended as permissible.

The vertical component of the passive resistance is such that resultant pressure is inclined from the normal to the back of the wall by the angle of wall friction.

Little is known about the location of the resultant passive pressure. It is probably affected by the rigidity of the wall. The usual assumption is that the distribution of lateral resistance is linear.

4-54. Pressures on Underground Structures. The vertical load reaction due to soil above a buried structure has been determined chiefly on culverts, although there are considerable data available from tunnel-construction experience. The relative rigidity or flexibility of surrounding earth and the intruding culvert, tunnel, or foundation has considerable influence on whether the load to be carried is greater than, equal to, or less than the weight of the soil directly above. A summary of general relationship is given in several reports of the Iowa State College Engineering Experiment Station, where a succession of fine work has been done on this problem from 1910 to the present. Generally, flexibility of the embedded structure reduces the load to be carried. Rigidity in the backfill has the same effect.

SHEETING AND BRACING OF EXCAVATIONS

The design of proper sheeting and bracing to hold excavation banks in safe position during a construction operation is often given too little attention. Legal responsibility for maintaining adjacent property and public streets with their buried utility services undamaged during and after construction is usually placed on the builder. Since the work to be done is temporary and is not a measurable item of completed production, the tendency is to minimize it and "take a chance." The results are usually expensive reconstruction, loss of valuable time, and lengthy litigation. Some cities require filing of a carefully prepared design for temporary sheeting.

4-55. Sheeting Design. Sheeting without special anchorage or buttresses can be used in several forms or combinations of shapes.

Any formula used for stresses in sheeting must be based on assumptions of the shape that the sheeting takes under load. In designing embedded poles subjected to unbalanced wire loading, the usual assumptions are:

1. The lateral passive resistance increases linearly with depth.
2. The point of inflection is at two-thirds the embedded length below the surface.

These assumptions may be safely used to determine depth of sheeting embedment for small structures, especially where reliable soil data are not available. On the assumption that the passive resistance increases linearly with depth and the maximum value at the bottom is the same on either side of sheeting and equals wK_pD, M. A. Drucker developed a formula that may be simplified to approximately

$$D = \frac{H}{3} \cdot \frac{1 + 3K_a}{1 - K_a} \tag{4-8}$$

where D = depth of embedment
H = height of retained fill
K_a = active pressure coefficient
K_p = passive-resistance coefficient (see Table 4-10)
(*Civil Engineering*, December, 1934).

Equation (4-8) should not be used where the coefficient of internal friction of the soil is greater than unity. When soil so hard is encountered (for values of ϕ greater than 45°) the formula to be used is

$$D = \frac{2K_a}{1 - 2K_a} H \tag{4-9}$$

These equations are applicable for cantilevered sheeting with no external supports other than the embedment below the level of excavation. Deeper types of steel sheeting can be used to depths of 20 ft or more and, even in saturated sands, when drained by well points, have permitted safe construction of foundations without interference from bracing. Where pile driving or other vibrations may occur, elimination of bracing is not advisable.

4-56. Bracing of Sheetpiles. Sheetpiling can be provided with additional supports by either bracing inside the cut or anchoring the top to buried "dead men" located outside. Where space is available, the latter procedure is more economical, taking the form of piles or buried masses of concrete, to which are attached adjustable lengths of cables or rods holding back continuous horizontal wales.

The reactions to be provided at each level of support cannot be definitely determined. The distribution of pressure on the sheeting depends on the flexibility of the sheeting, which in turn is affected by the relative stiffness of the supports. A liberal allowance at each level is therefore necessary, and the total of all the resistances should be considerably above the total pressure. However, unit stresses in the wales, struts, and ties can be taken at much higher values than for permanent construction; one-half the elastic limits of the respective materials is quite proper.

If the final shape of the sheeting is known, graphical determination of moments and pressures is possible. The Baumann solution given in *Transactions of the ASCE*, 1935, is such a method. Standard sheetpile handbooks usually give the Blum-Lohmeyer method for load evaluation at the various supports.

Normally, water pressure need not be considered, since the excavation must be kept dry, and any hydrostatic pressures on the back of the sheeting can be easily relieved. However, if water can accumulate in back of the sheeting, such pressure must be considered. Equally important, saturation will reduce considerably the soil-anchorage reaction at the bottom of the sheeting and increase the upper reactions. Saturation of the bottom will also require longer penetration of the sheeting to avoid blowouts or boils along the sheeting, either of which will prove disastrous. Drainage is a cheap investment for safety.

No bracing system is any stronger than the resistance provided at the bottom of the struts. Unless very good soil is found, the struts should be provided with screw jacks to develop the necessary resistances in the sills and to permit continuous adjustments.

Bracing against previously completed footings is permissible only if the footings are properly braced against undisturbed soil or have enough lateral strength. In evaluating lateral strength, the frictional resistance on the base of footings (disregarding any cohesive forces) may be added to the passive resistance of the soil against vertical faces.

Lateral resistance of piles, for movements of ¼ in., is taken by the U.S. Corps of Engineers in the design of river works as: 4 tons per wood pile, 5 tons per concrete pile, and 6 tons per steel pile, when driven to the usual bearing values of each. For substantial movements laterally, the values should be reduced by 30%.

4-57. Anchorage of Sheeting. Where exterior anchorages are used, the pull-out resistance varies with the type of soil and size of the buried anchor. Relative values of pull-out resistance for soils are 1.0 for compact and stiff clay and 0.5 for fairly soft clay and loose sand.

For equal vertical depth of embedment, resistance increases with the slope of the pull from the vertical. If we assign a value of 1.0 for vertical pull, then the resistance is 1.5 at 1:1 slope; 2.0 at 1:2 slope; 2.1 at 1:3 slope; and 2.3 at 1:4 slope.

Where embedment is larger than the maximum dimension of the anchor, the resistances to vertical pull in rammed average soil in pounds per square foot of anchor face at various depths are:

Resistance, Psf	*Depth, Ft*
800	1.0
1,000	1.5
1,900	2.0
3,000	3.0
5,400	4.0
8,000	5.0

With the techniques developed for posttensioned concrete, sheeting may be anchored by means of tendons inserted in diagonally drilled holes outside the sheeted excavation, tensioned against the wales to desired reactions, and grouted. This type of anchorage may be used instead of interior bracing frames. The grouted end of the tendon must be located beyond any possible slip plane in the soil body. Usually, the anchorage is in rock; some successful operations, however, have anchored the tendons or high-strength rods in soil. Tendons are allowed the same stress values as in posttensioned concrete. They should be pretested to 10 to 15% above the desired reaction before the holding anchorage is locked on the wales. Special sloping reaction blocks are needed on the wales, and are usually made of double beams; or else, the wales are set on a slope to provide a right-angle bearing surface for the tendon blocks.

Legal clearance from adjacent property owners and public lands is necessary to permit the subsurface encroachment by such anchorage.

Where excavations are necessarily kept open for long periods, say more than a year, some provision may be necessary to avoid possible corrosion loss of the metal in the tendons. The usual wet environment, often with stray electric currents, may cause some serious loss in original strength of the anchorages. Fully grouting the drilled holes in rock or the cased holes in earth after the tendons have been tensioned can reduce the hazard of corrosion.

Sheeting should not be removed until all backfill is properly consolidated. Some additional fill and compaction are necessary as sheeting is removed to avoid internal slip and possible damage to adjacent pavements.

SETTLEMENT AND FAILURES

4-58. Unwanted Foundation Movements. Structures that show unwanted vertical or lateral movements as a result of foundation settlements, heaves, or displacements are examples where the foundation design is a failure. Rarely does a structural failure occur in the footing; the usual trouble is in the soil and a consequence of the assumption that movements either will not occur or will be uniform.

Failures can be classified as unwanted upward, downward, and lateral movements.

That many structures are affected by the heave of fine-grained soils is now common knowledge. Certain bentonitic clay soils undergo large volume changes with seasonal variation in moisture content—especially when lightly loaded. The phenomenon is further complicated by the shading effect, both under and toward the northerly side of the buildings, causing an inequality in the drying out of the soil.

Tidal intrusion of water along shore lines has caused upward movements of light frame structures.

Evaporation of moisture taken out by trees adjacent to foundations has caused sufficient volume change to tip and crack walls.

Consolidation of soil layers from drainage induced by sewer and other subsurface trenches, as well as the normal slower consolidation from normal pressure of overburden, can cause tipping of foundations. Even pile foundations have been known to settle and move laterally where unconsolidated layers—usually not uniform in thickness—start to shrink after the trigger action of pile driving or change in loading conditions.

A very common failure is the dragging down of an existing wall by new construction built in intimate contact; an open gap or sliding plane must be provided between any previously settled-in-place structure and new work. A similar trouble is the sympathetic settlement of existing structures on a stable foundation underlain by soft strata, which are compressed and bent by new loading, even if the two structures do not touch.

Settlements can be easily observed, and the trouble can often be recognized from the shape of the masonry cracks. Every design should be examined to see where and how foundation movements can occur, and design modifications should be made to eliminate the effects on other structures and to equalize movements, in amount and direction, in the new structure.

For example, a large structure designed by the Los Angeles County Flood Control

District was to be founded partly on natural shale and partly on a blended compacted fill. The formula for the blending of fills was fixed by getting a mixture of decomposed granite and broken shale with the same triaxial shear test value as the natural shale. The computed expected settlement was 0.014 ft. Under 90% of the design load, the completed structure showed an actual settlement of 0.012 ft. Procedures of this type are available and should be used to avoid failures of the foundation design.

Settlements are possible for pile foundations or even footings on rock, and any assumption made that no settlement will occur can only lead to trouble. Pile foundations will settle as much as the supporting soils under the distributed loads. Certain rocks, of the shale and schist groups, will disintegrate under continuous loading. Unless bottoms of footings are sealed, any percolation of ground water, especially if it contains sewage or other wastes, will soften the contact layer sufficiently to permit lateral flow.

Compact hard dry clays do not remain in that condition if water is present. Backfill for basement walls with soils that shrink on dehydration—often caused by suction of the soil water into the concrete—show voids in the fill immediately adjacent to the wall. Such conditions have been known to be the cause of complete collapse immediately after a heavy rainfall, when enough water enters the voids to provide full hydrostatic pressure for which the wall was not designed.

Localized footing failures within buildings are often caused by the pipe trenches dug after the footings are completed, without regard to relative depths of excavations. Similar troubles often ocur when new trunk sewers or even transit tunnels are built in public streets without proper underpinning of all footings above the surface of influence from the new excavation. Such work will also affect the normal ground-water table, and together with pumping for industrial and air-conditioning requirements, has lowered the water level sufficiently to cause rotting of untreated wood piles. This difficulty is so prevalent that many cities restrict the use of untreated piles for any area where ground water may not remain stable. Proximity of natural water areas is no proof of such stability; the ground-water level has been shown to be at 35 ft below sea level at only short distances from the bulkhead lines where industrial pumping was carried to excess.

A serious foundation failure of a heavy factory bearing on compact fine sand was caused by the decision to sink some wells within the building and get the process water from the subsoil. The underlying layers were reduced in thickness and a dish-shaped floor resulted.

Local overload on ground-floor slabs, transmitted to some footings, explains the odd-shaped roofs on many single-story warehouses.

All these troubles can be controlled, but only by a frank admission in the design that foundations will settle if the soil support is altered and that every new foundation will settle as the load is applied.

ADOBE AND SOIL HOUSES

Use of local soil for building material must be as old as the first human who imitated an earth-boring animal and provided shelter. From that earth cave, it is only a short step to shelters made of sod, used extensively after World War II in the reclamation areas in Europe, and of mud bricks, called adobe when they are sun-cured. The oldest known adobe house is at Sialk, an oasis in Iran, dated before 4000 B.C. Many of the houses in Damascus with walls and roof of local mud applied to woven willow mats are reputed to date to early Biblical times.

4-59. Types of Soil Walls. The use of soil for wall construction is in five different processes:

1. Cajob, where soil wall panels are supported in wood or concrete frames.
2. Mud concrete, or poured adobe, is a fluid mix poured into full-height forms or into movable forms, which are lifted as the work progresses.
3. English Cob, which is a stiff mud piled up by hand to form a wall without any forms or structural skeleton.

4. Rammed earth, or "pise de terre," a damp mix placed between sturdy wood forms, in layers of about 4 in. which are rammed to about 2½ in.

5. Sun-baked soil bricks, adobe, where shrinkage occurs before using in the wall, requiring 2 to 3 months' storage. A good brick will have compressive strength of 300 to 500 psi and tensile value of 50 psi.

4-60. Adobe Blocks. Machine-made adobe blocks are available with admixtures such as cement (really a cement-soil concrete), emulsified asphalt, resins, silicates, soaps, and similar water-repellent ingredients. Rammed earth in mass or in blocks is fire-resistant, verminproof, and decay-proof and has excellent thermal insulation.

However, to reduce its susceptibility to moisture penetration, and to increase resistance to shrinkage, cracking, and surface deterioration, some soil-blending admixtures or ingredients become essential. These not only immediately increase the cost of materials, but require some skilled help in the preparation of the mixtures.

The mechanical resistance of the product is directly proportional to the care and skill in blending and preparing the blocks or walls. The soil either should be compressed or rammed to a density of at least 2.10, or 132 lb per cu ft—greater density than is normally accomplished by any compaction on a road or airport operation.

4-61. Materials for Soil Blocks or Walls. The soil mix should not have any gravel of 1 in. or larger size, and the total gravel must be less than 40% by weight. More gravel will cause the mix to coagulate, set badly on drying, and give fragile bricks. The sand should be clean and may require electrostatic treatment to be so ionized that the surfaces wet easily. Clay should make up 20 to 35% of the mix, with silt sizes eliminated as far as possible. If marl is to be used, it must not be too calcareous and should be broken up into a size range between 1/16 and ⅛ in.

Since optimum moisture conditions must be maintained, topsoils with vegetation should be avoided since they require too much water and shrink too much.

In France, binder stabilizers are found necessary. Those used are fat lime or hydraulic lime (at least 170 lb per cu yd), or portland cement (170 to 300 lb per cu yd). Very good results are obtained with a blend of 1 part hydraulic lime and 2 parts cement, by weight. A good report of strengths and densities possible with different percentages of cement and varying mixes of sand and clay sizes, based on work in Bogotá, Colombia, was given by Ralph Stone (*Civil Engineering,* December, 1952, p. 29). Considerable work has been done by and many reports are available from the United Nations, both in the Technical Assistance Administration and in the Housing, Town and Country Planning Section of the Department of Social Affairs.

Section 5

Concrete Construction

PAUL F. RICE

Technical Director,
Concrete Reinforcing Steel Institute, Chicago, Ill.

AND

EDWARD S. HOFFMAN

Chief Structural Engineer,
The Engineers Collaborative, Ltd., Chicago, Ill.

Economical, durable construction with concrete requires a thorough knowledge of its properties and behavior in service, of approved design procedures, and of recommended field practices. Not only is such knowledge necessary to avoid disappointing results, especially when concrete is manufactured and shaped on the building site, but also to obtain maximum benefits from its unique properties.

To provide the needed information, several organizations promulgate standards, specifications, recommended practices, and reports. Where these organizations are referred to in this section, they will be designated by their initials: PCA for Portland Cement Association, ACI for American Concrete Institute, CRSI for Concrete Reinforcing Steel Institute, and ASTM for American Society for Testing and Materials. In particular, for brevity, "Building Code Requirements for Reinforced Concrete," ACI 318-71, will be referred to either as ACI 318-71 or the ACI Building Code.

This code contains the following basic definitions:

Concrete is a mixture of portland cement, fine aggregate, coarse aggregate and water.

Admixture is a material other than portland cement, aggregate, or water that is added to concrete to modify its properties.

In this section, unless indicated otherwise, these definitions apply to the terms *concrete* and *admixture*.

CONCRETE AND ITS INGREDIENTS

5-1. Cements. (See also Art. 2-2.) Portland cements (ASTM C150) or air-entraining portland cements (ASTM C175) are available in types (I to V and IA to IIIA, respectively) for use under different service conditions. Portland-type cements also include portland blast-furnace slag cement (ASTM C205) and portland-pozzolan cement (ASTM C340), which may be employed in concrete under the ACI Building Code, and may therefore be included under its definition of concrete. Proprietary "shrinkage-compensating" portland-type cements are also available for general use in concrete.

Although all the preceding cements can be employed for concrete, they are *not* interchangeable. (See "Physical Requirements for Portland Cement," and long-time studies of performance, Art. 2–2.) Note that both tensile and compressive

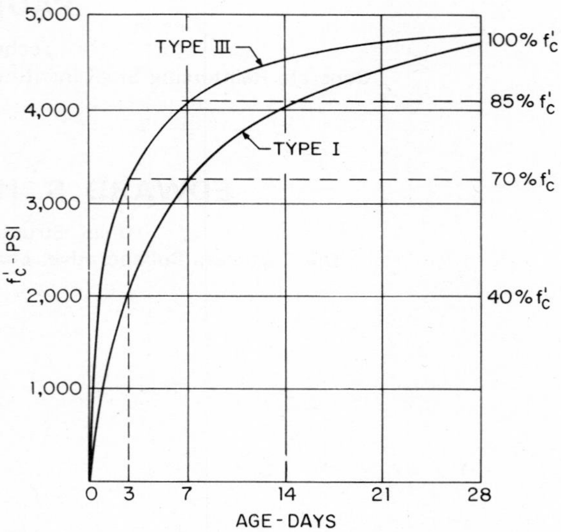

Fig. 5-1. Typical strength-gain rate with standard curing of non-air-entrained concrete with water-cement ratio of 0.50 and Chicago-area aggregates. (*Materials Service Corp., Chicago, Ill.*)

strengths vary considerably, at early ages in particular, even for the five types of basic portland cement. Consequently, although contract specifications for concrete strength are usually based on a standard 28-day age for the concrete, the proportions of ingredients required differ for each type. For the usual building project, where the load-strength relationship is likely to be critical at a point in strength gain equivalent to 7-day standard curing (Fig. 5-1), substitution of a different type (sometimes brand) of cement without reproportioning the mix may be dangerous.

The accepted specifications (ASTM) for cements do not regulate cement temperature nor color. Nevertheless, in hot-weather concreting, the temperature of the fresh concrete and therefore of its constituents must be controlled. Cement temperatures above 170°F are not recommended ("Recommended Practice for Hot Weather Concreting," ACI 305-72).

For exposed architectural concrete, not intended to be painted, control of color is desirable. For uniform color, the water-cement ratio and cement content must be kept constant, because they have significant effects on concrete color. Bear in mind that because of variations in the proportions of natural materials used, cements from different sources differ markedly in color. A change in brand of cement therefore can cause a change in color. Color differences also provide a convenient check for substitution of types (or brands) of cement different from those used in trial batches made to establish proportions to be employed for a building.

5-2. Aggregates. (See also Arts. 2-13 to 2-16.) Only material conforming to specifications for normal-weight aggregate (ASTM C33) or lightweight aggregate for structural concrete (ASTM C330) is accepted under the ACI Building Code without special tests. When an aggregate for which no experience record is available is considered for use, the modulus of elasticity and shrinkage as well as the compressive strength should be determined from trial batches of concrete made with the aggregate. In some localities, aggregates acceptable under C33 or C330 may impart abnormally low ratios of modulus of elasticity to strength (E_c/f'_c), or high-shrinkage to concrete. Such aggregates should not be used.

5-3. Proportioning Concrete Mixes. Principles for proportioning concrete to achieve a prescribed compressive strength after a given age under standard curing are simple.

1. The strength of a hardened concrete mix depends on the water-cement ratio. The water and cement form a paste. If the paste is made with more water, it becomes weaker (Fig. 5-2).

2. The ideal minimum amount of cement paste is that which will coat all aggregate particles and fill all voids.

3. For practical purposes, fresh concrete must possess workability sufficient for the placement conditions. For a given strength and with given materials, the cost of the mix increases as the workability increases. Additional workability is provided by more fine aggregate and more water, but more cement must also be added to keep the same water–cement ratio.

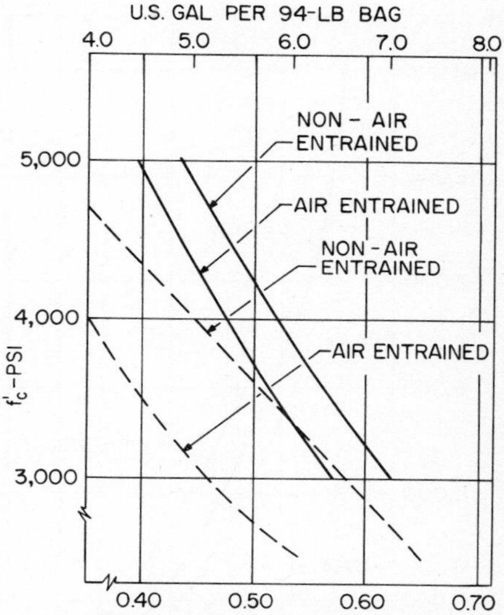

Fig. 5-2. Variation of 28-day compressive strength of normal-weight concrete with water-cement ratio. Solid lines for air-entrained and non-air-entrained concretes indicate average results of tests by Materials Service Corp., Chicago, Ill. Dash lines indicate relationship given in the ACI Building Code (ACI 318-71) for maximum permissible water-cement ratio and specified 28-day strengths.

Because of the variations in material constituents, temperature, and workability required at job sites, theoretical approaches for determining ideal mix proportions usually do not give satisfactory results on the job. Most concrete therefore is proportioned empirically, in accordance with results from trial batches made with the materials to be used on the job. Small adjustments in the initial basic mix may be made as a job progresses; the frequency of such adjustments usually depends on the degree of quality control.

When new materials or exceptional quality control will be employed, the trial-batch method is the most reliable and efficient procedure for establishing proportions.

In determination of a concrete mix, a series of trial batches (or past field experience) is used to establish a curve relating the water-cement ratio to the strength and ingredient porportions of concrete, including admixtures if specified, for the range of desired strengths and workability (slump). Each point on the curve should represent the average of test results on at least three specimens, and the curve should be determined by at least three points. Depending on anticipated quality control, a demonstrated or expected coefficient of variation or standard deviation is assumed for determination of minimum average strength of test specimens (Art. 5-9). Mix proportions are selected from the curve to produce this average strength.

For any large project, significant savings can be made through use of quality control to reduce the overdesign otherwise required by a building code (law). When the owner's specifications include a minimum cement content, however, much of the economic incentive for the use of quality control is lost. See Fig. 5-3 for typical water-cement ratios.

Note that separate procedures are required for selecting proportions when lightweight aggregates are used, because their water-absorption properties differ from those of normal-weight aggregates.

("Building Code Requirements for Reinforced Concrete," ACI 318-71; "Recommended Practice for Selecting Proportions for Normal Weight Concrete," ACI 211.1-70; "Recommended Practice for Evaluation of Compressive Test Results of Field Concrete," ACI 214; "Lightweight Concrete," SP29, American Concrete Institute.)

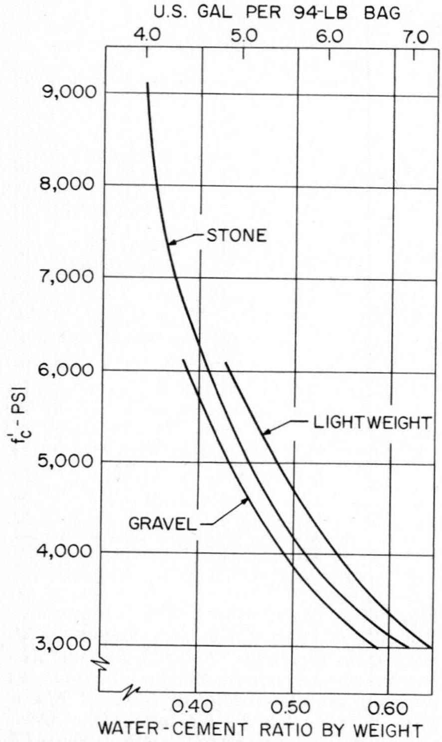

Fig. 5-3. Variation of 28-day compressive strength of concrete with type of aggregate and water-cement ratio, except that strengths in excess of 7,000 psi were determined at 56 days. All mixes contained a water-reducing agent and 100 lb per cu yd of fly ash, and were non-air-entrained. In calculation of the water-cement ratio, two-thirds of the weight of the fly ash was added to the cement content.

5-4. Yield Calculation. Questions often arise between concrete suppliers and buyers regarding "yield," or volume of concrete supplied. A major reason for this is that often the actual yield may be less than the yield calculated from the volumes of ingredients. For example, if the mix temperature varies, less air may be entrained; or if the sand becomes drier and no corrections in batch weights are made, the yield will be under that calculated.

If the specific gravity (sp. gr.) and absorption (abs.) of the aggregates have been determined in advance, accurate yield calculations can be performed as often as necessary to adjust the yield for control of the concrete.

Example. Yield of Non-air-entrained Concrete. The following material properties were recorded for materials used in trial batches: fine aggregate (sand) sp. gr. = 2.65, abs. = 1%; coarse aggregate (gravel) sp. gr. = 2.70, abs. = 0.5%; and cement, sp. gr. = 3.15 (typical). These properties are not expected to change significantly as long as the aggregates used are from the same source. The basic mix proportions, for 1 cu yd of concrete, selected from the trial batches are:

Cement: 564 lb (6 bags)
Surface-dry sand: 1,170 lb
Surface-dry gravel: 2,000 lb
Free water: 300 lb (36 gal) per cu yd

Check the yield.

$$\text{Cement volume} = \frac{564}{3.15 \times 62.4} = 2.86 \text{ cu ft}$$
$$\text{Water volume} = 300/62.4 = 4.81 \text{ cu ft}$$
$$\text{Sand volume} = \frac{1{,}170}{2.65 \times 62.4} = 7.08 \text{ cu ft}$$
$$\text{Gravel volume} = \frac{2{,}000}{2.70 \times 62.4} = 11.87 \text{ cu ft}$$
$$\text{Total volume of solid constituents} = 26.62 \text{ cu ft}$$

Volume of entrapped air = 27 − 26.62 = 0.38 cu ft (1.4%)
Total weight, lb per cu yd = 564 + 300 + 1,170 + 2,000 = 4,034
Total weight, lb per cu ft = 4,034/27 = 149.4
Weight of standard 6 × 12 in. cylinder (0.1965 cu ft) = 29.3 lb

These results indicate that some rapid field checks should be made. Total weight, lb, divided by the total volume, cu yd, reported on the trip tickets for truck mixers should be about 4,000 on this job, unless a different slump was ordered and the proportions adjusted accordingly. If the specified slump for the basic mix was to be reduced, the weight, lb per cu yd, should be increased, because less water and cement would be used and the cement paste (water plus cement) weighs 864/7.67 = 113 lb per cu ft $<$ 149.4 lb per cu ft. If the same batch weights are used for all deliveries, and the slump varies erratically, the yield also will vary. For the same batch weights, a lower slump is associated with underyield, a higher slump with overyield. With a higher slump, overyield batches are likely to be understrength, because some of the aggregate has been replaced by water.

The basic mix proportions in terms of weights may be based on surface-dry aggregates or on oven-dry aggregates. The surface-dry proportions are somewhat more convenient, since absorption then need not be considered in calculation of free water. Damp sand and gravel carry about 5 and 1% free water, respectively. The total weight of this free water should be deducted from the basic mix weight of water (300 lb per cu yd in the example) to obtain the weight of water to be added to the cement and aggregates. The weight of water in the damp aggregates also should be added to the weights of the sand and gravel to obtain actual batch weights, as reported on truck-mixer delivery tickets.

5.5. Properties and Tests of Fresh (Plastic) Concrete. About 2½ gal of water can be chemically combined with each 94-lb sack of cement for full hydration and maximum strength. Water in excess of this amount will be required, however, to provide necessary workability.

Although concrete technologists define and measure workability and consistency separately and in various ways, the practical user specifies only one—slump (technically a measure of consistency). The practical user regards workability requirements simply as provision of sufficient water to permit concrete to be placed and consolidated without honeycomb or excessive water rise; to make concrete "pumpable" if it is to be placed by pumps; and for slabs, to provide a surface that can be finished properly. These workability requirements vary with the job and the placing, vibration, and finishing equipment used.

Slump is tested in the field very quickly. An open-ended, 12-in.-high, truncated metal cone is filled in three equal-volume increments and each increment is consolidated separately, all according to a strict standard procedure (ASTM C143, "Slump of Portland Cement Concrete"). Slump is the sag of the concrete, in., after the cone is removed. The slump should be measured to the nearest ¼ in., which is about the limit of accuracy reproducible by expert inspectors.

Unless the test is performed exactly in accordance with the standard procedure, the results are not comparable and therefore are useless.

The slump test is invalidated if: the operator fails to anchor the cone down by standing on the base wings; the test is performed on a wobbly base, such as forms carrying traffic or a piece of metal on loose pebbles; the cone is not filled by inserting material in small amounts all around the perimeter, or filled and tamped in three equal increments; the top two layers are tamped deeper than their depth plus about 1 in.; the top is pressed down to level it; the sample has been transported and permitted to segregate without remixing; unspecified operations, such as tapping the cone, occur; the cone is not lifted up smoothly in one movement; the cone tips over due to filling from one side or pulling the cone to one side, or if the measurement of slump is not made to the center vertical axis of the cone.

Various penetration tests are quicker and more suitable for untrained personnel than the standard slump test. In each case, the penetration of an object into a flat surface of fresh concrete is measured and related to slump. For convenience, the measuring device is usually calibrated in inches of slump. These tests include use of the patented "Kelley ball" (ASTM C360, "Ball Penetration in Fresh Portland Cement Concrete") and a simple, standard tamping rod with a bullet nose marked with equivalent inches of slump.

Air Content. A field test frequently required measures the air entrapped and entrained in fresh concrete. Various devices (air meters) that are available give quick, convenient results. In the basic methods, the volume of a sample is measured, then the air content is removed or reduced under pressure, and finally the remaining volume is measured. The difference between initial and final volume is the air content. (See ASTM C138, C173, and C231.)

Cement Content. Tests on fresh concrete sometimes are employed to determine the amount of cement present in a batch. Although performed more easily than tests on hardened concrete, tests on fresh concrete nevertheless are too difficult for routine use and usually require mobile laboratory equipment.

For yield tests on fresh concrete, see Arts. 5-10, 5-11, and 5-12.

5-6. Properties and Tests of Hardened Concrete. The principal properties of concrete with which designers are concerned and symbols commonly used for some of these properties are:

$'_c$ = specified compressive strength, psi, determined in accordance with ASTM C39 from standard 6 × 12 in. cylinders under standard laboratory curing. Unless otherwise specified, f'_c is based on tests on cylinders 28 days old

E_c = modulus of elasticity, psi, determined in accordance with ASTM C469; usually assumed as $E_c = w^{1.5}(33)\sqrt{f'_c}$, or for normal-weight concrete (about 145 lb per cu ft), $E_c = 57{,}000\sqrt{f'_c}$

w = weight, lb per cu ft, determined in accordance with ASTM C138 or C567

f_t = direct tensile strength, psi

f_{ct} = average splitting tensile strength, psi, of lightweight-aggregate concretes determined by the split cylinder test (ASTM C496)

f_r = modulus of rupture, psi, the tensile strength at the extreme fiber in bending (commonly used for pavement design) determined in accordance with ASTM C78

Other properties, frequently important for particular conditions, are: durability to resist freezing and thawing when wet and with deicers, color, surface hardness, impact hardness, abrasion resistance, shrinkage, behavior at high temperatures (about 500°F), insulation value at ordinary ambient temperatures, insulation at the high temperatures of a standard fire test, fatigue resistance, and for arctic construction, behavior at cold temperatures (−60 to −75°F). For most of the research on these properties, specially devised tests were employed, usually to duplicate or simulate the conditions of service anticipated. (See "Index to Proceedings of the American Concrete Institute.")

In addition to the formal testing procedures specified by ASTM and the special procedures described in the research references, some practical auxiliary tests, precautions in evaluating tests, and observations that may aid the user in practical applications follow.

Compressive Strength, f'_c. The standard test (ASTM C39) is used to establish the quality of concrete, as delivered, for conformance to specifications. Tests of companion field-cured cylinders measure the effectiveness of the curing (Art. 5-13).

Core tests (ASTM C42) of the hardened concrete in place, if they give strengths higher than the specified f'_c or an agreed-on percentage of f'_c (often 85%), can be used for acceptance of material, placing, consolidation, and curing. If the cores taken for these tests show unsatisfactory strength but companion cores given accelerated additional curing show strengths above the specified f'_c, these tests establish acceptance of the material, placing, and consolidation, and indicate the remedy, more curing, for the low in-place strengths.

For high-strength concretes, say above 5,000 psi, care should be taken that the capping material is also high strength. Better still, the ends of the cylinders should be ground to plane.

Indirect testing for compressive strength includes surface-hardness tests (impact hammer). Properly calibrated, these tests can be employed to evaluate field curing. (See also Art. 5-20.)

Modulus of Elasticity E_c. This property is used in all design, but it is seldom determined by test, and almost never as a regular routine test. For important projects, it is best to secure this information at least once, during the tests on the

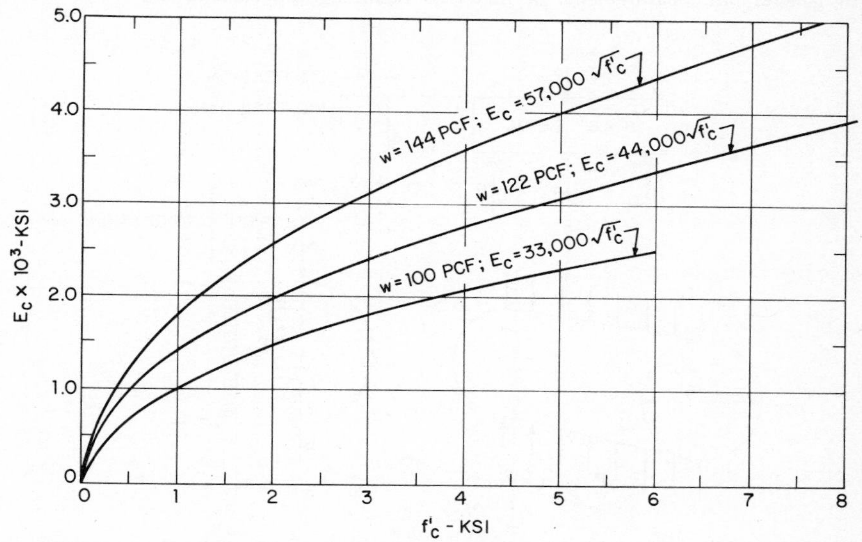

Fig. 5-4. Relationship between modulus of elasticity in compression and 28-day compressive strength of concrete. (*ACI 318-71.*)

trial batches at the various curing ages. An accurate value will be useful in prescribing camber or avoiding unusual deflections. An exact value of E_c is invaluable for long-span, thin-shell construction, where deflections can be large and must be predicted accurately for proper construction and for timing the removal of forms. Figure 5-4 is a plot of $E_c = w^{1.5}(33)\sqrt{f'_c}$.

Tensile strength. The standard splitting test is a measure of almost pure uniform tension f_{ct}. The beam test (Fig. 5-5*a*) measures bending tension f_r on extreme surfaces (Fig. 5-5*b*), calculated for an assumed perfectly elastic, triangular stress distribution.

The split-cylinder test (Fig. 5-5*c*) is used for structural design. It is not sensitive to minor flaws or the surface condition of the specimen. The most important application of the splitting test is in establishment of design values for reinforcing-steel development length, shear in concrete, and deflection of structural lightweight-aggregate concretes.

The values f_{ct} (Fig. 5-5*d*) and f_r bear some relationship to each other, but are not interchangeable. The beam test is very sensitive, especially to flaws on the surface of maximum tension and to the effect of drying-shrinkage differentials, even between the first and last of a group of specimens tested on the same day. The value

f_r is widely used in pavement design, where all testing is performed in the same laboratory and results are then comparable.

5-7. Measuring and Mixing Concrete Ingredients. Methods of measuring the quantities and mixing the ingredients for concrete, and the equipment available, vary greatly. For very small projects where mixing is performed on the site, the materials are usually batched by volume. Under these conditions, accurate proportioning is very difficult. To achieve a reasonable minimum quality of concrete, it is usually less expensive to prescribe an excess of cement than to employ quality control. The same conditions make use of air-entraining cement preferable to separate admixtures. This practical approach is preferable also for very small projects to be supplied with ready-mixed concrete. Economy with excess cement will be achieved whenever volume is so small that the cost of an additional sack of cement per cubic yard is less than the cost of a single compression test.

For engineered construction, some measure of quality control is always employed. In general, all measurements of materials including the cement and water should

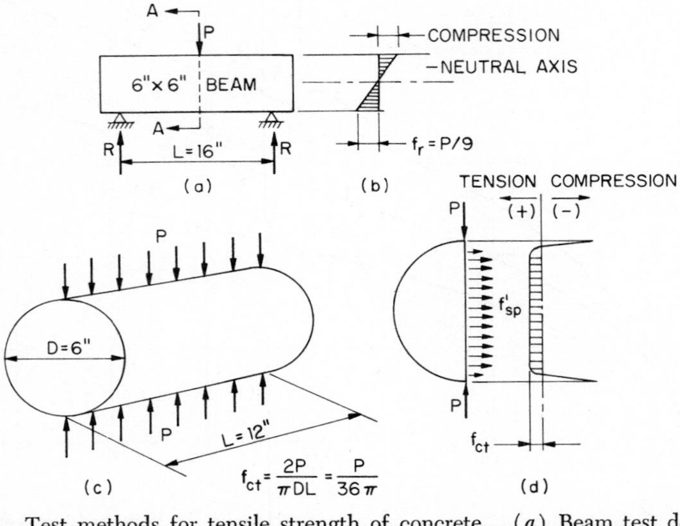

Fig. 5-5. Test methods for tensile strength of concrete. (*a*) Beam test determines modulus of rupture f_r. (*b*) Stress distribution assumed for calculation of f_r. (*c*) Split-cylinder test measures internal tension f_{ct}. (*d*) Stress distribution assumed for f_{ct}.

be by weight. The ACI Building Code provides a sliding scale of overdesign for concrete mixes that is inversely proportional to the degree of quality control provided. In the sense used here, such overdesign is the difference between the specified f'_c and the actual average strength as measured by tests.

Mixing and delivery of structural concrete may be performed by a wide variety of equipment and procedures:

Site mixed, for delivery by chute, pump, truck, conveyor, or rail dump cars. (Mixing procedure for normal-aggregate concretes and lightweight-aggregate concretes to be pumped are usually different, because the greater absorption of some lightweight aggregates must be satisfied before pumping.)

Central-plant mixed, for delivery in either open dump trucks or mixer trucks.

Central-plant batching (weighing and measuring), for mixing and delivery by truck. ("Dry-batched" ready mix.)

Complete portable mixing plants are available and are commonly used for large building or paving projects distant from established sources of supply.

Generally, drum mixers are used. For special purposes, various other types of mixers are required. These special types include countercurrent mixers, in which the blades revolve opposite to the turning of the drum, usually about a vertical axis, for mixing very dry, harsh, nonplastic mixes. Such mixes are required for concrete masonry or heavy-duty floor toppings. Dry-batch mixers are used for dry shotcrete (sprayed concrete), where water and the dry-mixed cement and aggregate are blended between the nozzle of the gun and impact at the point of placing.

("Recommended Practice for Measuring, Mixing, and Placing Concrete," ACI 614.)

5-8. Admixtures. These are materials other than portland cement, aggregate, or water that are added to concrete to modify its properties.

Air Entrainment. Air-entraining admixtures (ASTM C260) may be interground as additives with the cement at the mill or added separately at the concrete mixing plant, or both. Where quality control is provided, it is preferable to add such admixtures at the concrete plant so that the resulting air content can be controlled for changes in temperature, sand, or job requirements.

Use of entrained air is recommended for all concrete exposed to weathering or deterioration from aggressive chemicals. The ACI Building Code requires air entrainment for all concrete subject to freezing temperatures while wet. Detailed recommendations for air content are available in "Recommended Practice for Selecting Proportions for Normal Weight Concrete," ACI 211.1, and "Recommended Practice for Selecting Proportions for Structural Lightweight Concrete," ACI 211.2.

One common misconception relative to air entrainment is the fear that it has a deleterious effect on concrete strength. Air entrainment, however, improves workability. This will usually permit some reduction in water content. For lean, low-strength mixes, the improved workability permits a relatively large reduction in water content, sand content, and water-cement ratio, which tends to increase concrete strength. The resulting strength gain offsets the strength-reducing effect of the air itself, and a net increase in concrete strength is achieved. For rich, high-strength mixes, the relative reduction in the water-cement ratio is lower and a small net decrease in strength results, about on the same order of the air content (4 to 7%). The improved durability and reduction of segregation in handling, due to the entrained air, usually make air entrainment desirable, however, in all concrete except extremely high-strength mixtures, such as for lower-story interior columns or heavy-duty interior floor toppings for industrial wear.

Accelerators. Calcium chloride for accelerating the rate of strength gain in concrete (ASTM D98) is perhaps the oldest application of admixtures. Old specifications for winter concreting or masonry work commonly required use of a maximum of 1 to 3% $CaCl_2$ by weight of cement for all concrete. Proprietary admixtures now available may include accelerators, but not necessarily $CaCl_2$. The usual objective for use of an accelerator is to reduce curing time by developing 28-day strengths in 7 days approximately (ASTM C494).

In spite of users' familiarity with $CaCl_2$, a number of misconceptions about its effect persist. It has been sold (sometimes under proprietary names) as an accelerator, a cement replacement, an "antifreeze," a "waterproofer," and a "hardener." It is simply an accelerator; any improvement in other respects is pure serendipity. Recent discoveries, however, indicate possible corrosion damage from indiscriminate use of chloride-containing material in concrete exposed to stray currents, containing dissimilar metals, containing prestressing steel subject to stress corrosion, or exposed to severe wet freezing or salt water. For further information, see "Admixtures for Concrete," by ACI Committee 212.

Retarders. Unless proper precautions are taken, hot-weather concreting may cause "flash set," plastic shrinkage, "cold joints," or strength loss. Admixtures that provide controlled delay in the set of a concrete mix without reducing the rate of strength gain during subsequent curing offer an inexpensive solution to many hot-weather concreting problems. These (proprietary) admixtures are usually combined with water-reducing admixtures that more than offset the loss in curing time due to delayed set (ASTM C494). See "Recommended Practice for Hot Weather Concreting," ACI 305, for further details on retarders, methods of cooling concrete materials, and limiting temperatures for hot-weather concreting.

Waterproofing. A number of substances, such as stearates and oils, have been employed as masonry-mortar and concrete admixtures for "waterproofing." Indiscriminate use of such materials in concrete without extremely good quality control usually results in disappointment. The various water-repellent admixtures are intended to prevent capillarity, but most severe leakage in concrete occurs at honeycombs, cold joints, cracks, and other noncapillary defects. Concrete containing water-repellent admixtures also requires extremely careful continuous curing, since it will be difficult to rewet after initial drying.

Waterproof concrete can be achieved by use of high-strength concrete with a low water-cement ratio, air entrained to reduce segregation, designed to minimize crack width, and with good quality control and inspection during the mixing, placing, and curing operations. Surface coatings can be used to improve resistance to water penetration of vertical or horizontal surfaces. For detailed information on surface treatments, see "Durability of Concrete in Service," ACI 201.

Cement Replacement. The term "cement replacement" is frequently misused in reference to chemical admixtures intended as accelerators or water reducers. Strictly, a cement replacement is a finely ground material, usually weakly cementitious (pozzolanic), which combines into a cementlike paste replacing some of the cement paste to fill voids between the aggregates. The most common applications of these admixtures are for low heat, low strength, mass concrete, or concrete masonry. In the former, they fill voids and reduce the heat of hydration; in the latter, they fill voids and help to develop the proper consistency to be self-standing as the machine head is lifted in the forming process. Materials commonly used are fly ash (Fig. 5-3), hydraulic lime, natural cement, and pozzolans. Fly ash and other pozzolans used as admixtures should conform to ASTM C618, "Specifications for Fly Ash and Raw or Calcined Natural Pozzolans for Use in Portland Cement Concrete."

Special-purpose Admixtures. The list of materials employed from earliest times as admixtures for various purposes includes almost everything from human blood to synthetic coloring agents.

Admixtures for coloring concrete are available in all colors. The oldest and cheapest is perhaps carbon black.

Admixtures causing expansion for use in sealing cracks or under machine bases, etc., include powdered aluminum and finely ground iron.

Special admixtures are available for use where the natural aggregate is alkali reactive, to neutralize this reaction.

Proprietary admixtures are available that increase the tensile strength or bond strength of concrete. They are useful for making repairs to concrete surfaces.

For special problems requiring concrete with unusual properties, detailed recommendations of "Admixtures for Concrete," by ACI Committee 212, and references it contains may be helpful.

For all of these special purposes, a thorough investigation of admixtures proposed is recommended. Tests should be made on samples containing various proportions for colored concrete. Strength and durability tests should be made on concrete to be exposed to sunlight, freezing, salt, or any other job condition expected; and special tests should be made for any special properties required, as a minimum precaution.

QUALITY CONTROL

5-9. Mix Design. Concrete mixes are designed with the aid of information obtained from trial batches or field experience with the materials to be used. In either case, the proportions of ingredients must be selected to produce, for at least three test specimens, an average strength f_{cr} greater than the specified strength f'_c.

The required excess, $f_{cr} - f'_c$, depends on the standard deviation expected σ. Strength data for determining the standard deviation can be considered suitable if they represent either a group of at least 30 consecutive tests representing materials and conditions of control similar to those expected or the statistical average for two groups totaling 30 or more tests. The tests used to establish standard deviation should represent concrete produced to meet a specified strength within 1,000 psi of that specified for the work proposed.

$$\sigma = \sqrt{\frac{(x_1 - \bar{x})^2 + (x_2 - \bar{x})^2 + (x_3 - \bar{x})^2 + \cdots + (x_n - \bar{x})^2}{n}} \tag{5-1}$$

where $x_1, x_2, \ldots, x_n$ = strength, psi, obtained in test of first, second, . . . , nth sample, respectively
n = number of tests
$\bar{x}$ = average strength, psi, of n cylinders.

Coefficient of variation is the standard deviation expressed as a percentage of the average strength. ("Recommended Practice for Evaluation of Compression Test Results of Field Concrete," ACI 214.)

The strength used as a basis for selecting proportions of a mix should exceed the required f'_c by at least the amount indicated in Table 5-1.

Table 5-1. Recommended Average Strengths of Test Cylinders for Selecting Proportions for Concrete Mixes

Range of standard deviation σ, psi	Average strength f_{cr}, psi
Under 300	$f'_c + 400$
300–400	$f'_c + 550$
400–500	$f'_c + 700$
500–600	$f'_c + 900$
Over 600	$f'_c + 1{,}200$

The values for f_{cr} in Table 5-1 are the larger of the values calculated from Eqs. (5-2) and (5-3).

$$f_{cr} = f'_c + 1.343\sigma \tag{5-2}$$
$$f_{cr} = f'_c + 2.326\sigma - 500 \tag{5-3}$$

For an established supplier of concrete, it is very important to be able to document the value of σ. This value is based on a statistical analysis in which Eq. (5-1) is applied to at least 30 consecutive tests. These tests must represent similar materials and conditions of control not stricter than those to be applied to the proposed project. The lower the value of σ obtained from the tests, the closer the average strength is permitted to be to the specified strength. A supplier is thus furnished an economic incentive, lower cement content, to develop a record of good control (low σ). A supplier who does maintain such a record can, in addition, avoid the expense of trial batches.

Where no such production record exists, for example, when new sources of cement or aggregate are supplied to an established plant, to a new facility, such as a portable plant on the site, or for the first attempt at a specified strength f'_c more than 1,000 psi above previous specified strengths, trial batches must be used as a basis for selecting initial proportions.

If the specified $f'_c \leq 5{,}000$ psi for non-air-entrained concrete or $f'_c \leq 4{,}500$ psi for air-entrained normal-weight concrete, trial batches need not be employed. Table 5-2 lists maximum water-cement ratios that may be used to select *initial* proportions for normal-weight concrete. These ratios are expected to provide $f_{cr} - f'_c \geq 1{,}200$ psi and are economical only for small projects where the cost of trial batches is not justified.

Note that the term "initial proportions" includes cement. The initial proportions can be used during progress of a project only as long as the strength-test results justify them. The process of quality control of concrete for a project requires maintenance of a running average of strength-test results and changes in the proportions whenever the actual degree of control (standard deviation σ) varies from that assumed for the initial proportioning. Equations (5-2) and (5-3) are applied for this analysis. With contract specifications based on the 1971 ACI Building Code, no minimum cement content is required; so good control during a long-time project is rewarded by permission to use a lower cement content than would be permitted with inferior control.

Regardless of the method (field experience, trial batches, or maximum water-cement ratio) used, the basic initial proportions must be based on mixes with both air content and slump at the maximum permitted by the specifications.

Table 5-2. Maximum Water-cement Ratios for Specified Concrete Strengths*

Specified compressive strength f'_c, psi†	Non-air-entrained concrete		Air-entrained concrete	
	Absolute ratio by weight	U.S. gal per 94-lb bag of cement	Absolute ratio by weight	U.S. gal per 94-lb bag of cement
2,500	0.65	7.3	0.54	6.1
3,000	0.58	6.6	0.46	5.2
3,500	0.51	5.8	0.40	4.5
4,000	0.44	5.0	0.35	4.0
4,500	0.38	4.3	0.30	3.4
5,000	0.31	3.5		

*As specified by ACI 318-71, Table 4.2.4.

†28-Day strengths for cements meeting strength limits of ASTM C150 Types I, IA, II, or IIA, and 7-day strengths for Types III or IIIA. With most materials, the water-cement ratios shown will provide average strengths greater than indicated.

Table 5-3. Required Air-entrainment in Concrete Exposed to Freezing and Thawing While Wet*

Nominal maximum size of coarse aggregate, in.	Total air content, vol %
3/8	6–10
1/2	5–9
3/4	4–8
1	3.5–6.5
1 1/2	3–6
2	2.5–5.5
3	1.5–4.5

*From ACI 318-71, Table 4.2.5.

Other ACI Building Code requirements for mix design are:

1. Concrete exposed to freezing and thawing while wet shall have air entrained within the limits in Table 5-3, and the water-cement ratio by weight should not exceed 0.53. If lightweight aggregate is used, f'_c shall be at least 3,000 psi.

2. For watertight, normal-weight concrete, maximum water-cement ratios by weight are 0.48 for fresh water and 0.44 for seawater. With lightweight aggregate, minimum f'_c = 3,750 psi for concrete exposed to fresh water and f'_c = 4,000 psi for seawater.

Although the Code does not distinguish between a "concrete production facility" with in-house control and an independent concrete laboratory control service, the distinction is important. Very large suppliers have in-house professional quality control. Most smaller suppliers do not. Where the records of one of the latter might indicate a large standard deviation, but an independent quality-control service is utilized, the standard deviation used to select $f_{cr} - f'_c$ should be based on the proven record of the control agency. Ideally, the overdesign should be based, in these cases, on the record of the control agency operating in the concrete plant used.

5-10. Check Tests of Materials. Without follow-up field control, all the statistical theory involved in mix proportioning becomes an academic exercise.

The complete description of initial proportions should include: cement analysis and source; specific gravity, absorption, proportions of each standard sieve size; fineness modulus; and organic tests for fine and coarse aggregates used, as well as their weights and maximum nominal sizes.

If the source of any aggregate is changed, new trial batches should be made. A cement analysis should be obtained for each new shipment of cement.

The aggregate gradings and organic content should be checked at least daily, or for each 150 cu yd. The moisture content (or slump) should be checked continuously for all aggregates, and suitable adjustments should be made in batch weights. When the limits of ASTM C33 or C330 for grading or organic content are exceeded, proper materials should be secured and new mix proportions developed, or until these measures can be effected, concrete production may continue on an emergency basis but with a penalty of additional cement.

5-11. At the Mixing Plant—Yield Adjustments. Well-equipped concrete producers have continuous measuring devices to record changes in moisture carried in the aggregates or changes in total free water in the contents of the mixer. The same measurements, however, may be easily made manually by quality-control personnel.

To illustrate: for the example in Art. 5-4, the surface-dry basic mix is cement, 564 lb; water, 300 lb; sand, 1,170 lb; and gravel, 2,000 lb. Absorption is 1% for the sand and 0.5% for the gravel. If the sand carries 5.5% and the gravel 1.0% total water by weight, the added free water becomes:

Sand: 1,170 (0.055 − 0.01) = 53 lb
Gravel: 2,000 (0.010 − 0.005) = 10 lb

Batch weights adjusted for yield become:

Cement: 564 lb
Water: 300 − 53 − 10 = 237 lb
Sand: 1,170 + 53 = 1,223 lb
Gravel: 2,000 + 10 = 2,010 lb

Note that the corrective adjustment includes adding to aggregate weights as well as deducting water weight. Otherwise, the yield will be low, and slump (slightly) increased. The yield would be low by about

$$\frac{53 + 10}{2.65 \times 62.4} = 0.381 \text{ cu ft per cu yd} = 1.4\%$$

5-12. At the Placing Point—Slump Adjustments. With good quality control, no water is permitted on the mixing truck. If the slump is too low (or too high) on arrival at the site, additional cement must be added. If the slump is too low (the usual complaint), additional water can also be added. After such additions, the contents must be thoroughly mixed, 2 to 3 min. at high speed. Because placing-point adjustments are inconvenient and costly, telephone or radio communication with the supply plant is desirable so that most such adjustments may be made conveniently at the plant.

Commonly, a lesser degree of control is accepted in which the truck carries water, the driver is on the honor system not to add water without written authorization from a responsible agent at the site, and the authorization as well as the amounts added are recorded on the record (trip ticket) of batch weights.

Note: If site adjustments are made, test samples for strength-test specimens should be taken only after all site adjustments. For concrete in critical areas, such as lower-floor columns in high-rise buildings, strictest quality control is recommended.

5-13. Strength Tests. Generally, concrete quality is measured by the specified compressive strength f'_c of 6 × 12 in. cylinders after 28 days of laboratory curing. The tests made for this purpose but performed after various periods of field curing are typically specified for tests of curing adequacy. For lightweight-aggregate concretes only, the same type of laboratory-cured test specimen is tested for tensile splitting strength f_{ct} to establish design values for deflection, development of reinforcing steel, and shear. The applicable ASTM specifications for these tests are:

C31, "Making and Curing Concrete Compressive and Flexural Strength Test Specimens in the Field."

C39, "Test for Compressive Strength of Molded Concrete Cylinders."

C496, "Test for Splitting Tensile Strength of Molded Concrete Cylinders."

The specifications for standard methods and procedures of testing give general directions within which the field procedures can be adjusted to job conditions. One difficulty arises when the specimens are made in the field from samples taken at the job site. During the first 48 hr after molding, the specimens are very sensitive to damage and variations from standard laboratory curing conditions, which can significantly reduce the strength-test results. Yet, job conditions may preclude sampling, molding, and field storage on the same spot.

If the fresh-concrete sample must be transported more than about 100 ft to the point of molding cylinders, some segregation occurs. Consequently, the concrete sample should be remixed to restore its original condition. After the molds for test cylinders have been filled, if the specimens are moved, high-slump specimens segregate in the molds; low-slump specimens in the usual paper or plastic mold are often squeezed out of shape or separated into starting cracks. Such accidental damage varies with slump, temperature, time of set and molding, and degree of carelessness.

If the specimen cylinders are left on the job site, they must be protected against drying and accidental impact from construction traffic. If a workman stumbles over a specimen less than 3 days old, it should be inspected for damage. The best practice is to provide a small, insulated, dampproofed, locked box on the site in which specimens can be cast, covered, and provided with 60 to 80°F temperature and 100% humidity for 24 to 72 hours. Then, they can be transported and subjected to standard laboratory curing conditions at the testing laboratory. When transported, the cylinders should be packed and handled like fresh eggs, since loose rattling will have about an equivalent effect in starting incipient cracks.

Similarly, conditions for field-cured cylinders must be created as nearly like those of the concrete in place as possible. Also, absolute protection against impact or other damage must be provided. Because most concrete in place will be in much larger elements than a test cylinder, most of the in-place concrete will benefit more from retained heat of hydration (Fig. 5-6). This effect decreases rapidly, because the rate of heat development is greatest initially. To insure similar curing conditions, field-cured test cylinders should be stored for the first 24 hr in the field curing box with the companion cylinders for laboratory curing. After this initial curing, the field-cured cylinders should be stored near the concrete they represent and cured under the same conditions.

Exceptions to this initial curing practice arise when the elements cast are of dimensions comparable to those of the cylinders, or the elements cast are not protected from drying or low temperatures, including freezing, or test cylinders are cured *inside* the elements they represent (patented system).

These simple, seemingly overmeticulous precautions will eliminate most of the unnecessary, expensive, job-delaying controversies over low tests. Both contractor and owner are justifiably annoyed when costly later tests on hardened concrete after an even more costly job delay indicate that the original fresh-concrete test specimens were defective and not the building concrete.

Many other strength tests or tests for special qualities are occasionally employed for special purposes. Those most often encountered in concrete building construction are strength tests on drilled cores and sawed beams (ASTM C42); impact tests (e.g., Schmidt hammer); penetration tests; determination of modulus of elasticity during the standard compression test; and deflection measurements on a finished building element under load (Chapter 20, ACI 318-71). (See also "Commentary on ACI 318-71" and the "Manual of Concrete Inspection," ACI SP-2.)

5-14. Test Evaluation. On small jobs, the results of tests on concrete after the conventional 28 days of curing may be valuable only as a record. In these cases, the evaluation is limited to three options: (1) accept results, (2) remove and replace faulty concrete, or (3) conduct further tests to confirm option (1) or (2) or for limited acceptance at a lower-quality rating. The same comment can be applied to a specific element of a large project. If the element supports 28 days' additional construction above, the consequences of these decisions are expensive.

Samples sufficient for at least five strength tests of each class of concrete should be taken at least once each day, or once for each 150 cu yd of concrete or each 5,000 sq ft of surface area placed. Each strength test should be the average for two cylinders from the same sample. The strength level of the concrete can be considered satisfactory if the averages of all sets of three consecutive strength-test

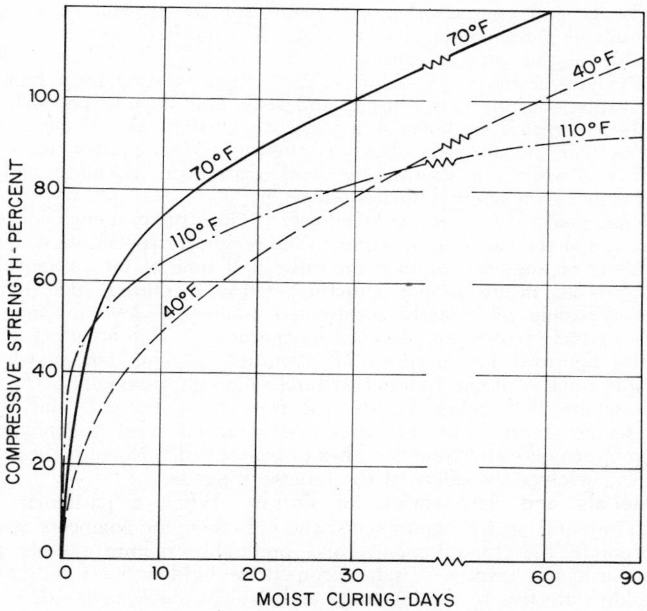

Fig. 5-6. Effect of curing temperatures on strength-gain rate of concrete, with 28-day strength as basis.

results equal or exceed the specified strength f'_c and no individual strength-test result falls below f'_c by more than 500 psi.

If individual tests of laboratory-cured specimens produce strengths more than 500 psi below f'_c, steps should be taken to assure that the load-carrying capacity of the structure is not jeopardized. Three cores should be taken for each case of a cylinder test more than 500 psi below f'_c. If the concrete in the structure will be dry under service conditions, the cores should be air-dried (temperature 60 to 80°F, relative humidity less than 60%) for 7 days before the tests and should be tested dry. If the concrete in the structure will be more than superficially wet under service conditions, the cores should be immersed in water for at least 48 hr and tested wet.

Regardless of the age on which specified design strength f'_c is based, large projects of long duration offer the opportunity for adjustment of mix proportions during the project. If a running average of test results and deviations from the average is maintained, then, with good control, the standard deviation achieved may be reduced significantly below the usually conservative, initially assumed standard deviation. In that case, a saving in cement may be realized from an adjustment corresponding to the improved standard deviation. If control is poor, the owner must be protected by an increase in cement. Specifications that rule out either adjustment are likely to result in less attention to quality control.

FORMWORK

For recommended over-all basis for specifications and procedures, see "Recommended Practice for Concrete Formwork," ACI 347. For materials, details, etc., for builders, see "Formwork for Concrete," ACI SP-4.

5-15. Responsibility for Formwork. The exact legal determination of responsibilities for formwork failures among owner, architect, engineer, general contractor, subcontractors, or suppliers can be determined only by a court decision based on the complete contractual arrangements undertaken for a specific project.

Generally accepted practice makes the following rough division of responsibilities:

Safety. The general contractor is responsible for the design, construction, and safety of formwork. Subcontractors or material suppliers may subsequently be held responsible to the general contractor. The term "safety" here includes prevention of any type of formwork failure. The damage caused by a failure always includes the expense of the form itself, and may also include personal injury or damage to the completed portions of a structure. Safety also includes protection of all personnel on the site from personal injury during construction. Only the supervisor of the work can control the workmanship in assembly and the rate of casting on which form safety ultimately depends.

Structural Adequacy of the Finished Concrete. The structural engineer is responsible for the design of the concrete structure. The reason for specification requirements that the architect or engineer approve the order and time of form removal, shoring, and reshoring is to insure proper structural behavior during such removal and to prevent overloading of recently constructed concrete below or damage to the concrete from which forms are removed prematurely. The architect or engineer should require approval for locations of construction joints not shown on plans to insure proper transfer of shear and other forces through these joints. Specifications should also require that debris be cleaned from form material and the bottom of vertical element forms, and that form-release agents used be compatible with appearance requirements and future finishes to be applied. None of these considerations, however, involves the safety of the formwork per se.

5-16. Materials and Accessories for Forms. When a particular design or desired finish imposes special requirements, and only then, the engineers' specifications should incorporate these requirements and preferably require sample panels for approval of finish and texture. Under competitive bidding, best bids are secured when the bidders are free to use ingenuity and their available materials ("Formwork for Concrete," ACI SP-4).

5-17. Pressure of Fresh Concrete on Vertical Forms. This may be estimated from

$$p = 150 + 9{,}000\,\frac{R}{T} \tag{5-4}$$

where p = lateral pressure, psf
R = rate of filling, ft per hr
T = temperature of concrete, °F

See Fig. 5-7*a*.

For columns, the maximum pressure $p_{\max}$ is 3,000 psf or $150h$, whichever is less, where h = height, ft, of fresh concrete above the point of pressure. For walls where R does not exceed 7 ft per hr, $p_{\max}$ = 2,000 psf or $150h$, whichever is less.

For walls with rate of placement $R > 7$,

$$p = 150 + \frac{43{,}400}{T} + 2{,}800\,\frac{R}{T} \tag{5-5}$$

where $p_{\max}$ = 2,000 psf or $150h$, whichever is less. See Fig. 5-7*b*.

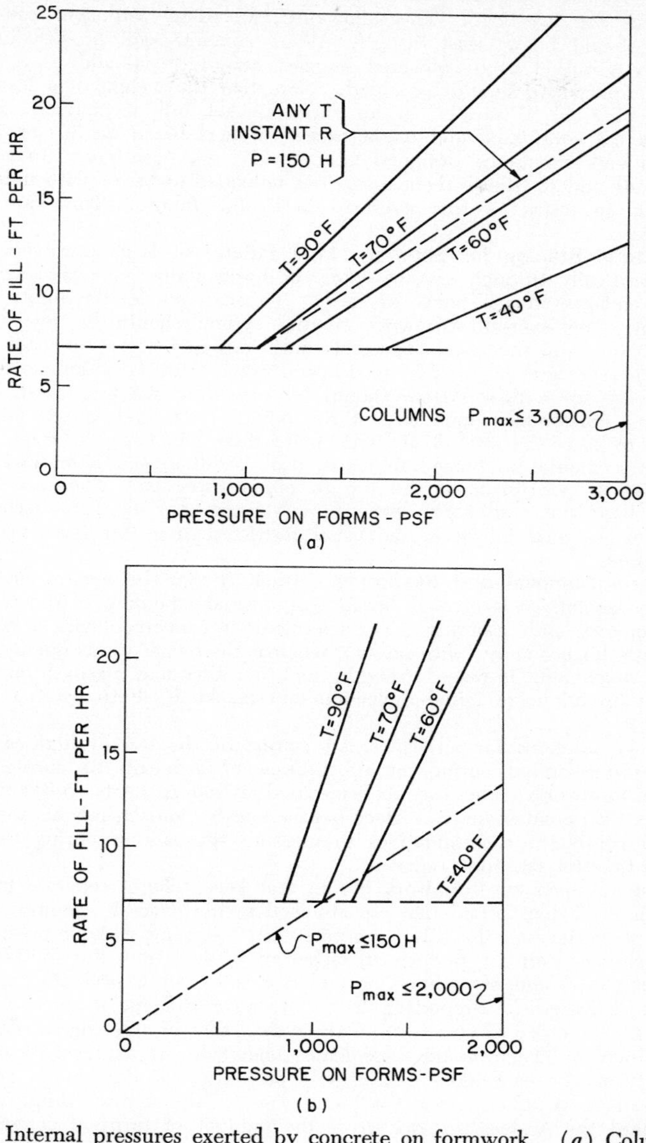

Fig. 5-7. Internal pressures exerted by concrete on formwork. (*a*) Column forms. (*b*) Wall forms.

The calculated form pressures should be increased if concrete weight exceeds 150 psf, cements are used that are slower setting than standard portland cement, slump is more than 4 in. (not recommended), retarders are used to slow set, or the concrete is revibrated full depth, or forms are externally vibrated. Under these conditions, a safe design assumes that the concrete is a fluid with weight w and $p_{max} = wh$ for the full height of placement.

5-18. Design Loads for Horizontal Forms. Best practice is to consider all known vertical loads, including the formwork itself, plus concrete, and to add an allowance for live load. This allowance, including workmen, runways, and equipment, should be at least 50 psf. When concrete will be distributed from overhead by a bucket or by powered buggies, an additional allowance of at least 25 psf for impact load should be added. Note that the weight of a loaded power buggy dropping off a runway, or an entire bucket full of concrete dropped at one spot, is not considered and might exceed designs based on 50- or 75-psf live load. Formwork should be designed alternatively, with continuity, to accept such spot overloads and distribute them to various unloaded areas, or with independently braced units to restrict a spot overload to a spot failure. The first alternative is preferable.

5-19. Lateral Bracing for Shoring. Most failures of large formwork are "progressive," vertically through several floors, or horizontally, as each successive line of shoring collapses like a house of cards. To eliminate all possibility of a large costly failure, the over-all formwork shoring system should be reviewed before construction to avoid the usual "house-of-cards" design for vertical loads only. Although it is not always possible to foresee exact sources or amounts of lateral forces, shoring for a floor system should be braced to resist at least 100 lb per lin ft acting horizontally upon any of the edges, or a total lateral force on any edge equal to 2% of the total dead loads on the floor, whichever is larger.

Wall forms should be braced to resist local building-code wind pressures, plus at least 100 lb per lin ft at the top in either direction. The recommendation applies to basement wall forms even though wind may be less, because of the high risk of personal injury in the usual restricted areas for form watchers and other workmen.

5-20. Form Removal and Reshoring. Much friction between contractors' and owners' representatives is created because of misunderstanding of the requirements for form removal and reshoring. The contractor is concerned with a fast turnover of form reuse for economy (with safety), whereas the owner wants quality, continued curing for maximum in-place strength, and an adequate strength and modulus of elasticity to minimize initial deflection and cracking. Both want a satisfactory surface.

Satisfactory solutions for all concerned consist of the use of high-early-strength concrete or accelerated curing, or substitution of a means of curing protection other than formwork. The use of field-cured cylinders (Arts. 5-6 and 5-13) in conjunction with nondestructive surface hardness tests (impact-hammer tests) enables owner and contractor representatives to measure the rate of curing to determine the earliest time for safe form removal.

Reshoring or ingenious formwork design that keeps shores separate from surface forms, such as "flying forms" that are attached to the concrete columns (Strickland Shoring System, Jacksonville, Fla.), permits early stripping without premature stress on the concrete. Properly performed, reshoring is ideal from the contractors' viewpoint. But the design of reshores several stories in depth becomes very complex. The loads delivered to supporting floors are very difficult to predict and often require a higher order of structural analysis than that of the original design of the finished structure. To evaluate these loads, knowledge is required of the modulus of elasticity E_c of each floor (different), properties of the shores (complicated in some systems by splices), and the initial stress in the shores, which is dependent on how hard the wedges are driven or the number of turns of screw jacks, etc. ("Formwork for Concrete," ACI SP-4). When stay-in-place shores are used, reshoring is simpler (because variations in initial stress, which depend on workmanship, are eliminated), and a vertically progressive failure can be averted.

One indirect measure is to read deflections of successive floors at each stage. With accurate measurements of E_c, load per floor can then be estimated by structural theory. A more direct measure (seldom used) is strain measurement on the shores, usable with metal shores only. On large important projects, where formwork cost and cost of failure justify such expense, both types of measurement can be employed.

5-21. Special Forms. Special formwork may be required for uncommon structures, such as folded plates, shells, arches, and posttensioned-in-place designs, or for special methods of construction, such as slip forming with the form rising on the finished concrete or with the finished concrete descending as excavation progresses, permanent forms of any type, preplaced-grouted-aggregate concreting, underwater concreting, and combinations of precast and cast-in-place concreting.

5-22. Inspection of Formwork. Inspection of formwork for a building is a service usually performed by the architect, engineer, or both, for the owner and, occasionally, directly by employees of the owner. Formwork should be inspected before the reinforcing steel is in place to insure that the dimensions and location of the concrete conform to design drawings (Art. 5-15). This inspection would, however, be negligent if deficiencies in the areas of contractor responsibility were not noted also.

(See "Recommended Practice for Concrete Formwork," ACI 347, and "Formwork for Concrete," ACI SP-4, for construction check lists, and "Manual of Concrete Inspection," ACI SP-2.)

REINFORCEMENT

5-23. Reinforcing Bars. The term *deformed steel bars for concrete reinforcement* is commonly shortened to *rebars*. The short form will be used in this section.

Standard United States Grade 60 rebars are available in 11 standard sizes, designated on design drawings and specifications by a size number (Table 5-4).

Table 5-4. Standard United States Rebar Sizes*

Bar size no.	Weight, lb per ft	Nominal diameter, in.	Cross-sectional area, sq in.	Nominal perimeter, in.
3	0.376	0.375	0.11	1.178
4	0.668	0.500	0.20	1.571
5	1.043	0.625	0.31	1.963
6	1.502	0.750	0.44	2.356
7	2.044	0.875	0.60	2.749
8	2.670	1.000	0.79	3.142
9	3.400	1.128	1.00	3.544
10	4.303	1.270	1.27	3.990
11	5.313	1.410	1.56	4.430
14	7.65	1.693	2.25	5.32
18	13.60	2.257	4.00	7.09

*Courtesy Concrete Reinforcing Steel Institute.

Standard billet-steel rebars are produced in three strength grades: 40, 60, and 75. Axle-steel rebars come in Grades 40 and 60, and rail-steel rebars in Grades 50 and 60. The grade number indicates yield strength, ksi.

The standard grade recommended in the CRSI "Manual of Standard Practice" for all building construction is Grade 60. It is also the standard grade on which the 1971 ACI Building Code requirements are based. Most of the United States rebar production is billet steel Grade 60. ASTM specifications for Grade 40 rebars provide for nine bar sizes, Nos. 3 through 11; and for Grade 75, three sizes, Nos. 11, 14, and 18.

Bending recommendations for Grade 60 rebars conforming to both ACI Building Code and industry practice are given in Table 5-5.

Standard rebars are not produced to any weldability requirements. The user intending any welding should follow procedures recommended by the American Welding Society (AWS D12.1), which are required by the ACI Building Code.

The standard specifications for rebars in the United States are ASTM A615 (billet steel), A616 (rail steel), and A617 (axle steel). Use the latest issue of these specifications, because properties involving yield strength and bendability are upgraded periodically.

5-24. Welded-wire Fabric (WWF). Welded-wire fabric is a grid made with two kinds of cold-drawn wire: smooth (or plain) or deformed. The wires can be spaced in each direction of the grid as desired, but for buildings, usually at

Table 5-5. Standard Hooks*

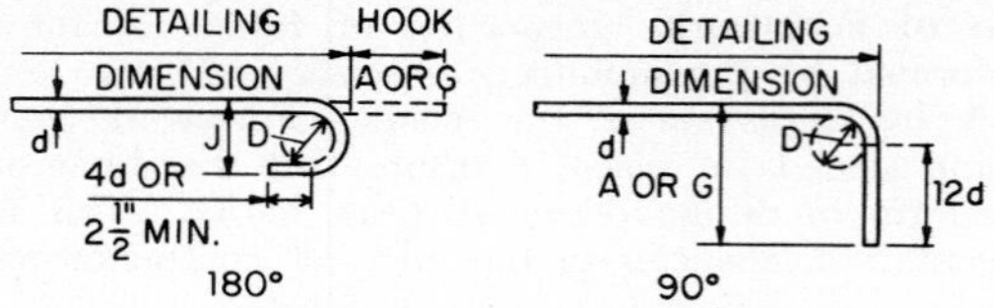

ACI Standard Hooks—All Grades

Bar size no.	*D*, in.	180° hooks		90° hooks
		A or *G*	*J*	*A* or *G*
3	2¼	5 in.	3 in.	6 in.
4	3	6 in.	4 in.	8 in.
5	3¾	7 in.	5 in.	10 in.
6	4½	8 in.	6 in.	1 ft 0 in.
7	5¼	10 in.	7 in.	1 ft 2 in.
8	6	11 in.	8 in.	1 ft 4 in.
9	9	1 ft 3 in.	11¼ in.	1 ft 7 in.
10	10	1 ft 5 in.	1 ft 0¾ in.	1 ft 10 in.
11	11¼	1 ft 7 in.	1 ft 2¼ in.	2 ft 0 in.
14	17	2 ft 2 in.	1 ft 8½ in.	2 ft 7 in.
18	22½	2 ft 11 in.	2 ft 3 in.	3 ft 5 in.

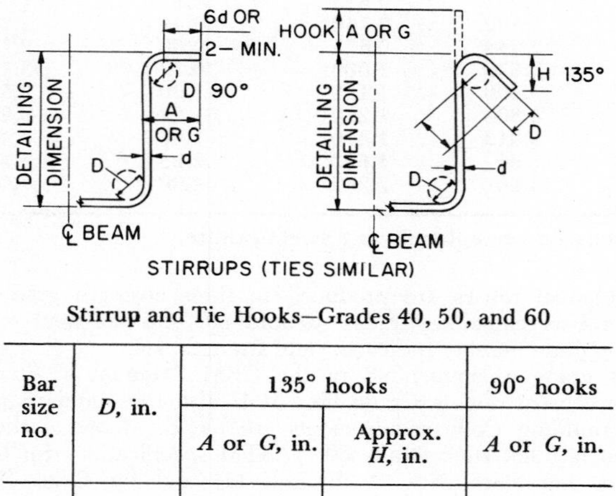

Stirrup and Tie Hooks—Grades 40, 50, and 60

Bar size no.	*D*, in.	135° hooks		90° hooks
		A or *G*, in.	Approx. *H*, in.	*A* or *G*, in.
3	1½	4	2½	4
4	2	4½	3	4½
5	2½	5½	3¾	6

*Courtesy Concrete Reinforcing Steel Institute.

12 in. maximum. Sizes of wires available in each type, with standard and former designations, are shown in Table 5-6.

Welded-wire fabric usually is designated WWF on drawings. Sizes of WWF are designated by spacing followed by wire sizes; for example, WWF 6 × 12, W11/W7, which indicates smooth wires, size W11, spaced at 6 in., and size W7, spaced at 12 in. WWF 6 × 12, D-11/D-7 indicates deformed wires of the same nominal size and spacing.

All WWF can be designed for Grade 60 material. Some of the smaller sizes are required to develop higher yield strengths measured at 0.2% offset, but limitations in the ACI Building Code on strain restrict general use to yield strength of 60 ksi.

Table 5-6. Standard Wire Sizes for Reinforcement

Deformed wire (A496) size	Plain wire (A82) size	Nominal dia, in.	Nominal area, sq in.	AS & W Steel-wire gages			
				Dia, in.	AS & W gage	Dia, in.	Area, sq in.
D-31	W31	0.628	0.310	1/2	. . .	0.5000	0.1964
D-30	W30	0.618	0.300	. . .	7/0	0.4900	0.1886
D-29	. . .	0.608	0.290	15/32	. . .	0.4688	0.1726
D-28	W28	0.597	0.280	. . .	6/0	0.4615	0.1673
D-27	. . .	0.586	0.270	7/16	. . .	0.4375	0.1503
D-26	W26	0.575	0.260	. . .	5/0	0.4305	0.1456
D-25	. . .	0.564	0.250	13/32	. . .	0.4062	0.1296
D-24	W24	0.553	0.240	. . .	4/0	0.3938	0.1218
D-23	. . .	0.541	0.230	3/8	. . .	0.3750	0.1104
D-22	W22	0.529	0.220	. . .	3/0	0.3625	0.1032
D-21	. . .	0.517	0.210	11/32	. . .	0.3438	0.09281
D-20	W20	0.504	0.200	. . .	2/0	0.3310	0.08605
D-19	. . .	0.491	0.190	5/16	. . .	0.3125	0.07670
D-18	W18	0.478	0.180	. . .	0	0.3065	0.07378
D-17	. . .	0.465	0.170	. . .	1	0.2830	0.06290
D-16	W16	0.451	0.160	9/32	. . .	0.2812	0.06213
D-15	. . .	0.437	0.150	. . .	2	0.2625	0.05412
D-14	W14	0.422	0.140	1/4	. . .	0.2500	0.04909
D-13	. . .	0.406	0.130	. . .	3	0.2437	0.04664
D-12	W12	0.390	0.120	. . .	4	0.2253	0.03987
D-11	. . .	0.374	0.110	7/32	. . .	0.2188	0.03758
D-10	W10	0.356	0.100	. . .	5	0.2070	0.03365
D-9	. . .	0.338	0.090	. . .	6	0.1920	0.02895
D-8	W8	0.319	0.080	3/16	. . .	0.1875	0.02761
D-7	W7	0.298	0.070	. . .	7	0.1770	0.02461
D-6	W6	0.276	0.060	. . .	8	0.1620	0.02061
. . .	W5.5	0.264	0.055	5/32	. . .	0.1562	0.01918
D-5	W5	0.252	0.050	. . .	9	0.1483	0.01727
. . .	W4.5	0.240	0.045	. . .	10	0.1350	0.01431
D-4	W4	0.225	0.040	1/8	. . .	0.1250	0.01227
. . .	W3.5	0.211	0.035	. . .	11	0.1205	0.01140
D-3	W3	0.195	0.030	. . .	12	0.1055	0.008715
. . .	W2.5	0.178	0.025	3/32	. . .	0.0938	0.006903
D-2	W2	0.159	0.020	. . .	13	0.0915	0.006576
. . .	W1.5	0.138	0.015	. . .	14	0.0800	0.005027
D-1	W1	0.113	0.010	. . .	15	0.0720	0.004072
. . .	W0.5	0.080	0.005	. . .	16	0.0625	0.003068

The standard specifications for wire in the United States are:
ASTM A82, Plain Wire
ASTM A496, Deformed Wire
ASTM A185, Plain Wire, WWF
ASTM A497, Deformed Wire, WWF

5-25. Prestressing Steel. Cold-drawn high-strength wires, singly or stranded, with ultimate tensile strengths up to 270 ksi, and high-strength, alloy-steel bars with ultimate tensile strengths up to 160 ksi, are used in prestressing. The applicable specifications for wire and strands are:
ASTM A416, Uncoated Seven-Wire Stress-relieved Strand.
ASTM A421, Uncoated Stress-Relieved Wire.
The ACI Building Code also permits use of other types of strand or wire, including galvanized wire or strand, that meet the minimum requirements of ASTM A416 and A421.

Single wires or strands are used for plant-made pretensioned, prestressed members. Posttensioned prestressing may be performed with the member in place, on a site fabricating area, or in a plant. In posttensioned applications, single wires are usually grouped into parallel-wire tendons. Posttensioned tendons may also consist of solid bars or strands.

High-strength, alloy-steel bars for posttensioning must be proof-stressed during manufacture to 85% of the minimum guaranteed tensile strength, and subsequently

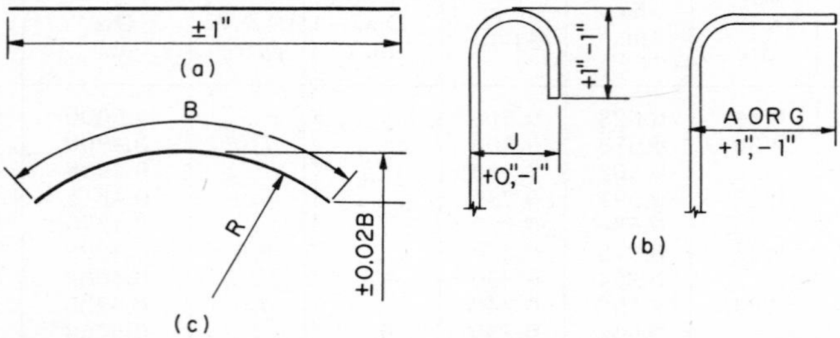

Fig. 5-8. Tolerance in fabrication of bar sizes Nos. 3 to 11. (*a*) Cutting tolerances. (*b*) Tolerances for standard hooks. (*c*) Radial bending tolerances.

given stress-relieving heat treatment. Final physical properties tested on full sections must conform to the following minimum requirements:

Yield strength (0.2% offset): 0.85 f_{pu} (specified ultimate).

Elongation in 20 diameters at rupture: 4%.

Reduction of area at rupture: 20%.

5-26. Fabrication and Placing of Rebars. Fabrication of rebars consists of cutting to length and required bending. The preparation of shop drawings and (field) placing drawings is termed *detailing*. Ordinarily, the rebar supplier details, fabricates, and delivers to the site, as required. In the far-western states, the rebar supplier also ordinarily places the bars. In the New York City area, fabrication is performed on the site by the same (union) workmen who place the reinforcement. ("Recommended Practice for Detailing Reinforced Concrete Structures," ACI 315.)

Fabrication Tolerances. The generally accepted industry tolerances on fabrication are the same regardless of varying contractual arrangements. For tolerances and shipping limits, see Figs. 5-8 through 5-10. ("Manual of Standard Practice," CRSI.)

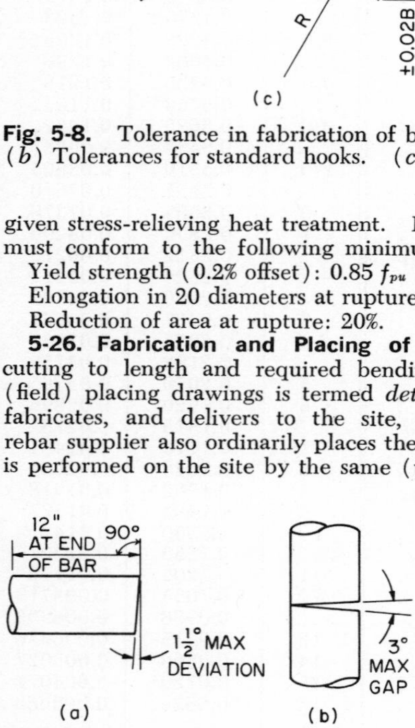

Fig. 5-9. Special fabrication tolerances. (*a*) End deviation at a cut. (*b*) Maximum gap at erected end-bearing splices.

Erection. For construction on small sites, such as high-rise buildings in metropolitan areas, delivery of materials is a major problem. Reinforcement required for each area to be concreted at one time is usually delivered separately. Usually, the only available space for storage of this steel is the formwork in place. Under such conditions, unloading time becomes important.

The bars for each detail length, bar size, or mark number are wired into *bundles* for delivery. A *lift* may consist of one or more bundles grouped together for loading or unloading. The maximum weight of a single lift for unloading is set by the job crane capacity. The maximum weight of a *shop lift* for loading is usually far larger, and so shop lifts may consist of several separately bundled *field lifts*. If the field equipment limits a field lift to less than 3,000 lb, the usual minimum, special arrangements for building are required. Where site storage is provided, the most economical unloading without an immediately available crane is by dumping or rolling bundles off the side. Unloading arrangements should be agreed on in advance, so that loading can be carried out in the proper order and bars bundled appropriately. ("Placing Reinforcing Bars," CRSI.)

Placement Tolerances. The ACI Building Code prescribes rebar placement tolerances applicable simultaneously to effective depth d and to concrete cover in all flexural members, walls, and columns as follows:

Where d is 8 in. or less, ±¼ in.; more than 8 in. but less than 24 in., ±⅜ in.; and 24 in. or more, ±½ in. These tolerances may not reduce cover more than one-third of that specified. The tolerance on longitudinal position of bends or ends of bars is ±2 in. At discontinuous ends, the specified cover cannot be reduced more than ½ in.

Bundling. Rebars may be placed in concrete members singly or in bundles (up to four No. 11 or smaller bars per bundle). This practice reduces rebar congestion or the need for several layers of single, parallel bars in girders. For columns, it eliminates many interior ties and permits use of No. 11 or smaller bars where small quantities of No. 14 or 18 are not readily available.

Only straight bars should be bundled ordinarily. Exceptions are bars with end hooks, usually at staggered locations, so that the bars are not bent as a bundle ("Placing Reinforcing Bars," CRSI).

A bundle is assembled by wiring the separate bars tightly in contact. If they are preassembled, placement in forms of long bundles requires a crane. Because cutoffs or splices of bars within a bundle must be staggered, it will often be necessary to form the bundle in place.

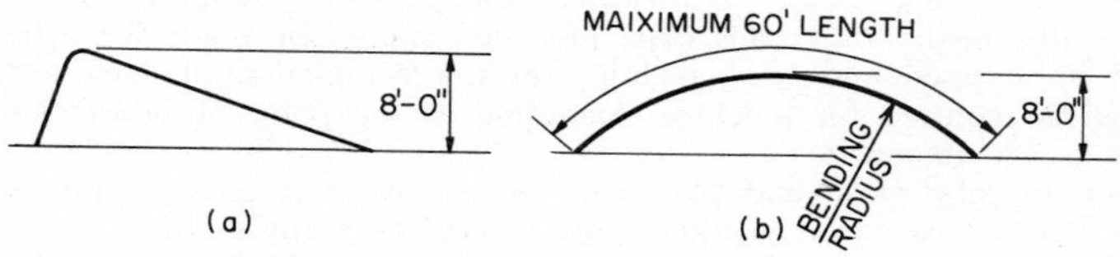

Fig. 5-10. Shipping limitations. (*a*) Height limit. (*b*) Length and height limits.

Bending and Welding Limitations. The ACI Code contains the following restrictions:

All bars must be bent without heating, except as permitted by the engineer.

Bars partly embedded in concrete may not be bent without permission of the engineer.

No welding of crossing bars (tack welding) is permitted without the approval of the engineer.

For unusual bends, heating may be permitted because bars bend more easily when heated. They are also less likely to crack if bent at less than the minimum radii permitted. If not embedded in thin sections of concrete, heating the bars to temperatures between 600 and 800°F facilitates bending, usually without damage to the bars or splitting of the concrete. If partly embedded bars are to be bent, heating controlled within these limits, plus the provision of a round fulcrum for the bend to avoid a sharp kink in the bar, are essential.

Tack welding creates a *metallurgical notch effect,* seriously weakening the bars. If different size bars are tacked together, the notch effect is aggravated in the larger bar. Tack welding therefore should never be permitted at a point where bars are to be fully stressed, and never for the assembly of ties or spirals to column verticals or stirrups to main beam bars.

When large, preassembled reinforcement units are desired, the engineer can plan the tack welding necessary as a supplement to wire ties at points of low stress or to added bars not required in the design.

5-27. Bar Supports. Bar supports are commercially available in four general types of material: wire, precast concrete, molded plastic, and asbestos concrete. Complete specifications for the various types of wire and precast-concrete bar supports, as well as recommended maximum spacings and details for use, are given in the CRSI "Manual of Standard Practice," and "Wire Bar Supports," Product Standard, U.S. Department of Commerce.

Wire bar supports are generally available in the United States in four classes of rust resistance: A, plain; B, with plastic-covered feet; C, stainless or with stainless-steel wire feet; and D, special stainless; listed in the order of increasing cost. Precast-block bar supports are commercially available in the far-western states in three types: plain block, block with embedded tie wire for bottom-bar support, and block with a hole for the leg of a vertical bar for top- and bottom-bar support.

Various proprietary molded plastic bar supports are available and have been used successfully in the United States. There is, however, no generally established set of standard types or specifications for the plastic material. British standards warn the user that the coefficient of temperature expansion for the molded plastic materials used can differ from that of concrete by as much as 10 to 1. Investigation of this property is advisable before use of plastic supports in concrete that will be exposed to high variations in temperature.

5-28. Inspection of Reinforcement. This involves approval of rebar material for conformance to the physical properties required, such as ASTM specifications for the strength grade specified; approval of the bar details and shop and placing drawings; approval of fabrication to meet the approved details within the prescribed tolerances; and approval of rebar placing.

Approvals of rebar material may be made on the basis of mill tests performed by the manufacturer for each heat from which the bars used originated. If samples are to be taken for independent strength tests, measurements of deformations, bending tests, and minimum weight, the routine samples may be best secured at the mill or the fabrication shop before fabrication. Occasionally, samples for check tests are taken in the field; but in this case, provision should be made for extra lengths of bars to be shipped and for schedules for the completion of such tests before the material is required for placing. Sampling at the point of fabrication, before fabrication, is recommended.

Inspection of fabrication and placement is usually most conveniently performed in the field, where gross errors would require correction in any event.

Under the ACI Building Code, the bars should be free of oil, paint, form coatings, and mud when placed. Rust or mill scale sufficiently loose to damage the bond is normally dislodged in handling.

If heavily rusted bars (which may result from improper storage for a long time exposed to rusting conditions) are discovered at the time of placing, a quick field test of suitability requires only scales, a wire brush, and calipers. In this test, a measured length of the bar is wire-brushed manually and weighed. If less than 94% of the nominal weight remains, or if the height of the deformations is deficient, the rust is deemed excessive. In either case, the material may then be rejected or penalized as structurally inadequate. Where space permits placing additional bars to make up the structural deficiency (in bond value or weight), as in walls and slabs, this solution is preferred, because construction delay then is avoided. Where job specifications impose requirements on rust more severe than the structural requirements of the ACI Code, for example, for decorative surfaces exposed to weather, the inspection should employ the special criteria required.

CONCRETE PLACEMENT

The principles governing proper placement of concrete are:

Segregation must be avoided during all operations between the mixer and the point of placement, including final consolidation and finishing.

The concrete must be thoroughly consolidated, worked solidly around all embedded items, and should fill all angles and corners of the forms.

Where fresh concrete is placed against or on hardened concrete, a good bond must be developed.

Unconfined concrete must not be placed under water.

The temperature of fresh concrete must be controlled from the time of mixing through final placement, and protected after placement.

("Mixing and Placing," ACI 318-71 and "Commentary"; "Specifications for Structural Concrete for Buildings," ACI 301; "Recommended Practice for Concrete Floor and Slab Construction," ACI 302; "Underwater Concreting," TRCS-3, The Concrete Society, Terminal House, Grosvenor Gardens, London, SWI, WOAJ.)

5-29. Methods of Placing. Concrete may be conveyed from a mixer to point of placement by any of a variety of methods and equipment, if properly transported to avoid segregation. Selection of the most appropriate technique for economy depends on job conditions, especially job size, equipment, and the contractor's experience. In building construction, concrete usually is placed with hand- or power-operated buggies; drop-bottom buckets with a crane; inclined chutes; flexible and rigid pipe by pumping; *shotcrete,* in which either dry materials and water are sprayed separately or mixed concrete is shot against the forms; and for underwater placing, tremie chutes (closed flexible tubes). For mass-concrete construction, side-dump cars on narrow-gage track or belt conveyors may be used. For pavement, concrete may be placed by bucket from the swinging boom of a paving mixer, directly by dump truck or mixer truck, or indirectly by trucks into a spreader.

A special method of placing concrete suitable for a number of unusual conditions consists of grout-filling preplaced coarse aggregate. This method is particularly useful for underwater concreting, because grout, introduced into the aggregate through a vertical pipe gradually lifted, displaces the water, which is lighter than the grout. Because of bearing contact of the aggregate, less than usual over-all shrinkage is also achieved.

5-30. Excess Water. Even within the specified limits on slump and water-cement ratio, excess water must be avoided. In this context, excess water is present for the conditions of placing if evidence of water rise (vertical segregation) or water flow (horizontal segregation) occurs. Excess water also tends to aggravate surface defects by increased leakage through form openings. The result may be honeycomb, sandstreaks, variations in color, or soft spots at the surface.

In vertical formwork, water rise causes weak planes between each layer deposited. In addition to the deleterious structural effect, such planes, when hardened, contain voids through which water may pass.

In horizontal elements, such as floor slabs, excess water rises and forms a weak laitance layer at the top. This layer suffers from low strength, low abrasion resistance, high shrinkage, and generally poor quality.

5-31. Consolidation. The purpose of consolidation is to eliminate voids of entrapped air and to insure intimate complete contact of the concrete with the surfaces of the forms and the reinforcement. Intense vibration, however, may also reduce the volume of desirable entrained air; but this reduction can be compensated by adjustment of the mix proportions.

Powered internal vibrators are usually used to achieve consolidation. For thin slabs, however, high-quality, low-slump concrete can be effectively consolidated, without excess water, by mechanical surface vibrators. For precast elements in rigid, watertight forms, external vibration (of the form itself) is highly effective. External vibration is also effective with in-place forms, but should not be used unless the formwork is specially designed for the temporary increase in internal pressures to full fluid head plus the impact of the vibrator ("Formwork for Concrete," ACI SP-4).

Except in certain paving operations, vibration of the reinforcement should be avoided. Although it is effective, the necessary control to prevent overvibration is difficult. Also, when concrete is placed in several lifts or layers, vibration of vertical rebars passing into partly set concrete below may be harmful. Note, however, that revibration of concrete before the final set, under controlled conditions, can improve concrete strength markedly and reduce surface voids (bugholes). This technique is too difficult to control for general use on field-cast vertical elements, but it is very effective in finishing slabs with powered vibrating equipment.

Manual spading is most efficient for removal of entrapped air at form surfaces. This method is particularly effective where smooth impermeable form material is used and the surface is inward sloping.

On the usual building project, different conditions of placement are usually encountered that make it desirable to provide for various combinations of the techniques described. One precaution generally applicable is that the vibrators not be used to move the concrete laterally.

("Consolidation of Concrete," ACI 609.)

5-32. Concreting Vertical Elements. The interior of columns is usually congested; it contains a large volume of reinforcing steel compared with the volume of concrete, and has a large height compared with its cross-sectional dimensions. Therefore, though columns should be continuously cast, the concrete should be placed in 2- to 4-ft-deep increments and consolidated with internal vibrators. These should be lifted after each increment has been vibrated. If delay occurs in concrete supply before a column has been completed, every effort should be made to avoid a cold joint. When the remainder of the column is cast, the first increment should be small, and should be vibrated to penetrate the previous portion slightly.

In all columns and reinforced narrow walls, concrete placing should begin with 2 to 4 in. of grout. Otherwise, loose stone will collect at the bottom and form honeycomb. This grout should be proportioned for about the same slump as the concrete or slightly more, but at the same or lower water-cement ratio. (Some engineers prefer to start vertical placement with a mix having the same proportions of water, cement, and fine aggregate, but with one-half the quantity of coarse aggregate, as in the design mix, and to place a starting layer 6 to 12 in. deep.)

When concrete is placed for walls, the only practicable means to avoid segregation is to place no more than a 24-in. layer in one pass. Each layer should be vibrated separately and kept nearly level.

For walls deeper than 4 ft, concrete should be placed through vertical, flexible trunks or chutes located about 8 ft apart. The trunks may be flexible or rigid, and come in sections so that they can be lifted as the level of concrete in place rises. The concrete should not fall free, from the end of the trunk, more than 4 ft or segregation will occur, with the coarse aggregate ricocheting off the forms to lodge on one side. Successive layers after the initial layer should be penetrated by internal vibrators for a depth of about 4 to 6 in., to insure complete integration at the surface of each layer. Deeper penetration can be beneficial (revibration), but control under variable job conditions is too uncertain for recommendation of this practice for general use.

The results of poor placement in walls are frequently observed: sloping layer lines; honeycombs, leaking, if water is present; and, if cores are taken at successive heights, up to a 50% reduction in strength from bottom to top. Some precautions necessary to avoid these ill effects are:

Place concrete in level layers through closely spaced trunks or chutes.

Do not place concrete full depth at each placing point.

Do not move concrete laterally with vibrators.

For deep, long walls, reduce the slump for upper layers 2 to 3 in. below the slump for the starting layer.

On any delay between placing of layers, vibrate the concrete thoroughly at the interface.

If concreting must be suspended between planned horizontal construction joints, level off the layer cast, remove any laitance and excess water, and form a straight, level construction joint, if possible, with a small cleat attached to the form on the exposed face (see also Art. 5-39).

5-33. Concreting Horizontal Elements. Concrete placement in horizontal elements follows the same general principles outlined in Art. 5-32. Where the surface will be covered and protected against abrasion and weather, few special precautions are needed.

For concrete slabs, careless placing methods result in horizontal segregation, with desired properties in the wrong location, the top consisting of excess water and fines with low abrasion and weather resistance, and high shrinkage. For a good surface in a one-course slab, low-slump concrete and a minimum of vibration and finishing are desirable. Immediate screeding with a power-vibrated screed is helpful in distributing low-slump, high-quality concrete. No further finishing should be undertaken until free water, if any, disappears. A powered, rotary tamping float can smooth very-low-slump concrete at this stage. Final troweling should be delayed, if necessary, until the surface can support the weight of the finisher.

When concrete is placed for deep beams that are monolithic with a slab, the beam should be filled first. Then, a short delay for settlement should ensue before slab concrete is cast. Vibration through the top slab should penetrate the beam concrete sufficiently to insure thorough combination.

When a slab is cast, successive batches of concrete should be placed on the edge of previous batches, to maintain progressive filling without segregation. For slabs with sloping surfaces, concrete placing should usually begin at the lower edge.

For thin shells in steeply sloping areas, placing should proceed downslope. Slump should be adjusted and finishing coordinated to prevent restraint by horizontal bars from causing plastic cracking in the fresh concrete.

5-34. Bonding to Hardened Concrete. The surface of hardened concrete should be rough and clean.

Vertical surfaces of planned joints may be prepared easily by wire brushing them, before complete curing, to expose the coarse aggregate. (The timing can be extended, if desired, by using a surface retarder on the bulkhead form.) For surfaces fully cured without earlier preparation, sandblasting, bush hammering, or acid washes (thoroughly rinsed off) are effective means of preparation for bonding new concrete. (See also Art. 5-32.)

Horizontal surfaces of previously cast concrete, for example, of walls, are similarly prepared. Care should be taken to remove all laitance and to expose sound concrete and coarse aggregate. (See also Art. 5-33. For two-course floors, see Art. 5-35.)

5-35. Heavy-duty Floor Finishes. Floor surfaces highly resistant to abrasion and impact are required for many industrial and commercial uses. Such surfaces are usually built as two-course construction, with a base or structural slab topped by a wearing surface. The two courses may be cast integrally or with the heavy-duty surface applied as a separate topping.

In the first process, which is less costly, ordinary structural concrete is placed and screeded to nearly the full depth of the floor. The wearing surface concrete, made with special abrasion-resistant aggregate, emery, iron filings, etc., then is mixed, spread to the desired depth, and troweled before final set of the concrete below.

The second method requires surface preparation of the base slab, by stiff brooming before final set to roughen the surface and thorough washing before the separate heavy-duty topping is cast. For the second method, the topping is a very dry (zero-slump) concrete, made with ⅜-in. maximum-size special aggregate. This topping should be designed for a minimum strength, $f'_c = 6{,}000$ psi. It must be tamped into place with powered tampers or rotary floats. (*Note:* If test cylinders are to be made from this topping, standard methods of consolidation will not produce a proper test; tamping similar in effect to that applied to the floor itself is necessary.) One precaution vital to the separate topping method is that the temperatures of topping and base slab must be kept compatible.

("Recommended Practice for Concrete Floor and Slab Construction," ACI 302.)

5-36. Concreting in Cold Weather. Frozen materials should never be used. Concrete should not be cast on a frozen subgrade, and ice must be removed from forms before concreting. Concrete allowed to freeze wet, before or during early curing, may be seriously damaged. Furthermore, temperatures should be kept above 40°F for any appreciable curing (strength gain).

Concrete suppliers are equipped to heat materials and to deliver concrete at controlled temperatures in cold weather. These services should be utilized.

In very cold weather, for thin sections used in building, the freshly cast concrete must be enclosed and provided with temporary heat. For more massive sections or in moderately cold weather, it is usually less expensive to provide insulated forms or insulated coverings to retain the initial heat and subsequent heat of hydration generated in the concrete during initial curing.

The curing time required depends on the temperature maintained and whether regular or high-early-strength concrete is employed. High-early-strength concrete may be achieved with accelerating admixtures (Art. 5-8) or with high-early-strength cement (Types III or IIIA) or by a lower water-cement ratio, to produce the required 28-day strength in about 7 days.

An important precaution in employing heated enclosures is to supply heat without drying the concrete or releasing carbon dioxide fumes. Exposure of fresh concrete to drying or fumes results in chalky surfaces. Another precaution is to avoid rapid temperature changes of the concrete surfaces when heating is discontinued. The heat supply should be reduced gradually, and the enclosure left in place to permit cooling to ambient temperatures gradually, usually over a period of at least 24 hr.

("Recommended Practice for Cold-weather Concreting," ACI 306.)

5-37. Concreting in Hot Weather. Mixing and placing concrete at a high temperature may cause *flash set* in the mixer, during placing, or before finishing can be completed. Also, loss of strength can result from casting hot concrete.

In practice, most concrete is cast at about $70 \pm 20°F$. Research on the effects of casting temperature shows highest strengths for concrete cast at 40°F and significant but practically unimportant increasing loss of strength from 40°F to 90°F. For higher temperatures, the loss of strength becomes important. So does increased shrinkage. The increased shrinkage is attributable not only to the high temperature, but also to the increased water content required for a desired slump as temperature increases. See Fig. 5-6.

For ordinary building applications, concrete suppliers control temperatures of concrete by cooling the aggregates and, when necessary, by supplying part of the mixing water as crushed ice. In very hot weather, these precautions plus sectional casting, to permit escape of the heat of hydration, may be required for massive foundation mats. Retarding admixtures are also used with good effect to reduce slump loss during placing and finishing.

("Recommended Practice for Hot Weather Concreting," ACI 305.)

5-38. Curing Concrete. Curing of concrete consists of the processes, natural and artifically created, that affect the extent and rate of hydration of the cement.

Many concrete structures are cured without artificial protection of any kind. They are allowed to harden while exposed to sun, wind, and rain. This type of curing is unreliable, because water may evaporate from the surface.

Various means are employed to cure concrete by controlling its moisture content or its temperature. In practice, curing consists of conserving the moisture within newly placed concrete by furnishing additional moisture to replenish water lost by evaporation. Usually, little attention is paid to temperature, except in winter curing and steam curing.

More effective curing is beneficial in that it makes the concrete more watertight and increases the strength.

Methods for curing may be classified as:

1. Those that supply water throughout the early hydration process and tend to maintain a uniform temperature. These methods include ponding, sprinkling, and application of wet burlap or cotton mats, wet earth, sawdust, hay, or straw.

2. Those designed to prevent loss of water but having little influence on maintaining a uniform temperature. These methods include waterproof paper and impermeable membranes. The latter is usually a clear or bituminous compound sprayed on the concrete to fill the pores and thus prevent evaporation. A fugitive dye in the colorless compound aids the spraying and inspection.

A white pigment that gives infrared reflectance can be used in a curing compound to keep concrete surfaces cooler when exposed to the sun.

The criterion for judging the adequacy of field curing provided in the ACI Building Code is that the field-cured test cylinders produce 85% of the strengths developed by companion laboratory-cured cylinders at the age for which strength is specified.

5-39. Joints in Concrete. Several types of joints may occur or be formed in concrete structures:

Construction joints are formed when fresh concrete is placed against hardened concrete.

Expansion joints are provided in long components to relieve compressive stresses that would otherwise result from a temperature rise.

Contraction joints (*control joints*) are provided to permit concrete to contract during a drop in temperature and to permit drying shrinkage without resulting uncontrolled random cracking.

Contraction joints should be located at places where concrete is likely to crack because of temperature changes or shrinkage. The joints should be inserted where there are thickness changes and offsets. Ordinarily, joints should be spaced 30 ft c to c or less in exposed structures, such as retaining walls.

To avoid unsightly cracks due to shrinkage, a dummy-type contraction joint is frequently used (Fig. 5-11). When contraction takes place, a crack occurs at this deliberately made plane of weakness. In this way, the crack is made to occur in a straight line easily sealed.

Control joints may also consist of a 2- or 3-ft gap left in a long wall or slab, with the reinforcement from both ends lapped in the gap. Several weeks after the wall or slab was concreted, the gap is filled with concrete. By that time, most of the shrinkage has taken place. A checkerboard pattern of slab construction can be used for the same reason.

In expansion joints, a filler is usually provided to separate the two parts of the structure. This filler should be a compressible substance, such as corkboard or premolded mastic. The filler should have properties such that it will not be squeezed out of the joint, will not slump when heated by the sun, and will not stain the surface of the concrete.

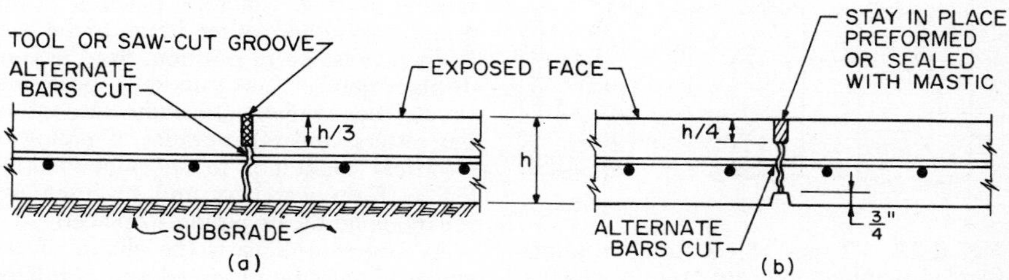

Fig. 5-11. Control joints for restraining shrinkage and temperature cracks. (*a*) Slab on grade. (*b*) Wall.

To be waterproof, a joint must be sealed. For this purpose, copper flashing may be used. It is usually embedded in the concrete on both sides of the joint, and folded into the joint so that the joint may open without rupturing the metal. The flashing must be strong enough to hold its position when the concrete is cast.

Proprietary flexible water stops and polysulphide calking compounds may also be used as sealers.

Open expansion joints are sometimes employed for interior locations where the opening is not objectionable. When exposed to water from above, as in parking decks, open joints may be provided with a gutter below to drain away water.

The engineer should show all necessary vertical and horizontal joints on design drawings. All pertinent details affecting reinforcement, water stops, and sealers should also be shown.

Construction joints should be designed and located if possible at sections of minimum shear. These sections will usually be at the center of beams and slabs, where the bending moment is highest. They should be located where it is most convenient to stop work. The construction joint is often keyed for shearing strength.

If it is not possible to concrete an entire floor in one operation, vertical joints preferably should be located in the center of a span. Horizontal joints are usually provided between columns and floor; columns are concreted first, then the entire floor system.

Various types of construction joints are shown in Fig. 5-12. The numbers on each section refer to the sequence of placing concrete.

If the joint is horizontal, as in Fig 5-12*a*, water may be trapped in the key of the joint. If the joint is vertical, the key is easily formed by nailing a wood strip to the inside of the forms. A raised key, as in Fig. 5-12*b*, makes formwork difficult for horizontal joints.

In the horizontal joint in Fig. 5-12*c*, the key is made by setting precast-concrete blocks into the concrete at intermittent intervals. The key in Fig. 5-12*d* is good if the shear acts in the directions shown.

The V-shaped key in Fig. 5-12*e* can be made manually in the wet concrete for horizontal joints.

The key is eliminated in Fig. 5-12*f*, reliance being placed on friction on the roughened surface. This method may be used if the shears are small, or if there are large compressive forces or sufficient reinforcement across the joint.

See also Arts. 5-32 to 5-34.

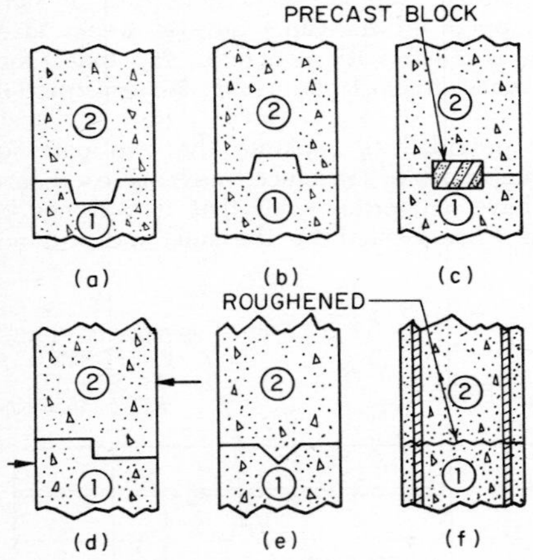

Fig. 5-12. Types of construction joints. Circled numbers indicate order of casting.

5-40. Inspection of Concrete Placement. Concrete should be inspected for the owner before, during, and after casting. Before concrete is placed, the formwork must be free of ice and debris and properly coated with bond-breaker oil. The rebars must be in place, properly supported to bear any traffic they will receive during concrete placing. Conduit, inserts, and other items to be embedded must be in position, fixed against displacement. Construction personnel should be available, usually carpenters, bar setters, and other trades, if piping or electrical conduit is to be embedded, to act as form watchers and to reset any rebars, conduit, or piping displaced.

As concrete is cast, the slump of the concrete must be observed and regulated within prescribed limits, or the specified strengths based on the expected slump may be reduced. If the inspector of placing is also responsible for sampling and making cylinders, he should test slump, entrained air, temperatures, and unit weights during concreting. He should control any field adjustment of slump and added water and cement. The inspector should also ascertain that handling, placing, and finishing procedures that have been agreed on in advance are properly followed, to avoid segregated concrete. In addition, he should insure that any emergency construction joints made necessary by stoppage of concrete supply, rain or other delays are properly located and made in accordance with procedures specified or approved by the engineer.

Inspection is complete only when concrete is cast, finished, protected for curing, and attains full strength.

(ACI "Manual of Concrete Inspection.")

ANALYSIS OF CONCRETE STRUCTURES

Under the ACI Building Code, concrete structures generally may be analyzed by elastic theory. When specific limiting conditions are met, certain approximate methods are permitted. For some cases, the Code recommends an empirical method.

5-41. Analyses of One-way Floor and Roof Systems. The ACI Building Code permits an approximate analysis for continuous systems in ordinary building if:

Components are not prestressed.

In successive spans, the ratio of the larger span to the smaller does not exceed 1.20.

The spans carry only uniform loads.

The ratio of live load to dead load does not exceed 3. This analysis determines the maximum moments and shears at faces of supports, and the midspan moments representing envelope values for the respective loading combinations. In this method, moments are computed from

$$M = CwL_n^2 \tag{5-6}$$

where C = coefficient, given in Fig. 5-13
w = uniform load
L_n = clear span for positive moment or shear and the average of adjacent clear spans for negative moment

For an elastic ("exact") analysis, the spans L of members that are not built integrally with their supports should be taken as the clear span plus the depth of slab or beam but need not exceed the distance between centers of supports. For spans of continuous frames, spans should be taken as the distance between centers of supports. For solid or ribbed slabs with clear spans not exceeding 10 ft, if built integrally with their supports, spans may be taken as the clear distance between supports.

If an elastic analysis is performed for continuous flexural members for each loading combination expected, calculated moments may be redistributed if the ratio ρ

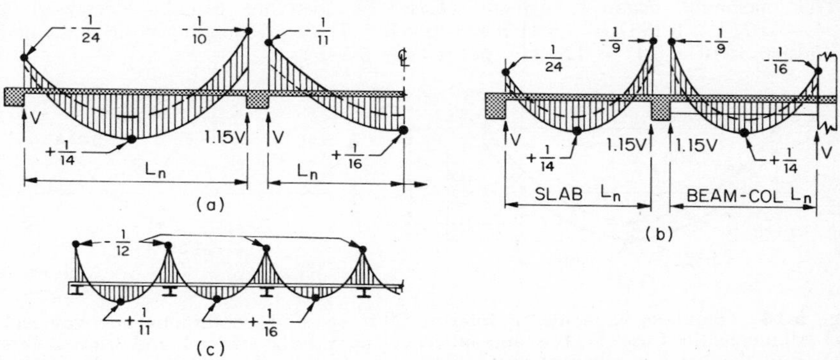

Fig. 5-13. Coefficients C for calculation of bending moments from $M = CwL_n^2$ in approximate analysis of beams and one-way slabs with uniform load w. For shears, $V = 0.5wL_n$. (*a*) More than two spans. (*b*) Two-span slab or beam. (*c*) Slabs—all spans, $L_n \leq 10$ ft.

of tension-reinforcement area to effective concrete area or ratio of $\rho - \rho'$, where ρ' is the compression-reinforcement ratio, to the balanced-reinforcement ratio ρ_b, lie within the limits given in the ACI Code. When ρ or $\rho - \rho'$ is not more than $0.5\rho_b$, computed negative moments may be increased or decreased a percentage

$$\gamma = 20\left(1 - \frac{\rho - \rho'}{\rho_b}\right) \tag{5-7}$$

Positive moments should be changed accordingly (increased by this percentage when negative moments are decreased) to reflect the moment redistribution property of underreinforced concrete in flexure.

For example, suppose a 20-ft interior span of a continuous slab with equal spans is made of concrete with a strength f'_c of 4 ksi and reinforced with bars having a yield strength f_y of 60 ksi. Dead load and live load are both 0.100 ksf. The design moments are determined as follows:

Maximum negative moments occur at the supports when this span and adjacent spans carry both dead and live loads. Call this loading Case 1. For Case 1 then, maximum negative moment equals

$$M = -(0.100 + 0.100)(20)^2/12 = -6.67 \text{ ft-kips per ft}$$

The corresponding positive moment at midspan is 3.33 ft-kips per ft.

Maximum positive moment in the interior span occurs when it carries full load but adjacent spans support only dead loads. Call this loading Case 2. For Case 2 then, the negative moment is $-(6.67 - 3.33/2) = -5.00$ ft-kips per ft, and the maximum positive moment is 5.00 ft-kips per ft. Figure 5-14*a* shows the maximum moments.

For the concrete and reinforcement properties given, the balanced-reinforcement ratio is, by Eq. (5-26),

$$\rho_b = \frac{0.85\beta_1 f'_c}{f_y} \frac{87}{87 + f_y} = 0.0285$$

Assume now that reinforcement ratios for the top steel and bottom steel are 0.00267 and 0.002, respectively. If alternate bottom bars extend into the supports, $\rho' = 0.001$. Substitution in Eq. (5-7) gives for the redistribution percentage

$$\gamma = 20\left(1 - \frac{0.00267 - 0.001}{0.0285}\right) = 18.8\%$$

The maximum negative moment (Case 1) therefore can be decreased to $M = -6.67(1 - 0.188) = -5.41$ ft-kips per ft. The corresponding positive moment at midspan is $10 - 5.41 = 4.59$ kips per ft (Fig. 5-14*b*).

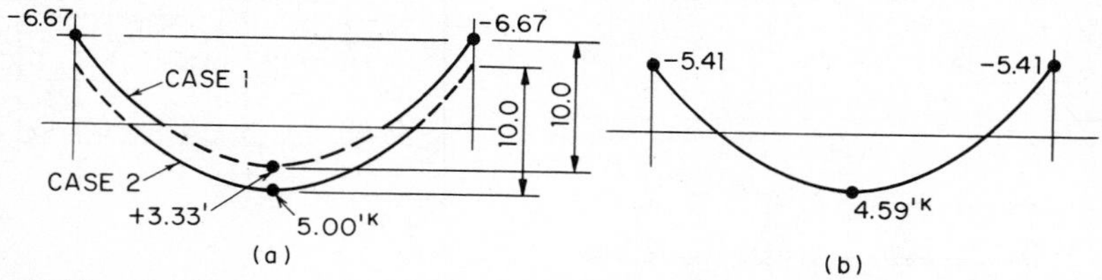

Fig. 5-14. Bending moments in interior 20-ft span of continuous one-way slab. (*a*) Moments for Case 1 (this and adjacent spans fully loaded) and Case 2 (this span fully loaded but adjacent spans with only dead load). (*b*) Case 1 moments after redistribution.

For Case 2 loading, if the negative moment is increased 18.8%, it becomes $-5.94 > 5.41$ ft-kips per ft. Therefore, the slab should be designed for the moments shown in Fig. 5-14*b*.

5-42. Two-Way Slab Frames. The ACI Building Code prescribes an *equivalent column* concept for use in analysis of two-way slab systems. This concept permits a three-dimensional (space-frame) analysis in which the "equivalent column" combines the flexibility (reciprocal of stiffness) of the real column and the torsional flexibility of the slabs or beams attached to the column at right angles to the direction of the bending moment under consideration. The method, applicable for all ratios of successive spans and of dead to live load, is an elastic ("exact") analysis called the "equivalent frame method."

An approximate procedure, the "direct design method," is also permitted (within limits of load and span). This method constitutes the direct solution of a one-cycle moment distribution. (See also Art. 5-58.)

(P. F. Rice and E. S. Hoffman, "Structural Design Guide to the ACI Building Code," Van Nostrand Reinhold Company, New York.)

5-43. Special Analyses. Space limitations preclude more than a brief listing of some of the special analyses required for various special types of reinforced concrete construction and selected basic references for detailed information. Further references to applicable research are available in each of the basic references.

Seismic-loading-resistant ductile frames: ACI 318-71; ACI 315 "Manual of Standard Practice for Detailing Reinforced Concrete Structures."

High-rise construction, frames, shear walls, frames plus shear walls, and tube concept: ASCE-IASBE Committee on High-rise Construction.

Nuclear containment structures: ASME-ACI Code for Nuclear Containment Structures.

Sanitary engineering structures: ACI Committee on Design of Sanitary Structures.

Bridges: "Criteria for Design of Reinforced Concrete Highway Bridges—Ultimate Design," U.S. Department of Transportation.

It should be noted that the ACI Building Code specifically provides for the acceptance of analyses by computer or model testing to supplement the manual calculations when required by building officials.

STRUCTURAL DESIGN OF FLEXURAL MEMBERS

5-44. Strength Design with Load Factors. Safe, economical strength design of reinforced concrete structures requires that their ultimate-load carrying capacity be predictable or known. The safe, or service-load carrying capacity can then be determined by dividing the ultimate-load carrying capacity by a factor of safety.

The ACI Building Code provides for strength design of concrete members by use of *design loads* (actual and specified loads multiplied by load factors). Design axial forces, shears, and moments in members are determined as if the structure were elastic. Strength-design theory is then used to design critical sections for these axial forces, shears, and moments.

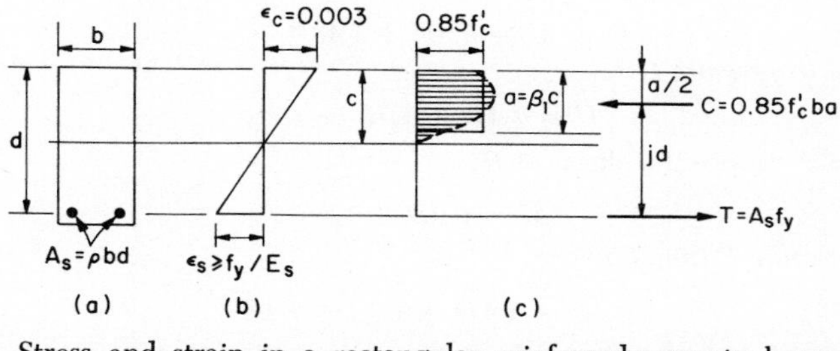

Fig. 5-15. Stress and strain in a rectangular reinforced-concrete beam, reinforced for tension only, at ultimate load. (*a*) Section. (*b*) Strain. (*c*) Stress.

Strength design of reinforced concrete flexural members (Art. 5-64) may be based on the following assumptions and applicable conditions of equilibrium and compatibility of strains:

1. Strain in the reinforcing steel and the concrete is directly proportional to the distance from the neutral axis (Fig. 5-15).

2. The maximum usable strain at the extreme concrete compression fiber equals 0.003 in. per in.

3. When the strain, in. per in., in reinforcing steel is less than f_y/E_s, where f_y = yield strength of the steel and E_s = its modulus of elasticity, the steel stress, psi, equals 29,000,000 times the steel strain. After the steel yield strength has been reached, the stress remains constant at f_y, though the strain increases.

4. Except for prestressed concrete (Art. 5-106) or plain concrete, the tensile strength of the concrete is negligible in flexure.

5. The shape of the concrete compressive stress distribution may be assumed to be a rectangle, trapezoid, parabola, or any other shape in substantial agreement with comprehensive strength tests.

6. For a rectangular stress block, the compressive stress in the concrete should be taken as $0.85f'_c$. This stress may be assumed constant from the surface of maximum compressive strain to a depth of $a = \beta_1 c$, where c is the distance to the neutral axis (Fig. 5-15). For f'_c = 4,000 psi, β_1 = 0.85; for greater concrete strengths, β_1 should be reduced 0.05 for each 1,000 psi in excess of 4,000.

(See also Art. 5-83 for columns.)

The ACI Code requires that the strength of a member based on strength design theory include a *capacity reduction factor* ϕ to provide for small adverse variations in materials, workmanship, and dimensions individually within acceptable tolerances. The degree of ductility, importance of the member, and the accuracy with which the member's strength can be predicted were considered in assigning values to ϕ:

ϕ should be taken as 0.90 for flexure and axial tension; 0.85 for shear and torsion; 0.70 for bearing on concrete; 0.65 for bending in plain concrete; and for axial compression combined with bending, 0.75 for members with spiral reinforcement, and 0.70 for other members.

For combinations of loads, a structure and its members should have the following strength U:

Dead load D and live load L, plus their internal moments and forces:

$$U = 1.4D + 1.7L \tag{5-8}$$

Wind load W:

$$U = 0.75\,(1.4D + 1.7L + 1.7W) \tag{5-9}$$

When D and L reduce the effects of W:

$$U = 0.9D + 1.3W \tag{5-10}$$

Earthquake forces E:

$$U = 0.75(1.4D + 1.7L + 1.87E) \tag{5-11}$$

When D and L reduce the effects of E:

$$U = 0.9D + 1.43E \tag{5-12}$$

Lateral earth pressure H:

$$U = 1.4D + 1.7L + 1.7H \tag{5-13}$$

When D and L reduce the effects of H:

$$U = 0.9D + 1.7H \tag{5-14}$$

Lateral pressure F from liquids:

$$U = 1.4D + 1.7L + 1.4F \tag{5-15}$$

Impact effects, if any, should be included with the live load L.

Where the effects of differential settlement, creep, shrinkage, or temperature change can be significant, they should be included with the dead load D, and the strength should be not less than

$$U = 0.75(1.4D + 1.7L) \tag{5-16}$$

5-45. Allowable-stress Design at Service Loads (Alternate Design Method). Nonprestressed-, reinforced-concrete flexural members (Art. 5-64) may be designed for flexure by the alternate design method of ACI 318-71 (working-stress design). In this method, members are designed to carry service loads (load factors and ϕ are taken as unity) under the straight-line (elastic) theory of stress and strain. (Because of creep in the concrete, only stresses due to short-time loading can be predicted with reasonable accuracy by this method.)

Working-stress design is based on the following assumptions:

1. A section plane before bending remains plane after bending. Strains therefore vary with distance from the neutral axis (Fig. 5-16*c*).

2. The stress-strain relation for concrete plots as a straight line under service loads within the allowable working stresses (Fig. 5-16*c* and *d*), except for *deep beams.*

3. Reinforcing steel takes all the tension due to flexure (Fig. 5-16*a* and *b*).

4. The modular ratio, $n = E_s/E_c$, where E_s and E_c are the moduli of elasticity of reinforcing steel and concrete, respectively, may be taken as the nearest whole number, but not less than 6 (Fig. 5-16*b*).

5. Except in calculations for deflection, n for lightweight concrete should be assumed the same as for normal-weight concrete of the same strength.

6. The compressive stress in the extreme surface of the concrete must not exceed $0.45f'_c$, where f'_c is the 28-day compressive strength of the concrete.

7. The allowable stress in the reinforcement must not be greater than the following:
Grades 40 and 50 .... 20,000 psi
Grade 60 or greater .... 24,000 psi
For ⅜-in. or smaller-diameter reinforcement in one-way slabs with spans not exceeding 12 ft, the allowable stress may be increased to 50% of the yield stress but not to more than .... 30,000 psi

8. For doubly reinforced flexural members, including slabs with compression reinforcement, an effective modular ratio of $2E_s/E_c$ should be used to transform the compression-reinforcement area for stress computations (Fig. 5-16*b* to an equivalent concrete area. (This recognizes the effects of creep.) The allowable stress in the compression steel may not exceed the allowable tension stress.

Because the strains in the reinforcing steel and the adjoining concrete are equal, the stress in the tension steel f_s is n times the stress in the concrete f_c. The total force acting on the tension steel then equals nA_sf_c. The steel area A_s therefore can be replaced in stress calculations by a concrete area n times as large.

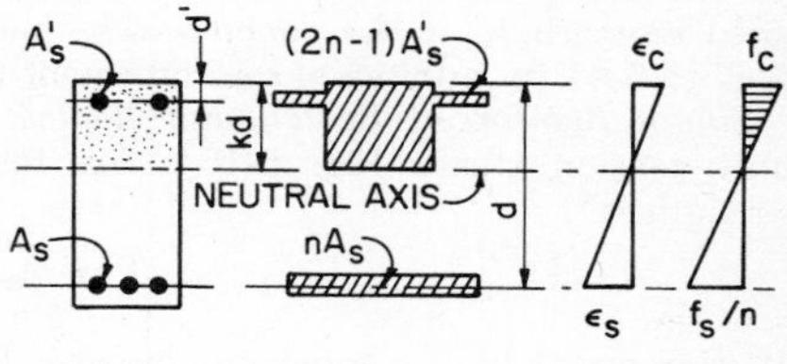

Fig. 5-16. Stresses and strains in a beam with compression reinforcement, as assumed for working-stress design. (*a*) Rectangular cross section of beam. (*b*) Transformed section with twice the steel area, to allow for effects of creep of concrete. (*c*) Assumed strains. (*d*) Assumed distribution of stress in concrete.

The transformed section of a concrete beam is a cross section normal to the neutral surface with the reinforcing replaced by an equivalent area of concrete (Fig. 5-16*b*). (In doubly reinforced beams and slabs, an effective modular ratio of $2n$ should be used to transform the compression reinforcement and account for creep and nonlinearity of the stress-strain diagram for concrete.) Stress and strain are assumed to vary with the distance from the neutral axis of the transformed section; that is, conventional elastic theory for homogeneous beams may be applied to the transformed section. Section properties, such as location of neutral axis, moment of inertia, and section modulus S, may be computed in the usual way for homogeneous beams, and stresses may be calculated from the flexure formula, $f = M/S$, where M is the bending moment at the section. This method is recommended particularly for T beams and doubly reinforced beams.

From the assumptions the following formulas can be derived for a rectangular section with tension steel only.

$$\frac{nf_c}{f_s} = \frac{k}{1-k} \tag{5-17}$$

$$k = \sqrt{2n\rho + (n\rho)^2} - n\rho \tag{5-18}$$

$$j = 1 - \frac{k}{3} \tag{5-19}$$

where $\rho = A_s/bd$ and b is the width and d the effective depth of the section (Fig. 5-16).

Compression capacity:

$$M_c = \tfrac{1}{2}f_ckjbd^2 = K_cbd^2 \tag{5-20a}$$

where $K_c = \tfrac{1}{2} f_ckj$

Tension capacity:

$$M_s = f_sA_sjd = f_s\rho jbd^2 = K_sbd^2 \tag{5-20b}$$

where $K_s = f_s\rho j$

Design of flexural members for shear, torsion, and bearing, and of other types of members, follows the strength design provisions of ACI 318-71, because allowable capacity by the alternate design method is an arbitrarily specified percentage of the strength.

5-46. Strength Design for Flexure. (See also Arts. 5-44 and 5-64.) The area A_s of tension reinforcement in a reinforced-concrete flexural member can be expressed as the ratio

$$\rho = \frac{A_s}{bd} \tag{5-21}$$

where b = beam width
d = effective beam depth = distance from the extreme compression surface to centroid of tension reinforcement

At ultimate strength of a critical section, the stress in this steel will be equal to its yield strength f_y, if the concrete does not first fail in compression. (See also Arts. 5-47 to 5-51 for additional reinforcement requirements.)

Singly Reinforced Rectangular Beams. For a rectangular beam, reinforced with only tension steel (Fig. 5-15), the total tension force in the steel at ultimate strength is

$$T = A_s f_y = \rho f_y bd \tag{5-22}$$

It is opposed by an equal compressive force

$$C = 0.85 f'_c b \beta_1 c \tag{5-23}$$

where f'_c = specified strength of the concrete
a = depth of equivalent rectangular stress distribution
c = distance from extreme compression surface to neutral axis
β_1 = a constant (given in Art. 5-44)

Equating the compression and tension at the critical section yields:

$$c = \frac{\rho f_y}{0.85 \beta_1 f'_c} d \tag{5-24}$$

The criterion for compression failure is that the maximum strain in the concrete equals 0.003 in. per in. In that case:

$$c = \frac{0.003}{f_s/E_s + 0.003} d \tag{5-25}$$

where f_s is the steel stress and E_s = 29,000,000 psi is the steel modulus of elasticity.

Under balanced conditions, the concrete will reach its maximum strain of 0.003 in. per in. when the steel reaches its yield strength f_y. Then, c as given by Eq. (5-25) will equal c as given by Eq. (5-24). Also, the steel ratio for balanced conditions in a rectangular beam with tension steel only becomes:

$$\rho_b = \frac{0.85 \beta_1 f'_c}{f_y} \frac{87{,}000}{87{,}000 + f_y} \tag{5-26}$$

All structures should be designed to avoid sudden collapse. Therefore, reinforcement should yield before the concrete crushes. Gradual yielding will occur if the quantity of tensile reinforcement is less than the balanced percentage determined by strength design theory. To avoid compression failures, the ACI Building Code therefore limits the steel ratio ρ to a maximum of $0.75\rho_b$.

The Code also requires that ρ for positive-moment reinforcement be at least $200/f_y$ to prevent sudden collapse when the design-positive moment strength is equal to or less than the cracking moment. This requirement does not apply, however, if the reinforcement area at every section of the member is at least one-third greater than that required by the design moment.

For underreinforced rectangular beams with tension reinforcement only (Fig.

5-15) and a rectangular stress block ($\rho \leq 0.75\rho_b$), the flexural design strength may be determined from:

$$M_u = 0.90bd^2\rho f_y\left(1 - \frac{0.59\rho f_y}{f'_c}\right) \quad (5\text{-}27a)$$

$$= 0.90A_sf_y\left(d - \frac{a}{2}\right) \quad (5\text{-}27b)$$

$$= 0.90A_sf_yjd \quad (5\text{-}27c)$$

where $a = A_sf_y/0.85f'_cb$ and $jd = d - a/2$. (See Fig. 5-17 for determination of j.)

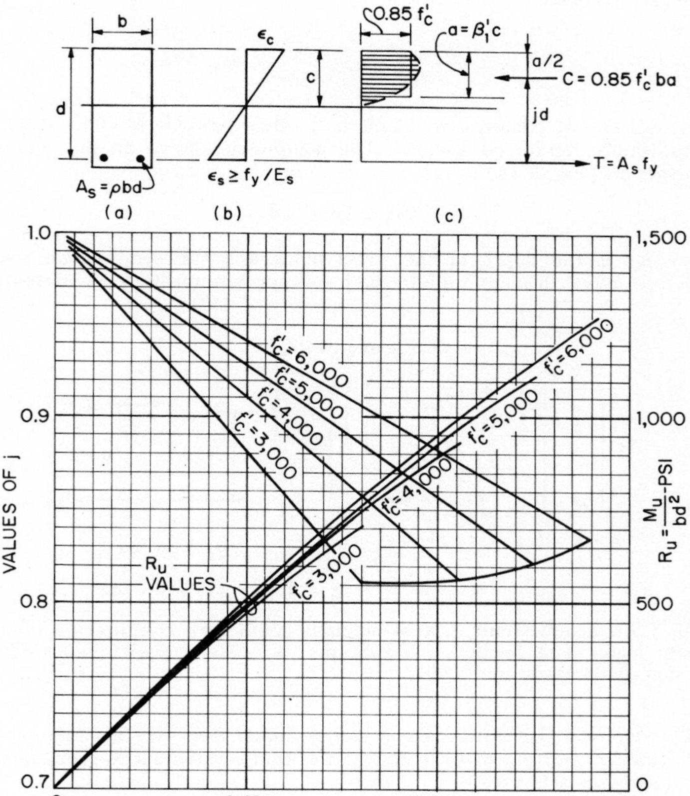

Fig. 5-17. For rectangular beam shown in (a) reinforced, in tension only, with ASTM A615, Grade 60 bars: (d) Moment-capacity chart based on assumed strain distribution in (b) and stress distribution in (c).

Tension-steel Limitations. For flexural members of any cross-sectional shape, with or without compression reinforcement, the tension steel is limited by the ACI Building Code so that A_sf_y does not exceed 0.75 times the total compressive force at balanced conditions. The total compressive force may be taken as the area of a rectangular stress block of a rectangular member; the strength of overhanging flanges or compression steel, or both, may be included.

Doubly Reinforced Rectangular Beams. For a rectangular beam with compression-steel area A'_s and tension-steel area A_s, the compression-steel ratio is

$$\rho' = \frac{A'_s}{bd} \tag{5-28}$$

and the tension-steel ratio is

$$\rho = \frac{A_s}{bd} \tag{5-29}$$

where b = width of beam
d = effective depth of beam
For design, ρ should not exceed $0.75\rho_b$.

$$\rho_b = \frac{0.85f'_c\beta_1}{f_y}\frac{87{,}000}{87{,}000 + f_y} + \rho'\frac{f'_s}{f_y} \tag{5-30}$$

where f'_s = stress in compression steel, and other symbols are the same as those defined for singly reinforced beams. The compression force on the concrete alone in a cross section (Fig. 5-18) is

$$C_{1b} = 0.85f'_cba \tag{5-31}$$

where $a = \beta_1 c$ is the depth of the stress block and the compression steel carries $A'_sf'_s$. Forces equal in magnitude to these but opposite in direction stress the tension

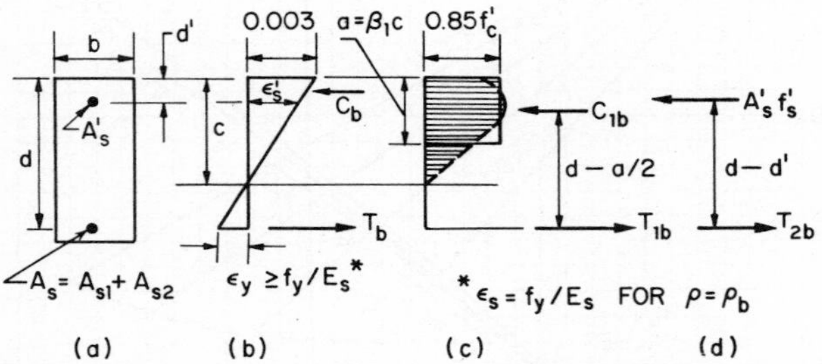

Fig. 5-18. Stress and strain in a rectangular beam with compression reinforcement at ultimate load. (*a*) Section. (*b*) Strain. (*c*) Compression block stress. (*d*) Compression steel stress.

steel. The depth to the neutral axis c can be found from the maximum compressive strain of 0.003 in. per in., or by equating the compression and tension forces on the section. (See also Art. 5-65.)

T Beams. When a T form is used to provide needed compression area for an isolated beam, flange thickness should be at least one-half the web width, and flange width should not exceed four times the web width.

When a T is formed by a beam cast integrally with a slab, only a portion of the slab is effective. For a symmetrical T beam, the effective flange width should not exceed one-fourth the beam span, nor should the width of the overhang exceed eight times the slab thickness nor one-half the clear distance to the next beam. For a beam having a flange on one side only, the effective flange width should not exceed one-twelfth the span, six times the slab thickness, nor one-half the clear distance to the next beam.

The overhang of a T beam should be designed to act as a centilever. Spacing of the cantilever reinforcing should not exceed 18 in. or five times the flange thickness.

In computing the moment capacity of a T beam, it may be treated as a singly reinforced beam with overhanging concrete flanges (Fig. 5-19). The compression force on the web (rectangular beam) is

$$C_w = 0.85f'_c b_w a \tag{5-32}$$

where b_w = width of web
The compression force on the overhangs is

$$C_f = 0.85f'_c(b - b_w)h_f \tag{5-33}$$

where h_f = flange thickness
b = effective flange width of T beam
Forces equal in magnitude to these but opposite in direction stress the tension steel:

$$T_w = A_{sw}f_y \tag{5-34}$$
$$T_f = A_{sf}f_y \tag{5-35}$$

where A_{sw} = area of reinforcing required to develop compression strength of web
A_{sf} = area of reinforcing required to develop compression strength of overhanging flanges

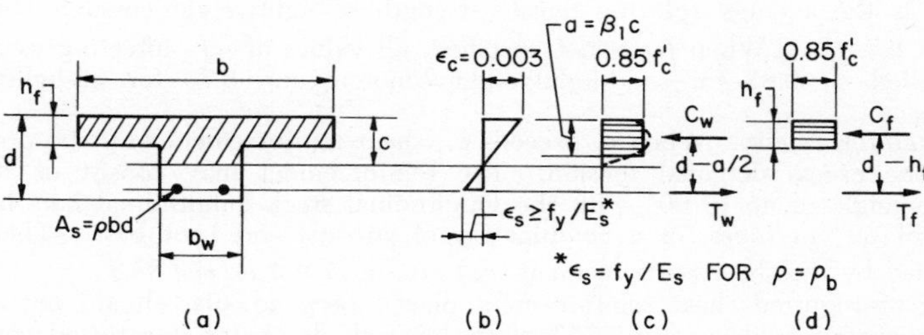

Fig. 5-19. Stress and strain in a T beam at ultimate load. (*a*) Section. (*b*) Strain. (*c*) Web compression block stress. (*d*) Flange compression block stress.

The steel ratio for balanced conditions is given by

$$\rho_b = \frac{b_w}{b}\left[\frac{0.85f'_c\beta_1}{f_y}\,\frac{87{,}000}{87{,}000 + f_y} + \frac{A_{sf}}{b_w d}\right] \tag{5-36}$$

The depth to the neutral axis c can be found in the same way as for rectangular beams.

5-47. Shear in Flexural Members. The nominal unit shear stress, psi, at a section of a reinforced-concrete flexural member with shear force V_u, lb, is

$$v_u = \frac{V_u}{\phi b_w d} \tag{5-37}$$

where ϕ = capacity reduction factor (given in Art. 5-44)
b_w = web width of rectangular or T beam, or diameter of circular section, in.
d = depth from extreme compression surface to centroid of reinforcement, in.
Except for brackets, deep beams, and other short cantilevers, the section for maximum shear may be taken at a distance d from the face of the support when the reaction in the direction of the shear introduces compression into the end region of the member.

For shear in two-way slabs, see Art. 5-58.

For nonprestressed flexural members of normal-weight concrete without torsion, the shear stress v_c carried by the concrete is limited to a maximum of $2\sqrt{f'_c}$, where

f'_c is the specified concrete strength, unless a more detailed analysis is made. In such an analysis, v_c should be obtained from

$$v_c = 1.9\sqrt{f'_c} + \frac{2{,}500\rho_w V_u d}{M_u} \leq 3.5\sqrt{f'_c} \tag{5-38}$$

where M_u = design bending moment occurring simultaneously with V_u at the section considered, but $V_u d/M_u$ must not exceed 1.0
$\rho_w = A_s/b_w d$
A_s = area of nonprestressed tension reinforcement

For one-way joist construction, the ACI Building Code allows these values to be increased 10%.

For members in which the torsional stress v_{tu} computed from Eq. (5-42) exceeds $1.5\sqrt{f'_c}$, the allowable shearing stress v_c that can be assigned to the concrete should be limited to a maximum value

$$v_c = \frac{2\sqrt{f'_c}}{\sqrt{1 + \left(\dfrac{v_{tu}}{1.2v_u}\right)^2}} \tag{5-39}$$

For lightweight concrete, v_c should be modified by substituting $f_{ct}/6.7$ for $\sqrt{f'_c}$, where f_{ct} is the average splitting tensile strength of lightweight concrete, but not more than $6.7\sqrt{f'_c}$. When f_{ct} is not specified, all values of $\sqrt{f'_c}$ affecting v_c should be multiplied by 0.85 for sand-lightweight concrete and 0.75 for all-lightweight concrete.

Shear Reinforcement. When v_u exceeds v_c, shear reinforcement must be provided to resist the excess diagonal tension. The reinforcement may consist of stirrups making an angle of 45 to 90° with the longitudinal steel, longitudinal bars bent at an angle of 30° or more, or a combination of stirrups and bent bars. The shear stress carried by the shear reinforcement, $v_u - v_c$, must not exceed $8\sqrt{f'_c}$.

Spacing of required shear reinforcement placed perpendicular should not exceed $0.5d$ for nonprestressed concrete, 75% of the over-all depth for prestressed concrete, or 24 in. Inclined stirrups and bent bars should be spaced so that at least one intersects every 45° line extending toward the supports from middepth of the member to the tension reinforcement. When $v_u - v_c$ is greater than $4\sqrt{f'_c}$, the maximum spacing of shear reinforcement should be reduced by one-half. (See Art. 5-111 for shear-strength design for prestressed concrete members.)

The area required in the legs of a vertical stirrup, sq in., is

$$A_v = \frac{v_u - v_c}{f_y} b_w s \tag{5-40a}$$

where s = spacing of stirrups, in.
f_y = yield strength of stirrup steel, psi

For inclined stirrups, the leg area should be at least

$$A_v = \frac{v_u - v_c}{(\sin\alpha + \cos\alpha) f_y} b_w s \tag{5-40b}$$

where α = angle of inclination with longitudinal axis of member

For a single bent bar or a single group of parallel bars all bent at an angle α with the longitudinal axis at the same distance from the support, the required area is

$$A_v = \frac{v_u - v_c}{f_y \sin\alpha} b_w d \tag{5-41}$$

in which $v_u - v_c$ should not exceed $3\sqrt{f'_c}$.

See also Art. 5-66.

5-48. Torsion in Concrete Members. Under twisting or torsional loads, a member develops normal (warping) and shear stresses. Torsional design stresses may be computed from

$$v_{tu} = \frac{3T_u}{\phi \Sigma x^2 y} \tag{5-42}$$

where T_u = design torsional moment
ϕ = capacity reduction factor = 0.85
x = shorter over-all dimension of the rectangular part of a cross section (Fig. 5-20)
y = longer over-all dimension of the rectangular part (Fig. 5-20)

For computation of v_{tu}, it is usually convenient to divide a cross section into n rectangles with sides x_r and y_r, where r ranges from 1 to n. Calculation of $\Sigma x^2 y$ for flanged sections depends on how the rectangles are chosen. They can be selected to yield the greatest value for $\Sigma x^2 y$. But they should not overlap.

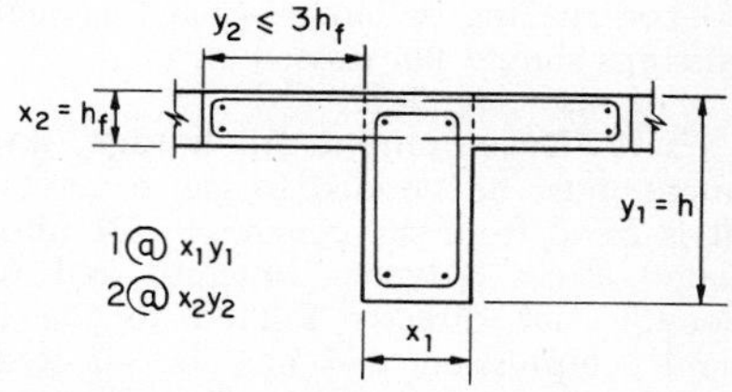

Fig. 5-20. Resolution of T beam into component rectangles for torsional shear calculations.

A rectangular box section may be taken as a solid section xy when wall thickness $h \geq x/4$. When $x/10 < h < x/4$, a solid section may be used, but $\Sigma x^2 y$ should be multiplied by $4h/x$. When $h < x/10$, divide the hollow section into separate rectangles. Fillets should be provided at all interior corners of box sections to reduce the effects of stress concentrations.

If v_{tu} computed from Eq. (5-42) does not exceed $1.5\sqrt{f'_c}$, where f'_c is the specified concrete strength, psi, torsional stresses need not be included with those from shear and bending.

The maximum allowable torsional stress is

$$v_{tu} = \frac{12\sqrt{f'_c}}{\sqrt{1 + (1.2v_u/v_{tu})^2}} \tag{5-43}$$

where v_u = nominal unit shear stress [Eq. (5-37)]

Members should be designed for the torsion at a distance d from the face of support, where d is the distance from extreme compression surface to centroid of tension steel of the members.

The portion of the torsion design stress carried by the concrete should not be taken larger than

$$v_{tc} = \frac{2.4\sqrt{f'_c}}{\sqrt{1 + (1.2v_u/v_{tu})^2}} \tag{5-44}$$

This is one-fifth of v_{tu} computed from Eq. (5-43).

Stirrups. Torsion reinforcement, when required, should be provided in addition to other reinforcement for flexure, axial forces, or shear. For this purpose, the required area A_t for one leg of a closed stirrup can be computed from

$$A_t = (v_{tu} - v_{tc}) \frac{s \Sigma x^2 y}{3\alpha_t x_1 y_1 f_y} \tag{5-45}$$

where $\alpha_t = 0.66 + 0.33y_1/x_1 \leq 1.50$
s = spacing of torsion reinforcement measured parallel to longitudinal reinforcement
x_1 = shorter center-to-center dimension of a closed rectangular stirrup
y_1 = larger center-to-center dimension of a closed rectangular stirrup
f_y = yield strength of stirrup steel

To control crack widths, insure development of the ultimate torsional strength of the member, and prevent excessive loss of torsional stiffness after cracking, the spacing of closed-stirrup torsion reinforcement should not exceed $(x_1 + y_1)/4$ or 12 in., whichever is smaller.

Longitudinal Reinforcement. Steel reinforcement parallel to the axis of the member should be provided in each corner of closed-stirrup torsion reinforcement to assist in development of the torsional design strength without excessive cracking. The amount of longitudinal reinforcement A_l, sq in., for resisting torsion should be the greater of

$$A_l = 2A_t(x_1 + y_1)/s \tag{5-46}$$

$$A_l = \left[\frac{400xs}{f_y}\left(\frac{v_{tu}}{v_{tu} + v_u}\right) - 2A_t\right]\frac{(x_1 + y_1)}{s} \tag{5-47}$$

except that A_l need not be greater than the values given by Eqs. (5-46) and (5-47) when $50b_ws/f_y$ is substituted for $2A_t$.

The spacing of longitudinal torsional reinforcement around the perimeter of closed stirrups should not exceed 12 in.

See also Art. 5-67.

5-49. Development, Anchorage, and Splices of Reinforcement. Steel reinforcement must be bonded to the concrete sufficiently so that the steel will yield before it is freed from the concrete. Despite assumptions made in the past to the contrary, bond stress between concrete and reinforcing bars is not uniform over a given length, not directly related to the perimeter of the bars, not equal in tension and compression, and may be affected by lateral confinement. ACI Building Code requirements therefore reflect the significance of average bond resistance over a length of bar or wire sufficient to develop its strength (**development length**).

The calculated tension or compression force in each reinforcing bar at any section must be developed on each side of that section by a development length l_d, or by end anchorage, or both. Hooks can be used to assist in the development of tension bars only.

The critical sections for development of reinforcement in flexural members are located at the points of maximum stress and where the reinforcement terminates or is bent.

The following requirements of ACI 318-71 for the development of reinforcement were developed to help provide for *shifts* in the location of maximum moment and for *peak stresses* that exist in regions of tension in the remaining bars wherever adjacent bars are cut off or bent. In addition, these requirements help minimize any loss of shear capacity or ductility resulting from flexural cracks that tend to open early whenever reinforcement is terminated in a tension zone.

All Flexural Reinforcement

Reinforcement should extend a distance of d or $12d_b$, whichever is larger, beyond the point where the steel is no longer required to resist stress, where d is the effective depth of the member and d_b is the reinforcement diameter. This requirement, however, does not apply at supports of simple spans and at the free end of cantilevers.

Continuing reinforcement should extend at least the development length l_d beyond the point where bent or terminated reinforcement is no longer required to resist tension.

Reinforcement should not be terminated in a tension zone unless *one* of the following conditions is satisfied:

1. Shear at the cutoff point does not exceed two-thirds of the shear permitted, including the shear strength of web reinforcement.

2. Stirrups with $(A_v/b_w)f_y \geq 60$ psi and having areas exceeding those required for shear and torsion are provided along each terminated bar over a distance from the termination point equal to $0.75d$. (A_v = cross-sectional area of stirrup leg, b_w = width of member, and f_y = yield strength of stirrup steel.) The spacing should not exceed $d/8\beta_b$, where β_b is the ratio of the area of the bars cut off to the total area of bars at the cutoff section.

3. For No. 11 bars and smaller, continuing bars provide double the area required

for flexure at the cutoff point, and the shear does not exceed three-fourths of that permitted.

Positive-moment Reinforcement

A minimum of one-third the required positive-moment reinforcement for simple beams should extend along the same face of the member into the support, and in beams, not less than 6 in.

A minimum of one-fourth the required positive-moment reinforcement for continuous members should extend along the same face of the member into the support, and in beams, at least 6 in.

For lateral-load-resisting members, the positive-moment reinforcement to be extended into the support in accordance with the preceding two requirements should be able to develop between the face of the support and the end of the bars the yield strength f_y of the bars.

Positive-moment tension reinforcement at simple supports and at points of inflection should be limited to a diameter such that the development length, in., computed for f_y with Eqs. (5-49) to (5-53) does not exceed

$$l_d = \frac{M_t}{V_u} + l_a \tag{5-48}$$

where M_t = theoretical moment strength of a section, in.-lb, assuming all reinforcement at the section stressed to f_y

V_u = applied design shear at the section, lb

l_a = embedment length, in., beyond center of support plus equivalent embedment length of any hook or mechanical anchorage, but not more than d or $12d_b$, whichever is greater

d = effective depth, in., of member

d_b = bar diameter, in.

Negative-moment Reinforcement

Negative-moment reinforcement in continuous, restrained, or cantilever members should be developed in or through the supporting member.

Negative-moment reinforcement should have sufficient distance between the face of the support and the end of each bar to develop its full yield strength.

A minimum of one-third of the required negative-moment reinforcement at the face of the support should extend beyond the point of inflection the greatest of d, $12d_b$, or one-sixteenth of the clear span.

Computation of Development Length. The basic development length l_d for deformed reinforcing bars and deformed wire **in tension** is:
For No. 11 or smaller bars:

$$l_d = \frac{0.04A_bf_y}{\sqrt{f'_c}} \geq 0.0004d_bf_y \tag{5-49}$$

For No. 14 bars:

$$l_d = \frac{0.085f_y}{\sqrt{f'_c}} \tag{5-50}$$

For No. 18 bars:

$$l_d = \frac{0.11f_y}{\sqrt{f'_c}} \tag{5-51}$$

For deformed wire:

$$l_d = \frac{0.03d_bf_y}{\sqrt{f'_c}} \tag{5-52}$$

For welded-wire fabric with deformed wires:

$$l_d = \frac{0.03d_b(f_y - 20{,}000n)}{\sqrt{f'_c}} \geq \frac{250A_w}{s_w} \tag{5-53}$$

where f'_c = specified concrete strength, psi
f_y = steel yield strength, psi
A_b = bar area, sq in.
n = number of cross wires within the development length that are at least 2 in. from the critical section
s_w = spacing of the wires, sq in.
A_w = area of one tension wire, sq in.

The basic development lengths for tension bars with various concrete strengths are given in Table 5-7.

Table 5-7. Development Length of Tension Reinforcement, l_d, In.*†

Bar size no.	f'_c = 3,000 psi		f'_c = 3,750 psi		f'_c = 4,000 psi		f'_c = 5,000 psi		f'_c = 6,000 psi	
	Top‡ bars	Other bars	Top‡ bars	Other bars	Top‡ bars	Other bars	Top‡ bars	Other bars	Top‡ bars	Other bars
3	13	12	13	12	13	12	13	12	13	12
4	17	12	17	12	17	12	17	12	17	12
5	21	15	21	15	21	15	21	15	21	15
6	27	19	25	18	25	18	25	18	25	18
7	37	26	33	24	32	23	29	21	29	21
8	48	35	43	31	42	30	38	27	34	25
9	61	44	55	39	53	38	48	34	43	31
10	78	56	70	50	67	48	60	43	55	39
11	96	68	86	61	83	59	74	53	68	48
14	130	93	117	83	113	81	101	72	92	66
18	169	120	151	108	146	104	131	93	119	85

*Courtesy Concrete Reinforcing Steel Institute.
†1. For bars enclosed in standard column spirals, use 0.75 l_d.
2. For bars, such as usual temperature bars, spaced 6 in. or more, use 0.8 l_d.
3. Longer embedments for lightweight concrete are generally required, depending on the tensile splitting strength f_{ct}.
4. Standard 90 or 180° end hooks may be used to replace part of the required embedment.
‡Horizontal bars with more than 12 in. of concrete below.

The basic development length in tension for top reinforcement, reinforcement with f_y greater than 60,000 psi, reinforced lightweight concrete, reinforcement with a spacing of 6 in. or greater, reinforcement in excess of that required, and bars enclosed within a spiral should be modified by multiplying the basic development length by the following factors:

For top reinforcement, horizontal bars with more than 12 in. of concrete below: 1.4
For bars with $f_y > 60{,}000$ psi: $2 - 60{,}000/f_y$
For reinforced, sand-lightweight concrete: 1.18
For reinforced all-lightweight concrete: 1.33

In lieu of the two preceding values for lightweight concrete: $6.7\sqrt{f'_c}/f_{ct} \geq 1.0$, where f_{ct} = average splitting tensile strength of the concrete, psi

When reinforcement is at least 3 in. from the side face of a member, is being developed in the length under consideration, and is spaced 6 in. or more c to c: 0.80

For reinforcement in excess of that required: (A_s required/A_s provided), where A_s = cross-sectional area of reinforcement

Basic development length for deformed bars **in compression** is

$$l_d = \frac{0.02 d_b f_y}{\sqrt{f'_c}} \geq 0.0003 d_b f_y \geq 8 \text{ in.} \tag{5-54}$$

When the reinforcement is enclosed by spirals at least ¼ in. in diameter and with not more than a 4-in. pitch, l_d can be reduced 25%. The basic development lengths for bars in compression are given in Table 5-8.

For **bundled bars,** the basic development length of each bar must be increased by 20% for a three-bar bundle and 33% for a four-bar bundle.

Hooks. For bars in tension, standard hooks (90 or 180°) can be used as part of the length required for development or anchorage of the bar. Table 5-9 gives the straight embedment length l_e equivalent to a hook for various reinforcing-bar sizes.

Table 5-10 lists the minimum tension embedment length E required with standard end hooks and Grade 60 bars to develop the specified yield strengths of bars.

Smooth Wire. The yield strength of smooth wire in tension can be considered

Table 5-8. Minimum Compression-dowel Development Lengths l_d, In.*†

Bar size no.	f'_c (Normal-weight concrete)			
	3,000 psi	3,750 psi	4,000 psi	Over 4,444 psi‡
3	8	8	8	8
4	11	10	10	9
5	14	12	12	11
6	17	15	14	14
7	19	17	17	16
8	22	20	19	18
9	25	22	22	20
10	28	25	24	23
11	31	28	27	25
14	37	33	32	31
18	50	44	43	41

* Courtesy Concrete Reinforcing Steel Institute.

† For embedments enclosed by spirals use 0.75 length given above but not less than 8 in.

‡ For $f'_c > 4{,}444$ psi, minimum embedment = $18d_b$.

developed if two cross wires are embedded and the closer wire is at least 2 in. from the critical section.

Web Reinforcement. Stirrups should be designed and detailed to come as close to the compression and tension surfaces of a flexural member as cover requirements and the proximity of other steel will permit. Ends of single-leg simple U stirrups or transverse multiple U stirrups must be anchored by one of the following means:

A standard hook plus half the development length l_d for tension reinforcement. The effective embedment length of a stirrup leg is the distance between middepth of the member ($d/2$ from the extreme compression surface) and the start of the hook (point of tangency).

Embedment, between middepth of the beam and the extreme compression surface, of the full l_d for tension reinforcement but not less than $24d_b$.

Bending around longitudinal reinforcement through at least 180° for stirrups inclined at an angle of at least 45° with longitudinal reinforcement.

Between anchored ends, each bend in the continuous portion of a simple U, or multiple U stirrup should enclose a longitudinal bar.

Stirrup Splices. Pairs of U stirrups or ties so placed as to form a closed unit can be considered properly spliced when the laps are $1.7l_d$. In members at least 18 in. deep, such splices can be considered adequate for No. 3 bars of Grade 60

Table 5-9. Equivalent Straight Embedment l_e, In., of Standard End Hooks for Grade 60 Bars*†

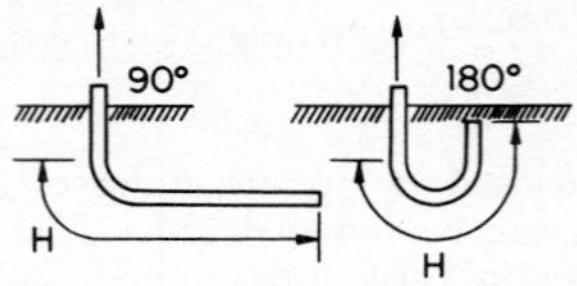

Bar size no.	For general use in enclosed members‡		In narrow members with no enclosure	
	Top bars	Other bars	Top bars	Other bars
3	6	6	4	4
4	8	8	6	6
5	10	10	7	7
6	11	12	8	10
7	12	17	9	13
8	15	22	11	17
9	19	28	14	22
10	24	32	18	24
11	29	34	23	26
14	37	37	28	28
18	32	32	24	24

*Courtesy Concrete Reinforcing Steel Institute.

†Equivalent to H in diagram above. Applicable for Nos. 14 to 18 bars for all concrete strengths f'_c, for Nos. 8 to 11 bars for $f'_c \leq 6{,}000$ psi, and for Nos. 3 to 7 bars for $f'_c \geq 3{,}000$ psi.

‡Massive elements or connections of principal members restrained against splitting by enclosure of reinforcement splices in external concrete or internal closed ties, spirals, or stirrups.

Table 5-10. Tension Embedment, E, In., with Standard End Hook for Grade 60 Bars*†

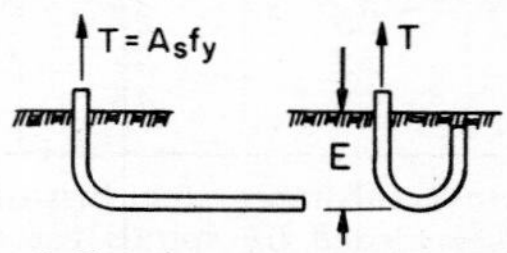

Bar size no.	f'_c = 3,000 psi		f'_c = 3,750 psi		f'_c = 4,000 psi		f'_c = 5,000 psi		f'_c = 6,000 psi	
	Top bars	Other bars	Top bars	Other bars	Top bars	Other bars	Top bars	Other bars	Top bars	Other bars
3	9	8	8	7	8	7	7	5	6	5
4	11	8	10	7	10	7	9	5	8	5
5	14	8	13	7	13	7	11	5	10	5
6	20	10	18	8	17	8	16	6	15	5
7	29	13	25	10	24	10	22	7	20	6
8	38	17	33	13	31	12	27	9	24	7
9	48	22	42	17	40	16	35	12	30	9
10	60	31	52	25	50	23	43	18	38	14
11	74	41	63	34	61	32	52	26	46	21
14	104	67	90	57	87	54	75	46	66	40
18	151	103	133	90	128	87	113	76	101	67

*Courtesy Concrete Reinforcing Steel Institute.

†For general use in enclosed members of normal-weight concrete (145 ± lb per cu ft).

NOTES: (1) For use in unconfined narrow sections subject to splitting, reduce the value of hook embedment (Table 5-9) and increase E tabulated above. (2) For lightweight-aggregate concrete, increase l_d by aggregate factor and add increase to value of E tabulated above.

and Nos. 3 and 4 bars of Grade 40 if the legs extend the full available depth of the member.

Compression Lap Splices. Minimum lengths of lap in splices of compression reinforcement vary with bar diameter d_b and yield strength of steel f_y for all values of f'_c greater than 3,000 psi. When f'_c is less than 3,000 psi, the length of lap should be one-third greater than the value computed from the following appropriate splice equation.

The ACI Building Code defines three classes of compression splices and requires the following minimum laps for each bar size No. 11 and smaller:

1. STANDARD SPLICE LAP l_s (about $30d_b$) equals the basic development length in compression [Eq. (5-54)], but not less than the largest of 12 in. or the values computed from Eqs. (5-55) and (5-56).

$$l_s = 0.0005 f_y d_b \qquad f_y \leq 60{,}000 \text{ psi} \tag{5-55}$$
$$l_s = (0.0009 f_y - 24) d_b \qquad f_y > 60{,}000 \text{ psi} \tag{5-56}$$

Splices confined by ties that have an effective area $A_v \geq 0.0015hs$, where h = over-all depth of member and s = tie spacing, should be lapped a minimum of 0.83 of the standard splice lap (about $25d_b$), but not less than 12 in.

Splices within column spirals must be lapped a minimum of 0.75 of the standard splice lap (about $23d_b$), but not less than 12 in.

2. COMPRESSION SPLICES BY BEARING. In members with closed ties, closed stirrups, or spirals, compressive stress in a rebar may be transmitted by bearing of square-cut ends held in concentric contact by a suitable device. The ends must be fitted within 3° of full bearing after assembly.

3. WELDED COMPRESSION SPLICES. The ACI Building Code requires butt welded compression splices to develop at least 125% of the specified yield strength of the steel.

Tension Lap Splices. Bars No. 11 or less in size can be spliced by lapping. Full (butt) welded splices or full positive connections may be used for bars of all sizes. These splices must have a capacity equal to at least 125% of the specified f_y.

Tension lap splices are classified in four groups with the minimum lap length l_s expressed as a multiple of the basic development length l_d of tension reinforcement.

CLASS A SPLICES. Splices at sections away from maximum stress and where the steel design stress is less than $0.5f_y$, if not more than 75% of the bars at these sections are spliced within one Class A splice length of the section.

$$l_s = l_d \tag{5-57}$$

CLASS B SPLICES. Splices where the design stress is less than $0.5f_y$ but where more than 75% of the bars at the section are spliced, or splices made at a section of maximum tensile stress where not more than 50% of the bars at the section are spliced.

$$l_s = 1.3 l_d \tag{5-58}$$

CLASS C SPLICES. Splices at a section of maximum tensile stress where more than 50% of the bars at the section are spliced.

$$l_s = 1.7 l_d \tag{5-59}$$

CLASS D SPLICES. Splices in tension tie members that are enclosed in a spiral and with 180° hooks on the ends of rebars larger in size than No. 4.

$$l_s = 2 l_d \tag{5-60}$$

Welded Tension Splices. These or other positive connections must be used for splices of Nos. 14 and 18 bars and may be used in lieu of lapped splices for No. 11 and smaller bars. Full (butt) welded splices or other full positive connections must develop at least 125% of the specified yield strength of the bars unless the splices are staggered at least 24 in. and located to develop at every section at least twice the calculated tensile force at the section, but not less than 20,000 psi on the total cross-sectional area of all bars at the section.

Wire Splices in Tension. The minimum splice length for deformed-wire reinforcement is

$$l_s = \frac{0.045 d_b f_y}{\sqrt{f'_c}} \geq 12 \text{ in.} \tag{5-61}$$

The minimum splice length for welded deformed-wire-fabric reinforcement is

$$l_s = \frac{0.045 d_b}{\sqrt{f'_c}} (f_y - 20{,}000n) \tag{5-62}$$

where n = number of cross wires in the splice
But the outermost cross wires in the splice must be lapped at least

$$l_s = \frac{A_w}{s_w}\left(360 - 8\frac{l_w}{d_b}\right) \tag{5-63}$$

where A_w = area of one wire
s_w = wire spacing
l_w = total lengths of wire extending beyond outermost cross wires, for each pair of spliced wires

Splices of welded plain-wire fabric must overlap between outermost cross wires of each fabric, but the lap should not be less than the spacing of cross wires plus 2 in. Splices of wires stressed to less than one-half the permissible stress should be made so that the overlap between the outermost cross wires is at least 2 in.

5-50. Crack Control. Because of the effectiveness of reinforcement in limiting crack widths, the ACI Building Code requires minimum areas of steel and limits reinforcement spacing, to control cracking.

Beams and One-way Slabs. If, in a structural floor or roof slab, principal reinforcement extends in one direction only, shrinkage and temperature reinforcement should be provided normal to the principal reinforcement, to prevent excessive cracking. The additional reinforcement should provide at least the ratios of reinforcement area to gross concrete area of slab given in Table 5-11, but not less than 0.0014.

Table 5-11. Minimum Shrinkage and Temperature Reinforcement

In slabs where Grade 40 or 50 deformed bars are used	0.0020
In slabs where Grade 60 deformed bars or welded-wire fabric, deformed or plain, are used (Table 5-16)	0.0018
In slabs reinforced with steel having a yield strength f_y exceeding 60,000 psi measured at a strain of 0.0035 in. per in.	$108/f_y$

This reinforcement should not be placed farther apart than five times the slab thickness or more than 18 in.

To control flexural cracking, tension reinforcement in beams and one-way slabs should be well distributed in zones of maximum concrete tension when the design yield strength of the steel f_y is greater than 40,000 psi. Spacing of principal reinforcement in slabs should not exceed 18 in. nor three times the slab thickness, except in concrete-joist construction.

Where slab flanges of beams are in tension, a part of the main reinforcement of the beam should be distributed over the effective flange width or a width equal to one-tenth the span, whichever is smaller. When the effective flange width exceeds one-tenth the span, some longitudinal reinforcement should be provided in the outer portions of the flange. Also, reinforcement for one-way joist construction should be uniformly distributed throughout the flange.

To control flexural cracking in beams, reinforcement should be so distributed that:
For interior exposures,

$$z \leq f_s \sqrt[3]{d_c A} \leq 175 \text{ kips per in.} \tag{5-64}$$

For exterior exposures,

$$z \leq f_s \sqrt[3]{d_c A} \leq 145 \text{ kips per in.} \tag{5-65}$$

where d_c = thickness, in., of concrete cover measured from extreme tension surface to center of bar

A = effective tension area, sq in., of concrete per rebar. It can be computed by dividing the concrete area surrounding the main tension reinforcing bars and having the same centroid as that reinforcement by the number of bars. If bar sizes differ, the number of bars should be computed as the total steel area divided by the area of the largest bar used

f_s = calculated stress in reinforcement at service loads, ksi, but may be taken as $0.60 f_y$ in lieu of such calculations

The numerical limitations on z of 175 and 145 kips per in. correspond to limiting crack widths of 0.016 and 0.013 in. for interior and exterior exposures, respectively. For watertight slabs or severe exposures, z should be smaller.

For one-way slabs, tests indicate that z should not exceed 155 kips per in. for interior exposures, or 129 kips per in. for exterior exposures.

The z values can be transformed into the following expressions for the maximum spacing of Grade 60 bars to control flexural cracking:

Interior Exposure	Maximum spacing, in.
Beams	$57.4/d_c^2$
One-way slabs	$39.9/d_c^2$
Exterior Exposure	Maximum spacing, in.
Beams	$32.6/d_c^2$
One-way slabs	$22.9/d_c^2$

See Tables 5-12 and 5-20, Art. 5-68.

Table 5-12. Maximum Spacing, In., of Grade 60 Bars for Control of Flexural Cracking

Bar size no.	Beams		One-way slabs					Two-way slabs			
	2-in. cover*		Interior exposure			Exterior exposure		Interior exposure		Exterior exposure	
	Interior exposure	Exterior exposure	Cover, in.			Cover, in.		Cover, in.		Cover, in.	
			¾	1	1½	1½	2	¾	1	1½	2
3	...	...	18.0	18.0	...	8.0	...	7.3	6.4	4.4	
4	...	...	18.0	18.0	...	7.5	...	8.1	7.2	5.0	
5	10.7	6.1	18.0	18.0	...	6.9	...	8.8	7.9	5.5	
6	10.2	5.8	18.0	18.0	...	...	4.1	9.4	8.4	...	5.2
7	9.7	5.5	18.0	18.0	...	...	3.9	9.9	8.9	...	5.6
8	9.2	5.2	18.0	17.7	...	...	3.7	10.2	9.4	...	5.9
9	8.7	5.0	18.0	16.3	...	...	3.5	10.6	9.7	...	6.2
10	8.3	4.7	18.0	14.9	...	...	3.3	11.0	10.1	...	6.5
11	7.8	4.5	17.9	13.7	...	...	3.1	11.3	10.4	...	6.7
14	7.1	4.0	...	...	7.2	...	...	...	...	...	7.2
18	5.9	...	...	...	5.8	...	...	...	...	...	7.9

*Cover to stirrups, if any, should be at least 1½ in.

Two-way Slabs. Flexural cracking in two-way slabs is significantly different from that in one-way slabs. For control of flexural cracking in two-way slabs, such as solid flat plates and flat slabs with drop panels, the ACI Building Code restricts the maximum spacing of tension bars to twice the overall thickness h of the slab. In waffle slabs or over cellular spaces, however, reinforcement should be the same as that for shrinkage and temperature in one-way slabs.

In the "Commentary" on ACI 318-71, Committee 224 on Cracking indicated that Eq. (5-66) can be used to predict the width of possible cracks w, in., that can develop in flexure.

$$w = K\beta f_s \sqrt{M_I} \tag{5-66}$$

where M_I = grid index = $d_{b1}s_2/\rho_{t1}$
f_s = 40% of design yield strength f_y, ksi
d_{b1} = diameter, in., of reinforcement closest to outer surface of concrete (direction 1)
s_2 = spacing, in., of perpendicular reinforcement (in direction 2)
ρ_{t1} = active steel ratio = A_{s1}/A_{t1}
A_{s1} = area, sq in., of tensile reinforcement in direction 1 per ft width of slab
A_{t1} = tensioned concrete area, sq in. per ft of width = $12(2C_1 + d_{b1})$
C_1 = clear concrete cover, in., for reinforcement in direction 1
K = experimentally determined constant, which can be taken as 2.8×10^{-5} for uniform loading
β = ratio of distance between extreme tension surface and neutral axis to distance between centroid of main reinforcement and neutral axis (generally $1.20 < \beta < 1.35$)

Crack width w computed from Eq. (5-66) is measured in direction 1.

For limiting crack widths of 0.016 in. for interior exposure and 0.013 for exterior exposure, as for one-way slabs, and with $K = 0.000028$ and $\beta = 1.3$, Eq. (5-66) can be transformed into Eqs. (5-67) and (5-68), which give the maximum spacing for Grade 60 bars in two-way slabs:

For interior exposures,

$$s = \sqrt{\frac{263d_{b1}}{2C_1 + d_{b1}}} \tag{5-67}$$

For exterior exposures,

$$s = \sqrt{\frac{173d_{b1}}{2C_1 + d_{b1}}} \tag{5-68}$$

See Table 5-12.

5-51. Deflection of Concrete Beams and Slabs. Reinforced-concrete flexural members must have adequate stiffness to limit deflection to an amount that will not adversely affect the serviceability of the structure under service loads.

Beams and One-way Slabs. Unless computations show that deflections will be small (Table 5-13), the ACI Building Code requires that the depth h of nonprestressed, one-way solid slabs, one-way ribbed slabs, and beams of normal-weight concrete—with Grade 60 reinforcement—be at least the fraction of the span L given in Table 5-14.

When it is necessary to compute deflections, calculation of short-time deflection may be based on elastic theory, but with an effective moment of inertia I_e.

For normal-weight concrete,

$$I_e = \left(\frac{M_{cr}}{M_a}\right)^3 I_g + \left[1 - \left(\frac{M_{cr}}{M_a}\right)^3\right] I_{cr} \leq I_g \tag{5-69}$$

where M_{cr} = cracking moment = $f_r I_g/y_t$
M_a = service-load moments for which deflections are being computed
I_g = gross moment of inertia of concrete section
I_{cr} = moment of inertia of cracked section transformed to concrete (for solid slabs, see Fig. 5-21)
f_r = modulus of rupture of concrete, psi = $7.5\sqrt{f'_c}$
f'_c = specified concrete strength, psi
y_t = distance from centroidal axis of gross section, neglecting the reinforcement, to the extreme surface in tension

When structural lightweight concrete is used, f_r in the computation of M_{cr} should be taken as $1.12f_{ct} \leq 7.5\sqrt{f'_c}$, where f_{ct} = average splitting tensile strength, psi, of the concrete. When f_{ct} is not specified, f_r should be taken as $5.6\sqrt{f'_c}$ for all-lightweight concrete and as $6.4\sqrt{f'_c}$ for sand-lightweight concrete.

Table 5-13. Maximum Allowable Ratios of Computed Deflection to Span L for Beams and Slabs

Type of member	Deflection to be considered	Deflection limitation
Flat roofs not supporting or attached to nonstructural elements likely to be damaged by large deflections	Immediate deflection due to the live load	$L/180$*
Floors not supporting or attached to nonstructural elements likely to be damaged by large deflections	Immediate deflection due to the live load	$L/360$
Roof or floor construction supporting or attached to nonstructural elements likely to be damaged by large deflections	That part of the total deflection that occurs after attachment of the nonstructural elements (the sum of the long-time deflection due to all sustained loads and the immediate deflection due to any additional live load)†	$L/480$‡
Roof or floor construction supporting or attached to nonstructural elements not likely to be damaged by large deflections		$L/240$§

*This limit is not intended to safeguard against ponding. Ponding should be checked by suitable calculations of deflection, including the added deflections due to ponded water, and considering long-time effects of all sustained loads, camber, construction, tolerances and reliability of provisions for drainage.

†The long-time deflection may be reduced by the amount of deflection that occurs before attachment of the nonstructural elements.

‡This limit may be exceeded if adequate measures are taken to prevent damage to supported or attached elements.

§But not greater than the tolerance provided for the nonstructural elements. This limit may be exceeded if camber is provided so that the total deflection minus the camber does not exceed the limitation,

Table 5-14. Minimum Depths h of Reinforced Concrete Beams and One-way Slabs*

	One-way solid slabs	Beams and one-way ribbed slabs
Cantilever	$L/10 = 0.1000L$	$L/8 = 0.1250L$
Simple span	$L/20 = 0.0500L$	$L/16 = 0.0625L$
Continuous:		
End span.	$L/24 = 0.0417L$	$L/18.5 = 0.0540L$
Interior span.	$L/28 = 0.0357L$	$L/21 = 0.0476L$

*For members with span L (Art. 5-41) not supporting or attached to partitions or other construction likely to be damaged by large deflections. Thinner members may be used if justified by deflection computations. For structural lightweight concrete of unit weight w, lb per cu ft, multiply tabulated values by $1.65 - 0.005w \geq 1.09$, for $90 < w < 120$. For reinforcement with yield strength $f_y > 60{,}000$ psi, multiply tabulated values by $0.4 + f_y/100{,}000$.

For deflection calculations for continuous spans, I_e may be taken as the average of the values obtained from Eq. (5-69) for the critical positive and negative moments.

Additional long-time deflection for both normal-weight and lightweight concrete flexural members can be estimated by multiplying the immediate deflection due to the sustained load by:

$$2 - 1.2\frac{A'_s}{A_s} \geq 0.6$$

where A'_s = area of compression reinforcement
A_s = area of tension reinforcement

The sum of the short-time and long-time deflections should not exceed the limits given in Table 5-13.

Two-way Slabs. Unless computations show that deflections will not exceed the limits listed in Table 5-13, thickness h for nonprestressed two-way slabs with a

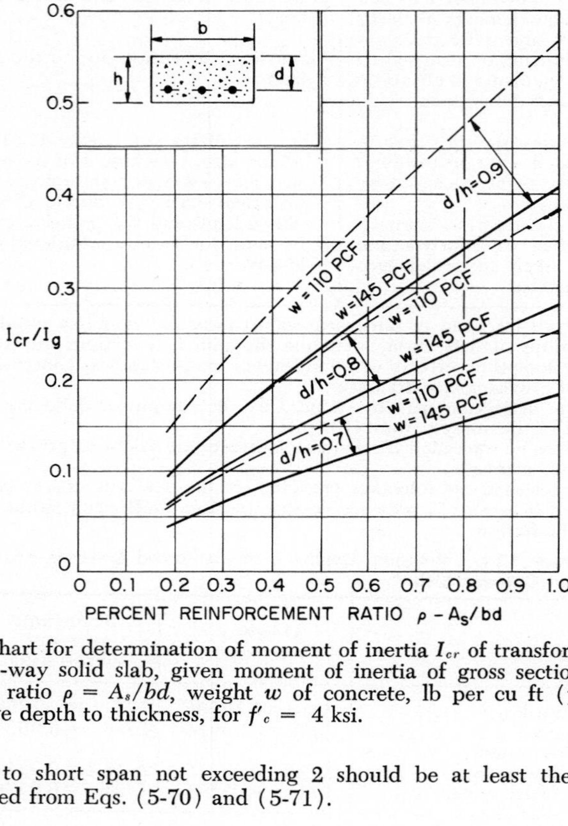

Fig. 5-21. Chart for determination of moment of inertia I_{cr} of transformed (cracked) section of one-way solid slab, given moment of inertia of gross section $I_g = bh^3/12$, reinforcement ratio $\rho = A_s/bd$, weight w of concrete, lb per cu ft (pcf), and ratio d/h of effective depth to thickness, for $f'_c = 4$ ksi.

ratio of long to short span not exceeding 2 should be at least the larger of the values computed from Eqs. (5-70) and (5-71).

$$h = \frac{L_n(800 + 0.005f_y)}{36{,}000 + 5{,}000\beta[\alpha_m - 0.5(1 - \beta_s)(1 + 1/\beta)]} \tag{5-70}$$

$$h = \frac{L_n(800 + 0.005f_y)}{36{,}000 + 5{,}000\beta(1 + \beta_s)} \tag{5-71}$$

But h need not be more than

$$h = \frac{L_n(800 + 0.005f_y)}{36{,}000} \tag{5-72}$$

where L_n = clear span in long direction, in.

α_m = average value of α for all beams along panel edges

α = ratio of flexural stiffness of beam section to flexural stiffness of a width of slab bounded laterally by the centerline of the adjacent panel, if any, on each side of the beam

β = ratio of clear span in long direction to clear span in short direction

β_s = ratio of length of continuous edges to total perimeter of slab panel

The thickness, however, should not be less than the following:

For slabs without beams or drop panels . . . . 5 in.
For slabs without beams, but with drop panels . . . . 4 in.
For slabs having beams on all four edges with a value of α_m at least equal to 2.0 . . . . 3½ in.

The minimum thicknesses for flat slabs with standard drop panels may be reduced 10%.

Unless edge beams with $\alpha \geq 0.8$ are provided at discontinuous edges, the minimum thicknesses of panels at those edges should be increased at least 10%.

The computed deflections of prestressed-concrete construction should not exceed the values listed in Table 5-13.

ONE-WAY REINFORCED-CONCRETE SLABS

A one-way reinforced-concrete slab is a flexural member that spans in one direction between supports and is reinforced for flexure only in one direction (Art. 5-64). If a slab is supported by beams or walls on four sides, but the span in the long direction is more than twice that in the short direction, most of the load will be carried in the short direction; hence, the slab can be designed as a one-way slab.

Table 5-15. Typical Fire Ratings for Concrete Members*

Fire rating, hr	Slabs and joists				Cover, in., for beams and columns 12 in. or larger in section
	Lightweight concrete		Normal-weight concrete		
	Depth h, in.	Cover, in.	Depth h, in.	Cover, in.	
1	3	¾	3½†	¾	1½
2	4	¾	4½	¾	1½
3	4½	1	5½	1	1½
4	5	1	6½	1	1½

*From Uniform Building Code, for Grade A concrete containing less than 40% quartz, chert, or flint.

†May be reduced to 3 in. when limestone aggregate is used.

One-way slabs may be solid, ribbed, or hollow. (For one-way ribbed slabs, see Arts. 5-54 to 5-57.) Hollow one-way slabs are usually precast (Arts. 5-98 to 5-105). Cast-in-place, hollow one-way slabs can be constructed with fiber or cardboard-cylinder forms, inflatable forms that can be reused, or precast hollow boxes or blocks. One-way slabs can be haunched at the supports for flexure or for shear strength.

5-52. Analysis and Design of One-way Slabs. Structural strength, fire resistance, crack control, and deflections of one-way slabs must be satisfactory under service loads.

Strength and Deflections. Approximate methods of frame analysis can be used with uniform loads and spans that conform to ACI Code requirements (see Art. 5-41). Deflections can be computed as indicated in Art. 5-51, or in lieu of calculations the minimum slab thicknesses listed in Table 5-14 may be used. In Fig. 5-21 is a plot of ratios of moments of inertia of cracked to gross concrete section for one-way slabs. These curves can be used to simplify deflection calculations.

Strength depends on slab thickness and reinforcement and properties of materials used. Slab thickness required for strength can be computed by treating a 1-ft width of slab as a beam (Arts. 5-44 to 5-46).

Fire Resistance. One-way concrete slabs, if not protected by a fire-resistant ceiling, must have a thickness and a concrete cover around reinforcement that conforms to the fire-resistance rating required by the building code. Table 5-15 gives typical slab thickness and cover around reinforcement for various fire-resistance ratings for normal-weight and structural-lightweight-concrete construction.

Reinforcement. Requirements for minimum reinforcement for crack control are

Table 5-16. Minimum and Maximum Reinforcement for One-way Reinforced-concrete Slabs

Slab thickness, in.	Minimum reinforcement*			Maximum reinforcement†		
	Area A_s, sq in.	Bar size and spacing, in.	Weight, psf	Area A_s, sq in.	Bar size and spacing, in.	Weight, psf
4	0.086	No. 3 @ 12‡	0.30	0.555	No. 5 @ 9½	1.89
4½	0.097	No. 3 @ 13½	0.33	0.655	No. 6 @ 8	2.24
5	0.108	No. 3 @ 12½	0.36	0.750	No. 6 @ 7	2.56
5½	0.119	No. 3 @ 11	0.41	0.845	No. 6 @ 6	2.99
6	0.130	No. 4 @ 18	0.45	0.931	No. 7 @ 7½	3.26
6½	0.140	No. 4 @ 17	0.48	1.025	No. 7 @ 7	3.50
7	0.151	No. 4 @ 15½	0.53	1.110	No. 8 @ 8½	3.80
7½	0.162	No. 4 @ 14½	0.56	1.208	No. 8 @ 7½	4.30
8	0.173	No. 4 @ 13½	0.61	1.291	No. 9 @ 9	4.54
8½	0.184	No. 4 @ 13	0.63	1.382	No. 9 @ 8½	4.80
9	0.194	No. 4 @ 12	0.68	1.482	No. 9 @ 8	5.10

* For Grade 60 reinforcement. Minimum area $A_s \geq 0.0018bh$, where b = slab width and h = slab thickness.

† For f'_c = 3,000 psi and no compression reinforcement.

‡ This spacing for a 4-in. slab is for flexure and can be increased to 15 in. for temperature and shrinkage reinforcement.

summarized in Art. 5-50. Table 5-16 lists minimum reinforcement when Grade 60 bars are used. Reinforcement required for flexural strength can be computed by treating a 1-ft width of slab as a beam (Arts. 5-44 to 5-46).

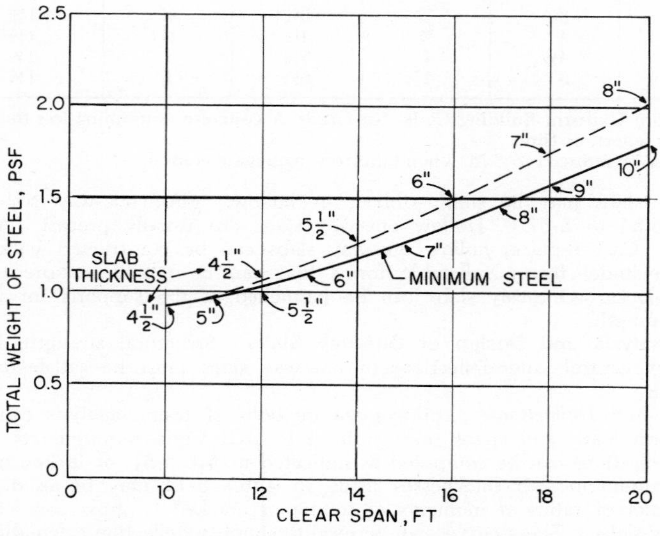

Fig. 5-22. Weight of reinforcing steel required for an interior span of a continuous, one-way solid slab of 3,000-psi concrete, weighing 150 lb per cu ft and carrying 100-psf live load (170-psf design live load).

Rebar weights, lb per sq ft of slab area, can be estimated from Fig. 5-22 for one-way, continuous, interior spans of floor or roof slabs made of normal-weight concrete.

One-way reinforced concrete slabs with spans less than 10 ft long can be reinforced with a single layer of draped welded-wire fabric for both positive and negative moments. These moments can be taken equal to $wL^2/12$, where w is the uniform load and L is the span, defined in Art. 5-41, if the slab meets ACI Code requirements for approximate frame analysis with uniform loads.

For development (bond) of reinforcement, see Art. 5-49.

Shear. This is usually not critical in one-way slabs, but the ACI Building Code requires that it be investigated (see Art. 5-47).

5-53. Embedded Pipes in One-way Slabs. Generally, embedded pipes or conduit, other than those merely passing through, should not be larger in outside dimension than one-third the slab thickness and should be spaced at least three diameters or widths on centers. Pipes that contain liquid, gas, or vapor should not be embedded in slabs if their temperature exceeds 150°F or the pressure is over 200 psi above atmospheric. Piping in solid one-way slabs is required to be placed between the top and bottom reinforcement unless it is for radiant heating or snow melting.

ONE-WAY CONCRETE-JOIST CONSTRUCTION

One-way concrete-joist construction consists of a monolithic combination of cast-in-place, uniformly spaced ribs (joists) and top slab (Fig. 5-23). (See also Art. 5-64.)

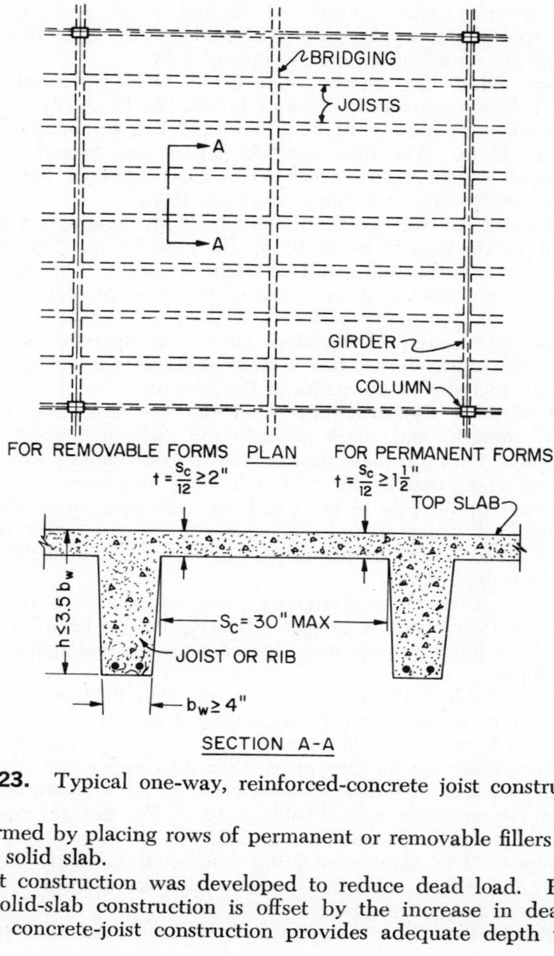

Fig. 5-23. Typical one-way, reinforced-concrete joist construction.

The ribs are formed by placing rows of permanent or removable fillers in what would otherwise be a solid slab.

One-way joist construction was developed to reduce dead load. For long spans, the utility of solid-slab construction is offset by the increase in dead load of the slab. One-way concrete-joist construction provides adequate depth with less dead load than for solid slabs, and results in smaller concrete and reinforcement quantities per square foot of floor area.

Uniform-depth floor and roof construction can be obtained by casting the joists integral with wide, supporting band beams of the same total depth as the joists. This design eliminates the need for interior beam forms.

5-54. Standard Sizes of Joists. One-way concrete-joist construction that exceeds the dimensional limitations of the ACI Building Code must be designed as slabs and beams. These dimensional limitations are:

Maximum clear spacing between ribs—30 in.

Maximum rib depth—3.5 times rib width.

Minimum rib width—4 in.

Minimum top-slab thickness with removable forms—2 in., but not less than one-twelfth the clear spacing of ribs.

Minimum top-slab thickness with permanent forms—1½ in., but not less than one-twelfth the clear spacing of ribs.

Removable form fillers can be standard, steel *pans* or hardboard, corrugated cardboard, fiberboard or glass reinforced plastic. Standard, removable, steel pans that conform to "Types and Sizes of Forms for One-way Concrete-joist Construction," NBS Voluntary Product Standard No. PS 16-69, include 20- and 30-in. widths and depths of 6, 8, 10, 12, 14, 16, and 20 in. Standard, steel, square-end pans are available in 36-in. lengths. Widths of 10 and 15 in. and tapered end fillers are available as special items. For forms 20 and 30 in. wide, tapered end forms slope to 16 and 25 in., respectively, in a distance of 3 ft.

Permanent form fillers are usually constructed from structural-clay floor tiles (ASTM C57) or hollow load-bearing concrete blocks (ASTM C90).

Structural-clay tiles are usually 12 in. square and can be obtained in thicknesses varying from 3 to 12 in. The titles are laid flat, or end to end, in rows between and at right angles to the joists. The usual clear distance between rows is 4 in., making the center-to-center spacing of the rows 16 in.

Hollow load-bearing concrete block are usually 16 in. long and 8 in. high. They can be obtained in thicknesses of 4, 6, 8, 10, and 12 in. The blocks are laid flat, end-to-end, in rows between and at right angles to the joists. The usual clear distance between rows is 4 in., making the center-to-center spacing of the rows 20 in.

If the clay-tile or concrete-block fillers have a compressive strength equal to that of the concrete in the joists, the vertical shells of the fillers in contact with the joist ribs can be included in the width of the joist rib.

5-55. Design of Joist Construction. One-way concrete joists must have adequate structural strength, and crack control and deflection must be satisfactory under service loads. Approximate methods of frame analysis can be used with uniform loads and spans that conform to ACI Code requirements (see Art. 5-41). Table 5-14 lists minimum depths of joists to limit deflection, unless deflection computations justify shallower construction (Table 5-13). Load tables in the Concrete Reinforcing Steel Institute "Design Handbook" indicate when deflections under service live loads exceed specified limits.

Economy can be obtained by designing joists and slabs so that the same-size forms can be used throughout a project. It will usually be advantageous to use square-end forms for interior spans and tapered ends for end spans, when required, with a uniform depth.

Fire Resistance. Table 5-17 gives minimum top-slab thickness and reinforcement cover for fire resistance when a fire-resistant ceiling is not used.

Temperature and Shrinkage Reinforcement. This must be provided perpendicular to the ribs and spaced not farther apart than five times the slab thickness, or 18 in. The required area of Grade 60 reinforcement for temperature and shrinkage is 0.0018 times the concrete area (Table 5-18). For flexural reinforcement, see Art. 5-56. For shear reinforcement, see Art. 5-57.

Embedded Pipes. Top slabs containing horizontal conduit or pipes that are allowed by the ACI Code (Art. 5-53) must have a thickness of at least 1 in. plus the depth of the conduit or pipe.

Bridging. Distribution ribs are constructed normal to the main ribs to distribute concentrated loads to more than one joist and to equalize deflections. These ribs are usually made 4 to 5 in. wide and reinforced top and bottom with one No. 4 or 5 continuous rebar. One distribution rib is usually used at the center of spans of up to 30 ft, and two distribution ribs are usually placed at the third points of spans longer than 30 ft.

Openings. These can be provided in the top slab of one-way concrete joist construction between ribs without significant loss in flexural strength. Header joists must be provided along openings that interrupt one or more joists.

Table 5-17. Required Top-slab Thickness and Cover for Fire Resistance of One-way Concrete Joist Construction*

Fire-resistance rating, hr	Normal-weight concrete		Structural lightweight concrete	
	Reinforcement cover, in.	Top-slab thickness, in.	Reinforcement cover, in.	Top-slab thickness, in.
1	¾	3½†	¾	3
2	1	4½	1	4
3	1¼	5½	1¼	4½

* From Uniform Building Code.
† Thickness can be reduced to 3 in. where limestones aggregate is used.

5-56. Reinforcement of Joists for Flexure. Reinforcement required for strength can be determined as indicated in Art. 5-46, by treating as a beam a section symmetrical about a rib and as wide as the spacing of ribs on centers.

Minimum Reinforcement. Positive-, or bottom, moment reinforcement, with a yield strength f_y, must have an area equal to or greater than $200/f_y$ times the concrete area of the rib $b_w d$, where b_w is the rib width and d = rib depth. Less bottom steel can be used, however, if the areas of both the positive and negative reinforcement at every section is one-third greater than the amount required by analysis. There is no minimum limitation on negative or top reinforcement. (See also Art. 5-55.)

Table 5-18. Temperature and Shrinkage Reinforcement for One-way Joist Construction

Top-slab thickness, in.	Required area of temperature and shrinkage reinforcement, sq in.	Reinforcement	Reinforcement weight, psf
2	0.043	WWF 4 × 12, W1.5/W1	0.19
2½	0.054	WWF 4 × 12, W2/W1	0.24
3	0.065	WWF 4 × 12, W2.5/W1	0.29
3½	0.076	No. 3 @ 17½ in.	0.26
4	0.086	No. 3 @ 15 in.	0.30
4½	0.097	No. 3 @ 13½ in.	0.33
5	0.108	No. 3 @ 12½ in.	0.36
5½	0.119	No. 3 @ 11 in.	0.41

If bottom bars in continuous joists are not continuous through the support, top bars should meet the requirements for shrinkage and temperature reinforcement.

Maximum Reinforcement. Positive- and negative-moment steel ratios must not be greater than three-quarters of the ratio that produces balanced conditions (Art. 5-46). The positive-moment reinforcement ratio is based on the width of the top flange, and the negative-moment reinforcement ratio is based on the width of the rib b_w.

Reinforcement for one-way concrete-joist construction may consist of one straight and one bent-up bar in each joist rib or straight top and bottom bars, cut off as required for moment. The straight-bar arrangement is preferred if crack control is essential at the top surface.

For top-slab reinforcement, see Art. 5-55 and Table 5-12. Straight top- and bottom-bar arrangements are more flexible in attaining uniform distribution of top bars to control cracking in the slab than straight and bent bars.

For development (bond) of reinforcement, see Art. 5-49.

Figure 5-24 shows rebar quantities, lb per sq ft of floor or roof area, for continuous,

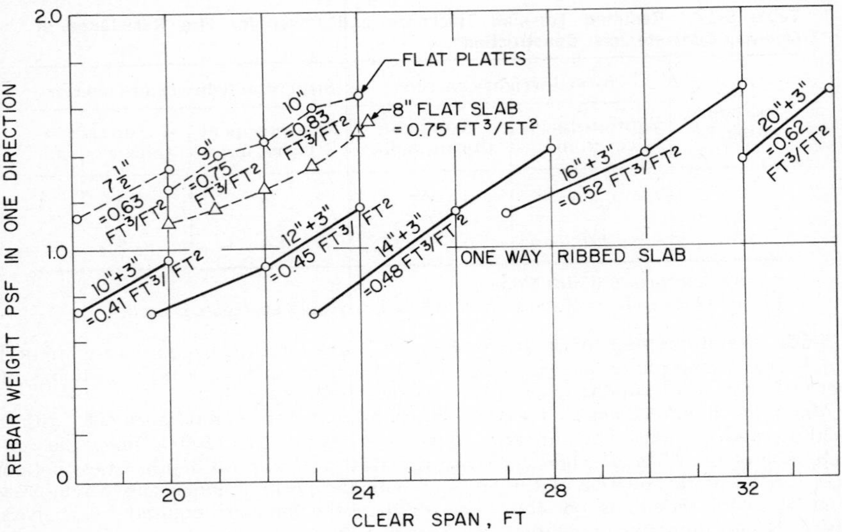

Fig. 5-24. Weights of steel and concrete in flat-plate, flat-slab, and one-way joist construction, for preliminary estimates. (*Concrete Reinforcing Steel Institute.*)

interior spans of one-way concrete-joist construction made with normal-weight concrete.

5-57. Shear in Joists. The average design shear stress v_u, psi, as a measure of diagonal tension is computed from

$$v_u = \frac{V_u}{\phi b_w d} \tag{5-73}$$

where V_u = design shear force at section, lb

ϕ = capacity reduction factor (Art. 5-44) = 0.85

d = distance, in., from extreme compression surface to centroid of tension steel

b_w = rib width, in.

The width b_w can be taken as the average of the width of joist at the compression face and the width at the tension steel. The slope of the vertical taper of ribs formed with removable steel pans can safely be assumed as 1 in 12. For permanent concrete block fillers, the shell of the block can be included as part of b_w, if the compressive strength of the masonry is equal to or greater than the concrete.

Based on satisfactory performance of joist construction, the ACI Building Code allows the average design shear stress v_c carried by concrete in joists to be taken 10% greater than that for beams or slabs.

If shear controls the design of one-way concrete-joist construction, tapered ends can be used to increase the shear capacity. The Concrete Reinforcing Steel Institute "Design Handbook" has comprehensive load tables for one-way concrete-joist construction that indicate where shear controls and when tapered ends are required for simple, end, and interior spans.

For joists supporting uniform loads, the critical section for shear stress at tapered ends is the narrow end of the tapered section. Shear need not be checked within the taper.

Reinforcement for shear must be provided when the average shear stress v_u exceeds that allowed on the concrete. The use of single-prong No. 3 stirrups spaced

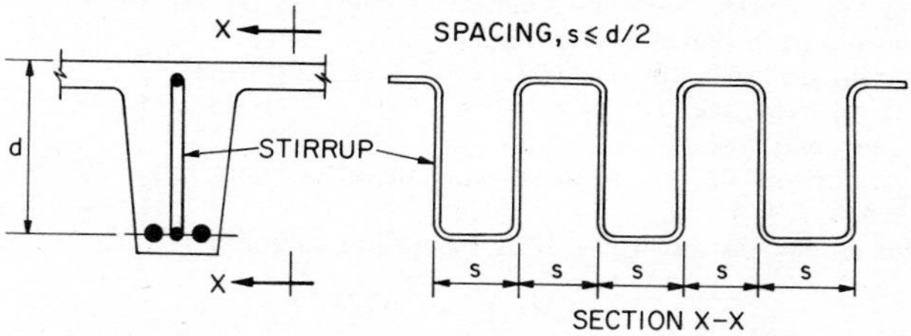

Fig. 5-25. Stirrups for concrete-joist construction.

at half depth, such as that shown in Fig. 5-25, is practical in narrow joists; they can be easily placed between two bottom bars.

TWO-WAY CONSTRUCTION

A two-way slab is a concrete panel reinforced for flexure in more than one direction. (See also Art. 5-64.) Many variations of this type of contruction have been used for floors and roofs, including flat plates, solid flat slabs, and waffle flat slabs. Generally, the columns that support such construction are arranged so that their center lines divide the slab into square or nearly square panels, but if desired, rectangular, triangular, or even irregular panels may be used.

5-58. Analysis and Design of Flat Plates. The flat plate is the simplest form of two-way slab—simplest for analysis, design, detailing, bar fabrication and placing, and formwork. A flat plate is defined as a two-way slab of uniform thickness supported by any combination of columns and walls, with or without edge beams, and without drop panels, column capitals, and brackets.

Shear and deflection limit economical flat-plate spans to under about 30 ft for light loading and about 20 to 25 ft for heavy loading. While use of reinforcing-steel or structural-steel shear heads for resisting shear at columns will extend these limits somewhat, their main application is to permit use of smaller columns. A number of other variations, however, can be used to extend economical load and span limits (Arts. 5-61 and 5-62).

The ACI Building Code permits two methods of analysis for two-way construction: *direct design,* within limitations of span and load, and *equivalent frame* (Art. 5-42). Limitations on use of direct design are:

A minimum of three spans continuous in each direction.

Rectangular panels with ratios of opposite sides less than 2.

Successive span ratios not to exceed 2:3.

Columns offset from center lines of successive columns not more than 0.10 span in either direction.

Specified ratio of live load to dead load not more than 3.

The procedure for either method of design begins with selection of preliminary dimensions for review, and continues with six basic steps.

Step 1. Select a plate thickness expected to be suitable for the given conditions of load and span. This thickness, unless deflection computations justify thinner plates, should not be less than h determined from Eq. (5-72). With Grade 60 reinforcement, minimum thickness is, from Eq. (5-72), for an interior panel

$$h = \frac{L_n}{32.7} \geq 5 \text{ in.} \qquad (5\text{-}74)$$

where L_n = clear span in direction moments are being determined

Also, as indicated in Art. 5-51, for discontinuous panels, h computed from Eq. (5-72) or (5-74) may have to be increased at least 10%.

Step 2. Determine for each panel the total static design moment

$$M_o = 0.125wL_2L_n^2 \tag{5-75}$$

where L_2 = panel width (center-to-center spans transverse to direction in which moment is being determined)
w = total design load, psf = $1.4D + 1.7L$, typically
D = dead load, psf
L = live load, psf

Step 3. Apportion M_o to positive and negative bending. In the direct-design method:

For interior spans, the negative design bending moment is

$$M_u = -0.65M_o \tag{5-76}$$

and the positive design bending moment is

$$M_u = 0.35M_o \tag{5-77}$$

For exterior spans (edge panels), the distribution of M_o is based on α_{ec}, the ratio of the flexural stiffness of the edge equivalent column K_{ec} to the flexural stiffness of the slab K_s.

$$\alpha_{ec} = \frac{K_{ec}}{K_s} \tag{5-78}$$

Equivalent column consists of actual columns above and below the slab plus an attached torsional member transverse to the direction in which moments are being determined and extending to the center lines on each side of the column of the bounding lateral panel. The torsional member should be taken as a portion of the slab as wide as the column in the direction in which moments are determined. (See also Art. 5-59.)

At the edge column, the negative design moment is

$$M_u = -0.65M_o \frac{\alpha_{ec}}{\alpha_{ec} + 1} \tag{5-79}$$

Also, in an edge panel, the design positive moment is

$$M_u = M_o \left(0.63 - 0.28 \frac{\alpha_{ec}}{\alpha_{ec} + 1}\right) \tag{5-80}$$

and at the first interior column, the negative design moment is

$$M_u = -M_o \left(0.75 - 0.10 \frac{\alpha_{ec}}{\alpha_{ec} + 1}\right) \tag{5-81}$$

Step 4. Distribute panel moments M_u to column and middle strips.

Column strip is a design strip with a width of $0.25L_2 \leq 0.25L_1$ on each side of the column center line, where L_1 is the center-to-center span in the direction in which moments are being determined (Fig. 5-26).

Middle strip is the design strip between two column strips (Fig. 5-26).

For flat plates without beams, the distributions of M_u become:

For positive moment, column strip 60%, middle strip 40%.

For negative moment at the edge column, column strip 100%.

For interior negative moments, column strip 75%, middle strip 25%.

A design moment may be modified up to 10% so long as the sum of the positive and negative moments in the panel in the direction being considered is at least that given by Eq. (5-75).

Step 5. Check for shear. Shear strength of slabs in the vicinity of columns or other concentrated loads is governed by the severer of two conditions: when the slab acts as a wide beam and when the load tends to punch through the slab. In the first case, a diagonal crack might extend in a plane across the entire width. Design for this condition is described in Art. 5-47. For the two-way action of the second condition, diagonal cracking might occur along the surface of a truncated cone or pyramid around the load.

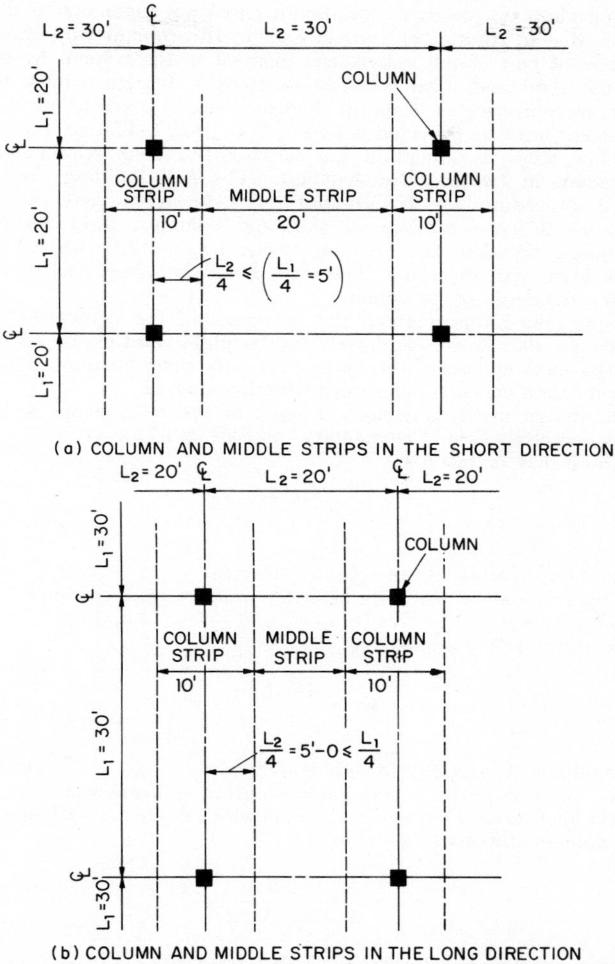

Fig. 5-26. Division of flat plate into column and middle strips.

The critical section for two-way action, therefore, should be taken perpendicular to the plane of the slab at a distance $d/2$ from the periphery of the column or load, where d is the effective depth of slab. The nominal shear stress for punching action can be computed from

$$v_u = \frac{V_u}{\phi b_o d} \tag{5-82}$$

where V_u = design shear force at critical section
ϕ = capacity reduction factor = 0.85
b_o = periphery of critical section

Unless shear reinforcement is provided, v_u should not exceed $v_c = 4\sqrt{f'_c}$, where f'_c is the specified concrete strength, psi. With shear reinforcement as for beams (Art. 5-47), v_u may be as high as $6\sqrt{f'_c}$. With steel-shape shearheads at interior columns meeting ACI Building Code requirements, v_u may be as high as $7\sqrt{f'_c}$.

Determine the maximum shear at each column for two cases: all panels loaded, and live load on alternate panels for maximum unbalanced moment to the columns. Combine shears due to transfer of vertical load to the column with shear resulting from the transfer of part of the unbalanced moment to the column by torsion. At this point, if the combined shear is excessive, steps 1 through 5 must be repeated with a larger column, thicker slab, or higher-strength concrete in the slab; or shear reinforcement must be provided where $v_u > v_c$ (Art. 5-47).

Step 6. When steps 1 through 5 are satisfactory, select reinforcement.

5-59. Stiffnesses in Two-way Construction. The ACI Building Code prescribes a sophisticated procedure for computation of stiffness to determine α_{ec} (Art. 5-58). Variations in cross sections of slab and columns, drop panels, capitals, and brackets should be taken into account. Columns should be treated as infinitely stiff within the joint with the slab. The slab should be considered to be stiffened somewhat within the depth of the column.

With the equivalent-frame method, the stiffnesses of the equivalent column K_{ec} (Art. 5-58) and the slab K_s are employed in a straightforward elastic analysis. With the direct-design method, use of the value of α_{ec} for determination of moments for edge spans comprises a one-step moment-distribution analysis.

In the direct-design method, neglect of some of the refinements in computation of stiffnesses is permitted (see "Commentary" on ACI 318-71).

Column stiffness may be taken as

$$K_c = \frac{4E_cI_c}{L_c} \qquad (5\text{-}83a)$$

where E_c = modulus of elasticity of column concrete
I_c = column moment of inertia based on uniform cross section
L_c = story height

Slab flexural stiffness may be taken as

$$K_s = \frac{4E_{cs}I_{cs}}{L_1} \qquad (5\text{-}83b)$$

where E_{cs} = modulus of elasticity of slab concrete
I_{cs} = moment of inertia of slab based on gross, uniform section
L_1 = center-to-center span in direction in which moments are being determined

Equivalent column stiffness is given by

$$K_{ec} = \frac{\Sigma K_c}{1 + (1/K_t)\,\Sigma K_c} \qquad (5\text{-}84)$$

where K_t = torsional stiffness of portion of slab (Art. 5-58)

The summation ΣK_c applies to the column above and the column below the slab.

$$K_t = \sum \frac{9E_{cs}C}{L_2(1 - c_2/L_2)^3} \qquad (5\text{-}85)$$

where $C = \Sigma(1 - 0.63x/y)x^3y/3$
c_2 = edge-column width transverse to direction in which moments are being determined
L_2 = center-to-center span transverse to L_1
x,y = dimensions of the section of flat plate in contact with side of column parallel to L_1, c_1 in width and h in depth ($x \leq y$)

In Eq. 5-85, the summation applies to the transverse spans of the slab on each side of the edge column.

Part of the unbalanced moment at the exterior-edge column is transferred to the column through torsional shear stresses on the slab periphery at a distance $d/2$ from the column, where d is the effective slab depth. The fraction of the unbalanced moment thus transmitted is

$$1 - \frac{1}{1 + \frac{2}{3}\sqrt{\frac{c_1 + d/2}{c_2 + d}}}$$

For preliminary design, with square columns flush at edges of the flat plate, a rapid estimate of the shear capacity to allow for effects of torsion can be made by using uniform vertical load w only, with allowable stresses for design load as follows:

For edge column, total shear $V_u = 0.5wL_2L_1$ and allowable shear, $v_c = 2\sqrt{f'_c}$

For first interior column, $V_u = 1.15wL_2L_1$ and allowable shear, $v_c = 4\sqrt{f'_c}$

where f'_c = specified concrete strength, psi

Use of this calculation in establishing a preliminary design is a short cut, which will often avoid the need for repeating steps 1 through 5 in Art. 5-58, because it gives a close approximation for final design.

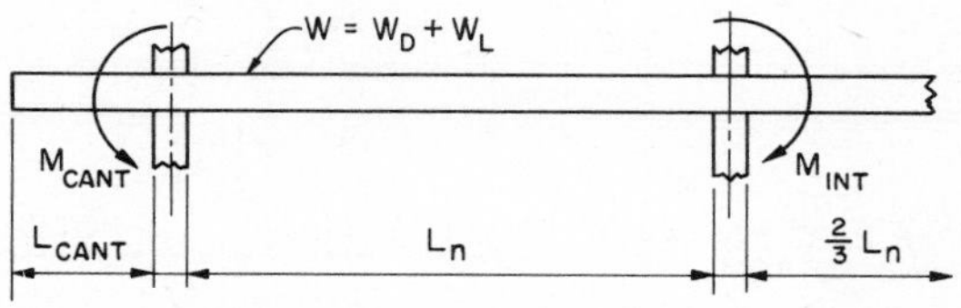

Fig. 5-27. Miminum cantilever span, for treatment of exterior column as interior column, equals $4L_n/15$, obtained by equating minimum cantilever moment at column to minimum negative moment at interior column.

Figure 5-27 shows the minimum cantilever edge span with which all columns can be considered interior columns and the direct-design method can be employed without tedious stiffness calculations.

5-60. Bar Lengths and Details for Flat Plates. The minimum bar lengths for reinforcing flat plates shown in Fig. 5-28, prescribed by the ACI Building Code, save development (bond) computations. The size of all top bars must be selected so that the development length l_d required for the bar size, concrete strength, and grade of the bar is not greater than the length available for development (see Tables 5-7 and 5-10).

The size of top bars at the exterior edge must be small enough that the hook plus straight extension to the face of the column E is less than that required for full embedment (Table 5-10).

Middle-strip bottom bars in Fig. 5-28 are shown extended into interior columns so that they lap, and one line of bar supports may be used. This anchorage, which exceeds ACI minimum requirements, usually insures ample development length and helps prevent temperature and shrinkage cracks at the center line.

Crack Control. The ACI Code prescribes formulas for determination of maximum bar spacings based on crack width considered suitable for interior and exterior exposure (Art. 5-50). These requirements apply only to one-way reinforced elements. For two-way slabs, similar procedures are recommended. Bar spacing at critical sections, however, should not exceed twice the slab thickness, except in the top slab of cellular or ribbed (waffle) construction, where requirements for temperature and shrinkage reinforcement govern.

With Grade 60 bars, the suggested maximum spacings for control of flexural cracking in two-way flat plates are given in Table 5-12.

5-61. Flat Slabs. A flat slab is a two-way slab generally of uniform thickness, but it may be thickened or otherwise strengthened in the region of columns by a drop panel, while the column top below the slab may be enlarged by a capital (round) or bracket (prismatic). If a drop panel is used to increase depth for negative reinforcement, the minimum side dimensions of this panel are $L_1/3$ and $L_2/3$, where L_1 and L_2 are the center-to-center spans in perpendicular directions. Minimum depth of a drop panel is $1.25h$, where h is the slab thickness elsewhere.

A *waffle flat slab* or *waffle flat plate* consists of a thin, two-way top slab and a grid of joists in perpendicular directions, cast on square dome forms. For strengthening around columns, the domes are omitted in the drop panel areas, to form a solid head, which also may be made deeper than the joists. Other variations of waffle patterns include various arrangements with solid beams on column center lines both ways. Standard sizes of two-way joist forms are given in Table 5-19.

The main complication introduced in the design of flat slabs and waffle flat plates is the change in cross section of the section considered effective for torsional stiffness K_t (Art. 5-59). The ACI Building Code permits this variation in depth to be neglected. This approximation results in an overestimation of flat-slab torsional

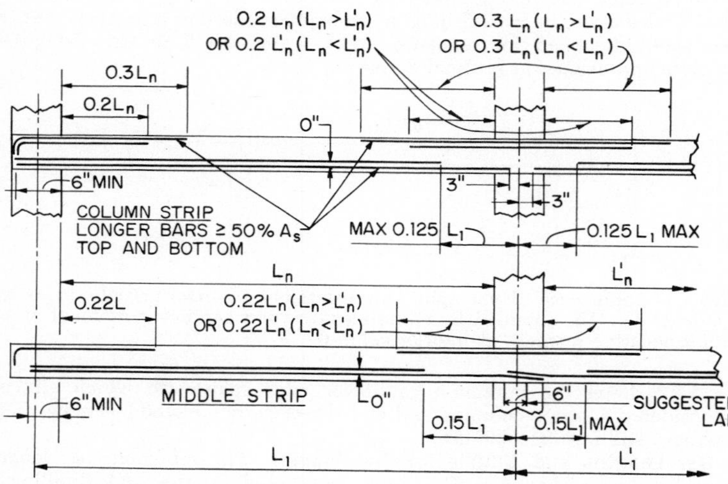

Fig. 5-28. Reinforcing details for column and middle strips of flat plates.

stiffness, making it equal to that of a flat plate with thickness equal to that of the flat-slab drop panel or waffle-flat-plate solid head. Except for the increase in negative moments at exterior columns, however, this assumption changes moments only slightly. Thus, the procedures outlined in Art. 5-58 can also be used for flat slabs.

The drop panel increases shear capacity. Hence, a solid flat slab can ordinarily be designed for concrete of lower strength than for a flat plate. Also, deflection of a flat slab is reduced by the added stiffness drop panels provide.

The depth of drop panels can be increased beyond $1.25h$ to reduce negative-moment reinforcement and to increase shear capacity when smaller columns are desired. If this adjustment is made, shear in the slab at the edge of the drop panel may become critical. In that case, shear capacity can be increased by making the drop panel larger, up to about 40% of the span. See Fig. 5-29 for bar details (column strip).

Waffle flat plates behave similarly to solid flat slabs with drop panels. Somewhat higher-strength concrete, to avoid the need of stirrups in the joists immediately around the solid head, is usually desirable. If required, however, such stirrups can be made in one piece as a longitudinal assembly, to extend the width of one dome between the drop head and the first transverse joist. For exceptional

Table 5-19. Standard Sizes of Two-way Joist Forms

Depth, in.	Volume, cu ft per dome	Weight of displaced concrete, lb per dome	3-in. top slab		4½-in. top slab	
			Equiv. slab thickness, in.	Weight,* psf	Equiv. slab thickness, in.	Weight,* psf
30-in. wide domes						
8	3.85	578	5.8	73	7.3	92
10	4.78	717	6.7	83	8.2	102
12	5.53	830	7.4	95	9.1	114
14	6.54	980	8.3	106	9.9	120
16	6.54	1116	9.1	114	10.6	133
20	9.16	1375	10.8	135	12.3	154
19-in. wide domes						
			2½-in. top slab		4½-in. top slab	
6	1.09	163	5.2	66	7.2	90
8	1.41	211	6.3	78	8.3	103
10	1.90	285	6.8	85	8.8	111
12	2.14	321	8.1	101	10.1	126

* Basis: w = 150 pcf.

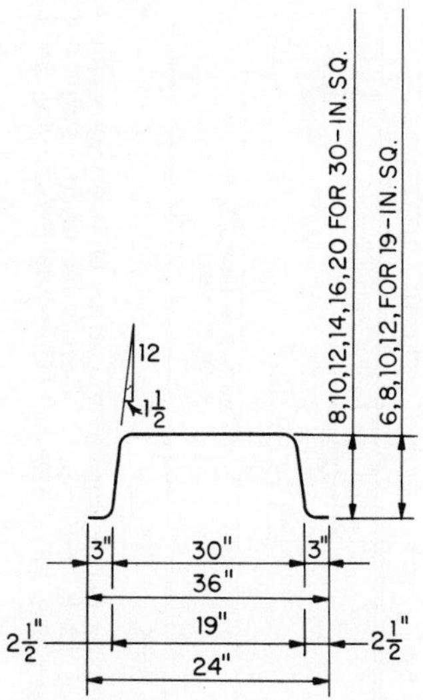

cases, such stirrups can be used between the second row of domes also. See Fig. 5-30 for reinforcement details.

5-62. Two-Way Slabs on Beams. The ACI Building Code provides for use of beams on the sides of panels, on column center lines. (A system of slabs and beams supported by girders, however, usually forms rectangular panels. In that case, the slabs are designed as one-way slabs.)

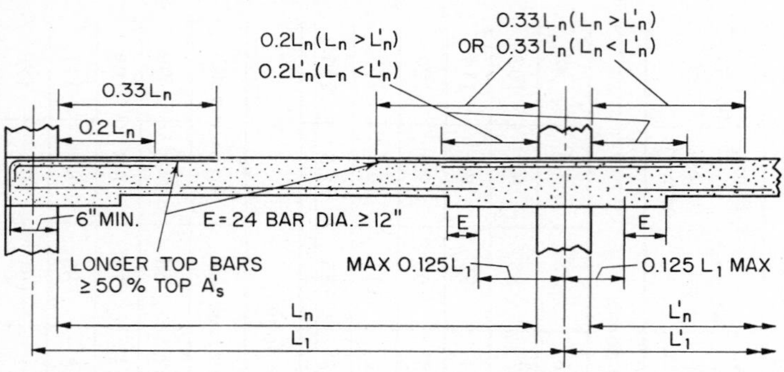

Fig. 5-29. Reinforcing details for column strips of flat slabs. Details for middle strips are the same as for middle strips of flat plates (Fig. 5-28).

Use of beams on all sides of a panel permits use of thinner two-way slabs, down to a minimum thickness $h = 3\frac{1}{2}$ in. A beam may be assumed to carry as much as 85% of the column-strip moment, depending on its stiffness relative to the slab (see Chapter 13, ACI 318-71). A secondary benefit, in addition to the direct advantages of longer spans, thinner slabs, and beam stirrups for shear, is that many local codes allow reduced live loads for design of the beams. These reductions are based on the area supported and the ratio of dead to live load. For

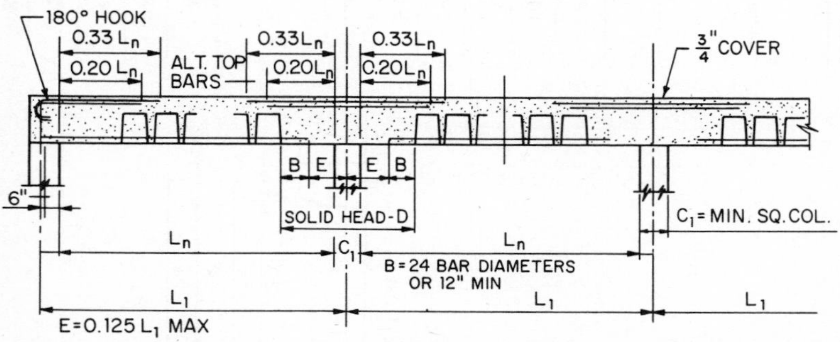

Fig. 5-30. Reinforcing details for column strips of two-way waffle flat plates. $B =$ 24 bar diameters or 12 in. minimum. Details for middle strips are the same as for middle strips of flat plates (Fig. 5-28). (*Concrete Reinforcing Steel Institute.*)

live loads up to 100 psf, such reductions are usually permitted to a maximum of 60%. Where such reductions are allowed, the reduced total panel moment M_o (Art. 5-58) and the increased effective depth to steel in the beams offer savings in reinforcement to offset partly the added cost of formwork for the beams.

5-63. Estimating Guide for Two-way Construction. Figure 5-31 can be used to estimate quantities of steel, concrete, and formwork for flat slabs, as affected by load and span. It also affords a guide to preliminary selection of dimensions for analysis, and can be used as an aid in selecting the structural system most appropriate for particular job requirements.

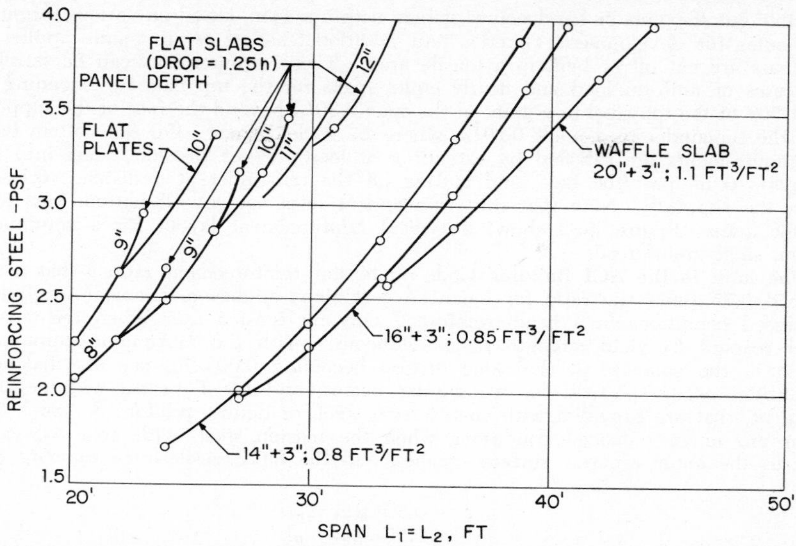

Fig. 5-31. Weight of reinforcing steel in square interior panels of flat plates, flat slabs, and waffle flat slabs of 4,000-psi concrete, with rebars of 60-ksi yield strength, carrying a design load of 200 psf, for preliminary estimates.

BEAMS

Most ACI Building Code requirements for design of beams and girders refer to *flexural members.* When slabs and joists are not intended, the Code refers specifically to *beams* and occasionally to *beams and girders,* and provisions apply equally to beams and girders. So the single term, *beams,* will be used in the following.

5-64. Definitions of Flexural Members. The following definitions apply for purposes of this section:

Slab—a flexural member of uniform depth supporting area loads over its surface. A slab may be reinforced for flexure in one or two directions.

Joist-slab—a ribbed slab with ribs in one or two directions. Dimensions of such a slab must be within the ACI Code limitations (see Art. 5-54).

Beam—a flexural member designed to carry uniform or concentrated line loads and to span in one direction. A beam may act as a primary member in beam-column frames, or may be used to support slabs or joist-slabs.

Girder—a flexural member used to support beams and designed to span between columns, walls, or other girders. A girder is always a primary member.

5-65. Flexural Reinforcement. Nonprestressed beams should be designed for flexure as explained in Arts. 5-44 to 5-46. For preliminary design, Fig. 5-17 is useful in estimating bending-moment capacity of beams reinforced for tension only.

If the beam size desired is adequate for the design moment with tension reinforcement only, beam size selected from Fig. 5-17 may be used for final design for flexure. If capacity must be increased without increasing beam size, additional capacity may be provided by addition of compression bars and more tension-bar area to match the compression forces developable in the compression bars (Fig. 5-18). (Shear, torsion, crack control, and deflection requirements must also be met to complete the design. See Arts. 5-47 to 5-51 and 5-66 to 5-68.) Deflection need not be calculated for ACI Code purposes if the total depth h of the beam, including top and bottom cover, is at least the fraction of the span L given in Table 5-14.

A number of interdependent complex requirements (Art. 5-49) regulate the permissible cutoff points or bend points of bars within a span, based on various formulas and rules for development (bond). An additional set of requirements applies if the bars are cut off or bent in a tensile area. These requirements can be satisfied for cases of uniform load and nearly equal spans for the top bars by extending at least 50% of the top steel to a point in the span $0.30L_n$ beyond the face of the support, and the remainder to a point $0.20L_n$, where L_n = clear span. For the bottom bars, all requirements are satisfied by extending at least 40% of the total steel into the supports 6 in. past the face, and cutting off the remainder at a distance $0.125L_n$ from the supports. Note that this arrangement does not cut off bottom bars in a tensile zone. Figure 5-32 shows a typical reinforcement layout for a continuous beam, singly reinforced.

The limit in the ACI Building Code on tension-reinforcement ratio ρ that it not exceed 0.75 times the ratio for balanced conditions applies to beams (Art. 5-46). Balanced conditions in a beam reinforced only for tension exist when the tension steel reaches its yield strength f_y simultaneously with the maximum compressive strain in the concrete at the same section becoming 0.003 in. per in. Balanced conditions occur similarly for rectangular beams, and for T beams with negative moment, that are provided with compression steel, or doubly reinforced. Such sections are under balanced conditions when the tension steel, with area A_s, yields just as the outer concrete surface crushes, and the total tensile-force capacity A_sf_y

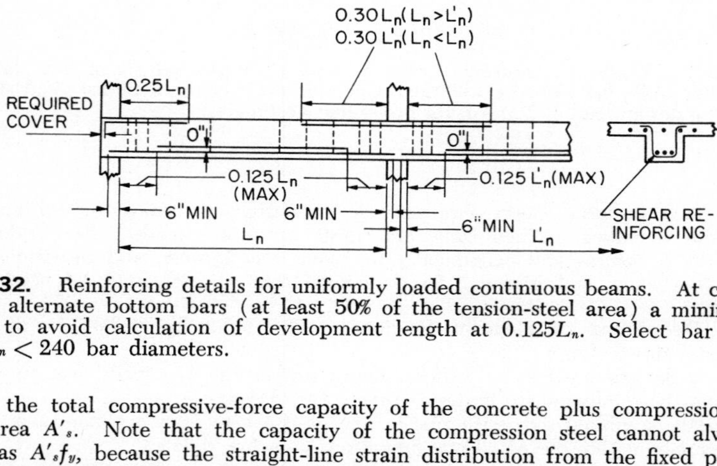

Fig. 5-32. Reinforcing details for uniformly loaded continuous beams. At columns, embed alternate bottom bars (at least 50% of the tension-steel area) a minimum of 6 in., to avoid calculation of development length at $0.125L_n$. Select bar size so that $L_n < 240$ bar diameters.

equals the total compressive-force capacity of the concrete plus compression steel, with area A'_s. Note that the capacity of the compression steel cannot always be taken as A'_sf_y, because the straight-line strain distribution from the fixed points of the outer concrete surface and centroid of the tension steel may limit the compression-steel stress to less than yield strength (Fig. 5-18).

For design of doubly reinforced beams, the force A_sf_y in the tension steel is limited to three-fourths the total compression force at balanced conditions. For a beam meeting these conditions in which the compression steel has not yielded, the design moment capacity is best determined by trial and error:

1. Assume the location of the neutral axis.
2. Determine the strain in the compression steel.
3. See if the total compressive force on the concrete and compression steel equals A_sf_y (Fig. 5-33).

Example. Design a T beam to carry a design negative bending moment of 225 ft-kips. The dimensions of the beam are shown in Fig. 5-33. Concrete strength f'_c = 4 ksi, the reinforcing steel has a yield strength f_y = 60 ksi, and capacity reduction factor ϕ = 0.90.

Need for Compression Steel. To determine whether compression reinforcement is required, first compute

$$R_u = \frac{M_u}{b_wd^2} = \frac{225{,}000 \times 12}{15(12.5)^2} = 1{,}150 \text{ psi}$$

A check with Fig. 5-17 indicates that, for the selected dimensions, compression reinforcement is needed, because the horizontal line through 1,150 does not intersect the line from the lower left representing $f'_c = 4{,}000$.

Compression on Concrete. Assume that the distance c from the neutral axis to the extreme compression surface is 5 in. The depth a of the portion of the concrete in compression then may be taken as $0.85c = 4.2$ in. (Art. 5-44). For

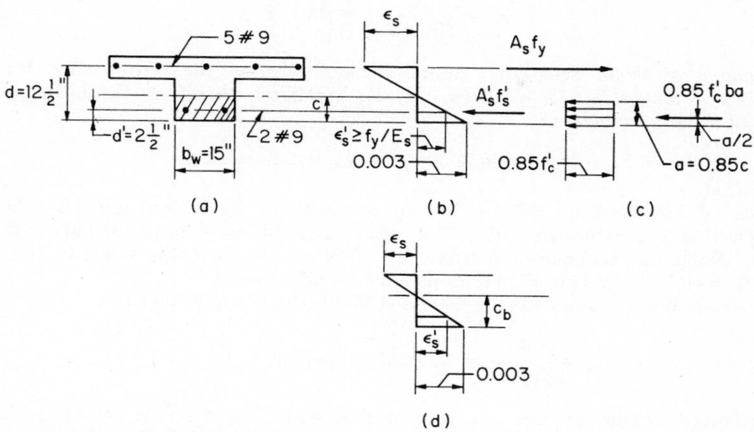

Fig. 5-33. Stress and strain in a T beam reinforced for compression. (*a*) Section. (*b*) Strain. (*c*) Stress. (*d*) Balanced strain.

a rectangular stress distribution over the concrete, the compression force on the concrete is

$$0.85f'_c b_w a = 0.85 \times 4 \times 15 \times 4.2 = 214 \text{ kips}$$

Selection of Tension Steel. To estimate the tension steel required, assume a moment arm $jd = d - a/2 = 12.5 - 4.2/2 = 10.4$ in. By Eq. (5-27*b*), the tension-steel force therefore should be about

$$A_s f_y = 60A_s = \frac{M_u}{\phi jd} = \frac{225 \times 12}{0.9 \times 10.4} = 288 \text{ kips}$$

from which $A_s = 4.8$ sq in. Select five No. 9 bars, supplying $A_s = 5$ sq in. This yields a tensile-steel ratio

$$\rho = \frac{5}{15 \times 12.5} = 0.0267$$

The bars can exert a force $A_s f_y = 5 \times 60 = 300$ kips.

Stress in Compression Steel. For a linear strain distribution, the strain ϵ'_s in the steel 2½ in. from the extreme compression surface can be found by proportion from the maximum strain of 0.003 in. per in. at that surface.

$$\epsilon'_s = \frac{5 - 2.5}{5} 0.003 = 0.0015 \text{ in. per in.}$$

With the modulus of elasticity E_s taken as 29,000 ksi, the stress in the compression steel is

$$f'_s = 0.0015 \times 29{,}000 = 43 \text{ ksi}$$

Selection of Compression Steel. The total compression force equals the 214-kip force on the concrete previously computed plus the force on the compression steel. If the total compression force is to equal the total tension force, the compression steel must resist a force

$$A'_s f'_s = 43A'_s = 300 - 214 = 86 \text{ kips}$$

from which the compression-steel area $A'_s = 2$ sq in. The compression-steel ratio then is

$$\rho' = \frac{2}{15 \times 2.5} = 0.0107$$

Adequacy of Beam Strength. Actual moment arm of the tensile steel for this beam is $d - d' = d - a/2 = 10$ in. By Eq. (5-27*b*), therefore, the beam section can resist a design moment

$$M_u = 0.9 \times 300 \times {}^{10}\!/_{12} = 225 \text{ ft-kips}$$

as required.

Check of Tension-steel Ratio. The tension reinforcement now must be checked to insure that ρ does not exceed $0.75\rho_b$, where ρ_b is the steel ratio for balanced conditions, which can be computed from Eq. (5-30). But first, the stress in the compression steel under balanced conditions must be determined:

The strain in the tension steel, which is at yield strength $f_y = 60$ ksi, is

$$\epsilon_s = \frac{60}{29{,}000} = 0.00207 \text{ in. per in.}$$

For balanced conditions, the concrete at this stage has a strain of 0.003 in. per in. at the extreme compression surface. For a linear strain distribution, by proportion, the distance of the neutral axis from the extreme compression surface is

$$c_b = \frac{0.003}{0.003 + 0.00207} 12.5 = 7.4 \text{ in.}$$

Also, by proportion, the stress in the compression steel is

$$f'_s = \frac{7.4 - 2.5}{12.5 - 7.4} 60 = 57.6 \text{ ksi}$$

Then, by Eq. (5-30),

$$\rho_b = \frac{0.85 \times 4{,}000 \times 0.85}{60{,}000} \frac{87{,}000}{87{,}000 + 60{,}000} + 0.0107 \frac{57.6}{60} = 0.0387$$

Hence, the tension-steel area is satisfactory because

$$\rho = 0.0267 < 0.75 \times 0.0387 = 0.0290$$

5-66. Reinforcement for Shear and Flexure. (See also Art. 5-47.) Minimum shear reinforcement is required in all beams with total depth greater than 10 in., or 2½ times flange (slab) thickness, or half the web thickness, except where the design shear stress v_u is less than half the allowable unit shear stress v_c on the concrete alone. Torsion stress must be combined with shear stress when the design loads cause a nominal torsion stress v_{tu}, psi, greater than $1.5\sqrt{f'_c}$, where f'_c = specified concrete strength, psi (Art. 5-48).

Shear stress v_u should be computed at critical sections in a beam from Eq. (5-37). Allowable shear stresses are given in Art. 5-47. Open or closed stirrups may be used as reinforcement for shear in beams; but closed stirrups are required for torsion. The minimum area of open or closed stirrups for vertical shear only, to be used where $0.5v_c \leq v_u \leq v_c$ and the torsional shear $v_{tu} \leq 1.5\sqrt{f'_c}$, should be calculated from

$$A_v = \frac{50 b_w s}{f_y} \qquad (5\text{-}86)$$

where A_v = area of all vertical legs in the spacing s, in., parallel to flexural reinforcement, sq in.
b_w = thickness of beam web, in.
f_y = yield strength of reinforcing steel, psi

Note that this minimum area provides a capacity for 50-psi shear on the cross section $b_w s$.

Where v_u exceeds v_c, the cross-sectional area A_v of the legs of open or closed vertical stirrups at each spacing s should be calculated from Eq. (5-40). A_v is the total area of vertical legs, two legs for a common open U stirrup or the total of all legs for a transverse multiple U. Note that there are three zones in which the required A_v may be supplied by various combinations of size and spacing of stirrups (Fig. 5-34):

1. Beginning 1 or 2 in. from the face of supports and extending over a distance d from each support, where d is the depth from extreme compression surface to centroid of tension steel (A_v is based on v_u at d from support).
2. Between the distance d from each support and the point where $v_u - v_c = 50$ psi (required A_v decreases from maximum to minimum).
3. Distance over which minimum reinforcement is required (minimum A_v extends from the point where $v_u - v_c = 50$ psi to the point where $v_u = 0.5v_c$).

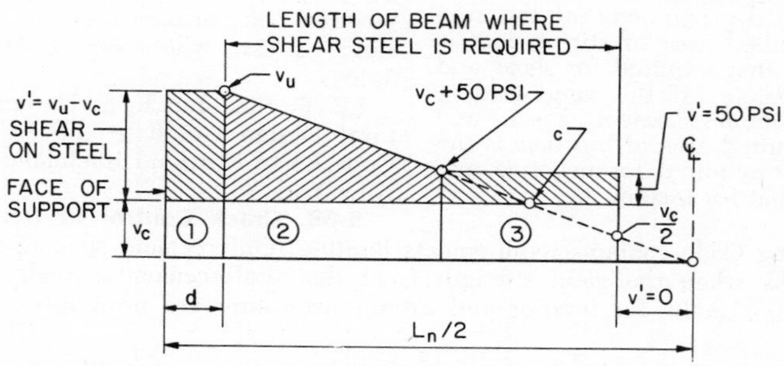

Fig. 5-34. Required shear reinforcement in three zones of a beam between supports and midspan is determined by cross-hatched areas.

5-67. Reinforcement for Torsion and Shear. (See also Art. 5-48.) Any beam supporting unbalanced loads transverse to the direction in which it is subjected to bending moments transmits an unbalanced moment to the supports and must be investigated for torsion. Generally, this requirement affects all spandrel and other edge beams, and interior beams supporting uneven spans or unbalanced live loads on opposite sides. The total unbalanced moment from a floor system with one-way slabs in one direction and beams in the perpendicular direction can often be considered to be transferred to the columns by beam flexure in one direction, neglecting torsion in the slab. The total unbalanced moment in the other direction, from the one-way slabs, can be considered to be transferred by torsional shear from the beams to the columns.

The torsional shear stress v_{tu} for rectangular or T beams should be computed from Eq. (5-42). If $v_{tu} \leq 1.5\sqrt{f'_c}$, where f'_c is the specified concrete strength, psi, v_{tu} may be neglected. The portion of the torsion design stress carried by the concrete should not exceed v_{tc} computed from Eq. (5-44).

The required area A_t of each leg of a closed stirrup for torsion should be computed from Eq. (5-45). Stirrup spacing should not exceed 12 in. or $(x_1 + y_1)/4$, where x_1 and y_1 are the dimensions center to center of a closed rectangular stirrup.

Torsion reinforcement also includes the longitudinal bars shown in each corner of the closed stirrups in Fig. 5-35, and the longitudinal bars spaced elsewhere inside the perimeter of the closed stirrups at not more than 12 in. At least one longitudinal bar in each corner is required. [For required areas of these bars, see Eqs. (5-46) and (5-47).] If a beam is fully loaded for maximum flexure and torsion simultaneously, as in a spandrel beam, the area of torsion-resisting longitudinal bars A_l should be provided in addition to flexural bars.

For interior beams, maximum torsion usually occurs with live load only on a slab on one side of the beam. Maximum torsion and maximum flexure cannot occur simultaneously. Hence, the same bars can serve for both.

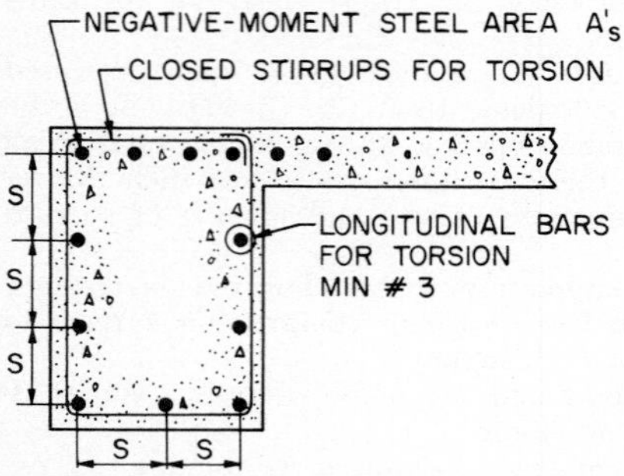

Fig. 5-35. Torsion reinforcement consists of closed stirrups and longitudinal bars within the perimeter of the stirrups. Required area of stirrup legs is the sum of that required for shear and that for torsion. If the same loading causes maximum moment, shear, and torsion, required area of top bars is the sum of that required for negative moment and that for torsion.

The closed stirrups required for torsion should be provided in addition to the stirrups required for shear, which may be the open type. Because the size of stirrups must be at least No. 3 and maximum spacings are established in the ACI Building Code for both shear and torsion stirrups, a closed-stirrup size-spacing combination can usually be selected for combined shear and torsion. Where maximum shear and torsion cannot occur under the same loading, the closed stirrups can be proportioned for the maximum combination of stresses or the maximum single stress, whichever is larger (Fig. 5-36).

(P. F. Rice and E. S. Hoffman, "Structural Design Guide to the ACI Building Code," Van Nostrand Rheinhold Company, New York.)

5-68. Crack Control in Beams. The ACI Building Code contains requirements limiting reinforcement spacing to regulate crack widths when the yield strength f_y of the reinforcement exceeds 40,000 psi (Art. 5-50). Limits for interior and exterior exposures are prescribed. Since the

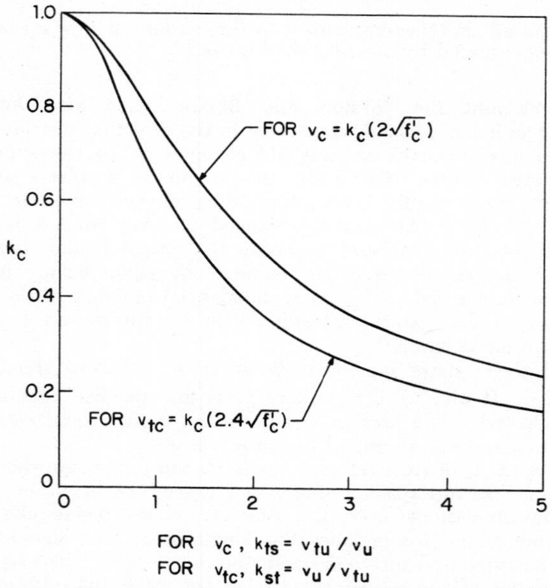

Fig. 5-36. Chart for determining torsional stress v_{tc} and shear stress v_c carried by concrete.

computed crack width is made a function of the tensile area of concrete tributary to and concentric with each bar, the required cover is important to crack control. (See Tables 5-12 and 5-20.)

WALLS

Generally, any concrete member whose length and height are considerably larger than the thickness may be treated as a wall. Walls subjected to vertical loads are called *bearing walls.* Walls subjected to no loads other than their own weight, such as panel or enclosure walls, are called *nonbearing walls.* Walls with a primary function of resisting horizontal loads are called *shear walls.* They also may serve as bearing walls. See Art. 5-91.

5-69. Bearing Walls. Reinforced concrete bearing walls may be designed as eccentrically loaded columns or by an empirical method given in the ACI Building Code. The empirical method may be used when the resultant of the applied load falls within the middle third of the wall thickness. This method gives the capacity of the wall as

Table 5-20. Minimum Clear Concrete Cover, In., for Beams and Girders

Bar size No.	Exterior exposure			Interior exposure		
	Cast-in-place concrete	Precast concrete	Prestressed concrete	Cast-in-place concrete	Precast concrete	Prestressed concrete
3, 4, and 5	1½	1¼	1½	1½	⅝	1½
6 through 11	2	1½	1½	1½	d_b*	1½
14 and 18	2	2	1½	1½	1½	1½
Stirrups	Same as above for each size				⅜	1

* d_b = nominal bar diameter, in.

$$P_u = 0.55\phi f'_c A_g \left[1 - \left(\frac{L_c}{40h}\right)^2\right] \tag{5-87}$$

where f'_c = specified concrete strength
ϕ = capacity reduction factor = 0.70
A_g = gross area of horizontal cross section of wall
h = wall thickness
L_c = vertical distance between supports

The allowable average design stress f_c for a wall is obtained by dividing P_u in Eq. (5-87) by A_g:

$$f_c = \frac{P_u}{A_g} = 0.385f'_c \left[1 - \left(\frac{L_c}{40h}\right)^2\right] \tag{5-88}$$

For given f'_c and ratio L_c/h, f_c can be obtained from Fig. 5-37.

Length. The effective length of wall for concentrated loads may be taken as the center-to-center distance between loads, but not more than the width of bearing plus four times the wall thickness.

Thickness. The minimum thickness of bearing walls for which Eqs. (5-87) and (5-88) are applicable is one-twenty-fifth of the least distance between supports at the sides or top, but at least 6 in. for the top 15 ft of wall height and 1 in. additional for each successive 25 ft, or fraction of it, downward. But bearing walls of two-story dwellings may be 6 in. thick throughout their height. Exterior basement walls, foundation walls, party walls, and fire walls should be at least 8 in. thick. Minimum thickness and reinforcement requirements may be waived, however, if justified by structural analysis.

Reinforcement. The area of horizontal steel reinforcement should be at least

$$A_h = 0.0025A_{wv} \tag{5-89}$$

where A_{wv} = gross area of the vertical cross section of wall

Area of vertical reinforcement should be at least

$$A_v = 0.0015A_{wh} \tag{5-90}$$

where A_{wh} = gross area of the horizontal cross section of wall
For Grade 60 bars, No. 5 or smaller, or for welded-wire fabric, these steel areas may be reduced to $0.0020A_{wv}$ and $0.0012A_{wh}$, respectively.

Walls 10 in. or less thick may be reinforced with only one rectangular grid of rebars. Thicker walls require two grids. The grid nearest the exterior wall surface should contain between one-half and two-thirds the total steel area required for the wall. It should have a concrete cover of at least 2 in., but not more than one-third the wall thickness. A grid near an interior wall surface should have a concrete cover of at least ¾ in., but not more than one-third the wall thickness. Minimum size of bars, if used, is No. 3. Maximum bar spacing is 18 in. (These requirements do not apply to basement walls, however. If such walls are cast against and permanently exposed to earth, minimum cover is 3 in. Otherwise, the cover should be at least 2 in. for bar sizes No. 6 and larger, and 1½ in. for No. 5 bars or ⅝-in. wire and smaller.)

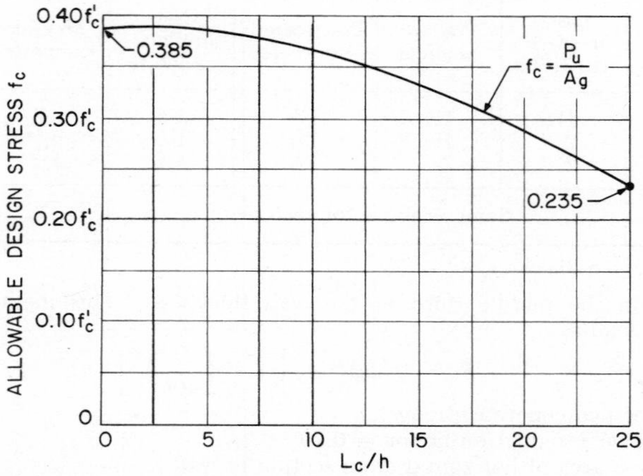

Fig. 5-37. Curve for determining allowable average design stress for bearing walls, given the ratio of unsupported height to thickness.

At least two No. 5 bars should be placed around all window and door openings. The bars should extend at least 24 in. beyond the corners of openings.

Design for Eccentric Loads. Bearing walls with bending moments sufficient to cause tensile stress must be designed as columns for combined flexure and axial load. Minimum reinforcement areas and maximum bar spacings are the same as for walls designed by the empirical method. Lateral ties, as for columns, are required for compression reinforcement and where the vertical bar area exceeds 0.01 times the gross horizontal concrete area of the wall. (For column capacity, see Arts. 5-83 and 5-85.)

Under the preceding provisions, a thin, wall-like (rectangular) column with a steel ratio less than 0.01 will have a greater carrying capacity if the bars are detailed as for walls. The reasons for this are: The effective depth is increased by omission of ties outside the vertical bars and by the smaller cover (as small as ¾ in.) permitted for vertical bars in walls. Furthermore, if the moment is low (eccentricity less than one-sixth the wall thickness), so that the wall capacity is determined by Eq. (5-87), the capacity will be larger than that computed for a column, except where the column is part of a frame braced against sidesway.

5-70. Nonbearing Wall. *Nonbearing*, reinforced-concrete walls, frequently classified as panels, partitions, or cross walls, may be precast or cast in place. Panels serving merely as exterior cladding, when precast, are usually attached to the columns or floors of a frame, supported on grade beams, or supported by and spanning between footings, serving as both grade beams and walls. Cast-in-place cross walls are most common in substructures. Less often, cast-in-place panels may be supported on grade beams and attached to the frame.

In most of these applications for nonbearing walls, stresses are low and alternative materials, such as unreinforced masonry, when supported by beams above grade, or panels of other materials, can be used. Consequently, unless aesthetic requirements dictate reinforced concrete, low-stressed panels of reinforced concrete must be designed for maximum economy. Minimum thickness, minimum reinforcement, full benefits of standardization for mass-production techniques, and design for double function as both wall and deep beam must be achieved.

Thickness of nonbearing walls of reinforced concrete should be at least one-thirtieth the distance between supports, but not less than 4 in.

The ACI Building Code, however, permits waiving all minimum arbitrary requirements for thickness and reinforcement where structural analysis indicates adequate strength and stability.

Where support is provided, as for a panel above grade on a grade beam, connections to columns may be detailed to permit shrinkage. Friction between base of panel and the beam can be reduced by an asphalt coating and omission of dowels. These provisions will permit elimination or reduction of horizontal shrinkage reinforcement. Vertical reinforcement is seldom required, except as needed for spacing the horizontal bars.

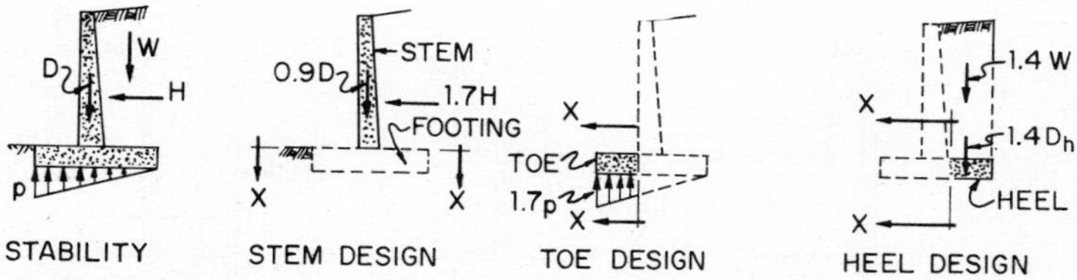

Fig. 5-38. Critical sections for design of cantilever retaining walls and design (factored) loads.

If a nonbearing wall is cast in place, reinforcement can be nearly eliminated except at edges. If the wall is precast, handling stresses will often control. Multiple pickup points with rigid-beam pickups will reduce such stresses. Vacuum pad pickups can eliminate nearly all lifting stresses.

Where deep-beam behavior or wind loads cause stresses exceeding those permitted on plain concrete, the ACI Code permits reduction of minimum tension-reinforcement [$A_s = 200bd/f_y$ (Art. 5-46)] if reinforcement furnished is one-third greater than that required by analysis. (For deep-beam design, see Art. 5-90.)

5-71. Cantilever Retaining Walls. Under the ACI Building Code, cantilever retaining walls are designed as slabs. Specific Code requirements are not given for cantilever walls, but when axial load becomes near zero, the Code requirements for flexure apply.

Minimum clear cover for bars in walls cast against and permanently exposed to earth is 3 in. Otherwise, minimum cover is 2 in. for bar sizes No. 6 and larger, and 1½ in. for No. 5 bars or ⅝-in. wire and smaller.

Two points requiring special consideration are analysis for load factors of 1.7 times lateral earth pressure and 1.4 times dead loads and fluid pressures, and provision of splices at the base of the stem, which is a point of maximum moment. The footing and stem are usually cast separately, and dowels left projecting from the footing are spliced to the stem reinforcement.

A straightforward way of applying Code requirements for strength design is illustrated in Fig. 5-38. Soil reaction pressure p and stability against overturning are determined for actual weights of concrete D and soil W and actual assumed lateral pressure of the soil H. The total cantilever bending moment for design of stem reinforcement is then based upon $1.7H$. The toe pressure used to determine the footing bottom bars is $1.7p$. And the top load for design of the top bars in the heel of the footing is $1.4(W + D_h)$, where D_h is the weight of the heel. The Code requires application of a factor of 0.9 to vertical loading that reduces the moment caused by H.

The top bars in the heel can be selected for the unbalanced moment between the design forces on the toe and the stem, but need not be larger than for the moment of the top loads on the footing (earth and weight of heel). For a footing proportioned so that the actual soil pressure approaches zero at the end of the heel, the unbalanced moment and the maximum moment in the heel due to the top loads will be nearly equal.

The possibility of an over-all sliding failure, involving the soil and the structure together, must be considered, and may require a vertical lug extending beneath the footing, tie backs, or other provisions. (See also Arts. 4-14 to 4-22.)

The base of the stem is a point of maximum bending moment and yet also the most convenient location for splicing the vertical bars and footing dowels. The ACI Code advises avoiding such points for the location of lap splices. But for cantilever walls, splices can be avoided entirely at the base of the stem only for low walls (8 to 10 ft high), in which L-shaped bars from the base of the toe

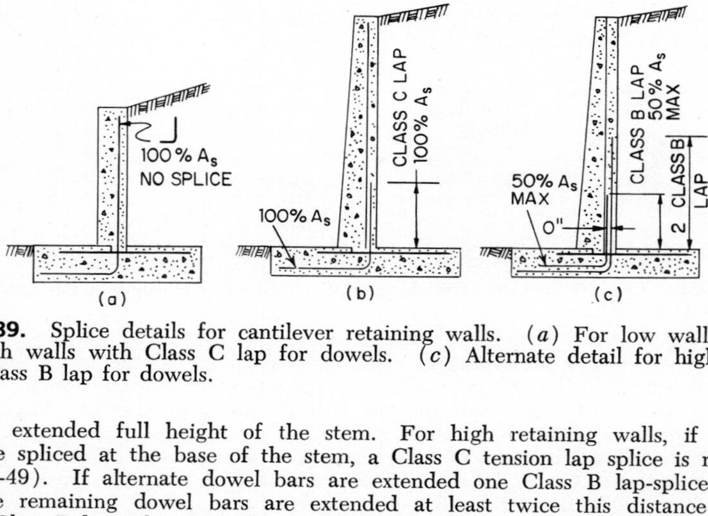

Fig. 5-39. Splice details for cantilever retaining walls. (*a*) For low walls. (*b*) For high walls with Class C lap for dowels. (*c*) Alternate detail for high walls, with Class B lap for dowels.

can be extended full height of the stem. For high retaining walls, if all the bars are spliced at the base of the stem, a Class C tension lap splice is required (Art. 5-49). If alternate dowel bars are extended one Class B lap-splice length and the remaining dowel bars are extended at least twice this distance before cutoff, Class B lap splices may be used. This arrangement requires that dowel-bar sizes and vertical-bar sizes be selected so that the longer dowel bars provide at least 50% of the steel area required at the base of the stem and the vertical bars provide the total required steel at the cutoff point of the longer dowels (Fig. 5-39).

5-72. Counterfort Retaining Walls. In this type of retaining wall, counterforts (cantilevers) are provided on the earth side between wall and footing to support the wall, which essentially spans as a continuous one-way slab horizontally. Counterfort walls seldom find application in building construction. A temporary condition in which basement walls may be required to behave as counterfort retaining walls occurs though, if outside fill is placed before the floors are constructed. Under this condition of loading, each interior cross wall and end basement wall can be regarded as a counterfort. It is usually preferable, however, to delay the fill operation rather than to design and provide reinforcement for this temporary condition.

The advantages of counterfort walls are the large effective depth for the cantilever reinforcement and concrete efficiently concentrated in the counterfort. For very tall walls, where an alternative cantilever wall would require greater thickness and larger quantities of steel and concrete, the savings in material will exceed the additional cost of forming the counterforts. Accurate design is necessary for economy in important projects involving large quantities of material and requires refinement of the simple assumptions in the definition of counterfort walls. The analysis becomes complex for determination of the division of the load between one-way horizontal slab and vertical cantilever action.

(F. S. Merritt, "Standard Handbook for Civil Engineers," McGraw-Hill Book Company, New York.)

5-73. Retaining Walls Supported on Four Sides. For walls more than 10 in. thick, the ACI Building Code requires two-way layers of bars in each face. Two-way-slab design of this reinforcement is required for economy in basement walls or subsurface tank walls supported as vertical spans by the floor above and the footing below, and as horizontal spans by stiff pilasters, interior cross walls, or end walls.

This type of two-way slab is outside the scope of the specific provisions in the ACI Code. Without an "exact" analysis, which is seldom justified because of the uncertainties involved in the assumptions for stiffnesses and loads, a realistic design can be based on the simple two-way-slab design method of Appendix A, Method 2, of the 1963 ACI Building Code, as follows:

Table 5-21. Moment Coefficients C for Two-way Slabs with Two Discontinuous Edges*

Location and type of moment	For short span when ratio of short span to long span is:						Long span, all values of m
	$m = 1.0$	$m = 0.9$	$m = 0.8$	$m = 0.7$	$m = 0.6$	$m = 0.5$	
Negative moment at continuous edge	0.049	0.057	0.064	0.071	0.078	0.090	0.049
Negative moment at discontinuous edge	0.025	0.028	0.032	0.036	0.039	0.045	0.025
Positive moment at midspan....	0.037	0.043	0.048	0.054	0.059	0.068	0.037

* Bending moments may be computed from $M = CwS^2$, where w = uniform load on slab and S = length of short span, the smaller of center-to-center distance between supports and clear span plus twice the slab thickness. From Appendix A, Method 2, "Building Code Requirements for Reinforced Concrete," ACI 318-63.

For the usual conditions of stiffness, the edges supported by a floor at the top of the wall and a footing below the wall can be considered "discontinuous." (Note that the following method provides for negative moments at "discontinuous" edges equal to 50 to 75% of those at "continuous" edges.) End walls or interior cross walls with separate footings can be considered continuous edge supports. Pilasters, unless very stiff and provided with separate footings, can be disregarded in the analysis for design of inner wall bars. Walls should be reinforced with at least a nominal amount of outer bars.

The wall panels should be divided into middle strips and column strips, each to be designed as one-way slabs (Fig. 5-26). Width of middle strips should be taken as half the panel width, in vertical and horizontal directions, respectively. Where the ratio of short span to long span is less than 0.5, however, the width of the middle strip in the short direction should be taken as the difference between the short and long spans.

Bending moments M at critical points in a two-way retaining wall may be computed with the aid of Table 5-21. It gives C for the computation of M for middle strips from

$$M = CwS^2 \tag{5-91}$$

where w = $1.7H$ or $1.4F$
H = lateral earth pressure
F = lateral fluid pressure
S = short span of wall

H can be converted into an equivalent uniform load, as indicated in Fig. 5-40, assuming that some load is resisted by cantilever action on outer-face dowels. The half-panel-wide column strips are assumed to resist two-thirds of the middle-strip moments. For the horizontal column strips one-quarter panel wide, the moment per foot may be taken as one-third the middle-strip moment.

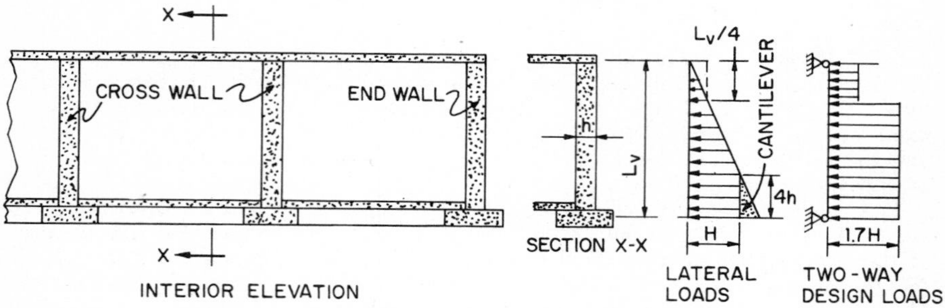

Fig. 5-40. Two-way retaining wall.

FOUNDATIONS

Building foundations should distribute wall and column loads to the underlying soil and rock within acceptable limits on resulting soil pressure and total and differential settlement. Wall and column loads consist of live load, reduced in accordance with the applicable general building code, and dead load, combined,

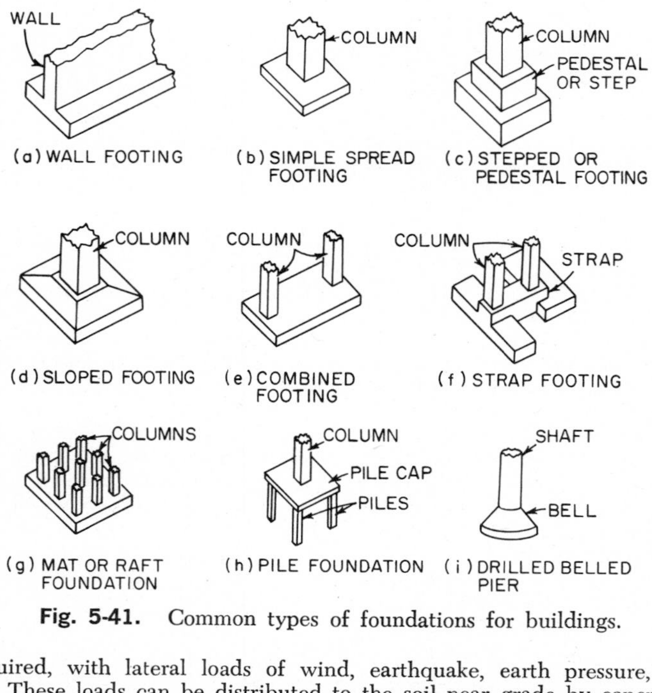

Fig. 5-41. Common types of foundations for buildings.

when required, with lateral loads of wind, earthquake, earth pressure, or liquid pressure. These loads can be distributed to the soil near grade by concrete spread footings, or to the soil at lower levels by concrete piles or drilled piers.

5-74. Types of Foundations. A wide variety of concrete foundations are used for buildings. Some of the most common types are illustrated in Fig. 5-41.

Spread wall footings consist of a plain or reinforced slab wider than the wall, extending the length of the wall (Fig. 5-41*a*). Plain- or reinforced-concrete, *individual-column spread footings* consist of simple, stepped, or sloped two-way concrete slabs, square or rectangular in plan (Fig. 5-41*b* to *d*). For two columns close together, or an exterior column close to the property line so that individual spread or pile-cap footings cannot be placed concentrically, a reinforced-concrete, spread *combined footing* (Fig. 5-41*e*) or a *strap footing* (Fig. 5-41*f*) can be used to obtain a nearly uniform distribution of soil pressure or pile loads. The strap footing becomes more economical than a combined footing when the spacing between the columns becomes larger, causing large bending moments in the combined footing.

For small soil pressures or where loads are heavy relative to the soil capacity, a reinforced-concrete *mat*, or *raft, foundation* (Fig. 5-41*g*) may prove economical. A mat consists of a two-way slab under the entire structure. Concrete cross walls or inverted beams can be utilized with a mat to obtain greater stiffness and economy.

Where sufficient soil strength is available only at lower levels, *pile foundations* (Fig. 5-41*h*) or *drilled-pier foundations* (Fig. 5-41*i*) can be used.

5-75. General Design Principles for Foundations. The area of spread footings, the number of piles, or the number of drilled piers are selected by a designer to support actual unfactored building loads without exceeding *settlement limitations*, a safe soil pressure q_a, or a safe pile or drilled-pier load. A factor of safety from 2 to 3, based on the ultimate strength of the soil and its settlement characteristics, is usually used to determine the safe soil pressure or safe pile or drilled-pier load.

For foundations not subjected to overturning moment, the area of spread footings or the number of piles or drilled piers is determined simply by dividing the actual unfactored wall or column loads by the safe soil pressure q_a, or the safe pile or drilled-pier load. The weight of the footing or drilled pier can be neglected in determining the dead load, if the safe soil pressure is a net pressure in excess of that at the same level due to the surrounding surcharge.

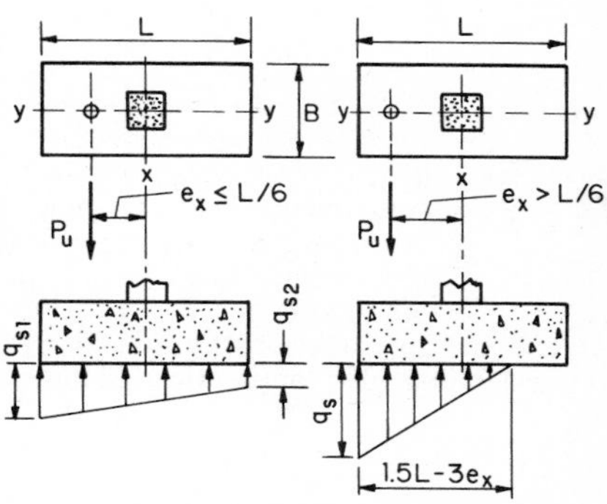

Fig. 5-42. Spread footing subjected to moment.

Soil Pressures. After the area of the spread footing or the number and spacing of piles or drilled piers has been determined, the spread footing, pile-cap footing, or drilled pier can be designed. The strength-design method of the ACI Building Code (Art. 5-44) uses factored loads of gravity, wind, earthquake, earth pressure, and fluid pressure to determine *design soil pressure* q_s and design pile or pier load. The design loadings are used in strength design to determine design moments and shears at critical sections.

For concentrically loaded footings, q_s is usually assumed as uniformly distributed over the footing area. This pressure is determined by dividing the concentric wall or column design load P_u by the area of the footing. The weight of the footing can also be neglected in determining q_s, because the weight does not induce design moments and shears. The design pile load for concentrically loaded pile-cap footings is determined in a similar manner.

When individual or wall spread footings are subjected to overturning moment about one axis, in addition to vertical load, as with a spread footing for a retaining wall, the pressure distribution under the footing is trapezoidal if the eccentricity e_x of the resultant vertical load P_u is within the kern of the footing, or triangular if beyond the kern, as shown in Fig. 5-42. Thus, when $e_x < L/6$, where L is the footing length in the direction of eccentricity e_x, the pressure distribution is trapezoidal (Fig. 5-42*a*) with a maximum

$$q_{s1} = \frac{P_u}{BL}\left(1 + \frac{6e_x}{L}\right) \tag{5-92}$$

and a minimum

$$q_{s2} = \frac{P_u}{BL}\left(1 - \frac{6e_x}{L}\right) \tag{5-93}$$

where B = footing width

When $e_x = L/6$, the pressure distribution becomes triangular over the length L, with a maximum

$$q_s = \frac{2P_u}{BL} \tag{5-94}$$

When $e_x > L/6$, the length of the triangular distribution decreases to $1.5L - 3e_x$ (Fig. 5-42*b*) and the maximum pressure rises to

$$q_s = \frac{2P_u}{1.5B(L - 2e_x)} \tag{5-95}$$

Reinforcement for Bearing. The bearing stress on the interface between a column and a spread footing, pile cap, or drilled pier should not exceed the allowable stress f_b given by Eq. (5-96), unless vertical reinforcement is provided for the excess.

$$f_b = 0.85\phi f'_c \sqrt{\frac{A_2}{A_1}} \qquad \frac{A_2}{A_1} \leq 4 \tag{5-96}$$

where ϕ = capacity reduction factor = 0.70
f'_c = specified concrete strength
A_1 = loaded area of the column or base plate
A_2 = supporting area of footing, pile cap, or drilled pier that is the lower base of the largest frustrum of a right pyramid or cone contained wholly within the footing, with A_1 the upper face, and with side slopes not exceeding 2 horizontal to 1 vertical (Fig. 5-43)

If the bearing stress on the loaded area exceeds f_b, reinforcement must be provided by extending the longitudinal column bars into the spread footings, pile cap, or

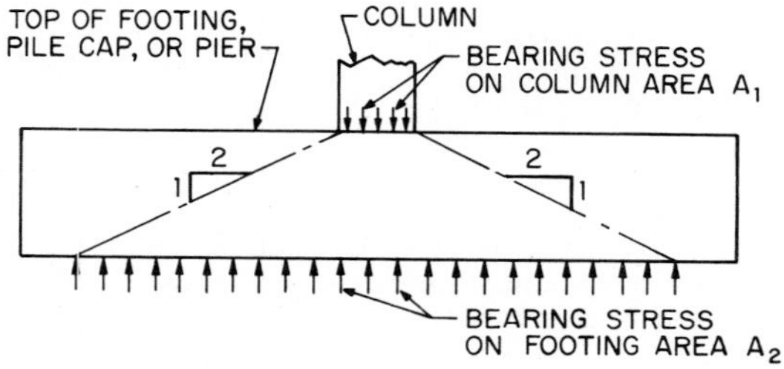

Fig. 5-43. Bearing stresses on column and footing.

drilled pier or by dowels. If so, the column bars or dowels required must have a minimum area of 0.005 times the loaded area of the column, and at least four bars must be provided.

The ACI Building Code requires that every column have a minimum tensile capacity in each face equal to one-fourth the area of the reinforcement in that face. This requirement applies throughout the full height of the column, from foundation up, and will sometimes control the dowel embedment for development of column verticals. If the same size and number of dowels are provided in the footing as rebars in the column and are fully developed in compression, they will usually develop the required minimum tension capacity required by the Code for the column.

Required compression-dowel embedment length cannot be reduced by hooks, nor by f'_c over 4,444 psi. Compression dowels can be smaller than column reinforcement. They may be one bar size larger than the column bars.

If the bearing stress on the loaded area of a column does not exceed $0.85\phi f'_c$, column compression bars or dowels do not need to be extended into the footing, pile cap, or pier, if they can be developed within three times the column dimension (pedestal height) above the footing (Art. 5-49). It is desirable, however, that a minimum of one No. 5 dowel be provided in each corner of a column.

Footing Thickness. The minimum thickness allowed by the ACI Building Code for footings is 8 in. for plain concrete footings on soil, 6 in. above the bottom reinforcement for reinforced-concrete footings on soil, and 12 in. above the bottom reinforcement for reinforced-concrete footings on piles. Plain-concrete pile-cap footings are not permitted.

Concrete Cover. The minimum concrete cover required by the ACI Code for reinforcement cast against and permanently exposed to earth is 3 in.

5-76. Spread Footings for Walls. (See also Art. 5-75.) The critical sections for shear and moment for spread footings supporting concrete or masonry walls are shown in Fig. 5-44*a* and *b*. Under the soil pressure, the projection of footing on either side of a wall acts as a one-way cantilever slab.

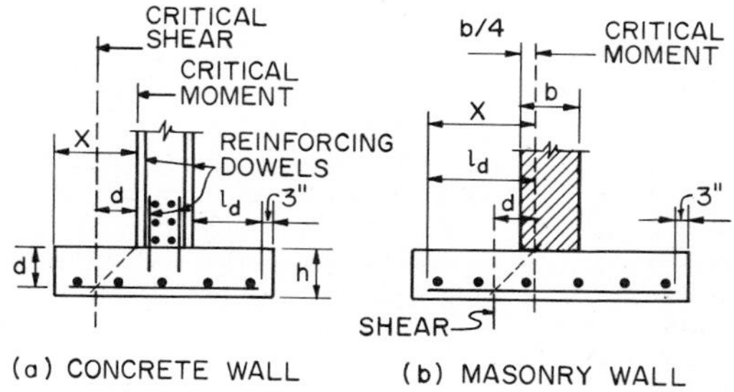

Fig. 5-44. Critical sections for shear and moment in wall footings.

Unreinforced Footings. For plain-concrete spread footings, the maximum permissible flexural tension stress, psi, in the concrete is limited by the ACI Building Code to $5\phi\sqrt{f'_c}$, where ϕ = capacity reduction factor = 0.65 and f'_c = specified concrete strength, psi. For constant-depth, concentrically loaded footings with uniform design soil pressure, the thickness h, in., can be calculated from

$$h = 0.08X\sqrt{\frac{q_s}{\sqrt{f'_c}}} \tag{5-97}$$

where X = projection of footing, in.

q_s = net design soil pressure, psf

Shear stress, $v_u = V_u/\phi b_w d$ (Art. 5-47), is not critical for plain-concrete wall footings. Tensile stress due to flexure controls the thickness.

Because it is usually not economical or practical to provide shear reinforcement in reinforced-concrete spread footings for walls, v_u is usually limited to the maximum value that can be carried by the concrete, $v_c = 2\sqrt{f'_c}$.

Flexural Reinforcement. Steel area can be determined from $A_s = M_u/\phi f_y jd$ as indicated in Art. 5-46. Sufficient length of reinforcement must be provided to develop the full yield strength of straight tension reinforcement. The critical development length is the shorter dimension l_d shown in Fig. 5-44*a* and *b*. The minimum lengths required to develop various bar sizes are tabulated in Table 5-7 (Art. 5-49). Reinforcement at right angles to the flexural reinforcement is usually provided as shrinkage and temperature reinforcement and to support and hold the flexural bars in position. The ACI Code does not limit the amount of this reinforcement.

5-77. Spread Footings for Individual Columns. (See also Art. 5-75.) Footings supporting columns are usually made considerably larger than the columns to keep soil pressure and settlement within reasonable limits. Generally, each column is also placed over the centroid of its footing to obtain uniform pressure distribution under concentric loading. In plan, the footings are usually square, but they can be made rectangular to satisfy space restrictions or to support rectangular columns or pedestals.

Under soil pressure, the projection on each side of a column acts as a cantilever slab in two perpendicular directions. The effective depth of footing d is the distance from the extreme compression surface of the footing to the centroid of the tension reinforcement.

Bending Stresses. Critical sections for moment are at the faces of square and rectangular concrete columns or pedestals (Fig. 5-45*a*). For round and regular-polygon columns or pedestals, the face may be taken as the side of a square having an area equal to the area enclosed within the perimeter of the column or pedestal. For steel columns with steel base plates, the critical section for moment may be taken halfway between the face of the column and the edges of the plate.

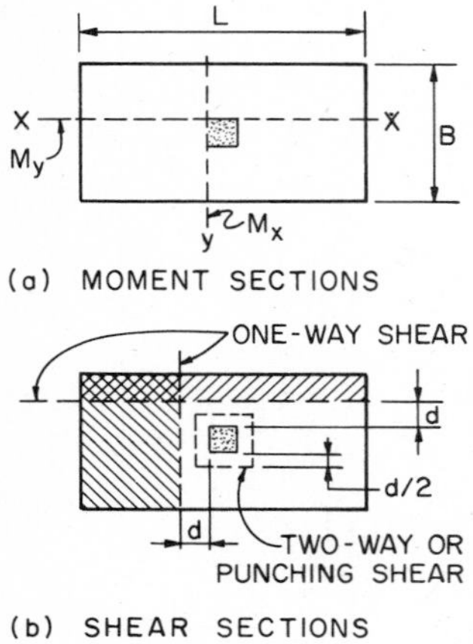

Fig. 5-45. Critical sections for shear and moment in column footings.

For plain-concrete spread footings, the flexural tensile stress must be limited to a maximum of $5\phi\sqrt{f'_c}$, where ϕ = capacity reduction factor = 0.65 and f'_c = specified concrete strength, psi. Thickness of such footings can be calculated with Eq. (5-97). Shear is not critical for plain-concrete footings.

Shear. For reinforced-concrete spread footings, shear is critical on two different sections:

Two-way or punching shear (Fig. 5-45*b*) is critical on the periphery of the surfaces at distance $d/2$ from the column, where d = effective footing depth. Shear stress may be computed from Eq. (5-82).

One-way shear, as a measure of diagonal tension, is critical at distance d from the column (Fig. 5-45*b*), and must be checked in each direction.

It is usually not economical or practical to provide shear reinforcement in column footings. So the shear stress v_c that can be carried by the concrete controls the thickness required.

$v_c = 2\sqrt{f'_c}$ for one-way shear for plain or reinforced concrete sections, but Eq. (5-38) may be used as an alternative for reinforced concrete

$v_c = 4\sqrt{f'_c}$ for two-way shear

Flexural Reinforcement. In square spread footings, reinforcing steel should be uniformly spaced throughout, in perpendicular directions. In rectangular spread footings, the ACI Building Code requires the reinforcement in the long direction to be uniformly spaced over the footing width. Also, reinforcement with an area $2/(\beta + 1)$ times the area of total reinforcement in the short direction should be uniformly spaced in a width that is centered on the column and equal to the short footing dimension, where β is the ratio of the long to the short side of the footing. The remainder of the reinforcing in the short direction should be uniformly spaced in the outer portions of the footing. To maintain a uniform spacing for the bars in the short direction for simplified placing, the theoretical number of bars required for flexure must be increased by about 15%. The maximum increase occurs when β is about 2.5.

The required area of flexural reinforcement can be determined as indicated in Art. 5-46. Maximum-size bars are usually selected to develop the yield strength by straight tension embedment without end hooks. The critical length is the shorter dimension l_d shown for wall footings in Figs. 5-44*a* and *b*.

5-78. Combined Spread Footings. (See also Arts. 5-74 and 5-75.) A combined spread footing under two columns should have a shape and location such that the center of gravity of the column loads coincides with the centroid of the footing area. It can be square, rectangular, or trapezoidal, as shown in Fig. 5-46.

Net design soil-pressure distribution can be assumed to vary linearly for most (rigid) combined footings. For the pressure distribution for flexible combined footings, see the report of ACI Committee 436, "Suggested Design Procedures for Combined Footings and Mats," *ACI Journal Proceedings,* Vol. 63, No. 10, Oct., 1966, pp. 1041–1058; and M. J. Haddadin, "Mats and Combined Footings—Analysis by the Finite Element Method," *ACI Journal Proceedings,* Vol. 69, No. 12, Dec. 1971, pp. 945–949. A computer program based on this method is available from the Cement and Concrete Research Institute, Portland Cement Association, Skokie, Illinois.

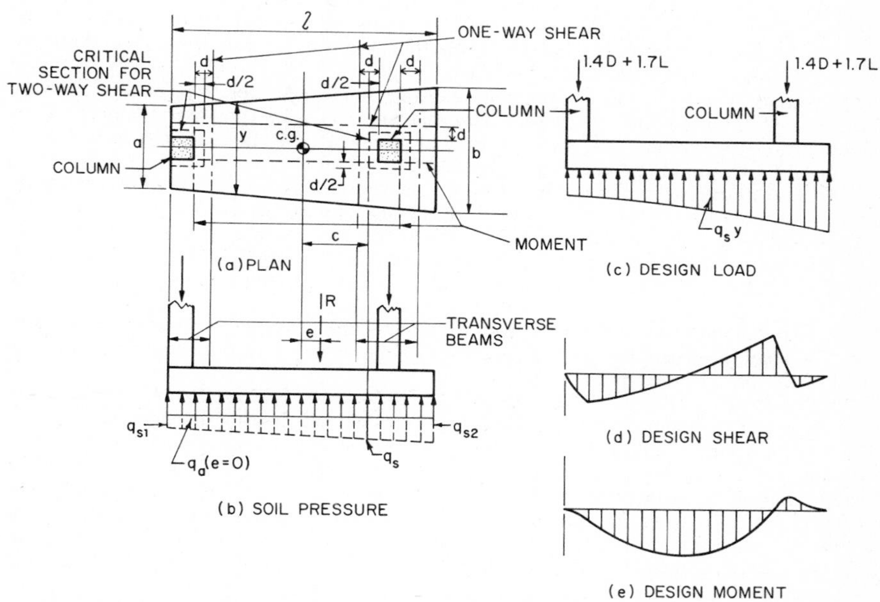

Fig. 5-46. Design conditions for combined spread footing of trapezoidal shape.

If the center of gravity of the unfactored column loads coincides with the centroid of the footing area, the net soil pressure q_a will be uniform.

$$q_a = \frac{R}{A_f} \tag{5-98}$$

where R = applied vertical load
A_f = area of footing

The factored design load on the columns, however, may change the location of the center of gravity of the loads. If the resultant is within the kern for footings with moment about one axis, the design soil pressure q_s can be assumed to have a linear distribution (Fig. 5-46*b*) with a maximum at the more heavily loaded edge:

$$q_{s1} = \frac{P_u}{A_f} + \frac{P_u ec}{I_f} \tag{5-99}$$

and a minimum at the opposite edge:

$$q_{s2} = \frac{P_u}{A_f} - \frac{P_u ec}{I_f} \tag{5-100}$$

where P_u = factored vertical load

e = eccentricity of load

c = distance from center of gravity to section for which pressure is being computed

I_f = moment of inertia of footing area

If the resultant falls outside the kern of the footing, the net design pressure q_s can be assumed to have a linear distribution with a maximum value at the more heavily loaded edge and a lengthwise distribution of three times the distance between the resultant and the pressed edge. The balance of the footing will have no net design pressure.

With the net design soil-pressure distribution known, the design shears and moments can be determined (Fig. 5-46*d* and *e*). Critical sections for shear and moment are shown in Fig. 5-46*a*. The critical section for two-way shear is at a distance $d/2$ from the columns, where d is the effective depth of footing. The critical section for one-way shear in both the longitudinal and transverse direction is at a distance d from the column face.

Maximum design negative moment, causing tension in the top of the footing, will occur between the columns. The maximum design positive moment, with tension in the bottom of the footing, will occur at the face of the columns in both the longitudinal and transverse direction.

Flexural reinforcement can be selected as shown in Art. 5-46. For economy, combined footings should be made deep enough to avoid the use of stirrups for shear reinforcement. In the transverse direction, the bottom reinforcement is placed uniformly in bands having an arbitrary width, which can be taken as the width of the column plus $2d$. The amount at each column is proportional to the column load.

Size and length of reinforcing must be selected to develop the full yield strength of the steel between the critical section, or point of maximum tension, and the end of the bar (Art. 5-49).

5-79. Strap Footings. (See also Arts. 5-74 and 5-75.) When the distance between two columns to be supported on a combined footing becomes large, cost increases rapidly, and a strap footing, or cantilever-type footing, may be more economical. This type of footing, in effect, consists of two footings, one under each column, connected by a strap beam (Fig. 5-47). This beam distributes the column loads to each footing to make the net soil pressure with unfactored loads uniform and equal at each footing, and with factored design loads uniform but not necessarily equal. The center of gravity of the actual column loads should coincide with the centroid of the combined footing areas.

The strap beam is usually designed and constructed so that it does not bear on the soil (Fig. 5-47*b*). The concrete for the beam is cast on compressible material. If the concrete for the strap beam were placed on compacted soil, the resulting soil pressure would have to be considered in design of the footing.

The strap beam, in effect, cantilevers over the exterior column footing, and the bending will cause tension at the top. The beam therefore requires top flexural reinforcement throughout its entire length. Nominal flexural reinforcement should be provided in the bottom of the beam to provide for any tension that could result from differential settlement.

The top bars at the exterior column must have sufficient length to develop their full yield strength. Hence, the distance between the interior face of the exterior column and the property-line end of the horizontal portion of the top bar, plus the value of the hook, must provide tension development length (Art. 5-49).

Strength-design shear reinforcement [Eq. (5-40*a*)] will be required when $v_u > v_c$, and minimum shear reinforcement [Eq. (5-86)] must be provided when $v_u > v_c/2$, where v_c = shear stress carried by the concrete (Art. 5-47).

For the strap footing shown in Fig. 5-47, the exterior column footing can be designed as a wall spread footing, and the interior column footing as an individual-column spread footing (Arts. 5-76 and 5-77).

5-80. Mat Foundations. (See also Arts. 5-74 and 5-75.) A mat or raft foundation is a single combined footing for an entire building unit. It is economical when building loads are relatively heavy and the safe soil pressure is small.

Weight of soil excavated for the foundation decreases the pressure on the soil under the mat. If excavated soil weighs more than the building, there is a net decrease in pressure at mat level from that prior to excavation.

When the mat is rigid, a uniform distribution of soil pressure can be assumed, and the design can be based on a statically determinate structure, as shown in Fig. 5-48. (See "Suggested Design Procedures for Combined Footings and Mats," by ACI Committee 426.)

If the centroid of the factored design loads does not coincide with the centroid of the mat area, the resulting nonuniform soil pressure should be used in the strength design of the mat.

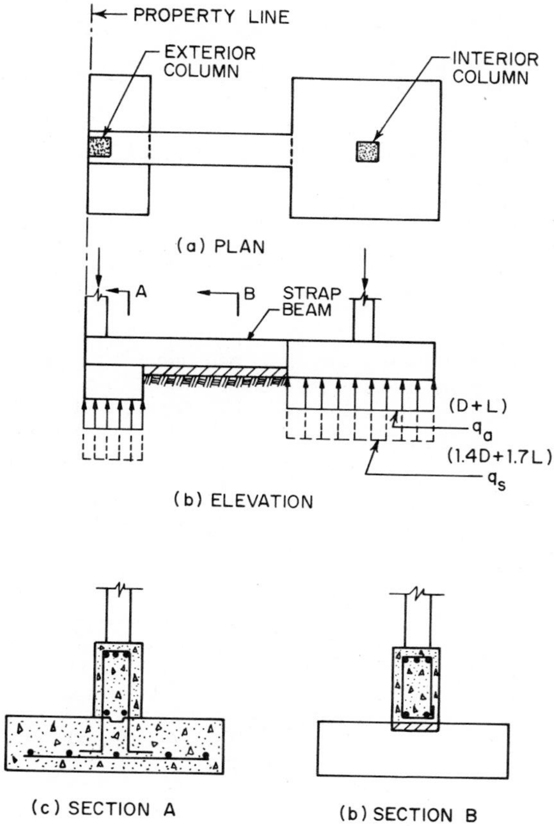

Fig. 5-47. Strap footing.

Strength-design provisions for flexure, one-way and two-way shear, development length, and serviceability should conform to ACI Building Code requirements (Arts. 5-58 to 5-60).

5-81. Pile Foundations. (See also Art. 5-75.) Building loads can be transferred to piles by a thick reinforced-concrete slab, called a pile-cap footing. The piles are usually embedded in the pile cap 4 to 6 in. They should be cut to required elevation after driving and prior to casting the footing. Reinforcement should be placed a minimum of 3 in. clear above the top of the piles. The pile cap is required by the ACI Building Code to have a minimum thickness of 12 in. above the reinforcement.

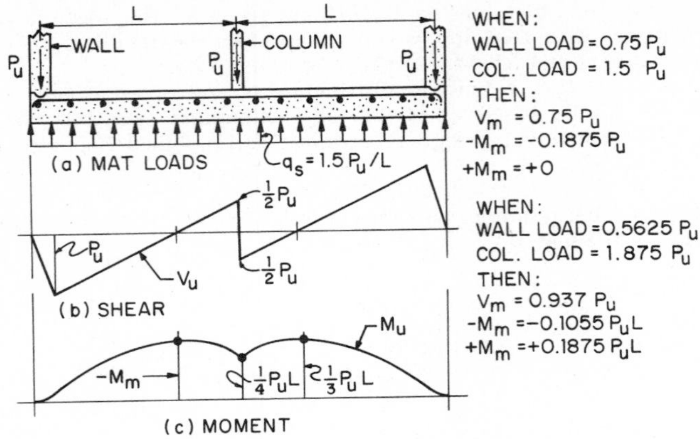

Fig. 5-48. Rigid mat footing.

Piles should be located so that the centroid of the pile cluster coincides with the center of gravity of the column design load. As a practical matter, piles cannot be driven exactly to the theoretical design location. A construction survey should be made to determine if the actual locations require modification of the original pile-cap design.

Pile-cap footings are designed like spread footings (Art. 5-77), but for concentrated pile loads. Critical sections for shear and moment are the same. Reaction from any pile with center $d_p/2$ or more inside the critical section, where d_p is the pile diameter at footing base, should be assumed to produce no shear on the section. The ACI Code requires that the portion of the reaction of a pile with center within $d_p/2$ of the section be assumed as producing shear on the section based on a straightline interpolation between full value for center of piles located $d_p/2$ outside the section and zero at $d_p/2$ inside the section.

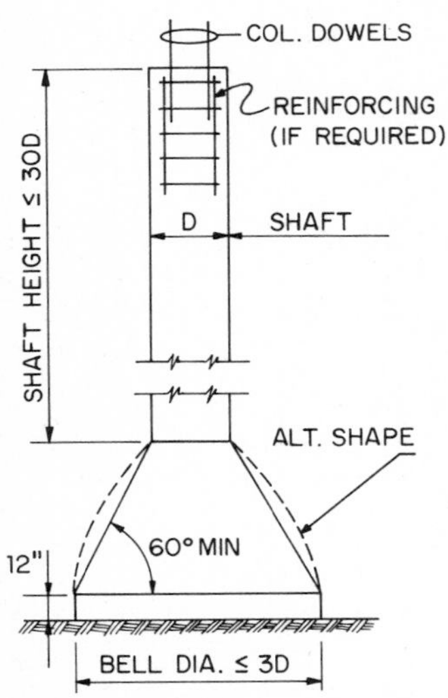

Fig. 5-49. Bell-bottom drilled pier.

For pile clusters without moment, the pile load P_u for strength design of the footing is obtained by dividing the design column load by the number of piles n. The design load equals $1.4D + 1.7L$, where D is the dead load, including the weight of the pile cap, and L is the live load. For pile clusters with moment, the design load on the pile ($n = 1$) is

$$P_{u(n=1)} = (1.4D + 1.7L)\left(\frac{1}{n} + \frac{eC_n}{\sum_1^n C_n^2}\right) \tag{5-101}$$

where e = eccentricity of resultant load with respect to neutral axis of pile group

C_n = distance between neutral axis of pile group and center of nth pile

5-82. Drilled-pier Foundations. (See also Art. 5-75.) A drilled-pier foundation is used to transmit loads to soil at lower levels through end bearing and, in some situations, side friction. It can be constructed in firm, dry earth or clay soil by machine excavating an unlined hole with a rotating auger or bucket with cutting vanes and filling the hole with plain or reinforced concrete. Under favorable conditions, pier shafts 12 ft in diameter and larger can be constructed economically to depths of 100 ft and more. Buckets with sliding arms can be used to form bells at the bottom of the shaft with a diameter as great as three times that of the shaft (Fig. 5-49).

Some building codes limit the ratio of shaft height to shaft diameter to a maximum of 30. They also may require the bottom of the bell to have a constant diameter for the bottom foot of height, as shown in Fig. 5-49.

The compressive stress permitted on plain-concrete drilled piers with lateral support from surrounding earth varies with different codes. ACI limits the computed bearing stress based on service loads to $0.30f'_c$, where f'_c is the specified concrete strength, and the computed bearing stress based on factored loads to $0.60f'_c$. ("Building Code Requirements for Structural Plain Concrete," ACI 322-72.)

Reinforced-concrete drilled piers can be designed as flexural members with axial load, as indicated in Art. 5-83.

Allowable unfactored loads on drilled piers with various shaft and bell diameters, supported by end bearing on soils of various allowable bearing pressures, are given in Table 5-22. For maximum-size bells (bell diameter three times shaft diameter) and a maximum concrete stress of $0.30f'_c$ for unfactored loads, the required concrete strength, psi, is 20.83% of the allowable soil pressure, psf.

COLUMNS

Column-design procedures are based on a comprehensive investigation reported by ACI Committee 105 ("Reinforced Concrete Column Investigation," *ACI Journal*, Feb., 1933) and followed by many supplemental tests. The results indicated that basically the total capacity for axial load can be predicted, over a wide range of steel and concrete strength combinations and percentages of steel, as the sum of the separate concrete and steel capacities.

5-83. Basic Assumptions for Strength Design of Columns. At maximum capacity, the load on the longitudinal reinforcement of a concentrically loaded concrete column can be taken as the steel area A'_s times steel yield strength f_y. The load on the concrete can be taken as the concrete area in compression times 85% of the compressive strength f'_c of the standard test cylinder. The 15% reduction from full strength accounts, in part, for the difference in size and, in part, for the time effect in loading of the column. Capacity of a concentrically loaded column then is the sum of the loads on the concrete and the steel.

The ACI Building Code applies a capacity reduction factor $\phi = 0.75$ for members with spiral reinforcement and $\phi = 0.70$ for other members. For small axial loads ($P_u \leq 0.10f'_cA_g$, where A_g = gross area of column), ϕ may be increased proportionately to as high as 0.90. Capacity of columns with eccentric load or moment may be similarly determined, but with modifications. These modifications introduce the assumptions made for strength design for flexure and axial loads.

The basic assumptions for strength design of columns can be summarized as follows:

1. Strain of steel and concrete is proportional to distance from neutral axis (Fig. 5-50*c*).
2. Maximum usable compression strain of concrete is 0.003 in. per in. (Fig. 5-50*c*).
3. Stress, psi, in longitudinal bars equals steel strain ϵ_s times 29,000,000 for strains below yielding, and equals the steel yield strength f_y, tension or compression, for larger strains (Fig. 5-50*f*).
4. Tensile strength of concrete is negligible.
5. Capacity of the concrete in compression, which is assumed at a maximum stress of $0.85f'_c$, must be consistent with test results. A rectangular stress distribution (Fig. 5-50*d*) may be used. Depth of the rectangle may be taken as $a = \beta_1 c$, where c is the distance from the neutral axis to the extreme compression surface and $\beta_1 = 0.85$ for $f'_c \leq 4{,}000$ psi and 0.05 less for each 1,000 psi that f'_c exceeds 4,000 psi.

In addition to these general assumptions, design must be based on equilibrium

Table 5-22. Allowable Service (Unfactored) Loads on Drilled Piers, Kips*

Shaft dia, ft	Shaft area, sq in.	$f'_c = 3{,}000$	$f'_c = 4{,}000$	$f'_c = 5{,}000$	$f'_c = 6{,}000$
1.5	254.5	229	305	382	458
2.0	452.4	407	543	679	814
2.5	706.9	636	848	1,060	1,272
3.0	1,017.9	916	1,221	1,527	1,832
3.5	1,385.4	1,247	1,662	2,078	2,494
4.0	1,809.6	1,629	2,172	2,714	3,257
4.5	2,290.2	2,061	2,748	3,435	4,122
5.0	2,827.4	2,545	3,393	4,241	5,099
5.5	3,421.2	3,079	4,105	5,132	6,158
6.0	4,071.5	3,664	4,886	6,107	7,329

Bell dia, ft	Bell area, sq ft	Safe allowable service-load bearing pressure on soil, psf					
		10,000	12,000	15,000	20,000	25,000	30,000
1.5	1.77	18	21	27	35	44	53
2.0	3.14	31	38	47	63	78	94
2.5	4.91	49	59	74	98	123	147
3.0	7.07	71	85	106	141	177	212
3.5	9.62	96	115	144	192	240	289
4.0	12.57	126	151	189	251	314	387
4.5	15.90	159	191	238	318	398	477
5.0	19.64	196	236	295	393	492	589
5.5	23.76	238	285	356	475	594	713
6.0	28.27	283	339	424	565	707	848
6.5	33.18	332	398	498	664	830	995
7.0	38.48	385	462	577	770	962	1,154
7.5	44.18	442	530	663	884	1,105	1,325
8.0	50.27	503	603	754	1,005	1,258	1,508
8.5	56.74	567	681	851	1,135	1,420	1,702
9.0	63.62	636	763	954	1,272	1,591	1,909
9.5	70.88	709	851	1,063	1,418	1,770	2,126
10.0	78.54	785	942	1,178	1,571	1,965	2,356
10.5	86.59	866	1,041	1,300	1,732	2,164	2,598
11.0	95.03	950	1,140	1,425	1,901	2,375	2,851
11.5	103.87	1,039	1,248	1,560	2,080	2,598	3,116
12.0	113.10	1,131	1,357	1,696	2,262	2,827	3,393
12.5	122.72	1,227	1,470	1,840	2,450	3,068	3,682
13.0	132.73	1,327	1,593	1,991	2,655	3,320	3,982
13.5	143.14	1,431	1,718	2,145	2,860	3,576	4,294
14.0	153.94	1,539	1,847	2,309	3,079	3,850	4,618
14.5	165.13	1,651	1,980	2,480	3,303	4,130	4,954
15.0	176.15	1,762	2,110	2,645	3,523	4,402	5,284
15.5	188.69	1,887	2,260	2,830	3,774	4,717	5,661
16.0	201.06	2,011	2,410	3,020	4,021	5,025	6,032
16.5	213.82	2,138	2,560	3,200	4,276	5,344	6,415
17.0	226.98	2,270	2,725	3,410	4,540	5,680	6,809
17.5	240.53	2,405	2,890	3,610	4,811	6,012	7,216
18.0	254.47	2,545	3,060	3,820	5,089	6,361	7,634

* $f_{c1} = 0.30 f'_c$ (ACI 322-72 Alternate Design Method).

NOTE: Bell diameter preferably not to exceed 3 times the shaft diameter. Check shear stress if bell slope is less than 2:1. (Courtesy Concrete Reinforcing Steel Institute.)

and strain compatibility conditions. No essential difference develops in maximum capacity between tied and spiral columns, but spiral-reinforced columns show far more toughness before failure. Tied-column failures have been relatively brittle and sudden, whereas spiral-reinforced columns that have failed have deformed a great deal and carried a high percentage of maximum load to a more gradual yielding failure. The difference in behavior is reflected in the higher value of ϕ assigned to spiral-reinforced columns.

Additional design considerations are presented in Arts. 5-84 to 5-89. Following is an example of the application of the basic assumptions for strength design of columns.

Example. Determine the capacity of the 20-in.-square reinforced-concrete column shown in Fig. 5-50*a*. The column is reinforced with four No. 18 bars, with $f_y = 60$ ksi, and lateral ties. Concrete strength is $f'_c = 6$ ksi. Assume the factored load P_u to have a minimum eccentricity of 2 in., and that slenderness can be ignored.

To begin, assume $c = 24$ in. Then, with $\beta_1 = 0.75$ for $f'_c = 6{,}000$ psi, the depth of the compression rectangle is $a = 0.75 \times 24 = 18$ in. This assumption can be checked by computing the eccentricity $e = 12M_u/P_u$, where M_u is the moment capacity, ft-kips.

Since the strain diagram is linear and the maximum compression strain is 0.003 in. per in., the strains in the reinforcing steel are found by proportion to be 0.00258 and 0.00092 in. per in. (Fig. 5-50*c*). The strain at yield is 60/29,000 = 0.00207 < 0.00258 in. per in. Hence, the stresses in the steel are 60 ksi and 0.00092 × 29,000 = 26.7 kips.

The maximum concrete stress, which is assumed constant over the depth a = 18 in., is $0.85f'_c = 0.85 \times 6 = 5.1$ ksi (Fig. 5-50*d*). Hence, the compression force on the concrete is 5.1 × 20 × 18 = 1,836 kips and acts at a distance 20/2 − 18/2 = 1 in. from the centroid of the column (Fig. 5-50*f*). The compression force on the more heavily loaded pair of reinforcing bars, which have a cross-sectional area of 8 sq in., is 8 × 60 less the force on concrete replaced by the steel 8 × 5.1, or 439 kips. The compression force on the other pair of bars is 8(26.7 − 5.1) = 173 kips (Fig. 5-50*f*).

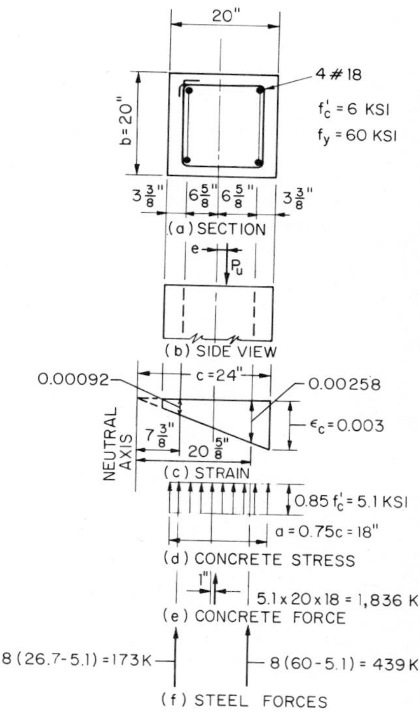

Fig. 5-50. Stress and strain in reinforced-concrete column.

Both pairs of bars act at a distance of 20/2 − 3.375 = 6.625 in. from the centroid of the column.

The capacity of the column for vertical load is the sum of the steel and concrete capacities multiplied by a capacity reduction factor $\phi = 0.70$.

$$P_u = 0.70(1{,}836 + 173 + 439) = 1{,}714 \text{ kips}$$

The capacity of the column for moment is found by taking moments of the steel and concrete capacities about the centroidal axis.

$$M_u = 0.70\left[1{,}836 \times \frac{1}{12} + (439 - 173)\,\frac{6.625}{12}\right] = 209 \text{ ft-kips}$$

The eccentricity for the assumed value of $c = 24$ in. is

$$e = \frac{209 \times 12}{1{,}710} = 1.46 < 2 \text{ in.}$$

If for a new trial, c is taken as 22.5 in., $P_u = 1{,}620$ kips, $M_u = 272$ ft-kips, and e checks out very close to 2 in. (See also Fig. 5-51.)

5-84. Design Requirements for Columns. The ACI Building Code contains the following principal design requirements for columns, in addition to the basic assumptions (Art. 5-83):

1. Columns must be designed for all bending moments associated with a loading condition.

2. For corner columns and other columns loaded unequally on opposite sides in perpendicular directions, biaxial bending moments must be considered.

3. All columns must be designed for a minimum eccentricity of at least 1 in., but not less than $e = 0.10h$ for tied columns and $0.05h$ for spiral-reinforced columns, where h is the overall column thickness in the direction of bending. Under special controls, precast-concrete columns may be designed for a minimum eccentricity of 0.6 in.

4. The minimum ratio of longitudinal-bar area to total cross-sectional area of column A_g is 0.01, and the maximum ratio is 0.08. For columns with a larger cross section than required by loads, however, a smaller A_g, but not less than half the gross area of the column, may be used for calculating both load capacity and minimum longitudinal bar area. This exception allows reuse of forms for larger-than-necessary columns, and permits longitudinal bar areas as low as 0.005 times the actual column area. At least four longitudinal bars should be used in rectangular reinforcement arrangements, and six in circular arrangements.

5. The ratio of the volume of spiral reinforcement to volume of concrete within the spiral should be at least

$$\rho_s = 0.45 \left(\frac{A_g - A_c}{A_c}\right) \left(\frac{f'_c}{f_y}\right) \qquad (5\text{-}102)$$

where A_g = gross cross-sectional area of concrete column, sq in.
A_c = area of column within outside diameter of spiral, sq in.
f'_c = specified concrete strength, psi
f_y = specified yield strength of spiral steel, psi (maximum 60,000 psi)

6. For tied columns, minimum size of ties is No. 3 for longitudinal bars that are No. 10 or smaller, and No. 4 for larger longitudinal bars. Minimum vertical spacing of sets of ties is 16 diameters of longitudinal bars, 48 tie-bar diameters, or the least thickness of the column. A set of ties should be composed of one round tie for bars in a circular pattern, or one tie enclosing four corner bars plus additional ties sufficient to provide a corner of a tie at alternate interior bars or at bars spaced more than 6 in. from a bar supported by the corner of a tie.

7. Splices of vertical bars from footing to roof must provide a minimum tensile capacity in each face of the column equal to 25% of the total vertical-bar area times the steel yield strength.

8. Minimum clear concrete cover required for reinforcement is listed in Table 5-23. See also Art. 5-86.

5-85. Design for Minimum Bending Moment. The requirement for design of columns for a minimum eccentricity (Art. 5-84) means that the maximum design load $P_u = 1.4D + 1.7L$, where D is the dead load and L the live load, for cast-in-place tied columns more than 10 in. thick is accompanied by a minimum bending moment $M_u = 0.10P_uh$.

Interior columns in buildings with reasonably uniform spans and usual loadings, when lateral deflection and moments due to wind are negligible or when lateral bracing is provided by walls or other elements, usually will be designed for this condition of maximum design axial load and minimum moment.

The universal column design chart in Fig. 5-51 gives the solution to this most common design problem directly for square columns with bars in four faces for

Table 5-23. Minimum Clear Cover, In., for Column Reinforcement*

Type of construction	Reinforcement	Interior exposure†	Exposed to Weather*
Cast-in-place	Longitudinal	1½	2
	Ties, spirals	1½	1½
Precast	Longitudinal	$5/8 \le d_b \le 1½$‡	1½
	Ties, spirals	⅜	1¼
Prestressed in-place	Longitudinal	1½	1½
	Ties, spirals	1	1½
Prestressed, precast	Longitudinal	$5/8 < d_b < 1½$‡	1½§
	Ties, spirals	⅜	1¼

* From ACI 318-71.
† See local code; fire protection may require greater thicknesses.
‡ d_b = nominal bar diameter, in.
§ Increase these covers 50% where tensile stress exceeds $6\sqrt{f'_c}$, where f'_c = specified concrete strength, psi.

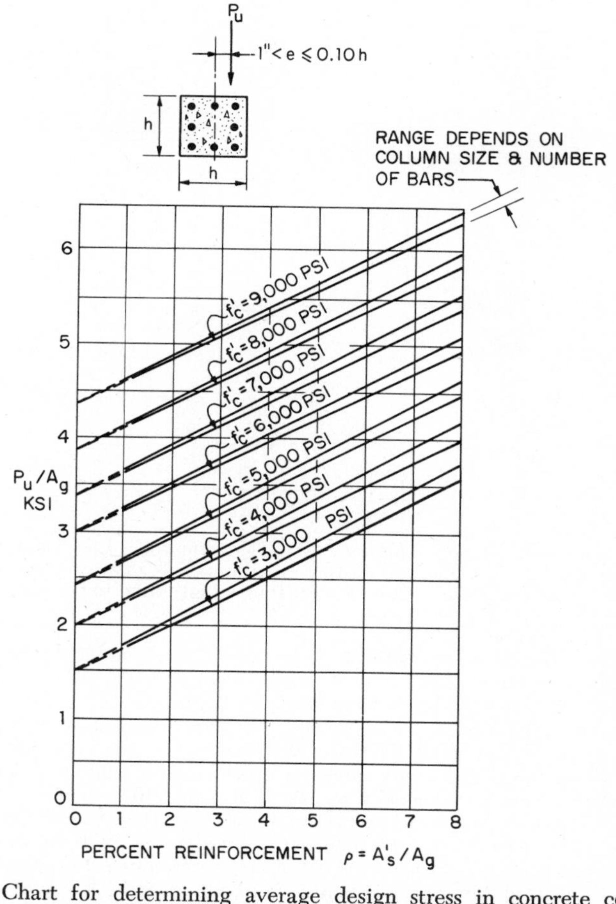

Fig. 5-51. Chart for determining average design stress in concrete column with minimum eccentricity e, capacity reduction factor ϕ of 0.70, and reinforcement with 60-ksi yield strength. (Courtesy Concrete Reinforcing Steel Institute.)

the practical range of concrete strengths and standard Grade 60 bars. The slight spread of the curves for each value of specified concrete strength f'_c is due to the variation in depth from face of concrete to centroid of various-size longitudinal bars and the practicable range of numbers of vertical bars. For rectangular columns with sides less than 2 to 1, round columns, or square columns with unknown, possibly larger bending moment, a preliminary size at a low steel ratio can be selected with confidence that the final design based on more accurate analysis can be completed for the size selected. Note that final design where large wind moment is a factor, where biaxial bending occurs, or where slenderness effects must be considered will usually require more steel and only occasionally a larger size ("CRSI Handbook," Concrete Reinforcing Steel Institute).

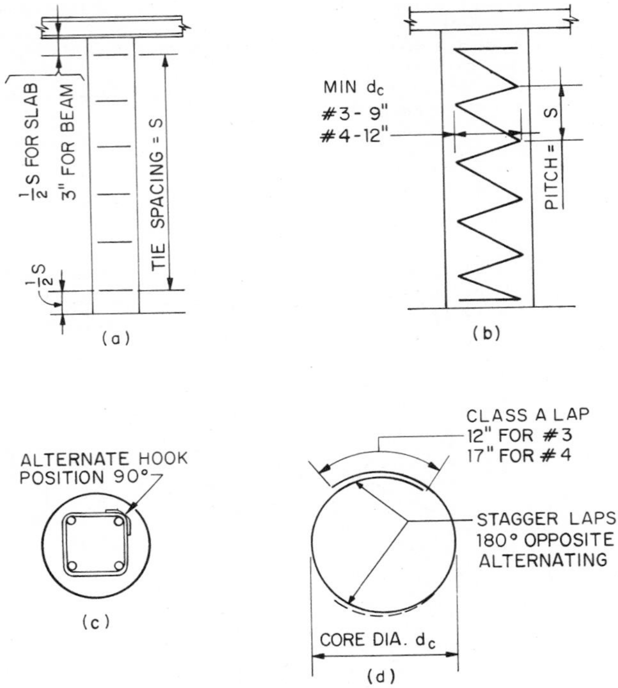

Fig. 5-52. Circular concrete column. (*a*) Tied column. Use separate ties when core diameter $d_c \leq s$. (*b*) Spiral-reinforced column. It may be used when $d_c > s$. (*c*) Rectangular tie for use in columns with four longitudinal bars. (*d*) Circular tie.

5-86. Column Ties and Tie Patterns. For full utilization, all ties in tied columns must be fully developed (for full tie yield strength) at each corner enclosing a vertical bar or, for circular ties, around the full periphery.

Splices. The ACI Building Code provides arbitrary minimum sizes and maximum spacings for column ties (Art. 5-84). No increase in size nor decrease in the spacings is required for Grade 40 materials. Hence, the minimum design requirements for splices of ties may logically be based on Grade 40 steel.

The ordinary closed, square or rectangular, tie is usually spliced by overlapping standard tie hooks around a longitudinal bar. Standard tie patterns require staggering of hook positions at alternate tie spacings, by rotating the ties 90 or 180°. ("Manual of Standard Practice," Concrete Reinforcing Steel Institute.) Two-piece ties are formed by lap splicing or anchoring the ends of U-shaped open ties. A Class C tension lap splice, with minimum lap of 12 in., must be provided, because it is not practicable to stagger such lap splices (Art. 5-49). Lapped bars should be securely wired together to prevent displacement during concreting.

Tie Arrangements. Commonly used tie patterns are shown in Figs. 5-52 to 5-54. In Fig. 5-53, note the reduction in required ties per set and the improvement

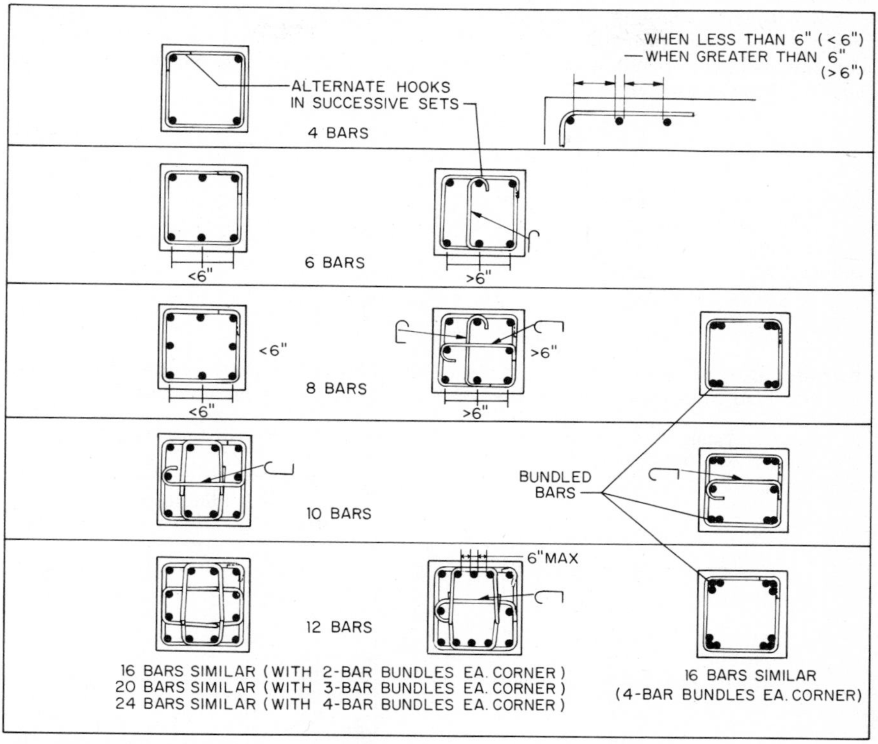

Fig. 5-53. Ties for square concrete columns. Additional single bars may be placed between any of the tied groups, but clear spaces between bars should not exceed 6 in.

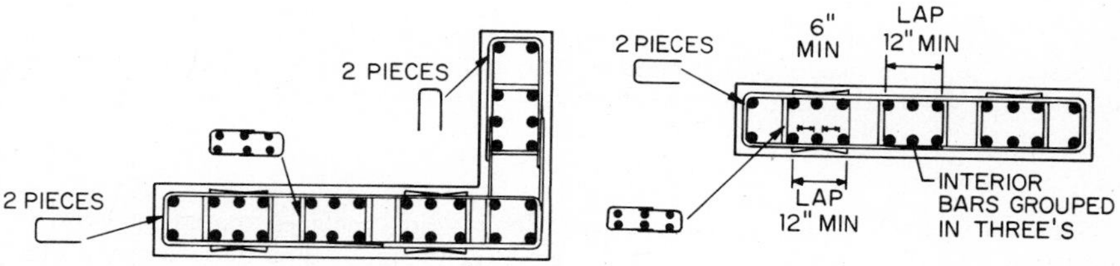

Fig. 5-54. Ties for wall-like columns. Spaces between corner bars and interior groups of three bars, and between interior groups may vary to accommodate average spacing not exceeding 6 in. A single additional bar may be placed in any of such spaces if the average spacing does not exceed 6 in.

in bending resistance about both axes achieved with the alternate bundled-bar arrangements. Bundles may not contain more than four bars, and bar size may not exceed No. 11.

Minimum sizes of ties and maximum spacings per set of ties are listed in Table 5-24.

Drawings. Design drawings should show all requirements for splicing longitudinal bars, that is, type of splice, lap length if lapped, location in elevation, and layout in cross section. On detail drawings, dowel erection details should be shown if special large longitudinal bars, bundled bars, staggered splices, or specially grouped bars are to be used.

5-87. Biaxial Bending of Columns. If column loads cause bending simultaneously about both principal axes of a column cross section, as for most corner columns, a biaxial bending analysis is required. For rapid preliminary design, Eq. (5-103) gives conservative results.

$$\frac{M_x}{M_{ox}} + \frac{M_y}{M_{oy}} \leq 1 \qquad (5\text{-}103)$$

where M_x, M_y = design moments about x and y axes, respectively
M_{ox}, M_{oy} = design capacities about x and y axes, respectively

For square columns with equal steel in all faces, $M_{ox} = M_{oy}$, and the relation reduces to:

$$\frac{M_x + M_y}{M_{ox}} \leq 1 \qquad (5\text{-}104)$$

Because $M_x = e_xP_u$ and $M_y = e_yP_u$, the safe biaxial capacity can be taken from uniaxial load-capacity tables for the load P_u and the uniaxial moment $M_u = (e_x + e_y)P_u$. Similarly, for round columns, the moment capacity is essentially equal in all directions, and the two bending moments about the principal axes may be combined into a single uniaxial design moment M_u which is then an exact solution.

Table 5-24. Minimum Sizes and Laps and Maximum Spacing, In., for Column Ties

Longitudinal-bar size No.	Minimum size No. 3	Minimum size No. 4	Minimum size No. 5
5	10		
6	12		
7	14		
8	16	16	
9	18	18	
10	18	20	
11	. . .	22	22
14	. . .	24	27
18	. . .	24	30
Lap splice, in.	15	20	26

$$M_u = \sqrt{M_x^2 + M_y^2} \qquad (5\text{-}105)$$

The linear solution always gives a safe design, but becomes somewhat overconservative when the moments M_x and M_y are nearly equal. For these cases, a more exact solution will be more economical for the final design. (Simple curves for direct solutions are available in the *"CRSI Handbook,"* Concrete Reinforcing Steel Institute.)

5-88. Slenderness Effects on Concrete Columns. The ACI Building Code requires that primary column moments be magnified to provide safety against buckling failure. Detailed procedures, formulas, and design aids are provided in the ACI Code and the "Commentary."

For most unbraced frames, an investigation will be required to determine the magnification factor to allow for the effects of sidesway and end rotation. The procedure for determination of the required increase in primary moments, after the determination that slenderness effects cannot be neglected, is complex. For direct solution, design aids are available (curves in "CRSI Handbook," Concrete Reinforcing Steel Institute; computer program, Portland Cement Association).

The ACI Code permits neglect of slenderness effects only for very short, braced columns, with the following limitations for columns with square or rectangular cross sections:

$L_u \leq 6.6h$ for bending in single curvature

$L_u \leq 10.2h$ for bending in double curvature with unequal end moments

$L_u \leq 13.8h$ for bending in double curvature with equal end moments

and for round columns, five-sixths of the maximum lengths for square columns, where L_u is the unsupported length and h the depth or overall thickness of column in the direction being considered.

These limiting heights are based on the ratio of the total stiffnesses of the columns to the total stiffnesses of the flexural members, $\Sigma K_c/\Sigma K_B = 50$, at the joint at each end of a column. As these ratios become less, the limiting heights can be increased. When the total stiffnesses of the columns and the floor systems are equal at each end of the column (a common assumption in routine frame analysis), the two ratios = 1.00, and the limiting heights increase about 30%. With this increase, the slenderness effects can be neglected for most columns in frames braced against sidesway.

A frame is considered braced when other structural elements, such as walls, provide stiffness resisting sidesway at least six times the sum of the column stiffnesses resisting sidesway in the same direction in the story being considered.

5-89. Economy in Column Design. Actual costs of reinforced-concrete columns in place per linear foot per kip carrying capacity vary widely. The following recommendations based on relative costs are generally applicable:

Formwork. Use of the same size and shape of column cross section throughout a floor and, for multistory construction, from footing to roof will permit mass production and reuse for economy. Within usual practicable maximum building heights, about 60 stories or 600 ft, increased speed of construction and saving in formwork will save more than the cost of the excess concrete volume over that for smaller column sizes in upper stories.

Concrete Strength. Use of the maximum concrete strength required to support the design loads with the minimum allowed steel area results in the lowest cost. The minimum size of a multistory column is established by the maximum concrete strength reliably available locally and the limit on maximum area of vertical bars. If the acceptable column size is larger than the minimum possible at the base of the multistory stack, the steel ratio can begin with less than the maximum limit (Art. 5-84). At successive stories above, the steel ratio can be reduced to the minimum, and thereafter, for additional stories, the concrete strength can be reduced. Near the top, as loads reduce further, a further reduction in the steel area to 0.005 may be made (Art. 5-84).

Steel. Grade 60 vertical bars offer the greatest economy. Minimum tie requirements can be achieved with four-bar or four-bundle (up to four bars per bundle) arrangements, or by placing an intermediate bar between tied corners not more than 6 in. (clear) from the corner bars. For these arrangements, no interior ties are required; only one tie per set is needed. (See Fig. 5-53 and Art. 5-86.) With no interior ties, low-slump concrete can be placed and consolidated more easily, and the cost and time for assembly of column reinforcement cages are greatly reduced. Note that, for small quantities, the local availability of Nos. 14 and 18 bars should be investigated before they are specified.

Details of Column Reinforcement. Where Nos. 14 and 18 bars are used in compression only, end-bearing splices usually save money. If the splices are staggered 50%, as with two-story lengths, the tensile capacity of the columns will also be adequate for the usual bending moments encountered. For unusually large bending moments, where tensile splices of No. 10 bars and larger are required, tension couplers are usually least expensive in place. For smaller bar sizes, lap splices, tensile or compressive, are preferred for economy. Some provision for staggered lap splices for No. 8 bars and larger may be required to avoid Class C tension splices (Art. 5-49).

Where butt splices are used, it will usually be necessary to assemble the column reinforcement cage in place. Two-piece interior ties or single ties with end hooks for two bars (see Art. 5-86) will facilitate this operation.

Where the vertical bar spacing is restricted and lap splices are used, even with the column size unchanged, offset bending of the bars from below may be required. However, where space permits, as with low steel ratios, an additional saving in fabrication and erection time will be achieved by use of straight column verticals, offset one bar diameter at alternate floors.

SPECIAL CONSTRUCTION

5-90. Deep Beams. The ACI Building Code defines deep beams as flexural members with over-all span-depth ratios less than 2.5 for continuous spans and 1.25 for simple spans. Some types of building components behave as deep beams and require analysis for nonlinear stress distribution in flexure. Some common examples are long, precast panels used as spandrel beams: below-grade walls, with or without openings, distributing column loads to a continuous slab footing or to end walls; and story-height walls used as beams to eliminate lower columns in the first floor area.

Shear. When the span-depth ratio is less than 5, beams are classified as deep for shear reinforcement purposes. Separate special requirements for shear apply when span-depth ratio is less than 2 or between 2 and 5. The critical section for shear should be taken at a distance from face of support of $0.15L_n \leq d$ for uniformly loaded deep beams, and of $0.50a \leq d$ for deep beams with concentrated loads, where a is the shear span, or distance from concentrated load to face of support, L_n the clear span, and d the distance from extreme compression surface to centroid of tension reinforcement. Shear reinforcement required at the critical section should be used throughout the span.

Allowable shear stress carried by the concrete can be taken as

$$v_c = 2\sqrt{f'_c} \tag{5-106}$$

where f'_c = specified concrete strength, psi

The ACI Building Code also presents a more complicated formula that permits the concrete to carry up to $6\sqrt{f'_c}$.

Maximum total shear when $L_n/d < 2$ should not exceed

$$v_u = 8\sqrt{f'_c} \tag{5-107}$$

and when L_n/d is between 2 and 5 should not exceed

$$v_u = \frac{2}{3}\left(10 + \frac{L_n}{d}\right)\sqrt{f'_c} \tag{5-108}$$

Required area of shear reinforcement should be determined from

$$\frac{A_v}{s}\left(\frac{1 + L_n/d}{12}\right) + \frac{A_{vh}}{s_2}\left(\frac{11 - L_n/d}{12}\right) = \frac{v_u - v_c}{f_y} b_w \tag{5-109}$$

where A_v = area of shear reinforcement perpendicular to main reinforcement within a distance s

s = spacing of shear reinforcement measured parallel to main reinforcement

A_{vh} = area of shear reinforcement parallel to main reinforcement within a distance s_2

s_2 = spacing of shear reinforcement measured perpendicular to main reinforcement

f_y = yield strength of shear reinforcement

b_w = width of beam web

Spacing s should not exceed $d/5$ or 18 in. Spacing s_2 should not exceed $d/3$ or 18 in. The area of shear reinforcement perpendicular to the main reinforcement should be a minimum of

$$A_v = 0.0015bs \tag{5-110}$$

where b = width of beam compression face
Area of shear reinforcement parallel to main reinforcement should be at least

$$A_{vh} = 0.0025bs_2 \tag{5-111}$$

When $b_w > 10$ in., shear reinforcement should be placed in each face of the beam. If the beam has a face exposed to the weather, between one-half and two-thirds of the total shear reinforcement should be placed in the exterior face. Bars should not be smaller than No. 3.

Bending. The area of steel provided for positive bending moment in a deep beam should be at least

$$A_s = \frac{200b_w d}{f_y} \tag{5-112}$$

where f_y = yield strength of flexural reinforcement, psi
This minimum amount can be reduced to one-third more than that required by analysis.

A safe assumption for preliminary design is that the extreme top surface in compression is $0.25h$ below the top of very deep beams for computation of a reduced effective depth d for flexure (Fig. 5-55).

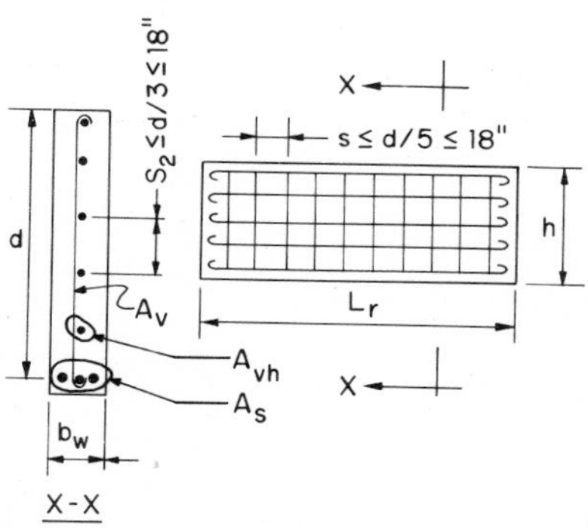

Fig. 5-55. Reinforcing for deep beams. When beam thickness exceeds 10 in., a layer of vertical steel should be placed near each face of the beam.

5-91. Shear Walls. Cantilevered shear walls used for bracing structures against lateral displacement (sidesway) are a special case of deep beams. They may be used as the only lateral bracing, or in conjunction with beam-column frames. In the latter case, the lateral displacement of the combination can be calculated with the assumption that lateral forces resisted by each element can be distributed to walls and frames in proportion to stiffness. For tall structures, the effect of axial shortening of the frames and the contribution of shear to lateral deformation of the shear wall should not be neglected.

Reinforcement required for flexure of shear walls as a cantilever should be proportioned as for deep beams (Art. 5-90). Shear reinforcement is usually furnished as a combination of horizontal and vertical bars distributed evenly in each story (for increment of load). For low shear stress ($v_u < 0.5v_c$, where v_c is the shear unit stress permitted on the concrete), the minimum shear reinforcement required and its location in a wall are the same as the minimum reinforcement for bearing walls (Art. 5-69). Maximum spacing of the horizontal shear reinforcement, however, should not exceed $L_w/5$, $3h$, or 18 in., where L_w is the horizontal length of wall and

h the overall wall thickness (Fig. 5-56). Maximum spacing of the vertical reinforcement should not exceed $L_w/3$, $3h$, or 18 in.

A minimum thickness $h = L_w/25$ is advisable for walls with high shear stress. Nominal shear stress is determined from

$$v_u = \frac{V_u}{\phi hd} \tag{5-113}$$

where V_u = design shear force (acting horizontally for shear walls)
ϕ = capacity reduction factor = 0.85
d = effective length of wall = $0.8L_w$

This stress at any section should not exceed $10\sqrt{f'_c}$, where f'_c is the specified concrete strength, psi.

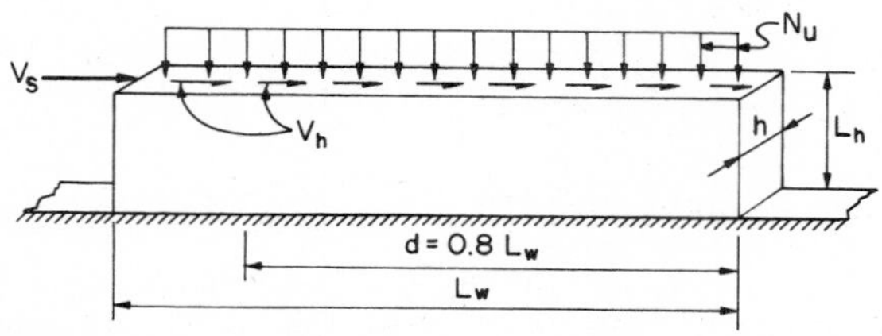

Fig. 5-56. Shear and normal forces on a section of a shear wall.

Shear stress carried by the concrete should not exceed the lesser of the values of v_c computed from Eqs. (5-114) and (5-115).

$$v_c = 3.3\sqrt{f'_c} + \frac{N_u}{4L_wh} \tag{5-114}$$

where N_u = design vertical axial load on wall acting with V_u, including tension due to shrinkage and creep (positive for compression, negative for tension)

$$v_c = 0.6\sqrt{f'_c} + \frac{L_w(1.25\sqrt{f'_c} + 0.2N_u/L_wh)}{M_u/V_u - L_w/2} \tag{5-115}$$

where M_u = design moment at section where V_u acts

Alternatively, $v_c = 2\sqrt{f'_c}$ may be used if N_u causes compression. Shear stress v_c computed for a section at a height above the base equal to $L_w/2$ or one-half the wall height, whichever is smaller, may be used for all lower sections.

When $v_u > v_c/2$, the area of horizontal reinforcement within a distance s required for shear is given by

$$A_h = \frac{(v_u - v_c)hs}{f_y} \geq 0.0025hL_h \tag{5-116}$$

where s = spacing of horizontal reinforcement (max $\leq L_w/5 \leq 3h \leq$ 18 in.)
f_y = yield strength of the reinforcement
L_h = wall height

Also, when $v_u > v_c/2$, the area of vertical shear reinforcement should be at least

$$A_{vh} = \left[0.0025 + 0.5\left(2.5 - \frac{L_h}{L_w}\right)\left(\frac{A_h}{hL_h} - 0.0025\right)\right]hL_w \geq 0.0025hL_w \tag{5-117}$$

but need not be larger than A_h computed from Eq. (5-116).

5-92. Reinforced-Concrete Arches. Arches are used in roofs for such buildings as hangars, auditoriums, gymnasiums, and rinks, where long spans are desired. An arch is essentially a curved beam with the loads, applied downward in its plane, tending to decrease the curvature. Arches are frequently used as the supports for thin shells that follow the curvature of the arches. Such arches are treated in analysis as two-dimensional, whereas the thin shells behave as three-dimensional elements.

The great advantage of an arch in reinforced concrete construction is that, if the arch is appropriately shaped, the whole cross section can be utilized in compression under the maximum (full) load. In an ordinary reinforced concrete beam, the portion below the neutral axis is assumed to be cracked and does not contribute to the bending strength. A beam can be curved, however, to make its axis follow the lines of thrust very closely for all loading conditions, thus virtually eliminating bending moments.

The component parts of a fixed arch are shown in Fig. 5-57. For a discussion of the different types of arches and the stress analyses required for each, see Arts. 3-86 to 3-89.

Because the depth of an arch and loading for maximum moments generally vary along the length, several cross sections must be chosen for design, such as the crown, springing, haunches, and the quarter points. Concrete compressive stresses and shear should be checked at each section, and reinforcement requirements determined. The sections should be designed as rectangular beams or T beams subjected to bending and axial compression, as indicated in Arts. 5-83 to 5-86.

When an arch is loaded, large horizontal reactions, as well as vertical reactions, are developed at the supports. For roof arches, tie rods may be placed overhead, or in or under the ground floor, to take the horizontal reaction. The horizontal

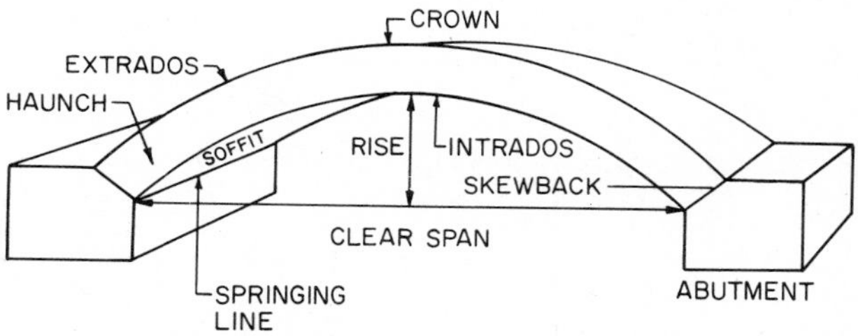

Fig. 5-57. Components of a fixed arch.

reaction may also be resisted externally by footings on sound rock or piles, by reinforced concrete buttresses, or by adjoining portions of the structure, for example a braced floor or roof at springing level.

Hinged arches are commonly made of structural steel or precast concrete. The hinges simplify the arch analysis and the connection to the abutment, and they reduce the indeterminant stresses due to shrinkage, temperature, and settlements of supports. For cast-in-place concrete, hingeless (fixed) arches are often used. They eliminate the cost of special steel hinges needed for hinged concrete arches and permit reduced crown thicknesses, to provide a more attractive shape.

Arches with spans less than 90 ft are usually constructed with ribs 2 to 4 ft wide. Each arch rib is concreted in a continuous operation, usually in 1 day. The concrete may be placed continuously from each abutment toward the crown, to obtain symmetrical loading on the falsework.

For spans of 90 ft or more, however, arch ribs are usually constructed by the alternate block, or voussoir, method. Each rib is constructed of blocks of such size that each can be completed in one casting operation. This method reduces the shrinkage stresses. The blocks are cast in such order that the formwork will settle uniformly. If blocks close to the crown section are not placed before blocks at the haunch and the springing sections, the formwork will rise at the crown, and placing of the crown blocks will then be likely to cause cracks in the haunch. The usual procedure is to cast two blocks at the crown, then two at the springing, and alternate until the complete arch is concreted.

In construction by the alternate block method, the block sections are kept separate by timber bulkheads. The bulkheads are kept in place by temporary struts between the voussoirs. Keyways left between the voussoirs are concreted later. Near piers and abutments where the top slopes exceed about 30° with the horizontal, top forms may be necessary, installed as the casting progresses.

If the arch reinforcement is laid in long lengths, settlement and deformation of the arch formwork can displace the steel. Therefore, depending on the curvature and total length, lengths of bars are usually limited to about 30 ft. Splicing should be located in the keyways. Splices of adjacent bars should be staggered (50% stagger), and located where tension is small.

Upper reinforcement in arch rings may be held in place with spacing boards nailed to props, or with wires attached to transverse timbers supported above the surface of the finished concrete.

Forms for arches may be supported on a timber falsework bent. This bent may consist of joists and beams supported by posts that are braced together and to solid ground. Wedges or other adjustment should be provided at the base of the posts so that the formwork may be adjusted if settlement occurs, and so that the entire formwork may be conveniently lowered after the concrete has hardened sufficiently to take its own load.

("Formwork for Concrete," American Concrete Institute.)

5-93. Reinforced-Concrete Thin Shells. Thin shells are curved slabs with thickness very small compared to the other dimensions. A thin shell possesses three-dimensional load-carrying characteristics. The best natural example of thin-shell behavior

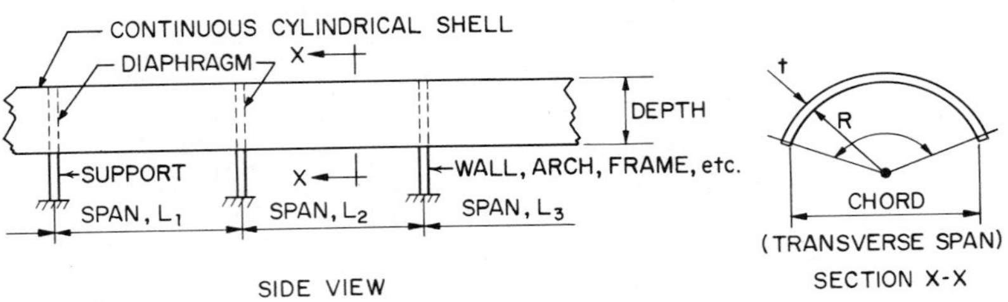

Fig. 5-58. Cylindrical thin shell.

is that of an ordinary egg, which may have a ratio of radius of curvature to thickness of 50. Loads are transmitted through thin shells primarily by direct stresses—tension or compression—called membrane stresses, which are almost uniform throughout the thickness. Reinforced-concrete thin-shell structures commonly utilize ratios of radius of curvature to thickness about five times that of an eggshell. Because concrete shells are always reinforced, their thickness is usually determined by the minimum thickness required to cover the reinforcement, usually 1 to 4 in. Shells are thickened near the supports to withstand localized bending stresses in such areas. (See also Arts. 3-90 to 3-97.)

Shells are most often used as roofs for such buildings as hangars, garages, theaters, and arenas, where large spans are required and the loads are light. The advantages of reinforced-concrete thin shells may be summarized as follows:

Most efficient use of materials.

Great freedom of architectural shapes.

Convenient accommodation of openings for natural lighting and ventilation.

Ability to carry very large unbalance of forces.

High fireproofing value due to lack of corners, thin ribs, and the inherent resistance of reinforced concrete.

Reserve strength due to many alternate paths for carrying load to the supports. One outstanding example withstood artillery fire punctures with only local damage.

Common shapes of reinforced-concrete thin shells used include cylindrical (barrel shells), dome, and hyperbolic paraboloid (saddle shape).

Cylindrical shells may be classified as long if the radius of curvature is shorter than the span, or as short (Fig. 5–58). Long cylindrical shells, particularly the continuous, multiple-barrel version which repeats the identical design of each bay (and permits reuse of formwork) in both directions, are advantageous for roofing rectangular-plan structures. Short cylindrical shells are commonly used for hangar roofs with reinforced-concrete arches furnishing support at short intervals in the direction of the span.

Structural analysis of these common styles may be simplified with design aids. ("Design of Cylindrical Concrete Shell Roofs," Manual No. 31, American Society of Civil Engineers; "Design Constants for Interior Cylindrical Concrete Shells," EB020D; "Design Constants for Ribless Concrete Cylindrical Shells," EB028D; "Coefficients for Design of Cylindrical Concrete Shell Roofs" (extension of ASCE

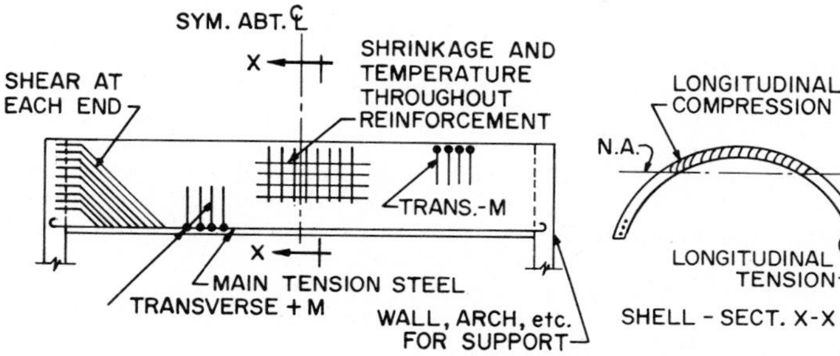

Fig. 5-59. Reinforcement in a long cylindrical shell. Folded plates are similarly reinforced.

Manual No. 31), EB035D; "Design of Barrel Shell Roofs," IS082D, Portland Cement Association.)

The ACI Building Code includes specific provisions for thin shells. It requires an elastic analysis and suggests model studies for important or unusual shapes, prescribes minimum steel, and prohibits use of the working-stress method for design, thus prescribing selection of all shear and flexural reinforcement by the strength-design method with the same load factors as for design of other elements. Figure 5-59 shows a typical reinforcing arrangement for a long cylindrical shell. (See also F. S. Merritt, "Standard Handbook for Civil Engineers," Sec. 8, "Concrete Design and Construction," McGraw-Hill Book Company, New York.)

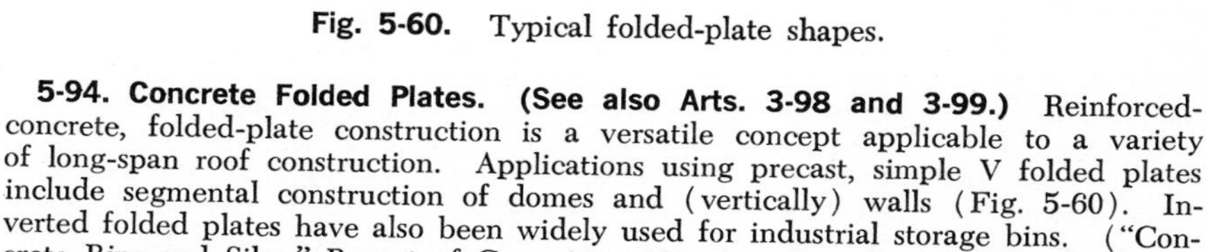

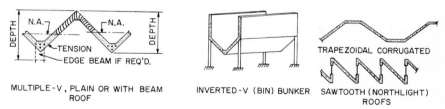

Fig. 5-60. Typical folded-plate shapes.

5-94. Concrete Folded Plates. (See also Arts. 3-98 and 3-99.) Reinforced-concrete, folded-plate construction is a versatile concept applicable to a variety of long-span roof construction. Applications using precast, simple V folded plates include segmental construction of domes and (vertically) walls (Fig. 5-60). Inverted folded plates have also been widely used for industrial storage bins. ("Concrete Bins and Silos," Report of Committee 313, American Concrete Institute.)

Formwork for folded plates is far simpler than that for curved thin shells. Precasting has also been a simpler process to save formwork, permit mass-production construction, and achieve sharp lines for exposed top corners (vees cast upside down) to satisfy aesthetic requirements. For very long spans, posttensioned, draped tendons have been employed to reduce the total depth, deflection, and reinforcing-steel requirements. The tendons may be placed in the inclined plates or, more conveniently, in small thickened edge beams. For cast-in-place, folded-plate construction, double forming can usually be avoided if the slopes are less than 35 to 40°.

Since larger transverse bending moments develop in folded plates than in cylindrical shells of about the same proportions, a minimum thickness less than 4 in. creates practical problems of placing the steel. A number of areas in the plates will require three layers of steel and, near the intersections of plates, top and bottom bars for transverse bending will be required. Ratios of span to total depth are similar to those for cylindrical shells, commonly ranging from 8 to 15. (See also, F. S. Merritt, "Standard Handbook for Civil Engineers," Sec. 8, "Concrete Design and Construction," McGraw-Hill Book Company, New York.)

5-95. Slabs on Grade. Slabs on ground are often used as floors in buildings. Special use requirements often include heavy-duty floor finish (Art. 5-35) and live-load capacity for heavy concentrated (wheel) loads or uniform (storage) loads, or both.

Although slabs on grade seem to be simple structural elements, actual analysis is extremely complicated. Design is usually based on rules of thumb developed by experience. For design load requirements that are unusually heavy and outside common experience, design aids are available. Occasionally, the design will be controlled by wheel loads only, as for floors in hangars, but more frequently by uniform warehouse loadings. ("Design of Floors on Ground for Warehouse Loadings," Paul F. Rice, *ACI Journal,* Aug., 1957, paper No. 54-7; "Design of Concrete Airport Pavement," EB050P, Portland Cement Association.)

A full uniform load over an entire area causes no bending moment if the boundaries of the area are simple construction joints. But actual loads in warehouse usage leave unloaded aisles and often alternate panels unloaded. As a result, a common failure of warehouse floors results from *uplift* of the slab off the subgrade, causing negative moment (top) cracking. In lieu of a precise analysis taking into account live-load magnitude, joint interval and detail, the concrete modulus of elasticity, the soil modulus, and load patterns, a quick solution to avoid uplift is to provide a slab sufficiently thick so that its weight is greater than one-fifteenth times the live load. Such a slab may be unreinforced, if properly jointed, or reinforced for temperature and shrinkage stresses only. Alternatively, for very heavy loadings, an analysis and design may be performed for the use of reinforcement, top and bottom, to control uplift moments and cracking. ("Design of Floors on Ground for Warehouse Loadings," Paul F. Rice, *ACI Journal,* Aug., 1957, paper No. 54-7.)

Shrinkage and temperature change in slabs on ground can combine effects adversely to create warping, uplift, and top cracking failures with no load. Closely spaced joint intervals, alternate panel casting sequence, and controlled curing will avoid these failures. Somewhat longer joint spacings can be employed if reinforcement steel with an area of about 0.002 times the gross section area of slab is provided in perpendicular directions.

With such reinforcement, warping will usually be negligible if the slab is cast in alternate lanes 12 to 14 ft wide, and provided with contraction joints at 20- to 30-ft spacings in the direction of casting. The joints may be tooled, formed by joint filler inserts, or sawed. One-half the bars or wires crossing the contraction joints should be cut accurately on the joint line. The warping effect will be aggravated if excess water is used in the concrete and it is forced to migrate in one direction to top or bottom of the slab, for example, when the slab has been cast on a vapor barrier or on a very dry subgrade. For very long slabs, continuous reinforcement, 0.006 times the gross area, has been used to eliminate transverse joints in more than 12,000 two-lane miles of highway and airport pavement. (Bulletin No. 1, "Design of Continuously Reinforced Concrete Pavement," and "Design and Construction," Continuously Reinforced Pavement Group, 180 N. LaSalle St., Chicago, Ill.; and "Suggested Specifications for Heavy-duty Concrete Floor Topping," IS021B; "Design of Concrete Floors on Ground," IS046B; "Suggested Specifications for Single-course Floors on Ground," IS070B, Portland Cement Association.)

5-96. Seismic-resistant Concrete Construction. The ACI Building Code contains special seismic requirements for design that apply only for areas where the probability of earthquakes capable of causing major damage to structures is high, and where ductility reduction factors for lateral seismic loads are utilized (ACI 318-71, Appendix A). The general requirements of ACI 318-71 for reinforced concrete provide sufficient seismic resistance for seismic zones where only minor seismic damage is probable and no reduction factor for ductility is applied to seismic forces. Designation of seismic zones is prescribed in general building codes, as are lateral force loads for design. (See also Art. 3-84.)

Special ductile-frame design is prescribed to resist lateral movements sufficiently to create "plastic" hinges and permit reversal of direction several times. These hinges must form in the beams at the beam-column connections of the ductile frame.

Shear walls used alone or in combination with ductile beam-column frames must also be designed against brittle (shear) failures under the reversing loads ("Commentary on ACI 318-71").

Ductility is developed in reinforced concrete by:

Conservative limits on the net tension steel ratio ($\rho < 0.50\,\rho_b$, where ρ_b is the steel ratio for balanced conditions, Art. 5-46) to insure underreinforced behavior.

Heavy confining reinforcement extending at joints through the region of maximum moment in both columns and beams, to include points where hinges may form. This confining reinforcement may consist of spiral reinforcing or heavy, closely spaced, well-anchored, closed ties with hooked ends engaging the vertical bars or the ties at the far face (ACI 315).

For complicated arrangements of vertical column bars that require several heavy ties per set at close set spacings, the resulting congestion and increased erection time have led to some use of square or rectangular spirals. For any particular application of the latter, the user is advised to investigate the feasibility of fabrication before detailing and specifying them ("Rectangular Spirals," Western Concrete Reinforcing Steel Institute, Burlingame, Calif. 94010).

5-97. Composite Flexural Members. Reinforced- and prestressed-concrete, composite flexural members are constructed from such components as precast members with cast-in-place flanges, box sections, and folded plates.

Composite structural-steel-concrete members are usually constructed of cast-in-place slabs and structural-steel beams. Interaction between the steel beam and concrete slab is obtained by natural bond if the steel beam is fully encased with a minimum of 2 in. of concrete on the sides of soffit. If the beam is not encased, the interaction may be accomplished with mechanical anchors (shear connectors). Requirements for composite structural-steel-concrete members are given in the AISC "Specification for the Design, Fabrication and Erection of Structural Steel for Buildings" (Art. 6-35).

The design strength of composite flexural members is the same for both shored and unshored construction. Shoring should not be removed, however, until the supported elements have the design properties required to support all loads and limit deflections and cracking. Individual elements should be designed to support all loads prior to the full development of the design strength of the composite member. Premature loading of individual precast elements can cause excessive deflections as the result of creep and shrinkage.

The design horizontal shear force for a composite member may be transferred between individual elements by contact stresses or anchored ties, or both. The horizontal unit shear stress at any section can be calculated from

$$v_{dh} = \frac{V_u}{\phi b_v d} \tag{5-118}$$

where V_u = total applied design shear force at section
b_v = width of section
ϕ = capacity reduction factor = 0.85
d = distance from extreme compression surface to centroid of tension steel

When $v_{dh} \le 80$ psi, the design shear force may be transferred by contact stresses without ties, if the contact surfaces are clean and intentionally roughened with a full amplitude of about ¼ in. Otherwise, fully anchored minimum ties [Eq. (5-86)], spaced not over 24 in. or four times the least dimension of the supported element, may be used.

When v_{dh} is between 80 and 350 psi, the design shear force may be transferred by shear-friction reinforcement placed perpendicular to assumed cracks. Shear stress v_{dh} should not exceed 800 psi or $0.2f'_c$, where f'_c is the specified concrete strength. Required reinforcement area is

$$A_{vf} = \frac{V_u}{\phi f_y \mu} \tag{5-119}$$

where f_y = yield strength of shear reinforcement
μ = coefficient of friction
= 1.4 for monolithic concrete
= 1.0 for concrete cast against hardened concrete
= 0.7 for concrete cast against as-rolled structural steel (clean and without paint)

PRECAST-CONCRETE MEMBERS

Precast-concrete members are assembled and fastened together on the job. They may be unreinforced, reinforced, or prestressed. Precasting is especially advantageous when it permits mass production of concrete units. But precasting is also beneficial because it facilitates quality control and use of higher-strength concrete. Form costs may be greatly reduced, because reusable forms can be located on a casting-plant floor or on the ground at a construction site in protected locations and convenient positions, where workmen can move about freely. Many complex thin-shell structures are economical when precast, but would be uneconomical if cast in place.

5-98. Design Methods for Precast Members. Design of precast-concrete members under the ACI Building Code follows the same rules as for cast-in-place concrete. In some cases, however, design may not be governed by service loads, because transportation and erection loads on precast members may exceed the service loads.

Design of joints and connections must provide for transmission of any forces due to shrinkage, creep, temperature, elastic deformation, gravity loads, wind loads, and earthquake motion.

("Suggested Design of Joints and Connections in Precast Structural Concrete," report by ACI Committee 512; "Precast Concrete Joists in Floor and Roof Construction," PA034B, Portland Cement Association, Skokie, Ill.)

5-99. Reinforcement Cover in Precast Members. Less concrete cover is required for reinforcement in precast-concrete members manufactured under plant control conditions than in cast-in-place members because the control for proportioning, placing, and curing is better. Minimum concrete cover for reinforcement required by ACI 318-71 is listed in Table 5-25.

For special, thin-section, precast-concrete members, less cover for small-size reinforcement is acceptable (Table 5-26).

For all sizes of reinforcement in precast-concrete wall panels, minimum cover of ¾ in. is acceptable at nontreated surfaces exposed to weather and ⅜ in. at interior surfaces. (Report of Committee 533, *ACI Journal,* April, 1970; "Minimum Requirements for Thin-section Precast Concrete Construction," ACI 525.)

5-100. Tolerances for Precast Construction. Dimensional tolerances for precast members and tolerances on fitting of precast members vary for type of member, type of joint, and conditions of use. See "PCI Design Handbook," Prestressed Concrete Institute; "Recommended Practice for Manufactured Reinforced Concrete Floor and Roof Units," ACI 512; "Precast Reinforced Concrete—Thin Sections," report of ACI Committee 525; ACI "Manual of Standard Practice for Detailing Concrete Structures."

5-101. Accelerated Curing. For strength and durability, precast concrete members require adequate curing. They usually are given some type of accelerated curing for economic reuse of forms and casting space. At atmospheric pressure, curing temperatures may be held between 125 and 185°F for 12 to 72 hr. Under pressure, autoclave temperatures above 325°F for 5 to 36 hr are applied for fast curing. Casting temperatures, however, should not exceed 90°F. See Fig. 5-6.

("Recommended Practice for Curing Concrete," report by ACI Committee 308; "Low-pressure Steam Curing," report by ACI Committee 517; "High-pressure Steam Curing: Modern Practice, Properties of Autoclaved Products," report by ACI Committee 516.)

5-102. Precast Floor and Roof Systems. Long-span, precast-concrete floor and roof units are usually prestressed. Short members, 30 ft or less, are often made with ordinary reinforcement. Types of precast units for floor and roof systems include solid or ribbed slabs, hollow-core slabs, single and double tees, rectangular beams, L-shaped beams, inverted-T beams, and I beams.

Table 5-25. Minimum Reinforcement Cover for Precast Members, In.

Concrete exposed to earth or weather:	
Wall panels:	
No. 14 and No. 18 bars	1½
No. 11 and smaller	¾
Other members:	
No. 14 and No. 18 bars	2
No. 6 through No. 11 bars	1½
No. 5 bars, ⅝-in. wire and smaller	1¼
Concrete not exposed to weather or in contact with the ground:	
Slabs, walls, joists:	
No. 14 and No. 18 bars	1¼
No. 11 and smaller	⅝
Beams, girders, columns:	
Principal reinforcement:	
Diameter of bar d_b but not less than ⅝ in. and need not be more than 1½ in.	
Ties, stirrups or spirals	⅜
Shells and folded-plate members:	
No. 6 bars and larger	⅝
No. 5 bars, ⅝-in. wire and smaller	⅜

Table 5-26. Minimum Reinforcement Cover for Thin-section Precast Members, In.

Concrete surfaces exposed to the weather:	
Main reinforcement in beams and columns	½
Slab reinforcement and secondary reinforcement in beams and columns	⅜
Concrete surfaces not exposed to weather, ground, or water:	
All reinforcement	⅜

Hollow-core slabs are usually available in normal-weight or structural lightweight concrete. Units range from 16 to 96 in. in width, and from 4 to 12 in. in depth. Hollow-core slabs may come with grouted shear keys to distribute loads to adjacent units over a slab width as great as one-half the span ("Load Distribution on Precast, Prestressed Hollow-core Slab Construction," *Journal of the Prestressed Concrete Institute,* Nov.-Dec. 1971). Manufacturers should be consulted for load and span data on hollow-core slabs, because camber and deflection often control the serviceability of such units, regardless of strength.

5-103. Precast Ribbed Slabs, Folded Plates, and Shells. Curved shells and folded plates have a thickness that is small compared to their other dimensions. Such structures depend on their geometrical configuration and boundary conditions for strength.

Thickness. With closely spaced ribs or folds, a minimum thickness for plane sections of 1 in. is acceptable.

Reinforcement. Welded-wire fabric with a maximum spacing of 2 in. may be used for slab portions of thin-section members, and for wide, thin elements 3 in. thick or less. Reinforcement should be preassembled into cages, using a template, and placed within a tolerance of +0 in. or −⅛ in. from the nearest face. The minimum clear distance between bars should not be less than 1½ times the nominal maximum size of the aggregate. For minimum cover of reinforcement, see Art. 5-99.

Compressive Strength. Concrete for thin-section, precast-concrete members protected from the weather and moisture and not in contact with the ground should have a strength of at least 4,000 psi at 28 days. For elements in other locations, a minimum of 5,000 psi is recommended.

Analysis. Determination of axial stresses, moments, and shears in thin sections is usually based on the assumption that the material is ideally elastic, homogeneous, and isotropic.

Forms. Commonly used methods for the manufacture of thin-section, precast-concrete members employ metal or plastic molds, which form the bottom of the slab and the sides of the boundary members. Forms are usually removed pneumatically or hydraulically by admitting air or water under pressure through the bottom form.

("Precast Reinforced Concrete—Thin Sections," Report by ACI Committee 525.)

5-104. Wall Panels. Precast-concrete wall panels include plain panels, decorative panels, natural stone-faced panels, sandwich panels, solid panels, ribbed panels, tilt-up panels, loadbearing and nonloadbearing panels, and thin-section panels. Prestressing, when used with such panels, makes it possible to handle and erect large units and thin sections without cracking ("Symposium on Precast Wall Panels," SP-11, American Concrete Institute).

Forms required to produce the desired size and shape of panel are usually made of steel, wood, concrete, vacuum-formed thermoplastics, fiber-reinforced plastics, or plastics formed into shape by heat and pressure, or any combination of these. For complicated form details, molds of plaster, gelatin, or sculptured sand can be used.

Glossy-smooth concrete finish can be obtained with forms made of plastic. But, for exterior exposure, this finish left untreated undergoes gradual and nonuniform loss of its high reflectivity. Textured surfaces or smooth but nonglossy surfaces obtained by early form removal are preferred for exterior exposure.

Exposed-aggregate monolithic finishes can be obtained with horizontal-cast panels by initially casting a thin layer containing the special surface aggregates in the forms and then casting regular concrete backup. With a thickness of exposed aggregate of less than 1 in., the panel can also be cast face up and the aggregate seeded over the fresh concrete or hand placed in a wet mortar. Variations of exposed surface can be achieved by use of set retardant, acid washes, or sandblasting. ("Jobsite Precast Concrete Panels," IS047A, IS048A, IS049.01A, IS061A; "Sandblasting of Concrete Surfaces," IS180T.)

Consolidation of the concrete in the forms to obtain good appearance and durability can be attained by one of the following methods.

External vibration with high-frequency form vibrators or a vibrating table.

Internal or surface vibration with a tamping-type or jitterbug vibrator.

Placing a rich, high-slump concrete in a first layer to obtain uniform distribution of the coarse aggregate and maximum consolidation, and then making the mix for the following layers progressively stiffer. This allows absorption of excessive water from the previous layer.

Tilt-up panels can be economical if the floor slab of the building can be designed for and used as the form for the panels. The floor slab must be level and smoothly troweled. Application of a good bond-breaking agent to the slab before concrete is cast for the panels is essential to obtain a clean lift of the precast panels from the floor slab.

If lifting cables are attached to a panel edge, large bending moments may develop at the center of the wall. For high panels, three-point pickup may be used. To spread pickup stresses, specially designed inserts are cast into the wall at pickup points.

Another method of lifting wall panels employs a vacuum mat—a large steel mat with a rubber gasket at its edges to contact the slab. When the air between mat and panel is pumped out, the mat adheres to the panel, because of the resulting vacuum, and can be used to raise the panel. The method has the advantage of spreading pickup forces over the mat area.

Panels, when erected, must be temporarily braced until other construction is in place to provide required permanent bracing.

Joints. Joint sealants for panel installations may be mastics or elastomeric materials. These are extensible and can accommodate the movement of panels. Guide lines for joint width are given in Table 5-27.

Recommended maximum joint widths and minimum expansions for the common sealants are listed in Table 5-28.

The joint sealant manufacturer should be asked to advise on backup material for use with his sealant and which shape factor should be considered. A good backup material is a rod of sponge material with a minimum compression of 30%, such as foamed polyethylene, polystyrene, polyurethane, polyvinyl chloride, or synthetic rubber.

5-105. Lift Slabs. Lift slabs are precast-concrete floor and roof panels that are cast on a base slab at ground level, one on top of the other, with a bond-breaking membrane between them. Steel collars are embedded in the slabs and fit loosely around the columns. After the slabs have cured, they are lifted to their final position by a patented jack system supported on the columns. The embedded steel collars then are welded to the steel columns to hold the lift slabs in place. This method of construction eliminates practically all formwork.

Table 5-27. Joint Widths for Precast Wall Panels*

Recommended maximum panel dimensions, ft	Design normal joint width, in.
5	3/8
18	1/2
30	3/4

* Report of ACI Committee 533, *ACI Journal,* April, 1970.

Table 5-28. Maximum Joint Widths for Sealants*

Type of sealant	Maximum joint width, in.	Maximum movement, tension and compression, %
Butyl	3/4	±10
Acrylic	3/4	±15–25
One-part polyurethane	3/4	±20
Two-part polyurethane	3/4	±25
One-part polysulfide	3/4	±25
Two-part polysulfide	3/4	±25

* Report of ACI Committee 533, *ACI Journal,* April, 1970.

PRESTRESSED-CONCRETE CONSTRUCTION

Prestressed concrete is concrete in which internal stresses have been introduced during fabrication to counteract the stresses produced by service loads. The prestress compresses the tensile area of the concrete to eliminate or make small the tensile stresses caused by the loads.

5-106. Basic Principles of Prestressed Concrete. In applying prestress, the usual procedure is to tension high-strength-steel elements, called **tendons,** and anchor them to the concrete, which resists the tendency of the stretched steel to shorten after anchorage and is thus compressed. If the tendons are tensioned before concrete has been placed, the prestressing is called **pretensioning.** If the tendons are tensioned after the concrete has been placed the prestressing is called **posttensioning.**

Prestress can prevent cracking by keeping tensile stresses small, or entirely avoiding tension under service loads. The entire concrete cross section behaves as an uncracked homogeneous material in bending. In contrast, in nonprestressed, reinforced-concrete construction, tensile stresses are resisted by reinforcing steel, and concrete in tension is considered ineffective. It is particularly advantageous with prestressed concrete to use high-strength concrete.

Loss of Prestress. The final compression force in the concrete is not equal to the initial tension force applied by the tendons. There are immediate losses, due to elastic shortening of the concrete, friction losses from curvature of the tendons, and slip at anchorages. There are also long-time losses, such as those due to shrinkage and creep of the concrete, and possibly relaxation of the steel. These losses should be computed as accurately as possible or determined experimentally. They are deducted from the initial prestressing force to determine the effective prestressing force to be used in design. (The reason that high-strength steels must be used for prestressing is to maintain the sum of these strain losses at a small percentage of the initially applied prestressing strain.) (See also Art. 5-107.)

Stresses. When stresses in prestressed members are determined, prestressing forces can be treated as other external loads. If the prestress is large enough to prevent cracking under design loads, elastic theory can be applied to the entire concrete cross section (Fig. 5-61).

Prestress may be applied to a beam by straight tendons or curved tendons. Stresses at midspan can be the same for both types of tendons, but the net stresses with the curved tendons can remain compressive away from midspan, whereas they become tensile at the top fiber near the ends with straight tendons. For a prestressing force P_s applied to a beam by a straight tendon at a distance e_1 below the neutral axis, the resulting prestress in the extreme surface throughout is

$$f = \frac{P_s}{A_c} \pm \frac{P_s e_1 c}{I_g} \qquad (5\text{-}120)$$

where P_s/A_c is the compressive stress on a cross section of area A_c, and $P_s e_1 c/I_g$ is the bending stress induced by P_s (positive for compression and negative for tension), as indicated in Fig. 5-61. If stresses $\pm Mc/I_g$ due to moment M caused by external gravity loads are superimposed at midspan, the net stresses in the extreme fibers can become zero at the bottom and compressive at the top. Because the stresses due to gravity loads are zero at the beam ends, the prestress is the final stress there and the top surface of the beam at the ends is in tension.

If the tensile stresses at the ends of beams with straight tendons are excessive, the tendons may be draped, or harped, in a vertical curve. Stresses at midspan will be substantially the same as with straight tendons (if the horizontal component of prestress is nearly equal to P_s), and the stresses at the beam ends will be compressive, because the prestressing force passes through or above the centroid of the end sections (Fig. 5-61). Between midspan and the ends, the cross sections will also be in compression.

5-107. Losses in Prestress. Assumption in design of total losses in tendon stress of 35,000 psi for pretensioning and 25,000 psi for posttensioning to allow for elastic shortening, frictional losses, slip at anchorages, shrinkage, creep, and relaxation of steel usually gives satisfactory results. Losses greater or smaller than these values have little effect on the design strength but can affect service-load behavior, such as cracking load, deflection, and camber.

Elastic Shortening of Concrete. In pretensioned members, when the tendons are released from fixed abutments and the steel stress is transferred to the concrete by bond, the concrete shortens under the compressive stress. The decrease in unit stress in the tendons equals $P_s E_s/A_c E_c = nf_c$, where E_s is the modulus of elasticity of the steel, psi; E_c the modulus of elasticity of the concrete psi; n the modular ratio, E_s/E_c; f_c the unit stress in the concrete, psi; P_s the prestressing force applied by the tendons; and A_c the cross-sectional area of the member.

In posttensioned members, the loss due to elastic shortening can be eliminated by using the members as a reaction in tensioning the tendons.

Frictional Losses. In posttensioned members, there may be a loss of prestress where curved tendons rub against their enclosure. The loss may be computed in terms of a curvature-friction coefficient μ. Losses due to unintentional misalignment may be calculated from a wobble-friction coefficient K (per lin ft). Since the coefficients vary considerably, they should, if possible, be determined experimentally. A safe range of these coefficients for estimates is given in the "Commentary on ACI 318-71."

Frictional losses can be reduced by tensioning the tendons at both ends, or by initial use of a larger jacking force which is then eased off to the required initial force for anchorage.

Slip at Anchorages. For posttensioned members, prestress loss may occur at the anchorages during the anchoring. For example, seating of wedges may permit some shortening of the tendons. If tests of a specific anchorage device indicate a

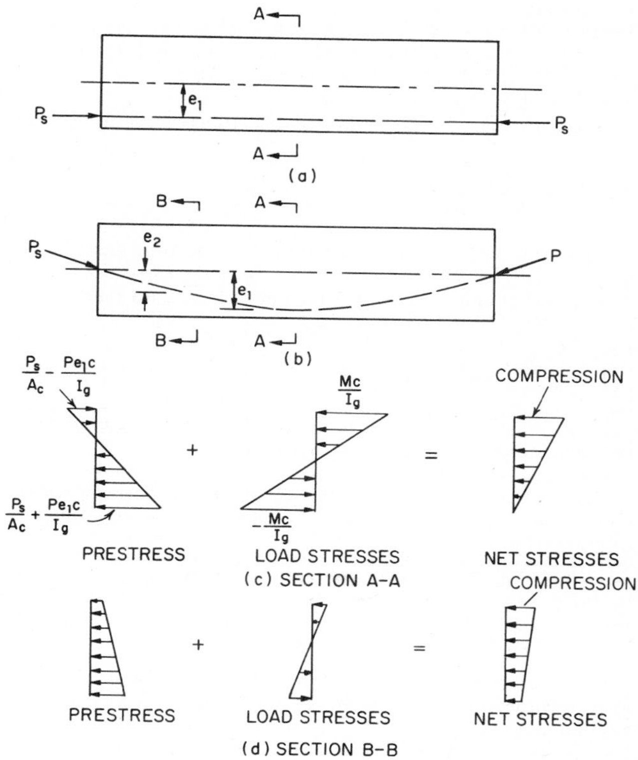

Fig. 5-61. Prestressed-concrete beam. (*a*) With straight tendons. (*b*) With curved tendons. (*c*) Midspan stresses with straight or curved tendons. (*d*) Stresses between midspan and supports with curved tendons. Net stresses near the supports become tensile with straight tendons.

shortening δl, the decrease in unit stress in the steel is equal to $E_s \delta l / l$, where l is the length of the tendon. This loss can be reduced or eliminated by overtensioning initially by an additional strain equal to the estimated shortening.

Shrinkage of Concrete. Change in length of a member due to concrete shrinkage results in a prestress loss over a period of time. This change can be determined from tests or experience. Generally, the loss is greater for pretensioned members than for posttensioned members, which are prestressed after much of the shrinkage has occurred. Assuming a shrinkage of 0.0002 in. per in. of length for a pretensioned member, the loss in tension in the tendons is $0.0002E_s = 0.002 \times 30 \times 10^6 = 6{,}000$ psi.

Creep of Concrete. Change in length of concrete under sustained load induces a prestress loss proportional to the load over a period of time depending greatly on the aggregate used. This loss may be several times the elastic shortening. An estimate of this loss may be made with an estimated creep coefficient C_{cr} equal to the ratio of additional long-time deformation to initial elastic deformation determined by test. The loss in tension for axial prestress in the steel is, therefore, equal to $C_{cr}nf_c$. Values ranging from 1.5 to 2.0 have been recommended for C_{cr}.

Relaxation of Steel. A decrease in stress under constant high strain occurs with some steels. Steel tensioned to 60% of its ultimate strength may relax and lose as much as 3% of the prestressing force. This type of loss may be reduced by temporary overtensioning, which artificially accelerates relaxation, reducing the loss that will occur later at lower stresses.

5-108. Allowable Stresses at Service Loads. At service loads and up to cracking loads. straight-line theory may be used with the following assumptions:

Strains vary linearly with depth through the entire load range.

At cracked sections, the concrete does not resist tension.

Areas of unbonded open ducts should not be considered in computing section properties.

Table 5-29. Allowable Stresses for Prestressed Concrete

Concrete:	
Temporary stresses after transfer of prestress but before prestress losses:	
Compression	$0.60f'_{ci}$
Tension in members without auxiliary reinforcement in tension zone	$3\sqrt{f'_{ci}}$*
Service-load stresses after prestress losses:	
Compression	$0.45f'_c$
Tension in precompressed tensile zone.	$6\sqrt{f'_c}$†
Prestressing steel:	
Due to jacking force	$0.80f_{pu}$‡
Pretensioning tendons immediately after transfer or posttensioning tendons immediately after anchoring	$0.70f_{pu}$‡

* Where the calculated tension stress exceeds this value, reinforcement should be provided to resist the total tension force on the concrete computed for assumption of an uncracked section.

† May be taken as $12\sqrt{f'_c}$ for members for which computations based on the transformed cracked section and on bilinear moment-deflection relationships show that immediate and long-time deflections do not exceed the limits given in Table 5-13.

‡ f_{pu} = ultimate strength of tendons, but not to exceed the maximum value recommended by the manufacturer of the tendons or the strength of anchorages.

The transformed area of bonded tendons and reinforcing steel may be included in pretensioned members and, after the tendons have been bonded by grouting, in posttensioned members.

Flexural stresses must be limited to insure proper behavior at service loads. Limiting these stresses, however, does not insure adequate design strength.

In establishing permissible flexural stresses, the ACI Building Code recognizes two service-load conditions, that before and that after prestress losses. Higher stresses are permitted for the initial state (temporary stresses) than for loadings applied after the losses have occurred.

Permissible stresses in the concrete for the initial load condition are specified as a percentage of f'_{ci}, the compressive strength of the concrete, psi, at time of initial prestress. This strength is used as a base instead of the usual f'_c, 28-day strength of concrete, because prestress is usually applied only a few days after concrete has been cast. The allowable stresses for prestressed concrete, as given in ACI 318-71, are tabulated in Table 5-29.

Bearing Stresses. Determination of bearing stresses at end regions around posttensioning anchorages is complicated, because of the elastic and inelastic behavior of the concrete and because the dimensions involved preclude simple analysis under the St. Venant theory of linear stress distribution of concentrated loads.

Lateral reinforcement may be required in anchorage zones to resist bursting, horizontal splitting, and spalling forces. Anchorages are usually supplied by a prestressing company and are usually designed by test and experience rather than theory.

Concrete in the anchorage region should be designed to develop the guaranteed ultimate tensile strength of the tendons. The ACI Code formula for bearing stresses [Eq. (5-96)] does not apply to posttensioning anchorages, but it can be used as a guide by substitution of f'_{ci}, the compressive strength of the concrete at the time of initial prestress, for f'_c.

5-109. Design Procedure for Prestressed-Concrete Beams. Beam design involves choice of shape and dimensions of the concrete member, positioning of the tendons, and selection of amount of prestress.

After a concrete shape and dimensions have been assumed, determine the geometrical properties—cross-sectional area, center of gravity, distances of kern and extreme surface from the centroid, moment of inertia, section moduli, and dead load of the member per unit of length.

Treat the prestressing forces as a system of external forces acting on the concrete.

Compute bending stresses due to service dead and live loads. From these, determine the magnitude and location of the prestressing force required at sections subject to maximum moment. The prestressing force must result in sufficient compressive stress in the concrete to offset the tensile stresses caused by the bending moments due to dead and live service loads (Fig. 5-61). But at the same time, the prestress must not create allowable stresses that exceed those listed in Table 5-29. Investigation of other sections will guide selection of tendons to be used and determine their position and profile in the beam.

After establishing the tendon profile, prestressing forces, and tendon areas, check stresses at critical points along the beam immediately after transfer, but before losses. Using strength-design methods (Art. 5-110), check the percentage of steel and the strength of the member in flexure and shear.

Design anchorages, if required, and shear reinforcement.

Finally, check the deflection and camber under service loads. The modulus of elasticity of high-strength prestressing steel should not be assumed equal to 29,000,000 psi, as for conventional reinforcement, but should be determined by test or obtained from the manufacturer as required by the ACI Building Code.

5-110. Flexural-strength Design of Prestressed Concrete. (See also Art. 5-109.) Flexural strength design should be based on factored loads and the assumptions of the ACI Building Code, as explained in Arts. 5-44 through 5-51. The stress f_{ps} in the tendons at design load ($1.4D + 1.7L$, where D is the dead load and L the live load), however, should not be assumed equal to the specified yield strength. High-strength prestressing steels lack a sharp and distinct yield point, and f_{ps} varies with the ultimate strength of the prestressing steel f_{pu}, the prestressing steel percentage ρ_p, and the concrete strength f'_c at 28 days. A stress-strain curve for the steel being used is necessary for stress and strain compatibility computations of f_{ps}. For unbonded tendons, successive trial-and-error analysis of tendon strain for strength design is straightforward but tedious. Assume a deflection at failure by crushing of the concrete (strain = 0.003 in. per in.). Determine from the stress-strain curve for the tendon steel the tendon stress corresponding to the total tendon strain at the assumed deflection. Proceed through successive trials, varying the assumed deflection, until the algebraic sum of the internal tensile and compressive forces equals zero. The moment of the resulting couple comprising the tensile and compressive forces times $\phi = 0.90$ is the design moment.

When such data are not available, and the effective prestress, after losses, is at least $f_{pu}/2$, the stress f_{ps} in unbonded tendons may be obtained from

$$f_{ps} = f_{pu}\left(1 - 0.5\rho_p \frac{f_{pu}}{f'_c}\right) \qquad (5\text{-}121a)$$

where $\rho_p = A_{ps}/bd$

A_{ps} = area of tendons in tension zone

b = width of compression face of member

d = distance from extreme compression surface to centroid of tendons and other tension reinforcement, if used and in unbonded tendons from

$$f_{ps} = f_{se} + 10{,}000 + \frac{f'_c}{100\rho_p} \leq f_{se} + 60{,}000 \tag{5-121b}$$

where f_{se} = effective prestress in tendons, psi, after prestress losses; but f_{ps} should not be taken more than the specified tendon yield strength.

Nonprestressed reinforcement conforming to ASTM A185, A615, A616, A617, and A496, in combination with tendons, may be assumed equivalent, at design-load moment, to its area times its yield point stress, but only if

$$\omega_p \leq 0.30 \tag{5-122}$$

$$(\omega + \omega_p - \omega') \leq 0.30 \tag{5-123}$$

$$\omega_w + \omega_{wp} - \omega'_w \leq 0.30 \tag{5-124}$$

where $\omega = \rho f_y / f'_c$

f_y = yield strength of nonprestressed reinforcement

$\omega' = \rho' f_y / f'_c$

$\omega_p = \rho_p f_{ps} / f'_c$

$\rho = A_s / bd$ = ratio of nonprestressed tension reinforcement

A_s = area of nonprestressed tension reinforcement

$\rho' = A'_s / bd$ = ratio of nonprestressed compression reinforcement

A'_s = area of nonprestressed compression reinforcement

$\omega_w, \omega_{wp}, \omega'_w$ = reinforcement indices for flanged sections, computed as for ω, ω_p, and ω', except that b is the web width, and the steel area is that required to develop the compressive strength of the web only

Strength in flexure (without the capacity reduction factor ϕ), as predicted by standard equations, does not correlate well with test results when ω exceeds 0.30.

Design and Cracking Loads. To prevent an abrupt flexural failure by rupture of the prestressing steel immediately after cracking without a warning deflection, the total amount of prestressed and nonprestressed reinforcement should be adequate to develop a design load in flexure of at least 1.2 times the cracking load, calculated on the basis of a modulus of rupture f_r. For normal-weight concrete, this modulus may be taken as

$$f_r = 7.5\sqrt{f'_c} \tag{5-125}$$

and for lightweight concrete as

$$f_r = 1.1 f_{ct} \leq 7.5\sqrt{f'_c} \tag{5-126a}$$

where f_{ct} = average splitting tensile strength of lightweight concrete

When the value for f_{ct} is not available, the modulus of rupture of lightweight concrete can be computed for sand-lightweight concrete from

$$f_r = 6.375\sqrt{f'_c} \tag{5-126b}$$

and for all-lightweight concrete from

$$f_r = 5.625\sqrt{f'_c} \tag{5-126c}$$

5-111. Shear-strength Design of Prestressed Concrete. (See also Art. 5-109.) The ACI Building Code requires that prestressed concrete beams be designed to resist diagonal tension by strength theory. There are two types of diagonal-tension cracks that can occur in prestressed-concrete flexural members: flexural-shear cracks initiated by flexural-tension cracks, and web-shear cracks caused by principal tensile stresses that exceed the tensile strength of the concrete.

The average shear stress v_u computed from Eq. (5-37) can be used as a measure of the diagonal-tension stress. The distance d from the extreme compression surface to the centroid of the tension reinforcement should not be taken less than 0.80 the over-all depth h of the beam.

When the beam reaction in the direction of the applied shear introduces compression into the end region of the member, the shear does not need to be checked within a distance $h/2$ from the face of the support.

Minimum Steel. The ACI Code requires that a minimum area of shear reinforcement be provided in prestressed-concrete members, except where the average diagonal-tension stress v_u is less $v_c/2$, where v_c is the shear stress that can be carried by the concrete; or where the depth h of the member is less than 10 in., 2.5 times the thickness of the compression flange, or one-half the thickness of the web; or where tests show that the required ultimate flexural and shear capacity can be developed without shear reinforcement.

When shear reinforcement is required, the amount provided perpendicular to the beam axis within a distance s should be not less than A_v given by Eq. (5-86). If, however, the effective prestress force is equal to or greater than 40% of the tensile strength of the flexural reinforcement, a minimum area A_v computed from Eq. (5-127) may be used.

$$A_v = \frac{A_{ps}}{80}\frac{f_{pu}}{f_y}\frac{s}{d}\sqrt{\frac{d}{b_w}} \tag{5-127}$$

where A_{ps} = area of tendons in tension zone
f_{pu} = ultimate strength of tendons
f_y = yield strength of nonprestressed reinforcement
s = shear reinforcement spacing measured parallel to longitudinal axis of member
d = distance from extreme compression surface to centroid of tension reinforcement
b_w = web width

The ACI Code does not permit the yield strength f_y of shear reinforcement to be assumed greater than 60,000 psi. The Code also requires that stirrups be placed perpendicular to the beam axis and spaced not farther apart than 24 in. or $0.75h$, where h is the over-all depth of the member.

The area of shear reinforcement required to carry the shear in excess of the shear that can be carried by the concrete can be determined from Eq. (5-40*a*).

Maximum Shear. For prestressed concrete members subjected to an effective prestress force equal to at least 40% of the tensile strength of the flexural reinforcement, the shear stress carried by the concrete is limited to that which would cause significant inclined cracking and, unless Eqs. (5-129) and (5-130) are used, can be taken as equal to the larger of $2\sqrt{f'_c}$ and

$$v_c = 0.60\sqrt{f'_c} + 700\frac{V_u d}{M_u} \leq 5\sqrt{f'_c} \tag{5-128}$$

where M_u = design-load moment at section
V_u = design shear force at section

The design moment M_u occurs simultaneously with the shear V_u at the section. The ratio $V_u d/M_u$ should not be taken greater than 1.0.

If the effective prestress force is less than 40% of the tensile strength of the flexural reinforcement, or if a more accurate method is preferred, the value of v_c should be taken as the smaller of the shear stress causing inclined flexure-shear cracking v_{ci} or web-shear cracking v_{cw}, but need not be smaller than $1.7f'_c$.

$$v_{ci} = 0.6\sqrt{f'_c} + \frac{V_d + V_l M_{cr}/M_{max}}{b_w d} \tag{5-129}$$

$$v_{cw} = 3.5\sqrt{f'_c} + 0.3f_{pc} + \frac{V_p}{b_w d} \tag{5-130}$$

where V_d = shear force at section due to dead load
V_l = shear force at section occurring simultaneously with M_{max}
M_{max} = maximum bending moment at the section due to externally applied design loads
M_{cr} = cracking moment based on the modulus of rupture (Art. 5-51)
b_w = width of web
f_{pc} = compressive stress in the concrete, after all prestress losses have occurred, at the centroid of the cross section resisting the applied loads, or at the junction of web and flange when the centroid lies within the flange
V_p = vertical component of effective prestress force at section considered

In a pretensioned beam in which the section $h/2$ from the face of the support is closer to the end of the beam than the transfer length of the tendon, the reduced prestress in the concrete at sections falling within the transfer length should be considered when calculating v_{cw}. The prestress may be assumed to vary linearly along the centroidal axis from zero at the beam end to the end of the transfer length. This distance can be assumed to be 50 diameters for strand and 100 diameters for single wire.

5-112. Bond, Development, and Grouting of Tendons. Three- or seven-wire pretensioning strand should be bonded beyond the critical section for a development length, in., of at least

$$l_d = (f_{ps} - \tfrac{2}{3} f_{se}) d_b \tag{5-131}$$

where d_b = nominal diameter of strand, in.
f_{ps} = calculated stress in tendons at design load, ksi
f_{se} = effective stress in tendons after losses, ksi

(The expression in parentheses is used as a constant without units.) Investigations for bond integrity may be limited to those cross sections nearest each end of the member that are required to develop their full strength under design load. When bonding does not extend to the end of the member, the bonded development length given by Eq. (5-131) should be doubled.

Minimum Bonded Steel. When prestressing steel is unbonded, the ACI Building Code requires that some bonded reinforcement be placed in the precompressed tensile zone of flexural members and distributed uniformly over the tension zone near the extreme tension surface. The amount of bonded reinforcement that should be furnished for beams, one-way slabs, and two-way slabs would be the larger of the values of A_s computed from Eqs. (5-132) and (5-133).

$$A_s = \frac{N_c}{0.5 f_y} \tag{5-132}$$

$$A_s = 0.004A \tag{5-133}$$

where N_c = tensile force in the concrete under actual dead load plus 1.2 times live load
A = area of that part of the cross section between the flexural tension face and centroid of gross section
f_y = yield strength of bonded reinforcement, but not more than 60,000 psi

The minimum amount of bonded reinforcement for two-way slabs with unbonded tendons is the same as for one-way slabs, except that the minimum amount may be reduced when there is no tension in the precompressed tension zone under service loads.

Grouting of Tendons. When posttensioned tendons are to be bonded, a cement grout is usually injected under pressure (80 to 100 psi) into the space between the tendon and the sheathing material of the cableway. The grout can be inserted in holes in the anchorage heads and cones, or through buried pipes. To insure filling of the space, the grout can be injected under pressure at one end of the member until it is forced out the other end. For long members, it can be injected at each end until it is forced out a vent between the ends.

Grout provides bond between the posttensioning tendons and the concrete member and protects the tendons against corrosion.

Members should be above 50°F in temperature at the time of grouting. This minimum temperature should be maintained for at least 48 hours.

Tendon Sheaths. Ducts for grouted or unbonded tendons should be mortar-tight and nonreactive with concrete, tendons, or filler material. To facilitate injection of the grout, the cableway or duct should be at least ¼ in. larger than the diameter of the posttensioning tendon, or large enough to have an internal area at least twice the gross area of the prestressing steel.

("Recommended Practice for Grouting Posttensioning Tendons," Prestressed Concrete Manufacturers Association of California and Western Concrete Reinforcing Steel Institute.)

5-113. Application and Measurement of Prestress. The actual amount of prestressing force applied to a concrete member should be determined by measuring the tendon elongation, also by checking jack pressure on a calibrated gage or load cell, or by use of a calibrated dynamometer. If the discrepancy in force determination exceeds 5%, it should be investigated and corrected. Elongation measurements should be correlated with average load-elongation curves for the particular prestressing steel being used.

5-114. Concrete Cover in Prestressed Members. The minimum thicknesses of cover required by the ACI Building Code for prestressed and nonprestressed reinforcement, ducts, and end fittings in prestressed concrete members are listed in Table 5-30.

Table 5-30. Minimum Concrete Cover, In., in Prestressed Members

Concrete cast against and permanently exposed to earth	3
Concrete exposed to earth or weather:	
Wall panels, slabs, and joists	1
Other members	1½
Concrete not exposed to weather or in contact with the ground:	
Slabs, walls, joists	¾
Beams, girders, columns:	
Principal reinforcement	1½
Ties, stirrups, or spirals	1
Shells and folded-plate members:	
Reinforcement ⅝ in. and smaller	⅜
Other reinforcement	d_b but not less than ¾

The cover for nonprestressed reinforcement in prestressed concrete members under plant control may be that required for precast members (Tables 5-25 and 5-26). When the general code requires fire-protection covering greater than required by the ACI Code, such cover should be used.

("PCI Design Handbook," Prestressed Concrete Institute.)

Section 6

Structural-Steel Construction

HENRY J. STETINA

Consulting Engineer, Jenkintown, Pa.,
Formerly Senior Regional Engineer,
American Institute of Steel Construction

Structural-steel construction embraces only the range of hot-rolled steel sections (or shapes) and plates, of thicknesses not less than about ⅛ in., together with such fittings as rivets, bolts, bracing rods, and turnbuckles.

The shop operation of cutting steel plates and shapes to final size, punching, drilling, and assembling the components into finished members ready for shipment is called **fabricating.** Some 400 companies comprise the structural-steel fabricating industry. The majority not only take contracts to furnish the fabricating services but also install or erect the steel—a branch of the industry called **erecting.**

The skilled laborers employed in the field to actually place and connect the steel in position are called **ironworkers.**

To promote uniformity in bidding practices, the American Institute of Steel Construction has adopted a more specific, although lengthy, definition for structural steel in the AISC "Code of Standard Practice." It is well for the owner and his engineer to understand fully just what the fabricator will supply when the latter submits a price for furnishing "structural steel." When it is desired that bids should include any other material, such as ornamental iron, open-web joists, and miscellaneous metalwork, a statement to that effect should appear in the bidding invitation.

6-1. Structural-steel Shapes. Steel mills have a standard classification for the many products they make, one of which is *structural shapes (heavy)*. By definition this classification takes in all shapes having at least one cross-sectional dimension of 3 in. or more. Shapes of lesser size are classified as *structural shapes (light)* or, more specifically, bars.

Shapes are identified by their cross-sectional characteristics—angles, channels, beams, columns, tees, pipe, tubing, and piles. For convenience, structural shapes are simply identified by letter symbols according to the following schedule:

Section	*Symbol*
Wide-flange shapes	W
Standard I shapes	S
Bearing-pile shapes	HP
Similar shapes that cannot be grouped in W, S, or HP	M
Structural tees cut from W, S, or M shapes	WT, ST, MT
American standard channels	C
All other channel shapes	MC
Angles	L

The industry recommended standard (adopted 1970) for indicating a specific size of beam or column-type shape on designs, purchase orders, shop drawings, etc., is listing of the symbol, depth, and weight, in that order. For example, W14 × 30 identifies a wide-flange shape whose nominal depth is 14 in. and which weighs 30 lb per lin ft. The ×, read as "by," is merely a separation. Similarly, plates are billed in the order of symbol (PL), followed by thickness and width, thus: PL¾ × 18.

Each shape has its particular functional use, but the work horse of building construction is the wide-flange W section. For all practical purposes, W shapes have parallel flange surfaces. The profile of a W shape of a given nominal depth and weight available from different producers is essentially the same, except for the size of fillets between web and flanges.

6-2. Tolerances for Structural Shapes. Mills are granted a tolerance because of variations peculiar to working of hot steel and wear of equipment. Limitations for such variations are established by Specification A6 of the American Society for Testing and Materials.

Wide-flange beams or columns, for example, may vary in depth by as much as ½ in.; i.e., ¼ in. over and under the nominal depth. The designer should always keep this in mind. Particularly, he should appreciate the fabricator's and erector's efforts to control these variations. Fillers, shims, and extra weld metal may not be desirable, but often they are the only practical solution.

Cocked flanges on column members are particularly troublesome to the erector, for it is not until the steel is erected in the field that the full extent of mill variations becomes evident. This is particularly true for a long series of spans or bays, where the accumulating effect of dimensional variation of many columns may require major adjustment. Fortunately, the average variation usually is negligible and nominal erection clearance allowed for by the fabricator will suffice.

Mill tolerances also apply to beams ordered from the mills cut to length. Where close tolerance may be desired, as sometimes required for welded connections, it may be necessary to order the beams long and then finish the ends in the fabricating shop to the precise dimensions.

6-3. Cambered Beams. Frequently, designers want long-span beams slightly arched (cambered) to offset deflection under load and to prevent a flat or saggy appearance. Such beams may be procured from the mills, the required camber being applied to cold steel. The AISC "Manual of Steel Construction" gives the maximum cambers that mills can obtain and their prediction of the minimum cambers likely to remain permanent. Smaller cambers than these minimums may be specified, but their permanency cannot be guaranteed. It should be observed that nearly all beams will have some camber as permitted by the tolerance for straightness, and advantage may be taken of such camber in shop fabrication.

A method of cambering, not dependent on mill facilities, is to employ heat. In welded construction, it is commonplace to flame-straighten members that have become distorted. By the same procedure, it is possible to distort or camber a beam to desired dimensions.

6-4. In-service Steel. Over the years, there have been numerous changes in the size of beam and column sections. The reasons are many: efficiency, popularity, stocking, standardization, simplification, and conservation. Occasionally, an engineer is called on to investigate an existing building, possibly because of increased floor loadings or because an additional story or two is to be added. Whatever the reason, he may be faced with the task of identifying sections that are no longer rolled and for which physical properties are not readily obtainable. To aid the designer in this problem, the American Institute of Steel Construction published in 1953 a complete compilation of steel and wrought-iron beams and columns that were rolled in this country during the period 1873–1952. Knowing the year of construction, consequently the approximate date of mill rolling, will enable the engineer to establish grade of the steel material and thus determine the yield and working stresses.

6-5. Structural-steel Specifications. Uniform quality is an outstanding characteristic of structural steel. This high degree of uniformity, regardless of the source of the steel, is achieved through universal acceptance of American Society for Testing and Materials specifications. There is no significant metallurgical or physical difference between the products rolled by United States structural mills.

With the variety of structural steels in current use, however, the problem of identifying steels with their wide range of strength levels as used in structures of recent origin will be more difficult. When construction records are not readily available or deemed unreliable, it may become necessary to take samples for laboratory analysis to determine the chemical and physical properties. This procedure is essential where welding is employed on existing steel, or where the strength of old steel is to be fully utilized. An alternate method, satisfactory for minor work, is to assume the old steel to be of the lowest grade in common use when the structure was built and make all connections with bolts.

In contrast to the single all-purpose structural steel (former ASTM A7) in general use during the first half of the twentieth century, many structural steels are now recommended by the American Institute of Steel Construction for building construction. While several are restricted to particular products, such as pipe, tubing, and strip steel for open-web steel joists, most of the specifications pertain to material for the plates and shapes generally used for principal members in steel structures. The most important characteristics of these steels are given in Table 6-1.

Weldability. All steels listed in Table 6-1 are of good welding quality.

Steel Strengths. In general, all steels of greater strength than A36 are more costly to buy from mills, but this does not mean they are too expensive to use. Their advantage of greater strength and possibly simpler fabrication often results in the most economical solution. Nevertheless, the desire for one grade of structural steel that is widely available and generally economical has made A36 the most commonly used steel.

Table 6-1 lists several steels with more than one level of tensile strength and yield stress, the levels being dependent on thickness of material. The listed thicknesses are precise for plates and nearly correct for shapes. To obtain the precise values for shapes, refer to the AISC "Steel Construction Manual" or to mill catalogs.

Weathering Steels. Strictly from a consideration of cost efficiency, corrosion-resistant A242 and A588 steels are not economical for ordinary usage. Their use in buildings, however, may be justified where steel will be exposed and perhaps left unpainted. On exposure to ordinary atmospheric conditions, these steels, called weathering steels, take on a thin rust film that inhibits further corrosion. Thus, they have been used without paint for building frameworks exposed to the elements.

Steel Identification. Because of the many steel types and strengths in use, ASTM specifications require each piece of hot-rolled steel to be properly identified with vital information, including heat numbers of the ingots. The American Institute of Steel Construction "Specification for the Design, Fabrication and Erection of Structural Steel for Buildings" requires fabricators to be prepared to demonstrate, by written procedure and by actual practice, the visible identification of all main stress-carrying elements, at least through shop assembly. Steel identification includes ASTM designation, heat numbers (if required), and mill test reports when specially ordered.

AISC Specifications. The AISC "Specification for the Design, Fabrication and

Erection of Structural Steel for Buildings," first promulgated in 1923, is widely used throughout the United States. The AISC revises the specification periodically to keep abreast of professional development, research findings, and availability of new materials.

AISC also issues the following supplementary specifications: "Specification for Architecturally Exposed Structural Steel," and in conjunction with the Steel Joist Institute, "Standard Specifications for Open Web Steel Joists, J- and H-Series,"

Table 6-1. Characteristics of Structural Steels

ASTM specification	Thickness, in.	Minimum tensile strength, ksi	Minimum yield stress,* ksi	Relative corrosion resistance†
		Carbon Steels		
A36	To 8 in. incl.	58–80¶	36	1‡
A529	To ½ in. incl.	60–85¶	42	1
	High-strength and High-strength, Low-alloy Steels			
A441	To ¾ incl.	70	50	2
	Over ¾ to 1½	67	46	2
	Over 1½ to 4 incl.	63	42	2
	Over 4 to 8 incl.	60	40	2
A572	Gr 42: to 4 incl.	60	42	1
	Gr 45: to 1½ incl.	60	45	1
	Gr 50: to 1½ incl.	65	50	1
	Gr 55: to 1½ incl.	70	55	1
	Gr 60: to 1 incl.	75	60	1
	Gr 65: to ½ incl.	80	65	1
A242	To ¾ incl.	70	50	4–8
	Over ¾ to 1½	67	46	4–8
	Over 1½ to 4 incl.	63	42	4–8
A588	To 4 incl.	70	50	4
	Over 4 to 5	67	46	4
	Over 5 to 8 incl.	63	42	4
		Heat-treated Low-alloy Steels		
A514	To ¾ incl.	115–135	100	1–4
	Over ¾ to 2½	115–135	100	1–4
	Over 2½ to 4 incl.	105–135	90	1–4

* Yield stress or yield strength, whichever shows in the stress-strain curve.

† Relative to carbon steels low in copper.

‡ A36 steel with 0.20% copper has a relative corrosion resistance of 2.

¶ Minimum tensile strength may not exceed the higher value.

"Standard Specifications for Longspan Steel Joists, LJ- and LH-Series," and "Standard Specifications for Deep Longspan Steel Joists, DLJ- and DLH-Series."

STEEL-FRAMING SYSTEMS

Steel construction may be classified into three broad categories: wall bearing, skeleton framing, and long-span construction. Depending on the functional requirements of the building itself and the materials used in constructing its roof, floors, and walls, one or more of these methods of framing may be employed in a single structure.

6-6. Wall-bearing Framing. Probably the oldest and commonest type of framing, wall-bearing framing (not to be confused with bearing-wall construction), occurs whenever a wall of a building, interior or exterior, is used to support ends of main structural elements carrying roof or floor loads. The walls must be strong enough to carry the reaction from the supported members and thick enough to insure stability against any horizontal forces that may be imposed. Such construction often is limited to relatively low structures, because load-bearing walls become massive in tall structures. Nevertheless, a wall-bearing system may be advantageous for tall buildings when designed with reinforcing steel (Sec. 11).

A common application of wall-bearing construction may be found in many one-family homes. A steel beam, usually 8 or 10 in. deep, is used to carry the interior

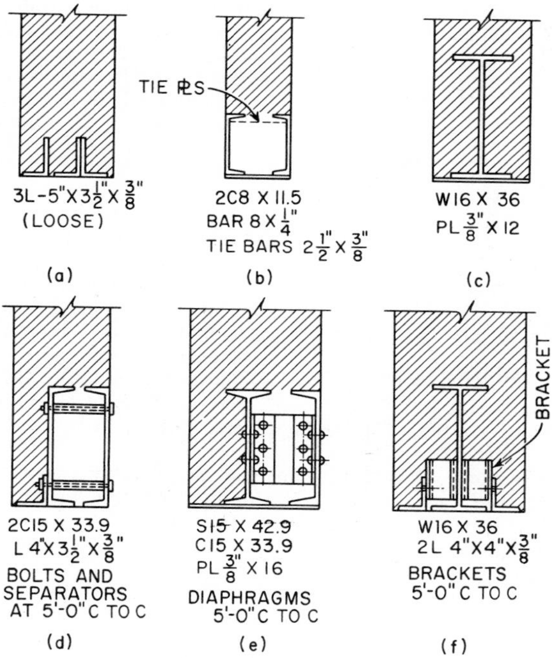

Fig. 6-1. Lintels supporting masonry.

walls and floor loads across the basement with no intermediate supports, the ends of the beam being supported on the foundation walls. The relatively shallow beam depth affords maximum headroom for the span. In some cases, the spans may be so large that an intermediate support becomes necessary to minimize deflection—usually a steel pipe column serves this purpose.

Another example of wall-bearing framing is the member used to support masonry over windows, doors, and other openings in a wall. Such members, called **lintels,** may be a steel angle section (commonly used for brick walls in residences) or, on longer spans and for heavier walls, a fabricated assembly. A variety of frequently used types is shown in Fig. 6-1. In types (*b*), (*c*), and (*e*), a continuous plate is used to close the bottom, or soffit, of the lintel, and to join the load-carrying beams and channels into a single shipping unit. The gap between the toes of the channel flanges in type (*d*) may be covered by a door frame or window trim, to be installed later. Pipe and bolt separators are used to hold the two channels together to form a single member for handling. (See also Art. 6-80.)

Bearing Plates. Because of low allowable pressures on masonry, bearing plates (sometimes called masonry plates) are usually required under the ends of all beams that rest on masonry walls, as illustrated in Fig. 6-2. Even when the pressure on the wall under a member is such that an area no greater than the contact portion of the member itself is required, wall plates are sometimes prescribed, if the member is of such weight that it must be set by the steel erector. The plates, shipped loose and in advance of steel erection, are then set by the mason to provide a satisfactory seat at the proper elevation.

Anchors. The beams are usually anchored to the masonry. Of the two common methods shown in the American Institute of Steel Construction "Manual of Steel Construction," **government anchors** as illustrated in Fig. 6-2 have been preferred.

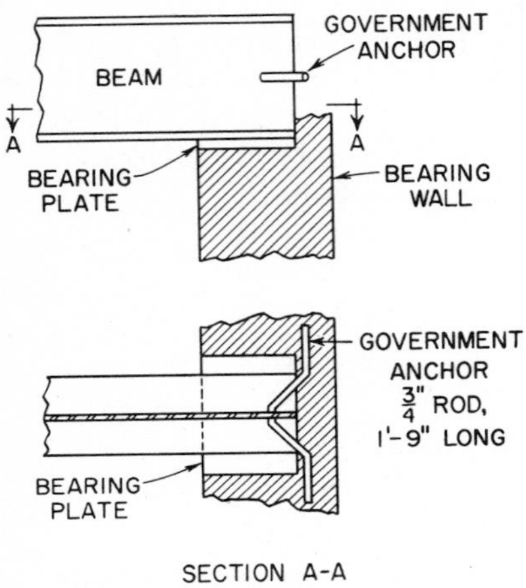

Fig. 6-2. Wall-bearing beam.

Nonresidential Uses. Another common application for the wall-bearing system is in one-story commercial and light industrial-type construction. The masonry side walls support the roof system, which may be rolled beams, open-web joists, or light trusses. Clear spans of moderate size are usually economical, but for longer spans (probably over 40 ft), wall thickness and size of buttresses (pilasters) must be built to certain specified minimum proportions commensurate with the span distance—a requirement of building codes to assure stability. Therefore, the economical aspect should be carefully investigated. It may cost less to introduce steel columns and keep wall size to the minimum permissible. On the other hand, it may be feasible to cut the span length in two by introducing intermediate columns and still retain the wall-bearing system for the outer end reactions.

Planning for Erection. One disadvantage of wall-bearing construction needs emphasizing: Before steel can be set by the ironworkers, the masonry must be built up to the proper elevation to receive it. When these elevations vary, as is the case at the end of a pitched or arched roof, then it may be necessary to proceed in alternate stages, progress of erection being interrupted by the work that must be performed by the masons, and vice versa. The necessary timing to avoid delays is seldom obtained. A few columns or an additional rigid frame at the end of a building may cost less than calling trades to a job on some intermittent and expensive arrangement. Remember, too, that labor-union regulations may prevent the trades from handling any material other than that belonging to their own craft. An economical rule may well be: Lay out the work so that the erector and his ironworkers can place and connect all the steelwork in one continuous operation (Bibliography, Art. 6-92).

6-7. Skeleton Framing. In skeleton framing all the gravity loadings of the structure, including the walls, are supported by the steel framework. Such walls are termed **nonbearing** or **curtain walls.** This system made the skyscraper possible. Steel, being so much stronger than all forms of masonry, is capable of sustaining far greater load in a given space, thus obstructing less of the floor area in performing its function.

With columns properly spaced to provide support for the beams spanning between them, there is no limit to the floor and roof area that can be constructed with this type of framing, merely by duplicating the details for a single bay. Erected tier upon tier, this type of framing can be built to any desired height. Fabricators refer to this type of construction as "beam and column"; a typical arrangement is illustrated in Fig. 6-3.

The spandrel beams, marked B1 in Fig. 6-3, are located in or under the wall

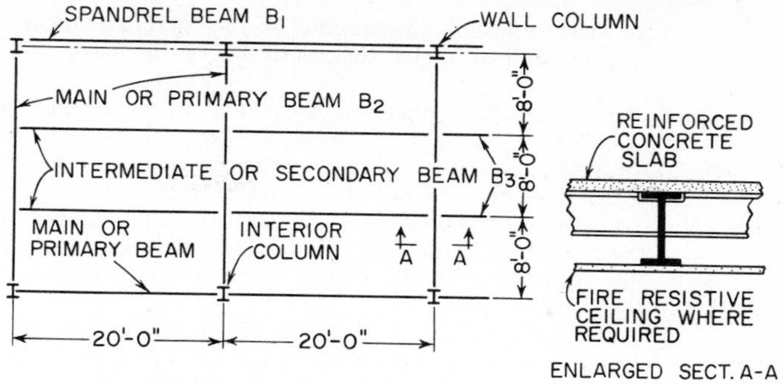

Fig. 6-3. Typical steel beam-and-column framing.

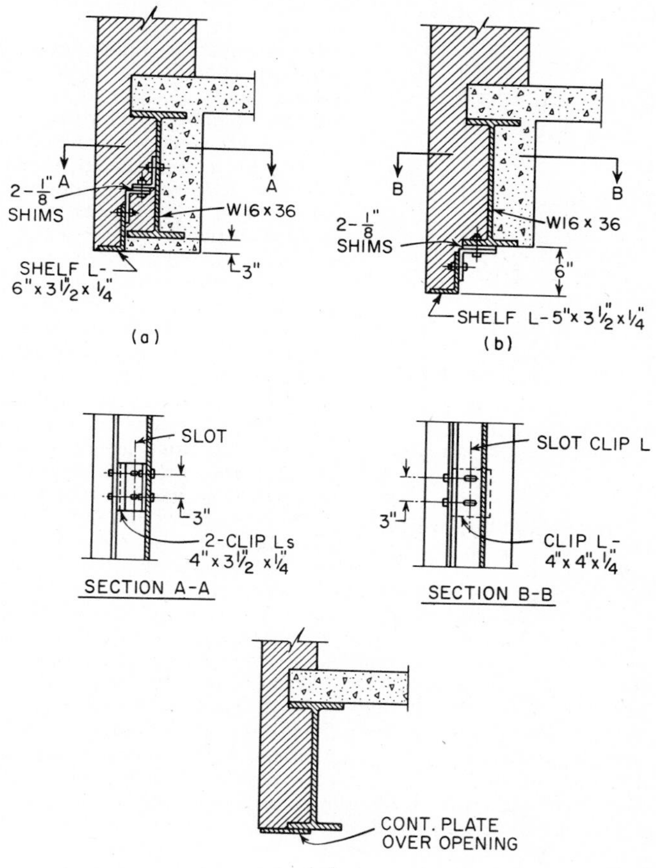

Fig. 6-4. Typical steel spandrel beams.

so as to reduce eccentricity caused by wall loads. Figure 6-4 shows two methods for connecting to the spandrel beam the shelf angle that supports the outer course of masonry over window openings 6 ft or more in width. In order that the masonry contractor may proceed expeditiously with his work, these shelf angles must be in alignment with the face of the building and at the proper elevation to match a masonry joint. The connection of the angles to the spandrel beams is made by bolting; shims are provided to make the adjustments for line and elevation (Art. 6-80).

Figure 6-4*a* illustrates a typical connection arrangement when the outstanding leg of the shelf angle is about 3 in. or less below the bottom flange of the spandrel beam; Fig. 6-4*b* illustrates the corresponding arrangement when the outstanding leg of the shelf angle is more than about 3 in. below the bottom flange of the spandrel beam. In the cases represented by Fig. 6-4*b*, the shelf angles are usually shipped attached to the spandrel beam; however, if the distance from the bottom flange to the horizontal leg of the shelf angle is greater than 10 in., a hanger may be required.

In some cases, as over door openings, the accurate adjustment features provided by Fig. 6-4*a* and *b* may not be needed. It may then be more economical to simplify the detail, as shown in Fig. 6-4*c*. The elevation and alignment will then conform to the permissible tolerances associated with the steel framework (Bibliography, Art. 6-92).

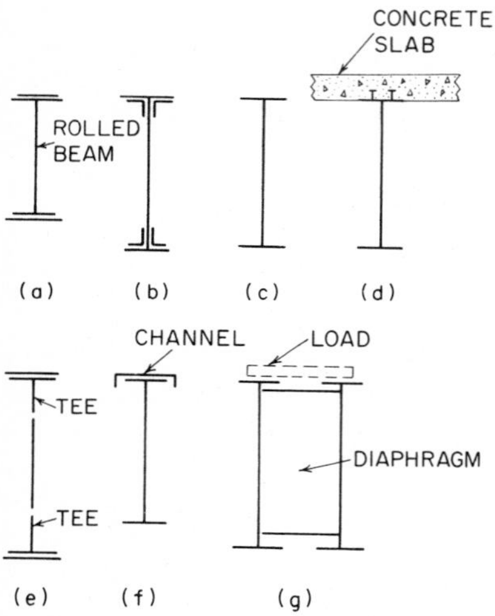

Fig. 6-5. Typical built-up girders.

6-8. Long-span Steel Framing. Large industrial buildings, auditoriums, gymnasiums, theaters, hangars, and exposition buildings require much greater clear distance between supports than can be supplied by beam and column framing. When the clear distance is greater than can be spanned with rolled beams, several alternatives are available. These may be classified as *girders, simple trusses, arches, rigid frames, cantilever-suspension spans,* and various types of space frames, such as *folded plates, curvilinear grids, thin-shell domes, two-way trusses, and cable networks.*

Girders are the usual choice where depths are limited, as over large unobstructed areas in the lower floors of tall buildings, where column loads from floors above must be carried across the clear area. Sometimes, when greater strength is required than is available in rolled beams, cover plates are added to the flanges (Fig. 6-5*a*) to provide the additional strength.

When depths exceed the limit for rolled beams, i.e., for spans exceeding about 67 ft (based on the assumption of a depth-span ratio of 1:22 with 36 in. deep Ws), the girder must be built up from plates and shapes. Welded girders usually are preferred to the old-type conventional riveted girders (Fig. 6-5*b*), composed of web plate, angles, and cover plates. Where this type may still be specified, all components are fastened with high-strength bolts rather than with rivets.

Welded girders generally are composed of three plates (Fig. 6-5*c*). This type offers the most opportunity for simple fabrication, efficient use of material, and least weight. Top and bottom flange plates may be of different size (Fig. 6-5*d*), an arrangement advantageous to composite construction, which integrates a concrete floor slab with the girder flange to function together.

Heavy girders may use cover-plated tee sections (Fig. 6-5*e*). Where lateral loads are a factor, as in the case of girders supporting cranes, a channel may be fastened to the top flange (Fig. 6-5*f*). In exceptionally heavy construction, it is not unusual to use a pair of girders diaphragmed together to share the load (Fig. 6-5*g*).

The availability of high-strength, weldable steels resulted in development of **hybrid girders.** For example, a high-strength steel, say A572 Grade 50, whose yield stress is 50 ksi, may be used in a girder for the most highly stressed flanges, and the lower-priced A36 steel, whose yield stress is 36 ksi, may be used for lightly stressed flanges and web plate and detail material. The AISC Specification requires that the top and bottom flanges at any cross section have the same cross-sectional area, and that the steel in these flanges be of the same grade. The allowable bending stress may be slightly less than for conventional homogeneous girders of the high-strength steel, to compensate for possible overstress in the web at the junction with the flanges. Hybrid girders are efficient and economical for heavy loading and long spans and, consequently, are frequently employed in bridgework.

Trusses. When depth limits permit, a more economical way of spanning long distances is with trusses, for both floor and roof construction. Because of their greater depth, trusses usually provide a greater stiffness against deflection when compared pound for pound with the corresponding rolled beam or plate girder than otherwise would be required. Six general types of trusses frequently used in building frames are shown in Fig. 6-6, together with modifications that can be made to suit particular conditions.

Trusses in Fig. 6-6*a* to *d* and *k* may be used as the principal supporting members in floor and roof framing; types *e* to *j* serve a similar function in the framing of symmetrical roofs having a pronounced pitch. As shown, types *a* to *d* have a top chord which is not quite parallel to the bottom chord. Such an arrangement is used to provide for drainage of flat roofs. Most of the connections of the roof beams (**purlins**), which these trusses support, can be identical, which would not be the case if the top chord were dead level and the elevation of the purlins varied. When used in floors, truss types *a* to *d* have parallel chords.

Properly proportioned, bow-string trusses (Fig. 6-6*j*) have the unique characteristic that the stress in their web members is relatively small. The top chord, which usually is formed in the arc of a circle, is stressed in compression, and the bottom chord in tension. In spite of the relatively expensive operation of forming the top chord, this type of truss has proved very popular in roof framing on spans of moderate lengths up to about 100 ft.

The Vierendeel truss (Fig. 6-6*k*) generally is shop welded to develop full rigidity of connections between the vertical and chords. It is useful where absence of diagonals is desirable to permit passage between the verticals.

Trusses also may be used for long spans, as three-dimensional trusses (space frames) or as grids. In two-way grids, one set of parallel lines of trusses is intersected at 90° by another set of trusses so that the verticals are common to both sets. Because of the rigid connections at the intersections, loads are distributed nearly equally to all trusses. Reduced truss depth and weight savings are among the apparent advantages of such grids.

Long-span joists are light trusses closely spaced to support floors and flat roofs. They conform to standard specifications (Art. 6-5) and to standard loading tables. Both Pratt and Warren types are used, the shape of chords and webs varying with the fabricator. Yet, all joists with the same designation have the same guaranteed load-supporting capacity. The standard loading tables list allowable loads for joists up to 72 in. deep and with clear span up to 144 ft. The joists may have parallel or sloping chords or other configuration.

Truss Applications. Cross sections through a number of buildings having roof trusses of the general type just discussed are shown diagrammatically in Fig. 6–7. Cross section (*a*) might be that of a storage building or a light industrial building. A Fink truss has been used to provide a substantial roof slope. Roofs of this type are often designed to carry little loading, if any, except that produced by wind and snow, since the contents of the building are supported on the ground floor. For light construction, the roof and exterior wall covering may consist of thin, cold-formed metal or asbestos-cement sheets, lapped at their seams so as to shed rain. Lighting and ventilation, in addition to that provided by windows in the vertical side walls, frequently are furnished by means of sash installed in the vertical side of a continuous monitor, framing for which is indicated by the dotted lines in the sketch.

Cross section (*b*) shows a scissors truss supporting the high roof over the nave of a church. This type of truss is used only when the roof pitch is steep, as in ecclesiastical architecture.

A modified Warren truss, shown in cross section (*c*), might be one of the main supporting roof members over an auditorium, gymnasium, theater, or other assembly-type building where large, unobstructed floor space is required. Similar trusses, including modified Pratt, are used in the roofs of large garages, terminal buildings, and airplane hangars, for spans ranging from about 80 up to 500 ft.

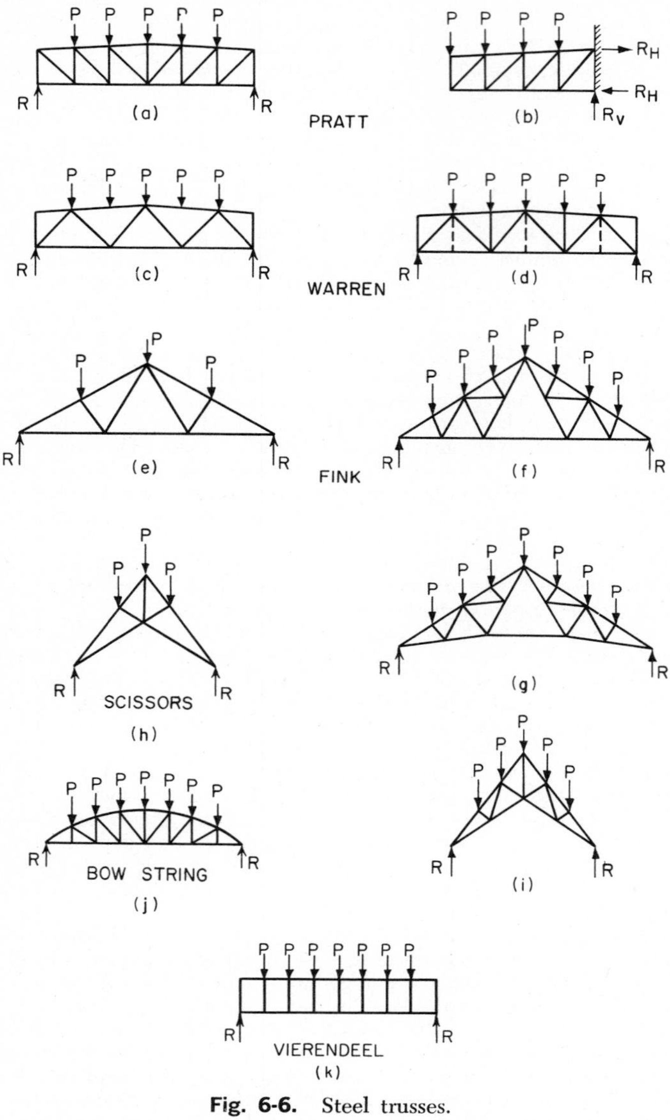

Fig. 6-6. Steel trusses.

The Pratt truss (Fig. 6-7*d*) is frequently used in industrial buildings, while (*e*) depicts a type of framing often used where overhead traveling cranes handle heavy loads from one point on the ground to another.

Arches. When very large clear spans are needed, the bent framing required to support walls and roof may take the form of solid or open-web arches, of the kind shown in Fig. 6-8. A notable feature of bents (*a*) and (*b*) is the heavy steel pins at points *A*, *B*, and *C*, connecting the two halves of the arch together at the crown and supporting them at the foundation. These pins are designed to carry all the reaction from each half arch, and to function in shear and bearing much as a single rivet is assumed to perform when loaded in double shear.

Use of hinge pins offers two advantages in long-span frames of the type shown in Fig. 6-8. In the first place, they simplify design calculations. Second, they simplify erection. All the careful fitting can be done and strong connections required to develop the needed strength at the ends of the arch can be made in the shop, instead of high above ground in the field. When these heavy members have been

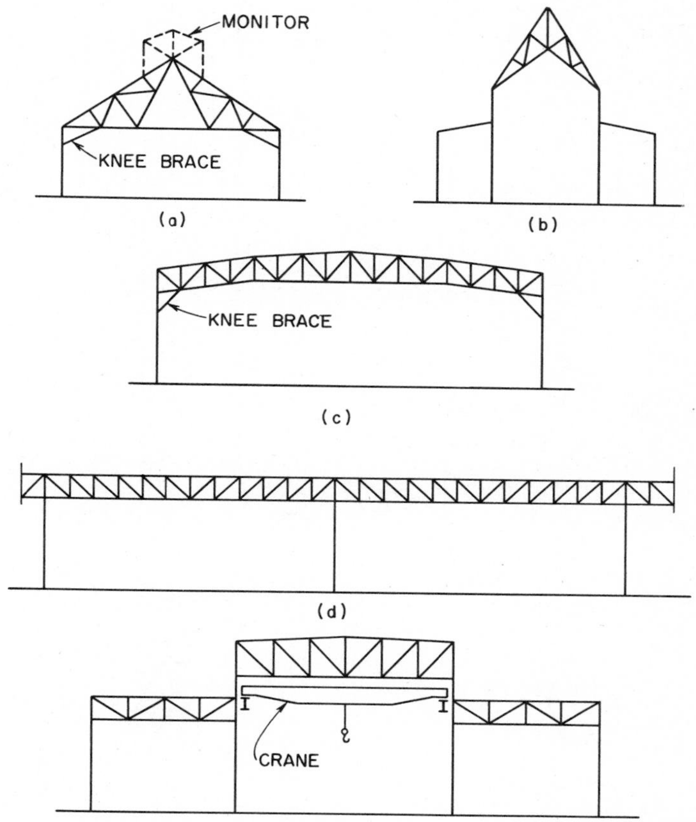

Fig. 6-7. Structures with trussed roofs.

raised in the field about in their final position, the upper end of each arch is adjusted, upward or downward, by means of jacks near the free end of the arch. When the holes in the pin plates line up exactly, the crown pin is slipped in place and secured against falling out by the attachment of keeper plates. The arch is then ready to carry its loading. Bents of the type shown in Fig. 6-8*a* and *b* are referred to as **three-hinged arches.**

When ground conditions are favorable and foundations are properly designed, and if the loads to be carried are relatively light, as, for example, for a large gymnasium, a **hingeless arch** similar to the one shown diagrammatically in Fig. 6-8*c* may offer advantage in over-all economy.

In many cases, the arches shown in Fig. 6-8*a* and *b* are designed without the pins at *B* (***two-hinged arch***). Then, the section at *B* must be capable of carrying the moment and shear present. Therefore, the section at *B* may be heavier than for the three-hinged arch, and erection will be more exacting for correct closure.

Rigid Frames. These are another type of long-span bent. In design, the stiffness afforded by beam-to-column connections is carefully evaluated and counted on

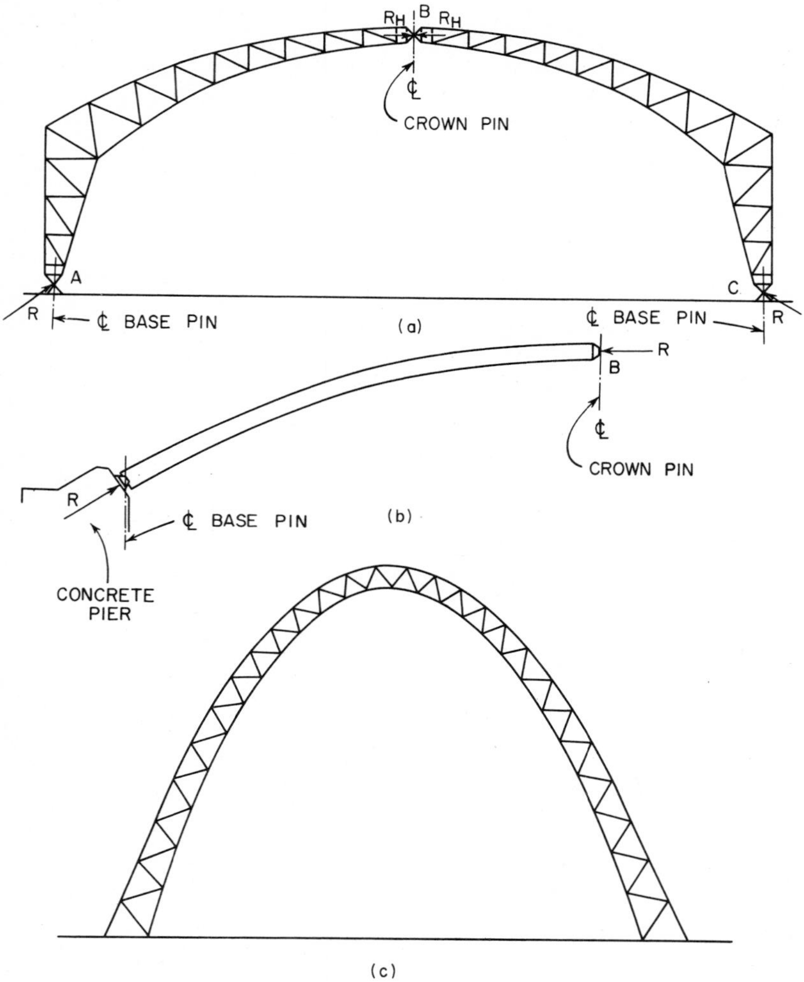

Fig. 6-8. Steel arches.

in the design to relieve some of the bending moment that otherwise would be assumed as occurring with maximum intensity at midspan. Typical examples of rigid frame bents are shown in Fig. 6-9. When completely assembled in place in the field, the frames are fully continuous throughout their entire length and height. A distinguishing characteristic of rigid frames is the absence of pins or hinges at the crown, or midspan.

In principle, single-span rigid-frame bents are either two-hinged or hingeless arches—the latter, when the column bases are fully restrained by large rigid foundations, to which they are attached by a connection capable of transmitting moment as well as shear. Since such foundations may not be economical, if at all possible, because of soil conditions, the usual practice is to consider the bents hinged at each reaction. However, this does not imply the necessity of expensive pin details; in most cases, sufficient rotation of the column base can be obtained with the ordinary flat-ended base detail and a single line of anchor bolts placed perpendicular to the span on the column center line. Many designers prefer to obtain a hinge effect by concentrating the column load on a narrow bar, as shown in Fig. 6-9*c*; this refinement is worthwhile in larger spans.

Regardless of how the frame is hinged, there is a problem in resisting the horizontal shear that the rigid frame imparts to the foundation. For small spans and light

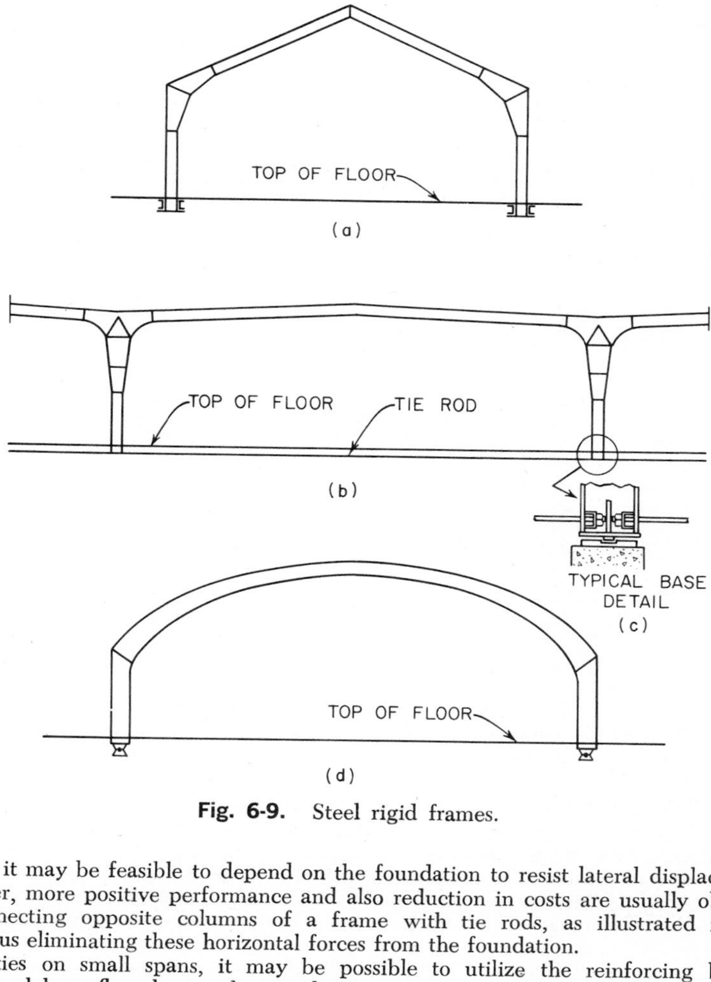

Fig. 6-9. Steel rigid frames.

thrusts, it may be feasible to depend on the foundation to resist lateral displacement. However, more positive performance and also reduction in costs are usually obtained by connecting opposite columns of a frame with tie rods, as illustrated in Fig. 6-9*b*, thus eliminating these horizontal forces from the foundation.

For ties on small spans, it may be possible to utilize the reinforcing bars in the floor slab or floor beams, by simply connecting them to the column bases. On larger spans, it is advisable to use tie rods and turnbuckles, the latter affording the opportunity to prestress the ties and thus compensate for elastic elongation of the rods when stressed. Prestressing the rod during erection to 50% of its value has been recommended for some major installations; but, of course, the foundations should be checked for resisting some portion of the thrust.

Single-story, welded rigid frames often are chosen where exposed steelwork is desired for such structures as churches, gymnasiums, auditoriums, bowling alleys, and shopping centers, because of attractive appearance and economy. Columns may be tapered, girders may vary in depth linearly or parabolically, haunches (knees) may be curved, field joints may be made inconspicuous, and stiffeners may simply be plates.

Field Splices. One problem associated with long-span construction is that of locating field splices compatible with the maximum sizes of members that can be shipped and erected. Field splices in frames are generally located at or near the point of counterflexure, thus reducing the splicing material to a minimum. In general, the maximum height for shipping by truck is 8 ft, by rail 10 ft. Greater over-all depths are possible, but these should always be checked with the carrier; they vary with clearances under bridges and through tunnels.

Individual shipping pieces must be stiff enough to be handled without buckling or other injury, light enough to be lifted by the raising equipment, and capable of erection without interference from other parts of the framework. This suggests a study of the entire frame to insure orderly erection, and to make provisions for temporary bracing of the members, to prevent jackknifing, and for temporary guying of the frame, to obtain proper alignment.

Hung-span Beams. In some large one-story buildings, an arrangement of cantilever-suspension (hung) spans (Fig. 6-10) has proved economical and highly efficient.

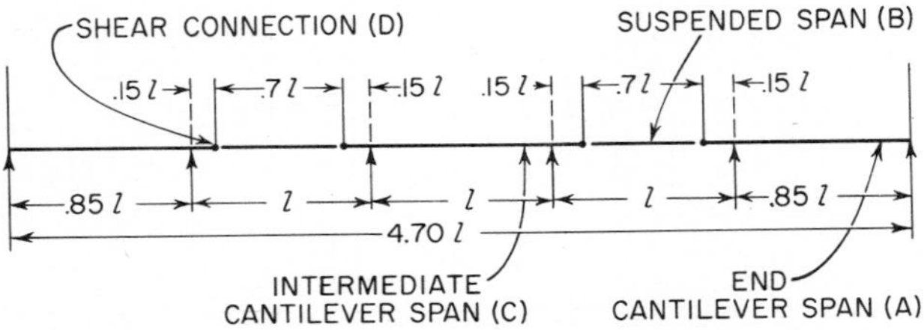

Fig. 6-10. Hung-span steel construction.

This layout was made so as to obtain equal maximum moments, both negative and positive, for the condition of uniform load on all spans. A minimum of three spans is required; that is, a combination of two end spans (*A*) and one interior span (*B*). The connection at the end of the cantilever (point *D*) must be designed as a shear connection only. If the connection is capable of transmitting moment as well as shear, it will change the design to one of continuity and the dimensions in Fig. 6-10 will not apply. This scheme of cantilever and suspended spans is not necessarily limited to one-story buildings.

As a rule, interior columns are separate elements in each story. Therefore, horizontal forces on the building must be taken solely by the exterior columns (Bibliography, Art. 6-92).

6-9. Steel and Concrete Framing. A fourth category or framing system could be added, wherein a partial use of structural steel has an important role, namely, the *reinforced-concrete–structural-steel combination framing.*

Composite or combination construction actually occurs whenever concrete is made to assist steel framing in carrying loads. The term composite, however, often is used for the specific cases in which concrete slabs act together with flexural members (Art. 6-15).

Reinforced-concrete columns of conventional materials when employed in tall buildings and for large spans become excessively large. One method of avoiding this objectionable condition is to use high-strength concrete and high-strength reinforcing bars. Another is to use a structural-steel column core. In principle, the column load is carried by both the steel column and the concrete that surrounds the steel shape; building codes usually contain an appropriate formula for this condition.

A number of systems employ a combination of concrete and steel in various ways. One method features steel columns supporting a concrete floor system by means of a steel grillage connected to the columns at each floor level. The shallow grillage is embedded in the floor slab, thus obtaining a smooth ceiling without drops or capitals.

Another combination system is the **lift-slab** method. In this system, the floor slabs are cast one on top of another at ground level. Jacks, placed on the permanent steel columns, raise the slabs, one by one, to their final elevation, where they are made secure to the columns. When fireproofing is required, the columns may be boxed in with any one of many noncombustible materials available for that purpose. The merit of this system is the elimination of formwork and shoring that are essential in conventional reinforced-concrete construction.

For high-rise buildings, structural-steel framing often is used around a central, loadbearing, concrete core, which contains elevators, stairways, and services. The thick walls of the core, whose tubular configuration may be circular, square, or rectangular, are designed to resist all the wind forces as well as gravity loads. Sometimes, the surrounding steel framing is cantilevered from the core, or the perimeter members are hung from trusses or girders atop the core and possibly also, in very tall buildings, at midheight of the core.

(Bibliography, Art. 6-92.)

FLOOR AND ROOF SYSTEMS

In most types of buildings, floor and roof systems are so intimately related to the structural frame that the two must be considered together in the development of a steel design. Both are important, but floors usually predominate in tier-type structures.

6-10. Factors Affecting Floor Design. Selection of a suitable and economical floor system for a steel-frame building involves many considerations: load-carrying capacity, durability, fire resistance, dead weight, over-all depth, facility for installing power, light, and telephones, facility for installing air conditioning, sound transmission, appearance, maintenance, and construction time.

Building codes set minimum design live loads for those structures within jurisdictional areas. In the absence of a code regulation, one may use "Minimum Design Loads in Buildings and Other Structures," American Standard A58.1, American National Standards Institute. See also Art. 3-2. Floors should be designed to support the actual loading or these minimum loads, whichever is larger. Most floors can be designed to carry any given load. However, in some instances, a building code may place a maximum load limit on particular floor systems without regard to calculated capacity.

Resistance to lateral forces is usually disregarded, except in areas of seismic disturbances or for perimeter windbents where floors may be employed as horizontal diaphragms to distribute lateral forces to walls or framing designed to transmit them to the ground.

Durability becomes a major consideration when a floor is subject to loads other than static or moderately kinetic types of forces. For example, a light joist system may be just the floor for an apartment or an office building but may be questionable for a manufacturing establishment where a floor must resist heavy impact and severe vibrations. Shallow floor systems deflect more than deep floors; the system selected should not permit excessive or objectionable deflections.

Fire resistance and fire rating are very important factors, because building codes, in the interest of public safety, specify the degree of resistance that must be provided. Many floor systems are rated by the codes or by fire underwriters for purposes of satisfying code requirements or basing insurance rates (see Arts. 6-86 to 6-91).

The dead weight of the floor system, including the framing, is an important factor affecting economy of construction. For one thing, substantial saving in the weight and cost of a steel frame may result with lightweight floor systems. In addition, low dead weight may also reduce foundation costs.

Sometimes, the depth of a floor system is important. For example, the height of a building may be limited for a particular type of fire-resistant construction or by zoning laws. The thickness of the floor may well be the determining factor limiting the number of stories that can be built. Also, the economy of a deep floor is partly offset by the increase in height of walls, columns, pipes, etc.

Another important consideration, particularly for office buildings and similar-type occupancies, is the need for furnishing an economical and flexible electrical wiring system. With the accent on movable partitions and ever-changing office arrangements, the readiness and ease with which telephones, desk lights, and electric-powered business machines can be relocated are of major importance. Therefore, the floor system that by its make-up provides large void spaces or cells for concealing wiring possesses a distinct advantage over competitive types of solid construction. Likewise, the problem of recessed lighting in the ceiling may disclose an advantage in one system over another. Furthermore, air conditioning and ventilation are a must in new office buildings; a competitive necessity in older structures, if at all possible. So location of ducts and method of attachment warrant study of several floor systems for optimum results.

Sound transmission and acoustical treatments are other factors that need to be evaluated. A wealth of data is available in reports published by the National Bureau of Standards. In general, floor systems of sandwich type with air spaces between layers will afford better resistance to sound transmission than solid systems, which do not interrupt sound waves. Although the ideal soundproof floor is impractical, costwise, there are several reasonably satisfactory systems. Much depends on type of occupancy, floor coverings, and ceiling finish—acoustical plaster or tile.

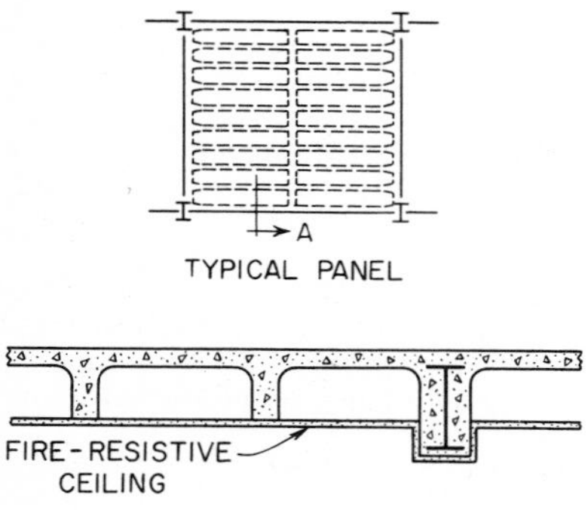

Fig. 6-11. Concrete joist floor.

Appearance and maintenance are also weighed by the designer and the owner. A smooth, neat ceiling is usually a prerequisite for residential occupancy; a less expensive finish may be deemed satisfactory for an institutional building.

Speed of construction is essential. Contractors are alert to the systems that enable the following-up trades to work immediately behind the erector and with unimpeded efficiency.

In general, the following constructions are commonly used in conjunction with steel framing: concrete arch, concrete joists (removable pan), steel joists, cellular steel and composite concrete-steel beams.

6-11. Concrete-arch Floors. The old-type floor used in office and industrial buildings consists of a reinforced-concrete slab, about 4 in. thick, supported on steel beams spaced about 8 ft apart. These beams are designated secondary or intermediate, whereas the beams or girders framing into columns and supporting secondary beams are the primary members (Fig. 6-3). It is called an "arch" floor because it evolved from earlier construction with thick brick and tile floors that were truly flat-top arches; some of the early reinforced-concrete floors also were cast on arch-shaped forms.

The arch system is so much heavier than the contemporary lightweight systems that it has lost most of its earlier appeal for light-occupancy construction. It ranks high, however, where heavy loads, durability, and rigidity are key factors. It is without competition wherever building codes restrict the use of other types. The system works well with the steel frame because all formwork can be supported directly, or by wire suspension, on the floor beams, thus obviating vertical shoring.

6-12. Concrete-pan System. Concrete floors cast on removable metal forms or pans, which form the joists, are frequently used with steel girders in certain areas. Since the joists span the distance between columns, intermediate steel beams are not needed (Fig. 6-11). This floor generally weighs less than the arch system but still considerably more than the lightest types.

There are a number of variations of the concrete-joist system, such as the "grid" or "waffle" system, where the floor is cast on small, square, removable pans, or domes, so that the finished product becomes a two-way joist system. Others such as the "Republic," "Schuster," and "Nassau" systems employ permanent filler blocks—usually a lightweight tile. Some of these variations fall in the heaviest floor classification; also the majority require substantial forms and shoring.

6-13. Open-web Joist Floors. The lightest floor system in common use is the open-web joist construction shown in Fig. 6-12. It is popular for all types of light occupancies, principally because of initial low cost.

Many types of open-web joists are on the market. Some employ bars in their make-up, while others are entirely of rolled shapes; they all conform to standards and good-practice specifications promulgated by the Steel Joist Institute and the American Institute of Steel Construction. All joists conform to the standard loading tables and carry the same size designation so that the designer need only indicate on his drawings the standard marking without reference to manufacturer, just as he would for a steel beam or column section.

Satisfactory joist construction is assured by adhering to SJI and AISC recommendations. Joists generally are spaced 2 ft c to c; they should be adequately braced

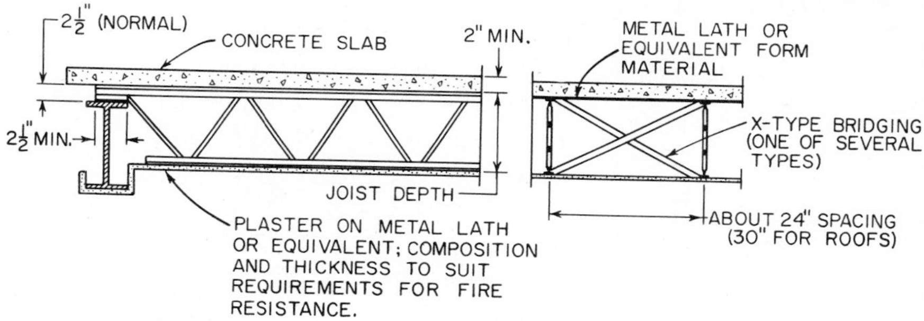

Fig. 6-12. Open-web steel joist.

(with bridging) during construction to prevent rotation or buckling; and to avoid "springy" floors, they should be carefully selected to provide sufficient depth.

One of the main appeals of this system is the elimination of falsework. Joists are easily handled, erected, and connected to supporting beams—usually by tack welding. Temporary coverage and working platforms are quickly placed. The open space between joists, and through the webs, may be utilized for ducts, cables, light fixtures, and piping. A thin floor slab is cast on steel lath, corrugated-steel sheets, or wire-reinforced paper lath laid on top of the joists. A plaster ceiling may be suspended or attached direct to the bottom flange of the joist.

Lightweight beams, or so-called "junior" beams, are also used in the same manner as the open-web joists, and with the same advantages and economy, except that the solid webs do not allow so much freedom in installation of utilities. Beams may be spaced according to their safe load capacity; 3- and 4-ft spacings are common. As a type, therefore, the lightweight-steel-beam floor is intermediate between concrete arches and open-web joists.

6-14. Cellular-steel Floors. Light-gage steel decking has rapidly advanced to the forefront of floor systems, particularly for modern office buildings. One type is illustrated in Fig. 6-13. Other manufacturers make similar cellular metal decks, the primary difference being in the shape of the cells. Often, decking with half cells is used. These are open ended on the bottom, but flat sheets close those cells, usually 5 to 6 ft apart, that incorporate services. Sometimes, cells are enlarged laterally to transmit air for air conditioning.

Two outstanding advantages held for cellular floors are rapidity of erection and the ease with which present and future connections can be made to telephone, light, and power wiring, each cell serving as a conduit. Each deck unit becomes a working platform immediately on erection, thus enabling the several finishing trades to follow right behind the steel erector.

While generally thought to be high in initial costs as compared with other floor systems, the cost differential can be narrowed to competitive position when equal consideration for electrical facility is imposed on the other systems; e.g., the addition of 4 in. of concrete fill to cover embedded electrical conduit on top of a concrete flat-slab floor.

In earlier floors of this type, the steel decking was assumed to be structurally independent. In that case, the concrete fill served only to provide fire resistance and a level floor. Most modern deckings are bonded or locked to the concrete, so that the two materials act as a unit. Therefore, the metal can be made thinner or the spans made longer. Usually, only top-quality stone concrete (ASTM C33 aggregates) is used, although lightweight concrete is a potential alternative.

Usage of cellular deck in composite construction is facilitated by economical attachment of shear connectors to both the decking and underlying beams. For example, when welded studs are used, a welding gun automatically fastens the studs through two layers of hot-dipped galvanized decking to the unpainted top flanges of the steel beams. This construction is similar to composite concrete-steel beams (Art. 6-15).

The total floor weight of cellular steel construction is low, comparable to open-web steel joists. Weight savings of about 50% are obtained in comparison with concrete

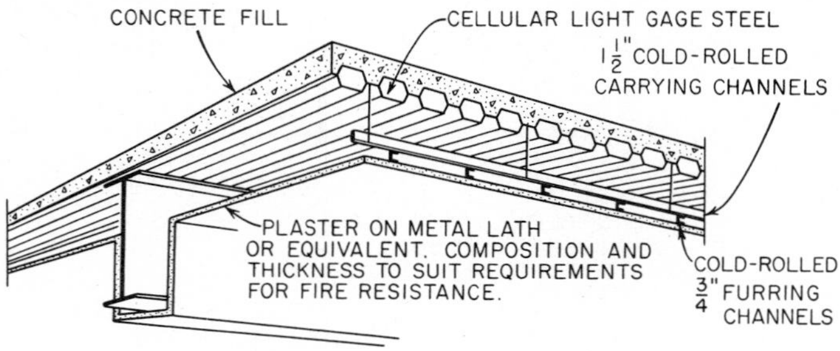

Fig. 6-13. Cellular steel floor.

slabs; 30% savings in over-all weight of the building. However, costwise, a big factor in a high-labor-rate area is the elimination of costly formwork needed for concrete slabs, since the decking serves as the form.

Fire resistance for any required rating is contributed by the fill on top of the cells and by the ceiling below (Fig. 6-13). Generally, removable panels for which no fire rating is claimed are preferred for suspended ceilings, and fireproofing materials are applied directly to the under side of the metal deck and all exposed surfaces of steel floor beams, a technique often called spray-on fireproofing (Art. 6-89).

6-15. Composite Concrete-steel Beams. In composite construction, the structural concrete slab is made to assist the steel beams in supporting loads. Hence, the concrete must be bonded to the steel to insure shear transfer. When the steel beams are completely encased in the concrete, the natural bond is considered capable of resisting horizontal shear. But that bond generally is disregarded when only the top flange is in contact with the concrete, and the slab acts much like a cover plate on the beam. (See also Art. 6-35.) Instead, shear connectors are used to resist the horizontal shear. Commonly used connectors are welded studs, hooked or headed, and short lengths of channels.

Usually, composite construction is most efficient for heavy loading, long spans, large beam spacing, and restricted depths. Because the concrete serves much like a cover plate, lighter steel beams may be used for given loads, and deflections are smaller than for noncomposite construction.

6-16. Effect of Intermediate Beams on Costs. The joist systems, either steel or concrete, require no intermediate support, since they are obtainable in lengths to meet the normal bay dimensions in tier building construction. On the other hand, the concrete arch and cellular steel floors are usually designed with one or two intermediate beams within the panel. The elimination of secondary beams does not necessarily mean over-all economy just because the structural-steel contract is less. These beams are simple to fabricate and erect and allow much duplication. An analysis of contract prices shows that the cost per ton of secondary beams will average 20% under the cost per ton for the whole steel structure; or viewed another way, the omission of secondary beams increases the price per ton on the balance of the steelwork by 3½% on the average. This fact should be taken into account when making a cost analysis of several systems.

6-17. Other Floor Systems. Aside from the basic floor systems of Arts. 6-11 to 6-15, there are numerous adaptations and proprietary systems, such as these:

Cofar (corrugated form and reinforcing), where concrete reinforcement is shop-welded to corrugated sheets, which serves as the form for the concrete slab. Clear spans up to 14 ft are possible.

Smooth ceilings, where the floor slab is supported with the aid of short cantilever steel beams, or grids, rigidly connected to the columns and embedded in the slab—a system that eliminates beams between columns.

Dox, a system of precast-concrete blocks tied together in the shop with steel rods to form beams or slabs.

Flexicore, a prefabricated, light, concrete floor plank, achieving weight saving through the use of circular, longitudinal, hollow spaces.

Battledeck or steel-plate floor, the concrete slab of the arch system being replaced with a steel plate. It requires close spacing of secondary beams.

And in addition, there are available a number of precast floor planks (or tiles) of concrete or gypsum, some featuring lightweight aggregates.

6-18. Roof Systems. These are similar in many respects to the floor types; in fact, for flat-top tier buildings, the roof may be just another floor. However, when roof loads are smaller than floor loads, as is usually the case, it may be economical to lighten the roof construction. For example, steel joists may be spaced farther apart. Where roof decking is used, the spacing of the joists is determined by the load-carrying ability of the applied decking and of the joists.

Most of the considerations listed for floors also are applicable to roof systems (Arts. 6-10 to 6-17). In addition, however, due thought should be given to weather resistance, heat conductance and insulation, moisture absorption and vapor barriers, and especially to maintenance.

Many roof systems are distinctive as compared with the floor types; for example, the corrugated sheet-metal roofing commonly employed on many types of industrial or mill buildings. The sheets rest on small beams, channels, or joists, called **purlins,** which in turn are supported by trusses. Similar members on the sidewalls are called **girts.**

ALLOWABLE DESIGN STRESSES

Structural-steel members are designed for any one, or any combination, of five possible stress conditions: bending, shearing, web crippling, direct or axial tension, and direct or axial compression. There are other conditions that must be investigated under special conditions: local buckling, excessive deflection, and torsion.

When elastic theory is used, design is based on allowable unit stresses, usually those given in local building codes. Since several grades of steel are available, it is expedient to relate all allowable unit stresses to the minimum yield stress specified for each grade.

When plastic theory is used, design is based on the ultimate strength of members. A safety factor, comparable to that established for elastic design, is applied to the design load to determine the ultimate-load capacity required of a member.

Formulas and allowable stresses in the following articles follow recommendations of the American Institute of Steel Construction "Specification for the Design, Fabrication and Erection of Structural Steel for Buildings," 1969 and Supplements 1 to 3. An accompanying commentary provides background and clarifies the intent of the specification. Both are incorporated in the AISC "Manual of Steel Construction."

6-19. Allowable Tension in Steel. Unit stress on the net section should not exceed $F_t = 0.60F_y$, where F_y is the minimum yield stress of the steel. But for pin holes in eyebars, pin-connected plates or built-up members, $F_t = 0.45F_y$. (See Table 6-2.)

Table 6-2. Allowable Tension on Net Section
(except at pin holes)

F_y, ksi	F_t, ksi	F_y, ksi	F_t, ksi
36.0	22.0	60.0	36.0
42.0	25.2	65.0	39.0
45.0	27.0	90.0	52.5*
50.0	30.0	100.0	57.5*
55.0	33.0		

* Limited to 0.5 tensile strength.

Net section for a member with a chain of holes extending along a diagonal or zigzag line is the product of the net width and thickness. To determine net width, deduct from the gross width the sum of the diameters of all the holes in the chain, then add, for each gage space in the chain, the quantity

$$\frac{s^2}{4g}$$

where s = longitudinal spacing (**pitch,** in.) of any two consecutive holes
g = transverse spacing (**gage,** in.) of the same two holes

The critical net section of the member is obtained from that chain with the least net width; but the net section through a hole should not be taken as more than 85% of the corresponding gross section.

6-20. Allowable Shear in Steel. Unit stress in shear on the gross section should not be greater than $F_v = 0.40F_y$, where F_y is the minimum yield point of the steel (Table 6-3).

Table 6-3. Allowable Shear on Gross Section

F_y, ksi	F_v, ksi	F_y, ksi	F_v, ksi
36.0	14.5	60.0	24.0
42.0	17.0	65.0	26.0
45.0	18.0	90.0	36.0
50.0	20.0	100.0	40.0
55.0	22.0		

Unit shear on a beam web may be computed by dividing the total shear at a section by the web area, taken as the product of web thickness and over-all beam depth. Except when the web has cutouts or holes, beams never fail because of high shear. Failure will be a form of buckling because of the complex pattern of stresses.

A special case occurs when a web lies in a plane common to intersecting members; for example, the knee of a rigid frame. Then, shear stresses generally are high. Such webs, in elastic design, should be reinforced when the web thickness is less than $32M/A_{bc}F_y$, where M is the algebraic sum of clockwise and counterclockwise moments (in ft-kips) applied on opposite sides of the connection boundary, and A_{bc} is the planar area of the connection web, sq in. (approximately the product of the depth of the member introducing the moment and the depth of the intersecting member). In plastic design, this thickness is determined from $23M_p/A_{bc}F_y$, where M_p is the plastic moment, or M times a load factor of 1.70. In this case, the total web shear produced by the factored loading should not exceed the web area (depth times thickness) capacity in shear. Otherwise, the web must be reinforced with diagonal stiffeners or a doubler plate.

For deep girder webs, allowable shear is reduced. The reduction depends on the ratio of clear web depth between flanges to web thickness and an aspect ratio of stiffener spacing to web depth. In practice, this reduction does not apply when the ratio of web depth to thickness is less than $380/\sqrt{F_y}$. Table 3 in the Appendix of the AISC "Specification for the Design, Fabrication and Erection of Structural Steel for Buildings" gives the allowable shear stresses in plate girder webs.

6-21. Allowable Compression in Steel. The allowable compressive stress on the gross section of axially loaded members is given by formulas determined by the effective slenderness ratios Kl/r of the members, where Kl is effective length (Art. 6-22) and r is the least radius of gyration. A critical value, designated C_c, occurs at the slenderness ratio corresponding to the maximum stress for elastic

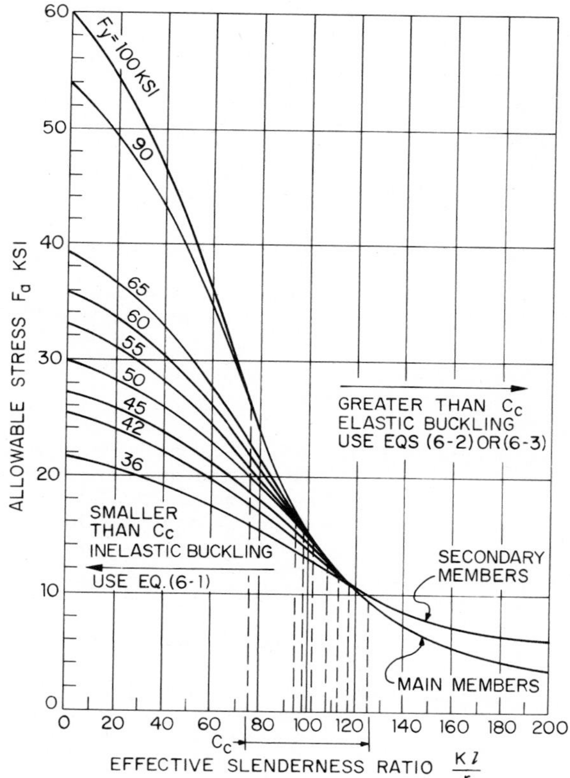

Fig. 6-14. Allowable stresses for axial compression.

buckling failure (Table 6-4). This is illustrated in Fig. 6-14. An important fact to note: When Kl/r exceeds $C_c = 126.1$, the allowable compressive stress is the same for A36 and all higher-strength steels.

$$C_c = \sqrt{2\pi^2 E/F_y}$$

where E = the modulus of elasticity of the steel = 29,000 ksi
F_y = the specified minimum yield stress, ksi

When Kl/r for any unbraced segment is less than C_c, the allowable compressive stress is

$$F_a = \frac{[1 - (Kl/r)^2/2C_c^2]F_y}{F.S.} \tag{6-1}$$

where $F.S.$ is the safety factor, which varies from 1.67 when $Kl/r = 0$ to 1.92 when $Kl/r = C_c$.

$$F.S. = \frac{5}{3} + \frac{3Kl/r}{8C_c} - \frac{(Kl/r)^3}{8C_c^3}$$

Table 6-4. Slenderness Ratio at Maximum Stress for Elastic Buckling Failure

F_y, ksi	C_c	F_y, ksi	C_c
36.0	126.1	60.0	97.7
42.0	116.7	65.0	93.8
45.0	112.8	90.0	79.8
50.0	107.0	100.0	75.7
55.0	102.0		

When Kl/r is greater than C_c:

$$F_a = \frac{12\pi^2 E}{23(Kl/r)^2} = \frac{149{,}000}{(Kl/r)^2} \qquad (6\text{-}2)$$

This is the Euler column formula for elastic buckling with a constant safety factor of 1.92 applied.

Increased stresses are permitted for bracing and secondary members with l/r greater than 120. (K is taken as unity.) For such members, the allowable compressive stress is

$$F_{as} = \frac{F_a}{1.6 - l/200r} \qquad (6\text{-}3)$$

where F_a is given by Eq. (6-1) or (6-2). The higher stress is justified by the relative unimportance of these members and the greater restraint likely at their end connections. The full unbraced length should always be used for l.

Tables giving allowable stresses for the entire range of Kl/r appear in the AISC "Manual of Steel Construction." Approximate values may be obtained from Fig. 6-14.

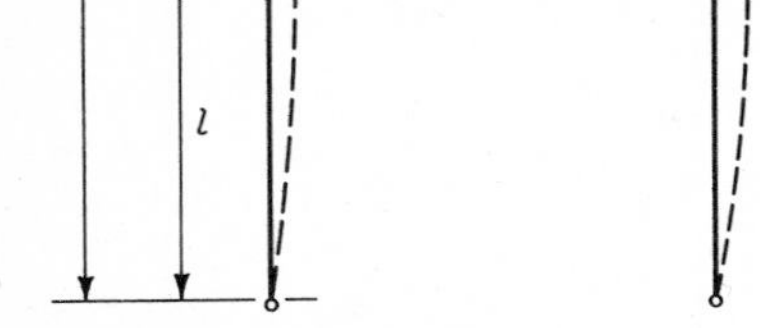

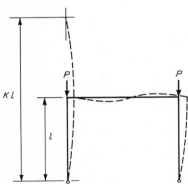

Fig. 6-15. Configurations of members of rigid frame due to sidesway.

6-22. Effective Column Length. The proper application of the column formulas, Eqs. (6-1) and (6-2) in Art. 6-21, is dependent on judicious selection of K. This term is defined as the ratio of effective column length to actual unbraced length.

For a pin-ended column with translation of the ends prevented, $K = 1$. But in general, K may be greater or less than unity. For example, consider the columns in the frame in Fig. 6-15. They are dependent entirely on their own stiffness for stability against sidesway. If enough axial load is applied to them, their effective length will exceed their actual length. But if the frame were braced to prevent sidesway, the effective length would be less than the actual length because of the resistance to end rotation provided by the girder.

Theoretical values of K for six idealized conditions in which joint rotation and translation are either fully realized or nonexistent are given in Fig. 6-16. Also noted are values recommended by the Column Research Council for use in design when these conditions are approximated. Since joint fixity is seldom fully achieved, slightly higher design values than theoretical are given for fixed-end columns.

Specifications do not provide criteria for sidesway resistance under vertical loading, because it is impossible to evaluate accurately the contribution to stiffness of the various components of a building. Instead, specifications cite the general conditions that have proven to be adequate.

Constructions that inhibit sidesway in building frames include substantial masonry walls; interior shear walls; braced towers and shafts; floors and roofs providing diaphragm action; that is, stiff enough to brace the columns to shear walls or bracing systems; frames designed primarily to resist large side loadings or to limit horizontal deflection; and diagonal X bracing in the planes of the frames. Compression members in trusses are considered to be restrained against translation at connections. Generally, for all these constructions, K may be taken as unity, but a value less than one is permitted if proven by analysis.

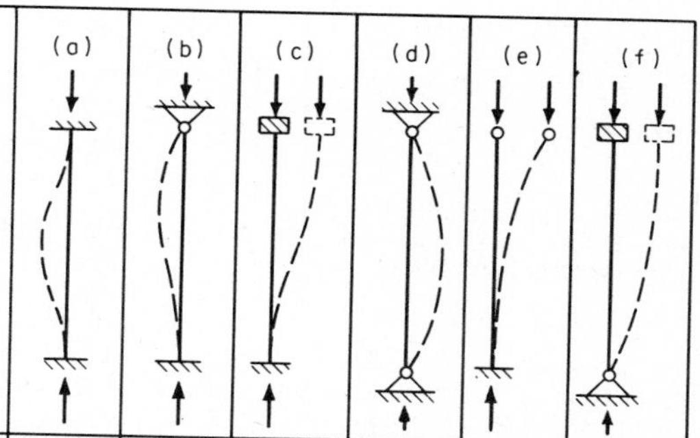

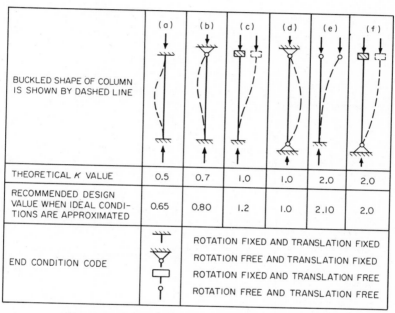

	(a)	(b)	(c)	(d)	(e)	(f)
BUCKLED SHAPE OF COLUMN IS SHOWN BY DASHED LINE						
THEORETICAL K VALUE	0.5	0.7	1.0	1.0	2.0	2.0
RECOMMENDED DESIGN VALUE WHEN IDEAL CONDITIONS ARE APPROXIMATED	0.65	0.80	1.2	1.0	2.10	2.0
END CONDITION CODE	ROTATION FIXED AND TRANSLATION FIXED ROTATION FREE AND TRANSLATION FIXED ROTATION FIXED AND TRANSLATION FREE ROTATION FREE AND TRANSLATION FREE					

Fig. 6-16. Values of K for idealized conditions.

When resistance to sidesway depends solely on the stiffness of the frames; for example, in tier buildings with light curtain walls or with wide column spacing, and with no diagonal bracing systems or shear walls, the designer may use any of several proposed rational methods for determining K. A quick estimate, however, can be made by using the alignment chart in the AISC "Manual." The effective length Kl of compression members, in such cases, should not be less than the actual unbraced length.

6-23. Web Crippling. Webs of rolled beams and welded plate girders should be so proportioned that the compressive stress, ksi, at the web toe of the fillets does not exceed

$$F_a = 0.75F_y$$

Web failure probably would be in the form of buckling caused by concentrated loading, either at an interior load or at the supports. The capacity of the web to transmit the forces safely should be checked.

Load Distribution. Loads are resisted not only by the part of the web directly under them but also by the parts immediately adjacent. A 45° distribution usually is assumed, as indicated in Fig. 6-17 for two common conditions. The distance k is determined by the point where the fillet of the flange joins the web; it is tabulated in the beam tables of the AISC "Manual of Steel Construction." F_a is applicable to the horizontal web strip of length $b + k$ at the end support or $b + 2k$ under an interior load. Bearing stiffeners are required when F_a is exceeded.

Bearing Atop Webs. The sum of the compression stresses resulting from loads bearing directly on or through a flange on the compression edge of a plate-girder web should not exceed the following:

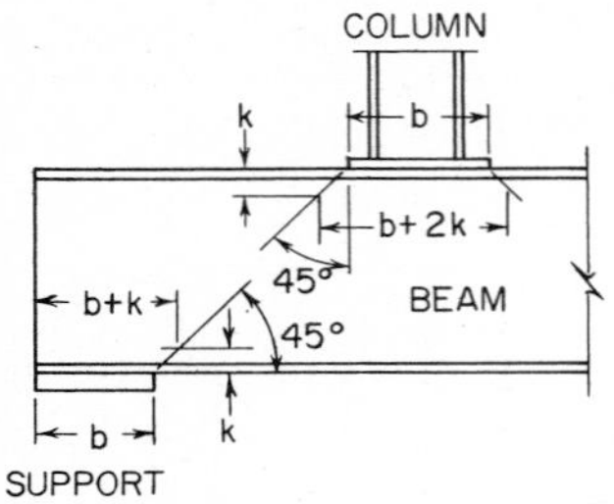

Fig. 6-17. Web crippling in a simple beam. Critical web section is assumed at fillet.

When the flange is restrained against rotation, the allowable compressive stress, ksi, is

$$F_a = \left[5.5 + \frac{4}{(a/h)^2}\right]\frac{10{,}000}{(h/t)^2} \qquad (6\text{-}4)$$

When the flange is not restrained against rotation,

$$F_a = \left[2 + \frac{4}{(a/h)^2}\right]\frac{10{,}000}{(h/t)^2} \qquad (6\text{-}5)$$

where a = clear distance between transverse stiffeners, in.
h = clear distance between flanges, in.
t = web thickness, in.

The load may be considered distributed over a web length equal to the panel length (distance between vertical stiffeners) or girder depth, whichever is less.

Web Stiffeners on Columns. The webs of columns may also be subject to crippling due to the thrust from the compression flange of rigidly connected beams, as shown

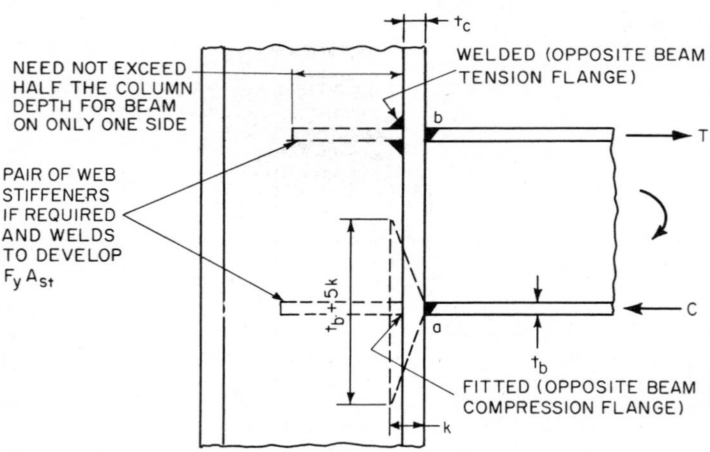

Fig. 6-18. Web crippling in column at joint with welded beam.

at point a in Fig. 6-18. Likewise, to insure full development of the beam plastic moment, the column flange opposite the tensile thrust at point b may require stiffening.

Therefore, web stiffeners are required on the column at point a when

$$t < \frac{C_1 A_f}{t_b + 5k} \qquad (6\text{-}6a)$$

or when

$$t \leq \frac{d_c \sqrt{F_y}}{180} \qquad (6\text{-}6b)$$

and when at *b*

$$t_c < 0.4\sqrt{C_1 A_f} \qquad (6\text{-}7)$$

where t = thickness of column web, in.
t_b = thickness of beam flange delivering concentrated load, in.
t_c = thickness of column flange, in.
A_f = area of beam flange delivering concentrated load, sq in.
k = distance from outer face of column to edge of web toe of fillet, in.
d_c = column web depth clear of fillets, in.
C_1 = ratio of beam flange yield stress to column yield stress

The area of one pair of stiffener plates, when required, should be at least

$$A_{st} = C_2[C_1 A_f - t(t_b + 5k)] \qquad (6\text{-}8)$$

where C_2 = ratio of column yield stress to stiffener yield stress.

When a rigidly connected beam is only on one side of the column, the length of the web stiffeners need not exceed one-half the column depth, provided the connecting weld to the column web develops $F_y A_{st}$.

6-24. Pipe and Tubular Columns. Pipe columns are particularly useful in low-height buildings where light loads prevail, and where attractive architectural effects are derived from exposed steel. Theoretically, the circular section is ideal because the slenderness ratio, l/r, is the same in every direction; however, the use of pipe columns for multistory buildings is limited by the cost of making beam connections.

Pipe meeting the requirements of American Society for Testing and Materials Specification A53, Types E and S, Grade B, is comparable to A36 steel, with F_y = 36 ksi. It comes in three weight classifications: standard, extra strong, and double extra strong, and in diameters varying from 3 to 12 in.

Several mills produce square and rectangular tubing in sizes ranging from 3 × 2 and 2 × 2 to 12 × 8 and 10 × 10 in., with wall thickness up to ⅝ in. These flat-sided shapes afford easier connections than pipes, not only for connecting beams but also for such items as window and door frames.

The main strength properties of several grades of steel used for pipe and tubular sections are summarized in Table 6-5.

Table 6-5. Characteristics of Pipe and Tubular Steels

ASTM spec.	Grade	Product	Min tensile strength, ksi	Min yield stress, ksi
A53	B	Pipe	60.0	35.0*
A500	A	Round	45.0	33.0
	A	Shaped	45.0	39.0
	B	Round	58.0	42.0
	B	Shaped	58.0	46.0
A501	...	All tubing	58.0	36.0
A618	I	All tubing	70.0	50.0
	II	All tubing	70.0	50.0
	III	All tubing	65.0	50.0

* Use 36.0 for purpose of design.

The concentric load capacity of pipes and tubes is determined from the column formulas of Art. 6-21. It is also given in safe-load tables in the AISC "Manual of Steel Construction." The formulas and tables are based on the assumption that the pipes and tubes are not filled with concrete.

6-25. Bending. Beams classified as **compact** are allowed a bending stress $F_b = 0.66\ F_y$ for the extreme surfaces in both tension and compression, where F_y is the specified yield stress, ksi. Such members have an axis of symmetry in the plane of loading, their compression flange is adequately braced to prevent lateral displacement, and they develop their full plastic moment (section modulus times yield stress) before buckling.

Compactness Requirements. To qualify as compact, beams must meet the following conditions:

1. The flanges must be continuously connected to the web or webs.

2. The width-thickness ratio of the unstiffened projecting elements of the compression flange must not exceed $65.0/\sqrt{F_y}$, where width b is one-half the full flange width of I-shaped sections, or the dimension from the free edge to the first row of fasteners (or welds) for projecting plates, or the full width of legs of angles, or the flanges of channels, or tee stems.

3. The depth-thickness ratio d/t_w of webs must not exceed $640(1 - 3.74f_a/F_y)/\sqrt{F_y}$ when f_a, the computed axial stress, is equal to or less than $0.16F_y$, or $257/\sqrt{F_y}$ when $f_a > 0.16F_y$.

4. The width-thickness ratio of stiffened compression flange plates in box sections and that part of the cover plates for beams and built-up members that is included between longitudinal lines of rivets, bolts, or welds does not exceed $190/\sqrt{F_y}$.

5. For the compression flange of members not box shaped to be considered supported, the unbraced length between lateral supports does not exceed $76.0b_f/\sqrt{F_y}$ or $20{,}000A_f/F_yd$, where b_f is the flange width, A_f the flange area, and d the web depth.

6. The unbraced length for rectangular box-shaped members with depth not more than six times the width and with flange thickness not more than two times the web thickness does not exceed $(1{,}950 + 1{,}200\ M_1/M_2)b/F_y$. The unbraced length in such cases, however, need not be less than $1{,}200b/F_y$. M_1 is the smaller and M_2 the larger of bending moments at points of lateral support.

Stresses for Compact Beams. Most sections used in building framing, including practically all rolled W shapes of A36 steel and most of those with $F_y = 50$ ksi, comply with these requirements for compactness, as illustrated in Fig. 6-19. Such sections, therefore, are designed with $F_b = 0.66F_y$. Excluded from qualifing are hybrid girders, tapered girders, and sections made from A514 steel.

Braced sections that meet the requirements for compactness, and are continuous over their supports or rigidly framed to columns, are also permitted a redistribution of the design moments. Negative gravity-load moments over the supports may be reduced 10%. But then, the maximum positive moment must be increased by 10% of the average negative moments. This moment redistribution does not apply to cantilevers, hybrid girders, or members made of A514 steel.

Stresses for Noncompact Beams. Many other beam-type members, including nearly compact sections that do not meet all six requirements, are accorded allowable bending stresses, some higher and some considerably lower than $0.66F_y$ depending on such conditions as shape factor, direction of loading, inherent resistance to torsion or buckling, and external lateral support. The common conditions and applicable allowable bending stresses are summarized in Fig. 6-20. In the formulas,

l = distance, in., between cross sections braced against twist or lateral displacement of the compression flange

r_T = radius of gyration, in., of a section comprising the compression flange plus one-third of the compression web area, taken about an axis in the plane of the web

A_f = area of the compression flange, sq in.

Lateral Support. In computation of allowable bending stresses in compression for beams with distance between lateral supports exceeding requirements, a range sometimes called laterally unsupported, the AISC formulas contain a moment factor C_b in recognition of the beneficial effect of internal moments, both in magnitude and direction, at the points of support. For the purpose of this summary, however, the moment factor has been taken as unity and the formulas simplified in Fig. 6-20. The formulas are exact for the case in which the bending moment at any point within an unbraced length is larger than that at both ends of this length. They are conservative for all other cases. Where more refined values are desired, reference should be made to the AISC Specification.

Limits on Width-thickness Ratios. For flexural members in which the width-thick-

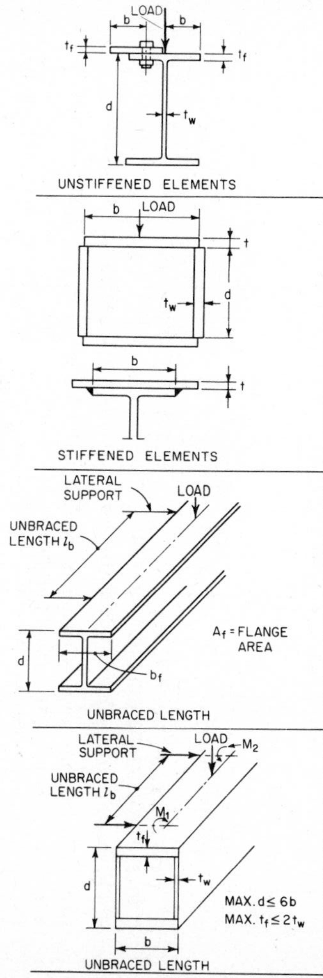

F_y =	36.0	42.0	45.0	50.0	55.0	60.0	65.0
$\frac{b}{t_f} \leq \frac{65.0}{\sqrt{F_y}}$	10.8	10.0	9.7	9.2	8.8	8.4	8.1
$\frac{d}{t_w} \leq \frac{640}{\sqrt{F_y}}(1-3.74\frac{f_a}{F_y})$ FOR $f_a/F_y \leq 0.16$	$107-11.1f_a$	$98.8-8.8f_a$	$95.4-7.9f_a$	$90.5-6.8f_a$	$86.3-5.9f_a$	$82.6-5.2f_a$	$79.4-4.6f_a$
$\frac{d}{t_w} \leq \frac{257}{\sqrt{F_y}}$ FOR $f_a/F_y > 0.16$	42.8	39.7	38.3	36.3	34.7	33.2	31.9
$\frac{b}{t} \leq \frac{190}{\sqrt{F_y}}$	31.7	29.3	28.3	26.9	25.6	24.5	23.6
$l_b \leq \frac{76.0b_f}{\sqrt{F_y}}$	12.7	11.7	11.3	10.7	10.2	9.8	9.4
$l_b \leq \frac{20{,}000\,A_f}{dF_y}$	$556\frac{A_f}{d}$	$476\frac{A_f}{d}$	$444\frac{A_f}{d}$	$400\frac{A_f}{d}$	$364\frac{A_f}{d}$	$333\frac{A_f}{d}$	$308\frac{A_f}{d}$
$l_b \leq \frac{b}{F_y}(1{,}950+1{,}200\frac{M_1}{M_2})$	—	—	—	—	—	—	
BUT NEED NOT BE LESS THAN $1{,}200\frac{b}{F_y}$	33.3b	28.6b	26.7b	24.0b	21.8b	20.0b	18.5b

Fig. 6-19. Requirements for laterally supported compact sections.

SECTION	REQUIREMENTS	F_b
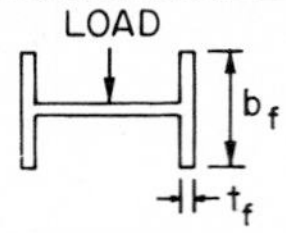 SYMMETRICAL ABOUT BOTH AXES	MEETS COMPACTNESS 1 & 2*	0.75 F_y FOR BOTH TENSION AND COMPRESSION
	MEETS COMPACTNESS 1*† $\frac{65.0}{\sqrt{F_y}} < \frac{b_f}{2t_f} < \frac{95.0}{\sqrt{F_y}}$	$F_y\left[1.075 - 0.005\left(\frac{b_f}{2t_f}\right)\sqrt{F_y}\right]$
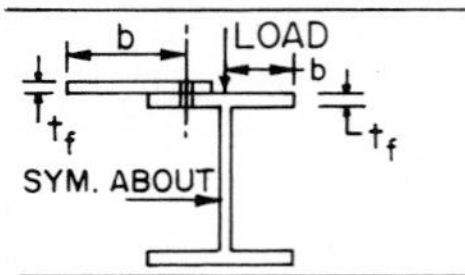	MEETS COMPACTNESS 1,3,4,5*† $\frac{65.0}{\sqrt{F_y}} < \frac{b}{t_f} < \frac{95.0}{\sqrt{F_y}}$	$F_y\left[0.79 - 0.002\left(\frac{b}{t_f}\right)\sqrt{F_y}\right]$
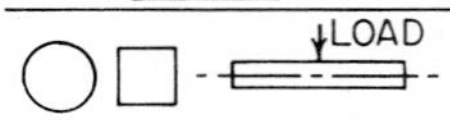	SOLID SQUARES AND ROUNDS SOLID FLAT ELEMENTS BENT ABOUT WEAKER AXIS	0.75 F_y FOR BOTH TENSION AND COMPRESSION
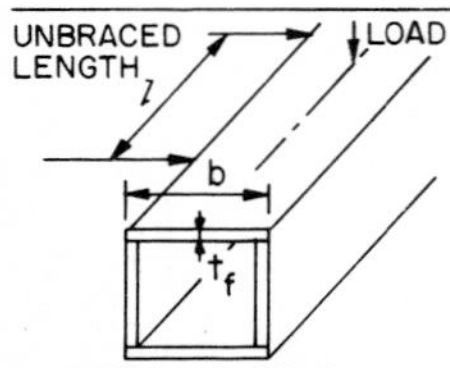	MEETS COMPACTNESS 1,3,6 $\frac{b}{t_f} \le \frac{238}{\sqrt{F_y}}$ OR $\le \frac{317}{\sqrt{F_y}}$ IF PERFORATED WITH ACCESS HOLES MIN. $l = \frac{1{,}200b}{F_y}$	0.60 F_y FOR BOTH TENSION AND COMPRESSION
TENSION FLANGE FLEXURAL SECTIONS NOT COVERED ABOVE		0.60 F_y
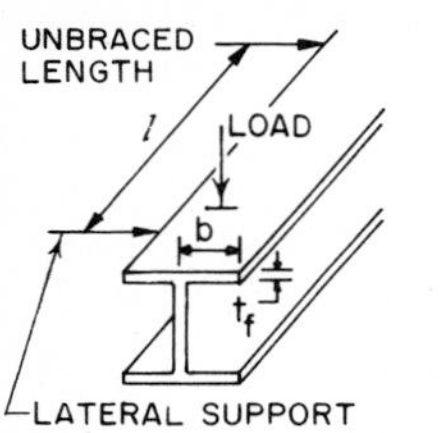	$\frac{b}{t_f} \le \frac{95.0}{\sqrt{F_y}}$	FOR COMPRESSION, USE LARGER OF FOLLOWING:
	WHEN $\frac{319}{\sqrt{F_y}} \le \frac{l}{r_t} \le \frac{714}{\sqrt{F_y}}$	$F_y\left[\frac{2}{3} - \frac{F_y(l/r_t)^2}{1{,}530{,}000}\right] \le 0.60F_y$
	WHEN $\frac{l}{r_t} \ge \frac{714}{\sqrt{F_y}}$	$\frac{170{,}000}{(l/r_t)^2} \le 0.60F_y$
FOR CHANNELS USE ONLY →	OR FOR SOLID RECTANGULAR FLANGE WITH AREA NOT LESS THAN TENSION FLANGE †	$\frac{12{,}000A_f}{ld} \le 0.60F_y$
COMPRESSION FLANGE FOR FLEXURAL SECTIONS NOT COVERED ABOVE	$\frac{b}{t_f} \le \frac{95.0}{\sqrt{F_y}}$ $l \le \frac{152b}{\sqrt{F_y}}$ FOR MAJOR-AXIS BENDING	0.60 F_y

* EXCEPT A514 STEEL † EXCEPT HYBRID GIRDERS

NOTE: FOR I-SHAPED SECTIONS $b_f = 2b$

Fig. 6-20. Allowable bending stresses for sections not qualifying as compact.

ness ratios of compression elements exceed the limits given in Fig. 6-20 and which are usually lightly stressed, appropriate allowable bending stresses are suggested in "Slender Compression Elements," Appendix C, AISC "Specification for the Design, Fabrication and Erection of Structural Steel for Buildings."

For additional discussion of lateral support, see Art. 6-44; also, additional information on width-thickness ratios of compression elements is given in Fig. 6-25.

The allowable bending stresses F_b, ksi, for values often used for various grades of steel are listed in Table 6-6.

6-26. Bearing. For bearing on contact surfaces, such as milled, sawed or accurately finished joints, bearing of pins in finished holes, and bearing stiffeners, the allowable stress is

$$F_p = 0.90F_y \tag{6-9}$$

Table 6-6. Allowable Bending Stresses, Ksi

F_y	$0.60F_y$	$0.66F_y$	$0.75F_y$
36.0	22.0	24.0	27.0
42.0	25.2	28.0	31.5
45.0	27.0	29.7	33.8
50.0	30.0	33.0	37.5
55.5	33.0	36.3	41.3
60.0	36.0	39.6	45.0
65.0	39.0	42.9	48.8
90.0	54.0		
100.0	60.0		

Table 6-7. Allowable Bearing on Milled Surfaces, Ksi

F_y	F_p	F_y	F_p
36.0	33.0	60.0	54.0
42.0	38.0	65.0	58.5
45.0	40.5	90.0	81.0
50.0	45.0	100.0	90.0
55.0	49.5		

where F_y is the specified minimum yield stress of the steel. When the parts in contact have different yield stresses, use the smaller F_y (Table 6-7).

The allowable bearing stress on expansion rollers and rockers, in lb per lin in., is

$$F_p = \frac{F_y - 13}{20} 0.66d \tag{6-10}$$

where d is the diameter of roller or rocker, in. (Table 6-8).

Table 6-8. Allowable Bearing on Expansion Rollers and Rockers, Kips per Lin In.

F_y	F_p	F_y	F_p
36.0	$0.76d$	60.0	$1.55d$
42.0	$0.96d$	65.0	$1.72d$
45.0	$1.06d$	90.0	$2.54d$
50.0	$1.22d$	100.0	$2.87d$
55.0	$1.39d$		

Allowable bearing stresses on masonry usually can be obtained from a local or state building code, whichever governs. In the absence of such regulations, however, the values in Table 6-9 can be used.

Table 6-9. Allowable Bearing on Masonry, Ksi

On sandstone and limestone	0.40
On brick in cement mortar	0.25
On the full area of concrete	$0.35f'_c$
On less than full concrete area	$0.35f'_c\sqrt{A_2/A_1} \le 0.7f'_c$

where f'_c = specified compressive strength, ksi, of the concrete
A_1 = bearing area
A_2 = concrete area

6-27. Combined Axial Compression and Bending. When a member is subjected to bending, the neutral axis is displaced from its original position a maximum distance Δ. If it also carries an axial load P, secondary bending moments, equal to the product of the load and the displacements of the axis, result. This often is referred as the P-Δ effect.

When the computed axial stress, f_a, is less than 15% of F_a, the stress that would be permitted if axial force alone were present, the influence of the secondary moment is so small that it may be neglected and a straight-line interaction formula may be used. Thus, when $f_a/F_a \leq 0.15$:

$$\frac{f_a}{F_a} + \frac{f_{bx}}{F_{bx}} + \frac{f_{by}}{F_{by}} \leq 1.0 \tag{6-11}$$

where subscripts x and y indicate, respectively, the major and minor axes of bending (if bending is about only one axis, then the term for the other axis is omitted), and

f_b = computed compressive bending stress, ksi, at point under consideration
F_b = compressive bending stress, ksi, that is allowed if bending alone existed

When $f_a/F_a > 0.15$, the effect of the secondary bending moment should be taken into account and the member proportioned to satisfy Eqs. (6-12) and (6-13):

$$\frac{f_a}{F_a} + \frac{C_{mx}f_{bx}}{[1 - f_a/F'_{ex}]F_{bx}} + \frac{C_{my}f_{by}}{[1 - f_a/F'_{ey}]F_{by}} \leq 1.0 \tag{6-12}$$

$$\frac{f_a}{0.60F_y} + \frac{f_{bx}}{F_{bx}} + \frac{f_{by}}{F_{by}} \leq 1.0 \tag{6-13}$$

where, as before, subscripts x and y indicate axes of bending

$$F'_e = \frac{12\pi^2 E}{23(Kl_b/r_b)^2}$$

E = modulus of elasticity, 29,000 ksi
l_b = actual unbraced length, in., in the plane of bending
r_b = corresponding radius of gyration, in.
K = effective-length factor in the plane of bending
C_m = reduction factor whose value is determined from the following conditions:

1. For compression members in frames subject to joint translation (sidesway), $C_m = 0.85$.

2. For restrained compression members in frames braced against joint translation and not subject to transverse loading between their supports in the plane of bending, $C_m = 0.6 - 0.4M_1/M_2$, but not less than 0.4, where M_1/M_2 is the ratio of the smaller to larger moments at the ends of that portion of the member unbraced in the plane of bending under consideration. M_1/M_2 is positive when the member is bent in reverse curvature, and negative when it is bent in single curvature.

3. For compression members in frames braced against joint translation in the plane of loading and subjected to transverse loading between their supports, the value of C_m may be determined by rational analysis. Instead, however, C_m may be taken as 0.85 for members whose ends are restrained, and 1.0 for ends unrestrained.

In wind and seismic design F'_e may be increased one-third, conforming with the increases covered in Art. 6-30.

Additional information, including illustrations of the foregoing three conditions for determining the value of C_m, is given in the "Commentary on the AISC Specification for the Design, Fabrication and Erection of Structural Steel for Buildings."

6-28. Combined Axial Tension and Bending. Members subject to both axial tension and bending stresses should be proportioned to satisfy Eq. (6-11), with f_b and F_b, respectively, as the computed and allowable bending tensile stress. But the compressive bending stresses must not exceed the values given in Art. 6-25.

6-29. Design of Beam Sections for Torsion. This is a special type of load application, since in normal practice eccentric loads on beams are counterbalanced to the point where slight eccentricities may be neglected. For example, spandrel beams supporting a heavy masonry wall may not be concentric with the load, thus inducing torsional stresses, but these will largely be canceled out by the equally eccentric loads of the floor, partitions, attached beams, and similar restraints. For this reason, one seldom finds any ill effects from torsional stresses.

It is during the construction phase that torsion may be in evidence, usually the result of faulty construction procedure. In Fig. 6-21 are illustrated some of the bad practices that have caused trouble in the field. When forms for concrete slabs are hung on one edge of a beam (usually the light secondary beam), the weight of the wet concrete may be sufficient to twist the beam. Figure 6-21*a* shows the correct method, which reduces torsional effect to the minimum. Likewise for spandrels, the floor ties, if any, forms, or the slab itself should be placed prior to the construction of the eccentric wall (Fig. 6-21*b*). Connectors for heavy roofing sheets when located on one side of the purlin may distort the section; the condition should be corrected by staggering, as indicated in Fig. 6-21*c*.

Equations for computing torsion stresses are given in Arts. 3-77 and 3-78. Also, see Bibliography, Art. 6-92.

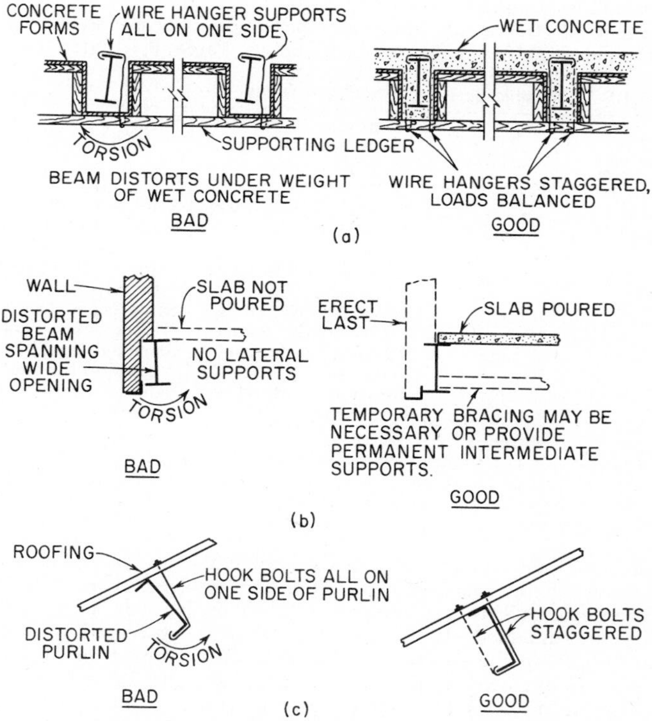

Fig. 6-21. Steel beams subjected to torsion.

6-30. Wind and Seismic Stresses. For wind or earthquake forces, acting alone or in combination with the design dead and live loads, allowable stresses may be increased one-third. But the resultant section should not be less than that required for dead and live loads alone without this one-third increase in stress. The increase is allowed because wind and seismic forces are of short duration.

6-31. Cyclic Loaded Members. Relatively few structural members in a building are ever subjected to large, repeated variations of stress or stress reversals (tension to compression, and vice versa) that could cause fatigue damage to the steel. Members need not be investigated for this possibility unless the number of stress cycles exceeds 20,000, which is nearly equivalent to two applications every day for 25 years.

Wind and seismic stresses of any appreciable magnitude occur too infrequently to cause fatigue damage. On the other hand, members such as cranes, crane girders, and supports for heavy machinery (e.g., presses and forges) are subjected to frequent stress variations. Allowable stresses given in the previous articles for static loadings have to be reduced to offset the lower resistance of steel to fatigue conditions. The subject is treated in considerable detail, both for members and their connections, including suggested maximum stress ranges, in "Fatigue," Appendix B, AISC Specification for the Design, Fabrication and Erection of Structural Steel for Buildings. An interesting point to note is that all steels (except for one condition favorable to A514 steel) are treated alike; i.e., fatigue damage is independent of steel strength. (See also, F. S. Merritt, "Structural Steel Designers' Handbook," McGraw-Hill Book Company, New York.)

6-32. Stresses for Welds. Allowable stresses for welds joining structural steel parts depend on the type of welded joint, strength of electrode, and strength of base material. Commonly used welded joints are fillet and groove. The latter is classified in accordance with penetration, as either complete or partial. Welded joints and definitions are illustrated in Fig. 6-37.

When proper electrodes are used for the grade of base steel, the allowable stresses for welded groove joints summarized in Table 6-10 may be used.

Table 6-10. Allowable Stresses for Groove Welds

Type of groove weld	Kind of stress*	Direction of stress	Allowable stress
Complete penetration	F_t and F_c	Parallel to axis of weld	Same as for base metal
	F_t and F_c	Normal to effective throat	Same as for base metal
	F_v	In plane of effective throat	Same as allowable shear for base metal
Partial penetration	F_t	Normal to weld axis,† on effective throat	Same as for shear on fillet welds, Table 6-11
	F_c	Normal to effective throat	Same as allowable compression for base metal
	F_v	In plane of effective throat	Same as allowable shear for base metal

* Subscripts *t, c, v* designate, respectively, tension, compression, and shear.

† Where partial penetration groove welds are used to join elements, such as girder flanges to webs, the stress in the elements parallel to the axis of the welds may be disregarded.

The most significant characteristic of fillet-welded joints (also illustrated in Fig. 6-37) is that all forces, regardless of direction, are resolved as shear on the effective throat. When fillet welds are used to join elements such as girder flanges to a web, they are designed for horizontal shear without regard to the tensile or compressive stresses in the elements. Allowable shear stresses for fillet welds based upon the strengths of electrodes and base steels are given in Table 6-11.

Allowable shear stress on the effective area of groove or fillet welds is limited to 0.3 the nominal strength of weld metal, ksi, except that the stress on base metal cannot exceed $0.40F_y$, where F_y is the yield strength, ksi, of the base metal. With the exception of allowable tensile stress for a complete-penetration groove weld, it is permissible to use weld metal whose strength is less than indicated as "matching" in Table 6-11. In general, weld metal one strength stronger than "matching," as given in Table 6-11, is permitted for all welds.

Specifications for all welding electrodes, promulgated by the American Welding Society (AWS), are identified as A5.1, A5.5, A5.17, etc., depending on the welding process (Art. 6-58). Electrodes for manual arc welding, often called stick electrodes, are designated by the letter E followed by four or five digits. The first two or three digits designate the strength level; thus, E70XX means electrodes having a minimum tensile strength of 70.0 ksi. Allowable shear stress on the deposited weld metal is taken as 0.30 times the electrode strength classification; thus, 0.30 times 70 for an E70 results in an allowable stress of 21.0 ksi. The remaining digits provide information on the intended usage, such as the particular welding positions and types of electrode coating.

(AISC "Specification for the Design, Fabrication and Erection of Structural Steel for Buildings", "Structural Welding Code," AWS D1.1.)

Table 6-11. Allowable Shear Stresses for Fillet Welds

Shear F_v	Electrode	Matching base steel
18.0 ksi	AWS A5.1, E60XX electrodes AWS A5.17, F6X-EXXX flux-electrode combination AWS A5.20, E60T-X electrodes	A500 Grade A
21.0 ksi	AWS A5.1 or A5.5, E70XX electrodes AWS A5.17, F7X-EXXX flux-electrode combination AWS A5.18, E70S-X or E70U-1 electrodes AWS A5.20, E70T-X electrodes	A36, A53 Grade B A242 A441, A500 Grade B A501, A529, A572 Grades 42 to 60, A588
24.0 ksi	AWS A5.5, E80XX electrodes Grade 80 submerged arc, gas metal-arc or flux-cored arc weld metal	A572 Grade 65
27.0 ksi	AWS A5.5, E90XX electrodes Grade 90 submerged arc, gas metal-arc or flux-cored arc weld metal	A514 over 2½ in. thick
30.0 ksi	AWS A5.5, E100XX electrodes Grade 100 submerged arc, gas metal-arc or flux-cored arc weld metal	A514 over 2½ in. thick
33.0 ksi	AWS A5.5, E110XX electrodes Grade 110 submerged arc, gas metal-arc or flux-cored arc weld metal	A514 2½ in. and less in thickness

6-33. Stresses for Rivets and Bolts. Allowable tension and shear stresses on rivets, bolts, and threaded parts (kips per sq in. of area of rivets before driving or unthreaded body area of bolts and threaded parts, except as noted) are given in Table 6-12.

The allowable bearing stress on the projected area of bolts in bearing type connec-

Table 6-12. Allowable Tension and Shear on Rivets, Bolts, and Threaded Parts, Ksi

Fastener	Tension F_t	Shear F_y
A502, Grade 1, hot-driven rivets . . .	20.0	15.0
A502, Grade 2, hot-driven rivets . . .	27.0	20.0
A307 bolts.	20.0*	10.0
Threaded parts structural-grade steel .	$0.60F_y$*	$0.30F_y$
A325-F bolts†	40.0‡	15.0
A325-N bolts†	40.0‡	15.0
A325-X bolts†	40.0‡	22.0
A490-F bolts†	54.0‡ §	20.0
A490-N bolts†	54.0‡ §	22.5
A490-X bolts†	54.0‡ §	32.0

* Applied to tensile stress area, sq in., equal to $0.7854(D - 0.9743/n)^2$, where D = major thread diameter, in.; n = number of threads per in.

† F = friction-type connection; N = bearing-type connection with threads included in shear plane; X = bearing-type connection with threads excluded from shear plane.

‡ Applied to the nominal bolt or shank area.

§ Static loading only.

tions (see Art. 6-62) and on rivets is given by $F_p = 1.35F_y$, where F_y is the specified yield stress of the connected material (not the fastener) (Table 6-13).

Table 6-13. Allowable Bearing, Ksi, on Projected Area of Bolts and Rivets

F_y	F_p	F_y	F_p
36.0	48.6	60.0	81.0
42.0	56.7	65.0	87.8
45.0	60.8	90.0	121.5
50.0	67.5	100.0	135.0
55.0	74.3		

Since connections do not fail in bearing, use of bearing stresses serves only as an index of the efficiency of net sections. The same index is valid for joints assembled with rivets or bolts, regardless of fastener shear strength or the presence or absence of threads in the bearing area. But bearing stress is not restricted in friction-type connections assembled with A325 and A490 bolts. Similarly, no distinction is made between bearing stresses for single and double, or enclosed, bearing; tests show no difference between these conditions.

6-34. Combined Shear and Tension on Rivets and Bolts. Tests indicate that allowable stress for concurrent shear and tension on fasteners can be obtained from an interaction formula representing an ellipse (Fig. 6-22). But for simplifica-

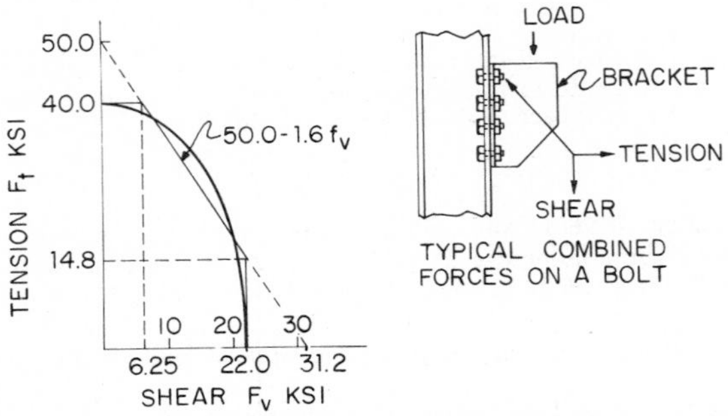

Fig. 6-22. Allowable combined stress in A325 bolts in bearing-type connections, with threads excluded from shear plane.

tion, use of three straight lines is recommended. This permits substantial concurrent stresses without a reduction. For example, as shown in Fig. 6-22 for A325 bolts in a bearing-type connection with threads excluded from shear planes, if a bolt is fully stressed in tension at the allowable of 40 ksi, it can also be stressed in shear up to 6.25 ksi. If it is stressed in shear at the allowable of 22 ksi, it can also be stressed in tension up to 14.8 ksi. Between these limiting conditions, the combined stresses are limited to those determined by the diagonal straight line. The ellipse, straight lines, and limits are different for different types of fasteners.

For combined shear and tension, with shear stress f_v produced by the same force as that producing the tension, but not exceeding the values in Table 6-12, the tension stress f_t in fasteners in bearing-type joints should not exceed the values indicated in the upper part of Table 6-14.

Similarly for combined tension and shear in high-strength bolts in friction-type joints, the shear stress f_v should not exceed the values indicated in the lower part of Table 6-14.

6-35. Allowable Stresses for Composite Design. In composite construction, steel beams are so interconnected with the concrete slab they support that beam and slab act together to resist compression. In one type (Fig. 6-23c), the concrete completely encases the steel beam; in another type (Fig. 6-23*a* and *b*), the slab rests on the top flange of the beam, and shear is transferred mechanically between the steel and concrete.

Encased Beams. Two design methods are approved for the encased type. In one method, stresses are computed on the assumption that the steel beam alone supports all the dead load applied prior to concrete hardening (unless the beam

Table 6-14. Allowable Combined Shear and Tension on Rivets and Bolts

Material	Allowable stress, ksi, for gravity loads	Allowable stress, ksi, increased for wind or seismic loads
	Bearing-type Joints	
A502, Grade 1 rivets	$F_t = 28.0 - 1.6f_v \leq 20.0$	$F_t = 37.3 - 1.6f_v \leq 26.7$
A502, Grade 2 rivets	$F_t = 38.0 - 1.6f_v \leq 27.0$	$F_t = 50.7 - 1.6f_v \leq 36.0$
A307 bolts (applied to stress area)	$F_t = 28.0 - 1.6f_v \leq 20.0$	$F_t = 37.3 - 1.6f_v \leq 26.7$
A325 bolts	$F_t = 50.0 - 1.6f_v \leq 40.0$	$F_t = 66.7 - 1.6f_v \leq 53.3$
A490 bolts	$F_t = 70.0 - 1.6f_v \leq 54.0$	$F_t = 93.3 - 1.6f_v \leq 72.0$
	Friction-type Joints	
A325 bolts	$F_v \leq 15.0(1 - f_tA_b/T_b)$	$F_v \leq 15.0(1.33 - f_tA_b/T_b)$
A490 bolts	$F_v \leq 20.0(1 - f_tA_b/T_b)$	$F_v \leq 20.0(1.33 - f_tA_b/T_b)$

F_t = allowable tensile stress when fastener is subjected to combined shear and tension
f_v = calculated shear stress, ksi
F_v = allowable shear stress when fastener is subjected to combined shear and tension
f_t = average tensile stress, ksi, due to direct load applied to all bolts in a connection
T_b = specified pretension load of bolt, kips
A_b = nominal area, sq in., of bolt body

is temporarily shored), and the composite beam supports the remaining dead and live loads. Then, for positive bending moments, the total stress, ksi, on the steel-beam bottom flange is

$$f_b = \frac{M_D}{S} + \frac{M_L}{S_t} \leq 0.66F_y \tag{6-14}$$

where F_y = specified yield stress of the steel, ksi
M_D = dead-load bending moment, in.-kips
M_L = live-load bending moment, in.-kips
S = section modulus of steel beam, in.^{3}
S_t = section modulus of transformed section, in.^{3} To obtain the transformed equivalent steel area, divide the effective concrete area by the modular ratio n (modulus of elasticity of steel divided by modulus of elasticity of concrete). In computation of effective concrete area, use effective width of concrete slab (Fig. 6-23*a* and *b*)

The stress $0.66F_y$ is allowed because the steel beam is restrained against lateral buckling.

The second method stems from a "shortcut" provision contained in many building codes. This provision simply permits higher bending stresses in beams encased in concrete. For example,

$$f_b = \frac{M_D + M_L}{S} \leq 0.76F_y \tag{6-15}$$

Of course, this higher stress would not be realized, because of composite action.

Beams with Shear Connectors. For composite construction where shear connectors transfer shear between slab and beam, the design is based on behavior at ultimate load. It assumes that all loads are resisted by the composite section, even if shores are not used during construction to support the steel beam until the concrete gains strength. For this case, the computed stress in the bottom flange for positive bending moment is

$$f_b = \frac{M_D + M_L}{S_t} \leq 0.66F_y \tag{6-16}$$

where S_t = section modulus, in.^{3}, of transformed section of composite beam. Figure 6-23*a* and *b* give the width of slab assumed effective in composite action. To prevent overstressing the bottom flange of the steel beam when temporary shoring is

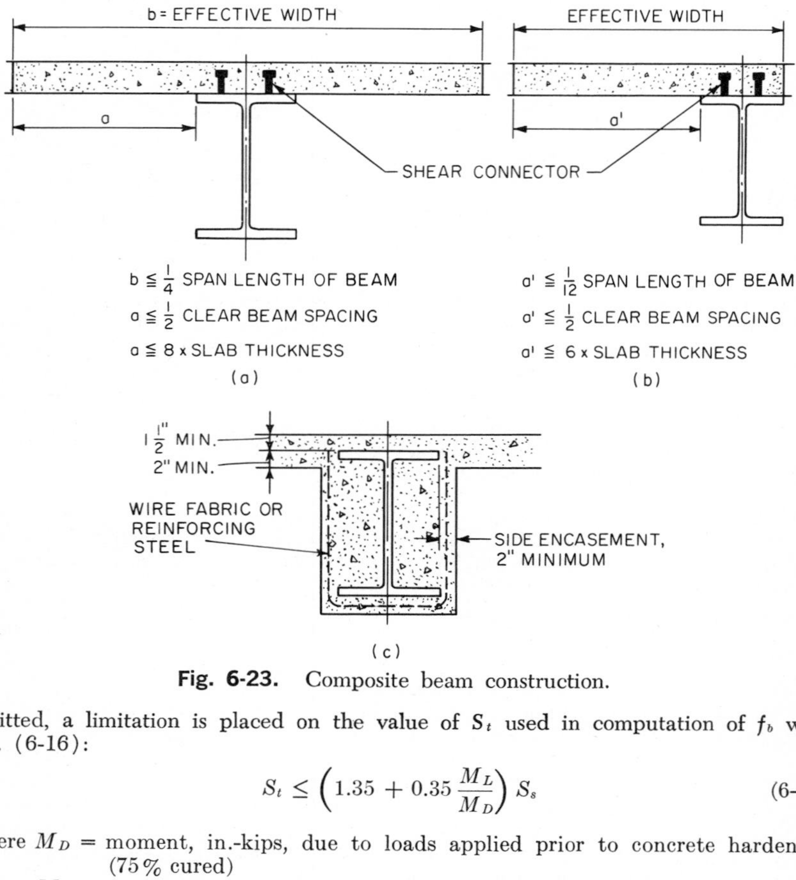

Fig. 6-23. Composite beam construction.

omitted, a limitation is placed on the value of S_t used in computation of f_b with Eq. (6-16):

$$S_t \leq \left(1.35 + 0.35\frac{M_L}{M_D}\right) S_s \tag{6-17}$$

where M_D = moment, in.-kips, due to loads applied prior to concrete hardening (75% cured)

M_L = moment, in.-kips, due to remaining dead and live loads

S_s = section modulus, in.^{3}, of steel beam alone relative to bottom flange

Shear on Connectors. Shear connectors usually are studs or channels. The total horizontal shear to be taken by the connectors between the point of maximum positive moment and each end of a simple beam, or the point of counterflexure in a continuous beam, is the smaller of the values obtained from Eqs. (6-18) and (6-19).

$$V_h = \frac{0.85f_c'A_c}{2} \tag{6-18}$$

$$V_h = \frac{A_sF_y}{2} \tag{6-19}$$

where f_c' = specified strength of concrete, ksi
A_c = actual area of effective concrete flange, as indicated in Fig. 6-23*a* and *b*, sq in.
A_s = area of steel beam, sq in.

These formulas represent the horizontal shear at ultimate load divided by 2 to approximate conditions at working load.

Number of Connectors. The minimum number of connectors N_1, spaced uniformly between the point of maximum moment and adjacent points of zero moment, is V_h/q, where q is the allowable shear load on a single connector, as given in Table 6-15. Values in this table, however, are applicable only to concrete made

Table 6-15. Allowable Horizontal-Shear Loads, q, for Connectors, Kips
(Applicable only to concrete made with ASTM C33 aggregates)

Connector	$f_c' = 3.0$	$f_c' = 3.5$	$f_c' \geq 4.0$
½-in. dia. × 2-in. hooked or headed stud	5.1	5.5	5.9
⅝-in. dia. × 2½-in. hooked or headed stud	8.0	8.6	9.2
¾-in. dia. × 3-in. hooked or headed stud	11.5	12.5	13.3
⅞-in. dia. × 3½-in. hooked or headed stud	15.6	16.8	18.0
3-in. channel, 4.1 lb	4.3w	4.7w	5.0w
4-in. channel, 5.4 lb	4.6w	5.0w	5.3w
5-in. channel, 6.7 lb	4.9w	5.3w	5.6w

w = length of channel, in.

with aggregates conforming to ASTM C33. For concrete made with rotary-kiln-produced aggregates conforming to ASTM C330 and with concrete weight of 90 pcf or more, the allowable shear load for one connector is obtained by multiplying the values in Table 6-15 by the factors in Table 6-16.

Table 6-16. Shear-load Factors for Connectors in Lightweight Concrete

Air dry weight, pcf, of concrete	90	95	100	110	115	120
Factors for $f'_c \leq 4.0$ ksi	0.73	0.76	0.78	0.83	0.86	0.88
Factors for $f'_c \geq 5.0$ ksi	0.82	0.85	0.87	0.91	0.96	0.99

If a concentrated load occurs between the points of maximum and zero moments, the minimum number of connectors required between the concentrated load and the point of zero moment is given by

$$N_2 = \frac{V_h}{q}\left(\frac{S_t M_c/M - S_s}{S_t - S_s}\right) \qquad (6\text{-}20)$$

where M = maximum moment, in.-kips
M_c = moment, in.-kips, at concentrated load $< M$
S_s = section modulus, in.^{3}, of steel beam relative to bottom flange
S_t = section modulus, in.^{3}, of transformed section of composite beam relative to bottom flange but not to exceed S_t computed from Eq. (6-17)

The allowable shear loads for connectors incorporate a safety factor of about 2.5 applied to ultimate load for the commonly used concrete strengths. Not to be confused with shear values for fasteners given in Art. 6-33, the allowable shear loads for connectors are applicable only with Eqs. (6-18) and (6-19).

Connector Details. Shear connectors, except those installed in ribs of formed steel decks, should have a minimum lateral concrete cover of 1 in. The diameter of a stud connector, unless located directly over the beam web, is limited to 2.5 times the thickness of the beam flange to which it is welded. Minimum center-to-center stud spacing is 6 diameters along the longitudinal axis, 4 diameters transversely. Studs may be spaced uniformly, rather than in proportion to horizontal shear, inasmuch as tests show a redistribution of shear under high loads similar to the stress redistribution in large riveted and bolted joints. Maximum spacing is 8 times the slab thickness.

Slab Compressive Stresses. These seldom are critical, but they should be investigated, especially when the slab lies on only one side of the steel beam (Fig.

6-23*b*). Usually, the effective width of the slab is determined by the criterion: width may be assumed up to eight times the thickness. Thickening the slab, particularly when there is a cover plate on the beam tension flange, generally is economical.

6-36. Criteria for Plate Girders. Allowable stresses for tension, shear, compression, bending, and bearing are the same for plate girders as those assigned in Arts. 6-19, 6-20, 6-23, 6-25, and 6-26. But reductions in allowable stress are required under some conditions, and there are limitations on the proportions of girder components.

Web Depth-thickness Limits. The ratio of the clear distance h between flanges, in., to web thickness t, in., is limited by

$$\frac{h}{t} \leq \frac{14{,}000}{\sqrt{F_y(F_y + 16.5)}} \qquad (6\text{-}21a)$$

where F_y is the specified yield stress of the compression flange steel, ksi (Table 6-17).

Table 6-17. Limiting Depth-Thickness Ratios for Plate-girder Webs

F_y, ksi	h/t Eq. (6-21*a*)	h/t Eq. (6-21*b*)	F_y, ksi	h/t Eq. (6-21*a*)	h/t Eq. (6-21*b*)
36.0	322	333	60.0	207	258
42.0	282	309	65.0	192	248
45.0	266	298	90.0	143	211
50.0	243	283	100.0	130	200
55.0	223	270			

When, however, transverse stiffeners are provided at spacings not exceeding 1.5 times h, the limit on h/t is increased to

$$\frac{h}{t} \leq \frac{2{,}000}{\sqrt{F_y}} \qquad (6\text{-}21b)$$

General Design Method. Plate girders may be proportioned to resist bending on the assumption that the moment of inertia of the gross cross section is effective. No deductions need be made for fastener holes, unless the holes reduce the gross area of either flange by more than 15%. When they do, the excess should be deducted.

Hybrid girders may also be proportioned by the moment of inertia of the gross section when they are not subjected to an axial force greater than 15% of the product of yield stress of the flange steel and the area of the gross section. At any given section, the flanges must have the same cross-sectional area and be made of the same grade of steel.

The allowable compressive bending stress F_b must be reduced from that given in Art. 6-25 when h/t exceeds $760/\sqrt{F_b}$. For greater values of this ratio, the allowable compressive bending stress, except for hybrid girders, becomes

$$F_b' \leq F_b\left[1 - 0.0005\frac{A_w}{A_f}\left(\frac{h}{t} - \frac{760}{\sqrt{F_b}}\right)\right] \qquad (6\text{-}22)$$

where A_w = the web area, sq in.

A_f = the compression flange area, sq in.

For hybrid girders, not only is the allowable compressive bending stress limited to that given by Eq. (6-22), but also the maximum stress in either flange may not exceed

$$F_b' \leq F_b\left[\frac{12 + (A_w/A_f)(3\alpha - \alpha^3)}{12 + 2(A_w/A_f)}\right] \qquad (6\text{-}23)$$

where α = ratio of web yield stress to flange yield stress.

Flange Limitations. The projecting elements of the compression flange must comply with the limitations for b/t given in Art. 6-39. The area of cover plates, where used, should not exceed 0.70 times the total flange area. Partial-length cover plates should extend beyond the theoretical cutoff point a sufficient distance to develop their share of bending stresses at the cutoff point. Preferably for welded plate girders, the flange should consist of a series of plates, which may differ in thickness and width, joined end to end with complete-penetration groove welds.

Bearing Stiffeners. These are required on girder webs at unframed ends. They may also be needed at concentrated loads, including supports. Set in pairs, bearing stiffeners may be angles or plates placed on opposite sides of the web, usually normal to the bending axis. Angles are attached with one leg against the web. Plates are welded perpendicular to the web. The stiffeners should have close bearing against the flanges through which they receive their loads, and should extend nearly to the edges of the flanges.

These stiffeners are designed as columns, with allowable stresses as given in Art. 6-21. The column section is assumed to consist of a pair of stiffeners and a strip of girder web with width 25 times web thickness for interior stiffeners and 12 times web thickness at ends. In computing the effective slenderness ratio Kl/r, use an effective length Kl of at least 0.75 the length of the stiffeners.

Intermediate Stiffeners. With properly spaced transverse stiffeners strong enough to act as compression members, a plate-girder web can carry loads far in excess of its buckling load. The girder acts, in effect, like a Pratt truss, with the stiffeners as struts and the web forming fields of diagonal tension. The following formulas for stiffeners are based on this behavior. Like bearing stiffeners, intermediate stiffeners are placed to project normal to the web and the bending axis, but they may consist of a single angle or plate. They may be stopped short of the tension flange a distance up to four times the web thickness. If the compression flange is a rectangular plate, single stiffeners must be attached to it to prevent the plate from twisting. When lateral bracing is attached to stiffeners, they must be connected to the compression flange to transmit at least 1% of the total flange stress, except when the flange consists only of angles.

The total shear force, kips, divided by the web area, sq in., for any panel between stiffeners should not exceed the allowable shear F_v given by Eqs. (6-24) and (6-25).

Except for hybrid girders, when C_v is less than unity:

$$F_v = \frac{F_y}{2.89}\left[C_v + \frac{1 - C_v}{1.15\sqrt{1 + (a/h)^2}}\right] \leq 0.4F_y \tag{6-24}$$

For hybrid girders or when C_v is more than unity or when intermediate stiffeners are omitted:

$$F_v = \frac{F_y C_v}{2.89} \leq 0.4F_y \tag{6-25}$$

where a = clear distance between transverse stiffeners, in.

h = clear distance between flanges within an unstiffened segment, in.

$C_v = \dfrac{45{,}000k}{F_y(h/t)^2}$ when C_v is less than 0.8

$= \dfrac{190}{h/t}\sqrt{\dfrac{k}{F_y}}$ when C_v is more than 0.8

t = web thickness, in.

$k = 5.34 + 4/(a/h)^2$ when $a/h > 1$

$= 4 + 5.34/(a/h)^2$ when $a/h < 1$

Stiffeners for an end panel or for any panel containing large holes and for adjacent panels should be so spaced that the largest average web shear f_v in the panel does not exceed the allowable shear given by Eq. (6-25).

Intermediate stiffeners are not required when h/t is less than 260 and f_v is less than the allowable stress given by Eq. (6-25). When these criteria are not satisfied, stiffeners should be spaced so that the applicable allowable shear, Eq.

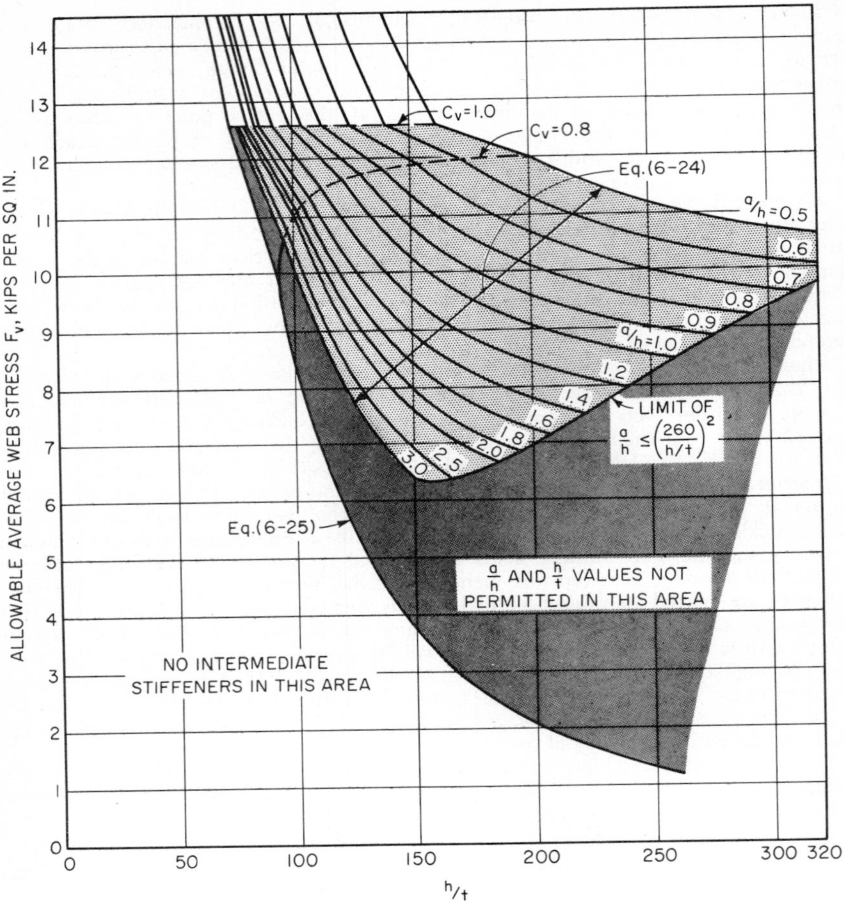

Fig. 6-24. Chart shows relationship between allowable web shears in plate girders, with $F_y = 36.0$, and web thickness, distance between flanges, and stiffener spacing.

(6-24) or (6-25), is not exceeded, and in addition, so that a/h is not more than $[260/(h/t)]^2$ or 3.

Solution of the preceding formulas for stiffener spacing requires assumptions of dimensions and trials. The calculations can be facilitated by using tables in the AISC "Manual of Steel Construction." Also, Fig. 6-24 permits rapid selection of the most efficient stiffener arrangement, for webs of A36 steel. Similar charts can be drawn for other steels.

If the tension field concept is to apply to plate girder design, care is necessary to insure that the intermediate stiffeners function as struts. When these stiffeners are spaced to satisfy Eq. (6-24), their gross area, sq in. (total area if in pairs) should be at least

$$A_{st} = \frac{1 - C_v}{2}\left[\frac{a}{h} - \frac{(a/h)^2}{\sqrt{1 + (a/h)^2}}\right] YDht \qquad (6\text{-}26)$$

where Y = ratio of yield stress of web steel to yield stress of stiffener steel
D = 1.0 for stiffeners in pairs
= 1.8 for single-angle stiffeners
= 2.4 for single-plate stiffeners

When the greatest shear stress f_v in a panel is less than F_v determined from Eq. (6-24), the gross area of the stiffeners may be reduced in the ratio f_v/F_v.

The moment of inertia of a stiffener or pair of stiffeners, about the web axis, should be at least $(h/50)^4$. The connection of these stiffeners to the web should be capable of developing shear, in kips per lin in. of single stiffener or pair, of at least

$$f_{vs} = h\sqrt{\left(\frac{F_y}{340}\right)^3} \tag{6-27}$$

where F_y is the yield stress of the web steel (Table 6-18). This shear also may be reduced in the ratio f_v/F_v, as above.

Table 6-18. Required Shear Capacity of Intermediate-Stiffener Connections to Girder Web

F_y, ksi	f_{vs}, kips per lin in.	F_y, ksi	f_{vs}, kips per lin in.
36.0	$0.034h$	60.0	$0.074h$
42.0	$0.043h$	65.0	$0.084h$
45.0	$0.048h$	90.0	$0.136h$
50.0	$0.056h$	100.0	$0.160h$
55.0	$0.065h$		

Combined Stresses in Web. A check should be made for combined shear and bending in the web where the tensile bending stress is approximately equal to the maximum permissible. When f_v, the shear force at the section divided by the web area, is greater than that permitted by Eq. (6-25), the tensile bending stress in the web should be limited to no more than $0.6F_y$ or $F_y(0.825 - 0.375f_v/F_v)$ where F_v is the allowable web shear given by Eq. (6-24) or (6-25). For girders with A514 steel flanges and webs, when the flange bending stress is more than 75% of the allowable, the allowable shear stress in the web should not exceed that given by Eq. (6-25).

Also, the compressive stresses in the web should be checked (see Art. 6-23).

6-37. Design of Trusses. A truss is a framework arranged so as to form a series of rigid triangles, the bending moments being translated into axial stresses. The assumptions that these stresses are applied along the center of gravity of each member and that these stress lines meet at a point are not always precisely true. Steel angles, for example, in bolted construction are located so that gage lines, not gravity axes, are on the stress lines. Naturally, for wide angles with two gage lines, the line closest to the gravity axis is centered on the stress line to minimize eccentricity. Secondary stresses caused by such eccentricity should be avoided or minimized.

As a general rule, members framing into trusses are located at **panel points**—the intersections of three or more members. When loads are introduced between panel points, the truss members subjected to them must be designed for combined axial stress and bending.

The design problems encountered with trusses are the same as the basic problems of designing for tension, compression, and combined stresses (Arts. 6-19, 6-21, 6-27, and 6-28) plus connections at the panel points.

DIMENSIONAL AND DEFLECTION LIMITATIONS

Allowable stresses are based on certain minimum sizes of structural members and their elements that make possible full development of strength before premature buckling occurs. The higher the allowable stresses the more stringent must be the dimensional restrictions to preclude buckling or excessive deflections.

6-38. Slenderness Ratios. (See also Art. 6-22.) The American Institute of Steel Construction "Specification for the Design, Fabrication and Erection of Structural Steel for Buildings" limits the effective slenderness ratio Kl/r to 200 for columns, struts, and truss members, where K is the ratio of effective length to actual unbraced length l, and r is the least radius of gyration.

A practical rule also establishes limiting slenderness ratios l/r for tension members:

For main members 240
For bracing and secondary members 300

But this does not apply to rods or other tension members that are drawn up tight (prestressed) during erection. The purpose of the rule is to avoid objectionable slapping or vibration in long, slender members.

6-39. Width-thickness Ratios. The AISC specifies several restricting ratios for compression members. One set applies to projecting elements subjected to axial compression or compression due to bending. Another set applies to compression elements supported along two edges.

Figure 6-25 lists maximum width-thickness ratios, b/t, for commonly used elements and grades of steel. Tests show that when b/t of elements normal to the direction of compressive stress does not exceed these limits, the member may be stressed close to the yield stress without failure by local buckling. Since the allowable

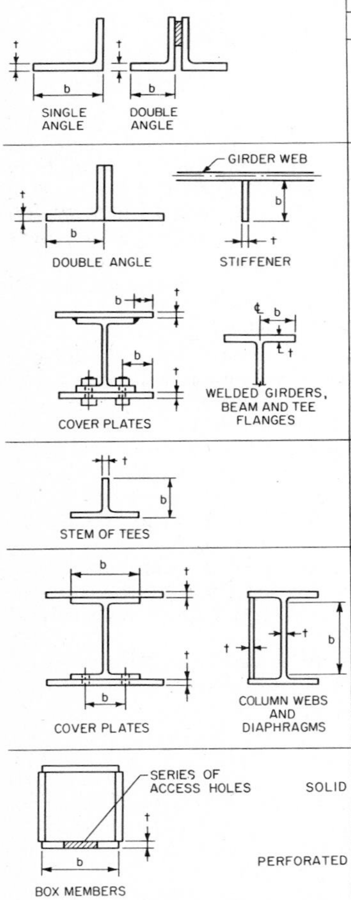

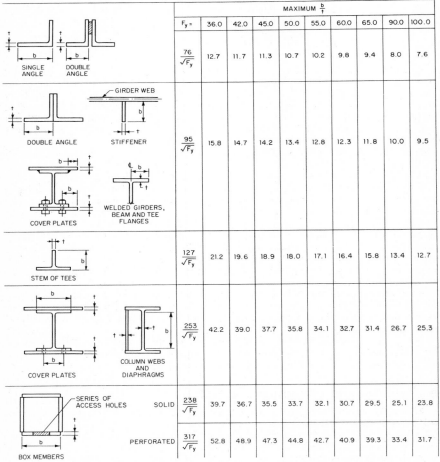

Element		Maximum $\frac{b}{t}$, F_y =	36.0	42.0	45.0	50.0	55.0	60.0	65.0	90.0	100.0
Single angle; double angle		$\frac{76}{\sqrt{F_y}}$	12.7	11.7	11.3	10.7	10.2	9.8	9.4	8.0	7.6
Double angle; stiffener; cover plates; welded girders, beam and tee flanges		$\frac{95}{\sqrt{F_y}}$	15.8	14.7	14.2	13.4	12.8	12.3	11.8	10.0	9.5
Stem of tees		$\frac{127}{\sqrt{F_y}}$	21.2	19.6	18.9	18.0	17.1	16.4	15.8	13.4	12.7
Cover plates; column webs and diaphragms		$\frac{253}{\sqrt{F_y}}$	42.2	39.0	37.7	35.8	34.1	32.7	31.4	26.7	25.3
Box members	SOLID	$\frac{238}{\sqrt{F_y}}$	39.7	36.7	35.5	33.7	32.1	30.7	29.5	25.1	23.8
Box members	PERFORATED	$\frac{317}{\sqrt{F_y}}$	52.8	48.9	47.3	44.8	42.7	40.9	39.3	33.4	31.7

Fig. 6-25. Maximum width-thickness ratios for compression elements.

stress increases with F_y, the specified yield stress of the steel, width-thickness ratios are less for higher-strength steels.

These b/t ratios should not be confused with the width-thickness ratios described in Art. 6-25. There, more restrictive conditions are set in defining compact sections qualified for higher allowable stresses.

6-40. Limits on Deflections. The AISC "Specification for the Design, Fabrication and Erection of Structural Steel for Buildings" restricts the maximum live-load deflection of beams and girders supporting plastered ceilings to ⅟₃₆₀ of the span. This rule is intended to prevent cracking of monolithic plaster; therefore, it is not necessarily applicable to members with unfinished ceilings or ceilings of other materials.

All other deflection problems are left to the designer; it is impracticable to cover with a simple specification all the possible variations in loads, occupancies, and tolerable amounts of movement.

Minimum Depth-span Ratios. As a guide, Table 6-19 lists suggested minimum depth-span ratios for various loading conditions and yield strengths of steel up to

Table 6-19. Suggested Minimum Depth-Span Ratios for Beams

Specific beam condition	F_y, ksi				Deflection at	
	36.0	42.0	45.0	50.0	$0.60F_y$	$0.66F_y$
Heavy shock or vibration	$\frac{1}{18}$	$\frac{1}{15.5}$	$\frac{1}{14.5}$	$\frac{1}{13}$	$\frac{l}{357}$	$\frac{l}{324}$
Heavy pedestrian traffic	$\frac{1}{20}$	$\frac{1}{17}$	$\frac{1}{16}$	$\frac{1}{14.5}$	$\frac{l}{320}$	$\frac{l}{291}$
Normal loading	$\frac{1}{22}$	$\frac{1}{19}$	$\frac{1}{18}$	$\frac{1}{16}$	$\frac{l}{290}$	$\frac{l}{264}$
Beams for flat roofs*	$\frac{1}{25}$	$\frac{1}{21.5}$	$\frac{1}{20}$	$\frac{1}{18}$	$\frac{l}{258}$	$\frac{l}{232}$
Roof purlins, except for flat roofs*	$\frac{1}{28}$	$\frac{1}{24}$	$\frac{1}{22}$	$\frac{1}{20}$	$\frac{l}{232}$	$\frac{l}{210}$

* Investigate for stability against ponding. l = span of beam.

F_y = 50.0 ksi. These may be useful for estimating or making an initial design selection. Since maximum deflection is a straight-line function of maximum bending stress f_b and therefore is nearly proportional to F_y, a beam of steel with F_y = 100.0 ksi would have to be twice the depth of a beam of steel with F_y = 50.0 ksi when each is stressed to allowable values and has the same maximum deflection.

Vibration of large floor areas that are usually free of physical dampeners such as partitions may occur in buildings, such as shopping centers and department stores, where pedestrian traffic is heavy. The minimum depth-span ratios suggested for "heavy pedestrian traffic" are intended to provide an acceptable solution.

One rule of thumb, often used to determine beam depth quickly, is that the depth, in., equals one-half the span length, ft. This, however, is accurate only for the case in which F_b = 24.0 ksi, which, in general, holds for A36 steel.

Ponding. Beams for flat roofs may require a special investigation to assure stability against water accumulation, commonly called ponding, unless there is adequate provision for drainage during heavy rainfall. The AISC "Specification for the Design, Fabrication and Erection of Structural Steel for Buildings" gives these criteria for stable roofs:

$$C_p + 0.9C_s \leq 0.25$$

$$I_d \geq \frac{25S^4}{10^6} \qquad \text{(6-28a)}$$

where $C_p = 32L_sL_p^4/10^7I_p$
$C_s = 32SL_s^4/10^7I_s$
L_p = column spacing in direction of girder, ft (length of primary members)
L_s = column spacing perpendicular to direction of girder, ft (length of secondary member)
S = spacing of secondary members, ft
I_p = moment of inertia for primary members, in.^{4}
I_s = moment of inertia for secondary member, in.^{4} Where a steel deck is supported on primary members, it is considered the secondary member. Use $0.85I_s$ for joists and trusses
I_d = moment of inertia of a steel deck supported on secondary members, in.^{4} per ft

Uniform-load Deflections. For the common case of a uniformly loaded simple beam loaded to the maximum allowable bending stress, the deflection in inches may be computed from

$$\delta = \frac{5}{24}\frac{F_b l}{Ed/l} \tag{6-28b}$$

where F_b = the bending stress, ksi
l = the span, in.
E = 29,000 ksi
d/l = the depth-span ratio

Camber. Trusses of 80-ft or greater span should be cambered to offset dead-load deflections. Crane girders 75 ft or more in span should be cambered for deflection under dead load plus one-half live load.

PLASTIC DESIGN OF STEEL

6-41. Plastic-design Criteria. The AISC "Specification for the Design, Fabrication and Erection of Structural Steel for Buildings" permits plastic design for simple or continuous beams and braced or unbraced planar rigid frames. The design method is applicable to all structural steels listed in Table 6-1 for which F_y does not exceed 65.0 ksi.

Plastic design is not recommended for parts of structures subjected to fatigue-producing loads, such as continuous crane-runway girders, although rigid frame supports for them may be plastic designed.

Load Factors. A distinguishing characteristic of plastic design is the absence of controlling allowable stresses, which are basic to elastic design. Instead, conventional safety factors are maintained through use of load factors by which the design loads are multiplied to determine the required capacity at ultimate load. (See example, Art. 6-42.)

The AISC Specification recommends two load factors:

1.70 for gravity loads (dead plus live).
1.30 for gravity loads plus wind or seismic forces.

Braced Frames. A moderate-height multistory framing system is considered braced when securely attached shear walls (interior and perimeter), floor slabs, and roof decks function together with the steel framework (Art. 6-48).

For taller buildings, the steel frame by itself must be braced to maintain stability under the effects of both gravity and horizontal loads. This is achieved either by a system of diagonal bracing or by a moment-resisting frame in which the beam-to-column connections are rigid (Art. 6-47). The members (columns, beams, diagonals) comprising the vertical bracing system can be analyzed as a simply connected vertical-cantilever truss for frame buckling and lateral stability. But in that case, the axial force in a member caused by the factored loading may not exceed $0.85P_y$, where P_y is the axial-load capacity at yield, the yield stress times the area of the member. Girders and beams in the bracing system must also satisfy Eq. (6-29*a*) for beam-columns, with P_{cr} taken as the maximum axial strength of the beam based on actual l/r between braced points in the plane of bending.

Unbraced Frames. Unbraced multistory frames are permitted when the effect of frame instability and column axial deformation is considered in the analysis. Such frames must be stable for both of the aforementioned factored loadings. In this case, the axial column force may not exceed $0.75P_y$.

Limits on Width-thickness Ratios. In plastic design, the limits on width-thickness ratios for compression elements are more restrictive than corresponding ratios for elastic design (Art. 6-39). The latter ratios insure safety against local buckling of compression elements up to the yield stress, whereas the ratios for plastic design are set to insure adequate hinge rotation at ultimate loads and throughout the plastic range. The maximum width-thickness ratios permitted for compression elements are listed in Table 6-20.

Table 6-20. Maximum Width-thickness and Depth-thickness Ratios for Plastic Design

	Yield stress F_y						
	36.0	42.0	45.0	50.0	55.0	60.0	65.0
b/t_f* for flanges of W beams and similar built-up single-web sections	8.5	8.0	7.4	7.0	6.6	6.3	6.0
b/t* for flange plates in box sections	31.7	29.3	28.3	26.9	25.6	24.5	23.6
d/t_w* for webs of members subjected to bending:							
When $P/P_y \geq 0.27$†	42.8	39.6	38.3	36.3	34.7	33.2	31.9
When $P/P_y = 0$‡	68.5	63.5	61.3	58.2	55.5	53.2	51.1

* b = projection, in., of compression flange of W and similar sections or width, in. (distance between longitudinal lines of fasteners or welds) of compression flange of box sections

t_f = thickness, in., of that flange in W and similar beams

t = thickness, in., of that flange in box sections

d = depth, in., of web

t_w = thickness, in., of web

† Depth-thickness ratios based on $257/\sqrt{F_y}$. P = applied axial load, kips, and P_y = yield stress, ksi, times cross-sectional area of member.

‡ $d/t_w = (1 - 1.4P/P_y)412/\sqrt{F_y}$ when $0 \leq P/P_y \leq 0.27$; use straight-line interpolation between values in table for $P/P_y = 0$ and $P/P_y \geq 0.27$.

Columns. The maximum strength of an axially loaded column, kips, is given by

$$P_{cr} = 1.7AF_a \tag{6-29a}$$

where A = gross area of column, sq in.

F_a = allowable stress, ksi, given by Eq. (6-1)

The factored axial load, kips, may not exceed P_{cr}.

For combined axial and bending forces, a column must satisfy the following interaction formulas:

$$\frac{P}{P_{cr}} + \frac{C_m M}{M_m(1 - P/P_e)} \leq 1.0 \tag{6-29b}$$

$$\frac{P}{P_y} + \frac{M}{1.18M_p} \leq 1.0M \leq M_p \tag{6-29c}$$

where M_p = plastic moment, in.-kips $= F_y Z$

F_y = steel yield stress, ksi

Z = plastic section modulus, in.³, of column

P = applied axial load, kips

$P_e = (23/12)AF'_e$

F'_e = stress defined for Eq. (6-12)

M = maximum applied moment, in.-kips, produced by applied loads

C_m = coefficient defined for Eq. (6-12)

M_m = critical moment in absence of axial load, in.-kips

= M_p for column braced in weak direction

= $M_p[1.07 - (l/r_y)\sqrt{F_y}/3{,}160] \leq M_p$ for column not braced in weak direction. Additional information is given in the "Commentary on the AISC Specification"

l/r_y = slenderness ratio about weak axis

See also following discussion of lateral bracing.

Shear. Webs of beams, girders, columns, and large connections (e.g., rigid-frame knees) should be proportioned for shear so that

$$V_u \leq 0.55F_y td \tag{6-30}$$

where V_u = shear, kips, for the factored loads
d = depth of member, in.
t = thickness of web, in.

Should the computed shear exceed V_u, the web may be reinforced with stiffeners or doubler plates.

Web Stiffeners. These are required at points of load application where a plastic hinge would form. Conditions governing web thickness and the need for stiffeners for members subjected to concentrated loads delivered by flanges of a member framing into it are summarized in Art. 6-23. The rules are applicable to both elastic and plastic design methods.

Beams. The plastic bending capacity of a beam is given by

$$M_p = ZF_y \tag{6-31}$$

where M_p = moment, in.-kips, at which a plastic hinge would form
F_y = specified minimum yield stress, ksi
Z = plastic section modulus, in.^{3}, about beam axis under consideration. The AISC "Manual of Steel Construction" lists Z for both the x and y axes of all rolled-beam sections

Lateral Bracing. An important consideration in plastic design of structural members is the adequacy of lateral bracing. To insure behavior as predicted by plastic theory, lateral bracing must be provided to resist lateral and torsional displacements at plastic-hinge locations. Also, the laterally unsupported distance l_{cr} between each hinge and an adjacent braced point must satisfy the criteria:

$$\frac{l_{cr}}{r_y} = \frac{1{,}375}{F_y} + 25 \qquad 1.0 > \frac{M}{M_p} > -0.5 \tag{6-32a}$$

$$\frac{l_{cr}}{r_y} = \frac{1{,}375}{F_y} \qquad -0.5 \geq \frac{M}{M_p} > -1.0 \tag{6-32b}$$

where r_y = radius of gyration, in., of member about its weak axis
M = lesser of the moments at ends of unbraced segment, in.-kips
M_p = moment, in.-kips, at which a hinge would form at end of member
M/M_p = end-moment ratio, positive when segment is bent in reverse curvature and negative when bent in single curvature

Members built into a masonry wall, with web perpendicular to the wall, are considered laterally supported.

Fasteners and Welds. Capacities of high-strength bolts, in both bearing and friction-type connections, A307 bolts, rivets, and welds for resisting forces computed for factored loading may be taken equal to 1.70 times the capacities determined with allowable unit stresses used for elastic design.

In general, other aspects of plastic design follow the criteria recommended for elastic design. Additional information is given in the AISC Specification and Commentary. (See also Bibliography, Art. 6-92.)

6-42. Comparison of Elastic and Plastic Design Methods. To illustrate the fundamental difference between elastic design, based on allowable stresses, and plastic design, based on load factors, the following simple example is presented in which a laterally supported beam with both ends fixed is designed by both methods.

Problem: Given total bending moment $WL/8$ = 234 ft-kips, where W is the total uniform gravity loading, kips, and L is the span, ft. Assume both ends are fixed. Select a compact beam section of A36 steel.

Solution by elastic theory: The maximum positive moment is

$$\frac{WL}{24} = \frac{1}{3} \times \frac{WL}{8} = \frac{234}{3} = 78 \text{ ft-kips}$$

The maximum negative moment is $WL/12 = ⅔ \times WL/8 = 156$ ft-kips (Fig. 6-26).

As noted in Art. 6-25, a limited redistribution of moments is permitted by the AISC specification. The negative end moments may be reduced 10% if the positive moment is increased by 10% of the average end moments. Thus, the adjusted end negative moments are $156 - 0.1 \times 156 = 140.4$, and the adjusted positive moment is $78 + 15.6 = 93.6$, since both end moments are equal. The negative moment, 140.4 ft-kips, governs. So with the allowable bending stress, $F_b = 24$ kips per sq in., the required elastic section modulus is

$$S_x = 140.4 \times {}^{12}\!/\!_{24} = 70.2$$

Since a W16 × 40 has $S_x = 64.6$, whereas a W16 × 45 has $S_x = 72.5$, use the W16 × 45.

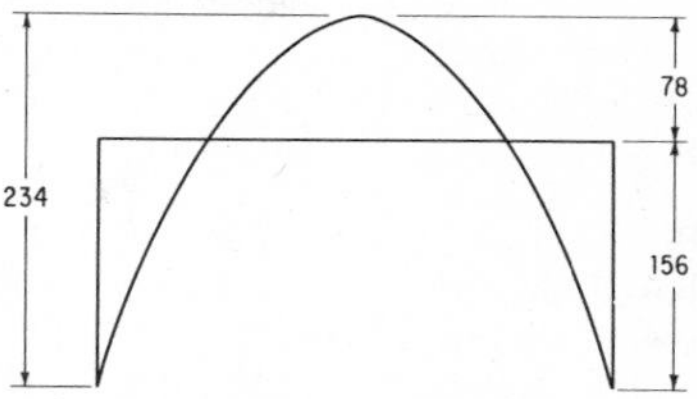

Fig. 6-26. Bending moment diagram for a fixed-end beam with $WL/8 = 234$.

Solution by plastic theory: Assume plastic hinges to form at both ends and at the midpoint. The plastic moment M_p is the same at all those points and thus must be equal to half the total, or simple-beam, bending moment. With the load factor for continuous beams of 1.70, $M_p = 1.70WL/16 = 1.70 \times 117$. Then, the required plastic section modulus is

$$Z_x = \frac{M_p}{F_y} = 1.70 \times 117 \times \frac{12}{36} = 66.3$$

Since a W16 × 36 has $Z_x = 63.9$ and a W16 × 40 has $Z_x = 72.7$, and other sections with a plastic modulus between those values are heavier, use the W16 × 40. The weight saving compared with elastic design is 11%.

BRACING

Bracing as it applies to steel structures includes secondary members incorporated into the system of main members to serve these principal functions:

1. Slender compression members, such as columns, beams, and truss elements, are braced, or laterally supported, so as to restrain the tendency to buckle in a direction normal to the stress path. The rigidity, or resistance to buckling, of an individual member, is determined from its length and certain physical properties of its cross section. Economy and size usually determine whether bracing is to be employed.

2. Since most structures are assemblies of vertical and horizontal members forming rectangular (or square) panels, they possess little inherent rigidity. Consequently, additional rigidity must be supplied by a secondary system of members or by rigid or semirigid joints between members. This is particularly necessary when the framework is subject to lateral loads, such as wind, earthquakes, and moving loads. Exempt from this second functional need for bracing are trusses, which are basically an arrangement of triangles possessing an inherent ideal rigidity both individually and collectively.

3. There is a need for bracing frequently to resist erection loads and to align or prevent overturning of trusses, bents, or frames during erection. Such bracing may be temporary; however, usually bracing needed for erection is also useful in supplying rigidity to the structure and therefore is permanently incorporated into the building. For example, braces that tie together adjoining trusses and prevent their overturning during erection are useful to prevent sway—even though the swaying forces may not be calculable.

6-43. Column Bracing. Interior columns of a multistory building are seldom braced between floor connections. Bracing of any kind generally interferes with occupancy requirements and architectural considerations. Since the slenderness ratio l/r in the weak direction usually controls column size, greatest economy is achieved by using only wide-flange column sections or similar built-up sections.

It is frequently possible to reduce the size of wall columns by introducing knee braces or struts in the plane of the wall, or by taking advantage of deep spandrels or girts that may be otherwise required. Thus the slenderness ratio for the weak and strong axis can be brought into approximate balance. The saving in column weight may not always be justified; one must take into account the weight of additional bracing and cost of extra details.

Column bracing is prevalent in industrial buildings because greater vertical clearances necessitate longer columns. Tall slender columns may be braced about both axes to obtain an efficient design.

Undoubtedly, heavy masonry walls afford substantial lateral support to steel columns embedded wholly or partly in the wall. The general practice, however, is to disregard this assistance.

An important factor in determining column bracing is the allowable stress for the column section (Art. 6-21). Column formulas for obtaining this stress are based on the ratio of two variables, effective length Kl and the physical property called radius of gyration r. For bracing for plastic design, see Art. 6-41.

The question of when to brace (to reduce the unsupported length and thus slenderness ratio) is largely a matter of economics and architectural arrangements; thus no general answer can be given.

Composite Columns. Concrete encasement (when used) is usually credited with partial lateral support. This is reflected in an increase in allowable load-carrying capacity. For example, the American Concrete Institute Building Code permits the strength of concrete columns with a structural-steel core (composite compression members) to be computed for the same limiting conditions applicable to ordinary reinforced-concrete members. Thus, for factored loads, the capacity of the column is determined by the yield stress of the structural steel, the compressive strength of the concrete ($0.85f'_c$), the yield stress of longitudinal reinforcing steel, critical buckling load, and slenderness ratio. But the minimum thickness of the concrete encasement around the core should be at least 1½ in. for precast and 2 in. for cast-in-place concrete. The concrete should have a strength f'_c of at least 2.5 ksi. Design yield stress for the structural steel core may not exceed 50 ksi. The concrete encasement must be spiral bound or laterally tied. In addition, longitudinal reinforcing bars with a total cross-sectional area between 1 and 8% of the net concrete section must be placed within the ties or spiral. (These bars may be taken into account in computing the area and moment of inertia of the core for spiral-bound columns, but not the moment of inertia for tied columns.) Then, the radius of gyration, in., of the composite member may be taken as

$$r = \sqrt{\frac{0.2E_cI_g + E_sI_t}{0.2E_cA_g + E_sA_t}} \tag{6-33a}$$

where E_c = modulus of elasticity of concrete, ksi
E_s = modulus of elasticity of structural steel core, ksi
I_g = moment of inertia, in.^{4}, of gross concrete section about centroidal axis, neglecting core and reinforcement
I_t = moment of inertia, in.^{4}, of structural steel about centroidal axis of member cross section
A_g = gross area of composite column, sq in.
A_t = area of structural steel, sq in.

In computation of the buckling load $P_c = \pi^2EI/(Kl)^2$, where K is the effective length factor and l the unbraced length, in., EI of the composite section may be taken as

$$EI = \frac{0.2E_cI_g + E_sI_t}{1 + \beta_d} \tag{6-33b}$$

where β_d = ratio of maximum design dead-load moment to maximum design total-load moment, always positive

6-44. Beam Bracing. Economy in size of member dictates whether laterally unsupported beams should have additional lateral support between end supports. Lateral support at intermediate points should be considered whenever the allowable stress obtained from the reduction formulas for large l/r_T given in Fig. 6-20 falls below some margin, say 25%, of the stress allowed for the fully braced condition.

There are cases, however, where stresses as low as 4.0 ksi have been justified, because intermediate lateral support was impractical.

The question often arises: When is a steel beam laterally supported? There is no fixed rule in specifications (nor any intended in this discussion) because the answer requires application of sound judgment based on experience. Tests and studies that have been made indicate that it takes rather small forces to balance the lateral thrusts of initial buckling.

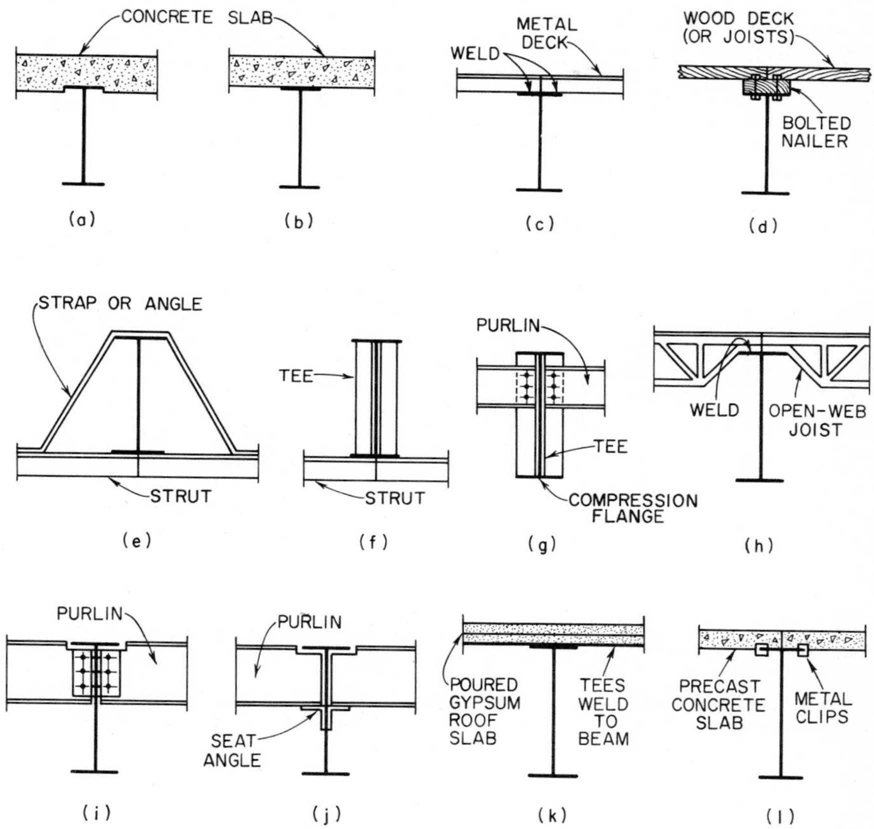

Fig. 6-27. Lateral support for beams.

In Fig. 6-27 are illustrated some of the common situations encountered in present-day practice. In general, positive lateral support is provided by:

(*a*) and (*b*). All types of cast-in-place concrete slabs (questionable for vibrating loads and loads hung on bottom flange).

(*c*). Metal and steel plate decks, welded connections.

(*d*). Wood decks nailed securely to nailers bolted to beam.

(*e*) and (*f*). Beam flange tied or braced to strut system, either as shown in (*e*) or by means of cantilever tees, as shown in (*f*); however, struts should be adequate to resist rotation.

(*g*) Purlins used as struts, with tees acting as cantilevers (common in rigid frames and arches). If plate stiffeners are used, purlins should be connected to them with high-strength bolts to insure rigidity.

(*h*). Open-web joists tack-welded (or the equivalent) to the beams; but the joists themselves must be braced together (bridging), and the flooring so engaged with the flanges that the joists, in turn, are adequately supported laterally.

(i). Purlins connected close to the compression flange.

(k). Tees (part of cast-in-place gypsum construction) welded to the beams.

Doubtful lateral support is provided by:

(j). Purlins seated on beam webs, where the seats are distant from the critical flange.

(l). Precast slabs not adequately fastened to the compression flange.

The reduction formulas for large l/r_T given in Fig. 6-20 do not apply to steel beams fully encased in concrete, even though no other lateral support is provided.

Introducing a secondary member to cut down the unsupported length does not necessarily result in adequate lateral support. The assumed thrust, or rather the resistive capacity of the member, must be traced through the system to ascertain its effectiveness. For example the system in Fig. 6-28*a* may be free to deflect laterally as shown. This can be prevented by a rigid floor system that acts as a diaphragm, or in the absence of a floor, it may be necessary to X brace the system as shown in Fig. 6-28*b*.

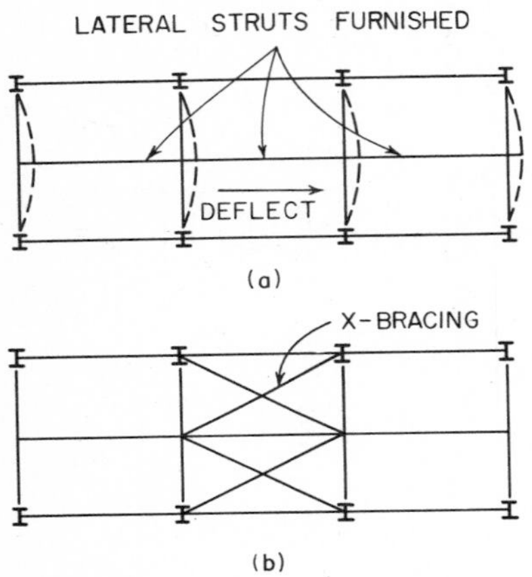

Fig. 6-28. Lateral bracing systems.

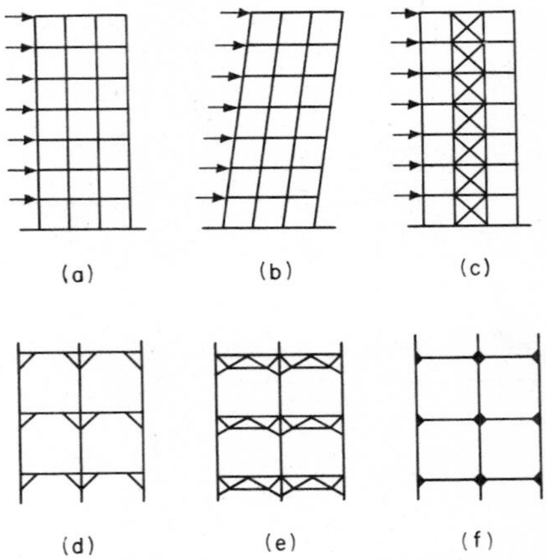

Fig. 6-29. Wind bracing for multistory buildings.

6-45. Capacity of Bracing. For an ideally straight, exactly concentrically loaded beam or column, only a small force may be needed to be supplied by an intermediate brace intended to reduce the unbraced length of a column or the unsupported length of the compression flange of a beam. But there is no generally accepted method of calculating that force.

The only function of a brace is to provide a node point in the buckled configuration. Since no force is considered present at the node, rigidity is the only requirement for the brace. But actual members do contain nonuniform residual stresses and slight initial crookedness and may be slightly misaligned, and these eccentricities create a thrust that must be resisted by the brace.

A rule used by some designers that has proved satisfactory is to design the brace for 2% of the axial load of columns, or 2% of the total compressive stress in beam flanges. Studies and experimental evidence indicate that this rule is conservative.

6-46. Lateral Forces on Building Frames. Design of bracing to resist forces induced by wind, seismic disturbances, and moving loads, such as those caused by cranes is not unlike, in principle, design of members that support vertical dead and live loads. These lateral forces are readily calculable. They are collected at points of application and then distributed through the structural system until delivered to the ground. Wind loads, for example, are collected at each floor level and distributed to the columns that are selected to participate in the system. Such loads are cumulative; that is, columns resisting wind shears must support at any floor level all the wind loads on the floors above the one in consideration.

6-47. Bracing Tall Buildings. If the steel frame of the multistory building in Fig. 6-29*a* is subjected to lateral wind load, it will distort as shown in Fig. 6-29*b*, assuming that the connections of columns and beams are of the standard type, for which rigidity (resistance to rotation) is nil. This can be visualized readily by assuming each joint is connected with a single pin. Naturally, the simplest method to prevent this distortion is to insert diagonal members—triangles being inherently rigid, even if all the members forming the triangles are pin-connected.

Braced Bents. Bracing of the type in Fig. 6-29*c*, called X bracing, is both efficient and economical. Unfortunately, X bracing is usually impracticable because of interference with doors, windows, and clearance between floor and ceiling. Architects of modern office buildings require clear floor areas; this offers flexibility of

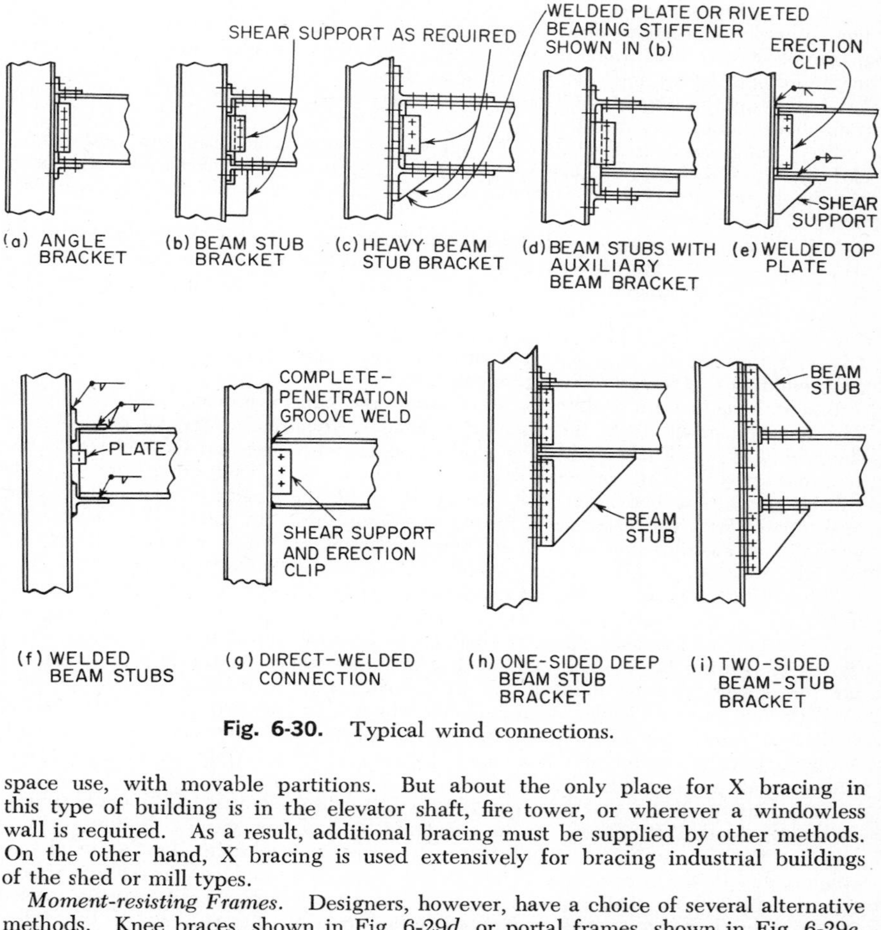

Fig. 6-30. Typical wind connections.

space use, with movable partitions. But about the only place for X bracing in this type of building is in the elevator shaft, fire tower, or wherever a windowless wall is required. As a result, additional bracing must be supplied by other methods. On the other hand, X bracing is used extensively for bracing industrial buildings of the shed or mill types.

Moment-resisting Frames. Designers, however, have a choice of several alternative methods. Knee braces, shown in Fig. 6-29*d*, or portal frames, shown in Fig. 6-29*e*, may be used in outer walls, where they are likely to interfere only with windows. However, the trend to window walls has decreased use of this type of bracing. The type that has emerged for almost uniform application is the bracket type (Fig. 6-29*f*). It consists simply of developing the end connection for the calculated wind moment. Connections vary in type, depending on size of members, magnitude of wind moment, and compactness needed to comply with floor-to-ceiling clearances.

Figure 6-30 illustrates a number of bracket-type wind-braced connections. The minimum type, represented in Fig. 6-30*a*, consists of angles top and bottom; they are ample for moderate-height buildings. Usually the outstanding leg (against the column) is of a size that permits only one gage line; a second line of fasteners would not be effective owing to the eccentricity. When greater moment resistance is needed, the type shown in Fig. 6-30*b* should be considered. This is the type that has become rather conventional in field-bolted construction. Figure 6-30*c* illustrates the maximum size with beam stubs having flange widths that permit additional gage lines, as shown. It is thus possible on larger wide-flange columns to obtain 16 fasteners in the stub-to-column connection.

The resisting moment of a given connection varies with the distance between centroids of the top and bottom connection piece. To increase this distance, thus increasing the moment, an auxiliary beam may be introduced as shown in Fig. 6-30*d*, provided it does not create an interference.

All the foregoing types may be of welded construction, rather than bolted. In fact, it is not unusual to find mixtures of both because of the fabricator's decision to shop-bolt and field-weld, or vice versa. Welding, however, has much to offer in simplifying details and saving weight, as illustrated in Fig. 6-30*e*, *f*, and *g*. The last represents the ultimate efficiency with respect to weight saving, and furthermore, it eliminates interfering details.

Deep wing brackets (Fig. 6-30*h* and *i*) are sometimes used for wall beams and spandrels designed to take wind stresses. Such deep brackets are, of course, acceptable for interior beam bracing whenever the brackets do not interfere with required clearances.

Not all beams need be wind-braced in tall buildings. Usually the wind is concentrated on certain column lines, called *bents*, and the forces are carried through the bents to the ground. For example, in a wing of a building, it is possible to concentrate the wind load on the outermost bent. To do so may require a stiff floor or diaphragmlike system capable of distributing the wind loads laterally. One-half these loads may be transmitted to the outer bent, and one-half to the main building to which the wing connects.

Braced bents are invariably necessary across the narrow dimension of a building. The question arises as to the amount of bracing required in the long dimension, since wind of equal unit intensity is assumed to act on all exposed faces of structures. In buildings of square or near square proportions, it is likely that braced bents will be provided in both directions. In buildings having a relatively long dimension, as compared with width, the need for bracing diminishes. In fact, in many instances, wind loads are distributed over so many columns that the inherent rigidity of the whole system is sufficient to preclude the necessity of additional bracing.

Why all the emphasis on beam-to-column connections and not on column-to-column joints, since both joints are subjected to the same wind forces? Columns are compression members and as such transmit their loads, from section above to section below, by direct bearing between finished ends. It is not likely, in the average building, for the tensile stresses induced by wind loads ever to exceed the compressive pressure due to dead loads. Consequently, there is no theoretical need for bracing a column joint. Actually, however, column joints are connected together with nominal splice plates for practical considerations—to tie the columns during erection and to obtain vertical alignment.

This does not mean that one may always ignore the adequacy of column splices. In lightly loaded structures, or in exceptionally tall but narrow buildings, it is possible for the horizontal wind forces to cause a net uplift in the windward column due to the overturning action. The commonly used column splices should then be checked for their capacity to resist the maximum net tensile stresses caused in the column flanges. This computation and possible heaving up of the splice material may not be thought of as bracing; yet, in principle, the column joint is being "wind-braced" in a manner similar to the wind-braced floor-beam connections.

Tubular Framing. Designs of some very tall buildings employ a structural concept that departs from the foregoing conventional method of wind-bracing steel frames. In these buildings, the perimeter columns and spandrel beams form a theoretical, vertical-cantilever hollow tube that resists the entire lateral load. The floors, through diaphragm action, transfer the lateral forces to the exterior columns. These members are closely spaced and either X-braced or rigidly connected to the spandrel beams to approximate an idealized solid-surface tube. Interior columns merely support their share of the gravity loads. The elimination of floorbeam-to-column moment connections and standardization of floorbeam sizes, freed from accumulative lateral forces, which are characteristic of conventional design, offer substantial savings.

6-48. Shear Walls. It is well known that masonry walls enveloping the steel frame, interior masonry walls, and perhaps some of the stiffer partitions can resist a substantial amount of lateral load. Rigid floor systems participate by distributing the shears induced at each floor level to the columns and walls. Yet, it is common design practice to carry wind loads on the steel frame, little or no credit being given to the substantial resistance rendered by the floors and walls. In the past, some engineers deviated from this conservatism by assigning a portion of the wind loads to the floors and walls; but even then, the steel frame carried the major share. The present trend in multistory building design, particularly for offices, is to walls of glass, thin metallic curtain walls, lightweight floors, and removable partitions. This construction imposes on the steel frame almost complete responsibility for transmittal of wind loads to the ground. Consequently, the importance of wind-bracing tall steel structures should not diminish.

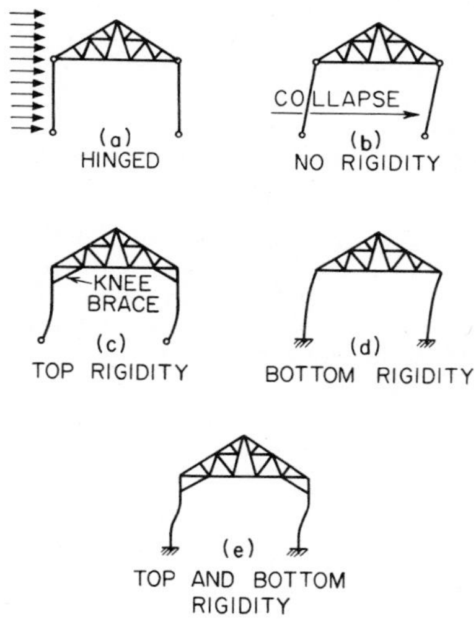

Fig. 6-31. Relative stiffness of mill-building frames depends on restraints on columns.

In fact, in tall, slender, partitioned buildings, such as hotels and apartments, the problem of preventing or reducing to tolerable limits the cracking of rigid-type partitions has been related to the wracking action of the frame caused by excessive deflection. One solution that may be used for exceptionally slender frames (those most likely to deflect excessively) is to supplement the normal bracing of the steel frame with shear walls. Acting as vertical cantilevers in resisting lateral forces, these walls, often constructed of reinforced concrete, may be arranged much like structural shapes, such as plates, channels, Ts, Is, or Hs. (See also Sec. 11.) Concrete walls normally needed for fire towers, elevator shafts, divisional walls, etc., may be extended and reinforced to serve as shear walls, and may relieve the steel frame of cumbersome bracing or uneconomical proportions.

6-49. Bracing Industrial-type Buildings. Bracing of low industrial buildings for horizontal forces presents fewer difficulties than bracing of multistory buildings, because the designer is virtually free to select the most efficient bracing without regard to architectural considerations or interferences. For this reason, conventional X bracing is widely used—but not exclusively. Knee braces, struts, and sway frames are used where needed.

Wind forces acting upon the frame shown in Fig. 6-31*a*, assuming hinged joints at the top and bottom of supporting columns, would cause collapse as indicated in Fig. 6-31*b*. In practice, the joints would not be hinged. However, a minimum-type connection at the truss connection and a conventional column base with anchor bolts located on the axis transverse to the frame would approximate this theoretical consideration of hinged joints. Therefore, the structure requires bracing capable of preventing collapse and undue deflection.

In the usual case, the connection between truss and column will be stiffened by means of knee braces (Fig. 6-31*c*). The rigidity so obtained may be supplemented by providing partial rigidity at the column base by simply locating the anchor bolts in the plane of the bent.

In buildings containing overhead cranes, the knee brace may interfere with crane operation. Then, the problem may be resolved by fully anchoring the column base so that the column may function as a vertical cantilever (Fig. 6-31*d*).

The method often used for very heavy industrial buildings is to obtain substantial rigidity at both ends of the column so that the behavior under lateral load will resemble the condition illustrated in Fig. 6-31*e*. In both (*d*) and (*e*), the footings must be designed for such moments.

A common assumption in wind distribution for the type of light mill building shown in Fig. 6-32 is that the windward column takes a large share of the load acting on the side of the building and delivers it directly to the ground. The remaining wind load on the side is delivered by the same columns to the roof systems, where it joins with the wind forces imposed directly on the roof surface. Then, by means of diagonal X bracing, working in conjunction with the struts and top chords of the trusses, the load is carried to the eave struts, thence to the gables and, through diagonal bracing, to the foundations.

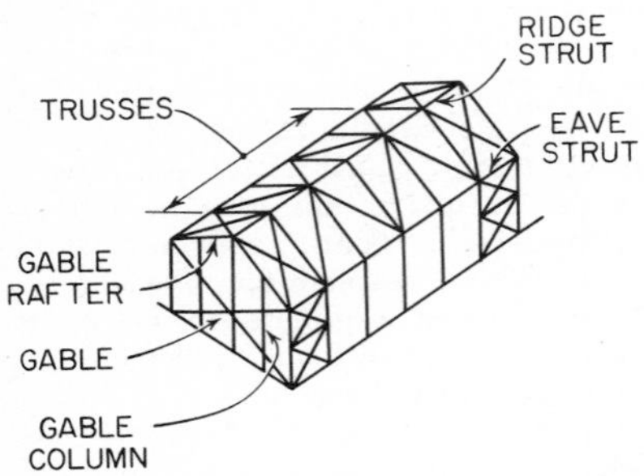

Fig. 6-32. Bracing in factory transmits wind load to ground.

Since wind may blow in all directions, the building also must be braced for the wind load on the gables. This bracing becomes less important as the building increases in length and conceivably could be omitted in exceptionally long structures. The stress path is not unlike that assumed for the transverse wind forces. The load generated on the ends is picked up by the roof system and side framing, delivered to the eaves, and then transmitted by the diagonals in the end side-wall bays to the foundation.

No distribution rule for bracing is intended in this discussion; bracing can be designed many different ways. Whereas the foregoing method would be sufficient for a small building, a more elaborate treatment is suggested for larger structures.

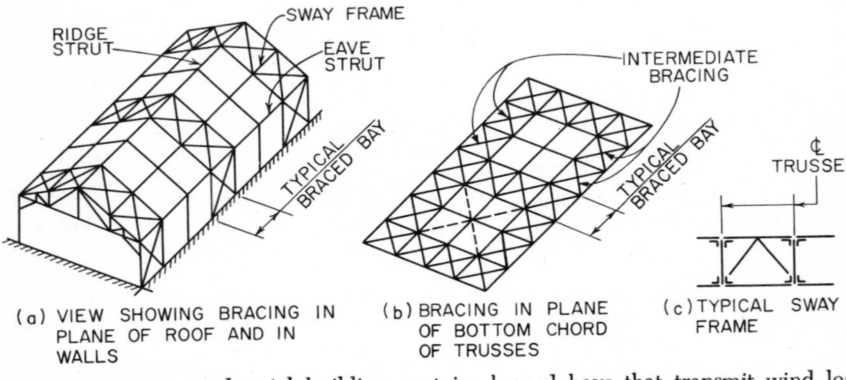

Fig. 6-33. Large industrial building contains braced bays that transmit wind load to ground.

Braced bays, or towers, are usually favored in the better-braced structures, such as in Fig. 6-33. There, a pair of transverse bents are connected together with X bracing in the plane of the columns, plane of truss bottom chords, plane of truss top chords, and by means of struts and sway frames. It is assumed that each such tower can carry the wind load from adjacent bents, the number depending on assumed rigidities, size, span, and also on sound judgment. Usually every third or fourth bent should become a braced bay. Participation of bents adjoining the braced bay can be assured by insertion of bracing designated "intermediate" in

Fig. 6-33*b*. This bracing is of greater importance when knee braces between trusses and columns cannot be used. When maximum lateral stiffness of intermediate bents is desired, it can be obtained by extending the X bracing across the span; this is shown with broken lines in Fig. 6-33*b*.

Buildings with flat or low-pitched roofs, shown in Fig. 6-7*d* and *e*, require little bracing because the trusses are framed into the columns. These columns are designed for the heavy moments induced by wind pressure against the building side. The bracing that would be provided, at most, would consist of X bracing in the plane of the bottom chords for purpose of alignment during erection and a line or two of sway frames for longitudinal rigidity. Alignment bracing is left in the structure since it affords a secondary system for distributing wind loads.

6-50. Bracing Crane Structures. All buildings that house overhead cranes should be braced for the thrusts induced by sidesway and longitudinal motions of the cranes. Bracing used for wind or erection may, of course, be assumed to sustain the lateral crane loadings. These forces are usually concentrated on one bent. Therefore, normal good practice dictates that adjoining bents share in the distribution. Most effective is a system of X bracing located in the plane of the bottom chords of the roof trusses.

In addition, the bottom chords should be investigated for possible compression, although the chords normally are tension members. A heavily loaded crane is apt to draw the columns together, conceivably exerting a greater compression stress than the actual tension stress obtainable under condition of zero snow load. This may indicate the need for intermediate bracing of the bottom chord.

6-51. Design of X Bracing. X bracing may consist of rods (probably the least expensive) or stiffer shapes, such as angles. Rods are unsuitable for compression, whereas an angle may be able to take some compression and thereby aid another diagonal member in tension. Generally, the rod or the angle is designed to take all the stress in tension.

One objection to rods is that they may rattle, particularly in buildings subjected to vibrations. Angle bracing is often larger than need be for the calculated stress. The American Institute of Steel Construction, to assure satisfactory service under normal operation, requires that tension members (except rods) have a minimum slenderness ratio l/r of 300.

6-52. Bracing Rigid Frames. Rigid frames of the type shown in Fig. 6-9*a* have enjoyed popular usage for gymnasiums, auditoriums, mess halls, and with increasing frequency, industrial buildings. The stiff knees at the junction of the column with the rafter imparts excellent transverse rigidity. Each bent is capable of delivering its share of wind load directly to the footings. Nevertheless, some bracing is advisable, particularly for resisting wind loads against the end of the building. Most designers emphasize the importance of an adequate eave strut; it usually is arranged so as to brace the inside flange (compression) of the frame knee, the connection being located at the mid-point of the transition between column and rafter segments of the frame. Intermediate X bracing in the plane of the rafters usually is omitted.

FASTENERS

There are three basic types of fasteners: rivets, bolts, and welds. Rivets, formerly widely used for main connections, are now seldom used in new construction.

Many variables affect the selection of a fastener. Included among these are economy of fabrication and of erection, availability of equipment, inspection problems, labor supply, and such design considerations as fatigue, size and type of connections, continuity of framing, reuse, and maintenance.

Most structural shops are sufficiently versatile to furnish all types of fasteners, but some are limited. For example, a shop arranged for all-welded construction may be short on punching and reaming capacity to furnish bolted construction.

It is not uncommon for steel framing to be connected with such combinations as shop and field welds, or shop welds and field bolts. Even for field fasteners, there may be a combination of welds for the main beam-to-column connections and bolts for the small fill-in beams. The variables affecting the decisions to use these fasteners and combinations of them are too numerous and often too controversial to permit establishment of fixed rules.

6-53. Hot-driven Rivets. Use of field riveting declined decades ago when trained riveters became scarce. Shop riveting, however, continued until the increasing volume of welded construction raised the cost of infrequent riveting to the point where it usually became more economical to substitute high-strength bolts, although the purchase price of bolts is higher. As a consequence, few structural shops now have the necessary equipment for riveted construction.

Available rivets in sizes up to 1½-in. diam conform to ASTM Specification A502, Grades 1 or 2. Grade 1 carbon-steel rivets are intended for use with the lower-strength carbon steels, while Grade 2 manganese-carbon rivets are more efficient for use with the higher-strength carbon and low-alloy, high-strength structural steels.

Allowable stresses for rivets are given in Table 6-12.

6-54. Unfinished Bolts. Known in construction circles by several names—ordinary, common, machine, or rough—unfinished bolts are characterized chiefly by the rough appearance of the shank. They are covered by ASTM Specification A307. They fit into holes 1/16 in. larger in diameter than the nominal bolt diameter. Symbols for bolts are shown in Fig. 6-34.

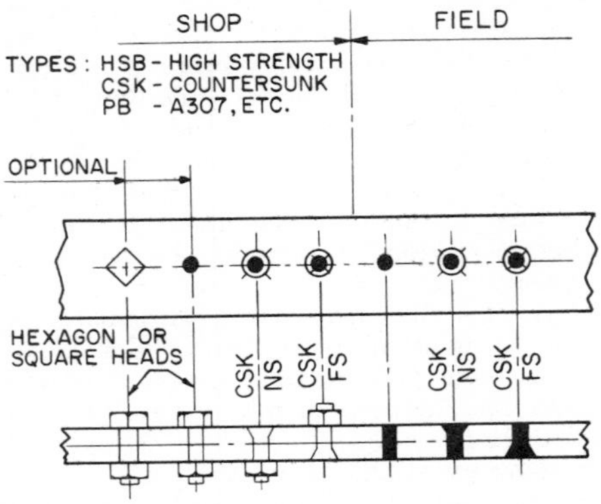

Fig. 6-34. Symbols for shop and field bolts.

Allowable stresses are given in Tables 6-12 and 6-14. The low allowable stress in shear takes into account the possibility of threads in shear planes. Thus, washers used to extend the bolt body are unnecessary.

One advantage of unfinished bolts is the ease of making a connection; only a wrench is required. On large jobs, however, erectors find they can tighten bolts more economically with a pneumatic-powered impact wrench. Power tightening probably yields greater uniformity of tension in the bolts and makes for a better-balanced connection.

While some old building codes restrict unfinished bolts to minor applications, such as small, secondary (or intermediate) beams in floor panels and in certain parts of one-story, shed-type buildings, the AISC, with a basis of many years of experience, permits A307 bolts for main connections on structures of substantial size. For example, these bolts may be used for beam and girder connections to columns in buildings up to 125 ft in height.

There is an economic relation between the strength of a fastener and that of the base material. So while A307 may be economical for connecting steel with a 36.0 ksi yield point, this type of bolt may not be economical with 50.0-ksi yield-point steel. The number of fasteners to develop the latter becomes excessive and perhaps impractical due to size of detail material.

A307 bolts should always be considered for use, even in an otherwise all-welded building, for minimum-type connections, such as for purlins, girts, struts, etc.

6-55. High-strength Bolts. Development of high-strength bolts is vested in the Research Council on Riveted and Bolted Structural Joints of the Engineering

Foundation. Its "Specification for Structural Steel Joints Using A325 or A490 Bolts" was adopted by the American Institute for Steel Construction.

Continuing research and experience have brought many changes since 1951, when this type of bolt was introduced. Initially, high-strength bolts were used as a substitute for rivets because of growing scarcity of trained riveters, and to overcome certain deficiencies inherent in rivets. The initial tension of cooled rivets often is insufficient to resist dynamic and fatigue-type loads; subsequently, rivets loosen and vibrate, eventually requiring replacement. But if that tension could be substantially increased and maintained, it could develop sufficient frictional resistance to slippage in joints. This is the principle utilized in friction-type connections with high-strength bolts. When high-strength bolts are used in a connection, they are highly tensioned by tightening of the nuts and thus tightly clamp together the parts of the connection.

For convenient computation of load capacity, the clamping force and resulting friction are resolved as shear. Bearing between bolt body and connected material

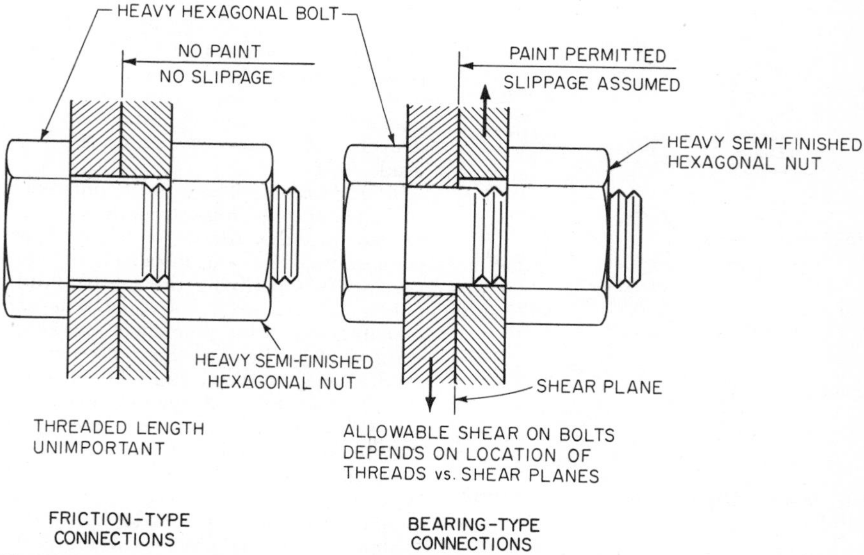

Fig. 6-35. Two main types of connections with high-strength bolts. Although, in general, no paint is permitted on faying surfaces in friction-type connections, the following exceptions are allowed: scored galvanized coatings, inorganic zinc-rich paint, metallized zinc or aluminum coatings.

is not a factor until loads become large enough to cause slippage between the parts of the connection. The bolts are assumed to function in shear following joint slippage into full bearing, but have much higher resistance than rivets because of the higher-strength steel used for these bolts.

The clamping and bearing actions lead to the dual concept: friction-type connections and bearing-type connections. For the latter, the allowable shear depends on the cross-sectional bolt area at the shear plane. Hence, two shear values are assigned, one for the full body area and one for the reduced area at the threads (Fig. 6-35). Allowable stresses are given in Tables 6-12 to 6-14.

Bolt Holes. Diameter of holes are 1/16 in. larger than the bolt diameter. For friction-type connections and when approved by the designer, however, hole diameters may be larger than bolt diameter as follows:

3/16 in. larger for bolts 7/8-in. diam. and under.
1/4 in. for 1-in. diam. bolts.
5/16 for bolts 1 1/8-in. diam. and over

Table 6-21. Washer Requirements for High-strength Bolts

Method of tensioning	A325 bolts	A490 bolts	
		Base material $F_y < 40.0$	Base material $F_y > 40.0$
Calibrated wrench	One washer under turned element	Two washers	One washer under turned element
Turn-of-the-nut	None	Two washers	One washer under turned element
Both methods, slotted and oversized holes	Two washers	Two washers	Two washers

Slotted holes, usually required for expansion or adjustment, are allowed, if their length does not exceed 2½ times bolt diameter and their width is not more than 1/16 in. greater than bolt diameter. In bearing-type connections, however, the force being resisted must be in the direction transverse to the slot length.

Washers. Washer requirements, depending on method of installation and diameter of holes, are summarized in Table 6-21.

Identification. There is no difference in appearance of high-strength bolts intended for either friction or bearing-type connections. To aid installers and inspectors in identifying the several available grades of steel, bolts and nuts are manufactured with permanent markings (Fig. 6-36). Symbols for bolts for use on drawings are shown in Fig. 6-34.

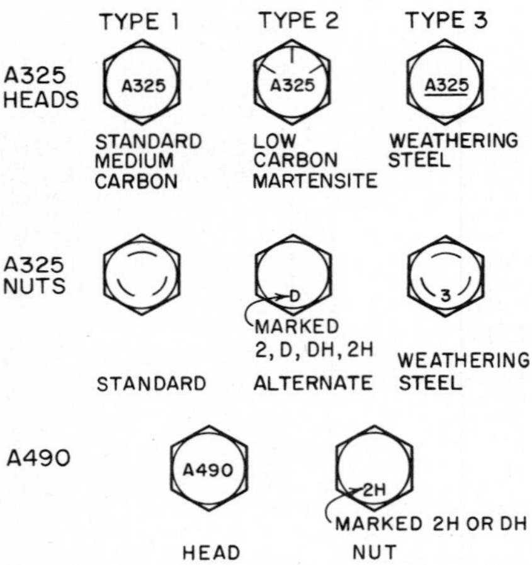

Fig. 6-36. High-strength-bolt identification.

Bolt Tightening. Specifications require all high-strength bolts to be tightened to 70% of their specified minimum tensile strength, which is nearly equal to the proof load (specified lower bound to the proportional limit) for A325 bolts, and within 10% of the proof load for A490 bolts. Tightening above these minimum tensile values does not damage the bolts, but it is prudent to avoid excessive uncontrolled tightening. The required minimum tension, kips, for bolt sizes is given in Table 6-22.

Table 6-22. Minimum Tightening Tension, Kips, for High-strength Bolts

Dia, in.	A325	A490
5/8	19	24
3/4	28	35
7/8	39	49
1	51	64
1 1/8	56	80
1 1/4	71	102
1 3/8	85	121
1 1/2	103	148

There are three methods for tightening bolts to assure the prescribed tensioning:

Turn-of-nut. By means of a manual or powered wrench, the head or nut is turned from an initial snug-tight position. The amount of rotation, varying from one-third to a full turn, depends on the ratio of bolt length (underside of head to end of point) to bolt diameter and on the disposition (normal or sloped not more than 1:20 with respect to the bolt axis) of the outer surfaces of bolted parts. Required rotations are tabulated in the "Specification for Structural Steel Joints Using A325 or A490 Bolts."

Calibrated wrench. By means of a powered wrench with automatic cutoff and calibration on the job. Control and test are accomplished with a hydraulic device equipped with a gage that registers the tensile stress developed.

Direct tension indicator. Special indicators are permitted on satisfactory demonstration of performance. One example is a hardened steel washer with protrusions on one face. The flattening that occurs on bolt tightening is measured and correlated with the induced tension.

6-56. Other Bolt-type Fasteners. *Interference body or bearing-type bolts* are characterized by a ribbed or interrupted-ribbed shank and a button-shaped head; otherwise, including strength, they are similar to the regular A325 high-strength bolts. The extreme diameter of the shank is slightly larger than the diameter of the bolt hole. Consequently, the tips of the ribs or knurlings will groove the side of the hole, assuring a tight fit. One useful application has been in high television towers, where minimum-slippage joints are desired with no more installation effort than manual tightening with a spud wrench. Nuts may be secured with lock washers, self-locking nuts, or Dardelet self-locking threads. The main disadvantage of interference body bolts is the need for accurate matching of truly concentric holes in the members being joined; reaming sometimes is necessary.

Huckbolts are grooved (not threaded) and have an extension on the end of the shank. When the bolt is in the hole, a hydraulic machine, similar to a bolting or riveting gun, engages the extension. The machine pulls on the bolt to develop a high clamping force, then swages a collar into the grooved shank and snaps off the extension, all in one quick operation.

6-57. Locking Devices for Bolts. Unfinished bolts (ASTM A307) and interference body type bolts (Art. 6-56) usually come with American Standard threads and nuts. Properly tightened, connections with these bolts give satisfactory service under static loads. But when the connections are subjected to vibration or heavy dynamic loads, a locking device is desirable to prevent the nut from loosening.

Locking devices may be classified according to the method employed: special threads, special nuts, special washers, and what may be described as field methods. Instead of conventional threads, bolts may be supplied with a patented self-locking thread called Dardelet. Sometimes, locking features are built into the nuts. Patented devices, the Automatic-Nut, Union-Nut, and Pal-Nut are among the common ones. Washers may be split rings or specially toothed. Field methods generally used include checking, or distorting, the threads by jamming them with a chisel or locking by tack welding the nuts.

6-58. Welding. The method of joining steel parts by fusion at high temperatures is widely used in steel construction, especially in fabricating shops, where conditions favorable to quality production are easily controlled. Field welding, however, depends on availability of skilled welders and reliability of inspection services, and therefore field bolting often is used instead.

An important advantage of welding is economy of material. Welding reduces the amount of detail connection material, especially in continuous and rigid-frame designs, where welded connections are relatively simple. In addition to economy, welding is quiet and therefore highly desirable for use in congested areas.

Any of several welding processes may be used: manual shielded metal-arc, submerged-arc, flux-cored arc, gas-metal arc, electro-gas, and electro-slag. They are not all interchangeable, however; each has its advantageous applications.

Many building codes accept the recommendations of the American Welding Society ("Structural Welding Code," AWS D1.1). The American Institute of Steel Construction "Specification for the Design, Fabrication and Erection of Structural Steel for Buildings" embodies many of this code's salient requirements. Symbols for welds are shown in Fig. 6-38.

Weld Types. Practically all welds used for connecting structural steel are of two types: fillet and groove.

Figure 6-37*a* and *b* illustrates a typical fillet weld. As stated in Art. 6-32, all stresses on fillet welds are resolved as shear on the effective throat. The normal throat dimension, as indicated in Fig. 6-37*a* and *b*, is the effective throat for all welding processes, except the submerged-arc method. The deep penetration characteristic of the latter process is recognized by increasing the effective throat dimension, as shown in Fig. 6-37*c*. Allowable stresses for fillet welds are given in Table 6-11.

Groove welds (Fig. 6-37*d*, *e*, and *f*) are classified in accordance with depth of solid weld metal as either complete or partial penetration. Most grooves,

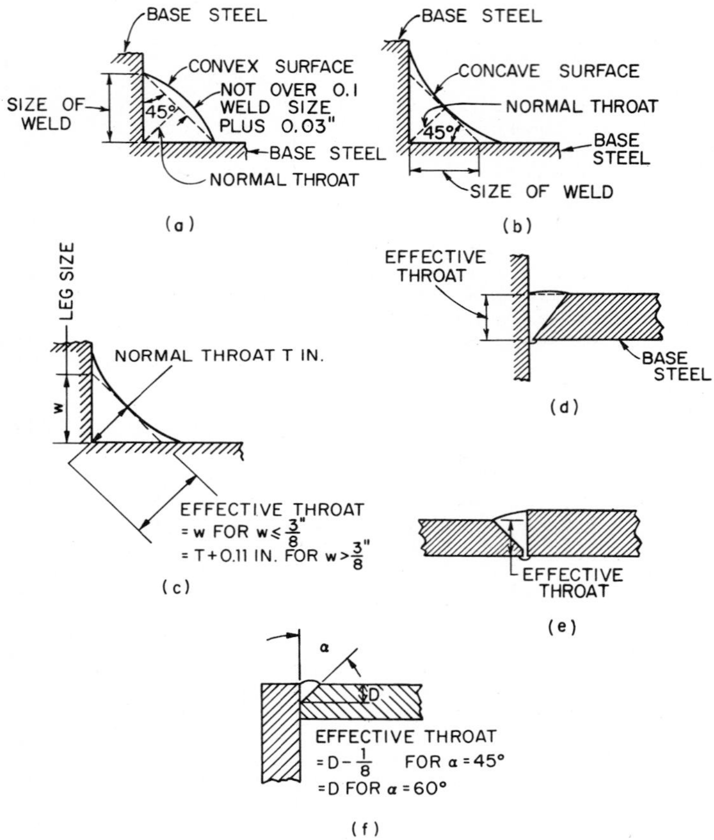

Fig. 6-37. Fillet and groove welds.

such as Fig. 6-37*d* and *e*, are made complete penetration by the workmanship requirements: Use back-up strips (Figs. 6-53 and 6-54) or remove slag inclusions and imperfections (step called back-gouging) on the unshielded side of the root weld. The partial-penetration groove weld shown in Fig. 6-37*f* is typical of the type of weld used for box-type members and column splice joints. Allowable stresses for groove welds are given in Table 6-10.

Base-metal Temperatures. An important requirement in the production of quality welds is the temperature of base metal. Minimum preheat and interpass temperature as specified by the AWS and AISC standards must be obtained within 3 in. of the welded joint before welding starts and then maintained until completion. Table 6-23 gives the temperature requirements based on thickness (thickest part of joint) and welding process for several structural steels. When base metal temperature is below 32°F, it must be preheated to at least 70°F and maintained at that

Table 6-23. Minimum Preheat and Interpass Temperatures for Base Metal to Be Welded

Steel*	Shielded metal-arc welding with other than low-hydrogen electrodes		Shielded metal-arc welding with low-hydrogen electrodes, gas metal-arc, and flux-cored arc welding	
	Thickness, in.	Temp, °F	Thickness, in.	Temp, °F
A36	To 1 in. incl. Over 1 to 1½ Over 1½ to 2½ Over 2½	32 150 225 300	To 1 in. incl. Over 1 to 2 Over 2 to 2½ Over 2½	32 50 150 225
A242 A441 A588 A572 to $F_y = 50$	Not permitted		To ¾ in. incl. Over ¾ to 1½ Over 1½ to 2½ Over 2½	32 70 150 225

* For temperatures for other steels, see AWS D1.1, "Structural Welding Code," American Welding Society.

temperature during welding. No welding is permitted when ambient temperature is below 0°F. Additional information, including temperature requirements for other structural steels, is given in AWS D1.1 and the AISC Specification.

Another quality-oriented requirement applicable to fillet welds is minimum leg size, depending on thickness of steel (Table 6-24). The thicker part governs,

Table 6-24. Minimum Sizes* of Fillet and Partial-penetration Welds

Base-metal Thickness, In.	*Weld Size, In.*
To ¼ incl.	⅛
Over ¼ to ½	3/16
Over ½ to ¾	¼
Over ¾ to 1½	5/16
Over 1½ to 2¼	⅜
Over 2¼ to 6	½
Over 6	⅝

* Leg dimension for fillet welds; minimum effective throat for partial-penetration groove welds.

except that the weld size need not exceed the thickness of the thinner part. This rule is intended to minimize the effects of restraint resulting from rapid cooling due to disproportionate mass relationships.

6-59. Combinations of Fasteners. The American Institute of Steel Construction "Specification for the Design, Fabrication and Erection of Structural Steel for Buildings" distinguishes between existing and new framing in setting conditions for connections of fasteners:

In new work, A307 bolts or high-strength bolts in bearing-type connections should not be considered as sharing the stress with welds. If welds are used, they should be designed to carry the entire stress in the connection. But when one leg of a connection angle is connected with one type of fastener and the other leg with a different type, this rule does not apply. The stress is transferred into the angle by one method, then taken out by another. Such a connection is commonly used, since one fastening method may be selected for shop work and a different method for field connection. But high-strength bolts may share stress with welds on the same leg in friction-type joints (Art. 6-55) if the bolts are tightened before the welds are made.

For connections in existing frames, existing rivets and high-strength bolts may be used for carrying stresses from existing dead loads and welds may be provided for additional dead loads and design loads. This provision assumes that whatever slip is likely to occur in the existing joint already has taken place.

6-60. Fastener Symbols. Fasteners are indicated on design, shop, and field erection drawings by notes and symbols. A simple note may suffice for bolts; for example: "⅞-in. A325 bolts, except as noted." Welds require more explicit information, since their location is not so obvious as that of holes for bolts.

Symbols are standard throughout the industry. Figure 6-34 shows the symbols for bolts, Fig. 6-38 the symbols for welds. The welding symbols (Fig. 6-38*a*)

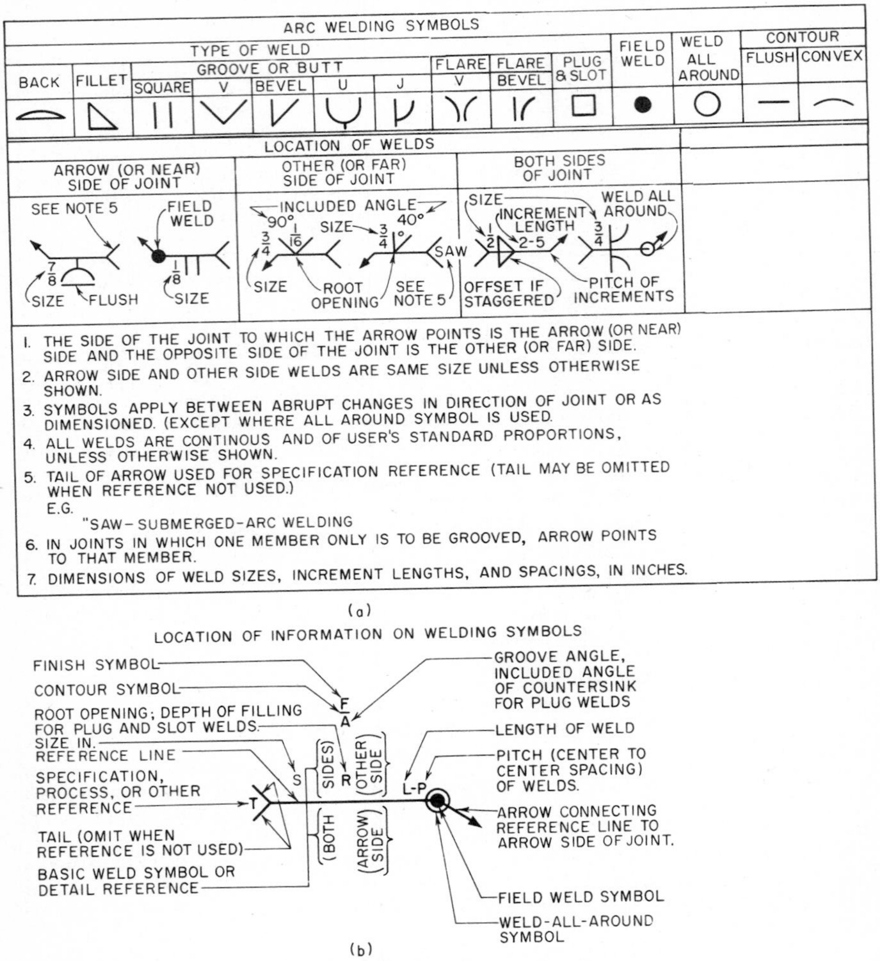

Fig. 6-38. Welding symbols.

together with the information key (Fig. 6-38*b*) are from the American Welding Society "Standard Welding Symbols," AWS 2.0.

6-61. Erection Clearances for Fasteners. All types of fasteners require clearance for proper installation in both shop and field. Shop connections seldom are a problem, since each member is easily manipulated and wholly under the control of the shop and drafting-room personnel. Field connections, however, require careful planning, because connections can be made only after all members are erected and aligned in final position. It becomes the responsibility of the shop draftsmen to arrange connection details in such manner that adequate clearance is provided in the field.

Clearances are necessary for two reasons: to permit entry, as in the case of bolts entering into holes, and to permit tightening of bolts with pneumatic bolting machines, and movement of "stick" electrodes in welding. Approximate clearances for installing high-strength bolts with powered impact wrenches are given in Fig. 6-39.

Somewhat less standardized, but equally important, are the clearances for welded joints. Ample room should be provided for the operator to observe the weld as it is deposited and to manipulate the welding electrode. In many respects, more latitude is afforded to the welder since he can shorten the electrode (12 to 18 in. initially) and even prebend it to accommodate a tight situation.

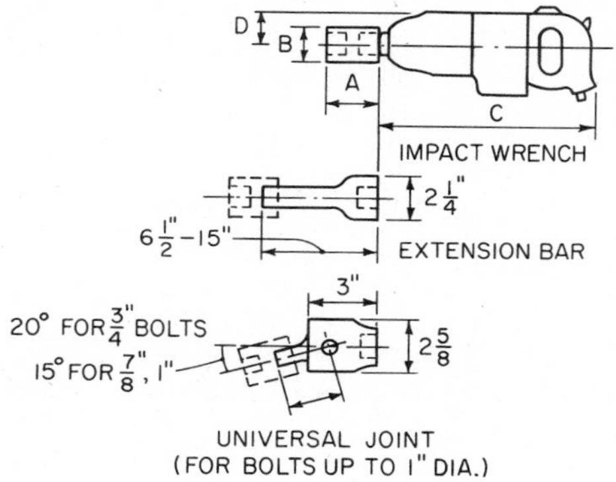

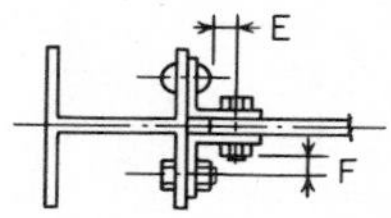

WRENCH DIMENSIONS, IN.

	BOLT SIZE	C	D
LIGHT WRENCHES	5/8 – 1	13 1/4 – 14	2 1/8
HEAVY WRENCHES	1 – 1 3/8	14 3/4 – 17 1/4	2 1/2

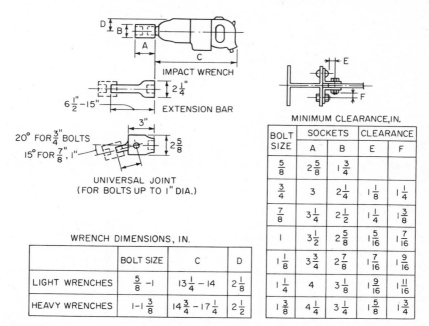

MINIMUM CLEARANCE, IN.

BOLT SIZE	SOCKETS		CLEARANCE	
	A	B	E	F
5/8	2 5/8	1 3/4		
3/4	3	2 1/4	1 1/8	1 1/4
7/8	3 1/4	2 1/2	1 1/4	1 3/8
1	3 1/2	2 5/8	1 5/16	1 7/16
1 1/8	3 3/4	2 7/8	1 7/16	1 9/16
1 1/4	4	3 1/8	1 9/16	1 11/16
1 3/8	4 1/4	3 1/4	1 5/8	1 3/4

Fig. 6-39. Approximate clearances for installing high-strength bolts with powered impact wrenches.

In general, the preferred position of the electrode for fillet welding in the horizontal position is in a plane forming an angle of 30° with the vertical side of the weld. Frequently, this angle must be varied to clear a projecting part, as illustrated in Fig. 6-40*a* and *b*. Satisfactory results can be obtained when the weld root is visible to the operator and when the clear distance to a projection that may interfere with the electrode is not less than one-half the projection height ($y/2$ in Fig. 6-40*b*).

Still another example of welding clearance is illustrated in Fig. 6-40*c*. There the problem is to weld an end-connection angle to a beam web, which is in a horizontal plane, as it would lie on shop skids. The 20° angle in the plan view is the minimum for satisfactory welding with a straight electrode. Assuming a ½-in. setback for the end of the beam and a ⅜-in. outside diameter of the electrode, the clearance should not be less than 1¼ in. for an angle whose width W is 3 in., or less than 1⅝ in. where W equals 4 in.

Although both of the foregoing examples deal with shop welding, the beams being laid in positions most favorable to the work, the limiting positions of the electrode are equally applicable to comparable situations that may arise with field connections.

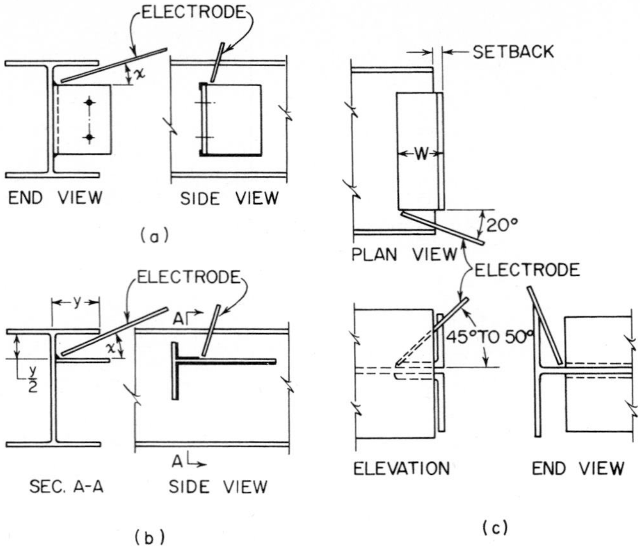

Fig. 6-40. Welding clearance near projecting flanges (*a* and *b*) and end-connection angles (*c*).

CONNECTIONS

Design of connections and splices is a specialized art of the structural-steel industry known as *detailing.* Usually, designers simply indicate the type and size of fastener and type of connection; for example, "⅞-in. A325 bolts in bearing-type joints, connections framed." For beams, the designer may furnish the reactions; if not, then the detailer would determine the reactions from the uniform-load capacity (tabulated in the AISC "Manual of Steel Construction"), giving due consideration to the effect of large concentrated loads near the connection.

For wind-resisting connections in multistory buildings, the wind moments are usually given in the design drawings and perhaps a sketch of a typical connection showing type of wind bracing desired.

Complete development of each connection becomes the responsibility of the fabricator's engineering organization.

6-62. Stresses in Bearing-type Connections. When some slip, although very small, may occur between connected parts, the fasteners are assumed to function in shear. The presence of paint on contact surfaces is therefore of no consequence. Fasteners may be rivets, A307 bolts, or high-strength bolts, or any other similar fastener not dependent on development of friction on the contact surfaces.

Single shear occurs when opposing forces act on a fastener as shown in Fig. 6-41*a*, the plates tending to slide on their contact surfaces. The body of the fastener resists this tendency; a state of shear then exists over the cross-sectional area of the fastener.

Double shear takes place whenever three or more plates act on a fastener as illustrated in Fig. 6-41*b*. There are two or more parallel shearing surfaces (one on each side of the middle plate in Fig. 6-41*b*). Accordingly, the shear strength of the fastener is measured by its ability to resist two or more single shears.

Bearing on Base Metal. This is a factor to consider; but, as stated in Art. 6-33, calculation of bearing stresses in most joints is useful only as an index of efficiency of the net section of tension members.

Edge Distances. For a single fastener (or two fasteners in line of stress) as shown in Fig. 6-41, however, there must be sufficient metal behind a fastener to prevent a tear-out. The AISC "Specification for the Design, Fabrication and

Erection of Structural Steel for Buildings" recommends minimum edge distances, center of hole to edge of connected part, as given in Table 6-25. In addition, special rules cover a case in which tear-out could be critical—tension members with not more than two fasteners in a line parallel to the direction of stress. In bearing-type connections, the edge distance, in., should be at least A_bC/t for single shear and twice that for double shear, where A_b is the nominal cross-sectional area, sq in., of the bolt, t the thickness, in., of the connected part, and C the

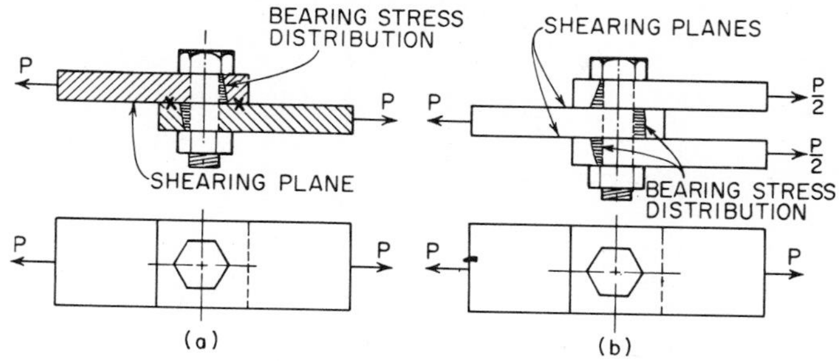

Fig. 6-41. Bolt connection—shear and bearing.

ratio of specified minimum tensile strength of the bolt, kips, to the specified minimum tensile strength of the connected part, kips. But this distance may be decreased in the proportion in which the fastener stress is less than the allowable, but not below that given in Table 6-25. The edge distance need not exceed 1½ times the transverse spacing of the fasteners.

Table 6-25. Minimum Edge Distance for Punched, Reamed, or Drilled Holes, In.

Fastener diameter, in.	At sheared edges	At rolled edges of plates, shapes or bars or gas-cut edges†
½	⅞	¾
⅝	1⅛	⅞
¾	1¼	1
⅞	1½*	1⅛
1	1¾*	1¼
1⅛	2	1½
1¼	2¼	1⅝
Over 1¼	1¾ × diameter	1¼ × diameter

* These may be 1¼ in. at the ends of beam connection angles.

† All edge distances in this column may be reduced ⅛ in. when the hole is at a point where stress does not exceed 25% of the maximum allowed stress in the element.

The maximum distance from the center of a fastener to the nearest edge of parts in contact should not exceed 6 in. or 12 times the part thickness.

Minimum Spacing. The AISC specification also requires that the minimum distance between centers of rivet and bolt holes not be less than 2⅔ times the rivet or bolt diameter. But at least three diameters is desirable.

Eccentric Loading. Stress distribution is not always as simple as for the joint in Fig. 6-41*a*, where the fastener is directly in the line of stress. Sometimes, the load is applied eccentrically, as shown in Fig. 6-42. For such connections, tests show that use of actual eccentricity to compute the maximum force on the extreme fastener is unduly conservative because of plastic behavior and clamping force generated by the fastener. Hence, it is permissible to reduce the actual eccentricity to a more realistic "effective" eccentricity.

For fasteners equally spaced on a single gage line, the effective eccentricity in inches is given by

$$l_{\text{eff}} = l - \frac{1 + 2n}{4} \tag{6-34a}$$

where l = the actual eccentricity
n = the number of fasteners

For the bracket in Fig. 6-42*b*, the reduction applied to l_1 is $(1 + 2 \times 6)/4 = 3.25$ in.

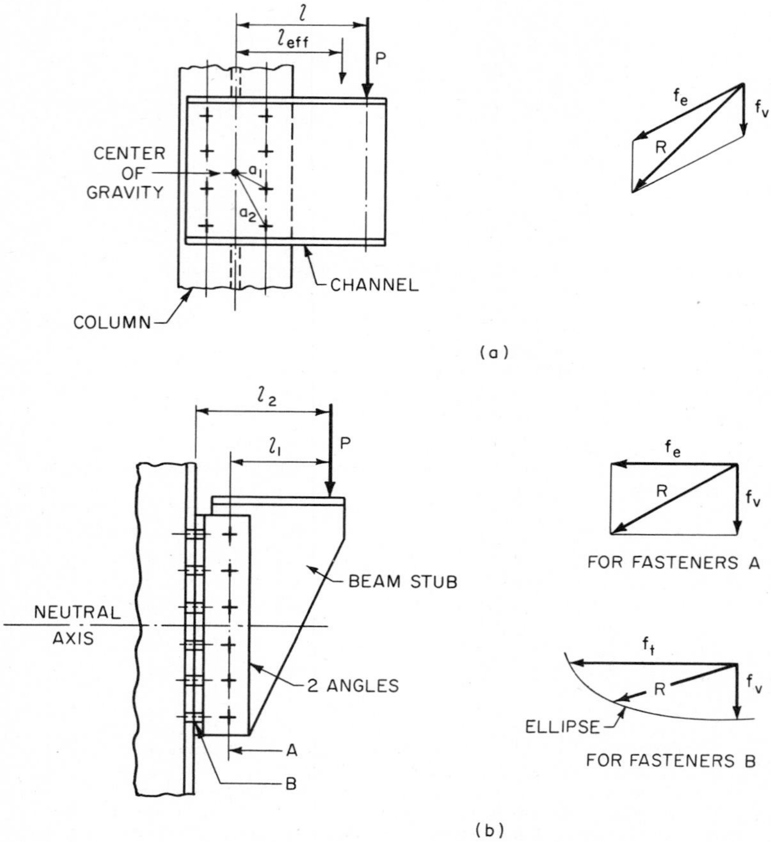

Fig. 6-42. Eccentrically loaded fastener groups.

For fasteners on two or more gage lines

$$l_{\text{eff}} = l - \frac{1 + n}{2} \tag{6-34b}$$

where n is the number of fasteners per gage line. For the bracket in Fig. 6-42*a*, the reduction is $(1 + 4)/2 = 2.5$ in.

In Fig. 6-42*a*, the load P can be resolved into an axial force and a moment: Assume two equal and opposite forces acting through the center of gravity of the fasteners, both forces being equal to and parallel to P. Then, assuming equal distribution on the fasteners, the shear on each fastener due to the force acting in the direction of P is $f_v = P/n$, where n is the number of fasteners.

The other force forms a couple with P. The shear stress f_e due to the couple is proportional to the distance from the center of gravity and acts perpendicular to the line from the fastener to the center. In determining f_e, it is convenient to first express it in terms of x, the force due to the moment Pl_{eff} on an imaginary fastener at unit distance from the center. For a fastener at a distance a from the center, $f_e = ax$, and the resisting moment is $f_e a = a^2x$. The sum of the moments equals Pl_{eff}. This equation enables x to be evaluated and hence, the various values of f_e. The resultant R of f_e and f_v can then be found; a graphical solution usually is sufficiently accurate. The stress so obtained must not exceed the allowable value of the fastener in shear (Arts. 6-33 and 6-34).

For example, in Fig. 6-42*a*, $f_v = P/8$. The sum of the moments is

$$4a_1^2x + 4a_2^2x = Pl_{\text{eff}}$$

$$x = \frac{Pl_{\text{eff}}}{4a_1^2 + 4a_2^2}$$

Then, $f_e = a_2x$ for the most distant fastener, and R can be found graphically as indicated in Fig. 6-42*a*.

Tension and Shear. For fastener group B in Fig. 6-42*b*, use actual eccentricity l_2 since these fasteners are subjected to combined tension and shear. Here too, the load P can be resolved into an axial shear force through the fasteners and a couple. Then, the stress on each fastener due to the axial shear is P/n, where n is the number of fasteners. The tensile forces on the fasteners vary with distance from the center of rotation of the fastener group.

A simple method, erring on the safe side, for computing the resistance moment of group B fasteners assumes that the center of rotation coincides with the neutral axis of the group. It also assumes that the total bearing pressure below the neutral axis equals the sum of the tensile forces on the fasteners above the axis. Then, with these assumptions, the tensile force on the fastener farthest from the neutral axis is

$$f_t = \frac{d_{\max}Pl_2}{\Sigma Ad^2} \tag{6-35}$$

where d = distance of each fastener from the neutral axis
$d_{\max}$ = distance from neutral axis of farthest fastener
A = nominal area of each fastener

The maximum resultant stresses f_t and $f_v = P/n$ are then plotted as an ellipse and R is determined graphically. The allowable stress is given as the tensile stress F_t as a function of the computed shear stress f_v. (In Table 6-14, allowable stresses are given for the ellipse approximated by three straight lines.)

Note that the tensile stress of the applied load is not additive to the internal tension (pretension) generated in the fastener on installation. On the other hand, the AISC Specification does require the addition to the applied load of tensile stresses resulting from prying action, depending on the relative stiffness of fasteners and connection material. Prying force Q (Fig. 6-43*b*) may vary from negligible to a substantial part of the total tension in the fastener. A lengthy method for computing this force is given in the AISC "Manual of Steel Construction" and "Structural Steel Detailing" for determining the prying forces, and also the thickness of the connection material. (See also F. S. Merritt, "Structural Steel Designers' Handbook," McGraw-Hill Book Company, New York.)

The old method for checking the bending strength of connection material ignored the effect of prying action. It simply assumed bending moment equal to P/n times e (Fig. 6-43). This procedure may be used for noncritical applications.

6-63. Stresses in Friction-type Connections. Design of this type of connection assumes that the fastener, under high initial tensioning, develops frictional resistance between the connected parts preventing slippage despite external load. Properly installed rivets and A307 bolts provide some friction, but since it is not dependable, it is ignored. High-strength steel bolts tightened nearly to their yield strengths, however, develop substantial, reliable friction. No slippage will occur at design loads if the contact surfaces are clean and free of paint or have only scored galvanized coatings, inorganic zinc-rich paint, or metallized zinc or aluminum coatings.

The AISC "Specification for the Design, Fabrication and Erection of Structural Steel for Buildings" lists allowable "shear" stresses for high-strength bolts in friction-type connections. Though there actually is no shear on the bolt shank, the shear concept is convenient for measuring bolt capacity.

Since most joints in building construction can tolerate tiny slippage, bearing-type joints, which are allowed much higher stresses for the same high-strength bolts when the threads are not in shear planes, may, for reasons of economy, lessen the use of friction-type joints.

The capacity of a friction-type connection does not depend on the bearing of the bolts against the sides of their holes. Hence general specification requirements for protection against high bearing stresses or bending in the bolts may be ignored.

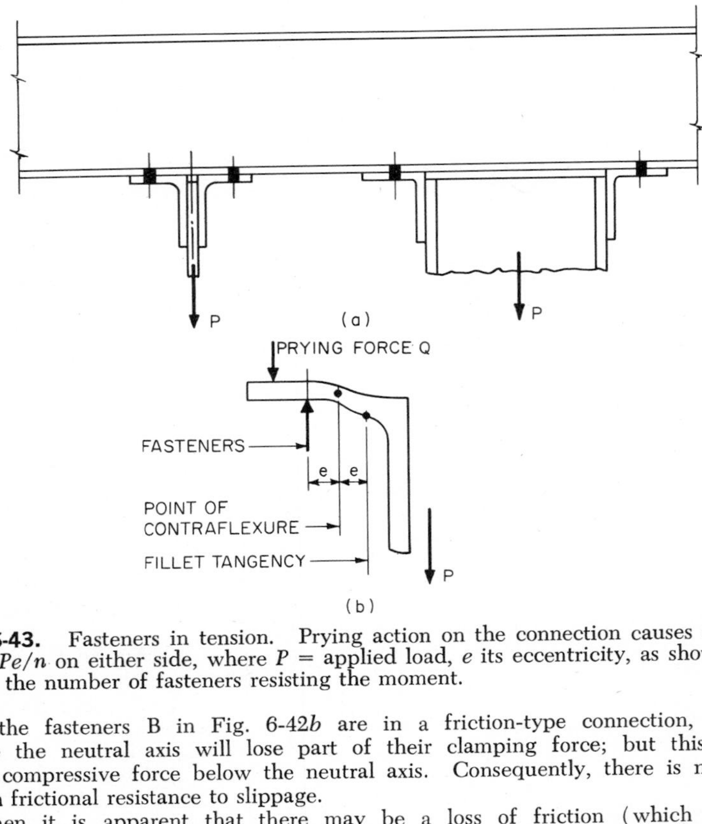

Fig. 6-43. Fasteners in tension. Prying action on the connection causes a moment $M = Pe/n$ on either side, where P = applied load, e its eccentricity, as shown above, and n the number of fasteners resisting the moment.

If the fasteners B in Fig. 6-42*b* are in a friction-type connection, the bolts above the neutral axis will lose part of their clamping force; but this is offset by a compressive force below the neutral axis. Consequently, there is no over-all loss in frictional resistance to slippage.

When it is apparent that there may be a loss of friction (which occurs in some type of brackets and hangers subject to tension and shear) and slip under load cannot be tolerated, the working value in shear should be reduced in proportion to the ratio of residual tension to initial tension.

Friction-type connections subjected to eccentric loading, such as that illustrated in Fig. 6-42, are analyzed in the same manner as bearing-type connections (Art. 6-62).

6-64. Stresses in Welded Connections. Welds are of two general types, fillet (Fig. 6-44*a*) and groove (Fig. 6-44*b*), with allowable stresses dependent on grade of weld and base steels. Since all forces on a fillet weld are resisted as shear on the effective throat (Art. 6-58), the strength of connections resisting direct tension, compression and shear are easily computed on the basis that a kip of fillet shear resists a kip of the applied forces. Many connections, some of which are shown in Fig. 6-45, are not that simple because of eccentricity of applied force with respect to the fillets. In designing such joints it is customary to take into account the actual eccentricity.

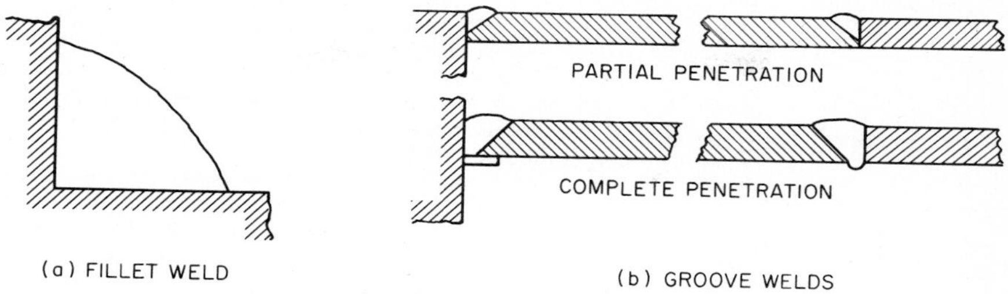

Fig. 6-44. Two main types of welds.

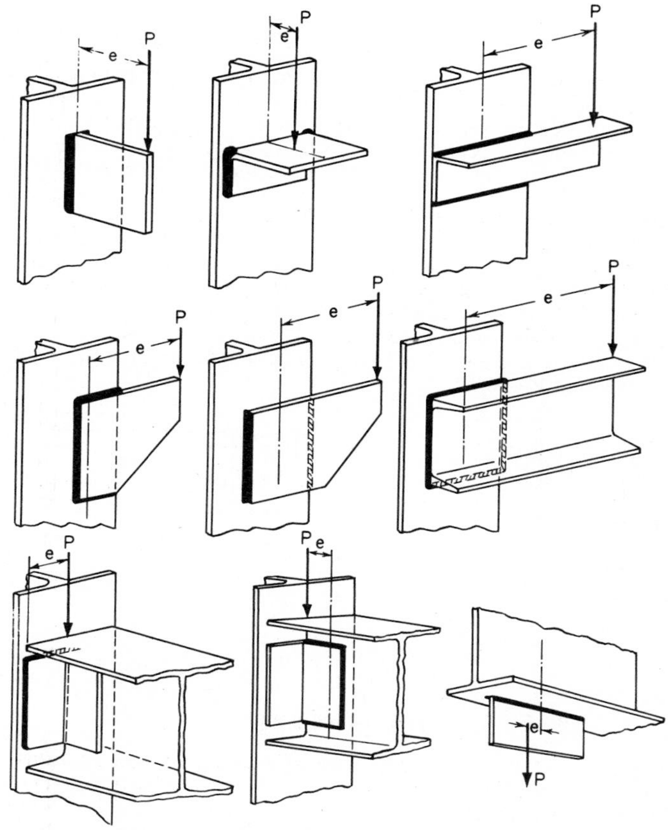

Fig. 6-45. Typical eccentric welded connections.

The underlying design principles for eccentric welded connections are similar to those for eccentric riveted or bolted connections (Art. 6-62). Consider the welded bracket in Fig. 6-46. The first step is to compute the center of gravity of the weld group. Then, the load P can be resolved into an equal and parallel load through the center of gravity and a couple. The load through the center of gravity is resisted by a uniform shear on the welds; for example, if the welds are all the same size, the shear per linear inch is $f_v = P/n$, where n is the total linear inches of weld. The moment Pl of the couple is resisted by the moment of the weld group. The maximum stress, which occurs on the weld element farthest from the center of gravity, may be expressed as $f_e = Pl/S$, where S is the polar section modulus of the weld group.

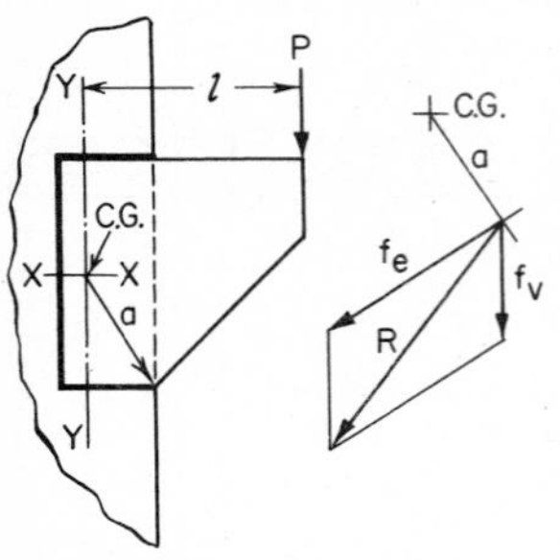

Fig. 6-46. Stresses on welds due to eccentricity.

To find S, first compute the moments of inertia I_X of the welds about the XX axis and I_Y about the perpendicular YY axis. (If the welds are all the same size, their lengths, rather than their relative shear capacities, can be conveniently used in all moment calculations.) The polar moment of inertia $J = I_X + I_Y$, and the polar section modulus $S = J/a$, where a is the distance from the center of gravity to the farthest weld element. The resultant R of f_v and f_e, which acts normal to the line from the center of gravity to the weld element for which the stress is being determined, should not exceed the capacity of the weld element (Art. 6-32).

6-65. Types of Beam Connections. In general, all beam connections are classified as either *framed* or *seated*. In the framed type, the beam is connected to the supporting member with fittings (short angles are common) attached to the beam web. With seated connections, the ends of the beam rest on a ledge or seat, in much the same manner as if the beam rested on a wall.

6-66. Bolted Framed Connections. When a beam is connected to a support, a column or a girder, with web connection angles, the joint is termed "framed." Each connection should be designed for the end reaction of the beam, and type, size, and strength of the fasteners, and bearing strength of base materials should be taken into account. To speed design, the AISC "Manual of Steel Construction" lists a complete range of suitable connections with capacities depending on these variables. Typical connections for beams or channels ranging in depth from 3 to 30 in. are shown in Fig. 6-47.

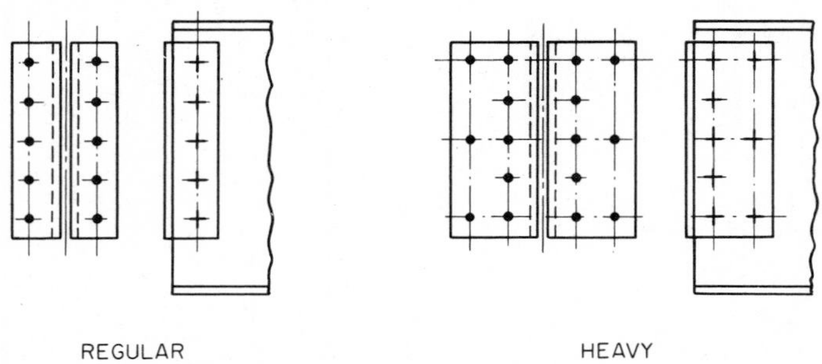

Fig. 6-47. Typical bolted framed connections.

To provide sufficient stability and stiffness, the length of connection angles should be at least half the clear depth of beam web.

For economy, select the minimum connection adequate for the load. For example, assume an 18-in. beam is to be connected. The AISC Manual lists three- and four-row connections in addition to the five-row type shown in Fig. 6-47. Total shear capacity ranges from a low of 26.5 kips for ¾-in. diam A307 bolts in a three-row regular connection to a high of 263.0 kips for 1-in. diam A325 bolts in a five-row heavy connection, bearing type. This wide choice does not mean that all types of fasteners should be used on a project, but simply that the tabulated data cover many possibilities, enabling an economical selection. Naturally, one type of fastener should be used throughout, if practical; but shop and field fasteners may be different.

Bearing stresses on beam webs should be checked against allowable stresses (Art. 6-33), except for friction-type connections, in which bearing is not a factor. Sometimes, the shear capacity of the field fasteners in bearing-type connections may be limited by bearing on thin webs, particularly where beams frame into opposite sides of a web. This could occur where beams frame into column or girder webs.

One side of a framed connection usually is shop connected, the other side field connected. The capacity of the connection is the smaller of the capacities of the shop or field group of fasteners.

In the absence of specific instructions in the bidding information, the fabricator should select the most economical connection. Deeper and stiffer connections, if desired by the designer, should be clearly specified.

6-67. Bolted Seated Connections. Sizes, capacities, and other data for seated connections for beams, shown in Fig. 6-48, are tabulated in the AISC "Manual of Steel Construction." Two types are available, stiffened seats (Fig. 6-48*a*) and unstiffened seats (Fig. 6-48*b*).

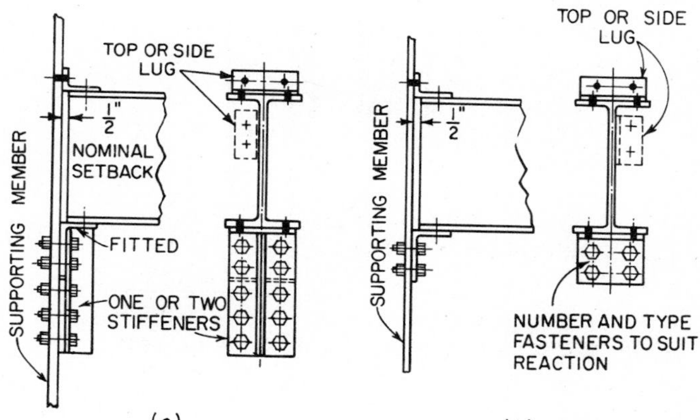

Fig. 6-48. Typical bolted seated connections. (*a*) Stiffened seat. (*b*) Unstiffened seat.

Unstiffened Seats. Capacity is limited by the bending strength of the outstanding, horizontal leg of the seat angle. A 4-in. leg 1 in. thick generally is the practical limit. An angle of A36 steel with these dimensions has a top capacity of 60.5 kips for beams of A36 steel, and 78.4 kips when $F_y = 50$ ksi for the beam steel. Therefore, for larger end reactions, stiffened seats are recommended.

The actual capacity of an unstiffened connection will be the lesser of the bending strength of the seat angle, the shear resistance of the fasteners in the vertical leg, or the bearing strength of the beam web. (See also Art. 6-23 for web crippling stresses.) Data in the AISC Manual make unnecessary the tedious computations of balancing the seat-angle bending strength and beam-web bearing.

The nominal setback from the support of the beam to be seated is ½ in. But tables for seated connections assume ¾ in. to allow for mill underrun of beam length.

Stiffened Seats. These may be obtained with either one or two stiffener angles, depending on the load to be supported. As a rule, stiffeners with outstanding legs having a width less than 5 in. are not connected together; in fact, they may be separated, to line up the angle gage line (recommended center line of fasteners) with that of the column.

The capacity of a stiffened seat is the lesser of the bearing strength of the fitted angle stiffeners or the shear resistance of the fasteners in the vertical legs. Crippling strength of the beam web usually is not the deciding factor, because of ample seat area. When legs larger than 5 in. wide are required, eccentricity should be considered, in accordance with the technique given in Art. 6-62. The center of the beam reaction may be taken at the midpoint of the outstanding leg.

Advantages. Seated connections offer some significant advantages. For economical fabrication, the beams merely are punched and are free from shop-fastened details. They pass from the punching machine to the paint shed, after which they are ready for delivery. In erection, the seat provides an immediate support for the beam while the erector aligns the connecting hole. The top angle is used to prevent accidental rotation of the beam. For framing into column webs, seated connections allow more erection clearance for entering the trough formed by column flanges than do framed connections. A framed beam usually is detailed to within ¹⁄₁₆ in. of the column web. This provides about ⅛ in. total clearance, whereas a seated beam is cut about ½ in. short of the column web, yielding a total clearance of about 1 in. Then, too, each seated connection is wholly independent, whereas

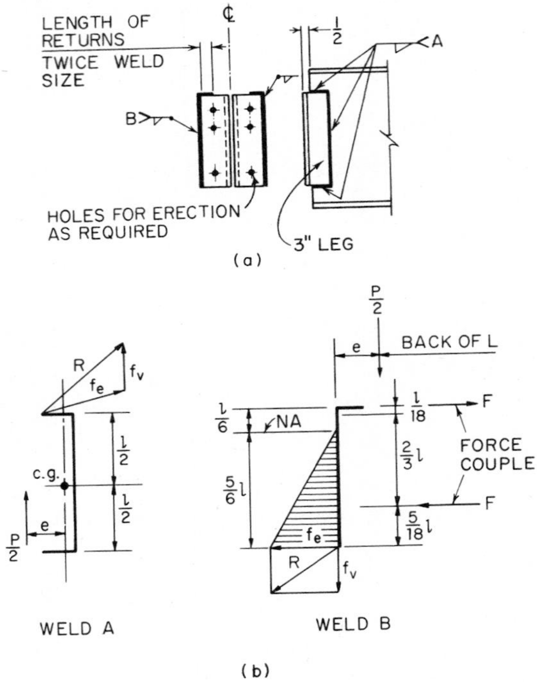

Fig. 6-49. Welded framed connections.

for framed beams on opposite sides of a web, there is the problem of aligning the holes common to each connection.

Frequently, the angles for framed connections are shop attached to columns. Sometimes, one angle may be shipped loose to permit erection. This detail, however, cannot be used for connecting to column webs, because the column flanges may obstruct entering or tightening of bolts. In this case, a seated connection has a distinct advantage.

6-68. Welded Framed Connections. The AISC "Manual of Steel Construction" tabulates sizes and capacities of angle connections for beams for three conditions: all welded, both legs (Fig. 6-49); web leg shop welded, outstanding leg for hole-type fastener; and web leg for hole-type fastener installed in shop, outstanding leg field welded. Tables are based on E70 electrodes. Thus, the connections made with A36 steel are suitable for beams of both carbon and high-strength structural steels.

Eccentricity of load with respect to the weld patterns causes stresses in the welds that must be considered in addition to direct shear. Assumed forces, eccentricities, and induced stresses are shown in Fig. 6-49*b*. Stresses are computed as in the example in Art. 6-64, based on vector analysis that characterizes elastic design. The capacity of welds A or B that is smaller will govern design.

If ultimate strength (plastic design) of such connections is considered, many of the tabulated "elastic" capacities are more conservative than necessary. Although AISC deemed it prudent to retain the "elastic" values for the weld patterns, recognition was given to research results on plastic behavior by reducing the minimum beam-web thickness required when welds A are on opposite sides of the web. As a result, welded framed connections are now applicable to a larger range of rolled beams than strict elastic design would permit.

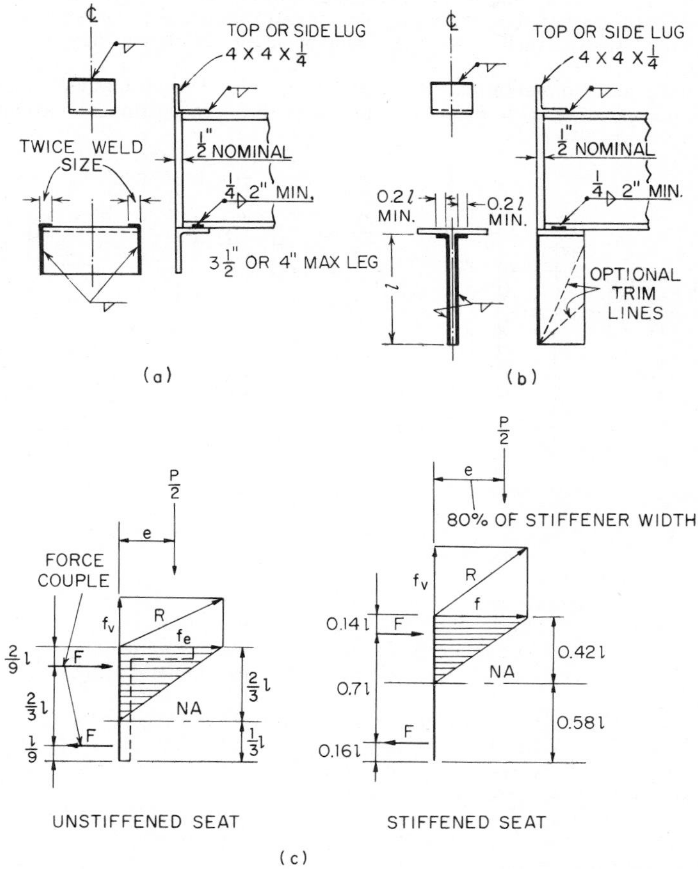

Fig. 6-50. Welded-seat connections. (*a*) Unstiffened seat. (*b*) Stiffened seat. (*c*) Stresses in welds.

Shear stresses in the supporting web for welds B should also be investigated, particularly when beams frame on opposite sides of the web.

6-69. Welded-seat Connections. Also tabulated in the AISC "Manual of Steel Construction," welded-seat connections (Fig. 6-50) are the welded counterparts of bolted-seat connections for beams (Art. 6-67). As for welded framed connections (Art. 6-68), the load capacities for seats, taking into account the eccentricity of loading on welds, are computed by "elastic" vector analysis. Assumptions and the stresses involved are shown in Fig. 6-50*c*.

An unstiffened seat angle of A36 steel has a maximum capacity of 60.5 kips for supporting beams of A36 steel, and 78.4 kips for steel with $F_y = 50$ ksi (Fig. 6-50*a*). For heavier loads, a stiffened seat (Fig. 6-50*b*) should be used.

Stiffened seats may be a beam stub, a tee section, or two plates welded together to form a tee. Thickness of the stiffener (vertical element) depends on the strengths of beam and seat materials. For a seat of A36 steel, stiffener thickness should be at least that of the supported beam web when the web is A36 steel, and 1.4 times thicker for web steel with $F_y = 50$ ksi.

When stiffened seats are on line on opposite sides of a supporting web of A36 steel, the weld size made with E70 electrodes should not exceed one-half the web thickness, and for web steel with $F_y = 50$ ksi, two-thirds the web thickness.

Although top or side lug angles will hold the beam in place in erection, it often is advisable to use temporary erection bolts to attach the bottom beam flange to the seat. Usually, such bolts may remain after the beam flange is welded to the seat.

6-70. End-plate Connections. The art of welding makes feasible connections that were not possible with older-type fasteners, e.g., end-plate connections (Fig. 6-51).

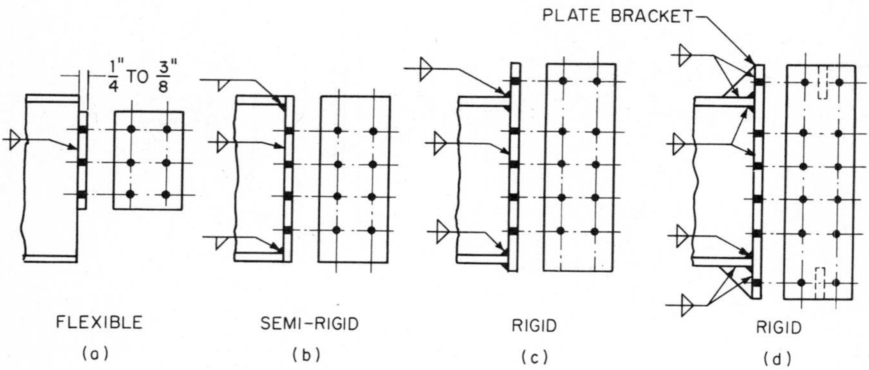

Fig. 6-51. End-plate connections.

Of the several variations, only the flexible type (Fig. 6-51*a*) has been "standardized" with tabulated data in the AISC "Manual of Steel Construction." Flexibility is assured by making the end plate ¼ in. thick wherever possible (never more than ⅜ in.). Such connections in tests exhibit rotations similar to those for framed connections.

The weld connecting the end plate to the beam web is designed for shear. There is no eccentricity. Weld size and capacity are limited by the shear strength of the beam web adjoining the weld. Effective length of weld is reduced by twice the weld size to allow for possible deficiencies at the ends.

As can be observed, this type of connection requires accurate cutting of the beam to length. Also the end plates must be squarely positioned so as to compensate for mill and shop tolerances.

The end plate connection is easily adapted for resisting beam moments (Fig. 6-51*b*, *c*, and *d*). One deterrent, however, to its use for tall buildings where column flanges are massive and end plates thick is that the rigidity of the parts may prevent drawing the surfaces into tight contact. Consequently, it may not be easy to make such connections accommodate normal mill and shop tolerances.

6-71. Special Connections. In some structural frameworks, there may be connections in which a standard type (Arts. 6-66 to 6-70) cannot be used. Beam centers may be offset from column centers, or intersection angles may differ from 90°, for example.

For some skewed connections the departure from the perpendicular may be taken care of by slightly bending the framing angles. When the practical limit for bent angles is exceeded, bent plates may be used (Fig. 6-52*a*).

Special one-sided angle connections, as shown in Fig. 6-52*b*, are generally acceptable for light beams. When such connections are used, the eccentricity of the fastener group in the outstanding leg should be taken into account. Length l may be reduced to the effective eccentricity (Art. 6-62).

Spandrel and similar beams lined up with a column flange may be conveniently connected to it with a plate (Fig. 6-52*c* and *d*). The fasteners joining the plate to the beam web should be capable of resisting the moment for the full lever arm l for the connection in Fig. 6-52*c*. For beams on both sides of the column with equal reactions, the moments balance out. But the case of live load on

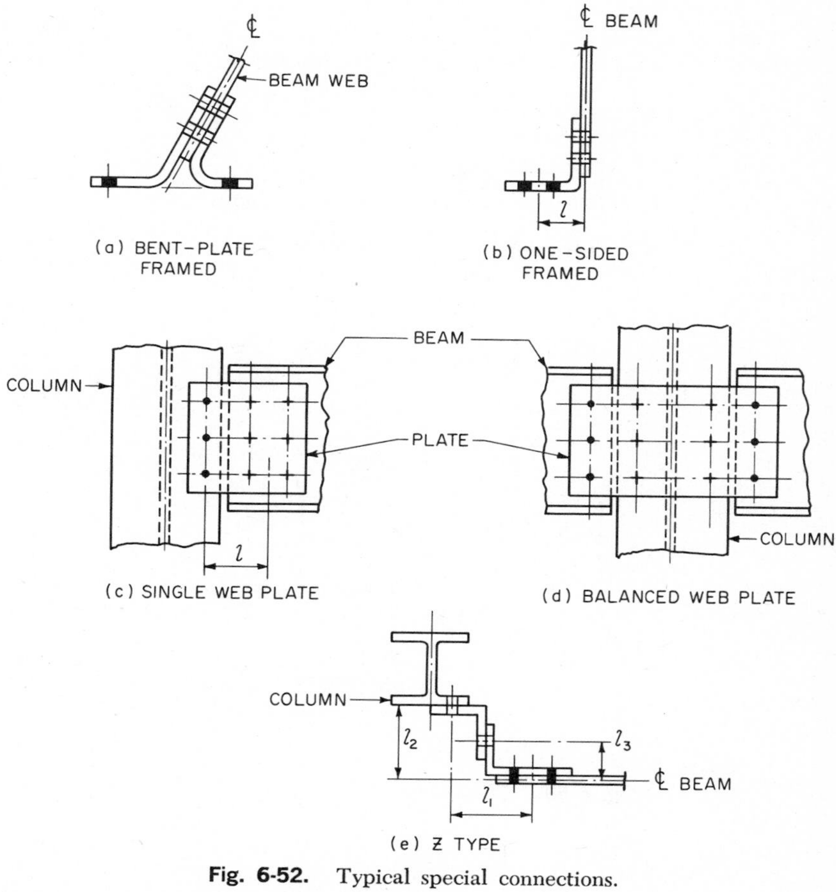

Fig. 6-52. Typical special connections.

one beam only must be considered. And bear in mind the necessity of supporting the beam reaction as near as possible to the column center to relieve the column of bending stresses.

When spandrels and girts are offset from the column, a Z-type connection (Fig. 6-52*e*) may be used. The eccentricity for beam-web fasteners should be taken as l_1, for column-flange fasteners as l_2, and for fasteners joining the two connection angles as l_3 when l_3 exceeds 2½ in.; smaller values of l_3 may be considered negligible.

6-72. Simple, Rigid, and Semirigid Connections. Moment connections are capable of transferring the forces in beam flanges to the columns. This moment transfer, when specified, must be provided for in addition to and usually independent of the shear connection needed to support the beam reaction. Framed, seated, and end-plate connections (Arts. 6-66 to 6-70) are examples of shear connections. Those shown in Fig. 6-30 are moment connections. In Fig. 6-30*a* to *g*, flange stresses are developed independently of the shear connections, whereas in *h* and *i*, the forces are combined and the entire connection resolved as a unit.

Moment connections may be classified according to their design function: those resisting moment due to lateral forces on the structure, and those needed to develop continuity, with or without resistance to lateral forces.

The connections generally are designed for the computed bending moment, which often is less than the beam's capacity to resist moment. A maximum connection is obtained, however, when the beam flange is developed for its maximum allowable stress.

The ability of a connection to resist moment depends on the elastic behavior of the parts. For example, the light lug angle shown connected to the top flange

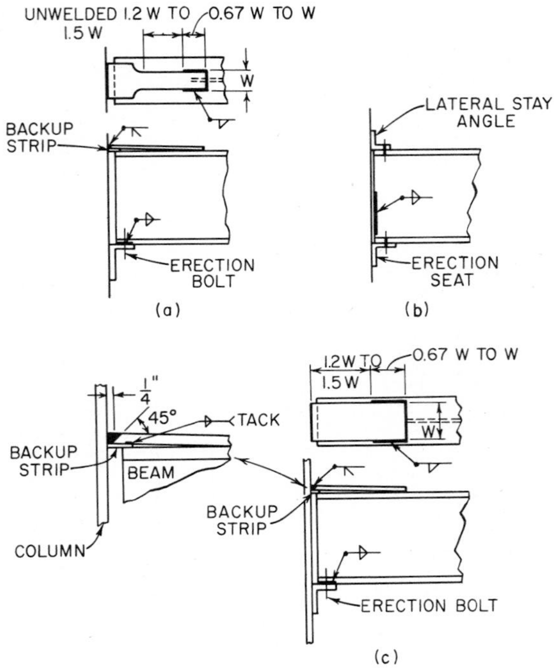

Fig. 6-53. Flexible welded connections.

of the beam in Fig. 6-53*b* is not designed for moment and accordingly affords negligible resistance to rotation. In contrast, full rigidity is expected of the direct welded flange-to-column connection in Fig. 6-54*a*. The degree of fixity, therefore, is an important factor in design of moment connections.

Fixity of End Connections. Specifications recognize three types of end connections: simple, rigid, and semirigid. The type designated *simple* (unrestrained) is intended to support beams and girders for shear only and leave the ends free to rotate under load. The type designated *rigid* (known also as rigid-frame, continuous, restrained frame) aims at not only carrying the shear but also providing sufficient rigidity to hold virtually unchanged the original angles between members connected. *Semirigid,* as the name inplies, assumes that the connections of beams and girders possess a dependable and known moment capacity intermediate in degree between the simple and rigid types. Figure 6-55 illustrates these three types together with the uniform-load moments obtained with each type.

Although no definite relative rigidities have been established, it is generally conceded that the simple or flexible type could vary from zero to 15% (some researchers recommend 20%) end restraint and that the rigid type could vary from 90 to 100%. The semirigid types lie between 15 and 90%, the precise value assumed in the design being largely dependent on experimental analysis. These percentages of rigidity represent the ratio of the moment developed by the connection, with no column rotation, to the moment developed by a fully rigid connection under the same conditions, multiplied by 100. (See also "Structural Steel Connections," Engineering Research Institute, University of Michigan.)

Framed and seated connections offer little or no restraint. In addition, several other arrangements come within the scope of simple-type connections, although they appear to offer greater resistance to end rotations. For example, in Fig. 6-53*a*, a top plate may be used instead of an angle for lateral support, the plate being so designed that plastic deformation may occur in the narrow unwelded portion. Naturally, the plate offers greater resistance to beam rotation than a light angle, but it can provide sufficient flexibility that the connection can be classified as a simple type. Plate and welds at both ends are proportioned for about 25% of the beam moment capacity. The plate is shaped so that the metal across the least width is at yield stress when the stresses in the wide portion, in the butt welds, and in the fillet welds are at allowable working values. The unwelded length is then made from 20 to 50% greater than the least width to assure ductile yielding. This detail can also be developed as an effective moment-type connection.

Another flexible type is the direct web connection in Fig. 6-53*b*. Figured for shear loads only, the welds are located on the lower part of the web, where the rotational effect of the beam under load is the least, a likely condition when the beam rests on erection seats and the axis of rotation centers about the seat rather than about the neutral axis.

Tests indicate that considerable flexibility also can be obtained with a properly proportioned welded top-plate detail as shown in Fig. 6-53*c* without narrowing it as in Fig. 6-53*a*. This detail is usually confined to wind-braced simple-beam designs. The top plate is designed for the wind moment on the joint, at the increased stresses permitted for wind loads.

The problem of superimposing wind bracing on what is otherwise a clear-cut simple beam with flexible connections is a complex one. Some compromise is usually effected between theory and actual design practice. Two alternates usually are permitted by building codes:

1. Connections designed to resist assumed wind moments should be adequate to resist the moments induced by the gravity loading and the wind loading, at specified increased unit stresses.

2. Connections designed to resist assumed wind moments should be so designed that larger moments, induced by gravity loading under the actual condition of restraint, will be relieved by deformation of the connection material.

Obviously, these options envisage some nonelastic, but self-limiting, deformation of the structural-steel parts. Innumerable wind-braced buildings of riveted, bolted, or welded construction have been designed on this assumption of plastic behavior and have proved satisfactory in service.

Fully rigid, bolted beam end connections are not often used because of the awkward, bulky details, which, if not interfering with architectural clearances, are often so costly to design and fabricate as to negate the economy gained by using smaller beam sections. In appearance, they resemble the types shown in Fig. 6-30 for wind bracing; they are developed for the full moment-resisting capacity of the beam.

Much easier to accomplish and more efficient are welded rigid connections (Fig. 6-54). They may be connected simply by butt-welding the beam flanges to the columns—the "direct" connection shown in Fig. 6-54*a* and *b*. Others may prefer the "indirect" method, with top plates, because this detail permits ordinary mill tolerance for beam length. Welding of plates to stiffen the column flanges, when necessary, is also relatively simple (Art. 6-23).

In lieu of the erection seat angle in Fig. 6-54*b*, a patented, forged hook-and-eye device, known as Saxe erection units, may be used. The eye, or seat, is shop welded to the column, and the hook, or clip, is shop welded to the under side of the beam bottom flange. For deep beams, a similar unit may be located on the top flange to prevent accidental turning over of the beams. Saxe units are capable of supporting normal erection loads and dead weight of members; but their contribution to the strength of the connection is ignored in computing resistance to shear.

A comparison of fixities intermediate between full rigidity and zero restraint in Fig. 6-55 reveals an optimum condition attainable with 75% rigidity; end and center-span moments are equal, each being $WL/16$, or one-half the simple-beam moment. The saving in weight of beam is quite apparent.

Perhaps the deterrent to a broader usage of semirigid connections has been the proviso contained in specifications: "permitted only upon evidence that the connections to be used are capable of resisting definite moments without overstress

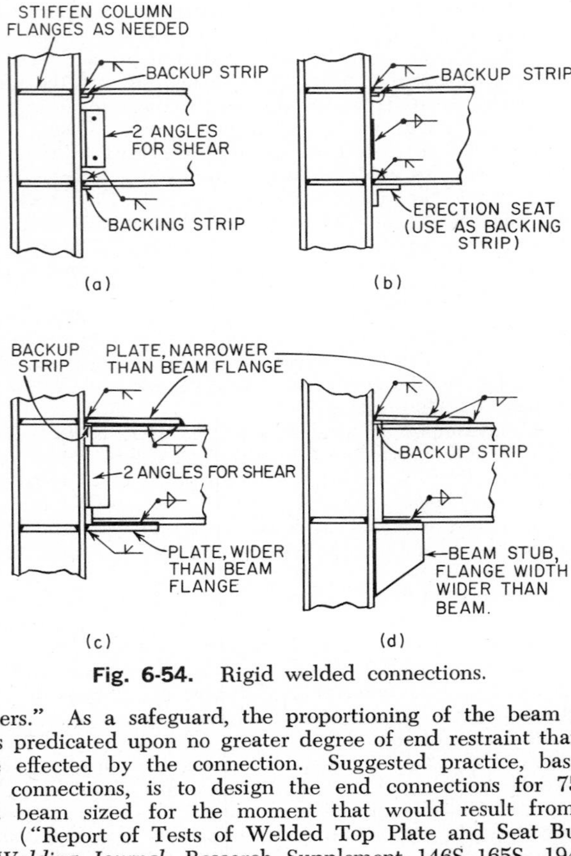

Fig. 6-54. Rigid welded connections.

of the fasteners." As a safeguard, the proportioning of the beam joined by such connections is predicated upon no greater degree of end restraint than the minimum known to be effected by the connection. Suggested practice, based on research with welded connections, is to design the end connections for 75% rigidity but to provide a beam sized for the moment that would result from 50% restraint; i.e., $WL/12$. ("Report of Tests of Welded Top Plate and Seat Building Connections," *The Welding Journal,* Research Supplement 146S–165S, 1944.) The type of welded connection in Fig. 6-53*c,* when designed for the intended rigidity, is generally acceptable.

End-plate connections (Fig. 6-51) are another means of achieving negligible, partial, and full restraint.

6-73. Column Splices. Column-to-column connections are usually determined by the change in section. In general, a change is made at every second floor level, where a shop or field splice is located. From an erection viewpoint, as well as for fabrication and shipment, splices at every third floor may be more economical because of the reduced number of pieces to handle. Naturally, this advantage is partly offset by extra weight of column material, because the column size is determined by loads on the lowest story of each tier, there being an excess of section for the story or two above.

Splices are located just above floor-beam connections, usually about 2 to 3 ft above the floor. Since column stresses are transferred from column to column by bearing, the splice plates are of nominal size, commensurate with the need for safe erection and bending moments the joint may be subjected to during erection. From the viewpont of moment resistance, a conventional column splice develops perhaps 20% of the moment capacity of the column.

Although column splices are not standardized, as are beam connections, they are generally uniform throughout the industry. (Splices shown in the Appendix, "Structural Steel Detailing," American Institute of Steel Construction, may be used by the fabricator or erector unless the designer indicates his preference.) Figure 6-56 illustrates the common types of column splices made with high-strength bolts. In Fig. 6-56*a* and *b*, the upper column bears directly on the lower column; filler

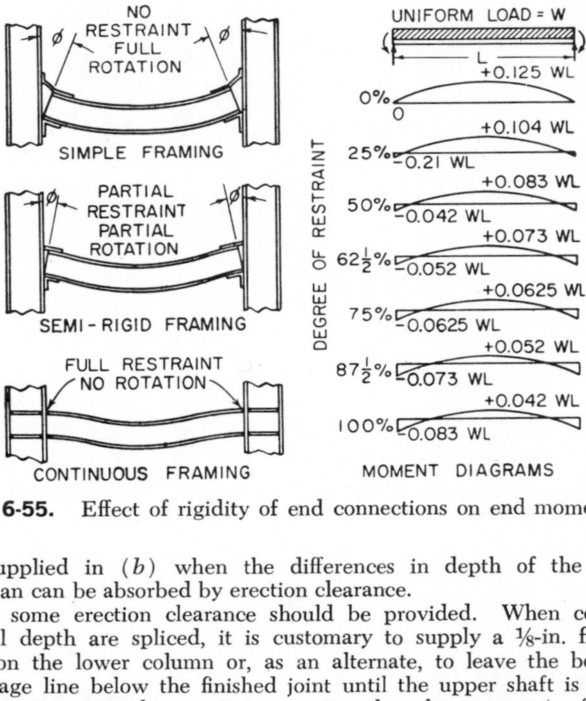

Fig. 6-55. Effect of rigidity of end connections on end moments.

plates are supplied in (*b*) when the differences in depth of the two columns are greater than can be absorbed by erection clearance.

As a rule, some erection clearance should be provided. When columns of the same nominal depth are spliced, it is customary to supply a ⅛-in. fill under each splice plate on the lower column or, as an alternate, to leave the bolt holes open on the top gage line below the finished joint until the upper shaft is erected. The latter procedure permits the erector to spring the plates apart to facilitate entry of the upper column.

When the upper column is of such dimension that its finished end does not wholly bear on the lower column, one of two methods must be followed: In Fig. 6-56*c*, stresses in a portion of the upper column not bearing on the lower column are transferred by means of flange plates that are finished to bear on the lower column. These bearing plates must be attached with sufficient single-shear bolts to develop the load transmitted through bearing on the finished surface.

When the difference in column size is pronounced, the practice is to use a horizontal bearing plate as shown in Fig. 6-56*d*. These plates, known as *butt plates,* may be attached to either shaft with tack welds or clip angles. Usually it is attached to the upper shaft, because a plate on the lower shaft may interfere with erection of the beams that frame into the column web.

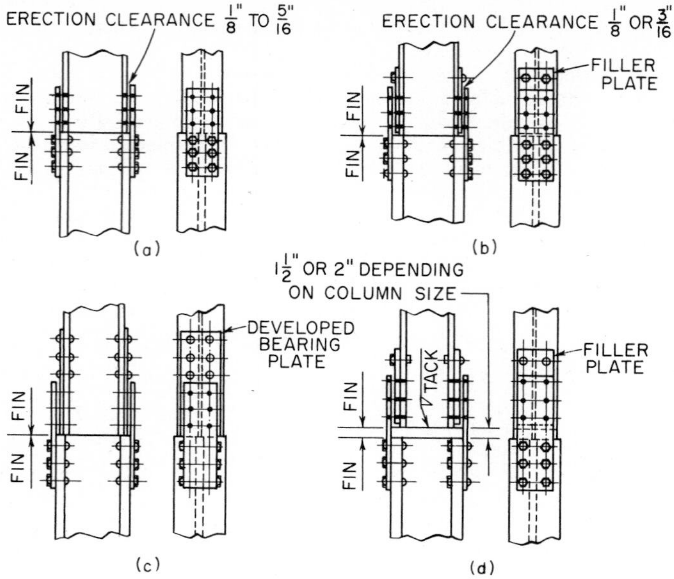

Fig. 6-56. Bolted column splices, friction-type connections.

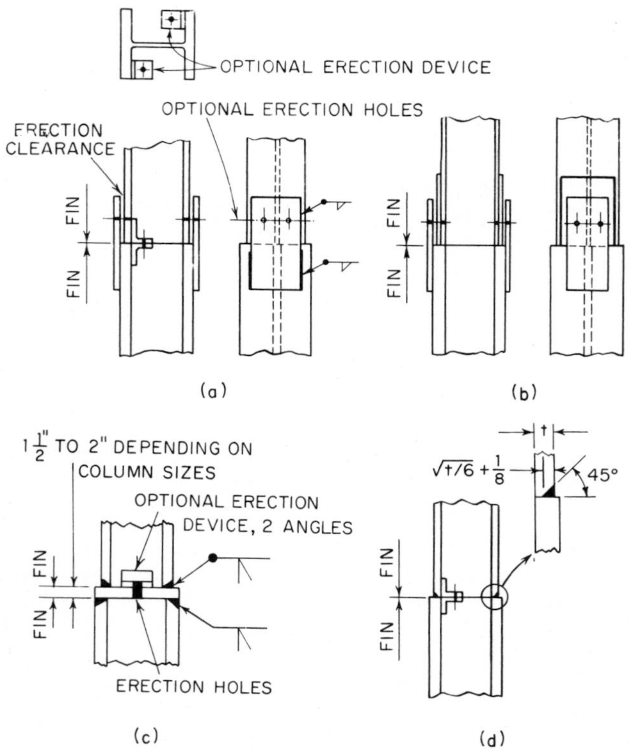

Fig. 6-57. Welded column splices.

Somewhat similar are welded column splices. In Fig. 6-57*a*, a common case, holes for erection purposes are generally supplied in the splice plates and column flanges as shown. Some, however, prefer to avoid drilling and punching of thick pieces, and use instead clip angles welded on the inside flanges of the columns, one pair at diagonally opposite corners, or some similar arrangement. Figure 6-57*b* and *c* corresponds to the bolted splices in Fig. 6-56*c* and *d*. The shop and field welds for the welded butt plate in Fig. 6-57*c* may be reversed, to provide erection clearance for beams seated just below the splice. The erection clip angles would then be shop-welded to the underside of the butt plate, and the field holes would pierce the column web.

The butt-weld splice in Fig. 6-57*d* is the most efficient from the standpoint of material saving. The depth of the bevel as given in the illustration is for the usual column splice, in which moment is unimportant. However, should the joint be subjected to considerable moment, the bevel may be deepened; but a ⅛-in. minimum shoulder should remain for the purpose of landing and plumbing the column. For full moment capacity, a complete-penetration welded joint would be required.

6-74. Beam Splices. These are required in rigid frames, suspended-span construction, and continuous beams. Such splices are usually located at points of counterflexure or at points where moments are relatively small. Therefore, splices are of moderate size. Flanges and web may be spliced with plates or butt-welded.

For one reason or another it is sometimes expedient to make a long beam from two short lengths. A welded joint usually is selected, because the beams can be joined together without splice plates and without loss of section due to bolt holes. Also, from the viewpoint of appearance, the welded joint is hardly discernible.

Usually, the joint must be 100% efficient, to develop the full section. Figure 6-58 illustrates such a detail. The back side of the initial weld is gouged or chipped out; access holes in the beam webs facilitate proper edge preparation and depositing of the weld metal in the flange area in line with the web. Such holes are usually left open, because plugs would add undesirable residual stresses to the joint (Bibliography, Art. 6-92).

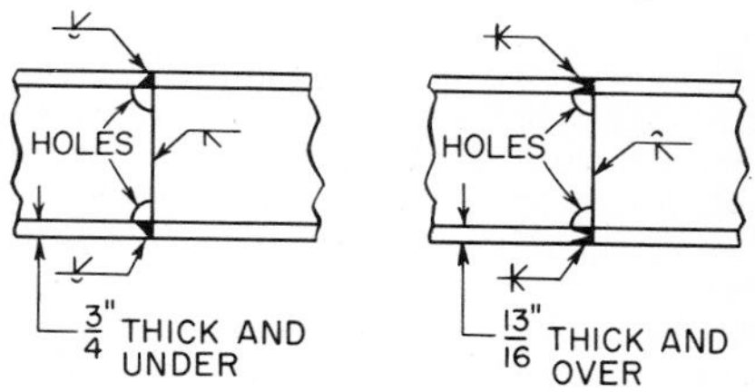

Fig. 6-58. Welded beam splices.

STEEL ERECTION

A clear understanding of what the fabricator furnishes or does not furnish to the erector, particularly on fabrication contracts that may call for delivery only, is all-important—and most fabricated steel is purchased on delivery basis only.

Purchasing structural steel has been greatly simplified by the industry's "Code of Standard Practice for Buildings and Bridges"—an American Institute of Steel Construction publication. A contract provision making the code part of the contract is often used, since it establishes a commonly accepted and well-defined line of demarcation between what is, and what is not, to be furnished under the contract. Lacking such a provision, the contract, to avoid later misunderstandings, must enumerate in considerable detail what is expected of both parties to the contract.

Under the code—and unless otherwise specifically called for in the contract documents—such items as steel sash, corrugated-iron roofing or siding, and open-web steel joists, and similar items, even if made of steel and shown on the contract design drawings, are not included in the category "structural steel." Also, such items as door frames are excluded, even when made of structural shapes, if they are not fastened to the structure in such way as to comply with "constituting part of the steel framing." On the other hand, loose lintels shown on design plans or in separate schedules are included.

According to the code, a fabricator furnishes with "structural steel," to be erected by someone else, the field bolts required for fastening the steel. The fabricator, however, does not furnish the following items unless specified in the invitation to bid: shims, fitting-up bolts, drift pins, temporary cables, welding electrodes, or thin leveling plates for column bases.

The code also defines the erection practices. For example, the erector does not paint field boltheads and nuts, field welds, or touch up abrasions in the shop coat, or perform any other field painting unless required in specifications accompanying the invitation to bid.

6-75. Erection Equipment. If there is a universal piece of erection equipment, it is the crane. Mounted on wheels or tractor treads, it is extremely mobile, both on the job and in moving from job to job. Practically all buildings are erected with this efficient raising device. The exception, of course, is the skyscraper whose height exceeds the reach of the crane. Operating on ground level, cranes have been used to erect buildings of about 20 stories, the maximum height being dependent on the length of the boom and width of building.

Cranes are also mounted on locomotive cars. Their use in building construction is limited to heavy plant construction serviced by railroads. By laying the rails early in the construction schedule, the erector can bring in the heavy steelwork on railway cars and use a heavy-duty locomotive crane for erection.

The guy derrick is a widely used raising device for erection of tall buildings. Its principal asset is the ease by which it may be "jumped" from tier to tier as erection proceeds upward. The boom and mast reverse position; each in turn serves to lift up the other. It requires about 2 hr to make a two-story jump.

Stiff-leg derricks and gin poles are two other rigs sometimes used, usually in the role of auxiliaries to cranes or guy derricks. Gin poles are the most elementary—simply a guyed boom. The base must be secure because of the danger of kicking out. The device is useful for the raising of incidental materials, for dismantling and lowering of larger rigs, and for erection of steel on light construction where the services of a crane are unwarranted.

Stiff-leg derricks are most efficient where they may be set up to remain for long periods of time. They have been used to erect multistory buildings but are not in popular favor because of the long time required to jump from tier to tier. Among the principal uses for stiff legs are (1) unloading steel from railroad cars for transfer to trucks, (2) storage and sorting, and (3) when placed on a flat roof, raising steel to roof level, where it may be sorted and placed within reach of a guy derrick.

Less time for "jumping" the raising equipment is needed for cranes mounted on steel box-type towers, about three stories high, that are seated on interior elevator wells or similar shafts for erecting steel. These tower cranes are simply jacked upward hydraulically or raised by cables, with the previously erected steelwork serving as supports. In another method, a stiff-leg derrick is mounted on a trussed platform, spanning two or more columns, and so powered that it can creep up the erected exterior columns. In addition to the advantage of faster jumps, these methods permit steel erection to proceed as soon as the higher working level is reached.

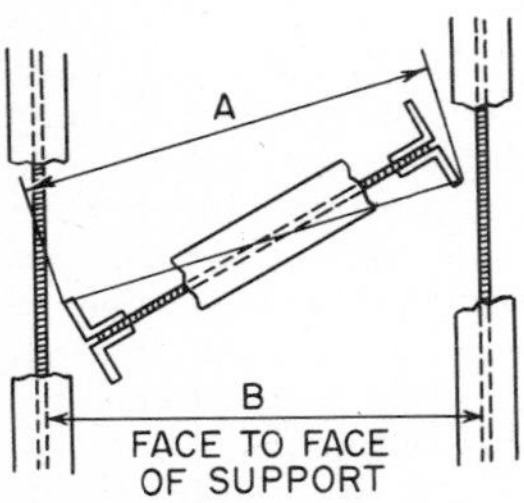

Fig. 6-59. Erection clearance for beams.

6-76. Clearance for Erecting Beams. Clearances for tightening bolts and welding are discussed in Arts. 6-61 and 6-73. In addition, designers also must provide sufficient field clearance for all members so as to permit erection without interference with members previously erected. The shop draftsman should always arrange the details so that the members can be swung into their final position without shifting the members to which they connect from their final positions. The following examples illustrate the type of problem most frequently encountered in building work:

In framed beam connections (Fig. 6-59) the slightly shorter distance out-to-out of connection angles ($B - 1/8$ in.), as compared with the face-to-face distance between supporting members, is usually sufficient to allow forcing the beam into position. Occasionally, however, because the beam is relatively short, or because heavy connection angles with wide outstanding legs are required, the diagonal distance *A* may exceed the clearance distance *B*. If so, the connection for one end must be shipped bolted to the framed beam to permit its removal during erection.

An alternate solution is to permanently fasten one connection angle of each pair to the web of the supporting beam, temporarily bolting the other angle to the same web for shipment, as shown in Fig. 6-60. The beam should be investigated for the clearance in swinging past permanently bolted connection angles. Attention must also be paid to possible interference of stiffeners in swinging the beam into place when the supporting member is a plate girder.

Another example is that of a beam seated on column-web connections (Fig. 6-61). The first step is to remove the top angles and shims temporarily. Then, while hanging from the derrick sling, the beam is tilted until its ends clear the edges of the column flanges, after which it is rotated back into a horizontal position

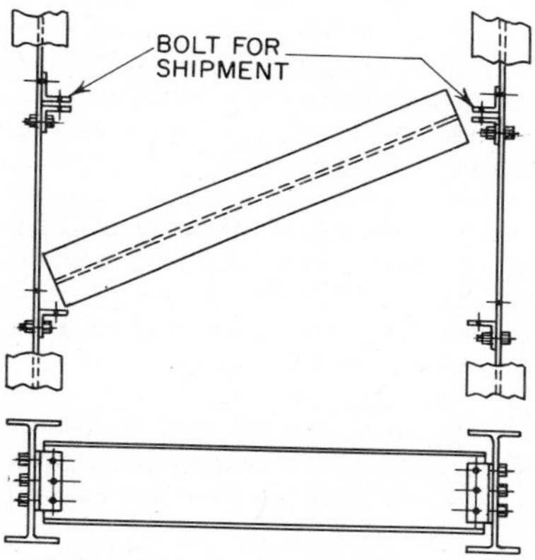

Fig. 6-60. Alternate method for providing erection clearance.

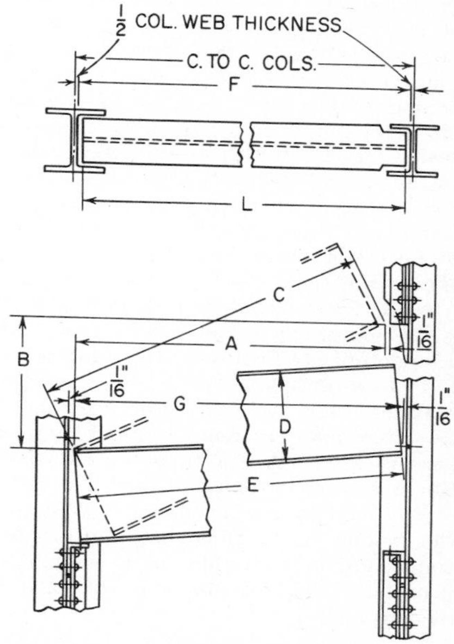

Fig. 6-61. Clearance for beams seated on column-web connections.

and landed on the seats. The greatest diagonal length *G* of the beam should be about ⅛ in. less than the face-to-face distance *F* between column webs. It must also be such as to clear any obstruction above; e.g., *G* must be equal to or less than *C*, or the obstructing detail must be shipped bolted for temporary removal. To allow for possible overrun, the ordered length *L* of the beam should be less than the detailing length *E* by at least the amount of the permitted cutting tolerance.

Frequently, the obstruction above the beam connection may be the details of a column splice. As stated in Art. 6-73, it may be necessary to attach the splice material on the lower end of the upper shaft, if erection of the beam precedes erection of the column in the tier above.

6-77. Erection Sequence. The order in which steel is to be fabricated and delivered to the site should be planned in advance so as not to conflict with the erector's methods or his construction schedule. For example, if steel is to be erected with derricks, the approximate locations at which the derricks will be placed will determine the shipping installments, or sections, into which the frame as a whole must be segregated for orderly shipment. When installments are delivered to the site at predetermined locations, proper planning will eliminate unnecessary rehandling. Information should be conveyed to the drafting room so that the shipping installments can be indicated on the erection plans and installments identified on the shipping lists.

In erecting multistory buildings with guy derricks, the practice is to hoist and place all columns first, spandrel beams and wall bracing next, and interior beams with filler beams last. More specifically, erection commences with bays most distant from the derrick and progresses toward the derrick, until it is closed in. Then, the derrick is jumped to the top and the process is repeated for the next tier. Usually, the top of the tier is planked over to obtain a working platform for the erectors and also to afford protection for the trades working below. However, before the derrick is jumped, the corner panels are plumbed; similarly when panels are erected across the building, cables are stretched to plumb the structure.

There is an established sequence for completing the connections. The raising gang connects members together with temporary fitting-up bolts. The number of bolts is kept to a minimum, just enough to draw the joint up tight and take care of the stresses caused by dead weight, wind, and erection forces. Permanent connections are made as soon as alignment is within tolerance limits. Usually, permanent bolting or welding follows on the heels of the raising gang. Sometimes, the latter moves faster than the gang making the permanent connections, in which case it may be prudent to skip every other floor, thus obtaining permanent connections as close as possible to the derrick—a matter of safe practice.

Some erectors prefer to use the permanent high-strength bolts for temporary fitting-up. According to AISC rules, A325 bolts that were tightened one-half turn of nut from initial snug-tight position may be loosened and retightened a second time. But such reuse is not permitted for A490 bolts under the same conditions.

6-78. Field Welding Procedures. The main function of a welding sequence is to control distortion due primarily to the effects of welding heat. In general, a large input of heat in a short time tends to produce the greatest distortion. Therefore, it is always advisable, for large joints, to weld in stages, with sufficient time between each stage to assure complete dispersal of heat, except for heat needed to satisfy interpass-temperature requirements (Art. 6-58). Equally important, and perhaps more efficient from the erector's viewpoint, are those methods that balance the heat input in such a manner that the distortional effects tend to cancel out.

Welding on one flange of a column tends to leave the column curled toward the welded side cooling, because of shrinkage stresses. A better practice for beams connecting to both sides of a column is to weld the opposite connections simultaneously. Thus the shrinkage of each flange is kept in balance and the column remains plumb.

If simultaneous welding is not feasible, then the procedure is to weld in stages. About 60% of the required weld might be applied on the first beam, then the joint on the opposite flange might be completely welded, and finally, welding on the first beam would be completed. Procedures such as this will go far to reduce distortion.

Experience has shown that it is good practice to commence welding at or near the center of a building and work outward. Columns should be checked frequently for vertical alignment, because shrinkage in the welds tends to shorten the distance between columns. Even though the dimensional change at each joint may be very small, it can accumulate to an objectionable amount in a long row of columns. One way to reduce the distortion is to allow for shrinkage at each joint, say, $\frac{1}{16}$ in. for a 20-ft bay, by tilting or spreading the columns. Thus, a spread of $\frac{1}{8}$ in. for the two ends of a beam with flanges butt-welded to the columns may be built in at the fabricating shop; for example, by increasing the spacing of erection-bolt holes in the beam bottom flange. Control in the field, however, is maintained by guy wires until all joints are welded.

Shortening of bays can become acute in a column row in which beams connect to column flanges, because the shrinkage shortening could possibly combine with the mill underrun in column depths. Occasionally, in addition to spreading the columns, it may be necessary to correct the condition by adding filler plates or building out with weld metal.

Some designers of large welded structures prefer to detail the welding sequence for each joint. For example, on one project, the procedure for the joint shown in Fig. 6-62 called for four distinct operations, or stages: first, the top 6 in. of the shear weld on the vertical connection was made; second, the weld on the top flange; third, the bottom-flange weld; and fourth, the remaining weld of the vertical connection. The metal was allowed to return to normal temperature before starting each stage. One advantage of this procedure is the prestressing benefits obtained in the connecting welds. Tensile stresses are developed in the bottom-flange weld on cooling; compressive stresses of equal magnitude consequently are produced in the top flange. Since these stresses are opposite to those caused by floor loads, welding stresses are useful in supporting the floor loads. Although this by-product assistance may be worth while, there are no accepted methods for resolving the alleged benefits into design economy.

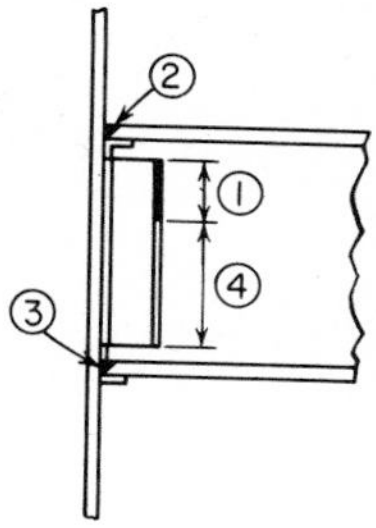

Fig. 6-62. Indication of sequence in welding connections.

Multistory structures erected with a guy derrick supported on the steelwork as it rises will be subjected by erection loads to stresses and strains. The resulting deformations should be considered in formulating a field-welding sequence.

6-79. Field Tolerances. Dimensional variations in the field often are a consequence of permissible variations in rolling of steel and in shop fabrication. Limits for mill variations are prescribed in the American Society for Testing and Materials Standard Specification A6, "General Requirements for Delivery of Rolled Steel Plates, Shapes, Sheet Piling, and Bars for Structural Use." For example, wide-flange beams are considered straight, vertically or laterally, if they are within ⅛ in. for each 10 ft of length. Similarly, columns are straight if the deviation is within ⅛ in. per 10 ft, with a maximum deviation of ⅜ in.

It is standard practice to compensate in shop details for certain mill variations. The adjustments are made in the field, usually with clearances and shims.

Shop-fabrication tolerance for straightness of columns and other compression members often is expressed as a ratio, 1:1,000, between points of lateral support. (This should be recognized as approximately the equivalent of ⅛ in. per 10 ft, and since such members rarely exceed 30 ft in length, between lateral supports, the ⅜-in. maximum deviation prevails.) Length of fabricated beams have a tolerance of 1/16 in. up to 30 ft and ⅛ in. over 30 ft. Length of columns finished to bear on their ends have a tolerance of 1/32 in.

Erected beams are considered level and aligned if the deviation does not exceed 1:500. Similarly, columns are plumb and aligned if the deviation of individual pieces, between splices in the usual multistory building, does not exceed 1:500. The total or accumulative displacement for multistory columns cannot exceed the limits prescribed in the American Institute of Steel Construction "Code of Standard Practice." For convenience, these are indicated in Fig. 6-63. Control is placed only on the exterior columns and those in the elevator shaft.

Field measurements to determine whether columns are plumb should always be made at night or on cloudy days, never in sunshine. Solar heat induces differential thermal stresses, which cause the structure to curl away from the sun by an amount that renders plumbing measurements useless.

If beam flanges are to be field welded (Fig. 6-54*a*) and the shear connection is a high-strength-bolted, friction-type joint, the holes should be made oversize or horizontal slotted (Art. 6-55), thus providing some built-in adjustment to accommodate mill and shop tolerances for beams and columns.

Similarly, for beams with framed connections (Figs. 6-47 and 6-49) that will be field bolted to columns, allowance should be made in the details for finger-type shims, to be used where needed for column alignment.

Because of several variables, bearing of column joints is seldom in perfect contact across the entire cross-sectional area. The AISC recommends acceptance if gaps between the bearing surfaces do not exceed 1/16 in. Should a gap exceed 1/16 in. and an engineering investigation shows need for more contact area, the gap may be filled with mild steel shims.

Tolerance for placing of machinery directly on top of several beams is another problem occasionally encountered in the field. The elevation of beam flanges will vary because of permissible variations for mill rolling, fabrication, and erection. This should be anticipated and adequate shims provided for field adjustments.

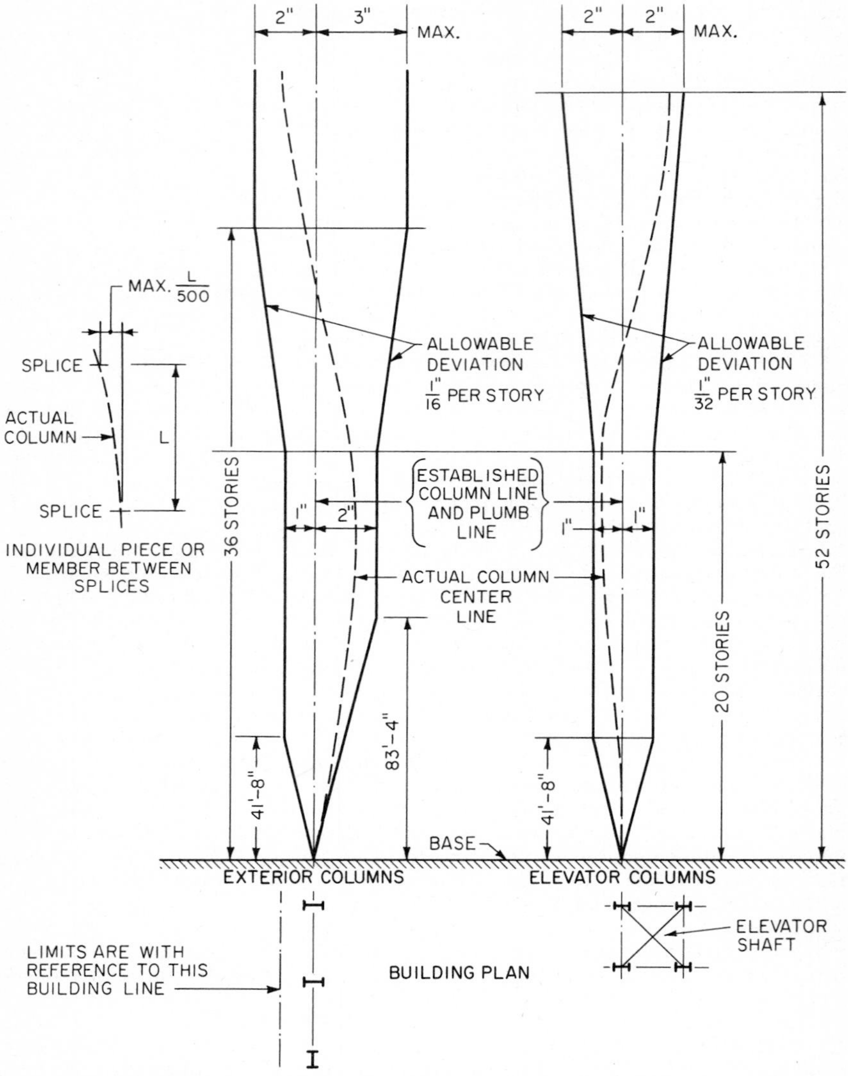

Fig. 6-63. Permissible deviations from plumb for columns.

6-80. Adjusting Lintels. Lintels supported on the steel frame (sometimes called shelf angles) may be permanently fastened in the shop to the supporting spandrel beam, or they may be attached so as to allow adjustment in the field (see Fig. 6-4, Art. 6-7). In the former case, the final position is solely dependent on the alignment obtained for the spandrel itself; whereas for the latter, lintels may be adjusted to line and grade independently of the spandrel. Field adjustment is the general rule for all multistory structures. Horizontal alignment is obtained by using slotted holes in the connection clip angles; vertical elevation (grade) is obtained with shims.

When walls are of masonry construction, a reasonable amount of variation in the position of lintels may be absorbed without much effort by masons. So the erector can adjust the lintels immediately following the permanent fastening of the spandrels to the columns. This procedure is ideal for the steel erector, since it allows him to complete his contract without costly delays and without interference with other trades. Subsequent minor variations in the position of the lintels, due to deflection or torsional rotation of the spandrel when subjected to dead weight of the floor slab, are usually absorbed without necessitating further lintel adjustment.

However, with lightweight curtain walls, the position of the lintels is important, because large paneled areas afford less latitude for variation. As a rule, the steel erector is unable to adjust the lintels to the desired accuracy at the time the main framework is erected. If he has contracted to do the adjusting, he must wait until the construction engineer establishes the correct lines and grades. In the usual case, floor slabs are concreted immediately after the steelwork is inspected and accepted. The floor grades then determined become the base to which the lintels can be adjusted. At about the same time, the wall contractor has his scaffolds in place, and by keeping pace with wall construction, the steel erector, working from the wall scaffolds, adjusts the lintels.

In some cases, the plans call for concrete encasement of the spandrel beams, in which case concreting is accomplished with the floor slab. Naturally, the construction engineer should see that the adjustment features provided for the lintels are not frozen in the concrete. One suggestion is to box around the details, thus avoid chopping out concrete. In some cases, it may be possible to avoid the condition entirely by locating the connection below the concrete encasement where the adjustment is always accessible.

The whole operation of lintel adjustment is one of coordination between the several trades; that this be carried out in an orderly fashion is the duty of the construction engineer. Furthermore, the desired procedure should be carefully spelled out in the job specifications so that erection costs can be estimated fairly.

Particularly irksome to the construction engineer is the lintel located some distance below the spandrel and supported on flexible, light steel hangers. This detail can be troublesome because it has no capacity to resist torsion. Avoid by developing the lintel and spandrel to act together as a single member.

PAINTING

Protection of steel surfaces has been, since the day steel was first used, a vexing problem for the engineers, paint manufacturers, and maintenance personnel. Over the years, there have been many developments, the result of numerous studies and research activities. However, it was not until 1950 that a concerted attempt was made to correlate all available data. Then, the Steel Structures Painting Council was organized, and an effort was made to determine and outline the best methods developed up to the present time, issue specifications covering practical and economical methods of surface preparation and painting steel structures, and engage in further research aimed at reducing or preventing steel corrosion.

Results are published in the "Steel Structures Painting Manual." This work is in two volumes—Vol. I, "Good Painting Practice," and Vol. II, "Systems and Specifications" (Steel Structures Painting Council, 4400 Fifth Avenue, Pittsburgh Pa. 15213). Each of the Council's paint systems covers the method of cleaning surfaces, types of paint to be used, number of coats to be applied, and techniques to be used in their applications. Each surface treatment and paint system is identified by uniform nomenclature; e.g., Paint System Specification SSPC-PS7.00-64T, which happens to be the identity of the minimum-type protection as furnished for most buildings. Undoubtedly, these paint systems will eventually become a national standard.

The Council, because of its broad representation, did not confine its work to bridges and buildings but has embraced the entire field of protection for all types of steel structures; viz., tanks, ships, pipelines, petroleum refineries, chemical plants, sewage works, coke ovens, etc., and under all kinds of service conditions. All kinds of paints, oils, lacquers, and metalizing methods are reviewed.

6-81. Corrosion of Steel. Ordinarily, steel corrodes in the presence of both oxygen and water, but corrosion rarely takes place in the absence of either. For instance, steel does not corrode in dry air, and corrosion is negligible when the relative humidity is below 70%, the critical humidity at normal temperature. Likewise, steel does not corrode in water that has been effectively deaerated. Therefore, the corrosion of structural steel is not a serious problem, except where water and oxygen are in abundance and where these primary prerequisites are supplemented with corrosive chemicals such as soluble salts, acids, cleaning compounds, and welding fluxes.

In ideal dry atmospheres, a thin transparent film of iron oxide forms. This layer of ferric oxide is actually beneficial, since it protects the steel from further oxidation.

When exposed to water and oxygen in generous amounts, steel corrodes at an average rate of roughly 5 mils loss of surface metal per year. If the surface is comparatively dry, the rate drops to about ½ mil per year after the first year, the usual case in typical industrial atmospheres. Excessively high corrosion rates occur only in the presence of electrolytes or corrosive chemicals; usually, this condition is found in localized areas of a building.

Mill scale, the thick layer of iron oxides that forms on steel during the rolling operations, is beneficial as a protective coating, provided it is intact and firmly adheres to the steel. In the mild environments generally encountered in most buildings, mill scale that adheres tightly after weathering and handling offers no difficulty. In buildings exposed to high humidity and corrosive gases, broken mill scale may be detrimental to both the steel and the paint. Through electrochemical action, corrosion sets in along the edges of the cracks in the mill scale and in time loosens the scale, carrying away the paint.

Galvanic corrosion takes place when dissimilar metals are connected together. Noble metals such as copper and nickel should not be connected to structural steel with steel fasteners, since the galvanic action destroys the fasteners. On the other hand, these metals may be used for the fasteners, because the galvanic action is distributed over a large area and consequently little or no harm is done. When dissimilar metals are to be in contact, the contacting surfaces should be insulated; paint is usually satisfactory.

6-82. Painting Steel Structures. Evidence obtained from dismantled old buildings and from frames exposed during renovation indicates that corrosion does not occur where steel surfaces are protected from the atmosphere. Where severe rusting was found and attributed to leakage of water, presence or absence of shop paint had no significant influence. Consequently, the American Institute of Steel Construction "Specification for the Design, Fabrication and Erection of Structural Steel for Buildings" exempts from one-coat shop paint, at one time mandatory, all steel framing that is concealed by interior finishing materials—ceilings, fireproofing partitions, walls, and floors.

Structures may be grouped as follows: (1) those that need no paint, shop or field; (2) those in which interior steelwork will be exposed, probably field painted; (3) those fully exposed to the elements. Thus, shop paint is required only as a primer coat before a required coat of field paint.

Group (1) could include such structures as apartment buildings, hotels, dormitories, office buildings, stores and schools, where the steelwork is enclosed by other materials. The practice of omitting the shop and field paint for these structures, however, may not be widely accepted because of tradition and the slowness of building-code modernization. Furthermore, despite the economic benefit of paint omission, clean, brightly painted steel during construction has some publicity value.

In group (2) are warehouses, industrial plants, parking decks, supermarkets, one-story schools, inside swimming pools, rinks, and arenas, all structures shielded from the elements but with steel exposed in the interior. Field paint may be required for corrosion protection or appearance or both. The severity of the corrosion environment depends on type of occupancy, exposure, and climatic conditions. The paint system should be carefully selected for optimum effectiveness.

In group (3) are those structures exposed at all times to the weather: crane runways, fire escapes, towers, exposed exterior columns, etc. When made of carbon steel, the members are painted after erection and therefore are primed with shop paint. The paint system selected should be the most durable one for the atmospheric conditions at the site. For corrosion-resistant steels, such as those meeting ASTM Specification A242 and A588, field painting may be unnecessary. On exposure, these steels acquire a relatively hard coat of oxide, which shields the surface from progressive rusting. The color, russet brown, has architectural appeal.

6-83. Paint Systems. The Steel Structures Painting Council has correlated surface preparations, primer, intermediate, and finish coats of paints into systems, each designed for a common service condition ("Steel Structures Painting Manual"). In addition, the Council publishes specifications for each system and individual specifications for surface preparations and paints. Methods for surface cleaning include solvent, hand-tool, power-tool, pickling, flame, and several blast techniques.

Surface preparation is directly related to the type of paints. In general, a slow-drying paint containing oil and rust-inhibitive pigments and one possessing good wetting ability may be applied on steel nominally cleaned. On the other hand, a fast-drying paint with poor wetting characteristics requires exceptionally good surface cleaning, usually entailing complete removal of mill scale. Therefore, in specifying a particular paint, the engineer should include the type of surface preparation, to prevent an improper surface condition from reducing the effectiveness of an expensive paint.

Paint selection and surface preparation are a matter of economics. For example, while blast-cleaned surfaces are conceded to be the best paint foundation for lasting results, its high cost is not always justified. Nevertheless, the Council specifies a minimum surface preparation by a blast cleaning process for such paints as alkyd, phenolic, vinyl, coal tar, epoxy, and zinc-rich.

As an aid for defining and evaluating the various surface preparations, taking into account the initial condition of the surface, an international visual standard is available and may be used. A booklet of realistic color photographs for this purpose can be obtained from the Council or American Society for Testing and Materials. The applicable standard and acceptance criteria are given in "Quality Criteria and Inspection Standards," published by the American Institute of Steel Construction.

The Council stresses the relationship between the prime coat (shop paint) and the finish coats. A primer that is proper for a particular type of field paint could be an unsatisfactory base for another type of field paint. Since there are numerous paint formulations, refer to Council publications when faced with a painting problem more demanding than ordinary.

In the absence of specific contract requirements for painting, the practice described in the American Institute of Steel Construction "Specification for the Design, Fabrication and Erection of Structural Steel for Buildings" may be followed. This method may be considered "nominal." The steel is brushed, by hand or power, to remove loose mill scale, loose rust, weld slag, flux deposit, dirt, and foreign matter. Oil and grease spots are solvent cleaned. The shop coat is a commercial-quality paint applied by brushing, dipping, roller coating, flow coating, or spraying to a 2-mil thickness. It affords only short-time protection. Therefore, finished steel that may be in ground storage for long periods or otherwise exposed to excessively corrosive conditions may exhibit some paint failure by the time it is erected, a condition beyond the control of the fabricator. Where such conditions can be anticipated, as for example, an overseas shipment, the engineer should select the most effective paint system.

6-84. Field-painting Steel. There is some question on the justification for protecting steelwork embedded in masonry or in contact with exterior masonry walls built according to good workmanship standards but not impervious to moisture. For example, in many instances, the masonry backing for a 4-in. brick wall is omitted to make way for column flanges. Very definitely, a 4-in. wall will not prevent penetration of water. In many cases, also, though a gap is provided between a wall and steelwork, mortar drippings fall into the space and form bridges over which water may pass, to attack the steel. The net effect is premature failure of both wall and steel. Walls have been shattered—sheared through the brick—by the powerful expansion of rust formations. The preventatives are: (1) coat the steel with suitable paint and (2) good wall construction (see Sec. 10).

A typical building code reads: "Special precautions shall be taken to protect the outer surfaces of steel columns located in exterior walls against corrosion, by painting such surfaces with waterproof paints, by the use of mastic, or by other methods or waterproofing approved by the building inspector."

In most structures an asphalt-type paint is used for column-flange protection. The proviso is sometimes extended to include lintels and spandrels, since the problem of corrosion is similar, depending on the closeness and contact with the wall. However, with the latter members, it is often judicious to supplement the paint with flashing, either metallic or fabric. A typical illustration, taken from an actual apartment-building design, is shown in Fig. 6-64.

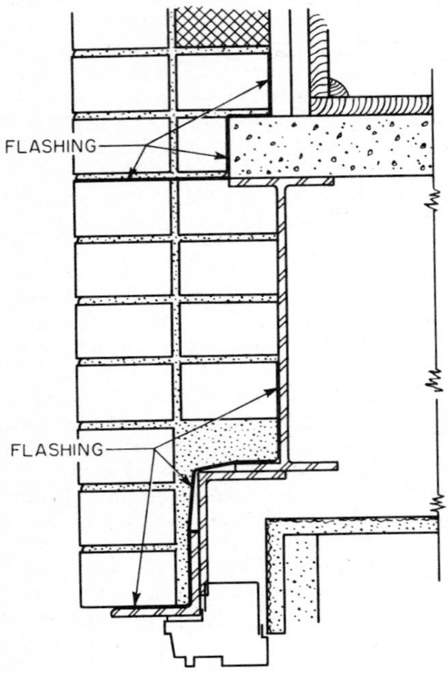

Fig. 6-64. Flashing at spandrels and lintels.

In general, building codes differ on field paint; either paint is stipulated or the code is silent. From a practical viewpoint, the question of field painting cannot be properly resolved with a single broad rule. For an enclosed building in which the structural members are enveloped, for example, a field coat is sheer wastage, except for exterior steel members in contact with walls. On the other hand, exposed steel subject to high-humidity atmospheres and to exceptionally corrosive gases and contaminants may need two or three field coats.

Manufacturing buildings should always be closely scrutinized, bearing in mind that original conditions are not always permanent. As manufacturing processes change, so do the corrosive environments stimulated by new methods. It is well to prepare for the most adverse eventuality.

Special attention should be given to steel surfaces that become inaccessible; e.g., tops of purlins in contact with roof surfaces. A three-coat job of particularly suitable paint may pay off in the long run, even though it delays placement of the roof covering.

6-85. Steel in Contact with Concrete. According to the "Steel Structures Painting Manual," Vol. I, "Good Painting Practice" (Steel Structures Painting Council, Pittsburgh, Pa.):

1. Steel that is embedded in concrete for reinforcing should not be painted. Design considerations require strong bond between the reinforcing and the concrete so that the stress is distributed; painting of such steel does not supply sufficient bond. If the concrete is properly made and of sufficient thickness over the metal, the steel will not corrode.

2. Steel that is encased with exposed lightweight concrete that is porous should be painted with at least one coat of good quality rust-inhibitive primer. When conditions are severe, or humidity is high, two or more coats of paint should be applied, since the concrete may accelerate corrosion.

3. When steel is enclosed in concrete of high density or low porosity, and when the concrete is at least 2 in. to 3 in. thick, painting is not necessary, since the concrete will protect the steel.

4. Steel in partial contact with concrete is generally not painted. This creates an undesirable condition, for water may seep into the crack between the steel

and the concrete, causing corrosion. A sufficient volume of rust may be built up, spalling the concrete. The only remedy is to chip or leave a groove in the concrete at the edge next to the steel and seal the crack with an alkali-resistant calking compound (such as bituminous cement).

5. Steel should not be encased in concrete that contains cinders, since the acidic condition will cause corrosion of the steel.

FIRE PROTECTION FOR STRUCTURAL STEEL

Structural steel is a noncombustible material. It is therefore satisfactory for use without protective coverage in many types of buildings where noncombustibility is sufficient, from the viewpoint of either building ordinances or owner's preference. When structural steel is used in this fashion, it is described as "exposed" or "unprotected." Naturally, unprotected steel may be selected wherever building codes permit combustible construction; the property of noncombustibility makes it more attractive than competitive materials not possessing this property.

Exposed or unprotected structural steel is commonly used for industrial-type buildings, hangars, auditoriums, stadiums, warehouses, parking garages, billboards, towers and low stores, schools, and hospitals. In most cases, these structures contain little combustible material. In others, where the contents are highly combustible, sprinkler or deluge systems may be incorporated to protect the steelwork.

6-86. Need for Fire Protection of Steel. Steel building frames and floor systems must be covered with fire-resistant materials in certain buildings to reduce the chances of fire damage. These structures may be tall buildings, such as offices, apartments, and hotels, or low-height buildings, such as warehouses, where there is a large amount of combustible content. The buildings may be located in congested areas, where the spread of fire is a strong possibility. So for public safety, as well as to prevent property loss, building codes regulate the amount of fire resistance that must be provided.

The following are some of the factors that enter into the determination of minimum fire resistance for a specific structure: height, floor area, type of occupancy (a measure of combustible contents), fire-fighting apparatus, sprinkler systems, and location in a community (fire zone), which is a measure of hazard to adjoining properties.

6-87. Fire-protection Engineering. At one time, most building codes designated all buildings as fireproofed or nonfireproofed, sometimes with an intermediate grade of semifireproofed, but usually without consideration of the combustible contents. However, it is just as logical to design a building for a fire resistance compatible with intended occupancy as it is to design the structure to support the live load for that occupancy. Fire-protection engineers are motivated by the desire to eliminate or reduce the waste in protection that was engendered by the older building codes.

Too little significance has been given to the meaning of "fireproofed." Some so-called "fireproof" buildings have been totally destroyed by fire, while others suffered no physical damage to their frame—like a good stove—but loss of life was great.

The modern concept of fire-resistive construction has followed two courses: (1) Buildings are designed to resist the potential fire severity of their combustible contents; (2) the fire-resistance ratings obtainable with various materials of construction are established by performance tests that are nationally recognized.

The first course is based on the evidence that there is a fairly definite relationship between the amount of combustible contents and resulting fire severity. Combustible contents may be readily expressed in terms of weight per square foot of floor area (see BMS 92 and 149).

Fire severity is a measure of the intensity and duration of a fire. It is expressed in terms of time of exposure equivalent to that in the standard furnace test—an internationally recognized standard (Fig. 6-65). For any stated period of duration of test, the rise in temperature must approximate the gradient shown by the curve; e.g., at 2 hr the structural assembly under test is exposed to a temperature of 1850°F, but in attaining this temperature, the furnace must record: 1000°F in

5 min, 1300°F in 10 min, 1550°F in 30 min; these being recorded on nine or more thermocouples distributed in the furnace near the test specimen. This performance test approximates the actual conditions encountered in real fires. Many fires, however, are of the smoldering type—of low intensity. Consequently, they could burn for hours in what may be not more than 1-hr resistive construction. Therefore, established ratings are a measure of time and intensity related together according to the standard curve in Fig. 6-65.

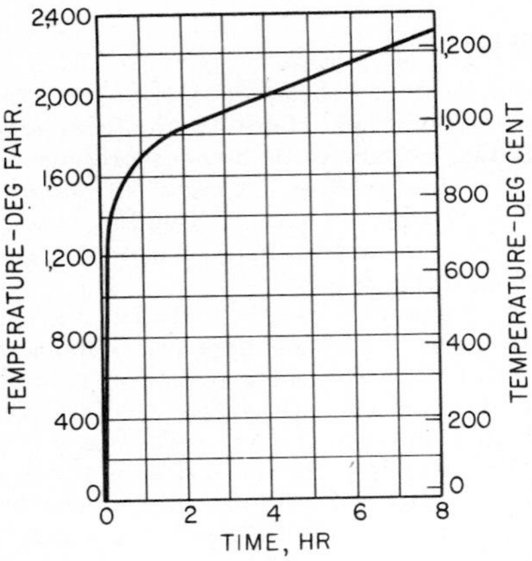

Fig. 6-65. Time-temperature curve for standard fire test.

The relationship of weight of combustibles to fire severity is given in Table 6-26. Included also is the type of occupancy commensurate with the amount of combustibles usually found. This realistic appraisal of fire load is not necessarily a recommendation. In authoring building-code regulations, it becomes necessary to evaluate other factors; e.g., height, area, fire zones, etc.

Of equal importance are the ratings assigned to the fire resistance of various building components—trusses, girders, floor systems, columns, and walls—based on a common standard (ASTM E119, "Standard Methods of Fire Tests of Building Construction and Material"). It is applicable to assemblies of masonry units and to composite assemblies of structural materials for buildings, including walls, partitions, columns, girders, beams, slabs, and composite slab-and-beam assemblies for floors and roofs.

The test specimens must be a certain minimum size; e.g., a floor assembly must have an area exposed to fire of not less than 180 sq ft, with neither dimension less than 12 ft. Furnace temperatures must adhere to the standard time-temperature relationship (Fig. 6-65). Critical end points for terminating the test are fixed

Table 6-26. Fire Severity of Various Occupancies

Average weight of combustibles, psf floor area	Fire severity, hr	Building type
5	½	Hospitals
7½	¾	Hotels, apartments, schools, department stores
10	1	
15	1½	Furniture storage, offices
20	2	Laundries
30	3	Heavy storage and warehouses
40	4½	
50	6	
60	7½	

for each type of construction. For example, one of the critical end points of a floor assembly is reached when the heat penetrating the floor raises the temperature on the unexposed surface more than 250°F above its initial temperature. Similarly, the critical end point of a steel-column test is reached when the average temperature of the steel across any one of four test levels reaches 1000 or 1200°F on any single reading. Corresponding limits for beams are 1100 and 1300°F.

6-88. Effect of Heat on Steel. A moderate rise in temperature of structural steel, say up to 500°F, is beneficial in that the strength is about 10% greater than the normal value. Above 500°F, strength falls off, until at 700°F it is approximately equal to the normal temperature strength. At a temperature of 1000°F, the compressive strength of steel is about the same as the maximum allowable working stress in columns. Therefore, it is permissible for tension and compression members to carry their maximum working stresses if the average temperature in the member does not exceed the temperature limits given in Art. 6-87 (for carbon and low-alloy steels; for other steels some adjustment may be necessary).

Unprotected steel members have a rating of about 15 min, based on tests of columns with cross-sectional areas of about 10 sq in. Heavier columns, possessing greater mass for dissipation of heat, afford greater resistance—20 min perhaps. Columns with reentrant space between flanges filled with concrete, but otherwise exposed, have likewise been tested; where the total area of the solid cross section approximates 36 sq in., the resistance is 30 min, and where the area is 64 sq in., the resistance is 1 hr.

The average coefficient of expansion for structural steel between the temperatures of 100 and 1200°F is given by the formula

$$C = 0.0000061 + 0.0000000019t \qquad (6\text{-}36)$$

in which C = coefficient of expansion per °F
t = temperature, °F

Below 100°F, the average coefficient of expansion is taken as 0.0000065.

The modulus of elasticity of structural steel, about 29,000 ksi at room temperature, decreases linearly to 25,000 ksi at 900°F. Then, it drops at an increasing rate at higher temperatures.

6-89. Materials for Improving Fire Resistance. Structural steel has been protected with many materials—brick, stone, concrete, tile, gypsum, mineral fibers, and various fire-resistant plasters.

Concrete insulation serves well for column protection, in that it gives additional stability to the steel section. Also, it is useful where abrasion resistance is needed. Concrete, however, is not an efficient insulating medium compared with fire-resistant plasters. Normally, it is placed completely around the columns, beams, or girders, with all reentrant spaces filled solid. Although this procedure contributes to the stability of columns and effects composite action in beams and slabs, it has the disadvantage of imposing great weight on the steel frame and foundations. For instance, full protection of a W12 column with stone concrete weighs about 355 psf, whereas plaster protection weighs about 40 psf. However, lightweight concretes may be used, made with such aggregates as perlite, vermiculite, expanded shale, expanded slag, pumice, pumicite, and sintered flyash.

Considerable progress has been made in the use of lightweight plasters with aggregates possessing good insulating properties. Two aggregates used extensively are perlite and vermiculite; they replace sand in the sanded-gypsum plaster mix. A 1-in. thickness weighs about 4 psf, whereas the same thickness of sanded-gypsum plaster weighs about 10 psf.

Typical details of lightweight plaster protection for columns are shown in Fig. 6-66. Generally, vermiculite and perlite plaster thicknesses of 1 to 1¾ in. afford protection of 3 and 4 hr, depending on construction details.

For buildings where rough usage is expected, a hard, dense insulating material such as concrete, brick, or tile would be the logical selection for fire protection. Concrete made with lightweight aggregates may be suitable.

For many buildings, finished ceilings are mandatory. It is therefore logical to employ the ceiling for protecting roof and floor framing. All types of gypsum plasters are used extensively for this dual purpose. Figure 6-67 illustrates typical installations. For 2-hr floors, ordinary sand-gypsum plaster ¾ in. thick is sufficient. Three and four-hour floors may be obtained with perlite gypsum and vermiculite gypsum in thickness range of ¾ to 1 in.

Instead of plastered ceilings, use may be made of fire-rated dry ceilings, acoustic tiles, or drop (lay-in) panels. But to qualify for the specified rating, such units must be individually secured to their suspended supports with hold-down clips in the same manner as when the units are fire tested.

Another alternative is to spray the structural steel mechanically (where it is not protected with concrete) with plasters of gypsum, perlite, or vermiculite, proprietary cementitious mixtures, or mineral fibers not deemed a health hazard during spraying (Fig. 6-68). In such cases, the fire resistance rating of the structural system is independent of the ceiling. Therefore, the ceiling need not be of fire-rated construction. Drop panels, if used, need not be secured to their suspended supports.

Still another sprayed-on material is the intumescent fire-retardant coating, essentially a paint. Tested in conformance with ASTM Specification E119, a $\frac{3}{16}$-in.-thick coat applied to a steel column has been rated 1 hr, a $\frac{1}{2}$-in.-thick coating 2 hr. As applied, the coating has a hard, durable finish, but at high temperatures, it puffs to many times its original thickness, thus forming an effective insulating blanket. Thus, it serves the dual need for excellent appearance and fire protection.

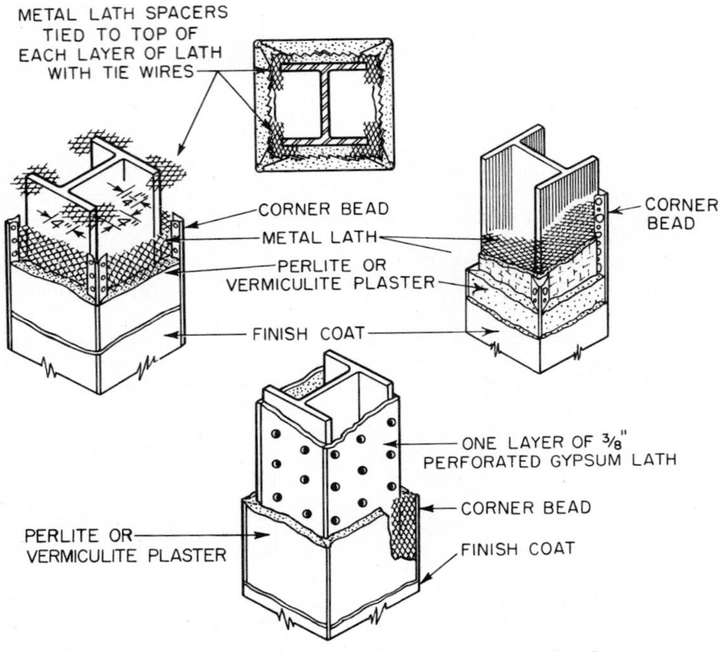

Fig. 6-66. Typical column fire protection with plaster.

Aside from dual functioning of ceiling materials, the partitions, walls, etc., being of incombustible material, also protect the structural steel, often with no additional assistance. Fireproofing costs, therefore, may be made a relatively minor expense in the over-all costs of a building through dual use of materials.

6-90. Pierced Ceilings and Floors. Some buildings require recessed light fixtures and air-conditioning ducts, thus interrupting the continuity of fire-resistive ceilings. A rule that evolved from early standard fire tests permitted 100 sq in. of openings for noncombustible pipes, ducts, and electrical fixtures in each 100 sq ft of ceiling area.

It has since been demonstrated, with over 100 fire tests that included electrical fixtures and ducts, that the fire-resistance integrity of ceilings is not impaired when, in general:

Recessed light fixtures, 2 by 4 ft, set in protected steel boxes, occupy no more than 25% of the gross ceiling area.

Air-duct openings, 30-in. maximum in any direction, are spaced so as not to occupy more than 576 sq in. each 100 sq ft of gross ceiling area. They must be protected with fusible link dampers against spread of smoke and heat.

These conclusions are not always applicable. Reports of fire tests of specific floor systems should be consulted.

A serious infringement of the fire rating of a floor system could occur when pipes, conduit, or other items pierce the floor slab, a practice called "poke-through." Failure to calk the openings with insulating material results in a lowering of fire ratings from hours to a few minutes.

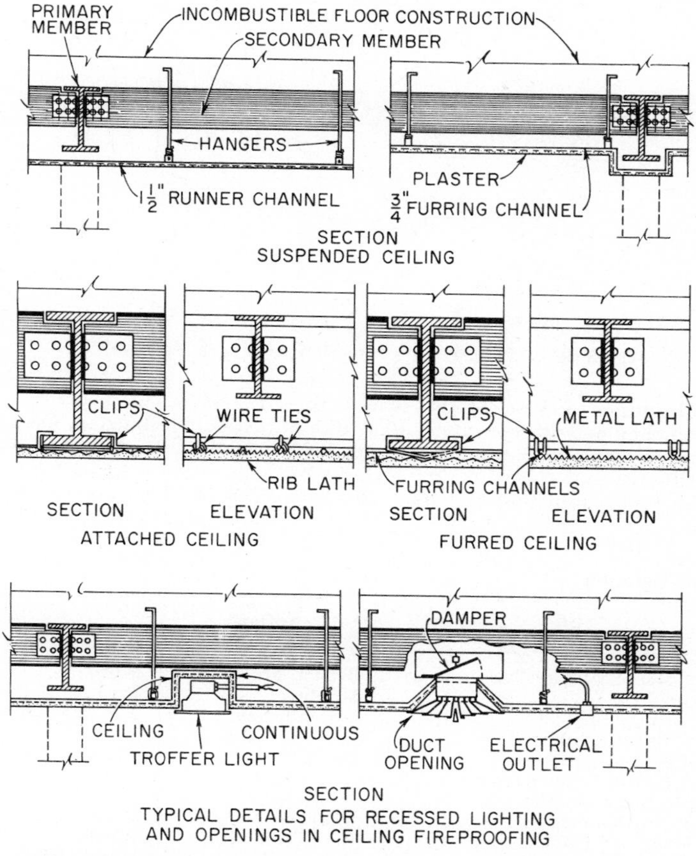

Fig. 6-67. Plaster-ceiling fire protection for floor framing.

6-91. Fire-resistance Ratings. Most standard fire tests on structural-steel members and assemblies have been conducted at one of two places—the National Bureau of Standards, Washington, D.C., or the Underwriters' Laboratories, Chicago. Fire-testing laboratories also are available at Ohio State University, Columbus, Ohio, and the University of California, Berkeley, Calif. Laboratory test reports form the basis for establishing ratings. Summaries of these tests, together with tabulation of recognized ratings, are published by a number of organizations listed below. The trade associations, for the most part, limit their ratings to those constructions employing the material they represent.

The American Insurance Association (formerly The National Board of Fire Underwriters), 85 John St., New York, N.Y. 10038.

The National Bureau of Standards, Washington, D.C. 20234.

Gypsum Association, 201 North Wells St., Chicago, Ill. 60606.

Metal Lath Association, W. Federal St., Niles, Ohio 44446.

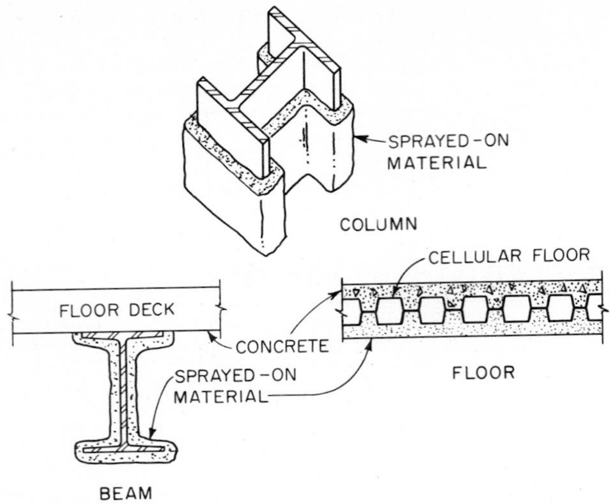

Fig. 6-68. Typical fire protection with sprayed material.

Perlite Institute, 45 West 45th St., New York, N.Y. 10036.
American Iron and Steel Institute, 1000 16th St., N.W., Washington, D.C. 20036.
American Institute of Steel Construction, 1221 Avenue of the Americas, New York, N.Y. 10020.
Acoustical and Insulating Materials Association, 205 W. Touhy Ave., Park Ridge, Ill. 60068.

6-92. Bibliography

Title	Author	Publisher
"Design of Steel Structures"	Bresler, Lin, Scalzi	Wiley
"Steel Structures"	McGuire	Prentice-Hall
"Erecting Structural Steel"	Oppenheimer	McGraw-Hill
"Structural Steel Detailing"	....	AISC
"Design of Welded Structures"	Blodgett	Lincoln Foundation
"Column Research Council Guide to Design Criteria for Metal Compression Members"	Johnston	Wiley
"Plastic Design in Steel"	Joint Committee	ASCE
"Applied Plastic Design in Steel"	Disque	Van Nostrand Reinhold
"Steel Design Manual"	Brockenbrough and Johnston	U.S. Steel Corp.
"Structural Steel Designers' Handbook"	Merritt	McGraw-Hill
"Manual of Steel Construction"	....	AISC
"Iron and Steel Beams, 1873–1952"	....	AISC
"Fundamentals of Welding"	....	AWS
"Structural Welding Code, D1.1"	....	AWS
"Quality Criteria and Inspection Standards"	....	AISC
"Steel Structures Painting Manual," Vols. I and II	Steel Structures Paint Council	SSPC
"Torsion Analysis of Rolled Steel Sections"	....	Bethlehem Steel Co.
"Fire-Resistant Steel-Frame Construction"	....	AISI

Section 7

Lightweight Steel Construction

F. E. FAHY, F.ASCE

The term lightweight steel construction, as used in this section, refers to structural components and assemblies that are made of steel but consist of and are produced from other than rolled structural-steel shapes and plates. Although many efficient, lightweight rolled structural shapes are available, they usually are considered structural steel, whereas this section deals with fabricated components made from basic plain-material forms, such as bars, sheet, and strip. These include:

Open-web steel joists, which are usually fabricated from relatively small bars, bar-size angles, and shapes formed from flat-rolled material.

Framing members formed from sheet and strip steel, either by roll-forming or in bending- or press-brake operations.

Deck and panel work in which steel sheets are shaped to provide both coverage and load-carrying capacity.

Many different kinds of lightweight structural components and building systems that employ them have been developed. The term lightweight does not mean that use of these components is limited to small or light-occupancy structures. Alone or in combination with structural steel, lightweight components are used in both custom-built and factory-produced buildings for all kinds of occupancies, from small and sometimes temporary structures to massive high-rise buildings.

This section describes a few forms that have become well established, and the general principles underlying their structural design and use.

OPEN-WEB STEEL JOISTS

Under "Standard Specifications for Open-web Steel Joists, J- and H-Series," adopted jointly by the Steel Joist Institute (SJI) and the American Institute of Steel Construction (AISC), open-web steel joists are relatively small, parallel-chord steel trusses.

They are suitable for direct support of floors and roof decks in buildings, when designed in accordance with those specifications and the standard load tables that accompany them.

Longspan steel joists, also covered by specifications adopted by SJI and AISC, are essentially structural steel trusses designed for somewhat heavier duty and substantially longer spans than regular open-web steel joists. They are not included in the following discussion.

As usually employed in floor construction, open-web steel joists are covered by a slab of concrete, 2 to 2½ in. thick, cast on permanent forms (centering). A wall-bearing, open-web steel-joist floor assembly, with ceiling, is illustrated in Fig. 7-1. Use of open-web steel joists in roofs is discussed in Art. 7-7.

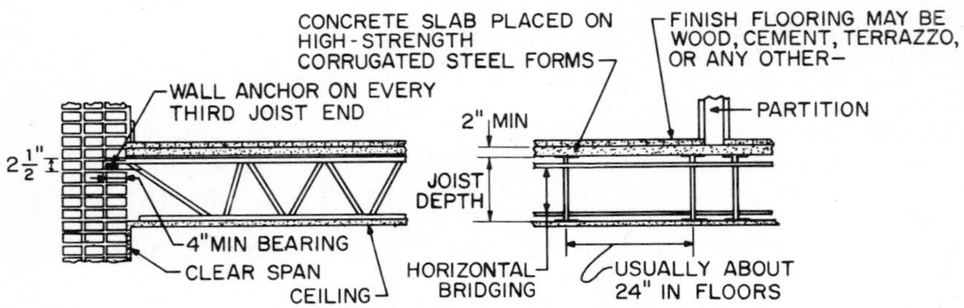

Fig. 7-1. Open-web steel-joist construction.

Open-web steel joists usually are supported either on structural-steel framing or masonry bearing walls. When used with structural steel, the joists are preferably welded to the supporting framework, but may be bolted or clipped. For wall-bearing applications, wall anchors are usually specified, as shown in Fig. 7-1.

In addition to light weight, one of the advantages of open-web steel joist construction is that the opening between chords provides space for electrical work, ducts, and piping.

7-1. Joist Design and Fabrication. Standardization of open-web steel joists under SJI-AISC specifications mentioned previously consists essentially of definition of product, specification of design basis, and specifications covering bridging and certain other details. Exact forms of the members and web systems, and methods of manufacture, are determined by individual producers. A number of proprietary designs have been developed. A few types are shown in Fig. 7-2. For the regular underslung end construction, the standard end depth is 2½ in., as shown. Square-end joists, which do not have underslung ends, have been supplied for special purposes.

SJI publishes load tables for standardized joists in depths of 8 to 30 in. in 2-in. increments. There are several different weights for each joist depth, except for the 8-in., which has only one standard weight.

J- and H-Series Joists. There are two strength groups, the J series and the H series. The J series is designed at a basic working stress of 22 ksi, based on material with a specified minimum yield strength of 36 ksi, as for A36 structural steel. The H series is designed at a basic working stress of 30 ksi, based on a minimum specified yield strength of 50 ksi. This strength is obtained by using

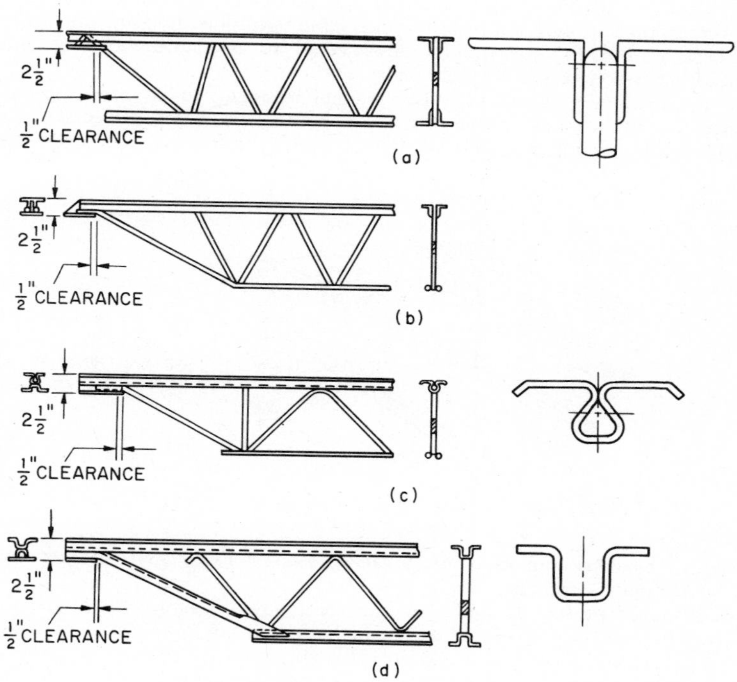

Fig. 7-2. Open-web steel joists.

either high-strength, low-alloy steel, or cold-formed sections of carbon steel with yield strength increased by the forming operation.

Specifications. Detailed rules for the structural design of open-web steel joists are given in SJI-AISC "Standard Specifications for Open-web Steel Joists, J- and H-Series," Steel Joist Institute, 2001 Jefferson Davis Highway, Arlington, Va. 22202. (See also Art. 7-2.)

Composite Construction. Although standard open-web steel joists are designed as simply supported, self-contained structural members, other joists have been designed with the top chord intended to act in a composite manner with the concrete slab that covers it. Composite action may be obtained by shaping the chord to provide shear anchorage within the slab, or by other means. Detailed information may be obtained from manufacturers' catalogs.

Fabrication. Open-web steel joists are different in one important respect from the fabricated structural-steel framing members commonly used in building construction: The joists are usually manufactured by production-line methods with special equipment designed for quantity production. Component parts are generally joined together by either resistance or electric-arc welding.

Shop Paint. Open-web steel joists are given a prime coat of paint prior to shipment. The SJI-AISC specifications (1970) call for shop paint that conforms to or meets the minimum performance requirements of Steel Structures Painting Council Specification 15-68T for Type I (red oxide) or Type II (asphalt), or of Federal Specification T-TP-636 (red oxide).

7-2. Design of Open-web Joist Floors. Open-web joists are designed primarily to be used under uniformly distributed loading, and at substantially uniform spacing. They can safely carry concentrated loads, however, if proper consideration is given to the effect of such loads. For example, good practice requires that heavy concentrated loads be applied at joist panel points. (The weight of an ordinary partition running crosswise of the joists usually is considered distributed by the floor slab to the extent that it will not cause appreciable local bending in the top chords of the joists.) But the joists must be selected to resist the bending moments, shears, and end reactions due to such loads.

Joists at Openings. Open-web joists are not designed to be used as individual framing members, although special joists have been so used under special conditions. Relatively small openings between joists may usually be framed with angle, channel,

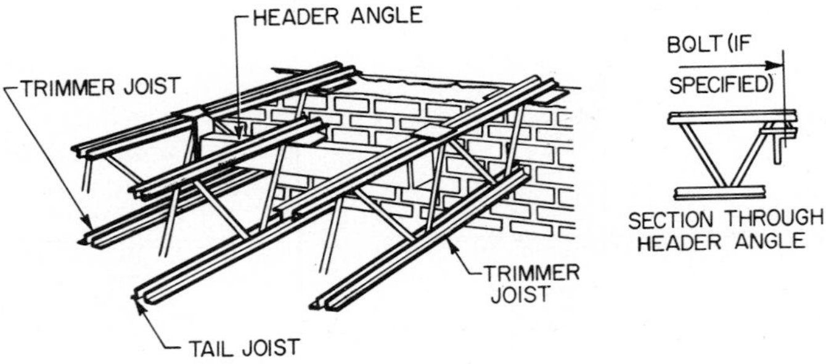

Fig. 7-3. Header construction at floor openings.

or Z-shaped headers supported on adjacent joists (Fig. 7-3). Larger openings should be framed in structural steel.

Headers should preferably be located so that they are supported at panel points of the trimmer joists. In any case in which that is not practicable and in which the header reaction exceeds 0.4 kip, consideration must be given to the bending stresses induced in the top chords of the trimmer joists by the concentrated loads from the header, or struts may be inserted in the joists at the points at which the header is supported ("Recommended Code of Standard Practice for Open-web and Longspan Steel Joists," Steel Joist Institute, May, 1968).

Joists at Heavy Loads. Open-web steel joists can be doubled and even tripled where necessary to support heavy loads.

Design Methods. The method of selecting joist sizes for any floor depends on whether or not the effect of any cross partitions or other concentrated loads must be considered. Under uniform loading only, joist sizes and spacings are most conveniently selected from a table of safe loads. The Steel Joist Institute standard load tables provide that equal concentrated top-chord loads, as in bulb-tee roof construction, may be considered uniformly distributed if the spacing of the concentrations along the joists does not exceed 33 in. Where other concentrated loads or nonuniform loads exist, calculate bending moments, end reactions, and shears, and select joists accordingly. Any case in which there may be local bending of either chord should be investigated by appropriate methods.

The deflection of open-web steel joists can be computed in the same way as for other types of beams. In computing moment of inertia of the cross section, however, the chords only should be considered, neglecting the web system. The resulting moment of inertia should be reduced 15% to account for that part of the deflection caused by web-member strain.

Effective moments of inertia of standard SJI-AISC joists are published by the Steel Joist Institute (Art. 7-1) or are available from manufacturers.

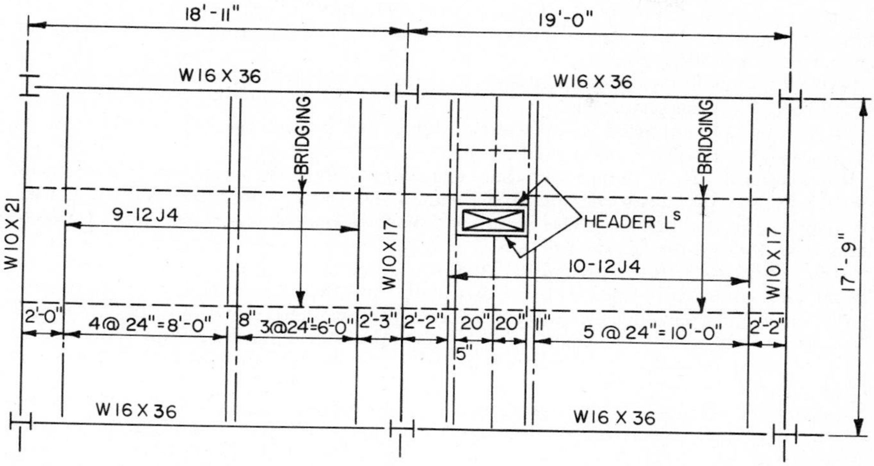

Fig. 7-4. Framing plan for part of a floor of open-web steel-joist construction.

Depth-span Limitations. The SJI-AISC specifications provide that the clear span of a joist shall not exceed 24 times its depth, except that for floors the clear span of an H-series joist shall not exceed 20 times its depth.

Figure 7-4 shows a portion of a framing plan for a typical open-web joist floor.

7-3. Extended Ends and Ceiling Extensions. Plastered ceilings attached directly to regular open-web joists are usually supported at the underslung ends by means of *ceiling extensions*. These may consist of an extension of the lower chord of the joist past the end bottom-chord panel point (Figs 7-1 and 7-5*a*), or they may consist of special attachments, depending on the length of the extension and the detailed design of the joist. For details, in a particular case, refer to the manufacturers' catalogs.

Open-web joists are frequently provided with extended ends overhanging the support (Fig. 7-5*b*). Depending on the length of the extension and the detailed design of the joist, the extended end may consist of a simple extension of the top chord of the joist, separate fabricated attachments, or a combination of both. The Steel Joist Institute publishes load tables for extended ends based on attachments that are separately fabricated from angle or channel sections. Since details may vary among different manufacturers, the individual manufacturer's practice should be followed in a particular case.

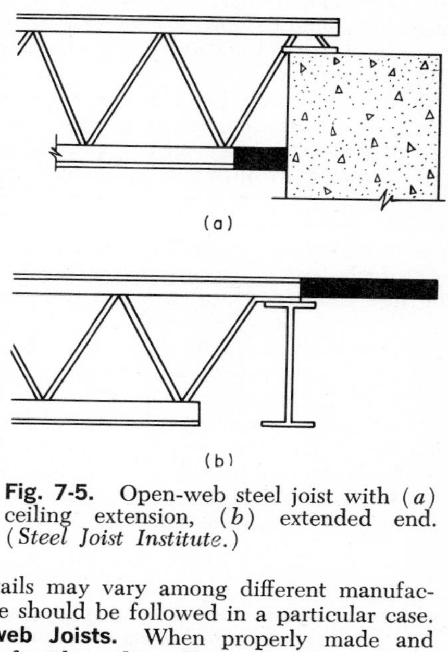

Fig. 7-5. Open-web steel joist with (*a*) ceiling extension, (*b*) extended end. (*Steel Joist Institute.*)

7-4. Bridging and Anchors for Open-web Joists. When properly made and installed, open-web steel joists are capable of withstanding all usual service requirements. The joists are, however, very flexible in the lateral direction. They are not intended to be used without proper lateral support. Consequently, bridging should be installed between joists as soon as possible after joists have been placed and before application of construction loads. The principal function of bridging is to provide lateral support during the construction period.

Bridging details vary with different manufacturers. The most commonly used type is composed of rods or angles fastened to the top and bottom chords of the joists in a vertical plane.

SJI-AISC specifications for oven-web joists contain provisions for the number of rows of bridging for each chord size.

Care must be exercised even after bridging has been installed, to avoid undue concentrations of construction loads.

Connections. It is important that masonry anchors be used on wall-bearing joists. Where the joists rest on steel beams, the joists should be welded, bolted, or clipped to the beams. The standard specifications of the Steel Joist Institute should be followed in all such details.

7-5. Formwork for Joist Construction. Formwork (centering) for concrete floors and roofs with open-web joists usually consists of a decking of corrugated steel sheets. Especially produced for this purpose and marketed under various tradenames, the sheets are usually placed on top of the joists and form a permanent part of the construction. Expanded metal rib lath or paper-backed welded-wire fabric has also been used.

Corrugated sheets can be fastened with self-tapping screws or welded to the joists. SJI-AISC specifications for open-web joists require that each attachment to the top chords of the joists be capable of resisting a lateral force of at least 0.3 kip. Spacing of attachments should not exceed 36 in. along the top chord.

7-6. Slab Reinforcement in Steel Joist Construction. When the usual cast-in-place floor slab is used, it is customary to install reinforcing bars in two perpendicular directions, or welded-wire fabric. No other reinforcement is usually considered necessary.

7-7. Open-web Joists in Roof Construction. Open-web joists frequently are used as purlins to support roofs at spacings determined by the load-carrying capacity of the decking. The joists should be properly braced. For steep-sloping roofs, sag rods should be provided, as for structural-steel purlins.

Open-web steel joists have also been used as rafters (inclined beams) to support sloping roofs.

When open-web joists are used for supporting flat roofs, consideration should be given to the problem of ponding, or increase in load due to the accumulation of rainwater in concavities caused by deflection of the roof structure. SJI-AISC specifications for open-web steel joists require that, unless a roof surface is provided with sufficient slope toward points of free drainage or adequate individual drains to prevent accumulation of rainwater, the roof system should be investigated to insure stability under ponding conditions.

A roof slope of ⅛ in. per ft is usually considered sufficient for normal conditions of free drainage. For a roof wholly or partly surrounded by parapets or other walls, and provided with individual drains, it is important that the drains be properly designed and maintained to prevent build-up of live load beyond safe limits. Such build-up, including excessive snow accumulation, can prove more damaging than the ponding that may be caused by deflection of the roof system.

The stability investigation may be made in the same way, and follow the same criteria, as for structural steel (Art. 6-40). Where open-web steel joists rest on relatively rigid supports, such as masonry bearing walls, an approximate solution for the required moment of inertia of the joists, derived from Eq. (6-28) with C_s considered negligible for bearing walls, is given by

$$I_{\min} = 0.0000128SL^4 \qquad (7\text{-}1)$$

where $I_{\min}$ = minimum allowable moment of inertia of each joist, in.^{4}
S = spacing of joists, ft
L = span of joists, ft

According to AISC, the ponding deflection contributed by a metal deck is usually such a small part of the total ponding deflection of a roof panel that it is sufficient to limit the moment of inertia of the deck, per foot of width normal to its span, to 0.000025 times the fourth power of its span length. ("Commentary on AISC Specification," 1969.) In any close or doubtful case, the stability against ponding of the entire roof structure should be investigated, as for structural steel.

For a detailed treatment of steel joist roofs under ponding loads, refer to "Structural Design of Steel Joist Roofs to Resist Ponding Loads," *Steel Joist Institute Technical Digest*, No. 3, May, 1971.

7-8. Fire Resistance of Open-web Steel Joist Construction. Any required degree of fire resistance can be obtained for steel joist construction with proper protection. Numerous fire-resistance tests have been made of floor and roof assemblies that include steel joists. Complete lists of such tests are published by the American Insurance Association (formerly the National Board of Fire Underwriters) and the Underwriters Laboratories, Inc. (See also "Fire Resistance Classification of Building Constructions," National Bureau of Standards, Report BMS 92.)

Table 7-1 lists a few typical ratings for steel joist construction:

Table 7-1. Fire Ratings for Floor Construction with Open-Web Joists

1 hr or 1½-hr fire resistance:

a. Top slab: 2-in. reinforced concrete or precast reinforced gypsum tile.
Ceiling: ¾-in. portland cement or sanded gypsum plaster on expanded metal lath.

b. Top slab: 2-in. reinforced concrete.
Ceiling: ⅝-in. Underwriters' Laboratories listed gypsum board.

2-hr fire resistance:

a. Top slab: 2½-in. reinforced concrete, or 2-in. reinforced gypsum tile with ¼-in. mortar finish.
Ceiling: ⅞-in. vermiculite plaster on expanded metal lath.

b. Top slab: 2-in. reinforced concrete.
Ceiling: ⅝-in. Underwriters' Laboratories listed acoustic tile.

c. Roof deck: 1-in. noncombustible insulation board over 18- to 22-gage steel roof deck.
Ceiling: ⅞-in. vermiculite plaster on expanded metal lath.

3-hr fire resistance:

Top slab: 2½-in. reinforced concrete, or 2-in. reinforced gypsum tile with ½-in. mortar finish.
Ceiling: 1-in. gypsum plaster, or ¾-in. gypsum-vermiculite plaster on expanded metal lath.

4-hr fire resistance:

Top slab: 2½-in. reinforced concrete, or 2-in. reinforced gypsum tile with ½-in. mortar finish.
Ceiling: 1-in. gypsum-vermiculite plaster on expanded metal lath.

Some additional ratings for open-web steel joists supporting steel roof deck are listed in Table 7-15.

COLD-FORMED SHAPES

Cold-formed shapes usually imply relatively small, thin sections made by bending sheet or strip steel in roll-forming machines, press brakes, or bending brakes. Because of the relative ease and simplicity of the bending operation and the comparatively low cost of forming rolls and dies, the cold-forming process lends itself well to the manufacture of special shapes for special purposes and makes it possible to use thin material shaped for maximum stiffness.

The use of cold-formed shapes for ornamental and other non-load-carrying purposes is commonplace. Door and window frames, metal-partition work, nonloadbearing studs, facing, and all kinds of ornamental sheet-metal work employ such shapes. The following deals with cold-formed shapes used for structural purposes in the framing of buildings.

There is no standard series of cold-formed structural sections, such as those for hot-rolled shapes, although groups of such sections have been designed ("Cold-formed Steel Design Manual," American Iron and Steel Institute). For the most part, however, cold-formed structural shapes are designed to serve a particular purpose. The general approach of the designer is therefore similar to that involved in the design of built-up structural sections; namely, to fit a particular application.

As a general rule, cold-formed shapes will cost considerably more per pound than hot-rolled sections. They will accordingly be found to be economical under the following circumstances:

1. Where their use permits a substantial reduction in weight relative to comparable hot-rolled sections. This will occur usually where relatively light loads are to

Table 7-2. Classification by Size of Flat-rolled Carbon Steel

a. Hot-rolled

Width, in.	Thicknesses, in.			
	0.2300 and thicker	0.2299–0.2031	0.2030–0.1800	0.1799–0.0449
To 3½ incl.	Bar	Bar	Strip	Strip[a]
Over 3½ to 6 incl	Bar	Bar	Strip	Strip[b]
Over 6 to 8 incl	Bar	Strip	Strip	Strip
Over 8 to 12 incl	Plate[c]	Strip	Strip	Strip
Over 12 to 48 incl	Plate[d]	Sheet	Sheet	Sheet
Over 48	Plate[d]	Plate[d]	Plate[d]	Sheet

b. Cold-rolled

Width, in.	Thicknesses, in.		
	0.2500 and thicker	0.2499–0.0142	0.0141 and thinner
To 12, incl.	Bar	Strip[e,f]	Strip[e]
Over 12 to 23 15/16, incl.	Sheet[g]	Sheet[g]	Strip[h]
Over 23 15/16	Sheet	Sheet	Black plate[i]

[a] 0.0255-in. minimum thickness.

[b] 0.0344-in. minimum thickness.

[c] Strip, up to and including 0.5000-in. thickness, when ordered in coils.

[d] Sheet, up to and including 0.5000-in. thickness, when ordered in coils.

[e] Except that when the width is greater than the thickness, with a maximum width of ½ in. and a cross-sectional area not exceeding 0.05 sq in., and the material has rolled or prepared edges, it is classified as flat wire.

[f] Sheet, when slit from wider coils and supplied with cut edge (only), in thicknesses 0.0142 to 0.0821 and widths 2 to 12 in., inclusive, and carbon content 0.25% maximum by ladle analysis.

[g] May be classified as strip when a special edge, a special finish, or single-strand rolling is specified or required.

[h] Also classified as black plate[i], depending on detailed specifications for edge, finish, analysis, and other features.

[i] Black plate is a cold-rolled, uncoated tin-mill product that is supplied in relatively thin gages.

be supported over short spans, or where stiffness rather than strength is the controlling factor in the design.

2. In a special case in which a suitable combination of standard rolled shapes would be heavy and uneconomical.

3. Where the quantities required are too small to justify the investment in equipment necessary to produce a suitable hot-rolled section.

4. In dual-purpose panel work, where both strength and coverage are desired.

7-9. Material for Cold-formed Shapes. Cold-formed shapes are usually made from sheet or strip steel. In the thicknesses in which it is available, hot-rolled material is usually used. Cold-rolled material, that is, material that has been cold-reduced to the desired thickness, is used for the thinner gages or where, for any reason, the surface finish, mechanical properties, or closer tolerances that result from cold-reducing are desired. Manufacture of cold-formed shapes from plates for use in building construction is infrequent.

Plate, Sheet, or Strip? The commercial distinction between steel plates, sheet, and strip is principally a matter of thickness and width of material, although in some sizes it depends on whether the material is furnished in flat form or in coils, whether it is carbon or alloy steel, and, particularly for cold-rolled material, on surface finish, type of edge, temper or heat treatment, chemical composition, and method of production. Although the manufacturers' classification of flat-rolled

Table 7-3. Principal Mechanical Properties of Structural-Quality Carbon and Low-Alloy Sheet and Strip Steel

ASTM specification	Material	Grade	Minimum yield point, ksi	Minimum tensile strength, ksi		Minimum elongation, %, in 2 in.	Bend test, 180°, ratio of inside dia to thickness
				Hot-rolled	Cold-rolled		
A570, A611	Hot-rolled sheet and strip, and cold-rolled sheet, carbon steel	A	25	45	42	*	Flat
		B	30	49	45	*	1
		C	33	52	48	*	1½
		D	40	55	52	*	2
		E (hot-rolled)	42	58	...	*	2½
		E (cold-rolled)	80	...	82	†	†
A446	Galvanized sheet	A	33	45		20	1½
		B	37	52		18	2
		C	40	55		16	2½
		D	50	65		12	†
		E	80	82		†	†
A607	High-strength, low-alloy columbium or vanadium sheet and strip	45	45	60		*	1
		50	50	65		*	1
		55	55	70		*	1½
		60	60	75		*	2
		65	65	80		*	2½
		70	70	85		*	3
A606	High-strength, low-alloy sheet and strip with improved corrosion resistance	Cut lengths	50	70	...	22	1
		Coils	45	65	...	22	1
		Annealed or normalized	45	65	...	22	1
		Cold-rolled	45	...	65	...	1

* Varies. See specifications.
† Not specified or required.

steel products by size is subject to change from time to time, and may vary slightly among different producers, that given in Table 7-2 for carbon steel is representative.

Carbon steel is generally used, although high-strength, low-alloy steel may be used where strength or corrosion resistance justify it. Stainless steel may be used for exposed work.

Mechanical Properties. Material to be used for structural purposes generally conforms to one of the standard specifications of the American Society for Testing and Materials (ASTM). Table 7-3 lists the ASTM specifications for structural-quality carbon and low-alloy sheet and strip, and the principal mechanical properties of the material described by them.

Stainless Steel Applications. Stainless-steel cold-formed shapes, although not ordinarily used in floor and roof framing, are widely used in exposed components, such as stairs; railings and balustrades; doors and windows; mullions; fascias, curtain walls and panel work; and other applications in which a maximum degree of corrosion resistance, retention of appearance and luster, or compatibility with other materials are primary considerations. Stainless-steel sheet and strip are available in several types and grades, with different strength levels and different degrees of formability, and in a wide range of finishes.

Types 302 and 304, in which chromium and nickel are the principal alloying elements, are the primary types used in architectural work. Type 316, which contains molybdenum in addition to chromium and nickel, offers even greater corrosion resistance. For detailed information regarding these and other types, refer to the appropriate publications of the American Iron and Steel Institute, ASTM Specifications A167 and A402, and publications of The International Nickel Company, Inc.

Table 7-4. Manufacturers' Standard Gage for Steel Sheets

[Thickness equivalents are based on 0.023912 in. per lb per sq ft (reciprocal of 41.820 psf per in. thick)]*

Manufacturers' standard gage no.	Weight, psf	Equivalent sheet thickness, in.	Manufacturers' standard gage no.	Weight, psf	Equivalent sheet thickness, in.
3	10.0000	0.2391	21	1.3750	0.0329
4	9.3750	0.2242	22	1.2500	0.0299
5	8.7500	0.2092	23	1.1250	0.0269
6	8.1250	0.1943	24	1.0000	0.0239
7	7.5000	0.1793	25	0.87500	0.0209
8	6.8750	0.1644	26	0.75000	0.0179
9	6.2500	0.1495	27	0.68750	0.0164
10	5.6250	0.1345	28	0.62500	0.0149
11	5.0000	0.1196	29	0.56250	0.0135
12	4.3750	0.1046	30	0.50000	0.0120
13	3.7500	0.0897	31	0.43750	0.0105
14	3.1250	0.0747	32	0.40625	0.0097
15	2.8125	0.0673	33	0.37500	0.0090
16	2.5000	0.0598	34	0.34375	0.0082
17	2.2500	0.0538	35	0.31250	0.0075
18	2.0000	0.0478	36	0.28125	0.0067
19	1.7500	0.0418	37	0.26562	0.0064
20	1.5000	0.0359	38	0.25000	0.0060

* The density of steel is ordinarily taken as 489.6 lb per cu ft, 0.2833 lb per cu in., or 40.80 psf per in. thick. However, since sheet weights are calculated on the basis of the specified width and length, with all shearing tolerances on the over side, and also since sheets are somewhat thicker at the center than they are at the edges, a further adjustment must be made to obtain a closer approximation for interchangeability between weight and thickness. This value for sheets has been found to be 41.820 psf per in. thick. ("Steel Products Manual," Carbon Steel Sheets, American Iron and Steel Institute.)

Coatings. Material for cold-formed shapes may be either *black* (uncoated) or galvanized. Because of its higher cost, galvanized material is used only where exposure conditions warrant paying for the increased protection afforded against corrosion.

Low-carbon sheets coated with vitreous enamel are frequently used for facing purposes, but not as a rule to perform a load-carrying function in buildings.

Selection of Grade. The choice of a grade of material, within a given class or specification, usually depends on the severity of the forming operation required to make the desired shape, strength desired (see also Art. 7-10), weldability requirements, and the economics of the situation. Grade C of ASTM A570 and A611, with a specified minimum yield point of 33 ksi, has long been popular for structural use. Some manufacturers, however, use higher-strength grades to good advantage.

Gage Numbers. Thickness of cold-formed shapes is sometimes expressed as the manufacturers' standard gage number of the material from which the shapes are formed. Use of decimal parts of an inch, instead of gage numbers, however, is preferable. Gage numbers are applicable only to sheets. Strip thicknesses are always given as decimal parts of an inch. The relationships between gage number, weight, and thickness for uncoated sheets are given in Table 7-4, and for galvanized sheets in Table 7-5.

7-10. Utilization of the Cold Work of Forming. When strength alone, particularly yield strength, is an all-important consideration in selecting a material or grade for cold-formed shapes, it is sometimes possible to take advantage of the strength increase that results from cold-working the material during the forming operation and thus use a lower-strength, more workable, and possibly more economical grade than would otherwise be required. This increase in strength is ordinarily most noticeable in relatively stocky, compact sections produced in the thicker gages. Cold-formed chord sections for open-web steel joists (Figs. 7-2*c* and *d*) are good examples. Over-all average yield strengths of more than 150% of the minimum specified yield strength of the plain material are obtained in such sections.

Table 7-5. Galvanized-sheet Gage Numbers, Unit Weights, and Thicknesses

Galvanized sheet gage no.	Weight, psf	Thickness equivalent,* in.
8	7.03125	0.1681
9	6.40625	0.1532
10	5.78125	0.1382
11	5.15625	0.1233
12	4.53125	0.1084
13	3.90625	0.0934
14	3.28125	0.0785
15	2.96875	0.0710
16	2.65625	0.0635
17	2.40625	0.0575
18	2.15625	0.0516
19	1.90625	0.0456
20	1.65625	0.0396
21	1.53125	0.0366
22	1.40625	0.0336
23	1.28125	0.0306
24	1.15625	0.0276
25	1.03125	0.0247
26	0.90625	0.0217
27	0.84375	0.0202
28	0.78125	0.0187
29	0.71875	0.0172
30	0.65625	0.0157
31	0.59375	0.0142
32	0.56250	0.0134

* Total thickness, in., including zinc coating. To obtain base metal thickness, deduct 0.0015 in. per oz coating class, or refer to ASTM A446.

The strengthening effect of the forming operation varies across the section, but is most pronounced at the bends and corners of a cold-formed section. Accordingly, the over-all strength of shapes in which bends and corners constitute a relatively large percentage of the whole section is increased more than that of shapes having a relatively large proportion of wide, flat elements that are not heavily worked in forming. For the latter type of shapes, the strength of the plain, unformed sheet or strip may be a controlling factor in the selection of a grade of material.

Full-section tests constitute a relatively simple, straightforward way of determining as-formed strength. They are particularly applicable to sections that do not contain any elements that may be subject to local buckling; that is, sections for which the form factor Q (defined in Art. 7-21) is unity. However, each case has to be considered individually in determining the extent to which cold-forming will produce an increase in utilizable strength. For further information, refer to the American Iron and Steel Institute "Specification for the Design of Cold-Formed Steel Structural Members" and its "Commentary."

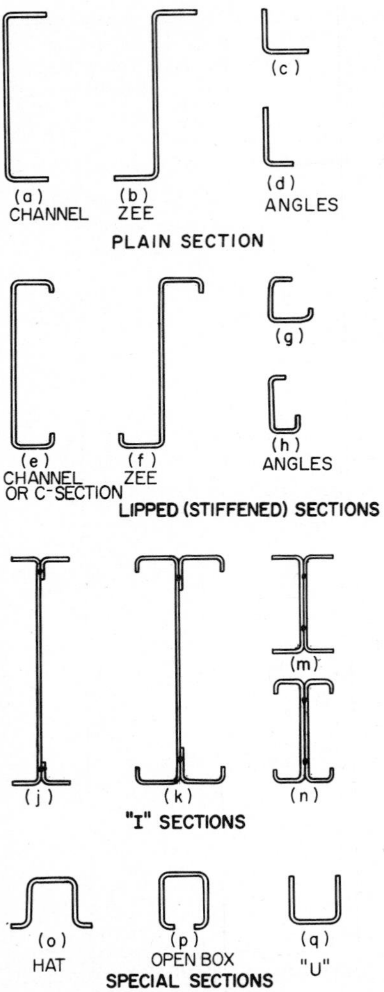

Fig. 7-6. Typical cold-formed steel structural sections.

7-11. Types of Cold-formed Sections. Many cold-formed shapes used for structural purposes are similar in their general configuration to hot-rolled structural sections. Channels, angles, and zees can be roll-formed in a single operation from one piece of material. I sections are usually made by welding two channels back to back or by welding two angles to a channel. All sections of this kind may be made with either plain flanges as in Fig. 7-6*a* to *d*, *j*, and *m*, or with flanges stiffened by lips at outer edges, as in Fig. 7-6*e* to *h*, *k*, and *n*.

In addition to these sections, which follow somewhat conventional lines and have their counterparts in hot-rolled structural sections, the flexibility of the forming process makes it relatively easy to obtain inverted U, or hat-shaped, sections and open box sections (Fig. 7-6*o* to *q*). These sections are very stiff in a lateral direction and can be used without lateral support where other more conventional types of sections would fail because of lateral instability.

Other special shapes are illustrated in Fig. 7-7. Some of these are nonstructural in nature; others are used for special structural purposes.

Figure 7-8 shows a few cold-formed stainless steel sections.

An important characteristic of cold-formed shapes is that thickness of section is substantially uniform. (A slight reduction in thickness may occur at bends but may be ignored for computing weights and section properties.) This means that, for a specified gage, the amount of flange material in a section, such as a channel, is almost entirely a function of the width of the section—except where additional flange has been obtained by doubling the material back on itself. Another distinguishing feature of cold-formed sections is that the corners are rounded on both the inside and outside of the bend, since the shapes are formed by bending flat material.

Sharp corners, such as can be obtained with hot-rolled structural channels, angles, and zees, cannot be obtained in cold-formed shapes, except with very soft material and in a coining or upsetting operation, as distinguished from a simple bending operation. This is not customary in the manufacture of structural cold-formed sections; and in proportioning such sections, the inside radius of bends should never be less, and should preferably be 33 to 100% greater, than specified for the relatively small ASTM bend-test specimens.

Deck and panel sections, such as are used for floors, roofs, and walls, are as a rule considerably wider, relative to their depth, than are the structural framing members shown in Figs. 7-6 to 7-8. Such shapes are illustrated and discussed in Arts. 7-31 to 7-42.

7-12. Design Principles for Cold-formed Sections. The structural behavior of cold-formed shapes follows the same laws of structural mechanics as does that of conventional structural-steel shapes and plates. Thus, design procedures commonly used in the selection of hot-rolled shapes are generally applicable to cold-

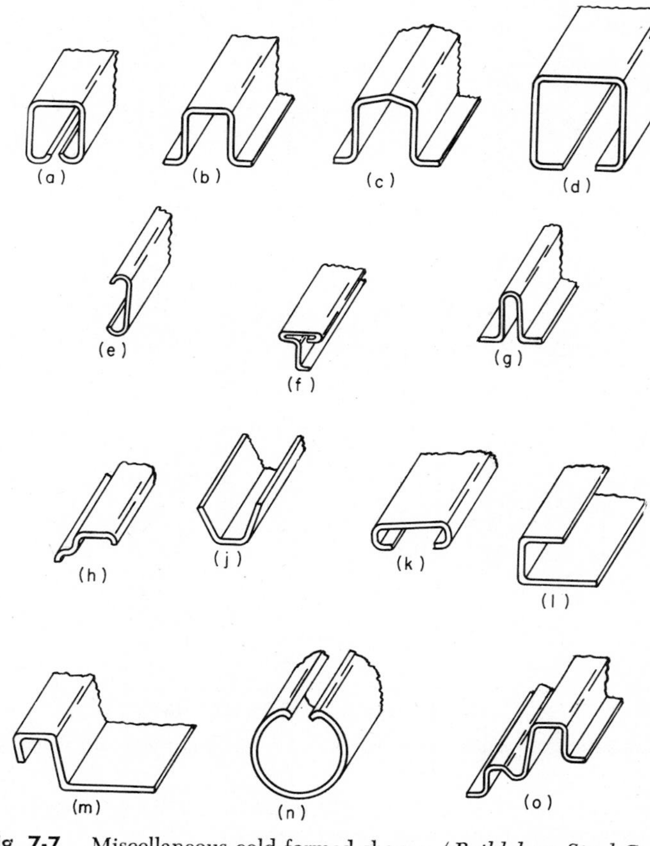

Fig. 7-7. Miscellaneous cold-formed shapes. (*Bethlehem Steel Corp.*)

formed sections. Although only a portion of a section, in some instances, may be considered structurally effective, computation of the structural properties of the effective portion follows conventional procedure.

The uniform thickness of most cold-formed sections, and the fact that the widths of the various elements comprising such a section are usually large relative to the thickness, make it possible to consider, in computing structural properties (moment of inertia, section modulus, etc.), that such properties vary directly as the first power of the thickness. So in most cases, section properties can be approximated by first assuming that the section is made up of a series of line elements, omitting the thickness dimension, and then multiplying the result by the thickness to obtain the final value. With this method, the final multiplier is always the first power of the thickness, and first-power quantities such as radius of gyration and those locating the centroid of the section do not involve the thickness dimension. The assumption that the area, moment of inertia, and section modulus vary directly as the first power of the thickness is particularly useful in determining the required thickness of a section after the lengths of the various elements that comprise it have been fixed. Although the method is sufficiently accurate for most practical purposes, it is well, particularly where the section is fairly thick relative to the widths of its elements, to check the final result through an exact method of computation.

Properties of thin elements are given in Table 7-6.

One of the distinguishing characteristics of light-gage cold-formed sections is that they are usually composed of elements that are relatively wide and thin,

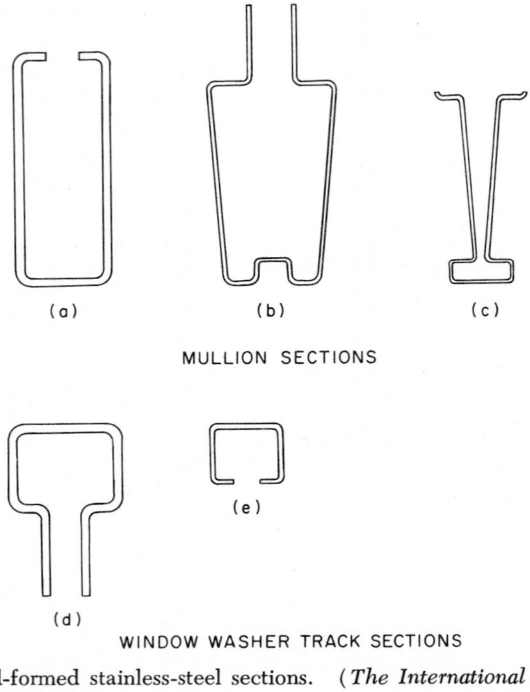

Fig. 7-8. Cold-formed stainless-steel sections. (*The International Nickel Co., Inc.*)

and as a result, attention must be given to certain modes of structural behavior customarily neglected in dealing with heavier material.

When wide thin elements are in axial compression—as in the case of a beam flange or a part of a column—they tend to buckle elastically at stresses below the yield point. This local elastic buckling is not to be confused with the general buckling that occurs in the failure of a long column or of a laterally unsupported beam. Rather local buckling represents failure of a single element of a section, and conceivably may be relatively unrelated to buckling of the entire member. In addition, there are other factors—such as shear lag, which gives rise to nonuniform stress distribution; torsional instability, which may be considerably more pronounced in thin sections than in thicker ones, requiring more attention to bracing; and other related structural phenomena customarily ignored in conventional structural work that sometimes must be considered with thin material. Means of taking care of these factors in ordinary structural design are described in the American Iron and Steel Institute (AISI) "Specification for the Design of Cold-Formed Steel Structural Members." The following discussion is based on the 1968 edition of

Table 7-6. Properties of Area and Line Elements

Area

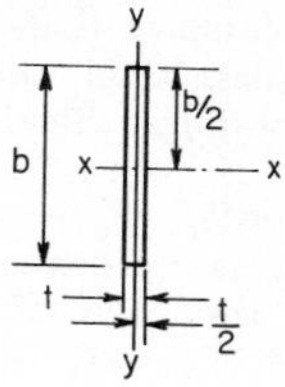

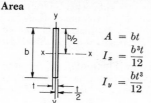

$$A = bt$$
$$I_x = \frac{b^3 t}{12}$$
$$I_y = \frac{bt^3}{12}$$

Rectangle

Line

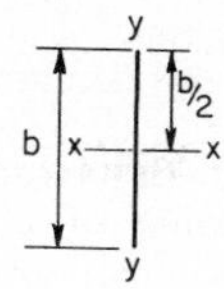

$$A_l = b$$
$$I_{lx} = \frac{b^3}{12}$$
$$I_{ly} = 0$$

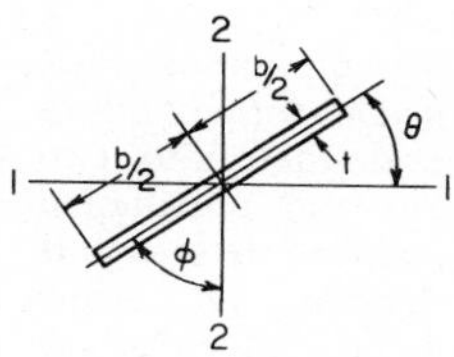

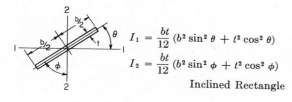

$$I_1 = \frac{bt}{12}(b^2 \sin^2 \theta + t^2 \cos^2 \theta)$$
$$I_2 = \frac{bt}{12}(b^2 \sin^2 \phi + t^2 \cos^2 \phi)$$

Inclined Rectangle

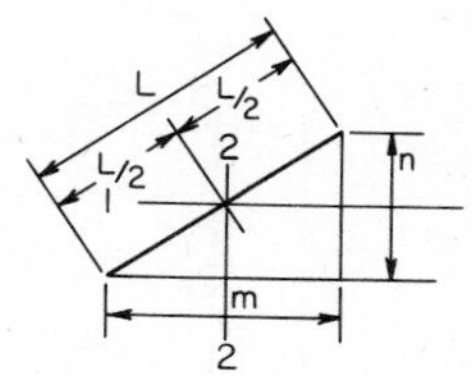

$$I_1 = \frac{Ln^2}{12}$$
$$I_2 = \frac{Lm^2}{12}$$

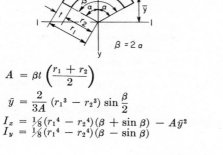

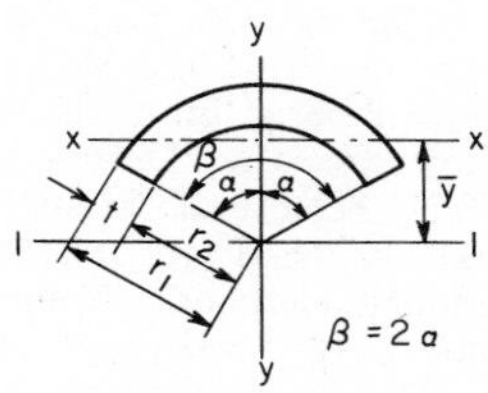

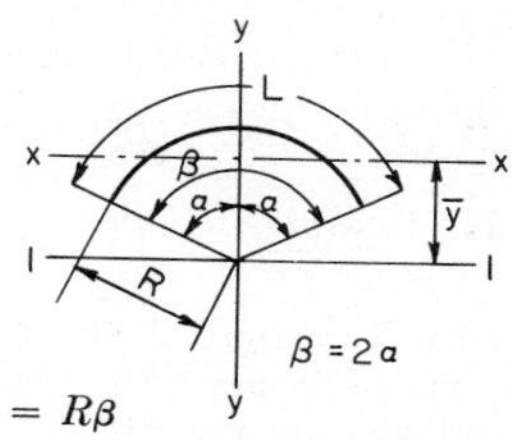

$$A = \beta t \left(\frac{r_1 + r_2}{2}\right)$$
$$\bar{y} = \frac{2}{3A}(r_1^3 - r_2^3) \sin \frac{\beta}{2}$$
$$I_x = \tfrac{1}{8}(r_1^4 - r_2^4)(\beta + \sin \beta) - A\bar{y}^2$$
$$I_y = \tfrac{1}{8}(r_1^4 - r_2^4)(\beta - \sin \beta)$$

$$L = R\beta$$
$$\bar{y} = 2R \frac{\sin \frac{\beta}{2}}{\beta}$$
$$I_x = R^3 \left(\frac{\beta + \sin \beta}{2} - \frac{4 \sin^2 \frac{\beta}{2}}{\beta}\right)$$
$$I_y = \tfrac{1}{2}R^3(\beta - \sin \beta)$$

Circular Arc

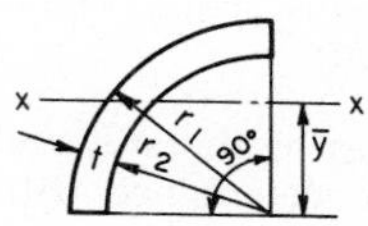

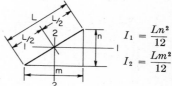

$$A = \frac{\pi}{4} t(r_1 + r_2)$$
$$= 0.7854t(r_1 + r_2)$$
$$\bar{y} = \frac{4}{3\pi} \frac{(r_1^3 - r_2^3)}{t(r_1 + r_2)}$$
$$= 0.424 \frac{(r_1^3 - r_2^3)}{t(r_1 + r_2)}$$
$$I_x = \frac{\pi}{16}(r_1^4 - r_2^4) - A\bar{y}^2$$
$$= 0.1964(r_1^4 - r_2^4) - A\bar{y}^2$$

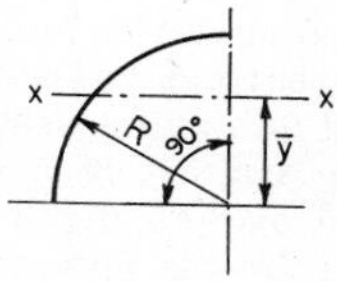

$$L = \frac{\pi R}{2}$$
$$\bar{y} = \frac{2}{\pi} R = 0.637R$$
$$I_x = 0.1488R^3$$

For small radii:

r_2	A	$\bar{y}$	I_x
$2t$	$3.927t^2$	$1.613t$	$2,549t^4$
$1.5t$	$3.142t^2$	$1.300t$	$1.369t^4$
t	$2.356t^2$	$0.990t$	$0.635t^4$
$0.75t$	$1.963t^2$	$0.838t$	$0.400t^4$
$0.5t$	$1.571t^2$	$0.690t$	$0.235t^4$

90° Circular Corner

that specification. It applies only to carbon and low-alloy steel. Although the same concepts are applicable to stainless steel, there is sufficient difference in stress-strain characteristics to require somewhat different detail treatment than for carbon and low-alloy material. For procedures applicable to stainless steel, see the AISI "Specification for the Design of Light-gage Cold-formed Stainless Steel Structural Members."

7-13. Structural Behavior of Flat Compression Elements. It is convenient, in discussing buckling of plates or flat compression elements in beams, columns, or other structural members, to use **flat-width ratio** w/t, the ratio of the width w of a single flat element, in., exclusive of edge fillets, to the thickness t of the element, in (Fig. 7-9).

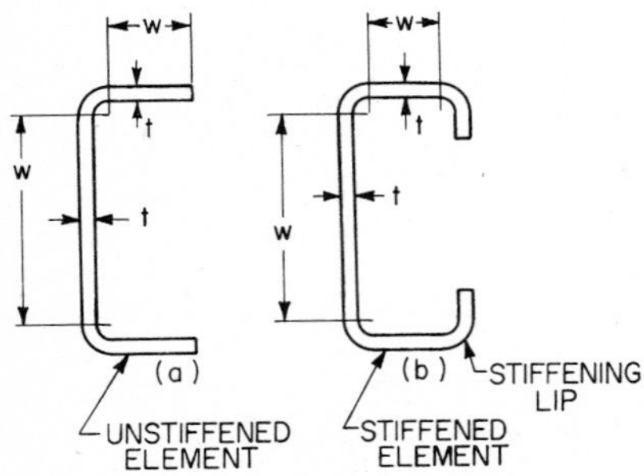

Fig. 7-9. Compression elements.

For structural design computations, flat compression elements of cold-formed structural members can be divided into two kinds—stiffened elements and unstiffened elements. **Stiffened compression elements** are flat compression elements; i.e., plane compression flanges of flexural members and plane webs and flanges of compression members, of which *both* edges parallel to the direction of stress are stiffened by a web, flange, stiffening lip, or the like (AISI "Specification for the Design of Cold-formed Steel Structural Members"). A flat element stiffened at only *one* edge parallel to the direction of stress is called an **unstiffened element.** If the sections shown in Fig. 7-9 are used as compression members, the webs are considered stiffened compression elements. The lips that stiffen the outer edges of the flanges are, in turn, unstiffened elements. Any section composed of a number of plane portions, or elements, can be broken down into a combination of stiffened and unstiffened elements.

Stiffener Requirements. According to the AISI Specification, in order that a compression element may qualify as a stiffened compression element, its edge stiffeners should comply with the following:

$$I_{\min} = 1.83t^4 \sqrt{\left(\frac{w}{t}\right)^2 - \frac{4{,}000}{F_y}} \tag{7-2}$$

but not less than $9.2t^4$

where w/t = flat-width ratio of stiffened element

$I_{\min}$ = minimum allowable moment of inertia of stiffener (of any shape) about its own centroidal axis parallel to stiffened element, in.^{4}

F_y = specified minimum yield point of the material, ksi

Where the stiffener consists of a simple lip bent at right angles to the stiffened element, the required over-all depth d, in., of such a lip may be determined with satisfactory accuracy from the following formula:

$$d = 2.8t \sqrt[6]{\left(\frac{w}{t}\right)^2 - \frac{4{,}000}{F_y}} \tag{7-3}$$

but not less than $4.8t$. A simple lip should not be used as an edge stiffener for any element having a flat-width ratio greater than 60.

The values of $I_{\min}$ and d from Eqs. (7-2) and (7-3) are plotted in Fig. 7-10 for $F_y = 33$ and $F_y = 50$ ksi. The influence of F_y is negligible for w/t ratios above about 25 or 30. For practical reasons, in the thinner gages, the depth d of a simple lip will frequently be larger than the required minimum.

Multiple-stiffened Elements. The discussion of stiffened compression elements in this section deals primarily with simple elements, such as those shown in Figs. 7-6 to 7-9. Multiple-stiffened elements, i.e., elements that have longitudinal stiffeners placed between webs or between a web and an edge, are sometimes encountered, particularly in deck and panel work. For methods of computing the effectiveness of such elements and for requirements for stiffeners for such elements, see the AISI "Specification for the Design of Cold-formed Steel Structural Members."

Buckling of Unstiffened Elements. Whether or not local buckling of a flat compression element must be considered depends on its flat-width ratio. For unstiffened elements, local buckling does not have to be considered unless the flat-width ratio exceeds $63.3/\sqrt{F_y}$. (See also Arts. 7-15 and 7-18.)

Buckling of Stiffened Elements. For stiffened elements, which behave in buckling in a substantially different manner from unstiffened elements, the flat-width ratio

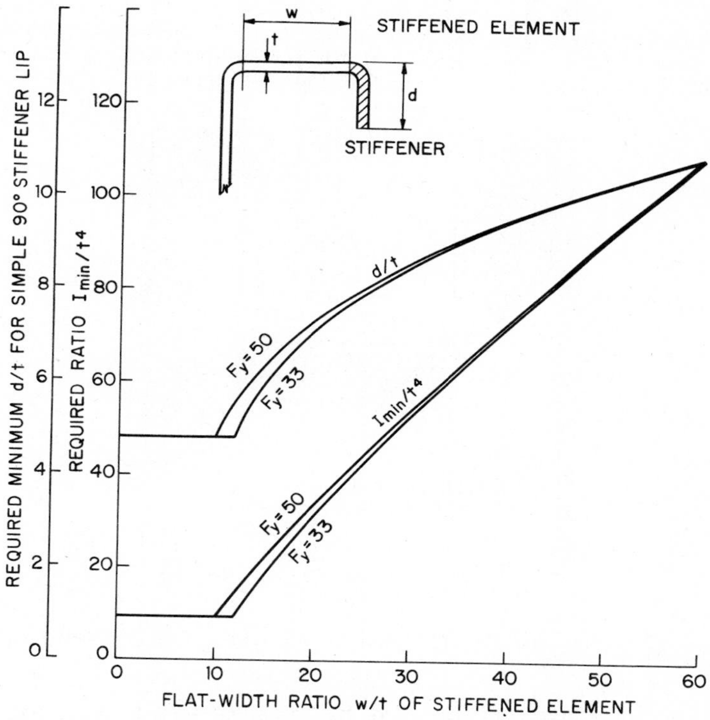

Fig. 7-10. Curves for determining minimum dimensions of edge stiffeners for stiffened compression elements.

beyond which allowance must be made for local buckling varies with the actual unit stress in the element. The effect of local buckling may be ignored for stiffened elements with flat-width ratios equal to or less than $(w/t)_{lim}$ in Eqs. (7-4) and (7-5). (These equations do not apply to the sides of square and rectangular structural tubing, for which a more liberal treatment is permissible. See the AISI Specification.)

For safe-load determination, i.e., in computing effective area and section modulus:

$$\left(\frac{w}{t}\right)_{lim} = \frac{171}{\sqrt{f}} \qquad (7\text{-}4)$$

where f = computed unit stress, ksi, in the element based upon effective width.

Equation (7-4) is based on a safety factor of about 1.67 against yield stress at the outer fiber of the section. For any other safety factor m, multiply the right-hand side of Eq. (7-4) by $\sqrt{1.67/m}$.

For deflection determinations, i.e., in computing moment of inertia to be used in deflection calculations or other calculations involving stiffness:

$$\left(\frac{w}{t}\right)_{\text{lim}} = \frac{221}{\sqrt{f}} \qquad (7\text{-}5)$$

See also Arts. 7-16 and 7-18.

Beam and Column Design. As long as the proportions of cold-formed sections are such that there are no unstiffened compression elements having net flat-width ratios greater than $63.3/\sqrt{F_y}$, and there are no stiffened compression elements having flat-width ratios greater than the limits established by Eq. (7-4), they can be treated, in their structural use as beams and columns, in the same general way as ordinary structural shapes, and without any modification to take into account local buckling. The maximum widths of flat compression elements that may be so used may be determined from the charts of Figs. 7-11 and 7-14.

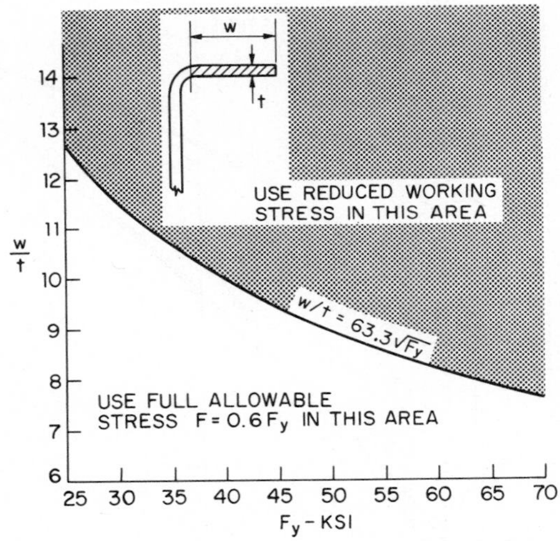

Fig. 7-11. Curve for determining maximum net width of unstiffened compression elements without stress reduction.

When the above limits are exceeded, the treatment in Arts. 7-15 and 7-16 is recommended. (AISI, "Cold-formed Steel Design Manual.")

Curves and Corners. The theoretical buckling strength of a curved plate carrying axial compression transverse to the direction of curvature is many times that of a flat plate, even if the curvature is very slight. Hence, ordinary bends and corners in cold-formed shapes may be relied upon to carry any stress up to the full yield strength of the material. The possibility of local buckling in transversely curved plates in compression need not be of concern unless the curve is very shallow relative to the width and thickness of the element. Such cases are rarely encountered in load-carrying elements in buildings. For doubtful cases that have not been proven by experience or tests, refer to aircraft practice or to appropriate texts dealing with elastic stability. The discussion of compression elements in this section deals only with flat elements.

7-14. Unit Stresses for Cold-formed Shapes. The basic working stress F, ksi, in tension and bending for cold-formed sections made of sheet and strip steel is usually specified as $F = 0.60F_y$, where F_y is the specified minimum yield point of the material, ksi. This corresponds to a safety factor of 1.67 applied to the yield point. As discussed in Art. 7-15, this basic stress must be reduced in dealing with wide unstiffened elements in compression to provide against local buckling.

An increase of 33⅓% in all allowable stresses is customary for combined wind, earthquake, and gravity loads, or in members subject only to the effect of wind and earthquake forces.

7-15. Design of Unstiffened Elements Subject to Compression. Unstiffened elements are defined in Art. 7-13. Compute section properties of the full section in the conventional manner, and use the basic working stress if the flat-width ratio w/t of any unstiffened element does not exceed $63.3/\sqrt{F_y}$, where F_y is the yield point, ksi, of the steel. (See Fig. 7-11.)

If w/t is greater than $63.3/\sqrt{F_y}$, the AISI "Specification for the Design of Cold-formed Steel Structural Members" prescribes that the allowable compression stress be reduced to a value F_c, ksi, given by Eqs. (7-6) to (7-8).

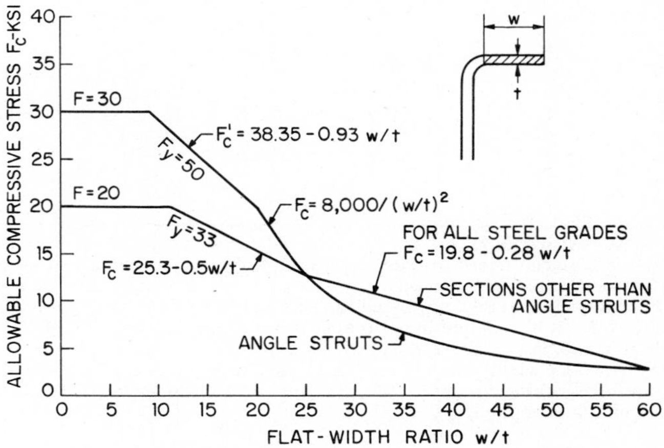

Fig. 7-12. Curves for allowable compressive stress in unstiffened elements for $F_y = 33$ and $F_y = 50$.

For the usual case in which F_y is 33 ksi or more, Eqs. (7-6) and (7-7) are applicable when the flat-width ratio w/t is less than 25.

For $63.3/\sqrt{F_y} < w/t \leq 144/\sqrt{F_y}$,

$$F_c = F_y \left[0.767 - 0.00264 \frac{w}{t} \sqrt{F_y} \right] \tag{7-6}$$

For $144/\sqrt{F_y} < w/t \leq 25$,

$$F_c = \frac{8{,}000}{(w/t)^2} \tag{7-7}$$

In the relatively infrequent case in which F_y is less than 33 ksi, use straight-line interpolation between $F_c = F$ (basic working stress) at $w/t = 63.3\sqrt{F_y}$, and $F_c =$ 12.8 ksi at $w/t = 25$.

For values of w/t from 25 to 60, inclusive, for all grades of steel,

$$F_c = 19.8 - 0.28 \frac{w}{t} . \tag{7-8}$$

except that for angle struts Eq. (7-7) applies.

These equations are plotted in Fig. 7-12 for yield points of 33 and 50 ksi. Note that Eq. (7-6) plots as a pair of straight lines drawn between $F_c = F$ when $w/t =$ $63.3/\sqrt{F_y}$, and $F_c = 8{,}000/(w/t)^2$ when $w/t = 144/\sqrt{F_y}$. Note also that, when $w/t \geq 25$, F_c is the same curve for all grades of steel.

Table 7-7. Allowable Compression Stresses in Unstiffened Elements

1	2	3	4	5	6
F_y, ksi	For $w/t \le 63.3/\sqrt{F_y}$		For $w/t \le 144/\sqrt{F_y}$ or 25, whichever is smaller		For $w/t > 144/\sqrt{F_y}$ or 25
	$63.3/\sqrt{F_y}$	$F_c = 0.60F_y$*	$144/\sqrt{F_y} \le 25$	F_c†	
25	12.7	15	25	12.8	If w/t is greater than $144/\sqrt{F_y}$ (column 4) but less than 25, $F_c = 8{,}000/(w/t)^2$
30	11.5	18	25	12.8	
33	11.0	20	25	12.8	
36‡	10.6	22	24	13.9	
37	10.4	22	23.7	14.2	
40	10.0	24	22.8	15.4	
42	9.8	25	22.2	16.2	
45	9.4	27	21.5	17.3	For all flat-width ratios greater than 25, use Eqs. (7-7) and (7-8), as applicable, to obtain F_c
50	9.0	30	20.4	19.2	
55	8.5	33	19.4	21.8	
60	8.2	36	18.6	23.1	
65	7.8	39	17.9	25.0	
70	7.6	42	17.2	27.0	
80	§	§			

* Rounded to nearest whole ksi.

† For w/t values between those in columns 2 and 4, F_c may be obtained by straight-line interpolation between columns 3 and 5.

‡ Material of this yield point is not listed in the ASTM sheet and strip specifications summarized in Table 7-3. It is included here because it is a popular standard grade of structural-steel plates, shapes, and bars as covered by ASTM A36.

§ Sheet steel of 80-ksi yield point (Grade E of ASTM A446 and A611) is rarely, if ever, used at an allowable working stress of $0.6F_y$, because deflection and stiffness usually govern the designs to which it is applied.

These limiting values and the corresponding allowable compression stresses are given in Table 7-7. Straight-line interpolation may be used between them.

Selection of Flat-width Ratios. For maximum economy of material in the use of unstiffened elements in compression, w/t values should be held to a maximum of $63.3/\sqrt{F_y}$, so that the full basic working stress may be used. With Eqs. (7-6) to (7-8), the maximum capacity of an element (unit stress times cross-sectional area) occurs at about $w/t = 35$ for F_y below about 40 ksi, and at $w/t = 144/\sqrt{F_y}$ for higher-strength grades. The reason for this is that as w/t increases beyond these values the buckling stress decreases faster than the area increases.

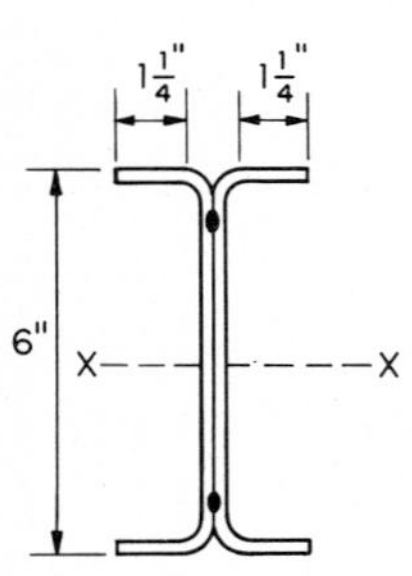

Fig. 7-13. Double-channel beam.

Example. As an example of the application of the equations, consider the double-channel section shown in Fig. 7-13, which is to be used as a simply supported beam about axis xx with the top flange in compression. The two legs of that flange are unstiffened compression elements. The section has a section modulus of about $27t$, where t is the thickness of the material, in. It is to be made of Grade C uncoated steel, $F_y = 33$ ksi.

If the thickness is 0.120 in., then the flat-width ratio of the unstiffened top-flange elements is $10.4 < (63.3/\sqrt{33} = 11.0)$. Thus, the section can be used at the basic working stress of 20 ksi under gravity loads. Its section modulus is $27 \times 0.120 = 3.2$ in.^{3}, and its resisting moment is $3.2 \times 20/12 = 5.3$ ft-kips.

If the thickness is 0.060 in., then the flat-width ratio of the unstiffened compression flange becomes $20.8 < (144/\sqrt{33} = 25)$. According to Eq. (7-6) or by interpolation in Table 7-7, the allowable stress must be reduced to about 15 ksi. The section modulus is 27 × 0.060 = 1.6 in.^{3}, and the resisting moment of the section is 1.6 × 15.1/12 = 2.0 ft-kips.

If the thickness is further reduced to 0.030 in., then the flat-width ratio of the unstiffened compression elements becomes 41.7, and the allowable stress about 8 ksi. The section modulus is reduced to 0.81 in.^{3}, and the resisting moment is only 0.54 ft-kips.

The above example illustrates the desirability of holding w/t for unstiffened compression elements to a maximum value of $63.3/\sqrt{F_y}$, unless larger values are required for reasons other than load-carrying capacity alone.

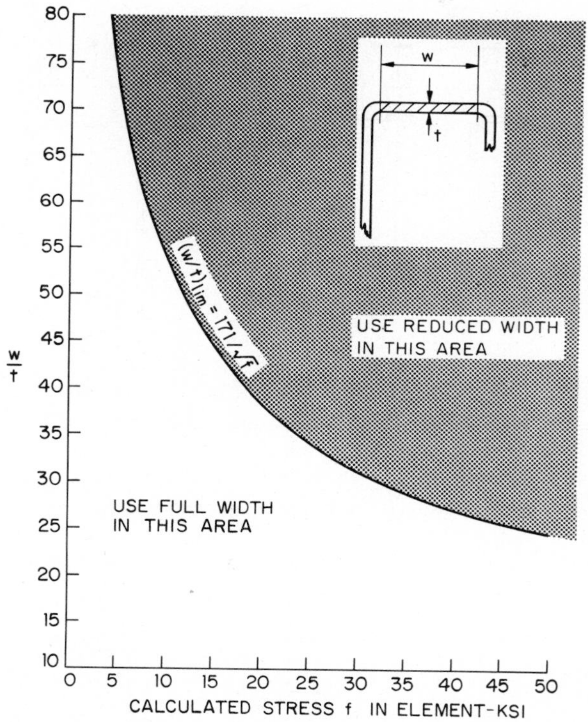

Fig. 7-14. Curve for determining maximum net width for full effectiveness of stiffened compression elements for load-carrying calculations.

7-16. Stiffened Elements Subject to Compression. As discussed in Art. 7-13, local buckling of stiffened elements in compression cannot be ignored when the flat-width ratio w/t is larger than $171/\sqrt{f}$ for safe-load determination (Fig. 7-14), or $221/\sqrt{f}$ for deflection calculations. In such cases, compute section properties based on an *effective width b,* in., of each stiffened compression element (Fig. 7-15). Determine b by means of Eqs. (7-9) and (7-10).

For safe-load determinations, i.e., in computing effective area and section modulus

$$\frac{b}{t} = \frac{253}{\sqrt{f}}\left[1 - \frac{55.3}{(w/t)\sqrt{f}}\right] \tag{7-9}$$

where f = unit stress, ksi, in the element computed on basis of reduced section
w = width of the element, in.
t = thickness, in.

For deflection determinations, i.e., in computing moment of inertia to be used in deflection calculations or in other calculations involving stiffness,

$$\frac{b}{t} = \frac{326}{\sqrt{f}}\left[1 - \frac{71.3}{(w/t)\sqrt{f}}\right] \tag{7-10}$$

with f, w, and t the same as for Eq. (7-9). [Equations (7-9) and (7-10) do not apply to the sides of square or rectangular structural tubing, for which a more liberal treatment is permissible. See the AISI "Specification for the Design of Cold-formed Steel Structural Members."]

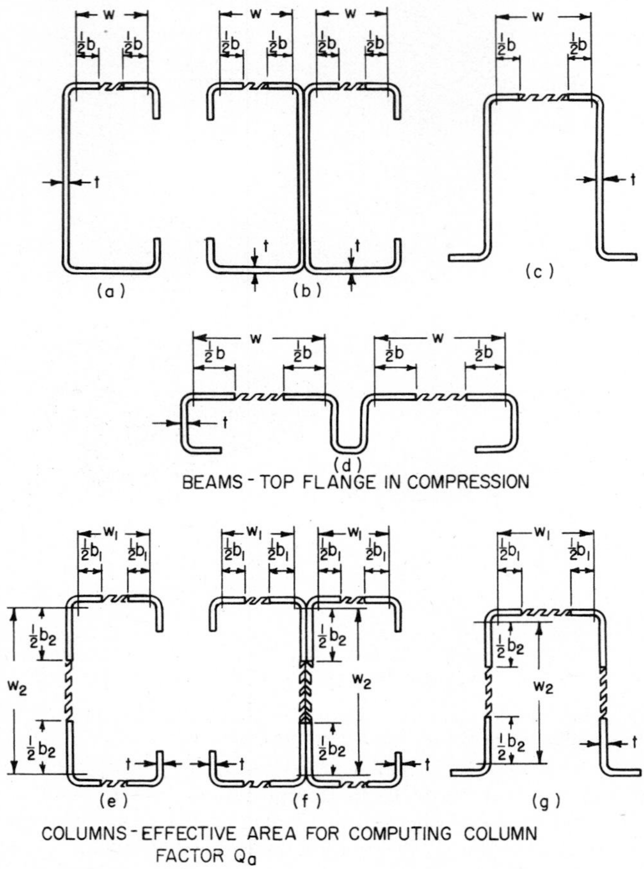

Fig. 7-15. Effective width of stiffened compression elements.

As shown in Fig. 7-15 that portion of the element not considered to be effective should be located symmetrically about the center line of the element.

The curves of Figs. 7-16 and 7-17 were plotted from Eqs. (7-9) and (7-10), and may be used to determine the ratio b/t for different values of w/t and unit stress.

The effective width b in Eqs. (7-9) and (7-10) is dependent on the unit stress f, and since reduced-section properties are a function of effective width, the reverse is also true. In the general case of flexural members, therefore, determination of effective width by these formulas requires successive approximations: First, a unit stress is assumed for the compression flange. With that stress, b is computed. Next, the section modulus and unit stress in the compression flange are calculated for the reduced section. Then the process is repeated if the effective width so computed should be significantly different from that for the stress first assumed.

However, this somewhat cumbersome procedure can be avoided and the correct value of b/t obtained directly from the formulas when f is known or is held to a specified maximum allowable value (such as 20 ksi for flexural members of Grade C uncoated steel), and the neutral axis of the section is closer to the tension flange than the compression flange so that the stress in the compression flange always controls. The latter condition holds for symmetrical channels, zees, and I sections used as flexural members about their major axes, such as (*e*), (*f*), (*k*), and (*n*) in Fig. 7-6. For hat and box sections, of the proportions shown in (*o*) and (*p*) in Fig. 7-6, or unsymmetrical channels, zees, and I sections, the

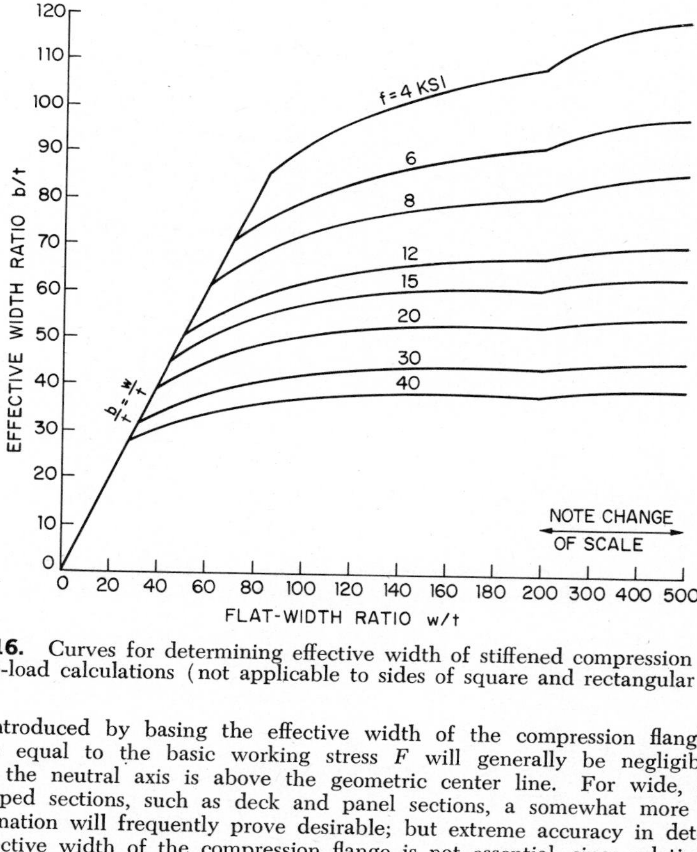

Fig. 7-16. Curves for determining effective width of stiffened compression elements for safe-load calculations (not applicable to sides of square and rectangular tubing).

error introduced by basing the effective width of the compression flange on an f value equal to the basic working stress F will generally be negligible, even though the neutral axis is above the geometric center line. For wide, inverted, pan-shaped sections, such as deck and panel sections, a somewhat more accurate determination will frequently prove desirable; but extreme accuracy in determining the effective width of the compression flange is not essential, since relatively large variations in effective width will have relatively little effect on section modulus and moment of inertia. Even for sections of this kind, a good first approximation can be made by basing b/t on the basic working stress F, and two approximations will almost always suffice.

In computing moment of inertia for deflection or stiffness calculations, properties of the full unreduced section can be used without significant error when w/t of the compression elements exceeds $(w/t)_{\lim}$ but does not exceed about 60 or 70.

Example. As an example of effective-width determination, load-carrying capacity is to be computed for the hat section in Fig. 7-18*a*, which is to be made of

Grade C uncoated steel. This shape is to be used as a simply supported beam with the top flange in compression, at a basic working stress of 20 ksi.

The top flange, 3 in. wide, is a stiffened compression element. If the thickness is ⅛ in., the flat-width ratio is 24. For a unit stress of 20 ksi, $(w/t)_{lim}$, according to

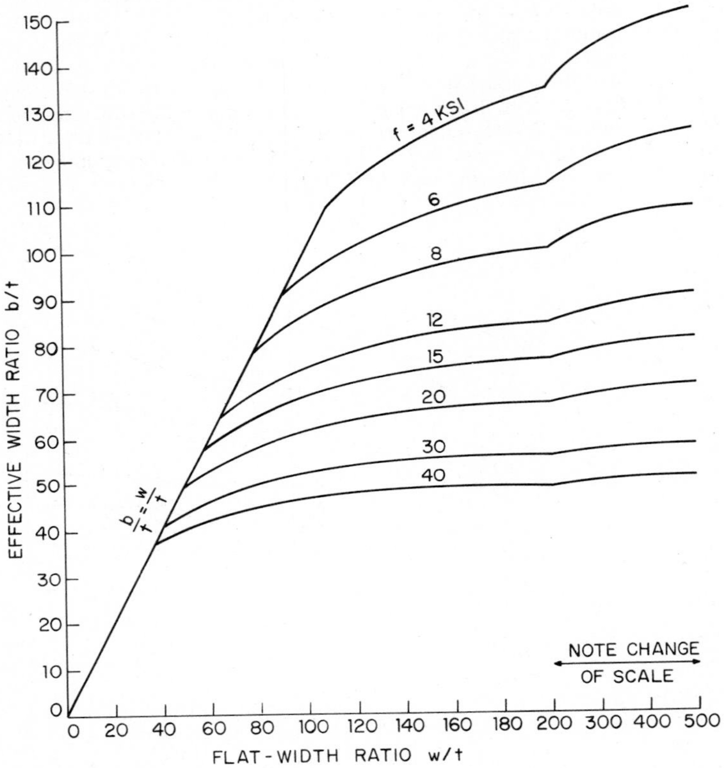

Fig. 7-17. Curves for determining effective width of stiffened compression elements for deflection calculations (not applicable to sides of square and rectangular tubing).

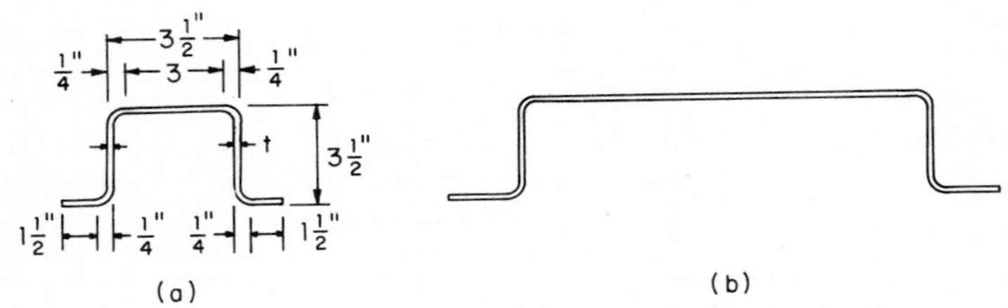

Fig. 7-18. Hat sections.

Eq. (7-4), is 38.2 > 24. Therefore, properties of the section may be computed in the usual manner, with the entire section considered structurally effective.

If, however, the thickness is 1/16 in., the flat-width ratio becomes 48, and Eq. (7-9) applies. For this value of w/t and f = 20 ksi, Eq. (7-9) gives b/t = 42. Thus, only 87½% of the top flange is considered effective. The neutral axis of the section will lie below the horizontal center line; so the compression-flange stress will control. Since that stress is limited to 20 ksi, the effective width may be accurately determined from Eq. (7-9) for 20 ksi.

For the section in Fig. 7-18*b*, in which the horizontal centroidal axis will be much nearer the compression than the tension flange, the stress in the tension flange controls. Determination of the unit stress in the compression flange and the effective width of that flange requires a trial-and-error procedure.

7-17. Light-gage Members with Unusually Wide Flanges on Short Spans. The phenomenon known as *shear lag*, resulting in nonuniform stress distribution in the flanges of a flexural member, requires recognition only in comparatively extreme cases of concentrated loads on spans that are short relative to the width of the member. The effects of shear lag can be taken care of by an effective-width procedure applicable to both tension and compression flanges. The AISI "Specification for the Design of Cold-formed Steel Structural Members" provides methods for handling this situation. This specification should be consulted in any case of concentrated loading in which the span of a beam is less than 30 times the half width of flange for I-beam and similar sections or thirty times half the distance between webs for box or U-type sections.

7-18. Maximum Flat-width Ratios for Light-gage Elements. When the flat-width ratio exceeds about 30 for an unstiffened element and 250 for a stiffened element, noticeable buckling of the element is likely to develop at relatively low stresses. Present-day practice is to permit buckles to develop in the sheet and to take advantage of what is known as the post-buckling strength of the section. The effective-width formulas [Eqs. (7-9) and (7-10)] are based on this practice. However, to avoid excessive deformations, over-all flat-width ratios, disregarding intermediate stiffeners and based on the actual thickness of the element, should not exceed the following:

Stiffened compression element having one longitudinal edge connected to a web or flange, the other to a simple right-angle lip	60
Stiffened compression element having both edges stiffened by other stiffening means than a simple right-angle lip	90
Stiffened compression element with both longitudinal edges connected to a web or flange element such as in a hat, U, or box type of section	500
Unstiffened compression elements	60

In special kinds of panel work, where the flat compression element may be stiffened by insulation board or other similar material fastened to it with an adhesive, elastic buckling does not occur with the same freedom as without such stiffening. The extent to which the stiffening effect of collateral materials may be utilized has to be determined by tests in each case. It will further depend on the extent to which the adhesive can be relied on to furnish continuous permanent bond between the two materials, and whether or not such bond is likely to be affected by fire or other external actions. In general, it is unwise to rely on collateral materials to prevent elastic buckling in the compression element.

7-19. Laterally Unsupported Light-gage Beams. In the relatively infrequent cases in which light-gage cold-formed sections used as beams are not laterally supported at frequent intervals, the unit stress must be reduced to avoid failure from lateral instability. The amount of reduction depends on the shape and proportions of the section and the spacing of lateral supports. For details, see the AISI "Specification for the Design of Cold-formed Steel Structural Members."

Because of the torsional flexibility of light-gage channel and zee sections, their use as beams without lateral support is not recommended. In the usual case in which one flange of such a section is connected to a deck or sheathing material, bracing of the other flange may or may not be needed to prevent twisting of the member, depending on: (1) whether or not the deck or sheathing material and its connection to the beam will effectively restrain the connected beam flange from lateral deflection and twist; (2) the dimensions of the member and the span; and (3) whether the unbraced flange is in compression or tension. In any doubtful case, a load test should be made to determine whether or not additional bracing is necessary. When bracing against twist is required, means of proportioning such bracing can be found in the AISI Specification. Bracing details will depend on the particular form of the sections involved. Bridging, such as is used with open-web steel-joist construction (Art. 7-4), can be effectively used.

In any case in which laterally unsupported beams must be used, or where lateral buckling of a flexural member is likely to be a problem, consideration should be given to use of relatively bulky sections having two webs, such as hat or box sections, as illustrated in Fig. 7-6*o* and *p*. Sections for which the moment of inertia about the vertical y axis is equal to or greater than that about the horizontal x axis do not fail because of lateral instability; but as the ratio I_y/I_x decreases, the section becomes more susceptible to failure by lateral buckling. Thus, double-web sections almost always are considerably more stable laterally than single-web sections of normal proportions. Hat sections (Fig. 7-6*o*) are especially suitable where lateral stiffness is necessary, and when the flat-width ratios of the individual elements are low enough to be fully effective structurally (Art. 7-13), design of the section is simple and straightforward. Sections of the kind shown in Fig. 7-7*c* having a top width of 2 in., depth of 2½ in., and bottom flanges ⅝ in.

Table 7-8. Allowable Shear Stresses in Webs, Ksi, and Maximum Depth-thickness Ratios*

1	2	3	4	5	6
F_y	For $F_v \leq [0.4F_y = (2/3)F]$		For F_v when $h/t \leq (547/\sqrt{F_y} = 424/\sqrt{F})$		For h/t less than the values in column 3, use F_v from column 2
	$0.4F_y$	Max. h/t	F_v	$547/\sqrt{F_y}$	
25	10	76	6.9	109	For h/t between the values in columns 3 and 5, $F_v = 152\sqrt{F_v}/(h/t) = 118\sqrt{F}/(h/t)$
30	12	69	8.3	100	
33	13.5	66	9.3	95	
37	14.5	63	10.3	90	
40	16	60	11.1	87	
42	17	59	11.6	85	
45	18	57	12.5	82	
50	20	54	13.9	77	
55	22	51	15.3	74	For h/t greater than the values in column 5, $F_v = 83{,}200/(h/t)^2$
60	24	49	16.7	71	
65	26	47	18.1	68	
70	28	45	19.4	65	

* F_v = allowable shear stress, ksi
F_y = minimum specified yield point, ksi
F = basic working stress in tension and bending, ksi
h = clear distance between flanges, in.
t = thickness of web, in. Where the web consists of two sheets, as in the case of two channels fastened back to back to form an I section, each sheet should be considered a separate web carrying its share of the shear

The values of F_v in this table are based upon the basic working stress values F from Table 7-7. Most of those in column 2 are rounded to agree with 1971 published values.

wide have been successfully used under floating loads without any lateral support on spans up to 7 ft, or 42 times the top width.

Even with hat sections, extreme proportions should be avoided. With very high narrow sections made of thin material, lateral instability may result from lack of stiffness in the webs, although the flat-width ratio of the compression flange may be low. It is desirable also to keep the flat-width ratio of the bottom flanges relatively low to avoid the necessity for using a low allowable stress (Art. 7-15) when the section is to be used as a continuous beam and the bottom flanges are in compression over the interior supports.

In pan construction, consisting of inverted hat sections or U sections, in which the narrow flanges are in compression, the lateral stability of the compression flanges depends on the rigidity of the webs and the bottom, or tension, flange. An approximate method of handling this case analytically is given in the AISI "Cold-formed Steel Design Manual," or tests may be made.

Table 7-9. Maximum End Reaction, or Concentrated Load on End of Cantilever, for Single Unreinforced Web of Light-gage Steel

[Grade C uncoated steel (F_y = 33), corner radii = thickness of material]

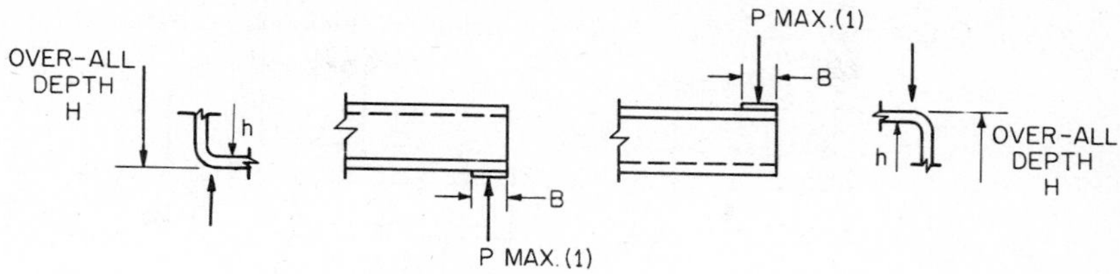

t = web thickness
h = clear distance between flanges
B = length of bearing
H = over-all depth

$$P \max (1) = 0.1t^2 \left(980 + 42\frac{B}{t} - 0.22\frac{B}{t}\frac{h}{t} - 0.11\frac{h}{t}\right) \tag{7-11}$$

NOTE: In solving this formula, B should not be assigned any value greater than h.

Gage and thickness	H, in.	P max (1), kips, for different values of B					
		1 in.	2 in.	3 in.	4 in.	6 in.	8 in.
No. 18, 0.0478 in.	1	0.39					
	2	0.38	0.53				
	3	0.36	0.50	0.62			
	4	0.34	0.45	0.57	0.67		
	6	0.29	0.36	0.43	0.50		
No. 16, 0.0598 in.	1	0.55					
	2	0.56	0.74				
	3	0.54	0.72	0.89			
	4	0.51	0.68	0.84	0.99		
	6	0.47	0.59	0.71	0.83		
	8	0.42	0.50	0.58	0.66		
No. 14, 0.0747 in.	2	0.82	1.05				
	3	0.80	1.05	1.26			
	4	0.77	1.00	1.23	1.43		
	6	0.73	0.91	1.10	1.28	1.62	
	8	0.68	0.82	0.96	1.10	1.39	
	10	0.64	0.73	0.83	0.93	1.12	
No. 12, 0.1046 in.	2	1.47	1.79				
	3	1.45	1.82	2.12			
	4	1.42	1.78	2.14	2.42		
	6	1.38	1.69	2.00	2.31	2.87	
	8	1.33	1.60	1.87	2.13	2.67	3.15
	10	1.28	1.51	1.73	1.96	2.40	2.85
	12	1.24	1.42	1.60	1.78	2.14	2.50
No. 10, 0.1345 in.	2	2.30	2.68				
	3	2.27	2.78	3.15			
	4	2.25	2.73	3.22	3.57		
	6	2.20	2.64	3.08	3.52	4.28	
	8	2.16	2.55	2.95	3.34	4.13	4.81
	10	2.11	2.46	2.81	3.16	3.86	4.56
	12	2.06	2.37	2.68	2.98	3.60	4.21
No. 8, 0.1644 in.	3	3.28	3.91	4.33			
	4	3.25	3.86	4.47	4.88		
	6	3.20	3.77	4.34	4.90	5.85	
	8	3.16	3.68	4.20	4.72	5.76	6.64
	10	3.11	3.59	4.06	4.54	5.50	6.45
	12	3.06	3.49	3.93	4.36	5.23	6.10

Table 7-10. Maximum Interior Reaction or Concentrated Load on Single Unreinforced Web of Light-gage Steel

[Grade C uncoated steel (F_y = 33), corner radii = thickness of material]

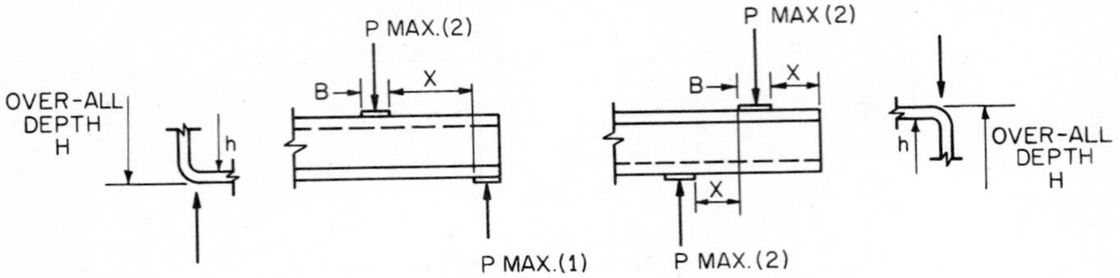

t = web thickness
h = clear distance between flanges
B = length of bearing
H = over-all depth

$$P \max (2) = 0.1t^2 \left(3{,}050 + 23\frac{B}{t} - 0.09\frac{B}{t}\frac{h}{t} - 5\frac{h}{t}\right) \qquad (7\text{-}12)$$

NOTE: In solving this formula, B should not be assigned any value greater than h. This table and formula apply only where the distance x is greater than $1.5h$. Otherwise Eq. (7-11) and Table 7-9 govern.

Gage and thickness	H, in.	P max (2), kips, for different values of B					
		1 in.	2 in.	3 in.	4 in.	6 in.	8 in.
No. 18, 0.0478 in.	1	0.77					
	2	0.74	0.83				
	3	0.71	0.79	0.87			
	4	0.68	0.75	0.83	0.90		
	6	0.61	0.67	0.73	0.78		
No. 16, 0.0598 in.	1	1.18					
	2	1.15	1.26				
	3	1.12	1.23	1.33			
	4	1.08	1.18	1.28	1.37		
	6	1.00	1.08	1.17	1.25		
	8	0.92	0.99	1.05	1.12		
No. 14, 0.0747 in.	2	1.79	1.92				
	3	1.74	1.89	2.01			
	4	1.69	1.83	1.97	2.09		
	6	1.60	1.72	1.84	1.96	2.18	
	8	1.51	1.61	1.71	1.81	2.02	
	10	1.42	1.50	1.58	1.67	1.83	
No. 12, 0.1046 in.	2	3.47	3.65				
	3	3.41	3.62	3.79			
	4	3.34	3.55	3.76	3.92		
	6	3.22	3.41	3.60	3.79	4.13	
	8	3.10	3.27	3.44	3.61	3.95	4.26
	10	2.98	3.13	3.28	3.43	3.74	4.04
	12	2.85	2.99	3.12	3.26	3.53	3.80
No. 10, 0.1345 in.	3	5.62	5.90	6.11			
	4	5.54	5.82	6.09	6.30		
	6	5.39	5.65	5.90	6.16	6.61	
	8	5.24	5.48	5.72	5.96	6.44	6.85
	10	5.08	5.31	5.53	5.75	6.19	6.64
	12	4.93	5.14	5.34	5.54	5.95	6.36

Table 7-10 continued

Gage and thickness	H, in.	P max (2), kips, for different values of B					
		1 in.	2 in.	3 in.	4 in.	6 in.	8 in.
No. 8, 0.1644 in.	3	8.38	8.73	8.97			
	4	8.29	8.63	8.98	9.21		
	6	8.10	8.43	8.76	9.08	9.63	
	8	7.92	8.23	8.54	8.85	9.47	9.98
	10	7.74	8.03	8.32	8.61	9.19	9.78
	12	7.56	7.83	8.10	8.38	8.92	9.47

Values above and to right of heavy lines should be checked to see that shear in web does not exceed allowable value recommended in Art 7-20.

7-20. Web Stresses in Cold-formed Shapes. Allowable shear stresses on the gross section of webs of cold-formed shapes may be assigned the same values, relative to yield point, as for structural steel. Since web stiffeners are not ordinarily used in such shapes, except possibly at reactions and points of concentrated loading, allowable shear stresses may be expressed in somewhat simpler terms than for structural-steel girders, where web stiffeners are customary. Shear stresses allowable under the 1968 AISI "Specification for the Design of Cold-formed Steel Structural Members" are summarized in Table 7-8. For any case in which intermediate stiffeners are used, as in plate girders, web design may be handled in the same way as for structural steel.

The use of unstiffened webs in which the depth-thickness ratio h/t exceeds 150 is not recommended unless stiffeners are provided at reactions and under concentrated loads. In such cases, h/t should not exceed 200.

Combined Shear and Bending. The buckling effect of combined shear and bending stresses in beam webs will rarely be critical. However, it may be when high h/t values and high-strength steel are being used, with high working stresses, and then only where high shear and high bending stresses occur at the same section. An interaction formula for investigating such a case can be found in the AISI Specification.

Web Crippling. Empirical formulas for design against web crippling also can be found in this specification. Those formulas are summarized in Tables 7-9 to 7-11 for Grade C uncoated steel and for sections where inside-corner radii are equal to the thickness of the material. Care must be taken not to extend the formulas beyond the tabular ranges of B/t. Tables 7-9 and 7-10 apply to all cases of unreinforced webs, including hat sections as well as channels and zees. Table 7-11 applies only where the web is backed up by an opposing web or portion of a web, such as in an I section made by welding two angles to a channel.

For other grades of steel or other corner radii adjust the tabular values as indicated by Eqs. (7-14*a*) and 7-14*b*).

Multiply the values in Table 7-9 by the quantity

$$k\left(1.15 - 0.15\frac{r}{t}\right)(1.33 - 0.33k) \qquad (7\text{-}14a)$$

Multiply the values in Table 7-10 by the quantity

$$k\left(1.06 - 0.06\frac{r}{t}\right)(1.22 - 0.22k) \qquad (7\text{-}14b)$$

Multiply the values in Table 7-11 by the ratio of 20 to the basic working stress, ksi, or by $33/F_y$, where F_y is the yield point of the steel, ksi.

In Eqs. (7-14*a*) and (7-14*b*),

k = specified minimum yield point, ksi, divided by 33
t = web thickness, in.
r = inside corner radius, in., if not greater than $4t$. For larger corner radii the web crippling strength should be determined by test

Table 7-11. Maximum Reactions and Concentrated Loads Bearing on Restrained Webs of Light-gage Steel

[Grade C uncoated steel ($F_y = 33$), corner radii = thickness of material]

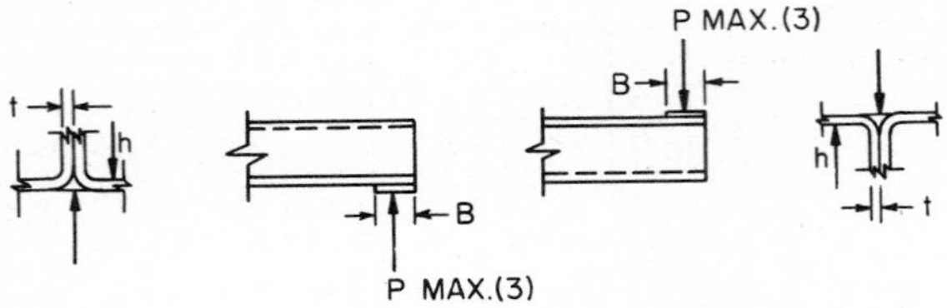

t = web thickness (each web sheet)
h = clear distance between flanges

$$P \max (3) = 20t^2 \left(7.4 + 0.93 \sqrt{\frac{B}{t}}\right) \qquad (7\text{-}14c)$$

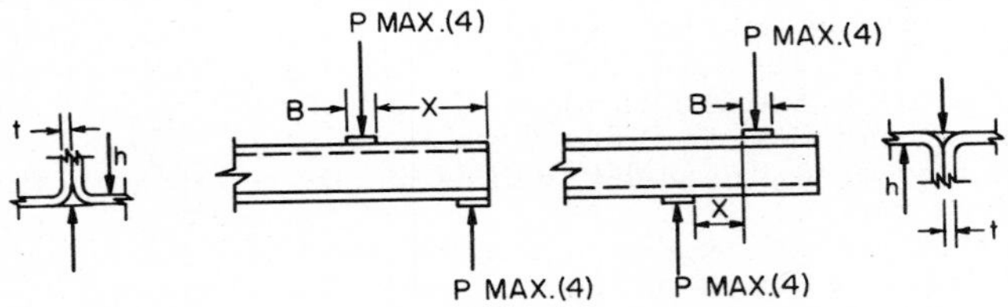

t = web thickness (each web sheet)
h = clear distance between flanges

$$P \max (4) = 20t^2 \left(11.1 + 2.41 \sqrt{\frac{B}{t}}\right) \qquad (7\text{-}15)$$

Values of P max (4) apply only where the distance x is greater than $1.5h$. Otherwise use P max (3).

The effective length of bearing B for substitution in the above formulas should not be taken as greater than h, and tabular values of P for values of B greater than h should not be used.

Gage and thickness	End bearing. P max (3), kips, for different values of B						Interior bearing. P max (4), kips, for different values for B					
	1 in.	2 in.	3 in.	4 in.	6 in.	8 in.	1 in.	2 in.	3 in.	4 in.	6 in.	8 in.
No. 18, 0.0478 in. . . .	0.53	0.61	0.68	0.72	....	....	1.01	1.22	1.38	1.51		
No. 16, 0.0598 in. . . .	0.80	0.91	1.00	1.08	....	....	1.50	1.79	2.01	2.20		
No. 14, 0.0747 in. . . .	1.20	1.37	1.49	1.59	1.76	....	2.22	2.63	2.94	3.21	3.64	
No. 12, 0.1046 in. . . .	2.24	2.51	2.71	2.88	3.16	3.40	4.06	4.73	5.26	5.69	6.42	7.04
No. 10, 0.1345 in. . . .	3.60	3.98	4.27	4.51	4.92	5.28	6.38	7.38	8.13	8.77	9.84	10.74
No. 8, 0.1644 in	5.24	5.76	6.14	6.48	7.03	7.51	9.21	10.54	11.57	12.42	13.87	15.09

Tabular values are for one web thickness t. Multiply by 2 for double-thickness webs.

7-21. Cold-formed Steel Columns. When cold-formed sections are used as compression members, conventional procedures may be followed without modification, except when a section falls into either of the following categories:

Class I. Sections containing elements that exceed the limits $w/t = 63.3/\sqrt{F_y}$ for unstiffened elements, and $w/t = 171/\sqrt{F} = 221/\sqrt{F_y}$ for stiffened elements, where F = basic working stress, ksi, and F_y = yield point, ksi, of the steel. When either of these limits is exceeded, requiring that provision be made against failure by local buckling, the reduced strength of the section can be taken into account by introducing a form factor or buckling factor into the column design formula. This is done merely by using QF_y instead of F_y in the basic formula, where the factor Q is determined as outlined later.

Class II. Open sections whose shear center and centroid do not coincide, and which are not braced against twisting. Under these conditions, because of low torsional resistance, failure of a thin-walled section may occur by torsional-flexural buckling instead of by simple column action. Hence, the torsional-flexural strength should be investigated as outlined later. This precaution applies to such sections as channels, hat sections, angles, tees, unsymmetrical I sections, and any other singly symmetrical or nonsymmetrical sections, if twisting can occur. It is not necessary for doubly symmetrical shapes, such as symmetrical I sections; closed sections, such as box or tubular sections; point-symmetric shapes such as equal-flange zee sections; or for any section which is held against twisting throughout its length, as by the attachment of wall sheathing.

Bear in mind, however, that members braced against twisting in the finished structure may not be so braced during construction.

The safe-load formulas recommended by the AISI "Specification for the Design of Cold-formed Steel Structural Members" for concentrically loaded columns are as follows:

A. Sections that do not fall into either Class I or II, i.e., sections not subject to either local buckling or twisting, and are at least 0.09 in. thick:

Same as for structural steel (see Sec. 6)

B. All sections that are less than 0.09 in. thick and all sections, regardless of thickness, that fall into Class I but do not fall into Class II, i.e., sections not subject to twisting but containing elements that may be subject to local buckling ($Q < 1.0$):

For $Kl/r < (\sqrt{2\pi^2E/QF_y} = 763/\sqrt{QF_y})$,

$$\frac{P}{A} = 0.522QF_y - \left(\frac{QF_y}{1{,}494}\right)^2 \left(\frac{Kl}{r}\right)^2 \qquad (7\text{-}16)$$

where P = total allowable axial load, kips
A = full, unreduced cross-sectional area of member, sq in.
F_y = specified minimum yield point, ksi
Kl = effective unsupported length of member, in. (For determination of the effective length factor K, follow structural-steel practice or refer to the AISI "Specification for the Design of Cold-formed Steel Structural Members" and its "Commentary.")
r = radius of gyration of full, unreduced cross section, in.
Q = a factor determined as described later

For $Kl/r \geq 763/\sqrt{QF_y}$,

$$\frac{P}{A} = \frac{12\pi^2E}{23(Kl/r)^2} = \frac{151{,}900}{(Kl/r)^2} \qquad (7\text{-}17)$$

as for structural steel columns.

Equation (7-16) represents a family of curves beginning at $P/A = 0.522QF_y$ for $Kl/r = 0$ and tangent to the Euler curve of Eq. (7-17) at $Kl/r = 763/\sqrt{QF_y}$, $P/A = 0.261QF_y$ (Fig. 7-19). An infinite number of curves may be drawn, each for a different value of QF_y.

Equation (7-16) reduces to Eq. (7-18) for Grade C uncoated steel ($F_y = 33$ ksi), and to Eq. (7-19) for a high-strength low-alloy steel for which $F_y = 50$ ksi.

For $F_y = 33$ and $Kl/r < 133/\sqrt{Q}$,

$$\frac{P}{A} = 17.2Q - 0.000488Q^2 \left(\frac{Kl}{r}\right)^2 \tag{7-18}$$

For $F_y = 50$ ksi and $Kl/r < 108/\sqrt{Q}$,

$$\frac{P}{A} = 26.1Q - 0.00112Q^2 \left(\frac{Kl}{r}\right)^2 \tag{7-19}$$

For Kl/r greater than the limiting values for Eqs. (7-18) and (7-19), for all grades of steel, and all values of Q, use Eq. (7-17).

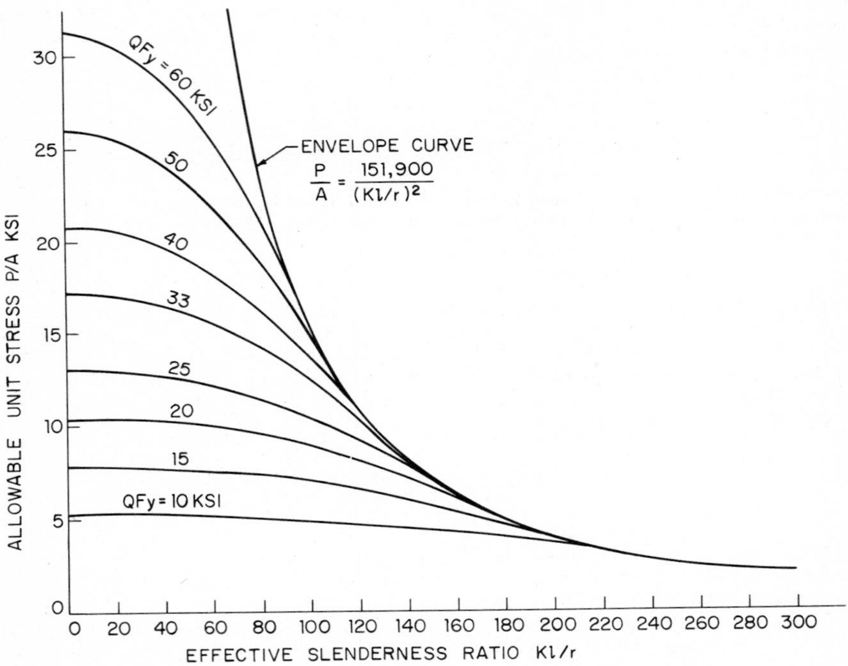

Fig. 7-19. Column design curves for sections less than 0.09 in. thick and sections subject to local buckling ($Q < 1$) but not twisting (group B sections).

C. Sections that fall into Class II:

1. For singly symmetrical shapes (shapes symmetrical about only one axis) that do not also fall into Class I; that is, shapes for which $Q = 1.0$, the AISI Specification requires that P/A be the smaller of the following:

a. P/A determined as for sections in group *A* or *B*, as applicable.

b. A function of the torsional-flexural buckling stress σ_{TFO}:

If $\sigma_{TFO} > 0.5F_y$,

$$\frac{P}{A} = 0.522F_y - \frac{F_y^2}{7.67\sigma_{TFO}} \tag{7-20}$$

If $\sigma_{TFO} \leq 0.5F_y$,

$$\frac{P}{A} = 0.522\sigma_{TFO} \tag{7-21}$$

Determination of σ_{TFO} is a complex procedure that involves the location of the shear center, the polar radius of gyration about the shear center, and the torsion and warping constants of the section. For thin-walled shapes of uniform thickness, these quantities are not mathematically difficult to determine, but the process is likely to prove tedious and laborious unless computerized. But the mere fact that a section or component falls in Class II does not necessarily mean that the torsional-flexural mode will control, i.e., that Eqs. (7-20) and (7-21) will yield a lower allowable stress than that given by the regular column formula. For a few common shapes of normal proportions, a preliminary appraisal of whether the torsional-flexural buckling stress may control the design can be made by means of the following approximate but conservative criteria (adapted from "Torsional-Flexural Buckling, Elastic and Inelastic, of Cold-Formed Thin-Walled Columns," by Alexander Chajes, Pen J. Fang, and George Winter, *Cornell Engineering Research Bulletin* 66-1, Cornell University, August, 1966, and the AISI "Cold-Formed Steel Design Manual"):

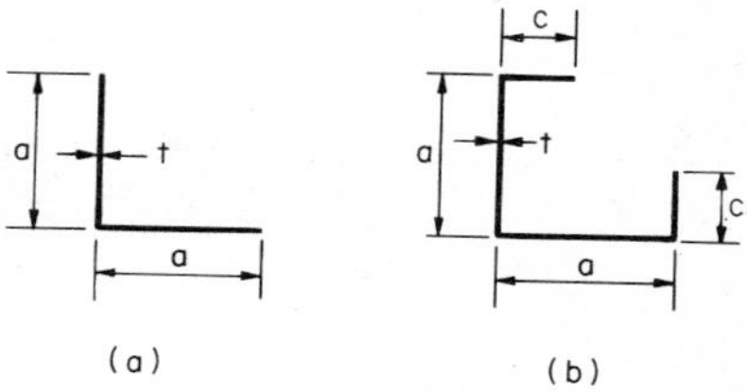

Fig. 7-20. Equal-legged angles.

For plain equal-leg angles (Fig. 7-20*a*), torsional-flexural buckling need not be investigated if

$$\frac{t}{a^2Kl} \geq 1.1 \tag{7-22}$$

where t = thickness of material, in.
a = leg width, in., as shown in Fig. 7-20*a*
Kl = effective length, in., between twist-restraining attachments (for open, singly symmetrical or nonsymmetrical components of built-up sections, the effective length between battens, diaphragms, or other bracing)

For equal-leg lipped angles for which $c/a \leq 0.4$ (Fig. 7-20*b*), torsional-flexural buckling need not be investigated if

$$\frac{t}{a^2Kl} \geq 1.1 + \frac{3c}{a} \tag{7-23}$$

For symmetrical channels and hat sections, see Figs. 7-21 to 7-23.

Torsional-flexural buckling should be investigated if $(t/a^2)Kl$ is less than the values given by Eqs. (7-22) and (7-23), or falls below and to the right of the curves in Figs. 7-21 to 7-23; if, for lipped angles (Fig. 7-20*b*), c/a is greater than 0.4; or if, for hat sections (Fig. 7-23), c/a is less than 0.2 or greater than 0.8. More precise methods of determining the failure mode, and methods of calculating the torsional-flexural buckling stress, are available in the AISI Specification and Design Manual.

2. For singly symmetrical shapes that also fall into Class I, i.e., sections for which Q is less than 1.0, either replace F_y by QF_y in Eqs. (7-20) and (7-21), or use tests to determine the critical load. General rules relating to such tests are given in the AISI Specification.

3. For nonsymmetric shapes, i.e., shapes that are not symmetric about any point or axis, the AISI Specification recommends either analysis or tests. A general method of analysis is given in the AISI Design Manual.

Note that torsional-flexural buckling must be considered for sections in group C only if a member or component is not restrained against twisting over any appreciable proportion of its length. The amount of restraint required to prevent twisting is small, perhaps of the same order of magnitude as that required to brace a column or to keep a wall stud from buckling in the plane of the wall (Art. 7-23). In any unproven or doubtful case, however, tests are advisable.

For bracing and secondary members with l/r greater than 120, the AISI Specifica-

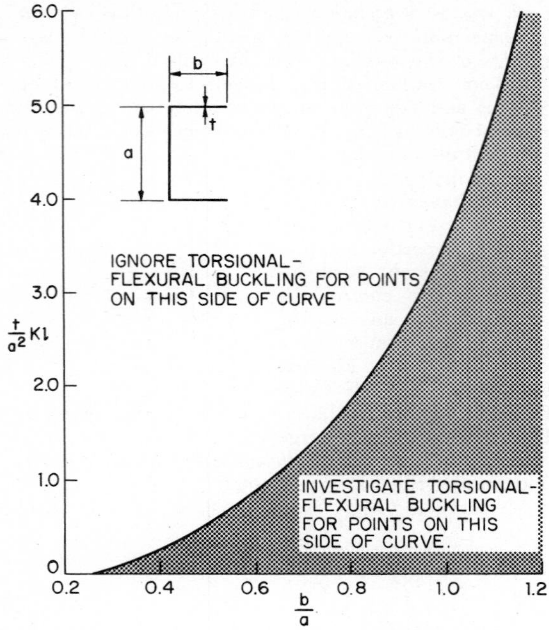

Fig. 7-21. Failure-mode criterion for plain channels under concentric compressive load.

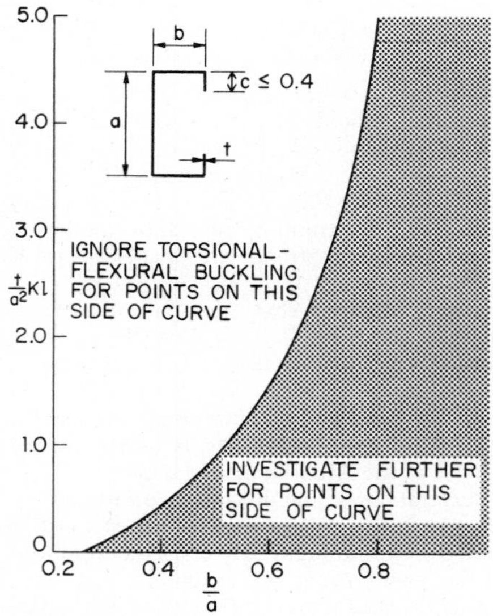

Fig. 7-22. Preliminary failure-mode criterion for lipped channels under concentric compressive load for $c/a \leq 0.4$.

tion recommends that the allowable P/A value determined as for sections in group *A, B,* or *C* be increased to

$$\left(\frac{P}{A}\right)_s = \frac{P/A}{1.3 - (L/r)/400} \tag{7-24}$$

where l is the actual unbraced length of the member. (The effective-length factor K is taken as unity for secondary members.) Kl/r should not exceed 200, except that during construction a value of 300 may be temporarily allowed.

Determination of Q. The form factor, or buckling factor, Q can never be greater than 1.0. For sections that do not contain any elements that exceed the lower w/t limits set for Class I, $Q = 1.0$ and disappears from the equation. For sections that do contain such elements, Q is determined as follows:

a. For members composed entirely of stiffened elements, Q is the ratio between the effective design area, as determined from the effective design widths of such

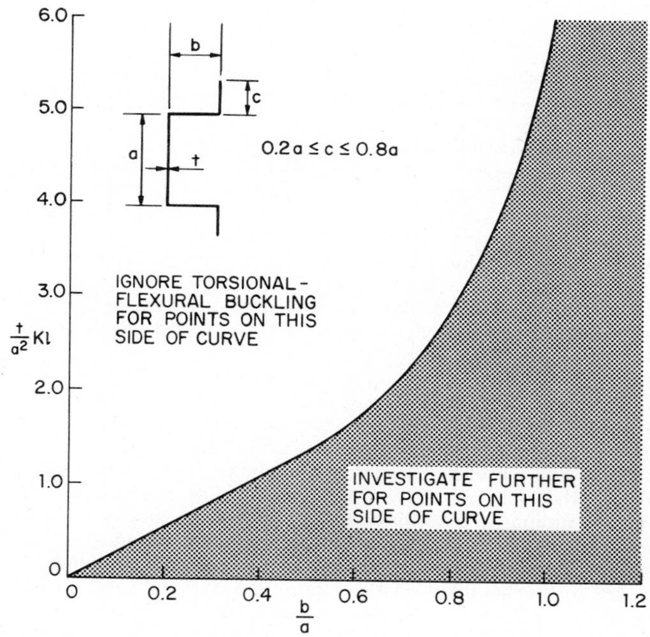

Fig. 7-23. Preliminary failure-mode criterion for hat sections under concentric compressive load for $0.2 \leq c/a \leq 0.8$.

elements, and the full or gross area of the cross section. The effective design area used in determining Q is to be based on the basic design stress F allowed in tension and bending.

b. For members composed entirely of unstiffened elements, Q is the ratio between the allowable compression stress for the weakest element of the cross section (the element having the largest flat-width ratio) and the basic design stress.

c. For members composed of both stiffened and unstiffened elements, the factor Q is to be the product of a stress factor Q_s computed as outlined in *b* above and an area factor Q_a computed as outlined in *a* above. However, the stress on which Q_a is to be based shall be that value of the unit stress f_c used in computing Q_s; and the effective area to be used in computing Q_a shall include the full area of all unstiffened elements.

7-22. Combined Axial and Bending Stresses. For eccentrically loaded members in groups *A* and *B* of Art. 7-21; i.e., members not subject to torsional-flexural buckling, the procedure for dealing with combined axial and flexural loads is substantially the same as for structural steel, and the same types of interaction equations may be used. The AISI "Specification for the Design of Cold-formed Steel Structural Members" prescribes the following criteria for compressive axial loading:

$$\frac{f_a}{F_{a1}} + \frac{C_{mx} f_{bx}}{\left(1 - \frac{f_a}{F'_{ex}}\right) F_{bx}} + \frac{C_{my} f_{by}}{\left(1 - \frac{f_a}{F'_{ey}}\right) F_{by}} \leq 1.0 \tag{7-25}$$

$$\frac{f_a}{F_{ao}} + \frac{f_{bx}}{F_{b1x}} + \frac{f_{by}}{F_{b1y}} \leq 1.0 \tag{7-26}$$

When $f_a/F_{a1} \leq 0.15$, Eq. (7-27), may be used instead of Eqs. (7-25) and (7-26).

$$\frac{f_a}{F_{a1}} + \frac{f_{bx}}{F_{bx}} + \frac{f_{by}}{F_{by}} \leq 1.0 \tag{7-27}$$

where F_{a1} = allowable axial stress P/A under concentric loading only

F_{ao} = allowable axial stress under concentric loading when effective column length $Kl = 0$

F_{b1x}, F_{b1y} = allowable bending stress in compression, about axis x and y, respectively when the possibility of lateral buckling is excluded. For sections containing unstiffened compression elements, F_{b1x} and F_{b1y} are limited to the allowable stresses in such elements as recommended in Art. 7-15

f_a = computed stress under concentric loading only

f_{bx}, f_{by} = computed bending stresses about axis x and y, respectively

C_m = reduction factor defined in Art. 6-27

Note that for members with unstiffened compression elements the bending stresses f_{bx} and f_{by} are to be computed on the basis of the effective width of such elements, as discussed in Art. 7-16.

For handling combined axial and bending loads on eccentrically loaded members in group *C* (Class II of Art. 7-21, i.e., members that may be subject to failure by torsional-flexural buckling), the procedure is more complex. The AISI "Cold-Formed Steel Design Manual" should be consulted in such cases.

7-23. Braced Wall Studs. A fairly common application of cold-formed shapes—and one that requires special treatment—is as load-bearing wall studs in light-occupancy buildings. Cold-formed steel studs are usually I, channel, or zee sections, to both flanges of which collateral materials are fastened. The AISI "Specification for the Design of Cold-formed Steel Structural Members" permits the strength of studs to be computed on the assumption that they are laterally supported in the plane of the wall when the wall material and its attachments to the studs comply with certain criteria that reduce to the following:

1. Wall sheathing is attached to both faces or flanges of the studs.
2. The spacing, in., of attachments of wall material to each face or flange of the stud does not exceed *a* in either Eq. (7-28) or (7-29):

$$a_{\max} = \frac{236{,}000}{F_y^2} \frac{r_2^2}{A} k \tag{7-28}$$

$$a_{\max} = \frac{L}{2} \frac{r_2}{r_1} \tag{7-29}$$

where L = length of stud, in.

A = full cross-sectional area of stud, sq in.

r_1 = radius of gyration of full cross section of stud about its axis parallel to wall, in.

r_2 = radius of gyration of full cross section of stud about its axis perpendicular to wall, in.

k = combined modulus of elastic support, or spring constant, of one attachment of wall material to stud and of wall material tributary to that attachment, kips per in.

F_y = specified minimum yield point of the stud material, ksi

For Grade C uncoated steel, $F_y = 33$ ksi, Eq. (7-28) becomes

$$a_{\max} = 220 \frac{r_2^2}{A} k \tag{7-30}$$

and for a steel with $F_y = 50$ ksi,

$$a_{\max} = 94 \frac{r_2^2}{A} k \tag{7-31}$$

For continuous attachment—by an adhesive, for example—a should be taken as unity in solving Eqs. (7-28), (7-30), and (7-31) for k.

One attachment and the wall material tributary to it should be able to exert on the stud a force, kips, in the plane of the wall at least equal to

$$F_{\min} = \frac{kLP}{82{,}200\sqrt{I_2(k/a)} - 240P} \tag{7-32}$$

where I_2 = moment of inertia of full cross section of stud about its axis perpendicular to wall, in.[4]

P = load on stud, kips

For continuous attachment, $a = 1$.

Values of k and F for any particular combination of stud, wall material, and attaching means can only be determined by test. The following table shows the range of values obtained at Cornell University on five different wallboards fastened to channel studs with 3/16-in.-diameter bolts and washers. (For background material, see Green, Winter, and Cuykendall, "Light-Gage Steel Columns in Wall-braced Panels," *Cornell University Engineering Experiment Station Bulletin* 35, Part 2, October, 1947, and Winter, "Lateral Bracing of Columns and Beams," *Transactions, American Society of Civil Engineers,* Vol. 125, 1960.)

Wallboard	Range of k values, kips per in.	Range of F values, kips
1/2 -in. standard-density wood and cane-fiber insulating boards	0.290–0.603	0.675–0.1575
1/2 -in. paper-base insulating board	0.915–1.460	0.162–0.227
3/8 -in. gypsum board sheathing	0.775–1.535	0.125–0.292
3/16 -in. medium-density compressed-wood fiberboard	2.010–4.560	0.140–0.256
5/32 -in. high-density compressed-wood fiberboard	3.960–7.560	0.435–0.600

The above values were obtained from relatively simple tension tests. Pieces of wall material are fastened to two short pieces of studs at about the contemplated spacing. Starting with a small initial load P_0 (50 lb or so), increments of load are applied, and the distance between studs is measured at each increment. The test should be carried to failure. A satisfactory value of k can be determined from the expression

$$k = \frac{0.75P - P_0}{0.5ny} \tag{7-33}$$

where P_{ult} = ultimate load, kips

P_0 = initial or zero load, kips

n = number of attachments

y = average change in length between attachments of each of the stud pieces from initial load P_0 to $0.75P_{ult}$

For continuous attachment, use n equal to four times the attached length along each faying surface. If it is desired to simulate conditions at a joint in the wallboard, the specimen can be altered accordingly.

For any particular assembly, it may be considered somewhat more direct to test in direct compression as a column a stud specimen to which collateral materials have been fastened. Thereby, it can be determined from the manner of failure whether or not the collateral materials and their attachments to the studs furnish sufficient lateral support in the plane of the wall. This procedure, however, should be used only under expert supervision. The instrumentation and technique necessary to perform and interpret such tests properly call for a high degree of skill. Among other precautions, care must be exercised to see that the collateral materials do not obscure the results by carrying part of the axial load on the stud.

It can be seen from the values of k and F from the Cornell tests, which are for one kind of stud with one kind of attachment, that a very wide variation can be expected in the ability of collateral materials to provide lateral support for studs. Tests have shown that even apparently minor changes in fasteners or technique can produce significant differences in the values of k and F (*Cornell University Bulletin* 35, Part 2). Fortunately, however, the degree of lateral support actually required to hold a column straight against buckling below its yield point is usually very small. With any reasonable spacing of attachments and strength of wall covering, sufficient lateral support will be provided.

7-24. Electric-arc Welding of Light-gage Steel. Electric-arc welding can be effectively used for making joints in light-gage steel construction. Care must be exercised not to use electrodes that are too large relative to the thickness of material being welded. Care also must be taken to avoid burning through the material because of the use of too much current. As with thicker material, the welding procedure and technique to be used depend on composition of material to be welded, conditions under which work is to be performed, and requirements of each case.

Butt welds and other edge-to-edge welds in thin material should be made only when proper alignment of the parts being joined and proper backup are provided.

For details, refer to texts dealing with welding techniques and procedures, such as "Welding Handbook," American Welding Society.

The AISI "Specification for the Design of Cold-Formed Steel Structural Members" prescribes the following allowable stresses in electric-arc welds in structural-quality sheet and strip:

Groove welds in tension or compression, including stresses due to eccentricity:

Same as for the base metal (or the lower grade if two grades are being joined) provided there is complete penetration and the yield strength of the filler metal is equal to or greater than that of the base metal.

Shear on throat of fillet or plug welds:

Specified minimum yield point F_y for lowest grade being joined	ASTM electrode classification	Allowable shear, ksi
36 ksi or lower	A233 E60	13.6
Greater than 36 ksi but not more than 50 ksi	A233 E60	13.6
	A233 E70	15.8
Greater than 50 ksi	A316 E70	15.8
	A316 E80	17.7

The throat of a fillet, plug, or slot weld should be considered as having the traditional thickness, i.e., $0.707t$ for a 45° fillet, where t is the thickness of the base material. As in structural-steel work, plug and slot welds should not be assigned any value in resistance to other than shear stresses.

7-25. Resistance Welding. Resistance welding is defined as "a group of welding processes wherein coalescence is produced by the heat obtained from resistance of the work to the flow of electric current in a circuit of which the work is a part, and by the application of pressure." ("Welding Handbook," American Welding Society.)

Although there are a number of different resistance-welding processes, **spot welding** is the simplest. It is commonly used in the manufacture of built-up or compound shapes of sheet and strip steel for structural purposes. In spot welding, the work is held under pressure between two electrodes through which an electric current passes, and the weld is made at contact interfaces between the pieces being joined (Figs. 7-24 and 7-25).

Resistance welding is essentially a shop process, because of the size of the equipment required for good work. Although portable resistance welders are available and will produce good results in very light work, electric-arc welding is generally more satisfactory for field use.

For structural design purposes, spot welds can be treated in the same way as rivets, except that no reduction in net section due to holes need be made. Table 7-12 and the curves of Figs. 7-26 and 7-27 show essential information for design purposes, for uncoated material, based upon "Recommended Practices for Resistance Welding," American Welding Society, 1966.

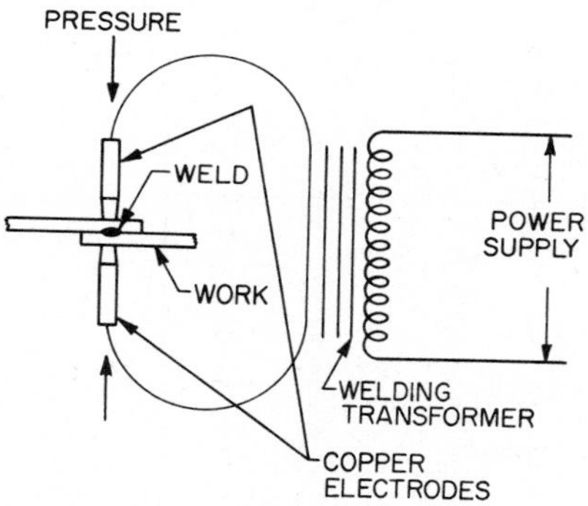

Fig. 7-24. Schematic diagram of spot-welding process.

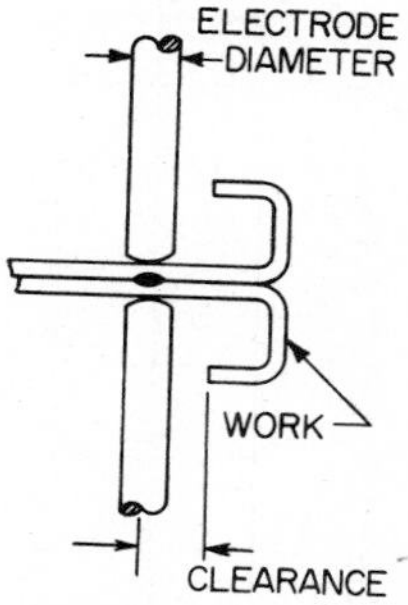

Fig. 7-25. Electrode clearance in spot welding.

Based on the shear strengths in Table 7-12 and a safety factor of about 2½, the allowable shear on spot welds for design purposes may be read from the curve of Fig. 7-26.

The thickest material listed in Table 7-12 for plain spot welding is ⅛ in. For welding material thicker than ⅛ in., variations of spot welding known as pulsation welding or projection welding are usually used unless large-capacity machines are available.

Projection welding is a form of spot welding wherein the weld is made through projections (usually consisting of buttons or protuberances in one or more of the pieces being joined) so as to localize the welding current and pressure. Thus, the welds can be made with somewhat lighter equipment than required for ordinary spot welding.

Pulsation welding, more accurately called multiple-impulse welding, can be defined as the making of spot welds by more than one impulse of current. Except where requirements for very small weld spacing or edge distance make it necessary to use projection welding, the choice between pulsation welding and projection welding is a matter for the fabricator to decide, based on the nature of the work, the number of pieces to be made, the capacity and limitations of the equipment, and other operating considerations.

Design values in Fig. 7-26 are based on the assumption that welding is to be performed in accordance with AWS "Recommended Practices for Resistance Welding," and with suitable procedure and controls. When it is, these values may be applied to any of the grades of carbon steel covered by ASTM A570 and A611 for uncoated structural-quality sheet and strip (Art. 7-9). In the absence of other information, the values in Fig. 7-26 may also be applied to higher-strength

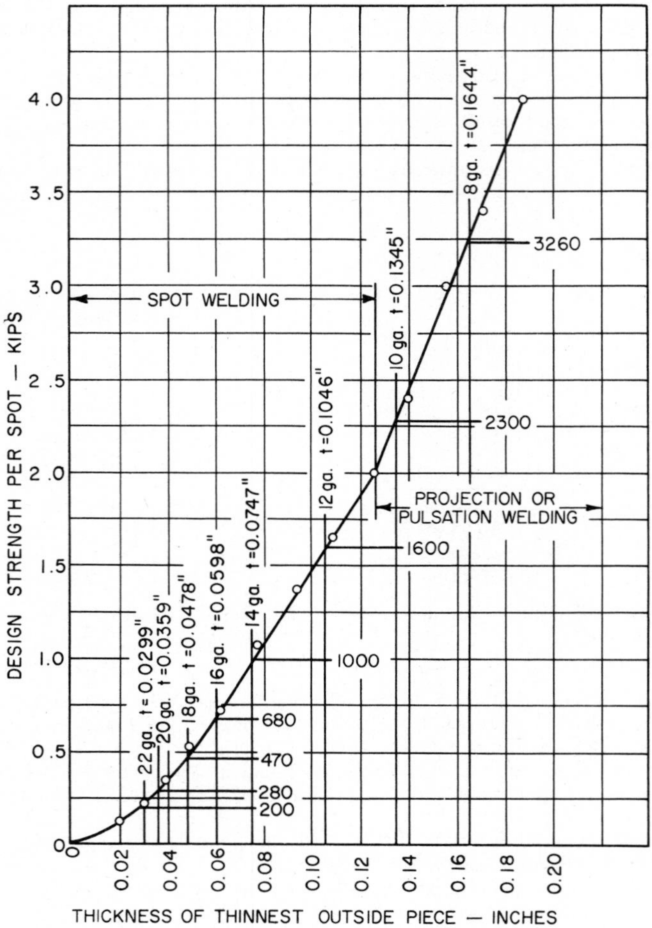

Fig. 7-26. Design strength of spot welds. For values for thicker material, divide the shear strengths given in Table 7-12 by 2.5 and use straight-line interpolation.

steel; but spot welds in higher-strength steel yield somewhat higher shear strengths than those in Table 7-12 and may require special welding conditions. For further information refer to AWS recommended practices.

Material for spot welding should be free from scale; therefore, either hot-rolled and pickled or cold-rolled material is usually specified. Steel containing not more than 0.15% carbon is more readily weldable than is higher-carbon material. But material of higher carbon content can be spot-welded successfully with the proper equipment and technique. Welding, by any usual process, of high-carbon material such as Grade D of ASTM A446, which may contain up to 0.40% carbon (heat analysis), is not, however, ordinarily recommended except with specialized procedures and due regard for the consequences of weld failure under service conditions. Designers should consider other means of joining when specifying such material.

It is important that the minimum contacting overlap shown in Table 7-12 be observed in detailing spot-welded joints. If smaller overlaps are used, the weld strengths will be difficult to maintain under shop conditions, or distortion of the lapping sheets may occur in welding. The minimum spacing specified should also be observed, so that no special precautions need be taken to compensate for the current-shunting effect of the adjacent welds. In detailing compound sections to

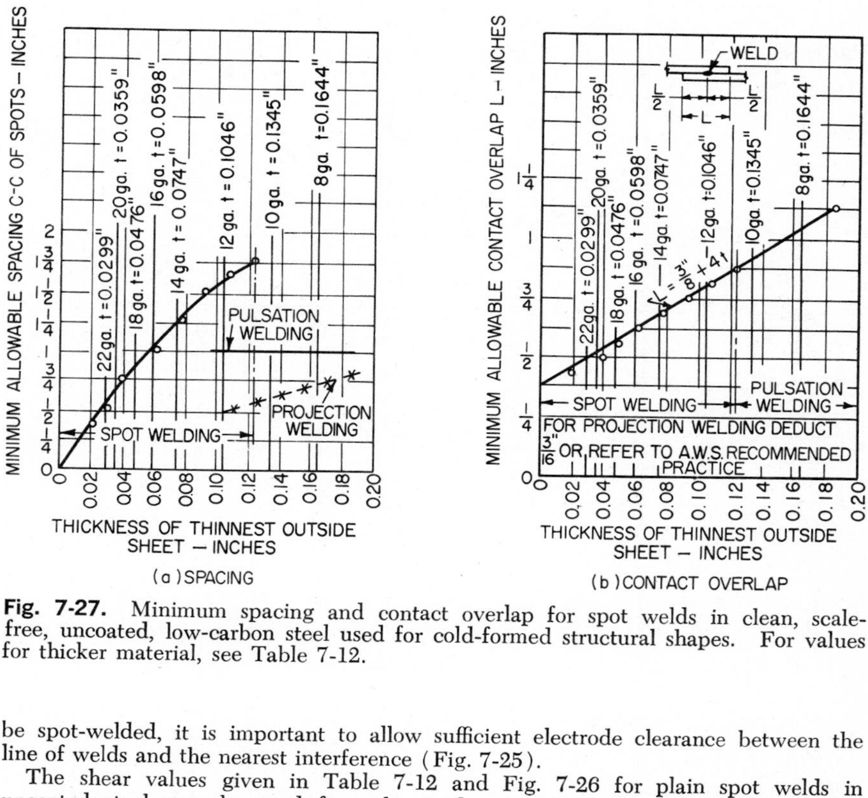

Fig. 7-27. Minimum spacing and contact overlap for spot welds in clean, scale-free, uncoated, low-carbon steel used for cold-formed structural shapes. For values for thicker material, see Table 7-12.

be spot-welded, it is important to allow sufficient electrode clearance between the line of welds and the nearest interference (Fig. 7-25).

The shear values given in Table 7-12 and Fig. 7-26 for plain spot welds in uncoated steel may be used for galvanized material up to and including ⅛ in. thick with commercial (1.25 oz per sq ft) or lighter coating weight. The electrode sizes and other dimensions recommended in Table 7-12 and Fig. 7-27 do not apply to galvanized steel, for which contacting overlap, minimum weld spacing, and other dimensional requirements must generally be larger than for uncoated material. For detailed information, refer to AWS "Recommended Practices for Resistance Welding Coated Low-Carbon Steel," and to other publications dealing with resistance welding.

7-26. Maximum Spacing of Welds. In addition to observing requirements for minimum spacing of welds (Table 7-12), it is also necessary to observe certain precautions with respect to the maximum spacing that may be permitted. These are as follows:

For Shear Transfer. The spacing should not exceed that required to transmit the shear between the connected parts. This refers, of course, to the primary function of the welds to act in the same way as rivets in structural connections.

For Column Action. The AISI "Specification for the Design of Cold-formed Steel Structural Members" recommends that, to avoid failure of a compression cover plate or sheet by column action (cylindrical buckling) between welds, weld spacing in line of stress, in., should not exceed $200t/\sqrt{f}$, where t is the thickness, in., of the cover plate, and f is the design stress, ksi, in the cover plate.

Table 7-12. Resistance Welding Data for Uncoated Low-carbon Steel

Thickness t of thinnest outside piece, in.	Min OD of electrode D, in. 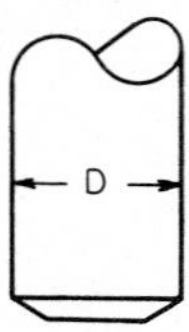	Min contacting overlap, in. 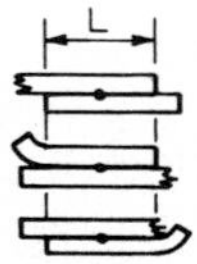	Min weld spacing c to c, in.	Approx diam of fused zone, in. 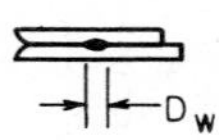	Min shear strength per weld, kips	Diam of projection D, in. 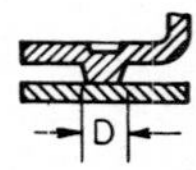
			Plain Spot Welding			
0.021	3/8	7/16	3/8	0.13	0.32	
0.031	3/8	7/16	1/2	0.16	0.57	
0.040	1/2	1/2	3/4	0.19	0.92	
0.050	1/2	9/16	7/8	0.22	1.35	
0.062	1/2	5/8	1	0.25	1.85	
0.078	5/8	11/16	1 1/4	0.29	2.70	
0.094	5/8	3/4	1 1/2	0.31	3.45	
0.109	5/8	13/16	1 5/8	0.32	4.15	
0.125	7/8	7/8	1 3/4	0.33	5.00	
			Pulsation Welding			
1/8	1	7/8	....	3/8	5	
3/16	1 1/4	1 1/8	....	9/16	10	
1/4	9/16	1 3/8	....	3/4	15	
			Projection Welding			
0.125	...	11/16	9/16	0.338	4.8	0.281
0.140	...	3/4	5/8	7/16	6.0	0.312
0.156	...	13/16	11/16	1/2	7.5	0.343
0.171	...	7/8	3/4	9/16	8.5	0.375
0.187	...	15/16	13/16	9/16	10.0	0.406
0.203	...	1	7/8	5/8	12	0.437
0.250	...	1 1/4	1 1/16	11/16	15	0.531

To Prevent Local Plate Buckling. The AISI Specification prescribes that the spacing, in line of stress, of welds connecting a compression cover plate or sheet to a stiffener or other element should not exceed three times the flat width of the narrowest unstiffened compression element (Art. 7-13) in that portion of the cover plate that is tributary to the connections. But the spacing need not be less than the following, unless closer spacing is required to transmit shear or prevent cylindrical buckling:

Value of F_c permitted in unstiffened element	*Minimum required spacing of connections, in.*
$0.54F_y$ or less	$228/\sqrt{F_y}$
Greater than $0.54F_y$	$190/\sqrt{F_y}$

where F_y = specified minimum yield point of the steel, ksi
t = thickness, in., of cover plate or sheet
F_c = allowable compressive stress in the unstiffened element, ksi

This reduces to the following, for $F_y = 33$ ksi and $F_y = 50$ ksi:

Allowable F_c, nearest ksi	Minimum required spacing, in., unless closer spacing is required to transmit shear or prevent cylindrical buckling	
	$F_y = 33$	$F_y = 50$
18 or less	$40t$	$32t$
Greater than 18, not more than 20	$33t$	$32t$
Greater than 20, not more than 27	. . .	$32t$
Greater than 27 . . .	. . .	$27t$

For intermittent fillet welds parallel to the direction of stress, the spacing should be taken as the clear distance, in., between welds plus ½ in. In all other cases, the spacing should be taken as the center-to-center distance between connections.

None of the above requirements regarding weld spacing applies to cover sheets which act only as sheathing material and are not considered load-carrying elements.

Between Two Channels Forming an I. The AISI Specification also contains recommendations for the maximum longitudinal spacing of welds connecting two channels to form an I section (Figs. 7-6*m* and 7-6*n*). For compression members, it is recommended that spacing, in., not exceed

$$s_{\max} = \frac{Lr_{cy}}{2r_y} \tag{7-34}$$

where r_{cy} is the radius of gyration, in., of one channel about its centroidal axis parallel to web, and L/r_y is the slenderness ratio of the built-up member about the axis that controls its load-carrying capacity; e.g., about the minor axis for free-standing columns, or about the major axis where that axis controls, as in the case of braced wall studs (Art. 7-23). For flexural members, the AISI provisions are somewhat more complicated, and involve the span of the beam, the tensile strength of each connection, the distance between rows of connections, the intensity of the load on the beam, and the location of the shear center of each channel. In most cases, any reasonably close spacing of the welds will be satisfactory, but it is always advisable to check the spacing against the AISI criteria.

7-27. Bolting Cold-formed Members. Bolting is often employed for making field connections in light-gage steel construction. With thin material, use of washers on both sides of joints is desirable.

The AISI "Specification for the Design of Cold-formed Steel Structural Members" prescribes that the distance, in., between bolt centers, in line of stress, and the distance, in., from bolt center to edge of sheet, in line of stress, shall not be less than 1½ times the bolt diameter, nor less than $P/0.6F_yt$, or $P/0.444F_ut$, whichever is smaller, where P is the load on the bolt, kips; t the thickness of thinnest connected sheet, in.; and F_u and F_y the tensile strength, ksi, and specified minimum yield point of material being joined, ksi.

Rules relating to maximum allowable spacing of welds (Art. 7-26) also apply to bolts.

Allowable stresses recommended by AISI for bolted joints are as follows:

Shear, on gross cross-section of bolts:
- ASTM A307 (unfinished) bolts . . . . 10 ksi
- ASTM A325 (high-strength) bolts
  - Threads excluded from shear plane . . . . 22 ksi
  - Threads not excluded from shear plane . . . . . . 15 ksi

Bearing:
$2.1F_y$ or $1.56F_u$, whichever is the smaller

Tension on net section:

$$F_t = \left(1.0 - 0.9r + \frac{3rd}{s}\right) 0.6F_y \leq 0.6F_y \leq 0.444F_u \qquad (7\text{-}35)$$

where d = nominal bolt diameter, in.
s = spacing, in., of bolts in direction normal to direction of stress being transferred through joint
= width, in., of sheet being connected for a connection with a single line of bolts in direction of stress
r = ratio of force transmitted by bolt or bolts at cross section under consideration to tension force in the member at that section

The ratio r should be computed on the assumption that each bolt in a joint transfers an equal part of the load, and the force in the member at any bolt, or transverse row of bolts, is the total force that has not been transmitted by preceding bolts, working inward from the outer bolt or row of bolts in the joint. If r is smaller than 0.2, it may be taken as zero.

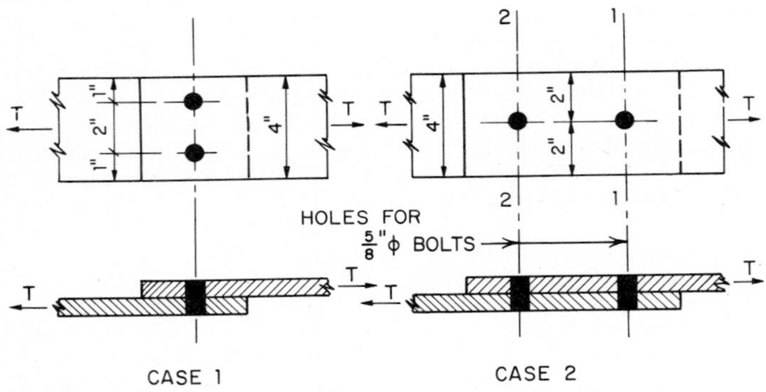

Fig. 7-28. Bolted connections with two bolts.

The foregoing requires a reduction of allowable net-section tensile stress, from the usual value of $0.6F_y$, only when the bolt spacing transverse to the direction of stress exceeds 3⅓ times the bolt diameter *and* r is greater than 0.2.

As examples of the application of Eq. (7-35), consider the upper sheet in each of the two-bolt arrangements shown in Fig. 7-28. In Case 1, the bolts are arranged in a single transverse row and the force transmitted by that row is equal to the total load T. The force in the sheet at the bolts is also considered to be T. Therefore, $r = T/T = 1$. With $s = 2$ in. and bolt diameter $d = 5/8$ in., $d/s = (5/8)/2 = 0.31$. The allowable net-section stress is, therefore, by Eq. (7-35),

$$F_t = (1.0 - 0.9 \times 1 + 3 \times 1 \times 0.31)0.6F_y = (1.03)0.6F_y$$

Since the quantity in parentheses is greater than 1, the allowable net-section stress is $0.6F_y$, or 100% of the basic allowable stress F, when $F_u > 1.35F_y$.

In Case 2, the two bolts are arranged in a single line in the direction of the applied load. The force being transmitted by the bolt at each section (1-1 and 2-2) is $T/2$. Still considering only the upper sheet, the force in that sheet at section 1-1 is taken as the total force T. At section 1-1, therefore, $r = (T/2)/T = \frac{1}{2}$. In this case, the spacing s is taken as the total sheet width, 4 in. Hence, $d/s = (\frac{5}{8})/4 = 0.16$. The allowable net-section stress in the upper sheet at section 1-1 is, therefore, by Eq. (7-35),

$$F_t = (1.0 - 0.9 \times \tfrac{1}{2} + 3 \times \tfrac{1}{2} \times 0.16)0.6F_y = (0.79)0.6F_y = 0.47F_y$$

Since F_t is less than $0.6F_y$, the allowable net-section stress is 79% of the basic allowable stress F, when $F_u > 1.35F_y$. At section 2-2, for the upper sheet, $r = 1$ because the force in the sheet at that section is $T/2$. The allowable stress at that section given by Eq. (7-35) is $(0.58)0.6F_y = 0.35F_y = 0.58F$, but this has no significance because the force in the upper sheet is only 50% of the total force T. For the lower sheet in Case 2, the allowable stresses at sections 1-1 and 2-2 are those computed for the upper sheet at sections 2-2 and 1-1, respectively.

Despite the stress reduction required in Case 2, the total force T that can be transferred through this particular joint on the basis of net-section tensile stress is about the same in both cases. This will not necessarily be true for other sizes and joint arrangements.

7-28. Riveting of Cold-formed Shapes. Although widely practiced in aircraft work and traditional applications such as drainage products, riveting seldom is used for load-carrying cold-formed members in buildings. When riveting is used for such purposes, ordinary structural practice may be followed as a guide in strength calculations, but, for maximum rivet spacing, rules for maximum allowable spacing of spot welds (Art. 7-26) should be observed. The provisions of Art. 7-27 for net-section stresses should also be observed, unless the results of carefully made tests or experience with a particular application dictate otherwise.

Care must be used in making use of the results of tests of joints in thin material, because of the variations in strength that can result from variations in the texture of the surfaces being joined, hole clearance, and tightness after driving.

7-29. Self-tapping Screws for Joining Light-gage Components. Self-tapping screws that are hardened so that their threads form or cut mating threads in one or both of the parts being connected are frequently used for making field joints in light-gage steel construction. Such screws provide a rapid and efficient means of making light-duty connections. The screws are especially useful for such purposes as fastening sheet-metal siding and roofing to structural steel; making attachments at joints, side laps, and closures in siding, roofing, and decking; fastening collateral materials to steel framing; and fastening steel studs to sill plates or channel tracks. The screws also may be used for fastening bridging to steel joists and studs, fastening corrugated decking to steel joists, and similar connections to secondary members.

A few types of self-tapping screws are illustrated in Fig. 7-29. Other types are available. There are many different head styles—slotted, recessed, hexagonal, flat, round, etc. Some types, called *Sems,* are supplied with preassembled washers under the heads. Other types are supplied with neoprene washers for making watertight joints in roofing.

All the types of screws shown in Fig. 7-29 require prepunched or predrilled holes. Self-drilling screws, which have a twist drill point that drills a hole of the proper size just ahead of threading, are especially suited for field work, because they eliminate a separate punching or drilling operation. Another type of self-drilling screw, capable of being used in relatively thin gages of material in situations where the parts being joined can be firmly clamped together, has a very sharp point that pierces the material until the threads engage.

Torsional-strength requirements for self-tapping screws have been standardized under American National Standards Institute B18.6.4, "Slotted and Recessed Head Tapping Screws and Metallic Drive Screws." Safe loads in shear and tension on such screws can vary considerably, depending on type of screw and head, tightening torque, and details of the assembly. They may be used for structural load-carrying purposes on the basis of tests of the type of assembly involved, manufacturers' recommendations, or experience with a particular application.

Essential body dimensions for some types of self-tapping screws are given in

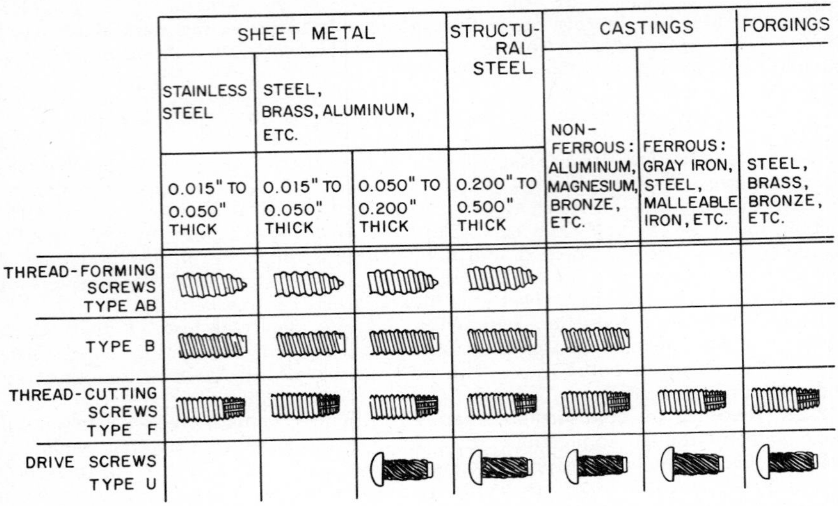

SHEET METAL
STRUCTURAL STEEL
CASTINGS
FORGINGS
STAINLESS STEEL
STEEL, BRASS, ALUMINUM, ETC.
NON-FERROUS: ALUMINUM, MAGNESIUM, BRONZE, ETC.
FERROUS: GRAY IRON, STEEL, MALLEABLE IRON, ETC.
STEEL, BRASS, BRONZE, ETC.
0.015" TO 0.050" THICK
0.015" TO 0.050" THICK
0.050" TO 0.200" THICK
0.200" TO 0.500" THICK
THREAD-FORMING SCREWS TYPE AB
TYPE B
THREAD-CUTTING SCREWS TYPE F
DRIVE SCREWS TYPE U

Fig. 7-29. Self-tapping screws. (*Extracted with permission from "P-K Selector Guide," USM Corporation, Parker-Kalon Division.*) For other types, head styles, detailed dimensions, and information on self-drilling screws, refer to manufacturers' literature.

Table 7-13. Average Diameters of Self-Tapping Screws, In.*

Number or size, in.	Types AB and B (Fig. 7-29)		Type F† (Fig. 7-29)	Type U (Fig. 7-29)
	Outside	Root	Outside	Outside
No. 4	0.112	0.084	0.110	0.114
No. 6	0.137	0.102	0.136	0.138
No. 8	0.164	0.119	0.161	0.165
No. 10	0.186	0.138	0.187	0.180
No. 12	0.212	0.161	0.213	0.209
No. 14‡ or 1/4	0.243	0.189	0.247	0.239‡
5/16	0.312	0.240	0.309	0.312
3/8 §	0.376	0.304	0.371	0.375

* Averages of standard maximum and minimum dimensions adopted under ANSI B18.6.4-1966.

† Type F has threads of machine-screw type approximating the Unified Thread Form (ANSI B1.1-1960). The figures shown are averages of those for two different thread pitches for each size of screw.

‡ Size No. 14 for Type U.

§ Does not apply to Type AB.

Table 7-13. Complete details of these and other types, and recommended hole sizes, may be found in ANSI B18.6.4, and in manufacturers' publications.

7-30. Special Fasteners for Light-gage Steel. Special fasteners such as tubular rivets, blind rivets (capable of being driven from one side only), special bolts used for "blind insertion," special studs, lock nuts, and the like, and even metal stitching, which is an outgrowth of the common office stapling device, are all used for special applications. When such a fastener is required, refer to manufacturers' catalogs, and base any structural strength that may be attributed to it on the results of carefully made tests or the manufacturers' recommendations.

STEEL ROOF DECK

The term steel roof deck usually refers to relatively long, ribbed steel sheets with overlapping or interlocking edges designed to serve primarily for the support of roof loads. A steel roof deck assembly, with insulation and roofing, is illustrated in Fig. 7-30.

7-31. Types of Steel Roof Deck. Most steel decking intended primarily to be used as roof deck has relatively narrow longitudinal ribs from 1½ to 2 in. deep and

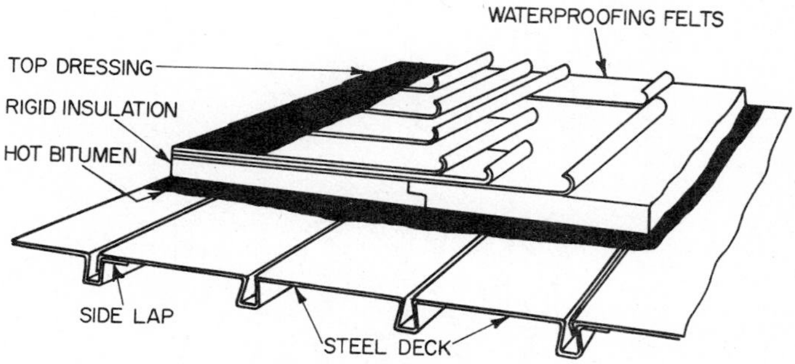

Fig. 7-30. Roof-deck assembly.

spaced about 6 in. c to c, as illustrated in Fig. 7-31. Some manufacturers produce shallower sections, with ribs from ⅞ to 1 in. deep and spaced about 4 in. c to c. Others produce deeper sections, with 8-in. rib spacing.

In addition to ribbed sections of the general type illustrated in Fig. 7-31, some manufacturers produce long-span roof-deck sections, from 3 to 7½ in. deep, some types of which are shown in Fig. 7-32. The cellular floor decks described in Arts. 7-38 to 7-40 may also be used in roofs.

The flat top portions of roof-deck sections may be either plain, as illustrated, or may have one or more longitudinal flutes, as shown for the long-span section in Fig. 7-32*a*; or they may have other patterns formed in them. Details of roof-deck sections vary with different manufacturers.

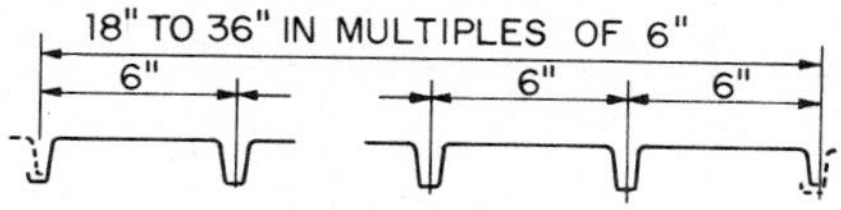

Fig. 7-31. Ribbed roof-deck section. Other widths and profiles are also available.

Roof-deck sections are also available with perforations in the ribs or top sheets to form acoustical ceilings. When the ribs are perforated (Figs. 7-33), they are placed in the normal position, with the flat portion up, and the rib space is filled with sound-absorbing material. If the top sheet is perforated, the deck is placed with ribs upstanding, and the space between them is filled with acoustical material. Closed cellular sections with perforated bottom sheets may also be used for acoustical ceilings in roof construction. (See also Art. 7-39 and Fig. 7-38.)

7-32. Materials for Steel Roof Deck. The usual material for steel roof deck is structural-quality sheet steel, either black or galvanized, conforming to Grade C of ASTM A611 for uncoated material, or to Grade A of A446 for galvanized material. Some manufacturers, however, produce decking from higher-strength grades also.

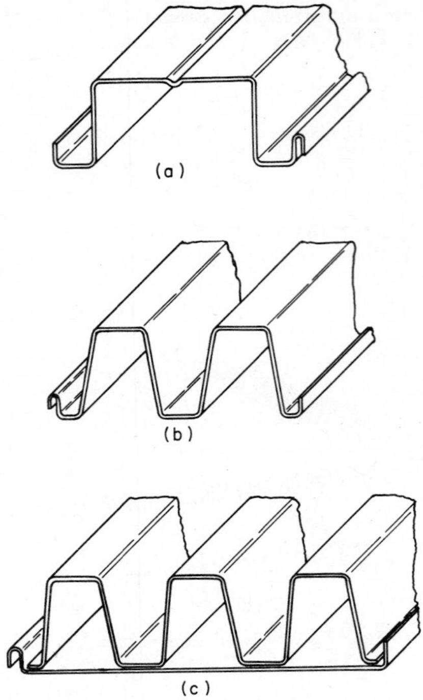

Fig. 7-32. Long-span roof-deck sections. See Figs. 7-36*d* and 7-38 for other sections used for long-span roof decks.

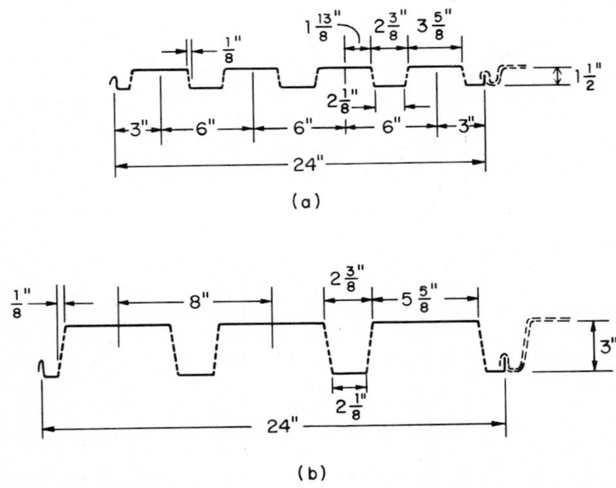

Fig. 7-33. Roof-deck sections with perforated ribs for acoustical control. (*H. H. Robertson Co.*) Cellular units with perforated bottom sheets (Fig. 7-38) also are used as roof deck with acoustic ceilings.

Common gages for sections of the general type shown in Fig. 7-31 are Nos. 18, 20, and 22, although some manufacturers offer these sections in heavier material. The very shallow, short-span sections with ribs spaced 4 in. c to c are usually produced in Nos. 24 and 26. The deep, long-span sections of Fig. 7-32 are furnished in gages as heavy as No. 12.

Steel deck made from uncoated material is given a prime coat of paint or other protective material prior to shipment. Galvanized material may or may not be painted.

Weight of the roof-deck sections illustrated in Fig. 7-31 varies, depending on the detailed design and dimensions of the section. With 20-gage material, sections 1½ in. deep will weigh between 2 and 3 psf. For structural design calculations, 3.0 psf is suggested for 18-gage deck, 2.5 psf for 20-gage, and 2.0 psf for 22-gage. The weight of the deeper types varies from less than 2 psf to more than 11 psf, depending on design and thickness.

7-33. Load-carrying Capacity of Steel Roof Deck. Standard load tables published by the Steel Deck Institute (SDI) apply only to decking of the general configuration shown in Fig. 7-31, with 6-in. rib spacing, 1½-in. minimum depth of ribs, three different rib widths, and three different thicknesses of material. (See Table 7-14.) Allowable uniformly distributed loads vary from 30 psf to well over 100 psf, depending on width of rib, span condition (simple or continuous), length of span c to c of supports, and whether allowable stress or deflection controls. Live-load deflection is limited to $\frac{1}{240}$ times the span. Safe-load and deflection calculations are based on the American Iron and Steel Institute "Specification for the Design of Cold-formed Steel Structural Members," following the procedures outlined in Arts. 7-12 to 7-30.

The SDI specifications require that steel deck units be anchored to the supporting framework to resist the following gross uplift forces, from which the dead load of the roof-deck construction may be deducted:

45 psf for eave overhangs.

30 psf for all other roof areas.

The SDI specifications further provide: "Suspended ceilings, light fixtures, ducts or other utilities shall not be supported by the steel deck."

Safe loads and useful spans for sections deeper than those covered by the SDI load table are substantially greater than those shown in Table 7-14, and for the deepest long-span sections, tabulated spans are as long as 35 ft for roof loads. For detailed information regarding such sections, refer to manufacturers' publications.

In addition to their normal function as roof panels to resist gravity loading, steel roof-deck assemblies can be used as shear diaphragms under lateral loads, such as wind and seismic forces. When steel roof deck is used for this purpose, special attention must be given to attachments between panels and attachments of panels to building frames. Detailed information is available from manufacturers. (See also A. H. Nilson, "Shear Diaphragms of Light-Gage Steel," *Journal of the Structural Division*, Vol. 86, No. ST11, November, 1960, American Society of Civil Engineers.)

7-34. Details and Accessories for Steel Roof Deck. As illustrated in Fig. 7-31, all roof-deck sections have lapping or interlocking edges. Most sections are designed so that the ends lap (Fig. 7-34*a*). Special filler or cover pieces are sometimes provided instead of telescoping end laps.

Special auxiliary shapes to fit various roofing details are furnished by most manufacturers of steel roof deck. These usually include side closure plates, ridge and valley plates, and cant strips. Although details and dimensions will vary with different manufacturers, the general shapes of these accessories are as shown in Fig. 7-34.

Steel roof decks are usually fastened to structural-steel supports by electric arc welding or by self-tapping screws. The Steel Deck Institute recommends that adjacent sections spanning more than 7 ft be fastened together at midspan. For further information see the SDI "Steel Roof Deck Design Manual." In a particular case, the details to be used depend on the circumstances surrounding the job and on the remmendations of the manufacturer.

Table 7-14. Allowable Total (Dead plus Live) Uniform Loads, Psf, on Steel Roof Deck*

NARROW RIB DECK

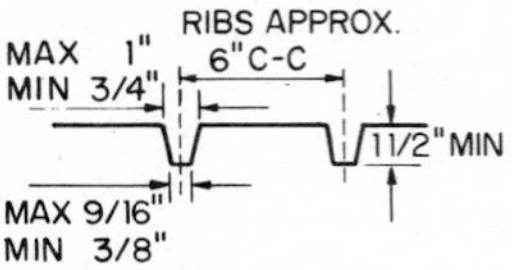

INTERMEDIATE RIB DECK

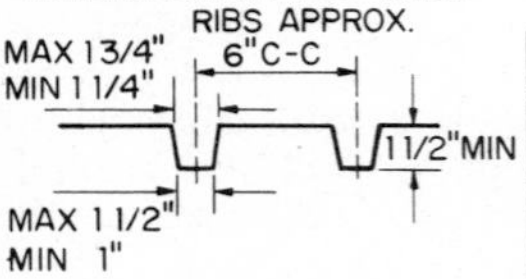

WIDE RIB DECK

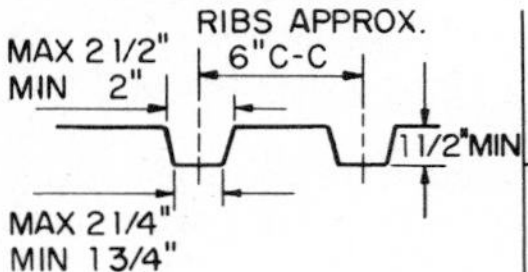

Deck	Span condition	Gage	5'-0"	5'-6"	6'-0"	6'-6"	6'-8"	7'-0"	7'-6"	8'-0"	8'-4"	8'-6"	9'-0"	9'-6"	10'-0"
Narrow rib deck	SIMPLE	22	47	39	32										
		20	57	47	40	34	32								
		18	77	64	54	46	44	40	34	30					
	2-SPAN	22	51	42	35	30									
		20	61	50	42	36	34	31							
		18	80	66	55	47	45	41	35	31					
	3 OR MORE	22	64	53	44	38	36	32							
		20	76	63	53	45	43	39	34	30					
		18	100	82	69	59	56	51	44	39	36	34			
Intermediate rib deck	SIMPLE	22	69	57	48	41	38	35	30						
		20	85	70	59	49	46	41	35	31					
		18	115	95	80	65	61	54	46	40	36	35	31		
	2-SPAN	22	76	63	53	45	43	39	34	30					
		20	90	75	63	54	51	46	40	35	33	31			
		18	119	98	82	70	67	61	53	46	43	41	37	33	30
	3 OR MORE	22	95	78	66	56	53	48	42	37	34	33			
		20	113	93	79	67	64	58	50	44	41	39	34	31	
		18	148	123	103	88	83	76	66	58	53	51	45	39	35
Wide rib deck	SIMPLE	22	94	73	58	48	45	41	35	30					
		20	117	90	72	59	55	49	42	36	33	32			
		18	163	125	99	80	75	66	55	47	43	41	36	32	
	2-SPAN	22	101	84	70	60	57	52	45	40	38	35	31		
		20	129	107	90	76	73	66	57	51	47	45	40	36	32
		18	182	150	126	107	102	93	81	71	65	62	54	47	42
	3 OR MORE	22	127	105	88	74	66	61	51	44	40	38	34	30	
		20	162	134	112	91	85	75	63	53	48	46	41	36	32
		18	227	188	158	126	118	103	86	72	65	62	54	47	42

(Spans are SPAN – C TO C JOISTS OR PURLINS.)

Load tables were calculated with sectional properties based on "Manufacturer's Standard Gage for Carbon Steel Sheets," as published February, 1964, in AISI Steel Products Manual."

Loads shown in tables are uniformly distributed total (dead plus live) loads, psf. Loads in shaded areas are governed by live-load deflection not in excess of $\frac{1}{240} \times$ span. The dead load included is 10 psf. All other loads are governed by the allowable flexural stress limit of 20 ksi for a 33-ksi minimum yield point.

Rib-width limitations shown are taken at the theoretical intersection points of flange.

Span length assumes c-to-c spacing of supports. Tabulated loads shall not be increased by assuming clear-span dimensions.

Bending moment formulas used for flexural stress limitation are: For simply supported and two-span decking, $M = wl^2/8$; for decking with three continuous spans or more, $M = wl^2/10$.

Deflection formulas for deflection limitation are: For simply supported decking, $\Delta = 5wl^4/384EI$; for two- and three-span decking, $\Delta = 3wl^4/384EI$.

Normal installations covered by these tables do not require midspan fasteners for spans of 7 ft or less.

† For decks having a fully effective top-flange stiffener, as defined in the AISI "Cold-Formed Steel Design Manual," the 1-in. minimum width of the rib bottom may be reduced by 1½ times the depth of the stiffener, but in no case shall the rib bottom be less than ⅝ in.

‡ For decks exceeding 1½ in. in depth but less than 2 in. deep, minimum rib widths may be reduced, but not more than twice the increase in depth.

7-35. Insulation of Steel Roof Deck. Although insulation is not ordinarily supplied by steel roof deck manufacturers, it is more-or-less standard practice to install insulation between the roof deck and the roof covering, as illustrated in Fig. 7-30. The Steel Deck Institute "Steel Roof Deck Design Manual" contains the following recommendations with respect to insulation:

"All steel roof decks shall be covered with a material of sufficient insulating value as to prevent condensation under normal occupancy conditions. Insulation shall be adequately attached to the steel deck by adhesives or mechanical fasteners. Insulation materials shall be protected from the elements at all times during storage and installation."

7-36. Fire Resistance of Steel Roof Deck. The Steel Deck Institute "Steel Roof Deck Design Manual," 1972, contains the fire resistance ratings given in Table 7-15. Additional fire ratings are available.

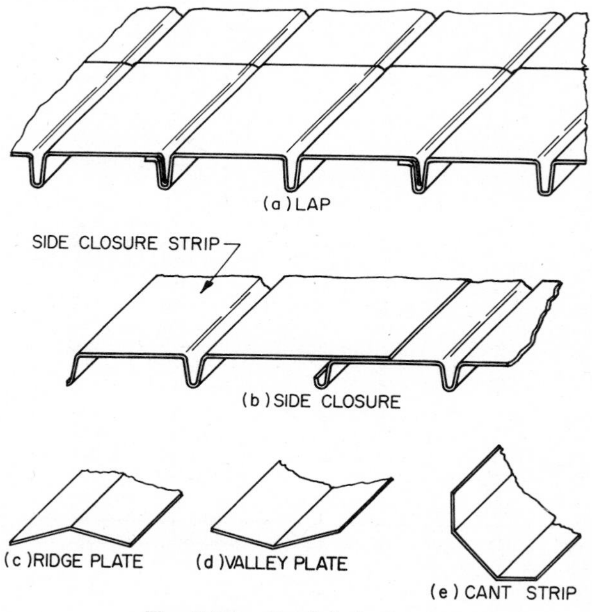

Fig. 7-34. Roof-deck details.

7-37. Other Forms of Steel Roof Decking. Forms of steel roof decking other than those described in Arts. 7-31 and 7-32 are also available. Corrugated sheets are an example. In addition to the traditional sinusoidal cross section, corrugated sheets are available in a variety of other cross-sectional configurations, sizes, and load-carrying capacities. Some of these have special venting slots, for use with gypsum roofs. Others have protrusions or indentations formed in them, or cross wires welded to them, to provide anchorage in concrete fill and function in a composite manner with the concrete. Some are made from full-hard material such as Grade E of ASTM A570, A611, and A446, with yield points of 80 ksi and higher.

Although ribbed roof-deck sections are usually placed with the flat portion up, as shown in Figs. 7-30 and 7-31, they may also be used with the ribs upstanding without the acoustical treatment mentioned in Art. 7-30. Roofs of this general kind have found numerous applications, especially for light industrial, agricultural, and relatively small storage structures. Sections for this purpose may be conventional, or they may have very wide, flat portions between upstanding ribs.

Table 7-15. Fire Resistance Ratings, Steel Roof Deck Construction

Hours	Roof construction	Insulation	Underside protection	Authority
2	Min 1½-in.-deep steel deck on steel joists, 6 ft c to c max	Min 1-in.-thick, UL-listed mineral-fiber board	⅞-in.-thick, light-weight-aggregate, gypsum plaster on metal lath	UL Design* P404 (RC1-2 hr)
	Min 1½-in.-deep steel deck on steel joists, 5 ft 6 in. c to c max	Min ¾-in.-thick, UL-listed glass-fiber insulation board	1-in.-thick vermiculite, gypsum plaster on metal-lath suspended ceiling	UL Design P409 (RC8-2 hr)
1½	Min 1½-in.-deep steel deck on steel beams	1-in.-thick, UL-listed mineral-fiber insulation board applied in two layers of ½-in. board, with joints staggered	Min 1⅜ to 1½-in.-thick, direct-applied, sprayed vermiculite plaster, UL listed	UL Design P703 (RC5-1½ hr)
1	Min 1½-in.-deep steel deck on steel joists, 7 ft c to c max	¾-in.-thick, UL-listed mineral-fiber board	Suspended ceiling, ⅝-in.-thick, UL-listed, acoustical, lay-in boards and UL-listed ceiling grid	UL Design P201 (RC7-1 hr)
1	Min 1½-in.-deep steel deck on steel beams	1-in.-thick, UL-listed mineral-fiber insulation board	Min 1⅜ to 1½-in.-thick, direct-applied, sprayed vermiculite plaster, UL listed	UL Design P701 (RC10-1 hr)
1	Min ¾-in.-deep steel deck on steel joists, 4 ft c to c max	1- to 3-in.-thick, UL-listed mineral-fiber insulation board	⅝-in.-thick, ceramic, acoustical, lay-in boards in ceiling grid	UL Designs P211 and P210 (RC-1 hr and RC4-1 hr)
1	Min 1⅜-in.-deep steel deck on 10-in. steel joists, 5 ft 8 in. c to c max	1-in.-thick, listed mineral-fiber insulation board	⅝-in.-thick, listed, acoustical, lay-in boards in listed ceiling grid	FM Design† FC37-1 hr

* Fire Resistance Index, Underwriters' Laboratories, Inc., January, 1972. Old numbers are in parentheses.
† Factory Mutual.

Manufacturers' literature contains a wealth of detailed information regarding all types of roof-deck sections and their applications.

CELLULAR STEEL FLOORS

Cellular steel floors are often used in many kinds of structures, including massive high-rise buildings for institutional, business, and mercantile occupancies.

7-38. Advantages of Cellular Decking. One of the principal advantages of cellular construction (Fig. 7-35) is that, in addition to acting as load-carrying components, they provide spaces for power and telephone wiring. With proper outlet fittings, changes in electrical services can be made quickly and easily. Cells of the proper size can also function as air ducts. In addition, use of cellular steel construction can result in a substantial saving in weight, particularly important in high-rise buildings.

From the standpoint of the contractor, cellular decking units are advantageous, because they can be quickly installed. Following structural-steel erection very closely, they can provide an immediate working platform for all trades.

Cellular steel floors are not to be considered, necessarily, light-load construction.

They are capable of carrying the heaviest loads likely to be encountered in any commercial building.

7-39. Types of Cellular Floor Units. Many different designs of cellular steel panels are available. Outline cross sections of some popular types are shown in

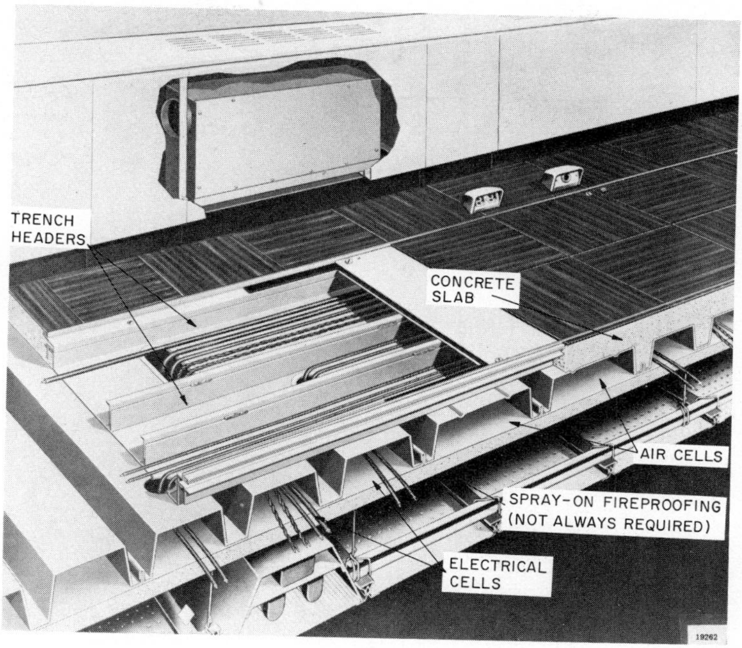

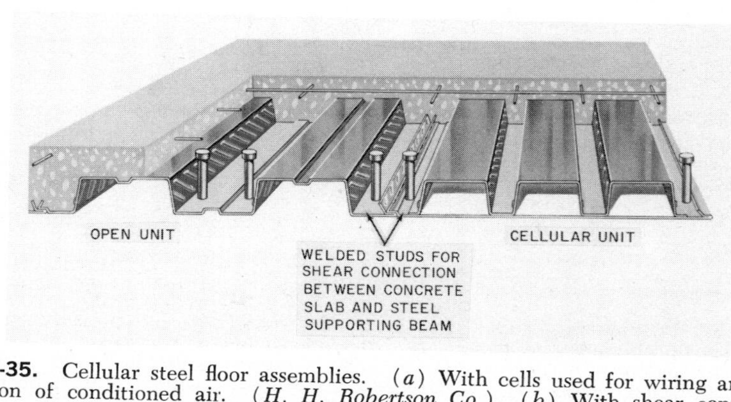

Fig. 7-35. Cellular steel floor assemblies. (*a*) With cells used for wiring and distribution of conditioned air. (*H. H. Robertson Co.*) (*b*) With shear connectors for composite action. (*Inland-Ryerson Construction Products Co.*)

Fig. 7-36. Cellular long-span roof deck sections of the type illustrated in Fig. 7-32*c* are also suitable for use in floor construction, although they have been generally superseded by those shown in Fig. 7-36.

Cellular steel units may be designed to function as self-contained structural components, carrying the entire floor load with the concrete acting only as fill, or they may be designed to function as a composite member with the concrete slab. Cellular units designed for composite action have indentations or protrusions (embossments) formed in them, as may be seen in Fig. 7-35*b*, to provide mechanical anchorage with the concrete. Most of the units used in new construction are of the composite type.

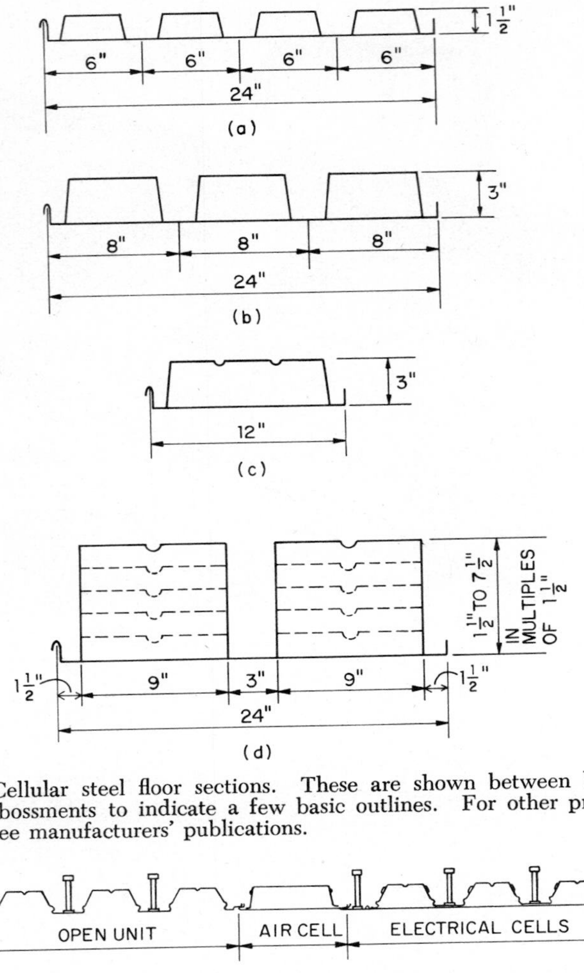

Fig. 7-36. Cellular steel floor sections. These are shown between basic indentations and embossments to indicate a few basic outlines. For other profiles, widths, and depths, see manufacturers' publications.

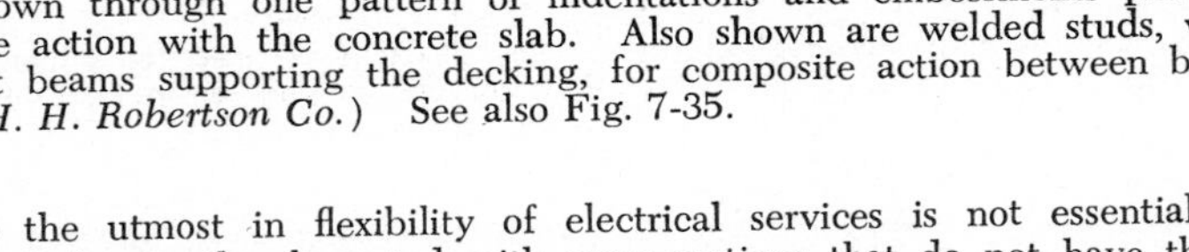

Fig. 7-37. Portion of blend system combining electrical cells, air cells, and open units, shown through one pattern of indentations and embossments providing for composite action with the concrete slab. Also shown are welded studs, which are placed at beams supporting the decking, for composite action between beams and slab. (*H. H. Robertson Co.*) See also Fig. 7-35.

Where the utmost in flexibility of electrical services is not essential, cellular sections are frequently alternated with open sections that do not have the bottom sheet (Fig. 7-35*b*), forming what is known as a *blend* system. Figure 7-37 shows a cross section of a portion of a blend system that combines closed cells for electrical raceways, closed cells for air ducts, and open sections.

The bottom sheets of closed cells can be perforated and acoustic pads inserted to provide acoustic ceilings. Some cells are available with specially designed pads to provide acoustic control and at the same time permit air to escape through the perforations, so that the cell acts as an air-diffusion chamber. The ceiling cavities in the deeper open sections (without the bottom sheet) may also be used to house fixtures for recessed lighting (Fig. 7-38).

7-40. Materials for Cellular Decking. Cellular steel floor and roof panels usually are made of structural-quality sheet steel conforming to the requirements for Grade C of ASTM A611 (uncoated) or Grade A of A446 (galvanized), both with a minimum specified yield point of 33 ksi. Thicknesses range from No. 12 to No. 22. Combinations of gages may be used, for example, 20-gage top and 18-gage bottom sheet.

The thickness of concrete over the tops of cellular units varies from about 2½ in. up, depending on job requirements and on whether the concrete functions simply as fill material or as a structural component, composite with the steel units. For composite action, concrete having a compressive strength of 3,000 psi is commonly used, although concrete of higher strength is sometimes specified. It may be either stone concrete, weighing about 145 lb per cu ft, or lightweight concrete, weighing about 110 lb per cu ft. (Concrete containing chloride salts should not be used with galvanized units.) Shrinkage reinforcement in the concrete usually consists of welded wire fabric. Wire gage depends on slab thickness.

Fig. 7-38. Cellular steel floor panel combining an acoustical ceiling, air-diffusion chamber, and recessed lighting.

7-41. Structural Design of Cellular Floors. Strength and deflection calculations of cellular floor units are usually based upon the AISI "Specification for the Design of Cold-formed Steel Structural Members," using the procedures discussed in Arts. 7-12 through 7-26. Structural design for composite action follows conventional practice. It may be based on either shored or unshored construction.

Usual spans for the popular 1½- and 3-in.-deep sections of the kind shown in Fig. 7-36*a* and *b* range from about 7 to 16 ft, depending on depth and thickness of section, thickness of concrete, and whether the design is noncomposite, unshored composite, or shored composite. The deeper sections, Fig. 7-36*d*, for example, can span up to 32 ft under floor loadings when used in shored composite construction.

As for other forms of construction, deflection limitations will sometimes control the design. These limitations depend to some extent on local building-code requirements. The traditional figure of 1/360 times the span is a common limit for live-load deflection. Regardless of deflection considerations, it is generally considered good practice to limit the span of plain (noncomposite) floor units to 25 times the total depth of the floor (the distance from the bottom of the floor units to the top of the structural concrete slab), and to limit the span of composite cellular floor construction to 32 times that depth.

In addition to normal floor-load service, cellular steel floors can also function as shear diaphragms under lateral loadings, such as wind or seismic forces.

Steel floor units usually are attached to supporting framework by electric-arc welding. Where the units are continuous over one or more supports, attachment to the intermediate supports is usually specified. Intermittent positive attachment the overlapping edges or by welding. Some units are made with continuously interlocking joints along the edges.

Welded-stud shear connectors can be installed in the troughs of cellular units, at steel beams, to provide for composite steel-concrete design of the beams, and incidentally to fasten the deck to the beams. Some cellular units are available with wide troughs designed to accommodate pairs of studs lengthwise of the beams (Fig. 7-35*b*). Units are also available with markedly trapezoidal troughs for additional concrete when required for shear transfer (Fig. 7-37).

Each manufacturer publishes specifications, load tables, and detailed recommendations for the use of his products. Manufacturers' recommendations should always be followed in designing any cellular-floor installation.

Table 7-16. Fire-resistance Ratings for Cellular Steel Floors without Underside Protection

Hours	Floor units	Concrete cover over top of units*	UL Design No.†
3	1½- or 3-in.-deep cellular or open	4 3/16 in. (lightweight)	D902 (225-3 hr)
2	1½- or 3-in.-deep cellular or open	3¼ in. (lightweight)	D840 (267-2 hr)
	1½- or 3-in.-deep cellular or open	4½ in. (stone)	D902 (300-2 hr)
	4½- to 7½-in.-deep cellular	3 in. or 4 in.	D903 (304-2 hr)
	4½- to 7½-in.-deep cellular, with perforated bottom sheets	3¼ in. or 4 in.	D903 (304-2 hr)
	4½- to 7½-in.-deep open	3½ in. or 4½ in.	D903 (304-2 hr)
1½	1½- or 3-in.-deep cellular or open	4 in. (stone)	D902 (32-1½ hr)
	4½- to 7½-in.-deep cellular or open	3 in. or 3½ in.	D903 (33-1½ hr)
	4½- to 7½-in.-deep cellular, solid bottom sheets	3¼ in. (stone)	D903 (33-1½ hr)
1	1½- or 3-in.-deep cellular or open	3½ in. (stone)	D902 (59-1 hr)
	4½- to 7½-in.-deep cellular, solid or perforated bottom sheets	2½ in. (lightweight)	D903 (60-1 hr)
	4½- to 7½-in.-deep open	2¾ in. (lightweight) or 3 in. (stone)	D903 (60-1 hr)

* Where two or three different types of units (solid bottom, perforated, or open) in a blend system require different thicknesses of concrete for the desired fire rating, the greatest concrete thickness required for any type of unit in the combination should be used throughout. Unless otherwise indicated, concrete may be lightweight or stone.

† Fire Resistance Index, Underwriters' Laboratories, Inc., January, 1972. Old numbers are in parentheses.

7-42. Details and Accessories for Cellular Steel Floors. Structural details of cellular units differ considerably among manufacturers. These details include the exact configurations and dimensions of the sections, the design of the edge laps or interlocks, and the pattern of the indentations or embossments that are formed in some units for composite action. Structural accessories required for all designs include closure pieces, installed at the ends of cell runs; flashing or cover strips at columns, beams and girders parallel to the cells, and elsewhere where the cellular construction is interrupted; and sealing material at the joints of abutting cells.

In addition to these structural accessories, other components include trench headers (Fig. 7-35*a*) and crossover cells for wiring; header ducts and inlet fittings for air cells (usually a part of the air-circulation system); shelf tees for supporting the translucent cover sheets of open units housing recessed lighting fixtures; hanger tabs for suspended ceilings; and other items required to produce a complete floor-ceiling system. Nonstructural accessories are not necessarily furnished by the manufacturer of the floor units. Some of them may be included in the electrical and air-conditioning contracts.

7-43. Fire Resistance of Cellular Floor Construction. Any required degree of fire protection for cellular steel floor and roof assemblies can be obtained with concrete topping and either rated membrane ceilings or direct application compounds (spray-on fireproofing). Under certain provisos with respect to the kind and thickness of concrete topping, cellular floors without any underside protection have been awarded fire-resistance ratings up to and including 3 hr (Table 7-16). A 2-hr rating requires less thickness of concrete than a 3-hr, and is sufficient to satisfy the fire-resistance requirements prescribed by practically all building codes for floor-ceiling assemblies for any occupancy in which cellular floors are likely to be used. Underside protection, however, is usually required for units underneath trench headers (Fig. 7-35*a*) and for the supporting steelwork. Under some conditions, it may prove economical to use a minimum of concrete cover and protect the entire underside with a spray-on compound. This depends on the economics of each job.

The listing in Table 7-16 is not intended to be all-inclusive. For other ratings, for ratings with underside protection, and for other information, see the Underwriters' Laboratories Fire Resistance Index. Additional details of fire tests and fire resistance ratings for cellular and noncellular steel floor-ceiling assemblies are available from the American Insurance Association, Factory Mutual Research Corp., and other fire-testing agencies, and from the manufacturers of steel floor and roof decking.

OTHER FORMS OF LIGHTWEIGHT STEEL CONSTRUCTION

7-44. Building Systems. Open-web sections of various kinds, unitary cold-formed framing members, and panels may be combined in a variety of ways to form complete building systems. A number of such systems are available, most of them including sheet-steel curtain-wall construction. (See Sec. 10.) They may or may not include structural-steel components. The systems are suitable for a variety of occupancies—industrial, commercial, educational, agricultural, and residential.

Modular construction, in which more-or-less standardized modules are put together to form buildings of various sizes, uses many different types of steel components.

Most of these building systems are of a proprietary nature. Refer to manufacturers' literature for detailed information in this rapidly changing field.

Section **8**

Wood Construction

MAURICE J. RHUDE

President, Sentinel Structures, Inc.
Peshtigo, Wis.

Wood is the greatest renewable resource in the building materials field. More trees can always be grown to provide the wood needed for construction. Trees often are planted with a purpose, stands are improved, and the trees are then harvested with a particular end product in mind.

As a consequence of its origin, wood as a building material has inherent characteristics with which users should be familiar. For example, although cut simultaneously from trees growing side by side in a forest, two boards of the same species and size most likely do not have the same strength. The task of describing this nonhomogeneous material, with its variable biological nature, is not easy. But it can be described accurately, and much better than was possible in the past. For research has provided much useful information on wood properties and behavior in structures.

Research has shown, for example, that a compression grade cannot be used, without modification, for the tension side of a deep bending member. Also, a bending grade cannot be used, unless modified, for the tension side of a deep bending member, or for a tension member. Experience indicates that typical growth characteristics are more detrimental to tensile strength than to compressive strength. Furthermore, research has made possible better estimates of engineering qualities of wood. No longer is it necessary to use only visual inspection, keyed to averages, for estimating the engineering qualities of a piece of wood.

Practice, not engineering design, has been the criterion in the past in home building. As a result, the strength of wood often was not fully utilized. With a better understanding of wood now possible, the availability of sound structural design criteria, and development of economical manufacturing processes, greater and more efficient use is being made of wood in home building.

Improvements in adhesives also have contributed to the betterment of wood construction. In particular, the laminating process, employing adhesives to build up thin boards into deep timbers, improves on nature. Not only are stronger structural members thus made available, but also higher grades of lumber can be placed in regions of greatest stress, and lower grades in regions of lower stress, for over-all economy. Despite variations in strength of wood, lumber can be transformed into glued-laminated timbers of predictable strength and with very little variability in strength.

8-1. Basic Characteristics and How to Use Them. Wood differs in several significant ways from other building materials. Its cellular structure is responsible, to a considerable degree, for this. Because of this structure, structural properties depend on orientation. While most structural materials are essentially isotropic, with nearly equal properties in all directions, wood has three principal grain directions—longitudinal, radial, and tangential. (Loading in the longitudinal direction is referred to as parallel to the grain, whereas transverse loading is considered across the grain.) Parallel to the grain, wood possesses high strength and stiffness. Across the grain, strength is much lower. (In tension, wood stressed parallel to the grain is 25 to 40 times stronger than when stressed across the grain. In compression, wood loaded parallel to the grain is 6 to 10 times stronger than when loaded perpendicular to the grain.) Furthermore, a wood member has three moduli of elasticity, with a ratio of largest to smallest as large as 150:1.

Wood undergoes dimensional changes from causes different from those for dimensional changes in most other structural materials. For instance, thermal expansion of wood is so small as to be unimportant in ordinary usage. Significant dimensional changes, however, occur because of gain or loss in moisture. Swelling and shrinkage from this cause vary in the three grain directions; size changes about 6 to 16% tangentially, 3 to 7% radially, but only 0.1 to 0.3% longitudinally.

Wood offers numerous advantages nevertheless in construction applications—beauty, versatility, durability, workability, low cost per pound, high strength-to-weight ratio, good electrical insulation, low thermal conductance, and excellent strength at low temperatures. It is resistant to many chemicals that are highly corrosive to other materials. It has high shock-absorption capacity. It can withstand large overloads of short time duration. It has good wearing qualities, particularly on its end grain. It can be bent easily to sharp curvature. A wide range of finishes can be applied for decorative or protective purposes. Wood can be used in both wet and dry applications. Preservative treatments are available for use when necessary, as are fire retardants. Also, there is a choice of a wide range of species with a wide range of properties.

In addition, a wide variety of wood framing systems is available. The intended use of a structure, geographical location, configuration required, cost, and many other factors determine the framing system to be used for a particular project.

Design Recommendations. The following recommendations aim at achieving economical designs with wood framing:

Use standard sizes and grades of lumber. Consider using standardized structural components, whether lumber, stock glued beams, or complex framing designed for structural adequacy, efficiency, and economy.

Use standard details wherever possible. Avoid specially designed and manufactured connecting hardware.

Use as simple and as few joints as possible. Place splices, when required, in areas of lowest stress. Do not locate splices where bending moments are large, thus avoiding design, erection, and fabrication difficulties.

Avoid unnecessary variations in cross section of members along their length.

Use identical member designs repeatedly throughout a structure, whenever practicable. Keep the number of different arrangements to a minimum.

Consider using roof profiles that favorably influence the type and amount of load on the structure.

Specify allowable design stresses rather than the lumber grade or combination of grades to be used.

Select an adhesive suitable for the service conditions, but do not overspecify. For example, waterproof resin adhesives need not be used where less expensive water-resistant adhesives will do the job.

Use lumber treated with preservatives where service conditions dictate. Such treatment need not be used where decay hazards do not exist. Fire-retardant treatments may be used to meet a specific flame-spread rating for interior finish, but are not necessary for large-cross-section members that are widely spaced and already a low fire risk.

Instead of long, simple spans, consider using continuous or suspended spans or simple spans with overhangs.

Select an appearance grade best suited to the project. Do not specify premium appearance grade for all members if it is not required.

Table 8-1 may be used as a guide to economical span ranges for roof and floor framing in buildings.

8-2. Standard Sizes of Lumber and Timber. Details regarding dressed sizes of various species of wood are given in the grading rules of agencies that formulate and maintain such rules. Dressed sizes in Table 8-2 are from the American Softwood Lumber Standard, Voluntary Product Standard PS20-70. These sizes are generally available; but it is good practice to consult suppliers before specifying sizes not commonly used to find out what sizes are on hand or can be readily secured.

8-3. Sectional Properties of Lumber and Timber. Table 8-3 lists properties of sections of solid-sawn lumber and timber.

8-4. Standard Sizes of Glued-laminated Timber. Standard finished sizes of structural glued-laminated timber should be used to the extent that conditions permit. These standard finished sizes are based on lumber sizes given in Voluntary Product Standard PS20-70. Other finished sizes may be used to meet the size requirements of a design, or to meet other special requirements.

Nominal 2-in.-thick lumber, surfaced to 1½ in. before gluing, is used to laminate straight members and curved members having radii of curvature within the bending-radius limitations for the species. (Formerly, a net thickness of 1⅝ in. was common for nominal 2-in. lumber and may be used, dependent on its availability.) Nominal 1-in.-thick lumber, surfaced to ¾ in. before gluing, may be used for laminating curved members when the bending radius is too short to permit use of nominal 2-in.-thick laminations, if the bending-radius limitations for the species are observed. Other lamination thicknesses may be used to meet special curving requirements.

8-5. Sectional Properties of Glued-laminated Timber. Table 8-4 lists properties of sections of glued-laminated timber.

8-6. Basic and Allowable Stresses for Timber. Testing of a species to determine average strength properties should be carried out from either of two viewpoints:

1. Tests should be made on specimens of large size containing defects. Practically all structural uses involve members of this character.

2. Tests should be made on small, clear specimens to provide fundamental data. Factors to account for the influence of various characteristics may be applied to establish the strength of structural members.

Tests made in accordance with the first viewpoint have the disadvantage that the results apply only to the particular combination of characteristics existing in the test specimens. To determine the strength corresponding to other combinations requires additional tests; thus, an endless testing program is necessary. The second viewpoint permits establishment of fundamental strength properties for each species, and application of general rules to cover the specific conditions involved in a particular case.

It is this second viewpoint that has been generally accepted. When a species has been adequately investigated under this concept, there should be no need for further tests on that species unless new conditions arise. ASTM Standard D143, "Standard Methods of Testing Small, Clear Specimens of Timber," gives the procedure for determination of fundamental data on wood species.

Basic stresses are essentially unit stresses applicable to clear and straight-grained defect-free material. These stresses, derived from the results of tests on small, clear specimens of green wood, include an adjustment for variability of material, length of loading period, and factor of safety. They are considerably less than the average of the species. They require only an adjustment for grade to become allowable unit stresses.

Allowable unit stresses are computed for a particular grade by reducing the basic stress according to the limitations on defects for that grade. The basic stress is multiplied by a strength ratio to obtain an allowable stress. This strength ratio

Table 8-1. Economical Span Range for Framing Members

Framing member	Economical span range, ft	Usual spacing, ft
Roof beams (generally used where a flat or low-pitched roof is desired):		
Simple span:		
Constant depth		
Solid-sawn	0–40	4–20
Glued-laminated	20–100	8–24
Tapered	25–100	8–24
Double tapered (pitched beams)	25–100	8–24
Curved beams	25–100	8–24
Simple beam with overhangs (usually more economical than simple span when span is over 40 ft):		
Solid-sawn	24	4–20
Glued-laminated	10–90	8–24
Continuous span:		
Solid-sawn	10–50	4–20
Glued-laminated	10–50	8–24
Arches (three-hinged for relatively high-rise applications and two-hinged for relatively low-rise applications):		
Three-hinged:		
Gothic	40–90	8–24
Tudor	30–120	8–24
A-frame	20–160	8–24
Three-centered	40–250	8–24
Parabolic	40–250	8–24
Radial	40–250	8–24
Two-hinged:		
Radial	50–200	8–24
Parabolic	50–200	8–24
Trusses (provide openings for passage of wires, piping, etc.):		
Flat or parallel chord	50–150	12–20
Triangular or pitched	50–90	12–20
Bowstring	50–200	14–24
Tied arches (where no ceiling is desired and where a long clear span is desired with low rise):		
Tied segment	50–100	8–20
Buttressed segment	50–200	14–24
Domes	50–350	8–24
Simple-span floor beams:		
Solid sawn	6–20	4–12
Glued laminated	6–40	4–16
Continuous floor beams	25–40	4–16
Roof sheathing and decking:		
1-in. sheathing	1–4	
2-in. sheathing	6–10	
3-in. roof deck	8–15	
4-in. roof deck	12–20	
Plywood sheathing	1–4	
Sheathing on roof joists	1.33–2	
Plank floor decking (floor and ceiling in one):		
Edge to edge	4–16	
Wide face to wide face	4–16	

Table 8-2. Nominal and Minimum Dressed Sizes of Boards, Dimension, and Timbers

Item	Thickness, in.			Face width, in.		
	Nominal	Minimum dressed		Nominal	Minimum dressed	
		Dry*	Green†		Dry*	Green†
Boards	1	3/4	25/32	2	1 1/2	1 9/16
	1 1/4	1	1 1/32	3	2 1/2	2 9/16
	1 1/2	1 1/4	1 9/32	4	3 1/2	3 9/16
				5	4 1/2	4 5/8
				6	5 1/2	5 5/8
				7	6 1/2	6 5/8
				8	7 1/4	7 1/2
				9	8 1/4	8 1/2
				10	9 1/4	9 1/2
				11	10 1/4	10 1/2
				12	11 1/4	11 1/2
				14	13 1/4	13 1/2
				16	15 1/4	15 1/2
Dimension	2	1 1/2	1 9/16	2	1 1/2	1 9/16
	2 1/2	2	2 1/16	3	2 1/2	2 9/16
	3	2 1/2	2 9/16	4	3 1/2	3 9/16
	3 1/2	3	3 1/16	5	4 1/2	4 5/8
				6	5 1/2	5 5/8
				8	7 1/4	7 1/2
				10	9 1/4	9 1/2
				12	11 1/4	11 1/2
				14	13 1/4	13 1/2
				16	15 1/4	15 1/2
	4	3 1/2	3 9/16	2	1 1/2	1 9/16
	4 1/2	4	4 1/16	3	2 1/2	2 9/16
				4	3 1/2	3 9/16
				5	4 1/2	4 5/8
				6	5 1/2	5 5/8
				8	7 1/4	7 1/2
				10	9 1/4	9 1/2
				12	11 1/4	11 1/2
				14	....	13 1/2
				16	....	15 1/2
Timbers	5 and thicker	...	1/2 in. less	5 and wider	....	1/2 in. less

* Dry lumber is defined as lumber seasoned to a moisture content of 19% or less.

† Green lumber is defined as lumber having a moisture content in excess of 19%.

represents that proportion of the strength of a defect-free piece that remains after taking into account the effect of strength-reducing features.

The principal factors entering into the establishment of allowable unit stress for each species include inherent strength of wood, reduction in strength due to natural growth characteristics permitted in the grade, effect of long-time loading, variability of individual species, possibility of some slight overloading, characteristics of the species, size of member and related influence of seasoning, and factor of safety. The effect of these factors is a strength value for practical-use conditions lower than the average value taken from tests on small, clear specimens.

Table 8-3. Properties of Sections of Solid-sawn Wood

Nominal size, in.	Standard dressed size, in. (S4S)	Area of section sq in.	Moment of inertia in.⁴	Section modulus, in.³	Weight, lb per lin ft, of piece when weight of wood, lb per cu ft, equals					
					25	30	35	40	45	50
1 × 3	¾ × 2½	1.875	0.977	0.781	0.326	0.391	0.456	0.521	0.586	0.651
1 × 4	¾ × 3½	2.625	2.680	1.531	0.456	0.547	0.638	0.729	0.820	0.911
1 × 6	¾ × 5½	4.125	10.398	3.781	0.716	0.859	1.003	1.146	1.289	1.432
1 × 8	¾ × 7¼	5.438	23.817	6.570	0.944	1.133	1.322	1.510	1.699	1.888
1 × 10	¾ × 9¼	6.938	49.466	10.695	1.204	1.445	1.686	1.927	2.168	2.409
1 × 12	¾ × 11¼	8.438	88.989	15.820	1.465	1.758	2.051	2.344	2.637	2.930
2 × 3*	1½ × 2½	3.750	1.953	1.563	0.651	0.781	0.911	1.042	1.172	1.302
2 × 4	1½ × 3½	5.250	5.359	3.063	0.911	1.094	1.276	1.458	1.641	1.823
2 × 6	1½ × 5½	8.250	20.797	7.563	1.432	1.719	2.005	2.292	2.578	2.865
2 × 8	1½ × 7¼	10.875	47.635	13.141	1.888	2.266	2.643	3.021	3.398	3.776
2 × 10	1½ × 9¼	13.875	98.932	21.391	2.409	2.891	3.372	3.854	4.336	4.818
2 × 12	1½ × 11¼	16.875	177.979	31.641	2.930	3.516	4.102	4.688	5.273	5.859
2 × 14	1½ × 13¼	19.875	290.775	43.891	3.451	4.141	4.831	5.521	6.211	6.901
3 × 1	2½ × ¾	1.875	0.088	0.234	0.326	0.391	0.456	0.521	0.586	0.651
3 × 2	2½ × 1½	3.750	0.703	0.938	0.651	0.781	0.911	1.042	1.172	1.302
3 × 4	2½ × 3½	8.750	8.932	5.104	1.519	1.823	2.127	2.431	2.734	3.038
3 × 6	2½ × 5½	13.750	34.661	12.604	2.387	2.865	3.342	3.819	4.297	4.774
3 × 8	2½ × 7¼	18.125	79.391	21.901	3.147	3.776	4.405	5.035	5.664	6.293
3 × 10	2½ × 9¼	23.125	164.886	35.651	4.015	4.818	5.621	6.424	7.227	8.030
3 × 12	2½ × 11¼	28.125	296.631	52.734	4.883	5.859	6.836	7.813	8.789	9.766
3 × 14	2½ × 13¼	33.125	484.625	73.151	5.751	6.901	8.051	9.201	10.352	11.502
3 × 16	2½ × 15¼	38.125	738.870	96.901	6.619	7.943	9.266	10.590	11.914	13.238
4 × 1	3½ × ¾	2.625	0.123	0.328	0.456	0.547	0.638	0.729	0.820	0.911
4 × 2	3½ × 1½	5.250	0.984	1.313	0.911	1.094	1.276	1.458	1.641	1.823
4 × 3	3½ × 2½	8.750	4.557	3.646	1.519	1.823	2.127	2.431	2.734	3.038
4 × 4	3½ × 3½	12.250	12.505	7.146	2.127	2.552	2.977	3.403	3.828	4.253
4 × 6	3½ × 5½	19.250	48.526	17.646	3.342	4.010	4.679	5.347	6.016	6.684
4 × 8	3½ × 7¼	25.375	111.148	30.661	4.405	5.286	6.168	7.049	7.930	8.811
4 × 10	3½ × 9¼	32.375	230.840	49.911	5.621	6.745	7.869	8.933	10.117	11.241
4 × 12	3½ × 11¼	39.375	415.283	73.828	6.836	8.203	9.570	10.938	12.305	13.672
4 × 14	3½ × 13½	47.250	717.609	106.313	8.203	9.844	11.484	13.125	14.766	16.406
4 × 16	3½ × 15½	54.250	1,086.130	140.146	9.418	11.302	13.186	15.069	16.953	18.837
6 × 1	5½ × ¾	4.125	0.193	0.516	0.716	0.859	1.003	1.146	1.289	1.432
6 × 2	5½ × 1½	8.250	1.547	2.063	1.432	1.719	2.005	2.292	2.578	2.865
6 × 3	5½ × 2½	13.750	7.161	5.729	2.387	2.865	3.342	3.819	4.297	4.774
6 × 4	5½ × 3½	19.250	19.651	11.229	3.342	4.010	4.679	5.347	6.016	6.684
6 × 6	5½ × 5½	30.250	76.255	27.729	5.252	6.302	7.352	8.403	9.453	10.503
6 × 8	5½ × 7½	41.250	193.359	51.563	7.161	8.594	10.026	11.458	12.891	14.323
6 × 10	5½ × 9½	52.250	392.963	82.729	9.071	10.885	12.700	14.514	16.328	18.142
6 × 12	5½ × 11½	63.250	697.068	121.229	10.981	13.177	15.373	17.569	19.766	21.962
6 × 14	5½ × 13½	74.250	1,127.672	167.063	12.891	15.469	18.047	20.625	23.203	25.781
6 × 16	5½ × 15½	85.250	1,706.776	220.229	14.800	17.760	20.720	23.681	26.641	29.601
6 × 18	5½ × 17½	96.250	2,456.380	280.729	16.710	20.052	23.394	26.736	30.078	33.420
6 × 20	5½ × 19½	107.250	3,398.484	348.563	18.620	22.344	26.068	29.792	33.516	37.240
6 × 22	5½ × 21½	118.250	4,555.086	423.729	20.530	24.635	28.741	32.847	36.953	41.059
6 × 24	5½ × 23½	129.250	5,948.191	506.229	22.439	26.927	31.415	35.903	40.391	44.878
8 × 1	7¼ × ¾	5.438	0.255	0.680	0.944	1.133	1.322	1.510	1.699	1.888
8 × 2	7¼ × 1½	10.875	2.039	2.719	1.888	2.266	2.643	3.021	3.398	3.776
8 × 3	7¼ × 2½	18.125	9.440	7.552	3.147	3.776	4.405	5.035	5.664	6.293
8 × 4	7¼ × 3½	25.375	25.904	14.802	4.405	5.286	6.168	7.049	7.930	8.811
8 × 6	7½ × 5½	41.250	103.984	37.813	7.161	8.594	10.026	11.458	12.891	14.323
8 × 8	7½ × 7½	56.250	263.672	70.313	9.766	11.719	13.672	15.625	17.578	19.531
8 × 10	7½ × 9½	71.250	535.859	112.813	12.370	14.844	17.318	19.792	22.266	24.740
8 × 12	7½ × 11½	86.250	950.547	165.313	14.974	17.969	20.964	23.958	26.953	29.948
8 × 14	7½ × 13½	101.250	1,537.734	227.813	17.578	21.094	24.609	28.125	31.641	35.156
8 × 16	7½ × 15½	116.250	2,327.422	300.313	20.182	24.219	28.255	32.292	36.328	40.365
8 × 18	7½ × 17½	131.250	3,349.609	382.813	22.786	27.344	31.901	36.458	41.016	45.573
8 × 20	7½ × 19½	146.250	4,634.297	475.313	25.391	30.469	35.547	40.625	45.703	50.781
8 × 22	7½ × 21½	161.250	6,211.484	577.813	27.995	33.594	39.193	44.792	50.391	55.990
8 × 24	7½ × 23½	176.250	8,111.172	690.313	30.599	36.719	42.839	48.958	55.078	61.198
10 × 1	9¼ × ¾	6.938	0.325	0.867	1.204	1.445	1.686	1.927	2.168	2.409
10 × 2	9¼ × 1½	13.875	2.602	3.469	2.409	2.891	3.372	3.854	4.336	4.818
10 × 3	9¼ × 2½	23.125	12.044	9.635	4.015	4.818	5.621	6.424	7.227	8.030
10 × 4	9¼ × 3½	32.375	33.049	18.885	5.621	6.745	7.869	8.993	10.117	11.241
10 × 6	9½ × 5½	52.250	131.714	47.896	9.071	10.885	12.700	14.514	16.328	18.142
10 × 8	9½ × 7½	71.250	333.984	89.063	12.370	14.844	17.318	19.792	22.266	24.740
10 × 10	9½ × 9½	90.250	678.755	142.896	15.668	18.802	21.936	25.069	28.203	31.337
10 × 12	9½ × 11½	109.250	1,204.026	209.396	18.967	22.760	26.554	30.347	34.141	37.934
10 × 14	9½ × 13½	128.250	1,947.797	288.563	22.266	26.719	31.172	35.625	40.078	44.531
10 × 16	9½ × 15½	147.250	2,948.068	380.396	25.564	30.677	35.790	40.903	46.016	51.128
10 × 18	9½ × 17½	166.250	4,242.836	484.896	28.863	34.635	40.408	46.181	51.953	57.726
10 × 20	9½ × 19½	185.250	5,870.109	602.063	32.161	38.594	45.026	51.458	57.891	64.323
10 × 22	9½ × 21½	204.250	7,867.879	731.896	35.460	42.552	49.644	56.736	63.828	70.920
10 × 24	9½ × 23½	223.250	10,274.148	874.396	38.759	46.510	54.262	62.014	69.766	77.517
12 × 1	11¼ × ¾	8.438	0.396	1.055	1.465	1.758	2.051	2.344	2.637	2.930
12 × 2	11¼ × 1½	16.875	3.164	4.219	2.930	3.516	4.102	4.688	5.273	5.859
12 × 3	11¼ × 2½	28.125	14.648	11.719	4.883	5.859	6.836	7.813	8.789	9.766
12 × 4	11¼ × 3½	39.375	40.195	22.969	6.836	8.203	9.570	10.938	12.305	13.672
12 × 6	11½ × 5½	63.250	159.443	57.979	10.981	13.177	15.373	17.569	19.766	21.962
12 × 8	11½ × 7½	86.250	404.297	107.813	14.974	17.969	20.964	23.958	26.953	29.948
12 × 10	11½ × 9½	109.250	821.651	172.979	18.967	22.760	26.554	30.347	34.141	37.934
12 × 12	11½ × 11½	132.250	1,457.505	253.479	22.960	27.552	32.144	36.736	41.328	45.920
12 × 14	11½ × 13½	155.250	2,357.859	349.313	26.953	32.344	37.734	43.125	48.516	53.906

Table 8-3. Properties of Sections of Solid-sawn Wood (*Continued*)

Nominal size, in.	Standard dressed size, in. (S4S)	Area of section, sq in.	Moment of inertia in.^{4}	Section modulus, in.^{3}	Weight, lb per lin ft, of piece when weight of wood, lb per cu ft, equals					
					25	30	35	40	45	50
12 × 16	11½ × 15½	178.250	3,568.713	460.479	30.946	37.135	43.325	49.514	55.703	61.892
12 × 18	11½ × 17½	201.250	5,136.066	586.979	34.939	41.927	48.915	55.903	62.891	69.878
12 × 20	11½ × 19½	224.250	7,105.922	728.813	38.932	46.719	54.505	62.292	70.078	77.865
12 × 22	11½ × 21½	247.250	9,524.273	885.979	42.925	51.510	60.095	68.681	77.266	85.851
12 × 24	11½ × 23½	270.250	12,437.129	1,058.479	46.918	56.302	65.686	75.069	84.453	93.837
14 × 2	13¾ × 1½	19.875	3.727	4.969	3.451	4.141	4.831	5.521	6.211	6.901
14 × 3	13¾ × 2½	33.125	17.253	13.802	5.751	6.901	8.051	9.201	10.352	11.502
14 × 4	13¾ × 3½	46.375	47.34	27.052	8.047	9.657	11.266	12.877	14.485	16.094
14 × 6	13½ × 5½	74.250	187.172	68.063	12.891	15.469	18.047	20.625	23.203	25.781
14 × 8	13½ × 7½	101.250	474.609	126.563	17.578	21.094	24.609	28.125	31.641	35.156
14 × 10	13½ × 9½	128.250	964.547	203.063	22.266	26.719	31.172	35.625	40.078	44.531
14 × 12	13½ × 11½	155.250	1,710.984	297.563	26.953	32.344	37.734	43.125	48.516	53.906
14 × 16	13½ × 15½	209.250	4,189.359	540.563	36.328	43.594	50.859	58.125	65.391	72.656
14 × 18	13½ × 17½	236.250	6,029.297	689.063	41.016	49.219	57.422	65.625	73.828	82.031
14 × 20	13½ × 19½	263.250	8,341.734	855.563	45.703	54.844	63.984	73.125	82.266	91.406
14 × 22	13½ × 21½	290.250	11,180.672	1,040.063	50.391	60.469	70.547	80.625	90.703	100.781
14 × 24	13½ × 23½	317.250	14,600.109	1,242.563	55.078	66.094	77.109	88.125	99.141	110.156
16 × 3	15¼ × 2½	38.125	19.857	15.885	6.619	7.944	9.267	10.592	11.915	13.240
16 × 4	15¼ × 3½	53.375	54.487	31.135	9.267	11.121	12.975	14.828	16.682	18.536
16 × 6	15½ × 5½	85.250	214.901	78.146	14.800	17.760	20.720	23.681	26.641	29.601
16 × 8	15½ × 7½	116.250	544.922	145.313	20.182	24.219	28.255	32.292	36.328	40.365
16 × 10	15½ × 9½	147.250	1,107.443	233.146	25.564	30.677	35.790	40.903	46.016	51.128
16 × 12	15½ × 11½	178.250	1,964.463	341.646	30.946	37.135	43.325	49.514	55.703	61.892
16 × 14	15½ × 13½	209.250	3,177.984	470.813	36.328	43.594	50.859	58.125	65.391	72.656
16 × 16	15½ × 15½	240.250	4,810.004	620.646	41.710	50.052	58.394	66.736	75.078	83.420
16 × 18	15½ × 17½	271.250	6,922.523	791.146	47.092	56.510	65.929	75.347	84.766	94.184
16 × 20	15½ × 19½	302.250	9,577.547	982.313	52.474	62.969	73.464	83.958	94.453	104.948
16 × 22	15½ × 21½	333.250	12,837.066	1,194.146	57.856	69.427	80.998	92.569	104.141	115.712
16 × 24	15½ × 23½	364.250	16,763.086	1,426.646	63.238	75.885	88.533	101.181	113.828	126.476
18 × 6	17½ × 5½	96.250	242.630	88.229	16.710	20.052	23.394	26.736	30.078	33.420
18 × 8	17½ × 7½	131.250	615.234	164.063	22.786	27.344	31.901	36.458	41.016	45.573
18 × 10	17½ × 9½	166.250	1,250.338	263.229	28.863	34.635	40.408	46.181	51.953	57.726
18 × 12	17½ × 11½	201.250	2,217.943	385.729	34.939	41.927	48.915	55.903	62.891	69.878
18 × 14	17½ × 13½	236.250	3,588.047	531.563	41.016	49.219	57.422	65.625	73.828	82.031
18 × 16	17½ × 15½	271.250	5,430.648	700.729	47.092	56.510	65.929	75.347	84.766	94.184
18 × 18	17½ × 17½	306.250	7,815.754	893.229	53.168	63.802	74.436	85.069	95.703	106.337
18 × 20	17½ × 19½	341.250	10,813.359	1,109.063	59.245	71.094	82.943	94.792	106.641	118.490
18 × 22	17½ × 21½	376.250	14,493.461	1,348.229	65.321	78.385	91.450	104.514	117.578	130.642
18 × 24	17½ × 23½	411.250	18,926.066	1,610.729	71.398	85.677	99.957	114.236	128.516	142.795
20 × 6	19½ × 5½	107.250	270.359	98.313	18.620	22.344	26.068	29.792	33.516	37.240
20 × 8	19½ × 7½	146.250	685.547	182.813	25.391	30.469	35.547	40.625	45.703	50.781
20 × 10	19½ × 9½	185.250	1,393.234	293.313	32.161	38.594	45.026	51.458	57.891	64.323
20 × 12	19½ × 11½	224.250	2,471.422	429.813	38.932	46.719	54.505	62.292	70.078	77.865
20 × 14	19½ × 13½	263.250	3,998.109	592.313	45.703	54.844	63.984	73.125	82.266	91.406
20 × 16	19½ × 15½	302.250	6,051.297	780.813	52.474	62.969	73.464	83.958	94.453	104.948
20 × 18	19½ × 17½	341.250	8,708.984	995.313	59.245	71.094	82.943	94.792	106.641	118.490
20 × 20	19½ × 19½	380.250	12,049.172	1,235.813	66.016	79.219	92.422	105.625	118.828	132.031
20 × 22	19½ × 21½	419.250	16,149.859	1,502.313	72.786	87.344	101.901	116.458	131.016	145.573
20 × 24	19½ × 23½	458.250	21,089.047	1,794.813	79.557	95.469	111.380	127.292	243.203	159.115
22 × 6	21½ × 5½	118.250	298.088	108.396	20.530	24.635	28.741	32.847	36.953	41.059
22 × 8	21½ × 7½	161.250	755.859	201.563	27.995	33.594	39.193	44.792	50.391	55.990
22 × 10	21½ × 9½	204.250	1,536.130	323.396	35.460	42.552	49.644	56.736	63.828	70.920
22 × 12	21½ × 11½	247.250	2,724.901	473.896	42.925	51.510	60.095	68.681	77.266	85.851
22 × 14	21½ × 13½	290.250	4,408.172	653.063	50.391	60.469	70.547	80.625	90.703	100.781
22 × 16	21½ × 15½	333.250	6,671.941	860.896	57.856	69.427	80.998	92.569	104.141	115.712
22 × 18	21½ × 17½	376.250	9,602.211	1,097.396	65.321	78.385	91.450	104.514	117.578	130.642
22 × 20	21½ × 19½	419.250	13,284.984	1,362.563	72.786	87.344	101.901	116.458	131.016	145.573
22 × 22	21½ × 21½	462.250	17,806.254	1,656.396	80.252	96.302	112.352	128.403	144.453	160.503
22 × 24	21½ × 23½	505.250	23,252.023	1,978.896	87.717	105.260	122.804	140.347	157.891	175.434
24 × 6	23½ × 5½	129.250	325.818	118.479	22.439	26.927	31.415	35.903	40.391	44.878
24 × 8	23½ × 7½	176.250	826.172	220.313	30.599	36.719	42.839	48.958	55.078	61.198
24 × 10	23½ × 9½	223.250	1,679.026	353.479	38.759	46.510	54.262	62.014	69.766	77.517
24 × 12	23½ × 11½	270.250	2,978.380	517.979	46.918	56.302	65.686	75.069	84.453	93.837
24 × 14	23½ × 13½	317.250	4,818.234	713.813	55.078	66.094	77.109	88.125	99.141	110.156
24 × 16	23½ × 15½	364.250	7,292.586	940.979	63.238	75.885	88.533	101.181	113.828	126.476
24 × 18	23½ × 17½	411.250	10,495.441	1,199.479	71.398	85.677	99.957	114.236	128.516	142.795
24 × 20	23½ × 19½	458.250	14,520.797	1,489.313	79.557	95.469	111.380	127.292	143.203	159.115
24 × 22	23½ × 21½	505.250	19,462.648	1,810.479	87.717	105.260	122.804	140.347	157.891	175.434
24 × 24	23½ × 23½	552.250	25,415.004	2,162.979	95.877	115.052	134.227	153.403	172.578	191.753

* For lumber surfaced 1⅝ in. thick, instead of 1½ in., the area, moment of inertia, and section modulus may be increased 8.33%.

Basic stresses for laminated timbers under wet service conditions are the same as for solid timbers; i.e., these stresses are based on the strength of wood in the green condition. When moisture content in a member will be low throughout its service, a second set of higher basic stresses, based on the higher strength of dry material, may be used. Technical Bulletin 479, U.S. Department of Agriculture, "Strength and Related Properties of Woods Grown in the United States," presents test results on small, clear, and straight-grained wood species in the green

Table 8-4. Properties of Sections of Glued-laminated Timber*

No. of laminations 1½-in.	No. of laminations ¾-in.	d	C_F	A	S	I	Vol.
2¼-in. Width							
2	4	3.00	1.00	6.8	3.4	5.1	0.05
	5	3.75	1.00	8.4	5.3	9.9	0.06
3	6	4.50	1.00	10.1	7.6	17.1	0.07
	7	5.25	1.00	11.8	10.3	27.1	0.08
4	8	6.00	1.00	13.5	13.5	40.5	0.09
	9	6.75	1.00	15.2	17.1	57.7	0.11
5	10	7.50	1.00	16.9	21.1	79.1	0.12
	11	8.25	1.00	18.6	25.5	105.3	0.13
6	12	9.00	1.00	20.2	30.4	136.7	0.14
	13	9.75	1.00	21.9	35.6	173.8	0.15
7	14	10.50	1.00	23.6	41.3	217.0	0.16
	15	11.25	1.00	25.3	47.5	267.0	0.18
8	16	12.00	1.00	27.0	54.0	324.0	0.19
	17	12.75	0.99	28.7	61.0	388.6	0.20
9	18	13.50	0.99	30.4	68.3	461.3	0.21
	19	14.25	0.98	32.1	76.1	542.6	0.22
10	20	15.00	0.98	33.8	84.4	632.8	0.23
3⅛-in. Width							
2	4	3.00	1.00	9.4	4.7	7.0	0.06
	5	3.75	1.00	11.7	7.3	13.7	0.08
3	6	4.50	1.00	14.1	10.5	23.7	0.10
	7	5.25	1.00	16.4	14.4	37.7	0.11
4	8	6.00	1.00	18.8	18.8	56.3	0.13
	9	6.75	1.00	21.1	23.7	80.1	0.15
5	10	7.50	1.00	23.4	29.3	109.9	0.16
	11	8.25	1.00	25.8	35.4	146.2	0.18
6	12	9.00	1.00	28.1	42.2	189.8	0.20
	13	9.75	1.00	30.5	49.5	241.4	0.21
7	14	10.50	1.00	32.8	57.4	301.5	0.23
	15	11.25	1.00	35.2	65.9	370.8	0.24
8	16	12.00	1.00	37.5	75.0	450.0	0.26
	17	12.75	0.99	39.8	84.7	539.8	0.28
9	18	13.50	0.99	42.2	94.9	640.7	0.29
	19	14.25	0.98	44.5	105.8	753.6	0.31
10	20	15.00	0.98	46.9	117.2	878.9	0.33
	21	15.75	0.97	49.2	129.2	1,017.4	0.34
11	22	16.50	0.97	51.6	141.8	1,169.8	0.36
	23	17.25	0.96	53.9	155.0	1,336.7	0.37
12	24	18.00	0.96	56.3	168.8	1,518.8	0.39
	25	18.75	0.95	58.6	183.1	1,716.6	0.41
13	26	19.50	0.95	60.9	198.0	1,931.0	0.42
	27	20.25	0.94	63.3	213.6	2,162.4	0.44

No. of laminations 1½-in.	No. of laminations ¾-in.	d	C_F	A	S	I	Vol.
6¾-in. Width							
4	8	6.00	1.00	40.5	40.5	121.5	0.28
	9	6.75	1.00	45.6	51.3	173.0	0.32
5	10	7.50	1.00	50.6	63.3	237.3	0.35
	11	8.25	1.00	55.7	76.6	315.9	0.39
6	12	9.00	1.00	60.8	91.1	410.1	0.42
	13	9.75	1.00	65.8	106.9	521.4	0.46
7	14	10.50	1.00	70.9	124.0	651.2	0.49
	15	11.25	1.00	75.9	142.4	800.9	0.53
8	16	12.00	1.00	81.0	162.0	972.0	0.56
	17	12.75	0.99	86.1	182.9	1,165.9	0.60
9	18	13.50	0.99	91.1	205.0	1,384.0	0.63
	19	14.25	0.98	96.2	228.4	1,627.7	0.67
10	20	15.00	0.98	101.3	253.1	1,898.4	0.70
	21	15.75	0.97	106.3	279.1	2,197.7	0.74
11	22	16.50	0.97	111.4	306.3	2,526.8	0.77
	23	17.25	0.96	116.4	334.8	2,887.3	0.81
12	24	18.00	0.96	121.5	364.5	3,280.5	0.84
	25	18.75	0.95	126.6	395.5	3,707.9	0.88
13	26	19.50	0.95	131.6	427.8	4,170.9	0.91
	27	20.25	0.94	136.7	461.3	4,670.9	0.95
14	28	21.00	0.94	141.8	496.1	5,209.3	0.98
	29	21.75	0.94	146.8	532.2	5,787.6	1.02
15	30	22.50	0.93	151.9	569.5	6,407.2	1.05
	31	23.25	0.93	156.9	608.1	7,069.5	1.09
16	32	24.00	0.93	162.0	648.0	7,776.0	1.12
	33	24.75	0.92	167.1	689.1	8,528.0	1.16
17	34	25.50	0.92	172.1	731.5	9,327.0	1.20
	35	26.25	0.92	177.2	775.2	10,174.4	1.23
18	36	27.00	0.91	182.3	820.1	11,071.7	1.27
	37	27.75	0.91	187.3	866.3	12,020.2	1.30
19	38	28.50	0.91	192.4	913.8	13,021.4	1.34
	39	29.25	0.91	197.4	962.5	14,076.7	1.37
20	40	30.00	0.90	202.5	1,012.5	15,187.5	1.41
	41	30.75	0.90	207.6	1,063.8	16,355.3	1.44
21	42	31.50	0.90	212.6	1,116.3	17,581.4	1.48
	43	32.25	0.90	217.7	1,170.1	18,867.4	1.51
22	44	33.00	0.89	222.8	1,225.1	20,214.6	1.55
	45	33.75	0.89	227.8	1,281.4	21,624.4	1.58
23	46	34.50	0.89	232.9	1,339.0	23,098.3	1.62
	47	35.25	0.89	237.9	1,397.9	24,637.7	1.65
24	48	36.00	0.88	243.0	1,458.0	26,244.0	1.69
	49	36.75	0.88	248.1	1,519.4	27,918.7	1.72
25	50	37.50	0.88	253.1	1,582.0	29,663.1	1.76
	51	38.25	0.88	258.2	1,645.9	31,478.7	1.79

No. of laminations 1½-in.	No. of laminations ¾-in.	d	C_F	A	S	I	Vol.
8¾-in. Width							
6	12	9.00	1.00	78.8	118.1	531.6	0.55
	13	9.75	1.00	85.3	138.6	675.8	0.59
7	14	10.50	1.00	91.9	160.8	844.1	0.64
	15	11.25	1.00	98.4	184.6	1,038.2	0.68
8	16	12.00	1.00	105.0	210.0	1,260.0	0.73
	17	12.75	0.99	111.6	237.1	1,511.3	0.77
9	18	13.50	0.99	118.1	265.8	1,794.0	0.82
	19	14.25	0.98	124.7	296.1	2,109.9	0.87
10	20	15.00	0.98	131.3	328.1	2,460.9	0.91
	21	15.75	0.97	137.8	361.8	2,848.8	0.96
11	22	16.50	0.97	144.4	397.0	3,275.5	1.00
	23	17.25	0.96	150.9	433.9	3,742.8	1.05
12	24	18.00	0.96	157.5	472.5	4,252.5	1.09
	25	18.75	0.95	164.1	512.7	4,806.5	1.14
13	26	19.50	0.95	170.6	554.5	5,406.7	1.18
	27	20.25	0.94	177.2	598.0	6,054.8	1.23
14	28	21.00	0.94	183.8	643.1	6,752.8	1.28
	29	21.75	0.94	190.3	689.9	7,502.5	1.32
15	30	22.50	0.93	196.9	738.3	8,305.7	1.37
	31	23.25	0.93	203.4	788.3	9,164.2	1.41
16	32	24.00	0.93	210.0	840.0	10,080.0	1.46
	33	24.75	0.92	216.6	893.3	11,054.8	1.50
17	34	25.50	0.92	223.1	948.3	12,090.6	1.55
	35	26.25	0.92	229.7	1,004.9	13,189.1	1.59
18	36	27.00	0.91	236.3	1,063.1	14,352.2	1.64
	37	27.75	0.91	242.8	1,123.0	15,581.7	1.69
19	38	28.50	0.91	249.4	1,184.5	16,879.6	1.73
	39	29.25	0.91	255.9	1,247.7	18,247.5	1.78
20	40	30.00	0.90	262.5	1,312.5	19,687.5	1.82
	41	30.75	0.90	269.1	1,378.9	21,201.3	1.87
21	42	31.50	0.90	275.6	1,447.0	22,790.7	1.91
	43	32.25	0.90	282.2	1,516.8	24,457.7	1.96
22	44	33.00	0.89	288.8	1,588.1	26,204.1	2.00
	45	33.75	0.89	295.3	1,661.1	28,031.6	2.05
23	46	34.50	0.89	301.9	1,735.8	29,942.2	2.10
	47	35.25	0.89	308.4	1,812.1	31,937.7	2.14
24	48	36.00	0.88	315.0	1,890.0	34,020.0	2.19
	49	36.75	0.88	321.6	1,969.6	36,190.9	2.23
25	50	37.50	0.88	328.1	2,050.8	38,452.2	2.28
	51	38.25	0.88	334.7	2,133.6	40,805.7	2.32
26	52	39.00	0.88	341.3	2,218.1	43,253.4	2.37
	53	39.75	0.88	347.8	2,304.3	45,797.1	2.42
27	54	40.50	0.87	354.4	2,392.0	48,438.6	2.46
	55	41.25	0.87	360.9	2,481.4	51,179.8	2.51

14	28	21.00	0.94	65.6	229.7	2,411.7	0.46
	29	21.75	0.94	68.0	246.4	2,679.5	0.47
15	30	22.50	0.93	70.3	263.7	2,966.3	0.49
	31	23.25	0.93	72.7	281.5	3,272.9	0.50
16	32	24.00	0.93	75.0	300.0	3,600.0	0.52
$5^1/_8$-in. Width							
3	6	4.50	1.00	23.1	17.3	38.9	0.16
	7	5.25	1.00	26.9	23.5	61.8	0.19
4	8	6.00	1.00	30.8	30.8	92.3	0.21
	9	6.75	1.00	34.6	38.9	131.3	0.24
5	10	7.50	1.00	38.4	48.0	180.2	0.27
	11	8.25	1.00	42.3	58.1	239.8	0.29
6	12	9.00	1.00	46.1	69.2	311.3	0.32
	13	9.75	1.00	50.0	81.2	395.8	0.35
7	14	10.50	1.00	53.8	94.2	494.4	0.37
	15	11.25	1.00	57.7	108.1	608.1	0.40
8	16	12.00	1.00	61.5	123.0	738.0	0.43
	17	12.75	0.99	65.3	138.9	885.2	0.45
9	18	13.50	0.99	69.2	155.7	1,050.8	0.48
	19	14.25	0.98	73.0	173.4	1,235.8	0.51
10	20	15.00	0.98	76.9	192.2	1,441.4	0.53
	21	15.75	0.97	80.7	211.9	1,668.6	0.56
11	22	16.50	0.97	84.6	232.5	1,918.5	0.59
	23	17.25	0.96	88.4	254.2	2,192.2	0.61
12	24	18.00	0.96	92.3	276.8	2,490.8	0.64
	25	18.75	0.95	96.1	300.3	2,815.2	0.67
13	26	19.50	0.95	99.9	324.8	3,166.8	0.69
	27	20.25	0.94	103.8	350.3	3,546.4	0.72
14	28	21.00	0.94	107.6	376.7	3,955.2	0.75
	29	21.75	0.94	111.5	404.1	4,394.3	0.77
15	30	22.50	0.93	115.3	432.4	4,864.7	0.80
	31	23.25	0.93	119.2	461.7	5,367.6	0.83
16	32	24.00	0.93	123.0	492.0	5,904.0	0.85
	33	24.75	0.92	126.8	523.2	6,475.0	0.88
17	34	25.50	0.92	130.7	555.4	7,081.6	0.91
	35	26.25	0.92	134.5	588.6	7,725.0	0.93
18	36	27.00	0.91	138.4	622.7	8,406.3	0.96
	37	27.75	0.91	142.2	657.8	9,126.4	0.99
19	38	28.50	0.91	146.1	693.8	9,886.6	1.01
	39	29.25	0.91	149.9	730.8	10,687.8	1.04
20	40	30.00	0.90	153.8	768.8	11,531.3	1.07
	41	30.75	0.90	157.6	807.7	12,417.9	1.09
21	42	31.50	0.90	161.4	847.5	13,348.9	1.12
	43	32.25	0.90	165.3	888.4	14,325.2	1.15
22	44	33.00	0.89	169.1	930.2	15,348.1	1.17
	45	33.75	0.89	173.0	972.9	16,418.5	1.20
23	46	34.50	0.89	176.8	1,016.7	17,537.6	1.23
	47	35.25	0.89	180.7	1,061.4	18,706.4	1.25
24	48	36.00	0.88	184.5	1,107.0	19,926.0	1.28

26	52	39.00	0.88	263.3	1,711.1	33,366.9	1.83
	53	39.75	0.88	268.3	1,777.6	35,329.2	1.86
27	54	40.50	0.87	273.4	1,845.3	37,367.0	1.90
	55	41.25	0.87	278.4	1,914.3	39,481.6	1.93
28	56	42.00	0.87	283.5	1,984.5	41,674.5	1.97
	57	42.75	0.87	288.6	2,056.0	43,947.2	2.00
29	58	43.50	0.87	293.6	2,128.8	46,301.0	2.04
	59	44.25	0.87	298.7	2,202.8	48,737.4	2.07
30	60	45.00	0.86	303.8	2,278.1	51,257.8	2.11
	61	45.75	0.86	308.8	2,354.7	53,863.7	2.14
31	62	46.50	0.86	313.9	2,432.5	56,556.4	2.18
	63	47.25	0.86	318.9	2,511.6	59,337.3	2.21
32	64	48.00	0.86	324.0	2,592.0	62,208.0	2.25

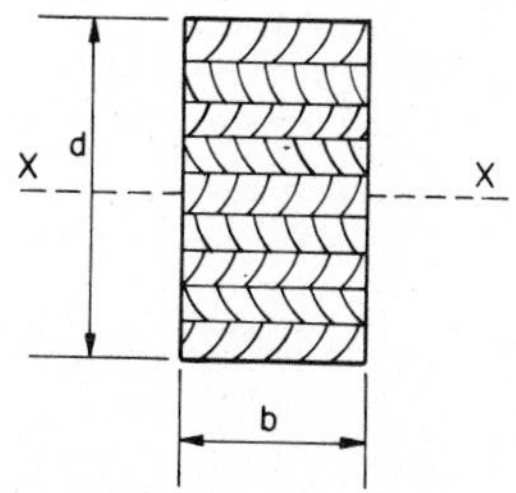

28	56	42.00	0.87	367.5	2,572.5	54,022.5	2.60
	57	42.75	0.87	374.1	2,665.2	56,968.6	2.55
29	58	43.50	0.87	380.6	2,759.5	60,019.8	2.64
	59	44.25	0.87	387.2	2,855.5	63,178.1	2.69
30	60	45.00	0.86	393.8	2,953.1	66,445.3	2.73
	61	45.75	0.86	400.3	3,052.4	69,823.3	2.78
31	62	46.50	0.86	406.9	3,153.3	73,313.8	2.83
	63	47.25	0.86	413.4	3,255.8	76,918.8	2.87
32	64	48.00	0.86	420.0	3,360.0	80,640.0	2.92
	65	48.75	0.86	426.6	3,465.8	84,479.4	2.96
33	66	49.50	0.85	433.1	3,573.3	88,438.7	3.01
	67	50.25	0.85	439.7	3,682.4	92,519.9	3.05
34	68	51.00	0.85	446.3	3,793.1	96,724.7	3.10
	69	51.75	0.85	452.8	3,905.5	101,055.0	3.14
35	70	52.50	0.85	459.4	4,019.5	105,512.7	3.19
	71	53.25	0.85	465.9	4,135.2	110,099.6	3.24
36	72	54.00	0.85	472.5	4,252.5	114,817.5	3.28
	73	54.75	0.85	479.1	4,371.4	119,668.3	3.33
37	74	55.50	0.84	485.6	4,492.0	124,653.9	3.37
	75	56.25	0.84	492.2	4,614.3	129,776.0	3.42
38	76	57.00	0.84	498.8	4,738.1	135,036.6	3.46
	77	57.75	0.84	505.3	4,863.6	140,437.4	3.51
39	78	58.50	0.84	511.9	4,990.8	145,980.4	3.55
	79	59.25	0.84	518.4	5,119.6	151,667.3	3.60
40	80	60.00	0.84	525.0	5,250.0	157,500.0	3.65
	81	60.75	0.84	531.6	5,382.1	163,480.4	3.69
41	82	61.50	0.83	538.1	5,515.8	169,610.3	3.74
	83	62.25	0.83	544.7	5,651.1	175,891.5	3.78
42	84	63.00	0.83	551.3	5,788.1	182,326.0	3.83

Table 8-4. Properties of Sections of Glued-laminated Timber* (*Continued*)

10¾-in. Width

No. of laminations 1½-in.	No. of laminations ¾-in.	d	C_F	A	S	I	Vol.
7	14	10.50	1.00	112.9	197.5	1,037.0	0.78
	15	11.25	1.00	120.9	226.8	1,275.5	0.84
8	16	12.00	1.00	129.0	258.0	1,548.0	0.90
	17	12.75	0.99	137.1	291.3	1,856.8	0.95
9	18	13.50	0.99	145.1	326.5	2,204.1	1.01
	19	14.25	0.98	153.2	363.8	2,592.2	1.06
10	20	15.00	0.98	161.3	403.1	3,023.4	1.12
	21	15.75	0.97	169.3	444.4	3,500.0	1.18
11	22	16.50	0.97	177.4	487.8	4,024.2	1.23
	23	17.25	0.96	185.4	533.1	4,598.3	1.29
12	24	18.00	0.96	193.5	580.5	5,224.5	1.34
	25	18.75	0.95	201.6	629.9	5,905.2	1.40
13	26	19.50	0.95	209.6	681.3	6,642.5	1.46
	27	20.25	0.94	217.7	734.7	7,438.8	1.51
14	28	21.00	0.94	225.8	790.1	8,296.3	1.57
	29	21.75	0.94	233.8	847.6	9,217.3	1.62
15	30	22.50	0.93	241.9	907.0	10,204.1	1.68
	31	23.25	0.93	249.9	968.5	11,258.9	1.74
16	32	24.00	0.93	258.0	1,032.0	12,384.0	1.79
	33	24.75	0.92	266.1	1,097.5	13,581.7	1.85
17	34	25.50	0.92	274.1	1,165.0	14,854.1	1.90
	35	26.25	0.92	282.2	1,234.6	16,203.7	1.96
18	36	27.00	0.91	290.3	1,306.1	17,632.7	2.02
	37	27.75	0.91	298.3	1,379.7	19,143.3	2.07
19	38	28.50	0.91	306.4	1,455.3	20,737.8	2.13
	39	29.25	0.91	314.4	1,532.9	22,418.4	2.18
20	40	30.00	0.90	322.5	1,612.5	24,187.5	2.24
	41	30.75	0.90	330.6	1,694.1	26,047.3	2.30
21	42	31.50	0.90	338.6	1,777.8	28,000.1	2.35
	43	32.25	0.90	346.7	1,863.4	30,048.1	2.41
22	44	33.00	0.89	354.8	1,951.1	32,193.6	2.46
	45	33.75	0.89	362.8	2,040.8	34,438.8	2.52
23	46	34.50	0.89	370.9	2,132.5	36,786.2	2.58
	47	35.25	0.89	378.9	2,226.3	39,237.8	2.63
24	48	36.00	0.88	387.0	2,322.0	41,796.0	2.69
	49	36.75	0.88	395.1	2,419.8	44,463.1	2.74
25	50	37.50	0.88	403.1	2,519.5	47,241.2	2.80
	51	38.25	0.88	411.2	2,621.3	50,132.8	2.86
26	52	39.00	0.88	419.3	2,725.1	53,139.9	2.91
	53	39.75	0.88	427.3	2,830.9	56,265.0	2.97
27	54	40.50	0.87	435.4	2,938.8	59,510.3	3.02
	55	41.25	0.87	443.4	3,048.6	62,878.1	3.08
28	56	42.00	0.87	451.5	3,160.5	66,370.5	3.14
	57	42.75	0.87	459.6	3,274.4	69,989.9	3.19
29	58	43.50	0.87	467.6	3,390.3	73,738.6	3.25
	59	44.25	0.87	475.7	3,508.2	77,618.8	3.30
30	60	45.00	0.86	483.8	3,628.1	81,632.8	3.36
	61	45.75	0.86	491.8	3,750.1	85,782.9	3.42
31	62	46.50	0.86	499.9	3,874.0	90,071.2	3.47
	63	47.25	0.86	507.9	4,000.0	94,500.2	3.53
32	64	48.00	0.86	516.0	4,128.0	99,072.0	3.58

12¼-in. Width

No. of laminations 1½-in.	No. of laminations ¾-in.	d	C_F	A	S	I	Vol.
8	16	12.00	1.00	147.0	294.0	1,764.0	1.02
	17	12.75	0.99	156.2	331.9	2,115.8	1.08
9	18	13.50	0.99	165.4	372.1	2,511.6	1.15
	19	14.25	0.98	174.6	414.6	2,953.8	1.21
10	20	15.00	0.98	183.8	459.4	3,445.3	1.28
	21	15.75	0.97	192.9	506.4	3,988.6	1.34
11	22	16.50	0.97	202.1	555.8	4,585.7	1.40
	23	17.25	0.96	211.3	607.5	5,239.7	1.47
12	24	18.00	0.96	220.5	661.5	5,953.5	1.53
	25	18.75	0.95	229.7	717.7	6,728.9	1.60
13	26	19.50	0.95	238.9	776.3	7,569.4	1.66
	27	20.25	0.94	248.1	837.2	8,476.5	1.72
14	28	21.00	0.94	257.2	900.4	9,453.9	1.79
	29	21.75	0.94	266.4	965.8	10,503.1	1.85
15	30	22.50	0.93	275.6	1,033.6	11,627.9	1.91
	31	23.25	0.93	284.8	1,103.6	12,829.5	1.97
16	32	24.00	0.93	294.0	1,176.0	14,112.0	2.04
	33	24.75	0.92	303.2	1,250.6	15,476.3	2.10
17	34	25.50	0.92	312.4	1,327.6	16,926.8	2.17
	35	26.25	0.92	321.6	1,406.8	18,464.1	2.23
18	36	27.00	0.91	330.8	1,488.4	20,093.1	2.30
	37	27.75	0.91	339.9	1,572.2	21,813.7	2.36
19	38	28.50	0.91	349.1	1,658.3	23,631.4	2.42
	39	29.25	0.91	358.3	1,746.7	25,545.7	2.49
20	40	30.00	0.90	367.5	1,837.5	27,562.5	2.55
	41	30.75	0.90	376.7	1,930.5	29,680.8	2.62
21	42	31.50	0.90	385.9	2,025.8	31,907.0	2.68
	43	32.25	0.90	395.1	2,123.4	34,239.7	2.74
22	44	33.00	0.89	404.2	2,223.4	36,685.7	2.81
	45	33.75	0.89	413.4	2,325.6	39,244.1	2.87
23	46	34.50	0.89	422.6	2,430.1	41,919.1	2.93
	47	35.25	0.89	431.8	2,536.9	44,712.7	3.00
24	48	36.00	0.88	441.0	2,646.0	47,628.0	3.06
	49	36.75	0.88	450.2	2,757.4	50,667.0	3.13
25	50	37.50	0.88	459.4	2,871.1	53,833.0	3.19
	51	38.25	0.88	468.6	2,987.1	57,127.8	3.25
26	52	39.00	0.88	477.8	3,105.4	60,554.8	3.32
	53	39.75	0.88	486.9	3,226.0	64,115.8	3.38
27	54	40.50	0.87	496.1	3,348.8	67,814.1	3.45
	55	41.25	0.87	505.3	3,474.0	71,651.5	3.51
28	56	42.00	0.87	514.5	3,601.5	75,631.5	3.57
	57	42.75	0.87	523.7	3,731.3	79,755.7	3.64
29	58	43.50	0.87	532.9	3,863.3	84,027.7	3.70
	59	44.25	0.87	542.1	3,997.7	88,449.1	3.76
30	60	45.00	0.86	551.2	4,134.4	93,023.4	3.83
	61	45.75	0.86	560.4	4,273.3	97,752.2	3.89
31	62	46.50	0.86	569.6	4,414.6	102,639.3	3.96
	63	47.25	0.86	578.8	4,558.1	107,685.9	4.02
32	64	48.00	0.86	588.0	4,704.0	112,896.0	4.08
	65	48.75	0.86	597.2	4,852.1	118,270.7	4.15
33	66	49.50	0.85	606.4	5,002.6	123,814.2	4.21

14¼-in. Width

No. of laminations 1½-in.	No. of laminations ¾-in.	d	C_F	A	S	I	Vol.
9	18	13.50	0.99	192.4	432.8	2,921.7	1.34
	19	14.25	0.98	203.1	482.3	3,436.2	1.41
10	20	15.00	0.98	213.8	534.4	4,007.8	1.48
	21	15.75	0.97	224.4	589.1	4,639.5	1.56
11	22	16.50	0.97	235.1	646.6	5,334.4	1.63
	23	17.25	0.96	245.8	706.7	6,095.4	1.71
12	24	18.00	0.96	256.5	769.5	6,925.5	1.78
	25	18.75	0.95	267.2	835.0	7,827.8	1.86
13	26	19.50	0.95	277.9	903.1	8,805.2	1.93
	27	20.25	0.94	288.6	973.9	9,860.7	2.00
14	28	21.00	0.94	299.3	1,047.4	10,997.4	2.08
	29	21.75	0.94	309.9	1,123.5	12,218.3	2.15
15	30	22.50	0.93	320.6	1,202.3	13,526.4	2.23
	31	23.25	0.93	331.3	1,283.8	14,924.6	2.30
16	32	24.00	0.93	342.0	1,368.0	16,416.0	2.38
	33	24.75	0.92	352.7	1,454.8	18,003.6	2.45
17	34	25.50	0.92	363.4	1,544.3	19,690.4	2.52
	35	26.25	0.92	374.1	1,636.5	21,479.4	2.60
18	36	27.00	0.91	384.8	1,731.4	23,373.6	2.67
	37	27.75	0.91	395.4	1,828.9	25,376.0	2.75
19	38	28.50	0.91	406.1	1,929.1	27,489.6	2.82
	39	29.25	0.91	416.8	2,032.0	29,717.4	2.89
20	40	30.00	0.90	427.5	2,137.5	32,062.5	2.97
	41	30.75	0.90	438.2	2,245.7	34,527.8	3.04
21	42	31.50	0.90	448.9	2,356.6	37,116.4	3.12
	43	32.25	0.90	459.6	2,470.1	39,831.1	3.19
22	44	33.00	0.89	470.3	2,586.4	42,675.2	3.27
	45	33.75	0.89	480.9	2,705.3	45,651.5	3.34
23	46	34.50	0.89	491.6	2,826.8	48,763.1	3.41
	47	35.25	0.89	502.3	2,951.1	52,012.9	3.49
24	48	36.00	0.88	513.0	3,078.0	55,404.0	3.56
	49	36.75	0.88	523.7	3,207.6	58,939.4	3.64
25	50	37.50	0.88	534.4	3,339.8	62,622.1	3.71
	51	38.25	0.88	545.1	3,474.8	66,455.0	3.79
26	52	39.00	0.88	555.8	3,612.4	70,441.3	3.86
	53	39.75	0.88	566.4	3,752.6	74,583.9	3.93
27	54	40.50	0.87	577.1	3,895.6	78,885.8	4.01
	55	41.25	0.87	587.8	4,041.2	83,350.0	4.08
28	56	42.00	0.87	598.5	4,189.5	87,979.5	4.16
	57	42.75	0.87	609.2	4,340.5	92,777.4	4.23
29	58	43.50	0.87	619.9	4,494.1	97,746.5	4.30
	59	44.25	0.87	630.6	4,650.4	102,890.1	4.38
30	60	45.00	0.86	641.3	4,809.4	108,211.0	4.45
	61	45.75	0.86	651.9	4,971.0	113,712.2	4.53
31	62	46.50	0.86	662.6	5,135.3	119,396.7	4.60
	63	47.25	0.86	673.3	5,302.3	125,267.7	4.68
32	64	48.00	0.86	684.0	5,472.0	131,328.0	4.75
	65	48.75	0.86	694.7	5,644.3	137,580.7	4.82
33	66	49.50	0.85	705.4	5,819.3	144,028.8	4.90
	67	50.25	0.85	716.1	5,997.0	150,675.2	4.97
34	68	51.00	0.85	726.8	6,177.4	157,523.1	5.05

33	66	49.50	0.85	532.1	4,390.0	108,653.3	3.70
	67	50.25	0.85	540.2	4,524.1	113,667.3	3.75
34	68	51.00	0.85	548.3	4,660.1	118,833.2	3.81
	69	51.75	0.85	556.3	4,798.2	124,153.3	3.86
35	70	52.50	0.85	564.4	4,938.3	129,629.9	3.92
	71	53.25	0.85	572.4	5,080.4	135,265.2	3.98
36	72	54.00	0.85	580.5	5,224.5	141,061.5	4.03
	73	54.75	0.85	588.6	5,370.6	147,021.1	4.09
37	74	55.50	0.84	596.6	5,518.8	153,146.2	4.14
	75	56.25	0.84	604.7	5,668.9	159,439.1	4.20
38	76	57.00	0.84	612.8	5,821.1	165,902.1	4.26
	77	57.75	0.84	620.8	5,975.3	172,537.4	4.31
39	78	58.50	0.84	628.9	6,131.5	179,347.3	4.37
	79	59.25	0.84	636.9	6,289.8	186,334.1	4.42
40	80	60.00	0.84	645.0	6,450.0	193,500.0	4.48
	81	60.75	0.84	653.1	6,612.3	200,847.4	4.54
41	82	61.50	0.83	661.1	6,776.5	208,378.4	4.59
	83	62.25	0.83	669.2	6,942.8	216,095.3	4.65
42	84	63.00	0.83	677.3	7,111.1	224,000.5	4.70
	85	63.75	0.83	685.3	7,281.4	232,096.1	4.76
43	86	64.50	0.83	693.4	7,453.8	240,384.5	4.82
	87	65.25	0.83	701.4	7,628.1	248,867.9	4.87
44	88	66.00	0.83	709.5	7,804.5	257,548.5	4.93
	89	66.75	0.83	717.6	7,982.9	266,428.8	4.98
45	90	67.50	0.83	725.6	8,163.3	275,510.8	5.04
	91	68.25	0.82	733.7	8,345.7	284,796.9	5.10
46	92	69.00	0.82	741.8	8,530.1	294,289.3	5.15
	93	69.75	0.82	749.8	8,716.6	303,990.5	5.21
47	94	70.50	0.82	757.9	8,905.0	313,902.4	5.26
	95	71.25	0.82	765.9	9,095.5	324,027.5	5.32
48	96	72.00	0.82	774.0	9,288.0	334,368.0	5.38
	97	72.75	0.82	782.1	9,482.5	344,926.3	5.43
49	98	73.50	0.82	790.1	9,679.0	355,704.5	5.49
	99	74.25	0.82	798.2	9,877.6	366,704.8	5.54
50	100	75.00	0.82	806.3	10,078.1	377,929.7	5.60

34	68	51.00	0.85	624.8	5,310.4	135,414.6	4.34
	69	51.75	0.85	633.9	5,467.7	141,476.6	4.40
35	70	52.50	0.85	643.1	5,627.3	147,717.8	4.47
	71	53.25	0.85	652.3	5,789.3	154,138.9	4.53
36	72	54.00	0.85	661.5	5,953.5	160,744.5	4.59
	73	54.75	0.85	670.7	6,120.0	167,535.1	4.66
37	74	55.50	0.84	679.9	6,288.8	174,515.4	4.72
	75	56.25	0.84	689.1	6,459.9	181,685.8	4.79
38	76	57.00	0.84	698.2	6,633.4	189,051.2	4.85
	77	57.75	0.84	707.4	6,809.1	196,611.7	4.91
39	78	58.50	0.84	716.6	6,987.1	204,372.5	4.98
	79	59.25	0.84	725.8	7,167.4	212,333.5	5.04
40	80	60.00	0.84	735.0	7,350.0	220,500.0	5.10
	81	60.75	0.84	744.2	7,534.9	228,871.8	5.17
41	82	61.50	0.83	753.4	7,722.1	237,454.4	5.23
	83	62.25	0.83	762.6	7,911.6	246,247.3	5.30
42	84	63.00	0.83	771.8	8,103.4	255,256.3	5.36
	85	63.75	0.83	780.9	8,297.4	264,480.7	5.42
43	86	64.50	0.83	790.1	8,493.8	273,926.5	5.49
	87	65.25	0.83	799.3	8,692.5	283,592.7	5.55
44	88	66.00	0.83	808.5	8,893.5	293,485.5	5.61
	89	66.75	0.83	817.7	9,096.7	303,603.8	5.68
45	90	67.50	0.83	826.9	9,302.3	313,954.1	5.74
	91	68.25	0.82	836.1	9,510.2	324,534.9	5.81
46	92	69.00	0.82	845.2	9,720.4	335,352.9	5.87
	93	69.75	0.82	854.4	9,932.8	346,406.5	5.93
47	94	70.50	0.82	863.6	10,147.6	357,702.7	6.00
	95	71.25	0.82	872.8	10,364.6	369,239.4	6.06
48	96	72.00	0.82	882.0	10,584.0	381,024.0	6.12
	97	72.75	0.82	891.2	10,805.6	393,054.2	6.19
49	98	73.50	0.82	900.4	11,029.6	405,337.6	6.25
	99	74.25	0.82	909.6	11,255.8	417,871.5	6.32
50	100	75.00	0.82	918.8	11,484.4	430,664.1	6.38
	101	75.75	0.82	927.9	11,715.2	443,712.2	6.44
51	102	76.50	0.81	937.1	11,948.3	457,024.1	6.51
	103	77.25	0.81	946.3	12,183.7	470,596.7	6.57
52	104	78.00	0.81	955.5	12,421.5	484,438.5	6.64
	105	78.75	0.81	964.7	12,661.5	498,545.9	6.70
53	106	79.50	0.81	973.9	12,903.8	512,927.8	6.76
	107	80.25	0.81	983.1	13,148.4	527,580.3	6.83
54	108	81.00	0.81	992.2	13,395.4	542,512.7	6.89
	109	81.75	0.81	1,001.4	13,644.5	557,720.6	6.95
55	110	82.50	0.81	1,010.6	13,896.1	573,213.9	7.02
	111	83.25	0.81	1,019.8	14,149.9	588,987.6	7.08
56	112	84.00	0.81	1,029.0	14,406.0	605,052.0	7.15

35	70	52.50	0.85	748.1	6,546.1	171,835.3	5.20
	71	53.25	0.85	758.8	6,734.5	179,305.0	5.27
36	72	54.00	0.85	769.5	6,925.5	186,988.5	5.34
	73	54.75	0.85	780.2	7,119.2	194,888.4	5.42
37	74	55.50	0.84	790.9	7,315.6	203,007.7	5.49
	75	56.25	0.84	801.6	7,514.6	211,349.5	5.57
38	76	57.00	0.84	812.3	7,716.4	219,916.7	5.64
	77	57.75	0.84	822.9	7,920.8	228,712.4	5.71
39	78	58.50	0.84	833.6	8,127.8	237,739.5	5.79
	79	59.25	0.84	844.3	8,337.6	247,001.0	5.86
40	80	60.00	0.84	855.0	8,550.0	256,500.0	5.94
	81	60.75	0.84	865.7	8,765.1	266,239.5	6.01
41	82	61.50	0.83	876.4	8,982.8	276,222.5	6.09
	83	62.25	0.83	887.1	9,203.3	286,451.9	6.16
42	84	63.00	0.83	897.8	9,426.4	296,930.8	6.23
	85	63.75	0.83	908.4	9,652.1	307,662.3	6.31
43	86	64.50	0.83	919.1	9,880.6	318,649.2	6.38
	87	65.25	0.83	929.8	10,111.7	329,894.6	6.46
44	88	66.00	0.83	940.5	10,345.5	341,401.5	6.53
	89	66.75	0.83	951.2	10,582.0	353,173.0	6.61
45	90	67.50	0.83	961.9	10,821.1	365,212.0	6.68
	91	68.25	0.82	972.6	11,062.9	377,521.5	6.75
46	92	69.00	0.82	983.3	11,307.4	390,104.5	6.83
	93	69.75	0.82	993.9	11,554.5	402,964.0	6.90
47	94	70.50	0.82	1,004.6	11,804.3	416,103.1	6.98
	95	71.25	0.82	1,015.3	12,056.8	429,524.8	7.05
48	96	72.00	0.82	1,026.0	12,312.0	443,232.0	7.12
	97	72.75	0.82	1,036.7	12,569.8	457,227.8	7.20
49	98	73.50	0.82	1,047.4	12,830.3	471,515.1	7.27
	99	74.25	0.82	1,058.1	13,093.5	486,097.1	7.35
50	100	75.00	0.82	1,068.8	13,359.4	500,976.6	7.42
	101	75.75	0.82	1,079.4	13,627.9	516,156.7	7.50
51	102	76.50	0.81	1,090.1	13,899.1	531,640.5	7.57
	103	77.25	0.81	1,100.8	14,173.0	547,430.7	7.64
52	104	78.00	0.81	1,111.5	14,449.5	563,530.6	7.72
	105	78.75	0.81	1,122.2	14,728.7	579,943.1	7.79
53	106	79.50	0.81	1,132.9	15,010.6	596,671.2	7.87
	107	80.25	0.81	1,143.6	15,295.1	613,718.0	7.94
54	108	81.00	0.81	1,154.3	15,582.4	631,086.2	8.02
	109	81.75	0.81	1,164.9	15,872.3	648,779.2	8.09
55	110	82.50	0.81	1,175.6	16,164.8	666,799.8	8.16
	111	83.25	0.81	1,186.3	16,460.1	685,151.2	8.24
56	112	84.00	0.81	1,197.0	16,758.0	703,836.1	8.31
	113	84.75	0.80	1,207.7	17,058.6	722,857.7	8.39
57	114	85.50	0.80	1,218.4	17,361.8	742,218.8	8.46
	115	86.25	0.80	1,229.1	17,667.8	761,922.8	8.54
58	116	87.00	0.80	1,239.8	17,976.4	781,972.3	8.61
	117	87.75	0.80	1,250.4	18,287.6	802,370.6	8.68
59	118	88.50	0.80	1,261.1	18,601.6	823,120.6	8.76
	119	89.25	0.80	1,271.8	18,918.2	844,225.2	8.83
60	120	90.00	0.80	1,282.5	19,237.5	865,687.6	8.91
	121	90.75	0.80	1,293.2	19,559.5	887,510.6	8.98
61	122	91.50	0.80	1,303.9	19,884.1	909,697.3	9.05
	123	92.25	0.80	1,314.6	20,211.4	932,250.8	9.13
62	124	93.00	0.80	1,325.3	20,541.4	955,174.0	9.20
	125	93.75	0.80	1,335.9	20,874.0	978,470.0	9.28
63	126	94.50	0.80	1,346.6	21,209.3	1,002,141.6	9.35
	127	95.25	0.79	1,357.3	21,547.3	1,026,192.0	9.43
64	128	96.00	0.79	1,368.0	21,888.0	1,050,624.2	9.50

* d = section depth, in.; C_F = size factor; A = cross-sectional area, sq in.; S = section modulus, in.^{3}; I = moment of inertia, in.^{4}; Vol. = volume, cu ft per lin ft.

state and in the 12%-moisture-content, air-dry condition. Technical Bulletin 1069, U.S. Department of Agriculture, "Fabrication and Design of Glued-laminated Structural Members," gives the basic stresses for clear, solid-sawn members and glued-laminated timbers in the wet condition, and basic stresses for clear, glued-laminated timbers in the dry condition.

Allowable unit stresses for commercial species of lumber, widely accepted for wood construction in economical and efficient designs, are given in "National Design Specification for Stress-grade Lumber and Its Fastenings," National Forest Products Association. Allowable unit stresses for glued-laminated timber are given in the American Institute of Timber Construction "Timber Construction Manual," John Wiley & Sons, Inc., and in the AITC "Standard Specifications for Structural Glued-laminated Timber," AITC 117.

See also Arts. 8-9 to 8-12, and 8-14.

8-7. Structural Grading of Wood. Strength properties of wood are intimately related to moisture content and specific gravity. Therefore, data on strength properties, unaccompanied by corresponding data on these physical properties, would be of little value.

The strength of wood is actually affected by many other factors, such as rate of loading, duration of load, temperature, direction of grain, and position of growth rings. Strength is also influenced by such inherent growth characteristics as knots, cross grain, shakes, and checks.

Analysis and integration of available data have yielded a comprehensive set of simple principles for grading structural timber (Tentative Methods for Establishing Structural Grades of Lumber, ASTM D245).

The same characteristics, such as knots and cross grain, that reduce the strength of solid timber also affect the strength of laminated members. However, there are additional factors, peculiar to laminated wood, that must be considered: Effect on strength of bending members is less from knots located at the neutral plane of the beam, a region of low stress. Strength of a bending member with low-grade laminations can be improved by substituting a few high-grade laminations at the top and bottom of the member. Dispersement of knots in laminated members has a beneficial effect on strength. With sufficient knowledge of the occurrence of knots within a grade, mathematical estimates of this effect may be established for members containing various numbers of laminations.

Allowable design stresses taking these factors into account are higher than for solid timbers of comparable grade. But cross-grain limitations must be more restrictive than for solid timbers, to justify these higher allowable stresses.

8-8. Weight and Specific Gravity of Commercial Lumber Species. Specific gravity is a reliable indicator of fiber content. Also, specific gravity and the strength and stiffness of solid wood or laminated products are interrelated. See Table 8-5 for weights and specific gravities of several commercial lumber species.

8-9. Moisture Content of Wood. Wood is unlike most other structural materials in regard to the causes of its dimensional changes. These are primarily from gain or loss of moisture, not change in temperature. It is for this reason that expansion joints are seldom required for wood structures to permit movement with temperature changes. It partly accounts for the fact that wood structures can withstand extreme temperatures without collapse.

A newly felled tree is green (contains moisture). In removing the greater part of this water, seasoning first allows free water to leave the cavities in the wood. A point is reached where these cavities contain only air, and the cell walls still are full of moisture. The moisture content at which this occurs, the fiber-saturation point, varies from 25 to 30% of the weight of the oven-dry wood.

During removal of the free water, the wood remains constant in size and in most properties (weight decreases). Once the fiber-saturation point has been passed, shrinkage of the wood begins as the cell walls lose water. Shrinkage continues nearly linearly down to zero moisture content (Table 8-6). (There are, however, complicating factors, such as the effects of timber size and relative rates of moisture movement in three directions, longitudinal, radial, and tangential to the growth rings.) Eventually, the wood assumes a condition of equilibrium, with the final moisture content dependent on the relative humidity and temperature of the ambient air. Wood swells when it absorbs moisture, up to the fiber-saturation point. The

Table 8-5. Weights and Specific Gravities of Commercial Lumber Species

Species	Specific gravity based on oven-dry weight and volume at 12% moisture content	Weight, lb per cu ft			Moisture content when green (avg), %	Specific gravity based on oven-dry weight and volume when green	Weight when green, lb per cu ft
		At 12% moisture content	At 20% moisture content	Adjusting factor for each 1% change in moisture content			
Softwoods:							
Cedar:							
Alaska	0.44	31.1	32.4	0.170	38	0.42	35.5
Incense	0.37	25.0	26.4	0.183	108	0.35	42.5
Port Orford	0.42	29.6	31.0	0.175	43	0.40	35.0
Western red	0.33	23.0	24.1	0.137	37	0.31	26.4
Cypress, southern	0.46	32.1	33.4	0.167	91	0.42	45.3
Douglas fir:							
Coast region	0.48	33.8	35.2	0.170	38	0.45	38.2
Inland region	0.44	31.4	32.5	0.137	48	0.41	36.3
Rocky Mountain	0.43	30.0	31.4	0.179	38	0.40	34.6
Fir, white	0.37	26.3	27.3	0.129	115	0.35	39.6
Hemlock:							
Eastern	0.40	28.6	29.8	0.150	111	0.38	43.4
Western	0.42	29.2	30.2	0.129	74	0.38	37.2
Larch, western	0.55	38.9	40.2	0.170	58	0.51	46.7
Pine:							
Eastern white	0.35	24.9	26.2	0.167	73	0.34	35.1
Lodgepole	0.41	28.8	29.9	0.142	65	0.38	36.3
Norway	0.44	31.0	32.1	0.142	92	0.41	42.3
Ponderosa	0.40	28.1	29.4	0.162	91	0.38	40.9
Southern shortleaf	0.51	35.2	36.5	0.154	81	0.46	45.9
Southern longleaf	0.58	41.1	42.5	0.179	63	0.54	50.2
Sugar	0.36	25.5	26.8	0.162	137	0.35	45.8
Western white	0.38	27.6	28.6	0.129	54	9.36	33.0
Redwood	0.40	28.1	29.5	0.175	112	0.38	45.6
Spruce:							
Engelmann	0.34	23.7	24.7	0.129	80	0.32	32.5
Sitka	0.40	27.7	28.8	0.145	42	0.37	32.0
White	0.40	29.1	29.9	0.104	50	0.37	33.0
Hardwoods:							
Ash, white	0.60	42.2	43.6	0.175	42	0.55	47.4
Beech, American	0.64	43.8	45.1	0.162	54	0.56	50.6
Birch:							
Sweet	0.65	46.7	48.1	0.175	53	0.60	53.8
Yellow	0.62	43.0	44.1	0.142	67	0.55	50.8
Elm, rock	0.63	43.6	45.2	0.208	48	0.57	50.9
Gum	0.52	36.0	37.1	0.133	115	0.46	49.7
Hickory:							
Pecan	0.66	45.9	47.6	0.212	63	0.60	56.7
Shagbark	0.72	50.8	51.8	0.129	60	0.64	57.0
Maple, sugar	0.63	44.0	45.3	0.154	58	0.56	51.1
Oak:							
Red	0.63	43.2	44.7	0.187	80	0.56	56.0
White	0.68	46.3	47.6	0.167	68	0.60	55.6
Poplar, yellow	0.42	29.8	31.0	0.150	83	0.40	40.5

relationship of wood moisture content, temperature, and relative humidity can actually define an environment (Fig. 8-1).

This explanation has been simplified. Outdoors, rain, frost, wind, and sun can act directly on the wood. Within buildings, poor environmental conditions may be created for wood by localized heating, cooling, or ventilation. The conditions

Table 8-6. Shrinkage Values of Wood Based on Dimensions When Green

Species	Dried to 20% MC*			Dried to 6% MC†			Dried to 0% MC		
	Radial, %	Tangential, %	Volumetric, %	Radial, %	Tangential, %	Volumetric, %	Radial, %	Tangential, %	Volumetric, %
Softwoods:‡									
Cedar:									
Alaska	0.9	2.0	3.1	2.2	4.8	7.4	2.8	6.0	9.2
Incense	1.1	1.7	2.5	2.6	4.2	6.1	3.3	5.2	7.6
Port Orford	1.5	2.3	3.4	3.7	5.5	8.1	4.6	6.9	10.1
Western red	0.8	1.7	2.3	1.9	4.0	5.4	2.4	5.0	6.8
Cypress, southern	1.3	2.1	3.5	3.0	5.0	8.4	3.8	6.2	10.5
Douglas fir:									
Coast region	1.7	2.6	3.9	4.0	6.2	9.4	5.0	7.8	11.8
Inland region	1.4	2.5	3.6	3.3	6.1	8.7	4.1	7.6	10.9
Rocky Mountain	1.2	2.1	3.5	2.9	5.0	8.5	3.6	6.2	10.6
Fir, white	1.1	2.4	3.3	2.6	5.7	7.8	3.2	7.1	9.8
Hemlock:									
Eastern	1.0	2.3	3.2	2.4	5.4	7.8	3.0	6.8	9.7
Western	1.4	2.6	4.0	3.4	6.3	9.5	4.3	7.9	11.9
Larch, western	1.4	2.7	4.4	3.4	6.5	10.6	4.2	8.1	13.2
Pine:									
Eastern white	0.8	2.0	2.7	1.8	4.8	6.6	2.3	6.0	8.2
Lodgepole	1.5	2.2	3.8	3.6	5.4	9.2	4.5	6.7	11.5
Norway	1.5	2.4	3.8	3.7	5.8	9.2	4.6	7.2	11.5
Ponderosa	1.3	2.1	3.2	3.1	5.0	7.7	3.9	6.3	9.6
Southern (avg.)	1.6	2.6	4.1	4.0	6.1	9.8	5.0	7.6	12.2
Sugar	1.0	1.9	2.6	2.3	4.5	6.3	2.9	5.6	7.9
Western white	1.4	2.5	3.9	3.3	5.9	9.4	4.1	7.4	11.8
Redwood (old growth)	0.9	1.5	2.3	2.1	3.5	5.4	2.6	4.4	6.8
Spruce:									
Engelmann	1.1	2.2	3.5	2.7	5.3	8.3	3.4	6.6	10.4
Sitka	1.4	2.5	3.8	3.4	6.0	9.2	4.3	7.5	11.5
Hardwoods:‡									
Ash, white	1.6	2.6	4.5	3.8	6.2	10.7	4.8	7.8	13.4
Beech, American	1.7	3.7	5.4	4.1	8.8	13.0	5.1	11.0	16.3
Birch:									
Sweet	2.2	2.8	5.2	5.2	6.8	12.5	6.5	8.5	15.6
Yellow	2.4	3.1	5.6	5.8	7.4	13.4	7.2	9.2	16.7
Elm, rock	1.6	2.7	4.7	3.8	6.5	11.3	4.8	8.1	14.1
Gum, red	1.7	3.3	5.0	4.2	7.9	12.0	5.2	9.9	15.0
Hickory:									
Pecan §	1.6	3.0	4.5	3.9	7.1	10.9	4.9	8.9	13.6
True	2.5	3.8	6.0	6.0	9.0	14.3	7.5	11.3	17.9
Maple, hard	1.6	3.2	5.0	3.9	7.6	11.9	4.9	9.5	14.9
Oak:									
Red	1.3	2.7	4.5	3.2	6.6	10.8	4.0	8.2	13.5
White	1.8	3.0	5.3	4.2	7.2	12.6	5.3	9.0	15.8
Poplar, yellow	1.3	2.4	4.1	3.2	5.7	9.8	4.0	7.1	12.3

* MC = moisture content, as a percent of weight of oven-dry wood. These shrinkage values have been taken as one-third the shrinkage to the oven-dry condition as given in the last three columns of this table.

† These shrinkage values have been taken as four-fifths of the shrinkage to the oven-dry condition as given in the last three columns of this table.

‡ The total longitudinal shrinkage of normal species from fiber saturation to oven-dry condition is minor. It usually ranges from 0.17 to 0.3% of the green dimension.

§ Average of butternut hickory, nutmeg hickory, water hickory, and pecan.

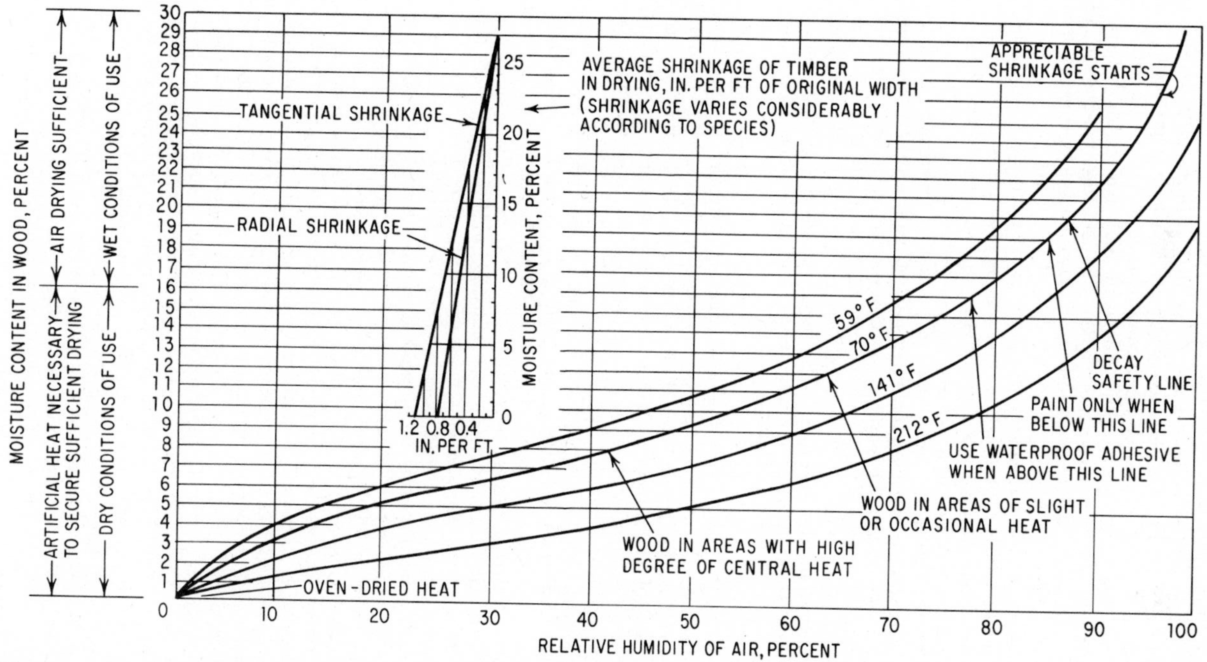

Fig. 8-1. Approximate relationship of wood equilibrium moisture content and temperature and relative humidity is shown by the curves. The triangular diagram indicates the effect of wood moisture content on shrinkage.

of service must be sufficiently well known to be specifiable. Then, the proper design stress can be assigned to wood and the most suitable adhesive selected.

Dry Condition of Use. Design stresses—Allowable unit stresses for dry condition of use are applicable for normal loading when the wood moisture content in service is less than 16%, as in most covered structures.

Adhesive selection—Dry-use adhesives are those that perform satisfactorily when the moisture content of wood does not exceed 16% for repeated or prolonged periods of service, and are to be used only when these conditions exist.

Wet Condition of Use. Design stresses—Allowable unit stresses for wet condition of use are applicable for normal loading when the moisture content in service is 16% or more. This may occur in members not covered or in covered locations of high relative humidity.

Adhesive selection—Wet-use adhesives will perform satisfactorily for all conditions, including exposure to weather, marine use, and where pressure treatments are used, whether before or after gluing. Such adhesives are required when the moisture content exceeds 16% for repeated or prolonged periods of service.

8-10. Checking in Timbers. Separation of grain, or checking, is the result of rapid lowering of surface moisture content combined with a difference in moisture content between inner and outer portions of the piece. As wood loses moisture to the surrounding atmosphere, the outer cells of the member lose at a more rapid rate than the inner cells. As the outer cells try to shrink, they are restrained by the inner portion of the member. The more rapid the drying, the greater will be the differential in shrinkage between outer and inner fibers, and the greater the shrinkage stresses. Splits may develop.

Checks affect the horizontal shear strength of timber. A large reduction factor is applied to test values in establishing allowable unit stresses, in recognition of stress concentrations at the ends of checks. Allowable unit stresses for horizontal shear are adjusted for the amount of checking permissible in the various stress grades at the time of the grading. Since strength properties of wood increase with dryness, checks may enlarge with increasing dryness after shipment, without appreciably reducing shear strength.

Cross-grain checks and splits that tend to run out the side of a piece, or excessive checks and splits that tend to enter connection areas, may be serious and may require servicing. Provisions for controlling the effects of checking in connection areas may be incorporated in design details.

To avoid excessive splitting between rows of bolts due to shrinkage during seasoning of solid-sawn timbers, the rows should not be spaced more than 5 in. apart, or a saw kerf, terminating in a bored hole, should be provided between the lines of bolts. Whenever possible, maximum end distances for connections should be specified to minimize the effect of checks running into the joint area. Some designers require stitch bolts in members, with multiple connections loaded at an angle to the grain. Stitch bolts, kept tight, will reinforce pieces where checking is excessive.

One of the principal advantages of glued-laminated timber construction is relative freedom from checking. Seasoning checks may, however, occur in laminated members for the same reasons that they exist in solid-sawn members. When laminated members are glued within the range of moisture contents set in AITC 103, "Standard for Structural Glued-laminated Timber," they will approximate the moisture content in normal-use conditions, thereby minimizing checking. Moisture content of the lumber at the time of gluing is thus of great importance to the control of checking in service. However, rapid changes in moisture content of large wood sections after gluing will result in shrinkage or swelling of the wood, and during shrinking, checking may develop in both glued joints and wood.

Differentials in shrinkage rates of individual laminations tend to concentrate shrinkage stresses at or near the glue line. For this reason, when checking occurs, it is usually at or near glue lines. The presence of wood-fiber separation indicates adequate glue bonds, and not delamination.

In general, checks have very little effect on the strength of glued-laminated members. Laminations in such members are thin enough to season readily in kiln drying without developing checks. Since checks lie in a radial plane, and the majority of laminations are essentially flat grain, checks are so positioned in horizontally laminated members that they will not materially affect shear strength. When members are designed with laminations vertical (with wide face parallel to the direction of load application), and when checks may affect the shear strength, the effect of checks may be evaluated in the same manner as for checks in solid-sawn members.

Seasoning checks in bending members affect only the horizontal shear strength. They are usually not of structural importance unless the checks are significant in depth and occur in the midheight of the member near the support, and then only if shear governs the design of the members. The reduction in shear strength is nearly directly proportional to the ratio of depth of check to width of beam. Checks in columns are not of structural importance unless the check develops into a split, thereby increasing the slenderness ratio of the columns.

Minor checking may be disregarded, since there is ample factor of safety in allowable unit stresses. The final decision as to whether shrinkage checks are detrimental to the strength requirements of any particular design or structural member should be made by a competent engineer experienced in timber construction.

8-11. Allowable Unit Stresses and Modifications for Stress-grade Lumber. (See also Arts. 8-6 to 8-10.) A wide range of allowable unit stresses for many species is given in the "National Design Specification for Stress-grade Lumber and Its Fastenings" (NDS). They apply to solid-sawn lumber under normal duration of load and under continuously dry service conditions, such as those in most covered structures. For continuously wet-use service conditions, apply the appropriate factor

Table 8-7. Allowable-stress Modification Factors for Moisture in Solid-Sawn Lumber

Service condition	Allowable unit stress				Modulus of elasticity E
	Extreme fiber in bending F_b or tension parallel to grain F_t	Compression parallel to grain F_c	Compression perpendicular to grain $F_{c\perp}$	Horizontal shear F_v	
At or above fiber-saturation point, or continuously submerged	1.00	0.90	0.67	1.00	0.91

from Table 8-7. NDS may be obtained from the National Forest Products Association, 619 Massachusetts Ave., N.W., Washington, D.C. 20036.

The modifications for duration of load, temperature, treatment, and size are the same as for glued-laminated timber (Art. 8-12).

8-12. Allowable Unit Stresses and Modifications for Structural Glued-laminated Timber. (See also Arts. 8-6 to 8-10.) The allowable unit stresses given in the AITC "Timber Construction Manual" and "Standard Specification for Structural Glued-laminated Timber," AITC 117, are for normal conditions of loading, assuming uniform loading, horizontal laminating, and a 12-in.-deep member with a span-to-depth ratio of 21:1. Tables are given for dry-use and for wet-use conditions. AITC specifications may be obtained from the American Institute of Timber Construction, 333 W. Hampden Ave., Englewood, Colo. 80110.

Allowable unit stresses for dry-use conditions are applicable when the moisture content in service is less than 16%, as in most covered structures. Allowable unit stresses for wet-use conditions are applicable when the moisture content in service is 16% or more, as may occur in exterior or submerged construction and in some structures housing wet processes or otherwise having constantly high relative humidities.

Allowable stresses for vertically laminated members made of combination grades of lumber are the weighted average of the lumber grades.

Requirements for limiting cross grain, type, and location of end joints, and certain manufacturing and other requirements, must be met for these allowable-unit-stress combinations to apply.

Species other than those referenced in these specifications may be used if allowable unit stresses are established for them in accordance with the provisions of U.S. Product Standard PS 56-73.

For laminated bending members 16¼ in. or more in depth, the outermost tension-side laminations representing 5% of the total depth of the member should have additional grade restrictions, as described in AITC 117, "Standard Specification for Structural Glued-laminated Timber of Douglas Fir, Southern Pine, Western Larch, and California Redwood."

Use of hardwoods in glued-laminated timbers is covered by AITC 119, "Standard Specification for Hardwood Glued-laminated Timber."

Lumber that is E-rated, as well as visually graded and then positioned in laminated timber, is covered by AITC 120, "Standard Specification for Structural Glued-laminated Timber Using E-rated and Visually Graded Lumber of Douglas Fir, Southern Pine, Hemlock, and Lodgepole Pine." The lumber is sorted into E grades by measuring the modulus of elasticity of each piece and positioning the lumber in the member according to stiffness, but still grading visually so as to meet the visual grade requirements.

Modifications for Duration of Load. Wood can absorb overloads of considerable magnitude for short periods; thus, allowable unit stresses are adjusted accordingly. The elastic limit and ultimate strength are higher under short-time loading. Wood members under continuous loading for years will fail at loads one-half to three-fourths as great as would be required to produce failure in a static-bending test when the maximum load is reached in a few minutes.

Normal load duration contemplates fully stressing a member to the allowable unit stress by the application of the full design load for a duration of about 10 years (either continuously or cumulatively).

When a member is fully stressed by maximum design loads for long-term loading conditions (greater than 10 years, either continuously or cumulatively), the allowable unit stresses are reduced to 90% of the tabulated values.

When duration of full design load (either continuously or cumulatively) does not exceed 10 years, tabulated allowable unit stresses can be increased as follows: 15% for 2-month duration, as for snow; 25% for 7-day duration; 33⅓% for wind or earthquake: and 100% for impact. These increases are not cumulative. The allowable unit stress for normal loading may be used without regard to impact, if the stress induced by impact does not exceed the allowable unit stress for normal loading. These adjustments do not apply to modulus of elasticity, except when it is used to determine allowable unit loads for columns.

Modifications for Temperature. Tests show that wood increases in strength as temperature is lowered below normal. Tests conducted at about −300°F indicate that the important strength properties of dry wood in bending and compression, including stiffness and shock resistance, are much higher at extremely low temperatures.

Some reduction of the allowable unit stresses may be necessary for members subjected to elevated temperatures for repeated or prolonged periods. This adjustment is especially desirable where high temperature is associated with high moisture content.

Temperature effect on strength is immediate. Its magnitude depends on the moisture content of the wood and, when temperature is raised, on the duration of exposure.

Between 0 and 70°F, the static strength of dry wood (12% moisture content) roughly increases from its strength at 70°F about ⅓ to ½% for each 1°F decrease in temperature. Between 70 and 150°F, the strength decreases at about the same rate for each 1°F increase in temperature. The change is greater for higher wood moisture content.

After exposure to temperatures not much above normal for a short time under ordinary atmospheric conditions, the wood, when temperature is reduced to normal, may recover essentially all its original strength. Experiments indicate that air-dry wood can probably be exposed to temperatures up to nearly 150°F for a year or more without a significant permanent loss in most strength properties. But its strength while at such temperatures will be temporarily lower than at normal temperature.

When wood is exposed to temperatures of 150°F or more for extended periods of time, it will be permanently weakened. The nonrecoverable strength loss depends on a number of factors, including moisture content and temperature of the wood, heating medium, and time of exposure. To some extent, the loss depends on the species and size of the piece.

Glued-laminated members are normally cured at temperatures of less than 150°F. Therefore, no reduction in allowable unit stresses due to temperature effect is necessary for curing.

Adhesives used under standard specifications for structural glued-laminated members, for example, casein, resorcinol-resin, phenol-resin, and melamine-resin adhesives, are not affected substantially by temperatures up to those that char wood. Use of adhesives that deteriorate at low temperatures is not permitted by standard specifications for structural glued-laminated timber. Low temperatures appear to have no significant effect on the strength of glued joints.

Modifications for Pressure-applied Treatments. The allowable stresses given for wood also apply to wood treated with a preservative when this treatment is in accordance with American Wood Preservers Association (AWPA) standard specifications. These limit pressure and temperature. Investigations have indicated that, in general, any weakening of timber as a result of preservative treatment is caused almost entirely by subjecting the wood to temperatures and pressures above the AWPA limits.

Highly acidic salts, such as zinc chloride, tend to hydrolyze wood if they are present in appreciable concentrations. Fortunately, the concentrations used in wood preservative treatments are sufficiently small that strength properties other than impact resistance are not greatly affected under normal service conditions. A significant loss in impact strength may occur, however, if higher concentrations are used.

None of the other common salt preservatives is likely to form solutions as highly acidic as those of zinc chloride. So, in most cases, their effect on the strength of the wood can be disregarded.

In wood treated with highly acidic salts, such as zinc chloride, moisture is the controlling factor in corrosion of fastenings. Therefore, wood treated with highly acidic salts is not recommended for use under highly humid conditions.

The effects on strength of all treatments, preservative and fire-retardant, should be investigated, to assure that adjustments in allowable unit stresses are made when required.

("Manual of Recommended Practice," American Wood Preservers Association.)

Modifications for Size Factor. When the depth of a rectangular beam exceeds 12 in., the tabulated unit stress in bending F_b should be reduced by multiplication by a size factor C_F.

$$C_F = \left(\frac{12}{d}\right)^{1/9} \tag{8-1}$$

where d = depth of member, in.

Table 8-4 lists values of C_F for various cross-sectional sizes.

Equation (8-1) applies to bending members satisfying the following basic assumptions: simply supported beam, uniformly distributed load, and span-depth ratio $L/d = 21$. C_F may thus be applied with reasonable accuracy to beams usually used in buildings. Where greater accuracy is required for other loading conditions or span-depth ratios, the percentage changes given in Table 8-8 may be applied

Table 8-8. Change in Size Factor C_F

Span-depth ratio	Change, %	Loading condition for simply supported beams	Change, %
7	+6.2	Single concentrated load	+7.8
14	+2.3	Uniform load	0
21	0	Third-point load	−3.2
28	−1.6		
35	−2.8		

directly to the size factor given by Eq. (8-1). Straight-line interpolation may be used for L/d ratios other than those listed in the table.

Modifications for Radial Tension or Compression. The radial stress induced by a bending moment in a member of constant cross section may be computed from

$$f_r = \frac{3M}{2Rbd} \tag{8-2}$$

where M = bending moment, in.-lb
R = radius of curvature at centerline of member, in.
b = width of cross section, in.
d = depth of cross section, in.

Equation (8-2) can also be used to estimate the stresses in a member with varying cross section. Information on more exact procedures for calculating radial stresses in curved members with varying cross section can be obtained from the American Institute of Timber Construction.

When M is in the direction tending to decrease curvature (increase the radius), tensile stresses occur across the grain. For this condition, the allowable tensile stress across the grain is limited to one-third the allowable unit stress in horizontal shear for southern pine for all load conditions, and for Douglas fir and larch for wind or earthquake loadings. The limit is 15 psi for Douglas fir and larch for other types of loading. These values are subject to modification for duration of load. If these values are exceeded, mechanical reinforcement sufficient to resist all radial tensile stresses is required.

When M is in the direction tending to increase curvature (decrease the radius), the stress is compressive across the grain. For this condition, the allowable stress is limited to that for compression perpendicular to grain for all species covered by AITC 117.

Modifications for Curvature Factor. For the curved portion of members, the allowable unit stress in bending should be modified by multiplication by the following curvature factor:

$$C_c = 1 - 2{,}000\left(\frac{t}{R}\right)^2 \tag{8-3}$$

where t = thickness of lamination, in.
R = radius of curvature of lamination, in.

t/R should not exceed $1/100$ for hardwoods and southern pine, or $1/125$ for softwoods other than southern pine. The curvature factor should not be applied to stress in the straight portion of an assembly, regardless of curvature elsewhere.

The recommended minimum radii of curvature for curved, structural glued-laminated members are 9 ft 4 in. for ¾-in. laminations, and 27 ft 6 in. for 1½-in. laminations. Other radii of curvature may be used with these thicknesses, and other radius-thickness combinations may be used.

Certain species can be bent to sharper radii, but the designer should determine the availability of such sharply curved members before specifying them.

Modifications for Lateral Stability. The tabulated allowable bending unit stresses are applicable to members that are adequately braced. When deep and slender members not adequately braced are used, allowable bending unit stresses must be reduced. For the purpose, a slenderness factor should be applied, as indicated in the AITC "Timber Construction Manual."

The reduction in bending stresses determined by applying the slenderness factor should not be combined with a reduction in stress due to the application of the size factor. In no case may the allowable bending unit stress used in design exceed the stress determined by applying a size factor.

8-13. Lateral Support of Wood Framing. To prevent beams and compression members from buckling, they may have to be braced laterally. Need for such bracing, and required spacing, depend on the unsupported length and cross-sectional dimensions of members.

When buckling occurs, a member deflects in the direction of its least dimension b, unless prevented by bracing. (In a beam, b usually is taken as the width.) But if bracing precludes buckling in that direction, deflection can occur in the direction of the perpendicular dimension d. Thus, it is logical that unsupported length L, b, and d play important roles in rules for lateral support, or in formulas for reducing allowable stresses for buckling.

For glued-laminated beams, design for lateral stability is based on a function of Ld/b^2. For solid-sawn beams of rectangular cross section, maximum depth-width ratios should satisfy the approximate rules, based on nominal dimensions, summarized in Table 8-9. When the beams are adequately braced laterally, the depth of

Table 8-9. Approximate Lateral-support Rules for Sawn Beams*

Depth-width ratio (Nominal Dimensions)	*Rule*
2 or less	No lateral support required
3	Hold ends in position
4	Hold ends in position and member in line, e.g., with purlins or sag rods
5	Hold ends in position and compression edge in line, e.g., with direct connection of sheathing, decking, or joists
6	Hold ends in position and compression edge in line, as for 5 to 1, and provide adequate bridging or blocking at intervals not exceeding 6 times the depth
7	Hold ends in position and both edges firmly in line

If a beam is subject to both flexure and compression parallel to grain, the ratio may be as much as 5:1 if one edge is held firmly in line, e.g., by rafters (or roof joists) and diagonal sheathing. If the dead load is sufficient to induce tension on the underside of the rafters, the ratio for the beam may be 6:1.

* "Wood Handbook," Agricultural Handbook 72, U.S. Forest Service, Forest Products Laboratory, Madison, Wis., 1955.

the member below the brace may be taken as the width. For glued-laminated arches, maximum depth-width ratios should satisfy the approximate rules, based on actual dimensions, in Table 8-10.

Table 8-10. Approximate Lateral-support Rules for Glued-laminated Arches

Depth-width ratio (Actual Dimensions)	*Rule*
5 or less	No lateral support required
6	Brace one edge at frequent intervals

Where joists frame into arches or the compression chords of trusses and provide adequate lateral bracing, the depth, rather than the width, of arch or truss chord may be taken as b. The joists should be erected with upper edges at least ½ in. above the supporting member, but low enough to provide adequate lateral support. The depth of arch or compression chord also may be taken as b, where joists or planks are placed on top of the arch or chord and securely fastened to them and blocking is firmly attached between the joists.

For glued-laminated beams, no lateral support is required when the depth does not exceed the width. In that case also, the allowable bending stress does not have to be adjusted for lateral instability. Similarly, if continuous support prevents lateral movement of the compression flange, lateral buckling cannot occur and the allowable stress need not be reduced.

When the depth of a glued-laminated beam exceeds the width, bracing must be provided at supports. This bracing must be so placed as to prevent rotation of the beam in a plane perpendicular to its longitudinal axis. And unless the compression flange is braced at sufficiently close intervals between the supports, the allowable stress should be adjusted for lateral buckling. All other modifications of the stresses, except for size factor, should be made.

When the buckling factor

$$C = \frac{L_e d}{b^2} \tag{8-4}$$

where L_e is the effective length, in., and b and d also are in in., does not exceed 100, the allowable bending stress does not have to be adjusted for lateral buckling. Such beams are classified as short beams.

For intermediate beams, for which C exceeds 100 but is less than

$$K = \frac{3E}{5F_b} \tag{8-5}$$

the allowable bending stress should be determined from

$$F'_b = F_b\left[1 - \frac{1}{3}\left(\frac{C}{K}\right)^2\right] \tag{8-6}$$

where E = modulus of elasticity, psi

F_b = allowable bending stress, psi, adjusted, except for size factor

For long beams, for which C exceeds K but is less than 2,500, the allowable bending stress should be determined from

$$F'_b = \frac{0.40E}{C} \tag{8-7}$$

In no case should C exceed 2,500.

The effective length L_e for Eq. (8-4) is given in terms of unsupported length of beam in Table 8-11. Unsupported length is the distance between supports

Table 8-11. Ratio of Effective Length to Unsupported Length of Glued-laminated Beams

Simple beam, load concentrated at center	1.61
Simple beam, uniformly distributed load	1.92
Simple beam, equal end moments	1.84
Cantilever beam, load concentrated at unsupported end	1.69
Cantilever beam, uniformly distributed load	1.06
Simple or cantilever beam, any load (conservative value)	1.92

or the length of a cantilever when the beam is laterally braced at the supports to prevent rotation and adequate bracing is not installed elsewhere in the span. When both rotational and lateral displacement are also prevented at intermediate points, the unsupported length may be taken as the distance between points of lateral support. If the compression edge is supported throughout the length of the beam and adequate bracing is installed at the supports, the unsupported length is zero.

Acceptable methods of providing adequate bracing at supports include anchoring the bottom of a beam to a pilaster and the top of the beam to a parapet; for a wall-bearing roof beam, fastening the roof diaphragm to the supporting wall or installing a girt between beams at the top of the wall; for beams on wood columns, providing rod bracing.

For continuous lateral support of a compression flange, composite action is essential between deck elements, so that sheathing or deck acts as a diaphragm. One example is a plywood deck with edge nailing. With plank decking, nails attaching the plank to the beams must form couples, to resist rotation. In addition, the planks must be nailed to each other, for diaphragm action. Adequate lateral support is not provided when only one nail is used per plank and no nails between planks.

(American Institute of Timber Construction, "Timber Construction Manual," John Wiley & Sons, Inc., New York; R. F. Hooley and B. Madsen, "Lateral Stability of Glued-laminated Beams," *Journal of the Structural Division,* No. ST3, June, 1964, American Society of Civil Engineers; "Western Woods Use Book," Western Wood Products Association, 1500 Yeon Building, Portland, Ore. 97204.)

8-14. Combined Stresses in Timber. Allowable unit stresses given in the "National Design Specification for Stress-grade Lumber and Its Fastenings" apply directly to bending, horizontal shear, tension parallel or perpendicular to grain, and compression parallel or perpendicular to grain. For combined axial stress P/A

and bending stress M/S, stresses are limited by

$$\frac{P/A}{F} + \frac{M/S}{F_b} \leq 1 \tag{8-8}$$

where F = allowable tension stress parallel to grain or allowable compression stress parallel to grain adjusted for unsupported length, psi

F_b = allowable bending stress, psi

When a bending stress f_x, shear stress f_{xy}, and compression or tension stress f_y perpendicular to the grain exist simultaneously, the stresses should satisfy the Norris interaction formula:

$$\frac{f_x^2}{F_x^2} + \frac{f_y^2}{F_y^2} + \frac{f_{xy}^2}{F_{xy}^2} \leq 1 \tag{8-9}$$

where F_x = allowable bending stress F_b modified for duration of loading but not for size factor

F_y = allowable stress in compression or tension perpendicular to grain $F_{c\perp}$ or $F_{b\perp}$ modified for duration of loading

F_{xy} = allowable horizontal shear stress F_v modified for duration of loading

The usual design formulas do not apply to sharply curved glued-laminated beams. Nor do they hold where laminations of more than one species, each with markedly different modulus of elasticity, comprise the same member.

8-15. Effect of Shrinkage or Swelling on Shape of Curved Members. Wood shrinks or swells across the grain, but has practically no dimensional change along the grain. Radial swelling causes a decrease in the angle between the ends of a curved member; radial shrinkage causes an increase in this angle.

Such effects may be of great importance in three-hinged arches that become horizontal, or nearly so, at the crest of a roof. Shrinkage, increasing the relative end rotations, may cause a depression at the crest and create drainage problems. For such arches, therefore, consideration must be given to moisture content of the member at time of fabrication and in service, and to the change in end angles that results from change in moisture content and shrinkage across the grain.

8-16. Fabrication of Structural Timber. Fabrication consists of boring, cutting, sawing, trimming, dapping, routing, planing, and otherwise shaping, framing, and finishing wood units, sawn or laminated, including plywood, to fit them for particular places in a final structure. Whether fabrication is performed in shop or field, the product must exhibit a high quality of workmanship.

Jigs, patterns, templates, stops, or other suitable means should be used for all complicated and multiple assemblies to insure accuracy, uniformity, and control of all dimensions. All tolerances in cutting, drilling, and framing must comply with good practice and applicable specifications and controls. At the time of fabrication, tolerances must not exceed those listed below, unless they are not critical and not required for proper performance. Specific jobs, however, may require closer tolerances.

Location of Fastenings. Spacing and location of all fastenings within a joint should be in accordance with the shop drawings and specifications, with a maximum permissible tolerance of ±1/16 in. The fabrication of members assembled at any joint should be such that the fastenings are properly fitted.

Bolt-hole Sizes. Bolt holes in all fabricated structural timber, when loaded as a structural joint, should be 1/16 in. larger in diameter than bolt diameter for 1/2-in. and larger-diameter bolts, and 1/32 in. larger for smaller-diameter bolts. Larger clearances may be required for other bolts, such as anchor bolts and tension rods.

Holes and Grooves. Holes for stress-carrying bolts, connector grooves, and connector daps must be smooth and true within 1/16 in. per 12 in. of depth. The width of a split-ring connector groove should be within +0.02 in. of and not less than the thickness of the corresponding cross section of the ring. The shape of ring grooves must conform generally to the cross-sectional shape of the ring. Departure from these requirements may be allowed when supported by test data. Drills and other cutting tools should be set to conform to the size, shape, and depth of holes, grooves, daps, etc., specified in the "National Design Specification for Stress-grade Lumber and Its Fastenings," National Forest Products Association.

Lengths. Members should be cut within ±1/16 in. of the indicated dimension when they are up to 20 ft long, and ±1/16 in. per 20 ft of specified length when they are over 20 ft long. Where length dimensions are not specified or critical, these tolerances may be waived.

End Cuts. Unless otherwise specified, all trimmed square ends should be square within 1/16 in. per ft of depth and width. Square or sloped ends to be loaded in compression should be cut to provide contact over substantially the complete surface.

8-17. Fabrication of Glued-laminated Timbers. Structural glued-laminated timber is made by bonding layers of lumber together with adhesive so that the grain direction of all laminations is essentially parallel. Narrow boards may be edge-glued, and short boards may be end-glued. The resultant wide and long laminations are then face-glued into large, shop-grown timbers.

Recommended practice calls for lumber of nominal 1- and 2-in. thicknesses for laminating. The lumber is dressed to 3/4- and 1 1/2-in. thicknesses, respectively, just prior to gluing. The thinner laminations are generally used in curved members.

Constant-depth members normally are a multiple of the thickness of the lamination stock used. Variable-depth members, due to tapering or special assembly techniques, may not be exact multiples of these lamination thicknesses.

Nominal width of stock, in	3	4	6	8	10	12	14	16
Standard member finished width, in .	2 1/4	3 1/8	5 1/8	6 3/4	8 3/4	10 3/4	12 1/4	14 1/4

Standard widths are most economical, since they represent the maximum width of board normally obtained from the stock used in laminating.

When members wider than the stock available are required, laminations may consist of two boards side by side. These edge joints must be staggered, vertically in horizontally laminated beams (load acting normal to wide faces of laminations), and horizontally in vertically laminated beams (load acting normal to the edge of laminations). In horizontally laminated beams, edge joints need not be edge-glued. Edge gluing is required in vertically laminated beams.

Edge and face gluings are the simplest to make, end gluings the most difficult. Ends are also the most difficult surfaces to machine into scarfs or finger joints.

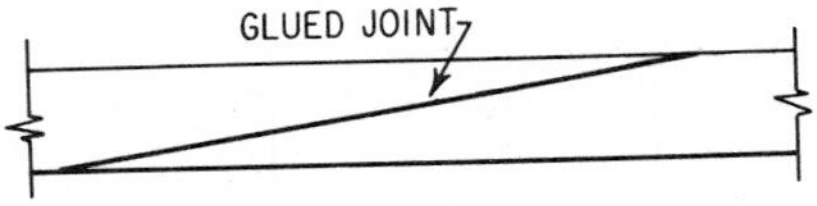

Fig. 8.2. Plane sloping scarf.

A plane sloping scarf (Fig. 8-2), in which the tapered surfaces of laminations are glued together, can develop 85 to 90% of the strength of an unscarfed, clear, straight-grained control specimen. Finger joints (Fig. 8-3) are less wasteful of lumber. Quality can be adequately controlled in machine cutting and in high-frequency gluing. A combination of thin tip, flat slope on the side of the individual fingers and a narrow pitch is desired. The length of fingers should be kept short for savings of lumber, but long for maximum strength.

The usefulness of structural glued-laminated timbers is determined by the lumber used and glue joint produced. Certain combinations of adhesive, treatment, and wood species do not produce the same quality of glue bond as other combinations, although the same gluing procedures are used. Thus, a combination must be supported by adequate experience with a laminator's gluing procedure.

The only adhesives currently recommended for wet-use and preservative-treated lumber, whether gluing is done before or after treatment, are resorcinol and phenol-resorcinol resins. The prime adhesive for dry-use structural laminating is casein. Melamine and melamine-urea blends are used in smaller amounts for high-frequency curing of end gluings.

Glued joints are cured with heat by several methods. R.F. (high-frequency) curing of glue lines is used for end joints and for limited-size members where there are repetitive gluings of the same cross section. Low-voltage resistance heating, where current is passed through a strip of metal to raise the temperature of a glue line, is used for attaching thin facing pieces. The metal may be left in the glue line as an integral part of the completed member. Printed electric circuits, in conjunction with adhesive films, and adhesive films impregnated on paper or on each side of a metal conductor placed in the glue line are other alternatives.

Preheating the wood to insure reactivity of the applied adhesive has limited application in structural laminating. The method requires adhesive application as a wet or dry film simultaneously to all laminations, and then rapid handling of multiple laminations.

Curing the adhesive at room temperature has many advantages. Since wood is an excellent insulator, a long time is required for elevated ambient temperature

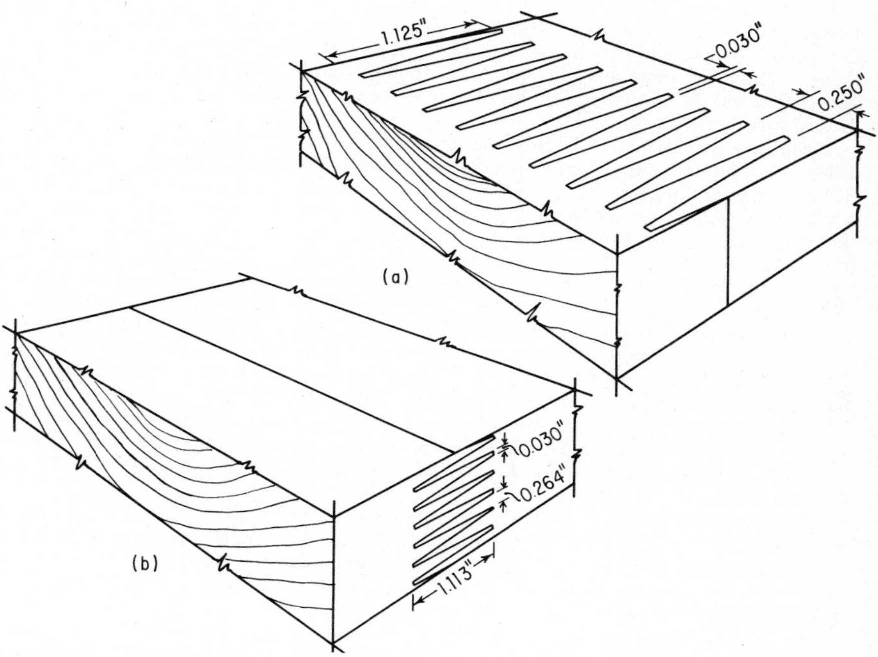

Fig. 8-3. Finger joint. (*a*) Fingers formed by cuts perpendicular to the wide face of the board. (*b*) Fingers formed by cuts perpendicular to the edges.

to reach inner glue lines of a large assembly. With room-temperature curing, equipment needed to heat the glue line is not required, and the possibility of injury to the wood from high temperatures is avoided.

8-18. Preservative Treatments for Wood. Wood-destroying fungi must have air, suitable moisture, and favorable temperatures to develop and grow in wood. Submerge wood permanently and totally in water, or keep the moisture content below 18 to 20%, or hold temperatures below 40°F or above 110°F, and wood remains permanently sound. If wood moisture content is kept below the fiber-saturation point (25 to 30%) when the wood is untreated, decay is greatly retarded. Below 18 to 20% moisture content, decay is completely inhibited.

If wood cannot be kept dry, a wood preservative, properly applied, must be used.

Wood members are permanent without treatment if located in enclosed buildings where good roof coverage, proper roof maintenance, good joint details, adequate flashing, good ventilation, and a well-drained site assure moisture content of the wood continuously below 20%. Also, in arid or semiarid regions where climatic conditions are such that the equilibrium moisture content seldom exceeds 20%, and then only for short periods, wood members are permanent without treatment.

Where wood is in contact with the ground or with water, where there is air and the wood may be alternately wet and dry, a preservative treatment, applied by a pressure process, is necessary to obtain an adequate service life. In enclosed buildings where moisture given off by wet-process operations maintains equilibrium moisture contents in the wood above 20%, wood structural members must be preservatively treated. So must wood exposed outdoors without protective roof covering and where the wood moisture content can go above 18 to 20% for repeated or prolonged periods.

Where wood structural members are subject to condensation by being in contact with masonry, preservative treatment is necessary.

To obtain preservatively treated glued-laminated timber, lumber may be treated before gluing, and the members then glued to the desired size and shape. Or the already-glued and machined members may be treated with certain treatments. When laminated members do not lend themselves to treatment because of size and shape, gluing of treated laminations is the only method of producing adequately treated members.

There are problems in gluing some treated woods. Certain combinations of adhesive, treatment, and wood species are compatible; other combinations are not. All adhesives of the same type do not produce bonds of equal quality for a particular wood species and preservative. The bonding of treated wood is dependent on the concentration of preservative on the surface at the time of gluing, and on the chemical effects of the preservative on the adhesive. In general, longer curing times or higher curing temperatures, and modifications in assembly times, are needed for treated than for untreated wood to obtain comparable adhesive bonds.

Each type of preservative and method of treatment has certain advantages. The preservative to be used depends on the service expected of the member for the specific conditions of exposure. The minimum retentions shown in Table 8-12 may be increased where severe climatic or exposure conditions are involved.

Creosote and creosote solutions have low volatility. They are practically insoluble in water, and thus are most suitable for severe exposure, contact with ground or water, and where painting is not a requirement or a creosote odor is not objectionable.

Oil-borne chemicals are organic compounds dissolved in a suitable petroleum carrier oil, and are suitable for outdoor exposure or where leaching may be a factor, or where painting is not required. Dependent on the type of oil used, they may result in relatively clean surfaces. While there is a slight odor from such treatment, it is usually not objectionable.

Water-borne inorganic salts are dissolved in water or aqua ammonia, which evaporates after treatment and leaves the chemicals in the wood. The strength of solutions varies to provide net retention of dry salt required. These salts are suitable where clean and odorless surfaces are required. The surfaces are paintable after proper seasoning.

When treating before gluing is required, water-borne salts, or oil-borne chemicals in mineral spirits, or AWPA P9 volatile solvent are recommended. When treatment before gluing is not required or desired, creosote, creosote solutions, or oil-borne chemicals are recommended.

8-19. Resistance of Wood to Chemical Attack. Wood is superior to many building materials in resistance to mild acids, particularly at ordinary temperatures. It has excellent resistance to most organic acids, notably acetic. However, wood is seldom used in contact with solutions that are more than weakly alkaline. Oxidizing chemicals and solutions of iron salts, in combination with damp conditions, should be avoided.

Wood is composed of roughly 50 to 70% cellulose, 25 to 30% lignin, and 5% extractives with less than 2% protein. Acids such as acetic, formic, lactic, and boric do not ionize sufficiently at room temperature to attack cellulose, and thus do not harm wood.

Table 8-12. Recommended Minimum Retentions of Preservatives, Lb per Cu Ft*

Preservatives	Sawn and laminated timbers		Laminations		Sawn and laminated timbers		Laminations	
	Western woods†	Southern pine	Western woods†	Southern pine	Western woods†	Southern pine	Western woods†	Southern pine
	Coastal waters				General use			
Creosote or creosote solutions:								
Coal-tar creosote	12	20	15	25	8‡	8	10	10
Creosote–coal-tar solution	Not recommended	20	Not recommended	25	8‡	8	10	10
Creosote–petroleum solution	Not recommended				8‡	8	10	10
Oil-borne chemicals:								
Pentachlorophenol, (5% in specified petroleum oil)	Not recommended				0.4§	0.4	0.5	0.5
Penta (water-repellent, moderate decay hazard)	Not recommended				0.2	0.2	0.25	0.25
	Ground contact				Above ground			
Water-borne inorganic salts:								
Chromated zinc arsenate (Boliden salt)	1.00	1.00	1.25	1.25	0.50	0.50	0.625	0.625
Acid copper chromate (Celcure)	1.00	1.00	1.25	1.25	0.50	0.50	0.625	0.625
Ammoniacal copper arsenite (Chemonite)	0.50	0.50	0.625	0.625	0.30	0.30	0.375	0.375
Chromated zinc chloride	1.00	1.00	1.25	1.25	0.75	0.75	0.94	0.94
Copperized chromated zinc chloride	1.00	1.00	1.25	1.25	0.75	0.75	0.94	0.94
Chromated copper arsenite (Greensalt, Erdalith)	0.75	0.75	0.94	0.94	0.35	0.35	0.44	0.44
Fluor chrome arsenate phenol (Tanalith, Wolman salt)	0.50	0.50	0.625	0.625	0.35	0.35	0.44	0.44
Fluor chrome arsenate phenol (Osmossar, Osmosalt)	0.50	0.50	0.625	0.625	0.35	0.35	0.44	0.44

* AITC 109 Treating Standard for Structural Timber Framing, American Institute of Timber Construction.

† Douglas fir, western hemlock, western larch.

‡ 10 lb for timber less than 5 in. thick.

§ 0.5 lb for timber less than 5 in. thick.

When the pH of aqueous solutions of weak acids is 2 or more, the rate of hydrolysis of cellulose is small and is dependent on the temperature. A rough approximation of this temperature effect is that, for every 20°F increase, the rate of hydrolysis doubles. Acids with pH values above 2, or bases with pH below 10, have little weakening effect on wood at room temperature, if the duration of exposure is moderate.

8-20. Designing for Fire Safety. Maximum protection of the occupants of a building and of the property itself can be achieved by taking advantage of the fire-endurance properties of wood in large cross sections, and by close attention to details that make a building fire-safe. Building materials alone, building features alone, or detection and fire-extinguishing equipment alone cannot provide maximum safety from fire in buildings. A proper combination of these three will provide the necessary degree of protection for the occupants and the property.

The following should be investigated:

Degree of protection needed, as dictated by the occupancy or operations taking place in the building.

Number, size, type (such as direct to the outside), accessibility of exitways (particularly stairways), and their distance from each other.

Installation of automatic alarm and sprinkler systems.

Separation of areas in which hazardous processes or operations take place, such as boiler rooms and workshops.

Enclosure of stairwells and use of self-closing fire doors.

Fire stopping and elimination, or proper protection, of concealed spaces.

Interior finishes to assure surfaces that will not spread flame at hazardous rates.

Roof venting equipment or provision of draft curtains where walls might interfere with production operations.

When exposed to fire, wood forms a self-insulating surface layer of char, which provides its own fire protection. Even though the surface chars, the undamaged wood beneath retains its strength and will support loads in accordance with the capacity of the uncharred section. Heavy-timber members have often retained their structural integrity through long periods of fire exposure and remained serviceable after refinishing of the charred surfaces. This fire endurance and excellent performance of heavy timber are attributable to the size of the wood members, and to the slow rate at which the charring penetrates.

The structural framing of a building, which is the criterion for classifying a building as combustible or noncombustible, has little to do with the hazard from fire to the building occupants. Most fires start in the building contents and create conditions that render the inside of the structure uninhabitable long before the structural framing becomes involved in the fire. Thus, whether the building is of combustible or noncombustible classification has little bearing on the potential hazard to the occupants. However, once the fire starts in the contents, the material of which the building is constructed can be of significant help in facilitating evacuation, fire fighting, and property protection.

The most important protection factors for occupants, firemen fighting the fire, and the property, as well as adjacent exposed property, are prompt detection of the fire, immediate alarm, and rapid extinguishment of the fire. Firemen do not fear fire fighting in buildings of heavy-timber construction as in buildings of many other types of construction. They need not fear sudden collapse without warning; they usually have adequate time, because of the slow-burning characteristics of the timber, to ventilate the building and fight the fire from within the building or on top.

With size of member of particular importance to fire endurance of wood members, building codes specify minimum dimensions for structural members, and classify buildings with wood framing as heavy-timber construction, ordinary construction, or wood-frame construction.

Heavy-timber construction is the type in which fire resistance is attained by placing limitations on the minimum size, thickness, or composition of all load-carrying wood members; by avoidance of concealed spaces under floors and roofs; by use of approved fastenings, construction details, and adhesives; and by providing the required degree of fire resistance in exterior and interior walls. (See AITC 108, Heavy Timber Construction, American Institute of Timber Construction.)

Ordinary construction has exterior masonry walls and wood-framing members of sizes smaller than heavy-timber sizes.

Wood-frame construction has wood-framed walls and structural framing of sizes smaller than heavy-timber sizes.

Dependent on the occupancy of a building or hazard of operations within it, a building of frame or ordinary construction may have its members covered with fire-resistive coverings. The interior finish on exposed surfaces of rooms, corridors, and stairways is important from the standpoint of its tendency to ignite, flame, and spread fire from one location to another. The fact that wood is combustible does not mean that it will spread flame at a hazardous rate. Most codes exclude the exposed wood surfaces of heavy-timber structural members from flame-spread requirements, because such wood is difficult to ignite and, even with an external source of heat, such as burning contents, is resistant to spread of flame.

Fire-retardant chemicals may be impregnated in wood with recommended retentions to lower the rate of surface flame spread and make the wood self-extinguishing if the external source of heat is removed. After proper surface preparation, the surface is paintable. Such treatments are accepted under several specifications, including Federal and military. They are recommended only for interior or dry-use service conditions or locations protected against leaching. These treatments are sometimes used to meet a specific flame-spread rating for interior finish, or as an alternate to noncombustible secondary members and decking meeting the requirements of Underwriters' Laboratories, Inc., NM 501 or NM 502, nonmetallic roof-deck assemblies in otherwise heavy-timber construction.

8-21. Mechanical Fastenings. Nails, spikes, screws, lags, bolts, and timber connectors, such as shear plates and split rings, are used for connections in wood construction. Joint-design data have been established by experience and tests, because determination of stress distribution in wood and metal fasteners is complicated.

Allowable loads or stresses and methods of design for bolts, connectors, and other fasteners used in one-piece sawn members also are applicable to laminated members.

Problems can arise, however, if a deep-arch base section is bolted to the shoe attached to the foundation by widely separated bolts. A decrease in wood moisture content and shrinkage will set up considerable tensile stress perpendicular to the grain, and splitting may occur. If the moisture content at erection is the same as that to be reached in service, or if the bolt holes in the shoe are slotted to permit bolt movement, the tendency to split will be reduced.

Fasteners subject to corrosion or chemical attack should be protected by painting, galvanizing, or plating. In highly corrosive atmospheres, such as in chemical plants, metal fasteners and connections should be galvanized or made of stainless steel. Consideration may be given to covering connections with hot tar or pitch. In such extreme conditions, lumber should be at or below equilibrium moisture content at fabrication to reduce shrinkage. Such shrinkage could open avenues of attack for the corrosive atmosphere.

Iron salts are frequently very acidic and show hydrolytic action on wood in the presence of free water. This accounts for softening and discoloration of wood observed around corroded nails. This action is especially pronounced in acidic woods, such as oak, and in woods containing considerable tannin and related compounds, such as redwood. It can be eliminated, however, by using zinc-coated, aluminum, or copper nails.

Hardened deformed-shank nails and spikes are made of high-carbon-steel wire and are headed, pointed, annularly or helically threaded, and heat-treated and tempered, to provide greater strength than common wire nails and spikes. But the same loads as given for common wire nails and spikes or the corresponding lengths are used with a few exceptions. The common nail and spike dimensions are shown in Table 8-13.

Nails should not be driven closer together than half their length, unless driven in prebored holes. Nor should nails be closer to an edge than one-quarter their length. When one structural member is joined to another, penetration of nails into the second or farther timber should be at least half the length of the nails. Holes for nails, when necessary to prevent splitting, should be bored with a diameter less than that of the nail. If this is done, the same allowable load as for the same-size fastener with a bored hole applies in both withdrawal and lateral resistance.

Nails or spikes should not be loaded in withdrawal from the end grain of wood. Also, nails inserted parallel to the grain should not be used to resist tensile stresses parallel to the grain.

Safe lateral loads and safe resistance to withdrawal of common wire nails are given in Table 8-14, and of spikes in Table 8-15. The tables provide a safety

Table 8-13. Nail and Spike Dimensions

Nails			Spikes		
Penny-weight	Length, in.	Wire diam, in.	Penny-weight	Length, in.	Wire diam, in.
6d	2	0.113	10d	3	0.192
8d	2½	0.131	12d	3¼	0.192
10d	3	0.148	16d	3½	0.207
12d	3¼	0.148	20d	4	0.225
16d	3½	0.162	30d	4½	0.244
20d	4	0.192	40d	5	0.263
30d	4½	0.207	50d	5½	0.283
40d	5	0.225	60d	6	0.283
50d	5½	0.244	5/16	7	0.312
60d	6	0.263	3/8	8½	0.375

Table 8-14. Strength of Wire Nails

Size of nail, pennyweight	6d	8d	10d	12d	16d	20d	30d	40d	50d	60d
Length of nail, in.	2	2½	3	3¼	3½	4	4½	5	5½	6
Safe lateral strength, lb (inserted perpendicular to the grain of the wood, penetrating 11 diameters), Douglas fir or southern pine	63	78	94	94	107	139	154	176	202	223
Safe resistance to withdrawal, lb per lin in. of penetration into the member receiving the point (inserted perpendicular to the grain of the wood):										
Douglas fir	29	34	38	38	42	49	53	58	63	68
Southern pine	42	48	55	55	60	71	76	83	90	97

factor of about 6 against failure. Joint slippage would be objectionable long before ultimate load is reached.

For lateral loads, if a nail or spike penetrates the piece receiving the point for a distance less than 11 times the nail or spike diameter, the allowable load is determined by straight-line interpolation between full load at 11 diameters, as given in the tables, and zero load for zero penetration. But the minimum penetration in the piece receiving the point must be at least 3⅔ diameters. Allowable lateral loads for nails or spikes driven into end grain are two-thirds the tabulated values. If a nail or spike is driven into unseasoned wood that will remain wet or will be loaded before seasoning, the allowable load is 75% of the tabulated values, except that for hardened, deformed-shank nails the full load may be used. Where properly designed metal side plates are used, the tabulated allowable loads may be increased 25%.

Wood Screws. The common types of wood screws have flat, oval, or round heads. The flat-head screw is commonly used if a flush surface is desired. Oval and round-headed screws are used for appearance or when countersinking is objectionable.

Wood screws should not be loaded in withdrawal from end grain. They should be inserted perpendicular to the grain by turning into predrilled holes, and should not be started or driven with a hammer. Spacings, end distances, and side distances must be such as to prevent splitting.

Table 8-15. Strength of Spikes

Size of spike, pennyweight	10d	12d	16d	20d	30d	40d	50d	60d	5/16 in.	3/8 in.
Length of spike, in.	3	3½	3½	4	4½	5	5½	6	7	8½
Safe lateral strength, lb (inserted perpendicular to the grain of the wood, penetrating 11 diameters), Douglas fir and southern pine	139	139	155	176	202	223	248	248	289	380
Safe resistance to withdrawal, lb per lin in. of penetration into the member receiving the point (inserted perpendicular to the grain of the wood):										
Douglas fir	49	49	53	58	63	68	73	73	80	96
Southern pine	71	71	76	83	90	97	104	104	115	138

Table 8-16 gives the allowable loads for lateral resistance at any angle of load to grain, when the wood screw is inserted perpendicular to the grain (into the side grain of main member) and a wood side piece is used. Embedment must be seven times the shank diameter into the member holding the point. For less penetration, reduce loads in proportion; penetration, however, must not be less than four times the shank diameter.

Table 8-16. Strength of Wood Screws

Gage of screw	6	7	8	9	10	12	14	16	18	20	24
Diam. of screw, in.	0.138	0.151	0.164	0.177	0.190	0.216	0.242	0.268	0.294	0.320	0.372
Allowable lateral load, lb (normal duration), Douglas fir or southern pine	75	90	106	124	143	185	232	284	342	406	548
Allowable withdrawal load, lb per in. of penetration (normal duration):											
Douglas fir	102	112	121	131	141	160	179	199	218	237	276
Southern pine	137	150	163	176	189	214	240	266	292	317	369

Table 8-16 also gives allowable withdrawal loads, lb per in., of penetration of the threaded portion of a screw into the member holding the point when the screw is inserted perpendicular to the grain of the wood.

When metal side plates rather than wood side pieces are used, the allowable lateral load, at any angle of load to grain, may be increased 25%. Allowable lateral loads when loads act perpendicular to the grain and the screw is inserted into end grain are two-thirds of those shown.

For Douglas fir and southern pine, the lead hole for a screw loaded in withdrawal should have a diameter of about 70% of the root diameter of the screw. For lateral resistance, the part of the hole receiving the shank should be about seven-eighths the diameter of the shank, and that for the threaded portion should be about seven-eighths the diameter of the screw at the root of the thread.

The loads given in Table 8-16 are for wood screws in seasoned lumber, for joints used indoors or in a location always dry. When joints are exposed to the weather, use 75%, and where joints are always wet, 67% of the tabulated loads. For lumber pressure-impregnated with fire-retardant chemicals, use 75% of the tabulated loads.

Lag Screws or Lag Bolts. Lag screws are commonly used because of their convenience, particularly where it would be difficult to fasten a bolt or where a nut on the surface would be objectionable. They range, usually, from about 0.2 to 1.0 in. in diameter, and from 1 to 16 in. in length. The threaded portion ranges from ¾ in. for 1- and 1¼-in.-long lag screws to half the length for all lengths greater than 10 in.

Lag screws, like wood screws, require prebored holes of the proper size. The lead hole for the shank should have the same diameter as the shank. The lead-hole diameter for the threaded portion varies with the density of the wood species. For Douglas fir and southern pine, the hole for the threaded portion should be 60 to 75% of the shank diameter. The smaller percentage applies to lag screws of smaller diameters. Lead holes slightly larger than those recommended for maximum efficiency should be used with lag screws of excessive length.

In determining withdrawal resistance, the allowable tensile strength of the lag screw at the net or root section should not be exceeded. Penetration of the threaded portion to a distance of about 7 times the shank diameter in the denser species and 10 to 12 times the shank diameter in the less dense species will develop approximately the ultimate tensile strength of a lag screw.

The resistance of a lag screw to withdrawal from end grain is about three-quarters that from side grain.

Table 8-17 gives the allowable normal-duration lateral and withdrawal loads for Douglas fir and southern pine. When lag screws are used with metal plates, the allowable lateral loads parallel to the grain are 25% higher than with wood side plates. No increase is allowed for load applied perpendicular to the grain.

Lag screws should preferably not be driven into end grain, because splitting may develop under lateral load. If lag screws are so used, however, the allowable loads should be taken as two-thirds the lateral resistance of lag screws in side grain with loads acting perpendicular to the grain.

Spacings, edge and end distances, and net section for lag-screw joints should be the same as those for joints with bolts of a diameter equal to the shank diameter of the lag screw.

For more than one lag screw the total allowable load equals the sum of the loads permitted for each lag screw, provided that spacings, end distances, and edge distances are sufficient to develop the full strength of each lag screw.

The allowable loads in Table 8-17 are for lag screws in lumber seasoned to a moisture content about equal to that which it will eventually have in service. For lumber installed unseasoned and seasoned in place, the full allowable lag-screw loads may be used for a joint having a single lag screw and loaded parallel or perpendicular to grain; or a single row of lag screws loaded parallel to grain; or multiple rows of lag screws loaded parallel to grain with separate splice plates for each row. For other types of lag-screw joints, the allowable loads are 40% of the tabulated loads. For lumber partly seasoned when fabricated, proportionate intermediate loads between 100 and 40% may be used.

For lumber pressure-impregnated with fire-retardant chemicals, allowable loads for lag screws are the same as those for unseasoned lumber. Where joints are to be exposed to weather, use 75%, and where joints are always wet, 67% of the tabulated allowable loads.

Bolts. Standard machine bolts with square heads and nuts are used extensively in wood construction. Spiral-shaped dowels are also used at times to hold two

Table 8-17. Strength of Lag Screws

Diam of lag screw, in.	1/4		5/16		3/8		7/16		1/2		9/16		5/8		3/4		7/8		1		1 1/8		1 1/4	
Allowable withdrawal load, lb per in., of penetration of the threaded part into the member holding the point (normal duration):																								
Douglas fir	232		274		313		352		389		425		460		528		593		655		716		774	
Southern pine	289		341		391		439		485		529		573		657		738		815		891		964	
	Allowable lateral load, lb per lag screw, in single shear with 1/2-in. metal side plates (Douglas fir or southern pine) for normal duration																							
Length of lag screw in main members, in.	Parallel to grain	Perpendicular to grain	Parallel to grain	Perpendicular to grain	Parallel to grain	Perpendicular to grain	Parallel to grain	Perpendicular to grain	Parallel to grain	Perpendicular to grain	Parallel to grain	Perpendicular to grain	Parallel to grain	Perpendicular to grain	Parallel to grain	Perpendicular to grain	Parallel to grain	Perpendicular to grain	Parallel to grain	Perpendicular to grain	Parallel to grain	Perpendicular to grain	Parallel to grain	Perpendicular to grain
3	210	160	265	180	320	245	370	210	415	215	455	225	490	235										
4	235	185	355	240	480	290	575	320	625	325	680	340	740	355										
5	...	...	375	255	535	325	710	405	850	440	930	460	1,005	480	1,190	525								
6	...	...	400	270	545	330	735	415	945	490	1,095	540	1,250	600	1,480	650								
7	...	...	...	...	555	340	750	425	970	505	1,210	600	1,460	700	2,030	890								
8	...	...	...	...	...	...	760	430	985	510	1,240	615	1,500	720	2,130	935	2,720	1,130						
9	...	...	...	...	...	...	...	...	990	515	1,250	620	1,510	725	2,160	950	2,880	1,200						
10	...	...	...	...	...	...	...	...	...	...	.....	...	1,540	740	2,190	965	2,960	1,230	3,710	1,485				
11	...	...	...	...	...	...	...	...	...	...	.....	...	.....	...	2,220	970	2,990	1,240	3,880	1,550				
12	...	...	...	...	...	...	...	...	...	...	.....	...	.....	...	.....	...	3,000	1,250	3,900	1,560	4,900	1,960		
13	...	...	...	...	...	...	...	...	...	...	.....	...	.....	...	.....	...	3,030	1,260	3,930	1,570	4,920	1,970		
14	...	...	...	...	...	...	...	...	...	...	.....	...	.....	...	.....	...	.....	.....	3,950	1,570	4,950	1,980	6,060	2,420
15	...	...	...	...	...	...	...	...	...	...	.....	...	.....	...	.....	...	.....	.....	3,960	1,580	4,980	1,990	6,110	2,450
16	...	...	...	...	...	...	...	...	...	...	.....	...	.....	...	.....	...	.....	.....	3,960	1,580	5,000	2,000	6,150	2,460

pieces of wood together; they are used to resist checking and splitting in railroad ties and other solid-sawn timbers.

Holes for bolts should always be prebored and should have a diameter that permits the bolt to be driven easily. Careful centering of holes in main members and splice plates is necessary.

Center-to-center distance along the grain between bolts acting parallel to the grain should be at least four times the bolt diameter. For a tension joint, distance from end of wood to center of nearest bolt should be at least seven times the bolt diameter for softwoods, and five times for hardwoods. For compression joints, the end distance should be at least four times the bolt diameter for both softwoods and hardwoods. If closer spacings are used, the loads allowed should be reduced proportionately.

Also for bolts acting parallel to the grain, the distance from center of a bolt to the edge of the wood should be at least 1½ times the bolt diameter. Usually, however, the edge distance is set at half the distance between bolt rows. In any event, the area at the critical section through the joint must be sufficient to keep unit stresses in the wood within the allowable.

The critical section is that section at right angles to the direction of the load that gives maximum stress in the member over the net area remaining after bolt holes at the section are deducted. For parallel-to-grain loads, the net area at a critical section should be at least 100% for hardwoods, and 80% for softwoods, of the total area in bearing under all the bolts in the joint.

For parallel- or perpendicular-to-grain loads, spacing between rows paralleling a member should not exceed 5 in. unless separate splice plates are used for each row.

For bolts bearing perpendicular to the grain, center-to-center spacing across the grain should be at least four times the bolt diameter if wood side plates are used. But if the design load is less than the bolt bearing capacity of the plates, the spacing may be reduced proportionately. For metal side plates, the spacing only need be sufficient to permit tightening of the nuts. Distance from end of wood to center of bolt should be at least four diameters. So should the distance between center of bolt and edge toward which the load is acting. The edge distance at the opposite edge is relatively unimportant.

A load applied to only one end of a bolt perpendicular to its axis may be taken as half the symmetrical two-end load.

Basic bolt bearing stresses for calculating allowable loads parallel and perpendicular to the grain are the same for Douglas fir and southern pine:

Parallel to grain	1,450 psi
Perpendicular to grain	320 psi

These basic stresses are for permanent loads. For short-duration loads, basic stresses may be modified as noted in Arts. 8-11 and 8-12. Allowable stresses for aligned bolts satisfying the spacing requirements previously given and with the load applied, parallel or perpendicular to grain, through metal plates to both ends of the bolt are calculated by multiplying the appropriate basic stress by a factor r (Table 8-18). For loads perpendicular to grain, an additional diameter factor v should be applied to the basic stresses (Table 8-19).

When bolts are properly spaced and aligned, the allowable load on a group of bolts may be taken as the sum of the individual load capacities.

To determine allowable bolt load parallel to grain: Select the basic bearing stress parallel to grain for the wood species. Adjust this stress for the service condition if the joint is to be used in other than a dry, inside location. Call the adjusted stress S_1. Calculate the bolt length-diameter ratio L/D. For this value, select from Table 8-18 the corresponding r factor. Multiply S_1 by r to obtain the allowable unit stress S_2, which is assumed to be uniformly distributed. Multiply S_2 by the projected area of the bolt to obtain the allowable load P_1 for one bolt when metal splice plates are used. For wood splice plates each of which is at least half the thickness of the main member, the allowable load is $0.8P_1$.

To determine allowable bolt load perpendicular to grain: Select the basic bearing stress perpendicular to grain for the wood species. Adjust it for the service condition if the joint is used in other than a dry, inside location. Call this stress S_1. Calculate the bolt length-diameter ratio L/D. For this value, select r from Table 8-18; then, pick v corresponding to the bolt diameter from Table 8-19. Multiply S_1 by r and v to obtain the allowable unit stress S_2. Finally, multiply S_2 by the projected area of the bolt to obtain the allowable load for the bolt loaded through wood or metal plates.

Table 8-18. Factor r for Adjusting Basic Stresses for Bolts

(Used in calculating allowable bearing stresses for common bolts when load is applied through metal splice plates)

Ratio of bolt length to diameter L/D	Bearing stress, % of basic stress for	
	Bolts bearing parallel to grain* when basic stress is 1,300 to 2,000 psi	Bolts bearing perpendicular to grain† when basic stress is 300 to 350 psi
1	100.0	100.0
2	100.0	100.0
3	99.0	100.0
4	92.5	100.0
5	80.0	100.0
6	67.2	100.0
7	57.6	97.3
8	50.4	88.1
9	44.8	76.7
10	40.3	67.2
11	36.6	59.3
12	33.6	52.0
13	31.0	45.9

* For wood splice plates, each of which is half the thickness of the main member, the allowable load should be taken as four-fifths that computed for metal splice plates.

† No reduction need be made when wood splice plates are used, except that the allowable load perpendicular to the grain should never exceed the allowable load parallel to the grain for any given size and quality of bolt and timber.

Table 8-19. Factor v for Adjusting Basic Stresses for Bolts

(For loads acting perpendicular to the grain and applied through metal plates)

Bolt diam, in.	*Diam factor v*
1/4	2.50
3/8	1.95
1/2	1.68
5/8	1.52
3/4	1.41
7/8	1.33
1	1.27
1 1/4	1.19
1 1/2	1.14
1 3/4	1.10
2	1.07
3 or over	1.00

Timber Connectors. These are metal devices used with bolts for producing joints with fewer bolts without reduction in strength. Several types of connectors are available. Usually, they are either steel rings that are placed in grooves in adjoining members to prevent relative movement or metal plates embedded in the faces of adjoining timbers. The bolts are used with these connectors to prevent the timbers from separating. The load is transmitted across the joint through the connectors.

Split rings are the most efficient device for joining wood to wood. They are placed in circular grooves cut by a hand tool in the contact surfaces. About half the depth of each ring is in each of the two members in contact. A bolt hole is drilled through the center of the core encircled by the groove.

A split ring has a tongue and groove split to permit simultaneous bearing of the inner surface of the ring against the core, and the outer surface of the ring against the outer wall of the groove. The ring is beveled for ease of assembly. Rings are manufactured in 2½- and 4-in. diam.

Shear plates are intended for wood-to-steel connections. But when used in pairs, they may be used for wood-to-wood connections, replacing split rings. Set with one plate in each member at the contact surface, they enable the members to slide easily into position during fabrication of the joint, thus reducing the labor needed to make the connection. Shear plates are placed in precut daps and are completely embedded in the timber, flush with the surface. As with split rings, the role of the bolt through each plate is to prevent the components of the joint from separating; loads are transmitted across the joint through the plates. They come in 2⅝- and 4-in. diam.

Shear plates are useful in demountable structures. They may be installed in the members immediately after fabrication and held in position by nails.

Toothed rings and spike grids sometimes are used for special applications. But split rings and shear plates are the prime connectors for timber construction.

Safe long-time working loads for split rings and shear plates have been derived from tests of full-scale joints and other basic information on strength and behavior of timber. Particular consideration has been given to the effects of short-time loading and allowances for variability in timber quality.

Working loads acting parallel to the grain of the wood were derived by applying a reduction factor to the ultimate loads observed in tests. A reduction factor of 4 kept working loads for split rings and shear plates within five-eighths of the test loads at the proportional limit.

Table 8-20. Allowable Loads for One Split Ring and Bolt in Single Shear
(Normal loading)

Split-ring diam, in.	Bolt diam, in.	No. of faces of piece with connectors on same bolt	Thickness (net) of lumber, in.	Loaded parallel to grain: Min edge distance, in.	Loaded parallel to grain: Allowable load per connector and bolt, lb	Loaded perpendicular to grain: Edge distance, in.: Unloaded edge, min	Loaded perpendicular to grain: Edge distance, in.: Loaded edge	Loaded perpendicular to grain: Allowable load per connector and bolt, lb
2½	½	1	1 min	1¾	2,270	1¾	1¾ min	1,350
							2¾ or more	1,620
			1½ or more	1¾	2,730	1¾	1¾ min	1,620
							2¾ or more	1,940
		2	1½ min	1¾	2,100	1¾	1¾ min	1,250
							2¾ or more	1,500
			2 or more	1¾	2,730	1¾	1¾ min	1,620
							2¾ or more	1,940
4	¾	1	1 min	2¾	3,510	2¾	2¾ min	2,030
							3¾ or more	2,440
			1½ or more	2¾	5,160	2¾	2¾ min	2,990
							3¾ or more	3,590
		2	1½ min	2¾	3,520	2¾	2¾ min	2,040
							3¾ or more	2,450
			2	2¾	4,250	2¾	2¾ min	2,470
							3¾ or more	2,960
			2½	2¾	5,000	2¾	2¾ min	2,900
							3¾ or more	3,480
			3 or more	2¾	5,260	2¾	2¾ min	3,050
							3¾ or more	3,660

Table 8-20 gives safe working loads for split rings and bolts. Table 8-21 gives safe working loads for shear plates and bolts.

In design for wind forces acting alone or with dead and live loads, the safe working loads for connectors may be increased by one-third. But the number and size of connectors should not be less than those required for dead and live loads alone.

Impact may be disregarded up to the following percentages of the static effect of the live load producing the impact:

Connector	*Impact allowance, %*
Split ring, any size, bearing in any direction	100
Shear plate, any size, bearing parallel to grain	66⅔
Shear plate, any size, bearing perpendicular to grain	100

One-half of any impact load that remains after disregarding the percentages indicated should be included with the dead and live loads in designing the joint.

The above procedures for increasing the allowable loads on connectors for suddenly applied and short-duration loads do not reduce the actual factor of safety of the joint. They are recommended because of the favorable behavior of wood under such forces. Different values are allowed for different types and sizes of connectors

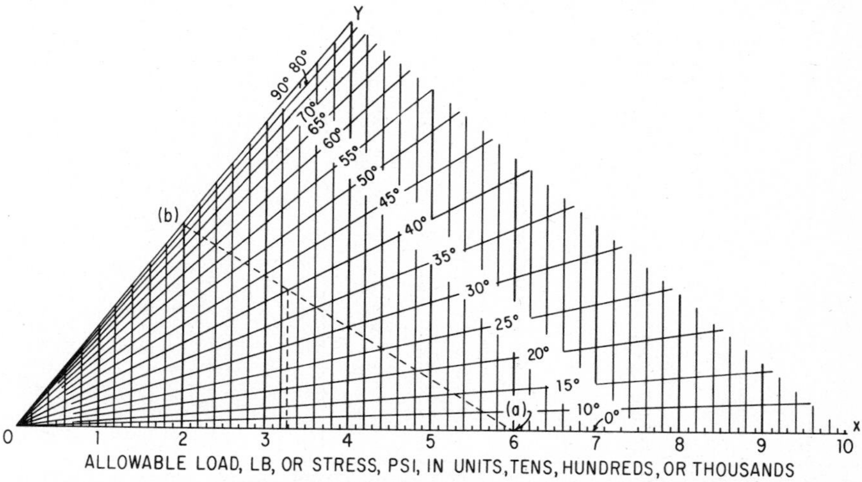

Fig. 8-4. Scholten nomograph for determining allowable bolt load on or bearing stress in wood at various angles to the grain.

and directions of bearing because of variations in the extent to which distortion of the metal, as well as the strength of the wood, affects the ultimate strength of the joint.

The tabulated loads apply to seasoned timbers used where they will remain dry. If the timbers will be more or less continuously damp or wet in use, two-thirds of the tabulated values should be used.

Safe working loads for split rings and shear plates for angles between 0° (parallel to grain) and 90° (perpendicular to grain) may be obtained from the Scholten nomograph (Fig. 8-4). This determines the bearing strength of wood at various angles to the grain. The chart is a graphical solution of the Hankinson formula:

$$N = \frac{PQ}{P \sin^2 \theta + Q \cos^2 \theta} \tag{8-10}$$

where N, P, and Q are, respectively, the allowable load, lb, or stress, psi, at inclination θ with the direction of grain, parallel to grain, and perpendicular to grain.

For example, given P = 6,000 lb and Q = 2,000 lb, find the allowable load at an angle of 40° with the grain. Connect 6 on the 0° line (point a) with the intersection of a vertical line through 2 and the 90° line (point b). N = 3,280 lb is found directly below the intersections of line ab with the 40° line.

Safe working loads are based on the assumption that the wood at the joint is clear and relatively free from checks, shakes, and splits. If knots are present

Table 8-21. Allowable Loads for One Shear Plate and Bolt in Single Shear
(Normal load duration and wood side plates*)

Shear-plate diam, in.	Bolt diam, in.	No. of faces of piece with connectors on same bolt	Thickness (net) of lumber, in.	Loaded parallel to grain: Min edge distance, in.	Loaded parallel to grain: Allowable load per connector and bolt, lb	Loaded perpendicular to grain: Edge distance, in.: Unloaded edge, min	Loaded perpendicular to grain: Edge distance, in.: Loaded edge	Loaded perpendicular to grain: Allowable load per connector and bolt, lb
2⅝	¾	1	1½ min	1¾	2,760	1¾	1¾ min	1,550
							2¾ or more	1,860
		2	1½ min	1¾	2,080	1¾	1¾ min	1,210
							2¾ or more	1,450
			2	1¾	2,730	1¾	1¾ min	1,590
							2¾ or more	1,910
			2½ or more	1¾	2,860	1¾	1¾ min	1,660
							2¾ or more	1,990
4	¾ or ⅞	1	1½ min	2¾	3,750	2¾	2¾ min	2,180
							3¾ or more	2,620
			1¾ or more	2¾	4,360	2¾	2¾ min	2,530
							3¾ or more	3,040
		2	1¾ min	2¾	2,910	2¾	2¾ min	1,680
							3¾ or more	2,020
			2	2¾	3,240	2¾	2¾ min	1,880
							3¾ or more	2,260
			2½	2¾	3,690	2¾	2¾ min	2,140
							3¾ or more	2,550
			3	2¾	4,140	2¾	2¾ min	2,400
							3¾ or more	2,880
			3½ or more	2¾	4,320	2¾	2¾ min	2,500
							3¾ or more	3,000

* Tabulated loads also apply to metal side plates, except that for 4-in. shear plates the parallel-to-grain (not perpendicular) loads for wood side plates shall be increased 11%; but loads shall not exceed: for all loadings, except wind, 2,900 lb for 2⅝-in. shear plates; 4,970 and 6,760 lb for 4-in. shear plates with ¾- and ⅞-in. bolts, respectively; or for wind loading, 3,870, 6,630, and 9,020 lb, respectively. If bolt threads are in bearing on the shear plate, reduce the preceding values by one-ninth.

Metal side plates, when used, shall be designed in accordance with accepted metal practices. For A36 steel, the following unit stresses, psi, are suggested for all loadings except wind: net section in tension, 22,000; shear, 14,500; bearing, 33,000. For wind, these values may be increased one-third. If bolt threads are in bearing, reduce the preceding shear and bearing values by one-ninth.

in the longitudinal projection of the net section within a distance from the critical section of half the diameter of the connector, the area of the knot should be subtracted from the area of the critical section. It is assumed that slope of the grain at the joint does not exceed 1 in 10.

The stress, whether tension or compression, in the net area, the area remaining at the critical section after subtracting the projected area of the connectors and the bolt from the full cross-sectional area of the member, should not exceed the safe stress of clear wood in compression parallel to the grain.

Tables 8-20 and 8-21 list the least thickness of member that should be used with the various sizes of connectors. The loads listed for the greatest thickness of member with each type and size of connector unit are the maximums to be used for all thicker material. Loads for members with thicknesses between those listed may be obtained by interpolation.

For connectors in a multiple joint, there are several rules to follow based on observed behavior of single-connector joints tested with variables that simulate those in a multiple joint:

When two or more connectors in the same face of a member are in line at right angles to the grain of the member and are bearing parallel to the grain, the clear distance between the connectors should not be less than ½ in.

When two or more connectors act perpendicular to the grain and are on a line at right angles to the length of the member, rules for width of member and edge distances used with one connector are applicable to width and edge distance for multiple connectors. The clear distance between the connectors should equal the clear distance between the edge of the timber toward which the load is acting and the connector nearest this edge.

Table 8-22. Anchor-bolt Load Values for Steel Base Shoes

Bolt diam, in.	Allowable shear load,* lb	Embedment in concrete, in.
½	1,500	4
⅝	2,200	4
¾	3,000	5
⅞	3,500	6
1	4,000	8
1¼	5,000	8
1½	7,000	9
1¾	9,000	9
2	11,000	10
2¼	13,500	10
2½	16,000	12
2¾	19,000	12
3	22,000	14

* Shear load based on 3,000-psi concrete and bolt yield strength of 50,000 psi.

In a joint with two or more connectors on a line parallel to the grain and with the load acting perpendicular to the grain, the clear spacing between adjacent connectors should not be less than 1 in. The total load used should equal the full load of one connector plus one-third this amount for each additional connector. In a joint of this type, somewhat more favorable results are obtained in tests if the connectors are staggered so that they do not act along the same line with respect to the grain of the transverse member.

Placement of connectors in joints with members at right angles to each other is subject to the limitations of either member. Since rules for alignment, spacing, and edge and end distance of connectors for all conceivable directions of applied load would be complicated, designers must rely on a sense of proportion and adequacy in applying the above rules to conditions of loading outside the specific limitations mentioned.

Anchor Bolts. To attach columns or arch bases to concrete foundations, anchor bolts are embedded in the concrete, with sufficient projection to permit placement of angles or shoes bolted to the wood. Sometimes, instead of anchor bolts, steel straps are embedded in the concrete with a portion projecting above for bolt attachment to the wood members. Table 8-22 lists allowable shear loads for anchor bolts and steel base shoes.

Washers. Bolt heads and nuts bearing on wood require metal washers to protect the wood and to distribute the pressure across the surface of the wood. Washers may be cast, malleable, cut, round-plate, or square-plate. When subjected to salt air or salt water, they should be galvanized or given some type of effective coating. Ordinarily, washers are dipped in red lead and oil prior to installation.

Setscrews should never be used against a wood surface. It may be possible, with the aid of proper washers, to spread the load of the setscrew over sufficient surface area of the wood that the compression strength perpendicular to grain is not exceeded.

Table 8-23 gives washer sizes capable of developing the capacity of A307 bolts.

Table 8-23. Allowable Loads for Bolts and Washers

Rod or bolt		Plate washers†		Cut washers			
Diam, in.	Tensile capacity,* lb	Side of square, in.	Thickness, in.	Outside diam, in.	Hole diam, in.	Thickness, in.	Max load,‡ lb
3/8	1,550	1 3/4	3/16	1	7/16	5/64	290
7/16	2,100	2 1/8	1/4	1 1/4	1/2	5/64	460
1/2	2,750	2 3/8	1/4	1 3/8	9/16	7/64	550
5/8	4,300	3	5/16	1 3/4	11/16	9/64	910
3/4	6,190	3 3/4	3/8	2	13/16	5/32	1,100
7/8	8,420	4 3/8	1/2	2 1/4	13/16	11/64	1,100
1	11,000	5	1/2	2 1/2	1 1/16	11/64	1,800
1 1/8	13,920	5 5/8	5/8	2 3/4	1 1/4	11/64	2,100
1 1/4	17,180	6 3/8	3/4	3	1 3/8	11/64	2,500
1 3/8	20,800	7	3/4	3 1/4	1 1/2	3/16	2,900
1 1/2	24,700	7 3/4	7/8	3 1/2	1 5/8	3/16	3,400
1 5/8	29,000	8 3/8	7/8	3 3/4	1 3/4	3/16	3,900
1 3/4	33,700	9	1	4	1 7/8	3/16	4,400
1 7/8	38,700	9 3/4	1	4 1/4	2	3/16	5,000
2	44,000	10 1/4	1 1/8	4 1/2	2 1/8	3/16	5,500
2 1/8	49,700	11	1 1/8				
2 1/4	55,700	11 3/4	1 1/4	4 3/4	2 3/8	7/32	6,000
2 3/8	62,000	12 1/2	1 1/4				
2 1/2	68,700	13	1 3/8	5	2 5/8	15/64	6,500
2 5/8	75,800	13 3/4	1 1/2				
2 3/4	83,200	14 1/2	1 1/2	5 1/4	2 7/8	1/4	6,800
2 7/8	90,900	15 1/4	1 5/8				
3	99,000	15 3/4	1 5/8	5 1/2	3 1/8	9/32	7,200

* Based on ASTM A36 steel and A307 bolts.

† Size required to develop capacity of rod or bolt in accordance with note ‡.

‡ Based on allowable strength of wood in compression perpendicular to grain of 450 psi.

Tie Rods. To resist the horizontal thrust of arches not buttressed, tie rods are required. The tie rods may be installed at ceiling height or below the floor. Table 8-24 gives the allowable loads on bars used as tie rods.

Hangers. Standard and special hangers are used extensively in timber construction. Stock hangers are available from a number of manufacturers, but most hangers are of special design. Where appearance is of prime importance, concealed hangers are frequently selected.

(National Design Specification for Stress-grade Lumber and Its Fastenings, National Forest Products Association, 1619 Massachusetts Ave., N.W., Washington, D.C. 20026; "Design Manual for TECO Timber Connector Construction," Timber Engineering Co., 1619 Massachusetts Ave., N.W., Washington, D.C. 20036; American Institute of Timber Construction "Timber Construction Manual," John Wiley &

Sons, Inc., New York; "Western Woods Use Book," Western Wood Products Association, 1500 Yeon Building, Portland, Ore. 97204.)

8-22. Glued Fastenings. Glued joints are generally between two pieces of wood where the grain directions are parallel (as between the laminations of a beam or arch). Or such joints may be between solid-sawn or laminated timber and plywood, where the face grain of the plywood is either parallel or at right angles to the grain direction of the timber.

It is only in special cases that lumber may be glued with the grain direction of adjacent pieces at an angle. When the angle is large, dimensional changes from changes in wood moisture content set up large stresses in the glued joint. Consequently, the strength of the joint may be considerably reduced over a period

Table 8-24. Allowable Tension on Round and Square Upset Bars

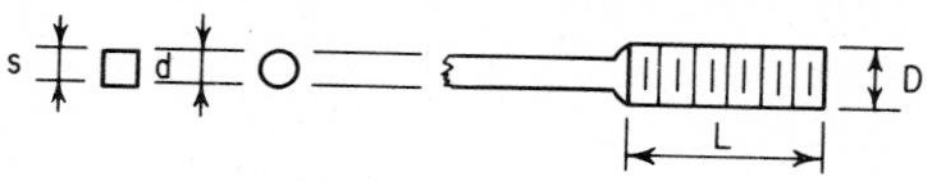

UNC and 4UN Class-2A Thread

Diam d or side s, in.	Round bars			Square bars		
	Capacity,* lb	Upset		Capacity,* lb	Upset	
		Diam D, in.	Length L, in.		Diam D, in.	Length L, in.
3/4	9,700	1	4	12,400	1 1/8	4
7/8	13,200	1 1/8	4	16,900	1 1/4	4
1	17,300	1 3/8	4	22,000	1 1/2	4
1 1/8	21,800	1 1/2	4	27,800	1 3/4	4
1 1/4	27,000	1 3/4	4	34,400	2	4 1/2
1 3/8	32,700	1 3/4	4	41,600	2	4 1/2
1 1/2	38,900	2	4 1/2	49,500	2 1/4	5
1 5/8	45,600	2 1/4	5	58,100	2 1/2	5 1/2
1 3/4	53,000	2 1/4	5	67,400	2 1/2	5 1/2
1 7/8	60,800	2 1/8	5 1/2	77,300	2 3/4	5 1/2
2	69,100	2 1/2	5 1/2	88,000	2 3/4	5 1/2
2 1/8	78,000	2 3/4	5 1/2	99,300	3	6
2 1/4	87,500	2 3/4	5 1/2	111,400	3 1/4	6 1/2
2 3/8	97,500	3	6	124,100	3 1/4	6 1/2
2 1/2	108,000	3 1/4	6 1/2	137,500	3 1/2	7
2 5/8	119,100	3 1/4	6 1/2	151,600	3 3/4	7

* Based on ASTM A36 steel.

of time. Exact data are not available, however, on the magnitude of this expected strength reduction.

In joints connected with plywood gusset plates, this shrinkage differential is minimized, because plywood swells and shrinks much less than does solid wood.

Glued joints can be made between end-grain surfaces. But they are seldom strong enough to meet the requirements of even ordinary service. Seldom is it possible to develop more than 25% of the tensile strength of the wood in such butt joints. It is for this reason that plane sloping scarfs of relatively flat slope (Fig. 8-2), or finger joints with thin tips and flat slope on the sides of the individual fingers (Fig. 8-3), are used to develop a high proportion of the strength of the wood.

Joints of end grain to side grain are also difficult to glue properly. When subjected to severe stresses as a result of unequal dimensional changes in the members due to changes in moisture content, joints suffer from severely reduced strength.

For the above reasons, joints between end-grain surfaces and between end-grain and side-grain surfaces should not be used if the joints are expected to carry load.

For joints made with wood of different species, the allowable shear stress for parallel-grain bonding is equal to the allowable shear stress parallel to the grain for the weaker species in the joint. This assumes uniform stress distribution in the joint. When grain direction is not parallel, the allowable shear stress on the glued area between the two pieces may be estimated from the Scholten nomograph (Fig. 8-4).

Adhesives used for fabricating structural glued-laminated timbers also are satisfactory for other glued joints. In selecting an adhesive, consideration should be given to initial and service-condition wood moisture content.

[AITC 103, Standard for Structural Glued-laminated Timber, American Institute of Timber Construction, 833 W. Hampden Ave., Englewood, Colo. 80110; Federal

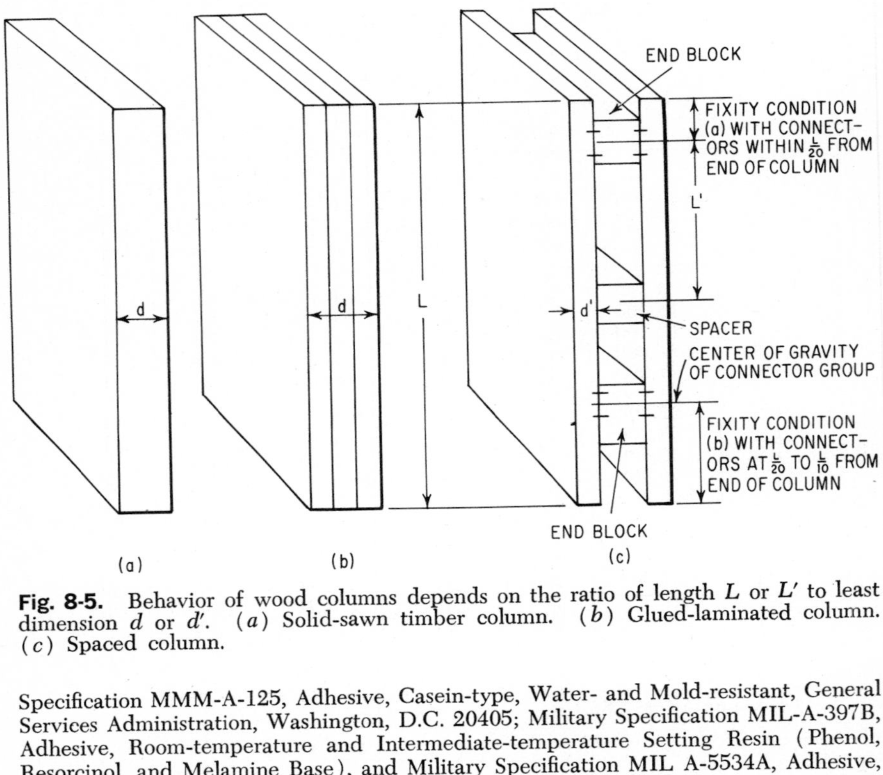

Fig. 8-5. Behavior of wood columns depends on the ratio of length L or L' to least dimension d or d'. (*a*) Solid-sawn timber column. (*b*) Glued-laminated column. (*c*) Spaced column.

Specification MMM-A-125, Adhesive, Casein-type, Water- and Mold-resistant, General Services Administration, Washington, D.C. 20405; Military Specification MIL-A-397B, Adhesive, Room-temperature and Intermediate-temperature Setting Resin (Phenol, Resorcinol, and Melamine Base), and Military Specification MIL A-5534A, Adhesive, High-temperature Setting Resin (Phenol, Melamine, and Resorcinol Base), U.S. Naval Supply Depot, Philadelphia, Pa. 19120.]

8-23. Wood Columns. Wood compression members may be a solid piece of timber (Fig. 8-5*a*) or laminated (Fig. 8-5*b*) or built up of spaced members (Fig. 8-5*c*). The latter comprise two or more wood compression members with parallel longitudinal axes. The members are separated at ends and midpoints by blocking, and joined to the end blocking with connectors with adequate shear resistance.

Columns with a ratio of unsupported length L, in., to least dimension d, in., less than 11 fail by crushing. The allowable concentric load for such members equals the cross-sectional area, sq in., times F_c, the allowable compression stress, psi, parallel to grain for the species, adjusted for service conditions and duration of load.

When the slenderness ratio L/d exceeds 11, wood columns generally fail by buckling. In that case, the allowable stress is determined from formulas that yield values less than F_c. The computed allowable stress must be adjusted for duration of load.

For rectangular, solid-sawn or glued-laminated columns, the allowable unit stress may not exceed F_c or

$$F'_c = \frac{0.30E}{(L/d)^2} \tag{8-11}$$

where E = modulus of elasticity, psi, adjusted for duration of load. In no case may L/d exceed 50. The formula was derived for pin-end conditions, but may also be used for square-cut ends.

For round columns, the allowable stress may not exceed that for a square column of the same cross-sectional area or

$$F'_c = \frac{3.619E}{(L/r)^2} \tag{8-12}$$

where r = radius of gyration of cross section, in.

For tapered columns, d may be taken as the sum of the least dimension plus one-third the difference between this and the maximum thickness parallel to this dimension.

For spaced columns, the allowable load is the smallest computed with F_c, F'_c calculated from Eq. (8-11) with the over-all dimensions of the column, and F''_c determined from Eq. (8-13*a*) or (8-13*b*) with the dimensions of the individual members. For the individual members, L/d' may not exceed 80, nor may L'/d' exceed 40, where L' is the spacing of the blocking and d' the least dimension. End blocks are required to insure spaced-column action when L/d' exceeds $\sqrt{0.30E/F_c}$. For smaller L/d', the individual members should be designed for allowable stresses computed from Eq. (8-11). Allowable stresses for spaced-column action depend on the fixity condition (Fig. 8-5*c*). For fixity condition (*a*),

$$F''_c = \frac{0.75E}{(L/d')^2} \tag{8-13a}$$

For fixity condition (*b*),

$$F''_c = \frac{0.90E}{(L/d')^2} \tag{8-13b}$$

Each member of a spaced column should be designed separately on the basis of its L/d'. The allowable load on each equals its cross-sectional area, sq in., times its allowable stress, psi, adjusted for duration of load. The sum of the allowable loads on the individual members should equal or exceed the total load on the column.

When a single spacer block is placed in the middle tenth of a spaced column, connectors are not required. They should be used for multiple spaced blocks. The distance between two adjacent blocks may not exceed half the distance between centers of connectors in the end blocks in opposite ends of the column.

For all types of columns, the distance between adequate bracing, including beams and struts, should be used as L in determining L/d. The largest L/d for the column or any component, whether it be computed for a major or minor axis, should be used in calculating the allowable unit stress.

For combined axial and bending stress, P/A divided by the allowable compression stress plus M/S divided by the allowable bending stress must not exceed unity. P is the axial load, lb; A the area of the column, sq in.; M the bending moment, in.-lb; and S the section modulus, in.^{3} Bending caused by transverse loads, wind loads, or eccentric loads, or any combination of these, should be taken into account.

The critical section of columns supporting trusses frequently exists at the connection of knee brace to column. Where no knee brace is used, or the column supports a beam, the critical section for moment usually occurs at the bottom of truss or beam. Then a rigid connection must be provided to resist moment, or adequate diagonal bracing must be provided to carry wind loads into a support.

Figure 8-6 shows typical column base anchorages, and Fig. 8-7 typical beam-to-column connections (AITC 104, Typical Construction Details, American Institute of Timber Construction).

(American Institute of Timber Construction, "Timber Construction Manual," John Wiley & Sons, Inc., New York; "Western Woods Use Book," Western Wood Products Association, 1500 Yeon Building, Portland, Ore. 97204; "Wood Structural Design Data," National Forest Products Association, 1619 Massachusetts Ave, Washington, D.C. 20036.)

8-24. Design of Timber Joists. Joists are relatively narrow beams, usually spaced 12 to 24 in. c to c. They generally are topped with sheathing and braced

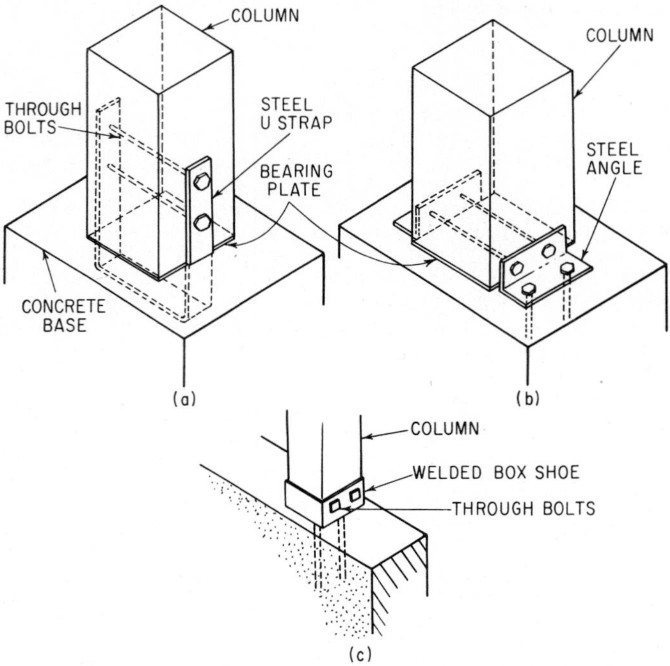

Fig. 8-6. Typical anchorages of wood column to base. (*a*) Wood column anchored to concrete base with U strap. (*b*) Anchorage with steel angles. (*c*) With a welded box shoe.

with diaphragms or cross bridging at intervals up to 10 ft. For joist spacings of 16 to 24 in. c to c, 1-in. sheathing usually is required. For spacings over 24 in., 2 in. or more of wood decking is necessary.

Standard beam formulas for bending, shear, and deflection may be used to determine joist sizes. Connections shown in Figs. 8-7 and 8-8 may be used for joists.

See also Art. 8-25.

8-25. Design of Timber Beams. Standard beam formulas for bending, shear, and deflection may be used to determine beam sizes. Ordinarily, deflection governs design; but for short, heavily loaded beams, shear is likely to control.

Figure 8-9 shows the types of beams commonly produced in timber. Straight and single- and double-tapered straight beams can be furnished solid-sawn or glued-laminated. The curved members can be furnished only glued-laminated. Beam names describe the top and bottom surfaces of the beam. The first part describes the top surface, the word following the hyphen the bottom. Sawn surfaces on the tension side of a beam should be avoided.

Table 8-25 gives the load-carrying capacity for various cross-sectional sizes of glued-laminated, simply supported beams.

Example. With the following data, design a straight, glued-laminated beam, simply supported and uniformly loaded: span, 28 ft; spacing, 9 ft c to c; live load, 30 psf; dead load, 5 psf for deck and 7.5 psf for roofing. Allowable bending stress of combination grade is 2,400 psi, with modulus of elasticity $E = 1{,}800{,}000$ psi. Deflection limitation is $L/180$, where L is the span, ft. Assume the beam is laterally supported by the deck throughout its length.

With a 15% increase for short-duration loading, the allowable bending stress F_b becomes 2,760 psi, and the allowable horizontal shear F_v, 230 psi.

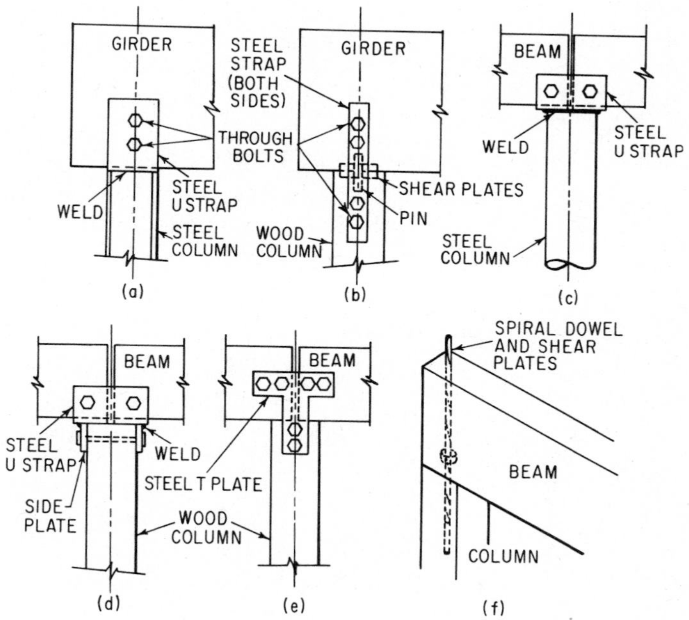

Fig. 8-7. Typical wood beam and girder connections to columns. (*a*) Girder to steel column. (*b*) Girder to wood column. (*c*) Beam to pipe column. (*d*) Beam to wood column, with steel strap welded to steel side plates. (*e*) Beam to wood column, with a T plate. (*f*) With spiral dowel and shear plates.

Assume the beam will weigh 22.5 lb per lin ft, averaging 2.5 psf. Then, the total uniform load comes to 45 psf. So the beam carries $w = 45 \times 9 = 405$ lb per lin ft.

The end shear $V = wL/2$ and the maximum shearing stress $= 3V/2 = 3wL/4$. Hence, the required area, sq in., for horizontal shear is

$$A = \frac{3wL}{4F_v} = \frac{wL}{306.7} = \frac{405 \times 28}{306.7} = 37.0 \text{ sq in.}$$

The required section modulus, in.^{3}, is

$$S = \frac{1.5wL^2}{F_b} = \frac{1.5 \times 405 \times 28^2}{2{,}760} = 172.6 \text{ in.}^3$$

If $D = 180$, the reciprocal of the deflection limitation, then the deflection equals

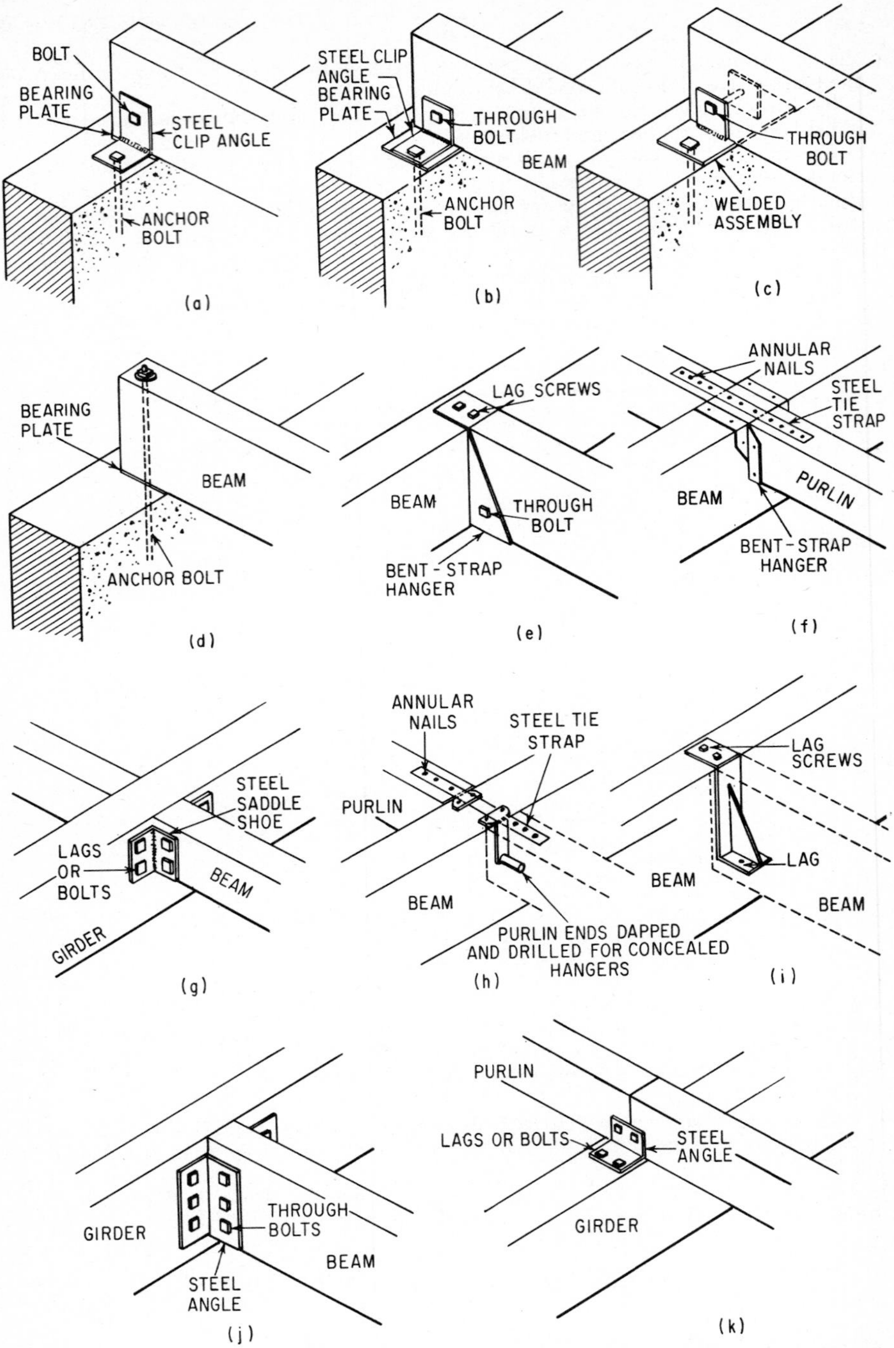

Fig. 8-8. Beam connections. (*a*) and (*b*) Wood beam anchored on wall with steel angles. (*c*) With welded assembly. (*d*) Beam anchored directly with bolt. (*e*) Beam supported on girder with bent-strap hanger. (*f*) Similar support for purlins. (*g*) Saddle connects beam to girder (suitable for one-sided connection). (*h*) and (*i*) Connections with concealed hangers. (*j*) and (*k*) Connections with steel angles.

$5 \times 1{,}728wL^4/384EI \leq 12L/D$, where I is the moment of inertia of the beam cross section, in.[4] Hence, to control deflection, the moment of inertia must be at least

$$I = \frac{1.875DwL^3}{E} = \frac{1.875 \times 180 \times 405 \times 28^3}{1{,}800{,}000} = 1{,}688 \text{ in.}^4$$

Assume that the beam will be fabricated with 1½-in. laminations. From Table 8-4, the most economical section satisfying all three criteria is 5⅛ × 16½, with $A = 84.6$, $S = 232.5$, and $I = 1{,}918.5$. But it has a size factor of 0.97%. So the

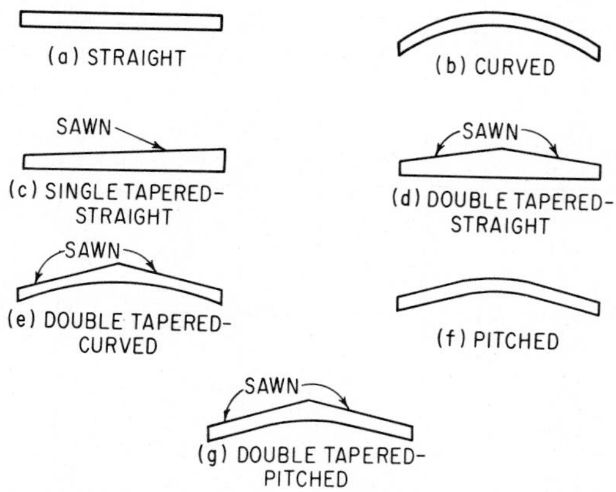

Fig. 8-9. Types of glued-laminated beams.

allowable bending stress must be reduced to $2{,}760 \times 0.97 = 2{,}677$ psi. And the required section modulus must be increased accordingly to $172.6/0.97 = 178$. The selected section still is adequate.

Cantilever systems may be comprised of any of the various types and combinations of beam illustrated in Fig. 8-10. Cantilever systems permit longer spans or larger loads for a given size member than do simple-span systems, if member size is not controlled by compression perpendicular to grain at the supports or by horizontal shear. Substantial design economies can be effected by decreasing the depths of the members in the suspended portions of a cantilever system.

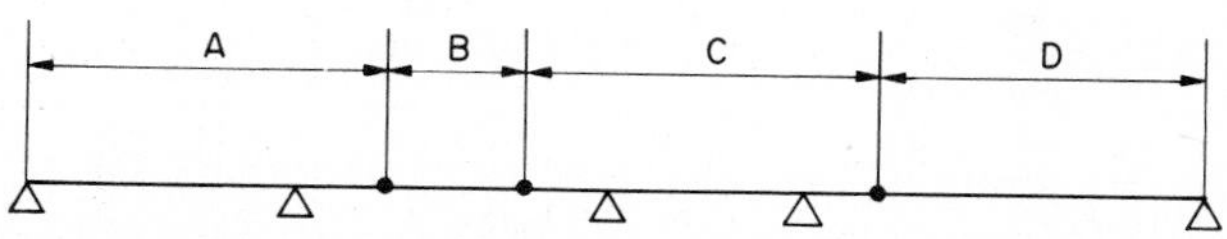

Fig. 8-10. Cantilevered-beam systems. *A* is a single cantilever. *B* is a suspended beam. *C* has a double cantilever, and *D* is a beam with one end suspended.

For economy, the negative bending moment at the supports of a cantilevered beam should be equal in magnitude to the positive moment.

Consideration must be given to deflection and camber in cantilevered multiple spans. When possible, roofs should be sloped the equivalent of ¼ in. per ft of horizontal distance between the level of drains and the high point of the roof to eliminate water pockets, or provisions should be made to insure that accumulation of water does not produce greater deflection and live loads than anticipated. Unbalanced loading conditions should be investigated for maximum bending moment, deflection, and stability.

(American Institute of Timber Construction, "Timber Construction Manual," John Wiley & Sons, Inc., New York; "Western Woods Use Book," Western Wood Products

Table 8-25. Load-carrying Capacity of Simple-span Laminated Beams*

Span, ft	Spacing, ft	Roof beam total load-carrying capacity						Floor beams total load
		30 psf	35 psf	40 psf	45 psf	50 psf	55 psf	50 psf
8	4	3⅛ × 4½	3⅛ × 4½	3⅛ × 6	3⅛ × 6	3⅛ × 6	3⅛ × 6	3⅛ × 6
	6	3⅛ × 4½	3⅛ × 4½	3⅛ × 6	3⅛ × 6	3⅛ × 6	3⅛ × 6	3⅛ × 6
	8	3⅛ × 4½	3⅛ × 4½	3⅛ × 6	3⅛ × 6	3⅛ × 6	3⅛ × 6	3⅛ × 7½
10	4	3⅛ × 4½	3⅛ × 4½	3⅛ × 6	3⅛ × 6	3⅛ × 6	3⅛ × 6	3⅛ × 7½
	6	3⅛ × 4½	3⅛ × 6	3⅛ × 6	3⅛ × 6	3⅛ × 6	3⅛ × 7½	3⅛ × 7½
	8	3⅛ × 6	3⅛ × 6	3⅛ × 7½	3⅛ × 7½	3⅛ × 7½	3⅛ × 7½	3⅛ × 9
	10	3⅛ × 6	3⅛ × 7½	3⅛ × 7½	3⅛ × 7½	3⅛ × 7½	3⅛ × 9	3⅛ × 9
12	6	3⅛ × 6	3⅛ × 6	3⅛ × 7½	3⅛ × 7½	3⅛ × 7½	3⅛ × 7½	3⅛ × 9
	8	3⅛ × 6	3⅛ × 7½	3⅛ × 9	3⅛ × 9	3⅛ × 9	3⅛ × 9	3⅛ × 10½
	10	3⅛ × 7½	3⅛ × 7½	3⅛ × 9	3⅛ × 9	3⅛ × 9	3⅛ × 10½	3⅛ × 10½
	12	3⅛ × 7½	3⅛ × 9	3⅛ × 9	3⅛ × 9	3⅛ × 10½	3⅛ × 10½	3⅛ × 12
14	8	3⅛ × 7½	3½ × 9	3⅛ × 9	3⅛ × 9	3⅛ × 10½	3⅛ × 10½	3⅛ × 12
	10	3⅛ × 9	3⅛ × 9	3⅛ × 10½	3⅛ × 10½	3⅛ × 10½	3⅛ × 12	3⅛ × 12
	12	3⅛ × 9	3⅛ × 10½	3⅛ × 10½	3⅛ × 10½	3⅛ × 12	3⅛ × 12	3⅛ × 13½
	14	3⅛ × 10½	3⅛ × 10½	3⅛ × 12	3⅛ × 12	3⅛ × 12	3⅛ × 13½	3⅛ × 13½
16	8	3⅛ × 9	3⅛ × 9	3⅛ × 10½	3⅛ × 10½	3⅛ × 12	3⅛ × 12	3⅛ × 13½
	12	3⅛ × 10½	3⅛ × 12	3⅛ × 12	3⅛ × 12	3⅛ × 13½	3⅛ × 13½	3⅛ × 15
	14	3⅛ × 12	3⅛ × 12	3⅛ × 13½	3⅛ × 13½	3⅛ × 15	3⅛ × 15	3⅛ × 15
	16	3⅛ × 12	3⅛ × 13½	3⅛ × 13½	3⅛ × 15	3⅛ × 15	3⅛ × 16½	3⅛ × 15
18	8	3⅛ × 9	3⅛ × 10½	3⅛ × 12	3⅛ × 12	3⅛ × 12	3⅛ × 13½	3⅛ × 15
	12	3⅛ × 12	3⅛ × 12	3⅛ × 13½	3⅛ × 13½	3⅛ × 15	3⅛ × 16½	3⅛ × 16½
	16	3⅛ × 13½	3⅛ × 15	3⅛ × 15	3⅛ × 16½	5⅛ × 13½	5⅛ × 13½	5⅛ × 15
	18	3⅛ × 15	3⅛ × 15	3⅛ × 16½	3⅛ × 18	5⅛ × 15	5⅛ × 15	5⅛ × 15
20	8	3⅛ × 12	3⅛ × 12	3⅛ × 13½	3⅛ × 13½	3⅛ × 13½	3⅛ × 15	3⅛ × 16½
	12	3⅛ × 13½	3⅛ × 13½	3⅛ × 15	3⅛ × 16½	3⅛ × 16½	5⅛ × 13½	5⅛ × 15
	16	3⅛ × 15	3⅛ × 16½	3⅛ × 18	3⅛ × 18	5⅛ × 15	5⅛ × 16½	5⅛ × 18
	18	3⅛ × 16½	3⅛ × 16½	3⅛ × 18	5⅛ × 15	5⅛ × 16½	5⅛ × 16½	5⅛ × 18
22	8	3⅛ × 13½	3⅛ × 13½	3⅛ × 13½	3⅛ × 15	3⅛ × 15	3⅛ × 16½	5⅛ × 15
	12	3⅛ × 15	3⅛ × 15	3⅛ × 16½	3⅛ × 18	3⅛ × 18	5⅛ × 15	5⅛ × 16½
	16	3⅛ × 16½	3⅛ × 18	5⅛ × 15	5⅛ × 16½	5⅛ × 16½	5⅛ × 18	5⅛ × 19½
	18	3⅛ × 18	5⅛ × 15	5⅛ × 16½	5⅛ × 16½	5⅛ × 18	5⅛ × 18	5⅛ × 19½
24	8	3⅛ × 13½	3⅛ × 15	3⅛ × 15	3⅛ × 16½	3⅛ × 16½	3⅛ × 18	5⅛ × 16½
	12	3⅛ × 16½	3⅛ × 16½	3⅛ × 18	5⅛ × 15	5⅛ × 16½	5⅛ × 16½	5⅛ × 18
	16	3⅛ × 18	5⅛ × 16½	5⅛ × 16½	5⅛ × 18	5⅛ × 18	5⅛ × 19½	5⅛ × 21
	18	5⅛ × 15	5⅛ × 16½	5⅛ × 18	5⅛ × 18	5⅛ × 19½	5⅛ × 21	5⅛ × 21
26	8	3⅛ × 15	3⅛ × 16½	3⅛ × 16½	3⅛ × 16½	3⅛ × 18	5⅛ × 16½	5⅛ × 18
	12	3⅛ × 18	3⅛ × 18	5⅛ × 16½	5⅛ × 16½	5⅛ × 18	5⅛ × 18	5⅛ × 19½
	16	5⅛ × 16½	5⅛ × 16½	5⅛ × 18	5⅛ × 18	5⅛ × 19½	5⅛ × 21	5⅛ × 22½
	18	5⅛ × 16½	5⅛ × 18	5⅛ × 18	5⅛ × 19½	5⅛ × 21	5⅛ × 21	5⅛ × 22½
28	8	3⅛ × 16½	3⅛ × 16½	3⅛ × 16½	3⅛ × 18	5⅛ × 16½	5⅛ × 16½	5⅛ × 19½
	12	3⅛ × 18	5⅛ × 16½	5⅛ × 18	5⅛ × 18	5⅛ × 18	5⅛ × 19½	5⅛ × 21
	16	5⅛ × 18	5⅛ × 18	5⅛ × 19½	5⅛ × 19½	5⅛ × 21	5⅛ × 22½	5⅛ × 24
	18	5⅛ × 18	5⅛ × 19½	5⅛ × 19½	5⅛ × 21	5⅛ × 22½	5⅛ × 24	5⅛ × 24
30	8	3⅛ × 18	3⅛ × 18	5⅛ × 16½	5⅛ × 16½	5⅛ × 18	5⅛ × 18	5⅛ × 21
	12	5⅛ × 16½	5⅛ × 18	5⅛ × 18	5⅛ × 19½	5⅛ × 19½	5⅛ × 21	5⅛ × 22½
	16	5⅛ × 18	5⅛ × 19½	5⅛ × 21	5⅛ × 21	5⅛ × 22½	5⅛ × 24	5⅛ × 25½
	18	5⅛ × 19½	5⅛ × 21	5⅛ × 21	5⅛ × 22½	5⅛ × 24	5⅛ × 25½	5⅛ × 27
32	8	3⅛ × 18	5⅛ × 16½	5⅛ × 18	5⅛ × 18	5⅛ × 18	5⅛ × 19½	5⅛ × 21
	12	5⅛ × 18	5⅛ × 19½	5⅛ × 19½	5⅛ × 21	5⅛ × 21	5⅛ × 22½	5⅛ × 24
	16	5⅛ × 19½	5⅛ × 21	5⅛ × 22½	5⅛ × 22½	5⅛ × 24	5⅛ × 25½	5⅛ × 27
	18	5⅛ × 21	5⅛ × 21	5⅛ × 22½	5⅛ × 24	5⅛ × 25½	5⅛ × 27	5⅛ × 28½
34	8	5⅛ × 16½	5⅛ × 18	5⅛ × 18	5⅛ × 19½	5⅛ × 19½	5⅛ × 21	5⅛ × 22½
	12	5⅛ × 19½	5⅛ × 19½	5⅛ × 21	5⅛ × 21	5⅛ × 22½	5⅛ × 24	5⅛ × 25½
	16	5⅛ × 21	5⅛ × 22½	5⅛ × 22½	5⅛ × 24	5⅛ × 25½	5⅛ × 27	5⅛ × 28½
	18	5⅛ × 22½	5⅛ × 22½	5⅛ × 24	5⅛ × 25½	5⅛ × 27	5⅛ × 28½	5⅛ × 28½
36	12	5⅛ × 19½	5⅛ × 21	5⅛ × 22½	5⅛ × 22½	5⅛ × 24	5⅛ × 25½	6¾ × 25½
	16	5⅛ × 22½	5⅛ × 24	5⅛ × 24	5⅛ × 25½	5⅛ × 27	5⅛ × 28½	6¾ × 27
	18	5⅛ × 22½	5⅛ × 24	5⅛ × 25½	5⅛ × 28½	5⅛ × 30	6¾ × 27	6¾ × 28½
	20	5⅛ × 24	5⅛ × 25½	5⅛ × 27	5⅛ × 30	6¾ × 27	6¾ × 28½	6¾ × 30
38	12	5⅛ × 21	5⅛ × 22½	5⅛ × 24	5⅛ × 24	5⅛ × 25½	5⅛ × 27	6¾ × 27
	16	5⅛ × 24	5⅛ × 24	5⅛ × 25½	5⅛ × 27	5⅛ × 28½	5⅛ × 30	6¾ × 28½
	18	5⅛ × 24	5⅛ × 25½	5⅛ × 27	5⅛ × 30	6¾ × 27	6¾ × 28½	6¾ × 30
	20	5⅛ × 25½	5⅛ × 27	5⅛ × 28½	6¾ × 27	6¾ × 28½	6¾ × 30	6¾ × 31½

Table 8-25. Load-carrying Capacity of Simple-span Laminated Beams* *(Continued)*

Span, ft	Spacing, ft	Roof beams total load-carrying capacity						Floor beams total load
		30 psf	35 psf	40 psf	45 psf	50 psf	55 psf	50 psf
40	12	5⅛ × 22½	5⅛ × 24	5⅛ × 24	5⅛ × 25½	5⅛ × 27	6¾ × 25½	6¾ × 28½
	16	5⅛ × 24	5⅛ × 25½	5⅛ × 27	5⅛ × 28½	6¾ × 27	6¾ × 28½	6¾ × 31½
	18	5⅛ × 25½	5⅛ × 27	5⅛ × 28½	6¾ × 27	6¾ × 28½	6¾ × 30	6¾ × 31½
	20	5⅛ × 27	5⅛ × 28½	6¾ × 27	6¾ × 28½	6¾ × 30	6¾ × 31½	6¾ × 33
42	12	5⅛ × 24	5⅛ × 24	5⅛ × 25½	5⅛ × 27	6¾ × 25½	6¾ × 25½	6¾ × 30
	16	5⅛ × 25½	5⅛ × 27	5⅛ × 28½	5⅛ × 30	6¾ × 28½	6¾ × 30	6¾ × 33
	18	5⅛ × 27	5⅛ × 28½	5⅛ × 30	6¾ × 28½	6¾ × 30	6¾ × 31½	6¾ × 33
	20	5⅛ × 28½	5⅛ × 30	6¾ × 28½	6¾ × 30	6¾ × 31½	6¾ × 33	6¾ × 34½
44	12	5⅛ × 24	5⅛ × 25½	5⅛ × 27	5⅛ × 27	6¾ × 25½	6¾ × 27	6¾ × 31½
	16	5⅛ × 27	5⅛ × 28½	5⅛ × 30	6¾ × 28½	6¾ × 30	6⅜ × 31½	6¾ × 33
	18	5⅛ × 28½	5⅛ × 30	6¾ × 28½	6¾ × 30	6¾ × 31½	6¾ × 33	6¾ × 34½
	20	5⅛ × 30	6¾ × 27	6¾ × 30	6¾ × 30	6¾ × 33	6¾ × 34½	6¾ × 36
46	12	5⅛ × 25½	5⅛ × 27	5⅛ × 28½	6¾ × 25½	6¾ × 27	6¾ × 28½	6¾ × 31½
	16	5⅛ × 28½	5⅛ × 30	6¾ × 28½	6¾ × 28½	6¾ × 31½	6¾ × 33	6¾ × 36
	18	5⅛ × 28½	6¾ × 28½	6¾ × 30	6¾ × 31½	6¾ × 33	6¾ × 34½	6¾ × 36
	20	5⅛ × 30	6¾ × 28½	6¾ × 31½	6¾ × 33	6¾ × 34½	6¾ × 36	8¾ × 34½
48	12	5⅛ × 27	5⅛ × 28½	5⅛ × 30	5⅛ × 30	6¾ × 28½	6¾ × 30	6¾ × 33
	16	5⅛ × 30	6¾ × 28½	6¾ × 30	6¾ × 30	6¾ × 31½	6¾ × 34½	6¾ × 37½
	18	5⅛ × 30	6¾ × 30	6¾ × 30	6¾ × 33	6¾ × 34½	6¾ × 36	8¾ × 34½
	20	6¾ × 28½	6¾ × 30	6¾ × 31½	6¾ × 34½	6¾ × 36	6¾ × 37½	8¾ × 36
50	12	5⅛ × 28½	5⅛ × 28½	5⅛ × 30	6¾ × 28½	6¾ × 30	6¾ × 31½	6¾ × 34½
	16	5⅛ × 30	6¾ × 30	6¾ × 30	6¾ × 31½	6¾ × 33	6¾ × 36	8¾ × 34½
	18	6¾ × 28½	6¾ × 30	6¾ × 31½	6¾ × 34½	6¾ × 36	8¾ × 33	8¾ × 36
	20	6¾ × 30	6¾ × 31½	6¾ × 33	6¾ × 36	6¾ × 37½	8¾ × 34½	8¾ × 37½
52	12	5⅛ × 28½	5⅛ × 30	6¾ × 28½	6¾ × 30	6¾ × 31½	6¾ × 31½	6¾ × 36
	16	6¾ × 28½	6¾ × 30	6¾ × 31½	6¾ × 33	6¾ × 34½	6¾ × 37½	8¾ × 36
	18	6¾ × 30	6¾ × 31½	6¾ × 33	6¾ × 34½	6¾ × 37½	6¾ × 39	8¾ × 37½
	20	6¾ × 31½	6¾ × 33	6¾ × 34½	6¾ × 37½	6¾ × 39	8¾ × 36	8¾ × 39
54	12	5⅛ × 30	6¾ × 28½	6¾ × 30	6¾ × 31½	6¾ × 33	6¾ × 33	6¾ × 37½
	16	6¾ × 30	6¾ × 31½	6¾ × 33	6¾ × 34½	6¾ × 36	6¾ × 37½	8¾ × 37½
	18	6¾ × 31½	6¾ × 33	6¾ × 34½	6¾ × 36	6¾ × 39	8¾ × 36	8¾ × 39
	20	6¾ × 33	6¾ × 34½	6¾ × 36	6¾ × 39	8¾ × 36	8¾ × 37½	8¾ × 40½
56	12	6¾ × 28½	6¾ × 30	6¾ × 31½	6¾ × 33	6¾ × 33	6¾ × 34½	8¾ × 36
	16	6¾ × 31½	6¾ × 33	6¾ × 34½	6¾ × 36	6¾ × 37½	8¾ × 34½	8¾ × 39
	18	6¾ × 33	6¾ × 34½	6¾ × 36	6¾ × 37½	8¾ × 34½	8¾ × 37½	8¾ × 40½
	20	6¾ × 33	6¾ × 36	6¾ × 37½	8¾ × 34½	8¾ × 37½	8¾ × 39	8¾ × 42
58	12	6¾ × 30	6¾ × 31½	6¾ × 31½	6¾ × 33	6¾ × 34½	6¾ × 36	8¾ × 37½
	16	6¾ × 31½	6¾ × 34½	6¾ × 36	6¾ × 37½	6¾ × 39	8¾ × 36	8¾ × 40½
	18	6¾ × 33	6¾ × 34½	6¾ × 37½	6¾ × 39	8¾ × 36	8¾ × 39	8¾ × 42
	20	6¾ × 34½	6¾ × 36	6¾ × 39	8¾ × 36	8¾ × 39	8¾ × 40½	8¾ × 43½
60	12	6¾ × 30	6¾ × 31½	6¾ × 33	6¾ × 34½	6¾ × 36	6¾ × 37½	8¾ × 39
	16	6¾ × 33	6¾ × 34½	6¾ × 36	6¾ × 39	8¾ × 36	8¾ × 37½	8¾ × 42
	18	6¾ × 34½	6¾ × 36	6¾ × 39	8¾ × 36	8¾ × 37½	8¾ × 39	8¾ × 43½
	20	6¾ × 36	6¾ × 37½	8¾ × 36	8¾ × 37½	8¾ × 40½	8¾ × 42	8¾ × 45

* This table applies to straight, simply supported, laminated timber beams. Other beam support systems may be employed to meet varying design conditions.

1. Roofs should have a minimum slope of ¼ in. per ft to eliminate water ponding.
2. Beam weight must be subtracted from total load carrying capacity. Floor beams are designed for uniform loads of 40-psf live load and 10-psf dead load.
3. Allowable stresses:
   Bending stress, F_b = 2,400 psi (reduced by size factor).
   Shear stress F_v = 165 psi.
   Modulus of elasticity, E = 1,800,000 psi.
   For roof beams, F_b and F_v were increased 15% for short duration of loading.
4. Deflection limits:
   Roof beams—1/180 span for total load.
   Floor beams—1/360 span for 40-psf live load only.
   For preliminary design purposes only.
   For more complete design information, see the AITC "Timber Construction Manual."

Association, 1500 Yeon Building, Portland, Ore. 97204; "Wood Structural Design Data," National Forest Products Association, 1619 Massachusetts Ave., N.W., Washington, D.C. 20036.)

8-26. Deflection and Camber of Timber Beams. The design of many structural systems, particularly those with long span, is governed by deflection. Strength calculations based on allowable stresses alone may result in excessive deflection. Limitations on deflection increase member stiffness.

Table 8-26 gives recommended deflection limits, as a fraction of the beam span, for timber beams. The limitation applies to live load or total load, whichever governs.

Table 8-26. Recommended Roof-beam Deflection Limitations, In.*

(In terms of span, l, in.)

Use classification	Live load only	Dead load plus live load
Roof beams:		
Industrial	$l/180$	$l/120$
Commercial and institutional:		
Without plaster ceiling	$l/240$	$l/180$
With plaster ceiling	$l/360$	$l/240$
Floor beams:		
Ordinary usage†	$l/360$	$l/240$
Highway bridge stringers	$l/200$ to $l/300$	
Railway bridge stringers	$l/300$ to $l/400$	

* Camber and Deflection, AITC 102, Appendix B, American Institute of Timber Construction.

† Ordinary usage classification is intended for construction in which walking comfort, minimized plaster cracking, and elimination of objectionable springiness are of prime importance. For special uses, such as beams supporting vibrating machinery or carrying moving loads, more severe limitations may be required.

Glued-laminated beams are cambered by fabricating them with a curvature opposite in direction to that corresponding to deflections under load. Camber does not, however, increase stiffness. Table 8-27 lists recommended minimum cambers for glued-laminated timber beams.

Table 8-27. Recommended Minimum Camber for Glued-laminated Timber Beams*

Roof beams†	1½ times dead-load deflection
Floor beams‡	1½ times dead-load deflection
Bridge beams: §	
Long span	2 times dead-load deflection
Short span	2 times dead load plus ½ applied-load deflection

* Camber and Deflection, AITC 102, Appendix B, American Institute of Timber Construction.

† The minimum camber of 1½ times dead-load deflection will produce a nearly level member under dead load alone after plastic deformation has occurred. Additional camber is usually provided to improve appearance or provide necessary roof drainage.

‡ The minimum camber of 1½ times dead-load deflection will produce a nearly level member under dead load alone after plastic deformation has occurred. On long spans, a level ceiling may not be desirable because of the optical illusion that the ceiling sags. For warehouse or similar floors where live load may remain for long periods, additional camber should be provided to give a level floor under the permanently applied load.

§ Bridge members are normally cambered for dead load only on multiple spans to obtain acceptable riding qualities.

8-27. Minimum Roof Slopes. Flat roofs have collapsed during rainstorms even though they were adequately designed on the basis of allowable stresses and definite deflection limitations. The failures were caused by ponding of water as increasing deflections permitted more and more water to collect.

Roof beams should have a continuous upward slope equivalent to ¼ in. per ft between a drain and the high point of a roof, in addition to minimum recommended camber (Table 8-26), to avoid ponding. When flat roofs have insufficient slope for drainage (less than ¼ in. per ft), the stiffness of supporting members should be such that a 5-psf load will cause no more than ½-in. deflection.

Because of ponding, snow loads or water trapped by gravel stops, parapet walls, or ice dams magnify stresses and deflections from existing roof loads by

$$C_p = \frac{1}{1 - W'L^3/\pi^4 EI} \tag{8-14}$$

where C_p = factor for multiplying stresses and deflections under existing loads to determine stresses and deflections under existing loads plus ponding
W' = weight of 1 in. of water on roof area supported by beam, lb
L = span of beam, in.
E = modulus of elasticity of beam material, psi
I = moment of inertia of beam, in.^{4}

(Kuenzi and Bohannan, "Increases in Deflection and Stresses Caused by Ponding of Water on Roofs," Forest Products Laboratory, Madison, Wis.)

8-28. Design of Wood Trusses. Type of truss and arrangement of members may be chosen to suit the shape of structure, the loads, and stresses involved. The types most commonly built are bowstring, flat or parallel chord, pitched, triangular or A type, camelback, and scissors (Fig. 8-11). For most construction other than houses, trusses usually are spaced 12 to 20 ft apart. For houses, very light trusses generally are erected 16 to 24 in. c to c.

Joints are critical in the design of a truss. Use of a specific truss type is often governed by joint considerations.

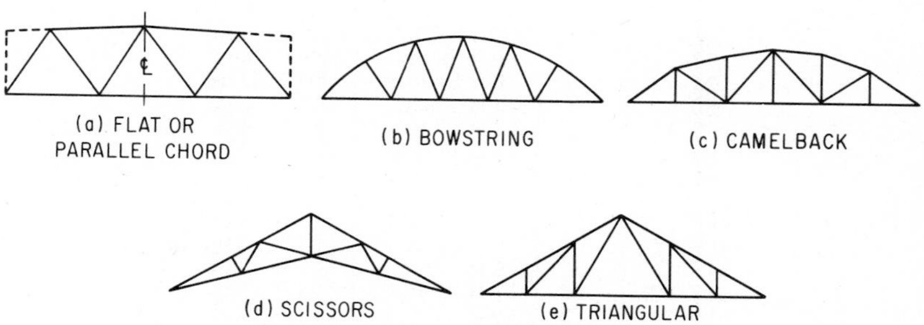

Fig. 8-11. Types of wood trusses.

Chords and webs may be single-leaf (or monochord), double-leaf, or multiple-leaf members. Monochord trusses and trusses with double-leaf chords and single-leaf web system are common. Web members may be attached to the sides of the chords, or web members may be in the same plane as the chords and attached with straps or gussets.

Individual truss members may be solid-sawn, glued-laminated, or mechanically laminated. Glued-laminated chords and solid-sawn web members are usually used. Steel rods or other steel shapes may be used as members of timber trusses if they meet design and service requirements.

The bowstring truss is by far the most popular. Spans of 100 to 200 ft are common, with single or two-piece top and bottom chords of glued-laminated timber, webs of solid-sawn timber, and metal heel plates, chord plates, and web-to-chord connections. This system is light weight for the loads that it can carry. It can be shop or field assembled. Attention to the top chord, bottom chord, and heel connections is of prime importance, since they are the major stress-carrying components. Since the top chord is nearly the shape of an ideal arch, stresses in chords are almost uniform throughout a bowstring truss; web stresses are low under uniformly distributed loads.

Parallel-chord trusses, with slightly sloping top chords and level bottom chords, are used less often, because chord stresses are not uniform along their length and web stresses are high. Hence, different cross sections are required for successive chords, and web members and web-to-chord connections are heavy. Eccentric joints and tension stresses across the grain should be avoided in truss construction, whenever possible, but particularly in parallel-chord trusses.

Triangular trusses and the more ornamental camelback and scissors trusses are used for shorter spans. They usually have solid-sawn members for both chords and webs where degree of seasoning of timbers, hardware, and connections are of considerable importance.

For joints, split-ring and shear-plate connectors are generally most economical. Sometimes, when small trusses are field-fabricated, only bolted joints are used. However, grooving tools for connectors can also be used effectively in the field.

Longitudinal sway bracing perpendicular to the plane of the truss is usually provided by solid-sawn X bracing. Lateral wind bracing may be provided by end walls or intermediate walls, or both. The roof system and horizontal bracing should be capable of transferring the wind load to the walls. Knee braces between trusses and columns are often used to provide resistance to lateral loads.

Table 8-28. Bowstring-truss Dimensions

Span range, ft	No. of panels	Avg truss height*	Avg arc length†
53–57	6	8 ft 7 in.	60 ft 8 in.
58–62	. . .	9 ft 3 in.	65 ft 11 in.
63–67	. . .	9 ft 11 in.	71 ft 5 in.
68–72	8	10 ft 7 in.	76 ft 5 in.
73–77	. . .	11 ft 3 in.	81 ft 8 in.
78–82	. . .	11 ft 11 in.	86 ft 10 in.
83–87	. . .	12 ft 7 in.	92 ft 1 in.
88–92	. . .	13 ft 3 in.	97 ft 4 in.
93–97	. . .	13 ft 11½ in.	102 ft 7 in.
98–102	. . .	14 ft 7½ in.	107 ft 10 in.
103–107	10	15 ft 5 in.	113 ft 2 in.
108–112	. . .	16 ft 1 in.	118 ft 5 in.
113–117	. . .	16 ft 9 in.	123 ft 8 in.
118–122	. . .	17 ft 5 in.	128 ft 11 in.
123–127	. . .	18 ft 1 in.	134 ft 2 in.
128–132	12	18 ft 11 in.	139 ft 6 in.
133–137	. . .	19 ft 7 in.	144 ft 9 in.

* The vertical distance from top of truss heel bearing plate to top of chord at midspan.

† Measured along the top of the top chord, wood to wood. Dimensions shown are for the longest of the span range. For smaller spans, lengths are proportionately shorter.

Horizontal framing between trusses consists of struts between trusses at bottom-chord level and diagonal tie rods, often of steel with turnbuckles for adjustment.

Table 8-28 gives typical bowstring-truss dimensions based on uniform loading conditions on the top chord, as normally imposed by roof joists.

For ordinary roof loads and spacing of 16 to 24 ft, vertical X bracing is required every 30 to 40 ft of chord length. This bracing is placed in alternate bays. Horizontal T-strut bracing should be placed from lower chord to lower chord in the same line as the vertical X bracing for the complete length of the building. If a ceiling is framed into the lower chords, these struts may be omitted.

Joists, spaced 12 to 24 in. c to c, usually rest on the top chords of the trusses and are secured there by toenailing. They may also be placed on ledgers attached to the sides of the upper chords or set in metal hangers, thus lowering the roof line.

Purlins, large-cross-section joists spaced 4 to 8 ft c to c, are often set on top of the top chords, where they are butted end to end and secured to the chords with clip angles. Purlins may also be set between the top chords on metal purlin

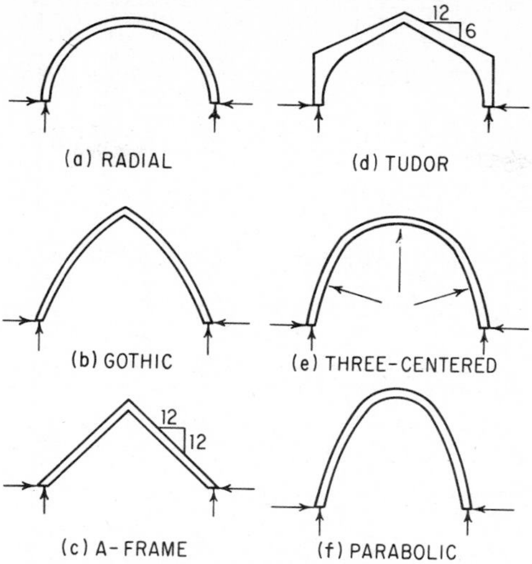

Fig. 8-12. Types of wood arches.

hangers. Roof sheathing, 1-in. D&M (dressed and matched) or ⅜- to ½-in.-thick plywood, is laid directly on joists. For purlin construction, 2-in. D&M roof sheathing is normally used.

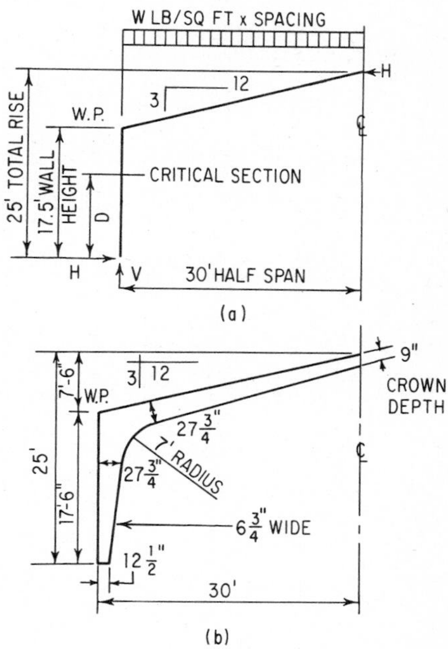

Fig. 8-13. Tudor arch.

("Design Manual for TECO Timber Connector Construction," Timber Engineering Co., 1619 Massachusetts Ave., N.W., Washington, D.C. 20036; AITC 102, Appendix A, "Trusses and Bracing," American Institute of Timber Construction, 333 W. Hampden Ave., Englewood, Colo., 80110.)

8-29. Design of Timber Arches. Arches may be two-hinged, with hinges at each base, or three-hinged, with a hinge at the crown. Figure 8-12 presents typical forms of arches.

Tudor arches are gabled rigid frames with curved haunches. Columns and pitched roof beam on each side of the crown usually are one piece of glued-laminated timber. This type of arch is frequently used in church construction with a high rise. Table 8-29 gives typical dimensions of three-hinged Tudor arches.

Preliminary sizes of key sections of a Tudor arch, for use in a detailed arch analysis, may be determined for uniform vertical loads as follows (for steep roofs, the effect of wind loads must be investigated): Assume an allowable bending stress F_b = 2,200 psi, after adjustment for short-duration loading, curvature, and size factor; horizontal shear stress F_v = 230 psi; live plus dead load of 45 psf; span L = 60 ft; rise from base to crown T = 25 ft; arch spacing = 16 ft; and haunch radius = 7 ft for ¾-in. laminations (Fig. 8-13*a*).

Table 8-29. Preliminary Design Dimensions, In., for Three-hinged Tudor Arches*

Loading	Roof pitch	Wall hgt, ft	30-ft span Width	30-ft span Base	30-ft span Lower tang.	30-ft span Upper tang.	30-ft span Crown	35-ft span Width	35-ft span Base	35-ft span Lower tang.	35-ft span Upper tang.	35-ft span Crown	40-ft span Width	40-ft span Base	40-ft span Lower tang.	40-ft span Upper tang.	40-ft span Crown	50-ft span Width	50-ft span Base	50-ft span Lower tang.	50-ft span Upper tang.	50-ft span Crown
Vertical dead + live load = 400 lb per ft	3:12	12	5⅛	7½	11	10¾	7½	5⅛	7½	12	12	7½	5⅛	7½	13¼	13	7½	5⅛	10½	14¾	14½	7½
		14	5⅛	7½	12	12	7½	5⅛	7½	13½	13¼	7½	5⅛	7½	14¾	14½	7½	5⅛	9½	16¾	16¼	7½
		16	5⅛	7½	13¼	13	7½	5⅛	7½	14¾	14½	7½	5⅛	7½	16	16	7½	5⅛	8¾	18¾	17½	7½
		18	5⅛	7½	14¼	14¼	7½	5⅛	7½	16	15¾	7½	5⅛	7½	17½	17¼	7½	5⅛	8	20½	19	7½
		20	5⅛	7½	15¼	15¼	7½	5⅛	7½	17	16¾	7½	5⅛	7½	18¾	18½	7½	5⅛	7½	22	20¾	7½
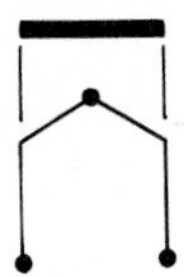	4:12	12	5⅛	7½	10¾	10¾	7½	5⅛	7½	11¾	11¾	7½	5⅛	7½	12¾	12½	7½	5⅛	9½	14½	13½	7½
		14	5⅛	7½	12	12	7½	5⅛	7½	13¼	13	7½	5⅛	7½	14¼	14¼	7½	5⅛	8¾	16¾	15	7½
		16	5⅛	7½	13¼	13	7½	5⅛	7½	14½	14½	7½	5⅛	7½	15¾	15½	7½	5⅛	8	18½	16½	7½
		18	5⅛	7½	14¼	14	7½	5⅛	7½	15¾	15½	7½	5⅛	7½	17	17	7½	5⅛	7½	20¼	18	7½
		20	5⅛	7½	15¼	15	7½	5⅛	7½	17	16¾	7½	5⅛	7½	18¼	18¼	7½	5⅛	7½	21¾	19½	7½
	6:12	12	5⅛	7½	10½	10½	7½	5⅛	7½	11½	11¼	7½	5⅛	7½	12¼	11½	7½	5⅛	8	14¾	11¾	7½
		14	5⅛	7½	11¾	11¾	7½	5⅛	7½	12¾	12¾	7½	5⅛	7½	13¾	13¼	7½	5⅛	7½	16½	12¾	7½
		16	5⅛	7½	13	12¾	7½	5⅛	7½	14	14	7½	5⅛	7½	15	15	7½	5⅛	7½	18¼	14¼	7½
		18	5⅛	7½	14	14	7½	5⅛	7½	15¼	15¼	7½	5⅛	7½	16½	16¼	7½	5⅛	7½	19¾	16	7½
	8:12	12	5⅛	7½	10¼	10	7½	5⅛	7½	11	10½	7½	5⅛	7½	12	10¼	7½	5⅛	7½	14¼	11½	7⅛
		14	5⅛	7½	11½	11¼	7½	5⅛	7½	12¼	12	7½	5⅛	7½	13½	11¾	7½	5⅛	7½	16	12¼	7½
		16	5⅛	7½	12¾	12½	7½	5⅛	7½	13½	13½	7½	5⅛	7½	14¾	13½	7½	5⅛	7½	17¾	13	7½
		18	5⅛	7½	13¾	13½	7½	5⅛	7½	14¾	14¾	7½	5⅛	7½	16	15	7½	5⅛	7½	19¼	14	7½
Vertical dead + live load = 600 lb per ft	3:12	12	5⅛	7½	12	12	7½	5⅛	8½	13½	13	7½	5⅛	10¾	15	13½	7½	5⅛	15½	17½	16	7½
		14	5⅛	7½	13½	13¼	7½	5⅛	7¾	15½	14¼	7½	5⅛	9¾	17½	14½	7½	5⅛	14	21	16½	7½
		16	5⅛	7½	14¾	14¾	7½	5⅛	7½	17¼	15¼	7½	5⅛	8¾	19¼	15¾	7½	5⅛	12¾	23½	17	7½
		18	5⅛	7½	16	16	7½	5⅛	7½	18¾	16¼	7½	5⅛	8	21¼	16¾	7½	6¾	9	20½	20¼	7½
		20	5⅛	7½	17¼	17	7½	5⅛	7½	20	17½	7½	5⅛	7½	22¾	18	7½	6¾	8¼	22	22	7½
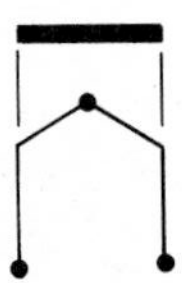	4:12	12	5⅛	7½	12	12	7½	5⅛	8	13¾	12¼	7½	5⅛	9¾	15¼	12¾	7½	5⅛	14	18	15¼	7½
		14	5⅛	7½	13½	13¼	7½	5⅛	7½	15½	13¾	7½	5⅛	9	17½	13¾	7½	5⅛	12¾	21	16	7½
		16	5⅛	7½	14¾	14¾	7½	5⅛	7½	17	15	7½	5⅛	8¼	19¼	15¼	7½	5⅛	11¾	23½	16½	7½
		18	5⅛	7½	16	16	7½	5⅛	7½	18½	16¼	7½	5⅛	7½	21	16½	7½	5⅛	10¾	25½	16¾	7½
		20	5⅛	7½	17¼	17	7½	5⅛	7½	20	17	7½	5⅛	7½	22¾	17¼	7½	6¾	7¾	21½	21½	7½
	6:12	12	5⅛	7½	12	11¼	7½	5⅛	7½	13¾	11¼	7½	5⅛	8½	15¼	11	7½	5⅛	11¾	18¼	14½	7¾
		14	5⅛	7½	13½	12¾	7½	5⅛	7½	15½	12½	7½	5⅛	7¾	17¼	12½	7½	5⅛	10¾	20¾	15¼	7½
		16	5⅛	7½	14¾	14	7½	5⅛	7½	17	14	7½	5⅛	7½	19	14	7½	5⅛	10	23	15¾	7½
		18	5⅛	7½	16	15¼	7½	5⅛	7½	18¼	15½	7½	5⅛	7½	20½	15¼	7½	5⅛	9½	25	16½	7½
	8:12	12	5⅛	7½	12	10½	7½	5⅛	7½	13½	10¼	7½	5⅛	7½	15¼	11	7½	5⅛	10	18	14	9½
		14	5⅛	7½	13¼	12¼	7½	5⅛	7½	15¼	11½	7½	5⅛	7½	17	11¾	7½	5⅛	9¼	20¼	15	8
		16	5⅛	7½	14½	13½	7½	5⅛	7½	16½	13¼	7½	5⅛	7½	18½	12¾	7½	5⅛	8¾	22¼	15¾	7½
		18	5⅛	7½	15¾	14¾	7½	5⅛	7½	18	14¼	7½	5⅛	7½	20¼	13¾	7½	5⅛	8¼	24¼	16¼	7½
Vertical dead + live load = 800 lb per ft	3:12	12	5⅛	8¾	13¾	12	7½	5⅛	11¼	15¾	12½	7½	5⅛	14	17¼	13¼	7½	5⅛	20½	20½	18½	12¼
		14	5⅛	7¾	15¾	12¾	7½	5⅛	10	18¼	13¼	7½	5⅛	12¾	20½	13¾	7½	5⅛	18½	24½	19	12¼
		16	5⅛	7½	17½	13¾	7½	5⅛	9¼	20¼	14	7½	5⅛	11½	22¾	14½	7½	6¾	13	22	18¾	12¼
		18	5⅛	7½	19¼	14¼	7½	5⅛	8¼	22¼	14¾	7½	5⅛	10½	25	15¼	7½	6¾	12	24	20	12¼
		20	5⅛	7½	20¾	14¾	7½	5⅛	7¾	24	15½	7½	6¾	7½	21	20¾	7½	6¾	11	25¾	21½	12¼
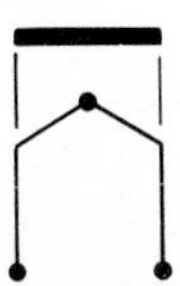	4:12	12	5⅛	8	14	11½	7½	5⅛	10½	16	11¾	7½	5⅛	12¾	17¾	13	7½	5⅛	18½	20¾	17½	12
		14	5⅛	7½	16	12½	7½	5⅛	9½	18¼	12¾	7½	5⅛	11¾	20½	13½	7½	5⅛	16¾	24¾	18¼	12
		16	5⅛	7½	17½	13½	7½	5⅛	8½	20¼	13¾	7½	5⅛	10¾	22¾	13¾	7½	6¾	12	22	17¾	12
		18	5⅛	7½	19	14½	7½	5⅛	7¾	22	14¾	7½	5⅛	9¾	25	14½	7½	6¾	11	23¾	19¼	12
		20	5⅛	7½	20¾	14¾	7½	5⅛	7½	23¾	15½	7½	6¾	7½	20¾	20½	7½	6¾	10¼	25¼	21	12
	6:12	12	5⅛	7½	14¼	10¾	7½	5⅛	9	16¼	10½	7½	5⅛	11	18	12½	7½	5⅛	15¼	21½	16½	11¼
		14	5⅛	7½	16	11¾	7½	5⅛	8¼	18¼	11¾	7½	5⅛	10	20¼	13¼	7½	5⅛	14¼	24½	17¼	11¼
		16	5⅛	7½	17½	13	7½	5⅛	7½	20	13	7½	5⅛	9¼	22½	13¾	7½	6¾	10¼	21½	16	11¼
		18	5⅛	7½	19	14	7½	5⅛	7½	21¾	14	7½	5⅛	8¾	24½	14¼	7½	6¾	9½	23	17½	11¼
	8:12	12	5⅛	7½	14¼	9¾	7½	5⅛	8	16	10¾	7½	5⅛	9½	17¾	12½	8	5⅛	13¼	21¼	16	11
		14	5⅛	7½	15¾	11¼	7½	5⅛	7½	18	11½	7½	5⅛	9	20	13½	7½	5⅛	12½	24	17	10½
		16	5⅛	7½	17¼	12½	7½	5⅛	7½	19¾	12	7½	5⅛	8¼	22	14	7½	6¾	9	20¾	16	10½
		18	5⅛	7½	18¾	13½	7½	5⅛	7½	21½	13	7½	5⅛	7¾	24	14½	7½	6¾	8½	22½	16½	10½

Vertical dead + live load = 1000 lb per ft 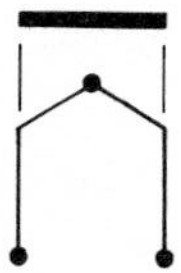	3:12	12	5⅛	10¾	15½	11¼	7½	5⅛	13¾	17½	13	7½	5⅛	17½	19	16	12¼	5⅛	25	25	22¼	12¼
		14	5⅛	9½	18	11¾	7½	5⅛	12½	20½	12¾	7½	5⅛	15¾	23	15¾	12¼	6¾	17½	22	18¾	12¼
		16	5⅛	8¾	20	12¼	7½	5⅛	11¼	23	12¾	7½	6¾	11	20½	17¼	12¼	6¾	16	24¾	19	12¼
		18	5⅛	8	22	12¾	7½	5⅛	10¼	25¼	13¼	7½	6¾	10	22¼	18½	12¼	6¾	14¾	27	19½	12¼
		20	5⅛	7½	23¾	13¼	7½	6¾	7½	20¾	19¾	7½	6¾	9¼	23¾	20	12¼	6¾	13¾	29	20¼	12¼
	4:12	12	5⅛	10	15¾	11	7½	5⅛	12¾	18	13	7½	5⅛	16	19¾	15¾	12	5⅛	22½	22½	21¾	12
		14	5⅛	9	18	11¾	7½	5⅛	11½	20¾	12¾	7⅛	5⅛	14½	23¼	15½	12	6¾	16	22¼	18	12
		16	5⅛	8¼	20	12¼	7½	5⅛	10½	23	12½	7½	6¾	10¼	20½	16¾	12	6¾	14¾	24¾	18½	12
		18	5⅛	7½	21¾	13¼	7½	5⅛	9¾	25¼	13	7½	6¾	9½	22	18¼	12	6¾	13¾	26¾	19	12
		20	5⅛	7½	23¾	13¼	7½	6¾	7½	20¾	19½	7½	6¾	8¾	23½	19¾	12	6¾	12¾	28¾	19¾	12
	6:12	12	5⅛	8¾	16¼	10	7½	5⅛	11	18¼	11¾	7½	5⅛	13¾	20½	14¼	11¼	5⅛	19	24	19	11¼
		14	5⅛	8	18	11¼	7½	5⅛	10¼	20¾	12¼	7½	5⅛	12½	23¼	14½	11¼	6¾	13½	22	17¼	11¼
		16	5⅛	7½	20	12	7½	5⅛	9½	22¾	12½	7½	5⅛	11¾	25¾	15	11¼	6¾	12¾	24¼	17¾	11¼
		18	5⅛	7½	21¾	12¾	7½	5⅛	8¾	25	13	7½	6¾	8¼	21¾	17¼	11¼	6¾	11¾	26¼	18½	11¼
	8:12	12	5⅛	7¾	16	9¾	7½	5⅛	9¾	18¼	12	7½	5⅛	12	20¼	14	10½	5⅛	16¼	24	17¾	12½
		14	5⅛	7½	18	10½	7½	5⅛	9	20¼	12¾	7½	5⅛	11	22¾	14¾	10½	6¾	11¾	21½	16¾	10½
		16	5⅛	7½	19¾	11½	7½	5⅛	8½	22½	13¼	7½	5⅛	10¼	25¼	15½	10½	6¾	11	23½	17¾	10½
		18	5⅛	7½	21½	12¼	7½	5⅛	8	24½	13½	7½	6¾	7½	21	16½	10½	6¾	10½	25¼	18½	10½
Vertical dead = 240 lb per ft Horizontal wind = 320 lb per ft 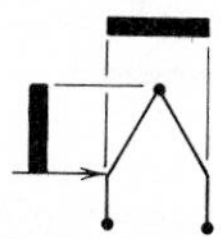	10:12	8	5⅛	7½	7½	11	7½	5⅛	7½	7½	12	10¼	5⅛	7½	8¼	13	12¾	5⅛	7½	9¼	15½	15½
		10	5⅛	7½	9¼	12½	12½	5⅛	7½	10	13¼	8	5⅛	7½	10¾	14¼	11¾	5⅛	7½	12	16¼	16¼
		12	5⅛	7½	11¼	13¾	13¾	5⅛	7½	12	14¾	7½	5⅛	7½	12¾	15¾	9¾	5⅛	7½	14¼	17½	16
	12:12	8	5⅛	7½	7½	12½	8¾	5⅛	7½	8¼	13½	12	5⅛	7½	8¾	14¾	14½	5⅛	7½	9¾	17½	17½
		10	5⅛	7½	9¾	13¾	9	5⅛	7½	10½	15	10¼	5⅛	7½	11	16¼	13¾	5⅛	7½	12½	18½	18½
		12	5⅛	7½	11¾	15¼	15¼	5⅛	7½	12½	16½	9	5⅛	7½	13¼	17½	11¾	5⅛	7½	14¾	20	18¼
	14:12	8	5⅛	7½	8¾	13¾	10½	5⅛	7½	9	15	13¾	5⅛	7½	9½	16¼	16¼	5⅛	7½	10¼	19½	19½
		10	5⅛	7½	11¼	15¼	8¼	5⅛	7½	11½	16½	12¼	5⅛	7½	12½	18	15¾	5⅛	7¾	13¼	20½	20½
		12	5⅛	7½	13¼	17	17	5⅛	7½	14	18	8¼	5⅛	7½	14½	19½	14¼	5⅛	8¼	16	22	20¾
	16:12	8	5⅛	7½	10	15	12½	5⅛	7½	10½	16¼	15¾	5⅛	7½	11	18	18	5⅛	9	11	21½	21½
		10	5⅛	7½	12¼	16½	9½	5⅛	7½	13	18	14½	5⅛	7¾	14	19½	18	5⅛	9¼	14¾	22¾	22¾
		12	5⅛	7½	14¾	18¼	7½	5⅛	7½	15½	19¾	11½	5⅛	8¼	16¼	21¼	16¾	5⅛	9¾	17½	24	23½
Vertical dead = 320 lb per ft horizontal wind = 320 lb per ft 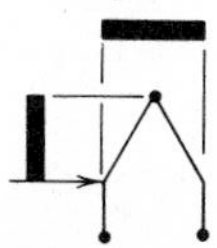	10:12	8	5⅛	7½	7½	10¾	9	5⅛	7½	8¼	11¾	11½	5⅛	7½	9	13	13	5⅛	8	10	15¾	15¾
		10	5⅛	7½	10	12	7½	5⅛	7½	10¾	13	10½	5⅛	7½	11½	14	13½	5⅛	7¾	13	16½	16½
		12	5⅛	7½	12	13¼	13¼	5⅛	7½	12¾	14¼	8½	5⅛	7½	13¾	15¼	12½	5⅛	7¾	15¾	17¼	17¼
	12:12	8	5⅛	7½	8¼	12¼	10¼	5⅛	7½	9	13¼	13¼	5⅛	7½	9½	14¾	14¾	5⅛	7¾	10½	17¾	17¾
		10	5⅛	7½	10½	13½	9	5⅛	7½	11¼	14¾	12¼	5⅛	7½	12	15¾	15¼	5⅛	7¾	13½	18¾	18¾
		12	5⅛	7½	12½	15	9	5⅛	7½	13¼	16	10¼	5⅛	7½	14¼	17¼	14¼	5⅛	7¾	16	19½	19½
	14:12	8	5⅛	7½	8¾	13½	12	5⅛	7½	9¼	14¾	14¾	5⅛	7½	10	16½	16½	5⅛	7¾	10½	19¾	19¾
		10	5⅛	7½	10¾	15	10¼	5⅛	7½	11¾	16¼	14¼	5⅛	7½	12½	17½	17¼	5⅛	7¾	13¾	20¾	20¾
		12	5⅛	7½	12¾	16½	8¼	5⅛	7½	13¾	17¾	12½	5⅛	7½	14¾	19¼	16½	5⅛	8	16½	21¾	21¾
	16:12	8	5⅛	7½	9½	15	13¾	5⅛	7½	9¾	16¼	16¼	5⅛	7½	10¼	18¼	18¼	5⅛	8¼	10¾	22	22
		10	5⅛	7½	11¾	16½	12¼	5⅛	7½	12¾	18	16¼	5⅛	7½	13	19¼	19¼	5⅛	8½	14	23	23
		12	5⅛	7½	14¼	18	7¾	5⅛	7½	14¾	19½	14½	5⅛	7½	15¾	21	18¾	5⅛	9	16¾	24¼	24¼
Vertical dead = 480 lb per ft horizontal wind = 320 lb per ft 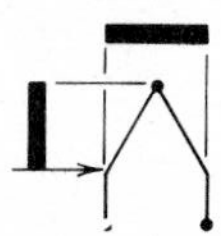	10:12	8	5⅛	7½	9	10½	10½	5⅛	7½	9¾	12	12	5⅛	8	10½	13½	13½	5⅛	10¼	11¼	16½	16½
		10	5⅛	7½	11¼	11½	10¼	5⅛	7½	12¼	12½	12½	5⅛	7¾	13¼	14	14	5⅛	10	15¼	17	17
		12	5⅛	7½	13½	12¾	8¾	5⅛	7½	14¾	13½	12½	5⅛	7¾	16	14¾	14¾	5⅛	9¾	18	17½	17½
	12:12	8	5⅛	7½	9¼	12	12	5⅛	7½	10¼	13½	13½	5⅛	7¾	11	15¼	15¼	5⅛	9¾	11¾	18½	18½
		10	5⅛	7½	11¾	13¼	11½	5⅛	7½	12¾	14¼	14¼	5⅛	7¾	13¾	15¾	15¾	5⅛	9¾	15½	19	19
		12	5⅛	7½	13¾	14½	10	5⅛	7½	15	15½	14	5⅛	7¾	16¼	16¾	16¾	5⅛	9½	18¼	19¾	19¾
	14:12	8	5⅛	7½	9¾	13¼	13¼	5⅛	7½	10½	15	15	5⅛	7¾	11¼	17	17	5⅛	9½	12	20½	20½
		10	5⅛	7½	12	14¾	13	5⅛	7½	13	16	16	5⅛	7¾	14	17¾	17¾	5⅛	9½	15½	21¼	21¼
		12	5⅛	7½	14¼	16	11½	5⅛	7½	15½	17½	15½	5⅛	7¾	16½	18¾	18¾	5⅛	9½	18½	22¼	22¼
	16:12	8	5⅛	7½	10	14¾	14¾	5⅛	7½	10¾	16¾	16¾	5⅛	7¾	11½	18¾	18¾	5⅛	9½	12	22¾	22¾
		10	5⅛	7½	12¼	16¼	14½	5⅛	7½	13¼	17¾	17¾	5⅛	7¾	14¼	19½	19½	5⅛	9½	15½	23½	23½
		12	5⅛	7½	14½	17¾	13¼	5⅛	7½	15¾	19¼	17½	5⅛	7¾	17	20¾	20¾	5⅛	9½	18¾	24½	24½

Table 8-29. Preliminary Dimensions, In., for Three-hinged Tudor Arches* (*Continued*)

Loading	Roof pitch	Wall hgt, ft.	60-ft span					70-ft span					80-ft span					90-ft span				
			Width	Base	Lower tang.	Upper tang.	Crown	Width	Base	Lower tang.	Upper tang.	Crown	Width	Base	Lower tang.	Upper tang.	Crown	Width	Base	Lower tang.	Upper tang.	Crown
Vertical dead + live load = 400 lb per ft	3:12	12	5⅛	14	16¼	16¾	7½	5⅛	17¾	17¾	20	7½	5⅛	21¾	21¾	23	7½	6¾	20	20	22¾	8
		14	5⅛	12¾	19½	17½	7½	5⅛	16¼	22	20¾	7½	5⅛	20	24	24	7½	6¾	18½	21½	24	7½
		16	5⅛	11¾	22	17¾	7½	5⅛	15	25	21¼	7½	6¾	14¼	22½	22	7½	6¾	17¼	24¾	24¾	7½
		18	5⅛	10¾	24	19¼	7½	6¾	10¾	22	21¼	7½	6¾	13¼	24½	22¾	7½	6¾	16	27	25¾	7½
		20	6¾	7¾	22	22	7½	6¾	10	23½	23¼	7½	6¾	12½	26½	23	7½	6¾	15	29¼	26	7½
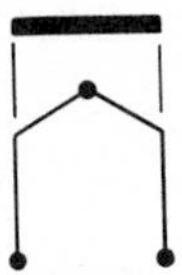	4:12	12	5⅛	12½	16¾	15¾	7½	5⅛	15¾	18½	18¾	7¾	5⅛	19¼	20	21½	9½	5⅛	22¾	22¾	24¼	12
		14	5⅛	11½	19½	16½	7½	5⅛	14½	22	19½	7½	5⅛	17¾	24¼	22¾	8¼	6¾	16¼	21½	22¼	8½
		16	5⅛	10½	21¾	17	7½	5⅛	13½	24½	20¼	7½	6¾	12¾	22	20¾	7½	6¾	15¼	24¼	23¼	7½
		18	5⅛	10	23¾	17¾	7½	6¾	9¾	21½	19½	7½	6¾	12	24	21¼	7½	6¾	14¼	26½	24¼	7½
		20	5⅛	9¼	25½	19¼	7½	6¾	9	23	21½	7½	6¾	11¼	25¾	22	7½	6¾	13½	28½	24¾	7½
	6:12	12	5⅛	10¼	16¾	14¼	8¼	5⅛	12¾	18¾	16¾	10½	5⅛	15¼	20½	19	13¼	5⅛	18	22¼	21¼	16
		14	5⅛	9½	19	15¼	7½	5⅛	12	21½	17¾	9	5⅛	14½	23¾	20¼	11½	6¾	13	21	19¾	12
		16	5⅛	9	21	16	7½	5⅛	11¼	23¾	18¾	7¾	6¾	10½	21¼	18¾	8½	6¾	12¼	23¼	21	10½
		18	5⅛	8½	23	16½	7½	6¾	8	20¾	17¼	7½	6¾	9¾	23	19½	7½	6¾	11¾	25¼	22	9¼
	8:12	12	5⅛	8¾	16½	13½	10	5⅛	10¾	18½	15½	12¾	5⅛	12¾	20¼	17¼	16	5⅛	15	22	19¼	19¼
		14	5⅛	8¼	18½	14½	8¾	5⅛	10¼	20¾	16¾	11¼	5⅛	12¼	23	18¾	14	5⅛	14¼	25¼	20¾	17
		16	5⅛	7¾	20½	15½	7½	5⅛	9½	23	17¾	9¾	5⅛	11½	25½	20	12½	6¾	10¼	22¼	19½	12¾
		18	5⅛	7½	22¼	16¼	7½	5⅛	9¼	25¼	18¾	8¾	6¾	8½	22	18½	9¼	6¾	10	24	20½	11½
Vertical dead + live load = 600 lb per ft	3:12	12	5⅛	20½	20½	20¼	7½	5⅛	26¼	26¼	24	12¼	6¾	25	25	24½	12¼	6¾	29¾	29¾	27½	12¼
		14	5⅛	18¾	24	21	7½	6¾	18½	22	22¼	12¼	6¾	23	23½	26	12¼	6¾	27½	27½	29¼	12¼
		16	6¾	13¼	22	19	7½	6¾	17¼	25	23	12¼	6¾	21¼	27½	26¾	12¼	6¾	25½	30	30¼	12¼
		18	6¾	12¼	24	20¼	7½	6¾	16	27¼	23½	12¼	6¾	19¾	30½	27¼	12¼	6¾	23¾	33½	31	12¼
		20	6¾	11½	25¾	21¾	7½	6¾	14¾	29½	24	12¼	6¾	18½	33	28	12¼	8¾	17½	30	28¼	12¼
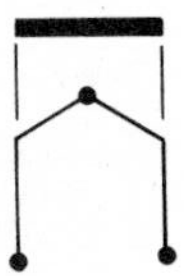	4:12	12	5⅛	18¼	20¼	19¼	7½	5⅛	23¼	23¼	22¾	12	6¾	21¾	21¾	23	12	6¾	25¾	25¾	25½	12¾
		14	5⅛	17	24¼	20	7½	6¾	16½	22	21¼	12	6¾	20¼	24¼	24¼	12	6¾	24	26	27¼	12
		16	6¾	12	21¾	18¼	7½	6¾	15½	24¾	21¾	12	6¾	18¾	27¼	25	12	6¾	22½	29¾	28¼	12
		18	6¾	11¼	23½	18¾	7½	6¾	14½	27	22½	12	6¾	17¾	30	26	12	6¾	21¼	32¾	29¼	12
		20	6¾	10½	25¼	20½	7½	6¾	13½	28¾	23	12	6¾	16¾	32¼	26½	12	8¾	15½	19¼	26¾	12
	6:12	12	5⅛	15¼	21	17½	10¼	5⅛	18¾	23¼	20½	13¼	5⅛	22¾	25½	23½	16¼	6¾	20½	22¼	22¾	17
		14	5⅛	14	24	18½	8¾	6¾	13½	21¾	19¼	11¼	6¾	16¼	23¾	22	12	6¾	19¼	26	24¼	14¾
		16	6¾	10¼	21¼	17¼	7½	6¾	12¾	24	20	11¼	6¾	15½	26½	23	11¼	6¾	18¼	29	25½	13
		18	6¾	9½	22¾	17¾	7½	6¾	12	26	21	11¼	6¾	14¾	28¾	23¾	11¼	6¾	17¼	31½	26¾	11¾
	8:12	12	5⅛	13	20¾	16½	12½	5⅛	16	23¼	19	16	5⅛	19	25½	21¼	19¾	6¾	17	22¼	20½	20¼
		14	5⅛	12¼	23¼	17¾	10¾	6¾	11½	21	18	12	6¾	13¾	23¼	20¼	14¾	6¾	16	25¼	22¼	18
		16	6¾	8¾	20½	16½	8	6¾	11	23	19¼	10½	6¾	13	25½	21½	13¼	6¾	15½	27¾	23¾	16
		18	6¾	8¼	22	17½	7½	6¾	10½	25	20	10½	6¾	12½	27½	22½	11¾	6¾	14¾	30¼	25	14½
Vertical dead + live load = 800 lb per ft	3:12	12	5⅛	27¼	27¼	23	12¼	6¾	26¾	26¾	24¼	12¼	6¾	32½	32½	27¾	12¼	6¾	38¾	38¾	31	12½
		14	6¾	19¼	22½	21¼	12¼	6¾	24½	24½	25¾	12¼	6¾	30	30	29½	12¼	8¾	28¼	28¼	29¾	12¼
		16	6¾	17½	25¾	21¾	12¼	6¾	22½	28¾	26¼	12¼	6¾	27¾	31¾	30¾	12¼	8¾	26¼	29	30¾	12¼
		18	6¾	16¼	28	22¼	12¼	6¾	21	31¾	26¾	12¼	8¾	20¼	29½	27¾	12¼	8¾	24½	32¼	31¾	12¼
		20	6¾	15	30¼	22½	12¼	8¾	15¼	28¼	24¾	12¼	8¾	19	31¾	28½	12¼	8¾	23	34¾	32¼	12¼
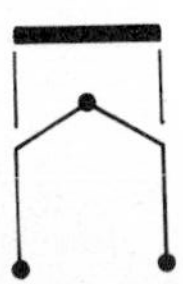	4:12	12	5⅛	24¼	24¼	22	12	6¾	23½	23½	23	12	6¾	28½	28½	26	12½	6¾	33¾	33¾	29¼	15¼
		14	6¾	17¼	22¾	20¼	12	6¾	21¾	25¼	24¼	12	6¾	26½	27½	28	12	6¾	31½	31½	31¼	12¾
		16	6¾	16	25½	21	12	6¾	20¼	28¾	25	12	6¾	24¾	31¾	28¾	12	8¾	23	29	29	12
		18	6¾	14¾	27¾	21½	12	6¾	19	31½	25¾	12	8¾	18¼	29	26½	12	8¾	21¾	31¾	29¾	12
		20	6¾	13¾	29¾	22	12	6¾	17¾	33¾	26¼	12	8¾	17	31	27¼	12	8¾	20½	34	30¾	12
	6:12	12	5⅛	20	24¼	20¼	12	6¾	19	22	21	13	6¾	23	23¾	23¾	16¼	6¾	27	27	26¼	19¾
		14	6¾	14¼	22½	18¾	11¼	6¾	17¾	25¼	22	11¼	6¾	21½	27¾	25	14¼	6¾	25½	30	28	17½
		16	6¾	13½	24¾	19¾	11¼	6¾	16¾	28	23	11¼	6¾	20¼	31	26¼	12½	6¾	24	33¾	29½	15½
		18	6¾	12¾	26¾	20½	11¼	6¾	16	30½	24	11¼	6¾	19¼	33¾	27½	11¼	8¾	17¾	30¼	27¼	11¾
	8:12	12	5⅛	17	24½	19	14¾	6¾	16	22	19¼	16	6¾	19¼	24	21½	19¾	6¾	22½	25¾	23¾	23¾
		14	6¾	12¼	22	18	10¾	6¾	15¼	24½	20¾	14	6¾	18¼	27¼	23¼	17½	6¾	21¼	29½	25¾	21
		16	6¾	11½	24	19	10½	6¾	14½	27	22	12¼	6¾	17¼	30	24¾	15½	6¾	20¼	32¾	27¼	19
		18	6¾	11	26	20	10½	6¾	13¾	29¼	23	11	6¾	16½	32½	26	13¾	8¾	15¼	29	25½	14½

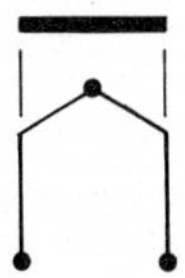

Vertical dead + live load = 1000 lb per ft	3:12	12	6¾	26	26	22½	12¼	6¾	32¾	32¾	26½	12¼	6¾	40½	40½	30¾	17	8¾	37¾	37¾	30¾	17
		14	6¾	23½	25	23¾	12¼	6¾	30	30	28½	12¼	6¾	37¼	37¼	32¾	17	8¾	35	35	32¾	17
		16	6¾	21¾	28¾	24¼	12¼	6¾	27¾	32¼	29¼	12¼	8¾	27	29¾	30½	17	8¾	32½	32½	34½	17
		18	6¾	20	31½	24¾	12¼	8¾	20¼	29¾	26½	12¼	8¾	25¼	33	31	17	8¾	30¼	36	35½	17
		20	8¾	14½	28	23½	12¼	8¾	18¾	31¾	27	12¼	8¾	23½	35½	31¾	17	8¾	28½	39	36	17
	4:12	12	6¾	23	23	21½	12	6¾	29	29	25¼	12	6¾	35¼	35¼	29¼	16¾	8¾	32¾	32¾	28¾	16¾
		14	6¾	21¼	25½	22½	12	6¾	26¾	28	27	12	6¾	33	33	31	16¾	8¾	30½	30½	31	16¾
		16	6¾	19¾	28¾	23¼	12	6¾	25	32¼	27¾	12	8¾	24	29¾	28½	16¾	8¾	28¾	32	32¼	16¾
		18	6¾	18¼	31¼	24	12	8¾	18¼	29¼	25½	12	8¾	22½	32½	29½	16¾	8¾	27	35½	33¼	16¾
		20	6¾	17¼	33¾	24¼	12	8¾	17¼	31¼	26	12	8¾	21¼	35	30¼	16¾	8¾	25½	38¼	34¼	16¾
	6:12	12	6¾	19	22	19¾	11½	6¾	23½	24¼	23¼	14¾	6¾	28½	28½	26¼	18½	6¾	33¼	33¼	29	22¾
		14	6¾	17¾	25¼	21	11¼	6¾	22	28¼	24½	13	6¾	26¾	31	28	16¼	8¾	24¾	28½	27¾	16¾
		16	6¾	16½	28	22	11¼	6¾	20¾	31½	25¾	11¼	8¾	19¾	29	26	15¾	8¾	23½	31½	29¼	15¾
		18	6¾	15¾	30½	22¾	11¼	8¾	15¼	28¼	23¾	11¼	8¾	18¾	31¼	27	15¾	8¾	22¼	34¼	30¼	15¾
	8:12	12	6¾	16¼	22	18¾	14	6¾	20	24½	21½	18	6¾	23¾	26¾	24¼	22¼	6¾	27¾	28¾	26¾	26¾
		14	6¾	15¼	24¾	20¼	12¼	6¾	18¾	27¾	23¼	15¾	6¾	22½	30¾	26	19¾	6¾	26½	33¼	28¾	24
		16	6¾	14¼	27¼	21¼	10¾	6¾	18	30½	24½	14	8¾	16¾	28	24¼	15	8¾	19¾	30½	27	18½
		18	6¾	13¾	29½	22¼	10½	6¾	17	33¼	25¾	12½	8¾	16	30	25¾	14¾	8¾	18¾	32¾	28½	16½

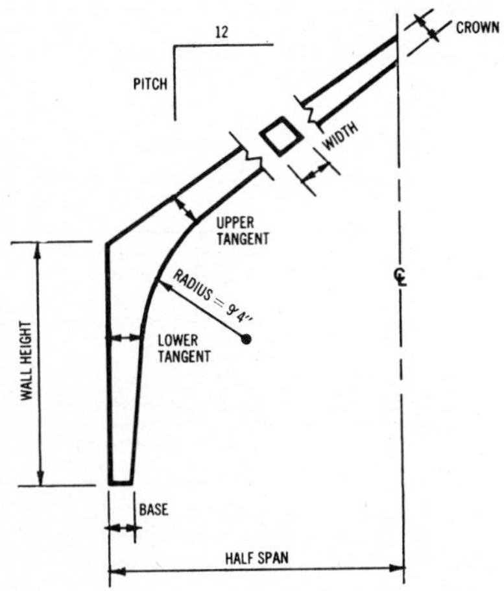

* Typical dimensions for three-hinged tudor arches for preliminary design purposes are given in Table 8-29. Sizes are based on Douglas fir laminated timber, developing an allowable shear stress of 165 psi and with a bending radius of 9 ft 4 in. For southern pine laminated timber, an allowable shear stress of 200 psi and a bending radius of 7 ft 2 in. may be used.

For roof pitches less than 10:12, the critical loading is generally the combined dead and live load on the horizontal projection of the full span. For roof pitches of 10:12 or greater, the critical loading is generally a combination of dead load and horizontal wind load. Sizes shown were determined for a uniformly distributed wind load applied on the vertical projection of the roof arm with a concentrated wind load equal to ½ the total wind load on the wall height acting at the haunch.

In the combined stress analysis, it was assumed that the bending portion of the loading exceeded the axial compression portion. The section sizes are based on the following design criteria:

1. Uniform loading.
2. Radius of curvature at the haunch = 9 ft 4 in.
3. Allowable stresses:
   Bending stress F_b = 2,400 psi (reduced by size factor and curvature factor when applicable).
   Shear stress F_v = 165 psi.
   Compression parallel to grain stress F_c = 1,500 psi (adjusted for $4d$ ratio).
   Modulus of elasticity E = 1,800,000 psi.
   These stresses were increased 15% for short duration of loading and 33⅓% for wind loading when applicable.
4. Deflection limits: $1/180$ for dead plus live load; $1/240$ for dead load only.
5. Vertical arch legs are laterally unsupported with tangent-point depth-to-width ratio not exceeding 5:1. (When vertical arch legs are laterally supported, tangent-point depth-to-ratio not exceeding 6:1 may be used.)

The total uniform vertical load $w = 45 \times 16 = 720$ lb per lin ft. Compute the horizontal thrust:

$$H = \frac{wL^2}{8T} = \frac{720 \times 60^2}{8 \times 25} = 12{,}960 \text{ lb}$$

For this shearing force, determine the base depth d for an assumed width b (from Table 8-29) of 6¾ in.

$$d = \frac{3H}{2F_v b} = \frac{3 \times 12{,}960}{2 \times 230 \times 6.75} = 12.6$$

Use $d = 12\frac{1}{2}$ in. Next compute the bending moment M at the critical section, which is a distance D from the base hinge. To find D, subtract the dimension

Table 8-30. Distance of Critical Section below Working Point

(WP in Fig. 8-13)

Pitch	1:12	2:12	3:12	4:12	5:12	6:12	7:12	8:12	9:12
Distance, ft	6.43	5.94	5.47	5.05	4.67	4.32	4.03	3.75	3.5

given in Table 8-30 for a 3:12 pitch from the wall height: 17.50 — 5.47 = 12.03 ft.

$$M = 12HD = 12 \times 12{,}960 \times 12.03 = 1{,}871{,}000 \text{ in.-lb}$$

The section modulus required at the critical section then is

$$S = \frac{M}{F_b} = \frac{1{,}871{,}000}{2{,}200} = 850 \text{ in.}^3$$

From Table 8-4, for ¾-in. laminations, select the most economical section with this section modulus: 6¾ × 27, with S = 866. Make the upper tangent the same size as the lower. Finally, determine the depth d at the crown. This depth should at least equal the width, and preferably, it should be one-third larger. Hence,

$$d = 1.33 \times 6.75 = 9 \text{ in.}$$

This may be increased for architectural reasons or to equal depth of purlins. Figure 8-13b summarizes the preliminary dimensions of the arch.

Table 8-31. Tie-rod Capacities Based on Tensile-stress Area*

Bar no.	Diam of rod, in.	Area of rod, sq in.	Stress area at root of thread, sq in.	Weight, lb per lin ft	Permissible horizontal thrust H			
					Threaded		Unthreaded	
					20,000 psi	32,000 psi	20,000 psi	32,000 psi
4	½	0.1964	0.1416	0.67	2,832	4,531	3,928	6,284
5	⅝	0.3068	0.2256	1.04	4,512	7,219	6,136	9,817
6	¾	0.4418	0.3340	1.50	6,680	10,688	8,836	14,138
7	⅞	0.6013	0.4612	2.04	9,224	14,758	12,026	19,242
8	1	0.7854	0.6051	2.67	12,102	19,363	15,708	25,133
9	1⅛	0.9940	0.7627	3.38	15,254	24,406	19,880	31,808
10	1¼	1.2272	0.9684	4.17	19,368	30,989	24,544	39,270
11	1⅜	1.4849	1.1538	5.05	23,076	36,922	29,698	47,517
12	1½	1.7671	1.4041	6.01	28,082	44,931	35,342	56,547
	1¾	2.4053	1.8983	8.18	37,966	60,746	48,106	76,970
	2	3.1416	2.4971	10.68	49,942	79,907	62,832	100,530
	2½	4.9087	3.9976	16.69	79,952	127,920	98,174	157,080
	3	7.0686	5.9659	24.03	119,320	190,910	141,370	226,200
	3½	9.6211	8.3268	32.71	166,540	266,460	192,420	307,880
	4	12.5660	11.0800	42.73	221,600	354,560	251,320	402,110

* To determine tie-rod size to resist horizontal thrust H, lb, of the tied segment arch: Compute thrust $H = 0.932wL$, where w = total load, lb per lin ft. Span L = radius. Select rod size with adequate permissible thrust H from above table.

Segmented arches are fabricated with overlapping lumber segments, nailed or glued-laminated. They generally are three-hinged, and they may be tied or buttressed. If an arch is tied, the tie rods, which resist the horizontal thrust (Table 8-31), may be above the ceiling or below grade, and simple connections may be used where the arch is supported on masonry walls, concrete piers, or columns (Fig. 8-14).

Segmented arches are economical because of the ease of fabricating them and simplicity of field erection. Field splice joints are minimized; generally there is only one simple connection, at the crown (Fig. 8-15*c*). Except for extremely long spans, they are shipped in only two pieces. Erected, they need not be concealed by false ceilings, as may be necessary with trusses. And the cross section is large enough for segmented arches to be classified as heavy-timber construction.

Figure 8-16 shows typical moment connections for arches.

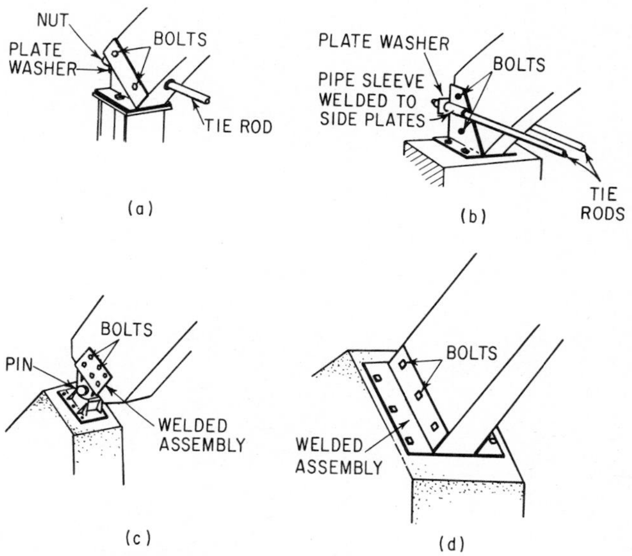

Fig. 8-14. Bases for segmented wood arches. (*a*) and (*b*) Tie rod anchored to arch shoe. (*c*) Hinge anchorage for large arch. (*d*) Welded arch shoe.

8-30. Wood Framing for Small Houses. Though skeleton framing may be used for one- and two-family dwellings, such structures up to three stories high generally are built with loadbearing walls. When wood framing is used, the walls are conventionally built with slender studs spaced 16 in. c to c. Similarly, joists and rafters, which are supported on the walls and partitions, are usually also spaced 16 in. c to c. Facings, such as sheathing and wallboard, and decking, floor underlayment, and roof sheathing are generally available in appropriate sizes for attachment to studs, joists, and rafters with that spacing. Sometimes, however, builders find use of 24-in. spacing with thicker attached materials more economical.

Wood studs are usually set in walls and partitions with wide faces perpendicular to the face of the wall or partition. The studs are nailed at the bottom to bear on a horizontal plank, called the bottom or sole plate, and at the top to a pair of horizontal planks, called the top plate. These plates often are the same size as the studs. Joists or rafters may be supported on the top plate or on a header, called a ribband, supported in a cutout in the studs.

Studs may be braced against racking by diagonals or horizontal blocking and facing materials, such as plywood or gypsum sheathing.

With 2×4 or 3×4-in. studs, 16 in. c to c, walls may be as tall as 14 ft; with

2 × 6-in. studs, up to 20 ft high. Angles at corners where stud walls or partitions meet should be framed solid so that no lath may extend from one room to another.

Three types of wood bearing wall construction generally are used in the United States: braced frame or eastern, balloon frame, and platform frame or western.

Braced-timber framing. This type originated in New England when the settlers were required to cut framing members from logs. It is the oldest type but it is also the strongest and most rigid.

In modern adaptations, sometimes called the **combination frame,** a braced-timber exterior frame consists of a wood sill resting on the foundation wall; comparatively heavy posts at all corners and some intermediate points, seated on the sill; horizontal

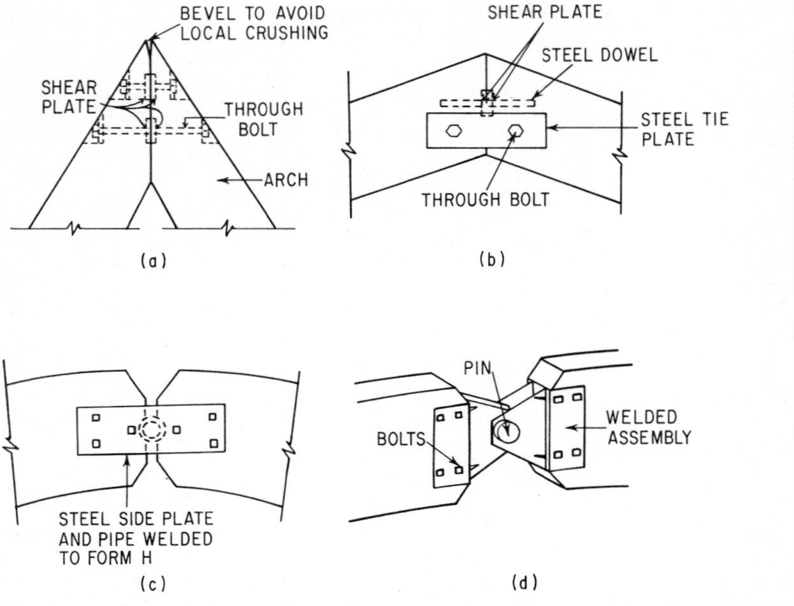

Fig. 8-15. Crown connections for arches. (*a*) For arches with slope 4:12 or greater, connection consists of pairs of back-to-back shear plates with through bolts or threaded rods counterbored into the arch. (*b*) For arches with flatter slopes, shear plates centered on a dowel may be used in conjunction with tie plates and through bolts. (*c*) and (*d*) Hinge at crown.

members, called girts, at second-floor level—"dropped girts" supporting ends of joists and "raised girts" paralleling the joists; a plate across the top of the frame, supporting roof rafters; and studs, 16 in. c to c between posts, extending from sill to girts and from girts to plate. Diagonal bracing is set between posts and sill, girts and plate. Posts are tenoned into sills, and girts are connected to posts with mortise-and-tenon joints.

Rough boards, ⅞ in. thick, generally are used to sheathe braced frames. Set horizontally, each board should be attached with at least two nails to each stud (three nails if boards are more than 6 in. wide).

Balloon framing (*Fig. 8-17a*). This type, built almost entirely of 2-in. lumber, is distinguished by two-story-high studs, to the faces of which joists at all levels are nailed. Ends of the upper-story joists rest on a false girt, or "ribband." It differs from the braced frame also because of the omission of diagonal bracing; instead the frame is stiffened by diagonal boards or sheathing. Fire stops—horizontal 2 × 4s—are placed between the studs at about mid-height and at floor level in each story.

Diagonal boards should be nailed to each stud with two or four nails. (Three nails are no better than two under racking action, because the center nail acts as a pivot and adds nothing to the strength.) Plywood or gypsum sheathing, applied in large sheets, also imparts considerable rigidity to the frame. It should be nailed with 8d common nails, 6 in. c to c, for ⅝-in. material and with 6d nails, 2½ in. c to c, for ¼-in. material.

Platform framing (*Fig. 8-17b and c*). This type, also known as western framing, is characterized by one-story-high walls supported by the floors at each level. At the bottom, joists rest on the sill and frame into headers of the same depth, which extend around the periphery of the building. Generally, before the superstructure is erected, the first-floor subflooring is laid on the joists to serve as a working platform. Then, a sole plate of the same size as the studs is set on top of the subfloor to form a base for the verticals. The studs extend to a plate at the next floor level. (Some builders assemble the wall framing in a horizontal position, then tilt it into place.) Joists are seated on the plate and frame into a header. After that, subflooring is placed over the joists, extending to the edge of the frame, another sole is laid, and the cycle is repeated. For this type of framing also, plywood or gypsum sheathing, well nailed, is highly desirable.

8-31. Timber Decking. Wood decking used for floor and roof construction may consist of solid-sawn planks with nominal thickness of 2, 3, or 4 in. Or it may be panelized or laminated. Panelized decking is made up of splined panels, usually about 2 ft wide.

Mechanically laminated decks are erected by setting square-edge dimension lumber on edge and fastening wide face to wide face. If side nails are used, they should be long enough to penetrate about 2½ lamination thickness for load transfer. Where supports are 4 ft c to c or less, side nails should be spaced not more than 30 in. c to c. They should be staggered one-third of the spacing in adjacent laminations. When supports are more than 4 ft c to c, nails should be spaced about 18 in. c to c alternately near top and bottom edges. These nails also should be staggered one-third of the spacing in adjacent laminations. Two side nails should be used at each end of butt-joined pieces. In addition, laminations should be toenailed to supports with nails 4 in. (20d) or longer. When supports are 4 ft c to c or less, alternate laminations should be nailed to alternate supports; for longer spans, alternate laminations at least should be nailed to every support.

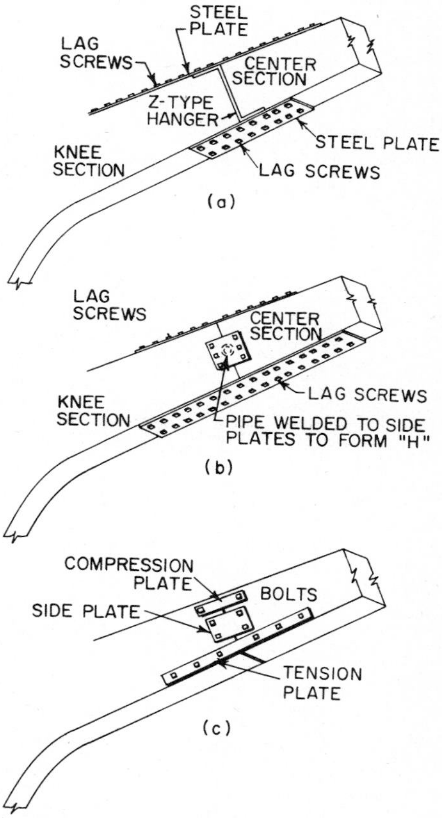

Fig. 8-16. Moment connections for an arch. (*a*) and (*b*) Connections with top and bottom steel plates. (*c*) Connections with side plates.

For glued-laminated decking, two or more pieces of lumber are laminated into a single decking member, usually with 2- to 4-in. nominal thickness.

Solid-sawn decking usually is fabricated with edges tongued and grooved, shiplap, or groove cut for splines, for transfer of vertical load between pieces. The decking may be end-matched, square end, or end-grooved for splines. As indicated in Fig. 8-18, the decking may be arranged in various patterns over supports.

In Type 1, the pieces are simply supported. Type 2 has a controlled random layup. Type 3 contains intermixed cantilevers. Type 4 consists of a combination of simple-span and two-span continuous pieces. Type 5 is two-span continuous.

In Types 1, 4, and 5, end joints bear on supports. For this reason, these types are recommended for thin decking, such as 2-in.

Type 3, with intermixed cantilevers, and Type 2, with controlled random layup, are used for deck continuous over three or more spans. These types permit some of the end joints to be located between supports. Hence, provision must be made for stress transfer at those joints. Tongue-and-groove edges, wood splines on each edge of the course, horizontal spikes between courses, and end matching or metal end splines may be used to transfer shear and bending stresses.

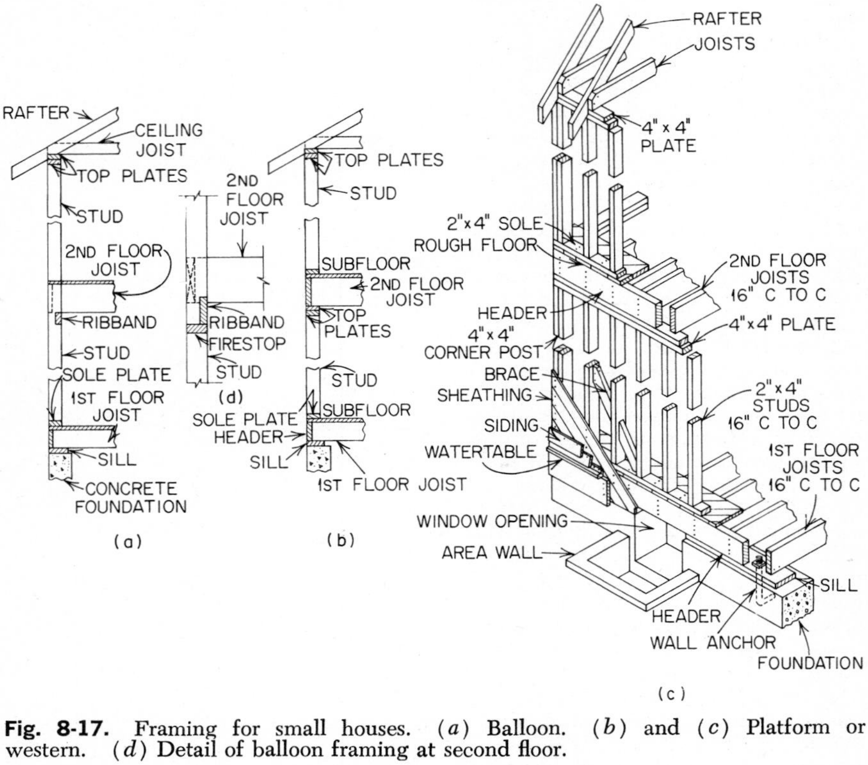

Fig. 8-17. Framing for small houses. (*a*) Balloon. (*b*) and (*c*) Platform or western. (*d*) Detail of balloon framing at second floor.

In Type 2, the distance between end joints in adjacent courses should be at least 2 ft for 2-in. deck and 4 ft for 3- and 4-in. deck. Joints approximately lined up (within 6 in. of being in line) should be separated by at least two courses. All pieces should rest on at least one support. And not more than one end joint should fall between supports in each course.

In Type 3, every third course is simple span. Pieces in other courses cantilever over supports and end joints fall at alternate quarter or third points of the spans. Each piece rests on at least one support.

To restrain laterally supporting members of 2-in. deck in Types 2 and 3, the pieces in the first and second courses and in every seventh course should bear on at least two supports. End joints in the first course should not occur on the same supports as end joints in the second course unless some construction, such as plywood overlayment, provides continuity. Nail end distance should be sufficient to develop the lateral nail strength required.

Heavy-timber decking is laid with wide faces bearing on the supports. Each piece must be nailed to each support. Each end at a support should be nailed to it. For 2-in. decking a 3½-in. (16d) toe and face nail should be used in each 6-in.-wide piece at supports, and three nails for wider pieces. Tongue-and-groove decking generally is also toenailed through the tongue. For 3-in. decking, each piece should be toenailed with one 4-in. (20d) spike and face-nailed with one 5-in. (40d) spike at each support. For 4-in. decking, each piece should be toenailed at each support with one 5-in. (40d) nail and face-nailed there with one 6-in. (60d) spike.

Courses of 3- and 4-in. double tongue-and-groove decking should be spiked to each other with 8½-in. spikes not more than 30 in. apart. One spike should not be more than 10 in. from each end of each piece. The spikes should be

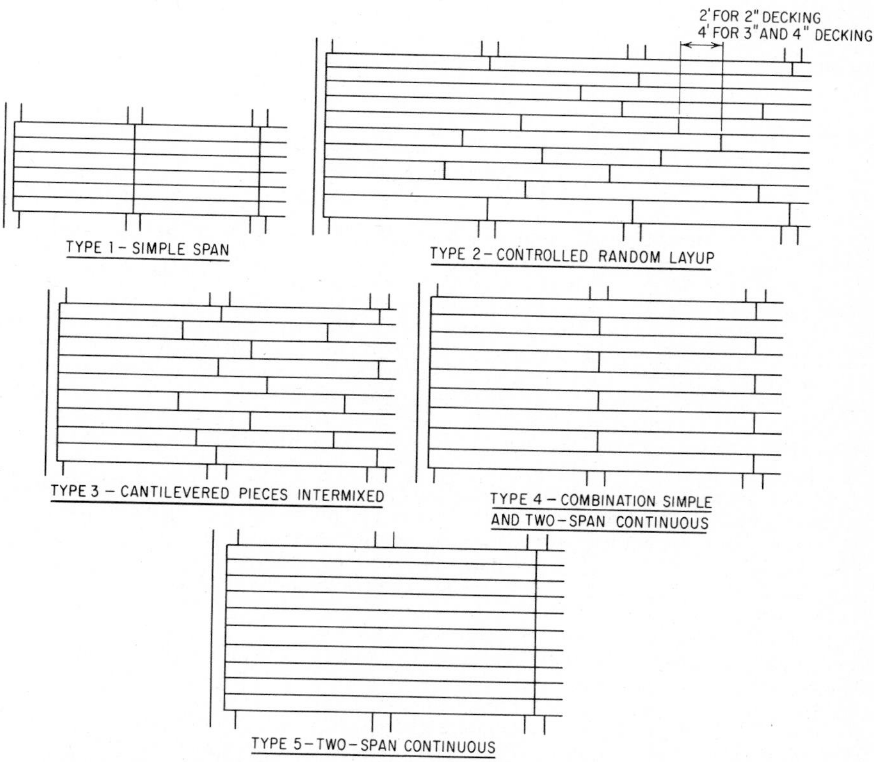

Fig. 8-18. Typical arrangement patterns for heavy-timber decking.

driven through predrilled holes. Two-inch decking is not fastened together horizontally with spikes.

Deck design usually is governed by maximum permissible deflection in end spans. But each design should be checked for bending stress. Deflection, in., may be calculated for a 12-in.-wide section from

$$\delta = \frac{wL^4}{K_1EI} \qquad (8\text{-}15)$$

where w = uniform load, psf
L = span, ft
E = modulus of elasticity, psi
I = moment of inertia, in.^{4}
K_1 = constant, obtained from Table 8-32

Bending stress, psi, for a 12-in.-wide section may be computed from

$$f_b = \frac{wL^2}{K_2 S} \tag{8-16}$$

where S = section modulus, in.^{3} Use the full cross section of the deck for Types 1, 4, and 5; use two-thirds for Types 2 and 3

K_2 = constant, obtained from Table 8-32

Table 8-32. Constants for Computing Deck Deflections and Bending Stresses

Deck type	K_1	K_2
1	0.0445	0.667
2	0.0566*	0.555
3	0.0608	0.555
4	0.0631	0.667
5	0.1071	0.667

* For 2-in. deck. Use 0.0671 for 3- and 4-in. deck.

Table 8-33 gives the maximum allowable, uniformly distributed total load for heavy-timber decking for various spans. The table is based on use of seasoned lumber, normal duration of loading, and loads applied normal to the deck. A limiting deflection ratio of $l/240$ and allowable bending stress of 1,200 psi have been assumed. For allowable bending stresses F_b other than 1,200, multiply tabulated values by $F_b/1{,}200$. For modulus of elasticity E other than 1,760,000, multiply tabulated values by $E/1{,}760{,}000$. For deflection limitations δ_A other than $l/240$, multiply tabulated values by $240\delta_A/l$; for example, for $l/360$, multiply by $^{240}/_{360}$.

(AITC 112, Standard for Heavy Timber Roof Decking, American Institute of Timber Construction, 333 W. Hampden Ave., Englewood, Colo., 80110; AITC "Timber Construction Manual," John Wiley & Sons, Inc., New York.)

8-32. Plywood. Plywood is made of softwood layers glued together; the grain of each layer is perpendicular to the grain of adjoining layers. Softwood plywood is manufactured in accordance with U.S. Product Standard PS1-66 for Softwood Plywood—Construction and Industrial. Plywood for exterior use should be made with an adhesive that is fully waterproof and suitable for wet service conditions and severe exposure.

PS1-66 classifies the softwoods according to stiffness:

Group 1. Douglas fir from Washington, Oregon, California, Idaho, Montana, Wyoming, British Columbia, and Alberta; western larch; southern pine (loblolly, longleaf, shortleaf, slash); yellow birch; tan oak.

Group 2. Port Orford cedar; Douglas fir from Nevada, Utah, Colorado, Arizona, and New Mexico; fir (California red, grand, noble, Pacific silver, white); western hemlock; red and white lauan; western white pine, red pine; Sitka spruce.

Group 3. Red alder, Alaska yellow cedar, jack pine, lodgepole and ponderosa pine, redwood, black, red, and white spruce.

Group 4. Incense and western red cedar, subalpine fir, eastern hemlock, sugar and eastern white pine, western poplar, Englemann spruce, paper birch, and bigtooth and quaking aspen,

Group 5. Balsam fir and balsam poplar.

Construction plywood is available under the standard in two grades made with exterior glue, Structural I and Structural II, and grades C-D, C-C, and Standard. Structural I is made only of Group I species. Structural II may be made of Group 1, 2, or 3 or of any combination of these species. Structural I and II are suitable for such applications as box beams, gusset plates, stressed-skin panels, and folded-plate roofs. Structural I A-C (face and back of veneer grade A and inner plies of grade C) is commonly used as an exterior type. Subflooring, wall sheathing, and roof decks can be made of grades C-D (plugged), standard and standard with exterior glue for interior use and of grades C-C and C-C (plugged) for exterior use. Structural I is stiffer than the other grades.

The standard also classifies plywood made for use as concrete forms in two grades. Plyform (B-B) Class I is limited to Group 1 species on face and back, with limitations on inner plies. Plyform (B-B) Class II permits Group 1, 2, or 3 for face and back, with limitations on inner plies. High-density overlay should be specified for both classes when highly smooth, grain-free concrete surfaces and maximum reuses are required. The bending strength of Plyform Class I is greater than that of Class II. Grades other than Plyform, however, may be used for forms.

Manufacturers stamp plywood to indicate group, grade, and Plyform class. They also include an identification index. This consists of a pair of numbers separated by a slant bar, on unsanded construction grades (Standard, Structural I and II, and C-C). The number on the left indicates the maximum recommended spacing,

Table 8-33. Allowable Total Uniform Load, Psf, for Heavy-timber Deck

Nominal deck thickness	Span ft	When bending governs F_b = 1,200 psi		When deflection governs E = 1,760,000 psi; $\delta_A = l/240$*				
		Types 1, 4, 5	Types 2 and 3	Type 1	Type 2	Type 3	Type 4	Type 5
2 in. (dressed thickness 1½ in.) single tongue and groove	7	73	61	39	50	53	55	93
	8	56	47	26	33	35	37	62
	9	44	37	18	24	24	26	43
	10	36	30	13	17	18	19	31
3 in. (dressed thickness 2½ in.) double tongue and groove	11	83	69	46	70	62	66	110
	12	70	58	35	54	48	51	85
	13	58	49	28	42	38	40	67
	14	51	43	22	34	31	32	54
	15	44	37	18	28	25	26	43
4 in. (dressed thickness 3½ in.) double tongue and groove	14	100	83	61	93	83	88	147
	15	87	73	50	75	68	71	120
	16	77	64	41	62	56	58	98
	17	68	56	34	52	46	48	81
	18	61	50	29	44	40	41	70
	19	54	45	24	37	33	35	57
	20	50	41	21	32	29	30	50

* l = span, in.

in., of supports when the panel is used for roof decking. The number on the right gives the maximum recommended spacing, in., of supports when the panel is used for subflooring. If a zero is stamped on the right, the panel should not be used for subflooring.

Appearance grades of plywood also are identified by a stamp. The group number indicates the species group of the veneer used on face and back. Letter symbols describe these veneer grades used in plywood:

N—Special-order natural-finish veneer. Selected all heartwood or all sapwood. Free of open defects. Allows some repairs.

A—Smooth and paintable. Neatly made repairs permissible. Also used for natural finish in less-demanding applications.

B—Solid-surface veneer. Circular repair plugs and tight knots permitted. Can be painted.

C—Minimum veneer permitted in exterior-type plywood. Knotholes up to 1 in. in diameter. (Occasional knotholes ½ in. larger permitted if total width of all knots and knotholes within a specified section does not exceed certain limits.) Limited splits permitted.

C plugged—Improved C veneer with splits limited to ⅛ in. in width and knotholes and borer holes limited to ¼ × ½ in.

D—Used only in interior-type plywood for inner plies and backs. Permits limited splits and knots and knotholes up to 2½ in. (½ in. larger under specified conditions).

Table 8-34 lists allowable loads and nail requirements for plywood roof sheathing. It also gives maximum support spacing and nail requirements for plywood subflooring. Table 8-35 presents similar data for plywood wall-sheathing. Table 8-36 gives the allowable uniform load to insure that the deflection of a plywood floor will not exceed $l/180$, where l is the joist spacing, in.

("The Plywood How to Book," "Plywood Construction Guide for Residential Building," "Plywood Construction Systems for Commercial and Industrial Buildings," "Plywood Concrete Forms," "Plywood Design Specification," "Plywood in Apart-

Table 8-34. Plywood Roof Sheathing and Subflooring Recommendations[a]

a. Roof Sheathing Laid with Face Grain Perpendicular to Joists[b]

Panel identification index	Plywood thickness, in.	Maximum span, in.[c]	Unsupported edge—max, length, in.[d]	Allowable roof loads, psf[e]										
				Spacing of supports, in., c to c										
				12	16	20	24	30	32	36	42	48	60	72
12/0	5/16	12	12	100 (130)										
16/0	5/16, 3/8	16	16	130 (170)	55 (75)									
20/0	5/16, 3/8	20	20	....	85 (110)	45 (55)								
24/0	3/8, 1/2	24	24	....	150 (160)	75 (100)	45 (60)							
30/12	5/8	30	26	....	....	145 (165)	85 (110)	40 (55)						
32/16	1/2, 5/8	32	28	....	....	....	90 (105)	45 (60)	40 (50)					
36/16	3/4	36	30	....	....	....	125 (145)	65 (85)	55 (70)	35 (50)				
42/20	5/8, 3/4, 7/8	42	32	....	....	....	....	80 (105)	65 (90)	45 (60)	35 (40)			
48/24	3/4, 7/8	48	36	....	....	....	....	....	105 (115)	75 (90)	55 (55)	40 (40)		
2-4-1	1 1/8	72	48	....	....	....	....	....	....	160 (160)	95 (95)	70 (70)	45 (45)	25 (30)
1 1/8 in. G 1 and 2	1 1/8	72	48	....	....	....	....	....	....	145 (145)	85 (85)	65 (65)	40 (40)	30 (30)
1 1/4 in. G3 and 4	1 1/4	72	48	....	....	....	....	....	....	160 (165)	95 (95)	75 (75)	45 (45)	25 (35)

[a] American Plywood Association.

[b] Applies to Standard, Structural I and II, and C-C grades only. Plywood continuous over two or more spans.

For applications where the roofing is to be guaranteed by a performance bond, recommendations may differ somewhat from these values. Contact American Plywood Association for bonded roof recommendations.

Use 2-in. (6d) common, smooth, ring-shank, or spiral-thread nails for ½ in. thick or less, and 2½-in. (8d) common, smooth, ring-shank, or spiral thread for plywood 1 in. thick or less (if ring-shank or spiral-thread nails have the same diameter as common). Use 2½-in. (8d) ring-shank or spiral-thread or 3-in. (10d) common, smooth-shank nails for 2-4-1, 1⅛-, and 1¼-in. panels. Space nails 6 in. at panel edges and 12 in. at intermediate supports, except that where spans are 48 in. or more, nails shall be 6 in. c to c at all supports.

[c] The spans shall not be exceeded for any load conditions.

[d] Provide adequate blocking, tongued-and-grooved edges, or other suitable edge support such as Ply-Clips when spans exceed indicated value. Use two PlyClips for 48-in. or greater spans, and one for lesser spans.

[e] Uniform load deflection limitation: 1/180 the span under live load plus dead load, 1/240 under live load only. Allowable live load shown without parentheses and allowable total load shown within parentheses.

Allowable loads were established by laboratory test and calculations assuming evenly distributed loads. Figures shown are not applicable for concentrated loads.

Table 8-34. Plywood Roof Sheathing and Subflooring Recommendations[a] (*Continued*)

b. Allowable Shears in Plywood Roof Diaphragms for Wind or Seismic Loads, Lb per Ft[f]
(Plywood and framing assumed already designed for perpendicular loads)

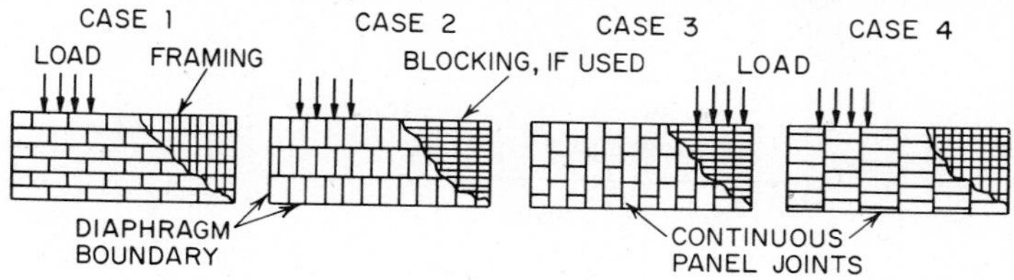

Plywood species and grade	Common nail length, in.	Min nail penetration into framing, in.	Min plywood thickness, in.	Min nominal width of framing member, in.	Blocked diaphragms: Nail spacing at diaphragm boundaries (all cases) and continuous panel edges parallel to load (cases 3 and 4)[g]: 4 / Nail spacing at other plywood panel edges[g]: 6	Blocked: $2\frac{1}{2}$ / 4	Blocked: 2 / 3	Unblocked diaphragms: Nails spaced 6 in. max at supported edges[g]: Load perpendicular to unblocked edges and continuous panel joints (case 1)	Unblocked: All other configurations (cases 2, 3, and 4)
Structural I	2 (6d)	$1\frac{1}{4}$	$\frac{5}{16}$ or $\frac{1}{4}$	2	250	375	420	167	125
				3	280	420	475	187	140
	$2\frac{1}{2}$ (8d)	$1\frac{1}{2}$	$\frac{3}{8}$	2	360	530	600	240	180
				3	400	600	675	267	200
	3 (10d)	$1\frac{5}{8}$	$\frac{1}{2}$	2	425	640	730	283	212
				3	480	720	820	320	240
Structural II	2 (6d)	$1\frac{1}{4}$	$\frac{5}{16}$ or $\frac{1}{4}$	2	167	250	280	111	84
				3	187	280	317	125	93
	$2\frac{1}{2}$ (8d)	$1\frac{1}{2}$	$\frac{3}{8}$	2	240	353	400	160	120
				3	267	400	450	178	133
	3 (10d)	$1\frac{5}{8}$	$\frac{1}{2}$	2	283	427	487	189	141
					320	480	547	213	160

[f] Design for diaphragm stresses depends on direction of continuous panel joints with reference to load, not direction of long dimension of plywood sheet.
[g] Space nails 12 in. c to c along intermediate framing members.

c. Plywood Subflooring and Subfloor Underlayment with Face Grain Perpendicular to Joists[h]

Panel identification index[i]	Plywood thickness, in.	Maximum span,[j] in.	Nail size and type	Nail spacing, in.: Panel edges	Nail spacing, in.: Intermediate
30/12	$\frac{5}{8}$	12[k]	$2\frac{1}{2}$-in. (8d) common	6	10
32/16	$\frac{1}{2}$, $\frac{5}{8}$	16[l]	$2\frac{1}{2}$-in. (8d) common[m]	6	10
36/16	$\frac{3}{4}$	16[l]	$2\frac{1}{2}$-in. (8d) common	6	10
42/20	$\frac{5}{8}$, $\frac{3}{4}$, $\frac{7}{8}$	20[l]	$2\frac{1}{2}$-in. (8d) common	6	10
48/24	$\frac{3}{4}$, $\frac{7}{8}$	24	$2\frac{1}{2}$-in. (8d) common	6	10
2-4-1 and $1\frac{1}{8}$-in. Groups 1 and 2	$1\frac{1}{8}$	48	3-in. (10d) common	6	6
$1\frac{1}{4}$-in. Groups 3 and 4	$1\frac{1}{4}$	48	3-in. (10d) common	6	6

[h] These values apply for Structural I and II, Standard sheathing, and C-C Exterior grades only, for application of $\frac{25}{32}$-in. wood-strip flooring or separate underlayment layer. Plywood continuous over two or more spans.
[i] Identification index appears on all panels, except $1\frac{1}{8}$- and $1\frac{1}{4}$-in. panels.
In some nonresidential buildings, special conditions may impose heavy concentrated loads and heavy traffic requiring subfloor constructions in excess of these minimums.
[j] Edges shall be tongued-and-grooved, or supported with blocking, unless underlayment is installed, or finish floor is $\frac{25}{32}$-in. wood strip. Spans limited to values shown because of possible effect of concentrated loads. At indicated maximum spans, panels will support uniform loads of at least 65 psf. For spans of 24 in. or less, panels will support loads of at least 100 psf.
[k] May be 16 in. if $\frac{25}{32}$-in. wood-strip flooring is installed at right angles to joists.
[l] May be 24 in. if $\frac{25}{22}$-in. wood-strip flooring is installed at right angles to joists.
[m] 2-in. (6d) common nail permitted if plywood is $\frac{1}{2}$ in. thick.

Table 8-35. Plywood Wall Sheathing and Siding Recommendations[a]

a. Plywood Wall Sheathing[b]

Panel identification index	Panel thickness, in.	Maximum stud spacing, in. — Exterior covering nailed to: Stud	Maximum stud spacing, in. — Exterior covering nailed to: Sheathing	Nail size[c]	Nail spacing, in. — Panel edges (when over framing)	Nail spacing, in. — Intermediate (each stud)
12/0, 16/0, 20/0	5/16	16	16[d]	2-in. (6d)	6	12
16/0, 20/0, 24/0	3/8	24	16 24[d]	2-in. (6d)	6	12
24/0, 32/16	1/2	24	24	2-in. (6d)	6	12

[a] American Plywood Association.

[b] The plywood may be installed horizontally or vertically. When plywood sheathing is used, building paper and diagonal wall bracing can be omitted. Plywood should be continuous over two or more spans.

[c] Common, annular, spiral-threaded, or T nails of the same diameter as common nails [0.113 in. for 2-in.-long (6d) nails] may be used. Staples also are permitted, but at reduced spacing.

[d] When sidings such as shingles are nailed only to the sheathing, apply the plywood with the face grain perpendicular to the supports.

b. Plywood Siding

Application	Plywood thickness, in.	Max spacing of supports, in. c to c	Nail length and type[e]	Nail spacing, in. — Panel edges	Nail spacing, in. — Intermediate
Panel siding[f]	3/8[g] 1/2 5/8 or thicker	16 24 24	2-in. (6d) casing or siding 2-in. (6d) casing or siding 2½-in. (8d) casing or siding	6 6 6	12 12 12
Lap siding	3/8[h] 1/2 5/8	16 20 24	2-in. (6d) casing or siding 2½-in. (8d) casing or siding 2½-in. (8d) casing or siding	One nail per stud along bottom edge	4 at vertical joint; 8 at studs (siding wider than 12 in.)
Bevel siding	9/16 min butt	16	2-in. (6d) casing or siding	Same as lap siding	
Soffits or ceilings	3/8 5/8	24 48	2-in. (6d) casing or siding	6 (or one nail at each support)	12 12

[e] Use galvanized, aluminum, or other noncorrosive nails. Nails may be color-coated.

[f] Battens, if used, can be applied with 2½-in.-long (8d) noncorrosive casing nails spaced 12 in. c to c (staggered).

[g] When separate sheathing is applied, 3/8-in. panel siding may be used over supports 24 in. on centers, 1/4-in. over supports 16 in. c to c.

[h] When separate sheathing is applied, 5/16-in. overlaid plywood lap siding may be used over supports 16 in. c to c.

Table 8-35. Plywood Wall Sheathing and Siding Recommendations[a] (*Continued*)

c. Allowable Shears in Plywood Walls, Lb/Lin In.[i]

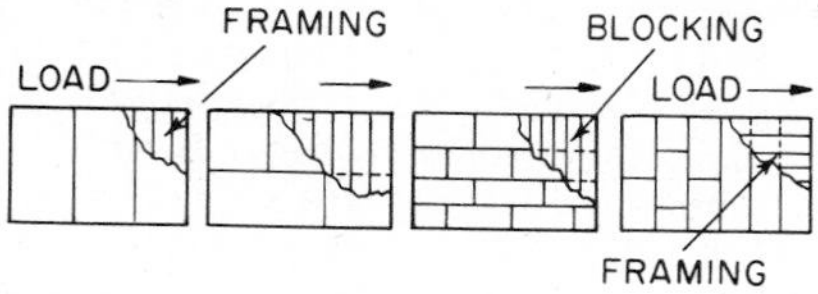

Plywood species	Min nominal plywood thickness, in.	Plywood sheathing direct to framing[j]				Plywood sheathing over 1/2-in. gypsum sheathing[k]				Plywood siding direct to framing[l]		Plywood siding over 1/2-in. gypsum sheathing[n]	
		Nail spacing, in., at plywood panel edges				Nail spacing, in., at plywood panel edges				Nail spacing, in., at plywood panel edges		Nail spacing, in., at plywood panel edges	
		6	4	2 1/2	2	6	4	2 1/2	2	6	4	6	4
Structural I	5/16	200	300	450	510	200	300	450	510				
	3/8	280	430	640	730	280	430	640	730	160 200[m]	240 300[m]	160 200[o]	240 300[o]
	1/2, 5/8	340	510	770	870	...	...	...	...	200	300	200	300
Structural II[p]	5/16	180	270	400	450	180	270	400	450				
	3/8	260	380	570	640	260	380	570	640	140 160[m]	210 240[m]	140 160[o]	210 240[o]
	1/2, 5/8	310	460	690	770	...	...	...	...	160	240	160	240

[i] All panel edges should be backed with framing, which should be 2-in. nominal or wider. The plywood may be installed horizontally or vertically. Space nails 12 in. c to c along intermediate framing members.

[j] Smooth, bright, common, or galvanized box nails—use 2-in. (6d) for 5/16-in. plywood, 2 1/2-in. (8d) for 3/8-in., and 3-in. (10d) for 1/2-in.

[k] Smooth, bright, common, or galvanized box nails—use 2 1/2-in. (8d) for 5/16-in. plywood and 3-in. (10d) for 3/8-in.

[l] Galvanized casing nails—use 2-in. (6d) or 2 1/2-in.[m] (8d) for 3/8-in. plywood and 2 1/2-in. (8d) for 1/2- and 5/8-in. For siding attached with galvanized box nails, use shears for sheathing with the same size nail. For 3/8- and 1/2-in. plywood, nails at panel edges should penetrate the full plywood thickness.

[m] Shear value for 2 1/2-in. (8d) nails.

[n] Galvanized casing nails—use 2 1/2-in. (8d) or 3-in.[o] (10d) for 3/8-in. plywood and 3-in. (10d) for 1/2- and 5/8-in. For siding attached with galvanized box nails, use shears for sheathing with the same size nail. For 3/8- and 1/2-in. plywood, nails at panel edges should penetrate the full plywood thickness.

[o] Shear value for 3-in. (10d) nails.

[p] Shears in this group may be increased 20% when using plywood one or more sizes thicker than minimum.

ments," "Plywood Truss Designs," "Plywood Components," American Plywood Association, 1119 A St., Tacoma, Wash. 98401.)

8-33. Composite Construction. The combination of concrete and steel is a prime example of composite construction. Steel serves a function, resisting tension, not handled as well by concrete. Wood and concrete do not complement each other in the same manner, since the strength properties of these materials do not differ so greatly. Nevertheless, some use has been made of concrete and wood in bridge decks. There are some materials, however, which, when combined with wood, yield improved constructions.

For example, an open-web joist has been fabricated with wood top and bottom chords and metal web members. Also, as an example, glued-laminated wood beams have been prestressed with high-strength steel strands. These were placed longitudinally in the tension zone of the beams and tensioned to apply a bending moment opposite to those that will be imposed in service. Results indicate that such beams are, on the average, about one-third stronger than nonprestressed beams, and variation in strength properties is about half (B. Bohannan, "Prestressed Laminated Wood Beams," U.S. Forest Service Research Paper, January, 1964, U.S. Forest Products Laboratory, Madison, Wis.). In another experiment, steel plates under tension were glued to the outer, tensile laminations of wood beams. Such beams were, on the average, 25% stiffer and 75% stronger then nonprestressed beams, and variability in strength properties decreased 70% (J. Peterson, "Wood Beams Prestressed with Bonded Tension Elements," *Journal of the Structural Division,* No. ST1, February, 1965, American Society of Civil Engineers).

8-34. Pole Construction. Wood poles are used for various types of construction, including flagpoles, utility poles, and framing for buildings. These employ preservatively treated round poles set into the ground as columns. The ground furnishes vertical and horizontal support and prevents rotation at the base (Table 8-37).

Isolated poles, such as flagpoles or signs, may be designed with lateral bearing values equal to twice these tabulated values.

In buildings, a bracing system can be provided at the top of the poles to reduce bending moments at the base and to distribute loads. Design of buildings supported by poles without bracing requires good knowledge of soil conditions, to eliminate excessive deflection or sidesway.

Bearing values under the base of poles should be checked. For backfilling the holes, well-tamped native soil, sand, or gravel may be satisfactory. But concrete or soil cement is more effective. They can reduce the required depth of embedment and improve bearing capacity by increasing the skin-friction area of the pole. Skin friction is effective in reducing uplift due to wind.

To increase bearing capacity under the base end of poles for buildings, concrete footings often are used. They should be designed to withstand the punching shear of the poles and bending moments. Thickness of concrete footings should be at least 12 in. Consideration should be given to use of concrete footings even in firm soils, such as hard dry clay, coarse firm sand, or gravel.

Calculation of required depth of embedment in soil of poles subject to lateral loads generally is impractical without many simplifying assumptions. While an approximate analysis can be made, the depth of embedment should be checked by tests, or at least against experience in the same type of soil.

Table 8-38 gives allowable stresses for treated poles, Table 8-39 lists standard dimensions for Douglas fir and southern pine poles, and Table 8-40 gives safe concentric column loads.

8-35. Timber Erection. Erection of timber framing requires experienced crews and adequate lifting equipment to protect life and property and to assure that the framing is properly assembled and not damaged during handling.

Each shipment of timber should be checked for tally and evidence of damage. Before erection starts, plan dimensions should be verified in the field. The accuracy and adequacy of abutments, foundations, piers, and anchor bolts should be determined. And the erector must see that all supports and anchors are complete, accessible, and free from obstructions.

Jobsite Storage. If wood members must be stored at the site, they should be placed where they do not create a hazard to other trades or to the members themselves. All framing, and especially glued-laminated members, stored at the site should be set above the ground on appropriate blocking. The members should be separated with strips, so that air may circulate around all sides of each member. The top and all sides of each storage pile should be covered with a moisture-resistant covering that provides protection from the elements, dirt, and jobsite debris. (Do not use clear polyethylene films, since wood members may be bleached by sunlight.)

Individual wrappings should be slit or punctured on the lower side to permit drainage of water that accumulates inside the wrapping.

Glued-laminated members of Premium and Architectural Appearance (and Industrial Appearance in some cases) are usually shipped with a protective wrapping of water-resistant paper. While this paper does not provide complete freedom

Table 8-36. Safe Uniform Floor Load for $l/180$ Deflection*

Joist spacing l, in.	Plywood thickness, in.	Safe load, psf
12	3/8	326
	1/2	435
	5/8	543
	3/4	653
16	3/8	193
	1/2	327
	5/8	408
	3/4	490
24	3/8	57
	1/2	111
	5/8	217
	3/4	311

* Exterior-type Douglas fir, C-C grade, applied with the face grain across supports.

Table 8-37. Allowable Lateral Passive Soil Pressure

Class of material	Allowable stress per foot of depth below natural grade, psf	Max allowable stress, psf
Good: Compact, well-graded sand and gravel, hard clay, well-graded fine and coarse sand (all drained so water will not stand)	400	8,000
Average: Compact fine sand, medium clay, compact sandy loam, loose sand and gravel (all drained so water will not stand)	200	2,500
Poor: Soft clay, clay loam, poorly compacted sand, clays containing large amounts of silt (water stands during wet season)	100	1,500

from contact with water, experience has shown that protective wrapping is necessary to insure proper appearance after erection. Though used specifically for protection in transit, the paper should remain intact until the roof covering is in place. It may be necessary, however, to remove the paper from isolated areas to make connections from one member to another. If temporarily removed, the paper should be replaced and should remain in position until all the wrapping is removed.

At the site, to prevent surface marring and damage to wood members, the following precautions should be taken:

Table 8-38. Stress Values for Treated Wood Poles*

(Normal duration of loading)

Species	Modulus of rupture, extreme fiber in bending, psi	Extreme fiber in bending F_b, psi	Modulus of elasticity E, psi
Cedar, northern white	4,000	1,080	800,000
Cedar, western red	6,000	1,620	1,000.000
Douglas fir	8,000	2,160	1,600,000
Hemlock, western	7,400	2,000	1,400,000
Larch, western	8,400	2,270	1,500,000
Pine, jack	6,600	1,780	1,100,000
Pine, lodgepole	6,600	1,780	1,000,000
Pine, ponderosa	6,000	1,620	1,000,000
Pine, red or Norway	6,600	1,780	1,200,000
Pine, southern	8,000	2,160	1,600,000

* See the latest edition of Standard Specifications and Dimensions for Wood Poles, ANSI 05.1.

Table 8-39. Dimensions of Douglas-fir and Southern-pine Poles*

Class		1	2	3	4	5	6	7	9	10
Min circumference at top, in.		27	25	23	21	19	17	15	15	12
Length of pole, ft	Ground-line distance from butt,† ft	Min circumference at 6 ft from butt, in.								
20	4	31.0	29.0	27.0	25.0	23.0	21.0	19.5	17.5	14.0
25	5	33.5	31.5	29.5	27.5	25.5	23.0	21.5	19.5	15.0
30	5½	36.5	34.0	32.0	29.5	27.5	25.0	23.5	20.5	
35	6	39.0	36.5	34.0	31.5	29.0	27.0	25.0		
40	6	41.0	38.5	36.0	33.5	31.0	28.5	26.5		
45	6½	43.0	40.5	37.5	35.0	32.5	30.0	28.0		
50	7	45.0	42.0	39.0	36.5	34.0	31.5	29.0		
55	7½	46.5	43.5	40.5	38.0	35.0	32.5			
60	8	48.0	45.0	42.0	39.0	36.0	33.5			
65	8½	49.5	46.5	43.5	40.5	37.5				
70	9	51.0	48.0	45.0	41.5	38.5				
75	9½	52.5	49.0	46.0	43.0					
80	10	54.0	50.5	47.0	44.0					
85	10½	55.0	51.5	48.0						
90	11	56.0	53.0	49.0						
95	11	57.0	54.0	50.0						
100	11	58.5	55.0	51.0						
105	12	59.5	56.0	52.0						
110	12	60.5	57.0	53.0						
115	12	61.5	58.0							
120	12	62.5	59.0							
125	12	63.5	59.5							

* See the latest edition of Standard Specifications and Dimensions for Wood Poles, ANSI 05.1.

† The figures in this column are intended for use only when a definition of ground line is necessary to apply requirements relating to scars, straightness, etc.

Lift members or roll them on dollies or rollers out of railroad cars. Unload trucks by hand or crane. Do not dump, drag, or drop members.

During unloading with lifting equipment, use fabric or plastic belts, or other slings that will not mar the wood. If chains or cables are used, provide protective blocking or padding.

Guard against soiling, dirt, footprints, abrasions, or injury to shaped edges or sharp corners.

Equipment. Adequate equipment of proper load-handling capacity, with control for movement and placing of members, should be used for all operations. It should be of such nature as to insure safe and expedient placement of the material. Cranes and other mechanical devices must have sufficient controls that beams, columns, arches, or other elements can be eased into position with precision. Slings, ropes, cables, or other securing devices must not damage the materials being placed.

The erector should determine the weights and balance points of the framing members before lifting begins, so that proper equipment and lifting methods may be employed. When long-span timber trusses are raised from a flat to a vertical

Table 8-40. Safe Concentric Column Loads on Douglas-fir and Southern-pine Poles, Lb*

Top diam, in.	8	7	6	5
ASA pole class	2	3	5	6
Unsupported pole length, ft (above ground line):				
0	65,000	50,000	37,000	25,000
10	65,000	50,000	28,000	14,000
12	65,000	41,000	26,000	12,000
14	52,000	31,000	18,000	9,000
16	40,000	24,000	14,000	7,000
18	33,000	20,000	11,000	6,000
20	28,000	17,000	10,000	5,000
25	19,000	12,000	7,000	
30	14,000	9,000		
35	11,000			

* See the latest edition of Standard Specifications and Dimensions for Wood Poles, ANSI 05.1.

position preparatory to lifting, stresses entirely different from design stresses may be introduced. The magnitude and distribution of these stresses depend on such factors as weight, dimensions, and type of truss. A competent rigger will consider these factors in determining how much suspension and stiffening, if any, is required and where it should be located.

Accessibility. Adequate space should be available at the site for temporary storage of materials from time of delivery to the site to time of erection. Material-handling equipment should have an unobstructed path from jobsite storage to point of erection. Whether erection must proceed from inside the building area or can be done from outside will determine the location of the area required for operation of the equipment. Other trades should leave the erection area clear until all members are in place and are either properly braced by temporary bracing or are permanently braced in the building system.

Assembly and Subassembly. Whether done in a shop or on the ground or in the air in the field, assembly and subassembly are dependent on the structural system and the various connections involved.

Care should be taken with match-marking on custom materials. Assembly must be in accordance with the approved shop drawings. Any additional drilling or dapping, as well as the installation of all field connections, must be done in a workmanlike manner.

Trusses are usually shipped partly or completely disassembled. They are assembled on the ground at the site before erection. Arches, which are generally shipped in half sections, may be assembled on the ground or connections may be made after the half arches are in position. When trusses and arches are assembled on the ground at the site, assembly should be on level blocking to permit connections to be properly fitted and securely tightened without damage. End compression joints should be brought into full bearing and compression plates installed where intended.

Prior to erection, the assembly should be checked for prescribed over-all dimensions, prescribed camber, and accuracy of anchorage connections. Erection should be planned and executed in such a way that the close fit and neat appearance of joints and the structure as a whole will not be impaired.

Field Welding. Where field welding is required, the work should be done by a qualified welder in accordance with job plans and specifications, approved shop drawings, and specifications of the American Institute of Steel Construction and the American Welding Society.

Hardware. All connections should fit snugly in accordance with job plans and specifications and approved shop drawings. All cutting, framing, and boring should be done in accordance with good shop practices. Any field cutting, dapping, or drilling should be done in a workmanlike manner, with due consideration given to final use and appearance.

Bracing. Structural elements should be placed to provide restraint or support, or both, to insure that the complete assembly will form a stable structure. This bracing may extend longitudinally and transversely. It may comprise sway, cross, vertical, diagonal, and like members that resist wind, earthquake, erection, acceleration, braking, and other forces. And it may consist of knee braces, cables, rods, struts, ties, shores, diaphragms, rigid frames, and other similar components in combinations.

Bracing may be temporary or permanent. Permanent bracing, required as an integral part of the completed structure, is shown on the architectural or engineering plans and usually is also referred to in the job specifications. Temporary construction bracing is required to stabilize or hold in place permanent structural elements during erection until other permanent members that will serve the purpose are fastened in place. This bracing is the responsibility of the erector and is normally furnished and erected by him. It should be attached so that children and other casual visitors cannot remove it or prevent it from serving as intended. Protective corners and other protective devices should be installed to prevent members from being damaged by the bracing.

In timber-truss construction, temporary bracing can be used to plumb trusses during erection and hold them in place until they receive the rafters and roof sheathing. The major portion of temporary bracing for trusses is left in place, because it is designed to brace the completed structure against lateral forces.

Failures during erection occur occasionally and regardless of construction material used. The blame can usually be placed on insufficient or improperly located temporary erection guys or braces, overloading with construction materials, or an externally applied force sufficient to render temporary erection bracing ineffective.

Structural members of wood must be stiff, as well as strong. They must also be properly guyed or laterally braced, both during erection and permanently in the completed structure. Large rectangular cross sections of glued-laminated timber have relatively high lateral strength and resistance to torsional stresses during erection. However, the erector must never assume that a wood arch, beam, or column cannot buckle during handling or erection.

Specifications often require that:

1. Temporary bracing shall be provided to hold members in position until the structure is complete.

2. Temporary bracing shall be provided to maintain alignment and prevent displacement of all structural members until completion of all walls and decks.

3. The erector should provide adequate temporary bracing and take care not to overload any part of the structure during erection.

While the magnitude of the restraining force that should be provided by a cable guy or brace cannot be precisely determined, general experience indicates that a brace is adequate if it supplies a restraining force equal to 2% of the applied load on a column or of the force in the compression flange of a beam. It does not take much force to hold a member in line; but once it gets out of alignment, the force then necessary to hold it is substantial.

Section 9

Windows

PHILIP M. GRENNAN

Consulting Engineer, Rockville Centre, N.Y.

Selection of a proper window for a building entails many considerations. Functional requirements are usually most important. However, size, type, arrangement, and materials very often establish the character of the building, and this external show of fenestration is an integral part of design and expression.

9-1. How Much Window Area? Distribution and control of daylight, desired vision of outdoors, privacy, ventilation control, heat loss, and weather resistance are all-important aspects of good design principles. Most building codes require that glass areas be equal to at least 10% of the floor area of each room. Nevertheless, it is good practice to provide glass areas in excess of 20% and to locate the windows as high in the wall as possible to lengthen the depth of light penetration. Continuous sash or one large opening in a room provides a more desirable distribution of light than separated narrow windows, and eliminates the dark areas between openings.

Location, type, and size of window are most important for natural ventilation. The pattern of air movement within a building depends to a great extent on the angle at which the air enters and leaves. It is desirable, particularly in summer, to direct the flow of air downward and across a room at a low level. However, the type and location of windows best suited to ventilation may not provide adequately for admission of light and clear vision, or perhaps proper weather protection. To arrive at a satisfactory relationship, it may be necessary to compromise with these functional requirements.

While building codes may establish a minimum percentage of glass area, they likewise may limit the use of glass and also require a fire rating in particular locations. In hazardous industrial applications subject to explosions, scored glass is an added precaution for quick release of pressure.

Heat loss is of economic importance and can also seriously affect comfort. Weather stripping or the use of windows with integral frame and trim can minimize air infiltration. Double glazing or sealed double glazing makes large glass areas feasible by materially reducing heat loss. Heat gain is accomplished mainly by the location of glass areas so that the rays of the sun can be admitted during the winter.

9-2. Window Materials. Window units are generally made of wood or metal. Plastics have been employed for special services.

Use of metal in window construction has been highly developed, with aluminum and steel most commonly used. Bronze, stainless steel, and galvanized steel are also extensively used for specific types of buildings and particular service. Use of metal windows is very often dictated by fire-resistance requirements of building codes.

9-3. Wood Windows. These can be used for most types of construction and are particularly adaptable to residential work. The most important factor affecting wood windows is weather exposure. However, with proper maintenance, long life and satisfactory service can be expected. Sash and frames have lower thermal conductivity than metal.

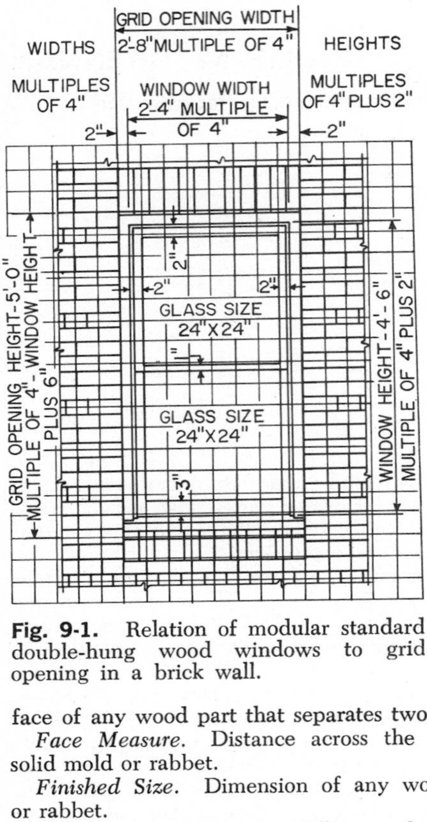

Fig. 9-1. Relation of modular standard double-hung wood windows to grid opening in a brick wall.

The kinds of wood commonly used to resist shrinking and warping for the exposed parts of windows are white pine, sugar pine, ponderosa pine, fir, redwood, cedar, and cypress. The stiles against which a double-hung window slides should be of hard pine or some relatively hard wood. The parts of a window exposed to the inside are usually treated as trim and are made of the same material as the trim. (See also Art. 9-4 and Figs. 9-1 to 9-3.)

9-4. Definitions for Wood Windows:

Sash. A single assembly of stiles and rails made into a frame for holding glass, with or without dividing bars. It may be supplied either open or glazed.

Window. One or more single sash made to fill a given opening. It may be supplied either open or glazed.

Window Unit. A combination of window frame, window, weather strip, balancing device, and at the option of the manufacturer, screen and storm sash, assembled as a complete and properly operating unit.

Measurements:

Between Glass. Distance across the face of any wood part that separates two sheets of glass.

Face Measure. Distance across the face of any wood part, exclusive of any solid mold or rabbet.

Finished Size. Dimension of any wood part over-all, including the solid mold or rabbet.

Wood Allowance. The difference between the outside opening and the total glass measurement of a given window or sash.

Stiles. Upright, or vertical, outside pieces of a sash or screen.

Rails. Cross, or horizontal, pieces of the framework of a sash or screen.

Check Rails. Meeting rails sufficiently thicker than the window to fill the opening between the top and bottom sash made by the check strip or parting strip in the frame. They are usually beveled and rabbeted.

Bar. Either vertical or horizontal member that extends the full height or width of the glass opening.

Muntin. Any short light bar, either vertical or horizontal.

Frame. A group of wood parts so machined and assembled as to form an enclosure and support for a window or sash.

Jamb. Part of frame that surrounds and contacts the window or sash the frame is intended to support.

Side Jamb. The upright member forming the vertical side of the frame.

Head Jamb. The horizontal member forming the top of the frame.

Rabbeted Jamb. A jamb with a rectangular groove along one or both edges to receive a window or sash.

Sill. The horizontal member forming the bottom of the frame.

Pulley Stile. A side jamb into which a pulley is fixed and along which the sash slides.

Casing. Molding of various widths and thicknesses used to trim window openings.

Parting Stop. A thin strip of wood let into the jamb of a window frame to separate the sash.

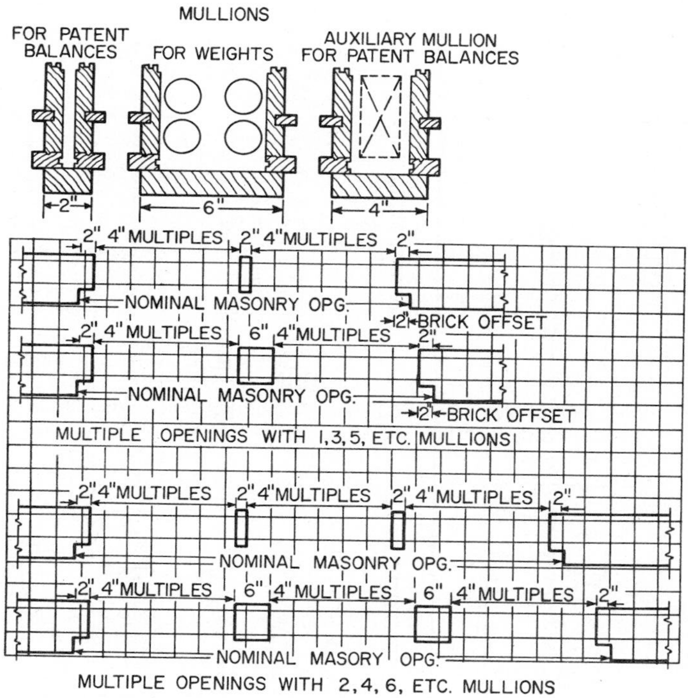

Fig. 9-2. Dimensions for multiple modular openings for double-hung wood windows.

Drip Cap. A molding placed on the top of the head casing of a window frame.

Blind Stop. A thin strip of wood machined so as to fit the exterior vertical edge of the pulley stile or jamb and keep the sash in place.

Extension Blind Stop. A molded piece, usually, of the same thickness as the blind stop, and tongued on one edge to engage a plow in the back edge of the blind stop, thus increasing its width and improving the weathertightness of the frame.

Dado. A rectangular groove cut across the grain of a frame member.

Jamb Liner. A small strip of wood, either surfaced four sides or tongued on one edge, which, when applied to the inside edge of a jamb, increases its width for use in thicker walls.

9-5. Steel Windows. These are, in general, made from hot-rolled, structural-grade new billet steel. However, for double-hung windows, the principal members are cold-formed from new billet strip steel. The principal manufacturers conform

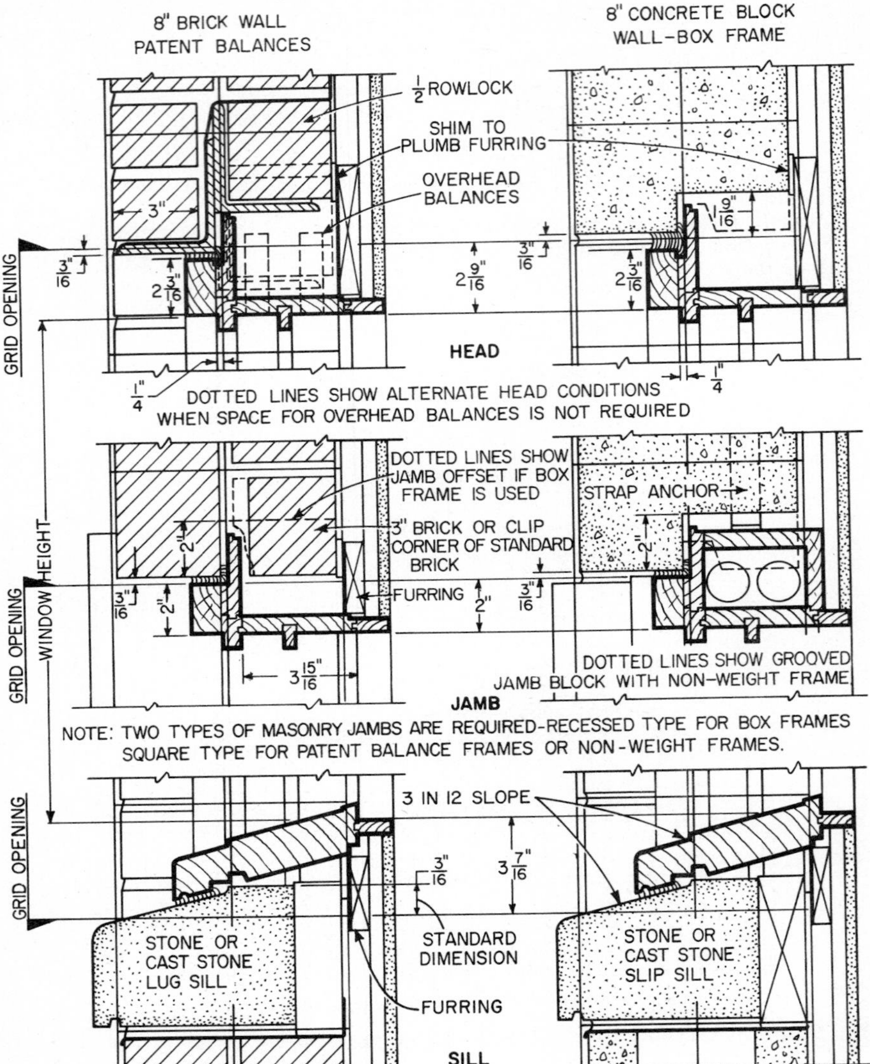

Fig. 9-3. Details of installation of double-hung wood windows. (*National Woodwork Manufacturers Association.*)

to the specifications of The Steel Window Institute, which has standardized types, sizes, thickness of material, depth of sections, construction, and accessories.

The life of steel windows is greatly dependent on proper shop finish, field painting, and maintenance. The more important aspects of proper protection against corrosion are as follows:

Shop Paint Finish. All steel surfaces should be thoroughly cleaned and free from rust and mill scale. A protective treatment of hot phosphate, cold phosphate-chromate, or similar method such as bonderizing or parkerizing is necessary to protect the metal and provide a suitable base for paint. All windows should then be given one shop coat of rust-inhibitive primer, baked on.

Field Painting. Steel windows should receive one field coat of paint either before or immediately after installation. A second coat should then be applied after glazing is completed and putty set.

Hot-dipped Galvanized. Assembled frames and assembled ventilators should be galvanized separately. Ventilators should be installed into frames after galvanizing and bonderizing. Specifications must guard against distortion by the heat of hot dipping, abrasion in handling, and damage to the zinc galvanizing by muriatic acid used to wash brickwork. Painting is optional.

9-6. Aluminum Windows. Specifications set up by the Aluminum Window Manufacturers' Association are classified as residential, commercial, and monumental. These specifications provide for a minimum thickness of metal and regulate stiffness of the component parts, as well as limiting the amount of air infiltration.

The manufacturers offer about the same window types that are available in steel. However, in addition to these standard types, substantial progress has been made in the development of aluminum windows as an integral part of the walls of a building. With the exception of ventilator sections, which are shop-fabricated, some windows are completely assembled at the job. The wide range of extruded aluminum sections now manufactured has made it feasible to treat large expanses of glass as window walls. Picture windows with many combinations of ventilating sash are also furnished in aluminum. Many features are available, such as aluminum trim and casing, weather stripping of stainless steel or Monel metal, combination storm sash and screens, sliding wicket-type screens, and metal bead glazing.

Aluminum windows should be protected for shipment and installation with a coating of clear methacrylate lacquer or similar material able to withstand the action of lime mortar.

9-7. Stainless-steel and Bronze Windows. These are high-quality materials used for durability, appearance, corrosion resistance, and minimum maintenance. Stainless steel is rustproof and extremely tough, with great strength in proportion to weight. Weather stripping of stainless steel is quite often used for wood and aluminum sash.

Bronze windows are durable and decorative and used for many fine commercial and monumental buildings.

9-8. Window Sizes. Manufacturers have contributed substantially toward standardization of window sizes and greater precision in fabrication. The American National Standards Institute has established an increment of sizing generally accepted by the building industry ("Basis for the Coordination of Building Materials and Equipment," A62.1).

The basis for this standardization is a nominal increment of 4 in. in dimension. However, it will be noted in Fig. 9-1, that for double-hung wood windows the 4-in. module applies to the grid dimensions, whereas the standard window opening is 4 in. less in width and 6 in. less in height than the grid opening. Also, to accommodate 2- and 6-in. mullions for multiple openings in brick walls, it is necessary to provide a 2-in. masonry offset at one jamb. (Modular Standard, National Woodwork Manufacturers Association.)

Use of modular planning effects maximum economies from the producer and simplification for the building industry. At the same time, a given layout does not confine acceptable products to such a limited range.

9-9. Window Types. The window industry has kept well apace with modern advances in building design and has made available a wide range of window types to satisfy the ever-increasing scope of requirements. Windows are basically functional, but at the same time, they are usually employed as a means of architectural expression. There are many materials, sizes, arrangements, details, and specific features from which to choose.

Types of windows (see symbols Fig. 9-4) include:

Pivoted windows, an economical, industrial window used where a good tight closure is not of major importance (Figs. 9-5 and 9-8). Principal members of metal units are usually 1⅜ in. deep, with frame and vent corners and muntin joints riveted. However, when frame and vent corners are welded, principal members may be 1¼ in. deep. Vents are pivoted about 2 in. above the center. These windows are adaptable to multiunits, both vertical and horizontal, with mechanical operation. Bottom of vent swings out and top swings in. No provision is made for screening. Putty glazing is placed inside.

Commercial projected windows, similar to pivoted windows (Figs. 9-6 and 9-8). These are an industrial-type window but also are used for commercial buildings where economy is essential. Vents are balanced on arms that afford a positive and easy control of ventilation and can be operated in groups by mechanical operators. Maximum opening about 35°. Factory provision is made for screens.

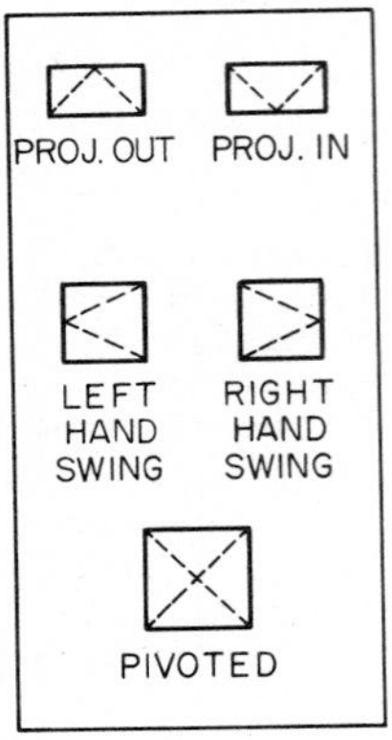

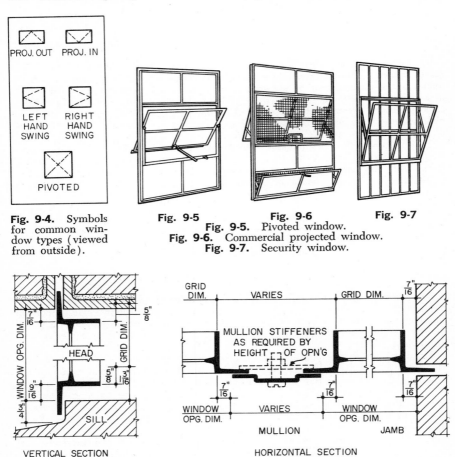

Fig. 9-4. Symbols for common window types (viewed from outside).

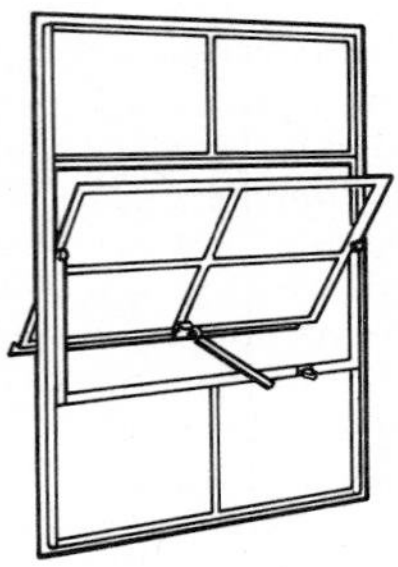

Fig. 9-5

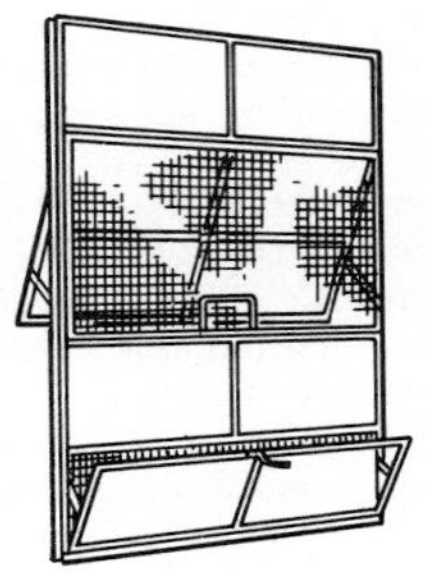

Fig. 9-6

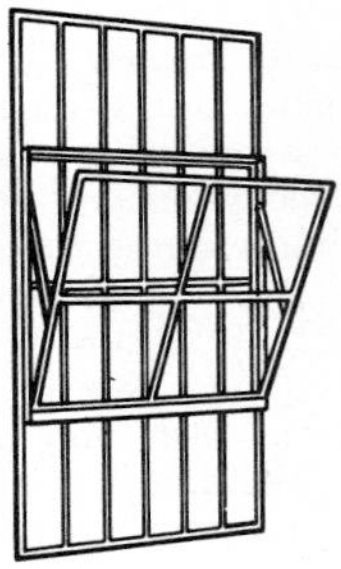

Fig. 9-7

Fig. 9-5. Pivoted window.
Fig. 9-6. Commercial projected window.
Fig. 9-7. Security window.

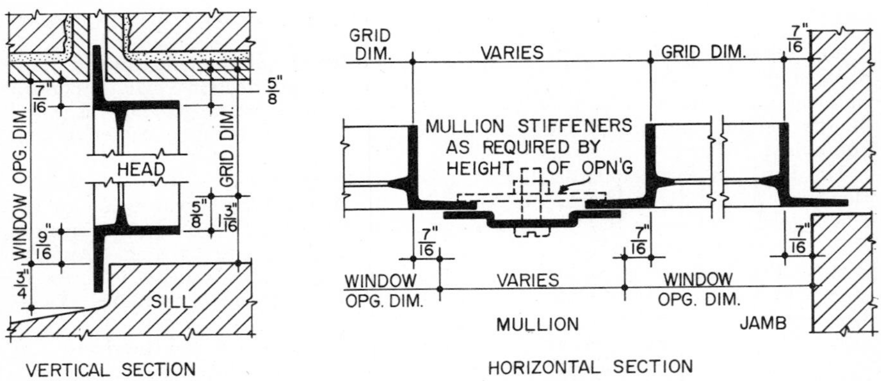

Fig. 9-8. Details of installation of commercial projected, pivoted, and security windows.

Security windows, an industrial window for use where protection against burglary is important, such as for factories, garages, warehouses, rear and side elevations of stores (Figs. 9-7 and 9-8). They eliminate the need for separate guard bars. They consist of a grille and ventilated window in one unit. Maximum grille openings are 7 × 16 in. The ventilating section, either inside or outside the grille, is bottom-hinged or projected. Muntins are continuous vertically and horizontally. Factory provision is made for screening. Glazing is placed from the inside.

Basement and utility windows, economy sash designed for use in basements, barracks, garages, service stations, areaways, etc. (Figs. 9-9 and 9-10). Ventilator opens inward and is easily removed for glazing or cleaning, or to provide a clear opening for passage of materials. Center muntin is optional. Screens are attached on the outside.

Architectural projected windows, a medium-quality sash used widely for commercial, institutional, and industrial buildings (Figs. 9-11 and 9-12). Ventilator arrangement permits cleaning from inside and provides a good range of ventilation control with easy operation. Steel ventilator sections are usually 1¼ in. deep, with corners welded. Frames may be 1⅜ in.* deep when riveted or 1¼ in. deep when welded. Screens are generally furnished at extra cost. Glazing may be either inside or outside.

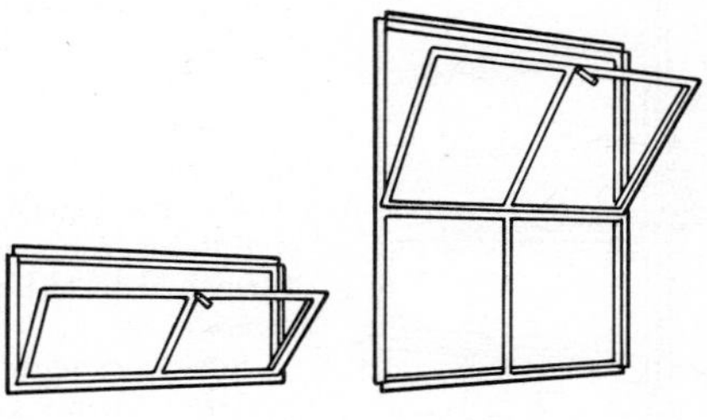

Fig. 9-9. Basement and utility windows.

Intermediate projected windows, high-quality ventilating sash used for schools, hospitals, commercial buildings, etc. (Figs. 9-13 and 9-14). When made of metal, corners of frames and ventilators are welded. Depth of frame and vent sections varies from "intermediate" to "heavy intermediate," depending on the manufacturer. Each vent is balanced on two arms pivoted to the frame and vent. The arms are equipped with friction shoes arranged to slide in the jamb sections. Screens are easily attached from inside. All cleaning is done from the inside. Glazing is set from the outside. Double insulating glass is optional.

Psychiatric projected windows, for use in housing for mental patients, to provide protection against exit but minimizing appearance of restraint (Fig. 9-15). Ventila-

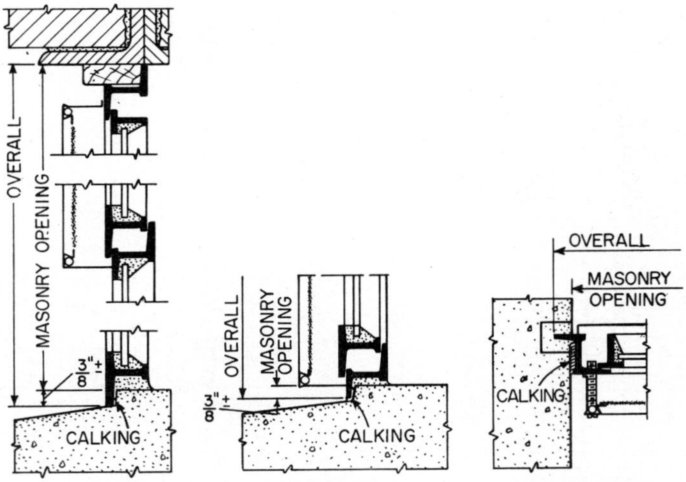

Fig. 9-10. Details of installation of basement and utility windows.

tors open in at the top, with a maximum clear opening of 5 to 6 in. Glazing is set outside. Screens also are placed on the outside but installed from inside. Metal casing can be had completely assembled and attached to window ready to install. Outside glass surfaces are easily washed from the inside.

Detention windows, designed for varying degrees of restraint and for different lighting and ventilating requirements. The guard type (Fig. 9-16) is particularly adaptable to jails, reform schools, etc. It provides security against escape through window openings. Ventilators are attached to the inner face of the grille and can be had with a removable key-locking device for positive control by attendants. Screens are installed from inside between vent and grille. Glazing is done from outside, glass being omitted from the grille at the vent section.

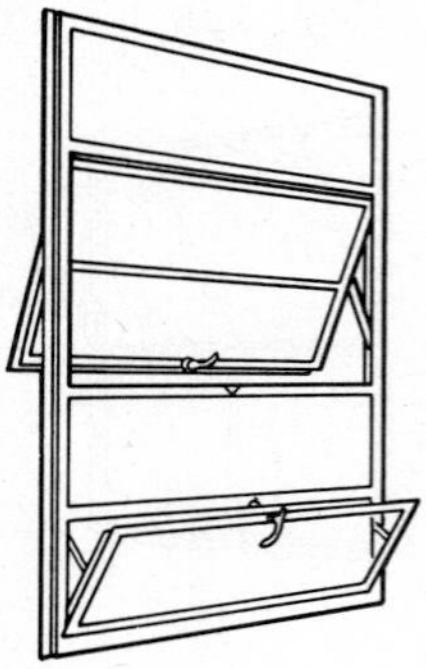

Fig. 9-11. Architectural projected window.

Residential double-hung windows (Figs. 9-1 to 9-3 and 9-17), available in different designs and weights to meet various service requirements for all types of buildings. When made of metal, the frame and ventilator corners are welded and weathertight. These windows are also used in combination with fixed picture windows for multiple window openings. They are usually equipped with weather stripping, which maintains good weathertightness. Screens and storm sash are furnished in either full or half sections. Glazing is done from outside.

Residence casements, available in various types, sizes, and weights to meet service requirements of homes, apartments, hotels, institutions, commercial buildings, etc. (Fig. 9-18). Rotary or lever operators hold the vent at the desired position, up to 100% opening. Screens and storm sash are attached to the inside of casement. Extended hinges on vents permit cleaning from inside. Glazing is done from outside.

Intermediate combination windows, with side-hinged casements and projected-in ventilators incorporated to furnish flexibility of ventilation control, used extensively for apartments, offices, hospitals, schools, etc., where quality is desired (Fig. 9-19). They are available in several weights with rotary or handle operation. When made of metal, corners of vents and frames are welded. Factory provision is made for screens. All cleaning is done from inside. Glazing is set from outside. Special glazing clips permit use of double insulating glass.

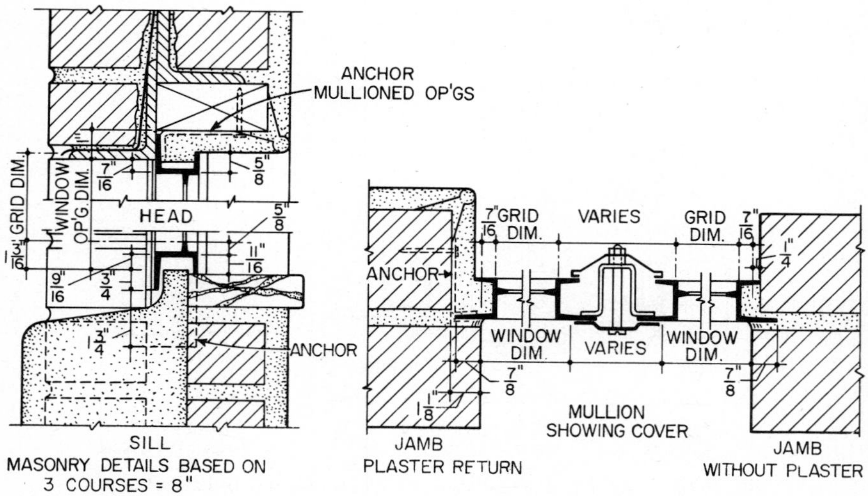

Fig. 9-12. Details of installation of architectural projected windows.

Intermediate casements, a heavier and better quality than residence casements, used particularly for fine residences, apartments, offices, institutions, and similar buildings (Fig. 9-20). Frames and ventilators of metal units have welded corners. Easy control of ventilation by rotary or handle operation. Extended hinges permit safe cleaning of all outside surfaces from inside. Screens are attached or removed from inside. Single or double glazing is set from outside.

Awning windows, suitable for residential, institutional, commercial, and industrial buildings (Fig. 9-21), furnishing approximately 100% opening for ventilation. Mechanical operation can provide for the bottom vent opening prior to the other vents, which open in unison. This is desirable for night ventilation. Manual operation can be had for individual or group venting. All glass surfaces are easily cleaned from inside through the open vents. Glazing is set from outside, storm sash and screens from inside.

Jalousie windows (Fig. 9-22) combine unobstructed vision with controlled ventilation and are used primarily for sunrooms, porches, and the like where protection from the weather is desired with maximum fresh air. The louvers can be secured in any position. Various kinds of glass, including obscure and colored, often are used for privacy or decoration. They do not afford maximum weathertightness but can be fitted with storm sash on the inside. Screens are furnished interchangeable with storm sash.

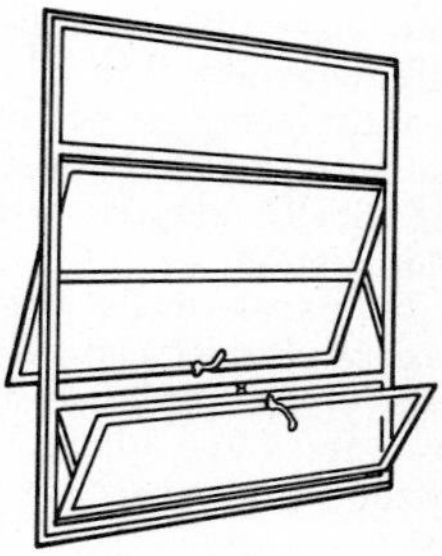

Fig. 9-13. Intermediate projected window.

Ranch windows, particularly suited to modern home design and also used effectively in other types of buildings where more light or better view is desired (Fig. 9-23). When made of metal, corners of frames and vents are welded. Depth of sections vary with manufacturers. These windows are designed to accommodate insulating glass or single panes. Screens are attached from inside.

Continuous top-hung windows, used for top lighting and ventilation in monitor and sawtooth roof construction (Fig. 9-24). They are hinged at the top to the structural-steel framing members of the building and swing outward at the bottom. Two-foot lengths are connected end to end on the job. Mechanical operators may be either manual or motor-powered, with lengths of runs as specified in Table 9-1. Sections may be installed as fixed windows. Glazing is set from outside.

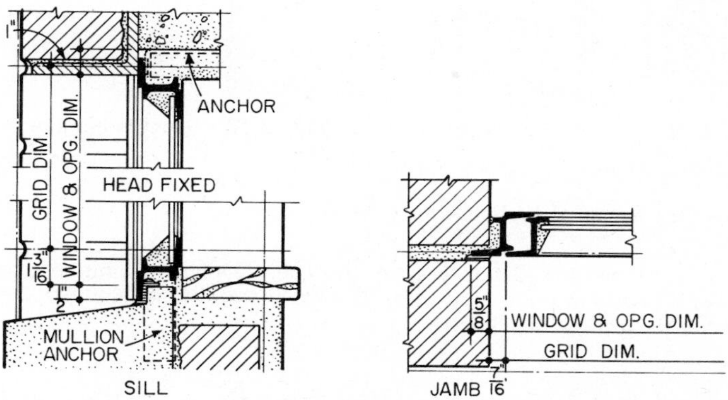

Fig. 9-14. Details of installation of intermediate projected window.

Additional types of windows (not illustrated) include:

Vertical pivoted, which sometimes are sealed with a rubber or plastic gasket.

Horizontal sliding sash usually in aluminum or wood for residential work.

Vertical folding windows, which feature a flue action for ventilation.

Double-hung windows with removable sash, which slide up and down, or another type which tilts for ventilation but does not slide.

Austral type, with sash units similar to double-hung but which operate in unison as projected sash. The sash units are counterbalanced on arms pivoted to the frame, the top unit projecting out at bottom and the bottom unit projected in at top.

Picture windows, often a combination of fixed sash with or without auxiliary ventilating units.

Store-front construction, usually semicustom-built of stock moldings of stainless steel, aluminum, bronze, or wood.

9-10. Windows in Wall-panel Construction. In the development of thin-wall buildings and wall-panel construction, windows have become more a part of the wall rather than units to fill an opening. Wall panels of metal and concrete incorporate windows as part of their general make-up; a well-integrated design will recognize this inherent composition.

Manufacturers of metal wall panels can furnish formed metal window frames specifically adapted to sash type and wall construction.

In panel walls of precast concrete, window frames can be cast as an integral part of the unit or set in openings as provided (Fig. 9-25). The cast-in-place method minimizes air infiltration around the frame and reduces installation costs on the job. Individual units or continuous bands of sash, both horizontal and vertical, can be readily adapted by the proper forming for head, jamb, and sill sections.

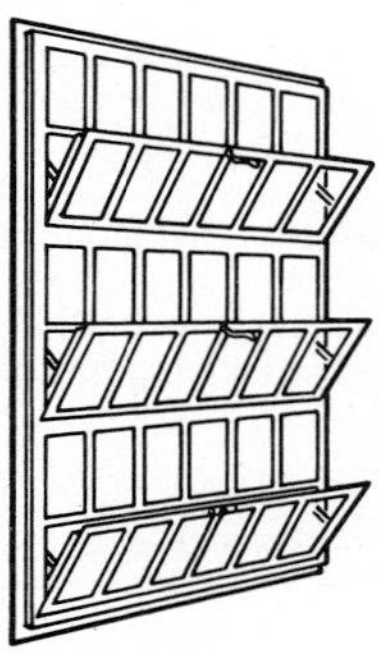

Fig. 9-15

Fig. 9-16

Fig. 9-15. Psychiatric projected window.
Fig. 9-16. Guard-type detention window.

Fig. 9-17. Double-hung windows.

Fig. 9-18

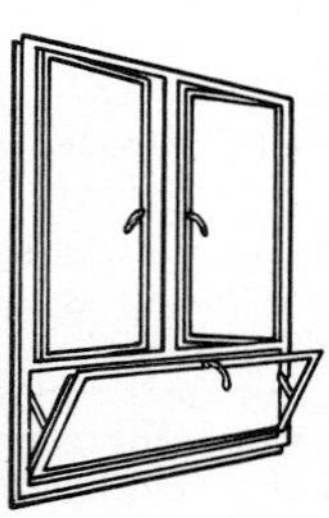

Fig. 9-19

Fig. 9-20

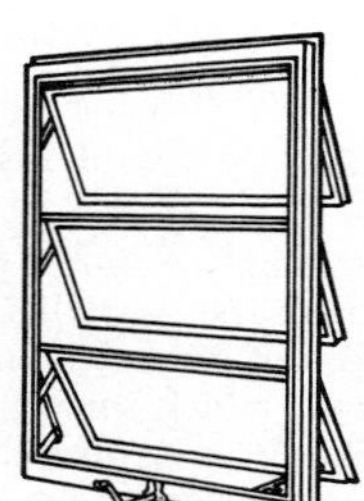

Fig. 9-21

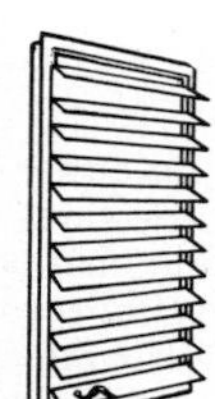

Fig. 9-22

Fig. 9-18. Residence casement window.
Fig. 9-19. Intermediate combination window.
Fig. 9-20. Intermediate casement window.
Fig. 9-21. Awning window.
Fig. 9-22. Jalousie window.

9-11. Mechanical Operators for Windows. Mechanical operation for pivoted and projected ventilators is achieved with a horizontal torsion shaft actuated by an endless hand chain or by motor power. Arms attached to the torsion shaft open and close the vents.

Two common types of operating arms are the lever and the rack and pinion (Fig. 9-26). The lever is used for manual operation of small groups of pivoted vents where rapid opening is desirable. The rack and pinion is used for longer runs of vents and can be motor-powered or manually operated. The opening and closing by rack and pinion are slower than by lever arm.

Usually one operating arm is furnished for each vent less than 4 ft wide and two arms for each vent 4 ft or more wide. Mechanical operators should be installed by the window manufacturer before glazing of the windows.

Continuous top-hung windows are mechanically controlled by rack-and-pinion or tension-type operators (Fig. 9-26). Maximum length of run is given in Table 9-1.

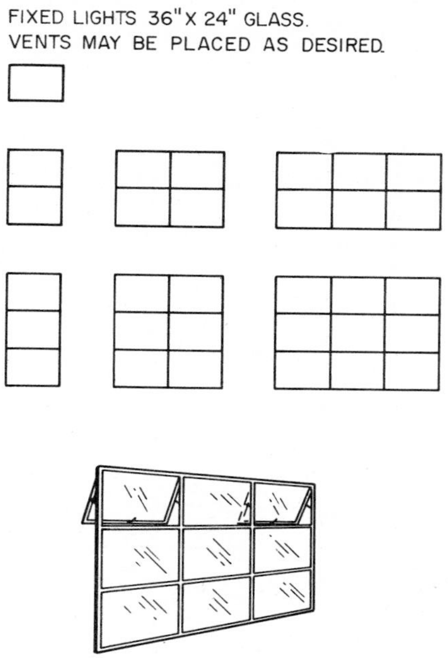

FIXED LIGHTS 36" X 24" GLASS.
VENTS MAY BE PLACED AS DESIRED.

Fig. 9-23. Ranch windows.

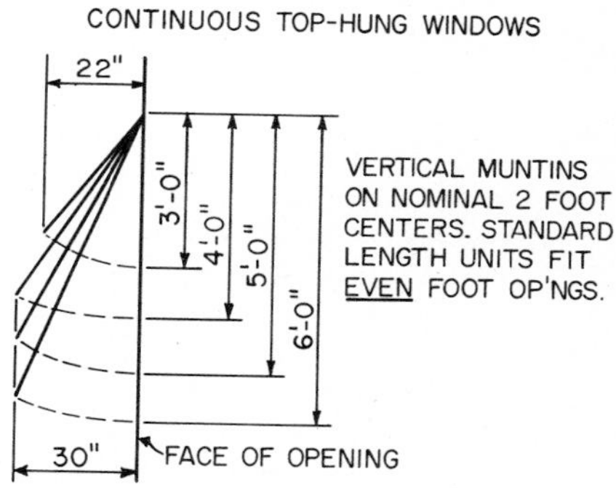

VERTICAL MUNTINS
ON NOMINAL 2 FOOT
CENTERS. STANDARD
LENGTH UNITS FIT
EVEN FOOT OP'NGS.

Fig. 9-24. Continuous top-hung windows.

9-12. Weather Stripping and Storm Sash. Weather stripping is a barrier against air leakage through cracks around sash. Made of metal or a compressible resilient material, it is very effective in reducing heat loss.

Storm sash might be considered an over-all transparent blanket that shields the window unit and reduces heat loss. Weather stripping and storm sash also reduce condensation and soot and dirt infiltration, as well as decreasing the amount of cold air near the floor.

Storm sash and screens are often incorporated in a single unit.

9-13. Window Glass. Various types and grades of glass are used for glazing:

Clear Window Glass. This is the most extensively used type for windows in all classes of buildings. A range of grades, as established by Federal Government Standard DD-G-451c, classifies quality according to defects. The more commonly used grades are A and B. A is used for the better class of buildings where appearance is important, and B is used for industrial buildings, some low-cost residences, basements, etc.

With respect to thickness, clear window glass is classified as "single-strength," about $\frac{3}{32}$ in. thick; "double-strength," about $\frac{1}{8}$ in. thick; and "heavy-sheet," up to $\frac{7}{32}$ in. thick. Maximum sizes are as follows: single-strength, 40×50 in.; double-strength, 60×80 in.; and heavy sheet, 76×120 in. Because of flexibility, single strength and double strength should never be used in areas exceeding 12 sq ft, and for appearance's sake areas should not exceed 7 sq ft.

Table 9-1. Maximum Run for Mechanical Operators, Ft

Operators for Pivoted Windows

Type of operator	Lever (1-in. pipe)				Rack and pinion (1-in. pipe)				
Control	Manual				Manual				Electric oil encl.
	Dust*		Oil†		Dust*		Oil†		
	Chain	Rod	Chain	Rod	Chain	Rod	Chain	Rod	
Sash open 16-in.	100	80	150	120	160	140	200	180	240

Operators for Projected Windows

Type of operator	Rack and pinion				
Control	Manual				Electric
	Dust*		Oil†		
	Chain	Rod	Chain	Rod	
	100	90	140	120	200

Operators for Continuous Top-hung Windows

Type of operator			Rack and pinion	Tension	
Control			Manual or electric	Manual	Electric
Limits	Sash	Height, ft	1-in. pipe	1-in. pipe	1-in. pipe
	Vertical	3, 4, 5, 6	150	200	300
	30° slope	3	120	160	240
		4	100	140	210
		5	80	120	180
		6	60	100	150

* Dust-protected—cast gearing.
† Oil-enclosed—cut gearing.

Plate and float glass have, in general, the same performance characteristics. They are of superior quality, more expensive, and have better appearance, with no distortion of vision at any angle. Showcase windows, picture windows, and exposed windows in offices and commercial buildings are usually glazed with polished plate or float glass. Thicknesses range from ⅛ to ⅞ in. There are two standard qualities, *silvering* and *glazing*, the latter being employed for quality glazing.

Processed glass and rolled figured sheet are general classifications of obscure glass. There are many patterns and varying characteristics. Some provide true obscurity with a uniform diffusion and pleasing appearance, while others may give a maximum transmission of light or a smoother surface for greater cleanliness. The more popular types include a clear, polished surface on one side with a pattern for obscurity on the other side.

Obscure wired glass usually is specified for its fire-retarding properties, although it is also used in doors or windows where breakage is a problem. It should not be used in pieces over 720 sq in. in area (check local building code).

Polished wired glass, more expensive than obscure wired glass, is used where clear vision is desired, such as in school or institutional doors.

There are also many special glasses for specific purposes:

Heat-absorbing glass reduces heat, glare, and a large percentage of ultraviolet rays, which bleach colored fabrics. It often is used for comfort and reduction of air-conditioning loads where large areas of glass have a severe sun exposure. Because of differential temperature stresses and expansion induced by heat absorption under severe sun exposure, special attention must be given to edge conditions. Glass having clean-cut edges is particularly desirable, because these affect the edge strength, which, in turn, must resist the central-area expansion. A resilient glazing material should be used.

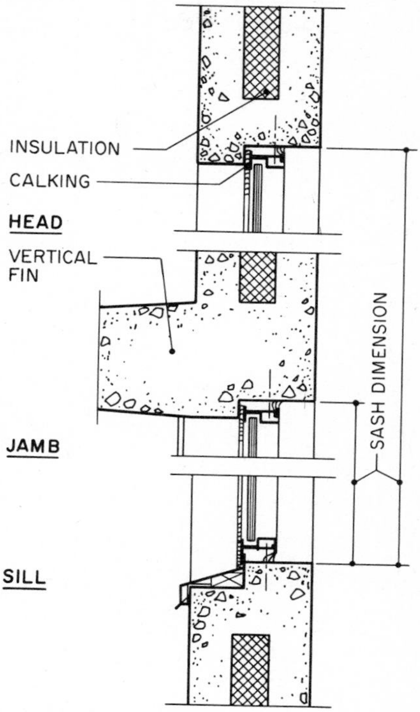

Fig. 9-25. Window installed in precast-concrete sandwich wall panel.

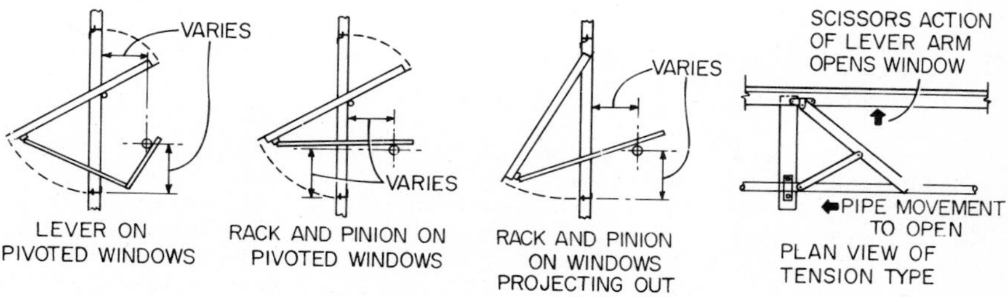

Fig. 9-26. Levers, rack-and-pinion, and tension-type window operators.

Corrugated glass, corrugated wired glass, and corrugated plastic panels are used for decorative treatments, diffusing light, or as translucent structural panels with color.

Laminated glass consists of two or more layers of glass laminated together by one or more coatings of a transparent plastic. This construction adds strength. Some types of laminated glass also provide a degree of security, sound isolation, heat absorption, and glare reduction. Where color and privacy are desired, fadeproof opaque colors can be included. When fractured, laminated glass tends to adhere to the inner layer of plastic and, therefore, shatters into small splinters, thus minimizing the hazard of flying glass.

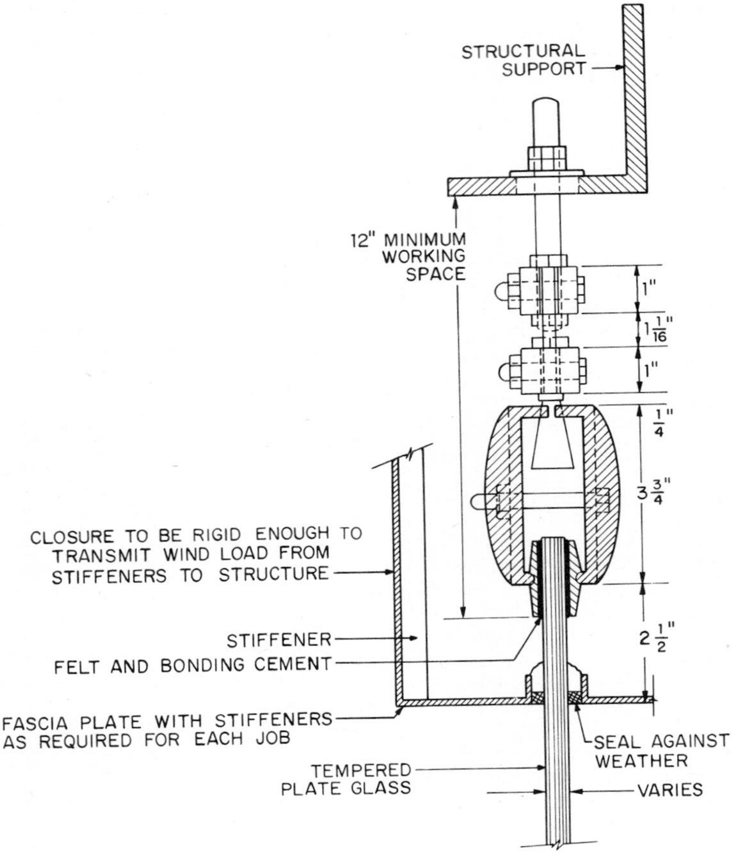

Fig. 9-27. Typical details of suspended glazing. (*F. H. Sparks Co., Inc., New York, N.Y.*)

Bullet-resisting glass is made of three or more layers of plate glass laminated under heat and pressure. Thicknesses of this glass vary from ¾ to 3 in. The more common thicknesses are 1³⁄₁₆ in., to resist medium-powered small arms; 1½ in., to resist high-powered small arms; and 2 in., to resist rifles and submachine guns. (Underwriters Laboratories lists materials having the required properties for various degrees of protection.) Greater thicknesses are used for protection against armor-piercing projectiles. Uses of bullet-resisting glass include cashier windows, bank teller cages, toll-bridge booths, peepholes, and many industrial and military applications. Transparent plastics also are used as bullet-resistant materials, and some of these materials have been tested by the Underwriters Laboratories. Thicknesses of 1¼ in. or more have met UL standards for resisting medium-powered small arms.

Tempered glass is produced by a process of reheating and sudden cooling that greatly increases strength. All cutting and fabricating must be done before tempering. Doors of ½- and ¾-in.-thick tempered glass are commonly used for commercial building. Other uses, with thicknesses from ⅛ to ⅞ in., include backboards for basketball, showcases, balustrades, sterilizing ovens, and windows, doors, and mirrors in institutions. Although tempered glass is 4½ to 5 times as strong as annealed glass of the same thickness, it is breakable, and when broken, disrupts into innumerable small fragments of more-or-less cubical shape.

Tinted and coated glasses are available in several types and for varied uses. As well as decor, these uses can provide for light and heat reflection, lower light transmission, greater safety, sound reduction, reduced glare, and increased privacy.

Transparent mirror glass appears as a mirror when viewed from a brightly lighted side, and is transparent to a viewer on the darker opposite side. This one-way-vision glass is available as a laminate, plate or float, tinted, and in tempered quality.

Plastic window glazing, made of such plastics as acrylic or polycarbonate, is used for urban school buildings and in areas where high vandalism might be anticipated. These plastics have substantially higher impact strength than glass or tempered glass. Allowance should be made in the framing and installation for expansion and contraction of plastics, which may be about eight times as much as that of glass. Note also that the modulus of elasticity (stiffness) of plastics is about one-twentieth that of glass. Standard sash, however, usually will accommodate the additional thickness of plastic and have sufficient rabbet depth.

Suspended glazing (Fig. 9-27) utilizes metal clamps bonded to tempered plate glass at the top edge, with vertical glass supports at right angles for resistance to wind pressure. These vertical supports, called stabilizers, have their exposed edges polished. The joints between the large plates and the stabilizers are sealed with a bonding cement. The bottom edge or sill is held in position by a metal channel, and sealed with resilient waterproofing. Suspended glazing offers much greater latitude in use of glass and virtually eliminates visual barriers.

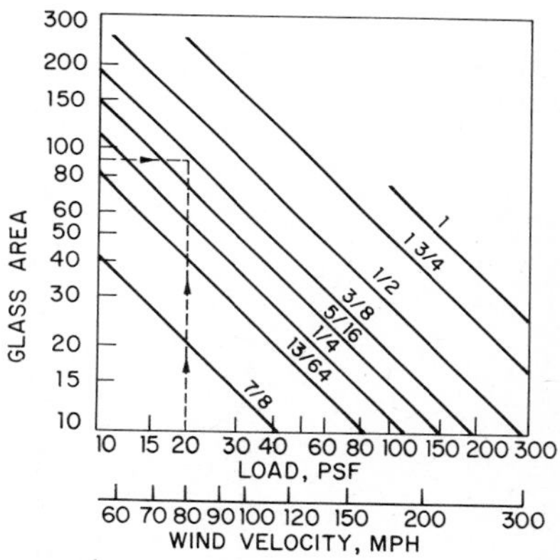

Fig. 9-28. Wind load chart indicates maximum glass area for various nominal thicknesses of plate or sheet glass to withstand specified wind pressures, with design factor of 2.5. (*Pittsburgh Plate Glass Co.*)

Special glazing. A governmental specification Z-97, adopted by many states, requires entrance-way doors and appurtenances glazed with tempered, laminated, or plastic material.

Thickness. Figure 9-28 can be used to determine thickness of plate, float, or sheet glass for a given glass area, or the maximum area for a given thickness to withstand a specified wind pressure. Based on minimum thickness permitted by Federal Specification DD-G-451c, the wind-load chart provides a safety factor of 2.5. It is intended for rectangular lights with four edges glazed in a stiff, weathertight rabbet.

For example, determine the thickness of a 108 × 120-in. (90-sq ft) light of polished plate glass to withstand a 20-psf wind load. Since the 20-psf and 90-sq ft ordinates intersect at the ⅜-in. glass thickness line, the thickness to use is ⅜ in.

The correction factors in Table 9-2 also allow Fig. 9-28 to be used to determine the thickness for certain types of fabricated glass products. The table, however, makes no allowance for the weakening effect of such items as holes, notches, grooves, scratches, abrasion, and welding splatter.

The appropriate thickness for the fabricated glass product is obtained by multiplying the wind load, psf, by the factor given in Table 9-2. The intersection of

Table 9-2. Resistance of Glass to Wind Loads

Product	*Factor**
Plate, float, or sheet glass	1.0
Rough rolled plate glass	1.0
Sand-blasted glass	2.5
Laminated glass†	2.0
Sealed double glazing:‡	
Metal edged, up to 30 sq ft	0.667
Metal edged, over 30 sq ft	0.556
Glass edged	0.5
Heat-strengthened glass	0.5
Fully tempered glass	0.25

* Enter Fig. 9-28 with the product of the wind load in psf multiplied by the factor.

† At 70°F or above, for two lights of equal thickness laminated to 0.015-in.-thick vinyl. At 0°F, factor approaches 1.

‡ For thickness, use thinner of the two lights, not total thickness.

the vertical line drawn from the adjusted load and the horizontal line drawn from the glass area indicates the minimum recommended glass thickness.

Table 9-3 gives the over-all thermal conductance, or U factor, and weight of glass for several thicknesses. Table 9-4 presents typical sound-reduction factors for glass, and for comparison, for other materials. As a sound barrier, glass, inch for inch, is about equal to concrete and better than most brick, tile, or plaster. Double glazing (Fig. 9-29) is particularly effective where high resistance to sound transmission is required. Unequal glass thicknesses minimize resonance in this type of glazing, and a nonreflective mounting reduces sound transmission. For optimum performance, sound-isolation glazing should be carefully detailed and constructed, and should be mounted in airtight, heavy walls with acoustically treated surfaces.

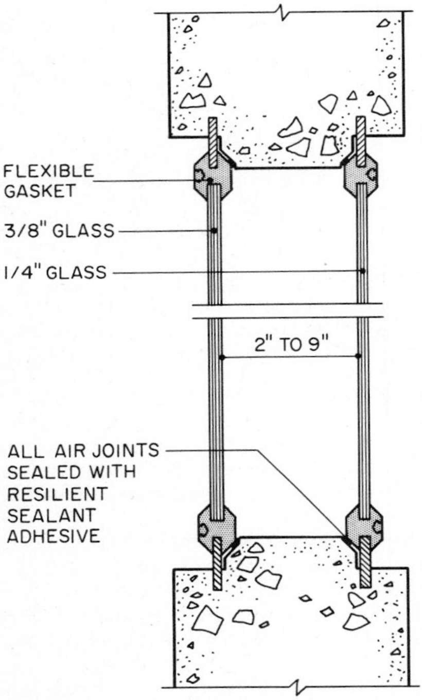

Fig. 9-29. Sound-isolation glazing.

9-14. Glazing Compounds. Putty, glazing compound, rubber, or plastic strips and metal or wood molds are used for holding glass in place in sash. Metal clips for metal sash and glazing points for wood sash also are employed for this purpose.

Putty and glazing compounds are generally classified in relation to the sash material and should be used accordingly for best results.

Bedding of glass in glazing compound is desirable, because it furnishes a smooth bearing surface for the glass, prevents rattling, and eliminates voids where moisture can collect (Commercial Standards, U.S. Department of Commerce). Glazing terminology relative to putty follows:

Face Puttying (Fig. 9-30). Glass is inserted in the glass rabbet and securely wedged where necessary to prevent shifting. Glazing points are also driven into the wood to keep the glass firmly seated. The rabbet then is filled with putty, the putty being beveled back against the sash and muntins with a putty knife.

Back Puttying (Fig. 9-30*b*). After the sash has been face-puttied, it is turned over, and putty is run around the glass opening with a putty knife, thus forcing putty into any voids that may exist between the glass and the wood parts.

Bedding (Fig. 9-30*c*). A thin layer of putty or bedding compound is placed in the rabbet of the sash, and the glass is pressed onto this bed. Glazing points are then driven into the wood, and the sash is face-puttied. The sash then is turned over, and the excess putty or glazing compound that emerged on the other side is removed.

Wood-stop or Channel Glazing (Fig. 9-30*d*). A thin layer of putty or bedding compound is placed in the rabbet of the sash, and the glass is pressed onto this

Table 9-3. Over-all Conductance and Weight of Glass

Thickness, in.	*U* factor	Weight, psf
1/8	1.14	1.65
1/4	1.13	3.29
3/8	1.12	4.94
1/2	1.11	6.58
1-in. insulating glass	0.55	7.00

Table 9-4. Sound Reduction—Glass and Other Materials*

Material thickness, in.	*Sound transmission classification rating*
Glass:	
0.087 (single strength)	26
0.118 (double strength)	29
1/8	29
3/16	29
1/4	27
5/16	29
3/8	30
1/2	33
1	32
1/4 laminated (up to 0.045 plastic layer)	33
1-in. sealed double glazing, 1/2-in. air space	27
Other materials:	
1/4 plywood	22
1/2 fiberboard	21
1/4 lead	45
2 solid vermiculite plaster on metal lath	30
Hollow wall, 2 × 4 studs, 1/2-in. gypsum plaster on metal lath on two sides	33

* Courtesy Libby-Owens-Ford Company, Toledo, Ohio.

bed. Glazing points are not required. Wood stops are securely nailed in place. The sash then is turned over, and the excess putty or glazing compound that emerged on the other side is removed.

Glazing Beads (Fig. 9-30*f*). These are designed to cover exterior glazing compound and improve appearance.

Continuous Glazing Angles (Fig. 9-30*g*). These angles or similar supports for glass usually are required for "labeled windows" by the National Board of Fire Underwriters.

Table 9-5 lists some commonly used glazing compounds and indicates the form and method in which they usually are used. Other compounds used for special service include silicone rubber, butadiene-styrene (GR-S) rubber, polyethylene, nitrile rubbers, polyurethane rubbers, acrylics, and epoxy.

Table 9-5. Commonly Used Glazing Compounds

Compound	Type of material	Final form	Type of glazing
Vegetable oil base (skin-forming type)	Gun, knife	Hard	Channel, bedding, face
Vegetable oil-rubber or nondrying oil blends (polybutene) etc. . . .	Gun, knife	Hard, plastic	Face, channel, bedding
Nondrying oil types.	Gun, knife, tacky tape	Plastic	Channel, bedding
Butyl rubber or polyisobutylene .	Tacky tape	Plastic, rubber	Bedding, channel
Polysulfide rubber	Gun	Plastic, rubber	Bedding, channel
Neoprene (polychloroprene). . . .	Gun, preformed gasket	Rubber	Channel
Vinyl chloride and copolymers . .	Preformed gasket	Hard, plastic	

9-15. Factory-Sealed Double Glazing. This is a factory-fabricated, insulating-glass unit composed of two panes of glass separated by a dehydrated air space. This type of sash is also manufactured with three panes of glass and two air spaces, providing additional insulation against heat flow or sound transmission. (See Fig. 9-30*e*).

Heat loss and heat gain can be substantially reduced by this insulated glass, permitting larger window areas and added indoor comfort. Heat-absorbing glass often is used for the outside pane and a clear plate or float glass for the inside.

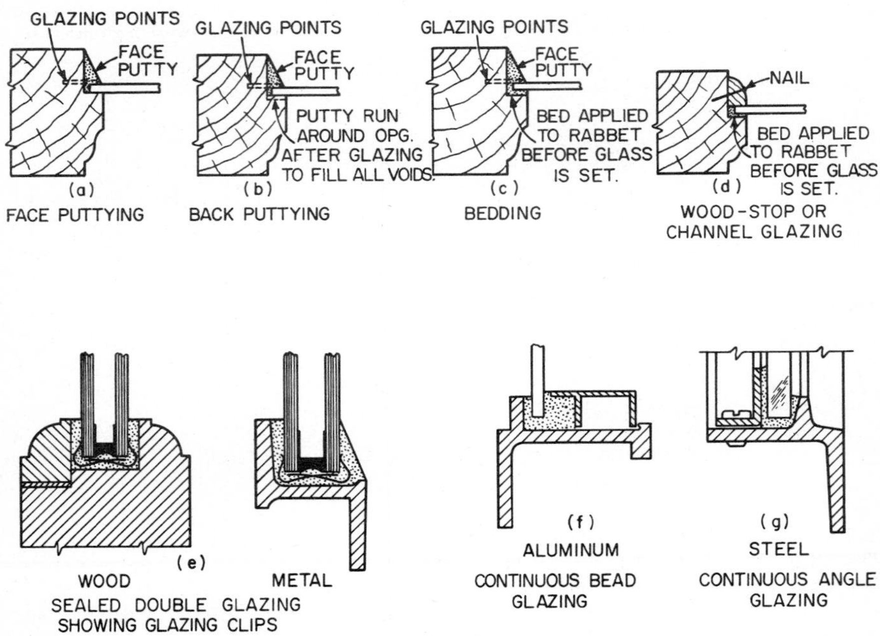

Fig. 9-30. Glazing details.

However, there are many combinations of glass available, including several patterned styles. Thickness of glass and air space between can be varied within prescribed limitations. In the selection of a window for double or triple glazing, accommodation of the over-all glass thickness in the sash is an important consideration.

9-16. Gaskets for Glazing. Structural gaskets, made of preformed and cured elastomeric material, may be used instead of sash for some applications to hold glass in place. Gaskets are extruded in a single strip, molded into the shape of the window perimeter and installed against the glass and window frame. A continuous locking strip of harder elastomer is then forced into one side of the gasket as a final

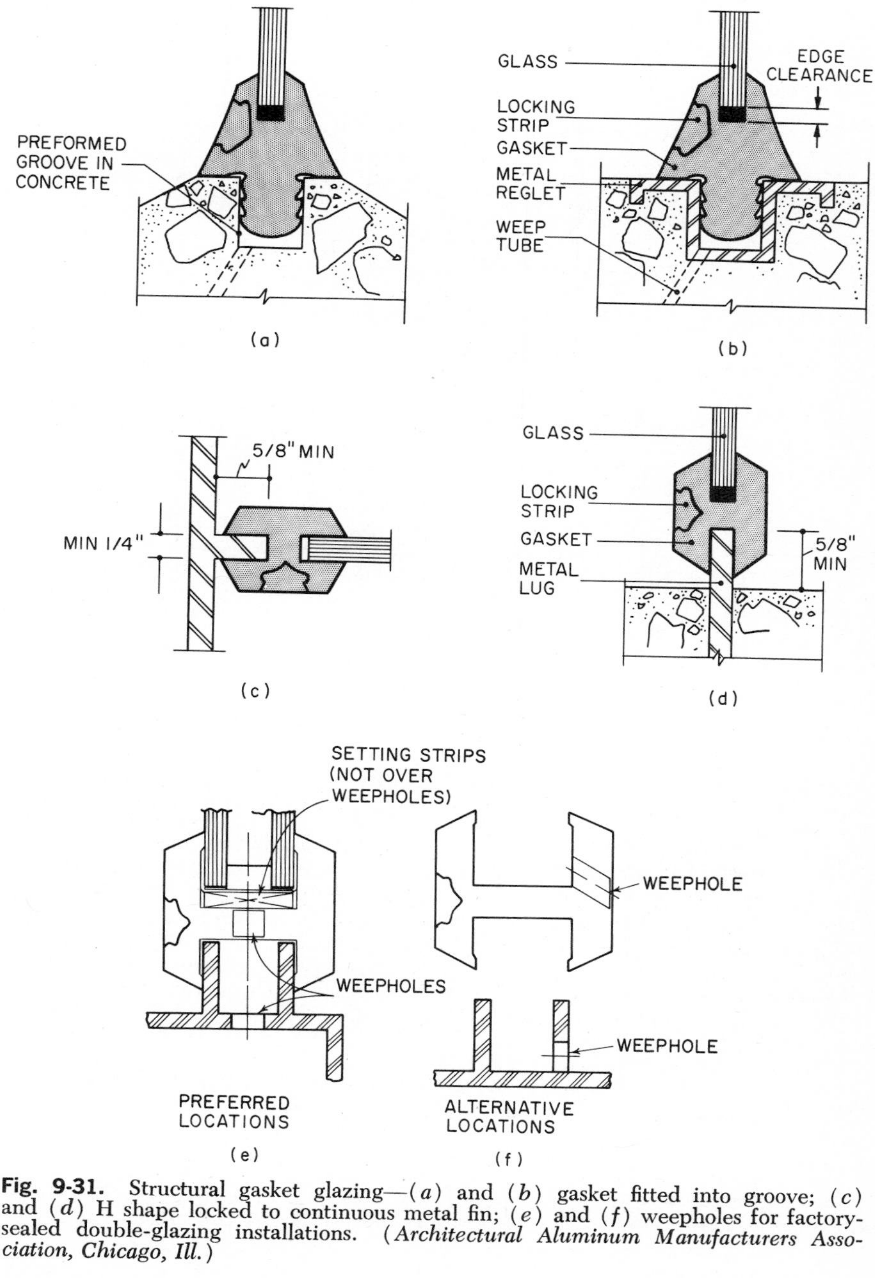

Fig. 9-31. Structural gasket glazing—(*a*) and (*b*) gasket fitted into groove; (*c*) and (*d*) H shape locked to continuous metal fin; (*e*) and (*f*) weepholes for factory-sealed double-glazing installations. (*Architectural Aluminum Manufacturers Association, Chicago, Ill.*)

sealing component. Fit and compression of the gasket determine weathertightness. With proper installation, calking is not ordinarily required; but should it be necessary, sealants are available that are compatible with the material of which the gasket is made.

There are many variations in the design of lockstrip gaskets, some of which are shown in Fig. 9-31. The two most commonly used types are the H and reglet types. The H gasket fits over a flange on the surrounding frame (Fig.

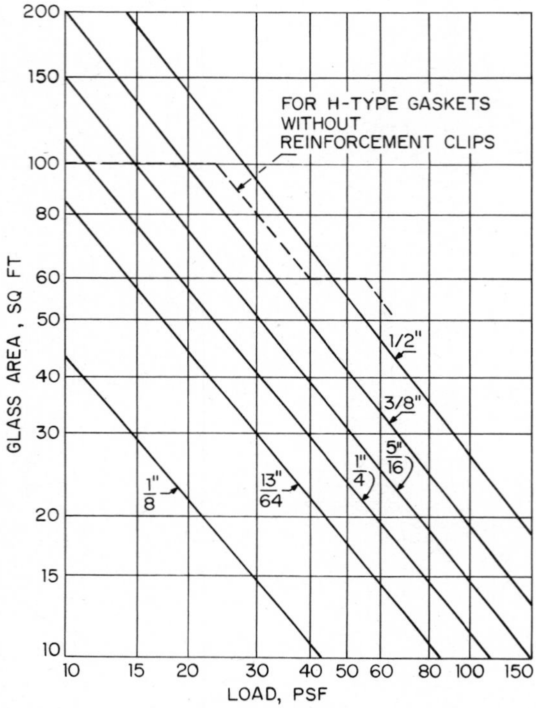

Fig. 9-32. Maximum permissible areas of plate, float, or sheet glass, with design factor of 2.5, for various wind loads on vertical windows supported on four sides and having standard H-type lockstrip gaskets. Loads apply to glass lights having a length-to-width ratio of 3 or less. (*Architectural Aluminum Manufacturers Association, Chicago, Ill.*)

9-31*c* and *d*), whereas the reglet gasket fits into a recess in the frame (Fig. 9-31*a* and *b*). When gaskets are used in concrete wall panels, the entire opening should be cast within one panel. In all cases, the smoothness, tolerance, and alignment of the contact surfaces are very important. The gaskets shown in Fig. 9-31*e* and *f* for use with double glazing require weepholes to allow water that may collect in the gaskets to drain.

The maximum glass area recommended for glazing with standard H-type lockstrip gaskets may be obtained from Fig. 9-32. Loads for the chart are based on minimum glass thicknesses allowed in Federal Specification DD-G-451c.

Section 10

Walls, Partitions, and Doors

FREDERICK S. MERRITT

Consulting Engineer, Syosset, N.Y.

MASONRY WALLS

Walls and partitions are classified as loadbearing and non-loadbearing. Different design criteria are applied to the two types.

Minimum requirements for both types of masonry walls are given in ANSI Standard Building Code Requirements for Masonry, A41.1 and ANSI Standard Building Code Requirements for Reinforced Masonry, A41.2, American National Standards Institute, and Building Code Requirements for Engineered Brick Masonry, Brick Institute of America.

Like other structural materials, masonry may be designed by application of engineering principles (Sec. 11). In the absence of such design, the empirical rules given in this section and adopted by building codes should be used.

10-1. Masonry Definitions. Following are some of the terms most commonly encountered in masonry construction:

Architectural Terra Cotta. (See Ceramic Veneer.)

Ashlar Masonry. Masonry composed of rectangular units usually larger in size than brick and properly bonded, having sawed, dressed, or squared beds. It is laid in mortar.

Bonder. (See Header.)

Brick. A rectangular masonry building unit, not less than 75% solid, made from burned clay, shale, or a mixture of these materials.

Buttress. A bonded masonry column built as an integral part of a wall and decreasing in thickness from base to top, though never thinner than the wall. It is used to provide lateral stability to the wall.

Ceramic Veneer. Hard-burned, non-loadbearing, clay building units, glazed or unglazed, plain or ornamental.

Chase. A continuous recess in a wall to receive pipes, ducts, conduits.

Column. A compression member with width not exceeding four times the thickness, and with height more than three times the least lateral dimension.

Concrete Block. A machine-formed masonry building unit composed of portland cement, aggregates, and water.

Coping. A cap or finish on top of a wall, pier, chimney, or pilaster to prevent penetration of water to masonry below.

Corbel. Successive course of masonry projecting from the face of a wall to increase its thickness or to form a shelf or ledge.

Course. A continuous horizontal layer of masonry units bonded together.

Cross-sectional Area. Net cross-sectional area of a masonry unit is the gross cross-sectional area minus the area of cores or cellular spaces. Gross cross-sectional area of scored units is determined to the outside of the scoring, but the cross-sectional area of the grooves is not deducted to obtain the net area.

Grout. A mixture of cementitious material, fine aggregate, and sufficient water to produce pouring consistency without segregation of the constituents.

Grouted Masonry. Masonry in which the interior joints are filled by pouring grout into them as the work progresses.

Header (Bonder). A brick or other masonry unit laid flat across a wall with end surface exposed, to bond two wythes.

Height of Wall. Vertical distance from top of wall to foundation wall or other intermediate support.

Hollow Masonry Unit. Masonry with net cross-sectional area in any plane parallel to the bearing surface less than 75% of its gross cross-sectional area measured in the same plane.

Masonry. A built-up construction or combination of masonry units bonded together with mortar or other cementitious material.

Mortar. A plastic mixture of cementitious materials, fine aggregates, and water.

Partition. An interior, nonbearing wall one story or less in height.

Pier. An isolated column of masonry. A bearing wall not bonded at the sides into associated masonry is considered a pier when its horizontal dimension measured at right angles to the thickness does not exceed four times its thickness.

Pilaster. A bonded or keyed column of masonry built as part of a wall, but thicker than the wall, and of uniform thickness throughout its height. It serves as a vertical beam, column, or both.

Rubble:

Coursed Rubble. Masonry composed of roughly shaped stones fitting approximately on level beds, well bonded, and brought at vertical intervals to continuous level beds or courses.

Random Rubble. Masonry composed of roughly shaped stones, well bonded and brought at irregular vertical intervals to discontinuous but approximately level beds or courses.

Rough or Ordinary Rubble. Masonry composed of nonshaped field stones laid without regularity of coursing, but well bonded.

Solid Masonry Unit. A masonry unit with net cross-sectional area in every plane parallel to the bearing surface 75% or more of its gross cross-sectional area measured in the same plane.

Stretcher. A masonry unit laid with length horizontal and parallel with the wall face.

Veneer. A wythe securely attached to a wall but not considered as sharing load with or adding strength to it.

Wall. Vertical or near-vertical construction, with length exceeding three times the thickness, for enclosing space or retaining earth or stored materials.

Bearing Wall. A wall that supports any vertical load in addition to its own weight.

Cavity Wall. (See *Hollow Wall* below.)

Curtain Wall. A non-loadbearing exterior wall.

Faced Wall. A wall in which the masonry facing and the backing are of different materials and are so bonded as to exert a common reaction under load.

Hollow Wall. A wall of masonry so arranged as to provide an air space within the wall between the inner and outer wythes. A cavity wall is built of masonry units or plain concrete, or of a combination of these materials, so arranged as to provide an air space within the wall, which may be filled with insulation, and in which inner and outer wythes are tied together with metal ties.

Nonbearing Wall. A wall that supports no vertical load other than its own weight.

Party Wall. A wall on an interior lot line used or adapted for joint service between two buildings.

Shear Wall. A wall that resists horizontal forces applied in the plane of the wall.

Spandrel Wall. An exterior curtain wall at the level of the outside floor beams in multistory buildings. It may extend from the head of the window below the floor to the sill of the window above.

Veneered Wall. A wall having a facing of masonry or other material securely attached to a backing, but not so bonded as to exert a common reaction under load.

Wythe. Each continuous vertical section of a wall one masonry unit in thickness.

Table 10-1. Types of Mortar

Mortar type	Parts by volume				Min avg compressive strength of three 2-in. cubes at 28 days, psi
	Portland cement	Masonry cement	Hydrated lime or lime putty	Aggregate measured in damp, loose condition	
M	1	1 (Type II)	....	Not less than 2¼ and not more than 3 times the sum of the volumes of the cements and limes used	2,500
	1	....	¼		
S	½	1 (Type II)	....		1,800
	1	....	Over ¼ to ½		
N	...	1 (Type II)	....		750
	1	....	Over ½ to 1¼		
O	...	1 (Types I or II)	....		350
	1	....	Over 1¼ to 2½		
K	1	....	Over 2½ to 4		75
PL	1	....	¼ to ½		2,500
PM	1	1	....		2,500

10-2. Quality of Materials for Masonry. Materials used in masonry construction should be capable of meeting the requirements of the applicable standard of the American Society for Testing and Materials.

Second-hand materials should be used only with extreme caution. Much salvaged brick, for example, comes from demolition of old buildings constructed of solid brick in which hard-burned units were used on the exterior and salmon units as backup. Because the color differences that guided the original masons in sorting and selecting bricks become obscured with exposure and contact with mortar, there is a definite danger that the salmon bricks may be used for exterior exposure and may disintegrate rapidly. Masonry units salvaged from chimneys are not recommended because they may be impregnated with oils or tarry material.

10-3. Mortar. For unit masonry, mortar should meet the requirements of the American Society for Testing and Materials Specifications C270 and C476. These define the types of mortar shown in Table 10-1. Each type is used for a specific purpose, as indicated in Table 10-2, based on compressive strength. However, it should not be assumed that higher-strength mortars are preferable to lower-strength mortars where lower strength is permitted for particular uses.

Mortars containing lime are generally preferred because of greater workability. Commonly used:

For concrete block, 1 part cement, 1 part lime putty, 5 to 6 parts sand.

For rubble, 1 part cement, 1 to 2 parts lime hydrate or putty, 5 to 7 parts sand.

For brick, 1 part cement, 1 part lime, 6 parts sand.

For setting tile, 1 part cement, ½ part lime, 3 parts sand.

Table 10-2. Mortar Requirements of Masonry

Kind of Masonry	*Types of Mortar*
Masonry in contact with earth:	
Footings	M or S
Walls of solid units	M, S, or N
Walls of hollow units	M or S
Hollow walls	M or S
Masonry above grade or interior:	
Piers of solid masonry	M, S, or N
Piers of hollow units	M or S
Walls of solid masonry	M, S, N, or O
Grouted masonry	PL or PM
Walls of hollow units; load-bearing or exterior, and hollow walls 12 in. or more in thickness	M, S, or N
Hollow walls less than 12 in. in thickness where assumed design wind pressure:	
a. Exceeds 20 psf	M or S
b. Does not exceed 20 psf	M, S, or N
Glass-block masonry	M, S, or N
Nonbearing partitions of fireproofing composed of structural clay tile or concrete masonry units	M, S, N, O, or gypsum
Gypsum partition tile or block	Gypsum
Firebrick	Refractory air-setting mortar
Linings of existing masonry	M or S
Masonry other than above	M, S, or N

10-4. Allowable Stresses in Masonry. In determining stresses in masonry, effects of loads should be computed on actual dimensions, not nominal. Except for engineered masonry, the stresses should not exceed the allowable stresses given in ANSI Standard Building Code Requirements for Masonry (A41.1), which are summarized for convenience in Table 10-3.

This standard recommends also that, in composite walls or other structural members composed of different kinds or grades of units or mortars, the maximum stress should not exceed the allowable stress for the weakest of the combination of units and mortars of which the member is composed.

Compressive strength of masonry depends to a great extent on workmanship and the completeness with which units are bedded. Tensile strength is a function of the adhesion of mortar to a unit and of the area of bonding (degree of completeness with which joints are filled). Hence, in specifying masonry work, it is important to call for a full bed of mortar, with each course well hammered down, and all joints completely filled with mortar.

10-5. Leakproof Walls. To minimize the entrance of water through a brick wall, follow the practices recommended in Art. 12-14.

In particular, in filling head joints, apply a heavy buttering of mortar on one end of the brick, press the brick down into the bed joint, and push the brick into place so that the mortar squeezes out from the top and sides of the head joint. Mortar should correspondingly cover the entire side of a brick before it is placed as a header. An attempt to fill head joints by slushing or dashing will not succeed in producing watertight joints. Partial filling of joints by "buttering" or "spotting" the vertical edge of the brick with mortar cut from the extruded bed joint is likewise ineffective and should be prohibited. Where closures are required, the opening should be filled with mortar so that insertion of the closure will extrude mortar both laterally and vertically.

Tooling of joints, if done properly, can help to resist penetration of water; but it is not a substitute for complete filling, or a remedy for incomplete filling of joints. A concave joint is recommended. Use of raked or other joints that provide horizontal water tables should be avoided. Mortar should not be too stiff at time of tooling, or compaction will not take place, nor should it be too fluid, or the bricks may move—and bricks should never be moved after initial contact with mortar. If a brick is out of line, it should be removed, mortar scraped off and fresh mortar applied before the brick is relaid.

Back-plaster, or parge, the back face of exterior wythes before backup units are laid. If the backup is laid first, the front of the backup should be parged. The

Table 10-3. Allowable Stresses in Unit Masonry

Construction and grade of unit	Allowable compressive stress on gross cross section, psi				
	Type M mortar	Type S mortar	Type N mortar	Type O mortar	Type PL or PM mortar
Solid masonry of brick or other solid units:					
8,000+ psi	400	350	300	200	
4,500 to 8,000 psi	250	225	200	150	
2,500 to 4,500 psi	175	160	140	100	
1,200 to 2,500 psi	125	115	100	75	
Grouted solid masonry of brick or other solid units:					
8,000+ psi	...	...	...	...	500
4,500 to 8,000 psi	...	...	...	...	350
2,500 to 4,500 psi	...	...	...	...	275
1,200 to 2,500 psi	...	...	...	...	225
Solid masonry of through-the-wall solid units:					
8,000+ psi	500	400	300		
4,500 to 8,000 psi	350	275	200		
2,500 to 4,500 psi	275	200	150		
1,200 to 2,500 psi	225	175	125		
Masonry of hollow units:					
1,000+ psi	100	90	85		
700 to 1,000 psi	85	75	70		
Piers of hollow units, cellular spaces filled	105	95	90		
Hollow walls (cavity or masonry bonded):*					
Solid units:					
4,500+ psi	180	170	140		
2,500 to 4,500 psi	140	130	110		
1,200 to 2,500 psi	100	90	80		
Hollow units:					
1,000+ psi	80	70	65		
700 to 1,000 psi	70	60	55		
Stone ashlar masonry:					
Granite	800	720	640	500	
Limestone or marble	500	450	400	325	
Sandstone or cast stone	400	360	320	250	
Rubble, coursed, rough or random	140	120	100	80	

* On gross cross section of wall minus area of cavity between wythes. The allowable compressive stresses for cavity walls are based on the assumption that floor loads bear on only one of the two wythes. Increase stresses 25% for hollow walls loaded concentrically.

mortar should be the same as that used for laying the brick and should be applied from ¼ to ⅜ in. thick.

10-6. Lateral Support for Masonry Walls. For nonengineered solid masonry walls or bearing partitions, the ratio of unsupported height to nominal thickness, or the ratio of unsupported length to nominal thickness, should not exceed 20. For hollow walls or walls of hollow masonry units, the ratio should be 18 or less. See ANSI Standard Building Code Requirements for Masonry, A41.1, American National Standards Institute.

In calculating the ratio of unsupported length to thickness for cavity walls, you can take the thickness as the sum of the nominal thicknesses of the inner and outer wythes. For walls composed of different kinds or classes of units or mortars, the ratio should not exceed that allowed for the weakest of the combinations. Veneers should not be considered part of the wall in computing thickness for strength or stability.

Cantilever walls and masonry walls in locations exposed to high winds should not be built higher than ten times their thickness unless adequately braced or designed in accordance with engineering principles. Backfill should not be placed against foundation walls until they have been braced to withstand horizontal pressure.

In determining the unsupported length of walls, existing cross walls, piers, or buttresses may be considered as lateral supports, if these members are well bonded or anchored to the walls and capable of transmitting the lateral forces to connected structural members or to the ground.

In determining the unsupported height of walls, floors and roofs may be considered as lateral supports, if provision is made in the building to transmit the lateral forces to the ground. Ends of floor joists or beams bearing on masonry walls should be securely fastened to the walls. When lateral support is to be provided by joists parallel to walls, anchors should be spaced no more than 6 ft apart and engage at least three joists, which should be bridged solidly at the anchors.

Anchors for floor beams bearing on walls should be provided at intervals of 4 ft or less. The anchors should be solidly embedded in mortar and extend into the walls a distance of at least half the wall thickness. Interior ends of anchored joists should be lapped and spiked, or the equivalent, so as to form continuous ties across the building.

Unsupported height of piers should not exceed ten times the least dimension. However, when structural clay tile or hollow concrete units are used for isolated piers to support beams or girders, unsupported height should not exceed four times the least dimension unless the cellular spaces are filled solidly with concrete or either Type M or S mortar (see Table 10-1).

10-7. Masonry-thickness Requirements. Walls should not vary in thickness between lateral supports. When it is necessary to change thickness between floor levels to meet minimum-thickness requirements, the greater thickness should be carried up to the next floor level.

Where walls of masonry hollow units or bonded hollow walls are decreased in thickness, a course of solid masonry should be interposed between the wall below and the thinner wall above, or else special units or construction should be used to transmit the loads between the walls of different thickness.

The following limits on dimensions of masonry walls should be observed unless the walls are designed by application of engineering principles (Sec. 11):

Bearing walls should be at least 12 in. thick for the uppermost 35 ft of their height. Thickness should be increased 4 in. for each successive 35 ft or fraction of this distance measured downward from the top of the wall. Rough or random or coursed rubble stone walls should be 4 in. thicker than this, but in no case less than 16 in. thick. However, for other than rubble stone walls, the following exceptions apply to masonry bearing walls:

Stiffened Walls. Where solid masonry bearing walls are stiffened at distances not greater than 12 ft by masonry cross walls or by reinforced-concrete floors, they may be made 12 in. thick for the uppermost 70 ft but should be increased 4 in. in thickness for each successive 70 ft or fraction of that distance.

Top-story Walls. The top-story bearing wall of a building not over 35 ft high may be made 8 in. thick. But this wall should be no more than 12 ft high and should not be subjected to lateral thrust from the roof construction.

Residential Walls. In dwellings up to three stories high, walls may be 8 in. thick (if not more than 35 ft high), if not subjected to lateral thrust from the roof construction. Such walls in one-story houses and one-story private garages may be 6 in. thick, if the height is 9 ft or less or if the height to the peak of a gable does not exceed 15 ft.

Penthouses and Roof Structures. Masonry walls up to 12 ft high above roof level, enclosing stairways, machinery rooms, shafts, or penthouses, may be made 8 in. thick. They need not be included in determining the height for meeting thickness requirements for the wall below.

Plain Concrete Walls. Such walls may be 2 in. less in thickness than the minimum requirements for other types of masonry walls, but in general not less than 8 in.—and not less than 6 in. in one-story dwellings and garages.

Hollow Walls. Cavity or masonry bonded hollow walls should not be more than 35 ft high. In particular, 10-in. cavity walls should be limited to 25 ft in height, above supports. The facing and backing of cavity walls should be at least 4 in. thick, and the cavity should not be less than 2 in. or more than 3 in. wide.

Faced Walls. Neither the height of faced (composite) walls nor the distance between lateral supports should exceed that prescribed for masonry of either of the types forming the facing and the backing. Actual (not nominal) thickness of material used for facings should not be less than 2 in. and in no case less than one-eighth the height of the unit.

In general, parapet walls should be at least 8 in. thick and the height should not exceed three times the thickness. The thickness may be less than 8 in., however, if the parapet is reinforced to withstand safely earthquake and wind forces to which it may be subjected.

Nonbearing exterior masonry walls may be 4 in. less in thickness than the minimum for bearing walls. However, the thickness should not be less than 8 in. except that where 6-in. bearing walls are permitted, 6-in. nonbearing walls can be used also.

Nonbearing masonry partitions should be supported laterally at distances of not more than 36 times the actual thickness of the partition, including plaster. If lateral support depends on a ceiling, floor, or roof, the top of the partition should have adequate anchorage to transmit the forces. This anchorage may be accomplished with metal anchors or by keying the top of the partition to overhead

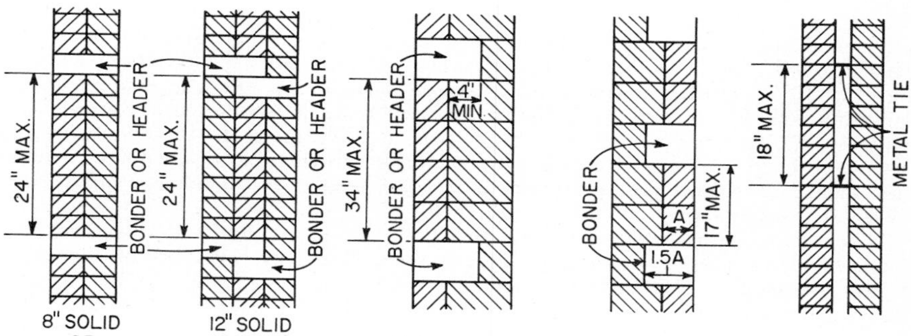

Fig. 10-1. Requirements for vertical spacing of bonding elements in masonry walls.

work. Suspended ceilings may be considered as lateral support if ceilings and anchorages are capable of resisting a horizontal force of 150 lb per lin ft of wall.

10-8. Bond in Masonry Walls. When headers are used for bonding the facing and backing in solid masonry walls and faced walls, not less than 4% of the wall surface of each face should be composed of headers, which should extend at least 4 in. into the backing. These headers should not be more than 24 in. apart vertically or horizontally (Fig. 10-1). In walls in which a single bonder does not extend through the wall, headers from opposite sides should overlap at 4 in. or should be covered with another bonder course overlapping headers below at least 4 in.

If metal ties are used for bonding, they should be corrosion-resistant. For bonding facing and backing of solid masonry walls and faced walls, there should be at least one metal tie for each 4½ sq ft of wall area. Ties in alternate courses should be staggered, the maximum vertical distance between ties should not exceed 18 in., and the maximum horizontal distance should not be more than 36 in.

In walls composed of two or more thicknesses of hollow units, stretcher courses should be bonded by one of the following methods: At vertical intervals up to 34 in., there should be a course lapping units below at least 4 in. (Fig. 10-1). Or at vertical intervals up to 17 in., lapping should be accomplished with units at least 50% thicker than the units below (Fig. 10-1). Or at least one metal tie should be incorporated for each 4½ sq ft of wall area. Ties in alternate courses should be staggered; the maximum vertical distance between ties should be 18 in. and maximum horizontal distance, 36 in. Full mortar coverage should be provided in both horizontal and vertical joints at ends and edges of face shells of the hollow units.

In ashlar masonry, bond stones should be uniformly distributed throughout the wall and form at least 10% of the area of exposed faces.

In rubble stone masonry up to 24 in. thick, bond stones should have a maximum spacing of 3 ft vertically and horizontally. In thicker walls, there should be at least one bond stone for each 6 sq ft of wall surface on both sides.

For bonding ashlar facing, the percentage of bond stones should be computed from the exposed face area of the wall. At least 10% of this area should be composed of uniformly distributed bond stones extending 4 in. or more into the backup. Every bond stone, and when alternate courses are not full bond courses every stone, should be securely anchored to the backup with corrosion-resistant metal anchors. These should have a minimum cross section of $\frac{3}{16} \times 1$ in. There should be at least one anchor to a stone and at least two anchors for stones more than 2 ft long or with a face area of more than 3 sq ft. Larger facing stones should have at least one anchor per 4 sq ft of face area of the stone, but not less than two anchors.

Cavity-wall wythes should be bonded with $\frac{3}{16}$-in.-diameter steel rods or metal ties of equivalent stiffness embedded in horizontal joints. There should be at least one metal tie for each $4\frac{1}{2}$ sq ft of wall area. Ties in alternate courses should be staggered, the maximum vertical distance between ties should not exceed 18 in. (Fig. 10-1), and the maximum horizontal distance, 36 in. Rods bent to rectangular shape should be used with hollow masonry units laid with cells vertical. In other walls, the ends of ties should be bent to 90° angles to provide hooks at least 2 in. long. Additional bonding ties should be provided at all openings. These ties should be spaced not more than 3 ft apart around the perimeter and within 12 in. of the opening.

When two bearing walls intersect and the courses are built up together, the intersections should be bonded by laying in true bond at least 50% of the units at the intersection. When the courses are carried up separately, the intersecting walls should be regularly toothed or blocked with 8-in. maximum offsets. The joints should be provided with metal anchors having a minimum section of $\frac{1}{4} \times 1\frac{1}{2}$ in. with ends bent up at least 2 in. or with cross pins to form an anchorage. Such anchors should be at least 2 ft long and spaced not more than 4 ft apart.

10-9. Grouted Masonry. Masonry units in the outer wythes of grouted masonry walls should be laid with full head and bed joints of Type M or S mortar (see Table 10-1). All interior joints should be filled with grout.

Masonry units in the interior wythes should be placed or floated in grout poured between the two outer wythes.

One of the outer wythes may be carried up not more than three courses before grouting, but the other should be carried up not more than one course above the grout. Each grout pour should be stopped at least $1\frac{1}{2}$ in. below the top and stirred. Grouted longitudinal vertical joints should be at least $\frac{3}{4}$ in. wide. Bonders should not be used.

High-lift grouting, sometimes called *reinforced-concrete and masonry construction,* is an alternative method for grouted masonry. A cavity wall is constructed with a space of at least $2\frac{1}{2}$ in. between the tied wythes. Vertical reinforcement is placed in the space. The cavity is then filled with grout in lifts not exceeding 4 ft and over lengths of not more than 25 ft.

10-10. Glass Block. For control of light that enters a building and for better insulation than obtained with ordinary glass panes, masonry walls of glass block are frequently used. These units are hollow, $3\frac{7}{8}$ in. thick by 6 in. square, 8 in. square, or 12 in. square (actual length and height $\frac{1}{4}$ in. less, for modular coordination). Faces of the units may be cut into prisms to throw light upward or the block may be treated to diffuse light.

Glass block may be used as in nonbearing walls and to fill openings in walls. The glass block so used should have a minimum thickness of $3\frac{1}{2}$ in. at the mortar joint. Also, surfaces of the block should be satisfactorily treated for mortar bonding.

For exterior walls, glass-block panels, when set in chases, should not have an unsupported area of more than 144 sq ft. They should be no more than 25 ft long, or more than 20 ft high between supports. With wall anchor construction, panel area should be limited to 100 sq ft, height to 10 ft, and length to 10 ft.

For interior walls, glass-block panels should not have an unsupported area of more than 250 sq ft. Neither length nor height should exceed 25 ft.

Exterior panels should be held in place in the wall opening to resist both internal and external wind pressures. The panels should be set in recesses at the jambs so as to provide a bearing surface at least 1 in. wide along the edges. Panels more than 10 ft long should also be recessed at the head. Some building codes, however, permit anchoring small panels in low buildings with noncorrodible perforated metal strips.

Steel reinforcement should be placed in the horizontal mortar joints of glass-block panels at vertical intervals of 2 ft or less. It should extend the full length of the joints but not across expansion joints. When splices are necessary, the reinforcement should be lapped at least 6 in. In addition, reinforcement should be placed in the joint immediately below and above any openings in a panel.

The reinforcement should consist of two parallel longitudinal galvanized-steel wires, No. 9 gage or larger, spaced 2 in. apart, and having welded to them No. 14 or heavier-gage cross wires at intervals of up to 8 in.

Mortar joints should be from ¼ to ⅜ in. thick. They should be completely filled with mortar.

Exterior glass-block panels should be provided with expansion joints at sides and top. These joints should be kept free of mortar and should be filled with resilient material (Fig. 10-2). An opening may be filled one block at a time, as in Fig. 10-2, or with a preassembled panel.

10-11. Openings, Chases, and Recesses in Masonry Walls. Masonry above openings should be supported by arches or lintels of metal or reinforced masonry, which should bear on the wall at each end at least 4 in. Stone or other nonreinforced masonry lintels should not be used unless supplemented on the inside of the wall with structural steel lintels, suitable masonry arches, or reinforced-masonry lintels carrying the masonry backing. Lintels should be stiff enough to carry the superimposed load with a deflection of less than $1/720$ of the clear span.

In plain concrete walls, reinforcement symmetrically disposed in the thickness of the wall should be placed not less than 1 in. above and 2 in. below openings. It should extend at least 24 in. on each side of the opening or be equivalently developed with hooks. Minimum reinforcement that should be used is one ⅝-in.-diameter bar for each 6 in. of wall thickness.

In structures other than low residences, masonry walls should not have chases and recesses deeper than one-third the wall thickness, or longer than 4 ft horizontally or in horizontal projection. There should be at least 8 in. of masonry in back of chases and recesses, and between adjacent chases or recesses and the jambs of openings.

Chases and recesses should not be cut in walls of hollow masonry units or in hollow walls but may be built in. They should not be allowed within the required area of a pier.

The aggregate area of recesses and chases in any wall should not exceed one-fourth of the whole area of the face of the wall in any story.

In dwellings not more than two stories high, vertical chases may be built in 8-in. walls if the chases are not more than 4 in. deep and occupy less than 4 sq ft of wall area. However, recesses below windows may extend from floor to sill and may be the width of the opening above. Masonry above chases or recesses wider than 12 in. should be supported on lintels.

Recesses may be left in walls for stairways and elevators, but the walls should not be reduced in thickness to less than 12 in. unless reinforced in some approved manner. Recesses for alcoves and similar purposes should have at least 8 in. of masonry at the back. They should be less than 8 ft wide and should be arched over or spanned with lintels.

10-12. Support Conditions for Walls. Provision should be made to distribute concentrated loads safely on masonry walls and piers. Heavily loaded members should have steel bearing plates under the ends to distribute the load to the masonry

Fig. 10-2. (*Top left*) First step in installing a glass-block panel is to coat the sill with an asphalt emulsion to allow for movement due to temperature changes. Continuous-expansion strips are installed at jambs and head. (*Top right*) Block are set with full mortar joints. (*Bottom left*) Welded-wire ties are embedded in the mortar to reinforce the panel. (*Bottom right*) When all the block are placed, joints tooled to a smooth concave finish, and the perimeter of the panel calked, the block are cleaned.

within allowable bearing stresses. Length of bearing should be at least 3 in. Lightly loaded members may be supported directly on the masonry if the bearing stresses in the masonry are within permissible limits and if length of bearing is 3 in. or more.

Masonry should not be supported on wood construction.

10-13. Corbeling. Except for chimneys, the maximum corbeled horizontal projection beyond the face of a wall should not exceed one-half the wall thickness. Maximum projection of one unit should not be more than one-half its depth or one-third the width (measured at right angles to the offset face).

Chimneys may not be corbeled more than 6 in. from the face of a wall more than 12 in. thick, and corbeling should not be permitted in thinner walls, unless there are equal projections on opposite sides of the wall. However, in the second story of two-story dwellings, corbeling of chimneys on the exterior of enclosing walls may equal the wall thickness. In every case, the corbeling should not be more than 1-in. projection for each course of brick projected.

10-14. Cold-weather Construction of Masonry Walls. Masonry should be protected against damage by freezing. The following special precautions should be taken:

Materials to be used should be kept dry. Tops of all walls not enclosed or sheltered should be covered whenever work stops. The protection should extend downward at least 2 ft.

Frozen materials must be thawed before use. Masonry units should be heated to at least 20°F. Mortar temperature should be between 40 and 120°F, and mortar should not be placed on a frozen surface. If necessary, protect the wall with heat and windbreaks for at least 48 hr. Use of mortars made with high-early-strength cement may be advantageous for cold-weather masonry construction.

10-15. Chimneys and Fireplaces. Minimum requirements for chimneys may be obtained from local building codes or the National Building Code recommended by the American Insurance Association. (See also H. C. Plummer, "Brick and Tile Engineering," Brick Institute of America, McLean, Va.)

In brief, chimneys should extend at least 3 ft above the highest point where they pass through the roof of a building and at least 2 ft higher than any ridge within 10 ft. They should be constructed of solid masonry units or reinforced concrete and lined with fire clay or other suitable refractory clay. In dwellings, thickness of chimney walls may be 4 in. In other buildings, the thickness of chimneys for heating appliances should be at least 8 in. for most masonry. Rubble stone thickness should be a minimum of 12 in.

Fireplaces should have backs and sides of solid masonry or reinforced concrete, not less than 8 in. thick. A lining of firebrick at least 2 in. thick or other approved material should be provided unless the thickness is 12 in.

Fireplaces should have hearths of brick, stone, tile, or other noncombustible material supported on a fireproof slab or on brick trimmer arches. Such hearths should extend at least 20 in. outside the chimney breast and not less than 12 in. beyond each side of the fireplace opening along the chimney breast. Combined thickness of hearth and supporting construction should not be less than 6 in. Spaces between chimney and joists, beams, or girders and any combustible materials should be fire-stopped by filling with noncombustible material.

The throat of the fireplace should be not less than 4 in. and preferably 8 in. above the top of the fireplace opening. A metal damper extending the full width of the fireplace opening should be placed in the throat. The flue should have an effective area equal to one-twelfth to one-tenth the area of the fireplace opening. For more details, see H. C. Plummer, "Brick and Tile Engineering," Brick Institute of America, McLean, Va.; C. G. Ramsey and H. R. Sleeper, "Architectural Graphic Standards," John Wiley & Sons, Inc., New York.

CURTAIN WALLS

With skeleton-frame construction, exterior walls need carry no load other than their own weight, and therefore their principal function is to keep wind and weather out of the building—hence the name curtain wall. Nonbearing walls may be supported on the structural frame of a building, on supplementary framing (girts, for example) in turn supported on the structural frame of a building, or on the floors.

10-16. Functional Requirements of Curtain Walls. Curtain walls do not have to be any thicker than required to serve their principal function. Many industrial buildings are enclosed only with light-gage metal. However, for structures with certain types of occupancies and for buildings close to others, fire resistance is an important characteristic. Fire-resistance requirements in local building codes often govern in determining the thickness and type of material used for curtain walls.

In many types of buildings, it is desirable to have an exterior wall with good insulating properties. Sometimes a dead air space is used for this purpose. Sometimes insulating material is incorporated in the wall or erected as a backup.

The exterior surface of a curtain wall should be made of a durable material, capable of lasting as long as the building. Maintenance should be a minimum; initial cost of the wall is not so important as the annual cost (amortized initial cost plus annual maintenance and repair costs).

To meet requirements of the owner and the local building code, curtain walls may vary in construction from a simple siding to a multilayer-sandwich wall. They may be job-assembled or be delivered to the job completely prefabricated.

Walls with masonry components should meet the requirements of Arts. 10-2 to 10-14 or Sec. 11.

("Comparative Ultimate Cost of Building Walls," Brick Institute of America, McLean, Va.)

10-17. Wood Walls. Either in the form of boards or plywood, wood is often used as sheathing behind various types of facing materials. It is generally covered on the outside face with a waterproofing material before the facing is placed. Wood sheathing is most often used with wood framing, as in residential construction.

Wood is also applied as an exterior finish in the form of siding, shingles, half timbers, or plywood sheets.

Siding may be drop, or novelty; lap, or clapboard; vertical boarding or horizontal flush boarding.

Drop siding can combine sheathing and siding in one piece. Tongued-and-grooved individual pieces are driven tightly up against each other when they are nailed in place to make the wall weathertight. It is not considered a good finish for permanent structures.

Lap siding or clapboard are beveled boards, thinner along one edge than the opposite edge, which are nailed over sheathing and building paper. Usually narrow boards lap each other about 1 in., wide boards more than 2 in. At the eaves, the top siding boards slip under the lower edge of a frieze board to make a weathertight joint.

When vertical or horizontal boards are used for the exterior finish, precautions should be taken to make the joints watertight. Joints should be coated with white lead in linseed oil just before the boards are nailed in place, and the boards should be driven tight against each other. Battens should be applied over the joints if the boards are squared-edged.

Half timbers may be used to form a structural frame of heavy horizontal, vertical, and diagonal members, the spaces between being filled with brick. This type of construction is sometimes imitated by nailing boards in a similar pattern to an ordinary sheathed frame and filling the space between boards with stucco.

Plywood for exterior use should be an exterior grade, with plies bonded with permanent waterproof glue. The curtain wall may consist of a single sheet of plywood or of a sandwich of which plywood is a component. Also, plywood may be laminated to another material, such as a light-gage metal, to give it stiffness.

10-18. Wall Shingles. Wood, asphalt, or asbestos cement are most frequently used for shingles over a sheathed frame. Shingles are made in a variety of forms and shapes and are applied in different ways. The various manufacturers make available instructions for application of their products.

10-19. Stucco. Applied like plaster, stucco is a mixture of sand, portland cement, lime, and water. Two coats are applied to masonry, three coats on metal lath. The finish coat may be tinted by adding coloring matter to the mix or the outside surface may be painted with a suitable material.

The metal lath should be heavily galvanized. It should weigh at least 2.5 lb per sq yd, even though furring strips are closely spaced. When supports are 16 in. c to c, it should weigh 3.4 lb per sq yd. The lath sheets should be applied with long dimensions horizontal and should be tied with 16-ga wire. Edges should be lapped at least 1 in., ends 2 in.

The first, or scratch, coat should be forced through the interstices in the lath so as to embed the metal completely. This coat may consist of 1 part quicklime putty or hydrated lime putty and 3 parts sand by volume, plus 1 bu of long hair or fiber per cu yd of sand. To increase the rate of hardening, some of the lime may be replaced by portland cement. A common mix is 1 part portland cement, 1 part lime putty, and 5 or 6 parts sand.

The second, or brown, coat may be based also on lime or portland cement. With lime, the mix may be 1 part quicklime putty or hydrated lime putty and 3 parts of sand, by volume. With cement, the mix may be 1 part portland cement to 3 parts sand, plus lime putty in amount equal to 15 to 25% of the volume of cement. The finish coat may have the same proportions as the brown coat.

For all mixes, the ingredients, except for the hair or fiber, are thoroughly mixed. A small amount of water may be added for this purpose, though the hydrated lime may be mixed dry. The mix should be left standing at least 24 hr. So enough water should be added to the mix before this waiting period to prevent it from drying out. The hair or fiber and more water should be added and the ingredients thoroughly mixed just prior to use.

The brown coat may be applied as soon as the scratch coat has hardened sufficiently to withstand pressure without breaking the keys, usually in 1 week to 10 days. For the finish coat it may be wise to wait several months to give the building a chance to settle and the base coats to shrink.

See also Art. 13-11.

10-20. Metal and Asbestos-cement Siding. Either in flat sheets or corrugated form, light-gage metal or asbestos cement may be used to form a lightweight enclosure. Corrugated sheets are stiffer than the flat. If the sheets are very thin, they should be fastened to sheathing or closely spaced supports.

When corrugated siding is used, details should be planned so that the siding will shed water. Horizontal splices should be placed at supporting members and the sheets should lap about 4 in. Vertical splices should lap at least 1½ corrugations. Sheets should be held firmly together at splices and intersections to prevent water from leaking through. Consideration should be given to sealing strips at openings where corrugated sheets terminate against plane surfaces. The bottommost girt supporting the siding should be placed at least 1 ft above the foundation because of the difficulty of attaching the corrugated materials to masonry. The siding should not be sealed in a slot in the foundation because the metal may corrode or the asbestos-cement crack.

When flat sheets are used, precautions should be taken to prevent water from penetrating splices and intersections. The sheets may be installed in sash like window glass, or the splices may be covered with battens. Edges of metal sheets may be flanged to interlock and exclude wind and rain.

Pressed-metal panels, mostly with troughed or boxed cross sections, are also used to form lightweight walls.

Provision should be made in all cases for expansion and contraction with temperature changes. Allowance for movement should be made at connections. Methods of attachment vary with the type of sheet and generally should be carried out in accordance with the manufacturer's recommendations. For typical details, see C. G. Ramsey and H. R. Sleeper, "Architectural Graphic Standards," John Wiley & Sons, Inc., New York.

10-21. Metal and Glass Facings. In contrast to siding in which a single material forms the complete wall, metal and glass are sometimes used as the facing, which is backed up with insulation, fire-resistant material, and an interior finish. The glass usually is tinted and is held in a light frame in the same manner as window glass. Metal panels may be fastened similarly in a light frame, attached to mullions or other secondary framing members, anchored to brackets at each floor level, or connected to the structural frame of the building. The panels may be small and light enough for one man to carry or one or two stories high, prefabricated with windows.

Provision for expansion and contraction should be made in the frames, when they are used, and at connections with building members. Metal panels should be shaped so that changes in surface appearance will not be noticeable as the metal expands and contracts. Frequently, light-gage metal panels are given decorative patterns, which also hide movements due to temperature variations ("canning") and stiffen the sheets. Flat sheets may be given a slight initial curvature and stiffened on the rear side with ribs, so that temperature variations will only change the curvature a little and not reverse it. Or flat sheets may be laminated to one or more flat stiffening sheets, like asbestos cement or asbestos cement and a second light-gage metal sheet, to prevent "canning."

It may be desirable in many cases to treat the metal to prevent passage of sound. Usual practice is to apply a sound-absorbing coating on the inside surface of the panel. Some of these coatings have the additional beneficial effect of preventing moisture from condensing on this face.

Metal panels generally are flanged and interlocked to prevent penetration of water. A good joint will be self-flashing and will not require calking. Care must be taken that water will not be blown through weep holes from the outside into the building. Flashing and other details should be arranged so that any water that may penetrate the facing will be drained to the outside. (See also Art. 10-22.)

10-22. Sandwich Panels. Walls may be built of prefabricated panels that are considerably larger in size than unit masonry and capable of meeting requirements of appearance, strength, durability, insulation, acoustics, and permeability. Such panels generally consist of an insulation core sandwiched between a thin lightweight facing and backing.

When the edges of the panels are sealed, small holes should be left in the seal. Otherwise, heat of the sun could set up sizable vapor pressure, which could cause trouble.

The panels could be fastened in place in a light frame, attached to secondary framing members, anchored to brackets at each floor level, or connected to the structural frame of the building. Because of the large size of the panels, special precautions should be taken to allow for expansion and contraction due to temperature changes. Usually such movements are provided for at points of support.

When light frames are not used to support the units, adjoining panels generally interlock, and the joints are calked and sealed with rubber or rubberlike material to prevent rain from penetrating. Flashing and other details should be arranged so that any water that comes through will be drained to the outside. For typical details, see C. G. Ramsey and H. R. Sleeper, "Architectural Graphic Standards," John Wiley & Sons, Inc., New York; and "Tilt-up Concrete Walls," PA079.01B, Portland Cement Association.

Metal curtain walls may be custom, commercial, or industrial type. Custom-type walls are those designed for a specific project, generally multistory buildings. Commercial-type walls are those built up of parts standardized by manufacturers. Industrial-type walls are comprised of ribbed, fluted, or otherwise preformed metal sheets in stock sizes, standard metal sash, and insulation.

Metal curtain walls may be classified according to the methods used for field installation:

Stick Systems. Walls installed piece by piece. Each principal framing member, with windows and panels, is assembled in place separately (Fig. 10-3*a*). This type of system involves more parts and field joints than other types and is not so widely used.

Mullion-and-panel Systems. Walls in which vertical supporting members (mullions) are erected first, and then wall units, usually incorporating windows (generally unglazed), are placed between them (Fig. 10-3*b*). Often, a cover strip is added to cap the vertical joint between units.

Panel Systems. Walls composed of factory-assembled units (generally unglazed) and installed by connecting to anchors on the building frame and to each other (Fig. 10-3*c*). Units may be one or two stories high. This system requires fewer pieces and fewer field joints than the other systems.

Ample provision for movement is one of the most important considerations in designing metal curtain walls. Movement continually occurs, owing to thermal expansion and contraction, wind loads, gravity, and other causes. Joints and connections must be designed to accommodate it.

When mullions are used, it is customary to provide for horizontal movement at each mullion location, and in multistory buildings, to accommodate vertical movement at each floor, or at alternate floors when two-story-high components are used. Common ways of providing for horizontal movements include use of split mullions, bellows mullions, batten mullions, and elastic structural gaskets. Split mullions comprise two channel-shaped components permitted to move relative to each other in the plane of the wall. Bellow mullions have side walls flexible enough to absorb wall movements. Batten mullions consist of inner and outer cap sections that clamp the edges of adjacent panels, but not so tightly as to restrict movement in the plane of the wall. Structural gaskets provide a flexible link between mullions and panels. To accommodate vertical movement, mullions are spliced with a telescoping slip joint.

When mullions are not used and wall panels are connected to each other along their vertical edges, the connection is generally made through deep flanges. With the bolts several inches from the face of the wall, movement is permitted by the flexibility of the flanges.

Slotted holes are unreliable as a means of accommodating wall movement, though

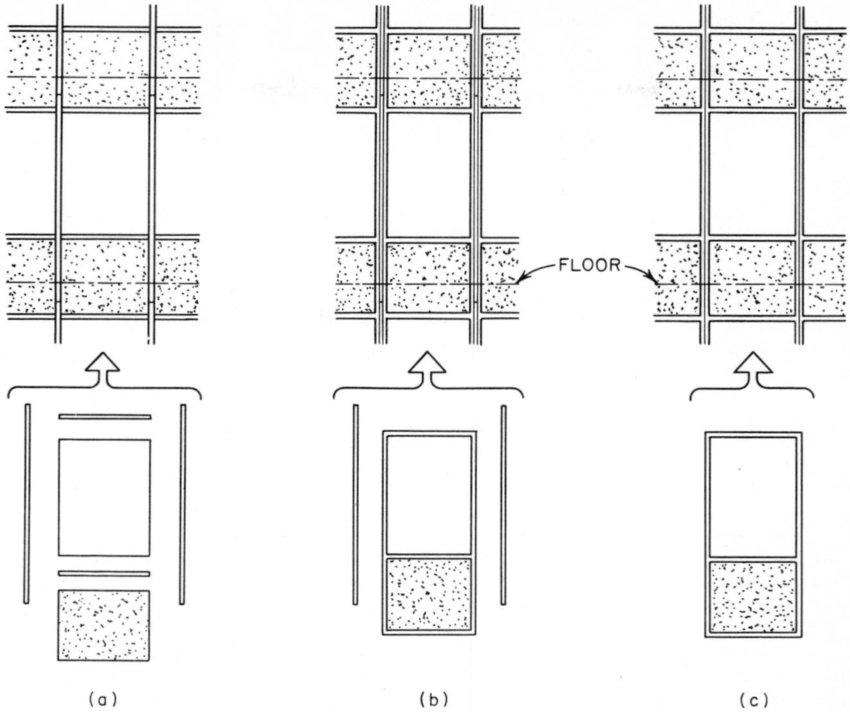

Fig. 10-3. Methods for field installation of metal curtain walls: (*a*) stick, (*b*) mullion and panel, and (*c*) panel system.

they are useful in providing dimensional tolerance in installing wall panels. Bolts drawn up too tightly or corrosion may prevent slotted holes from functioning as intended. If slotted holes are used, the connections should be made with shoulder bolts or sleeves and Bellville or nylon washers, to provide light but positive pressure and prevent rattling.

With metal curtain walls, special consideration also must be given to prevention of leakage, since metal and glass are totally nonabsorptive. It is difficult to make the outer face completely invulnerable to water penetration under all conditions; so a secondary defense in the form of an internal drainage system must be provided. Any water entering the wall must be drained to the exterior. Bear in mind that water can penetrate a joint through capillary action, reinforced by wind pressure, and running down a vertical surface, water can turn a corner to flow along a horizontal surface, defying gravity, into a joint.

Since metals are good transmitters of heat, it is particularly important with metal curtain walls to avoid thermal short circuits and metallic contacts between inner and outer wall faces. When, for example, mullions project through the wall, the inner face should be insulated, or each mullion should comprise two sections separated by insulation.

For more details on curtain walls, see W. F. Koppes, "Metal Curtain Wall Specifications Manual," National Association of Architectural Metal Manufacturers, 1010 West Lake St., Oak Park, Ill. 60301; "Curtain Wall Handbook," U.S. Gypsum Co., Chicago, Ill. 60606.

PARTITIONS

Partitions are dividing walls one story or less in height used to subdivide the interior space in buildings. They may be bearing or nonbearing walls.

10-23. Types of Partitions. Bearing partitions may be built of masonry or concrete or of wood or light-gage steel studs. These materials may be faced with plaster, wallboard, plywood, wood boards, plastic, or other materials that meet functional and architectural requirements. Masonry partitions should satisfy the requirements of Arts. 10-2 to 10-14 or Sec. 11.

Nonbearing partitions may be permanently fixed in placed, temporary (or movable) so that the walls may be easily shifted when desired, or folding. Since the principal function of these walls is to separate space, the type of construction and materials used may vary widely. They may be opaque or transparent; they may be louvered or hollow or solid; they may extend from floor to ceiling or only partway; and they may serve additionally as cabinets or closets or as a concealment for piping and electrical conduit.

Fire resistance sometimes dictates the type of construction. If a high fire rating is desired or required by local building codes, the local building official should be consulted for information on approved types of construction or the fire ratings of the American Insurance Association should be used.

When movable partitions may be installed, the structural framing should be designed to support their weight wherever they may be placed.

Acoustics also sometimes affects the type of construction of partitions. Thin construction that can vibrate like a sounding board should be avoided. Depending on functional requirements, acoustic treatment may range from acoustic finishes on partition surfaces to use of double walls separated completely by an air space or an insulating material.

Light-transmission requirements may also govern the selection of materials and type of construction. Where transparency or translucence is desired, the partition may be constructed of glass, or of glass block or plastic, or it may contain glass windows.

For data on installation of facings of ceramic wall tiles, see ANSI Standard Specifications for Glazed Ceramic Wall Tile, Ceramic Mosaic Tile, Quarry Tile and Pavers Installed in Portland Cement Mortars, A108.1, A108.2, A108.3, and A108.5, American National Standards Institute. Information also may be obtained from the Tile Council of America, Inc., Princeton, N.J.

Consideration should also be given to the necessity for concealing pipes, conduits, and ducts in partitions.

10-24. Structural Requirements of Partitions. Bearing partitions should be capable of supporting their own weight and superimposed loads in accordance with recommended engineering practice and should rest in turn on adequate supports that will not deflect excessively. Masonry partitions should meet the requirements of Arts. 10-2 to 10-14 or Sec. 11.

Nonbearing partitions should be stable laterally between lateral supports or additional lateral supports should be added. Since they are not designed for vertical loads other than their own weight, such partitions should not be allowed to take loads from overhead beams that may deflect and press down on them. Also, the beams under the partition should not deflect to the extent that there is a visible separation between bottom of partition and the floor or that the partition cracks.

Folding partitions, in a sense, are large doors. Depending on size and weight, they may be electrically or manually operated. They may be made of wood, light-gage metal, or synthetic fabric on a light collapsible frame. Provision should be made for framing and supporting them in a manner similar to that for large folding doors (Art. 10-35).

ORDINARY DOORS

Many types of doors are available to serve the primary purpose of barring or permitting access to a structure and between its interior spaces. They may be hinged on top or sides to swing open and shut; they may slide horizontally or vertically; or they may revolve about a vertical axis in the center of the opening.

A large variety of materials also is available for door construction. Wood, metal, glass, plastics, and combinations of these materials with each other and with other materials, in the form of sandwich panels, are in common use.

Selection of a type of door and door material depends as much on other factors as on the primary function of serving as a barrier. Cost, psychological effect, fire resistance, architectural harmony, and ornamental considerations are but a few of the factors that must be taken into account.

10-25. Fire and Smokestop Doors. Building codes require fire-resistant doors in critical locations to prevent passage of fire. Such doors are required to have a specific minimum fire-resistance rating, and are usually referred to as fire doors. The codes also may specify that doors in other critical locations be capable of preventing passage of smoke. Such doors, known as smokestop doors, need not be fire rated.

Fire protection of an opening in a wall or partition depends on the door frame and hardware, as well as on the door. All these components must be "labeled" or "listed" as suitable for the specific application. Bear in mind that fire doors are tested as an assembly of these components, and hence only approved assemblies should be specified.

Table 10-4. Typical Fire Ratings Required for Doors

Door use	*Rating, hr**
Doors in 3- or 4-hr fire barriers	3†
Doors in 2- or 1½-hr fire barriers	1½
Doors in 1-hr fire barriers	¾
Exit doors	1½ ‡
Doors to stairs and exit passageways	¾
Doors in 1-hr corridors	¾
Other corridor doors	0§
Smokestop doors	0¶

* Self-closing, swinging doors. Normally kept closed.

† Some codes require two 1½-hr opening protectives, with one protective installed on each face of a fire barrier.

‡ No rating required for street-floor exit doors with an exterior separation of more than 15 ft, or for exit doors of one- or two-family houses.

§ Should be noncombustible or 1¾-in. solid-core wood doors. Some codes do not require self-closing for doors in hospitals, sanitariums, nursing homes, and similar occupancies.

¶ Should be metal, metal-covered, or 1¾-in. solid-core wood doors (1⅜-in. in buildings less than three stories high), with 600-sq in. or larger, clear, wire-glass panels in each door.

SOURCE: Based on New York City Building Code.

All fire doors should be self-closing or should close automatically when a fire occurs. In addition, they should be self-latching, so that they remain closed. Push-pull hardware should not be used. Exit doors for places of assembly for more than 100 persons usually must be equipped with fire-exit (panic) hardware capable of releasing the door latch when pressure of 15 lb, or less, is applied to the device in the direction of exit. Combustible materials, such as flammable carpeting, should not be permitted to pass under a fire door.

Fire-door assemblies are rated, in hours, according to ability to withstand a standard fire test, such as that specified in American Society for Testing and Materials Standard E152. They may be identified as products qualified by tests by a UL label, provided by Underwriters' Laboratories, Inc.; an FM symbol of approval, authorized by Factory Mutual Research Corp.; or by a self-certified label, provided by the manufacturer (not accepted by National Fire Protection Association and some code officials).

Openings in walls and partitions that are required to have a minimum fire-resistance rating must have protection with a corresponding fire-resistance rating. Typical requirements are listed in Table 10-4.

This table also gives typical requirements for fire resistance of exit doors, doors to stairs and exit passageways, corridor doors, and smokestop doors.

In addition, some building codes also limit the size of openings in fire barriers. Typical maximum areas, maximum dimensions, and maximum percent of wall length occupied by openings are given in Table 10-5.

Table 10-5. Maximum Sizes of Openings in Fire Barriers

Protection of adjoining areas	Max area, sq ft	Max dimension, ft
Unsprinklered	120*	12†
Sprinklers on both sides	150*	15*
Building fully sprinklered	Unlimited*	Unlimited*

* But not more than 25% of the wall length or 56 sq ft per door if the fire barrier serves as a horizontal exit.
† But not more than 25% of the wall length.
SOURCE: Based on New York City Building Code.

Smokestop doors should be of the construction indicated in the footnote to Table 10-4. They should close openings completely, with only the amount of clearance necessary for proper operation.

["Standard for Fire Doors and Windows," NFPA No. 80; Life Safety Code, NFPA No. 101; "Fire Tests of Door Assemblies," NFPA No. 252, National Fire Protection Association, 60 Batterymarch St., Boston, Mass. 02110.

"Fire Tests of Door Assemblies," Standard UL 10(b); "Fire Door Frames," Standard UL 63; "Building Materials List" (annual, with bimonthly supplements), Underwriters' Laboratories, Inc., 207 East Ohio St., Chicago, Ill. 60611.

"Factory Mutual Approval Guide," Factory Mutual Research Corp., 1151 Boston-Providence Turnpike, Norwood, Mass. 02062.

"Hardware for Labeled Fire Doors," National Builders Hardware Association, 1290 Avenue of the Americas, New York, N.Y. 10019.]

10-26. Traffic Flow and Safety. Openings in walls and partitions must be sized for their primary function of providing entry and exit to or from a building or its interior spaces, and doors must be sized and capable of operating so as to prevent or permit such passage, as required by the occupants of the building. In addition, openings must be adequately sized to serve as an exit under emergency conditions. In all cases, traffic must be able to flow smoothly through the openings.

To serve these needs, doors must be properly selected for the use to which they are to be put, and properly arranged for maximum efficiency. In addition, they must be equipped with suitable hardware for the application. (See also Art. 10-25.)

Safety. Exit doors and doors leading to exit passageways should be so designed and arranged as to be clearly recognizable as such and to be readily accessible at all times. A door from a room to an exit or to an exit passageway should be the swinging type, installed to swing in the direction of travel to the exit.

Because of the heavy flow of traffic at building entrances, safety provisions at entrance doors are an important design consideration. Account must be taken of the location of such doors in the building faces, flow of outdoor traffic, and type and volume of traffic generated by the building. Following are some design recommendations:

Arc of a door swing should exceed 90°.

An entrance should always be set back from the building face.

Hinge jambs of swinging doors should be located at least 6 in. from a wall perpendicular to the building face.

If hinge jambs for two doors have to be placed close together, there should be enough distance between them to permit the doors to swing through an arc of 110°.

If several doors swinging in the same direction are placed close together in the same plane, they should be separated by center lights and should also have sidelights, to enable the doors to swing through 110° arcs.

If doors hung on center pivots are arranged in pairs, they should be hinged at the side jambs and not at the central mullion.

The floor on both sides of an exit door should be substantially level for a distance on each side equal to at least the width of the widest single leaf of the door. If, however, the exit discharges to the outside, the level outside the door may be one step lower than inside, but not more than 7½ in. lower.

Code Limitations on Door Sizes. To insure smooth, safe traffic flow, building codes generally place maximum and minimum limits on door sizes. Typical restrictions are as follows:

No single leaf in an exit door should be less than 28 in. wide or more than 48 in. wide. Minimum nominal width of opening should be at least:

36 in. for single corridor or exit doors.

32 in. for each of a pair of corridor or exit doors with central mullion.

48 in. for a pair of doors with no central mullion.

32 in. for doors to all occupiable and habitable rooms.

44 in. for doors to rooms used by bedridden patients and single doors used by patients in such buildings as hospitals, sanitariums, and nursing homes.

32 in. for toilet-room doors.

Jambs, stops, and door thickness when the door is open should not restrict the required width of opening by more than 3 in. for each 22 in. of width.

Nominal opening height for exit and corridor doors should be at least 6 ft 8 in. Jambs, stops, sills, and closures should not reduce the clear opening to less than 6 ft 6 in.

Opening Width Determined by Required Capacity. Width of an opening used as an exit is a measure of the traffic flow that the opening is permitted to accommodate. Capacities of exits and access facilities generally are measured in units of width of 22 in., and the number of persons per unit of width is determined by the type of occupancy. Thus, the number of units of exit width for a doorway is found by dividing by 22 the clear width of the doorway when the door is in the open position. (Projections of stops and hinge stiles may be disregarded.) Fractions of a unit of width less than 12 in. should not be credited to door capacity. If, however, 12 in. or more is added to a multiple of 22 in., one-half unit of width can be credited. Table 10-6 lists capacities in persons per unit of width that may be assumed for various types of occupancy.

Table 10-6. Capacity of Doors, Persons per 22-in. Unit of Door Width

Occupancy type	To outdoors at grade	Other doors
High hazard	50	40
Storage	75	60
Mercantile	100	80
Industrial	100	80
Business	100	80
Educational	100	80
Institutional		
For detention	50	40
For handicapped	30	30
Hotels, motels, apartments	50	40
	From assembly place	**From safe area**
Assembly		
Theaters	50	100
Concert halls	80	125
Churches	80	125
Outdoor structures	400	500
Museums	80	125
Restaurants	50	100

SOURCE: Based on New York City Building Code.

Every floor of a building should be provided with exit facilities for its occupant load. The number of occupants for whom exit facilities must be provided is determined by the actual number of occupants for which the space is designed, or by dividing the net floor area by the net floor area per person specified in the local building code, as, for example, in Table 10-7.

Table 10-7. Typical Occupant Load Requirements for Buildings

Occupancy	*Net floor area per occupant, sq ft*
Billiard rooms	50
Bowling alleys	50
Classrooms	20
Dance floors	10
Dining spaces (nonresidential)	12
Exhibition spaces	10
Garages and open parking structures	250
Gymnasiums	15
Habitable rooms	140
Industrial shops	200
In schools	30
Institutional sleeping rooms	
Adults	75
Children	50
Infants	25
Kindergartens	35
Kitchens (nonresidential)	200
Laboratories	50
Preparation rooms	100
Libraries	25
Locker rooms	12
Offices	100
Passenger terminals or platforms	1.5*C**
Sales areas (retail)	
First floor or basement	25
Other floors	50
Seating areas (audience) in places of assembly	
Fixed seats	*D*†
Movable seats	10
Skating rinks	15
Stages	*S*‡
Standing room (audience) in places of assembly	4
Storage rooms	200

* C = capacity of all passenger vehicles that can be unloaded simultaneously.

† D = number of seats or occupants for which space is to be used.

‡ S = 75 persons per unit of width of exit openings serving a stage directly, or one person per 15 sq ft of performing area plus one person per 50 sq ft of remaining area plus number of seats that may be placed for an audience on stage.

SOURCE: Based on New York City Building Code.

10-27. Water Exclusion at Exterior Doors. Exterior doors are subjected to all the effects of natural forces, as are walls and windows, including solar heat, rain, and wind. For ordinary installations, closed doors cannot be expected to completely exclude water or stop air movement under all conditions. One important reason for this is that clearances must be provided around each door. These are necessary to permit easy operation and thermal expansion and contraction of doors.

Exterior doors, however, can be made less vulnerable to water penetration by setting them back from the building face, or by providing overhead protection, such as a canopy, marquee, or balcony. These measures also will help reduce collection of snow and ice at door thresholds. In addition, provision of an entrance vestibule is desirable, because it can serve as a weather barrier. Also, weatherstripping around doors assists in preventing passage of water and air past closed doors.

10-28. Structural Requirements for Openings and Doors. A wall or partition above a door or window opening must be adequately supported by a structural member. In design of such a member, stiffness as well as strength must be taken into account. Excessive deflection could interfere with door operation.

In design of door framing, wind loads are generally more critical than dead loads imposed by the wall or partition above. Wind load should be taken as at least 15 psf. Under this loading, maximum deflection of framing members should not exceed ¾ in. or $1/175$ the clear span. The design should take into account the fact that wind forces acting outward may be larger than those acting inward. Door framing should generally be made independent of other framing.

("Entrance Manual," National Association of Architectural Metal Manufacturers, 228 North La Salle St., Chicago, Ill. 60601.)

10-29. Control of Air Movements at Entrances. In tall buildings, there is likely to be a difference in air pressure between the inside and outside at entrances. This pressure results from air movements through stairways and shafts. When the building is heated in cold weather, warm air rises in the vertical passageways. When the building is cooled in warm weather, cool air flows downward through these passageways. The resulting pressure differential varies with building height and difference between indoor and outdoor temperatures. In many cases, this stack effect causes large air flows through entrance doors, significantly increases heating and cooling loads, and may make entrance-door operation difficult.

A commonly used means of reducing such loads is provision of an entrance vestibule with one set of doors leading outdoors and another set leading to the inside. In some cases, however, traffic flow is so great that at least one door of both sets of doors may be partly open simultaneously, permitting air to flow through the vestibule between the inside and outside of the building. Thus, the effectiveness of the vestibule is decreased. The loss of effectiveness may be reduced, however, by venting the vestibule to outdoors and providing compensating heating or cooling.

Another means of reducing air movements due to stack effect is installation of revolving doors (Art. 10-33). While in continuous contact with its enclosure, a revolving door permits entry and exit without extra force to offset the differential pressure between indoors and outdoors. (Building codes, however, require some swinging doors in conjunction with revolving doors.)

Where stack effects make operation of swinging doors difficult, operation may be made easier by hanging the doors on balanced pivots or equipping the doors with automatic operators. In some cases, it may be advantageous to replace the doors with automatic, horizontally sliding doors.

10-30. Swinging Doors. These are doors hinged near one edge to rotate about a vertical axis. Swinging doors are hung on butts or hinges. The part of a doorway to which a door is hinged and against which it closes is known as the door frame. It consists of two verticals, commonly called jambs, and a horizontal member, known as the header (Figs. 10-4 and 10-5). **Single-acting doors** can swing 90° or more in only one direction; **double-acting doors** can swing 90° or more in each of two directions.

To stop drafts and passage of light, the header and jambs have a stop, or projection, extending the full height and width, against which the door closes. The projection may be integral with the frame, or formed by attaching a stop on the surface of the frame, or inset slightly.

Door frames for swinging wood doors generally are fastened to rough construction members known as rough bucks, and the joints between the frame and the wall are covered with casings, or trim. With metal construction, the trim is often integral with the frame, which is attached to the wall with anchors. ("Recommended Standard Details, Steel Doors and Frames," SDI 111, Steel Door Institute, 2130 Keith Building, Cleveland, Ohio 44115.)

At the bottom of a door opening for an exterior door is a sill (Fig. 10-4), which forms a division between the finished floor on the inside and the outside construction. The sill generally also serves as a step; for the door opening usually is raised above exterior grade to prevent rain from entering. The top of the sill is sloped to drain water away from the interior. Also, it may have a raised section

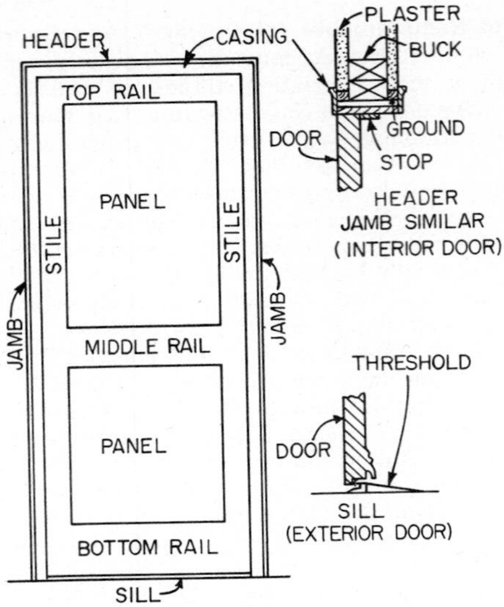

Fig. 10-4. Typical paneled wood door.

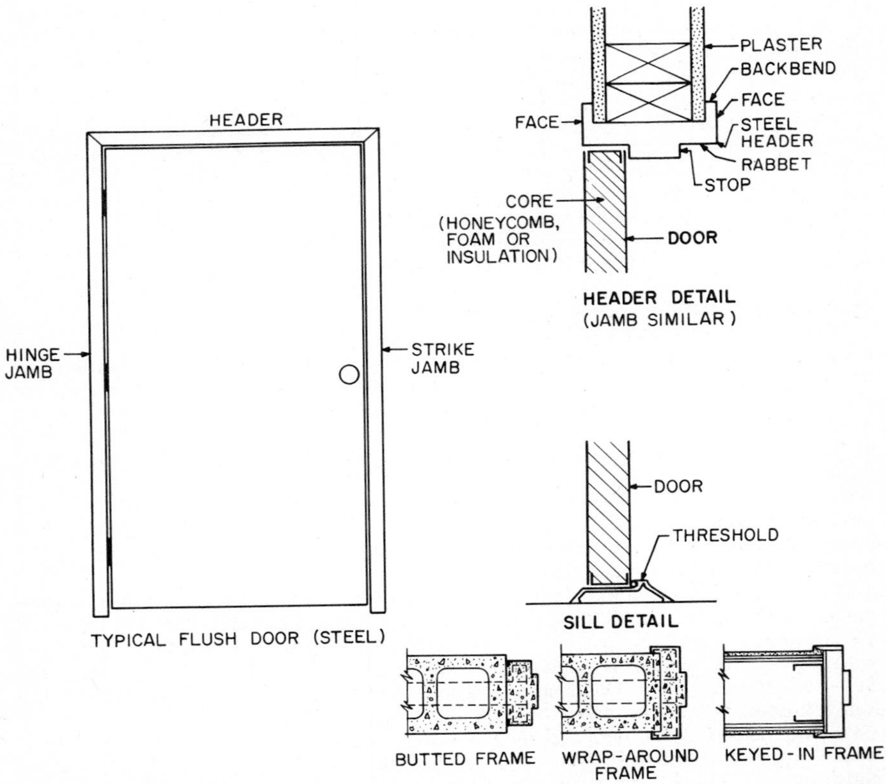

Fig. 10-5. Typical steel flush door. (*Courtesy V. C. Braun, Amweld Building Products, Niles, Ohio.*)

in the plane of the door or slightly to the rear, so that water dripping from the door will fall on the slope. The raised section may be integral with the sill or a separate threshold. In either case, the rear portion covers the joint between sill and floor. In addition, all joints should be sealed.

To weatherproof the joint between the bottom of a wood door and the threshold, a weatherstrip in the form of a hooked length of metal often is attached to the underside of the door. When the door is closed, the weatherstrip locks into the threshold to seal out water. Other types of weatherstripping, including plastic gaskets, generally are used for steel doors and may be installed on header and jambs ("Recommended Weatherstripping for Standard Steel Doors and Frames," SDI 111-E, Steel Door Institute, 2130 Keith Building, Cleveland, Ohio, 44115).

The direction of swing, or hand, of each door must be known and specified in ordering such hardware as latches, locks, closers, and panic hardware, because the hand determines the type of operation required of them. The hand is determined with respect to the outside or key, or locking, side, following the conventions illustrated in Fig. 10-6.

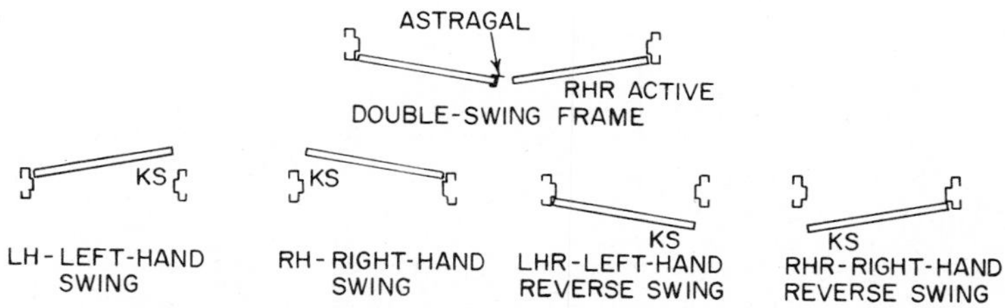

Fig. 10-6. Swing of doors. *KS* indicates key, or locking, side.

Swinging doors generally are available in thicknesses of 1⅜ or 1¾ in., opening heights of 6 ft 8 in. or 7 ft, and widths of 30, 36, or 42 in. Nonstandard sizes are obtainable on special order, but certain precautions should be observed: Size and number of butts or hinges and offset pivots should be suitable for the door size. For rail-and-stile doors, check with the manufacturer to insure that face areas will not be excessive for the stile width. Large doors should not be used where they will be exposed to strong winds. To prevent spreading of stiles of tall doors, push bars or other intermediate bar members should be attached to both stiles of doors over 8 ft high.

Package entrance doors are available from several manufacturers. The package includes a single door or pair of doors, door frame, and all hardware. Such doors offer the advantages of integrated design, assumption by the manufacturer of responsibility for satisfactory performance, usually quick delivery, and often cost savings.

("Entrance Manual," National Association of Architectural Metal Manufacturers, 228 North La Salle St., Chicago, Ill. 60601.)

10-31. Horizontally Sliding Doors. This type of door may roll on a track in the floor and have only guides at the top, or the track and rollers may be at the top and guides at the bottom. Some doors fold or collapse like an accordion, to occupy less space when open. A pocket must always be provided in the walls on either or both sides, to receive rigid-type doors; with the folding or accordion types, a pocket is optional.

Horizontally sliding doors are advantageous for unusually wide openings, and where clearances do not permit use of swinging doors. Operation of sliding doors is not hindered by windy conditions or differences in air pressure between indoors and outdoors.

10-32. Vertically Sliding Doors. This type of door may rise straight up, may rise up and swing in, or may pivot outward to form a canopy. Sometimes, the door may be in two sections, one rising up, the other dropping down. Generally, all types are counterweighted for ease of operation.

In designing structures to receive sliding doors, the clearance between the top of the doors and the ceiling or structural members above should be checked—especially the clearance required for vertically sliding doors in the open position. Also, the deflection of the construction above the door opening should be investigated to be certain that it will not cause the door to jam.

To keep out the weather, the upper part of a sliding door either is recessed into the wall above, or the top part of the door extends slightly above the bottom of the wall on the inside. Similarly, door sides are recessed into the walls or lap them and are held firmly against the inside. Also, the finished floor is raised a little above outside grade. Very large sliding doors require special study (Arts. 10-35 and 10-36).

10-33. Revolving Doors. This type of door is generally selected for entranceways carrying a continuous flow of traffic without very high peaks. They offer the advantage of keeping interchange of inside and outside air to a relatively small amount compared with other types of doors. They usually are used in combination with swinging doors because of the inability to handle large groups of people in a short period of time.

Revolving doors consist of four leaves that rotate about a vertical axis inside a cylindrical enclosure. Diameter of the enclosure generally is at least 6 ft 6 in., and the opening to the enclosure usually is between 4 and 5 ft.

Building codes prohibit use of revolving doors for some types of occupancies; for example, for theaters, churches, and stadiums, because of the limited traffic flow in emergencies. Where they are permitted as exits, revolving doors have limitations imposed on them by building codes. The National Fire Protection Association "Life Safety Code" allows only one-half unit of width per revolving door, but some codes permit one unit per door. Also, revolving doors may not provide more than 50% of the required exit capacity at any location. The remaining capacity must be supplied by swinging doors within 20 ft of the revolving doors. Rotation speed must be controlled so as not to exceed 15 rpm. Each wing should be provided with at least one push bar and should be glazed with tempered glass at least $\frac{7}{32}$ in. thick. Some codes also require the doors to be collapsible.

10-34. Door Materials. *Wood* is used in several forms for doors. When appearance is unimportant and a low-cost door is required, it may be made of boards nailed together. When the boards are vertical and held together with a few horizontal boards, the door is called a batten door. Better-grade doors are made with panels set in a frame or with flush construction.

Paneled doors consist of solid wood or plywood panels held in place by verticals called stiles and horizontals known as rails (Fig. 10-4). The joints between panels and supporting members permit expansion and contraction of the wood with atmospheric changes. If the rails and stiles are made of a single piece of wood, the paneled door is called solid. When hardwood or better-quality woods are used, the doors generally are veneered; rails and stiles are made with cores of softwood sandwiched between the desired veneer.

Tempered glass or plastic may be used for panels. In exterior doors, the lights must be installed to prevent penetration of water, especially in veneered doors. One way is to insert under the glass a piece of molding that extends through the door and is turned down over the outside face of the door to form a drip. Another way is to place a sheet-metal flashing under the removable outer molding that holds the glass. The flashing is turned up behind the inside face of the glass and down over the exterior of the door, with only a very narrow strip of the metal exposed.

Flush doors also may be solid or veneered. The veneered type has a core of softwood, while the flat faces may be hardwood veneers. When two plies are used for a face, they are set with grain perpendicular to each other.

Flush doors, in addition, may be of the hollow-core type. In that case, the surfaces are made of plywood and the core is a supporting grid. Edges of the core are solid wood boards.

Metal doors generally are constructed in one of three ways: cast as a single unit or separate frame and panel pieces; metal frame covered with sheet metal; and sheet metal over a wood or other type of insulating core. (See "ANSI Standard Nomenclature for Steel Doors and Steel Door Frames," A123.1, and "Recommended

Standard Details, Steel Doors and Frames," Steel Door Institute, 2130 Keith Building, Cleveland, Ohio, 44115.)

Cast-metal doors are relatively high-priced. They are used principally for monumental structures.

Hollow metal doors may be of flush or panel design, with steel faces having a thickness of at least 20 gage. Flush doors incorporate steel stiffeners or polyurethane foam, polystyrene foam, or honeycomb core as a lightweight support for the faces (Fig. 10-5). Voids between stiffeners may be filled with lightweight insulation. Panel doors may be of stile-and-rail or stile-and-panel construction with insulated panels. Also, a light-duty, 22-gage, steel-faced door with an insulated core is available for use in houses.

Metal-clad (Kalamein) doors are of the swinging type only. They may be of flush or panel design, with metal-covered wood cores for stiles and rails and insulated panels covered with steel 24 gage or lighter.

Other Materials. Doors may be made wholly or partly transparent or translucent. Lights may be made of tempered glass or plastic. Doors made completely of glass are pivoted at top and bottom because the weight makes it difficult to support them with hinges or butts.

Sliding doors of the collapsible accordion type generally consist of wood slats or a light steel frame covered with textile. Plastic coverings frequently are used.

SPECIAL-PURPOSE DOORS

Large-size doors, such as those for hangars, garages, and craneway openings in walls and for subdividing gymnasiums and auditoriums, often have to be designed

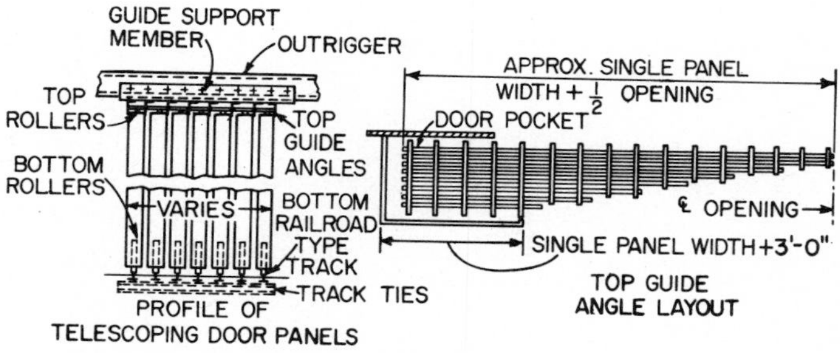

Fig. 10-7. Typical horizontal-sliding, telescoping door.

individually. As distinguished from those made on a production basis, special-purpose doors require careful design of the main structure to support loads and to allow space for the open door and its controls.

Manufacturers classify special-purpose doors as horizontal-sliding, vertical-sliding, swing, and top or horizontal-hinge.

10-35. Horizontal-sliding Doors. Door leaves in the horizontal-sliding type are equipped with bearing-type bottom wheels and ride rails in the floor while top rollers operate in overhead guides. Two variations are in common use—telescoping and folding.

Telescoping doors (Fig. 10-7) are frequently used for airplane hangars. Normally composed of 6 to 20 leaves, they generally are center parting. They are built of wood, steel, or a combination of the two. Open doors are stacked in pockets at each end of the opening.

Telescoping doors are frequently operated by two motors located in the end pockets. The motors drive an endless chain attached to tops of center leaves. Remaining leaves are moved by a series of interconnecting cables attached to the powered leaf and arranged so that all leaves arrive at open or closed position simultaneously. Motor size ranges from 1 to 10 hp, travel speed of leaves from 45 to 160 fpm.

The weight of the leaves must be taken by footings below the rails. Provision also must be made to take care of wind loads transmitted to the top guide channels by the doors and to carry the weight of these channels.

Folding doors (Fig. 10-8) are commonly used for subdividing gymnasiums, auditoriums, and cafeterias and for hangars with very wide openings. This type of door is made up of a series of leaves hinged together in pairs. Leaves fold outward, and when the door is shut, they are held by automatic folding stays. Motors that operate the door usually are located in mullions adjacent to the center of the opening. The mullions are connected by cables to the ends of the opening, and when the door is to be opened, the mullions are drawn toward the ends, sweeping the leaves along. Travel speed may vary from 45 to 160 fpm.

Chief advantage of folding over telescoping types is that only two guide channels are required, regardless of width. Thus, less metal is required for guide channels and rails, and less material for the supporting member above. Also, since wind loading is applied to leaves that are always partly in folded position, the triangular configuration gives the door considerable lateral stiffness. Hence, panel thickness

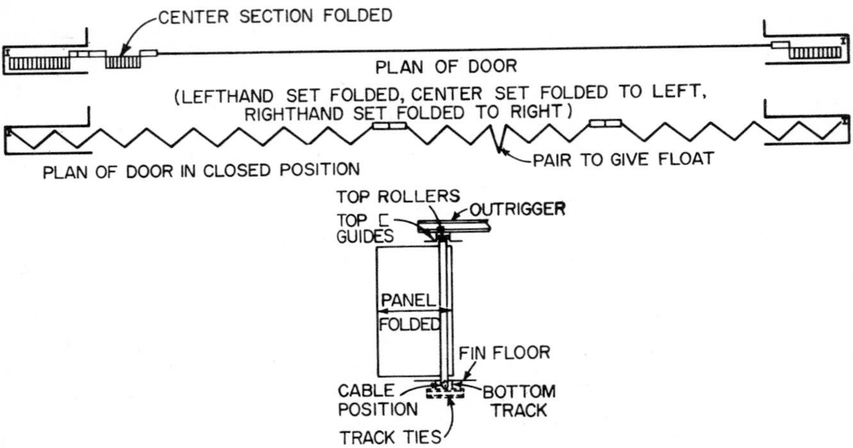

Fig. 10-8. Typical folding horizontal-sliding door.

may be less for folding than for telescoping doors, and a lighter load may be used for designing footings.

10-36. Vertical-sliding Doors. When space is available above and below an opening into which door leaves can be moved, vertical-sliding doors are advantageous. They may be operated manually or electrically. Leaves may travel at 45 to 60 fpm. About 1½ ft in excess of leaf height must be provided in the pockets into which the leaves slide. So the greater the number of leaves the less space needed for the pockets.

Vertical-sliding doors normally are counterweighted. About 15 in. of clear space back of the jamb line is needed for this purpose, but on the idler side only 4 in. is required.

Vertical loads are transmitted to the door guides and then by column action to the footings. Lateral support is provided by the jamb and building walls.

10-37. Swing Doors. When there is insufficient space around openings to accommodate sliding doors, swing types may be used. Common applications have been for firehouses, where width-of-building clearance is essential, and railway entrances, where doors are interlocked with the signal system.

Common variations include single-swing (solid leaf with vertical hinge on one jamb), double-swing (hinges on both jambs), two-fold (hinge on one jamb and another between folds and leaves), and four-fold (hinges on both jambs and between each pair of folds).

The more folds, the less time required for opening and the smaller the radius needed for swing. Tighter swings make doors safer to open and allow material to be placed closer to supports.

10-38. Horizontal-hinge Doors. Effective use of horizontal-hinge doors is made in such applications as craneway entrances to buildings. Widths exceeding 100 ft at or near the top of the building can be opened to depths of 4 to 18 ft. Frequently, horizontal-sliding doors are employed below crane doors to increase the opening. If so, the top guides are contained in the bottom of the crane door; so the sliding door must be opened before the swing door.

Top-hinged swing doors are made of light materials, such as structural steel and exterior-grade plywood. The panel can be motor-operated to open in 1 min or less.

10-39. Radiation-Shielding Doors. These are used as a barrier against harmful radiation and atomic particles across openings for access to "hot" cells, and against similar radioactive-isotope handling arrangements and radiation chambers of high-energy X-ray machines or accelerators. Usually, they must protect not only personnel but also instruments even more sensitive to radiation than people.

Shielding doors usually are much thicker and heavier than ordinary doors, because density is an important factor in barring radiation. Generally, these special-purpose doors are made of steel plates, steel-sheathed lead, or concrete. To reduce thickness, concrete doors may be of medium-heavy (240 lb per cu ft) or heavy (300 lb per cu ft) concrete, often made with iron-ore aggregate.

The heavy doors usually are operated hydraulically or by electric motor. Provision must be made, however, for manual operation if the mechanism should break down.

Common types of shielding doors include hinged, plug, and overlap. The hinged type is similar to a bank vault door. The plug type, flush with the walls when closed, may roll on floor-mounted tracks or hang from rails. Overlap doors, surface-mounted, also may roll or hang from rails. In addition, vertical-lift doors sometimes are used.

Section 11

Engineered Brick Construction

ALAN H. YORKDALE

Director of Engineering and Research, Brick Institute of America, McLean, Va.

Design of loadbearing brick structures may be based on rational engineering analysis instead of the empirical requirements for minimum wall thickness and maximum wall height contained in building codes and given in Sec. 10. Those requirements usually make bearing-wall construction for buildings higher than three to five stories uneconomical, and encourage use of other methods of support (steel or concrete skeleton frame). Since 1965, engineered brick buildings 10 or more stories high, with design based on rational structural analysis, have been built in the United States. This construction was stimulated by the many loadbearing brick buildings exceeding 10 stories in height constructed during the preceding two decades in Europe.

Design requirements for engineered brick structures given in this section are taken from "Building Code Requirements for Engineered Brick Masonry," promulgated by the Structural Clay Products Institute (now the Brick Institute of America, 1750 Old Meadow Road, McLean, Va. 22101).

11-1. Design Standards. Working with a highly qualified Engineering Advisory Committee, the Structural Clay Products Institute developed in 1966 a rational design standard for engineered brick construction, based on laboratory research and historical performance data. The Institute, in collaboration with the Engineering Advisory Committee, also developed and published a second-generation standard, "Building Code Requirements for Engineered Brick Masonry," August, 1969. This newer edition incorporated the findings of additional research and the experience of architects and engineers in use of the 1966 standard.

Provisions of the first or second editions of the standard have been adopted by reference or incorporated in the masonry chapters of the BOCA Basic Building Code, Southern Standard Building Code, National Building Code, and Uniform Building Code.

In addition to adoption by the model building codes, many local political subdivisions have also adopted, by reference or in principle, these provisions and requirements for engineered brick construction.

General Requirements of Design Standard. The 1969 standard, "Building Code Requirements for Engineered Brick Masonry," provides minimum requirements predicated on a general analysis of the structure and based on generally accepted engineering analysis procedures.

The standard contains a requirement for architectural or engineering inspection of the workmanship to ascertain, in general, if the construction and workmanship are in accordance with the contract drawings and specifications. Frequency of inspections should be such that an inspector can inspect the various stages of construction and see that the work is being properly performed. The standard requires reduced allowable stresses and capacities for load when such architectural or engineering inspection is not provided.

Structural Requirements of Design Standard. Engineered brick, bearing-wall structures do not require new techniques of analysis and design, but merely application of engineering principles used in the analysis and design of other structural systems. The method of analysis depends on the complexity of the building with respect to height, shape, wall location, and openings in the wall. A few conservative assumptions, however, accompanied by proper details to support them, can result in a simplified and satisfactory solution for most bearing-wall structures up to 12 stories high. More rigorous analysis for bearing-wall structures beyond this height may be required to maintain the economics of this type of construction.

11-2. Definitions of Design Terms. Where terms are not defined in this section, they may be assumed to have generally accepted meanings. Unless otherwise noted in this section, terms listed in this article have the meanings given here. (For definitions of commonly used masonry terms, see Art. 10-1.)

Bearing Wall. (See Loadbearing Wall.)

Brick. A masonry unit having approximately the shape of a rectangular prism, made from burned clay or shale or a mixture of these.

Column. (See Art. 10-1.)

Eccentricity. The normal distance between the centroidal axis of a member and the component of resultant load parallel to that axis.

Effective Height. The height of a member to be assumed for calculating the slenderness ratio.

Effective Thickness. The thickness of a member to be assumed for calculating the slenderness ratio.

Lateral Support. Members such as cross walls, columns, pilasters, buttresses, floors, roofs, or spandrel beams that have sufficient strength and stability to resist horizontal forces transmitted to them may be considered lateral supports.

Loadbearing Wall. A wall that supports any vertical load in addition to its own weight.

Prism. An assemblage of brick and mortar for the purpose of laboratory testing for design strength, quality control of materials, and workmanship. Minimum height for prisms is 12 in., and the slenderness ratio should lie between 2 and 5.

Slenderness Ratio. Ratio of the effective height of a member to its effective thickness.

Solid Masonry Unit. (See Art. 10-1.)

Solid Masonry Wall. A wall built of solid masonry units laid contiguously, with joints between units filled with mortar or grout.

Virtual Eccentricity. The eccentricity of resultant axial loads required to produce axial and bending stresses equivalent to those produced by applied axial and transverse loads.

11-3. Materials for Engineered Brick Construction. Strength (compressive, shearing, and transverse) of brick structures is affected by the properties of the brick and the mortar in which they are laid. In compression, strength of brick has the greater effect. Although mortar is also a factor in compressive strength, its greater effect is on the transverse and shearing strengths of masonry. For these reasons, there are specific design requirements for and limitations on materials used in engineered brick structures.

Brick. These units must conform to the requirements for grade MW or SW, ASTM "Standard Specifications for Building Brick," C62. In addition, brick used in loadbearing or shear walls must comply with the dimension and distortion tolerances specified for type FBS of ASTM "Standard Specifications for Facing Brick," C216. Bricks that do not comply with these tolerance requirements may be used if the ultimate compressive strength of the masonry is determined by prism tests. Used or salvaged bricks are not permitted in engineered brick masonry construction.

Mortar. Most of the test data on which allowable stresses for engineered brick masonry are based were obtained for specimens built with portland cement-hydrated lime mortars. Three mortar types are provided for: M, S, and N, as described in ASTM C270 (see Table 10-1), except that the mortar must consist of mixtures of portland cement (type I, II, or III), hydrated lime (type S, non-air-entrained), and aggregate when the allowable stresses specified in "Building Code Requirements for Engineered Brick Masonry" are used. This standard provides, however, that "Other mortars . . . may be used when approved by the Building Official, provided strengths for such masonry construction are established by tests . . ."

11-4. Determination of Masonry Compressive Strength. Allowable stresses are based, for the most part, on the ultimate compressive strength f'_m, psi, of masonry used. This strength may be determined by prism testing, or may be assumed based on the compressive strength of the bricks and the type of mortar used.

Prism Tests. When the compressive strength of the masonry is to be established by preliminary tests, the tests should be made in advance of construction with prisms built of similar materials, assembled under the same conditions and bonding arrangements as for the structure. In building the prisms, the moisture content of the units at the time of laying, consistency of the mortar, thickness of mortar joints, and workmanship should be the same as will be used in the structure.

Prisms. Prisms should be stored in an air temperature of not less than 65°F and aged, before testing, for 28 days in accordance with the provisions of ASTM "Standard Method of Test for Compressive Strength of Masonry Assemblages," E447. Seven-day test results may be used if the relationship between 7- and 28-day strengths of the masonry has been established by tests. In the absence of such data, the 7-day compressive strength of the masonry may be assumed to be 90% of the 28-day compressive strength.

The value of f'_m used in the standard, "Building Code Requirements for Engineered Brick Masonry," is based on a height-thickness ratio h/t of 5. If the h/t of prisms tested is less than 5, which will cause higher test results, the compressive strength of the specimens obtained in the tests should be multiplied by the appropriate correction factor given in Table 11-1. Interpolation may be used to obtain intermediate values.

Table 11-1. Strength Correction Factors for Short Prisms

Ratio of height to thickness h/t	2.0	2.5	3.0	3.5	4.0	4.5	5.0
Correction factor	0.73	0.80	0.86	0.91	0.95	0.98	1.00

Brick Tests. When the compressive strength of masonry is not determined by prism tests, but the brick, mortar, and workmanship conform to all applicable requirements of the standard, allowable stresses may be based on an assumed value of the 28-day compressive strength f'_m computed from Eq. (11-1) or interpolated from the values in Table 11-2.

$$f'_m = A(400 + Bf'_b) \tag{11-1}$$

where A = coefficient (⅔ without inspection and 1.0 with inspection)

B = coefficient (0.2 for type N mortar, 0.25 for type S mortar, and 0.3 for type M mortar)

f'_b = average compressive strength of brick, psi $\leq$ 14,000 psi

When there is no engineering or architectural inspection to insure compliance with the workmanship requirements of the standard, the values in Table 11-2 under "Without inspection" should be used.

Compressive strength tests of brick should be conducted in accordance with ASTM "Standard Methods of Sampling and Testing Brick," C67.

11-5. Allowable Stresses for Brick Construction. Allowable stresses; compressive f_m, tensile f_t, and shearing f_v for brick construction should not exceed the values shown in Tables 11-3 and 11-4. Allowable shears on bolts and anchors are given in Table 11-9. For allowable loads on walls and columns, see Art. 11-6.

Table 11-2. Assumed Compressive Strength of Brick Construction, Psi

Compressive strength of brick, psi	Without inspection			With inspection		
	Type N mortar	Type S mortar	Type M mortar	Type N mortar	Type S mortar	Type M mortar
14,000 plus	2,140	2,600	3,070	3,200	3,900	4,600
12,000	1,870	2,270	2,670	2,800	3,400	4,000
10,000	1,600	1,930	2,270	2,400	2,900	3,400
8,000	1,340	1,600	1,870	2,000	2,400	2,800
6,000	1,070	1,270	1,470	1,600	1,900	2,200
4,000	800	930	1,070	1,200	1,400	1,600
2,000	530	600	670	800	900	1,000

Table 11-3. Allowable Stresses for Nonreinforced Brick Construction, Psi

Description		Without inspection	With inspection
Axial compression			
Walls	f_m	$0.20 f'_m$	$0.20 f'_m$
Columns	f_m	$0.16 f'_m$	$0.16 f'_m$
Flexural compression			
Walls	f_m	$0.32 f'_m$	$0.32 f'_m$
Columns	f_m	$0.26 f'_m$	$0.26 f'_m$
Flexural tension			
Normal to bed joints			
M or S mortar	f_t	24	36
N mortar	f_t	19	28
Parallel to bed joints			
M or S mortar	f_t	48	72
N mortar	f_t	37	56
Shear			
M or S mortar	v_m	$0.5\sqrt{f'_m} \leq 40$	$0.5\sqrt{f'_m} \leq 80$
N mortar	v_m	$0.5\sqrt{f'_m} \leq 28$	$0.5\sqrt{f'_m} \leq 56$
Bearing			
On full area	f_m	$0.25 f'_m$	$0.25 f'_m$
On one-third area or less	f_m	$0.375 f'_m$	$0.375 f'_m$
Modulus of elasticity	E_m	$1{,}000 f'_m \leq 2{,}000{,}000$ psi	$1{,}000 f'_m \leq 3{,}000{,}000$ psi
Modulus of rigidity	E_v	$400 f'_m \leq 800{,}000$ psi	$400 f'_m \leq 1{,}200{,}000$ psi

For wind, blast, or earthquake loads combined with dead and live loads, the allowable stresses in brick construction may be increased by one-third, if the resultant section will not be less than that required for dead loads plus reduced live loads alone. Where the actual stresses exceed the allowable, the designer should specify a larger section or reinforced brick masonry.

11-6. Allowable Loads on Walls and Columns. Two stress-reduction factors are used in calculating allowable loads on walls and columns: slenderness coefficient C_s and eccentricity coefficient C_e. The eccentricity coefficient is used to reduce the allowable axial load in lieu of performing a separate bending analysis. The slenderness coefficient is used to reduce the allowable axial load to prevent buckling.

To determine these two factors, three constants are needed: end eccentricity ratio e_1/e_2, ratio of maximum virtual eccentricity to wall thickness e/t, and slenderness ratio h/t.

Eccentricity Ratio. At the top and bottom of any wall or column, a virtual eccentricity (Art. 11-2) of some magnitude (including zero) occurs. e_1/e_2 is the ratio of the smaller virtual eccentricity to the larger virtual eccentricity of the loads acting on a member. By this definition, the absolute value of the ratio

Table 11-4. Allowable Stresses for Reinforced Brick Construction, Psi

Description	Without inspection	With inspection
Axial compression		
Walls . . . . f_m	$0.25 f'_m$	$0.25 f'_m$
Columns . . . . f_m	$0.20 f'_m$	$0.20 f'_m$
Flexural compression		
Walls and beams . . . . . . f_m	$0.40 f'_m$	$0.40 f'_m$
Columns . . . . f_m	$0.32 f'_m$	$0.32 f'_m$
Shear		
No shear reinforcement		
Flexural members . . . . v_m	$0.7\sqrt{f'_m} \leq 25$	$0.7\sqrt{f'_m} \leq 50$
Shear walls . . . . v_m	$0.5\sqrt{f'_m} \leq 50$	$0.5\sqrt{f'_m} \leq 100$
With shear reinforcement taking entire shear		
Flexural members v	$2.0\sqrt{f'_m} \leq 60$	$2.0\sqrt{f'_m} \leq 120$
Shear walls v	$1.5\sqrt{f'_m} \leq 75$	$1.5\sqrt{f'_m} \leq 150$
Bond		
Plain bars . . . . u	53	80
Deformed bars . . . . u	107	160
Bearing		
On full area . . . . f_m	$0.25 f'_m$	$0.25 f'_m$
On one-third area or less . . f_m	$0.375 f'_m$	$0.375 f'_m$
Modulus of elasticity . . . . . E_m	$1{,}000 f'_m \leq 2{,}000{,}000$ psi	$1{,}000 f'_m \leq 3{,}000{,}000$ psi
Modulus of rigidity . . . . . . E_v	$400 f'_m \leq 800{,}000$ psi	$400 f'_m \leq 1{,}200{,}000$ psi

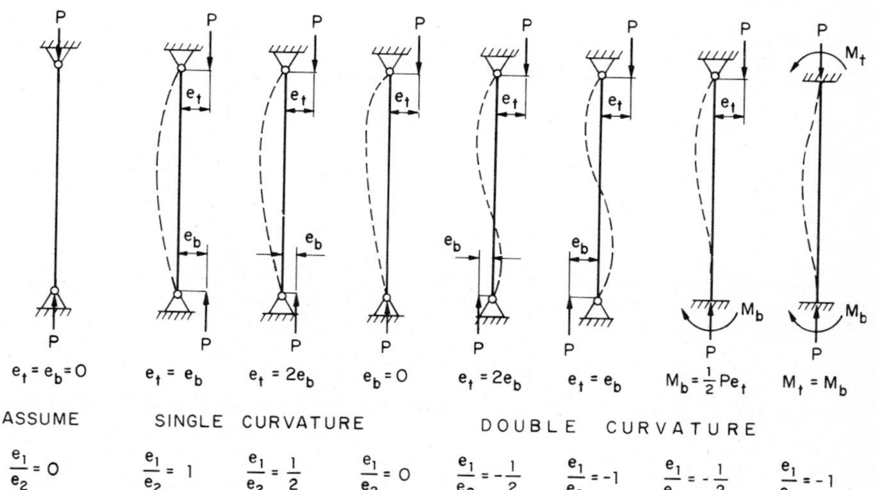

Fig. 11-1. Axis of compression member with positive eccentricity ratio e_1/e_2 has single curvature, and with negative eccentricity ratio, double curvature.

is always less than or equal to 1.0. Where e_1 or e_2, or both, are equal to zero, e_1/e_2 is assumed to be zero. Where the member bent in single curvature (top and bottom virtual eccentricities occurring on the same side of the centroidal axis of a wall or column), e_1/e_2 is positive. Where the member is bent in double curvature (top and bottom virtual eccentricities occurring on opposite sides of a wall or column centroidal axis), e_1/e_2 is negative (Fig. 11-1).

Eccentricity-thickness Ratio. The ratio of maximum virtual eccentricity to wall thickness e/t is used in selecting the eccentricity coefficient C_e. Design of a nonreinforced member requires that e/t be less than or equal to ⅓. If e/t is greater than ⅓, the designer should specify a larger section, different bearing details for load transfer to the masonry, or reinforced brick construction.

Slenderness Ratio. This is the ratio of the unsupported height h to the wall thickness t. It is used in selecting the slenderness coefficient C_s.

The unsupported height h is the actual distance between lateral supports. Not always the floor-to-floor height, h may be taken as the distance from the top of the lower floor to the bearing of the upper floor where these floors provide lateral support.

The effective thickness t for nonreinforced solid masonry is the actual wall thickness, except for metal-tied cavity walls. In cavity walls, each wythe is considered to act independently, thus producing two different walls. When the cavity is filled with grout, the effective thickness becomes the total wall thickness.

Table 11-5. Eccentricity Coefficients C_e

e/t	e_1/e_2								
	-1	-3/4	-1/2	-1/4	0	1/4	1/2	3/4	1
0–0.05	1.00	1.00	1.00	1.00	1.00	1.00	1.00	1.00	1.00
0.10	0.86	0.86	0.85	0.84	0.84	0.83	0.83	0.82	0.81
0.15	0.78	0.77	0.76	0.75	0.73	0.72	0.71	0.70	0.68
0.167	0.77	0.75	0.74	0.72	0.71	0.69	0.68	0.66	0.65
0.20	0.74	0.72	0.71	0.68	0.66	0.64	0.62	0.60	0.59
0.25	0.69	0.66	0.64	0.61	0.59	0.56	0.54	0.51	0.49
0.30	0.64	0.61	0.58	0.55	0.52	0.48	0.45	0.42	0.39
0.333	0.61	0.57	0.54	0.50	0.47	0.43	0.40	0.36	0.32

Eccentricity Coefficients. C_e may be selected from Table 11-5, or calculated from Eqs. (11-2) to (11-4). Linear interpolation is permitted within the table.

$$C_e = 1.0 \qquad 0 < \frac{e}{t} \leqq 0.05 \tag{11-2}$$

$$C_e = \frac{1.3}{1 + 6e/t} + \frac{1}{2}\left(\frac{e}{t} - \frac{1}{20}\right)\left(1 - \frac{e_1}{e_2}\right) \qquad 0.05 < \frac{e}{t} \leqq 0.167 \tag{11-3}$$

$$C_e = 1.95\left(\frac{1}{2} - \frac{e}{t}\right) + \frac{1}{2}\left(\frac{e}{t} - \frac{1}{20}\right)\left(1 - \frac{e_1}{e_2}\right) \qquad 0.167 < \frac{e}{t} \leqq 0.333 \tag{11-4}$$

Slenderness Coefficient. C_s may be selected from Table 11-6, or calculated from Eq. (11-5). Linear interpolation is permitted within the table.

$$C_s = 1.20 - \frac{h/t}{300}\left[5.75 + \left(1.5 + \frac{e_1}{e_2}\right)^2\right] \leq 1.0 \tag{11-5}$$

Allowable Axial Loads. Allowable loads, lb, on brick walls and columns can be computed from

$$P = C_e C_s f_m A_g \tag{11-6}$$

where C_e = eccentricity coefficient
C_s = slenderness coefficient
f_m = allowable axial compressive stress, psi (Table 11-3 or 11-4)
A_g = gross cross-sectional area, sq in.

For shear walls with eccentric loading, see Art. 11-16.

11-7. Shear-wall Action. The general design concept of a bearing wall should take into account the combined structural action of floor and roof systems with the walls. The floor and roof systems carry vertical loads and, acting as a diaphragm, lateral loads to the walls for transfer to the foundation. Some walls, called shear walls, are designed specifically to resist the components in their plane of lateral forces of wind and earthquake. These shear walls, by their resistance to shear and overturning, transfer the lateral loads to the foundation (Fig. 11-2).

Distribution of the lateral-load components to the shear walls parallel to them depends on the rigidities of the shear walls and the building components, such

Table 11-6. Slenderness Coefficients C_s

h/t	e_1/e_2								
	-1	-3/4	-1/2	-1/4	0	1/4	1/2	3/4	1
5.0	1.00	1.00	1.00	1.00	1.00	1.00	1.00	1.00	1.00
7.5	1.00	1.00	1.00	1.00	1.00	0.98	0.96	0.93	0.90
10.0	1.00	0.99	0.98	0.96	0.93	0.91	0.88	0.84	0.80
12.5	0.95	0.94	0.92	0.90	0.87	0.83	0.79	0.75	0.70
15.0	0.90	0.88	0.86	0.83	0.80	0.76	0.71	0.66	0.60
17.5	0.85	0.83	0.81	0.77	0.73	0.69	0.63	0.57	0.50
20.0	0.80	0.78	0.75	0.71	0.67	0.61	0.55	0.48	0.40
22.5	0.75	0.73	0.69	0.65	0.60	0.54	0.47	0.39	
25.0	0.70	0.67	0.64	0.59	0.53	0.47	0.39		
27.5	0.65	0.62	0.58	0.53	0.47	0.39			
30.0	0.60	0.57	0.52	0.47	0.40				
32.5	0.55	0.52	0.47	0.41					
35.0	0.50	0.46	0.41						
37.5	0.45	0.41							
40.0	0.40								

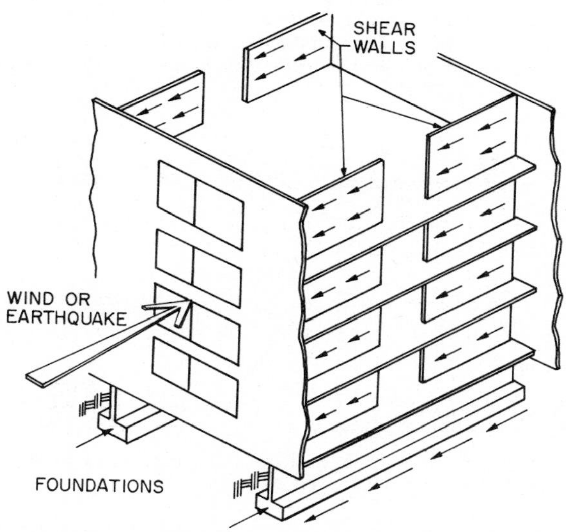

Fig. 11-2. Vertical shear walls resist horizontal forces parallel to their plane.

as floors and roofs, acting as diaphragms, that transmit the lateral loads to the shear walls (Art. 11-13).

11-8. Diaphragms. Horizontal distribution of lateral forces to shear walls is achieved by the floor and roof systems acting as diaphragms (Fig. 11-3).

To qualify as a diaphragm, a floor or roof system must be able to transmit the lateral forces to the shear walls without exceeding a horizontal deflection that would cause distress to any vertical element. The successful action of a diaphragm also requires that it be properly tied into the supporting shear walls. Designers should insure this action by appropriate detailing at the juncture between horizontal and vertical structural elements of the building.

Diaphragms may be considered analogous to horizontal (or inclined, in the case of some roofs) plate girders. The roof or floor slab constitutes the web; the joists, beams, and girders function as stiffeners; and the walls or bond beams act as flanges.

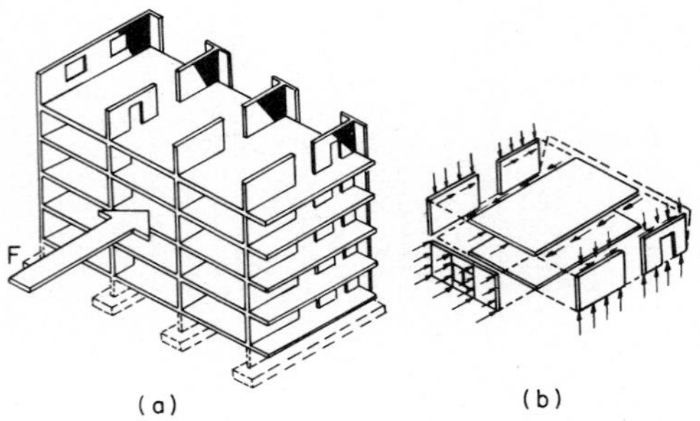

Fig. 11-3. Floors of building distribute horizontal loads to shear walls (diaphragm action).

Diaphragms may be constructed of structural materials, such as concrete, wood, or metal in various forms. Combinations of such materials are also possible. Where a diaphragm is made up of units, such as plywood, precast-concrete planks or steel decking, its characteristics are, to a large degree, dependent on the attachments of one unit to another and to the supporting members. Such attachments must resist shearing stresses due to internal translational and rotational actions.

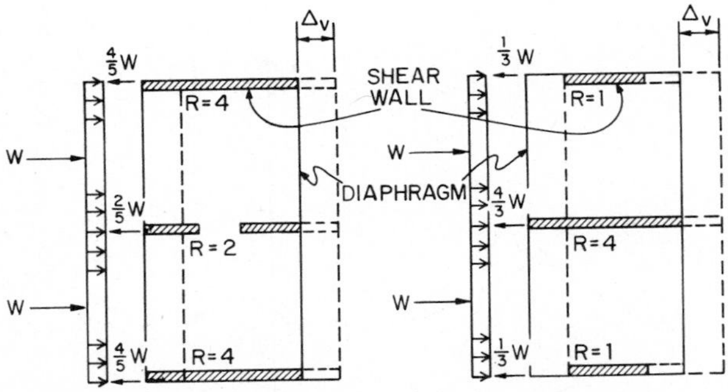

Fig. 11-4. Horizontal section through shear walls connected by a rigid diaphragm. R = relative rigidity and Δ_v = shear-wall deflection.

The stiffness of a horizontal diaphragm affects the distribution of the lateral forces to the shear walls. For the purpose of analysis, diaphragms may be classified into three groups; rigid, semirigid or semiflexible, and flexible, although no diaphragm is actually infinitely rigid or infinitely flexible.

A rigid diaphragm is assumed to distribute horizontal forces to the vertical resisting elements in proportion to the relative rigidities of these elements (Fig. 11-4).

Semirigid or semiflexible diaphragms are diaphragms that deflect significantly under load, but have sufficient stiffness to distribute a portion of the load to the vertical elements in proportion to the rigidities of these elements. The action is analogous to a continuous beam of appreciable stiffness on yielding supports (Fig. 11-5). Diaphragm reactions are dependent on the relative stiffnesses of diaphragm and vertical resisting elements.

A *flexible diaphragm* is analogous to a continuous beam or series of simple beams spanning between nondeflecting supports. Thus, a flexible diaphragm is

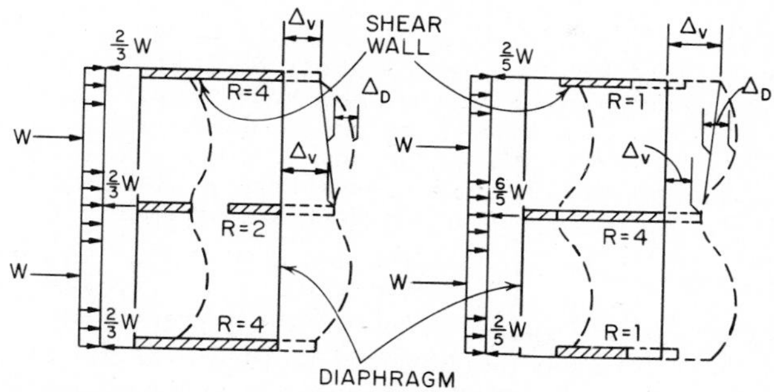

Fig. 11-5. Horizontal section through shear walls connected by a semirigid diaphragm. Δ_D = diaphragm horizontal deflection.

considered to distribute the lateral forces to the vertical resisting elements in proportion to the wall tributary areas (Fig. 11-6).

A rigorous analysis of lateral-load distribution to shear walls is sometimes very time-consuming, and frequently unjustified by the results. Therefore, in many cases, a design based on reasonable limits may be used. For example, the load may be distributed by first considering the diaphragm rigid, and then by considering it flexible. If the difference in results is not great, the shear walls can then be safely designed for the maximum applied load. (See also Art. 11-9.)

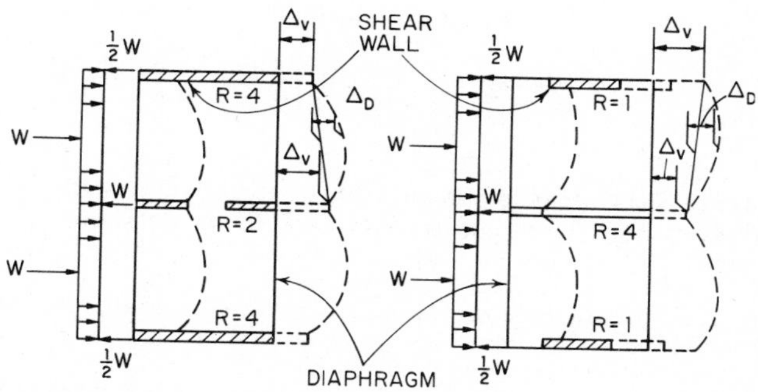

Fig. 11-6. Horizontal section through shear walls connected by a flexible diaphragm.

11-9. Torque Distribution to Shear Walls. When the line of action of the resultant of lateral forces acting on a building does not pass through the center of rigidity of a shear-wall system, distribution of the rotational forces must be considered as well as distribution of the translational forces. If rigid or semirigid diaphragms are used, the designer may assume that torsional forces are distributed to the shear walls in proportion to their relative rigidities and their distances from the center of rigidity. A flexible diaphragm should not be considered capable of distributing torsional forces. See also Arts. 11-10 and 11-13.

11-10. Example of Torque Distribution to Shear Walls. To illustrate load-distribution calculations for shear walls with rigid or semirigid diaphragms, Fig. 11-7 shows a horizontal section through three shear walls *A*, *B*, and *C* taken above a rigid floor. Wall B is 16 ft from wall A, and 24 ft from wall C. Rigidity of A is 0.33, of B 0.22, and of C 0.45 (Art. 11-13). A 20-kip horizontal force acts at floor level parallel to the shear walls and midway between A and C.

The center of rigidity of the shear walls is located, relative to wall A, by taking moments about A of the wall rigidities and dividing the sum of these moments by the sum of the wall rigidities, in this case 1.00.

$$x = 0.22 \times 16 + 0.45 \times 40 = 21.52 \text{ ft}$$

Thus, the 20-kip lateral force has an eccentricity of 21.52 − 20 = 1.52 ft. The eccentric force may be resolved into a 20-kip force acting through the center of rigidity and not producing torque, and a couple producing a torque of 20 × 1.52 = 30.4 ft-kips.

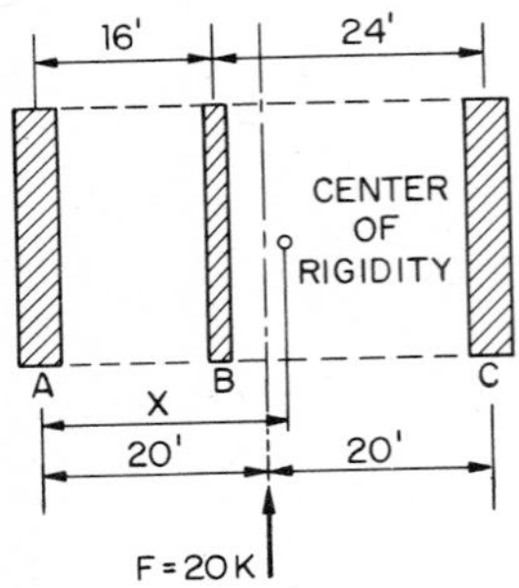

Fig. 11-7. Rigid diaphragm distributes 20-kip horizontal force to shear walls A, B, and C.

The nonrotational force is distributed to the shear walls in proportion to their rigidities:

$$\text{Wall } A\text{: } 0.33 \times 20 = 6.6 \text{ kips}$$
$$\text{Wall } B\text{: } 0.22 \times 20 = 4.4 \text{ kips}$$
$$\text{Wall } C\text{: } 0.45 \times 20 = 9.0 \text{ kips}$$

For distribution of the torque to the shear walls, the equivalent of moment of inertia must first be computed:

$$I = 0.33(21.52)^2 + 0.22(5.52)^2 + 0.45(18.48)^2 = 313$$

Then, the torque is distributed in direct proportion to shear-wall rigidity and distance from center of rigidity and in inverse proportion to I.

$$\text{Wall } A\text{: } 30.4 \times 0.33 \times 21.52/313 = 0.690 \text{ kips}$$
$$\text{Wall } B\text{: } 30.4 \times 0.22 \times 5.52/313 = 0.118 \text{ kips}$$
$$\text{Wall } C\text{: } 30.4 \times 0.45 \times 18.48/313 = 0.808 \text{ kips}$$

The rotational forces should be added to the nonrotational forces acting on Walls A and B, whereas the rotational force on Wall C acts in the opposite direction to the nonrotational force. For a conservative design, the rotational force on Wall C should not be subtracted. Hence, the walls should be designed for the following forces:

$$\text{Wall } A\text{: } 6.6 + 0.7 = 7.3 \text{ kips}$$
$$\text{Wall } B\text{: } 4.4 + 0.1 = 4.5 \text{ kips}$$
$$\text{Wall } C\text{: } 9.0 \text{ kips}$$

11-11. Relative Deflections of Shear Walls. When shear walls are connected by rigid diaphragms (Art. 11-8) and horizontal forces are distributed to the vertical resisting elements in proportion to their relative rigidities, the relative rigidity of a shear wall is dependent on shear and flexural deflections. For the proportions of shear walls used in most high-rise buildings, however, flexural deflections greatly exceed shear deflections. In such cases, only flexural rigidity need be considered in determination of relative rigidity of the shear walls (Art. 11-13).

Deflections can be determined by treating a shear wall as a cantilever beam. For a shear wall with a solid rectangular cross section, the cantilever flexural deflection under uniform loading may be computed from

$$\delta_m = \frac{wH^4}{8E_mI} = 0.0015\,\frac{wH}{f'_mt}\left(\frac{H}{L}\right)^3 \tag{11-7}$$

where w = uniform lateral load (Fig. 11-8)
H = height of wall
E_m = modulus of elasticity = $1{,}000f'_m$
f'_m = 28-day compressive strength of the masonry wall
I = moment of inertia of wall cross section = $tL^3/12$
t = wall thickness
L = length of wall

The cantilever shear deflection under uniform loading may be computed from

$$\delta_v = \frac{0.6wH^2}{E_vA} = 0.0015\,\frac{wH}{f'_{m}t}\left(\frac{H}{L}\right) \tag{11-8}$$

where E_v = modulus of rigidity of wall cross section = $400f'_m$
A = cross-sectional area of wall = tL

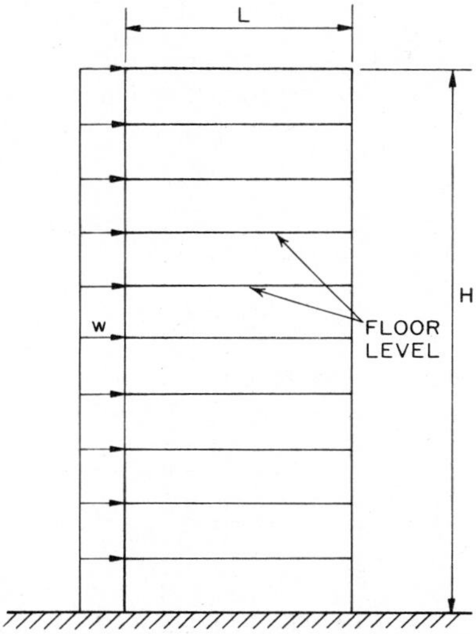

Fig. 11-8. Shear wall acts as vertical cantilever under uniformly distributed wind load w.

The total deflection then is

$$\delta = \delta_m + \delta_v = 0.0015\,\frac{wH}{f'_{m}t}\left[\left(\frac{H}{L}\right)^3 + \frac{H}{L}\right] \tag{11-9}$$

Table 11-7 lists relative flexural and shear deflections (δ_m/δ and δ_v/δ) for several height-length ratios.

Table 11-7. Relative Deflections of Shear Walls

Wall ratio H/L	Relative flexural deflection δ_m/δ	Relative shear deflection δ_v/δ
1	0.50	0.50
2	0.80	0.20
3	0.90	0.10
4	0.94	0.06
5	0.96	0.04

For a cantilever wall subjected to a concentrated load P at the top (Fig. 11-9a), the flexural deflection is

$$\delta_m = \frac{PH^3}{3E_mI} \tag{11-10}$$

The shear deflection under P for the cantilevered wall is

$$\delta_v = \frac{1.2PH}{E_v A} \tag{11-11}$$

If E_m is taken as 3,000 ksi, E_v as $0.4E_m$, t as 12 in., and P as 1,000 kips, the total cantilever deflection is given by

$$\delta = 0.1112\left(\frac{H}{L}\right)^3 + 0.0833\,\frac{H}{L} \tag{11-12}$$

For a wall fixed at the top against rotation and subjected to a concentrated load P at the top (Fig. 11-9*b*), the flexural deflection is

$$\delta_m = \frac{PH^3}{12E_m I} \tag{11-13}$$

The shear deflection for the fixed wall is also given by Eq. (11-11). For the same values assumed for Eq. (11-12), the total deflection for the fixed wall is

$$\delta = 0.0278\left(\frac{H}{L}\right)^3 + 0.0833\,\frac{H}{L} \tag{11-14}$$

11-12. Diaphragm-deflection Limitations. As indicated in Art. 11-8, horizontal deflection must be considered in designing a horizontal diaphragm. Diaphragm

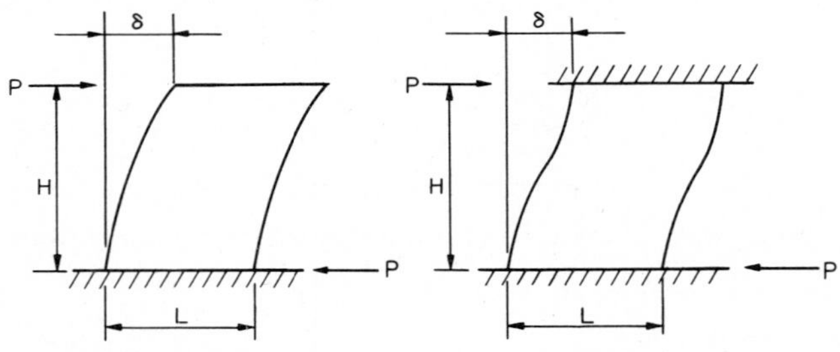

Fig. 11-9. Deflections δ of shear walls caused by horizontal load P applied at the top.

deflection should be limited to prevent excessive stresses in walls perpendicular to shear walls. Equation (11-15) was suggested by the Structural Engineers Association of Southern California for allowable story deflection, in., of masonry or concrete building walls.

$$\Delta = \frac{h^2 f}{0.01Et} \tag{11-15}$$

where h = height of wall between adjacent horizontal supports, ft
t = thickness of wall, in.
f = allowable flexural compressive stress of wall material, psi
E = modulus of elasticity of wall material, psi

This limit on deflection must be applied with engineering judgment. For example, continuity of wall at floor level is assumed, and in many cases is not present because of through-wall flashing. In this situation, the deflection may be based on the allowable compressive stress in the masonry, assuming a reduced cross section of wall. The effect of reinforcement, which may be present in a reinforced brick masonry wall or as a tie to the floor system in a nonreinforced or partly reinforced masonry wall, was not considered in development of Eq. (11-15). Note also that the limit on wall deflection is actually a limit on differential deflection between two successive floor, or diaphragm, levels.

Maximum span-width or span-depth ratios for diaphragms are usually used to control horizontal diaphragm deflection indirectly. Normally, if the diaphragm is designed with the proper ratio, the diaphragm deflection will not be critical. Table 11-8 may be used as a guide for proportioning diaphragms.

Table 11-8. Maximum Span-width or Span-depth Ratios for Diaphragms—Roofs or Floors*

Diaphragm construction	Masonry and concrete walls	Wood and light steel walls
Concrete	Limited by deflection	
Steel deck (continuous sheet in a single plane) .	4:1	5:1
Steel deck (without continuous sheet)	2:1	2½:1
Cast-in-place reinforced gypsum roofs	3:1	4:1
Plywood (nailed all edges)	3:1	4:1
Plywood (nailed to supports only—blocking may be omitted between joists).	2½:1†	3½:1
Diagonal sheathing (special)	3:1†	3½:1
Diagonal sheathing (conventional construction).	2:1†	2½:1

* From California Administrative Code, Title 21, Public Works.

† Use of diagonal sheathed or unblocked plywood diaphragms for buildings having masonry or reinforced concrete walls shall be limited to one-story buildings or to the roof of a top story.

11-13. Shear-wall Rigidity. Where shear walls are connected by rigid diaphragms (Art. 11-8) so that they must deflect equally under horizontal loads, the proportion of total horizontal load at any level carried by a shear wall parallel to the load depends on the relative rigidity, or stiffness, of the wall in the direction of the load. Rigidity of a shear wall is inversely proportional to its deflection under unit horizontal load. This deflection equals the sum of the shear and flexural deflections under the load (Art. 11-11).

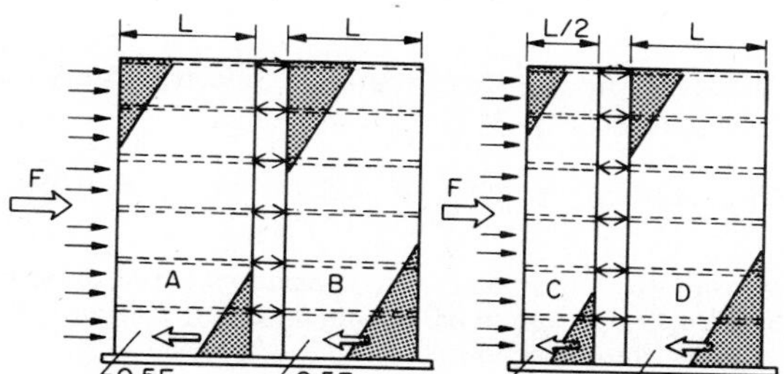

Fig. 11-10. Distribution of horizontal load to parallel shear walls. (*a*) Walls with same length and rigidity share the load equally. (*b*) Wall half the length of another carries less than one-eighth the load.

Where a shear wall contains no openings, computations for deflection and rigidity are simple. In Fig. 11-10*a*, each of the shear walls has the same length and rigidity. So each takes half the total load. In Fig. 11-10*b*, length of wall *C* is half that of wall *D*. By Eq. (11-7), *C* therefore receives less than one-eighth the total load.

Walls with Openings. Where shear walls contain openings, such as doors and windows, computations for deflection and rigidity are more complex. But approximate methods may be used.

For example, the wall in Fig. 11-11, subjected to a 1,000-kip load at the top, may be treated in parts. The wall is 8 in. thick, and its modulus of elasticity $E_m = 2{,}400$ ksi. Its height-length ratio H/L is $12/20 = 0.6$. The wall is perforated by two, symmetrically located, 4-ft-square openings.

Deflection of this wall can be estimated by subtracting from the deflection it would have if it were solid the deflection of a solid, 4-ft-deep, horizontal midstrip, and then adding the deflection of the three coupled piers *B*, *C*, and *D*.

Deflection of the 12-ft-high solid wall can be obtained from Eq. (11-12) modified, by proportion, for actual thickness *t* and E_m.

$$\delta = [0.1112(0.6)^3 + 0.0833 \times 0.6] \frac{3{,}000}{2{,}400} \times \frac{12}{8} = 0.138 \text{ in.}$$

Rigidity of the solid wall then is

$$R = \frac{1}{0.138} = 7.22$$

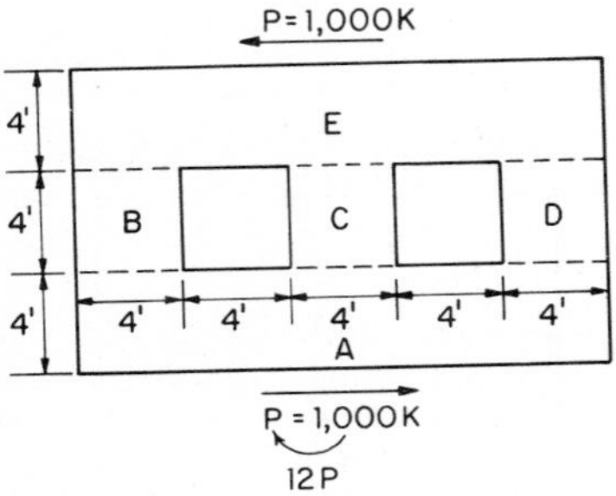

Fig. 11-11. Shear wall, 8 in. thick, with openings.

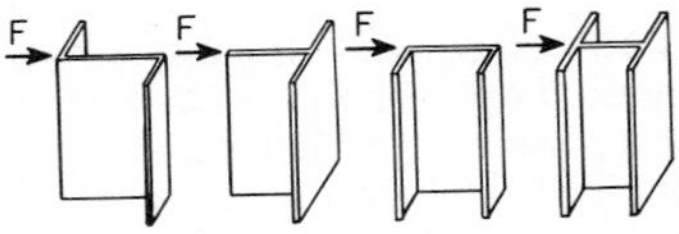

Fig. 11-12. Shear walls with flange walls have high resistance to lateral loads parallel to the web walls.

Similarly, the deflection of the 4-ft-deep solid midstrip can be computed from Eq. (11-12), with $H/L = 4/20 = 0.20$.

$$\delta = [0.1112(0.2)^3 + 0.0833 \times 0.2] \frac{3{,}000}{2{,}400} \times \frac{12}{8} = 0.033 \text{ in.}$$

Deflection of the piers, which may be considered fixed top and bottom, can be obtained from Eq. (11-14) modified for actual *t* and E_m, with $H/L = 4/4 = 1$. For any one of the piers, the deflection is

$$\delta' = [0.0278(1)^3 + 0.0833 \times 1] \frac{3{,}000}{2{,}400} \times \frac{12}{8} = 0.208 \text{ in.}$$

The rigidity of a single pier is $1/0.208 = 4.81$, and of the three piers, $3 \times 4.81 = 14.43$. Therefore, the deflection of the three piers when coupled is

$$\delta = \frac{1}{14.43} = 0.069 \text{ in.}$$

The deflection of the whole wall, with openings, then is approximately

$$\delta = 0.138 - 0.033 + 0.069 = 0.174 \text{ in.}$$

And its rigidity is

$$R = \frac{1}{0.174} = 5.74$$

11-14. Effects of Shear-wall Arrangements. To increase the stiffness of shear walls and thus their resistance to bending, intersecting walls or flanges may be used. Often in the design of buildings, Z-, T-, U-, L-, and I-shaped walls in plan develop as natural parts of the design (Fig. 11-12). Shear walls with these shapes have better flexural resistance than a single, straight wall.

In "Building Code Requirements for Engineered Brick Masonry," a limit is placed on the effective flange width that may be used in calculating flexural stresses in shear walls. For symmetrical T or I sections, the effective flange width may not exceed one-sixth the total wall height above the level being analyzed. For unsymmetrical L or C sections, the width considered effective may not exceed one-sixteenth the total wall height above the level being analyzed. In either case, the overhang for any section may not exceed six times the flange thickness (Fig. 11-13).

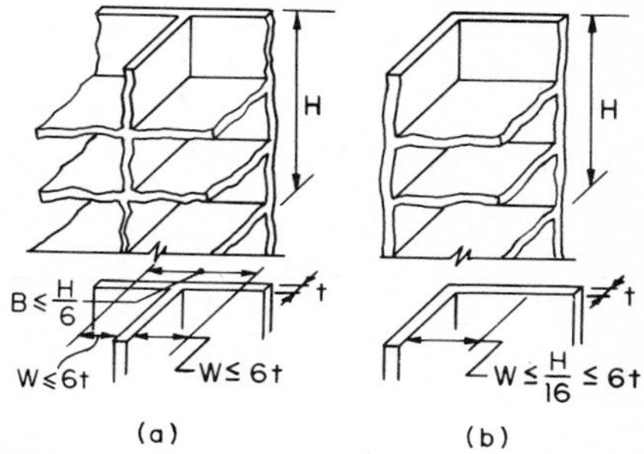

Fig. 11-13. Effective flange width of shear walls may be less than actual width. (*a*) Limits for flanges of I and T shapes. (*b*) Limits for C and L shapes.

The shear stress at the intersection of the walls should not exceed the permissible shear stress. The allowable stress depends on the method used in bonding the two walls together. If the intersection is bonded with masonry, the allowable shear stress is that given in Table 11-3 or Table 11-4; if with metal anchors, the allowable stress is given in Table 11-9.

Table 11-9. Allowable Shear on Bolts and Anchors*

Dia, in.	Minimum embedment, in.†	Allowable shear, lb	
		Without inspection	With inspection
1/4	4	180	270
3/8	4	270	410
1/2	4	370	550
5/8	4	500	750
3/4	5	730	1,100
7/8	6	1,000	1,500
1	7	1,230	1,850
1 1/8	8	1,500	2,250

* From "Building Code Requirements for Engineered Brick Masonry," Brick Institute of America.

† Bolts and anchors should be solidly embedded in mortar.

See also Art. 11-15.

11-15. Coupled Shear Walls. Another method than that described in Art. 11-14 for increasing the stiffness of a bearing-wall structure and reducing the possibility of tension developing in shear walls under lateral loads is coupling of coplanar shear walls.

Figures 11-14 and 11-15 indicate the effect of coupling on stress distribution in a pair of walls under horizontal forces parallel to the walls. A flexible connection between the walls is assumed in Figs. 11-14*a* and 11-15*a*, so that the walls act as independent vertical cantilevers in resisting lateral loads. In Figs. 11-14*b* and 11-15*b*, the walls are assumed to be connected with a more rigid member, which is capable of shear and moment transfer. A rigid-frame-type action results. This can be accomplished with a steel, reinforced concrete, or reinforced brick masonry coupling. A plate-type action is indicated in Figs. 11-14*c* and 11-15*c*. This assumes an extremely rigid connection between walls, such as full story-height walls or deep rigid spandrels.

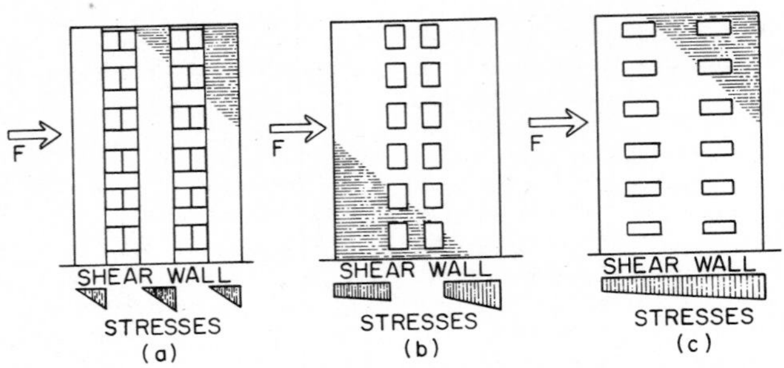

Fig. 11-14. Stress distribution in end shear walls (*a*) with flexible coupling, (*b*) with rigid-frame-type action, and (*c*) with plate-type action.

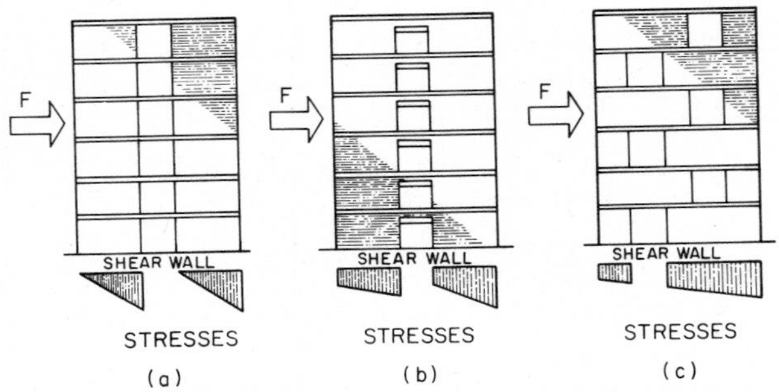

Fig. 11-15. Stress distributions in interior shear walls (*a*) with flexible coupling, (*b*) with rigid-frame-type action, and (*c*) with plate-type action.

11-16. Eccentrically Loaded Shear Walls. "Building Code Requirements for Engineered Brick Masonry" also provides a basis for design of eccentrically loaded shear walls. The standard requires that, in a nonreinforced shear wall, the virtual eccentricity e_L about the principal axis normal to the length L of the wall not exceed an amount that will produce tension. In a nonreinforced shear wall subject to bending about both principal axes, $e_tL + e_Lt$ should not exceed $tL/3$, where e_t = virtual eccentricity about the principal axis normal to the thickness t of the shear wall. Where the virtual eccentricity exceeds the preceding limits, shear walls should be designed as reinforced or partly reinforced walls.

Consequently, for a planar wall the virtual eccentricity e_L, which is found by dividing the overturning moment about an axis normal to the plane of the wall by the axial load, should not exceed $L/6$. In theory, any virtual eccentricity exceeding $L/6$ for planar shear walls will result in development of tensile stresses. If, however, intersecting walls form resisting flanges, the classic approach may be used where the bending stress Mc/I is combined with the axial stress P/A. If the bending stress exceeds the axial stress, then tensile stresses are present. (Note that some building codes require a safety factor against overturning. It is usually 1.5 for nonreinforced walls. But the codes state that, if the walls are vertically reinforced to resist tension, the safety factor does not apply.)

For investigation of biaxial bending, the eccentricity limitation may, for convenience, be placed in the form:

$$\frac{e}{t} = \frac{e_t}{t} + \frac{e_L}{L} \leq \frac{1}{3} \tag{11-16}$$

11-17. Allowable Stresses for Shear Walls. Allowable vertical loads on shear walls may be computed from Eq. (11-6). The effective height h for computing C_s should be taken as the minimum vertical or horizontal distance between lateral supports.

Allowable shearing stresses for shear walls should be taken as the sum of the allowable shear stress given in Table 11-3 or 11-4 and one-fifth the average compressive stress produced by dead load at the level being analyzed, but not more than the maximum values listed in these tables.

In computation of shear resistance of a shear wall with intersecting walls treated as flanges, only the parts serving as webs should be considered effective in resisting the shear.

When nonloadbearing shear walls are required to resist overturning moment only by their own weight, the design can become critical. Consequently, positive ties should be provided between shear walls and bearing walls, to take advantage of the bearing-wall loads in resisting overturning. The shear stress at the connection of the shear wall to the bearing wall should be checked. The designer should exercise judgment in assumption of the distribution of the axial loads into the nonloadbearing shear walls.

11-18. Bearing Walls. Walls whose function is to carry only vertical loads should be proportioned primarily for compressive stress. Allowable vertical loads are given by Eq. (11-6). Effects of all loads and conditions of loading should be investigated. For application of Eq. (11-6), each loading should be converted to a vertical load and a virtual eccentricity.

In the lower stories of a building, the compressive stress is usually sufficient to suppress development of tensile stress. But in the upper stories of tall buildings, where the exterior walls are subject to high lateral wind loads and small axial loads, the allowable tensile stress may be exceeded occasionally and make reinforcement necessary.

When lateral forces act parallel to the plane of bearing walls that act as loadbearing shear walls, they must meet the requirements for both shear walls and biaxial bending (Arts. 11-16 and 11-17). Such walls must be checked, to preclude development of tensile stress or excessive compressive or shearing stresses. If analysis indicates that tension requirements are not satisfied, the size, shape, or number of shear walls must be revised or the wall must be designed as reinforced masonry.

11-19. Floor-Wall Connections. Bending moments caused in a wall by floor loading depend on such factors as type of floor system, detail of floor-wall connections and sequence of construction. Because information available on their effects is limited, engineers must make certain design assumptions when providing for such moments. Conservative assumptions that may be used in design are discussed in the following.

For a floor system that acts hinged at the floor-wall connection, such as steel joists and stems from precast-concrete joists, a triangular stress distribution can be assumed under the bearing (Fig. 11-16). The moment in the wall produced by dead and live loads is then equal to the reaction times the eccentricity resulting from this stress distribution.

For precast-concrete-plank floor systems, which deflect and rotate at the time of placing, a triangular stress distribution similarly can be assumed to result from the dead load of the plank and to induce a moment in the wall. When the topping is placed as each level is constructed, a triangular stress distribution can still be assumed. The moment resulting from the dead load of the floor system, including the topping, then is that due to the eccentric loading. If, however, the topping is placed after the wall above has been built and the wall clamps the plank end in place, creating a restrained end condition, the moment in the wall will then be the sum of the moment due to the eccentric load of the plank itself and the fixed-end moment resulting from the superimposed loads of topping weight and live load (Fig. 11-17).

The degree of fixity and the resulting magnitude of the restrained end moments usually must be assumed. Full fixity of floors due to the clamping action of a wall under large axial loads in the lower stories of high- and medium-rise buildings appears a logical assumption. The same large axial loads that provide the clamping

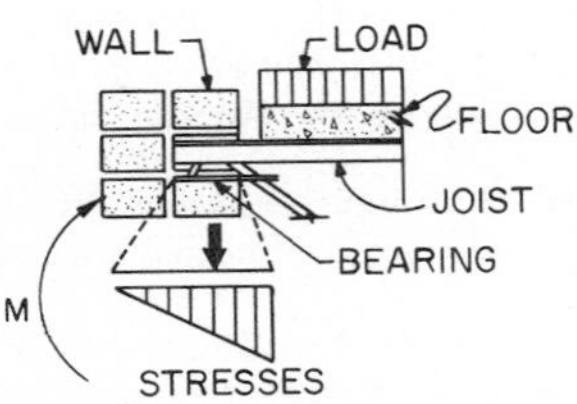

Fig. 11-16. Stress distribution in wall from joist reaction.

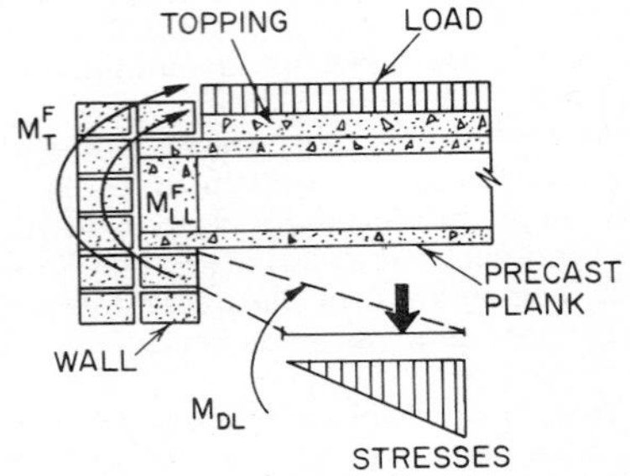

Fig. 11-17. Stress distribution in wall from plank reaction.

action in the lower stories also act to suppress development of tensile stresses in the wall at the floor-wall connection. Because axial loads are smaller in upper stories, however, the degree of fixity may be assumed reduced, with occurrence of slight rotation and elevation of the extreme end of the slab. Based on this assumption, slight, local stress-relieving in connections in upper stories could take place. Regardless of the assumption, the maximum moment transferred to the wall can never be greater than the negative-moment capacity of the floor system.

When full fixity is assumed, the magnitude of the moment in the wall will be approximately the distribution factor times the initial fixed-end moment of the slab at the face of the wall. As an approximation for precast plank with uniform load w and span L, $wL^2/36$ may be conservatively assumed as the wall moment. [Preliminary test results have indicated about 80% moment transfer from the slab into the wall sections (40% to the upper and 40% to the lower wall section) with flat, precast plank penetrating the full wall thickness.]

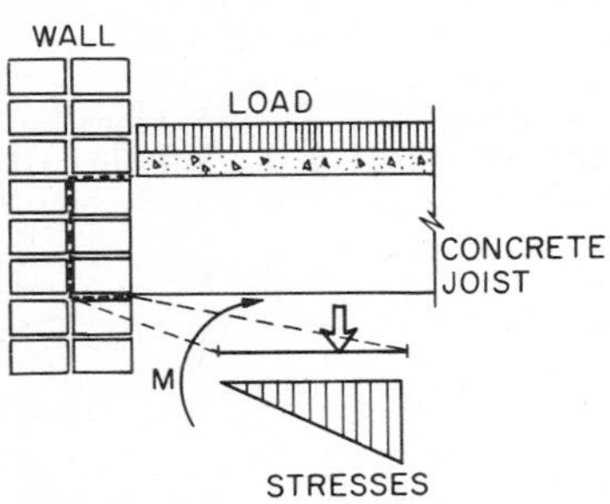

Fig. 11-18. Stress distribution in wall with only joist stem embedded.

For a cast-in-place concrete slab, a fixed-end moment may be assumed for both dead and live loads, because usually the wall above the slab will be built before removal of shoring.

Because restrained end moments in a wall can become large, reduction of the eccentricity of the floor reaction is advantageous in limiting the moment in the wall. This may be accomplished by projecting only the stems of cast-in-place or precast-concrete systems into the wall (Fig. 11-18). In such cases, a bearing pad should be placed immediately under each stem.

11-20. Example—End Shear Wall, Concrete Joist Floors. Design a nonreinforced, loadbearing, shear wall between the second and third floors (Fig. 11-19) of a 10-story building, given the following data:

Nominal thickness is 8 in. Actual thickness $t = 7.5$ in.

Unsupported height $h_u = 8$ ft.

Precast-concrete-joist floor system with 4-in. minimum bearing.

Story height = 8 ft 10 in.

Intersecting flange wall 3 ft long. Ratio of cross-sectional shear-wall area to total wall area = 16.6/18.6 = 0.892.

	Floor	Roof
Live loads LL, psf	40	30
Dead loads DL, psf	60	50
Total loads TL, psf	100	80

Maximum live-load reduction permitted is

$$R = \frac{100TL}{4.33LL} = \frac{100 \times 100}{4.33 \times 40} = 58\%$$

Lateral wind load = 20 psf.
Parallel shear load = 22.5 kips.
Parallel wind moment = 900 ft-kips.
Wall dead load = 120 lb per cu ft = 74.4 lb per sq ft.

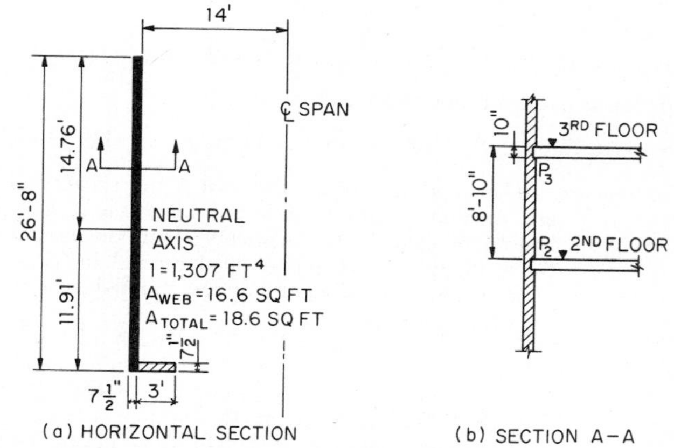

Fig. 11-19. Properties of end shear wall of 10-story building.

Required Masonry Strength. The allowable compressive stress for nonreinforced walls is $f_m = 0.20f'_m$, where f'_m is the 28-day compressive strength of the masonry, psi. From Eq. (11-6), the required masonry strength is

$$f'_m = \frac{P}{0.20A_gC_eC_s} \qquad (11\text{-}17)$$

When gravity loads and wind loads are combined, f'_m given by Eq. (11-17) may be multiplied by 1/1.33.

Loads at Third Floor. Gravity loading per linear foot of wall from floors or roof is contributed by half the floor or roof span, a length of 14 ft. Loading from floors above is shared with the flange wall, so that the shear wall carries 0.892 of this loading.

The dead load just below the third floor equals the weight of eight stories, or 8 × 8.333 = 70.7 ft, of masonry, plus roof dead load, plus dead load of eight floors, seven of which are shared by the flange wall.

$$P_{DL} = 70.7 \times 74.4 + (14 \times 50 + 7 \times 14 \times 60)0.892 + 14 \times 60 = 11{,}970 \text{ lb per ft}$$

Similarly, the live load, without reduction, is

$$P_{LL} = (14 \times 30 + 7 \times 14 \times 40)0.892 + 14 \times 40 = 4{,}430 \text{ lb per ft}$$

The contributory floor area is large enough that the maximum reduction may be applied. Hence, the wall just below the third floor must carry 100 − 58 = 42% of the total live load from the floors. The reduced live load is

$$P_{RLL} = (14 \times 30 + 0.42 \times 7 \times 14 \times 40)0.892 + 0.42 \times 14 \times 40 = 2{,}080 \text{ lb per ft}$$

Loads at Second Floor. The dead load just above the second floor equals the weight of nine stories, or 9 × 8.33 = 79.5 ft, of masonry, plus roof dead load, plus dead load of eight floors.

$$P_{DL} = 79.5 \times 74.4 + (14 \times 50 + 8 \times 14 \times 60)0.892 = 12{,}555 \text{ lb per ft}$$

Similarly, the live load, without reduction, is

$$P_{LL} = (14 \times 30 + 8 \times 14 \times 40)0.892 = 4{,}370 \text{ lb per ft}$$

and the reduced live load is

$$P_{RLL} = (14 \times 30 + 0.42 \times 8 \times 14 \times 40)0.892 = 2{,}055 \text{ lb per ft}$$

Summation of Loads at Second Floor. The total dead load just above the second floor is

$$P'_{DL} = 14 \times 26(50 + 8 \times 60) + 79.5 \times 74.4(26.67 + 3.0) = 370{,}400 \text{ lb}$$

The total unreduced live load plus dead load is

$$P'_{TL} = 14 \times 26(80 + 8 \times 100) + 79.5 \times 74.4(26.67 + 3.0) = 497{,}700 \text{ lb}$$

Source of Moments. The bending moments to which the shear wall is subjected are moments acting about the longitudinal axis of the wall, caused by floor loads; moments about the longitudinal axis of the wall, caused by perpendicular wind loads acting between floor levels; and moments acting parallel to the wall, caused by wind loads.

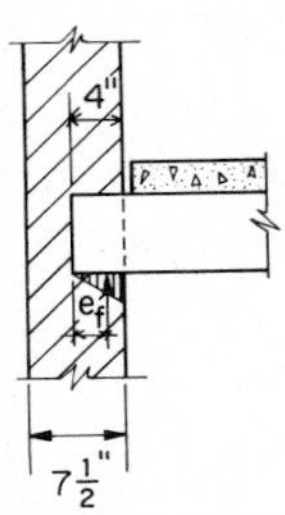

Fig. 11-20. Bearing of concrete joist on end wall.

A conservative approach in selection of loading conditions to produce maximum stresses is to assume loadings such that total-load moment occurs at the top of the wall (third floor) and dead-load moment at the bottom (second floor). This arrangement makes e_1/e_2 small, thus giving conservative C_e and C_s values for use in Eq. (11-17). Note that end walls normally go into double curvature when moments are caused by floor loads (eccentricity ratio e_1/e_2 is negative).

Moments from Floor Loads. Assume 4-in. bearing of joists on the wall and a triangular stress distribution under the bearing (Fig. 11-20). The eccentricity of the joist reaction then is

$$e_f = \frac{7.5}{2} - \frac{4}{3} = 2.42 \text{ in.}$$

The dead-load and live-load moments due to this eccentricity are distributed in equal proportions to the wall sections above and below the floor, and equal

$$M_{DL} = \tfrac{1}{2} \times 40 \times 14 \times 2.42 = 675 \text{ in.-lb per ft}$$
$$M_{LL} = \tfrac{1}{2} \times 60 \times 14 \times 2.42 = 1{,}015 \text{ in.-lb per ft}$$
$$\text{Total moment} = M_{TL} = 1{,}690 \text{ in.-lb per ft}$$

Perpendicular Wind Moment. For the 20-psf wind load, with the wall assumed to act as a fixed-end beam between floors, the bending moments at the floors are

$$M = \tfrac{1}{12} \times 20(8.83)^2 12 = 1{,}560 \text{ in.-lb per ft}$$

Parallel Wind Moment at Second Floor. This was given initially as $M' = 900$ ft-kips.

Overturn Check. To determine if tensile stress develops in the shear wall from overturning moment, the wall should be considered to be carrying only dead loads in addition to the wind parallel to the wall. The location of the neutral axis of the wall is found, relative to the flanged end, by dividing the sum of the moments of the wall areas about that end by the wall area $A = 18.6$ sq ft.

$$x = \frac{26.67}{2} \times 16.6 \times \frac{1}{18.6} = 11.91 \text{ ft}$$

The moment of inertia of the wall about this axis is $I = 1{,}307 \text{ ft}^4$.

The allowable eccentricity e_L, which is the maximum for which no tension occurs on the wall cross section, is given by I/cA, where c is the maximum distance of a wall end from the neutral axis.

$$e_L = \frac{1{,}307}{(26.67 - 11.91)18.6} = 4.76 \text{ ft}$$

The actual eccentricity is

$$e_L = \frac{M'}{P'_{DL}} = \frac{900{,}000}{370{,}400} = 2.43 < 4.76 \text{ ft}$$

Therefore, no tensile reinforcement is required.

Loading Conditions. Five cases are considered for checking eccentricities and stresses in the wall under gravity and wind loads:

Case I. Total loads on floors and roof plus wind parallel to the wall.
Case II. Dead loads on floors and roof plus wind parallel to the wall.
Case III. Total loads on floors and roof plus wind perpendicular to the wall.
Case IV. Dead loads on floors and roof plus wind perpendicular to the wall.
Case V. Dead loads plus reduced live loads.

Case I (TL + Parallel Wind). The wall is subjected to biaxial bending. Eccentricities must satisfy Eq. (11-16) and strength f'_m must satisfy Eq. (11-17).

The maximum eccentricity of floor load at the third floor is

$$e_t = \frac{1{,}690}{4{,}370 + 12{,}555} = 0.0997 \text{ in.}$$

The virtual eccentricity of the 900-ft-kip overturning moment is

$$e_L = \frac{M'}{P'_{TL}} = \frac{900{,}000}{497{,}700} = 1.81 \text{ ft}$$

In accordance with Eq. (11-16) for biaxial bending,

$$\frac{e}{t} = \frac{0.0997}{7.5} + \frac{1.81}{26.67} = 0.081 < \frac{1}{3}$$

The eccentricity of floor load at the top of the wall is computed for floor total-load moment.

$$e_{\text{top}} = \frac{M}{P} = \frac{1{,}690}{4{,}430 + 11{,}970} = \frac{1{,}690}{16{,}400} = 0.103 \text{ in.}$$

The eccentricity of floor load at the bottom of the wall for floor dead-load moment is

$$e_{\text{bot}} = \frac{1{,}015}{4{,}370 + 12{,}555} = \frac{1{,}015}{16{,}925} = 0.060 \text{ in.}$$

The eccentricity ratio, negative because of double curvature of the wall, then is

$$\frac{e_1}{e_2} = -\frac{0.060}{0.103} = -0.582$$

For use in determining the slenderness coefficient C_s, the slenderness ratio is

$$\frac{h}{t} = \frac{8.0 \times 12}{7.5} = 12.8$$

From Eq. (11-5) or Table 11-6, $C_s = 0.92$. From Table 11-5, with $e/t = 0.081$, the eccentricity coefficient $C_e = 0.90$. For use in Eq. (11-17), the area of a linear foot of wall is $A_g = 7.5 \times 12 = 90$ sq in. From Eq. (11-17), the required masonry strength of Case I, with the one-third increase in stress allowed for combined gravity and wind loads, is

$$f'_m = \frac{16{,}926}{0.20 \times 90 \times 0.90 \times 0.92} \frac{1}{1.33} = 855 \text{ psi}$$

Case II (DL + Parallel Wind). This case is similar to Case I, but dead load is used instead of total load.

The eccentricity of the floor load at the bottom of the wall is

$$e_{\text{bot}} = \frac{1{,}015}{12{,}555} = 0.081 \text{ in.}$$

The virtual eccentricity of the 900-ft-kip overturning moment is

$$e_L = \frac{M'}{P'_{DL}} = \frac{900{,}000}{370{,}400} = 2.43 \text{ ft}$$

A check with Eq. (11-16) yields

$$\frac{e}{t} = \frac{0.081}{7.5} + \frac{2.43}{26.67} = 0.102 < \frac{1}{3}$$

The eccentricity of the floor dead load at the top of the wall is

$$e_{\text{top}} = \frac{1{,}015}{11{,}970} = 0.085 \text{ in.}$$

The eccentricity ratio for the floor loading is

$$\frac{e_1}{e_2} = -\frac{0.081}{0.085} = -0.950$$

With $h/t = 12.8$ as in Case I, $C_s = 0.94$. From Table 11-5, with $e/t = 0.102$, $C_e = 0.86$. From Eq. (11-17), the required masonry strength for Case II is

$$f'_m = \frac{12{,}555}{0.20 \times 90 \times 0.86 \times 0.94} \frac{1}{1.33} = 650 \text{ psi}$$

Case III (TL + Perpendicular Wind). This case differs from Case I in that the wall acts as a one-way slab, supported by floors and roof, in resisting the wind load. Bending due to wind and bending due to floor loads act about the same axis at the top and bottom of the wall in each story. The eccentricity, at the top, of the total floor loads and the perpendicular wind is

$$e_{\text{top}} = \frac{1{,}690 + 1{,}560}{16{,}400} = \frac{3{,}250}{16{,}400} = 0.198 \text{ in.}$$

A check of the eccentricity-thickness ratio yields

$$\frac{e}{t} = \frac{0.198}{7.5} = 0.026 < \frac{1}{3}$$

Because the transverse loading exceeds 10 psf, e_1/e_2 is taken as unity. With $h/t = 12.8$, $C_s = 0.70$. From Table 11-5, with $e/t = 0.026$, $C_e = 1.0$. From

Eq. (11-17), the required masonry strength for Case III is

$$f'_m = \frac{16{,}925}{0.20 \times 90 \times 1.0 \times 0.70} \frac{1}{1.33} = 1{,}010 \text{ psi}$$

Case IV (DL + Perpendicular Wind). This case is similar to Case II, but dead load is used instead of total load.

$$e_{\text{top}} = \frac{1{,}015 + 1{,}560}{11{,}970} = \frac{2{,}575}{11{,}970} = 0.215 \text{ in.}$$

A check of the eccentricity-thickness ratio yields

$$\frac{e}{t} = \frac{0.215}{7.5} = 0.029 < \frac{1}{3}$$

Because the transverse loading exceeds 10 psf, e_1/e_2 is taken as unity. With $h/t = 12.8$, $C_s = 0.70$. From Table 11-5, with $e/t = 0.029$, $C_e = 1.0$. From Eq. (11-17), the required masonry strength for Case IV is

$$f'_m = \frac{12{,}555}{0.20 \times 90 \times 1.0 \times 0.70} \frac{1}{1.33} = 750 \text{ psi}$$

Case V (DL + RLL). No wind is included in this case. The eccentricity at the top of the wall is

$$e_{\text{top}} = \frac{1{,}690}{11{,}970 + 2{,}080} = \frac{1{,}690}{14{,}050} = 0.120 \text{ in.}$$

The eccentricity at the bottom of the wall is

$$e_{\text{bot}} = \frac{1{,}015}{12{,}555 + 2{,}055} = \frac{1{,}015}{14{,}610} = 0.069 \text{ in.}$$

A check of the eccentricity-thickness ratio yields

$$\frac{e}{t} = \frac{0.120}{7.5} = 0.016 < \frac{1}{3}$$

The eccentricity ratio, with the wall in double curvature, is

$$\frac{e_1}{e_2} = -\frac{0.069}{0.120} = -0.575$$

With $h/t = 12.8$, $C_s = 0.93$. From Table 11-5, with $e/t = 0.016$, $C_e = 1.0$. From Eq. (11-17), the required masonry strength for Case V is

$$f'_m = \frac{14{,}610}{0.20 \times 90 \times 1.0 \times 0.93} = 885 \text{ psi}$$

Shear Check. Assume $f'_m = 1{,}010$ psi and Type N mortar, without inspection. The allowable stress (Art. 11-17) is

$$v_m = 0.5\sqrt{f'_m} + \frac{f_{DL}}{5} = 0.5\sqrt{1{,}015} + \frac{1}{5}\frac{370{,}400}{18.6 \times 144} = 44 > 28 \text{ psi}$$

Use 28 psi. The actual average shear stress over the wall web is

$$v_m = \frac{V}{A} = \frac{22{,}500}{16.6 \times 144} = 9.5 < 28 \text{ psi}$$

The 7.5-in.-thick, nonreinforced wall is satisfactory between the second and third floors if masonry with $f'_m \geq 1{,}010$ psi is used.

11-21. Example—End Shear Wall, Concrete Plank Floors. Design a shear wall for the same conditions as for the example in Art. 11-20, except that the floors are flat, precast-concrete plank with topping. All loadings remain the same for purposes of this example.

Sources of Moments. Assume 4-in. bearing of the plank on the wall for dead loads (Fig. 11-21). The eccentricity of the dead-load reaction then is 2.42 in., as in Art. 11-20. For uniform live load w_{LL}, because of the restraint of the floor end imposed by the clamping action of the wall, $w_{LL}L^2/36$ is conservatively assumed as the live-load moment induced in the wall section above or below the floor. For the loads used in Art. 11-20, the moments in the wall are

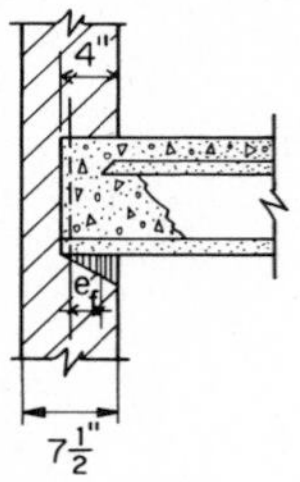

Fig. 11-21. Bearing of concrete plank on end wall.

$$M_{DL} = \tfrac{1}{2} \times 60 \times 14 \times 2.42 = 1{,}015 \text{ in.-lb per ft}$$
$$M_{LL} = \tfrac{1}{36} \times 40(28)^2 12 = 10{,}400 \text{ in.-lb per ft}$$
$$\text{Total moment} = M_{TL} = 11{,}415 \text{ in.-lb per ft}$$

Case I (TL + Parallel Wind). This case is similar to Case I of Art. 11-20. The maximum eccentricity at the second floor is

$$e_t = \frac{11{,}415}{16{,}925} = 0.673 \text{ in.}$$

As in Art. 11-20, the virtual eccentricity of the 900-ft-kip overturning moment is $e_L = 1.81$ ft. In accordance with Eq. (11-15) for biaxial bending,

$$\frac{e}{t} = \frac{0.673}{7.5} + \frac{1.81}{26.67} = 0.158 < \frac{1}{3}$$

The eccentricity of floor load at the top of the wall (third floor) is computed for the floor total-load moment.

$$e_{\text{top}} = \frac{11{,}415}{16{,}400} = 0.698 \text{ in.}$$

The eccentricity of floor load at the bottom of the wall (second floor) for floor dead-load moment is

$$e_{\text{bot}} = \frac{1{,}015}{16{,}925} = 0.060 \text{ in.}$$

The eccentricity ratio, negative because of double curvature of the wall, then is

$$\frac{e_1}{e_2} = -\frac{0.060}{0.698} = -0.086$$

With $h/t = 12.8$, Table 11-6 gives $C_s = 0.88$. From Table 11-5, with $e/t = 0.158$, $C_e = 0.72$. From Eq. (11-17), the required masonry strength for Case I, with the one-third increase in stress allowed for combined gravity and wind loads, is

$$f'_m = \frac{16{,}925}{0.20 \times 90 \times 0.72 \times 0.88} \frac{1}{1.33} = 1{,}115 \text{ psi}$$

Case II (DL + Parallel Wind). This case is the same as Case II in Art. 11-20, because there are no live-load moments.

Case III (TL + Perpendicular Wind). This case is similar to Case III, Art. 11-20.

$$e_{\text{top}} = \frac{11{,}415 + 1{,}560}{16{,}400} = \frac{12{,}975}{16{,}400} = 0.790 \text{ in.}$$

A check of the eccentricity-thickness ratio yields

$$\frac{e}{t} = \frac{0.790}{7.5} = 0.105 < \frac{1}{3}$$

Because the transverse loading exceeds 10 psf, e_1/e_2 is taken as unity. With $h/t = 12.8$, $C_s = 0.70$. From Table 11-5, with $e/t = 0.105$, $C_e = 0.80$. From Eq. (11-17), the required masonry strength for Case III is

$$f'_m = \frac{16{,}925}{0.20 \times 90 \times 0.80 \times 0.70}\frac{1}{1.33} = 1{,}270 \text{ psi}$$

Case IV (*DL* + *Perpendicular Wind*). This case is the same as Case IV, Art. 11-20, because there are no live-load moments.

Case V (*DL* + *RLL*). No wind is included in this case. The eccentricity at the top of the wall is

$$e_{\text{top}} = \frac{11{,}415}{14{,}050} = 0.815 \text{ in.}$$

The eccentricity at the bottom of the wall is

$$e_{\text{bot}} = \frac{1{,}015}{14{,}610} = 0.069 \text{ in.}$$

A check of the eccentricity-thickness ratio yields

$$\frac{e}{t} = \frac{0.815}{7.5} = 0.119 < \frac{1}{3}$$

The eccentricity ratio, negative with the wall in double curvature, is

$$\frac{e_1}{e_2} = -\frac{0.069}{0.815} = -0.085$$

With $h/t = 12.8$, $C_s = 0.88$. From Table 11-5, with $e/t = 0.119$, $C_e = 0.79$. From Eq. (11-17), the required masonry strength for Case V is

$$f'_m = \frac{14{,}610}{0.20 \times 90 \times 0.79 \times 0.88} = 1{,}170 \text{ psi}$$

The nonreinforced wall is satisfactory between the second and third floors if masonry with $f'_m \geq 1{,}270$ psi is used.

11-22. Example—Interior Shear Wall, Concrete Joist Floors. Design a nonreinforced, loadbearing, interior shear wall between the second and third floors of a 10-story building, loaded on both sides of the wall, given the same data as in the example in Art. 11-20, except with the changes shown in Fig. 11-22 and the following.

Parallel wind moment = 1,500 ft-kips.

Parallel shear = 37 kips.

For an interior wall, there is no perpendicular wind load.

Distribution of Axial Loads. The axial loads, as in the examples in Art. 11-20 and 11-21, are assumed to spread over the total cross-sectional area of the wall, which in this example is I-shaped.

Loads at the Third Floor. Gravity loading per linear foot of wall from floors or roof is contributed by half the floor or roof span on either side of the wall, lengths of 11.5 and 14.5 ft. Loading from floors above is shared with the flange walls, so that the shear wall carries 15.6/21.9 = 0.713 of this loading.

The dead load just below the third floor, from eight stories of masonry, roof dead load, and dead load of eight floors, is

$$\begin{aligned} P_{DL} &= 70.7 \times 74.4 + [(11.5 + 14.5)50 + 7 \times 26 \times 60]0.713 + 26 \times 60 \\ &= 15{,}550 \text{ lb per ft} \end{aligned}$$

Similarly, the live load, without reduction, is

$$P_{LL} = (26 \times 30 + 7 \times 26 \times 40)0.713 + 26 \times 40 = 6{,}800 \text{ lb per ft}$$

With the maximum permissible reduction, the wall need carry only 42% of the floor live load (Art. 11-20). The reduced live load is

$$P_{RLL} = (26 \times 30 + 0.42 \times 7 \times 26 \times 40)0.713 + 0.42 \times 26 \times 40$$
$$= 3{,}180 \text{ lb per ft}$$

Loads at the Second Floor. The dead load just above the second floor, from nine stories of masonry, roof dead load, and dead load of eight floors, is

$$P_{DL} = 79.5 \times 74.4 + (26 \times 50 + 8 \times 26 \times 60)0.713 = 15{,}760 \text{ lb per ft}$$

Similarly, the live load, without reduction, is

$$P_{LL} = (26 \times 30 + 8 \times 26 \times 40)0.713 = 6{,}500 \text{ lb per ft}$$

and the reduced live load is

$$P_{RLL} = (26 \times 30 + 0.42 \times 8 \times 26 \times 40)0.713 = 3{,}050 \text{ lb per ft}$$

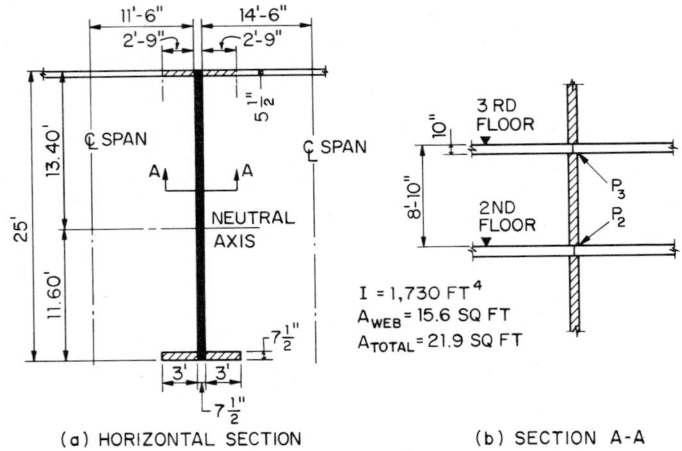

Fig. 11-22. Properties of interior shear wall of 10-story building.

Summation of Load at the Second Floor. The total dead load just above the second floor is

$$P'_{DL} = 79.5 \times 120 \times 21.9 + 23.83 \times 26(30 + 8 \times 40) = 423{,}900 \text{ lb}$$

The total unreduced live load plus dead load is

$$P'_{TL} = 79.5 \times 120 \times 21.9 + 23.83 \times 26(80 + 8 \times 100) = 752{,}300 \text{ lb}$$

Source of Moments. To produce maximum total-load and reduced-live-load moments, assume checkerboard loading of floors and roof, with the live load occurring on the long span at the top of the wall (second floor) and on the short span at the bottom of the wall (third floor). This arrangement of loads will normally place the wall in single curvature, thus giving conservative (positive) values of e_1/e_2.

Moments from Floor Loads. Assume 4-in. bearing of joists on the wall and a triangular stress distribution under each bearing (Fig. 11-23). The eccentricity of the joist reaction then is

$$e_f = \frac{7.5}{2} - \frac{4}{3} = 2.42 \text{ in.}$$

The dead-load and total-load moments due to this eccentricity are distributed in equal proportions to the wall sections above and below the floor.

At the top of the wall:

$$M_{DL} = -\tfrac{1}{2} \times 60 \times 11.5 \times 2.42 = -835 \text{ in.-lb per ft}$$
$$M_{TL} = \tfrac{1}{2} \times 100 \times 14.5 \times 2.42 = 1{,}750 \text{ in.-lb per ft}$$
$$\text{Moment in wall} = M = 915 \text{ in.-lb per ft}$$

At the bottom of the wall:

$$M_{DL} = -\tfrac{1}{2} \times 60 \times 14.5 \times 2.42 = -1{,}055 \text{ in.-lb per ft}$$
$$M_{TL} = \tfrac{1}{2} \times 100 \times 11.5 \times 2.42 = 1{,}390 \text{ in.-lb per ft}$$
$$\text{Moment in wall} = M = 335 \text{ in.-lb per ft}$$

Parallel Wind Moment. For the second floor, this was given initially for this example as $M' = 1{,}500$ ft-kips.

Overturn Check. To determine if tensile stress develops in the shear wall from overturning moment, the wall should be considered to be carrying only dead loads in addition to the wind parallel to the wall. The location of the neutral axis of the wall is found relative to the narrow flange, by dividing the sum of the moments of the wall areas about that end by the wall area $A = 21.9$ sq ft.

$$x = \left[\frac{0.625(25)^2}{2} + 0.625 \times 6 \times 24.69 + \frac{0.46(5.5)^2}{2}\right]\frac{1}{21.9} = 13.40 \text{ ft}$$

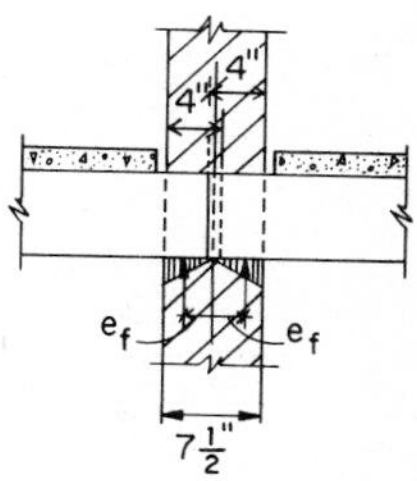

Fig. 11-23. Bearing of concrete joists on interior wall.

The moment of inertia of the wall about this axis is $I = 1{,}730 \text{ ft}^4$.

The allowable eccentricity e_L, which is the maximum for which no tension occurs on the wall cross section, is given by I/cA, where c is the maximum distance of a wall end from the neutral axis.

$$e_L = \frac{I}{cA} = \frac{1{,}730}{13.40 \times 21.9} = 5.90 \text{ ft}$$

The actual eccentricity is

$$e_L = \frac{M'}{P'_{DL}} = \frac{1{,}500{,}000}{423{,}900} = 3.55 < 5.90 \text{ ft}$$

Therefore, no tensile reinforcement is required.

Loading Conditions. Three cases are considered for checking eccentricities and stresses in the wall under wind and gravity loads:

Case I. Total load on floors and roof plus wind parallel to the wall.

Case II. Dead load on floors and roof plus wind parallel to the wall.

Case III. Dead load plus reduced live load.

Case I (TL + Parallel Wind). This case involves biaxial bending and is similar to Case I, Art. 11-20. The maximum eccentricity at the second floor is

$$e_t = \frac{915}{22{,}260} = 0.041 \text{ in.}$$

The virtual eccentricity of the 1,500-ft-kip overturning moment is

$$e_L = \frac{M'}{P'_{TL}} = \frac{1{,}500{,}000}{752{,}300} = 2.00 \text{ ft}$$

In accordance with Eq. (11-16) for biaxial bending,

$$\frac{e}{t} = \frac{0.041}{7.5} + \frac{2.00}{25.0} = 0.086 < \frac{1}{3}$$

The eccentricity of the floor load at the top of the wall (third floor) is computed for live load only on the long span.

$$e_{top} = \frac{915}{22{,}350} = 0.041 \text{ in.}$$

The eccentricity of floor load at the bottom of the wall (second floor) for live load only on the short span is

$$e_{bot} = \frac{335}{22{,}260} = 0.015 \text{ in.}$$

The eccentricity ratio, positive because of single curvature of the wall, is

$$\frac{e_1}{e_2} = \frac{0.015}{0.041} = 0.375$$

For use in determining the slenderness coefficient C_s, the slenderness ratio $h/t = 12.8$. From Eq. (11-5) or Table 11-6, $C_s = 0.81$. From Table 11-5, with $e/t = 0.086$, the eccentricity coefficient is $C_e = 0.87$. From Eq. (11-17), the required masonry strength for Case I, with the one-third increase in allowable stress for combined gravity and wind loads, is

$$f'_m = \frac{22{,}260}{0.20 \times 90 \times 0.87 \times 0.81} \frac{1}{1.33} = 1{,}320 \text{ psi}$$

Case II (DL + Parallel Wind). This case is similar to Case I, but dead load is used instead of total load.

The eccentricity of the floor load at the bottom of the wall is

$$e_{bot} = \frac{1{,}055 - 835}{15{,}760} = \frac{220}{15{,}760} = 0.014 \text{ in.}$$

The virtual eccentricity of the 1,500-ft-kip overturning moment is

$$e_L = \frac{M'}{P'_{DL}} = \frac{1{,}500{,}000}{423{,}900} = 3.55 \text{ ft}$$

A check with Eq. (11-16) yields

$$\frac{e}{t} = \frac{0.014}{7.5} + \frac{3.55}{25.0} = 0.144 < \frac{1}{3}$$

The eccentricity of the floor load at the top of the wall is

$$e_{top} = \frac{220}{15{,}550} = 0.014 \text{ in.}$$

The eccentricity ratio, negative for the wall in double curvature, is

$$\frac{e_1}{e_2} = -\frac{0.014}{0.014} = -1$$

With $h/t = 12.8$, $C_s = 0.95$. From Table 11-5, with $e/t = 0.144$, $C_e = 0.80$. From Eq. (11-17), the required masonry strength for Case II is

$$f'_m = \frac{15{,}760}{0.20 \times 90 \times 0.80 \times 0.95} \frac{1}{1.33} = 860 \text{ psi}$$

Case III ($DL + RLL$). No wind is included in this case. The eccentricity at the top of the wall is, for live load only on the long span,

$$e_{top} = \frac{915}{15,550 + 3,180} = \frac{915}{18,730} = 0.049 \text{ in.}$$

The eccentricity at the bottom of the wall is, for live load only on the short span,

$$e_{bot} = \frac{335}{15,760 + 3,050} = \frac{335}{18,810} = 0.018 \text{ in.}$$

A check of the eccentricity-thickness ratio yields

$$\frac{e}{t} = \frac{0.049}{7.5} = 0.007 < \frac{1}{3}$$

The eccentricity ratio, with the wall in single curvature, is

$$\frac{e_1}{e_2} = \frac{0.018}{0.049} = 0.368$$

With $h/t = 12.8$, $C_s = 0.81$. From Table 11-5, with $e/t = 0.007$, $C_e = 1.0$. From Eq. (11-17), the required masonry strength for Case III is

$$f'_m = \frac{18,810}{0.20 \times 90 \times 1 \times 0.81} = 1,300 \text{ psi}$$

Shear Check. Assume $f'_m = 1,320$ psi and Type N mortar, without inspection. The allowable stress (Art. 11-17) is

$$v_m = 0.5\sqrt{f'_m} + \frac{f_{DL}}{5} = 0.5\sqrt{1,320} + \frac{1}{5}\frac{423,900}{21.9 \times 144} = 46 > 28 \text{ psi}$$

Use 28 psi. The actual average shear stress over the wall web is

$$v_m = \frac{37,000}{15.6 \times 144} = 16.5 < 28 \text{ psi}$$

The nonreinforced wall is satisfactory between the second and third floors if masonry with $f'_m \geq 1,320$ psi is used.

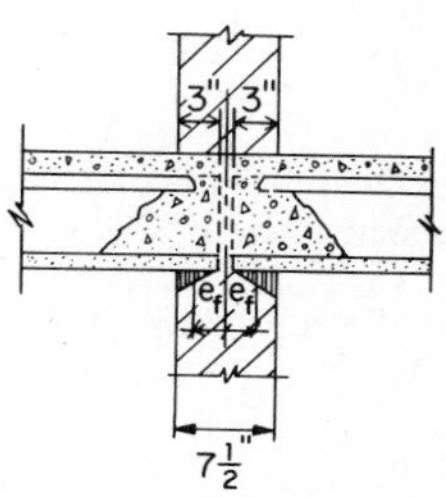

Fig. 11-24. Bearing of concrete planks on interior wall.

11-23. Example—Interior Shear Wall, Concrete Plank Floors. Design a shear wall for the same conditions as given for the example in Art. 11-22, except that the floor is flat, precast-concrete plank with topping.

Moments from Floors. Assume conservatively that the wall above or below a floor receives a live-load moment from the floor of $w_{LL}L^2/36$, as in Art. 11-22. Because precast planks cannot lap each other, the bearing is assumed to be 3 in. (Fig. 11-24), and the eccentricity of the plank reaction is

$$e_f = \frac{7.5}{2} - \frac{3}{3} = 2.75 \text{ in.}$$

With live load only on the long span, moments at the top of the wall are

$$\begin{aligned} M_{DL} &= -\tfrac{1}{2} \times 60 \times 11.5 \times 2.75 &&= -950 \text{ in.-lb per ft} \\ M_{TL} &= \tfrac{1}{2} \times 60 \times 14.5 \times 2.75 + 40(29)^2 12/36 &&= 12,400 \text{ in.-lb per ft} \\ \text{Moment in wall} &= M &&= 11,450 \text{ in.-lb per ft} \end{aligned}$$

With live load only on the short span, moments at the bottom of the wall are

$$\begin{aligned} M_{DL} &= -\tfrac{1}{2} \times 60 \times 14.5 \times 2.75 &&= -1,200 \text{ in.-lb per ft} \\ M_{TL} &= \tfrac{1}{2} \times 60 \times 11.5 \times 2.75 + 40(23)^2 12/36 &&= 8,000 \text{ in.-lb per ft} \\ \text{Moment in wall} &= M &&= 6,800 \text{ in.-lb per ft} \end{aligned}$$

Case I (TL + Parallel Wind). This case is similar to Case I, Art. 11-22. The maximum eccentricity at the second floor is

$$e_t = \frac{11{,}450}{22{,}260} = 0.514 \text{ in.}$$

As in Art. 11-22, the virtual eccentricity of the 1,500-ft-kip overturning moment is $e_L = 2.00$ ft. In accordance with Eq. (11-16) for biaxial bending,

$$\frac{e}{t} = \frac{0.514}{7.5} + \frac{2.00}{25.0} = 0.148 < \frac{1}{3}$$

The eccentricity of floor load at the top of the wall is computed for live load only on the long span.

$$e_{\text{top}} = \frac{11{,}450}{22{,}350} = 0.512 \text{ in.}$$

The eccentricity of floor load at the bottom of the wall for live load only on the short span is

$$e_{\text{bot}} = \frac{6{,}800}{22{,}260} = 0.302 \text{ in.}$$

The eccentricity ratio, positive because of single curvature, is

$$\frac{e_1}{e_2} = \frac{0.302}{0.512} = 0.590$$

With $h/t = 12.8$, $C_s = 0.77$. From Table 11-5, with $e/t = 0.148$, $C_e = 0.71$. From Eq. (11-17), the required masonry strength for Case I, with the one-third increase in stress allowed for combined gravity and wind loads, is

$$f'_m = \frac{22{,}260}{0.20 \times 90 \times 0.71 \times 0.77}\,\frac{1}{1.33} = 1{,}700 \text{ psi}$$

Case II (DL + Parallel Wind). This case is the same as Case II, Art. 11-22, because there are no live-load moments.

Case III (DL + RLL). No wind is included in this case. The eccentricity at the top of the wall is

$$e_{\text{top}} = \frac{11{,}450}{18{,}730} = 0.613 \text{ in.}$$

The eccentricity at the bottom of the wall is

$$e_{\text{bot}} = \frac{6{,}800}{18{,}810} = 0.360 \text{ in.}$$

A check of the eccentricity-thickness ratio yields

$$\frac{e}{t} = \frac{0.613}{7.5} = 0.082 < \frac{1}{3}$$

The eccentricity ratio, positive with the wall in single curvature, is

$$\frac{e_1}{e_2} = \frac{0.360}{0.613} = 0.592$$

With $h/t = 12.8$, $C_s = 0.77$. From Table 11-5, with $e/t = 0.082$, $C_e = 0.87$. From Eq. (11-17), the required masonry strength for Case III is

$$f'_m = \frac{18{,}810}{0.20 \times 90 \times 0.87 \times 0.77} = 1{,}560 \text{ psi}$$

The nonreinforced wall is satisfactory between the second and third floors if masonry with $f'_m \geq 1{,}700$ psi is used.

11-24. Bibliography. The following publications are available from the Brick Institute of America, McLean, Va.:

J. G. Gross, R. D. Dikkers, and J. C. Grogan, "Recommended Practice for Engineered Brick Masonry."

H. C. Plummer, "Brick and Tile Engineering."

"Building Code Requirements for Engineered Brick Masonry."

"Technical Notes on Brick and Tile Construction"—a series.

Section 12

Water Permeability of Masonry Structures

CYRUS C. FISHBURN

Formerly, Division of Building Technology, National Bureau of Standards

This section discusses the permeability of concrete and masonry structures subjected to water on exposed faces. It gives information on the construction of water-resistant above-grade and below-grade structures and the use of admixtures and coatings.

12-1. Definitions of Terms Related to Water Resistance

Permeability. Quality or state of permitting passage of water and water vapor into, through, and from pores and interstices, without causing rupture or displacement.

Terms used in this section to describe the permeability of materials, coatings, structural elements, and structures follow in decreasing order of permeability:

Pervious or Leaky. Cracks, crevices, leaks, or holes larger than capillary pores, which permit a flow or leakage of water, are present. The material may or may not contain capillary pores.

Water-resistant. Capillary pores exist that permit passage of water and water vapor, but there are few or no openings larger than capillaries that permit leakage of significant amounts of water.

Water-repellent. Not "wetted" by water; hence, not capable of transmitting water by capillary forces alone. However, the material may allow transmission of water under pressure and may be permeable to water vapor.

Waterproof. No openings are present that permit leakage or passage of water and water vapor; the material is impervious to water and water vapor, whether under pressure or not.

These terms also describe the permeability of a surface coating or a treatment against water penetration, and they refer to the permeability of materials, structural members, and structures whether or not they have been coated or treated.

12-2. Permeability of Concrete. Concrete contains many interconnected voids and openings of various sizes and shapes, most of which are of capillary dimensions. If the larger voids and openings are few in number and not directly connected with each other, there will be little or no water penetration by leakage and the concrete may be said to be water-resistant.

Concrete in contact with water not under pressure ordinarily will absorb it. The water is drawn into the concrete by the surface tension of the liquid in the wetted capillaries.

The tendency for the water to rise above its level is known as the **capillary potential.** The potential capillary rise is termed the **equivalent negative hydraulic head.** The surface forces in the capillaries and the potential capillary rise increase inversely with the bore of the capillaries.

Water under pressure and in contact with the exposed sides and bottom of a concrete foundation floor or wall will tend to fill all capillaries in the concrete below the height of the hydrostatic pressure head. The water will tend to rise above this level to heights equivalent to the negative hydraulic head of the capillaries. The rise, however, may be greatly reduced if the rate of evaporation from the interior surfaces exceeds the rate of flow through the capillaries.

A slow percolation may occur from any capillary containing water at a hydrostatic pressure greater than the potential of the capillary. Some of the capillaries in concrete may be permeable to the flow of water under a low hydraulic head. Tests of permeability, under pressure, of sand-and-gravel concrete, however, indicate that the rate at which water under pressure passes through capillaries in concrete is very small when compared with the possible rate of leakage through large interconnected voids, subjected to like pressure.

If a capillary in concrete is made repulsive to water with a water-repellent admixture, the capillary potential is of the same value but is reversed in sign. Such a capillary will not permit absorption of water that is not under pressure. If the water is under a pressure head greater than that of the capillary potential, the water will enter the concrete and will tend to rise to a height equal to the difference between the water pressure head and the equivalent positive hydraulic head of the capillary. This height will increase with increase in capillary bore and will equal the height of the pressure head in voids or openings having zero capillary potential.

Water-resistant concrete for buildings should be a properly cured, dense, rich concrete containing durable, well-graded aggregate. The water content of the concrete should be as low as is compatible with workability and ease of placing and handling. Information on the kind of concrete to be used in floors and walls is given in Arts. 12-7, 12-8, 12-9, and 12-11. More detailed information on the proportioning, mixing, placing, and curing of water-resistant concrete is given in Sec. 5. See also "Recommended Practice for Selecting Proportions for Normal Weight Concrete," ACI 211.1, and "Specifications for Structural Concrete for Buildings," ACI 301, American Concrete Institute.

12-3. Water-repellent Admixtures for Concrete. These are the most effective and widely used of the so-called integral "waterproofers" or "dampproofers" for concrete. They tend to keep the surfaces of voids and capillaries in concrete from being wetted by water.

A water-repellent admixture usually reduces the strength of concrete. For a given repellent, the strength reduction tends to increase with increase in repellency. The amount of water repellent commonly used is limited, rarely exceeding 0.2% of the weight of the cement.

Water-repellent concrete is permeable to water vapor. If a vapor-pressure gradient is present, moisture may penetrate from the exposed face to an inner face. The concrete is not made waterproof (in the full meaning of the term) by the use of an integral water repellent.

Whether or not a capillary is repellent, water under a sufficient pressure head will enter and pass through any capillary whose potential is less than that of the pressure head. Water will not enter a repellent capillary whose potential exceeds the applied water pressure head.

Water repellents may not make concrete impermeable to penetration of water under pressure, but they may reduce absorption of water by the concrete. The reduction in absorption resulting from the use of a water-repellent admixture may be a good measure of the probable effectiveness of the admixture against the capillary rise of water not under pressure.

The "soaps" or salts of fatty acids, such as calcium, aluminum, and ammonium stearates and oleates, are commonly used water-repellent admixtures for concretes. When used in amounts equal to about 0.2% by weight of cement, they may reduce the absorption of a concrete to roughly 70% of that of the plain untreated concrete. These repellents, however, have little effect on the permeability of concrete subjected to water under pressure.

Butyl stearate used in amounts equal to 0.5 to 1.0% by weight of cement reduced absorption of concrete test specimens to about 35 and 55%, respectively, of that of plain concrete. Used in the form of an emulsion, the butyl stearate may become well distributed throughout the mass of the concrete without excessive foaming and possible reduction in strength of concrete. Permeability of concrete containing butyl stearate may not differ much from that of standard plain concrete, but butyl stearate is markedly superior to the "soaps" (salts of fatty acids) in reducing absorption.

Heavy mineral oil used in amounts equivalent to about 4% by weight of cement and containing 12% of stearic acid significantly reduces the permeability of concrete subjected to a water pressure of 20 psi. The absorption of treated concrete is about 45% of that of plain concrete, but the compressive strength is 85% of that of the standard. Some engineers recommend that the viscosity of the oil should not be less than SAE 60, and that the oil should not saponify or emulsify with alkali and seriously impair the strength of the concrete.

Asphalt emulsions have been used as water-repellent admixtures and are claimed to be effective. Repellency of treated concrete may be increased if the concrete is dried sufficiently to permit the emulsion to "break."

Coal-tar pitch cut with benzene, added to concrete in the amount of 1% of the pitch, by weight of cement, increases absorption, reduces compressive strength to about 85% of that of the standard, and decreases permeability, under pressure, of the concrete.

Powered iron, with an oxidizing agent such as ammonium chloride, is not usually considered as an admixture for concrete. However, it is occasionally added to cementitious grouts and mortar that are applied as coatings for concrete and masonry. Oxidation of the iron begins when water is added and results in an expansion of the volume occupied by the iron particles. This expansion is claimed to control, reduce, or eliminate shrinkage of the grout or mortar.

Directions for the use of the iron may vary with different producers of the material. Equal parts by weight of portland cement and the chemically prepared iron are sometimes used in a grout. A trowel-applied mortar coat of portland cement, iron, and sand may contain as much as 25% of iron by weight of cement. Usually the grout coat is followed by one or more parge coats of a mortar. It is difficult to obtain a continuous coating with the iron alone and cement, or cement and sand, are usually added to obtain a workable mixture.

In permeability tests of surface coatings for concrete, several coatings containing powdered iron and an oxidizing agent slightly reduced the absorption of cylindrical specimens of concrete immersed in water. Other data, some of which are unpublished, indicate the need for using portland cement, or portland cement and sand, with the iron to insure application of a continuous coating on highly permeable masonry walls. Although the data are meager, they indicate that trowel-applied mortar coatings are highly water-resistant whether or not they contain iron.

Many contractors specializing in preventive measures and remedial treatments against water penetration use chemically prepared iron powder in coatings applied to subgrade concrete and masonry surfaces. Use of iron may have some advantages, but it should be noted that coatings containing iron are permeable to the penetration of water by capillarity, whether or not the water is under hydrostatic head. Such a coating is not waterproof in the full sense of the term.

("Admixtures for Concrete," ACI Manual of Concrete Practice, Part 1, American Concrete Institute.)

12-4. Permeability of Unit-masonry Walls. Most masonry units will absorb water. Some are highly pervious under pressure. The mortar commonly used in masonry will absorb water but usually contains few openings permitting leakage.

Masonry walls usually leak at the joints between the mortar and the units, however. Excepting single-leaf walls of highly pervious units, leakage at the joints results from failure to fill them with mortar and poor bond between the masonry unit and mortar. As with concrete, rate of capillary penetration through masonry walls is small compared with the possible rate of leakage.

12-5. Permeability of Concrete and Masonry Structures. Many above- and below-grade structures, built without surface coatings, membranes, and integral waterproofings, have given adequate protection against leakage under severe exposure conditions and have not become objectionably damp from the penetration of capillary moisture. Such structures contain few, if any, large openings permitting leakage of water in quantity.

Capillary penetration of moisture through above-grade walls that resist leakage of wind-driven rain is usually of minor importance. Such penetration of moisture into well-ventilated subgrade structures may also be of minor importance if the moisture is readily evaporated. However, long-continued capillary penetration into some deep, confined subgrade interiors frequently results in an increase in relative humidity, a decrease in evaporation rate, and objectionable dampness.

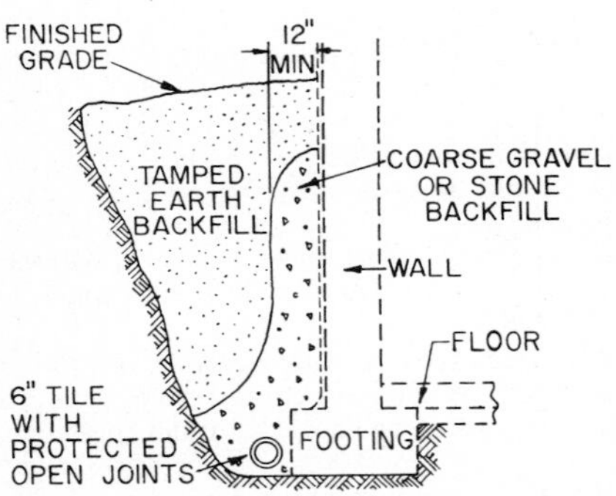

Fig. 12-1. Drainage of basement wall with drain tile along the footing and gravel fill.

12-6. Drainage for Subgrade Structures. Subgrade structures located above ground-water level in drained soil may be in contact with water and wet soil for periods of indefinite duration after long-continued rains and spring thaws. Drainage of surface and subsurface water may greatly reduce the time during which the walls and floor of a structure are subjected to water, may prevent leakage through openings resulting from poor workmanship and reduce the capillary penetration of water into the structure. If subsurface water cannot be removed by drainage, the structure must be made waterproof or highly water-resistant.

Surface water may be diverted by grading the ground surface away from the walls and by carrying the runoff from roofs away from the building. The slope of the ground surface should be at least ¼ in. per ft for a minimum distance of 10 ft from the walls. Runoff from high ground adjacent to the structure should also be diverted.

Proper subsurface drainage of ground water away from basement walls and floors requires a drain of adequate size, sloped continuously, and, where necessary, carried around corners of the building without breaking continuity. The drain should lead to a storm sewer or to a lower elevation that will not be flooded and permit water to back up in the drain.

Drain tile should have a minimum diameter of 6 in. and should be laid in gravel or other kind of porous bed at least 6 in. below the basement floor. The open joints between the tile should be covered with a wire screen or building paper to prevent clogging of the drain with fine material. Gravel should be laid above the tile, filling the excavation to an elevation well above the top of the footing. Where considerable water may be expected in heavy soil, the gravel fill should be carried up nearly to the ground surface and should extend from the wall a distance of at least 12 in. (Fig. 12-1).

12-7. Concrete Floors on Ground. These should preferably not be constructed in low-lying areas that are wet from ground water or periodically flooded with surface water. The ground should slope away from the floor. The level of the finished floor should be at least 6 in. above grade. Further protection against ground moisture and possible flooding of the slab from heavy surface runoffs may be obtained with subsurface drains located at the elevation of the wall footings.

All organic material and topsoil of poor bearing value should be removed in preparation of the subgrade, which should have a uniform bearing value to prevent unequal settlement of the floor slab. Backfill should be tamped and compacted in layers not exceeding 6 in. in depth.

Except where the minimum winter air temperatures are above 35°F, slabs on ground of residences should be insulated at the edges. A mineral insulating material, at least 1 in. thick, that will not absorb water should be used. For best protection, the insulation should be placed along the outside edges and under the bottom of the slab for a distance of at least 18 in. from the edges. In mild climates, the insulation may be placed along the outside edges downward to a distance of 12 in. below the bottom of the slab.

Where the subgrade is well-drained, as where subsurface drains are used or are unnecessary, floor slabs of residences should be insulated either by placing a granular fill over the subgrade or by use of a lightweight-aggregate concrete

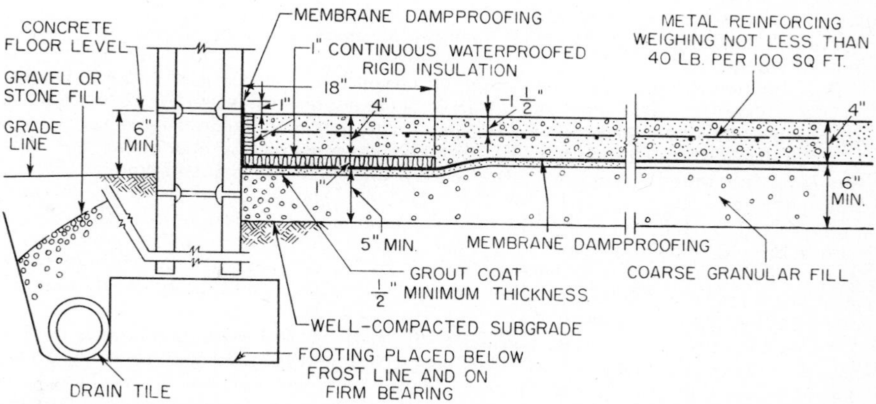

Fig. 12-2. Concrete slab on ground insulated and waterproofed with bituminous membrane.

slab covered with a wearing surface of gravel or stone concrete. The granular fill, if used, should have a minimum thickness of 5 in. and may consist of coarse slag, gravel, or crushed stone, preferably of 1-in. minimum size. A layer of 3-, 4-, or 6-in.-thick hollow masonry building units is preferred to gravel fill for insulation and provides a smooth, level bearing surface.

Moisture from the ground may be absorbed by the floor slab, thereby increasing the thermal conductivity of lightweight-aggregate concrete. A waterproof membrane is effective in preventing such absorption only if it is continuous and not punctured or damaged during the concreting. It is desirable, therefore, that the concrete be low absorptive.

Insulation also retards the capillary rise of ground moisture.

If the subgrade is not properly drained and the ground-water level should reach the bottom of the slab, the insulation value of granular fill, hollow masonry units, or lightweight-aggregate concrete will be greatly reduced. In such cases, a waterproof layer of insulation may be needed beneath the slab.

Floor coverings, such as oil-base paints, linoleum, and asphalt tile, acting as a vapor barrier over the slab, may be damaged by a rise of capillary moisture and water vapor from the ground. If such floor coverings are used and where a complete barrier against the rise of moisture from the ground is desired, a two-ply bituminous membrane waterproofing should be placed beneath the slab and over the insulating concrete or granular fill (Fig. 12-2). The top of the lightweight-aggregate concrete, if used, should be troweled or brushed to a smooth level surface for the membrane. The top of the granular fill should be covered with a grout coating, similarly finished. (The grout coat, ½ to 1 in. thick, may consist of a 1:3 or a 1:4 mix by volume of portland cement and sand. Some ⅜- or ½-in. maximum-sized coarse aggregate may be added to the grout if desired.) After the top surface of the insulating concrete or grout coating has hardened and dried, it should be mopped with hot asphalt or coal-tar pitch and covered before cooling with a lapped layer of 15-lb bituminous saturated felt. The first ply of felt then should be mopped with hot bitumen and a second ply of felt laid and mopped on its top surface. Care should be exercised not to puncture the membrane, which should preferably be covered with a coating of mortar, immediately after its completion. If properly laid and protected from damage, the membrane may be considered to be a waterproof barrier.

Where there is no possible danger of water reaching the underside of the floor, a single layer of 55-lb smooth-surface asphalt roll roofing or a similar product of 55-lb minimum weight may be substituted for the membrane. Joints between the sheets should be lapped and sealed with bituminous mastic. Great care should be taken to prevent puncturing of the roofing layer during concreting operations. When so installed, asphalt roll roofing provides a low-cost and adequate barrier against the movement of excessive amounts of moisture by capillarity and in the form of vapor.

If a lightweight-aggregate insulating concrete slab is used, it is placed directly on the subgrade. The slab should have a minimum thickness of 4 in. and a minimum compressive strength of 1,500 psi. The coarse aggregate should not exceed 1 in. in size. If the insulating slab is to be covered with a membrane, the slab should be permitted to dry out after curing and before the membrane is placed. The wearing surface applied over the membrane should be at least 2½ in. and preferably 3 in. thick of 3,000 psi concrete and should be reinforced with welded wire fabric weighing at least 40 lb per square. A 6×6 mesh of No. 6 wire is suitable. If the concrete wearing slab is to be covered with an oil-base paint or a floor covering that is a vapor barrier, the slab should be in a thoroughly dry condition before it is covered.

A concrete slab placed over a granular fill, with or without a membrane, should have a minimum thickness of 4 in. The concrete should have a minimum compressive strength of 3,000 psi, should contain durable aggregate of not more than 1-in. maximum size, and should be reinforced with welded wire fabric of 40-lb minimum weight.

("Concrete Floor Construction," Portland Cement Association, Old Orchard Road, Skokie, Ill. 60076.)

12-8. Basement Floors in Drained Soil. Where a basement is to be used as living quarters or for the storage of things that may be damaged by moisture, the floor should be insulated and should preferably contain the membrane waterproofing described in Art. 12-7. In general, the design and construction of such basement floors are similar to those of floors on ground discussed in Art. 12-7.

If the ground adjacent to the structure is properly drained, the insulation obtained from a granular fill, hollow masonry units, or lightweight-aggregate concrete should suffice. No edge insulation of the slab is needed if the basement floor is more than 2 ft beneath the ground surface. Where insulation is needed and the exposure conditions are severe, for example, where the floor slab is in contact with water or wet ground for extended periods, a 1-in.-thick layer of a nonabsorptive and waterproof mineral insulating material is preferred to granular fill and lightweight concrete.

If passage of moisture from the ground into the basement is unimportant or can be satisfactorily controlled by air conditioning or ventilation, the waterproof membrane need not be used. The concrete slab should have a minimum thickness of 4 in. and need not be reinforced, but should be laid on a granular fill or other insulation placed on a carefully prepared subgrade. The concrete in the slab should have a minimum compressive strength of 2,000 psi and may contain an integral water repellent.

A bituminous-filled joint between the basement walls and the floor will prevent leakage into the basement of any water that may occasionally accumulate under the slab. Space for the joint may be provided by use of beveled siding strips, which are removed after the concrete has hardened. After the slab is properly cured, it and the wall surface should be in as dry a condition as is practicable before the joint is filled to insure a good bond of the filler and to reduce the effects of slab shrinkage on the permeability of the joint. Hot asphalt or coal-tar pitch may be placed in the joint after its sides have been primed with a suitable bitumen.

12-9. Monolithic Concrete Basement Walls. These should have a minimum thickness of 6 in. Where insulation is desirable, as where the basement is used for living quarters, lightweight aggregate, such as those prepared by calcining or sintering blast-furnace slag, clay, or shale that meet the requirements of American Society for Testing and Materials Standard C330, may be used in the concrete. The concrete should have a minimum compressive strength of 2,000 psi.

Wall form ties of an internal-disconnecting type are preferred to twisted-wire ties. Entrance holes for the form ties should be sealed with mortar after the forms are removed. If twisted-wire ties are used, they should be cut a minimum distance of 1½ in. from the face of the wall and the holes filled with mortar.

The concrete should be placed in the forms in level layers, preferably not more than 18 in. thick. The concrete should be either vibrated or carefully tamped in the forms. Construction joints, where necessary, should be horizontal. Before concreting is resumed at a construction joint, the top surface of the wall should be cleaned of soft material to expose the aggregate. The cleaned top surface should be dampened and covered with a thin coating of a 1:2 mix of cement-sand grout, followed at once and before drying out with the next layer of concrete. The wall should be moist-cured for at least 7 days.

The resistance of the wall to capillary penetration of water in temporary contact with the wall face may be increased by the use of a water-repellent admixture. The water repellent may also be used in the concrete at and just above grade to reduce the capillary rise of moisture from the ground into the superstructure walls.

Where it is desirable to make the wall resistant to passage of water vapor from the outside and to increase its resistance to capillary penetration of water, the wall face may be treated with a bituminous coating. The continuity and the resultant effectiveness in resisting moisture penetration of such a coating is dependent on the smoothness and regularity of the concrete surface and on the skill and technique used in applying the coating to the dry concrete surface. Some bituminous coatings that may be used are listed below in increasing order of their resistance to moisture penetration:

Spray- or brush-applied asphalt emulsions.

Spray- or brush-applied bituminous cutbacks.

Trowel coatings of bitumen with organic solvent, applied cold.

Hot-applied asphalt or coal-tar pitch, preceded by application of a suitable primer.

Cementitious brush-applied paints and grouts and trowel coatings of a mortar increase moisture resistance of monolithic concrete, especially if such coatings contain a water repellent. However, in properly drained soil, such coatings may not be justified unless needed to prevent leakage of water through openings in the concrete resulting from segregation of the aggregate and bad workmanship in casting the walls. The trowel coatings may also be used to level irregular wall surfaces in preparation for the application of a bituminous coating.

12-10. Basement Walls of Masonry Units. Water-resistant basement walls of masonry units should be carefully constructed of durable materials to prevent leakage and damage due to frost and other weathering exposure. Frost action is most severe at the grade line and may result in structural damage and leakage of water. Where wetting followed by sudden severe freezing may occur, the masonry units should meet the requirements of the following specifications:

Building brick (solid masonry units made from clay or shale), American Society for Testing and Materials Standard C62, Grade SW.

Facing brick (solid masonry units made from clay or shale), ASTM Standard C216, Grade SW.

Structural clay load-bearing wall tile, ASTM Standard C34, Grade LBX.

Hollow load-bearing concrete masonry units, ASTM Standard C90, Grade N.

For such exposure conditions, the mortar should be a Type S mortar (Table 10-1) having a minimum compressive strength of 1,800 psi when tested in accordance with the requirements of ASTM Standard C270. For milder freezing exposures and where the walls may be subjected to some lateral pressure from the earth, the mortar should have a minimum compressive strength of 1,000 psi.

Rise of moisture, by capillarity, from the ground into the superstructure walls may be greatly retarded by use of an integral water-repellent admixture in the mortar. The water-repellent mortar may be used in several courses of masonry located at and just above grade.

Both the vertical and horizontal joints in brick masonry foundation walls should be completely filled with mortar. Walls of hollow masonry units laid with the cells vertical should be bedded with mortar under all face shells of the units. Mortar should also be spread on the cross webs, which serve as the bearing for the face shell of the units above, as where the wall contains two tiers, or wythes, of units. Mortar for head joints should be buttered in the exposed faces of the wall a distance equal to the thickness of the face shells. Side bearing units should be laid in full mortar beds. Both the bed and head joints of walls of solid concrete masonry units should be filled with mortar.

Thickness of basement walls should be as great as that of the wall they support and should not be less than one-twelfth of the vertical distance between lateral supports. Any requirement for the thermal conductance of the wall will also affect its thickness under usual conditions of service. It is likely that, if the thermal conductance C of the wall does not exceed 0.4, or possibly 0.5 Btu per hr per sq ft per °F, there will be no surface condensation on the interior face. This usually requires that masonry walls of hollow units be at least 12 in. thick. If additional insulation is needed in the wall, insulation material or a nonabsorptive mineral type should be used.

All-brick basement walls containing two or three tiers of units can be built to be highly water-resistant (Arts. 12-14 and 12-16). However, chief reliance against leakage of water through masonry basement walls, particularly those of hollow units, should be placed on use of trowel-applied mortar coats of pargings on the exposed faces of the walls. The mortar coats should also be used on the outside faces of masonry foundation walls of basementless structures located where freezing temperature may occur, to prevent leakage and the subsequent freezing of water trapped in the masonry.

Two trowel coats of a mortar containing 1 part portland cement to 3 parts sand by volume should be applied to the outside faces of basement walls built of hollow masonry units. One trowel coat may suffice on the outside of all-brick and of brick-faced walls.

The wall surface and the top of the wall footing should be cleaned of dirt and soil, and the masonry should be thoroughly wetted with water. While still damp, the surface should be covered with a thin scrubbed-on coating of portland cement tempered to the consistency of thick cream. Before this prepared surface has dried, a ⅜-in.-thick trowel-applied coating or mortar should be placed on the wall and over the top of the footing; a fillet of mortar may be placed at the juncture of the wall and footing.

Where a second coat of mortar is to be applied, as on hollow masonry units, the first coat should be scratched to provide a rough bonding surface. The second coat should be applied at least 1 day after the first, and the coatings should be cured and kept damp by wetting for at least 3 days. A water-repellent admixture in the mortar used for the second or finish coat will reduce the rate of capillary penetration of water through the walls. If a bituminous coating is not to be used, the mortar coating should be kept damp until the backfill is placed.

The bituminous coatings described in Art. 12-9 may be applied to the plaster coating if resistance to penetration of water vapor is desired. The plaster coating should be dry and clean before the bituminous coating is applied over the surfaces of the wall and the top of the footing. Unless backed with a tier of masonry or a self-supporting layer of concrete, the bituminous coatings may fail, in time, to bond properly to the inner face of the masonry. They should not be used on the inner faces of the walls unless properly supported in position.

Expansion joints in both concrete and masonry foundation walls should be of bellows type made of 16-oz copper sheet, which should extend a minimum distance of 6 in. on each side of the joint. The sheet should be embedded between wyths of masonry units or faced with a 2-in.-thick cover of mortar reinforced with welded-wire fabric. The copper bellows may be backed with bituminous saturated fabric. The outside face of the expansion joint should be filled flush with the wall face with bituminous plastic cement conforming with the requirements of Federal Specification SS-C-153.

12-11. Basements Subjected to Hydrostatic Pressure, without Membranes. Basement walls and floors subjected to water under pressure should be designed and built to resist it. Where exposure conditions are severe, both walls and floors should be made of reinforced concrete. Expansion joints should be in the same vertical plane throughout walls and floors. These joints should be waterproof and of the bellows type, composed of noncorrosive metal sheet of sufficient strength and ductility to resist the anticipated maximum pressure and movement. The exposed face of the expansion joint should be sealed with plastic cement (Federal Specification SS-C-153).

Addition of an integral water repellent to concrete continuously subjected to water under pressure slightly increases permeability of the concrete. The increase has been partly attributed to difficulty in properly mixing the concrete, frothing of the mix, and entrapment of air in large-sized voids. Use of an integral water repellent in basement concrete subjected for long periods to hydrostatic pressure may not, therefore, be advisable.

If slow seepage of moisture by capillarity and passage of water vapor into a basement is not objectionable, or if the effects of such moisture penetration are controlled by ventilation or air conditioning, the walls and floor may not require the protection afforded by a bituminous membrane.

The concrete should be dense and should contain a well-graded, low-absorptive aggregate. The cement factor should be not less than 6 bags of portland cement per cubic yard and the water-cement ratio preferably should not exceed 0.45 by weight. The minimum 28-day compressive strength should be 3,000 psi. For maximum density, the concrete should be carefully handled and vibrated into place. If not vibrated, the concrete should have a maximum slump of 5 in. Construction joints in wall footings and walls should be keyed and should otherwise be made as described in Art. 12-9. The walls and floors should be carefully cured to avoid cracking due to shrinkage. Furthermore, the walls should be kept damp until after they have been backfilled.

The use of shotcrete or trowel-applied mortar coatings, ¾ in. or more in thickness, to the outside faces of both monolithic concrete and masonry unit walls greatly increases their resistance to penetration of moisture. Such coatings cover and seal construction joints and other vulnerable joints in the walls against leakage. When applied in a thickness of 2 in. or more, they may be reinforced with welded-wire fabric. The reinforcement may reduce the incidence of large shrinkage cracks in the coating. However, the cementitious coatings do not protect the walls against leakage if the walls, and subsequently the coatings, are badly cracked as a result of unequal foundation settlement, excessive drying shrinkage, and thermal changes.

12-12. Bituminous Membrane. This is a waterproof barrier providing protection against the penetration of water under hydrostatic pressure and water vapor. To resist hydrostatic pressure, the membrane should be made continuous in the walls and floor of a basement. It also should be protected from damage during building operations and should be laid by experienced workmen under competent supervision. It usually consists of three or more alternate layers of hot, mopped-on asphalt or coal-tar pitch and plies of bituminous saturated felt or woven cotton fabric. The number of moppings exceeds the number of plies by one. An alternative is a ¹⁄₁₆- to ⅛-in.-thick layer of butyl rubber secured with a compatible adhesive.

Bituminous saturated cotton fabric is stronger and is more extensible than bituminous saturated felt but is more expensive and more difficult to lay. At least one or two of the plies in a membrane should be of saturated cotton fabric to provide strength, ductility, and extensibility to the membrane. Where vibration, temperature changes, and other conditions conducive to displacement and volume changes in the basement are to be expected, the relative number of fabric plies may be increased.

The minimum weight of bituminous saturated felt used in a membrane should be 13 lb per 100 sq ft. The minimum weight of bituminous saturated woven cotton fabric should be 10 oz per sq yd.

Although a membrane is held rigidly in place, it is advisable to apply a suitable primer over the surfaces receiving the membrane and to aid in the application of the first mopped-on coat of hot asphalt or coal-tar pitch.

Materials used in the membrane should meet the requirements of the following current American Society for Testing and Materials standards:

Creosote primer for coal-tar pitch—D43.

Primer for asphalt—D41.

Coal-tar pitch—D450.

Asphalt—D449.

Woven cotton fabric, bituminous saturated—D173.

Coal-tar saturated felt—D227.

Asphalt saturated felt—D226.

The number of plies of saturated felt or fabric increases with the hydrostatic head to which the membrane is to be subjected. Five plies is the maximum commonly used in building construction, but 10 or more plies have been recommended for pressure heads of 35 ft or greater. The thickness of the membrane crossing the wall footings at the base of the wall should be no greater than necessary to reduce possible settlement of the wall due to plastic flow in the membrane materials.

The amount of primer to be used may be about 1 gal per 100 sq ft. The amount of bitumen per mopping should be at least 4½ gal per 100 sq ft. The thickness of the first and last moppings is usually slightly greater than the thickness of the moppings between the plies.

The surfaces to which the membrane is to be applied should be smooth, dry, and at a temperature above freezing. Air temperature should be not less than 50°F. The temperature of coal-tar pitch should not exceed 300°F and asphalt, 350°F.

If the concrete and masonry surfaces are not sufficiently dry, they will not readily absorb the priming coat, and the first mopping of bitumen will be accompanied by bubbling and escape of steam. Should this occur, application of the membrane should be stopped and the bitumen already applied to damp surfaces should be removed.

The membrane should be built up ply by ply, the strips of fabric or felt being laid immediately after each bed has been hot-mopped. One of several methods of laying the plies is the solid or continuous method, in which a portion of each strip of fabric or felt is in contact with the supporting base of the membrane. The lap of succeeding plies or strips over each other is dependent on the width of the roll and the number of plies. In any membrane there should be some lap of the top or final ply over the first, initial ply. The American Railway Engineering Association requires a minimum distance of 2 in. for this lap (Fig. 12-3). End laps should be staggered at least 24 in., and the laps between succeeding rolls should be at least 12 in.

The floor membrane should be placed over a concrete base or subfloor whose top surface is troweled smooth and which is level with the tops of the wall footings. The membrane should be started at the outside face of one wall and extend over the wall footing, which may be keyed. It should cover the floor and tops of other footings to the outside faces of the other walls, forming a continuous horizontal waterproof barrier. The plies should project from the edges of the floor membrane and lap into the wall membrane.

The loose ends of felt and fabric must be protected; one method is to fasten them to a temporary vertical wood form about 2 ft high, placed just outside the wall face. Immediately after the floor membrane is laid, its surface should be protected and covered with a layer of portland-cement concrete, at least 2 in. thick.

After its completion, a wall membrane should be protected against damage and should be held in position by a facing of brick, tile, or concrete block. A brick facing should have a minimum thickness of 2½ in. Facings of asphalt plank, asphalt block, or mortar require considerable support from the membrane itself and give protection against abrasion of the membrane from lateral forces only. Protection against downward forces such as may be produced by settlement of the backfill is given only by the self-supporting masonry walls.

The kind of protective facing may have some bearing on the method of constructing the membrane. The membrane may be applied to the face of the wall after its construction, or it may be applied to the back of the protective facing before the main wall is built. The first of these methods is known as the outside application; the second is known as the inside application (Fig. 12-4).

For the inside application, a protective facing of considerable stiffness against lateral forces must be built, especially if the wall and its membrane are to be used as a form for the pouring of a main wall of monolithic concrete. Again, the inner face of the protecting wall must be smooth or else leveled with mortar to provide a suitable base for the membrane. The completed membrane should be covered with a ⅜-in.-thick layer of mortar to protect it from damage during construction of the main wall.

Application of wall membranes should be started at the bottom of one end of the wall and the strips of fabric or felt laid vertically. Preparation of the surfaces and laying of the membrane proceed much as they do with floor membranes. The surfaces to which the membrane is attached must be dry and smooth, which may require that the faces of masonry walls be leveled with a thin coat of grout or mortar. The plies of the wall membrane should be lapped into those of the floor membrane.

If the outside method of application is used and the membrane is faced with masonry, the narrow space between the units and the membrane should be filled with mortar as the units are laid. The membrane may be terminated at the grade line by a return into the superstructure wall facing.

Contraction joints in walls and floors containing a bituminous membrane should be the usual metal-bellows type. The membrane should be placed on the exposed face of the joint and it may project into the joint, following the general outline of the bellows.

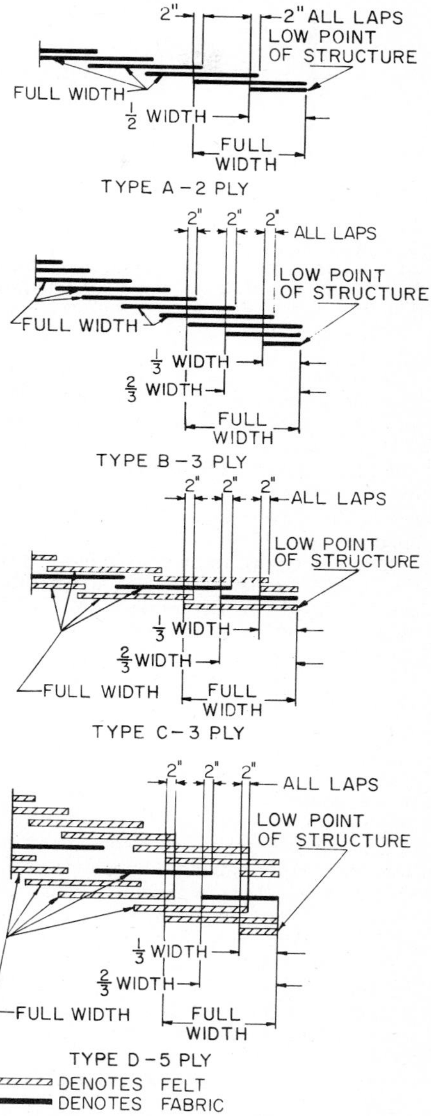

Fig. 12-3. Lapping of membranes recommended by American Railway Engineering Association.

The protective facing for the membrane should be broken at the expansion joint and the space between the membrane and the line of the facing filled with a bituminous plastic cement.

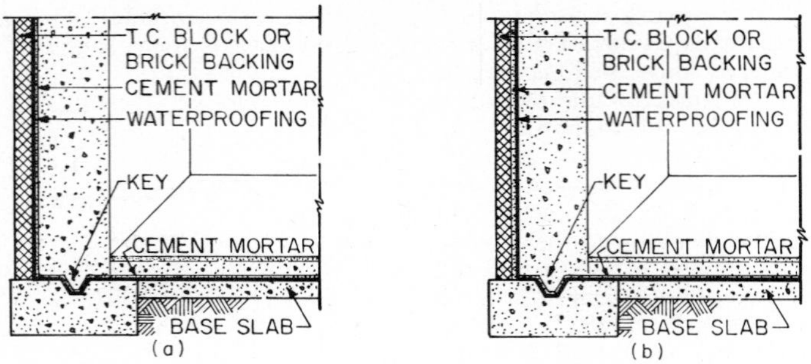

Fig. 12-4. Two methods of applying waterproofing membranes to walls. (*a*) Outside application (membrane applied to the wall, then protected by the backing). (*b*) Inside application (membrane applied to the backing, after which the wall is built).

Details at pipe sleeves running through the membrane must be carefully prepared. The membrane should be reinforced with additional plies and may be calked at the sleeve. Steam and hot-water lines should be insulated to prevent damage to the membrane.

12-13. Above-grade Water-resistant Concrete Walls. The rate of moisture penetration through capillaries in above-grade walls is low and usually of minor importance. However, such walls should not permit leakage of wind-driven rain through openings larger than those of capillary dimension.

Monolithic, cast-in-place concrete walls are usually of sufficient thickness to permit readily the proper placing of a workable concrete. Construction joints should be carefully made, as described in Art. 12-9. Cracking of the concrete due to volume changes resulting from a fluctuation in temperature and moisture content of the concrete may be controlled by the use of steel reinforcement.

Walls of cellular concrete, made of aggregates containing no fines, are highly permeable and require surface coatings of grout or mortar to prevent leakage of wind-driven rain.

Precast-concrete panels are usually made of dense highly water-resistant reinforced concretes. However, the walls made of these panels are vulnerable to leakage at the joints. In precast construction, edges of the panels may be recessed and the interior of vertical joints filled with grout after the panels are aligned.

Calking compound is commonly used as a facing for the joints. Experience has shown that calking compounds often weather badly; their use as a joint facing creates a maintenance problem and does not prevent leakage of wind-driven rain after a few years' exposure.

The amount of movement to be expected in the vertical joints between precast-concrete panels is a function of the panel dimension and the seasonal fluctuation in temperature and moisture content of the concrete. For such wall construction, it may be more feasible to use an interlocking water-resistant joint, faced on the weather side with mortar and backed with either a compressible premolded strip or calking.

("Guide to Joint Sealants for Concrete Structures," Report by Committee 504, American Concrete Institute.)

12-14. Above-grade Water-resistant Brick Walls. Brick walls 4 in. or more in thickness can be made highly water-resistant. The measures that need to be taken to insure there will be no leakage of wind-driven rain through brick facings are not extensive and do not require the use of materials other than those commonly used in masonry walls. The main factors that need to be controlled are the rate of "suction" of the brick at the time of laying and filling of all joints with mortar (Art. 10-5).

In general, the greater the number of brick leaves, or wythes, in a wall, the more water-resistant the wall. Walls of hollow masonry units are usually highly permeable, and brick-faced walls backed with hollow masonry units are greatly dependent upon the water resistance of the brick facing to prevent leakage of wind-driven rain.

Facing brick and building brick made of clay or shale should meet the current requirements of American Society for Testing and Materials Standards C216 and C62 respectively. Methods of sampling and testing the brick are given in ASTM Standard C67. Brick meeting the requirements for Grades SW and MW of Standards C62 and C216 should be satisfactory for use in water-resistant brick walls.

The water absorption of the brick during complete immersion from a dry condition is of minor importance compared with the rate of absorption of the brick at the time of laying. This rate can be reduced by wetting the brick before laying. Medium absorptive brick may need to be thoroughly soaked with water. Highly absorptive brick may require total immersion in water for some time before brick "suction" is reduced to the low limits needed. The test for the rate of absorption of brick is described in ASTM Standard C67. In this absorption test, the brick, in a flat position, is partly immersed in water to a depth of ⅛ in. for 1 min, and the absorption is calculated for a net equivalent area of 30 sq ft. For water-resistant masonry, the suction (rate of absorption) of the brick should not exceed 0.35, 0.5, and 0.7 oz, respectively, for properly constructed all-brick walls or facings of nominal 4-, 8-, and 12-in. thickness.

In general, method of manufacture and surface texture of brick do not greatly affect the permeability of walls. However, water-resistant joints may be difficult to obtain if the brick are deeply scored, particularly if the mortar is of a dry consistency. Loose sand should be brushed away or otherwise removed from brick that are heavily sanded.

Mortar to be used in above-grade, water-resistant brick-faced and all-brick walls should be of as wet a consistency as can be handled by the mason and meet requirements of ASTM Standard C270, Type N. Water retention of the mortar should be less than 75%, and preferably 80% or more. For laying absorptive brick that contain a considerable amount of absorbed water, the mortars having a water retention of 80% or more may be used without excessive "bleeding" at the joints and "floating" of the brick.

The mortar may contain a masonry cement meeting the requirements of ASTM Standard C91, except that water retention should not be less than 75 or 80%. Excellent mortar may also be made with portland cement and hydrated lime, mixed in the proportion of 1:1:6 parts by volume of cement, lime, and loose damp sand. The hydrated lime should be highly plastic. Type S lime conforming with the requirements of ASTM Standard C207 is highly plastic, and mortar containing it, in equal parts by volume with cement, will probably have a water retention of 80% or more.

Since capillary penetration of moisture through concrete and mortar is of minor importance, particularly in above-grade walls, the mortar need not contain an integral water repellent. However, if desired, water-repellent mortar may be advantageously used in a few courses at the grade line to reduce capillary rise of moisture from the ground into the masonry. To secure a strong water-resistant bond of the mortar to the brick, the mortar should be mixed to as wet a consistency as possible for the mason to handle. The mortar should be of a type that does not stiffen rapidly on the board, except through loss of moisture by evaporation.

The amount of wetting the brick will require to control rate of absorption properly when laid should be known or determined by measurements made before the brick are used in the wall. Some medium absorptive brick may only require frequent wetting in the pile; others may need to be totally immersed for an hour or more. While the immersion method is more costly than hosing in the pile, it insures that all brick are more or less saturated when removed from immersion. A short time interval may be needed before the brick are laid; but the brick are likely to remain on the scaffold, in a suitable condition to lay, for some time. Brick on the scaffold should be inspected and moisture condition checked several times a day.

Mortar should be retempered frequently if necessary to maintain as wet a consistency as is practically possible for the mason to use. At air temperatures below 80°F, mortar should be used or discarded within 3½ hr after mixing; for air temperatures of 80°F or higher, unused mortar should be discarded after 2½ hr.

The resistance of a brick facing to the penetration of rain may be increased if the mortar in the joints is compacted by tooling the joints concave with a rounded iron bar. The tooling should be done after the mortar has begun to stiffen but care should be exercised not to displace the brick.

12-15. Above-grade Water-resistant Concrete Masonry Walls. Exterior concrete masonry walls, without facings of brick, may be 8 or 12 in. thick and are usually highly permeable to leakage of wind-driven rain. A water-resistant concrete masonry wall is the exception rather than the rule, and protection against leakage is obtained by facing the walls with a cementitious coating of paint, stucco, or shotcrete.

For walls of rough-textured units, a portland cement–sand grout provides a highly water-resistant coating. The cement may be either white or gray.

Factory-made portland-cement paints containing a minimum of 65%, and preferably 80%, portland cement may also be used as a base coat on concrete masonry. Application of the paint should conform with the requirements of American Concrete Institute Standard 616. The paints, stuccos, and shotcrete should be applied to dampened surfaces. Shotcrete should conform with the requirements of ACI Standard 506.

Pneumatically applied coatings thinner than those discussed above and which contain durable-low absorptive aggregates are also highly water-resistant.

Shrinkage of concrete masonry due to drying and a drop in temperature may result in cracking of a wall and its cementitious facing. Such cracks readily permit leakage of wind-driven rain. The chief factor reducing incidence of shrinkage cracking is the use of dry block. When laid in the wall, the block should have a low moisture content, preferably one that is in equilibrium with the driest condition to which the wall will be exposed.

The block should also have a low potential shrinkage. See moisture-content requirements in ASTM C90 and method of test for drying shrinkage of concrete block in ASTM C426.

Formation of large shrinkage cracks may be controlled by use of steel reinforcement in the horizontal joints of the masonry and above and below wall openings. Where there may be a considerable seasonal fluctuation in temperature and moisture content of the wall, high-yield-strength, deformed-wire joint reinforcement should be placed in at least 50% of all bed joints in the wall.

Use of control joints faced with calking compound has also been recommended to control shrinkage cracking; however, this practice is marked by frequent failures to keep the joints sealed against leakage of rain. Steel joint reinforcement strengthens a concrete masonry wall, whereas control joints weaken it, and the calking in the joints requires considerable maintenance.

12-16. Water-resistant Cavity Walls. Cavity walls, particularly brickfaced cavity walls, may be made highly resistant to leakage through the wall facing. However, as usually constructed, facings are highly permeable, and the leakage is trapped in the cavity and diverted to the outside of the wall through conveniently located weep holes. This requires that the inner tier of the cavity be protected against the leakage by adequate flashings, and weep holes should be placed at the bottom of the cavities and over all wall openings.

Flashings should preferably be hot-rolled copper sheet of 10-oz minimum weight. They should be lapped at the ends and sealed either by solder or with bituminous plastic cement. Mortar should not be permitted to drop into the flashings and prevent the weep holes from functioning. The weep holes may be formed by the use of sash-cord head joints or ⅜-in.-diameter rubber tubing, withdrawn after the wall is completed.

12-17. Water-resistant Surface Treatments for Above-grade Walls. Experience has shown that leakage of wind-driven rain through masonry walls, particularly those of brick, ordinarily cannot be stopped by use of an inexpensive surface treatment or coating that will not alter the appearance of the wall. Such protective devices either have a low service life or fail to stop all leakage.

Both organic and cementitious pigmented coating materials, properly applied as a continuous coating over the exposed face of the wall, do stop leakage. Many of the organic pigmented coatings are vapor barriers and are therefore unsuitable for use on the outside, "cold" face of most buildings. If vapor barriers are used on the cold face of the wall, it is advisable to use a better vapor barrier on the warm face to reduce condensation in the wall and behind the exterior coating.

In Arts. 12-18 to 12-22, coatings for masonry are described and arbitrarily divided into five groups, as follows: (1) colorless coating materials; (2)cementitious coatings; (3) pigmented organic coatings; (4) joint treatment for brick masonry; and (5) bituminous coatings.

12-18. Colorless Coating Materials. The colorless "waterproofings" are often claimed to stop leakage of wind-driven rain through permeable masonry walls. Solutions of oils, paraffin wax, sodium silicate, chlorinated rubber, silicone resins, and salts of fatty acids have been applied to highly permeable test walls and have been tested at the National Bureau of Standards under exposure conditions simulating a wind-driven rain. Most of these solutions contained not more than 10% of solid matter. These treatments reduced the rate of leakage but did not stop all leakage through the walls. The test data show that colorless coating materials applied to permeable walls of brick or concrete masonry may not provide adequate protection against leakage of wind-driven rain.

Solutions containing oils and waxes tended to seal the pores exposed in the faces of the mortar joints and masonry units, thereby acting more or less as vapor barriers, but did not seal the larger openings, particularly those in the joints.

Silicone water-repellent solutions greatly reduced leakage through the walls as long as the treated wall faces remained water-repellent. After an exposure period of 2 or 3 hr, the rate of leakage gradually increased as the water repellency of the wall face diminished.

Coatings of the water-repellent, breather type, such as silicone and "soap" solutions, may be of value in reducing absorption of moisture into the wall surface. They may be of special benefit in reducing the soiling and disfiguration of stucco facings and light-colored masonry surfaces. They may be applied to precast-concrete panels to reduce volume changes that may otherwise result from changes in moisture content of the concretes. However, it should be noted that a water-repellent treatment applied to the surface may cause water, trapped in the masonry, to evaporate beneath the surface instead of at the surface. If the masonry is not water-resistant and contains a considerable amount of soluble salts, as evidenced by efflorescence, application of a water repellent may cause salts to be deposited beneath the surface, thereby causing spalling of the masonry. The water repellents therefore should be applied only to walls having water-resistant joints.

Emulsions of waxes in water have also been used as water-resistant coatings for permeable masonry walls. In tests of two emulsions at the National Bureau of Standards, one with 30% wax did not reduce leakage much, and the other, with 45% wax, was effective and stopped leakage through brick walls when tested soon after application. However, the treated walls were no longer water-resistant and leaked badly when again tested after exposure outdoors for a few months at Washington, D.C.

Claims for the effectiveness of colorless solutions and emulsions as waterproofings for masonry are often based on tests of small prisms of brick, stone, or concrete that have been treated and immersed in water. The treated specimens are found to absorb only a small percentage of the moisture taken up by similar but untreated specimens. Since highly permeable masonry walls, particularly brick masonry walls, leak at the joints and since the colorless solutions are known to fail in sealing such openings against leakage of wind-driven rain, the claims based on absorption tests are frequently fallacious. Furthermore, application of a colorless material makes the treated face of the masonry water-repellent and may prevent the proper bonding of a cementitious coating which could otherwise be used to stop leakage.

12-19. Cementitious Coatings. Coatings of portland-cement paints, grouts, and stuccos, and of pneumatically applied mortars are highly water-resistant. They are preferred above all other types of surface coatings for use as water-resistant base coatings on above-grade concrete masonry (see also Art. 12-15). They may also be applied to the exposed faces of brick masonry walls that have not been built to be water-resistant.

The cementitious coatings absorb moisture and are of the breather type, permitting passage of water vapor. Addition of water repellents to these coatings does not greatly affect their water resistance but does reduce the soiling of the surface from the absorption of dirt-laden water. If more than one coating is applied, as in a two-coat paint or stucco facing job, the repellent is preferably added only to the finish coat, thus avoiding the difficulty of bonding a cementitious coating to a water-repellent surface.

The technique used in applying the cementitious coatings is highly important. The backing should be thoroughly dampened. Paints and grouts should be scrubbed into place with stiff fiber brushes and the coatings should be properly cured by wetting. Properly applied, the grouts are highly durable; some grout coatings applied to concrete masonry test walls were found to be as water-resistant after 10 years out-of-doors exposure as when first applied to the walls.

12-20. Pigmented Organic Coatings. These include textured coatings, mastic coatings, conventional paints, and aqueous dispersions. The thick-textured and mastic coatings are usually spray-applied but may be applied by trowel. Conventional paints and aqueous dispersions are usually applied by brush or spray. Most of these coatings are vapor barriers but some textured coatings, conventional paints, and aqueous dispersions are breathers. Excepting the aqueous dispersions, all the coatings are recommended for use with a primer.

Applied as a continuous coating, without pinholes, the pigmented organic coatings are highly water-resistant. They are most effective when applied over a smooth backing. When they are applied with paintbrush or spray by conventional methods to rough-textured walls, it is difficult to level the surface and to obtain a continuous water-resistant coating free from holes. A scrubbed-on cementitious grout used as a base coat on such walls will prevent leakage through the masonry without the use of a pigmented organic coating.

The pigmented organic coatings are highly decorative but may not be so water-resistant, economical, or durable as the cementitious coatings.

12-21. Joint Treatments for Masonry. Leakage of wind-driven rain through the joints in permeable brick masonry walls can be stopped by either repointing or grouting the joints. It is usually advisable to treat all the joints, both vertical and horizontal, in the wall face. Some "tuck-pointing" operations in which only a few, obviously defective joints are treated may be inadequate and do not necessarily insure that the untreated joints will not leak.

Repointing consists of cutting away and replacing the mortar from all joints to a depth of about ⅝ in. After the old mortar has been removed, the dust and dirt should be washed from the wall and the brick thoroughly wetted with water to near saturation. While the masonry is still very damp but with no water showing, the joints should be repointed with a suitable mortar. This mortar may have a somewhat stiff consistency to enable it to be tightly packed into place, and it may be "prehydrated" by standing for 1 or 2 hr before retempering and using. Prehydration is said to stabilize the plasticity and workability of the mortar and to reduce the shrinkage of the mortar after its application to the joints.

After repointing, the masonry should be kept in a damp condition for 2 or 3 days. If the brick are highly absorptive, they may contain a sufficient amount of water to aid materially in curing.

Weathering and permeability tests described in C. C. Fishburn, "Effect of Outdoor Exposure on the Water Permeability of Masonry Walls," *National Bureau of Standards BMS Report* 76 indicate that repointing of the face joints in permeable brick masonry walls was the most effective and durable of all the remedial treatments against leakage that did not change the appearance of the masonry.

Joints are *grouted* by scrubbing a thin coating of a grout over the joints in the masonry. The grout may consist of equal parts by volume of portland cement and fine sand, the sand passing a No. 30 sieve.

The masonry should be thoroughly wetted and in a damp condition when the grout is applied. The grout should be of the consistency of a heavy cream and should be scrubbed into the joints with a stiff bristle brush, particularly into the juncture between brick and mortar. The apparent width of the joint is slightly increased by some staining of the brick with grout at the joint line. Excess grout may be removed from smooth-textured brick with a damp sponge, before the grout hardens; care should be taken not to remove grout from between the edges of the brick and the mortar joints. If the brick are rough-textured, staining may be controlled by the use of a template or by masking the brick with paper masking tape.

Bond of the grout to the joints is better for "cut" or flush joints than for tooled joints. If the joints have been tooled, they should preferably not be grouted until after sufficient weathering has occurred to remove the film of cementing materials from the joint surface, exposing the sand aggregate.

Grouting of the joints has been tried in the field and found to be effective on leaky brick walls. The treatment is not so durable and water-resistant as a repointing job but is much less expensive than repointing. Some tests of the water resistance of grouted joints in brick masonry test walls are described in *National Bureau of Standards BMS Report* 76.

The cost of either repointing or grouting the joints in brick masonry walls probably greatly exceeds the cost of the additional labor and supervision needed to make the walls water-resistant when built.

12-22. Bituminous Coatings. Bituminous cutbacks, emulsions, and plastic cements are usually vapor barriers and are sometimes applied as "dampproofers" on the inside faces of masonry walls. Plaster is often applied directly over these coatings, the bond of the plaster to the masonry being only of a mechanical nature. Tests show that bituminous coatings applied to the inside faces of highly permeable masonry walls, not plastered, will readily blister and permit leakage of water through the coating. It is advisable not to depend on such coatings to prevent the leakage of wind-driven rain unless they are incorporated in the masonry or held in place with a rigid self-sustaining backing.

Even though the walls are resistant to wind-driven rain, but are treated on their inner faces with a bituminous coating, water may be condensed on the warm side of the coating and damage to the plaster may result, whether the walls are furred or not. However, the bituminous coating may be of benefit as a vapor barrier in furred walls, if no condensation occurs on the warm side.

Section 13

Plaster and Gypsumboard

FREDERICK S. MERRITT

Consulting Engineer, Syosset, N.Y.

Plaster construction may consist of materials partly or completely prepared in the field, or of prefabricated sheets or tiles (dry-type construction). Factors such as initial cost, cost of maintenance and repair, fire resistance, sound control, decorative effects, and speed of construction must be considered in choosing between them.

When field-prepared materials are used, the plaster finish generally consists of a base and two or more coats of plaster. When dry-type construction is used, one or more plies of prefabricated sheet or tile may be combined to achieve desired results. For fire-resistance and sound-transmission ratings of plaster construction, see "Design Data—Fire Resistance," Gypsum Association, 201 North Wells St., Chicago, Ill. 60606.

13-1. Plaster Construction Terms.

Accelerator. Any material added to gypsum plaster that speeds the set.

Acoustical Plaster. A finishing plaster that corrects sound reverberations or reduces noise intensity.

Adhesives:

Contact. An adhesive that forms a strong, instantaneous bond between two plies of gypsumboard when the two surfaces are brought together.

Laminating. An adhesive that forms a slowly developing bond between two plies of gypsumboard or between masonry or concrete and gypsumboard.

Stud. An adhesive suitable for attaching gypsumboard to framing.

Admixture. Any substance added to a plaster component or plaster mortar to alter its properties. (See also Dope.)

Arris. A sharp edge forming an external corner at the junction of two surfaces.

Back Blocking. Support provided for gypsumboard butt joints that fall between framing members.

Back Plastering. Application of plaster to one face of a lath system after application and hardening of plaster applied to the opposite face (used for solid plaster partitions and curtain walls).

Band. A flat molding.

Base Coat. The plaster coat or combination of coats applied before the finish coat.

Bead. A strip of sheet metal usually with a projecting nosing, to establish plaster grounds, and two perforated or expanded flanges, for attachment to the plaster base, for use at the perimeter of a plaster membrane as a stop or at projecting angles to define and reinforce the edge.

Beaded Molding. A cast plaster string of beads set in a molding or cornice.

Beading. (See Ridging.)

Bed Mold or Bed. A flat area in a cornice in which ornamentation is placed.

Blisters. Protuberances on the finish coat of plaster caused by application over too damp a base coat, or troweling too soon.

Bond Plaster. A plaster formulated to serve as a first coat applied to monolithic concrete.

Boss. A Gothic ornament set at the intersection of moldings.

Brown Coat. Coat of plaster directly beneath the finish coat. In two-coat work, brown coat refers to the base-coat plaster applied over the lath. In three-coat work, the brown coat refers to the second coat applied over a scratch coat.

Buckles. Raised or ruptured spots that eventually crack, exposing the lath. Most common cause for buckling is application of plaster over dry, broken, or incorrectly applied wood lath.

Bull Nose. This term describes an external angle that is rounded to eliminate a sharp corner and is used largely at window returns and door frames.

Butterflies. Color imperfections on a lime-putty finish wall, caused by lime lumps not put through a screen, or insufficient mixing of the gaging.

Capital or Cap. The ornamental head of a column or pilaster.

Case Mold. Plaster shell used to hold various parts of a plaster mold in correct position. Also used with gelatin and wax molds to prevent distortions during pouring operation.

Casing Bead. A bead set at the perimeter of a plaster membrane or around openings to provide a stop or separation from adjacent materials.

Casts. (See Staff.)

Catface. Flaw in the finish coat comparable to a pock mark.

Ceilings:

Coffered. Ornamental ceilings composed of recessed panels between ribs.

Contact. Ceilings attached in direct contact with the construction above, without use of runner channels or furring.

Cross Furred. Ceilings applied to furring members attached at right angles to the underside of main runners or other structural supports.

Furred. Ceilings applied to furring members attached directly to the structural members of the building.

Suspended. Ceilings applied to furring members suspended below the structural members of the building.

Chamfer. A beveled corner or edge.

Chase. A groove in a masonry wall to provide for pipes, ducts, or conduits.

Check Cracks. Cracks in plaster caused by shrinkage, but the plaster remains bonded to its base.

Chip Cracks. Similar to check cracks, except the bond is partly destroyed. Also referred to as fire cracks, map cracks, crazing, fire checks, and hair cracks.

Corner Bead. A strip of sheet metal with flanges and a nosing at the junction of the flanges; used to protect arrises.

Corner Cracks. Cracks in joint of intersecting walls or walls and ceilings.

Corner Floating. (See Floating Angles.)

Cornerite. Reinforcement for plaster at a reentrant corner.

Cornice. A molding, with or without reinforcement.

Cove. A curved concave, or vaulted, surface.

Cure. Treatment, usually of a portland-cement plaster, to insure hydration after application.

Dado. The lower part of a wall usually separated from the upper by a molding or other device.

Darby. A flat wood tool with handles about 4 in. wide and 42 in. long; used to smooth or float the brown coat; also used on finish coat to give a preliminary true and even surface.

Dentils. Small rectangular blocks set in a row in the bed mold of a cornice.

Dope. Additives put in any type of mortar to accelerate or retard set.

Double-up. Applications of plaster in successive operations without a setting and drying interval between coats.

Dry Out. Soft chalky plaster caused by water evaporating before setting.

Efflorescence. White fleecy deposit on the face of plastered walls, caused by salts in the sand or backing; also referred to as "whiskering" or "saltpetering."

Egg and Dart. Ornamentation used in cornices consisting of an oval and a dart alternately.

Eggshelling. Plaster chip-cracked concave to the surface, the bond being partly destroyed.

Enrichments. Any cast ornament that cannot be executed by a running mold.

Expanded Metals. Sheets of metal that are slit and drawn out to form diamond-shaped openings.

Fat. Material accumulated on a trowel during the finishing operation of plaster and used to fill in small imperfections. Also denotes a mortar that is not too stiff, too watery, or oversanded.

Feather Edge. A beveled-edge wood tool used to straighten reentrant angles in the finish plaster coat. Also denotes a thin edge formed by tapering joint compound at a joint to blend with adjacent gypsumboard surface.

Fines. Aggregate capable of passing through a No. 200 sieve.

Finish Coat. Last and final coat of plaster.

Fisheyes. Spots in finish coat about ¼ in. in diameter caused by lumpy lime due to age or insufficient blending of material.

Float. A tool shaped like a trowel, with a handle braced at both ends and wood base for blade, used to straighten, level, and texture finish plaster coats.

Floating Angles (Corner Floating). Unrestrained surfaces intersecting at about 90°, usually with fasteners omitted near the intersection (Fig. 13-1).

Furring. Strips that are nailed over studs, joists, rafters, or masonry to support lath. This construction permits free circulation of air behind the plaster.

Gaging. Mixing of gaging plaster with lime putty to acquire the proper setting time and initial strength. Also denotes type of plaster used for mixing with the putty.

Green Plaster. Wet or damp plaster.

Grounds. A piece of wood, metal, or plaster attached to the framing to indicate the thickness of plaster to be applied.

Gypsum Base. Gypsum lath used as a base for veneer plasters.

Gypsumboard. A noncombustible board with gypsum core enclosed in tough, smooth paper.

Hardwall. Gypsum neat base-coat plaster.

Joint Compound. Material used to conceal fasteners, joints, and surface indentations in gypsumboard construction.

Joint Treatment. Concealing of gypsumboard joints, usually with tape and joint compound.

Joints:

Butt. Joints in which gypsumboard ends with core exposed (usually in the direction of the board width) are placed together.

Crown. Protrusion of joint compound from gypsumboard surface at a joint.

Keene's Cement. A dead-burned gypsum product that yields a hard, high-strength plaster.

Lamination:

Sheet. A ply of gypsumboard attached to another ply with adhesive over the entire surface to be bonded.

Strip. A ply of gypsumboard attached to another ply by parallel strips of adhesive, usually 16 or 24 in. apart.

Lath. A base to receive plaster.

Lime. Oxide of calcium produced by burning limestone. Heat drives out the carbon dioxide leaving calcium oxide, commonly termed "quicklime." Addition of water to quicklime yields hydrated or slaked lime.

Lime Plaster. Base-coat plaster consisting essentially of lime and aggregate.

Lime Putty. Thick paste of water and slaked quicklime or hydrated lime.

Marezzo. An imitation marble formed with Keene's cement to which colors have been added.

Neat Plaster. A base-coat plaster to which sand is added at the job.

Niche. A small recess in a wall.

Ogee. A curved section of a molding, partly convex and partly concave.

Pinhole. A small hole appearing in a cast because of excess water.

Putty Coat. A smooth, troweled-finish coat containing lime putty and a gaging material.

Quicklime. (See Lime.)

Relief. Ornamental figures above a plane surface.

Retarder. Any material added to gypsum plaster that slows its set.

Return. The terminal of a cornice or molding that takes the form of an external miter and stops at the wall line.

Reveal. The vertical face of a door or window opening between the face of the interior wall and the window or door frame.

Ridging. A linear surface protrusion along treated joints.

Scagliola. An imitation marble, usually precast, made with Keene's cement.

Scratch Coat. First coat of plaster in three-coat work.

Screeds. Long, narrow strips that serve as guides for plastering.

Skim Coat. (See Finish Coat.)

Slaking. Adding water to hydrate quicklime into a putty.

Soffit. The underside of an arch, cornice, bead, or other construction.

Splay Angle. An angle of more than 90°.

Staff (Casts). Plaster casts made in molds and reinforced with fiber; usually wired or nailed into place.

Stucco. Plaster applied to the exterior of a building.

Sweat Out. Soft, damp wall area caused by poor drying conditions.

Temper. Mixing of plaster to a workable consistency.

Template. A gage, pattern, or mold used as a guide to produce arches, curves, and various other shapes.

Veneer Plasters. Gypsum plasters meeting requirements of ASTM C587, and that may be applied in one or more coats to a maximum thickness of ¼ in.

Wadding. The act of hanging staff by fastening wads made of plaster of paris and excelsior or fiber to the casts and winding them around the framing.

Wainscot. The lower 3 or 4 ft of an interior wall when it is finished differently from the remainder of the wall.

White Coat. A gaged lime-putty troweled-finish coat.

13-2. Isolation and Control Joints. Plaster construction has low resistance to stresses induced by structural movements. It also is subject to dimensional changes because of changes in temperature and humidity. If proper provision is not made in the application of plaster systems for these conditions, cracking or chipping may occur or fasteners may pop out.

Generally, such defects may be prevented by applying plaster systems so that movement is not restrained. Thus, lath and plaster surfaces should be isolated by control joints, floating angles, or other means where a partition or ceiling abuts any structural element, except the floor, or meets a dissimilar wall or partition assembly, or other vertical penetration. Isolation also is advisable where the construction changes in the plane of the partition or ceiling.

In long walls or partitions, control joints should be inserted no more than 30 ft apart. In large ceiling areas, control joints should not be spaced more than 50 ft apart. At control joints, continuity of both lath and plaster or of gypsumboard should be interrupted. Control joints may be conveniently located where door frames extend from floor to ceiling or along lighting fixtures, heating vents, or air-conditioning diffusers, where continuity would ordinarily be broken.

A floating angle is desirable at intersections of ceilings and walls or partitions

(Fig. 13-1). It is constructed by installing gypsum lath or gypsumboard first in the ceiling. The first line of ceiling fasteners should be spaced about 8 in. from the wall. Normal fastener spacing should be used in the rest of the ceiling away from the wall. Gypsum lath or gypsumboard should then be applied to the wall with firm contact at the ceiling line to support the edges of the ceiling panels. The first line of fasteners for the wall panels should be spaced about 8 in. from the ceiling.

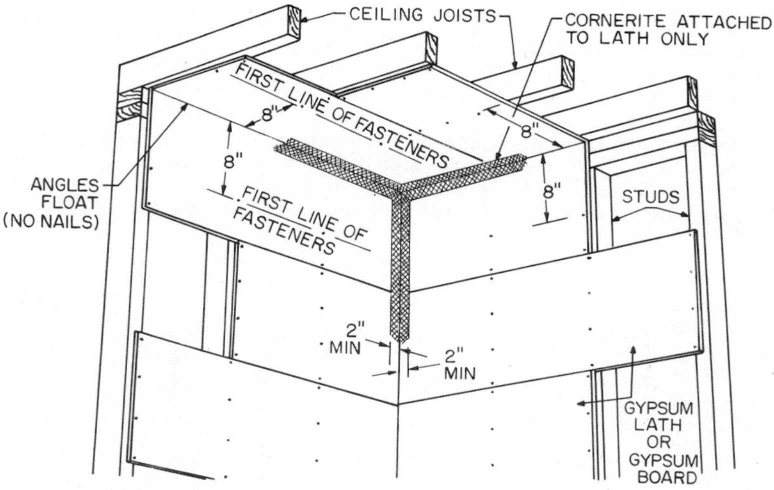

Fig. 13-1. Floating-angle construction at intersection of walls and ceiling.

WET-TYPE CONSTRUCTION

13-3. Gypsum Bases. One commonly used base for plaster is gypsum lath. This is a noncombustible sheet generally 16 × 48 in. by ⅜ or ½ in. thick, or 16 × 96 in. by ⅜ in. thick. It is composed principally of calcined gypsum that has been mixed with water, hardened, dried, and then sandwiched between two paper sheets (ASTM C37). Perforated gypsum lath is produced, during the manufacturing process, by punching ¾-in.-diam holes through the lath 4 in. c to c in perpendicular directions. Such lath has high fire resistance, because it provides both mechanical and chemical bond with the base coat of plaster. Insulating gypsum lath is made by cementing a sheet of shiny aluminum to the back of plain lath. It is used for vapor control, and for insulation against heat loss or gain.

When nailed to wood members, ⅜-in. lath should be attached with four nails, and ½-in. lath with five nails, to each framing member covered. Nails should be blued gypsum lath nails, made of 13-gage wire, 1⅛ in. long, with a 19/64-in.-diam flat head. They should be driven until the head is just below the paper surface without breaking the paper. Lath also may be attached to wood framing with four or five 16-gage staples, 7/16 in. wide, with ⅜-in. divergent legs. With metal framing, screws should be used, as recommended by the lath manufacturer. Clips, however, are a suitable alternative for use with wood or metal framing. Fasteners should be driven at least ⅜ in. away from ends and edges. Clips must secure the lath to framing at each intersection with the framing.

Studs or ceiling members supporting lath may be spaced up to 16 in. c to c with ⅜-in. gypsum lath, and up to 24 in. c to c with ½-in. lath.

The lath should be applied to studs with long dimension horizontal, and vertical joints should be staggered (Fig. 13-2). In ceilings, the long sides should span supports. Ends should rest on or be nailed to framing, headers, or nailing blocks.

Each lath should be in contact with adjoining sheets, but if spaces more than ½ in. wide are necessary, the plaster should be reinforced with self-furring metal lath stapled or tied with wire to the gypsum lath.

Except at intersections that are to be unrestrained, reentrant corners should be reinforced with cornerite stapled or tied with wire to the gypsum lath. Exterior-angle corners should be finished with corner beads set to true grounds and nailed or tied with wire to the structural frame or to furring. Casing beads should be used around wall openings and at intersections of plaster with other finishes and of lath and lathless construction.

Gypsum Base for Veneer Plasters. Special gypsum lath meeting requirements of ASTM C588 is required as a base for veneer plasters. It is formulated to provide the strength and absorption necessary for proper application and performance of these thin coatings. The lath comes in thicknesses of ⅜ (for two-coat systems), ½, and ⅝ in., the last permitting 24-in. spacing of wood framing. In general, gypsum base should be applied first to the ceiling, then to the walls. Maximum spacing of nails is 7 in. on ceilings, and 8 in. on walls. Screw spacing should not exceed 12 in. for wood framing 24 in. c to c or for steel framing or for wood ceiling framing, or 16 in. for wood studs spaced 16 in. c to c.

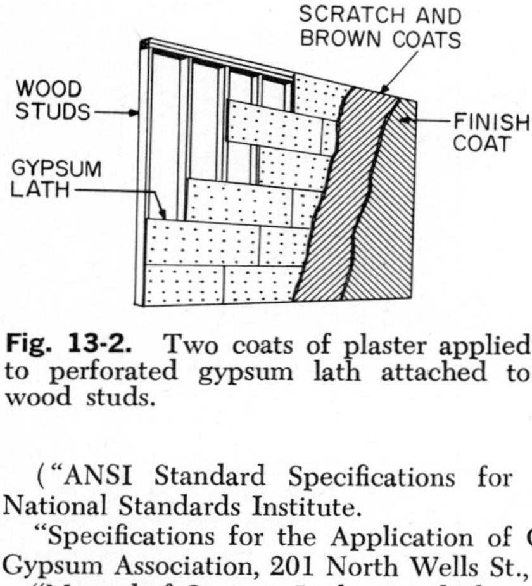

Fig. 13-2. Two coats of plaster applied to perforated gypsum lath attached to wood studs.

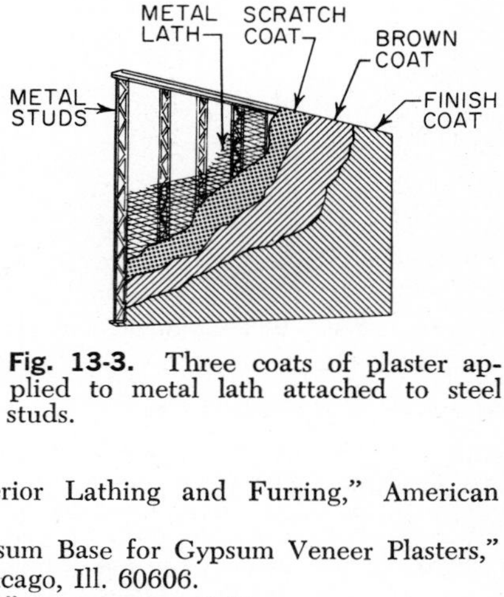

Fig. 13-3. Three coats of plaster applied to metal lath attached to steel studs.

("ANSI Standard Specifications for Interior Lathing and Furring," American National Standards Institute.

"Specifications for the Application of Gypsum Base for Gypsum Veneer Plasters," Gypsum Association, 201 North Wells St., Chicago, Ill. 60606.

"Manual of Gypsum Lathing and Plastering," Gypsum Association.

"Red Book of Lathing and Plastering," United States Gypsum Company, 101 South Wacker Drive, Chicago, Ill. 60606).

13-4. Metal Lath. A metal base often is used in plaster construction because it imparts strength and resists cracking. Plaster holds to metal lath by mechanical bond between the initial coat of plaster and the metal. So it is important that the plaster completely surrounds and embeds the metal.

Basic types commonly used are expanded-metal, punched sheet-metal, and paper-backed welded-wire lath. Woven-wire lath may be used as a supplemental reinforcement over solid plaster bases, but not as the primary base for gypsum plaster. Wire lath should be made of galvanized, copper-bearing steel. The other types should be made of copper-bearing steel with a protective coat of paint or of galvanized steel.

Expanded-metal lath is fabricated by slitting sheet steel and expanding it to form a mesh. Several types are available:

Diamond-mesh lath, with more than 11,000 meshes per sq yd, is an all-purpose lath, suitable as a base for flat or curved plaster surfaces (Fig. 13-3). The small meshes are helpful in reducing droppings of plaster during plastering. A self-furring type also is available. When it is attached to a backing, it is separated from the backing by at least ¼ in. Thus, self-furring lath is convenient for use as exterior stucco bases and bases for column fireproofing and replastering over old surfaces.

Flat-rib expanded-metal lath comes with smaller openings than diamond mesh, and it has ribs parallel to the length of the sheet that make it more rigid. Flat-rib lath is generally preferred for nailing to wood framing and tying to framing for flat ceilings, but it is not suitable for contour lathing.

High-rib expanded-metal lath is used when greater rigidity is desired, for example, for spacing supports up to 24 in. c to c and for solid, studless plaster partitions. The lath has a herringbone mesh pattern and V-shaped ribs running the length of each sheet. For ⅜-in. rib lath, ⅜-in.-deep ribs are spaced 4½ in. c to c, alternating with inverted ³⁄₁₆-in. ribs. For ¾-in. rib lath, ¾-in.-deep ribs are spaced 6 in. c to c. This type of high-rib lath may be used as a form and reinforcement

Table 13-1. Limiting Spans for Metal Lath, In.

Type of lath	Min wt., lb per sq yd	Vertical supports			Horizontal supports	
		Wood	Metal		Wood or concrete	Metal
			Solid partitions*	Others		
Diamond mesh (flat expanded) metal lath	2.5	16	16	12	0	0
	3.4	16	16	16	16	13½
Flat- (⅛-in.) rib expanded metal lath. . .	2.75	16	16	16	16	12
	3.4	19	24	19	19	19
⅜-in. rib expanded metal lath†	3.4	24	. . .	24	24	19
	4.0	24	. . .	24	24	24
¾-in. rib expanded metal lath.	5.4	. . .	†	24	36‡	36‡
Sheet-metal lath†	4.5	24	. . .	24	24	24
Welded-wire lath	1.4§	16	16	16	16	16
	1.95¶	24	24	24	24	24

* For paper-backed lath, only absorbent, perforated, or slotted paper separator should be used.

† Permitted for studless solid partitions.

‡ Permitted only for contact or furred ceilings.

§ Welded 16-gage wire, paper-backed lath.

¶ Paper-backed lath with welded wire, face wires 16 gage, every third back wire parallel to line dimension of lath 11 gage.

SOURCE: Based on Standard A42.4, American National Standards Institute.

for concrete slabs, but its thickness makes it generally unsuitable for plaster construction. The ⅜-in. rib lath may also be used as a concrete form, but its rigidity makes it unsuitable for contour lathing.

Welded-wire lath should be made of wire 16 gage or thicker, forming 2×2-in. or smaller meshes, stiffened continuously parallel to the long dimension of the sheet at intervals not exceeding 6 in. The paper backing should comply with Federal Specification UU-B-790, "Building Paper, Vegetable Fiber (Kraft, Waterproofed, Water Repellant, and Fire Resistant)." Acting as a base to which plaster can adhere while hardening about the wire, the backing should permit full embedment in at least ⅛ in. of plaster of more than half the total length and weight of the wires.

Table 13-1 lists limiting spans for various types and weights of metal lath for ceilings and walls.

Tying and Nailing. Attachments of metal lath to supports should not be spaced farther apart than 6 in. along the supports.

When metal framing or furring is used, metal lath should be tied to it with 18-gage, or heavier, galvanized soft-annealed wire. Rib lath, however, should be attached to open-web steel joists with single loops of 16-gage, or heavier, wire, or double loops of 18-gage wire, with the ends of each loop twisted together. Also, rib lath should be tied to concrete joists with loops of 14-gage, or heavier, wire or with wire hangers not less than 10 gage.

With wood supports, diamond-mesh, flat-rib, and welded-wire lath should be attached to horizontal framing with 1½-in., 11-gage, 7⁄16-in.-head, barbed, galvanized, or blued roofing nails, driven full length. For vertical wood supports the following may be used: 4d common nails; 1-in., 14-gage wire staples driven full length; and 1-in. roofing nails driven at least ¾ in. into the supports. Common nails should be bent over to engage a rib or at least three strands of lath. Alternatives of equal strength also may be used.

Metal lath should be applied with long sides of sheets spanning supports. Each sheet should underlap or overlap adjoining sheets on both sides and ends. Expanded-metal and sheet-metal lath should be lapped ½ in. along the sides, or have edge ribs nested, and 1 in. along the ends. Welded-wire lath should be lapped one mesh at sides and ends. All side laps of metal lath should be fastened to supports and tied between supports at intervals not exceeding 9 in.

Wherever possible, end laps should be staggered, and the ends should be placed at and fastened to framing. If end laps fall between supports, the adjoining ends should be laced or securely tied with 18-gage, galvanized, annealed steel wire.

Normally, metal lath should be applied first to ceilings. Flexible sheets may be carried down 6 in. on walls and partitions. As an alternative, preferable for more rigid sheets, sides and ends of the lath may be butted into horizontal reentrant angles and the corner reinforced with cornerite. But for large ceilings (length exceeding 60 ft in any direction, or more than 2,400 sq ft in area) and other cases in which restraint should be avoided, and for portland-cement plaster ceilings, cornerite should not be used. Instead, the abutting sides and ends should terminate at a casing bead, control joint, or similar device that will isolate the ceiling lath and plaster from the walls and partitions.

Similar considerations govern installation of metal lath at vertical reentrant corners. Between partitions, flexible sheets may be bent around vertical corners and attached at least one support away from them; more rigid sheets may be butted and the corner reinforced with cornerite. But where restraint is undesirable at reentrant corners, for example, where partitions meet structural walls or columns, or where loadbearing walls intersect, cornerite should not be used. Instead, the walls, partitions, and structural members should be isolated from each other, as described for ceiling-wall corners.

("ANSI Standard Specifications for Interior Lathing and Furring," American National Standards Institute.

Technical Bulletins, Metal Lath Association, West Federal St., Niles, Ohio 44446.

"Red Book of Lathing and Plastering," United States Gypsum Company, 101 South Wacker Drive, Chicago, Ill. 60606.)

13-5. Masonry Bases. Gypsum partition tile has scored faces to provide a mechanical bond as well as the natural bond of gypsum to gypsum. The 12 × 30-in. faces of the tile present an unwarped plastering surface because the tile is dried without burning; so a mechanic can lay a straighter wall than with other types of units. The Gypsum Association's "Manual of Gypsum Lathing and Plastering" recommends that only gypsum plaster be applied to gypsum partition tile, since lime and portland cement do not bond adequately. Also, only gypsum mortar should be used for tile erection.

Brick and clay tile can be used as a plaster base if they are not smooth-surfaced or of a nonporous type. If the surface does not provide sufficient suction, it should offer a means for developing a mechanical bond, such as does scored tile.

Plaster should not be applied directly to exterior masonry walls because dampness may damage the plaster. It is advisable to fur the plaster at least 1 in. in from the masonry.

Properly aged concrete block may serve as a plaster base, but for block ceilings, a bonding agent or a special bonding plaster should be applied first.

For precast or cast-in-place concrete with smooth dense surfaces, a bonding agent or a special bonding plaster should be used first. But if a plaster thickness of more than ⅜ in. is required for concrete ceilings, or ⅝ in. for concrete walls, metal lath should be secured to the concrete before plastering, in which case sanded plaster can be used.

13-6. Plaster Base Coats. The base coat is the portion of the plaster finish that is applied to masonry or lath bases and supports the finish coat. For typical thicknesses, see Art. 13-7.

Except for veneer plasters, plaster applications may be three-coat (Fig. 13-3) or two-coat (Fig. 13-2). The former consists of (1) a scratch coat, which is applied directly to the plaster base, cross-raked after it has "taken up," and allowed to set and partly dry; (2) a brown coat, which is surfaced out to the proper grounds, darbied (float-finished), and allowed to set and partly dry; and (3) the finish coat. Three-coat plaster is required over metal lath, ½-in. gypsum lath spanning horizontal supports more than 16 in. c to c, all gypsum lath attached by clips providing only edge support, and ⅜-in. perforated gypsum lath on ceilings.

The two-coat application is similar, except that cross-raking of the scratch coat is omitted and the brown coat is applied within a few minutes to the unset scratch coat. Three coats are generally preferred, because the base coat thus produced is stronger and harder.

Veneer plaster applications, 1⁄16 in. thick, may be one-coat or two-coat, both applied to special gypsum base (Art. 13-3). The single coat is composed of a scratch coat without cross-raking and an immediately applied finish.

See also Art. 13-13, and "Manual of Gypsum Lathing and Plastering," Gypsum Association.

13-7. Plaster Grounds. Except for veneer plasters, thickness of base-coat plaster should be controlled with grounds—wood or metal strips applied at the perimeter of all openings and at baseboards or continuous strips of plaster applied at intervals along a wall or ceiling, to serve as screeds. Plaster screeds should be used on all plaster surfaces of large area.

Grounds should be set to provide a minimum plaster thickness of ½ in. over gypsum lath and gypsum partition tile; ⅝ in. over brick, clay tile, or other masonry; and ⅝ in. from the face of metal lath. A thickness of 1⁄16 in. should be allowed for the finish coat.

13-8. Plaster Ingredients. Plasters other than veneer plasters are generally composed of portland cement, or gypsum and lime; an aggregate (sand, vermiculite, perlite); and water. Lime mortar also may be used. See also Arts. 13-11 to 13-13.

Hair or sisal fibers may be added to some scratch-coat plasters for application to metal lath, to limit the amount of plaster that passes through the lath meshes to what is needed for good bond. Fibering, however, adds no strength.

Sand should comply with ASTM C35. It should be clean, free of organic material, more than about 5% clay, silt, or other impurities, and should not contain salt or alkali. The proportion of sand in the plaster has an important bearing on the characteristics of the product. Oversanding results in considerable reduction in strength and hardness. A mix as lean as 4:1 by weight should never be used.

The Gypsum Association suggests that a 1-cu-ft measuring box be used for preparing mixes. Some plasterers use a No. 2 shovel, which holds about 16 lb of moist sand, for maintaining proper proportions. Thus, with each 100-lb bag of plaster, a 1:2 mix requires 12 shovels of sand and a 1:3 mix 18 shovels ("Manual of Gypsum Lathing and Plastering").

Water should be clean and free of substances that might affect the rate of set of the plaster. It is not advisable to use water in which plasterers' tools have been washed because it might change the set. Excessive water is undesirable in the mix, because when the water evaporates, it leaves numerous large voids, which decrease the strength of the plaster. Hence, manufacturers' recommendations should be observed closely in determining water requirements.

Perlite and vermiculite are manufactured lightweight aggregates that are used to produce a lightweight plaster with relatively high fire resistance for a given thickness. Both aggregates should conform with ASTM C35, and the mix should be prepared strictly in accordance with manufacturers' recommendations.

13-9. Mixing Plaster. A mechanical mixer disperses the ingredients of a mix more evenly and therefore is to be preferred over box mixing. Recommended practice is as follows: (1) Place the anticipated water requirements in the mixer;

(2) add about half the required sand (or all required perlite or vermiculite); (3) add all the plaster; (4) add the rest of the sand; (5) mix at least 30 sec, but not more than 3 min, adding water, if necessary, to obtain proper workability; and (6) dump the entire batch at once.

The mixer should be thoroughly cleaned when it is not in use. If partly set material is left in it, the set of the plaster might be accelerated. For this reason also, tools should be kept clean.

For hand mixing, first sand and plaster should be mixed dry to a uniform color in a mixing box, water added, and the plaster hoed into the water immediately and thoroughly mixed. Undermixed plaster is difficult to apply and will produce soft and hard spots in the plastered surface.

Plaster should not be mixed more than 1 hr in advance. Nor should a new mix, or gaging, be mixed in with a previously prepared one. And once plaster has started to set, it should not be remixed or retempered.

13-10. Plaster Drying. A minimum temperature of 55°F should be maintained in the building where walls are to be plastered when outdoor temperatures are less than 55°F, and held for at least 1 week before plaster is applied and 1 week after the plaster is dry.

In hot, dry weather, precautions should be taken to prevent water from evaporating before the plaster has set. Plastered surfaces should not be exposed to drafts, and openings to the outside should be closed off temporarily. After the plaster sets, the excess moisture it contains evaporates. Hence, the room should be adequately ventilated to allow this moisture to escape.

13-11. Portland-cement Plaster and Stucco. Portland cement, mixed with lime putty and sand, is used when a hard, sand-finish, fire-retarding, non-water-absorbent plaster is desired, and for all coats of stucco. (Portland-cement plaster and stucco are essentially the same. Stucco is usually associated with exterior applications, plaster with interior applications.) Portland-cement plaster also is used for scratch and brown coats under Keene's cement finish, and for scratch and brown coats that serve as a backing for tile wainscots. A typical mix for a coat consists of 1 part portland cement, ¼ part lime putty or hydrated lime, and 3 parts sand by weight. See also Art. 10-19.

("Guide to Portland Cement Plastering," report of Committee 524, American Concrete Institute; "Plasterer's Manual," Portland Cement Association.)

13-12. Gypsum Base-coat Plasters. (See also Art. 13-6.) Two types of gypsum base-coat plasters are in general use: gypsum neat plaster and gypsum ready-mixed plaster. Also, veneer plasters may be used in thin one-coat or two-coat systems. In one-coat systems, the finish veneer plaster is also the base-coat plaster. In both veneer-plaster systems, a base-coat veneer plaster should be applied to a special gypsum base (Art. 13-3) to a thickness of 1/16 to 3/32 in., and left with a rough surface to receive the finish coat. Veneer plasters should meet the requirements of ASTM C587. (See "Recommended Specifications for Gypsum Plastering" and "Specifications for the Application of Gypsum Veneer Plaster," Gypsum Association.)

Gypsum neat plaster, sometimes called hardwall or gypsum-cement plaster, is sold in powder form and mixed with an aggregate and water at the construction site. Mixed with no more than 3 parts sand by weight, it makes a strong base coat at low cost. Scratch coats generally consist of 1 part plaster powder to 2 parts sand by weight, fibered or unfibered; the base coat in two-coat work usually is a 1:2½ mix; brown coats are 1:3 mixes. With perlite or vermiculite instead of sand, a 1:2 mix may be used.

Gypsum ready-mixed plaster requires the addition only of water at the site, since it is sold in bags containing the proper proportions of aggregate and plaster. It is specified when good plastering sand is high cost or not available, or to avoid the possibility of oversanding. It is equal in strength to gypsum neat plaster, but costs a little more because of the extra cost of transporting the sand.

The water ratio for base-coat plasters should be such that slump does not exceed 4 in. when tested with a 2 × 4 × 6-in. cone at the mixer, for mixes with sand proportions not exceeding those given for gypsum neat plaster.

Total thickness of base coat for other than veneer plasters should be at least the values given in Art. 13-7. The scratch coat (Figs. 13-2 and 13-3) applied to lath should be laid on with enough pressure to form a strong clinch or key. (The lath should be wetted before the scratch coat is applied, or strength will be reduced.) The coat should cover the lath to a thickness of ¼ in. For two-coat systems, the brown coat is applied immediately. For three-coat systems, after the surface has been trued, the scratch coat should be scratched horizontally and vertically with a toothed tool to form a good bonding surface, then left to dry partly. When the surface is so hard that the edges of the scoring do not yield easily under the pressure of a thumbnail, the brown coat may be applied. Hardening may take at least 1 day, and sometimes as long as 1 week, depending on the weather and the amount of lime in the mix.

The brown coat not only forms the base for the finish coat, but is also the straightening coat. When plaster grounds are used, the plaster should be laid on with a steel float, trued, and finally roughened, in preparation for the finish coat. With plaster screeds, the brown coat should be applied until flush with the screeds. It is put on with a darby, a two-handed wood float. The brown coat should be cross-raked or scratched to receive the finish coat.

13-13. Finish-coat Plasters. Several types of plasters are available for the finish coat (Figs. 13-2 and 13-3). In two of the most common—made with gaging plaster or Keene's cement—lime is an important ingredient, because it gives plasticity and bulk to the coat.

The lime is prepared first, being slaked to a smooth putty, then formed on the plasterer's board into a ring with water in the center. Next, gaging plaster or Keene's cement is gradually sifted into the water. Then, aggregates, if required, are added. Finally, all ingredients are thoroughly mixed and kneaded.

Gaging plasters are coarsely ground gypsum plasters, which are available in quick-setting and slow-setting mixtures; so it is not necessary to add an accelerator or a retarder at the site. Gaging plasters also are supplied as white gaging plaster and a slightly darker local gaging plaster. Finish coats made with these plasters are amply hard for ordinary usages and are the lowest-cost plaster finishes. However, they are not intended for ornamental cornice work or run moldings, which should be made of a finer-ground plaster. Gypsum gaging plasters should conform with ASTM C28.

Typical mixes consist of 3 parts lime putty to 1 part gaging plaster, by volume. If a harder surface is desired, the gaging content may be increased up to 1 part gaging to 2 parts lime putty.

The lime-gaging plaster should be applied in at least two coats, when the brown coat is nearly dry. The first coat should be laid on very thin, with sufficient pressure to be forced into the roughened surface of the base coat. After the first coat has been allowed to draw a few minutes, a second or leveling coat, also thin, should be applied, and after this has been allowed to draw, the third coat, if desired, added.

The base coat draws the water from the finish coats; so the finished surface should be moistened with a wet brush as it is being troweled. Pressure should be exerted on the trowel to densify the surface and produce a smooth hard finish. Finally, the surface should be dampened with the brush and clean water. It should be allowed to stand at least 30 days before oil paints are applied.

When a harder surface is required, or one with greater resistance to dampness than can be obtained with lime-gaging plaster, such as for gymnasiums, school corridors, trucking areas, and bathrooms, **Keene's cement** may be used with lime. The regular cement sets initially in about 1½ hr and has a final set of about 4 to 6 hr, but a quicker-setting variety is available with a 2-hr set. Hardness of the surface depends on the lime content and the vigor with which the surface is troweled. Keene's cement should conform with ASTM C61.

The base coat must be dry before a Keene's-cement finish is applied. If suction is too great, however, the surface should be moistened with a wet brush. The plaster should be applied in thin coats in the same manner as with lime-gaging plasters, except that Keene's-cement finishes may require additional troweling.

Prepared gypsum trowel finishes also are available that require only addition of water at the site. The resulting surface may be harder than obtainable with

Keene's cement and may be decorated as soon as dry. The plaster is applied in the same manner as lime-gaging plaster, but as with Keene's cement finishes, the base coat should be dry. However, the plaster has a moderately fast set and should be troweled before it sets. For best results, three very thin coats should be applied and water should be used sparingly.

Sand float finishes are similar to gypsum trowel finishes, except that these float finishes contain a fine aggregate to yield a fine-textured surface and the final surface is finished with a float. The base coat should be firm and uniformly damp when the finish coat is applied. These finishes have high resistance to cracking.

Molding plaster, intended for ornamental work, is made with a finer grind than other gaging plasters. It produces a smooth surface, free from streaks or indentations as might be obtained with coarser-ground materials. Equal parts of lime putty and molding plaster are recommended by the Gypsum Association for cornice moldings.

Veneer plasters, applied to a thickness of only $\frac{1}{16}$ to $\frac{3}{32}$ in., develop hard, strong (3,000-psi compressive strength) surfaces that can be decorated the day after application. Factory prepared, these plasters are easy to work, have high plasticity, and provide good coverage. They may be applied in one-coat systems over special gypsum base (Arts. 13-3 and 13-12), or in two-coat systems over a veneer-plaster or sanded gypsum base coat. They require only addition of water on the job. Veneer plasters should meet the requirements of ASTM C587.

("Manual of Lathing and Plastering," "Recommended Specifications for Gypsum Plastering," and "Specifications for the Application of Gypsum Veneer Plaster," Gypsum Association, 201 North Wells St., Chicago, Ill. 60606.

"Red Book of Lathing and Plastering," United States Gypsum Company, 101 South Wacker Drive, Chicago, Ill. 60606.)

DRY-TYPE CONSTRUCTION

To avoid construction delays due to the necessity of waiting for plaster to dry or to obtain different types of finishes, dry-wall construction may be used. Rigid or semirigid boards, some of which require no additional decorative treatment, are nailed directly to studs or masonry or to furring.

Joints may be concealed or accentuated according to the architectural treatment desired. The boards may interlock with each other, or battens, moldings, or beads may be applied at joints.

When the boards are thinner than conventional lath and plaster, framing members should be furred out to conventional thicknesses so that stock doors and windows can be used.

13-14. Plywood Finishes. Available with a large variety of surface veneers, including plastics, plywood is nailed directly to framing members or furring. Small finish nails, such as 4d, should be used, with a spacing not exceeding 6 in. All edges should be nailed down, intermediate blocking being inserted if necessary. The plywood should also be nailed to intermediate members. When joints are not to be covered, the nailheads should be driven below the surface (set). With wood battens, ordinary flat-headed nails can be used. Water-resistant adhesive may be applied between plywood and framing members for additional rigidity.

13-15. Asbestos-cement Boards. Like other types of boards, asbestos-cement can be nailed directly to framing members or furring. Thin sheets usually are backed with plywood or insulation boards to increase resistance to impact. Joints generally are covered with moldings or beads.

13-16. Fiber and Pulp Boards. Fabricated with a wide variety of surface effects, fiber and pulp boards also are available with high acoustic and thermal insulation values. Application is similar to that for plywood. Generally, adjoining boards should be placed in moderate contact with each other, not forced.

13-17. Gypsumboard. Composed of a gypsum core enclosed in smooth, tough paper or other sheet material, gypsumboard generally comes in widths of 4 ft, lengths of 8 to 12 ft, and thickness of $\frac{1}{4}$ to $\frac{5}{8}$ in. Boards $\frac{1}{2}$ in. thick are adequate for 16- to 24-in. spacing of supports.

Several different types of gypsumboard are available. They include various types of wallboard, backing board, coreboard, fire-resistant gypsumboard, water-resistant gypsumboard, gypsum sheathing, and gypsum formboard.

Gypsum Wallboard. This type is used for the surface layer on interior walls and ceilings. Regular gypsum wallboard comes with gray liner paper on the back and a special paper covering, usually cream-colored, on facing side and edges. This covering provides a smooth surface suitable for decoration. Insulating gypsum wallboard has a sheet of aluminum foil bonded to the liner paper. Predecorated gypsum wallboard does not require decorative treatment after installation because it comes with a finished surface, often a decorative vinyl or paper sheet. Wallboard should conform with ASTM C36.

Wallboard usually is available in the following thicknesses:

¼ in.—for covering and rehabilitating old walls and ceilings.

⅜ in.—mainly for the outer face in two-layer wall systems.

½ in.—for single-layer new construction with supports 16 to 24 in. c to c.

⅝ in.—for better fire resistance and sound control than ½ in. provides.

Standard edges are rounded, beveled, tapered, or square.

Backing Board. This type is used as a base layer in multi-ply construction, where several layers of gypsumboard are desired for high fire resistance, sound control, and strength in walls. It has gray liner paper on front and back faces. Insulating backing board comes with aluminum foil bonded to the back face. Gypsum backing board should conform with ASTM C442.

Gypsum Coreboard. To save space, this type is used as a base in multi-ply construction of self-supporting (studless) gypsum walls. Coreboard may be supplied as 1-in.-thick, solid backing board or as two factory-laminated, ½-in.-thick layers of backing board.

Type X Gypsumboard. For use in fire-rated assemblies, Type X may be a gypsum wallboard, backing board, or coreboard with core made more fire resistant by addition of glass fiber or other reinforcing materials.

Water-resistant Gypsum Backing Board. This type comes with a water-resistant gypsum core and water-repellant face paper. It may be used as a base for wall tile in baths, showers, and other areas subject to wetting. The board should conform with ASTM C630.

Gypsum Sheathing. This type is used as fire protection and bracing of exterior frame walls. It must be protected from the weather by an exterior facing. Sheathing should conform with ASTM C79.

Fire-block Gypsumboard. This type is used under combustible roof coverings to protect rafters and other internal construction from roof fires.

Gypsum Formboard. This type is used as a permanent form in the casting of gypsum-concrete roof decks.

Applications. Gypsumboard can be applied over existing firm, flat surfaces, or to wood or metal framing or furring. Wallboard can be applied directly to masonry or concrete. Generally, however, attachment of the wallboard to furring fastened to the masonry or concrete is advisable for protection against possible dampness. Gypsumboards also can be used to construct self-supporting, solid or semisolid partitions, attached to floor and ceiling.

Gypsumboards may be used in single-ply construction (Fig. 13-6) or combined in multi-ply systems (Fig. 13-9). The latter are preferred for greater sound control and fire resistance.

When outdoor temperatures are less than 55°F, the temperature of the building interior should be maintained at a minimum of 55°F for at least 24 hours before installation of gypsumboards. Heating should be continued until a permanent heating system is in operation, or outdoor temperatures stay continuously above 55°F. In warm or cold weather, gypsumboards should be protected from the weather.

Green lumber should not be used for framing or furring gypsumboard systems. Moisture content of the lumber should not exceed 19%, to avoid defects caused by shrinkage as the wood dries. For the same reason, dry lumber should be kept dry during storage and erection and afterward.

See also Arts. 13-18 to 13-21.

("ANSI Specifications for Application and Finishing of Gypsum Wallboard," A97.1, American National Standards Institute.

"Design Data—Gypsum Wallboard" and "Recommended Guide for Gypsum Drywall Construction," Gypsum Association.)

13-18. General Installation Procedures. Many building components may be affected by the decision to use gypsumboard construction. For example, window and door frames should have the appropriate depth for the wall thickness resulting from use of wallboard. Therefore, walls and partitions and related installations, including mechanical and electrical equipment, should be carefully planned and coordinated.

Placement. Wallboard preferably should be applied to studs with long dimension horizontal, and vertical joints should be staggered (Fig. 13-4). In ceilings, the long sides should span supports (often referred to as horizontal application). Board ends and edges parallel to framing or furring should be supported on those members, except for face layers of two-ply systems. Otherwise, back blocking should be used to reinforce the joints.

Ceiling panels should be installed first, then the walls. Adjoining boards should be placed in contact, but not forced against each other. Tapered edges should be placed next to tapered edges, square ends in contact with square ends. (Joints formed by placing square ends next to tapered edges are difficult to conceal.)

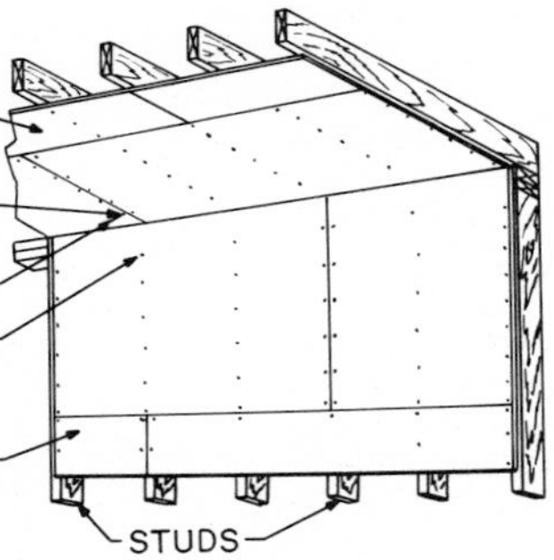

Fig. 13-4. Floating-angle construction with horizontal application of wallboard to wall and ceiling.

To prevent damage from structural movements or dimensional changes, control joints should be provided and floating-angle construction used (Fig. 13-4), as described in Art. 13-2.

Furring. Supplementary framing, or furring, should be used when framing spacing exceeds the maximum spacing recommended by the gypsumboard manufacturer for the thickness of board to be used (Table 13-2), or when the surface of framing or base layer is too far out of alignment.

At gypsumboard joints, wood furring should be at least 1½ in. wide, and metal furring 1¼ in. wide, to provide adequate bearing surface and space for attachment of the gypsumboard. The furring should be aligned to receive the board, and securely fastened to framing or masonry or concrete backing. For rigidity, lumber used for furring should be at least 2 × 2-in. when nails are used and 1 × 3-in. when screws are used for attachment of gypsumboard. Furring on masonry or concrete may be as small as ⅝ × 1½-in.

Fastening. Gypsumboards may be held in place with various types of fasteners or adhesives, or both. Special nails, staples, and screws are required for attachment of gypsumboards, because ordinary fasteners may not hold the boards tightly in place or countersink neatly, to be easily concealed. Clips and staples may be used only to attach the base layer in multi-ply construction.

To avoid damaging edges or ends, fasteners should be placed no closer to them than ⅜ in. (Figs. 13-7 and 13-8). With a board held firmly in position, fastening should start at the middle of the board and proceed outward toward the edges or ends. Nails should be driven with a crown-headed hammer until a uniform depres-

Table 13-2. Maximum Framing Spacing for Application of Gypsumboard

Base layer thickness, in.	Face layer thickness, in.	Location	Application*	Maximum c to c spacing, in.: Single-ply	Two-ply: Fasteners only	Two-ply: Adhesive between plies
3/8	....	Ceilings	Horizontal	16	16	16
3/8	3/8	Ceilings	Horizontal	†	16	16
3/8	3/8	Ceilings	Vertical	†	‡	16
1/2	....	Ceilings	Horizontal	24	24	24
1/2	....	Ceilings	Vertical	16	16	16
1/2	3/8	Ceilings	Horizontal	†	16	24
1/2	3/8	Ceilings	Vertical	†	‡	24
1/2	1/2	Ceilings	Horizontal	†	24	24
1/2	1/2	Ceilings	Vertical	†	16	24
5/8	....	Ceilings	Horizontal	24	24	24
5/8	....	Ceilings	Vertical	16	16	24
5/8	3/8	Ceilings	Horizontal	†	16	24
5/8	3/8	Ceilings	Vertical	†	‡	24
5/8	1/2 or 5/8	Ceilings	Horizontal	†	24	24
5/8	1/2 or 5/8	Ceilings	Vertical	†	16	24
1/4	....	Walls	Vertical	‡	16	16
1/4	3/8	Walls	‡	†	‡	‡
1/4	1/2 or 5/8	Walls	Horizontal or vertical	†	16	16
3/8	....	Walls	Horizontal or vertical	16	16	24
3/8	3/8, 1/2, or 5/8	Walls	Horizontal or vertical	†	16	24
1/2 or 5/8	....	Walls	Horizontal or vertical	24	24	24
1/2 or 5/8	3/8, 1/2, or 5/8	Walls	Horizontal or vertical	†	24	24

* Horizontal application places gypsumboard with long sides perpendicular to supports. Vertical application sets long sides parallel to supports.
† Not applicable.
‡ Not recommended.

sion, or dimple, not more than 1/32 in. deep is formed around the nail head. Care should be taken in driving not to tear the paper or crush the gypsum core.

Nails. For use in attachment of gypsumboard, nails should conform with ASTM C380 and C514. (Some fire-rated systems, however, may require special nails recommended by the gypsumboard manufacturer.) For easy concealment, nail heads should be flat or slightly concave and thin at the rim. Diameter of nail heads should be 1/4 or 5/16 in. Length of smooth-shank nails should be sufficient for penetration into wood framing of at least 7/8 in., and length of annular-grooved nails, for 3/4-in. penetration.

Screws. Three types of screws, all with cupped Phillips heads, are used for attachment of gypsumboard. Type W, used with wood framing, should penetrate at least 5/8 in. into the wood. Type S, used for sheet metal, should extend at least 3/8 in. beyond the inner gypsumboard surface. Type G, used for solid gypsum construction, should penetrate at least 3/8 in. into supporting gypsumboard. These screws, however, should not be used to attach wallboard to 3/8-in. backing board, because sufficient holding power cannot be developed. Nails or longer screws should be driven through both plies.

Staples. Used for attaching base ply to wood framing in multi-ply systems (Fig. 13-5), staples should be made of flattened, galvanized, 16-gage wire. The crown should be at least 7/16 in. wide. Legs should be long enough to permit penetration into supports of at least 5/8 in., and should have spreading points.

Adhesives. These may be used to attach gypsumboard to framing or furring, or to existing flat surfaces. Nails or screws may also be used to provide supplemental support.

Adhesives used for bonding wallboard may be classified as stud, laminating, or contact or modified contact.

Stud adhesives are used to attach wallboard to wood or steel framing or furring. They should conform with ASTM C557. They should be applied to supporting members in continuous, or nearly so, beads.

Laminating adhesives are used to bond gypsumboards to each other, or to suitable masonry or concrete surfaces. They are generally supplied in powder form, and water is added on the site. Only as much adhesive should be mixed at one time as can be applied within the period specified by the manufacturer. The adhesive may be spread over the entire area to be bonded, or in parallel beads or a pattern of large spots, as recommended by the manufacturer. Supplemental fasteners or temporary support should be provided the face boards until sufficient bond has been developed.

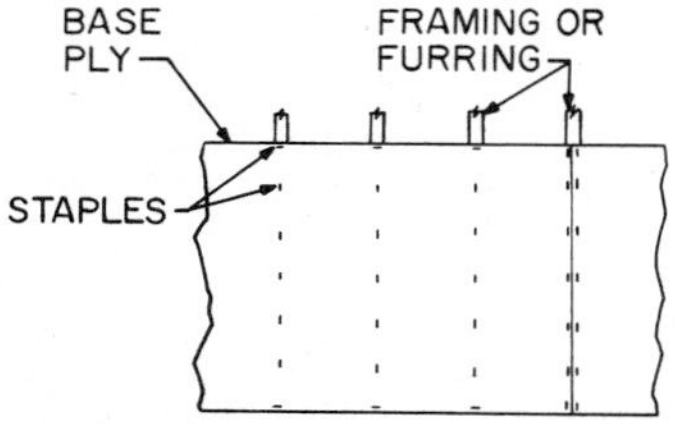

Fig. 13-5. Wallboard base ply attached to wood supports with staples, spaced 7 in. c to c when face ply is to be laminated, and 16 in. c to c when face ply is to be nailed.

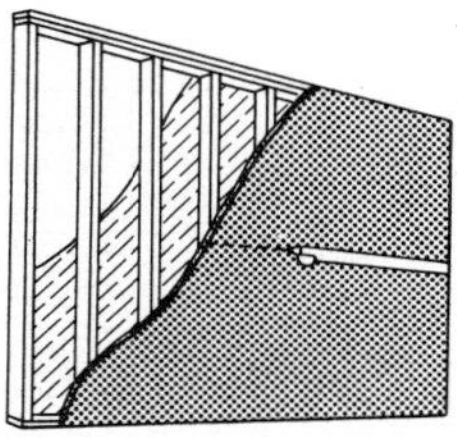

Fig. 13-6. Single-ply application of wallboard.

Contact adhesives are used to laminate gypsumboards to each other, or to attach wallboard to metal framing or furring. A thin, uniform coat of adhesive should be applied to both surfaces to be joined. After a short drying time, the face board should be applied to the base layer and tapped with a rubber mallet, to insure over-all adhesion. Once contact has been made, it may not be feasible to move or adjust the boards being bonded.

Modified contact adhesives, however, permit adjustments, often for periods of up to ½ hr after contact. Also, they generally are formulated with greater bridging ability than contact adhesives. Modified adhesives may be used for bonding wallboard to all kinds of supporting construction. See also Arts. 13-19 to 13-21.

("ANSI Specifications for Application and Finishing of Gypsum Wallboard," A97.1, American National Standards Institute.

"Using Gypsumboard for Walls and Ceilings," Gypsum Association, 201 North Wells St., Chicago, Ill. 60606.

"Drywall Construction Handbook," United States Gypsum Company, 101 South Wacker Drive, Chicago, Ill. 60606.

"Gypsum Wallboard Construction," National Gypsum Company, Buffalo, N.Y. 14202.)

13-19. Single-ply Gypsumboard Construction. (See also Art. 13-18.) A single-ply system consists of one layer of wallboard attached to framing, furring, masonry, or concrete (Fig. 13-6). This type of system is usually used for residential construction and where fire-rating and sound-control requirements are not stringent.

Nail Attachment. Spacing of nails generally is determined by requirements for fire resistance and for firmness of contact between wallboard and framing necessary to avoid surface defects. Spacing normally used depends on whether single nailing or double nailing is selected. Double nailing provides tighter contact, but requires more nails. In either method, one row of nails is driven along each support crossed by a board or on which a board end or edge rests, and the spacing applies to the center-to-center distance between nails in each row.

In the single-nailing method, nails should be spaced not more than 7 in. apart for ceilings, and 8 in. apart for walls (Fig. 13-7).

In the double-nailing method, pairs of nails are driven 12 in. c to c, except at edges or ends, in a special sequence (Fig. 13-8). First, one nail of each pair is placed, starting at the middle of the board and then proceeding toward edges and ends. Next, the second nail is driven 2 in. from the first. Finally, the first

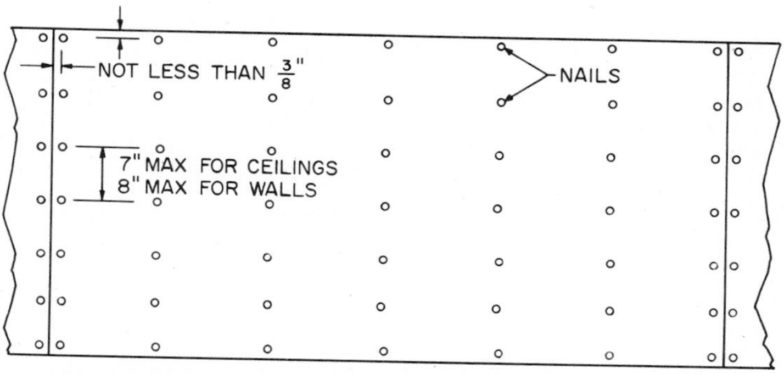

Fig. 13-7. Wallboard attached to wood framing with single nailing.

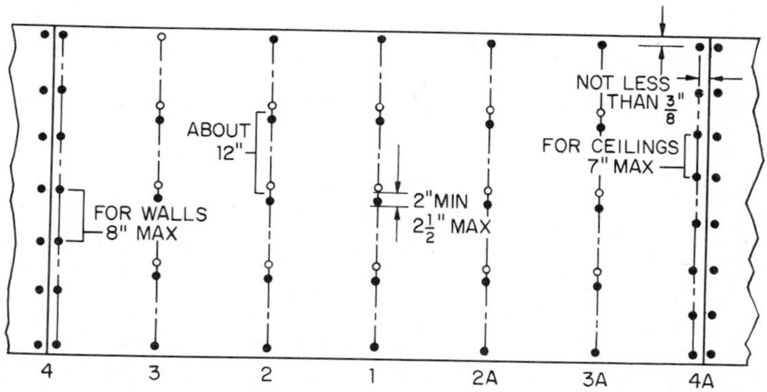

Fig. 13-8. Wallboard attached to wood framing with double nailing. One man nails rows 1 to 4, and a second man, rows 2A to 4A (solid dots, starting at middle of board). They then drive the second set of nails (open dots, again starting at middle of board). Finally, they reseat first set of nails with a blow on each.

set of nails placed should be given an extra hammer blow to reseat them firmly. Single-nailing spacing should be used for edges or ends at supports.

Screw Attachment. Fewer screws than nails are needed for attachment of wallboard. With wood framing, Type W screws should be spaced 12 in. c to c for ceilings. Spacing should not exceed 16 in. for studs 16 in. c to c, or 12 in. for studs 24 in. c to c. With metal framing, spacing of Type S screws should not exceed 12 in. for walls or ceilings.

Adhesive Nail-on Attachment. With this method, at least 50% fewer nails are required in the middle of the wallboard than with conventional nailing. Also, a stiffer assembly results.

The first step is to apply stud adhesive to framing or furring in beads about ¼ in. in diameter. Each bead should contain enough adhesive so that it will spread to an average width of 1 in. and thickness of 1/16 in. when wallboard is pressed against it, but there should not be so much adhesive that it will squeeze out at joints. If joints are to be treated, an undulating or zigzag bead should be applied to supports under joints. If joints are to be untreated, as is likely to be the case for predecorated wallboards, two parallel beads should be applied near each edge of the framing. Along other supports a single, continuous, straight bead is adequate. Adhesive, however, should not be applied to members, such as diagonal bracing, blocking, and plates, not required for wallboard support.

The number and spacing of supplemental fasteners needed with adhesives depend on adhesive properties. Stud adhesives generally require only perimeter fasteners for walls. One fastener should be driven wherever each edge or end crosses a stud. For edges or ends bearing on studs, fasteners should be spaced 16 in. c to c. The same perimeter fastening should be used for ceilings, but in addition, fasteners should be spaced 24 in. c to c along all framing crossed by the wallboard.

Where perimeter fasteners cannot be used, for example, with predecorated wallboard for which joint treatment is unnecessary or undesirable, prebowing or temporary bracing should be used to keep the wallboard pressed against the framing until the adhesive develops full strength. Bracing should be left in place for at least 24 hours. Prebowing bends a wallboard so that the finish side faces the center

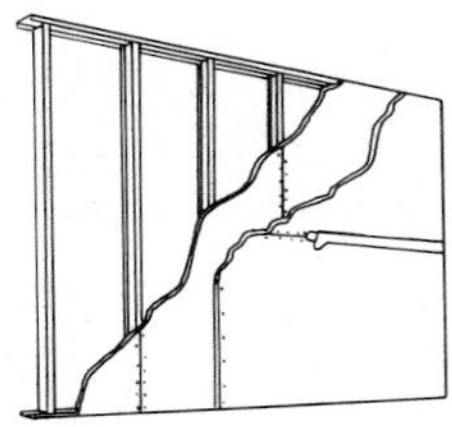

Fig. 13-9. Double-ply application of wallboard.

of curvature. The arc is employed to keep the board in tight contact with the adhesive as the board is pressed into place starting at one end.

See also Art. 13-21.

13-20. Multi-ply Gypsumboard Construction. (See also Art. 13-18.) A multi-ply system consists of two or more layers of gypsumboard attached to framing, furring, masonry, or concrete (Fig. 13-9). This type of system has better fire resistance and sound control than can be achieved with single-ply systems, principally because of greater thickness, but also because better insulation can be used for the base layer.

Face layers usually are laminated (glued), but may be nailed or screwed, to a base layer of wallboard, backing board, or sound-deadening board. Backing board, however, often is used for economy. When adhesives are used for attaching the face ply, some fasteners generally are also used to insure bond.

Maximum support spacing depends principally on the thickness of the base ply and its orientation relative to the framing, as indicated in Table 13-2, and is the same for wood or metal framing or furring.

Nails, screws, or staples may be used to attach base ply to supports. For wood framing, when the face ply is to be laminated to the base ply, staples should be spaced 7 in. c to c (Fig. 13-5), and nails and screws should be single-nailed as recommended for single-ply construction (Art. 13-19). If the face ply is to be nailed, nails and staples should be driven 16 in. c to c, and screws 24 in. c to c. For metal framing and furring, Type S screws should be spaced 12 in. c to c if the face ply is to be laminated, and 16 in. c to c if the face ply is to be attached with screws.

For attachment of the face ply with adhesive, sheet, strip, or spot lamination may be used. In **sheet lamination,** the entire back of the face ply is covered with adhesive, usually applied with a notched spreader. In **strip lamination,** adhesive is applied by a special spreader in grouped parallel beads, with groups spaced

16 to 24 in. c to c. In **spot lamination,** adhesive is brushed or daubed on the back of the face ply at close intervals. Partitions with strip or spot lamination provide better sound control than those with sheet lamination.

Supplemental fasteners or temporary bracing is required to insure complete bond between face and base plies. If a fire rating is not required, temporary fasteners may be placed at 24-in. intervals. For fire-rated assemblies, permanent fasteners generally are required, and spacing depends on that used in the assembly tested and rated. Nails should penetrate at least 1⅛ in. into supports. With sound-deadening base plies, fastener spacing should be as recommended by the base manufacturer.

In placing face ply on base ply, joints in the two layers should be offset at least 10 in. Face ply may be applied horizontally or vertically. Horizontal application (long sides of sheets perpendicular to supports) usually provides fewer joints, but vertical application may be preferred for predecorated wallboard that is to have joints trimmed with battens.

See also Art. 13-21.

13-21. Finishing Procedures. The finish surface of wallboard may come predecorated or may require decoration. Predecorated wallboard may require no further treatment other than at corners, or may need treatment of joints and concealment of fasteners. Corner and edge trim are applied for appearance and protection, and battens often are used for decorative concealment of flush joints. Other types of wallboard require preparation before decoration can be applied.

While trim can be applied to undecorated wallboard, as for predecorated panels, usually, instead, joints are made inconspicuous.

Joint-treatment Products. Materials used for treatment of wallboard joints to make them inconspicuous should meet the requirements of ASTM C475. These materials include:

Joint tape, a strip of strong paper reinforcement with feathered edges, for embedment in joint compound. Sometimes supplied with small perforations to improve the bond, the tape usually is about 2 in. wide and 1⁄16 in. thick.

Joint compound, and adhesive, with or without fillers, for bonding and embedding the tape. Two types may be used. One, usually referred to as joint, or taping, compound, is applied as an initial coat for filling depressions at joints and fasteners and for adhering joint tape. The second, called topping, or finishing, compound, is used to conceal the tape and for final smoothing and leveling at joints and fasteners. As an alternative, an all-purpose compound that combines the features of taping and topping compounds may be used for all coats. Compounds may be supplied premixed, or may require addition of water on the job.

Flush Joints and Fasteners. For concealing fastener heads and where wallboards in the same plane meet, at least three coats of joint compound should be used. The first coat, of taping compound, should be used for adhering tape at edges and ends and to fill all depressions over fastener heads and at tapered edges. Joint tape should be centered over each joint for the length or width of the wall or panel and pressed into the compound, without wrinkling, with a tape applicator, a broad knife. Excess compound should be redistributed as a skim coat over the tape. A skim coat of joint compound applied immediately after tape embedment reduces the possibility of edge wrinkling or curling, which may result in edge cracking.

A second, or bedding, coat should be applied with topping compound at fastener heads and joints after the taping compound has dried. The waiting period may be 24 hr if regular taping compound is used, and about 3 hr if a quick-setting compound is used. This bedding coat should be thin. At joints, it should completely cover the tape and should be feathered out 1 to 2 in. beyond each edge of the tape at tapered edges, and over a width of 10 to 12 in. at square edges.

When the coat has dried, a second, thin coat of the topping compound should be applied for finish at fasteners and joints. At joints, it should completely cover the preceding coat and should be feathered out 1 to 2 in. beyond its edges at tapered edges, and over a width of 18 in. at square edges. Irregularities may be removed by light sanding after each coat has hardened. But if joint treatment is done skillfully, such sanding should not be necessary.

Reentrant Corners. Treatment of interior corners differs only slightly from that of flush joints. Joint compound should first be applied to both sides of each joint. Joint tape should be creased along its center, then embedded in the compound to form a sharp angle. After the compound has dried, finishing compound should be applied, as for flush joints.

Trim. Exposed corners and edges should be protected with a hard material. Wood or metal casings may be used for protection and to conceal the joints at door and window frames. Baseboard may be attached to protect wallboard edges at floors. Corner beads may be applied at exterior corners when inconspicuous protection is desired. Casing beads may be similarly used in other places.

At least three-coat finishing with joint compound is required to conceal fasteners and obtain a smooth surface with trim shapes.

Decoration. Before wallboard is decorated, imperfections should be repaired. This may require filling with joint compound and sanding. Joint compound should be thoroughly dry before decoration starts.

The first step should be to seal or prime the wallboard surface. (Glue size, shellac, and varnish are not suitable sealers or primers.) An emulsion sealer blocks the pores and reduces suction and temperature differences between paper and joint compound. If wallpaper is used over such a sealer, the paper can be removed later without damaging the wallboard surface and will leave a base suitable for redecorating. A good primer provides a base for paint and conceals color and surface variations. Some paints, such as high-quality latex paints, often have good sealing and priming properties.

("Using Gypsumboard for Walls and Ceilings," Gypsum Association, 201 North Wells St., Chicago, Ill. 60606.

"Fireproof Gypsum Wallboard Application Instructions" and "Gypsum Drywall Joint Treatment Guide," National Gypsum Company, Buffalo, N.Y. 14202.

"Drywall Construction Handbook," United States Gypsum Company, 101 South Wacker Drive, Chicago, Ill. 60606.)

Section 14

Floor Coverings

THOMAS H. BOONE

Formerly, Building Research Division,
National Bureau of Standards

AND

JOHN E. FITZGIBBONS

Chemical Engineer, Standardization Division,
Naval Construction Battalion Center
Davisville, R.I.

For proper selection of floor coverings, designers should take into account many factors, including use of lightweight-concrete slabs, subfloors in direct contact with the ground, radiant heating, air conditioning; possible necessity for decontamination dustlessness; traffic loads, and maintenance costs—all of which have an important bearing on flooring selection. Consideration must be given to current standards of styling, comfort, color, and quietness.

The primary consideration of the designer of a flooring system, is to select a floor covering that can meet the maximum standards at reasonable cost. To avoid the dissatisfaction that would arise from failure to select the proper flooring, designers must consider all the factors relevant to flooring selection.

This section contains information that can provide a guide toward this end. It summarizes the characteristics of the major types of nontextile or resilient floor coverings, and describes briefly the methods for the proper installation of these materials.

14-1. Asphalt Tiles. These tiles are intended for use on rigid subfloors, such as smooth-finished or screeded concrete, structurally sound plywood, or hardboard floors not subject to excessive dimensional changes or flexing. The tiles can be satisfactorily installed on below-grade concrete subject to slight moisture from the ground. (See also Art. 14-8.)

Low cost and large selection of colors and designs make asphalt tile an economically desirable flooring.

Asphalt tile is composed of asbestos fibers, mineral coloring pigments, and inert fillers bound together. For dark colors the binder is Gilsonite asphalt; for intermediate and light colors, the binder may consist of resins of the cumarone indene type or of those produced from petroleum. Tiles most commonly used are ⅛ in. thick.

Simplified Practice Recommendation R225 established four color classes: A, B, C, and D, graded from black and dark red (A) to cream, white, yellow, blue, and bright red (D). A color comparison chart showing commercial equivalents of various manufacturers' color lines, for use as a guide, may be obtained from the Resilient Tile Institute, 101 Park Avenue, New York, N.Y. 10017.

To avoid permanent indentations in asphalt tiles, contact surfaces of furniture or equipment should be smooth and flat to distribute the weight. This is particularly necessary for installations over radiant-heated floors and on areas near windows exposed to sun.

Never use on asphalt tiles waxes containing benzene, turpentine, or naphtha-type solvents and free fat or oils. Avoid strong detergents or cleaning compounds containing abrasives or preparations not readily soluble in water. These may soften the tiles and cause colors to bleed. Grease, oils, fats, vinegar, and fruit juices allowed to remain in contact with asphalt tile will stain and soften them. Because of these restrictions, asphalt tiles are not recommended for use in kitchens or bathrooms.

14-2. Cork Tiles. Cork flooring is intended for use on rigid subfloors, such as smooth-finished or screeded concrete supported above grade and free of moisture, or on structurally sound plywood or hardboard. Cork tile is not recommended for application below grade. When it is installed at grade, moisture-free conditions must be insured. (See also Art. 14-8.)

Cork tile is manufactured by baking cork granules with phenolic or other resin binders under pressure. Four types of finishes are produced: natural, factory-prefinished wax, resin-reinforced wax, and vinyl cork tile (Art. 14-8). The tiles are ⅛, 3/16, 5/16, and ½ in. thick.

Natural cork tile must be sanded (to level), sealed, and waxed immediately after installation.

Unless the exposed surface of cork floors is maintained with sealers and protective coatings, permanent stains from spillage and excessive soiling by heavy traffic will result.

Cork tiles are particularly suitable for areas where quiet and comfort are of paramount importance.

14-3. Vinyl Flooring. Flooring of this type is unbacked. It is intended for use on rigid subfloors, such as smooth-finished or screeded concrete supported above grade, or structurally sound plywood or hardboard floors. Vinyl floors are not recommended for use below grade. They must be applied with an alkaline, moisture-resistant adhesive when used at grade. (See also Art. 14-8.)

Vinyl mats or runners may be laid without adhesive over relatively smooth surfaces. Large mats generally are installed in a recess in the concrete floor at building entrances. The mats are ribbed or perforated for drainage.

Vinyl flooring consists predominantly of polyvinyl chloride resin as a binder, plasticizers, stabilizers, extenders, inert fillers, and coloring pigments. Because of its unlimited color possibilities and opaqueness to transparent effects, it is widely used. Common thicknesses are 0.080, 3/32, and ⅛ in.

Since vinyl resins are tough synthetic polymers, vinyl flooring can withstand heavy loads without indentation, and yet is resilient and comfortable under foot. It is practically unaffected by grease, fat, oils, household cleaners, or solvents. But it is easily scratched and scuffed. The lustrous surface of new tile is not retained under constant foot traffic unless adequately protected with a floor polish.

14-4. Vinyl-asbestos Tiles. These are intended for use on rigid subfloors, such as smooth-finished or screeded concrete, or structurally sound plywood or hardboard subfloors not subjected to excessive dimensional change or flexing. Like asphalt tiles, vinyl-asbestos tiles can be installed satisfactorily on below-grade concrete subject to slight moisture from the ground. (See also Art. 14-8.)

Vinyl-asbestos tiles are composed of asbestos fibers, mineral coloring pigments, and inert fillers, bound together with vinyl resin and plasticizers. They have brighter and clearer colors than asphalt tiles, which have a similar composition except for the binder. A color comparison chart showing commercial equivalents of various manufacturers' color lines, for use as a guide, may be obtained from the Asphalt and Vinyl Asbestos Tile Institute, 101 Park Ave., New York, N.Y. 10017. Common thicknesses are 1/16 and 1/8 in.

Vinyl-asbestos tile has a duller and harder surface than vinyl tile, and therefore its appearance is not damaged as much by foot traffic. Resistance to oils, grease, and household reagents is a little less than that of vinyl.

14-5. Backed Vinyl. The family of backed-vinyl flooring comprises vinyl wearing surfaces from 0.002 to 0.050 in. thick, laminated to many different backing materials. In some products, the vinyl surfaces are unfilled transparent films placed over a design on paper, cork, or degraded vinyl. Filled vinyl surfaces with a 34% vinyl resin binder are placed over rag felt or asbestos felt. The asbestos felt backing is intended for use in moist areas. Foamed rubber or plastic is incorporated in some of these materials to increase comfort and decrease impact noise.

Asbestos-backed vinyl materials may be applied with a moisture-resistant adhesive on concrete at or below grade. Installation of other backed vinyl flooring is similar to that of linoleum. (See also Art. 14-8.)

14-6. Rubber. Rubber flooring is intended for use on rigid subfloors, such as smooth-finished or screeded concrete supported above grade, or on structurally sound plywood or hardwood subfloors. Rubber is not recommended for use below grade. When used at grade, it must be applied with an alkaline, moisture-resistant adhesive. (See also Art. 14-8.)

Rubber mats or runners may be laid without adhesive over relatively smooth surfaces. Large mats generally are installed in a recess in the concrete floor at building entrances. The mats are ribbed or perforated for drainage.

Most rubber flooring is produced from styrene-butadiene rubber. Reclaimed rubber is added to some floorings. The flooring also contains mineral pigments and mineral fillers, such as zinc oxide, magnesium oxide, and various clays. Another synthetic-rubber flooring, chlorosulfonated polyethylene (Hypalon), also is available.

Rubber floorings can be obtained in thicknesses from 3/32 to 1/4 in. They have excellent resistance to permanent deformation under load. Yet they are resilient and quiet under foot.

14-7. Linoleum. Since the burlap and rag felt backings of linoleum are susceptible to moisture and fungus attack, and the wearing surface is damaged by alkaline solutions, linoleum should never be used on floors where moisture can accumulate between the flooring and subfloor. Like wood and cork floors, linoleum should not be used on concrete at or below grade.

Linoleum basically is made from drying oils, such as linseed, natural and synthetic resins, a filler, and pigments similar to those used in paints. The filler is largely wood flour. The flooring usually has a backing or carrier of burlap or rag felt.

With burlap backing, thickness is 1/8 or 3/16 in. With felt backing, over-all thickness ranges from 0.065 to 0.125 in., and the thickness of the linoleum wearing surface from 0.022 to 0.085 in.

Linoleum is available as sheet or tile. When properly maintained, it performs outstandingly in residential and commercial buildings.

14-8. Installation of Thin Coverings. Most manufacturers and trade associations make available instructions and specifications for installation and maintenance of their floorings. (See Art. 14-16.)

The most important requirement for a satisfactory installation of a thin floor covering is a dry, even, rigid, and clean subfloor.

Protection from moisture is a prime consideration in applying a flooring over concrete. Moisture within a concrete slab must be brought to a low level before installation begins. Moisture barriers, such as 6-mil polyethylene, 55-lb asphalt-saturated and coated roofing felt, or 1/32-in. butyl rubber, should be placed under concrete slabs at or below grade, and a minimum of 30 days' (90 in some cases) drying time should be allowed after placement of concrete before installing the flooring.

Lightweight concrete always has a higher gross water requirement than ordinary concrete. Lightweight concrete, therefore, takes longer to dry. So a longer drying period should be allowed before installing flooring. Flooring manufacturers provide advice and sometimes also equipment to test for moisture.

An indication of dryness at any given time is no assurance that a concrete slab at or below grade will always remain dry. Therefore, protection from moisture from external sources must be given considerable attention. (See also Arts. 12-1 to 12-8.)

Some concrete curing agents containing oils and waxes may cause trouble with adhesive-applied flooring. Commercial curing agents that have been shown to be satisfactory include styrene-butadiene copolymer or petroleum-hydrocarbon resin

Table 14-1. Adhesives Used to Install Flooring*

Flooring	Concrete below grade	Concrete on grade	Concrete above grade	Plywood or hardboard
Asphalt and vinyl-asbestos tiles	Asphalt, cutback Asphalt, emulsion	Asphalt, cutback Asphalt, emulsion	Asphalt, cutback Asphalt, emulsion	Asphalt, cutback Asphalt, emulsion
Rubber and vinyl	Chemical set Latex	Latex	Latex	Latex
Linoleum, cork, and vinyl backed with felt or cork	Do not install	Do not install	Linoleum paste	Linoleum paste
Vinyl backed with asbestos felt	Latex	Latex	Latex Linoleum paste	Linoleum paste
Laminated wood block	Do not install	Asphalt, hot melt Asphalt, cutback Rubber base	Asphalt, hot melt Asphalt, cutback Rubber base	Asphalt, hot melt Asphalt, cutback Rubber base
Solid unit wood block	Do not install	Asphalt, hot melt Asphalt, cutback	Asphalt, hot melt Asphalt, cutback	Asphalt, hot melt Asphalt, cutback

* The adhesives listed are intended for each type of flooring; i.e., asphalt adhesives used for asphalt and vinyl-asbestos tiles are not the same as asphalt adhesives used for wood blocks. There are a number of adhesives having special properties and characteristics, such as heat resistance, resistance to water spillage, and brush-on types not included in this table.

dissolved in a hydrocarbon solvent. These products dry to a hard film in 24 hr. Parting agents, used as slab separators in lift-slab and tiltup construction, may cause bond failure of adhesive-applied flooring if they contain nonvolatile oils or waxes. Concrete surfaces that have been treated with or have come in contact with oils, kerosene, or waxes must be cleaned as recommended by the adhesive manufacturer.

All concrete surfaces to receive adhesive-applied, thin flooring must be smooth, and free from serious irregularities that would "telegraph" through the covering and be detrimental to appearance and serviceability. For rough or uneven concrete floors, a troweled-on underlayment of rubber latex composition or asphalt mastic is recommended. It can be applied from a thickness of ¼ in. to a featheredge. Small holes, cracks, and crevices may be filled with a reliable cement crack filler.

Wood subfloors, of sufficient strength to carry intended loads without deflection, should be covered with plywood or hardboard underlayments. These should be nailed 6 in. c to c in perpendicular directions over the entire area with ringed or barbed nails. The nails should be driven flush with the underlayment surface without denting it around the nail head.

Adhesives commonly used to attach flooring to concrete and underlayment are listed in Table 14-1.

14-9. Ceramic Tiles. Three main types are used as floor covering: mosaic, paver, and quarry.

Ceramic mosaic tile has a dense, fully vitrified body. It is made either from natural clays or a blended clay-flint-feldspar mixture. The area of each tile is less than 6 sq in. For easy handling, the small mosaic tiles are assembled at the factory in desired patterns and fastened together. One method of holding a pattern, for example, is to glue the top face of the tiles to a sheet of paper, which can be removed easily when the tiles are in place on the floor.

Paver tiles are similar to the ceramic mosaic floor tile in composition and properties. They differ mainly in size. The face area of pavers is greater than 6 sq in. These tiles usually range in size from 3×3 in. to 6×6 in.

Quarry tile has a very dense body. Very durable, it is resistant to freezing, abrasion, and moisture.

For information on selection of bed, joint, and tiles and on methods of installation, see American National Standard "Specifications for Ceramic Mosaic Tile, Quarry Tile and Pavers Installed in Portland Cement Mortars," A108.2, A108.3, and A108.5, American National Standards Institute. Also, specification work sheets are published by the American Institute of Architects for ceramic and quarry tile.

Besides portland cement mortars, organic-adhesive thin setting beds may be used to install tile over smooth, rigid, and dry subfloors. Mosaic tiles in 9×9 in. units, bonded in a preformed rubber or vinyl grid, also are applied with adhesive. These adhesives, however, are not recommended for use in floors subject to heavy traffic, concentrated loads, or excessive amounts of water. Table 14-2 lists characteristics of chemical-resistant setting beds and grouts for ceramic tiles.

Attractive appearance and tough serviceability make ceramic tiles suitable for flooring for kitchens and baths.

14-10. Terrazzo. A Venetian marble mosaic, with portland cement matrix, terrazzo is composed of two parts marble chips to one part portland cement. Color pigments may be added. Three methods of casting in place portland-cement terrazzo atop structural concrete floor slabs are commonly used: sand cushion, bonded, and monolithic.

Sand cushion (floating) terrazzo is used where structural movement that might injure the topping is anticipated from settlement, expansion, contraction, or vibration. This topping is at least 3 in. thick. The concrete slab is covered with a ¼- to ½-in. bed of dry sand. Over this is laid a membrane, then wire-mesh reinforcing. The terrazzo underbed is installed to ⅝ in. below the finished floor line. Next, divider strips are placed and finally, the terrazzo topping.

Bonded terrazzo has a minimum thickness of 1¾ in. After the underlying concrete slab has been thoroughly cleaned and soaked with water, the surface is slushed with neat portland cement to insure a good bond with the terrazzo. Then, the underbed is laid, divider strips are installed, and terrazzo is placed.

Monolithic terrazzo is constructed by placing a ⅝-in. topping as an integral part of a green-concrete slab. Adhesive-bonded monolithic terrazzo with an epoxy resin adhesive also has been used successfully, with a topping thickness of only ⅜ in.

Terrazzo may be precast. It generally is used in this form for treads, risers, platforms, and stringers on stairs.

Because of the large variety of color and surface textures that can be attained with terrazzo, it is used extensively as an exterior and interior decorative flooring. Portland-cement terrazzo, however, should not be used in areas subject to spillage, such as might be encountered in kitchens.

Details on selecting the proper type of terrazzo, marble-chip sizes, methods of applying, and finishing of the surface may be obtained from the National Terrazzo and Mosaic Association, 716 Church St., Alexandria, Va. 22314. Also, specification sheets for terrazzo are published by the American Institute of Architects.

Other matrix materials used with marble chips include rubber latex, epoxy, and polyesters. Suppliers should be consulted for installation details.

14-11. Concrete Floors. A concrete topping may be applied to a concrete structural slab before or after the base slab has hardened. Integral toppings usually are ½ in. thick; independent toppings about 1 in.

It is well to remember in specifying or installing a concrete floor that the difference between good and poor concrete lies in selection and grading of aggregates, proportioning of the mix, and the care with which the vital operations of placing, finishing, and curing are carried out. Useful information on concrete floors may be obtained from the Portland Cement Association. (See also Sec. 5.)

14-12. Wood Floors. Both hardwoods and softwoods are used for floors. Hardwoods most commonly used are maple, beech, birch, oak, and pecan. Softwoods are yellow pine, Douglas fir, and western hemlock. The hardwoods are more resistant to wear and indentation than softwoods.

Hardwood strip floorings are available in thicknesses of $^{11}/_{32}$, $^{15}/_{32}$, and $^{25}/_{32}$ in. and in widths of $1\frac{1}{2}$, 2, $2\frac{1}{4}$, and $3\frac{1}{4}$ in.

Softwood strip flooring usually is $^{25}/_{32}$ in. thick and can be obtained in widths of $2\frac{3}{8}$, $3\frac{1}{4}$, and $5\frac{3}{16}$ in.

Solid-unit wood blocks for floors are made from two or more units of strip-wood flooring fastened together with metal splines or other suitable devices. A block usually is square. Tongued and grooved, either on opposite or adjacent sides, it is held in place with nails or an asphalt adhesive (Table 14-1). Blocks $^{25}/_{32}$ in. thick are made up of multiples of $1\frac{1}{2}$- or $2\frac{1}{4}$-in. strips; blocks $\frac{1}{2}$ in. thick are composed of multiples of $1\frac{1}{2}$- or 2-in. strips.

A laminated block is formed with plywood comprising three or more plies of wood glued together. The core or cross bonds are laid perpendicular to the face and back of the block. Usually square, the block is tongued and grooved on either opposite or adjacent sides. The most common thickness is $\frac{1}{2}$ in., but other thicknesses used are $\frac{3}{8}$, $\frac{7}{16}$, $\frac{5}{8}$, and $\frac{13}{16}$ in. Laminated blocks are installed with adhesives (Tables 14-1).

Average moisture content of wood flooring at time of installation should be 6% in dry southwestern states, 10% in damp southern coastal states, and 7% in the rest of the United States ("Moisture Content of Wood in Use," *U.S. Forest Products Laboratory Publication No.* 1655, Madison, Wis.). The U.S. Forest Products Laboratory also recommends heating the building before installing any type of wood flooring, except when outdoor temperatures are high. Bundles of flooring should be opened and stored in the building before installation. Leave at least 1 in. of expansion space at walls and columns.

Specifications for wood floors on concrete, gymnasium floors, or other special designs or conditions may be obtained from associations of wood flooring manufacturers and installers. (See Art. 14-16.)

14-13. Industrial Floors. The primary purpose of a floor covering in an industrial building is to protect the structural floor from foot and truck traffic and from corrosive effluents. The flooring also must provide a durable surface that will not be detrimental to plant operations.

A good-quality concrete floor is adequate if factory conditions are dry and the surface has to withstand only foot and truck traffic. A cast-iron grid filled with concrete on $\frac{1}{4}$-in.-thick steel plates may be used in areas where heavy equipment or materials are dragged continually. End-grain wood blocks may be used in areas of heavy trucking of heavy castings, to protect both the structure and castings from impact damage. Asphalt mastic floors also are used in areas with heavy truck traffic.

If the factory process is wet and corrosive, the designer may have to choose between a less-resistant flooring that has to be replaced periodically and a flooring with low maintenance costs but with higher initial cost. Table 14-2 rates the relative resistance to liquids of some floor toppings and bedding and jointing materials for clay tiles and bricks. In addition to selecting the most suitable materials, the designer should provide adequate slope and drainage, a liquid-tight layer below the finished floor as a second line of defense, and a nonslip surface. Selection and laying of chemical-resistant materials are a specialized operation on which specialists should be consulted.

14-14. Conductive Flooring. Where explosive vapors are present, sparks resulting from accumulation of static electricity constitute a hazard. The most effective of several possible ways of mitigating this hazard is to keep electrical resistance so low that dangerous voltages never are attained.

Most objects normally rest or move on the floor and therefore can be electrically connected through the floor. Flooring of sufficiently low electrical resistance (conductive flooring) thus is of paramount importance in elimination of electrostatic

Table 14-2. Resistance of Floor Toppings and of Bedding and Jointing Material to Various Liquids*

Material	Water	Organic solvents	Oils	Acids	Alkalies
Floor topping:					
Portland cement concrete	VG	G	F	VP	G
High-alumina cement concrete	G	G	F-G	VP	F
Pitch (coal tar) mastic	VG	F	F	F-G	G
Asphalt mastic	VG	VP	P	F-G	G
Epoxy	VG	G	G	G	G
Polyester	VG	G	G	G	F
Bedding and jointing materials for clay tiles and bricks:					
Portland cement mortar	As above for concrete				
High-alumina cement mortar	As above for concrete				
Silicate cement	F	G	G	G	VP
Sulfur cement	VG	G	F	G	G
Epoxy	As above				
Furan	VG	VG	VG	VG	VG

VG = very good; G = good; F = fair; P = poor; VP = very poor.

* This table is intended for broad comparisons, as the items under various headings may need qualifications for particular conditions, such as temperature, and concentration of the liquids and acids which are oxidizing agents.

hazards. But the electrical resistance must be high enough to eliminate electric shock from faulty electrical wiring or equipment. ("Code for the Use of Flammable Anesthetics," NFPA No. 56, National Fire Protection Association, 60 Batterymarch Street, Boston, Mass. 02110.)

Electrical conductivity of flooring depends on the presence of acetylene black (carbon). In ceramic, linoleum, rubber, and vinyl floors, the carbon is finely dispersed in the material during manufacture; in latex terrazzo, concrete terrazzo, and setting-bed cement for ceramic tile, the carbon is uniformly dispersed in the dry powder mixes, placed in containers, and shipped for on-the-job composition ("Conductive Flooring for Hospital Operating Rooms," Monograph 11, National Bureau of Standards, Washington, D.C. 20234).

For linoleum flooring, brass seam connectors with projecting points are used for electrical intercoupling between sheets. For vinyl tiles, copper foil is placed between the adhesive and tile for electrical intercoupling.

When flammable gases are in use, everyone and everything must be electrically intercoupled via a static conductive floor at all times. To insure this, constant vigilance, inspection by testing, and a high standard of housekeeping are required. Wax or dirt accumulation on the floor and grounding devices can provide high resistance to electrical flow and cancel the effect of conductive flooring.

14-15. Specifications and Standards for Flooring. Federal specifications are written to establish minimum standards for purposes of competitive bidding and to insure that government agencies will receive quality material. Some flooring materials, however, cannot be classed under these specifications because only one manufacturer makes the products. Prices of the Federal or interim Federal specifications listed in Table 14-3 may be obtained from the General Services Administration, Federal Supply Service, Standardization Division, Washington, D.C. 20025, or the nearest GSA regional office.

Commercial Standards are specifications that establish quality levels for manufactured products in accordance with the principal demands of the trade. The Office of Commodity Standards, National Bureau of Standards, Washington, D.C. 20234, will supply a price list. Standards concerned with flooring include Water-Resistant Organic Adhesives for Installation of Clay Tile, 181-52; Stripoak Flooring, 56-60; Laminated Hardwood Block Flooring, 233-60; and Hardwood Stair Treads and Risers, 89-40.

Table 14-3. Federal Specifications

Title	Remarks	Specification number*
Tile, floor, asphalt	⅛- and 3/16-in.	SS-T-312A
Tile, floor, vinyl-asbestos	⅛- and 1/16-in.	SS-T-312A
Adhesive, asphalt, cutback type	For asphalt and vinyl-asbestos tile	MMM-A-110A
Adhesive, asphalt, emulsion type	For asphalt and vinyl-asbestos tile	MMM-A-115A
Tile, floor, vinyl	0.050-in.	SS-T-312A
Tile, floor, rubber	⅛- and 3/16-in.	SS-T-312A
Floor coverings, linoleum	0.125-, 0.090-, 0.070-, and 0.065-in.	LLL-F-1238A
Floor covering, vinyl wearing surface, with backing	0.085-, 0.065-, and 0.055-in.	L-F-475A
Paste, linoleum	....	MMM-A-137
Flooring, vinyl, plastic	⅛- and 3/16-in.	L-F-00450
Floor covering, translucent or transparent vinyl surface, with backing	0.085-, 0.065-, and 0.055-in.	L-F-001641
Wall base, rubber and vinyl	For coves and straights between walls and floors	SS-W-40A
Block, floor, wood	Solid unit and laminated hardwood	NN-B-350
Federal Test Method Standard	Sampling and testing method for resilient flooring	501A

* As of 1974.

14-16. Flooring Trade Associations. Detailed information on installation and maintenance of flooring materials may be obtained from the following organizations:

Resilient Tile Institute, 101 Park Avenue, New York, N.Y. 10017.

Ceramic Tile Institute, 3415 West 8th Street, Los Angeles, Calif. 90005.

American Plywood Association, 1119 A Street, Tacoma, Wash. 98402.

Hardwood Plywood Manufacturers Association, P. O. Box 6246, Arlington Station, Arlington, Va. 22208.

Maple Flooring Manufacturers Association, 424 Westington Ave., Oshkosh, Wis. 54901.

National Oak Flooring Manufacturers Association, 814 Sterick Building, Memphis, Tenn. 38103.

National Terrazzo and Mosaic Association, 716 Church St., Alexandria, Va. 22314.

Portland Cement Association, Old Orchard Rd., Skokie, Ill. 60076.

Rubber and Vinyl Flooring Council Division, Rubber Manufacturers Association, Inc., 444 Madison Avenue, New York, N.Y. 10022.

Tile Contractors Association of America, Inc., 112 North Alfred St., Alexandria, Va. 22314.

Tile Council of America, 800 Second Avenue, New York, N.Y. 10017.

Wood Flooring Institute, 201 North Wells St., Chicago, Ill. 60606.

Section 15

Roof Coverings

J. C. GUDAS

Editor-Publisher, American Roofer & Building Improvement Contractor, Downers Grove, Ill.

The term *roofing* or *roof covering* is generally considered to refer to the covering material installed in a building over the roof deck, which in turn is erected over the building frame. Roofing does not normally include rafters, sheathing, nor purlins, but it does include the roofing felt placed over the roof deck. The type of covering used on a building depends on the roofing system specified to weatherproof the structure properly.

Primarily, design of a roofing system is the responsibility of an architect or engineer. The designer often relies on specifications developed by roofing trade associations and materials manufacturers.

It is of extreme importance that plans and specifications call for a stable, structurally sound building, with a minimum of movement. Roofing materials cannot withstand the three-way movement of a poorly designed building, which often results from a misguided attempt to reduce costs.

The importance of proper design of roofs was expressed by an architect who stated that the roof, representing 2% of the total construction cost of a new building, creates 98% of the structural problems.

15-1. Roof Slopes. Roofs are constructed in a variety of shapes, of which the more common are gable or shed, hip, gambrel, mansard, sawtooth, and flat. The slope of a roof is generally referred to as the **pitch** and is expressed as the ratio of the rise of the roof to the horizontal span.

If a roof is dead flat, there is no rise. If a roof that has a constant slope and spans a building 20 ft wide has the peak 10 ft above the eaves, the pitch is 10/20 or **half-pitch** (also expressed as 6 in. on 12 in., or 6:12). Common roof pitches are 1/3, 1/2, and occasionally 5/8 or 3/4.

The greater the pitch, the more roofing material is required to cover the roof. With added allowances for waste, pitches result in the following increase in roofing-material area over a flat area: 1/4, 12%; 1/3, 20%; 1/2, 42%; 5/8, 60%; and 3/4, 80%. (See also Table 15-1.)

15-2. Bituminous and Plastic Roofing Materials. Expensive errors in roof application have resulted from failure of architects, engineers, builders, roofing contractors, and property owners to agree on trade terminology. As many as three or four different meanings have been conveyed by a term or phrase in a contract. The generally accepted definitions of bituminous materials are as follows:

Table 15-1. Variation of Roof Area with Slope

Classification	Incline		Inclined area per sq ft horizontal area	Percentage increase in area over flat roof
	In. per ft horizontal	Angle with horizontal		
Flat roofs	1/8	0°36′	1.000	0.0
	1/4	1°12′	1.000	0.0
	3/8	1°47′	1.000	0.0
	1/2	2°23′	1.001	0.1
	5/8	2°59′	1.001	0.1
	3/4	3°35′	1.002	0.2
	1	4°46′	1.003	0.3
	1 1/8	5°21′	1.004	0.4
	1 1/4	5°57′	1.005	0.5
	1 1/2	7°8′	1.008	0.8
	1 3/4	8°18′	1.011	1.1
	2	9°28′	1.014	1.4
Steep roofs	2 1/4	10°37′	1.017	1.7
	2 1/2	11°46′	1.021	2.1
	2 3/4	12°54′	1.026	2.6
	3	14°2′	1.031	3.1
	3 1/4	15°9′	1.036	3.6
	3 1/2	16°16′	1.042	4.2
	3 3/4	17°21′	1.048	4.8
	4	18°26′	1.054	5.4
	4 1/4	19°30′	1.061	6.1
	4 1/2	20°34′	1.068	6.8
	5	22°37′	1.083	8.3
	6	26°34′	1.118	11.8
	7	30°16′	1.158	15.8
	8	33°42′	1.202	20.2
	9	36°52′	1.250	25.0
	10	39°48′	1.302	30.2
	11	42°31′	1.356	35.6
	12	45°0′	1.414	41.4
Extra-steep roofs	14	49°24′	1.537	53.7
	16	53°8′	1.667	66.7
	18	56°19′	1.803	80.3
	20	59°2′	1.943	94.3
	22	61°23′	2.088	108.8
	24	63°26′	2.235	123.5

Bitumen. A generic term used to indicate either asphalt or coal-tar pitch.

Asphalt. A by-product of the refining processes of petroleum oils.

Coal-tar Pitch. A by-product of crude tars derived from coking of coal. The crude tars are distilled to produce coal-tar pitch.

Asphalt and coal-tar pitch serve in a variety of waterproofing and water-shedding applications. In their solid state coal-tar pitch and asphalt resemble each other in appearance. There are, however, major differences: Coal-tar pitch has a lower melting point, and as a result is self-sealing and melts faster. Because of the low melting point, the top layer of a coal-tar-pitch roof must be restrained by a covering of gravel or slag from flowing caused by solar heat. Asphalt usually does not require such a covering, but it does not have the self-sealing features of coal-tar pitch. Nevertheless, the range of use of asphalt is greater, because it can be used on steeper slopes.

Both asphalt and pitch are used as a saturant to produce the felts and prepared roofings in common usage. These products can be classified into four basic groups: saturated felts, coated felts, prepared roofing, and plastic roll materials.

For the first three categories, two types of felts are used in the manufacturing process: organic felts, manufactured from rags, paper, wood pulp, or a combination of these; and inorganic felts, composed of asbestos or glass fibers. Saturated and coated felts, delivered in rolls, have bitumen on both sides and are surfaced with a mineral release agent on either side or both sides, to prevent adhesion in the roll. Prepared roofings have a surfacing material, such as ceramic granules, sand, mica, or talc, added on the top side for decorative purposes.

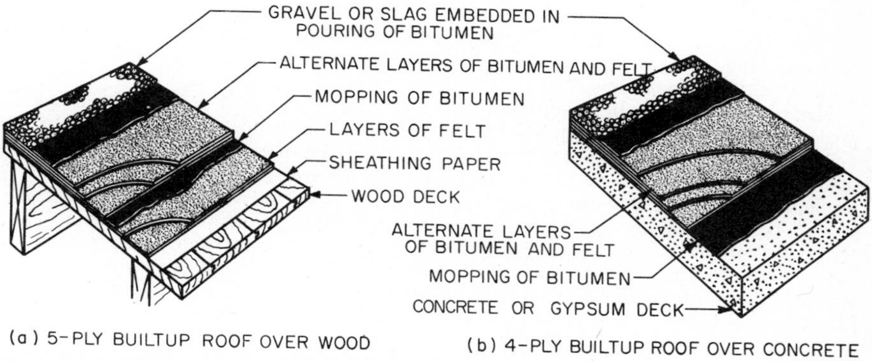

Fig. 15-1. Built-up roofing is applied with bitumen mopping between each two layers.

The fourth category, plastic roll materials, includes sheets of such plastics as polyethylene, vinyl, hypalon, and rubber. These materials are used as roof coverings, as well as for flashing, expansion joints, and vapor barriers.

15-3. Types of Roof Coverings. While a multitude of materials are utilized as roof coverings, there are basically two main groups into which they are classified, single-unit roof coverings and multiunit roof coverings.

Single-unit roof coverings are any roof surfaces, flat or sloped, that become, after application, a single entity. Included in this category are:

Built-up roofs (often referred to as *flat roofs*), with little or no slope, consisting of a number of layers (plies) of roofing felts bonded together with asphalt or pitch (Fig. 15-1).

Flat-seam metal roofings.

Synthetic roof coverings, which may include rigid vinyl sheet roofing or flexible sheet roofing.

Fluid-applied products, which are sprayed or brushed on and are manufactured from various components, such as plastic, rubber, epoxy compounds, and bitumen emulsions.

Multiunit roof coverings are materials suitable for application on steep roofs (slopes of 4 in. or more) and consist of many pieces installed individually, usually partly overlapped. This category includes such products as roofing shingles manufactured from asphalt-coated felts; tiles and shingles of plastic or glass compounds; asbestos-cement, aluminum, steel, and wood shingles; and clay, concrete, and slate tiles.

Attachment to Decks. For monolithic roof decks, such as cast-in-place concrete and poured gypsum, felts can be cemented directly to the roof deck. For roof decks made of separate units that permit nailing, the first two plies of felt are nailed to the deck, sealing the joints between the units from leakage and tying the roof covering to the deck. For multiunit roof decks that do not permit nailing, such as precast-concrete slabs, the first ply of felt is cemented to the deck by a strip-mopping operation, the deck surface being mopped with bitumen to within 4 in. of its perimeter.

Surface Coatings. Where the slope of a roof deck is 2 in. on 12 in. or more, it is difficult to retain slag or gravel surfacing applied to bitumen roofing to prevent flowing. The surfacing has a tendency to slide or to be blown off the roof. This has resulted in specification of top-layer, or cap, felts with a mineral surfacing embedded in the exposed face of the felts. Minimum weight of mineral-surfaced cap sheets is 55 lb per square (100 sq ft). Maximum specifications require the cap sheets to be applied over two 15-lb felts, one mopped and the first one nailed.

Use of smooth-surfaced roofs on suitable low slopes has resulted in specifications that eliminate the heavy surface coating in favor of a 30-lb felt nailed to the deck and followed by three 15-lb felts and a final coat of asphalt, applied at the rate of 10 lb per square. The thin coating is preferable to a heavy coating, because the thin film is not so subject to **alligatoring** (a wrinkling of membranes with an alligator-skin appearance). Also, loss of volatile oils under solar radiation is less than with thick films.

Slag, however, can be used on roofs with slopes up to 4 in. per ft if special precautions are taken in its application.

Mineral-surfaced Cap Sheets. These also may be used on slopes up to 4 in. per ft. These cap sheets are furnished in three forms. The simplest is a heavy sheet of saturated felt with its upper surface entirely covered with granules. The second type is similar, except for an unsurfaced lap portion along one side. The third has about one-half its width surfaced with granules, and is applicable to two-ply construction in the final assembly. Known as double-coverage or 19-in. selvage-edge roofing, this type should be applied so that each sheet laps the unsurfaced portions, which can be nailed to the roof deck or to wood nailers embedded in the deck.

Flat Roofs. Because of the tendency of a flat deck to retain water, application of built-up roofing on a level roof requires greater mechanical skill and judgment than for steep roofs. Gravity helps even a badly constructed steep roof to shed water.

15-4. Selection of Roof Coverings. The type of roofing material used to waterproof a structure depends largely on shape and slope of the roof, compatibility of materials to be installed together, geographical location, end use of the building, desired esthetic effect, and cost.

For steep roofs, a multiunit roof covering generally is advisible. In selection of the type, location of the building is an important factor, because climatic differences dictate the choice. For example, in the southern United States, clay or concrete tiles are usually selected to provide insulation from the semitropical heat and to protect the deck from rot and mildew. In Alaska, extreme temperature changes create severe stress on built-up bituminous roofs. Hence, an elastomeric coating might be advisable, even though costlier initially. Asphalt shingles are a practical, inexpensive choice for North American houses. Slate is used extensively in England, but less frequently in the United States because of high cost.

Apartment, commercial (stores and offices), and industrial (manufacturing and storage) buildings, schools, and hospitals are generally covered with a flat, single-unit roof. Occasionally, combinations of single-unit and multiunit roofings are specified, for example, for a mansard roof, flat on top, with a part of the top exterior of the building framed out to form a base for decorative shingles.

Roofs with sweeping curves, thin-shell domes, hyperbolic-paraboloid shells, folded-plate roofs, and other unconventional surfaces call for a single-unit covering that can be applied over difficult-to-waterproof shapes. Elastomeric or plastic fluid-applied compounds are often specified for the purpose.

Compatibility of materials used in the construction of a roof is of extreme importance and requires careful study before a decision is made in the selection of products. Chemical reaction between two incompatible products can create undesirable softening, hardening, bleeding, cracking, or peeling. Differing thermal contraction or expansion of roof deck, insulation, and covering can result in splitting and cracking of the roof membrane. For example, foamed-in-place polyurethane insulation, with its exceptionally high coefficient of expansion, when used with the wrong coatings, will crack and split, permitting moisture to enter.

Structures such as laundries, with a high interior moisture content, or cold-storage buildings, with a sharp contrast between interior and exterior temperatures, require vapor barriers to safeguard the roof membrane against infiltration of moisture and eventual deterioration of the roof covering.

SINGLE-UNIT COVERINGS

15-5. Built-up Roofs. The term built-up roof describes the roof covering that is applied on a deck with relatively little or no slope and that is composed of

Fig. 15-2. Five-ply built-up roofing.

a sandwich of alternate layers (or plies) of saturated felt and moppings of bitumen. An optional top pouring of pitch or asphalt, in which slag, or gravel may be embedded, completes the built-up roof. The felt membranes may be of the organic or inorganic type, and may be over an insulated or uninsulated deck.

The roof sandwich may consist of a few or many of the following products, depending on the specifications: vapor barrier; asbestos base sheet or coated base sheet; sheathing paper (over wood-fiber or wood decks); primer (over concrete and precast-concrete decks); insulation (board, urethane foam, or insulating concrete); coated or saturated felts, No. 15, 20, 30, or 40, or a combination, according to specifications; steep asphalt; low-slope asphalt; slag or gravel topping; nails or fasteners; No. 11 glass-fiber sheet; glass-fiber cap sheet or mineral cap sheet; glass-fiber tape to seal insulation; flashings.

Manufacturers' or roof-association specifications, as selected by the architect or engineer, must be closely observed to insure a serviceable roof covering.

The five-ply built-up roof specification is the oldest of all built-up roofing specifications. With the introduction of cast-in-place roof decks, the four-ply specification came into wide use, because it requires one ply of felt fewer than a wood-sheathed deck or a metal deck that has a multiplicity of side and end joints.

While one ply of felt may be dispensed with for a cast-in-place concrete deck, one extra layer of bitumen is required to compensate for omission of nailing of the first two felts. Thus, a four-ply built-up roof over concrete is equal to a five-ply over wood. (See Figs. 15-1 and 15-2.)

All subsequent specifications are modifications of the standard specifications for four- and five-ply roofing to take care of varying conditions. Minimum specifications have been achieved by reducing these standards by one or two plies of felt and one or two layers of bitumen, or equal.

Many factors must be considered to obtain a successful built-up roof application. For example: In multiacre-sized roofs, restraint of relative movement of roofing creates stresses never meant for the roofing to withstand. Dimensionally unstable insulation can cause ridges that weaken and split the membrane. Water in insulation not properly stored and handled can vaporize and blister the roof. A concrete deck, shrinking 1 in. per 100 ft in 16 years, will split a roof covering that otherwise would have performed for 25 years. Equipment, such as heating and air-conditioning units installed on top of a building, creates an additional source of roof-covering problems.

15-6. Dependence of Built-up Roofing on Roof Decks. *The roof deck is the foundation or base on which the whole roofing system depends.* The deck must satisfy considerably more than the engineering requirement that the deck be designed to support live loads (rain, snow, equipment, and workmen) and dead loads (weight of the deck itself and the roofing assembly). The roof deck must also provide proper drainage of water from the roof with enough slope so that water does not collect in the low areas between roof-forming members, and so that the roof area is completely dry within about 48 hr after it stops raining. The deck must, in addition, provide a foundation that is smooth, free of humps, depressions, and offsets at joints (Fig. 15-4), and that is stiff enough to support the equipment and materials needed to apply and maintain the roof system.

Many roof failures can be traced to poorly designed roof decks, faulty preparation of the deck to receive the roofing, and roofing applied before the concrete or gypsum deck has properly cured and dried. The following test may be used to determine deck dryness before start of a roofing application:

Pour on the surface of the deck 1 pt of bitumen that is specified for membrane heated to 350 to 400°F. If the bitumen foams, the deck is not dry enough to roof. After the bitumen has cooled, try to strip it from the deck. If the bitumen strips clean from the deck, the deck is not dry enough to roof.

15-7. Types of Roof Decks. Following are brief descriptions of commonly used roof decks.

Steel Decks. This type is composed of a number of pieces of sheet metal formed with ribs, to give strength and rigidity. The sheets are attached by welding or clips to the roof framing system.

Precast, Prestressed Concrete Decks. Individual units of concrete may be cast away from the building site and erected after they harden. They may be T-shaped, double tees, hollow-core slabs, flat plates, or channel slabs. Prestress applied with high-strength steel offsets in some part stresses that will be imposed on the unit by live and dead loads. Camber differentials between adjacent members after erection must not be more than ¼ in. Lightweight fill or grout over the slab or between adjacent members may be used to correct differentials in the deck before the roofing application begins.

Structural Cement-fiber Roof Decks. These are assembled of panels made with treated wood fibers bonded with cementitious binders. The panels are rated noncombustible and possess insulating qualities. Additional insulation may be required when metal supports are within ½ in. of the deck top.

Wood Decks. A wood deck should be of sound, well-seasoned lumber, connected by tongue and groove, or shiplapped or splined together at the side joints. Because of the moisture content of lumber, it is important that a slip sheet be installed between the deck and roofing membrane to permit dimensional changes to take place in the wood.

Gypsum Concrete Decks. This type is comprised of slabs or planks made of a mixture of gypsum, wood chips or shavings, mineral aggregate, and water, and reinforced with steel, or a galvanized woven-wire fabric. Manufacturers' specifications must be adhered to rigidly. Gypsum concrete must not be poured in rain or in freezing temperatures. If material freezes before it is cured, it must be replaced completely before roofing commences. This type of deck will not prove satisfactory if the building will have a high-moisture occupancy, such as frozen-food storage, laundry, or restaurant.

Insulating-Concrete Decks. These are made of lightweight aggregates, such as vermiculite or perlite, mixed with portland cement and water, or by foaming the concrete with chemical agents. If this material is used with galvanized steel decks, underside venting should be used. This type of deck will not give satisfactory service in areas or over surfaces and forms where complete evaporation of moisture is not possible.

Reinforced Concrete Decks. These are made of stone or crushed gravel and sand mixed with portland cement and water and reinforced with steel bars or welded-steel fabric.

Thermosetting Insulating Fill. This is a mixture of asphalt and vermiculite or perlite compacted in place, to provide insulation and slope for drainage. The fill must be applied over a rigid subbase that will not deflect under pressure from the compacting roller. If the building will have a high-moisture occupancy, venting precautions should be taken, and a vapor barrier should be installed on the deck under the insulation. If moisture from rain or other sources wets the fill before the roofing is applied, it is essential to thoroughly dry the fill.

15-8. Vapor Barriers for Built-up Roofing. Unless precautions are taken, water vapor in the interior of a building, especially if it has a high-moisture occupancy, will condense on the cold underside of roofing felts, saturating the insulation and reducing its effectiveness, or will drip back into the building, staining the ceiling or wetting the floor. In some cases, the moisture will remain in the felts, freezing in cold weather and rupturing the roof membrane.

A solid, unbroken vapor barrier placed below insulation installed on the underside of the roofing felts can, however, prevent such condensation.

The following general rules will assist in determining whether or not to install a vapor seal under built-up roofing:

A vapor seal should generally be applied on all roof decks in cold climates, and on all decks of heated buildings where the midwinter temperature may fall below 45°F.

Vapor barriers should be used on all roof decks in temperate climates wherever conditions of high inside humidity exist, such as in textile mills, laundries, canning factories, creameries, and breweries.

Vapor barriers should be used on all roof decks where the roof deck itself contains an appreciable amount of moisture. A vapor barrier should be used in all cases on top of concrete and poured gypsum decks.

Perm Ratings. The effectiveness of a vapor barrier is measured by its perm rating, which is a measure of porosity of material to passage of water vapor. Perm ratings are established by ASTM procedures. To be classified as a vapor barrier, the material should have a perm rating between 0.00 and 0.50 perms.

A perm rating for a material is the number of grains of water vapor (7,000 grains equal 1 lb) that will pass through 1 sq ft of the material in 1 hr when the vapor-pressure differential between the two sides of the material equals 1 in. of mercury (0.49 psi).

Vapor-barrier Materials. Coated base sheets, saturated felts, laminated kraft papers, and polyvinyl chloride (PVC) sheets are commonly used as roof vapor barriers.

PVC sheets are generally specified when a noncombustible roof rating is required. PVC has a very low perm rating, about 0.10 to 0.50 perms. The sheets are attached to the deck with mechanical fasteners that pass through the insulation (on top of the vapor barrier), or with a compatible adhesive.

Coated base sheets may be used with lightweight, insulating roof fills, such as foamed, cellular, perlite, or vermiculite concretes, which have a high moisture content. Vents, for example, a vent stack or edge venting, should be provided to release moisture from the deck itself. Coated base sheets have proved to be a satisfactory vapor barrier when the material has been properly stored and handled before application, and if the installation includes venting.

Kraft laminates, vapor-barrier sheets made up of two layers of special heavy-duty kraft paper laminated together, have a low perm rating, about 0.25 perm. Used with compatible adhesives or mechanical fasteners through the insulation, these laminates will result in a UL or FM acceptable fire-retardant system.

Saturated felts are sometimes used as vapor barriers, but coated base sheets are preferable.

15-9. Insulation with Built-up Roofing. Insulation is a material or product that retards transfer of heat through itself. In a roof system, heat is transferred primarily by conduction, the movement of heat from a warm surface to a cold surface of a material. Insulation usually is placed below the roof deck, but it may instead be placed above (Art. 15-13).

Efficiency of an insulation material is measured by its thermal conductivity K or thermal conductance C.

The K factor of a material is the amount of heat, Btu, that will pass through 1 in. of the material over an area of 1 sq ft in 1 hr when the temperature difference is 1°F. Thermal conductance C is similar to K, but applies to nonhomogeneous material or a specific thickness of a material. R values are used to indicate the thermal resistance of a material, and are the reciprocals of the K or C values.

The principal types of roof insulations are:

Cellular-glass insulation. It has a C value of 0.39 for 1 in. thickness. Because it has a perm rating of 0.0, a vapor barrier is not required.

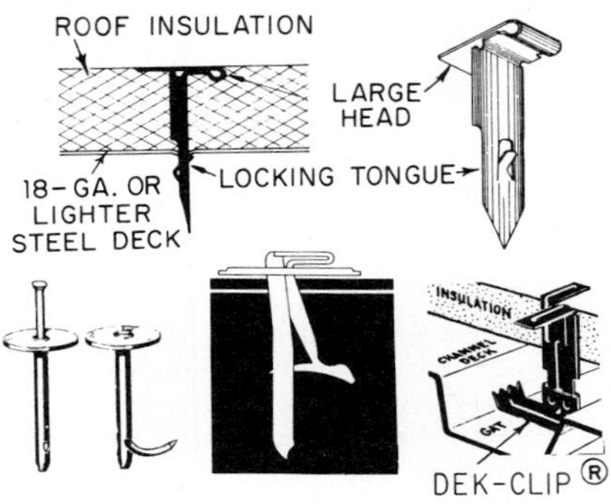

Fig. 15-3. Types of fasteners used to anchor insulation to roof deck.

Glass-fiber insulation. It has a nominal K value of 0.25.

Perlite insulation. Consisting of expanded volcanic glass combined with mineral fibers and binders, it has a C value of 0.36 for a 1-in. nominal thickness.

Cork insulation. It has a K value of 0.26 for densities of 6.4 lb per cu ft.

Fiberboard insulation. Made primarily of wood fibers and a binder, it has a K value of 0.36.

Urethane rigid-board insulation or urethane foamed-in-place insulation. These are made of a closed-cell plastic foamed under pressure. The rigid-board insulation is sometimes sandwiched between two layers of coated roofing felts. Foamed-in-place urethane (a term used interchangeably with polyurethane) is sprayed directly on the roof. The board insulation is usually covered with a standard built-up roof, whereas the foamed-in-place insulation requires a liquid roof coating formulated for the purpose. Urethane insulation has a C value of about 0.15 for a 1-in. nominal thickness.

Installation of Insulation. Insulation should be protected from adverse weather at all times, including transit to the job. All material installed in a day's work must be covered and exposed edges protected. Care should be taken to avoid damage to the material. For some roofing, a taping of joints is recommended by manufacturers, particularly with glass-fiber roof insulation.

Insulation may be attached to a roof deck by any of several possible methods, depending on type of roof deck, type of insulation, and whether a vapor barrier has been specified (Fig. 15-3).

Adhesives or mechanical fasteners should be used where a vapor barrier is installed. Nailing is usually the method with wood decks. Adhesives are often used as a fastening agent with steel or concrete decks. But the time-tested method, mopping with hot asphalt, generally is the preferred installation method.

Often, installation of insulation starts with a "sprinkle" of hot asphalt on the deck into which insulation boards are laid. (Channel mopping is another term used to describe the application. Solid mopping refers to covering of the entire roof surface with hot, steep asphalt.)

15-10. Built-up Roofing Failures. The primary cause of roofing failures may be summarized as follows:

Faulty design of the entire structure.
Faulty design of the roof deck.
Poorly designed flashing details.
Improperly installed flashings.
Application of roofing materials during inclement weather.
Improper storage of roofing materials before installation.
Improper use of materials.
Faulty and careless workmanship. (See, for example, Fig. 15-4.)

Movement of roofing, which often causes roofing failures, is difficult to avoid, because of differences in dimensional changes that occur in connected materials

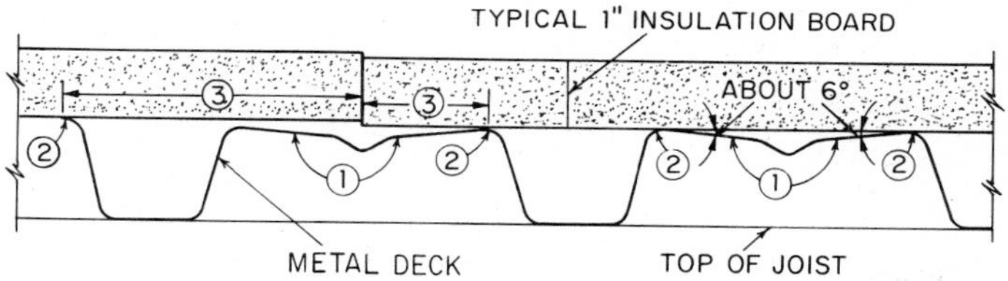

Fig. 15-4. Poor installation of insulation on metal deck. Irregularities (1) in the top flanges of the deck prevent adherence of the insulation. Top bends (2) of the ribs provide the only support for the insulation. The end joint (3) of the insulation board is cantilevered, because the top flange is not level, as a result of which the board may fracture under foot traffic.

as temperatures change or moisture content varies (Table 15-2). These failures result basically from a design error caused by any of the following factors:

Deck or building movement.
Thermal movement of the substrate, such as insulation.
Thermal movement of the built-up roofing.

Movement of the built-up roofing due to alternate wetting and drying as condensation occurs in the roofing plies and evaporates. Prevention requires careful attention to detail in preparing the specifications, with full knowledge of the properties of materials to be installed.

See also Art. 15-12.

15-11. Temporary Roof Coverings. When weather conditions are likely to be adverse to application of built-up roofing, or on roofs where trades other than roofers have further work, a temporary roof covering may be applied and application of the permanent roof covering delayed to a date when more favorable conditions prevail. Use of temporary coverage helps extend the construction season through the full year; allows use of the roof as a storage area during construction; permits installation on the roof of equipment, such as ventilating fans, signs, air conditioners, and heating units; and allows foot traffic on the roof.

A temporary roof covering may consist of a coated base sheet with 4-in. side and end laps. If extraheavy traffic is expected, double coverage should be employed.

The temporary roof is installed exactly as a conventional roof, but will have served its purpose when the final complete roof is installed. In most instances, the temporary covering should be removed, especially if extensive damage has occurred to it. If damage is light, the temporary roof can be repaired, to serve as a vapor barrier under the roof insulation or as the deck for the new application.

Table 15-2. Moisture and Thermal Expansions of Roof Materials

Material	Expansion on wetting of dry sample		Expansion for 100°F temperature rise	
	%	In. in 10 ft	%	In. in 10 ft
Reinforced polyester	0.001	0.001	0.10	1/8
Dense stone (marble)	0.001	0.001	0.03	1/32
Brick	0.007	1/128	0.03	1/32
Concrete	0.03	3/64	0.06	5/64
Steel	.....	.....	0.06	5/64
Sandstone	0.07	5/64	0.07	5/64
Wood				
across grain	2.5	3	*	
along grain	0.05	1/16	*	
Felt (wood fiber)				
Coal-tar pitch				
along grain	0.5	5/8	0.2	1/4
across grain	2.0	2 1/2	0.3	23/64
Asphalt				
along grain	0.2	1/4	0.1	1/8
across grain	1.5	2	0.2	1/4
Wood fiberboard	0.5	5/8	0.03	1/32
Glass fiberboard	0.05	1/16	*	
Perlite board	0.26	5/16	*	
Polystyrene	0.02	1/32	0.4	1/2
Polyurethane	0.9	1	0.6	3/4
Foamed glass	*	.....	0.05	1/16

* No figures available.

15-12. Good-practice Recommendations for Built-up Roofing. Careful adherence to the basic conditions for good application of a built-up roofing system will result in an economical long life for the roof. Often, however, when the simple rules are overlooked, expensive failures requiring premature roof replacement occur. The basic requirements for good roofing are:

Structurally sound buildings and stable roof decks.

Insulation with acceptable thermal dimensional stability.

Acceptable vapor barrier, whenever required.

Dry insulation and roofing materials.

Uniform and positive adhesion between materials, including adhesion of the vapor barrier to the deck, insulation to the vapor barrier, the first ply of roofing felt to the insulation, and between all plies of felt.

Application of asphalt or pitch in quantities not less than those established in the specifications.

Prevention of moisture in any form from being sealed into the roofing system during its application.

The roofer should exercise good quality control and insist on good workmanship throughout the roofing application.

The roofing contractor must have the cooperation of the architect or engineer, general contractor, and roofing materials manufacturers, so that the application of the roofing system may be coordinated under the most favorable weather and working conditions.

The architect and the general contractor should avoid the lowest bid price if it appears to be far below the price structure within a particular area. An abnormally low price for roofing initially establishes the primary reasons for a roof failure before construction of the project is even started.

15-13. Inverted Built-up Roofing System. Also known as upside-down roof assembly, an inverted built-up roofing system consists of several layers installed in a sequence different from the usual one in which insulation is placed below the roof deck. First, standard built-up roofing is applied directly to the deck. Then, rigid insulation that is impervious to moisture, such as extruded polystyrene foam, is bonded to the top of the built-up roofing with a mopping of steep asphalt (Fig. 15-5). A layer of ¾-in. crushed stone (1,000 lb per square) or paving blocks or structural concrete on top of the insulation completes the assembly. Gravel or slag should not be used, because the sharp edges would damage the bare insulation underneath.

The theory is that the insulation, set above the roofing, both insulates the building and protects the built-up roofing from the harmful effects of thermal cycling, ultraviolet degradation, weathering, and roof traffic. Most common defects caused by these elements, such as blistering, ridging, cracking, alligatoring, and wrinkling, are virtually eliminated.

The roofing can be installed over a variety of decks: wood, structural wood-fiber boards, cast-in-place concrete, precast concrete, poured gypsum, lightweight fills, or light-gage metal (with a rigid board installed as a base).

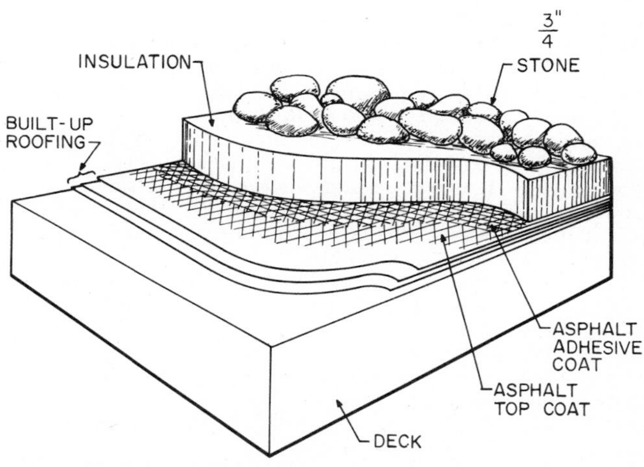

Fig. 15-5. Inverted built-up roofing (patented), with insulation placed above the deck and roofing.

Note that the inverted system specifies twice as much stone per square as usually applied. Thus, this roof system is about 300 to 400 lb heavier per square, and consequently requires a stronger roof deck.

15-14. Foamed-in-place Polyurethane Roof System. Urethane (or polyurethane) board insulation, factory produced, is often included in built-up roofing. Foamed-in-place urethane, however, is usually specified when a roof is difficult to waterproof with conventional built-up roofing. Foamed-in-placed urethane insulation is produced on the job site and foamed directly onto the roof deck. The urethane immediately conforms to the shape of the substrate and seals all cracks and crevices. Flashing becomes unnecessary.

Chemically, polyurethane foam is not a single material, but a family of plastics based on urethane polymer. Flexible, rigid, and semirigid foams are produced by the reaction of a number of components when mixed together in differing ratios, depending on the end use of the resulting product. Roofing insulation is usually foamed with truck-mounted, portable equipment. Materials used include a polyisocyanate and a polyhydroxyl, which are reacted under carefully controlled conditions at a low viscosity. Other ingredients include a blowing agent and surfactants to control cell structure, and a catalyst to control speed of reaction. Formulations may also include flame retardants, fillers, extenders, and pigments.

Flexible foam derives its physical properties from cells that are broken and open. Semirigid or rigid foams are produced by adding functional resins to conventional foam formulations. The purpose is to keep the cells closed, retaining the gas and giving the foam a rigid or semirigid structure.

The insulation produced in this manner rises (somewhat like bread dough) to 30 times its original volume. Then, it sets into a plastic compound of uniform cross-linked cells (more than 200,000 cells per cu ft). Actually, the material is 97% gas and is better insulation than still air.

The foam, uniformly sprayed over the surface to be insulated, rises to its full height and hardens within a few minutes. A protective coating should be applied as soon as possible thereafter, to avoid moisture from rain or dew condensing on the surface and preventing good adherence of a top covering.

Except in cold weather, deck preparation is minimal. All that is required is that the deck be clean, dry, and free of debris. On structures being reroofed, major patching should be done in advance. Because the foam adheres to any surface, it is self-flashing. Urethane foam has been used successfully on sloping decks and on particularly troublesome decks with unusual design features.

Cold-weather applications present special problems. Although it is possible to heat the isocyanate and the resin and to insulate spraying lines, the substrate to be sprayed is still cold. Adhesion to the deck will be poor, and the entire roof may crack and shear away from its base. Foamed-in-place urethane may be used, however, if the roof is temporarily enclosed and the enclosure is heated.

15-15. Metal Roof Coverings. One of the most important factors in selection of metal roof coverings is their ability to conform readily to a wide variety of roof shapes and slopes. Principal metals used include steel, aluminum, copper, lead, and alloys, such as terneplate, stainless steels, and Monel metal. In addition, multimetal systems, new alloys, dead-soft metals, together with fast installation methods, such as those used for lock-in or snap-in panel systems and the zipperlike zip-rib roofing, offer many design possibilities.

Multimetal systems employ two or more metals to obtain properties not available in a single metal. The metals are joined in a permanent molecular bond by heat and pressure. Usually, one metal is bonded to both sides of a base metal that provides over 50% of the thickness of the system. Cladding, however, need be done only on one side for some applications, with any ratio of thickness desired.

The multimetals used in roof coverings include:

Terne-coated stainless steel, which is a nickel-chrome stainless steel clad on both sides with a terne alloy (80% lead, 20% tin).

Copper-clad stainless steel, fully annealed and provided with a bright surface. More oxide-free than commercial copper, the surface can be lacquered to preserve its brilliance. Stainless steel clad on only one side with copper is frequently laminated to a stiff substrate, such as plywood. This product has a low coefficient of thermal expansion.

Brass-clad stainless steel consists usually of 85-15 brass clad to a 400-series, ferritic stainless steel.

Galvanized steel.

Nickel stainless steel.

Roofing panels are manufactured from either 24- or 26-gage steel, usually with a protective coating on both sides. Color pigment is often added by manufacturers, who offer 15- and 20-year guarantees on the color fastness.

Dead-soft Stainless Steel. This metal roofing handles easily, conforming to any desired shape with little or no springback. It is easy to solder and can be joined by all the standard methods. It also may be used for flashing, expansion joints, fascia, and copings.

Corrosion Prevention. Corrosion or gradual deterioration is a phenomenon encountered in the use of practically all commercial metals. This disintegration occurs because each metal possesses a potential force that tends to make it revert to the more stable state of chemical combination in which it is frequently found in nature.

Dissimilar metals, when in contact or connected by an electrolyte, set up galvanic action, resulting in the destruction of the metal lower in the electromotive-force series. Aluminum, high in the series, for example, will attack zinc, next lowest in the series. Roofing metals, in the order of highest potential, list as follows: copper, lead, tin-nickel (Monel), iron (steel), aluminum, and zinc. Because current flows from copper to steel, for example, these metals should not be in contact.

A metal lower in the series, however, may be used for sacrificial protection of another metal. For example, an unbroken zinc coating (galvanizing) will protect steel against corrosion.

Metals near each other in the series suffer little damage from each other's contact. The farther apart two metals in the series, the greater the damage may be from contact. For this reason, a copper roof should not be applied directly over a wood roof deck fastened with steel nails. Instead, the copper should be separated from the deck by insertion of an asphalt felt to insulate the copper from attack by the steel nails in the roof deck. Also, aluminum sheets should not rest directly on steel purlins, but should be insulated from the steel with felt strips.

As far as possible, nails or other fasteners used in securing metal roof coverings should be of the same metal or alloy as the roofing.

Thermal Movements. All building materials are in constant, although imperceptible, state of motion because of temperature changes, and metal sheets move much more than other roofing materials. Table 15-3 indicates the relative thermal movements of several roofing metals.

Table 15-3. Change, In., in 8-Ft Sheet for 150°F Temperature Change

Steel	6/64
Monel metal	7/64
Copper	9/64
Aluminum	12/64
Lead	15/64
Zinc	16/64

Because the amount of expansion and contraction is proportional to the dimensions, large roofing sheets undergo larger movements than small sheets. A metal roof laid in large sheets that are permanently fastened to the deck, consequently, would buckle and split at seams and joints.

To allow for expansion and contraction, four types of metal-roof installations are often used, batten seam, standing seam, flat seam, and corrugated metal.

Corrugated-metal roofing, as the name implies, consists of sheets with corrugations, which absorb thermal movements in the direction across the undulations. In the other direction, connections between sheets must permit thermal movements.

Standing-seam construction employs narrow panels interconnected along their length by raised folded edges that provide for expansion and contraction, while loose-locked seams allow longitudinal motion. (This is also true of batten-seam construction, a type of roofing not used as frequently as standing seam.)

Batten-seam roofing has a metal cap, or batten, installed over raised seams.

Flat-seam construction employs a series of small sheets that are interconnected by solder on all four edges. The sheets are so small that expansion and contraction are negligible and are amply provided for by a slight buckling in the center of each sheet.

Except for corrugated metal, metal roofing sheets should not be fastened directly to structural framing, because the metal must be free to move. Connections to structural framing may be made with small metal clips for standing-seam and flat-seam roofing. In all three cases, the clips should be locked into the laps or folds of the roofing sheets.

Where metal sheets are fastened directly to the framing, although provisions are made for expansion and contraction, the consequent movement of the sheets makes it virtually impossible for any of the fastening devices used to remain tight. This has led to use of lead or Neoprene washers, which take up the slack caused by loosening of bolts, friction of moving sheets, and enlargement of holes drilled for fasteners. Note that interior and exterior air pressures on this type of roof should also be reckoned with in providing for sheet movement.

Special caps shaped to the angle of the roof should be used where metal-roofing sheets meet at a ridge. These caps should have pockets on both sides to house the ends of the sheets.

15-16. Cold-process Roof-covering Systems. These are used primarily for maintenance and repair rather than for new construction, but they are not limited to these applications. Many systems are available, some involving chemical combinations with bituminous products. One cold-process roofing system requires a heavy-duty glass mesh to be embedded in a high-viscosity, asphalt-based emulsion which is brushed or sprayed on a roof deck, and a finish coat is added for a smooth surface. Another system combines polyurethane with coal tar to produce a sprayed-on waterproofing membrane.

To produce workable liquidity, bituminous materials are thinned with solvents or emulsified with water.

Cutbacks. Solvent-thinned bitumens are known as cutbacks. Solvents include kerosene, in the manufacture of cutback asphalt, and toluene, in the manufacture of cutback tar. With cutbacks, the cementing process of a roof is achieved by evaporation of the solvents, which leaves a film of bituminous material.

Emulsions. These are produced by melting bitumen and dispersing it into tiny particles suspended in water by means of soap or a chemical emulsifying agent. On a roof, the water evaporates, leaving a solid film of bitumen on the surface.

Felts and Fabrics. Plies used with cold-process bitumen differ from those used with hot. Since cementing takes place on evaporation of solvents or emulsion water, it is necessary that the drying-out process be speeded. For the purpose, porous mats made of glass fibers often are used, because they have the advantage that the open weave allows water vapor to travel through it, a breathing process that eliminates blisters. Glass fibers, however, require a thick mastic-type emulsion, because light films are not strong enough to hold the fibers in place.

Application. Cold-process materials are applied by brush or spray. But first, the roof surface must be prepared, to eliminate cracks, ruptures, holes, fish mouths, and other undesirable characteristics of weathered roof decks.

In selection of cold-processing roofing materials, designers should beware of unprincipled producers who promote doubtful products by making extravagant advertising claims. Cold-process roofing materials, when properly selected and applied, can add years of life to roofs, but they are not cure-alls and will not add much life to roof coverings that have disintegrated. They will not bridge gaps in surfaces and will prolong life only in materials in which life exists.

15-17. Elastomeric Roof Coverings. The term elastomer refers to a synthetic polymer with rubberlike properties. This type of material gained rapid acceptance for use on free-form roof shapes, such as large domes, barrel roofs, folded plates, and hyperbolic paraboloids. Also, elastomers are advantageous for cable-supported roof decks, which require a roofing membrane with minimal weight and an elasticity that withstands continuous movement. Furthermore, liquid-applied Neoprene-hypalon systems weighing only 20 to 50 lb per square are often specified for very large industrial buildings rather than conventional built-up roofs weighing up to 600 lb per square.

Elastomeric roofing systems fall into two general categories:

Fluid-applied systems, applied by roller, pressure-fed roller, spray gun, brush, or squeegee in single or multiple coats.

Sheet-applied systems of the one-ply type or those requiring a top coat for color.

The fluid-applied system has the advantages of conforming readily to any shape of substrate without material waste due to pattern-cutting requirements, as is necessary with the sheet system, and better adhesion. Fluid application also forms a continuous membrane, free of joints. The elastomer will not, however, fill voids and requires continuous substrate junctures and expansion joints. Insulation-covered decks present special problems, except for sprayed-in-place polyurethane insulation, which offers a continuous unbroken surface to be covered by the elastomer.

One-ply sheet systems often have material, such as asbestos, laminated to the back. The single ply is provided in roll form and is installed with an adhesive applied to the roof deck and at the joints where rolls lap. The product is lightweight, easy to install, and forms a long-lasting watertight roofing, with excellent expansion-contraction properties. The one-ply system also helps to resist the high stress caused by roof-deck movements when air conditioning is used in the building interior. Furthermore, this roofing will withstand light foot traffic. Repairs are made with patches, cutouts, or sealants.

Sheet systems should not be used for steep roofs or roofs with intricate designs that would involve extensive cutting, fitting, and seaming. Sheet systems will, however, tolerate greater deck defects. If large thermal movements may occur, sheet elastomers can be permitted to "float," remaining unbonded over some areas.

A number of compounds fall into the elastomer family, including:

Silicone. This is a group name for polymers with a structure of alternate silicone and oxygen atoms, with various organic groups attached to the silicone. The elastomer is compounded with a suitable filler and vulcanizing agent to produce a liquid-applied film. This coating has higher dirt retention than most other elastomers.

Butyl Rubber. This is a synthetic rubber, a liquid formed by copolymerization of isobutylene with a small proportion of isoprene or butadiene. It is compounded in black only, and is somewhat difficult to adhere under normal field conditions.

Polyurethane Rubber. This elastomer should not be confused with polyurethane insulation, which is an entirely different product. The polyurethane rubber formulation is produced by addition of a diisocyanate to polyesters. Fire retardants can be added to increase fire resistance. This rubber can be made with a wide range of physical properties. It has low vapor transmission and excellent resistance to thermal exposure and weathering. Fewer coats are required in application than for other elastomers, thus reducing labor costs.

Neoprene. This is a family of elastomers made by polymerization of chloroprene and formed into liquid, sheet, or puttylike consistency. The product will not support combustion and has excellent adhesive properties. It is normally compounded in dark colors.

Hypalon. This is made of chlorosulfonated polyethylene, in liquid, sheet, or puttylike consistency. It will not support combustion, and has even greater resistance to thermal movement and weathering than Neoprene. It is formulated in a variety of subtle colors.

Roof Decks. To be successful, elastomeric roofing must be applied over a satisfactory roof deck, carefully selected and specified. Lightweight-aggregate concrete decks or foamed-concrete fills, for example, should not be used with liquid elastomeric coatings. Air and moisture trapped in such substrates will eventually blister the elastomeric film. Structural concrete, properly finished and cured for a minimum of 30 days, makes an excellent foundation for liquid roofing. Plywood provides an excellent surface for Neoprene and hypalon coatings. Closed-cell polyurethane insulation serves as an excellent substrate for elastomers, and where cost is not an obstacle, elastomers are the preferred covering for the foamed-in-place product.

Application. Manufacturers' specifications should be followed carefully in all applications. A typical application of the liquid elastomer over a concrete deck might call for (1) a neoprene primer, (2) two body coats of elastomer, and (3) two color finishing coats of elastomer.

A sheet system might require (1) a water-dispersed deck adhesive; (2) the laminate roofing, cut to size and shape according to the deck design and then rerolled before application; (3) brooming or pushing out of air blisters and pressing the material into place within 15 min after the adhesive has been spread; (4) "flashing tape," caps, and seals for field joints, vents, and roof edges; and (5) often, a top coating of liquid synthetic rubber to cover the tapes and provide a seamless surface.

15-18. Corrugated Asbestos-cement Sheets. These are applied somewhat like corrugated metal. The same type of fastener is employed. Because asbestos cement is a more easily punctured material, however, the sheets are placed over welded studs on purlins, and the corrugations are malleted until the studs pierce the material. Rubber mallets are used to deliver enough force without damaging the sheets or the studs.

Asbestos-cement roofing differs in shape from metal roofing. Asbestos-cement sheets are molded so that the valleys of the first and last corrugations lie downward, to allow the minimum side lap to be one corrugation instead of 1½.

Two methods of application are employed, the step method and the miter method. Both prevent occurrence of four-thickness end and side laps, a feature of corrugated-metal roofing that would be unsightly as well as difficult to waterproof with the thicker asbestos-cement roofing. In the step method, alternate courses of the sheets are started with half sheets. In the miter method, no effort is made to break joints, but with a miter at one corner of each sheet, end and side laps have uniform thickness.

15-19. Rigid Vinyl Roofing. Polyvinyl chloride in flat, corrugated, and ribbed sheets is available for roofing in standard and special lengths. Standard lengths range from 8 to 20 ft. Longer sheets can be ordered, however, to reduce the number of end laps required.

The sheets are fastened to structural framing with machine screws, bolts, or wood screws, depending on whether the frame is metal or wood. Fastening may be made less laborious by use of fasteners that seal both top and underside when thrust through a hole drilled from the top. These fasteners include "pop" rivets and machine screws with a Neoprene cap and jacket.

Vinyl sheets can be cut quickly on the job with portable power saws with abrasive cutting wheels. Where possible on large jobs, sawing, cutting, and drilling of vinyl sheets should be done in the shop. The sheets can be nested and drilled and cut a half-dozen at a time. In addition, much of the sawing can be speeded with stationary machines.

As with all sheet roofing material, application starts at the eaves opposite the prevailing wind direction. While the sheeter has no jurisdiction over purlin spacing, the sheeting should be arranged to end over a purlin. Overhangs at eaves should not exceed 6 in. Each of the second and succeeding sheets placed alongside each other should cover one corrugation of the previous sheets and should be fastened along the lap (in the valley) at 12-in. intervals. Sheets laid at the ends of these, and succeeding courses, should cap the lower ones at least 8 in. At a ridge, corrugated closures should be placed to insure a neat finish and adequate sealing. Where full weatherproofing is required, filler strips may be inserted at eaves, and calking cement employed between end laps.

For economy, sheets should extend from eaves to ridge. This cuts down the number of fasteners, amount of accessories, and amount of material, because end laps are eliminated. For many arched roofs, however, sufficiently long sheets are not available, not because manufacturers cannot produce them, but because of shipping difficulties.

MULTIUNIT ROOF COVERINGS

15-20. Asphalt Shingles. A typical asphalt shingle is produced from five major compatible components: a smooth base felt, saturating asphalt, coating asphalt reinforced with mineral stabilizers, back coating for dimensional stability and as a barrier against moisture from the building interior, and opaque mineral granules embedded in the heavy coating of asphalt for texture and color.

Asphalt shingles are widely used in preference to traditional slate, tile, or thatched roofs. Many types of asphalt shingle are available. They range from a low-cost, lightweight type to a heavyweight shingle weighing from 230 to 350 lb per square. Giant individual shingles and square-butt strip shingles, with or without cutouts for design tabs, provide labor-saving advantages. Self-sealing shingles, with a factory-applied adhesive activated by the sun's rays (roof temperature 60°F or over), offer resistance to wind uplift. Another design features a slit or tab that matches an engaging device in the adjacent shingle and interlocks the shingles into a secure roof covering.

To prevent wind damage, self-sealing or interlocking shingles should be used on roofs with less than 4 on 12 pitch in all wind zones. In medium-wind areas, roofs with slopes of 4 on 12 or steeper should have sealed tabs if the shingle weight is less than 275 lb per square. Low-wind zones require secured tabs in areas where winds reach 70 mph or more.

Also, a large variety of asphalt roofing styles is available, in a wide range of colors, textures, and patterns. The style of the building will reflect on the type of shingles selected to complement it.

Application. With all asphalt-shingle materials, the first step in application is to fasten an underlay of No. 15 asphalt felt to the roof deck, with a starting strip at the low end of the roof from which work proceeds. The felt should be lapped 6 in. at the end and sides.

Because many backup leaks at eaves are caused by moisture traveling upward from the eaves beneath the shingles, through capillary attraction, it is a wise precaution to cement the starter material to the roof deck, rather than nail it. In winter, water may dam up in a gutter and backing up under the shingles, follow a nail puncture through the roof into the building interior.

Faulty shingle applications on wood roof decks can be traced in part to too few nails, but many defective roofs can also be attributed to nailing shingles too high. If the shingle is nailed ½ in. above the slots, as in the case of square-butt shingles, each shingle will be held to the roof not just by the four nails called for in specifications, but by eight nails—four driven to attach the shingle and four others driven to attach the shingle on top. Each nail travels through two thicknesses of shingle. If the shingle is nailed 2 in. or more above the slots, however, it is held by only four nails at the top, and it is more likely to be pulled loose by winds.

Nails should be at least 1¼ in. long on new work, and 1¾ in. long on reroofing. Hot-dipped, zinc-coated iron or steel nails, with barbed stems, usually should be used. But ordinary steel nails give good service, and if the stems rust slightly, the effect will be beneficial, because the holding power of a rusty nail is excellent.

While shingle tabs are not required to be fastened, it is a useful precaution in high-wind areas to place a small dab of asphalt cement under each tab, which then should be pressed firmly onto the concealed portion of the underlying shingle.

Special Treatments. Color and form may be used imaginatively in special applications to dramatize or accent a roof, or to stress the horizontal portions of a building (Fig. 15-6). Special treatments most frequently employed are ribbon courses, two-tone rib courses, single shingle applications, and complementary color courses.

Ribbon Courses. A rugged, horizontal roof texture can be achieved by applying a triple thickness of asphalt shingles every third, fourth, or fifth course. This presents eye appeal and texture and is especially suited for the large roof areas of a long one-story building.

Two-tone Rib Courses. A colorful approach to roof styling, similar in appearance to ribbon courses, utilizes two colors or tones on the roof. The boldness of the design depends on the colors of the shingles selected for the entire roof.

Every three or four courses, one course of 9-in. black starter strip is applied. The next course is applied with only 1 in. of starter strip exposed to the weather. A black band or rib results. The lighter the main roof color, the greater the contrast with the black and the bolder the styling. This design is particularly effective on square one-story buildings.

Single Shingle Applications. Visual appeal can be achieved by application of heavyweight shingles to obtain a rough-textured individual shingle look. Application is simple with a no-tab, two-tab, or three-tab shingle.

The first course is applied the standard way. Shingle sections 4, 6, and 8 in.-wide are cut from additional bundles of shingles and are nailed over the cutouts and the butted ends of the shingles. For a three-tab shingle, a 4-in. section is cut off each end, and each remaining section is cut in half. If no-tab shingles are used, the cut sections are placed at random intervals between butted joints. The sections are extended about ¼ in. below the shingle course, depending on the amount of texture desired. Care should be used so that no two sections of the same size are applied next to one another.

Complementary Color Courses. Complementary color courses take advantage of the variety of colors asphalt shingles offer. This application is suited to commercial buildings, such as restaurants and motels. Different-color shingles are used in alternate courses up the roof. The shingles can be a contrasting or complementary color. For example, the different courses may be dark green and a lighter green, or red and brown, or slate and black.

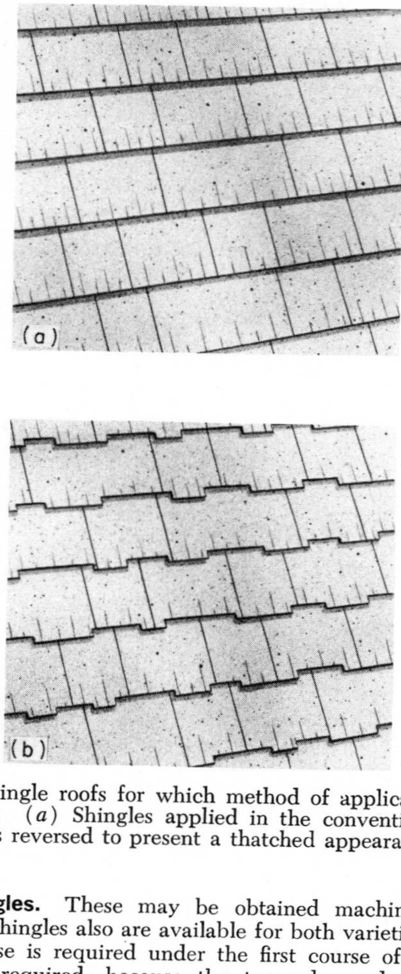

Fig. 15-6. Asphalt-shingle roofs for which method of application of shingles determines the appearance. (*a*) Shingles applied in the conventional straight-line way. (*b*) The same shingles reversed to present a thatched appearance. (*Courtesy Globe Industries, Inc.*)

15-21. Wood Shingles. These may be obtained machine-sawn or hand-split. Fire-retardant-treated shingles also are available for both varieties.

An undereaves course is required under the first course of shingles on the roof; but no cant strip is required, because the tapered wood shingles fit closely to the roof.

Because of water drip at the eaves, eaves shingles receive more abuse than the other shingles on the roof. To prevent any possibility of leakage at eaves or cornice, a triple layer of shingles at those areas is a wise and comparatively inexpensive precaution, adding little in material and labor to the average job. If a triple layer is used, the joints in all three layers should be laid so as not to coincide.

The standard exposure of wood shingles is 5 in. But depending on the length of shingles, they may be laid at exposures ranging from 3½ to 12 in.

As the shingles are applied in alternate courses, with joints broken in relation to courses above and below, the desired exposure is maintained, except where it is adjusted to line up with dormers or with the ridge line.

Because a sharp hatchet and saw are used for instant trimming, there is no necessity for special shingles on wood shingle roofs. At gables, joints are broken by use of wide and narrow shingles in alternate courses. To prevent water dripping down the gables, it is a good plan to nail a length of beveled siding from the eaves to ridge along the gable and nail shingles over it. This forces water onto the main body of the roof, and the slight elevation is imperceptible from the ground.

Although it is difficult to avoid lining up an occasional joint, matching joints of shingles in any three courses should be avoided. This is important because a leak is unavoidable if a shingle splits on a line between the joints in alternate courses.

To avoid such splits, nails should never be driven into the center of a shingle, no matter how wide the shingle. There is the additional hazard that a center-nailed shingle may have a joint above it, exposing the nail to the weather, to corrosion, and to an eventual leak.

Only the widest shingles should be used on valley sides. For ridges and hips, the shingles themselves are employed, the best construction being a modified Boston

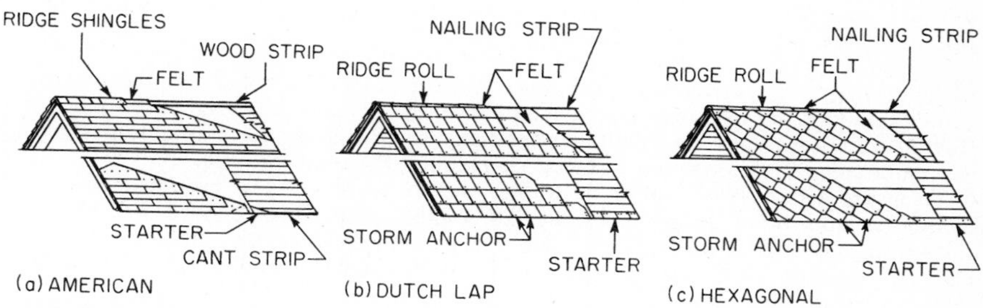

Fig. 15-7. Application of asbestos-cement shingles begins at the eaves with a cant or starter strip. Ridge shingles or ridge rolls can be used at the peak of sloping roofs.

hip. Hip and ridge shingles should be at least 5 in. wide. To avoid waste, shingles of this width should be selected from the random widths available. Hip and ridge shingles should be applied horizontally and allowed to project above the ridge shingle on the other side of the roof, the projection being chipped and shaved off with the shingler's hatchet. The hip shingles should be applied so that the shingles are woven in a weathertight pattern.

15-22. Asbestos-cement Shingles. Composed of 15 to 25% asbestos fiber and 75 to 85% portland cement, asbestos-cement products might be described as lightweight sheets of reinforced concrete. They are produced in a variety of shapes and sizes for numerous specific applications. Principal products are roofing shingles, flat sheets, siding shingles, and corrugated sheets.

Basically, there are four standard types of asbestos-cement roofing shingles, known as American-method, multiple-unit, Dutch-lap, and hexagonal.

American-method shingles (Fig. 15-7*a*) are so called because of their shape and finish. Laid in a rectangular pattern and having a simulated wood-grain, they present an appearance similar to conventional wood shingles.

Multiple-unit shingles are produced in large units, each of which covers an area equal to that of two to five standard-size shingles. Made in styles and sizes that vary with manufacturers, they retain the same general appearance, when installed, as the small units.

Dutch-lap shingles (Fig. 15-7*b*), sometimes called *Scotch-lap,* are larger than conventional shingles and are lapped at the top and one side, a method that effects savings in material and labor.

Hexagonal shingles (Fig. 15-7*c*), also known as shingles laid in the *French method,* are nearly square. They are laid in a diamond pattern with an overlap at the top and bottom. The effect is a hexagonal pattern.

Sizes of asbestos-cement roof shingles vary from 16 × 16 in. to 14 × 30 in.

Asbestos-cement shingles are highly resistant to water. But if, before installation, they are not stored under tarpaulins or under cover during inclement weather, seeping moisture may discolor them.

Special eaves starters are manufactured to aid in application of the first course of shingles (Fig. 15-7). A cant strip or furring strip is required along the eaves. Furring strips are also required at hips and ridges to enable ridge and hip shingles to cover adjacent material at the correct angle. Specially designed hip and ridge shingles and terminals are manufactured and should be used for finishing these areas.

Most asbestos-cement shingles require a minimum roof pitch of 5 in. per ft. American-method shingles, however, can be applied (because of the small exposure) on roofs with a 4-in. pitch.

15-23. Roofing Tiles. Clay roofing tiles offer considerable latitude in design and range of choice. The tiles are manufactured from special clays. Shale, a claylike rock, which when fired becomes as enduring as stone, is often preferred.

While a variety of patterns is available, all tiles fall into two classifications, roll and flat. Roll tiles are furnished in semicircular shape, reverse-curve S shape, pan and cover, and flat-shingle design.

To lay tiles successfully, vertical and horizontal chalk lines should be struck as guide lines. In laying out these lines, it must not be assumed that the manufacturer's standard tile measurements coincide with the tile size delivered to the job. Although variations may be small, contractions occurring in the burning of the tile can cause enough difference to warrant checking of the delivered tile.

To insure a full-length tile at the ridge, and to avoid needless cutting at dormers, chimneys, and other projections, the horizontal guide lines should be adjusted, always in favor of increasing the lap. With interlocking tile, however, because the tile exposure and lap are governed by the locking device, it is impossible to make any adjustments.

Except for Spanish mission tile, an undereaves course plus a cant strip are required before the first course of tile is laid. Tile application generally starts from the bottom right-hand corner of a roof and proceeds from right to left. Tiles adjacent to chimneys, vents, and other projections are left loose so that flashing material can be inserted.

Tile joints are broken, as with other types of roofing. But because there is an element of risk in fastening a half tile at such exposed locations as gables, manufacturers make special tile-and-a-half units. These not only save time by eliminating cutting, but make the gable windtight.

These large-width tiles can be used effectively at valleys and hips. Where cutting must be done, as around chimneys, the tile may be trimmed on a slater's stake, the cutting line being tapped with a hammer until the unwanted portion breaks off.

Tiles that verge along the hips must be fitted tight against the hip board with mastic or cement colored to match the tile.

15-24. Roofing Slate. This is one of the older mechanical arts, one in which skills are disappearing as craftsmen become scarce.

Slate roofs may be divided into two classifications, standard commercial slate and textural slate, or random slating. The latter is the older form, in which slates are delivered to the job in a variety of sizes and thicknesses, to be sorted by slaters. The longer and heavier slates are placed at the eaves, medium sized at the center, and the smallest at the ridge.

Standard commercial slating results from grading of slates at the quarries and costs less in place. Slates are graded not only by length and width but also by thickness, the latter being about ¼ in.

In manufacture, slate is split, resulting in a smooth and a rough side. The slates are applied on the roof with the smooth side down.

A slate should be laid with a lap of at least 3 in. over the second slate beneath it. Slate exposure, or margin, is found by deducting the lap from the total length of the slate and dividing the result by 2. For example, the exposure of a 20-in. slate with a 3-in. lap is $(20 - 3)/2 = 8\frac{1}{2}$ in.

Application starts with an undereaves course of slate fastened over a batten. The purpose of the batten is to cant the first course of slate to coincide with the angular projection of the other courses.

Slate is applied so that all joints are broken. It is nailed through countersunk holes with noncorroding nails, which are not driven home but level with the slate. Slating nails, with a large, thin head designed to fit within the countersunk holes, should be used.

15-25. Roof Finishing Details. The juncture of chimneys, walls, and roof projections with the edge of a roofing membrane is the most vulnerable area of the whole structure, because of the possibility of poor flashing. Flashings usually join two different materials, roof to wall, with different thermal characteristics. Resulting relative movement is likely to cause tears or cracks in the fabric or felt, and even in the roof membrane.

Whenever a sharp angle is introduced into roof or wall construction—in fact, at almost all intersecting planes—flashing is necessary to waterproof the angle. Because movement is to be expected at intersecting planes, flashing should be made of fairly elastic material, or fashioned so that there is room for movement.

All flashing material is affected by the expansion and contraction of the material itself, and by similar forces in the material with which it is in contact. These forces tend to pull the flashing away from the wall, leaving a gap through which water can leak past the flashing into the wall.

Bituminous flashings have the ability to hug tight against the wall when applied properly. Metal flashings, on the other hand, require added protection in the into a mortar line between the nearest row of bricks above the base flashing, If the flashing pulls away from the wall, there will be no gap through which water can pass to the rear. Plastic flashing sheets are useful in sealing the junctions of vents and pipes with roof decks.

On flat roof decks, installation of flashing is fairly simple, base flashing of bituminous materials or metal being applied so that at least 6 in. of the material is secured to the roof deck and 8 in. to the wall. Counterflashing should be installed above it. Not less than $1\frac{1}{2}$ in. of the counterflashing should enter a raggle cut into a mortar line between the nearest row of bricks above the base flashing, and not less than 4 in. of the flashing should hang vertically down the wall.

On steep roofs, recourse must be made to step flashing. This entails cutting short pieces of flashing metal, bending them at right angles, and sandwiching one flange between every roofing unit adjacent to the intersecting surface, the remaining flange being in contact with the vertical plane. In application of the flashing, care must be taken that each unit laps the unit below by at least 2 in. Thus, water traveling downward will be conducted from one unit to another, and should water seep under the shingles or other roofing materials, the sandwiched metal will conduct it safely back to the surface. Counterflashing is installed above these base flashings in a series of steps dictated by the slope of the roof and the mortar-line raggles into which the counterflashing is inserted.

Crickets. Chimneys projecting through one side of a sloping roof require the building of a cricket, or flashing saddle, behind the chimney. Ordinary methods of flashing are not suitable in such locations, where water propelled by a 15-mph wind can build up to a depth of several inches within minutes.

Saddle is an apt description of the sheet-metal flashing designed for chimney backs. It takes the form of an equestrian saddle or a miniature roof with a ridge and two valleys built to divide the streams of water traveling down the roof and conduct them away from the chimney, instead of pocketing behind the chimney.

The saddle should be securely soldered to a base step flashing behind the chimney. Because of the severe punishment they take, saddles should be fabricated of heavy-gage metals, highly resistant to corrosion.

Valleys. Where roofs intersect, special precautions are necessary to design a waterproof roof. It is obvious that, where the intersecting planes meet at the aptly named valley, the area receives twice as much rainwater as flows down the main roof area.

Bituminous valley materials should be laid in two plies at this vulnerable point, an 18-in.-wide base sheet followed by a 35-in.-wide mineral-surfaced top sheet.

Valley metal should be at least 12 in. wide, preferably much wider, and should bridge all junctures to make the intersection watertight. To guard against water building up in the valley, metal should be folded back 1 in. to act both as a barrier to infiltrating water and as a device in which fastening clips can be hooked. Metal valleys, because of expansion and contraction, should not be fastened directly to the deck, but through the clips.

Section 16

Builders' Hardware

RAYMOND V. MILLER

Director of Construction, Rider College, Lawrenceville, N.J.

AND

RICHARD A. HUDNUT

Product Standards Coordinator, Builders Hardware Manufacturers Association, New York, N.Y.

Hardware is a general term covering a wide variety of fastenings and gadgets. By builders' hardware we usually mean items made of metal or plastics ordinarily used in building construction. By common usage, the term builders' hardware generally covers only finishing hardware, but some rough hardware is discussed in this section.

In point of cost, the finishing hardware for a building represents a relatively small part of the finished structure. But the judicious selection of suitable items of hardware for all the many conditions encountered in construction and use of any building can mean a great deal over the years in lessened installation and maintenance costs and general satisfaction.

To make the best selection of hardware requires some knowledge of the various alternates available and the operating features afforded by each type. In this section, pertinent points relating to selecting, ordering, and installing some of the more commonly used builders' hardware items are discussed briefly.

16-1. Finishing and Rough Hardware. Finishing hardware consists of items that are made in attractive shapes and finishes and that are usually visible as an integral part of the completed building. Included are locks, hinges, door pulls, cabinet hardware, door closers and checks, door holders, exit devices, and lock-operating trim, such as knobs and handles, escutcheon plates, strike plates, and knob rosettes. In addition, there are push plates, push bars, kick plates, door stops, and flush bolts.

Included in rough hardware are utility items not usually finished for attractive appearance, such as fastenings and hangers of many types, shapes, and sizes—nails, screws, bolts, studs secured by electric welding guns, studs secured by powder-actuated cartridge guns, inserts, anchor bolts, straps, expansion bolts and screws, toggle bolts, sash balances and pulleys, casement and special window hardware, sliding-door and folding-door supports, and fastenings for screens, storm sash, shades, venetian blinds, and awnings.

16-2. When to Select Hardware. An important point to bear in mind in connection with selection of hardware for a new building is that, in many instances, it is one of the earliest decisions that should be made, particularly when doors and windows are to be of metal. This is true despite the fact that the finishing hardware may not actually be applied until near the end of the construction period.

Metal door bucks and window frames, for example, are among the first things that are needed on the job after construction gets above the first floor. These bucks and frames cannot be designed until the exact hardware to be applied to them has been decided on in detail, and precise template information has been obtained from the hardware manufacturer. Even after all the shop drawings for these items have been completely detailed, revised, and approved by the architect, it usually takes a considerable period of time to have them manufactured and delivered to the building site. Furthermore, the particular type of finishing hardware selected may also require considerable time to manufacture. Yet, if the frames are not available on the job during the early stages, work on the exterior walls and interior partitions cannot proceed. The delay would, of course, set back progress on the entire building. So it is wise to select the hardware and design the bucks and frames as soon as possible.

A good deal of time and expense can sometimes be saved on these metal frames by specifying those of standard design. If any fancy trim is desired, it is often advisable to have this separately made and applied later. Such trim may well be of wood; its fabrication will not delay delivery of the frames that must be built into the walls and partitions as they progress.

16-3. Function of Hardware Consultants. In view of the wide variety of types, styles, materials, and finishes found in the numerous items comprising finishing hardware, it is advisable to call in a hardware consultant or manufacturer's representative before specifications are made in too great detail. He is in the best position to streamline the requirements, to obtain the desired result in the simplest, most economical and satisfactory manner. In this way, one can often avoid unnecessary delay and complications.

It is well to remember that what was standard a short time ago may now be out of regular production. A hardware consultant should be able to advise on the selection of stock items.

Advance study with a representative of the manufacturer usually has another distinct advantage; he can often point out how definite savings can be achieved by making slight changes in the specifications. In many instances, these changes may mean little or nothing to the owner of the building and may be just as acceptable to him as what was originally specified. To the manufacturer, however, these minor changes may mean a great deal in what he can contribute to faster progress and lower ultimate cost of the completed work.

16-4. Factors Affecting Selection of Hardware. The operating characteristics of hardware items, of course, govern their selection, according to the particular requirements in each case. Then, the question of material, such as plastics, brass, bronze, aluminum, or steel, can be settled, as well as the finish desired. Selection of material and finish depends on the architectural treatment and decorative scheme:

Wrought, cast, and stamped parts are available for different items. Finishes include polished, satin, and oxidized. When solid metal is desired rather than a surface finish that is plated on metal, the order should definitely so specify.

National standards defining characteristics, sizes, dimensions, and spacings of holes, materials, and finishes for many items can assist greatly in identification and proper fit of hardware items to the parts on which they will be mounted.

16-5. Template and Nontemplate Hardware. For hardware items that are to be fastened to metal parts, such as jambs or doors, so-called template hardware is used. Template items are made to a close tolerance to agree exactly with drawings furnished by the manufacturer. The sizes, shapes, location, and size of holes in this type of hardware are made to conform so accurately to the standard drawings that the ultimate fit of all associated parts is assured. In the case of hinges or butts, holes that are template-drilled usually form a crescent-shaped pattern (Fig. 16-4).

Hardware for stock may be nontemplate; however, certain lines are all template-made, whether for stock or for a specific order. Nontemplate items may vary somewhat and may not fit into a template cutout.

For use on wood or metal-covered doors, template drilling is not necessary; nontemplate hardware of the same type and finish can be used. Generally, there is no price difference between template and nontemplate items.

FINISHING HARDWARE

16-6. Hinges and Butts. A hinge is a device permitting one part to turn on another. In builders' hardware, the two parts are metal plates known as leaves. They are joined together by a pin, which passes through the knuckle joints.

When the leaves are in the form of elongated straps, the device is usually called a hinge (Figs. 16-1 and 16-2). This type is suitable for mounting on the surface of a door.

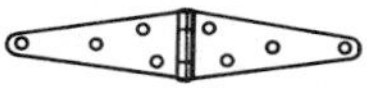

Fig. 16-1.

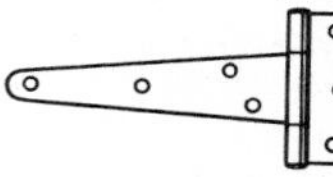

Fig. 16-2.

Fig. 16-1. Heavy strap hinge. (*C. Hager & Sons Hinge Mfg. Co.*)
Fig. 16-2. Heavy tee hinge. (*C. Hager & Sons Hinge Mfg. Co.*)

When the device is to be mounted on the edge of a door, the length of the leaves must be shortened, because they cannot exceed the thickness of the door. The leaves thus retain only the portion near the pin, or butt end, of the hinge (Figs. 16-3 to 16-5). Thus, hinges applied to the edge of a door have come to be known as butts, or butt hinges.

Butts are mortised into the edge of the door. They are the type of hinge most commonly used in present-day buildings.

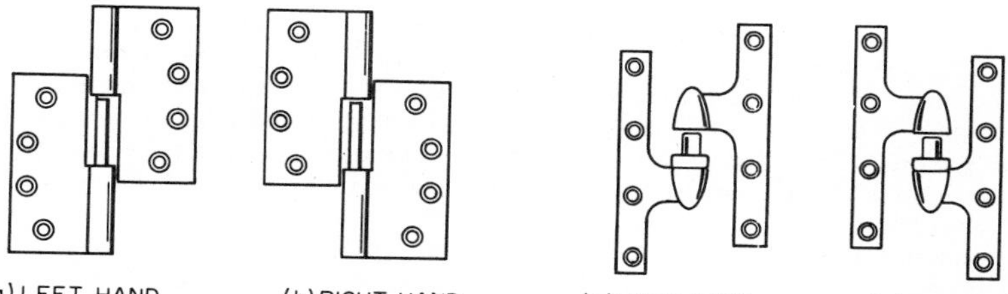

Fig. 16-3. Special types of butt hinges that are handed. For hand of doors, see Fig. 16-13.

Sizes of butt hinges vary from about 2 to 6 in., and sometimes to 8 in. Length of hinge is usually made the same as the width; but they can be had in other widths. Sometimes, on account of projecting trim, special sizes, such as $5 \times 4\frac{1}{2}$ in., are used.

For the larger, thicker, and heavier doors receiving high-frequency service, and for doors requiring silent operation, bearing butts (Fig. 16-4) or butts with Oilite bearings or other antifriction surfaces are generally used. It is also customary to use bearing butts wherever a door closer is specified.

Plain bearings (Fig. 16-5) are recommended for residential work. The lateral thrust of the pin should bear on hardened steel.

Unusual conditions may dictate the use of extra-heavy hinges or ball bearings where normal hinges would otherwise be used. One such case occurred in a group of college dormitories where many of the doors developed an out-of-plumb condition that prevented proper closing. It was discovered that the students had been using the doors as swings. Heavier hinges with stronger fastenings eliminated the trouble.

Two-bearing and four-bearing butt hinges should be selected, as dictated by weight of doors, frequency of use, and need for maintaining continued floating, silent operation. Because most types of butt hinges may be mounted on either right-hand or left-hand doors, it should be remembered that the number of bearing units actually supporting the thrust of the vertical load is only one-half the bearing units available. With a two-bearing butt, for example, only one of the bearings carries the vertical load, and with a four-bearing butt, only two carry the load. It should be noted, however, that some hinges, particularly olive knuckle hinges (Fig. 16-3*c* and *d*), which are frequently used on high-grade work, are "handed" and must be specified for use on either a right-hand or left-hand door. The loose-joint hinge (Fig. 16-3*a* and *b*) is also "handed."

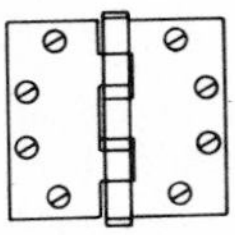

Fig. 16-4.

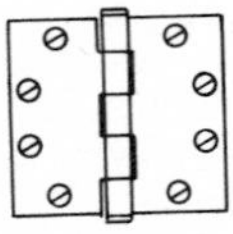

Fig. 16-5.

Fig. 16-4. Bearing butt hinge. (*Builders Hardware Manufacturers Association.*)

Fig. 16-5. Plain bearing butt hinge. (*Builders Hardware Manufacturers Association.*)

When butts are ordered for metal doors and jambs, "all machine screws" should be specified.

16-7. Location of Hinges. One rule for locating hinges is to allow 5 in. from rabbet of head jamb to top edge of top hinge and 10 in. from finished floor line to bottom of bottom hinge. The third hinge should be spaced equidistant between top and bottom hinges.

16-8. Types of Hinge Pins. A very important element in the selection of hinges is the hinge pin. It may be either a loose pin or a fast pin.

Loose pins are generally used wherever practicable, because they simplify the hanging of doors. There are four basic pin types:

1. Ordinary loose pins.
2. Nonrising loose pins.
3. Nonremovable loose pins.
4. Fast (or tight) pins.

The **ordinary loose pin** can be pulled out of the hinge barrel so that the leaves may be separated. Thus, the leaves may be installed on the door and jamb independently, with ease and "mass-production" economy. However, these pins have a tendency—with resulting difficulties—of working upward and out of the barrel of the hinge. This "climbing" is caused by the constant twisting of the pin due to opening and closing the door. Present-day manufacture of hinges has tended to drift away from this type of pin, which is now found only in hinges in the lowest price scale.

The **nonrising loose pin** (or self-retaining pin) has all the advantages of the ordinary loose pin; but at the same time, the disadvantage of "climbing" is eliminated. The method of accomplishing the nonrising features varies with the type of hinge and its manufacture.

The **nonremovable loose pin** is generally used in hinges on entrance doors, or doors of locked spaces, which open out and on which the barrel of the hinge is therefore on the outside of the door. If such a door were equipped with ordinary loose-pin hinges or nonrising loose-pin hinges, it would be possible to remove the pin from the barrel and lift the door out of the frame and in so doing overcome the security of the locking device on the door. In a nonremovable loose-pin hinge, however, a setscrew in the barrel fits into a groove in the pin, thereby preventing its removal. The setscrew is so placed in the barrel of the hinge that it becomes inaccessible when the door is closed. This type of hinge offers the advantage of the ordinary loose-pin type plus the feature of security on doors opening out.

The **fast (or tight) pin** is permanently set in the barrel of the hinge at the time of manufacture. Such pins cannot be removed without damaging the hinge. They are regularly furnished in hospital- or asylum-type hinges. The fact that the leaves of this type of hinge cannot be separated, however, makes the installation more difficult and costly. However, the difficulty is not too great, because with this type of hinge it is only necessary to hold the door in position while the screws for the jamb leaf are being inserted.

Ends of pins are finished in different ways. Shapes include flat-button, ball, oval-head, modern, cone, and steeple. They can be chosen to conform with type of architecture and decoration. Flat-button tips are generally standard and are supplied unless otherwise specified.

16-9. How to Select Hinges. One hinge means one pair of leaves connected with a pin. The number of hinges required per door depends on the size and weight of the door, and sometimes on conditions of use. A general rule recommends two butt hinges on doors up to 60 in. high; three hinges on doors 60 to 90 in. high; and four hinges on doors from 90 to 120 in. high.

Table 16-1 gives general recommendations covering the selection of suitable hinges. Figure 16-6 indicates how hinge width is determined when it is governed by clearance.

For transoms, butt hinges can be either on the top or the bottom. The transom lifter or adjusting device may be any one of several types, as desired. Orders should definitely specify which location and which type.

The proper operating clearance between the hinged edge of a door and the jamb is taken care of in the manufacture of the hinges by "swaging" or slightly bending the leaves of the hinge near the pin. Since the amount of such bending required is determined by whether one or both leaves are to be mortised or surface-mounted, it is important in ordering hinges to specify the type of hinge needed to satisfy mounting conditions. If hinge leaves are to be mortised into both the edge of the door and the jamb, a full-mortise hinge is required (Fig. 16-7*a, c,* and *h*). If leaves are to be surface-applied to both the side of the door and the face of the jamb, a full-surface hinge is needed (Fig. 16-7*d*). If one leaf is to be surface-applied and the other mortised, the hinge is a half-surface hinge or a half-mortise hinge, depending on how the leaf for the door is to be applied—half-surface if applied to door surface (Fig. 16-7*b, e,* and *f*) and half-mortise if mortised in the door end (Fig. 16-7*g*).

Where door-closing devices are not required and where quiet is demanded, as in hospitals, friction hinges or door-control hinges may be used to prevent slamming. They will cause the door to remain in any position. Only slight hand pressure is required to close it.

Exterior doors should have butts of nonferrous metal. Although chromium plating does not tarnish, it is not considered to be satisfactory on steel for exterior use. Interior doors in rooms where dampness and steam may occur should be of nonferrous metal and should also have butts of nonferrous metal. Butts for other interior work may be of ferrous metal.

Ferrous-metal butts should be of hardened cold-rolled steel. Where doors and door frames are to be painted at the job site, butts should be supplied with a prime-coat finish. For doors and trim that are to be stained and varnished, butts are usually plated.

Other types of hinges include some with a spring that keeps the door shut. They may be either single- or double-acting. The spring may be incorporated in a hinge mounted on the door in the usual manner, or it may be associated with a pivot at the bottom of the door. In the latter case, the assembly may be of the type that is mortised into the bottom of the door, or it may be entirely below the floor.

16-10. Door-closing Devices. These include overhead closers, either surface-mounted or concealed, and floor-type closers. These are some of the hardest-worked items in most buildings. To get the most satisfactory operation at low first cost and low maintenance cost, each closer should be carefully selected and installed to suit the particular requirements and conditions at each door.

Most of these devices are a combination of a spring—the closing element—and an oil-cushioned piston, which dampens the closing action, inside a cylinder (Fig.

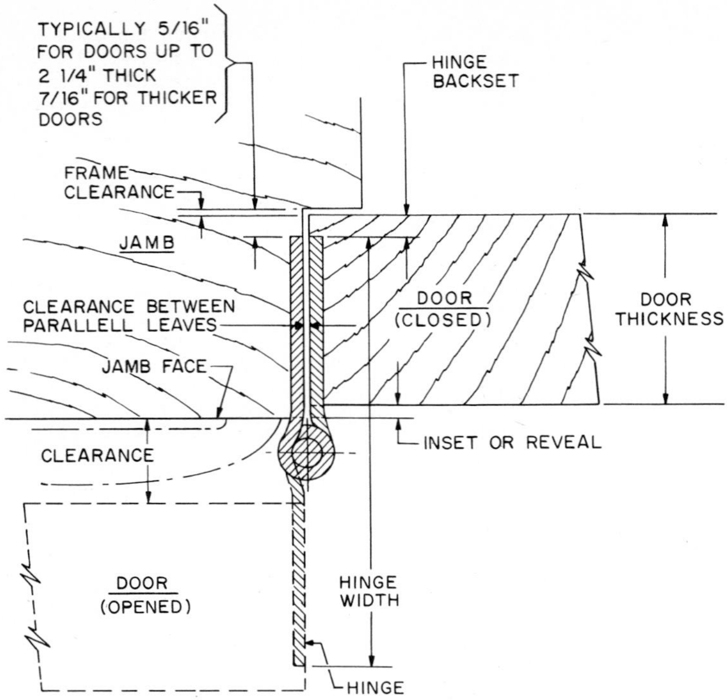

Fig. 16-6. When determined by clearance, width of door hinge equals inset (typically ⅛ in.), plus hinge backset (typically ¼ in. for doors up to 2¼ in. thick, and ⅜ in. for thicker doors), plus twice the door thickness, plus clearance.

Table 16-1. Hinges for Doors

Door thickness, in.	Door width, in.	Minimum hinge height, in.
⅞ or 1	Any	2½
1⅛	To 36	3
1⅜	To 36	3½
1⅜	Over 36	4
1¾	To 41	4½
1¾	Over 41	4½ heavy
1¾ to 2¼	Any	5 heavy*

* To be used for heavy doors subject to high-frequency use or unusual stress.

16-8). The piston operates with a crank or a rack-and-pinion action. It displaces the fluid through ports in the cylinder wall, which are closed or open according to the position of the piston in the cylinder. Opening of the door energizes the spring, thus storing up closing power. Adjustment screws are provided to change the size of the ports, controlling flow of fluid. This arrangement makes the closer extremely responsive to the conditions of service at each individual door and permits a quiet closure, which at the same time insures positive latching of the door.

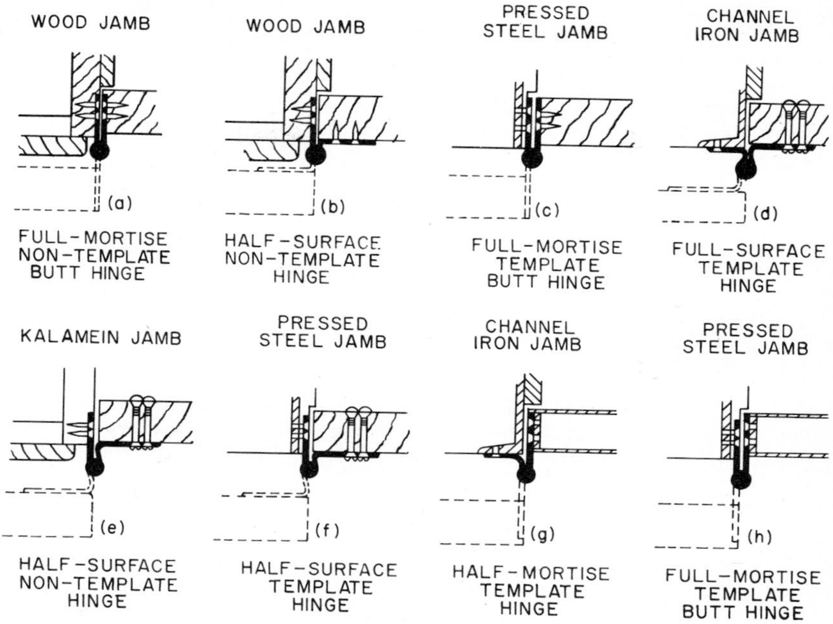

Fig. 16-7. Hinge-mounting classification depends on where and how hinge is fastened: (*a*), (*c*), and (*h*), full mortise; (*d*) full surface; (*b*), (*e*), and (*f*), half surface; and (*g*) half mortise.

While the fluid type of closer is preferred, pneumatic closers are also used, particularly for light doors, like screen doors.

Overhead door closers are installed in different ways, on the hinge side of the door or on the top jamb on the stop side of the head frame or on a bracket secured to the door frame on the stop side. Various types of brackets are available for different conditions. Also, when it is desired to install a closer between two doors hung from the same frame, or on the inside of a door that opens out, an arrangement with a parallel arm makes this possible. Other types of closers may be mortised into the door or housed in the head above the door.

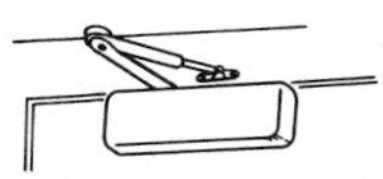

Fig. 16-8. Door closer—spring closes the door, while hydraulic mechanism (cylinder with piston) keeps the door from slamming. (*Builders Hardware Manufacturers Association.*)

Closers may be semiconcealed or fully concealed. Total concealment greatly enhances appearance but certain features of operation are limited.

An exposed-type closer should be mounted on the hinge side or stop side of the door unless there is real need for a bracket mounting.

The bracket mounting that allows a closer to do the best job is a soffit bracket. A soffit bracket, however, may obstruct the head of the opening more than is desirable. In that event, a corner bracket may be used. It places the closer nearer to the hinge edge of the door.

Whereas the use of brackets reduces headroom and may become a hazard, a parallel arm closer mounted on the door rides out with the door, leaving the opening entirely clear.

When surface-applied door closers are used, careful consideration should be given to the space required in order that doors may be opened at least 90° before the closers strike an adjacent wall or partition.

Semiconcealed door closers are recommended for hollow metal doors. These closers are mortised into the upper door rail.

Various hold-open features also are available in different closer combinations to meet specific requirements. A fusible link closer is a type that is used to close a door automatically in case of fire. It acts like an ordinary hold-open arm closer until fire-heated air enters the doorway. At a specified temperature, the link breaks, and the closer shuts the door. Another available feature is delayed action. This allows plenty of time to push a loaded vehicle through the opening before the door closes.

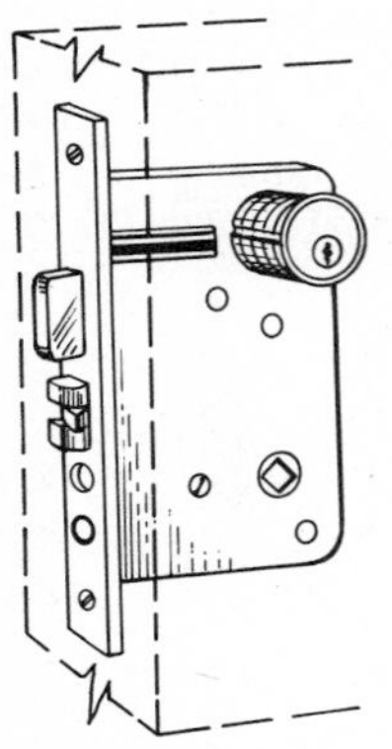

Fig. 16-9. Mortise lock. (*Builders Hardware Manufacturers Association.*)

A hydro-hinge is a hinge in which a door-closing device is incorporated. A spring is placed in the top hinge, a checking device in the bottom hinge. It is used for half doors, gates, or partition doors where closers of conventional type cannot be installed because of physical conditions.

When floor-type checking and closing devices are used, floor conditions should be carefully determined in order that there will be sufficient unobstructed depth available for their installation. Closers concealed in the floor tend to foul up faster than overhead types (because of scrub water and floor dirt) and are more expensive to install.

To get maximum performance from any door closer it must be of ample size to meet the conditions imposed on it. If abnormal conditions exist, such as drafts or severe traffic, a closer of larger than the normal capacity should be employed. Installing too small a closer is an invitation to trouble. Manufacturers' charts should be used to determine the proper sizes and types of closers to suit door sizes and job conditions.

It is very important that door closers be installed precisely as recommended by the manufacturer. Experience has shown that a large percentage of troubles with closers results from disregard of mounting instructions.

16-11. Locks and Latches. The function of locks and latches is to hold doors in a closed position. Those known as rim locks or latches are fastened on the surface of the door. The ones that are mortised into the edge of the door are known as mortise locks (Fig. 16-9) or latches.

When the locking bolt is beveled, the device is usually referred to as a latch bolt; such a bolt automatically slides into position when the door is closed. A latch is usually operated by a knob or lever. Sometimes it may be opened with a key on the other side. Night latches came to be so called because they are generally used at night with other ordinary locks to give additional security.

Latches must of course take into account the hand of the door so the bevel will be right. A large percentage of latches are "reversible"—that is, they may be used on a right- or left-hand door (Art. 16-12). It is well, however, when ordering any lock to specify the hand of the door on which it is to be used.

When the locking bolt is rectangular in shape, it does not slide into position automatically when the door is closed; it must be projected or retracted by a thumb turn or key. This type of bolt is referred to as a dead bolt, and the lock as a dead lock. It may be worked with a key from one or both sides. Such a bolt is often used in conjunction with a latch. When latch bolts and dead bolts are combined into one unit, it is known as a lock.

For keyed locks that do not have dead bolts, it is desirable that the latch bolt be equipped with a dead latch (Fig. 16-12). This is a small plunger or an auxiliary dead latch (Fig. 16-10) that is held depressed when the door is closed and "deadlocks" the latch bolt so that it cannot be retracted by a shim, card, or similar device inserted between latch and door frame.

Various combinations of latches, dead bolts, knobs and keys, and locking buttons are applied to all types and kinds of doors. The exact combination most suitable for any given door is determined by a careful analysis of the use to which the door is to be put.

A point to bear in mind is that a uniform size should be selected, if practicable, for a project, no matter what the individual functions of the different locks may be. This makes possible the use of standard-size cutouts or sinkages on each installation. When this is done, not only is the cost of installation reduced, but any changes that may be made in the drawings as the job progresses, or any changes that may have to be made later are simplified, and special hardware is avoided.

Unit locks (Fig. 16-10) are complete assemblies that eliminate most of the adjustments during installation that would otherwise be necessary. These locks have merely to be slipped into a standard notch cut into a wood door or formed in a metal door.

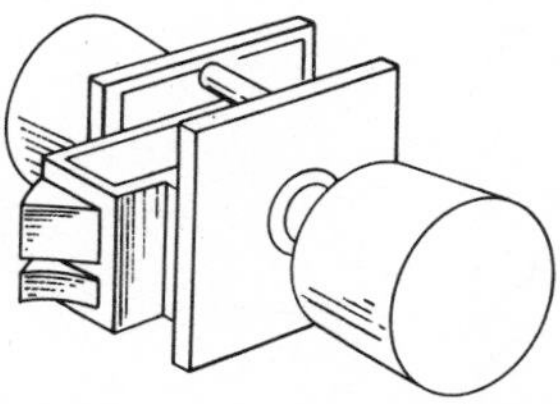

Fig. 16-10. Unit lock. (*Builders Hardware Manufacturers Association.*)

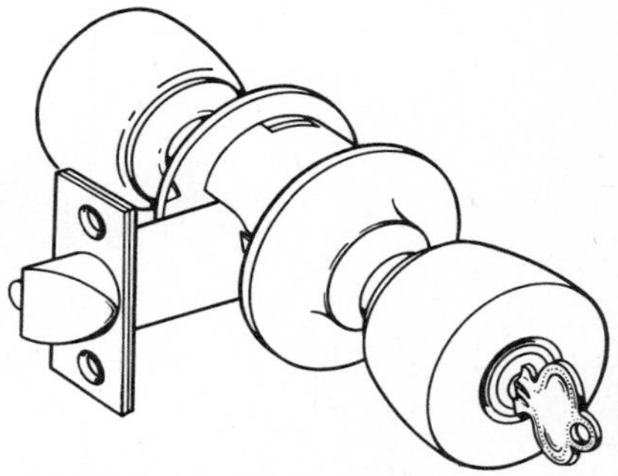

Fig. 16-11. Cylindrical lock. (*Builders Hardware Manufacturers Association.*)

Bored-in locks are another type that can be installed by boring standard-size holes in wood doors or by having uniform circular holes formed in metal doors. These bored-in locks are often referred to as tubular-lock sets or cylindrical-lock sets depending on how the holes have to be bored to accommodate them.

Tubular locks have a tubular case extending horizontally at right angles to the edge of the door. This type of case requires a horizontal hole of small diameter to be bored into the door at right angles to the vertical edge of the door; another small hole is required at right angles to the first hole to take care of the locking cylinder.

Cylindrical locks (Fig. 16-11), the other type of bored-in lock, have a cylindrical case requiring a relatively large-diameter hole in the door, bored perpendicular to the face of the door. This hole accommodates the main body of the lock. A hole of smaller diameter to take the latch bolt must be bored at right angles to the edge of the door.

When bored-in locks are used in hollow metal doors, a reinforcing unit is required in the door. This unit is generally supplied by the door manufacturer.

Exit devices are a special series of locks required by building codes and the National Fire Protection Association "Life Safety Code" on certain egress doors in public buildings. On the egress side, there is a horizontal bar running a minimum of two-thirds the width of the door. When pressed against, the bar releases the locking or latching mechanism, allowing the door to open. These devices are required to be labeled for safe egress by a nationally recognized independent testing laboratory. When used on fire doors, they must carry an additional label showing that the devices have also been investigated for fire. They then bear the name "fire exit hardware." These devices are available with various functions, including arrangements for having them locked from the outside and openable with a key. For all functions, however, they must be openable from the inside by merely depressing the cross bar.

Keys. Locks are further classified according to the type of key required to retract the bolts. On all but the cheaper installations the principal type of key used is the cylinder key, which operates a pin-tumbler cylinder.

In the preceding locks, the locking cylinder, which is the assembly that supplies the security feature of the lock, is a cylindrical shell with rotatable barrel inside. The barrel has a longitudinal slot or keyway formed in it. The cross section of the keyway for every lock has a shape requiring a similarly designed key. Several holes (usually five or six) are bored through the shell and into the barrel at

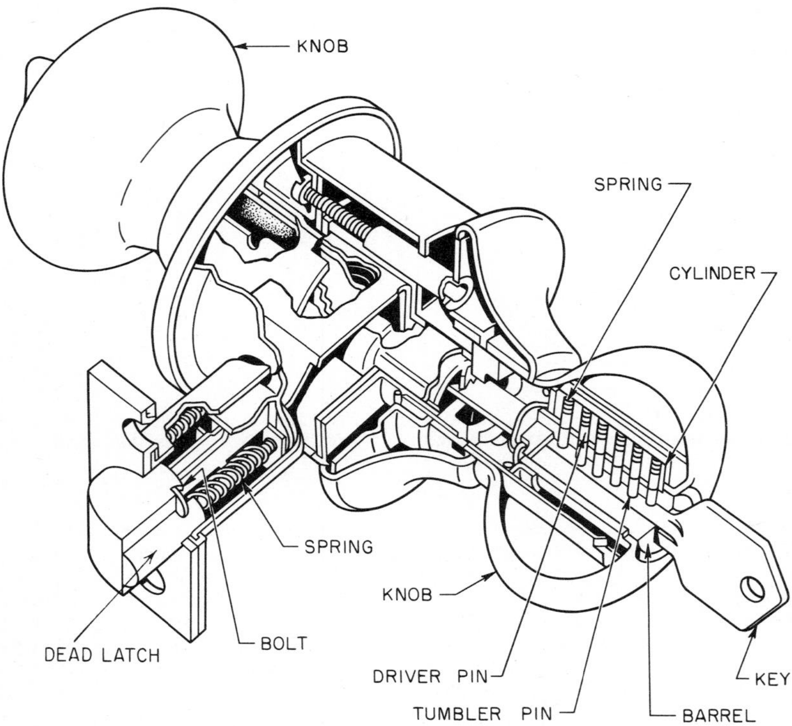

Fig. 16-12. Mechanism of cylindrical lock. (*Kwikset Locks.*)

right angles to the axis of the barrel (Fig. 16-12). In each hole is placed a pair of pins—a driver pin in the shell end and a tumbler pin in the barrel end. Pins vary in length, and the combination of lengths differs from that of pins in other locks. A spring mounted in the shell end of each hole behind each driver pin forces these pins into the hole in the barrel, so that normally they are partly in the barrel portion of the hole and partly in the shell portion. Thus, the barrel is prevented from rotating in the shell to operate the bolt, and the door remains locked.

Keys have notches spaced along one side to correspond with the spacing and length of the pins. When a key is inserted in the slot in the barrel, each notch forces a tumbler pin back, against pressure from the spring. When the proper key is inserted in the slot, each notch pushes its corresponding pair of pins just far enough back into the shell to bring the junction point of that pair exactly at the circumference of the barrel. The barrel then can be rotated within the shell by merely turning the key.

Turning the key operates a cam attached to the end of the barrel that withdraws the bolt from its locked position.

The security feature of the cylinder lock results from the fact that only one series of notches in the key will correspond exactly to the respective lengths of the several individual pins. With as many as five or six tumblers, it is apparent that many combinations are available.

By using split pins, different keyways, and various sizes and arrangements of the pins, master keying and grand master keying of cylinder locks is made possible. Thus one master key may be made to open all the separate locks on each floor of a building, while a grand master key will open any lock in any floor of that building. In case there are a number of such buildings in a group, a great-grand master key can be made that will open any door on any building in the group.

Security Precautions. The security of a building from the standpoint of unauthorized entry and life safety is affected by the proper selection of hardware. Although locks are important for security, the total system, including the building design,

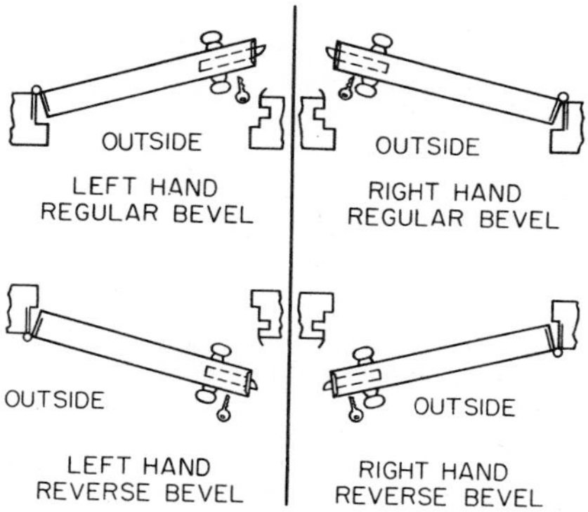

Fig. 16-13. Conventional method for determining hand of doors and bevel.

must be considered. A lock installed in a door that is loosely fitted in the frame may be ineffective. Hollow metal and aluminum frames that are not reinforced and anchored properly are easily "spread," and undetectable entry can result.

In high levels of master keying, more split pins must be used, and thus more "shear lines" are established. This makes the cylinder easier to pick. Hence, unnecessarily complicated master key systems should be avoided.

A good security lock is available with a dead bolt and a deadlocking latch bolt. Both bolts are retractable in one operation, merely by turning the inside knob. This satisfies requirements for security against unauthorized entry and life safety.

Hardware requirements for life safety and fire doors are covered by two National Fire Protection Association Standards, "Life Safety Code" and "Fire Doors and Windows." Both are incorporated in many building codes.

16-12. Hand of Doors. As you look at a door, the hinges can be either on your right or left, and the door can open either away from or toward you. Because of the fact that it is from the outside of a door that the lock is operated by a key, it is customary to speak of the hand of a door as viewed from the outside—the side on which protection is needed against unauthorized entry. In a series of connecting rooms, the outside of each door is considered to be the side you first approach as you proceed into the rooms from the entrance.

If when you are standing outside the door the hinges are to your right and the door opens away from you, it is a right-hand, regular-bevel door; if the door opens toward you, it is a right-hand, reverse-bevel door. Similarly, if when you are outside the door, the hinges are to your left and the door swings away from you, it is a left-hand, regular-bevel door; if the door opens toward you, it is a left-hand, reverse-bevel door (Fig. 16-13).

In ordering hardware, the hand of the door must be given; also, whether the beveled face of the latch bolt has to be regular-beveled or reverse-beveled. In some cases, the face is reversible, but it is best to state the bevel definitely anyway. For closet and cabinet doors, which never open away from you, it is sufficient to state merely whether the hand of these doors is right or left and the fact that they are closet or cabinet doors. It is then understood that reverse-beveled latch bolts are required.

For casement sash, the hand is stated a little differently; it is safest to specify whether the hinges are on the right or on the left as viewed from the room side, and also whether the sash opens in or out.

16-13. Casement-sash Operators. There are many types of devices for operating casement sash. Some of these require that screens have openings to make the operating handle accessible. Other types, which operate independently of the screens, have an interior crank that actuates a worm gear. Various arrangements of arms actuated by the gear are used to move the sash.

16-14. Sliding Hardware. Hardware for interior and exterior sliding doors and screens ranges from light to heavy. Hangers, tracks, and wheels are of various designs to serve different requirements. Ball bearings and composition or nylon wheels are generally used. The tracks and hangers are often made of aluminum, especially for the lighter installations. Because of the wide choice, a careful examination of the features peculiar to each design is well worthwhile before a decision is made.

16-15. Folding-door Hardware. Folding doors require special supports and tracks because of their wide range of movement and their various weights. Hardware is available for folding doors weighing from 30 to 250 lb each.

16-16. Locking Bolts (Doors and Windows). Various types of bolts and rods are fastened to doors and windows for the purpose of securing them in closed position or to other doors and windows. Flush bolts and surface bolts, manual or automatic, are most often used.

Top and bottom vertical bolts operated by a knob located at convenient height between them are known as **cremorne bolts.**

ROUGH HARDWARE

16-17. Window Hardware. Sash balances are commonly used with double-hung windows as counterweights instead of weights and pulleys. Some balances have tape or cable with clock-type springs, which coil and uncoil as the sash is raised and lowered. Another type employs torsion springs with one end fixed to the side jamb of the window and the other arranged to turn as the sash goes up or down. The turning device in this type of balance may be a slide working in a rotatable spiral tube, or it may be a slotted bushing attached to the free end of the spring and fitted around a vertical rod attached to the sash. This rod (Fig. 16-14) is a flat piece of metal twisted into a spiral shape. The up-and-down movement of the sash causes the slotted bushing to revolve on its spiral sliding rod, thus winding and unwinding the spring. Still another type utilizes a vertical tension spring of the ordinary coil variety. One end of the spring is fixed and the other is fastened to the sash; the spring stretches or compresses in a vertical direction as the sash is moved up or down.

Some sash balances combine weather stripping with the balances. Others have friction devices to hold the sash in the desired position.

One patented type of sash balance known as the Unique sash balance incorporates a clever counterbalancing feature by making a change in the degree of pitch of the spiral rod from top to bottom, thereby controlling the increase and decrease of spring tension (Fig. 16-14). The pitch varies from 30 to 80°. As the spring turns, the changing spring tension is automatically compensated for by the variable pitch of the spiral rod sliding through the slotted bushing on the end of the spring. Thus the tension is equalized at every point of operation. This automatically prevents the sash from creeping or dropping of its own accord, without the necessity of introducing friction devices that interfere with easy operation.

Two sash balances are used per sash (one on each side) or four balances per double-hung window.

Other window hardware—locks, sash pulls, sash weights, and pulleys—are simple items, supplied as standard items with specific windows.

16-18. Inserts, Anchors, and Hangers. Metal inserts of various types are cast into concrete floor slabs to serve as hangers or connectors for other parts of the structure that will be supported from the floor system. These inserts include electric conduit and junction boxes, supports for hung ceilings, slots for pipe hangers, fastenings for door closers to be set under the floor, and circular metal shapes for vertical pipe openings.

Metal anchors of various types are used in building construction. Each type is specifically shaped, according to the purpose served. These include anchors for securing stone facings to masonry walls, anchors for fastening marble slabs in place, column anchor bolts set in foundations, and anchor bolts for fastening sills to masonry.

Joist hangers are used for framing wood joists into girders and for framing headers into joists around stairwells and chimneys.

16-19. Nails. **Wire nails,** made of mild steel, are commonly used for most nailing purposes. Cylindrical in shape, they are stronger for driving than cut nails and are not so liable to bend when driven into hardwoods.

Cut nails, sheared from steel plate, are flat and tapered. They have holding power considerably greater than wire nails. They are usually preferred to wire nails for fastening wood battens to plaster; also in places where there is danger that the nails may be drawn out by direct pull. They are frequently used for driving into material other than wood. They are generally used for fastening flooring.

When driven with width parallel to grain of wood, they have less tendency to split wood than wire nails.

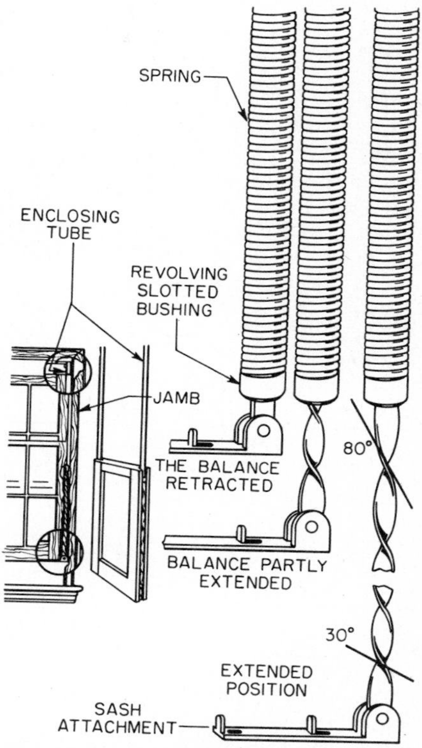

Fig. 16-14. Constant-tension sash balance. (*Unique Balance Co.*)

The length of cut and wire nails is designated by the unit "pennies." Both cut and wire nails come in sizes from 2-penny (2d) which are 1 in. long, to 60-penny (60d), which are 6 in. long. For each penny above 2, the length increases by ¼ in. up to and including the 12d, measuring 3¼ in. Above 12 some of the numbers are omitted, and the sizes increase by ½ in. for each designation from 16d up. Thus, 16d is 3½ in.; 20d, 4 in.; 30d, 4½ in.; 40d, 5 in.; 50d, 5½ in.; and 60d, 6 in.

Above 6 in., the fasteners are called **spikes.** They run in 7-, 8-, 9-, 10-, and 12-in. lengths.

Various gages or thicknesses of nails, and different sizes and shapes of heads and points (Figs. 16-15 and 16-16) are available in both wire and cut nails. Certain types have distinguishing names. For examples, the term **brads** is applied to nails with small heads, suitable for small finish work.

Common brads are the same thickness as common nails but have different heads and points. **Clout nails** have broad flat heads. They are used mostly for securing gutters and metalwork.

Casing nails are about half a gage thinner than common wire nails of the same length; **finishing nails**, in turn, are about half a gage thinner than casing nails of the same length. **Shingle nails** are half a gage to a full gage thicker than common nails of the same length.

Certain manufacturers have developed nails for special purposes that hold tighter and longer. Among these are threaded nails (Fig. 16-16), which combine the ease of driving of the ordinary nail with much greater holding power.

One type is a spiral-threaded flooring nail. These nails turn as they drive, minimizing splitting of the tongues of the floor boards. These nails are said to actually grip more firmly with the passage of time. A nailing machine is available for driving these nails. In one operation, it starts the nail, drives the joint between the flooring strips tight, and drives and sets the nail. With this machine, mashed tongues and marred edges of the wood are avoided.

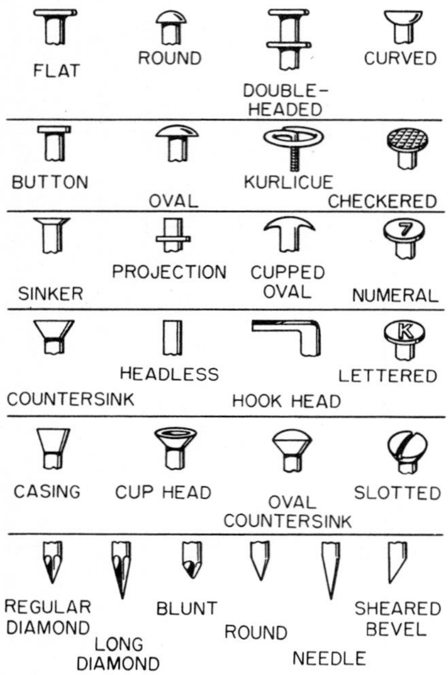

Fig. 16-15. Nail heads and points.

Nails of aluminum and stainless steel are particularly useful in exposed locations where rust or corrosion of steel nails might cause unsightly stains to form on exterior finished surfaces. These nails are now made in most of the usual sizes, including special spiral-threaded nails.

Galvanized nails are used for fastening slate and shingles. These nails are sometimes used in exposed locations as protection against corrosion.

To satisfy the special requirements and varying conditions in each case, the proper nails to use for fastening various sizes and types of materials must be carefully and intelligently chosen by men experienced in this kind of work.

16-20. Screws. These are used for applying hardware of all descriptions; also for panel work, cabinet work, and all types of fine finish work. Parts of electric and plumbing fixtures are applied with screws. They have greater holding power than nails and permit easy removal and replacement of parts without injury to the wood or finish. Screws avoid danger of splitting the wood or marring the finish, when the screw holes are bored with a bit.

Screws are made in a large variety of sizes and shapes to suit different uses (Fig. 16-17) and they are made of different metals to match various materials. A much-used type of head, other than the ordinary single-slot type, is the **Phillips head,** which has two countersunk slots at right angles to each other. The head keeps the screw driver exactly centered during driving and also transmits greater driving power to the screw, while holding the screw driver firmly on the head. Phillips heads are smoother at the edges, because the slots in the head do not extend to the outer circumference.

Steel screws for wood vary in length from ¼ to 6 in. Each length is made in a variety of thicknesses. Heads may be ordinary flat (for countersinking), round, or oval.

Parker screws for securing objects to thin metal are self-threading when screwed into holes of exactly the correct size.

Sizes of screws are given in inches of length and in the gage of the diameter. Lengths vary by eighths of an inch up to 1 in., by quarters from there up to 3 in., and by halves from 3 to 5 in. Unlike wire gages, the smallest diameter of a screw gage is the lowest number; the larger the number, the greater is the diameter in a screw gage. Gage numbers range from 0 to 30.

Lag screws are large, heavy screws used for framing timber and ironwork. Lengths vary from 1½ to 12 in. and diameters from 5⁄16 to 1 in. Two holes should

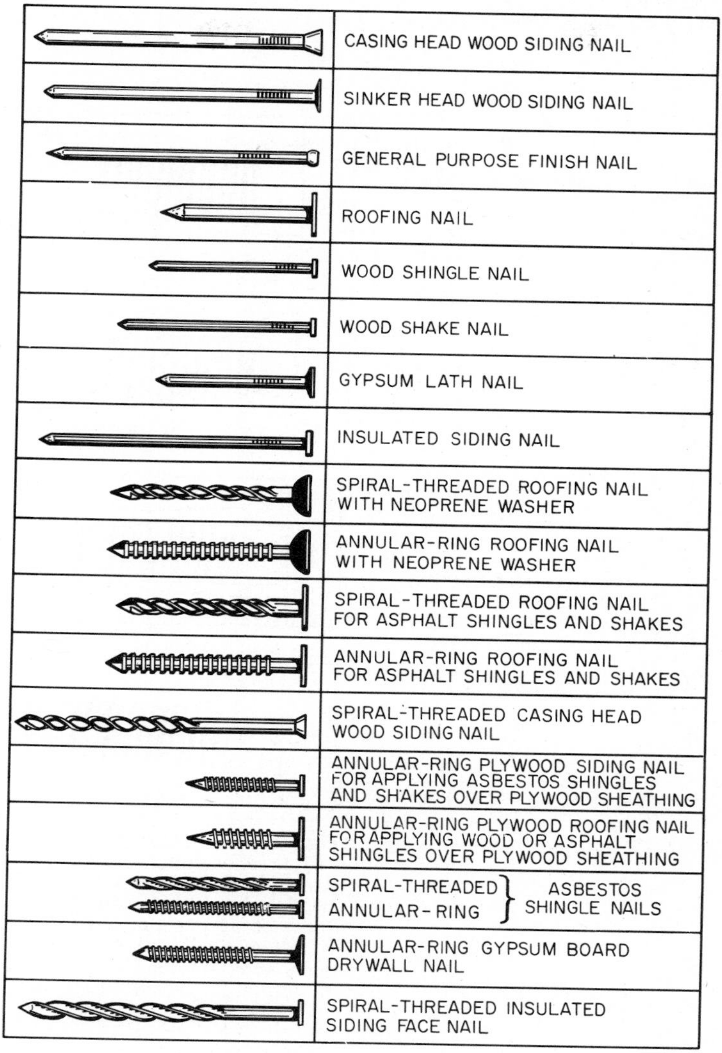

Fig. 16-16. Typical nails.

be bored for lag screws, one to take the unthreaded shank without binding and the other (a smaller hole) to take the threaded part. This smaller hole is usually somewhat shorter in length than the threaded portion. Lag screws usually have square heads and are tightened with a wrench.

16-21. Welded Studs. Studs electrically welded to the steel framework of a building are often used as the primary element for securing corrugated siding and roofing, insulation, metal window frames, ornamental metal outer skins, anchorages for concrete, and other items. The welded studs thus form an integral part of the basic structure.

Many types of studs or fasteners are available, each one being designed for a particular purpose. Most of the studs have threads formed on them, either externally or internally. Some of the studs are designed to have the material impaled over them and riveted to them.

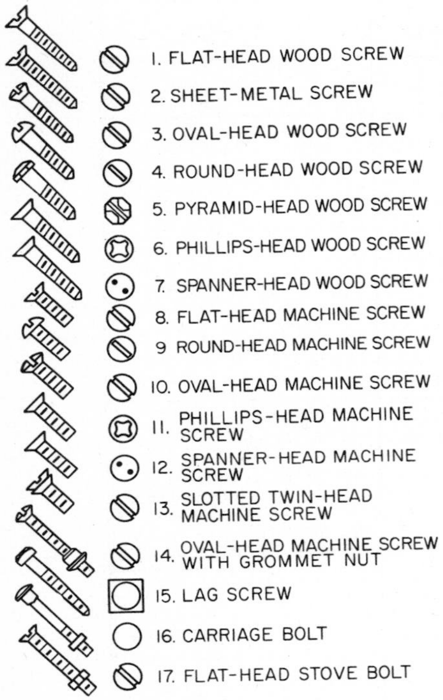

Fig. 16-17. Typical screws.

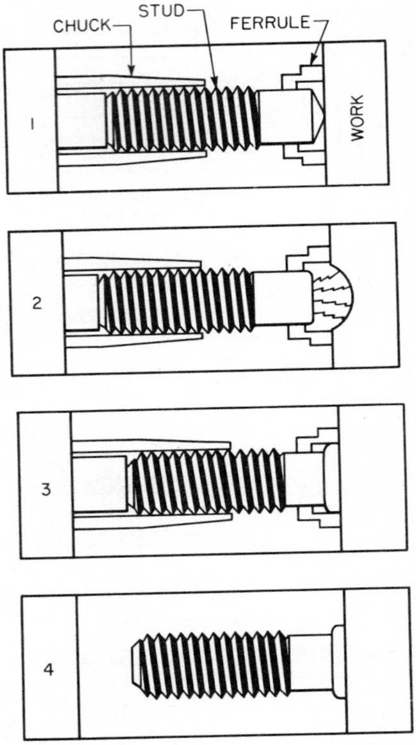

Fig. 16-18. Steps in stud welding: (1) press welding end of gun against work plate; (2) press trigger, creating an electric arc between stud and plate and melting portions of each; (3) stud is automatically plunged into molten pool; and (4) remove gun and knock off ferrule.

The studs are designed so as to project the exact distance desired after they have been welded. Special sealing washers and nuts are usually placed on each stud over the flat sheet of material being fastened. Tightening the nut or expanding the head of the stud with a riveting hammer then makes the fastening complete, weathertight, and secure.

Studs are usually mild steel, cadmium plated, or stainless steel. The latter is recommended for corrosive atmospheric conditions. Flux to assure a good weld is contained in the center of each stud at the welding end.

Equipment required for stud welding includes a stud-welding gun, a control unit for adjusting the amount of welding current fed to the gun, and a power source. The source of welding current may be a direct-current generator, a rectifier, or a battery unit. When a welding generator is used, the minimum National Electrical Manufacturers Association rating should be 400 amp.

The welding gun usually has a chuck for holding the stud in position for welding (Fig. 16-18), and a leg assembly holding an adjustable-length extension sleeve into which the necessary arc-shielding ferrule is inserted. Expendable ceramic arc ferrules are generally used to confine the arc and control the weld fillet. After each weld, the arc ferrule is broken and removed by a light tap with any convenient metal object. In some cases where the required finished stud is short, an extra length is provided on the stud as furnished, for proper chucking in the gun. A groove is provided so the extra length can be easily broken off after the stud is welded.

The operation of fastening corrugated or flat asbestos-cement roofing and siding to steel girts is typical of how these stud fastenings are installed on buildings. First, the sheet of asbestos cement is placed in the correct position on the girts or purlins. If the sheets have not already been predrilled, a carboloy-tip drill is then used to drill a hole of proper diameter through the sheet. Next, an end-mill drill is inserted through the hole in the material to remove any scale, rust, paint, or dirt and to provide a clean welding surface. Then, the stud is inserted in the gun chuck, the ceramic arc shield or ferrule is put into place in the adjustable-length device, and the tubular assembly is inserted through the hole in the asbestos-cement siding. The gun is held perpendicular to and firmly against the purlin or girt. The stud is end-welded in a fraction of a second by pressing the gun trigger. Finally, the ferrule is knocked off.

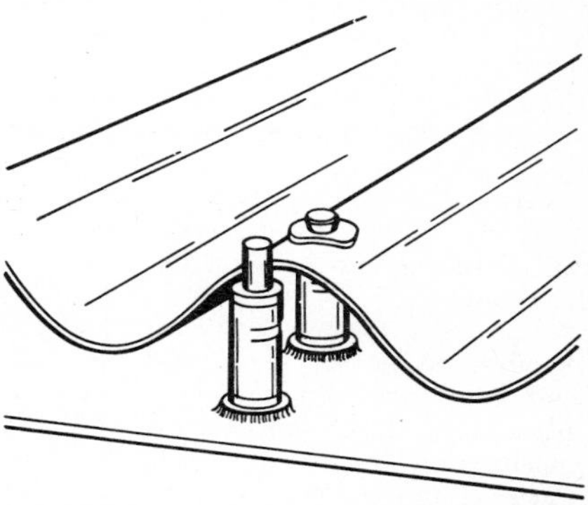

Fig. 16-19. Corrugated metal impaled on and fastened in place with welded stud.

The sealing collar and nut are then placed on the stud and tightened sufficiently to make a positive and weathertight seal between the collar and the asbestos-cement sheet. A similar fastening, but with corrugated-metal sheet, is shown in Fig. 16-19.

The above method permits application of siding and roofing entirely from the outside. A man is not required to be on the inside because there are no clips or fasteners on the interior. The expense of an interior scaffold is thus saved.

In securing corrugated metal, it is sometimes desirable to weld the studs to the steel frame in advance of placing predrilled sheets. In these cases, templates for quickly marking sheet and stud locations may be employed, as desired. Stud welding is applicable to any steel frame composed of standard structural steel. Steels of the high-carbon variety such as rerolled rail stock are not suitable for stud welding.

The manufacturer's recommendations should be followed in selecting the best type and size of stud for each specific installation. The leading manufacturers have direct field representatives in all areas who can supply valuable advice as to the best procedures to follow.

16-22. Powder-driven Studs. For many applications requiring the joining of steel or wood parts, or fastenings to concrete, steel, and brick surfaces, powder-actuated stud drivers are found to decrease costs because of their simplicity and speed. These drivers use a special powder charge to drive a pin or stud into relatively hard materials. The key to their efficiency is the proper selection of drive pins and firing charges. Because of the high velocity, the drive pin, in effect, fuses to the materials and develops the holding power.

Pull-out tests of these driven studs prove remarkable holding power. Average pull-out in 3,500-psi concrete exceeds 1,200 lb for 10-gage studs, and 2,400 lb for ¼-in.-diameter studs. In steel plate the pull-out resistance is still greater.

All that is required is a stud driver, the correct stud, and the correct cartridge, as recommended by the manufacturer for each specific set of fastening conditions. There are about a half dozen different strengths of cartridge, each identified by color, and some two dozen varieties of studs. Some studs have external threads, others internal threads. A plastic coating protects the threads from damage while driving.

The drivers will force studs into steel up to about ½ in. thick, into concrete through steel plates up to about ¼ in. thick, into concrete through various thicknesses of woods, and into steel through steel up to ¼ in. thick.

Powder-driven studs should never be driven into soft materials or into very hard or brittle materials, such as cast iron, glazed tile, or surface-hardened steel. Neither should they be used in face brick, hollow tile, live rock, or similar materials.

In driving studs a suitable guard must be used for each operation, and the driver must be held squarely to the work. If the driver is not held perpendicular, a safety device prevents firing of the charge. There also must be sufficient backup material to absorb the full driving power of the charge. Studs cannot safely be driven closer than ½ in. from the edge of a steel surface, or closer than 3 in. from the outside edge of concrete or brick surfaces.

16-23. Miscellaneous Anchoring Devices. Various types of expansion bolts, screw anchors, and toggle bolts are available for securing fixtures, brackets, and equipment to solid material, such as masonry, brick, concrete, and stone. Anchors also can be had for fastening to hollow walls, such as plaster on metal lath in furred spaces and hollow tile. The best device to use depends on the requirements in each case.

For use with practically any materials, including soft and brittle ones, such as composition board, glass, and tile, fiber screw anchors with a hollow metal core find a universal application. These plugs with braided metal cores possess many advantages: They can be used with wood screws and with lag screws. The flexible construction permits the plug to conform to any irregularities. Because of this elastic compression, the fibers are compressed as the screw enters. Screws can be unscrewed and replaced. Shock and vibration have no effect on gripping power. The plugs come in about 40 sizes to fit anything from a No. 6 screw to a ⅝-in. lag screw. In practice, a hole is drilled first, the plug is driven into it, and then the screw is inserted, expanding the plug against the sides of the hole.

For some fastenings, one-piece drive bolts are hammered like a nail into prepared holes in masonry or concrete. Other types of expansion bolts have expansion shields or are calked into place. The expansion shield is expanded in the hole by a tapered sleeve forced against a cone-shaped nut by the tightening of a bolt threaded into the cone. These types are not recommended for soft or brittle materials.

For thin hollow walls, toggle bolts equipped with spring-actuated wings are used. The wings will pass into the hole in folded position. After entering the hollow space, the wings open out, thereby obtaining a secure hold.

For fastening lightweight materials to nailable supports, stapling machines are extensively used. In one patented system, for example, staples secure acoustic tile to wood furring strips. In this system, invisible fastenings are obtained by using a full-spline suspension for the kerfed pieces of tile and a special stapling machine adjusted to function at the proper distance below the furring strips. The machine staples the splines (at the joints of the tiles) to the supporting strips, giving a speedy, economical, and secure fastening.

16-24. Bolts. The number of sizes and varieties of this basic type of fastening found in buildings is great. One of the more significant applications is in making field connections for structural-steel framing. The bolts are known as high-tensile-strength bolts (Art. 6-55). For other types of bolts used with structural steel, see Arts. 6-54 and 6-56.

In timber construction, bolts generally are used with timber connectors—patented metal rings, grids, shear plates, framing anchors, and clamping plates, which increase the strength of joints (Art. 8-21).

Section 17

Acoustics

LYLE F. YERGES

Consulting Engineer, Downers Grove, Ill.

Acoustics, derived from a Greek word meaning to hear, refers to the entire science of sound, including its generation, transmission, absorption, and control.

Acoustics has a long history. In Roman times, Vitruvius wrote extensively on the acoustics of rooms, particularly theaters and performing spaces. Some of his observations were correct, some incorrect. During the latter part of the nineteenth century, excellent scientific work in acoustics was done by Helmholz and Lord Rayleigh. But it was the pioneer work of Wallace Clement Sabine at Harvard University during the first two decades of the twentieth century that founded the modern science of acoustics.

Sabine established the relationship between reverberation and intelligibility within a room. He also discovered the relationship of reverberation time to the acoustical absorption present within a room.

Although acoustics remains part science and part art, modern studies are stripping away much of its mystique. As a result, it is possible to plan and predict the acoustics of finished spaces with reasonable certainty.

17-1. Sound Production and Transmission. Sound is a vibration in an elastic medium. It is a simple form of mechanical energy, and can be described by the mathematics associated with the generation, transmission, and control of energy.

Almost any moving, vibrating, oscillating, or pulsating object is a potential sound source. Usually, though, sound sources of interest to us radiate enough energy to be audible to humans or felt by them.

An understanding of the nature of "elastic matter" is essential to an understanding of the mechanism of sound transmission. If a single molecule of a quantity of matter subject to sound motion were observed, it would be found to move first to one side of its normal, or rest, position to some maximum displacement, then back through normal to an opposite maximum position, and back to normal (Fig. 17-1.) If the motion is regular and repetitive, it is called a **vibration.**

While each molecule or particle moves only an infinitesimal distance to either side of its rest position, it bumps the particles adjacent to it and imparts motion and energy to these particles. Thus, the impulse or disturbance progresses rapidly and travels great distances, but the medium in which it travels only vibrates. Figure 17-2 shows what the molecules in a medium look like when a sound wave passes through the medium. The impulse travels in the same direction as the movement of the particles; in other words, it is a longitudinal or "squeeze" wave motion, not a transverse wave like light.

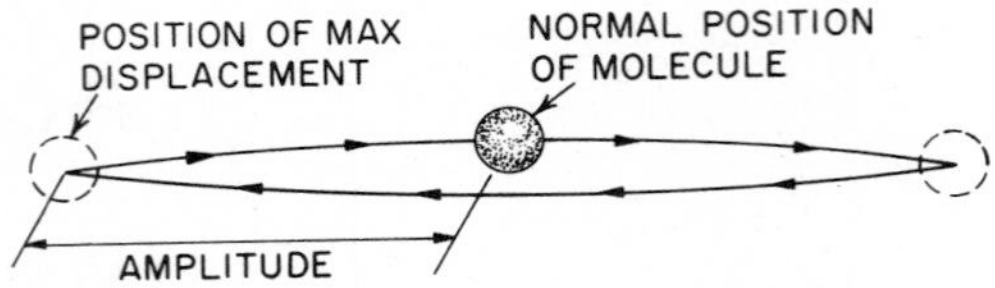

Fig. 17-1. Motion of a molecule during a single cycle of vibration.

Sound waves in air (or other gases or fluids) travel outward from the source like a rapidly expanding soap bubble, although any particular group of molecules behaves like a pulsating balloon, moving only slightly, while the wave progresses swiftly to great distances.

17-2. Nomenclature for Analysis of Sound.

Cycle. A complete single excursion of a vibrating molecule (Fig. 17-1).

Frequency. The number of cycles of vibration in a given unit of time, usually cycles per second (cps) or hertz (Hz).

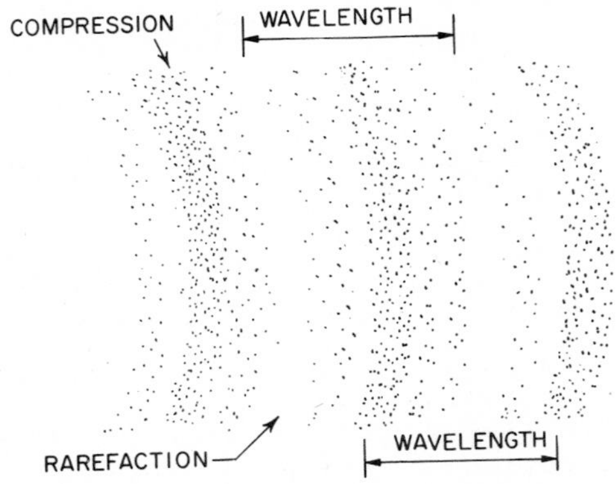

Fig. 17-2. An elastic medium subjected to sound.

Sound Wave. The portion of a sound between two successive compressions or rarefactions.

Wavelength. The distance between two successive rarefactions or compressions in a sound wave (Fig. 17-2).

Amplitude. Maximum displacement, beyond its normal, or rest, position, of a vibrating element. In most audible sounds, these excursions are very small, although low-frequency sound may cause large excursions (as would be observed in the motion of a loudspeaker cone reproducing very low-frequency sounds at audible level). Amplitude of motion is related to the increased pressure created in the medium, and to the intensity of the energy involved.

Velocity. Speed at which a sound impulse travels (not the speed of movement of any particular molecule). In a given medium, under fixed conditions, sound velocity is a constant. Therefore, the relationship between velocity, frequency, and wavelength can be expressed by the equation:

$$\text{Velocity} = \text{frequency} \times \text{wavelength} \qquad (17\text{-}1)$$

Because velocity is constant in air, low-frequency sound has long wavelengths, and high-frequency sound has short wavelengths. This is important to remember in any acoustical design considerations.

In addition to the longitudinal (compression or "squeeze") waves by which sound travels, there is another type of vibrational motion to which most building materials and construction systems are subjected, a **transverse wave** (Fig. 17-3). This is the familiar motion of a vibrating string or reed. This type of wave, too, transmits energy that can be felt or heard, or both.

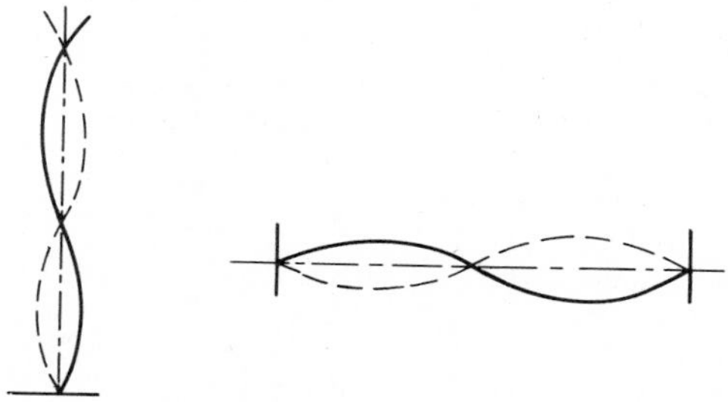

Fig. 17-3. Transverse waves.

Sheets or panels, studs and joists, hangers and rods, and similar slender and somewhat flexible members are particularly apt to vibrate in a transverse mode and to transmit sound energy along their length, as well as radiating it from their surfaces to the surrounding air.

17-3. Elastic Media. The elasticity and the mass of all materials determine their acoustical performance. All matter has mass; that is, inertia and weight. (For most purposes, we can think of weight and mass as interchangeable in our gravitational system.)

Sound does not travel faster in dense media than in less dense media. In fact sound velocity is inversely proportional to the density of the medium, as can be seen from Eq. (17-2).

$$\text{Velocity} = k\sqrt{\frac{\text{modulus of elasticity}}{\text{density}}} \qquad (17\text{-}2)$$

where k is a constant. Usually, however, dense materials often have a much larger modulus of elasticity than less dense materials, and the effect of the modulus more than offsets the effects of density. Table 17-1 lists the velocity of sound in some common materials.

Another characteristic of material, very important to its acoustical performance, is acoustical **impedance.** This value is determined by multiplying the velocity of

Table 17-1. Velocity of Sound in Various Media

Material	*Approximate Sound Velocity, ft per sec*
Air	1,100
Wood	11,000
Water	4,500
Aluminum	16,000
Steel	16,000
Lead	4,000

sound in the material by the density of the material. An examination of the units resulting from this operation will show that this is a way of measuring the rate at which the material will accept energy. Table 17-2 lists the acoustical impedance of several typical materials used in building construction.

Table 17-2. Acoustical Impedance of Various Materials

Material	*Acoustical Impedance, Psi per Sec*
Rubber	100
Cork	165
Pine	1,900
Water	2,000
Concrete	14,000
Glass	20,000
Lead	20,500
Cast iron	39,000
Copper	45,000
Steel	58,500

Sound must travel via some path—the elastic media. Hard, rigid, dense materials make excellent transmission paths, because they can accept much energy readily.

17-4. Hearing. In most cases of interest, a human is the "receiver" in the source-path-receiver chain. Hearing is the principal subjective response to sound. Within certain limits of frequency and energy levels, sound creates a sensation within the auditory equipment of humans and most animals.

At very low frequencies or at very high-energy levels, additional sensations, ranging from pressure in the chest cavity to actual pain in the ears, are experienced.

Our sense of feeling is also affected by vibrations in building structures, a subject closely related to ordinary building acoustics but outside the scope of this section.

From a strictly mechanical standpoint, the ear responds in a relatively predictable manner to physical changes in sound. Table 17-3 relates the objective characteristics of sound to our subjective responses to those characteristics.

Table 17-3. Subjective Responses to Characteristics of Sound

Objective (Sound)	*Subjective (Hearing)*
Amplitude, Pressure, Intensity	Loudness
Frequency	Pitch
Spectral distribution of acoustical energy	Quality

Loudness is the physical response to sound pressure and intensity. It is influenced somewhat by the frequency of the sound.

Pitch is the physical response to frequency. Low frequencies are identified as low in pitch, high frequencies as high in pitch. Middle C on the piano, for example, is 261 Hz; 1 octave below is 130 Hz; 1 octave above is 522 Hz.

An **octave** represents a 2:1 ratio of frequencies.

For practical purposes, humans can be considered to have a hearing range from about 16 Hz to somewhat less than 20,000 Hz, and to be significantly deaf to low frequencies and very high frequencies. Human loudness response to sounds of identical pressure but varying frequency plots something like the curve in Fig. 17-4.

Any measure of loudness must, in some way, specify frequency as well as pressure or intensity to have any real significance in human acoustics. As will be shown later, measuring equipment and scales used are all modified to recognize this. Table 17-4 lists some of the significant frequency ranges.

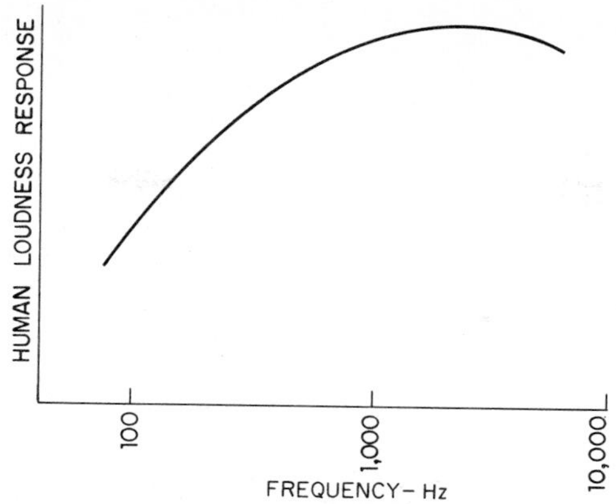

Fig. 17-4. Human response to sounds of identical level but varying frequency.

Low frequencies do not affect humans very strongly, while those from 500 to 5,000 Hz are very important.

Sounds generated by mechanical equipment may encompass the entire frequency range of human hearing. Jet aircraft, for example, have significant output throughout the entire range, and large diesel engines produce substantial energy from 30 to 10,000 Hz.

Wanted sound (*signal*), whatever its nature, communicates information to us. *Unwanted sound,* whatever its nature, is *noise.* When the signal becomes sufficiently louder than the noise, the signal can be detected and the information becomes available. Thus, the distinction between noise and communication is completely subjective, completely a matter of human desires and needs of the moment.

Table 17-4. Significant Frequency Ranges

	Approximate Frequency Range, Hz
Range of human hearing	16–20,000
Speech intelligibility (containing the frequencies most necessary for understanding speech)	600–4,800
Speech privacy range (containing speech sounds that intrude most objectionably into adjacent areas)	250–2,500
Typical small table radio	200–5,000
Male voice	350*
Female voice	700*

* Frequency at about which energy output tends to peak.

While very weak sounds can be heard in a very quiet background, the level of wanted sound may be increased to override the unwanted background (or noise); but there is a significant noise level above which even loud (shouting) speech is scarcely intelligible. Such a level, if long continued, can prove damaging to the hearing mechanism. At still higher levels, pain can be experienced, and immediate physical damage (often irreversible) may occur.

Complete silence (if such a thing could occur) would be most unpleasant. Humans would quickly become disoriented and distressed. Somewhere between silence and the uncomfortably loud, however, is a wide range that humans find acceptable.

17-5. Force and Energy Concepts. In acoustics, the fundamental concept is the well-known:

$$\text{Force} = \text{mass} \times \text{acceleration} \tag{17-3}$$

All materials possess mass. This mass, when moved with change in direction and velocity, which any elastic, oscillating motion must involve, must be accelerated. This process requires force; and force, acting over displacement, is energy. Thus, for sound, the amount of motion of the particles in the medium depends on the sound pressure (force) or the intensity (amount) of the sound energy involved.

The energy concepts of importance are kinetic energy (energy of motion), potential energy (stored energy), and energy conversion.

Transmission (flow), or prevention of transmission of sound, and conversion of sound energy to a nonaudible form are the functions of so-called **acoustical materials.**

17-6. Units and Dimensions for Sound Measurements. Generally, absolute numbers obtained from measurements have little significance in acoustics. Instead, measurements are almost always compared with some base or reference, and they are usually quoted as **levels** above or below that reference level. The levels are usually ratios of observed values to the reference level, such as 2:1, 1:2, 1:10, because rarely are simple, linear relationships found between stimuli and effects in humans.

Zero level of sound pressure is not a true physical zero; that is, the absence of any sound pressure at all. Rather, zero level is something of an average threshold of human hearing, of sound at about 1,000 Hz. The physical pressure associated with this threshold level is very small, 0.0002 dynes per sq cm (often referred to as a microbar, one-millionth of normal standard atmospheric pressure).

Changes in human response tend to occur according to a ratio of the intensity of the stimuli producing the response. In acoustics, the ratio of 10:1 is called a bel, and one-tenth of a bel, a decibel (dB). Thus, power and intensity levels, in dB, are computed from

$$\text{Level} = 10 \log_{10} \frac{\text{quantity measured}}{\text{reference quantity}} \tag{17-4}$$

Intensity level *IL*, dB, for example, represents the ratio of the intensity being measured to some reference level, and is given by

$$IL = 10 \log_{10} \frac{I}{I_o} \tag{17-5}$$

where I = intensity measured, watts per sq cm

I_o = reference intensity = 10^{-16} watts per sq cm

Since intensity varies as the square of the pressure, intensity level also is given by

$$IL = 10 \log_{10} \frac{p^2}{p_o^2} = 20 \log_{10} \frac{p}{p_o} \tag{17-6}$$

where p = pressure measured, dynes per sq cm

p_o = reference pressure = 0.0002 dynes per sq cm

Sound pressure level, dB, to correspond with intensity level, is defined by

$$SPL = 20 \log_{10} \frac{p}{p_o} \tag{17-7}$$

Sound power level refers to the power of a sound source relative to a reference power of 10^{-12} watts. (*Note:* At one time, 10^{-13} watts was used; thus, it is imperative that the reference level always be explicitly stated.)

The ear responds in a roughly logarithmic manner to changes in stimulus intensity, but approximately as shown in Table 17-5.

Another comparison, which gives more meaning to various levels, is shown in Table 17-6.

Table 17.5. Subjective Effect of Changes in Sound Characteristics

Change in Sound Level, dB	*Change in Apparent Loudness*
3	Just perceptible
5	Clearly noticeable
10	Twice as loud (or ½)
20	Much louder (or quieter)

Table 17-6. Comparison of Intensity, Sound Pressure Level, and Common Sounds

Relative intensity	*SPL,* dBA*	Loudness
100,000,000,000,000	140	Jet aircraft and artillery fire
10,000,000,000,000	130	Threshold of pain
1,000,000,000,000	120	Threshold of feeling
100,000,000,000	110	
10,000,000,000	100	Inside propellor plane
1,000,000,000	90	Full symphony or band
100,000,000	80	Inside automobile at high speed
10,000,000	70	Conversation, face-to-face
1,000,000	60	
100,000	50	Inside general office
10,000	40	Inside private office
1,000	30	Inside bedroom
100	20	Inside empty theater
10	10	
1	0	Threshold of hearing

* *SPL* as measured on A scale of standard sound level meter.

17-7. Sound Measurements and Measurement Scales. Most measurements or evaluations of sound intensity or level are made with an electronic instrument that measures the sound pressure. The instrument is calibrated to read pressure levels in decibels (rather than volts). It can measure the over-all sound pressure level throughout a frequency range of about 20 to 20,000 Hz, or within narrow frequency bands (such as an octave, third-octave, or even narrower ranges). Usually, the sound level meter contains filters and circuitry to bias the readings so that the instrument responds more like the human ear—"deaf" to low frequencies and most sensitive to the midfrequencies (from about 500 to 5,000 Hz). Such readings are called A-scale readings. Most noise level readings (and, unless otherwise specifically stated, most sound pressure levels with no stated qualifications) are A-scale readings (often expressed as dBA). This means that actual sound pressure readings have been modified electrically within the instrument to give a readout corresponding somewhat to the ear's response (Fig. 17-4).

For various measurements and evaluations of performance for materials, constructions, systems, and spaces, see Art. 17-17.

17-8. Sound and Vibration Control. This is a process, not a product. It consists of:

A. Acoustical analysis

1. Determining the use of the structure—the subjective needs.
2. Establishing the desirable acoustical environment in each usable area.
3. Determining noise and vibration sources inside and outside the structure.
4. Studying the location and orientation of the structure and its interior spaces with regard to noise and noise sources.

B. Acoustical design

1. Designing shapes, areas, volumes, and surfaces to accomplish what the analysis indicates.
2. Choosing materials, systems, and constructions to achieve the desired result.

Sound and vibration sources are usually speech and sounds of normal human activity—music, mechanical equipment sound and vibration, traffic, and the like. Characteristics of these sound sources are well known or easily determined. Therefore, the builder or designer is usually most interested in the transmission paths for sound and vibration. These are gases (usually air); denser fluids (water, steam, oil, etc.); and solids (building materials themselves).

The mechanism for energy transmission via these paths is discussed in Art. 17-1. During this transmission process, some of the sound energy is absorbed or dissipated, some is reflected from various surfaces or constructions, and some is transmitted through the constructions.

Sound control is accomplished by means of barriers and enclosures, acoustically absorbent materials, and other materials, constructions, and systems properly shaped and assembled. Vibration control is accomplished by means of various resilient materials and assemblies, and by damping materials (viscoelastic materials of various types).

Air-borne and structure-borne energy are controlled by somewhat different techniques, described in the following articles.

17-9. Air-borne Sound Transmission. A sound source in a room sets the air into vibration. The vibrating air causes any barrier it touches (partitions, floors, ceilings, etc.) to vibrate; the vibrating barrier, in turn, sets into vibration the air on the opposite side of the barrier (Fig. 17-5).

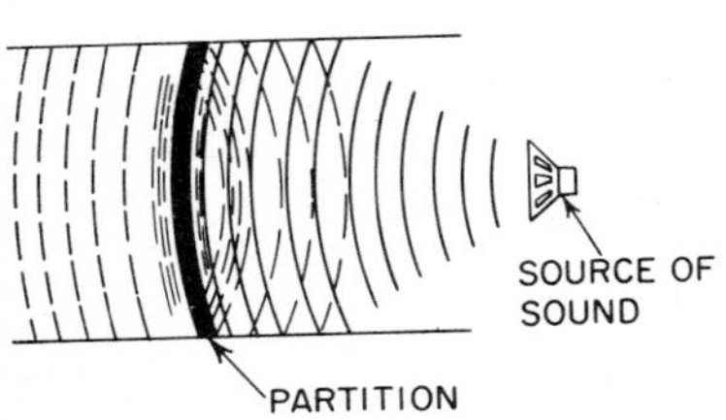

Fig. 17-5. Sound transmission through a barrier.

The barrier, like any other body, resists motion because of its inherent inertia. Because it takes more force, and therefore more energy, to move a heavier (more massive) barrier, sound transmission through the barrier depends directly on the mass of the barrier.

The loss in energy level between the original signal striking the barrier and the level of the energy transmitted to the opposite side is called the **sound transmission loss** of the barrier. The more effective the barrier, the greater the sound transmission loss.

Figure 17-6 shows graphically how a truly limp barrier of infinite width and height responds to sound energy of varying frequency (the *mass law,* straight-line variation). The straight line rises at the rate of 6 dB per octave. At 800 Hz, for example, such a barrier moves back and forth twice as fast as at 400 Hz (1 octave lower), and one-half as fast as at 1,600 Hz (1 octave higher). The moving barrier has kinetic energy *KE,* which is related to mass *M* and velocity v by

$$KE = \tfrac{1}{2}Mv^2 \tag{17-8}$$

As the velocity of movement (frequency) doubles (1 octave higher), the energy of motion quadruples. Therefore, the energy or intensity level increases by 6 dB, inasmuch as Eq. (17-5) gives $10 \log_{10} 4 \approx 6$ dB.

Likewise, as the mass of the barrier doubles, the force or pressure required to maintain the same frequency doubles, and because energy is proportional to the square of the pressure, the energy level quadruples. By Eq. (17-6), the increase in intensity level is $10 \log_{10} 4 \approx 6$ dB.

In practice, however, this simple mathematical relationship rarely exists. No barrier is truly limp; a wall, floor, or ceiling always has a finite stiffness. Therefore, at low frequencies, barriers tend to have higher sound transmission losses than the mass law predicts. In some region of the spectrum, however, barriers tend to have lower sound transmission losses (pass sound more readily) than the mass law indicates (see curve for solid panel in Fig. 17-6).

This latter phenomenon results because the barrier, in addition to its back-and-forth motion, as shown in Fig. 17-5, moves in a simultaneous shear wave, like a rope being shaken (Fig. 17-3). At some frequency, the velocity of this shear wave in the barrier coincides with the velocity of the impinging sound wave in the air. Then, the partition is quite transparent to sound, and a deep coincidence dip in the sound transmission loss curve occurs (see curve for double wall in Fig. 17-6). At frequencies above the coincidence frequency, the curve tends to recover its 6 dB per octave slope.

These are few practical limp materials useful for ordinary building panels. Soft sheet lead is occasionally used for specialized barriers, as described in Art. 17-16, or is applied to other panel material to increase the mass of the composite without increasing its stiffness. In addition, damping materials (Art. 17-12) can be applied to the panels to damp out their vibrations quickly and to increase the energy loss as the panels vibrate.

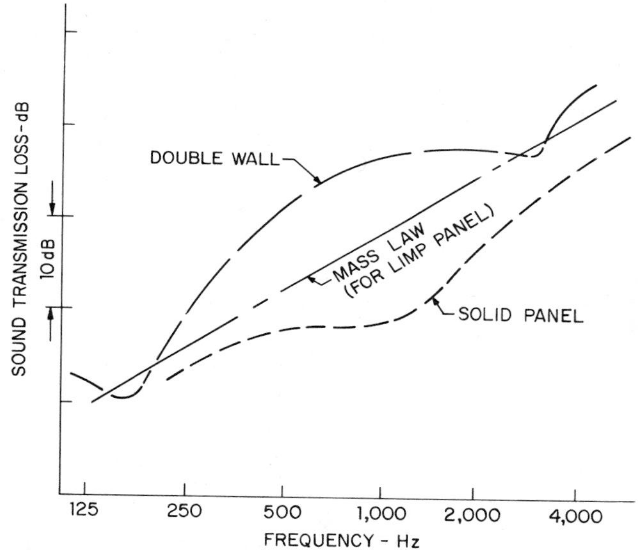

Fig. 17-6. Sound transmission loss of barriers.

Figure 17-6 shows another characteristic of barriers that further complicates their performance. While single, solid panels have somewhat smaller transmission losses than the mass law predicts, a double wall, a barrier comprising separated wythes or leaves, of the same total weight produces greater transmission losses than the mass law predicts.

If the mass of a single, solid panel is divided into two separate, unconnected layers, the only energy transfer between them occurs via the air between the layers. Air is not stiff; it can sustain little of the shear wave, and it is not an efficient transmitter of the back-and-forth motion of one layer to the other (except at certain resonant frequencies). As a result, particularly in the midfrequencies, the sound transmission loss actually exceeds the mass-law values, often considerably. The greater the distance between layers, the better the performance. Theoretically, the sound transmission loss should increase about 6 dB for each doubling of the width of the air space. In practice, however, the increase in decibels is somewhat less than this.

If a porous, sound-absorbent blanket is inserted in the void between layers, the standing waves in the air between layers are minimized. Furthermore, such a blanket has the effect of reducing the stiffness of the air, as well as absorbing some of the energy of the air as it pumps back and forth through the blanket. The result is an additional increase in sound transmission loss.

The performance of various barriers, and rating systems for their performance, are shown in Table 17-7, p. 17-18.

17-10. Bypassing of Sound Barriers. Rarely is a sound barrier the sole transmission path for the acoustic energy reaching it. Some energy invariably travels via the connecting structures (floors, ceilings, etc.), or through various openings in or around the barrier.

Structural flanking via edge attachments and junctions of walls, partitions, floors, and ceilings can seriously degrade the performance of a barrier. Figure 17-7 shows some typical flanking paths and how to avoid such flanking.

Even trivial openings through a sound barrier seriously degrade its performance. An opening with an area of only about 1 sq in. transmits as much acoustical

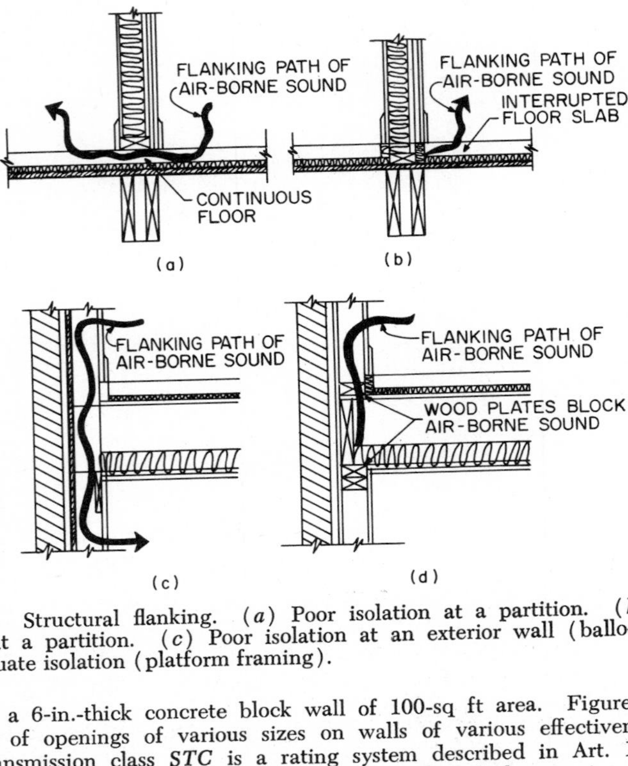

Fig. 17-7. Structural flanking. (*a*) Poor isolation at a partition. (*b*) Adequate isolation at a partition. (*c*) Poor isolation at an exterior wall (balloon framing). (*d*) Adequate isolation (platform framing).

energy as a 6-in.-thick concrete block wall of 100-sq ft area. Figure 17-8 shows the effect of openings of various sizes on walls of various effectiveness. (*Note:* Sound transmission class *STC* is a rating system described in Art. 17-16.) The better the wall, the more serious the effect of openings or leaks.

The more common points of leakage through or around barriers include perimeter of pipes, ducts, or conduits penetrating the barriers; relief grilles for return air; perimeter of doors or glazing; shrinkage or settlement cracks at partition heads or sills; joints between partitions and exterior curtain-wall mullions; joints around or openings through back-to-back electrical outlet boxes, medicine cabinets, etc.; common supply or return ducts with short, unlined runs between rooms; and operable windows opening to a common court. Such openings should be avoided, when possible, and all cracks, joints, and perimeters should be calked and sealed.

In some spaces (an open-plan school, "landscaped" office, etc.), screens, or partial-height barriers are used. Their effectiveness is much less than that of full-height partitions or walls. When partial-height partitions are contemplated in building design, the services of experienced acoustical consultants should always be used, because experts are needed to deal with the many factors that enter into the acoustical design of open-plan spaces.

See also Art. 17-15.

17-11. Structure-borne Sound Transmission. Acoustical energy transmission from sound sources to distant parts of the structure via the structure itself is often a major problem in building acoustics. One reason, as discussed in Art. 17-10, is that structural flanking can seriously degrade the performance of sound barriers. Additional transmission paths include pipes, ducts, conduits, and almost any solid, continuous, rigid member in the building. Impacts and vibrations can be transmitted through floor-ceiling assemblies quite readily, if proper precautions are not taken.

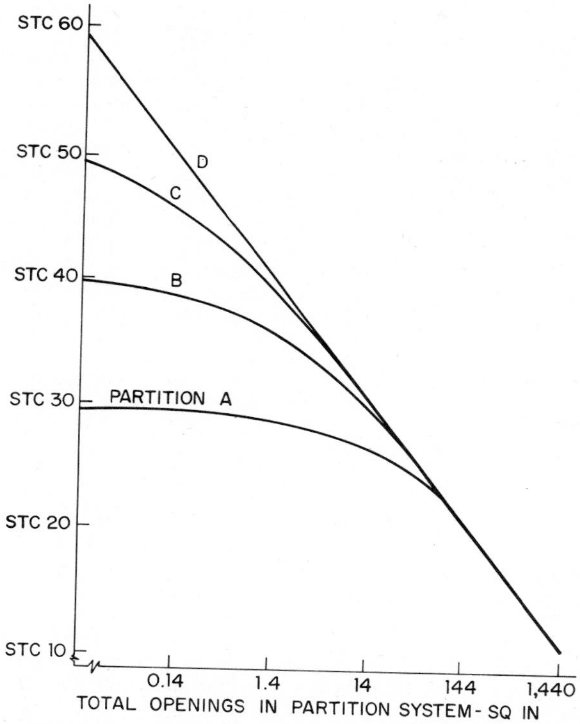

Fig. 17-8. Effect of leaks on *STC* of partitions (for 100-sq-ft test wall).

Normally, in ordinary construction, a good carpet over a good pad will provide sufficient impact isolation against the sounds of footfalls, heel clicks, and dropped objects. Where carpet is not feasible, or in critical spaces, special floor-ceiling assemblies can be used. (See Art. 17-16, for various constructions and their impact isolation performance.)

Isolation against vibration or impact produced by machinery or other vibrating equipment is usually provided by use of special resilient mounts or mounting systems. Springs, elastomeric pads, and other devices are used in such work. For a detailed discussion of such isolation, refer to the current ASHRAE "Guide," American Society of Heating, Refrigerating and Air-Conditioning Engineers, or other references on vibration and shock isolation.

It is important to remember that any moving, rotating, oscillating, or vibrating equipment, rigidly attached to the building structure, will transmit some of its energy to the structure. Isolating against such transmission is always advisable, even in ordinary construction, and is imperative in critical buildings and where the equipment is large or of high power.

Noise and vibration transmitted via plumbing and heating pipes, ducts, and conduits can be objectionable unless proper precautions are taken. Resilient connectors for rigid members, acoustically absorptive duct linings, and similar approaches are described in detail in the ASHRAE "Guide."

17-12. Damping of Vibrations. As a panel or object vibrates, it radiates acoustical energy to the air surrounding it and to solid surfaces touching it or attached to it. If the energy of vibration could be dissipated, the radiation would be reduced and the sound and vibration levels lowered. One way to do this is to attach firmly to a vibrating panel certain "lossy" substances (those with high internal friction or poor connections between particles) or viscoelastic materials (neither elastic nor completely viscous, such as certain asphaltic compounds). These damp the vibrations by absorbing the energy and converting it to heat.

In the assembly of barriers, damping can be accomplished with proper connections and attachments, use of viscoelastic adhesives, proper attachment of insulating materials, and similar means. Special viscoelastic materials for adhesive attachment or brush or spray application to panels are available.

17-13. Sound Absorption. The best-known acoustical materials are acoustical absorbents (although actually all materials are acoustical). Generally, these absorbents are lightweight, porous, "fuzzy" types of boards, blankets, and panels.

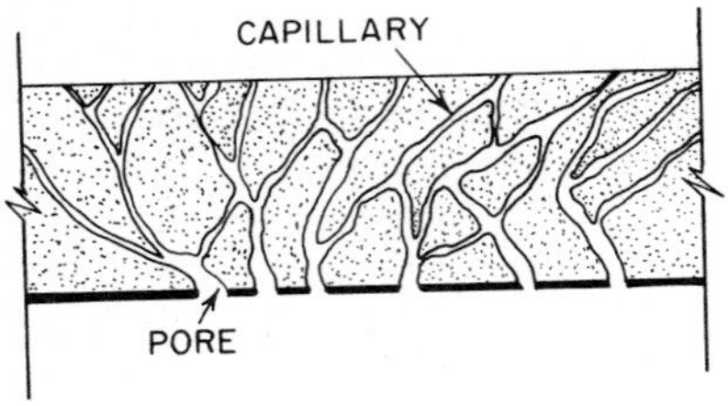

Fig. 17-9. Cross section of sound absorbent (greatly enlarged).

Acoustical absorbents act as energy transducers, converting the mechanical energy of sound into heat. The conversion mechanism involves either pumping of air contained within the porous structure of the material, or the flexing of thin panels or sheets. Most materials employ the first principle.

The internal construction of most absorbents consists of a random matrix of fibers or particles, with interconnected pores and capillaries (Fig. 17-9). It is necessary that the air contained within the matrix be able to move sufficiently to create friction against the fibers or capillaries. Nonconnected-cell or closed-cell porous materials are not effective absorbents.

Tuned chambers, with small openings and a restricted neck into the chambers, also are used for sound absorption; but they are somewhat specialized in design and function, and their use requires expert design in most instances.

The surface of absorbents must be sufficiently porous to permit the pressures of impinging sound waves to be transferred to the air within the absorbent. Very thin, flexible facings (plastic or elastomeric sheets) stretched over panels and blankets do not interfere significantly with this pressure transfer, but thick, rigid, heavy coatings (even heavy coatings of paint) may seriously restrict the absorption process. Perforated facings, if sufficiently thin and with sufficient closely spaced openings, do not appreciably degrade the performance of most absorbents.

The visible surface of absorbent panels and tiles may be smooth or textured, fissured or perforated, or decorated or "etched" in many ways. Figures 17-10 and 17-11 show typical commercial tile products.

Absorbents are normally produced from vegetable or mineral fibers, porous or granular aggregates, foamed elastomers, and other products, employing either added binders or their own structure to provide the structural integrity required. Because of their lightweight, porous structure, most such products are relatively fragile and must be installed where they are not subject to abuse; or they may be covered with sturdy perforated or porous facings to protect them.

Absorbents are chosen for their appearance, fire resistance, moisture resistance, strength, maintainability, and similar characteristics. Their performance ratings in respect to these characteristics are usually published in advertising materials and in various bulletins and releases of trade associations.

The acoustical property of absorbents most important to designers and builders is absorptive efficiency. [Sound transmission through most absorbents takes place readily. They are very poor in this respect and should never be used to attempt to improve the air-borne sound isolation of a barrier. Sound transmission over the top of a partition, through a lightweight, mechanically suspended acoustical panel ceiling, is frequently a serious problem in buildings (Art. 17-15).] The absorptivity of a tile or panel is usually expressed as the fraction or percentage of acoustical energy absorbed from an impinging plane wave. If a perfectly absorptive plane surface represents 1.00 or 100%, the ratio of absorption of a given product to this perfect absorber is called the **sound absorption coefficient** of the product.

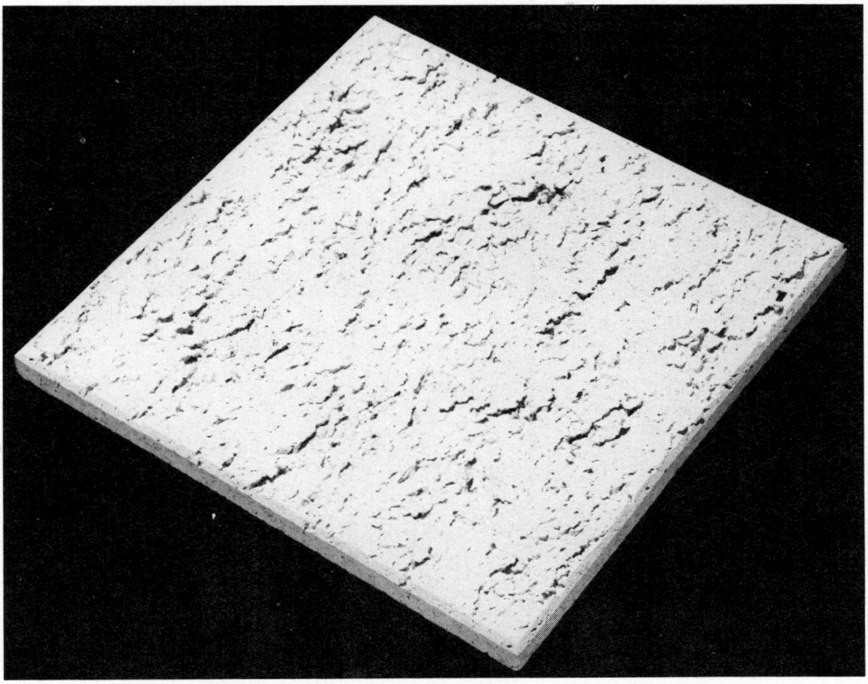

Fig. 17-10. Fissured-surface acoustic tile with the appearance of travertine stone.

The absorptivity of a material varies with its thickness, density, porosity, flow resistance, and other characteristics. Further, absorptivity varies significantly with the frequency of the impinging sound. Usually, very thick layers of absorptive material are required for good absorption of low-frequency sound, while relatively thin layers are effective at higher frequencies. Little or no absorption, however, can be obtained with thin layers or flocked surfaces, textured but nonporous surfaces, or other products often mistakenly thought to be absorptive or claimed to break up the sound.

Generally, there is an optimum density and flow resistance for any particular family of materials (particularly fibrous materials). Usually, absorptivity increases with thickness of the material.

Performance data for absorbent materials are usually readily available from manufacturers and trade associations. Performance of some typical products is shown in Table 17-10, p. 17-19. See also Arts. 17-14 to 17-16.

17-14. Control of Reflection and Reverberation. Normally, acoustical absorbents are used to prevent or minimize reflections of sound from the surfaces of rooms or enclosures. Distinct reflections—**echoes**—are usually objectionable in any occupied space. Rapid, repeated, but still partly distinguishable echoes, such as occur between parallel sidewalls of a corridor—**flutter**—are also objectionable.

Reverberation comprises very rapid, repeated, jumbled echoes, blending into an indistinct but continuing sound after the source that created them has ceased. Usually, reverberation is one of the major causes of poor intelligibility of speech within a room; but, within limits, it may actually enhance the sound of music

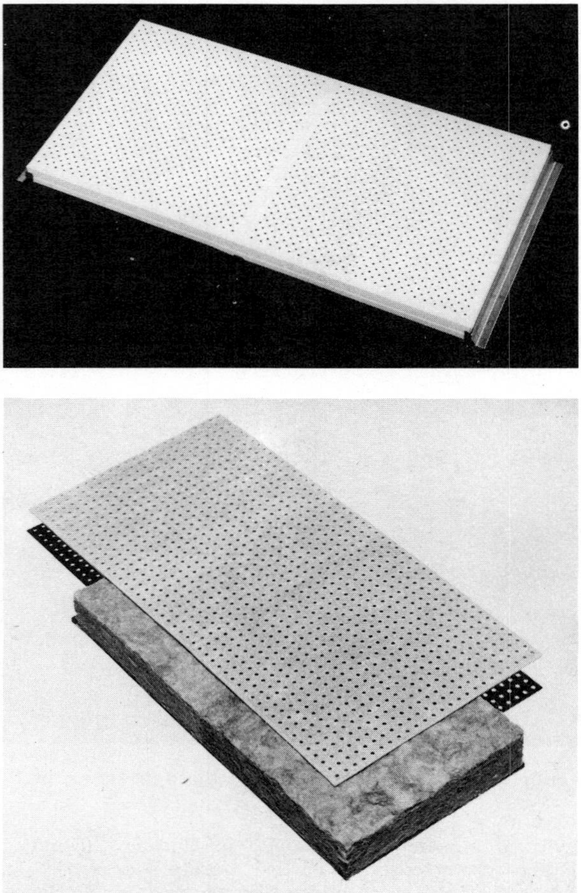

Fig. 17-11. Typical perforated acoustical tiles.

within a space. Reverberation control is a necessary and important aspect of good acoustic design, but it is often greatly overemphasized. Good room proportions and configuration, control of echoes, and absorption of noise usually assure an acceptable reverberation time within a space. Where careful determination and control of reverberation are required in a room, the services of a competent acoustical consultant are always advisable.

Reflections from strategically located and properly shaped room surfaces may be highly desirable, because such reflections may strongly enhance the source signal. But excessively delayed or highly persistent reflections are usually undesirable. (For most purposes, and within the normal frequency range of importance to human hearing, it can be assumed that sound waves, like light waves, reflect from a surface at the same angle as the angle of incidence. Because of the enormously longer wavelength of sound compared with light, this assumption is inexact but acceptable for most design work.)

In most rooms, absorption of most of the acoustical energy impinging on many of the surfaces (the floor, distant walls, etc.) is desirable to prevent build-up or increase of unintelligible or useless sound. For this purpose, sound absorbents may be placed on some or all of the surfaces. The difference in sound pressure level (or noise level) caused within a space by the introduction of absorbents can be calculated simply. From such a calculation, it is possible to determine how effective such treatment will be.

Noise reduction *NR*, dB, provided by adding acoustical absorbents in a space can be determined from

$$NR = 10 \log_{10} \frac{A_o + A_a}{A_o} \tag{17-9}$$

where A_o = original acoustical absorption present
A_a = added acoustical absorption

Acoustical absorption equals the sum of the products of each area (in consistent units) in the space times the absorption coefficient of the material constituting the surface of the area; for example, the floor area, sq ft × its absorption coefficient, plus ceiling area, sq ft × its absorption coefficient, plus total wall area, sq ft × its absorption coefficient.

[*Note:* An anomalous but useful term is often used in advertising data, the **noise reduction coefficient** *NRC*. This is the arithmetic average of the sound absorption coefficients of a material as determined at 250, 500, 1,000, and 2,000 Hz (Art. 17-13). Since these frequencies include the most significant speech and intelligibility ranges, such a figure is a reasonably good means of comparing similar materials; that is, materials with absorption characteristics not differing widely from one another within this frequency range. Often, *NRC* is used, instead of the absorption coefficients at various frequencies, to determine an average noise reduction from Eq. (17-9).]

Equation (17-9) indicates that the more absorption present originally the less the improvement provided by added absorption. Thus, in a very "hard," bare room, addition of acoustical (sound absorbent) tile to a full ceiling significantly reduces the noise level. But addition of the same ceiling tile in a room with a thick carpet, upholstered furniture, and heavy draperies would make little change.

Heavy carpet, upholstered furniture, heavy draperies, and similar materials are very effective absorbers. In residences, for example, rarely is additional acoustical absorption required in bedrooms, living rooms, and similar spaces. In kitchens, bathrooms, or recreation rooms, with normal hard floors and few additional furnishings or fabrics, however, an acoustical tile ceiling is helpful. This is equally true of offices and similar spaces. In theaters and auditoriums, the large expanse of upholstered seating and aisle carpets is normally adequate for most noise control, but added absorption on some wall surfaces may be required for control of echoes.

Reverberation Time. The reverberation within a space is usually expressed as the time required for a sound pulse to decay 60 dB (to one-millionth of its original level). For most purposes, reverberation time *T*, sec, can be calculated from the simple Sabine formula:

$$T = \frac{0.049V}{A} \tag{17-10}$$

where V = volume of the space, cu ft
A = total acoustical absorption in the space

Equation (17-10) assumes a smooth, steady, logarithmic decay; random distribution of sound within the room, with the wave front striking every surface quickly and within the decay time; and no standing waves between surfaces that could support a persistent mode. These are idealized conditions and never exist, but the formula is sufficiently accurate for most purposes.

Because the absorption of an absorbent material varies with frequency of sound, it is necessary to calculate T for each significant frequency. For most reverberation calculations, determinations at 500 Hz are adequate. For concert halls and critical spaces, calculations are usually made 2 octaves above and 2 octaves below 500 Hz as well.

Optimum reverberation time for a room is a subjective determination, governed by speech intelligibility and the fullness and richness of musical sound desired. Figure 17-12 shows, within the shaded area, the acceptable range of reverberation times for normal spaces of varying volume. For critical spaces (radio studios, concert halls, auditoriums, etc.), it is advisable to obtain the advice of competent acoustical consultants, because experts are needed to evaluate and provide for many additional factors that are equally important.

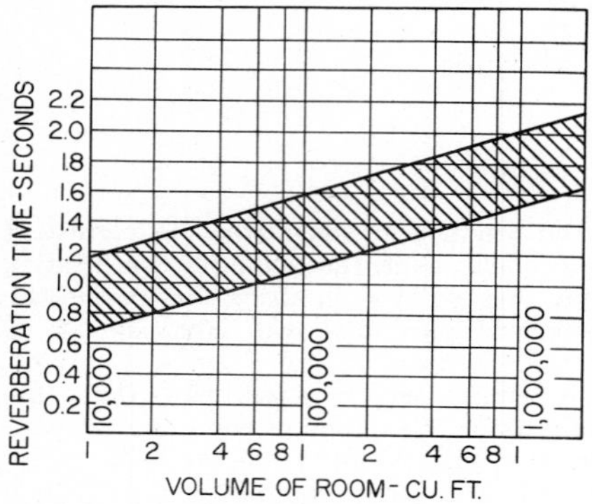

Fig. 17-12. Recommended reverberation time, indicated by shaded area, varies with size of room.

It is imperative that designers understand that a reverberation-time determination is not an acoustical analysis, and that, in many instances, reverberation time is a trivial part of an acoustical study.

17-15. Installation of Absorbents. The amount, location, and installation method for absorbents are all significant in a room. As discussed in Art. 17-14, the material is located on surfaces from which reflections are undesirable and where it is reasonably free from damage. The amount of material required can be determined from noise-reduction and reverberation-time calculations. (For most simple spaces, a full ceiling treatment is acceptable.)

Acoustical absorbents can be cemented directly to a smooth, solid surface, nailed or stapled to furring strips, or suspended by any of a number of mechanical systems, such as that shown in Fig. 17-13.

While acoustical absorbents are usually installed as, or on, flat, horizontal surfaces, various other configurations are used. Coffers, grids of hanging panels, and similar arrangements are often employed. The acoustical performance of each configuration must be tested to determine its effectiveness.

Sprayed-on and trowel-applied acoustical products are also employed, although their use is limited.

The structural, fire-resistance, and acoustical performance of most acoustical panels and tiles are significantly affected by the installation method. It is imperative that performance data be explicitly related to the specific installation method.

Partition Bypassing. When partitions or other sound barriers are constructed to (but not through) a suspended acoustical tile or panel ceiling, sound transmission through the tile from one space, over the top of the partition, into an adjacent space can be a serious problem. Figure 17-14 shows how sound can bypass a partition, and alternative methods of preventing this.

Similarly, structure-borne, flanking transmission along the continuous metal runners of some suspension systems may seriously degrade the performance of an acoustical ceiling as a sound barrier. Performance data for most typical commercial ceilings are published in various bulletins and advertising matter.

17-16. Performance Data. To simplify and standardize evaluation of the acoustical performance of materials, systems, and constructions, various rating systems have been adopted. The best known and most widely used are those published by the American Society for Testing and Materials (ASTM) and the Acoustical and Insulating Materials Association (AIMA).

Partitions, Floor-Ceiling Assemblies, and Barriers. Insulation (or isolation) of air-borne sound provided by a barrier is usually expressed as its **sound transmission class** STC. For a specific construction, STC is determined from a sound-transmission-

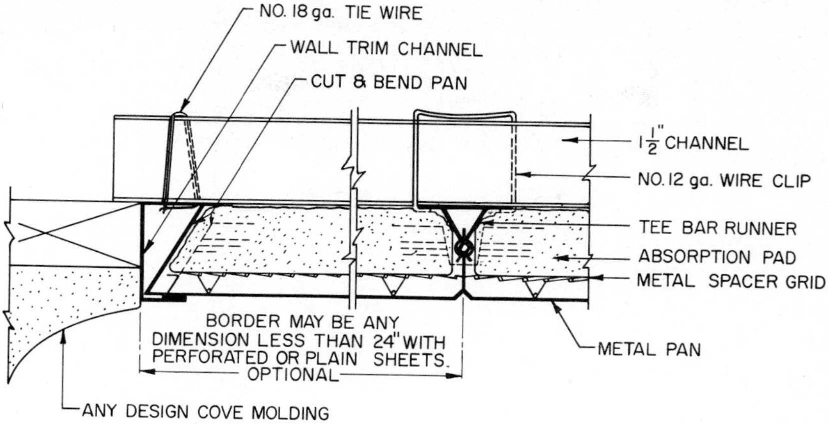

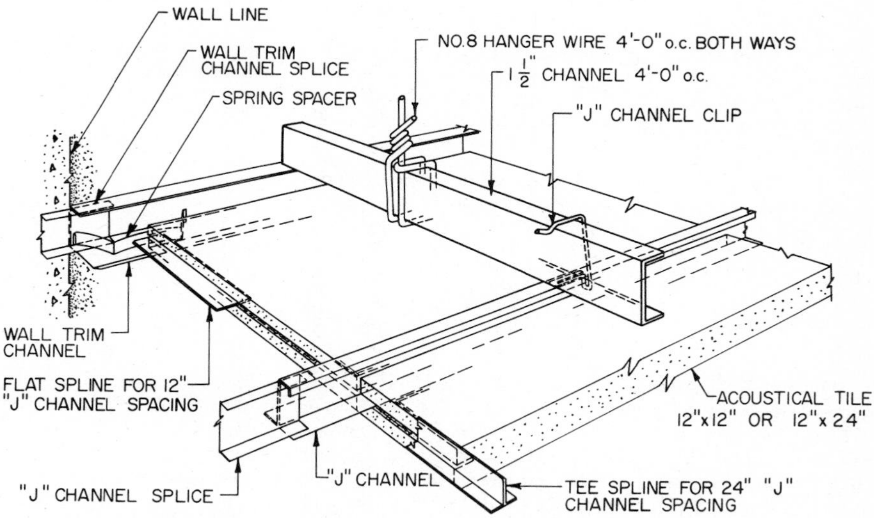

Fig. 17-13. Typical mechanically suspended acoustical tiles.

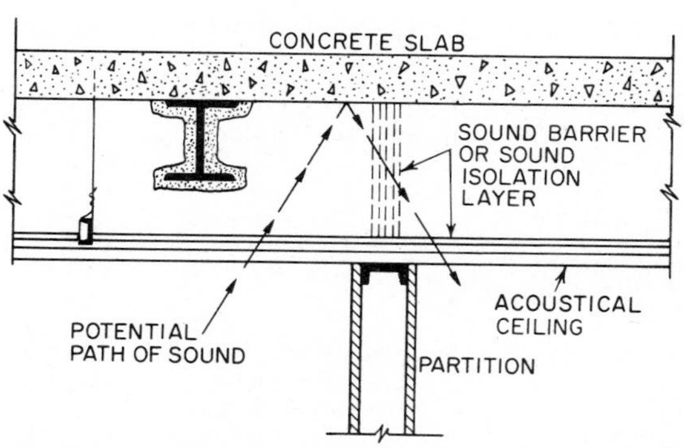

Fig. 17-14. Prevention of sound transmission over partitions.

loss curve obtained from a standardized test of a large-scale specimen in a recognized laboratory. This curve is compared with a standard contour, and a numerical rating is assigned to the specimen (ASTM E90 test procedure and ASTM E413. Determination of Sound Transmission Class).

Table 17-7 lists typical *STC* ratings of several partition, wall, and floor-ceiling components or assemblies. Published data for almost any type of construction can be obtained from various sources.

Table 17-7. ***STC* of Various Constructions**

Construction	*STC*
¼-in. plate glass	26
¾-in. plywood	28
½-in. gypsum board, both sides of 2 × 4 studs	33
¼-in. steel plate	36
6-in. concrete block wall	42
8-in. reinforced-concrete wall	51
12-in. concrete block wall	53
Cavity wall, 6-in. concrete block, 2-in. air space, 6-in. concrete block	56

It is important to remember that a difference of one or two points between two similar constructions is rarely significant. Normally, constructions tend to fall into groups or classes with their median values about five points apart. In Table 17-8, the boldface number represents the median of a performance group which includes the numbers on either side of the median.

Table 17-8. ***STC* Performance Groups**

30	31	**32**	33	34
35	36	**37**	38	39
40	41	**42**	43	44
45	46	**47**	48	49
50	51	**52**	53	54
55	56	**57**	58	59

The impact insulation (or isolation) provided by floor-ceiling assemblies is usually expressed as their impact noise rating *INR* or impact insulation class *IIC*. Like *STC, INR* and *IIC* values are obtained by comparing the curve of the sound spectrum obtained in a test with a standard contour (except that for *INR* and *IIC* the sound pressure level is measured in the room below the noise source). The entire procedure is controversial and far from widely accepted; but its use is so widespread that designers and builders should be aware of it. (See ASTM RM 14-4.)

Table 17-9 lists the impact isolation provided by several types of construction.

Table 17-9. ***IIC* of Various Floor Constructions**

Construction	*IIC*
Oak flooring on ½-in. plywood subfloor, 2 × 10 joists, ½-in. gypsum board ceiling	23
With carpet and pad	48
8-in. concrete slab	35
With carpet and pad	57
2½-in. concrete on light metal forms, steel bar joists	27
With carpet and pad	50

Note particularly the enormous effect of certain floor coverings on the performance of the construction. (While *IIC* values are shown in Table 17-9, they can be converted to *INR* values by subtracting 51 points; for example, *IIC* 60 = *INR* 9, and *IIC* 45 = *INR* −6, etc.)

Acoustical Absorbents. Sound absorption coefficients and noise reduction coefficients of acoustical absorbents, including carpets and other furnishings, are usually readily available, particularly from the Acoustical and Insulating Materials Association, 205 W. Touhy Ave., Park Ridge, Ill. 60068. They are normally obtained from laboratory tests of panel specimens of about 72- to 80-sq-ft area, tested according to ASTM C423. It is imperative that the test specimens be as nearly identical as possible, in construction system and detail, with actual field installations, since construction details enormously affect acoustical performance.

Table 17-10 lists performance ranges of typical absorbent materials. For specific data on specific materials or systems, always refer to specific tests by accredited test agencies.

Table 17-10. Performance of Commonly Used Sound Absorbents

Absorbent	Thickness, in.	Density, lb per cu ft	Noise reduction coefficient
Mineral or glass fiber blankets	½–4	½–6	0.45–0.95
Molded or felted tiles, panels, and boards	½–1⅛	8–25	0.45–0.90
Plasters (porous)	⅜–¾	20–30	0.25–0.40
Sprayed-on fibers and binders	⅜–1⅛	15–30	0.25–0.75
Foamed, open-cell plastics, elastomers, etc	½–2	1–3	0.35–0.90
Carpets	Varies with weave, texture, backing, pad, etc.		0.30–0.60
Draperies	Varies with weave, texture, weight, fullness		0.10–0.60

Absorbent	Absorption coefficient per sq ft of floor area at frequencies, Hz:					
	125	250	500	1,000	2,000	4,000
Seated audience	0.60	0.75	0.85	0.95	0.95	0.85
Unoccupied upholstered (fabric) seats	0.50	0.65	0.80	0.90	0.80	0.70

The sound transmission loss through an acoustical ceiling (up and over a barrier) when the ceiling is used as a continuous membrane, is often an important rating. The *Bulletin of the Acoustical and Insulating Materials Association* normally lists values for various constructions, as obtained from test procedures developed by the Association. As might be expected, the effectiveness of acoustical absorbents as acoustical barriers is limited, and supplementary barriers or isolation are frequently required in normal construction.

Other Acoustical Materials. Performance ratings of damping materials, duct linings, vibration isolation materials and devices, etc., are available from various sources. Use of such materials is somewhat complex and specialized.

17-17. Acoustical Criteria. Increasingly, acoustical performance criteria and environmental specifications are becoming a part of building contract documents. Governmental agencies, lending institutions, owners, and tenants frequently require objective standards of performance.

To a large degree, acoustical criteria are subjective or are based on subjective response to acoustic parameters. This complicates attempts to provide objective specifications, but long experience has permitted acoustical experts to determine broad classes or ranges of criteria and standards that will produce satisfaction in most instances.

It is important to remember that one-point differences are normally insignificant in acoustical criteria. Usually, a tolerance of ±2½ points from a numerical value is acceptable in practice.

Tables 17-11 to 17-13 list some of the more common criteria for ordinary building spaces. They should, however, be used only as a guide; always attempt to obtain specific data or requirements whenever possible.

Acceptable Background Noise Levels. Steady, constant, unobtrusive sound levels that normally occur in typical rooms and that are acceptable as background noise are indicated in Table 17-11. For specific applications and in acoustically critical spaces, specific requirements should always be determined.

Table 17-11. Typical Acceptable Background Levels

Space	*Background Level, dBA*
Recording studio	25
Suburban bedroom	30
Theater	30
Church	35
Classroom	35
Private office	40
General office	50
Dining room	55
Computer room	70

[*Note:* Levels are given in dBA, easily obtained numbers with simple measuring equipment (Art. 17-7). When noise criterion *NC* values are specified, they can be determined by subtracting about 7 to 10 points from dBA values; that is, dBA 50 = *NC* 40 to *NC* 43.]

Sound-transmission-loss Requirements. Acoustical performance requirements of sound barriers separating various occupancies may vary widely, depending on the particular needs of the particular occupants. Typical requirements for common occupancies in normal buildings are shown in Table 17-12.

Table 17-12. Sound Isolation Requirements Between Rooms

Between: Room	and Adjacent area	Sound isolation requirement, *STC*
Hotel bedroom	Hotel bedroom	47
Hotel bedroom	Corridor	47
Hotel bedroom	Exterior	42
Normal office	Normal office	33
Executive office	Executive office	42
Bedroom	Mechanical room	52
Classroom	Classroom	37
Classroom	Corridor	33
Theater	Classroom	52
Theater	Music rehearsal	57

Table 17-13. Impact Isolation Requirements between Rooms

Between: Room	and Room below	Impact isolation requirement, *IIC*
Hotel bedroom	Hotel bedroom	55
Public spaces	Hotel bedroom	60
Classroom	Classroom	47
Music room	Classroom	55
Music room	Theater	62
Office	Office	47

In highly critical spaces, or where a large number of spaces of identical use are involved (as in a large hotel or multiple-dwelling building), it is always advisable to obtain expert acoustical advice.

Impact Isolation Requirements. Because the only standard test method available is controversial, impact isolation performance specifications are only broad, general suggestions, at best. The values listed in Table 17-13, however, are reasonably safe and economically obtainable with available construction systems.

Acoustical Absorption Requirements. It is ironic that requirements for the most widely used acoustical materials are the least well defined. So-called optimum reverberation time requirements (Fig. 17-12) are a reasonably safe specification for most rooms, but many additional requirements are often involved.

In general, where noise control is the significant requirement, the equivalent absorption of a full ceiling of acoustical tile providing a noise-reduction coefficient of about 0.65 to 0.70 is adequate. This absorption may be provided by carpet, furnishings, or other materials, as well as by acoustical tile. In many instances, however, no absorption at all should be applied to the ceiling of a room, because the ceiling may be a necessary sound reflector.

For important projects, always obtain the advice of competent acoustical consultants.

Table 17-14. Available Options in Noise Control

Objectives of sound-control efforts	Sound-control procedures*					
	Quiet the source†	Barriers or enclosures	Vibration isolation or damping	Absorption	Masking	Personal protection
Reduce the general noise level to:						
Improve communication	X		X	X‡		
Increase comfort	X		X	X‡		
Reduce risk of hearing damage	X		X	X‡		
Reduce extraneous, intruding noise to:						
Increase privacy		X‡			X	
Increase comfort		X‡	X			
Improve communication		X‡	X			
Protect many persons against localized source producing damaging levels	X	X‡	X	X		X
Protect many persons against many distributed sources producing damaging levels	X	X	X	X		X‡
Protect one person against localized source producing damaging levels	X	X§	X			X‡
Protect a few persons against many distributed sources producing damaging levels		X§				X‡
Eliminate echoes and flutter				X¶		
Reduce reverberation				X‡		
Eliminate annoying vibration			X‡			

* No mention has been made of another option, reinforcement, because its purpose is to increase levels. It is the only available option when the signal level must be increased.

† Always the best and simplest means of eliminating noise, if practical and economical.

‡ Indicates the most likely solution(s).

§ A closed booth or small room for the person(s) is often feasible.

¶ Assumes that the configuration of the reflecting surface(s) cannot be modified.

Table 17-15. Typical Acoustical Problems and Likely Solutions

Problem	Possible causes	Solution
"It's so noisy, I can't hear myself think"	High noise levels Excessive reverberation Excessive transmission Excessive vibration Focusing effects	Absorption Absorption Sound isolation Vibration isolation Eliminate cause of focusing effects
"Speech and music are fuzzy, indistinct"	Excessive reverberation	Absorption
"Little sounds are most distracting"	Background level too low Room too "dead"	Masking Optimum reverberation
"There's an annoying echo"	Echo Flutter Focusing effects Excessive reverberation	Proper room shape Proper room shape Eliminate cause of focusing effects Absorption
"I can hear everything the fellow across the office says"	Room too "dead" Background level too low Focusing or reflection	Optimum reverberation Masking Eliminate focusing or reflection
"It doesn't sound natural in here"	Flutter Distortion caused by improper sound system Selective absorption Room too "dead", low reverberation time	Alter room shape Add absorption Proper sound system Proper type and amount of absorption
"It feels oppressive"	Reverberation time too low Background sound level too low	Proper amount of absorption Use of background and masking sound
"It's not loud enough at the rear of the room"	Room too large Improper shape Lack of reflecting surfaces Poor distribution Too much absorption	Electronic amplification Alter room shape Add reflecting surfaces Eliminate absorption on surfaces needed for reflection
"There are dead spots in the room"	Poor distribution Improper shape Echoes	Reflecting surfaces Eliminate cause of focusing effects Alter room shape
"Sound comes right through the walls of these offices"	Sound leaks Sound transmission Vibration Receiving room too quiet Poor room location	Eliminate leaks Sound isolation Vibration isolation Masking Proper room layout
"I can hear machine noises and people walking around upstairs"	Vibration Sound transmission Poor room location	Vibration isolation Sound isolation Proper room layout
"Outside noises drive me crazy"	Poor room location Sound leaks Sound transmission	Proper acoustical environment Eliminate leaks Sound isolation

17-18. Helpful Hints for Noise Control. A building encloses a myriad of activities and the equipment used in those activities. It is imperative that designers and builders consider the control of noise and vibration associated with such activities. While it is not practical to give here detailed solutions to the many potential acoustical problems that may arise in the design of even one building, the more common problems encountered and the most likely solutions to such problems are indicated in Tables 17-14 and 17-15.

Source-Path-Receiver. Sound originates at a source and travels via a path to a **receiver.** Sound control consists of modifying or treating any or all of these three elements in some manner.

The most effective control measures often involve eliminating noise at the source. For example:

Balancing moving parts, lubricating bearings, improving aerodynamics of duct systems, etc.

Modifying parts or processes.

Changing to a different, less noisy process.

The most common sound control measures usually involve acoustical treatment to absorb sound; but equally important are:

Use of barriers to prevent air-borne sound transmission.

Interruption of the path with carefully designed discontinuities.

Use of damping materials to minimize radiation from surfaces.

Another important approach involves reinforcing the direct sound with controlled reflections from properly designed reflective surfaces.

Often, the most simple and effective approach involves protecting the receiver, enclosing the person within adequate barriers, or equipping the person with personal protection devices (ear plugs or muffs), rather than trying to enclose or modify huge sources or an entire room or building.

Table 17-14 indicates the usual options available in noise control, and the most likely solutions to problems.

In existing buildings, identifying problems, determining their causes, and deciding on the most economical solutions are frequently the tasks of designers and builders. Table 17-15 indicates typical acoustical problems, their possible causes, and likely solutions to the problems.

17-19. Bibliography.

L. L. Beranek, "Music, Acoustics, and Architecture," John Wiley & Sons, Inc., New York.

L. L. Beranek, "Noise and Vibration Control," McGraw-Hill Book Company, New York.

P. D. Close, "Sound Control and Thermal Insulation of Buildings," Van Nostrand Reinhold Company, New York.

M. D. Egan, "Concepts in Architectural Acoustics," McGraw-Hill Book Company, New York.

C. M. Harris, "Handbook of Noise Control," McGraw-Hill Book Company, New York.

P. M. Morse and K. U. Ingard, "Theoretical Acoustics," McGraw-Hill Book Company, New York.

L. F. Yerges, "Sound, Noise and Vibration Control," Van Nostrand Reinhold Company, New York.

Section 18

Thermal Insulation

E. B. J. ROOS
Partner

AND

T. H. QUINLAN
Senior Associate,
Seelye, Stevenson, Value & Knecht,
New York, N.Y.

Certain materials are used in building construction specifically because they offer high resistance to passage of heat. They are called thermal insulation.

Such materials may be used in buildings to retain heat; for example, to decrease heat losses through a building enclosure in cold weather or reduce loss of heat from hot-water pipes. The same insulation may also serve to exclude heat; for example, to decrease heat gains through a building enclosure in hot weather or reduce heat gains through air-conditioning ducts carrying cooling air. In addition, insulation may be useful in preventing undesirable conditions, such as "sweating" of cold pipes. On the other hand, improper installations of insulation may create undesirable conditions, such as unwanted condensation of water vapor on or within building assemblies.

A knowledge of insulation characteristics, behavior, and installation therefore is important in securing maximum economic advantages from heat conservation and satisfactory performance in service.

18-1. Effect of Insulation. There is a very simple way of evaluating insulation without going into details and theoretical considerations, if the following basic principles, which will be discussed later, are accepted here at their face value:

1. All materials offer some resistance to flow of heat. The resistance is directly proportional to thickness.

2. Insulations, as used in construction, are considerably more resistant to heat flow than structural materials, such as stone, brick, concrete, steel, and wood, and for the purpose of providing resistance to heat flow are considerably more economical.

The amount of heat that flows through a material is called **conductance.** Many books contain tables that give the conductance for various materials, as well as for built-up walls, roofs, slabs, etc. ("Handbook of Fundamentals," American Society of Heating, Refrigerating and Air-Conditioning Engineers.)

Consider, for example, a wall in which we are thinking of adding, say, 1 in. of a certain insulation. From tables, we can get the heat conductance of the insulation and the over-all conductance of the wall. If, for example, the conductance of the insulation is one-third that of the wall, then the over-all conductance of the wall with the insulation added is one-fourth of what it would be without insulation. The heat flow stopped is three-fourths.

Table 18-1 will make this simple method very clear. (All terms are relative heat conductances, with the heat conductance of the uninsulated wall taken as unity.)

Table 18-1. Effect of Insulating a Wall

Relative conductance of insulation added	Resultant conductance of insulated wall	Decrease in heat flow
$1/2$	$1/3$	$2/3$
$1/3$	$1/4$	$3/4$
$1/4$	$1/5$	$4/5$
$1/5$	$1/6$	$5/6$
$1/n$	$1/(n+1)$	$n/(n+1)$

It can be seen that the savings in heat are considerable. Nevertheless, increasing the thickness of insulation provides diminishing returns. Considering the effect of insulation alone, if the thickness is doubled, say, from 1 in. to 2 in., the heat flow becomes 50%, or 50% is saved. If another 1 in. is added, the heat flow compared with the original 1-in. thickness is 33%. With another 1-in. addition, the heat flow is reduced to 25%. So doubling the original thickness reduces the heat flow by 50%; tripling the thickness reduces heat flow another 17%; and quadrupling the thickness reduces heat flow only another 8%. So for each increase in the thickness of insulation, the effect is less, and hence an economical balance can be calculated (Art. 18-11).

In any case, once installed, insulation not subject to abuse needs no repair or maintenance. Hence, first cost is the final cost, which in a new structure is offset by a considerable reduction in the required heating or cooling plant. Furthermore, in any structure, insulation makes possible a continuous reduction in operating expense for heating or cooling. These savings can generally pay for the cost of insulation in anywhere from 3 to 10 years, depending on the structure. Any investment having a return of 10 to 30% is attractive. Also, added comfort due to the resulting more uniform inside temperatures is another advantage.

18-2. Types of Insulation. All materials have some resistance to heat flow, but modern construction and the cost and weight of structural materials do not permit or require walls of stone, brick, or other structural materials thick enough to provide sufficient resistance to heat flow for economical heating and cooling. For this reason, new and more economical insulations for structures have been and are being developed. These insulating materials generally add no strength to the structure; they are employed only for the purpose of resisting flow of heat. In every case, the cost and weight of these materials are less than those of a sufficient thickness of structural materials of the same resistance to heat flow.

Modern building insulations are made of many different materials—glass fibers, glass foam, mineral fibers, wood or other organic fibers, metal foil, foamed plastics, lightweight concrete. Proper selection depends generally first on cost and then on the physical characteristics.

Except for metal foils, these materials utilize still air, which is an excellent insulator, to resist heat flow. Some, such as cork, cellular glass, or foamed plastics, enclose small particles of air in cells. Granular materials like pumice, vermiculite, or perlite trap air in relatively large enclosures. Fibrous materials employ the principle that thin films of air cling persistently to all surfaces and serve as a heat barrier.

Insulating value of these materials depends on (1) ability of each component of the material to conduct heat, (2) ability of the components to transmit heat to each other, (3) length of path of heat flow, and (4) size of cells in which air is trapped or closeness of adjacent air films on the fibers or granules.

Metal foils involve a different principle; they are known as reflective insulation. Just as a mirror reflects light, shiny metals, such as aluminum, reflect heat. They also have the properties of conducting heat very rapidly and emitting it very slowly from the surface away from the heat source. Thus, if heat is radiated to a bright aluminum foil, 95% will be reflected back. If it receives heat by conduction, it will lose only 5% by radiation from the opposite face.

To use reflective insulation properly, always provide an enclosed air space on at least one side of the material (Fig. 18-1*a*). A gap of ¾ to about 2 in. is

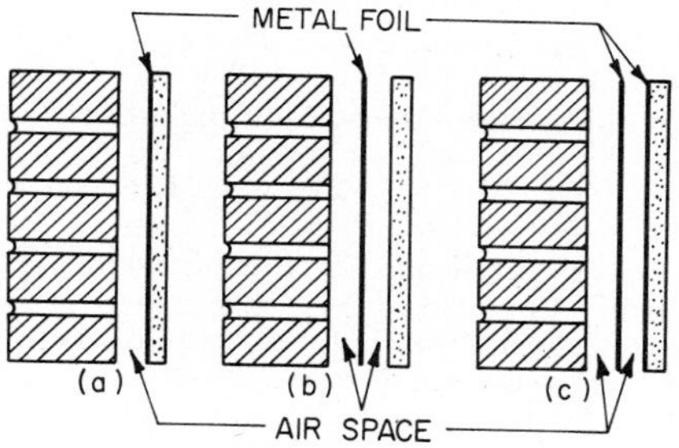

Fig. 18-1. Metal foil serving as reflective insulation is used in conjunction with an air space. (*a*) With foil placed against interior facing, only one air space is provided. (*b*) Splitting the gap with the foil improves insulation by forming two air spaces. (*c*) Adding a second foil on the warm side increases insulating value even more and prevents condensation.

most effective. If a foil is placed inside a gap, two air spaces are formed (Fig. 18-1*b*), thereby increasing resistance to heat flow. To prevent condensation troubles it is wise to use at least two reflective surfaces separated by an air space (Fig. 18-1*c*).

Supports for the foil should have high resistance to heat flow, so as not to carry off heat. In particular, aluminum foil should not be placed in contact with other metals, or it may be subjected to galvanic corrosion, and it should not be exposed to the alkalies in wet plaster or cement. A foil should not be placed on the cold side of a construction, unless a better vapor barrier is provided close to the warm side—to prevent condensation.

Calculation of heat transmission through a construction with reflective insulation is simplified by use of data compiled by the National Bureau of Standards (Housing Research Paper No. 32, U.S. Government Printing Office, Washington, D.C. 20402).

18-3. Heat Flow across an Air Space. The best practical heat insulation at our disposal is still air. An air space of about ¾ in. has been found to be best. Air spaces, however, have a limitation due to inability to keep the air motionless.

Surfaces of all materials exposed to air retain a film of air held fast by molecular attraction. This film also is an excellent insulation, but its thickness and effectiveness are diminished by any air motion wiping the surface.

In a vertical air space, such as may be constructed in a wall, with a higher temperature on one side than on the other, the air within the gap in contact with the warm side becomes heated. The hot air expands, becomes lighter, and tends to rise. On the cooler side, the air is cooled, contracts, becomes heavier and sinks. A circulation is set up, the warm air rising and being replaced with cooler air. The result is that the cooler air comes in contact with the warm side of the air space and absorbs heat, and the warm air comes in contact with the cooler side of the air space, giving up heat. Thus the air motion transfers heat from the warm side to the cold side. At the same time, its motion reduces the thickness of the surface air film and so decreases the film's resistance to heat.

This type of air motion is called **convection.** It is one of the means whereby heat flows. The amount of heat transferred is the same for the same temperature difference, regardless of which side is warm or cold.

Consider now a horizontal air space, such as may exist between a flat roof and a furred ceiling. In summer sunshine, the roof slab becomes very hot. The air in contact, in the space below, becomes heated, expands, and becomes lighter. On the other hand, the air in contact with the cooler ceiling below the air space is cooled, contracts, and becomes heavier. Since the light air is at the top of the air space it cannot rise, and the cooler, heavier air at the bottom of the air space cannot sink. Hence air motion due to temperature differences, namely, convection, is at a minimum, and so also is transfer of heat.

This condition is desirable for summer. But in winter the situation is less desirable. The roof slab is cold, and the air below is cooled, tending to fall—which it is free to do. The air above the ceiling is warmed and tends to rise—which it is free to do. So a thermal circulation is set up. This again is convection, a conveyor belt taking heat from the ceiling below and transporting it to the roof slab above.

Air spaces in pitched roofs fall somewhere in between vertical and horizontal air spaces.

Insulations that depend on air spaces for their resistance to heat flow incorporate within them a great many tiny cells. Not only do they contain a large number of air films in the path of heat flow, but the minute size of the air spaces tends to keep the air still and thus considerably reduces convection.

18-4. How Heat Flows. Heat always flows from a region of high temperature to a region of lower temperature.

Heat may be transmitted by three different methods—conduction, convection, and radiation.

When heat flows by **conduction** through a material, it is transmitted much like electricity through a conductor. All materials conduct heat. Some, such as metals, conduct heat very well; others, such as cork, have comparatively great resistance to heat flow. In every case, some heat will flow if a difference in temperature exists.

Convection, strictly speaking, is not true flow. It is a combination of flow by conduction from a warm surface to cooler air in contact with it, or from warm air to a cooler surface, the heated air tending to rise and the cooled air to fall. This rising and falling make possible a physical transportation of heat (Art. 18-3).

The amount of heat transmitted by conduction or convection is proportional to the temperature difference. For example, take the wall of a building in winter, with an inside temperature of 70°F. When the outside temperature is 10°F, there is a temperature difference of 60°F. The heat flow out (in this case a heat loss) is twice as much as occurs when the outside temperature is 40°F, the temperature difference being 30°F.

Radiation is the flow of heat through space from a warm surface to a cooler one. This type of flow is like that of light and is of practical importance in connection with reflective insulation (Art. 18-2).

18-5. Measurement of Heat. To have practical and economical design with insulation, it is necessary to be able to measure heat. The unit of heat generally employed is the Btu (British thermal unit). For practical purposes, **1 Btu** is the amount of heat required to raise the temperature of 1 pound of water 1 degree Fahrenheit.

The basic unit used in measuring heat flow is **thermal conductivity** *K,* defined as the number of Btu that will flow through a material 1 ft square and 1 in. thick due to a temperature difference of 1°F, in 1 hr.

Another basic unit is **thermal conductance,** ***C*****,** defined as the heat flow through a given thickness of 1-ft-square material with a 1°F temperature differential. It is useful for comparing standard structural materials, such as cinder block, hollow tile, brick, window glass, and plywood, which have standard thicknesses not equal to 1 in., as well as air spaces.

Note that these basic units do not include an allowance for air films but consider only flow from surface to surface.

Since most construction—walls, roofs, floor slabs, etc.—is built up of several different materials, possibly including air spaces, the desired end result in heat-flow measurement is the over-all heat flow through the structure including air films. This factor, *U*, is termed the **over-all conductance.** It is defined as the number of Btu that will flow through 1 sq ft of the structure from air to air due to a temperature difference of 1°F, in 1 hr.

The basic factors—*K*, *C*, and *U*—have been determined by experiment for most structural materials, as well as for most commonly used built-up walls, roofs, slabs, for easy reference, so that little calculation is required ("Handbook of Fundamentals," American Society of Heating, Refrigerating and Air-Conditioning Engineers).

With these factors, it is a simple matter to calculate the heat flow (heat loss in winter, heat gain in summer) for any building.

For example, let us consider a wall. All that is required is to find in a table the over-all conductance factor *U* corresponding to the type of wall selected for the building. This factor is then multiplied by the wall area in square feet (less window and door area) and by the temperature difference existing between outside design conditions and desired inside conditions. The result is the number of Btu gained or lost through the wall. A similar calculation must be made for the glass area, door area, roof, floor, partition, etc., wherever temperature differences will exist.

It is of interest to note that the maximum over-all conductance *U* encountered is 1.5 Btu per hr per sq ft per °F. This would occur with a sheet-metal wall. The metal has, for practical purposes, no resistance to heat flow. The *U* value of 1.5 is due entirely to the resistance of the inside and outside air films. Most types of construction have *U* factors considerably less than 1.5.

The minimum *U* factor generally found in standard construction with 2 in. of insulation is about 0.10.

Since the *U* factor for single glass is 1.13, it can be seen that windows are a large source of heat gain, or heat loss, compared with the rest of the structure. For double glass, the *U* factor is 0.45. For further comparison, the conductivity *K* of most commercial insulations varies from about 0.24 to about 0.34.

18-6. Calculation of Heat Flow. It is not always possible to find in tables the *U* factor corresponding to the actual type of construction and choice of materials desired. In that case, it is necessary to calculate *U* by first finding the thermal conductivity *K* or thermal conductance *C* for each material. These are found in tables, such as those in the "Handbook of Fundamentals" (American Society of Heating, Refrigerating and Air-Conditioning Engineers). The thickness in inches must also be known for each material.

For example, the thermal conductivity *K* for brick is 9.2. The conductance of a 4-in. thickness of brick veneer is then 9.2/4, or 2.3. Note that the value 2.3 might be listed in tables as thermal conductance *C* for 4-in. veneer brick, in which case no calculation is needed.

Before the thermal characteristics of any wall, roof, etc., can be calculated it is necessary to know the heat conductance of the air films and possible air spaces encountered. These are generally accepted as follows:

	Btu per Hr
Outside air film f_o (15-mph wind)	6.00
Inside air film f_i (still air)	1.65
Air space ¾ in. or more	1.10

In all tables, conductivities or conductances are listed, but to find the over-all conductance of, say, a built-up wall, it is necessary to compute the total resistance, which is the sum of the resistances R corresponding to the thickness of each material employed.

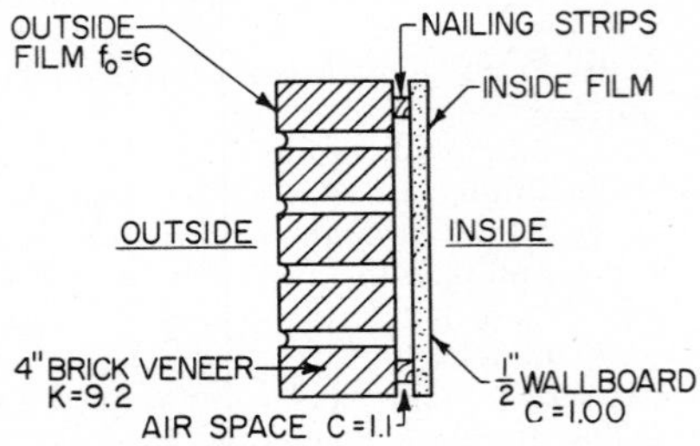

Fig. 18-2. Section through a brick-veneer wall.

Resistance is the reciprocal of conductance. If, for example, the heat conductivity of 1 in. of insulation is 0.25, then the resistance is 1/0.25 or 4.

There are formulas for calculating U, but a simpler method is as follows: Assume the hypothetical wall in Fig. 18-2.

Item	K	Thickness, in.	C	$R = 1/C$
Outside film	...	...	6	0.166
Brick	9.2	4	2.30	0.434
Air space	...	...	1.10	0.910
Wallboard	...	½	1.00	1.000
Inside film	...	...	1.65	0.606
Total resistance	...	...	...	3.116

$$\text{Over-all conductance } U = \frac{1}{3.116} = 0.32$$

Assume we wish to find out what the over-all conductance would be for the wall in Fig. 18-2 with 1-in. insulation ($K = 0.25$).

Item	K	Thickness, in.	C	$R = 1/C$
Wall	...	...	0.32	3.116
Insulation	0.25	1	0.25	4.000
Total resistance	...	...	...	7.116

$$\text{Over-all conductance } U = \frac{1}{7.116} = 0.14$$

The last calculation indicates that, if the over-all conductance of a construction is known and it is desired to add to or subtract a material from this wall, it is not necessary to recalculate completely, but as above, use the existing resistance and add or subtract the resistance of the material to be varied. Then, the reciprocal is the conductance sought.

A simpler device for finding the over-all conductance resulting from the addition of insulation to a known wall is as follows:

Using the above example, we have 0.32 for the over-all conductance of the wall, and the conductivity of the desired 1-in. insulation is 0.25. We then multiply the two conductances together, and divide by their sum:

$$\frac{0.32 \times 0.25}{0.32 + 0.25} = \frac{0.08}{0.57} = 0.14$$

If we wish to find the over-all conductance of the above wall by the addition of 2 in. of insulation ($C = 0.25/2$) instead of 1 in., we have

$$\frac{0.32 \times 0.125}{0.32 + 0.125} = \frac{0.04}{0.445} = 0.09$$

18-7. Danger of Condensation. If reduction of heat flow were all that had to be considered in the structural application of insulation, the whole matter of design would be merely a simple matter of economics—comparing the cost of insulation with the cost saved in heating or cooling, as well as the reduction in size of heating or cooling plant in new structures.

Unfortunately for insulation, air almost always contains water vapor. When insulation is not correctly installed, this water vapor may, under certain conditions, destroy the insulating value, as well as the structure itself. This by-product nuisance of insulation is the greatest single cause of insulation failure. It is essential, therefore, to the proper installation of insulation that the behavior of water vapor be investigated, and then its effect on insulation. Here it can truthfully be stated: "It is not the heat but the humidity."

The problem involved is more easily grasped if the following two statements are kept in mind:

1. For practical purposes, water vapor is a gas and behaves exactly like air. It goes wherever air goes, penetrating voids, porous materials, etc.
2. Generally, water vapor always flows from a region of high temperature to a region of lower temperature.

When a certain volume of air contains all the water vapor it can, it is saturated, and the so-called relative humidity is 100%. The higher the temperature, the more vapor it can contain. Conversely, the lower the temperature, the less vapor it can hold.

Whatever the temperature, if the air is saturated, any drop in temperature will immediately cause some of the water to condense. Since this occurs in nature in the formation of dew, this temperature is called the **dew point.**

Assume air containing a certain definite amount of vapor less than saturation. As the temperature drops, it takes less and less vapor to saturate the air, until a temperature is reached at which the air is saturated with vapor. It has reached the dew point. Any further cooling will condense out moisture.

Charts are available that give the dew point for any conditions of temperature and relative humidities ("Handbook of Fundamentals," American Society of Heating, Refrigerating and Air-Conditioning Engineers).

Fortunately, the air around us is seldom saturated. The amount of vapor actually in the air can be expressed as a percentage of the amount this air would contain if saturated at the same temperature; and this percentage is for practical purposes the **relative humidity.**

Assume a room with air at 70°F and 30% relative humidity. The dew point is approximately 38°F. If the outside temperature is, say, 0°F, the temperature in the outside wall of the room will vary from 0°F at the exterior air film up to 70°F at the inside air film. Therefore, at some point within the wall the temperature will be 38°F, the dew point.

Since almost all building materials are porous, the vapor permeating the pores will tend to condense into water near the section where the dew point occurs, and the higher temperature inside the room will cause more vapor to penetrate the wall, repeating the process.

The dew point often occurs within the insulation and may in time saturate it, drastically reducing its insulating value. Furthermore, the moisture may rot or rust the structure, or stain interior finishes. Under severe conditions, the moisture may freeze, and in expanding, as ice always does, it may cause cracks in the structure.

Condensation within a structure is generally a winter problem. The greater the relative humidity existing within the building, the more serious the problem becomes.

The relative humidity maintained within a structure depends on the occupancy of the building. If the conditions are intermittent, for example, as in a church, condensation that may occur for a short period will reevaporate in between such periods. On the other hand, an indoor swimming pool will maintain a continuous comparatively high inside relative humidity; so unless condensation is prevented, the process is accumulative during any cold weather.

Buildings often are maintained at lower temperatures (50 to 60°F) during the night when outside temperatures are at their minimum. The entire wall or roof is then at an even lower temperature. So condensation may occur at night, even if not during the day. On the other hand, though condensation may occur at night, it may not exceed the amount that would evaporate during daytime occupancy.

Generally, there is always danger of condensation within a wall or roof, etc., if the temperature on the cold side is below the dew point of the air on the warm side. Whenever condensation occurs on a window, it can be assumed that it is also occurring within the structure, although to a lesser degree.

18-8. Use of Vapor Barriers to Prevent Condensation. Whenever insulation is installed in a wall, roof, or slab, its resistance to the flow of heat is so much greater than that of the other elements of the construction that the dew point and resulting condensation may occur within the insulation.

Since water vapor flows from regions of high temperature to regions of low temperature, a simple solution to condensation is to stop the flow of water vapor by means of some surface material impervious to moisture—provided this surface is always maintained above the dew point. Such a surface is called a **vapor barrier.** It must always be applied on the warm side.

Because condensation is generally most severe during the heating season, all vapor barriers should be installed on the interior side of walls and roofs. From a practical standpoint, this means that the vapor barrier should be next to and part of the insulation.

One of the best and most economical vapor barriers is aluminum foil. Some insulations come equipped with this foil attached to one surface. However, unless reinforced with kraft paper or some other strong material, the foil is easily ripped, torn, or punctured, and so is of little value as a barrier.

Since vapor behaves as a gas, a vapor barrier, to be effective, must be airtight, or as nearly so as possible. But this is often an impractical requirement. For example, consider a roof with the insulation above the deck and between a vapor barrier and waterproof roofing. Unless the insulation is of a firm material, such as cellular glass, the heat of the sun will cause the air within the insulation to expand, forming bubbles under the waterproofing. During the coolness of the night, the bubbles will contract. After a series of sunny days and cool nights, the bending back and forth of the surface may destroy the roofing. One way to prevent this is to side-vent the roof insulation so the contained air can freely expand and contract. The side vents must, however, be protected from driving rain.

Vapor barriers can be made of other materials besides aluminum foil. There are aluminum paints, plastic paints, some plastic films, asphalt paints, rubber-base paints, asphalt, and foil-laminated papers. It must be remembered that water-repellent surfaces are not necessarily vapor barriers, that is, airtight.

To evaluate a vapor barrier, a unit known as the **perm** is used. It is defined as a vapor-transmission rate of 1 grain of water vapor through 1 square foot of material per hour when the vapor-pressure difference is equal to 1 inch of mercury (7,000 grains equal 1 pound). A material having a vapor-transmission rate of 1 perm or less is considered a good vapor barrier. The corresponding unit for permeance of 1-in. thickness is **perm-inch.**

Resistance to vapor transmission is the reciprocal of the permeance.

Since vapors flow from the warm side of a wall or roof to the cold side, the exterior surface should be as porous as possible or vented and yet offer protection against penetration of rain. This is particularly important with "blown-in" insulation as applied to frame houses, for which a vapor barrier generally cannot be installed.

This type of insulation also involves another principle, which, if ignored, frequently is the cause of peeling of paint and leads to unnecessary repair of rain gutters that do not leak.

"Blown-in" insulation is sprayed into the spaces between the studs of frame construction. The interior surface is generally lath and plaster, or wall-board—both porous. The exterior is generally wood sheathing, with shingles, clapboards, or stucco. The heat resistance of the insulation is such that during the winter the location of the dew point falls within the insulation. Theoretically, the resulting condensation should occur within the insulation. This, however, does not occur. Condensation, when it does happen, does not form at the location of the dew point within the insulation, but on the inside surface of the sheathing.

The principle involved is this: Whenever the dew point occurs within a material, condensation will not occur until the flow of water vapor encounters the surface of another material of greater resistance to the flow of water vapor. That is, as long as the air can keep on moving, it will carry the moisture along with it and will not deposit the moisture until it reaches a surface that resists its flow and is colder than the dew point.

The problem inherent in blown-in insulation can be solved by "cold-side venting." In applying blown-in insulation, an opening usually is drilled through the exterior wall surface between each pair of studs. These holes should never be sealed, only covered with porous water-repellent material for protection against the weather. Then, whatever water vapor flows through the inside porous finish can escape to the cold air outside without condensing. With clapboard construction, "toothpick" wedges may be driven under the lower edge of each clapboard to provide the required openings for breathing.

To sum up: Vapor barriers, or as much resistance as possible to vapor flow (or air) should be provided on the warm side of walls and roofs. Openings or porous materials—as little resistance as possible to vapor flow—should be provided on the cold side.

If vapor barriers were perfect, cold-side venting would not be required. Unfortunately, vapor barriers are not perfect; therefore, cold-side venting is worthwhile insurance against failure of insulation in all cases.

18-9. Condensation during Summer Cooling. The discussions in Arts. 18-7 and 18-8 of winter condensation seem to contradict summer requirements when the warm and cold sides of a construction are the reverse of what they are in winter. In most parts of the United States, however, cooling seldom results in maintenance of inside temperatures more than 15°F below outside conditions, whereas in winter, inside temperatures are generally maintained at 60 to 75°F above outside conditions. So in winter, the prevailing maximum temperature differences are from four to five times what they are in summer. Furthermore, in summer very little cooling is required during the night. Hence, as far as insulation is concerned, summer condensation is so intermittent that it can be completely disregarded for the average structure and average occupancy.

It should be mentioned, however, that in low-temperature work, such as cold-storage rooms and low-temperature test cells, special conditions arise for which it is best to refer to a specialist.

18-10. Ventilation to Get Rid of Condensation. In addition to cold-side venting as described in Art. 18-7, ventilation is used in other circumstances to prevent damage from condensation. For example, where insulated ceilings exist below an attic, or an air space below a roof, louvers or some other provision for admission of outside air should be provided.

Under winter conditions, outside air, being very cold, has little capacity for holding moisture. Yet, outside air is the only source for inside air, and inside any structure there are many sources from which moisture is absorbed—people, shower baths, kitchens.

Hence, an occupied and heated building contains air with more moisture than in the outside air. The moisture tends to flow from the higher temperatures inside to the lower temperatures in the attic or air space under the roof. If these spaces are not ventilated with air capable of absorbing the moisture, it will condense and cause rust, rot, leaks, stains, etc.

In other words, any unheated space above heated spaces must be ventilated. Where prevailing winds are not dependable, one opening should be high, and another low, to take advantage of the tendency of heated air to rise.

As a general rule, the upward movement of heated air should be provided for in all air spaces above insulation placed over warmer spaces.

Ventilation is required because of the lack of a perfect vapor barrier and is workable only because the drier outside air is capable of absorbing more moisture. In general, vent area should total about 1/300th of the horizontal projection of the roof area.

18-11. Economics of Insulation. Generally, construction materials almost always require some insulation. The question, therefore, is not whether or not to insulate but rather what thickness and type of insulation to use.

Select the type of insulation by the installed cost involved. This should be the one whose prevailing cost per square foot divided by its resistance R_i gives the lowest figure, all values being based on 1-in. thickness.

Figure the total wall area to be insulated for each building exposure, as for example A_n (north wall), A_s (south wall), A_e (east wall), A_w (west wall), sq ft. Calculate winter heat loss and summer heat gain for each wall exposure without insulation. Do not neglect summer solar gains.

Determine the effect of adding the resistance of 1 in. of insulation to the resistance of the uninsulated wall for each case: $R_w + R_{1i}$, where R_{1i} is the resistance of 1 in. of insulation. The insulated wall heat gain or loss will be that of the uninsulated wall times $R_w/(R_w + R_{1i})$, and for 2-in. insulation, the heat gain or loss of the uninsulated wall times $R_w/(R_w + R_{2i})$. The effect of 3 or more inches of insulation can be similarly determined.

As an example, the resistance R_w of the wall shown in Fig. 18-2 is 3.116. Addition of 1 in. of insulation with a resistance of 4.0 ($R = 1/C = 1/0.25$) increases the total resistance to 7.116. Thus, the reduction in heat gain or loss due to 1 in. of insulation is $R_w/(R_w + R_{1i}) = 3.116/7.116 = 0.438$, or 43.8%.

With 2 in. of insulation, the total resistance is $3.116 + 2 \times 4.0 = 11.116$. Therefore, reduction in heat gain or loss is $3.116/11.116 = 0.28$, or 28%.

The preceding calculations can be continued to cover the reductions with 3, 4, 5, and 6 in. of insulation. With costs of fuel and power for heating and cooling known, the designer can then compare the resulting savings in operating cost with investment in installed cost of insulation. The savings in operating cost due to insulation are worthwhile if they prove greater than interest and amortization charges of the added cost of the insulation that brings about the savings.

Thickness of insulation is generally limited in residential use to 3 in. in walls and 6 in. in roof or ceilings, because of the depth available between 2 × 4 studs and 2 × 6 rafters. In commercial and industrial construction, thickness of insulation is usually limited by the distance columns and beams or trusses extend within the structure. In any case, the preceding calculations indicate the optimum thickness of insulation for any type of structure.

As an adjunct to the above considerations, the savings due to double or triple fenestration should not be overlooked.

Section **19**

Heating and Air Conditioning

RALPH TOROP

Chief Engineer, Forman Air Conditioning Company, Inc., New York, N.Y.

HEATING

In the design and installation of a heating system for a building, the first objective is to determine the size of the heating plant required and the proper distribution of the heat to the various rooms or zones of the structure.

19-1. General Procedure for Sizing a Heating Plant. The basic procedure used in sizing a heating plant is as follows: We isolate the part of the structure to be heated. To estimate the amount of heat to be supplied to that part, we must first decide on the design indoor and outdoor temperatures. For if we maintain a temperature of, say, 70°F inside the structure and the outside temperature is, say, 0°F, then heat will be conducted and radiated to the outside at a rate that can be computed from this 70° temperature difference. If we are to maintain the design inside temperature, we must add heat to the interior by some means at the same rate that it is lost to the exterior.

Recommended design inside temperatures are given in Table 19-1. Recommended design outdoor temperatures for a few cities are given in Table 19-2. (More extensive data are given in the "ASHRAE Guide," American Society of Heating, Refrigerating and Air-Conditioning Engineers.)

Note that the recommended design outdoor winter temperatures are not the lowest temperatures ever attained in each region. For example, the lowest temperature on record in New York City is −14°F, whereas the design temperature is 0°F. If the design indoor temperature is 70°F, we would be designing for $(70 - 0)/(70 + 14)$, or 83.3%, of the capacity we would need for the short period that a record cold of −14°F would last.

Once we have established for design purposes a temperature gradient (indoor design temperature minus outdoor design temperature) across the building exterior, we obtain the heat-transmission coefficients of the various building materials in

Table 19-1. Recommended Design Indoor Winter Temperatures

Type of Building	*Temp, °F*
Schools:	
Classrooms	72
Assembly rooms, dining rooms	72
Playrooms, gymnasiums	65
Swimming pool	75
Locker rooms	70
Hospitals:	
Private rooms	72
Operating rooms	75
Wards	70
Toilets	70
Bathrooms	75
Kitchens and laundries	66
Theaters	72
Hotels:	
Bedrooms	70
Ballrooms	68
Residences	72
Stores	68
Offices	72
Factories	65

Table 19-2. Recommended Design Outdoor Winter Temperatures

State	City	Temp, °F	State	City	Temp, °F
Ala.	Birmingham	10	Miss.	Vicksburg	10
Ariz.	Flagstaff	−10	Mo.	St. Louis	0
Ariz.	Phoenix	25	Mont.	Helena	−20
Ark.	Little Rock	5	Nebr.	Lincoln	−10
Calif.	Los Angeles	35	Nev.	Reno	−5
Calif.	San Francisco	35	N.H.	Concord	−15
Colo.	Denver	−10	N.J.	Trenton	0
Conn.	Hartford	0	N.Mex.	Albuquerque	0
D.C.	Washington	0	N.Y.	New York	0
Fla.	Jacksonville	25	N.C.	Greensboro	10
Fla.	Miami	35	N.Dak.	Bismarck	−30
Ga.	Atlanta	10	Ohio	Cincinnati	0
Idaho	Boise	−10	Okla.	Tulsa	0
Ill.	Chicago	−10	Ore.	Portland	10
Ind.	Indianapolis	−10	Pa.	Philadelphia	0
Iowa	Des Moines	−15	R.I.	Providence	0
Kans.	Topeka	−10	S.C.	Charleston	15
Ky.	Louisville	0	S.Dak.	Rapid City	−20
La.	New Orleans	20	Tenn.	Nashville	0
Maine	Portland	−5	Tex.	Dallas	0
Md.	Baltimore	0	Tex.	Houston	20
Mass.	Boston	0	Utah	Salt Lake City	−10
Mich.	Detroit	−10	Vt.	Burlington	−10
Minn.	Minneapolis	−20	Va.	Richmond	15

the exterior construction for computation of the heat flow per square foot (Arts. 18-5 and 18-6). These coefficients may be obtained from the manufacturers of the materials or from tables, such as those in the "ASHRAE Handbook of Fundamentals." Next, we have to take off from the plans the areas of exposed walls, windows, roof, etc., to determine the total heat flow, which is obtained by adding the sum of the products of the area, temperature gradient, and heat-transmission coefficient for each item (see Art. 19-7 for an application of these coefficients).

19-2. Heat Loss through Basement Floors and Walls. Although heat-transmission coefficients through basement floors and walls are available, it is generally not practicable to use them because ground temperatures are difficult to determine owing to the many variables involved. Instead, the rate of heat flow can be estimated, for all practical purposes, from Table 19-3. This table is based on ground-water temperatures, which range from about 40 to 60°F in the northern sections of the United States and 60 to 75°F in the southern sections. (For specific areas, see "ASHRAE Guide.")

19-3. Heat Loss from Floors on Grade. Attempts have been made to simplify the variables that enter into determination of heat loss through floors set directly on the ground. The most practical method breaks it down to a heat flow in Btu per hour per linear foot of edge exposed to the outside. With 2 in. of edge insulation, the rate of heat loss is about 50 in the cold northern sections of the United States, 45 in the temperate zones, 40 in the warm south. Corresponding rates for 1-in. insulation are 60, 55, and 50. With no edge insulation the rates are 75, 65, and 60.

Table 19-3. Below-grade Heat Losses

Ground water temp, °F	Basement floor loss,* Btu per hr per sq ft	Below-grade wall loss, Btu per hr per sq ft
40	3.0	6.0
50	2.0	4.0
60	1.0	2.0

* Based on basement temperature of 70°F.

19-4. Heat Loss from Unheated Attics. Top stories with unheated attics above require special treatment. To determine the heat loss through the ceiling, we must calculate the equilibrium attic temperature under design inside and outside temperature conditions. This is done by equating the heat gain to the attic via the ceiling to the heat loss through the roof:

$$U_cA_c(T_i - T_a) = U_rA_r(T_a - T_o) \tag{19-1}$$

where U_c = heat-transmission coefficient for ceiling
U_r = heat-transmission coefficient for roof
A_c = ceiling area
A_r = roof area
T_i = design room temperature
T_o = design outdoor temperature
T_a = attic temperature

Thus

$$T_a = \frac{U_cA_cT_i + U_rA_rT_o}{U_cA_c + U_rA_r} \tag{19-2}$$

The same procedure should be used to obtain the temperature of other unheated spaces, such as cellars and attached garages.

19-5. Air Infiltration. When the heating load of a building is calculated, it is advisable to figure each room separately, to ascertain the amount of heat to be supplied to each room. Then, compute the load for a complete floor or building and check it against the sum of the loads for the individual rooms.

Once we compute the heat flow through all exposed surfaces of a room, we have the heat load if the room is perfectly airtight and the doors never opened. However, this generally is not the case. In fact, windows and doors, even if weather-stripped, will allow outside air to infiltrate and inside air to exfiltrate. The amount of cold air entering a room depends on crack area, wind velocity, and number of exposures, among other things.

Attempts at calculating window- and door-crack area to determine air leakage usually yield a poor estimate. Faster and more dependable is the air-change method, which is based on the assumption that cold outside air is heated and pumped into the premises to create a static pressure large enough to prevent cold air from infiltrating.

The amount of air required to create this static pressure will depend on the volume of the room.

If the number of air changes taking place per hour N are known, the infiltration Q in cubic feet per minute can be computed from

$$Q = \frac{VN}{60} \tag{19-3}$$

where V = volume of room, cu ft

The amount of heat q in Btu per hour required to warm up this cold air is given by

$$q = 1.08QT \tag{19-4}$$

where Q = cfm of air to be warmed
T = temperature rise

19-6. Choosing Heating-plant Capacity. Total heat load equals the heat loss through conduction, radiation, and infiltration.

If we provide a heating plant with a capacity equal to this calculated heat load, we shall be able to maintain design room temperature when the design outside temperature prevails, if the interior is already at design room temperature. However, in most buildings the temperature is allowed to drop to as low as 55°F during the night. Thus, theoretically, it will require an infinite time to approach design room temperature. It, therefore, is considered good practice to add 20% to the heating-plant capacity for morning pickup.

The final figure obtained is the minimum heating-plant size required. Consult manufacturer's ratings and pick a unit with a capacity no lower than that calculated by the above method.

On the other hand, it is not advisable to choose a unit too large, because then operating efficiency suffers, increasing fuel consumption.

With a plant of 20% greater capacity than required for the calculated heat load, theoretically after the morning pickup, it will run only 100/120, or 83⅓%, of the time. Furthermore, since the design outdoor temperature occurs only during a small percentage of the heating season, during the rest of the heating season the plant would operate intermittently, less than 83⅓% of the time. Thus it is considered good practice to choose a heating unit no smaller than required but not much larger.

If the heating plant will be used to produce hot water for the premises, determine the added capacity required. (See Art. 21-33.)

19-7. Heating-load-calculation Example. As an example of the method described in Arts. 19-1 to 19-6 for sizing a heating plant, let us take the building shown in Fig. 19-1.

A design outdoor temperature of 0°F and an indoor temperature of 70°F are assumed. The wall is to be constructed of 4-in. brick with 8-in. cinder-block backup. Interior finish is metal lath and plaster (wall heat-transmission coefficient $U = 0.25$).

The method of determining the heat load is shown in Table 19-4.

Losses from the cellar include 4 Btu per sq ft per hr through the walls [column (4)] and 2 Btu per sq ft per hr through the floors. Multiplied by the corresponding areas, they yield the total heat loss in column (6). In addition, some heat is lost because of infiltration of cold air. One-half an air change per hour is assumed, or 71.2 cfm [column (3)]. This causes a heat loss, according to Eq. (19-4), of

$$1.08 \times 71.2 \times 70 = 5{,}400 \text{ Btu per hr}$$

To the total for the cellar, 20% is added to obtain the heat load in column (7).

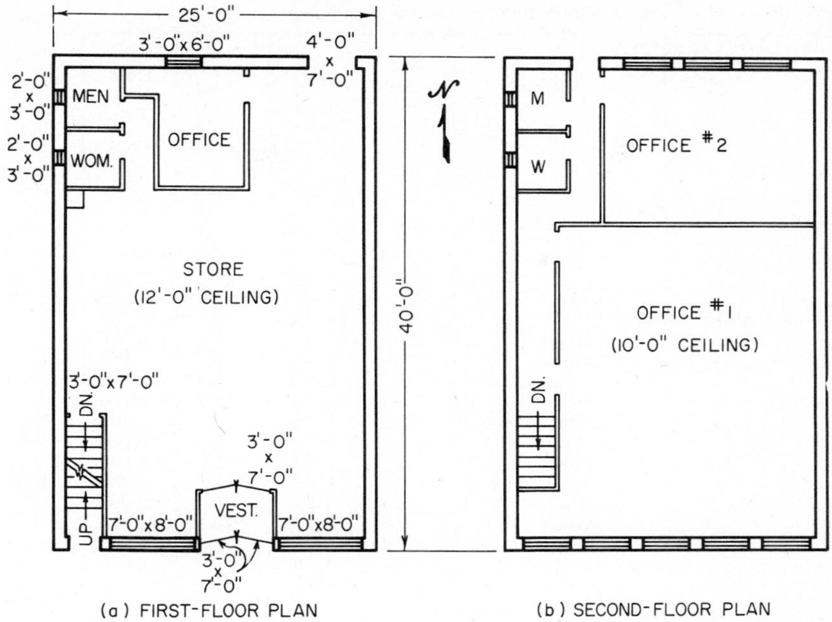

Fig. 19-1. First- and second-floor plans of a two-story building.

Similarly, heat losses are obtained for the various areas on the first and second floors. Heat-transmission coefficients [column (4)] were obtained from the "Handbook of Fundamentals," American Society of Heating, Refrigerating and Air-Conditioning Engineers. These were multiplied by the temperature gradient (70 — 0) to obtain the heat losses in column (6).

The total for the building, plus 20%, amounts to 144,475 Btu per hr. A heating plant with approximately this capacity should be selected.

19-8. Warm-air Heating. A warm-air heating system supplies heat to a room by bringing in a quantity of air above room temperature, the amount of heat added by the air being at least equal to that required to counteract heat losses.

A gravity system (without a blower) is rarely installed because it depends on the difference in density of the warm-air supply and the colder room air for the working pressure. Air-flow resistance must be kept at a minimum with large ducts and very few elbows. The result usually is an unsightly duct arrangement.

A forced warm-air system can maintain higher air velocities, thus requires smaller ducts, and provides much more sensitive control. For this type of system,

$$q = 1.08Q(T_h - T_i) \qquad (19\text{-}5)$$

where T_h = temperature of air leaving grille
T_i = room temperature
q = heat added by air, Btu per hr
Q = cfm of air supplied to room

Equation (19-5) indicates that the higher the temperature of the discharge air T_h the less air need be handled. In cheaper installations, the discharge air may be as high as 170°F and ducts are small. In better systems, more air is handled with a discharge temperature as low as 135 to 140°F. With a room temperature of 70°F, we shall need (170 — 70)/(135 — 70) = 1.54 times as much air with the 135°F system as with a 170°F system.

It is not advisable to go much below 135°F with discharge air, because drafts will result. With body temperature at 98°F, air at about 100°F will hardly seem

Table 19-4. Heat-load Determination for Two-story Building

Space (1)	Heat-loss source (2)	Net area or cfm infiltration (3)	U or coefficient (4)	Temp gradient, °F (5)	Heat loss, Btu per hr (6)	Total plus 20% (7)
Cellar	Walls	1,170	4		4,680	
	Floor	1,000	2		2,000	
	½ air change	71.2	1.08	70	5,400	14,500
First-floor store	Walls	851	0.25	70	14,900	
	Glass	135	1.13	70	10,680	
	Doors	128	0.69	70	1,350	
	1 air change	170	1.08	70	12,850	47,600
Vestibule	Glass	48	1.13	70	3,800	
	2 air change	12	1.08	70	905	5,650
Office	Walls	99	0.25	70	1,730	
	Glass	21	1.13	70	1,660	
	½ air change	9	1.08	70	680	4,900
Men's room	Walls	118	0.25	70	2,060	
	Glass	6	1.13	70	48	
	½ air change	3	1.08	70	226	2,800
Ladies' room	Walls	60	0.25	70	1,050	
	Glass	6	1.13	70	48	
	½ air change	3	1.08	70	226	1,590
Second-floor office No. 1	Walls	366	0.25	70	6,400	
	Glass	120	1.13	70	9,500	
	Roof	606	0.19	70	8,050	
	½ air change	50	1.08	70	3,786	33,300
Office No. 2	Walls	207	0.25	70	3,620	
	Glass	72	1.13	70	5,700	
	Roof	234	0.19	70	3,120	
	½ air change	19	1.08	70	1,435	16,650
Men's room	Walls	79	0.25	70	1,380	
	Glass	6	1.13	70	475	
	Roof	20	0.19	70	266	
	½ air change	3	1.08	70	226	2,820
Ladies' room	Walls	44	0.25	70	770	
	Glass	6	1.13	70	475	
	Roof	20	0.19	70	266	
	½ air change	3	1.08	70	226	2,080
Hall	Walls	270	0.25	70	4,720	
	Door	21	0.69	70	1,015	
	Roof	75	0.19	70	1,000	
	2 air change	40	1.08	70	3,020	12,585
						144,475

warm. If we stand a few feet away from the supply grille, 70°F room air will be entrained with the warm supply air and the mixture will be less than 98°F when it reaches us. We probably would complain about the draft.

Supply grilles should be arranged so that they blow a curtain of warm air across the cold, or exposed, walls and windows. (See Fig. 19-2 for a suggested arrangement.) These grilles should be placed near the floor, since the lower-density warm air will rise and accumulate at the ceiling.

Return-air grilles should be arranged in the interior near unexposed walls—in foyers, closets, etc.—and preferably at the ceiling. This is done for two reasons:

1. The warm air in all heating systems tends to rise to the ceiling. This creates a large temperature gradient between floor and ceiling, sometimes as high as 10°F. Taking the return air from the ceiling reduces this gradient.

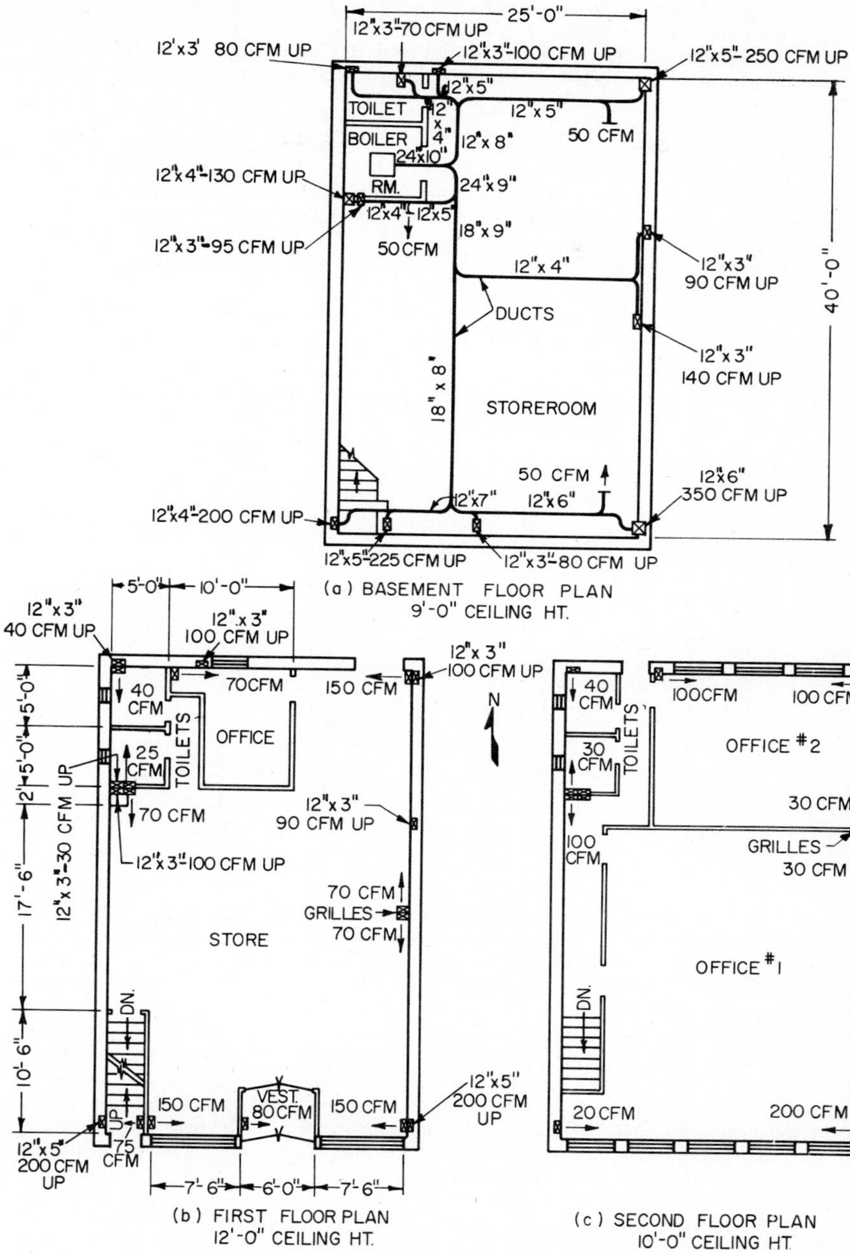

Fig. 19-2. Layout of a duct system for warm-air heating of the basement, first floor, and second floor of the building shown in Fig. 19-1. The boiler room is in the basement.

2. Returning the warmer air to the heating plant is more economical in operation than using cold air from the floor.

19-9. Duct Design. After discharge grilles and the warm-air heater are located, it is advisable to make a single-line drawing showing the air quantities each branch and line must be able to carry (Fig. 19-2).

Of the methods of duct design in use, the equal-friction method is the most practical. It is considered good practice not to exceed a pressure loss of 0.15 in. of water per 100 ft of ductwork due to friction. Higher friction will result in large power consumption for air circulation. It is also considered good practice to stay below a starting velocity in main ducts of 900 fpm in residences; 1,300 fpm in schools, theaters, and public buildings; and 1,800 fpm in industrial buildings. Velocity in branch ducts should be about two-thirds of these and in branch risers about one-half.

Too high a velocity will result in noisy and panting ductwork. Too low a velocity will require uneconomical, bulky ducts.

Table 19-5. Diameters of Circular Ducts in Inches Equivalent to Rectangular Ducts

Side	4	8	12	18	24	30	36	42	48	60	72	84
3	3.8	5.2	6.2									
4	4.4	6.1	7.3									
5	4.9	6.9	8.3									
6	5.4	7.6	9.2									
7	5.7	8.2	9.9									
12	...	10.7	13.1									
18	...	12.9	16.0	19.7								
24	...	14.6	18.3	22.6	26.2							
30	...	16.1	20.2	25.2	29.3	32.8						
36	...	17.4	21.9	27.4	32.0	35.8	39.4					
42	...	18.5	23.4	29.4	34.4	38.6	42.4	45.9				
48	...	19.6	24.8	31.2	36.6	41.2	45.2	48.9	52.6			
60	...	21.4	27.3	34.5	40.4	45.8	50.4	54.6	58.5	65.7		
72	...	23.1	29.5	37.2	43.8	49.7	54.9	59.6	63.9	71.7	78.8	
84	...	....	....	39.9	46.9	53.2	58.9	64.1	68.8	77.2	84.8	91.9
96	...	....	....	....	49.5	56.3	62.4	68.2	73.2	82.6	90.5	97.9

The shape of ducts usually installed is rectangular, because dimensions can easily be changed to maintain the required area. However, as ducts are flattened, the increase in perimeter offers additional resistance to air flow. Thus a flat duct requires an increase in cross section to be equivalent in air-carrying capacity to one more nearly square.

A 12 × 12-in. duct, for example, will have an area of 1 sq ft and a perimeter of 4 ft, whereas a 24 × 6-in. duct will have the same cross-sectional area but a 5-ft perimeter and thus greater friction. Therefore, a 24 × 7-in. duct is more nearly equivalent to the 12 × 12. Equivalent sizes can be determined from tables, such as those in the "ASHRAE Guide" (American Society of Heating, Refrigerating and Air-Conditioning Engineers), where rectangular ducts are rated in terms of equivalent round ducts (equal friction and capacity). Table 19-5 is a shortened version.

Charts also are available in the Guide giving the relationship between duct diameter in inches, air velocity in feet per minute, air quantity in cubic feet per minute, and friction in inches of water pressure drop per 100 ft of duct. Table 19-6 is based on data in the Guide.

In the equal-friction method, the equivalent round duct is determined for the required air flow at the predetermined friction factor.

As an example, let us compute the duct sizes for the structure for which the heat load was calculated in Art. 19-7.

Table 19-6. Sizes of Round Ducts for Air Flow*

Friction, in. per 100 ft	0.05		0.10		0.15		0.20		0.25		0.30	
Air flow, cfm	Diam, in.	Velocity, fpm	Diam, in.	Velocity, fpm	Diam, in.	Velocity, fpm	Diam, in.	Velocity, fpm	Diam, in.	Velocity, fpm	Diam, in.	Velocity, fpm
50	5.3	350	4.6	450	4.2	530	3.9	600	3.8	660	3.7	710
100	6.8	420	5.8	550	5.4	640	5.1	720	4.8	780	4.7	850
200	8.7	480	7.6	650	6.9	760	6.6	860	6.3	940	6.1	1,020
300	10.2	540	8.8	730	8.2	850	7.7	960	7.3	1,050	7.1	1,120
400	11.5	580	9.8	770	9.0	920	8.5	1,040	8.2	1,130	7.8	1,200
500	12.4	620	11.8	820	9.8	970	9.3	1,080	8.8	1,160	8.6	1,270
1,000	15.8	730	13.7	970	12.8	1,140	12.0	1,280	11.5	1,400	11.2	1,500
2,000	20.8	870	18.0	1,150	16.6	1,370	15.7	1,520	15.0	1,660	14.5	1,780
3,000	24.0	960	21.0	1,280	19.7	1,500	18.3	1,680	17.5	1,850		
4,000	26.8	1,050	23.4	1,360	21.6	1,600	20.2	1,800				
5,000	29.2	1,100	25.5	1,460	23.7	1,700	22.2	1,900				
10,000	37.8	1,310	33.2	1,770	30.3	2,000						

* Based on data in "ASHRAE Guide," American Society of Heating, Refrigerating and Air-Conditioning Engineers.

Table 19-4 showed that a heating unit with 144,475 Btu per hr capacity is required. After checking manufacturers' ratings of forced warm-air heaters, we choose a unit rated at 160,000 Btu per hr and 2,010 cfm.

If we utilized the full fan capacity and supply for the rated 160,000 Btu per hr and applied Eq. (19-5), the temperature rise through the heater would be

$$\Delta T = \frac{q}{1.08Q} = \frac{160{,}000}{1.08 \times 2{,}010} = 73.6°\text{F}$$

If we adjusted the flame (oil or gas) so that the output was the capacity theoretically required, the temperature rise would be

$$\Delta T = \frac{q}{1.08Q} = \frac{144{,}475}{1.08 \times 2{,}010} = 66.5°\text{F}$$

In actual practice, we do not tamper with the flame adjustment in order to maintain the manufacturer's design balance. Instead, the amount of air supplied to each room is in proportion to its load. (See Table 19-7, which is a continuation

Table 19-7. Air-supply Distribution in Accordance with Heat Load

Space	Load, Btu per hr	% of Load	Cfm (% × 2,010)
Cellar	14,500	10.05	200
First-floor store	47,600	33.00	660
Vestibule	5,650	3.91	80
Office	4,900	3.39	70
Men's room	2,800	1.94	40
Ladies' room	1,590	1.10	25
Second-floor office No. 1	33,300	23.01	460
Office No. 2	16,650	11.51	230
Men's room	2,820	1.95	40
Ladies' room	2,080	1.44	30
Hall	12,585	8.70	175
	144,475	100.00	2,010

of Table 19-4 in the design of a forced-air heating system.) Duct sizes can then be determined for the flow indicated in the table (see Fig. 19-2). In this example, minimum duct size for practical purposes is 12 × 3 in. Other duct sizes were obtained from Table 19-6, assuming a friction loss per 100 ft of 0.15, and equivalent rectangular sizes were obtained from Table 19-5 and shown in Fig. 19-2.

19-10. Humidification in Warm-air Heating. A warm-air heating system lends itself readily to humidification. Most warm-air furnace manufacturers provide a humidifier that can be placed in the discharge bonnet of the heater. This usually consists of a pan with a ball float to keep the pan filled.

Theoretically, a building needs more moisture when the outside temperature drops. During these colder periods, the heater runs more often, thus vaporizing more water. During warmer periods less moisture is required and less moisture is added because the heater runs less frequently.

Some manufacturers provide a woven asbestos-cloth frame placed in the pan to materially increase the contact surface between air and water. These humidifiers have no control.

Where some humidity control is desired, a spray nozzle connected to the hot-water system with an electric solenoid valve in the line may be actuated by a humidistat.

With humidification, well-fitted storm windows must be used, for with indoor conditions of 70°F and 30% relative humidity and an outdoor temperature of 0°F, condensation will occur on the windows. Storm windows will cut down the loss of room moisture.

19-11. Control of Warm-air Heating. The sequence of operation of a warm-air heating system is usually as follows:

When the thermostat calls for heat, the heat source is started. When the air chamber in the warm-air heater reaches about 120°F, the fan is started by a sensitive element. This is done so as not to allow cold air to issue from the supply grilles and create drafts.

If the flame size and air quantity are theoretically balanced, the discharge air will climb to the design value of, say, 150°F and remain there during the operation of the heater. However, manual shutoff of grilles by residents, dirty filters, etc., will cause a reduction of air flow and a rise in air temperature above design. A sensitive safety element in the air chamber will shut off the heat source when the discharge temperature reaches a value higher than about 180°F. The heat source will again be turned on when the air temperature drops a given amount.

When the indoor temperature reaches the value for which the thermostat is set, the heat source only is shut off. The fan, controlled by the sensitive element in the air chamber, will be shut off after the air cools to below 120°F. Thus, most of the usable heat is transmitted into the living quarters instead of escaping up the chimney.

With commercial installations, the fan usually should be operated constantly, to maintain proper air circulation in windowless areas during periods when the thermostat is satisfied. If residential duct systems have been poorly designed, often some spaces may be too cool, while others may be too warm. Constant fan operation in such cases will tend to equalize temperatures when the thermostat is satisfied.

Duct systems should be sized for the design air quantity of the heater. Insufficient air will cause the heat source to cycle on and off too often. Too much air may cool the flue gases so low as to cause condensation of the water in the products of combustion. This may lead to corrosion, because of dissolved flue gases.

19-12. Warm-air Perimeter Heating. This type of heating is often used in basementless structures, where the concrete slab is laid directly on the ground. The general arrangement is as follows: The heater discharges warm air to two or more under-floor radial ducts feeding a perimeter duct. Floor grilles or baseboard grilles are located as in a conventional warm-air heating system, with collars connected to the perimeter duct.

To prevent excessive heat loss to the outside, it is advisable to provide a rigid waterproof insulation between the perimeter duct and the outside wall.

19-13. Hot-water Heating Systems. A hot-water heating system consists of a heater or furnace, radiators, piping systems, and circulator.

A gravity system without circulating pumps is rarely installed. It depends on a difference in density of the hot supply water and the colder return water for working head. Air-flow resistance must be kept to a minimum, and the circulating piping system must be of large size. A forced circulation system can maintain higher water velocities, thus requires much smaller pipes and provides much more sensitive control.

Three types of piping systems are in general use for forced hot-water circulation systems:

One-pipe system (Fig. 19-3). This type has many disadvantages and is not usually recommended. It may be seen in Fig. 19-3 that radiator No. 1 takes

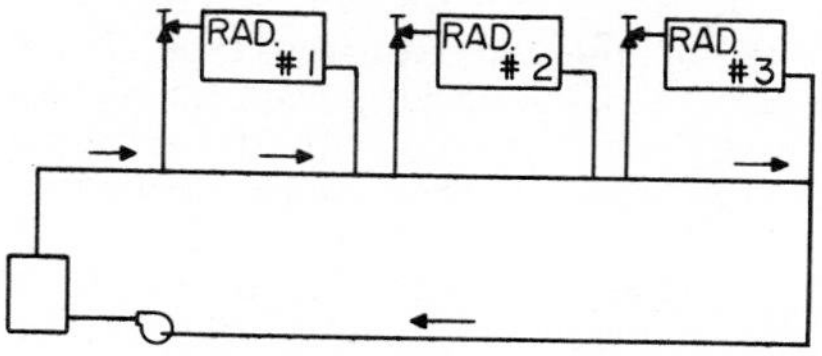

Fig. 19-3. One-pipe hot-water heating system.

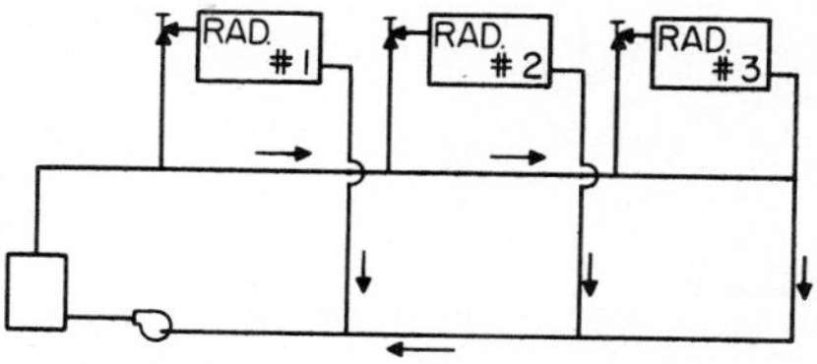

Fig. 19-4. Two-pipe direct-return hot-water heating system.

hot water from the supply main and dumps the colder water back in the supply main. This causes the supply-water temperature for radiator No. 2 to be lower, requiring a corresponding increase in radiator size. (Special flow and return fittings are available to induce flow through the radiators.) The design of such a system is very difficult, and any future adjustment or balancing of the system throws the remainder of the temperatures out.

Two-pipe direct-return system (Fig. 19-4). Here all radiators get the same supply-water temperature, but the last radiator has more pipe resistance than the first. This can be balanced out by sizing the pump for the longest run and installing orifices in the other radiators to add an equivalent resistance for balancing.

Two-pipe reversed-return system (Fig. 19-5). The total pipe resistance is about the same for all radiators. Radiator No. 1 has the shortest supply pipe and the longest return pipe, while radiator No. 3 has the longest supply pipe and the shortest return pipe.

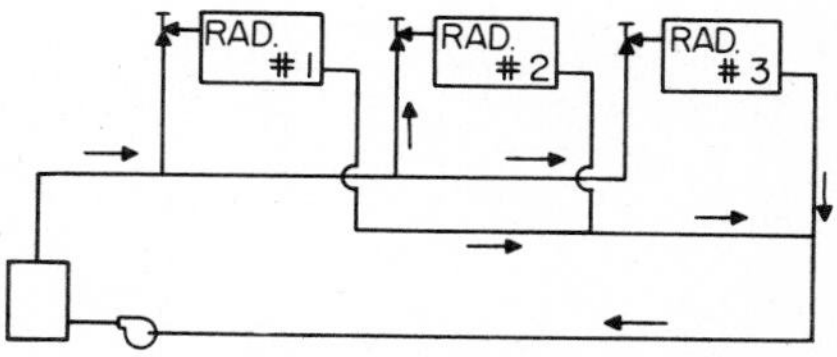

Fig. 19-5. Two-pipe reversed-return hot-water heating system.

Supply design temperatures usually are 180°F, with a 20°F drop assumed through the radiators; thus the temperature of the return riser would be 160°F.

When a hot-water heating system is designed, it is best to locate the radiators, then calculate the water flow in gallons per minute required by each radiator. For a 20°F rise

$$q = 10{,}000Q \tag{19-6}$$

where q = amount of heat required, Btu per hr

Q = flow of water, gpm

A one-line diagram showing the pipe runs should next be drawn, with gallons per minute to be carried by each pipe noted. The piping may be sized, using friction-flow charts and tables showing equivalent pipe lengths for fittings, with water velocity limited to a maximum of 4 ft per sec. (See, for example, Fig. 21-15 or "ASHRAE Guide," American Society of Heating, Refrigerating, and Air-Conditioning Engineers.) Too high a water velocity will cause noisy flow; too low a velocity will create a sluggish system and costlier piping.

The friction should be between 250 and 600 milinches per ft (1 milinch = 0.001

in.). It should be checked against available pump head, or a pump should be picked for the design gallons per minute at the required head.

It is very important that piping systems be made flexible enough to allow for expansion and contraction. Expansion joints are very satisfactory but expensive. Swing joints as shown in Fig. 19-6 should be used where necessary. In this type of branch take-off, the runout pipe, on expansion or contraction, will cause the threads in the elbows to screw in or out slightly, instead of creating strains in the piping.

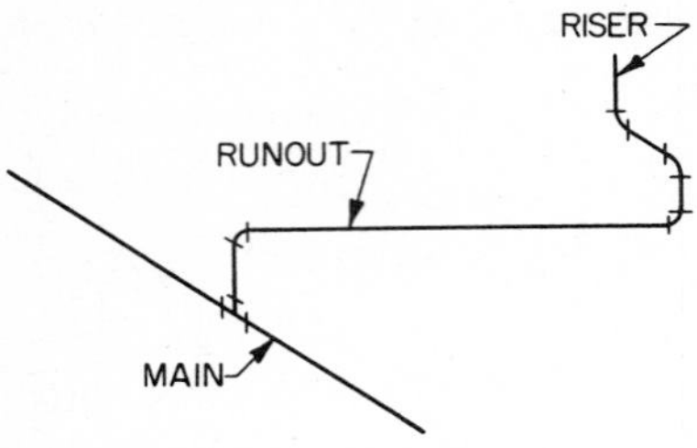

Fig. 19-6. Swing joint permits expansion and contraction of piping.

19-14. Hot-water Radiators. Radiators, whether of the old cast-iron type, finned pipe, or other, should be picked for the size required, in accordance with the manufacturer's ratings. These ratings depend on the average water temperature. For 170°F average water temperature, 1 sq ft of radiation surface is equal to 150 Btu per hr.

19-15. Expansion Tanks for Hot-water Systems. All hot-water heating systems must be provided with an expansion tank of either the open or closed type.

Figure 19-7 shows an open-type expansion tank. This tank should be located at least 3 ft above the highest radiator and in a location where it cannot freeze.

The size of tank depends on the amount of expansion of the water. From a low near 32°F to a high near boiling, water expands 4% of its volume. Therefore, an expansion tank should be sized for 6% of the total volume of water in radiators, heater, and all piping. That is, the volume of the tank, up to the level of its overflow pipe, should not be less than 6% of the total volume of water in the system.

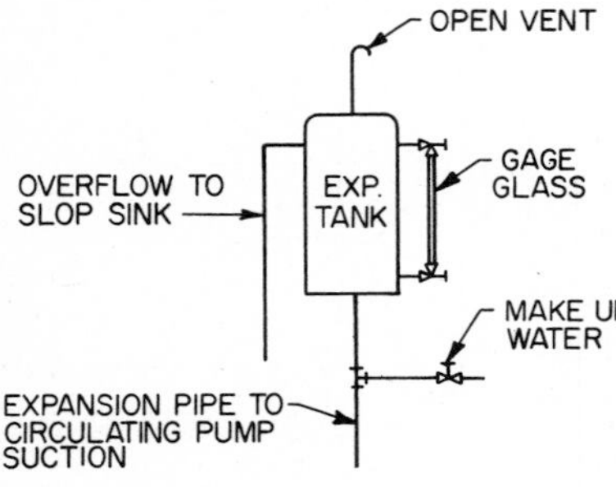

Fig. 19-7. Open-type expansion tank.

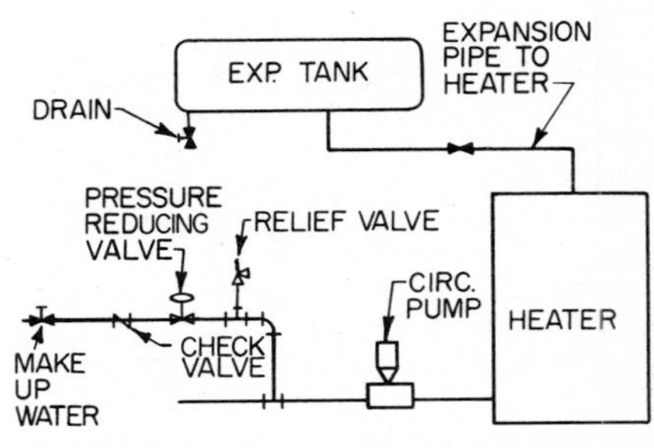

Fig. 19-8. Closed-type expansion tank.

Figure 19-8 is a diagram of the hookup of a closed-type expansion tank. This tank is only partly filled with water, creating an air cushion to allow for expansion and contraction. The pressure-reducing valve and relief valve are often supplied as a single combination unit.

The downstream side of the reducing valve is set at a pressure below city water-main pressure but slightly higher than required to maintain a static head in the highest radiator. The minimum pressure setting in pounds per square inch is equal to the *height in feet* divided by 2.31.

The relief valve is set above maximum possible pressure. Thus, the system will automatically fill and relieve as required.

19-16. Precautions in Hot-water Piping Layout. One of the most important precautions in a hot-water heating system is to avoid air pockets or loops. The pipe should be pitched so that vented air will collect at points that can be readily vented either automatically or manually. Vents should be located at all radiators.

Pipe traps should contain drains for complete drainage in case of shutdown. Zones should be valved so that the complete system need not be shut down for repair of a zone. Multiple circulators may be used to supply the various zones at different times, for different temperature settings and different exposures.

Allow for expansion and contraction of pipe without causing undue stresses. All supply and return piping should be insulated.

In very high buildings, the static pressure on the boiler may be too great. Heat exchangers may be installed as indicated in Fig. 19-9. The boiler temperature and lowest zone would be designed for 200°F supply water and 180°F return. The second lowest zone and heat exchanger can be designed for 170°F supply water and 150°F return, etc.

19-17. Control of Hot-water Systems. The control system is usually arranged as follows: An immersion thermostat in the heater controls the heat source, such as an oil burner or gas solenoid valve. The thermostat is set to maintain design heater water temperature (usually about 180°F). When the room thermostat calls for heat, it starts the circulator. Thus, an immediate supply of hot water is available for the radiators. A low-limit immersion stat, usually placed in the boiler and wired in series with the room stat and the pump, is arranged to shut off the circulator in the event that the water temperature drops below 70°F. This is an economy measure; if there is a flame failure, water will not be circulated unless it is warm enough to do some good. If the boiler is used to supply domestic hot water via an instantaneous coil or storage tank, hot water will always be available for that purpose. It should be kept in mind that the boiler must be sized for the heating load plus the probable domestic hot-water demand. (See Art. 21-33.)

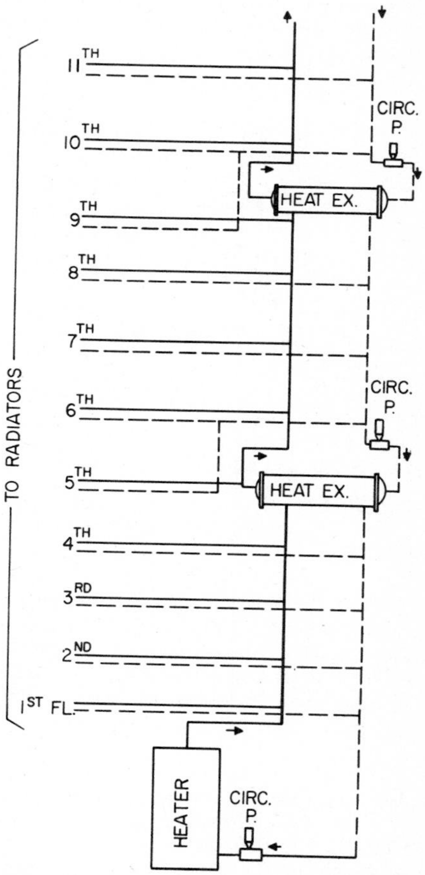

Fig. 19-9. Piping layout for tall building, including heat exchangers.

19-18. High-temperature, High-pressure Hot-water Systems. Some commercial and industrial building complexes have installed hot-water heating systems in which the water temperature is maintained well above 212°F. This is made possible by subjecting the system to a pressure well above the saturation pressure of the water at the design temperature. Such high-temperature hot-water systems present some inherent hazards. Most important is the danger of a leak, because then the water will flash into steam. Another serious condition occurs when a pump's suction strainer becomes partly clogged, creating a pressure drop. This may cause steam to flash in the circulating pump casing and vapor bind the pump.

These systems are not generally used for heating with radiators. They are mostly used in conjunction with air-conditioning installations in which the air-handling units contain a heating coil for winter heating. The advantage of high-temperature hot-water systems is higher rate of heat transfer from the heating medium to the air. This permits smaller circulating piping, pumps, and other equipment.

19-19. Steam-heating Systems. A steam-heating system consists of a boiler or steam generator and a piping system connecting to individual radiators or convectors.

A *one-pipe heating system* (Fig. 19-10) is the simplest arrangement. The steam-supply pipe to the radiators is also used as a condensate return to the boiler. On start-up, as the steam is generated, the air must be pushed out of the pipe and radiators by the steam. This is done with the aid of thermostatic air valves in the radiators. When the system is cold, a small vent hole in the valve is open. After the air is pushed out and steam comes in contact with the thermostatic element, the vent hole automatically closes to prevent escape of steam.

Where pipe runouts are extensive, it is necessary to install large orifice air vents to eliminate the air quickly from the piping system; otherwise the radiators near the end of the runout may get steam much later than the radiators closest to the boiler.

Air vents are obtainable with adjustable orifice size for balancing a heating system. The orifices of radiators near the boiler are adjusted smaller, while radiators far from the boiler will have orifices adjusted for quick venting. This helps balance the system.

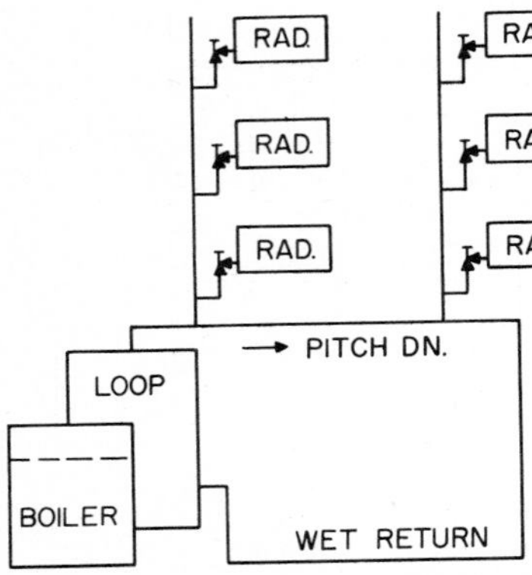

Fig. 19-10. One-pipe steam-heating system. Condensate returns through supply pipe.

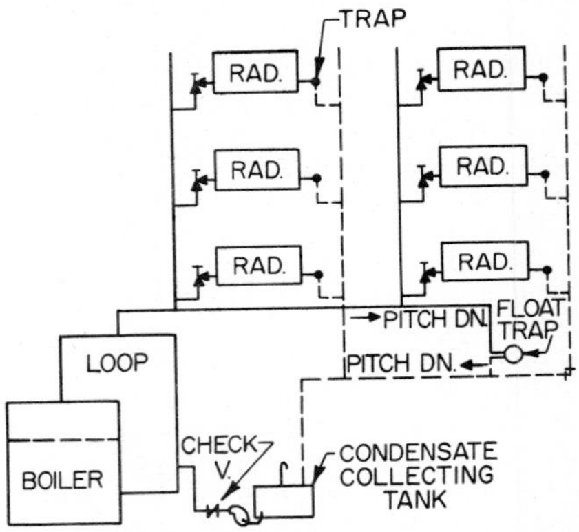

Fig. 19-11. Two-pipe steam-heating system. Different pipes are used for supply and return.

The pipe must be generously sized so as to prevent gravity flow of condensate from interfering with supply steam flow. Pipe capacities for supply risers, runouts, and radiator connections are given in the "ASHRAE Guide" (American Society of Heating, Refrigerating and Air-Conditioning Engineers). Capacities are expressed in square feet of **equivalent direct radiation (EDR)**:

$$1 \text{ sq ft EDR} = 240 \text{ Btu per hr}$$

Where capacities are in pounds per hour, 1 lb per hr = 970 Btu per hr.

Valves on radiators in a one-pipe system must be fully open or closed. If a valve is throttled, the condensate in the radiator will have to slug against a head of steam in the pipe to find its way back to the boiler by gravity. This will cause water hammer.

A *two-pipe system* is shown in Fig. 19-11. The steam supply is used to deliver steam to the supply end of all radiators. The condensate end of each radiator is connected to the return line via a thermostatic drip trap (Fig. 19-12). The trap is adjusted to open below 180°F and close above 180°F. Thus when there is enough condensate in the radiator to cool the element, it will open and allow the condensate to return to the collecting tank. A float switch in the tank starts the pump to return this condensate to the boiler against boiler steam pressure.

There are many variations in combining the two-pipe system and one-pipe system to create satisfactory systems.

In the *two-pipe condensate pump return system,* the pressure drop available for pipe and radiator loss is equal to boiler pressure minus atmospheric pressure, the difference being the steam gage pressure.

A wet-return system, shown in Fig. 19-13, will usually have a smaller head available for pipe loss. It is a self-adjusting system depending on the load. When steam is condensing at a given rate, the condensate will pile up in the return main above boiler level, creating a hydraulic head that forces the condensate into the boiler. The pressure above the water level on the return main will be less than the steam boiler pressure. This pressure difference—boiler pressure minus the pressure that exists above the return main water level—is available for pipe friction and radiator pressure drop.

On morning start-up when the air around the radiators is colder and the boiler is fired harder than during normal operation in an effort to pick up heat faster, the steam side of the radiators will be higher in temperature than normal. The air-side temperature of the radiators will be lower. This increase in temperature differential will create an increase in heat transfer causing a faster rate of condensation and a greater piling up of condensate to return the water to the boiler.

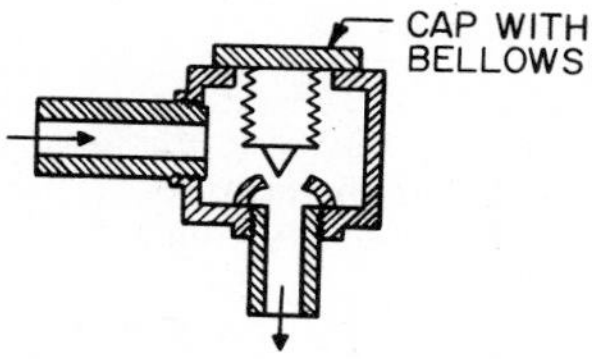

Fig. 19-12. Drip trap for condensate end of steam radiator.

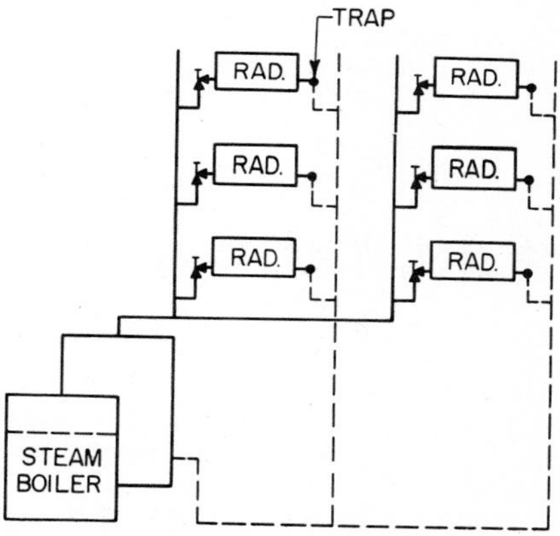

Fig. 19-13. Wet-return two-pipe steam-heating system.

In laying out a system, the steam-supply runouts must be pitched to remove the condensate from the pipe. They may be pitched back so as to cause the condensate to flow against the steam or pitched front to cause the condensate to flow with the steam.

Since the condensate will pile up in the return pipe to a height above boiler water level to create the required hydraulic head for condensate flow, a check of boiler-room ceiling height, steam-supply header, height of lowest radiator, height of dry return pipe, etc., is necessary to determine the height the water may rise in the return pipe without flooding these components.

The pipe may have to be oversized where condensate flows against the steam (see "ASHRAE Guide"). If the pitch is not steep enough, the steam may carry the condensate along in the wrong direction, causing noise and water hammer.

A **vacuum heating system** is similar to a steam pressure system with a condensate return pump. The main difference is that the steam pressure system can eliminate air from the piping and radiators by opening thermostatic vents to the atmosphere. The vacuum system is usually operated at a boiler pressure below atmospheric. The vacuum pump must, therefore, pull the noncondensables from the piping and radiators for discharge to atmosphere. Figure 19-14 shows diagrammatically the operation of a vacuum pump.

This unit collects the condensate in a tank. A pump circulates water through an eductor (1), pulling out the noncondensable gas to create the required vacuum.

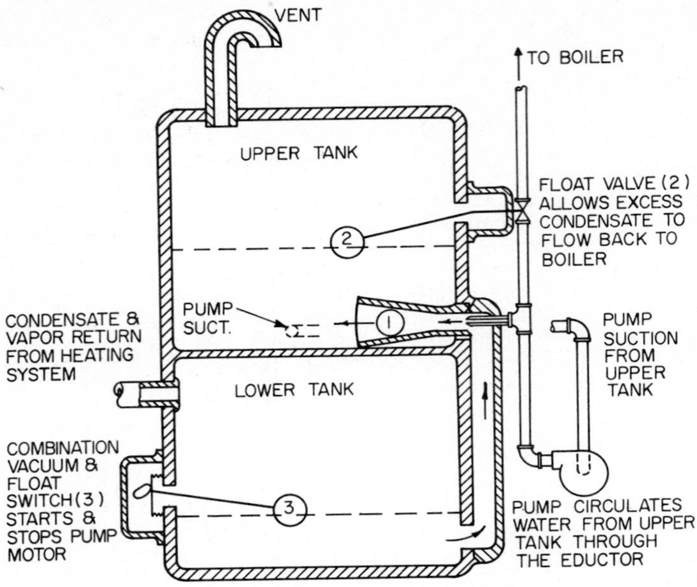

Fig. 19-14. Diagram of a vacuum pump for a vacuum steam-heating system. The eductor (1) maintains the desired vacuum in the lower tank.

The discharge side of the eductor nozzle is above atmospheric pressure. Thus, an automatic control system allows the noncondensable gas to escape to atmosphere as it accumulates, and the excess condensate is returned to the boiler as the tank reaches a given level.

Vacuum systems are usually sized for a total pressure drop varying from ¼ to ½ psi. Long equivalent-run systems (about 200 ft) will use ½ psi total pressure drop to save pipe size.

Some systems operate without a vacuum pump by eliminating the air during morning pickup by hard firing and while the piping system is above atmospheric pressure. During the remainder of the day, when the rate of firing is reduced, a tight system will operate under vacuum.

19-20. Unit Heaters. Large open areas, such as garages, showrooms, stores, and workshops, are usually best heated by unit heaters placed at the ceiling.

Figure 19-15 shows the usual connections to a steam unit-heater installation. The thermostat is arranged to start the fan when heat is required. The surface thermostat strapped on the return pipe prevents the running of the fan when insufficient steam is available. Where hot water is used for heating, the same arrangement is used, except that the float and thermostatic trap are eliminated and an air-vent valve is installed. Check manufacturers' ratings for capacities in choosing equipment.

Where steam or hot water is not available, direct gas-fired unit heaters may be installed. However, an outside flue is required to dispose of the products of combustion properly (Fig. 19-16). Check with local ordinances for required flues from direct gas-fired equipment. (Draft diverters are usually included with all gas-fired heaters. Check with the manufacturer when a draft diverter is not included.) For automatic control, the thermostat is arranged to start the fan and open the gas solenoid valve when heat is required. The usual safety pilot light is included by the manufacturer.

Gas-fired unit heaters are often installed in kitchens and other premises where large quantities of air may be exhausted. When a make-up air system is not provided, and the relief air must infiltrate through doors, windows, etc., a negative pressure must result in the premises. This negative pressure will cause a steady downdraft through the flue pipe from the outdoors into the space and prevent proper removal of the products of combustion. In such installations, it is advisable to place the unit heater in an adjoining space from which air is not exhausted in large quantities and deliver the warm air through ducts. Since propeller fans on most unit heaters cannot take much external duct resistance, centrifugal blower unit heaters may give better performance where ductwork is used. Sizes of gas piping and burning rates for gas can be obtained from charts and tables in the "ASHRAE Guide" for various capacities and efficiencies.

The efficiency of most gas-fired heating equipment is between 70 and 80%.

19-21. Panel Heating. Panel heating, or radiant heating as it is sometimes referred to, consists of a warm pipe coil embedded in the floor, ceiling, or walls. The most common arrangement is to circulate warm water through pipe under the floor. Some installations with warm-air ducts, steam pipes, and electric-heating elements have been installed.

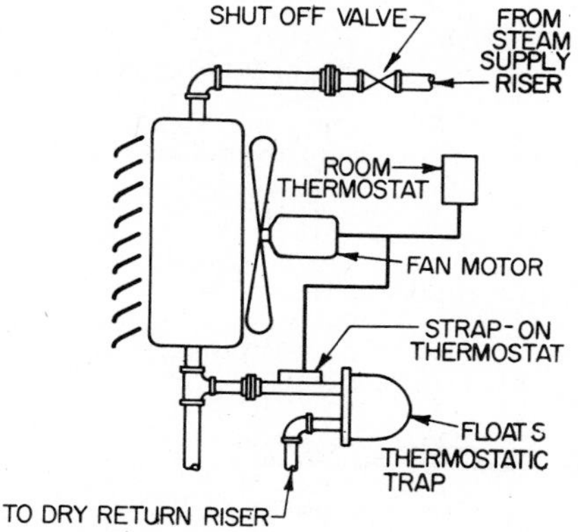

Fig. 19-15. Steam unit heater. Hot-water heater has air vent, no float and trap.

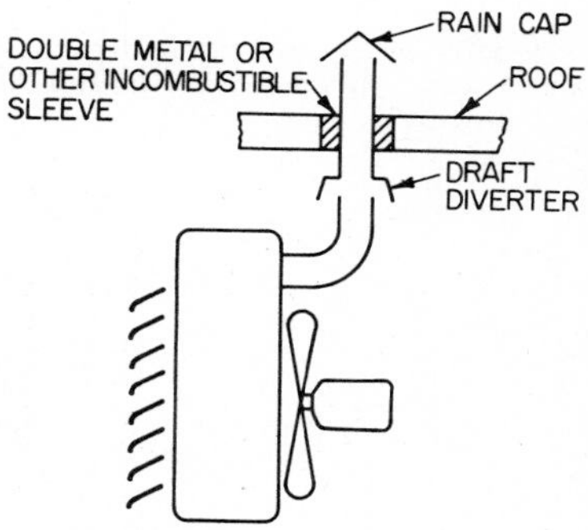

Fig. 19-16. Connections to outside for gas-fired unit heater.

Warm-air ducts for panel heating are not very common. A modified system normally called the perimeter warm-air heating system circulates the warm air around the perimeter of the structure before discharging the air into the premises via grilles (Art. 19-12).

Pipe coils embedded in concrete floor slabs or plaster ceilings and walls should not be threaded. Ferrous pipe should be welded, while joints in nonferrous metal pipe should be soldered. Return bends should be made with a pipe bender instead of fittings to avoid joints. All piping should be subjected to a hydrostatic test of at least three times the working pressure, with a minimum of 150 psig. Inasmuch as repairs are costly after construction is completed, it is advisable to adhere to the above recommendations.

Construction details for ceiling-embedded coils are shown in Figs. 19-17 to 19-20. Floor-embedded coil construction is shown in Fig. 19-21. Wall-panel coils may be installed as in ceiling panels.

Electrically heated panels are usually prefabricated and should be installed in accordance with manufacturer's recommendations and local electrical codes.

The piping and circuiting of a hot-water panel heating system are similar to hot-water heating systems with radiators or convectors, except that cooler water is used. However, a 20°F water-temperature drop is usually assumed. Therefore, charts used for the design of hot-water piping systems may be used for panel heating, too. (See Fig. 21-15 or "ASHRAE Guide," American Society of Heating, Refrigerating and Air-Conditioning Engineers.) In panel heating, the panel coils

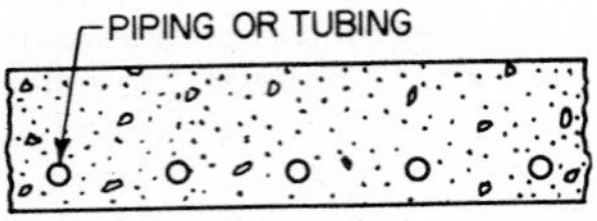

Fig. 19-17. Pipe coil embedded in a structural concrete slab for radiant heating.

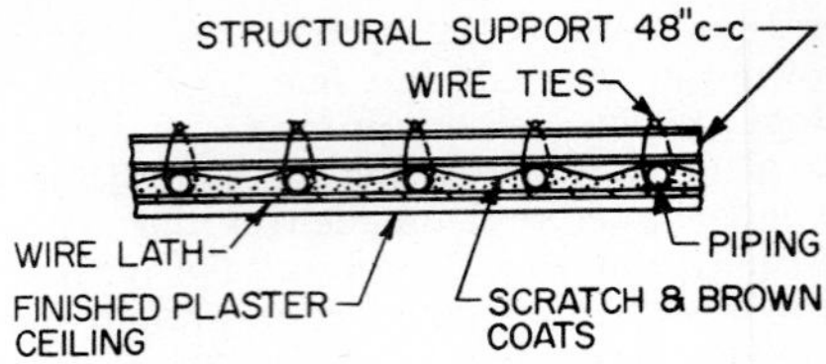

Fig. 19-18. Pipe coil embedded in a suspended plaster ceiling for radiant heating.

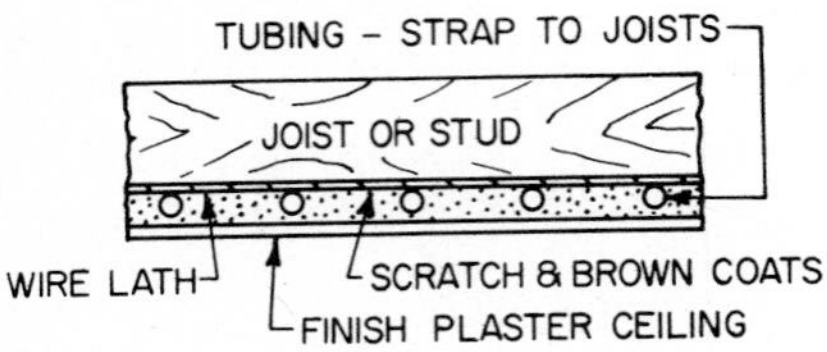

Fig. 19-19. Pipe coil attached to joists or studs and embedded in plaster for radiant heating.

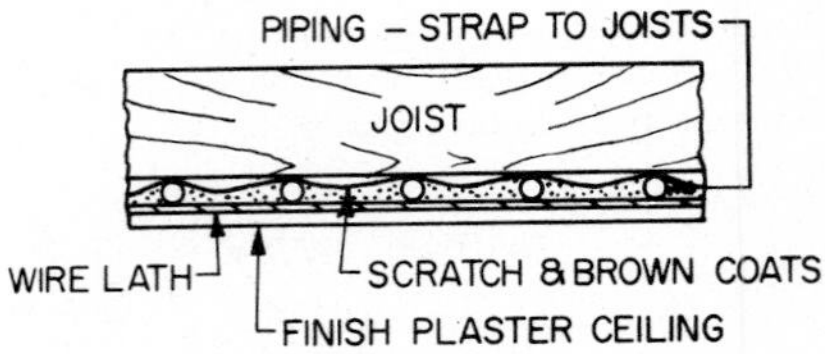

Fig. 19-20. Pipe coil embedded in plaster above lath for radiant heating.

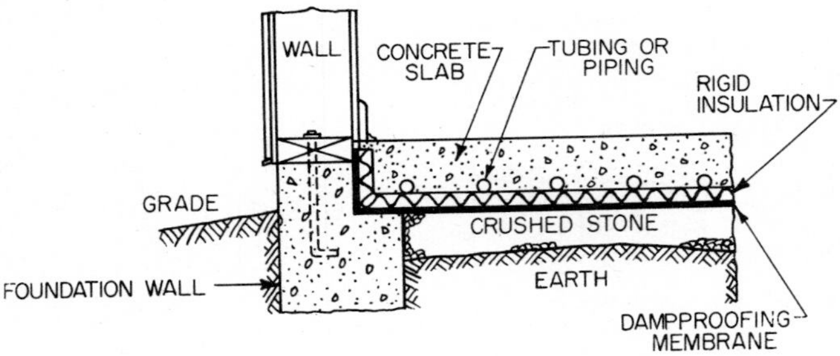

Fig. 19-21. Pipe coil embedded in a floor slab on grade for radiant heating.

replace radiators. Balancing valves should be installed in each coil, as in radiators. One wise precaution is to arrange coils in large and extensive areas so as not to have too much resistance in certain circuits. This can be done by using high-resistance continuous coils (Fig. 19-22) or low-resistance grid (Fig. 19-23) or a combination of both (Fig. 19-24).

One of the advantages of this system is its flexibility; coils can be concentrated along or on exposed walls. Also, the warmer supply water may be routed to

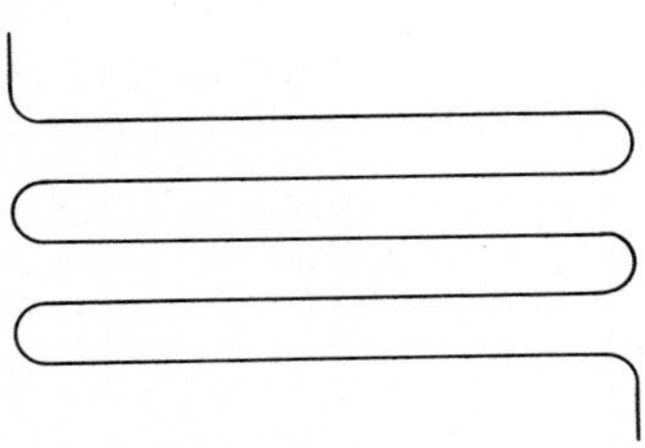

Fig. 19-22. Continuous pipe coil.

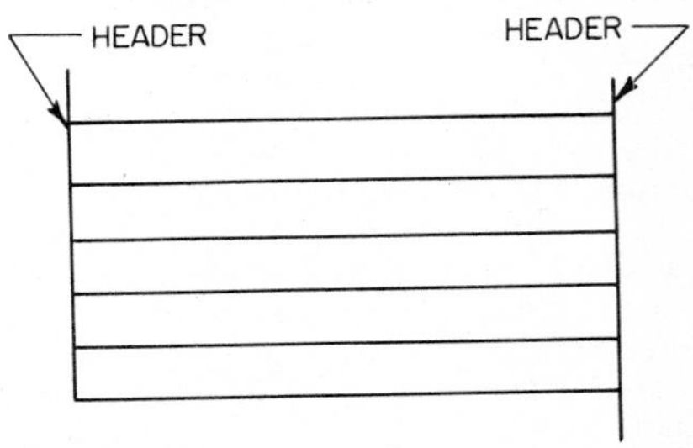

Fig. 19-23. Piping arranged in a grid for radiant heating.

the perimeter or exposed walls and the cooler return water brought to the interior zones.

Panel and Room Temperatures. Heat from the embedded pipes is transmitted to the panel, which in turn supplies heat to the room by two methods: (1) convection and (2) radiation. The amount of heat supplied by convection depends on the temperature difference between the panel and the air. The amount of heat supplied by radiation depends on the difference between the fourth powers of the absolute temperatures of panel and occupants. Thus, as panel temperature is increased, persons in the room receive a greater percentage of heat by radiation than by convection. Inasmuch as high panel temperatures are uncomfortable, it is advisable to keep floor panel temperatures about 85°F or lower and ceiling panel temperatures 100°F or lower. The percentage of radiant heat supplied by a panel at 85°F is about 56% and by one at 100°F about 70%.

Most advocates of panel heating claim that a lower than the usual design inside temperature may be maintained because of the large radiant surface comforting the individual; i.e., a dwelling normally maintained at 70°F may be kept at an air temperature of about 65°F. The low air temperature makes possible a reduction in heat losses through walls, glass, etc., and thus cuts down the heating load. However, during periods when the heating controller is satisfied and the water circulation stops, the radiant-heat source diminishes, creating an uncomfortable condition due to the below-normal room air temperature. It is thus considered good practice to design the system for standard room temperatures (Table 19-1) and the heating plant for the total capacity required (Arts. 19-6 and 19-7).

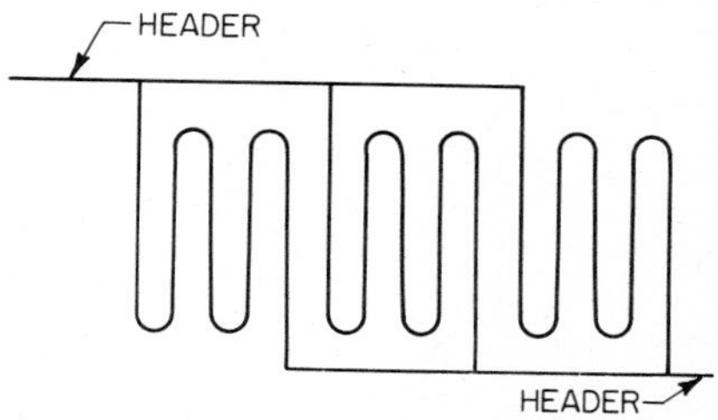

Fig. 19-24. Combination of grid arrangement and continuous pipe coil.

Design of Panel Heating. Panel output, Btu per hour per square foot, must be estimated to determine panel-heating area required. Panel capacity is determined by pipe spacing, water temperature, area of exposed walls and windows, infiltration air, insulation value of structural and architectural material between coil and occupied space, and insulation value of structural material preventing heat loss from the reverse side and edge of the panel. For other than ordinary panel-heating design, it is best to leave such design to a specialist or refer to "Systems, ASHRAE Guide and Data Book." For the usual installation with pipe spacing between 6 and 9 in. and installed as indicated in Figs. 19-19, 19-20, and 19-21, the curves shown in Fig. 19-25 may be used.

Example. Figure 19-26 illustrates a room 10 × 10 ft with two walls exposed, two walls unexposed, ceiling exposed, and floor panel heated.

First, obtain the heating-load requirements for this room, as discussed in Arts. 19-1 to 19-6. Assume storm windows with an over-all glass factor of 0.65 Btu per hr per sq ft per °F are used. Assume also that exposed walls and ceiling are insulated to provide an over-all factor of 0.1 Btu per hr per sq ft per °F. Then, the heating load comes to 7,320 Btu per hr. The panel output for a floor panel with 85°F average water temperature, from Fig. 19-25, is 28.5 Btu per hr per sq ft.

$$\text{Required panel area} = \frac{\text{heat load}}{\text{panel output}} = \frac{7{,}320}{28.5} = 256 \text{ sq ft}$$

Inasmuch as the total floor area is only 100 sq ft, this panel output will be insufficient to maintain comfortable conditions. If this panel were to be designed for 95°F average water temperature (not recommended), the panel capacity would be 54 Btu per hr per sq ft (Fig. 19-25). The panel area required would be 135 sq ft, again exceeding 100 sq ft. Thus, the total heating job in this room cannot be done with panel heating alone. The floor, however, can provide

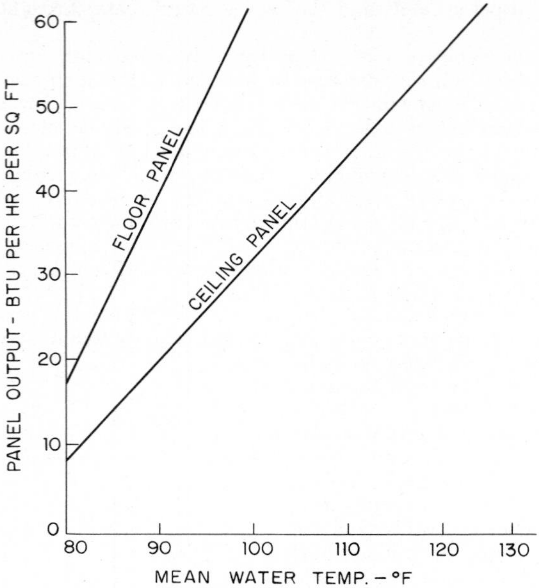

Fig. 19-25. Output of a radiant-heating panel for various panel temperatures and mean radiant temperatures of premises. (*Reprinted from "ASHRAE Guide," American Society of Heating, Refrigerating and Air-Conditioning Engineers, with permission.*)

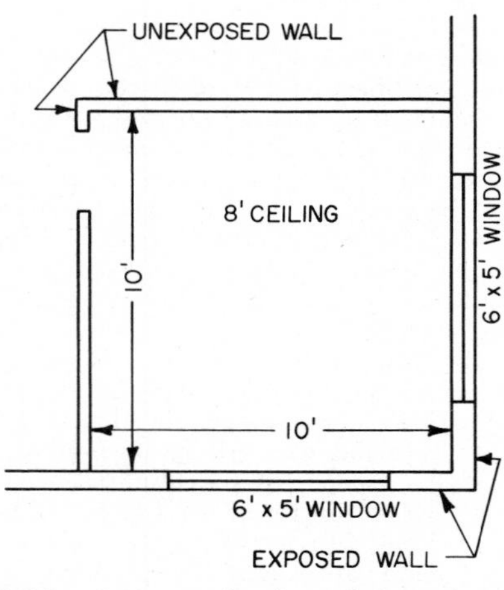

Fig. 19-26. Room to be heated by radiant panel.

$100 \times 28.5 = 2{,}850$ Btu per hr with panel heating. The remainder, 4,470 Btu per hr, can be taken care of with radiators set under the large windows.

Where conditions are such that only part of the floor area is required, the pipe grid should be placed near the exposed windows and the interior area left blank.

19-22. Snow Melting. Design of a snow-melting system for sidewalks, roads, parking areas, etc., involves the determination of a design amount of snowfall, sizing and layout of piping, and selection of heat exchanger and circulating medium. The pipe is placed under the wearing surface, with enough cover to protect it

Table 19-8. Water Equivalent of Snowfall

City	In. per hr per sq ft	City	In. per hr per sq ft
Albany, N.Y.	0.16	Evansville, Ind.	0.08
Asheville, N.C.	0.08	Hartford, Conn.	0.25
Billings, Mont.	0.08	Kansas City, Mo.	0.16
Bismarck, N.Dak.	0.08	Madison, Wis.	0.08
Boise, Idaho	0.08	Minneapolis, Minn.	0.08
Boston, Mass.	0.16	New York, N.Y.	0.16
Buffalo, N.Y.	0.16	Oklahoma City, Okla.	0.16
Burlington, Vt.	0.08	Omaha, Nebr.	0.16
Caribou, Maine	0.16	Philadelphia, Pa.	0.16
Chicago, Ill.	0.06	Pittsburgh, Pa.	0.08
Cincinnati, Ohio	0.08	Portland, Maine	0.16
Cleveland, Ohio	0.08	St. Louis, Mo.	0.08
Columbus, Ohio	0.08	Salt Lake City, Utah	0.08
Denver, Colo.	0.08	Spokane, Wash.	0.16
Detroit, Mich.	0.08	Washington, D.C.	0.16

Table 19-9. Heat Output and Circulating-fluid Temperatures for Snow-melting Systems

Rate of snowfall, in. per hr per sq ft of water equivalent		Air temp, 0°F			Air temp, 10°F			Air temp, 20°F			Air temp, 30°F		
		Wind, velocity, mph			Wind velocity, mph			Wind velocity, mph			Wind velocity, mph		
		5	10	15	5	10	15	5	10	15	5	10	15
0.08	Slab output, Btu per hr per sq ft	151	205	260	127	168	209	102	128	154	75	84	94
	Fluid temp, °F	108	135	162	97	117	138	85	97	110	70	75	79
0.16	Slab output, Btu per hr per sq ft	218	273	327	193	233	274	165	191	217	135	144	154
	Fluid temp, °F	142	169	198	120	149	170	117	129	142	100	105	109
0.25	Slab output, Btu per hr per sq ft	292	347	401	265	305	346	235	261	287	203	212	221
	Fluid temp, °F	179	206	234	165	186	206	151	163	176	134	139	144

NOTE: This table is based on a relative humidity of 80% for all air temperatures and construction as shown in Fig. 19-27.

against damage from traffic loads, and a heated fluid is circulated through it.

If friction is too high in extensive runs, use parallel loops (Art. 19-21). All precautions for drainage, fabrication, etc., hold for snow-melting panels as well as interior heating panels.

Table 19-8 gives a design rate of snowfall in inches of water equivalent per hour per square foot for various cities.

Table 19-9 gives the required slab output in Btu per hour per square foot at a given circulating-fluid temperature. This temperature may be obtained once we determine the rate of snowfall and assume a design outside air temperature and wind velocity. The table assumes a snow-melting panel as shown in Fig. 19-27.

Once the Btu per hour per square foot required is obtained from Table 19-9 and we know the area over which snow is to be melted, the total Btu per hour needed for snow melting is the product of the two. It is usual practice to add 40% for loss from bottom of slab.

The circulating-fluid temperature given in Table 19-9 is an average. For a 20°F rise, the fluid temperature entering the panel will be 10°F above that found in the table, and the leaving fluid temperature will be 10°F below the average. The freezing point of the fluid should be a few degrees below the minimum temperature ever obtained in the locality.

Check manufacturers' ratings for antifreeze solution properties to obtain the gallons per minute required and the friction loss to find the pumping head.

When ordering a heat exchanger for a given job, specify to the manufacturer the steam pressure available, fluid temperature to and from the heat exchanger, gallons per minute circulated, and physical properties of the antifreeze solution.

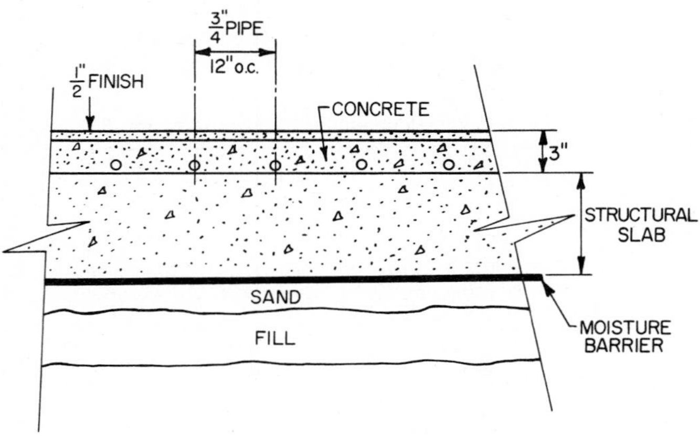

Fig. 19-27. Pipe coils embedded in outdoor concrete slab for snow melting.

19-23. Radiators and Convectors. In hot-water and steam-heating systems, heat is released to the spaces to be warmed by radiation and convection. The percentage transmitted by either method depends on the type of heat-dispersal unit used.

A common type of unit is the tubular radiator. It is composed of a series of interconnected sections, each of which consists of vertical tubes looped together. Steam-radiator sections are attached by nipples only at the base, whereas hot-water sections are connected at both top and bottom. Steam radiators should not be used for hot-water heating because of the difficulty of venting.

Pipe-coil radiators are sometimes used in industrial plants. The coils are usually placed on a wall under and between windows and are connected at the ends by branch tees or manifolds. Sometimes, finned-pipe coils are used instead of ordinary pipe. The fins increase the area of heat-transmitting surface.

Since radiators emit heat by convection as well as radiation, any enclosure should permit air to enter at the bottom and leave at the top.

Convectors, as the name implies, transmit heat mostly by convection. They usually consist of finned heating elements placed close to the floor in an enclosure that has openings at bottom and top for air circulation.

Baseboard units consist of continuous heating pipe in a thin enclosure along the base of exposed walls. They may transmit heat mostly by radiation or by convection. Convector-type units generally have finned-pipe heating elements. Chief advantage is the small temperature difference between floor and ceiling.

19-24. Vent Connections to Chimney. Heaters in which fuel is burned must be connected in some manner to the outside so that combustion products can be removed.

A vent connection to the building chimney must be equal to the size of the boiler or heater flue. The run must be as short as possible and pitch up toward the chimney at least ¼ in. per ft. Where the vent connection must be long and there is danger that the flue gases will cool to 212°F or below before entering the chimney, be sure to insulate the vent pipe to prevent condensation of combustion products. Dissolved flue gases in the condensate could cause corrosion.

Gas-fired heaters and boilers are usually provided with an approved draft hood. This should be installed as per manufacturer's recommendations. Oil-fired heaters and boilers should be provided with an approved draft stabilizer in the vent pipe. The hoods and stabilizers are used to prevent snuffing out of the flame in extreme cases and pulling of excessive air through the combustion chamber when the chimney draft is above normal, as in extremely cold weather.

19-25. Combustion-air Requirements for Heaters. All fuel-burning heaters must have sufficient air for combustion. Every boiler room, utility room, cellar, etc., must have adequate air available from the outdoors. An area—open window or duct—equal to about twice the size of the flue connection should suffice. Too small an area will result in exhaustion of boiler-room air. The resulting lowering of the boiler-room static pressure will cause higher-pressure atmospheric air to flow through the chimney back to the boiler, causing flare-backs and smoke in the boiler room. In some cases, where boiler-room walls were constructed of light gypsum block and where the ceilings were high, complete collapse of the wall resulted because of the difference in static pressure from the outside to the inside of the wall.

AIR CONDITIONING

To determine the size of cooling plant required in a building or part of a building, we determine the heat transmitted to the conditioned space through the walls, glass, ceiling, floor, etc., and add all the heat generated in the space. This is the cooling load. The unwanted heat must be removed by supplying cool air. The total cooling load is divided into two parts—sensible and latent.

19-26. Sensible and Latent Heat. The part of the cooling load that shows up in the form of a dry-bulb temperature rise is called sensible heat. It includes heat transmitted through walls, windows, roof, floor, etc.; radiation from the sun; and heat from lights, people, electrical and gas appliances, and outside air brought into the air-conditioned space.

Cooling required to remove unwanted moisture from the air-conditioned space is called latent load, and the heat extracted is called latent heat. Usually, the moisture is condensed out on the cooling coils in the cooling unit.

For every pound of moisture condensed from the air, the air-conditioning equipment must remove about 1050 Btu. Instead of rating items that give off moisture in pounds or grains per hour, common practice rates them in Btu per hour of latent load. These items include gas appliances, which give off moisture in products of combustion; steam baths, food, beverages, etc., which evaporate moisture; people; and humid outside air brought into the air-conditioned space.

19-27. Design Temperatures for Cooling. Before we can calculate the cooling load, we must first determine a design outside condition and the conditions we want to maintain inside.

For comfort cooling, indoor air at 80°F dry bulb and 50% relative humidity is usually acceptable.

Table 19-10 gives recommended design outdoor summer temperatures for various cities. Note that these temperatures are not the highest ever attained; for example, in New York City, the highest dry-bulb temperature recorded exceeds 105°F, whereas the design outdoor dry-bulb temperature is 95°F. Similarly, the wet-bulb temperature is sometimes above the 75°F design wet-bulb for that area.

19-28. Heat Gain through Enclosures. To obtain the heat gain through walls, windows, ceilings, floors, etc., when it is warmer outside than in, the heat-transfer

Table 19-10. Recommended Design Outdoor Summer Temperatures

State	City	Dry-bulb temp, °F	Wet-bulb temp, °F	State	City	Dry-bulb temp, °F	Wet-bulb temp, °F
Ala.	Birmingham	95	78	Miss.	Vicksburg	95	78
Ariz.	Flagstaff	90	65	Mo.	St. Louis	95	78
Ariz.	Phoenix	105	75	Mont.	Helena	95	67
Ark.	Little Rock	95	78	Nebr.	Lincoln	95	78
Calif.	Los Angeles	90	70	Nev.	Reno	95	65
Calif.	San Francisco	85	65	N.H.	Concord	90	73
Colo.	Denver	95	65	N.J.	Trenton	95	78
Conn.	Hartford	95	75	N.Mex.	Albuquerque	95	70
D.C.	Washington	95	78	N.Y.	New York	95	75
Fla.	Jacksonville	95	78	N.C.	Greensboro	95	78
Fla.	Miami	95	79	N.Dak.	Bismarck	95	73
Ga.	Atlanta	95	76	Ohio	Cincinnati	95	75
Idaho	Boise	95	65	Okla.	Tulsa	100	77
Ill.	Chicago	95	75	Ore.	Portland	90	68
Ind.	Indianapolis	95	75	Pa.	Philadelphia	95	78
Iowa	Des Moines	95	78	R.I.	Providence	95	75
Kans.	Topeka	100	78	S.C.	Charleston	95	78
Ky.	Louisville	95	78	S.Dak.	Rapid City	95	70
La.	New Orleans	95	80	Tenn.	Nashville	95	78
Maine	Portland	90	73	Tex.	Dallas	100	78
Md.	Baltimore	95	78	Tex.	Houston	95	78
Mass.	Boston	95	75	Utah	Salt Lake City	95	65
Mich.	Detroit	95	75	Vt.	Burlington	90	73
Minn.	Minneapolis	95	75	Va.	Richmond	95	78

coefficient is multiplied by the surface area and the temperature gradient (Arts. 19-1 and 19-7).

Radiation from the sun through glass is another source of heat. It can amount to about 200 Btu per hr per sq ft for a single sheet of unshaded common window glass facing east and west, about three-fourths as much for windows facing northeast and northwest, and one-half as much for windows facing south. For most practical applications, however, the sun effect on walls can be neglected, since the time lag is considerable and the peak load is no longer present by the time the radiant heat starts to work through to the inside surface. Also, if the wall exposed to the sun contains windows, the peak radiation through the glass also will be gone by the time the radiant heat on the walls gets through.

Radiation from the sun through roofs may be considerable. For most roofs, total equivalent temperature differences for calculating heat gain through sunlit roofs is about 50°F.

19-29. Roof Sprays. Many buildings have been equipped with roof sprays to reduce the sun load on the roof. Usually the life of a roof is increased by the spray system, because it prevents swelling, blistering, and vaporization of the volatile components of the roofing material. It also prevents the thermal shock of thunderstorms during hot spells. Equivalent temperature differential for computing heat gain on sprayed roofs is about 18°F.

Water pools 2 to 6 in. deep on roofs have been used, but they create structural difficulties. Furthermore, holdover heat into the late evening after the sun has set creates a breeding ground for mosquitoes and requires algae-growth control. Equivalent temperature differential to be used for computing heat gain for water-covered roofs is about 22°F.

Spray control is effected by the use of a water solenoid valve actuated by a temperature controller whose bulb is embedded in the roofing. Tests have been carried out with controller settings of 95, 100, and 105°F. The last was found to be the most practical setting.

The spray nozzles must not be too fine, or too much water is lost by drift. For ridge roofs, a pipe with holes or slots is satisfactory. When the ridge runs north and south, two pipes with two controllers would be practical, for the east pipe would be in operation in the morning and the west pipe in the afternoon.

19-30. Heat Gains from Interior Sources. Electric lights and most other electrical appliances convert their energy into heat.

$$q = 3.42W \tag{19-7}$$

where q = Btu per hr developed
W = watts of electricity used

Where fluorescent lighting is used, add 25% of the lamp rating for the heat generated in the ballast. Where electricity is used to heat coffee, etc., some of the energy is used to vaporize water. Tables in the "ASHRAE Guide," American Society of Heating, Refrigerating, and Air-Conditioning Engineers, give an estimate of the Btu per hour given up as sensible heat and that given up as latent heat by appliances.

Heat gain from people can be obtained from Table 19-11.

Table 19-11. Rates of Heat Gain from Occupants of Conditioned Spaces*

Degree of activity	Typical application	Total heat, adults, male, Btu per hr	Total heat adjusted,† Btu per hr	Sensible heat, Btu per hr	Latent heat, Btu per hr
Seated at rest	Theater, matinee	390	330	180	150
	Theater, evening	390	350	195	155
Seated, very light work	Offices, hotels, apartments	450	400	195	205
Moderately active office work	Offices, hotels, apartments	475	450	200	250
Standing, light work; or walking slowly	Department store, retail store, dime store	550	450	200	250
Walking; seated	Drug store				
Standing; walking slowly	Bank	550	500	200	300
Sedentary work	Restaurant‡	490	550	220	330
Light bench work	Factory	800	750	220	530
Moderate dancing	Dance hall	900	850	245	605
Walking 3 mph; moderately heavy work	Factory	1000	1000	300	700
Bowling¶	Bowling alley				
Heavy work	Factory	1500	1450	465	985

NOTE: Tabulated values are based on 80°F room dry-bulb temperature. For 78°F room dry-bulb, the total heat remains the same, but the sensible heat values should be increased by approximately 10%, and the latent heat values decreased accordingly.

* Reprinted from "ASHRAE Guide," Chap. 13, 1957, American Society of Heating, Refrigerating and Air-Conditioning Engineers, with permission.

† *Adjusted total heat gain* is based on normal percentage of men, women, and children for the application listed, with the postulate that the gain from an adult female is 85% of that for an adult male, and that the gain from a child is 75% of that for an adult male.

‡ Adjusted total heat value for *sedentary work, restaurant,* includes 60 Btu per hr for food per individual (30 Btu sensible and 30 Btu latent).

¶ For *bowling* figure one person per alley actually bowling, and all others as sitting (400 Btu per hr) or standing (550 Btu per hr).

19-31. Heat Gain from Outside Air. The sensible heat from outside air brought into a conditioned space can be obtained from

$$q_s = 1.08Q(T_o - T_i) \tag{19-8}$$

where q_s = sensible load due to outside air, Btu per hr
Q = cfm of outside air brought into conditioned space
T_o = design dry-bulb temperature of outside air
T_i = design dry-bulb temperature of conditioned space

The latent load due to outside air in Btu per hour is given by

$$q_l = 0.67Q(G_o - G_i) \tag{19-9}$$

where Q = cfm of outside air brought into conditioned space
G_o = moisture content of outside air, grains per lb of air
G_i = moisture content of inside air, grains per lb of air

The moisture content of air at various conditions may be obtained from a psychrometric chart.

19-32. Miscellaneous Sources of Heat Gain. In an air-conditioning unit, the fan used to circulate the air requires a certain amount of brake horsepower depending on the air quantity and the total resistance in the ductwork, coils, filters, etc. This horsepower will dissipate itself in the conditioned air and will show up as a temperature rise. Therefore, we must include the fan brake horsepower in the air-conditioning load. For most low-pressure air-distribution duct systems, the heat from this source varies from 5% of the sensible load for smaller systems to 3½% of the sensible load in the larger systems.

Where the air-conditioning ducts pass through nonconditioned spaces, the ducts must be insulated. The amount of the heat transmitted to the conditioned air through the insulation may be calculated from the duct area and the insulation heat-transfer coefficient.

19-33. Cooling Measured in Tons. Once we obtain the cooling load in Btu per hour, we convert the load to tons of refrigeration by

$$\text{Load in tons} = \frac{\text{load in Btu per hr}}{12{,}000} \tag{19-10}$$

A ton of refrigeration is the amount of cooling that can be done by a ton of ice melting in 24 hr.

19-34. Example of Cooling Calculation. Consider the building shown in Fig. 19-2. The first and second floors only will be air-conditioned. The design outdoor condition is assumed to be 95°F DB (dry-bulb) and 75°F WB (wet-bulb). The design indoor condition is 80°F DB and 50% relative humidity.

The temperature gradient across an exposed wall will be 15°F (95°F − 80°F). The temperature gradient between a conditioned and an interior nonconditioned space, such as the cellar ceiling, is assumed to be 10°F.

Exterior walls are constructed of 4-in. brick with 8-in. cinder backup; interior finish is plaster on metal lath. Partitions consist of 2 × 4 studs, wire lath, and plaster. First floor has double flooring on top of joists; cellar ceiling is plaster on metal lath.

Lights may be assumed to average 4 watts per sq ft. Assume 50 persons will be in the store, 2 in the first-floor office, 10 in the second-floor office No. 1, and 5 in second-floor office No. 2.

Air Requirements for First-floor Store (Table 19-12A).

$$\text{Load in tons} = \frac{77{,}249}{12{,}000} = 6.45 \text{ tons}$$

If supply air is provided at 18°F differential, with fresh air entering the unit through an outside duct, the flow required for sensible heat is [from Eq. (19-8)]:

$$Q = \frac{49{,}279}{1.08 \times 18} = 2{,}530 \text{ cfm}$$

Table 19-12A. First-floor Store Load

	Area, Sq Ft		$U(T_o - T_i)$, °F	Heat Load, Btu per Hr
Heat gain through enclosures:				
North wall	92	×	0.25 × 15	344
North door	28	×	0.46 × 15	193
South wall	412	×	0.25 × 15	1,540
South glass	154	×	1.04 × 15	2,405
East wall	480	×	0.25 × 15	1,800
West wall	228	×	0.25 × 15	854
Partition	263	×	0.39 × 10	1,025
Partition door	49	×	0.46 × 10	225
Floor	841	×	0.25 × 10	2,100
Ceiling	112	×	0.25 × 10	280
Load due to conduction				10,766
Sun load on south glass, 154 sq ft × 98				15,100
Occupants (sensible heat) from Table 19-12, 50 × 200				10,000
Lights [Eq. (19-7)] 841 sq ft × 4 watts per sq ft × 3.42				11,500
				47,366
Fan hp (4% of total)				1,913
Total internal sensible heat load				49,279
Fresh air (sensible heat) at 10 cfm per person [Eq. (19-8), 500 cfm × 1.08 × 15]				8,100
Total sensible load				57,379
Occupants (latent load) from Table 19-12, 50 × 250				12,500
Fresh air (latent heat) Eq. (19-9), 500 × 0.67 (99-77)				7,360
Total latent load				19,860
Total load				77,239

If no fresh-air duct is provided and all the infiltration air enters directly into the premises, the flow required is computed as follows:

$$\text{Store volume} = 841 \times 12 = 10{,}100 \text{ cu ft}$$

$$\text{Infiltration} = \frac{10{,}100}{60} \times 1.5 = 252 \text{ cfm}$$

Sensible load from the infiltration air [Eq. (19-8)]

$$= 252 \times 1.08 \times 15 = 4{,}090 \text{ Btu per hr}$$

$$\text{Internal sensible load plus infiltration} = 49{,}279 + 4{,}090 = 53{,}369 \text{ Btu per hr}$$

$$\text{Flow with no fresh-air duct} = \frac{53{,}369}{1.08 \times 18} = 2{,}750 \text{ cfm}$$

Air Requirements for First-floor Office (*Table 19-12B*).

$$\text{Load in tons} = \frac{3{,}947}{12{,}000} = 0.33 \text{ ton}$$

$$\text{Supply air required with fresh-air duct} = \frac{2{,}546}{1.08 \times 18} = 131 \text{ cfm}$$

$$\text{Supply air required with no fresh-air duct} = \frac{3{,}064}{1.08 \times 18} = 158 \text{ cfm}$$

Air Requirements for Second-floor Office No. 1 (*Table 19-12C*).

$$\text{Load in tons} = \frac{40{,}759}{12{,}000} = 3.4 \text{ tons}$$

$$\text{Supply air required with fresh-air duct} = \frac{33{,}139}{1.08 \times 18} = 1{,}700 \text{ cfm}$$

$$\text{Supply air required with no fresh-air duct} = \frac{36{,}049}{1.08 \times 18} = 1{,}850 \text{ cfm}$$

Table 19-12B. First-floor Office Load

	Btu per Hr	
Heat gain through enclosures:		
North wall, 102 × 0.25 × 15	382	
North glass, 18 × 1.04 × 15	281	
Partition, 30 × 0.39 × 10	117	
Floor, 79 × 0.25 × 10	198	
	978	
2 occupants (sensible load) at 195	390	
Lights, 79 × 4 × 3.42	1,080	
	2,448	
Fan hp 4%	98	
Total internal sensible load		2,546
Fresh air (sensible load) at 2 changes per hour, 32 × 1.08 × 15		518
Total sensible load		3,064
Occupants (latent load), 2 at 205	410	
Fresh air (latent load), 32 × 0.67(99 − 77)	473	
Total latent load		883
Total load		3,947 Btu per hr

Table 19-12C. Second-floor Office No. 1

	Btu per Hr	
Heat gain through enclosures:		
South wall, 130 × 0.25 × 15	388	
South glass, 120 × 1.04 × 15	1,870	
East wall, 260 × 0.25 × 15	975	
West wall, 40 × 0.25 × 15	150	
Partition, 265 × 0.39 × 10	1,035	
Door in partition, 35 × 0.46 × 10	161	
Roof, 568 × 0.19 × 54	5,820	
	10,399	
Sun load on south glass, 120 × 98	11,750	
Occupants (sensible load), 10 × 195	1,950	
Lights, 568 × 4 × 3.42	7,770	
	31,869	
Fan hp 4%	1,270	
Total internal sensible load		33,139
Fresh air (sensible load), 180 × 1.08 × 15		2,910
Total sensible load		36,049
Occupants (latent load), 10 × 205	2,050	
Fresh air (latent load), 180 × 0.67(99 − 77)	2,660	
Total latent load		4,710
Total load		40,759

Air Requirements for Second-floor Office No. 2 (Table 19-12D).

$$\text{Load in tons} = \frac{12{,}799}{12{,}000} = 1.07 \text{ tons}$$

$$\text{Supply air required with fresh-air duct} = \frac{9{,}384}{1.08 \times 18} = 483 \text{ cfm}$$

$$\text{Supply air required with no fresh-air duct} = \frac{10{,}634}{1.08 \times 18} = 550 \text{ cfm}$$

Cooling and supply-air requirements for the building are summarized in Table 19-13.

Table 19-12D. Second-floor Office No. 2

	Btu per Hr	
Heat gain through enclosures:		
North wall, 103 × 0.25 × 15	386	
North glass, 72 × 1.04 × 15	1,110	
East wall, 132 × 0.25 × 15.	495	
Partition, 114 × 0.39 × 10.	445	
Door in partition, 18 × 0.46 × 10.	83	
Roof, 231 × 0.19 × 54.	2,370	
	4,889	
Occupants (sensible load), 5 × 195	975	
Lights, 231 × 4 × 3.42	3,160	
	9,024	
Fan hp 4%	360	
Total internal sensible load		9,384
Fresh air (sensible load), 77 × 1.08 × 15.		1,250
Total sensible load		10,634
Occupants (latent load), 5 × 205	1,025	
Fresh air (latent load), 77 × 0.67(99 – 77).	1,140	
Total latent load		2,165
Total load		12,799

Table 19-13. Cooling-load Analysis for Building in Fig. 19-2

Space	Tons	Cfm with fresh-air duct	Cfm with no fresh-air duct
First-floor store	6.45	2,545	2,750
First-floor office	0.33	131	158
Second-floor office No. 1	3.40	1,700	1,850
Second-floor office No. 2	1.07	483	550
Total	11.25	4,859	5,308

19-35. Basic Cooling Cycle. Figure 19-28 shows the basic air-conditioning cycle of the *direct-expansion type.* The compressor takes refrigerant gas at a relatively low pressure and compresses it to a higher pressure. The hot gas is passed to a condenser where heat is removed and the refrigerant liquefied. The liquid is then piped to the cooling coil of the air-handling unit and allowed to expand to a lower pressure (suction pressure). The liquid vaporizes or is boiled off by the relatively warm air passing over the coil. The compressor pulls away the vaporized refrigerant to maintain the required low coil pressure with its accompanying low temperature.

In some systems, water is chilled by the refrigerant and circulated to units in or near spaces to be cooled (Fig. 19-29), where air is cooled by the water.

In formerly used water-cooled and belt-driven air-conditioning compressors, motor winding heat was usually dissipated into the atmosphere outside the conditioned space (usually into the compressor rooms). For average air-conditioning service 1 hp could produce about 1 ton of cooling. When sealed compressor-motor units came into use, the motor windings were arranged to give off their heat to the refrigerant suction gas. This heat was therefore added to the cooling load of the compressor. The result was that 1 hp could produce only about 0.85 ton of cooling.

Because of water shortages, many communities restrict the direct use of city water for condensing purposes. As an alternative, water can be cooled with cooling towers and recirculated. But for smaller systems, cooling towers have some inherent disadvantages. As a result, air-cooled condensers have become common for small- and medium-sized air-conditioning systems. Because of the higher head pressures

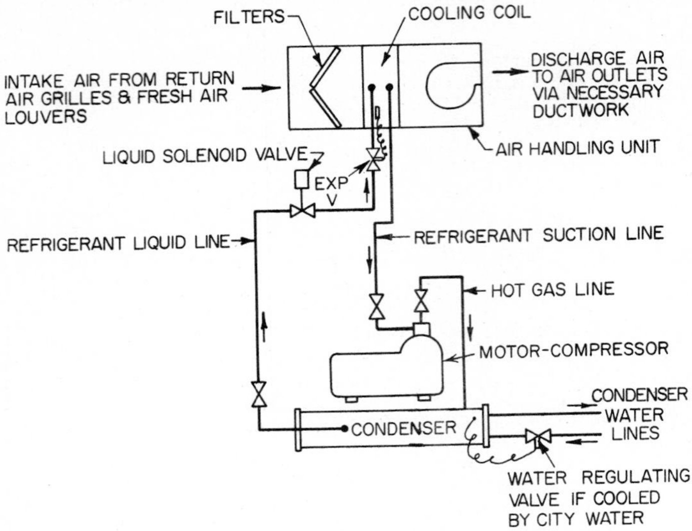

Fig. 19-28. Direct-expansion air-conditioning cycle.

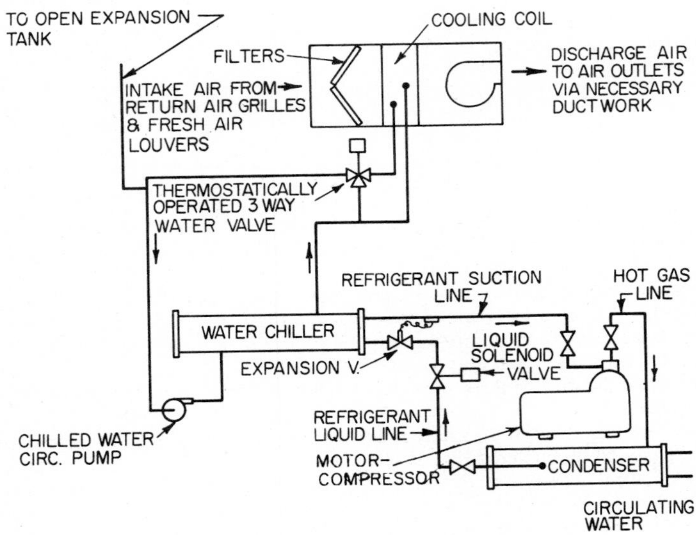

Fig. 19-29. Chilled-water air-conditioning cycle.

resulting from air-cooled condensers, each horsepower of compressor-motor will produce only about 0.65 ton of cooling. Because of these developments, air-conditioning equipment manufacturers rate their equipment in Btu per hour output and kilowatts required for the rated output, at conditions standardized by the industry, instead of in horsepower.

19-36. Air-distribution Temperature for Cooling. The air-distribution system is the most critical part of an air-conditioning system. If insufficient air is circulated, proper cooling cannot be done. On the other hand, handling large quantities of air is expensive in both initial cost and operation. The amount of cool air required increases rapidly the closer its temperature is brought to the desired room temperature.

If, for example, we wish to maintain 80°F DB (dry bulb) in a room and we introduce air at 60°F, the colder air when warming up to 80°F will absorb an amount of sensible heat equal to q_s. According to Eq. (19-8), $q_s = 1.08Q_1(80 - 60) = 21.6Q_1$, where Q_1 is the required air flow in cubic feet per minute. From Eq. (19-8), it can be seen also that, if we introduce air at 70°F, with a temperature rise of 10°F instead of 20°F, $q_s = 10.8Q_2$, and we shall have to handle twice as much air to do the same amount of sensible cooling.

From a psychrometric chart, the dew point of a room at 80°F DB and 50% relative humidity is found to be 59°F. If the air leaving the air-conditioning unit is 59°F or less, the duct will sweat and will require insulation. Even if we spend the money to insulate the supply duct, the supply grilles may sweat and drip. Therefore, theoretically, to be safe, the air leaving the air-conditioning unit should be 60°F or higher.

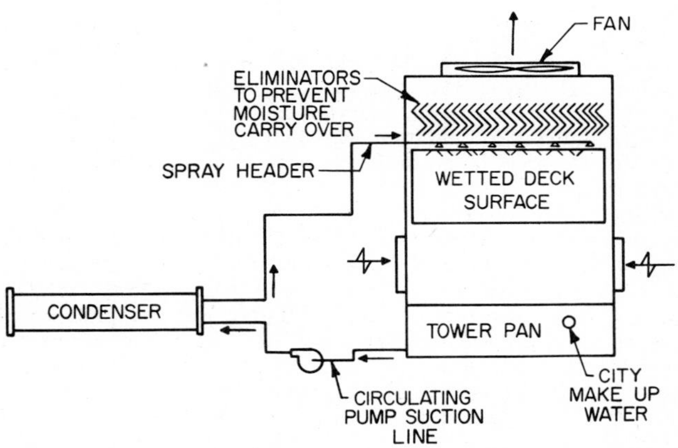

Fig. 19-30. Water-tower connection to condenser of an air-conditioning system.

Because there are many days when outside temperatures are less than design conditions, the temperature of the air supplied to the coil will fluctuate, and the temperature of the air leaving the coil may drop a few degrees. This will result in sweating ducts. It is good practice, therefore, to design the discharge-air temperature for about 3°F higher than the room dew point.

Thus, for 80°F DB, 50% RH, and dew point at 59°F, the minimum discharge air temperature would be 62°F as insurance against sweating. The amount of air to be handled may be obtained from Eq. (19-8), with a temperature difference of 18°F.

19-37. Condensers. If a water-cooled condenser is employed to remove heat from the refrigerant, it may be serviced with city water, and the warm water discharged to a sewer. Or a water tower (Fig. 19-30) may be used to cool the condenser discharge water, which can then be recirculated back to the condenser.

Where practical, the water condenser and tower can be replaced by an evaporative condenser as in Fig. 19-31.

The capacity of water savers, such as towers or evaporative condensers, depends on the wet-bulb temperature. The capacity of these units decreases as the wet-bulb temperature increases.

Such equipment should be sized for a wet-bulb temperature a few degrees above that used for sizing air-conditioning equipment.

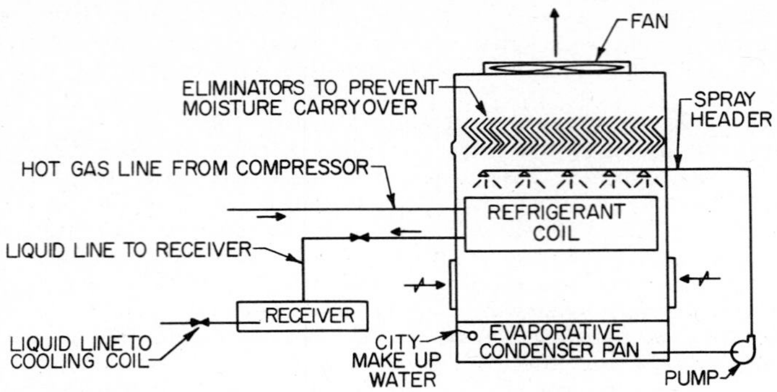

Fig. 19-31. Evaporative condenser for an air-conditioning system.

As an example, consider an area where the design wet-bulb temperature is 75°F. If we size the air-conditioning equipment for this condition, we shall be able to maintain design inside conditions when the outside conditions happen to be 75°F WB. There will be a few days a year, however, when the outside WB may even attain 79 or 80°F WB. During the higher wet-bulb days, with the air-conditioning equipment in operation, we shall balance out at a relative humidity above design. For example, if the design relative humidity is 50%, we may balance out at 55% or higher.

However, if the water towers and evaporative condensers are designed for 75°F WB, then at a wet-bulb temperature above 75, the water-saver capacity would be decreased and the compressor head pressure would build up too high, overload the motor, and kick out on the overload relays. So we use 78°F WB for the design of water savers. Now, on a 78°F WB day, the compressor head pressure would be at design pressure, the equipment will be in operation, although maintaining room conditions a little less comfortable than desired. At 80°F WB, the compressor head pressure will build up above design pressure and the motor will be drawing more than the design current but usually less than the overload rating of the safety contact heaters.

Water condensers are sized for 104°F refrigerant temperature and 95°F water leaving. When condensers are used with city water, these units should be sized for operation with the water at its warmest condition.

The amount of water in gallons per minute required for condensers is equal to

$$Q = \frac{\text{tons of cooling} \times 30}{\text{water-temperature rise}} \tag{19-11}$$

This equation includes the heat of condensation of the compressor when used for air-conditioning service. In many cities, the water supplied is never above 75°F. Thus 75°F is used as a condenser design condition.

When high buildings have roof tanks exposed to the sun, it will be necessary to determine the maximum possible water temperature and order a condenser sized for the correct water flow and inlet-water temperature.

When choosing a condenser for water-tower use, we must determine the water temperature available from the cooling tower with 95°F water to the tower. Check the tower manufacturer for the capacity required at the design WB, with 95°F water to the sprays, and the appropriate wet-bulb approach. (Wet-bulb approach is equal to the number of degrees the temperature of the water leaving the tower is above the wet-bulb temperature. This should be for an economic arrangement about 7°F.) Thus, for 78°F, the water leaving the tower would be 85°F, and the condenser would be designed for a 10°F water-temperature rise; i.e., for water at 95°F to the tower and 85°F leaving the tower.

Evaporative condensers should be picked for the required capacity at a design wet-bulb temperature for the area in which they are to be installed. Manufacturers ratings should be checked before the equipment is ordered.

For small- and medium-size cooling systems, air-cooled condensers are available for outdoor installation with propeller fans or for indoor installation usually with centrifugal blowers for forcing outdoor air through ducts to and from the condensers.

19-38. Compressor-motor Units. Compressor capacity decreases with increase in head pressure (or temperature) and fall in suction pressure (or temperature). Therefore, in choosing a compressor-motor unit, one must first determine design conditions for which this unit is to operate; for example:

Suction temperature—whether at 35°F, 40°F, 45°F, etc.

Head pressure—whether serviced by a water-cooled or air-cooled condenser. Manufacturers of compressor-motor units rate their equipment for Btu per hour output and kilowatt and amperage draw at various conditions of suction and discharge pressure. These data are available from manufacturers, and should be checked before ordering a compressor-motor unit.

Although capacity of a compressor decreases with the suction pressure (or back pressure), it is usual practice to rate compressors in tons or Btu per hour at various suction temperatures.

For installations in which the latent load is high, such as restaurants, bars, and dance halls, where a large number of persons congregate, the coil temperature will have to be brought low enough to condense out large amounts of moisture. Thus, we find that a suction temperature of about 40°F will be required. In offices, homes, etc., where the latent load is low, 45°F suction will be satisfactory. Therefore, choose a compressor with capacity not less than the total cooling load and rated at 104°F head temperature, and a suction temperature between 40 and 45°F, depending on the nature of the load.

Obtain the brake horsepower of the compressor at these conditions, and make sure that a motor is provided with horsepower not less than that required by the compressor.

A standard NEMA motor can be loaded about 15 to 20% above normal. Do not depend on this safety factor, for it will come in handy during initial pull-down periods, excess occupancy, periods of low-voltage conditions, etc.

Compressor manufacturers have all the above data available for the asking so that one need not guess.

Sealed compressors used in most air-conditioning systems consist of a compressor and a motor coupled together in a single housing. These units are assembled, dehydrated, and sealed at the factory. Purchasers cannot change the capacity of the compressor by changing its speed or connect to it motors with different horsepower. It therefore is necessary to specify required tonnage, suction temperature, and discharge temperature so that the manufacturer can supply the correct, balanced, sealed compressor-motor unit.

19-39. Air-handling Units. For an economical installation, the air-handling units should be chosen to handle the least amount of air without danger of sweating ducts and grills. In most comfort cooling, an 18°F discharge temperature below room temperature will be satisfactory.

If a fresh-air duct is used, the required fresh air is mixed with the return air, and the mix is sent through the cooling coil; i.e., the fresh-air load is taken care of in the coil, and discharge air must take care of the internal load only. Then, from Eq. (19-8), the amount of air to be handled is

$$Q = \frac{q_s}{1.08\,\Delta T}$$

where q_s = internal sensible load

ΔT = temperature difference between room and air leaving coil (usually 18°F)

See Art. 19-36.

If a fresh-air duct is not installed, and the outside air is allowed to infiltrate into the premises, we use Eq. (19-8) with the sensible part of the load of the infiltration air added to q_s, because the outside air infiltrating becomes part of the internal load.

Once the amount of air to be handled is determined, choose a coil of face area such that the coil-face velocity V would be not much more than 500 cfm. Some coil manufacturers recommend cooling-coil-face velocities as high as 700 cfm. But there is danger of moisture from the cooling coil being carried along the air stream at such high velocities.

$$V = \frac{Q}{A_c} \qquad (19\text{-}12)$$

where Q = air flow, cfm
A_c = coil face area, sq ft
V = air velocity, fpm

The number of rows of coils can be determined by getting the manufacturer's capacity ratings of the coils for three, four, five, six, etc., rows deep and choosing a coil that can handle not less than the sensible and latent load at the working suction temperature.

Multizone air-handling units (Fig. 19-32) are used to control the temperature of more than one space without use of a separate air-handling unit for each zone.

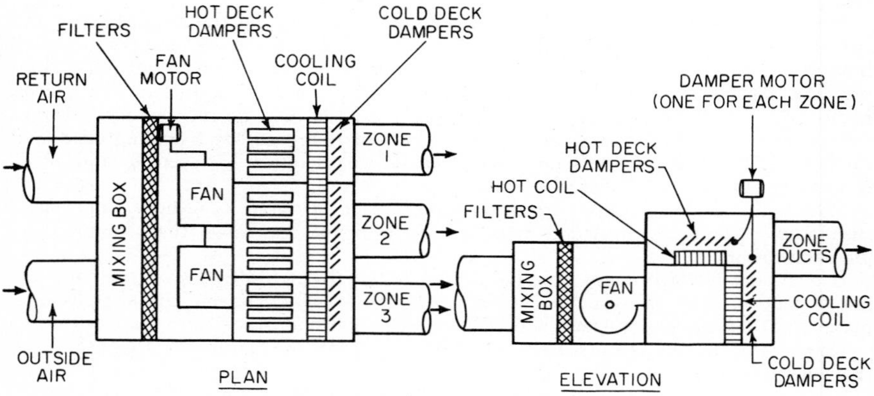

Fig. 19-32. Multizone air-handling unit.

When a zone thermostat calls for cooling, the damper motor for that zone opens the cold deck dampers and closes or throttles the warm deck dampers. Thus, the same unit can provide cooling for one zone while it can provide heating for another zone at the same time.

19-40. Filters for Air-conditioning Units. All air-handling units must be provided with filter boxes. Removal of dust from the conditioned air not only lowers building maintenance costs and creates a healthier atmosphere but prevents the cooling and heating coils from becoming blocked up.

Air filters come in a number of standard sizes and thicknesses. The filter area should be such that the air velocity across the filters does not exceed 350 fpm for low-velocity filters nor 550 fpm for high-velocity filters. Thus, the minimum filter area in square feet to be provided equals the air flow, in cubic feet per minute, divided by the maximum air velocity across the filters, in feet per minute.

Most air filters are of either the throwaway or cleanable type. Both these types will fit a standard filter rack.

Electrostatic filters are usually employed in industrial installations, where a higher percentage of dust removal must be obtained. Check with manufacturers' ratings for particle-size removal, capacity, and static-pressure loss; also check electric service required. These units generally are used in combination with regular throwaway or cleanable air filters, which take out the large particles, while the charged electrostatic plates remove the smaller ones.

19-41. "Package" Air-conditioning Units. To meet the great demand for cheaper air-conditioning installations, manufacturers produce "package," or preassembled, units. These vary from 4,000 to 30,000 Btu per hr for window units and 9,000 to 600,000 Btu per hr for commercial units. Less field labor is required to install them than for custom-designed installations.

Package window units operate on the complete cycle shown in Fig. 19-28, but the equipment is very compactly arranged. The condenser is air-cooled and projects outside the window. The cooling coil extends inside. Both the cooling-coil fan and condenser fan usually are run by the same motor.

These package units require no piping, just an electric receptacle of adequate capacity. Moisture that condenses on the cooling coil runs via gutters to a small sump near the condenser fan. Many manufacturers incorporate a disk slinger to spray this water on the hot condenser coil, which vaporizes it and exhausts it to the outside. This arrangement serves a double purpose:

1. Gets rid of humidity from the room without piping.
2. Helps keep the head pressure down with some evaporative cooling.

Floor-type and ceiling-type package units also contain the full air-conditioning cycle (Fig. 19-28). The air-cooled package units are usually placed near a window to reduce the duct runs required for the air-cooled condenser.

Roof-type package units are available for one-story structures or top-floor areas such as industrial spaces and supermarkets. These units contain a complete cooling cycle (usually with an air-cooled condenser) and a furnace. All controls are factory prewired, and the refrigeration cycle is completely installed at the factory. Necessary ductwork, wiring, and gas piping are supplied in the field.

A combination package unit including an evaporative condenser is also available.

Most packaged units are standardized to handle about 400 cfm of air per ton of refrigeration. This is a good average air quantity for most installations. For restaurants, bars, etc., where high latent loads are encountered, the unit handles more than enough air; also, the sensible capacity is greater than required. However, the latent capacity may be lower than desired. This will result in a somewhat higher relative humidity in the premises.

For high sensible-load jobs, such as homes, offices, etc., where occupancy is relatively low, air quantities are usually too low for the installed tonnage. Latent capacity, on the other hand, may be higher than desired. Thus, if we have a total load 5 tons—4½ tons sensible and ½ ton latent—and we have available a 5-ton unit with a capacity of 4 tons sensible and 1 ton latent, it is obvious that during extreme weather the unit will not be able to hold the dry-bulb temperature down, but the relative humidity will be well below design value.

Under such conditions, these units may work out satisfactorily. In the cases in which the latent capacity is too low and we have more than enough sensible capacity, we can set the thermostat below design indoor dry-bulb temperature and maintain a lower dry-bulb temperature and higher relative humidity to obtain satisfactory comfort conditions. Where we have insufficient sensible capacity, we have to leave the thermostat at the design setting. This will automatically yield a higher dry-bulb temperature and lower relative humidity—and the premises will be comfortable if the total installed capacity is not less than the total load.

As an example of the considerations involved in selecting package air-conditioning equipment, let us consider the structure in Fig. 19-2 and the load analysis in Table 19-13.

If this were an old building, it would be necessary to do the following:

First-floor store (load 6.45 tons). Inasmuch as a 5-ton unit is too small, we must choose the next larger size—a 7½-ton package unit. This unit has a greater capacity than needed to maintain design conditions. However, many people would like somewhat lower temperature than 80°F dry bulb, and this unit will be capable of maintaining such conditions. Also, if there are periods when more than 50 occupants will be in the store, extra capacity will be available.

First-floor office (load 0.33 tons). Ordinarily a 4,000-Btu-per-hr window unit would be required to do the job. But because the store unit has spare capacity, it would be advisable to arrange a duct from the store 7½-ton unit to cool the office.

Second-floor offices (No. 1, load 3.40 tons; No. 2, load 1.07 ton). A 5-ton unit is required for the two offices. But it will be necessary to provide a fresh-air connection to eliminate the fresh-air load from the internal load, to reduce air requirements to the rated flow of the unit. Then, the 2,000 cfm rating of the 5-ton package unit will be close enough to the 2,183 cfm required for the two offices.

Inasmuch as the total tonnage is slightly above that required, we will balance out at a slightly lower relative humidity than 50%, if the particular package unit selected is rated at a sensible capacity equal to the total sensible load.

A remote air-conditioning system would be more efficient, because it could be designed to meet the needs of the building more closely. A single air-handling unit for both floors can be arranged with the proper ductwork. However, local ordinances should be checked, since some cities have laws preventing direct-expansion systems servicing more than one floor. A chilled water system (Fig. 19-29) could be used instead.

19-42. Package Chillers. These units consist of a compressor, water chiller, condenser, and all automatic controls. The contractor has to provide the necessary power and condensing water, from either a cooling tower, city water, or some other source such as well or river. The controls are arranged to cycle the refrigeration compressor to maintain a given chilled-water temperature. The contractor need provide only insulated piping to various chilled-water air-handling units.

Package chillers also are available with air-cooled condensers for complete outdoor installation of the unit, or with a condenserless unit for indoor installation connected to an outdoor air-cooled condenser.

The larger packaged chillers, 50 to 1,000 tons, are generally powered by centrifugal compressors. When the units are so large that they have to be shipped knocked down, they are usually assembled by the manufacturer's representative on the job site, sealed, pressure-tested, evacuated, dehydrated, and charged with the proper amount of refrigerant.

19-43. Absorption Units for Cooling. Absorption systems for commercial use are usually arranged as package chillers. In place of electric power, these units use a source of heat to regenerate the refrigerant. Gas, oil, or low-pressure steam is used.

In the absorption system used for air conditioning, the refrigerant is water. The compressor of the basic refrigeration cycle (Art. 19-35) is replaced by an absorber, pump, and generator. A weak solution of lithium bromide is heated to evaporate the refrigerant (water). The resulting strong solution is cooled and pumped to a chamber where it absorbs the cold refrigerant vapor (water vapor) and becomes dilute.

Condensing water required with absorption equipment is more than that required for electric-driven compressors. Consult the manufacturer in each case for the water quantities and temperature required for the proper operation of this equipment.

The automatic-control system with these package absorption units is usually provided by the manufacturer and generally consists of control of the amount of steam, oil, or gas used for regeneration of the refrigerant to maintain the chilled water temperature properly. In some cases, condenser-water flow is varied to control this temperature.

When low-cost steam is available, absorption systems may be more economical to operate than systems with compressors. In general, steam consumption is about 20 lb per hr per ton of refrigeration. Tower cooling water required is about 3.7 gpm per ton with 85°F water entering the absorption unit and 102°F water leaving it.

19-44. Ducts for Air Conditioning. In designing a duct system for air conditioning, we must first determine air-outlet locations. If wall grilles are used, they should be spaced about 10 ft apart to avoid dead spots. Round ceiling outlets should be placed in the center of a zone. Rectangular ceiling outlets are available that blow in either one, two, three, or four directions.

Manufacturers' catalog ratings should be checked for sizing grilles and outlets. These catalogs give the recommended maximum amount of air to be handled by an outlet for the various ceiling heights. They also give grille sizes for various lengths of blows. It is obvious that the farther the blow, the higher must be the velocity of the air leaving the grille. Also the higher the velocity, the higher must be the pressure behind the grille.

When grilles are placed back to back in a duct as in Fig. 19-33, be sure that grille A and grille B have the same throw; for if the pressure in the duct is large enough for the longer blow, the short-blow grille will bounce the air off the opposite wall, causing serious drafts. But if the pressure in the duct is just enough for the short blow, the long-blow grille will never reach the opposite wall. Figure 19-34 is recommended for unequal blows because it allows adjustment of air and build-up of a higher static pressure for the longer blow.

In some modern buildings perforated ceiling panels are used to supply conditioned air to the premises. Supply ductwork is provided in the plenum above the suspended ceiling as with standard ceiling outlets. However, with perforated panels, less acoustical fill is used to match the remainder of the hung ceiling.

After all discharge grilles and the air-handling unit are located, it is advisable to make a single-line drawing of the duct run. The air quantities each line and branch must carry should be noted. Of the few methods of duct-system design in use, the equal-friction method is most practical (Art. 19-9). For most comfort

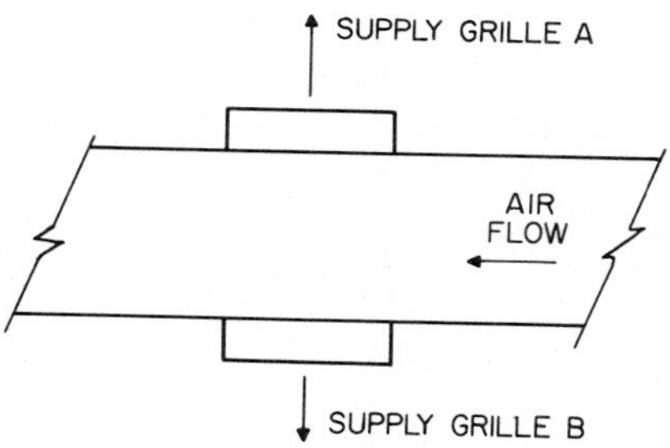

Fig. 19-33. Grilles placed back to back on an air-supply duct with no provisions for adjustment of discharge.

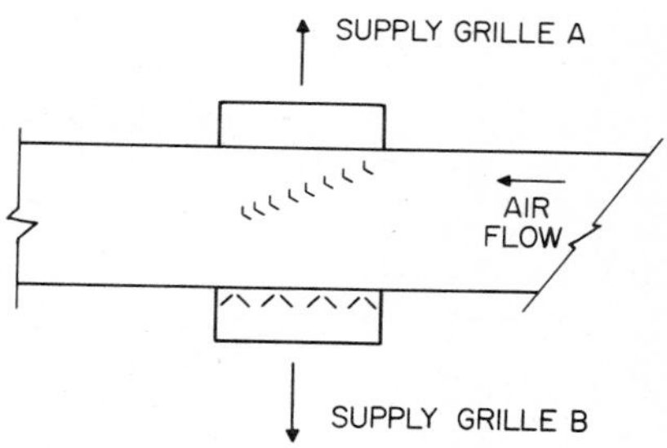

Fig. 19-34. Vanes placed in duct to control air flow to back-to-back grilles, useful when air throws are unequal.

cooling work, it is considered good practice not to exceed 0.15 in. friction per 100 ft of ductwork. It is also well to keep the air below 1,500 fpm starting velocity.

If a fresh-air duct is installed, the return-air duct should be sized for a quantity of air equal to the supply air minus the fresh air.

It is advisable, where physically possible, to size the fresh-air duct for the full capacity of the air-handling unit. For example, a 10-ton system handling 4,000 cfm of supply air—1,000 cfm fresh air and 3,000 cfm return air—should have the fresh-air duct sized for 4,000 cfm of air. A damper in the fresh-air duct will throttle the air to 1,000 cfm during the cooling season; however, during an intermediate season, when the outside air is mild enough, cooling may be obtained by operating only the supply-air fan and opening the damper, thus saving the operation of the 10-ton compressor-motor unit.

As an example of the method for sizing an air-conditioning duct system, let us determine the ductwork for the first floor of the building in Fig. 19-35. Although a load analysis (see Art. 19-34 for an example) shows that the air requirement is 2,676 cfm, we must design the ducts to handle the full capacity of air of the package unit we supply. Handling less air will unbalance the unit, causing a drop in suction temperature, and may cause freezing up of the coil. If a 7½-ton package unit is used, for example, the ducts should have a capacity at least equal to the 3,000 cfm at which this unit is rated.

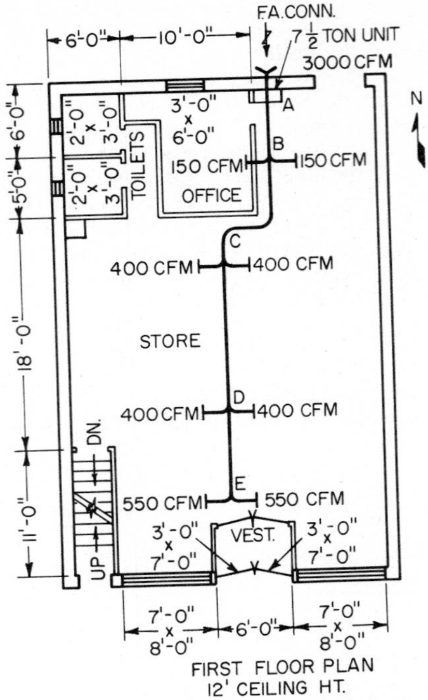

Fig. 19-35. Ductwork for cooling a store.

Table 19-14 shows the steps in sizing the ducts. The 3,000 cfm is apportioned to the various zones in the store in proportion to the load from each, and the flow for each segment of duct is indicated in the second column of the table. Next, the size of an equivalent round duct to handle each air flow is determined from Table 19-6 with friction equal to 0.15 in. per 100 ft. The size of rectangular duct to be used is obtained from Table 19-5.

Table 19-14. Duct Calculation for 7½-ton Package Unit for Store (Fig. 19-35)

Duct	Cfm	Equivalent round duct (friction = 0.15 in. per 100 ft, max velocity = 1,500 fpm), diam, in.	Rectangular duct, in.
A-B	3,000	19.7	28 × 12
B-C	2,700	18.4	24 × 12
C-D	1,900	16.1	18 × 12
D-E	1,100	13.2	12 × 12

The above example of ductwork design and the example in Art. 19-9 fall into the category of low-pressure duct systems. This type of design is used for most air-distribution systems that are not too extensive, such as one- or two-floor systems, offices, and residences. In general, the starting air velocity is below 2,000 fpm, and the fan static pressure is below 3 in. of water.

For large multistory buildings, high-velocity air-distribution duct systems often are used. These systems operate at duct velocities well above 3,000 fpm and above 3-in. static pressure. Obvious advantages include smaller ducts and lower building cost, since smaller plenums are needed above hung ceilings. Disadvantages are high power consumption for fans and need for an air pressure-reducing valve and sound attenuation box for each air outlet, resulting in higher power consumption for operation of the system.

Some of the more elaborate heating and air-conditioning installations consist of a high-pressure warm-air duct system and a high-pressure cold-air duct system. Each air outlet is mounted in a sound attenuation box with pressure-reducing valves and branches from the warm- and cold-air systems (Fig. 19-36). Room temperature is controlled by a thermostat actuating two motorized volume dampers. When cooling is required, the thermostat activates the motor to shift the warm-air damper to the closed or throttled position and the cold-air damper to the open position.

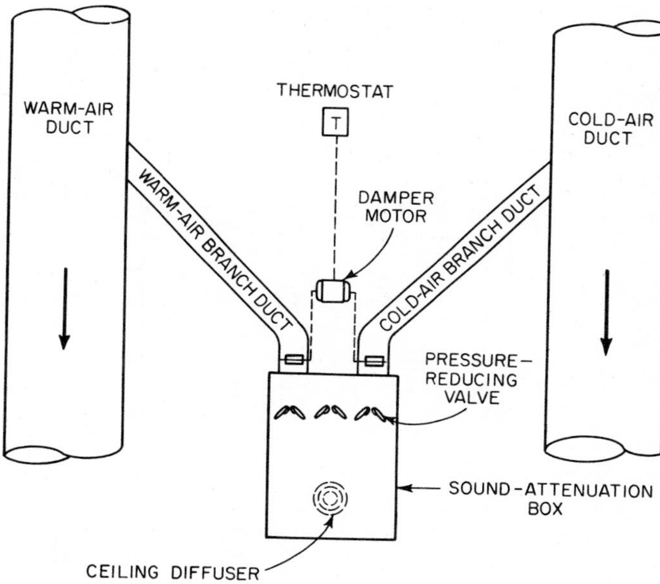

Fig. 19-36. Double-duct air-distribution system.

19-45. Control Systems for Air Conditioning. One commonly used system has a thermostat wired in series with the compressor holding-coil circuit (Fig. 19-37). Thus the compressor will stop and start as called for by room conditions. The high-low pressure switch in series with the thermostat is a safety device that stops the compressor when the head pressure is too high and when the suction pressure approaches the freezing temperature of the coil. A liquid solenoid will shut off the flow of refrigerant when the compressor stops, to prevent flooding the coil back to the compressor during the off cycle. This valve may be eliminated when the air-handling unit and compressor are close together.

A second type of control is the pump-down system (Fig. 19-38). The thermostat shuts off the flow of the refrigerant, but the compressor will keep running. With the refrigerant supply cut off, the back pressure drops after all the liquid in the coil vaporizes. Then, the high-low pressure switch cuts off the compressor.

Either of these two systems is satisfactory. However, the remainder of this discussion will be restricted to the pump-down system.

Additional safety controls are provided on package units to reduce compressor burnouts and increase the average life of these units. The controls include crankcase heaters, motor-winding thermostats, and nonrecycling timers. These controls are usually prewired in the factory. The manufacturer supplies installation and wiring

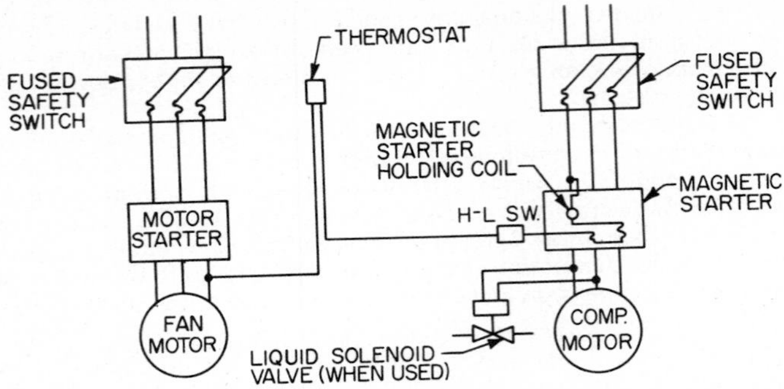

Fig. 19-37. Control circuit for an air-conditioning system in which the thermostat controls compressor operation.

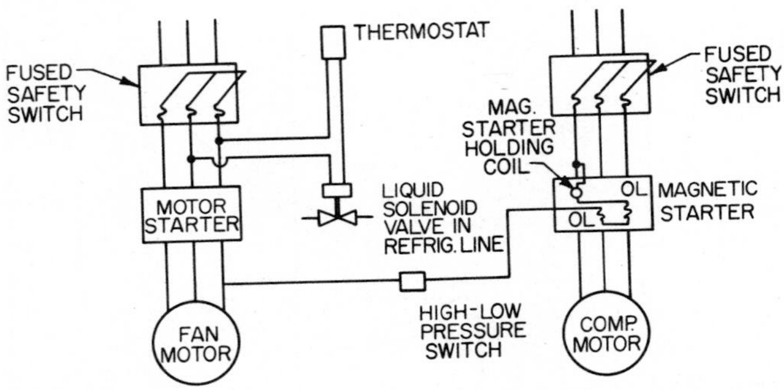

Fig. 19-38. Pump-down system for control of air conditioning.

instructions for interconnecting the various components of the air-conditioning system.

Where city water is used for condensing purposes, an automatic water-regulator valve is supplied (usually by the manufacturer). Figure 19-39 shows a cross

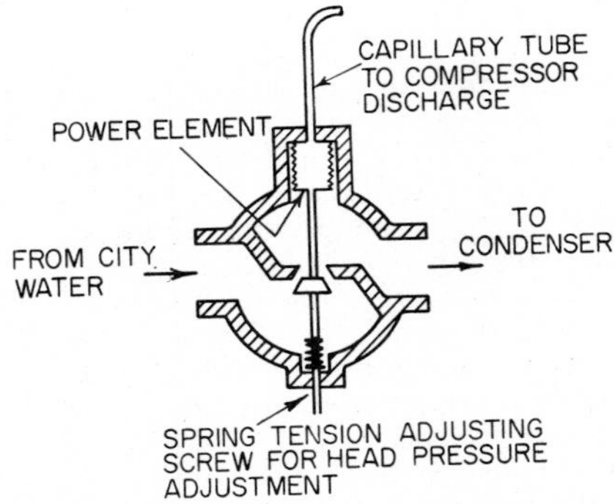

Fig. 19-39. Water-regulating valve for condensing in an air-conditioning system.

section of such a valve. The power element is attached to the hot-gas discharge of the compressor. As the head pressure builds up, the valve is opened more, allowing a greater flow of water to the condenser, thus condensing the refrigerant at a greater rate. When the compressor is shut off, the head pressure drops, the flow of water being cut off by the action of a power spring working in opposition to the power element. As the water temperature varies, the valve responds to the resulting head pressure and adjusts the flow automatically to maintain design head pressure.

When a water tower is used, the automatic water-regulating valve should be removed from the circuit, because it offers too much resistance to the flow of water. It is not usually necessary to regulate the flow of water in a cooling-tower system. For when the outside wet-bulb temperature is so low that the tower yields too low a water temperature, then air conditioning generally is not needed.

Figure 19-40 shows a simplified wiring diagram for a tower system. Because of the interconnecting wiring between the magnetic starters, the tower fan cannot run unless the air-conditioning fan is in operation. Also, the circulating pump cannot run unless the tower fan is in operation, and the compressor cannot run unless the circulating pump is in operation.

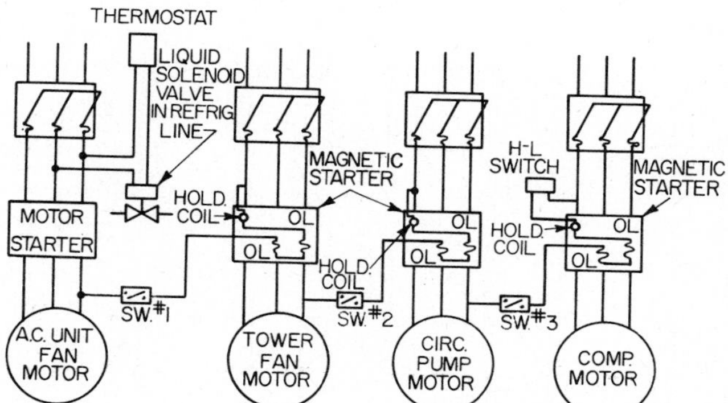

Fig. 19-40. Control circuit for a cooling-tower-type air-conditioning system.

When the system is started, the air-conditioning fan should be operated first, then the tower via switch No. 1, the pump via switch No. 2, and the compressor via switch No. 3. When shutting down, switch No. 3 should be snapped off, then switch No. 2, then switch No. 1, and then the air-conditioning fan.

In buildings where one tower and pump are used to provide water for many units, pressure switches may be used in series with the holding coil of each compressor motor starter. No interconnecting wiring is necessary; for the compressor will not be able to run unless the pump is in operation and provides the necessary pressure to close the contacts on the pressure switch.

Some city ordinances require that indirect-expansion, or chilled-water, systems be installed in public buildings, such as theaters, night clubs, hotels, and depots. This type of system is illustrated in Fig. 19-29. The refrigerant is used to cool water and the water is circulated through the cooling coil to cool the air. The water temperature should be between 40 and 45°F, depending on whether the occupancy is a high latent load or a low latent load. Because the cost of equipment is increased and the capacity decreased as water temperature is lowered, most designers use 45°F water as a basis for design and use a much deeper cooling coil for high-latent-load installations.

The amount of chilled water in gallons per minute to be circulated may be obtained from

$$Q = \frac{24 \times \text{tons of refrigeration}}{\Delta T} \tag{19-13}$$

where ΔT is the temperature rise of water on passing through the cooling coil, usually 8 or 10°F. The smaller the temperature change, the more water to be circulated and the greater the pumping cost, but the better will be the heat transfer through the water chiller and cooling coil.

Coil manufacturers' catalogs usually give the procedure for picking the number of rows of coil necessary to do the required cooling with the water temperatures available.

The water chiller should be sized to cool the required flow of water from the temperature leaving the cooling coil to that required by the coil at a given suction temperature.

Assuming 45°F water supplied to the cooling coil and a temperature rise of 8°F, the water leaving the coil would be at 53°F. The chiller would then be picked to cool the required flow from 53 to 45°F at 37°F suction. The lower the suction temperature, the smaller will be the amount of heat-transfer surface required in the chiller. Also, the lower will be the compressor capacity. In no case should the suction temperature be less than 35°F, since the freezing point of water—32°F—is too close. A frozen and cracked chiller is expensive to replace.

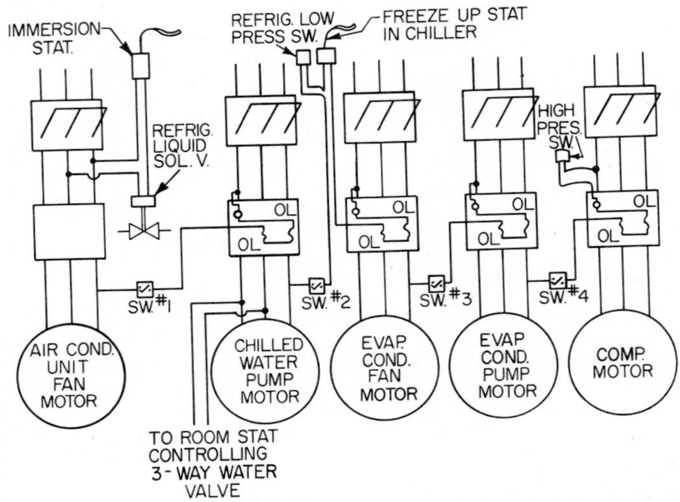

Fig. 19-41. Control circuit for a chilled-water system with evaporative condenser.

A simplified control system for a chilled-water system with an evaporative condenser is shown in Fig. 19-41.

19-46. Heating and Air Conditioning. Most manufacturers of air-conditioning equipment allow space in the air-handling compartment for the installation of a humidification unit and a heating coil for hot water or steam. These make it possible to humidify and heat the air in cold weather (Art. 19-10). Before a decision is reached to heat through the air-conditioning duct system, it is important to consider the many pitfalls present:

When a unit provides air conditioning in a single room, a heating coil may do the job readily, provided the Btu-per-hour rating of the coil is equal to or greater than the heating load. The fact that the supply grilles are high is usually no disadvantage; since winter heating is designed for the same amount of air as summer cooling, a small temperature rise results, and the large air-change capacity of the air-handling unit creates enough mixing to prevent serious stratification. Difficulties usually are encountered, however, in a structure with both exposed and interior zones. In the winter, the exposed zones need heating, while the interior zones are warm. If the heating thermostat is located in the exposed zone, the interior zones will become overheated. If the thermostat is located in the interior zone, the exposed zones will be too cold. Where some heat is generated in the interior zones, the system may require cooling of the interior and heating of the exterior zones at the same time. Thus it is impossible to do a heating and cooling job with a single system in such structures.

Where heating and cooling are to be done by the same duct system, the air-handling equipment should be arranged to service individual zones—one or more units for the exposed zones, and one or more units for the interior zones.

When a heating system is already present and an air-conditioning system is added, a heating coil may be used to temper the outside air to room temperature. A duct-type thermostat may be placed in the discharge of the unit to control the steam or hot-water valve. When a room thermostat is used, the spare coil capacity may be used for quick morning pickup. Later in the day, the system may be used for cooling the premises with outside air if the building-heating system or other internal heat sources overheat the premises. In buildings with large window areas, it is advisable to place under the windows for use in cold weather some radiation to supply heat in addition to that from the heating coil, to counteract the down draft from the cold glass surfaces.

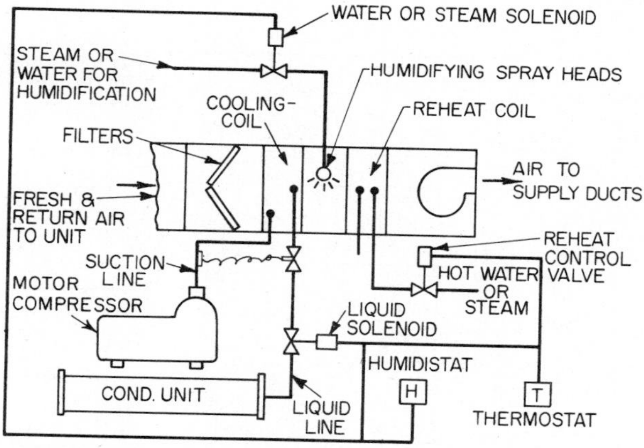

Fig. 19-42. Circuit for close control of temperature and humidity.

The heating-coil size should be such that its face area is about equal to the cooling-coil face area. The number of rows deep should be checked with manufacturers' ratings. When the unit is to be used for tempering only, the coil need be sized for the fresh-air load only [Eq. (19-8)].

When large quantities of outside air are used, it is usual practice to install a preheat coil in the fresh-air duct and a reheat coil in the air-conditioning unit. Install the necessary filters before the preheat coil to prevent clogging.

19-47. Industrial Air Conditioning. Certain manufacturing processes call for close control of temperature and relative humidity. A typical control system is shown in Fig. 19-42. The humidistat may be of the single-pole double-throw type.

On a rise in room humidity, the refrigeration compressor cuts in for the purpose of dehumidification. Since the sensible capacity of the cooling coil usually is much higher than the sensible load, whereas the latent capacity does not differ too much from the latent load, undercooling results. The room thermostat will then send enough steam or hot water to the reheat coil to maintain proper room temperature.

A small drop in room relative humidity will cut off the compressor. A further drop in room relative humidity will energize the water or steam solenoid valve to add moisture to the air. The thermostat is also arranged to cut off all steam from the reheat coil when the room temperature reaches the thermostat setting. However, there may be periods when the humidistat will not call for the operation of the refrigeration compressor and the room temperature will climb above the thermostat setting. An arrangement is provided for the thermostat to cut in the compressor to counteract the increase in room temperature.

When calculating the load for an industrial system, designers should allow for the thermal capacity and moisture content of the manufactured product entering and leaving the room. Often this part of the load is a considerable percentage of the total.

19-48. Chemical Cooling. When the dew point of the conditions to be maintained is 45°F or less, the method of controlling temperature and humidity described in Art. 19-47 cannot be used because the coil suction temperature must be below freezing and the coil will freeze up, preventing flow of air. For room conditions with dew point below 45°F, chemical methods of moisture removal, such as silica gel or lithium chloride, may be used. This equipment is arranged to have the air pass through part of the chemical and thus give up its moisture, while another part of the system regenerates the chemical by driving off the moisture previously absorbed. The psychrometric process involved in this type of moisture removal is adiabatic or, for practical purposes, may be considered as being at constant wet-bulb temperature. For example, if we take room air at 70°F and 30% relative

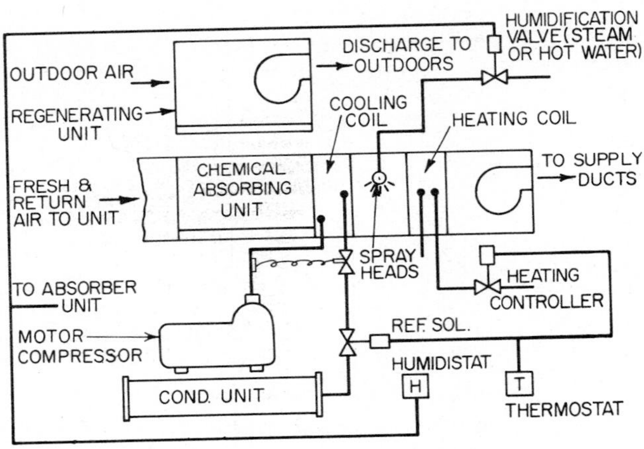

Fig. 19-43. Controls for cooling with a chemical moisture absorber.

humidity and pass the air through a chemical moisture absorber, the air leaving the absorber will be at a wet-bulb temperature of 53°F; i.e., the conditions leaving may be 75°F and 19% RH or 80°F and 11% RH, etc.

It may be noticed that the latent load here is converted to sensible load, and a refrigeration system will be necessary to do sensible cooling. See Fig. 19-43 for the control of such a system.

19-49. Year-round Air Conditioning. Electronic-computor rooms and certain manufacturing processes require temperature and humidity to be closely controlled throughout the year. Air-conditioning systems for this purpose often need additional controls for the condensing equipment.

When city water is used for condensing and the condensers are equipped with a water regulating valve, automatic throttling of the water by the valve automatically compensates for low water temperatures in winter. When cooling towers or evaporative condensers are used, low wet-bulb temperatures of outside air in winter will result in too low a head pressure for proper unit operation. A controller may be used to recirculate some of the humid discharge air back to the evaporative-condenser suction to maintain proper head pressure. With cooling towers, the water temperature is automatically controlled by varying the amount of air across the towers, or by recirculating some of the return condenser water to the supply. Figure 19-44 shows the piping arrangement.

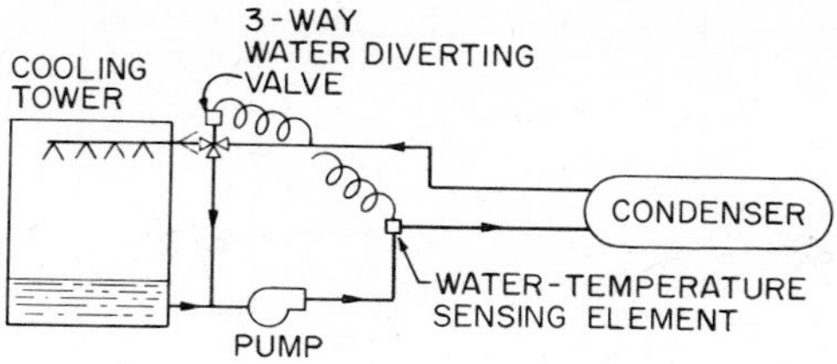

Fig. 19-44. Piping layout for control of temperature of water from a cooling tower.

When air-cooled condensers are used in cold weather, the head pressure may be controlled in one of the following ways: for a single-fan condenser, by varying the fan speed; for a multifan condenser, by cycling the fans or by a combination of cycling and speed variation of fans for closer head-pressure control.

Some head-pressure control systems are arranged to flood refrigerant back to the condenser coil when the head pressure drops. Thus, the amount of heat-transfer surface is reduced, and the head pressure is prevented from dropping too low.

19-50. Heat Pumps. A heat-pump cycle is a sequence of operations in which the heat of condensation of a refrigerant is used for heating. The heat required to vaporize the refrigerant is taken from ambient air at the stage where the normal refrigeration cycle (Art. 19-35) usually rejects the heat. In the summer cycle, for cooling, the liquid refrigerant is arranged to flow to the cooling coil through an expansion valve, and the hot gas from the compressor is condensed as in the standard refrigeration cycle (Fig. 19-28). During the heating season, the refrigerant gas is directed to the indoor (heating) coil by the use of multiport electric valves. The condensed liquid refrigerant is then directed to the "condenser" via an expansion valve and is evaporated. This method of heating is competitive with fuel-burning systems in warmer climates where the cooling plant can provide enough heat capacity during the winter season and where electric rates are low. In colder latitudes, the cooling plant, when used as a heat pump, is not large enough to maintain design indoor temperatures and is therefore not competitive with fuel-burning plants.

Other heat sources for heat pumps are well water or underground grid coils. These installations usually call for a valve system permitting warm condenser water to be piped to the air-handling unit for winter heating and the cold water from the chiller to be pumped to the air-handling unit for summer cooling. In winter, the well water or the water in the ground coil is pumped through the chiller to evaporate the liquid refrigerant, while in the summer this water is pumped through the condenser. Obviously this system is more efficient than air-to-air but must be tailor-made for each installation and is noncompetitive except for large buildings.

19-51. Ventilation. There are many codes and rules governing minimum standards of ventilation. All gravity or natural-ventilation requirements involving window areas in a room as a given percentage of the floor area or volume are at best approximations. The amount of air movement or replacement by gravity depends on prevailing winds, temperature difference between interior and exterior, height of structure, window-crack area, etc. For controlled ventilation, a mechanical method of air change is recommended.

Where people are working, the amount of ventilation air required will vary from one air change per hour where no source of heat or offensive odors are generated to about 60 air changes per hour.

At best, a ventilation system is a dilution process, by which the rate of odor or heat removal is equal to that generated in the premises. Occupied areas below grade or in windowless structures require mechanical ventilation to give occupants a feeling of outdoor freshness. Without outside air, a stale or musty odor may result. The amount of fresh air to be brought in depends on the number of persons occupying the premises, type of activity, volume of the premises, and amount of heat, moisture, and odor generation. Table 19-15 gives the recommended minimum amount of ventilation air required for various activities.

Table 19-15. Minimum Ventilation Air for Various Activities

Type of Occupancy	*Ventilation Air, Cfm per Person*
Inactive, theaters	5
Light activity, offices	10
Light activity with some odor generation, restaurant	15
Light activity with moderate odor generation, bars	20
Active work, shipping rooms	30
Very active work, gymnasiums	50

The amount of air to be handled, obtained from the estimate of the per person method, should be checked against the volume of the premises and the number of air changes per hour given in Eq. (19-14).

$$\text{Number of air changes per hr} = \frac{60Q}{V} \qquad (19\text{-}14)$$

where Q = air supplied, cfm
V = volume of ventilated space, cu ft

When the number of changes per hour is too low (below one air change per hour), the ventilation system will take too long to create a noticeable effect when first put into operation. Five changes per hour are generally considered a practical minimum. Air changes above 60 per hour usually will create some discomfort because of air velocities that are too high.

Toilet ventilation and locker-room ventilation are usually covered by local codes—50 cfm per water closet and urinal is the usual minimum for toilets and six changes per hour minimum for both toilets and locker rooms.

Removal of heat by ventilation is best carried out by locating the exhaust outlets as close as possible to the heat source. Where concentrated sources of heat are present, canopy hoods will remove the heat more efficiently.

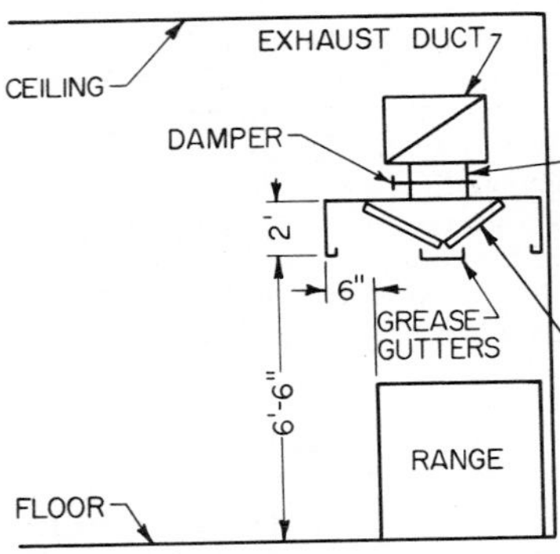

Fig. 19-45. Canopy hood for exhausting heat from a kitchen range.

Figure 19-45 shows a canopy-hood installation over a kitchen range. Grease filters reduce the frequency of required cleaning. When no grease is vaporized, they may be eliminated.

Greasy ducts are serious fire hazards and should be cleaned periodically. There are on the market a number of automatic fire-control systems for greasy ducts. These systems usually consist of fusible-link fire dampers and a means of flame smothering—CO_2, steam, foam, etc.

Figure 19-46 shows a double hood. This type collects heat more efficiently; i.e., less exhaust air is required to collect a given amount of heat. The crack area is arranged to yield a velocity of about 1,000 fpm.

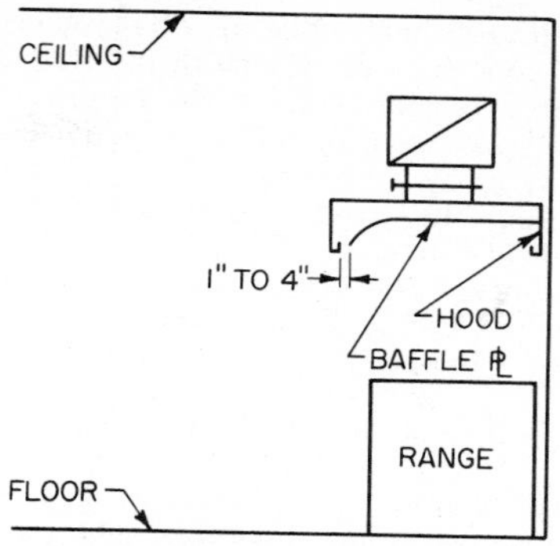

Fig. 19-46. Double hood for exhausting heat.

A curtain of high-velocity air around the periphery of the hood catches the hot air issuing from the range or heat source. Canopy hoods are designed to handle about 50 to 125 cfm of exhaust air per square foot of hood. The total amount of ventilation air should not yield more than 60 changes per hour in the space.

Where hoods are not practical to install and heat will be discharged into the room, the amount of ventilation air may be determined by the following method:

Determine the total amount of sensible heat generated in the premises—lights, people, electrical equipment, etc. (Arts. 19-30 to 19-32). This heat will cause a temperature rise and an increase in heat loss through walls, windows, etc. To maintain desired temperature conditions, ventilation air will have to be used to remove heat not lost by transmission through enclosures (Arts. 19-2 to 19-5 and 19-28).

$$q_v = 1.08Q(T_i - T_o) \qquad (19\text{-}15)$$

where q_v = heat, Btu per hr, carried away by ventilation air
Q = flow of ventilation air, cfm
T_i = indoor temperature to be maintained
T_o = outdoor temperature

With Eq. (19-15), we can calculate the amount of ventilation air required by assuming a difference between room and outdoor temperatures or we can calculate this temperature gradient for a given amount of ventilation air.

The same method may be used to calculate the air quantity required to remove any objectionable chemical generated. For example, assume that after study of a process we determine that a chemical will be evolved in vapor or gas form at the rate of X lb per min. If Y is the allowable concentration in pounds per cubic foot, then $Q = X/Y$, where Q is the ventilation air needed in cubic feet per minute.

Where moisture is the objectionable vapor, the same equation holds, but with X as the pounds per minute of moisture vaporized, Y the allowable concentration of moisture in pounds per cubic foot above outdoor moisture concentration.

Once the amount of ventilation air is determined, a duct system may be designed to handle it, if necessary.

Ventilation air may be provided by installing either an exhaust system, a supply system, or both.

In occupied areas where no unusual amounts of heat or odors are generated, such as offices and shipping rooms, a supply-air system may be provided, with grilles or ceiling outlets located for good distribution. When the building is tight, a relief system of grilles or ducts to the outside should be provided. But when the relief system is too extensive, an exhaust fan should be installed for a combination supply and exhaust system.

In spaces where concentrated heat, odor, or other objectionable vapors are generated, it is more practical to design an exhaust distribution system with hoods. The closer the hood to the source, the more efficient the system.

All air exhausted from a space must be replaced by outside air either by infiltration through doors and windows or by a fresh-air make-up system. Make-up air systems that have to operate during the winter season are often equipped with heating coils to temper the cold outside air.

19-52. Integrated Systems. The proliferation of mechanical installations in new structures has resulted in the use of hitherto dead space for wireways, pipe runs, air supply or air return, etc. Raised floor space (as in computer rooms), hung-ceiling space, hollow structural columns (round, rectangular, or square tubing) may be used. The basic principles in the design of air requirements for the various spaces do not change.

Once the air quantities have been determined, it is then necessary to check the areas available to move these air quantities and the required horsepower for the job. Utilizing otherwise unused spaces will require reference to basic design manuals for relationship of velocity, friction factors, etc., and should be left to a specialist.

Section 20

Fire Protection

HUGH B. KIRKMAN

Consulting Engineer,
Charleston, S.C.

There are two distinct aspects of fire protection: life safety and property protection. Although providing for one aspect generally results in some protection for the other, the two goals are not mutually inclusive. A program that provides for prompt notification and evacuation of occupants meets the objectives for life safety, but provides no protection for property. Conversely, it is possible that adequate property protection might not be sufficient for protection of life.

Absolute safety from fire is not attainable. It is not possible to eliminate all combustible materials or all potential ignition sources. Thus, in most cases, an adequate fire protection plan must assume that unwanted fires will occur despite the best efforts to prevent them. Means must be provided to minimize the losses caused by the fires that do occur.

The first obligation of designers is to meet legal requirements while providing the facilities required by the client. In particular, the requirements of the applicable building code must be met. The building code will contain fire safety requirements, or it will specify some recognized standard by reference. Many owners will also require that their own insurance carrier be consulted—to obtain the most favorable insurance rate, if for no other reason.

20-1. Fire-protection Standards. The standards most widely adopted are those published by the National Fire Protection Association (NFPA), 60 Batterymarch St., Boston, Mass. 02110. The NFPA "National Fire Codes" comprise 10 volumes containing more than 200 standards, updated annually. (These are also available separately.) The standards are supplemented by the NFPA "Fire Protection Handbook," which contains comprehensive and detailed discussion of fire problems and much valuable statistical and engineering data.

Underwriters Laboratories, Inc. (UL), 207 E. Ohio St., Chicago, Ill. 60611, publishes testing laboratory approvals of devices and systems in its "Fire Protection

Equipment List," updated annually and by bimonthly supplements. The publication outlines the tests that devices and systems must pass to be listed. The UL "Building Materials List" describes and lists building materials, ceiling-floor assemblies, wall and partition assemblies, beam and column protection, interior finish materials, and other pertinent data. UL also publishes lists of "Accident Equipment," "Electrical Equipment," "Electrical Construction Materials," "Hazardous Location Equipment," "Gas and Oil Equipment," and others.

Separate standards for application to properties insured by the Factory Mutual System are published by the Factory Mutual Engineering Corporation (FM), Norwood, Mass. 02062. The "Handbook of Industrial Loss Prevention," prepared by the FM staff and published by McGraw-Hill Book Company, New York, treats many industrial fire problems in detail. FM also publishes a list of devices and systems it has tested and approved.

The General Services Administration, acting for the Federal government, has developed many requirements that must be considered, if applicable. Also, the Federal government encourages cities to adopt some uniform code.

The Federal Occupational Safety and Health Act of 1970 (OSHA) sets standards for protecting the health and safety of nearly all employees. It is not necessary that a business be engaged in interstate commerce for the law to apply. OSHA defines **employer** as "a person engaged in a business affecting commerce who has employees, but does not include the United States or any State or political subdivision of a State."

An employer is required to "furnish to each of his employees employment and a place of employment which are free from recognized hazards that are causing or are likely to cause death or serious physical harm to his employees." He is also required to "comply with occupational safety and health standards promulgated under the Act." Regulation and enforcement are the responsibility of the Department of Labor.

There are also other OSHA requirements, such as record keeping. Every employer must permit Labor Department investigators to enter his establishment "without delay and at reasonable times."

Part 1910 of the Act contains all the standards that have been adopted. Subpart E, "Means of Egress," adopts NFPA 101, "Life Safety Code" as the standard. Subpart L, "Fire Protection," adopts the appropriate NFPA standard, where any is given.

Many states have codes for safety to life in commercial and industrial buildings, administered by the Department of Labor, the State Fire Marshal's Office, the State Education Department, or the Health Department. Some of these requirements are drastic and must always be considered.

Obtaining optimum protection for life and property can require consultation with the owner's insurance carrier, municipal officials, and the fire department. If the situation is complicated enough, it can require consultation with a specialist in all phases of fire protection and prevention. In theory, municipal building codes are designed for life safety and for protection of the public, whereas insurance-oriented codes (except for NFPA 101, "Life Safety Code") are designed to minimize property fire loss. Since about 70% of any building code is concerned with fire protection, there are many circumstances that can best be resolved by a fire protection consultant.

20-2. Fire Loads and Resistance Ratings. The nature and potential magnitude of fire in a building are directly related to the amount and physical arrangement of combustibles present, as contents of the building or as materials used in its construction. Because of this, all codes classify buildings by *occupancy* and *construction*, because these features are related to the amount of combustibles.

The total amount of combustibles is called the *fire load* of the building. Fire load is expressed in pounds per square foot (psf) of floor area, with an assumed calorific value of 7,000 to 8,000 Btu per lb. (This Btu content applies to organic materials similar to wood and paper. Where other materials are present in large proportion, the weights must be adjusted accordingly. For example, for petroleum products, fats, waxes, alcohol, and similar materials, the weights are taken at twice their actual weights, because of the Btu content.)

National Bureau of Standards burnout tests presented in NBS Report BMS92 indicate a relation between fire load and fire severity as shown in Table 20-1.

Table 20-1. Relation between Weight of Combustibles and Fire Severity*

Average Weight of Combustibles, Psf	*Equivalent Fire Severity, Hr*
5	½
7½	¾
10	1
15	1½
20	2
30	3
40	4½
50	6
60	7½

* Based on National Bureau of Standards Report BMS92, "Classifications of Building Constructions," U.S. Government Printing Office, Washington, D.C. 20402.

The temperatures used in standard fire tests of building components are indicated by the internationally recognized time-temperature curve shown in Fig. 20-1. Fire resistance of construction materials, determined by standard fire tests, is expressed in hours. The UL "Building Materials List" tabulates fire rating for materials and assemblies it has tested.

Every building code specifies required fire-resistance ratings for structural members, exterior walls, fire divisions, fire separations, ceiling-floor assemblies, and any other constructions for which a fire rating is necessary. (Fire protection for structural steel is discussed in Arts. 6-86 to 6-91. Design for fire resistance of open-web steel joist construction is treated in Art. 7-8, of steel roof deck in Art. 7-36, and of cellular floor construction in Art. 7-43. Design for fire safety with wood construction is covered in Art. 8-20.)

Building codes also specify the ratings required for interior finish of walls, ceilings, and floors. These are classified as to flame spread, fuel contributed, and smoke developed, determined in standard tests performed according to ASTM E84 or ASTM E119. Ratings of specific products are given in the UL "Building Materials List."

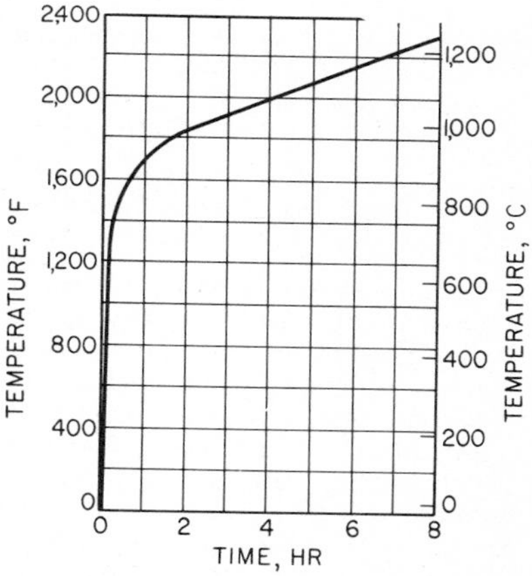

Fig. 20-1. Time-temperature curve for standard fire test.

20-3. Fire-resistance Classification of Buildings. Although building codes classify buildings by occupancy and construction, there is no universal standard for number of classes of either occupancy or construction. The National Building Code (NBC), sponsored by the American Insurance Association (formerly National Board of Fire Underwriters), lists nine occupancy classes and seven construction classes. The Building Officials and Code Administrators International, Inc. (BOCA) code lists 17 occupancy classes and 10 construction classes. The New York City code lists 10 occupancy classes with a total of 19 subclasses, and 2 construction classes with a total of 10 subclasses. The nine occupancy classes of the National Building Code are tabulated with approximate fire loads in Table 20-2. This table should be used only as a guide. For a specific project refer to the applicable local code. Codes do not relate life-safety hazards to the actual fire load, but deal with them through requirements for exit arrangements, interior finishes, and ventilation.

The seven construction classes of the National Building Code are:

1. Fire-resistive construction—Type A.
2. Fire-resistive construction—Type B.
3. Protected noncombustible construction.
4. Unprotected noncombustible construction.
5. Heavy-timber construction.
6. Ordinary construction.
7. Wood-frame construction.

Table 20-2. Approximate Fire Loads for Various Occupancies*

Occupancy Class	*Typical Average Fire Load Including Floors and Trim, Psf*
Assembly	10.0
Business	12.6
Educational	7.6
High hazard	†
Industrial	25.0
Institutional	5.7
Mercantile	15–20
Residential	8.8
Storage	30.0

* From National Bureau of Standards Report BMS92, "Classifications of Building Constructions," U.S. Government Printing Office, Washington, D.C. 20402.

† Special provisions are made for this class, and hazards are treated for the specific conditions encountered, which might not necessarily be in proportion to the actual fire load.

The required fire resistance varies from 4 hr for exterior bearing walls and interior columns in the Fire Resistive—Type A class to 1 hr for walls and none for columns in the Wood-frame Construction class.

Class of construction affects fire-protection-system design through requirements that combustible structural members as well as contents of buildings be protected.

Height and Area Restrictions. Limitations on heights and floor areas included between fire walls in any story of a building are given in every building code and are directly related to occupancy and construction. From the standpoint of fire protection, these provisions are chiefly concerned with safety to life, and endeavor to insure this through requirements determining minimum number of exits, proper location of exits, and maximum travel distance (hence escape time) necessary to reach a place of refuge. The limitations are also aimed at limiting the size of fires.

Unlimited height and area are permitted for construction in Class 1. Permissible heights and areas are decreased with decrease in fire resistance of construction. Area permitted between fire walls in any story reduces to 6,000 sq ft for a one-story, wood-frame building.

Installation of automatic sprinklers increases permissible heights and areas in all classes except those allowed unlimited heights and areas.

Permissible unlimited heights and areas in fire-resistive buildings considered generally satisfactory in the past may actually not be safe. A series of fires involving loss of life and considerable property damage opened the fire safety of such construction to question. As a result, some cities have made more stringent the building-code regulations applicable to high-rise buildings.

The New York City code prohibits floor areas of unlimited size unless the building is sprinklered. Without automatic sprinklers, floor areas must be subdivided into fire-wall-protected areas of from 7,500 to 15,000 sq ft, and the enclosing fire walls must have 1- or 2-hr fire ratings, depending on occupancy and construction.

Smoke Elimination. Because smoke inhalation has been the cause of nearly all fatalities in high-rise buildings, New York City building regulations require that a smoke venting system be installed and made to function independently of the air-conditioning system. Also, smoke detectors must be provided to actuate exhaust fans and at the same time warn the fire department and the building's control center. The control center must have two-way voice communication, selectively, with all floors and be capable of issuing instructions for occupant movement to a place of safety.

Sprinklers and Risers. The New York City code also includes conditions for connecting sprinkler systems to the standpipe riser, eliminating duplicate supplies. Also, automatic sprinklers with a discharge rate smaller than usual are permitted. This combination of use of a single riser both to supply automatic sprinklers and serve as a standpipe, and installation of sprinkler systems with a calculated rate of discharge suitable for office occupancies, can result in substantial savings and improved protection, especially in high-rise buildings. (NFPA 13, "Installation of Sprinkler Systems" and NFPA 14, "Standpipes" permit the combined use when risers are 6 in. in diameter or larger.) Economy results not only because a complete duplicate riser and water supply are eliminated, but also because the largest horizontal pipe size may be as small as 3 in. and most of the horizontal piping need not be larger than 2½ in. in a building 200 × 300 ft.

20-4. Classes of Fires. For convenience in defining effectiveness of extinguishing media, UL has developed a classification that separates combustible materials into four types:

1. *Class A fires* involve ordinary combustibles and are readily extinguishable by water or cooling, or by coating with a suitable chemical powder.
2. *Class B fires* involve flammable liquids where smothering is effective and where a cooling agent must be applied with care.
3. *Class C fires* are those in live electrical equipment where the extinguishing agent must be nonconductive. Since a continuing electrical malfunction will keep the fire source active, circuit protection must operate, after which an electrically conductive agent can be used with safety.
4. *Class D fires* involve metals that burn, such as magnesium, sodium, and powdered aluminum. Special powders are necessary for such fires, as well as special training for operators. These fires should never be attacked by untrained personnel.

AUTOMATIC SPRINKLER AND STANDPIPE SYSTEMS

20-5. Applicability of Sprinklers. The most widely used apparatus for fire protection in buildings is the automatic sprinkler system. In one or more forms, automatic sprinklers are effective protection in all nine of the National Building Code occupancy classes (Art. 20-3). Special treatment and use of additional extinguishing agents, though, may be required in many high-hazard, industrial, and storage occupancies.

Sprinkler systems are suitable for extinguishing all Class A fires and, in many cases, also Class B and C fires. For Class B fires, a sealed (fusible) head system may be used if the flammable liquid is in containers or is not present in large quantity. Sprinklers have a good record for extinguishing fires in garages, for example. An oil-spill fire can be extinguished or contained when the water is applied in the form of spray, as from a sprinkler head. When an oil spill or process-pipe rupture can release flammable liquid under pressure, an open-head (deluge) system may be required to apply a large volume of water quickly and to keep surrounding equipment cool.

For Class C fires, water can be applied to live electrical equipment if it is done in the form of a nonconducting foglike spray. This is usually the most economical way to protect outdoor oil-filled transformers and oil circuit breakers.

20-6. Sprinkler System Design. Fire protection should be based on complete coverage of the building by the sprinkler system. Partial coverage is rarely advisable, because extinguishing capacity is based on detecting and extinguishing fires in their incipiency, and the system must be available at all times in all places. Systems are not designed to cope with fires that have gained headway after starting in unsprinklered areas.

Basically, a sprinkler system consists of a network of piping installed at the ceiling or roof and supplied with water from a suitable source. On the piping at systematic intervals are placed heat-sensitive heads, which discharge water when a predetermined temperature is reached at any head. A gate valve is installed in the main supply, and drains are provided. An alarm can be connected to the system so that local and remote signals can be given when the water flows.

The most widely accepted standard for sprinkler system design is NFPA 13, "Installation of Sprinkler Systems," supplemented when necessary by NFPA 24, "Outside Protection," NFPA 15, "Water Spray Fixed Systems," and NFPA 20, "Centrifugal Fire Pumps," National Fire Protection Association (Art. 20-1). Other Standards will be required for special hazards. These standards must be used with discretion when they are incorporated in specifications, because of the many references to the "authority having jurisdiction." Thus, before these standards are incorporated in a specificaton, rulings must be obtained from the "authority having jurisdiction" over the project at hand. The owner's insurance carrier can furnish the necessary information.

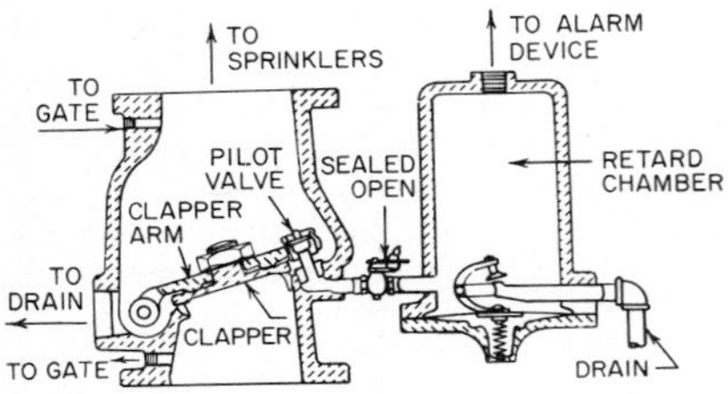

Fig. 20-2. Alarm-check valve.

If the owner's insurance carrier is the Factory Mutual System, the FM standards will apply, and Factory Mutual will be the authority to be consulted (Art. 20-1).

In many cases the local building code will be the governing factor, and an insurance requirement a secondary matter. In such cases, conflicts must be reconciled to avoid a penalty for the owner and ambiguity for the construction bidders.

20-7. Types of Sprinkler Systems. The types of system that should be used depends chiefly on temperature maintained in the building, damageability of contents, expected propagation rate of a fire, and total fire load.

Wet-pipe systems are so called because there is always water in the piping. Consequently, building temperature must be maintained above freezing. Wet-pipe systems are the simplest and most economical type that can be used. Other than a gate valve and an alarm valve, there are no other devices between the water supply and the sprinkler heads. Except for periodic tests of the alarm, no maintenance is required.

An alarm can be provided to indicate when a sprinkler head has started to operate and water is flowing in the piping. This is generally done by installing, in the main supply pipe, an alarm-check valve designed to distinguish between a momentary pressure surge and a steady flow. An alarm-check valve (Fig. 20-2) is essentially a swing check valve with an interior orifice that will admit water to a retard chamber that must fill completely to open a valve to the alarm supply pipe. The retarding chamber can also admit water to a pressure-actuated switch that can transmit a local or remote alarm electrically. In cases in which only a local alarm is considered adequate, water is admitted directly to a water-motor-driven gong (Fig. 20-3), which requires no outside energy.

Where no water-motor gong is needed or required, a vane-type water-flow indicator can be inserted in the supply pipe to give notification electrically of water flow. This device (Fig. 20-4) is equipped with an electric retard that prevents alarms on pressure surges. To avoid damage to the vane on sudden pressure surges, a swing check valve should be installed immediately upstream. The check valve should be drilled and tapped above and below the clapper, for pressure gages.

Pipe sizing for wet-pipe systems, as well as for all other types of systems, is based on the degree of hazard covered. Pipe-sizing schedules are provided in

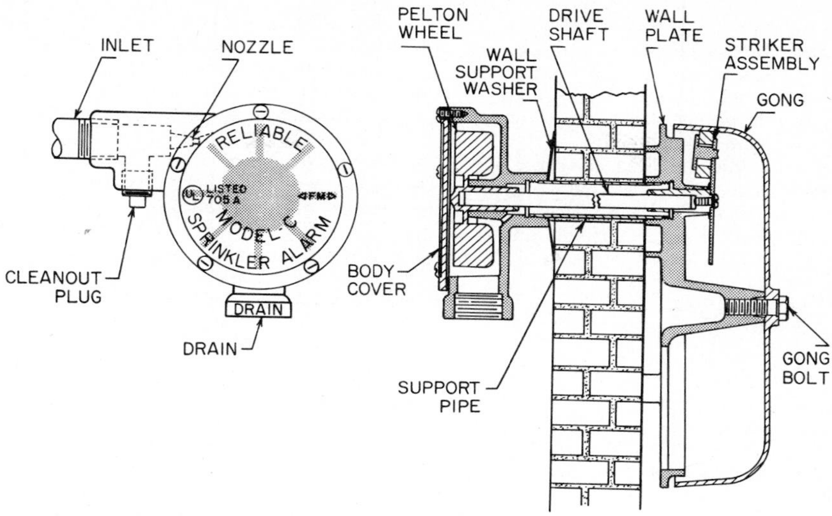

Fig. 20-3. Typical water-motor alarm. (*The Reliable Automatic Sprinkler Co., Inc., Mount Vernon, N.Y.*)

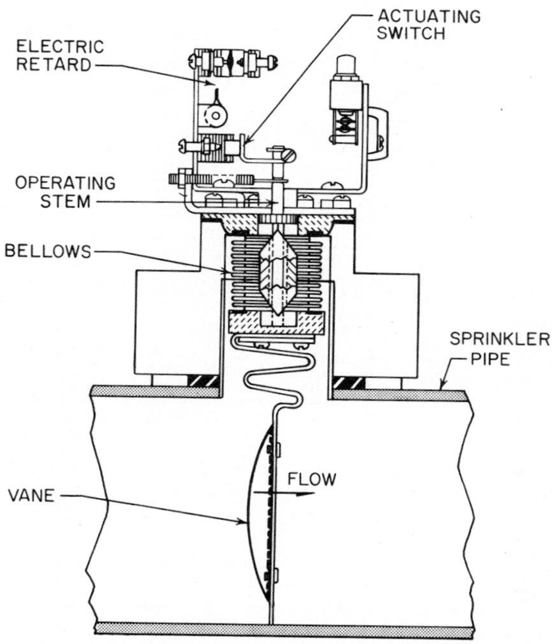

Fig. 20-4. Vane-type water-flow indicator. Flow in sprinkler pipe deflects vane, thus causing the operating stem to tilt and close the actuating switch.

"Installation of Sprinkler Systems," National Fire Protection Association, for light-hazard, ordinary-hazard, and extra-hazard systems. These schedules apply to all systems having sealed heads. A pipe-sizing schedule is also given for deluge systems, but this is only a guide. Determination of the schedule to be used must be made by the insurance or code authority. Frequently, a single building can require the use of all the schedules.

For buildings where the fire load is unusually heavy, or where stock is piled higher than 12 ft, or where deluge systems are used, the authority can require that the schedule be disregarded and the sizing be calculated for a specific rate of discharge over a specific area. In this case, the authority will indicate the total area over which a specified discharge rate must be made. The area indicated will have no relation to the total area of the building or the total number of heads subject to the same fire. For example, in a 25,000-sq ft building the authority might consider that the sprinkler system will bring a fire under control within an area of 2,000 sq ft with a water application rate (density) of 0.35 gpm per sq ft of floor area. This means that the maximum anticipated discharge of the sprinkler system would be 2,000 × 0.35, or 700 gpm. With heads spaced for a coverage of 100 sq ft each, this means that each head would have to discharge 35 gpm. With the discharge rate per head known, and the total area known, the pressure required to deliver this volume of water at the most distant point from the water supply can be calculated. Calculations are made in the manner described in NFPA 13, and NFPA 15, "Water Spray Fixed Systems."

In determining the area and density required, the authority considers the contents, type of building construction, heat venting, and type of system, among other factors. If high piling of stock is involved, the authority takes into account the pile size, pile height, aisle width, and combustibility of contents. Based on full-scale fire tests, a standard, NFPA 231C, describes the protection required for stock piling from 12 to 20 ft, in racks. Storage heights less than 12 ft are covered by NFPA 13.

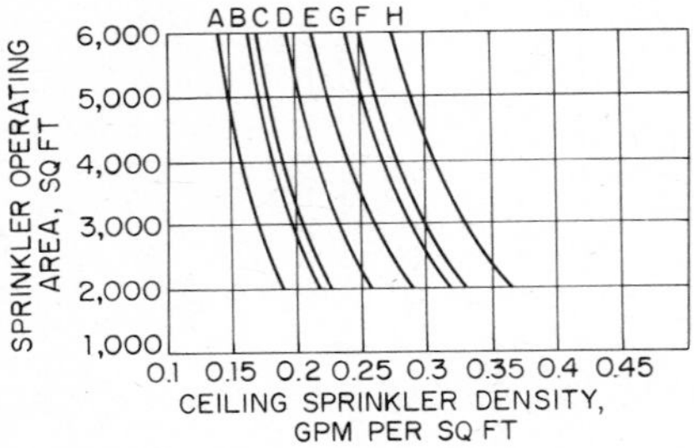

Fig. 20-5. Curves determine sprinkler discharge for design purposes, for buildings containing 20-ft-high rack storage of Class I commodities, in double-row racks, with conventional pallets. Curve application: *A*, 8-ft aisles with 286°F ceiling sprinklers and 165°F in-rack sprinklers; *B*, 8-ft aisles with 165°F ceiling sprinklers and 165°F in-rack sprinklers; *C*, 4-ft aisles with 286°F ceiling sprinklers and 165°F in-rack sprinklers; *D*, 4-ft aisles with 165°F ceiling sprinklers and 165°F in-rack sprinklers; *E*, 8-ft aisles with 286°F ceiling sprinklers; *F*, 8-ft aisles with 165°F ceiling sprinklers; *G*, 4-ft aisles with 286°F ceiling sprinklers; *H*, 4-ft aisles with 165°F ceiling sprinklers. *(Reproduced with permission from NFPA 231C, "Rack Storage of Materials," 1972, copyright National Fire Protection Association, Boston, Mass. 02110.)*

Figure 20-5 presents sprinkler system design curves for 20-ft-high rack storage, with conventional pallets and with Class I commodities (noncombustible products on wood pallets or in ordinary paper cartons on wood pallets, such as noncombustible foodstuffs or beverages).

Dry-pipe systems are used in locations and buildings where it is impractical to maintain sufficient heat to accommodate a wet-pipe system. The system of piping is the same, but it is normally empty, containing only air under pressure. A normally high water pressure is held by a normally low air pressure by use of a *dry-pipe valve.* This valve generally employs a combined air and water clapper (Fig. 20-6) where the area under air pressure is about 16 times the area subject to water pressure. When a sprinkler head operates, air is released to the atmosphere, allowing the water to overcome the pressure differential and enter the piping. Air pressure must be checked periodically, but devices are available to maintain pressure automatically and are equipped with an alarm that goes off in case of a malfunction allowing air pressure to drop too low.

In dry-pipe systems of large capacity, the relatively slow drop in air pressure when a single head or a few heads operate is overcome by use of an accelerator or exhauster. The former is a device, installed near the dry-pipe valve, to sense a small drop in pressure and transmit the system air pressure to a point under the air clapper. The water then can open the clapper quickly and push its way into the system without requiring a further pressure drop. An exhauster accomplishes the opening by sensing a small pressure drop and exhausting the pressurized air to the atmosphere through a large opening.

The dry-pipe valve must be installed in a heated area, or in a heated enclosure, because there is water in the piping up to the valve, and priming water in the valve itself.

Deluge systems are those in which all heads open to discharge water simultaneously. The water is controlled by a *deluge valve*, which is operated by a tempera-

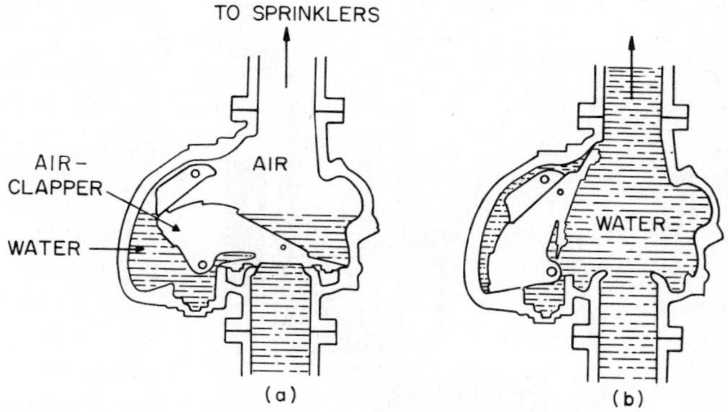

Fig. 20-6. Differential dry-pipe valve. (*a*) Air pressure keeps clapper closed. (*b*) Venting of air permits clapper to open and water to flow.

ture-sensitive detection system installed throughout the same area as the piping and heads. The detecting devices may be electrical, pneumatic, or mechanical, and may be connected to the deluge valve by wiring, tubing, or piping. Some valves are adaptable to all three means of operation.

A deluge system may be used for a few heads for a small, isolated industrial hazard, or in an entire, huge structure, such as an airplane hangar requiring several thousand heads. Deluge systems are also used to advantage in chemical plants where process vessels and tanks contain flammable materials. The purposes of such systems are to isolate a fire and to cool exposed equipment not originally involved in the fire.

Piping in deluge systems must be sized by calculation. Consideration must be given to the probability of simultaneous operation of adjacent systems, if any are involved. In addition, authorities frequently require that provision be made in the water supply system for operation of two or more fire-hose streams while the deluge system is operating. In a multisystem installation, each system must be of manageable size, and good practice requires that no system be larger than one 6-in. deluge valve can supply.

An electrically controlled system may employ fixed-temperature thermostats, rate-of-rise units, smoke detectors, combustion products detectors, or infrared or ultraviolet detectors. The nature and extent of the hazard determine the kind of detector required.

Operation of a hydraulically or pneumatically controlled deluge valve can be by means of a pilot line of small-diameter pipe on which are spaced automatic sprinkler heads at suitable intervals. These heads can be augmented, when necessary, by use of a mechanical air- or water-release device, which operates on the rate-of-rise principle as well as the fixed temperature of the sprinkler heads. Other pneumatic means include small copper air chambers, sensitive to rate-of-rise conditions, connected by small-diameter copper tubing to the release mechanism of the valve.

Most deluge valves are essentially check valves with a clapper that is normally latched in the closed position, as in Fig. 20-7. The actuating system unlatches the valve, allowing water to enter the system and flow out the heads.

One deluge valve design (Fig. 20-8) employs a single differential diaphragm

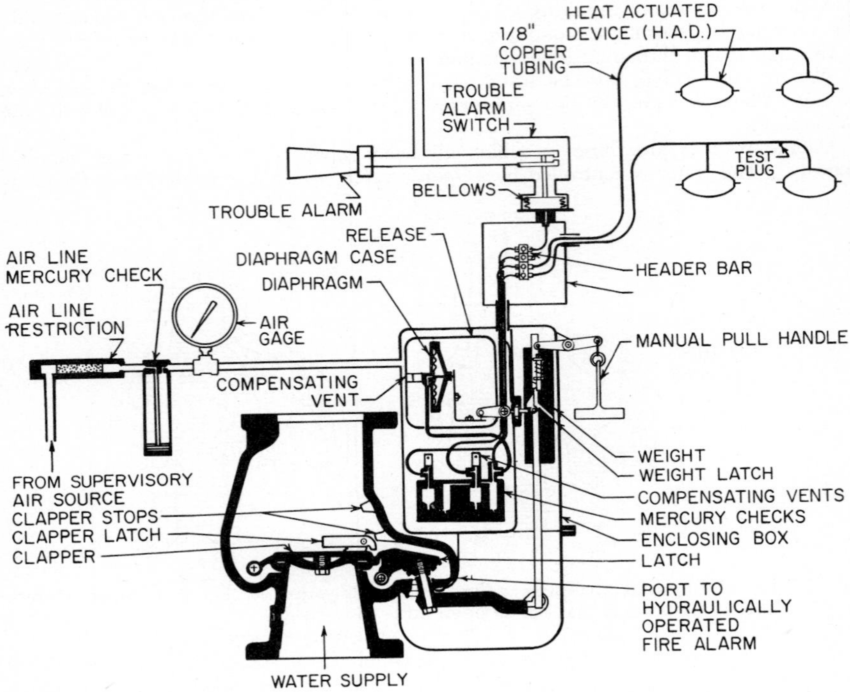

Fig. 20-7. Automatic deluge valve. (*"Automatic" Sprinkler Corporation of America, Cleveland, Ohio.*)

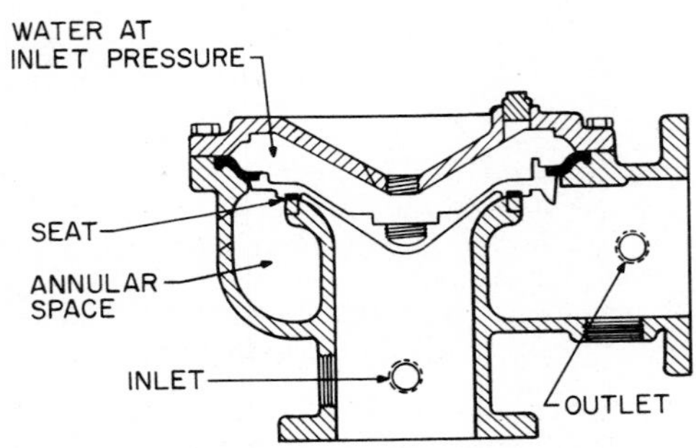

Fig. 20-8. Deluge valve with single differential diaphragm, in set position. When the diaphragm is raised, water fills the annular space and discharges through the outlet. (*The Viking Corporation, Hastings, Mich.*)

in which the water pressure bears on both sides, while the top, large-area side adjoins a closed chamber. The actuating system opens the closed chamber, allowing the water to push the diaphragm up and off the water seat, to release water to the system.

A modification of this valve (Fig. 20-9) employs a pressure regulator that maintains on the outlet side any predetermined pressure less than the available system pressure. This allows the system to discharge at a constant rate. When a deluge system is supplied by a pressure tank, the high initial pressure is reduced to the pressure actually required, thus conserving the limited water supply.

For application of water from a deluge system, a variety of nozzles may be

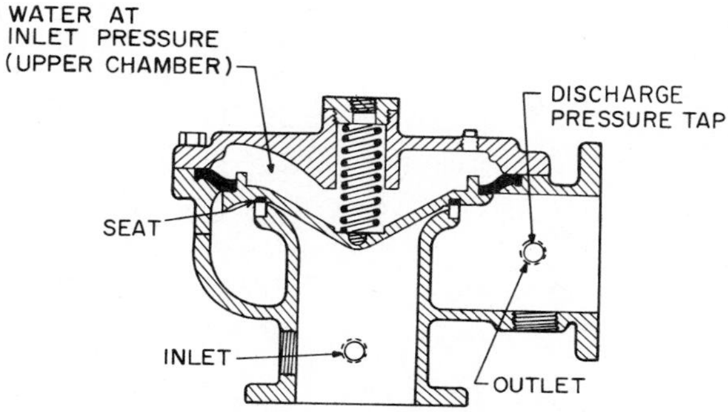

Fig. 20-9. Deluge valve with outlet-pressure regulator, in set position. When the diaphragm is raised, water fills the annular space and discharges through the outlet. (*The Viking Corporation, Hastings, Mich.*)

needed. If general area coverage is required, a standard sprinkler head, without a fusible element, may be used. A standard sprinkler orifice is ½ in., but larger or smaller orifices may be required. For specific equipment protection, orifices may be directional, or with a wide or narrow included angle, and with varying degrees of droplet size, including atomizing nozzles.

Preaction systems employ the same components as deluge systems, but the sprinkler heads are sealed and the pipe sizing is in accordance with wet- and dry-pipe systems. A separate heat-detecting system with sealed heads is employed, because the detecting system is designed to operate before a head opens and at the same time to sound an alarm. If the fire can be extinguished before any heads open, the total damage from fire and water may be less. In addition, if any part of the piping is damaged, water will not issue unless heat is present also. The sealed-head piping system is usually supervised by air at low pressure, 1 or 2 psi, the loss of which will sound a trouble alarm.

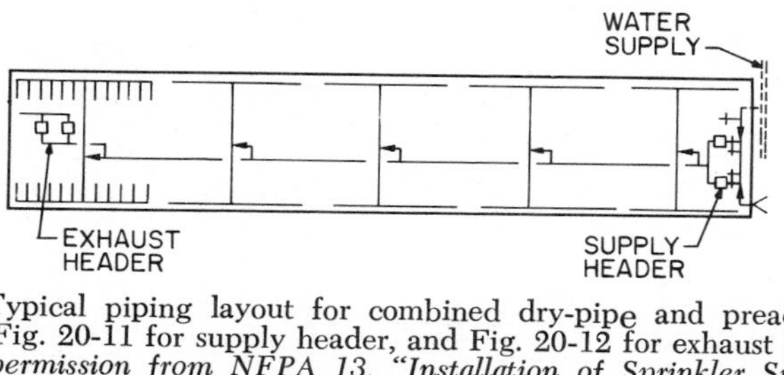

Fig. 20-10. Typical piping layout for combined dry-pipe and preaction sprinkler systems. See Fig. 20-11 for supply header, and Fig. 20-12 for exhaust header. (*Reproduced by permission from NFPA 13, "Installation of Sprinkler Systems," 1972, copyright National Fire Protection Association, Boston, Mass. 02110.*)

This system lends itself to use where highly damageable contents are involved. It is also a substitute for a dry-pipe system when the additional expense of the detecting system is justified.

Combined dry-pipe and preaction systems are suitable for large-area structures, such as piers, where more than one dry-pipe system would be required and where a water supply pipe would be required in an unheated area.

In this arrangement (Fig. 20-10), the heat-detecting system is the same as for a standard preaction system, but two dry-pipe valves in parallel are used instead of deluge valves. The system uses compressed air, as in a dry-pipe system. Valve arrangement at the supply end is shown in Fig. 20-11, and the exhauster system at the outboard end in Fig. 20-12.

When a fire is detected, the detecting system sounds an alarm and operates the exhausters at the valves and at the outboard end. The exhausters hasten

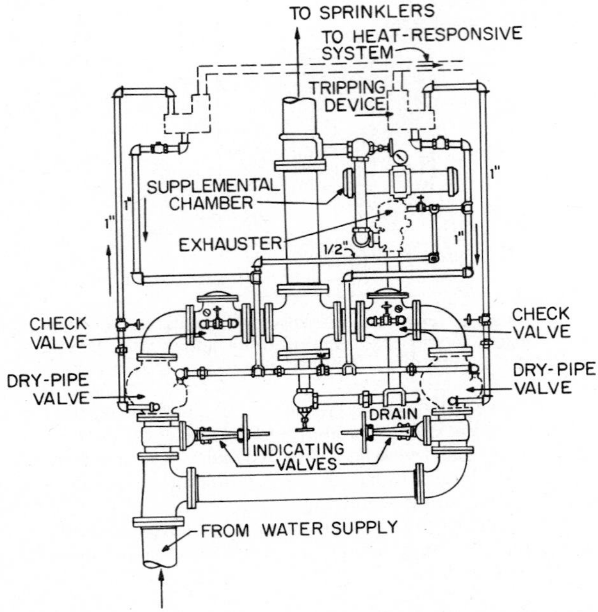

Fig. 20-11. Supply header for sprinkler system in Fig. 20-10. Valve trimmings are not shown. (*Reproduced by permission from NFPA 13, "Installation of Sprinkler Systems," 1972, copyright National Fire Protection Association, Boston, Mass. 02110.*)

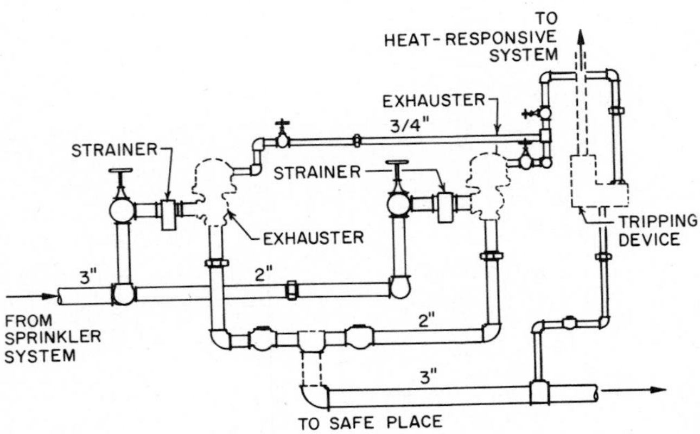

Fig. 20-12. Exhaust header for sprinkler system in Fig. 20-10. (*Reproduced by permission from NFPA 13, "Installation of Sprinkler Systems," 1972, copyright National Fire Protection Association, Boston, Mass. 02110.*)

the entry of water into the sprinkler pipes. Rubber-faced check valves on each subsystem avoid the need for exhausting air in the subsystems that are not involved in the fire. If there is a failure in the detecting system, it will still operate as a dry-pipe system, and if both dry-pipe valves fail, at least an alarm has been given.

In a large system, a number of dry-pipe valves, each with expensive heated enclosures, can be eliminated, as well as wrapping and heat-tracing of the supply line and the consequent maintenance of these items.

Firecycle system is a recycling sprinkler system (patented by Viking Corporation, Hastings, Mich.) that will detect a fire, extinguish it with standard sprinklers, and turn off the water automatically when the fire is out (Fig. 20-13). If the fire is not completely out, or if it rekindles, the system will start again and repeat the performance as many times as necessary. This system can be valuable when there is highly damageable stock, or if it is likely to be a long time before there is a response to the alarm.

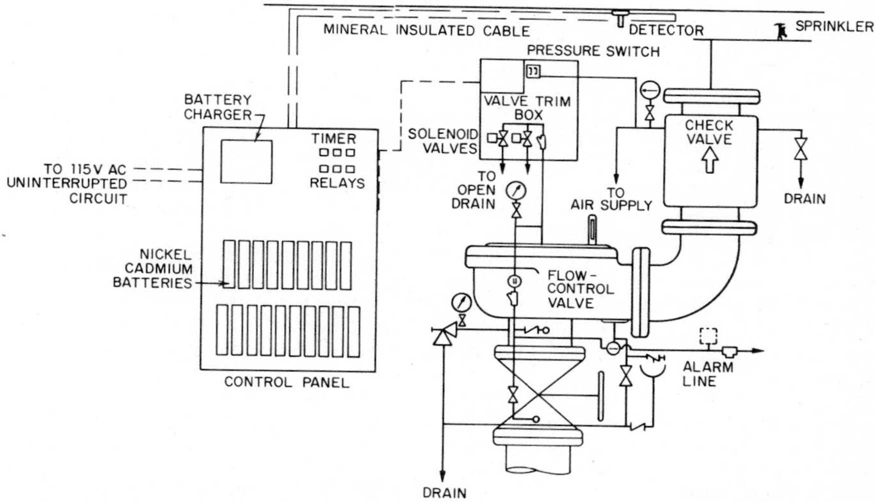

Fig. 20-13. General arrangement of Firecycle system. (*The Viking Corporation, Hastings, Mich.*)

The basic design of the system is similar to the preaction system. It employs self-resetting, rate-of-rise, electrically operated detectors connected with conductors not subject to fire damage. The detector circuit operates a flow-control valve that is a modified form of the Viking deluge valve. As soon as the detecting system senses the fire, an alarm is actuated and the valve is opened, admitting water to the system. As ceiling temperature increases, sprinklers operate and discharge water over the fire area. This discharge will continue until the ceiling temperature is reduced to below the operating temperature of the detector. At that time, the detector circuit closes and a timing cycle starts. At the end of the timing cycle, the flow-control valve closes, and the system is ready for a repeat operation if the ceiling temperature increases again.

Outside sprinklers for exposure protection may be installed outside a building on any side exposed to an adjacent combustible structure or other potential fire hazard. Such systems have open nozzles directed onto the wall, windows, or cornices to be protected. The water supply may be taken from a point below the inside-sprinkler-system control valve if the building is sprinklered, otherwise from any other acceptable source, with the controlling valve accessible at all times. The system is usually operated manually by a gate valve but can be made automatic by use of a deluge valve actuated by suitable means on the exposed side of the building. The distribution piping is usually installed on the outside of the wall, with nozzles provided in sufficient numbers to wet the entire surface to be protected.

20-8. Water Supplies for Fire Protection. These must be of the most reliable type obtainable. An example of a satisfactory supply is a connection to a reliable municipal water system. Where this is not possible, storage or pumping facilities must be provided. An elevated water tank is acceptable for this purpose and requires no outside energy source. An automatic fire pump supplied from a storage reservoir is usually used as a secondary supply, augmenting a municipal water connection or a gravity tank, but such a supply is generally acceptable as a primary supply if it is supervised by a central-station system (Art. 20-10). A hydropneumatic storage tank is also acceptable as a primary supply. It is generally augmented by an automatic fire pump. National Fire Protection Association standard NFPA 20, applies to fire pumps, NFPA 22 to water tanks.

The absolute minimum water supply requirement is to provide for a pressure of 15 psi at the highest line of sprinklers concurrent with a flow at the base of the riser of 250 gpm for light-hazard occupancy, and 500 gpm for ordinary-hazard occupancy. This requirement is based on the availability of a municipal fire department that will respond to alarms set off by a system in the building. Water supply requirements are given in municipal building codes. If, however, NFPA 13, "Installation of Sprinkler Systems," is the governing standard, the rules are indefinite and the local authority must be consulted.

Where the occupancy requires a system with an unusual water demand, this must be met. In a calculated system, the maximum demand is known within limits, and at most must be supplemented with a reasonable flow from hose streams. Where there are deluge systems, attention must be given to the possible simultaneous operation of more than one system.

A property located so that buildings cannot be reached by a fire department with 250 ft of hose may require installation of a private, underground water distribution system. If a municipal water system is available within a reasonable distance, this can form the primary water source. Water demand may be such that this would have to be supplemented by a private elevated tank, or automatic fire pump, or both.

Under such circumstances, the distribution system would have to provide private hydrants to reach all sides of all buildings. Supplies for automatic sprinklers and standpipes would be taken directly from this underground system. Standard for the installation of such systems is NFPA 24.

20-9. Standpipes. Hoses supplied with water from standpipes are the usual means of manual application of water to interior building fires. Standpipes are usually designed for this use by the fire department, but they can be used by building fire fighters also.

Standpipes are necessary in buildings higher than those that ground-based fire department equipment can handle effectively. The National Building Code requires standpipes in buildings higher than 50 ft. The New York City building-code requirement starts at six stories or 75 ft, except where there are more than 10,000 sq ft per floor, in which case they are required for two or more stories.

Riser Sizes. The National Building Code permits the use of 4-in.-diameter risers up to 75 ft high, but requires 6-in. risers for greater height. The New York City code permits 4-in. up to 150 ft, but requires 6-in. for greater heights. The National Fire Protection Association standard for standpipes, NFPA 14, permits 4-in. risers up to 100 ft high, but requires 6-in. for greater heights.

Maximum Pressure. All codes require vertical zoning in riser sections of about 300 ft, to prevent excessive pressures.

Number of Risers. Most codes require that any part of any floor be within 130 ft of a standpipe outlet valve.

Hose-valve Sizes. Where standpipes are required for fire department use, 2½-in. hose valves should be provided. Where it is intended and permitted that building fire fighters use hose, 1½-in. hose may be specified. Some codes provide for both, in which case the valves should be 2½-in. and equipped with easily removable adapters.

Water Supply. Standpipes can be supplied from municipal water mains when adequate pressure and volume are available. When necessary, this source can be supplemented with water from a tank or automatic fire pump.

NFPA requires that the minimum supply be 500 gpm for one riser, and an additional 250 gpm for each additional riser, up to a maximum of 2,500 gpm. A pressure of 65 psi is required at the highest outlet for all risers.

Standpipes with no water supply other than a fire-department pumper connection are permitted in open-air parking garages.

20-10. Central-station Supervisory Systems. Any mechanical device or system is more reliable if it is tested and supervised regularly. Sprinkler systems are designed to be rugged and dependable, as shown by their satisfactory performance record of 96.5% over a period exceeding 75 years. Performance, however, is improved where systems have been connected to an approved central-station supervisory service that provides continuous monitoring of alarms, gate valves, pressures, and other pertinent functions.

A central station supplies the necessary attachments to a system or water supply and transmits appropriate signals over supervised circuits to a constantly monitored location. When any signal is received at the station, at whatever hour, suitable action should be taken by dispatch of investigators, notification of the fire department, or whatever is necessary. Notification of a closed gate valve, for example, should be investigated immediately and thus avoid a possible loss. Any special extinguishing system, such as a carbon dioxide system, can also be supervised as to pressure, operation, or other vital condition.

Such services are available in most cities and are arranged by contract, usually with an installation charge and an annual maintenance charge. Requirements for such systems are in National Fire Protection Association standard NFPA 71. Where no such service is available, a local or proprietary substitute can be provided when such a facility is desirable. Standards for such installations are NFPA 72A, "Local Protective Signaling Systems," NFPA 72B, "Auxiliary Protective Signaling Systems," NFPA 72C, "Remote Station Protective Signaling Systems," and NFPA 72D, "Proprietary Protective Signaling Systems."

CHEMICAL EXTINGUISHING SYSTEMS

20-11. Foam. A foamed chemical, mostly a mass of air- or gas-filled bubbles, formed by chemical or mechanical means, may be used to control fires in flammable liquids. Foam is most useful in controlling fires in flammable liquids with low flash points and low specific gravity, such as gasoline. The mass of bubbles forms a cohesive blanket that extinguishes fire by excluding air and cooling the surface.

Foam clings to horizontal surfaces and can also be used on vertical surfaces of process vessels to insulate and cool. It is useful on fuel-spill fires, to extinguish and confine the vapors.

For fire involving water-soluble liquids, such as alcohol, a special foam concentrate must be used. Foam is not suitable for use on fires involving compressed gases, such as propane, nor is it practical on live electrical equipment. Because of the water content, foam cannot be used on fires involving burning metals, such as sodium, which reacts with water. It is not effective on oxygen-containing materials.

Three distinct types of foam are suitable for fire control: chemical foam, air foam (mechanical foam), and high-expansion foam.

Chemical foam was the first foam developed for fire fighting. It is formed by the reaction of water with two chemical powders, usually sodium bicarbonate and aluminum sulfate. The reaction forms carbon dioxide, which is the content of the bubbles. This foam is the most viscous and tenacious of the foams. It forms a relatively tough blanket, resistant to mechanical or heat disruption. The volume of expansion may be as much as 10 times that of the water used in the solution.

Chemical foam is sensitive to the temperature at which it is formed, and the chemicals tend to deteriorate during long storage periods. It is not capable of being transported through long pipe lines. For these reasons, it is not used as much as other foams. National Fire Protection Association standard NFPA 11 covers chemical foam.

Air foam (mechanical foam) is made by mechanical mixing of water and a protein-based chemical concentrate. There are several methods of combining the components, but essentially the foam concentrate is induced into a flowing stream of water through a metering orifice and a suitable device, such as a venturi. The volume of foam generated is from 16 to 33 times the volume of water used. Several kinds of mixing apparatus are available, choice depending on volume required, availability of water, type of hazard, and characteristics of the protected area or equipment.

Air foam can be conducted through pipes and discharged through a fixed chamber mounted in a bulk fuel storage tank, or it can be conducted through hoses and discharged manually through special nozzles. This foam can also be distributed through a sprinkler system of special design to cover small equipment, such as process vessels, or in multisystem applications, over an entire airplane hangar. The standard for use and installation of air foam is NFPA 11, and for foam-water sprinkler systems, NFPA 16.

High-expansion foam was developed for use in coal mines, where its extremely high expansion rate allowed it to be generated quickly in sufficient volume to fill mine galleries and reach inaccessible fires. This foam can be generated in volumes of from 100 to 1,000 times the volume of water used, with the latter expansion in most general use. The foam is formed by passage of air through a screen constantly wetted by a solution of chemical concentrate, usually with a detergent base. The foam can be conducted to a fire area by ducts, either fixed or portable, and can be applied manually by small portable generators. Standard for equipment and use of high-expansion foam is NFPA 11A.

High-expansion foam is useful for extinguishing fires by totally flooding indoor confined spaces, as well as for local application to specific areas. It extinguishes by displacing air from the fire and by the heat-absorbing effect of converting the foam water content into steam. The foam forms an insulating barrier for exposed equipment or building components.

High-expansion foam is more fragile than chemical or air foam. Also, it is not generally reliable when used outdoors where it is subject to wind currents. High-expansion foam is not toxic, but it has the effect of disorienting people who may be trapped in it.

20-12. Carbon Dioxide. This gas is useful as an extinguishing agent, particularly on surface fires, such as those involving flammable liquids in confined spaces. It is nonconductive and is effective on live electrical equipment. Because carbon dioxide requires no clean-up, it is desirable on equipment such as gasoline or diesel engines. The gas can be used on Class A fires. But when a fire is deep-seated, an extended discharge period is required to avoid rekindling.

Carbon dioxide provides its own pressure for discharge and distribution and is nonreactive with most common industrial materials. Because its density is 1½ times that of air, carbon dioxide tends to drop and to build up from the base of a fire.

Extinguishment of a fire is effected by reduction of the oxygen concentration surrounding a fire. Carbon dioxide may be applied to concentrated areas or machines by hand-held equipment, either carried or wheeled. Or the gas may be used to flood totally a room containing a hazard. The minimum concentrations for total flooding for fires involving some commercial liquids are listed in Table 20-3. Standard for design and installation of carbon dioxide systems is NFPA 12.

Carbon dioxide is not effective on fires involving burning metals, such as magnesium, nor is it effective on oxygen-containing materials, such as nitrocellulose. Hazard to personnel is involved to the extent that a concentration of 9% will cause suffocation in a few minutes, and concentrations of 20% can be fatal. When used in areas where personnel are present, a time delay before discharge is necessary to permit evacuation.

For use in total flooding systems, carbon dioxide is available in either high-pressure or low-pressure equipment. Generally, it is more economical to use low-pressure equipment for large volumes, although there is no division point applicable in all cases.

Table 20-3. Minimum Carbon Dioxide Concentrations for Extinguishment*

Material	Theoretical min CO_2 concentration, %	Min design CO_2 concentration, %
Acetylene	55	66
Acetone	26†	31
Benzol, benzene	31	37
Butadiene	34	41
Butane	28	34
Carbon disulfide	55	66
Carbon monoxide	53	64
Coal or natural gas	31†	37
Cyclopropane	31	37
Dowtherm	38†	46
Ethane	33	40
Ethyl ether	38†	46
Ethyl alcohol	36	43
Ethylene	41	49
Ethylene dichloride	21	25
Ethylene oxide	44	53
Gasoline	28	34
Hexane	29	35
Hydrogen	62	74
Isobutane	30†	36
Kerosene	28	34
Methane	25	30
Methyl alcohol	26	31
Pentane	29	35
Propane	30	36
Propylene	30	36
Quench, lube oils	28	34

* Reproduced by permission from "Standard on Carbon Dioxide Extinguishing Systems," NFPA 12, 1972, copyright National Fire Protection Association, Boston, Mass. 02110.

† Calculated from accepted residual oxygen values. Other theoretical minimum concentrations were obtained from "Limits of Flammability of Gases and Vapors," Bulletin 503, Bureau of Mines, available from U.S. Government Printing Office, Washington, D.C. 20402.

High-pressure systems use the gas compressed in standard cylinders, available in 50-, 75-, or 100-lb sizes. These can be used in banks, with appropriate manifold and distribution piping to nozzles located in the protected area. Automatic-discharge mechanisms can be triggered by any appropriate detection system, electric, pneumatic, or mechanical. While there is no theoretical limit to the size of a high-pressure system, storage space required for large numbers of cylinders is large and can limit system size. Also, the required semiannual weighing of cylinders is a disadvantage.

Low-pressure systems are available in units as small as 750 lb and as large as 52 tons. In these systems, the gas is stored at 300 psi and maintained in insulated, cylindrical tanks, refrigerated automatically to 0°F. A low-pressure system can be discharged automatically, and a single supply tank can provide protection for a number of separate areas, each operated by a separate control valve. This system is valuable for large areas when a multiton discharge is required in a few minutes. Capacity is usually provided for a second application, manually operated, to be used when and if the first application is not totally effective.

20-13. Halon 1301. This gas is one of a series of halogenated hydrocarbons, bromotrifluoromethane ($CBrF_3$), used with varying degrees of effectiveness as fire-extinguishing agents. Its use, in general, is the same as that of carbon dioxide (Art. 20-12). It also needs no clean-up. Its action is quicker than carbon dioxide, however, because Halon 1301 does not have to displace the air from around the fire.

Table 20-4. Halon-1301 Design Concentrations for Flame Extinguishment

(In air at atmospheric pressure and 70°F)*

Material	*Minimum Design Concentration*† *% by Volume*
Commercial denatured alcohol	4.0
n-Butane	2.9
Isobutane	3.3
Carbon disulfide	12.0
Carbon monoxide	1.0
Ethane	3.3
Ethyl alcohol	4.0
Ethylene	7.2
n-Heptane	3.7
Hydrogen	20.0
Methane	2.0
Propane	3.2
Kerosene	2.8
Petroleum naphtha	6.6

* Reproduced by permission from "Standard on Halogenated Fire Extinguishing Agent Systems—Halon 1301," NFPA 12A, 1972, copyright National Fire Protection Association, Boston, Mass. 02110.

† Includes a safety factor of 10% minimum above experimental threshold values. For other temperatures or pressures, specific test data should be obtained.

At room temperatures, Halon 1301 is a colorless, odorless gas with a density five times that of air. The gas is stored in pressure vessels, similar to those for carbon dioxide. These vessels can be manifolded for large systems. Because the vapor pressure varies widely with temperature of Halon 1301, the containers are superpressurized with dry nitrogen to 600 psig or 360 psig.

Action of the gas as an extinguisher is considered to be that of a chemical inhibitor, interfering with the chain reaction necessary to maintain combustion. The action is almost instantaneous. Required concentrations are substantially less than those for carbon dioxide. Concentrations for extinguishment of surface fires involving various industrial materials are given in Table 20-4.

Because of the low concentration provided, and the density of Halon 1301, the discharge system must provide a thorough mixing of the agent with the air surrounding the fire. Flame extinguishment is very rapid. This characteristic is aided by quick detection of fires, indicating the need for an automatically actuated system.

If rapid egress is possible, there should be no prohibition on use of Halon 1301 in normally occupied areas in concentrations less than 7%. It is considered by Underwriters Laboratories to be about half as toxic as carbon tetrachloride, which was used for many years as a fire-extinguishing agent. The decomposition products of Halon 1301, however, are irritating, and prolonged exposure of the gas to temperatures above 900°F will cause the formation of toxic fluorides and bromides.

20-14. Dry Chemical Extinguishing Agents. The powder used originally to extinguish Class B fires was described as a dry chemical extinguishing agent. It consisted of a sodium bicarbonate base with additives to prevent caking and to improve fluid flow characteristics. Later, multipurpose dry chemicals effective on Class A, B, and C fires were developed. These chemicals are distinctly different from the dry powder extinguishing agents used on combustible metals (Art. 20-15).

Dry chemicals are effective on surface fires, especially on flammable liquids. When used on Class A fires, they do not penetrate into the burning material. So when a fire involves porous or loosely packed material, water is used as a backup. The major effect of dry chemicals is due almost entirely to ability to break the chain reaction of combustion. A minor effect of smothering is obtained on Class A fires.

Fires that are likely to rekindle are not effectively controlled by dry chemicals. When these chemicals are applied to machinery or equipment at high temperatures, caking can cause some difficulty in cleaning up after the fire.

Dry chemicals can be discharged in local applications by hand-held extinguishers, wheeled portable equipment, or nozzles on hose lines. These chemicals can also be used for extinguishing fires by total flooding, when they are distributed through a piped system with special discharge nozzles. The expellant gas is usually dry nitrogen.

20-15. Dry Powder Extinguishing Agents. Powders effective in putting out combustible-metal fires are called dry powders. There is no universal extinguisher that can be used on all fires involving combustible metals. Such fires should never be fought by untrained personnel.

There are several proprietary agents effective on several metals, but none should be used without proper attention to the manufacturer's instructions and the specific metal involved. For requirements affecting handling and processing of combustible metals, reference should be made to National Fire Protection Association standards NFPA 48 and 652 for magnesium, NFPA 481 for titanium, NFPA 482M for zirconium, and NFPA 65 and 651 for aluminum.

FIRE DETECTION

Every fire-extinguishing activity must start with detection. To assist in this, many types of automatic detectors are available, with a wide range of sensitivity. Also, a variety of operations can be performed by the detection system. It can initiate an alarm, local or remote, visual or audible; notify a central station; actuate an extinguishing system; start or stop fans or processes, or perform any other operation capable of automatic control.

There are five general types of detectors, each employing a different physical means of operation. The types are designated *fixed-temperature, rate-of-rise, photoelectric, combustion-products,* and *ultraviolet* or *infrared* detectors.

A wide variety of detectors has been tested and reported on by Underwriters Laboratories, Inc. (UL). See Art. 20-1.

20-16. Fixed-temperature Detectors. In its approval of any detection device, UL specifies the maximum distance between detectors to be used for area coverage. This spacing should not be used without competent judgment. In arriving at the permitted spacing for any device, UL judges the response time in comparison with that of automatic sprinkler heads spaced at 10-ft intervals. Thus, if a device is more sensitive than a sprinkler head, the permitted spacing is increased until the response times are nearly equal. If greater sensitivity is desired, the spacing must be reduced.

With fixed-temperature devices, there is a thermal lag between the time the ambient temperature reaches rated temperature and the device itself reaches that temperature. For thermostats having a rating of 135°F, the ambient temperature can reach 206°F.

Disk thermostats are the cheapest and most widely used detectors. The most common type employs the principle of unequal thermal expansion in a bimetallic assembly to operate a snap-action disk at a preset temperature, to close electrical contacts. These thermostats are compact. The disk, ½ in. in diameter, is mounted on a plastic base 1¾ in. in diameter. The thermostats are self-resetting, the contacts being disconnected when normal temperature is restored.

Thermostatic cable consists of two sheathed wires separated by a heat-sensitive coating which melts at high temperature, allowing the wires to contact each other. The assembly is covered by a protective sheath. When any section has functioned, it must be replaced.

Continuous detector tubing is a more versatile assembly. This detector consists of a small-diameter Inconel tube, of almost any length, containing a central wire, separated from the tube by a thermistor element. At elevated temperatures, the resistance of the thermistor drops to a point where a current passes between the wire and the tube. The current can be monitored, and in this way temperature changes over a wide range, up to 1,000°F, can be detected. The detector can be assembled to locate temperature changes of different magnitudes over the same length of detector. It is self-restoring when normal temperature is restored. This detector is useful for industrial applications, as well as for fire detection.

Fusible links are the same devices used in sprinkler heads and are made to operate in the same temperature range. They are used to restrain operation of a fire door, electrical switch, or similar mechanical function, such as operation of dampers. Their sensitivity is substantially reduced when installed at a distance below a ceiling or other heat-collecting obstruction.

20-17. Rate-of-rise Detectors. Detectors and detector systems are said to operate on the rate-of-rise principle when they function on a rapid increase in temperature, whether the initial temperature is high or low. The devices are designed to operate when temperature rises at a specified number of degrees, usually 10 or 15°F, per min. They are not affected by normal temperature increases and are not subject to thermal lag, as are fixed-temperature devices.

A rate-of-rise system widely used to operate deluge and preaction systems employs a hollow copper bulb connected to a suitably encased diaphragm by small-diameter copper tubing. Increased ambient temperature raises the temperature and hence the pressure of air in the bulb. The increased pressure is transmitted to the diaphragm through the tubing. The expanded diaphragm starts a lever train that performs any function required. A calibrated vent, however, permits slowly developed air pressure to escape. A limited number of bulbs can be placed on a tubing circuit, but any number of circuits can be used to operate a single system.

A related system uses copper tubing as the detector, with a specified minimum length exposed to a given protected area. Pressure from heated air is transmitted to one or two suitably encased diaphragms, which are connected to electrical contacts.

A self-contained pneumatic detector employs an air chamber about the size and shape of half a tennis ball. The ball is mounted on a plastic base containing a flexible diaphragm connected to electrical contacts. Pressure from heated air in the chamber bulges the diaphragm directly, and closes or opens electrical contacts. A fixed-temperature element is also incorporated in the device.

Rate-of-rise operation also is provided by a device consisting of a sealed, stainless-steel cylinder ⅝ in. in diameter, 3 in. long, enclosing two struts in compression, with electrical contacts. Rapid increase in temperature causes the cylinder to expand faster than the struts, relieving the compression and causing the contacts to close. Fixed-temperature operation is obtained at a predetermined maximum temperature, when all components have expanded to a point where the contacts close.

Similar action is obtained with a different device used to operate a pilot air or water line controlling a deluge system. An aluminum tube 1½ in. in diameter, 19 in. long, encloses an air or water valve. Rapid rise in temperature reaches the outer case first, expanding it. The expanding case pulls a rod, opening the valve. Fixed-temperature operation is obtained by installing a standard sprinkler head in an opening provided in the base.

A rate-of-rise detector employing thermocouples is used by one manufacturer. A parabolic mirror directs radiant or convected heat to the thermopile for maximum effect.

20-18. Photoelectric Detectors. These indicate a fire condition by detecting the smoke. Sensitivity can be adjusted to operate when obscuration is as low as 0.4% per ft. In these devices, a light source is directed so that it does not impinge on a photoelectric cell. When sufficient smoke particles are concentrated in the chamber, their reflected light reaches the cell, changing its resistance and initiating a signal.

These detectors are particularly useful when a potential fire is likely to generate a substantial amount of smoke before appreciable heat and flame erupt. A fixed-temperature, snap-action disk is usually included in the assembly.

20-19. Combustion-products Detectors. Two physically different means, designated ionization type and resistance-bridge type, are used to operate combustion-products detectors.

The ionization type, most generally used, employs ionization of gases by alpha particles emitted by a small quantity of radium or americum. The detector contains two ionization chambers, one sealed and the other open to the atmosphere, in electrical balance with a cold cathode tube or transistorized amplifier. When sufficient combustion products enter the open chamber, the electrical balance is upset, and the resulting current operates a relay.

The resistance-bridge type of detector operates when combustion products change the impedance of an electric bridge grid circuit deposited on a glass plate.

Combustion-products detectors are designed for extreme early warning, and are most useful when it is desirable to have warning of impending combustion when combustion products are still invisible. These devices are sensitive in some degree to air currents, temperature, and humidity, and should not be used without consultation with competent designers.

20-20. Flame Detectors. These discriminate between visible light and the light produced by combustion reactions. Ultraviolet detectors are responsive to flame having wavelengths up to 2,850 angstroms. The effective distance between flame and detectors is about 10 ft for a 5-in.-diam pan of gasoline, but a 12-in.-square pan fire can be detected at 30 ft.

Infrared detectors are also designed to detect flame. These are not designated by range of wavelength because of the many similar sources at and above the infrared range. To identify the radiation as a fire, infrared detectors usually employ the characteristic flame flicker to identify it, and also have a built-in time delay to eliminate accidental similar phenomena.

OTHER SAFETY MEASURES

20-21. Smoke and Heat Venting. In extinguishment of any building fire, the heat-absorption capacity of water is the principal medium of reducing the heat release from the fire. When, however, a fire is well-developed, the smoke and heat must be released from confinement to make the fire approachable for final manual action. If smoke and heat venting is not provided in the building design, holes must be opened in the roof or building sides by the fire department. In many cases, it has been impossible to do this, with total property losses resulting.

Large-area, one-story buildings can be provided with venting by use of monitors, or a distribution of smaller vents. Multistory buildings present many problems, particularly since life safety is the principal consideration in these buildings.

Ventilation facilities should be provided in addition to the protection afforded by automatic sprinklers and hose stations.

Large One-Story Buildings. For manufacturing purposes, low buildings are frequently required to be many hundreds of feet in each horizontal dimension. Lack of automatic sprinklers in such buildings has proven to be disastrous where adequate smoke and heat venting has not been provided. Owners generally will not permit fire division walls, because they interfere with movement and processing of materials. With the whole content of a building subject to the same fire, fire protection and venting are essential to prevent large losses in windowless buildings, underground structures, and buildings housing hazardous operations.

There is no accepted formula for determining the exact requirements for smoke and heat venting. Establishment of guide lines is the nearest approach that has been made to venting design, and these must be adapted to the case at hand. Consideration must be given to quantity, shape, size, and combustibility of contents.

Venting Ratios. It is generally accepted that the ratio of effective vent opening to floor area should be:

Low heat-release contents	1:150
Moderate heat-release contents	1:100
High heat-release contents	1:30–1:50

Venting can be accomplished by use of monitors, continuous vents, unit-type vents, or sawtooth skylights. In moderate-sized buildings exterior-wall windows may be used if they are near the eaves.

Monitors must be provided with operable panels or other effective means of providing openings at the required time.

Continuous gravity vents are continuous narrow slots provided with a weather hood above. Movable shutters can be provided and should be equipped to open automatically in a fire condition.

Vent Spacing. Unit-type vents are readily adapted to flat roofs, and can be installed in any required number, size, and spacing. They are made in sizes from 4×4 ft to 10×10 ft, with a variety of frame types and means of automatic opening. In arriving at the number and size of vents, preference should be given to a large number of small vents, rather than a few large vents. Because it is desirable to have a vent as near as possible to any location where a fire can start, a limit should be placed on the distance between units.

The generally accepted maximum distance between vents is:

Low heat-release contents	150 ft
Moderate heat-release contents	120 ft
High heat-release contents	75–100 ft

Releasing Methods. Roof vents should be automatically operated by means that do not require electric power. They also should be capable of being manually operated. Roof vents approved by Underwriters Laboratories, Inc. (Art. 20-1) are available from a number of manufacturers.

Refer to National Fire Protection Association standard NFPA 204 in designing vents for large, one-story buildings. Tests conducted prior to publication of NFPA 231C indicated that a sprinkler system designed for adequate density of water application will eliminate the need for roof vents, but the designers would be well advised to consider the probable speed of fire and smoke development in making a final decision. NFPA 231C covers the rack storage of materials as high as 20 ft.

High-rise Buildings. Building codes vary in their definition of high-rise buildings, but the intent is to define buildings in which fires cannot be fought successfully by ground-based equipment and personnel. Thus, ordinarily, high-rise means buildings 100 ft or more high. In design for smoke and heat venting, however, any multistory building presents the same problems.

Because the top story is the only one that can be vented through the roof, all other stories must have the smoke conducted through upper stories to discharge safely above the roof. A separate smoke shaft extending through all upper stories will provide this means. It should be provided with an exhaust fan and should be connected to return-air ducts with suitable damper control of smoke movement, so that smoke from any story can be directed into the shaft. The fan and dampers should be actuated by smoke detectors installed in suitable locations at each inlet to return-air ducts. Operation of smoke detectors also should start the smoke-vent-shaft fan and stop supply-air flow. Central-station supervision (Art. 20-10) should be provided for monitoring smoke-detector operation. Manual override controls should be installed in a location accessible under all conditions.

Windows with fixed sash should be provided with means for emergency opening by the fire department.

Openings between floors for pipes, ducts, wiring, and other services should be sealed with the equal of positive fire stops. Partitions between each floor and a suspended ceiling above are not generally required to be extended to the slab above unless this is necessary for required compartmentation. But smoke stops should be provided at reasonable intervals to prevent passage of smoke to noninvolved areas.

Pressurizing stair towers to prevent the entrance of smoke is highly desirable but difficult to accomplish. Most standpipe connections are presently located in stair towers, and it is necessary to open the door to the fire floor to advance the hose stream toward the fire. A more desirable arrangement would be to locate the riser in the stair tower, if required by code, and place the hose valve adjacent to the door to the tower. Some codes permit this, and it is adaptable to existing buildings.

20-22. Safety during Construction. Most building codes provide specific measures that must be taken for fire protection during construction of buildings. But when they do not, fundamental fire-safety precautions must be taken. Even those structures that will, when completed, be noncombustible contain quantities of forming and packing materials that present a serious fire hazard.

Multistory buildings should be provided with access stairways and, if applicable, an elevator for fire department use. Stairs and elevator should follow as close as possible the upward progress of the structure and be available within one floor of actual building height. In buildings requiring standpipes, the risers should be placed in service as soon as possible, and as close to the construction floor as practicable. Where there is danger of freezing, the water supply can consist of a Siamese connection for fire department use.

In large-area buildings, required fire walls should be constructed as soon as possible. Competent watchman service also should be provided.

The greatest source of fires during construction is portable heaters. Only the safest kind should be used, and these safeguarded in every practical way. Fuel supplies should be isolated and kept to a minimum.

Welding operations also are a source of fires. They should be regulated in accordance with building-code requirements.

Control of tobacco smoking is impossible during building construction; so control of combustible materials is necessary. Good housekeeping should be provided, and all combustible materials not necessary for the work should be removed as soon as possible.

Construction offices and shanties should be equipped with adequate portable extinguishers. So should each floor in a multistory building.

Section 21

Water Supply, Plumbing, Sprinkler, and Waste-Water Systems

TYLER G. HICKS
International Engineering Associates,
New York, N.Y.

WATER SUPPLY AND PURIFICATION

Enough water to meet the needs of occupants must be available for all buildings. Further water needs for fire protection, heating, air conditioning, and possibly process use must also be met. This section provides specific data on all these water needs except the last—process use. Water needs for process use must be computed separately because the demand depends on the process served.

21-1. Water Supply. Sources of water for buildings include public water supplies, ground water, and surface water. Each requires careful study to determine if sufficient quantities are available for the building being designed.

Water for human consumption must be of suitable quality to meet local requirements. Public water supplies generally furnish suitably treated water to a building, eliminating the need for treatment in the building. However, ground and surface waters may require treatment in the building prior to distribution for human consumption. Useful data on water treatment are available from the American Water Works Association.

Useful data on water supply for buildings are available in the following publications: American Society of Civil Engineers, "Glossary—Water and Sewage Control Engineering"; E. W. Steel, "Water Supply and Sewerage," McGraw-Hill Book Company, New York; G. N. Fair, J. C. Geyer, and D. A. Okun, "Water and Wastewater Engineering," John Wiley & Sons, Inc., New York; E. Nordell, "Water Treatment for Industrial and Other Uses," Van Nostrand-Reinhold, New York; American

Society for Testing and Materials, "Manual on Industrial Water." The last-mentioned publication contains extensive data on process-water and steam requirements of a variety of industries. Data on water for fire protection are available from the National Board of Fire Underwriters (NBFU).

Transmission and distribution of water for buildings is generally in pipes, which may be run underground or aboveground. Useful data on pipeline sizing and design are given in: H. E. Babbitt, J. J. Doland, and J. L. Cleasby, "Water Supply Engineering," and C. E. Davis and K. E. Sorenson, "Handbook of Applied Hydraulics," both McGraw-Hill Book Company, New York. NBFU has a series of publications on water storage tanks for a variety of services.

21-2. Characteristics of Water. Physical factors to be considered in raw water are temperature, turbidity, color, taste, and odor. All but temperature are characteristics to be determined in the laboratory by qualified technicians working on carefully procured samples.

Turbidity, a condition due to fine, visible material in suspension, is usually due to presence of colloidal soils. It is expressed in parts per million (ppm) of suspended solids. It may vary widely with discharge of relatively small streams of water. Larger streams or rivers tending to be muddy are generally muddy all the time. The objection to turbidity in potable supplies is its ready detection by the drinker; 10-ppm turbidity shows up as cloudiness in a glass of water. Therefore, the usual limit is 5 ppm.

Color, also objectionable to the drinker, is preferably kept down to 10 ppm or less. It is measured, after all suspended matter (turbidity) has been centrifuged out, by comparison with standard hues.

Tastes and odors due to organic matter or volatile chemical compounds in the water should of course be removed completely from drinking water. But slight, or threshold, odors due to very low concentrations of these compounds are not harmful—just objectionable. Perhaps the most common source of taste and odor is due to decomposition of algae.

Chemical Content. Chemical constituents commonly found in raw waters and measured by laboratory technicians if the water is to be considered for potable use include hardness, pH, iron, and manganese, as well as total solids. Total solids should not exceed 1,000 ppm and in some places are kept below 500 ppm. Other limits on individual chemical constituents generally are:

	Ppm
Lead	0.1
Arsenic	0.05
Copper	0.2
Zinc	5.0
Magnesium	125
Iron	0.3
Chlorides	250
Sulfates	250
Phenol (in compounds)	0.001
Fluorine	1.0 (optimum concentration for prevention of tooth decay)

Hardness, measured as calcium carbonate, may be objectionable as a soap waster with as little as 150 ppm of $CaCo_3$ present. But use of synthetic detergents decreases its significance and makes even much harder waters acceptable for domestic uses. Hardness is of concern, however, in waters to be used for boiler feed, where boiler scale must be avoided. Here, 150 ppm would be too much hardness; the water would require softening (treatment for decrease in hardness).

Hydrogen-ion concentration of water, commonly called pH, can be a real factor in corrosion and encrustation of pipe and in destruction of cooling towers. A pH under 7 indicates acidity; over 7 indicates alkalinity; 7 is neutral. Color indicators commonly used can measure pH to the nearest tenth, which is near enough.

Iron and manganese when present in more than 0.3-ppm concentrations may discolor laundry and plumbing. Their presence and concentration should be determined. More than 0.2 ppm is objectionable for most industrial uses.

Organic Content. Bacteriological tests of water must be made on carefully taken and transported samples. A standard sample is five portions of 10 cu cm, each sample a different dilution of the water tested. A competent laboratory will use standard methods for analyses.

Organisms other than bacteria, such as plankton (free-floating) and algae, can in extreme cases be real factors in the design of water treatment; therefore, biological analyses are significant. Microscopic life, animal and vegetable, is readily tracked down under a high-powered microscope.

21-3. Water Treatment. To maintain water quality within the limits given in Art. 21-2, water supplied to a building usually must undergo some treatment. Whether treatment should be at the source or after transmission to the point of consumption is usually a question of economics, involving hydraulic features, pumping energies and costs, and possible effects of raw water on transmission mains.

In general, treatment, in addition to disinfection, should be provided for water averaging over 20 coliform group organisms per 100 ml, if the water is to be used for domestic purposes. Treatment methods include screening, plain settling, coagulation and sedimentation, filtration, disinfection, softening, and aeration. It is the designer's problem to choose the method that will take the objectionable elements out of the raw water in the simplest, least expensive manner.

Softening of water is a process that must be justified by its need, depending upon the use to which the water will be put. It has been stated that, with a hardness in excess of about 150 ppm, cost of softening will be offset partly by the saving in soap. When synthetic detergents are used instead of soap, this figure may be stretched considerably. But when some industrial use of water requires it, the allowable figure for hardness must be diminished appreciably.

Since corrosion can be costly, corrosive water must often be treated in the interest of economics. In some cases, it may be enough to provide threshold treatment that will coat distribution lines with a light but protective film of scale. But in other cases—boiler-feed water for high-pressure boilers is one—it is important to have no corrosion or scaling. Then, deaeration and pH control may be necessary. (The real danger here is failure of boiler-tube surfaces because of overheating due to scale formation.)

(American Water Works Association, "Water Quality and Treatment," McGraw-Hill Book Company, New York; G. M. Fair, J. C. Geyer, and D. A. Okun, "Elements of Water Supply and Wastewater Disposal," John Wiley & Sons, Inc., New York.)

WASTE PIPING

Human, natural, and industrial wastes resulting from building occupancy and use must be disposed of in a safe, quick manner if occupant health and comfort are to be safeguarded. So design of an adequate plumbing system requires careful planning and adherence to state or municipal regulations governing these systems.

The National Plumbing Code, published by ASME, gives basic goals in environmental sanitation and is a useful aid in designing plumbing systems for all classes of buildings. It is also helpful for areas having antiquated codes or none at all. Much of the material in this section is based on this code.*

21-4. Plumbing-system Elements. The usual steps in planning a plumbing system are: (1) secure a sewer or waste-disposal plan of the site; (2) obtain preliminary architectural and structural plans and elevations of the building; (3) tabulate known and estimated occupancy data, including the number of persons, sex, work schedules, and pertinent details of any manufacturing process to be performed in the building; (4) obtain copies of the latest edition of the applicable code; (5) design the system in accordance with code recommendations; and (6) have the design approved by local authorities before construction is begun.

The typical plumbing layout in Fig. 21-1 shows the major elements necessary in the usual plumbing system. *Fixtures* (lavatories, water closets, bathtubs, showers, etc.) are located as needed on each floor of the structure. Table 21-1 lists the

* Extracted with permission of the publisher, The American Society of Mechanical Engineers, 345 East 47th Street, New York, N.Y. 10017.

Table 21-1. Minimum Facilities Recommended by National Plumbing Code

Type of building or occupancy[b]	Water closets	Urinals	Lavatories	Bathtubs or showers	Drinking fountains[c]
Dwelling or apartment houses[d]	1 for each dwelling or apartment unit	....	1 for each apartment or dwelling unit	1 for each apartment or dwelling unit	
Schools:[e]	Male | Female				
Elementary	1 per 100 | 1 per 35	1 per 30 male	1 per 60 persons		1 per 75 persons
Secondary	1 per 100 | 1 per 45	1 per 30 male	1 per 100 persons		1 per 75 persons
Office or public buildings	No. of persons | No. of fixtures 1–15 | 1 16–35 | 2 36–55 | 3 56–80 | 4 81–110 | 5 111–150 | 6 1 fixture for each 40 additional persons	Wherever urinals are provided for men, one water closet less than the number specified may be provided for each urinal installed[b] except that the number of water closets in such cases shall not be reduced to less than ⅔ of the minimum specified	No. of persons | No. of fixtures 1–15 | 1 16–35 | 2 36–60 | 3 61–90 | 4 91–125 | 5 1 fixture for each 45 additional persons		1 for each 75 persons
Manufacturing, warehouses, workshops, loft buildings, foundries and similar establishments[f]	No. of persons | No. of fixtures 1–9 | 1 10–24 | 2 25–49 | 3 50–74 | 4 75–100 | 5 1 fixture for each additional 30 employees	Same substitution as above	1–100 persons, 1 fixture for each 10 persons Over 100, 1 for each 15 persons[g,h]	1 shower for each 15 persons exposed to excessive heat or to skin contamination with poisonous, infectious, or irritating material	1 for each 75 persons
Dormitories[i]	Male: 1 for each 10 persons Female: 1 for each 8 persons Over 10 persons, add 1 fixture for each 25 additional males and 1 for each 20 additional females	1 for each 25 men Over 150 persons, add 1 fixture for each additional 50 men	1 for each 12 persons. (Separate dental lavatories should be provided in community toilet rooms. Ration of dental lavatories for each 50 persons is recommended.) Add 1 lavatory for each 20 males, 1 for each 15 females	1 for each 8 persons. In the case of women's dormitories, additional bathtubs should be installed at the ratio of 1 for each 30 females Over 150 persons, add 1 fixture for each 20 persons	1 for each 75 persons

Table 21-1. Minimum Facilities Recommended by National Plumbing Code (*Continued*)

Type of building or occupancy[b]	Water closets			Urinals		Lavatories		Bathtubs or showers	Drinking fountains[c]
	No. of persons	No. of fixtures		No. of persons	No. of fixtures	No. of persons	No. of fixtures		
		Male	Female						
Theaters, auditoriums	1–100	1	1	(Male) 1–200	1	1–200	1		1 for each 100 persons
	101–200	2	2	201–400	2	201–400	2		
	201–400	3	3	401–600	3	401–750	3		
	Over 400, add 1 fixture for each additional 500 males and 1 for each 300 females			Over 600; 1 for each additional 300 males		Over 750, 1 for each additional 500 persons			

[a] The figures shown are based upon one fixture being the minimum required for the number of persons indicated or any fraction thereof.

[b] Building category not shown on this table will be considered separately by the administrative authority.

[c] Drinking fountains shall not be installed in toilet rooms.

[d] Laundry trays—one single compartment tray for each dwelling unit or 2 compartment trays for each 10 apartments. Kitchen sinks—1 for each dwelling or apartment unit.

[e] This schedule has been adopted (1945) by the National Council on Schoolhouse Construction.

[f] As required by the American Standard Safety Code for Industrial Sanitation in Manufacturing Establishments (ANSI Z4.1-1935).

[g] Where there is exposure to skin contamination with poisonous, infectious, or irritating materials, provide 1 lavatory for each 5 persons.

[h] 24 lin in. of wash sink or 18 in. of a circular basin, when provided with water outlets for such space, shall be considered equivalent to 1 lavatory.

[i] Laundry trays, 1 for each 50 persons. Slop sinks, 1 for each 100 persons.

General. In applying this schedule of facilities, consideration must be given to the accessibility of the fixtures. Conformity purely on a numerical basis may not result in an installation suited to the need of the individual establishment. For example, schools should be provided with toilet facilities on each floor having classrooms.

Temporary workingmen facilities:
1 water closet and 1 urinal for each 30 workmen
24-in. urinal trough = 1 urinal. 48-in. urinal trough = 2 urinals.
36-in. urinal trough = 2 urinals. 60-in. urinal trough = 3 urinals. 72-in. urinal trough = 4 urinals.

Courtesy of American Society of Mechanical Engineers.

minimum facilities required by the National Plumbing Code for various classes of buildings. In industrial and commercial buildings, the number of fixtures on a given floor may be extensive because, besides including the above sanitary fixtures, there may be other units, such as central locker and washrooms, process-machine drains, and water coolers.

Each fixture is served by a *soil* or *waste stack,* a *vent* or *vent stack,* and a *trap,* Fig. 21-1. Vertical soil stacks conduct waste solids and liquids from one or more fixtures to a sloped *house* or *building drain* in the building cellar, crawl space, or wall. Each vent stack extends to a point above the building roof and may or may not have *branch vents* connected to it. Vents and vent stacks permit the entrance of fresh air to the plumbing system, diluting any gases present and balancing the air pressure in various branches.

Traps on each fixture provide a water seal, which prevents sewer gases from entering the working and living areas. The house drain delivers the discharge from the various stacks to the *house* or *building trap* (Fig. 21-1), which is generally provided with a separate vent. Between the house trap and *public sewer,* or other main sewer pipe, is the *house sewer.* The house sewer is outside the building structure, while the house trap is just inside the building foundation wall.

Where the house drain is below the level of the public sewer line, some arrangement for lifting the sewage to the proper level must be provided. This can be

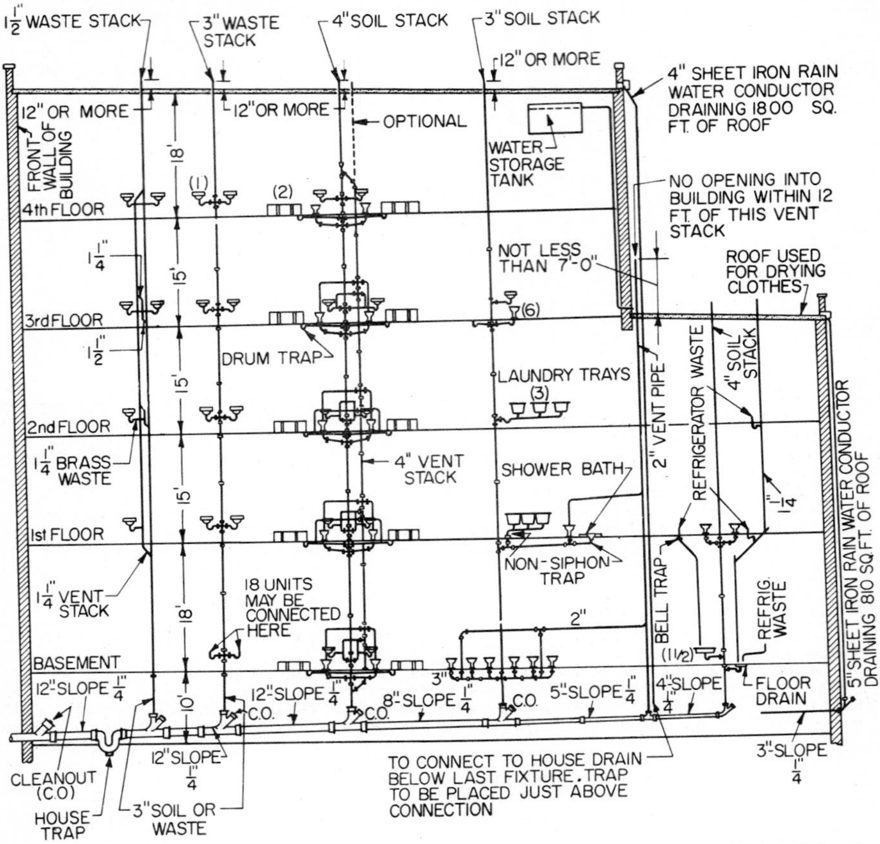

Fig. 21-1. Typical plumbing layout for a residential building. (*Babbitt, "Plumbing," McGraw-Hill Book Company, New York.*)

done by allowing the house drain to empty into a suitably sized *sump pit.* The sewage is discharged from the sump pit to the public sewer by a pneumatic ejector or motor-driven pump.

(V. T. Manas, "National Plumbing Code Handbook," McGraw-Hill Book Company, New York.)

21-5. Plumbing Piping. Cast iron is the most common piping material for systems in which extremely corrosive wastes are not expected. In recent years, copper has become increasingly popular because it is lightweight and easy to install. Vitrified-clay, steel, brass, wrought-iron, and lead pipe are also used in plumbing systems.

No matter what material is used for plumbing-system piping, it should conform to one or more of the accepted standards approved by the National Plumbing Code, or the code used in the area in which the building is located. Figure 21-2 shows a number of different traps used in plumbing piping systems.

In cast-iron pipe, the fitting joints are calked with oakum or hemp and filled with molten lead not less than 1 in. deep. This provides some flexibility without the danger of leaks. Clay pipe is calked with oakum, cement grout, and mortar. Copper pipe is commonly brazed, while steel and wrought-iron pipe have screwed or flanged connections.

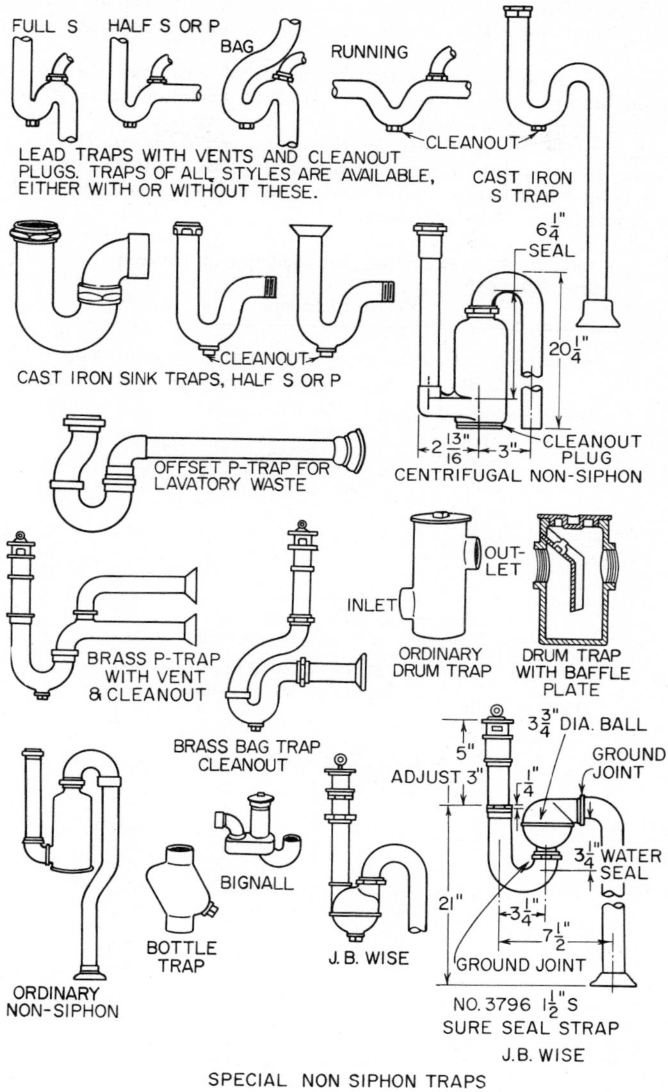

Fig. 21-2. Types of traps used on plumbing fixtures. (*Babbitt, "Plumbing," McGraw-Hill Book Company, New York.*)

When planning a plumbing system, check with the applicable code before specifying the type of joint to be used in the piping. Joints acceptable in some areas may not be allowed in others.

See also Art. 21-7.

21-6. Rain-water Drainage. Exterior sheet-metal *leaders* for rain-water drainage are not normally included in the plumbing contract. Interior leaders or storm-water drains, however, are considered part of the plumbing work. Depending on the size of the building and the code used in the locality, rain water from various roof areas may or may not be led into the house drains. Where separate rain-water leaders or storm drains are used, the house drains are then called *sanitary drains* because they convey only the wastes from the various plumbing fixtures in the building.

Interior storm-water drain pipes may be made of cast iron, steel, or wrought iron. All joints must be tight enough to prevent gas and water leakage. It is common practice to insert a cast-iron running trap between the storm drain and the house drain to maintain a seal on the storm drain at all times. However, not every code requires this type of protection. Use of storm drains does not eliminate the need for separate drains and vents for sewage. All codes prohibit use of storm drains for any type of sewage.

Table 21-2. Sizes of Vertical Leaders and Horizontal Storm Drains*

Vertical Leaders

Size of leader or conductor,† in.	Max projected roof area, sq ft
2	720
2½	1,300
3	2,200
4	4,600
5	8,650
6	13,500
8	29,000

Horizontal Storm Drains

Diam of drain, in.	Max projected roof area for drains of various slopes, sq ft		
	⅛-in. slope	¼-in. slope	½-in. slope
3	822	1,160	1,644
4	1,880	2,650	3,760
5	3,340	4,720	6,680
6	5,350	7,550	10,700
8	11,500	16,300	23,000
10	20,700	29,200	41,400
12	33,300	47,000	66,600
15	59,500	84,000	119,000

* From National Plumbing Code, ASME.

† The equivalent diameter of square or rectangular leader may be taken as the diameter of that circle which may be inscribed within the cross-sectional area of the leader.

Water falling on the roof may be led either to a gutter, from where it flows to a leader, or it may be directed to a leader inlet by means of a slope in the roof surface. Many different designs of leader inlets are available for different roof constructions and storm-water conditions. When vertical leaders are extremely long, it is common practice to install an expansion joint between the leader inlet and the leader itself.

Drains for building yards, areaways, and floors may be connected to the inlet side of an interior rain-leader trap. Where this is not possible, these drains may be run to a municipal sewer, or to a dry well. When a dry well is used, only the drains may discharge to it, other sewage being prohibited by most codes.

21-7. Plumbing-pipe Sizes. There are two ways of specifying the pipe size required for a particular class of plumbing service: (1) directly in terms of area served, as in roof-drainage service (Table 21-2) and (2) in terms of "fixture units" (Table 21-3).

Table 21-3. Sizes of Building Drains and Sewers*

Diam of pipe, in.	Max number of fixture units that may be connected to any portion† of the building drain or the building sewer			
	Fall per ft			
	1/16 in.	1/8 in.	1/4 in.	1/2 in.
2	....	....	21	26
2½	....	....	24	31
3	....	20‡	27‡	36‡
4	....	180	216	250
5	....	390	480	575
6	....	700	840	1,000
8	1,400	1,600	1,920	2,300
10	2,500	2,900	3,500	4,200
12	3,900	4,600	5,600	6,700
15	7,000	8,300	10,000	12,000

* From National Plumbing Code, ASME.
† Includes branches of the building drain.
‡ Not over two water closets.

As can be seen from these tables, the capacity of a leader or drain varies with the pitch of the installed pipe. The greater the pitch per running foot of pipe, the larger the capacity allowed, in terms of either the area served or the number of fixture units. This is because a steeper pitch increases the static head available to produce flow through the pipe. As static head increases, so does the amount of liquid which the pipe can handle.

Fixture Unit. This is the average discharge, during use, of an arbitrarily selected fixture, such as a lavatory or toilet. Once this value is established, the discharge rates of other types of fixtures are stated in terms of the basic fixture. For example, when the basic fixture is a lavatory served by a 1¼-in. trap, the average flow during discharge is 7.5 gpm. So a bathtub that discharges 15 gpm is rated as two fixture units (2×7.5). Thus, a tabulation of fixture units can be set up, based on an assumed basic unit.

Table 21-4 gives fixture-unit values recommended by the National Plumbing Code. These values designate the relative load weight to be used in estimating the total load carried by a soil waste pipe. They are to be used in connection with Tables 21-2 and 21-3 and similar tabulations for soil, waste, and drain pipes for which the permissible load is given in terms of fixture units.

Fixtures not listed in Table 21-4 should be estimated in accordance with Table 21-5. For a continuous or semicontinuous flow into a drainage system, such as from a pump, pump ejector, air-conditioning system, or similar device, two fixture units should be used for each gallon per minute of flow. When additional fixtures are to be installed in the future, the drain size should be based on the ultimate load, not on the present load. Provide plugged fittings at the stack for connection of future fixtures.

21-8. Branches and Vents. The fixture unit (Art. 21-7) is also used for sizing vents and vent stacks (Table 21-6). In general, the diameter of a branch vent or vent stack is one-half or more of the branch or stack it serves, but not less than 1¼ in. Smaller diameters are prohibited, because they might restrict the venting action.

Fixture branches connecting one or more fixtures with the soil or waste stack are usually sized on the basis of the maximum number of fixture units for a given size of pipe or trap (Table 21-7). Where a large volume of water or other liquid may be contained in a fixture, such as in bathtubs, slop sinks, etc., an oversize branch drain may be provided to secure more rapid emptying.

Table 21-4. Fixture Units per Fixture or Group*

Fixture type	Fixture-unit value as load factors	Min size of trap, in.	
1 bathroom group consisting of water closet, lavatory and bathtub or shower stall	Tank water closet, 6 Flush-valve water closet, 8		
Bathtub† (with or without overhead shower)	2		1½
Bathtub†	3		2
Bidet	3	Nominal	1½
Combination sink and tray	3		1½
Combination sink and tray with food-disposal unit	4	Separate traps	1½
Dental unit or cuspidor	1		1¼
Dental lavatory	1		1¼
Drinking fountain	½		1
Dishwasher, domestic	2		1½
Floor drains‡	1		2
Kitchen sink, domestic	2		1½
Kitchen sink, domestic, with food-waste grinder	3		1½
Lavatory¶	1	Small P.O.	1¼
Lavatory¶	2	Large P.O.	1½
Lavatory, barber, beauty parlor	2		1½
Lavatory, surgeon's	2		1½
Laundry tray (1 or 2 compartments)	2		1½
Shower stall, domestic	2		2
Showers (group) per head	3		
Sinks:			
Surgeon's	3		1½
Flushing rim (with valve)	8		3
Service (trap standard)	3		3
Service (P trap)	2		2
Pot, scullery, etc.	4		1½
Urinal, pedestal, siphon jet, blowout	8	Nominal	3
Urinal, wall lip	4		1½
Urinal stall, washout	4		2
Urinal trough (each 2-ft section)	2		1½
Wash sink (circular or multiple) each set of faucets	2	Nominal	1½
Water closet, tank-operated	4	Nominal	3
Water closet, valve-operated	8		3

* From National Plumbing Code, ASME.
† A shower head over a bathtub does not increase the fixture value.
‡ Size of floor drain shall be determined by the area of surface water to be drained.
¶ Lavatories with 1¼- or 1½-in. trap have the same load value; larger P.O. (plumbing orifice) plugs have greater flow rate.

21-9. Gutters. Semicircular gutters are sized on the basis of the maximum projected roof area served. Table 21-8 shows how gutter capacity varies with diameter and pitch. Where maximum rainfall is either more than, or less than, 4 in. per hr, refer to the National Plumbing Code for suitable correction factors.

21-10. How to Use Fixture Units. The steps in determining pipe sizes by means of fixture units (Art. 21-7) are: (1) list all fixtures served by one stack or branch; (2) alongside each fixture, list its fixture unit; (3) take the sum of the fixture units and enter the proper table (Tables 21-3, 21-6, 21-7) to determine the pipe size required for the stack or the branch.

21-11. Traps. Separate traps are required by almost every code on most fixtures not fitted with an integral trap. The trap should be installed as close as possible to the unit served. More than one fixture may be connected to a trap if certain

Table 21-5. Other Fixture-unit Values*

Fixture Drain or Trap Size, In.	*Fixture-unit Value*
1¼ in. and smaller	1
1½	2
2	3
2½	4
3	5
4	6

* From National Plumbing Code, ASME.

Table 21-6. Size and Length of Vents*

Size of soil or waste stack, in.	Fixture units connected	Diam of vent required, in.								
		1¼	1½	2	2½	3	4	5	6	8
		Max length of vent, ft								
1¼	2	30								
1½	8	50	150							
1½	10	30	100							
2	12	30	75	200						
2	20	26	50	150						
2½	42		30	100	300					
3	10		30	100	200	600				
3	30			60	200	500				
3	60			50	80	400				
4	100			35	100	260	1,000			
4	200			30	90	250	900			
4	500			20	70	180	700			
5	200				35	80	350	1,000		
5	500				30	70	300	900		
5	1,100				20	50	200	700		
6	350				25	50	200	400	1,300	
6	620				15	30	125	300	1,100	
6	960					24	100	250	1,000	
6	1,900					20	70	200	700	
8	600						50	150	500	130
8	1,400						40	100	400	120
8	2,200						30	80	350	110
8	3,600						25	60	250	80
10	1,000							75	125	100
10	2,500							50	100	50
10	3,800							30	80	35
10	5,600							25	60	25

* From National Plumbing Code, ASME.

code regulations are observed. For specific requirements, refer to the governing code. Table 21-4 lists the minimum trap size recommended by the National Plumbing Code for various fixtures.

A water seal of at least 2 in., and not more than 4 in., is generally required in most traps. Traps exposed to freezing should be suitably protected to prevent ice formation in the trap body. Cleanouts of suitable size are required on all traps except those made integral with the fixture or those having a portion which is easily removed for cleaning of the interior body. Most codes prohibit use of traps in which a moving part is needed to form the seal. Double trapping is also usually prohibited.

Table 21-7. Horizontal Fixture Branches and Stacks*

Diam of pipe, in.	Max number of fixture units that may be connected to			
	Any horizontal† fixture branch	One stack of 3 stories in height or 3 intervals	More than 3 stories in height	
			Total for stack	Total at one story or branch interval
1¼	1	2	2	1
1½	3	4	8	2
2	6	10	24	6
2½	12	20	42	9
3	20‡	30¶	60¶	16‡
4	160	240	500	90
5	360	540	1,100	200
6	620	960	1,900	350
8	1,400	2,200	3,600	600
10	2,500	3,800	5,600	1,000
12	3,900	6,000	8,400	1,500
15	7,000			

* From National Plumbing Code, ASME.
† Does not include branches of the building drain.
‡ Not over two water closets.
¶ Not over six water closets.

Table 21-8. Size of Gutters*

Diam of gutter,† in.	Max projected roof area for gutters of various slopes, sq ft			
	1/16-in. slope	⅛-in. slope	¼-in. slope	½-in. slope
3	170	240	340	480
4	360	510	720	1,020
5	625	880	1,250	1,770
6	960	1,360	1,920	2,770
7	1,380	1,950	2,760	3,900
8	1,990	2,800	3,980	5,600
10	3,600	5,100	7,200	10,000

* From National Plumbing Code, ASME.
† Gutters other than semicircular may be used provided they have an equivalent cross-sectional area.

21-12. Cleanouts. In horizontal drainage lines, at least one cleanout is required per 50 ft of 4-in. or smaller pipe. For larger sizes, a cleanout is required for each 100 ft of pipe. In underground drainage lines, the cleanout must be extended to the floor or grade level to allow easier cleaning. Cleanouts should open in a direction opposite to that of the flow in the pipe, or at right angles to it.

In pipes up to 4 in., the cleanout should be the same size as the pipe itself. For pipes larger than 4 in., the cleanout should be at least 4 in. in diameter but may be larger, if desired. When underground piping over 10 in. in diameter is used, a manhole is required at each 90° bend and at intervals not exceeding 150 ft.

21-13. Interceptors. These are devices installed to separate and retain grease, oil, sand, and other undesirable matter from the sewage, while permitting normal liquid wastes to discharge to the sewer or sewage-treatment plant.

Grease interceptors are used for kitchens, cafeterias, restaurants, and similar establishments where the amount of grease discharged might obstruct the pipe or interfere with disposal of the sewage. Oil separators are used where flammable liquids or oils might be discharged to the sewer. Sand interceptors are used to remove sand or other solids from the liquid sewage before it enters the building sewer. They are provided with large cleanouts for easy removal of accumulated solids.

Other types of plants in which interceptors are usually required include laundries, beverage-bottling firms, slaughterhouses, and food-manufacturing establishments.

21-14. Plumbing Fixtures. To insure maximum sanitation and health protection, most codes have rigid requirements regarding fixtures used in plumbing systems. These usually include items like materials of construction, connections, overflows, installation, prevention of backflow, flushing methods, types of fixtures allowed, and inlet and outlet sizes. So before specifying or purchasing a given fixture, check the governing code carefully to see that the proposed units are acceptable. Figure 21-3 shows a typical connection diagram for a group of fixtures. Figure 21-4 shows the designs of typical water closets.

21-15. Pipe Hangers. To insure adequate support of piping used in plumbing systems, the National Plumbing Code allows the following *maximum* distance between supports in vertical pipes: cast-iron soil pipe, one story; screwed pipe, every other story; copper tubing 1½ in. and over, each story; copper tubing 1¼ in. and under, 4-ft intervals; lead pipe, 4-ft intervals.

For horizontal pipes, the following maximum distances are allowed: cast-iron soil pipe, 5-ft intervals; screwed pipe, 12-ft intervals; copper tubing 1½ in. and smaller, 6-ft intervals; copper tubing 2 in. and larger, 10-ft intervals; lead pipe is to be supported by strips or other means for its entire length. Pipes run in the ground should have firm support for their complete length.

Note that the above allowances are for the *maximum* distances between supports. Where possible, supports that are closer together than the allowable maximum distance should be used.

The hangers and anchors used for plumbing piping should be metal, and strong enough to prevent vibration. Each should be designed and installed to carry its share of the total weight of the pipe.

21-16. Piping for Indirect Wastes. Certain wastes like those from food-handling, dishwashing (commercial), and sterile-materials machines should be discharged through an indirect waste pipe; i.e., one that is not directly connected with the building drainage pipes but handles waste liquids by discharging them into a plumbing fixture or receptacle from where they flow directly to the building drainage pipes. Indirect-waste piping is generally required for the discharge from rinse sinks, laundry washers, steam tables, refrigerators, egg boilers, iceboxes, coffee urns, dishwashers, stills, sterilizers, etc. It is also required for units that must be fitted with drip or drainage connections but are not ordinarily regarded as plumbing fixtures.

An air gap is generally required between the indirect-waste piping and the regular drainage system. The gap should have at least twice the effective diameter of the drain it serves. A common way of providing the required air gap is by leading the indirect-waste line to a floor drain, slop sink, or similar fixture that is open to the air and is vented or trapped in accordance with the governing code. To provide the necessary air gap, the indirect-waste pipe is terminated above the flood level of the fixture.

Where a device discharges only clear water, such as engine-cooling jackets, water lifts, sprinkler systems, or overflows, an indirect-waste system must be used. Or, if desired, devices discharging only clear water may be emptied by leading their waste lines to the building roof. The roof provides the air gap and the waste flows off it into the building storm drains.

Hot water above 140°F and steam pipes usually must be arranged for indirect connection into the building drainage system or into an approved interceptor.

To prevent corrosion of plumbing piping and fittings, any chemicals, acids, or

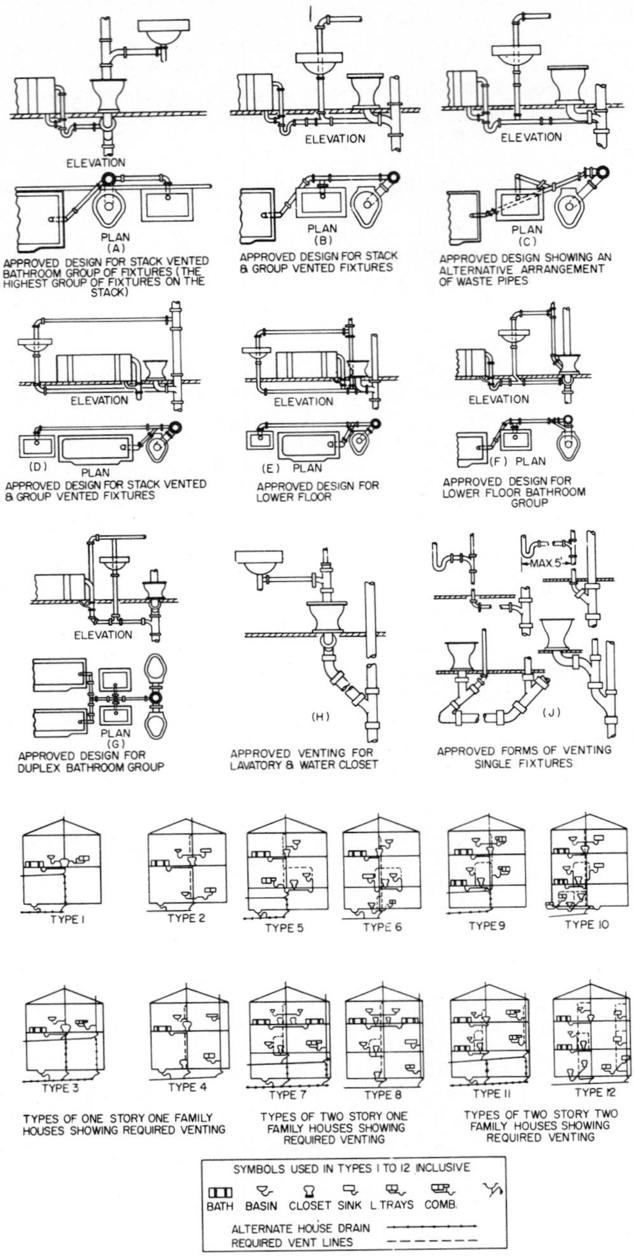

Fig. 21-3. Typical plumbing installations. (*Babbitt, "Plumbing," McGraw-Hill Book Company, New York.*)

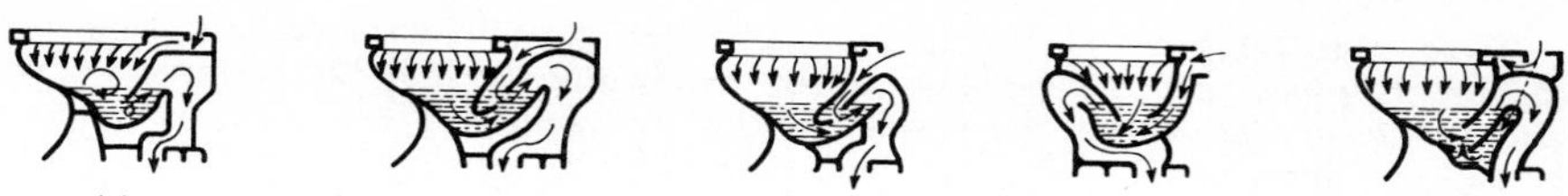

Fig. 21-4. Typical water closets. (*a*) Siphon-vortex—water from rim washes bowl, creates vortex, becomes a jet, and discharges by siphon action. (*b*) Siphon-jet—water from rim becomes jet in the upleg of discharge, then is siphoned out in downleg. (*c*) Reverse trap—similar to siphon-jet, but the water closet is smaller. (*d*) Washdown—pressure buildup causes upleg to overflow, creating discharge siphon. (*e*) Blowout—used with direct-flush valve, strong jet into upleg causes discharge. (See also C. G. Ramsey and H. R. Sleeper, "Architectural Graphic Standards," and W. J. McGuinness and B. Stein, "Mechanical and Electrical Equipment for Buildings," John Wiley & Sons, Inc., New York.)

corrosive liquids are generally required to be automatically diluted or neutralized before being discharged into the plumbing piping. Sufficient fresh water for satisfactory dilution, or a neutralizing agent, should be available at all times. A similar requirement is contained in most codes for liquids that might form or give off toxic or noxious fumes.

21-17. Plumbing-system Inspection and Tests. Plans for plumbing systems must usually be approved before construction is started. After installation of the piping and fixtures is completed, both the new work and any existing work affected by the new work must be inspected and tested. The plumber or plumbing contractor is then informed of any violations. These must be corrected before the governing body will approve the system.

Either air or water may be used for preliminary testing of drainage, vent, and plumbing pipes. After the fixtures are in place, their traps should be filled with water and a final test made of the complete drainage system.

When testing a system with water, all pipe openings are tightly sealed, except the highest one. The pipes are then filled with water until overflow occurs from the top opening. Using this method, either the entire system or sections of it can be tested. In no case, however, should the head of water on a portion being tested be less than 10 ft, except for the top 10 ft of the system. Water should be kept in the system for at least 15 min before the inspection starts. During the inspection, piping and fixtures must be tight at all points; otherwise approval cannot be granted.

An air test is made by sealing all pipe outlets and subjecting the piping to an air pressure of 5 psi throughout the system. The system should be tight enough to permit maintaining this pressure for at least 15 min without the addition of any air.

The final test required of plumbing systems uses either smoke or peppermint. In the smoke test, all traps are sealed with water and a thick, strong-smelling smoke is injected into the pipes by means of a suitable number of smoke machines. As soon as smoke appears at the roof stack outlets, they should be closed and a pressure equivalent to 1 in. of water should be maintained throughout the system for 15 min before inspection begins. For the peppermint test, 2 oz of oil of peppermint are introduced into each line or stack.

GAS PIPING

As with piping and fixtures for plumbing systems, gas piping and fixtures are subject to code regulations in most cities and towns. Natural and manufactured gases are widely used in stoves, water heaters, and space heaters of many designs. Since gas can form explosive mixtures when mixed with air, gas piping must be absolutely tight and free of leaks at all times. Usual codes cover every phase of gas-piping size, installation, and testing. The local code governing a particular building should be carefully followed during design and installation.

21-18. Gas Supply. The usual practice is for the public-service gas company to run its pipes into the building cellar, terminating with a brass shutoff valve and gas meter inside the cellar wall. From this point the plumbing contractor or gas-pipe fitter runs lines through the building to the various fixture outlets. When the pressure of the gas supplied by the public-service company is too high for the devices in the building, a pressure-reducing valve can be installed near the point where the line enters the building. This valve is usually supplied by the gas company.

Besides municipal codes governing the design and installation of gas piping and devices, the gas company serving the area will usually have a number of regulations that must be followed. In general, gas piping should be run in such a manner that it is unnecessary to locate the meter near a boiler, under a window or steps, or in any other area where it may be easily damaged. Where multiple meter installations are used, the piping must be plainly marked by means of a metal tag showing which part of the building is served by the particular pipe. When two or more meters are used in a building to supply separate consumers, there should be no interconnection on the outlet side of the meters.

Table 21-9. Heat Input to Common Appliances

Unit	*Approx Input, Btu per Hr*
Water heater, side-arm or circulating type	25,000
Water heater, automatic instantaneous:	
4 gpm	150,000
6 gpm	225,000
8 gpm	300,000
Refrigerator	2,500
Ranges, domestic:	
4 top burners, 1 oven burner	62,500
4 top burners, 2 oven burners	82,500

Table 21-10. Typical Heating Values of Commercial Gases

Gas	*Net Heating Value, Btu per Cu Ft*
Natural gas (Los Angeles)	971
Natural gas (Pittsburgh)	1,021
Coke-oven gas	514
Carbureted water gas	508
Commercial propane	2,371
Commercial butane	2,977

21-19. Gas-pipe Sizes. Gas piping must be designed to provide enough gas to appliances without excessive pressure loss between them and the meter. It is customary to size gas piping so the pressure loss between the meter and any appliance does not exceed 0.3 in. of water during periods of maximum gas demand. Other factors influencing the pipe size include maximum gas consumption anticipated, length of pipe and number of fittings, specific gravity of the gas, and the diversity factor.

21-20. Estimating Gas Consumption. Use the manufacturer's Btu rating of the appliances and the heating value of the gas to determine the flow required in cubic feet per hour. When Btu ratings are not immediately available, the values in Table 21-9 may be used for preliminary estimates. The average heating value of gas in the area can be obtained from the local gas company, but when this is not immediately available, the values in Table 21-10 can be used for preliminary estimates.

Example. A building has two 8 gpm hot-water heaters and 10 domestic ranges, each of which has four top burners and one oven burner. What is the maximum gas consumption that must be provided for if gas with a net heating value of 500 Btu per cu ft is used?

Solution. Heat input, using values from Table 21-9, is

$$2(300,000) + 10(62,500) = 1,225,000 \text{ Btu per hr}$$

Maximum gas consumption is therefore 1,225,000/500 = 2,450 cu ft per hr. The supply piping would be sized for this flow, even though all appliances would rarely operate at the same time.

21-21. Gas-pipe Materials. Materials recommended for gas piping in buildings are wrought iron or steel complying with American Standard B36.10. Malleable-iron or steel fittings should be used, except for stopcocks and valves. Above 4-in. nominal size, cast-iron fittings may be used.

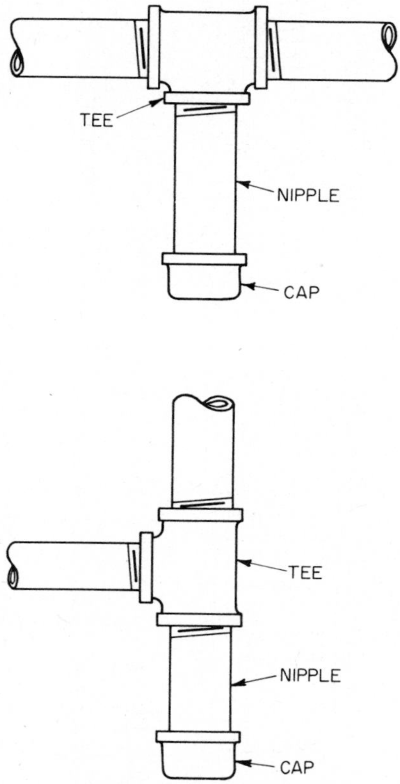

Fig. 21-5. Drips for gas piping.

Some local codes permit the use of brass or copper pipe of the same sizes as iron pipe if the gas handled is not corrosive to the brass or copper. Threaded fittings are generally required with these two materials.

Since the usual gas supplied for heating and domestic cooking generally contains some moisture, all piping should be installed so it pitches back to the supply main, or drips should be installed at suitable intervals. Neither unions nor bushings are permitted in gas pipes because of the danger of leakage and moisture trapping. To permit moisture removal, drips are installed at the lowest point in the piping, at the bottom of vertical risers, and at any other places where moisture might accumulate. Figure 21-5 shows typical drips for gas piping.

21-22. Testing Gas Piping. Once installation of the gas piping in a building is complete, it must be tested for leaks. Portions of the piping that will be enclosed in walls or behind other permanent structures should be tested before the enclosure is completed.

The usual code requirement for gas-pipe testing stipulates a test pressure of about 10 psig for 15 min, using air, city gas, or some inert gas. Use of oxygen in pressure tests is strictly prohibited. Prior to connecting any appliances, the piping should be purged, using the regular gas supply.

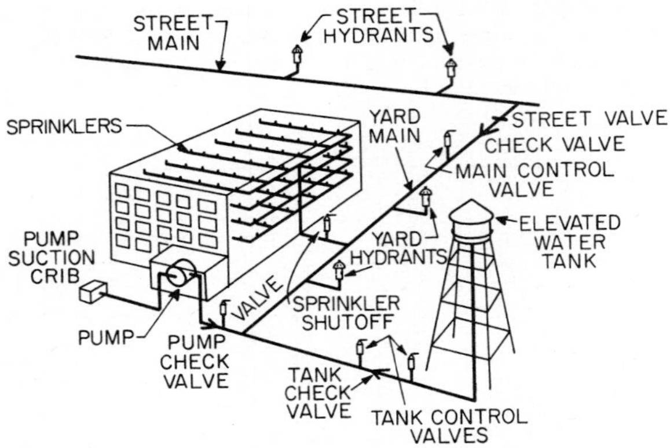

Fig. 21-6. Water-supply piping for sprinklers.

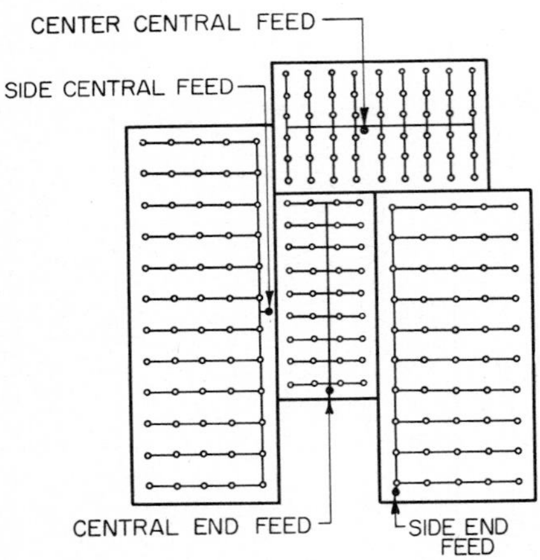

Fig. 21-7. Typical layouts of sprinkler piping.

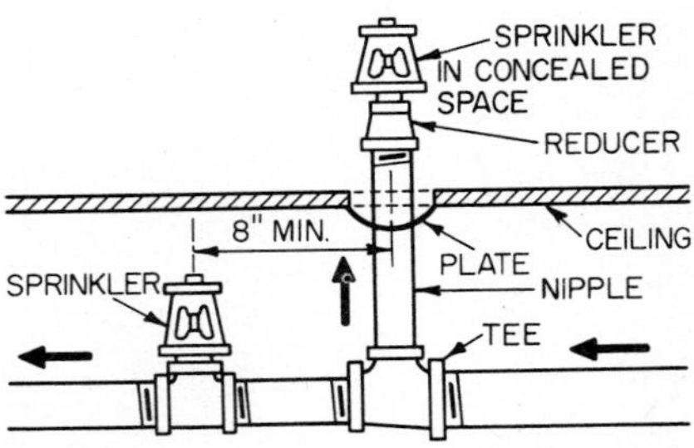

Fig. 21-8. Sprinklers below and above ceiling.

SPRINKLER SYSTEMS

Well-proved as effective in preventing the spread of fires in buildings of all types, sprinkler systems for fire protection are in wide use. Essentially (Fig. 21-6), they consist of a series of horizontal parallel pipes placed near the building ceiling. Vertical risers (Fig. 21-7) supply the liquid—usually water—used for extinguishing fires. The horizontal pipes contain sprinklers (Fig. 21-8) preset to open automatically and discharge liquid when the temperature of the air in the room reaches a certain predetermined level—usually 135 to 160°F. No manual control of the sprinklers is required; so they provide continuous protection day and night.

21-23. Approvals of Sprinkler Systems. Four common types of automatic sprinkler systems are in use today: (1) wet-pipe, (2) dry-pipe, (3) preaction, and (4) deluge. The type of system to be used depends on a number of factors, including occupancy classification, local code requirements, and the requirements of the building fire underwriters.

Prior to installing or remodeling any sprinkler system, the applicable drawings should be submitted to the local municipal agency and the underwriter of the building. Since beneficial reductions in insurance rates may be obtained by suitable installation of sprinkler systems, it is important that the underwriter have sufficient time for a full review of the plans before construction begins. Likewise, municipal approval of the sprinkler-system plans is usually necessary before the structure can be occupied. In actual construction, the piping contractor installing the sprinkler system generally secures the necessary municipal approval. However, existence of the approval should always be checked before construction is begun.

21-24. Water Supply for Sprinklers. At least one automatic water supply having sufficient capacity and pressure is required for every sprinkler system. Factors influencing the type of supply include the area, height, and value of the structure protected, occupancy, type of construction, type and availability of public fire protection, and in some instances, the exposures of the structure. A secondary supply may be required, depending on a number of factors, including those mentioned above. For a single, or a primary supply, a connection from a reliable waterworks system having adequate pressure and capacity is preferred.

A good secondary supply can be obtained by use of one or more fire pumps, provided they have suitable drivers, sufficient capacity and reliability, and an adequate suction supply and are properly located in the premises. Underwriters may accept a motor-driven automatically controlled fire pump supplied from a water main or taking its suction under pressure from a storage system having sufficient capacity to meet the requirements of the structure protected.

For light-hazard occupancy, the pump should have a capacity of at least 250 gpm; when supplying sprinklers and hydrants, the capacity of the pump should be at least 500 gpm. Where the occupancy is classed as an ordinary hazard, the capacity should be at least 500 gpm, or 750 gpm, depending on whether or not hydrants are supplied in addition to sprinklers. For extra-hazard occupancy, consult the underwriter and local fire-protection authorities.

Light hazards are those met in apartment houses, asylums, clubhouses, colleges, churches, dormitories, hospitals, hotels, libraries, museums, office buildings, and schools. **Ordinary hazards** include mercantiles, warehouses, manufacturing, and occupancies not classed as light or extra-hazardous. **Extra-hazard occupancies** include only those buildings or portions of buildings where the inspection agency having jurisdiction determines the hazard is severe.

Other sources of water supply include gravity tanks, pressure tanks, penstocks, flumes, and fire-department connections suitable for pumping from the outside. In general, gravity tanks must have a capacity of at least 5,000 gal with the tank bottom not less than 35 ft above the underside of the roof for light- and ordinary-hazard occupancies. Pressure tanks must have at least 2,000 and 3,000 gal of water available, respectively, for these occupancies. Where penstocks, flumes, rivers, or lakes supply the water, approved strainers or screens must be provided to prevent solids from entering the piping. Figure 21-9 shows a fire-department tie-in pit.

See also Art. 20-8.

21-25. Sprinkler Piping. The local fire-prevention authorities and the fire underwriters usually have specific requirements regarding the type, material, and size of pipe used in sprinkler systems. They should be consulted prior to the design of any system. Table 21-11 lists pipe-size schedules for the three occupancy classifications met in various structures (Art. 21-24). Figure 21-10 shows a typical drain connection for a sprinkler riser pipe.

21-26. Area Protected by Sprinklers. In structures of fire-resistive construction and light-hazard occupancy (Art. 21-24), the area protected by one sprinkler should not exceed 196 sq ft; the distance c to c of the sprinkler pipes and between the sprinkler heads should not exceed 14 ft. In this type of construction with ordinary-hazard occupancy, the area protected per sprinkler should not ordinarily exceed 100 sq ft; the distance between pipes and between sprinklers should not be more than 12 ft. For extra-hazard occupancy, the area protected by each sprinkler should not exceed 90 sq ft; distance between pipes and between sprinklers should not be more than 10 ft. Local fire-protection codes and underwriters' require-

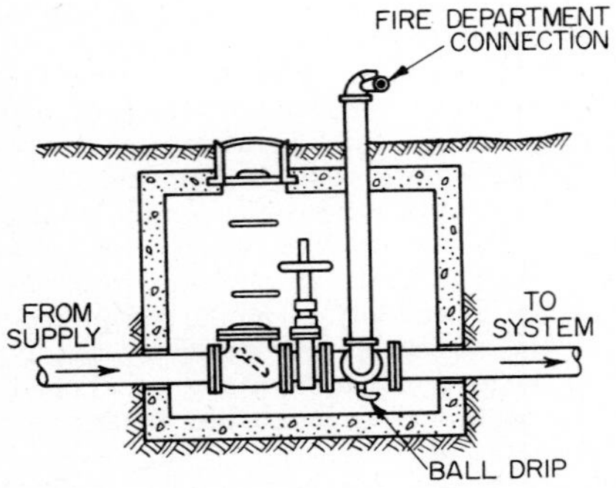

Fig. 21-9. Fire-department tie-in pit for tapping a water supply.

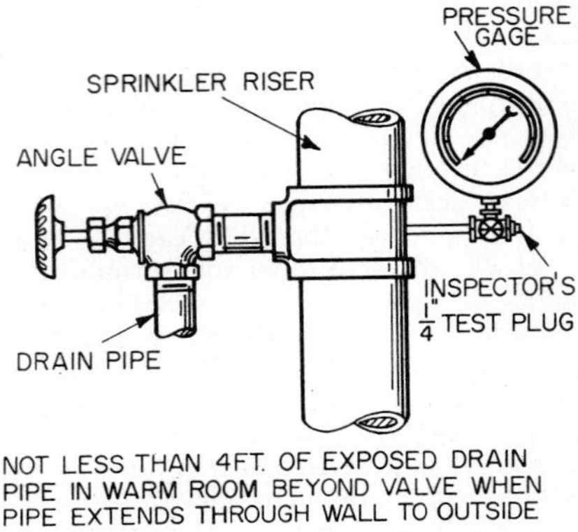

Fig. 21-10. Drain connection for sprinkler riser.

Table 21-11. Pipe-size Schedule for Typical Sprinkler Installations.

Occupancy and Pipe Size, In.	*No. of Sprinklers*
Light Hazard	
1	2
1¼	3
1½	5
2	10
2½	40
3	No limit
Ordinary Hazard	
1	2
1¼	3
1½	5
2	10
2½	20
3	40
3½	65
4	100
5	160
6	250
Extra Hazard	
1	1
1¼	2
1½	5
2	8
2½	15
3	27
3½	40
4	55
5	90
6	150

ments cover other types of construction, including mill, semimill, open-joist, and joist type with a sheathed or plastered ceiling.

21-27. Sprinkler Position. In all usual applications, the sprinklers should be installed in the upright position, where they are protected from damage by vehicles and other objects passing beneath them. When it is necessary to install sprinklers in a pendent position, they must be of a design approved for that position.

Sprinkler deflectors should be parallel to the building ceilings and roof, or the incline of stairs when installed in a stairwell. Deflectors for sprinklers in the peak

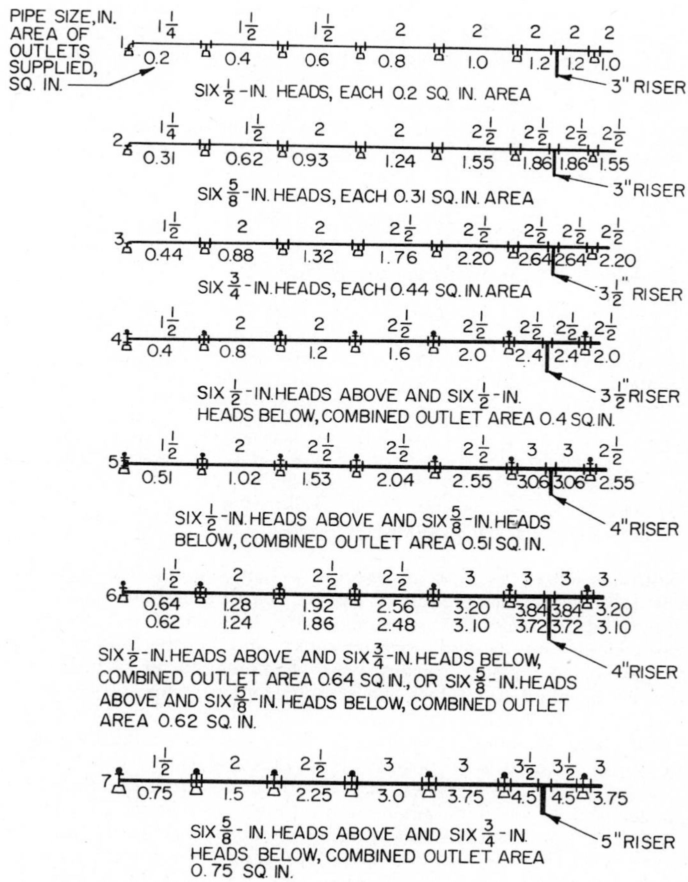

Fig. 21-11. Sprinkler piping for outside areas.

of a pitched roof should be horizontal. The distance between the deflectors and ceiling should not exceed 14 in. in fire-resistive structures; in mill and other smooth construction, the distance should not be more than 10 in.

21-28. Sprinkler Alarms. These are often required, depending on what other types of alarms or fire-notification devices are installed in the structure. The usual sprinkler alarm is designed to sound automatically when the flow from the system is equal to or greater than that from a single automatic sprinkler. Dry-pipe systems use a water-actuated alarm; preaction and deluge systems use electric alarm attachments actuated independently of the water flow in the system.

21-29. Outside Sprinklers. For protection of structures against exposure fires, outside sprinklers may be used. They can be arranged to protect cornices, windows, side walls, ridge poles, mansard roofs, etc. They are also governed by underwriters' requirements. Figure 21-11 shows the pipe sizes used for sprinklers protecting outside areas of a building, including cornices, windows, side walls, etc.

See also Art. 20-7.

HOT- AND COLD-WATER SUPPLY

Most plumbing codes contain sections dealing with hot- and cold-water distribution within buildings. Table 21-12 lists the minimum fixture-supply pipe sizes recommended by the National Plumbing Code for various types of plumbing fixtures requiring a water supply. These sizes apply to both hot- and cold-water fixtures.

21-30. Water Pressure at Fixtures. No fixture should have a water-supply pressure less than 8 psi. Furthermore, for flushometer-type water closets, a minimum pressure of 15 psi is required. There also are units for which the manufacturer requires a higher pressure.

The minimum required pressure should be available at the fixture, with sufficient allowance for pressure loss in the piping and fittings between the supply source and the fixture. Where there is a possibility that the minimum pressure in the system will fall below that required by the local plumbing code, some means must be provided to raise the pressure to the desired level. Devices used for this purpose include gravity and pressure tanks, and booster pumps. Where there is danger of excessive pressure causing water hammer, an air chamber or approved device must be installed to protect the piping from pressure surges and to reduce noise.

Table 21-12. Minimum Sizes for Fixture-supply Pipes*

Type of Fixture or Device	*Pipe Size, In.*	*Type of Fixture or Device*	*Pipe Size, In.*
Bath tubs	½	Shower (single head)	½
Combination sink and tray	½	Sinks (service, slop)	½
Drinking fountain	⅜	Sinks, flushing rim	¾
Dishwasher (domestic)	½	Urinal (flush tank)	½
Kitchen sink, residential	½	Urinal (direct flush valve)	¾
Kitchen sink, commercial	¾	Water closet (tank type)	⅜
Lavatory	⅜	Water closet (flush valve type)	1
Laundry tray, 1, 2 or 3 compartments	½	Hose bibs	½
		Wall hydrant	½

* From National Plumbing Code, ASME.

21-31. Hot- and Cold-water Piping. All water-supply and distribution piping must be designed so there is no possibility of backflow at any time. The minimum required air gap—distance between the fixture outlet and the flood-level rim of the receptacle—should be maintained at all times. Table 21-13 shows minimum air gaps generally used. Before any potable-water piping is put into use it must be disinfected with chlorine, using a procedure approved by the local code.

Pipes and tubing for water distribution may be made of copper, brass, lead, cast iron, wrought iron, or steel, provided they are approved by the local code. Threaded fittings are usually required to be galvanized or cement-coated. When choosing materials for potable-water piping, care should be taken to see there is no possibility of chemical action, or any other action that might cause a toxic condition. Figure 21-12 shows a typical piping arrangement for hot- and cold-water supply to plumbing fixtures.

21-32. Water-pipe Sizing. Since hot- and cold-water demand and usage vary considerably from one type of building to another, much care is necessary in pipe sizing. In domestic installations, each person will use a total of 20 to 80 gal per day, depending on family size, garden requirements, etc. For apartment houses, a consumption of 50 gal per person per day is generally assumed in system design, while 40 gal per person per day is often used for housing projects.

Table 21-14 shows the water pressure and flow required for various plumbing fixtures of usual design. In some instances, where special designs are used, a higher pressure or flow rate may be required.

As with drain and vent piping, the required flow for cold-water fixtures is usually computed in terms of fixture units. Table 21-15 lists the demand weight of fixtures in fixture units. Once the number of fixtures of various types is known, the load for a given type is determined by multiplying the number in the building or on a particular riser by the demand weight for this type. Figure 21-13 gives the

Table 21-13. Minimum Air Gaps for Generally Used Plumbing Fixtures*

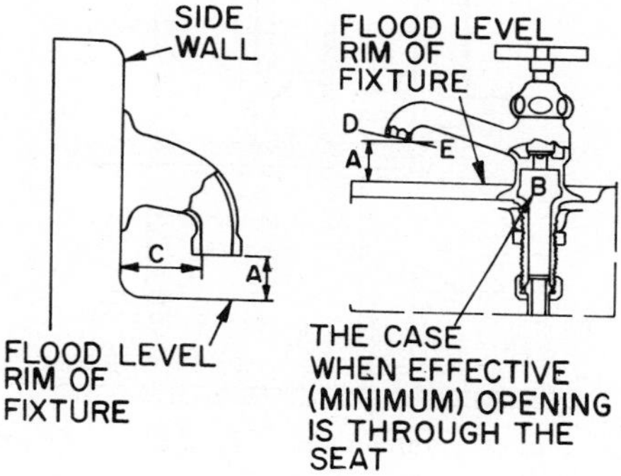

Fixture	Minimum air gap	
	When not affected by near wall†	When affected by near wall‡
Lavatories with effective openings not greater than ½-in. diameter	1.0	1.50
Sink, laundry trays, and goose neck bath faucets with effective openings not greater than ¾-in. diameter	1.5	2.25
Overrim bath fillers with effective openings not greater than 1-in. diameter	2.0	3.00
Effective openings greater than 1-in.	¶	§

* From National Plumbing Code, ASME.

† Side walls, ribs, or similar obstructions do not affect the air gaps when spaced from inside edge of spout opening a distance greater than three times the diameter of the effective opening for a single wall, or a distance greater than four times the diameter of the effective opening for two intersecting walls (see figure).

‡ Vertical walls, ribs, or similar obstructions extending from the water surface to or above the horizontal plane of the spout opening require a greater air gap when spaced closer to the nearest inside edge of spout opening than specified in note † above. The effect of three or more such vertical walls or ribs has not been determined. In such cases, the air gap shall be measured from the top of the wall.

¶ 2 × effective opening.

§ 3 × effective opening.

demand of any group of fixtures in terms of gallons of water per minute. Where fixtures other than those listed in Table 21-15 must also be supplied, such as air-conditioning units, garden hoses, process devices, etc., the required flow in gallons per minute must be computed and added to the fixture demand of the riser supplying the unit or units.

The National Plumbing Code recommends the following steps in sizing cold-water distribution piping systems: (1) Sketch all the proposed risers, horizontal mains, and branch lines, indicating the number and the type of fixtures served, together with the required water flow. (2) Compute the demand weights of the fixtures,

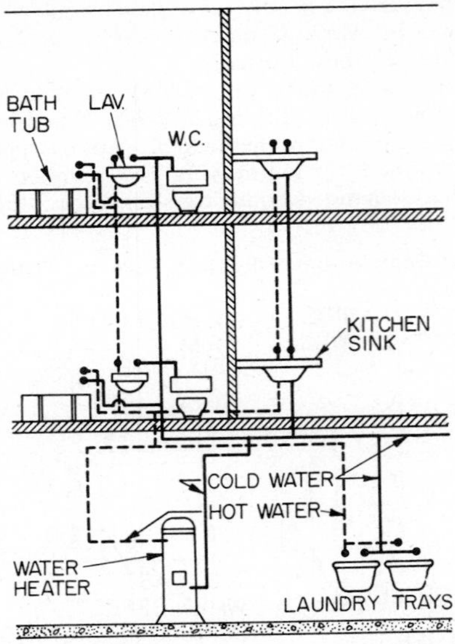

Fig. 21-12. Domestic piping arrangement. (*Babbitt, "Plumbing," McGraw-Hill Book Company, New York.*)

in fixture units, using Table 21-15. (3) Using Fig. 21-13 or 21-14, and the total number of fixture units, determine the water demand in gallons per minute. (4) Compute the equivalent length of pipe for each stack in the system, starting from the street main. (5) Obtain by test or from the water company the average

Table 21-14. Rate of Flow and Required Pressure During Flow of Different Fixtures*

Fixture	Flow pressure,† psi	Flow rate, gpm
Ordinary basin faucet	8	3.0
Self-closing basin faucet	12	2.5
Sink faucet, ⅜ in.	10	4.5
Sink faucet, ½ in.	5	4.5
Bathtub faucet	5	6.0
Laundry-tub cock, ½ in.	5	5.0
Shower	12	5.0
Ball cock for closet	15	3.0
Flush valve for closet	10–20	15–40‡
Flush valve for urinal	15	15.0
Garden hose, 50 ft and sill cock	30	5.0

* From National Plumbing Code, ANSI A40.8-1955.

† Flow pressure is the pressure in the pipe at the entrance to the particular fixture considered.

‡ Wide range due to variation in design and type of flush-valve closets.

minimum pressure in the street main. Determine the minimum pressure needed for the highest fixture in the system. (6) Compute the pressure loss in the piping, using the equivalent length found in item 4. (7) Choose the pipe sizes from a chart like that in Fig. 21-15, or from the charts given in the National Plumbing Code. (See also V. T. Manas, "National Plumbing Code Handbook," McGraw-Hill Book Company, New York.)

21-33. Hot-water Heaters. Two types of heaters are in common use. *Storage heaters* (Fig. 21-16) generally use steam as the heating medium, though hot water, gas, or electricity also find some use. The tank used with this type of heater stores hot water for future use. In general, the less uniform the demand for

Table 21-15. Demand Weight of Fixtures in Fixture Units*,†

Fixture or group‡	Occupancy	Type of supply control	Weight in fixture units¶
Water closet	Public	Flush valve	10
Water closet	Public	Flush tank	5
Pedestal urinal	Public	Flush valve	10
Stall or wall urinal	Public	Flush valve	5
Stall or wall urinal	Public	Flush tank	3
Lavatory	Public	Faucet	2
Bathtub	Public	Faucet	4
Shower head	Public	Mixing valve	4
Service sink	Office, etc.	Faucet	3
Kitchen sink	Hotel or restaurant	Faucet	4
Water closet	Private	Flush valve	6
Water closet	Private	Flush tank	3
Lavatory	Private	Faucet	1
Bathtub	Private	Faucet	2
Shower head	Private	Mixing valve	2
Bathroom group	Private	Flush valve for closet	8
Bathroom group	Private	Flush tank for closet	6
Separate shower	Private	Mixing valve	2
Kitchen sink	Private	Faucet	2
Laundry trays (1–3)	Private	Faucet	3
Combination fixture	Private	Faucet	3

* From National Plumbing Code, ASME.

† For supply outlets likely to impose continuous demands, estimate continuous supply separately and add to total demand for fixtures.

‡ For fixtures not listed, weights may be assumed by comparing the fixture to a listed one using water in similar quantities and at similar rates.

¶ The given weights are for total demand. For fixtures with both hot and cold water supplies, the weights for maximum separate demands may be taken as three-fourths the listed demand for supply.

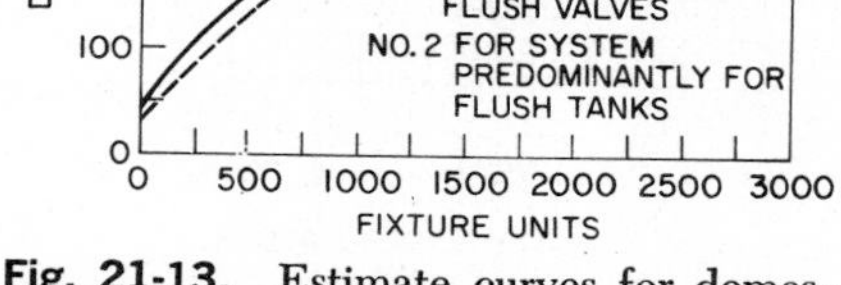

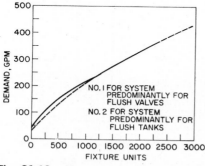

Fig. 21-13. Estimate curves for domestic water-demand load. (*National Plumbing Code, ASME.*)

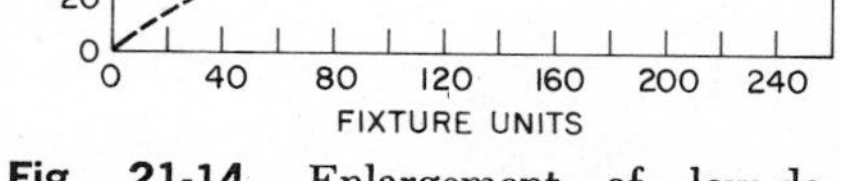

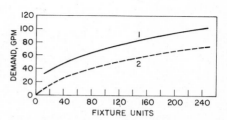

Fig. 21-14. Enlargement of low-demand portion of Fig. 21-13. (*National Plumbing Code, ASME.*)

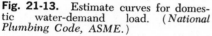

hot water, the greater the storage capacity needed in the tank. When the water demand is fairly uniform, a smaller storage capacity is allowable, but the capacity of the heating coil must be greater. However, a large storage capacity is used whenever possible, because this reduces the load on the boiler and allows use of a smaller heating coil. Industrial, office, and school-building hot-water demands are usually nonuniform with peak loads occurring during very short periods. Loads in hotels, apartment houses, and hospitals are usually more uniform.

Instantaneous heaters (Fig. 21-17) have V-shaped or straight tubes to heat the water. Steam or hot water is the usual heating medium. The water for the building is heated as needed; there is no storage section in the heater. The heater coils may be installed in a boiler instead of a tank.

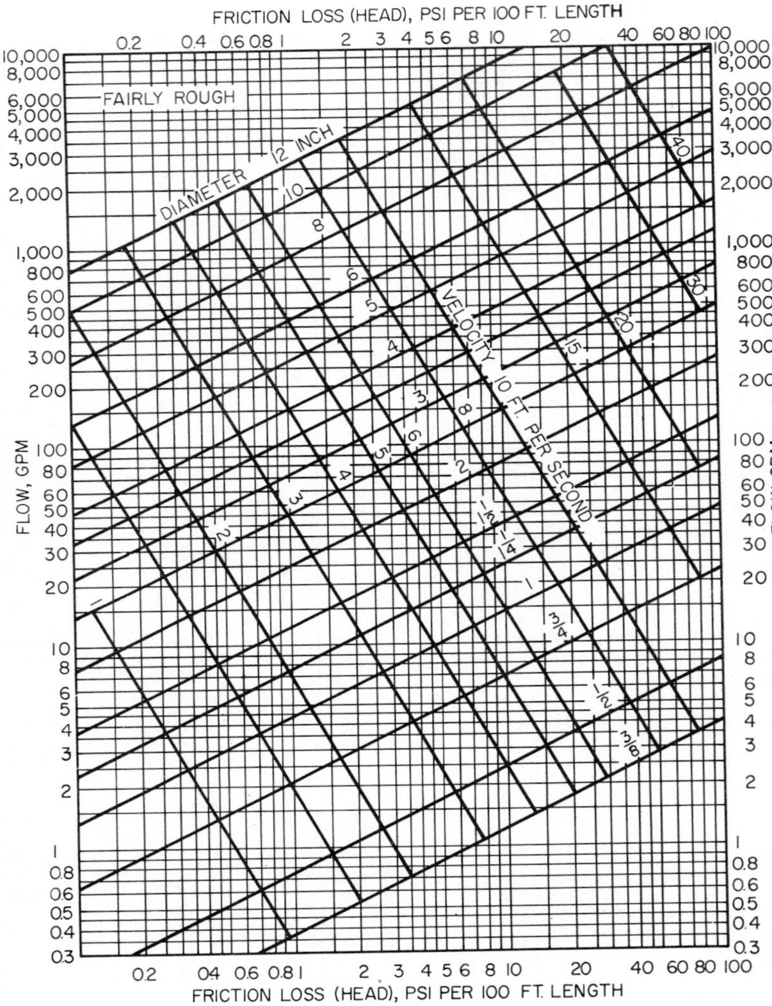

Fig. 21-15. Chart for selecting water-pipe size for various flows.

The hot-water load for a given building is computed in a manner similar to that described in Art. 21-32 but with Table 21-16 and the tabulated demand factor for the particular building type. The heating-coil capacity of the heater must at least equal the maximum probable demand for hot water.

For storage-type heaters, the storage capacity is obtained by multiplying the maximum probable demand by a suitable factor, which usually varies from 1.25

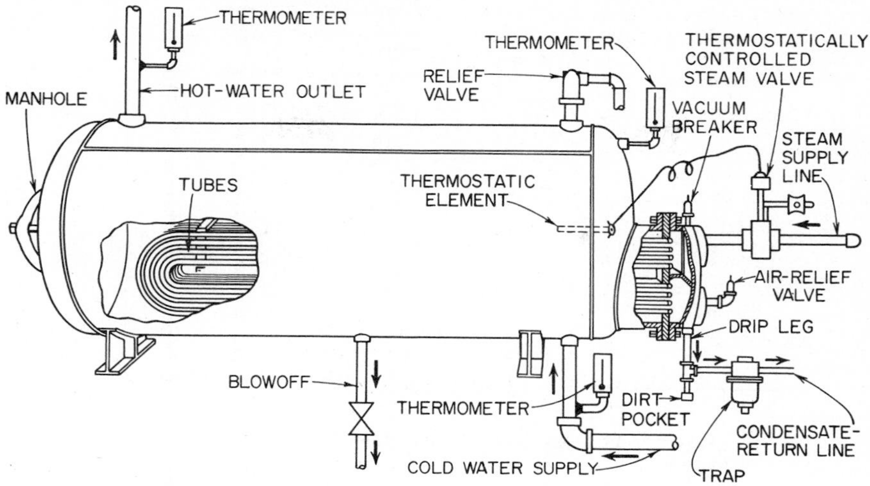

Fig. 21-16. Storage heater for domestic hot water.

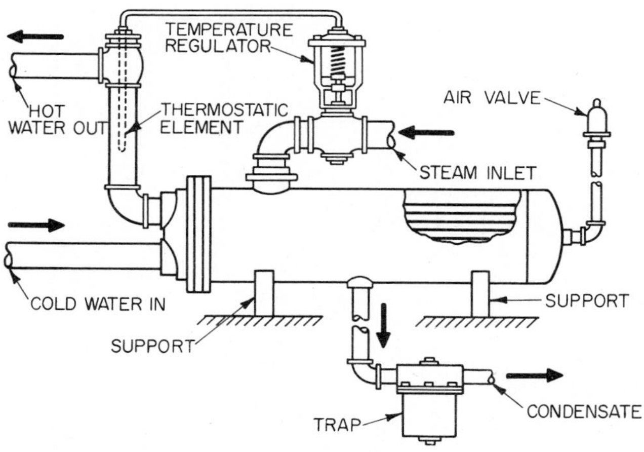

Fig. 21-17. Instantaneous heater for domestic hot water.

for apartment houses to 0.60 for hospitals. Table 21-17 lists hot-water temperatures for various services.

Example. An industrial plant has 9 showers, 200 private lavatories, 20 slop sinks, 20 public lavatories, and 1 dishwasher. What is the hourly water consumption and what storage and heating capacities are required for this load?

Using Table 21-16:

9 showers at 225 gph	=	2,025 gph
200 private lavatories at 2 gph	=	400
20 slop sinks at 20 gph	=	400
20 public lavatories at 12 gph	=	240
1 dishwasher at 250 gph	=	250
Total hourly demand	=	3,315 gph

Table 21-16. Hot-water Demand per Fixture for Various Building Types
(Based on average conditions for types of buildings listed, gallons of water per hour per fixture at 140°F)

Type of fixture	Apartment House	Hospital	Hotel	Industrial plant	Office building
Basins, private lavatories	2	2	2	2	2
Basins, public lavatories	4	6	8	12	6
Showers	75	75	75	225	
Slop sinks	20	20	30	20	15
Dishwashers (per 500 people)	250	250	250	250	250
Pantry sinks	5	10	10		
Factors					
Demand factor	0.30	0.25	0.25	0.40	0.30
Storage factor	1.25	0.60	0.80	1.00	2.00

Table 21-17. Hot-water Temperatures for Various Services, °F

Cafeterias (serving areas)	130
Lavatories and showers	130
Slop sinks (floor cleaning)	150
Slop sinks (other cleaning)	130
Cafeteria kitchens	130 + steam

Using the demand factor from Table 21-16, maximum probable demand is

$$0.40 \times 3{,}315 = 1{,}326 \text{ gph}$$

Heating-coil capacity must at least equal this maximum probable demand. Storage capacity should be $1.00 \times 3{,}315 = 3{,}315$ gal.

21-34. Hot-water Piping System. Upfeed or downfeed systems (Fig. 21-18) can be used to supply hot water to the building fixtures. To supply hot water continuously, circulating piping is often employed. Unused water is returned through this piping to the heater. Here, its temperature is raised to the desired level and the water is returned to the risers.

WASTE-WATER DISPOSAL

Waste water is comprised of domestic sewage and industrial waste. Where a public sewage system is available, the problem of domestic waste-water disposal is usually minor, except perhaps for the largest buildings. But industrial waste may present special problems because (1) the flow volume may be beyond the public sewer capacity, and (2) local regulations may prohibit the discharge of industrial waste into public sewers. Further, many recent pollution regulations prohibit discharge of industrial waste into streams, lakes, rivers, and tidal waste without suitable prior treatment. Treatment methods for a variety of industrial wastes are discussed in E. B. Besselievre, "The Treatment of Industrial Wastes," McGraw-Hill Book Company, New York. Specific design procedures for sewers, drains, and sewage treatment, with accompanying numerical examples, are given in T. G. Hicks, "Standard Handbook of Engineering Calculations," McGraw-Hill Book Company, New York.

21-35. Sewers. A sewer is a conduit for water carriage of wastes. For the purpose of this section any piping for waste water inside a building will be considered plumbing or process piping; outside the building, waste-water lines are called sewers.

Sewers carry **sewage.** And a system of sewers and appurtenances is **sewerage.** (See "Glossary—Water and Sewage Control Engineering," American Society of Civil Engineers.)

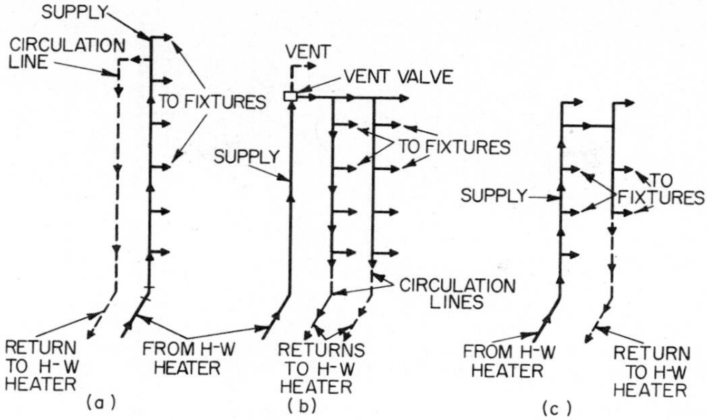

Fig. 21-18. Hot-water piping systems.

Sanitary sewers carry domestic wastes or industrial wastes. Where buildings are located on large sites or structures with large roof areas are involved, a storm sewer is used for fast disposal of rain and is laid out to drain inlets located for best collection of runoff.

For figuring rates of runoff to determine storm-sewer requirements, the rational method may be used. The so-called **rational method** for determination of runoff involves the formula $Q = AIR$, where Q is the maximum rate of runoff in cubic feet per second; A the watershed area in acres; I the imperviousness ratio of the watershed area; and R the rate of rainfall on the area in inches per hour. The imperviousness ratio I is the ratio of water running off the land to water precipitating on it. It will range from 0.7 to 0.9 for built-up areas and paved surfaces; 0.05 to 0.30 for unpaved surfaces, depending on slopes. In storm-sewer design, however, it is necessary to know not only rate of runoff and total runoff, but also at what point in time after the start of a storm the rate of runoff reaches its peak. It is this peak runoff for which pipe must be sized and sloped. It should be noted that the water necessary to fill the drainage system—storm sewers and manholes—is, in effect, storage. Such storage is a sort of buffer tending to decrease slightly the magnitude of theoretical maximum flow. Its effect can be approximated by figuring about how much flow it would take to fill the system before it can begin to flow as designed (Fig. 21-19).

21-36. Determining Sewer Size. Sanitary sewers or lines carrying exclusively industrial wastes from a building to disposal must be sized and sloped according to best hydraulic design. The problem is generally one of flow in a circular pipe. (C. V. Davis and K. E. Sorenson, "Handbook of Applied Hydraulics," McGraw-Hill Book Company, New York.) Gravity flow is to be desired; but pumping is sometimes required.

Pipe should be straight and of constant slope between manholes, and manholes should be used at each necessary change in direction, slope, elevation, or size of pipe. Manholes should be no farther apart than 500 ft, and preferably as close as 200 ft where pipe used is of a diameter too small to enter.

The sewer from a building must be sized to carry out all the water carried in by supply mains or other means. Exceptions to this are the obvious cases where losses might be appreciable, such as an industrial building where considerable water is consumed in a process or evaporated to the atmosphere. But, in general, water out about equals water in, plus all the liquid and water-borne solid wastes produced in the building.

Another factor to consider in sizing a sewer is infiltration. Sewers, unlike water mains, often flow at less pressure than that exerted by ground water around them. Thus they are more likely to take in ground water than to leak out sewage. An infiltration rate of 2,000 to 200,000 gal per day per mile might be expected. It

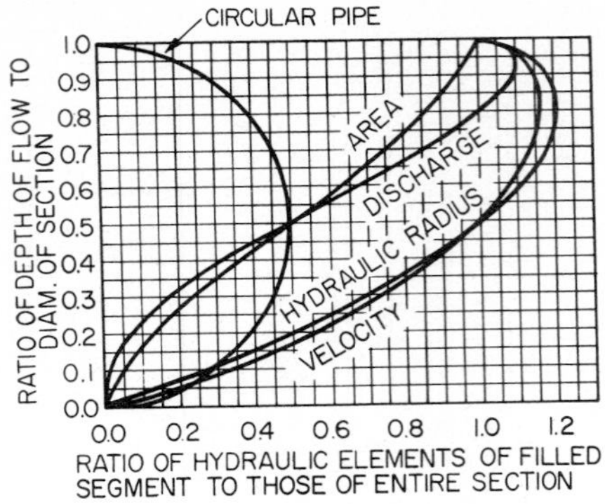

Fig. 21-19. Hydraulic properties of circular pipe for various depths of flow.

all depends on diameter of pipe (which fixes length of joints), type of soil, groundwater pressure—and, of course, workmanship.

In an effort to keep infiltration down, sewer-construction contracts are written with a specified infiltration rate. Weir tests in a completed sewer can be used to check the contractor's success in meeting the specification; but unless the sewer is large enough to go into, prevention of excessive infiltration is easier than correction. In addition to groundwater infiltration through sewer-pipe joints, the entry of surface runoff through manhole covers and thus into sewers is often a factor. Observers have gaged as much as 150 gpm leaking into a covered manhole.

Size and slope of a sanitary sewer depend lastly—and vitally—on a requirement that velocity when flowing full be kept to at least 2 fps to keep solids moving and prevent clogging. In general, no drain pipe should be less than 6 in. in diameter; an 8-in. minimum is safer.

21-37. Sewer-pipe Materials. Vitrified clay, concrete, asbestos cement, cast iron, bituminized fiber, and steel may be used for pipe to carry sewage and industrial wastes. Vitrified clay up to about 36 in. diameter and concrete in larger sizes are used most generally; asbestos cement is growing in use for the smaller diameters.

Choice of pipe material depends on required strength to resist load or internal pressure; corrosion resistance, which is especially vital in carrying certain industrial wastes; erosion resistance in sewers carrying coarse solids; roughness factor where flat slopes are desirable; and cost in place.

Vitrified-clay pipe requires some care in handling to minimize breakage. Concrete pipe must be made well enough or protected to withstand effects of damaging sewer gas (hydrogen sulfide) or industrial wastes. Asbestos-cement pipe comes in greater lengths, therefore has fewer joints. It is smooth, noncorrosive, and easily jointed. Cast-iron sewer pipe is good under heavy loads, exposed as on bridges, in inverted siphons, or in lines under pressure. Bituminized-fiber pipe in smaller diameters (to 8 in.) for drains is lighter than but about as strong as clay or concrete. It may deteriorate in sunlight and must be handled with reasonable care. Steel is used chiefly for its strength or flexibility. Corrugated-steel pipe with protective coatings is made especially for sewer use; its long lengths and light weight ease handling and jointing.

Section 22

Electric Power and Lighting

CHARLES J. WURMFELD

Consulting Engineer, New York, N.Y.

22-1. Design Standards. The National Electrical Code is the basis for all municipal electric wiring ordinances and the basic standard for electrical design in the United States. Some of the larger cities, however, have elaborated on this code and the local code may contain more restrictive provisions. The local ordinance should always be consulted.

The National Electrical Code, sponsored by the National Fire Protection Association, has been adopted by the American Insurance Association (formerly the National Board of Fire Underwriters) and approved by the American National Standards Institute. Copies can be obtained from the NFPA, 60 Batterymarch Street, Boston, Mass. 02110. (See also F. Stetka, "NFPA Handbook of the National Electrical Code," McGraw-Hill Book Company, New York.)

The American Insurance Association sponsors the Underwriters Laboratories, Inc., which passes on electrical material and equipment in accordance with standard test specifications. The UL also issues a semiannual List of Inspected Electrical Appliances, which can be obtained from the UL at 207 East Ohio Street, Chicago, Ill. 60611.

Electrical codes and ordinances are written primarily to protect the public from fire and other hazards to life. They represent minimum safety standards. Strict application of these codes will not, however, guarantee satisfactory or even adequate performance. Correct design of an electrical system, over these minimum safety standards, to achieve a required level of performance, is the responsibility of the electrical designers.

22-2. Definitions. **Electric current** I is the rate at which electricity flows through a conductor or circuit. The practical unit is the ampere, which is a current of one coulomb per second. A coulomb is a basic quantity of electricity.

Electric currents are classified as:

1. **Direct current** (dc) which always flows in the same direction.
2. **Alternating current** (ac) which reverses direction at regular intervals.

Resistance R is the opposition offered by a material to the flow of an electric current in it. The unit of electrical resistance is the ohm.

Electromotive force, potential difference, or voltage E is the force that makes electrons move in a circuit. It is measured in volts.

$$E = IR \qquad \text{(Ohm's law)} \tag{22-1}$$

By convention, electric current is assumed to flow from a positive (+) to a negative (—) terminal.

Electric power W is the rate of doing electrical work. The unit is the watt or kilowatt (1,000 watts). 746 watts = 0.746 kw = 1 hp. Direct-current power is given in watts by

$$W = EI = I^2R \tag{22-2}$$

where E = voltage drop
I = current, amp
R = resistance, ohms

Power in ac circuits may not be equal to the product of the current and voltage. In single-phase ac circuits, for example, power generally is the product of current, voltage, and a power factor which measures the effect of a phase difference between current and voltage.

Kilowatthour (kwhr) is the energy expended if work is done for 1 hr at the rate of 1 kilowatt.

Efficiency is the ratio of output to input. Output is the useful energy delivered by a machine. Input is the energy supplied.

A cycle is a complete set of positive and negative values through which an alternating current repeatedly passes.

Frequency is the number of cycles of alternating current per second.

Phase is the time relationship between a voltage and a current with the same frequency. If the current and voltage pass through zero, maximum, or minimum values simultaneously, they are in phase. If current and voltage are represented by a sine curve, one may lead or lag the other by up to 360°. If, for example, the maximum of a sinusoidal current occurs 90° before the maximum of the voltage, the current may be considered to lead the voltage by 90° or lag by 270°. Phase differences also may exist between two currents or two voltages. For example, a three-phase power supply delivers three alternating currents 120° out of phase with one another.

Power factor is the ratio of true power to apparent power. It usually is expressed as a percentage. It reaches 100% only when current and voltage are in phase.

Apparent power comprises an energy component and a wattless (inductance or capacitance) component. The smaller the power factor the larger the wattless component. As a result, larger equipment and conductors are needed to deliver the required power. Low power factors often may be corrected by installing synchronous motors, which, when overexcited, have the property of neutralizing wattless components of current, or by connecting static condensers across the line.

Inductance, L, opposes the change in magnitude of an alternating current, makes the current lag the voltage. The unit of measure is the henry. Inductive reactance, X_L, impedes the flow of current. It is measured in ohms.

Capacitance, C, the ratio of charge to potential difference between two points of a circuit, measures the ability of the circuit to store electrical energy in the form of restrained charge. A condenser or capacitor, consisting of two conductors separated by an insulator, has this ability to store energy when the circuit carries alternating current. Capacitance is measured in farads. Capacitive reactance, X_C, which impedes the flow of current, makes the voltage lag the current. The unit of measure is the ohm.

Impedance, Z, measured in ohms, is the total opposition to flow of alternating current due to resistance, R; inductive reactance, X_L; and capacitive reactance, X_C. It is the vector sum of these components:

$$Z^2 = R^2 + (X_L - X_C)^2 \tag{22-3}$$

A **series circuit** has components connected in sequence. Hence, the current at every point is the same, the total resistance is the sum of the individual resistances, and the voltage across each component varies with the load. A disadvantage of this type of circuit is that current will not flow if there is a break at any point in the circuit. The most important commercial application is in street lighting. But if one lamp in a group goes out, all go out.

Multiple, parallel, or shunt circuits have components with common terminals. Voltage across the components is the same, and the current, varying with the load, divides among them. Parallel circuits generally are used for distribution of electricity for power and lighting in buildings.

Circular mil is the area of a circle 0.001 in. in diameter.

Square mil is the area of a square with sides 0.001 in. long.

22-3. Building Wiring Systems. The electrical load in a building is the sum of the loads, in kilowatts (kw), for lighting, motors, and appliances.

In computing the size of feeders for some types of buildings, the designer may apply a demand factor to the lighting load, since all lights may not be on at the same time. Table 22-1 gives recommended minimum lighting loads based on floor areas and the allowed demand factors for several types of buildings. These minimum loads should be used only when the actual lighting-fixture installation is not known; for example, in apartment building design. For other building types, such as factories or commercial buildings, where illumination is designed, the installed-fixture load should be used.

Motor and appliance loads usually are taken at full value. Household and kitchen appliances, however, are exceptions. The National Electrical Code (Art. 22-1) lists demand factors for household electric ranges, ovens, and clothes dryers. Some municipal codes allow the first 3,000 watts of apartment appliance load to be included with lighting load and therefore to be reduced by the demand factor applied to lighting.

For factories and commercial buildings, the electrical designer should obtain from the mechanical design the location and horsepower of all blowers, pumps, compressors, and other electrical equipment, as well as the load for elevators, boiler room, and other machinery. The load in amperes for running motors is given in Tables 22-2A and 22-2B.

Plans. Electrical plans should be drawn to scale, traced or reproduced from the architectural plans. Architectural dimensions may be omitted except for such rooms as meter closets or service space, where the contractor may have to detail his equipment to close dimensions. Floor heights should be indicated if full elevations are not given. Locations of windows and doors should be reproduced accurately, and door swings shown, to facilitate location of wall switches. For estimating purposes, feeder or branch runs may be scaled from the plans with sufficient accuracy.

Indicate on the plans by symbol the location of all electrical equipment (Table 22-3). Show all ceiling outlets, wall receptacles, switches, junction boxes, panel boxes, telephone and interior communication equipment, fire alarms, television master-antenna connections, etc.

A complete set of electrical plans should include a diagram of feeders, panel lists, service entrance location, and equipment. Before these can be shown on the plans, however, wire sizes must be computed in accordance with procedures outlined in Arts. 22-5 to 22-7 and 22-11.

Where there is only one panel box in an area, and it is clear that all circuits in that area connect to that box, it is not necessary to number the panel other than to designate it as, for example, "apartment panel." In larger areas, where two or more panel boxes may be needed, each should be labeled for identification and location; for example, L.P. 1-1, L.P. 1-2 . . . for all panels on the first floor; L.P. 2-1, L.P. 2-2 . . . for panels on the second floor.

Branch Circuits. It is good practice to limit branch runs to a maximum of 50 ft by installing sufficient panels in efficient locations.

Connect each outlet with a branch circuit and show the home runs to the panel, as indicated in Table 22-3. General lighting branch circuits with a 15-amp fuse or circuit breaker in the panel usually are limited to 6 to 8 outlets, although most

Table 22-1. Code Demand Factors for Lighting*

Type of occupancy	Unit load per sq ft, watts	Load to which demand factor applies, watts	Demand factor, %
Armories and auditoriums	1	Total wattage	100
Banks	2	Total wattage	100
Barber shops and beauty parlors	3	Total wattage	100
Churches	1	Total wattage	100
Clubs	2†	Total wattage	100
Court rooms	2	Total wattage	100
Dwellings (other than hotels)	3†	3,000 or less	100
		Next 117,000	35
		Over 120,000	25
Garages—commercial (storage)	½	Total wattage	100
Hospitals	2	50,000 or less	40‡
		Over 50,000	20
Hotels and motels, including apartment houses without provisions for cooking by tenants	2†	20,000 or less	50‡
		Next 80,000	40
		Over 100,000	30
Industrial commercial (loft) buildings	2	Total wattage	100
Lodge rooms	1½	30,000 or less	100
		Over 30,000	70
Office buildings	5	Total wattage	100
Restaurants	2	Total wattage	100
Schools	3	Total wattage	100
Stores	3	12,500 or less	100
		Over 12,500	50
Warehouses, storage	¼		
In any of above occupancies except single-family dwellings and individual apartments of multifamily dwellings:			
Assembly halls and auditoriums	1	Total wattage as specified for the specific occupancy	
Halls, corridors	½		
Closets, storage spaces	¼		

* From "National Electric Code," 1968.

In view of the trend toward higher-intensity lighting systems and increased loads due to more general use of fixed and portable appliances, each installation should be considered as to the load likely to be imposed and the capacity should be increased to ensure safe operation.

Where electric-discharge lighting systems are to be installed, high power-factor type should be used, or the conductor capacity may need to be increased.

† For general illumination in dwelling occupancies, it is recommended that one 15-amp branch circuit be installed for each 375 sq ft of floor area.

‡ For subfeeders to areas in hospitals and hotels where entire lighting is likely to be used at one time, e.g., in operating rooms, ballrooms, dining rooms, a demand factor of 100% shall be used.

codes permit 12. No more than two outlets should be connected in a 20-amp appliance circuit.

It is good practice to use wire no smaller than No. 12 in branch circuits, though some codes permit No. 14. Special-purpose individual branch circuits for motors or appliances should be sized to suit the connected load.

22-4. Electric Services. For economy, alternating current is transmitted long distances at high voltages and then changed to low voltages by step-down transformers at the point of service.

Small installations, such as one-family houses, usually are supplied with three-wire service. This consists of a neutral (transformer mid-point) and two power wires with current differing 180° in phase. From this service, the following types of

Table 22-2A. Protection of Single-phase Motors and Circuits

Size of motor: Horsepower	Size of motor: Ampere rating	Branch-circuit protection*: Maximum size fuse permitted by the Code	Branch-circuit protection*: Time-delay-cartridge or low-peak fuse that can be used	Motor-running protection†, size of time-delay-cartridge or low-peak fuse: Ordinary service	Motor-running protection†, size of time-delay-cartridge or low-peak fuse: Heavy service	Motor-running protection†, size of time-delay-cartridge or low-peak fuse, Maximum size: 40°C motor	Motor-running protection†, size of time-delay-cartridge or low-peak fuse, Maximum size: All other motors	Size of fused switch or fuse holder: Maximum size switch that can be used	Size of fused switch or fuse holder: Size that can be used with time-delay-cartridge or low-peak fuses	Minimum size of starter: NEMA size	Minimum size and type of wire: AWG or MCM	Minimum size of conduit: diameter, in.
115 volts												
1/6	4.4	15	7	4½	5	5 6/10	5 6/10					
1/4	5.8	20	10	5 6/10	6¼	8	7	30	30	00	14 R	½
1/3	7.2	25	12	7	8	9	9					
1/2	9.8	30	17½	10	12	12	12	30	30	0	14 R	½
3/4	13.8	45	25	15	17½	17½	17½	60	30			
1	16	50	30	17½	20	20	20	60	30			
1½	20	60	30	20	25	25	25	60	30	0	12 R	½
2	24.0	80	40	25	30	30	30	100	60	1	10 R	¾
230 volts												
1/6	2.2	15	3½	2¼	2½	2 8/10	2 8/10					
1/4	2.9	15	5	2 8/10	3 2/10	4	3½	30	30	00	14 R	½
1/3	3.6	15	6	3½	4	4½	4½					
1/2	4.9	15	8	5	5 6/10	6¼	6¼					
3/4	6.9	25	12	7	8	9	8	30	30	00	14 R	½
1	8	25	15	8	9	10	10					
1½	10	30	17½	10	12	12	12	30	30			
2	12	40	20	12	15	15	15	60	30	0	14 R	½
3	17	60	25	17½	20	20	20	60	30	1	10 R	¾
5	28	90	45	30	35	35	35	100	60	2	8 R	¾
7½	40	125	60	40	45	50	50	200	60	2	6 R	1
10	50	150	80	50	60	70	60	200	100	3	4 R	1¼

* These do not give motor-running protection.
† On normal installations these also give branch-circuit protection.

Table 22-2B. Protection of Three-phase 208-volt Motors and Circuits

Size and class of motor			Branch circuit protection*		Motor-running protection†				Size of fused switch or fuse holder		Mini-mum size of starter: NEMA size	Mini-mum size and type of wire: AWG or MCM	Mini-mum size of conduit: diame-ter, in.
					Size of time-delay-cartridge or low-peak fuse								
							Maximum size						
Horse-power	Ampere rating	Class	Maximum size fuse permitted by the Code	Time-delay-cartridge or low-peak fuse that can be used	Ordinary service	Heavy service	40°C motor	All other motors	Maximum size switch that can be used	Size that can be used with time-delay-cartridge or low-peak fuses			
½	2.1	Any	15	5	2¼	2½	2 8/10	2½	30	30	00	14 R	½
¾	3	Any	15	8	3 2/10	3½	4	3½					
1	3.7	Any	15	8	4	4½	5	4½					
1½	5.3	Any	15	10	5 6/10	6¼	7	6¼					
2	6.9	1	25	12	7	8	9	8	30	30	0	14 R	½
		2	20	12	7	8	9	8					
		3–4	15	12	7	8	9	8					
3	9.5	1	30	15	10	12	12	12	30	30	0	14 R	½
		2	25	15	10	12	12	12					
		3	20	15	10	12	12	12					
		4	15	15	10	12	12	12					
5	15.9	1	50	25	17½	20	20	20	60	30	1	12 R	½
		2	40	25	17½	20	20	20	60	30			
		3	35	25	17½	20	20	20	60	30			
		4	25	25	17½	20	20	20	30	30			
7½	23.3	1	80	35	25	30	30	30	100	60	1	10 R	¾
		2	60	35	25	30	30	30	60	60			
		3	50	35	25	30	30	30	60	60			
		4	40	53	25	30	30	30	60	60			

10	28.6	1	90	45	30	35	40	35	100	60			
		2	70	45	30	35	40	35	100	60			
		3	60	45	30	35	40	35	60	60	2	8 R	¾
		4	45	45	30	35	40	35	60	60			
15	42.3	1	125	60	45	50	50	50	200	60			
		2	110	60	45	50	50	50	200	60			
		3	90	60	45	50	50	50	100	60	2	6 R	1
		4	70	60	45	50	50	50	100	60			
20	55	1	175	90	60	70	70	70	200	100			
		2	150	90	60	70	70	70	200	100			
		3	110	90	60	70	70	70	200	100	3	4 R	1¼
		4	90	90	60	70	70	70	100	100			
25	68	1	225	100	70	80	90	80	400	100			
		2	175	100	70	80	90	80	200	100			
		3	150	100	70	80	90	80	200	100	3	2 R	1¼
		4	110	100	70	80	90	80	200	100			
30	83	1	250	125	90	100	110	100	400	200			
		2	225	125	90	100	110	100	400	200			
		3	175	125	90	100	110	100	200	200	3	1 R	1½
		4	125	125	90	100	110	100	200	200			
40	110	1	350	175	110	125	150	150	400	200			
		2	300	175	110	125	150	150	400	200			
		3	225	175	110	125	150	150	400	200	4	0 RH	2
		4	175	175	110	125	150	150	200	200			
50	132	1	400	200	150	175	175	175	400	200			
		2	350	200	150	175	175	175	400	200			
		3	300	200	150	175	175	175	400	200	4	00 RH	2
		4	200	200	150	175	175	175	200	200			
60	159	1	500	250	175	200	200	200	600	400			
		2	400	250	175	200	200	200	400	400			
		3	350	250	175	200	200	200	400	400	5	000 RH	2
		4	250	250	175	200	200	200	400	400			

Table 22-2B. Protection of Three-phase 208-volt Motors and Circuits (*Continued*)

Size and class of motor			Branch circuit protection*		Motor-running protection†				Size of fused switch or fuse holder				
					Size of time-delay-cartridge or low-peak fuse								
							Maximum size						
Horse-power	Ampere rating	Class	Maximum size fuse permitted by the Code	Time-delay-cartridge or low-peak fuse that can be used	Ordinary service	Heavy service	40°C motor	All other motors	Maximum size switch that can be used	Size that can be used with time-delay-cartridge or low-peak fuses	Minimum size of starter: NEMA size	Minimum size and type of wire: AWG or MCM	Minimum size of conduit: diameter, in.
75	196	1	600	300	200	225	250	250	600	400			
		2	500	300	200	225	250	250	600	400			
		3	400	300	200	225	250	250	400	400	5	250 RH	2½
		4	300	300	200	225	250	250	400	400			
100	260	1		400	250	300	350	300		400			
		2		400	250	300	350	300		400			
		3	600	400	250	300	350	300	600	400	5	400 RH	3
		4	400	400	250	300	350	300	400	400			
125	328	1–3		500	350	400	450	400		600		2 sets‡	
		4	500	500	350	400	450	400	600	600	6	0000 RH	2½‡
150	381	1–3		600	400	450	500	450		600		2 sets‡	
		4	600	600	400	450	500	450	600	600	6	250 RH	2½‡

* These do not give motor-running protection.
† On normal installations these also give branch-circuit protection.
‡ Indicates two sets of multiple conductors and two runs of conduit.

Table 22-3. Electrical Symbols*

Wall	*Ceiling*	
⊢○	○	Outlet
-Ⓑ	Ⓑ	Blanked outlet
	Ⓓ	Drop cord
-Ⓔ	Ⓔ	Electrical outlet—for use only when circle used alone might be confused with columns, plumbing symbols, etc.
-Ⓕ	Ⓕ	Fan outlet
-Ⓙ	Ⓙ	Junction box
-Ⓛ	Ⓛ	Lamp holder
-Ⓛ PS	Ⓛ PS	Lamp holder with pull switch
-Ⓢ	Ⓢ	Pull switch
Ⓥ	-Ⓥ	Outlet for vapor-discharge lamp
-ⓧ	ⓧ	Exit-light outlet
-Ⓒ	-Ⓒ	Clock outlet (specify voltage)
⊖		Duplex convenience outlet
⊖ 1,3		Convenience outlet other than duplex. 1 = single, 3 = triplex, etc.
⊖ WP		Weatherproof convenience outlet
⊖ R		Range outlet
⊖ S		Switch and convenience outlet
⊖ [R]		Radio and convenience outlet
(△)		Special-purpose outlet (designated in specifications)
⊙		Floor outlet
S		Single-pole switch
S_2		Double-pole switch
S_3		Three-way switch
S_4		Four-way switch
S_D		Automatic door switch
S_E		Electrolier switch
S_K		Key-operated switch

Table 22-3. Electrical Symbols (*Continued*)

Wall *Ceiling*	
S_P	Switch and pilot lamp
S_{CB}	Circuit breaker
S_{WCB}	Weatherproof circuit breaker
S_{MC}	Momentary-contact switch
S_{RC}	Remote-control switch
S_{WP}	Weatherproof switch
S_F	Fused switch
S_{WF}	Weatherproof fused switch
⊖$_{a,b,c}$ etc. ○$_{a,b,c}$ etc. $S_{a,b,c}$ etc.	Any standard symbol as given above with the addition of a lower-case subscript letter may be used to designate some special variation of standard equipment of particular interest in a specific set of architectural plans. When used they must be listed in the key of symbols on each drawing and if necessary further described in the specifications
■	Lighting panel
▨	Power panel
——	Branch circuit; concealed in ceiling or wall
—·—	Branch circuit; concealed in floor
- - - -	Branch circuit; exposed
—▸▸	Home run to panel board. Indicate number of circuits by number of arrows. NOTE: Any circuit without further designation indicates a two-wire circuit. For a greater number of wires indicate as follows: —///— (3 wires), —////— (4 wires), etc.
——	Feeders. NOTE: Use heavy lines and designate by number corresponding to listing in feeder schedule
≡▭≡	Underfloor duct and junction box. Triple system. NOTE: For double or single systems eliminate one or two lines. This symbol is equally adaptable to auxiliary-system layouts
Ⓖ	Generator
Ⓜ	Motor
Ⓘ	Instrument
Ⓣ	Power transformer (or draw to scale)
⊠	Controller
▭┘	Isolating switch

Table 22-3. Electrical Symbols (*Continued*)

Wall	*Ceiling*	
		Push button
		Buzzer
		Bell
		Annunciator
		Outside telephone
		Interconnecting telephone
		Telephone switchboard
		Bell-ringing transformer
		Electric door opener
		Fire-alarm bell
		Fire-alarm station
		City fire-alarm station
		Fire-alarm central station
		Automatic fire-alarm device
		Watchman's station
		Watchman's central station
		Horn
		Nurse's signal plug
		Television antenna outlet
		Radio outlet
		Signal central station
		Interconnection box
		Battery
—·—·—		Auxiliary-system circuits. NOTE: Any line without further designation indicates a 2-wire system. For a greater number of wires designate with numerals in manner similar to 12—No. 18W-¾″-C., or designate by number corresponding to listing in schedule.
□a,b,c		Special auxiliary outlets. Subscript letters refer to notes on plans or detailed description in specifications

* National Electrical Code, National Fire Protection Association, Boston, Mass. 02110.

Table 22-4. Properties of Conductors*

Size AWG M cm	Area, cm	Concentric lay stranded conductors		Bare conductors		D-c resistance, Ω/1,000 ft at 25°C, 77°F		
						Copper		Aluminum
		No. wires	Diam. each wire, in	Diam., in	Area,† in²	Bare cond.	Tin'd. cond.	
18	1,624	Solid	0.0403	0.0403	0.0013	6.51	6.79	10.7
16	2,583	Solid	0.0508	0.0508	0.0020	4.10	4.26	6.72
14	4,107	Solid	0.0641	0.0641	0.0032	2.57	2.68	4.22
12	6,530	Solid	0.0808	0.0808	0.0051	1.62	1.68	2.66
10	10,380	Solid	0.1019	0.1019	0.0081	1.018	1.06	1.67
8	16,510	Solid	0.1285	0.1285	0.0130	0.6404	0.659	1.05
6	26,250	7	0.0612	0.184	0.027	0.410	0.427	0.674
4	41,740	7	0.0772	0.232	0.042	0.259	0.269	0.424
3	52,640	7	0.0867	0.260	0.053	0.205	0.213	0.336
2	66,370	7	0.0974	0.292	0.067	0.162	0.169	0.266
1	83,690	19	0.0664	0.332	0.087	0.129	0.134	0.211
0	105,500	19	0.0745	0.373	0.109	0.102	0.106	0.168
00	133,100	19	0.0837	0.418	0.137	0.0811	0.0843	0.133
000	167,800	19	0.0940	0.470	0.173	0.0642	0.0668	0.105
0000	211,600	19	0.1055	0.528	0.219	0.0509	0.0525	0.0836
250	250,000	37	0.0822	0.575	0.260	0.0431	0.0449	0.0708
300	300,000	37	0.0900	0.630	0.312	0.0360	0.0374	0.0590
350	350,000	37	0.0973	0.681	0.364	0.0308	0.0320	0.0505
400	400,000	37	0.1040	0.728	0.416	0.0270	0.0278	0.0442
500	500,000	37	0.1162	0.814	0.520	0.0216	0.0222	0.0354
600	600,000	61	0.0992	0.893	0.626	0.0180	0.0187	0.0295
700	700,000	61	0.1071	0.964	0.730	0.0154	0.0159	0.0253
750	750,000	61	0.1109	0.998	0.782	0.0144	0.0148	0.0236
800	800,000	61	0.1145	1.031	0.835	0.0135	0.0139	0.0221
900	900,000	61	0.1215	1.093	0.938	0.0120	0.0123	0.0197
1,000	1,000,000	61	0.1280	1.152	1.042	0.0108	0.0111	0.0177
1,250	1,250,000	91	0.1172	1.289	1.305	0.00863	0.00888	0.0142
1,500	1,500,000	91	0.1284	1.412	1.566	0.00719	0.00740	0.0118
1,750	1,750,000	127	0.1174	1.526	1.829	0.00616	0.00634	0.0101
2,000	2,000,000	127	0.1255	1.631	2.089	0.00539	0.00555	0.00885

* SOURCE: Table 17-12 from W. T. Stuart, Wiring Design—Commercial and Industrial Buildings, in D. G. Fink and J. M. Carroll (eds.), "Standard Handbook for Electrical Engineers," 10th ed., McGraw-Hill Book Co., 1968. Used by permission.

† Area given is that of a circle having a diameter equal to the overall diameter of a stranded conductor.

The values given in the table are those given in *NBS Circ.* 31 except that those shown in the eighth column are those given in Specification B33 of the American Society for Testing and Materials.

The resistance values given in the last three columns are applicable only to direct current. When conductors larger than No. 4/0 are used with alternating current, the multiplying factors in Table 17-11 should be used to compensate for skin effect.

interior branch circuits are available:

Single-phase two-wire 230-volt—by tapping across the phase wires.

Single-phase two-wire 115-volt—by tapping across one phase wire and the neutral.

Single-phase three-wire 115/230-volt—by using both phase wires and the neutral.

For larger installations, the service most widely used is the 120/208-volt three-phase four-wire system. This has a neutral and three power wires carrying current differing 120° in phase. From this service, the following types of interior branch circuits are available:

Single-phase two-wire 208-volt—by tapping across two phase wires.

Single-phase two-wire 120-volt—by tapping across one phase wire and the neutral.

Two-phase three-wire 120/208-volt—by using two phase wires and the neutral.

Three-phase three-wire 208-volt—by using three phase wires.

Three-phase four-wire 120/208-volt—by using three phase wires and the neutral.

22-5. Circuit and Conductor Calculations. The current in a conductor may be computed from the following formulas, in which

I = conductor current, amp
W = power, watts
f = power factor, as a decimal
E_p = voltage between any two phase legs
E_g = voltage between a phase leg and neutral, or ground

Single-phase two-wire circuits:

$$I = \frac{W}{E_p f} \quad \text{or} \quad I = \frac{W}{E_g f} \tag{22-4}$$

Single-phase three-wire (and balanced two-phase three-wire) circuits:

$$I = \frac{W}{2E_g f} \tag{22-5}$$

Three-phase three-wire (and balanced three-phase four-wire) circuits:

$$I = \frac{W}{3E_g f} \tag{22-6}$$

When circuits are balanced in a three-phase four-wire system, no current flows in the neutral. So when a three-phase four-wire feeder is brought to a panel from which single-phase circuits will be taken, the system should be designed so that under full load the load on each phase leg will be nearly equal.

22-6. Voltage-drop Calculations. Voltage drop in a circuit may be computed from the following formulas, in which

V_d = voltage drop between any two phase wires, or between phase wire and neutral when only one phase wire is used in the circuit
I = current, amp
L = one-way run, ft
R = resistance, ohms per mil-ft
c.m. = circular mils

Single-phase two-wire (and balanced single-phase three-wire) circuits:

$$V_d = \frac{2RIL}{\text{c.m.}} \tag{22-7}$$

Balanced two-phase three-wire, three-phase three-wire, and balanced three-phase four-wire circuits:

$$V_d = \frac{\sqrt{3}\,RIL}{\text{c.m.}} \tag{22-8}$$

Equations (22-7) and (22-8) contain a factor R that represents the resistance in ohms to direct current of 1 mil-ft of wire. The value of R may be taken as 10.7 for copper and 17.7 for aluminum. For small wire sizes, up to No. 3, resistance is the same for alternating and direct current. But above No. 3, ac resistance is larger, and a correction factor should be applied (Tables 22-4 and 22-5).

Voltage drops used in design may range from 1 to 5% of the service voltage. Some codes set a maximum for voltage drop of 2.5% for combined light and power circuits from service entry to the building to point of final distribution at branch panels.

When this voltage drop is apportioned to the various parts of the circuit, it is economical to assign the greater part, say 1.5 to 2%, to the smaller, more numerous feeders, and only 0.5 to 1% to the heavy main feeders between the service and main distribution panels. Table 22-6 gives the maximum allowable current for each wire size for copper wire; use 84% of these values for aluminum. Table 22-4 gives the area, in circular mils, to be used in the voltage-drop formulas.

First, select the minimum-size wire allowed by the Code, and test it for voltage drop. If this drop is excessive, test a larger size, until one is found for which the voltage drop is within the desired limit. This trial-and-error process can be

Table 22-5. Ac/Dc Resistance Ratio*

Size	Multiplying factor			
	For nonmetallic sheathed cables in air or nonmetallic conduit		For metallic sheathed cables or all cables in metallic raceways	
	Copper	Aluminum	Copper	Aluminum
Up to 3 AWG	1	1	1	1
2	1	1	1.01	1.00
1	1	1	1.01	1.00
0	1.001	1.000	1.02	1.00
00	1.001	1.001	1.03	1.00
000	1.002	1.001	1.04	1.01
0000	1.004	1.002	1.05	1.01
250 M cmils	1.005	1.002	1.06	1.02
300 M cmils	1.006	1.003	1.07	1.02
350 M cmils	1.009	1.004	1.08	1.03
400 M cmils	1.011	1.005	1.10	1.04
500 M cmils	1.018	1.007	1.13	1.06
600 M cmils	1.025	1.010	1.16	1.08
700 M cmils	1.034	1.013	1.19	1.11
750 M cmils	1.039	1.015	1.21	1.12
800 M cmils	1.044	1.017	1.22	1.14
1,000 M cmils	1.067	1.026	1.30	1.19
1,250 M cmils	1.102	1.040	1.41	1.27
1,500 M cmils	1.142	1.058	1.53	1.36
1,750 M cmils	1.185	1.079	1.67	1.46
2,000 M cmils	1.233	1.100	1.82	1.56

* SOURCE: Table 17-11 from W. T. Stuart, Wiring Design—Commercial and Industrial Buildings, in D. G. Fink and J. M. Carroll (eds.), "Standard Handbook for Electrical Engineers," 10th ed., McGraw-Hill Book Co., 1968. Used by permission.

shortened by first assuming the desired voltage drop, and then computing the required wire area with Eq. (22-7) and (22-8). The wire size can be selected from Table 22-4.

For circuits designed for motor loads only, no lighting, the maximum voltage drop may be increased to a total of 5%. Of this, 1% can be assigned to branch circuits and 4% to feeders.

Tables 22-7 to 22-9 give dimensions of trade sizes of conduit and tubing and permissible numbers of conductors that can be placed in each size.

22-7. Wiring for Motor Loads. Motors have a high starting current that lasts a very short time. But it may be two to three times as high as the rated current when running. Although motor windings will not be damaged by a high current of short duration, they cannot take currents much greater than the rated value for long periods without excessive overheating and consequent breakdown of the insulation.

Overcurrent protective devices, fuses and circuit breakers, should be selected to protect motors from overcurrents of long duration, and yet permit short-duration starting currents to pass without disconnecting the circuit. For this reason, the National Electrical Code permits the fuse or circuit breaker in a motor circuit to have a higher ampere rating than the allowable current-carrying capacity of the wire. Table 22-10 gives the overcurrent protection for motors allowed by the Code. Time-delay fuses are available that permit smaller fuse holders for a given-size motor than with standard fuses (Table 22-2).

The National Electrical Code requirements for motor circuit conductors and overcurrent protection are as follows:

Branch Circuits (One Motor). Conductors shall have an allowable current-carrying capacity not less than 125% of the motor full-load current. Overcurrent protection, fuses or circuit breakers, must be capable of carrying the starting current of the motor. Maximum rating of such protection is given in Table 22-10, which shows that the maximum rating varies with the type, starting method, and locked-rotor current of the motor. For the great majority of motor applications in buildings,

Table 22-6. Allowable Current-carrying Capacities of Copper Conductors, Amperes*
(Not more than three conductors in raceway or cable)
Based on room temperature of 30°C (86°F)

Size, AWG or Mcm	Rubber type R, type RW, type RU, type RUW (14-2), type RH-RW† / Thermoplastic type T, type TW	Rubber type RH, type RH-RW† type RHW	Paper / Thermoplastic Asbestos type TA / Var-Cam type V / Asbestos Var-Cam type AVB MI cable	Asbestos Var-Cam type AVA, type AVL	Impregnated Asbestos type AI (14-8), type AIA	Asbestos type A (14-8), type AA
14	15	15	25	30	30	30
12	20	20	30	35	40	40
10	30	30	40	45	50	55
8	40	45	50	60	65	70
6	55	65	70	80	85	95
4	70	85	90	105	115	120
3	80	100	105	120	130	145
2	95	115	120	135	145	165
1	110	130	140	160	170	190
0	125	150	155	190	220	225
00	145	175	185	215	230	250
000	165	200	210	245	265	285
0000	195	230	235	275	310	340
250	215	255	270	315	335	
300	240	285	300	345	380	
350	260	310	325	390	420	
400	280	335	360	420	450	
500	320	380	405	470	500	
600	355	420	455	525	545	
700	385	460	490	560	600	
750	400	475	500	580	620	
800	410	490	515	600	640	
900	435	520	555			
1000	455	545	585	680	730	
1250	495	590	645			
1500	520	625	700	785		
1750	545	650	735			
2000	560	665	775	840		

Correction Factor for Room Temperatures over 30°C (86°F)

°C °F						
40 104	.82	.88	.90	.94	.94	
45 113	.71	.82	.85	.90	.92	
50 122	.58	.75	.80	.87	.89	
55 131	.41	.67	.74	.83	.86	
60 140	...	.58	.67	.79	.83	.91
70 158	...	.35	.52	.71	.76	.87
75 167	...	...	.43	.66	.72	.86
80 176	...	...	.30	.61	.69	.84
90 194	...	...	...	.50	.61	.80
100 212	...	...	...	...	.51	.77
120 248	...	...	...	...	...	.69
140 284	...	...	...	...	...	.59

* From "Standard Handbook for Electrial Engineers," edited by A. E. Knowlton; 9th ed., copyright © 1957, McGraw-Hill Book Company. Used by permission.
Allowable current-carrying capacities of aluminum wire are 84% of the values in this table.
† For RH-RW in wet locations use column 2, for dry locations use column 3.

Table 22-7. Smallest Trade Size of Conduit or Tubing, In., for Specific Numbers of Conductors*

(Rubber-covered types RF-2, RFH-2, etc., R, RH, RW, RH-RW, RU, RUH, and RUW thermoplastic types TF, T, and TW, one to nine conductors)

Size, AWG or Mcm	Number of conductors in one conduit or tubing								
	1	2	3	4	5	6	7	8	9
18	½	½	½	½	½	½	½	¾	¾
16	½	½	½	½	½	½	¾	¾	¾
14	½	½	½	½	¾	¾	1	1	1
12	½	½	½	¾	¾	1	1	1	1¼
10	½	¾	¾	¾	1	1	1	1¼	1¼
8	½	¾	¾	1	1¼	1¼	1¼	1½	1½
6	½	1	1	1¼	1½	1½	2	2	2
4	½	1¼	*1¼	1½	1½	2	2	2	2½
3	¾	1¼	1¼	1½	2	2	2	2½	2½
2	¾	1¼	1¼	2	2	2	2½	2½	2½
1	¾	1½	1½	2	2½	2½	2½	3	3
0	1	1½	2	2	2½	2½	3	3	3
00	1	2	2	2½	2½	3	3	3	3½
000	1	2	2	2½	3	3	3	3½	3½
0000	1¼	2	2½	3	3	3	3½	3½	4
250	1¼	2½	2½	3	3	3½	4	4	4½
300	1¼	2½	2½	3	3½	4	4	4½	4½
350	1¼	3	3	3½	3½	4	4½	4½	5
400	1½	3	3	3½	4	4	4½	5	5
500	1½	3	3	3½	4	4½	5	5	6
600	2	3½	3½	4	4½	5	6	6	6
700	2	3½	3½	4½	5	5	6	6	
750	2	3½	3½	4½	5	6	6	6	
800	2	3½	4	4½	5	6	6		
900	2	4	4	5	6	6	6		
1000	2	4	4	5	6	6			
1250	2½	4½	4½	6	6				
1500	3	5	5	6					
1750	3	5	6	6					
2000	3	6	6						

* From "Standard Handbook for Electrical Engineers," edited by A. E. Knowlton, 9th ed., copyright © 1957, McGraw-Hill Book Company. Used by permission.
See also tables in Chap. 9 of "NFPA Handbook of the National Electrical Code," edited by Frank Stetka, McGraw-Hill Book Company.

conductor and fuse protection may be selected more readily from Table 22-2.

Feeder Circuits (More than One Motor on a Conductor). The conductor shall have an allowable current-carrying capacity not less than 125% of the full-load current of the largest motor plus the sum of the full-load currents of the remaining motors on the same circuit. The rating of overcurrent protection, fuses or circuit breakers, shall not be greater than the maximum allowed by the code for protection of the largest motor plus the sum of the full-load currents of the remaining motors on the circuit.

If the allowable current-carrying capacity of the conductor or the size of the computed overcurrent device does not correspond to the rating of a standard-size fuse or circuit breaker, the next larger standard size should be used.

Amp-Traps and Hi-Caps are high-interrupting-capacity current-limiting fuses used in service switches and main distribution panels connected near service switches. This type of fuse is needed here because this part of the wiring system in large buildings consists of heavy cables or buses and large switches that have very little resistance. If a short circuit occurs, very high currents will flow, limited only by the interrupting capacity of the protective device installed by the utility company on its own transformers furnishing the service. Ordinary fuses cannot interrupt this current quickly enough to avert damage to the building wiring and connected

Table 22-8. Maximum Number of Conductors in Conduit or Tubing*

Size AWG or M cm	Maximum number of conductors in conduit or tubing (based upon % conductor fill, Table 22-9, for new work)											
	½ in	¾ in	1 in	1¼ in	1½ in	2 in	2½ in	3 in	3½ in	4 in	5 in	6 in
18	7	12	20	35	49	80	115	176				
16	6	10	17	30	41	68	98	150				
14	4	6	10	18	25	41	58	90	121	155		
12	3	5	8	15	21	34	50	76	103	132	208	
10	1	4	7	13	17	29	41	64	86	110	173	
8	1	3	4	7	10	17	25	38	52	67	105	152
6	1	1	3	4	6	10	15	23	32	41	64	93
4	1	1	1	3†	5	8	12	18	24	31	49	72
3	.	1	1	3	4	7	10	16	21	28	44	63
2	.	1	1	3	3	6	9	14	19	24	38	55
1	.	1	1	1	3	4	7	10	14	18	29	42
0	.	..	1	1	2	4	6	9	12	16	25	37
00	.	..	1	1	1	3	5	8	11	14	22	32
000	.	..	1	1	1	3	4	7	9	12	19	27
0000	.	..	..	1	1	2	3	6	8	10	16	23
250	.	..	..	1	1	1	3	5	6	8	13	19
300	.	..	..	1	1	1	3	4	5	7	11	16
350	.	..	..	1	1	1	1	3	5	6	10	15
400	.	..	..	..	1	1	1	3	4	6	9	13
500	.	..	..	..	1	1	1	3	4	5	8	11
600	.	..	..	..	..	1	1	1	3	4	6	9
700	.	..	..	..	..	1	1	1	3	3	6	8
750	.	..	..	..	..	1	1	1	3	3	5	8
800	.	..	..	..	..	1	1	1	2	3	5	7
900	.	..	..	..	..	1	1	1	1	3	4	7
1,000	.	..	..	..	..	1	1	1	1	3	4	6
1,250	.	..	..	..	..	..	1	1	1	1	3	5
1,500	.	..	..	..	..	..	...	1	1	1	3	4
1,750	.	..	..	..	..	..	...	1	1	1	2	4
2,000	.	..	..	..	..	..	...	1	1	1	1	3

* SOURCE: Table 17-8 from W. T. Stuart, Wiring Design—Commercial and Industrial Buildings, in D. G. Fink and J. M. Carroll (eds.), "Standard Handbook for Electrical Engineers," 10th ed., McGraw-Hill Book Co., 1968. Used by permission.

† Where an existing service run of conduit or electrical metallic tubing does not exceed 50 ft in length and does not contain more than the equivalent of two quarter bends from end to end, two No. 4 insulated and one No. 4 bare conductors may be installed in 1-in. conduit or tubing.

electrical equipment. The interrupting-capacity value needed can be obtained from the utility company.

Fuses in service switches and connected main panels should have current-time characteristics that will isolate only the circuit in which a short occurs, without permitting the short-circuit current to pass to other feeders and interrupt those circuits too. The electrical designer should obtain data from manufacturers of approved fusing devices on the proper sequence of fusing.

22-8. Service-entrance Switch and Metering Equipment. Fused switches or circuit breakers must be provided near the entrance point of electrical service in a building for shutting off the power. The National Electrical Code requires that each incoming service in a multiple-occupancy building be controlled near its entrance by not more than six switches or circuit breakers.

Many types of service-entrance switches are available to meet the requirements of utility companies. They may be classified as fuse pull switch, externally operated safety switch, bolted pressure contact-type switch, and circuit breaker.

Metering equipment consists of a meter pan, meter cabinet, current transformer cabinet, or a combination of these cabinets, depending on the load requirements and other characteristics of the specific project. The meters and metering trans-

Table 22-9. Dimensions and Percent Area of Conduit or Tubing*

Trade size	Internal diameter, in	Area, in								
			Not lead-covered			Lead-covered				
		Total 100%	1 cond. 31%	2 cond. 31%	3 cond. and over 40%	1 cond. 55%	2 cond. 30%	3 cond. 40%	4 cond. 38%	Over 4 cond. 35%
½	0.622	0.30	0.16	0.09	0.12	0.17	0.09	0.12	0.11	0.11
¾	0.824	0.53	0.28	0.16	0.21	0.29	0.16	0.21	0.20	0.19
1	1.049	0.86	0.46	0.27	0.34	0.47	0.26	0.34	0.33	0.30
1¼	1.380	1.50	0.80	0.47	0.60	0.83	0.45	0.60	0.57	0.53
1½	1.610	2.04	1.08	0.63	0.82	1.12	0.61	0.82	0.78	0.71
2	2.067	3.36	1.78	1.04	1.34	1.85	1.01	1.34	1.28	1.18
2½	2.469	4.79	2.54	1.48	1.92	2.63	1.44	1.92	1.82	1.68
3	3.068	7.38	3.91	2.29	2.95	4.06	2.21	2.95	2.80	2.58
3½	3.548	9.90	5.25	3.07	3.96	5.44	2.97	3.96	3.76	3.47
4	4.026	12.72	6.74	3.94	5.09	7.00	3.82	5.09	4.83	4.45
5	5.047	20.00	10.60	6.20	8.00	11.00	6.00	8.00	7.60	7.00
6	6.065	28.89	15.31	8.96	11.56	15.89	8.67	11.56	10.98	10.11

* SOURCE: Table 17-10 from W. T. Stuart, Wiring Design—Commercial and Industrial Buildings, in D. G. Fink and J. M. Carroll (eds.), "Standard Handbook for Electrical Engineers," 10th ed., McGraw-Hill Book Co., 1968. Used by permission.

formers for recording current consumed are furnished by the utility company. Unless otherwise permitted by the utility company, meters must be located near the point of service entrance. Sometimes, the utility company permits one or more tenant meter rooms at other locations in the cellar of an apartment house to suit economical building wiring design. Tenant meter closets on the upper floors, opening on public halls, also may be permitted. The most common form of tenant meters used is the three-wire type, consisting of two phase wires and the neutral, taken from a 208/120-volt three-phase four-wire service.

The service switch and metering equipment may be combined in one unit, or the switch may be connected with conduit to a separate meter trough. For individual metering, the detachable-socket-type meter with prongs that fit into the jaws of the meter-mounting trough generally is used.

22-9. Switchboards. A switchboard is defined in the National Electrical Code as a large single panel, frame, or assembly of panels, on which are mounted, on the face or back or both, switches, overcurrent and other protective devices, buses, and usually instruments. Switchboards are generally accessible from the rear as well as from the front and not intended to be installed in cabinets (see also Art. 22-10).

Switchboards are commonly divided into the following types:

1. Live-front.
2. Dead-front.
3. Safety enclosed switchboards.
   a. Unit or sectional.
   b. Draw-out.

Live-front switchboards have the current-carrying parts of the switch equipment mounted on the exposed front of the vertical panels and are usually limited to systems not exceeding 600 volts. They generally are installed in restricted areas.

Dead-front switchboards have no live parts mounted on the front of the board and are used in systems limited to a maximum of 600 volts for dc and 2,500 volts for ac.

Unit safety-type switchboard is a metal-enclosed switchgear consisting of a completely enclosed self-supporting metal structure, containing one or more circuit breakers or switches.

Draw-out-type switchboard is a metal-clad switchgear consisting of a stationary housing mounted on a steel framework and a horizontal draw-out circuit-breaker structure. The equipment for each circuit is assembled on a frame forming a self-contained and self-supporting mobile unit.

Metal-clad switchgear consists of a metal structure completely enclosing a circuit breaker and associated equipment such as current and potential transformers, interlocks, controlling devices, buses and connections.

22-10. Panelboards. A panelboard is a single panel or a group of panel units designed for assembly in the form of a single panel in which are included buses and perhaps switches and automatic overcurrent protective devices for control of light, heat, or power circuits of small capacity. It is designed to be placed in a cabinet or cutout box placed in or against a wall or partition and accessible only from the front. In general, panelboards are similar to but smaller than switchboards (see Art. 22-9).

A panelboard consists of a set of copper mains from which the individual circuits are tapped through overload protective devices or switching units.

Panelboards are designed for dead-front or live-front construction. In dead-front construction, no live parts are exposed when the door of the panelboard is opened. In the live-front type, the current-carrying parts of the fuses or switches are exposed when the door of the panelboard is opened. Panelboards preferably should be dead-front construction.

Panelboards are designed for flush, semiflush, or surface mounting. They fall into two general classifications, those designed for medium loads, usually required for lighting systems, and those for heavy-duty industrial-power-distribution loads.

Lighting panelboards are generally used for distribution of branch lighting circuits and fall into the following types: plug-fuse branches; single-pole tumbler switch, with one-fuse branches, plug or cartridge type; double-pole tumbler switch, with two-fuse branches, plug or cartridge type; single-pole quick-lag circuit breaker branches; single-pole quick-make and quick-break circuit-breaker branches.

Panelboards are designed with mains for distribution systems consisting of:

1. Three-wire, single-phase 120/240-volt, solid-neutral, alternating current.
2. Three-wire, 120/240-volt, solid-neutral, direct current.
3. Four-wire, three-phase, 120/208-volt, solid-neutral, alternating current.

Distribution panelboards are designed to distribute current to lighting panelboards and power loads and panelboards.

The mains in the panelboard may be provided with lugs only, fuses, switch and fuses, or circuit breakers.

Power panelboards fall into the following types:

1. Live-front, fused cutouts in branches.
2. Live-front, fused knife switches in branches.
3. Safety dead-front, pull switch in branches.
4. Safety dead-front, circuit breaker in branches.

Since motors fed from power panelboards vary in sizes, the switches and breakers in a power panelboard are made of several different sizes corresponding to the rating of the equipment.

The following items should be taken into consideration in determining the number and location of panelboards:

1. No lighting panelboard should exceed 42 single-pole protective devices.
2. Panelboard should be located as near as possible to the center of the load it supplies.
3. Panelboard should always be accessible.
4. Length of run from panelboard to the first outlet should not exceed 100 ft.
5. Panelboard should be located so that the feeder is as short as possible and have a minimum number of bends and offsets.
6. Spare circuit capacity should be provided at the approximate rate of one spare to every five circuits originally installed.
7. At least one lighting panelboard should be provided for each floor of a building.

22-11. Sample Calculations for Apartment-building Riser. A diagram of a light and power riser for a nine-story apartment building is shown in Fig. 22-1.

1. Typical Meter Branch to Apartment Panel. Note that the meters are three-wire type, and three apartments are connected to the same neutral. Under balanced

Table 22-10. Overcurrent Protection for Motors*

Full-load current rating of motor, A	For running protection of motors		Maximum allowable rating or setting of branch-circuit protective devices							
			With code letters: Single-phase, squirrel-cage, and synchronous. Full voltage, resistor or reactor starting, code letters F to V inclusive Without code letters: Same as above		With code letters: Single-phase, squirrel-cage, and synchronous. Full voltage, resistor or reactor start, code letters B to E inclusive. Autotransformer, start, code letters F to V inclusive Without code letters: (Not more than 30 A) Squirrel-cage and synchronous, autotransformer start, high-reactance squirrel-cage†		With code letters: Squirrel-cage and synchronous autotransformer start, code letters B to E inclusive Without code letters: (More than 30 A) Squirrel-cage and synchronous autotransformer start, high-reactance squirrel-cage†		With code letters: All motors code letter A Without code letters: D-c and wound-rotor motors	
	Maximum rating of nonadjustable protective devices, A	Maximum setting of adjustable protective devices, A	Fuses	Circuit breakers (nonadjustable overload trip)	Fuses	Circuit breakers (nonadjustable overload trip)	Fuses	Circuit breakers (nonadjustable overload trip)	Fuses	Circuit breakers (nonadjustable overload trip)
1	2	1.25	15	15	15	15	15	15	15	15
2	3	2.50	15	15	15	15	15	15	15	15
3	4	3.75	15	15	15	15	15	15	15	15
4	6	5.0	15	15	15	15	15	15	15	15
5	8	6.25	15	15	15	15	15	15	15	15
6	8	7.50	20	15	15	15	15	15	15	15
7	10	8.75	25	20	20	15	15	15	15	15
8	10	10.0	25	20	20	20	20	20	15	15
9	12	11.25	30	30	25	20	20	20	15	15
10	15	12.50	30	30	25	20	20	20	15	15
11	15	13.75	35	30	30	30	25	30	20	20
12	15	15.00	40	30	30	30	25	30	20	20
13	20	16.25	40	40	35	30	30	30	20	20
14	20	17.50	45	40	35	30	30	30	25	30
15	20	18.75	45	40	40	30	30	30	25	30
16	20	20.00	50	40	40	40	35	40	25	30

17	25	21.25	60	50	45	40	35	40	30	30
18	25	22.50	60	50	45	40	40	40	30	30
19	25	23.75	60	50	50	40	40	40	30	30
20	25	25.00	60	50	50	40	40	40	30	30
22	30	27.50	70	70	60	50	45	50	35	40
24	30	30.00	80	70	60	50	50	50	40	40
26	35	32.50	80	70	70	70	60	70	40	40
28	35	35.00	90	70	70	70	60	70	45	50
30	40	37.50	90	100	80	70	60	70	45	50
32	40	40.00	100	100	80	70	70	70	50	50
34	45	42.50	110	100	90	70	70	70	60	70
36	45	45.00	110	100	90	100	80	100	60	70
38	50	47.50	125	100	100	100	80	100	60	70
40	50	50.00	125	100	100	100	80	100	60	70
42	50	52.50	125	125	110	100	90	100	70	70
44	60	55.00	125	125	110	100	90	100	70	70
46	60	57.50	150	125	125	100	100	100	70	70
48	60	60.00	150	125	125	100	100	100	80	100
50	60	62.50	150	125	125	100	100	100	80	100
52	70	65.00	175	150	150	125	110	125	80	100
54	70	67.50	175	150	150	125	110	125	90	100
56	70	70.00	175	150	150	125	125	125	90	100
58	70	72.50	175	150	150	125	125	125	90	100
60	80	75.00	200	150	150	125	125	125	90	100
62	80	77.50	200	175	175	125	125	125	100	100
64	80	80.00	200	175	175	150	150	150	100	100
66	80	82.50	200	175	175	150	150	150	100	100
68	90	85.00	225	175	175	150	150	150	110	125
70	90	87.50	225	175	175	150	150	150	110	125
72	90	90.00	225	200	200	150	150	150	110	125
74	90	92.50	225	200	200	150	150	150	125	125
76	100	95.00	250	200	200	175	175	175	125	125
78	100	97.50	250	200	200	175	175	175	125	125
80	100	100.00	250	200	200	175	175	175	125	125
82	110	102.50	250	225	225	175	175	175	125	125
84	110	105.00	250	225	225	175	175	175	150	150
86	110	107.50	300	225	225	175	175	175	150	150
88	110	110.00	300	225	225	200	200	200	150	150
90	110	112.50	300	225	225	200	200	200	150	150
92	125	115.00	300	250	250	200	200	200	150	150

Table 22-10. Overcurrent Protection for Motors* *(Continued)*

Full-load current rating of motor, A	For running protection of motors		Maximum allowable rating or setting of branch circuit protective devices							
			With code letters: Single-phase, squirrel-cage, and synchronous. Full voltage, resistor or reactor starting, code letters F to V inclusive Without code letters: Same as above		With code letters: Single-phase, squirrel-cage, and synchronous. Full voltage, resistor or reactor start, code letters B to E inclusive. Autotransformer, start, code letters F to V inclusive Without code letters: (Not more than 30 A) Squirrel-cage and synchronous, autotransformer start, high-reactance squirrel-cage†		With code letters: Squirrel-cage and synchronous autotransformer start, code letters B to E inclusive Without code letters: (More than 30 A) Squirrel-cage and synchronous autotransformer start, high-reactance squirrel-cage†		With code letters: All motors code letter A Without code letters: D-c and wound-rotor motors	
	Maximum rating of nonadjustable protective devices, A	Maximum setting of adjustable protective devices, A	Fuses	Circuit breakers (nonadjustable overload trip)	Fuses	Circuit breakers (nonadjustable overload trip)	Fuses	Circuit breakers (nonadjustable overload trip)	Fuses	Circuit breakers (nonadjustable overload trip)
94	125	117.50	300	250	250	200	200	200	150	150
96	125	120.00	300	250	250	200	200	200	150	150
98	125	122.50	300	250	250	200	200	200	150	150
100	125	125.00	300	250	250	200	200	200	150	150
105	150	131.50	350	300	300	225	225	225	175	175
110	150	137.50	350	300	300	225	225	225	175	175
115	150	144.00	350	300	300	250	250	250	175	175
120	150	150.00	400	300	300	250	250	250	200	200
125	175	156.50	400	350	350	250	250	250	200	200
130	175	162.50	400	350	350	300	300	300	200	200
135	175	169.00	450	350	350	300	300	300	225	225
140	175	175.00	450	350	350	300	300	300	225	225
145	200	181.50	450	400	400	300	300	300	225	225
150	200	187.50	450	400	400	300	300	300	225	225
155	200	194.00	500	400	400	350	350	350	250	250
160	200	200.00	500	400	400	350	350	350	250	250

165	225	206.00	500	500	450	350	350	350	250	250
170	225	213.00	500	500	450	350	350	350	300	300
175	225	219.00	600	500	450	350	350	350	300	300
180	225	225.00	600	500	450	400	400	400	300	300
185	250	231.00	600	500	500	400	400	400	300	300
190	250	238.00	600	500	500	400	400	400	300	300
195	250	244.00	600	500	500	400	400	400	300	300
200	250	250.00	600	500	500	400	400	400	300	300
210	250	263.00	800	600	600	500	450	500	350	350
220	300	275.00	800	600	600	500	450	500	350	350
230	300	288.00	800	600	600	500	500	500	350	350
240	300	300.00	800	600	600	500	500	500	400	400
250	300	313.00	800	700	800	500	500	500	400	400
260	350	325.00	800	700	800	600	600	600	400	400
270	350	338.00	1000	700	800	600	600	600	450	500
280	350	350.00	1000	700	800	600	600	600	450	500
290	350	363.00	1000	800	800	600	600	600	450	500
300	400	375.00	1000	800	800	600	600	600	450	500
320	400	400.00	1000	800	800	700	800	700	500	500
340	450	425.00	1200	. . .	1000	700	800	700	600	600
360	450	450.00	1200	. . .	1000	800	800	800	600	600
380	500	475.00	1200	. . .	1000	800	800	800	600	600
400	500	500.00	1200	. . .	1000	800	800	800	600	600
420	600	525.00	1600	. . .	1200	. . .	1000	. . .	800	700
440	600	550.00	1600	. . .	1200	. . .	1000	. . .	800	700
460	600	575.00	1600	. . .	1200	. . .	1000	. . .	800	700
480	600	600.00	1600	. . .	1200	. . .	1000	. . .	800	800
500	. . .	625.00	1600	. . .	1600	. . .	1000	. . .	800	800

* SOURCE: Table 17-21 from W. T. Stuart, Wiring Design—Commercial and Industrial Buildings, in D. G. Fink and J. M. Carroll (eds.), "Standard Handbook for Electrical Engineers," 10th ed., McGraw-Hill Book Co., 1968. Used by permission.

† High-reactance squirrel-cage motors are those designed to limit the starting current by means of deep-slot secondaries or double-wound secondaries and are generally started on full voltage.

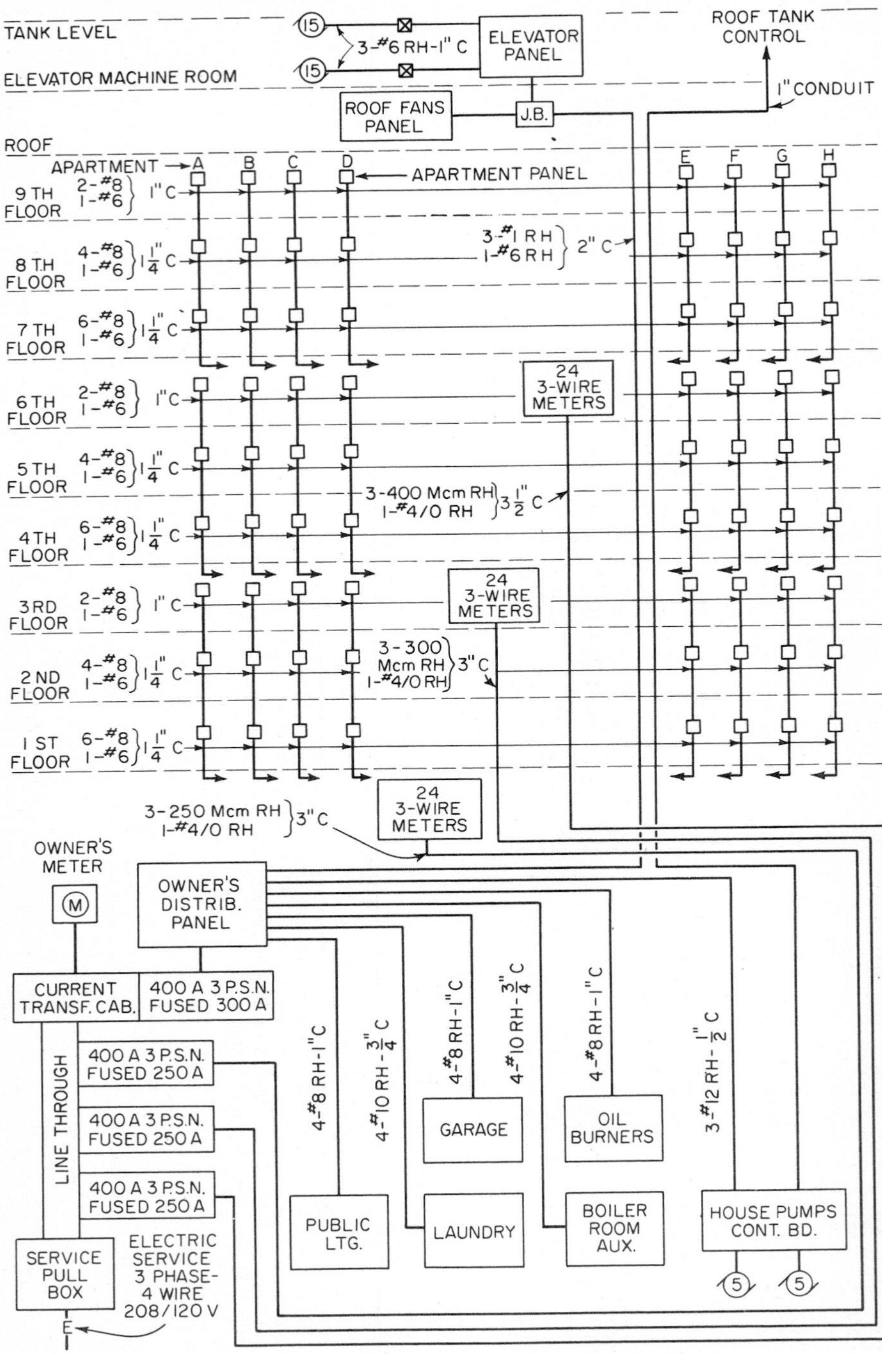

Fig. 22-1. Typical diagram of an apartment-building electrical riser.

conditions, when each of the three identical apartments is under full load, no current will flow in the neutral. But at maximum unbalance, current in the neutral may be twice the current in the phase wire for any one apartment. The neutral wire must be sized for this maximum current, though the usual practice is to compute voltage drops for the balanced condition.

Assume that the apartment area is 900 sq ft. The one-way run from meter to apartment panel (apartment A) is 110 ft.

Apartment lighting load	= 900 × 3 watts per sq ft	= 2,700 watts
Apartment appliance load		= 3,000
Total		= 5,700 watts

The electric service is three-phase four-wire 208/120 volts. Thus, the voltage between phase wires is 208, and between one phase and neutral 120 volts. Assume a 90% power factor. From Eq. (22-5):

$$\text{Current per phase} = \frac{5{,}700}{2 \times 120 \times 0.9} = 26.4 \text{ amp}$$

The local electrical code requires the minimum size of apartment feeder to be No. 8 wire. According to Table 22-6, the allowable current in No. 8 RH wire is 45 amp; so this wire would be adequate for the current, but it still must be checked for voltage drop. The neutral must be sized for the maximum unbalance, under which condition the current in the neutral will be 2 × 26.4 = 52.8 amp. This will require No. 6 wire.

The voltage drop between phase wires can be obtained from Eq. (22-8), with the area of No. 8 wire taken as 16,510 circular mils (Table 22-10) and length of wire as 110 ft:

$$V_d = \frac{\sqrt{3} \times 10.7 \times 26.4 \times 110}{16{,}510} = 3.24 \text{ volts}$$

$$\%\text{ voltage drop} = \frac{3.24 \times 100}{208} = 1.56\%$$

2. *Feeder to Meter Bank on Sixth Floor.*

Total load (24 apartments)	= 24 × 5,700	= 136,800 watts
Demand load:	first 15,000 watts at 100%	= 15,000
	Balance, 121,800 watts at 50%	= 60,900
Total demand load		= 75,900 watts

Assume a power factor of 90% and apply Eq. (22-6):

$$\text{Current per phase} = \frac{75{,}900}{3 \times 120 \times 0.9} = 234 \text{ amp}$$

According to Table 22-6, minimum-size type RH wire for this current is 250 Mcm. This has to be tested for voltage drop. For a one-way run of 150 ft from service switch to sixth-floor meter bank and an area of 250,000 circular mils, Eq. (22-8) gives:

$$V_d = \frac{\sqrt{3} \times 10.7 \times 234 \times 150}{250{,}000} = 2.60 \text{ volts}$$

From Table 22-5, the correction factor for ac resistance of 250-Mcm wire is 1.06. Application of this factor yields a corrected voltage drop of 1.06 × 2.60 = 2.76 volts.

$$\%\text{ voltage drop} = \frac{2.76 \times 100}{208} = 1.33\%$$

Then, the total voltage drop from service switch to apartment panel *A* is

$$1.56 + 1.33\% = 2.89\%,$$

which exceeds the 2.5% maximum voltage drop allowed by the local code. It is necessary, therefore, to increase the size of the meter bank feeder over the minimum size required for the current.

Table 22-11. Dimensions of Rubber-covered and Thermoplastic-covered Conductors*

Size, AWG Mcm	Types RF-2, RFH-2, R, RH, RHH, RHW, RH-HW, RW		Types TF, T, THW, TW, RU, RUH, RUW		Types THHN, THWN		Types FEP, FEPB			
	Approx. diam., in	Approx. area, in²	Approx. diam., in	Approx. area, in²	Approx. diam., in	Approx. area, in²	Approx. diam., in		Approx. area, in²	
18	0.146	0.0167	0.106	0.0088						
16	0.158	0.0196	0.118	0.0109						
14	$^{2}/_{64}$ in 0.171	0.0230	0.131	0.0135	0.105	0.0087	0.105	0.105	0.0087	0.0087
14	$^{3}/_{64}$ in 0.204	0.0327								
14	....	......	0.162	0.0206						
12	$^{2}/_{64}$ in 0.188	0.0278	0.148	0.0172	0.122	0.0117	0.121	0.121	0.0115	0.0115
12	$^{3}/_{64}$ in 0.221	0.0384								
12	....	......	0.179	0.0251						
10	.... 0.242	0.0460	0.168	0.0224	0.153	0.0184	0.142	0.142	0.0159	0.0159
10	....	......	0.199	0.0311						
8	.... 0.311	0.0760	0.228	0.0408	0.201	0.0317	0.189	0.169	0.0280	0.0225
8	....	......	0.259	0.0526						
6	0.397	0.1238	0.323	0.0819	0.257	0.0519	0.244	0.302	0.0467	0.0716
4	0.452	0.1605	0.372	0.1087	0.328	0.0845	0.292	0.350	0.0669	0.0962
3	0.481	0.1817	0.401	0.1263	0.356	0.0995	0.320	0.378	0.0803	0.1122
2	0.513	0.2067	0.433	0.1473	0.388	0.1182	0.352	0.410	0.0973	0.1316
1	0.588	0.2715	0.508	0.2027	0.450	0.1590				
0	0.629	0.3107	0.549	0.2367	0.491	0.1893				
00	0.675	0.3578	0.595	0.2781	0.537	0.2265				
000	0.727	0.4151	0.647	0.3288	0.588	0.2715				
0000	0.785	0.4840	0.705	0.3904	0.646	0.3278				
250	0.868	0.5917	0.788	0.4877	0.716	0.4026				
300	0.933	0.6837	0.843	0.5581	0.771	0.4669				
350	0.985	0.7620	0.895	0.6291	0.822	0.5307				
400	1.032	0.8365	0.942	0.6969	0.869	0.5931				
500	1.119	0.9834	1.029	0.8316	0.955	0.7163				
600	1.233	1.1940	1.143	1.0261						
700	1.304	1.3355	1.214	1.1575						
750	1.339	1.4082	1.249	1.2252						
800	1.372	1.4784	1.282	1.2908						
900	1.435	1.6173	1.345	1.4208						
1,000	1.494	1.7531	1.404	1.5482						
1,250	1.676	2.2062	1.577	1.9532						
1,500	1.801	2.5475	1.702	2.2748						
1,750	1.916	2.8895	1.817	2.5930						
2,000	2.021	3.2079	1.922	2.9013						

* SOURCE: Table 17-2 from W. T. Stuart, Wiring Design—Commercial and Industrial Buildings, in D. G. Fink and J. M. Carroll (eds.), "Standard Handbook for Electrical Engineers," 10th ed., McGraw-Hill Book Co., 1968. Used by permission.

To bring the total drop down to 2.5% the meter bank feeder drop must be reduced to 0.94% or 1.96 volts. The required wire size may be found by proportion:

$$\text{Required area} = \frac{2.76}{1.96} \times 250{,}000 = 352{,}000 \text{ circular mils}$$

According to Table 22-6, the nearest larger wire size is 400 Mcm.

For computing the size of the neutral for carrying 234 amp, the local code allows a demand factor of 70% on lighting loads over 200 amp. Hence, the net current in the neutral is $200 + 34 \times 0.70 = 223.8$ amp. From Table 22-6, minimum-size wire is No. 4/10 RH. And with Table 22-11, the size of conduit required for the feeder is computed as follows:

Three 400-Mcm RH wires	$= 3 \times 0.8365$	$= 2.5095$ sq in.
One 0000 RH wire$_g$		$= 0.4840$
Total area		$= 2.9935$ sq in.

According to Table 22-9, permissible raceway fill for four conductors is 40%. Hence the minimum area required for the conduit is $2.9935/0.40 = 7.484$ sq in. And from Table 22-9, the required conduit size is 3½ in.

3. *Feeder to Elevator and Roof Fans Panel.* The total load on this feeder is

8 roof fans at ½ hp	= 16.8 amp per phase
8 roof fans at ¼ hp	= 17.4
2 elevators at 15 hp	= 84.6
Total	= 118.8 amp per phase

The minimum current-carrying capacity of the feeder must be 125% of the rated full-load current of the largest motor, an elevator motor, 42.3 amp, plus the sum of the rated load currents of the other motors. Thus, the capacity must be:

$$1.25 \times 42.3 + 42.3 + 16.8 + 17.4 = 129.4 \text{ amp}$$

From Table 22-6, the minimum-size wire that can be used is No. 1 RH, with an area of 83,690 circular mils.

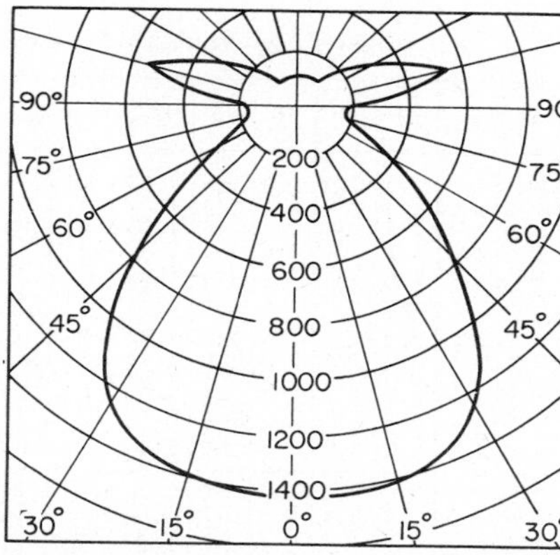

Fig. 22-2. Candlepower distribution curve.

Check for voltage drop with a run of 150 ft:

$$V_d = \frac{10.7 \times 118.8 \times 150}{83{,}690} = 2.09 \text{ volts}$$

$$\% \text{ voltage drop} = \frac{2.09 \times 100}{208} = 1\%$$

22-12. Lighting Design. The following terms are basic in illumination design for buildings:

Lumen, *L*, is the unit of light quantity.

Foot-candle (ft-c) is the unit of light intensity. Numerically, it equals the number of lumens per square foot on an area.

Candlepower (cp) is the unit of intensity of the light source. A light source of one candlepower will produce a light intensity of one foot-candle on a surface one foot away from the light source. The intensity, in foot-candles, on the surface will vary inversely as the square of the distance from the light source.

A theoretical point source has equal distribution of light in all directions. Filaments of lamps are not true point sources. Furthermore, the base of the lamp stops light from passing. As a result, candlepower distribution curves for lamps are irregular. The intensity usually is a maximum directly under the center of the lamp and diminishes as the angle increases away from the vertical. Figure 22-2 shows a typical candlepower distribution curve.

Average illumination on a horizontal working plane in a uniformly lighted room may be calculated by the lumen method, which is based on the following formulas:

$$\text{Ft-c} = \frac{LUM}{A} \tag{22-9}$$

$$L = \frac{A\ \text{Ft-c}}{UM} \tag{22-10}$$

where L = amount of light, lumens
M = maintenance factor
A = room area, sq ft
U = coefficient of utilization

The coefficient of utilization depends on a factor called room ratio, which takes into account shape of room, length and width, and height of light source above the floor. Room ratios are given in Table 22-12. Typical coefficients of utilization are presented in Table 22-13. Note that in this table the coefficient of utilization also varies with the reflective property of ceiling and walls.

Table 22-12. Room Ratios*†

Room W, ft	Room L, ft	Height of light source‡ above floor, ft															
		7	8	9	10	11	12	13	15	17	19	23	27	33	43	53	63
35	40	4.2	3.4	2.9	2.5	2.2	2.0	1.8	1.5	1.3	1.1	0.9	0.8	0.6	0.5		
	60	4.9	4.0	3.4	2.9	2.6	2.3	2.1	1.8	1.5	1.3	1.1	0.9	0.7	0.6		
	80	5.4	4.4	3.7	3.2	2.9	2.6	2.3	1.9	1.7	1.5	1.2	1.0	0.8	0.6	0.5	
	100	...	4.7	4.0	3.4	3.1	2.7	2.5	2.1	1.8	1.6	1.3	1.1	0.9	0.6	0.5	
	120	...	4.9	4.2	3.6	3.2	2.8	2.6	2.2	1.9	1.7	1.3	1.1	0.9	0.7	0.5	
	140	...	5.1	4.3	3.7	3.3	2.9	2.7	2.2	1.9	1.7	1.4	1.1	0.9	0.7	0.6	
40	40	4.4	3.6	2.4	2.7	2.4	2.1	1.9	1.6	1.4	1.2	1.0	0.8	0.7	0.5		
	60	5.3	4.4	3.7	3.2	2.8	2.5	2.3	1.9	1.7	1.5	1.2	1.0	0.8	0.6	0.5	
	80	...	4.9	4.1	3.6	3.2	2.8	2.5	2.1	1.8	1.6	1.3	1.1	0.9	0.7	0.5	
	100	...	5.2	4.4	3.8	3.4	3.0	2.7	2.3	2.0	1.7	1.4	1.2	0.9	0.7	0.6	0.5
	120	...	5.5	4.6	4.0	3.5	3.2	2.8	2.4	2.1	1.8	1.5	1.2	1.0	0.8	0.6	0.5
	140	...	...	4.8	4.1	3.7	3.3	3.0	2.5	2.1	1.9	1.5	1.3	1.0	0.8	0.6	0.5
50	50	...	4.6	3.8	3.3	3.0	2.6	2.4	2.0	1.7	1.5	1.2	1.0	0.8	0.6	0.5	
	70	...	5.3	4.5	3.9	3.4	3.1	2.8	2.3	2.0	1.8	1.4	1.2	1.0	0.7	0.6	0.5
	100	...	...	5.1	4.4	3.9	3.5	3.2	2.7	2.3	2.0	1.6	1.4	1.1	0.8	0.7	0.5
	140	...	...	...	4.9	4.3	3.9	3.5	2.9	2.5	2.2	1.8	1.5	1.2	0.9	0.7	0.6
	170	...	...	...	5.1	4.6	4.1	3.7	3.1	2.7	2.4	1.9	1.6	1.3	1.0	0.8	0.6
	200	...	...	...	5.3	4.7	4.2	3.8	3.2	2.8	2.4	2.0	1.6	1.3	1.0	0.8	0.6
60	60	...	5.5	4.6	4.0	3.5	3.2	2.8	2.4	2.1	1.8	1.5	1.2	1.0	0.8	0.6	0.5
	80	...	...	5.3	4.6	4.0	3.6	3.3	2.7	2.4	2.1	1.7	1.4	1.1	0.9	0.7	0.5
	100	...	...	...	5.0	4.4	3.9	3.6	3.0	2.6	2.3	1.8	1.5	1.2	0.9	0.8	0.6
	140	...	...	...	...	5.0	4.4	4.0	3.4	2.9	2.6	2.1	1.7	1.4	1.0	0.8	0.7
	170	...	...	...	...	5.2	4.7	4.2	3.5	3.1	2.7	2.2	1.8	1.5	1.1	0.9	0.7
	200	...	...	...	...	5.5	4.9	4.4	3.7	3.2	2.8	2.3	1.9	1.5	1.2	0.9	0.7
80	80	...	...	...	5.3	4.7	4.2	3.8	3.2	2.8	2.4	2.0	1.6	1.3	1.0	0.8	0.6
	140	...	...	...	...	...	5.3	4.8	4.1	3.5	3.1	2.5	2.1	1.7	1.3	1.0	0.8
	200	...	...	...	...	...	...	5.4	4.6	3.9	3.5	2.8	2.3	1.9	1.4	1.1	0.9
100	100	...	...	...	...	...	5.2	4.8	4.0	3.4	3.0	2.4	2.0	1.6	1.2	1.0	0.8
	150	...	...	...	...	...	...	...	4.8	4.1	3.7	2.9	2.5	2.0	1.5	1.2	1.0
	200	...	...	...	...	...	...	...	5.3	4.6	4.1	3.3	2.7	2.2	1.7	1.3	1.1
120	120	...	...	...	...	...	...	...	4.8	4.1	3.7	2.9	2.5	2.0	1.5	1.2	1.0
	160	...	...	...	...	...	...	...	5.5	4.7	4.2	3.4	2.8	2.3	1.7	1.4	1.1
	200	...	...	...	...	...	...	...	...	5.2	4.6	3.7	3.1	2.5	1.9	1.5	1.2

* From "Standard Handbook for Electrical Engineers," edited by A. E. Knowlton, 9th ed., copyright © 1957, McGraw-Hill Book Company. Used by permission.

† As computed for a limited range of room sizes; use formula in text for other rooms. Since coefficients of utilization are not as a rule substantially increased for ratios greater than 5.0, the table has not been extended.

‡ Equals ceiling height when semi- or totally indirect luminaires are used.

Table 22-13. Coefficients of Utilization* for Six Typical Luminaires†

Typical distribution	Luminaires	Ceiling	80%			70%			50%			30%		
		Walls	50%	30%	10%	50%	30%	10%	50%	30%	10%	50%	30%	10%
		Room ratio	Coefficients of utilization floor 10%											
MF 0.60 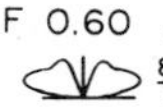 85 / 5	 INDIRECT	0.6 (J)	.27	.21	.16	.24	.19	.14	.17	.14	.11	.12	.09	.07
		0 8 (I)	.34	.28	.22	.30	.25	.20	.22	.18	.15	.15	.12	.09
		1.0 (H)	.39	.33	.28	.35	.30	.25	.26	.22	.18	.17	.14	.12
		1.25 (G)	.45	.39	.33	.40	.34	.29	.30	.26	.22	.20	.17	.14
		1.5 (F)	.49	.43	.38	.43	.38	.33	.32	.28	.24	.22	.19	.16
		2.0 (E)	.55	.49	.44	.48	.43	.39	.36	.32	.29	.24	.22	.19
		2.5 (D)	.58	.53	.48	.52	.47	.43	.38	.35	.32	.26	.23	.21
		3.0 (C)	.61	.56	.52	.54	.50	.46	.40	.37	.34	.27	.25	.23
		4.0 (B)	.65	.61	.57	.57	.54	.50	.43	.40	.37	.28	.26	.25
		5.0 (A)	.68	.64	.61	.59	.56	.53	.44	.42	.39	.29	.27	.26
MF 0.60  60 / 20	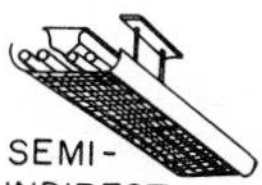 SEMI-INDIRECT	0.6 (J)	.26	.20	.16	.23	.19	.15	.19	.16	.13	.15	.12	.10
		0.8 (I)	.32	.27	.22	.29	.24	.20	.24	.20	.17	.18	.16	.13
		1.0 (H)	.36	.31	.27	.33	.29	.25	.27	.24	.21	.21	.18	.16
		1.25 (G)	.41	.36	.32	.38	.33	.29	.31	.27	.24	.24	.21	.19
		1.5 (F)	.45	.40	.36	.41	.36	.33	.33	.30	.27	.26	.23	.21
		2.0 (E)	.50	.45	.43	.46	.42	.38	.37	.34	.31	.28	.26	.24
		2.5 (D)	.53	.49	.45	.49	.45	.41	.39	.36	.33	.30	.28	.26
		3.0 (C)	.55	.52	.48	.51	.47	.44	.40	.38	.35	.31	.29	.28
		4.0 (B)	.58	.55	.50	.53	.50	.47	.42	.40	.38	.33	.30	.30
		5.0 (A)	.59	.57	.54	.54	.52	.50	.43	.42	.40	.34	.32	.31
MF 0.65  40 / 40	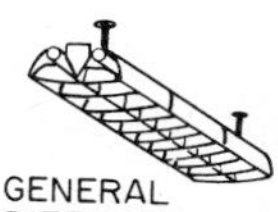 GENERAL DIFFUSER	0.6 (J)	.27	.22	.18	.26	.21	.17	.23	.19	.16	.20	.18	.15
		0.8 (I)	.34	.28	.24	.32	.27	.24	.29	.25	.22	.26	.22	.20
		1.0 (H)	.39	.34	.30	.37	.32	.28	.33	.29	.26	.29	.26	.23
		1.25 (G)	.44	.39	.35	.42	.37	.33	.38	.34	.30	.33	.30	.27
		1.5 (F)	.48	.43	.39	.46	.41	.37	.41	.37	.34	.36	.33	.30
		2.0 (E)	.53	.49	.45	.51	.46	.43	.45	.42	.39	.39	.37	.35
		2.5 (D)	.57	.53	.49	.54	.50	.47	.48	.45	.42	.42	.39	.37
		3.0 (C)	.60	.56	.52	.56	.53	.50	.49	.47	.44	.43	.41	.39
		4.0 (B)	.63	.60	.53	.59	.56	.54	.52	.50	.47	.45	.44	.42
		5.0 (A)	.65	.62	.59	.61	.58	.56	.53	.52	.50	.47	.45	.44
MF 0.65  25 / 55	 SEMI-DIRECT	0.6 (J)	.33	.28	.24	.32	.27	.24	.30	.26	.23	.29	.25	.22
		0.8 (I)	.40	.35	.31	.39	.34	.30	.37	.32	.30	.34	.31	.28
		1.0 (H)	.46	.41	.37	.44	.40	.36	.42	.38	.35	.39	.36	.33
		1.25 (G)	.51	.46	.42	.50	.45	.41	.46	.43	.40	.43	.40	.38
		1.5 (F)	.55	.50	.46	.53	.49	.45	.50	.46	.43	.46	.44	.41
		2.0 (E)	.60	.56	.52	.58	.54	.51	.54	.51	.48	.50	.48	.46
		2.5 (D)	.63	.59	.56	.61	.57	.55	.57	.54	.52	.53	.50	.49
		3.0 (C)	.66	.62	.59	.63	.60	.57	.59	.56	.54	.54	.53	.51
		4.0 (B)	.69	.66	.63	.66	.63	.61	.61	.59	.57	.56	.55	.53
		5.0 (A)	.71	.68	.66	.68	.66	.64	.64	.61	.59	.58	.57	.55
MF 0.70  0 / 60	 DIRECT	0.6 (J)	.35	.30	.28	.34	.30	.27	.34	.30	.27	.33	.30	.27
		0.8 (I)	.41	.37	.34	.41	.37	.34	.40	.36	.34	.39	.36	.34
		1.0 (H)	.45	.41	.38	.45	.41	.38	.44	.41	.38	.43	.40	.38
		1.25 (G)	.49	.45	.43	.49	.45	.43	.48	.45	.42	.47	.44	.42
		1.5 (F)	.52	.48	.46	.51	.48	.46	.50	.48	.45	.49	.47	.45
		2.0 (E)	.55	.52	.50	.54	.52	.50	.53	.51	.49	.52	.50	.49
		2.5 (D)	.57	.54	.52	.56	.54	.52	.55	.52	.52	.54	.52	.51
		3.0 (C)	.58	.56	.54	.58	.56	.54	.56	.55	.53	.55	.54	.53
		4.0 (B)	.60	.58	.56	.59	.58	.56	.58	.57	.55	.57	.56	.55
		5.0 (A)	.61	.59	.58	.60	.59	.58	.59	.58	.57	.58	.57	.56
MF 0.70 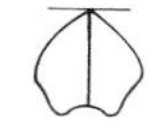 0 / 100	 DIRECT	0.6 (J)	.53	.46	.42	.53	.46	.42	.52	.46	.42	.51	.46	.41
		0.8 (I)	.64	.57	.52	.63	.57	.52	.62	.56	.52	.61	.56	.52
		1.0 (H)	.72	.65	.60	.71	.65	.60	.70	.64	.60	.68	.64	.60
		1.25 (G)	.78	.72	.68	.78	.72	.68	.76	.71	.68	.75	.70	.67
		1.5 (F)	.83	.77	.73	.82	.77	.73	.81	.76	.72	.80	.76	.72
		2.0 (E)	.89	.84	.80	.88	.84	.80	.87	.83	.80	.85	.82	.79
		2.5 (D)	.93	.88	.85	.92	.88	.84	.90	.86	.84	.88	.86	.83
		3.0 (C)	.95	.92	.88	.94	.91	.88	.92	.90	.87	.91	.88	.86
		4.0 (B)	.99	.95	.93	.97	.94	.92	.95	.93	.91	.94	.92	.90
		5.0 (A)	1.01	.98	.96	1.00	.97	.95	.98	.96	.94	.96	.94	.92

* Coefficients of utilization are based on the lumen output of reflectorized lamps. Coefficients greater than 1.0 are the result of interreflections of light by the floor, walls and other surfaces. This builds up the footcandle level beyond the initial lamp lumens per square foot of work-plane area.

† From "Standard Handbook for Electrical Engineers," edited by A. E. Knowlton, 9th ed., copyright © 1957, McGraw-Hill Book Company. Used by permission.

Table 22-14. Recommended Levels of Illumination*

Area	Footcandles on Tasks†
Assembly:	
Rough easy seeing	30
Rough difficult seeing	50
Medium	100
Fine	500‡
Extra fine	1,000‡
Cloth products:	
Cloth inspection	2,000‡
Cutting	300‡
Sewing	500‡
Pressing	300‡
Inspection:	
Ordinary	50
Difficult	100
Highly difficult	200‡
Very difficult	500‡
Most difficult	1,000‡
Machine shops	
Rough bench- and machine work	50
Medium bench- and machine work, ordinary automatic machines, rough grinding, medium buffing and polishing	100
Fine bench- and machine work, fine automatic machines, medium grinding, fine buffing and polishing	500‡
Extra-fine bench- and machine work, grinding, fine work	1,000‡
Offices:	
Cartography, designing, detailed drafting	200
Accounting, auditing, tabulating, bookkeeping, business-machine operation, reading poor reproductions, rough layout drafting	150
Regular office work, reading good reproductions, reading or transcribing handwriting in hard pencil or on poor paper, active filing, index references, mail sorting	100
Reading or transcribing handwriting in ink or medium pencil on good-quality paper, intermittent filing	70
Reading high-contrast or well-printed material; tasks and areas not involving critical or prolonged seeing such as conferring, interviewing, inactive files, and washrooms	30
Corridors, elevators, escalators, stairways	20§
Schools:¶	
Reading printed material	30
Reading pencil writing	70
Spirit duplicated material:	
Good	30
Poor	100
Drafting, benchwork	100‡
Lip reading, chalkboards, sewing	150‡
Woodworking:	
Rough sawing and benchwork	30
Sizing, planing, rough sanding, medium-quality machine- and benchwork, gluing, veneering, cooperage	50
Fine bench- and machine work, fine sanding and finishing	100

* SOURCE: Table 19-21 from W. T. Stuart, Wiring Design—Commercial and Industrial Buildings, in D. G. Fink and J. M. Carroll (eds.), "Standard Handbook for Electrical Engineers," 10th ed., McGraw-Hill Book Co., 1968. Used by permission.

† Minimum on the task at any time.

‡ Obtained with a combination of general lighting plus specialized supplementary lighting. Care should be taken to keep within the recommended brightness ratios. These seeing tasks generally involve the discrimination of fine detail for long periods of time and under conditions of poor contrast. To provide the required illumination, a combination of the general lighting indicated plus specialized supplementary lighting is necessary. The design and installation of the combination system must provide for not only a sufficient amount of light but also the proper direction of light, diffusion, and eye protection. As far as possible, it should eliminate direct and reflected glare as well as objectionable shadows.

§ Or not less than one-fifth the level in adjacent areas.

¶ Tasks are listed here, rather than areas.

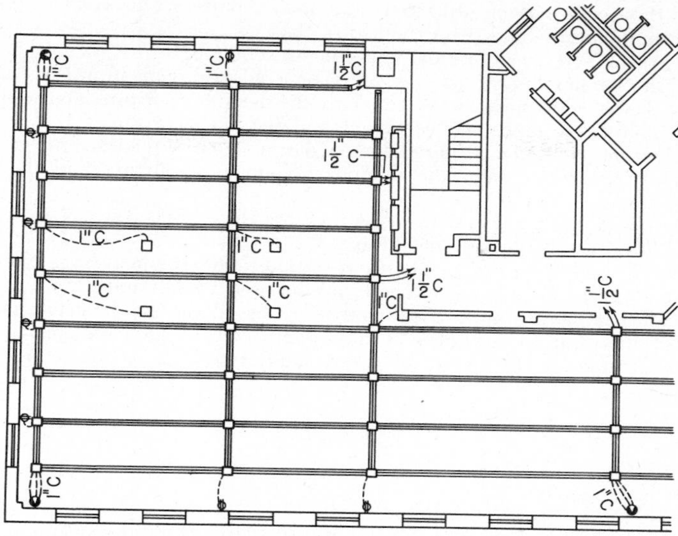

Fig. 22-3. Underfloor distribution of electric power through ducts.

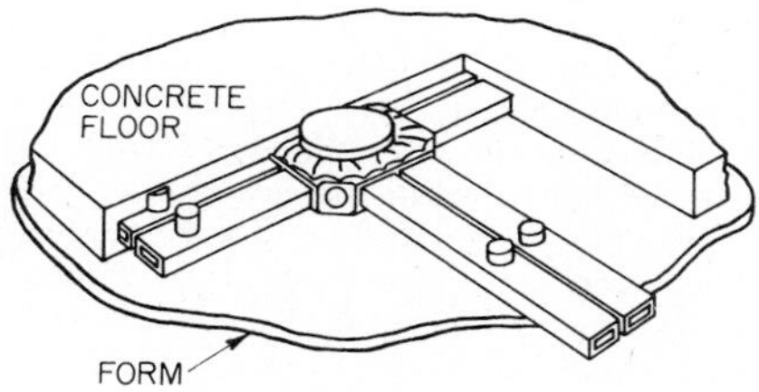

Fig. 22-4. Electric ducts in a concrete floor.

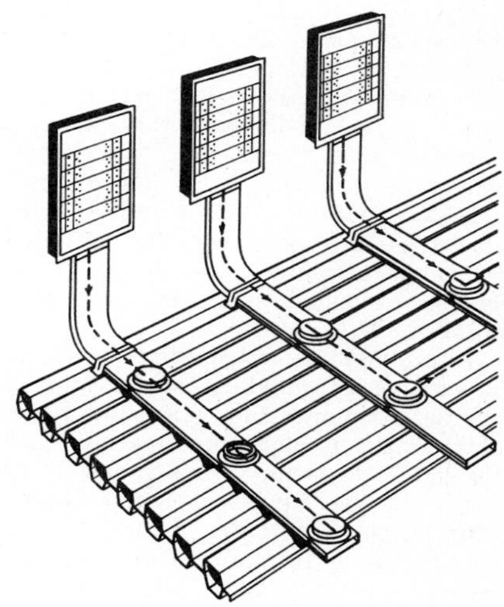

Fig. 22-5. Cellular steel decking serving as underfloor electric ducts. Wires in headers distribute power to wires in the cells.

The maintenance factor allows for such conditions as depreciation of lamp output with prolonged use and cleaning programs.

Recommended foot-candle levels for use in design for various tasks are shown in Table 22-14.

Steps in design by the lumen method are as follows:

1. Determine the foot-candles of illumination from Table 22-14.
2. Find the room ratio from Table 22-12.

3. Select the coefficient of utilization and maintenance factor from Table 22-13 or from a manufacturer's catalog for the fixture chosen.
4. Using Eq. (22-10), compute the total lumens required.
5. Establish the number of fixtures required. A good general rule is to make the distance between fixtures about equal to the height of fixture above the floor and to allow half this distance between a wall and first line of fixtures.
6. Select fixtures with the required lumen output from a manufacturer's catalog. For fluorescent fixtures, specify the number of lamps per fixture and the size of the lamps in watts.
7. On each 15-amp fixture circuit, allow no more than 1,500 watts of incandescent lighting or 1,250 watts of fluorescent lighting.

The lumen method is approximate, since it is used to determine average illumination in the room. The point-by-point method may be used to determine more accurately the illumination at any point. This method is based on the inverse-square law, which states that the foot-candles at any point equals the candlepower of the light source divided by the square of the distance from the source to point being investigated.

$$\text{Ft-c} = \frac{\text{cp}}{D^2} \qquad (22\text{-}11)$$

where cp = candlepower
D = distance from light source

Fluorescent lamps produce more lumens per watt than incandescent lamps. Output of a fluorescent lamp ranges from 42 lumens per watt for a 25-watt lamp to 58 lumens per watt for a 75-watt lamp. In comparison, output of incandescents ranges from 10 lumens per watt for a 25-watt lamp to 16 lumens per watt for the 75-watt size. Hence, fluorescent lamps generally are preferred for air-conditioned buildings.

Lighting adds 3.416 Btu per watt in heat gain to a room. This is an important factor in air conditioning the space.

In computing heat gain, watts consumed by the ballast for fluorescent lamps should be included. This power can add 25% to the lamp rating in watts.

22-13. Underfloor Distribution Systems. An underfloor duct system provides electrical distribution with the flexibility and capacity required for buildings where use of space is subject to change, more telephones may be added, intercommunication systems expanded, and electrically operated business machines of greater complexity installed.

Advantages of distributing electrical wiring under the floor include: improved appearance of building, less chance of accidental injury to wires, elimination of overhead conduit and wires and thus of interference from these with movement of materials to and from machines, and less interference with efficient layout of the lighting system. Underfloor duct has greater wiring capacity than conduit. The wires can be tapped at regular and short intervals (Fig. 22-3). The electrical distribution system can be installed without waiting for the layout of partitions, equipment, or furniture. There is no need to tear up floors to install conduit and boxes. The underfloor ducts will meet the building's need for a long time, no matter how often electrical requirements may change.

The ducts can be installed easily in almost all types of floors (Figs. 22-4 and 22-5).

Cellular-steel deck construction (Fig. 22-5) is economical because it also provides cells that serve as underfloor ducts. It has very high wire-carrying capacity. And outlets can be installed on the floor at almost any point.

Section **23**

Vertical Transportation

FREDERICK S. MERRITT

Consulting Engineer, Syosset, N.Y.

Vertical circulation of traffic in a multistory building is the key to successful functioning of the design, both in normal use and in emergencies. In fact, location of elevators or stairs may determine the floor plan. So in the design of a building, much thought should be given to the type of vertical circulation to be provided, number of units needed, and their location, arrangement, and design.

Traffic may pass from level to level in a multistory building by ramps, stairs, elevators, or escalators. The powered equipment is always supplemented by stairs for use when power is shut off, or there is a mechanical failure, or maintenance work is in process, or in emergencies. In addition to conventional elevators, other types of man lifts are occasionally installed in residences, factories, and garages. For moving small packages or correspondence between floors, dumbwaiters or vertical conveyors also are installed in certain types of buildings.

23-1. Ramps. When space permits, a sloping surface, or ramp, can be used to provide an easy connection between floors. In some garages, to save space, every floor serves as a ramp; each one is split longitudinally, each section sloping gradually in opposite directions to meet the next level above and below.

Ramps are especially useful when large numbers of people or vehicles have to be moved from floor to floor. Therefore, they are frequently adopted for public buildings, such as railroad stations, stadiums, and exhibition halls. They are often used in garages. And they are sometimes incorporated in special-purpose buildings, such as schools for physically handicapped children. In all cases, ramps should be constructed with a nonslip surface.

Ramps generally have been built with slopes up to 15% (15 ft in 100 ft), but 10% is a preferred maximum. Some idea of the space required for a ramp may be obtained from the following: With the 10% maximum slope and a story height of, say, 12 ft, a ramp connecting two floors would have to be 120 ft long. The ramp need not be straight for the whole distance, however. It can be curved, zigzagged, or spiraled. Level landings, with a length of at least 44 in. in the direction of travel, should be provided at door openings and where ramps change slope or direction abruptly. Ramps and landings should be designed for a live load of at least 100 psf. Railings should be designed for a horizontal thrust of 50 lb per ft at top of rail. Enclosures should be designed for 50 lb per ft applied 42 in. above the floor.

Minimum width of pedestrian ramps is 30 in. for heights between landings not exceeding 12 ft. Minimum width is 44 in. for greater height when the slope does not exceed 10%. Landings should be at least as wide as the ramps.

Powered ramps, or moving walks, carrying standing passengers, may operate on slopes up to 8° at speeds up to 180 fpm, and on slopes up to 15° at speeds up to 140 fpm. Construction is similar to that of escalators, except that the treadway is continuous.

("Life Safety Code," National Fire Protection Association, Boston, Mass. 02110; "American National Standard Safety Code for Elevators, Dumbwaiters, Escalators, and Moving Walks," A17.1, American National Standards Institute.)

23-2. Stairs. Less space is required for stairs than for ramps, because steeper slopes can be used. Maximum slope of stairs for comfort is estimated to be about 1 on 2 (27°), but this angle frequently is exceeded for practical reasons. Exterior stairs generally range in slope from 20 to 30°, interior stairs from 30 to 35°.

Among the principal components of a stairway are:

Flight. A series of steps extending from floor to floor, or from a floor to an intermediate landing or platform.

Landings (Platforms). Are used where turns are necessary or to break up long climbs. Landings should be level, as wide as the stairs, and at least 44 in. long in the direction of travel.

Step. Combination of a riser and the tread immediately above.

Rise. Distance from floor to floor.

Run. Total length of stairs in a horizontal plane, including landings.

Riser. Vertical face of a step. Its height is generally taken as the vertical distance between treads.

Tread. Horizontal face of a step. Its width is usually taken as the horizontal distance between risers.

Nosing. Projection of a tread beyond the riser below.

Soffit. Underside of a stair.

Header. Horizontal structural member supporting stair stringers or landings.

Carriage. Rough timber supporting the steps of wood stairs.

Stringers. Inclined members along the sides of a stairway. The stringer along a wall is called a wall stringer. Open stringers are those cut to follow the lines of risers and treads. Closed stringers have parallel top and bottom, and treads and risers are supported along their sides or mortised into them. In wood stairs, stringers are placed outside the carriage to provide a finish.

Railing. Framework or enclosure supporting a handrail and serving as a safety barrier.

Balustrade. A railing composed of balusters capped by a handrail.

Handrail. Protective bar placed at a convenient distance above the stairs for a handhold.

Baluster. Vertical member supporting the handrail in a railing.

Newel Post. Post at which the railing terminates at each floor level.

Angle Post. Railing support at landings or other breaks in the stairs. If the angle post projects beyond the bottom of the stringers, the ornamental detail formed at the bottom of the post is called the **drop.**

Winders. Steps with tapered treads in sharply curved stairs.

Headroom. Minimum clear height from a tread to overhead construction, such as the ceiling of the next floor, ductwork, or piping.

Vertical Clearance. Ample headroom should be provided not only to prevent tall people from injuring their heads, but to give a feeling of spaciousness. A person of average height should be able to extend his hand forward and upward without touching the ceiling above the stairs. Minimum vertical distance from the nosing of a tread to overhead construction should never be less than 6 ft 8 in. and preferably not less than 7 ft.

Stair Width. Width of a stair depends on its purpose and the number of persons to be accommodated in peak hours or emergencies. Generally, the minimum width that can be used is specified in the local building code. For example, for interior stairs, clear width may be required to be at least 20 in. in one- and two-family dwellings, 36 in. in hotels, motels, apartment buildings, and industrial buildings, and 44 in. for other types of occupancy.

Number of Stairways Required. This is usually controlled by local building codes. This control may be achieved through a restriction on the maximum horizontal distance from any point on a floor to a stairway, or on the maximum floor area contributory to a stairway. In addition, codes usually have special provisions for public buildings, such as theaters and exhibition halls. Restrictions might also be placed on the maximum capacity of a stairway. For example, assigned capacity, the number of persons that may be served by stairs per floor per 22-in. unit of stair width, might be 15 for such buildings as hospitals and nursing homes; 30 for other institutional and residential buildings; 45 for storage buildings; 60 for mercantile, business, educational, and industrial buildings, theaters, and restaurants; 80 for churches, concert halls, and museums, and 320 for stadiums and amusement structures.

Step Sizes. Risers and treads generally are proportioned for comfort, although sometimes space considerations control or the desire to achieve a monumental effect, particularly for outside stairs of public buildings. Treads should be 10 to 13 in. wide, exclusive of nosing. Treads less than 10 in. wide should have a nosing of about 1 in. The most comfortable height of riser is 7 to 7½ in. Risers less than 6 in. and more than 8 in. high should not be used. The steeper the slope of the stairs, the greater the ratio of riser to tread. Among the more common simple formulas generally used with the preceding limits are:

1. Product of riser and tread must be between 70 and 75.
2. Riser plus tread must equal 17 to 17.5.
3. Sum of the tread and twice the riser must lie between 24 and 25.5.

In designing stairs, account should be taken of the fact that there is always one less tread than riser per flight of stairs. No flight of stairs should contain less than three risers.

Curved Stairs. Winders should be avoided when possible, because the narrow width of tread at the inside of the curve may cause accidents. Sometimes, instead, **balanced steps** can be used. Instead of radiating from the center of the curve, like winders, balanced steps, though tapered, have the same width of tread along the line of travel as the straight portion of the stairs. (Line of travel in this case is assumed to be about 20 in. from the rail on the inside of the curve.) With balanced steps, the change in angle is spread over a large portion of the stairs.

Loading. Stairs and landings should be designed for a live load of 100 psf or a concentrated load of 300 lb placed to produce maximum stresses.

Railings. Handrails generally are set 2 ft 6 in. to 2 ft 10 in. above the intersections of treads and risers at the front of the steps. At landings, railings usually are from 2 ft 10 in. to 3 ft high, although lower railings can be used safely if the parapet is very wide. Low, wide railings usually are used for monumental stairs.

Interior stairs more than 88 in. wide should have intermediate handrails that divide the stairway into widths of not more than 88 in., preferably into a nominal multiple of 22 in. Handrails along walls should have a clearance of at least 1½ in.

Railings should be designed for a horizontal force of 40 lb per ft and a vertical force of 50 lb per ft, applied at top of rail. But neither force should be taken as less than 200 lb.

Types of Stairs. Generally, stairs are of the following types: straight, circular, curved, or spiral, or a combination.

Straight stairs are stairs along which there is no change in direction on any flight between two successive floors. There are several possible arrangements of straight stairs. For example, they may be arranged in a **straight run** (Fig. 23-1*a*), with a single flight between floors, or a series of flights without change in direction (Fig. 23-1*b*). Also, straight stairs may permit a change in direction at an intermediate landing. When the stairs require a complete reversal of direction (Fig. 23-1*c*), they are called **parallel stairs.** When successive flights are at an angle to each other, usually 90° (Fig. 23-1d), they are called **angle stairs.** In addition, straight stairs may be classified as **scissors stairs** when they comprise a pair of straight runs in opposite directions and are placed on opposite sides of a fire-resistive wall (Fig. 23-1*e*).

Circular stairs when viewed from above appear to follow a circle with a single center of curvature and large radius

Curved stairs when viewed from above appear to follow a curve with two or more centers of curvature, such as an ellipse.

Spiral stairs are similar to circular stairs except that the radius of curvature is small and the stairs may be supported by a center post. Over-all diameter of such stairs may range from 3 ft 6 in. to 8 ft. There may be from 12 to 16 winder treads per complete rotation about the center.

See also Arts. 23-3 to 23-7.

23-3. Fire Exits. In many types of buildings, exit stairs must be enclosed with walls having a high fire-resistance rating to prevent spread of smoke and flames. Walls between these stairs and the building interior should not have any openings. Wired glass windows, however, are permitted in exterior walls of the stair tower that are not exposed to severe fire hazards. Access to such smokeproof towers

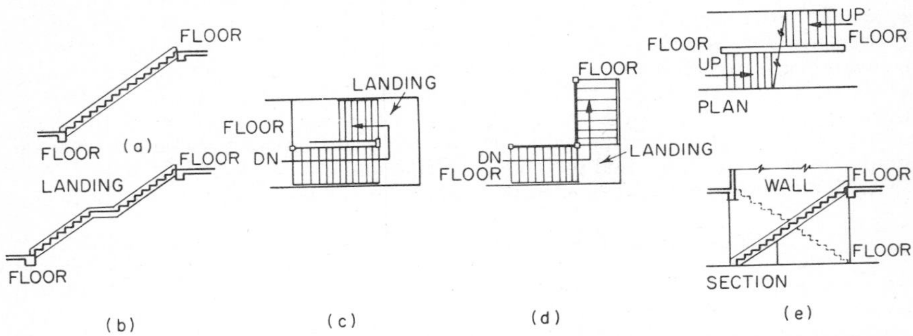

Fig. 23-1. Arrangements of straight stairs. (a) A single flight between floors. (b) A series of flights without change in direction. (c) Parallel stairs. (d) Angle stairs. (e) Scissors stairs.

should be provided in each story through vestibules open to the outside on an exterior wall, or from balconies on an exterior wall, neither exposed to severe fire hazards. Doors should be at least 40 in. wide, self-closing, and provided with a viewing window of clear, wired glass not exceeding 720 sq in. in area. It also is wise to incorporate some means of opening the top of the shaft, either with a thermally operated device or with a skylight, to let escape any heat that might enter the tower from a fire.

In public buildings, there should be more than one smokeproof tower, and these towers should be as far apart as possible. Exits at the bottom of smokeproof towers should be directly to the outdoors, where people can remove themselves quickly to a safe distance from the building.

Stairs outside a building are acceptable as a required fire exit instead of inside stairs, if they satisfy all the requirements of inside stairs. Where enclosure of inside stairs is required, however, outside stairs should be separated from the building interior by fire-resistant walls with fire doors or fixed wired glass windows protecting openings. Some building codes limit the height of outside stairs to a maximum of six stories or 75 ft.

Fire escapes, outside metal-grating stairs, and landings attached to exterior walls with unprotected openings were acceptable at one time as required exits, but usually now are not accepted as such for new construction.

23-4. Wood Stairs. (See also Art. 23-2.) In wood-frame buildings, low nonfireproof buildings, and one- and two-family houses, stairs may be constructed of wood. They may be built in place or shop fabricated.

Construction of a built-in-place stair starts with cutting of carriages to the right size and shape to receive the risers and treads. Next, the lower portion of the wall stringer should be cut out at least ½ in. deep to house the steps (Fig. 23-2). The stringer should be set in place against the wall with the housed-out profile fitted to the stepped profile of the top of the carriage. Then, treads and risers should be firmly nailed to the carriages, tongues at the bottom of the risers fitting into grooves at the rear of the treads. Nosings are generally finished on the underside with molding.

If the outer stringer is an open stringer it should be carefully cut to the same profile as the steps, mitered to fit corresponding miters in the ends of risers, and nailed against the outside carriage. Ends of the treads project beyond the open stringer.

If the outer stringer is a curb or closed stringer it should be plowed out in the same way as the wall stringer to house the steps. Ends of the treads and risers should be wedged and glued into the wall stringer.

(American Institute of Architects and Ramsey and Sleeper, "Architectural Graphic Standards," John Wiley & Sons, Inc., New York; J. D. Wilson and S. O. Werner, "Simplified Stair Layout," Delmar Publishers, Albany, N.Y.)

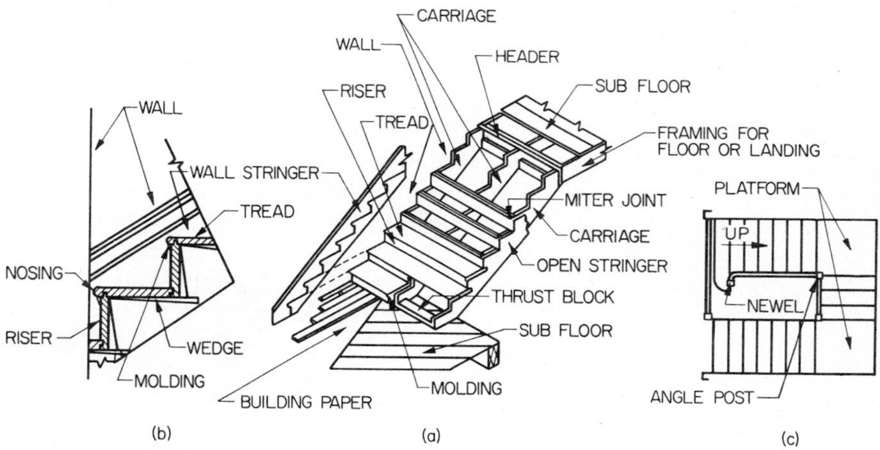

Fig. 23-2. Typical construction for wood stairs.

23-5. Steel Stairs (See also Art. 23-2.) Pressed-sheet steel stairs generally are used in fire-resistant buildings. They may be purchased from various manufacturers in stock patterns.

The steel sheets are formed into risers and subtreads or pans, into which one of several types of treads may be inserted (Fig. 23-3). Stringers usually are channel-shaped. Treads may be made of stone, concrete, composition, or metal. Most types are given a nonslip surface.

("Metal Stairs Manual," National Association of Architectural Metal Manufacturers, 1010 West Lake St., Oak Park, Ill. 60301.)

23-6. Concrete Stairs. (See also Art. 23-2.) Depending on the method of support provided, concrete stairs may be designed as cantilevered or inclined beams and slabs. The entire stairway may be cast in place as a single unit, or slabs or T beams formed first and the steps built up later. Soffits formed with plywood or hardboard forms may have a smooth enough finish to make plastering unnecessary. Concrete treads should have metal nosings to protect the edges. Stairs also may be made of precast concrete.

23-7. Escalators. Escalators, or powered stairs, are used when it is necessary to move large numbers of people from floor to floor. They have the advantage of continuous operation without the need for operators. They have large capacity with low power consumption. Large department stores provide vertical-transportation facilities for one person per hour for every 20 to 25 sq ft of sales area above the entrance floor, and powered stairs generally carry 75 to 90% of the traffic, elevators the remainder.

In effect, an escalator is an inclined bridge spanning between floors. It includes a steel trussed framework, handrails, and an endless belt with steps (Fig. 23-4). At the upper end are a pair of motor-driven sprocket wheels and a worm-gear driving machine. At the lower end is a matching pair of sprocket wheels. Two precision-made roller chains travel over the sprockets pulling the endless belt of steps around.

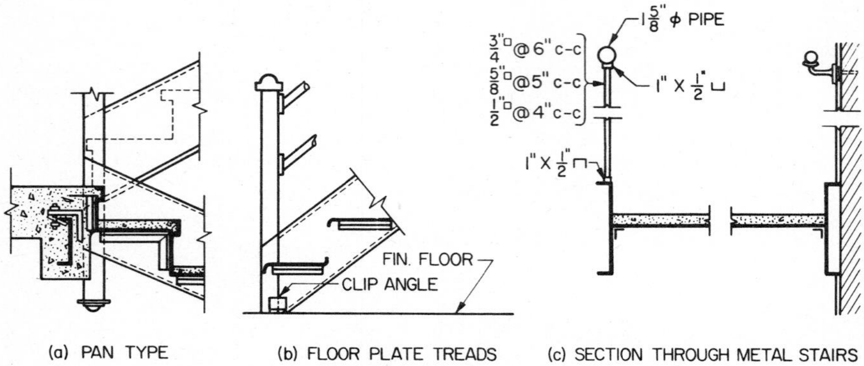

Fig. 23-3. Metal-stair details.

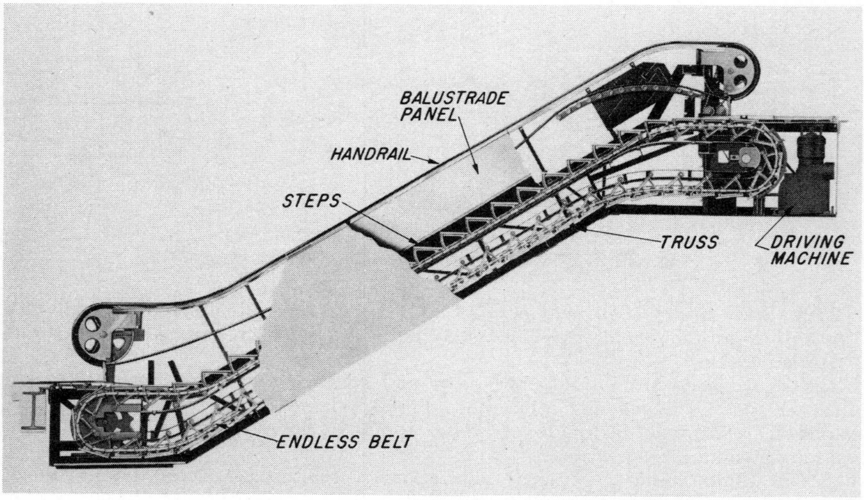

Fig. 23-4. Escalator details. (*Courtesy of Otis Elevator Co.*)

The steps move on an accurately made set of tracks attached to the trusses, with each step supported on four resilient rollers.

Escalators usually can operate at 90 or 120 fpm, as needed for peak traffic and are reversible in direction. Slope of the stairs is standardized at 30°.

For a given speed of travel, width of step determines the capacity of the powered stairs. Standard widths are 32 and 48 in. between handrails, with corresponding capacities at 90 fpm of 5,000 and 8,000 persons per hr. At 120 fpm, a 48-in. escalator can carry as many as 10,000 persons per hr.

Planning for Escalators. Location of powered stairs should be selected only after a careful study of traffic flow in the building. They should be installed where traffic is heaviest and where convenient for passengers, and their location should be obvious to people approaching them. In stores, escalators should lead to strategic sales areas.

In the design of a new building, adequate space should be allotted for powered stairs. Structural framing should be made adequate to support them.

For an escalator installation in an existing building, careful study should be made to determine the necessary alterations to assure adequate space and supports. When the driving machine cannot be housed inside the trussed framework, building beams must be cut to permit passage of the drive chain; these beams must be reinforced.

Foor-to-floor height should be taken into account in determining loads on supporting members. If the total rise is within certain limits, the stairs need be supported only at top and bottom; otherwise, they may have to be supported at intermediate points also. A structural frame should be installed around the stair well to carry the floor and wellway railing. The stairway should be independent of this frame.

Installation. Design of escalators permits a vertical variation of ½ in. in the level of the supporting beams from the specified floor-to-floor height. The escalator is shimmed to bring it level. If variations in elevation exceed ½ in., installation is difficult and much time may be lost. Also, when upper and lower beams are too far apart horizontally, there will be delay, because an extension must be added to the trussed framework.

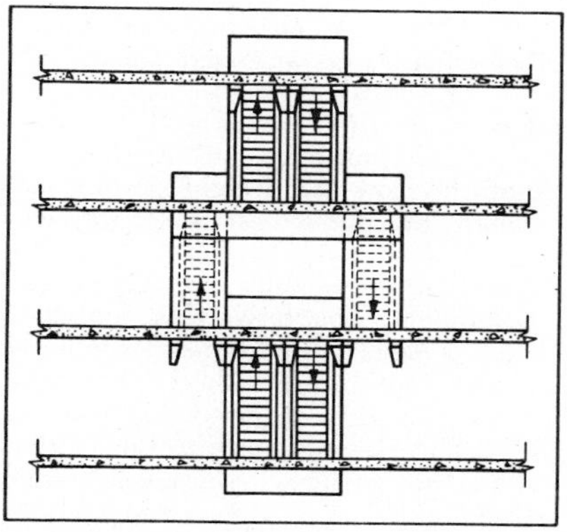

Fig. 23-5. Parallel arrangement of escalators.

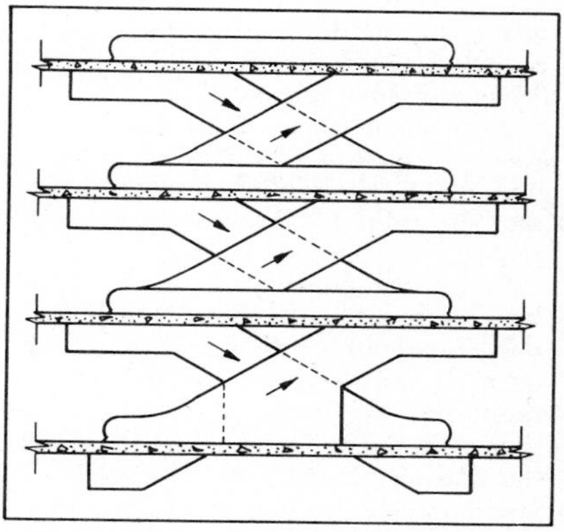

Fig. 23-6. Crisscross arrangement of escalators.

Trusses generally are brought to the job in three sections. There, they are assembled and raised into position with chain hoists, either through an elevator shaft or on the outside of the building.

Either the owner or the manufacturer may furnish the exterior treatment of the balustrade and wellway railings. Often, plaster is specified, and the owner's forces apply it.

Escalator Arrangements. Escalators usually are installed in pairs—one for carrying traffic up and the other for moving traffic down. The units may be placed parallel to each other in each story (Fig. 23-5), or crisscrossed (Fig. 23-6). Crisscrossed stairs generally are preferred because they are more compact, reducing walking distance between stairs at various floors to a minimum.

Fire Protection. Escalators may be acceptable as required exits if they comply with the applicable requirements for exit stairs (Art. 23-3). Such escalators must be enclosed in the same manner as exit stairs. Escalators capable of reversing direction, however, may not qualify as required exits.

An escalator not serving as a required exit must have its floor openings enclosed or protected as required for other vertical openings. Acceptable protection, as an alternative, is afforded in buildings completely protected by a standard supervised sprinkler system by any of the following:

Sprinkler-vent method, a combination of an automatic fire or smoke detection system, automatic air-exhaust system, and an automatic water curtain.

Spray-nozzle method, a combination of an automatic fire or smoke detection system and a system of high-velocity water-spray nozzles.

Rolling shutter method, in which an automatic, self-closing, rolling shutter is used to enclose completely the top of each escalator.

Partial enclosure method, in which kiosks, with self-closing fire doors, provide an effective barrier to spread of smoke between floors.

Escalator trusses and machine spaces should be enclosed with fire-resistant materials. But ventilation must be provided for machine and control spaces.

("Life Safety Code," National Fire Protection Association, Boston, Mass. 02110; "American National Standard Safety Code for Elevators, Dumbwaiters, Escalators, and Moving Walks," A17.1, American National Standards Institute; G. R. Strakosch, "Vertical Transportation: Elevators and Escalators," John Wiley & Sons, Inc., New York.)

23-8. Elevators. Elevators are desirable in all multistory buildings for providing vertical transportation of passengers or freight, and may be required by local building codes for buildings more than four stories high. They are not, however, usually accepted as required means of exit. But they should be arranged for use by firemen in emergencies.

Two major types of elevators are in general use—electric traction and hydraulic. Electric traction elevators are used exclusively in tall buildings and in most low buildings. Hydraulic elevators are usually used for low-rise freight service and may be used for low-rise passenger service where low initial cost is sought, with rises up to about six stories.

Major components of an electric traction installation include the car or cab, hoist wire ropes, driving machine, control equipment, counterweights, hoistway, rails, penthouse, and pit (Fig. 23-7). The car is a cage of light metal supported on a structural frame, to the top of which the wire ropes are attached. The ropes raise and lower the car in the shaft. They pass over a grooved motor-driven sheave and are fastened to the counterweights. The driving machine that drives the sheave consists of an electric motor, brakes, and auxiliary equipment, which are mounted, with the sheave, on a heavy structural frame. The counterweights, consisting of blocks of cast iron in a frame, are needed to reduce power requirements.

The paths of both the counterweights and the car are controlled by separate sets of T-shaped guide rails. The control and operating machinery may be located in a penthouse above the shaft or in the basement. Safety springs or buffers are placed in the pit, to bring car or counterweight to a safe stop if either passes the bottom terminal at normal operating speed.

Elevators and related equipment, such as machinery, signal systems, ropes, and guide rails, are generally supplied and installed by the manufacturer. The general contractor has to guarantee the dimensions of the shaft and its freedom from encroachments. The owner's architect or engineer is responsible for the design and construction of components needed for supporting the plant, including buffer supports, machine-room floors, trolley beams, and guide-rail bracket supports. Magnitudes of loads generally are supplied by the manufacturer with a 100% allowance for impact.

For design of machinery and sheave beams and floor systems, unit stresses should not exceed 80% of those allowed for static loads in the design of usual building structural members. Where stresses due to loads, other than elevator loads, supported on the beams or floor system exceed those due to the elevator loads, 100% of the allowable unit stresses may be used. Unit stresses in a guide rail or its reinforcement, due to horizontal forces, calculated without impact, should not exceed 15,000 psi, and deflection should not exceed ¼ in. Guide-rail supports should be capable of resisting horizontal forces with a deflection of not more than ⅛ in.

For elevators serving more than three floors, means should be provided for venting smoke and hot gases from the hoistways to the outer air in case of fire. (Windows in the walls of hoistway enclosures are prohibited.) Vents may be located in the penthouse or in the enclosure just below the uppermost floor, with direct openings to the outside or with noncombustible duct connections to the outside. Vent area should be at least 3.5% of the hoistway cross-sectional area, but not less than 2 sq ft per elevator car. At least one-third of the vent area should be permanently open or automatically opened by a damper.

See also Arts. 23-9 to 23-22. Controls are described in Arts. 23-9 and 23-21. Hydraulic elevators are discussed in Art. 23-22.

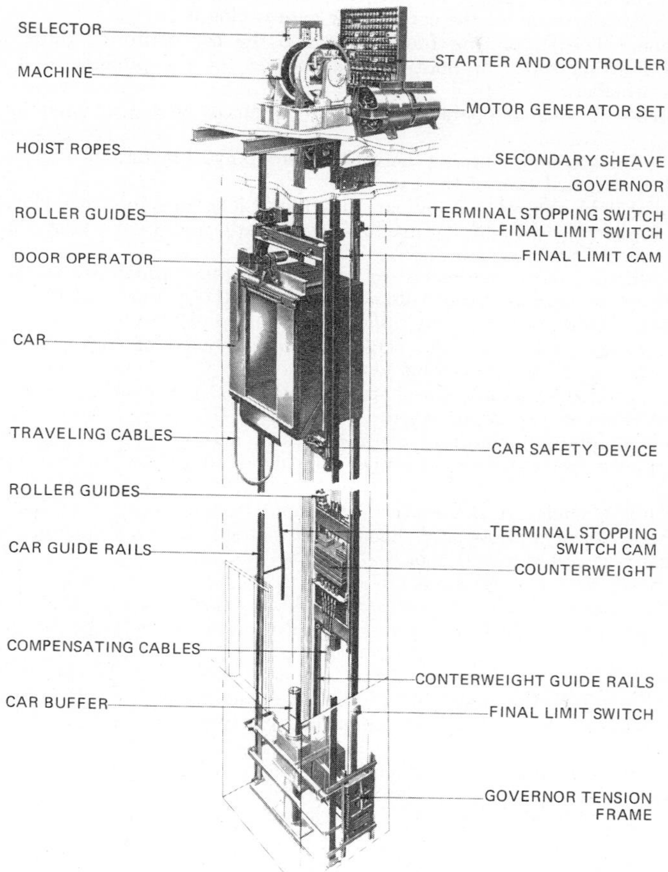

Fig. 23-7. Typical passenger-elevator installation. (*Courtesy of Otis Elevator Co.*)

("Life Safety Code," National Fire Protection Association, Boston, Mass. 02110; "American National Standard Safety Code for Elevators, Dumbwaiters, Escalators, and Moving Walks," A17.1, American National Standards Institute; G. R. Strakosch, "Vertical Transportation: Elevators and Escalators," John Wiley & Sons, Inc., New York.)

23-9. Definitions of Elevator Terms.

Annunciator. An electrical device that indicates, usually by lights, the floors at which an elevator landing signal has been registered.

Buffer. A device for stopping a descending car or counterweight beyond its bottom terminal by absorbing and dissipating the kinetic energy of the car or counterweight. The absorbing medium may be oil, in which case the buffer may be called an **oil buffer,** or a spring, in which case the buffer may be referred to as a **spring buffer.**

Bumper. A device other than a buffer for stopping a descending car or counterweight beyond its bottom terminal by absorbing the impact.

Car. The load-carrying element of an elevator, including platform, car frame, enclosure, and car door or gate.

Car-door Electric Contact. An electrical device for preventing normal operation of the driving machine unless the car door or gate is closed.

Car Frame. The supporting frame to which the car platform, guide shoes, car safety, and hoisting ropes or hoisting-rope sheaves, or the plunger of a hydraulic elevator are attached.

Car Switch. A manual operating device in a car by which an operator actuates the control.

Control. The system governing the starting, stopping, direction of motion, acceleration, speed, and retardation of the car.

Generator-field control employs an individual generator for each elevator, with voltage applied to the driving-machine motor adjusted by varying the strength and direction of the generator field.

Multivoltage control impresses successively on the armature of the driving-machine motor various fixed voltages, such as those that might be obtained from multicommutator generators common to a group of elevators.

Rheostatic control varies the resistance or reactance of the armature or the field circuit of the driving-machine motor.

Single-speed, alternating-current control governs a driving-machine induction motor that runs at a specified speed.

Two-speed alternating-current control governs a two-speed driving-machine induction motor, with motor windings connected to obtain various numbers of poles.

Dispatching Device. A device that operates a signal in a car to indicate when the car should leave a designated floor or to actuate the car's starting mechanism when the car is at a designated floor.

Driving Machine. See Machine.

Emergency Stop Switch. A car-located device that, when operated manually, causes the car to be stopped by disconnecting electric power from the driving-machine motor.

Hoistway. A shaft for travel of one or more elevators. It extends from the bottom of the pit to the underside of the overhead machine room or the roof. A **blind hoistway** is the portion of the shaft that passes floors or other landings without providing a normal entrance.

Hoistway Access Switch. A switch placed at a landing to permit car operation with both the hoistway door at the landing and the car door open.

Hoistway-door Electric Contact. An electrical device for preventing normal operation of the driving machine unless the hoistway door is closed.

Hoistway-door Locking Device. A device for preventing the hoistway door or gate from being opened from the landing side unless the car has stopped within the landing zone.

Leveling Device. A mechanism for moving a car that is within a short distance of a landing toward the landing and stopping the car there. An **automatic maintaining, two-way, leveling device** will keep the car floor level with the landing during loading and unloading.

Machine (Driving Machine). The power unit for raising and lowering an elevator car.

Electric driving machines include an electric motor and brake, driving sheave or drum, and connecting gearing, belts, or chain, if any. A **geared-drive machine** operates the driving sheave or drum through gears. A **traction machine** drives the car through friction between suspension ropes and a traction sheave. A **gearless traction machine** has the traction sheave and the brake drum mounted directly on the motor shaft. A **winding-drum machine** has the motor geared to a drum on which the hoisting ropes wind. A **worm-geared machine** operates the driving sheave or drum through worm gears.

Hydraulic driving machines raise or lower a car with a plunger or piston moved by a liquid under pressure in a cylinder.

Nonstop Switch. A device for preventing a car from making registered landing stops.

Operating Device. The car switch, push button, lever, or other manual device used to actuate the control.

Operation. The method of actuating the control.

Automatic operation starts the car in response to operating devices at landings, or located in the car and identified with landings, or located in an automatic starting mechanism, and stops the car automatically at landings. **Group automatic operation** starts and stops two or more cars under the coordination of a supervisory control system, including automatic dispatching means, with one button per floor in each car and up and down buttons at each landing. **Nonselective collective automatic operation** has one button per floor in each car and only one button at each landing. **Selective collective automatic operation** is a form of group automatic operation in which car stops are made in the order in which landings are reached in each direction of travel after buttons at those landings have been actuated. **Single automatic operation** has one button per floor in each car and only one button per landing, so arranged that after any button has been actuated, actuation of any other button will have no effect on car operation until response to the first button has been completed.

Car-switch operation starts and stops a car in response to a manually operated car switch or continuous-pressure buttons in a car.

Preregister operation is one in which signals to stop are registered in advance by buttons in a car or at landings and then, at the proper point in car travel, are given to an operator in the car, who initiates the stop, which is completed automatically.

Signal operation starts and stops a car automatically as landings are reached, in response to actuation of buttons in cars or at landings, irrespective of direction of car travel or sequence in which buttons are actuated; but the car can be started only by a button or starting switch in the car.

Parking Device. A device for opening from the landing side the hoistway door at any landing when the car is within the landing zone.

Pit. Portion of a hoistway below the lowest landing.

Position Indicator. Device for showing the location of a car in the hoistway.

Rise. See Travel.

Rope Equalizer. A device installed on a car or counterweight to equalize automatically the tensions in the hoisting ropes.

Runby. The distance a car can travel beyond a terminal landing without striking a stop.

Safety. A mechanical device attached to the counterweight or to the car frame or an auxiliary frame to stop or hold the counterweight or the car, whichever undergoes overspeed or free fall, or if the hoisting ropes should slacken.

Safety Bulkhead. In a cylinder of a hydraulic elevator, a closure, at the bottom of the cylinder but above the cylinder head, with an orifice for controlling fluid loss in case of cylinder-head failure.

Signal Registering Device. A button or other device in a car or at a landing that causes a stop signal to be registered in a car.

Signal Transfer Device. A device used in automatic elevator operation for automatically transferring a signal registered in one car to the next car if the former passes, without stopping, a floor for which a signal has been registered.

Signal Transfer Switch. A manually operated switch for accomplishing the same function as a signal transfer device.

Slack-rope Switch. A device that automatically disconnects electric power from the driving machine when the hoisting ropes of a winding-drum machine become slack.

Starter's Control Panel. An assembly of devices with which an elevator starter can control the way in which one or more elevators function.

Terminal Speed-limiting Device (Emergency). A device for reducing automatically the speed of a car approaching a terminal landing, independently of the car-operating device and the normal terminal stopping device if the latter should fail to slow the car as intended.

Terminal Stopping Device. Any device for slowing or stopping a car automatically at or near a terminal landing, independently of the car-operating device. A **final terminal stopping device,** after a car passes a terminal landing, disconnects power from the driving apparatus, independently of the operating device, normal terminal stopping device, or emergency terminal speed-limiting device. A **stop-motion switch,** or **machine final terminal stopping device,** is a final terminal stopping device operated directly by the driving machine.

Transom. One or more panels that close an opening above a hoistway entrance.

Travel (Rise). The vertical distance between top and bottom terminal landings.

Traveling Cable. A cable containing electrical conductors for providing electrical connections between a car and a fixed outlet in a hoistway.

Truck Zone. A limited distance above a landing within which the truck-zoning device permits movement of a freight-elevator car with its door or the hoistway door open.

Truck-zoning Device. A device that permits a car operator to move, within a specified distance above a landing, a freight-elevator car with its door or the hoistway door open.

23-10. Roping for Elevators. For a given weight of car and load, the method of "roping up" an elevator has a considerable effect on the loading of the hoisting ropes, machine bearings, and building members.

Driving machines usually are winding-drum or traction type, depending on whether the ropes are wound on drums on the drive shaft or are powered by a drive sheave. The traction type is usually used. It may be classified as double wrap or single wrap.

For the double wrap—to obtain sufficient traction between the ropes and the driving sheave, which has U-shaped or round-seat grooves—a secondary or idler sheave is used (Fig. 23-8*a*).

In the single-wrap type, the ropes pass over the traction or driving sheave only once; so there is a single wrap, or less, of the ropes on the sheave (Fig. 23-8*d*). The traction sheave has wedge-shaped or undercut grooves that grip the ropes by virtue of the wedging action between the sides of the grooves and the ropes. With this type of installation, the sheave has one-half the loading obtained with the double-wrap machine for the same weight of car and counterweight. As a result, the single-wrap design is lighter than the double-wrap. But rope life may be shorter because of the pinching action of the sheave grooves.

In most buildings, elevator machines are located in a penthouse. When a machine must be installed in the basement (Fig. 23-8*b*), the load on the overhead supports is increased, rope length is tripled, and additional sheaves are needed, adding substantially to the cost. Other disadvantages include higher friction losses and larger number of rope bends, requiring greater traction between ropes and driving sheaves for the same elevator loads and speeds; higher power consumption; more rope wear; and consequently, greater operating expenses.

When heavy loads are to be handled and speed is not important, a 2-to-1 roping may be used (Fig. 23-8*c*), in which case the car speed is only one-half that of the rope. Ends of the rope are dead-ended to the overhead beams, instead of being attached to car and counterweights, as for 1-to-1 roping. With this setup, the anchorages carry one-half the weight of car and counterweights. So the loading on the traction and secondary sheaves is only about one-half that for the 1-to-1 machine. Because of the higher speed of the ropes with 2-to-1 roping, a higher-speed and therefore a less costly motor can be used.

(G. R. Strakosch, "Vertical Transportation: Elevators and Escalators," John Wiley & Sons, Inc., New York.)

23-11. Passenger Elevators. (See also Arts. 23-8 and 23-9.) The number of passenger elevators needed to serve a building adequately depends on their capacity, the volume of traffic, and the interval between cars.

Traffic is measured by the number of persons handled in 5-min periods. For proposed buildings, peak traffic generally can be estimated from comparisons with existing buildings of the same class and occupancy located in the same or a similar neighborhood. Or the peak traffic may be estimated in terms of the probable population. The maximum 5-min traffic flow can be counted (in existing buildings) or estimated from the population of the building, which in turn may be computed from the net rentable area (see Arts. 23-13 to 23-20).

Dividing the peak 5-min traffic flow by the 5-min handling capacity of an elevator gives the minimum number of elevators required. The 5-min handling capacity of an elevator is determined from the round-trip time.

In addition to handling capacity, there is another factor to be considered in determining the number of elevators for adequate service—**interval,** the average time between elevators leaving the ground floor. As an indication of the length of time passengers must wait for an elevator, it is a significant measure of good service.

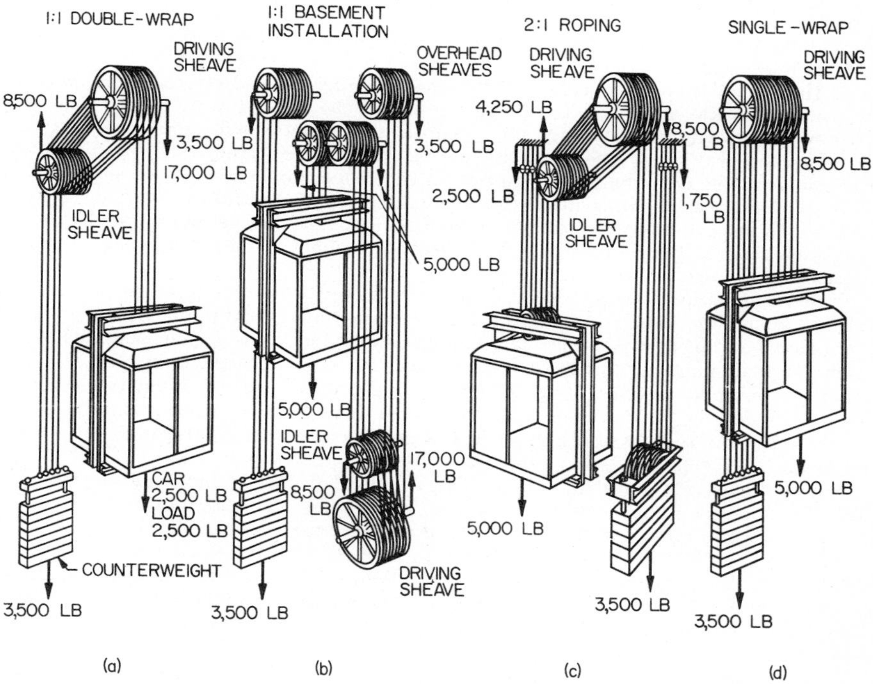

Fig. 23-8. Types of roping for elevators driven by traction machines.

After the number of elevators has been computed on the basis of traffic flow, the interval should be checked (see also Arts. 23-13 and 23-16). It is obtained by dividing the round-trip time by the number of elevators.

Round-trip time is composed principally of the time for a full-speed round-trip run without stops, time for accelerating and decelerating per stop, time for leveling at each stop, time for opening and closing gates and doors, time for passengers to move in and out, reaction time of operator, lost time due to false stops, and standing time at top and bottom floors.

Opening and closing of doors may contribute materially to lost time unless the doors are properly designed. A 3-ft 6-in. opening is good, because two passengers may conveniently enter and leave a car abreast. A slightly wider door would be of little advantage. Department stores, hospitals, and other structures served by large passenger elevators (4,000 lb and over) usually require much bigger openings.

Center-opening doors, preferred for power operation, are faster than either the single or two-speed type of the same width. The impact on closing is smaller with the center-opening door; hence, there is less chance of injuring a passenger. Also, transfer time is less since passengers can move out as the door starts to open.

Another factor affecting passenger-transfer time is the shape of the car. The narrower and deeper a car, the greater is the time required for passengers to leave likely to be. Platform sizes should conform to the standards of the National Elevator Industry, Inc.

23-12. Automatic Elevators. (See also Arts. 23-8, 23-9, and 23-21.) Elevator operation may be either car-switch (operator-controlled) or automatic. With dual operation, the car may be operated by a car switch or by push buttons under the control of an operator, the buttons at each floor signaling but not operating the car. Several types of automatic elevators are available.

The simplest type is the **single automatic.** The car responds to the first button pressed, ignoring all future calls until it arrives at the destination. Used principally in apartment houses and hospitals, these elevators have the disadvantage of not being able to store calls.

With collective controls, calls can be stored. When the **selective-collective type** is used, calls are answered in the direction of the car's travel; when an "up" button is pressed at a landing, for example, the elevator will stop there only if it is on the way up. The controls can be extended to groups of cars, so that the call is answered by the first car to pass in the proper direction.

In addition, the elevators may also be dispatched automatically. A timer signals the cars to leave the terminal at predetermined intervals. Since automatic dispatching often reduces the round-trip time, it can bring about an increase in the carrying capacity of the elevators.

Fully automatic systems (operatorless) are widely used for passenger elevators. These systems are capable of adjusting to varying traffic conditions. For example, for office buildings, when office employees report for work in the morning and traffic is predominantly up, the supervisory control sets itself for the up peak. Cars leave the bottom terminal floor either at the end of a fixed interval or when loaded to 80% of capacity, as indicated by a load-weighing device in the floor of the cab. As soon as the highest call is answered, each car returns to the lower terminal.

For very heavy up traffic, the building is divided into a high and low zone, and elevators are automatically assigned to each zone. This results in a decrease in the round-trip time, which improves the carrying capacity of the elevators.

At the close of business in the evening, the system adjusts to a down peak. The building again is divided into two zones. Waiting time at the bottom terminal is eliminated. Cars start down as soon as they reach the highest call, and when fully loaded, automatically bypass remaining floor calls. High-zone cars meanwhile answer all up calls. For very heavy down traffic, each zone is again divided in two.

Between peak periods, when traffic is moderate, multiple zoning may be used to control automatically the location of cars throughout a building. The objectives are to minimize passenger waiting time and eliminate unnecessary car travel. For the purpose, the building is divided into vertical zones, usually one zone per car. When traffic does not require use of a car, it is parked within a zone it will serve. Thus, when a call comes from that zone, the parked car will be near the call. As traffic grows heavier—for example, around lunch hour—cars leave parked positions at more rapid intervals, and when necessary, a shift may be made to down-peak operation.

During off hours—nights, Sundays, holidays—cars park at the bottom terminal with motor generators shut off. When a call is registered, the motor-generator set powering one car starts up automatically.

Since the elevators are operatorless, several safety devices are incorporated in these automatic elevators in addition to those commonly carried by manually operated cars—automatic load weigher to prevent overcrowding, buttons in car and starter station to stop the doors from closing and to hold them open, lights to indicate floor stops pressed, two-way loudspeaker system for communication with the starter station, and auxiliary power systems if the primary power and supervisory systems should fail. Safety devices also prevent the doors from closing when a passenger is standing in the doorway, and of course, the elevators cannot move when the doors are open.

23-13. Elevators in General-purpose Buildings. (See also Arts. 23-8 to 23-12.) For a proposed diversified-tenancy, or general-purpose, office building, peak traffic may be estimated from the probable population computed from the net rentable area (usually 65 to 70% of the gross area). Net rentable area per person may range from 90 to 135 sq ft. However, when several floors are occupied by a single organization with a large clerical staff, the population density may be much higher.

Diversified-tenancy office buildings usually have important traffic peaks in the morning, at noon, and in the evening, and very little interfloor traffic. The 5-min morning peak generally is the controlling factor, because if the elevators can handle that peak satisfactorily, they can also deal with the others. In a well-diversified office building, the 5-min peak will be about one-ninth of the population; but it can rise to one-eighth if the tenancy is not sufficiently diversified.

For busy, high-class office buildings in large cities, time intervals between elevators may be classified as follows: 20 sec, excellent; between 20 and 25 sec, good; between 25 and 30 sec, fair; between 30 and 35 sec, poor; and over 35 sec, unsatisfactory. In small cities, however, intervals of 30 sec and longer may be satisfactory.

When elevators are divided into local and express service, it is desirable to equalize the intervals. Often, this cannot be done because the necessary number of express elevators would not be economically justified. Intervals for high-use elevators of 30 to 35 sec, therefore, may be considered acceptable.

Car speeds used vary with height of building: 4 to 10 stories, 200 to 500 fpm; 10 to 15 stories, 600 to 700 fpm; 15 to 20 stories, 700 to 800 fpm; 20 to 50 stories, 900 to 1,200 fpm; and over 50 stories, 1,200 to 1,800 fpm.

Elevators should be easily accessible from all entrances to a building. For maximum efficiency, they should be grouped near the center. Except in extremely large buildings, two banks of elevators located in different parts of the structure should not serve the same floors. Generally, traffic cannot be equalized on the two banks, and the bank near the principal exit or street will handle most of the traffic.

Elevators should not serve two lower terminal floors. The extra stop increases the round-trip time and decreases the handling capacity. If there is sufficient traffic between the two lower floors, escalators should be installed.

In laying out an elevator group, not more than four elevators should be placed in a straight line. Elevators placed in alcoves off the main corridor make a good arrangement. Such a layout eliminates interference between elevator and other traffic, makes possible narrow corridors, saves space in the upper floors, decreases passenger-transfer time, and reduces walking distances to the individual elevators. Width of alcove may be 10 to 12 ft.

It is necessary to divide elevator groups into local and express banks in tall buildings, especially those with setbacks and towers, and in low buildings with large rental areas. In general, when more than eight elevators are needed, consideration should be given to such an arrangement. One or two of the elevators of the high-use express bank should be provided with openings at all floors for Sunday and holiday service. In addition to improving service, the division into local and express banks has the advantage that corridor space on the floors where there are no doors can be used for toilets, closets, and stairs.

Unless the building is very large (500,000 sq ft of net rentable area), separate service elevators are not needed for a diversified-tenancy office building. One of the passenger elevators can be detached from the bank for freight work when passenger traffic is light.

23-14. Elevators in Single-purpose Buildings. (See also Arts. 23-8 to 23-12.) Elevator requirements and layouts are similar in general for both single-purpose and diversified-tenancy office buildings (Art. 23-13), but several different factors should be taken into consideration: Single-purpose buildings are occupied by one large organization. Generally, the floors that are occupied by the clerical staff are not subdivided into many offices; the net rentable area is about 80% of the gross area. Population densities are higher than for general-purpose buildings. Depending on the kind of business to be carried on, population density varies from 40 sq ft per person for some life-insurance companies to about 100 for some telephone companies.

While traffic peaks occur at the same periods as in the diversified-tenancy type, the morning peak is very high, unless working hours are staggered. The maximum 5-min periods may be one-eighth to one-half the population, depending on the type of occupancy. Because of this volume of traffic, elevators should have a capacity of 4,000 lb or more, to keep the number required to a minimum. Furthermore, interfloor traffic is large in this type of building and must be considered in designing the elevator system. Usually, so many elevators are required for handling traffic that the interval is satisfactory.

23-15. Professional-building Elevators. (See also Arts. 23-8 to 23-12.) Population cannot be used as the sole basis for determining the number of elevators needed for buildings occupied by doctors, dentists, and other professional men, because of the volume of patient and visitor traffic. Peaks may occur in the forenoon and midafternoon, with the maximum taking place when reception hours coincide. Traffic studies indicate that the maximum peak varies from two to six persons per doctor per hour up and down.

Since crowding of incapacitated patients is inadvisable, elevators should be of at least 3,000-lb capacity. If the building has a private hospital, then one or two of the elevators should have stretcher-size platforms, 5 ft 8 in. by 8 ft 8 in.

23-16. Hotel Elevators. (See also Arts. 23-8 to 23-12.) Hotels with transient guests average 1.3 to 1.5 persons per sleeping room. They have pronounced traffic peaks in morning and early evening. The 5-min maximum occurs at the check-out hour in the evening and is about 10% of the estimated population, with traffic moving in both directions.

Ballrooms and banquet rooms should be located on lower floors and served by separate elevators. Sometimes it is advisable to provide an express elevator to serve heavy roof-garden traffic. Passenger elevators should be of 3,000-lb capacity or more to allow room for bellboys with baggage. Intervals for passenger elevators should not exceed 50 sec.

Service elevators are very important in hotels. They should have a capacity of at least 3,500 lb, with platforms about 7 ft by 6 ft 2 in. The number required usually is determined by comparison with existing hotels.

23-17. Hospital Elevators. (See also Arts. 23-8 to 23-12.) Traffic in a hospital is of two types: (1) medical staff and equipment and (2) transient traffic, such as patients and visitors. Greatest peaks occur when visitor traffic is combined with regular hospital traffic. Waiting rooms should be provided at the main floor and only a limited number of visitors should be permitted to leave them at one time, so that the traffic peaks can be handled in a reasonable period and to keep corridors from getting congested. In large hospitals, however, pedestrian and vehicular traffic should be separated.

For vehicular traffic or a combination of vehicular and pedestrian traffic, hospital elevators should be of stretcher size—5 ft 8 in. to 6 ft 4 in. wide and 8 ft 8 in. to 10 ft deep, with a capacity of 4,000 to 5,000 lb. Speeds vary from 50 to 700 fpm, depending on height of building and load. The elevators should be self-leveling, to avoid bumping patients on and off. For staff, visitors, and other pedestrian traffic, passenger-type elevators, with wide, shallow platforms, such as those used for office buildings, should be selected.

Elevators should be centrally located and readily accessible from the main entrance. Service elevators can be provided with front and rear doors and, if desired, so located that they can assist the passenger elevators during traffic peaks.

23-18. Apartment-building Elevators. (See also Arts. 23-8 to 23-12.) Multistory residential buildings do not have peaks so pronounced as other types of buildings. Generally, the evening peak is the largest. Traffic flow at that time may be 6 to 8% of the building population in a 5-min period. Building population may be estimated at 2.6 persons per apartment.

One elevator rated at 2,000 lb and 200 fpm usually is sufficient for a six-story apartment building with 50 to 75 units. For taller buildings, two elevators operating at 400 to 500 fpm are generally adequate. One of these elevators can be used for service work at times.

23-19. Elevators in Government Buildings. (See also Arts. 23-8 to 23-12.) Municipal buildings, city halls, state office buildings, and other government office buildings may be treated the same as single-purpose office buildings (Art. 23-14). Population density often may be assumed as one person per 100 sq ft of net area. The 5-min maximum peak occurs in the morning, and may be as large as one-third the population.

23-20. Department-store Elevators. (See also Arts. 23-8 and 23-9.) Department stores should be served by a coordinated system of escalators and elevators. The required capacity of the vertical-transportation system should be based on the transportation or merchandising area and the maximum density to which it is expected to be occupied by shoppers.

The **transportation area** is all the floor space above or below the first floor to which shoppers and employees must be moved. Totaling about 80 to 85% of the gross area, it includes the space taken up by counters, showcases, aisles, fitting rooms, public rooms, restaurants, credit offices, and cashiers' counters but does not include kitchens, general offices, accounting departments, stockrooms, stairways, elevator shafts, or other areas for utilities.

The **transportation capacity** is the number of persons per hour that the vertical-transportation system can distribute from the main floor to the other merchandising floors. The ratio of the peak transportation capacity to the transportation area is called the density ratio. This ratio is about 1 to 20 for a busy department store. So the required hourly handling capacity of a combined escalator and elevator system is numerically equal to one-twentieth, or 5%, of the transportation area. The elevator system generally is designed to handle about 10% of the total.

The maximum peak hour usually occurs from 12 to 1 P.M. on weekdays and between 2 and 3 P.M. on Saturdays.

The type of elevator preferred for use with moving stairs is one with 3,500-lb capacity or more. It should have center-opening, solid-panel, power-operated car and hoistway doors, with a 4-ft 6-in. opening and a platform 8 ft by 5 ft 6 in.

23-21. Freight Elevators. (See also Arts. 23-8 and 23-9.) In low-rise buildings, freight elevators may be of the hydraulic type, but in taller buildings (higher than about 50 ft) electric elevators (Fig. 23-9) generally will be more practical. In selecting freight elevators, the following should be considered:

1. Building characteristics, including the travel, number of floors, floor heights, and openings required on two sides of a car. Also, structural conditions that may influence the size, shape, or location of the elevator should be studied.
2. Units to be carried on the elevator—weight, size, type, and method of loading.
3. Number of units to be handled per hour.
4. Probable cycle of operation and principal floors served during the peak of the cycle.
5. Will the elevator handle passengers during periods of peak passenger traffic?

For low-rise, slow-speed applications, especially where industrial trucks will be used, rugged hydraulic freight elevators generally will be more economical than electric freight elevators. Hydraulic freight elevators are discussed in Art. 23-22.

Electric-elevator Controls. Two types of control equipment are generally used for electric elevators—multivoltage and ac rheostatic.

With **multivoltage**, the hoisting motor is dc operated. A motor-generator set is provided for each elevator, and the speed and direction of motion of the car are controlled by varying the generator field. This type of elevator permits the most accurate stops, the most rapid acceleration and deceleration, and minimum power consumption and maintenance for an active elevator. Automatic leveling to compensate for cable stretch or other variations from floor level is an inherent part of multivoltage equipment. Multivoltage is generally used for passenger elevators.

Standard speeds with multivoltage geared freight elevators are 75, 100, 150, and 200 fpm, with 75 fpm a popular standard for 10,000-lb cars. Generally recommended are 100 fpm for 2 or 3 floors, 150 fpm for 4 or 5 floors, and 200 fpm for 6 to 10 floors. If one of the stories exceeds 20 ft, however, the next higher speed should be used.

The **ac rheostatic type** is often chosen to keep initial cost down when the elevator is to be used infrequently (less than five trips per hour on a normal business day). If the ordinary landing accuracy of this type will not be acceptable for the loading methods expected, either of two leveling devices may be installed:

1. Inching, which may be added to the controller for a single-speed ac motor at nominal cost. It permits an operator to move the car with continuous-pressure switches to floor level within a limited zone (about 9 in.) with the doors open.

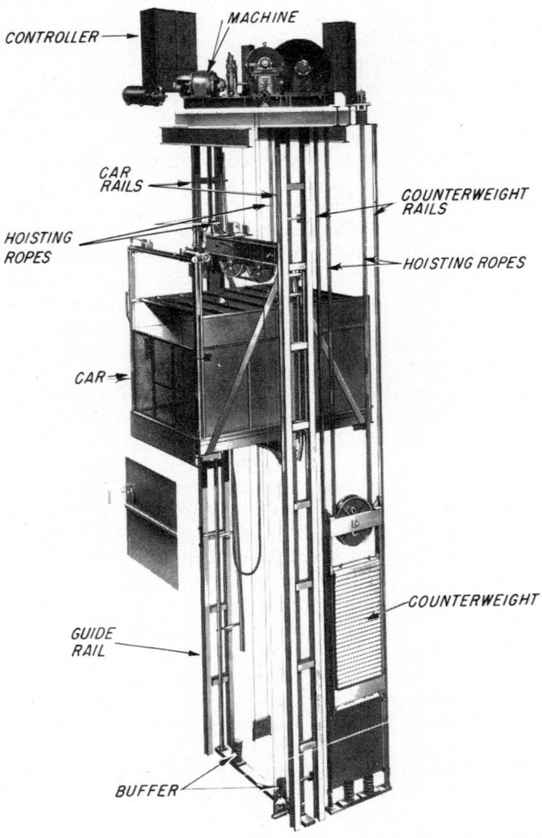

Fig. 23-9. Typical freight-elevator installation. (*Courtesy of Otis Elevator Co.*)

2. Automatic leveling, which is achieved with a two-speed motor and control apparatus. It is available only with selective-collective or single-automatic operation (Art. 23-12) at speeds up to 100 fpm, for capacities up to 8,000 lb, and at 75 fpm, for 10,000-lb elevators.

Standard speeds for ac rheostatic-type geared freight elevators are 50, 75, and 100 fpm. Generally recommended are 50 fpm for two floors, 75 fpm for three or four floors, and 100 fpm for five to eight floors. If one of the stories is more than 20 ft high, the next higher speed should be used.

Car Capacity. The size of car to be used for a freight elevator is generally dependent on the dimensions of the freight package to be carried per trip and the weight of the package and loading equipment. Power trucks, for example, impose severer strains on the entire car structure and the guide rails than do hand trucks. As a power truck with palletized load enters an elevator, most of the weight of the truck and its load are concentrated at the edge of the platform, producing heavy eccentric loading. Maximum load on an elevator should include most of the truck weight as well as the load to be lifted, since the truck wheels are on the elevator as the last unit of load is deposited.

The carrying capacity per hour of freight elevators is determined by the capacity or normal load of the elevator and the time required for a round trip. Round-trip time is composed of the following elements:

1. Running time, which may be readily calculated from the rated speed, with due allowance for accelerating and decelerating time (about 2¼ sec for ac rheostatic with inching, 1¾ sec for multivoltage), and the distance traveled.

2. Time for operation of the car gate and hoistway doors (manual 16 sec, power 8 sec).

3. Loading and unloading time (hand truck 25 sec, power truck 15 sec). Wherever practical, a study should be made of the loading and unloading operations for a similar elevator in the same type of plant.

Operation. The most useful and flexible type of operation for both multivoltage and ac rheostatic freight elevators is selective-collective with an annunciator. When operated without an attendant, the car automatically answers the down calls as approached when moving down and similarly answers up calls when moving up. The elevator attendant, when present, has complete control of the car and can answer calls indicated by the annunciator by pressing the corresponding car button.

The single-automatic type, available for both multivoltage and ac rheostatic controls, is a somewhat simpler, less expensive type for operation without an attendant. But it has the disadvantage that corridor calls cannot be registered when the car is in motion, nor can the car respond to more than one call at a time.

If an attendant will always operate the elevator, use may be made of car-switch releveling, which is a simple, inexpensive means of operating with variable-voltage control. The car will travel in either direction, or stop, in response to the movement of a switch, leveling automatically at each floor.

Also, a car-switch type of operation with ac rheostatic control may be used for freight elevators. The car is leveled to a mark on the hoistway door by the attendant using the normal car switch.

The least expensive type of operation is the continuous-pressure type. It is available for light-service ac rheostatic elevators. Buttons are installed in the corridor and car and, when pressed, will cause the car to move in a selected direction. The car will stop when the button is released. This control is not recommended for travel of more than 70 ft.

The standard hoistway door is the vertical bi-parting metal-clad wood type. For active elevators and openings wider than 8 ft, doors should be power-operated.

When passengers are carried on a freight elevator, the most important safeguards are: (1) a capacity rating equal to that of a passenger elevator with the same net area; (2) interlocks on hoistway doors, instead of locks and contacts; (3) interlocking of the car gate and the counterbalanced hoistway door, so that the hoistway door opens first and the car gate closes first to eliminate the tripping hazard. This third requirement is met preferably by use of power-operated doors and gates with sequence operation. The door time with sequence operation is about the same as for manual doors.

Automatic freight elevators can be integrated into material-handling systems for multistory warehouse or production facilities. On each floor, infeed and outfeed horizontal conveyors may be provided to deliver and remove loads, usually palletized, to and from the freight elevator. The elevator may be loaded, transported to another floor, and unloaded—all automatically.

23-22. Hydraulic Elevators. For low-rise elevators, hydraulic equipment may be used to supply the lift. The car sits atop a plunger, or ram, which operates in a pressure cylinder (Fig. 23-10). Oil serves as the pressure fluid and is supplied through a motor-driven positive-displacement pump, actuated by an electric-hydraulic control system.

To raise the car, the pump is started, discharging oil into the pressure cylinder and forcing the ram up. When the car reaches the desired level, the pump is stopped. To lower the car, oil is released from the pressure cylinder, returning to a storage tank.

Single-bearing cylinders (Fig. 23-11a) are a simple type that operate like a hydraulic jack. They are suitable for elevator and sidewalk lifts where the car is guided at top and bottom, preventing eccentric loading from exerting side thrust on the cylinder bearing. A cylinder of heavy steel usually is sunk in the ground as far as the load rises. The ram, of thick-walled steel tubing polished to a mirror finish, is sealed at the top of the cylinder with compression packing. Oil is admitted under pressure near the top of the cylinder, while air is removed through a bleeder.

A different cylinder design should be used where the car or platform does not operate in guides. One type capable of taking off-balance loads employs a two-bearing plunger (Fig. 23-11b). The bearings are kept immersed in oil.

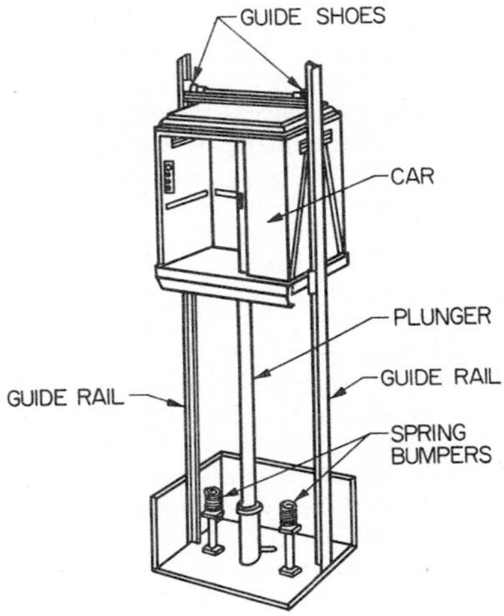

Fig. 23-10. Hydraulic elevator.

Another type, suitable for general industrial applications, has a movable bearing at the lower end of the ram to give support against heavy eccentric loads (Fig. 23-11c). At the top of the cylinder, the plunger is supported by another bearing.

For long-stroke service, a cage-bearing type can be used (Fig. 23-11d). The cage bearing is supported by a secondary cylinder about 3 ft below the main cylinder head. Oil enters under pressure just below the main cylinder head, passes down through holes in the bearing, and lifts the plunger.

When the car or platform is not heavy enough to insure gravity lowering, a double-acting cylinder may be used (Fig. 23-11e). To raise the plunger, oil is admitted under pressure below the piston; to lower it, oil is forced into the cylinder near the top, above the piston, and flows out below. Jack plunger sizes for the various types range from 2½ in. in diameter for small low-capacity lifts to 18 in. for large lifts, operating at 150 to 300 psi.

Hydraulic elevators have several advantages over electric elevators: They are simpler. The car and its frame rest on the hydraulic ram that raises and lowers them. There are no wire ropes, no overhead equipment, no penthouse. Without heavy overhead loads, hoistway columns and footings can be smaller. Car safeties or speed governors are not needed, because the car and its load cannot fall faster than normal speed. Speed of the elevators is low; so the bumpers need be only heavy springs.

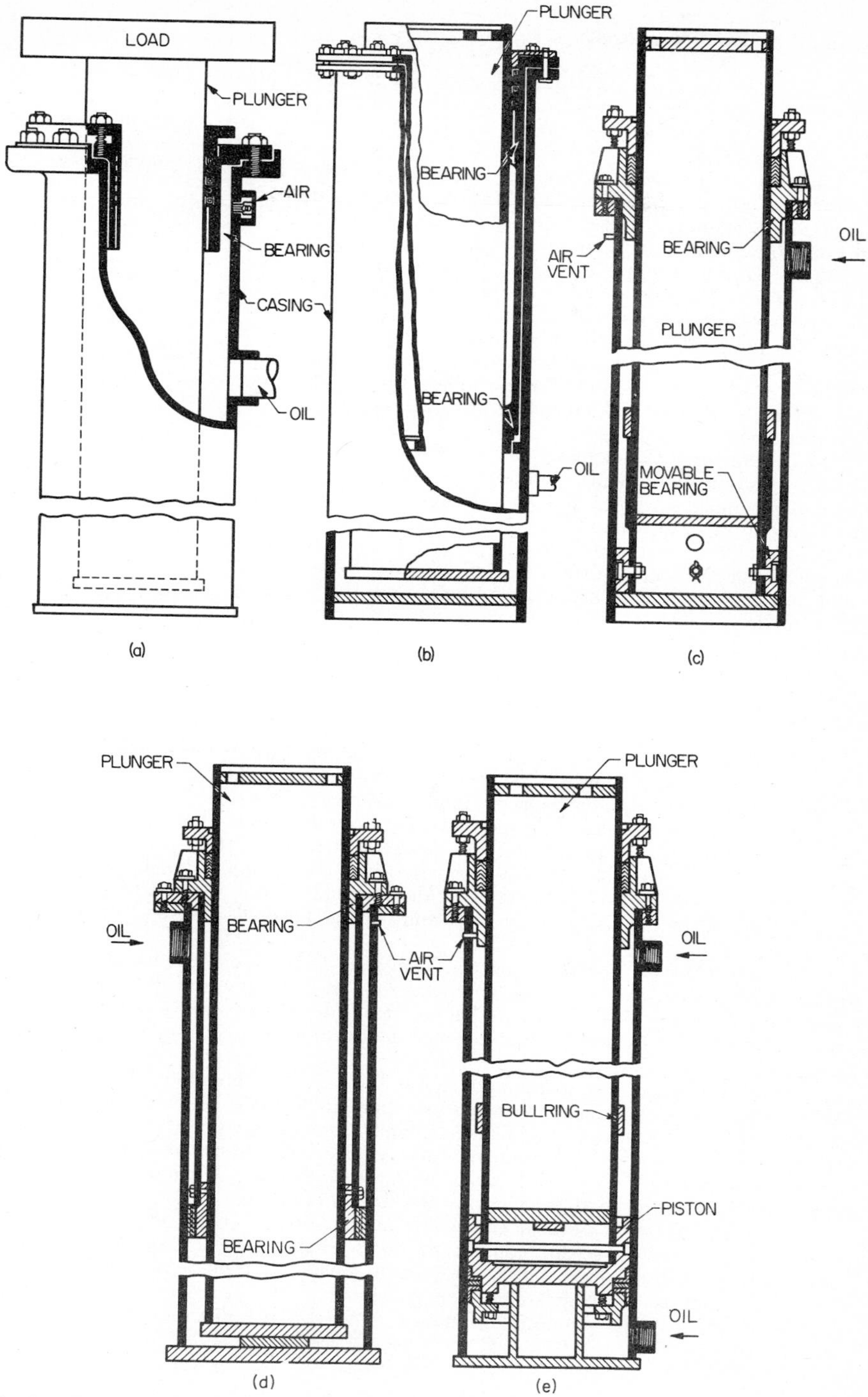

Fig. 23-11. Jacks commonly used for hydraulic elevators: (*a*) single-bearing plunger for guided loads; (*b*) two-bearing plunger for off-balance loads; (*c*) movable-bearing plunger for heavy service; (*d*) cage bearing for long-stroke service; (*e*) double-acting plunger.

Capacity of hydraulic passenger elevators ranges from 1,000 to 4,000 lb at speeds from 40 to 125 fpm. With gravity lowering, down speed may be 1.5 to 2 times up speed. So the average speed for a round trip can be considerably higher than the up speed.

Capacity of standard freight elevators ranges from 2,000 to 20,000 lb at 20 to 85 fpm, but they can be designed for much greater loads.

(G. R. Strakosch, "Vertical Transportation: Elevators and Escalators," John Wiley & Sons, Inc., New York.)

23-23. Dumbwaiters. For small loads, dumbwaiters may be used in multistory buildings to transport items between levels. These are cars, generally too small for an operator or passenger, which are raised or lowered like elevators. They may be powered—controlled by push buttons—or manually operated by pulling on ropes. Powered dumbwaiters available can automatically handle from 100 to 500 lb at speeds from 45 to 150 fpm. They also are available with special equipment for automatic loading and unloading and also designed for floor-level loading suitable with cart-type conveyances.

23-24. Vertical Conveyors. When there is a continuous flow of materials to be distributed throughout a multistory building, vertical conveyors may be the most economical means. In some installations, 200 lb or more of paper work and light supplies are circulated per minute.

A typical conveyor installation is similar to powered stairs (Art. 23-7). A continuous roller chain is driven by an electric motor. Engaging sprockets at top and bottom, the chain extends the height of the building, or to the uppermost floor to be served. Carriers spaced at intervals along it transport trays from floor to floor at a speed of about 72 fpm.

Like elevators, however, vertical conveyors are enclosed in fire-resistant shafts. Generally, the only visual evidences of the existence of a machine are the wall cutouts for receiving and dispatching trays at each floor. Short gravity-control conveyor sections may lead from the conveyor receiving station into a work area.

In event of fire, vertical sliding doors, released by fusible links, will snap down over the openings, sealing off the conveyor shaft at each floor.

Operation of vertical conveyors is simple. When the trays are ready for dispatch, the attendant sets the floor-selector dial or presses a button alongside the dispatch cutout. As the trays are placed on the loading station, they are automatically moved into the path of the traveling carriers. Each tray rides up and around the top sprocket and is automatically discharged on the downward trip at the preselected floor. It takes only 4½ min for a 26-story delivery of interoffice correspondence.

The best place to install a vertical conveyor is in a central location, next to other vertical shafts, to minimize horizontal runs in collecting and distributing correspondence at each level. An installation alongside a stairway is advantageous in providing access to the rear maintenance platform of the conveyor, for servicing equipment.

Vertical conveyors are made and shipped in sections about 10 ft long, which are assembled in the shaft. Principal components include the head sprocket and drive unit, the foot sprocket and take-up units, and intermediate sections consisting of braced steel shapes and tubular track guides. Generally, one or more walls of the shaft are left open until the conveyor is installed. A removable floor plate is provided above the shaft for access to the drive machinery.

In choosing between vertical conveyors and pneumatic tubes, the designer's first consideration should be the number of floors to be served. An arterial system of pneumatic tubes would satisfy the requirements of a predominantly horizontal building, whereas a vertical conveyor generally would be more advantageous in a tall building. The amount of paper work to be distributed is a second consideration.

Section 24

Surveying for Building Construction

REGINALD S. BRACKETT

Formerly Architectural Superintendent,
Smith, Smith, Haines, Lundberg & Waehler,
New York, N.Y.

Surveying for building construction can be divided into three classifications:

1. *Preliminary.* So the architect can prepare plans for the building, information should be obtained on:
   - *a.* Boundary lines and general topography (Art. 24-2).
   - *b.* Streets, curbs, and pavements.
   - *c.* Utilities—sewers, water mains, gas lines, electric and steam service.
   - *d.* Buildings on and adjacent to the site.
   - *e.* Subsurface conditions—from test holes and borings.
   - *f.* Soil tests.
   - *g.* Quantity survey (engineers' estimate).
2. *Construction.* Condition survey of adjacent structures and establishing lines and grades. Needed are:
   - *a.* Survey of conditions of adjacent structures that are subject to damage from new construction (Art. 24-3).
   - *b.* Elevation readings on existing structures, including utility appurtenances and street pavements (Art. 24-4).
   - *c.* General layout for construction—base lines, offsets, and bench marks (Art. 24-5).
3. *Possession Survey.* Made after the building is completed.

24-1. Units of Measurement. In land surveying, both horizontal and vertical distances are measured in feet and decimals of a foot. In building construction, however, horizontal distances are given in feet and inches, while as a rule, elevations are noted in feet and decimals of a foot. Conversion is easily made by remembering that there are 96 eighth inches in 1 ft, or ⅛ in. equals 0.01 ft (more accurately 0.0104) and 1 in. equals 0.0833 ft.

24-2. Survey of General Topography. For a preliminary survey, property lines can be established by any competent person, provided the locality is properly

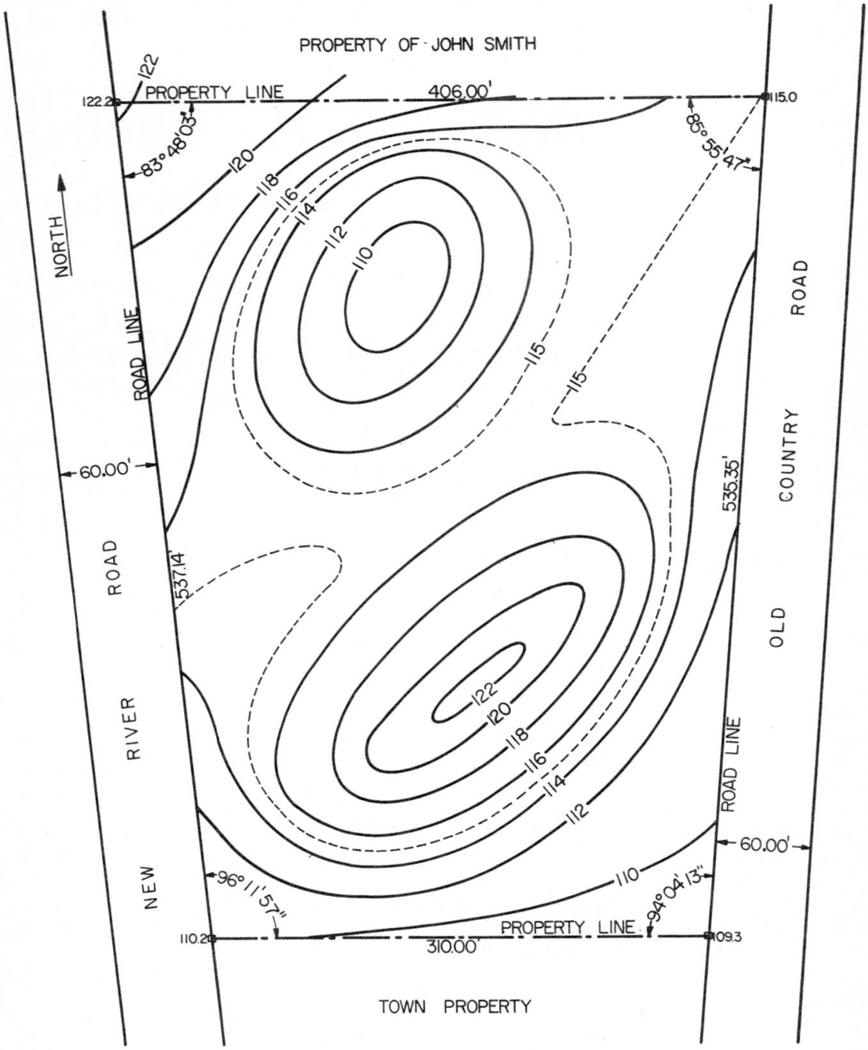

Fig. 24-1. Building site in open country.

monumented. In many cities and towns where monuments are not established, the local surveyor is the only one who has the knowledge of landmarks referred to in deeds for land in the vicinity. In such cases, it is always advisable to have the main corners of property staked out by a local licensed surveyor.

One or two bench marks are also required to establish the datum, or reference point for elevation. It is a good policy when the excavation goes below ground-water level to call water level +100, so as to eliminate minus elevations.

The new survey should show any difference between the present standard and that used in the original survey. Surveys will differ when a city has been monumented long after the buildings are in place. The differences should be noted so that the architect can make corrections before the new plans are started. It is confusing to find too late that 100 ft in the deed is only 99.85 ft.

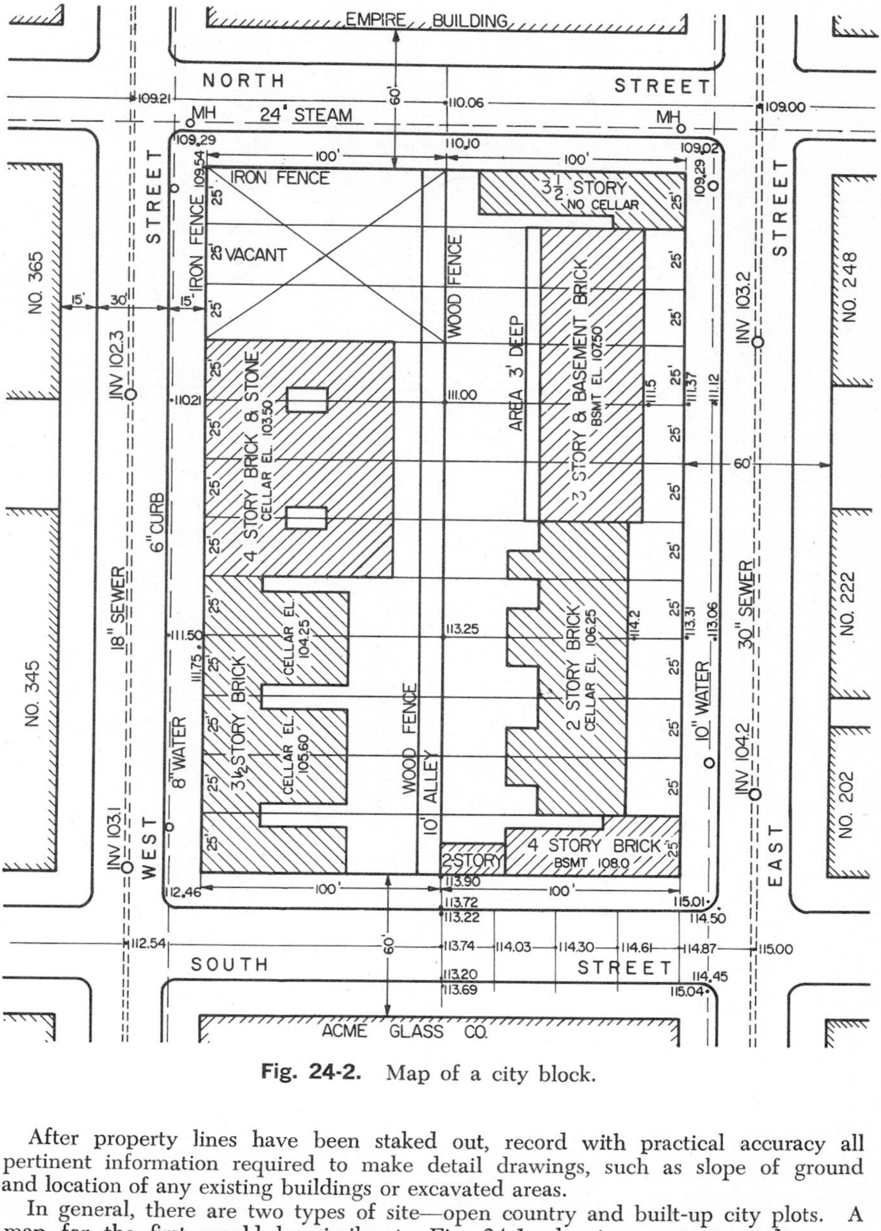

Fig. 24-2. Map of a city block.

After property lines have been staked out, record with practical accuracy all pertinent information required to make detail drawings, such as slope of ground and location of any existing buildings or excavated areas.

In general, there are two types of site—open country and built-up city plots. A map for the first would be similar to Fig. 24-1, showing contours, and a map for the second would be like Fig. 24-2, showing existing buildings. The contours can be drawn using either a cross-section grid system (Fig. 24-3) or the stadia method, laying off angles and distances (Fig. 24-4). The latter is perhaps a little quicker but probably less accurate. (R. E. Davis, F. S. Foote, and J. W. Kelly, "Surveying: Theory and Practice," McGraw-Hill Book Company, New York.)

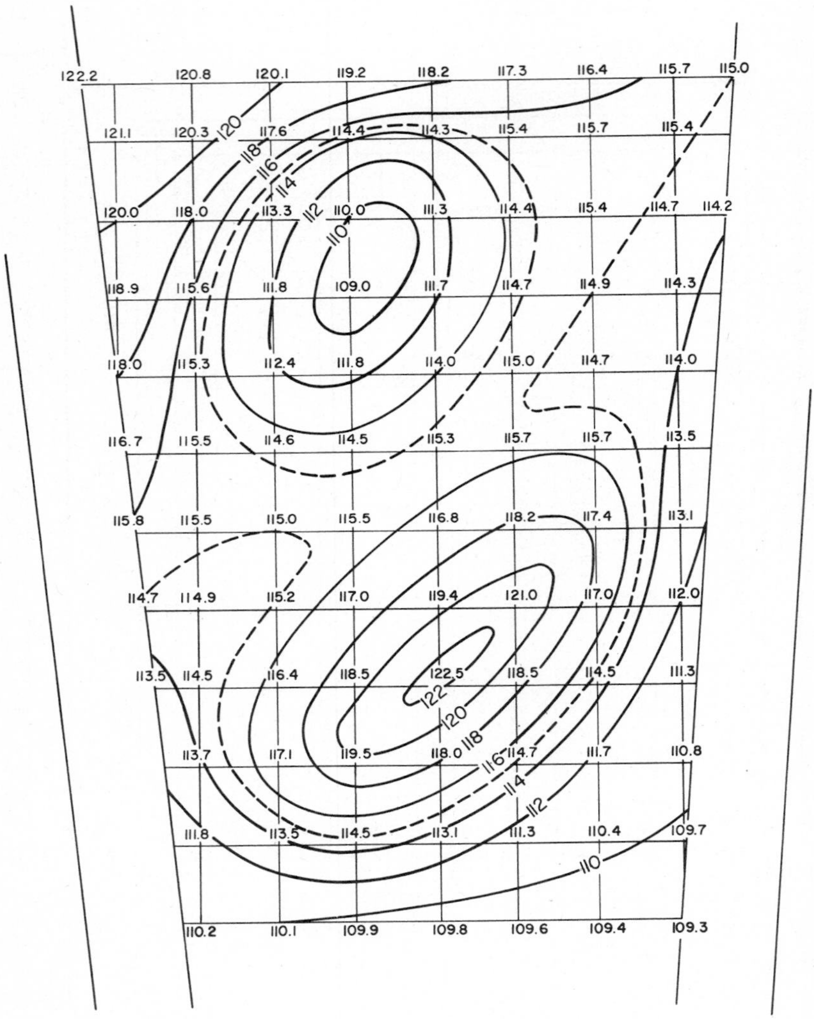

Fig. 24-3. Contours plotted from a grid system.

The type of building to be constructed and its details will determine the frequency with which elevations and dimensions are required. Walls of existing buildings should be plumbed when on the property line.

For street curbs and pavements in this preliminary work, only the location, elevation, and condition are required.

Sizes and condition of sewers should be recorded, giving invert elevations at manholes. Other utilities can be arrived at from hydrants and other manhole covers.

Existing buildings on the site should be shown. The number of stories, kind

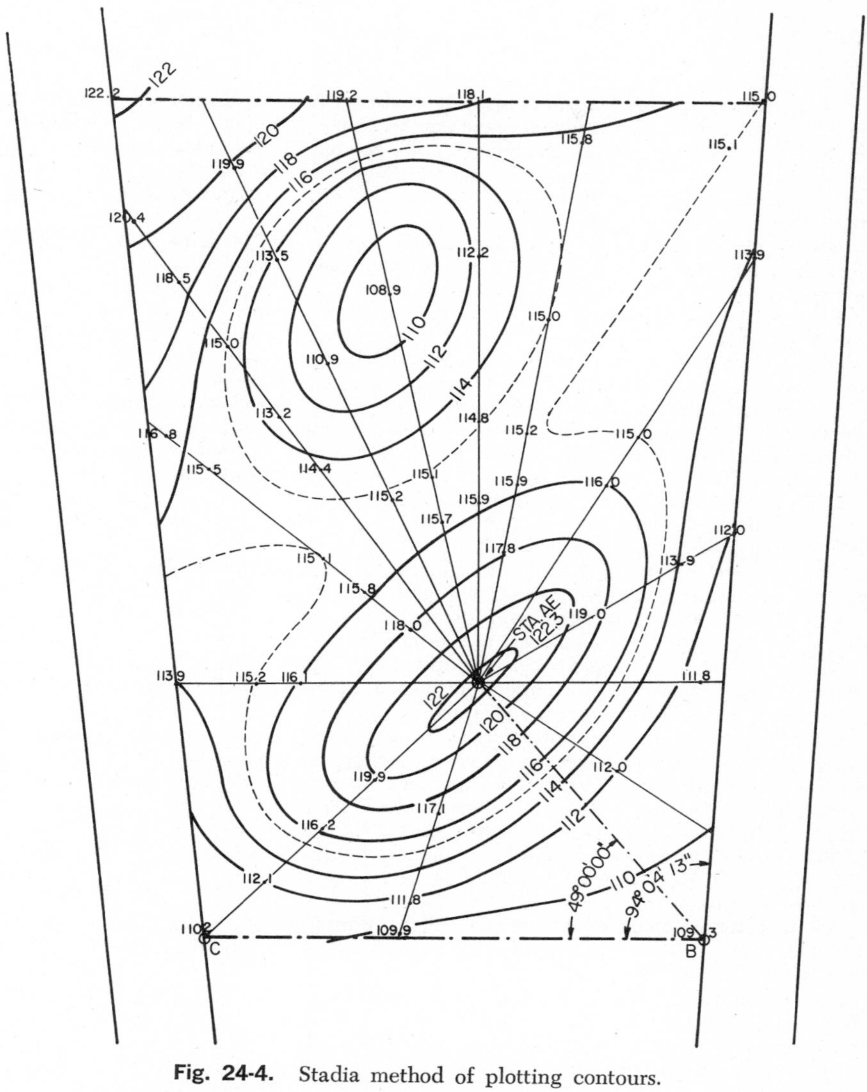

Fig. 24-4. Stadia method of plotting contours.

of material, and whether there is a cellar or basement should be noted as in Fig. 24-2.

Unless the character of the soil on which the foundation is to rest is known, test holes or borings should be made (Sec. 4).

An estimate should be made of excavation quantities. This involves plotting cross sections from elevations obtained from a survey and grading for the new building. Volumes can be figured by either the end-area formula [Eq. (24-1)] or the prismoidal formula [Eq. (24-2)].

$$V = \tfrac{1}{2}L(A_1 + A_2) \tag{24-1}$$
$$V = \tfrac{1}{6}L(A_1 + 4A_m + A_2) \tag{24-2}$$

where V is the volume in cubic feet of a prismoid with end bases or cross sections having areas in square feet A_1 and A_2 and a midsection of A_m and with a length in feet of L.

24-3. Survey of Nearby Buildings. It is always possible that adjacent buildings might be damaged by construction operations for a new building. So an examination, both inside and out, of all buildings close enough to the new construction to be affected should be made. The survey should be carried out, prior to start of construction, by a party representing the architect, contractor, and owners of the building to be inspected.

The record of this survey should be made accurately and in detail, to be available for comparison, if needed, to prove or disprove any change in appearance. The exterior can be examined, where beyond range of unaided vision, with high-powered field glasses. Record all defects or irregularities of masonry, which if noted later could be attributed to construction of the new building. It is a good policy to mention all sides where exposed, whether defective or not. If various elements of existing structures are in good condition, just state no comment.

For example, the record of an examination could read as follows: "Empire Building on the north side of North St.—South elevation, granite watertable requires pointing, no cracks in stone, a few spalls at corners; limestone above to 3rd floor, no comment. The entablature at 3rd floor has a large crack, starting at the top of the 5th column from the west corner of the building and extending diagonally upward through the architrave and frieze to the cornice; open at the top about the thickness of a mortar joint. Above the 3rd floor . . ."

The interior inspection should cover all rooms from the lowest floor to the roof. Cracks developing from settlement will generally carry from floor to floor. If a crack in a wall does not appear in rooms above and below it, it is safe to say that it did not come from settlement of the foundation.

In describing rooms, describe floor, walls, and ceiling. Thus: "Room 203—Floor covered with asphalt tile. South wall has a diagonal crack in plaster, starting in the upper left hand corner and running down to a point one foot above the floor and 3 ft from the right corner. Starts as 1/32 in. closing to zero. East wall has one vertical crack in center, extending from floor to ceiling, open 1/16 in. for full height. North wall—no comment. West wall has some hairline crazing near ceiling. Ceiling is acoustical gypsum tile, two pieces missing. Door to room binds at top on the lock side. Door trim split at top on the hinge side . . ."

Sewers and pavement should be examined by the architect, contractor, and a city representative. As with the building survey, photographs contribute an important part of the record.

24-4. Establishment of Elevations before Construction. The examination described in Art. 24-3 must be supported by readings of elevations at the foundations of the various buildings to determine any settlement or lateral movement. The number and type of such reference marks will be controlled by the kind of new foundation. In general, reference points should be located and recorded in such a way that settlement or horizontal motion can be detected.

A reference grid can be laid out on street pavements and sidewalks. (Spikes driven into the pavement will aid in making future readings.) The water table, window sills, or mortar joints on existing buildings can be used as reference points to observe settlement. These have an advantage over marks placed on walls at some convenient arbitrary elevation, inasmuch as readings on them will show up any irregular previous settlement.

Establish levels inside the buildings too, if they are close to the excavation. Readings on utility castings, hydrants, valve boxes, etc., will tie in subsurface pipelines.

It is generally accepted that most buildings settle during construction as the loading increases; but it also must be remembered that settlement continues for many years. If any of the adjacent buildings are comparatively new, then differential settlement would have to be watched for. After work starts, keep a record of readings taken at regular intervals. Let the readings give warning; do not wait for cracks to appear.

24-5. General Layout for Construction. This differs, in general, for private dwellings, low industrial buildings, and large commercial buildings. However, they all require building lines at the start as a reference for measurements.

Small Buildings. Because it is not practical to work to a masonry line without disturbing it, offset lines are used. Distances to control points—building corners, wall faces, footings, etc.—are measured from these lines. The simplest layout (Fig. 24-5) requires principally setting of batter or corner boards nailed to posts or other supports. The boards should be long enough to enable the building and offset lines to be marked, and the top should be set to a known elevation, such as level of the first floor. The posts should be driven first and grade nails set

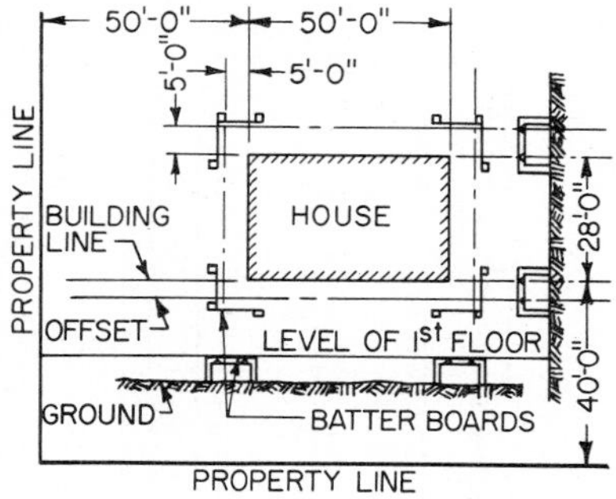

Fig. 24-5. Offset line for building layout determined with batter boards.

so as to determine the elevation of the boards before the lines are marked. A piece of shiplap, laid with the tongue projecting up, is generally used for the board. The step in the shiplap top is set at the desired elevation, and a saw cut the depth of the tongue allows a string to rest on the step at the correct level.

Layouts for low industrial buildings follow the same pattern, except that the corner boards are farther apart, and additional intermediate batter boards are needed. The position of the batter boards will vary with the depth and method used in excavation.

Large Buildings. The layout for large buildings, although basically the same as for small buildings, has to be expanded. Offset lines have to be established with the accuracy of a base line and will be called base lines in the following discussion.

Temperature correction for tape measurements is very important for layouts of large buildings; so it is necessary to have a standardized tape, which is kept in the field office and is used for this kind of work only. (The temperature at which any tape is standard can be found by comparison with a standard tape. For this purpose, the tape may be sent to the National Bureau of Standards, Washington, D.C.) A 100-ft steel tape will change ¾ in. when the temperature changes 100°F.

Work that requires extreme accuracy should be done on a cloudy day, in the early morning or evening. It is very difficult to determine the temperature of a steel tape in the sun. One expedient is to measure two lines, each 100 ft long—one that will remain in the sun and the other in the shade all day, where they will be readily accessible for checking a tape at any time. This will automatically give the correction to be applied to tape measurements.

Figure 24-6 shows a typical layout of main reference lines. The offset traverse or base line, which is wholly outside the building site, can vary in distance from the building line, but 6 ft is good because the line will clear the conventional swing scaffold. The sidewalk bridge, which is necessary for protection, prevents sighting above it, unless the transit is set up outside the curb line at either end. The

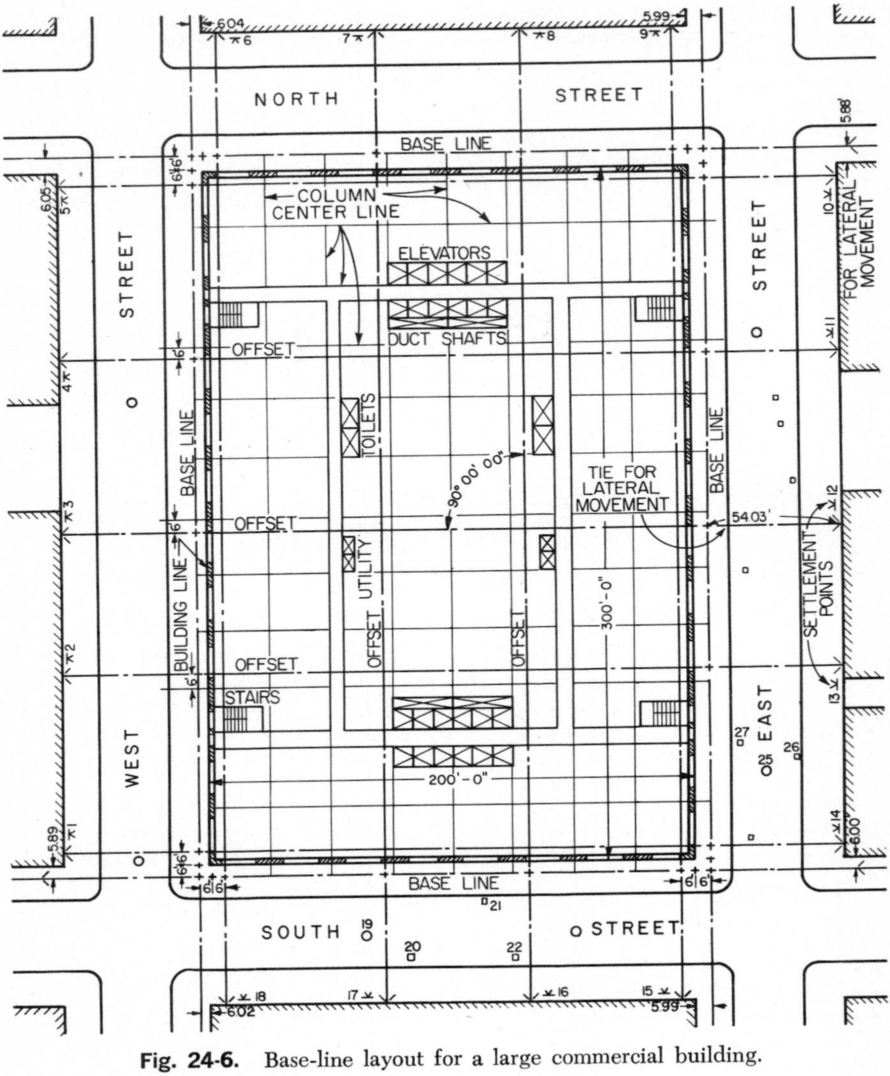

Fig. 24-6. Base-line layout for a large commercial building.

base line, therefore, must be produced well beyond the limits of the sidewalk bridge so as to be of use above it.

After the base line is finished and checked, all points of use in layout should be projected to a permanent location on nearby existing buildings. Control points for base lines can in many cases be obtained by sighting on some architectural ornamentation a short distance away. Those for the inside building-line offset and column offset lines should be located as high as possible, so as to be seen from the bottom of the excavation. The roof cornice in Fig. 24-7 is ideal. Targets at such locations can be used to establish column lines by "bucking in" or double centering between control points on opposite sides.

The inside building line is useful in aligning foundation walls after the general excavation is completed. Placing removable windows in the sidewalk fence, which is part of the sidewalk bridge, permits transit lines to be dropped into the excavation.

Steel Framing. The center lines of columns can be marked on batter boards set with the top a given number of feet above subgrade, footings, or pile caps. If piles are used, they can be staked out with a tape by measuring from the center lines. Before columns are set, transit lines can be run on the center lines, but after they are in place, the offset lines have to be used.

There is a certain satisfaction in being able to sight directly on the line to be used. As soon as some concrete is placed, get a permanent bench mark set in it. In a steel-frame building, an anchor bolt may be used; if not, set a special rod for this purpose.

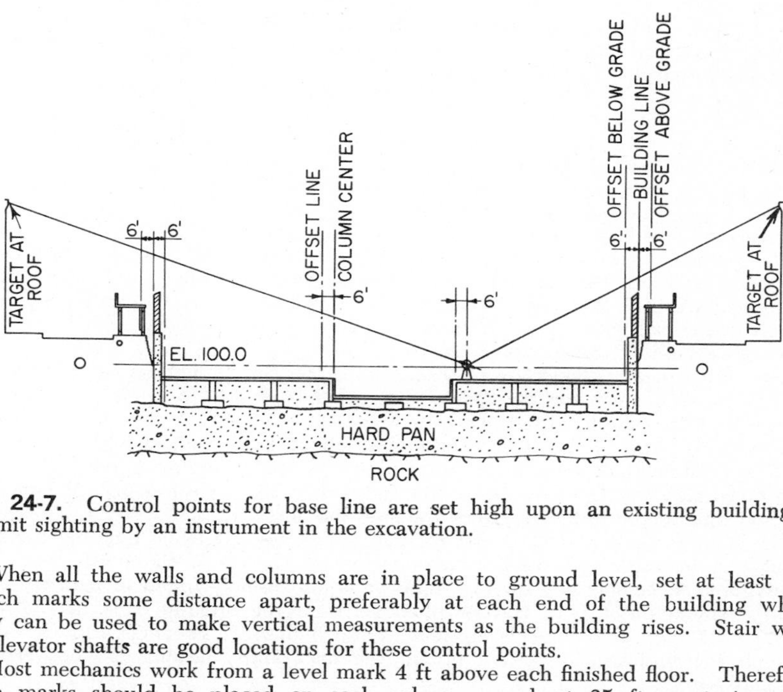

Fig. 24-7. Control points for base line are set high upon an existing building to permit sighting by an instrument in the excavation.

When all the walls and columns are in place to ground level, set at least two bench marks some distance apart, preferably at each end of the building where they can be used to make vertical measurements as the building rises. Stair wells or elevator shafts are good locations for these control points.

Most mechanics work from a level mark 4 ft above each finished floor. Therefore, such marks should be placed on each column, or about 25 ft apart, in every story. They must be tied in accurately between floors, which is the reason for the vertical base lines. Two marks must be seen from the same level setup before the 4-ft marks are established on the columns. With an extension-leg tripod, set the horizontal hair of the level directly on the taped 4-ft marks. On the original run do not set above or below the mark and read a rule. Between setting the rule and reading the backsight, and reading the rule to set the foresight, errors can be made, especially when about ½ in. from it. If the rule is turned the wrong way, the error will be twice the amount of the reading.

Since the building starts to settle shortly after loading, do not try to check with an outside bench mark. Hold and watch the original vertical base lines, and from time to time take a quick check on the preestablished 4-ft marks on the lower floors to spot a differential settlement. If there is any, set the 4-ft marks on the upper floors so that, when a particular column that is slow or fast in settling catches up, all the marks will be level. In some factory buildings, settlement effects are unimportant, but when the interior finish is marble, trouble may be encountered if settlement is not taken care of in advance.

Extreme care must be used in setting columns around elevator shafts. Check each column independently with a heavy plumb bob. (It can consist merely of a 4-in. pipe, 8 in. long, filled with lead and suspended by piano wire.) Swinging can be reduced by suspending the bob in a pail of oil.

As the building rises above ground level, the exterior base line is the most important control line. The interior column offset lines, when used, must always be checked with the building line.

Accuracy of alignment of a steel-framed building stems from the anchor bolts and billets. It is important to have all horizontal measurements correct so that the lower section will go together without difficulty. Each tier of framing above depends on the accuracy of shop fabrication.

The elevation of the column base plates controls for the full height of the building; so it is obvious that extreme care must be used in setting them. If this is done, column splices, as a rule, will be as close to the true elevation as if established with a tape. This is very important so that the beams and girders will be in the correct vertical location. Clearances for mechanical work also could be affected by variation from the true location.

As columns are loaded, a shortening takes place. So vertical measurements must be made and tied in before this happens. Any total over-all measuring done after loading will have to be corrected by calculated deformations based on the modulus of elasticity of the column material.

Fireproofing and the rough floor are located with respect to the steel. This is done because it is not practical to use an instrument on the frame until it is braced by some kind of deck, whether concrete or steel.

The frame is plumbed by the steel erector with a heavy bob and checked by the engineer using the base line produced beyond the sidewalk bridge. When the temperature in the field differs from that of the fabrication shop, the frame will be longer or shorter, an effect to be considered when checking. (Shop temperature is always a little above tape standard.)

After the structural floors are in place, the center line is marked on each exterior column in each story for the mason and other trades to work from. This also is done from the base-line extension.

A straight piece of 1 × 1-in. white pine, about 8 ft long, should be used for an offset rod, with a 6-in. piece fastened to the end, forming a tee. Two nails are driven into the head of the tee to sight on. The transit man will know the rod is at right angles to his line when the nails line up.

The column marks should be set accurately, but the mechanic building the walls should be instructed to set his masonry line ¼ in. inside the finish line. This is advisable to compensate for the slight variations usually found in masonry work.

The column offset lines can be plumbed from the opposite side of the street to the floor edge at both ends or sides of the building. The lines can be carried through the building by setting up on the floor and "bucking in" between the end marks.

Concrete Framing. A concrete frame requires a layout for each floor. A level mark to work from is put on a column splice rod 1 ft above the finish floor. Exterior column centers are set from the base line. Intermediate columns may be located by taping.

After column forms are stripped, 4-ft marks are established as is done with a steel frame. Column centers for masonry also are established as for a steel frame. The mason uses the 4-ft marks to determine window head and sill heights.

A competent mason foreman checks to the level mark on the floor above before he gets there.

24-6. Surveying Instruments. For survey and layout work for building construction, generally transits and levels, supplemented by tapes and stadia and level rods, can perform all required tasks satisfactorily. Surveys for remote sites and very large lots, especially over rough terrain, can be expedited with electronic devices such as godimeters, which use the speed of light, tellurometers, which employ microwaves, and electrotapes, which utilize radiofrequency signals.

Transits. These instruments principally provide a telescope, a scale for reading horizontal angles, a scale for reading vertical angles, and a plumb bob attachment for positioning the telescope over a point below, plus a tripod support. The telescope contains two perpendicular hairs, which when the instrument has been leveled become horizontal and vertical, for determination of line of sight, and two additional hairs on opposite sides of and parallel to the horizontal hair, for stadia measurement of distance. Transits are used mainly for property surveys, setting base lines and other reference lines, alignment, and other observations requiring measurement of horizontal or vertical angles. In choosing a transit, always select one with a telescope supported by standards high enough to permit a 360° vertical swing of the telescope and with a prism eyepiece in place. A transit equipped with an optical plummet is a time saver in high-wind setups.

Levels. These instruments consist principally of a telescope and a leveling device, a means for insuring that the line of sight is horizontal. The telescope contains a central hair, which is horizontal when the instrument has been leveled. Levels are used mainly to determine or set elevations.

A hand level consists of a telescope designed to be held in the hand of an observer or supported on a rod. It is suitable for rough readings and for sights up to about 50 ft.

An engineer's level consists of a telescope with a level-indicating bubble and leveling screws, supported on a tripod. It may be of the wye or Dumpy type. Either is capable of measuring elevations to within 0.01 ft. When the instrument has been leveled, the telescope rotates only about a vertical axis, and the lines of sight always lie in the same plane.

For faster, more accurate sighting, a tilting level or a self-leveling level may be used. With a conventional engineer's level, the bubble must be checked before each sight. Care must also be taken to observe the bubble length, which changes with temperature. This is automatically compensated for with tilting or self-leveling levels.

A tilting level is a Dumpy level with a telescope that can be tilted a few degrees through a vertical angle by a micrometer screw at the eyepiece end. When a tilting level is to be set up at a new station, a preliminary setup is first made with the aid of a bulls-eye level, similar to that used on some cameras. When a reading is taken, the telescope bubble is reflected by three mirrors and observed through a small window at the left side of the telescope. The mirrors are arranged so as to split the image and enable the observer to see both ends of the bubble simultaneously. The observer, while viewing the split image, can line up the bubble ends with the tilting screw and thus level the telescope.

A self-leveling level, like a tilting level, has a bulls-eye level for making the initial setup. With the instrument nearly level, a self-leveling device takes over. The line of sight through the telescope is kept horizontal with the aid of gravity and three prisms. One or more of these prisms are suspended as a pendulum in the telescope. Even if the telescope is not level, gravity tends to make the pendulums plumb, a braking arrangement stopping their swing. With the pendulums plumb, the line of sight becomes horizontal. Functioning of the device can be checked by turning one of the bulls-eye screws and observing if the line of sight moves up and down. These instruments can also be obtained with stadia hairs and an azimuth scale.

(R. E. Davis, F. S. Foote, and J. W. Kelly, "Surveying: Theory and Practice," McGraw-Hill Book Company, New York.)

Section 25

Estimating Building Construction Costs

E. D. LOWELL

**Chief Estimator, Kaiser Engineers,
Oakland, Calif.**

Scoping, planning, scheduling, estimating, and economic evaluation of complex engineering and construction projects require the specialized skills of many persons. Estimators, planners, and schedulers, along with architects, engineers, economists, managers, and other professionals, are very important members of the team.

Because building construction involves millions of dollars, accurate estimating is critical to the success of these projects. A poor estimate by a construction company can inflict serious losses on the company and could well threaten its solvency.

Plant capital-cost estimating has developed specialty application; that is, an estimator of commercial facilities, such as banks, hospitals, and office buildings, tends to stay in this field and further expand his knowledge and expertise in commercial construction. The broad categories of specialization are generally:

Commercial facilities.
Industrial complexes.
Chemical and refinery plants.
Power plants—fossil and nuclear.
Dams, tunnels, and highways.

The following discussions, illustrations, and applications are addressed to all of these fields of specialization, with recognition, however, of the fact that each field has its own unique methods, procedures, and presentation relative to capital-cost estimating.

25-1. Estimating Before, During, and After Design. Project cost estimates are prepared by different firms for different purposes and are of varying degrees of accuracy. A general contractor may prepare lump-sum bid estimates from completed or semicompleted plans and specifications. His methods and techniques are designed for this type of estimating, which, in many cases, depends on his having firm lump-sum quotations from specialty subcontractors on 85 to 100% of the facility cost. At the other end of the spectrum, an owner's management may decide that a new facility having a certain approximate area or annual capacity is needed. Based on prior plant construction-cost data, a reasonable unit price for the area or capacity may be obtained, from which a total value for the plant, including engineering, administrative, financing, and start-up costs, can be calculated. Between these two extremes lies a broad band of estimating methods and techniques.

The general contractor bidding a lump sum usually has the advantage of completed drawings and specifications, as well as lump-sum quotations for most of the work. The architect-engineer, on the other hand, relies on his in-house conceptual estimating capability to estimate construction costs as design proceeds.

Monitoring Estimates. An architect-engineer estimator is involved in a project from its inception. The scope of his activities encompasses the entire project. During the development of a project and through detail design and construction, the estimator is involved in making continuous *monitoring estimates* to assure design-cost compliance and, if necessary, to evaluate alternative methods. (The trade-off study, value engineering, is a natural outgrowth of this. So are other evaluation techniques, which include complete reviews of total cost, scope, plan, and schedule, at least at the one-third and two-thirds points of design progress. Participants in these reviews should include the owner and the planning, management, and engineering staff, as well as the estimator.) Consequently, these monitoring estimates should be in sufficient depth to assure understanding of the project by all concerned, and should yield reliable costs commensurate with the data available and properly correlated with original concepts.

Cost Control. A related aspect of estimating, cost control of a project, requires particular attention during the conceptual and detail design stages to assure compliance with the amount appropriated for construction and with the engineering and architectural requirements of the project for quality and quantity of product. It is less costly and time-consuming to revise a "runaway" project during the detail design phase than after construction has commenced, when few alternatives or options are left. Estimating during the detail design phase is critical, and is dependent on the expertise of the conceptual estimators.

Planning-stage Estimating. From the inception of a facility, financial studies should be continually developed, evaluated, and reevaluated on the basis of increasing in-depth studies. During these stages, for one reason or another, projects may be estimated, reestimated, and re-reestimated; many "trade-off" studies are required as the scope of the project continually changes. This is the period in which the reliability of the estimate is a direct function of the architect-engineer conceptual estimator's knowledge and experience, over and above the quality and quantity of information available.

The form of the estimates and the estimating methods used are generally unique to the individual estimator, facility, and conditions. The estimator must spend a great deal of time in determining the scope and general parameters of the job and then translating these into meaningful, measurable quantities with reasonable units of cost that are all-encompassing and defendable. For example, for concrete construction, instead of estimating item by item the form costs, the rebar density and cost, and the concrete purchase, placing, and finishing costs, the estimator may choose to use an all-inclusive unit cost, such as $85 per cu yd. The reason is that at this stage it is more important to determine the quantity of concrete accurately than to detail unit prices.

The estimator must involve himself in the over-all project plan, its economics, its phases, and its schedule, to evaluate properly the intangibles that must always be reflected in construction costs. The scope and schedule of the project are so closely related to the estimate that isolated or inadequate analysis of scope or schedule can negate the reliability of the estimate.

25-2. Types of Estimates. Estimates may be defined by at least two criteria, quality and end use.

INFORMATION REQUIRED	TYPES OF ESTIMATES: DOOR-KNOB	MAGNITUDE	PRELIM'RY.	DEFINITIVE	ENGINEER'S	BID
	TYPE 1	TYPE 2	TYPE 3	TYPE 4	TYPE 5	TYPE 6
PRODUCT, CAPACITY, & LOCATION	●	●	●	●	●	●
FACILITY DESCRIPTION	●	●	●	●	●	●
PLANT LAYOUT	●	●	●	●	●	●
TIME FOR ESTIMATE PREPARATION	●	●	●	●	●	●
LIST OF EQUIPMENT, MAJOR EQUIPMENT PRICED		●	●	●	●	●
GENERAL ARRANGEMENTS, APPROVED BY CLIENT		●	●	●	●	●
OUTLINE SCOPE, GENERAL PLANT FEATURES		●	●	●	●	●
ELECTRICAL MOTOR LIST, WITH HP			●	●	●	●
PIPE & INSTRUMENT DIAGRAMS			●	●	●	●
SINGLE-LINE ELECTRICAL DRAWINGS			●	●	●	●
DRAWINGS OF PIPING-SYSTEM RUNS				●	●	●
PRELIMINARY DESIGN DRAWINGS				●	●	●
DETAIL EQUIPMENT, LIST PRICED				●	●	●
DETAIL # SCOPE OF WORK				●	●	●
DETAIL CONSTRUCTION DRAWINGS					●	●
DETAIL SPECIFICATIONS					●	●
SUBCONTRACT AND VENDOR-FIRM LUMP-SUM QUOTES						●

(a)

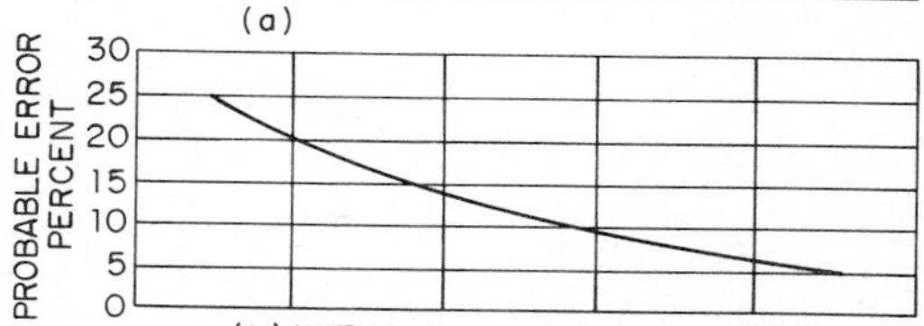

(b) WITHOUT HISTORIC COST DATA

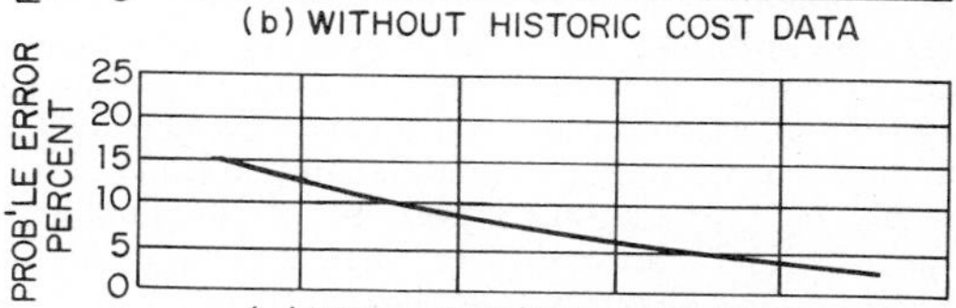

(c) WITH HISTORIC COST DATA

Fig. 25-1. (*a*) Types of estimates and information required for preparation of each type (indicated by dots). Basic data needed include craft wage rates and fringe benefits; payroll taxes and insurance; local sales, use, and other taxes; design and construction schedule; and insurance requirements. (*b*) Probable contingency of each type of estimate when prepared without historic cost data. (*c*) Probable contingency with historic cost data.

Quality. The quality of a cost estimate is directly proportional to the quality and quantity of the engineering available, the expertise of the estimators, and the methods used for estimate preparation. The simplified dot chart (Fig. 25-1*a*) is an approach to quality definition of a cost estimate. This chart indicates that estimates may range from a doorknob estimate (Type 1), which is based on facility quantity and unit price, to an engineer's estimate (Type 5), which is an in-house cost estimate based on completed detail plans and specifications. A bid estimate (Type 6) is the same as an engineer's estimate, except that it incorporates an evaluation of specialty contractors' competitive bids.

The degree of accuracy of these estimates is indicated in Fig. 25-1*b* and *c* by the magnitude of the required contingency allowance, which may range from a probable 25% for Type 1 estimates to 5% for Type 6. No one estimate conforms completely with one type, because usually an estimate is a montage of varying qualities.

End Use. Types of cost estimate differ in accordance with the end use of the estimate. Doorknob estimates, for example, have been used for studying the economic feasibility of a project, as well as for funding or financing a project. Any quality of estimate may be prepared for any one or more of the following end uses:

Studies.
Comparative appraisals.
Checking.
Budgeting purposes.
Control.
Feasibility studies.
Financing.
Proposals.
Bids.
Value engineering.
Scoping efforts.

Doorknob and Magnitude Estimates. Figure 25-1*a* indicates that the data available for preparing Type 1 and 2 estimates are very general in nature. These estimates therefore cannot be prepared by the accepted methods of quantity take-off and pricing. These types of estimates are normally prepared by a highly skilled conceptual estimator, who from his own knowledge and experience applies ratio and comparison techniques to determine the cost of the major features. The techniques and methods employed are unique to the estimator and unique for the project.

Often, such estimates are merely the proposed plant's annual (or daily) capacity times a unit cost, or ratios of total plant costs from prior job histories factored to the proposed capacity, by use of the sixth-tenth factor or some other method of cost-capacity adjustments. This type of estimating is extremely demanding of an estimator and important in the early feasibility study stage.

Preliminary, Definitive, and Engineer's Estimates. Type 3, 4, and 5 estimates may be further subdivided according to the progress of the project. For instance, an engineer's estimate, Type 5, may be made when detail design is 30% complete, 60% complete, and 100% complete. These estimates are usually prepared by architect-engineer conceptual estimators and are based on quantity takeoffs and pricing of partly or fully designed portions of the facility to be built. As design progresses, the margin of error due to quantity variances decreases and consequently estimate reliability increases, while the contingency requirement decreases. From 50 to 70% of the time consumed in preparing these estimates is spent on quantity takeoff, with the balance for pricing, summarizing, and comparing.

Bid Estimates. Estimates for preparation of a competitive bid, Type 6, which are based on completed construction drawings, technical specifications, and general and special conditions, require the estimator to takeoff and price only the work to be performed by the general contractor. Specialty subcontractor's competitive quotations are received for the balance of the work, which may cover from 80 to 100% of the job. The proper evaluation of local conditions, productivity, competitiveness of the market, volume of work the general contractor has underway, as well as the quality and completeness of the plans and specifications, play a significant role in determination of the final price of the facility. An award to a low bidder does not relieve the architect-engineer of the responsibility for informing the client of the probable final cost based on his best judgment. This estimate may be in the form of the low bid plus a contingent reserve for change orders, changes in conditions, and other unforeseen events best evaluated by the estimator.

25-3. Preliminary Steps. Certain basic information must be established for each cost estimate. A standard checklist should be available to form a guide and reminder, and should include at least the following:

Abstract of the job.
Local craft union agreements and wages.
Climatic data.
Labor availability and relative productivity.
Insurance and tax requirements.
Proposed schedule for preparation, review, and submittal of the estimate.
Man-hours assigned for estimate preparation.
Approximate engineering and construction schedule.
Outline and summary page format for the estimate.

Prior to preparing a cost estimate, the estimator, to plan and schedule the estimate properly, should establish a list of job task and manpower requirements. A general estimator who estimates all the work other than mechanical and electrical and assembles the completed estimate should be allocated sufficient time to prepare the estimate.

There are at least two good bases for establishing manpower requirements for estimating:

1. By actual drawing count and, from past experience, the average man-hour effort per drawing. The latter generally ranges from 3 to 6 hr per drawing. Examples of drawing distribution expressed as percentage of the total drawings might be as follows:

	Commercial construction, %	Industrial construction, %
Pipe drawings	25	15
Electrical drawings	25	35
Other drawings	50	50
Total drawings	100	100

2. By relative dollar value of the major disciplines or crafts; for example:

	Commercial construction, %	Industrial construction, %
General construction	60	40
Piping	20*	10
Electrical	10	10
Engineered equipment	10	40
Total	100	100

* Includes plumbing, heating, ventilation, air conditioning, etc.

Whatever method is used, note that electrical and mechanical takeoffs represent from 20 to 50% of the effort, and should be allocated adequate time for proper preparation of the estimate.

The estimate summary format, with general outlines of each division, should be established so that takeoff sheets, pricing sheets, and supporting data can be properly organized. In summarizing an estimate, it is advisable to adopt the organization of one of the following:

Job specifications.
Outline of prior estimate, if one or more exists.
Standard specifications, if job specifications do not exist.
Process flow of industrial-type plants.

25-4. Quantity Takeoffs. Clearly defined, accurate, and well-organized quantities required for a project are prerequisite to good estimates. Special forms usually are used for recording takeoff. They vary with type of material being surveyed, individuality of the particular contractor, architect, or engineer, and requirements of the particular job. Figure 25-2 shows a sample form applicable to general construction items of work. Figure 25-3 is a piping takeoff and pricing sheet developed for manual pipe estimating and incorporating an extensive checklist across the column headings. Figure 25-4 is a takeoff and pricing sheet generally used

KAISER ENGINEERS

QUANTITY SHEET

JOB NO. ______ SUBJECT ______ REF. NO. ______

TITLE ______ SHEET ______ OF ______

DRG. ______ -R ______ DATE ______

BY ______

	DESCRIPTION	NO. PCS.	LENGTH	WIDTH	HEIGHT OR WEIGHT	QUANTITY	TOTAL QUANTITY

Fig. 25-2. Sample form for quantity takeoff of general construction items.

KAISER ENGINEERS

PIPE-QUANTITY & ESTIMATE SHEET

BY ______ DATE ______ JOB NAME ______ JOB NO. ______

DRG. REF. ______ LOCATION ______ ACC'T NO. ______

SYSTEM ______ SHEET ______ OF ______

DESCRIPTION	SIZE	SCHEDULE	SPECS.	PIPE LENGTH	FITTINGS NO. PCS.	VALVES NO. PCS.	INSUL. LENGTH	HANGERS & SUPPORTS QUANTITY	OTHER QUANTITY	UNIT M	UNIT L	TOTAL MAT. & SUBS.	TOTAL LABOR HOURS
		AVERAGE			LIN. FT.				TOTAL	MAT. LIST	LABOR STD		
MATERIAL X FACTOR													
LABOR X FACTOR													

Fig. 25-3. Sample takeoff and pricing sheet for piping estimating.

by electrical estimators. The quantities must be clearly defined with proper references to their location (drawing number, specification reference, facility name, project title, date of takeoff, and name of person doing the takeoff), and recorded in clean, clear, legible lettering and readily understandable computation sequences.

The extent of architectural and engineering data available will determine the technique best suited for quantity surveying. Estimators working only with preliminary scope information generally do not have the time or information available for takeoff of detail quantities. These must be developed by one of the following means:

Ratio basis (for example, past experience may indicate that 1½ cu yd of concrete is needed per ton of installed equipment, or 150 lb of reinforcing steel per cu yd of concrete).

Comparison basis (for example, records may show that a certain-size pump required 150 cu yd of concrete for foundations; thus, the pump for the project being estimated, being smaller, might take only 100 cu yd).

Use of good judgment and experience.

Engineers' estimates based on 75% design completion, however, require definite quantities to justify the quality of the estimate. Detailed quantity surveying may represent 50 to 70% of the estimating effort.

Hurried quantity surveys with inadequate checks and poor summarization and definition are the basic cause of poor-quality estimates.

For industrial construction, properly prepared and priced *capital equipment lists* are essential. Generally, a process engineer will define equipment requirements during the conceptual stages of a program and will obtain either judgment prices or telephone-quote prices. As the facility becomes better defined, more refined

PRICING SHEET

JOB ________ SHEET ____ OF ____

WORK ________ DATE ________

Estimate by ____ Priced by ____ Mat'l. Extended by ____ Labor Extended by ____ Checked by ____ Estimate No. ____

DESCRIPTION	QUANTITY	MATERIAL UNIT PRICE	Per	MATERIAL COST	LABOR UNIT	Per	LABOR HOURS

Fig. 25-4. Sample takeoff and pricing sheet for electrical estimating.

practices are employed. In such cases, the engineer defines each piece of equipment and solicits reliable quotations. Figure 25-5 is a form that minimizes the repetitive writing effort by simultaneously serving for equipment definition, equipment pricing, and equipment installation estimating.

Recording Requirements. There are four basic requirements for quantity surveys:

1. **Neatness.** Write clearly and simply and do not crowd your work.

2. **Exactness.** Define the item being surveyed and give the location and drawing reference.

3. **Accuracy.** Check all computations and check the quantities for rationality.

4. **Utility.** Be sure quantities taken off are the proper quantities for the pricing structure of the job.

25-5. Unit Prices. Pricing techniques vary with the type of estimate, estimator, construction company, and data available.

Material Costs. Current material unit prices often can be obtained by telephone. Otherwise, material prices based on prior job records, properly adjusted to reflect current costs, competitive nature of the market, and specific job locations and quantities, have to be extracted from company records. Such prices are probably the easiest to obtain and the most reliable for preliminary-type estimates. Piping and electrical material prices are available as published list prices which must be adjusted to reflect market experience.

Process Equipment Unit Prices. These are important inasmuch as equipment may represent 40% of the estimated value of an industrial project. Sizing and pricing of equipment are the most critical steps in preparation of an industrial-type estimate; hence, care and good judgment must be used. The best sources of price data are the vendors of the equipment. They are usually agreeable to giving, for estimating purposes, budget prices based on minimum specifications.

It is important that all material and equipment prices be quoted FOB job site and, if required, include applicable sales, use, and other taxes.

EQUIPMENT LIST AND COST ESTIMATE

TITLE ____________ JOB NO ________ SHEET ______

FACILITY ____________ AREA NO ________

DATE ________ OF ______

REVISIONS							NOTES (BY LINE NO)
NO	DATE						

LINE NO	EQUIP. NO.	P.O. NO.	SPEC. NO.	QUAN.	EQUIPMENT DESCRIPTION	LOCATION	EQUIPMENT DATA									COST ESTIMATE											COST ACCOUNT	
							UTIL REQ'D	CAPACITY	SPEED	DRIVE	ENCL	TORQUE	VOLTS	SHAFT BASE	WEIGHT	LABOR 00	BURDEN 10	EQUIP USAGE 20	MATERIAL 30	SUBCONTR 40	BASIS FOR EQUIPMENT PRICES			TOTAL EQUIP 60	TOTAL ESTIMATE		
																					PRICE	TAX	FREIGHT				
1																												1
2																												2
3																												3

Fig. 25-5. Sample form for equipment listing, pricing, and installation estimating.

Labor Costs. Estimating labor unit prices is another major task of estimators. It can be done by various methods, such as:

Crew size and experienced productivity.
Experienced man-hours per unit of work.
Experienced unit costs.

In all cases, the costs should be well defined and may include all or a portion of the following:

Bare craft wage rate.
Craft fringe benefits.
Statutory taxes and insurance.
Small tools and consumable supply allowances.
Construction equipment costs.
Contractors' overhead and profit.

In the crew system of estimating, a first step is establishment of the crew size and crew rate. For example:

One each carpenter foreman:	8 hr per day @ $7.45 =	$ 59.60 per day
Eight each carpenter journeymen:	64 hr per day @ $7.00 =	448.00 per day
Four each construction laborers:	32 hr per day @ $5.45 =	174.40 per day
Crew rate:	104 hr @ $6.56	= $682.00 per day

From job experience, the estimator knows that this crew will build and erect, complete with screeds, headers, keyways, and the like, about 500 sq ft of average industrial construction forms in one 8-hr day, at about 80 to 85% productivity. If the job calls for 1,000 sq ft of forms, building and erecting them will take 2 days and cost 2 × $682 = $1,364 for labor only.

If the method of experienced man-hours per units were to be used, the estimator would refer to prior job records. For the preceding example, the records may show that this type of form takes 0.21 hr per sq ft to build and erect. Based on the previously calculated average hourly rate of $6.56, the cost for 1,000 sq ft of forms in place would be, for labor only: 0.21 × 1,000 × $6.56 = $1,377.60.

The unit-cost method is very broad, and is used for preliminary estimating only. For instance, in unit-cost estimating for the preceding example, the estimator would have to know that the concrete forms would cost $1.40 per sq ft for labor only. Then, the cost of 1,000 sq ft of forms in place would be 1,000 × $1.40 = $1,400 for labor only.

While estimators hope for as much detail regarding equipment and materials as is possible, often the quantities and units are on a much broader scale. For instance: What is the cost of an integrated steel plant with modest finishing facilities for 1,000,000 short tons annually? Records would indicate that construction cost of the plant should range from $350 to $400 per ton, or from $350,000,000 to $400,000,000, plus the owner's costs, which could include some or all of the following:

Escalation.
Spare parts.
Land.
Owners' engineering and administrative costs.
Interest during construction.
Technical assistance and training.
Operating inventory.
Working capital.
Start-up expenses.
Financing costs.

Quantities and unit prices, be they for a cubic yard of concrete or a complete facility, must be clearly defined and limited so that costs based on them represent as accurate a prediction of future costs as is possible to obtain. Simplicity and clarity of presentation cannot be overstressed.

25-6. Allowances. The alternative to detailed quantities and unit prices (Arts. 25-4 and 25-5) is use of allowances. These are generally used in lieu of detailed unit prices when the scope or definition of the work is incomplete. But their use should be held to a minimum and established by experienced estimators who have developed a feel for costs. For instance, the electrical subcontract value, excluding equipment, for an industrial plant may be from 10 to 13% of the total costs; process and utility piping may be 6%. During the early stages of an estimate, this method may be more realistic than an attempt to detail the quantities.

Before an allowance is used for any element in the estimate, all sources of information should be explored, such as data on similar installations in other plants, or perhaps a sketch by process engineers as to the intent of the undefined element. In presenting the estimate to management and the client, the estimator should identify and list the value of each allowance.

25-7. Estimate Details. The details of an estimate vary with the type of estimate, information available, and time allocated for preparing the estimate. A magnitude estimate (Type 2), such as the one for the 1,000,000-ton integrated steel mill in Art. 25-5, may require only a one-page tabulation of major plant facilities, listing annual capacity of each facility and an all-inclusive experienced unit cost per annual unit of capacity. On the other hand, a definitive estimate (Type 4) for a process plant provides more details and is further subdivided into plant facilities and subfacilities. The definitive estimate would have construction quantity and pricing detail for all elements of the work. Figure 25-6 is a form for preparing such an estimate. The depth of detail varies with the item being estimated, the amount of engineering available, and the time available. Care should be taken not to over-detail the item of work being estimated.

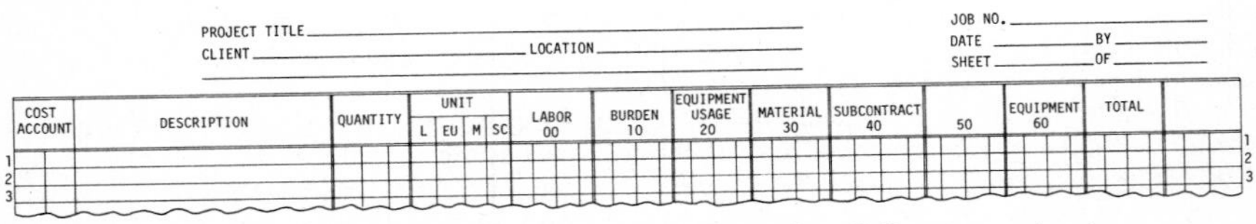

Fig. 25-6. Sample form for preparation of a definitive estimate.

Bid (Type 6) estimates, which are prepared by estimators for general contractors and specialty subcontractors, are usually based on completed plans and specifications. Such estimates are developed in greater detail than definitive estimates, and are prepared only for the work the contractor intends to do himself. The estimate outline and resulting summarization will invariably conform to the construction specification divisions.

The cost estimate, as presented, does not include all the supporting studies and detail that an estimator develops to establish his unit prices and quantities. It is good practice, however, for the estimator to assemble an appendix to the estimate, incorporating worksheets properly classified and identified, to complete the file of detail support information. The extra time and effort required to obtain, create, and organize all papers used in the preparation of the estimate are well spent, because the data may be invaluable at a later date when the inevitable extra-work orders and claims develop.

25-8. Evaluations of General Cost Items. After quantities have been taken off and priced (Arts. 25-4 and 25-5), the engineered equipment has been defined and priced for purchase and installation, and all such costs have been properly summarized, the next step involves judgment evaluation of escalation, contingency, fee, overhead, and engineering costs.

The labor, material, and equipment prices used in an estimate are usually defined as those in effect at the date of the estimate. Future price increases, either known or evaluated, must be taken into account to develop a valid estimate.

Escalation. This applies to increases in costs for material and labor and does not take into consideration changes in craft-labor productivity, design innovations, and improved construction techniques. These should be reflected in the body of the estimate, or as suggested alternative schemes.

One method of evaluating escalation is shown for illustrative purposes in Fig. 25-7, and consists of four steps:

The first step involves preparation of a schedule showing the design and construction periods in simple bar chart form. The time scale is selected to match that of physical progress curves drawn later below the bar chart.

Next, the craft wage rates for the five major crafts (carpenters, laborers, ironworkers, pipefitters, and electricians) are summarized directly underneath the bar chart using the same time scale. The X to the right of each craft indicates the expiration date of the wage contract with the craft union, and the dollar amount shows the wage and benefit amounts defined by the contract for the period indicated.

The third step involves preparation of curves indicating physical progress of the project. The construction-progress curve is based on on-site construction-labor

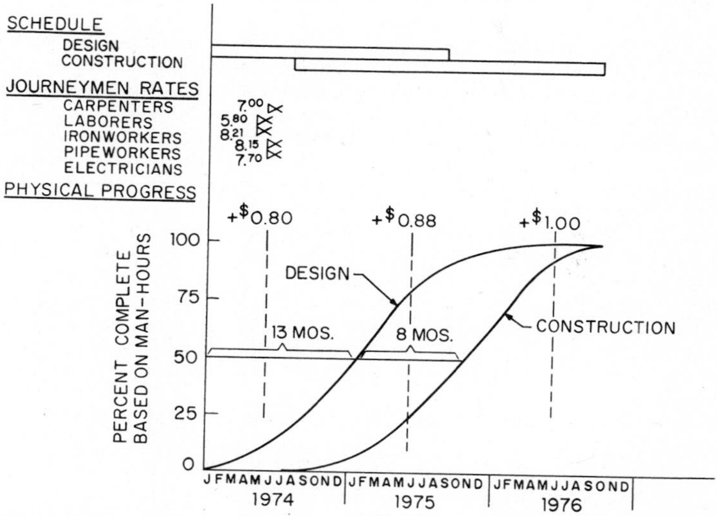

Fig. 25-7. Chart illustrating evaluation of cost escalation for an industrial building. Bar chart at the top illustrates schedule for design and construction. Journeymen hourly rates are given immediately below, with date of expiration of wage contract with each union indicated by an *X* at the right. The two curves at the bottom indicate, respectively, design and construction progress. Vertical lines show when union contracts expire and are labeled with anticipated amount of hourly wage increase. Cost escalation with data from the chart is computed as follows:

Labor, including fringe benefits (10% annual increase):	
3,200,000 hr @ 100% @ \$0.80	= \$2,560,000
3,200,000 hr @ 75% @ \$0.88	= 2,110,000
	\$4,670,000
Payroll taxes and insurance (15% total increase):	\$ 700,000
Material and other field costs (5% annual increase, 21-month period):	
\$16,500,000 × 5%/12 × 21	= \$1,450,000
Engineered equipment:	
Firm quotation: \$20,630,000 @ 0%	....
Balance (5% annual rate): \$8,700,000 × 5%/12 × 21	= \$ 780,000
Engineering: \$5,300,000 @ 7%	= \$ 370,000
Total escalation exposure	= \$7,970,000

man-hours and represents an average indicated by experience. The vertical lines with +0.80, +0.88, and +\$1.00 indicate actual or estimated craft wage increase at appropriate effective dates.

The final step involves calculating the expected cost escalation, which is divided into five areas:

1. Labor. For this size job and based on the cost estimate, the estimator determines that there will be 3,200,000 on-site construction man-hours. Since the cost estimate is based on wage rates in effect at the time of the estimate, it is necessary to allow for future increases. Because construction did not commence until mid-

1974, all man-hours (100%) received the $0.80 estimated increase for mid-1974, and 75% received the $0.88 increase. The curve indicates that the $1.00 increase will have no effect if the project stays on schedule.

2. Payroll taxes and burden. An estimated 15% is added to the labor increase to allow for the increase in statutory wage costs.

3. Material. The estimate indicates the value of material purchases required. The example in Fig. 25-7 shows $16,500,000 of such purchases. Records show that material costs have been and probably will increase at an annual rate of 5 or 6%. The center of these purchases is taken at the 50% physical completion date, which is 21 months from the date of the estimate. This results in the calculation shown in Fig. 25-7 for this group of costs.

4. Engineered equipment. An analysis of the project indicates that $20,630,000 of equipment is based on firm quotations, leaving $8,700,000 subject to escalation. With the same formula for equipment as was used for material, the estimator can calculate the escalation exposure for equipment.

5. Engineering. The estimate allows $5,300,000 for engineering. An estimated 7% is added for the probable increase in these costs over the duration of the project.

Total. The sum of labor and other escalation is the total estimated escalation exposure for the project.

Contingency. This is an evaluation based on experience, and is included as part of the estimated cost to compensate for accidental events; errors; imperfections in estimating; quality and quantity of engineering data available; omissions; and unknown events. With the addition of an evaluated contingency, the estimator intends that the resulting estimate precisely represents the final cost of the facility. Man is not perfect, however, and the best estimator usually is only able to keep the variation between total estimated cost and actual cost figures within a very small percentage.

An example of a contingency evaluation for an estimate might be as follows:

Labor costs	$ 9,297,000 @ 15% =	$1,395,000
Construction equipment	1,932,000 @ 15% =	290,000
Material purchase	6,125,000 @ 10% =	612,000
Subcontracts		
Firm lump sum	13,250,000 @ 2% =	265,000
Unit price	3,073,000 @ 10% =	307,000
Estimated	2,150,000 @ 15% =	322,000
Engineered equipment		
Firm lump sum	22,351,000 @ 3% =	670,000
Estimated	5,625,000 @ 10% =	563,000
Field overhead		
General conditions	3,632,000 @ 15% =	545,000
Escalation	8,250,000 @ 15% =	1,238,000
Total construction cost	$75,685,000	
Total contingency		$6,207,000

This calculation indicates that the average contingency for the job is about 8%. The estimator who prepared the estimate is the best judge of contingent requirements, because he is intimately associated with all facets of the estimate and the detail of information available.

Overhead. Contractor's field overhead or job general conditions include project costs not readily or equitably assignable to individual direct-cost items of work. There are many standard check sheets available for use in defining the costs to be included in this category. Individual company accounting practices, nature of work, and estimating-department definitions should be the bases for a standard and comprehensive checklist that reflects the individuality of the company and gives continuity of estimate definition from job to job.

Contractor's Fee. Profit or fee evaluation is management's prerogative. Estimators, however, should be aware of their management's considerations and evaluating techniques, so that the proper bare figures for computation are readily available. Individual company policies vary so widely in the definition of profit and fees that a discussion would require extreme detail analysis. Each company should establish clear definitions compatible with management's policies.

25-9. Computer Estimating. For estimates that are prepared from minimum engineering data, where basic quantities are nonexistent and where maximum use of experience, ratios, and factors, as well as the knowledge of a conceptual estimator are required, computerization is improbable, because the methods and techniques used are not amenable to strict mathematical solution. Where design has progressed sufficiently to permit reasonable quantity surveying, there is a definite place in estimating for the use of computers, especially where large volumes of data are used, as in piping and electrical estimating. The cost of maintaining up-to-date material prices, however, is a major cost factor that must be considered, because an outdated material data base will adversely affect the estimate.

Computer equipment availability and cost are also important considerations in developing and using estimating systems. Programs developed for an IBM 360/50, for example, require availability of such equipment on a continuing basis. Service-center operation with, if need be, terminal equipment at the home office is practical and indeed desirable. But establishment of files and procedures for using the computer at outlying job locations or under bidding conditions requires planning and careful implementation.

Because of the large volume of detail unit material and labor prices in pipe estimating, an outline of a computer approach to this subject is given in Art. 25-10.

25-10. Pipe Estimating by Computer. Development and implementation of a detailed piping estimating program require an in-depth study of all possible alternatives and requirements. For instance, the program should be able to sort and summarize according to various items and subjects, such as takeoff sheet number, bill of material number, commodity number, cost-account number, drawing number, line or system number, and specification number. These may not all be developed and used in the initial program, but consideration should be given to them and space allotted for them during the establishment of the program, because they may be developed at a later date.

Data to be included in the computer memory bank must be defined in detail to assure consistency from estimate to estimate. Unit material costs may be defined as list prices; unit labor man-hours may be defined as craft journeyman-hours; and unit subcontract costs for insulation, painting, etc., may be the current competitive subcontract unit price for medium-size jobs. The weight of each item, while not essential, does provide useful information for checking over-all costs and determining shipping weights and discounts.

Introduction of adjustment factors permits the estimator to make reasonable judgment decisions to revise the standard units in the files, and is necessary for any practical program. Adjustment factors are generally classified in three major groups, as follows:

Labor wage-rate adjustments, including fringe benefits, statutory insurances, and taxes, which have to be inserted for each project due to the variation in these costs from time to time and from place to place.

Standard unit man-hours adjustments to take into account the individuality of the particular system and circumstances. These adjustments compensate for completeness of drawings, adequacy of specifications, craft supervisory requirements, availability and quality of crafts, area productivity, weather, adequacy of working space, job size, complexity of installation, crew size, area congestion, and job schedule.

Adjustments to list prices of materials, for area pricing and available discounts.

Test runs on piping estimates indicate that little cost savings may be realized by use of a computer. It does, however, reduce the estimator's time by 30% or more, thus freeing him for other activities for which he is skilled.

Human judgment is still the prime ingredient of good estimates. But with continuing development, study, and objective approach, computer estimating will improve and incorporate more sophisticated judgment factoring. Figures 25-8 and 25-9 contain samples of quantity input sheets and computer output sheets for pipe estimates.

25-11. Cost Estimate and Construction Schedule. The cost of a project is very much dependent on the project schedule. The relationship is best described by means of an optimum time-cost curve (Fig. 25-10). This curve indicates that

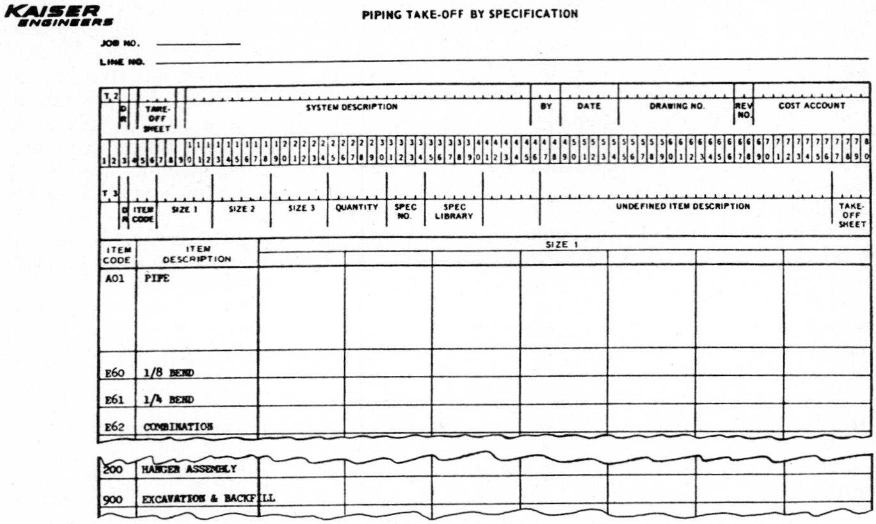

Fig. 25-8. Sample input sheet for computerized estimating of piping costs.

```
FILE UPDATE NO.         1           ***********************************      PAGE        1
FILE REVISE DATE        13JAN2      *   ESTIMATE BY TAKE-OFF SHEET    *      DATE   13JAN2
                                    ***********************************

JOB TITLE               ROCK PLANT BRIDGEPORT                                JOB NO. 3001

TAKE-OFF DESCRIPTION    SCREENING PLANT PLANT WATER                          TAKE-OFF SHEET    2
BY                      EDL
DRAWING NO./REVIS
COST ACCOUNT            1363.635
TAKE-OFF DATE           31AUG1
```

SPEC CODE	ITEM	SIZE	DESC CODE	SCH	RATING PRES WALL	QUANT	UN	MATERIAL $ STD	MATERIAL $ ADJ	LABOR MH STD	LABOR MH ADJ	SUBCON $ STD	WEIGHT LBS
	SPRAY NOZZLES	1/4	*Z01			672	EA	0	0	54	67	0	0
	DRILL HOLES	1/4	*Z02			672	EA	0	0	67	84	0	0
A CELEST	PIPE	2	120A	40		143	LF	117	70	30	38	0	522
A CELEST	PIPE	3	53XA	40		34	LF	73	44	8	10	0	257
A CELEST	HANGER ASSEMBLY	16	Z203			1	EA	29	18	3	4	0	33
* SUMMARY *						TOTALS		4286	2572	540	673	0	8218

```
TOTAL COST (ADJ)        $  2572
TOTAL PIPE LENGTH          419 LF
EQUIV PIPE DIAMETER        5.67 INCHES
TOTAL COST/LENGTH       $  6.14/LF
TOTAL LBS/LENGTH          19.61/LF

STD MH/(TOTAL LENGTH X EQUIV DIAM)             0.23
STD MATERIAL $/(TOTAL LENGTH X EQUIV DIAM)     1.81
```

Fig. 25-9. Sample computer output for computer estimate of piping costs.

construction cost for a project is at a minimum for a specific construction time, increases rapidly when construction is speeded, and also rises, although not so fast, when project time is lengthened. The cost estimate, therefore, is affected by the schedule, which should be well thought out to obtain an optimum project cost. It should be recognized, however, that an optimum project schedule resulting in best costs may not be economically sound as far as the over-all project is concerned. For example, if an increase of $100,000 in construction cost will bring the plant on-line 2 months earlier, the additional product profit may far outweigh the construction cost penalty. A cost estimate prepared for too short or too long a construction period also simulates the time-cost curve in reflecting project value.

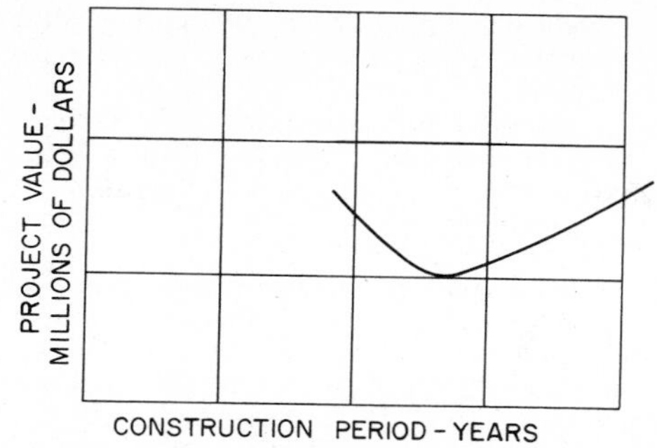

Fig. 25-10. Curve indicates the variation in construction cost of a project with construction duration.

Preparation of a cost estimate should start with a plan and schedule of accomplishment indicating major milestones. These schedules range from a detailed bar chart to a relatively complex critical-path-method (CPM) diagram. First, the estimator should ascertain the time available and the manpower required for preparing the estimate. With this information, a simple bar chart indicating estimating activities and dates can be prepared. The chart should allow ample time for management review and approval. Because most estimates are due at a specific time and date, the plan for accomplishing the major tasks should establish well-defined milestone dates for easy monitoring.

Methods for scheduling construction vary from a simple bar chart showing design, procurement, and construction times to complex CPM diagrams requiring computer solutions. To support and define the estimate properly, the estimator should establish and submit a schedule of project activities with the cost estimate.

If the project is in the design stage, adequate detail of design accomplishment should be included as restraints on procurement and construction activities. The scheduling of the detail design effort is an extremely complex task with many interrelations and dependencies. The drawing release schedule, however, could be the basis for determining initiation of procurement activities and commencement of field activities.

There are various rules of thumb for predicting duration of design and construction activities. The relatively simple, but sometimes unreliable, relationship between dollars and time offers a basic check of schedule. Table 25-1, which applies to industrial construction projects, may be used as a guide:

Table 25-1. Construction Times for Industrial Buildings

Construction value, millions of dollars	Design period, months	Construction period, months	Over-all period, months
5	12	13	20
10	12	15	20
20	14	18	24
30	16	20	26
50	20	24	30
75	20	27	36
100	24	30	36

Detail Control. With development of the detailed scope of work and the definitive cost estimate, a detailed project schedule showing the interrelationship between design, procurement, and construction, with proper resource allocation and reasonable durations, should be established. This schedule, when reviewed and approved, should be updated periodically and used for monitoring physical accomplishment relative to dollars expended, as a check against the estimate and as a method of highlighting cost variances.

How much detail is required for proper schedule definition? The answer is not simple. A small job requires fewer activities than a large job, and a complex job requires more activities than a simple job. The detail required therefore should be only that needed to assure proper definition and sequencing of the salient job activities.

Monitoring. A project manager uses project schedules and cost estimates for evaluating his team accomplishments. The estimate and schedule should be developed so that periodic updatings are as free from partial or slanted input data as possible. All input should be factual. It is the responsibility of the project manager and his staff to interpret these facts and initiate the required action.

25-12. Presentation of the Estimate. The time consumed in and cost of preparation of cost-estimate details certainly justify allowing adequate time for assembling the worksheets and data that support the estimate.

During the assembly phase of an estimate, two very important checks should be made:

1. Mathematical computations should be verified by a competent clerk knowledgeable in the estimating field. Mathematical errors are inexcusable.

2. The plans, specifications, scope, equipment list, and other supporting details should be reviewed thoroughly to make sure that all cost items are included and are in their proper place.

Table 25-2. Typical Summary-page Format for Industrial Building

1.0	*Construction Costs*
1.1	Direct costs
1.11	Yard improvements and development
1.12	Buildings and structures
1.13	Crushing and grinding
1.14	Screening and classifying
1.15	Concentrating
1.16	Finish-product processing
1.17	Storage and shipping
1.18	Escalation
	Total direct costs (1.1)
1.2	Indirect costs (general conditions)
1.21	Contractor's field overhead
1.22	Construction plant and equipment
	Total indirect costs (1.2)
	Total construction costs (1.0)
2.0	*Engineering, Supervision, and Procurement*
3.0	*Engineer-contractor Fee*
	Subtotal
4.0	*Contingency*
	Total project

The estimate in its entirety should then include at least the following documents: summary, criteria sheet, scope or job abstract, comparisons with other estimates, and supporting quantity and pricing detail.

Summary. The summary page should be consistent with prior jobs. One method of accomplishing this is adoption of a standard summary-page format for presentation of all cost estimates. Standard summary pages consistent with the type of estimating being done should be established within a company for ready understanding and ease of recognizing project elements. The summary-page format in Table 25-2 is typical for a turnkey industrial project.

Another approach for summarization would be to follow the outline of the specifications. As a substitute in the case of commercial work, the format established by the Construction Specifications Institute, Inc. for building specifications is excellent (Art. 27-7). Estimate summaries should be confined to a single page, and present as tersely as possible the basic dollar facts about the job.

Criteria Sheet. Probably one of the most significant documents that can be prepared is the estimate criteria sheet (Table 25-3), which outlines in estimator's language the over-all financial limitations of the estimate. A criteria sheet should be prepared for each job to include at least the following:

The schedule expressed in starting and completing months and years for both design and field construction.

The work shift—standard or otherwise—on which the cost estimate is based.

The date for labor and material costs and an indication of the inclusion or exclusion of evaluated escalation exposure.

Table 25-3. Example of Estimate Criteria for an Industrial Building March 10, 1974

Estimate Criteria

The estimate of cost for the XYZ Plant is based on the following assumptions:

1. The estimate is based on the construction program starting January, 1974, and being completed in 14 months.
2. The estimate is based on labor, material, and equipment prices as of October, 1973. An escalation allowance has been included as a line item for future increases.
3. The demolition of existing structures that are in the way of construction has been included.
4. Rough grading of the plant site has been included.
5. Prices for certain pieces of equipment to be furnished by the owner have been obtained from the owner and are included.
6. The estimate does not include additional monies, if required, for hiring and training of minority workers.
7. The cost of ecological studies and the costs due to these studies, such as added work or construction delays, are not included.
8. Added costs, if any, due to the Occupational Safety and Health Act of 1970 are not included.
9. The estimate is based on engineering and construction being performed on a one-shift standard work week.
10. The Arizona Gross Receipts Tax (sales tax) is not included.
11. The cost of spare parts is not included.
12. Special tools or equipment, such as zinc furnace and vulcanizer, are not included.
13. Rerouting of utility lines within the plant area is by others.
14. The costs for plant start-up are not included.
15. Owner's engineering, administration, financing, and other such costs are not included.

The statement that added costs, if required, have or have not been included for: hiring and training of special workers; recruiting and expediting costs due to shortage of skilled craftsmen or material; and implementation of certain new or unusual governmental regulatory acts.

The statement that costs have or have not been included for any or all of the following: land, owner's engineering and administration, training of operating personnel, furniture and furnishings, spare parts, interest during construction, production materials, and financing.

A list of major process or facility exclusions, if any, that may include owner-furnished material or equipment.

It should be stressed that the criteria sheet is not to be worded as a legal document, but is the medium through which an estimator, in his own language, outlines major considerations in the estimate.

Scope. Inclusion of a general scope or descriptive abstract of the facility with a general arrangement plan of the plant and, if available, a material-flow diagram, will aid considerably in defining the estimate limitations and afford easy reference during review meetings with management, client, and financial representatives. The statements should be clear, concise, and factual for the plant description and for the engineering data available.

Estimate Comparisons. A page should be included in the estimate package with data from the cost estimate for comparison with prior estimates for the same facility with a brief analysis of the differences, if any, and comparisons with costs for similar installations that have been constructed or estimated. For instance, if an estimate of $14,000,000 for a 8,000-short-ton-per-day copper concentrator can be compared with four or five similar plants within the range of $1,500 to $2,200 per short ton per day, it would appear that the estimate being reviewed is within reason. Comparisons in detail or in totals are good checks and a basis for evaluating the estimate at hand with other experience data. Well-documented and readily accessible historical information makes such comparisons easy and reliable.

Details. Estimate details should be organized in a loose-leaf binder with proper tabs, so that answers to questions and supporting documents are readily accessible. Neat, orderly worksheets may not improve the estimate quality (many people believe that they do, however), but they do improve the ability to sell the cost estimate. Estimating is the application of personal knowledge, experience, and understanding on a judgment basis to the logical flow of plant and material. Clarity of thought and logical presentation of detail will materially enhance the quality of the estimate.

After the estimate with its details has served its purpose, it should be given an identification symbol, placed in fixed binders, and properly filed for future references.

25-13. Reference Material. Job cost data derived from experience must be properly analyzed and organized to be readily available for estimators' use. There are many fine books available for aiding in detail and over-all cost estimating of mechanical, electrical, and brick-and-mortar activities to supplement job records. The sources of prices on engineered mechanical equipment are the manufacturers, who are generally willing to give prices that range from list prices to firm quotations. The following is a sampling of the many good cost guide books available. Be sure to read and understand the introductions of these books, to determine the particular qualifications and areas of specialization of each:

"Marshall Valuation Service," Marshall and Swift Publication Company, Los Angeles, Calif., updated periodically.

"Building Construction Cost Data," Robert Snow Means Company, Inc., Duxbury, Mass., annual editions.

"Construction Pricing and Scheduling Manual," McGraw-Hill Information Systems Company, New York, annual editions.

"Commercial-Industrial Estimating & Engineering Standards," Richardson Engineering Service, Inc., Solana Beach, Calif., 3 volumes, updated periodically.

Series of publications by John S. Page and James G. Nation, published by Gulf Publishing Company: "Estimator's Piping Man-Hour Manual"; "Estimator's Man-Hour Manual on Heating, Air Conditioning, Ventilating and Plumbing"; "Estimator's General Construction Man-Hour Manual"; "Estimator's Electrical Man-Hour Manual."

25-14. Cost Indices. These are invaluable for updating past estimates and prices, as well as for plotting trends to aid in forecasting future costs. The composition and intent of an index should be studied and thoroughly understood before it is used.

The various indices published periodically by *Engineering News-Record* are probably the most widely known. Other indices that may be helpful are the building cost index published in *Architectural Record;* the "Nelson Refinery Construction Cost (Inflationary and True Cost) Index," and the "Chemical Engineering Plant Cost Index." These represent only a few of the indices available. It is sometimes advantageous, however, for the estimator to develop his own index for specific building types.

Section 26

Construction Management

ROBERT F. BORG

President, Kreisler Borg Florman Construction Company
Scarsdale, N.Y.

Construction management of a building project encompasses organizing the field forces and backup personnel in administrative and engineering positions necessary for supervising labor, awarding subcontracts, purchasing materials, record keeping, and financial and other management functions to insure profitable and timely performance of the job. The combination of managerial talents required presupposes training and experience, both in field and office operations of a construction job. Proper construction management will spell the difference between a successful building or contracting organization and a failure.

This section outlines practical considerations in construction management based on the operations of a functioning general contracting organization. Wherever possible, in illustrations given, the forms are from actual files for specific jobs. These forms, therefore, not only illustrate various management techniques, but also give specific details as they apply to particular situations.

26-1. Types of Construction Companies. The principles of construction management, as outlined in this article, apply equally to those engaged in subcontracting and those engaged in general contracting.

Small Renovation Contractors. These companies generally work on jobs requiring small amounts of capital and the type of work that does not require too great a knowledge of estimating or construction organization. They usually perform home alterations or small commercial and office work. Many small renovation contractors have their offices in their homes and perform the "paper work" at night or on weekends after working with the tools of their trade during the day. The ability to grow from this type of contractor to a general contractor depends mainly on the training and business ability of the individual. Generally, if one is intelligent enough to be a good small renovation contractor, that person may be expected to eventually move into the field of larger work.

General Contractors. These companies often are experts in either new buildings or alteration work. Many building contractors subcontract a major portion of their work, while alteration contractors generally perform many of the trades with their own forces.

Some general contractors specialize in public works. Others deal mainly with private and commercial work. Although a crossing of the lines by many general contractors is common, it is often in one or another of these fields that many general contractors find their niche.

Owner-builder. The company that acts as an owner-builder is not a contractor in the strict sense of the word. Such a company builds buildings only for its own ownership, either to sell on completion or to rent and operate. Examples of this type of company include giants in the industry, and many of them are listed on the various stock exchanges. Many owner-builders, on occasion, act in the capacity of general contractor or as construction manager (see below) as a sideline to their main business of building for their own account.

Consultant Construction Manager. A construction manager may be defined as a company, an individual, or a group of individuals who perform the functions required in building a project for an owner, as the agent of the owner, as if the owner were performing the job with his own employees. The construction management organization usually supplies all the personnel required for the various duties the owner would need done were he doing them himself. Such personnel include construction superintendents, expediters, project managers, and accounting personnel.

The manager sublets the various portions of the construction work in the name of the owner and does all the necessary office administration, requisitioning, paying of subcontractors, payroll reports, and other work on the owner's behalf. Generally, construction management is performed without any risk of capital to the construction manager. All the financial obligations are contracted in the name of the owner by the construction manager.

See also Art. 1-22.

Package (Turn-key) Builders. Such companies take on a contract for both design and construction of a building. Often these services, in addition, include acquisition of land and financing of the project. Firms that engage in package building usually are able to show prospective clients prototypes of similar buildings completed by them for previous owners. From an inspection of the prototype and discussion of possible variations, or features to be included, an approximate idea is gained by the prospective owner of the cost and function of the proposed building.

Package builders often employ their own staff of architects and engineers, as well as construction personnel. Some package builders subcontract the design portion to independent architects or engineers. It is important to note that, when a package builder undertakes design as part of the order for a design-construction contract, the builder must possess the necessary professional license for engineering or architecture, which is required in most states for those performing that function.

See also Arts. 1-23 to 1-26.

Sponsor-builder. In the field of government-aided or subsidized building, particularly in the field of housing, a sponsor-builder may be given the responsibility for planning, design, construction, rental, and maintenance. A sponsor guides a project through the government processing and design stages. The sponsor employs attorneys to deal with the various government agencies, financial institutions, and real estate consultants, to provide the know-how in land acquisition and appraisal. On signing the contract for construction of the building, the sponsor assumes the building role, and in this sense functions very much as an owner-builder would in building for its own account.

26-2. Construction Company Organization. How a construction company organizes for its work depends on number and size of projects, project complexity, and geographical distribution of the work.

Sole Proprietor. This is a simple form of organization for construction contractors. It is often used by subcontractors, including those licensed in plumbing or electrical work. The advantage of operating as a sole proprietorship is that taxes on profits are much lower for individual owners. But there is the disadvantage of having the sole exposure to potential debts associated with a disastrous job.

Partnership. This is the joint ownership and operation of a company by two or more persons. Each partner, however, is personally liable for all the debts of the partnership. Profits and losses are shared in some manner predetermined by the partners. A partnership comes to an end with the death of one of the partners. (For typical provisions to be included in a partnership agreement, see Richard H. Clough, "Construction Contracting," John Wiley & Sons, Inc., New York.)

Corporation. This is the most common form of organization used by general contractors. A corporation is an entity that has the power to act as a separate body and enter into contracts. It has perpetual life and is owned by stockholders, each of whom has a share in the profits and losses of the corporation. An important advantage of the corporate form of ownership for general contractors is the absence of personal liability of the stockholders. This is desirable because of the risks of the contracting business, and is more than recompense for the additional burden of taxes that those taking part in corporate ownership must bear. (Small corporations can obtain some relief from Federal taxes, however.)

Corporations formed in one state must obtain, as a foreign corporation, a certificate of authority to do business in other states. This is important when bidding jobs in locations other than the home state of the general contractor.

Some general contractor corporations are large enough to find it advantageous to raise capital by becoming public corporations, with shares sold over the counter or on the various stock exchanges. Such corporations publish financial reports yearly for the benefit of the stockholders, as required by law. A study of such reports is often helpful for those engaged in the contracting business.

Joint Venture. Often when an individual job is too large to be undertaken by one company, or the risks involved are too great for one company to want to assume (although it may be capable of doing so), a joint venture is formed. This is an association between two or more contracting firms for a particular project. It joins the resources of the venturers, who share the financing and management of the job and the profits or losses in some predetermined manner.

Generally, there are specific reasons for the formation of a joint venture between specific companies. For example, one may possess the equipment and the other the know-how for a particular job. Or one may possess the financing and the other the personnel required to perform the contract. Joint ventures do not bind the members to any debts of the coventurers other than for those obligations incurred for the particular jobs undertaken.

Staff Organization. An organization chart for a typical medium-size general building contracting company is shown in Fig. 26-1. The organization shown is for a company that subcontracts most of its work and is engaged mainly in new construction. For an organization with district offices, see the chart in Paul G. Gill, "Systems Management Techniques for Builders and Contractors," McGraw-Hill Book Company, New York.

26-3. Sources of Business. For continuity of operation, a construction organization needs a supply of new projects to build. After a company has been in existence for a long time and built up a reputation, new business may come to it with little or no effort. But most companies must work hard at obtaining new jobs. Furthermore, work that happens to come in may not be of a type that the organization prefers. To find that type requires serious skillful efforts.

To be successful, a contracting company should have a person specifically assigned to attract new business. This person might be the proprietor of the construction company. In large firms a complex organization with sales and public relations personnel, backed up by engineers and cost estimators, is used. The organization should be geared to follow up on all possible sources of new business.

Public Works. The following sources for leads to new jobs and submitting proposals can be used for public work bids:

Dodge Bulletin, or other construction industry newsletters.

Engineering News-Record,—"Pulse" and "Official Proposals" sections.

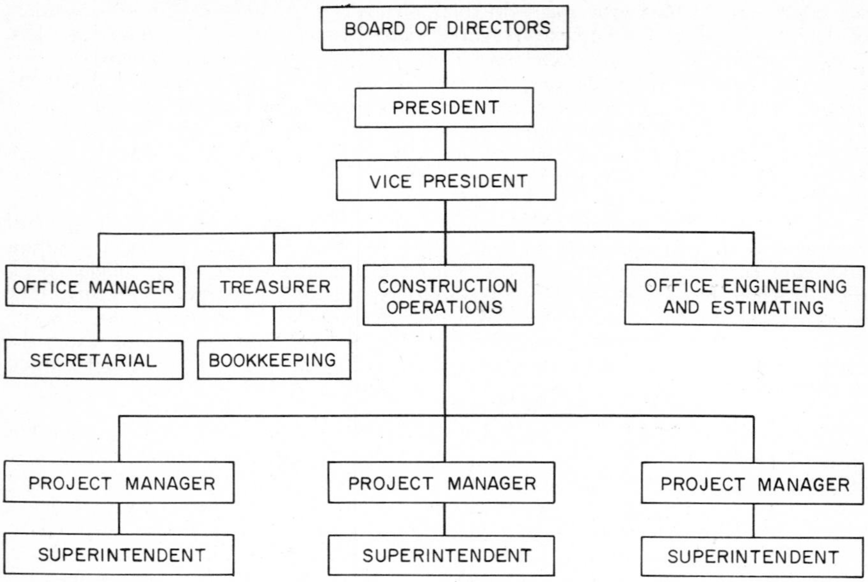

Fig. 26-1. Organization chart for a medium-size general contracting company.

Bid invitations, as a result of requests to be placed on bid-invitation mailing lists of various government agencies.

Newspaper announcements and articles.

Official publications of government agencies that contain advertisements of contracts to be bid.

Private Contracts. All the sources for public works.

Contacts with and letters to architects.

Contacts with and letters to owners.

Personal recommendation.

Sponsorship. Applying for sponsorship of any of the following:

Government-encouraged housing programs.

Urban renewal.

Purchase of land, with financing of building construction to be provided by various government programs.

Owner-builder.

Construction and rental of apartment buildings.

Construction and rental of commercial and office facilities.

Construction and leasing of post offices and other government buildings.

Construction Manager.

Application to city and state agencies or large corporations awarding this type of contract.

All the sources for obtaining private contracts.

Uses of Dodge Bulletin. From a typical *Dodge Bulletin* (McGraw-Hill Information Systems Company, New York), a subscriber to this daily information bulletin can gain the following information:

Contracts for which general contract and prime bids for mechanical and electrical work, etc., are being requested. A contractor interested in any of these types of work can obtain the plans and specifications from owners or designers, whose names and addresses are given, and submit a bid.

For contracts awarded, lists of names of contractors and amounts of contracts. Subcontractors or material suppliers who are interested in working for the contractors can communicate with those who have received the awards.

Lists of jobs being planned and estimates of job costs. Contractors and subcontractors who are interested in jobs in those locations and of the sizes indicated can communicate directly with the owner.

Additional information that may be obtained from the *Dodge Bulletin* includes lists of subcontractors and suppliers being employed by general contractors on other jobs that are already under way, and tabulation of the low bidders on jobs bid and publicly opened.

26-4. Plans and Specifications. (See also Art. 1-10.) To the construction manager, the most important thing to be alert to regarding plans and specifications is: What constitutes the plans and specifications and other contract documents that pertain to the project?

It is surprising how often this question is neglected by those entering into construction contracts and by attorneys and others concerned with signing of contracts. A clear understanding of what constitutes the contract documents and the revisions, if any, is one of the most urgent aspects of construction management. Furthermore, a precise list of what constitutes the contractor's obligation under the contract is essential to proper performance of the contract by the contractor.

In general, the contract documents should be identified and agreed to by both parties. A listing of the contract documents should be included as part of every contract. Contract documents generally include the following:

Plans (list each plan and revision date of each plan, together with title.)

Specifications, properly identified, a copy initialed by each party and in the possession of each party.

The agreement.

General conditions.

Soil borings.

Existing site plan.

Special conditions.

Original proposal (if it contains alternates and unit prices, and these are not repeated in the contract).

Invitation (if it contains data on completion dates or other information that is not repeated in the contract).

Addenda (if any).

The contractor should repeat and list the contract documents in each subcontract and purchase order (Arts. 26-8 and 26-9).

It is essential to have a properly drawn and understood list of the contract documents agreed to by all parties if construction management is to proceed smoothly and if changes and disputes are to be handled in an orderly manner (Art. 26-16). To prevent misunderstandings and doubts as to which documents are in the possession of various subcontractors and suppliers for estimating, a properly drafted transmittal form should accompany each transmission of such documents (Fig. 26-2).

26-5. Estimating, Bidding, and Costs. (See also Sec. 25.) It is advisable to have the routine to be followed in preparing cost estimates and submitting bids well established in a contractor's organization.

Particular attention should be given to the answers to the question: For whom is the estimate being prepared and for what purpose? The answers will influence the contractor as to the amount of time and effort that should be expended on preparation of the estimate, and also indicate how serious the organization should be about attempting to negotiate a contract at the figure submitted. Decision on the latter should be made at an early date, even before the estimate is prepared, so that the type of estimate can be decided.

Bid Documents. The documents should be examined for completeness of plans and specifications, and for the probable accuracy that an estimate will yield from the information being furnished. For example, sometimes contract documents are sent out for bid when they are only partly complete and the owner does not seriously intend to award a contract at that stage but merely wishes to ascertain whether construction cost will be within his budget.

Preparation of the Top Sheet. This is usually based on an examination of the specifications' table of contents. If there are no specifications, then the contractor

TRANSMITTAL

Kreisler Borg Florman
CONSTRUCTION COMPANY
97 Montgomery Street, Scarsdale, New York 10583/Telephone (914) SC 5-4600

TO: Lipsky & Rosenthal, Inc.
155 Utica Avenue
Brooklyn, N.Y. 11213

DATE February 25, 1972

ATT: Mr. D. G.
Mr. A. H.

RE: J. 47 Combined School and Apartment, New York, N.Y.

WE ARE SENDING YOU THIS DATE THE FOLLOWING:

- [X] BLUEPRINTS
- [ ] SPECS
- [ ] SAMPLES
- [ ] BOOKLETS
- [ ] CATALOGUE CUTS
- [ ] LETTERS
- [ ] TEST REPORTS
- [ ]

SENT FOR THE FOLLOWING REASON:

- [ ] YOUR APPROVAL
- [ ] YOUR FINAL APPROVAL
- [ ] APPROVED
- [ ] APPROVED AS NOTED
- [ ] DISAPPROVED
- [ ] RESUBMIT
- [X] YOUR INFORMATION
- [ ] YOUR USE
- [ ]

NUMBER COPIES	DRAWING NUMBER	PREPARED BY	DESCRIPTION
1 Set	A-1 thru A-10 Dated 2/18/72	Carl Puchall & Assoc.	School Preliminary Plans
1	7 of 12 Rev. 10/18/71	" " "	Typical Apartment Floor Plan
1	9 of 12 Rev. 1/17/72	" " "	Sub. Cellar Garage

REMARKS: We would appreciate your budget figure for referenced project, as soon as possible. Price must be broken down to School portion and Residential portion.

Thank you.

Kreisler Borg Florman
CONSTRUCTION COMPANY

BY ____________________

TITLE Estimator

Fig. 26-2. Form for transmittal of contract documents to subcontractors and suppliers.

should use as a guide top sheets (summary sheet showing each trade) from previous estimates for jobs of a similar nature, or checklists.

Subcontractor Prices. Decide on which trades subbids will be obtained, and solicit prices from subcontractors and suppliers in those trades. These requests for prices should be made by postcard, telephone, or personal visit.

Decide on which trades work will be done by the contractor's own forces, and prepare a detailed estimate of labor and material for those trades.

Pricing. Use either unit prices arrived at from the contractor's own past records, estimates made by the members of the contractor's organization, or various reference books that list typical unit prices ("Building Construction Cost Data," Robert Snow Means Co., Inc., P.O. Box G, Duxbury, Mass. 02332).

Hidden Costs. Carefully examine the general conditions of the contract and visit the site, so as to have a full knowledge of all the possible hidden costs, such as special insurance requirements, portions of site not yet available, etc.

Final Steps. Receive prices for materials and subcontracts.

Review the estimate and carefully go over exclusions and exceptions in each subcontract bid and in material quotations. Fill in with allowances or budgets those items or trades for which no prices are available.

Decide on the markup. This is an evaluation that should be made by the contractor, weighing factors such as the amount of extras that may be expected, the reputation of the owner, the need for work on the part of the general contractor, and a gage as to the contractor's overhead.

Finally, and most importantly, the estimate must be submitted in the form requested by the owner. The form must be filled in completely, without any qualifying language or exceptions, and must be submitted at the time and place specified in the invitation to bid.

26-6. Types of Bids and Contracts. Contractors usually submit bids for a lump-sum contract or a unit-price-type contract, in either case based on complete plans and specifications.

When plans and specifications have not yet advanced to a stage where a detailed estimate can be made, the type of contract usually resorted to is a cost-plus-fee type. The bid may be based on either a percentage markup the contractor will require over and above costs, or may include a lump-sum fee that the contractor proposes to charge, over and above costs. Sometimes an incentive fee is also incorporated, allowing owner and contractor to share, in an agreed-on ratio, the cost savings achieved by the contractor, or to reward the contractor for completing the project ahead of schedule. Evaluation of bids by an owner may take into consideration the experience and reputation of the contractor, as a result of which awards may not be made on a strictly low-bid basis.

Budget Estimate. Often, an owner in preparing plans and specifications will want to determine the expected construction cost while the plans are still in a preliminary stage. He will then ask contractors for a cost estimate for budget purposes. If the estimate from a specific contractor appears to be satisfactory to an owner, and if the owner is desirous of establishing a contractual relationship with the contractor early in the planning stage so as to benefit from the contractor's suggestions and guidance, a contract may be entered into after the submission of the estimate.

On the other hand, the owner may refrain from formally entering into the contract, but may treat the contractor as the "favored contractor." When requested, this contractor will assist the architect and engineers with advice and cost estimates and will expect to receive the contract for construction on completion of plans and specifications, if the cost of the project will lie within the budget estimate when plans and specifications have been completed.

Separate Prime Contracts. Sometimes an owner has the capability for managing construction projects and will take on some of the attributes of a general contractor. One method for an owner to do this is to negotiate and award separate prime contracts to the various trades required for a project. Administration of these trades will be done either by the owner's own organization or by a construction manager hired by the owner (Art. 26-1). See also Arts. 1-20 and 27-10.

Sale Lease-back. This is a method used by some owners and government agencies to obtain a constructed project. Prospective builders are asked to bid not only on cost of construction, but also on supplying a completed building and leasing it to the prospective user for a specified time. This type of bid requires a knowledge of real estate analysis and financing, as well as construction. Contractors who bid may have to associate with a real estate firm to prepare such a bid.

Sponsor-builder. In this type of arrangement (Art. 26-1), the contractor may not only have to prepare a construction-cost estimate but may also require knowledge of real estate, to be prepared to act as owner of the completed project, in accordance with the terms of a sponsor-builder agreement with a government agency, or government-assisted neighborhood or nonprofit group.

The following types of contracts are used for general construction work:

Letter of Intent. This is used where a quick start is necessary and where there is not sufficient time for drafting a more detailed contract. A letter of intent also may be used where an owner wishes material ordered before the general contract is started, or where the commitment of subcontractors requiring extensive lead time must be secured immediately.

Lump-sum Contract. (For example, Document A-1, American Institute of Architects, 1735 New York Ave., N.W. Washington, D.C. 20006.) Basis of payment is a stipulated sum. Progress payments, however, are made during the course of construction.

Cost-plus-fixed-fee Contract. (For example, A-111, American Institute of Architects.) This type of agreement is used generally with an "up-set" price. The contractor guarantees that the total cost plus a fee will not exceed a certain sum. (See types of bids, preceding.) Generally, there are provisions for auditing of construction costs by the owner.

Cost-plus-percentage-of-fee Contract. Similar to cost-plus-fixed-fee contracts, but the fee paid, instead of being a lump sum in addition to the cost, is a percentage of the costs.

Unit-price Contract. This type of agreement is used where the type of work involved is subject to variations in quantities and it is impossible to ascertain the total amount of work when the job is started. Bids will be submitted by the contractors on the basis of estimated quantities for each classification of work involved. On the basis of the unit prices submitted and the estimated quantities, a low bidder will be chosen for award of the contract. After the contract has been completed, the final amount paid to the contractor will be the sum of the actual quantities encountered for each class of work multiplied by the unit prices bid for that class.

Design-construction Contracts. This type of agreement is used for turnkey projects and by package builders (Art. 26-1). Cost estimates must be prepared from preliminary plans or from similar past jobs. Preparation of these estimates requires high skill and knowledge of construction methods and costs, because the usual methods for preparing cost estimates do not apply.

Management Contracts (Art. 26-1). This type of contract is generally administered through subcontracts and purchase orders awarded by the contractor for the owner in the owner's name. Labor performed by the contractor will generally also be paid for by the owner directly.

26-7. Contract Administration. Administration of construction contracts requires an intimate knowledge of the relationship of the various skills required for construction, which involves labor, material suppliers, and subcontractors. Feeding into the job are all of the life-supplying services. This is illustrated in the "Borg wheel of construction progress" (Fig. 26-3). Like any wheel, it is dependent on all the spokes being in place if it is to move. Whether the contractor combines one or more of the jobs shown in more than one person in his organization is immaterial. But all the spokes shown in the wheel are necessary.

In the upper left of Fig. 26-3, the task of the contractor, principal or partner, shows his relationship to the wheel of construction progress. As indicated, this person must be familiar with and has responsibility for legal, bonding, insurance, and banking requirements of the firm. He feeds into the job necessary organization and policy decisions. This contribution, when added to what is fed into the job by the project manager (progress), bookkeeping (money), superintendent (progress), clerical (correspondence and records), architect and engineers (plans and approvals), building department (approvals and inspection), and the owner (money), is essential for job progress.

Planning of the job is dealt with in Art. 26-13. Profit and loss of the job are controlled in the manner described in Art. 26-17.

26-8. Subcontracts. General contractors usually obtain subcontract bids as well as material-price solicitations during the general-contract bidding stage. Some-

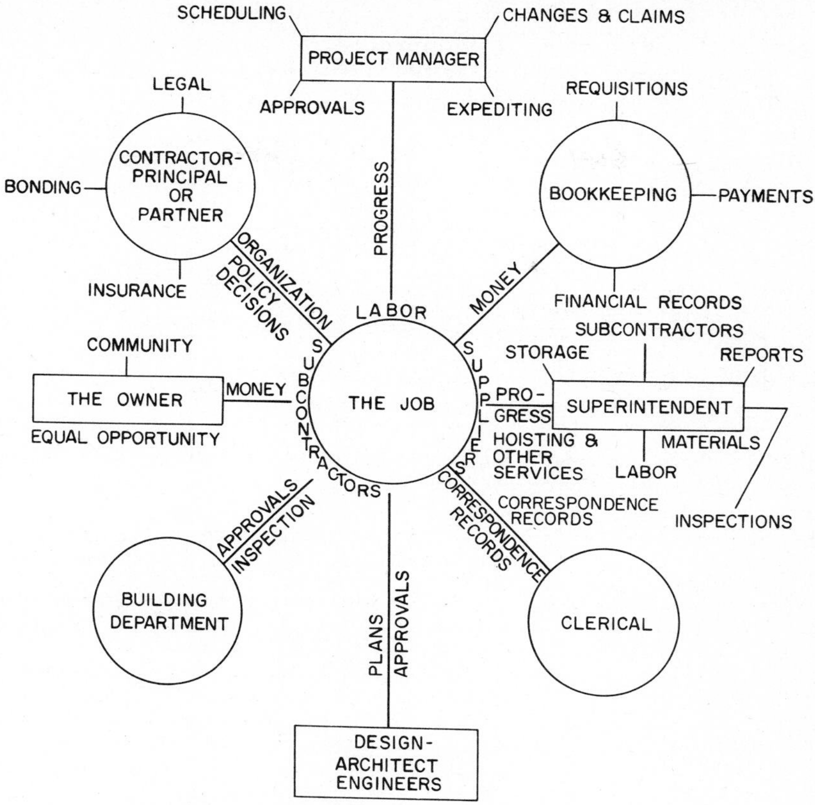

Fig. 26-3. Borg wheel of construction progress.

times, however, general contractors continue shopping after award of the general contract to attain budget goals for the work that may have been exceeded during the initial bidding stages. In such cases, additional bids from subcontractors are solicited after the award of contract.

Purchasing Index. Contractors would find it advantageous to approach purchasing of subcontracts with a purchasing index (Fig. 26-4). The index should list everything necessary to be purchased for the job, together with a budget for each of the items. As subcontracts are awarded, the name of the subcontractor is entered, and the amount of the subcontract is noted in the appropriate column. Then, the profit or loss on the purchase is later entered in the last column; thus, a continuous tabulation is maintained of the status of the purchases.

Priority numbers are given to the various items, in order of preference in purchasing. The contractor should concentrate his efforts on those subcontracts that must be awarded first. Those that follow in due course are given priority numbers that are appropriate. These priorities are indicated in Fig. 26-4 by the numbers in the left column.

Bid Solicitation. Bids are generally solicited through notices in trade publications, such as *Dodge Bulletin,* or from lists of subcontractors that the contractor maintains. Solicitation also can be by telephone call, letter, or postcard to those invited to bid. Where the owner or the law requires use of specific categories of subcontractors, bids have to be obtained from qualified members of such groups.

After subcontractor bids have been received, careful analysis and tabulation are needed for the contractor to compare bids fairly (Fig. 26-5).

KREISLER BORG FLORMAN CONSTRUCTION CO.
97 Montgomery Street
Scarsdale, New York 10583

PRIORITY NO.		ITEM	ESTIMATE	SUBCONTRACTOR	AMOUNT	PROFIT
1	2A	Excavation	77,000			
1	2B	Foundations	INCL.			
1	2H	Piling	INCL.			
1	3A	Concrete Superstructure	242,000			
2	4	Masonry	104,400			
3	4B	Cast Stone	INCL.			
3	4C	Ext. Cement Ash Sills	INCL.			
6	12A	Venetian Blinds	2,200			
2	14A	Elevators	29,600			

Fig. 26-4. Purchasing index. Numbers in first column indicate priority in purchasing subcontracts, materials, and equipment.

KREISLER BORG FLORMAN CONSTRUCTION CO.
97 Montgomery St., Scarsdale, N.Y. 10583

FOR: C.I. 17
MISC. IRON SUB CHECK SHEET - SEC. 5A

ITEMS	Lieb Iron Works	Premier Structural Steel	Ment Bros.	REMARKS
Plans & Specs. C.I. 17 - Base Bid	13,180.	13,987.	14,900.	
No Sales Tax	✓	✓	✓	
Shop Drawings & Samples	✓	✓	✓	
Painting - one coat	✓	✓	✓	
field	NO	NO	NO	
two coat	NO	NO	shelf angle only	
Attached Lintels - F & I	✓	✓	✓	
Loose Lintels - Furnish	✓	✓	✓	
Site Bridge (no crane)	+ 2,350.	NO	NO	
Exterior Saddles furn. & install	+ 1,360.	NO	NO	
Alts.				
Site Bridge Crane	+ 1,000.	—	—	

Fig. 26-5. Analysis of subcontractor bids.

In a complicated trade, such as Miscellaneous Iron, shown in Fig. 26-5, it is necessary to tabulate, from answers obtained by questioning each of the bidders, the exact items that are included and excluded. In this way, an evaluation can intelligently be made, not only of the prices submitted but also as to whether or not the subcontractors are offering a complete job of the section of work being solicited. Where an indication in the subcontractor's proposal or in Fig. 26-5 shows that a portion of the work is being omitted, it is necessary to cross-check the specifications and other trades to be purchased to ascertain that the missing items are covered by other subcontractors.

Subcontract Forms. Various subcontract forms are available for the written agreements. A commonly used form is the standard form of agreement between contractor

Kreisler Borg Florman
CONSTRUCTION COMPANY (INC.)
97 Montgomery Street, Scarsdale, New York 10583
Telephone (914) SC 5-4600 DATE August 31, 1975

TO: Brisk Waterproofing Co Inc
720 Grand Avenue
Ridgefield, New Jersey

PROJECT: East Midtown Plaza Stage II, Project #HRB 66-14B
24th & 25th Sts betw. 1st & 2nd Aves, New York, NY

SUBJECT TO CONDITIONS ON REVERSE SIDE HEREOF, YOU ARE AUTHORIZED TO PROCEED WITH THE WORK AS DESCRIBED BELOW IN CONNECTION WITH THE ABOVE PROJECT, THE COST OF WHICH IS TO BE $ 4,000.00

FOUR THOUSAND DOLLARS

DESCRIPTION

HYDROLITHIC IRON TYPE WATERPROOFING (Specification Division 5, Section 35) and all such work shown on the Plans, Specification, Specification Addenda 1, 2, 3 and 4, Construction Contract, General Conditions of the Contract, Invitation, Contractor's Loan Agreement, Bid Form, Performance and Payment Bond, and Subcontract Rider. If Performance and Payment Bond is required premium is to be paid by Contractor.
As a condition precedent to the duty of performance of the work hereunder the Contractor shall be awarded and execute a contract for the work of the project.
Subcontractor agrees to include at no additional cost any further details or corrections to the plans and specifications resulting from requirements of the Building Department or Housing and Development Administration and agrees to be bound by final plans and specifications when they are completed and approved.
This work includes the H.I.T. waterproofing of two elevator pits, 1 ejector pit and laundry troughs and curbs as indicated on plans. In connection with this work, light, water, heat, pumping and power to be furnished to us without charge. Floor slabs to be left raked by others.

PLEASE SIGN EXTRA COPY OF THIS ORDER AND RETURN TO KREISLER-BORG AT ONCE.
PLEASE SEND US CERTIFICATES OF INSURANCE FOR WORKMEN'S COMPENSATION, PUBLIC LIABILITY AND PROPERTY DAMAGE INSURANCE.

ALL OF THE ABOVE MATERIALS TO BE DELIVERED OR THE WORK COVERED BY THIS ORDER ARE TO BE COMPLETED IN ACCORDANCE WITH PLANS, SPECIFICATIONS AND ALL CONTRACT REQUIREMENTS BETWEEN US AND THE OWNER, BY ALL OF WHICH YOU AGREE TO BE BOUND UPON ACCEPTANCE OF THIS ORDER.

Kreisler Borg Florman (CONTRACTOR)
CONSTRUCTION COMPANY (INC.)

BY Robert F. Borg, President

ACCEPTED:
BRISK WATERPROOFING CO INC
SUBCONTRACTOR

BY R. W. Ehrenberg, V.P.

DATE: 9/6/75

(over)

Fig. 26-6. (*a*) Front side of short form for subcontract.

and subcontractor (Contractor-Subcontractor Agreement, A401, American Institute of Architects). A short form of subcontract with all the information appearing on two sides of one sheet is shown in Fig. 26-6. Changes may be made on the back of the printed form with the permission of both parties to the agreement.

Important, and not to be neglected, is a subcontract rider (Fig. 26-7), which is tailored for each job. Only one page of the rider is shown in Fig. 26-7. The rider takes into account modifications required to adapt the standard form to the specific project. The rider, dealing with such matters as options, alternates, completion dates, insurance requirements, and special requirements of the owner or lending agency, should be attached to all copies of the subcontract and initialed by both parties.

CONDITIONS OF CONTRACT

Within five days after the date of this contract and before commencement of the work, the subcontractor agrees to furnish the contractor with a certificate showing that he is properly covered by Workmen's Compensation Insurance as required by the law of the State where the work is to be performed and with such other insurance that may be required by Contractor, the specifications and terms of contract between Contractor and the Owner.

Where the order covers the furnishing of labor and materials on a time and material basis, it is distinctly understood the subcontractor will furnish daily vouchers for verification and signature to an authorized representative of Contractor showing labor used and materials installed in the work. A copy of these signed vouchers to be presented with invoice, together with duplicate bills for materials furnished. Contractor shall have the right to examine all records of the subcontractor relating to said charges.

Where this order is issued to a subcontractor, and purports to cover labor and materials in addition to the original contract, it is given with the express understanding that, should it subsequently be proven that the work covered herein is in the subcontractor's original contract, this order becomes null and void.

Should the subcontractor or material dealer at any time refuse or neglect to supply an adequate number of properly skilled workmen or sufficient materials of the proper quality, or fail in any respect to prosecute the work with promptness and diligence, the Contractor shall in its exclusive opinion be at liberty after three days' notice to the subcontractor or material dealer to provide any such labor or materials in accordance with such notice and to deduct the cost thereof from any money then due or to become due to the subcontractor or material dealer under this contract, any excess cost to Contractor will be immediately paid by the subcontractor or material dealer.

Where this order covers materials only, it is agreed that the materials will be delivered F. O. B. job unless otherwise ordered, in such quantity and at such times as may be authorized by this company's representative. It is further understood that in all cases, quantities of materials mentioned herein are approximate only and the dealer agrees that deliveries will be based upon actual needs and requirements of the work.

It is understood that no claims for extra work performed or additional materials furnished shall be made unless ordered in writing by an officer or authorized representative of the Contractor.

TERMS OF PAYMENT: Payments to subcontractors will be made monthly to the amount of 85% of the value of the work and materials incorporated in the building during the previous calendar month. Final balance to become due and payable within sixty days after the subcontractor has completed his work to the entire satisfaction of the architects, engineers, other representatives of the Owner and Contractor.

All payments covering subcontractor's work and/or material shall be payable only after the Contractor has received corresponding payments from the Owner.

The subcontractor hereby accepts exclusive liability for the payment of all taxes now or which may hereafter be enacted covering the labor and material to be furnished hereunder, and any contributions under the New York State Unemployment Insurance Act, The Federal Social Security Act, and all legislation enacted either Federal, State or Municipal, upon the payroll of employees engaged by him or materials purchased for the performance of this contract, and agrees to meet all the requirements specified under the aforesaid acts or legislation, or any acts or legislation which may hereafter be enacted affecting said labor and/or materials. The subcontractor will furnish to the Contractor, any records the Contractor may deem necessary to carry out the intent of said acts or legislation and hereby authorizes the Contractor to deduct the amount of such taxes and contributions from any payments due the subcontractor and to pay same direct or take any such precaution as may be necessary to guarantee payment.

Samples and details are to be submitted to Contractor, if requested, and approval must be secured before proceeding with the work.

If inferior work or material is installed or furnished and allowed to remain, the Owner and/or Contractor at its option, may reject such work and/or material or are to be allowed the difference in value between cost of work and/or material installed or furnished and cost of work and/or material specified or ordered.

The subcontractor will furnish all labor, materials, tools, scaffolds, rigging, hoists, etc. required to carry on the work in the best and most expeditious manner and furnish protection for his and other work, and will do all necessary cutting and patching and also remove and replace any interfering work, for the proper installation of his work. The subcontractor will remove all rubbish from premises in connection with his work. The subcontractor agrees to perform work in a safe and proper manner and save the Owner and Contractor harmless against all penalties for violation of governing ordinances and all claims for damages resulting from negligence of the subcontractor, or his employees, or accidents in carrying on his work.

The subcontractor agrees to repair, replace or make good any damages, defects or faults resulting from defective work, that may appear within one year after acceptance of work or for such additional period as may be required by Owner or by the specifications relating to same.

Time is of the essence of this agreement.

All labor employed shall be Union labor of such type and character as to cause no Union or jurisdictional disputes at the site of the work.

The subcontractor shall furnish all labor, material and equipment and permits and pay all fees and furnish all shop drawings, templates, and field measurements incidental to the work hereby let to it that the Contractor is required to perform and furnish under the General Conrtact, and whatever the Contractor is required to do or is by the General Contractor bound in and about the work hereby let to the Subcontractor shall be done by the Subcontractor without any extra charge.

Any controversy or claim arising out of or relating to this contract, or breach thereof, shall be settled by arbitration in the City of New York in accordance with the Rules of the American Arbitration Association, and judgment upon the award rendered by the Arbitrator(s) may be entered in any Court having jurisdiction thereof.

The Subcontractor expressly covenants and agrees to file no lien of any nature or kind for any reason whatsoever arising out of this contract for matters and things related thereto and does hereby expressly and irrevocably constitute the Contractor as its agent to discharge as a public record any lien which may have been filed by it for any reason or cause whatsoever.

Fig. 26-6. (*b*) Back side of short form for subcontract.

Also, the list of drawings (Fig. 26-8) must not be neglected. The content of the exact contract drawings should never be left in doubt. Without a dated list of the drawings that both parties have agreed to have embodied in the subcontract, disputes may later arise.

See also Arts. 26-9 and 26-16.

26-9. Purchase Orders. Issuance of a purchase order differs from issuance of a subcontract (Art. 26-8). A purchase order is issued for material on which no labor is expected to be performed in the field. A subcontract, on the other hand, is an order for a portion of the work for which the subcontractor is expected not only to furnish materials but also to perform labor in the field. An example of a purchase order form, front and back, is shown in Fig. 26-9.

For the specific project, a purchase order rider (like Fig. 26-7) and list of contract drawings (Fig. 26-8) must be appended to the standard purchase order form. The rider describes special conditions pertaining to the job, options or alternates, informa-

EAST MIDTOWN PLAZA STAGE II
24th & 25th Sts betw. 1st & 2nd Aves
New York City, New York

1. Several options are available to the General Contractor. Subcontractor agrees to accommodate himself to whichever options are exercised at no additional cost, except as specifically noted.

The Owner has accepted bid alternates a, b, and d
The work of these alternates are included in this agreement. No additional amounts will be paid to the Subcontractor for the Subcontractor's work in connection with same.

The Subcontractor has had its attention specifically called to the following Articles in the Contract: Article 50, Abnormal Foundations; Article 51, Alternate Prices; Article 52, Street Crossing Matters, and agrees to comply with the appropriate portions of same at no additional cost.

2. The Subcontractor understands that the date of completion of the General Contract under this Agreement is 7/1/76 . Any delays directly or indirectly attributable to the Subcontractor which may cause the imposition of liquidated damages or additional cost to the Contractor under his Agreement with the Sponsor will be backcharged to the Subcontractor at the rate of damages or additional cost.

3. Wherever the word "Owner" appears in the Standard Form of Subcontract it shall be construed to mean "Sponsor" or "Housing and Development Administration" as their interest may appear.

Wherever any statements, documents or data of any sort, nature, or description are required under this contract to be submitted to the Architect or Owner, or both, duplicates of such statements, documents or data shall be likewise submitted to the HDA. This approval of the HDA shall be a condition precedent to any action being taken or not taken hereunder or a prerequisite to the exercise by the parties here of any right or rights hereunder, including the right to receive payment under this contract.

4. The following provisions are made a part of this Agreement as required by the Construction Contract:
"In the event that the right of the Contractor to proceed with the work under his contract with the Owner is terminated in whole or in part, and that the Owner, by contract or otherwise, or the Contractor's surety, or the Lender, replaced the Contractor and continues with the performance of that Contract, then the Subcontractor, if required, agrees to complete the work under this subcontract then remaining unfinished in accordance with the terms and conditions in place and stead of the Contractor the Owner, another Contractor engaged by the Owner, the surety, or the Lender, as if they, or any of them, were originally parties to the subcontract.

"It is understood and agreed that the Contractor under his Contract with the Owner is subject to audit as directed by the Administration of Pay-

Fig. 26-7. Subcontract rider.

tion pertaining to shop drawings, or sample submissions and other particular requirements of the job.

Material-price solicitations are handled much in the same manner as subcontract-price solicitations (Art. 26-8). Material bids should be analyzed for complicated trades in the same manner as for subcontracts (Fig. 26-5).

To properly administer both the subcontract and the purchase orders, which may number between 40 and 60 on an average building job, it is necessary to have a purchasing log (Fig. 26-10) in which is entered every subcontract and purchase order after it has been sent to the subcontractor or vendor. The entry in the log is copied from the file copy of the typed document, and an initial in the upper right-hand corner of the file copy indicates that entry has been made. The log serves as a ready cross reference, not only to names of subcontractors

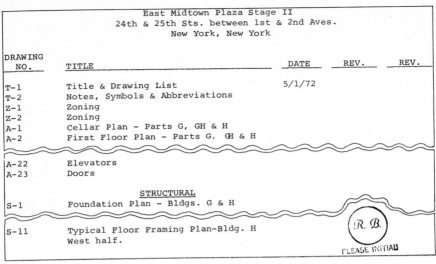

East Midtown Plaza Stage II
24th & 25th Sts. between 1st & 2nd Aves.
New York, New York

DRAWING NO.	TITLE	DATE	REV.	REV.
T-1	Title & Drawing List	5/1/72		
T-2	Notes, Symbols & Abbreviations			
Z-1	Zoning			
Z-2	Zoning			
A-1	Cellar Plan - Parts G, GH & H			
A-2	First Floor Plan - Parts G, GH & H			
A-22	Elevators			
A-23	Doors			
	STRUCTURAL			
S-1	Foundation Plan - Bldgs. G & H			
S-11	Typical Floor Framing Plan-Bldg. H West half.			

R. B.
PLEASE INITIAL

Fig. 26-8. List of drawings for construction of a project.

and vendors but also to the amounts of their orders and the dates the orders were sent.

In negotiating and awarding either a subcontract or a material purchase, the contractor must take into account the scope of the work, list inclusions properly, note exceptions or exclusions, and, where practicable, record unit prices for added or deleted work. Consideration must be given for the time of performance of units of work and availability of men and materials, or equipment for performing the work. Purchase orders should contain a provision for field measurements by the vendor, if this is required, and should indicate whether delivery and transportation charges and sales taxes are included in the prices.

26-10. Bonding. (See also Sec. 28.) From the construction management point of view, the most important question involving bonding is: What avenues of business are open to the contractor who lacks sufficient bonding capacity to do bonded work?

In Art. 26-3 various sources of business are described, and in Art. 26-1 various types of construction companies are discussed. Many of these types of business and construction companies do not require bonds for their work. For example, it is very rare that a bond is required in a construction management contract. When a contractor lacks capacity for bonding, it is well to pursue the lines of work described in those articles for which a bond will not be required.

The second question confronting a construction manager is whether or not to require a bond of subcontractors. The problem arises when a contemplated subcontractor award is to a firm that is deficient in either financial capability or experience for the job on hand. Usually, the question answers itself when it arises. If a firm has questionable financial capability and questionable experience, so that the general contractor believes that additional protection afforded by a subcontract bond is necessary, then generally the bonding companies will look with disfavor on issuing a bond to such a subcontractor. In general, therefore, if the financial capability or experience of a subcontractor is sufficiently doubtful as to require bonding, the job should not be awarded to that company. Exceptions to this can be made to assist young companies in starting and gaining experience.

There are alternatives to subcontract bonds. These alternatives include the following:

Personal guarantees by the principals of the subcontracting company.

KBF № 3002 **PURCHASE ORDER**

Kreisler Borg Florman

97 Montgomery Street, Scarsdale, New York 10583 Telephone (914) SC 5-4600 CONSTRUCTION COMPANY, INC.
(herein referred to as "Purchaser")

TO
Construction Products Co.
Route #7
Brookfield, Connecticut
Att: Mr. Alan Fishkin

(referred to herein as "Vendor")

DATE 180
JOB NO.

Please enter our order for the following materials, subject to the terms and conditions and all of the provisions set forth herein and in accordance with all pertainent provisions of the contract dated 6/9/71 entered into between purchaser and Construction for Progress, Inc. (referred to herein as "The Owner") for the construction of: (Name of project and shipping address)

13-Story Apartment Building, 170th Street to 172nd Street and 93rd Avenue, Jamaica New York

(Which contract, together with all plans, specifications and all provisions and documents incorporated or referred to therein is referred to herein as the "Principal Contract").

QUANTITY	DESCRIPTION	PRICE (Unit or Lump Sum)
	REFUSE CHUTE HOPPER DOORS, ACCESS DOOR AND SPARK ARRESTOR (Specification Division 23, Section 10), furnished, fabricated, and delivered, and all such work shown on the Plans, Specifications, AIA General Conditions, Modifications to AIA General Conditions, General Conditions, Contract and Rider.	
	This order is based on the following:	
	Twelve (12) 15"x18" hopper doors and frames	
	One (1) 12"x12", 1-1/2 hr. fireproof self-closing access door and sleeve	
	One (1) explosive vent	
	Five (5) hoppers to have sprinkler heads installed.	
	The price for this material FOB job site will be	$1,350.00
	Shop drawings and sample hopper door to be submitted immediately.	

The price(s) set forth above shall constitute payment in full for the prompt and proper furnishing of the materials hereunder in accordance with the terms of this Purchase Order. If unit prices are set forth above, Vendor agrees to be bound by the quantities of such items which may be certified by the Owner as having been properly furnished under the terms of the Principal Contract, it being understood in such event that unit quantities set forth above are estimated and used only for convenience in determining the approximate amount of this Purchase Order and that the actual quantities required may substantially vary, upward or downward from the estimated quantity.

If no price is set forth above, it is agreed that Vendor's price will be the lowest prevailing market price and in no event is the order to be filled at higher prices than last previously quoted or charged without Purchaser's consent..

This order shall not become effective unless within ten (10) days from the date of execution hereof by Purchaser as set forth below, Purchaser receives the acknowledgment copy hereof, unconditionally executed by an officer of Vendor. If Purchaser does not so receive the executed acknowledgment hereof, the offer contained herein is withdrawn and shall thereafter be renewed only by written statement to such effect made by Purchaser.

TIME OF DELIVERY: approximately six to eight weeks.

THE ABOVE IS HEREBY ACCEPTED
VENDOR CONSTRUCTION PROD. CO. INC
BY J S Clarke VICE PRES
DATE 9/28/71
(Name and Title)

TERMS OF PAYMENT: Regular
Kreisler Borg Florman CONSTRUCTION COMPANY, INC.
BY R Borg
(Name and Title) President

1. ACKNOWLEDGMENT MUST BE SIGNED AND RETURNED BEFORE INVOICES WILL BE PAID. 2. PURCHASE ORDER NUMBER MUST APPEAR ON ALL INVOICES. 3. THERE MUST BE A SEPARATE INVOICE FOR EACH PURCHASE ORDER.

VENDOR

Fig. 26-9. (*a*) Front side of purchase order.

Personal guarantees of other individuals of substantial worth unconnected with the subcontracting company.

Posting of a sum of money or of a security until performance of the subcontractor's work has been completed by the subcontractor.

26-11. Insurance. (See also Sec. 28.) For a construction manager, the limits of insurance that should be carried by the contractor and the limits that should be carried by the subcontractors are of fundamental importance.

Usually, the contractor's limits are set forth in the agreement with the owner. When the owner's insurance limits are not high enough to afford the contractor full protection for the exposure the contractor anticipates, it is often worth the

TERMS AND CONDITIONS

The purchase orders is subject to the following terms and conditions and by accepting this Purchase Order, Vendor agrees to be bound by each and every of said terms and conditions.

1. Acceptance of this Purchase Order by Vendor in time and manner hereinbefore specified shall constitute the only possible manner of acceptance hereof and Purchaser will in no way be responsible or indebted to Vendor for any materials or samples thereof furnished by Vendor or for any work or services performed by Vendor in the absence of such acceptances, it being understood that Vendor shall in such event make no claim against the Purchaser.

2. This Purchase Order may only be accepted by Vendor unconditionally and exactly as written. Any additional or different terms or conditions stated by Vendor are hereby objected to and rejected and the statement by Vendor of such additional terms and conditions shall be deemed to constitute a rejection of the offer contained therein, even if accompanied by the ackknowledgment copy hereof executed, by Vendor.

3. The principal contract has been, is and will be available for review by Vendor at Purchaser's office during normal business hours. By accepting this order, Vendor acknowledges that it has read the Principal Contract and is fully aware of each and every of its provisions which in any way bear upon or relate to the materials to be furnished hereunder and Vendor understands and agrees that all of such provisions of the Principal Contract are applicable hereto and made a part hereof as if expressly set forth herein in full, including but not limited to the general provisions of the Principal Contract dealing with the subject project as a whole as well as the special provisions and parts thereof specially relating to the materials to be furnished hereunder and Vendor agrees to furnish said materials as if it were the contractor named in the principal contract, with Vendor hereby assuming the same obligations, undertakings, warranties, guarantees, risk and liabilities to the Purchaser as Purchaser has as sumed to the Owner in addition to any other obligations, undertakings, warranties, guarantees, risks and liabilities provided for herein or otherwise provided as a matter of Law. In the event of any conflict between the Provisions of this Purchase Order and of the Principal Contract, this Purchase Order shall control. It is understood that if the Principal Contract affords to the Owner the right to approve a Vendor, then the Purchase Order is conditioned upon such approval and in the event the Owner does not so approve, then this Purchase Order shall be null and void and of no force or effect and Vendor shall have no rights whatsoever against Purchaser and of all the materials hereunder and/or any samples thereof, if any, which have previously been furnished by Vendor, may be returned to Vendor at Vendor's sole risk and expense and Vendor shall have no claim of any sort against Purchaser.

4. All materials hereunder shall be furnished in strict accordance with the Principal Contract and to the satisfaction of both Purchaser and the Owner. Both the Owner and Purchaser have the right to inspect the materials hereunder at Vendor's place of manufacture and/or at any time prior to delivery to the job and/or after delivery to the job and/or after installation in the project and in the event that the materials are rejected in whole or in part by the Owner and/or the Purchaser as not conforming to the requirements hereof, such materials may be returned to Vendor at Vendor's sole risk and expense. In such event all costs incurred by Purchaser in testing and/or inspecting the materials shall be for the account of Vendor. Purchaser reserves the right to elect to keep non-conforming materials or some portion thereof rather than returning them to Vendor. In such event, in addition to all other liabilities provided for hereunder and under the law otherwise, Vendor shall be charged for the difference between the value of the delivered non-confirming materials and the value of such materials would have had had they conformed to the requirements of this purchase order. Notwithstanding anything to the contrary hereunder or under the Law otherwise, Purchaser is not required to inspect and/or reject the materials hereunder within any particular period of time and delay in such inspection and/or rejection shall in no way limit Vendors obligation to furnish materials which strictly conform to the requirements hereof and shall in no way constitute an acceptance of non-conforming goods, nor shall payment to Vendor for or utilization by Purchaser of any or all of the materials hereunder be deemed an acceptance thereof. Vendor shall, at its own expense, arrange for an assist in all testing of ma- terials hereunder as required under the Principal Contract or when requested by Purchaser and/or the Owner.

5. Unless hereinbefore expressly stated to the contrary, Purchaser shall, within ten (10) days after eceiving payment therefor from the Owner, pay to the Vendor eighty-five (85%) percent of the amount due in accordance with the prices set forth above, for materials furnished and delivered by Vendor and Vendor shall not be so entitled to said payment or any portion thereof unless and until Purchaser is paid by the Owner for such materials. The remaining fifteen (15%) percent of said payment shall be paid to Vendor within ten (10) days after Purchaser receives payment from the Owner of all retained percentages under the terms of the Principal Contract. Before any partial of final payment, Vendor shall furnish written proof satisfactory to Purchaser that all labor and all suppliers of material, equipment and/or services in connection with the materials hereunder have been paid in full and that there are no claims outstanding against it in connection with the materials hereunder.

6. Vendor shall be entitled to no price increases or escalators of any kind in connection with the furnishing of the materials hereunder nor shall Vendor be entitled to reimbursement for boxing or packing costs or cartage charges or taxes of any type (including any Federal, State or local taxes and including but not limited to Sales or Use Taxes), all of such costs and expenses and all other costs and/or expenses which might be incurred by Vendor in connection with the furnishing of the materials hereunder being included in the consideration set forth on the reverse side hereof, payment of which con- sideration shall constitute payment in full for the furnishing of the materials hereunder. Vendor represents that the price charged for the items or services covered by this Purchase Order, is the lowest price charged by Vendor to buyers of a class similar to Purchaser under conditions to those specified in this Purchase Order and that prices comply with applicable Government regulations in effect at time of quotation, sale or delivery. Vendor agrees that any price reduction made in merchandise covered by this order subsequent to the placement of this Purchase Order will be applicable to the Purchase Order.

7. The number of this Purchase Order must appear on all invoices and if more than one purchase order has been issued by Purchaser to Vendor, there must be separate invoices for each order. Written evidence (such as delivery receipts) of delivery of the materials hereunder to Purchaser must be attached to all invoices.

8. Unless a specific delivery date or period of delivery is expressly specified herein, Vendor shall within five (5) days after being notified so to do by Purchaser, deliver the subject material or any portion thereof in such amounts and at such times as Purchaser may direct. It is understood that such time of delivery is of the essence and shall be determined by the requirements of the Principal Contract and that Purchaser shall be the sole judge thereof. Delivery shall be made FOB job site. Vendor shall unload, uncrate and stock pile in the places and manner requested by Purchaser.

9. Purchaser is under no obligation to accept any delivery exceeding the quantity specified in this Purchase Order and reserves the right to return, at Vendor's expense, part or all of any amount so de- livered. Unless the entire order can be filled, Purchaser is under no obligation to accept delivery of any part thereof, unless previously consented to in writing and Purchaser reserves the right to return, at Vendor's expense, part or all of any such delivery. Material must be delivered as per instructions and if not so delivered, any extra handling costs incurred by Purchaser will be billed back to Vendor.

10. Vendor shall prepare and submit all samples and drawings and any other back-up data when and as required by the Principal Contract and/or when and as otherwise required by Purchaser or Owner and Vendor shall perform any and all other matters necessary in order to have the materials hereunder approved by the Owner. All of the preparation, submissions and other work, materials and/or services under or in connection with this paragraph shall be at the sole expense of Vendor. Unless otherwise agreed in writing, all special dies, molds, patterns, jigs, fixtures, and any other property furnished to the Vendor by the Purchaser, or specifically paid for by the Purchaser for use in the performance of this Purchase Order, shall be and remain the property of Purchaser, shall be subject to removal upon Purchaser's instructions; shall be used only in filling orders from Purchaser; shall be held at Vendor's risk, and shall be kept insured by Vendor at the Vendor's expense while in its custody or control in an amount equal to the replacement cost thereof, with loss payable to Purchaser. Copies of policies or certificates of such insurance will be furnished to Purchaser on demand.

11. Vendor agrees to employ workmen and to use materials and work practices and delivery methods which shall not in any manner cause or contribute to strikes or other labor disturbances by workmen employed by Purchaser and/or by subcontractors of Purchaser or by any other contractors involved in the subject project. Union labels shall be affixed to the materials hereunder if Purchaser so requires. Vendor will not discriminate against any employee or applicant for employment because of race, creed, color, or national origin. Vendor will take affirmative action to insure that applicants are employed, and that employees are treated during employment, without regard to their race, creed, color, or national origin.

12. If the Owner makes or causes to be made changes or issues or causes to be issued orders or directions, any of which effect the materials to be furnished hereunder, this Purchase Order shall be deemed to be modified accordingly. If such modification causes an increase or a decrease in the cost of furnishing the materials hereunder, then in lieu of all claims and demands which vendor might other- wise have or make, Vendor shall accept in full satisfaction and discharge of all such claims and demands, the amounts, if any, as may be allowed by the Owner to the Purchaser on account of Vendor's claims and demands. If such modification causes a decrease in the cost of furnishing the materials hereunder, Vendor shall accept and be bound by the Owner's determination as to the amount of money to be deducted on account thereof. Vendor agrees to accept the Owner's determination, as aforesaid, in full and complete discharge and release of Purchaser's liability to Vendor by reason of such modification and agrees to be bound and concluded thereby and agrees never to look to Purchaser on account thereof, except for such monies, if any, as the Owner may pay to Purchaser in satisfaction of Vendor's claims and demands. Vendor agrees to immediately comply with all orders and directions of Purchaser in any way bearing upon the material to be furnished hereunder, including but not limited to, orders or directions modifying the materials hereunder, irrespective of whether Vendor shall dispute same in any particular (including but not limited to Vendor claiming additional reimbursement therefor), it being understood that the furnishing of the materials hereunder is not to be held up pending the resolution of such dispute.

13. In addition and without prejudice to all other warranties expressed or implied by Law, Vendor warrants that all materials covered by this Purchase Order will strictly conform to the requirements set forth in the Principal Contract and will be of the best quality and workmanship and free from defects and fit for the intended purposes. All warranties, both expressed and implied, aslo constitute conditions and shall survive inspection, acceptance and payment and shall inure to Purchaser and the Owner. Without limitation of any rights by reason of any breach of warranty or otherwise, material or articles which are not as warranted may at any time be returned to Vendor at Vendor's expense for credit, correction or replacement, as Purchaser may direct. This warranty shall remain in effect for one (1) year from the date of acceptance of all work under the Principal Contract unless a longer period of warranty is contained in the Principal Contract, in which event such longer period shall govern.

14. Vendor shall observe and comply with all Federal, State, Municipal and Local Laws, orders, rules, ordinances and regulations and those of all pertient bureaus, agencies and departments, which in any way relate to the furnishing of the materials hereunder.

15. Vendor undertakes and agrees to defend at Vendor's own expense and to indemnify and hold Purchaser, its bonding company and the Owner harmless from and in connection with all claims, suits, actions, and proceedings in which Purchaser, and/or Purchaser's bonding company and/or the Owner are made defendants for actual or alleged infringement of any U.S. or foreign letters patent or copy- right or trade secret or unfair competition resulting from the use or sale of the materials purchased hereunder and Vendor further agrees to pay and discharge any and all judgments or decrees which may be rendered in any such suit, action or proceedings against such defendants. The provisions of this paragraph shall apply whether or not Purchaser furnishes any specification to Vendor.

16. Vendor shall promptly pay for all labor, materials, equipment, services and all other charges it incurs in connection with the furnishing of the materials hereunder. All payments received by Vendor hereunder shall be held by Vendor as trust funds for the payment of such labor, materials, equipment, services and all other charges incurred by Vendor in connection with the furnishing of the materials hereunder and shall not be applied to any other purposes until all of Vendor's obligations in connection with the materials hereunder are satisfied. Vendor agrees that it shall file no lien in connection herewith and that it shall file no notice or claims with the Owner or with Purchaser's bonding company. Vendor further agrees that if a lien is filed by any third person or entity for whom Vendor obtains labor, materials, equipment or services in connection with the materials to be furnished hereunder, then Vendor shall at its own expense and immediately after being notified by Purchaser of the existence of such lien, take all steps necessary to discharge or vacate such lien including but not limited to, furnishing a lien bond if necessary to discharge such lien. If such lien is filed by a third party and is not discharged or vacated within five (5) days after Vendor is notified as to its existence, Purchaser may take the steps necessary to discharge or vacate such lien and charge to Vendor all costs in connection therewith, including attorney's fees. Further to the above, Vendor agrees to assume the entire responsibility and liability for and defense of and to indemnify and hold Purchaser and Purchaser's bonding company harmless from any and all demands, claims, liens and/or charges which may be filed, made or asserted against Purchaser and/or its bonding company and/or the Owner, for or in connection with the materials hereunder, which demands, claims, liens and/or charges may be filed and/or asserted by any third person or entity, including but not limited to demands, claims, liens and/or charges for payment for labor, equipment, materials or services as well as those for personal injury (including death) and/or property damage in any way related to the materials hereunder. If at any time there should be any such lien, demand, claim or charge, Purchaser shall have, in addition to all other rights, the right to retain out of any payment then due or which thereafter might become due to Vendor, an amount sufficient in Purchaser's opinion to fully indemnify it against any loss in connection therewith. If Purchaser so elects, it may make payment directly to such third persons or entities in order to resolve such third party claim and such payment shall be for the account of Vendor.

17. Vendor agrees that it shall not delegate or sublet or assign this Purchase Order, or any of its obligations or rights hereunder or any portion of such right or obligations, without prior written per- mission of Purchaser. Any such delegation, subletting or assignment without prior written permission of Purchaser shall vest no rights against Purchaser and any third party.

18. In the event that any provisions of the Bankruptcy Act of the United States of America shall be invoked by or against Vendor, or in the event Vendor shall become insolvent, or make an assignment for the benefit of creditors, or a Turstee or Receiver shall be appointed of his property, or any judgment is taken against Vendor and execution issued thereon, or any writ of attachment or the like is served upon Purchaser because of any pending litigation of Vendor, or if the Vendor shall in the opinion of Purchaser violate any of the terms of this Purchase Order, then upon the happening of any of the fore- going, Purchaser shall have the right to terminate the employment of Vendor under this Purchase Order, and in such event, may either employ its won forces or those of others to complete the work remaining to be performed under this Purchase Order either in whole or in part, and shall be entitled to have and use without cost to it, either through its own forces or through any other persons so employed by it, any plant, equipment or materials, whether on the premises or otherwise provided by Vendor in the performance of the work called for herein and the completion thereof (the right to possession of which plant, equipment and materials Vendor hereby assings and transfers to Purchaser), and Vendor shall receive no further payment until the work under the Purchase Order shall be finished, completed, accepted, and final payment made by the Owner. Then, if the unpaid balance that would be due under this Purchase Order shall exceed the cost to Purchaser of completing same, including Purchaser's overhead plus any damages or loss which the Purchaser may sustain by reason of Vendor's fault such excess shall be paid to Vendor, but if such cost plus all such damages shall exceed such unpaid balance, Vendor shall pay the difference to Purchaser. In addition to having the right to terminate this Purchase Order for cause, as aforesaid, Purchaser may at any time, terminate this Purchase Order in whole or in part, for Purchaser's own convenience. Upon such termination for convenience, the principals contained in Section 8-706 of the Armed Service Procurement Regulations relating to termination of subcontracts (as such is in effect as of the date of execution by Purchaser hereof) shall determine the rights of Vendor and Vendor shall have no other claims or rights against Purchaser.

19. None of the rights and remedies conferred upon or reserved by Purchaser under this Purchase Order or the liabilities or obligations hereby imposed or assumed by Vendor, shall be deemed exclusive, or impose any limitation upon or be in derogation of any right or remedy of Purchaser now existing or which hereafter may exist, at Law, in equity or by statute. Each and every right and remedy of Purchaser under this Purchase Order shall be cumulative and in addition to all of the rights and remedies of the Purchaser under this Purchase Order now existing, or hereafter to exist, at Law, in equity or statute. In the event that Vendor fails to perform any of its obligations hereunder, then in addition to all other obligations and/or liabilities imposed upon Vendor hereunder and/or under the Law otherwise, Vendor shall hold Purchaser harmless from and indemnify Purchaser for any and all losses, costs, damages and expenses (including, but not limited to, all incidental and consequential damages due to delay or otherwise) incurred by Purchaser as a result of any of such failures, it being understood by Vendor that any failure on its part may effect Purchaser to an extent in excess of the dollar amount of this Purchase Order.

20. Failure on the part of Vendor to perform any of the obligations called for hereunder shall be deemed to be a substantial breach of the within Purchase Order. It is understood that this Purchase Order constitutes an entire and indivisible agreement. No waiver by Purchaser of any breach of this Purchase Order shall be deemed to be a waiver of any other or any subsequent breach. Acquiescense by Pur- chaser in a course of performance by Vendor contrary to the terms of this Purchase Order shall not be construed as a waiver or modification of any term hereof. A waiver or modification of any term hereof may be accomplished only be a writing signed by Purchaser in which Purchaser so states its intent.

21. It is understood and agreed that each and every provision of this Purchase Order, including any alleged breach thereof, shall be interpreted in accordance with the Law of the State of New York.

22. This Purchase Order constitutes the entire agreement between Purchaser and Vendor and shall supersede and merge all prior understandings, whether oral or in form of correspondence heretofore had between said parties in connection with any and all matters pertaining to this purchase order and the performance thereof, and it is specifically agreed that the within Purchase Order shall be subject to modification only by a writing signed by an officer of Purchaser. Nothing contained in this Purchase Order is intended to inure to the benefit of any third party.

23. Purchase Order valid only if signed by an authorized officer of the Purchaser.

Fig. 26-9. (*b*) Back side of purchase order.

additional cost of increasing these limits. For example, insurance with limits of \$500,000 to \$1,000,000 for public liability and \$500,000 for property damage often may be less expensive in the long run than lower limits that may be the maximum required by the owner. Furthermore, if the general contractor requires subcontractors to carry insurance with the same limits, it may be provident for the general contractor to pay the additional cost to the subcontractor for the increase of limits above what the subcontractor normally carries. When insurance limits are raised from \$100,000 to \$500,000, or from \$50,000 to \$500,000, the increase in cost is not proportional to the increase in limits.

PURCHASING LOG

KREISLER BORG FLORMAN CONSTRUCTION CO.
97 Montgomery Street, Scarsdale, N. Y.
FOR: 170 Street, Jamaica, N.Y. – TURNKEY PROJECT

SPEC	ITEM	NAME	DATE	AMOUNT
4.00	Excavation and Grading	Dedona Contracting	4/27/71	$25,000.-
5.00	Concrete Foundations	AD&M General Cont.	4/26/71	28,000.-
30.00	Heating & Ventilation	John A. Jones	6/11/71	150,000.-
29.18, 19 &22	Asphalt Paving & Steel Curb	Strada Contracting	6/26/72	4,500.-
22.0	Lath & Plaster	A. Palmese & Son	7/20/72	1,600.-
12.00	Resilient Flooring	Staples Floorcraft	7/20/72	20,000.-

Fig. 26-10. Purchasing log for recording subcontract and purchasing orders.

Contractors should also be alert to instances in which excessive costs result from overinsurance. For example, in a builder's risk policy with coverage on the completed-value basis, there are certain exclusions from the policy that may be included in the contract price but that should not be included in the amount of insurance purchased. For instance, in a $1,000,000 contract for a new building, the demolition work, portions of the foundation below the basement slab, and certain site and landscaping work are not covered in the builder's risk insurance policy and therefore should be deducted from the amount of insurance purchased. Consequently, insurance coverage should be substantially less than $1,000,000.

26-12. Scheduling and Expediting. Two common methods of scheduling construction projects are by means of the Gantt (bar) chart (Fig. 26-11) and by means of the critical path method (Fig. 26-12).

Bar Chart. The bar chart is preferred by many contractors because of its simplicity, ease in reading, and ease of revision. The bar chart can show a great deal of information besides expected field progress. It can also show actual field progress; dates for required delivery; fabrications and approvals; percent of completion, both planned and actual; and time relationships of the various trades. Copies of bar charts may be distributed to subcontractors, trade foremen, and in many cases laymen, with the expectation that it will be easily understood.

Steps in preparation of a bar chart should include the following:

1. On a rough freehand sketch, lay out linearly and to scale horizontally, the amount of time contemplated for the total construction of the job, based on either the contract or past experience.

2. List in the first column all the major trades and items of work to be performed by the contractor for the job.

3. Based on past experience or on previous bar charts that give actual times of completion for portions of past jobs, block in the amount of time that will be needed for each of the major trades and items of work, and indicate their approximate starting and completion date in relationship to the other trades on the job.

4. On completion of Step 3, reexamine the chart in its entirety to ascertain whether the total amount of time being allocated for completion is realistic.

5. After adjusting various trades and times of completion and starting and completion dates on the chart, work backward on dates for those trades requiring fabrication of material off the site and for those trades that require approval and submission of shop drawings, samples, or schedules.

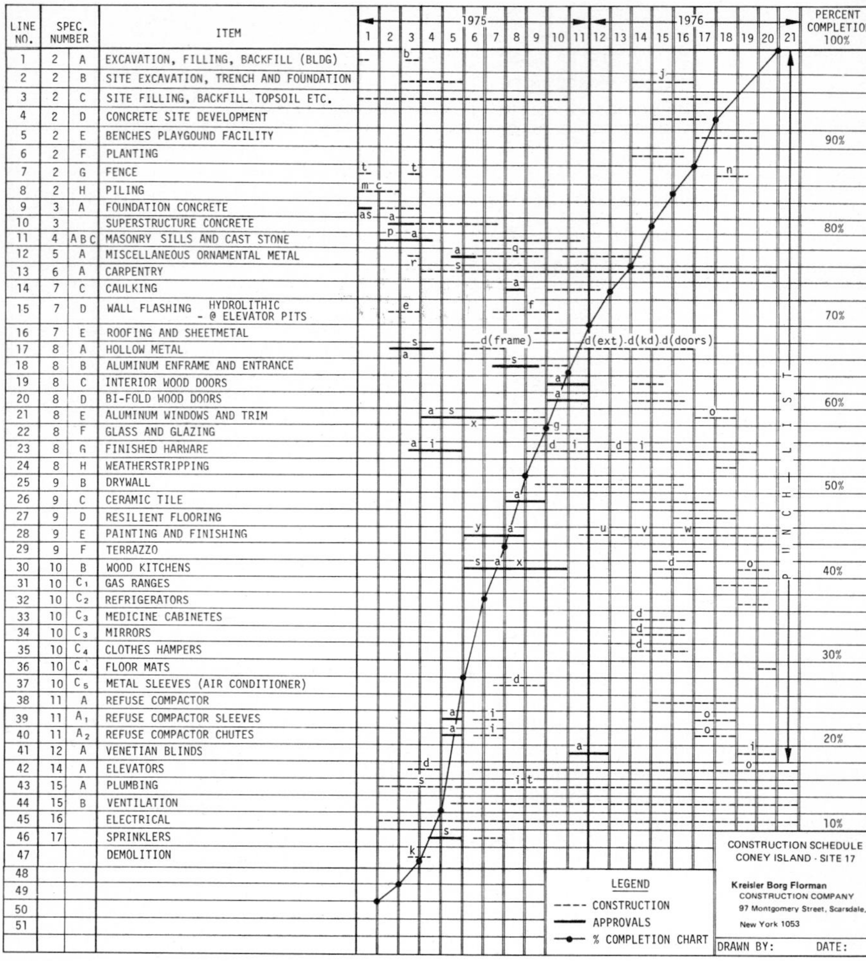

Fig. 26-11. Bar chart for construction scheduling. Letters associated with bars indicate: *a*, approvals; *b*, backfill; *c*, contract piles; *d*, delivery; *e*, elevator pits, hydrolithic; *f*, spandrel waterproofing; *g*, glazing; *h*, handrails; *i*, installation; *j*, site work; *k*, taxpayers; *l*, hardware schedule; *m*, load test; *n*, permanent; *o*, adjustments; *p*, brick panels; *q*, lintel; *r*, saddles, wedge inserts; *s*, shop drawing; *t*, temporary; *u*, tape; *v*, prime, ceiling spray; *w*, finish; *x*, fabrication; *y*, finish schedule.

6. Block in the length of time necessary for fabrication and approval of items requiring those steps.

7. Using the contractor's trade-payment breakdown or schedule of payments for each trade and month-by-month analysis of which trades will be on the job, sketch in the percent-completion graph across the face of the chart.

8. After the chart has been completely checked and reviewed for errors, draw the chart in final form.

Critical Path Method (*CPM*). This is favored by some owners and government agencies because it provides information on the mutually dependent parts of a construction project and the effect that each component has on the over-all completion of the project and scheduling of other components. CPM permits a more realistic analysis of the daily problems that tend to delay work than does the customary bar chart. Strict adherence to the principles of CPM will materially aid builders to reduce costs substantially and to enhance their competitive position in the industry.

Briefly, CPM involves detailing, in normal sequence, the various steps to be taken by each trade, from commencement to conclusion. The procedure calls for the coordination of these steps with those of other trades with contiguous activities or allied or supplementary operations, to insure completion of the tasks as scheduled, so as not to delay other work. And CPM searches out those trades that control the schedule. This knowledge enables the contractor to put pressure where it will do the most good to speed a project and to expedite the work at minimum cost.

An important, but not essential, element of CPM is a chart consisting of a network of arrows and nodes. Each arrow represents a step or task for a particular trade. Each node, assigned a unique number, represents the completion of the steps or tasks indicated by the arrows leading to it and signifies the status of the project at that point. The great value of this type of chart lies in its ability to indicate what tasks can be done concurrently and what tasks follow in sequence.

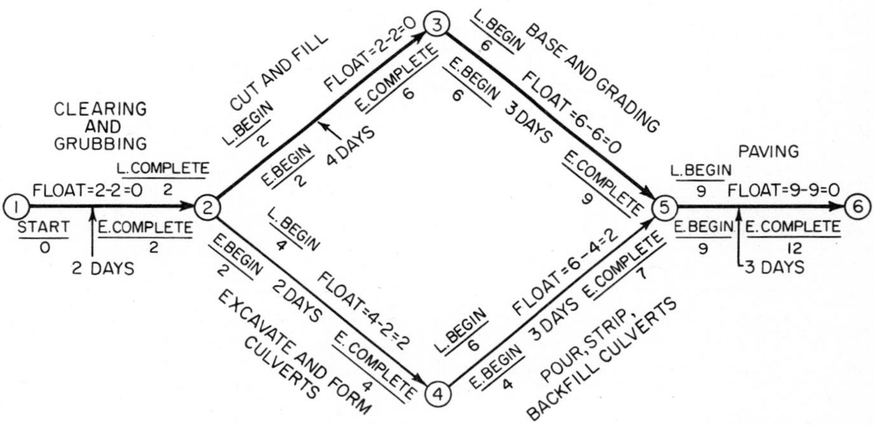

Fig. 26-12. Network for critical path method of scheduling comprises arrows representing steps, or tasks, and numbered nodes representing completion of those tasks. Heavy line marks critical path, the sequence of steps taking longest to complete.

This information facilitates expediting, and produces a warning signal for future activities.

The actual critical path in the network is determined by the sequence of operations requiring the most time or that would be considered the most important parts of the contract on which other trades would depend. This path determines the total length of time for construction of the project. Usually, it is emphasized by heavier or colored lines, to remind the operating staff that these tasks take precedence (see Fig. 26-12). Also, it is extremely important that the contractor be guided by the knowledge that a task leaving a node cannot be started until all tasks entering that node have been completed.

Another benefit of CPM scheduling is the immediate recognition of float time. Associated with noncritical activities, float time is the difference between time required and time available to execute a specific item of work. This information often enables a supervisor to revise his thinking and scheduling to advantage.

Float is determined in two steps, a forward and a backward pass over the network. The forward pass starts with the earliest begin time for the first activity. Addition of the duration of this task to the begin time yields the earliest complete time. This also is the earliest begin time for the next task or tasks. The forward pass continues with the computation of earliest complete times for all subsequent tasks. At nodes where several arrows meet, the earliest begin time is the largest of the earliest complete times of those tasks. The backward pass starts with the earliest complete time of the final task. Subtraction of the duration of that activity yields the latest allowable begin time for it. The backward pass continues with the computation of the latest allowable begin times for all preceding tasks. At nodes from which several arrows take off, the latest allowable complete time is the smallest of the latest allowable begin times, and the latest allowable begin times of preceding tasks are found by subtracting their duration from it. Float is the difference between earliest and latest begin (or complete) times for each task.

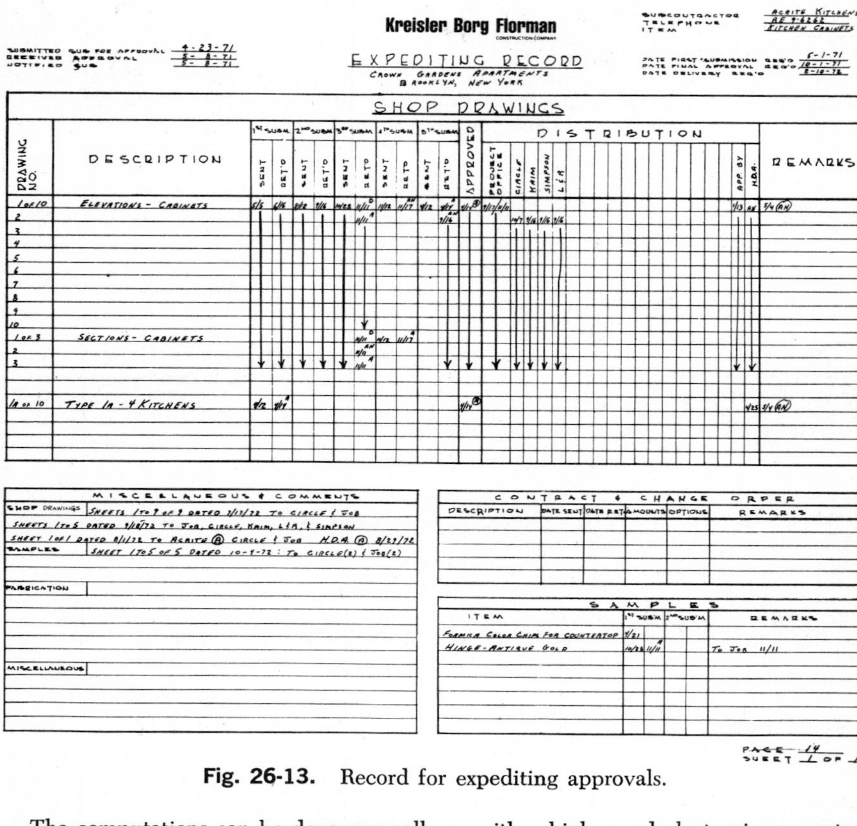

Fig. 26-13. Record for expediting approvals.

The computations can be done manually or with a high-speed electronic computer. The latter is desirable for projects involving a large number of tasks, or for which frequent updating of the network is required.

Expediting. The task of keeping a job on schedule should be assigned to an expediter. The expediter must be alert to and keep on schedule the following major items: letting of subcontracts; securing of materials; and expediting of shop drawings, sample approvals, fabrication and delivery of materials, and building department and government agency submissions and approvals.

Records must be kept by the expediter of all these functions, so that the work can be properly administered. One method of record keeping for the expediting of shop drawings, change orders, samples, and other miscellaneous approvals is illustrated in Fig. 26-13. Information entered includes name of subcontractor, description of work being done, and dates of submissions and approvals received. Space is provided for shop-drawing and sample submissions, approvals and distributions, and other miscellaneous information, as well as contract and change orders. A page similar to the one shown in Fig. 26-13 should be used for every subcontractor and every supplier on the job.

Expediting requires detail work, follow-up work, and awareness of what is happening and what will be needed on the job. Its rewards are a completed job on or ahead of schedule.

26-13. Project Management. In a small contracting organization, project management is generally the province of the proprietor. In larger organizations, an individual assigned to project management will be responsible for one large job or several small jobs.

A project manager is responsible in whole or in part for the following:

Progress schedule (Fig. 26-11 or 26-12).

Purchasing (Arts. 26-8 and 26-9).

Arranging for surveys and layout.

Obtaining permits from government agencies from start through completion of the job.

Familiarity with contract documents and their terms and conditions.

Submission of and obtaining approvals of shop drawings and samples, and material certifications.

Conducting job meetings with the job superintendent and subcontractors and following up decisions of job meetings. Job meetings should result in assignments to various individuals for follow-up of the matters discussed at the meetings. Minutes kept of the plans of action discussed at each meeting should be distributed as soon as possible to all attending. At the start of the following job meeting, these minutes can be checked to verify that follow-up has been properly performed. Figure 26-14 illustrates typical job-meeting minutes.

26-14. Field Supervision. A field superintendent has the most varied duties of anyone in the construction organization. His responsibilities include the following:

Field office (establishment and maintenance); fencing and security; watchmen; familiarity with contract documents; ordering out, receiving, storing, and installing materials; ordering out and operation of equipment and hoists; daily reports; assisting in preparation of the schedule for the project; maintenance of the schedule; accident reports; monitoring extra work; drafting of back-charges; dealing with inspectors, subcontractors, and field labor; punch-list work; and safety.

Familiarity with contract documents and ability to interpret the plans and specifications are essential for performance of many of the superintendent's duties. (The importance of knowing the contract documents is discussed in Art. 26-4.) Should the superintendent detect that work being required of him by the architect, owner, or inspectors exceeds the requirements of the contract documents, he should alert the contractor's office. A claim for pay for extra work, starting with a change-order proposal, may result (Art. 26-16).

The daily reports from the superintendent are a record that provides much essential information on the construction job. From these daily reports, the following information is derived: names of men working and hours worked; cost code amounts; subcontractor operations and description of work being performed; materials received; equipment received or sent; visitors to the job site; other remarks; temperature and weather; accidents or other unusual occurrences. Figure 26-15 shows a typical daily report.

Back-charges. Frequently, either at the request of a subcontractor or because of the failure of a subcontractor to perform, work must be done on behalf of a subcontractor and his account charged. If the work performed and the resultant back-charge is at the request of the subcontractor, then obtaining the information and the agreement of all parties to a back-charge order (Fig. 26-16) is easy.

If, however, there is a dispute as to whether or not the work is part of the obligation of the subcontractor, then the task becomes more complicated. Back-charges in this situation should be used sparingly. Sending a back-charge to a subcontractor under circumstances that are controversial is at best only a self-serving declaration, and in all likelihood will be vigorously disputed by the subcontractor.

26-15. Who Pays for the Unexpected? A well-drafted construction contract between contractor and owner should always contain a changed-conditions clause in its general conditions. (See "General Conditions of the Contract for Construc-

K B F JOB MINUTES

Kreisler Borg Florman — Meeting # 14

CONEY ISLAND SITE 17 HOUSES — Date: 1-12-73

Project #182

PRESENT: MESSRS. CARNICELLI - LOUIE - WALLACE

Spec. Section	Item of Work	Remarks	To Follow
2A	EXCAVATION	Con Ed Vault Work	H.L.
2	SITE WORK	Filling/Surcharge N&S Ends to Commence	H.L.
		Sanitary Lines Complete	J.E.C.
		Planting	
		Fencing	
		Benches	
2B	FOUNDATIONS		
2H	PILING		
3A	CONC. SUPERSTRUCTURE	Rat Patching - Clean-up	H.L.
4	MASONRY	Top Out - 2 Week Schedule	H.L.
4B	CAST STONE	Follow on Deliveries	H.L.
4C	EXTRUDED CEMENT ASH SILLS	Top Out - 2 Week Schedule	H.L.
5A	MISC. & ORNAMENTAL METAL	(Top Out - 2 Week Schedule Follow)	H.L.
5B	LOUVERS	(on Elev. Bm Installation)	
5C			
6A-1	CARPENTRY	Protection - Dirt Chute	H.L.
6A-2	MILLWORK	Approvals - Hardware - Schedule	H.L.
7C	CAULKING		J.E.C.
7D	SPANDREL WATERPROOFING	Top Out - 2 Week Schedule	H.L.
7E	ROOFING & SHEET METAL	Delivery of Materials	H.L.
7G			
8A	HOLLOW METAL	Delivery Schedules	H.L.
8B	ALUM. ENFRA. & ENTRANCES	Mezz. Mgr. Office to Complete	H.L.
8C	INTERIOR WOOD DOORS	Approvals	J.E.C.
8D	BI-FOLD DOORS	Approval of Hardware	J.E.C.
8E	ALUM. WINDOWS & TRIM	Top Out - 2 Week Schedule - Delivery Sliders	H.L.
8F	GLAZING	Expediting to Commence w/i 2 Weeks	J.E.C.
8G	HARDWARE	Follow on Deliveries	H.L.
8H	WEATHERSTRIPPING		
8I			
9B	DRYWALL	2 Week Schedule - Int. Part. 2-3-4	H.L.
9C	CERAMIC TILE		
9D	RESILIENT TILE		
9E-1	DUSTPROOFING		
9E	PAINTING	Lintels	H.L.
9F	TERRAZZO	Shop Drawings	J.E.C.
9G	AC TILE		
10B	WOOD KITCHEN CABINETS		
10C-1	GAS RANGES		
10C-2	REFRIGERATORS		
10C-3	MEDICINE CABINETS		
10C-4	MIRRORS		
10C-5	CLOTHES HAMPERS		
10C-6	METAL SLEEVES A/C	Order Complete	-
10C-7	METAL GRAB BARS	Back Plate Installation	H.L.
10C-8	VET. PROVISIONS		
10E	TOILET PARTITIONS		J.E.C.
11A	REFUSE COMPACTOR		
	Sleeves)	Complete except Roof Area	
	Chutes)	Top Out - 2 Week Schedule	
12A	VENETIAL BLINDS		
14A	ELEVATORS	On Strike	
15A	PLUMBING & Hot Water	On Schedule	
15B	VENTILATION-~~HOT WATER~~	Follow on Delivery of Roof Fans On Schedule	J.E.C.
16	ELECTRICAL & HEAT	On Schedule - Follow on Telephone System complete	J.E.C.
17	SPRINKLERS		

Fig. 26-14. Minutes of a job meeting.

DAILY REPORT

Kreisler Borg Florman

CONSTRUCTION COMPANY (INC.)
97 Montgomery Street, Scarsdale, New York 10583
Telephone (914) SC 5-4600

JOB No. 182 SHEET 1 OF 1
JOB NAME Coney Island Site 17 WEATHER Fair
SUPTS NAME H. L. TEMP. A.M. 38° P.M. 49°

KREISLER BORG FLORMAN OPERATIONS

	CLASS	EMPLOYEE'S NAME	HRS.	COST CODE	EXPLANATION	TOTAL HOURS	RATE	AMOUNT
1	Proj. Mgr.	J. C.						
2	Secty	M. E.						
3	Secty	P. W.			UTSCO deliv. 500' of .018 S/S base flashing			
4	Supt.	H. L.			for Sepia			
5	Asst.	D. W.						
6	Lab.	J. F.	7		Sweeping & scraping 10th & 11th fl.			
7	Lab.	V. C.	7		Sweeping & cleaning floors for bathroom layout			
8	Lab.	B. M.	8		on the 17th & 18th fls.light salamanders and			
9	Lab.	C. T.	7		change propane cylinder on 2nd,3rd & 4th fls.			
10	Lab.	E. M. (1)	7		Cleaning "C" type window for installation of			
					hopper frame on the 6th & 7th floors			

SUBCONTRACTOR OPERATIONS

SUB'S NAME & WORK FORCE	WORK PERFORMED
Island Security (4) 4 Watchmen	2 shifts, 3PM to 5AM
Alwinseal 2 Ironworkers	setting 19th floor windows;installing hopper frames on 6th fl.
Lieb 2 Ironworkers(1)	Setting roof level lintels;Stair platform frame at Mech.Rm. level Bldg "C"
Lipsky 1-15(1)-3(1)-17(1) Plb.	C.I.test 13th fl. thru roof Bldg B;installing gas risers 16th,
	17th fls Blds.A,B,C; setting tubs 16th fl;maint.temp. water
Lefferts Waterproofer	Spandrel waterproofing south end "C" east side of B & A at
	roof level & mech.room slab level at bldg C.
Public Improvement 2-12(1)-8(3) elec.	Installing conduits & corrective work 2nd & 3rd fl. working with
	sheetrock crew;misc.corrective work on 6th flr; pulling wire
	on 10th flr. bldgs B & C; nippling on 14th & 15th fl; layout
	for nippling on 15th & 16th flr; fabrication work;maint. temp.
	lights
Star-Circle 2-26(4) Carp	Sheetrock 2nd flr; framing stairs 2 to 3 & 3 to 4 flr; framing
	kit.3 fl; sheetrock bathrms 3 fl; installing insul & sheetrock
	back of stairs 4 fl; fire cod'g duct risers 5 flr & framing
	5 flr.apts; layout of bathrooms on 17 & 18 flrs.
Bafill 26(7)-2-27(2) Brklyrs	Laying face brick and backup block to roof level at Building A West side of Building B
1-19(7) M.Tender 1 Hst.Egr.	Brickwk. at bulkhead to mech.rm. slab level at building C
	Block work at elev.shafts 7 & 8 flr. of buildings A,B & c
Dic 2 carp. 4 lab.	Framing & grouting pipe & duct openings on 10th flr; cleaning inserts at roof level.

REMARKS

Reid - 3 engr. Took settlement readings with Raamot's Engr.
Active - 2 carp. Installing storefront frames at 1st flr. of Bldgs B & C
Tropey Continue excavating for Con Edison vault at 24th St near Bldgs B & C

MATERIAL RECEIVED	VISITORS
	F. B - UDC
153 men on jobsite	F. F - Bafill
40 minority	

DATE Jan. 16, 1973 DAY Tuesday
PROJ. MGR. INITIAL ____ REPORT No. 266 OVER ☐ SUPT. SIGNATURE Superintendent

Fig. 26-15. Daily report, prepared by superintendent.

BACK CHARGE ORDER

PAGE 1 OF 1

Kreisler Borg Florman
CONSTRUCTION COMPANY
97 Montgomery Street, Scarsdale, New York 10583
Telephone (914) SC 5-4600

NAME OF COMPANY TO BE BACKCHARGED Universal Ductwork Corporation

LOCATION Bronx Municipal Hospital Center Pelham Parkway So & Eastchester Rd, Bronx, NY (JOB NAME) JOB NO. 173

PERFORMED BY General Contractor (INSERT GENERAL CONTRACTOR OR NAME OF SUBCONTRACTOR) DATE January 22, 1975

LABOR

NAME OF EMPLOYEE	OCCUPATION	HOURS WORKED	RATE OF PAY	AMOUNT
John M	Laborer	7		

MATERIALS, SUPPLIES AND/OR EQUIPMENT

QUANTITY	DESCRIPTION	UNIT PRICE	AMOUNT
2	Bullpoints	$7.50	

COMPLETE DESCRIPTION OF WORK DONE THIS DATE

Chopped holes in existing wall for new ducts, Room 301

(SIGNED) ____ (SUBCONTRACTOR'S SUPERINTENDENT OR TRADE FOREMAN)

THIS BACK CHARGE WORK HAS BEEN PERFORMED AND THE COST HAS BEEN DEDUCTED FROM THE MONEY DUE YOU ON THIS JOB

UNIVERSAL DUCTWORK CORP. (NAME OF SUBCONTRACTOR)

BY ____ (SIGNATURE OF K-B REPRESENTATIVE)

January 22, 1975 (DATE) Superintendent (TITLE)

AT YOUR REQUEST ☒
BECAUSE OF FAULTY WORK BY YOU ☐
BECAUSE OF YOUR FAILURE TO DO THE WORK DESCRIBED ☐
OTHER, AS FOLLOWS: ☐

NUMBER 173-6

UNLESS WE RECEIVE YOUR WRITTEN PROTEST WITHIN 3 DAYS IT WILL BE ASSUMED THAT YOU ACCEPT THE CHARGE.

Fig. 26-16. Back-charge order, for work done by general contractor on behalf of a subcontractor.

tion," AIA A201, American Institute of Architects, 1735 New York Ave., NW, Washington, D.C. 20006.) This clause should answer the question:

If the contractor encounters subsurface conditions different from what might normally be expected or is described in plans, specifications, or other job information provided by the owner or his agents, resulting in increased cost, should the contractor absorb the increased cost or should the owner pay it?

A properly drafted changed-conditions clause should contain the following elements:

A requirement that the owner pay for the unexpected.

Arrangements so that the owner will not be the arbiter of whether the unexpected has occurred.

Recognition that the contract documents have been based on an assumed, described set of facts.

Indication that a changed condition can exist because of unanticipated difficulty of performance.

A requirement that the owner be made aware of a changed condition when it occurs.

Indication that changed conditions can include obstructions.

A requirement that the contractor stop work in the area of a changed condition until ordered to proceed.

Indication that either party can claim changed conditions.

Recognition that the method of procedure for handling changed conditions is provided for in the contract.

An arbitration clause or effective means of appeal other than to the courts.

Because of the comprehensiveness of the elements and because of the lack of uniformity in changed-conditions clauses, the American Society of Civil Engineers Committee on Contract Administration drafted the following recommended changed-conditions clause:

"The contract documents indicating the design of the portions of the work below the surface are based upon available data and the judgment of the Engineer. The quantities, dimensions, and classes of work shown in the contract documents are agreed upon by the parties as embodying the assumptions from which the contract price was determined.

"As the various portions of the subsurface are penetrated during the work, the Contractor shall promptly, and before such conditions are disturbed, notify the Engineer and Owner, in writing, if the actual conditions differ substantially from those which were assumed. The Engineer shall promptly submit to Owner and Contractor a plan or description of the modifications which he proposes should be made in the contract documents. The resulting increase or decrease in the contract price, or the time allowed for the completion of the contract, shall be estimated by the Contractor and submitted to the Engineer in the form of a proposal. If approved by the Engineer, he shall certify the proposal and forward it to the Owner with recommendation for approval. If no agreement can be reached between the Contractor and the Engineer, the question shall be submitted to arbitration as provided elsewhere herein. Upon the Owner's approval of the Engineer's recommendation, or receipt of the ruling of the arbitration board, the contract price and time of completion shall be adjusted by the issuance of a change order in accordance with the provisions of the sections entitled, 'Changes in the Work' and 'Extensions of Time.'"

A contractor would be well advised not to enter into agreement for construction without a changed-conditions clause.

26-16. Changes, Extras, and Claims. Most contracts provide means by which the owner can order changes in the work or require extra work. These changes or extras may be priced in any of the following ways:

Unit Prices. At the time of either the bid or the signing of the contract, unit prices are listed by the contractor for various classes of work that may be subject to change. Usually, unit prices are easily administered for such trades as excavation, concrete, masonry, and plastering. The task of the purchaser is to obtain unit prices from subcontractors for various classes of work for the same trades that are in the contract. Although usually the same unit price is agreed for both added and deducted work, occasionally the unit prices for deducted work will be agreed to be 10% less than those for added work.

Cost of Labor and Materials, plus Markup. Another method of computing the value of changes or extra work is by use of actual certified costs, as derived from record keeping as the project proceeds. Rates for wages and fringe benefits must be verified, and the amount of percentage markup must be agreed to either in the contract or before the work is started. Usually, the general contractor is allowed a markup over and above subcontractors' costs and markup, but the general contractor's markup is less than the subcontractor's allowance in such cases.

When work is done on a cost-plus-markup basis, the contractor must maintain an accurate daily tabulation of all field costs. This document should be agreed to and signed by all parties responsible for the record keeping for the change. It will form an agreed-on certification that the work has been performed and of the quantities of labor and material used. A daily work-report certification is shown in Fig. 26-17.

Negotiation by Lump Sum. If the owner desires to have changes or extra work performed and does not want it done on a cost-plus or unit-price basis, owner and contractor may negotiate a lump-sum payment. In this situation, a cost estimate is prepared by the contractor or subcontractor involved, and a breakdown of costs is submitted, together with the estimate total. If the owner accepts the lump sum, he or the architect writes a change order, and the work is performed. Such a change order is shown in Fig. 26-18 as issued to a subcontractor.

Kreisler Borg Florman
CONSTRUCTION COMPANY

97 Montgomery Street, Scarsdale, New York 10583
Telephone (914) SC 5-4600

DESCRIPTION OF WORK relocation site drainage line — NO. 3 PAGE 1 OF 1
LOCATION Apartment House, 170th Street, Jamaica, New York (JOB NAME) — JOB NO.
PERFORMED BY General Contractor (INSERT GENERAL CONTRACTOR OR NAME OF SUBCONTRACTOR) — DATE January 23, 1975

LABOR

NAME OF EMPLOYEE	OCCUPATION	HOURS WORKED	RATE OF PAY	AMOUNT
George M	Laborer	7	$7.50	

MATERIALS, SUPPLIES AND/OR EQUIPMENT

QUANTITY	DESCRIPTION
60 LF	4" Clay Drainage Pipe

COMPLETE DESCRIPTION OF WORK DONE THIS DATE

Relocated site drainage to catch basin B as shown on revised sketch A-301 dated January 10, 1973

THE INFORMATION CONTAINED ABOVE IS CORRECT. ALL WORK SUBJECT TO TERMS OF SUBCONTRACT AND PRIME CONTRACT.

Kreisler Borg Florman CONSTRUCTION COMPANY
BY
TITLE Superintendent

SUBCONTRACTOR (NAME)
BY
TITLE

OWNER'S OR [redacted] REPRESENTATIVE CONSTRUCTION FOR PROGRESS, INC. (COMPANY)
BY
TITLE Engineer

Fig. 26-17. Daily work-report certification by general contractor for changes made or extra work performed.

Extra Work. An order from either the architect or the owner's representative may sometimes result in work that is not part of the original contract documents (Art. 26-4). Usually, there is acknowledgment on the part of the person requesting the work that it is not the contractor's obligation to perform it. This acknowledgment will generally be in the form of an authorization to proceed with the extra work or with a change order to the contractor.

Claims. Sometimes, however, a dispute may arise as to whether or not the work is part of the contract documents and hence the obligation of the contractor. Such a dispute may result in a claim on the part of the contractor. In making a claim, the contractor may request that a change order be issued prior to proceeding with the work. Or the contractor may nevertheless proceed with the work so as not to delay the job, but request that a change order be issued to him. Such a request is made in the form of a change-order proposal (Fig. 26-19). Depending on the size of the claim and the attitude of the owner or architect, the contractor

CHANGE ORDER

Kreisler Borg Florman
CONSTRUCTION COMPANY
97 Montgomery Street, Scarsdale, New York 10583
Telephone (914) SC 5-4600

DATE: March 2, 1972

TO: National Tile & Marble Corporation
300 West 102nd Street
New York, New York 10025

RE: 13-Story Apartment Building
170th Street, Jamaica, New York

WE ARE INCREASING ~~(DECREASING)~~ YOUR SUBCONTRACT AMOUNT BY THE SUM OF $1100.00 FOR

Change tile in lobby and vestibule in accordance with letter from Clarence Lilien and Associates dated January 26, 1972.

Reference your proposal dated February 15, 1972.

ALL WORK MUST COMPLY WITH THE PLANS, SPECIFICATIONS AND ALL OTHER CONDITIONS AND REQUIREMENTS OF THE GENERAL CONTRACT. ALL TERMS AND CONDITIONS OF THE ORIGINAL SUBCONTRACT BETWEEN US SHALL APPLY AND REMAIN IN FULL FORCE.

KINDLY SIGN AND RETURN ONE COPY OF THIS ORDER.

ACCEPTED:
NATIONAL TILE & MARBLE CORPORATION
BY ______ DATE ______
OWNER'S CHANGE ORDER NO. 4
BASED ON CHANGE ORDER PROPOSAL NO. ______

Kreisler Borg Florman
CONSTRUCTION COMPANY
BY ______
TITLE Vice-President

Fig. 26-18. Change order.

may decide whether to continue with the extra work or to press for a decision on the claim, through either arbitration (Art. 26-22), or some other remedy available to him either under the contract or at law.

26-17. Cost Records. Segregation of costs for each job is essential for proper construction management. Only with such records can profits or losses for each job be calculated and predictions made for future work costs. After award of subcontracts, in order for records to be up to date as to any extras or credits that are being claimed by or awarded to the subcontractors for changes (Art. 26-16), a monthly update of all anticipated costs for each subcontract is essential. This can be done by means of an Application for Payment Form (Fig. 26-20) from each subcontractor.

On this form, the approved extras are shown on the second line on the front of the sheet under Amount (Fig. 26-20*a*). Also, all extras claimed by the subcontractor must be listed on the back (Fig. 26-20*b*), as shown on the portion of the form Extras to Date.

Payrolls. From the daily reports received from the field (Art. 26-14), weekly payrolls can be prepared for each job. Labor for each job must be segregated and tabulated in a payroll report (Fig. 26-21). This report also provides statistics on tax withholding to the government, as well as other payroll information. In addition, the report gives the gross wages week by week.

Monthly Cost Report. This report summarizes subcontracts and extras, material purchases, and labor costs encountered to date and expected to be encountered until completion. A monthly report prepared in a manner similar to Fig. 26-22 will yield information to the contractor long before the job has been completed for calculating anticipated profit for the job, and will offer a method of monitoring job progress and costs to ascertain whether the anticipated profit is being maintained.

CHANGE ORDER PROPOSAL NO.

Kreisler Borg Florman
CONSTRUCTION COMPANY
97 Montgomery Street, Scarsdale, New York 10583
Telephone (914) SC 5-4600

TO Bond-Ryder Associates, Inc.
101 Central Park North
New York, New York 10026

DATE: January 15, 1973

RE: Lionel Hampton Houses
UDC #29
New York, New York

WE SUBMIT OUR ESTIMATE IN THE AMOUNT OF $ 325.40 + Mark up FOR PERFORMING THE FOLLOWING extra WORK:

Furnishing and installing labor and material to remove and or eliminate receptacles that are directly over kitchen sinks and install blank cover plates in Buildings "A", "B", and "C".

Attached find back-up information from Meyerbank Electric Co., Inc., dated 12/13/72 and letter of authorization to proceed from Mr. R. Germano, Project Manager, Urban Development Corporation dated December 11, 1972.

This proposal is our request for a Change Order to reimburse us for Change Order No. 56 issued to Meyerbank Electric Co., Inc., a copy of which we enclose.

This work is not part of the completed plans dated September 1, 1972, nor is it part of the Addenda, as agreed to by contractor. We are proceeding with this work so as not to delay the job.

~~BEFORE STARTING ANY WORK ON THIS CHANGE WE ARE AWAITING YOUR DECISION.~~ ~~IF YOU WISH US TO PROCEED WITH THIS WORK~~ PLEASE SIGN ONE COPY OF THIS PROPOSAL INDICATING YOUR APPROVAL. ALL OTHER CONDITIONS OF THE CONTRACT BETWEEN US SHALL REMAIN IN FULL FORCE AND EFFECT. ~~IN ACCORDANCE WITH THE APPLICABLE PROVISIONS OF THE CONTRACT WE REQUEST AN EXTENSION OF OUR CONTRACT COMPLETION DATE BECAUSE OF THIS CHANGE. A PROMPT DECISION IS REQUESTED.~~

APPROVED AS CHANGE ORDER NO.__________ TO THE CONTRACT

BY __________________ DATE:__________

DISAPPROVED; PROCEED IN ACCORDANCE WITH CONTRACT PLANS & SPECIFICATIONS.

BY __________________ DATE:__________

CC: Urban Development Corporation

Kreisler Borg Florman
CONSTRUCTION COMPANY

BY __________________

TITLE Project Manager

Fig. 26-19. Change order proposal.

26-18. Safety. Responsibility for job safety rests initially with the superintendent. Various safety manuals are available giving recommended practice for all conceivable types of construction situations; for example, see "Manual of Accident Prevention in Construction," Associated General Contractors of America, Inc., Washington, D.C.

Because of wide diversity in state safety laws, the Federal government in 1970 passed the Occupational Safety and Health Act (OSHA) (Title 29—Labor Code

Kreisler Borg Florman

CONSTRUCTION COMPANY (INC.)

97 Montgomery Street, Scarsdale, New York 10583

Telephone (914) SC 5-4600

APPLICATION FOR PAYMENT ON ACCOUNT OF CONTRACT

ALL REQUISITIONS MUST BE MAILED AND IN THIS OFFICE ON OR BEFORE THE LAST DAY OF THE MONTH.
BOTH SIDES OF THIS REQUISITION MUST BE COMPLETED AND SIGNED.

SUBCONTRACTOR Modern Pollution Control, Inc. PERIOD FROM Start TO September 30, 1975

FOR 170th St @92rd Ave, Jamaica, New York WORK REQ. No. 1

PROJECT Apartment Building

ITEM	AMOUNT	DO NOT WRITE IN THIS SPACE
AMOUNT OF ORIGINAL CONTRACT	9,825	
APPROVED EXTRAS TO DATE (LIST ON REVERSE SIDE)	250*	
TOTAL CONTRACT AND EXTRAS	10,075	
CREDITS TO DATE (LIST ON REVERSE SIDE)	nil	
NET CONTRACT TO DATE	10,075	
VALUE OF WORK PERFORMED TO DATE	TOTAL	
LESS RESERVE OF ____% AS PER CONTRACT	1,475	
BALANCE	8,600	
LESS PREVIOUS PAYMENTS	nil	
AMOUNT OF THIS REQUISITION		

DO NOT WRITE IN THIS SPACE

JOB ________ DISTRIBUTION ________

CHECKED: QTY. ________ PRICE ________ EXT. ________

AMOUNT PAID ________ DATE ________

CHECK NO. ________ FOLIO NO. ________

ENTER ________ PAY ________

(OVER)

Fig. 26-20. (*a*) Front side of application for payment submitted by subcontractor.

of Federal Regulations, Chapter XVII, Part 1926, U.S. Government Printing Office). Compared with state safety laws of the past, the Federal law had much stricter requirements. For example, in the past, a state agency had to take the contractor to court to penalize him for illegal practices. In contrast, the Occupational Safety and Health Administration can impose fines on the spot for violations, despite the fact that inspectors ask the employers to correct their deficiencies. OSHA enforcement, however, may eventually be taken over by the states if they develop state regulations as strict as those of OSHA.

THE FOLLOWING IS A FULL AND COMPLETE LIST OF ANY AND ALL PERSONS, FIRMS OR ENTITIES OF EVERY NATURE OR KIND AND DESCRIPTION WHO FURNISHED WORK, LABOR (OTHER THAN ON DIRECT PAYROLL OF APPLICANT), SERVICE AND/OR MATERIAL (INCLUDING BUT NOT LIMITED TO INSURANCE PREMIUMS, WATER, GAS, POWER, LIGHT, HEAT, OIL, GASOLINE, TELEPHONE SERVICE OR RENTAL OF EQUIPMENT AS WELL AS UNION WELFARE, PENSION AND OTHER FRINGE BENEFIT PAYMENTS AND FEDERAL, STATE & LOCAL TAXES) IN EXCESS OF $100.00 ARISING OUT OF OR IN CONNECTION WITH THE JOB FOR WHICH THIS PAYMENT IS REQUESTED TOGETHER WITH ANY AND ALL AMOUNTS NOW DUE AND OWING TO THEM AS WELL AS THE DATE THAT THEY LAST PERFORMED ANY WORK, LABOR, SERVICE OR DELIVERED ANY MATERIAL:

NAME	ADDRESS	DATE	AMOUNT DUE
NONE DUE AND OWING			

ALL CLAIMS FOR EXTRAS NOT HEREIN LISTED ARE WAIVED BY THE SUBCONTRACTOR

EXTRAS TO DATE				CREDITS TO DATE			
Order No.	Description	Date	Amount	Order No.	Description	Date	Amount
	Damages on electrical control box		$250.00				

THE FOREGOING REPRESENTATIONS ARE MADE IN ORDER TO INDUCE YOU TO MAKE AN INTERIM OR FINAL PAYMENT TO US, WELL KNOWING THAT YOU ARE RELYING ON THE TRUTH THEREOF

MODERN POLLUTION CONTROL INC	
APPLICANT'S NAME	SIGNATURE
36-50 38th St, L.I.C., New York	President
ADDRESS	TITLE

Fig. 26-20. (*b*) Back side of application for payment.

The most frequently cited violations of OSHA, in order, have been: guardrails, handrails, covers; scaffolding; ladders; gas welding and cutting; grounding and bonding; cranes and derricks; and housekeeping. Also cited have been handling of flammable and combustible liquids; general electrical installation and maintenance; fire protection; trenching; motor vehicles; head protection; and material hoists, personnel hoists, and elevators. Other violations include medical services and first aid; stairways; general requirements for hand and power tools; excavation; sanitation; personal protective equipment; eye and face protection; arc welding and cutting; and safety nets.

An essential aspect of job safety is fire prevention. In aiding in this endeavor, superintendents and project managers often have the advice of insurance companies who perform inspection of the jobs, free of charge. Such inspections by insurance

PAYROLL

Kreisler Borg Florman

CONSTRUCTION COMPANY (INC.)

97 MONTGOMERY STREET, SCARSDALE, N. Y., 10583, SC 5-4600

PROJECT TWIN PARKS NORTHWEST

PERIOD - FROM 1/17/75 TO 1/23/75

SHEET NO. 1 OF 1

	Badge Number	Name of Employee	Position		Hours Worked Daily							Total Hours	Rate	Amount Earned		Gross Wages	W.T. Exemp.	Deductions				Other	Net Payment
					Wed.	Thu.	Fri.	Sat.	Sun.	Mon.	Tue.			Straight	Premium			O.A.B.	W.T.	N.Y.S. W.T.	N.Y.C. W.T.		
1		J.R.		S.T.																			
				O.T.												200.00		11.70	26.30	6.90	2.30		152.80
2		V.L.		S.T.																			
			Lab.	O.T.												262.50		15.36	46.40	12.60	4.00		184.14
3		P.F.		S.T.																			
			Lab.	O.T.												262.50		15.36	30.50	8.70	3.05		204.89
4				S.T.																			
				O.T.																			
5				S.T.																			
				O.T.																			
6				S.T.																			
				O.T.																			
7				S.T.																			
				O.T.																			
8				S.T.																			
				O.T.																			
9				S.T.																			
				O.T.																			
10				S.T.																			
				O.T.																			
11				S.T.																			
				O.T.																			
12				S.T.																			
				O.T.																			
13				S.T.																			
				O.T.																			
14				S.T.																			
				O.T.																			
15				S.T.																			
				O.T.																			
16				S.T.																			
				O.T.																			
17				S.T.																			
				O.T.																			
18				S.T.																			
				O.T.																			
			TOTAL	S.T.																			
				O.T.												725.00		42.42	103.20	28.20	9.35		541.83

CODE:
A—Absent; not paid for
B—Absent; paid for
H—Holiday
V—Vacation with pay
P—Vacation without pay
S—Sick

DEPT. HEAD APPROVAL RB

AMOUNT OF PREVIOUS WEEK'S PAYROLL $755.00

SIGNATURE OF SUPERINTENDENT

Fig. 26-21. Payroll report.

KREISLER BORG FLORMAN CONSTRUCTION CO.

JOB: ________ PERIOD ENDING FEB. 28, 1974
LOCATION: ________ PREPARED BY S. F.
ARCHITECT: ________ CHECKED BY J. F.

Cost Analysis

NO	ITEM	NAME	COST OR COMMITMENT TO DATE	COST TO COMPLETE	ANTICIPATED TOTAL COST	REMARKS
	SUBCONTRACTS & MAJOR PURCHASE ORDERS					
1	EXCAVATION	AAA EXCAVATORS	23000.00	* 500.00	23500.00	*MAINTAINING TEMPORARY ROADS
2	CONCRETE-MATERIAL	BBB READY-MIX	8349.21	4650.79	*13000.00	*MAXIMUM
3	REINFORCING-MATERIAL	CCC STEEL CO.	2100.00	–	2100.00	PURCHASE AMOUNT
4	" -INSTALLATION	DDD ERECTORS	1850.00	–	1850.00	
5	STEEL & MISC. IRON	EEE FABRICATORS	42000.00	* 600.00	42600.00	*CLAIM, CORRECTING BASE PLATES
6	MASONRY	FFF MASON CO.	73000.00	1100.00	74100.00	OPTION CATCH BASINS-1100.*
7	LATH & PLASTER	GGG PLASTERING CO.	9000.00	–	9000.00	
8	HOLLOW METAL	HHH STEEL DOOR	1750.00	–	1750.00	
9	ROOFING	III ROOFING	12000.00	* 200.00	12200.00	*CLAIM-FLASHING EXTRA OPENINGS
10	WINDOWS	JJJ CO.	16000.00	–	16000.00	
11	GLAZING	NOT LET	–	* 8300.00	8300.00	*ANTICIPATED COST
12	ROUGH LUMBER	LLL LUMBER CO.	6914.63	2085.37	* 9000.00	*MAXIMUM PURCHASE AMOUNT
13	MILLWORK	MMM MILLWORK	28500.00	* (260.00)	28240.00	*BACKCHARGE REPAIRING CABINETS
14	HARDWARE	NNN CO.	9100.00	400.00	9500.00	*ALLOWANCE 9500.*
15	ASPHALT TILE	OOO FLOOR CO.	8400.00	* 500.00	8900.00	*OPTION WAXING 500.*
16	PAINTING	NOT LET	–	*23000.00	23000.00	*ANTICIPATED COST
17	PLUMBING & HEATING	QQQ HEATING CORP.	110000.00	–	110000.00	
18	ELECTRICAL	RRR ELECTRICAL CO.	49000.00	–	49000.00	
	TOTAL SUBCONTRACTING & MAJOR P.O.		400963.84	41076.16	442040.00	ALTERNATE STAGE LIGHTING 2200.*
	MISCELLANEOUS MATERIALS & EQUIPMENT					
19	CONCRETE ACCESSORIES, ETC.		743.21	200.00	943.21	
20	NAILS & ROUGH HARDWARE		612.73	400.00	1012.73	
21	CRANE & PUMP RENTAL		1350.00	500.00	1850.00	
	TOTAL MISC. MATERIALS & EQUIPMENT		2705.94	1100.00	3805.94	
	GENERAL CONDITIONS					
22	INSURANCE & WELFARE		8353.20	8920.00	17273.20	APPROXIMATELY 20% OF LABOR
23	SURVEYS		350.00	–	350.00	
24	SHANTIES & SUPPLIES		642.15	500.00	1142.15	
25	TEMPORARY FACILITIES		2315.10	600.00	2915.10	
26	GLASS BREAKAGE		–	200.00	200.00	
27	RUBBISH TRUCKING		160.00	120.00	280.00	
28	TRAVEL & MISC. EXPENSE		629.32	750.00	1379.32	
29	GLASS & ALUMINUM CLEANING		–	1000.00	1000.00	
	TOTAL GENERAL CONDITIONS		12449.77	12090.00	24539.77	
	LABOR					
30	SUPERVISION		6755.00	9000.00	15755.00	
31	HAND EXCAVATION		5787.65	600.00	6387.65	
32	CONCRETE		9930.36	10000.00	19930.36	
33	CARPENTRY		16355.34	22000.00	38355.34	
34	CLEANUP & MISC.		2937.66	3000.00	5937.66	
	TOTAL LABOR		41766.01	44600.00	86366.01	
	SUMMARY					
35	SUBCONTRACTS & MAJOR PURCHASE ORDERS		400963.84	41076.16	442040.00	
36	MISC. MATERIALS & EQUIPMENT		2705.94	1100.00	3805.94	
37	GENERAL CONDITIONS		12449.77	12090.00	24539.77	
38	LABOR		41766.01	44600.00	86366.01	
39	FUTURE COSTS ON APPROVED CHANGES		–	2100.00	2100.00	
40	CONTINGENCIES		–	2000.00	2000.00	
	ANTICIPATED TOTAL COST		457885.56	102966.16	560851.72	
	CONTRACT PRICE				593252.00	
	APPROVED CHANGE ORDERS				6941.16	
	PENDING CHANGES FOR WORK DONE				1159.74	
	ADJUSTED CONTRACT PRICE				601352.90	
	ANTICIPATED TOTAL COST				560851.72	
	ANTICIPATED GROSS PROFIT				40501.18	

Fig. 26-22. Monthly cost report. (*Courtesy Samuel C. Florman.*)

companies often result in reports with advice on fire-prevention procedures. Contractors will benefit from adoption of these recommendations.

To most effectively deal with safety in the contractor's organization, the contractor should assign one man responsibility for safety. He should be familiar with all Federal and state regulations in the contractor's area. This man should also instruct superintendents and foremen in safety requirements and, on his visits to job sites. be constantly alert for violations of safety measures. A file containing all the necessary records relative to government regulations should also be the responsibility of this person. He should obtain a copy of Record-Keeping Requirements under the Occupational Safety and Health Act, Occupational Safety and Health Administration (U.S. Department of Labor, Washington, D.C.). Management should hold frequent conferences with this individual and with the insurance company to review the safety record of the firm and to obtain advice for improving this safety record.

26-19. Community Relations. Community concerns with the results of new construction or disturbances from construction operations materially affect the construction industry. In the past, many communities merely helped shape projects that were being planned for construction in their environs. This aid consisted of recommendations from community advisory boards and localization of planning. More recently, communities have assumed a more vigorous role in regulating construction, including the power of veto or costly delay over many projects.

Some of the areas of community relations that must be dealt with in construction management are discussed in the following:

Employment of Local Labor. Because many construction projects are built in inner city, or core areas, where there is much unemployment, communities may insist on utilization of unemployed local labor. This may be done in accordance with local plans or in the form of an Equal Opportunity Program of nondiscrimination, or by recruitment of local labor for employment on the job site.

Utilization of Local Subcontractors. Various government agencies may require or give preference to Federal, state, or municipal employment of local subcontractors for construction work. In many cases, general contractors may be required to enter into written understandings with a government agency specifying goals to be set for local subcontractor employment on a job. As a result of such actions, poorly capitalized subcontractors have been able to make initial employment gains in fields requiring small capital investment, such as watchmen services and painting. Payrolls for such subcontractors, however, in a number of these trades do present difficulties. In many instances, it may be necessary for the general contractor to make special arrangements for interim payments to these subcontractors, prior to the regular payment date, for work performed.

Among the routes used to bring about employment of subcontractors short of capital on construction jobs are the following:

Awarding of a subcontract and orders to such firms.

Subdividing of work into manageable-size subcontracts.

Encouragement of subcontractors to enter into joint ventures with better-financed subcontractors.

Awarding of sub-subcontracts to subcontractors by better-financed subcontractors who hold a large subcontract.

Awarding of a pilot contract for a small job to a subcontractor, for example, tiling of one or two bathrooms, just to create an opportunity to begin to function.

Recruitment of local community labor and local subcontractors for a project requires maintenance by the contractor of an active program for the purpose. When the job is started, if there is a community group strongly organized and vocal in the area, the leaders of this group should be approached. If a request is made by the community group for employment of a member of the group as a community liaison man or organizer, this request should be given earnest consideration. With a salaried liaison between the contractor's organization and the community, many pitfalls can be avoided.

Up-to-date lists of community subcontractors should be maintained by the contractor's office. These lists should be frequently updated, or they will rapidly become obsolete as these small subcontractors either expand or phase out. The contacts thus made with local firms are important, because an acquaintanceship with local conditions is essential in obtaining and executing contracts and dealing with communities.

Public Interest Groups. These also express their opinions and ask for a voice in planning and construction of proposed projects. They can promote projects they favor or seriously delay or cause to be canceled projects they oppose, by lobbying, court actions, presentation of arguments at public hearings, or influencing local officials in a position to regulate construction.

Environmental Impact Statements. An environmental impact statement is an analysis of the effect that proposed construction will have on the environment of the locality in which the project is to be built. The statement should take into consideration, among other things, the following factors: effect on traffic; potential noise, sound and air pollution; effect on wild life and ecology; effect on population and community growth; racial characteristics; economic factors; and esthetics and harmony with the appearance of the community.

For each project, contractors should ascertain whether an environmental impact statement is required. If such a statement is required, it should be begun early in the construction planning stage, if it is required to be drafted by the contractor. If the impact statement for a project is to be drafted by a government agency, the contractor should ascertain that the agency has drafted the statement and that it has been filed and approved.

26-20. Relations with Public Agencies in Executing Construction Operations. A contractor must deal with numerous public agencies. In some localities, to obtain the necessary permits for start and approval of completed construction, a contractor, for example, may have business with the following agencies: building department, highway department, fire department, police department, city treasurer or controller, sewer and water department, and various government agencies providing the financing, such as Federal Housing Administration, State Division of Housing, or a public-interest corporation formed by the state or municipality for undertaking construction work.

The contractor must be familiar with the organization of the agency and the division of functions, that is, which portion of the agency does design, which does construction-cost approval, and which does inspection. The contractor should also determine whether financing is provided for completed construction, or merely by providing or guaranteeing a mortgage. In addition, the contractor should be knowledgeable on construction-code enforcement; source of payments, whether through a capital construction budget or merely a building mortgage; permits required, methods of obtaining them, and fees; record keeping; and methods of obtaining information from the agency's files.

The contractor should do the following to deal most effectively with public agencies. First, various members of the contractor's staff should concentrate their efforts and become experts on dealing with one or more agencies. For instance, the person in charge of field operations should be the one to deal with the building department and building inspectors. He should make it his business to be familiar with the organizational structure of the building department and to know the inspectors. Others in the organization should be versed in dealing with city treasurers or comptrollers, and still others with state or Federal agencies.

Second, up-to-date files and information should be kept and segregated on agency regulations and procedures. For example, building department codes and regulations should be obtained, and revisions of these should be maintained in the contractor's offices.

Also, it is very important that a personal relationship be established between the contractor's personnel and the members of the agencies with which the personnel deal. The contractor too should visit the agencies and introduce himself. He should explain his problems to agency personnel. Most of the personnel are willing to help when approached on a frank and open basis.

26-21. Labor Relations. Proper labor relations on a construction job involve many facets of a contractor's ability. Such relations often are affected by the type of labor organization involved.

Most craft labor on jobs in large cities and in the industrialized portions of the country is unionized. Most unionized employees on construction are members of American Federation of Labor building crafts unions. Open-shop contractors, however, often are able to perform work on a nonunion basis. In a few cases, local labor unions of an industrialized type perform construction work.

Strikes. Construction labor strikes after expiration of a labor contract can place a contractor in a difficult position. If the contractor is a member of a contractors' association, the association will do the bargaining for him. If not an association member, he has one or two alternatives. He can sign up as an independent contractor on condition that the final terms of the agreement settled at the conclusion of the strike will apply to his agreement retroactively. Or he can continue working around the affected trade and hope that the job will not be delayed or advanced too far from the normal job sequence.

Jurisdictional Disputes. Jurisdictional disputes between two crafts often make a contractor an innocent by-stander. Appeals may be made to the national headquarters of the crafts involved, and machinery exists in such cases for resolution of such disputes in the National Joint Board for the Settlement of Jurisdictional Disputes and the National Appeals Board in Washington, D.C.

Standby Pay. In many union contracts, a requirement is imposed on contractors for standby pay for men who may not actually be engaged in installing materials or performing work because of a requirement that members of the union be assigned to stand by for certain purposes. Examples of standby pay are pay for temporary water (plumbers), for temporary electric standby to maintain temporary power and lighting (electricians), and for steamfitters or electricians to maintain temporary heat. Standby pay is usually considered a normal part of the construction process, if in the union agreement, and must be allowed for by the contractor in his cost estimates.

Prevailing Wages. Contracts being done for city, state, or Federal agencies often require that the labor on a job be paid prevailing wages. These are defined as the wages received by persons normally performing that trade in the locality where the job is being performed. Prevailing wages in cities or localities where there is a strong unionized work force usually are the union wages paid to labor in that area. Also considered part of prevailing wages are all fringe benefits and allowances usually paid in addition to hourly wages. If a contractor enters into an agreement that provides for payment of prevailing wages, he may also be responsible for a subcontractor who fails to pay prevailing wages to his labor.

Labor Recruitment. A contractor's labor recruitment takes many forms. Most importantly, the work force comes from a following of labor that has previously worked with the contractor. These men are summoned by telephone or by word of mouth when they are needed on the job site. In instances in which the contractor's usual labor force must be vastly expanded, he can resort to advertising in newspapers or recruitment by advising local labor-union officials that men are needed.

Men are assigned to tasks on the basis of their trade. In strong union jurisdictions, crossing-over by labor from one trade to another is prohibited when the union labor is organized along craft lines.

In times of labor shortage, when other contractors may be competing for his labor, the contractor may have to resort to overtime work to attract and hold capable men, with the expectation that overtime pay will prevent others from hiring the labor from him. In instances in which a great many jobs, or one tremendous job, will be under construction in a locality with a relatively small labor force, other problems may arise. Labor may have to be brought in from the outside, and to do this the contractor may find it necessary to pay travel time or living allowances for such labor.

Termination. When a project ends, generally all labor is discharged, except the contractor's key personnel, who may be moved to another job. Subcontractors, however, frequently, through judicious timing, are able to employ entire labor groups by moving them from one project to another.

Efficiency. Keeping production efficiency high is generally the chief task of the labor superintendent or foreman. The means used may involve careful scheduling and planning of materials and equipment availability, weeding out of inefficient men, training and instruction of those who may not be totally familiar with the work, and most importantly, skilled supervision.

26-22. Arbitration. One of the favored ways of settling disputes in the construction industry is by arbitration. Parties to a contract, either at the time of entering into the contract or when a dispute arises, agree to submit the facts of the dispute to impartial third parties who will hear a presentation of the claims by both parties and render a decision.

If the parties have agreed to submit a matter to arbitration, the decision of the arbitrators is binding on both parties. The decision can be enforced in any court of law having jurisdiction over the party against whom a claim is made.

Because of the large number of construction arbitration cases filed, the American

Arbitration Association has, in conjunction with representatives from the construction industry, drafted special construction-industry arbitration rules ("Construction Contract Disputes—How They May Be Resolved under the Construction Industry Arbitration Rules," American Arbitration Association, 140 West 51st St., New York, N.Y. 10020). These rules provide information for proceeding with construction arbitration.

Parties may agree beforehand to submit disputes to arbitration by inclusion of the standard arbitration clause in their agreement:

"Any controversy or claim arising out of or relating to this contract, or the breach thereof, shall be settled by arbitration in accordance with the Construction Industry Arbitration Rules of the American Arbitration Association, and judgment upon the award rendered by the Arbitrator(s) may be entered in any Court having jurisdiction thereof."

To the standard arbitration clause, it is sometimes best to add the words "in the city of ________" immediately after "shall be settled by arbitration," to specify where the arbitration hearings should be held.

26-23. Social and Environmental Concerns in Construction. Construction and project managers should seriously consider the social and environmental effects of their construction operations, job safety operations (Art. 26-18), and for proper community relations, use of environmental impact statements and the opinions of public interest groups (Art. 26-19). How the constructor operates and whether or not he is prepared to act on these possible effects depend to a great extent on his awareness and knowledge of how his operations interfere with or become the concern of others. It is well therefore that builders be familiar with operations that have resulted in criticism and restraints, so that they can avoid pitfalls and operate within desirable guidelines.

The Committee on Social and Environmental Concerns of the Construction Division of the American Society of Civil Engineers has defined the three main areas of concern for constructors as follows:

Social. These areas cover:

Land usage, such as the visual aspect of the construction project, including housekeeping and security; avoidance of landscape defacement, such as needless removal of trees; prevention of earth cuts and borrow pits that would deface certain areas for a long time; protection of wildlife, vegetation, and other ecological systems; and visual protection of surrounding residential areas through installation of proper fencing, plantings, etc.

Historical and archaeological, including preservation of historical and archaeological items of an irreplaceable nature.

Crime. The construction process often creates temporary negative impacts on a community, resulting in a crime increase. This can include local crime as well as fraud and bribing of public officials.

Economics, including impact of a project on the economics of a region, such as a rapidly increased demand for labor far in excess of supply, with a negative effect on the wage structure in the area and economic harm to the area after construction has been completed.

Community involvement, including hiring practices and dealing with the leadership in the community, whether they be of different ethnic groups, income levels, or organizational affiliations (Art. 26-19); union hiring practices; and training programs and foreign language programs.

Safety (Art. 26-18).

Physical Media. The effects of construction on land, air, water, and of release of pollutants and toxics are the next broad area of concern. Water is often altered in its purity and temperature, and wildlife often is destroyed on land and water by construction of such projects as dams, power plants, and river and harbor facilities.

Energy Conservation, Vibration, and Noise. Vibration has become of increasing concern through the increasingly frequent use in buildings of light construction materials. These are usually flexible and prone to vibrate. In addition, construction machinery has become larger and more powerful, with the result that vibration of this machinery requires strict control.

Identification of noise-producing construction operations and equipment, and control of building construction noise must also be the concern of contractors. Noise-abatement codes for construction exist in many municipalities. Unless the provisions of these codes are properly understood and enforced, they may result in prohibiting of two- or three-shift construction work and delaying of work that requires overtime. Also, some Federal agencies have promulgated regulations requiring noise readings on projects. In accordance with such readings, local officials may place a construction project in one of the following categories:

Unacceptable: Noise levels exceed 80 dB for 1 hr or more per 24 hr, or 75 dB for 8 hr per 24 hr.

Normally unacceptable (discretionary): Noise exceeds 65 dB for 8 hr per 24 hr, or loud repetitive noises on site.

Normally acceptable (discretionary): Noise does not exceed 65 dB for more than 30 min per 24 hr.

Acceptable: Noise does not exceed 45 dB for more than 30 min per 24 hr.

Equipment may not be permitted on Federal building projects and generally should not be used on other types of projects if it produces a noise level 50 ft away exceeding the limits in Table 26-1.

Table 26-1. Limits on Noise Levels of Construction Equipment

Equipment	*Maximum Noise Level at 50 Ft, dB(A)*
Earthmoving	
Front loader	75
Backhoes	75
Dozers	75
Tractors	75
Scrapers	80
Graders	75
Trucks	75
Pavers	80
Materials handling	
Concrete mixers	75
Concrete pumps	75
Cranes	75
Derricks	75
Stationary	
Pumps	75
Generators	75
Compressors	75
Impact	
Pile drivers	95
Jackhammers	75
Rock drills	80
Pneumatic tools	80
Other	
Saws	75
Vibrators	75

Energy conservation in construction projects is part of the over-all problem of the conservation of energy resources of the nation as a whole. Contractors should be alert to and aware of any ways to bring this about.

To effectuate all of these social and environmental concerns, the U.S. Public Building Service, General Services Administration, has developed a chart, called the Environmental Interaction Model (Table 26-2) to be applied to planning, design, construction, and use of office buildings. It can, however, be extended to other types of building construction. The *wants of mankind,* which are extensions of basic needs, are shown on the left. On the top are listed the *environmental resources* that can be drawn on to satisfy these wants. Quantity ratings or cost ratings can be inserted in the various spaces. Through a comparison of alternative costs, an evaluation can be made of the step that will finally be taken.

26-24. Systems Building. The term systems building is used to define a method of construction in which use is made of integrated structural, mechanical, electrical,

Table 26-2. Environmental Interaction Model

ACTIVITY:

ENVIRONMENTAL RESOURCES

PHYSICAL AND CULTURAL STATES, CONDITIONS, AND INFLUENCES IMMEDIATELY AFFECTED BY THE CONSTRUCTION OF A FEDERAL OFFICE BUILDING

PHYSICAL

BASIC NEEDS	COMPLIMENTARY / EXTENDED HUMAN WANTS		1 AGRICULTURAL LAND	2 RESIDENTIAL LAND	3 PARK LAND	4 INDUSTRIAL LAND	5 WATERSHED	6 FOREST LAND	7 UNIQUE GEOLOGICAL FORMATIONS	8 HISTORICALLY SIGNIFICANT SITES	9 LAKES	10 PONDS	11 BOGS	12 SWAMPS	13 RIVERS	14 STREAMS	15 TREES / WOOD / PAPER	16 GROUND COVER	17 GRAVEL	18 TOPSOIL	19 GASOLINE	20 NATURAL GAS	21 OIL	22 ELECTRICITY	23 COAL	24 DIESEL FUEL	25 NATURAL RESERVOIRS	26 NATURAL BUILDING MATERIALS	27 REGIONAL POPULATION	28 LOCAL POPULATION
FOOD	URBAN LOCATION / RURAL LOCATION	A																												
	ACCESSIBILITY / ISOLATION	B																												
	EATING FACILITIES / SELF PROVISION	C																												
	SPECIALTIES / STAPLES	D																												
	NEW LAND / RECYCLED LAND	E																												
AIR	AIR POLLUTION / FILTRATION	F																												
	AIR CONDITIONING / NATURAL VENTILATION	G																												
WATER	NATURAL DRAINAGE / STORM SEWERS	H																												
	LANDSCAPE / PAVEMENT	I																												
PROTECTION	WATER POLLUTION / SEWAGE TREATMENT	J																												
	INSTITUTIONAL PROTECTION / SELF PROTECTION	K																												
	NOISE / QUIET	L																												
	PUBLIC / PRIVATE	M																												
	OPEN SPACE / COVERED SPACE	N																												
	SUNNY AREAS / SHADED AREAS	O																												
	WINDY AREAS / CALM AREAS	P																												
MOBILITY	FLOWS / CONTAINERS	Q																												
	INDIVIDUAL / MASS	R																												
	FAST / SLOW	S																												
	PROXIMITY / DISTANCE	T																												
	LOCAL / REGIONAL	U																												
	PEDESTRIAN / VEHICLE	V																												
COMMUNICATION	WRITTEN / SPOKEN	W																												
	CONCOURSE / TARGET	X																												
	CERVICE-FILLER / PRIME PURPOSE	Y																												
	MEETING PLACE / REMOTE SYSTEM	Z																												
	IDENTITY / CONFORMITY	AA																												
SELF-ESTEEM	EMPLOYMENT / LEISURE	BB																												
	VISTAS / INTIMATE SPACE	CC																												
	COMMUNITY / PRIVACY	DD																												
	INVOLVEMENT / DETACHMENT	EE																												
	CONTROL / FREEDOM	FF																												
	VARIETY / UNIFORMITY	GG																												
	SOCIAL / ANTI-SOCIAL	HH																												
	OPEN / TERRITORIAL	II																												
	HAPPINESS / FRUSTRATION	JJ																												

CULTURAL

BASIC NEEDS	COMPLIMENTARY / EXTENDED HUMAN WANTS		29 PUBLIC USER	30 OCCUPANT USER	31 MAINTENANCE USER	32 CONSTRUCTION PERSONNEL	33 SERVICE PERSONNEL	34 PLANNING TIME	35 CONSTRUCTION TIME	36 TIME OF USE	37 TRAVEL TIME	38 POLITICAL ATTITUDES	39 RELIGIOUS ATTITUDES	40 ECONOMIC ATTITUDES	41 EDUCATIONAL ATTITUDES	42 EDUCATIONAL FACILITIES	43 PLACES OF WORSHIP	44 LOW-MIDDLE INCOME HOUSING	45 HIGH COST HOUSING	46 NURSING HOMES	47 DAY CARE CENTERS	48 HOSPITALS	49 MEDICAL CLINICS	50 DRUG TREATMENT CENTERS	51 SAFETY	52 CORRECTIONAL FACILITIES	53 POLICE STATIONS	54 FIRE STATIONS	55 LOCAL GOVERNMENT LOCATION	56 COMPREHENSIVE PLANS	57 RETAIL STORES
FOOD	URBAN LOCATION / RURAL LOCATION	A																													
	ACCESSIBILITY / ISOLATION	B																													
	EATING FACILITIES / SELF PROVISION	C																													
	SPECIALTIES / STAPLES	D																													
	NEW LAND / RECYCLED LAND	E																													
AIR	AIR POLLUTION / FILTRATION	F																													
	AIR CONDITIONING / NATURAL VENTILATION	G																													
WATER	NATURAL DRAINAGE / STORM SEWERS	H																													
	LANDSCAPE / PAVEMENT	I																													
PROTECTION	WATER POLLUTION / SEWAGE TREATMENT	J																													
	INSTITUTIONAL PROTECTION / SELF PROTECTION	K																													
	NOISE / QUIET	L																													
	PUBLIC / PRIVATE	M																													
	OPEN SPACE / COVERED SPACE	N																													
	SUNNY AREAS / SHADED AREAS	O																													
	WINDY AREAS / CALM AREAS	P																													
MOBILITY	FLOWS / CONTAINERS	Q																													
	INDIVIDUAL / MASS	R																													
	FAST / SLOW	S																													
	PROXIMITY / DISTANCE	T																													
	LOCAL / REGIONAL	U																													
	PEDESTRIAN / VEHICLE	V																													
COMMUNICATION	WRITTEN / SPOKEN	W																													
	CONCOURSE / TARGET	X																													
	CERVICE-FILLER / PRIME PURPOSE	Y																													
	MEETING PLACE / REMOTE SYSTEM	Z																													
	IDENTITY / CONFORMITY	AA																													
SELF-ESTEEM	EMPLOYMENT / LEISURE	BB																													
	VISTAS / INTIMATE SPACE	CC																													
	COMMUNITY / PRIVACY	DD																													
	INVOLVEMENT / DETACHMENT	EE																													
	CONTROL / FREEDOM	FF																													
	VARIETY / UNIFORMITY	GG																													
	SOCIAL / ANTI-SOCIAL	HH																													
	OPEN / TERRITORIAL	II																													
	HAPPINESS / FRUSTRATION	JJ																													

BASIC NEEDS	COMPLIMENTARY / EXTENDED HUMAN WANTS		58 WHOLESALE OUTLETS	59 SHOPPING CENTERS	60 THEATERS	61 PROFESSIONAL SERVICES	62 BANKS	63 LIBRARIES	64 RECREATIONAL FACILITIES	65 SEWERS	66 RESERVOIRS	67 POWER SUB STATIONS	68 WELLS	69 RECYCLE AND DEPOSIT CENTERS	70 RAPID TRANSIT STATIONS	71 GAS STATIONS	72 TAXICAB AVAILABILITY	73 FOOT PATHS	74 BICYCLE PATHS	75 ROADS	76 FREEWAYS	77 PROPERTY TAX	78 BUSINESS & OCCUPATION TAX	79 CITY TAX	80 SALES TAX	81 ZONING ORDINANCES	82 URBAN RENEWAL PLANS	83 SPECIAL DISTRICT PLANS	84 MODEL CITIES PLANS	85 SEWER & WATER DISTRICT PLANS	86 FIRE DISTRICT PLANS
FOOD	URBAN LOCATION / RURAL LOCATION	A																													
	ACCESSIBILITY / ISOLATION	B																													
	EATING FACILITIES / SELF PROVISION	C																													
	SPECIALTIES / STAPLES	D																													
	NEW LAND / RECYCLED LAND	E																													
AIR	AIR POLLUTION / FILTRATION	F																													
	AIR CONDITIONING / NATURAL VENTILATION	G																													
WATER	NATURAL DRAINAGE / STORM SEWERS	H																													
	LANDSCAPE / PAVEMENT	I																													
PROTECTION	WATER POLLUTION / SEWAGE TREATMENT	J																													
	INSTITUTIONAL PROTECTION / SELF PROTECTION	K																													
	NOISE / QUIET	L																													
	PUBLIC / PRIVATE	M																													
	OPEN SPACE / COVERED SPACE	N																													
	SUNNY AREAS / SHADED AREAS	O																													
	WINDY AREAS / CALM AREAS	P																													
MOBILITY	FLOWS / CONTAINERS	Q																													
	INDIVIDUAL / MASS	R																													
	FAST / SLOW	S																													
	PROXIMITY / DISTANCE	T																													
	LOCAL / REGIONAL	U																													
	PEDESTRIAN / VEHICLE	V																													
COMMUNICATION	WRITTEN / SPOKEN	W																													
	CONCOURSE / TARGET	X																													
	CERVICE-FILLER / PRIME PURPOSE	Y																													
	MEETING PLACE / REMOTE SYSTEM	Z																													
	IDENTITY / CONFORMITY	AA																													
SELF-ESTEEM	EMPLOYMENT / LEISURE	BB																													
	VISTAS / INTIMATE SPACE	CC																													
	COMMUNITY / PRIVACY	DD																													
	INVOLVEMENT / DETACHMENT	EE																													
	CONTROL / FREEDOM	FF																													
	VARIETY / UNIFORMITY	GG																													
	SOCIAL / ANTI-SOCIAL	HH																													
	OPEN / TERRITORIAL	II																													
	HAPPINESS / FRUSTRATION	JJ																													

ENVIRONMENTAL INTERACTION MODEL
"MAN, CONSTRUCTION AND THE ENVIRONMENT"
GENERAL SERVICES ADMINISTRATION

and architectural systems. (See also Art. 1-19). Application of these systems should be controlled by an engineer-construction management firm rather than by use of prevailing contract building-management procedures.

Building systems have usually been utilized in two general areas: school construction and housing production, including that under the impetus of the Federal Housing and Urban Development Department's "Operation Breakthrough" in the early 1970s, and various types of industrialized housing. The ultimate goal is integration of planning, designing, programming, manufacturing, site operation, scheduling, financing, and management into a disciplined method of mechanized production of buildings. It has been difficult, however, to bring all these together into a single entity.

Housing Production. The greatest concentration of effort has occurred in the realm of structural framing, leading to development of mass production methods. These have, in general, been of the following types:

Panel type, consisting of floors and walls that are precast on site or at a factory and stacked in a house-of-cards fashion to form a building.

Volumetric type, consisting of boxes of precast concrete or preassembled steel, aluminum, plastic, or wood frames, or combinations of these, which are erected on the site after being produced in a factory.

Component type, consisting of individual members of precast-concrete beams and columns or prefabricated floor elements, which are brought to the site in volume, or mass produced.

These systems have displayed inherent disadvantages that came about through lack of opportunity to use the entire systems-building process, and because the systems often were not able to attain sufficient volume production to pay for the many fixed costs and start-up costs for the component factories that were built.

The lessons learned from past experiences offer the following guidelines for construction managers:

Early commitment must be made to use of a specific system in the design stage.

The systems builder must control the design.

Shop drawings should be started early and refined during the design.

Preparations of schedules, cost estimate, and bids must involve all the members of the team, that is, the designers, owner, and contractor, because there will be many last-minute proposed changes and details that will have to be challenged and modified.

Construction must follow an industrialized-building sequence, rather than a conventional bar chart or CPM diagram. Unlike the procedure for conventional construction, the method of scheduling and monitoring for industrialized building requires that all trades closely follow the erection sequence and that every trade match the speed of erection. Thus, if the erection speed is eight apartments per day, every trade must automatically fall on the critical path. It is necessary that each of the subcontractors work at the rate of eight apartments per day, otherwise the scheduled date of occupancy will not be met.

School Construction. The second major area in which systems building has been used appreciably is school construction. In this case, 8 or 10 prefabricated subsystems have been brought together for assembly on the site. The technique has been most effective in California school construction.

Types of Systems. In general, there are two kinds of system in use.

A *closed system* comprises one set of major components designed to integrate exclusively with each other. Such a system requires linked bids and coordinated product research and development by the manufacturers. In a closed system, the buyer has no alternative but to purchase the components specified. The California school construction system, for example, is virtually a closed system.

An open system is one in which many different components manufactured by different suppliers are compatible. Where the closed system will accept only a specific item, a specified ceiling lighting component for example, the ceiling lighting components in an open system can be met by three, four, or more components manufactured by different companies.

("Building Systems Information Clearing House Newsletter," Systems Division, School Planning Laboratory, 770 Pampas Lane, Stanford, Cal. 94305.)

Section 27

Specifications

JOSEPH F. EBENHOEH, JR.

Architect, Senior Associate
Chief, Architectural Specifications Division,
Albert Kahn Associates, Inc., Architects and Engineers,
Detroit, Mich.

27-1. Role of Specifications in Construction. Architects or engineers prepare contract documents for a project for the purpose of obtaining competitive bids for construction and to guide the successful contractor during construction. To the extent these documents, such as The Agreement between the Owner and the Contractor and certain aspects of the general conditions, involve legal matters, the role of the architects and engineers is to assist the owner's attorney. The contract documents consist of

The Agreement between the Owner and the Contractor.
General Conditions (and Special Conditions).
Drawings and Specifications.

The specifications are a written description and the drawings a diagrammatic presentation of the construction project. The drawings and specifications are complementary. For example, drawing notes are short, general, and describe a type of construction, its location and quantity required, whereas specifications expand on the characteristics of the materials involved and the workmanship desired in installing them. A note on the drawings may read "Suspended acoustic ceiling." The specifications will completely describe the ceiling and establish such details as whether the acoustic material is prefinished perforated metal with sound-absorbing pads, mineral fiber, or vegetable fiber; the size of the units; whether the metal grid supporting the acoustic units is concealed or exposed, and numerous other conditions.

Specifications are addressed to the prime contractor. They present a written description of the project in an orderly and logical manner. They are organized into divisions and sections representing, in the opinion of the specification writer, the trades that will be involved in construction (Art. 27-7). Proper organization of the specifications will facilitate estimating and aid in preparation of bids.

Workmanship required should be detailed in the specifications. Contractors study specifications to determine the sequence of work, quality of workmanship, and appearance of end product. From this information, they determine costs of the various skills and labor required. If workmanship is not determined properly, unrealistic costs will result.

For example, specifications will describe workmanship for architectural concrete in terms of forms, surface finish, curing methods, and weather protection. These items, representing large costs, will affect the bids substantially. Contractors will not include in their bids the cost of any item that is not specified or install it. If the form material (wood, metal, or plastic) and finished surface of the concrete are not defined in the specifications, contractors will select the form material that will give what they consider a satisfactory finish with the greatest number of reuses. The end product may or may not satisfy the building designer. A clearly written and comprehensive specification is the only means of indicating to contractors what must be done to produce an acceptable building.

Whenever a dispute arises, the contracting parties will scrutinize the specifications for instructions or requirements spelled out in writing before consulting any other part of the construction documents. So specifications must be specific and accurate in describing the requirements of the project.

A good specification will expand or clarify drawing notes, define quality of materials and workmanship, establish the scope of work, and spell out the responsibilities of the prime contractor.

27-2. Types of Specifications. These include performance, descriptive, reference, proprietary, and base-bid specifications.

Performance specifications define the work by the results desired. For example, curtain wall specifications will establish: that drawings will govern the design, character, and arrangement of the various components; materials and finishes; wind loads the assembled units will have to resist; that provisions should be made for expansion and contraction; acceptable minimum air infiltration; tests that must be conducted; acceptable heat-transmission factors; method of operation for ventilators (projected, pivoted, sliding); type of glazing (gaskets, metal bead with dry-seal gaskets, or putty glazing); and that all connections shall be made with concealed fasteners.

This type of specification gives the contractor complete freedom to employ his knowledge and experience to provide the itemized results. He will engineer the end product, design internal reinforcing and bracing, and assemble the various components to comply with the specification.

As another example: The engineer knows that numerous manufacturers are capable of producing fans with various characteristics. So he will write a performance specification that will establish the rotation, discharge, type, size, and capacity of the fan that will meet his requirements. Also, he will require the manufacturer to submit test data certifying the fan's performance. Under the performance specification, the manufacturer is responsible for establishing the physical and chemical properties of the metal used to build the fan and housing and the dimensions of the various parts, to insure that the fan will perform as specified.

The terms of the contract should obligate each prime contractor to execute and deliver to the architect or engineer and owner a written guarantee that all labor and materials furnished and the work performed are in accordance with the requirements of the contract. Should any defects develop from use of inferior materials, equipment, or workmanship during the guarantee period (one year or more from the date of final completion of the contract or full occupancy of the building by the owner, whichever is earlier), the contractor should be required by the contract to put all guaranteed work in satisfactory condition. Also, he should make good all damage resulting from the inferior work and restore any work, material, equipment, or contents disturbed in fulfilling the guarantee.

The prime contractor should get guarantees from fabricators, manufacturers, and subcontractors who furnish products. Therefore, these firms would have to make good any defects, even to the extent of furnishing an entirely new product, to comply with performance defined in the specifications.

Descriptive specifications describe the components of a product and how they are assembled. The specification writer specifies the physical and chemical properties of the materials, size of each member, size and spacing of fastening devices, exact relationship of moving parts, sequence of assembly, and many other requirements. The contractor has the responsibility of constructing the work in accordance with this description. The architect or engineer assumes total responsibility for the function and performance of the product. Usually, architects and engineers do not have the resources, laboratory, or technical staff capable of conducting research on the specified materials or products. Therefore, unless the specification writer is very sure the assembled product will function properly, he should avoid the use of descriptive specifications.

Reference specifications employ standards of recognized authorities to specify quality. Among these authorities are the American Society for Testing and Materials, American National Standards Institute, National Bureau of Standards, Underwriters' Laboratories, Inc., American Association of State Highway Officials, American Institute of Steel Construction, and American Concrete Institute.

Here is an example of a reference specification: Cement shall be portland cement conforming to ASTM C150-71, "Specification for Portland Cement," using Type I or Type II for general concrete construction.

Reputable companies state in their literature that their products conform to specific recognized standards and will furnish independent laboratory reports supporting their claims. The buyer is assured that the products conform to minimum requirements and that he will be able to use them repeatedly and expect the same end result.

Reference specifications generally are used in conjunction with one or more of the other types of specifications.

Proprietary specifications call for materials, equipment, and products by trade name, model number, and manufacturer. This type of specification simplifies the specification writer's task because commercially available products set the standard of quality acceptable to the architect or engineer.

Use of these specifications is dangerous, because manufacturers reserve the right to change their products without notice, and the product that is incorporated in the project may not be what the writer thought he would get.

For example, on one project, a product was specified by model number and manufacturer, in accordance with usual practice. A catalog showed that a door on the product was mounted on a stainless-steel piano hinge. But after the specification was completed, the manufacturer's agent delivered a new catalog. This one showed the door mounted on three hinges. The specification writer asked to see a sample and discovered that the door actually was mounted on three bright metal hinges that distracted from the esthetic value of the product. Yet, this was the standard product identified by model number of the specified manufacturer that, because of the proprietary specification, became the standard of quality for the project. The specification was promptly changed to instruct the manufacturer to provide a continuous stainless-steel piano hinge, which was still available but at extra cost.

A disadvantage of proprietary specifications is that they may permit the use of alternative products that are not equal in every respect. Therefore, the specifier should be familiar with the products and their past performance under similar use and should know whether they have had a history of satisfactory service. He should also take into consideration the reputation of the manufacturers or subcontractors for giving service and their attitude toward making good on defective or inferior work on previous projects.

Under a proprietary specification, the architect or engineer is responsible to the owner for the performance of the material or product specified and for checking the installation to see that it conforms with the specification. The manufacturer of the product specified by model number has the responsibility of providing the performance promised in his literature.

In general, the specification writer has the responsibility of maintaining competition between manufacturers and subcontractors, to keep costs in line. Naming only one supplier may result in a high price. Two or more names should be supplied for each product to insure competition.

Use of "or equal" should be avoided. It is not satisfactory in controlling quality of materials and equipment, though it saves time in preparing the specification, since only one or two products need be investigated and research time on other products is postponed.

Base-bid specifications establish acceptable materials and equipment by naming one or more (three preferred) manufacturers and fabricators. The bidder is required to prepare his proposal with prices submitted from these suppliers. Usually, base-bid specifications will permit the bidder to submit substitutions or alternates for the specified products. When this is done, the bidder should state in his proposal the price to be added to or deducted from the base bid and include the name, type, manufacturer, and descriptive data for the substitutions. Final selection rests with the owner.

Base-bid specifications provide the greatest control of quality of materials and equipment. But there are many pros and cons for the various types of specifications, and there are many variations of them.

27-3. Components of Specifications. A properly arranged specification comprises four major parts—bidding requirements, general conditions (and special conditions), contract forms, and technical specifications. Each major part has a function of grouping all pertinent information for easy reference.

27-4. Bidding Requirements. These inform contractors of all provisions for submission of proposals.

The usual documents prepared for bidders are called advertisement for bids, invitation to bid, instructions to bidders, and the proposal form.

The advertisement for bids generally is prepared by the specification writer for projects financed with public funds. But it may also be used on private projects if the owner elects to solicit bids in this manner. The advertisement for bids is a printed notice in newspapers or other periodicals. It is used to give public notice that proposals will be received for construction and completion of a structure.

The architect or engineer selects newspapers and other periodicals most likely to reach the largest number of contractors and other interested parties. Statute will determine the number of consecutive times the advertisement must appear to satisfy the requirements of public notice.

The advertisement for bids will give:

Name of owner.

Location of project.

Name of the prime contract or contracts.

Time and place for receiving bids.

Brief description of the project.

Time and place the contract documents (which actually are not contract documents officially until an agreement is signed by both parties, but for simplicity they will be called contract documents here) can be obtained.

Amount of deposit for the documents.

Place where interested parties may examine them.

Information regarding bid guarantees.

Information governing performance and payment bonds.

An invitation to bid contains practically the same information as an advertisement for bids. Both are addressed to bidders and signed by the owner or his duly authorized representative.

Invitations to bid generally are prepared for projects financed with private funds, but they may be used for public work when permitted by statute. The invitations are sent directly to qualified contractors selected for experience, qualifications, and financial ability to perform under the terms of the contract. The document usually is prepared in letter form.

Instructions to bidders describe conditions governing submission of proposals that must be established to protect the financial interests of the owner and the bidders. Each has a sizable sum of money invested in the project by the time bids are opened. The conditions usually are in accordance with acceptable industry standards mutually established by such organizations as the American Institute of Architects, Consulting Engineers Council, and Associated General Contractors. Some of these conditions must be in accordance with statutes.

The instructions to bidders will describe in detail all requirements governing:

Preparation and submission of bids.

Receipt and opening of bids.

Withdrawal and rejection of bids.

Interpretation of and addenda to the drawings and specifications.

Bidders' visits to the site to ascertain pertinent local conditions.

In addition, for public works or projects where the owner elects to advertise for bids, the specification writer will have to include other requirements. These will describe the qualification of bidders, submission and return of bidders' guarantee, taxes, and wage rates. After careful review of the project, the specification writer should add other pertinent information. For example, he may wish the bidders to submit a list of subcontractors and material suppliers. If so, he should define the terms of acceptance or rejection of these.

The proposal form is a means of obtaining uniformity in presentation of bidders' proposals. Preparation of this form by the architect or engineer is a widely accepted procedure and recommended in the interest of good practice.

The bidder is requested to complete the proposal form by filling in all blank spaces with the appropriate information. The proposal is addressed to the owner and signed by the bidder. When he signs, the bidder acknowledges having familiarized himself with the general and special conditions, drawings, and specifications; examined the site, and received addenda, written instructions affecting the work to be performed. Further, the bidder agrees to furnish all materials and perform the work for the stated price, to complete the work in the stipulated time, and to execute a contract for the project if the proposal is accepted within the established time.

The proposal form should provide blank space for prices that may be added to or deducted from the sum quoted for the complete project if the owner elects to make the changes described under the heading of "alternates." The general conditions will establish the rules governing changes in the contract. They also will itemize various methods that may be employed to determine the value of changes in the work. Usually, among the items included in the list of allowed costs is the contractor's fee stated as a percentage. Constituting all his charges for supervision, overhead, and profit, this fee varies, depending on whether the work is performed by his own forces or by a subcontractor. The proposal form should have blank spaces for the bidder to fill in the fees to be applied to the estimated cost of labor, materials, and equipment.

Another method used to determine the costs of changes is to compute the value of the work on the basis of unit prices stated in the contract. The usual method of obtaining unit prices is to require the bidder to state them in his proposal. The proposal should describe the work to which the unit price will apply and the unit of measure, and the form should provide blank spaces in which the bidders can place the prices.

Each paragraph containing blank spaces should define the scope of the item being quoted in clear, concise, and simple language.

The proposal form should be as simple and short as possible. It should not repeat general-condition and technical specification items, unless necessary for identification of quoted prices. Other items than those previously mentioned, however, can be included. Some of these are subcontractors' names and prices; subdivision of the bid total into prices for different types of work, or, if the project has more than one building, into prices for each of the buildings; statement on bidders' security; and space for the bidder to itemize substitutions.

The proposal form should have at the end blank spaces where the bidder can insert his full firm name and business address, his official signature, typewritten or printed name of signing party or parties, and his business form—individual, partnership, or corporation.

Proposals that are not signed by the individuals submitting them should be accompanied by a power of attorney establishing the authority of the signer to do so. Proposals made by a partnership should be signed by all the partners or an agent. If an agent signs, the partners should execute a power of attorney. Proposals submitted by a corporation should have the correct corporate name and the signature of the president or other authorized officer written below the corporate name following "By———" and the impression of the corporate seal. If an official other than the president signs, a certified copy of a resolution of the Board of Directors authorizing the official to do so should be attached to the proposal.

27-5. General Conditions and Special Conditions. The general conditions contain provisions that should be included in the contract documents for every project. Primarily, these provisions define the relationship and duties of the owner and the contractor.

The standard form, General Conditions of the Contract for Construction, American Institute of Architects, is general enough to apply to all building projects. This document should be used either by reference or by physical inclusion (binding) in the specifications. Experienced designers and some owners, however, may prepare a set of general conditions of their own. These often will serve better the practice of the designer or owner.

The specification writer should always use the general conditions adopted by his employer and never make revisions unless all parties concerned concur. This principle insures that all members of the office and field staff will be thoroughly familiar with all provisions and that contractors will be familiar with all provisions governing costs, a factor tending to reduce estimating costs and thus to yield better prices to the owner.

The standard AIA general conditions comprise many articles assembled under 14 headings. The author's office has found through long experience that additional articles are necessary to protect the interests of all parties concerned. Certain of these articles, such as use of the site by the contractor, are modified to suit the individual project.

When general conditions prepared by owners or agencies are used on a project, the author's office prepares a document titled special conditions. These modify and supplement the general conditions to the extent the architects and engineers consider necessary to protect the parties concerned.

27-6. Contract Forms. Since all parties executing a contract should be familiar with all provisions of the contract, the contract forms should be included in the specifications to give bidders the opportunity to review them before submitting their proposals. Then, when a bidder is awarded a contract, he will be acquainted with its provisions.

Usual documents comprising the contract forms are the form of agreement, bid bond, labor and material payment bond, performance bond, sworn statements, waiver of lien, guarantee, arbitration agreement, and form of subcontract.

Standard contract forms, such as those prepared by the American Institute of Architects and the American Consulting Engineers Council, can be used with confidence. They have been prepared under the direction of counsel and with the help of consultants thoroughly familiar with the subject.

Generally, the architect or engineer will assist the owner's attorney in the final formulation of the general conditions and contract forms. Government agencies, institutions, and large corporations, however, will prepare contract forms best suited to serve their interests, because the standard forms are too general.

27-7. Technical Specifications. Difficult and time-consuming to prepare, the technical specifications give a written description of the project, lacking only a portrayal of its physical shape and its dimensions. The specifications describe in detail every piece of material, whether concealed or exposed, in the project and every piece of fixed equipment needed for the normal functioning of the project. If they are properly prepared, well organized, comprehensive, and indexed, any type of work, kind of material, or piece of equipment in a project can be easily located.

Usually, the specification writer will receive a set of prints for a project when the drawings are from 50 to 75% completed. The ideal, but not always practical, time to start the specifications is at the time when drawings are 95% or more complete.

The technical specifications cover the major types of work—architectural, civil, structural, mechanical, and electrical. Each of these types is further divided and subdivided in the specifications and given a general title that describes work performed by building tradesmen or technicians, such as plasterers, tile setters, plumbers, and sheet metal workers.

The specification writer studies the drawings, reports of meetings, and correspondence, and then assigns all pertinent work to each trade best qualified to perform it. Though he should make a sincere effort to assign work to the subcontractors for these trades, there is no rule compelling him to do so. The prime contractor has the responsibility to perform all work and furnish all materials to the best of his ability, and to complete the project as rapidly as possible. He, therefore, has the right to select subcontractors or do the work with his own forces. Each specification, either in the general conditions or special conditions, should contain a statement informing the contractor that, regardless of the subdivision of the technical specifications, he shall be responsible for allocation of the work to avoid delays due to conflict with local customs, rules, and union jurisdictional regulations and decisions.

Standard forms for technical specifications can be obtained from the Construction Specifications Institute and from Associated General Contractors. The following format has been developed by the CSI:

CSI Format for Building Specifications*

Bidding Requirements (see Art. 27-4).
Contract Forms (see Art. 27-6).
General Conditions and Supplementary Conditions (see Art. 27-5).
Specifications.

Division 1. *General Requirements*

Includes most requirements that apply to the job as a whole or to several of the technical sections, and especially those requirements sometimes referred to as special conditions.

General conditions and supplementary conditions are not included within the divisions of the specifications.

Division 2. *Site Work*

Includes most subjects dealing with site preparation and development. Site utilities in Divisions 15 and 16 must be coordinated with these divisions.

Division 3. *Concrete*

Includes most items traditionally associated with concrete work. Exceptions: Paving, piles, waterproofing, terrazzo.

Division 4. *Masonry*

Includes most materials traditionally installed by masons. Exceptions: Paving, interior flooring.

Division 5. *Metals*

Includes most structural metals and those metals not falling under the specific provisions of other divisions. Exceptions: Reinforcing steel, curtain walls, roofing, piles, doors, and windows.

Division 6. *Wood and Plastics*

Includes most work traditionally performed by carpenters. Exceptions: Wood fences, concrete formwork, doors, windows, finish hardware.

Division 7. *Thermal and Moisture Protection*

Includes most items normally associated with insulation and preventing the passage of water or water vapor. Exceptions: Paint, waterstops and joints installed in concrete or masonry, and gaskets and sealants.

Division 8. *Doors and Windows*

Includes hardware, doors, windows, and frames; metal and glass curtain walls; transparent and translucent glazing. Exceptions: Glass block and glass mosaics.

Division 9. *Finishes*

Includes interior finishes not traditionally the work of the carpentry trade.

Division 10. *Specialties*

Includes factory-assembled, prefinished items.

Division 11. *Equipment*

Includes most items of specialized equipment.

Division 12. *Furnishings*

Includes most items placed within the finished building.

Division 13. *Special Construction*

Includes on-site construction that consists of items that normally would fall under several other divisions but require the control that can be attained only by including all parts in a single section.

Division 14. *Conveying Systems*

Includes those systems that utilize power to transport people or materials.

Division 15. *Mechanical*

Includes most items that have been traditionally associated with the mechanical trades.

Division 16. *Electrical*

Includes most items that have been traditionally associated with the electrical trades.

The CSI format lists typical subjects for inclusion in each division. For example:

Division 9. *Finishes*

Lath and plaster.
Gypsum drywall.
Tile.
Terrazzo.
Veneer stone.
Acoustical treatment.

Note that the sections comprising a division are each a basic unit of work. The CSI format is based on the principle of placing sections together in related groups.

In preparing technical specifications, the specifier relies heavily on background information developed from years of experience. This information is recorded in specifications written for previous projects, check lists, master specifications, interoffice correspondence, letters replying to contractors' inquiries, and research notes. Inexperienced specification writers should use as a guide CSI specifications accepted by many professionals. These basic documents set the framework for the trade sections, but the initiative of the specifier still is required to complete the technical specifications.

The form of each section should be based on the work to be described. But there is a similarity in structure of sections. Here are examples taken from two separate projects:

Ceramic tile	*Movable metal partitions*
Special note	Special note
Work required	Work required
Work included under other sections	Work included under other sections
Samples	Shop drawings
Manufacturers	Manufacturers
Materials	Types of partitions
Standard specifications	Partitions
Setting methods	Doors and frames
Mortar	Hardware
Tile work—general	Metal base
Application of wall tile	Metal ceilings
Application of floor tile	Hose cabinets
Protection	Glass and glazing
Completion	Finish
Extra tile for maintenance	Protection

The Special-note paragraph will make the general conditions (and special conditions) part of the technical specifications. The Work-required paragraph will clearly describe the scope of work. These paragraphs should lead each section.

Items of work not to be performed by the trade contractor will be described in the Work-included-under-other-sections paragraph.

The paragraph on shop drawings will state that these are to be provided. Since general-condition items should not be repeated in the trade sections, it should refer to the applicable articles in the general and special conditions that describe the provisions governing the submission of the drawings.

The other subheadings are general. They were selected because they represent classes or subdivisions for which requirements usually would be presented in these sections.

When preparing technical specifications, the specifier should place himself in the role of the estimator, general contractor, fabricator, architect's project representative, and subcontractor, so as to provide information required by each in the performance of his work.

Language should be clear and concise. Good specifications contain as few words as necessary to describe the materials and the work.

Use "shall" when specifying the contractor's duties and responsibilities under the contract. Use "will" to specify the owner's or architect's responsibilities. (See also Art. 27-8.)

Do not use "as directed by the architect," ". . . to the satisfaction of the architect," or ". . . approved by the architect." The specification should be comprehensive and adequate in scope to eliminate the necessity of using these phrases. "Approved by the architect" may be used, however, if it is accompanied by a specification that can be used to determine what the architect would consider in his evaluation. "By others" is not clear or definite and, when used, has resulted in many extras.

Never use "any" when "all" is meant.

27-8. Automated Specifications. This refers to a process that manipulates filed and new text and prints the text for a project specification. The process requires master specifications, data processing equipment, and editing techniques.

The master specification is a file of text covering every possible condition that would be needed. This text is stored in the memory bank of the data processing system, ready to be used on command. Use of master specifications gives the continued assurance that comprehensive and accurate specifications are produced.

The project specification is the written description of a specific project.

To be useful, the master specification file must be periodically updated to include supplemental text for new and revised building components and systems and text reflecting changing office experience. The automated specification process must permit updating without affecting text that specification writers intend to be preserved.

Preservation of master text during the manipulation of text required during the production of project specifications is another demand placed on the process.

Data processing equipment performing these operations include the automatic typewriter and the electronic computer.

Automatic Typewriter. This is a machine resembling a standard typewriter, except for a few modifications consistent with its function. The automatic typewriter, manually operated, produces a typed original and simultaneously records on tape the material being typed. Later, when the need arises, the machine can be used to prepare automatically a typed copy of the text that was recorded on tape. Changes can be made by stopping the running of the tape and typing out the change. The machine simultaneously records the revised text on a second tape. The second tape then becomes the file copy of a project specification, or a new master specification.

A specification writer converts the master text into a project specification by reading each line of the printed copy taken off the automatic typewriter and editing as required. Changes are indicated by drawing a line through the text, or using a caret indicating the location of text written in the margins. Thus, the edited copy serves as instructions to be followed by the operator of the automatic typewriter. This system is well suited for small and some medium-size offices.

Computer. The most advanced data processing system employed in the automated specification process, the computer has a tape or disk storage bank, core memory bank, and high-speed printer, all run by complex integrated mechanical, electrical, and electronic systems. The tape or disk storage bank holds the master specifications. The core memory bank holds among other things the program containing the basic instructions for operating the computer to manipulate or edit the text in the tape or disk storage bank. The high-speed printer produces a copy of the text brought up from the storage bank and edited by the program. The computer records the revised text in the form of a project specification which is preserved and later printed on media acceptable for reproduction. At the same time, the master text is preserved and may be used for the production of another project specification.

Computers are available through direct in-house and time-sharing plans.

The direct in-house plan allows for the leasing or purchasing of computer equipment and the purchasing of programs available from numerous sources.

The time-sharing plan is a method of computer usage that allows two or more firms to use a computer system in such a way that each can operate as if the entire system were under the firm's exclusive control. The computer system is usually located at a data center remote from the user's location. The user has in the office a teletype or typewriter terminal that is tied to the remote data center through a telephone circuit. Master and project specifications are prepared, edited, and printed out at the user's terminal. Many offices find this plan economical. Normally, the printing of the specification is done at the console of the teletype, which is a slow printer. But high-speed printers are available to the user at the data center. The user, however, must establish a technique for delivery of the specification printout by United States mail or special messenger.

Programs. Success of the computer system depends on a program and a file of master specifications. A variety of programs, though, is available. The program and the editing techniques adapted are closely related.

The program is the set of provisions or instructions in the core memory unit that directs the computer to operate in a predetermined manner on input data to produce required output data. The simplest of programs for specifications translates instructions into the action of deleting lines or inserting text.

A more advanced program permits selection of large blocks of text related to common subject matter from the master specification. This program directs the computer to write a project specification from a large volume of master specification text. The specification printed as a result of the automatic editing process resembles very closely the final project specification. This project specification is read, and additional editing, if needed, is done by line or word deletions and insertions. Hand editing is kept to a minimum, thus releasing the text for early reproduction.

The advanced program requires development of editing techniques. It also requires that text be assigned in the master specification to common related blocks.

This is done by giving each block a group name (corresponding to a family name) and each subdivision of that group a member name (like the given name of a family member). Also, phrases that change frequently can be assigned a number, and optional phrases can be given an alternate number. The advanced program is able to identify the options and select or delete them from the master text.

Because the goal of the advanced program is to select automatically from the file text applicable to a project, a system of editing techniques to decide which option is to be included must be developed by the office. Thus, when the responsible parties have selected the basic materials and systems to be included in the project, the decision is recorded in terms recognized by the program and master file. One method is to use questionnaires related to basic materials and systems, with questions requiring a simple yes or no answer. The computer is used to translate each yes answer into action, which consists of calling up from the master file the optional text identified with the selected materials and systems and printing copy on the high-speed printer. This project specification can be further edited by adding or deleting text.

When final approval is given to the project specification, the computer uses another program to assign section numbers and page numbers, format text, drop out all the optional names assigned to blocks of text, and omit notes to the specification writer that are customarily recorded in the master file.

27-9. Covers and Binding for Specifications. The cover page, title page, table of contents, and binding also are the responsibility of the specification writer.

The cover page should be made of heavy durable paper or plastic. Usually, the title of the project, location, name of owner or agency, name of architect or engineer, date of issue, and project number are printed on the cover page or title page or both.

The first page after the cover or title page should be the table of contents. It should list all parts, divisions, and sections included in the book.

Binding at the left side with staples, screw bolts, plastic binders, or equivalent fasteners is permitted. The binding method allowing the book to be laid flat is preferred. Some agencies, for legal reasons, require the stapling method.

27-10. Construction Contracts. The specification writer must know, before starting the specifications, what construction contracts will be awarded on any given project. The kind and number of contracts will be based on size of project, usual practice of owner or architect, time allowed to complete the project, necessity of purchasing long-delivery items, and many other considerations.

An owner may award a single contract for construction of a project or several contracts, each covering portions of the work. When several contracts are awarded, each will have a complete set of contract documents. Usual practice is to divide the project into the major classes—civil, architectural, structural, mechanical, and electrical.

Construction contracts may be awarded for the following:

Architectural work.
Structural steel.
Plumbing and heating.
Ventilating and air conditioning.
Fire-protection work.
Electrical work.
Site preparation work.
Paving.
Foundations.
Transportation equipment.
Elevators.
Primary and secondary switchgear.
Precast concrete work.
Track work.
Demolition.
Landscaping.
Compressors.
Kitchen equipment.
Hardware.
Stage equipment.
Laboratory equipment.
Boilers.
Interior decorating.

Individual preference will dictate whether a single contract or separate contracts will be awarded for the major classes of work. Some architects and engineers prefer to receive separate proposals for each class but to award one contract. They inform the contractor submitting a proposal for the architectural work to include in his proposal a fee for managing the work of the contractors selected by the owner. In submitting a proposal, the contractor for the architectural work agrees to sign contracts with the other contractors. Similarly, they, on submitting a proposal, agree to sign with the contractor selected for the architectural work. The owner reviews all bids and accepts them independently.

The total contract amount with separate contracts is the sum total of the individual bids accepted, including a fee for managing the work. The fee may be a lump sum or a percentage of the total of accepted bids.

Reasons frequently given favoring separate contracts for architectural, mechanical, and electrical work include: (1) Elimination of bid shopping for the major subcontracts. (Bid shopping is a practice of some contractors to force subcontractors to submit a lower bid by informing them of lower prices submitted by others.) (2) Owner retains control of selection of major contractors. (3) Some construction may be started while contract documents are being prepared for other work. This results in an early occupancy date. See also Arts. 1-20 and 26-6.

A single contract usually is preferred because it permits the general contractor to obtain bids from and award contracts to subcontractors with whom he knows he can work efficiently. Also, he can manage all the work, including requests for approval of materials and equipment, shop drawings, tests, changes in the work, application of payments, establishing a detailed construction schedule, and expediting the work.

When separate contracts are awarded, it is very important that the specifications define the responsibilities of each contractor. For example, who will employ the excavator when the work requires excavating, backfilling, compacting, and grading for building foundations, underground piping, and underground electrical service? Building foundations usually are specified as part of the architectural work; piping is part of the mechanical, and electrical service obviously is the responsibility of the electrical contractor. The specifications must tell each of the bidders and subsequently the contractors who will employ the excavator to complete the earth work.

Also, the specifications will tell each of the contractors how the work constructed under one separate contract is to be prepared to receive and connect to work performed under other separate contracts. For example, consider the case of a wall with face brick set on a ledge on a concrete foundation wall and anchored to concrete backup, a steel column anchored to the concrete, and dampproofing laid under the brick, down the foundation wall and up the backup specified distances. Assume that contract documents will be prepared for foundations, structural steel, and superstructure work. Then, the foundations specifications will establish that the work includes construction of foundations and concrete backup; column anchor bolts will be furnished by the structural steel contractor and installed by the foundation contractor; face brick, dampproofing, and metal anchors will be furnished and installed by the superstructure contractor; and dovetail slots to receive the metal anchors will be furnished and installed by the foundation contractor. The specifications for structural steel and for superstructure work will each establish the responsibilities of the contractors concerned in accordance with the assignment of work in the foundations specifications.

27-11. Revisions of Contract Documents. Revisions made after issuance of contract documents to the bidders must be in writing.

Revisions of the contract documents made before receipt of bids should be issued to all parties concerned as addenda. Revisions issued after award of contract to the successful bidder should be described in a bulletin, with or without accompanying drawings, and made part of the contract on execution of a change order. The specification writer prepares the written description for addenda and bulletins, sometimes also the change orders, but rarely the field, or work, orders.

Addenda inform bidders of revisions and modifications of work that will affect the quotations. Bulletins require the contractor to state a price for described revisions and modifications, but do not authorize a change in the contract. The contract is changed only when the price quoted is accepted by the owner, and contractor and owner sign the change order. Field orders describe a change in work that will take effect immediately after signing by architect or engineer, owner, and contractor.

27-12. Aids for Specification Writers. In the course of preparing specifications, specification writers have to review all drawings pertaining to the work, determine the various components detailed, then prepare specifications for each of these components, and supplement the drawings by specifying work that is not detailed. Each experienced specification writer has his own system of recording information pertaining to each trade section thus collected. The more experienced the writer, the fewer the notes that will be jotted down.

The specifier is expected to establish order out of the confusion resulting from numerous instructions and complex drawings. His usual procedure for accomplishing this is to prepare an outline. Some suggestions follow:

The writer should purchase a spiral notebook, or staple together several blank pages at the left-hand margin, to serve as a specification notebook or outline. The first page should be a list of divisions and trade sections he has assigned to the project. This list also serves as an index for the notebook.

Place the title of each division and section at the top of a blank sheet. Then, review the drawings and correspondence for the project. Decide, as each bit of information, item of material or equipment is discovered, under what trade the item should be specified. Use a short note to record the item under the selected section. Verbal instructions received also should be recorded. Assignment of materials and work to trade sections should continue until all drawings and correspondence have been reviewed. After that, commence writing the sections.

If the outline is complete and thorough, the trade sections will be comprehensive and can be readily prepared. Following is an example taken from an outline:

Movable Metal Partitions

Report of meeting 5/6, p. 4.
Type—flush, with recessed base, ceiling high
Dwg. Sht. 4—Sliding doors
Dwg. Sht. 18—Metal acoustic ceiling over shop offices
Dwg. Sht. 18—Metal filler above partition—Section A8
Dwg. Sht. 14—6-in.-high metal base, detail 58/14. Is this by MMP supplier?
Specify dry-seal glazing
Fire hose cabinets
Sliding-door hardware by MMP. Other hardware specified under hardware section

The final specification for movable partitions was prepared, a few weeks after the notes were recorded, from a master specification. The specifier had only to delete paragraphs that were not applicable and to modify others.

When personal experience with performance of materials and equipment is lacking, or it is necessary to specify unknown materials and equipment, the specification writer can rely on products conforming with the minimum standards established by recognized authorities.

One such authority is the American Society for Testing and Materials. Its committees have developed over 3,000 standard specifications and methods of test for materials. Each specification writer should have available those that are frequently used in preparing construction specifications.

Many industries produce construction materials complying with the physical and chemical properties specified in ASTM standards. Lack of an ASTM standard for a material should be a warning to a specifier that the material may be new, special (made on order only), or supplied by one manufacturer and more costly than alternatives.

Another important authority is the American National Standards Institute. It promulgates ANSI Standards, which define products, processes, or procedures with reference to one or more of the following: nomenclature, composition, construction, dimensions, tolerances, safety, operating characteristics, performance, quality, rating, certification, testing, and the service for which defined. Like ASTM specifications, these standards are subject to periodic review. They are reaffirmed or revised to meet changing economic conditions and technological progress. Users of ASTM specifications and ANSI Standards should avail themselves of the latest editions.

Federal specifications are the most comprehensive and competition-encouraging specifications available. They establish minimum requirements for various types of work.

The amended Federal Property and Administrative Services Act of 1949 made use of Federal specifications and standards mandatory for all Federal agencies. And often, manufacturers will submit test reports certifying that their product complies with a certain Federal specification. Therefore, specifiers will find it helpful to maintain a file of Federal specifications pertaining to construction. (See Index of Federal Specifications and Standards, and monthly supplements, Superintendent of Documents, U.S. Government Printing Office, Washington, D.C. 20402.)

27-13. Trade Associations and Societies. The following trade associations and societies publish reliable information and standards useful to specification writers:

Architectural Aluminum Manufacturers Association, 410 N. Michigan Ave., Chicago, Ill. 60611.

Architectural Woodwork Institute, 5055 S. Chesterfield Road, Arlington, Va. 22206.

Asphalt Institute, Asphalt Institute Building, College Park, Md. 20740.

Brick Institute of America, 1750 Old Meadow Road, McLean, Va. 22101.

Building Research Institute, 2101 Constitution Ave., N.W., Washington, D.C. 20418.

Concrete Reinforcing Steel Institute, 180 N. LaSalle St., Chicago, Ill. 60601.

Construction Specifications Institute, Suite 300, 1150 17th St., N.W., Washington, D.C. 20036.

Copper Development Association, Inc., 57th Floor, Chrysler Building, 405 Lexington Ave., New York, N.Y. 10017.

Facing Tile Institute, 111 E. Wacker Drive, Chicago, Ill. 60601.

Factory Insurance Association, 85 Woodland Street, Hartford, Conn. 06105.

Factory Mutual System, 1151 Boston-Providence Turnpike, Norwood, Mass. 02062.

Forest Products Research Society, 2801 Marshall Court, Madison, Wis. 53705.

Gypsum Association, 201 N. Wells St., Chicago, Ill. 60606.

Indiana Limestone Institute of America, Stone City National Bank Building, Suite 400, Bedford, Ind. 47421.

Industrial Fasteners Institute, 1505 East Ohio Building, 1717 E. 9th St., Cleveland, Ohio 44114.

International Masonry Institute, Inc., 823 15th St., N.W., Washington, D.C. 20005.

Maple Flooring Manufacturers Association, 424 Washington Ave., Oshkosh, Wis. 54901.

Marble Institute of America, 1984 Chain Bridge Road, McLean, Va. 22101.

Metal Lath Association, W. Federal St., Niles, Ohio 44446.

National Association of Architectural Metal Manufacturers, 1033 South Blvd., Oak Park, Ill. 60302.

National Builders' Hardware Association, 1290 Avenue of the Americas, New York, N.Y. 10019.

National Building Granite Quarries Association, N. State St., Box 44, Concord, N.H. 03301.

National Bureau for Lathing and Plastering, 938 K St., N.W., Washington, D.C. 20001.

National Concrete Masonry Association, 1800 N. Kent, Arlington, Va. 22209.

National Electrical Manufacturers Association, 155 E. 44th St., New York, N.Y. 10017.

National Fire Protection Association, 60 Batterymarch St., Boston, Mass. 02110.

National Forest Products Association, 1619 Massachusetts Avenue, N.W., Washington, D.C. 20036.

National Terrazzo and Mosaic Association, Inc., 716 Church St., Alexandria, Va. 22314.

National Woodwork Manufacturers Association, Inc., 400 W. Madison Ave., Chicago, Ill. 60606.

Perlite Institute, Inc., 45 W. 45th Street, New York, N.Y. 10036.

Plumbing Fixtures Manufacturers Association, 1145 19th Street, N.W., Washington, D.C. 20036.

Porcelain Enamel Institute, Inc., 1900 L Street, N.W., Washington, D.C. 20036.

Portland Cement Association, Old Orchard Road, Skokie, Ill. 60076.

Sheet Metal and Air Conditioning Contractors National Association, Inc., 1611 N. Kent Street, Arlington, Va. 22209.

Steel Deck Institute, Box 270, Westchester, Ill. 60153.

Steel Door Institute, 2130 Keith Building, Cleveland, Ohio 44115.

Steel Joist Institute, 2001 Jefferson Davis Highway, Arlington, Va. 22202.

Steel Structures Painting Council, 4400 Fifth Ave., Pittsburgh, Pa. 15213.

Superintendent of Documents, U.S. Government Printing Office, Washington, D.C. 20402.

Sweet's Division, McGraw-Hill Information Systems Company, 1221 Avenue of the Americas, New York, N.Y. 10020.

Tile Council of America, Inc., P.O. Box 326, Princeton, N.J. 08540.

Underwriters' Laboratories, Inc., 207 E. Ohio St., Chicago, Ill. 60611.

Vermiculite Institute, 52 Executive Park South, Atlanta, Ga. 30329.

Western Wood Products Association, 1500 Yeon Building, Portland, Ore. 97204.

Wire Reinforcement Institute, 5034 Wisconsin Ave., N.W., Washington, D.C. 20016.

Zinc Institute, Inc., 292 Madison Ave., New York, N.Y. 10017.

Section 28

Insurance and Bonds

C. S. COOPER

Retired, Former Resident Vice President,
Long Island Branch,
Fireman's Fund American Insurance Companies,
Garden City, N.Y.

Insurance policies are contracts under which an insurance company agrees to pay the insured, or a third party on behalf of the insured, should certain contingencies arise.

The importance of insurance protection cannot be overemphasized. No business is immune to loss resulting from ever-present risks. It is imperative, therefore, that a sound insurance program be designed and that it be kept up to date.

Because few businesses can afford the services of a full-time insurance executive, it is important that a competent agent or broker be selected to: (1) prepare a program that will provide complete coverage against the hazards peculiar to the construction business, as well as against the more common perils; (2) secure insurance contracts from qualified insurance companies; (3) advise about limits of protection; and (4) maintain records necessary to make continuity of protection certain. While a responsible executive of the business should interest himself in the subject of insurance, much of the detail can be eliminated by utilizing the services of a competent agent or broker.

There are three kinds of insurance carriers—stock, mutual, and reciprocal. Nevertheless, policy forms are rather well standardized. Rates are established by the rating organizations that serve the member companies of different groups of carriers, but generally, basic rates are very nearly the same.

It is necessary, of course, that the responsibility for providing protection be placed upon an insurance company whose financial strength is beyond doubt.

Another important point for the buyer of insurance is the service that the company selected may be in a position to render. Frequently, construction operations are conducted at a considerable distance from city facilities. It is necessary that the company charged with the responsibility of protecting the construction operations be in a position to render "on-the-job" service from both a claim and an engineering standpoint.

The interests of contractors and building owners are very closely allied. Particular attention should be given to the definition of these respective interests in all insurance policies. Where the insurable interest lies may depend upon the terms of the contract. Competent advice is frequently needed in order that all policies protect all interests as required.

The forms of protection purchased and the adequacy of limits are of great importance to a prime or general contractor. It is also of great importance to him to see that the insurance carried by subcontractors is written at adequate limits and is broad enough to protect against conditions that might arise as a result of their acts. Also, the policies should include the interests of the prime or general contractor in so far as such interests should be protected.

This section merely outlines those forms of insurance that may be considered fundamental. It includes brief, but not complete, descriptions of coverages without which a contractor should not operate. It is not intended to take the place of the advice of experienced insurance personnel.

28-1. Fire Insurance. Fire insurance policies are now well standardized. They insure buildings, contents, and materials on job sites against direct loss or damage by fire or lightning. They also include destruction that may be ordered by civil authorities to prevent advance of fire from neighboring property. Under such a policy, the fire insurance company agrees to pay for the direct loss or damage caused by fire or lightning and also to pay for removal of property from premises that may be damaged by fire.

Careful attention should be given to selection of the amount of insurance to be applied to property exposed to loss. In addition, the cost of debris removal must be taken into consideration if the property could be subject to total loss.

It is important to bear in mind that under no circumstances will the amount paid ever exceed the amount stated in the policy. Insurance on dwellings, however, includes extension of the policy to provide that 10% of the amount of insurance on the dwelling shall be additional insurance applicable to both the rental value of the dwelling and private structures appurtenant to it when these structures are on the same premises. If the fire insurance policy has a co-insurance clause, the problem of valuation and adequate amount of insurance becomes even more important.

A form of fire insurance particularly applicable in the construction industry is that known as Builders Risk Insurance. The purpose of this form is to insure an owner or contractor, as their interests may appear, against loss by fire while buildings are under construction. Such buildings may be insured under the Builders Risk form by the following methods:

1. The reporting form, under which values are reported monthly. Reports must be made regularly and accurately. If so, the form automatically covers increases in value.

2. The completed value form under which insurance is written for the actual value of the building when it is completed. This is written at a reduced rate because it is recognized that the full amount of insurance is not at risk during the entire term of the policy. No reports are necessary in connection with this form.

3. Automatic Builders Risk Insurance, which insures the contractor's interest automatically in new construction, pending issuance of separate policies for a period not exceeding 30 days. This form generally is used for contractors who are engaged in construction at a number of different locations.

There are certain hazards which, though not quite so common as fire and lightning, are nevertheless real. These should be insured by endorsement. A few of the available endorsements are:

1. Extended coverage endorsement, which insures the property for the same amount as the basic fire policy against loss or damage caused by windstorm, hail, explosion, riot, civil commotion, aircraft, vehicles, and smoke.

2. Vandalism and malicious mischief endorsement, which extends the protection

of the policy to include loss caused by vandalism or malicious mischief. There is a special extended coverage form that provides coverage on an all-risk basis and includes the peril of collapse.

Other forms that extend and make fire insurance protection more complete also are available. Special conditions and specific exposures should be studied to determine the advisability of the purchase of certain of these additional protections.

Note that completed-value and reporting forms treat foundations of a building in the course of construction as a part of the Builders Risk Value for insurance purposes. Builders Machinery and Equipment must be specifically insured as a separate item if coverage is desired under the policy.

28-2. Insurance Necessitated by Contractor's Equipment. The Inland Marine insurance market is the place to look for many coverages needed by contractors. From the insurance point of view, each contractor's problem is considered separately. The type of operation, the nature of equipment, the area in which the contractor works, and other pertinent factors are all points considered by an Inland Marine underwriter in arriving at a final form and rate.

Obvious contractors' equipment—mechanical shovels, hoists, bulldozers, ditchers, and all other mobile equipment not designed for highway use—is the primary subject of the Contractors Equipment Floater. Such protection is necessary because of the size of the investment in such equipment and of the multiplicity of perils to which the equipment is exposed.

Some companies will write the Contractors Equipment Floater Policy only on a named-perils basis, which ordinarily includes fire, collision, or overturning of a transporting conveyance, and sometimes theft of an entire piece of equipment. Other companies will write certain kinds of contractors' equipment on the so-called all-risk basis. Certain perils, such as collision during use, are subject to a deductible fixed amount. The rate for this broader insurance generally is higher than that for the named-perils form. In the all-risk form, the customary exclusions, such as wear and tear, the electrical exemption clause, strikes, riots, and other similar exclusions, are present. Because of increased exposure to nuclear hazards, all Inland Marine policies that insure against fire must carry a nuclear exclusion clause that provides that the company shall not be liable for loss by nuclear reaction or radiation or radioactive contamination, whether controlled or uncontrolled and whether such loss is direct or indirect, approximate or remote.

In addition to the equipment, there is a need to provide protection for the materials and supplies en route to or from the site. If these materials and supplies are transported at the risk of a contractor and are in the custody of a common carrier, a Transportation Floater should be obtained. If, on the other hand, these materials and supplies are moved on the contractor's own trucks, a "Motor Truck Cargo-owners Form" should be obtained. The premium for the transportation form is usually based on the value of shipments coming under the protection of the policy. The coverage is usually on an all-risk basis. The Motor Truck Cargo-owners Form is generally on a named-perils basis at a flat rate applied against the limit of liability required by the insured's needs.

Occasionally, a contractor may be responsible for machinery, tanks, and other property of that nature until such time as they are completely installed, tested, and accepted. Exposures of this kind are usually covered under an "Installation Floater," which would provide insurance to the site as well.

The Installation Floater Form is generally on a named-perils basis, including perils of loading and unloading, at a rate for the exposure deemed adequate by the underwriter. Large contractors should have such insurance written on a monthly reporting form to reflect increasing values as installation progresses. Small contractors generally can provide for the coverage under a stated amount on an annual basis subject to coinsurance.

There is also a form of policy known as a "Riggers Floater" that is designed for contractors doing that type of work. This policy is usually a named-perils form at rates based on the nature of the rigging operation.

Neither the forms nor rates in any of these classes are standard among the companies writing them, although in general they are all similar.

Some contractors building bridges might be required to take a Bridge Builders

Risk Form. This is an exception to most of the insurance provided to contractors, in that it is required to be rated in accordance with forms and rates filed in most of the states and administered by a licensed rating bureau.

28.3 Motor-vehicle Insurance. Loss and damage caused by or to motor vehicles should be separately insured under specific policies designed to cover hazards resulting from the existence and operation of such vehicles. Bodily injury or property damage sustained by the public as a result of the operation of contractor's motor vehicles or other self-propelled equipment is insured under a standard policy of insurance. To secure complete protection, a Comprehensive Automobile policy should be obtained to provide protection, in addition to the preceding, for hired cars, employers nonownership, and any newly acquired motor vehicles or self-propelled equipment during the term of the policy. Damage to owned vehicles can be added to the same policy on an automatic basis to provide Comprehensive Coverage and Collision Insurance.

It is important that all vehicles be covered, and it is also extremely important that high limits be carried.

Protection furnished by automobile liability and property damage insurance serves in two ways: (1) the insurance company agrees to pay any sum for bodily injury and property damage for which the insured is legally liable; (2) the policy agrees to defend the insured. It is important, therefore, that the limits be adequate to guarantee that the insured obtain full advantage of the services available. For example, suppose $25,000 and $50,000 limits are carried, and action is brought against the contractor in the amount of $100,000. It may be necessary that he employ an attorney to safeguard his interests for the amount of the action that exceeds the limits of the policy.

It is presumed that no contractor would operate any type of motor vehicle without insurance. It is important, therefore, that automatic coverage be provided to include all motor vehicles owned or acquired. Too great emphasis cannot be placed on adequacy of limits since the cost for highest available limits is reasonable.

Damage to owned motor vehicles may be insured under a fire, theft, and collision policy. Comprehensive motor-vehicle protection covers physical damage sustained by motor vehicles because of fires, theft, and other perils, including glass breakage. Collision insurance insures against loss from collision or upset. While the latter is available on a full-coverage basis, it is generally written on a deductible basis (loss less a fixed sum).

To cover a contractor for liability arising from the use by employees of their own automobiles while on his business, nonownership or contingent liability coverage is necessary. This may be included in the policy by endorsement.

Frequently, contractors have occasion to hire trucks or other vehicles. Liability and property damage insurance to cover the contractor's liability when using hired vehicles should be included at the time automobile insurance is arranged.

While few contractors have occasion to operate horse-drawn vehicles, the occasional necessary use should be covered under a team's liability policy. It is usually provided under a general liability policy covering the contractor's operations (Art. 28-4).

Maintenance and use of boilers and other types of pressure vessels and machinery require the protection provided by boiler and machinery insurance. These policies cover loss resulting from accidents to boilers or machinery, and in addition, cover contractor's liability for damage to the property of others. Policies may also include liability arising from bodily injuries sustained by persons other than employees. This may be of importance because of the exposure that many contractors have as a result of the interest of the public in construction work.

The service rendered by boiler and machinery companies is of great importance and cannot be overemphasized. Nearly all insurance companies that write this form have staffs of competent and experienced inspectors whose job it is to see that boilers, pressure vessels, and other machinery are adequately maintained.

28-4. Liability Policies Covering Contractor's Operations. Anyone who suffers bodily injury or whose property is damaged as a result of the negligence of another person can recover from that person, if the latter is legally liable. Every business should protect itself against claims and suits that may be brought against it because of bodily injuries or property damage suffered by third parties.

Maintenance of an office or yard, as well as the conduct of a construction job, presents exposures to the public. There may be no negligence, and consequently no legal liability on the part of the contractor, but should claim be brought or suit instituted for an injury, the contractor will require trained personnel to investigate the claim and negotiate a settlement or defend a lawsuit if the claim goes to court.

Public-liability insurance is expressly designed to serve contractors by providing insurance that will pay for bodily injuries and property damage suffered by third parties if the contractor is legally liable, but the policy will also serve by defending the interests of the contractor in court. Sometimes the problem in court is one of amount of damage; but frequently, it is one in which the contractor being sued is not legally liable for the injuries or damage. It is fundamental, therefore, that a policy be obtained at substantial limits for both bodily injuries and property damage, that the policy cover all existing exposures and also provide for protection against exposures that may not exist or be contemplated on the inception date of the policy. The scope of operations conducted by most contractors is such that it is frequently difficult to visualize all the hazards that may exist or come about simply by being in the construction business.

The changes that have taken place in public-liability protection have been very rapid and in the best interests of the insurance-buying public. While at one time it was necessary to obtain several policies to cover the numerous public-liability hazards, it is possible now to obtain a comprehensive policy that will include the legal liability resulting from known operations and hazards, as well as those from operations and hazards not contemplated at the time the policy is written. The comprehensive general-liability policy that covers all liability of the insured, except that resulting from the use of automobiles, is a standard form available in all states at rates regulated by law.

The comprehensive general-liability policy protects the contractor under one insuring clause and with one limit against claims that formerly had to be covered specifically under a schedule liability policy or by a number of different policies and endorsements. Blanket coverage is provided by this policy at a premium based upon actual exposures disclosed by an audit at the end of the policy term. Because of the value of this policy, every contractor should make provisions for maintaining accurate records of payrolls, value of sublet work, dollar amount of sales, and other factors that will be important at the time an audit is made.

Under the bodily injury liability clause of this policy the insurance company agrees to pay on behalf of the insured "all sums that the insured shall become legally obligated to pay as damages because of bodily injury, sickness or disease, including death at any time resulting therefrom, sustained by any person and caused by an occurrence." This is a very broad insuring clause. It obviously includes the entire business operations of the insured.

Property-damage liability is an optional coverage. If it is written, and no contractor should be without property-damage insurance, the coverage will apply to the entire risk with one or two exceptions.

There are certain exclusions in the policy that should be noted. The policy does not include:

1. Ownership, maintenance, or use (including loading or unloading) of water craft away from premises owned, rented, or controlled by the insured; automobiles while away from the premises or the ways immediately adjoining; and aircraft under any condition. However, this exclusion does not apply to operations performed by independent contractors or to liability assumed by any contract covered by the policy.

2. Bodily injury sustained by employees while engaged in the employment of the insured.

3. Liability for damage to property occupied, owned, or rented to the insured or in his care, custody, or control.

Of particular importance to contractors is the provision of the comprehensive policy pertaining to contractual liability. The policy automatically provides coverage for the following written contracts: lease of premises; easement agreement, except in connection with construction or demolition operations on or adjacent to a railroad;

undertaking to indemnify a municipality; side-track agreement; or elevator maintenance agreement. A premium is charged for such agreements as may be disclosed by audit.

There is no protection for the liability assumed in some very common types of agreements that include service, delivery, and work contracts. Many of these contracts are signed without full realization of the liability assumed. Each such agreement should be submitted to the insurance company at the time the policy is written in order that a premium charge may be computed and the agreement covered under the policy.

Another very important coverage, from the standpoint of contractors, is provided by products-liability coverage. The comprehensive general-liability policy includes complete and automatic products-liability insurance, including completed-operations protection. The one exception to this complete coverage is that the policy does not include liability for damage to the work or to the goods themselves, such as the obligation of the contractor to repair or replace if there are defects. While the policy provides coverage, it is in fact an optional protection, which may be deleted. However, it should be stressed that, because of the completed-operations protection provided under products-liability insurance, every contractor should take advantage of this coverage.

Most building contractors use elevators or hoists during construction. The comprehensive policy automatically covers elevators, hoists, and other such hazards. Escalators may be covered for an additional premium.

The breadth of public-liability protection available, the numerous hazards to which a contractor may be exposed, both known and unknown, and the necessity for having complete coverage at all times further emphasize the need for the advice of trained insurance representatives.

28-5. Workmen's Compensation. Every state requires an employer to secure a policy of workmen's compensation to provide for an injured employee the benefits of the workmen's compensation law of that state. It is not necessary to emphasize the need for this insurance. It is, however, necessary to point out that an insurance company that has had extensive experience in the workmen's compensation field is best suited to meet the requirements of most contractors.

The problem of acquiring and keeping labor may be troublesome. It is in the employer's, as well as the employee's, best interests to see that the company entrusted to provide workmen's compensation insurance is equipped to provide loss-prevention service, prompt first aid, and to settle compensation claims fairly and speedily.

Some states impose on contractors liability for injury to subcontractors and their employees unless insurance is specifically provided for subcontractors and their employees. Check the law of the state in which construction is to be performed.

The law of the state in which the work is done is the one that will control.

Subcontractors' insurance should be carefully examined and made complete where it is deficient. Certificates of insurance should be required. Because many complex situations arise, it is important that the advice of a qualified agent or broker be obtained.

28-6. Construction Contract Bonds. The United States government, state, county, and municipal governments generally obtain competitive bids on all construction. Awards are made to the lowest responsible bidder, who is required to furnish bond provided by qualified corporate surety. Also, more and more private owners are requiring bonds of contractors to whom they award construction contracts.

Generally, bidders are required to post a certified check or furnish a bid bond. A bid bond assures that, if a contract is awarded to the contractor, he will, within a specified time, sign the contract and furnish bond for its performance. If the contractor fails to furnish the performance bond, the measure of damage is the smaller of the following: the penalty of the bid bond or the amount by which the bid of the lowest bidder found to be responsible and to whom the contract is awarded exceeds the bid of the principal.

It should be noted that most surety companies follow the practice of authorizing a bid bond only after the performance bond on a particular contract has been underwritten and approved. For this reason, contractors are well advised against depositing a certified check with a bid unless there are assurances that the performance bond on that particular contract has been underwritten and approved.

There is no standard form of construction contract bond. The Federal government and each state, county, or municipal government has its own form. Private owners generally use the bond form recommended and copyrighted by the American Institute of Architects. Surety companies have developed a very broad form of bond, which is available to owners of private construction. Whatever form is used, the surety generally has a twofold obligation:

1. To indemnify the owner against loss resulting from the failure of the contractor to complete the work in accordance with the contract.

2. To guarantee payment of all bills incurred by the contractor for labor and materials.

Sometimes two bonds are furnished, one for the protection of the owner, and another to protect exclusively those who perform labor or furnish materials. If one bond is furnished, the owner has prior rights.

Sureties underwrite construction contract bonds carefully. They are interested in determining whether a contractor has the capital to meet all financial obligations, the equipment to handle the physical aspects of the particular undertaking, and the construction experience to fulfill the terms of the contract.

It is essential that contractors understand all the information that a surety will require and that is necessary to the underwriting of a contract bond. It is also important that a contractor take advantage of the services of a competent agent or broker who has close affiliations with a surety company that has the capacity to meet all the contractor's needs. Handling of construction contract bonds is rather highly specialized; consequently, agents and brokers must be experienced in the insurance requirements of contractors and they can be of valuable assistance in many ways to principal and obligee.

The importance of maintaining books and records is pointed out in Art. 28-4. Here again, the importance of complete records cannot be stressed too highly. Accurate information about the amount of work on hand, value of equipment, value of materials, and records of past performances will prove necessary.

To be more specific, there are outlined below several items of information that will be required by the surety:

1. A complete balanced financial statement with schedules of the principal items. This is a condition precedent to the approval of any contract bond. Sureties have forms on which contractors may furnish financial information. However, when such forms are used, they should be completed by the individual responsible for the financial operations of the company and the data taken directly from the company's books. It is preferable to have the financial statement prepared and certified by a public accountant. Contractors will find that they will save time and money if the services of qualified accountants are obtained.

2. A report regarding the contractor's organization. The surety is interested in knowing the length of time the contractor has been in business, whether he operates as an individual, a partnership, or a corporation, and certain specific details depending upon the form of organization.

3. A report of the technical qualifications and experience of the individuals who will be in charge of work to be performed.

4. A report of the type of work undertaken in the past, together with information regarding jobs successfully completed.

5. An inventory of equipment, noting value and age of each piece and any existing encumbrance. An inventory of materials will also be helpful.

Contractors should be fully acquainted with the penalties of liquidated damages that attach in the event any difficulties arise to prevent completion of a contract. Sometimes things occur over which the contractor has no control—sometimes mistakes are made that prevent completion—whatever the contingency it is far better to have qualified with a corporate surety bond than to have undertaken a job without the protection of a bond.

No contractor should feel that the requirement of a surety bond is a reflection upon him. Rather, a contractor should feel that a sound financial condition, a successful past, and the ability to carry out a contract are something of which to be proud. Furthermore, the cost factor is an item included in the gross contract price. To be qualified by a corporate surety is a stamp of approval.

28-7. Money and Securities Protection. Nearly every contractor has cash, securities, and a checking account that are vulnerable to attack by dishonest people, both on and off the payroll. The same hazards present in every business are present also in the construction business. And no contractor is immune to dishonesty, robbery of payroll, burglary of materials, or forgery of his signature on a check.

Employee dishonesty may be covered on a blanket basis, either under a Primary Commercial or Blanket Position Form of Bond. The fact that contractors generally entrust the maintenance of payroll records and the payment of employees to subordinates demonstrates the necessity for blanket dishonesty protection.

While many contractors maintain an organization on a year-round basis, some may not. For those contractors who do have a permanent staff, the bond may be written on a 3-year basis at a saving. It is important that adequate limits be purchased. A blanket bond in an amount equal to 5% of the gross sales is desirable.

General funds, securities, and payroll funds should be covered on the broadest basis now available that protects against burglary, robbery, mysterious disappearance, and destruction, on and away from any premises. The general funds may be covered in an amount sufficient to protect against the maximum single exposure. Payroll funds may be insured specifically and in a different amount.

Contractors who maintain inventories of materials should insure them against burglary and theft. It should be noted, however, that insurance companies are not willing to insure against loss by burglary or theft unless materials are under adequate protection. Insurance is not available to cover property on open sites or in yards, but only while within buildings that are completely secured when not open for business.

Forgery and alteration of checks are very common crimes. Forgeries are cleverly committed by "gentlemen" who have devoted their lives to acquiring the ability to duplicate the signature of another. Every business that maintains a checking account, however small the balance may be, should insure against loss caused by the forgery of the maker's name or by the forgery of an endorsement of checks issued.

The modern policy to cover all these hazards is the comprehensive dishonesty, disappearance, and destruction policy. Its several insuring agreements include employee-dishonesty coverage, broad-form money and securities protection, on and off the premises, and forgery. Other coverages to provide burglary and theft protection for merchandise and materials may be added by endorsement. Optional coverages available are numerous, and the contract may be designed specifically for all a contractor's exposures.

28-8. Employee Group Benefits. Interest in group benefits for employees has grown appreciably in recent years. There are many forms of group insurance. Included in the following is a brief description of those forms of "group insurance benefits" that are of greatest current interest.

Group Life Insurance. A form of term life insurance written to cover in a specified amount a group of employees of a single employer. Frequently, the amount of insurance is related to an employee's earnings and increases as the earnings increase. However, the amount of insurance may be a flat sum. Usually, a specified number of employees must participate.

Group Disability Insurance. If an employee is away from work as a result of a nonoccupational disability, this insurance provides a continuing income during his period of absence. The following states have adopted compulsory disability laws: Rhode Island, California, New Jersey, and New York. Other states are studying this insurance.

The amount of benefit is usually a percentage of earnings. Compulsory laws generally provide a benefit of 50% of the weekly earnings, subject to a maximum of $30. Under other plans, the benefit may range from 50 to 70%, with a maximum weekly benefit of from $40 to $60.

The period during which benefits will be paid may vary from 13 to 26 weeks, but in plans that are administered by employers the term during which benefits will be payable may be related to the length of service.

Group Hospitalization and Surgical Benefits. These forms of group protection are designed to protect an employee from the results of high hospital expense or the expense of a surgical operation. The protection may be written to cover an employee solely or it may be written to cover an employee and dependents.

The allowance for hospital room and board charges may be set at a figure intended to provide semiprivate care. Certain extras, such as the use of operating room, drugs, X rays, and electrocardiograms, may be included either on a service basis or on a basis that provides a maximum amount that will be paid for such service.

The period during which benefits are payable will vary depending on the contract and may run as long as 70 days. Some plans provide for payment of full charges for a comparatively short period and half charges for a longer specified period.

Surgical benefits are usually based on a schedule of amounts set forth in the policy. The maximum provided will, of course, depend upon the premium.

Group Coverage for Major Medical Expenses. Increased medical expense has brought about a demand for protection against catastrophic medical costs. This is one of the newest forms of group insurance. Rates vary greatly, as do forms of policy. The coverage is usually provided on a deductible basis (medical expense less a fixed sum). Frequently, a co-insurance feature is included so that the individual protected by such insurance bears part of the cost.

Index